THE PELAGIC DICTIONARY OF

NATURAL HISTORY

OF THE BRITISH ISLES

THE PELAGIC DICTIONARY OF
NATURAL HISTORY
OF THE BRITISH ISLES

PETER J. JARVIS

PELAGIC PUBLISHING

Published by Pelagic Publishing
PO Box 874
Exeter
EX3 9BR
UK

www.pelagicpublishing.com

The Pelagic Dictionary of Natural History of the British Isles

ISBN 978-1-78427-194-7   Hardback
ISBN 978-1-78427-195-4   ePub
ISBN 978-1-78427-196-1   PDF

A CIP record for this book is available from the British Library

Cover illustrations by Hennie Haworth © 2020

Printed and bound in India by Replika Press Pvt. Ltd.

MIX
Paper from
responsible sources
FSC® C016779

In memory of
Nicky Jarvis 1975–2010

Contents

Preface

The dictionary provides information on the animals, plants, fungi and algae of the British Isles – Great Britain, Ireland, the Isle of Man and the Channel Islands. It covers the taxonomy, habitat, distribution, abundance, status and, for animals, diet and size of individual species or species groups, with entries on those species that have English (common) names, and on many of the families, orders, classes, etc.

Even in the age of Internet search engines and hyperlinked web pages there is great value in an 'old-fashioned' dictionary relating to the natural history of a region. This goes a bit further: it is an annotated dictionary which provides information that, together, is not found on single web sites. It goes beyond definitions, but it is not an encyclopedia. The book does not compete with field guides so there are neither verbal nor pictorial descriptions to aid identification.

And in some ways information on natural history – or, rather, the words used in natural history – is fighting a bit of a rearguard action. The 2007 edition of the *Oxford Junior Dictionary*, for instance, deleted a number of previous entries describing nature, examples being acorn, adder, ash, beech, bluebell, buttercup, catkin, conker, cowslip, cygnet, dandelion, fern, hazel, heather, heron, ivy, kingfisher, lark, mistletoe, nectar, newt, otter, pasture and willow. (Admittedly, room was understandably needed for such recent additions to common vocabulary as blog, broadband, bullet point, celebrity, chat room, committee, cut-and-paste, MP3 player and voicemail.)

This compendium provides an annotated snapshot of the biodiversity of the British Isles in the early twenty-first century. The region's biodiversity is changing dramatically: the order of magnitude of species is probably much the same as in the previous millennium but there has been a pronounced turnover in the British and Irish flora and fauna, with extinctions offset by arrivals through introduction and natural dispersal. Also, distributions and balances of numbers have shifted as species respond to human activities and (maybe much the same thing) climatic change.

This dictionary is in two main parts. The smaller Part I looks at various terms that people interested in natural history may come across. It consists of explanations of words and phrases that users of the main (organism) section might find useful. Inevitably it echoes information that is available in a number of other natural history dictionaries, but the terms included and their definitions are here geared to the organism entries, and on many of the families, orders, classes, etc.

Part II is on the organisms. Information is given on individual species or species groups, with entries on those with English names, and on selected families, orders, classes, etc. There are very few, if any, common or readily identifiable species that do not have an English name, so what is excluded are generally those that are taxonomically difficult or rare. In the case of marine organisms, entries are given for intertidal and subtidal inverte-brate species, and generally speaking for fish and marine mammal species that might be observed inshore. A few entries are for diseases rather than the disease organism, for instance acute oak decline, ash dieback, Dutch elm disease, Lyme disease and myxomatosis, since these are the words an interested user would be looking for.

Entries are arranged alphabetically by primary name rather than by any qualifying adjectival name: thus all gobies are grouped together and collected under G for 'goby' rather than separated under, for instance, B for 'black goby' or C for 'common goby'. Alphabetically arranged entries in the organism section also include the names of many families, all orders and all classes in Latin, and synonyms of common names.

There is information on around 10,000 species, including natives, intro-ductions, and animals that are regular migrants and visitors. All British **vertebrates** are included – mammals (101 species), birds (590), herpeto-fauna (30), freshwater and estuarine fishes (56), together with some coastal fishes. There are around 2,800 entries for **invertebrates** – some two-thirds being for insects. Most invertebrates do not have common names, so for instance there are entries for only 84 of the over 250 species of bee found in the British Isles. All **vascular plants** are covered (2,950 species), as are all **mosses** (around 760), and **liverworts and hornworts** (288). About 1,000 of the 1,700 known **lichen** species are included, 1,410 of the 12,000 or so **fungi**, and some 50 **macroalgae** (i.e. seaweeds).

Each organism entry gives the Latin binomial and family (and sometimes subfamily) name. For vascular plants there is usually information on growth form (e.g. herb, shrub, tree) and life cycle (e.g. annual, biennial, perennial). Similarly, mosses are described as acrocarps (forming short cushions) or pleurocarps (free-branching), lichens as crustose, squamulose (scaly), foliose or fruticose, and liverworts as thallose (flattened, leafless) or foliose.

For all groups indication is usually given on distribution as well as whether a species is common, scarce or something in between – inevitably rather vague, but arguably better than nothing. To the geographer, 'Great Britain' refers to England (including the Scilly Isles), Wales and Scotland. The term 'United Kingdom' refers to England, Wales, Scotland and Northern Ireland. 'The British Isles' refers to the United Kingdom together with the Republic of Ireland, the Isle of Man and the Channel Islands. 'Britain', strictly, has no geographical (or political) meaning, but is in this dictionary somewhat guiltily used as a short-hand for Great Britain.

Distribution maps for the British Isles can be useful, but data varies according to survey effort, which in turn depends on where species are found (e.g. with less effort put into montane habitat survey than, say, in chalk grassland), taxonomic difficulty, and the number of 'experts' undertaking survey work. It is true that maps are available for many species, for example those produced by the National Biodiversity Network and the Biological Records Centre, but these are often at a scale to give little more than a general geographical impression, and crucially often include all records, including those from some time back, and therefore no guarantee of current presence. In any case, there is also simply not the space in a book of this size and taxonomic scope for distribution maps at any scale, useful or otherwise.

For introduced species there is an indication of the region of origin, and often the period or, where known, date of introduction. Comments are generally given on whether such species have become naturalised (having self-sustaining populations) or appear as casuals (not persisting in the wild, and therefore dependent on repeated reintroductions).

Comments are given for plants, animals and fungi on the habitat(s) where they are usually found.

For animals there are generally comments on size and, for birds and mammals, weight. For some species a note is made of status, population size and trends, diet, and feeding and reproductive behaviour. Population figures are inevitably estimates that vary in accuracy (depending on sample counts and a variety of statistical techniques) and recentness, and any single figure by definition cannot take into account the seasonal and annual variations that are responses to changes in weather. The emphasis of most population studies therefore involves identifying population trends rather than total numbers. The most accurate data are probably for birds, yet even here the effort involved in identifying trends is immense, and subject to often sizeable limits of confidence.

Comments are made where appropriate on etymology, both of the English name and the binomial components, though sometimes this is uncertain or unknown. Dialect terms were originally also included but added too many words to the already generous 600,000 word limit.

So while this dictionary lacks the detail of more taxonomically targeted publications, it does possess a breadth otherwise unobtainable on the printed page.

Acknowledgements

First thanks must go to Paul Cox, who provided a prompt that was the seed of this book and greatly encouraged its germination.

Many thanks to Nigel Massen (Pelagic Publishing) for his faith in this venture, and Chris Reed (BBR Design) for his editing and proofreading skills, and computer-tech advice. Ian Trueman has also been most encouraging and supportive.

As has my wife, Kat, who has provided patience, cake and tea during my hours in front of the keyboard, computer screen and an array of natural history books over a gestation period lasting well over three years, a length of time reflecting both the range of material consulted, and a shift from touch-typing efficiency to post-stroke single-digit hunt-and-peck.

Abbreviations and acronyms

BAP	Biodiversity Action Plan
BBS	Breeding Bird Survey, co-ordinated by the BTO, in partnership with the JNCC and the RSPB
BCE	Before Christian Era (formerly BC)
BTO	British Trust for Ornithology
CE	Christian Era (formerly AD)
c.f.:	compare (Latin *conferre*)
DEFRA	Department for the Environment, Food and Rural Affairs
DNA	Deoxyribonucleic acid
EA	Environment Agency
EN	English Nature
f.	*forma* = 'form'
fl.	*floreat*, i.e. let flourish/was active
GWCT	Game & Wildlife Conservation Trust
IUCN	International Union for Nature Conservation
JNCC	Joint Nature Conservation Committee
LNR	Local Nature Reserve
NERC Act	Natural Environment and Rural Communities Act (2006) – Species of Principal Importance in England
NNR	National Nature Reserve
OSPAR	The Convention for the Protection of the Marine Environment of the North-East Atlantic (The OSPAR Convention); List of threatened and/or declining species and habitats (Region II – Greater North Sea)
p.a.	each year, by the year (Latin *per annum*)
PTES	People's Trust for Endangered Species
q.v.	which see (Latin *quod vide*)
RBBP	Rare Breeding Birds Panel
RDB	Red Data Book – RDB 1 = Endangered; RDB 2 = Vulnerable; RDB 3 = Rare; Near Threatened (JNCC Species Status Assessment project)
RDL	Red Data List – JNCC Species Status Assessment project; see RDB
RNA	Ribonucleic acid
RSPB	Royal Society for the Protection of Birds
SMP	Seabird Monitoring Programme
SSSI	Site of Special Scientific Interest
UK BAP	UK Biodiversity Action Plan
WCA	Wildlife and Countryside Act (1981)
WWT	Wildfowl & Wetlands Trust

Part I: Terms

A

ABAPTATION The process by which the present-day match between an organism and its environment has been determined by evolutionary forces acting on its ancestors.

ABAXIAL Relating to the side away from the axis of a plant's lateral organ, for example leaf, usually the underside; opposite of adaxial.

ABDOMEN In insects, spiders and crustaceans, the section of the body furthest away from the head and attached to the thorax, and containing digestive and reproductive organs. In vertebrates, it contains the reproductive and alimentary systems.

ABIOTIC Non-living.

ABMIGRATION A northward summer migration of birds without a corresponding southward migration the previous autumn; movement of an individual from one breeding area to another by pairing in a winter flock with a bird from the new area and travelling there with it on spring migration, examples being found where winter flocks contain birds from different breeding grounds mixed together, particularly seen in dabbling ducks.

ABRASION Wear of feathers, which may change the appearance of birds by reducing tail and wing length, or by revealing a different colour from that at the tip.

ABSCISSION LAYER Cork layer forming at base of leaf stalk of deciduous tree in autumn, causing leaf to fall (abscission).

ACCESSORY FRUIT See pseudocarp.

ACCIDENTAL See vagrancy.

ACCLIMATISATION Adaptation of an introduced species to new conditions so that it can breed and spread.

ACCRESCENT Botanical name for the description of an organ, like the calyx, which continues to grow after flowering.

ACHENE Dry fruit that does not split open to release the single seed.

ACICLE The slenderest prickle of a bramble, often not more than a stiff bristle.

ACICULAR Pointed, like a needle, for example of a leaf.

ACIDIC With a pH below 7; see alkaline and pH.

ACORN Nut of the oak tree.

ACROCARP Type of moss in which the female sex organs and hence the capsules are borne at the tips of stems or branches. Acrocarps are usually unbranched and erect, forming a mounded colony. See pleurocarp.

ACTINOMORPHIC The shape of sea anemones and many flowers that are radially symmetrical.

ACULEATE Pointed or with prickles. In insects, having a sting, the Aculeata comprising ants, bees and stinging wasps.

ACUMINATE Leaf shapes that taper to a point.

ADAPTATION An anatomical, physiological or behavioural feature that makes a living organism better suited to its particular way of life.

ADAPTIVE RADIATION The evolution of two or more distinct species from a single original stock, each of the new species being adapted to a new environment.

ADAXIAL Relating to the side towards the axis of a plant's lateral organ, for example leaf, usually the upper side; opposite of abaxial.

ADIPOSE TISSUE Loose connective tissue whose main role is to store energy in the form of fat. **Brown adipose tissue** or **brown fat** is used in thermoregulation and is particularly important in newborn and hibernating mammals. See also subcutaneous fat.

ADNATE Botanical description for an organ which is joined to another organ, like a leaf joined to a stem, stamens to petals, or fungal gills to the stem.

ADNEXED GILLS Fungal gills with a narrow attachment to the stem.

ADRENALINE A hormone that prepares an animal's body for fight or flight in response to danger.

ADULT Having reached the final developmental stage.

ADVENTITIOUS Formed accidentally or in an unusual anatomical position, for example roots developing from aerial tissue.

ADVENTITIOUS COLORATION Superficial staining of (part of) a body, as in swans when their necks may be stained a rusty colour by the presence of iron in the water.

ADVENTIVE Non-native.

AEROBIC Requiring or living in the presence of air or oxygen.

AEROBIC RESPIRATION Chemical process that uses oxygen to release energy from food.

AESTIVAL Relating to summer.

AESTIVATION State of dormancy in animals involving a period of deep and prolonged sleep, or torpor, that occurs in the summer or dry season in response to heat and drought.

AFFORESTATION Planting stands of trees on previously unforested or non-woodland areas, usually for commercial and/ or recreational purposes.

AFTERSHAFT A supplementary small feather growing from the underside of the base of the shafts of certain feathers in birds.

AGARIC Fungus with a cap (pileus) and stalk (stipe), with gills on the underside of the cap.

AGONISTIC BEHAVIOUR Often used as a synonym of 'antagonistic behaviour' and refers to aggressive activity such as fighting, threatening, displaying and feigning, or avoidance and escaping, connected with conflict between social animals of the same species. However, this also includes 'synagonistic display', which is non-aggressive social behaviour, for example courtship.

AGRESTAL Plants growing wild in arable land.

AGROSTOLOGY The branch of systematic botany dealing with grasses.

AIGRETTE An elongated ornamental plume or crest that appears on the head of grey herons and egrets.

AIPHYLLUS Synonym of evergreen, particularly in a horticultural context.

AIR SAC An air-filled cavity in a bird's body that forms a connection between the lungs or respiratory system and bone cavities, helping breathing and temperature regulation – birds have at least five. Also a thin-walled extension of an insect's tracheae which increases the efficiency of gaseous exchange.

ALAR, ALATE Relating to wings; used of ants, aphids and other insects which have both winged and wingless forms.

ALBINISM, ALBINO Lack of melanin pigment in the body, showing white body parts and pink eyes.

ALBUMEN The 'white' of an egg, being a protein store for the embryo.

ALEVIN A newly hatched fish, still attached to the yolk, particularly used of trout and salmon. Once the yolk has been eaten, the alevin become fry, emerge from the gravel, move towards the light and start to feed on tiny insects in the water.

ALGAL BLOOM See bloom (3) and eutrophication.

ALGAL MAT A covering of cyanobacteria on the surface of sediments in shallow water, or on the water itself.

ALGIVOROUS Feeding on algae.

ALGOLOGY The study of algae, synonymous with phycology.

ALIEN A species introduced to a region through deliberate or accidental human activity.

ALIMENTARY SYSTEM A long tube, called the gut, or alimentary canal, along which food passes and is gradually digested.

ALKALINE With a pH above 7; see acidic and pH.

ALLELOPATHY Release by an organism, usually a plant, of a chemical substance that inhibits the germination or growth of other organisms.

ALLEN'S RULE The contention that within an animal species the extremities of the body tend to be longer in the warmer regions of its distribution and shorter in the cooler regions.

ALLOCHTHONOUS Moved from an original location or region.

ALLOGAMY Cross-fertilisation; fertilisation of a flower by pollen from another flower, especially one on a different plant.

ALLOGENIC A change in plant succession resulting from a change in the abiotic environment, in contrast to autogenic change.

ALLOMETRY Growth of one part of an organism faster or slower than that of the whole organism.

ALLOMONE A chemical substance produced and released by an individual of one species that affects the behaviour of a member of another species to the benefit of the originator. Production of allomones is a common form of defence, especially by plant species against insect herbivores, but is also used for example by the bee orchid to produce a scent that attracts its bee pollinators, which in turn receive a food reward.

ALLOPARENTING Care of young by a non-related individual.

ALLOPATRIC Refers to different species or subspecies with areas of distribution that do not overlap; see sympatric.

ALLOPOLYPLOID An interspecific hybrid which has doubled its chromosomes and so is able to breed; most hybrids are sterile.

ALLOPREENING Preening of one individual by another, often seen during courtship, and helping in pair-bonding.

ALOPECOID Resembling a fox (c.f. vulpine).

ALTITUDINAL MIGRATION Movement of upland animals, particularly birds, down to a lower level for the winter and back to higher levels for breeding.

ALTRICIAL Animals that hatch in an undeveloped state, often blind and completely dependent on their parent(s); the opposite of precocial.

ALTRUISM Behaviour detrimental to an individual but favouring the survival or spread of that individual's genes, for example by benefiting a relative.

ALULA A structure consisting of three to five feathers growing on a bird's 'thumb', also known as the 'bastard wing'. It controls the airflow over the leading edge of the wing, reducing stalling when landing or flying at slow speeds. Also a lobe on the trailing edge of a fly's wing.

AMBERGRIS Aromatic substance found in the intestines of the sperm whale, which can be seen cast up on the shore or floating on the sea. Used in the cosmetic industry for fixing and intensifying perfumes.

AMBIENT Relating to the immediate environment, for example ambient temperature.

AMBIVALENCE Showing fear and aggression simultaneously, as in territory holders that both attack and retreat when displaying.

AMENTACEOUS Bearing catkins.

AMINO ACIDS The building blocks of proteins. There are 20 common types of amino acid. They join together in special sequences to form different proteins.

AMMOCOETES Larval form of lampreys.

AMPLEXICAUL A leaf with its base clasping or surrounding the stem, for example seen in teasel.

AMPLEXUS Where a male frog or toad clasps the back of a female when mating, during which he fertilises eggs released into the water.

ANADROMOUS, ANADROMY Behaviour of animals such as eels and salmon which live in the sea as adults and migrate into fresh water to breed. The opposite of catadromous.

ANAEROBIC Living in the absence of air or oxygen; see anoxic. **Anaerobic respiration**, a chemical process that releases energy from food without the need of oxygen.

ANAMORPH The asexual reproductive stage of a fungus, often mould-like.

ANASTOMOSING Name given to the veins on a leaf when each loops round to reconnect with the next one, so forming a network.

ANATOMY The study of the bodily structure of animals. The study of the functioning of the body is 'physiology'.

ANCIENT Habitat such as woodland or pasture with semi-natural character that has existed since 1600 CE or before.

ANDRODIOECIOUS A dioecious plant in which hermaphrodite and male flowers occur on different plants.

ANDROECIOUS Plants bearing only male flowers.

ANDROECIUM The group of male parts of a flower; all the stamens.

ANDROMONOECIOUS A monoecious plant in which hermaphrodite and male flowers occur on the same plant.

ANECIC Of earthworms, species that make permanent vertical burrows in soil.

ANEMOCHORY Dispersal of organisms by wind.

ANEMOPHILY Pollination of flowers by the wind.

ANGEL Use of radar to track flocks of birds. The traces of the flocks are known as angels.

ANIMAL A multicellular living organism that lives by taking in food. Animals (Animalia) make up one of the five kingdoms of nature, the others being Monera, Protista, Fungi and Plantae.

ANIMALCULE Tiny, usually microscopic animals.

ANISODACTYL The commonest arrangement of toes in birds, with 3 toes forward and one back.

ANNUAL (PLANT) A plant that completes its life cycle in one year.

ANNUAL RING Ring of wood laid down in stem and branches

of tree or shrub (and in shells) during one growing season. See also growth ring.

ANNULUS 1) A 'ring' around the stipe (stem) of an agaric fungus, a remnant of the partial veil. 2) In some ferns, a ring of cells around the sporangium.

ANOESTRUS A period of sexual inactivity in females between breeding, particularly in mammals; outside the breeding season.

ANOINTING See self-anointing.

ANOXIC Lacking oxygen.

ANTAGONISTIC DISPLAY A type of behaviour involving aggression, such a display being well seen in the defence of territory. This is also commonly known as 'agonistic display'.

ANTENNA One of the paired sense organs, sometimes known as 'feelers', on the head of an invertebrate or anthropod. An animal uses its antennae to feel or to taste, as well as sensing vibrations and smells.

ANTHER The part of the stamen that produces pollen, usually at the end of a stalk (filament).

ANTHERIDIUM The male sex organ in algae, bryophytes, pteridophytes and fungi.

ANTHESIS The flowering time (strictly, pollen-shedding time) of a plant.

ANTHOCYANIN The red, blue and purple pigments in the sap of plants, which gives us the autumn colours of the dying leaves.

ANTHRACNOSE A fungal disease responsible for diseases on many woody plant species. Infected plants develop dark, water soaked lesions on stems, leaves or fruit.

ANTHROPOCHORY Dispersal of organisms, including seeds, eggs, etc. by human activity.

ANTHROPOMORPHISM The practice of ascribing human qualities to animals other than mankind.

ANTING Where birds self-anoint by rubbing ants on to their feathers using the formic acid to rid themselves of parasites.

ANTLER Bone extensions of the skull grown by male deer as objects of sexual attraction and as weapons in fights for control of harems. They are shed and regrow each year. As they grow they are covered with skin richly supplied with blood vessels – 'velvet' – that is shed or rubbed off when growth is complete. A mature red deer stag may have 12–15 branches (tines or points). Reindeer are the only species where females also have (smaller) antlers. See switch.

ANVIL A stone or hard object used by song thrush to smash open the shells of snails to eat the soft insides.

APERTURE An opening in some kinds of mollusc shell where the body emerges for locomotion, feeding, etc.

APEX Tip of a stem or other organ.

APEX PREDATOR A predator at the 'top' of the food chain, itself without a predator.

APIARY A place where bees are kept; an apiarist is a bee-keeper.

APICAL MERISTEM See meristem.

APICULATE Leaf with a small abrupt point at the apex.

APOCARPOUS A plant in which the gynoecium has multiple, distinct (free, unfused) carpels.

APOGAMY In ferns, apomictic reproduction in which a gametophyte develops directly into a sporophyte without fertilisation.

APOMATIC, APOMICTIC (noun APOMIXIS) Relating to a plant whose seeds are formed (or in ferns the whole plant) without normal fertilisation; progeny are all female and genetically identical to the mother plant.

APOSEMATIC COLORATION From the Greek apo +

sema = 'away' + 'sign', bright colours that warn predators that an animal is dangerous or toxic if attacked, for example the black and yellow stripes of bees and wasps.

APOTHECIUM A wide, open, saucer-shaped or cup-shaped fruiting body of a fungus or lichen.

APPENDAGE A projection from the body, such as a limb or a crest.

APPETITIVE BEHAVIOUR Activity increasing the chances of successfully achieving a goal, for example a hungry animal looking for food. Consummatory behaviour follows achievement of the goal.

APPRESSED A botanical description of organs that are pressed together but not joined, as in sepals to petals, or hairs lying flat against a stem.

APTERY, APTEROUS Wingless, for example in insects such as fleas and worker ants.

AQUATIC Relating to water as opposed to land (which is terrestrial).

AQUILINE Of or like an eagle; (of a nose) curved like an eagle's beak.

AQUIPRATA Communities of plants where surface water is an important factor, for example wet meadows.

ARABLE Ploughed land, sown with cereal and other crops.

ARACHNOLOGY The scientific study of Arachnida – eight-legged invertebrates, such as spiders, scorpions, ticks, mites or harvestmen.

ARANEOLOGY The scientific study of spiders.

ARBOREAL Living in trees.

ARBORETUM A place where (usually) rare trees or shrubs are cultivated.

ARCHAEOPHYTE An introduced plant associated with human activity and present in the British Isles before around 1500; see neophyte.

ARCTIC-ALPINE Relating to species that occur at high latitude and high elevation.

ARENICOLOUS Growing, living or burrowing in sand.

ARENOPHILOUS Plants adapted to living in sandy areas.

ARIL Fleshy cup formed from fused cone scales surrounding seeds.

ARISTATE In plants, having a pointed, bristle-like tip such as the awn of a grass; in insects (particularly in true flies), having a lateral bristle on an antenna.

ARISTOTLE'S LANTERN The dental apparatus of a sea urchin's mouth on the underside of its body, used for biting and rasping, named after the description Aristotle gave it from its resemblance to a horn lantern.

ARRAY Collective noun for hedgehogs.

ARRESTED MOULT When the moulting of a bird is suspended for a period to be resumed later, for example the common tern which completes its moult after migration.

ASCENDENT MOULT While primary moult usually follows a descendent (outward) order, a few bird species have an ascendent (inward) moult. See also descendent moult.

ASCOCARP The fruiting body (sporocarp) of an ascomycete fungus, with different shapes: see apothecium (the commonest shape), cleistothecium, gymnothecium and perithecium.

ASCUS The sac-like organ in which ascomycete fungi produce sexual spores.

ASEXUAL REPRODUCTION Reproduction in which there is only one parent; includes parthenogenesis (q.v.). The young have exactly the same chromosomes as the parent, and therefore the same characteristics.

ASPERGILLOSIS A number of diseases caused by a fungal mould (*Aspergillus*), particularly affecting the respiratory system, and found especially in wildfowl and gallinaceous birds.

ASSOCIATION A stable, broad plant community within a particular geographical unit.

ASYMMETRICALLY POSITIONED EARS As in nocturnal owls, one ear being positioned higher than the other for good directional hearing.

ASYNCHRONOUS Occurring at different times, as in birds laying eggs at intervals to ensure survival of first born.

ATLANTIC CLIMATE Wet and mild, thus influenced by the Atlantic Ocean.

ATLASSING Collecting data for use in compiling an atlas showing the distribution of animals, particularly birds, in a particular area.

ATROPHY Partial or complete wasting away of a body part or, loosely, reduction in size for example the swift's legs and feet have become atrophied as their lifestyle has become aerial.

ATTIRE Deer antlers.

AURICLE 1) In botany, an ear-shaped lobe, appendage or extension at the base of a leaf (especially in grasses) or petal. 2) The external mammalian ear.

AUSTRAL Southern.

AUSTRINGERS Old falconry term for keeping hawks, particularly goshawks.

AUTECOLOGY The study of the ecology of a single species.

AUTOCHORY Dispersal of seeds or spores by the parent.

AUTOCHTHONOUS Taxa originating in its current location; not allochthonous (q.v.).

AUTOGAMY Self-fertilisation.

AUTOGENIC A self-generated change in plant succession; see also allogenic.

AUTOLYCISM Making non-parasitic or non-commensal use of activities of another animal, as in birds taking advantage of human activities, for instance using buildings for nesting and arable fields for feeding.

AUTOTOMY Spontaneous severance a limb or tail (that later regrows) as an anti-predator defence mechanism, for example a lizard snapping off its tail when caught by a predator.

AUTOTROPH An organism that can synthesise complex organic compounds from simple substances, for example plants synthesising carbohydrates by photosynthesis, or microbial organisms using chemosynthesis. See primary producer.

AVIAN Of birds.

AVICULTURE The keeping and breeding of birds in captivity.

AVIFAUNA The birds found in a particular area, or a description of them.

AWN Stiff bristle-like appendage on the tip of a plant, particularly barley, oats and a number of grasses.

AXI The angle between the stem and a branch, leaf or bract.

AXILLA Body part equivalent to the underarm or armpit.

AXILLARIES Feathers covering the 'armpit' of a bird, where the underwing joins the body.

AXILLARY 1) In plants, arising from an axil. 2) In birds, feathers growing from the axilla.

B

BACKBONE Spine; a flexible chain of bones running from the head to the tail in a vertebrate.

BACTERIOLOGY The scientific study of bacteria.

BACULUM The penis bone which aids sexual reproduction by maintaining stiffness during sexual penetration, found in bats, rodents, insectivores, cats, dogs, foxes, seals and mustelids.

BALEEN PLATES The fringed plates that hang from the roof of the mouth of the largest types of non-toothed whale. Baleen plates filter zooplankton from seawater.

BALLISTOSPORE A fungal spore that is discharged explosively.

BALLOONING A method of dispersal in spiderlings, which climb to a raised point paying out a length of silk until the wind catches them, lifting and transporting the animals often long distances.

BAND Collective name for a group of jays.

BARACHORY Dispersal of seeds or spores by gravity (i.e. falling to the ground).

BARB A horizontal branch from the shaft (rachis) of a feather, also called a 'ramus'. It carries the 'barbules' which are so arranged that those on one barb interlock with those of its neighbours, being held by tiny hooks called 'barbicels' or 'hamuli'.

BARBEL A whisker-like sensory organ near the side of the mouth of such fish as barbel, carp and catfish, which house the taste buds and aid the search for food in muddy water.

BARBICEL, BARBULE In feathers, see barb.

BARK The tough outer woody layer of a tree or shrub, the inner bark consisting of living tissue, and the outer bark including dead tissue on the surface.

BARREN(S) 1) Old-fashioned name for heathland, or more generally low-nutrient soil. 2) In animals, incapable of bearing offspring.

BASAL BRISTLE Small feather at the base of a bird's beak.

BASAL KNOB See knob.

BASE A substance that, in solution, can bind hydrogen ions and has a pH >7, reacting with an acid to yield a salt and water.

BASE STATUS Base-rich soils are rich in basic elements such as lime, strictly with a pH above 7 but in practice generally taken to be a pH of 5.5 or above. Base-poor soils typically have a pH below 4.5.

BASIDIUM A microscopic group of spore-forming cells found on the fruiting bodies of basidiomycete fungi. The presence of basidia is one of the main characteristic features of the Basidiomycota. A single toadstool may have a million basidia.

BAST FIBRE Fibre collected from the phloem of economic crops such as flax and hemp, and certain trees, for instance lime.

BASTARD WING See alula.

BATESIAN MIMICRY See mimicry.

BATOLOGY The study of brambles, this being a large apomictic group producing a large number of microspecies.

BEAK 1) Multifunctional projection from the head (also called the bill or rostrum), particularly in birds, used for eating, feeding young or partner, manipulation, grooming, capturing or killing prey, courtship and fighting. 2) A rigid projection from a fruit tip. 3) In orchid flowers, a projection separating an anther from the surface of the stigma beneath it.

BEE BREAD The food of bee larvae, a mixture of pollen and honey, and sometimes plant oils and gland exudates.

BEE DANCE Movement by honeybees to communicate direction and distance of (usually food) resources. A **round dance** occurs for nearby resources, typically <10–20 m from the nest; as the distance to the resource increases, the round dance transforms into a **waggle dance**, akin to a figure-of-eight movement.

BEESWAX Wax (mainly esters of fatty acids and various long-chain alcohols) excreted by glands of the worker honeybee, used to create cells for storing honey and protecting larvae and pupae.

BENTHOS The fauna and flora of the bottom of a lake or the sea.

BERR A fleshy fruit produced from a single flower with hard-coated seeds.

BEVY Collective name for a flock of larks, a group of partridge or quail, or a group of roe deer.

BIANNUAL Occurring twice a year.

BIENNIAL A plant that completes its life cycle every two years, flowering in the second year.

BIFID Divided into two.

BILATERAL SYMMETRY Arrangement of the parts of a body, flower, etc. such that an imaginary central line or plane divides it into mirror halves.

BILL See beak.

BILLING A pair of birds gently touching bills or caressing with their bills, also called 'nebbing'.

BI-LOBED PETALS Partly split petals.

BINDING-TO A falconry term, grasping prey by a raptor in mid-air, often at the end of a stoop.

BINOCULAR VISION Vision using two eyes to gauge depth, with a three-dimensional effect.

BINOMIAL NOMENCLATURE International, universally accepted system of naming species. Invented by Linnaeus, the system uses two Latin names, the genus name and the species name.

BIOCOENOSIS A community of interacting organisms inhabiting a habitat or biotope.

BIODIVERSITY The number of species or amount of genetic variation in a particular area.

BIOFILM A group of microorganisms in which cells stick to each other on a living or non-living surface, often embedded in what is technically an extracellular polymeric substance, more informally referred to as slime.

BIOGEOCHEMICAL CYCLE Movement of a chemical element from the physical environment through the living environment back to the physical environment. See also nutrient cycling.

BIOLOGICAL CLOCK An internal mechanism in organisms that controls the periodicity of particular activities, for example metabolic changes, sleep cycles or photosynthesis. Daily rhythms are called circadian. Also, in animals, a clock that measures ageing throughout the body.

BIOLOGICAL CONTROL Use of predators, parasites, diseases or competitors to control or suppress a pest population.

BIOLOGY The academic study of living organisms in both its general and particular aspects. Biology is basically divided into 'botany', the study of plants, 'zoology', the study of animals, and 'microbiology', the study of microscopic organisms. It also has various more generalised divisions, such as ecology, physiology and taxonomy.

BIOLUMINESCENCE Production of light by a living thing;

BIOLUMINESCENCE

also known as phosphorescence. Particularly associated with some marine life but also evident in, for example, some fungi and insects such as glow-worms.

BIOMAGNETISM Magnetic fields produced by living organisms.

BIOMASS The mass of living organisms in a defined area or habitat, usually measured as total weight or weight per unit area.

BIOME Ecosystems (biotic communities together with their general physical environment) that extend over large geographical areas, taking their name from the dominant, climatic climax vegetation, and broadly corresponding to a climatic region. Although the landscape would naturally incorporate other vegetation types, and indeed now contains little if any natural vegetation, the biome of the British Isles is temperate deciduous forest.

BIOMETRY Application of mathematical or statistical methods in biology; concerned with the accurate measurement of the characteristics of living things such as dimensions and weights.

BIOSPHERE All the parts of the Earth that are inhabited by living things, including the land, earth and the air.

BIOTA All the species of living plants and animals (and microorganisms) occurring within a certain area or region.

BIOTIC FACTOR The influence or effect of a living thing on a habitat or ecosystem.

BIOTOPE An area of uniform environmental conditions providing a living place for a particular biological community; see habitat.

BIOTYPE A group of animals or plants of fixed genetic make-up, adapted to some usually environmental condition.

BIPEDALISM A form of terrestrial locomotion in which an animal moves by means of its two rear limbs or legs.

BIPINNATE A pinnate leaf with leaflets that are themselves pinnate.

BIRKS Vernacular Yorkshire name for birchwood.

BISERRATE Serrate, with each 'tooth' toothed again.

BISEXUAL Possessing both male and female reproductive organs, for example a flower with both stamens and pistils; hermaphroditic.

BIVOLTINE Animals producing two broods in a year.

BLANKET BOG An ombrogenous bog, or 'flow', the peat accumulating on (near) flat terrain above the water table, relying on rainfall.

BLASTOCHORY Plant dispersal using offshoots or runners.

BLAZE In birds, a coloured patch at the base of the beak.

BLOOD A complex bodily fluid in animals that delivers substances such as nutrients and oxygen to the cells and transports metabolic waste products away.

BLOODSUCKER An animal that sucks blood, for example leech, mosquito or horsefly.

BLOOM 1) One or more flowers on a plant. 2) Waxy, easily removed, often whitish film or covering on fruit, leaves, etc. 3) Rapid algal growth in response to rising temperature and abundant nutrients. 4) Collective noun for jellyfish.

BLOSSOM The flowers of fruit trees, emerging in spring profusion; insects provide the cross-pollination necessary for fruit production.

BLOW SPOUT Seen during expiration of large whales when the surface: water vapour-rich air in the lungs is forced out, expands and cools, causing the vapour to condense and become visible. Whale spouts are therefore not of water.

BLOWHOLE Breathing hole, equivalent to nostrils, on a dolphin or whale, located on the top of its head.

BOG The habitat and plant community associated with wet areas where acid conditions reduce the decomposition rate of plant material, which accumulates to form peat. Types of bog in Britain are blanket, raised, valley, soligenous (also mire or fen) and quaking bog (see all).

BOLE The unbranched trunk or stem of a tree.

BOLLING Trunk or stump of pollarded tree.

BONE 1) A unit of the vertebrate skeleton. 2) The dense connective tissue from which the skeleton is made, mainly of calcium and phosphate salts and collagen. Bones support and protect the body's organs, produce red and white blood cells, store minerals, provide structure and support for the body, and enable mobility.

BOOK LUNGS Respiratory organs found in arachnids, over which haemolymph passes to allow oxygen absorption, probably evolved from **book gills** which are external equivalents in horseshoe crabs.

BOOTED Where no scales are present on a bird's leg, the legs being instead covered by a continuous skin, for example in the thrushes (Turdidae).

BOREAL Northern.

BORON A micronutrient essential for maintaining cell wall integrity, though excess leads to necrosis. Boron probably plays a number of essential roles in animals, but the physiology is little understood.

BOSCUS The wood or undergrowth produced by coppicing.

BOTANY The scientific study of plants.

BOTULISM Severe food poisoning caused by toxins produced by *Clostridium botulinum* bacteria, found in stagnant water and affecting waterfowl (and humans).

BOUQUET Collective noun for pheasants.

BRACE A pair of animals, particularly of game birds or waterfowl.

BRACHYPTERY In insects, a species or a sex with both sets of wings reduced (i.e. have become vestigial) so that the animal is flightless. Absence of wings is called aptery (q.v.).

BRACKISH Slightly salty; a mixture of fresh water and salt water, for example in estuaries and lagoons, with a fauna that includes some salt-tolerant freshwater species, a few marine taxa, and some species limited to the brackish conditions.

BRACT (Usually) modified leaf associated with a reproductive structure such as a flower or inflorescence.

BRACT SCALE Thin, papery, seed bearing structure rising from cone axis above cone scales.

BRACTEOLE A small, secondary bract; above a bract or on a lateral branch.

BRAILING A method of temporary flight constraint by tying a bird's wing making it unable to fully open the wing.

BRAKE An area covered by scrub or bracken.

BRANCH A more or less lateral stem that arises from the main stem (trunk) or another branch.

BRASHING Removing the lower (especially dead) branches of trees in a plantation, usually up to 2 m height. The removed material is the brash.

BREAST BAND A broad zone of colour along a bird's breast also known as a 'pectoral band' or a 'gorget'. The ring ousel has a white band, for example.

BRECK Medieval term for an area of heathland broken up for cultivation before being allowed to revert to grass-heath habitat.

BRECKLAND A multi-habitat region, but characteristically heather- and gorse-dominated sandy heathland in northern Suffolk and southern Norfolk. The Breckland Biodiversity Audit (2011) identified 12,845 species which included 28% of species in th UK Biodiversity Action Plan (including 72 species found nowhere else in the British Isles).

BREEDING Producing offspring by mating. In birds and mammals, breeding also means raising the young.

BREEDING CYCLE The complete sequence of reproductive activity, for example in many birds from initial courtship and pair formation through nesting to the final independence of the young.

BROAD A shallow lake formed in a depression caused by medieval peat digging. Broads are mainly found in north-east Norfolk and Ireland, and fringing reed habitats are valuable for wetland birds, including rarities such as bittern and bearded tit.

BROAD-LEAVED TREES Trees with wide leaves, in contrast with the needle-like leaves of conifers, and usually deciduous.

BROC, BROCC, BROCK Traditional name for a badger, the first being Old Irish, Manx and Scottish Gaelic.

BROOD The young hatched from a single clutch of eggs; collective name for pheasants and hens.

BROOD NEST 1) Built by parent birds for the young to rest up in; it may or may not be the nest in which it was hatched. 2) In bee-keeping, the part of the hive box containing the honeycomb.

BROOD PARASITE Birds and insects using hosts of the same or different species to raise their young. Intraspecific brood parasitism is seen in some duck species, interspecific parasitism as with the cuckoo and cuckoo bees which lay their eggs in the nests of other species.

BROOD PATCH A featherless area on a bird's underside which develops at the time of year when the eggs need incubating. Superficial blood vessels facilitate tranfer of heat to the eggs. Feathers are generally shed naturally but some ducks and geese may pluck the feathers (and use them as nest lining).

BROODING Warming or shading of the eggs or young by the parent bird sitting over them, depending on ambient temperature.

BROWN ADIPOSE TISSUE See adipose tissue.

BROWN EARTH A soil type generally developing under deciduous woodland and associated with mull (q.v.) humus.

BROWSE, BROWSING To feed by continual nibbling on twigs, leaves and other vegetation above ground level.

BRUMATION Dormancy in amphibians and reptiles, similar to hibernation (q.v.) in mammals but differing in the metabolic processes involved. Similar to hibernation brumation is triggered by lowering temperatures and a decrease in the hours of daylight during autumn.

BRYOLOGY The scientific study of mosses, liverworts and hornworts.

BUCCAL Of the mouth or, more specifically, the cheek.

BUCK The male of the fallow, roe and muntjac deer.

BUD 1) In botany, an undeveloped shoot, usually found on an axil or stem tip, from which a leaf, flower or new stem might emerge. 2) In zoology, an outgrowth that could develop into a new individual, an example of asexual reproduction.

BUD SCALES In many woody plants, hard, often sticky protective modified leaves that cover and protect the delicate bud. See also cataphyll.

BUILDING Collective noun for rooks.

BULB Swollen underground food storage organ formed of succulent leaves or scales.

BULBIL A small bud or tuber-like structure formed at the base of a leaf in place of a flower, which breaks off and grows into a new plant; common in lilies.

BULLATE In plants, with the surface raised into swellings resembling blisters.

BUR(R) 1) A seed or dead fruit with hooks or bristles that catch on the fur of passing animals allowing dispersal. 2) A large, often fairly hemispherical outgrowth on a tree-trunk.

BUSH General name for low woody plant or vegetation; interchangeable term with shrub.

BUSKING The aggressive territorial defensive display of a male mute swan in which he advances over the water towards an intruder with his neck drawn back and wings arched, proceeding with a jerky movement because, as opposed to normal practice, he paddles with both feet in unison.

BUTTERFLY FLIGHT Seen for example in ringed plover which has a slow, fluttering and often erratic type of flight display.

BUZZ POLLINATION Also known as sonication, where bees, on landing on a flower, rapidly contract their indirect flight muscles to produce strong vibrations (generating forces of up to 30 G) that expel pollen from the flower's anthers. Around 8% of flowering plants have poricidal (q.v.) anthers, where pollen is hidden inside a tube-like anther, accessible through a small pore at the tip. Most of these species do not offer nectar: the only reward for the bee is the pollen that is expelled through buzz pollination.

C

CACHE, CACHING Storage or hoarding of food by animals.

CADUCOUS Dropping off or shedding at an early stage of development, for example amphibian gills, or leaves, sepals and stipules of certain plants.

CAESPITOSE Growing in dense tufts.

CALAMISTRATED WEB Spider web formed from silk produced by the spinning organ (cribellum), enlarged versions called cobwebs.

CALAMISTRUM A row of specialised leg bristles used by spiders to comb out bands of silk with a woolly texture.

CALAMUS Part of a quill; in a feather the hollow, unfeathered part of the rachis which inserts into a skin follicle.

CALCAR Outgrowth from the heel of a bat, partly supporting the wing membrane round the tail.

CALCAREOUS Soils or rocks containing calcium carbonate.

CALCICOLE, CALCIPHILE A plant needing or preferring soils or water high in calcium carbonate.

CALCIFICATION The process of hardening when calcium carbonate is laid down by cells.

CALCIFUGE A plant that grows better in acid soils and avoids calcareous conditions.

CALCIUM 1) An element essential for plant growth, strengthening cell walls and involved in stomatal behaviour and root development. 2) In vertebrate animals, it is important in the cell physiology and biochemistry, and vital in bone formation and in the health of the muscular, circulatory, and digestive systems. 3) Some invertebrates use calcium compounds for building their exoskeleton (shells and carapaces) or endoskeleton (e.g. echinoderm plates and poriferan calcareous spicules).

CALLOW Arthropod lacking hardness and adult colour just after ecdysis or emergence from pupation (in insects in the teneral phase).

CALLUNETUM Plant community dominated by heath or heather.

CALYPTRA A protective cap on the capsule of a moss or liverwort.

CALYX All the sepals of a flower.

CAMBIUM A tissue layer in the stems, branches and roots of vascular plants lying between the phloem and xylem.

CAMOUFLAGE Use of colour, pattern and shape by animals for disguise or concealment, making them hard to see (see crypsis), allowing them to blend in with their surroundings. See also mimicry and disruptive coloration.

CAMPANULATE Shaped like a bell.

CAMPODEIFORM Body shape of typically predatory insect larvae that have well-developed legs, antennae and a flattened body, for example lacewings and many species of beetles.

CANINE 1) Of or like a dog or other canid. 2) A pointed tooth between the incisors and premolars of a mammal, often enlarged in carnivores.

CANOPY The uppermost layer of branches in a woodland (tree crowns).

CANT See coupe.

CAPILLARY Hair-like (Latin *capillus* = 'hair'), in botany, zoology and soil science coming to mean a tube which has an internal diameter of hair-like thinness, for example the narrowest type of blood vessel (one cell thick).

CAPITATE Having a head or headlike (Latin *caput* = 'head').

CAPITULUM A condensed head of flowers, especially of Asteraceae, the inflorescence made up of closely packed flowers reaching the same level on a flattened axis.

CAPSULE 1) Dry fruit which splits open to release its seeds. 2) In mosses and liverworts, the spore-bearing part on top of the seta (q.v.).

CARAPACE Dorsal section of the exoskeleton or shell of tortoises and crustaceans.

CARAVAN Young shrews travel by forming a 'caravan' behind the mother, each carrying the tail of the sibling in front in its mouth.

CARBOHYDRATE A group of compounds containing only carbon, hydrogen and oxygen that includes sugars and starch, cellulose and chitin.

CARINA See sternum.

CARNASSIAL TEETH Modified molars or premolars (usually the fourth upper premolar and the first lower molar) in some mammalian carnivores (e.g. red fox, stoats and weasels) that allow teeth in the upper and lower jaws to move against each other in a shearing action.

CARNIVORE 1) An animal that eats other animals. 2) Carnivorous plants such as sundew that trap and digest insects and other arthropods to supplement nutrients in nutrient-poor habitats such as bogs.

CARPEL The basic unit of the reproductive part of the female flower, containing the unfertilised seeds; may be solitary to many, fused or separate.

CARPUS, CARPAL The group of small bones between the main part of the forelimb and the metacarpus in terrestrial vertebrates. In birds, the outermost wrist joint, forming the bend in the wing.

CARR From Old Norse *kjarr* = 'swamp', a type of waterlogged wooded vegetation, often a successional stage between swamp and woodland. See also fen carr.

CARRION The decaying flesh of dead animals, eaten by scavenging animals, insect larvae, etc.

CARTILAGE Flexible, resilient, skeletal tissue that covers and protects the ends of long bones at the joints.

CARUNCLE 1) In plants, a reduced, often brightly coloured aril, or an elaiosome (q.v.). 2) A fleshy appendage, for example a bird wattle. *Caruncula* is Latin for 'fleshiness'.

CARYOPSIS An achene in which the ovary wall is fused with the seed coat, typical in grasses and cereals.

CAST Collective noun for falcons and hawks.

CASTE One of three or more functionally specialised forms, distinguished morphologically, making up a population among social insects, for example in honeybees usually queen, drone (male) and worker.

CASTING See pellet.

CASUAL A non-native plant or animal species of which populations do not persist in the wild, and therefore dependent on repeated reintroductions.

CATADROMOUS, CATADROMY Fish that spend most of their life in fresh water and migrate to the sea to reproduce, the young then migrating back, usually to their natal river. The opposite of anadromous.

CATAFLEXISTYLY A morphological change whereby flowers function first as a female then as a male.

CATAPHYLL A reduced, small scale-like leaf often covering a dormant bud, often containing resin. Many plants have both 'true leaves' (euphylls) which perform the majority of photosynthesis, and cataphylls that are modified for other specialised functions.

CATERPILLAR The wingless larva of moths, butterflies and sawflies.

CATKIN Elongated cluster of single-sex flowers bearing scaly bracts and usually lacking petals. Many trees bear catkins, including willows, poplars, alder, hazel, birches and oaks. In many species only the male flowers form catkins, and the female flowers are single (hazel, oak), a cone (alder) or other types (mulberry). In other plants (e.g. willows and poplars) both male and female flowers are borne in catkins, known as dioecy (q.v.). Wind carries pollen from male to female catkins.

CAUDAL Of the tail.

CAUDAL GLAND See subcaudal gland and supracaudal gland.

CAULDRON Collective noun for raptors.

CAULINE Growing on a stem, especially on the upper part, usually referring to leaves.

CAVERNICOLOUS Living in caves or caverns.

CECIDIUM A plant gall produced, for example, by an insect or a fungus.

CELL The living unit that make up most organisms, consisting of jelly-like cytoplasm held in by a membrane. At the centre of most cells is a nucleus.

CELL (WING) The area of an insect's wing enclosed by a number of veins. An open cell is partly bordered by the wing margin.

CELLULOSE An organic compound that is the most important structural component of the cell wall of green plants and algae.

CENTRAL NERVOUS SYSTEM See nervous system.

CEPHALOTHORAX The fused head and the thorax of spiders, mites and some crustaceans.

CERCI Paired appendages on the rearmost segments of a number of arthropods, for example bristletails, earwigs and mayflies, serving as pinching weapons, sensory organs or mating structures.

CERE From Latin *cera* = 'wax', a structure made of skin situated at the base of the upper mandible and containing the nostrils, found for example in raptors (providing a sexual signal of fitness), owls and skuas.

CEREAL A grass cultivated for the edible components of its grain, for example wheat, barley, oat and rye.

CERNUOUS Pendulous or drooping, of a bud, flower or fruit, or the capsule of a moss.

CERVIX, CERVICAL 1) (Of the) neck. 2) (Of the) the narrow neck-like passage forming the lower end of the uterus.

CHAEOTAXY The arrangement of bristles on an insect or mite, or the taxonomy based on their size and position.

CHAETA 1) Stiff hair or bristle on an insect. 2) Chitinous bristles on the outside of an earthworm or polychaete used in gripping the burrow.

CHAMAEPHYTE From Greek *chamai* + *phyton* = 'dwarf' or 'on the ground' + 'plant', a plant where the perennating bud is at or near the ground.

CHARM Collective noun for finches.

CHATTERING Collective noun for starlings.

CHELICERAE Appendages at the front of the mouth (biting jaws) of spiders and other arachnids, in some species also used for defence and digging.

CHEMOSYNTHESIS Synthesis of organic compounds by bacteria or other living organisms using energy derived from reactions involving inorganic chemicals, typically in the absence of sunlight.

CHEMOTAXIS Change in the direction of movement (towards or away from) in response to a chemical stimulus.

CHIPPING The breaking of the eggshell prior to hatching.

CHITIN A tough carbohydrate found in the external skeleton of insects, crustaceans and spiders, and in the cell wall of fungi.

CHLOROPHYLL The green pigment in plants which absorbs sunlight as part of the process of photosynthesis.

CHLOROPLAST A plastid in green plant cells which contains chlorophyll and in which photosynthesis takes place.

CHLOROSIS A mineral deficiency or disease in plants which inhibits chlorophyll production, turning leaves yellow or pale green.

CHROMOSOME From Greek *chroma* + *soma* = 'colour' + 'body', a DNA molecule – a thread-like structure of nucleic acids and protein – found in the nucleus of most living cells, carrying the information in the form of genes needed to control the structure and development of the cell. Most cells have two copies of each chromosome.

CHRYSALIS The hard protective pupal case of butterfly or some moths when undergoing metamorphosis. See pupa and cocoon. The term is derived from the metallic gold coloration found in the pupae of many butterflies, from Greek *chrysos* = 'gold'.

CHURRING, CHIRRING A prolonged trilling noise made by some birds (e.g. nightjar) and insects (e.g. crickets).

CILIA Hair-like structures on a plant or microscopic animal. In animals cilia can beat to make a cell move or to move things nearby.

CILIUM A minute thread-like organ that moves in unison with other cilia to move a protozoan organism or provide with food.

CIRCADIAN From Latin *circa* + *dies* = 'around' + 'day', any biological process (e.g. metabolic or behavioural) that displays an endogenous rhythm of about 24 hours. These **circadian rhythms** are driven by a **circadian clock**, and are widely seen in all living organisms.

CIRCALITTORAL ZONE That part of the sublittoral zone below about 5 m, dominated by sessile animals.

CIRCINATE An arrangement of bud scales or young leaves in a shoot bud (see vernation), for example seen in ferns.

CIRCUMSCISSILE Fruits that shed their seed by splitting along a circumference with the top coming off like a lid.

CLADODE A flattened green stem resembling a leaf.

CLAMOUR Collective noun for rooks.

CLASS Highest subdivision of a phylum, itself subdivided into orders.

CLASSIFICATION The arrangement of (extinct and) living organisms in taxonomic groups according to observed similarities, with taxonomy (q.v.) the branch of biology concerned with classification.

CLAUSILIUM A calcareous structure found in the door snails (Clausiliidae), a spoon- or tongue-shaped 'door' which can slide down to close the shell aperture.

CLAVATE Club-shaped, often applied to insect antennae.

CLAW 1) In plants, the narrow basal part of a sepal or petal. 2) In animals, a sharp, usually curved, nail on the foot used to seize, tear, scratch, dig, etc.

CLEAVES Two halves of a deer's foot.

CLEIDONIC EGG An egg enclosed by a shell that isolates it from the surrounding environment, as laid by reptiles and birds.

CLEISTOCARPOUS 1) In mosses, having the capsule opening

irregularly without an operculum. 2) In fungi, having or forming cleistothecia.

CLEISTOGAMY Describing specialised, usually very small flowers which self-pollinate and set seed without opening.

CLEISTOTHECIUM An ascocarp that is spherical rather than the usual cup shape.

CLIMATIC CLIMAX A stable plant community in a state of dynamic equilibrium characterised as the theoretical end point of succession (q.v.) in a particular climatic region. This was formerly viewed as the only kind of climax possible (monoclimax theory), but is now seen as one of a number of possible outcomes (see polyclimax theory).

CLIMAX The final stage of succession (q.v.) in relation to a key environmental control such as climate, fire, extreme topography or human activity, stable but remaining in a state of dynamic equilibrium. See also climatic climax, plagioclimax, subclimax and polyclimax theory.

CLIMBER A plant that uses another for support.

CLINE Gradual character changes between animal or plant populations over a large area, usually in response to an environmental gradient.

CLITELLUM The saddle on an earthworm or leech, a swollen region near the head that as a viscid sac secretes mucus that facilitates copulation, later secreting a cocoon that covers the animal and into which the eggs are laid.

CLOACA A bifunctional urogenital cavity at the lower end of the alimentary system, through which urine and excreta are passed, and providing access to the reproductive system, into which sperm is passed and out of which eggs are laid. All amphibians, birds, reptiles and a few mammals (none native to Britain) have this opening.

CLONE A cell, cell product or organism that is genetically identical to the unit or individual from which it was derived. In nature this occurs through asexual reproduction.

CLOVEN HOOF A hoof split into two toes (cleaves), found on members of the mammalian order Artiodactyla, for example deer, cattle, goats, sheep and pigs. The two cleaves are homologous to the third and fourth fingers of the hand.

CLUTCH A single batch or hatch of eggs.

CLYPEUS Lowest part of an insect's face, above the labrum (q.v.).

COARSE (FISH) Freshwater fish other than salmonids, which are game fish.

COB Male swan.

COBWEB From Anglo-Saxon and Old English words *kobbe* = 'spider' + *webb* = 'to weave', an enlarged calamistrated web (q.v.) incorporating strands of proteinous silk of great strength and high adhesive quality, and of various designs.

COCOON A silk case that protects certain invertebrates – most moths (but not butterflies), bees, fleas, spiders and worms – as they change from larvae to adults.

COELOM The principal body cavity in most animals, fluid-filled, located between the intestinal canal and the body wall, providing space for the growth of internal organs and acting to absorb shock. In earthworms it helps movement by allowig muscles to contract and expand; also **coelomic fluid** is excreted as protection against drying out or predation.

COHORT A group of same-aged individuals.

COITION Alternative name for copulation.

COLD-BLOODED ANIMAL See ectotherm and poikilotherm. The term is misleading: the blood temperature of the animal varies with ambient environmental temperature, and so need not be 'cold'.

COLD STRATIFICATION Seeds of some perennial plants require cold moist conditions, giving this term, before dormancy is broken and germination is triggered.

COLEOPTILE From Greek *coleos* + *ptilon* = 'sheath' + 'feather', a sheath surrounding the apical meristem of monocotyledon seeds that protects the shoot's growing tip as it pushes through the soil to the surface.

COLLAGEN A fibrous protein providing strength and elasticity, a major component of the connective tissue of an animal.

COLLET The meeting point between stem and root.

COLONISATION The invasion of a 'new' site by **colonists** leading to the establishment of a viable population.

COLONY, COLONIAL Conspecific individuals living and reproducing in close association with, or connected to, one another. Some animals are obligatory colonial organisms, for example slime moulds, corals and honeybees.

COLUMELLA The central spiral in the shell of a gastropod, and a similar axial structure in corals, and the sporangia of some fungi, mosses and hornworts.

COMB See wattle.

COMMENSALISM A relationship between two animals of benefit to one (the commensal), for example acquiring nutrients, protection or locomotion (see phoresis), but with a neutral impact on the other (the host).

COMMENSURAL POINT See gape.

COMMUNITY A functionally interacting assemblage of plants and animals.

COMPARTMENT Area of managed woodland.

COMPETITION Interaction between individuals of the same species (**intraspecific competition**) or different species (**interspecific competition**) for a limited resource such as food or space. Males can display to compete for females in courtship, involving **intersexual competition**. Male–male combat, to acquire females, is **intrasexual competition**. **Exploitation competition** occurs where a resource is in limited supply, with the competitively superior individual or species more likely to be successful. **Interference competition** occurs where one competitor denies the other access to a limited resource by excluding it from the resource, for example by occupying the space in which the resource is located.

COMPOUND 1) In plants, of a leaf or flower composed of a number of parts. 2) In animals, where a number of individuals independently perform some vital functions but are connected to form a united colony, for example corals and bryozoans.

COMPOUND EYE An eye consisting of a number of small visual units (ommatidia), as found in insects, arachnids and crustaceans.

CONCHOLOGY Study of mollusc shells. See also malacology.

CONDUPLICATE An arrangement of bud scales or young leaves in a shoot bud (see vernation).

CONE In conifers the reproductive structure that in males (usually herbaceous and inconspicuous) carries pollen, and in females (on woody scales) the ovules which when fertilised produce seeds.

CONGENERIC Belonging to the same genus.

CONGREGATION Collective noun for plovers.

CONGRESS Collective noun for ravens.

CONIDIA The asexual, non-motile spores of a fungus.

CONK The woody fruiting body of bracket fungi.

CONKER The shelled seed of horse-chestnut, an abbreviation of the children's game 'Conquerors'.

CONNATE The condition where similar organs of a plant (e.g. sepals and petals) have united. In yellow-wort, for instance, leaves appear in opposite pairs but are fused together around the stem.

CONSERVATION See nature conservation.

CONSPECIFIC Belonging to the same species.

CONSUMER In ecology, a living thing that eats other living things. Apart from plants, which make their own food (i.e. are producers), most living things are consumers. **Primary consumers** are herbivores, **secondary consumers** predators or omnivores. See heterotroph.

CONTOUR FEATHER Any of the feathers which form the outline of an adult bird's plumage, helping in streamlining, insulation, waterproofing, etc.

CONVERGENT EVOLUTION The process by which two separate taxa evolve similar structures through adapting to similar environments and living in a similar way.

CONVOCATION Collective noun for eagles.

CONVOLUTE An arrangement of bud scales or young leaves in a shoot bud (see vernation).

CONY Rabbit and rabbit fur.

COPPER 1) A micronutrient required by plants for photosynthesis and the activation of several enzyme, assisting in the metabolism of carbohydrates and proteins; deficiency leads to chlorosis and leaf necrosis. Deficiency is problematic in wheat, oat, carrot and spinach production. 2) In animals copper is critical in the functioning of organs and metabolic processes. Copper deficiency is associated with livestock diseases: swayback in sheep and goats (lack of limb co-ordination in lambs and kids) and falling disease (sudden heart failure) in cattle.

COPPICE Woodland management in which shrubs or trees are cropped by cutting them down to a stump (stool), allowing multi-stem regrowth into poles that are harvested, usually at intervals of 12–15 years. Coppice is normally divided into compartments or coupes (q.v.), also known as fells, cants or haggs. **Coppice-with-standards** is coppice within which 'standard' trees are retained within the coupe to reach maturity and provide timber.

COPROPHAGY The eating of animal dung.

COPROPHILOUS Dung-loving; living in or growing on dung.

COPSE A small group of trees, the term being a contraction of 'coppice' dating from the late sixteenth century.

CORBICULUM The pollen basket on the hind legs of many bees, formed by stiff hairs.

CORDATE Heart-shaped.

CORIACEOUS From Latin *corium* = 'leather', leaves and other plant organs with a tough leathery texture.

CORM A swollen, flattened, underground stem covered by papery scales.

COROLLA All the petals of a flower.

CORONA A series of petal-like structures, either outgrowths from the petals or as modified stamens, for example seen in daffodils.

CORTEX The outermost layer of an organ; in plants, found between the epidermis and vascular bundles.

CORTICOLOUS Growing in or on bark, for example many species of lichen.

CORTINA A web-like veil in fungi, for instance found in *Cortinarius*.

CORYMB A raceme with the lower pedicels longer than the upper so that the flowers are all borne at more or less the same time and create a flat-topped inflorescence.

COSTA A ridge, or the central vein of a leaf.

COSTAL MARGIN 1) In vertebrates, the lower edge of the chest or thorax. 2) In insects, the front margin of a wing immediately behind the leading edge.

COTYLEDON The first leaves of a germinating plant (one in monocotyledons, two in dicotyledons, and two or more in gymnosperms), usually different in appearance from subsequent leaves.

COUMARIN Aromatic substance described as the scent of new-mown hay, found plants such as meadowsweet, woodruff and sweet vernal-grass.

COUNTERSHADING A method of camouflage with dark colour on the back (dorsal side) of a body normally exposed to daylight and light colour (often white) on the ventral side which would be shaded, giving the body a more uniform shade and a perceived lack of depth.

COUPE A clear-felled or rotationally coppiced area of woodland (also known as fells, cants or haggs). Coupes are normally at least 0.5 ha, but under the UK Woodland Assurance Standard certification standard no more than 10% of semi-natural woodlands >10 ha in size should be felled in any five-year period. Large coupes have the advantage of deterring deer from entering.

COURTSHIP Often ritualised display behaviour that prompts and facilitates mate selection (usually by the female) or that forms a bond between a male and female before mating.

COURTSHIP FEEDING An example of a nuptial gift (q.v.).

COVER Collective noun for coots.

COVERT 1) Small wood where game animals can hide. 2) A feather covering the base of a main flight or tail feather of a bird, helping to smooth airflow, or as in ear coverts concealing the ear opening.

COVEY A group (usually a family group) of ptarmigans, quails or partridges.

COXA The base of the insect leg, articulating with the body.

CRAW A bird's crop, or more generally an animal's stomach.

CREMASTER Cluster of hooks gripping a silk pad at the hind end of a butterfly or moth pupa, supporting the dependent pupa.

CRENATE, CRENULATE With blunt, rounded or scalloped teeth.

CREPUSCULAR Active at twilight.

CREST A tuft of feathers on the upper part of the head, also called a 'horn', seen for example on lapwing, skylark and hoopoe.

CRIBELLUM The silk-spinning organ possessed by some spiders.

CRITICALLY ENDANGERED With a very high risk of extinction, categorised in the IUCN Red List as the most severe conservation status for wild populations.

CROP An expanded, muscular pouch near the gullet or throat of a bird, part of the digestive tract, used to store food prior to digestion in the gizzard.

CROSS-BREEDING, CROSS-POLLINATION Allogamy; reproduction by parents of different genotypes, and in plants transfer of pollen from the anthers of one plant to the stigma of another.

CROWN 1) The totality of an individual plant's above-ground parts, including stems, leaves and reproductive structures. 2) More specifically, in trees the branches, leaves and reproductive structures extending from the trunk or main stem.

CROZIER The tightly coiled unopened leaf of a fern.

CRUSTOSE Crust-like, for example in describing some lichen and coralline algae.

CRYPSIS, CRYPTIC COLORATION From Greek *crypsis* = 'hiding', a colour or pattern using camouflage or mimicry that allows an animal to avoid detection by making it difficult to see against its background, either to facilitate predation, for example crab spider, or to reduce the chances of predation, for example ground-nesting birds and species of moth such as peppered moth.

CRYPTOBIOSIS A physiological state in which an organism's metabolic activity is reduced to an undetectable level to survive a period of adverse environmental conditions, for example freezing or desiccation. See, for example, tardigrade.

CRYPTOFAUNA The fauna of concealed habitats or micro-habitats.

CRYPTOGAM A plant that reproduces by spores, not seeds, for example algae, bryophytes and pteridophytes.

CRYPTOPHYTE A plant whose perennating buds are below ground level or the water surface.

CRYPTOZOA Small but non-microscopic invertebrates that live in darkness and under conditions of high relative humidity, for example in the wet soil underneath rocks and in leaf litter.

CUCKOO-SPIT White froth seen on stems or shoots of a variety of plants in late spring, secreted by the young of the froghopper nymph, the froth protecting their bodies from desiccation. These bugs are therefore also known as spittlebugs and cuckoo-spit insects.

CULL Selective killing of animals to reduce numbers.

CULM The hollow, jointed stem of a grass, especially that bearing the flower.

CULMEN The ridge along the top of a bird's upper mandible.

CULTIVAR A plant variety that has been selectively bred for desired properties.

CUNEATE Wedge-shaped.

CUPULE 1) In botany, in the Fagaceae, a structure protecting the fruit of an oak, beech or sweet chestnut: in oak holding the acorn in a cup, in beech and sweet chestnut completely enclosing the nut(s), splitting open when mature. The cupule is covered by scales, which in sweet chestnut take the form of sharp spines, giving the nut protection from seed predators. 2) A cup-shaped outgrowth of the thallus of some liverworts. 3) The fruiting body of certain cup fungi. 4) In entomology, a sucker on the feet of some species of fly, or a structure on the head of some aquatic beetles.

CURRAG Manx and Irish for acidic boggy ground.

CURSORIAL Adapted for running.

CUSHION PLANT A compact plant with leaves growing close to the ground, characteristically creating a hummock, as an adaptation to windy conditions on the coast (e.g. thrift) or in mountain habitats.

CUSPIDATE In leaves, abruptly narrowed to a point.

CUTICLE A thin waxy layer that protects a plant, for example from water loss, or forming the exoskeleton of an invertebrate.

CYANOBIONT See photobiont.

CYGNET A young swan, dating to late Middle English, from the Anglo-Norman French *cignet*, a diminutive of Old French *cigne* = 'swan', based on Latin *cycnus*, in turn from Greek *cycnos*.

CYME An inflorescence formed by a spiral terminating in a flower, subsequent flowers being produced below it so that the oldest flower is at the top (or centre) of the inflorescence.

CYST A wall secreted around the resting stages in the development of many unicellular organisms; a membrane or bladder enclosing a parasitic larva, maintaining it in a temporarily dormant state.

CYTOLOGY The study of plant and animal cells.

CYTOPLASM The gel-like material within a cell, excluding the nucleus. See also protoplasm.

D

DABBLING DUCK One feeding along the surface of the water or by tipping head first into the water to graze on aquatic plants, vegetation and insects, for example mallard, teal, gadwall and garganey.

DACTYLY The arrangement of digits (fingers and toes). For birds, see anisodactyl, pamprodactyly, syndactyly and zygodactyly. For ungulate mammals, see under Artiodactyla (even-toed) and Perissodactyla (odd-toed) in Part II.

DARWINISM Theory of the evolution of species by natural selection advanced by Charles Darwin (1809–82).

DECEIT Collective noun for lapwings, derived from the birds' habit of feigning injury and moving away from the nest site as anti-predator behaviour.

DECIDUOUS An often seasonal non-persistence, for example in botany, shedding leaves in autumn or petals dropping off after anthesis, and in zoology relating, for instance, to deer antlers and milk teeth.

DECOMPOSER A saprobe which, by feeding on dead matter, breaks it down into simpler substances. These are heterotrophs that facilitate the process of **decomposition**, which is a part of nutrient cycling.

DECUMBENT With leaves or stems lying along the ground (procumbent) but turning upwards or becoming erect at the tip.

DECURRENT Extended downwards, as with a leaf extending to form a wing on a plant stem, or gills running down the stem of a fungus.

DECUSSATE Arranged in alternating pairs, each pair at right angles to the next.

DEFLEXED Used of leaves, sepals or other plant organs which are markedly bent downwards.

DEFOLIATION Removal of leaves from a plant.

DEFORESTATION Removal of forest by felling or burning.

DEHISCENT Opening naturally, for example fruit that splits open to release the seeds when mature.

DELIQUESCENT GILLS Fungal gills that dissolve as they mature, releasing an inky liquid containing black spores, peculiar to the genus *Coprinus*.

DELTOID Shaped like the Greek letter *delta* (Δ).

DEMERSAL Living on or close to the bottom of the sea or a lake.

DEN (Maternity) shelter of a fox and other animals.

DENDROLOGY From Greek *dendron* = 'tree', the scientific study of trees.

DENTATE, DENTICULATE Toothed and finely toothed.

DEPOSIT FEEDING Ingestion of sediment by a benthic animal from which it extracts nutrients.

DESCENDENT MOULT The usual sequence of moult in birds in which the primary feathers are replaced from the carpal joint outwards, being much more common than the ascendant.

DESCENT Collective noun for woodpeckers.

DETERMINATE LAYING Where the female bird usually produces a fixed number of eggs in a clutch.

DETRITIVORY Feeding on organic detritus.

DEWLAP Fold of loose skin hanging from the neck or throat of an animal, for example in chickens, dogs and cattle.

DEXTRAL Right-handed or turning to the right, for example in snails which have the shell aperture on the right-hand side (dextral coil); opposite of sinistral.

DIADROMOUS, DIADROMY Of fish that spend their life cycles partially in fresh water, partially in salt water. See anadromous and catadromous.

DIALECT Local variation in the song of a bird population.

DIAPAUSE A state of dormancy or quiescence during the development of insects and other invertebrates when they are subjected to unfavourable environmental conditions.

DIASPORE A propagule (seed or spore, together with any attached tissue, for example elaiosome or samara) that mainly functions in a plant's dispersal, but in different species may also provide protection, seed positioning and orientation, moisture adsorption and regulation of seed respiration.

DIASTATAXIC Having no secondary feather corresponding to the fifth feather of the great wing coverts, examples including grebes, owls, pigeons and some geese. If a secondary is present, the arrangement is called 'eutaxic'.

DICHASIUM A cyme in which the first flower is flanked by two younger ones, which in turn are flanked by two even younger ones, and so on.

DICHOGAMY Where male and female parts of a flower mature at different times, preventing self-fertilisation.

DICHOTOMOUS BRANCHING Repeated division, each time forking into two equal branches.

DICLINOUS Having stamens and pistils in separate flowers on the same plant or on different plants.

DICOTYLEDON A flowering plant with a seed that germinates to bear two cotyledons (leaves).

DIEBACK A common symptom or name of disease, especially of woody plants, characterised by progressive death of leaves, twigs, branches or roots, starting at the tips. See, for example, ash dieback.

DIGESTION The breakdown of food (carbohydrates, proteins and fats) into soluble substances that can be absorbed into the blood.

DIGITALIN A group of stimulatory drugs obtained from foxglove leaves containing cardiac glycosides for the treatment of heart conditions such as congestive heart failure, though with a narrow therapeutic window, and an excess of digitalis product becomes toxic.

DIGITATE Leaves with finger-like lobes (c.f. palmate).

DIGITIGRADE Walking on toes and not touching the ground with the heel, for example bird, fox and cat. See also plantigrade and unguligrade.

DIMORPHISM The existence of two distinctive forms within a species, but not regarded as constituting separate subspecies.

DIOECIOUS, DIOECY In plants, having male and female flowers growing on separate plants.

DIPLOID An organism having two sets of matching chromosomes, that is with the chromosomes paired in the cell nucleus.

DISC In flowers, an outgrowth from the stamens or receptacle, often producing nectar.

DISC FLORET One of the florets in the disc-like centre of a capitulum flower head.

DISPERSAL The movement of individual animals or the propagules of plants (e.g. seed dispersal) or sessile animals, using a variety of mechanisms and strategies, and comprising three stages: departure, transfer and settlement (oikesis). **Postnatal dispersal** occurs when young birds and mammals, often just the males, (are forced to) leave their place of birth on reaching maturity, generally to reduce the chances of inbreeding.

DISPLACEMENT ACTIVITY In animal behaviour, the performance of a behaviour pattern out of the particular functional context of behaviour to which it is normally related as a result, for example of frustration or conflicting stimuli, for example non-feeding ground-pecking or 'false preening' in birds.

DISPLAY A usually stereotypical method of communication in animals, which involves either showing off certain conspicuous physical features, for example the tail of a peacock, or the red breast of a robin, and/or the ritual performance of certain actions, such as bowing or turning of the head, and/or the uttering of certain sounds, in particular bird song. Display includes threat, territorial and courtship display.

DISRUPTIVE COLORATION Skin pattern that disrupts an animal's outline when viewed against a typical background, making it more difficult to see. See also camouflage.

DISSECTED Deeply divided.

DISSIMULATION Collective noun for a group of birds.

DISTAL At the end furthest away from a point of attachment; opposite of proximal.

DISTICHOUS Leaves, flowers and other organs, arranged in two rows.

DISTRIBUTION The geographical range of a species, or the disposition of individuals in a particular area.

DIURNAL Active during the day, or at daily intervals.

DIVARICATE 1) To stretch or spread apart. 2) Of stalks and other plant organs, to diverge at a wide angle.

DIVERGENCE, DIVERGENT Where two or more organs have ends further apart than their bases.

DIVERSITY See genetic and species diversity.

DIVING BIRDS In the British avifauna these include divers, grebes, auks, cormorants and some ducks, which dive from the surface of the water, and gannets, terns and kingfishers which dive from the air.

DIVISION A major grouping in botanical classification, but in zoology falling out of favour, equivalent to (and generally being replaced by) 'phylum'.

DNA Deoxyribonucleic acid, a double helix chain which contains each living organism's genetic material.

DOE Female deer, hare or rabbit.

DOMAIN The highest taxonomic rank, based on comparison of ribosomal RNA, and divided into Bacteria, Archaea and Eukaryota.

DOMESTICATION The process of adapting wild plants and animals for human use Animals are tamed and kept by humans for food, work and other resources, and cultivated plants have characteristics that reflect human control in effecting 'improvements' in desired properties, for example nutrition, yield or attractiveness.

DOMINANT An individual or species that exerts the greatest influence on others, for example through size, abundance or aggression.

DORMANCY, DORMANT A resting condition; a period in an organism's life when growth, development, and in animals physical activity temporarily cease, minimising metabolic activity and helping to conserve energy.

DORSAL Relating to the back or upper surface of a body, opposite (referring to the lower surface) to ventral.

DOWN A layer of fine feathers found under the tougher exterior feathers of birds, and the first feather covering in the young of some species.

DOWNLAND Lowland calcareous grassland found over chalk or limestone and maintained by grazing by stock and/or rabbits.

DRAG The resistance to motion that occurs when an animal travels through air or water.

DRAKE 1) Male duck. 2) In angling, a large mayfly (*Ephemera danica*).

DREAD FLIGHT Panic flights by kittiwakes made early in the breeding season, with spontaneous, silent and synchronous flight away from nesting cliffs, probably as a form of anti-predator behaviour.

DREY Squirrel's nest, typically a mass of twigs between forked branches in a tree.

DRONE Male bee, wasp or ant that comes from an unfertilised egg, its function being to mate with the queen.

DRUMMING 1) Rapid tapping by the bill of a woodpecker on a tree-trunk or dead branch. 2) The sound of air passing through the spread tail feathers of a snipe during summer aerial display.

DRUPE A fleshy fruit with stony seed(s), for example cherry and plum.

DRUPELET The individual small drupes of some aggregate fruits, for example blackberry.

DULE Collective noun for doves.

DUN The newly emerged winged subadult (subimago) stage of a mayfly, effective at escaping the water, with a hydrophobic surface, but they are clumsy fliers. Within a day or two they have a final moult into imagos (see spinner).

DUNE A coastal hill of loose sand created by wind, more or less bound by sand couch-grass, marram and other specialist plants as part of a psammosere (q.v.). **Dune slacks** are low-nutrient, seasonally flooded habitats often found between dune ridges.

DUST BATHING, DUSTING Forcing fine, dry material (such as dust or fine soil) into the plumage by squatting on the ground and moving the body, wings and legs. Like anting and smoke bathing, dust presumably discourages ectoparasite infestation.

DWARFISM Abnormally small size in a plant (e.g. as a response to cold or windy conditions) or animal.

DYNAMIC EQUILIBRIUM A state within a community in which species and ecosystem diversity, function and structure have reached a point where they remain relatively stable (are in equilibrium), but which can respond to environmental change (behaving as allogenic factors). Climax vegetation is an example.

DYSTROPHIC Of a lake or pond with brown acidic water that is low in oxygen, especially at lower depths, and supporting little life owing to high levels of dissolved humus.

E

EARLY WOOD See spring wood.

EARTH Fox's den.

ECDYSIS The moulting process in crustaceans, insects, etc. (of the exoskeleton) or of skin (e.g. sloughing in snakes).

ECESIS See oecesis.

ECHOLOCATION A method of sensing objects by emitting high-pitched sounds then listening to the echoes that return from nearby objects. Bats, dolphins and some whales use echolocation to 'see' in the dark or in water.

ECLIPSE PLUMAGE Ducks moult all their flight feathers at the same time and for about a month cannot fly, making them vulnerable to predators. To provide some protection for the coloured drakes (males), the moult starts with their bright body feathers, replaced by drab brown ones – so-called eclipse plumage.

ECOLOGY The study of the interaction between living things and their biotic and abiotic environment.

ECOSYSTEM The community of interacting organisms and their physical environment in a given area.

ECOTONE The transition zone between two types of habitat, for example a woodland edge is an ecotone between open country and forest habitats. Containing some species from each of the contributing habitat, and commonly with species found only in the ecotone itself, ecotones are often especially species-rich (see edge effect).

ECOTYPE A genetically distinct form or race of a plant or animal species occupying and adapted to a particular habitat.

ECTOHYDRIC, ECTOHYDROUS In mosses, liverworts and lichens, taking up water from all over the leaf or thallus surface, with no internal water transport system. See endohydric.

ECTOMYCORRHIZA From the Greek *ectos* + *myces* + *rhiza* = 'outside' + 'fungus' + 'root', a mycorrhiza in which the fungal hyphae do not penetrate the root of the host plant.

ECTOPARASITE A parasite living on the outside of the host's body.

ECTOSKELETON See exoskeleton.

ECTOTHERM From Greek *ectos* + *thermos* = 'outside' + 'hot', an organism that relies on external heat sources to maintain internal body temperature. Fluctuating ambient temperatures may lead to variation in body temperature (see poikilotherm), such variation being reduced by behavioural means, for example basking or using shade. These animals are often called 'cold blooded', but blood temperature actually varies with ambient environmental temperature, so may not be 'cold'.

EDAPHIC Relating to soil.

EDAPHOLOGY The study of how soil influences plants and other living organisms.

EDGE EFFECT Where two habitats overlap, the overlap area (edge) will contain (some) species from both habitats, and often species found only in the overlap area, so will tend to be species-rich, for example woodland edge (with hedgerows as an analogue of this habitat).

EFT The juvenile terrestrial stage of a newt.

EGG The fertilised ovum and developing embryo, enclosed inside a protective membrane. When laid outside water there is an impermeable shell that protects against drying, as seen in reptiles and birds.

EGG TOOTH A small horny growth at the tip of a hatching bird's upper mandible or reptile's jaw used to break through the egg shell.

ELAIOSOME From Greek *elaion* + *soma* = 'oil' + 'body', variously shaped fleshy structures attached to the seeds of many plant species, rich in lipids and proteins, that often attract ants which take the seed to their nest as larval food, thereby dispersing the seed.

ELUVIAL, ELUVIATION In soil, movement of chemicals or material in suspension or solution in water from upper horizons to be redeposited in lower horizons (see illuvial).

ELVER Young eel.

ELYTRA (singular ELYTRON) The hard front wings on insects such as beetles and earwigs.

EMARGINATE 1) Leaf or other plant organ slightly notched at tip. 2) A feather with one web, which produces a tapering effect.

EMBRYO A very young plant or animal, still inside the mother (in mammals), its eggs (in birds and reptiles), or its seeds (in plants).

EMERGENT A plant with root in water and stems rising above (not floating on) above the water surface.

ENDANGERED At risk of extinction, categorised in the IUCN Red List as the second most severe conservation status for wild populations after Critically Endangered.

ENDEMIC Confined to a particular region, for example the Scottish crossbill is the only endemic bird in the British Isles, and Lundy cabbage is only found on that island.

ENDOCRINE GLAND A hormone-producing gland, also known as the 'ductless gland'. All glands form the 'endocrine system'.

ENDOGEIC Of earthworms, species that live in and feed on the soil, making horizontal burrows.

ENDOGENOUS Of or produced from within the body.

ENDOHYDRIC, ENDOHYDROUS In mosses and liverworts, possessing internal water-conducting tissues, efficient enough to draw water into the plant directly from the substrate. See ectohydric.

ENDOMYCORRHIZA From Greek *endos* + *myces* + *rhiza* = 'inside' + 'fungus' + 'root', a mycorrhiza in which the fungal hyphae penetrate the root of the host plant.

ENDOPARASITE A parasite living inside the host's body.

ENDOPTERYGOTE An insect in which the wings develop inside the young body, emerging during metamorphosis in a pupal stage, followed by emergence of the winged adult.

ENDOSKELETON A skeleton entirely inside in an animal's body.

ENDOTHERM An animal (bird or mammal) that can maintain a constant temperature using internal mechanisms, for example by shivering, panting and sweating. See also homoiotherm and warm-blooded.

ENDOZOOCHORY Seed or spore dispersal by being ingested and excreted by animals.

ENSIFORM Sword-shaped.

ENTIRE 1) In plants, a leaf (or organ) without lobes, teeth or other indentations in the margin. 2) In animals, an uncastrated male mammal.

ENTOMOCHORY Seed or spore dispersal by insects.

ENTOMOLOGY The scientific study of insects.

ENTOMOPHILY Pollination by insects. Plants adapted to this are entomophilous.

ENVIRONMENT An organism's surroundings, which

includes non-living matter, such as air and water, as well as other living organisms.

ENVIRONMENTAL GRADIENT A gradual spatial change in abiotic factors, for example related to altitude, temperature or water depth, with species and their abundance usually changing in parallel.

ENZYME Biological catalysts, mostly proteins, which control and accelerate the chemical reactions inside living organisms.

EPHEMERAL A short-lived plant (**ephemerophyte**); an organism completing its life cycle in a short time.

EPIBIONT, EPIBIOTIC A plant or animal growing on the surface of another living organism. See epiphyte and epizoic.

EPICALYX An additional whorl similar to, but outside, the true calyx, for example in teasel and mallows.

EPICHILE Front part of lip or labellum of certain orchids.

EPICORMIC Growing from buds on a large stem or shoot, for example new shoots growing directly from a tree-trunk.

EPIDERMIS The outermost layer of tissue, in plants and many invertebrates having a thickness of a single cell.

EPIFAUNA Benthic animals that live on the surface of the bottom sediment of a water body (with infauna living within the sediment).

EPIGAMIC A secondary sexual characteristic that serves to attract the opposite sex during courtship, for example colour or song.

EPIGEAL 1) A plant creeping or growing on or close to the ground. 2) Where germination takes place above the ground.

EPIGEIC 1) Species that live in leaf litter on the surface of the soil, for example non-burrowing earthworms. 2) An animal that cannot burrow, swim or fly, therefore always in contact with a surface.

EPIGLOTTIS A flap made of elastic cartilage covered with a mucous membrane, attached to the entrance of the larynx. In whales this closes off the blowhole.

EPIGYNE The genital opening of the female spider.

EPIGYNOUS In flowers, having the ovary enclosed in the receptacle, with the stamens and other floral parts situated above.

EPILIMNION The warm upper water layer of a thermally stratified lake, above the colder hypolimnion (q.v.).

EPINEUSTON From Greek *epi* + *neustos* = 'upon' + 'able to swim', the organisms living on a water surface.

EPIPELIC An organism that moves over the surface of mud or sediment or lives at the sediment/water interface.

EPIPHRAGM A temporary disc of calcium phosphate and mucilage with which a snail seals its shell aperture to avoid desiccation.

EPIPHYTE A plant which grows on another plant, using it for support but not for nutrition.

EPIPSAMMIC An organism that moves over the surface of sand particles or lives at the sand/water interface.

EPIZOIC An animal or plant non-parasitically living on or attached to the body of another animal, for instance many mosses and lichens, and barnacles on whales.

EPIZOOCHORE A propagule such as a spore or a seed that is carried on the body of an animal.

EPIZOOTIC A large-scale out break of disease, being the equivalent of an 'epidemic' in humans.

EROSE Leaves or other plant organs with an irregular margin, looking as if they have been bitten.

ERUCIFORM Caterpillar-like: an insect larva with a more or less cylindrical body and stumpy prolegs on the hind region, as well as true legs on the thorax.

ERUPTION Mass movement of birds, occurring at irregular intervals.

ERYTHRISM Colour variation in which black or brown-black pigment is replaced by orange, red-brown or yellow-brown pigment.

ESCAPE 1) In botany, a plant having spread outside a cultivated habitat (usually a garden) vegetatively or by seed dispersal 2) In animals, a species or individual that has escaped or been liberated from captivity.

ETHNOBOTANY The scientific study of the human uses of plants, or the relationships between people and plants.

ETHOLOGY The comparative study of the behaviour of animals.

ETIOLATION A process in flowering plants grown in partial or complete absence of light, characterised by long, weak stems, smaller leaves and a pale yellow colour (chlorosis).

EUCARPIC In fungi, having only part of the thallus transformed into a sporangium (fruiting body). See holocarpic.

EUKARYOTA The domain that consists of the kingdoms Animalia, Fungi, Plantae and Protista.

EULITTORAL ZONE Or midlittoral, the intertidal zone extending from the spring high tideline, which is rarely covered by the tide, to the spring low tideline.

EUSOCIAL Co-operative care of the young, where only one female in a group produces offspring, and all other members care for her and her offspring, for example in ants, and social bees and wasps.

EUTAXIC With a secondary feather present. See diastataxic.

EUTROPHIC Naturally or artificially rich in nutrients, used in the context of freshwater habitats. See also mesotrophic and oligotrophic.

EUTROPHICATION Nutrient enrichment of a freshwater environment, especially from human activity, for example discharge of sewage, or run-off from fertiliser-treated agricultural soil. Usually involving nitrates and, especially, phosphates the productivity of the ecosystem increases, often leading to excessive plant growth and blooms of algae, the death and decomposition of which by oxygen-using bacteria leads to oxygen depletion (hypoxia) and mortality in sensitive animal species.

EVERGREEN A plant that retains its leaves throughout the year.

EVOLUTION Change in the heritable characteristics of populations and species over successive generations, leading to new species (speciation). The mechanism driving evolution is natural selection.

EXALTATION Collective noun for larks.

EXOGENOUS Produced from outside the body.

EXOPTERYGOTE Insect such as a grasshopper without a pupal stage, in which the wings develop gradually on the outside of the body; the young insect is known as a nymph.

EXOSKELETON 1) A hard, outer covering that surrounds the body of an animal without an internal skeleton, for example in insects and crustaceans. 2) Also used in the context of the calcium carbonate covering of stoneworts (q.v.).

EXOTIC Describing an introduced species in its non-native range.

EXOZOOCHORY Seed or spore dispersal by transportation on the outside of an animal.

EXSERTED Protruding, for example stamens that protrude from the mouth of a two-lipped or tubular flower.

EXTERNAL DIGESTION When enzymes are secreted to digest food outside of an organism, which then absorbs or consumes the soluble nutrients, as undertaken by many spider species, and insects such as true flies and assassin bugs.

EXTINCTION The disappearance of a species locally, regionally or globally.

EXTRAFLORAL NECTARY A nectary not produced on a flower, the nectar being used as a reward, for example for insects whose presence deters other insects that might otherwise feed on the plant. Laurel, for instance, has nectaries located at the base of the leaf blade close to the petiole that attract bees and wasps. Bracken and guelder-rose are other examples.

EXTRALIMITAL Occurring only outside the boundaries of the area under consideration.

EXUDATE A fluid emerging from an organism, for instance pus in animals and, in plants, sap, resin and, arguably, nectar.

EXUVIA The cast-off outer coat of an arthropod after a moult.

EYAS A young hawk or falcon, especially (in falconry) an unfledged nestling taken from the nest for training.

EYE-STRIPE Distinctively coloured stripe of feathers or fur leading back from or through the eye.

EYRIE The nest or nest site of a raptor such as the golden eagle.

F

FACET One of the hexagonal units making up a compound eye, for example in insects.

FACIAL DISC The slightly concave forward-facing part of the head seen in some owls, notably barn owl.

FACULTATIVE Capable of existing under different environmental conditions or by assuming various behaviours, or optionally living under particular conditions, in contrast to obligative behaviour.

FAECAL SAC A white mucous membrane serving as an 'envelope' in which the faeces of nestling birds are enclosed, easily removed from the nest by a parent.

FAIRY RING A circle or arc of fungi, usually in grassland but also in woodland, and often associated with a necrotic zone (dead grass). The fungi infect grass roots then expand from this centre point as the nutrients are exhausted, but eventually stabilising in size. Rings found in woods are 'tethered', because they are formed by mycorrhizal fungi symbiotic with trees; those in grassland are 'free' because they are not connected with other organisms.

FALCATE Sickle- or scythe-shaped.

FALL Collective noun for woodcock.

FAMILY In taxonomic classification, the rank between order and genus.

FANG A long, sharp tooth. Venomous snakes have hollow fangs that inject venom into their prey. Carnivorous mammals normally have two pairs of fangs (canines). Spiders have external fangs (chelicerae).

FASCICLE 1) In plants, a bundle or bunch of leaves or branches. 2) In animals, a bundle of nerve or muscle fibres.

FASTIGIATE A tree with nearly vertical branches, for example Lombardy poplar.

FAT A family of energy-rich chemicals that do not mix with water. The group includes solids (fats) and liquids (oils).

FAUNA A collective term for the animals of a given place or time (c.f. flora for plants). See also avifauna, epifauna, ichthyofauna, infauna, macrofauna, megafauna, meiofauna and mesofauna.

FAUNISTICS The taxonomic classification of animals.

FAWN First-year young deer.

FEATHER Keratin outgrowths from the skin of a bird that form the outer covering (plumage), providing insulation and waterproofing, and helping flight. Vaned feathers (including contour feathers) have a central shaft (rachis) with barbs on either side. Under these lie the down feathers which provide thermal insulation. Feather colour and pattern aid in camouflage and display. See also calamus.

FEATHERED MAIDEN (TREE) See maiden.

FECUNDITY The maximum possible number of offspring (maximum capacity) of an individual animal or population, measured as the number of eggs produced in a female's lifetime. The actual number of offspring is fertility (q.v.).

FELINE Animal of the cat family.

FELL See coupe.

FEMUR Thigh bone; in insects and spiders, the third and usually largest leg segment.

FEN A tract of low marshy ground containing peat, relatively rich in mineral salts, neutral or alkaline in reaction.

FEN CARR Wet woodland growing over peat, particularly found in East Anglia.

FERAL Domesticated species, populations or individuals where they or their antecedents have reverted to the wild.

FERTILITY 1) the ability to produce offspring. 2) The actual number of offspring that an individual animal or population produces in its lifetime. See also fecundity.

FIBRILLOSE With thread-like fibres.

FIBROUS ROOT Root system with no main axis, unlike a tap root.

FIBULA The calf bone, the smaller bone in the lower leg (hind leg in tetrapods).

FIELD LAYER A 'layer' of herbs and growing trees and shrubs, in a woodland found between the ground and shrub layers.

FILAMENT 1) The 'stalk' of a stamen. 2) Algal cells that are joined end-to-end. 3) A long protein chain such as in hair or muscle.

FILIFORM Thread-like.

FILTER-FEEDING Feeding by sieving suspended food particles from the water.

FIMBRIATE Leaves and other organs with fringed margins; in fungi, with prominent projecting hairs on the cap margin.

FIMICULOUS Growing on and feeding in dung.

FISTULA Stems and stalks that are tubular and hollow.

FITNESS Or **Darwinian fitness**, the balance between inherited advantages and disadvantages that determines an individual's survival and reproductive capacity, usually measured by the number of offspring or close kin that themselves survive to reproductive age. This is Darwin's 'survival of the fittest' as a mechanism of natural selection.

FIXED ACTION PATTERN An instinctive behavioural sequence in response to an external sensory stimulus (sign stimulus or releaser). These are complex, species-specific, stereotypical and independent of experience.

FLAGELLUM 1) A long, whip-like projection on a cell that beats to make the cell move. Sperm cells use flagella to swim. 2) In insects the part of the antenna furthest from the head, beyond the second segment.

FLEDGE To rear a young bird to the point that its wing feathers are sufficiently developed for flight.

FLEDGLING A young bird that has grown feathers and able to leave the nest.

FLEHMEN From the German for 'to bare the (upper) teeth', where a mammal curls back its upper lip and inhales through its mouth, holding this for several seconds and facilitating the transfer of pheromones and other scents into the vomeronasal organ (or Jacobson's organ) located above the roof of the mouth. The main function seems to be the acquisition of intraspecific information.

FLIGHT 1) Action of flying. 2) Collective noun for cormorants, doves, goshawks, pigeons and swallows.

FLOCCOSE Covered with soft white hairs to give a woolly appearance.

FLOCK 1) A group of birds, of the same species or in a mixed flock, that come together for example to feed, for safety or to fly. 2) Collective noun for pigeons and swifts, sheep and goats.

FLOOD MEADOW See meadow.

FLORA A collective term for the plants of given time or place (c.f. fauna for animals).

FLORAL FORMULA A representation of the structure of

a flower using numbers, letters and various symbols, presenting information about the flower in a compact manner.

FLORET A small flower which forms part of a compound head as in the family Asteraceae, or in a compact or open inflorescence as in sedges or grasses. See disc and ray florets.

FLORISTICS The branch of phytogeography concerned with the types, numbers and distribution of plant species in a particular area.

FLORULA Flora of small areas, for instance of a single crag or pond.

FLOW COUNTRY An area of peatland and wetland in Caithness and Sutherland covering about 4,000 km², the largest expanse of blanket bog in Europe. Much of the area was damaged from 1979 to 1987 from investment-orientated coniferous afforestation together with extensive drainage.

FLOWER The sexual reproductive structure in angiosperms.

FLOWER HEAD Or pseudanthium, a compact mass of flowers at the top of a stem that, as a group, resembles a single flower, for example in members of the Asteraceae.

FLUSH 1) To cause a bird or any game animal to take flight. 2) Collective noun for mallards. 3) In plants, a growth spurt.

FOETUS The embryo of a mammal in the later stages of development.

FOLIAGE The mass of leaves on a plant.

FOLIOSE Leaf-like.

FOLLICLE 1) In plants, a dry, usually multi-seeded fruit which splits open on one side only. 2) In animals, a small spherical group of cells containing a cavity, for example hair follicles.

FOOD CHAIN Organisation of food passing along a chain of living things, reflecting the transfer of energy from one trophic level to the next. In a simple food chain, food passes from a plant to a herbivore to a predator.

FOOD WEB A diagram, essentially a set of interlinked food chains, that shows how food, and therefore energy, passes from one living thing to another in the ecosystem, from primary producers (plants) to primary consumers (herbivores) to secondary consumers (e.g. predators).

FORAGING 1) Animal behaviour associated with searching for, acquiring and consuming food. 2) In plants, vegetative reproduction by the production of runners.

FORB A non-woody dicotyledonous plant.

FORCIPATE Forked or pincer-like.

FOREST In general use, synonymous with woodland, often where the trees are planted and managed for commercial purposes. In Norman law, royal forest was land owned by the monarch to be used for hunting and with strict laws of usage, and included non-wooded areas, for example the extensive areas of heathland in the New Forest.

FORESTRY The generally commercial growing of timber. See silviculture.

FORM A shallow depression in the ground used by a hare for shelter.

FOSSIL The remains or traces of an animal or a plant that has been preserved in rock.

FOSSORIAL Animals that dig, and are adapted for such behaviour, for example moles digging burrows, and birds such as kingfisher and puffin.

FRASS The powdery wood produced by the activity of boring insects, or their excrement.

FRAYING Damage to trees by deer cleaning velvet off of their antlers.

FREE GILLS Fungal gills with no stem attachment.

FROND The compound leaf of a fern or palm.

FRUGIVOROUS, FRUGIVORY Fruit-eating.

FRUIT The ripe, fertilised ovary that contains one or a number of seeds.

FRUITING BODY Part of a fungus that supports the cells needed for sexual reproduction (i.e. that produces spores).

FRY Trout and salmon hatch as alevins; once the yolk has been eaten, alevins become fry, emerge from the gravel redd, move towards the light and start to feed on tiny insects in the water, then grow into the next stage, parr.

FUGACIOUS Petals, leaves and other organs which fall soon after they are fully developed.

FULMAR OIL In fulmars, a stomach oil made of wax esters and triglycerides which can be sprayed out of the mouth as a defence against predators and as an energy-rich food source for chicks.

FUNDAMENTAL NICHE The niche (q.v.) that a species would theoretically occupy in the absence of competition for resources; the realised niche (q.v.) is found when competition exists.

FURCULA 1) The pectoral girdle (wishbone) of a bird. 2) The forked appendage folded beneath the body at the end of the abdomen in a springtail which, when tension is released, allows the animal to jump.

FUSIFORM Spindle-shaped; elongated, tapering at both ends.

G

GAGGLE A group of geese on the ground, a flying flock being called a 'skein'.

GALLINACEOUS Belonging to the order Galliformes (grouse, pheasants, etc.).

GAME Of salmonid fish (other freshwater fish are coarse fish) and some birds and mammals, hunted, caught or shot for sport or food.

GAMETOPHYTE In plants, the haploid generation that bears the sex organs that produce the gametes: in angiosperms and gymnosperms, the pollen (male) and embryo sac (female); in pteridophytes, the prothallus (q.v.).

GAMOPETALOUS Of petals which are fused completely or at the base, so creating a corolla tube.

GANDER Male goose.

GANG The collective noun for stoats.

GAPE In the vertebrate mouth, the space between the upper mandible and the lower. The angle of the gape is called the commensural point.

GASTER The part of the abdomen in Hymenoptera such as wasps that lies behind the 'waist'.

GEKKERING Chattering vocalisation made by fox cubs, often accompanying play-fighting or excitement.

GENE The basic unit of heredity, passed on from parent to offspring. A gene is made up of a length of DNA inside a cell. Each gene controls the assembly of a particular protein. The **gene pool** is the total gene complement in a population. **Gene flow** is the movement of genes within an interbreeding group.

GENETIC DIVERSITY The total number of genetic characteristics in a species' genetic make-up.

GENETICS The study of genes and heredity.

GENICULATE Bent like a knee.

GENOTYPE The genetic constitution of an organism. See also phenotype.

GENTLE Angler's term for maggot.

GENUS (plural GENERA) In taxonomy, the rank between family and species. The genus name is incorporated into the scientific name of all species, for example *Turdus* in *Turdus merula* (blackbird).

GEOCOLE An organism that spends a part of its life cycle in the soil.

GEOGRAPHICAL RANGE Or **geographical distribution**, the area within which a species occurs.

GEOPHYTE A perennial plant whose perennating buds overwinter beneath the surface of the soil.

GEOTROPISM See tropism.

GERMINATION The process in which a seed or similar structure starts to develop as a response to water, oxygen and warmth (though cold stratification may also be a requirement).

GESTATION (PERIOD) Pregnancy (carrying an embryo or foetus) and its length.

GILL 1) In fish, aquatic insect larvae, etc., a breathing organ, often containing feathery structures that extract oxygen from water for the purpose of respiration. 2) In fungi, a blade-like structure, usually hanging downwards, on which spores are produced.

GIRTH Measurement of the circumference of a tree, in the UK at 1.3 m above the ground (1.4 m in the USA).

GIZZARD A chamber in an animal's alimentary canal where food may be ground up, often facilitated by grit and small stones swallowed for this purpose.

GLABROUS Smooth and without hairs.

GLADE A woodland clearing.

GLAND An organ in an animal's body that synthesises substances such as hormones for release into the bloodstream (**endocrine gland**), for example pituitary and adrenal glands, or into cavities inside the body or its outer surface (**exocrine gland**), for example sweat and oil glands. Glands are also used by mammals in scent (q.v.) marking.

GLAUCOUS From Greek *glaucos* = 'grey', 'silvery' or 'gleaming', a grey-blue, grey-green or whitish colour; in plants, often seen as a waxy film on the surface of a leaf or stem of this colour, serving to reduce water loss.

GLEANING A feeding method in birds by which they catch invertebrate prey by plucking them from leaves or the ground, from crevices, and, for example with ticks and lice, from living animals.

GLEBA The fertile tissue in the fungal group that includes puffballs, stinkhorns and earthstars.

GLEY A sticky, waterlogged soil lacking oxygen.

GLIDING Or **soaring**, the ability of some birds, especially raptors and seabirds, to use air movement to fly without wing flapping, thus saving energy, the wings deflecting air downward to cause a lift force that supports the bird in the air.

GLUCOSE A type of sugar as a product of photosynthesis and used by most living organisms for energy.

GLUME A small bract at the base of a grass or sedge flower.

GONAD A gland in which sperm or eggs are created.

GONYS A ridge along the tip of the lower mandible of a bird's bill at the junction of the two joined halves, especially evident in gulls.

GOSLING Young goose.

GOSSAMER Filmy cobweb.

GRAFTING Artificial implantation of the aerial parts of one plant with that of another, usually a vigorous rootstock.

GRAMINIVOROUS Feeding on grass.

GRAMINOID Grass-like.

GRANIVORY Feeding on grain or seeds.

GRASSLAND Vegetation characterised by grasses and forbs. Managed grassland includes pasture (used for grazing livestock) and meadow (used for cutting hay). See also downland.

GRAVID Pregnant; carrying eggs or developing young.

GRAZE, GRAZING To eat plants, usually grass or other low-growing vegetation. See also browse and herbivore.

GRILSE A salmon which has spent one winter at sea before returning to its natal river.

GROUND LAYER A 'layer' of low-growing plants, including mosses, in a woodland found between soil/litter and the field layer.

GROVE A small group of trees with little or no undergrowth.

GROWTH RING The rings inside a tree, and also evident in shells and some corals, indicative of annual or other periodic changes in growth rate.

GRUB An immature (larval) insect, usually referring to beetles.

GUANO Accumulated dried bird and bat droppings, usually seabird manure, often collected as a phosphate-rich fertiliser.

GUILD A group of species using similar techniques to exploit similar resources, for example feeding guilds such as insectivores or detritivores.

GULAR Relating to or situated on the throat of an animal, especially a bird, reptile or fish.

GULAR FLUTTERING A means of thermoregulation in birds, counteracting overheating by rapidly vibrating the gular skin.

GULAR POUCH An extensible throat sac used for short-term storage of food, for instance seen in rooks and cormorants.

GULLET The muscular passage through which food passes from mouth to stomach (the oesophagus).

GYMNOTHECIUM An ascocarp that is globose or pear-shaped rather than the usual cup shape.

GYNANDROMORPHIC With both male and female characteristics.

GYNODIOECIOUS With female and bisexual (hermaphrodite) flowers on separate plants.

GYNOECIUM All of the female parts of a plant; all the carpels.

GYNOMONOECIOUS With female and bisexual (hermaphrodite) flowers occurring separately on the same monoecious plant.

GYPSOBELUM See love dart.

H

HABITAT The area (or arena) used by a particular species for its survival, defined in terms of its biotic and abiotic environmental characteristics.

HABITUATION A form of learning with the reduction or disappearance of an innate response to a frequently repeated stimulus when that stimulus experientially has no consequence, for example animals being initially frightened by but becoming habituated to human presence.

HAEMOGLOBIN A red protein responsible for transporting oxygen in the blood of vertebrates.

HAEMOLYMPH A fluid equivalent to blood in most invertebrates.

HALLUX The innermost digit of the hind foot of vertebrates.

HALOPHYTE A plant adapted to tolerate salty conditions.

HALOSERE Plant succession in a saline environment such as saltmarsh.

HALTERES Small, club-shaped organs (modified hind wings) in flies functioning as gyroscopes, maintaining balance by informing the insect about rotation of the body during flight.

HAMULUS A small hook or hook-like projection, especially one linking the fore- and hindwings of a bee or wasp.

HANGER Generally beech-dominated chalk woodland in southern England.

HAPLOID With a single set of chromosomes in each cell nucleus.

HARDWOOD The timber of a broad-leaved tree, or the tree itself.

HARDY Tolerant of adverse climatic conditions, usually relating to low temperature.

HAREM A group of female animals sharing and controlled by a single male.

HASTATE Shaped like a spear head, with two lower lobes.

HATCH 1) Breakage of an egg to allow the emergence of a chick or young reptile. 2) The mass emergence of adult mayflies.

HAUL-OUT Behaviour associated with seals (and other pinnipeds), temporarily moving from sea to shore for purposes of breeding (giving birth, mating, male fighting) or rest.

HAULM The stem of a cultivated plant after harvesting, often used for animal bedding.

HAUSTELLATE Adapted for sucking liquids rather than biting solids.

HAUSTELLUM The sucking organ or proboscis of an insect or crustacean.

HAUSTORIUM 1) The modified root or stem of a parasitic plant such as dodder or mistletoe 2) The outgrowth from a fungal hypha that penetrates a host plant to extract nutrients.

HAW Fruit of the hawthorn.

HAWKING Catching prey in the air.

HAY Grass or other herbaceous plants that have been cut, dried and stored, particularly for use as winter fodder.

HEARTWOOD Dead wood consisting of several annual rings at centre of tree-trunk or branch, no longer water-conducting tissue but providing structural support.

HEATHLAND A plant community, often a plagioclimax, dominated by low-growing shrubs of heather, heaths and gorse, associated with free-draining, nutrient-poor, acidic soils (generally podzolic).

HEAVY METAL A metal of relatively high density or of high relative atomic weight which may be intrinsically toxic, for example lead and mercury, or become so at high concentrations even if essential at trace levels, for example copper, iron and zinc.

HEDGE(ROW) A line of closely spaced shrubs and tree species, planted and trained to form a barrier to stock or to mark a boundary between fields or bordering a track or roadside, an ecologically important ecotone habitat that can serve as a wildlife corridor.

HELIOTROPE A flower or sessile animal that moves or grows towards sunlight.

HELOPHYTE A macrophyte in which the perennating bud lies in mud under water, but with emergent stems and flowers, evident in many marsh species.

HELOTISM A symbiotic relation of plants or animals in which one organism functions as a 'slave' (= helot) to the other, that is, forced to perform tasks for another but for mutual benefit, for example an alga and fungus in lichen, or with slave-making ants using black ants.

HEMELYTRON A forewing of a bug (heteropteran) having a leathery basal portion and a thinner membranous tip.

HEMICRYPTOPHYTE A plant with its overwintering perennating buds at ground level, with above-ground stems that die back.

HEMIMETABOLOUS In insects, having an incomplete metamorphosis, with developing young (nymphs) but no pupal stage, for example dragonflies and grasshoppers; the opposite of holometabolous.

HEMIPARASITE A plant without (or with poorly developed) roots but which contains chlorophyll via haustoria using a host for support, water and nutrients other than carbon.

HEPATIC 1) Relating to the liver. 2) A liverwort.

HERB, HERBACEOUS 1) (Relating to) a flowering plant with non-woody stems, usually dying down in autumn. 2) A plant used in cooking, pharmacology/medicine or perfumery.

HERBICIDE A chemical compound toxic to plants.

HERBIVORE An animal that feeds on plants.

HERD 1) A large group of (especially hoofed) mammals that live together or are kept together as livestock. 2) Collective noun for swans, cranes and wrens.

HERMAPHRODITE An animal or plant that has both male and female reproductive organs.

HERONRY A breeding colony of herons, characteristically in trees.

HERPETOLOGY From Greek *herpein* = 'to creep', the study of reptiles and amphibians.

HETEROBLASTY A series of morphological changes during a plant's development, especially seen with juvenile, transitional and mature leaves. See also heterophylly. The opposite is homoblasty.

HETEROCHROSIS In birds, an abnormal plumage coloration, for example albinism, melanism, intensification of the normal colours, or rarely the introduction of other colours.

HETEROGAMOUS Having two different types of flower on the same plant, for example in the capitulum of some Asteraceae the disc florets are bisexual, the ray florets female.

HETEROPHYLLY Where different plants of the same species have differently shaped leaves. In **seasonal heterophylly**, early and late growth in a season are morphologically different. The opposite of isophylly.

HETEROSPORY Seen in all gymnosperms and angiosperms, and some pteridophytes, with two kinds of spore – female megaspores and male microspores.

HETEROSTYLY A polymorphism in which different plants of the same species have two forms of flower, for example with anthers and styles of different lengths, ensuring cross-pollination.

HETEROTROPH An organism (primary or secondary consumer) that cannot manufacture its own food and obtains its food and energy by taking in living or dead organic substances. All animals, protozoans and fungi, and most bacteria, are heterotrophs.

HIBERNATION State of rest of a mammal during winter in which the metabolic and respiration rates are reduced, and body temperature greatly lowered. Hibernation may be interrupted repeatedly. British hibernators are hedgehogs, bats and dormice. The amphibian and reptile equivalent is brumation (q.v.).

HILUM A scar on a seed marking its point of attachment to the plant. On a bean seed, for example, this is called the 'eye'.

HISPID Having large, coarse hairs or bristles.

HISTOLOGY The study of cells or tissues.

HOLDFAST Rootlike structure that anchors aquatic sessile organisms, for example seaweeds, stalked crinoids, benthic cnidarians, and sponges, to the substrate.

HOLOCARPIC Of a fungus in which the whole thallus divides to form one or several sporangia (fruiting bodies). See eucarpic.

HOLOMETABOLOUS In insects, having a complete metamorphosis from juvenile to adult via a pupal stage, the opposite of hemimetabolous.

HOLOMICTIC LAKE One in which, at least once a year, there is a physical mixing of the surface and the deep waters. See also meromictic lake.

HOLOPARASITE An obligately parasitic plant, totally dependent on a host.

HOLT 1) An otter burrow or den. 2) A wood or wooded hill.

HOME RANGE The area within which an animal lives, moves and finds food and other resources, in some species an identical area to a defended territory (q.v.), in many species a rather larger area.

HOMOBLASTY An absence of morphological changes during a plant's development, especially in its leaves. The opposite is heteroblasty.

HOMOIOTHERM, HOMEOTHERM Generally, an endotherm (bird or mammal) that maintains its body temperature at a more or less constant level (thermoregulation), usually above that of the ambient temperature, by its metabolic activity. Some homoiotherms (e.g. lizards) are also ectotherms, maintaining a constant body temperature through behavioural mechanisms alone (behavioural thermoregulation). The opposite is poikilotherm.

HOMOLOGOUS 1) Of organs, similar in position, structure and evolutionary origin but not necessarily in function, for example a bird wing and a fox foreleg, or a human arm and a seal flipper. 2) Of chromosomes, with identical linear sequences of genes.

HOMOSPORY Having all spores of the same sort, the case with most pteridophytes.

HOMOSTYLY Not heterostylous (q.v.).

HONEYDEW A sticky, sugary excretion of some sap-sucking aphids, bugs and scale insects. To obtain sufficient protein from the sap these insects must excrete a large amount of surplus sugar.

HOOD An area of contrasting plumage covering much of a bird's head.

HORIZON A layer in a soil profile.

HORMONE An internally secreted chemical produced by an endocrine gland which triggers changes in body functions, regulating physiology and behaviour.

HOST 1) An organism that harbours a parasite, or a mutual or a commensal symbiont, typically providing nourishment and shelter. In parasitism, a **primary host** is one in which the parasite reaches maturity and reproduces sexually. A **secondary host** or **intermediate host** is a one that harbours the parasite for a transitional period during which some developmental stage is usually completed. 2) Collective noun for sparrows.

HOST-SPECIFIC Living only on or in one particular species of host.

HUMUS Amorphous organic matter in soil formed by the decomposition of leaves and other plant material by soil microorganisms.

HYALINE Transparent, translucent or glassy.

HYBRID Offspring of two plants or animals (cross fertilisation) of different but closely related species.

HYDROCHORY Dispersal of seeds or spores by water.

HYDROID 1) A type of cell found in some bryophytes that transport water and minerals taken from the soil. 2) Animals in the class Hydrozoa.

HYDROLOGY The study of the movement of water, especially fresh water in a natural state (in lakes, rivers, etc.), and in groundwater.

HYDROPHYTE A plant that is submerged in or floating on water.

HYDROSERE In plant succession a series of stages initially characterised by aquatic plant communities, from colonisation of open water, through marsh and swamp, then if the substrate continues to dry out, into woodland, all via autogenic processes.

HYDROTAXIS, HYDROTROPISM Movement or growth of an organism (or organ) towards or away from water.

HYGROPHANOUS In fungi, usually referring to the caps of agarics drying from the centre.

HYGROPHILY Inhabiting wet or moist environments.

HYMENIUM In fungi, the tissue layer in the fruiting body that contains cells that will develop into spore-bearing structures (asci or basidia).

HYPANTHIUM In plants, an enlargement of the receptacle creating a cup-shaped or tubular structure in which the sepals, petals and stamens are fused at the base to surround the gynoecium and fruit, and diagnostic of families such as the Rosaceae, Grossulariaceae and Fabaceae.

HYPERMETAMORPHOSIS With the Greek prefix *hyper* = 'above' or 'beyond', a life history with two or more different kinds of larva, for example oil beetles: the young louse-like stages of British examples parasitise solitary bees, are carried to the host's nests, detach themselves, eat bee eggs, pollen and nectar, and undergo a series of moults at each of which becoming increasingly grub-like, with soft bodies and smaller legs, before emerging as adults.

HYPERPARASITE A parasite whose host is another species of parasite.

HYPHA(E) The branching threads that make up the mycelium of a fungus.

HYPOCHILE From Greek *hypo* + *cheilos* = 'beneath' or 'below' + 'margin' or 'brim', the rear part of the lip or labellum of the flowers of certain orchids.

HYPOCOTYL Incorporating Greek *cotyle* = 'cup-shaped', in plants that part of the stem of an embryo beneath the stalks of the seed leaves (cotyledons) and directly above the root.

HYPOGEAL Underground; in germination, with the cotyledons emerging below the ground.

HYPOLIMNION The cold lower layer of water in a stratified lake, below the warm epilimnion (q.v.).

HYPONEUSTON From Greek *hypo* + *neustos* = 'beneath' + 'able to swim', organisms that live on the underside of a water surface.

HYPONOME See mantle.

HYPOPHLOEODAL Incorporating Greek *phloios* = 'bark', organisms living beneath tree bark.

HYPOXIA Oxygen deficiency.

I

ICHTHYOFAUNA The fish fauna of a particular area.

ICHTHYOLOGY The scientific study of fishes.

ILLUVIAL, ILLUVIUM In soil, movement of chemicals or material in suspension or solution in water eluviation (see eluvial) to be redeposited in a lower horizon, for example iron sesquioxide redeposited as an iron pan.

IMAGO (plural IMAGINES) The adult stage in the life history of a (usually winged) insect.

IMBRICATE Overlapping like roof tiles.

IMMATURITY A juvenile or non-adult stage; not fully grown or mature. A recently fledged bird will have juvenile plumage.

IMMUNITY An inherited or acquired resistance to a pathogen.

IMPARIPINNATE Leaf with the leaflets paired except for a single terminal one.

IMPERFECT FLOWER One that has a single set of reproductive organs, that is, is either male or female.

IMPRINTING A form of learning in which a very young animal fixes its attention on the first object with which it has visual, auditory, tactile or olfactory experience, and subsequently follows that object. In nature the object is usually a parent, so this has adaptive value. It is particularly evident in precocial birds, for example ducks and geese, but has also been found in mammals such as sheep and goats, and fishes (where olfactory imprinting leads to catadromy).

INBREEDING Mating between close relatives, especially over a number of generations, often leading to **inbreeding depression** – a decline in vigour or biological fitness because of the expression of recessive (non-dominant) genes.

INCISOR The front teeth in most mammals, with a cutting edge. Rodents are gnawing animals, and only the front surface of the chisel-like incisors has enamel, the back being softer dentine. This allows the continuously growing rootless incisors to remain sharp as they are worn down.

INCLUDED FLOWER A flower where the corolla encloses the stamens.

INCUBATION 1) in birds, the process of sitting on eggs to keep them warm, and sometimes to regulate humidity. Egg thermoregulation is also seen in some reptiles, often by laying in a warm substrate such as compost. 2) The medical term for the time between being exposed to infection and showing first symptoms.

INDICATOR SPECIES A species with a narrow range of tolerance for a particular environmental factor, or a set of these, so its presence or absence is indicative of these conditions.

INDIGENOUS Organisms that are native to the geographical area under consideration.

INDUMENTUM The hairs or down on a plant.

INDUSIUM The covering of the heaps of spores on the underside of a fern frond.

INFAUNA Benthic animals that live within the bottom sediment of a water body (with epifauna living on the sediment).

INFLORESCENCE The arrangement of a group of flowers on a plant, including the associated bracts and bracteoles (see both).

INFRALITTORAL ZONE That part of the sublittoral to a depth of around 5 m below low tide, characterised by macroalgae.

INGLUVIES A dilation or pouch in the oesophagus of certain animals that receives food prior to the main stomach, especially a bird's craw or the first stomach of a ruminant, but also found in some insects (e.g. in honeybees this is the fermentation site where nectar is processed into honey) and molluscs.

INQUILINE An animal that lives commensally in the dwelling-place of another animal.

INSECTIVORE 1) An insect-eating animal or a carnivorous plant. 2) As Insectivora, the now generally disused order name for shrews, hedgehog and mole, now placed in the Eulipotyphla.

INSTAR A stage between moults or moult and pupation of an insect nymph or larva, or the equivalent in other invertebrates.

INSTINCT Inborn (non-learned) complex but generally fixed patterns of behaviour in animals as a response to particular stimuli.

INSULATION Reduction of heat loss by a body layer such as fur, blubber or feathers.

INTERCALARY MERISTEM In grasses, the section where the blade and sheath meet, an area of cell division where new growth begins.

INTERNODE Section of stem between two nodes or leaf junctions.

INTERSEXUAL COMPETITION See competition.

INTERSEXUAL SELECTION See sexual selection.

INTERSPECIFIC Occurring between two or more species.

INTERTIDAL The littoral zone of the seashore exposed between high spring tide and low spring tide.

INTRASEXUAL COMPETITION See competition.

INTRASEXUAL SELECTION See sexual selection.

INTRASPECIFIC Occurring within a species or between individuals of a single species.

INTRAVAGINAL 1) In grasses, within a sheath, building shoot density in the immediate vicinity of the parent shoots. 2) In animals, within the vagina.

INTRODUCTION Non-native species brought into a country or region by deliberate or accidental human activity, subsequently established and naturalised.

INTROGRESSION The acquisition by one species or population of genetic characters originating in another closely related one, usually by means of hybridisation or repeated back-crossing, an example being Spanish bluebell hybridising with the native bluebell.

INVASION The sudden mass arrival of a species, usually birds or insects, not usually seen in an area.

INVASIVE SPECIES A particularly successful introduction that has spread to the extent that it causes change or damage to the environment.

INVERTEBRATE An animal without a backbone or vertebrate column.

INVOLUCEL A secondary involucre, or ring of small leaves or bracts at the base of each segment of a cluster.

INVOLUCRE Whorl or rosette of leaf-like bracts beneath an inflorescence or capitulum, as in the Asteraceae. In the Fagaceae, this is a term sometimes misused for the cupule surrounding developing nuts.

INVOLUTE An arrangement of bud scales or young leaves in a shoot bud (see vernation), or of other organs.

IRIDESCENCE Shimmering colours caused by the diffraction of light as the angle of view or the angle of illumination changes, evident for example on some butterfly wings, beetle carapaces and bird plumage.

IRIDOCYTES A cell that occurs in the skin of some fishes and reptiles, appears iridescent from containing guanine.

IRON An essential micronutrient, with iron proteins found in nearly all living organisms. The colour of blood in vertebrates is due to the iron-containing, oxygen-transporting protein haemoglobin. Haemoglobin and haemoglobin-like molecules are also found in many invertebrates, fungi, and plants.

IRREGULAR Describes flowers which are bilaterally symmetrical – can be cut into two halves in only one plane; zygomorphic.

IRRUPTION A sudden, rapid increase in population number, often leading to long-distance mass movement.

ISOCHRONE, ISOCHRONAL LINE A line drawn on a map connecting points at which something occurs or arrives at the same time, for example a line drawn across a migration map joining up those places where arrival and departure dates for a particular species are the same, showing the progress of migration across a particular region.

ISOMEROUS Having parts that are similar in number or position, for example when the different parts of a flower are the same in number, as in five petals and five sepals.

ISOPHYLLY Having leaves that are all the same shape on the same plant; the opposite of heterophylly.

ITEROPARITY Having more than one reproductive cycle in an organism's lifetime.

J

JACOBSON'S ORGAN Vomeronasal organ. See flehmen.

JIZZ The overall impression which a bird gives an observer, incorporating such features as shape, posture, flying style, other movements, size and colour, enabling an experienced bird-watcher at least to suspect its identity.

JUGGING The roosting of a covey of partridges, the group in the roost being called the 'jug'.

JUVENILE Young or pre-adult; a young bird in its first covering of feathers (juvenile plumage), which have replaced its initial down.

K

KEEL A sharply ridged structure or longitudinal ridge on an organ; a flat, sharp extension along part of the body formed by hairs (e.g. water shrew tail) or skin (along the outer side of the spur in some bats). Also in plants, the two fused lower petals of members of the pea family. Also see sternum.

KERATIN The fibrous protein that makes up the outer layer of skin, hair, fur, feathers, claws and hooves.

KERATINOPHILIC Growing on feathers, for example some fungi.

KERNEL The softer, usually edible part of a nut, seed or fruit stone contained inside its shell.

KEY A winged achene; see samara.

KID A young goat or roe deer.

KIDNEY-SHAPED Leaves with a single leaf blade that are either lobed or unlobed, with or without a tooth margin.

KINGDOM The taxonomic rank between the highest, domain, and phylum. The UK distinguishes five kingdoms: Animalia, Plantae, Fungi, Protista (or Protoctista, equivalent to the Protista together with their multicellular descendants) and Monera (Prokaryota). The USA identifies six, with Monera replaced by Archaea/Archaeabactera and Bacteria/Eubacteria. Some recent classifications have abandoned the term 'kingdom' because they are not monophyletic (not consisting of all the descendants of a common ancestor).

KIT 1) The young of some mammals, for example wildcat, rabbit, mink and beaver. 2) A group of flying pigeons.

KLEPTOPARASITISM The stealing of food (including stored food) by one species from another, also known as 'piracy' or 'food piracy'. Skuas are notable examples. The term also covers stealing nest material.

KNOB 1) A **basal knob** is a prominent bump, swelling or projection at the base of a bird's bill, typically positioned on the upper mandible, larger in males, probably serving as an indicator of health or sexual maturity, particularly during the mating season. It is evident on mute swans, shelducks, king eider and black scoters. 2) Collective noun for wigeons.

KNOT Collective noun for toads.

KNOTHOLE A hole in a piece of wood where a knot has been, depending on size often used by bats, hole-nesting birds, insects, snails, etc.

KYPE The hook on the lower jaw of salmon or trout.

L

LABELLUM The lip of an orchid's flower that attracts and serves as a landing platform for insects, which then pollinate the flower.

LABIAL PALPS Moustache-like scaly mouthparts of adult butterflies and moths (and some other insects) found on each side of the proboscis, covered with sensory hairs and scales, and which test whether something is food.

LABIATE Resembling or possessing a lip or labium, for example in the flowers of Lamiaceae (formerly Labiatae).

LABIUM The lower 'lip' of an insect's mouthpart, the corresponding 'upper lip' being the labrum. In dragonfly and damselfly nymphs the labium folds beneath the head and thorax, and can flick out to capture prey then carry it back to the head. In honeybees, the labium is elongated to form a tube and tongue. See also labial palps.

LABOUR Collective noun for moles, though it should be noted that adult moles are solitary animals (apart from mating pairings).

LABRUM The upper 'lip' of an insect's mouthpart, the corresponding 'lower lip' being the labium.

LACINIATE Of leaves, etc., bordered with a fringe; divided into deep narrow irregular segments; having toothed margins cut into deep irregular, usually pointed lobes.

LAEVIGATE Having a smooth surface.

LAGG A wet zone on the edge of a raised bog where water from the adjacent upland collects and flows slowly around the main body of peat.

LAMELLA 1) In flowers, an erect scale or blade inserted in some corollas and forming a part of their corona or crown. 2) A membrane on a fungal gill. 3) Miniature ridges inside the bills of water-feeding ducks (e.g. shoveler) serving as filters during feeding. 4) Flanges or teeth-forming barriers inside the aperture of a snail's shell.

LAMENTATION One of the collective nouns for swans.

LAMINA A sheet-like structure, for example a leaf blade.

LAMMAS GROWTH Additional, abnormal growth by some tree species around Lammas Day (1 August).

LANCEOLATE Shaped like a lance head, narrow, oval and tapering to a point at each end.

LARDER A collection of prey items impaled on thorns or barbed wire by shrikes.

LARVA The young stage of an insect or other invertebrate, or of an amphibian, of a different morphology from its parent (having to metamorphose), for example a tadpole is the larva of a frog, a maggot that of a fly.

LATERAL ROOT A shallow root running out sideways from stem.

LAWN 1) An area of short, regularly mown grass in a garden or park. 2) A grassed area in a wood created by deer or cattle grazing (also termed **laund**).

LAYERING Development of a new individual plant from a branch that produces adventitious roots when it touches the ground.

LAYING (LAYERING) The rejuvenation of hedgerows by encouraging them into new growth and helping to improve their overall structure. The upright stems are cut almost right through so that they can be bent over. Higher up, each stem is then cut at an angle to grow back vigorously with vertical shoots.

LEACHING The loss of water-soluble plant nutrients from soil through the percolation of rainwater.

LEADER, LEADING SHOOT Main shoot that develops from the terminal bud at the top of a tree each year.

LEAF In vascular plants, a thin flattened growth emerging from a stipule and petiole as the principal lateral appendage of a stem, or needle-like growths in conifers, used in photosynthesis and so characteristically green.

LEAF-ROSETTE A radiating cluster of leaves, often at the base of a stem.

LEAFLET A separate segment of a leaf, which often resembles a leaf, but has no associated bud; a division of a compound leaf.

LEK From the Swedish for 'play', an arena within which males gather for competitive display in order to attract females for mating, seen in capercaillie, black grouse, ruff and snipe, and in insects such as some hymenopterans and flies.

LEMMA The lower of the two scales that surround or enclose the stamens and ovary of a grass floret.

LENTIC WATER Non-flowing fresh water.

LENTICEL 1) Lens-shaped raised pores in the stems of woody plants and roots of herbaceous plants that allow air to reach underlying tissue. 2) Also seen as specks on some fruit, for example apples and pears, and on potato tubers.

LENTICULAR Shaped like a lens.

LEPTOCEPHALUS The flat, transparent larva of eels.

LEUCISM, LEUCISTIC A mutation in some animals leading to partial loss of pigmentation and resulting in white, pale or patchy coloration of skin, hair, feathers or scales, but not the eyes. It is caused by a reduction in multiple types of pigment, not just melanin as with albinism.

LEVERET Young hare.

LEY Grassland planted for a single season or for just a few years on otherwise arable land, used for pasture.

LICHENICOLOUS FUNGUS A parasitic fungus that only lives on lichen as the host. This is not the same as a fungal component of a lichen, which is known as a **lichenised fungus**.

LICHENOLOGY The scientific study of lichens.

LIFE CYCLE The series of changes in morphology and function in the life of an organism, including reproduction.

LIGAMENT A strong, flexible fibrous connective tissue that holds two bones together, or bone and cartilage, to allow articulation in joints.

LIGNEOUS Made of wood.

LIGNICOLOUS Growing on or in dead wood.

LIGNIN A organic polymer that provide important structural materials in the support tissues of vascular plants, especially important in the formation of cell walls in wood and bark because they lend rigidity.

LIGULE 1) A membranous scale or fringe of hairs where the leaf blade and sheath meet in most grasses and sedges. 2) Strap-like extension of the corolla of the floret in some Asteraceae.

LIMB An arm, leg, wing or flipper, or a tree branch.

LIMNOLOGY The study of freshwater ecosystems.

LINGULATE Tongue-shaped.

LINKS Coastal dune grassland, often managed as golf courses, associated with nutrient-poor conditions (contrasting with machair) in Scotland.

LIRELLAE Elongated, narrow, often branched apothecia that often have hard, black outer margins, said to resemble Arabic script, for example seen in handwriting lichen.

LITHOPHYTE A plant growing on rock or stone.

LITHOSERE Plant succession based on a rocky or stony substrate.

LITTER Dead plant material, for example leaves, needles and twigs, fallen to the ground, before decomposition.

LITTORAL Of the shore of a lake or the sea.

LOAM A soil with a mixture of sand, silt and clay (e.g. sandy loam) that can retain moisture and nutrients.

LOBE A roundish and flattish projecting or hanging part of something, for example a rounded indentation on a leaf margin, or the lobate webbing on a coot's feet.

LOBED In plants, a leaf having lobes or divisions extending less than halfway to the middle of the base.

LOCHAN Small inland (freshwater) loch in Scotland.

LOOMERY A breeding colony of guillemots.

LORE The region between nostril and eye in birds, reptiles and amphibians.

LOTIC WATER Flowing fresh water.

LOVE DART A calcium carbonate 'dart' (or made of chitin in the Vitrinidae) that some hermaphroditic species of slug or snail reciprocally 'fire' into each other just before physical mating, its mucus coating containing hormones that enhance sperm survival.

LUMPER Scientist working in the field of taxonomy who tends to emphasise the similarities between species and group them together in the same genus where possible. The opposite kind of taxonomist is a splitter, who emphasises the differences between species and increases the number of genera.

LUNATE With the shape of a crescent moon.

M

MACHAIR From the Scottish Gaelic for 'fertile plain', a species-rich grassland that has developed on stabilised sand behind coastal dunes, typical of some western Scottish and Irish coasts.

MACROFAUNA Benthic or soil organisms which are retained by a 0.5 mm sieve.

MACRONUTRIENT An essential element required in relatively large amounts, for example carbon, hydrogen, nitrogen, oxygen, calcium and phosphorus. See also micronutrient.

MACROPHYTE An aquatic plant visible non-microscopically.

MAGNESIUM An essential nutrient present in every cell type in every organism. 1) In plants, a micronutrient that contributes to chlorophyll, and therefore photosynthesis and growth, and membrane structure, as well as a number of enzyme reactions, with deficiency leading to chlorosis. 2) In animals, it is again important in many biochemical reactions including those associated with cellular functions, and again deficiency leads to disease. Ruminants are particularly vulnerable to magnesium deficiency in grassland, with symptoms including muscle weakness and loss of balance.

MAIDEN Or **feathered maiden**, in nursery terms a young fruit tree usually 1–2 m tall growing on vigorous rootstock, to be planted out to produce a standard.

MALACOLOGY The scientific study of molluscs, with conchology the study of their shells.

MAMMALOLOGY The scientific study of mammals.

MAMMARY GLAND The milk-producing organ of a female mammal.

MANDIBLE 1) The lower jaw of a mammal. 2) In birds, the term usually refers to both parts of the bill (i.e. upper and lower mandible). 3) In insects and other arthropods, one of a pair of external biting mouthparts.

MANGANESE A micronutrient important in photosynthesis, and in skeletal development, enzyme regulation and reproduction in vertebrates, but excessive amounts are deleterious to plants and animals, with neurodegenerative effects noted in vertebrates.

MANTLE 1) In molluscs, the dorsal body wall which covers the visceral mass (the digestive and reproductive organs and body parts associated with movement). In many species, the epidermis of the mantle secretes calcium carbonate and conchiolin, thereby creating a shell, adding to it to increase its size and strength as the animal grows. The mantle is highly muscular: in cephalopods contraction of the mantle is used to force water through a tubular siphon, the **hyponome**, thus propelling the animal through the water. In gastropods it is used as a 'foot' for locomotion. 2) The shoulder feathers and upper back of a bird. 3) Of a bird of prey on the ground or perching, its wings and tail spread so as to cover captured prey.

MARITORIES Underwater territories held by male seals near female haul-out sites.

MARSH Wetland with herbaceous plants on a mineral soil (by contrast, fen develops on peat) under water for much of the year. Marsh may follow swamp (which is permanently waterlogged) as part of a hydrosere.

MASK A dark patch around the eye, for example seen on badgers and shrikes.

MAST Fruit crop of certain trees, especially Fagaceae, for example beech and oak. 'Mæst' is Old English for the nuts of forest trees that have accumulated on the ground, especially those used as food for fattening pigs. A **mast year**, with particularly abundant mast production, occurs periodically as a method of predator satiation.

MATING The action of a male or female animal (or hermaphrodites), or a number of these, coming together to breed.

MAXILLA 1) In vertebrates, the upper jaw, which holds the upper teeth, the **premaxilla** in the front, sometimes fused with the maxilla, holding the incisors. 2) In birds, including the premaxilla, often referred to as the upper mandible. 3) In insects and other arthropods, a pair of mouthparts behind the mandibles used for ingestion.

MEADOW A general name for (long) grassland, more strictly a field used for producing hay. **Flood meadows** are naturally seasonally flooded by rivers. **Water meadows** are flooded annually using sluices and controlled irrigation. Working water meadows largely disappeared by the mid-twentieth century, but the channels have often remained, providing valuable habitat. Flood and water meadows are particularly nutrient-rich because of the deposition of river silt.

MEALY With a floury texture; generally refers to the surface of leaves or stems.

MEDULLA 1) The soft internal tissue or pith of a plant. 2) A loosely woven mass of fungal hyphae.

MEDULLARY RAYS Channels that connect the outer and inner layers of a tree-trunk, extending vertically and perpendicular to the growth rings.

MEDUSA The umbrella-shaped, free-swimming stage in the life cycle of the jellyfish and certain other cnidarians.

MEGAFAUNA The large animals of a particular region. A common threshold is a weight >40 kg, so includes such examples as red deer and seals.

MEGASPORE The female spores that give rise to female gametophytes in heterosporous plants.

MEIOFAUNA Benthic animals that can pass unharmed through a 0.5–1 mm mesh but are retained by one of 30–45 µm.

MELANISM Excess of black pigment (melanin) in the body, producing darker individuals than normal, in mammals evident in skin and hair.

MEMBRANE A very thin flexible sheet made up of a double layer of lipids and proteins that surrounds all cells.

MERE A pool or shallow lake.

MERICARP A one-seeded portion created by splitting up a two- or many-seeded fruit (schizocarp), seen for example in Apiaceae, Geraniaceae and Malvaceae.

MERISTEM Plant tissue that contains cells which continue to divide, generating new growth. **Apical meristems** occur at the tips of stems and roots.

MERMAID'S PURSE The empty egg-cases of thornback ray, commonly washed up on shore.

MEROMICTIC LAKE One in which part of the water body is permanently stratified. See also holomictic lake.

MESOFAUNA Macroscopic soil invertebrates, for example earthworms and arthropods.

MESOTROPHIC Fresh water having a moderate concentration of dissolved nutrients. See also eutrophic and oligotrophic.

METABOLISM The chemical reactions that occur inside a living organism in order to maintain life.

METAMORPHOSIS A major change in the shape or structure of an animal that happens as the animal grows. In insects, **incomplete metamorphosis** occurs when young animals are similar to adults and change gradually into the adult form, for example in grasshoppers; **complete metamorphosis** occurs when there is a profound change between larval and adult forms, for example in butterflies and moths, from caterpillar to adult via a pupal stage, and frogs, from tadpole via gradual change into the adult.

METAPLEURAL GLAND Secretory glands in ants that produce antibiotic substances which inhibit growth of bacteria and fungal spores on the ants themselves and inside their nest.

MICROBIOLOGY The study of microorganisms.

MICROCLIMATE The climate of a small area where this differs from the surrounding climate, for example because of topography, or associated with a particular habitat, for example inside a woodland.

MICROFAUNA The microscopic animals found in a particular habitat.

MICROHABITAT Where organisms are found within a larger habitat that differs in its characteristics, for example a rock pool in the intertidal zone, or the litter in woodland.

MICRONUTRIENT Or **trace element**, an essential element required in relatively small amounts, for example boron, copper, iron, manganese and zinc, deficiency leading to inhibition of growth, etc. (In excess, however, these can become toxic as heavy metals.) See also macronutrient.

MICROORGANISM A living thing that can be seen only by using a microscope, for example a virus or bacterium.

MIDRIB The thick structure running longitudinally through the centre of a leaf (serving as a vein) or thallus, providing structural support.

MIGRATION Movement of an animal (**migrant**) outward from an area (**emigration** or outmigration) or into a new area (**immigration** or in-migration), as a result of environmental or population pressures, or a periodic or cyclical movement between locations as a result of seasonal environmental changes in temperature and/or precipitation.

MILT The seminal fluid of fish, molluscs and other aquatic animals, containing sperm, and sprayed over eggs to fertilise them.

MIMESIS, MIMETIC Visual mimicry of inanimate objects, for example early instars of swallowtail caterpillars resembling bird faeces, or equivalent objects, for example stick insects. Also used with reference to sound, for instance starlings imitating the vocalisation of other birds or the sounds of inanimate objects.

MIMICRY Imitation of one living thing by another in appearance, behaviour, sound or scent, with some survival advantage, usually to gain protection or hide from predators. **Batesian mimicry** (named after the English naturalist Henry Walter Bates) occurs when a non-dangerous, non-toxic or indeed palatable species mimics the appearance of a dangerous or unpalatable one, for example hoverfly mimicry of bees. **Müllerian mimicry** (named after the German naturalist Fritz Müller) describes two or more species that have similar warning coloration, for example yellow and black caterpillars. In **Wasmannian mimicry** (named after the Austrian entomologist Erich Wasmann), the mimic resembles a model that it lives along with in a nest or colony, seen in some ants and social bees. See also mimesis.

MIRE Peat wetland or bog.

MOBBING Co-operative behaviour by potential prey in harassing a would-be predator.

MODER A humus intermediate between mor and mull, for instance found under some upland grassland.

MONOCARPIC, MONOCARPOUS 1) Plants that flower once, set seed, then die; the opposite of polycarpic. 2) Plants in which the gynoecium has a single carpel.

MONOCLIMAX THEORY See climatic climax.

MONOCOTYLEDON Plants with one seed leaf (cotyledon) in the embryo contained in the seed.

MONOECIOUS Bearing male and female flowers as separate organs but on the same plant.

MONOGAMY Having a sexual relationship with one partner during any one mating period, in long-lived animals sometimes for life.

MONOLECTIC An insect (especially a bee) that collects pollen from the flowers of only one plant species; see also oligolectic and polylectic.

MONOMICTIC LAKE One in which the water freely circulates during only one season.

MONOPHAGY Eating only one species or kind of food.

MONOSPECIFIC Relating to or (of a genus) consisting of only one species.

MONOTYPIC A taxon having only one species, with no subspecies. See also polytypic.

MOOR(LAND) (Usually) high open ground over acid soil or peat, often with heather or (if grazing is heavy) grass.

MOR Raw humus over an acid soil with slow decomposition and few worms or microorganisms other than fungi, particularly associated with coniferous woodland.

MORPHOLOGY The study of the shape or form of an organism.

MOSS Northern word for a bog. For moss plants, see also entry in Part II.

MOULT(ING) The process of shedding skin, feathers or fur to make way for new growth.

MOUSTACHIAL STRIPE In birds, a streak of contrasting feathers running back from the base of the bill, for example in bearded tit and reed bunting.

MUCILAGE A complex of carbohydrates produced by a number of plants, viscous, gelatinous or slimy when wet, hard when dry, used in storage of water and food, seed germination, and thickening membranes. The carnivorous sundews use this as a sticky trap for insects. Some microorganisms, for example protists, also produce mucilage for use in locomotion.

MUCRONATE Having a tiny bristle-like tip (a mucro) on a leaf or sepal.

MUCUS A slimy, slippery viscous colloid secreted by a mucous gland and covering a mucous membrane. It protects epithelial cells (tube and cavity linings) in the respiratory, gastro-intestinal, urogenital, visual and auditory systems in mammals; the epidermis in amphibians; and the gills in fish. Fish also have mucus over their scales. Slugs and snails use mucus to facilitate movement. A major function of mucus is to protect against infections.

MUDFLAT An area of soft intertidal mud not covered with vegetation. Its ecological importance, for example for feeding waders, has made this a priority habitat in the UK BAP.

MUIRBURN In Scotland, regular burning of areas of moorland to benefit grouse, permitted only between 1 October and 15 April inclusive. The Muirburn Code came into force in 2002 and has been updated several times, most recently in 2011 (with ongoing consultation).

MULCH A (usually) organic layer of material applied to the soil surface to conserve moisture, reduce the impact of heavy rain and extreme temperatures, improve soil fertility and health, and reduce weed growth.

MULL Mild (non-acid) humus, with rapid decomposition by earthworms and microorganisms, associated with deciduous woodland and grassland.

MÜLLERIAN MIMICRY See mimicry.

MULTIPAROUS An animal that produces more than one young at birth, or which has experienced more than one birth.

MURDER Collective noun for crows.

MURMURATION Collective noun for a group of seagulls and especially starlings, where in places numbers can reach many thousands.

MUSTER Collective noun for crows.

MUTAGEN A physical or chemical agent that changes the DNA of an organism, thus increasing the frequency of mutations above the natural background level.

MUTATION A sudden permanent change in a gene or group of genes: technically, alteration of the nucleotide sequence of the genome resulting from errors during DNA replication or other types of damage to DNA, which may then be subject to error-prone repair. This affects heritable structure, function and fitness of the organism, and is significant in evolution. Mutations may be harmful or beneficial, or have neutral impact.

MUTUALISM An association between two species of mutual benefit, for example plants and mycorrhizal fungi, specific plant-pollinator relationships, and hermit crab and parasitic anemone. Mutualisms can be obligatory or facultative. See also symbiosis and commensalism.

MYCELIUM The branching, thread-like hyphae ('root' system) of fungi which spread out and feed on decomposing matter. The mycelia send up fruiting bodies.

MYCOLOGY The scientific study of fungi.

MYCORRHIZA A mutualism in which a specialised root is 'infected' with fungus. Fungi supply mineral salts and water for growth of the plant.

MYOGLOBIN An iron- and oxygen-binding protein found in vertebrate muscle tissue. It is the primary oxygen-carrying pigment of muscle tissues, for example a high abundance allowing diving mammals to stay underwater for longer.

MYRMECHORY Dispersal of seeds and spores by ants.

MYRMECOPHILY A mutualism whereby an organism provides accommodation or food for ants.

N

NACREOUS Iridescent, with a pearly sheen like mother-of-pearl.

NAIAD 1) Immature dragonfly or mayfly. 2) A water plant (see entry in Part II).

NAIL A horny plate-like structure, shaped like a shield, found at the tip of the upper mandible of ducks, geese and swans, sometimes bent at the tip like a hook, used (depending on diet) to extract food out of mud, plants or shells.

NALOSPI The distance between the forward edge of a bird's nostril and the tip of the bill.

NANISM Dwarfism.

NARES The nostrils on birds, placed towards the base of the upper mandible. Cormorants and gannets have primitive external nares as nestlings but these close soon after fledging, and adults breathe through their mouths.

NASTIC MOVEMENT In plants, non-directional responses to such stimuli as temperature, humidity and light, for example see nictinasty and photonasty.

NATAL Of, relating to or at (the time of) birth.

NATANT Floating.

NATIVE A plant or animal belonging naturally to a country or region; having colonised by natural means.

NATURAL A community of native organisms or a landscape or ecosystem essentially untouched by human activity. See also semi-natural.

NATURAL SELECTION The process by which many different factors, from climate to food supply, steer the course of evolution. Organisms best suited to their environment have more surviving offspring (differential survival and reproduction), which carry on their successful characteristics (see fitness).

NATURALISED An introduced (non-native) plant or animal that has become established and self-perpetuating in a country or region.

NATURE CONSERVATION Management aimed at the protection of species, habitats and landscapes.

NATURE RESERVE An area in which management aims to protect and conserve habitat and species, often with some kind of statutory protection, and ranging in geographic scale and in importance from Local Nature Reserves to National Nature Reserves. These complement areas designated by local authorities as important for nature conservation (see SINCs).

NEAP TIDE A tide occurring just after the first or third quarters of the moon when there is least difference between high and low water.

NECROPHAGY Eating of carrion or other dead matter.

NECTAR A liquid, up to 60% sugar (itself 55% sucrose, 24% glucose and 21% fructose), secreted by a nectary as a reward for pollinating animals or those providing anti-herbivore protection, and which is the source of sugar for honey.

NECTARIFEROUS Nectar-bearing.

NECTARY A glandular body within the flower (**floral nectary**) or outside the flower (**extrafloral nectary**) which secretes nectar. The floral nectary is usually located at the base of the perianth, so pollinators brush the stamen and pistil while accessing the nectar, acquiring and depositing pollen, and so fertilising the plant.

NEEDLE The long, narrow, pointed leaf of a coniferous tree.

NEKTON From Greek for 'swimming', animals that are active in water, in contrast to passively moving plankton.

NEMATOCYST A specialised stinging cell in the tentacles of some cnidarians (sea anemones, jellyfish, some corals) containing a barbed or venomous coiled thread that, triggered by touch, is projected in self-defence or to capture prey.

NEONATE A newborn animal.

NEOPHYTE A plant associated with human activity and introduced into the British Isles after around 1500; see archaeophyte.

NEOTENY Retention of juvenile features in an adult animal, seen in Strepsiptera and other insects where only the female remains flightless, and also in dogs and pet mice. Such animals can reproduce even though the body is otherwise not fully developed.

NERVE A bundle of specialised cells that carry signals rapidly around the body of an animal.

NERVOUS SYSTEM The network of nerve cells transmitting signals to and from parts of an animal's body. The **central nervous system** is the complex of nerve tissues that controls the activities of the body, in vertebrates comprising the brain and spinal cord.

NEST A structure built by some animals to hold and often protect eggs, offspring and/or the animal itself. Material and structure are often species-specific.

NESTLING A bird too young to leave its nest.

NEUSTON From Greek *neustos* = 'able to swim', the organisms inhabiting the surface layer or moving on the surface film of a body of water. See epineuston and hyponeuston.

NEUTRAL Of soil, neither acid nor alkaline, with a pH value of 7.

NEWCOMER'S GLAND Caterpillars of many lycaenid butterflies have a gland on the seventh abdominal segment that produces honeydew and is called the 'dorsal nectary gland', also Newcomer's gland. In the large, silver-studded, common, Adonis and chalk hill blues and brown argus these secretions attract ants which provide food (large blue) and protection from parasites and predators.

NICHE 1) A crevice or platform on a wall or rock face. 2) An **ecological niche** is described by the role and position a species has in its environment – how it survives, feeds and reproduces, and how it interacts with the biotic and abiotic factors of its environment. One view of the niche is as an n-dimensional hypervolume, where the dimensions are those environmental conditions and resources that define the requirements of an individual or species to survive. No two species can occupy an identical niche, and there are various ways in which **niche separation** is effected, for example resource partitioning. See fundamental niche and realised niche.

NICTINASTY (sometimes NICTONASTY) A diurnal plant movement, for example flowers opening by day and closing at night.

NICTITATING MEMBRANE From Latin *nictare* = 'to blink', a translucent inner membrane ('third eyelid') found in many sharks, reptiles and birds, and in some mammals (for example seals), attached to the corner of the eye and drawn across the eye to reduce illumination, and moisten, clean and protect the surface.

NIDICOLOUS A young bird that remains in the nest until able to fly.

NIDIFUGOUS Precocial birds leaving the nest almost immediately after hatching, although remaining protected by parents.

NITROGEN FIXATION Conversion of gaseous nitrogen gas into ammonia, then into other compounds that roots can absorb, most commonly through the action of nitrogen-fixing bacteria.

NOCTURNAL Active at night.

NODE Stem joint bearing a leaf, flower or lateral stem.

NOMENCLATURE A system of applying a scientific name to each organism or group of them, using binomial nomenclature (q.v.) for species and trinomial for subspecies.

NUCIFEROUS Bearing nuts.

NUPTIAL Connected with mating.

NUPTIAL GIFT Or **courtship feeding**, food items or sperm packets (or sometimes inedible tokens) that are transferred to females by males during courtship or copulation. They are used by a number of insect species (e.g. some butterflies, grasshoppers and fruit flies), some arachnids, and a few birds (e.g. shrikes).

NUPTIAL PAD A thickened pad on the inner hand or inner toes of the front leg of a male frog or toad which helps to grasp the female during mating.

NUPTIAL PLUMAGE Or **breeding plumage**, assumed by male birds during the breeding season, sometimes with a rather extravagent result, for example in ruffs.

NUT Dry fruit containing a single seed within a woody shell.

NUTANT Of flowers: drooping, nodding or bent downwards.

NUTRIENT CYCLING A biogeochemical cycle (q.v.) in which the element being transferred and exchanged, whether in an ecosystem or regionally or globally, is used by living organisms as a nutrient.

NUTRIENTS Elements used as food by a living organism to sustain life. See also macronutrient and micronutrient.

NYCTIGAMOUS Of a flower that opens at night.

NYE Collective noun for pheasants.

NYMPH 1) The young stage of an exopterygote insect such as a grasshopper, that looks like a small version of its parents, but which as yet has no wings. 2) The aquatic phase of insects such as mayflies, stoneflies, damsel flies and dragonflies.

O

OBLIGATE Restricted to a particular way of life, an organism that is obliged to live under particular conditions, for example a host-specific parasite or a monophagous animal, in contrast to facultative.

OBOVATE Shaped like an egg, for instance in a leaf, with a broad end at the top, a narrower end at the base (petiole).

OCELLUS A single-lens simple eye in an insect and some other invertebrates. Spiders have several pairs of simple eyes with each pair adapted for a specific task or tasks.

OCHREA A delicate sheath-like organ formed from stipules encircling the stem just above the leaf.

OECESIS The process of settlement in a new location (see also dispersal).

OESOPHAGUS The muscular tube along which food passes between mouth and stomach, also called the gullet.

OESTRUS CYCLE A regularly occurring, hormone-regulated cycle of sexual receptivity (oestrus) in sexually mature female mammals (other than humans and most other primates). It is interrupted by non-breeding phases or by pregnancies.

OIL GLAND Technically the **uropygial gland**, informally also known as the **preen gland**, is a sebaceous gland found in most birds, located dorsally at the base of the tail. See preen gland/preening.

OLD-FIELD SUCCESSION Successional changes taking place on abandoned fields.

OLFACTION, OLFACTORY (Relating to) the sense of smell, used to identify pheromones, food, danger, etc. In mammals, especially carnivores and ungulates, olfactory cues provide information on identity, and social and reproductive status. Most mammals and reptiles have a main olfactory system, which detects volatile stimuli, and an accessory olfactory system (vomeronasal), which detects fluid-phase stimuli (see flehmen). Insects use olfactory neurons in their antennae in finding mates (for example moths), hunting (for example wasps), etc., and many are drawn towards carbon dioxide (for example Diptera, including mosquitoes). Plant tendrils can be sensitive to airborne volatile organic compounds. Plant parasites such as dodder use this ability to locate their hosts. Emission of volatile compounds is detected when foliage is browsed and plants can then take defensive chemical measures, for example increasing the concentration of tannins to the leaves, reducing their palatability.

OLIGOLECTIC An insect (especially a bee) that collects pollen from the flowers of only a few unrelated plants; see also monolectic and polylectic.

OLIGOPHAGY Eating only a few kinds of food.

OLIGOTROPHIC Of fresh water, poor in nutrients. See also eutrophic and mesotrophic.

OMBROGENOUS BOG A peat-forming bog found above the water table, depending on precipitation for both water and mineral nutrients, in other words is **ombrotrophic**. The resulting lack of dissolved bases leads to strongly acidic conditions and only specialised vegetation, characterised by sphagnum, will grow. See raised bog and blanket bog.

OMMATIDIUM Individual hexagonal visual receptors in a compound eye, sending a mosaic image to the brain.

OMNIVORE An animal that eats both plant and animal material.

OOLOGY The study of eggs.

OOTHECA An egg-case produced by some insects (particularly cockroaches) and some molluscs.

OPERCULUM A structure that closes or covers an aperture. Examples include: 1) A flap of the sporangium of a moss, covering the peristome. 2) A plate that closes the aperture of a gastropod shell when the animal is retracted. 3) Plates at the top of a barnacle. 4) A flexible flap of skin that covers the gills of most fish. 5) A membranous or cartilaginous flap covering the nostrils (nares) of diving birds, keeping water out of the nasal cavity.

OPISTHOSOMA The posterior part of an arachnid's body, behind the cephalothorax.

OPPORTUNISTIC SPECIES One with a low level of specialisation capable of adapting to varied living conditions or that gives priority to reproduction over survival, and so able to colonise a newly available or transitory habitat rapidly when conditions are suitable.

OPPOSITE Used of buds or leaves that are arranged in pairs on the stem or twig.

ORBIT The eye cavity or socket of the vertebrate skull.

ORBITAL RING A ring of small feathers (or of bare skin) immediately surrounding a bird's eye, often decorative, its colour contrasting with adjacent plumage, and conveying such information as sexual maturity and health.

ORDER Subdivision of a class, itself divided into families.

ORGAN A collection of tissues joined in a structural unit to serve a common function.

ORGANELLE A tiny cellular structure that performs specific functions within a cell.

ORGANIC Material containing carbon compounds; derived from or in some way connected with living things.

ORNITHOLOGY The scientific study of birds.

ORNITHOPHILY Pollination by birds.

OSMOREGULATION The maintenance of constant osmotic pressure in the fluids of an organism by the control of water and salt concentrations.

OSMOSIS The movement of water or another solvent from a region of low solute concentration to one of higher concentration across a partially permeable membrane, tending to equalise the concentration on the two sides. This movement, for example, allows plant roots to draw water from the soil.

OSTENTATION Collective noun for peacocks.

OUTCROSSING Reproduction involving non-related individuals, increasing genetic diversity and avoiding inbreeding.

OVARY The part of an animal or plant (here the basal part of the gynoecium) that produces the reproductive egg cells.

OVATE See ovoid.

OVIDUCT The tube that leads from the ovary to the outside of the body.

OVIPARY The method of reproduction in which eggs are laid in which the embryos develop outside the mother's body, each egg eventually hatching.

OVIPOSITOR Egg-laying organ used by insects and a few fish (e.g. bitterling). In bees, wasps and ants it is modified into a sting.

OVOID Egg-shaped, being widest towards the base.

OVOVIVIPARITY The method of reproduction in which the embryo develops in the maternal body but hatches only after the egg has been laid, e.g. in the adder and in sharks.

OVULE The part of a flower that develops into a seed after pollination. Each ovule contains an egg cell.

OVUM An egg cell (female gamete).

OXYGEN A gas that makes up 20.8% of the atmosphere. Oxygen is a product of photosynthesis, and most living things take in oxygen from the air or water (aerobic respiration) and use it to release energy from food.

P

PACK Collective noun for grouse.

PADDLING 1) Trampling the surface of the ground, for example done by gulls, the vibrations believed to encourage earthworms (but not marine worms) to rise to the surface thinking it is rain. 2) Collective noun for ducks on water.

PAEDOGENESIS Breeding while in the larval or other pre-adult stage.

PALAEONTOLOGY The study of fossils, including **palaeo-botany** – that of fossil plants.

PALE, PALEA One of the bracts in a grass flower.

PALMATE(D) Shaped like a hand, with spread fingers. 1) In plants, composed of more than three leaflets or lobes arising from a central point. 2) In birds and salamanders it means the front toes are webbed. 3) In fallow deer, antlers in which the angles between the tines are partly filled in to form a broad flat surface.

PALPS A pair of elongated segmented appendages near the mouth of an insect, spider or other arthropod, usually concerned with touch and taste. See also pedipalps.

PALUDAL Relating to a marsh.

PALYNOLOGY The study of extant or fossil pollen and spores.

PAMPRODACTYLY In birds, an arrangement where all four toes may point forward, or birds may rotate the outer two toes backward, a characteristic of swifts.

PAN A dense iron-rich mineral layer in a heathland or conif-erous woodland soil (podzol).

PANICLE A compound or branched inflorescence.

PAPILIONACEOUS Having flowers shaped like a butterfly, seen in plants in the pea family.

PAPILLA 1) A small rounded projection, for example on a tongue or from a leaf. 2) In birds, a brightly coloured, fleshy protuberance in the mouth of a nestling, combining with the colours of the gape to guide the parents in placing the food. 3) In some fishes, small lumps of dermal tissue found in the mouth which hold taste buds, for example in rays and sharks, or are ornamental, for example in thick-lipped mullet. 4) In some fishes, the **genital papilla** is a small, fleshy tube behind the anus from which sperm or eggs are released.

PAPPUS The modified calyx that surrounds the base of the corolla tube in the flower heads of Asteraceae, composed of bristles, feathery structures, awns or scales. In examples such as dandelion, it functions as a 'parachute' which enables the seed to be carried by the wind.

PARASITE, PARASITISM (Relating to) an organism which lives inside (**endoparasite**) or on (**ectoparasite**) another organism (the host), benefiting by deriving nutrients at the other's expense. See also hyperparasite and parasitoid.

PARASITOID An insect that as a larva lives on or in a host which it eventually kills. Most are found in the Hymenoptera. In many species eggs are laid within the bodies of other organisms. The eggs hatch and the larvae initially feed on non-essential organs of the host, but eventually either feed on vital organs, killing the host, or pupate within the host to emerges as an adult, killing the host.

PARENCHYMA 1) In plants, the cellular tissue found chiefly in the softer parts of leaves, pulp of fruits, bark and pith of stems, etc. 2) Cellular tissue lying between the body wall and the organs of invertebrate animals lacking a coelom, for example flatworms. 3) More generally, the functional (not the structural or connective) parts of an organ in the body.

PARIPINNATE LEAF A pinnate leaf with all the leaflets in pairs (i.e. with no single leaflet at the tip).

PARITY The number of times a female has given birth. See iteroparity and semelparity.

PARLIAMENT Collective noun for owls.

PARR A trout or salmon having developed from a fry to leave the hatch site, remaining in fresh water for 1–4 years, before migrating to sea as a smolt.

PARTHENOGENESIS Asexual reproduction in which the offspring develops from unfertilised eggs.

PARTIAL VEIL Thin, web- or skin-like tissue covering the gills or pores of an immature fungal fruit body.

PARTY Collective noun for jays.

PASSAGE A **bird of passage** – one that migrates seasonally.

PASSERINE A bird in the order Passeriformes, distinguished by the arrangement of the toes, three pointing forward and one back, which facilitates perching.

PASTURE Grassland used for grazing stock.

PATAGIUM 1) In a number of moths, a sensory organ at the base of the lower wing, or a lobe that covers the wing joint. 2) In birds, the skin in which the wing feathers are embedded. 3) A membrane or fold of skin between the fore- and hindlimbs of a bat.

PATHOGEN A microorganism capable of causing disease, usually after exceeding a tipping point in its population size.

PEAT Slowly decomposing organic material supporting and consisting of bog vegetation, including moss such as sphagnum, accumulating under waterlogged, mainly anaerobic conditions, for example in various kinds of bog, wet heath, etc.

PECTEN 1) Any of a number of comb-like structure found in animal bodies. 2) A kind of scallop.

PECTINATE(D) Having narrow ridges or projections close together like the teeth of a comb. Birds with long necks preen their necks and heads using pectinated toes. The pectinated claws of the middle toes of herons are used for combing slime and scales off fish.

PECTORAL Relating to the breast or chest. 1) In fishes and whales, the **pectoral fins** are a pair of fins situated on either side just behind the head, helping to control the direction of movement during swimming. 2) In birds, the **pectoral girdle** is the wishbone (furcula).

PEDATE 1) In animals, having feet, or a structure that resembles feet. 2) In plants, describing a palmate leaf with divided lobes.

PEDICEL 1) in plants, an individual flower-stalk. 2) In fungi, a cylindrical appendage on the spores of some species. 3) In insects, the second segment of the antenna, or to the 'waist' at the front of an ant's abdomen.

PEDIPALPS The second pair of appendages of chelicerates, including spiders and scorpions (the 'claw'), which use them to capture prey.

PEDOLOGY The study of the formation, description and classification of soils.

PEDUNCLE 1) Stalk of an inflorescence, for instance as seen in pedunculate oak. 2) In animals, a stalk-like part by which an organ is attached to a body, or by which sedentary animals, for example barnacles, are attached to a substrate.

PEEP A small wader such as stint or dunlin, and a collective noun for chickens.

PELAGE The fur, hair or wool covering a mammal's body.

PELAGIC Living in the open ocean, including birds that

are found at sea apart from when nesting, for example fulmar and kittiwake.

PELLET The mass of undigested food regurgitated by some bird species, for example owls, hawks, eagles, falcons, crows, gulls and shrikes.

PELORIA From the Greek *peloros* = 'monstrous', an abnormal regularity of structure occurring in normally irregular flowers, for example foxgloves may produce a flat, saucer-shaped peloric flower at the tip of the spike.

PELTATE Shield-shaped; of a more or less circular leaf with the stalk (pedicel) attached to a point on the underside.

PEN Female swan.

PENNATE 1) Having a feather-like structure. 2) A **pennate diatom** has bilateral symmetry.

PERENNATING Of a plant or part of a plant, living through a number of years, usually with an annual quiescent period. Thus a **perennating bud** remains dormant until favourable growing conditions return, for instance in spring. **Perennating organs** include storage structures such as bulbs, corms and tubers.

PERENNIAL A plant living for more than two years and usually flowering annually.

PERFECT FLOWER One with both male and female parts (stamens and carpels), that is, bisexual or hermaphroditic.

PERFOLIATE Of a stalkless leaf or bract, extended at the base to encircle the node, so that the stem appears to pass through it.

PERIANTH The sepals and petals together, especially when not distinguishable as calyx and corolla (see petal).

PERICARP 1) The soft or hard outer covering of a fruit, often made up of the **epicarp** (outermost layer), **mesocarp** and **endocarp** (the inner layer surrounding the ovary or the seeds). 2) A membranous envelope around the **cystocarp** (fruiting structure) of red algae.

PERISTOME From Greek *peri* + *stoma* = 'around' + 'mouth'. 1) In mosses and some fungi, a specialised structure in the sporangium that allows for gradual spore discharge; a fringe of small projections around the mouth of a capsule in mosses and certain fungi. 2) The aperture margin of the shell lip in some gastropods.

PERITHECIUM In fungi, a round or flask-shaped fruiting body (ascocarp) with a pore through which spores are discharged.

PERSISTENT CALYX Sepals remaining attached, and often present around fruit.

PEST A subjective, often emotive word used of an animal that adversely affects an economic activity in some way, for example reducing the yield of a crop, not necessarily for what they do as such, but how many of them are doing it (a population size threshold). See also weed.

PETAL Each of the segments of the corolla of a flower, which are modified leaves and are typically coloured; one segment of a whorl of floral leaves (perianth).

PETIOLATE Having or attached by a petiole. In petiolate leaves, the leaf stalk (petiole) may be long, as in the leaves of celery and rhubarb, short, or completely absent, in which case the blade attaches directly to the stem (is sessile).

PETIOLE 1) in plants, the stalk of a leaf. 2) In insects, the 'waist' of bees, wasps and other Hymenoptera (known as the pedicel in ants).

pH An abbreviation of 'potential of hydrogen', a measure of the acidity or alkalinity of an aqueous solution on a scale of 0–14 (7 is neutral, <7 acid, >7 alkaline or basic), calculated as the logarithm of the reciprocal of the concentration of hydrogen ions in the solution.

PHANEROPHYTE A plant whose perennating buds are above ground.

PHARYNX The cavity behind the nose and mouth, connecting them to the oesophagus.

PHENOLOGY The study of (the timing of) seasonal changes in plants and animals from year to year, for example in flowering, emergence of insects and migration of birds, especially in relationship to weather conditions.

PHENOPHASE A stage in the annual life cycle of a plant or animal that has an identifiable start and finish, for example the period over which open flowers are present on a plant.

PHENOTYPE The set of observable characteristics of an individual resulting from the interaction of its genotype (q.v.) with the environment.

PHEROMONE A chemical secreted or excreted by an animal that triggers off a behavioural or physiological response in another of the same species, commonly as a sexual attractant. See also gland and scent (1).

PHLOEM In vascular plants, the living tissue that transports the soluble organic compounds made during photosynthesis (photosynthate), in particular sucrose, to parts of the plant where needed. In trees, the phloem is the innermost layer of the bark.

PHORESIS, PHORESY The transport of one organism by another, for example mites carried by insects, a form of commensalism.

PHOSPHORUS 1) In plants, a macronutrient involved in the storage, transport and release of energy in photosynthesis, in the synthesis of most carbohydrates, and in cell division. A deficiency leads to leaf discoloration and necrosis. 2) In vertebrates, it is one of the most important minerals in nutrition, with 80% of phosphorus being found in bones and teeth, the remainder in body fluids and soft tissue.

PHOTOBIONT One of the symbiotic fungal partners in a lichen, either a green alga (phycobiont) or cyanobacteria (cyanobiont).

PHOTOBLASTIC Seeds that germinate in response to stimulation by light.

PHOTONASTY Response of a plant organ to light.

PHOTOPERIODISM The developmental responses of plants and animals to seasonal changes in the relative lengths of day and night (**photoperiod**), for example in plants generally provoking flowering. In vertebrates, changes in photoperiod trigger hormone production that affects timing of such behaviour as courtship and mating, migration and hibernation.

PHOTOSYNTHESIS The process by which green plants (also some bacteria and protistans) use the energy from sunlight absorbed by chlorophyll to produce glucose from carbon dioxide and water. The soluble organic compounds are known as **photosynthate**.

PHOTOTROPISM See tropism.

PHREATIC ZONE The waterlogged region at and below the water table.

PHREATOPHYTE A plant with long roots that can reach the water table.

PHYCOBIONT The algal symbiont in a lichen.

PHYCOLOGY The study of algae, synonymous with algology.

PHYLLODE Green leaf stalk, flattened to look like a leaf.

PHYLOGENY The evolutionary history of a taxon. A **phylogenetic tree** or evolutionary tree is a branching diagram showing the presumed evolutionary relationships between different taxa based on similarities and differences in their physical or genetic characteristics.

PHYLUM The taxonomic rank below kingdom and above class.

PHYSIOLOGY The science of the interacting functions, activities and mechanisms of living things and their components, including physical and chemical processes.

PHYTOGEOGRAPHY The study of the geographical distribution of plants. See also floristics.

PHYTOPHAGOUS Feeding on plants.

PHYTOPLANKTON Microscopic photosynthesising plants in freshwater and marine systems, including diatoms and dinoflagellates.

PICTURED In insects, used to describe wings, especially in flies, that are darkly mottled.

PIGEON MILK Or **crop milk**, a secretion created by sloughing off fluid-filled cells in the crop lining of parent birds that is regurgitated as food to young pigeons.

PILEUM The top of a bird's head from the base of the beak to the nape.

PILEUS The 'cap' of the fruiting body of a fungus.

PILIFEROUS Bearing or producing hairs.

PILOSE Covered with fine soft hairs.

PINION The outer part of a bird's wing including the flight feathers.

PINNA 1) One of the leaflets in a pinnate leaf. 2) The external mammalian ear, more commonly called auricle.

PINNATE 1) A compound leaf, composed of several pairs of leaflets each side of a common stalk or rachis. 2) Leaf venation in which the principal veins are found on either side of a midrib.

PIONEER PLANT A species in the forefront of the colonisation of open ground or new habitat opportunities, at the start of succession, for example lichen and mosses on rock, or at a later successional stage, for example birch trees on heathland.

PISCICULTURE The raising or farming of fish.

PISCIVOROUS Feeding on fish.

PISTIL The female parts of a flower. It may consist of one carpel, with its ovary, style and stigma, or several carpels joined together with a single ovary.

PITEOUSNESS, PITYING Collective nouns for doves.

PITH 1) Tissue that stores food at the centre of the stem of a non-woody plant. 2) Spongy white tissue lining the rind of citrus fruits.

PLACENTA 1) In plants, the fused margins of the carpel to which the ovules are attached. 2) In mammals, the organ that connects the developing foetus to the uterine wall, allowing nutrient uptake, thermoregulation, waste elimination, and gas exchange via the mother's blood supply.

PLAGIOCLIMAX From Greek *plagios* = 'oblique', a plant community or ecosystem where succession has been 'deflected' from how it might naturally have progressed by direct or indirect human activity, reaching a rather different state of stable equilibrium, for example cleared woodland becoming heathland and maintained by burning or chalk grassland maintained by grazing. Deflection is not the same as an arresting factor, which leads to a subclimax (q.v.).

PLANKTON From Greek *planctos* = 'wanderer' or 'drifter', floating and drifting microorganisms (phytoplankton and zooplankton) that live in the surface layers of fresh or marine waters. See also pleuston, nekton, phytoplankton and zooplankton.

PLANT Mainly multicellular eukaryotes of the kingdom Plantae, **green plants** absorbing water and inorganic substances through roots, and synthesising nutrients in the leaves by photosynthesis.

PLANTATION Planted commercial timber woodland using one or a few coniferous or broadleaf tree species.

PLANTIGRADE An animal that walks with toes and metatarsals flat on the ground, for example rodents, lagomorphs, hedgehogs and humans.

PLASMODIUM 1) Vegetative body of slime fungi. 2) A genus of a parasitic protozoan that causes malaria.

PLASTIC Varying in form or structure as a response to environmental (not genetic) conditions.

PLASTRON Lower part of shell of a terrapin, tortoise or turtle.

PLEACHER The stem half-cut and bent in hedge-laying.

PLEUROCARP A type of moss in which the female sex organs and capsules are carried on short lateral branches, tending to form spreading carpets. See acrocarp.

PLEUSTON Organisms that live in the thin surface layer at the air-water interface of a body of water.

PLICATE Wrinkled or folded; an arrangement of bud scales or young leaves in a shoot bud (see vernation).

PLUMAGE The layer of feathers that covers a bird, as well as their pattern and colour, which differ between species and subspecies, with age and sometimes with season. See also nuptial plumage.

PLUMOSE Having many fine feathers or filaments with a feathery appearance. In insects, applied especially to antennae, for example of moths.

PNEUMATOPHORE 1) A gas-filled sac serving as a float in some colonial marine hydrozoans, such as the Portuguese man-of-war. 2) A specialised respiratory root in certain aquatic plants that protrudes above water or mud into the air.

POD 1) A long dry cylinder fruit, which splits open releasing the seeds. 2) A group of dolphins.

PODZOL (PODSOL) From the Russian *pod* + *zola* = 'under ash', giving the more accurate spelling, a zonal group of acid soils having surface organic layers and thin, organic mineral horizons above illuvial dark brown horizons, developed under coniferous or mixed forests or heath vegetation.

POIKILOTHERM A so-called cold-blooded animal that cannot regulate its body temperature except by behavioural means, for example basking or burrowing, and otherwise reflects the surrounding environmental temperature. The opposite is homoiotherm.

POLLARD 1) A tree where the upper branches have been removed, promoting a dense head of foliage and branches, usually at a height of around 2 m (beyond the reach of browsing animals). See also coppice. 2) A deer which has shed its antlers. 3) De-horned (polled) livestock.

POLLEN A powdery substance comprising **pollen grains** which are male microgametophytes of seed plants, and used to fertilise seed plants.

POLLEN BASKET See corbiculum.

POLLINATION The transfer of pollen from the male anther of a flower to the female stigma so that fertilisation can occur.

POLLINIUM A coherent aggregation of pollen that is transported together, for example in orchids.

POLLUTION The introduction of contaminants into the environment or taken in by organisms that cause adverse change. These substances may be intrinsically harmful, for example the heavy metals mercury and lead, or are harmful above certain critical levels or concentrations, for example the otherwise essential micronutrients copper and zinc. **Thermal pollution** results from discharge of heated wastewater.

POLYANDRY A mating system involving one female mating

with more than one male during a single breeding season, simultaneously or sequentially, for example seen in red-necked phalarope.

POLYCARPIC A plant that flowers and sets seeds many times during its lifetime (iteroparous), the opposite of monocarpic.

POLYCLIMAX THEORY The theory that a stable plant community characterised as the theoretical end point of succession can be in a state of dynamic equilibrium with any controlling environmental factor, not just climate (see monoclimax theory and climatic climax), for example topography, natural fire or human activity, and subsumes subclimaxes and plagioclimaxes (see both).

POLYETHISM Specialisation in different members of a colony of social insects, leading to a division of labour. The different functions may be undertaken by individuals with different morphologies (**caste polyethism**) or of different ages (**age polyethism**).

POLYGAMY Mating with more than one individual during a single breeding attempt.

POLYGYNY A mating system involving one male mating with more than one female during a single breeding season, for instance as part of a harem.

POLYLECTIC An insect (especially a bee) that collects pollen from the flowers of a variety of unrelated plants; see also monolectic and oligolectic.

POLYMICTIC Of a (shallow) lake with continuous water circulation (i.e. with no seasonal thermal stratification).

POLYMORPHISM 1) The existence of two or more genetically distinct forms within an interbreeding population, or in the life cycle of an individual organism, for example most two-spot ladybirds are red with two black spots but some are black with red spots. 2) In genetics, the presence of genetic variation within a population, upon which natural selection can operate.

POLYP An individual anthozoan cnidarian (coral or sea anemone), which secretes a limestone casing as it matures, having a cylindrical body and a ring of tentacles around the mouth. Hydrozoan cnidarians have 2 growth forms, polyp and medusa.

POLYPHAGY Eating many different kinds of food.

POLYTYPIC Containing more than one taxonomic category of the next lower rank. A **polytypic genus** therefore contains two or more different species. See also monotypic.

POME A fleshy fruit in the family Rosaceae consisting of an outer thickened fleshy layer and a tough central core with usually five seeds enclosed in a capsule, for example apple, pear and quince.

POPULATION All the individuals of the same group or species which live in a particular geographical area, and have the ability to interbreed.

PORE 1) In biology, a minute opening in a surface, especially the skin of an organism, through which gases, liquids, or microscopic particles may pass. 2) In soils, the open space between soil particles through which gases, water and small organism can pass, and affected by soil texture.

PORICIDAL In plants, relating to small holes or pores. In poricidal anthers, pollen is packed tightly into the anther and released through a small pore at the tip of a tube via buzz pollination. Poricidally dehiscent fruits, for example those of poppies, use a 'shaker' mechanism to disperse the seeds. Seed capsules sway in wind, shaking seeds out through the pores.

PORRECT Extending horizontally.

POSTNUPTIAL Taking place after breeding.

POTASSIUM 1) In plants, a micronutrient involved in photosynthesis (especially carbon dioxide uptake via stomata), leaf movement, and control of the water potential of cells. Leaves normally require a potassium content of 1–2%; symptoms of deficiency include chlorosis, stunted growth and necrosis. Potassium is essential in the synthesis of most carbohydrates and in cell division. 2) Potassium is the major cation inside animal cells, sodium the major cation outside cells: the difference between the concentrations of these particles causes a difference in the ability of cells to produce electrical discharge – critical for body functions such as muscle contraction and heart function.

PRECOCIAL An animal born or hatching in an advanced state of development, able to walk, feed itself and thermoregulate, and quickly becoming independent, in contrast to altricial.

PREDATION A feeding interaction in which a carnivorous or omnivorous **predator** hunts or otherwise catches, usually kills, then feeds on other animals (**prey**).

PREDATOR SATIATION, PREDATOR SATURATION An anti-predator adaptation: populations of potential prey briefly emerge or reproduce in high numbers, swamping predator numbers which have remained the same and can only eat so much, thereby reducing the probability of an individual organism being eaten, for instance seen with mayflies. An analogous strategy relating to seed 'predators' is also used by some trees; see mast.

PREEN GLAND, PREENING See oil gland. A bird typically transfers preen oil to its body during preening by rubbing its beak and head against the gland opening, then rubbing the oil on the feathers of the body and wings, and on the skin of the feet and legs. Functions include maintenance of body and feather integrity, waterproofing, anti-parasitic effects, and in some species possibly pheromone production.

PREHENSILE Able to wrap around and grasp objects.

PREMAXILLA See maxilla.

PREMORSE Having the end irregularly truncated as if bitten or broken off, for example used of leaves or roots which end abruptly as is the case with devil's-bit scabious roots.

PRENUPTIAL Before breeding.

PREVERNAL Before or in early spring.

PREY An animal that is killed and eaten by another animal.

PRICKLE A small thorn (q.v.), for example seen on roses.

PRIMARY (FEATHER) One of the largest, outermost 'fingertip' wing feathers.

PRIMARY CONSUMER An animal that feeds on plants (primary producers); a herbivore or functionally, in part, an omnivore.

PRIMARY PRODUCER A photosynthetically or chemosynthetically active organism that can synthesise complex organic compounds from simple substances (i.e. is an autotroph).

PRIMARY PRODUCTION, PRIMARY PRODUCTIVITY The amount of biomass produced by primary producers (autotrophs) using photosynthesis or chemosynthesis, with **productivity** being the rate of production.

PRIMARY SUCCESSION Plant succession that begins on bare ground, for example on rock, sand or mud. See succession and secondary succession.

PRIMARY WOODLAND Woodland that has been continuously present since the last Ice Age and regenerating naturally. Wistman's Wood, Dartmoor, may be such a woodland: a high-altitude moss-rich oakwood (remnant of the much more extensive ancient forest cleared in the Mesolithic), even if rather altered in structure during historical times.

PRISERE A collection of seres in primary succession (vegetative development of an area from a non-vegetated surface to a climax community).

PROBOSCIS A long, flexible snout or mouthpart, in invertebrates usually used for feeding and sucking.

PROCUMBENT Trailing along the ground.

PRODUCER See primary producer.

PROLEG One of the stumpy non-jointed legs on the rear part of a caterpillar. Caterpillars have up to four pairs of prolegs, together with an extra pair – claspers – at the tail end. Larvae of geometrid moths have no prolegs, having only true legs at the front end and the claspers, and so move forward by bending in a loop, hence 'loopers' (q.v.) and the family name, Geometridae (earth measurers). Prolegs can also be found on other larval insects, for example sawflies and some true flies.

PRONOTUM A prominent plate-like structure that covers all or part of the thorax of insects such as grasshoppers and cockroaches.

PROPAGULE Plant or fungal material used in propagating an organism to the next stage in its life cycle – seed or spore, or (in vegetative reproduction) a bulb, etc., and in horticulture a leaf, stem or root cutting.

PROPOLIS A resinous mixture produced by honeybees by mixing saliva and beeswax with plant exudate (e.g. from tree buds or sap flow) to reinforce the structural stability of the hive, seal unwanted crevices, prevent diseases and parasites from entering the hive, and inhibit fungal and bacterial growth.

PROSOMA The fused anterior (front) segment of an invertebrate body, especially or synonymous with the cephalothorax of an arachnid.

PROSTRATE Lying close to or along the ground.

PROTANDRY 1) The condition in which an organism begins life as a male and subsequently changes into a female, seen in oysters which, during their first year, as males release sperm into the water, but as they grow over the next two or three years develop greater energy reserves, and spawn as females by releasing eggs. See also protogyny. 2) In botany, the maturation of male parts of a flower (anthers) before female parts (carpels), thus preventing self-fertilisation.

PROTEIN An essential organic polymer composed of amino acids and used for metabolism, as structural elements (e.g. keratin), and creation of genes, hormones, enzymes, etc.

PROTHALLUS The small gametophyte generation of a plant that bears the true sex organs, mostly used with reference to pteridophytes.

PROTOGYNY 1) The condition in which an organism begins life as a female and subsequently changes into a male, seen in wrasse. See also protandry. 2) In botany, the maturation of female parts of a flower (carpels) before male parts (anthers), thus preventing self-fertilisation but enhancing cross-pollination, for example seen in apples, pears and figworts.

PROTOPLASM The material comprising the living part of a cell.

PROXIMAL At the end near a point of attachment; opposite of distal.

PROXIMATE CAUSE, PROXIMATE FACTOR Something immediately responsible for causing an observed result, for example variations in day length (photoperiod), or a female bird selecting a particular male with which to mate based on his display. The **ultimate cause** (q.v.) may be the evolution of a particular preference for display plumage and display behaviour, since offspring will likely also possess these preferred characteristics.

PSAMMOPHYTE A plant thriving in sand, such as on dunes, for example marram and sand couch.

PSAMMOSERE Plant succession in a sandy habitat.

PSEUDANTHIUM See flower head.

PSEUDOCARP Or **accessory fruit**, one in which some of the fruit combines the ripened ovary with another structure, often the receptacle, for example a strawberry or mulberry.

PTEROCHORY Wind dispersal of winged seeds, for example of sycamore.

PTERYLOSIS The arrangement of feathers on a bird's body and how they are implanted in the skin.

PTILOSIS The properties of bird feathers.

PUBESCENT 1) Covered with down or soft short hair. 2) In humans, relating to the approach of puberty.

PULLUS A nestling bird.

PULMONARY Relating to the lungs.

PULSE 1) A member of the Fabaceae, or the edible seeds of such a plant. 2) In medicine, the near-surface arterial palpation of a heartbeat.

PULVERULENT Consisting of or covered with fine particles like dust or powder.

PULVINUS In some plants, an enlarged section at the base of a leaf stalk subject to changes of rigidity, leading to movements of the leaf.

PUNCTATE Marked with dots, pores, small excrescences or pitted with tiny depressions.

PUNGENT Having a sharp taste or smell.

PUPA In an insect subject to complete metamorphosis at the stage of development between a larva and an adult. Insects at the pupal stage do not move or feed, and the larval body is rebuilt into that of the adult, for example seen in Lepidoptera. See chrysalis and cocoon.

PUPARIUM The hardened, barrel-shaped case in which the pupa of many fly species is concealed, created from the larval skin.

PUPIPAROUS Some parasitic flies and a few other insects which give birth to a fully grown larva that pupates almost immediately.

PYGOSTYLE Fused caudal vertebrae forming a triangular plate supporting a bird's tail feathers.

PYRENE The stone within a drupe, consisting of a seed surrounded by hard endocarp tissue, for example in a cherry or plum.

PYRIFORM Pear shaped, including describing the shape of those eggs of birds that are tapered towards one end.

Q

QUADRAT A circumscribed area, typically of one square metre, selected (usually at random) to act as part of a sampling effort for assessing the presence or abundance of plants or (generally invertebrate) animals. Also the portable frame, typically with an internal grid, used to mark out a quadrat.

QUADRUPED An animal with four feet.

QUAGMIRE Soft, wet, boggy ground which quakes or yields underfoot.

QUAKING BOG Or *Schwingmoor*, a peat bog found in the wetter parts of some valley bogs and raised bogs. The vegetation, mostly sphagnum moss often anchored by sedges, forms a floating mat on the surface of water, soft mud or very wet peat, and shakes when walked upon.

QUARRY SPECIES A legally defined game bird, waterfowl or leporid mammal that can be shot within a defined open season.

QUARTERING Methodically covering an area of ground, for example by a hunting raptor.

QUEEN The (primary) reproductive female in a colony of eusocial insects, who also generally initiates the colony.

QUESTING Stimulated by chemicals such as carbon dioxide as well as heat and movement, ticks climb up and hold on to leaves and grass by their third and fourth pair of legs, with the first pair of legs outstretched. Subsequently, the ticks climb on to a potential host which brushes against their extended front legs.

QUICKSET HEDGE A type of hedge created by planting live cuttings, usually of hawthorn, directly into the earth.

QUILL A moulted flight feather (usually a primary) of a large bird, for example goose or swan.

R

RACE A geographically isolated breeding population that shares certain characteristics in higher frequencies than other populations of that species, but has not become reproductively isolated from other populations.

RACEME An inflorescence formed by a stem producing a flower then growing further before producing the next, so that the youngest flowers are at the top of the stem; all pedicels are of equal length. See corymb and thyrse.

RACHIS 1) In birds, the main shaft of a feather, its lower end inserted into a skin follicle. Also known as the shaft, its lower end, beyond the level of the lowest barb, is called the calamus or quill. 2) In plants, the central stalk (more accurately, axis) of a compound leaf or of an inflorescence.

RADIAL SYMMETRY The arrangement of parts of an organism around a central axis, so that the organism can be divided into similar halves by any plane that contains this axis. In animals such as echinoderms, ctenophores, cnidarians, and many sponges and sea anemones, this allows them to respond to stimuli from all directions without moving.

RADIALLY SYMMETRICAL FLOWERS Those with more than six petals with a capitulum made up of tiny florets, as one flower head. See actinomorphic.

RADICAL Of leaves, etc., growing from the roots or a stem base.

RADULA A rasping 'tongue' consisting of a chitinous band in the mouth of all molluscs except bivalves, set with numerous minute teeth (denticles) and moved back and forwards over the floor of the mouth to break up food.

RAFT A closely packed flock of birds on water, applied to some types of seabird and duck.

RAISED BOG An acidic, wet, nutrient-poor, sphagnum-rich ombrogenous bog, so called because it rises in height over time as a result of peat accumulation, forming a dome up to 12 m above the water table, the principal supply of water and nutrients coming from rainfall rather than ground water. Some 94% of the UK's lowland raised bog has been destroyed or severely degraded, with perhaps 13,000 ha remaining, most being in Scotland, mainly in the Solway and Galloway areas, and in Moray and Aberdeenshire. In Ireland, raised bogs are mostly found in central areas, but also in the Bann River Valley in Northern Ireland, near Omagh, Co. Tyrone, and in East Clare and North Limerick on either side of the mouth of the River Shannon, occurring on land <130 m with rainfall of 800–900 mm p.a.

RAMET An individual member of a clone.

RANGE See geographical range.

RAPTOR A predatory bird such as a hawk, eagle, osprey, kite, falcon and harrier (all diurnal) and owl (nocturnal).

RARITY The status of a species or other taxon, it being very uncommon or infrequently encountered. A rare species is not necessarily endangered or threatened with (local) extinction, though this is commonly the case.

RASORIAL An animal, usually a bird, that scratches the ground for food.

RAY FLORET Strap-shaped and typically sterile florets in the Asteraceae. In plants such as dandelions the flower head is composed entirely of ray florets.

REALISED NICHE The actual (rather than maximum theoretical) niche space that a species occupies in the presence of competition. See also fundamental niche.

RECEPTACLE The flat, domed or dished part of the stem to which the floral parts are attached.

RECTRICES (singular RECTRIX) Flight feathers on the tail (c.f. remiges on the wing).

RECURVED Turned or bent back.

RED, AMBER AND GREEN LIST Conservation status, for example for birds in the UK, is described as red, amber or green. The **Red List** critera include a historical population decline in UK during 1800–1995; at least a 50% decline in the UK breeding population over the last 25 years, at least a 50% contraction of the UK breeding range over last 25 years. The **Amber List** criteria include a historical population decline during 1800–1995, but recovering; population size has more than doubled over the last 25 years; moderate (25–49%) decline in the UK breeding population or breeding range over the last 25 years, rare breeder (1–300 breeding pairs in the UK); rare non-breeders (<900 individuals). The IUCN also produces a **Red List** of species at a global scale, ranging from 'least concern' to 'extinct'.

REDD Depression made in gravel on the bed of the river by salmon and trout for egg-laying.

REDHEAD Female smew.

REELING A continuous, high-pitched monotonous trill made by the grasshopper warbler.

REFLEXED Sharply turned back upon itself.

REGENERATION 1) Regrowth or re-creation of lost or damaged tissues, organs or limbs. 2) Regrowth or re-establishment of a habitat or plant community after damage or destruction.

REGULAR FLOWERS Flowers that are radially symmetrical, for example apple and lilies; actinomorphic.

REGURGITATION Ejection of partially digested food from the oesophagus or gizzard, mainly for the feeding of young, seen in a number of bird species with altricial young, and carnivores.

RELEASER A simple feature of a complex stimulus that can elicit a fixed action pattern in an animal.

RELIC 1) A population or taxon that was more widespread in the past. 2) A fragment of a formerly more widespread habitat or community.

REMIGES (singular REMEX) Flight feathers on the wings (c.f. rectrices on the tail).

RENDZINA From Polish rędzina via Russian, a shallow humus-rich soil overlying chalk or limestone, typically under calcareous grassland.

RENIFORM Kidney-shaped.

REPRODUCTION The process by which new individual organisms are produced from their 'parents' (see sexual reproduction) or without the involvement of another organism (see asexual reproduction).

RESIDENT A species of which examples can be found in a particular area at any time of the year, although they may not be the same individuals.

RESILIN An elastic protein that can be used to store energy. It enables insects of many species to jump or pivot their wings efficiently. In fleas, resilin pads in the legs are compressed like springs which, when released, allow the flea to jump.

RESIN A sticky substance, insoluble in water and solidifying when exposed to air, exuded by some (especially coniferous) trees and other plants after injury as protection from insects and pathogens.

RESORPTION Assimilation into the womb, seen in foetuses in some mammals, for example rabbits, when non-viable, or in response to stress, food scarcity, overpopulation, etc.

RESOURCE PARTITIONING A form of niche separation, the use of resources by ecologically similar species or by different sexes of the same species in ways that avoid competition, for example by using different microhabitats, acquiring different food items, or taking the same items using different behaviours.

Woodpeckers, treecreepers and nuthatches, for example, find food on tree-trunks using (literally) different approaches.

RESPIRATION A process involving the production of energy, typically with an intake of oxygen and release of carbon dioxide from the oxidation of complex organic substances. It is not the same as breathing, more accurately called ventilation.

RESUPINATE 1) Upside down. 2) A fungal fruit body that grows flat against the substrate without producing a cap.

RETICULATE With a net pattern, forming or covered with a network.

RETINA A tissue in the back of an animal's eye that detects light, sending images to the brain.

RETRIX (plural RETRICES) A tail feather.

REVERSED SEXUAL DIMORPHISM See size dimorphism.

REVOLUTE Rolled back or down, for example of leaves; an arrangement of bud scales or young leaves in a shoot bud (see vernation).

RHEOTAXIS Movement or directional change as a response to water movement, for example fish turning to face an oncoming current.

RHEOTROPISM Growth of a plant or sessile animal in the direction of a water current.

RHINARIUM Bare part of a mammal's muzzle in which the nostrils are located.

RHIZOID 1) In fungi, tightly spun hyphae. 2) In mosses and liverworts, a filamentous outgrowth that anchors the plant to the substrate, and which can absorb water and mineral nutrients. 3) Root hairs in higher plants.

RHIZOME An underground or ground-level creeping (often swollen) stem producing roots and shoots.

RHIZOMORPH A rootlike thread produced by fungi.

RHIZOSPHERE The region of soil within which plant roots are found, influenced by root secretions and associated soil microorganisms.

RIB 1) In plants, the midvein of a leaf (costa). 2) One of the long curved bones which form the rib cage in tetrapod vertebrates, protecting the internal organs. In some other animals, especially snakes, ribs may provide support and protection for the whole body. Fishes often have two sets of ribs attached to the vertebral column, dorsal and ventral.

RIBOSOMAL RNA Ribonucleic acid (rRNA) is the RNA component of the ribosome, essential for protein synthesis in living organisms. The ribosome contains roughly 60% rRNA and 40% protein by weight.

RIBOSOME A complex molecule made of ribosomal RNA molecules and proteins that form a factory for protein synthesis in cells.

RICTAL BRISTLE One of a group of stiff, hair-like feathers around the eyes and base of the bill, particularly evident in insectivorous birds. A suggested role in securing prey after in-flight capture is undemonstrable. They may serve a similar purpose to eyelashes and vibrissae (whiskers) in mammals.

RINGTAIL A young or female hen harrier or Montagu's harrier, from the barred tail pattern.

RIPARIAN Of or on the banks or margins of a river or stream.

RIPENING A process in fruit, generally associated with a colour change, in which as they age they become softer and sweeter and therefore more palatable.

RITUAL An action or set of actions performed by an animal in a fixed sequence, often as a means of communication (display or other social interaction).

RITUALISATION Occurs when an action or behaviour pattern in an animal has lost its original function but has been retained for its role in display or other social interaction.

RNA Ribonucleic acid, a substance which helps to make proteins, essential in a number of biological roles in coding, decoding, regulation and expression of genes.

RODENT RUN The practice by birds of fleeing from a predator by crouching and running with head down through ground vegetation mimicking a mammal.

RODING 1) The courtship flight of the male woodcock at dusk in spring when he flies over a regular course giving two types of call, a low croak and a high 'tsiwick'. 2) Clearing a fenland dyke of silt and vegetation using a long-handled scythe and rake, usually undertaken in winter. Also known as dydling.

ROOKERY 1) Roosting and breeding site in trees for rooks. 2) Breeding ground for a colony of seals.

ROOST A place where birds regularly settle to rest at night, or where bats congregate to rest in the day.

ROOT The part of a vascular plant which anchors it to the ground typically underground, conveying and storing water and nutrients for the rest of the plant.

ROOT HAIR Fine structure at the tip of young roots, through which water and mineral salts are absorbed from the soil. See rhizoid.

ROOTSTOCK In grafting, a plant which has an established root system, on to which a cutting or a bud from another plant is grafted.

ROSETTE 1) a circular arrangement of leaves, all at a similar height, on a short stem, for example daisy and dandelion. 2) A hemicryptophyte in which leaves only develop at the stem base.

ROSTRUM A beak-like projection, especially a stiff snout or anterior prolongation of the head in a cetacean, crustacean or insect, in the last case applied especially to the piercing mouthparts of bugs (Hemiptera) and weevils.

ROUND DANCE See bee dance.

ROYAL A stag with 12 points on antlers (six on each) with three of the antlers forming a crown on top of each antler.

ROYAL JELLY A honeybee secretion from worker bee glands, used in the nutrition of larvae and adult queens. After three days, drone and worker larvae are no longer so fed, but when workers decide to make a new queen, they choose several small larvae and feed them royal jelly, which triggers the development of queen morphology.

RUDERAL A plant that thrives in situations of high-intensity disturbance and low-intensity stress. Such plants are fast-growing, rapidly complete their life cycles, and generally produce large amounts of seeds; they are often found colonising disturbed waste or agricultural land, and are often annuals.

RUFF A thick band of feathers or fur round the neck of a bird or mammal. (For the species, see entry in Part II.)

RUGOSE With a rough, wrinkled or puckered surface.

RUMINANT An ungulate (deer, cattle, sheep) that chews the cud regurgitated from its rumen.

RUNCINATE Pinnately lobed leaves whose lobes point backwards; saw-toothed.

RUNNER A stolon or creeping stem that produces roots at its apex from which a daughter plant grows. See blastochory.

RUNT The smallest and weakest animal in a litter.

RUT An annual period of sexual activity in deer (and, in the wild, sheep and goats), during which males fight each other for dominance and thereby access to the females.

S

SACCATE Plant parts that are shaped like a pouch or sack.

SADDLE See clitellum.

SAGITTAL CREST A bony ridge on the top of some mammal skulls (less commonly in reptiles) to which the jaw muscles are attached, indicating great jaw muscle strength, for instance seen in badgers.

SAGITTATE Arrow-shaped.

SALINE Having a high salt content. Slightly salty water is 'brackish'.

SALTING A coastal area regularly covered by the tide.

SALTMARSH A community of salt-tolerant plants (halophytes) growing on coastal mud covered by tides, often showing zonation reflecting the extent of tidal submersion.

SAMARA From the Latin for 'seed of an elm', dry fruit on which a flattened wing of fibrous, papery tissue has developed from the ovary wall, enabling wind to transport the seed further, for example in sycamore.

SAP A fluid that transports water and nutrients in plants via xylem cells or phloem.

SAPLING A young and slender tree, conventionally 3–15 cm in diameter at breast height.

SAPROBE, SAPROPHAGE, SAPROPHYTE, SAPRO-TROPH Prefix from the Greek *sapros* = 'putrid', a plant, fungus or microorganism (more accurately called myco-heterotroph because it actually parasitises fungi, rather than dead organic matter directly) which obtains nutrients from dead or decaying organic material. Some insect larvae also feed on dead material.

SAPWOOD Living wood (xylem tissue) consisting of the outer annual rings in a tree-trunk, between the bark and dead heartwood, through which sap, containing water and nutrients from the soil, is conducted.

SAURIAN Relating to or resembling a lizard.

SAXICOLOUS Growing on or among rocks and stones.

SCABRID, SCABROUS Scaly or rough to the touch, often because of bristly hairs or tiny prickles.

SCALE 1) small, overlapping plates that protect the skin. 2) A papery or woody flap of tissue, usually a much reduced leaf (= scale leaf), for example typical of conifers.

SCANDENT Climbing.

SCAPE 1) in plants, a leafless pedicel, or a flower-stalk arising from the rootstock where all the leaves are basal. The scapes of chives, for example, are used as a vegetable. 2) In insects, the first segment of the antenna.

SCAPULA The shoulder blade or wing bone that connects the humerus to the clavicle. In fish, the scapular blade is attached to the upper surface of the articulation of the pectoral fin.

SCARIOUS Thin, dry and membranous (papery texture), for example as in some bracts.

SCAT Animal faeces or droppings.

SCAVENGER An animal that feeds on the remains of dead animals (carrion) or plants.

SCENT 1) A (distinctive) smell, often deliberately deposited (**scent marking**) via a **scent gland** (e.g. a subcaudal gland) to provide information about an individual's social or sexual status, or to leave a perceptible **scent trail**. See also pheromone. 2) Noxious scents are produced by some animals as an anti-predator

defence. 3) In plants, floral scent consists of volatile organic compounds produced to attract pollinators or deter herbivores.

SCHIZOCARP A dry fruit that breaks into one-seeded portions or mericarps when ripe.

SCHIZOPETALOUS FLOWER As seen in bell heather in which the bell corolla is split into four equal segments, giving it the appearance of a totally different plant.

SCHOOL An assemblage of fish, marine mammals, etc., the animals swimming close together as communicative, predatory or anti-predatory behaviour.

SCHWINGMOOR See quaking bog.

SCION A cutting grafted onto another plant (the stock).

SCLERENCHYMA Woody tissue in a partly or mainly non-woody organ providing mechanical stiffness and strength.

SCLERITE One of the hard sections of an arthropod's exoskeleton, an internal spicule of some sponges and soft corals, or found in combination with other structures, for example molluscan radula ('teeth').

SCLEROPHYLL Plants with stiff leathery, evergreen leaves that can reduce water loss and resist heat.

SCLEROTIUM A storage organ consisting of a mass of hyphal threads with a black protective rind produced by some fungi, containing nutrients, and which can remain dormant for long periods, enabling the fungus to grow in unfavourable environments.

SCOPA A bee's pollen-collecting apparatus, either the corbicola (pollen basket) on the leg, or a brush of abdominal hairs.

SCORPIOID Curved or coiled like the tail of a scorpion.

SCRAMBLER A plant that climbs and sprawls over other plants or physical supports.

SCRAPE 1) The simplest nest construction – a shallow depression in soil or vegetation, usually in open habitat, typically with a rim sufficiently deep to keep the eggs from rolling away. 2) A shallow water-filled hollow used by waterbirds for feeding, often created and maintained as part of conservation management.

SCRUB Habitat or vegetation characterised by shrubs or low trees (commonly hawthorn), often as an intermediate stage in succession from grassland or heathland to woodland.

SCUT The short white tail of a rabbit, hare or deer, visible in flight to warn nearby conspecifics of danger, or to indicate to a predator that it has been seen.

SCUTELLATE Covered or protected with scutella (small scales or plates, being Latin for 'small shield'); shield-shaped. The term is used in describing the pattern of leg scales on birds such as ducks where these form shield-like overlapping structures. Where the scales do not overlap the pattern is 'reticulate' or 'netlike', as found for example on swans.

SCUTELLUM In insects such as shieldbugs the hardened extension of the thorax over the abdomen.

SEAWASH BALLS Clumps of empty common whelk egg capsules detached from rocks and washed up on the beach.

SECONDARY CONSUMER A carnivore that preys on herbivores (which are primary consumers).

SECONDARY FEATHER An inner flight feather on the trailing edge of the wing, connected to the ulna.

SECONDARY SUCCESSION Succession that is initiated, often on land disturbed by fire, landslide, abandoned farmland or clear-felling, where there is already some kind of soil (in contrast to primary succession).

SECONDARY WOODLAND Woodland growing on a site that has not been wooded continuously throughout history, for example has previously been clear-felled then allowed to regrow – an example of secondary succession.

SECUND Arranged, curving or drooping to one side.

SEDENTARY Resident, non-migrating, especially used in relation to birds.

SEED A fertilised ovule (produced by sexual reproduction) produced as a tough protected package containing a plant embryo and food reserves for it to use. Following germination, each seed can develop into a new plant.

SEED BANK 1) The accumulated dormant seeds found in a soil. 2) A store where seeds are held to preserve genes or genetic diversity, and as a means of conserving rare or endangered plants.

SEGETAL From Latin *segetalis* = 'growing in cereal field', describing weeds growing among crops.

SELF-ANOINTING Where an animal smears substances over itself, often the secretions, parts or entire bodies of other animals, for example anting (q.v.). Hedgehogs cover their spines in saliva containing a range of toxic and irritating substances such as toad skin or poison glands, possibly to mask their own scent or as vicarious toxic protection against predation.

SELF-COMPATIBLE/SELF-INCOMPATIBLE Able/unable to self-fertilise, the latter preventing inbreeding and promoting outcrossing.

SELF FERTILISATION Autogamy; fertilisation of female gametes by male gametes from the same plant (i.e. **self-pollination**).

SEMELPARITY Characterised by a single reproductive episode in an organism's lifetime.

SEMI-NATURAL Vegetation or habitat which although modified by human activity has a rather natural character, for example ancient woodland and unploughed meadowland.

SEMI-PALMATE In birds such as avocet, with the front three toes partially webbed.

SEPAL One segment of the whorl of floret parts (usually green) outside the petals (the set of sepals provides the calyx).

SEPTUM A wall or membrane dividing the ovary of a fruit into cells.

SERE, SERAL COMMUNITY The group of plants at a particular stage in succession. A number of seral stages will occur before a climax community is reached. See also prisere, halosere, hydrosere, lithosere and psammosere.

SEROTINAL Referring to the latter part of the summer season.

SEROTINOUS, SEROTINY From Latin *serotinus* = 'belated' or 'deferred', the retention of seeds on a tree for several years until an environmental shock such as fire triggers their release.

SERRATE, SERRULATE Sharply toothed like a saw, finely toothed.

SESSILE 1) In plants, without a stalk. 2) In animals, attached to one place and unable to move.

SETA 1) In mosses and liverworts, the stalk supporting the capsule and supplying it with nutrients. 2) In fungi, thick-walled hairs on the cap, gills or stem. 3) In earthworms, bristles on each segment that help anchor and control the animal when moving through earth, preventing backsliding during peristaltic motion. 4) Scale-like structures on the legs of krill and other small crustaceans that help them gather phytoplankton.

SETACEOUS Like a bristle (seta), in insects especially on the antenna.

SETT Badger burrow complex.

SEXUAL DIMORPHISM The differences in appearance between males and females of the same species, for example brighter colour and different pattern of male bird plumage which aids display (intersexual competition) or the larger size of male mammals to support fighting (intrasexual competition). See also size dimorphism (which includes reversed sexual dimorphism).

SEXUAL REPRODUCTION Production of new living organisms by combining genetic information from individuals of different sexes; the fusion of male and female gametes.

SEXUAL SELECTION Involves **intrasexual selection**, or competition between members of the same sex (usually males) for access to mates, and **intersexual selection**, where members of one sex (usually females) choose members of the opposite sex.

SHARMING A combination of grunts and squeals produced by water rail especially over disputes over the feeding area.

SHEARING Behaviour by shearwaters when tipping from side to side when gliding over the water, first with one wing tip and then the other.

SHEATH 1) In general terms, a structure in living tissue which closely envelops another. 2) The base of a (grass) leaf, enclosing the stem. 3) In a mycorrhizal relationship, a layer of fungal hyphae that surrounds the root surface of the host plant, preventing direct contact between the root and the soil. 3) The outer protective covering of a non-human and non-marine mammalian penis.

SHODDY See wool shoddy.

SHRUB A much-branched, woody plant, not a tree, the branches arising at or immediately above soil level.

SHRUB LAYER The vegetation layer in some woodlands between the tree and field layers.

SHUNTING 1) Birds on (generally) autumn migration that fitfully follow the coastline instead of striking out to sea. 2) Birds moving between winter feeding sites.

SIEGE Collective noun for herons and bitterns.

SIGN STIMULUS A simple component of a complex stimulus that can elicit a specific behavioural response (see fixed action pattern) in animals.

SILICULA Seed-pod which is at least as broad or broader than long, with a central partition which persists after the seeds are released, characteristic of some members of the cabbage family.

SILIQUA Like a silicula but longer than broad, characteristic of the remaining members of the cabbage family.

SILK An elastic natural protein fibre. Spiders make their webs from silk, and some insect larvae make silk cocoons, for example mulberry silkworm, which is reared for silk fabric production.

SILVICULTURE Woodland or forest management controlling the establishment, growth, composition, health and quality of trees to meet a range of needs, often but not exclusively involving commercial production of wood or timber. See forestry.

SIMPLE Undivided or not compound, usually applied to a leaf which is not divided into leaflets.

SINC, SLINC, SNCI Site of Importance for Nature Conservation (SINC), Site of Local Importance for Nature Conservation (SLINC) and Site of Nature Conservation Interest (SNCI) are designations used by local authorities in England for sites of substantive local nature conservation value.

SINISTRAL Left-handed or turning to the left, for example in physid snails which have the shell aperture on the left-hand side; opposite of dextral.

SINUATE Wavy, for example of a leaf margin.

SIZE DIMORPHISM Where the two sexes of the same species are of different sizes, the male often being larger (to support intrasexual competition), but in some cases it is the female (**reversed size/sexual dimorphism**), for example in some birds of prey where enhanced flight performance and greater hunting skills are required before egg-laying.

SKEIN A small flock of flying geese usually referring to a ragged line or V formation. Such a flock on the ground forms a 'gaggle'.

SLACK See dune.

SLIME A general name that may refer to slime mould, biofilm or various kinds of mucus (see all).

SLINC See SINC.

SLOT The individual foot mark of a deer.

SLOUGHING Skin shedding by reptiles and insects.

SMELL One of the senses: the faculty of perceiving odours or scents via the nasal organs. See scent.

SNCI See SINC.

SNEAK Collective noun for weasels.

SNOUT Elongated part of an animal's head including the mouth and nose.

SOARING See gliding.

SOBOLE Underground sucker or creeping stem that produces roots and buds.

SOCIAL HIERARCHY Ranking of individuals in a group, established by fighting or display, with dominant and subordinate individuals.

SODIUM 1) In plants, an important micronutrient in controlling osmotic pressure, and which can substitute for potassium in some metabolic functions. It is also important in carbon dioxide fixation in some plants. 2) In animals, sodium regulates blood volume, blood pressure, osmotic equilibrium and nerve activity.

SOFTWOOD Conifer timber, or the tree itself.

SOIL A mixture of minerals, organic material (including living and dead organisms), gases and liquids found on much of the Earth's land surface, which serves to store water, as a medium for plant growth, and as a habitat for organisms. Its chemical, physical and biological make-up is a function of parent material (e.g. underlying rock), climate, topography, vegetation, animals, human activity, and time, all reciprocally interacting. **Edaphology** is concerned with the influence of soils on living things. **Pedology** involves the formation, description and classification of soils.

SOLIGENOUS BOG A bog that depends on water from surface run-off, lateral water movement or rising water table rather than precipitation.

SOLUM The surface and subsoil layers that have evidence of continuing soil-forming processes, containing most of the plant roots and animal and microbial activity, down to the relatively unweathered parent material.

SONICATION Synonym of buzz pollination.

SORD Collective noun for mallards, or ducks generally.

SORUS In ferns and fungi, a group of sporangia.

SOUNDER A group of wild boar, more particularly a group of females and young led by an old matriarch; young males leave their sounder aged 8–15 months.

SPADIX 1) In botany, a spike of minute sessile flowers closely arranged round a fleshy axil and typically enclosed in a spathe. 2) In zoology, a part or organ in certain invertebrates, roughly conical in shape, for example a group of connected tentacles in a nautiloid.

SPAT A juvenile form of a bivalve mollusc such as an oyster.

SPATHE Large bract, sometimes leaf-like, which encloses a spadix.

SPATULATE Spoon- or paddle-shaped.

SPAWN 1) The jelly-like mass of eggs laid by fish, amphibians, molluscs and crustacea. 2) A fungal mycelium on a growth medium such as sawdust, used by mushroom growers.

SPECIATION The evolutionary process by which potentially interbreeding populations evolve sufficient genetic differences to become non-interbreeding and thereby distinct species.

SPECIES The basic unit of biological classification and a taxonomic rank.; a group of living organisms which can produce fertile offspring with each other, but cannot breed successfully with (are reproductively isolated from) another species (biological species concept).

SPECIES DIVERSITY The number of different species in an area or community, incorporating the ideas of **species richness** (the total number of species) and **species evenness** (how close in numbers each species in an environment is).

SPECULUM In birds, a bright, often iridescent patch of colour on the secondary feathers, for example in ducks such as mallard, teal, gadwall and pintail.

SPICATE Resembling a spike, in particular arranged as a simple inflorescence with (nearly) sessile flowers.

SPIGGIN An adhesive glycoprotein produced by the kidneys of male three-spined sticklebacks excreted to be used as a glue to bind together the plant and inorganic material used in building his nest. The word comes from *spigg*, the Swedish name for this fish species.

SPIKE See spicate.

SPINE 1) In plants, a sharp modified leaf (c.f. thorn). 2) In vertebrates, the vertebral column or backbone housing the spinal canal, which encloses and protects the nervous tissue of the spinal cord.

SPINNER Final stage (imago) of the mature adult mayfly.

SPINNERET The organ through which the silk, gossamer or thread of spiders, silkworms, and some other insects is produced.

SPINNEY A small area of woodland, often where dominated by hawthorn.

SPINNING The action of some waterbirds, for example red-necked plalarope, swimming in tight circles and in so doing forcing water away from itself on the surface, and causing an upward flow which carries up small animals on which it feeds.

SPIRACLE 1) In insects, an opening in the body which lets air in and out of the tracheae. 2) A breathing hole (a vestigial gill slit) behind the eye of some cartilagenous fishes.

SPLITTER A taxonomist who argues the case for separating taxa into different species in contrast to lumpers (q.v.) who group taxa together often despite morphological and other differences. Splitting has recently dominated arguments because of new analytical techniques in genetics that detect previously hidden differences.

SPORANGIUM The part of a vascular plant (pollen sacs and ovules in gymnosperms and angiosperms) or fungus that produces spores.

SPORE The haploid package of cells produced on the sporophyte, and developing into the gametophyte.

SPOROPHYTE The main body of a vascular plant; the diploid generation of a plant that bears the sporangia.

SPRAINT Faeces of mink and otter, which are used as scent markers to define territory.

SPRING 1) A season (in Britain, notionally March–May). 2) A stream emerging where the water table meets the ground surface. 3) Collective noun for teal, which have a rapid, almost vertical take-off when disturbed.

SPRING TIDE A tide occurring just after the second or fourth quarters of the moon (just after a new or full moon) when there is the greatest difference between high and low water. The sun and the moon lie directly in line with the Earth so that their combined gravity 'pulls' seawater.

SPRING WOOD Or **early wood**, in trees the inner part of an annual ring, formed early in the growing season, consisting of thin-walled vessels for conducting water. See also summer wood.

SPUR 1) A projection from a plant organ, especially a **nectar spur** from a flower. 2) A short side branch (**stem spur**) bearing fruit.

SQUAB A young nestling pigeon or dove, fed at least partly on pigeon milk. An older nestling is called a **squeaker**.

SQUAMULOSE Describing lichens which have small, often overlapping 'scales' (squamules).

SQUARROSE Rough or scaly.

SSSI Site of Special Scientific Interest, an area of particular conservation interest because of its fauna, flora or geology, designated by Natural England, Natural Resources Wales, Scottish Natural Heritage, and, (Isle of Man) the Department of Environment, Food and Agriculture. Owners and occupiers of SSSIs must obtain consent to undertake or permit any activity listed in the notification of status.

STADIUM The time interval between successive moults.

STAMEN In angiosperms, the male reproductive organ bearing pollen, consisting of an anther and a stalk called a filament.

STAMINODE A sterile stamen, often modified into a petal or nectary.

STANDARD In forestry, a timber tree in coppice wood; in horticulture, a tree >2 m in height.

STANDARD PETAL The posterior petal of a pea flower. See also keel.

STARCH A carbohydrate used by some plants, for example potatoes, wheat and rice, as an energy store.

STELLATE Star-shaped.

STEM In vascular plants, the above-ground central axis which provides support and is used for the transport and storage of water and nutrients, and from which emerge leaf and flower nodes.

STENOTOPIC Organism with restricted range of tolerance for one or a number of environmental factors.

STERNUM 1) The breastbone – a flat shield-shaped bone in the mid-front of the rib cage of mammals and birds to which the lower ends of the ribs are attached. It has a 'keel' or 'carina' along its lower edge to which the pectoral muscles are anchored. In amphibians and reptiles it is typically composed of cartilage. 2) In arthropods, the ventral part of a segment of a thorax or abdomen, or the cephalothorax in arachnids.

STIGMA 1) The female part of the flower on the end of the style receptive to pollen. 2) A dark area on the wing of a moth, dragonfly or damselfly.

STIPE The stalk of a fungus or seaweed.

STIPULE Leaf-like organ at the base of a leaf or its petiole.

STOCK 1) The base of a plant from which the stems arise. 2) A plant onto which a scion is grafted.

STOLON Or runner, a creeping overground stem that forms adventitious roots at the nodes, and daughter plants from the buds. See blastochory.

STOMA (plural STOMATA) A small pore used for gas exchange, generally found on the under-surface of plant leaves.

STOOL A newly cut coppice stump.

STOOP The rapid steep dive of the hunting peregrine, often 240 km/h and not uncommonly 300 km/h (390 km/h has been recorded outside Britain), the fastest speed of any bird species.

STRATIFICATION See cold stratification.

STRIATED Ridged, furrowed or streaked, as a pattern or morphology.

STRIDULATION A mechanical sound produced by some insects usually for mating purposes. In grasshoppers and male crickets stridulation involves rubbing one part of the body containing a series of pegs or ridges (the 'file') over another (the 'scraper') to produce acoustic vibrations. The scraper is on the wing, the file on the leg (in grasshoppers) or on the opposite wing (in crickets). The sound comes in pulses of varying duration, volume and pitch depending on species. Male water boatmen rub their hairy front legs against a ridge on the face or, in one species, with a sound level of up to 99.2 dB, the penis. Screech beetles rub a sharp part of their abdominal plate against a part of the elytra when disturbed. Other beetles that stridulate include some weevils and tiger, scarab, bark and longhorn beetles. A number of bugs and velvet ant species also stridulate. So too can some spider species: male noble false widow spiders, for instance, stridulate during courtship, by scraping 10–12 teeth on the abdomen against a file on the rear of the carapace.

STRIGOSE Plants with stiff, closely appressed hairs, generally aligned in direction.

STROBILUS The cone of a conifer, or a cone-like structure such as a hop or alder flower.

STYLE 1) In plants, a usually elongated part of the pistil connecting the ovary and stigma. 2) In insects, a slender bristle rising from the tip of the antenna.

STYLET 1) A needle-like object. 2) Part of an insect's piercing mouthpart. 3) Part of a bee or wasp sting.

SUBALPINE Associated with environmental conditions in uplands or on mountains just below the treeline.

SUBCAUDAL GLAND Found under the tail on badger, voles, etc.; see supracaudal gland.

SUBCLIMAX A plant community or ecosystem where 'normal' succession has been arrested by a natural environmental factor, for example topography or natural fire, or human activity (but not 'deflected' to a different route: see plagioclimax), and not allowed to progress. In polyclimax theory, this has equivalent status to a climatic climax as a form of dynamic equilibrium.

SUBCUTANEOUS FAT A layer of fat stored just underneath the skin to provide padding, some thermoregulation and, importantly, stored energy. In birds, this is often called **pre-migratory fat**, being important to build up before migration, and in mammals it is a crucial requirement before entering hibernation. See also adipose tissue.

SUBIMAGO In mayflies, the winged animal (also known as a dun) that emerges from the nymphal skin; the subimago moults again (the only example of a winged insect undergoing a moult) to reveal the imago.

SUBLITTORAL ZONE The zone below the intertidal (i.e. low spring tide) and so permanently covered in seawater, but reached by sunlight and therefore highly productive and species-rich. It may be divided into the infralittoral and circalittoral zones (see both).

SUBULATE Awl-shaped, narrow, pointed and more or less flattened.

SUCCESSION A sequential change in plants in terms of both species and communities, influenced by environmental conditions and availability of species, but crucially resulting from autogenic processes: the environment alters from the activities of the plants themselves, for example in effecting soil development, or drying out or stabilising the substrate. See primary and secondary succession, and sere.

SUCCULENT Having thick fleshy leaves or stems adapted to storing water.

SUCKER A shoot growing from the roots or stem base of a tree or shrub, initially dependent on the parent but with the potential to become an independent plant.

SUFFRUTICOSE In plants, woody and perennial at the base but herbaceous above.

SUFFUSED Gradually spreading through, for example seen in some flowers or plumage.

SULCATE With parallel surface ridges or grooves.

SULPHUR 1) An element taken up in roots, required by plants as a macronutrient, though in low amounts, in the creation of some enzymes and proteins, water equilibrium and help in photosynthesis, a deficiency leading to chlorosis. 2) In sheep and cattle it is essential for growth and production of wool and milk.

SUMMER WOOD In trees, the outer part of an annual ring, formed during the middle and later part of growing season, consisting of thick-walled vessels for conducting water up the stem. See also spring wood.

SUPERCILIARY STRIPE, SUPERCILIUM A streak of contrasting feathers above a bird's eye, often useful in helping to distinguish between species such as warblers.

SUPRACAUDAL GLAND Found on the upper tail surface of foxes, and domestic dogs and cats, being modified sweat and sebaceous glands used for intraspecific signalling and scent marking. Badgers and voles are among mammals that possess a subcaudal gland in the same fashion.

SWALING A West Country term for the controlled burning of moorland, for example on Dartmoor, to replace old plant growth with fresh shoots to benefit of game birds and other wildlife, and for grazing livestock.

SWAMP Wetland habitat waterlogged throughout the year, characterised by a mixture of open water and emergent plants, for example as reed swamp. Swamp may precede marsh (which is waterlogged only for part of the year) as part of a hydrosere. (In America 'swamp' incorporates 'marsh'.)

SWAN-UPPING Ownership of mute swans in the Thames is shared by the Crown, the Vintners' Company and the Dyers' Company. Swan-upping is the capture, inspection and marking (ringing) of mute swans, carried out on this river during the third week in July under the direction of the Queen's Swan Warden. See also under swan in Part II.

SWIM BLADDER An internal gas-filled bladder that contributes to bony fish controlling their buoyancy, thereby remaining at the water depth of choice without having to waste energy in swimming.

SWIMMERET An abdominal limb or appendage in crustaceans, usually adapted for swimming, carrying eggs, moving water over the gills, or burrowing.

SWITCH A red deer antler with no tines.

SYLVAN Relating to woodland and trees.

SYMBIOSIS From the Greek for 'living together', an interaction between two different species living in close physical association, often to the advantage of both. It can take the form of an **obligative symbiosis**, where survival requires the relationship, or **facultative symbiosis**, where each species can survive independently. With physical attachment, where the organisms have bodily union, this is **conjunctive symbiosis**, for example in lichen. See also commensalism and mutualism.

SYMPATRIC Living in the same or overlapping geographical area. The converse is 'allopatric'.

SYNANTHROPIC Organisms that live near and benefit from an association with humans and the environmentally modified habitats created by humans, excluding domestic animals and horticultural or sylvicultural plants.

SYNCARPOUS Having carpels fused along their edges, for example tomato and tulip.

SYNCHRONOUS EMERGENCE, SYNCHRONOUS HATCHING Emergence at roughly the same time of, for instance, adult mayflies from water and chicks of colonially nesting seabirds as examples of predator satiation (q.v.) as an anti-predator strategy.

SYNDACTYLY Having the inner toe and middle toe fused for part of its length, for example found in kingfisher.

SYRINX The lower larynx or voice organ in birds, particularly well developed in songbirds.

T

TADPOLE Aquatic limbless tailed amphibian larva.

TAGMA (plural TAGMATA) In segmented invertebrates, a morphologically distinct part of the body, usually consisting of several adjoining segments, for example the head, thorax and abdomen of insects.

TALON A sharp, hooked claw on the toe of a raptor.

TANNIN A bitter-tasting organic substance made of derivatives of gallic acid present in some galls, bark, and other plant tissues, making the tissue unpalatable.

TAP ROOT The main descending root of plant, which can nevertheless throw out lateral roots.

TARSE Falconry term for a male falcon (see tiercel).

TARSUS 1) In tetrapods, the articulating bones of the hind foot between the tibia and fibula and the metatarsus (ankle bone). 2) In birds, the tarsus has fused with neighbouring bones to form a single **tarsometatarsus**, effectively giving the leg a third segment. 3) In insects and arachnids, the leg segment furthest from the body.

TAXIS An innate change in the direction of movement in response to an external stimulus.

TAXON A taxonomic group of any rank.

TAXONOMY The branch of biology concerned with the scientific classification, description and nomenclature of (extinct and) living organisms, in general into the ranks of domain, kingdom, phylum, class/division, order, family, genus and species.

TEAM Group name given to ducks in flight.

TEGMEN (plural TEGMINA) 1) The inner protective layer of a seed. 2) The leathery front wing of a grasshopper, cockroach, etc.

TELEOMORPH The sexual reproductive stage, typically a fruiting body, of a fungus.

TENDON The strong connective tissue of fibrous collagen that connects muscle to bone.

TENDRIL A thread-like modified stem, leaf or petiole used by climbing plants for support and attachment.

TENERAL In an insect, the phase immediately after moult or emergence from pupation when it is soft and lacking adult coloration.

TENTACLE A slender, flexible structure or organ, especially around the mouth of an invertebrate, used for grasping or moving about, and/or bearing sense organs.

TEPAL One of the outer parts of a flower (perianth) when these parts cannot easily be divided into sepals and petals, for example in lilies.

TERATOLOGY Scientific study of congenital physiological abnormalities in animals and plants, and their morphological expression.

TERCEL, TERCELET See tiercel.

TERETE Rounded or circular in section, for example smooth (unridged or ungrooved) stems.

TERNATE A compound leaf with three leaflets.

TERRESTIAL Of land as opposed to water.

TERRITORY An area claimed by animals which is defended against members of their own species, especially when breeding.

TEST The hard protective skeletal body of sea urchins, heart urchins, etc.

TESTA The outer coat of a seed.

TETRAMEROUS Having flower parts in four.

TETRAPOD From Greek *tetra* + *podos* = 'four' + 'legs', a vertebrate animal with four limbs (amphibians, reptiles, birds and mammals), or having evolved from such (i.e. snakes are tetrapods).

TETRASPOROPHYE Diploid phase in the life cycle of red algae, the site of meiosis (specifically, the **tetrasporangia** where haploid tetraspores are produced).

THALLUS A plant body that is not differentiated into stem and leaves and lacks true roots and a vascular system, typical of algae (e.g. seaweed fronds), fungi, lichens, and some liverworts.

THERMOREGULATION The ability of an organism through physiological mechanisms or behaviour to keep its body temperature within certain boundaries, even with a very different ambient (surrounding) temperature.

THERMOTROPISM See tropism.

THEROPHYTE An annual or ephemeral plant that completes its life cycle in a short period when conditions are favourable and survive harsh conditions, for instance overwintering, as seeds.

THICKET Vegetation dominated by a dense growth of small trees or shrub.

THINNING Removal of some plants to enhance the growth or survival of others. Plant populations can self-thin. Thinning is common in horticulture and forestry.

THORAX 1) In insects, the middle section of the body between the head and the abdomen, bearing the legs and wings. 2) In vertebrates, the part of the body between the neck and the abdomen.

THORN See also spine and prickle. 1) hard, rigid extensions or modifications of shoots, having sharp ends that serve to deter herbivores. 2) A general name for a spiny shrub such as hawthorn and blackthorn.

THYRSE A dense inflorescence consisting of a central raceme with cymose lateral branches (see cyme), for example seen in lilac and horse-chestnut.

TIBIA 1) In tetrapods, the larger bone (shin bone) in the lower hind leg. 2) In invertebrates, the fourth segment of leg, between the femur and tarsus.

TIDINGS Collective noun for magpies.

TIERCEL In falconry, a male falcon (the female is simply a 'falcon'), dating from Middle English, from Old French, based on Latin *tertius* = 'third', possibly from the belief that the third egg of a clutch (alternatively one egg in three) produced a male, because the male is a third smaller in size than the female.

TILLER In grasses, a lateral shoot emerging from the main stem at or near ground level.

TINE The points on a deer antler.

TIPPET The black frill at the base of the head of great crested grebe in breeding plumage.

TISSUE The material intermediate between cells and a complete organ of which the body of an organism is composed. A particular tissue consists of a mass of similar cells which give it distinctive characteristics.

TOMENTOSE In plants, covered with dense, white woolly hair (pubescence).

TOMIAL TOOTH A sharp protrusion on the tip of the mandible of falcons, hawks and kites used for killing prey.

TOOTH 1) In plants, one of a series of small, regular points on a leaf margin. 2) In animals, a small, calcified structure set into the jaws (or mouths) of many vertebrates, used for tearing and chewing food, for hunting and/or for defence.

TOP PREDATOR An animal at the top of a food chain.

TOPOGRAPHY 1) The science of the shape and features of the surface of the Earth. 2) A description of the external parts of a bird.

TORPOR A state of decreased physiological activity in an animal, usually by a reduced body temperature and reduced metabolic rate, in order to survive adverse (usually low temperature) conditions, in this way saving the energy that would normally be used to maintain a high body temperature. Many small birds, rodents and bats enter daily torpor at night. Closely related to aestivation and hibernation.

TOTIPALMATE In aquatic birds, having all four toes connected by a web instead of the most usual arrangement (palmate) where the hallux is free of webbing. In birds with totipalmate feet the 'hind' toe points forward like the others, for example in gannet, cormorant and shag.

TOXIN A poisonous substance.

TRACE ELEMENT See micronutrient.

TRACHEA 1) In insects, one of the small tubes that carries air into the body, open to the air at the spiracles. 2) In vertebrates, the tube (windpipe) leading from the throat to the lungs.

TRACHEID An elongated cell in the xylem of vascular plants that serves in the transport of water and mineral nutrients.

TRAGUS A fleshy and often hairy projection which covers the entrance of the mammalian ear. In many bats, this skin flap is important in echolocation, and therefore in both prey capture and obstacle avoidance.

TRANSPIRATION The process by which water is carried through plants from roots to leaf stomata, where it changes to vapour and released to the atmosphere.

TREE A large perennial woody plant (notionally >10 m in height) usually with a single trunk that supports a number of branches.

TREE LINE The altitude above which trees do not naturally grow on a mountainside.

TREE RING See annual ring and growth ring.

TRIBE A taxonomic rank above genus (grouping genera with particularly shared features), but below family and subfamily.

TRICHOME A small hair or other outgrowth from a plant's epidermis.

TRIFOLIATE Composed of 3 leaflets.

TRIGONOUS Triangular in cross section; stems with three sides.

TRIP Collective noun for dotterels.

TRIQUETROUS Triangular in cross section, with sharp edges and concave sides, seen for example in some stems and seeds.

TRIUNGULIN The first larval stage of oil beetles, which have three-clawed feet used to hitch a ride on host bees (phoresy) to be ferried to the bee nest where they develop.

TROPHIC Related to feeding (Greek *trophe* = 'food' or 'nutrition').

TROPHIC LEVEL A 'level' or stage in the transfer of energy (as food) or nutrients through a food chain, from primary producers (plants, first trophic level), through primary consumers (herbivores, second) and secondary consumers (carnivores, third and higher); the position held by an organism in a food chain.

TROPISM From Greek *tropos* = 'a turn', a movement by a plant or plant part towards (**positive tropism**) or away from (**negative tropism**) the source of an environmental stimulus, for example light (**phototropism**), temperature (**thermotropism**) or gravity (**geotropism**).

TRUNCATED Abruptly ending; in plants, a leaf with a sharply truncated base.

TUBER In plants an underground nutrient storage organ derived from a swollen stem or root, allowing overwintering survival, providing energy and nutrients for regrowth during the next growing season, and as a means of asexual reproduction, for example potato (**stem tuber**) and sweet potato (**root tuber**).

TUBERCLE A small rounded protuberance on the surface of an animal or plant, or on a bone.

TUNIC The brownish paper covering of a bulb.

TURION 1) A bud capable of growing into a complete plant that falls off an aquatic plant into the mud of a pond or river and overwinters. 2) A young shoot emerging from a rhizome, for example an asparagus shoot.

TURLOUGH In Ireland, a temporary lake in an area of limestone, filled by rising groundwater during the rainy winter season.

TUSSOCK Bunched growth typical of sedges, rushes and some grasses.

TWACK Collective noun for ducks.

TWIG A small thin terminal branch of a woody plant.

TWITCHER, TWITCHING The follower/pursuit of a rare bird, usually to 'tick them' on a list as having been seen.

TYMBAL The membranous resonating cavity of a male cicada, one on each side of the abdomen which vibrates through muscular activity to produce a monotonous call.

U

UBIQUIST A plant or animal species that, with a broad range of ecological tolerance, is found in several kinds of community, ecosystem or biome.

ULNA In tetrapods, the longer, thicker bone of the forelimb.

ULTIMATE CAUSE, ULTIMATE FACTOR Something explaining behavioural or physiological traits in terms of evolutionary forces acting on them, for example availability of food, or that, while a female bird selects a particular male with which to mate based on his display is a **proximate cause** (q.v.), the ultimate cause may be the evolution of a particular preference for display plumage and display behaviour, since offspring will likely also possess these preferred characteristics.

UMBEL An umbrella-shaped inflorescence with all the pedicels arising from a common point.

UMBO 1) The raised boss in the centre of a fungal cap (the adjective being **umbilicate**). 2) The highest part of each valve of a mollusc shell, formed when the animal was a juvenile.

UNCINATE Having a hooked shape.

UNDERSTOREY Layer of vegetation beneath a woodland canopy, incorporating the shrub, field and ground layers.

UNDULATE Wavy, for example along a leaf margin.

UNGULIGRADE Walking on hooves, like an ungulate.

UNIMPROVED Farmland or habitat which has not been recently cleared, drained or cultivated, often allowing secondary succession.

UNIPAROUS An animal that produces only one egg or one offspring at a time.

UNISEXUAL Refers to an organism that is either male or female, for example of a flower having either stamens or pistils but not both.

UNIVALVE Former name for a gastropod.

UNIVERSAL VEIL Thin web- or skin-like structure covering an immature fungal fruit body, splitting as the fruit body grows.

UNIVOLTINE An animal that has one generation a year.

UNKINDNESS Collective noun for ravens.

UPENDING Ducks and other waterbirds submerging the head and foreparts in order to feed, with the tail being raised in the air.

UPLAND Areas of Britain that are generally over 300 m (1,000 feet).

UPPER PARTS The mantle, back, rump and upper tail coverts of a bird.

URCEOLATE Shaped like a flask or pitcher, for example corollas which are more or less round in outline, but contracted at the mouth.

UROPYGIUM A bird's rump.

UROPYIAL GLAND See oil gland.

UTERUS Womb; the organ in female mammals in which the embryo and foetus develop before birth.

UTRICLE 1) A small sac, for example an air-filled cavity in a seaweed. 2) A thin bladder-like pericarp or seed vessel.

V

VAGRANCY, VAGRANT Relating to an animal that appears well outside its normal geographical range, for example a bird that has flown or been blown away from its usual range or migratory route.

VALLEY BOG A peat bog that develops on acidic substrates in gently sloping valley bottoms and depressions with poor drainage, and reliant on ground water, with good examples within New Forest heathland.

VARIANT, VARIETY Variation of a species or other taxon arising in the wild, usually differing in only one characteristic, for example colour or leaf shape.

VASCULAR Relating to the veins or conducting tissue of a body or organ.

VASCULAR BUNDLE A discrete bundle of vascular tissue as part of a plant's transport system; a strand of conducting vessels in stem or leaves, typically with phloem on the outside and xylem on the inside.

VASCULAR PLANT A plant with vascular bundles (angiosperms, gymnosperms and pteridophytes).

VECTOR A carrier of disease or parasites.

VEGETATION The sum total of plants found in a particular habitat.

VEGETATIVE PROPAGATION, REPRODUCTION Reproduction by cuttings, layering and grafting; or asexual reproduction.

VEIL In fungi, a thin web- or skin-like tissue that covers and protects the fruit body (universal veil) or just the gills or pores (partial veil).

VEIN 1) In plants, a strand of vascular tissue comprising one or more vascular bundles. 2) In higher animals, a tube carrying mainly oxygen-depleted blood towards the heart. 3) In insects, a series of often cross-connecting structures in wings carrying haemolymph and providing structural strength.

VELUTINOUS Covered in short soft hairs, like velvet.

VELVET 1) In deer, the skin richly supplied with blood vessels that covers the growing antler, subsequently shed or rubbed off. 2) In invertebrates, the term used for some animals covered with dense, matted hair (c.f. velvet ant under ant, velvet swimming crab under swimming crab).

VENATION The arrangement or pattern of veins in a leaf or in the wing of an insect.

VENOM A toxin secreted by an animal via fangs (some spiders and snakes) or stings (for example bees and wasps). The difference between venom and other poisons is the method of delivery.

VENTRAL On or relating to an animal's underside, the opposite (referring to the upper surface) to dorsal.

VERGE A grass strip bordering a path, road or field.

VERMICULATED Marked with a pattern of sinuous or wavy lines, especially used of bird plumage, for example male teal, male scaup in winter and red-breasted merganser.

VERMIN From Latin *vermis* = 'worm' and originally used for the worm-like larvae of certain foodstuff pests, now extended to include rodents and adult insects such as cockroaches, a subjective and emotive word (c.f. pest).

VERNAL Of or relating to spring.

VERNATION The arrangement of bud scales or young leaves in a shoot bud: **convolute** (rolled with one side round the other), **involute** (with the sides rolled towards the centre of the underside), **revolute** (as involute but on the upper side), **conduplicate** (folded, each leaf clasping those adjacent inside the fold), **plicate** (pleated lengthways) or **circinate** (rolled lengthways).

VERRUCOSE Covered in small, warty outgrowths.

VERTEBRATE An animal that has a backbone, or vertebral column (fish, amphibians, reptiles, birds and mammals), from **vertebrae**, the bone segments that make up the spine.

VERTICILLATE In one or more whorls.

VESTIGIAL A small remnant; of an organ or part of the body, present in a much reduced form, possibly atrophied, having become functionless in the course of evolution.

VIBRISSAE 1) Long stiff hairs (whiskers) growing around a mammalian mouth and elsewhere that stimulate sensory nerves and are used for tactile purposes. 2) Coarse bristle-like feathers (see rictal bristle) growing around the gape of some insectivorous bird.

VICARIANCE A process by which the geographical range of a taxon or a whole biota is split into discontinuous parts by the creation of a physical or biotic barrier to gene flow or dispersal.

VILLOUS In botany, shaggy or covered with soft hair.

VIOLET GLAND Alternative term for the supracaudal gland.

VISCERA The internal organs in the main cavities of a body, especially those in the abdomen, for instance intestines.

VIVIPAROUS, VIVIPARY 1) Refers to a plant that produces seed that germinate inside the fruit while still on the plant, or which reproduces vegetatively. 2) An animal in which the embryo develops within the body and is born active.

VIXEN Female fox.

VOLVA A cup-like structure at the base of the stipe of an agaric fungus, a remnant of the universal veil.

VOMERONASAL ORGAN See flehmen.

VULNERABLE A species with small breeding population, categorised by the IUCN as likely to become endangered unless the circumstances threatening its survival and reproduction improve.

VULPINE Of or resembling a fox (c.f. alopecoid).

W

WADER A (usually) long-legged, long-billed bird of coasts, wetlands and moors.

WAGGLE DANCE See bee dance.

WALLOW Of large mammals, to roll in water or mud to keep cool or as protection from biting insects; a location where such behaviour takes place.

WARM-BLOODED See endotherm.

WARNING COLORATION Or aposematic coloration (q.v.) – bright colours that warn predators that an animal is dangerous or toxic if attacked, for example the black and yellow stripes of bees and wasps.

WARREN 1) A series of underground tunnels and chambers where rabbits live. 2) In medieval times, an enclosed area maintained to raise rabbits for meat and fur.

WASMANNIAN MIMICRY See mimicry.

WATCH Collective noun for nightingales.

WATER MEADOW See meadow.

WATER POTENTIAL The potential energy of water per unit volume relative to pure water, quantifying the tendency of water to move from one area to another through osmosis, capillary action, gravity or mechanical pressure, for example the internal water potential of a plant cell is more negative than pure water, causing water to move from the soil into plant roots via osmosis. See also potassium.

WATER TABLE The upper boundary of the saturated zone of the substrate.

WATERFOWL Or wildfowl, a general name for birds associated with freshwater habitats, for example ducks, geese and swans, generally with short, webbed feet and a broad, flat bill.

WATTLE A bare fleshy appendage (see caruncle) hanging from various parts of the head or neck in several groups of birds and mammals (goats and some pigs), and functionally include **combs**, which are more associated with the top of the head, for instance in chickens. Males in the grouse family have wattles above the eye, and some pheasant species have them below the eye, particularly conspicuous in the mating season. As well as indicating sexual maturity and fitness, wattles function in heat regulation: they are thick with capillaries and veins for overheated blood to be air-cooled as it passes through, a process responsible for the bright pink or red colour commonly suffusing these features.

WEB 1) Membranes between the toes of waterbirds, frogs, etc. For birds, see palmate, semi-palmate, totipalmate and lobed. 2) A net spun by spiders; see cobweb and calamistrated web.

WEDGE Name given to swans flying in a V formation.

WEED A subjective, often emotive word used of a plant that adversely affects an economic activity in some way, for example competing with a crop for resources and reducing the yield, not necessarily for what they do as such, but how many of them are doing it (a population size threshold). See also pest.

WETLAND Defined by the Ramsar Convention on Wetlands of International Importance (1971) as 'all areas of marsh, fen, peatland or water, whether natural or artificial, permanent or temporary, with water that is static or flowing, fresh, brackish or salt, including areas of marine water, the depth of which at low tide does not exceed 6 m'.

WHIFFLING Behaviour in which a bird rapidly descends with a zigzagging movement, sometimes briefly with its body turned upside down but with neck and head twisted 180°, that is, in a 'normal' position.

WHITENESS Collective noun for swans.

WHORL A pattern of spirals or concentric circles, for example in plants, structures such as buds and leaves arising three or more at a time around a stem at the same point, or one of the coils of a snail shell.

WILDERNESS An extensive, more or less uninhabited area of uncultivated land, essentially undisturbed by at least recent human activity, together with naturally developed biotic communities. There is no true wilderness in Britain comparable to, say, Alaskan or Antarctic wilderness, but it is a term sometimes used to describe, for instance, the more remote parts of Dartmoor or the Cairngorms.

WILDFOWL Aquatic game birds; waterfowl.

WILDWOOD 1) An uncultivated woodland that has been allowed to grow naturally. 2) The postglacial (mixed) woodland, reaching a maximum extent in Britain around 6,000 years ago.

WING 1) In plants, the lateral petals of a pea or orchid flower. 2) Flattened appendages on seeds facilitating aerial dispersal (see samara). 3) In birds, bats and insects, a modified forelimb allowing flight, and in some birds swimming.

WING BAR A line of contrasting colour along the tips of the coverts of a bird's wing. A wing bar generally crosses the wing when folded and is seen down the centre of the wing in flight, while a **wing stripe** is evident along the edge of the wing.

WINNOWING Rapid wing beating through a shallow arc, characteristic of some falcons, for example hobby and peregrine, generally alternating with gliding flight.

WISP A small flock of snipe.

WITCH'S (or WITCHES') BROOM Masses of densely branched small twigs found among the branches of trees, said to resemble a witch's broom. They can be induced by various parasites including the fungus birch besom which affects birch trees, but these growths can be caused by a number of different types of organism, including other fungi, oomycetes, insects, mistletoe, dwarf mistletoes, mites, nematodes, phytoplasmas and viruses.

WOLD The open, grassy chalk or limestone hills with few trees, for example in Lincolnshire and East Yorkshire (Yorkshire Wolds), and limestone in the Cotswolds and east Midlands. Previously, however, it referred to wooded landscape, from the Old English, and cognate with the Dutch *woud* and German *Wald*. It is also the origin of the formerly much wooded **Weald**, an area in south-east England between the North and South Downs.

WOOD(LAND) A tree-covered landscape, or its plant community.

WOOD PASTURE Woodland, often containing ancient trees, grazed by livestock or deer which control sapling and shrub growth to maintain a semi-open habitat. Also, the dung enhances invertebrate and fungal diversity.

WOODY PLANTS Trees and shrubs (and some lianas: perennial plants whose stems and larger roots are reinforced with wood produced from secondary xylem). See also lignin.

WOOL ALIEN, WOOL SHODDY A seed or bur of a non-native plant can attach to the wool of a sheep; after shearing the impurities are discarded as 'wool shoddy' along with the attached seed, which might then germinate. Wool aliens are typically found on waste ground, or in arable fields where the shoddy has been spread as a dressing.

WORKER The caste in eusocial insects (sterile female ants, bees and wasps) that maintains the nest, tends the eggs and larvae, and forages for food.

WORT 1) An old term for a herb, often used in combination to refer to a specific plant, for example St John's-wort. 2) The liquid extracted from the mashing process while brewing beer or whisky.

WRECK An incident involving a large number of seabird deaths over a short period of time due to (usually winter) storms, for example see little auk under auk.

X

XANTHISM, XANTHOCHROISM An aberrant condition involving the presence of an excessive amount of yellow pigment in (part of) an animal, for example in bird feathers.

XEROMORPHIC Adapted to withstand dry conditions.

XEROPHYTE A drought-tolerant plant.

XEROSERE A stage in the succession of plants on dry substrates such as dunes.

XYLEM Plant tissue that mainly conducts water but also some dissolved nutrients from roots to leaves.

XYLOPHAGOUS Referring to organisms that feed on wood.

Z

ZINC 1) In plants, an essential micronutrient involved in the production of auxins, activation of enzymes in protein synthesis, creation of starch and other carbohydrates and formation of chlorophyll. It is also involved in plant and seed maturation, with deficiency leading to stunting. An excess translates the element into a toxic heavy metal. 2) In animals and microorganisms, this is also an important dietary component, involved with protein and enzyme production.

ZOEA Free-swimming planktonic larvae of many decapod crustaceans, especially crabs.

ZOOCHORY Dispersal of seeds and spores by animals.

ZOOGEOGRAPHY The study of the geographical distribution of animals.

ZOOID An animal arising from another by budding or division, especially each of the individuals making up a colonial organism, often with different forms and functions. They may be directly connected by tissue, for example corals, or share a common exoskeleton, for example bryozoans.

ZOOLOGY The scientific study of animals.

ZOONOSIS A disease capable of transmission from animals to humans, for example bird flu and rabies.

ZOOPHYTE A plantlike animal, for example sponge, coral or sea anemone.

ZOOPLANKTON Microscopic floating or weakly swimming animals in freshwater and marine systems.

ZOOSPORE A motile spore, possessing one or more flagella, in certain algae, fungi and protozoans.

ZUGUNRUHE A German word combining *Zug* + *Unruhe* = 'migration' + 'anxiety' or 'restlessness', describing animals, particularly birds, that demonstrate restless behaviour immediately before migration, or during the usual migration period if movement is constrained.

ZYGODACTYLY In birds, with two toes facing forward and two facing backwards, for example seen in woodpeckers.

ZYGOMORPHIC In flowers, bilaterally symmetrical in one plane; see irregular.

ZYGOTE The cell formed by the union of male and female sex cells at fertilisation.

Part II: Organisms

A

AARON'S ROD Alternative name for great mullein, less commonly agrimony and goldenrod. The tall raceme suggests the rod of Levi on which Aaron's name was written (*Numbers*, xvii: 3–8).

ABALONE Species of *Haliotis*, the only British representative being the edible green ormer. The thick inner layer of the shell is composed of nacre (mother-of-pearl). The name dates from the mid-nineteenth century, from American Spanish *abulones*, from *aulón* (American Indian).

ABELE Alternative name for white poplar, from Latin *albellus* = 'little white (one)', referring to the white underside of the leaf.

ABRAHAM-ISAAC-JACOB *Trachystemon orientalis* (Boraginaceae), a rhizomatous perennial neophyte herb from the Caucasus and Turkey, grown in gardens as a shade-tolerant, weed-smothering, ground cover ornamental, naturalised on banks and in damp woodland in Britain as far north as central Scotland. The English name may reflect the changing colour of the flowers as they age, with three shades often appearing on the same plant at the same time: Abraham, Isaac and Jacob were the founders of Judaism. This name has in the past been given to other such flowering plants, e.g. lungwort and common comfrey. Greek *trachys* + *stemon* = 'rough' + 'stamen'.

ACAENA Spineless acaena *Acaena inermis* (Rosaceae), a mat-forming perennial New Zealand neophyte, widespread and very scattered as a naturalised garden escape on often bare ground by streams and on banks, roadsides and waste ground. Greek *acaina* (*acantha*) = 'thorn', and Latin *inermis* = 'spineless'.

  Two-spined acaena *A. ovalifolia*, a mat-forming perennial South American neophyte, a local naturalised garden escape in woodland, and on roadsides, moorland and dunes, along tracks and paths and in other sparsely vegetated places, scattered in western Britain and Ireland. *Ovalifolia* is Latin for 'oval leaf'.

ACANTHACEAE Family name of bear's-breeches, derived from the Greek for this plant, *acanthos*.

ACANTHOCEPHALA A small phylum of endoparasitic worms known as spiny-headed or thorny-headed. Most are parasitic on aquatic vertebrates and use crustaceans as the intermediate host. Others are terrestrial and use insects as the host. Six of the nine avian acanthocephalids parasitise waterbirds. *Acanthos* is Greek for the spiny plant bear's-breech + *cephali* = 'spiny head'.

ACARI The subclass (in some classifications, order) containing ticks and mites, of which there are at least 1,700 British species grouped into five orders (or suborders) – four of mites (Astigmata, Mesostigmata, Oribatida and Trombidiformes) and one of ticks (Ixodida).

ACCENTOR Birds in the family Prunellidae; see dunnock.
Alpine accentor *Prunella collaris* is an occasional vagrant. The name dates from the early nineteenth century, from Latin *accentor* (*ad* + *cantor*) = 'one who sings with another'.

ACCIPITER, ACCIPITRIDAE The descriptive and family names of hawks. *Accipiter* is Latin for hawk or bird of prey.

ACERACEAE Former family name for maples and sycamore, now (together with horse-chestnuts) called Sapindaceae. *Acer*

is Latin for 'hard' or 'sharp': the wood is hard, and was used in making pikes and lances.

ACONITE Winter aconite *Eranthis hyemalis* (Ranunculaceae), a herbaceous tuberous perennial introduced from Europe by at least Tudor times, with a scattered distribution as a garden escape, naturalised, sometimes in large numbers, in open woodland, grassland and scrub associated with habitation, and also in gardens and on verges, most records coming from England south of the Midlands and from the east coast. 'Aconite' dates from the mid-sixteenth century, ultimately from Greek *aconiton* (see also monk's-hood and wolf's-bane). Greek *er* + *anthos* = 'spring' + 'flower', and Latin *hyemalis* = 'wintry', both reflecting the plant's early flowering.

ACORACEAE Family name of sweet-flag. *Acor* is Latin for 'sour taste'.

ACORN BARNACLE See barnacle.

ACRIDIDAE Grasshopper family (Greek *acrida* = 'grasshopper') comprising 12 native mainland species, together with blue-winged and Jersey grasshoppers, found only in the Channel Islands. There are also three occasional migrant species (desert locust *Schistocerca gregaria*, Italian locust *Calliptamus italicus* and migratory locust *Locusta migratoria*), and one occasional casual introduction (Egyptian grasshopper *Anacridium aegyptium*).

ACROCEPHALIDAE Family name for some warblers (Greek *acros* + *cephali* = 'end' + 'head').

ACTEON Lathe acteon *Acteon tornatilis* (Acteonidae), a carnivorous sea snail whose aperture accounts for two-thirds of the shell length of 2–3 cm, burrowing in sand from the intertidal to 250 m depth, feeding on infaunal polychaetes. Actaeon was a huntsman in Greek mythology, turned into a deer for viewing a naked bathing Artemis. *Tornatus* is Latin for 'turned'.

ACTINIIDAE Family name of sea anemones, with 14 species in 6 genera in British waters. See sea anemone and glaucous pimplet. *Actinos* is Greek for 'ray'.

ACTINOBACTERIA Phylum of gram-positive bacteria found in marine, freshwater and soil environments, often the commonest organisms, with an important role in organic material decomposition. *Actinos* is Greek for 'ray'.

ACTINOMYCETALES From Greek *actinos* + *myces* = 'ray' + '(shaped like) fungus', an order of Actinobacteria, either rod-shaped or filamentous.

ACULEATA Ants, bees and stinging wasps, a subclade of the Hymenoptera characterised by modification of the ovipositor into a stinger, though not all members of the group can sting. *Aculeata* is Latin for 'pointed'.

ACUTE OAK DECLINE A disease mainly affecting largely mature native oaks, making its first presence in Britain in the 1980s, mainly in south-east England and the Midlands, but having reached the Welsh Borders. Affected trees have weeping fissures that seep black fluid down the trunk. A lesion is formed in the live tissue beneath the bleeds, a sign of tissue decay. Some trees die four to six years after onset of symptoms. The larval galleries of oak jewel beetle are usually found in association with the lesions and various species of bacteria have also been isolated. High co-occurrence of the beetle and the bacteria suggest that both play a role in this disease.

ADDER *Viperus berus* (Viperidae), with a distinctive black zigzag dorsal marking, males up to 60 cm long, females 75 cm, locally common and widespread in Britain, though protected under the WCA, usually found in open heathland in southern regions, but also in sunny open areas within woodland, and woodland edge. Males appear from hibernation in early spring, females following about a month later. Reproduction tends to be every other year, mating occurring in late April and the first half of May, when males are territorial and duel or 'dance' as a show of strength. Young are born (an example of ovoviviparity) in late summer. This is Britain's only venomous snake, the bite rarely fatal to humans, though medical attention should be sought. The diet is mainly of small rodents, but also lizards, frogs, newts, and the young of ground-nesting birds. *Viper* is Latin for this species, derived from Latin *vivus + pario* = 'alive' + 'to produce'.

ADDER'S-TONGUE Members of the fern family Ophioglossaceae. Specifically, *Ophioglossum vulgatum*, common and widespread, associated with mostly lowland habitats, especially old grassland, woodland margins and dune slacks, and under bracken. The binomial is Greek for 'snake' + 'tongue', and Latin for 'common'.

Least adder's-tongue *O. lusitanicum*, a rarity found in Guernsey and Scilly in short coastal turf. *Lusitanicum* is Latin for 'Portuguese'.

Small adder's-tongue *O. azoricum*, with a very scattered distribution around the western British Isles, found in grassy coastal habitats on damp sandy soils and peat. *Azoricum* is Latin for 'from the Azores'.

ADELIDAE Family name for long-horn moths, from Greek *adelos* = 'unseen', from the larva being 'unseen' in its portable case.

ADEPHAGA From Greek for 'gluttonous', the suborder of beetles which contains carnivorous beetles in the families Carabidae, Dytiscidae, Gyrinidae, Haliplidae, Hygrobiidae and Noteridae.

ADERIDAE Family name for beetles (1.5–2.5 mm length) in the superfamily Tenebrionoidea, with three British species in three genera, all probably breeding in the heartwood of trees, especially oak, attacked by red rot. Larvae are found under bark, in decomposing wood and litter.

ADMIRAL (BUTTERFLY) Members of the family Nymphalidae.

Red admiral *Vanessa atalanta* (Nymphalinae), wingspan 45–50 mm, primarily a migrant, although sightings of individuals and immature stages in the first few months of the year, especially in southern England, mean that a few individuals are now resident. Because the number of adults seen in any one year depends on the number of migrants, numbers fluctuate: in some years this butterfly can be widespread and common, in others local and scarce. Peak flight period is from June to October. It is found in a range of habitats and in 'good' years is common in gardens. Larvae favour common nettle, creating and living within a tent formed by folding the edges of a leaf together, emerging only to feed. Adults have a variety of nectar sources, including ivy, bramble and buddleia, as well as feeding on rotting fruit and honeydew. The binomial's origin lies in Jonathan Swift's poem Cadenus and Vanessa, and the huntress Atalanta in Greek mythology.

White admiral *Limenitis camilla* (Limenitidinae), male wingspan 56–64 mm, females 58–66 mm, a deciduous woodland species of south, east and central England. Having declined in the earlier part of the twentieth century it has now more or less recovered its former range (though with a 14% decline between 2005 and 2014), and numbers continue to drop (by 45% in 2005–14). Global warming may have allowed it to re-establish at sites that had become too cool and cessation of coppicing may have benefited this species which depends on honeysuckle growing in shady woodland for successful larval development. Adults, flying in June–August, feed on honeydew and sap, the nectar of bramble and other plants, and salts and minerals from moist earth and animal droppings. It remains a priority species for conservation. The binomial is from Greek *limenitis* =

'harbour-keeping' (no entomological significance), and a princess mentioned by Virgil in the *Aeniad*.

ADOXACEAE Family name of moschatel (Greek *adoxos* = 'insignificant').

AEGITHALIDAE Family name of long-tailed tit. *Aegithalos* was used by Aristotle for some species of tit.

AEOLIS Sea slugs of the shallow sublittoral in the families Dendronotidae, Flabellinidae and Proctonotidae.

AESHNIDAE From New Latin *aeschna*, erroneously from Greek *aischros* = 'ugly', dragonfly family with three British genera: hawkers *Aeshna*, emperors *Anax* and hairy dragonfly *Brachytron pratense*.

AESOP Slim aesop *Facelina auriculata* (Facelinidae), a widespread sea slug up to 3.8 cm long, under boulders in the intertidal and sublittoral, feeding on hydroids. *Auriculata* is Latin for 'eared'.

AGARIC Fungus with a cap and stalk, with gills on the underside of the cap. 'Agaric' is late Middle English, originally denoting various bracket fungi with medicinal or other uses (Greek *agaricon* = 'tree fungus').

Clouded agaric Alternative name for clouded funnel.

Dingy agaric Alternative name for coalman.

Fly agaric *Amanita muscaria* (Agaricales, Amanitaceae), common and widespread, poisonous (though rarely fatal) and hallucinogenic, containing the psychoactive chemical compounds ibotenic acid and muscimol, found in late summer and autumn on infertile, often sandy soils, especially under birch and pine. Greek *amanitai* = 'amanitas', and Latin *muscaria* = 'relating to flies', a reference to a tradition of using this fungus to kill flies.

AGARICOMYCETES Class of fungi, more or less equivalent to the homobasidiomycetes (Greek *agaricon* and *myces*, both meaning fungus).

AGELENIDAE Family name for funnel-web spiders, from Greek *agele* = 'herd', here meaning living gregariously, with 13 species in the genera *Agelena* (labyrinth spider), *Tegenaria* (11 species) and *Textrix*.

AGNATHA From Greek for 'no jaws', a superclass of fish including lampreys.

AGRIMONY *Agrimonia eupatoria* (Rosaceae), a perennial herb of basic and neutral soils, widespread and common except in north Scotland, found in hedge banks, field margins and grassland, and on woodland margins, roadsides and railway banks, sometimes in waste ground. 'Agrimony' comes from the fourteenth-century *egremounde*, from Greek *argemon*, a white speck on the cornea for which the plant was prescribed. Mithridates Eupator was King of Pontus around 115 BCE.

Bastard agrimony *Aremonia agrimonioides*, a European neophyte, widespread but especially in central Scotland, naturalised in woodland and shaded roadsides.

Fragrant agrimony *Ag. procera*, widespread but scattered in hedge banks, field margins and grassland, and on woodland margins, roadsides, railway banks and waste ground. *Procera* is Latin for 'tall'.

AGRIOLIMACIDAE Family name for field and marsh slugs (Greek *agrios* = 'living in fields' and Latin *limax* = 'slug').

AGROMYZIDAE Family name (Latin *ager* = 'field', Greek *myzao* = 'to suck') for flies in the superfamily Opomyzoidea, with 394 species in 18 genera. Adults are usually 2–3 mm long. Larvae are generally leaf miners (some are gall-formers).

AIPTASIIDAE From Greek *aiptos* = 'one unable to fly', family name of sea anemones with two species in two genera. See trumpet anemone.

AIZOACEAE Family name, from Greek for 'evergreen', for dewplants, annual or semi-woody perennial neophytes occasionally naturalised on walls and hedge banks as garden escapes in southern England and the Scilly and Channel Islands.

AJOWAN *Trachyspermum ammi* (Apiaceae), an annual Indian or eastern Mediterranean neophyte scattered in England and central Scotland, casual on waste ground, originating from birdseed. Ajowan is from Hindi *ajvāyn*. The genus name is a Greek-Latin compound meaning 'rough-seeded'.

AKE-AKE *Olearia traversii* (Asteraceae), an evergreen neophyte shrub introduced from the Chatham Islands by 1887, used in hedges in the Channel Islands and the Scillies, and naturalised as a garden escape in rough grassland in Cornwall, Isle of Man and south and west Ireland. *Olearia* honours the German botanist Johann Ölschläger (1635–1711), his name latinised to Olearius; *traversii* is from William Travers or his son, Henry Travers, early explorers in New Zealand.

AKIRAHO *Olearia paniculata* (Asteraceae), an evergreen New Zealand neophyte shrub introduced by 1866, grown in Alderney and Guernsey hedges, naturalised in quarries, hedgerows and waste ground in Cornwall, Somerset and Aberdeenshire. *Olearia* honours the German botanist Johann Ölschläger (1635–1711), his name latinised to Olearius; *paniculata* refers to the plant's flower clusters (panicles).

ALAUDIDAE Family name of larks (Latin *alauda*).

ALBATROSS **Black-browed albatross** *Thalassarche melanophris* (Diomedeidae), wingspan 2–2.4 m, occasionally arriving in Britain during summer, the most famous being 'Albert Ross', which appeared on the Bass Rock in 1967, then subsequently at Hermanness, Shetland almost annually from 1972 to 1995. Greek *thalassa* + *arche* = 'sea' + 'beginning', and *melanos* + *ophris* = 'black' + 'eyebrow'.

ALCIDAE Family name of auks (Latin *alca*, but also note Icelandic *álka*).

ALDER Specifically, *Alnus glutinosa* (Betulaceae), also known as **black alder**, a common and widespread deciduous tree of damp, basic to moderately acidic soils, found by rivers, streams, canals, and ditches, and in flood plains, fens and bogs, carr and dune slacks. It can rapidly colonise open sites, producing even-aged stands, but seedlings are shade- and drought-sensitive, so regeneration in woodland is often poor. It is also widely planted. The alder-specific fungus *Phytophthora alni*, first noted in Britain in 1993, is now widespread: the number of trees affected has increased steadily since the mid-1990s and 20% of trees are now affected. Disease incidence is highest in south-east England; heavy tree losses are occurring in the Welsh Borders; and alders on Scottish river systems are suffering. 'Alder' comes from Old English *alor* and *aler*, possibly from Old German *elawer* = 'reddish-yellow', the colour of the freshly cut wood. *Alnus* is Latin for 'alder', *glutinosa* for 'sticky'.

Grey alder *A. incana*, a widespread deciduous European neophyte tree, present by 1780, widely planted in parks, on roadsides, reclaimed tips and riverbanks. Being very hardy and tolerant of poor, wet soils make it useful in amenity planting in the north. It sometimes becomes naturalised on waste ground and railway embankments. *Incana* is Latin for 'grey'.

Italian alder *A. cordata*, a deciduous neophyte tree introduced in 1820, widespread if commoner in England, naturalised in Somerset, but planted on roadsides, in urban parks and in amenity areas. Unlike other alders, it thrives on nutrient-poor, dry soils. Street trees commonly flower and fruit; seeds are wind-dispersed and seedlings are commonly found by pavements and on waste ground. *Cordata* is Latin for 'heart-shaped', referring to the leaves.

Red alder *A. rubra*, a rapidly growing deciduous North American neophyte tree, introduced before 1880, with a scattered distribution, found in large gardens and parks, occasionally planted for amenity in damp areas, regenerating naturally at sites in Argyll. It is light-demanding, with best growth on soils of poor to medium nutrient status and fresh to wet soil moisture. *Rubra* is Latin for 'red'.

ALDER FLY Members of the order Megaloptera (Sialidae) (Greek *sialis* = a kind of bird), with three British species, *Sialis lutaria* being common and widespread. Eggs are laid by still or slow-moving water, larvae moving into this medium as predators over winter, pupating, then emerging in May–June.

ALDER MOTH *Acronicta alni* (Noctuidae, Acronictinae), wingspan 33–8 mm, nocturnal, widespread but local in broadleaf woodland and scrub throughout much of England, Wales and Ireland. Adults fly in May–June. Young larvae resemble bird dropping, but when mature are black and yellow with club-like hairs. They feed on a number of trees, particularly alder, downy birch and goat willow. Greek *acronyx* = 'nightfall', Latin *alnus* = 'alder'.

ALDER TONGUE *Taphrina alni* (Taphrinales, Taphrinaceae), a fungal pathogen that causes a gall on alder catkins, recorded in Cornwall in the 1940s, then in Northumberland, Ayrshire and Skye, mainly since the 1990s, and now quite common and widespread. Greek *taphros* = 'ditch'.

ALDERCAP Some 44 species of inedible fungus in the genus *Naucoria* (Strophariaceae). The family name comes from Greek *strophos* = 'belt', a reference to the stem rings. **Ochre aldercap** *N. escharioides* and **striate aldercap** *N. striatula* are both widespread and locally common, in late summer and autumn on damp soil, usually with alder, sometimes birch and willow.

ALEXANDERS *Smyrnium olusatrum* (Apiaceae), a herbaceous perennial archaeophyte present since Roman times, widely cultivated until displaced by celery in the fifteenth century, naturalised, often near the sea, in hedge banks, on cliffs, at the base of walls, and on grassy roadsides, pathsides and waste ground, scattered, but absent from the north and west Scottish coast. 'Alexanders' comes from Latin *petroselinum Alexandrinum* = 'rock parsley of Alexandria'. Greek *smyrna* = 'myrrh', referring to its scent, and Latin *olus* = 'vegetable'.

Perfoliate alexanders *S. perfoliatum*, a south European neophyte, scattered mostly in southern England, naturalised in grassy habitats.

ALEXIIDAE From Greek *alexo* = 'to protect', beetle family represented by the uncommon fungivore *Sphaerosoma pilosum*, found in leaf litter and decaying wood.

ALEYRODIDAE Family name of whiteflies (Greek *aleuro* = 'flour').

ALFALFA Alternative name for lucerne, a Spanish word from Arabic *al fasfasah*.

ALGA(E) A polyphyletic group of simple aquatic photosynthetic organisms, ranging from the microscopic (microalgae), for example diatoms, to large seaweeds (macroalgae), for example giant kelp. The most complex freshwater forms are Charophyta, a division of green algae which includes stoneworts. See also cyanobacteria (blue-green algae), Phaeophycea (brown algae), Chlorophyta and Charales.

ALISMATACEAE Family name for water-plantain. *Alisma* is the Greek name for an unidentified water plant.

ALISON Herbaceous plants in the family Brassicaceae. For etymology, see small alison.

Golden alison *Aurinia saxatilis*, a perennial European neophyte cultivated by 1710, a garden escape found as a casual on walls, rocks and dry waste ground, scattered in Britain. Latin *aureus* = 'golden', referring to flower colour, *saxatilis* = 'growing among rocks'.

Hoary alison *Berteroa incana*, a biennial, sometimes annual or perennial European neophyte cultivated by 1640, scattered in England, occasionally naturalised in the south on waste ground, and mainly found as a grain impurity casual in arable fields, on waste ground, around docks and in newly sown grass or clover leys. Carlo Bertero (1789–1831) was an Italian physician and botanist. *Incana* is Latin, here meaning 'hairy'.

Small alison *Alyssum alyssoides*, a casual annual European neophyte, present by 1740, appearing in arable fields. This poor competitor needs disturbance to provide open soil for seedlings. It is now more or less confined to a few sites in Suffolk. Greek *a* + *lyssa* = 'without' + 'rage', referring to the plant's use to cure rabies.

Sweet alison *Lobularia maritima*, an annual, biennial or perennial European neophyte, present by 1722, naturalised on sea-cliffs, dunes and open ground and walls near the sea, and inland as a casual on waste ground. *Lobularia* is Latin for 'small pod'.

ALKANET *Anchusa officinalis* (Boraginaceae), a herbaceous European neophyte perennial present in gardens since at least 1200 and formerly cultivated for fodder, scattered but scarce outside England, on rough ground, hedgerows and railway banks, originating from gardens or from birdseed, occasionally naturalised. Medieval dyers gave the name 'alkanet' to the imported red dyestuff obtained from dyer's alkanet *Alkanna tinctoria*, from Spanish *alcaneta*, in turn from Arabic *al-henna*, relating to henna dye. *Anchusa* was the name used by the Romans for a plant used as a rouge; *officinalis* is Latin meaning 'used in pharmacology or medicine'.

False alkanet *Cynoglottis barrelieri*, a herbaceous south-east European neophyte perennial, an occasional garden escape in England. Greek *cyon* + *glossa* = 'dog' + 'tongue'. The specific name honours the French botanist Jacques Barrelier (1606–73).

Green alkanet *Pentaglottis sempervirens*, a common and widespread perennial south-west European neophyte herb present before 1700, often grown in but also viewed as an invasive weed in gardens, naturalised in lightly shaded habitats, including waste ground, roadside and stream banks, hedgerows, scrub and woodland. Greek *pente* + *glossa* = 'five' + 'tongue', from the five tongue-like scales on the corolla, and Latin *sempervirens* = 'evergreen'.

ALL-HEAL Alternative name for selfheal and common valerian, the roots being used in herbal medicine, for instance in treating epilepsy.

ALLIACEAE The onion family name, including chives, daffodil, garlic, leek, ramsons, snowdrop, snowflake and starflower. *Allium* is Latin for onion, garlic, leek, etc.

ALLSEED *Radiola linoides* (Linaceae), an annual herb, scattered and now mainly found near coastal regions apart from the east coasts of England and Ireland, on damp, bare, infertile, peaty or sandy ground in acid grasslands and heathland, by ponds, and in woodland rides. Near the coast it occurs in dune slacks, sandy grassland and machair. *Linoides* is Latin for 'flax'.

Four-leaved allseed *Polycarpon tetraphyllum* (Caryophyllaceae), an annual found in south-west England, Dorset, the Scillies and the Channel Islands, casual in a few other places, including Moray and Co. Wexford, in sunny sites that are summer-droughted and relatively frost-free in winter, on waste ground and compacted shingle and sand, in bulb fields and gardens. The binomial comes from Greek for 'many-fruited' and 'four-leaved'.

ALMOND *Prunus dulcis* (Rosaceae), a deciduous tree long-introduced from south-west Asia via Europe, scattered in England and Wales on waste ground, planted on roadsides and in parks. 'Almond' comes from Old French *amande*, itself from Low Latin *amandula*, Latin *amygdala*, Greek *amygdale*, and Hebrew *megedh el* (= 'sacred fruit'). The binomial is Latin for 'plum tree' and 'sweet'.

ALOE-MOSS Species of *Aloina* (Pottiaceae), the genus name possibly reflecting the fleshy characteristics of the aloe-like leaves. Species are associated with chalk and limestone habitats.

ALPINE-SEDGE Rare species of *Carex* (Cyperaceae). Latin *carex* = 'sedge'.

Black alpine-sedge *C. atrata*, rhizomatous, with a few records from south Scotland, Cumbria and Snowdonia, on wet calcareous rock ledges, growing in short vegetation, among tall herbs, or in dwarf willow scrub. Latin *atra* = 'black'.

Close-headed alpine-sedge *C. norvegica*, on wet ledges and grass over basic rock with late snow-lie. *Norvegica* is Latin for 'Norwegian'.

Scorched alpine-sedge *C. atrofusca*, on stony, calcareous mountain flushes and bogs. Latin *atra* + *fusca* = 'very dark' + 'dusky'.

ALTAR-LILY *Zantedeschia aethiopica* (Araceae), a rhizomatous South African neophyte perennial, scattered in (especially southern) England and south Wales, naturalised as an escape from cultivation as an ornamental, persistent and spreading in damp hedgerows, scrub and ditches. Francesco Zantedeschi (1797–1873) was an Italian priest and scientist.

ALYDIDAE A shieldbug family with a single British representative, *Alydus calcaratus*, wingspan 10–12 mm, widespread in England and parts of Wales but local to dry heathland together with brownfield sites in the Thames Gateway. Nymphs are ant-shaped, possibly allowing them to raid ant nests.

AMANITA Fungi in the genus *Amanita* (Amanitaceae), *amanitai* being Greek for this group. Rarities include **bearded amanita** *A. ovoidea* (Wiltshire and the Isle of Wight), **fragile amanita** (or **alder amanita**) *A. friabilis* (one site in each of Devon, Kent and Lancashire, and two in Monmouthshire), **jewelled amanita** *A. gemmata* (lethally poisonous, containing the psychoactive chemical compounds ibotenic acid and muscimol), and **solitary amanita** *A. echinocephala* (poisonous, in southern England and the Yorkshire Wolds).

Grey spotted amanita *A. excelsa*, undistinctively edible, widespread and locally common in summer and autumn on soil in broadleaf or mixed woodland. Latin *excelsa* = 'high'.

Grey veiled amanita *A. porphyria* var. *spissa*, inedible, possibly toxin-containing, widespread but uncommon in summer and autumn on acidic soil associated with coniferous trees, occasionally heathland. *Porphyros* is Greek for 'purple', *spissa* Latin for 'dense', referring to the closely spaced gills.

Warted amanita *A. strobiliformis*, probably edible, possibly hallucinogenic, rare, scattered, found in summer and autumn on preferably alkaline soil in broadleaf woodland. *Strobiliformis* comes from Latin meaning having overlapping scales (cone-like), from the bract-like projections arranged in a spiral pattern.

AMANITACEAE Family name for homobasidiomycete fungi in the order Agaricales, generally in the genus *Amanita*, including a number of Britain's most poisonous fungi, for example deathcap, destroying angel, fly agaric, jewelled agaric and panthercap. See also grisette and blusher.

AMARANTH Annual neophytes in the genus *Amaranthus* (Amaranthaceae) (Greek *amaranton* = 'everlasting flowers'). A number of species are occasional casuals introduced from the Americas, found as garden escapes and from waste products such as wool shoddy and birdseed.

Common amaranth *A. retroflexus*, from North America, in gardens by 1729, scattered mainly in England, escapes reinforced by introductions via wool shoddy, birdseed and soyabean waste, on disturbed, nutrient-rich waste ground, waysides and cultivated land, usually as a casual. *Retroflexus* is Latin for 'bent backwards'.

Green amaranth *A. hybridus*, a semi-tropical American plant in gardens by 1656, scattered, mainly in England, rare in Scotland and Ireland, on disturbed, nutrient-rich waste ground, waysides and arable fields, usually casual, occasionally naturalised, originating from wool shoddy, birdseed, etc.

Purple amaranth *A. cruentus*, probably a domesticated form of common amaranth, scattered in England as a casual, occasionally naturalised. *Cruentus* is Latin for 'blood-coloured'.

AMARANTHACEAE From Greek *amaranton* = 'everlasting

flowers' for the goosefoot family, formerly known as Cheno-podiaceae.

AMARYLLIDACEAE Former name of the family that includes snowdrops and daffodils, now subsumed into the Alliaceae. Meaning 'sparkling' in Greek, Amaryllis was also a shepherdess in Greek mythology.

AMAUROBIIDAE Family name of tangled nest (funnel-web) spiders, with three species of *Amaurobius* and two of *Coelotes*. See lace-weaver spider. *Amauros* is Greek for 'the dark'.

AMBER SNAIL Members of the terrestrial pulmonate family Succineidae in the genera *Oxyloma* (Greek *oxys* + *loma* = 'sharp' + 'fringe') and *Succinella* (Latin *succinum* = 'amber').

Large amber snail *S. putris*, shell length 15–22 mm, width 7–12 mm, common and widespread in England, Wales and Ireland, locally common in south Scotland, in damp well-vegetated sites, often on the stems of waterside plants. *Putris* is Latin for 'putrid'.

Pfeiffer's amber snail *O. elegans*, shell length 9–15 mm, common and widespread on plants, generally near the base, in damp habitats, such as marsh, water meadow and pond banks, sometimes on floating plants. Ludwig Pfeiffer (1805–77) was a German botanist and conchologist.

Slender amber snail *O. sarsii*, scarce and scattered, shell length 10–14 mm, found on mud or low vegetation on lake and river margins.

Small amber snail *S. oblonga*, shell up to 8 mm length, width 4.5, scattered and uncommon in Britain, locally common in Ireland, declining throughout, in sparsely vegetated sites often on damp sand or calcareous sediment and on lake shores.

AMBLYOMMIDAE Family name (Greek *amblys* + *omma* = 'blunt' + 'appearance') of six British tick species in five genera.

AMBROSIA FUNGUS Ascomycete fungal symbiont of ambrosia beetles.

AMELETIDAE Mayfly family (Greek *ameletes* = 'not worthy of attention') represented by upland summer mayfly.

AMMODYTIDAE Family name of sand eels (Greek *ammos* + *dytis* = 'sand' + 'diver').

AMOEBA A single-celled eukaryotic protoplasmic organism within a number of taxonomic groups, including protozoa, fungi and algae. Amoebae change shape by extending and contracting pseudopods which, with protoplasmic flow, allows movement. They reproduce by binary fission (splitting into two). Some amoebas are predatory, feeding on bacteria and other protists; others are detritivores. Several species occur in the mud and water of ponds and ditches.

AMPHIBIA Class that includes the order Caudata (newts) and Anura (frogs and toads), animals with both aquatic and terrestrial life stages.

AMPHIPODA Order of shrimp-like crustaceans (class Malacostraca) with no carapace and generally with laterally compressed bodies. Greek *amphi* + *podi* = 'both kinds' + 'foot', a reference to the two different kinds of leg, used for swimming and walking. They are mainly marine, but contain some fresh and brackish water species, feeding as detritivores or scavenging. See woodhopper, sand hopper and shrimp.

ANACARDIACEAE Family name for sumaches and smoke-tree, from Greek *anei* + *cardia* = 'without' + 'heart', because the pulp of the fruit has the nut growing from the end of it rather than having the seed enclosed.

ANATIDAE Family name of swans, ducks and geese (Latin *anas* = 'duck').

ANCHUSA **Garden anchusa** *Anchusa azurea* (Boraginaceae), a herbaceous south European neophyte perennial, an occasional casual on waste ground. Greek *agchousa* = 'alkanet', and Latin *azurea* = 'azure'.

ANDREAEOPSIDA Class of mosses including the order Andreaeales, the name honouring the German apothecary J.G.R. Andreae (1724–93).

ANDRENIDAE From Greek *andrene* = 'wasp', family name for solitary, ground-nesting mining bees.

ANEMONE (ANIMAL) See sea anemone.

ANEMONE (PLANT) Perennial herbs in the genus *Anemone* (Ranunculaceae), from Greek *anemos* = 'wind', with a feminine derivation meaning 'daughter of the wind' (Pliny claimed it only flowered when the wind blew), giving the alternative name wind flower.

Balkan anemone *A. blanda*, tuberous, introduced from the Balkans in 1898, with a scattered distribution as a naturalised garden throw-out in woodland, on verges, under park trees and in churchyards in England and south Scotland. *Blanda* is Latin for 'smooth'.

Blue anemone *A. apennina*, rhizomatous, introduced from Mediterranean Europe by 1724, widespread and naturalised as a garden escape favouring light shade in woodland, scrub and churchyards, and near former habitation. *Apennina* refers to the Apennine mountains.

Wood anemone *A. nemorosa*, widespread, rhizomatous, in broadleaf woodland and hedgerows in (semi-)shade in moist, moderately fertile soil. Seeds are rarely fertile, so spread is vegetative and very slow, making this a good indicator of ancient woodland. It is also found on hedge banks, heathy grassland, moorland, scree and limestone pavement. This is the 'county' flower of Middlesex as voted by the public in a 2002 survey by Plantlife. *Nemorosa* is Latin for 'of woodland'.

Yellow anemone *A. ranunculoides*, spring-flowering, rhizomatous, introduced from Europe by Tudor times, naturalised and scattered in England and south Scotland as a garden escape or throw-out in shady places such as woodland, and along paths. *Ranunculoides* is Latin meaning 'resembling buttercups'.

ANGEL **Destroying angel** *Amanita virosa* (Agaricales, Amaritaceae), an infrequent fungus, with a scattered generally upland British distribution, commoner in Scotland. It is lethal, containing amatoxins, a complex group of substances that initially cause gastro-intestinal disorders, then generally irreversible damage to kidneys and liver, followed by coma and death. It is found in summer and autumn on soil in broadleaf and mixed woodland. *Amanita* is Greek for this kind of fungus; *virosa* is Latin for 'rank' or 'poisonous'.

ANGELICA Herbaceous perennials in the genus *Angelica* (Apiaceae). The medieval Latin name was *herba angelica* = 'angel's plant', reflecting the tradition that an angel (or specifically Raphael, hence *archangelica*) had revealed the efficacy of this plant against cholera and the plague.

Garden angelica *A. archangelica*, a monocarpic European neophyte present by the mid-sixteenth century, scattered and naturalised, especially in England, on riversides, roadsides and in waste ground, or as a casual.

Wild angelica *A. sylvestris*, common and widespread on base-enriched soils in a variety of habitats, for instance damp woodland and carr, damp neutral grassland, marsh, tall-herb fen and ungrazed montane grassland. *Sylvestris* is Latin for 'wild'.

ANGELICA-TREE Deciduous neophyte shrubs or small trees in the genus *Aralia* (Araliaceae), etymology unknown, occasionally naturalised garden escapes in hedgerows, field edges, roadsides and waste ground. **Chinese angelica-tree** *A. chinensis* was introduced in 1830, **Japanese angelica-tree** *A. elata* in 1865. *Elata* is Latin for 'tall'.

ANGEL'S WINGS *Pleurocybella porrigens* (Agaricales, Marasmiaceae), an inedible fungus common and widespread in the northern half of Scotland, with a few scattered records elsewhere, on dead wood in coniferous woodland. Greek *pleuro* + *cybella* =

'side' + 'small cap' (the stalk being on the side of the cap), and Latin *porrigens* = 'stretching out'.

ANGIOSPERM A flowering plant that produce seeds within a carpel, from Greek *angeion* + *sperma* = 'case' + 'seed'. See also gymnosperm.

ANGLE Tawny-barred angle *Macaria liturata* (Geometridae, Ennominae), wingspan 22–7 mm, a widespread and common nocturnal moth found in coniferous woodland and gardens. Adults fly in June–July, with another brood in August in the south. Larvae feed on Scots pine and other conifers. Greek *macaria* = 'happiness', and Latin *litura* = 'blot'.

ANGLER'S CURSE Mayflies in the genus *Caenis* (Caenidae), Caenis being a figure in Greek mythology. *Caenis rivulorum* is the commonest and most widespread, though more localised in south-east England. Nymphs live in pools, stream margins, lakes and canals, adapted for moving in and on mud and silt, feeding by collecting fine particulate organic detritus from the sediment. Adults emerge in May–September. *Rivulorum* is Latin for 'of a stream'.

ANGUIDAE Family name of slow worm (Latin *anguis* = 'snake').

ANGUILLIDAE Family name of freshwater eels (Latin *anguillis* = 'eel').

ANGUILLIFORMES The order to which (true) eels belong.

ANISOPODIDAE Family name (Greek *aniso* + *podi* = 'unequal' + 'leg') of wood gnats, generally associated with moist habitats, including woodland. There are four species in one genus, *Sylvicola*, including the common and widespread *S. fenestralis* (6–8 mm length), often found near buildings and sewage-works, larvae feeding on dung and decaying organic matter. Latin *sylva* + *colo* = 'forest' + 'to inhabit', *fenestralis* for 'window'.

ANISOPTERA Suborder of dragonflies (Greek *aniso* + *ptero* = 'unequal' + 'wing'), with five families: Aeshnidae, Cordalegasteridae, Corduliidae, Gomphidae and Libellulidae.

ANNELIDA From Latin *anellus*, diminutive of *anulus* = 'ring', the phylum of ringed or segmented worms that includes earthworms, leeches, ragworms, tubeworms and fanworms. Classes include Polychaeta, Oligochaeta and Hirudinea, or (according to some taxonomists) Clitellata, divided into the subclasses Oligochaeta and Hirudinea.

ANNULET *Charissa obscurata* (Geometridae), wingspan 27–32 mm, a night-flying moth (July–August) well-distributed around the coast of Britain and inland on the heaths and downs of southern England and Wales. In Ireland it occurs on the coast. Also recorded on the Isle of Man and the Channel Islands. Adults feed on a variety of herbaceous plants; larvae favour sea campion, thrift and rock-rose on the coast, bird's-foot-trefoil and heather elsewhere. Colour tends to reflect habitat: pale grey in limestone and chalk, darker in peaty areas. An annulet is a small ring. Greek *charis* = 'grace', and Latin *obscurus* = 'dark'.

Irish annulet *Gnophos dumetata*, wingspan 26 mm, nocturnal, first discovered in Ireland in 1991, and now fairly well established in the Burren area of Co. Clare. Adults fly in August. Larvae feed on blackthorn. Greek *gnophos* = 'darkness' and Latin *dumetum* = 'thicket'.

Scotch annulet *G. obfuscata*, wingspan 41–6 mm, scarce, predominantly night-flying (July–August), on rocky moorland and mountain hillsides, local in the northern half of Scotland and parts of west Ireland. Larvae feed on ericaceous plants, stonecrops, saxifrages and petty whin. *Obfuscus* is Latin for 'dusky'.

ANOMALOUS (MOTH) *Stilbia anomala* (Noctuidae, Oncocnemidinae), wingspan 29–36 mm, local on moorland in Scotland, northern England, Isle of Man, Dartmoor, Exmoor and the New Forest, and through much of Wales and Ireland. With numbers declining by 97% over 1968–2007, it is of conservation concern

in the UK BAP. Adults fly in August–September. Larvae feed on grasses, especially wavy and tufted hair-grass. Greek *stilbo* = 'glisten' (from the glossy forewings) and *anomalos* = 'anomalous' (it looking more like a pyralid than a noctuid moth).

ANOMIIDAE Family name (Greek *anomos* = 'irregular') for some marine bivalve molluscs. See saddle oyster, and small oyster under oyster.

ANOSTRACA From Greek *anostracos* = 'anarchist', an order in the Crustacea, class Branchiopoda, that includes fairy shrimp.

ANSERIFORMES Order of ducks, geese and swans (Latin *anser* = 'goose').

ANT 1) Eusocial insect in the order Hymenoptera, family Formicidae, typically having a sting and living in a colony with one or more breeding queens. It is wingless except for fertile adults which form large mating swarms. *Formica* (Latin for 'ant') and *Lasius* (Greek *lasios* = 'hairy') are common genera. The word derives from Old English *æmete*.

Black ant or **small black ant** *L. niger*, queens 9 mm long, males 3.5–4.5 mm, workers 3–5 mm, common and widespread in southern Britain, locally common in Scotland, occasional in Ireland, found in dry open habitats including post-industrial sites, gardens and pavements. Nuptial flights take place in July–August. Nests are often large, located under stones, paving slabs, pieces of wood, plastic or metal places that can be warmed by the sun. Where there are no suitable sites, soft mound nests are sometimes found. Colonies have one queen and from several hundred up to 10,000 workers that forage in the open. They are aggressive and will attack other ants. They scavenge and prey on small invertebrates. They tend aphids and 'milk' them for honeydew. Early in the season, when food sources are scarce, they may enter houses, attracted to sweet substances. *Niger* is Latin for 'black'.

Brown ant *L. brunneus*, a tree-dwelling species recorded from south and central England, closely associated with a few species of ant-tended Homoptera. Flight is in June–July. Nests are usually within mature trees, especially oak, but also in stumps, hedgerows and timber-framed buildings. Common signs of a tree being inhabited are piles of frass resulting from excavations; tunnels can be found by removing bark but the nest itself can be deep within the tree. Workers are rarely seen away from their host tree or even on its surface. Most activity occurs in bark crevices or tunnels where the ants tend aphids. Aphid excreta form the majority of the ant diet, though they also take other small insects found on or under the bark. *Brunneus* is Latin for 'brown'.

Erratic ant *Tapinoma erraticum*, in dry, sunny habitats with short vegetation and patches of bare (including recently burned) ground on heathland, particularly in the New Forest and Surrey. It scavenges or preys on small invertebrates. Nests are shallow constructions in soil, occasionally under stones, on banks or in dry peat or clumps of moss. Ants may raise solaria of soil and vegetation fragments, often around supporting vegetation. Greek *tapenoma* = 'humility'. Latin *erraticum* = 'wandering'.

Grass ant *Myrmecina graminicola*, locally common in southern England and the south Midlands, and south and west Wales. Nests are under stones or fallen wood, in tree stumps, in soil, and under moss, and have also been found in the nests of other ant species, in pasture, dry open woodland and gardens. Workers scavenge and prey on small invertebrates on the ground. When disturbed, females curl into balls and appear dead. Sexuals develop during late summer, with mating taking place on the ground. *Myrmecina* is a diminutive of Greek *myrmex* = 'ant'; *graminicola* is Latin for 'living in grass'.

Hairy wood ant *F. lugubris*, found in northern England, north Wales and the Highlands in deciduous and coniferous woodland, extending into open, often rocky areas. Mating flights occur in June–July. Mound nests are composed of vegetable litter, some existing in isolation, but large groups of interconnecting nests often occur. Foraging workers follow pheromone trails from the nest mound, leading to proven nectar sites or to groups of aphids from which the ants collect honeydew and the aphids themselves. Workers also prey on other arthropods, including insect pests living in the tree canopy. *Lugubris* is Latin for 'doleful'.

Indolent ant *Ponera indica*, locally common in southern England, extending into south Wales in warm, sheltered habitats, including waste ground and gardens, with flight in August–September. Workers forage individually in soil, litter and moss, preying on (largely developmental stages of) ground-dwelling invertebrates. Nests comprise simple passageways running through the soil and with one or more small chambers; colonies are small, generally with up to 40 workers. Greek *poneros* = 'vicious'. Latin *indica* = 'Indian'.

Jet ant *L. fuliginosus*, locally common in southern England and Wales, contracting in range in north and central England and southern Ireland. It is a semi-social parasite of a semi-social parasite, establishing colonies in those of the *L. umbratus* group, which are in turn founded in colonies of the *L. niger/flavus* species complexes. The low chance of success for a founding queen under these circumstances may be the reason for the localised distribution of colonies. Workers forage on trails leading some distance from the nest to trees where they tend aphids for honeydew. Nests are made from wood fragments cemented together with saliva, usually in the hollow of a partially rotted tree, log or stump, or within a hedge bank or wall. *Fuliginosus* is Latin for 'sooty'.

Long-spined ant *Temnothorax interruptus*, recently recorded only from heathland in Dorset and the New Forest, and from Dungeness. Workers forage individually, scavenging on and hunting small arthropods. Nests are constructed under moss, lichen or small stones, among old heather or grass roots, or in peat, where an entrance hole leads several centimetres into a nest chamber. During warm weather, the brood is moved to the surface, where it is covered with small plant fragments. The genus name is from Greek *temno* = 'to cut' + thorax.

Narrow-headed ant *F. exsecta* has a highly disjunct population, Endangered in RDB 3. In southern England there are two foci: across the lowland heaths of Dorset and the New Forest, and around the lowland heaths of the Bovey Valley, east Devon. Most Scottish records are from the Caledonian forests. Nests are found in open heathland or moorland, as well as woodland rides and clearings, often built around a grass tussock. Nest mounds, created using fine vegetation fragments and generally half the size of a football, are usually found in isolation but occasionally occur in small groups. Workers forage from these to low-growing shrubs and trees for aphid honeydew and to capture or scavenge insects. *Exsecta* is Latin for 'upstanding'.

Negro ant *F. fusca* is common in lowland England and Wales, but has a more localised distribution in northern England, Ireland and Scotland. Winged queens and males appear in July–August. They nest under stones and in tree stumps under loose bark, and occasionally build earth nests. Workers forage individually, preying on other insects. They also feed on aphid honeydew. *Fusca* is Latin for 'dark'.

Red ant *Myrmica rubra*, widespread and locally common, especially in England and Wales, favouring sunny, warm, damp habitats such as meadows and riverbanks, but found in many urban gardens and parks, agricultural areas and woodland edges. Nuptial flights are usually in August–September. Ants feed on honeydew from aphids and scale insects and drink nectar from flowers. Nests are in the ground, in tufts of grass, under stones and in rotten wood. Colonies are usually polygynous with an average of 15 queens and around a thousand workers. This species is aggressive and stings with little provocation. *Myrmex* is Greek for 'ant', *rubra* Latin for 'red'.

Red barbed ant *F. rufibarbis*, very rare and endangered, the only well-substantiated records being from Surrey and on St Martins, Isles of Scilly, and the subject of a Species Recovery Programme funded by English Nature. The nest habitat is usually a south-facing bank with sparse vegetation on dry heathland, but some nests are in purple moor-grass/bristle bent grassland. Foraging is above ground; aphids are tended and dead arthropods are carried into the nest. *Rufibarbis* is Latin for 'red-bearded'.

Red wood ant *F. rufa*, locally common in the southern half of England and in Wales in woodland, occasionally spreading into scrubby heath, open rides and verges. Large nest mounds, built using plant fragments, may contain up to 400,000 individuals. Occasionally, several nests are interconnected, forming a super-colony. These ants are major predators and scavengers of woodland insects and feed extensively on aphid honeydew. Their colonies also support a wide range of often rare myrmecophilous arthropods. *Rufa* is Latin for 'red'.

Scottish wood ant *F. aquilonia*, restricted to the Highlands, associated with Caledonian pine forest and mature birch groves. These ants form trails to trees bearing honeydew-producing Homoptera. Mating flights are in June–July. They build large mound nests from plant fragments, which are often linked by long trails to neighbouring mounds to form one huge colony. Nests are usually built on free-draining soils or slopes with some exposure to the sun. *Aquilonia* is Latin for 'northern'.

Shining guest ant *Formicoxenus nitidulus*, uncommon, with a scattered distribution in England and Scotland. It lives only within the nest mounds of its ant hosts, usually wood ants. Within the host nest mound, this species nests in hollow twigs and stems, and wood fragments or the earth floor. A single wood ant mound may have several of these nests. Workers forage singly, leaving their nest via narrow galleries connecting with the interior of the host mound, obtaining food from host workers by intercepting regurgitated food or by soliciting. *Nitidulus* is the diminutive of Latin *nitidus* = 'shining'.

Slave-making ant *F. sanguinea* has a disjunct British distribution, predominating in southern England and central Scotland, with scattered records on the Welsh Borders. Its southern distribution is aligned to sandy soils. In Scotland it is mainly found around the Spey and Dee valleys. It largely occurs on heathland or in open woodland. Mating flight is in July–August. The species makes 'slave raids' on black ants such as *F. fusca* and *F. lemani*, hence its common name. Larvae and pupae of the captured ants are taken back to the nest where they mature and behave as if they were *F. sanguinea* individuals. Slaves carry out various tasks in the nest including foraging, brood rearing, nest maintenance and even assisting with slave raids themselves. Nests are usually in a stump or log, under stones or in a bank, and are founded by colony fission or by social parasitism in which mated queens rejoin their own colony or secure acceptance in the nest of a congeneric, usually of the *F. fusca* group. Workers prey and scavenge on other invertebrates; they also tend Homoptera for honeydew. They usually forage individually although nest-mates may be recruited to a large food source, particularly the nests of other ants which are often raided. *Sanguinea* is Latin for 'bloody'.

Slender ant *Leptothorax acervorum*, widespread and generally abundant. In woodland and farmland, nests are often found in dead trees, fallen branches or stumps, under bark and in old fence-posts. In upland moorland and lowland heathland, nests are often in the ground around the base of heather, in exposed dried peat, under stones or in rock crevices. Workers forage individually, mainly looking for carrion, though they also prey on small, weak invertebrates. They also take scraps of food, corpses and nest rubbish from around the colonies of other, larger ant species. *Leptos* is Greek for 'slender', *acervorum* Latin for 'of a heap'.

Turf ant *Tetramorium caespitum*, on rocky areas with short grass from the north Wales coast fairly continuously round to Norfolk; some records from Berwick; and also in south and south-east Ireland. There are also inland records (heathland) from southern England and Breckland. Workers prey and scavenge on other arthropods, feed on honeydew from root aphids, and seeds are taken into their nests, which are found in sandy or peaty ground, under stones and in rock crevices. At coastal sites, nests are usually found on crumbling, rocky cliffs down to the high water line. Greek *tetra* + *morion* = 'four' + 'part', and Latin *caespes* = 'turf'.

Westwood's ant *Stenamma westwoodii*, occasionally recorded from open habitats in southern England, with nests under stones, workers scavenging individually in leaf litter, and sometimes preying on small invertebrates. Greek *stenos* + *amma* = 'narrow' + 'knot'. John Obadiah Westwood (1805–93) was an English entomologist.

Wood ant Mound-nesting members of the genus *Formica*. See hairy, red and Scottish wood ants above. Also, *F. cunicularia* (frequent in southern England), *F. lemani* (found in south-west and north England, Wales and Scotland, occasionally in Ireland), and *F. picea* (on Dorset and Hampshire heaths and lowland raised bogs in south Wales).

Woodlouse ant Alternative name for grass ant.

Yellow meadow ant *L. flavus*, common and widespread, especially in England and Wales, in open sunny habitats including grassland, parkland, gardens, and woodland clearings and edges, with nuptial flights in July–August. Nests are built in soil, sometimes started under stones, and a large mound can be raised that may last for many years. Nests are also established in undisturbed grassland. Workers feed on small insects and the honeydew of root-feeding aphids. *Flavus* is Latin for 'yellow'.

2) **Velvet ant** Nest parasitoid wasps in the family Mutillidae. Females lack wings and have coarse setae that cover most of their body, making them resemble hairy ants.

Black-headed velvet ant *Myrmosa atra*, found throughout England, Wales and southern Ireland, mainly in sandy habitats, but also recorded from chalky banks, wooded slopes and roadsides. Males are found on flowers of bramble and umbellifers. Hosts are various ground-nesting crabonid wasps and halictine bees. This ant species may be a kleptoparasite feeding on the food stores of its host, or a parasitoid feeding on the host itself. Greek *myrmos* = 'ant', and Latin *atra* = 'black'.

Large velvet ant *Mutilla europaea*, locally common in southern England, especially on heathland but also on chalk grassland and in deciduous woodland. Males take nectar from flowers such as wild parsnip. The queen parasitises bumblebees, occasionally honey bees, ovipositing inside bee cocoons containing prepupae or young pupae. Larvae eat these immature stages then spin cocoons within those of the host. *Mutilla* is probably from Latin *mutilus* = 'maimed'.

Least velvet ant Alternative name for black-headed velvet ant.

Small velvet ant *Smicromyrme rufipes*, locally common in England from Dorset to Kent, north to Oxfordshire, Bedfordshire, Cambridgeshire and Norfolk, in sandy areas in warm, sunny places on the coast, for example dunes, and inland, for example heathland. Males have been recorded from umbellifers and ragworts. As a parasitoid of a number of ground-nesting wasps and bees, the female enters the host's burrow and bites open a cell. If a mature larva is present she inserts her gaster into the cell and lays an egg. On reaching maturity, the ant ectoparasite spins a cocoon within that of its host. *Smicros* is Ionic Greek for 'small', *myrmos* = 'ant', and *rufipes* Latin for 'red-footed'.

ANTHICIDAE Family of antlike flower beetles (2–5 mm length), with 13 British species in 6 genera, largely ground-dwelling and superficially resembling ants. They are mainly found as saprophages on decaying plant material, especially in saltmarsh or compost, but adults are also often found on flowers, hence the family name from Greek *anthos* = 'flower'.

ANTHOATHECATA From Greek *anthos* + *athece* = 'flower' + 'uncased', an order of Hydrozoa; see hydroid.

ANTHOCEROTOPHYTA From Greek *anthos* + *ceras* + *phyto* = 'flower' + 'horn' + 'plant', the phylum of hornworts.

ANTHOMYIIDAE Family name (Greek *anthos* + *myia* = 'flower' + 'fly') for flies in the suborder Bracycera, superfamily Muscoidea, with 245 species in 29 genera. Adults are small or medium in size. Larvae include root-maggots (the general English name for members of this family is root-maggot flies), leaf miners and parasites. A number, especially in the genus *Delia*, are agricultural pests.

ANTHOMYZIDAE Family name (Greek *anthos* + *myzo* = 'flower' + 'to suck'), for flies in the suborder Bracycera, superfamily Opomyzoidea, with 21 species in 10 genera. Adults are 1.3–4.5 mm long, found in a range of habitats. Larvae are found in decaying plants, leaf sheaths of grasses and rushes, and in fungi.

ANTHOZOA From Greek *anthos* + *zoo* = 'flower' + 'animal', a class of cnidarian marine invertebrates that includes sea anemones, corals (soft corals being in the order Alcyonacea, hard corals Scleractinia), sea fans and sea pens.

ANTHRAX From Greek for 'coal' or 'carbon', an infection by the bacterium *Bacillus anthracis*, usually transmitted from animals, persistent in soil, and lethal in humans if not treated with antibiotics.

ANTHRIBIDAE From Greek *anthos* = 'flower' + an element of unknown origin, the family name of fungus weevils, with nine species in eight genera found in England and Wales. See weevil.

ANTIRRHINUM See snapdragon.

APHID Members of the family Aphididae (Sternorryncha), sap-sucking bugs, including those commonly known as greenfly, with around 630 species in 160 genera, many of which are widespread and common pests of crops and trees. Some species are 'farmed' by ants for the sugary secretion honeydew. The *Oxford English Dictionary* comments that the 'least improbable' etymological explanation is that 'aphid' derives from the plural Greek *apheides*, meaning 'unsparing, lavishly bestowed', referring either to the 'prodigious rate of production' of the insects or their voracity.

Widespread and locally common on their host plants are **downy birch aphid** *Euceraphis punctipennis*, **elder aphid** *Aphis sambuci*, **giant willow aphid** *Tuberolachnus salignus*, **hellebore aphid** *Macrosiphum hellebori*, **maple aphid** *Periphyllus testudinaceus*, **nettle aphid** *Microlophium carnosum*, **rose aphid** *Macrosiphum rosae*, **silver birch aphid** *E. betulae* and **sycamore aphid** *Drepanosiphum platanoidis*.

Black bean aphid or **blackfly** *Aphis fabae*, common and widespread, overwintering as eggs on such trees as spindle, guelder-rose and mock orange, then in spring moving to a range of plants including those of economic value where heavy infestation will cause damage. *Faba* is Latin for 'bean'.

Potato aphid *Macrosiphum euphorbiae*, common and widespread, highly polyphagous in summer, feeding on over >200 plant species in >20 plant families, on spring emergence initially preferring Chenopodiaceae before favouring Solanaceae such as potato and tomato, and becoming a pest species.

Woolly apple aphid *Eriosoma lanigera*, a common pest of apple trees, pear, mountain-ash, pyracantha and cotoneaster. Adults suck sap from beneath the bark and secrete a waxy coating which gives them a woolly appearance. It was fifth in the Royal Horticultural Society's list of worst garden pests of 2017. *Lanigera* is Latin for 'wool-bearing' (*lana* = 'wool').

APIACEAE Name (Latin *apium* = 'celery') of the umbelliferous carrot family (formerly Umbelliferae), including a number of casual neophytes.

APIDAE Family name for some bees (Latin *apis* = 'bee'), with 30 genera in the British list.

APIONIDAE Family name (Greek *apion* = 'pear') for the 1.5–3.5 mm length seed weevils, with 87 British species in 34 genera. They are largely plant-feeding, many species restricted to a single host species (though several species may feed on the same plant species). See weevil (3, seed weevils).

APOCRITA The larger of the suborders of Hymenoptera, with adults narrowed at the waist, and ovipositors not being serrated. Subdivided into the Parasitica, or ichneumons and other parasitic wasps, and the Aculeata, or bees, ants and wasps. *Apocrita* is Greek for 'responder'.

APOCYNACEAE Family name of periwinkle plants, from Greek *apo* + *cyon* = 'away with' + 'dog', referring to the former use of the North American dogbane *Apocynum cannabinum* as a dog poison.

APODIDAE, APODIFORMES Family and order names of swifts (Greek *a* + *podi* = 'without feet'). The feet are small and weak, though with the aid of sharp claws they can cling to their nests and vertical surfaces.

APPLE *Malus pumila* (Rosaceae), a common and widespread small archaeophyte tree of garden origin, in hedgerows, scrub, roadsides and waste ground, usually occurring as single trees, as well as in orchards and gardens. *Malus* is Latin for 'apple tree', *pumila* = 'small'.

    Crab apple *M. sylvestris*, a small tree of hedgerows, scrub, copses, roadsides and rough ground, usually occurring as single trees. *Sylvestris* is Latin for 'wild'.

APPLE-MOSS Mosses with apple-shaped capsules in the genera *Bartramia* (after the American botanist John Bartram, 1699–1777), *Philonotis* (Greek *philein* + *notis* = 'to love' + 'moisture', reflecting the habitat) and *Plagiopus* (Greek *plagios* = 'oblique') all Bartramiaceae.

    Uncommon upland species, mainly from the Highlands, are **Haller's apple-moss** *B. halleriana* and **stiff apple-moss** *B. ithyphylla* (both also Lake District and Wales), **Oeder's apple-moss** *Pl. oederianus*, **spiral apple-moss** *Ph. seriata*, **swan-necked apple-moss** *Ph. cernua*, **upright apple-moss** *B. stricta* and **woolly apple-moss** *Ph. tomentella*.

    Arnell's apple-moss *Ph. arnellii*, scattered but locally common on wet soil in rock crevices and ledges, by streams, in woodland rides, and on quarry floors. Hampus Arnell (1848–1932) was a Swedish bryologist.

    Common apple-moss *B. pomiformis*, common and widespread if scattered in central and eastern England and in Ireland, on acidic or mildly base-rich rock, especially on cliffs, and on hedge banks and walls. *Pomiformis* is Latin for 'apple-shaped'.

    Fountain apple-moss *Ph. fontana*, common and widespread, but scattered in the Midlands, East Anglia and central Ireland, frequent in marsh and montane flushes and on wet cliff ledges, and by streams and lakes. *Fontana* is Latin for 'fountain'.

    Rigid apple-moss *Ph. rigida*, local in south-west England, Wales, Isle of Man, west Scotland and west and south-west Ireland, in crevices on damp cliffs and on riverbanks, often in coastal areas.

    Thick-nerved apple-moss *Ph. calcarea*, scattered in upland regions, but also in the New Forest and parts of East Anglia, in wet, base-rich sites. *Calcarea* is Latin for 'of chalky land'.

    Tufted apple-moss *Ph. caespitosa*, widespread but scattered on damp soil and rocks, usually where the ground is flushed. *Caespitosa* is Latin for 'tufted'.

APPLE-OF-PERU *Nicandra physalodes* (Solanaceae), an annual Peruvian neophyte, casual on waste and cultivated ground, especially in England and Wales, sometimes introduced with wool shoddy or birdseed. The genus name is that of Nikander of Colophon, a second-century BCE Greek poet and physician. *Physalodes* is from Greek meaning 'like a bladder'.

APPLE TWIG CUTTER *Involvulus caeruleus* (Rhynchitidae), a tooth-nosed snout weevil, 2.5–4 mm length, scattered mainly in the southern third of England. A local pest of apple, but other commercial fruit tree species are also attacked, adults feeding on leaves, larvae inside shoots. *Involvare* is Latin for 'wrap', referring to the larvae wrapping themselves in leaves, *caeruleus* for 'blue'.

APTERYGOTA Subclass of wingless insects, often living in soil or decaying vegetation (Greek *a* + *pteryga* = 'without' + 'wing').

AQUIFOLIACEAE Family name for holly (Latin *aquifolia*).

ARABIS **Garden arabis** *Arabis caucasica* (Brassicaceae), a herbaceous perennial, mat-forming south European neophyte species of rock-cress, introduced by 1798. It was first recorded in the wild in 1855 (Kew, Surrey) and persists in many places, especially in England and Wales, usually not far from habitation, and naturalised on walls, rocks and cliffs. *Arabis* comes from medieval Latin, in turn from Greek (feminine of *araps*), though possibly from Greek *Arabia* (Arabia), perhaps referring to the ability of this plant to grow in rocky or sandy soils. See also rock-cress.

ARACEAE Family name (Greek *aron* = 'arum') for lords-and-ladies, skunk-cabbage, and more recently including duckweeds.

ARACHNACTIDAE Incorporating Greek *arachni* = 'spider', family name of some sea anemones, with two species in two genera.

ARACHNIDA The class of eight-legged invertebrates that includes the subclass Acari (mites and ticks), and the orders Araneae (spiders), Opiliones (harvestmen), Scorpiones and Pseudoscorpiones. *Arachni* is Greek for spider.

ARALIACEAE Family name (etymology unknown) for ivy.

ARANEAE The order that contains spiders (Latin *araneus* = 'spider').

ARANEIDAE From Latin *araneus* = 'spider', family name of orb-web-weaving spiders, with 32 British species in 16 genera.

ARAUCARIACEAE Family name for monkey-puzzle, named after the Araucarian Indians of Patagonia.

ARCHANGEL **Yellow archangel** *Lamiastrum galeobdolon* ssp. *montanum* (Lamiaceae), a stoloniferous perennial herb, common and widespread in England and Wales, local and possibly introduced elsewhere, found in moist woodland, hedges and roadsides, usually on heavy soils, an indicator of ancient woods and wood-relic hedges, particularly in eastern England. Subspecies *argentatum*, possibly introduced in the 1960s, is now widespread, and naturalised on woodland edges, verges and tracksides. Subspecies *galeobdolon* is scarce, with a scattered distribution, mainly in eastern England. 'Archangel' comes from medieval Latin *archangelica*, a name given to a number of dead-nettles, as if nettles had been angelically deprived of their sting. *Lamiastrum* is Latin for 'resembling *Lamium*' (dead-nettles). *Galeobdolon* is Greek for 'smelling like a weasel'.

ARCHES Nocturnal moths in the family Noctuidae, including the genera *Anaplectoides* (subfamily Noctuinae) derived in three stages from Greek *a* + *plectos* + *eidos* = 'not' + 'folded' + 'in appearance' (i.e. *Aplectoides* similar to the genus *Aplecta*, prefixed by *ana*, indicating not *Aplectoides* but related to it); *Polia* (Hadeninae), from Greek *polios* = 'grey'; and *Apamea* (Xyleninae), from a town in Asia Minor. See also black arches (Nolidae) and buff arches (Drepanidae).

    Dark arches *Ap. monoglypha*, wingspan 45–55 mm, widespread, common and often abundant, in gardens, hedgerows, verges, woodland, heathland and moorland. Adults fly in July–August. Larvae feed in spring on the bases and stems of grasses. Greek *monos* + *glyphis* = 'single' + 'notch', describing such a feature on the forewing.

    Green arches *An. prasina*, wingspan 40–50 mm, widespread, common in broadleaf woodland. Population levels increased by 154% over 1968–2007. The marbled greenish colour of the adults, which are active in June–July, affords excellent camouflage on moss- or lichen-covered trees and fences. Larvae feed on a number of plants, including sallows, docks and bramble. *Prasina* is Latin for 'leek green', from the forewing colour.

Grey arches *P. nebulosa*, wingspan 45–55 mm, widespread, common in broadleaf woodland and gardens. Adults fly in June–July. Larvae feed in autumn on herbaceous plants, then after hibernation in spring on buds and leaves of trees such as birch and willow. *Nebulosa* is Latin for 'cloudy', from the forewing colour.

Light arches *Ap. lithoxylaea*, wingspan 43–50 mm, widespread, common in habitats such as dry pasture, calcareous grassland, woodland rides and gardens. Adults fly from June to August. Larvae feed on grasses. Greek *lithos* + *xylon* = 'stone' + 'wood', describes grain-like markings on a stone-grey forewing.

Northern arches *Ap. exulis*, wingspan 43–50 mm, scarce, on moorland and grassland in parts of mainland Scotland (ssp. *assimilis*) and on Shetland (ssp. *marmorata*, also known by the name exile). Adults fly in July–August. Larvae probably feed on grasses. *Exul* is Latin for exile.

Reddish light arches *Ap. sublustris*, wingspan 42–8 mm, local on chalk and limestone grassland, dunes and shingle beaches throughout England and parts of Ireland. Adults fly in June–July. Larvae feed on grasses. *Sublustris* is Latin for 'glimmering faintly'.

Silvery arches *P. hepatica*, wingspan 43–52 mm, local in wooded heathland, moorland and open woodland scattered throughout Britain, commonest on the heaths of south-east England, Cumbria and central Scotland. Adults fly in June–July. Larvae feed on birches, hawthorn, sallows, bog-myrtle and honeysuckle. *Hepatica* is Latin here meaning 'liver-coloured'.

ARDEIDAE Family name of herons, egrets and bitterns (Latin *ardea* = 'heron').

ARGENT (MOTH) Members of the genus *Argyresthia* (Argyresthiidae), the English name being the heraldic term for 'silver', and the genus name from Greek *argyros* + *esthes* = 'silver' + 'dress', from the metallic sheen on the forewings. Species are nocturnal.

Rarities are **bronze argent** *A. arceuthina* and **ochreous argent** *A. praecocella* (both on downland in southern England and East Anglia, and woodland in northern Scotland where juniper is present), and **triple-barred argent** *A. trifasciata* (first recorded in Britain in 1982 in London but since seen as far north as Lancashire, and in Wales around conifers).

Blackthorn argent *A. spinosella*, wingspan 9–11 mm, common in hedgerows throughout much of England and Wales, more scattered and local in Scotland and Ireland. Adults fly in June–July. Larvae mine buds and flowering shoots of blackthorn *Prunus spinosa*, giving the specific *spinosella*.

Brindled argent *A. curvella*, wingspan 10–12 mm, local in gardens and orchards throughout England and Wales, extending to Perthshire and the Western Isles. Adults fly in June–July. Larvae mine buds and shoots of apple, in places becoming a serious pest. *Curvella* is Latin for 'small curve', referencing the curved fascia (band of colour) on the forewing.

Brown rowan argent *A. semifusca*, wingspan 12 mm, common in open woodland and hedgerows, widespread if more numerous in the north. Adults fly in June–September. Larvae mine young shoots of hawthorn (more so in the south) or rowan during early summer. *Semifuscus* is Latin for 'half dark', forewings being half dark, half white.

Gold-four argent *A. ivella*, wingspan 10–12 mm, local in old orchards and on isolated wild apple trees, in England and Wales, as far north as Northumberland. Adults fly in July–August. Larvae feed inside young shoots of apple and hazel. *Ivella* comes from the markings on the forewing which approximate the shape IV.

Gold-ribbon argent *A. brockeella*, wingspan 9–12 mm, widespread and common in open (especially birch) woodland and freshwater margins. Adults fly in June–July. Larvae mine young shoots and catkins of birches and alder. The specific name honours the nineteenth-century German entomologist J.K. Brock.

Gold rowan argent *A. sorbiella*, wingspan 12 mm, widespread if local in open woodland, more numerous in the west, but rare in Ireland. Adults fly in June–July. Larvae mine buds and shoots of rowan and whitebeam, species of *Sorbus* giving the specific name.

Golden argent *A. goedartella*, wingspan 10–13 mm, widespread and common in open woodland, heathland, gardens and freshwater margins. Adults fly at night and on sunny afternoons in June–August. Larvae mine young shoots and catkins of birches and alder. Jan Gödart (1620–68) was a Dutch entomologist.

Hawthorn argent *A. bonnetella*, wingspan 9–11 mm, widespread and common in hedgerows, gardens, downland and scrub. Adults fly in June–August. Larvae mine hawthorn bark. Charles Bonnet (1720–93) was a Swiss entomologist.

Juniper argent *A. dilectella*, wingspan 7–9 mm, found in downland, open woodland and gardens, widespread, but rare in the northern half of Britain and in Ireland. Adults fly in July–August. Larvae mine shoots of juniper and cypress. *Dilectus* is Latin for 'beloved'.

Larch-boring argent *A. laevigatella*, wingspan 9–13 mm, local in plantations throughout Britain. Adults fly from May to July. Larvae feed inside terminal shoots of larch, in places becoming a serious pest. Latin *laevigatus* = 'smooth', from the glossy forewing texture.

Large beech argent *A. semitestacella*, wingspan 11–14 mm, local in deciduous woodland where beech is present, widespread though with fewer records from Scotland and Ireland. Adults fly in August–September. Caterpillars mine beech shoots in May–June. Latin *semi* + *testa* = 'half' + 'brick', from the brown (brick-coloured) and white forewing.

Netted argent *A. retinella*, wingspan 9–10 mm, afternoon- and night-flying, widespread and common in open woodland. Adults fly in June–July. Larvae mine catkins and shoots of birch. *Retinella* comes from Latin *rete* = 'net', from the reticulate forewing.

Oak-bark argent *A. glaucinella*, wingspan 9 mm, local in open woodland, especially of oak, throughout Britain and (less so) in southern Ireland. Adults fly in June–July. Larvae feed beneath the surface of the bark of, especially, oaks, but also that of horse-chestnut and birches. Latin *glaucus* = 'blue-grey', from the forewing colour.

Purple argent *A. albistria*, wingspan 9–11 mm, widespread and common in hedgerows and scrub. Adults fly in July–August. Caterpillars mine flowering shoots of blackthorn, overwintering as small larvae and feeding again in spring. *Albistria* is Latin for 'white streak', seen on the forewing.

Sallow argent *A. pygmaeella*, wingspan 10–13 mm, widespread and common in damp woodland, scrub, gardens and hedgerows. Adults fly in June–August. Larvae mine young shoots and catkins of sallow and willows. *Pygmaeella* in Latin for 'pygmy', though this moth is not smaller than many congeners.

Spruce argent *A. glabratella*, wingspan 9–13 mm, in coniferous woodland and plantations scattered throughout Britain as far north as Aberdeenshire. Adults fly in May–June. Larvae feed inside shoots of Norway spruce, in places becoming a pest. Latin *glabrus* = 'smooth', from the glossy forewing texture.

ARGENT & SABLE *Rheumaptera hastata* (Geometridae, Larentiinae), wingspan 30–8 mm, a day-flying moth (May–July) with a widely scattered if local distribution, commoner in west Scotland. A smaller race, *nigrescens*, is a moorland species whose caterpillars feed on bog-myrtle; the slightly larger and darker race, *hastata*, feeds on birch. Larvae spin together the leaves of their food plant, forming a cocoon. The heraldic terms argent and black refer to 'silver' and 'black', colours found on both wings and body. Greek *rheuma* + *pteron* = 'stream' + 'wing', Latin *hasta* = 'spear', from shapes on the wing.

Small argent & sable *Epirrhoe tristata*, wingspan 24–6 mm, widespread and relatively common in moorland, upland limestone grassland, upland woodland and other upland habitats. Adults fly in sunshine and at night from May to July. Larvae feed on heath bedstraw. The binomial is Greek for 'flood' (from the wavy forewing pattern), and Latin *tristis* = 'sad' (the black and white coloration suggesting mourning).

ARGIDAE Family name of some sawflies (Greek *arges* = 'bright').

ARGUS 1) Butterflies in the genus *Aricia* (Lycaenidae, Polyommatinae). Aricia was an ancient town in Latium (central-west Italy).

    Brown argus *A. agestis*, wingspan 25–31 mm, common in southern England on chalk and limestone grassland but also found in other warm open habitats, for example heathland and open woodland south and east of a line from south Wales to Yorkshire, and also in north Wales. The main larval food plant is common rock-rose. Larvae are often found in the presence of ants, which may offer them protection against parasites and predators. Ants are attracted from secretions exuded from the Newcomer's gland at the tail end of the larva. Adults, flying from May to September, feed on nectar of white clover, thyme, ragwort and marjoram. This butterfly extended its range by 115% over 1976–2014, though overall numbers fell by 25%. *Agestis* may be a typographical error for Latin *agrestis* = 'field', or for Argestes, the god of the north-west wind.

    Northern brown argus *A. artaxerxes*, wingspan 25–31 mm, found as two subspecies in northern England and parts of Scotland on limestone grassland, sheltered coastal valleys and quarries, and limestone pavement, these habitats favouring growth of the larval food plant, common rock-rose. Adults, flying in June–August, feed on nectar of thyme. Range contracted by 27% between 1976 and 2014 (and by 36% over 2005–14). Population levels have also shown an ongoing decline (by 52% over 1976–2014, though with a slight upturn of 6% in 2005–14), mainly because of inappropriate grazing and, in Scotland, afforestation of suitable habitat. This is a priority species for conservation effort. Artaxerxes was the name of several Persian kings.

    2) **Scotch argus** *Erebia aethiops* (Nymphalidae, Satyrinae), male wingspan 44–8 mm, female 46–52 mm, a 'brown' butterfly flying in June–September in sheltered damp sites, including acid grassland and bogs, in much of Scotland, and two sites on sheltered limestone grassland in Cumbria. Larvae feed on purple moor-grass, adults on nectar of heather, scabious, knapweeds, ragwort and thistles. Although experiencing a long-term decline in distribution (by 17% over 1976–2014), numbers – admittedly from a low base level – increased by 170% between 1976 and 2014 and 24% between 2005 and 2014. The binomial is from Erebos, the region of darkness between Earth and Hades, and Greek for African, both terms indicating the dark colour.

ARGYRESTHIIDAE Family name for argent moths (formerly Yponomeutidae), from Greek *argyros* + *esthes* = 'silver' + 'dress', from the metallic sheen on the forewings.

ARIONIDAE Family name for round-backed slugs. Arion was a musician of Lesbos rescued from drowning by a dolphin.

ARISTOLOCHIACEAE Family name for birthworts, from Greek *aristos* + *lochia* = 'best' + 'delivery', referring to the medicinal qualities of the plant in helping childbirth.

ARK Milky ark *Striarca lactea* (Noetiidae), a bivalve, shell size up to 2 cm, attached by a strap-like byssus to the underside of stones in the intertidal and sublittoral from Pembrokeshire to Kent. Latin *stria* + *arca* = 'channel' + 'box' or 'ark', and *lactea* = 'milk'.

ARK-SHELL *Arca tetragona* (Arcidae), a bivalve with a boat-shaped shell up to 4 cm long, attached to hard substrates by a stalked byssus in the intertidal and over the shelf around Britain except the southern North Sea and the eastern part of the English Channel, commonest off west Scotland, Orkney and Shetland. Latin *arca* = 'ark', Greek *tetragonon* = 'four-angled'.

ARROWGRASS Rhizomatous herbaceous perennials in the genus *Triglochin* (Juncaginaceae) (Greek *treis* + *glochis* = 'three' + 'pointed', referring to the projections on the carpel).

    Marsh arrowgrass *T. palustris*, widespread and locally common, though declining in southern England, in damp, grassy or marshy habitats, often on calcareous substrates, including wet heathland, fens, saltmarsh fringes, and river shingle in upland areas. *Palustris* is Latin for 'relating to marsh'.

    Sea arrowgrass *T. maritima*, common and widespread in coastal and estuarine saltmarshes, coastal rocks and cliff edges subject to sea spray, and the banks of tidal rivers. Inland it occurs in brackish pasture, and increasingly by salt-treated roads. *Maritima* is Latin for 'maritime'.

ARROWHEAD *Sagittaria sagittifolia* (Alismataceae), a herbaceous perennial, common and widespread in much of England, scattered in Wales and Ireland, and introduced in central Scotland, in shallow, still or slowly flowing, calcareous and eutrophic water. In larger rivers it may grow only as submerged leaves, but in ditches, lakes, ponds and canals it often produces emergent leaves and flowers. **Canadian arrowhead** *S. rigida* and **narrow-leaved arrowhead** *S. subulata* are both rare introductions. *Sagitta* is Latin for 'arrow'.

ARTEMIS Filter-feeding bivalves in the genus *Dosinia* (Veneridae), shell size up to 6 cm, common and widespread as a burrower in sand and shell-gravel in the low intertidal down to 100 m.

ARTHROPOD(A) Phylum of invertebrates that have an exoskeleton, segmented body, and jointed appendages/limbs. There are 12 classes, of which the most important are the Diplopoda (millipedes), Chilopoda (centipedes), Insecta, Crustacea, and Arachnida (spiders and mites). The name is from Greek *arthro* + *podi* = 'joint' + 'foot'.

ARTICHOKE Jerusalem artichoke *Helianthemum tuberosus* (Asteraceae), a herbaceous North American neophyte perennial, cultivated by 1617 for the edible tubers that develop from the rhizomes, and now much grown in gardens. The tubers can survive as garden throw-outs, or plants can germinate from birdseed, and it is occasionally recorded on waste ground. 'Artichoke' dates to the mid-sixteenth century, from Italian *articiocco*, in turn from Arabic *al-kkarsufa*; 'Jerusalem' is an early seventeenth-century alteration of the Italian *girasole* = 'sunflower'. Greek *helios* = anthos* = 'sun' + 'flower', and Latin *tuberosus* = 'tuberous'. (This species is unrelated to **globe artichoke** *Cynara cardunculus* var. *scolymus*, a kind of thistle.)

ARTIODACTYLA The order of even-toed ungulates (hoofed animals with their weight borne about equally by the third and fourth toes) that includes the families Bovidae (cattle, sheep and goat), Cervidae (deer) and Suidae (wild boar), and the infraorder Cetacea (whales, dolphins and porpoises). Greek *artios* + *dactylos* = 'even' + 'toe'.

ARUM Bog arum *Calla palustris* (Araceae), a rhizomatous herbaceous neophyte perennial in cultivation by 1738, very scattered growing as floating mats at the margins of lakes and ponds, or rooted in marshes and wet alder woodland, favouring shade. 'Arum' is Late Middle English, from Greek *aron* for this plant. Greek *callos* = 'beautiful', *palustris* Latin for 'marshy'.

    Wild arum: alternative name for lords-and-ladies.

ARUM LILY Alternative name for altar-lily.

ASARABACCA *Asarum europaeum* (Aristolochiaceae), a perennial herb introduced from Europe, naturalised since at least the seventeenth century, rare (having been declining since the 1930s) and scattered, found in shaded places, including woodland and hedges. It can reproduce by seed, at least in southern England, but generally spreads using surface rhizomes to form a dense mass. *Asaron* was the name given by Dioscorides for this plant; the English name compounds this with *baccharis*, probably a cyclamen.

ASCIDIACEA, ASCIDIAN A class in the subphylum Tunicata that includes sea squirts, with three main types: solitary ascidians; social ascidians that form clumped communities by attaching at their bases; and compound ascidians that consist of many small individuals (each individual called a zooid) forming colonies up to several meters in diameter. They are sac-like marine invertebrate filter feeders, mostly sublittoral. Greek *ascidion*, diminutive of *ascos* = 'wineskin'. See also sea squirt, tunicata and sea grapes.

Star ascidian *Botryllus schlosseri* (Styelidae), widespread as flat or fleshy colonies with zooids 2–4 mm across arranged in star-shaped systems. It grows on macroalgae and artificial substrates, including docks, mainly in the intertidal and shallow subtidal, but has been found at 200 m.

ASCOMYCETE, ASCOMYCOTA Class and division names for a large group of fungi, sometimes known as sac fungi because they carry their spores in a sac or ascus (Greek *ascos* = 'sac'). As the spores mature within the ascus, increasing fluid pressure builds up until eventually the top bursts off, rapidly releasing the spores. Formerly referred to as discomycetes and Discomycota.

ASCOT HAT *Hortiboletus bubalinus* (Boletales, Boletaceae), a very rare fungus, with only 11 records in 2006–12, most in south-east England, but others from Somerset, Gloucestershire, Oxfordshire and Yorkshire, generally on soil in grass. *Boletus* comes from Greek *bolos* = 'lump of clay', *bubalinus* from Latin for 'buff'.

ASH *Fraxinus excelsior* (Oleaceae), a recently common and widespread deciduous tree of woodland, scrub and hedgerows, especially on moist, basic soils, but also frequent on rock scars and cliffs, stabilised scree and the grikes of limestone pavement. It can tolerate periodically waterlogged soils, and is a rapid coloniser of waste ground, disused quarries and railway banks. It can become dominant on steep slopes in limestone areas, notably the Mendips, southern Pennines (Peak District), western Yorkshire and northern Lancashire. Ancient upland mixed ash woodlands include 900 ha in the Peak District National Park. Ash, however, has recently become endangered because of ash dieback disease, and has been predicted to lose perhaps 95% of its numbers in Britain. The English name derives from Old English *æsc*. *Fraxinus* is Latin for this tree, *excelsior* = 'noble' or 'exalted'.

Manna ash *F. ornus*, a deciduous European neophyte tree planted in streets and parks, growing on road, railway and canal banks, mainly in England. Both binomial parts are Latin for 'ash'.

Narrow-leaved ash *F. angustifolia*, an introduced deciduous tree planted in streets and parks, in a few places spreading to road, railway and canal banks. *Angustifolia* is Latin for 'narrow-leaved'.

ASH DIEBACK A disease of ash trees caused by the fungus *Hymenoscyphus fraxineus* with symptoms of leaf loss, crown dieback and bark lesions. The disease is usually fatal either directly, or indirectly by weakening the tree so that it becomes vulnerable to attacks by other pests or pathogens, especially honey fungus. It was first recorded in Britain in a Buckinghamshire nursery in February 2012. Local spread may be by wind. Over longer distances the risk of spread is most likely to be through movement of diseased trees. In October 2012, a small number of cases in Norfolk and Suffolk were confirmed, including in established woodland. Many further finds have since been confirmed, mostly on the eastern side of England and Scotland. In May 2013 the first case was found in (south-west) Wales. By February 2016, 26% of the 10 km survey grid squares in the UK had reported at least one infection, though none was in Northern Ireland. Possibly 90% of Britain's ash trees are at risk.

ASILIDAE From *asilus*, a Latin word for 'fly', the family name for robber flies in the suborder Bracycera, superfamily Asiloidea, with 29 species in 16 genera.

ASPARAGACEAE From Greek *aspharagos*, the family name for such plants as asparagus, bluebell, butcher's-broom, grape-hyacinth, hyacinth, lily-of-the-valley, Solomon's-seal and squill.

ASPARAGUS *Asparagus prostratus* (Asparagaceae), a dioecious, rhizomatous herbaceous perennial, very local and declining in the Channel Islands, south-west England, south Wales and south-east Ireland on free-draining soil. On sea-cliffs, it often grows through a dense mat of red fescue, usually in very rocky soils. On dunes, it grows in turf, often by paths. The name derives from *sperage*, from medieval Latin *sparagus*, in turn derived from Greek *aspharagos*, itself from Persian *asparag* = 'sprout'. Asparagus has been corrupted at times and in places to 'sparrow grass'. The specific name is Latin for 'prostrate'.

Garden asparagus *A. officinalis*, widely cultivated as a vegetable, scattered mainly in England and Wales, naturalised largely by bird-dispersed seed on dry sandy soils on sparse grassland, grassy heathland (in East Anglia) and dunes. *Officinalis* Latin for 'having medical or pharmacological use'.

ASPEN *Populus tremula* (Salicaceae), a common and widespread tree of both clay and sandy soils in mixed broad-leaved woodland, hedgerows, heathland, and former clay- and sandpits. In the north and west, it grows on rocky outcrops and riverbanks, often as a shrub. It suckers to form thickets, and readily colonises bare ground. The English name dates from the sixteenth century, from Old English *æspe*. The binomial is Latin for 'poplar', and 'trembling' from the action of the leaves in wind.

ASPEN TONGUE *Taphrina johansonii* (Taphrinales, Taphrinaceae), a fungal pathogen that causes a gall on aspen catkins, with a few records from Shropshire, East Anglia and the Highlands.

ASPERGILLOSIS A number of respiratory diseases caused by a fungal mould (*Aspergillus*) in wild animals, found especially in ducks and game birds. *Aspergare* is Latin for 'to spray'.

ASPHODEL 1) Rare, hardly naturalised species of *Asphodelus* (Xanthorrhoeaceae), from Greek *asphodelos*, flowers found in the Elysian Fields.

2) **Bog asphodel** *Narthecium ossifragum* (Nartheciaceae), a rhizomatous herbaceous perennial common and widespread, though absent from most of central and eastern England, in wet, moderately basic to strongly acidic mineral soils and peat in mire and in wet heathland and flushes. This is the county flower of Ross as voted by the public in a 2002 survey by Plantlife. Although slightly toxic, plants in the uplands are sometimes grazed, causing livestock death. In 2012, for example, lambs died in the Lake District, ears shrivelling and dropping off, wool falling out and animals becoming blind. *Narthecium* is from Greek *narthex* = 'hollow stem or cane'. *Ossifragum* ('bone breaker') reflects the belief that the bones of sheep became brittle from eating this plant; this, however, was caused by the calcium-deficient vegetation found in the habitats in which bog asphodel grows.

3) **Scottish asphodel** *Tofieldia pusilla* (Tofieldiaceae), a rhizomatous perennial herb, very local in the Highlands, with records also from Upper Teesdale, growing by streams and in calcareous flushes. Thomas Tofield (1730–79) was an English botanist. *Pusilla* is Latin for 'small'.

ASSASSIN BUG Predatory members of the family Reduviidae (Heteroptera), including *Reduvius personatus*, which lives around humans, feeding on bed bugs, psocids, flies, etc. They can give a painful stab using their large curved proboscis. *Reduvia* is Latin for 'hangnail' or 'remnant', possibly prompted by the lateral flanges on the abdomen of many species; *personatus* is Latin for 'masked'.

Heath assassin bug *Coranus subapterus*, length 9–12 mm, Britain's commonest assassin bug but with a scattered distribution, local on heathland and dune in England and Wales.

The binomial is from Greek *coris* = 'bug' and Latin *subapterus* = 'nearly wingless'.

ASTARTE English and genus name of near sublittoral filter-feeding marine bivalves in the family Astartidae. Astarte was the Phoenician goddess of fertility and sexual love.

ASTER Species of *Aster* (from Greek, then Latin, *astro* = 'star'), *Cosmos* (Greek *cosmicos* = 'universe') and *Callistephus* (Greek for 'beautiful crown') (Asteraceae), all names describing the flower head. **China aster** *Ca. chinensis* and **Mexican aster** *Co. bipinnatus* are annual Far Eastern neophytes, occasionally casual on waste ground.

Goldilocks aster *A. linosyris*, a herbaceous perennial found very locally in south-west England, south-west and north Wales, and Cumbria on shallow soil in open, grassy habitats on limestone sea-cliffs, clifftop grassland and low heathland overlying limestone. It is usually intolerant of heavy grazing, but in Pembrokeshire is found in sheep-grazed, clifftop grassland. The specific name may in part indicate a resemblance in the leaves to 'flax' (Latin *linum*).

Saltmarsh aster *A. squamatus*, an annual or biennial Central and South American neophyte, naturalised on waste ground in south Hampshire and the Isle of Wight since 2003, and at Plymouth Docks, Devon, and Co. Dublin (both 2006), likely to increase its range. *Squamatus* is Latin for 'scaly'.

Sea aster *A. tripolium*, a herbaceous perennial growing round most coastlines in ungrazed or lightly grazed saltmarsh, especially along creeksides, tidal riverbanks and in brackish ditches. In western Britain and in Ireland it also grows among rocks and on sea-cliffs. It is also found locally in inland saltmarsh, and recently has been recorded on salt-treated roadsides. *Tripolium* is Latin for 'from Tripoli' (i.e. North Africa).

ASTERACEAE The daisy family, from Greek *astro* = 'star', reflecting the shape of the flower head.

ASTEROIDEA Class that includes starfish (Greek *asterias* = 'starfish').

ASTIGMATA The order (or suborder in the order Sarcoptiformes) of mites, ranging in body size from 0.2–1.5 mm, found in a range of terrestrial habitats. They can be parasitic or predatory, or feed on decaying material, and some species are economically important, including dust mites, feather mites, and pests of stored products. Greek *astigma* = 'without markings'.

ASTRANTIA *Astrantia major* (Apiaceae), a herbaceous perennial European neophyte present by the sixteenth century, scattered in Britain on waste ground as a garden escape or an introduction in partially shaded habitats, usually near habitation. *Astrantia* may come from Greek *astro* = 'star', referencing the flower's star-like umbels.

ATTELABIDAE Family name of leaf-rolling weevils, with two monospecific genera in the British list. See leaf-roller. *Atellabos* is Greek for 'imperfect'.

ATYPIDAE Family name of purse web spider (Greek *atypos* = 'untypical').

AUBRETIA *Aubretia deltoidea* (Brassicaceae), a mat-forming herbaceous perennial European neophyte, occasionally found, especially in the south, as a naturalised garden escape on walls and stony banks. Claude Aubriet (1665–1742) was a French botanical artist. *Deltoidea* is from the Greek letter *delta*, meaning D- or triangle-shaped.

AUCHENORRHYNCHA A suborder of true bugs (Hemiptera), with membraceous wings that are uniform in texture and held like a tent over the abdomen, and with short, bristle-like antennae. The group includes leafhoppers, planthoppers, treehoppers and spittlebugs. The name comes from Greek *auchen* + *rhynchos* = 'neck' + 'nose', referring to the position of the mouthparts, which lie at the back of the head near the neck.

Families found in the British Isles are froghoppers or spittlebugs, leafhoppers, planthoppers, treehoppers and cicada.

AUK Member of the seabird family Alcidae, from Icelandic *álka*, itself from Old Norse *alka* for these birds.

Little auk *Alle alle* (Alcidae), wingspan 44 cm, length 18 cm, weight 170 g, a winter visitor to mostly east coast waters in small numbers each year. Some birds enter the North Sea in autumn and large numbers may be seen passing offshore during gales, with 'wrecks' of hundreds of birds being recorded. 'Unprecedented' numbers were reported, centred on the Moray Firth, in January 2016. The previous highest wreck was in November 2007, when 29,000 birds were recorded around the Farne Islands. Swedish *alle* means 'seabird'.

AUNT-ELIZA *Crocosmia paniculata* (Iridaceae), a cormous rhizomatous herbaceous South African neophyte perennial, widely scattered and naturalised in rough and waste ground. The binomial is from Greek meaning 'smelling of saffron', and Latin referencing the flower clusters (panicles).

AUTOSTICHIDAE Family name for the obscure moths, presumably from Greek *autostichidos* = 'self-styled'.

AVENS Herbaceous perennials in the genus *Geum* (Latin for these plants) and a shrub in the genus *Dryas* (Greek name for a nymph), both in the family Rosaceae, the English name dating from Middle English from Old French *avence*.

Large-leaved avens *G. macrophyllum*, a neophyte from North America and north-eastern Asia present in gardens by 1804, scattered in Britain, naturalised in woodland and on roadsides and riverbanks. *Macrophyllum* is Greek for 'big leaf'.

Mountain avens *D. octopetala*, a dwarf procumbent and creeping shrub, very local on base-rich ledges and rock crevices on mountains, for example in north-west Wales and the Lake District, in upland calcareous grassland in the Yorkshire Dales, on coastal shell sand in north Scotland, and on limestone pavement in the Burren, west Ireland. It is listed as Vulnerable in England and Endangered in Wales. *Octopetala* is Latin for 'eight petals'.

Water avens *G. rivale*, rhizomatous, widespread but only locally common in central and south-east England, on mildly acidic to calcareous, slow-draining or wet soils, in streamsides and flushes in deciduous woodland, carr, herb-rich hay meadows, montane willow scrub and tall-herb communities on mountain ledges. *Rivale* is Latin indicating growing by streams.

Wood avens *G. urbanum*, also known as herb bennet, a common, widespread perennial herb of free-draining, mildly acidic to calcareous soils, found in moderate shade in deciduous woodland (especially in disturbed sites in secondary woodland), scrub and hedgerows, and in more disturbed and open habitats, often growing as a street or garden weed, hence the specific name.

AVIAN POX A viral disease, symptoms including tumour-like growths on the skin, typically on the head, wings or legs, and spread by contact between birds or indirectly via contaminated surfaces, and by biting insects such as mosquitoes. Known for decades in such species as house sparrow, dunnock, starling and wood pigeon, a severe form was first noted in great tit in 2006, which has spread across much of England and Wales.

AVOCET *Recurvirostra avosetta* (Recurvirostridae), wingspan 78 cm, length 44 cm, weight 280 g, found on coastal lagoons on the east coast in summer and the Exe Estuary in winter. At least 95 sites with 1,629 pairs were reported in 2010, with 7,500 wintering individuals and 1,700 further birds as passage migrants. Numbers increased by 12% over 2010–15. Having been extinct in Britain since the 1840s, its 'return' in the 1940s (when coastal marshes in East Anglia were flooded as a defence against possible invasion by the Germans) and subsequent increase in numbers represent a major conservation success, and it has been adopted as the emblem of the RSPB. Nests are in scrapes on bare or thinly vegetated ground, clutch size is 3–4, incubation takes 23–5 days, and the precocial chicks fledge in 35–42 days. Feeding is on invertebrates and small fish, caught by sweeping the upcurved bill from side to side, prey located by touch. 'Avocet' dates from the late seventeenth century, derived from French *avocette*, in turn from Venetian *avosetta*, which provides the specific name. Latin *recurvus* + *rostris* = 'bent back' + 'bill'.

AWLWORT *Subularia aquatica* (Brassicaceae), an annual generally submerged aquatic plant, very local in Wales, north-west England, Scotland and western Ireland, growing on gravel or stony substrates in acidic, oligotrophic lakes, usually in water <1 m in depth. Latin *subula* = 'awl', from the leaf shape.

AZALEA Members of the Ericaceae, the name from Greek *azaleos* = 'dry', referring to the native habitat of the original species *A. procumbens*: deciduous species of *Rhododendron* (Greek *rhodon* + *dendros* = 'rose' + 'tree'), commonly planted in parks and gardens, only yellow azalea having become truly naturalised; also a species of *Kalmia* (named in honour of the Swedish naturalist Pehr Kalm, 1716–79).

Trailing azalea *K. procumbens*, a procumbent, calcifugous dwarf shrub locally frequent in the Highlands, rare on Orkney and Shetland, on dry, exposed, stony mountain moorland.

Yellow azalea *R. luteum*, an east European neophyte shrub present by the end of the eighteenth century, scattered and locally naturalised on acidic soils on heathland and moorland and in open woodland. *Luteum* is Latin for 'yellow'.

B

BABY'S-BREATH *Gypsophila paniculata* (Caryophyllaceae), a tap-rooted herbaceous perennial European neophyte, present in gardens by 1759, scarce and scattered in Britain, naturalised in grassy habitats, especially on light soils, and as a casual on tips. Greek *gypsos* + *philos* = 'gypsum' + 'loving', referring to the plant's chalk- or lime-loving nature. *Paniculata* is Latin referring to the flower clusters (panicles).

BABY'S TEARS Alternative name for mind-your-own-business.

BACHELOR'S-BUTTON Alternative name for a number of plants, often in a garden context, especially feverfew, cornflower and kerria. The name comes from a tradition that button flowers were picked by girls, given names of eligible men, and placed under aprons: the flower that opened first indicated the bachelor to marry.

BACILLACEAE Rod-shaped bacteria that form endospores, with two main subdivisions, the anaerobic spore-forming bacteria in the genus *Clostridium* (Latin for 'thread'), and the aerobic or facultatively anaerobic spore-forming bacteria in the genus *Bacillus* (Latin diminutive for 'rod').

BACILLARIOPHYTA Depending on the taxonomic classification used, the division or class name for diatoms.

BACILLUS From the Latin diminutive for 'rod', a genus of gram-positive, rod-shaped bacteria, including the causative agent of anthrax (*B. anthracis*).

BACKSWIMMER Waterbugs of the family Notonectidae (Greek *notos* + *nectos* = 'back' + 'swimming') and Pleidae (possibly Greek *pleon* = 'sail') that swim upside down through the water, often near the surface where they grab insects that have fallen in to the water film; they also feed on prey as large as tadpoles and small fish.

Widespread but local species include **mottled backswimmer** *Notonecta maculata*, **pygmy backswimmer** *Plea minutissima*, *N. obliqua* and *N. viridis*.

Common backswimmer *N. glauca*, widespread and abundant in ponds, ditches and canals. Sometimes incorrectly called greater water boatman. *Glauca* is Latin for 'glaucous'.

BACTERIA A domain of microscopic single-celled organisms, typically a few microns long, found in all environments, including the human body where many are benign or indeed essential. See also actinobacteria.

BADGER *Meles meles* (Mustelidae), head-body length 75–90 cm, tail 15–20 cm, weight 8–9 kg in spring, 11–12 kg in autumn, common in England though rarer to the north and east; thinly distributed in Scotland, common throughout most of Ireland, but absent from the Isle of Man and most other islands. They favour deciduous and mixed woodland, especially where near pasture, where earthworms are a favoured food. They also eat small animals, from rodents to insects, and blackberries, apples, cereals, fungi and roots when available. Badgers are increasingly evident in towns, and are especially drawn to peanuts. They increase their body fat in autumn, but enter a state of torpor, not hibernation, over winter. Badgers are nocturnal, spending the day in their sett, which comprises underground tunnels and nesting chambers excavated using large digging claws. The estimated pre-breeding population in 1997 was 310,000 in >50,000 social groups; in 2014 the number of social groups in England had risen to 71,600, more than doubling in 25 years. Estimates for 2014 suggest a population in England of 424,000 in England, and 61,000 in Wales. The Protection of Badgers Act (1992) consolidated previous legislation and, as well as protecting the badger itself, made it an offence to damage or obstruct the setts.

Badgers are vulnerable to bovine tuberculosis (bTB), a disease retransmitted to cattle mainly via urine and faeces, causing great financial loss to farmers. Despite scientific studies showing the long-term non-effectiveness of culling badgers in reducing bTB in cattle (indeed exacerbating it, since it encourages non-diseased badgers to disperse into vacated diseased areas) the British Government has initiated a series of local (county) culls since 2013, though targets have not been met. It is likely that bTB would be reduced most effectively using a combination of culling, vaccination and, above all, ensuring that movements of cattle to markets are of certified healthy animals. Road traffic accidents are a major cause of badger death. Mating takes place between February and May, with implantation delayed until late winter. Only one female in a social group normally breeds. Litters of two to three cubs are born around February, appearing above ground at about eight weeks. Weaning usually takes 12 weeks. By late summer cubs are feeding independently. 'Badger' may come from 'badge' + -*ard*, referring to the white mark on its forehead, and may date to the early sixteenth century. The binomial is Latin for badger.

BAETIDAE Named from the Spanish river Baetis, a mayfly family with four British genera: *Baetis* (nine species, only **large dark olive** *B. rhodani* being common and widespread) and **small spurwing** *Centroptilum leuteolum*, nymphs preferring running water, adults with hindwings; and *Cloeon* (two species) and *Procloeon* (two), nymphs preferably swimming in ponds or slow-moving water, adults without hindwings.

BAGWORM Members of the moth family Psychidae, so called because the larvae create and live in portable cases or 'bags'.

BALITORIDAE Family name of stone loach, possibly from Greek *balios* = 'swift' or 'spotted'.

BALLERINA **Pink ballerina** Alternative name for pink waxcap.

BALLOON FLY Alternative name given to some dagger flies (Empididae). Specifically, can refer to *Hilara maura*, 7 mm length, widespread and locally common in Britain, often seen swarming over water and marshy areas in summer. Greek *hilaros* = 'cheerful', and *mauros* = 'dark'.

BALLOONWORT Rare thallose liverworts in the genus *Sphaerocarpos* (Sphaerocarpaceae) (Greek *sphaera* + *carpos* = 'ball' + 'fruit'). **Micheli's balloonwort** *S. michelii* (for the Italian priest and botanist Pier Micheli, 1679–1737) and **Texan balloonwort** *S. texanus*, are both found on non-calcareous or neutral substrates in southern England and especially East Anglia in gardens and arable fields, rarely on roadsides and waste ground.

BALM *Melissa officinalis* (Lamiaceae), a lemon-scented herbaceous perennial neophyte cultivated for over a millennium, used in cooking and for teas, much grown in gardens where it seeds prolifically, often becoming a weed and escaping into nearby habitats, and so found in much of England, Wales and southeast Ireland (scarce elsewhere) on banks and verges close to habitation, and on waste ground. The English name derives from Latin *balsamum*, Greek *balsamon* and Hebrew *basam* = 'scent' or 'balsam'. The binomial comes from the Greek for 'honey bee' and Latin for 'of pharmacological use'.

Bastard balm *Melittis melissophyllum*, a herbaceous perennial, locally common in southern and south-west England and south-west Wales, in woodland, hedge banks and scrub on base-rich soils. In the New Forest it is indicative of ancient woodland. It prefers light shade and is often abundant in cleared or coppiced woodland, but is intolerant of grazing, and in the New Forest and Dorset has recently declined as a result of overshading and pony grazing. The binomial comes from the Greek for 'honey bee' and 'bee leaf'.

BALM-OF-GILEAD *Populus x jackii* (*P. balsamifera* x *P. deltoides*) (Salicaceae), a deciduous neophyte tree present by

1773, scattered in Britain, mostly planted in damp woods and by rivers and ponds, but increasingly in parks and by roadsides. Only female plants are known here; they often become naturalised by suckering. The English name refers to Jeremiah 8:22, 'Is there no balm in Gilead', from the balsam-like scent of the leaf buds and leaves as they unfold.

BALSAM Annual members of the genus *Impatiens* (Balsaminaceae). The English name derives from Latin *balsamum*, Greek *balsamon* and Hebrew *basam* = 'scent' or 'balsam'. The genus name is Latin for 'impatient', from the way the ripe seed pods burst open when touched.

Himalayan balsam Alternative name for Indian balsam.

Indian balsam *I. glandulifera*, a common and widespread Himalayan neophyte annual introduced in 1839, perniciously invasive on the banks of waterways, but also established in damp woodland, flushes and mires. The Environment Agency has estimated that the weed occupies >13% of riverbanks in England and Wales. The tallest annual in Britain, its rapid growth to 2–3 m can shade out other species and form monospecific stands. Each plant can produce up to 800 seeds which are dispersed widely as the ripe seedpods shoot the seeds up to 7 m away, which may then be carried downstream. With the existing physical and chemical control measures of 2014, the EA estimated it would cost up to £300 million to eradicate Himalayan balsam from the UK. In that year, however, a host-specific variety of a rust fungus new to science (*Puccinia komarovii* var *glanduliferae*) released at sites in Berkshire, Cornwall and Middlesex seemed to be effective in reducing the weed's vigour, and by 2017 some 33 sites were being used in this trial programme. *Glandulifera* is Latin for 'bearing little glands'.

Orange balsam *I. capensis*, a North American neophyte probably introduced in the early nineteenth century, locally common in England as far north as Yorkshire, naturalised by rivers and canals. Seeds are ejected from their capsules to a distance of a few metres, and can be dispersed by water. *Capensis* is Latin for 'from the cape'.

Small balsam *I. parviflora*, an East Asian neophyte present by 1823, naturalised in Britain in semi-natural woodland and plantations, especially along tracks, and also along shaded riverbanks, and in east Scotland on unshaded river shingle. *Parviflora* is Latin for 'small flower'.

Touch-me-not balsam *I. noli-tangere*, rare and local, probably native in central Wales and the Lake District and casual elsewhere in Britain as far north as central Scotland, found on nutrient-rich soils in damp woodland, on streamsides and in valley-side seepages. Latin *noli-tangere* = 'do not touch'.

BALSAM-POPLAR Species of *Populus* (Salicaceae), all deciduous trees. See also poplar.

Eastern balsam-poplar *P. balsamifera*, a North American neophyte introduced in 1689, widespread and scattered, planted in hedgerows, amenity areas and on roadsides, and in small quantities for timber. It can reproduce prolifically by suckering.

Hybrid balsam-poplar *Populus* 'Balsam Spire' (*P. balsamifera* x *P. trichocarpa*), a fastigiate neophyte planted in parks and amenity areas as screens or windbreaks, and in plantations. This female hybrid, raised in North America in the 1930s, was released for planting in Britain 1957, and is now one of the most planted poplars in the UK, being resistant to *Marssonina* leaf fungi and *Melampsora* leaf rusts.

Western balsam-poplar *P. trichocarpa*, a North American neophyte, introduced in 1892, often planted for amenity or timber. Most trees are male, which cannot reproduce, but which occasionally spread by suckers. Greek *trichos* + *carpos* = 'hair' + 'seed'.

BALSAMINACEAE Family name of balsam, from Latin *balsamum*, Greek *balsamon* and Hebrew *basam* = 'scent' or 'balsam'.

BAMBOO Woody members of the Poaceae, tribe Bambuseae, usually with a strong rhizome system flowering at best occasionally, with a few examples locally casual or naturalised as garden escapes. 'Bamboo' comes from Dutch *bamboes*, in turn from Portuguese *mambu*, itself from Malay *mambu*.

BANE See vampire's bane.

BANEBERRY *Actaea spicata* (Ranunculaceae), a perennial herb found in northern England in the grikes of limestone pavements, on rock ledges and in deciduous woodland, all sites having the same characteristics of shade and low competition. It is more widespread as a naturalised garden escape. Berries are poisonous, hence the English name. Actaea is a name from Ancient Greece; Latin *spicata* = 'spiked'.

BAR (MOTH) Members of the family Geometridae.

Grey scalloped bar *Dyscia fagaria* (Ennominae), wingspan 30–40 mm, found in moorland, heathland and peat bogs, day-flying in May–August, locally distributed mainly in the north of England and Wales, in Scotland and in Ireland, and occasionally in the very south of England. Larvae feed on heathers and cross-leaved heath. The binomial comes from Greek *dyo* + *scia* = 'two' + 'shadow' (from the dark blotches on the forewing), and Latin *fagus* = 'beech' (though this is not a food plant).

Purple bar *Cosmorhoe ocellata* (Larentiinae), wingspan 20–5 mm, nocturnal, widespread and common in hedgerows, scrub, woodland, calcareous grassland, heathland, dunes and gardens. Adults fly from May to August, usually with two generations in the south, but a single brood further north. Larvae feed on bedstraws. Greek *cosmos* + *rhoe* = '(good) order' + 'stream', and Latin *ocellatus* = 'eyed', from the forewing spots.

BARBASTELLE *Barbastella barbastellus* (Vespertilionidae), head-body length 40–55 mm, wingspan 260–90 mm, weight 6–13 g, a very rare bat found in southern and central England and Wales: breeding roosts are known from parts of Exmoor and the Quantock Hills (Devon and Somerset), Mottisfont (Hampshire), Ebernoe Common (West Sussex), and Paston Great Barn (Norfolk). There was an estimated total population in 1995 of 5,000. It roosts in buildings and trees, and mainly feeds on moths, and can drink from ponds or lakes. The usually single young are generally born in July. Juveniles can fly at three weeks, and become independent at six weeks. Hibernation is in buildings, tree holes and caves. Echolocation is best heard at around 32 kHz. The name is late eighteenth-century from French, in turn from Italian *barbastello*.

BARBEL *Barbus barbus* (Cyprinidae), length up to 90 cm, weight 12 kg, found in the middle reaches of clear oxygen-rich rivers and streams with sandy or stony beds, originally only in the Thames and Trent systems but introduced for anglers elsewhere in lowland England and Wales. A bottom-feeder, its underslung mouth makes it adapted for feeding on benthic crustaceans, insect larvae and molluscs. Spawning is in May–July, often with upstream migration in a search for suitable gravel beds. The name comes from the four whisker-like filaments located at the corners of the mouth, which are used to locate food.

BARBERRY Spiny introduced (neophyte) shrubs in the genus *Berberis* (Berberidaceae), specifically *B. vulgaris*, a locally common and widespread deciduous shrub, found in hedgerows and coppices, and on banks, cliffs and waste ground. Possibly an archaeophyte (recorded from the Neolithic Grimes Graves, Norfolk). It might be native in England and Wales, though was cultivated in medieval times. The English name possibly references the 'barb' of the plant's spines, but it has been suggested that *Berberis* is the Latinised form of the Arabic name. *Vulgaris* is Latin for 'common'.

Often widespread, locally naturalised neophytes from China are **Chinese barberry** *B. julianae* (introduced in 1900), **clustered barberry** *B. aggregata* (1908), **Gagnepain's barberry** *B. gagnepainii* (1904) and **Mrs Wilson's barberry** *B. wilsoniae* (1904); from the Himalayas **great barberry** *B. glaucocarpa* (1832); and from South America **box-leaved barberry** *B. buxifolia* (1830) and **Darwin's barberry** *B. darwinii* (1849).

Hedge barberry (or **golden barberry**) *B.* x *stenophylla*, an evergreen hybrid (*B. darwinii* x *B. empetrifolia*), arising in around 1860 in Sheffield, and now commonly planted in parks, roadsides and hedges, and occasionally naturalised. *Stenophylla* is Greek for 'narrow leaves'.

Thunberg's barberry *B. thunbergii*, a deciduous introduction in 1883 from Japan, widespread and commonly found as a garden escape in urban habitats, and often bird-sown, in waste ground, churchyards and hedgerows, and on roadsides, railway and canal banks. Karl von Thunberg (1743–1828) was a Swedish botanist.

BARK BEETLE Small weevils, mostly species of *Scolytus* (Curculionidae, Scolytinae), often associated with dead wood, many becoming forestry pests, and large elm and small elm bark beetles the vectors of Dutch elm disease. *Scolytos* is Greek for 'bark beetle'. See also Colydiidae, and flat bark beetle under beetle.

Birch bark beetle *S. ratzeburgi*, length 5–6 mm, mostly recorded from Cumbria and the Highlands, larvae feeding on phloem in galleries in birch, it being a vector of fungal disease. Ratzeburg is a town in Schleswig-Holstein, Germany.

Eight-toothed spruce bark beetle *Ips typographicus*, 4.2–5.5 mm length, with a few records from the north Midlands and East Anglia, first recorded from imported timber in 1997. This pest often builds up numbers in wind-blown, damaged and recently felled trees before attacking adjacent live trees where the female lays her eggs in galleries in the wood, in which the larvae feed. A 2014 survey suggests it has been controlled and is now absent. *Ips* is Greek for a worm (here a larva) that eats through wood, and *typos* + *grapheia* = 'form' + 'writing'.

Gorse bark beetle *Hylastinus obscurus*, length 2–3 mm, with a scattered distribution in England and Wales. *Hyla* is Greek for 'a wood', *obscurus* Latin for 'dark'.

Great spruce bark beetle *Dendroctonus micans*, 6–8 mm length, from North America, first discovered in Britain in 1982, now with scattered records mainly from Wales, the west Midlands and south Scotland, a pest that damages spruce trees by tunnelling into the bark (but not the wood itself) to lay eggs; larvae feed on the inner woody layers. *Micans* is Latin for 'sparkling'.

Large elm bark beetle *S. scolytus*, length 3–6 mm, with a scattered distribution in England and Wales, occasional in Scotland, larvae feeding in tunnels in elm trees, and a significant vector of Dutch elm disease.

Large fruit bark beetle *S. mali*, length 3.5–4.5 mm, with a scattered distribution in England and south Wales, a vector of fungal disease in fruit trees such as apple (giving the specific name, the Latin for 'apple tree' being *malus*).

Lineated bark beetle *Trypodendron lineatum*, 3.5 mm length, occasional, with a scattered distribution over Britain, boring into dead conifer trees and timber to create galleries, but also introducing ambrosia fungus (it is also known as striped ambrosia beetle) as adult and larval food. Greek *trypao* + *dendron* = 'to bore' + 'tree', and Latin *lineatum* = 'lined'.

Oak bark beetle *S. intricatus*, length 2.5–3.5 mm, common and widespread in England and Wales, found on oaks and, with larvae feeding on phloem in tunnels, a potential vector of oak wilt. *Intricatus* is Latin for 'complex'.

Pygmy elm bark beetle *S. pygmaeus*, with only three records from the Suffolk coast.

Small elm bark beetle *S. multistriatus*, length 2–3.5 mm, with a scattered distribution in the southern two-thirds of England and in Wales, a vector of Dutch elm disease (though not as significant as large elm bark beetle). *Multistriatus* is Latin for 'many-lined'.

Small fruit bark beetle *S. rugulosus*, length 2–3 mm, with a scattered distribution in the southern half of England and in Wales, a vector of fungal disease in fruit trees. *Rugulosus* is Latin for 'finely wrinkled'.

BARKBUG See flatbug.

BARKFLY Outdoor members of the order Psocoptera, small, winged soft-bodied insects, with 68 species, many of which live on or under bark, but also in leaf litter and under stones, feeding on algae, fungi or pollen.

BARKSPOT Inedible fungi in the genus *Diatrype* (Xylariales, Diatrypaceae) (Greek *dia* + *trypa* = 'through' + 'hole'), including **beech barkspot** *D. disciformis* and **willow barkspot** *D. bullata*, both common and widespread in woodland on dead branches.

BARLEY Species of *Hordeum* (Latin for 'barley'), the common cultivated plant being two-rowed barley, together with *Hordelymus* (wood barley), all Poaceae.

Introduced annuals occasionally found as casuals on waste ground, often originating from wool shoddy or as relics of cultivation, are **Antarctic barley** *H. pubiflorum*, **Argentine barley** *H. euclaston*, **little barley** *H. pusillum*, **Mediterranean barley** *H. geniculatum*, **six-rowed barley** *H. vulgare* and **two-rowed barley** *H. distichon*.

Foxtail barley *H. jubatum*, a perennial North American neophyte introduced as a fodder grass in 1782 and recorded from the wild since 1890, locally frequent on roadsides, resown grassland and waste ground as a casual from wool shoddy and birdseed, and as a contaminant of grass-seed mixtures. It is increasingly found by salt-treated roads. *Jubatum* is Latin for 'crested'.

Meadow barley *H. secalinum*, a perennial common in England south of a line from the Severn to the Humber estuaries (but scarce in south-west England), scattered elsewhere, in meadow, pasture and roadsides, often in valley floodplains, with a preference for clay. In coastal areas it is often abundant in grazing marsh grassland. *Secalinum* is Latin for 'resembling rye'.

Sea barley *H. marinum*, an annual of lightly vegetated coastal habitat, for example on the margins of dried-up pools and ditches in grazing saltmarsh, on tracks and sea walls, and in the upper saltmarsh, as well as occasionally inland by salt-treated roads. *Marinum* is Latin for 'marine'.

Wall barley *H. murinum*, an annual archaeophyte locally common in lowland England and Wales, eastern Scotland and east Ireland, growing on fertile, disturbed ground, roadsides, pavements, walls, railway banks and rough grassland, often in urban areas. Subspecies *leporinum* and *glaucum* are of Mediterranean origin, occasional casuals on waste ground originating from wool shoddy. *Murus* is Latin for 'wall'.

Wood barley *Hordelymus europaeus*, a perennial found locally in England and Wales on calcareous soils, especially in sheltered beech woodland and along medieval boundary banks and old hedgerows. The genus name conflates *hordeum* + *elymus* = 'barley' + 'wild rye'.

BARN SWALLOW See swallow.

BARNACLE Marine crustacean suspension-feeder in the class Maxillopoda, infraclass Cirripedia, with a swimming larval phase, and an adult stage fixed to a hard substrate in the intertidal and sublittoral. 'Barnacle' dates from the late sixteenth century, derived from medieval Latin *bernaca*, of unknown origin. In Middle English the term denoted the barnacle goose.

Common and widespread species include the **acorn barnacles** *Balanus balanus*, *B. crenatus* (Balanidae) and *Semibalanus balanoides* (Archaeobalanidae); **goose barnacle** *Lepas anatifera* (Lepadidae); **Montagu's stellate barnacle** *Chthamalus montagui* and **Poli's stellate barnacle** *C. stellatus* (Chthamalidae); and **wart barnacle** *Verruca stroemia* (Verrucidae).

The New Zealand/Australian *Austrominius modestus* (Austrobalanidae) probably arrived between 1940 and 1943 on ship hulls, though was first recorded in Chichester Harbour in 1946, and is now found around most coasts of England and Wales, a few areas of Scotland and some Scottish islands including the Outer Hebrides. Its rapid spread is attributed to pelagic larval dispersal and further transport by hull fouling, and it has become the dominant barnacle at a number of sites.

Crab hacker barnacle *Sacculina carcini* (Sacculinidae) is parasitic on crabs, and recorded from a few sites off the British coast; in 2010, *The Guardian* launched an annual competition for the public to suggest common names for previously unnamed species, and in 2012 this name was selected.

BARRED (MOTH) **Silver barred** *Deltote bankiana* (Noctuidae, Eustrotiinae), wingspan 24–8 mm, flying from May to July, nocturnal but active at day when disturbed. It is found in the Cambridgeshire fens and at a coastal marsh in south-east Kent. It also occurs as a probable migrant in the south-east and central-south of England. In Ireland it is restricted to peat bogs in Co. Cork and Co. Kerry. Preferred habitat is marsh, fen, damp grassland and boggy heathland, where larvae feed on such grasses as purple moor-grass and smooth meadow-grass. Conservation status is Rare (RDB 3). Greek *deltotos* means shaped like the capital Greek letter *delta*, from the triangular forewing. Sir Joseph Banks (1743–1820) was an English naturalist and collector.

BARTRAMIACEAE Family name of apple-mosses, named to honour the American botanist John Bartram (1699–1777).

BARTSIA Species in the family Orobanchaceae. Johann Bartsch (1709–38) was a Prussian botanist.

Alpine bartsia *Bartsia alpina*, a rhizomatous perennial found in the south-west Highlands in grassy habitats on base-rich soils and on rocky ledges, and locally in northern England on drier hummocks in flushes in damp upland pasture, and on steep, flushed, species-rich banks.

French bartsia *Odontites jaubertianus*, an annual neophyte naturalised in 1965 on gravelly rough ground near RAF Aldermaston, Berkshire, but reported as nearly gone in 1986; in the 1970s in a lay-by at RAF Greenham Common, also Berkshire; in an old airfield at Spanhoe, Northamptonshire, in 2005; and on calcareous grassland on an old airfield in north Essex in 2006. This is a French endemic, and these airfield records may have originated with material carried from France on muddy aircraft towards the end of the war. The genus name is from Latin *odontitis* (used by Pliny as a plant good for curing toothache), in turn from Greek *odous* = 'tooth'. Hippolyte Jaubert (1798–1874) was a French botanist.

Red bartsia *O. vernus*, a common and widespread annual root-hemiparasite found on short, often trampled grassland, tracks, waste ground, arable field margins, and (ssp. *litoralis* in Scotland and Cumbria) gravelly and rocky seashores, and saltmarsh. *Vernus* is Latin for the season 'spring'.

Yellow bartsia *Parentucellia viscosa*, a hemiparasitic annual, scattered but especially common in southern and south-western England in damp, open grassy disturbed habitats on sandy soils, often by tracks. It usually grows in drier dune slacks and in reclaimed heath-pasture, but is also found on pathsides, rough and scrubby grassland and field borders, as well increasingly in reseeded amenity grasslands and waste places. The binomial honours Pope Nicholas V. Pontifex Maximus, formerly Tomaso Parentucelli (1397–1455), plus the Latin for 'sticky'.

BASIDIOMYCETE Any fungus of the phylum Basidiomycota (Greek *mycetes* = 'fungus') in which spores are produced in basidia (the spore-bearing structure), and including puffballs, rusts, smuts and many so-called mushrooms.

BASIL **Wild basil** *Clinopodium vulgare* (Lamiaceae), a herbaceous rhizomatous perennial of hedges, common and widespread in Britain, decreasing northwards, and introduced and scarce in Ireland, growing in woodland edges, scrubby grassland, coastal cliffs and dunes, waste ground, old quarries, and railway banks, typically on dry calcareous soils. Greek *cline* + *podi* = 'bed' + 'foot', and Latin *vulgare* = 'common'.

BASKET SHELL *Corbula gibba* (Corbulidae), a widespread clam with shell size to 1.5 cm, from the low intertidal and deeper,

burrowing in muddy mixed sediments. *Corbula* is Latin for 'basket', *gibba* for 'hump'.

BASS See sea bass.

BASTARD-TOADFLAX *Thesium humifusum* (Santalaceae), a perennial with herbaceous shoots from a woody rootstock, which is hemiparasitic on the roots of various herbaceous plants. It is found in short, usually grazed, species-rich calcareous grassland, occasionally on clay or calcareous sandy soil near the coast, very local in England as far north as Lincolnshire. The mythological Theseus was king of Athens. *Humifusum* is Latin for 'prostrate' or 'sprawling'.

BAT Members of the order Chiroptera, in the families Rhinolophidae (horseshoe bats in the genus *Rhinolophus*, from Greek *rhinos* + *lophos* = 'nose' + 'crest') and Vespertilionidae (Latin *vespertilio* = 'bat', from *vesper* = 'evening', in the genera *Eptesicus*, *Myotis*, *Nyctalus* and *Plecotus*). They are small, (in Britain) insectivorous mammals that fly and hunt using echolocation, different species using different frequencies, with 18 species (17 breeding) native to Britain. Vagrant species are: **greater mouse-eared**, **Kuhl's pipistrelle**, **northern** and **parti-coloured bats**. See also noctule and serotine.

Alcathoe bat *M. alcathoe*, head-body length 40 mm, wingspan 200 mm, weight 3.5–5.5 g, confirmed as a resident British species in 2010 following records the previous year from woodland cave swarming sites in Ryedale (North York Moors) and the South Downs in Sussex. It favours deciduous woodland for roosting and hunting insects in the canopy and above water. It has a high-pitched echolocation call, with a frequency that falls from 120 kHz at the beginning of the call to 43 kHz at the end. Alcathoe is a figure from Greek mythology, turned into a bat when she refused the advances of the god Dionysus. Greek *myotis* = 'mouse-ear'.

Bechstein's bat *M. bechsteinii*, head-body length 43–53 mm, wingspan 250–300 mm, weight 7–13 g, very rare, found in parts of southern England and south Wales in (preferably old) deciduous woodland, which is used for roosting and hibernation (both, for example, in woodpecker tree holes) and foraging for insects in the canopy. With probably no more than 1,000 individuals it is a species of conservation priority in the UK BAP, and it is listed in the EC Habitats Directive, which requires designation of Special Areas of Conservation (SACs) to promote important populations; in 2006 there were six SACs designated specifically for this bat. Mating occurs in autumn and spring, with maternity colonies forming in April and May. Females gather in colonies of 10–30 bats (occasionally up to 100), with young born from late June to early July. Echolocation is mostly at around 50–60 kHz. Johann Bechstein (1757–1822) was a German naturalist.

Brandt's bat *M. brandtii*, head-body length 38–50 mm, wingspan 210–40 mm, weight 4.5–9.5 g, only separated from whiskered bat in 1970, found throughout England and Wales. The British population was estimated in 1995 as 30,000 (22,500 in England, 7,000 in Wales, 500 in Scotland). Roost surveys in Britain up to 2015 (including whiskered bat) indicate a 30.6% increase on a 1999 base year value, equivalent to an annual increase of 1.7%. They roost in crevices in buildings, especially older ones with stone walls and slate roofs, emerging within half an hour after sunset and hunting throughout the night along hedgerows and in tree canopies for invertebrates, particularly moths. Mating usually takes place in autumn, but has been observed throughout winter. Adult females form maternity colonies in summer, giving birth to their single young in June or early July. Young can fly by three weeks; after six weeks they can forage independently. Hibernation is in caves and tunnels. They echolocate at 33–89 kHz, being loudest at 45 kHz. Johann von Brandt (1802–79) was a German zoologist.

Brown long-eared bat *Plecotus auritus*, head-body length 37–52 mm, wingspan 230–85 mm, weight 6–12 g, widespread and locally common in woodland, parks and gardens. This is the commonest bat in Britain after common and soprano pipistrelles. The British population was estimated in 1995 as 200,000 (155,000 in England, 17,500 in Wales, 27,500 in Scotland). A Northern Ireland survey in 1999 indicated a further 45,000 animals. Roost counts in Britain up to 2015 indicate a 28.2% increase on a 1990 base year value, equivalent to an annual increase of 1.8%. Summer roosts of up to 50 males and females are in buildings and trees. They catch flying insects at night but are also gleaners, flying slowly to pick insects off leaves, bark and spider webs, and they also hunt from perches. Their broad wings and tail allow slow, highly manoeuvrable, hovering flight, and for catching non-flying prey they need powerful hearing, hence their large ears. Mating takes place in autumn and to an extent throughout winter. Maternity colonies are established in late spring, with one young born in late June to mid-July, able to fly at three weeks and independent at six weeks. Hibernation is in caves and tunnels, occasionally houses and trees. Echolocation calls in the range 25–50 kHz, peaking at 35 kHz. Greek *plecein* + *otis* = 'to twist, or weave' + 'ear', and Latin *auritus* = 'eared'.

Common long-eared bat Alternative name for brown long-eared bat.

Daubenton's bat *M. daubentonii*, head-body length 45–55 mm, wingspan 240–75 mm, weight 7–12 g, widespread and locally common. The British population was estimated in 1995 as 150,000 (95,000 in England, 15,000 in Wales, 40,000 in Scotland). Waterway surveys in Britain up to 2015 indicate a 4.6% increase on a 1993 base year value, equivalent to an annual increase of 0.3%. Increases in parts of its range may be associated with a greater number of reservoirs and flooded gravel-pits and quarries. A 1999 survey found 410,000 individuals in Northern Ireland. It is the second commonest bat species recorded in Ireland. In England and Wales, most summer roosts are in tunnels or under bridges over canals and rivers (in Scotland greater use is made of buildings), usually feeding within 6 km of the roost. They usually take insects from close to the water but can take prey directly from the water using their feet as a gaff or the tail membrane as a scoop. Mating takes place in autumn and to an extent in winter. Maternity roosts are used from late spring to October. Young are suckled for several weeks, becoming independent at 6–8 weeks. Hibernation is in tunnels, caves and mines. Echolocation in the range 35–85 kHz, loudest at 45–50 kHz. The English and specific names refer to the French naturalist Louis-Jean-Marie Daubenton (1716–1800).

Geoffroy's bat *M. emarginatus*, head-body length 41–58 mm, wingspan 220–50 mm, weight 7–15 g, a relatively sedentary continental species, but discovered in September 2012 in the South Downs, West Sussex, and near Bath in September 2013, so possibly extending its range into England. It roosts and hibernates underground. It hunts over grassland and scrub, feeding on insects, typically plucking its prey off surfaces such as leaves, but also capturing insects during flight, generally within 5 m of the ground. Echolocation when gleaning is in the range 52–124 kHz. Etienne Geoffroy Saint-Hilaire (1772–1844) was a French naturalist.

Greater horseshoe bat *Rhinolophus ferrumequinum*, head-body length 57–71 mm, tail length 35–43 mm, wingspan 350–400 mm, weight up to 30 g, rare and endangered in south-west England (east to the Isle of Wight) and south Wales. The population had declined by >90% over the previous 100 years, but survey suggested a population of around 6,600 in 2005, and an index of change in roost numbers from 1999 to 2015 showed a 126% increase, equivalent to an annual increase of 5.2%. It has a complex horseshoe-shaped noseleaf which is related to its particular echolocation system. Summer roosting is often in buildings, especially roofs, or caves. It hunts cockchafers, dung beetles, noctuid moths, craneflies and caddis flies. Mating mainly occurs in autumn. Maternity colonies begin to gather in May and reach peak numbers in June–July when most breeding females return to the maternity roost. Young are born in mid-July. In winter it hibernates in caves, disused mines, cellars and tunnels. It has an almost constant frequency call at about 82 kHz. The specific name is Latin for 'horseshoe'.

Greater mouse-eared bat *M. myotis*, head-body length 65–80 mm, wingspan 365–450 mm, weight 24–40 g, Britain's largest bat. Always rare since its discovery in 1958, it was declared extinct in 1990, but an emaciated female was found dying in Bognor Regis in January 2001, and a solitary juvenile male had been hibernating in a tunnel near Chichester, West Sussex, since 2002. This species uses echolocation (22–86 kHz, with most energy at 37 kHz) for spatial orientation, but while emitting ultrasound calls when approaching its arthropod prey, it also catches them by gleaning off the ground.

Grey long-eared bat *P. austriacus*, head-body length 41–58 mm, wingspan 255–300 mm, weight 7–12 g, found in a few colonies in Sussex, Hampshire, the Isle of Wight, Dorset, Devon and Somerset, with a total fragmented and declining population of around 1,000 individuals, the decline linked to the disappearance of lowland unimproved meadows, its main foraging habitat. Feeding is on moths, flies (mainly craneflies) and small beetles caught in flight over open ground. Summer and winter roosts are mainly in roofs, though cellars, caves and tunnels are also used for hibernation. Breeding colonies roost during April–May. A single young is born in late June/early July and can fly by August. Echolocation is at 29–62.5 kHz. Latin *austriacus* = 'southern'.

Hairy-armed bat Alternative name for Leisler's bat.

Leisler's bat *Nyctalus leisleri*, head-body length 50–70 mm, wingspan 260–320 mm, weight 12–20 g, widespread and scattered in Britain as far north as central Scotland, widespread and the third commonest bat in Ireland (>20,000 estimated in 2016, Europe's largest population), where it is also the largest bat. The UK population in 1999 was tentatively estimated as 28,000, and considered to be stable, but a 2016 estimate was only 9,500 in England. It roosts in buildings and tree holes, and forages for insects in woodland, pasture and riparian habitats, along woodland margins, and around street lamps. Mating occurs from late summer to mid-autumn. Gestation begins in spring and lasts up to 75 days. In summer, maternity colonies of females gather in tree holes, sometimes in buildings, particularly in Ireland where colonies may number 1,000. The usually single young are born in mid-June, are weaned after six weeks, but capable of flight after four weeks. Hibernation is in tree holes and in cavities in buildings. Echolocation calls in the range 15–45 kHz, peaking at 25 kHz. Johann Leisler (1771–1813) was a German physician and naturalist. *Nyctalus* is from Greek *nyx* = 'night', Nyx also being the goddess of the night.

Lesser horseshoe bat *R. hipposideros*, head-body length 35–45 mm, wingspan 200–50 mm, weight 5–9 g, rare and endangered, found in west England, Wales and west Ireland. The British population was estimated at 50,000 in 2008, and seems to be increasing: roost surveys up to 2015 indicate a 76.2% increase on a 1999 base year value, equivalent to an annual increase of 3.6%. The Irish population is around 14,000. Summer colonies are usually found in roofs with nearby caves or tunnels to use in poor weather. There are peaks of activity at dusk and dawn but bats are active all night throughout the breeding season, feeding on flies (mainly midges), small moths, caddis flies, lacewings, beetles and wasps. Mating takes place during autumn or early winter. Maternity roosts are almost always in buildings in April–May. Half to two-thirds of the females in the nursery roost give birth to a single young between mid-June and mid-July. Lactation lasts four to five weeks. Young are completely independent at six weeks. Hibernation is from September/October until April, often into May, using caves, mines, tunnels and cellars. It has a complex horseshoe-shaped noseleaf which is related to their particular echolocation system. They have an almost constant frequency call at about 110 kHz. Greek *hipposideros* = 'horseshoe'.

Lesser noctule Alternative name for Leisler's bat.

Natterer's bat *M. nattereri*, head-body length 40–50 mm, wingspan 245–300 mm, weight 7–12 g, widespread but scarce. The British population was estimated in 1995 as 100,000 (70,000 in England, 12,500 in Wales, 17,500 in Scotland). Roost surveys in Britain up to 2015 indicate an 11.2% decrease on a 2002 base year value, equivalent to an annual decrease of 0.9%. A 1999 survey found 48,000 in Northern Ireland, yet a 1994 survey found only 400–500 in the Republic where it is classified as Threatened in the Irish RDB 2. Most summer colonies are in old stone buildings, with roosts in crevices and gaps in beams. Their flight, sometimes over water, is usually among trees, the broad wings giving them good manoeuvrability at slow speed. They hunt at night generally at heights of <5 m. Much of the insect prey is taken from foliage. Mating mainly occurs in autumn. Maternity colonies form from May to June through to July. The female gives birth to a single young at the end of June or in early July. For the first three weeks the young feeds on milk and is left in a crèche inside the roost when its mother leaves to hunt. Within six weeks it is independent. The bats arrive at their hibernation sites, usually entrances to caves, tunnels and mines, from December, individually finding crevices, and generally leaving by early March. Echolocation frequency range is 35–80 kHz, peaking at 50 kHz. Johann Natterer (1787–1843) was an Austrian naturalist.

Northern bat *Eptesicus nilssonii*, head-body length 78–80 mm, wingspan 240–80, weight 8–16 g. A single vagrant specimen was found in a hibernaculum in Surrey in 1987. *Eptesicus* comes from Greek *epten* + *oikos* = 'I fly' + 'house'. Sven Nilsson (1787–1883) was a Swedish naturalist.

Parti-coloured bat *Vespertilio murinus*, head-body length 48–64 mm, wingspan 260–330 mm, weight 11–24 g, a vagrant continental species with one or two records in most years since 2000, from the south coast to Shetland, roosting in trees, buildings and rock fissures. They hunt for insects after dusk at heights of 20–40 m. Echolocation uses a wide range of ultrasonic sounds, but especially at 25–7 kHz. The English name comes from the two colours of the fur: the dorsal side is red or dark brown, the ventral side white or grey. *Murinus* comes from Latin *mus* = 'mouse'.

Whiskered bat *M. mystacinus*, head-body length 35–48 mm, wingspan 210–40 mm, weight 4–8 g, only separated from Brandt's bat in 1970, found throughout England, Wales, south Scotland and Ireland. The British population was estimated in 1995 as 40,000 (30,500 in England, 8,000 in Wales, 1,500 in Scotland), with a further 24,000 in Northern Ireland, though numbers in the Republic have been estimated as <1,000. Roost surveys in Britain up to 2015 (including Brandt's bat) indicate a 30.6% increase on a 1999 base year value, equivalent to an annual increase of 1.7%. They roost in crevices in buildings, especially older ones with stone walls and slate roofs, emerging within half an hour after sunset and hunting throughout the night along hedgerows and in tree canopies for invertebrates, particularly moths. Mating usually takes place in autumn, but has been observed throughout winter. Adult females form maternity colonies in summer, giving birth to their single young in June or early July. Young can fly by three weeks; after six weeks they can forage independently. Hibernation is in caves and tunnels. They echolocate in the range 32–89 kHz, being loudest at 45 kHz. *Mystacinus* comes from the Greek for 'moustached'.

BAT BUG *Cimex pipistrelli* (Cimicidae), with scattered and local records from England and Wales, feeding on bat blood. *Cimex* is Latin for 'bug'; *pipistrelli* refers to pipistrelle bats.

BAY *Laurus nobilis* (Lauraceae), an evergreen neophyte tree or shrub introduced from Mediterranean Europe possibly in Roman times, widely planted as a culinary and ornamental plant of garden or shrubbery, naturalised (often bird-sown) in scrub, woodland, cliffs and dunes particularly in the south. 'Bay' derives from Old French *baie*, in turn from Latin *baca* = 'berry'; 'bay tree' is therefore 'berry tree', the berries having supposed medicinal properties. *Laurus* is Latin for 'laurel' or 'bay', *nobilis* for 'notable'.

BAYBERRY *Myrica pensylvanica* (Myricaceae), a deciduous neophyte North American shrub, cultivated by 1727, occasionally planted for cover, naturalised on heathland and in wet woodland in Hampshire, casual at a few other English sites. *Myrica* may come from Greek *myrtos*, in turn from ancient Semitic (Accadian) *murru* = 'bitter'.

BEACON (FUNGUS) Inedible saprobic rarities in the genus *Mitrula* (Helotiaceae). The prefix *mitr-* is a reference to a mitre, cap or headdress (by implication indicating the shape of the cap). **Bog beacon** *M. paludosa* and **burning beacon** *M. sclerotipus* are both found in damp conditions.

BEAD-MOSS **Bog bead-moss** *Aulacomnium palustre* (Aulacomniaceae), common and widespread in bogs and wet heathland, and also recorded in chalk grassland and woodland clearings. Greek *aula* + *mnium* = 'furrow' + 'moss'; *palustre* is from Latin for 'marshy'.

BEAD WEED **Orkney bead weed** *Lomentaria orcadensis* (Lomentariaceae), a red seaweed with fronds 4 cm long, widespread if patchily so on the east coast, found on rock and kelp holdfasts in silty lower intertidal pools and subtidally, on exposed shores. *Lomentum* is Latin for 'bean flour', *orcadensis* = 'pertaining to Orkney'.

BEAK-SEDGE Species of *Rhynchospora* (Cyperaceae) (Greek *rhynchos* + *sporos* = 'beak' + 'seed').

Brown beak-sedge *R. fusca*, a rhizomatous perennial, locally common in the New Forest, west Scotland and Ireland on wet heathland, and the margins of acidic mires, favouring bare peat. *Fusca* is Latin for 'brown'.

White beak-sedge *R. alba*, a herbaceous perennial, locally common in southern and south-west England, Wales, the Lake District, western Scotland and Ireland, in acidic bogs, wet heathland and mires, often in association with sphagnum. It is commonly found on bare wet peat, sometimes in shallow standing water. *Alba* is Latin for 'white'.

BEAN **French bean** *Phaseolus vulgaris* and **runner bean** *P. coccineus* (Fabaceae), long-cultivated South/Central American vegetables, scattered and widespread casuals on waste ground throughout lowland regions. *Phaseolus* is Latin for 'bean', literally meaning 'small boat', describing the shape of the pods. *Vulgaris* is Latin for 'common', *coccineus* for 'scarlet' (from the flower colour).

Broad bean *Vicia faba* (Fabaceae), an erect annual, seeds of which have been found in Iron Age deposits at Glastonbury. Widely grown as a vegetable and increasingly as a fodder crop, it is frequently found as an escape from cultivation on waste ground. Most populations are casual. The binomial is Latin for 'vetch' and 'bean'.

BEAR See woolly bear.

BEAR LEEK Alternative name for ramsons.

BEARBERRY *Arctostaphylos uva-ursi* (Ericaceae), a procumbent shrub found on peaty, often rocky upland heathland and moorland, generally over well-drained soil, locally common in the Highlands (this is the county flower of Aberdeenshire as voted by the public in a 2002 survey by Plantlife) and west and north-west Ireland, and occasionally on upland as far south as the Derbyshire Pennines. In the Burren it grows in heathy grasslands on limestone. The binomial is from Greek *arcto* + *staphyle* = 'bear' + 'grapes', and the Latin equivalent, these in turn leading to 'bearberry'.

Arctic bearberry *A. alpinus*, a shrub found on acidic mineral soils or peat on exposed upland heathland, and also on drier blanket bog in the northern Highlands and Islands. This is the county flower of Orkney as voted by the public in a 2002 survey by Plantlife.

BEARD-GRASS **Annual beard-grass** *Polypogon monspeliensis* (Poaceae), a scarce annual mostly found in southern and south-east England, but scattered as far north as central Scotland, on thinly vegetated coastal habitats, cattle-poached marsh, the margins of dried-up brackish pools, and upper saltmarsh, and inland as a casual derived from wool shoddy and birdseed. Greek *poly* + *pogon* = 'many' + 'beard', referring to the hairy panicle; Latin *monspeliensis* = 'from Montpelier', France.

BEARD LICHEN See lichen.

BEARD-MOSS Species of *Barbula* (Latin for 'little beard'), *Bryoerythrophyllum* (Greek *bryo* + *erythros* + *phyllo* = 'moss' + 'red' + 'leaf'), *Didymodon* (Greek *didyma* = 'in pairs'), *Leptobarbula* (Greek *leptos* + *barbula* = 'slender' + 'little beard'), *Leptodontium* (Greek *leptos* + *donti* = 'slender' + 'tooth'), *Paraleptodontium* ('nearly' *Leptodontium*) and *Pseudocrossidium* ('false' *Crossidium*), all in the family Pottiaceae. The English name reflects the long, thread-like stems in a tangled mass that often hang from trees and other substrates.

Rarities include **drooping-leaved beard-moss** *Paraleptodontium recurvifolium* (oceanic-montane), **glaucous beard-moss** *D. glaucus* (disused chalk-pit near Swindon, Wiltshire), **Irish beard-moss** *D. maximus* (limestone cliffs in the Dartry Mountains, Ireland), **rufous beard-moss** *Bryoerythrophyllum ferruginascens*, **sausage beard-moss** *D. tomaculosus* (recognised as a distinct species in 1981) and **Scottish beard-moss** *Bryoerythrophyllum caledonicum* (western Highlands).

Bent-leaved beard-moss *Leptodontium flexifolium*, locally common and widespread on well-drained, peaty, acidic soil on heathland and moorland, and in open woodland. *Flexifolium* is Latin for 'flexibly leaved'.

Beric beard-moss *Leptobarbula berica*, scattered in England and Wales on limestone rock in woodland, and on limestone stonework in churchyards and mortar.

Bird's-claw beard-moss *Barbula unguiculata*, common and widespread on neutral or base-rich, disturbed and open habitats, including gardens, fields and old walls. *Unguiculata* is Latin for 'clawed'.

Brown beard-moss *D. spadiceus*, scattered in upland areas on stones and tree-trunks, often by base-rich streams and seepages. *Spadiceus* is Latin for 'dark brown'.

Cylindric beard-moss *D. insulanus*, common and widespread, usually on soil, also on walls, concrete, tarmac, pavements, damp rocks and wood. *Insulanus* is Latin for 'islander'.

Dusky beard-moss *D. luridus*, common and widespread in lowland England, Wales and south-east Scotland, scattered elsewhere, on base-rich, often moist rock and walls, and in dune slacks. *Luridus* is Latin for 'wan' or 'dull yellow'.

Fallacious beard-moss *D. fallax*, common and widespread, mainly in lowland areas, on bare, usually base-rich soil over rock and walls, by roadsides, on disturbed ground, in quarries and pits, and by watercourses. *Fallax* is Latin for 'fallacious'.

Hornschuch's beard-moss *Pseudocrossidium hornschuchianum*, widespread and locally common in England and Wales, less frequent in northern England, Scotland and Ireland, on calcareous substrates and compacted soil at the edge of paths, sometimes on old walls. Christian Hornschuch (1793–1850) was a German bryologist.

Lesser bird's-claw beard-moss *Barbula convoluta*, common and widespread in disturbed, open habitats, including gardens, fields and walls. *Convoluta* is Latin for 'rolled-up'.

Nicholson's beard-moss *D. nicholsonii*, common and widespread in much of lowland England, Wales and south Ireland in stream flood zones on stones, tree roots and wooden stumps, walls and the bottoms of bridges, as well as moist tarmac, gravel paths and concrete. William Nicholson (1866–1945) was an English bryologist.

Olive beard-moss *D. tophaceus*, common and widespread in lowland areas on damp, base-rich stones, walls, coastal cliffs and soil by tracks, and in pits and quarries. *Tophaceus* is Latin for 'gritty'.

Pointed beard-moss *D. acutus*, widespread and scattered

on unshaded, well-drained, calcareous, soil or in thin turf on chalk and limestone banks, quarries and fixed dunes. *Acutus* is Latin for 'pointed'.

Red beard-moss *Bryoerythrophyllum recurvirostrum*, common and widespread on stony, calcareous sites, on walls and thin soil and in rock crevices. *Recurvirostrum* is Latin for 'bent beak'.

Revolute beard-moss *Pseudocrossidium revolutum*, widespread and locally common in England and Wales, less frequent in Scotland and Ireland, on limestone outcrops and quarries, occasionally on stony ground and wall mortar. *Revolutum* is Latin for 'rolled back'.

Rigid beard-moss *D. rigidulus*, common and widespread, especially in the lowlands, on sheltered, base-rich walls, concrete, paving stones, and rocks. *Rigidulus* is Latin for 'somewhat rigid'.

Rusty beard-moss *D. ferrugineus*, widespread and scattered on well-drained, bare, calcareous soil and rock, for example in old quarries and chalk-pits, on scree, in calcareous dunes, and on machair. *Ferrugineus* is Latin for 'rust-coloured'.

Shady beard-moss *D. umbrosus*, scattered in (mainly south and east) England and in Ireland in moist, shaded, calcareous sites, for instance on mortar at the base of walls. *Umbrosus* is Latin for 'shady'.

Soft-tufted beard-moss *D. vinealis*, locally common in England and Wales, scattered elsewhere, on dry walls and unshaded, base-rich rock, occasionally stony calcareous ground, short dry turf, roof tiles and concrete. *Vinealis* is Latin for 'vine-like'.

BEAR'S-BREECH(ES) *Acanthus mollis* (Acanthaceae), a herbaceous Mediterranean neophyte perennial present since at least the mid-sixteenth century, scattered but commonest in south-west and south England, and absent from much of Scotland and Ireland, persisting where dumped from gardens on roadsides, railway banks and waste ground, and in woodlands. The Greek name for this plant was *acanthos*. *Mollis* is Latin for 'soft'. 'Breech' means 'rump', possibly reflecting the broad and soft leaves.

Spiny bear's-breech *A. spinosus*, a herbaceous Mediterranean neophyte perennial scattered in England, persisting where dumped from gardens on roadsides, railway banks and waste ground.

BEAUTY (BUTTERFLY) See Camberwell beauty.

BEAUTY (MOTH) 1) Members of the family Noctuidae.

Marbled beauty *Bryophila domestica* (Bryophilinae), wingspan 20–5 mm, nocturnal, common in orchards, woodland and gardens and around old buildings, dry-stone walls and sea-cliffs throughout Britain, though less so in Scotland, and in east Ireland. Population levels increased by 297% between 1968 and 2007. Adults fly in July–August. Larvae feed on lichens on walls or tree-trunks. The binomial is Greek for 'moss-loving' and Latin for 'native'.

Pine beauty *Panolis flammea* (Hadeninae), wingspan 30–3 mm, nocturnal, widespread and common in coniferous and mixed woodland, plantations, parks and gardens. Adults fly in March–April. Larvae feed on pine needles, causing sufficient damage to be a serious pest in some areas. Greek *pan* + *oloos* = 'all' + 'destructive', and Latin *flammea* = 'flame-coloured'.

2) Mostly nocturnal members of the family Geometridae, subfamily Ennominae (Greek *ennomos* = 'lawful', indicating that this subfamily contains the archetypal geometrids).

Belted beauty *Lycia zonaria*, wingspan 27–30 mm, a rare moth (RDB 3) on sandhills in a few scattered coastal sites in Cheshire, north Wales, Scotland (including machair) and parts of Ireland. Adults fly in afternoon sunlight and from the early evening in March–April. Larvae feed on a range of low-growing plants. The genus name is probably from Greek *lyceios* = 'relating to a wolf', from the hairy abdomen. *Zonaria* is Latin for 'belted', from the reddish abdominal margins.

Bordered beauty *Epione repandaria*, wingspan 25–30 mm,

generally common in damp woodland clearings, scrub, hedgerows and gardens, less so in Wales and Scotland. Adults fly in July–September. Larvae feed on sallows, black poplar, alder and hazel. Epione was the wife of the Ancient Greek physician Aesculapius. *Repandus* is Latin for 'bent back', from sharply angled lines on the forewing.

Brindled beauty *L. hirtaria*, wingspan 35–45 mm, commonly in broadleaf woodland, scrub, hedgerows, parks and gardens throughout England, Wales, south Scotland and the western half of Ireland. With numbers declining by 87% over 1968–2007, this is a species of conservation concern in the UK BAP. Adult males fly in March–April. Larvae feed on a number of deciduous trees. *Hirtaria* is from Latin *hirtus* = 'hairy', from the texture of the thorax and abdomen.

Dark-bordered beauty *E. vespertaria*, wingspan 25–30 mm, a rarity (RDB 3) in scrubby wet heathland at three locations in Moray and Aberdeenshire, where larvae mainly feed on aspen, and Strensall Common, near York. Numbers at the latter site fell by >90% following an accidental heath fire in 2009 that destroyed much of the larval food plant, creeping willow, exacerbated by increased sheep grazing (intended to maintain the heathland), with only 50–100 adult moths present in 2016. Conservation measures include planting new willow and fencing off sheep. Males fly from dawn and at dusk in July–August. *Vesper* is Latin for 'evening'.

Feathered beauty *Peribatodes secundaria*, wingspan 38–44 mm, a recent colonist and immigrant found in coniferous woodland in parts of south-east England, first recorded in 1981 in Orlestone Forest in Kent, and subsequently recorded in Essex, West Sussex and Surrey. Adults fly in July. Larvae feed on Norway spruce. Greek *peri* + *batodes* = 'around' + 'covered in thorns', possibly describing a habitat.

Great oak beauty *Hypomecis roboraria*, wingspan 40–50 mm, in ancient woodland and mature oak woodland throughout much of the southern half of England. Adults fly in June–July. Larvae, greatly resembling twigs, feed on pedunculate oak, giving the specific name *roboraria*.

Lilac beauty *Apeira syringaria*, wingspan 38–42 mm, local in broadleaf woodland, scrub and gardens throughout much of England, Wales and Ireland, rare in Scotland. Adults fly in June–July. Larvae feed on privet, honeysuckle and ash. Greek *apeiros* = 'without end' (from a line pattern that does not reach the wing edge), and Latin *syringaria* = 'lilac', another food plant.

Mottled beauty *Alcis repandata*, wingspan 30–45 mm, widespread and common in woodland, hedgerows, moorland, heathland, calcareous grassland, parks and gardens. Adults fly in June–July. Larvae feed on a variety of woody plants. Alcis was the surname of Athena, or a daughter of Aegyptus in Ancient Greece. *Repandus* is Latin for 'bent', from a line shape on the hindwing.

Oak beauty *Biston strataria*, wingspan 40–50 mm, common in oak woodland, gardens and areas with scattered trees throughout England and Ireland, more local in Wales and Scotland. Adult males fly in March–April. Larvae feed on oak, alder, hazel, aspen and elm. In Greek mythology, Biston was a son of Mars. *Strataria* comes from Latin *stratum* = 'quilt', reflecting the embroidered pattern on the forewing.

Pale brindled beauty *Phigalia pilosaria*, wingspan 35–40 mm, widespread and common in broadleaf woodland, gardens, and areas with scattered trees. Adult males fly from January to March, when the wingless females are crawling on tree-trunks. Larvae feed in late spring on deciduous trees. Phigalia was a city in Ancient Greece. *Pilosus* is Latin for 'hairy', describing the abdominal hair-scales.

Pale oak beauty *H. punctinalis*, wingspan 46–55 mm, common in broadleaf woodland, scrub, parks and gardens throughout much of southern England and, especially, East Anglia. Adults fly from May to July. Larvae feed on oaks and other trees. *Punctinalis* is from Latin *punctus* = 'point', from the speckling on the wings.

Rannoch brindled beauty *L. lapponaria*, wingspan 31–4 mm,

a scarce day-flying (March–May) moth of moorland and bog in central and northern Scotland (particularly around Rannoch and Upper Speyside), and of priority conservation status. Larvae feed on moorland plants, with preference for bog-myrtle, but also taking heathers, heaths and bilberry. Latin *lapponaria* = 'occurring in Lapland'.

Satin beauty *Deileptenia ribeata*, wingspan 30–40 mm, common mainly in coniferous woodland. From a low base, population levels increased by 2,925% between 1968 and 2007. Adults fly in June–August. Larvae feed in spring on yew and a number of coniferous trees. Greek *deile* + *ptynos* = '(late) afternoon' + 'winged' (i.e. 'dusk-flying'). *Ribeata* refers to currant (*ribes*), mistaken as a food plant.

Small brindled beauty *Apocheima hispidaria*, wingspan 28–35 mm, local in oak woodland and mature hedgerows throughout much of England and Wales, less commonly in the north. Adult males fly in February–March, when the wingless females are crawling on tree-trunks. Larvae mainly feed on oak, but are also found on other broadleaf trees. Greek *apo* + *cheimon* = 'after' + 'winter', from its appearance in late winter/early spring. Latin *hispidus* = 'hairy', describing the abdominal hair-scales.

Willow beauty *Peribatodes rhomboidaria*, wingspan 30–8 mm, widespread and common in woodland, scrub, hedgerows, parks and gardens. Adults fly in July–August, in the south sometimes again in September. Larvae feed on a variety of woody plants. The specific name describes the rhomboid-shaped spot on the forewing.

BEAUTY (PLANT) See spring beauty.

BEAVER **European beaver** *Castor fiber* (Castoridae), head-body length 80–100 cm, tail length 25–50 cm, weight 11–30 kg (average 18 kg), a rodent that became extinct in Wales in the twelfth century and Scotland in the sixteenth century, overhunted for its fur, meat and castoretum (a glandular secretion used for scent), as well as suffering from destruction of forest habitat. They are semi-aquatic, living in lakes, ponds, rivers and streams, usually in forested areas but also in marshes and swamps, constructing dams using branches and mud to back up water. Dams can retain water in upland areas, reducing flood volumes and creating new habitats for wildlife. They prefer herbaceous foods, but mainly eat woody vegetation in winter, preferring willow, aspen and birch with diameters <10 cm, which they can fell using powerful incisors. Winter food is cached in lodges which are entered via underwater openings. Gestation is 60–128 days (average 107 days), with 2–6 kits per litter (average 3), and a single litter in a year. Starting with a captive population of Norwegian animals in Kent in 2001, beavers have been the subject of a number of reintroduction projects, and are present in a number of enclosures in wildlife parks. Escaped beavers were noted in Tayside in 2001, have since spread widely, and in 2015 were estimated to number >150. In Knapdale Forest, Argyll, 11 animals were deliberately released in 2009, with 5 further releases in 2010, and are also breeding successfully. Three beavers were noted on the River Otter in 2013, producing three kits the following year. In 2015, following capture and testing, they were found to be free of disease and of Eurasian descent, so were allowed by Natural England to continue to live in the wild. 'Beaver' comes from Old English *beofor* or *befor*, ultimately from an Indo-European root meaning 'brown'. The binomial reflects Latin for 'beaver' and 'material with lignin/cellulose'.

BED BUG Members of the family Cimicidae (Heteroptera), which feed exclusively on blood. Common bed bug *Cimex lectularius* (Latin *cimex* + *lectula* = 'bug' + 'bed') is synanthropic, parasitic on humans.

BEDEGUAR GALL Alternative name for Robin's pincushion; also see rose bedeguar gall wasp under gall wasp.

BEDELLIDAE Family name for bindweed bent-wing.

BEDSTRAW Mostly herbaceous perennial species of *Galium*

(Greek *galion* = 'milk', with some species, for instance lady's bedstraw, having been used to curdle milk in cheese-making) together with the introduced *Coprosma* (Greek *copros + osme* = 'dung' + 'smell', referring to the plant's foetid smell), both Rubiaceae. Bedstraw was so named in the sixteenth century for its use in mattresses, though also believed to have been one of the herbs in the hay in the manger at Bethlehem.

Common marsh-bedstraw *G. palustre*, common and widespread in wetland habitats, for example wet meadows, marshes, fens, ditches, ponds and lakesides, growing on a range of soil types though with a preference for non-calcareous substrates. *Palustre* is from Latin for 'marsh'.

Fen bedstraw *G. uliginosum*, on base-rich marsh and fen, usually in relatively dry and calcareous habitats, widespread and scattered, though absent from north Scotland and north-east and southern Ireland. *Uliginosum* is from Latin for 'swamp'.

Heath bedstraw *G. saxatile*, common and widespread on infertile acidic soils in grassland, heathland, rocky habitats and open woodland, and locally on disturbed or waste ground. It remains abundant in upland areas, and is an indicator of unimproved hill grassland, but it has locally declined in the lowlands through habitat destruction. *Saxatile* is Latin for 'growing in rocks'.

Hedge bedstraw *G. album*, stoloniferous, on well-drained, base-rich soils on rough grassland, railway banks, roadsides, hedge banks, woodland edges, scrub and waste ground, common and widespread in England and Wales, more scattered in Wales and (where probably introduced) Ireland. *Album* is Latin for 'white', from the flower colour.

Lady's bedstraw *G. verum*, stoloniferous, common and widespread on well-drained, relatively infertile neutral or calcareous soil. Habitats include hay meadow, pasture, downland, dune grassland and machair, roadsides and railway embankments. The procumbent var. *maritimum* occurs widely in coastal clifftops and dunes. *Verum* is Latin for 'true' (i.e. the 'true bedstraw').

Limestone bedstraw *G. sterneri*, uncommon in Britain north of a line from the Severn to Humber estuaries, and in west and north Ireland, on short grassland overlying limestone and basic igneous rocks and mica-schist, and also on vegetated scree, rock ledges and limestone pavement. *Sterneri* may honour the English horticulturalist Frederick Stern (1884–1967).

Northern bedstraw *G. boreale*, widespread and locally common in northern England, parts of Wales and Ireland, and especially in Scotland, on damp, usually base-enriched substrates, in rocky habitats, montane grassland flushes, river shingle and stony lake shores, and on stabilised dunes. Greek *boreale* = 'northern'.

Slender bedstraw *G. pumilum*, rare and declining in southern England on short chalk and limestone grassland, limestone spoil, and anthills. *Pumilum* is Latin for 'dwarf'.

Smooth bedstraw Alternative name for crosswort.

Tree bedstraw *C. repens*, an evergreen New Zealand neophyte shrub planted as a windbreak in the Scilly Isles, naturalised in places. *Repens* is Latin for 'spreading'.

Wall bedstraw *G. parisiense*, an annual of old walls, sandy banks and bare ground on calcareous or neutral substrates, intolerant of competition, and susceptible to nutrient enrichment, with a scattered distribution in southern and south-west England and East Anglia, an occasional casual elsewhere in England. Latin *parisiense* = 'Parisian'.

Yellow bedstraw Alternative name for lady's bedstraw.

BEE There are over 250 species of bee on the British list, with 225 species of solitary bee, 25 bumblebee (q.v.) species (with three recently extinct), and one honeybee (q.v.) species. See also carder, carpenter, cuckoo, flower, furrow and nomad bees (Apidae), leaf-cutter and mason bees (Megachilidae) and mining bees (Andrenidae and Halictidae).

Bellflower (blunthorn) bee *Melitta haemorrhoidalis* (Melittidae), scattered in England, especially the south, the Isle of Man and south-east Scotland, mainly on calcareous grassland and open rides in broadleaf woodland on chalk, flying in July–August, using bellflowers and harebell for pollen, and meadow crane's-bill for nectar. Nests are burrows in soil. Greek *melissa* (*melitta*) = 'honeybee', and *haema + rein* = 'blood' + 'to flow'.

Clover (blunthorn) bee *Melitta leporina*, locally common in the southern half of England and in Wales in open grassland, flying in June–August, preferentially feeding on white clover but also using a number of other plants. Nesting burrows are in soil, often in loose aggregations. *Leporina* is Latin for 'like a hare'.

Common mourning bee *Melecta albifrons* (Apidae), a cuckoo bee that, as a social parasite, lays its eggs in the nests of *Anthophora* solitary bees, larvae feeding on pollen stored by the host, found in southern England, especially the south-east, in gardens and on coastal cliffs, visiting a variety of flowers from April to June. *Melecta* comes from Greek for 'honey collector', *albifrons* Latin for 'white-fronted'.

Common yellow face bee *Hylaeus communis* (Apidae), widespread, especially in England except for the north; sporadic in Wales, known mainly from the south coast; in Ireland, recorded in Counties Wicklow, Meath, Dublin and Kildare. In 2007 specimens were reared from the Black Wood of Rannoch, Scotland. Flying in summer, it is found in many habitats, for example open woodland, grassland and coastal sites, and is common in gardens, most often seen on sunlit surfaces and visiting umbellifer flowers. It is an opportunistic cavity nester, using existing burrows, particularly in dead wood and woody stems, but also in the soil and crevices in mortar joints. In Greek mythology, Hylaeus was a centaur who was killed by Meleager after trying to rape Atalanta.

Elusive knapweed bee In 2010, *The Guardian* launched an annual competition for the public to suggest common names for previously unnamed species. In 2012, this name was selected, though it actually already had a common name: see downland furrow bee under furrow bee.

Ivy bee *Colletes hederae* (Colletidae), described as new to science in 1993, first recorded in England in Dorset in 2001, now common in southern England, and found as far north as north Norfolk and Shrewsbury/south Staffordshire. Active in August–October, this is a ground-nester in sandy soil and soft grassy slopes, verges and gardens. There is a preference for ivy pollen and nectar, but it also visits other flowers. *Colletes* is Greek for one who 'glues or fastens', *hedera* Latin for 'ivy'.

Large yellow face bee *H. signatus*, local and scattered throughout southern and central Britain, flying in summer in open habitats such as calcareous grassland, quarries, coastal marsh, beaches, post-industrial sites and gardens. For pollen it is oligolectic on Resedaceae, especially visiting wild mignonette and weld. Nests have been found in burrows in the vertical faces of compacted soil, dead woody stems, in vacated nests of *Colletes*, and in crevices in mortar joints. *Signatus* is Latin for 'sealed'.

Long-horned bee *Eucera longicornis* (Apidae), locally abundant in southern England and south Wales, in coastal grasslands (including cliffs and landslips), deciduous woodland rides and, occasionally, heathlands, flying in May–July. Pollen is taken from Lamiaceae, nectar from a range of flowers. Nest burrows are generally found in aggregations in sparsely vegetated or eroded soil. Greek *eu + ceras* and Latin *longus + cornus* both mean 'long-horned'.

Plasterer bee Alternative name for colletes.

Red bartsia (blunthorn) bee *Melitta tricincta*, locally common and increasing in south England, nesting in burrows in chalk grassland and occasionally broadleaf woodland on chalky soils, relying on pollen of red bartsia, flying in July–September. *Tricincta* is Latin for 'three-banded'.

Red mortar bee Alternative name for red mason bee.

Sainfoin (blunthorn) bee *Melitta dimidiata*, recorded only from the three principal army ranges on Salisbury Plain, found in June–August, feeding on sainfoin. No nests have been found. *Dimidiata* is Latin for 'divided'.

Short-horned yellow face bee *H. brevicornis*, common throughout much of lowland England and into south-west Scotland, plus a few records from Wales, and in Ireland it is widely distributed, mainly in the eastern and southern counties. It is found in woodland, fen, calcareous grassland, heathland and coastal habitats, flying in May–September. Pollen and nectar are taken from a variety of flowers. Nesting is in dead woody stems, for example of bramble. *Brevicornis* is Latin for 'short-horned'.

Sweat bee or furrow bees (q.v.), species of *Halictus* (Halictidae), from New Latin *halictus* for these bees, with four British species, so called because many species are attracted to the salt in human sweat.

Yellow face bee Species of *Hylaeus*; see examples above.

Yellow loosestrife bee *Macropis europaea* (Melittidae), length <15 mm, locally common south of a line from the Wash to the Severn estuary, Rare in RDB 3, in fen, bog and alongside rivers and canals feeding on yellow loosestrife, the species being unique in Britain for the females lining their burrow nests with fatty floral oils as well provisioning them with pollen. They are active in July–September. *Macropis* may come from Greek *macros* + *pous* = 'large' + 'foot'.

BEE-EATER *Merops apiaster* (Meropidae), wingspan 46 cm, length 28 cm, weight 61 g, a brightly coloured scarce visitor from the continent, with a few records of breeding, for example a pair at Bishop Middleham Quarry (Co. Durham) in 2002, a pair near Hampton Bishop (Herefordshire) in 2005, two pairs in the Isle of Wight in 2014, two pairs in Cumbria in 2015, and an unsuccessful pair at a quarry at East Leake, Nottinghamshire, in 2017. The tunnel nests contain 6–7 eggs which are incubated for 20 days. Food is predominantly of insects, especially bees, wasps, and hornets. The binomial combines Greek and Latin for 'bee-eater'.

BEE-FLY Members of the family Bombyliidae (incorporating Latin *bombus* = 'bee'). Adults generally feed on nectar and pollen; larvae are usually parasitoids of other insects.

Dark-edged bee-fly *Bombylius major*, up to 18 mm length, wingspan 25 mm, common and widespread, especially in the south, often seen in gardens and hedgerows, with a furry body, and a long proboscis for drinking nectar. Larvae prey on the larvae of bees and wasps in their nests.

Dotted bee-fly *B. discolor*, up to 10 mm length, scarce if widespread in southern England and south Wales, with strongholds in Somerset, Dorset, the Isle of Wight and Kent. It underwent a dramatic decline and contraction of range in the 1960s and 1970s, but recently has been expanding its range northwards. Adults feed on nectar. Larvae are parasitic on solitary bee eggs and larvae. *Discolor* is Latin for 'two-coloured'.

Flea bee-fly *Phthiria pulicaria*, 4–6 mm length, widespread but local on coastal dune from Cumbria round to Lincoln, and around the Moray Firth, occasionally on inland sandy heathland. Adults feed on nectar, mainly from Asteraceae, in June–August. Larvae are parasitoids of caterpillars of micro-moths, especially of the family Gelechiidae. Greek *phtheir* = 'louse', and Latin *pulex* = 'flea'.

Heath bee-fly *B. minor*, 7–8.5 mm length, scarce and declining in range, now mainly found in east Dorset heathlands in July–August, feeding on nectar using their long tongues. Females collect dust in a 'basket' under their abdomens, and lay dust-coated eggs while hovering, flicking them at small nest holes of solitary bees of the genus *Colletes*, located in sandy banks. The dust helps to camouflage the emerging larva, which finds its way into the bee's burrow to parasitise bee larvae, or possibly feeding on the pollen store.

Mottled bee-fly *Thyridanthrax fenestratus*, 7–9 mm length, declining in range and currently restricted to Dorset, the New Forest, West Sussex, the Weald, Surrey and Berkshire. Adults feed on nectar. Larvae are parasitoids of the heath sand wasp or of the caterpillars gathered by the wasp for its own larvae. Greek *thyra* + *anthrax* = 'door' + 'coal', and Latin *fenestra* = 'window' or 'opening'.

Western bee-fly *B. canescens*, 6.5–7.5 mm length, scarce, with a scattered distribution in west and south-west England, Wales, east Ireland and central-eastern Scotland, especially evident in open habitats in June–July feeding on nectar. Larvae are parasitic on the eggs and larvae of solitary bees. *Canescens* is Latin for 'greyish'.

BEE WOLF *Philanthus triangulum* (Crabronidae), up to 17 mm long, a solitary wasp once a rarity but which has undergone a sizeable increase in both range and abundance since the late 1980s, and now locally common in an increasing number of sites in south England, with records also as far north as Yorkshire and north Wales, usually on dunes and lowland heathland, but nesting aggregations have been found in a park in Ipswich, Suffolk, and by Battersea Bridge, Greater London. A number of plants are used for nectar. Nests are dug in both level sandy exposures and in vertical soil faces, with some aggregations numbering 15,000 burrows. The main burrow can be 1 m in length, with 3–34 short lateral burrows at the end, each terminating in a cell which is provisioned with honey bees and other bees. Greek *philos* + *anthos* = 'loving' + 'flower'.

BEECH (TREE) *Fagus sylvatica* (Fagaceae), a late colonist tree of Britain after the last glacial period, restricted as a native to southern England and south-east Wales, now widely planted (becoming self-sown) elsewhere. It is found on a variety of base-rich to acidic, free-draining soils, growing in pure woodlands (for example, in hangers on chalk and limestone escarpments), as standard trees, or as pollards in wood pasture. The binomial is Latin for 'beech' (from Greek *phigos*) and 'of woodland' or 'wild'. See also southern beech.

BEECH (SEAWEED) See sea beech.

BEEFSTEAK FUNGUS *Fistulina hepatica* (Agaricales, Fistulinaceae), a common, widespread, annual, poorly edible, large, blood-red bracket fungus, resembling raw meat and shaped like a broad tongue, found in summer and autumn, parasitic in limited tiers usually on the lower parts of the trunks of oak and sweet chestnut, and causing brown rot. Oak timber stained by this parasite is valued by wood-turners and cabinetmakers, who sometimes refer to it as 'brown oak'. Cutting through the flesh gives the appearance of a steak. The binomial comes from the Latin for 'little pipes' (the walls of each tube being separate, not shared with neighbouring tubes), and a reference to the liver-like appearance of the mature brackets.

BEET Species of *Beta* (Latin for beet) in the family Amaranthaceae.

Caucasian beet *B. trigyna*, a perennial European neophyte, present by 1796, naturalised on a road verge in Cambridgeshire, possibly elsewhere, and scarce and scattered mainly in southern England as a relict on tips and waste ground. *Trigyna* is Greek for 'three ovaries'.

Fodder beet See root beet below.

Foliage beet *B. vulgaris* ssp. *cicla*, a biennial neophyte herb of garden origin, widespread as a casual garden escape on waste ground, and as a relic of cultivation. Includes var. *cicla* (leaf or spinach beet) and var. *flavescens* (Swiss chard). Latin *vulgaris* = 'common', *flavescens* = 'becoming yellow'.

Leaf beet See foliage beet above.

Root beet *B. vulgaris* ssp. *vulgaris*, an annual to biennial herb with an edible root, of garden origin, casual on waste ground and as a relic of cultivation, mainly in England and south-east Ireland. Includes beetroot, fodder beet, sugar beet and mangel-wurzel.

Sea beet *B. vulgaris* ssp. *maritima*, a herbaceous perennial, widespread except for the north and (apart from the Firth of Forth) east coasts of Scotland, on coastal rocks and cliffs, saltmarsh drift lines, sea walls, and on sand and shingle beaches, favouring nutrient-enriched sites such as seabird cliffs. It also occurs on waste ground near the sea, and occasionally inland as a casual of tips and roadsides.

Spinach beet See foliage beet above.

Sugar beet See root beet above.

BEET BUG Members of the family Piesmatidae (Heteroptera) (Greek *piesmos* = 'pressure', possibly *piestos* = 'compressed'), represented in Britain by *Piesma maculatum* (length 2–3 mm) with a wide but scattered distribution in England and Wales, feeding on goosefoot and other chenopods, occasionally beet and spinach; and *P. quadratum* (2–3 mm), widespread but local to saltmarsh, feeding on oraches and sea-purslane. Latin *maculatum* = 'spotted', *quadratum* = 'four-sided'.

BEETLE Members of the insect order Coleoptera, with a main pair of forewings especially hardened to form protective covers (elytra) for the underwings which they use for flying. There are 4,072 species of beetle in 1,265 genera in 103 families currently recognised/found in Britain, Ireland and the surrounding islands (*Checklist of beetles of the British Isles*, 2nd edition, 2012), maximum numbers since this includes a few definite and probable extinctions.

Families (common names, number of genera/species in Britain excluding rare migrants and occasional adventives) are: **Aderidae** (3/3), **Alexiidae** (1/1), **Anthicidae** (antlike flower beetles, 6/13), **Anthribidae** (fungus weevils, 8/9), **Apionidae** (seed weevils, 34/87), **Attelabidae** (leaf-rolling weevils, 2/2), **Biphyllidae** (false skin beetles, 2/2), **Bothrideridae** (3/4), **Byturidae** (fruitworm beetles, 1/2), **Carabidae** (ground beetles, 83/362), **Cerambycidae** (longhorn beetles, 47/63), **Cerylonidae** (3/5), **Chrysomelidae** (seed and leaf beetles, 62/273), **Ciidae** (minute tree fungus beetles, 7/22), **Cleridae** (chequered and ham beetles, 4/8), **Coccinellidae** (ladybird beetles, 25/46), **Colydiidae** (cylindrical bark beetles and narrow timber beetles, 7/10), **Corylophidae** (5/12), **Cryptophagidae** (silken fungus beetles, 11/103), **Cucujidae** (some flat bark beetles, 1/2), **Curculionidae** (weevils, 155/480), **Dascillidae** (orchid beetle, 1/1), **Dasytidae** (some soft-winged flower beetles, 3/9), **Dryophthoridae** (2/4), **Dytiscidae** (diving beetles, 29/118), **Endomychidae** (handsome fungus beetles, 5/8), **Erirhinidae** (wetland weevils, 7/13), **Erotylidae** (pleasing fungus beetles, 4/8), **Georissidae** (1/1), **Gyrinidae** (whirligig beetles, 2/12), **Haliplidae** (crawling water beetles, 3/19), **Helophoridae** (some water scavenger beetles, 1/20), **Histeridae** (hister and clown beetles, 21/47), **Hydrochidae** (some water scavenger beetles, 1/7), **Hydrophilidae** (some water scavenger beetles, 18/70), **Hygrobiidae** (screech beetle, 1/1), **Kateretidae** (short-winged flower beetles, 3/9), **Laemophloeidae** (lined flat bark beetles, 4/11), **Latridiidae** (minute brown scavenger beetles, 8/56), **Limnichidae** (1/), **Lymexylidae** (timberworm beetles, 2/2), **Malachiidae** (some soft-winged flower beetles, 11/17), **Megalopodidae** (1/3), **Melandryidae** (false darkling beetles, 9/17), **Meloidae** (oil beetles, 3/10), **Monotomidae** (root-eating beetles, 3/23), **Mordellidae** (tumbling flower beetles, 5/17), **Mycetophagidae** (hairy fungus beetles, 6/15), **Mycteridae** (1/1, probably extinct), **Nanophyidae** (2/2), **Nemonychidae** (1/1), **Nitidulidae** (sap and pollen beetles, 16/91), **Noteridae** (noterus, 1/2), **Oedemeridae** (false blister beetles, 4/10), **Orsodacnidae** (1/2), **Phalacridae** (3/15), **Platypodidae** (1/1), **Ptilodactylidae** (1/1), **Ptinidae** (wood-borer beetles, 25/56), **Pyrochroidae** (cardinal beetles, 2/3), **Pythidae** (1/1), **Raymondionymidae** (1/1), **Rhynchitidae** (tooth-nosed snout weevils, 9/18), **Ripiphoridae** (wedge-shaped beetles, 1/1), **Salpingidae** (6/11), **Scraptiidae** (false flower beetles. 2/14), **Silphidae** (sexton, burying and carrion beetles, 7/20), **Silvanidae** (some flat bark beetles and flat grain beetles, 10/12), **Sphaeritidae** (1/1), **Sphaeriusidae** (mud beetle, 1/1), **Sphindidae** (2/2), **Staphylinidae** (rove beetles, 278/1122), **Tenebrionidae** (darkling beetles, 32/45), **Tetratomidae** (polypore fungus beetles, 2/4), **Trogossitidae** (5/5).

As well as the species given below, see separate entries

for bark, borer, cardinal, carpet, carrion, click, darkling (under Tenebrionidae), diving, flea, flour, flower, ground, jewel, larder, leaf, longhorn, malachite, moss, mud, oil, pollen, pot, reed, riffle, sailor, seed, sexton, soldier, spider, stag, tiger, tortoise and willow beetles.

Ambrosia beetle *Xyleborus dryographus* (Curculionidae, Scolytinae), so named because it introduces ambrosia fungi to a dead or decaying host tree, with a scattered distribution in the southern half of England. Adults excavate tunnels in dead trees in which they cultivate fungal gardens, for themselves and their larvae, grazing on the mycelia (their only source of nutrition) on the gallery walls. Greek *xyle* + *boros* = 'wood' + 'food', and *drys* + *graphe* = 'tree' + 'description'.

Antirrhinum beetle *Brachypterolus pulicarius* and *B. vestitus* (see Kateretidae).

Asparagus beetle *Crioceris asparagi* (Chrysomelidae, Criocerinae), adults 6–8 mm long, common and widespread, especially south-east of a line from the Humber to the Severn estuaries and parts of south Wales. Adults emerge from overwintering sites in May–June. There are two or three generations between May and September. Both adults and larvae eat asparagus leaves, and gnaw bark from the stems, halting growth. Larvae feed for a few weeks before pupating in the soil around the plant base. Adults overwinter in soil, leaf litter, or hollow stems of asparagus. Greek *crios* + *ceras* = 'ram' + 'horn'.

Auger beetle Member of the wood-boring (woodworm) family Bostrichidae.

Bee beetle *Trichius fasciatus* (Scarabaeidae, Cetoniinae), up to 25 mm length, a black and yellow bee mimic, rare but widespread and particularly found in Wales and the Highlands, close to or in woodland. Adults are seen on a variety of flowers, especially those of thistles and bramble, from May to August; larvae live in and feed on dead wood, especially birch. *Fasciata* is Latin for 'in bundles'.

Biscuit beetle *Stegobium paniceum* (Ptinidae), up to 3.5 mm long, a scavenger and pest of stored products with a scattered distribution in England. Greek *stegein* = 'to cover', and Latin *panis* = 'bread'.

Black clock beetle *Pterostichus madidus* (Carabidae, Carabinae), length 15–20 mm, very common and widespread in woodland, moorland and other habitats under stones, loose bark and grass tussocks. It is very common in gardens and on arable land, particularly in summer. It is largely predatory on ground-living invertebrates, including caterpillars and slugs, but the adult also eats some plant material. Greek *pteron* + *stichos* = 'wing' + 'line', and Latin *madidus* = 'wet'.

Black snail beetle *Silpha atrata* (Silphidae, Silphinae), length 10–15 mm, common and widespread, particularly evident in January–May in many different habitats including woodland, meadows and gardens, sometimes resting under logs and debris. Adults are specialist nocturnal snail predators and have a narrow head adapted to this role. Larvae also feed on pulmonate snails as well as earthworms. Greek *silphe* = an insect producing an unpleasant smell, and Latin *atrata* = 'dressed in black'.

Bloodsucker beetle Alternative name for common red soldier beetle on account of its red colour.

Bloody-nosed beetle *Timarcha tenebricosa* (Chrysomelidae, Chrysomelinae), 18–23 mm long, widespread in England (apart from the north-east) and Wales, and occasional in south Ireland, in grassland, heathland, woodland edges and hedgerows. The flightless adults (the elytra are fused) are evident in April–June, both adults and larvae feeding on lady's bedstraw and congenerics mainly at night, hiding in moss or under stones during the day. Its common name comes from the visual deterrent of bright red fluid (haemolymph) – that also has an unpleasant taste – produced from the mouth as an anti-predator defence known as reflex bleeding. *Tenebricosa* is Latin for 'dark'.

Blue mint beetle *Chrysolina coerulans* (Chrysomelide, Chrysomelinae), length 7–10 mm, first found breeding in the Britain in 2011 in Kent. Since 2012 it has been found in several locations in Cambridgeshire and Lincolnshire. Of widespread occurrence in Europe it is expected spread further in the UK. Larvae and adults feed on mint leaves in spring and summer. Greek *chrysos* = 'golden', and Latin *coeruleus* = 'blue'.

Blue pepper-pot beetle *Cryptocephalus punctiger* (Chrysomelidae, Cryocephalinae), 2.4–3.5 mm length, with a very scattered distribution in Britain, scarce and continuing to decline in status, and a species of conservation concern in the UK BAP. Found in broadleaf woodland and wasteland, adults and larvae feed on birch leaves. Its English name was the winning suggestion when, in 2010, *The Guardian* launched a competition for the public to suggest common names for previously unnamed species. Greek *cryptos* + *cephale* = 'hidden' + 'head', and Latin *punctiger* = 'punctured'.

Bog beetle Alternative name for mud beetle.

Bombardier beetle *Brachinus crepitans* (Carabidae, Brachininae), length 7–10 mm, found mainly in dry, sunny habitats, often under stones, in calcareous grasslands, arable field margins and chalk quarries, south and east of a line from the Severn estuary to the Wash, with most modern records from coastal areas. Both this and streaked bombardier beetle have the ability to expel a 100°C jet of chemicals from the anus to deter predators, produced when hydroquinone and hydrogen peroxide from separate glands are mixed: the resulting explosion can be heard as an audible crackle, and *crepitans* is Latin for 'crackle'. See also streaked bombardier beetle below.

Bread beetle Alternative name for biscuit beetle.

Burying beetle **Black burying beetle** is an alternative name for black sexton beetle, **common burying beetle** for common sexton beetle.

Cellar beetle Alternative name for churchyard beetle.

Chequered beetle Member of the family Cleridae.

Churchyard beetle *Blaps mucronata* (Tenebrionidae), flightless, 30 mm length, with a scattered distribution in (especially south) England and Wales. It used to be fairly common in houses, favouring dark, damp habitats such as cellars (an alternative name is cellar beetle), below floorboards and in attics, and also in barns, sheds, stables and caves, scavenging on food particles and feeding nocturnally on plant material. Greek *blapsis* = 'an injuring', and Latin *mucronata* = 'pointed'.

Cigarette beetle *Lasioderma serricorne* (Ptinidae), length 2–3 mm, a rare pest of stored products in south-east England and East Anglia. Greek *lasios* + *derma* = 'hairy' + 'skin', and Latin *serra* + *cornu* = 'a saw' + 'horn'.

Clown beetle Members of the family Histeridae.

Colorado (potato) beetle *Leptinotarsa decemlineata* (Chrysomelidae, Chrysomelinae), 12 mm length, a major pest of potato and other crops, occasionally reported from but not established in the UK, and a notifiable quarantine pest. Introduction is prohibited under the EC Single Market Protected Zone arrangements for Plant Health. These beetles are most commonly intercepted in spring and early summer on a range of plant produce from continental Europe. The binomial is Latin for 'slender flat surface' and 'ten-striped', from the five stripes along the length of each elytra.

Darkling beetle Member of the family Tenebrionidae (q.v.), named for the dark elytra of most species.

Death-watch beetle *Xestobium rufovillosum* (Ptinidae), adults 7 mm long, common and widespread in England and Wales, inhabiting the dead wood of several hardwood species. For the xylophagous (wood-eating) larvae (up to 11 mm long) to flourish, the heartwood usually has to have been modified by fungal decay, making the timber more palatable. Tunnelling can cause major damage to wood beams and floors in buildings and to wooden furniture. Adults emerge during spring. The common name refers to the repeated ticking sound produced by the adults as they bang their heads against the wood, possibly to attract a mate. The male ejaculates the human equivalent of 14 litres of sperm. Greek *xestos* = 'polished by scraping', and Latin *rufus* + *villosus* = 'red' + 'hairy'.

Dor beetle Member of the family Geotrupidae; also known as lousy watchman and dumbledore.

Dried bean beetle Alternative name for bean weevil.

Dried fruit beetle *Carpophilus hemipterus* (Nitidulidae), length 2–4 mm, a pest of stored products with a few records in south-east England and the Midlands. Greek *carpos* + *philos* = 'fruit' + 'loving', and *hemi* + *pteryx* = 'half' + 'wing'.

Dung beetle Mostly in the family Scarabaeidae, subfamily Aphodiinae which has 55 species, 49 of which are in the genus *Aphodius*.

False blister beetle Member of the family Oedemeridae.

False click beetle Member of the family Eucnemidae. Specifically, *Eucnemis capucina*, with a few scattered records from England. Greek *eu* + *cnemis* = 'good' + 'knee'.

False darkling beetle Member of the family Melandryidae.

False flower beetle Member of the family Scraptiidae.

False skin beetle Member of the family Biphyllidae.

Featherwing beetle Member of the family Ptiliidae, the English name coming from the distinctive structure of the wings, which often protrude from beneath the elytra when at rest.

Flat bark beetle Member of the family Silvanidae or Cucujidae; also **lined flat bark beetle**, Laemophloeidae.

Flat grain beetle *Cryptolestes pusillus* (Laemophloeidae), a pest of stored grain, with a very few records from Kent and East Anglia. Greek *cryptos* + *lestes* = 'hidden' + 'robber', and Latin *pusillus* = 'small'.

Fruitworm beetle Member of the family Byturidae. See raspberry beetle below.

Furniture beetle *Anobium punctatum* (Ptinidae), adults 2.7–4.5 mm long, a woodland species now more commonly evident as a major pest of timber and furniture, particularly widespread in England and Wales. Adults (4–6 mm length) do not feed but larvae bore into wood and feed on cellulose. Adults emerge from infected timber in spring, leaving a small round hole 1–2 mm in diameter. *Anobium* comes from Greek for 'lifeless', reflecting the adult's habit of playing dead when disturbed; *punctatum* is Latin for 'spotted'.

Grain beetle Member of the family Silvanidae. But see also flat grain beetle above.

Graveyard beetle *Rhizophagus parallelocollis* (Monotomidae), body length 1.5–2 mm, with a scattered distribution in England and Wales, in basements, deep layers of compost piles, mammal burrows and sap, particularly associated with cemeteries where it swarms in graves and on corpses in 10–24 month-old coffins. Greek *rhiza* + *phago* = 'rot' + 'to eat', and Latin *parallelocollis* = 'parallel neck'.

Green dock beetle *Gastrophysa viridula* (Chrysomelidae, Chrysomelinae), 4–6 mm length, common and widespread, with at least two generations each year, possibly up to six, with the last brood hibernating as an adult. Found in a wide range of habitats, both adults and larvae feeding on dock and sorrel leaves, and sometimes on other Polygonaceae and buttercups. Greek *gaster* + *physis* = 'stomach' + 'growth', and Latin *viridula* = 'greenish'.

Green mint beetle See leaf beetle.

Heather beetle *Lochmaea suturalis* (Chrysomelidae, Galerucinae), 4.3–6 mm length, locally common and widespread on heathland and moorland where there is a damp understorey of moss or leaves for egg-laying and pupation. Adults feed on young shoots of heather, larvae on leaves and stems. *Lochmoios* is from Greek for 'bush', Latin *suturalis* = 'sewing'.

June beetle Alternative name for summer chafer.

Large pine-shoot beetle *Tomicus piniperda* (Curculionidae, Scolytinae), length 3–5 mm, widespread but scattered, in galleries of conifers such as Scots pine, damaged trees releasing a monoterpene scent that attracts the female. Greek *tomicos* = 'cutting'.

Mab's lantern beetle *Philorhizus quadrisignatus* (Carabidae), 3.5–4 mm length, very local, predatory in broadleaf and pasture woodland on trees, under bark, or among dead twigs and litter on slightly damp ground. There were single records in 1978, 1986 and 1987 from Sherwood Forest and Bushey Park, London, and none since. Its English name was the winning suggestion when, in 2010, *The Guardian* launched a competition for the public to suggest common names for previously unnamed species, and reflects its 'feminine' shape, Mab being the fairies' queen or midwife. Greek *philos* + *rhiza* = 'loving' + 'root', and Latin *quadrus* + *signatos* = 'fourfold' + 'marked'.

Minotaur beetle *Typhaeus typhoeus* (Geotrupidae), 12–20 mm length, in sandy grassland and heathland, widespread if scarce in England and Wales, Britain's only member of the 'roller' guild of dung beetles. Both sexes roll rabbit droppings and other dung into their deep (1–1.5 m) nest burrows, using their strong front legs, mainly doing so at night, but they are more often located by the tunnelling spoil heaps. The dung provides food for the larvae. Males possess 'horns' (prompting 'minotaur' from the bull-headed man of the Cretan Labyrinth in Greek mythology) which are used to defend the nest. Typhoeus, or Typhon, was a monstrous storm or volcano giant in Greek mythology.

Minute bog beetle See mud beetle (1).

Minute tree fungus beetle Members of the family Ciidae. Most species are associated with particular genera of bracket fungi. Many are widespread and common, including the non-native *Cis bilamellatus*, an Australian species first found in Britain at Kew Gardens in the nineteenth century, and now widespread in England, Wales and the central Scottish lowlands, and recently reported from Ireland. *Cis* here is Greek for 'woodworm'.

Moccas beetle *Hypebaeus flavipes* (Malachiidae), a false soldier beetle known only from Moccas Park NNR, Herefordshire, RDL Vulnerable, requiring ancient oak trees and flowering hawthorn to complete its life cycle. Larvae live in galleries made by wood borers in the red-rotten heartwood of oaks. Greek *hypo* + *baios* = 'under' + 'poor', and Latin *flavipes* = 'yellow-footed'.

Museum beetle *Anthrenus museorum* (Dermestidae), adult body length 2.7–3 mm, larvae 4–5 mm. Adults are pollen grazers, but larvae, sometimes called 'woolly bears', feed on stored goods and animal fibres such as wool, feathers and fur, often becoming household pests.

Musk beetle *Aromia moschata* (Cerambycidae), 13–35 mm length, with a scattered distribution in England, Wales and south-west Scotland. Adults are generally found in July–August on willow (and other tree) flowers and leaves, larvae in the wood. Named after its ability when threatened to produce a pleasant musky scent from thoracic glands (monoterpenes that include rose oxide, an important fragrance in perfumery), it can also produce a sound when handled. The binomial is Greek and Latin for 'aroma' and 'musk-scented'.

Mustard beetle *Phaedon tumidulus* (Chrysomelidae, Chrysomelinae), 3–4 mm length, common and widespread in England and Wales, more scattered in Scotland and Ireland in open woodland, parks, wasteland and gardens. Adults and larvae feed on various brassicas, including garlic mustard and water-cress, and take pollen from gorse. Greek *phaios* = 'dark' or 'grey', and Latin *tumidus* = 'swollen'.

Net-winged beetle Member of the family Lycidae.

Orchid beetle *Dascillus cervinus* (Dascillidae), 8–12 mm, widespread and locally common in unimproved grassland, especially in upland areas and chalk downland in England and Wales. Adults are seen, often on flowers, in May–July. Larvae live in the soil and eat roots, especially of grasses. *Cervinus* is Latin for 'tawny' (like a deer).

Pine-shoot beetle See large pine-shoot beetle above.

Plum beetle *Tetrops praeustus* (Cerambycidae), a longhorn beetle common and widespread in England and Wales in forest edges, hedgerows and orchards, larvae associated with rosaceous trees such as plum and apple. Greek *tetra* + *ops* = 'four' + 'eyes', and Latin *praeustus* = '(scorched) brown'.

Powderpost beetle *Lyctus brunneus* and *Stephanopachys*

substriatus (Bostrichidae), wood-boring (woodworm) species 4–7 mm long, where damage usually originates in timber yards or storerooms, but these beetles can also damage furniture, sports equipment, wood block floors and joinery.

Raspberry beetle *Byturus tomentosus* (Byturidae), 3.2–4 mm long, a common and widespread pest of raspberries and other cane fruits, adults eating leaves, larvae flower buds and developing fruit. Greek *bytouros* = 'tusk', and Latin *tomentosus* = 'densely covered with short hair'.

Rhinoceros beetle *Sinodendron cylindricum* (Lucanidae), a 15–18 mm cylindrical beetle, widespread (if rarely abundant) in England and Wales, local in Scotland and Ireland, in woodland and parkland and along hedgerows. Adults, active in summer, are strong fliers. Larvae live in and feed on old trees and rotting wood, particularly of beech; adults feed on tree sap. A projection on the male head looks like a rhino horn, giving it its common name; the female has a small bump rather than a horn. Greek *sinos* + *dendron* = 'damage' + 'tree'.

Root-eating beetle Members of the family Monotomidae, which despite the name are predatory.

Rosemary beetle *Chrysolina americana* (Chrysomelidae, Chrysomelinae), 6.7–8.1 mm length, a southern European introduction first recorded in Surrey in 1963, but recorded in greater numbers since the 1990s, initially mainly in London gardens but soon spreading rapidly, and now widespread throughout England and Wales. It is established in Scotland, local in Ireland. Adults and larvae both feed on leaves and flowers of aromatic herbs such as rosemary, lavender, sage and thyme. Adults are present year-round, though particularly active from August (when mating and egg-laying take place) to April when most damage occurs. It was eighth in the Royal Horticultural Society's list of most serious pests in 2017. The genus name incorporates Greek *chrysos* = 'gold', from the contribution of this colour to the multicoloured elytral stripes.

Sap beetle Members, with pollen beetles, of the family Nitidulidae.

Sawyer beetle Alternative name for tanner beetle.

Screech beetle *Hygrobia hermanni* (Hygrobiidae), 9 mm length, in ponds and ditches, often in soft mud, common in England and Wales, more scattered in Ireland. Both adults and larvae are predatory. The English name comes from its ability to stridulate loudly when disturbed. Greek *hygros* = 'wet'. Johann Hermann (1738–1800) was a German entomologist.

Shoot beetle See large pine-shoot beetle above.

Small bloody-nosed beetle *Timarcha goettingensis* (Chrysomelidae, Chrysomelinae), 8–13 mm length, found in open habitats with a scattered distribution in England, possibly declining outside the south, and occasional in Ireland. Adults and larvae mainly feed on bedstraw. The specific name refers to the German city of Göttingen.

Snail beetle See black snail beetle above.

Spruce shortwing beetle *Molorchus minor* (Cerambycidae), an introduced longhorn beetle, 6–16 mm length, widespread in conifer plantations in England and Wales. Adults frequent flowers of species such as hawthorn. Larvae live in spruce, fir and other conifers.

Streaked bombardier beetle *Brachinus sclopetas* (*sclopetum* is Latin for 'gun'), length 5–7 mm, has always been rare and now may only be found on a handful of brownfield sites in London, though most such sites have been subjected to development schemes and some beetle translocations have been tried. In 2012 Buglife reported a new population in Newham on a single mound due to make way for Olympics infrastructure. Redesigning the project meant the beetles remained safe. This mound is now the only known non-translocated colony of the species in the UK. Larvae feed on *Amara* and *Harpalus* ground beetles.

Striped ambrosia beetle Alternative name for lineated bark beetle.

Swollen-thighed beetle *Oedemera nobilis* (Oedemeridae),

8–10 mm length, a locally common false blister beetle widespread (and expanding its range) in the southern half of England and in Wales, in flower meadows, waste ground and gardens. Larvae feed on dry stems, for example of thistle, adults on pollen and nectar of a range of plants in April–September. Greek *oidema* = 'swelling', and Latin *nobilis* = 'notable'.

Tanner beetle *Prionus coriarius* (Cerambycidae) the largest of Britain's longhorn beetles (25–45 mm length, females larger than males), very rare, thinly scattered across southern England (on sandy soils from Dorset to East Anglia) and Wales, larvae and the short-lived adults feeding on decaying tree stumps and logs, usually of oak. Adults stridulate during mating. Greek *prion* = 'saw', and Latin *coriarius* = 'leathery'.

Timber beetle, narrow timber beetle Members of the family Colydiidae.

Timberman beetle *Acanthocinus aedilis* (Cerambycidae), a longhorn beetle, up to 20 mm in length, with a scattered distribution in Britain, adults predominantly found in March–September in Highlands pine forest, though there are also records from the Pennines, as well as from timber yards throughout the country. Larvae live and feed in rotting wood of Scots pine. The antennae of the adult males are three to four times their body length, and females two times, and extend in a manner resembling calipers when they land on felled tree-trunks, seeking dead wood, leading to their common name. Greek *acantha* + *cineo* = 'thorn' + 'to move', and Latin *aedilis* = a Roman magistrate.

Tumbling flower beetle Members of the family Mordellidae, with 17 species in 5 genera, with a southern British distribution. Adults, 2–9 mm long, characteristically curl up in a C shape on flowers, where they feed on pollen. When disturbed, they make a series of tumbling movements (prompting their common name) as a consequence of the efforts to get themselves back into take-off position for flight having been in a lateral or dorsal position.

Viburnum beetle *Pyrrhalta viburni* (Chrysomelidae, Galerucinae), 4.5–6.5 mm length, common and widespread in England, more scattered in Wales and Ireland. Both larvae (present in April–June) and adults (which emerge in July–August) feed on leaves and stems of viburnums such as guelder-rose and wayfaring-tree, overwintering as eggs in cavities chewed into stems and covered by a hard, dried cap of mucus and faeces or regurgitated food. In 2011 the Royal Horticultural Society placed this beetle first in its top ten pest species, though by 2017 it was ninth. Greek *pyrrhos* = 'flame-coloured'.

Wasp beetle *Clytus arietis* (Cerambycidae), a common longhorn beetle, 9–18 mm length, that mimics the markings and movements of wasps, with a widespread distribution in England and Wales, and occasionally found in southern Scotland. Adults are active in May–August in woodland and hedgerows, visiting flowers for pollen and nectar, basking on sunlit leaves and logs, and breeding in decaying wood. Greek *clytos* = 'glorious', and Latin *aries* = 'ram'.

Wasp-nest beetle *Metoecus paradoxus* (Ripiphoridae), 10–12 mm length, uncommon, with a scattered distribution in England and the Welsh Borders. Eggs are laid in the autumn on rotting wood, and larvae hitch a lift on wasps which visit to collect wood scrapings. Back at the nest, the larva drops off and finds a wasp grub, which it parasitises and eats before emerging as an adult in midsummer. Greek *metoicos* = 'one who changes home', and Latin *paradoxus* = 'strange'.

Water-cress beetle Alternative name for mustard beetle.

Water-lily leaf beetle *Galerucella nymphaeae* (Chrysomelidae, Galerucinae), 4–8 mm length, with a widespread but scattered distribution in wetland habitat, feeding on water-lily species, adults hibernating under leaf sheaths of reedmace. The binomial is New Latin for this beetle, possibly from *galerum* + *eruca* = 'helmet' + 'caterpillar', and from *Nymphaea* (water-lily).

Water-penny beetle *Eubria palustris* (Psephenidae), 2–3 mm long, scarce and endangered with a scattered distribution in England and Wales, found in the splash zone of fast-flowing

water. Adults are short-lived; more commonly found are the aquatic larvae, known as 'water-pennies' because of their bronze colour and flattened, rounded appearance as they cling on to rocks and logs, where they eat algae. The binomial may come from Greek *eu* + *briao* = 'good' + 'to be strong', and Latin for 'marsh'.

Water scavenger beetle Species in the families Hydrophilidae, Hydrochidae and Helophoridae.

Weaver beetle *Lamia textor* (Cerambycidae), an endangered longhorn beetle, 15–30 mm length, found locally in Avon, southern England and East Anglia, larvae particularly associated with willows. In Greek mythology Lamia was a monster that fed on human flesh. *Textor* is Latin for 'weaver'.

Wedge-shaped beetle Member of the family Ripiphoridae, with one British species, wasp-nest beetle.

Whirligig beetle Members of the family Gyrinidae, often found as groups of small dark elliptical beetles swimming in circles on the open water of ponds and streams. Scarcer species live among emergent vegetation, rather than circling openly by day. All are carnivorous, feeding on animals caught in the surface film. Larvae breathe through gills, rather than from the surface, which may explain the group's predilection for deeper water bodies.

BEETROOT See root beet under beet.

BEGGARTICKS *Bidens frondosa* (Asteraceae), an American annual neophyte cultivated in 1710, recorded as a casual in 1918 but not known as a naturalised plant until found on a canal towpath near Birmingham in 1952. It is now frequently naturalised by canals in the Midlands, along the Thames, Severn and Trent, on damp and waste ground, and on tips in urban areas and ports. The English name, also used for the congeneric black-jack, may come from the resemblance of the seed pods to ticks, or from the prickly fruits which stick to fur and material. *Bidens* comes from Latin *bis* + *dens* = 'twice' + 'tooth' (referring to the bristles on the achenes), and *frondosa* = 'leafy'. See also bur-marigold.

Fern-leaved beggarticks *B. ferulifolia*, an annual Mexican neophyte, with self-sown escapes from containers occasionally casual in England, Wales and the Isle of Man. *Ferulifolia* is Latin for 'fennel-leaved'.

BELL (FUNGUS) Members of the genus *Galerina* (Agaricales, Strophariaceae), inedible, characteristic of often acidic mossy areas in grassland, moorland and woodland. *Galerina* is Latin for 'like a helmet'.

Locally common and widespread are **bog bell** *G. paludosa*, **dwarf bell** *G. pumila*, **funeral bell** *G. marginata* (poisonous, containing amatoxins) and **moss bell** *G. hypnorum*.

Rarities are **hairy leg bell** *G. vittiformis* and **snowbed bell** *G. harrisonii* (ten records between 1981 and 2010 from the Highlands).

Ribbed bell *G. clavata*, widespread and common in summer and autumn in damp grassy habitats, including lawns. *Clavata* is Latin for 'club-shaped'.

Turf bell *G. graminea*, common and widespread in England and Wales, found from summer to early winter in woodland and short grassland, especially on calcareous soil. *Graminea* is Latin for 'grass-like'.

BELL (MOTH) Members of the family Tortricidae, subfamily Olethreutinae, including many species of *Epiblema* (Greek for a covering, possibly describing the dark thorax and forewing base, suggesting a cloak when the wings are folded), *Epinotia* (a genus name without cogent explanation derived from Greek *epi* + *noton* = 'upon' + 'the back') and *Eucosma* (Greek *eucosmos* = 'graceful'). Wing lengths are around 10–17 mm, except where otherwise indicated.

Bearberry bell *Epinotia nemorivaga*, in high moorland and mountains in north Scotland, the Orkneys and west Ireland. Adults fly in sunny afternoons in June–July. Larvae mine alpine bearberry leaves. *Nemorivaga* comes from Latin *nemus* + *vagus*

= 'woodland glade' + 'wandering', though this does not reflect the habitat.

Bilberry bell *Rhopobota myrtillana*, wingspan 9–12 mm, widespread, especially in the north, but scarce, on moorland, heathland and mountains. Adults fly during late afternoon and early evening in May–June. Larvae feed on leaves of bilberry and bog-bilberry. Greek *rhops* + *bosco* = 'shrub' + 'to eat'. The specific name refers to bilberry *Vaccinium myrtillus*.

Birch bell *Epinotia demarniana*, wingspan 12–22 mm, nocturnal, local in woodland, heathland, fen and bog in parts of England and Wales. Adults fly in June–July. Larvae feed inside catkins of birch, alder and goat willow. Dr Demarne was a nineteenth-century German entomologist.

Black-brindled bell *Epinotia signatana*, local in woodland, hedgerows and scrub in England and Wales, mostly in the south and east, plus one record from Co. Offaly. Adults fly in sunny afternoons in June–July. Larvae mine leaves of blackthorn, wild cherry and bird cherry in spring. *Signatus* is Latin for 'marked', from a forewing mark.

Bright bell *Eucosma hohenwartiana*, wingspan 11–23 mm, widespread and common in dry open areas and grassland. Adults fly in June–August. Larvae mine the flower heads of common knapweed and saw-wort. Sigismund von Hohenwarth (1745–1822) was a German prelate and entomologist.

Brown alder bell *Epinotia sordidana*, wingspan 18–23 mm, widespread but local in marsh and ditches, and on riverbanks. Adults fly in August–October. Larvae feed on alder leaves. *Sordidus* is Latin for 'dirty' or 'dingy', reflecting the greyish forewing.

Brown elm bell *Epinotia abbreviana*, nocturnal, widespread and common in hedgerows and woodland, especially in lowland regions. Adults fly in June–July. In spring, larvae feed on the buds of elm and sometimes field maple. *Abbreviata* is Latin for 'shortened', from an abbreviated line on the forewing.

Butterbur bell *Epiblema turbidana*, wingspan 16–22 mm, widespread but scarce and local in damp pasture, riverbanks and ditches. Adults fly from late afternoon into the evening in June–July. Larvae mine butterbur roots from August to April. *Turbidus* is Latin for 'confused' or 'muddy', perhaps reflecting the indistinctly marked forewing.

Chalk-hill bell *Pelochrista caecimaculana*, wingspan 13–20 mm, nocturnal, local on dry pasture, downland and waste ground in parts of southern England and south Wales. Adults fly in July. Larvae mine the roots of knapweed and saw-wort from September onwards. Greek *pelos* + *christos* = 'clay' + 'anointed', from markings on the forewing, and Latin *caecus* + *macula* = 'blind' + 'spot', reflecting the wing's eyespot.

Chalk rose bell *Notocelia incarnatana*, wingspan 14–20 mm, nocturnal, scarce in open woodland, downland, coastal limestone and dunes in parts of England, Wales, south-west Scotland and (particularly on dunes) Ireland, most numerous on the coast. Adults fly in June–September. Larvae mine the leaves of a number of wild roses, especially burnet rose. The genus name is obscure, possibly conjoining Greek and Latin for 'the back' and 'heaven'. *Incarnatana* is Latin for 'flesh-coloured'.

Coast spurge bell *Acroclita subsequana*, scarce, on shingle and sandy beaches in parts of southern England from Devon to Sussex. With two generations, adults fly in May–June and July–September. Larvae feed on the leaves of sea spurge and Portland spurge. Greek *acron* + *clitus* = 'point' + 'slope', from the abrupt tapering of the antennae, and Latin *subsequens* = 'following', referring to a sequential listing of the species.

Cock's-head bell *Zeiraphera isertana*, wingspan 13–18 mm, nocturnal, widespread and commonly found in oak woodland. Adults fly in July–August. Larvae feed on oak leaves in May–June. Greek *zeira* + *phero* = 'loose garment' + 'carry', possibly referring to nets spun by the larvae. Paul Isert (1756–87) was a German-born physician.

Colt's-foot bell *Epiblema sticticana*, wingspan 15–20 mm, nocturnal, widespread, local in waste ground, open areas, dunes and cliffs. Adults fly in May–June. Larvae mine the roots and later the lower stems of colt's-foot. Greek *stictos* = 'spotted', describing wing markings.

Common birch bell *Epinotia immundana*, widespread and common in woodland, heathland and marshes and on riverbanks. Adults fly in April–June and, in the south, again in August–September. Larvae feed on alder and birch, the first generation in catkins, the midsummer generation on leaves. Strictly, *immundus* is Latin for 'unclean', but here means 'plain'.

Common rose bell *Notocelia rosaecolana*, wingspan 15–20 mm, nocturnal, common in gardens and hedgerows throughout much of southern Britain and south Ireland. Adults fly in June–July. Larvae mine the leaves of a wild and cultivated roses, especially sweet-briar (eglantine). Latin *rosa* + *colo* = 'rose' + 'inhabit'.

Common spruce bell *Epinotia tedella*, nocturnal, common in coniferous woodland. Adults fly in May–June. Larvae mine Norway spruce needles. *Tedella* is from Latin *t(a)eda* = 'pinewood'.

Crescent bell *Epinotia bilunana*, nocturnal, widespread and common in birch woodland, heathland and gardens. Adults fly in June–July. Larvae mine birch catkins. Latin *bi* + *luna* = 'twice' + 'moon', with two crescent shapes evident on the forewing.

Dark aspen bell *Epinotia maculana*, wingspan 17–23 mm, nocturnal, scattered and local in woodland in much of Britain. Adults fly in August–October. Larvae feed on leaves of aspen and possibly other poplar species. Latin *macula* = 'spot', referencing a number of wing spots.

Dark spruce bell *Epinotia subsequana*, with a very scattered distribution in woodland, gardens and parks in parts of England, Wales and Co. Cork, Ireland. Larvae mine the needles of conifers such as Norway spruce and silver-fir. *Subsequana* comes from Latin *subsequens* = 'following', referring to the sequence in *Lepidoptera Britannica* (1803 et seq.).

Dingy spruce bell *Epinotia pygmaeana*, with a scattered distribution in coniferous woodland, plantations and gardens in parts of Britain. Adults fly in April–June. Larvae mine needles of Norway spruce, silver-fir and other conifers. Latin *pygmaeana* refers to the small size.

Downland bell *Rhopobota stagnana*, widespread but scarce in damp woodland and pastures (larvae feeding on devil's-bit scabious) and downland (on small scabious). With two generations, adults fly in April–June and August–September. *Stagnana* comes from Latin *stagna* = 'swamp', one of the habitats.

Eyed bell *Eucosma pupillana*, wingspan 13–19 mm, a scarce nocturnal moth of dry pasture and waste ground in a few places in England, as far north as Derbyshire, and Wales. Adults fly in July–August. Larvae mine wormwood stems and roots. *Pupillana* is from Latin *pupilla* = 'pupil', from the eye mark on the forewing.

Fir bell *Epinotia nigricana*, endangered, recorded (perhaps only with old records) from coniferous woodland and plantations in a few parts of England and Northern Ireland. Adults fly in afternoon sunshine during summer. Larvae mine buds of silver-fir. Latin *nigricans* = 'blackish', referring to the dark forewing.

Fleabane bell *Epiblema cnicicolana*, nocturnal, local in damp grassland and on riverbanks in southern England and Wales. Adults fly in May–July, usually during early evening. Larvae mine common fleabane stems. *Cnicus* is Latin for 'safflower', a food plant on the continent.

Four-spot bell *Eriopsela quadrana*, in woodland, dry pasture and downland at a few sites in England and on the Burren, Ireland. Adults fly during late afternoon and at night in May–June. Larvae feed on goldenrod leaves. Greek *erion* + *psallo* = 'wool' + 'touch', from the whitish wing colour, and Latin *quadra* = 'square', from the forewing markings.

Fulvous bell *Eucosma fulvana*, wingspan 19–25 mm, confused with *Eucosma hohenwartiana* so distribution and status are unclear, but associated with dry open areas and grassland where its food plant is found. Adults fly in June–July. Larvae mine flower heads of greater knapweed. Latin *fulvana* = 'fulvous', describing the tawny wing colour.

Goldenrod bell *Eucosma aspidiscana*, wingspan 13–19 mm, in woodland and on downland, waste ground and cliffs in parts of southern England, the Midlands and west Ireland. Adults fly in April–June. Larvae mine flowers and later the stems of goldenrod. Greek *aspis* + *discos* = 'shield' + 'round plate', from the wing eyespot.

Great bell *Epiblema grandaevana*, wingspan 21–30 mm, scarce, on waste ground and cliffs occasionally in parts of north-east England (mostly coastal, possibly imported with ballast from the Baltic and more recently from the east Midlands and East Anglia. Adults fly in May–July. Larvae mine the roots of colt's-foot and butterbur from August to May. *Grandaevus* is Latin for 'of great age'.

Grey aspen bell *Epinotia cinereana*, nocturnal, recently separated from *E. nisella* so distribution and status are unclear. Adults fly in July–August. Larvae mine aspen leaves. Latin *cinereus* = 'ash-coloured', referring to the greyish forewing.

Grey poplar bell *Epinotia nisella*, nocturnal, widespread and common in woodland, parks, marshes and freshwater margins. Adults fly in July–August. Larvae mine catkins of sallow and poplar species in spring. *Nisus* is Latin for 'sparrowhawk', perhaps reflecting the colour.

Highland bell *Epinotia crenana*, rare, on heathland, moorland, riverbanks and mountains in parts of north Scotland. Adults fly in August–October. Larvae feed on leaves of rusty willow, eared willow and sallows. Latin *crena* = 'notch' or 'serration', referring to the wing edge.

Hoary bell *Eucosma cana*, wingspan 15–23 mm, widespread and common in rough ground and flower meadows. Adults fly from late afternoon into the night in June–August. Larvae mine the flower heads of thistles and common knapweed. *Cana* is Latin for 'grey' or 'hoary'.

Honeysuckle bell *Eucosmomorpha albersana*, local in open woodland and woodland clearings in parts of mainly southern England and Wales. Adults fly during the afternoon and at night in May–June. Larvae feed on honeysuckle. Greek *eucosmos* + *morphe* = 'graceful' + 'form'. Johann Albers (1772–1821) was a German physician and entomologist.

Knapweed bell *Epiblema cirsiana*, wingspan 12–23 mm, nocturnal, widespread and common in a range of habitats, including damp sheltered meadows and woodland rides and clearings. The adult resembles a bird dropping and flies in May–June. Larvae mine the roots and later the lower stems of marsh thistle and common knapweed. *Circiana* comes from Latin *circium* = 'thistle'.

Large birch bell *Epinotia brunnichana*, wingspan 18–22 mm, nocturnal, common in birch woodland and heathland throughout much of Britain, more frequently in the north. Adults fly in July–August. Larvae feed on birch, sallow and hazel leaves. Morten Brünnich (1737–1827) was a Danish entomologist.

Large sallow bell *Epinotia caprana*, wingspan 16–22 mm, nocturnal, widespread but local on moorland, fen, damp heathland and mosses. Adults fly in July–August, sometimes into October. Larvae feed on shoots of bog-myrtle and sallows (the specific name referring to goat willow *Salix capraea*).

Lemon bell *Thiodia citrana*, wingspan 16–21 mm, a scarce nocturnal moth found in dry pasture and waste ground throughout much of England, especially East Anglia, and Wales. Adults fly in July. Larvae mine flower heads of yarrow, field wormwood and stinking camomile. Greek *theiodea* = 'brimstone-like', and Latin *citrus*, both refer to the yellowish colour.

Marbled bell *Eucosma campoliliana*, nocturnal, widespread and common in dry pasture, dunes and waste ground, if most abundant in coastal districts. Adults fly in June–July. Larvae mine the flower heads and stems of common ragwort. Latin *campus* + *lilium* = 'field' + 'lily', reflecting the moth's beauty (as in 'consider the lilies of the field', Matthew 6:28–9).

Maritime bell *Eucosma lacteana*, local in saltmarsh in England, from Sussex (last recorded in 1981) to Lincolnshire. The 2005 records at Titchwell, Norfolk, were the first sightings since 1966; and the last Suffolk records were in 2001. Adults fly in July–August. Larvae mine the flower of sea wormwood. *Lactis* is Latin for 'milk', from the colour.

Moorland bell *Epinotia mercuriana*, local on moorland and in mountains in parts of northern England, Scotland and west and south-west Ireland. Adults fly during the afternoon and early evening in July–September. Larvae mine leaves and shoots of heather, bilberry and other moorland plants. *Mercuriana* comes from the forewing marking that resembles the symbol of the planet (and god) Mercury.

Mugwort bell *Eucosma metzneriana*, wingspan 18–22 mm, recorded in Britain for the first time in Cambridgeshire in July 1977. Subsequent records include: Southsea, Hampshire, 1982 and on the Isle of Wight in 2008; East Sussex, 1989; the Isle of Grain, Kent, 1998; and six Essex records from 1994 to 2015. Adults fly in June–July. Larvae mine the stems of mugwort and wormwood in autumn and spring. Herr Metzner (d.1861) was a German collector.

Northern bell *Rhopobota ustomaculana*, scarce, on moorland and mountains in parts of northern England, north Wales and the Highlands. Adults fly during the afternoon and evening in June–July. Larvae feed on cowberry leaves. *Ustomaculana* is from Latin *ustus* + *macula* = 'burnt' + 'spot', from a wing marking.

Obscure bell *Eucosma aemulana*, nocturnal, local in woodland and on chalk downland and cliffs in parts of southern England. Adults fly in July–August. Larvae feed on goldenrod seeds. Latin *aemulus* = 'rivalling' (i.e. congenerics).

Pale lettuce bell *Eucosma conterminana*, wingspan 15–19 mm, nocturnal, local in dry open often chalky or stony habitats, and in gardens, allotments, etc. in parts of southern England. Adults fly in July–August. Larvae feed on the flower heads of various kinds of lettuce. *Conterminana* comes from Latin, referencing the (pale) end of the forewing.

Pale saltern bell *Eucosma rubescana*, wingspan 16–20 mm, nocturnal, local in saltmarsh in Kent, Essex and Suffolk. Adults fly in June–July. Larvae mine sea aster flowers. *Rubescens* is Latin for 'red'.

Pine bell *Epinotia rubiginosana*, wingspan 13–15 mm, local in pine woodland throughout much of Britain, rarer in Ireland. Adults fly in sunshine in the late afternoon and at night in June. Larvae mine needles of Scots and stone pine in September–October. *Rubiginosus* is Latin for 'rusty', from the colour of part of the forewing.

Ragwort bell *Epiblema costipunctana*, wingspan 13–18 mm, nocturnal, widespread and common in rough meadow and on waste ground, if more local in Scotland. Adults are bivoltine in the south, flying in May–July and late July–September. In Scotland there is one generation, in June–July. Larvae mine stems and roots of common ragwort from September to April, and in the south from June to July. Latin *costa* + *punctum* = 'side' + 'spot', describing wing markings.

Red-headed bell *Zeiraphera rufimitrana*, nocturnal, local in pine woodland in parts of southern England, possibly no longer present (for instance, last recorded in Norfolk in 1960). Adults fly in August. Larvae feed on Scots pine needles. Latin *rufus* + *mitra* = 'red' + 'mitre', from the red head.

Saltmarsh bell *Eucosma tripoliana*, local in saltmarsh throughout much of southern England and south Wales, with records north to Lancashire. Adults fly in the late afternoon and at night in July–August. Larvae mine flowers of sea aster *Aster tripolium* from August to October, giving the specific name.

Scarce rose bell *Phaneta pauperana*, nocturnal, on chalky ground, sea-cliffs and other rough ground in parts of south-east England and the Midlands. Adults fly in April–May. Larvae feed on flowers and seeds of dog-rose. *Pauper* is Latin for 'poor', suggesting an unattractive species.

Silver-barred bell *Epinotia fraternana*, scarce, in coniferous woodland in parts of Britain, with one record from south-west Ireland. Adults fly in the afternoon in July. Larvae mine needles of species of fir. Latin *fraternus* = 'brotherly', meaning related to other species.

Small birch bell *Epinotia ramella*, nocturnal, widespread and common in birch woodland. Adults fly in July–October. Larvae mine birch and sallow catkins, and *ramella* comes from Latin *ramus* = 'branch'.

Small spruce bell *Epinotia nanana*, widespread but rare in coniferous woodland, plantations, parks and gardens, absent in north Scotland, and with only one (pre-1970) Irish record. Adults fly in June–August. Larvae feed on Sitka and Norway spruce needles. Latin *nanus* = 'dwarf'.

Southern bell *Crocidosema plebejana*, a naturalised nocturnal moth discovered in south Devon in 1900, expanding its range and found locally in gardens and on waste ground in parts of southern England and East Anglia. Adults fly in July–October. Larvae mine the seed capsules and shoots of tree-mallow and related species. Greek *crocidos* + *sema* = 'woollen nap' + 'mark', from a hindwing mark. *Plebeius* is Latin here meaning 'common', reflecting its native distribution.

Square-barred bell *Epinotia tetraquetrana*, widespread and common in a range of habitats, especially light woodland. Adults fly in April–May. Larvae feed inside the stems (creating galls) then the leaves of birch and alder. *Tetraquetrus* is Latin for 'four-sided', reflecting part of the pattern on the forewing.

Square-spot bell *Notocelia tetragonana*, nocturnal, local in woodland and scrub in parts of Britain, probably declining, and with one record from west Ireland. Adults fly in July. Larvae mine the flower buds and leaves of wild roses. *Tetragonum* is Latin for 'quadrangle', from the shape of the blotch on the forewing.

Summer rose bell *Notocelia roborana*, wingspan 18–21 mm, nocturnal, widespread and common in gardens and hedgerows. Adults fly in June–August. Larvae mine the leaves of roses, despite which the specific name references oak *Quercus robur*.

Thistle bell *Epiblema scutulana*, wingspan 18–23 mm, nocturnal, widespread and common in waste ground and rough meadows. Adults fly in May–July. Larvae mine the roots and especially the lower stems of spear and musk thistle in August–April. *Scutulana* is Latin, either *scutula* = 'diamond shape' or *scutulum* = 'small shield', describing a blotch on the forewing.

Triple-blotched bell *Notocelia trimaculana*, nocturnal, widespread and common in hedgerows, gardens and woodland. Adults fly in June–July. Larvae mine hawthorn leaves. *Trimaculana* is Latin for 'three-spotted', describing forewing markings.

Two-coloured bell *Eucosma obumbratana*, wingspan 14–20 mm, nocturnal, local on farmland and along the margins of rough meadows throughout much of England and Wales. Adults fly in July–August. Larvae mine the seed heads of perennial sowthistle in autumn. *Obumbratus* is Latin for 'overshadowed' or 'darkened', reflecting part of the forewing colour.

Variable bell *Epinotia solandriana*, wingspan 16–21 mm, nocturnal, common in woodland and scrub over much of Britain. Adults fly in July–August. Larvae feed on the leaves of birch, goat willow and hazel. Daniel Solander (1733–82) was a Swedish-born naturalist.

White-blotch bell *Epinotia trigonella*, wingspan 16–21 mm, crepuscular, widespread and common on damp heathland, mosses and in oak woodland. Adults fly in August–September. Larvae feed on birch leaves. Latin *trigonus* = 'triangular', from the shape of a spot formed on the wings when folded.

White-foot bell *Epiblema foenella*, wingspan 17–26 mm, nocturnal, common in rough and waste ground throughout England and Wales and occasional in south Scotland. Adults fly during the late afternoon into the evening in June–August. Larvae mine the roots and the lower stems of mugwort in August–May. *Foenum* is Latin for 'hay', probably from a hayfield habitat.

White sallow bell *Epinotia subocellana*, nocturnal, widespread

and common in deciduous woodland and damp areas. Adults fly in May–July. Larvae feed on sallow leaves. Latin *sub* + *ocellus* = 'under' + 'eye', from the wing marking.

Yellow-faced bell *Notocelia cynosbatella*, wingspan 16–22 mm, nocturnal, widespread and common in gardens, hedgerows and scrub. Adults fly in May–July. Larvae mine the flower buds and leaves of a number of wild and cultivated roses. Greek *cynosbatos* = 'dog-rose'.

BELLADONNA Alternative name for deadly nightshade.

BELLCAP Alternative name for some bonnet fungi. **Bleeding bellcap** Alternative name for burgundydrop bonnet. **Lilac bellcap** Alternative name for lilac bonnet.

BELLE (MOTH) Species of *Aspitates* (Geometridae, Ennominae), from a misspelling of Greek *aspilates*, for an Arabian precious stone mentioned by Pliny, and *Scotopteryx* (Larentiinae), from Greek *scotos* + *pteryx* = 'darkness' + 'wing'.

July belle *S. luridata*, wingspan 32–8 mm, day-flying in June–August, fairly widespread on heathland, moorland, downland, shingle beaches and scrubby places. Larvae feed on gorse, petty whin and dyer's greenweed. *Luridata* is Latin for 'wan' or 'pale'.

Lead belle *S. mucronata*, wingspan 30–8 mm, day- and night-flying in May–June, locally common, especially in south-west England and in Wales, but scattered throughout the British Isles on heathland, moorland and some scrub. Larvae feed on gorse, broom, petty whin and dyer's greenweed. *Mucronatus* is Latin for 'pointed', from the acute forewing apex.

Straw belle *A. gilvaria*, wingspan 25–30 mm, scarce, day- and night-flying in June–September, on chalk and limestone grassland, with a declining distribution and separate subspecies on the North Downs of Kent and Surrey and on the Burren, west Ireland. Larvae feed on herbaceous plants such as bird's-foot-trefoil, cinquefoils and thyme. An RDB species of priority conservation concern. *Gilvus* is Latin for 'yellow'.

Yellow belle *A. ochrearia*, wingspan 28–36 mm, with two generations a year, adults flying on sunny days but more characteristically by night in May–June and August–September. Local throughout the south of England and Wales on dry grassland, sandy habitats, waste ground and especially a range of coastal habitats. Larvae feed on low-growing plants such as wild carrot, restharrow and buck's-horn plantain. Latin *ochra* = 'ochre'.

BELLFLOWER Herbaceous perennials in the family Campanulaceae, mostly in the genus *Campanula* (Latin for 'little bell'), with one species of *Wahlenbergia* (named after the Swedish botanist Georg Wahlenberg, 1780–1851).

Rarities include **broad-leaved bellflower** *C. rhomboidalis* (introduced from the Jura/Alps in 1775, formerly naturalised by River Esk in Dumfries and Galloway, and found on a sunken lane in Cumbria in 1983), **chimney bellflower** *C. pyramidalis* (Mediterranean, naturalised on walls in Guernsey and west Kent, casual elsewhere), **Cornish bellflower** *C. alliariifolia* (Near Eastern in southern England and west Yorkshire/east Lancashire), **giant bellflower** *C. latifolia* (in southern England and north Scotland, introduced in Ireland and the Isle of Man), and **milky bellflower** *C. lactiflora* (scattered). See also Canterbury-bells and harebell.

Adria bellflower *C. portenschlagiana*, a Mediterranean neophyte, scattered in the British Isles, locally common in England and Wales, naturalised on walls and rocky banks. Adria references the Adriatic Sea. Franz Portenschlag-Ledermeyer (1772–1822) was an Austrian naturalist.

Clustered bellflower *C. glomerata*, scattered in Britain on chalk and oolitic limestone, in calcareous grassland, scrub, open woodland, cliffs and dunes, and as a casual garden escape on roadsides and waste ground. This is the county flower of Rutland as voted by the public in a 2002 survey by Plantlife. *Glomerata* is Latin for 'clustered'.

Creeping bellflower *C. rapunculoides*, a rhizomatous European neophyte present by mid-Tudor times, widespread but scattered in Britain and east Ireland on rough ground, verges and railway banks, occasionally in woodland and field borders. Latin *rapunculoides* = 'resembling a small turnip'.

Ivy-leaved bellflower *W. hederacea*, locally common in south-west England and Wales, scattered in southern England and west Ireland, in damp, wet or boggy places on acidic soils, on heathland, heathy pasture, moorland, open woodland and willow carr, and by streams and in flushes. *Hederacea* is Latin for 'resembling ivy'.

Nettle-leaved bellflower *C. trachelium*, scattered mainly in England as a native on dry, base-rich, usually calcareous soils in woodland, scrubby grassland and hedge banks. In Ireland it is also recorded from riverbanks and swamp woodland. As a garden escape, it has become naturalised on a wider range of soils and habitats. Greek *trachelos* = 'neck'.

Peach-leaved bellflower *C. persicifolia*, a rhizomatous European neophyte, present in Tudor times, now scattered as a casual or naturalised on waste ground and roadsides, and in hedgerows and woodland, mostly in Britain, but with records also from Co. Dublin and Co. Waterford. Latin *persicifolia* = 'peach-leaved'.

Rampion bellflower *C. rapunculus*, an archaeophyte, recorded in the wild in 1597, occasionally found in England and south Scotland as naturalised in rough grassland and on roadsides, railway banks and in quarries. It also occurs as a relic of cultivation, having once been frequently grown in gardens for ornament and its edible parsnip-like roots (*rapunculus* is Latin for 'little turnip') and its leaves used in salads, but falling out of favour as a vegetable around 1700 and consequently declining.

Spreading bellflower *C. patula*, biennial, local in England as far north as Lincolnshire, and up to mid-Wales, in dry, well-drained, sunny sites on infertile sandy or gravelly soils in open woodland, woodland edge and hedge banks, and on roadsides and railway banks, but declining because of the cessation of coppicing and increased use of herbicides. *Patula* is Latin for 'spreading'.

Trailing bellflower *C. poscharskyana*, a Mediterranean neophyte, widespread and scattered, commoner in England and Wales, naturalised on walls and rocky banks. Gustav Poscharsky (1832–1915) was a German horticulturalist.

BELLS Poplar bells *Schizophyllum amplum* (Agaricales, Scizophyllaceae), a scarce inedible fungus, Near Threatened in the RDL (2006), mainly recorded from south-central England and East Anglia, on dead wood of poplar and willow. Greek *schizo* + *phyllon* = 'split' + 'leaf' (i.e. gills), and Latin *amplum* = 'large'.

BENT Grasses in the genus *Agrostis* (Greek *agros* = 'grass'), together with *Polypogon* (Greek *poly* + *pogon* = 'many' + 'beard', from the hairy panicle), both Poaceae.

Black bent *A. gigantea*, a perennial archaeophyte, common in England but scattered elsewhere, in grassy habitats and waste ground, found as a weed in cereal and neglected arable land, particularly on light, sandy soils, spreading by seed and by rhizomes. In wetter habitats, though rarer, it can persist in tall closed vegetation using rhizomes.

Bristle bent *A. curtisii*, a perennial locally common in south-west and southern England, and south Wales, on drier parts of sandy and peaty heathland, especially where there is impeded drainage. In closed communities it occurs as scattered plants, but it can seed prolifically into burnt or disturbed ground and create an almost continuous cover. It also grows in open acidic woodland over gravel and sand.

Brown bent *A. vinealis*, a widespread but scattered rhizomatous perennial, locally common on dry or free-draining, acidic, sandy or peaty soils on heathland, in acidic grass-heath and in open woodland. On some lowland heaths it may grow in damp but not waterlogged soils. It is also grown as a drought-resistant lawn grass. *Vinealis* is Latin for 'like a vine'.

Common bent *A. capillaris*, a common and widespread rhizomatous, occasionally stoloniferous, perennial of neutral to acidic, dry to moist soils. Habitats include lowland and hill pasture, hay meadow, heathland, open woodland, scrub, dunes, and a range of ruderal habitats including sites contaminated by heavy metals. It is much used on lawns. *Capillaris* is Latin for 'hair-like'.

Creeping bent *A. stolonifera*, a common and widespread stoloniferous perennial found in many habitats on heavy and light, moist and dry soils, for example in permanent grassland, upper saltmarsh, dunes and dune slacks, bog, springs, flushes and ditches. It also grows on spoil tips and a number of open and disturbed habitats, and is a weed of arable fields. *Stolonifera* is Latin for 'bearing runners (stolons)'.

Highland bent *A. castellana*, a perennial Mediterranean neophyte first recorded, as a casual, in 1924, scattered throughout Britain, increasingly recorded from roadsides, amenity grassland, temporary leys and cultivated land, often sown in grass-seed mixtures, and is also introduced with wool shoddy. Latin *castellum* = 'village' or 'fort'.

Rough bent *A. scabra*, an often annual North American neophyte, found on rough ground, and by roads and railways, often naturalised, originating from wool shoddy and grain. *Scabra* is Latin for 'rough'.

Small bent *A. hyemalis*, a generally annual North American neophyte, occasional on waste ground in dockyards, and by roads and railways, originating as wool shoddy and a grain contaminant. *Hiemalis* is Latin for 'of winter'.

Velvet bent *A. canina*, a widespread stoloniferous perennial locally common on infertile, acidic, peaty soils in bog, wet heathland, fen and fen-meadow, springs, swamp and water margins. *Canina* is Latin for 'relating to dogs'.

Water bent *Polypogon viridis*, an annual or perennial neophyte naturalised in the Channel Islands on roadsides and by pools. In England it grows on tips and damp waste ground, and is spreading as a weed of nurseries, gardens and pavements. *Viridis* is Latin for 'green'.

BENT-MOSS *Campylostelium saxicola* (Ptychomitriaceae), uncommon, widespread and scattered mainly on damp sandstone near streams in hilly districts, but also on rocks rich in heavy metals. Greek *campylos* + *stele* = 'curved' + 'pillar' (referring to the curved seta), and Latin *saxicola* = 'rock dweller'.

BENT-WING Leaf-mining moths, wingspan around 6–9 mm unless otherwise indicated, in the families Bedelliidae (*Bedellia*, honouring the British entomologist George Bedell, 1805–77), Bucculatricidae (*Bucculatrix*, from Latin *buccula* = 'little cheek', here meaning 'visor', from the large antennal eye caps), Gracillariidae subfamily Phyllocnistinae (*Phyllocnistis*, from Greek *phyllo* + *cnizo* = 'leaf' + 'scrape', from the leaf mine), Lyonetiidae subfamily Cemiostominae (*Leucoptera*, from Greek *leucos* + *pteron* = 'white' + 'wing') and Opostegidae (*Opostega*, from Greek *ops* + *stege* = 'face + 'roof', referencing the large antennal eye caps, and *Pseudopostega*, meaning 'false' *Opostega*).

Alder bent-wing *Buc. cidarella*, widespread on alder as far north as Inverness-shire, and on bog-myrtle on Anglesey and Borth Bog in north-west Wales, Co. Kerry in Ireland, and Arne NNR on the Isle of Purbeck, Dorset. Adults fly in May–June, sometimes with a second generation in August. Larvae are active in August–September. Latin *cidaris* = 'headdress', from a tuft conspicuous against the dark wings.

Bindweed bent-wing *Bedellia somnulentella*, local in hedgerows and on waste ground throughout much of southern Britain, occurring with less frequency as far north as Co. Durham, and recorded from a few sites in Ireland. Adults occur in two generations, evident in August and overwintering in October–May. Larvae mine leaves of bindweeds and morning-glory in July–August and September. *Somnulentus* is Latin for 'sleepy', either referring to overwintering or to the dusky colour.

Birch bent-wing *Buc. demaryella*, widespread but local in open woodland, heathland, moorland and bogs. Adults fly in May–June. Larvae are active in August, mining leaves of birch, occasionally hazel and sweet chestnut. M. Demary was a nineteenth-century French entomologist.

Broom bent-wing *L. spartifoliella*, nocturnal, local throughout Britain, with records also from the south and east of Ireland. Adults fly in June–July. Larvae feed under the bark of broom and dyer's greenweed. Latin *spartifoliella* references the leaves of Spanish broom.

Buckthorn bent-wing *Buc. frangutella*, found in parts of East Anglia, south and south-west England, south Wales and the Burren, west Ireland in fen, open woodland and hedgerows. Adults fly in June–July. Larvae mine buckthorn in drier habitats and alder buckthorn on damper ones, often several to a leaf, in August–September. After the initial mining phase, larvae feed externally on the leaf underside. The specific name is a misprinted derivation of *Frangula* (alder buckthorn).

Crested bent-wing *Buc. cristatella*, scattered and widespread in parts of England and Ireland on downland, grassland, waste ground and verges. Adults fly in two generations in May–June and July–August. Larvae feed on yarrow, initially in a mine then externally on the leaves, being active in April–May and in July. Latin *cristatus* = 'crest', describing the erect scales on the head.

Daisy bent-wing *Buc. nigricomella*, widely distributed on embankments, waste ground and rough grassland. Adults have two generations, flying in April–May and August. Larvae initially mine oxeye daisy, then feed externally on the leaves. Latin *niger* + *coma* = 'black' + 'plume', from the tuft on the head.

Elm bent-wing *Buc. albedinella*, local in hedgerows and woodland margins throughout much of southern England, though with records up to the Humber Estuary. Adults fly in June. Larvae feed on elm leaves. Latin *albidinellus* = 'whiteness', from the forewing colour.

Hawthorn bent-wing *Buc. bechsteinella*, local in scrub, hedgerows and woodland edges throughout much of England and the Welsh Border, with some Scottish records. Adults fly in evening sunshine in May–June, and again in July–August. Larvae, evident from July to October, mine leaves of hawthorn, apple, pear, rowan and wild service-tree. Johan Bechstein (1757–1810) was a German entomologist.

Kent bent-wing *P. xenia*, nocturnal, on riverbanks and in damp woodland. First discovered in Kent in 1974, subsequently rapidly spreading throughout south and south-east England and East Anglia. Adults fly in July–September in two generations, the second brood probably overwintering as an adult. Larval mines in grey and white poplars can be found from June to August or September. Greek *xenos* = 'gift'.

Lime bent-wing *Buc. thoracella*, common in ancient woodland, woodland rides, parks and gardens throughout much of south England and Wales, north to Lincolnshire, plus a couple of records from east Ireland. Adults fly in June, sometimes August in the south of their range. Larvae mine leaves of small-leaved and occasionally common lime. *Thoracella* comes from Latin *thorax*, from the conspicuous yellow thorax.

Little bent-wing *L. lotella*, wingspan 5–6 mm, a nocturnal rarity found on dry downland and in marsh and other damp areas in south-east England, Yorkshire and west Ireland. Adults fly in June–July. Larvae, mainly seen in June–September, mine leaves of bird's-foot-trefoil, species of *Lotus* that prompt the specific epithet.

Mint bent-wing *Pseudopostega crepusculella*, wingspan 8–12 mm, widespread though relatively local and generally scarce, in damp habitats such as fen and alder carr. Adults fly in June–July. Larvae mine leaves of mint.

Oak bent-wing *Buc. ulmella*, common and widespread in England and Wales, less so in Scotland, associated with mature oak trees, the leaves of which are mined by the larvae throughout much of the year but particularly in August–November. With two generations, adults fly in April–May and in July–September. Latin *ulmus* = 'elm', in error of the food plant.

Poplar bent-wing *Ph. unipunctella*, nocturnal, common in woodland, areas with scattered trees and gardens throughout much of southern England, East Anglia and south Wales, with records north to Cheshire. With two generations, adults fly in July–August and September then overwinter. Larvae mine leaves of poplar, especially in August–October. Latin *unus* + *punctum* = 'one' + 'spot', describes the mark on the forewing.

Saltern bent-wing *Buc. maritima*, found in a number of saltmarshes in England, Wales and south and east Ireland. Adults have two generations and fly in the late afternoon in June and August. Larvae mine sea aster. The specific name refers to the habitat.

Scarce bent-wing *L. lathyrifoliella* form *orobi* (or *L. orobi*), a rarity found on moorland, hillsides and open woodland in north-east Yorkshire, Co. Durham and the Aviemore district, and in Ireland from the Burren in Co. Clare, and Ballynahinch, Co. Galway. Larvae mine leaves of bitter-vetchling (*Lathyrus*), referenced in the specific epithet. *Orobus* is the former name of another *Lathyrus* food-plant species.

Sorrel bent-wing *O. salaciella*, wingspan 9–12 mm, widespread but local in Britain in open habitats such as heathland and dry pasture where the larval food plant, sheep's sorrel, is found. Adults fly in June–July. *Salaciella* is from Latin *salax* = 'lustful', possibly because the type specimen was captured while mating, or it is from Greek *salax* = 'miner's sieve'.

Southern bent-wing *L. lathyrifoliella*, confined to a few coastal counties in southern England and in Wales, occupying cliffs and similar habitats where the larval food plant occurs, narrow-leaved everlasting-pea (*Lathyrus*, giving the specific epithet). Adults fly in May–July.

Willow bent-wing *P. saligna*, nocturnal, with a local and scattered distribution in fen, marsh, riverbanks and other damp areas throughout southern England, the Midlands, East Anglia and Wales. With two generations, adults fly in July and September then overwinter into April. Larvae mine leaves of willows, especially purple and crack willows, being particularly active in June and August–September. *Saligna* conflates Latin *salix* + *lignum* = 'willow' + 'wood'.

BERBERIDACEAE Family name for members of the genera *Berberis* (spined) and *Mahonia* (without spines). See barberry and Oregon-grape.

BERMUDA-BUTTERCUP *Oxalis pes-caprae* (Oxalidaceae), a perennial neophyte herb, despite its name introduced from South Africa (1757), now naturalised as an agricultural weed, especially in bulb fields, in mild climates. Outside the Channel Islands and the Scillies, populations are usually casual. Greek *oxalis* = 'acid', referring to oxalic acid in the leaves and roots, and Latin *pes-caprae* = 'goat's foot', from the leaflet shape.

BERMUDA-GRASS *Cynodon dactylon* (Poaceae), a probably introduced perennial scattered in (especially southern) England and south Wales, naturalised on eroding fore-dunes in dry sandy places and short coastal grassland, and inland on lawns and verges. It is also found around docks, on tips and as a casual originating from wool shoddy. Greek *cyon* + *odonti* = 'dog' + 'tooth', referring to the sharp hard scales of the rhizomes and stolons, and *dactylon* = 'finger'. **African Bermuda-grass** *C. incompletus*, from South Africa, is a casual wool alien recorded from a few sites in England on waste ground.

BERYTIDAE Possibly from Greek *beris* = 'feast' (or *berys* = 'bear'), the family name of stilt bugs (Heteroptera), long, slender-legged insects with nine British species in four genera.

BETONY *Betonica officinalis* (Lamiaceae), a herbaceous perennial common and widespread in England and Wales, scarce elsewhere, in hedge banks, grassland, heath, open woodland and woodland edges. It is occasionally found in clifftop grassland, sometimes as the genetically dwarf var. *nana*. It favours but is not limited to mildly acidic soils. It has locally declined in England and Ireland as a result of the loss or improvement of permanent pasture, ploughing of fields to the edge of woods, and shading of woodland grassland following a decline in coppicing. 'Betony' comes from Middle English from Old French *betoine*, in turn from Latin *betonica*, from the name of an Iberian tribe (Vettones) and as a variant of *vettonica*, a plant found in Spain. *Officinalis* indicates a pharmacological use, including as a general tonic.

BETULACEAE Family name for birch (Latin *betula*), alder, hazel and hornbeam.

BIB Alternative name for pout.

BIBIONIDAE Family name of flies, with 14 species of *Bibio*, four of *Dilophus*, most species associated with grassy habitats, the larvae feeding on dead plant material or roots. See fever fly and St Mark's fly.

BILBERRY *Vaccinium myrtillus* (Ericaceae), a common and (apart from central and eastern England) widespread low shrub, locally dominant in well-drained heathland and moorland, especially in upland areas, and as an understorey in acid woodland of birch, oak and pine; also on hummocks in peat bogs in the north and west of its range. 'Bilberry' dates from the late sixteenth century, probably of Scandinavian origin. The genus name is Latin for this genus; *myrtillus* = 'small myrtle'.

Bog bilberry *V. uliginosum*, a low shrub, locally common in central and north Scotland, Orkney and Shetland, scattered elsewhere in Scotland and northern England, and naturalised on Haddon Hill, Exmoor (recorded in 1994), on podzolic or peaty acidic soils in upland dwarf-shrub heaths and blanket bog, occasionally in grass-sedge heathland, and rarely in calcareous mountain avens communities on montane ledges. *Uliginosum* is Latin for 'growing in bogs'.

BINDWEED Herbaceous perennials in the genera *Convolvulus* (Latin *convolvere* = 'to wrap') and *Calystegia* (Greek *calyx* + *stegon* = 'calyx' + 'cover', referring to the bracts that conceal the calyx), both in the Convolvulaceae. See also black-bindweed.

Field bindweed *Con. arvensis*, trailing or climbing, common and widespread in lowland regions on waste or cultivated ground, waysides, railway banks, open scrub and rough grassland. Latin *arvum* = 'farmland'.

Hairy bindweed *Cal. pulchra*, a widespread but scattered climbing (possibly) Asian neophyte, in cultivation since 1823, naturalised in hedges and on waste ground, usually close to habitation. Latin *pulcher* = 'beautiful'.

Hedge bindweed *Cal. sepium*, a common and widespread climber, found in hedges, scrub, woodland edge, tall-herb fen, open willow and alder carr, and on railway banks and waste ground. It also occurs in artificial habitats in built-up areas. Subspecies *roseata* (first described in 1967) has a scattered distribution, but is absent from Yorkshire to north Scotland, in brackish habitats at the upper edges of saltmarsh, in reedbeds, and in grassy waste places in coastal regions, as well as inland on riverbanks, in fen and on rough ground. *Sepium* is Latin for 'fence' or 'hedge'.

Large bindweed *Cal. silvatica*, a climbing or scrambling Mediterranean neophyte present since 1815, widespread and common except in upland Scotland, on hedges, fences, gardens and waste ground, spreading into semi-natural habitats away from habitation. *Silvatica* is from Latin for 'woodland'.

Sea bindweed *Cal. soldanella*, trailing, absent from north Scotland but elsewhere commonly found on dunes, and above the strand-line on sand and, sometimes, shingle beaches. *Soldanella* is Latin for a small coin, a reference to rounded leaves.

BIPHYLLIDAE Family name of false skin beetles, with two species, both found in woodland: *Biphyllus lunatus* in King Alfred's cakes fungus, particularly on ash, and *Diplocoelus fagi* mainly beneath beech bark and on sycamore.

BIRCH (MOTH) Grey birch *Aethalura punctulata* (Geometridae, Ennominae), nocturnal, wingspan 30–5 mm, common in birch woodland and heathland throughout Britain, most

numerous in East Anglia. Adults fly in May–June. Larvae feed on birches, occasionally alder. Greek *aithalos* + *ouron* = 'smoke' + 'boundary', and Latin *punctulum* = 'small spot', from the wing pattern.

BIRCH (TREE) Deciduous trees in the genus *Betula* (Betulaceae), 'birch' derived from Old English *bierce* and *birce*, related to German *Birke*. *Betula* is Latin for birch.

Downy birch *B. pubescens*, common and widespread, usually in mixed woodland or as isolated trees on roadsides and field boundaries, growing on a range of soils though preferring more acidic, wetter, peatier soils, especially in the uplands. It can rapidly colonise open, unshaded ground, particularly burned areas, especially on heathland. It is also widely planted, becoming established in urban areas. *Pubescens* is Latin for 'downy'.

Dwarf birch *B. nana*, a low-growing shrub of upland heaths and blanket bogs of highland Scotland (also north-west Yorkshire and south Northumberland), usually on acidic peat but occasionally rooted into rock crevices. It grows on both moderately dry, sloping sites and on waterlogged flat ground. Seedlings are vulnerable to grazing. While reaching 860 m in Glen Cannich (east Ross), it is found as low as 120 m in west Sutherland. *Nana* is Latin for 'dwarf'.

Grey birch *B. populifolia*, a North American neophyte, increasingly planted in parks and by roadsides, regenerating in places. *Populifolia* is Latin for 'having leaves like a poplar'.

Himalayan birch *B. utilis*, a neophyte increasingly planted in parkland and by roadsides, possibly regenerating in places. *Utilis* is Latin for 'useful'.

Paper-bark birch *B. papyrifera*, a North American neophyte increasingly planted in parkland and by roadsides, possibly regenerating in places. *Papyrifera* is Latin for 'bearing paper'.

Silver birch *B. pendula*, common and widespread in woodland on a range of light, well-drained acidic soils. It can rapidly colonise open ground, especially burned areas, and can invade open heathland. It is also widely planted on roadsides and in parkland. *Pendula* is Latin for 'hanging', from the disposition of the leaves.

BIRCH BESOM *Taphrina betulina* (Taphrinales, Taphrinaceae), a common and widespread fungal pathogen, one of the causative agents of witch's broom. *Taphrina* is Greek for 'small trench'.

BIRCH CATKIN BUG *Kleidocerys resedae* (Lygaeidae), locally common throughout England, Wales and south Ireland, associated with birch. Greek *cleidos* + *ceras* = 'key' + 'horn', and Latin *reseda* = 'mignonette'.

BIRD Warm-blooded animal in the class Aves with, among other characteristics, feathers, toothless beaked jaws, the laying of hard-shelled eggs, a high metabolic rate, and a lightweight but strong skeleton. All British examples can fly.

BIRD-IN-A-BUSH *Corydalis solida* (Papaveraceae), a tuberous perennial herb introduced from Europe, scattered in England and Wales, rare in Scotland, a garden escape naturalised since the eighteenth century in woodland, hedgerows, roadsides and rough grassland. *Corydalis* is Greek for lark and like this bird the flower has a spur or crest, probably giving it its English as well as genus name. *Solida* is Latin for 'dense' or 'solid'.

BIRD'S-FOOT *Ornithopus perpusillus* (Fabaceae), a locally common winter-annual, absent from much of northern England, Scotland and Ireland, on short grassland on free-draining acidic sand and gravel, around rock outcrops, on dunes, bare patches on dry heathland and beside tracks, and colonising bare areas after fire and other disturbances. Greek *ornis* + *pous* = 'bird' + 'foot', and Latin *perpusillus* = 'very small'.

BIRD'S-FOOT-TREFOIL Generally perennial herbs in the genus *Lotus* (Fabaceae), from Greek *lotos*, used for a variety of plants. 'Trefoil' refers to the three-lobed leaf.

Common bird's-foot-trefoil *L. corniculatus*, common and widespread in grassland, including well-drained meadows, chalk and limestone downland, hill pasture and montane rock ledges; also on coastal clifftops, shingle and dunes. *Corniculatus* is Latin for 'small-horned'.

Greater bird's-foot-trefoil *L. pedunculatus*, common and widespread in rushy pasture, wet meadow, marsh, ditches, lake and pond margins, and damp road verges, more frequent on acidic than calcareous soils. *Pedunculatus* is Latin for having a flower-stalk.

Hairy bird's-foot-trefoil *L. subbiflorus*, an annual or short-lived perennial largely found in southern and south-west England around rock outcrops and in dry, open grassland on coastal cliffs. It grows best on shallow, drought-prone neutral to moderately acidic soils. In Dorset, Hampshire and Co. Wexford it occurs inland in open, sandy grassland and on verges. *Subbifloris* is Latin for 'nearly two flowers'.

Narrow-leaved bird's-foot-trefoil *L. tenuis*, found mainly in England on coastal grazing marsh and sea walls; inland it occurs as a casual or recent colonist in rough grassland, and on railway banks and verges. It is found on a range of neutral to calcareous soils, preferring estuarine silts and heavy clays. *Tenuis* is Latin for 'slender' or 'narrow'.

Slender bird's-foot-trefoil *L. angustissimus*, an annual found locally in southern England on thin, drought-prone soils on sea-cliffs, growing on sunny banks by tracks, and in open areas among scrub. In Devon, it occurs in similar habitats inland. Sites in Hampshire and Kent are associated with sand and gravel extraction. *Angustissimus* is Latin for 'very narrow'.

BIRD'S NEST (FUNGUS) Inedible fungi in the genera *Crucibulum* (Latin for 'in the form of a crucible') and *Cyathus* (Greek *cyath* = 'cup-shaped') (Agaricaceae) in which a cup-shaped fruit body has a woolly membrane covering the top; once mature the membrane fall away revealing egg-like containers (peridioles) that contain the spores. The 'eggs' are attached to the base of the 'nest' by fine threads that break when raindrops knock the eggs from the nest, dispersing the spores.

Common bird's nest *Cr. laeve*, infrequent, scattered but widespread, found from summer to early winter on organic debris, and may be abundant on winter cereal stubble. *Laeve* is Latin for 'smooth', referencing the smooth inner surfaces of the 'nests'.

Dung bird's nest *Cy. stercoreus*, Near Threatened in the RDL (2006), on with dead marram and rabbit dung at coastal sites in north Devon and Wales. *Stercoreus* is Latin for 'filthy'.

Field bird's nest *Cy. olla*, widespread and locally common especially in England from spring to autumn on rotting wood, woodchip mulch, manured soil, horse dung, and other organic debris, especially dead Asteraceae and (by the coast) marram stems. *Olla* is Latin for a 'pot'.

Fluted bird's nest *Cy. striatus*, fairly common and widespread on soil and dead wood in open woodland, and also on woodchip mulch in parks and gardens, fruit bodies evident from early summer to early winter. *Striatus* is Latin for 'striped', referring to the striated cup sides.

BIRD'S-NEST (MOTH) *Tinea trinotella* (Tineidae, Tineinae), wingspan 12–18 mm, widespread and common in bird nests, larvae creating a portable case and feeding on nest detritus. Adults fly at night between May and August, possibly with two generations. *Tinea* is Latin for a chewing worm (i.e. larva), *trinotella* for 'three marks', from the stigmata on the forewing.

BIRD'S NEST (PLANT) Yellow bird's-nest *Hypopitys monotropa* (Ericaceae), a saprophytic (and chlorophyll-less) herbaceous perennial getting nutrition through parasitism on fungi, scattered and local, growing from leaf litter in shaded woodland, especially under beech and hazel on calcareous substrates, and under pine on more acidic soils. It also grows in damp dune slacks, often associated with creeping willow. Greek *hypo* + *pitys* = 'under' + 'pine', referring to the habitat, and Latin *monotropa* = 'one turn'.

BIRD'S WING *Dypterygia scabriuscula* (Noctuidae, Xyleninae), wingspan 32–7 mm, a nocturnal moth local in broadleaf woodland, heathland and scrub throughout the southern half of England. Adults fly in June–July, with an occasional second brood in the south. Larvae feed on docks, sorrels and knotgrass. Greek *di* + *pteryx* = 'two' + 'wing' (from a bird's wing pattern on the forewing, also giving the English name), and Latin *scabrius* = 'rougher'.

BIRTHWORT *Aristolochia clematitis* (Aristolochiaceae), a scrambling rhizomatous perennial herb introduced from Europe, with a scattered and declining distribution as an escape or relic of cultivation in rough ground and in woodland in central and southern England and south Wales. As a medicinal herb it was claimed to be useful as an aromatic stimulant in treating rheumatism and gout, and for relieving obstructions after childbirth, hence both its English and genus names, the latter from Greek *aristo* + *lochia* = 'best' + 'delivery'. Greek *clematitis* means related to a climbing plant.

Breckland birthwort *A. hirta*, introduced from Turkey, found under conifers on heathland in Suffolk since 1969. *Hirta* is Latin for 'hairy'.

BISTORT Herbaceous perennials in the genus *Persicaria* (Polygonaceae). 'Bistort' dates from the early sixteenth century, from French *bistorte* and Latin *bistorta*, from *bis* + *torta* (past participle of *torquere*) = 'twice' + 'twisted', describing the twisted root. *Persicaria* is the medieval name derived from the similarity of the leaves to a peach tree.

Alpine bistort *P. vivipara*, local in the Highlands and Southern Uplands and northern England, scarce in north Wales and north-west and south-west Ireland, usually on base-rich substrates, less frequently in acidic sites, on wet rock, and in montane grassland and flushes. Reproduction is usually via bulbils at the base of the inflorescence, often transported by streams. *Vivipara* is Latin for 'bearing live young'.

Amphibious bistort *P. amphibia*, a common and widespread floating aquatic in lakes, ponds, canals, slow-flowing rivers and ditches, or a terrestrial plant of damp sites on streamsides and in marsh, wet meadow and dune slacks, and as a weed of cultivated land. *Amphibia* indicates it grows both in and out of water.

Common bistort *P. bistorta*, widespread, particularly common in north-west England, on base-poor soils in grassy habitats such as damp pasture, hay meadows and tall-herb communities in river valleys, and on riverbanks and roadsides. Many populations derive from garden escapes or throw-outs.

Red bistort *P. amplexicaulis*, a widespread Himalayan neophyte introduced in 1826, expanding its range, naturalised on roadsides, hedge banks and waste ground, invasive and persistent in gardens. Latin *amplexi* + *caulis* = 'clasping' + 'stem'.

BITHYNIA Freshwater snails in the genus *Bithynia* (Bythiniidae). Bithynia is a former country in Anatolia.

Common bithynia or **mud bithynia** *B. tentacula*, shell height 12–15 mm, width 5–7 mm, common and widespread, in lentic or slow-flowing calcium-rich water, grazing on algae on the substrate, and filtering suspended algae. *Tentaculum* is Latin for 'feeler'.

Leach's bithynia *B. leachii*, shell height 4–8 mm, width 3–7 mm, common and widespread in England, Wales, central Scotland, and central and south-east Ireland, found in clean, calcium-rich, slow-running and thickly weeded water. William Leach (1791–1836) was an English naturalist.

BITTER-CRESS Species of *Cardamine* (Brassicaceae). Plants are edible as a rather bitter herb, giving the English name. Greek *cardamon* refers to the spice.

Hairy bitter-cress *C. hirsuta*, a common and widespread winter-annual which sometimes flowers and seeds again in autumn, especially found as a weed of cultivation and in other ruderal habitats, but also growing on rock outcrops, by streams and in woodland. *Hirsuta* is Latin for 'hairy'.

Large bitter-cress *C. amara*, a perennial winter-green herb locally common, though absent from south-west England, west Wales, west Scotland and all but north-east Ireland, found on streamsides, marsh, wet meadow and damp woodland, often growing in slow-moving or still water, preferring acidic conditions, and tolerant of shade. *Amara* is Latin for 'bitter'.

Narrow-leaved bitter-cress *C. impatiens*, a locally abundant biennial in (especially ash) woodland, on limestone rocks and scree, by rivers and on damp roadsides, occasionally as a garden escape. It does not like competition, but is often invasive in recently disturbed habitats. *Impatiens*, Latin for 'impatient', refers to the seed pod bursting open when ripe.

Wavy bitter-cress *C. flexuosa*, a common and widespread winter- or summer-annual (rarely a short-lived perennial), most frequent in open, moist, shaded vegetation in marsh, by rivers and streams, and in gardens. It prefers mildly basic soil, and avoids strongly acidic soil. It commonly colonises disturbed, fertile habitats. Latin *flexuosa* can mean 'wavy'.

BITTER-VETCH *Lathyrus linifolius* (Fabaceae), a perennial herb, common and widespread except for the east Midlands, East Anglia and central Ireland, on moist, infertile neutral and acidic soils in dry meadows, lightly grazed pasture, grassy banks, scrub and open woodland, and on upland stream banks and rock ledges. The binomial is Greek for 'pea' or 'pulse', and Latin for 'flax-like leaves'.

Wood bitter-vetch *Vicia orobus* (Fabaceae), a perennial herb of often stony and fertilised grassy or scrub habitats, on banks and field edges, mostly in the western British Isles, and declining in status. The binomial comprises Latin and Greek for 'vetch'.

BITTERLING *Rhodeus amarus* (Cyprinidae), length generally no more than 5 cm in England, a fish introduced from central Europe in the early twentieth century, mainly found in Lancashire but also in Cambridgeshire since the 1970s, and around Llandrindod Wells in Wales, generally eating plant material, with some small animals also taken. Spawning is in clear, slow-running or still water, often with a muddy bottom, eggs being laid in the shell of a swan mussel for protection, which it uses as a nursery. Bitterling were once used in human pregnancy testing. The common name reflects the bitter taste. Greek *rhodeos* = 'rose' and Latin *amarus* = 'bitter'.

BITTERN *Botaurus stellaris* (Ardeidae), wingspan 130 cm, length 75 cm, male weight 1.5 kg, females 1 kg, a thickset heron found in reedbeds. The often polygamous males produce a distinctive booming vocalisation from late January into spring to establish territories and attract mates. Bitterns became absent as a breeding British species between the 1870s and 1911. Following a slow recovery, and having reached a peak of about 80 booming males in the 1950s, the species started to decline again shortly afterwards. With concern over a possible second UK extinction in the 1990s, a conservation programme involving appropriate habitat management, restoration and re-creation led to sizeable population increases, from a low of 11 males in 1997 (numbers halving from the previous year because of a cold winter) to 153 in 2015, 162 at 78 sites in 2016, and at least 164 birds at 71 sites in 2017. East Anglia had more than 80 booming males, and remains a stronghold not only on the Suffolk coast (e.g. at Minsmere) and in the Norfolk Broads, but also increasingly in a newly created habitat in the Fens. Following restoration and creation of appropriate wetlands, Somerset had 49 booming males in 2017. Only one male remained at Leighton Moss, Lancashire, from 2014 to 2016. Other locations were the RSPB Old Moor reserve in South Yorkshire and the Blacktoft Sands reserve on the Humber. It is on the Red List. Clutch size is 4–6, incubation (by the female only) takes 25–6 days, and the altricial young fledge in 50–5 days. Diet is of fish, amphibians and large insects. 'Bittern' comes from late Middle English *bitore*, ultimately (like the genus name) from Latin *butio* + *taurus* = 'bittern' + 'bull' (from

its call). Latin *stellaris* = 'starred', from *stella* = 'star', a reference to the speckled plumage.

Little bittern *Ixobrychus minutus* (Ardeidae), wingspan 55 cm, length 36 cm, weight 150 g, generally a scarce visitor, but there was a breeding success (three young raised) in South Yorkshire in 1984, and following the restoration of Avalon Marshes in Somerset, breeding has been attempted every year since 2010 at the Ham Wall RSPB reserve. A male was sighted at Lakenheath Fen, Suffolk in May 2016. *Ixobrychus* is from Greek *ixias*, a reedlike plant and *bruchomai* = 'to bellow'; *minutus* is Latin for 'small'.

BITTERSWEET *Solanum dulcamara* (Solanaceae), common and widespread except in upland Scotland, a scrambling, woody perennial growing in woodland, hedgerows, ditches, walls and, as var. *marinum*, on shingle beaches. It often grows in moist habitats and is common in swamps and tall-herb fens, and beside rivers and lakes, where it sometimes grows in shallow water. *Solanum* is Latin for 'nightshade', itself derived from the narcotic properties of some species. Latin *dulcis* + *amara* = 'sweet' + 'bitter'.

BIVALVE, BIVALVIA Class of freshwater and intertidal/sublittoral molluscs, with a shell that consists of two similar parts (valves), as left and right surfaces (in contrast to the upper and lower surfaces of brachiopods) joined along one edge by a flexible ligament that forms a hinge. They include mussels, clams, oysters, cockles and scallops. Most are filter feeders.

BLACK (MOTH) Waved black *Parascotia fuliginaria* (Erebidae, Boletobiinae), wingspan 18–28 mm, scarce, in damp woodland and suburban habitats in parts of the southern half of England. Adults fly at night from June to August. Larvae feed on bracket fungi and decomposing wood. Greek *para* = 'other than' + *Scotia*, another genus, and Latin *fuliginaria* = 'sooty'.

BLACK (SOLDIER FLY) Members of the family Stratiomyidae. Rarities are **pine black** *Zabrachia tenella*, with recent records only from Surrey and Co. Galway; **scarce black** *Eupachygaster tarsalis*, with modern records for only four localities in woodland and parkland in Berkshire (Windsor Forest), Cambridgeshire, Hampshire (New Forest), and Dorset; and **silver-strips black** *Neopachygaster meromelas*, mainly in southern and south-east England, but extending to Yorkshire.

Dark-winged black *Pachygaster atra*, common and widespread in England and (especially south) Wales, with a gradient in abundance from south-east to north-west, and also recorded from Ireland. Larvae are found under bark and in dead wood but have also been found in garden compost, decaying vegetation and leaf litter. Adults are found from June to August in fen, coastal landslips, woodland and old hedgerows. Greek *pachys* + *gaster* = 'thick' + 'stomach', and Latin *ater* = 'black'.

Yellow-legged black *Pachygaster leachii*, common and widespread in England (as far north as Yorkshire) and Wales. Adults frequent woodland, hedgerows and scrub, from June to August. Larvae occur in umbellifer roots and in decaying deciduous trees.

BLACK ARCHES 1) *Lymantria monacha* (Erebidae, Lymantriinae), wingspan 30–50 mm, a nocturnal moth local in (especially oak) woodland throughout England and Wales, commoner in the south. Population levels increased by 164% over 1968–2007. Adults fly in July–August. Larvae feed on oaks, occasionally becoming a defoliating pest. Greek *lymanter* = 'destroyer', and Latin *monacha* = 'nun', from the black and white wing colour.

2) Nocturnal moths in the family Nolidae, subfamily Nolinae. Nola is a town in Campania, Italy.

Kent black arches *Meganola albula*, wingspan 18–24 mm, scarce, on heathland, saltmarsh, shingle beaches, sandy beaches, chalk downland and woodland clearings in south-east England, predominantly coastal, but increasing in recent years inland in central-southern England and East Anglia. Adults fly in June–August. Larvae feed on raspberry, bramble and wild strawberry. *Albula* is Latin for 'whitish', the wing colour.

Least black arches *Nola confusalis*, wingspan 16–18 mm, widespread but local in woodland, hedgerows, parks, orchards and gardens. Population levels increased by 198% between 1968 and 2007. Adults fly in May–June. Larvae feed on lime, downy birch, blackthorn and other trees and shrubs. Latin *confusus* refers to the original uncertainty in this moth's classification.

Small black arches *Meganola strigula*, wingspan 18–24 mm, scarce, in mature oak woodland in the southern half of England. Adults fly in June–July. Larvae feed on pedunculate oak leaves. *Strigula* is Latin for 'small line', referring to a forewing marking.

BLACK-BANDED (MOTH) *Polymixis xanthomista* (Noctuidae, Xylininae), wingspan 38–44 mm, nocturnal, scarce, on rocky coastlines of the Scilly Isles, Cornwall, Devon, Pembrokeshire, Anglesey, Isle of Man and Co. Cork. Adults fly in August–September. Larvae feed on thrift flowers. Greek *polys* + *mixis* = 'many' + 'mixing', from the mixed colouring, and *xanthos* + *meistos* = 'yellow' + 'least', from the sparse yellow scales on the forewing.

BLACK-BINDWEED *Fallopia convolvulus* (Polygonaceae), an annual archaeophyte, a twining arable weed since the Neolithic, once the major contaminant of agricultural seed in Britain, remaining common and widespread in arable fields, gardens and waste ground, and on roadsides. Gabriello Fallopia (1523–62) was an Italian anatomist. *Convolvulus* is from Latin *convolvere* = 'to wrap'.

BLACK-EYE Coastal black-eye *Miktoniscus patiencei* (Trichoniscidae), up to 3 mm long, an uncommon woodlouse found along the south coasts of England and Ireland with outlying populations as far north as east Scotland, exclusively coastal, mainly in damp peaty soil in supralittoral habitats such as saltmarsh, thinly vegetated shingle banks and at the base of sea-cliffs.

BLACK-EYED SUSAN (LICHEN) *Bunodophoron melanocarpum* (Sphaerophoraceae), locally common in western Scotland, scattered in other upland areas, in humid ancient woodland on tree-trunks or on mossy, mainly siliceous rocks. Greek *bonnos* + *hodos* + *phoros* = 'hill' + 'way' + 'carrying', and *melas* + *carpos* = 'black' + 'fruit'.

BLACK-EYED-SUSAN (PLANT) *Rudbeckia hirta* (Asteraceae), a herbaceous North American perennial neophyte in cultivation by 1714, naturalised as a garden escape or originating from wool shoddy, in rough and waste ground, occasional in England and south Wales. The genus name honours the Swedish botanists Olof Rudbeck (1630–1702) and his son, also Olof (1660–1740). *Hirta* is Latin for 'hairy'.

BLACK-GRASS *Alopecurus myosuroides* (Poaceae), an annual archaeophyte, common (though declining) south of a line from the Humber to the Severn estuaries, scattered elsewhere, on waste ground, neglected grassland and arable fields, often a cereal weed. Greek *alopex* + *ouros* = 'fox' + 'tail' (this is one of the foxtail grasses), and *myosuroides* = 'resembling plants in the genus *Myosurus*'.

BLACK-JACK *Bidens pilosa* (Asteraceae), an annual South American neophyte, with a few records in England as a casual wool alien. Latin *bis* + *dens* = 'twice' + 'tooth' (referring to the bristles on the achenes), and *pilosa* = 'covered in long hair'.

BLACKBERRIES AND CUSTARD *Pyrenula hibernica* (Pyrenulaceae), a crustose lichen known at three sites in the western Highlands, from a few oceanic woodland sites in Co. Kerry, Co. Killarney and the Burren, and discovered in Wales in 2005 at Trawsfynydd, Meirionnydd. It is found on nutrient-rich bark of holly and hazel. Greek *pyren* = 'kernel', and Latin *hibernica* = 'Irish'.

BLACKBERRY The fruit of bramble, sometimes referring to the plant itself.

BLACKBIRD *Turdus merula* (Turdidae), wingspan 36 cm, length 24 cm, weight 100 g, a common and widespread member of the thrush family, found in a range of habitats, including

gardens, with 4.9 million pairs in Britain in 2009, a 22% increase in numbers between 1995 and 2015, following a decline over 1970–95, and with essentially no change in numbers over 2010–15. Perhaps 1.8 million pairs breed in Ireland. Some 10–15 million birds are likely in winter, numbers augmented by Scandinavian migrants. The cup-shaped nest in hedges, bushes, etc. usually has 3–4 eggs, incubation lasts 13–14 days, and the altricial young fledging in 14–16 days. Feeding is on earthworms, insects, berries and other fruit. The 'black bird' is the male; females are brown. The binomial combines Latin for 'thrush' and 'blackbird'.

BLACKCAP *Sylvia atricapilla* (Sylviidae), wingspan 22 cm, length 13 cm, weight 21 g, mainly an insectivorous summer migrant to woodland, parks and gardens, with 1.2 million territories in Britain in 2009. An increasing number of birds overwinter, though most winter birds are visitors from Central Europe rather than British breeders who have stayed on. Numbers increased by 288% over 1970–2015, 151% over 1995–2015, and 21% over 2010–15. The cup-shaped nest made of plant material and lined with hair is built in bramble, scrub or trees. Clutch size is 4–5, incubation lasts 13–14 days, and fledging of the altricial young takes 11–12 days. The English name describes the black 'cap' on the male's head; the female's cap is brown. Latin *Sylvia* = a woodland sprite, in turn from *silva* = 'wood', and *ater* + *capillus* = 'black' + 'hair'.

BLACKCOCK The (male) black grouse.

BLACKFLY Members of the family Simuliidae; also see black bean aphid under aphid.

BLACKHEAD Fungi in the genus *Diatrypella* (Xylariales, Diatrypaceae) (Greek *dia* + *trypa* = 'through' + 'hole'), **birch blackhead** *D. favacea* on the dead wood of birch and hazel, **oak blackhead** *D. quercina* on that of oak, both widespread in woodland.

BLACKNECK *Lygephila pastinum* (Erebidae, Toxocampinae), wingspan 37–42 mm, a common moth flying in daylight and from dusk (June–September), in open grassland, marsh, damp woodland rides and wood edges, widespread in parts of south England, south Wales, the south Midlands, Yorkshire and East Anglia. The main larval food is tufted vetch. Greek *lyge* + *phileo* = 'darkness' + 'to love' (from the crepuscular behaviour), and Latin *pastinum* = ground that has been prepared for planting, referencing 'furrow' markings on the forewings.

Scarce blackneck *L. craccae*, wingspan 40–6 mm, a rare RDB nocturnal moth of sea-cliffs and rocky coasts in north Cornwall, north Devon and Somerset. Adults fly in July–August. Larvae feed on vetches, including tufted vetch *Vicia cracca*, which gives the specific epithet.

BLACKTHORN *Prunus spinosa* (Rosaceae), a common and widespread deciduous shrub or small tree of open woodland, scrub, hedgerow, scree and cliff slope; a prostrate form also occurs on shingle beaches. It grows on a wide variety of soils. It is a common hedging plant because of its thick thorny growth habit. An alternative name is sloe. *Prunus* is Latin for plum tree, *spinosa* = 'spined'.

BLACKWOOD Australian blackwood *Acacia melanoxylon* (Fabaceae), an evergreen neophyte tree or shrub from Australia grown in gardens since 1808, naturalised on cliffs and coastal scrub in south Devon and in woods on Tresco, Scilly Isles. Greek *acis* = 'sharp point', and *melas* + *xylo* = 'black' + 'wood'.

BLACKWORT *Southbya nigrella* (Amelliaceae), a foliose liverwort known only from thin soil over limestone on Portland (Dorset) and the Isle of Wight. Anthony Southby (1799–1883) was an English botanist. *Nigrella* is from Latin *niger* = 'black'. **Green blackwort** *S. tophacea* is scarce in calcareous habitats in Cornwall, Wales and Ireland. Latin *tophaceus* = 'gritty'.

BLADDER-FERN Members of the genus *Cystopteris* (Woodsiaceae) (Greek *cysti* + *pteri* = 'bladder' + 'fern'). Sori (clusters of sporangia) are usually covered in an inflated bladder-like protective cover.

Brittle bladder-fern *C. fragilis*, on shaded rocks and basic woodland soils, and in wall mortar, common in northern and western Britain, scattered elsewhere, though locally frequent in Ireland. *Fragilis* is Latin for 'brittle'.

Diaphanous bladder-fern *C. diaphana*, discovered as native in east Cornwall in 2001, probably introduced elsewhere in south-west England and Co. Cork, on rocks, walls and banks on acidic soil.

Mountain bladder-fern *C. montana*, rare, on dripping, shaded, basic rocks and scree and in gullies in central Scotland.

BLADDER-MOSS Species of *Physcomitrium* (Funariaceae). Greek *physa* + *mitrion* = 'bladder' + 'little turban', referring to the urn-like calyptra that protects the sporophyte-containing capsule. **Norfolk bladder-moss** *P. eurystomum* is confined to two seasonal ponds in Breckland.

Common bladder-moss *P. pyriforme*, common and widespread, especially in England and Wales in arable fields and other recently disturbed soil, particularly in cattle-poached pastures and next to streams. *Pyriforme* is Latin for 'pear-shaped', from its capsule morphology.

Dwarf bladder-moss *P. sphaericum*, scattered in Britain mainly in the north Midlands, Lancashire and west Yorkshire on mud at the margins of water bodies. *Spaericum* is Latin for 'spherical', from the capsule shape.

BLADDER-SEDGE *Carex vesicaria* (Cyperaceae), a locally common and widespread perennial of mesotrophic or slightly basic habitats where the water table lies close to or above the soil surface, by lakes, ponds and streams, in marsh and swamp, ditches, wet meadow and pasture, and in wet woodland. Latin *carex* = 'sedge', *vesica* = 'bladder'.

BLADDER-SENNA *Colutea arborescens* (Fabaceae), a deciduous neophyte shrub introduced from Europe by Tudor times, naturalised extensively since 1900, spreading along railway embankments, then becoming more widespread in verges, waste ground and rough grassland, commonest in south-east England. *Colutea* is Greek for this genus, *arborescens* Latin for 'tree-like'.

BLADDER SNAIL Sinistral freshwater members of the pulmonate snail family Physidae.

Acute bladder snail *Physella acuta*, North American, shell height 8–16 mm, width 4–5 mm, introduced as an aquarium and pond species, escaped into slow-flowing rivers, streams, lakes, ponds and swamps, including stagnant conditions, feeding on dead plant and animal material and detritus. Widespread but local in Ireland, spreading rapidly since its discovery in the Ards Peninsula, Co. Down in 2000. *Physella* is a diminutive of Greek for 'bladder'.

Common bladder snail *Physa fontinalis*, shell height 7–12 mm, width 4–7 mm, widespread in streams, ditches, canals, ponds and lakes, usually on leaves or stems of water weeds. Greek *physa* = 'bladder', *fontinalis* Latin for 'from a spring'.

Moss bladder snail *Aplexa hypnorum*, shell length 9–13 mm, widespread but sporadic and declining, Vulnerable in the RDB, in temporary waters or the margins of seasonally flooded water bodies. Greek *a* + *plexis* = 'not' + 'stroke', and *hypnos* = 'sleep'.

Tadpole bladder snail *Physella gyrina*, North American, shell height 12–22 mm, width 3.5–16 mm, introduced as an aquarium and pond species, escaping into shallow lake waters, canals and similar habitats, spreading since the 1990s. In Ireland it is particularly common in Lough Neagh.

BLADDERNUT *Staphylea pinnata* (Staphyleaceae), a deciduous European neophyte tree introduced by the Tudor period, scarce and scattered as a garden escape in woodland, hedgerows and on waste ground, mostly in England. The seed pods are bladder-shaped, hence the English name. Greek *staphyle* =

'cluster', referring to the flower cluster, and Latin *pinnata* = 'feather-shaped'.

BLADDERSEED *Physospermum cornubiense* (Apiaceae), a rhizomatous herbaceous perennial of arable, light woodland, gorse scrub on heathland, rough grassy slopes and shaded roadside banks, very local and declining in east Cornwall and south Devon, and long naturalised in woodland at Burnham Beeches, Buckinghamshire. The English name reflects the inflated seed shape, as does the genus name from Greek *physa* + *sperma* = 'bladder' + 'seed'. *Cornubiense* is Latin for 'from Cornwall'.

BLADDERWORT Species of perennial insectivorous herbs in the genus *Utricularia* (Lentibulariaceae), trapping prey via their modified-leaf 'bladders'; *utriculus* is Latin for a small leather bottle. Specifically, *U. australis*, widespread and scattered, mainly found in acidic water – lakes, ponds, slow-flowing streams, ditches, canals, and swampy ground over mineral or peaty soil – but also occasionally in moderately calcareous sites. *Australis* is Latin for 'southern'.

Greater bladderwort *U. vulgaris*, scattered but locally common in south-west and eastern England and north Ireland, in still or slow-moving oligotrophic and mesotrophic, base-rich waters, including limestone lakes, ponds, ditches and pools in calcareous fen and marsh, and flooded clay-, marl- and gravel-pits. *Vulgaris* is Latin for 'common'.

Intermediate bladderwort *U. intermedia*, found in the Lake District, north and west Scotland, and north and west Ireland, especially in shallow, oligotrophic water in acidic and peaty sites, occasionally growing in calcareous waters.

Lesser bladderwort *U. minor*, widespread and scattered, commonest in south-west England, the New Forest, Wales, Scotland and Ireland in nutrient-poor, acidic, though sometimes base-rich, shallow water in bog pools and abandoned peat cuttings, on lake edges, in ditches and small ponds, and in fen.

New Forest bladderwort *U. bremii*, discovered in the 1990s (and still present in 2014) in a small pond and adjacent large puddle at Shatterford, Hampshire.

Nordic bladderwort *U. stygia*, in still or slow-moving acidic water in peaty bog in north-west Scotland and Co. Tyrone. *Stygia* is Greek for 'dark', from the River Styx in Greek mythology.

Pale bladderwort *U. ochroleuca*, in still or slow-moving acidic water in peaty bog, locally frequent in north-west Scotland and Co. Tyrone. *Ochroleuca* is Greek for 'yellowish-white'.

BLAEBERRY Scottish name for bilberry, from *blae* = 'blue-black'.

BLANKETFLOWER *Gaillardia* x *grandiflora* (Asteraceae), an annual or short-lived perennial of garden origin (North American parents), occasional in England, sometimes naturalised on coastal sand or shingle, and with a few sites on rough ground. Gaillard de Marentoneau was an eighteenth-century French patron of botany.

BLASTOBASIDAE Family name for dowd moths, all recent introductions from Madeira (Greek *blastos* + *basis* = 'shoot' + 'step', from the shape of the antenna).

BLEAK *Alburnus alburnus* (Cyprinidae), length 18–25 cm, a fish found in shoals in slow-flowing lowland rivers and in some lakes in lowland England and Wales, feeding on small molluscs, floating insects, insect larvae, worms, small shellfish and plant detritus. Spawning is April–June in shallow water over gravel. 'Bleak' dates from the late fifteenth century, from Old Norse *bleikja*. Latin *alburnus* = 'whitish'.

BLEEDING-HEART *Dicentra formosa* (Papaveraceae), a rhizomatous perennial herb introduced from North America by 1796, commonly grown in gardens but naturalised in shaded woodland and roadside habitats, especially by streams. It was first recorded in Ireland in 1995 in Co. Tyrone. Greek *dis* +

centron = 'twice' + 'spur', referring to the flower shape, and Latin *formosa* = 'beautiful'.

BLENNIIDAE Family name for most blennies.

BLENNY Intertidal and subtidal fishes in the families Blenniidae (see also shanny) and Tripterygiidae. Latin *blennius*, itself from Greek *blennos* = 'mucus', refers to a mucous coating.

Black-faced blenny *Tripterygion delaisi* (Tripterygiidae), length up to 6 cm, recorded from only three locations on the south English coast (Portland Harbour, Roundham Head, Plymouth Sound) in rock crevices, pier pilings and other habitats with reduced light in shallow coastal waters ranging 0–40 m depth, feeding on crustaceans. The genus and family names come from the Greek for 'three fins'.

Montagu's blenny *Coryphoblennius galerita* (Blenniidae), length up to 85 mm, found from Pembrokeshire to Dorset, and in south and south-west Ireland, in the rocky intertidal most often in pools with coralline algae, feeding on barnacles. George Montagu (1753–1815) was a British soldier and naturalist. Greek *corypho* + *blennos* = 'top' + 'mucus' (from the mucus coating), and Latin *galerita* = 'fur hat', referencing a fleshy cap on the forehead.

BLEWIT Fungi in the genus *Lepista* (Agaricales, Tricholomataceae). The English name may come from the blueish gills of some species, especially when young. *Lepista* is Latin for a goblet, describing the concave mature cap.

Field blewit *L. saeva*, locally common, edible and tasty, absent from the northern half of Scotland, found in autumn and winter on often calcareous soil in pasture and other grassy habitats, occasionally broadleaf woodland. *Saeva* is Latin for 'savage', here probably simply 'wild'.

Flowery blewit *L. irina*, uncommon, poorly edible, scattered mainly on calcareous soil in southern England in summer and autumn in deciduous woodland, especially under beech. *Irina* is Latin for 'pertaining to irises', here in relation to their scent, which also gives the English name.

Sordid blewit *L. sordida*, common and widespread, edible, in summer and autumn on soil in plant debris in hedgerows, scrub, parkland and gardens, including compost heaps. *Sordida* is Latin for 'dingy'.

Wood blewit *L. nuda*, common and widespread, edible and tasty, in autumn and winter, one of the few large agarics that continue to emerge after the first frosts, on soil and leaf litter in mixed woodland, hedgerows, parks and gardens, and also in dune slacks under dwarf willow. *Nuda* is Latin for 'naked'.

BLIGHT Various fungal diseases that cause rotting, wilting, browning or chlorosis of plant tissue.

Box blight Caused by the fungus species *Cylindrocladium buxicola* (the more damaging) and *Pseudonectria buxi*, often found together, causing box leaves to brown and fall, the former also causing dieback of young stems. Identified as a new species in 1998, *Cylindrocladium* is now widespread in the UK (and found in Ireland in 2001), not only in gardens but found at Box Hill, Surrey (where box is native) in 2000. It was second in the Royal Horticultural Society's list of most serious diseases in 2015. Greek *cylindros* + *clados* = 'cylinder' + 'branch'; Latin *buxicola* = 'box dweller'.

Potato blight A disease of potato foliage and tubers (and of tomato) associated with humid weather, caused by the pathogenic fungus *Phytophthora infestans* (Peronosporales), responsible for the Irish Famine that began in 1845. Greek *phyto* + *phthora* = 'plant' + 'destruction', and Latin *infestans* = 'infested'.

Quince leaf blight *Diplocarpon mespili* (Dermateaceae), causing defoliation. It may also affect other Rosaceae such as hawthorn and medlar. Greek *diploos* + *carpos* = 'double fruit', and Latin *mespilus* = 'medlar'.

BLINDIA Dwarf blindia *Blindia caespiticia* (Seligeriaceae), a rare and endangered moss on rock at a few Highland sites. *Caespiticia* derives from Latin for 'tufted'.

Sharp-leaved blindia *B. acuta*, common in upland regions on rock or gravel in or by streams and flushes. *Acuta* is Latin for 'sharp'.

BLINKS *Montia fontana* (Montiaceae), an annual or perennial, widespread and scattered, especially in western regions, and absent from much of central England and central Ireland, in acidic or neutral, wet habitats such as springs and flushes (often in moss-rich communities), the margins of lakes, streams, and damp tracks. 'Blinks' dates from the late seventeenth century, in the sense of 'a brief gleam of light', referring to the small white flowers. Giuseppe Monti (1682–1760) was an Italian botanist. *Fontana* is Latin here meaning growing in water.

BLISTER Willow blister *Cryptomyces maximus* (Rhytismatales, Rhytismataceae), a very rare fungus, found in only five sites in the world, four being in Wales, on live and dead wood in willow scrub, Vulnerable in the RDL (2006). Greek *cryptos* + *mycis* = 'hidden' + 'fungus', and Latin *maximus* = 'greatest'.

BLOOD-DROP-EMLETS *Mimulus luteus* (Phrymaceae), a creeping mat-forming perennial South American neophyte herb introduced in around 1826, scattered, uncommon and naturalised in wet habitats such as marsh, and on riversides and river shingle. Its common name reflects the red spots on the yellow flowers. The binomial is from Greek for 'little mimic' and Latin for 'yellow'.

BLOOD SPOT *Ophioparma ventosa* (Ophioparmaceae), a crustose lichen, mainly in upland Britain and north-west and south-east Ireland on siliceous rocks. Blood red discs prompt its English name. Greek *ophis* = 'snake' and Latin *parma* = 'shield', and Latin *ventosa* = 'puffed up' (literally 'full of wind').

BLOOD-VEIN *Timandra comae* (Geometridae, Sterrhinae), wingspan 23–8 mm, a nocturnal moth common in marsh, riverbanks, water meadows and gardens throughout England and Wales, local and scattered in Scotland and south and east Ireland. This is a species of conservation concern in the UK BAP. The bivoltine adults fly in May–July and August–September. Larvae feed on low-growing herbaceous plants such as docks, sorrels and knotgrass. The adult rests with wings held so that the reddish cross-lines of the fore and hind wings form a continuous band, giving the English name. In Ancient Greece Timandra was the daughter of Leda, sister of Clyemnestra. *Comae* comes from Latin for 'hairs'.

Small blood-vein *Scopula imitaria*, wingspan 26–9 mm, nocturnal, common in a range of habitats including gardens, waste ground, hedgerows and broadleaf woodland throughout England and Wales, especially in the south, and south and east Ireland. Adults fly in July–August. Larvae feed on woody plants such as privet and honeysuckle. Latin *scopula* = 'small broom' (from scale tufts on the tibia), and *imitor* = 'imitate' (from similarity with other species).

BLOODWORM Larvae of both bristle worms and non-biting midges.

BLOTCH MINER Alfalfa blotch miner *Agromyza frontella* (Agromyzidae), a scarce fly, with records from Essex, Kent, Surrey, Warwickshire, Cambridgeshire, Glamorgan and Pembrokeshire. Larvae mine leaves of lucerne and red clover with at least two generations between June and September.

BLOW FLY Metallic blue, green or black flies of the family Calliphoridae, the name coming from an old English term for meat that had eggs laid on it, which was said to be fly-blown.

BLOWN-GRASS *Agrostis avenacea* (Poaceae), an annual Australasian neophyte scattered in England as a casual on waste ground, roadsides and railway land, originating from wool shoddy. Greek *agros* = 'field', and Latin *avena* = 'oats'.

BLUE (BUTTERFLY) Members of the family Lycaenidae, subfamily Polyommatinae.

Adonis blue *Polyommatus bellargus*, wingspan 30–40 mm,

double-brooded, flying in May–July and August–September, in warm sheltered locations on chalk downland in southern England where its sole larval food plant, horseshoe vetch occurs. It is found where turf is closely cropped, possibly because it provides a higher temperature for the immature stages or because this is a requirement for the ant species that attend the larvae and pupae, offering them protection against parasites and predators. Ants are attracted from secretions exuded from the Newcomer's gland at the tail end of the larva. This butterfly has suffered from both loss of habitat and inappropriate habitat management, and is a species of conservation concern: over 1995–2005 distribution did increase by 30% and numbers by 33%, but data reviewing 2005–14 showed a decline in distribution by 12% and in numbers by 43%. In Greek mythology Adonis was a beautiful youth beloved of Aphrodite. Greek *polyommatos* = 'many-eyed', refers to the underwing eyespots. Latin *bellus* = 'beautiful' + Argus, who in Greek mythology had a hundred eyes, again referencing the eyespots.

Chalk hill blue *Po. coridon*, wingspan 33–40 mm, on chalk downland and some limestone grassland in southern England, especially on warm south-facing slopes, where the larval food plant, horseshoe vetch, may grow. Larvae are often found in the presence of ants, which may offer them protection against parasites and predators. Ants are attracted from secretions exuded from the Newcomer's gland at the tail end of the larva. Adults feed on the nectar of such plants as bird's-foot-trefoil, field scabious, thistles, thyme and selfheal. Males will also visit moist earth or animal droppings to gather salts and minerals. Flying peaks in August. There was a 50% decrease in distribution from 1976 to 2014 (though numbers actually increased by 20%), but over 2005–14 there was an increase in distribution by 5% and in abundance by 55%. Corydon was a shepherd named in Virgil's *Eclogues*.

Common blue *Po. icarus*, wingspan 29–36 mm, the most widespread lycaenid in the British Isles, absent only from mountainous Wales and Scotland, and some central parts of Ireland, found in a variety of grasslands, for example downland, coastal dunes, undercliffs, verges, acid grassland and woodland clearings, as well as urban wasteland and churchyards. Flight is in May–September, peaking in August. The primary larval and adult food plant is bird's-foot-trefoil, but a variety of other plants are also used. Larvae are often found in the presence of ants, which may offer them protection against parasites and predators. Ants are attracted from secretions exuded from the Newcomer's gland at the tail end of the larva. Despite a general decline in numbers and distribution (both by 17% between 1976 and 2014), this butterfly remains widespread and is not a species of conservation concern, with 2005–14 data showing a very slight increase (by 1%) in population size and range. In Greek mythology Icarus flew too close to the sun, causing the wax attaching his wings to melt, and he fell into the sea and drowned.

Holly blue *Celastrina argiolus*, wingspan 26–34 mm, common and widespread in England, Wales and Ireland, spreading northwards and with a few sites now in Scotland. It is found in many types of habitat, including gardens, churchyards, woodland, parks, and anywhere its food plants and nectar sources occur. Flower buds of holly (in spring) and ivy (in summer) are the main larval foods. Adults feed on honeydew and nectar from various plants. Males will also take salts and minerals from damp mud and animal waste. Adults fly in April–September, peaking at different times in different years because of variations in weather. Numbers therefore fluctuate from year to year, though range increased by 36% since the 1970s into the twenty-first century. However, data over 2005–14 show a population decline by 61%. Greek *celastra* = 'holly' (food plant), and *argiolus* is the diminutive of Argus, who in Greek mythology had a hundred eyes (referencing underwing eyespots).

Large blue *Maculinea arion*, male wingspan 38–48 mm, females 42–52 mm, the largest of Britain's blue butterflies, always

having been rare and becoming extinct in 1979 owing to habitat loss, subsequently reintroduced in 1984 using Swedish stock. A major factor in its rarity and vulnerability lies in its life cycle – a highly specific double mutualistic behaviour: larvae initially feed on the flower heads of wild thyme on limestone or coastal acidic grassland (though the first instar is also cannibalistic) but at the fourth instar stage the larva drops to the ground in the hope of being found by a red ant, in which case the ant taps the larva causing it to secrete a sugary droplet from its Newcomer's gland. After 30 minutes to 4 hours, the larva distorts its body by rearing up on its prolegs, to give the illusion of being an ant grub. The ant picks the larva up in its jaws and carries it back to its nest where it lives alongside the ant grubs which then form its diet. Survival is highest within nests of *Myrmica sabuleti*, much poorer using *M. scabrinodis*. Adults, flying in June–July, feed on honeydew together with nectar from carline thistle, bugle and thyme. Most reintroduction sites are in south-west England, notable ones being Green Down, the Polden Hills (Somerset), Dartmoor, and Daneway Banks (Gloucestershire). The number of adults flying in 2006 was around 10,000 on eleven sites, the largest number seen in the British Isles for over 60 years, and in 2016, 10,000 individuals were reported at just two of these sites. Nevertheless data for 2005–14 had shown a decline in numbers by 20%. Latin *macula* + *linea* = 'spot' + 'line', from the upper forewing markings. Arion was a Greek poet and musician of the seventh century BCE.

Silver-studded blue *Plebejus argus*, wingspan 26–32 mm, on heathland and to a lesser extent on calcareous grassland, particularly in south and south-west England but with near-coastal colonies elsewhere, especially in East Anglia and north Wales, and inland at Prees Heath, Shropshire. Larvae feed on heather, heaths, gorse, vetches and bird's-foot-trefoil. They are often found in the presence of ants, especially the black ant *Lasius niger*, which may offer the larva protection against parasites and predators. Ants are attracted from secretions exuded from the Newcomer's gland at the tail end of the larva. Adults feed on the nectar of heathers and heaths, particularly bell heather. Long-term declines in distribution (by 64% in 1976–2014) seem to have stabilised and there are a few signs that the situation is improving (an increase in range by 19% in 2005–14), but the species has suffered from increasingly fragmented habitat. It flies in June–August, but often moves less than 20 m in its life, making it difficult for it to colonise even nearby sites. This is therefore a priority species for conservation efforts, with numbers over 2005–14 declining by 9%. Its English name comes from the light blue scales found on the wing undersides, most evident when reflecting light. Latin *plebeius* = 'plebeian'. Argus in Greek mythology had a hundred eyes, here referencing the underwing eyespots.

Small blue *Cupido minimus*, wingspan 16–17 mm, Britain's smallest butterfly, mainly localised in south and central England, south Wales, north-east Scotland and parts of coastal Ireland, often confined to small patches of sheltered grassland where its sole larval food plant, kidney vetch, is found. Adults fly in May–September, consuming nectar from vetches and bird's-foot-trefoil. Distribution declined by 44% from 1976 to 2014, and it was last noted in Northern Ireland in 2001, so it is a priority species for conservation. Overall British numbers did increase by 31% over 1995–2005, but declined by 27% over 2005–14. Cupid was the Roman god of love. Latin *minimus* = 'smallest'.

BLUE (MAYFLY) Northern iron blue and **southern iron blue**: see mayfly.

BLUE-EYED-GRASS *Sisyrinchium bermudiana* (Iridaceae), a probably native herbaceous perennial, local in south-west Ireland in wet meadows, ditches and stony ground by lake shores. Greek *sys* + *rynchos* = 'pig' + 'snout', referring to the roots being eaten by swine.

American blue-eyed-grass *S. montanum*, a herbaceous North

American neophyte perennial naturalised in grassy habitats and rough ground, scattered in Britain as far north as central Scotland.

BLUE-EYED MARY *Omphalodes verna* (Boraginaceae), a widespread but scattered creeping perennial European neophyte in cultivation by 1633, now a garden escape naturalised in woodland and along lanes. Greek *omphalos* = 'navel' (referring to the shape of the fruit), and Latin *verna* = 'spring'.

BLUE-GREEN ALGAE Former name for cyanobacteria.

BLUE-SOWTHISTLE Herbaceous perennial species of *Cicerbita* (Asteraceae). 'Cicerbita' was possibly the medieval name for sowthistle.

Rarities are **alpine blue-sowthistle** *C. alpina* (four sites at Lochnagar, Aberdeenshire), **hairless blue-sowthistle** *C. plumieri* (Pyrenean, scattered in England and Scotland), and **Pontic blue-sowthistle** *C. bourgaei* (Near Eastern, occasionally naturalised in England and Wales on waste ground).

Common blue-sowthistle *C. macrophylla*, a neophyte introduced from the Urals in 1823, widespread and locally common in Britain, scattered in Ireland, a garden escape naturalised on roadsides and waste ground. *Macrophylla* is Greek for 'large-leaved'.

BLUEBELL *Hyacinthoides non-scripta* (Asparagaceae), a common and widespread, generally lowland, bulbous herbaceous perennial found, sometimes abundantly, in deciduous woodland and hedgerows and, especially in western and upland areas, in meadows, under bracken and on cliffs. It also occurs as a naturalised garden escape. It has suffered in recent years from collecting and, especially, from hybridisation with Spanish bluebell. The genus name is Greek for resembling *Hyacinthus*, this in turn referring to a name mentioned by Homer. *Non-scripta* is Latin for 'no marks'.

Italian bluebell *H. italica*, a bulbous south-west European neophyte, occasionally naturalised in southern and central England and central Scotland in neglected woodland.

Spanish bluebell *H. hispanica*, a bulbous south-west European neophyte, common and widespread in lowland regions, in woodland, hedgerows, and shady roadsides, and rough and waste ground. It is the commonest cultivated bluebell in gardens: discarded bulbs become naturalised and hybridisation (introgression) with the native bluebell occurs.

BLUEBERRY Highbush blueberry *Vaccinium corymbosum* (Ericaceae), a deciduous North American neophyte shrub present by 1760, naturalised as a bird-sown escape on heathland and in pine plantations in Dorset, Hampshire and a few other places. *Vaccinium* is Latin for bilberry, *corymbosum* Greek meaning full of corymbs (flat-topped flower heads).

BLUEBOTTLE Large blow flies in the family Calliphoridae with a dark metallic blue body. Adults feed on nectar or detritus, larvae on carrion. The commonest and most widespread are *Calliphora vicina* and *C. vomitoria* (both 10–14 mm length). Females lay up to 300 eggs on fresh carcasses or in open wounds. They are important in forensic pathology because they lay eggs on corpses at a consistent time after death, age of larvae helping to determine time of death. Greek *callos* + *phorein* = 'beautiful' + 'carry', and Latin *vicina* = 'neighbour'.

BLUET (INSECT) Alternative name for some damselflies in the genera *Coenagrion* and *Enallagma*. **Azure bluet**: see azure damselfly under damselfly. **Spearhead bluet**: see northern damselfly under damselfly. **Variable bluet**: see variable damselfly under damselfly. 'Bluet' dates from the eighteenth century, a diminutive of French *bleu* = 'blue'.

BLUET (PLANT) Some members of the genus *Centaurea* or some genera in the Rubiaceae. (Note spelling: not to be confused with blewet fungi.)

BLUETHROAT *Luscinia svecica* (Turdidae), wingspan 22 cm, length 14 cm, weight 20 g, a scarce breeder (1995 and 1996, plus an unpaired male in Norfolk over summer 2010) and passage migrant to and from Europe and North Africa, with 85–600 birds (2013 was a peak year) seen along the south and east coast in spring and autumn in scrub and grassy habitats. The English name describes the bird's blue throat, with a white or chestnut central throat spot depending on subspecies. The binomial is Latin for 'nightingale' and 'Swedish'.

BLUSH (MOTH) See maiden's blush.

BLUSHER *Amanita rubescens* (Agaricales, Amanitaceae), a widespread, common fungus, poisonous when raw, edible if well boiled, found in summer and autumn on acidic soil in (particularly coniferous) woodland. The English and specific names (Latin for reddening) refer to the colour change from white to pink-red when cut or damaged flesh of the cap or stem is exposed to air. Greek *amanitae* = 'mushrooms'.

BOAR See wild boar.

BOAT BUG *Enoplops scapha* (Coreidae), length 11–12 mm, a squash bug local on dunes and dry open cliff sites from north Wales to Kent, feeding on scentless mayweed and other composites. Greek *enoplos* = 'armed', and *scaphe* = 'boat'.

BOATMAN See water boatman.

BOG-LAUREL *Kalmia polifolia* (Ericaceae), an evergreen North American neophyte shrub cultivated since 1767, naturalised on Chobham Common, Surrey, since at least 1910, and a garden escape or relic on damp, acidic waste ground and moorland at a few sites in northern England and Scotland. Pehr Kalm (1716–79) was a Swedish traveller and naturalist. Greek *polios* + Latin *folia* = 'grey' + 'leaves'.

BOG-MOSS Species in the genus *Sphagnum* (Sphagnaceae), from Greek *sphagnos*, a type of moss (used as an astringent).

Rarities include **Baltic bog-moss** *S. balticum* (mid-Wales through northern England to the Highlands), **cleft bog-moss** *S. riparium* (mainly Highlands), **olive bog-moss** *S. majus* (mostly northern Scotland with one remaining site in Northumberland), and **Skye bog-moss** *S. skyense* (Hebrides, and west and south Ireland).

Austin's bog-moss *S. austinii*, recorded from Dartmoor, north Wales, northern England, Scotland and Ireland (though not in the south), on undisturbed raised and blanket bog. Coe Austin (1831–80) was an American botanist.

Blunt-leaved bog-moss *S. palustre*, common and widespread, less so in the east and south Midlands, shade-tolerant, forming large mats or hummocks on sites moderately enriched with nutrients, for example wet woodland, ditches, stream margins and flushes. Latin *palustre* = 'marsh-loving'.

Blushing bog-moss *S. molle*, widespread but scattered, most frequent in north-west Scotland and the Hebrides, on wet heathland, boggy grassland and peaty streamsides. *Mollis* is Latin for 'soft'.

Compact bog-moss *S. compactum*, widespread except for parts of the Midlands and East Anglia, particularly common in Wales, the Highlands and west Ireland, in wet heathland and other acidic, sometimes disturbed, wet sites. A poor competitor, it is usually found in open vegetation with some bare ground.

Cow-horn bog-moss *S. denticulatum*, common and widespread except for parts of the Midlands and East Anglia, in boggy pools, and acid flushes, springs and ditches usually in nutrient-poor habitats, but occasionally where habitats are moderately nutrient-enriched. *Denticulatum* is Latin for 'fine-toothed'.

Feathery bog-moss *S. cuspidatum*, widespread except for much of the Midlands and East Anglia, the most aquatic British species of *Sphagnum*, in very acidic pools and depressions in bogs, and ditches on moorland. *Cuspidatum* is Latin for 'sharp-pointed'.

Fine bog-moss *S. angustifolium*, widespread if scattered in mineral-rich sites, generally quite base-demanding. *Angustifolium* is Latin for 'narrow-leaved'.

Five-ranked bog-moss *S. quinquefarium*, locally common

and widespread in south-west and northern England, Wales, Scotland and (non-central) Ireland on well-drained ground in heavy rainfall areas, but absent from bogs and flushes.

Flat-leaved bog-moss *S. platyphyllum*, scattered, mostly in Wales, the Lake District and the Highlands, requiring base-rich conditions in flushes, fens and pools. *Platyphyllum* is Greek for 'broad-leaved'.

Flat-topped bog-moss *S. fallax*, common and widespread except for parts of the Midlands and East Anglia, in permanently damp habitats, including nutrient-poor fen, and pools and runnels on bogs. *Fallax* is Latin for 'false'.

Flexuous bog-moss *S. flexuosum*, widespread if scattered, locally common in wet heathland in Dorset, the New Forest and Surrey, and in Wales, and in slightly mineral-rich sites such as poor fen and wet woodland.

Fringed bog-moss *S. fimbriatum*, common and widespread except in parts of the south Midlands, more scattered in Ireland, shade-tolerant, in damp sites moderately enriched with nutrients, especially among birch, willow and purple moor-grass. It is also found in more open wet sites such as stream sides, ditches, lake margins, fen and rush bog. *Fimbriatum* is Latin for 'fringed'.

Girgensohn's bog-moss *S. girgensohnii*, found in Wales, north England and Scotland, scattered elsewhere in England and in Ireland, a shade-demanding species at lower altitudes, in damp woodland, wet heathland, and among rushes in marshes, and also found in marginal parts of slightly base-enriched mires. Gustav Girgensohn (1786–1872) was an American botanist.

Golden bog-moss *S. pulchrum*, scarce, scattered, commonest in the Dorset heaths, in pools and hollows in bogs and valley mires, and in acidic flushes. *Pulcher* is Latin for 'beautiful'.

Imbricate bog-moss *S. affine*, locally common in upland north Wales, Lake District, south-west Scotland, Highlands and south-west and west Ireland, in fen, marsh, bog, base-rich seepages and ditches. *Affine* is Latin for 'similar'.

Lesser cow-horn bog-moss *S. inundatum*, common and widespread except in parts of the Midlands and East Anglia, in slightly base-rich habitats such as fen, swamp, pools, ditches and at the edges of flushes. *Inundatum* is Latin for 'flooded'.

Lindberg's bog-moss *S. lindbergi*, locally common in nutrient-poor flushes and springs in the Highlands, and at low altitude in and around pools on blanket bogs in Sutherland, and in boggy depressions in Shetland. Sextus Lindberg (1835–89) was a Swedish physician and botanist.

Lustrous bog-moss *S. subnitens*, common and widespread except for parts of the Midlands and eastern England, in boggy grassland, marsh, fen, flushes, ditches, wet woodland, damp rocky banks under heather, and in northern and western regions also on blanket bog. *Subnitens* is Latin for 'slightly shiny'.

Magellanic bog-moss *S. magellanicum*, widespread except for the Midlands, East Anglia and south-east Ireland, forming mats and low cushions on raised and blanket bogs, and in valley mires. Where hummocks and hollows are well-developed it tends to occupy a position intermediate between the top and base.

Pale bog-moss *S. strictum*, mainly found in north-western Scotland, including the Hebrides, and north, west and south-west Ireland in areas with high rainfall, often among purple moor-grass in wet heathland, boggy grassland, shallow blanket bog, and on humid, rocky slopes. *Strictum* is Latin for 'erect'.

Papillose bog-moss *S. papillosum*, widespread and fairly frequent though less common in the Midlands and eastern England, forming mats and hummocks on raised and blanket bogs, in valley mires and beside flushes.

Recurved bog-moss *S. recurvum* var *mucronatum*, scattered, mainly in parts of southern England, Wales, the north Midlands, north-west England and north-east Ireland, locally common in a range of damp nutrient-poor habitats.

Red bog-moss *S. capillifolium*, common and widespread except for the Midlands and peripheral regions, on bogs and heathland, in wet woodland, on well-drained mineral soils

and shallow peat in humid sites. *Capillifolium* is Latin for 'hairy-leaved'.

Rigid bog-moss *S. teres*, scattered, mostly in Wales, northern England and Scotland, in or bordering moderately base-rich flushes. *Teres* is Latin for 'cylindrical'.

Russow's bog-moss *S. russowii*, locally common in Wales, northern England and Scotland, scattered in Ireland, on sites moderately nutrient-enriched, often in wooded mires, flushed grassy and rocky banks, moorland, woodland and occasionally bogs. The name honours the Estonian botanist E. Russow.

Rusty bog-moss *S. fuscum*, found in northern England, the Southern Uplands and Highlands, and much of Ireland, on undisturbed parts of raised or blanket bogs, and on peaty slopes >500 m altitude, occasionally as hummocks within flushes. *Fuscum* is Latin for 'brown'.

Slender cow-horn bog-moss *S. subsecundum*, widespread in south and west Britain, scarce in Ireland, in moderately nutrient-enriched flushes and fens, ditches and swamps. *Subsecundum* is Latin for 'less secund' (arranged on one side only).

Soft bog-moss *S. tenellum*, common and widespread except for much of the Midlands and East Anglia, on acidic, bare, peaty ground, especially in hollows on bogs, and also on wet heathland and on well-drained, but humid, heathery slopes in areas with high rainfall. *Tenellum* is Latin for 'tender'.

Spiky bog-moss *S. squarrosum*, common and widespread, scattered in the south and east Midlands and in Ireland, in nutrient-enriched swampy ground such as wet woodland, or among sedges, rushes and purple moor-grass, for instance on stream banks and in moorland flushes. *Squarrosum* is Latin for having scale-like overlapping leaves or bracts.

Twisted bog-moss *S. contortum*, found in upland Britain, but also southern England and north Norfolk, and scattered in Ireland, in fens and flushes in base-rich conditions.

Warnstorf's bog-moss *S. warnstorfii*, widespread in Wales, northern England and Scotland, local and scattered in Ireland, in base-rich flushes. Carl Warnstorf (1837–1921) was a German bryologist.

BOG-MYRTLE *Myrica gale* (Myricaceae), a generally widespread shrub though common only in western regions and locally in southern England, and absent from much of central and eastern England. It suckers to form dense thickets, growing in organic soils with moving groundwater in base-poor bog and moorland, lowland raised bog, wet heaths and acid carr, tolerating light shade. Some lowland populations have been lost to peat extraction and reclamation; others have disappeared as woodland develops on drained bogs. *Myrica* may come from Greek *myrtos*, in turn from ancient Semitic (Accadian) *murru* = 'bitter'. *Gale* probably comes from the Old English name for this shrub, 'gagel'.

BOG-ROSEMARY *Andromeda polifolia* (Ericaceae), a dwarf shrub, locally common in the northern half of England, Wales, south and central Scotland, and central and west Ireland, on moist to wet acidic peaty ground, most abundant in lowland raised bogs but with scattered records from upland peatlands. This is the county flower of Cardiganshire, Kirkcudbright and Co. Tyrone as voted by the public in a 2002 survey by Plantlife. In Greek mythology Andromeda was chained to a rock, saved from a sea monster by Perseus. Greek *polios* + Latin *folia* = 'grey' + 'leaves'.

BOG-RUSH Perennial species of *Schoenus* (Cyperaceae) (Greek *schoinos* = 'rush').

Black bog-rush *S. nigricans*, locally common along the west and south coasts of Britain, East Anglia, and much of Ireland, on fen, marsh, bog, serpentine heathland, dune slacks, sea-cliff seepages and upper saltmarsh. In west Ireland, it is also frequent on acid blanket bog. *Nigricans* is Latin for 'black'.

Brown bog-rush *S. ferrugineus*, known from four sites (including two as transplants) in highland Perthshire in calcareous grassland seepages. *Ferrugineus* is Latin for 'rust-coloured'.

BOG-SEDGE Rhizomatous perennials in the genus *Carex* (Cyperaceae), Latin for 'sedge'. Specifically, *C. limosa*, is found in wet blanket bog and valley mires and peaty pool edges, often associated with sphagnum moss and growing in standing water. Most sites are acidic and oligotrophic, often where there is also some mineral enrichment. Latin *limosa* = 'marshy'.

Mountain bog-sedge *C. rariflora*, found at a few sites in the Highlands on wet, base-poor peaty substrates, mainly in flush bogs on gentle slopes, often in areas of late snow-lie. *Rariflora* is Latin for 'scattered flowers'.

Tall bog-sedge *C. magellanica*, scarce, scattered in parts of upland Wales, north England, Scotland and north-east Ireland in wet habitats, pools and hummocks in sphagnum bogs, or on the margins of gently sloping mires where there is slight lateral water movement.

BOGBEAN *Menyanthes trifoliata* (Menyanthaceae), a locally common and widespread rhizomatous perennial emergent at the shallow edge of lakes, pools and slow-flowing rivers, and in swamp, flushes and dune slacks, tolerating a wide range of water chemistry, but intolerant of shade. This is the county flower of Renfrewshire as voted by the public in a 2002 survey by Plantlife. *Menyanthes* is Greek for 'moon flower', *trifoliata* Latin for 'three-leaved'.

BOLETE Fungi in the order Boletales, many in the genera *Boletus* (Greek *bolos* = 'lump of clay') and *Leccinum* (an old Italian word meaning fungus) (Boletaceae) and *Suillus* (Latin for 'porcine', a reference to the greasy nature of the caps) (Suillaceae). Boletes have a fleshy cap (pileus) and stalk (stipe) fruit body with pores and soft tubes rather than gills under the cap.

Rarities include **bearded bolete** *S. bresadolae* var. *flavogriseus*, **chestnut bolete** *Gyroporus castaneus* (especially found in the New Forest), **constant bolete** *B. immutatus* (recently elevated to species rank, probably endemic, found at a single site in Berkshire), **cornflower bolete** *G. cyanescens*, **devil's bolete** *B. satanas* (lethally poisonous, particular recorded from Somerset), **foxy bolete** *L. vulpinum* (Highlands), **ghost bolete** *L. holopus*, **golden bolete** *Buchwaldoboletus sphaerocephalus*, **golden gilled bolete** *Phylloporus pelletieri*, **goldenthread bolete** *Xerocomus chrysonemus*, **hollow bolete** *S. flavidus* (Highlands), **inkstain bolete** *B. pulverulentulus*, **jellied bolete** *S. flavidus*, **matt bolete** *B. pruinatus*, **oak bolete** *B. appendiculatus*, **oldrose bolete** *B. rhodopurpureus*, **pine bolete** *B. pinophilus* (Highlands), **poplar bolete** *X. silwoodensis*, **royal bolete** *B. regius* (two accepted records from Hampshire), **sepia bolete** *B. porosporus*, **slate bolete** *L. duriusculum*, **slippery white bolete** *S. placidus* (five records from Kent), **velvet bolete** *S. variegatus*, **wood bolete** *Buchwaldoboletus lignicola* and **yellow bolete** *B. pseudosulphureus* (recorded from 13 locations between 1963 and 2012).

Alder bolete *Gyrodon lividus* (Paxillaceae), uncommon, edible, scattered in England, Wales and Ireland, Near Threatened in the RDL (2006), found in summer and early autumn in damp deciduous woodland and parkland under alder. Greek *gyros* + *donti* = 'turn' + 'tooth', and Latin *lividus* = 'lead-coloured'.

Bay bolete *B. badius*, edible and tasty, common and widespread, on soil in spruce and pine forests, occasionally under oak, beech and sweet chestnut, in late summer and autumn. *Badius* is Latin for 'chestnut'.

Bitter beech bolete *B. calopus*, inedible, widespread but only locally common in summer and early autumn on calcareous soil usually under broadleaf trees, especially beech. Greek *cala* + *pous* = 'beautiful' + 'foot', from the graduated yellow-to-red colouring of the stem.

Bitter bolete *Tylopilus felleus* (Boletaceae), widespread but uncommon, inedible, in August–September in deciduous woodland, ectomycorrhizal on oak, beech and other hardwoods. Greek *tylos* + *pilos* = 'bump' + 'hat' (referencing the swollen cap), and *felleus*, from Latin *fel* = 'bile' (a reference to the bitter taste).

Blackening bolete Alternative name for inkstain bolete.

Bovine bolete *S. bovinus*, common and widespread, poor tasting, found in August–November, usually ectomycorrhizal on Scots pine but also other pines and sometimes other conifers, often by woodland paths and in clearings. The English and specific names indicate a similar colour to a Jersey cow.

Bronze bolete *B. aereus*, edible and tasty, very scattered, most records coming from south England in late summer and early autumn on soil under broadleaf trees, especially beech and oak. *Aereus* is Latin for the colour bronze.

Brown birch bolete *L. scabrum*, common and widespread, edible, found in summer and autumn on woodland soil under birch, with which it is ectomycorrhizal. Latin *scabrum* = 'rough', describing the stem.

Crimson bolete *Rubinoboletus rubinus*, inedible, scattered and uncommon in England, found in summer and autumn on soil and grass in deciduous woodland under oak. Latin *rubino* and *rubinus* = 'ruby'.

Deceiving bolete *B. queletii*, inedible, widespread (mostly in southern England) but rare, found in late summer and early autumn on calcareous soil under broadleaf trees. Lucien Quélet (1832–99) was a French mycologist.

Dusky bolete *Porphyrellus porphyrosporus* (Boletaceae), widespread and scattered, inedible, found in July–September on soil in deciduous woodland. The English name and binomial refer to the dark purplish-brown colours.

False yellow bolete A variety (var. *discolor*) of scarletina bolete, often found under oak.

Gilded bolete *Aureoboletus gentilis* (Boletaceae), uncommon, scattered in southern England on soil and leaf litter in deciduous woodland, mainly associated with oak, but also other hardwoods. *Aureus* is Latin for 'golden', *gentilis* for 'belonging to a tribe'.

Hazel bolete *L. pseudoscabrum*, edible, scattered, commonest in south-east England in summer and autumn in woodland soil and below hedges, under hornbeam, with which it is ectomycorrhizal, and less commonly hazel. Latin *pseudo* + *scaber* = 'false' + 'rough', from the rough scaly surface of the stem.

Iodine bolete *B. impolitus*, edible, more widespread in southern England, local and scattered elsewhere and generally infrequent or rare, found in summer and early autumn on soil under broadleaf trees, especially oak, and often on mown grass. *Impolitus* is Latin for 'rough'.

King bolete Alternative name for cep.

Larch bolete *S. grevillei*, common and widespread, edible, found in July–October in grassland, ectomycorrhizal on larch. Robert Greville (1794–1866) was a Scottish mycologist.

Lurid bolete *B. luridus*, edible when cooked, widespread but not especially common, found in summer and autumn on often calcareous soil under broadleaf trees, especially beech, but also oak and occasionally lime, and also on open grassland. In the Burren, it has been noted growing with the lime-loving mountain avens with which it may form a mycorrhizal association.

Mottled bolete *L. variicolor*, widespread but uncommon, edible, found in summer and autumn on damp mossy soil in woodland, under birch, with which it is ectomycorrhizal. Latin *variicolor* and the English name refer to the mottled and variable cap colour.

Orange birch bolete *L. versipelle*, edible, common and widespread in Britain, rare in Ireland, found in summer and autumn on acid soil in woodland edge, scrub and heathland, under birch, with which it is ectomycorrhizal. *Versipelle* refers to the changing nature of the cap surface, from Latin *vertere* + *pellis* = 'turn' + 'skin'.

Orange oak bolete or **orange bolete** *L. aurantiacum*, scattered and locally common, edible, found in summer and autumn on soil, often under poplars, including aspen, with which it is ectomycorrhizal, and also other hardwoods such as oak and birch. *Aurantiacum* is Latin for orange, the colour of the cap.

Pale bolete *B. fechtneri*, edible, occasional in southern England, Cumbria and south-east Wales, found in late summer and autumn on chalky soil beneath beech, oak and sweet chestnut, rarely under conifers (usually spruce). František Fechtner (1883–1967) was a Czech mycologist.

Parasitic bolete *Pseudoboletus parasiticus*, inedible, widespread but uncommon in England and Wales, rare in Scotland and Ireland, found in August–October, only in association with common earthball, most likely on calcareous soil.

Peppery bolete *Chalciporus piperatus*, peppery-tasting but edible when well cooked, widespread and locally common in late summer and early autumn on soil under pine forest and broadleaf woodland, often under birch. Greek *chalcos* + *porous* = 'copper' + 'pores', and Latin *piperatus* = 'peppery'.

Red cracking bolete *B. chrysenteron*, effectively inedible, probably common and widespread (there has been confusion with identification), found in summer and autumn, usually beneath conifers, though under beech has also been recorded. Greek *chrysos* + *enteron* = 'gold' + 'intestine'.

Rooting bolete *B. radicans*, inedible, widespread (mostly in southern England) and fairly common in summer and autumn, mainly on alkaline or neutral soil, often on verges and beneath oak, occasionally beech. It also forms mycorrhizal relationships with rock-rose. *Radix* is Latin for 'root'.

Ruby bolete *B. rubellus*, infrequent, with a scattered distribution mainly in the southern half of Britain, on soil on woodland edges, in clearings and in low-nutrient grassland in parks and gardens beneath deciduous trees, especially oaks. *Rubellus* is Latin for 'ruby-red'.

Saffron bolete *L. crocipodium*, edible, with a scattered distribution in England and Wales, commoner in the south and south-east, found in summer and autumn on soil in woodland clearings and edges and beside woodland paths, under oak, sometimes hornbeam, with which it is ectomycorrhizal. Greek *crocos* + *podi* = 'crocus' + 'foot'.

Satan's bolete Alternative name for devil's bolete.

Scarletina bolete *B. luridiformis*, edible after cooking, widespread and common, found in late summer and early autumn on acid soil under conifers, especially spruce, occasionally broadleaf trees such as beech and (var. *discolor* = false yellow bolete) oak. *Luridus* is Latin for 'pale'.

Sticky bolete *S. viscidus*, widespread but generally uncommon, of no culinary value, found from August to October in coniferous plantations and parkland, ectomycorrhizal on larch.

Suede bolete *B. subtomentosus*, edible but tasteless, widespread but infrequent, found in late summer and early autumn, mainly under broadleaf trees in woodland and parks, and on the edges of conifer plantations where there are birch and willow. Latin *subtomentosus* = 'quite downy', describing the appearance of the young caps, and reflecting the English name.

Summer (king) bolete *B. reticulatus*, widespread, scattered and infrequent, edible and tasty, evident in summer and autumn on soil under broadleaf trees, particularly beech and oak. Latin *reticulatus* refers to the reticulated (netlike) pattern on the stem.

Weeping bolete *S. granulatus*, untasty but edible, widespread if more common in England, south Wales and north-east Ireland, found in July–November in coniferous woodland, ectomycorrhizal on Scots pine. The pores of young specimens release milky droplets, giving its common name. Latin *granulatus* describes the granular surface of the upper part of the stem.

Yellow cracking bolete Alternative name for suede bolete.

BOMBYCILLIDAE Family name of waxwing (Greek *bombycos* = 'silk' and Latin *cilla* = 'tail').

BOMBYLIIDAE From Greek *bombilios* = 'a buzzing insect', family name of bee-flies and villas in the superfamily Asiloidea, with nine species in four genera. Adults are bee mimics (affording some protection from predators) and generally feed on nectar and pollen in spring. Females of some species coat their eggs in dust or sand, then flick these at solitary bee or wasp nests while

hovering. Emerging larvae seek out the burrows of their hosts, and develop as ectoparasites on the host.

BONNET Inedible fungi in the genera *Hemimycena* (with 20 species), *Mycenella* (6 species) and *Mycena* (146 species) (Agaricales, Mycenaceae), the genus names being variants of Greek *myces* = 'fungus'.

Common and widespread inedible species on stumps, litter and woodland debris are **angel's bonnet** *Mycena arcangeliana* (especially on beech and ash), **bark bonnet** *M. speirea*, **beechleaf bonnet** *M. capillaris* (beech), **blackedge bonnet** *M. pelianthina* (mostly in southern and south-west England and the Welsh Borders, on chalk and limestone under beech), **bleeding bonnet** *M. sanguinolenta*, **bulbous bonnet** *M. stylobates*, **clustered bonnet** (or **oak bonnet**) *M. inclinata* (mostly oak), **coldfoot bonnet** *M. amicta*, **common bonnet** *M. galericulata* (mostly oak), **dripping bonnet** *M. rorida*, **frosty bonnet** *M. adscendens*, **grooved bonnet** *M. polygramma* (mostly oak), **iodine bonnet** *M. filopes* (mostly conifers), **mealy bonnet** *M. cinerella* (mostly conifera), **nitrous bonnet** *M. leptocephala* (mostly conifers), **orange bonnet** *M. acicula*, **pinkedge bonnet** *M. capillaripes* (mostly conifers), **rush bonnet** *M. bulbosa*, and **yellowleg bonnet** *Mycena epipterygia* (conifers).

Rarities include **beautiful bonnet** *Mycena renati*, **cryptic bonnet** *M. picta*, **dewdrop bonnet** *Hemimycena tortuosa*, **fanvault bonnet** *H. mairei*, **golden edge bonnet** *M. aurantiomarginata*, **milky bonnet** *H. lactea*, **olive edge bonnet** *M. viridimarginata*, **pelargonium bonnet** *M. septentrionalis*, **reed bonnet** *M. belliae* (mainly in East Anglia), **rooting bonnet** *M. megaspora*, **scarlet bonnet** *M. adonis*, **sideshoot bonnet** *M. latifolia*, **violet bonnet** *M. urania* (in Caledonian pinewood) and **willow bonnet** *Mycenella salicina*.

Black milking bonnet See milking bonnet below.

Brownedge bonnet *M. olivaceomarginata*, saprobic, widespread and locally common in Britain, rare in Ireland, found in autumn in short grassland and roadsides. The specific name is Latin for 'olive (colour) margined'.

Burgundydrop bonnet *M. haematopus*, common and widespread, found in late summer and autumn on stumps and trunks of broadleaf trees, especially oak. Its English name comes from the blood-red coloured latex that oozes when cut, seen also in the specific name from Greek *haema* + *pous* = 'blood' + 'foot'.

Drab bonnet *M. aetites*, widespread, local in late summer and autumn in short grassland leaf litter. Latin *aetites* may refer to aetite, a geode stone used to promote childbirth; it is also called an eagle-stone, derived from Greek *aetos* = 'eagle' (found in the eagle's stomach or neck it was said to possess properties warding off premature birth and miscarriage).

Ferny bonnet *M. pterigena*, scattered and uncommon, on often dead stems of bracken. *Pterigena* is from Greek referring to 'fern' (*ptero* = 'feather').

Ivory bonnet *M. flavoalba*, common and widespread, saprobic, found from August to late November on mossy lawns, in dune slacks, and on grassy woodland edges and clearings. Latin *flava* + *alba* = 'yellow' + 'white', or 'ivory-coloured'.

Lilac bonnet *M. pura*, common and widespread, saprobic, containing a small amount of the poison muscarine, found in summer and autumn on woodland soil in leaf and needle litter, occasionally in grassland. Some caps are lilac, but they can also be whitish, yellow or blue. *Purum* is Latin for 'pure' or 'clean'.

Milking bonnet *M. galopus*, common and widespread, saprobic, found in summer and autumn trooping in leaf litter and in grass in woodland and hedgerows. Var. *candica* is entirely white, while the less common var. *nigra* has a blackish cap and stem. Some authorities give the latter full species status as *M. leucogala*. Its English name comes from the white latex that oozes from the cut stem, reflected in the specific epithet, from Greek *gala* + *pous* = 'milk' + 'foot'.

Pink bonnet *M. rosella*, uncommon, saprobic, scattered in

Britain and east Ireland, on needle litter in coniferous woodland. *Rosella* is Latin for 'pinkish'.

Purple edge bonnet *M. purpureofusca*, scarce in England, locally common in the Highlands, on dead wood in coniferous forest, usually under pine. Latin *purpureofusca* = 'purple' + 'brown'.

Rancid bonnet *M. olida*, widespread but only locally common in England, found on dead wood of broadleaf trees, especially beech. *Olida* is Latin for 'smelly'.

Red edge bonnet *M. rubromarginata*, scattered and uncommon, found in autumn in woodland litter and on dead wood. Latin *rubromarginata* = 'red-margined'.

Rosy bonnet *M. rosea*, toxic (contains muscarine), saprobic, widespread and locally common in south and west-central England, scattered elsewhere, found from August to November in leaf litter under deciduous trees, especially beech, occasionally on acidic soil in coniferous woodland. *Rosea* is Latin for 'pink' or 'rosy'.

Saffrondrop bonnet *M. crocata*, scattered in (especially southeast) England and south Wales in late summer and autumn on twigs and litter in broadleaf woodland, favouring beech. *Crocata* is Latin for 'saffron-coloured'.

Scotch bonnet See champignon. (Not to be confused with the chili pepper *Capsicum chinense*.)

Snapping bonnet *M. vitilis*, common and widespread, found in summer and autumn on branches and detached bark of broadleaf trees. When its stem is pulled lengthways it breaks with a sharp snapping sound, hence its name. However, Latin *vitilis* is from *vitis* = 'vine', possibly referring to the stem's lateral flexibility.

White milking bonnet See milking bonnet above.

BOOKLICE Indoor members of the order Psocoptera, small, wingless or short-winged, soft-bodied insects, 30 species having been recorded in the British Isles, feeding on starch and mould found on paper, wallpaper, etc.

BOOTLACE WORM *Lineus longissimus* (Lineidae), Britain's commonest nemertean, a ribbon worm found round the coast of Britain (scarce from Hampshire to Yorkshire) and north and north-east Ireland, on the lower shore, often in rock pools, coiled in knots beneath boulders and on muddy sand. In sublittoral areas, it occurs on muddy, sandy, stony or shelly substrates. It is usually 5–15 m in length but can be >30 m, and usually 5 mm in width. (A specimen found in Scotland recorded as 55 m long, and reported as Britain's longest animal, was probably stretched.)

BOOTLACES Dead man's bootlaces Alternative name for the seaweed mermaid's tresses.

BOOTLEG (FUNGUS) Golden bootleg *Phaeolepiota aurea* (Agaricaceae), widespread, scattered and infrequently found, saprobic, containing hydrogen cyanide which can cause stomach upset, found from September to November beneath deciduous and coniferous trees on disturbed ground, often associated with nettles. Greek *phae-* + *lepis* = 'dusky' + 'scales', and Latin *aurea* = 'golden'.

BORAGE *Borago officinalis* (Boraginaceae), an annual European neophyte cultivated since at least 1200 as a herb and as an oil-seed crop, now widespread and scattered, more frequent in England and south-east Ireland, as a casual garden escape on roadsides and waste ground. It also originates from birdseed and as a relic of cultivation. It is increasingly being planted as a nectar source for bees. Also known as starflower: starflower oil contains gamma linolenic acid, an anti-inflammatory fatty acid helping to regulate the body's immune system and improving symptoms of rheumatoid arthritis, premenstrual tension, etc. 'Borage' dates from Middle English, from Old French *bourrache*, from medieval Latin *borrago*, perhaps from Arabic *abū 'arak* ('father of sweat', from its use in driving out fever) or *abū ḥurāš* (= 'father of roughness', referring to the leaves). Latin *officinalis* indicates pharmacological use.

Slender borage *B. pygmaea*, a herbaceous Mediterranean neophyte perennial, naturalised on heath in the Channel Islands, and a casual on rough ground in a few places in southern England and south Wales. *Pygmaea* is Latin for 'dwarf'.

BORDER (MOTH) Nocturnal moths in the family Geometridae.

Clouded border *Lomaspilis marginata* (Ennominae). wingspan 30–8 mm, widespread and commonly found in woodland, marshes, riverbanks and heathland. Adults fly from May to July. Larvae feed on sallows, hazel, aspen and other poplars. Greek *loma* + *spilos* = 'hem' + 'spot', and Latin *marginata* = 'margined'.

Dotted border *Agriopis marginaria* (Ennominae), wingspan 27–32 mm, widespread and commonly found in woodland, scrub, hedgerows, gardens, heathland and moorland. Adult males fly from February to April, when the flightless females climb up tree-trunks. Larvae feed on a number of broadleaf tree species, including fruit trees such as apple and plum. Greek *agrios* + *ops* = 'wild' + 'face' (from rough scales on the face), and Latin *marginis* = 'border', from the spotted forewing border.

Lace border *Scopula ornata* (Sterrhinae), wingspan 21–4 mm, day- and night-flying, with two generations flying in May–June then from late July to September. It has a declining population, now only found locally in south-east England, habitats including North Downs grassland and railway embankments. Larvae feed on thyme, marjoram and other herbs. Latin *scopula* = 'small broom', from tufts on the tibia of the males of some species, and *ornata* = 'ornate'.

BORER (BIVALVE) Mainly wood-boring shipworms (Teredinidae), but also rock-boring taxa.

Wrinkled rock borer *Hiatella arctica* (Hiatellidae), shell size up to 4 cm, common and widespread in crevices on hard substrates, among horse mussel beds and in kelp holdfasts, from the low intertidal to the shelf edge.

BORER (BEETLE), BORING BEETLE Mainly members of the families Ceramycidae and Ptinidae.

Clover root borer Alternative name for gorse bark beetle.

Ivy boring beetle *Ochina ptinoides* (Ptinidae), scattered but locally common in England and Wales, larvae boring into ivy stems.

Large poplar borer Longhorn beetle; see poplar borer below.

Maple wood-boring beetle *Gastrallus immarginatus* (Ptinidae), found very locally in southern England. Greek *gaster* = 'stomach', and Latin *immarginatus* = 'unmargined'.

Oak pinhole borer *Platypus cylindrus* (Platypodidae), 6–8 mm length, the only representative of the family in the British list, though only recently separated from the Curculionidae, and one of the 'ambrosia beetles', so named because they introduce ambrosia fungi to a dead or decaying host oak tree, cultivating fungal gardens which are the beetle's sole source of nutrition (they are wood borers, not wood feeders). It was a rarity until the Great Storm of 1987, when the large amount of extra dead oak that became available to the beetle saw it become abundant and widespread in the southern half of England and in Wales, a situation now promoted by oak dieback (see acute oak decline) and its spread into timber yards. Greek *platys* + *pous* = 'broad' + 'foot'.

Old house borer *Hylotrupes bajulus* (Cerambycidae), a scarce longhorn beetle, adults 15–25 mm in length, larvae (up to 30 mm long) found in softwood, particularly roofing timbers. Formerly a more widespread pest in houses, hence its English name, at present it is found only in England south-west of London (mainly in Surrey) where building regulations protect structural timbers and prevent further spread. Small inactive infestations are common in buildings over 100 years old in London. Isolated infestations in other parts of England usually come from imported infested packing cases. With large infestations, larval feeding may be audible as a scraping noise. Greek *hylotrupes* = 'house-' or 'wood-borer', Latin *bajulus* = 'carrier'.

Pine-stump borer *Asemum striatum* (Cerambycidae), a

longhorn beetle 8–23 mm length, with a scattered distribution in Britain, larvae feeding in dead wood and stumps of pine. *Striatus* is Latin for 'grooved'.

Pinhole borer See oak pinhole borer above.

Poplar borer *Saperda carcharias* (Cerambycidae), a longhorn beetle with a scattered distribution in (especially eastern) England and north-east Scotland. Larvae live and feed in dead wood, almost exclusively of poplar species. Greek *carcharos* = 'point' or 'shark'.

Shothole borer Alternative name for large and small fruit bark beetles.

Small poplar borer *Saperda populnea* (Cerambycidae), a longhorn beetle with a widespread distribution in England and south Wales. Larvae feed on aspen and other poplars, becoming a potential forestry pest by attacking the trunks of even young trees, creating a spindle-shaped gall. *Populnea* is Latin for 'relating to poplar'.

Tanbark borer *Phymatodes testaceus* (Cerambycidae), a longhorn beetle, 8–13 mm length, with a widespread but local distribution in England (scarcer north of the Midlands) and Wales, nocturnally active in open woodland mainly from May to July. Larvae develop under the bark of standing dead timber or recently cut trunks, particularly of oak. Latin *testa* = 'shell'.

Two-spot wood borer Alternative name for oak jewel beetle.

Wharf borer *Nacerdes melanura* (Oedemeridae), 9–13 mm length, a false blister beetle found around the coast in England and Wales, and occasionally inland in the Midlands, for example by Birmingham canals. Larvae bore into timber wharf pilings, harbour and dock timbers and into wood generally around the high tide mark. They also bore into ship timbers: 2% of stored waterlogged timbers from the Mary Rose were infested by these larvae. Adults feed on nectar and pollen from a variety of flowers from April to October. Greek *melas* + *ura* = 'black' + 'tail'.

Willow borer Alternative name for poplar borer.

BORER (MOTH) See also corn-borer.

Currant shoot-borer *Lampronia capitella* (Prodoxidae), wingspan 14–17 mm, rare, local in woodland and commercial fruit habitat in parts of south and south-east England, the west Midlands, coastal East Anglia, a few sites in Wales, and the Cairngorms. Adults fly in May–June. Larvae initially feed in the fruit of currants or gooseberry, often becoming a serious pest. They then overwinter, and in spring burrow into a new bud, a behaviour which gives it its English name. It is a proposed future RDB species and a priority species under the UK BAP. Greek *lampros* = 'bright' (referring to the wing colours), and Latin *caput* = 'head'.

Peach twig borer *Anarsia lineatella* (Gelechiidae, Anacampsinae), wingspan 11–14 mm, a nocturnal moth added to the British list in 1957 when reared from an imported apricot, followed by a number of similar records. It is now established in parks and gardens as a local resident in south and particularly south-east England. Indications of the potential for its ability to spread more widely have been noted with a record in Glamorgan in 1997, Dorset in 2003 and Suffolk in 2013. Adults fly in June–August. Larvae feed inside and damage fruit of peach and plum. Greek *anarsios* = 'incongruous' (from other gelechiids), and Latin *lineatus* = 'lined' (from lines on the forewing).

BOSTON-IVY *Parthenocissus vitacea* (Vitaceae), a perennial neophyte climber introduced from North America, grown for its autumn colour, now naturalised in a few places in England, Wales, Jersey, and Co. Dublin. In North America it covers the outer walls of many older buildings, particularly in Boston, hence its common name; it is the plant from which the term 'Ivy League' derives. Greek *parthena* = 'virgin' + Latin *cissos* = 'ivy', and Latin *vitacea* = 'like the vine'. The specific epithet *tricuspidata* is also used, being Latin for 'three-pointed', referring to the leaf shape.

BOSTRICHIDAE Family name for powderpost and auger beetles, with five British species, all boring into and living in rotting wood and dried plant material.

BOT FLY Members of the family Oestridae, whose larvae penetrate the skin of mammals, including cattle and deer. 'Bot' here is another word for maggot or fly larva.

BOTHRIDERIDAE Beetle family, with four extant species (1–3 mm length), scattered and scarce in England and Wales. *Anommatus diecki*, *A. duodecimstriatus* and *Oxylaemus variolosus* (this last species associated with ancient oak forest) are subterranean. *Teredus cylindricus* is a predator found in hollow oaks in association with Anobiidae or *Lasius brunneus* ants.

BOVIDAE Family name for cattle, sheep and goats (Latin *bos* = 'ox' or 'cow').

BOX *Buxus sempervirens* (Buxaceae), an evergreen shrub or small tree native to some woodlands and thickets on steep slopes on chalk, and in scrub on chalk downland, including Box Hill (Surrey), Boxley (east Kent) and possibly elsewhere, but it has been widely cultivated since Roman times and the limits of its native range are uncertain. It is popular for hedging in gardens (but see box tree moth under moth) and is commonly planted in woodlands, often becoming naturalised. The binomial is Latin for 'box tree' and 'evergreen'.

Carpet box *Pachysandra terminalis* (Buxaceae), a dwarf stoloniferous shrub introduced from Japan in 1882, widely planted in gardens, parks and amenity areas, sometimes naturalised, for example in Kent, Buckinghamshire and Berkshire. Greek *pachys* + *andras* = 'thick' + 'male', referring to the stamens, and Latin *terminalis* = 'terminal (position)'.

BOX BUG *Gonocerus acuteangulatus* (Coreidae), length 11–14 mm, an RDB squash bug historically limited to Box Hill (Surrey), but recently expanding its food plants from only box to include yew, hawthorn, buckthorn and plum, and now occurring widely in south-east England. In 2011 it was noted in Birmingham, and in 2014 in Leicestershire.

BRACHYCENTRIDAE From Greek *brachys* + *centron* = 'short' + 'point', a caddisfly family. See grannom.

BRACHYCERA Suborder of Diptera with reduced antenna segmentation (Greek *brachys* + *ceras* = 'short' + 'horn').

BRACKEN *Pteridium aquilinum* (Dennstaedtiaceae), a common and widespread fern of open habitats such as heathland, moorland and acid grassland. 'Bracken' dates from Middle English usage, of Scandinavian origin (c.f. Swedish *bräken*). Greek *pteris* = 'wing' or 'feather', and Latin *aquilinum* = 'like an eagle'.

BRACKEN BUG *Monalocoris filicis* (Miridae), length 2–3 mm, widespread, feeding on the sporangia of ferns, particularly bracken; this and fern bug are the only bug species with this diet.

BRACKET FUNGUS Basidiomycete fungi in the orders Polyporales and Hymenochaetales that form woody fruiting bodies (conks) with pores or tubes on the underside, usually found as inedible brackets, often tiered, on living or more usually dead tree-trunks or branches, and important agents of wood decay.

Scarce species include **aspen bracket** *Phellinus tremulae* (Hymenochaetaceae) (found in the Highlands), **benzoin bracket** *Ischnoderma benzoinum* (Fomitopsidaceae) (benzoin being the balsamic-like resin often found dripping from the edge of the cap), **bicoloured bracket** *Gloeoporus dichrous* (Meruliaceae), **brownflesh bracket** *Coriolopsis gallica* (Polyporaceae) (mostly from south-east England), **clustered bracket** *Inonotus cuticularis* (Hymenochaetaceae), **greasy bracket** *Aurantiporus fissilis* (Polyporaceae), **lacquered bracket** *Ganoderma lucidum* (Ganodermataceae), **pine bracket** *Porodaedalea pini* (Hymenochaetaceae) (mostly Highlands), **poplar bracket** *Oxyporus populinus* (Agaricomycetes), **red-belted bracket** *Fomitopsis pinicola* (Fomitopsidaceae), **rose bracket** *F. rosea*, **salmon bracket** *Erastia salmonicolor* (Polyporaceae) (only 12 records from 1963 to 2013), **soft bracket**

Leptoporus mollis (Polyporaceae), and **tufted bracket** *Fuscoporia torulosa* (Hymenochaetaceae).

Alder bracket *Mensularia radiata* (Hymenochaetaceae), widespread, locally common, annual, found year-round, sporulating in late summer and autumn, on trunks of dead or dying hardwoods, usually alder, and causing white rot. Latin *radiata* probably describes the radial wrinkles often found on the upper surfaces of mature brackets.

Artist's bracket *Ganoderma applanatum*, common and widespread, perennial, evident throughout the year on trunks of broadleaf trees, especially beech. The common name comes from the fact that the creamy white underside can be scratched with a sharp point to leave brown marks to produce 'artistic images'. Greek *ganos + derma* = 'glaze' + 'skin', a reference to the lacquered appearance of the cap, Latin *applanatum* = 'having a flattened shape'.

Beeswax bracket *Ganoderma pfeifferi*, uncommon, perennial, favouring southern and south-east England, evident throughout the year near the base of trunks of beech, occasionally oak. In late winter and spring the pore surface secretes a yellow, sweet-smelling waxy substance, hence its common name. Ludwig Pfeiffer (1805–77) was a German botanist.

Big smoky bracket *Bjerkandera fumosa* (Meruliaceae), annual, fairly widespread in England, local elsewhere, evident throughout the year, sporulating in late summer and autumn, on dead wood of broadleaf trees such as beech, sycamore and ash. Clas Bjerkander (1735–95) was a Swedish naturalist. Latin *fumosa* = 'smoky'.

Bitter bracket *Postia stiptica* (Fomitopsidaceae), annual, locally common and widespread, found all year but sporulating in late autumn, on dead pine wood, occasionally that of hardwoods. *Postia* is Latin for 'post', *stiptica* for 'styptic' (causing bleeding to stop).

Blueing bracket *Postia subcaesia*, annual, scattered and widespread, found all year, sporulating in autumn, on beech, ash and other hardwoods. *Subcaesia* is Latin for 'nearly sky blue', referring to the blueish colour of the mature cap.

Blushing bracket *Daedaleopsis confragosa* (Polyporaceae), widespread but only locally common, found throughout the year, sporulating in late summer and autumn, with kidney-shaped brackets on dead hardwood. *Daedaleopsis* means 'having the appearance of *Daedalea*' (the genus that includes oak mazegill); in Greek mythology Daedalus constructed the Minotaur labyrinth at Knossos, Crete. *Confragosa* is Latin for 'rough', referring to the wrinkled cap surface.

Cinnamon bracket *Hapalopilus nidulans* (Polyporaceae), neurotoxic, widespread, though only occasionally recorded from Ireland, on dead wood in deciduous and mixed woodland, under birch and other hardwoods. Latin *hapalus + pilus* = 'soft-boiled' + 'hair', and *nidulans* = 'nesting'.

Conifer blueing bracket *Postia caesia*, common and widespread, annual, evident all year, sporulating in late autumn, in damp shaded woodland on dead wood of pine and spruce, occasionally well-rotted hardwood. *Caesia* is Latin for 'sky blue', referring to the blue-grey colour of the mature cap.

Cushion bracket *Phellinus pomaceus*, widespread and locally common, on living and dead wood of mainly blackthorn. *Pomaceus* is Latin for 'apple-like'.

Fragrant bracket *Trametes suaveolens* (Polyporaceae), scattered and uncommon, evident throughout the year, sporulating in autumn, on living and dead hardwoods, especially willow and poplar. *Trametes* is Greek meaning 'one who is thin' (in cross section); *suaveolens* is Latin for 'sweet-smelling', referring to the aniseed scent given off by fresh young specimens when cut.

Giant elm bracket *Rigidoporus ulmarius* (Meripilaceae), a pathogenic fungus found all year, sporulating in late summer and autumn, scattered in England and Wales on the base of trunks and on stumps of elm and other hardwoods but having greatly declined following Dutch elm disease in the 1970s. A fruit body found at Kew in 2003 was (until 2011) the largest known fungal fruit body ever discovered, with a diameter of 150 x 133 cm, and a circumference of 425 cm. Latin *rigidoporus* = 'rigid' pores, *ulmarius* = 'relating to elm'.

Greyling bracket *Postia tephroleuca*, annual, scattered and widespread, commoner in England, found all year, sporulating in autumn, on dead wood of deciduous, occasionally coniferous trees. Greek *tephros + leucos* = 'ash-coloured' + 'white'.

Hairy bracket *Trametes hirsuta*, common and widespread, evident in late summer and autumn, sometimes persisting throughout the year, found on many hardwood trees, especially beech, and causing white rot. *Hirsuta* is Latin for 'hairy'.

Hazel bracket *Skeletocutis nivea* (Polyporaceae), common and widespread, an annual evident throughout the year, sporulating in late summer and autumn, in deciduous woodland on dead wood, especially of hazel and ash, causing white rot. Latin *sceleto + cutis* = 'skeleton' or 'withered' + 'skin', and *nivea* = 'snow white'.

Lumpy bracket *Trametes gibbosa*, common and widespread, perennial, sporulating in late summer and autumn, found on many hardwoods, especially beech, forming brackets on standing timber and more often rosettes on the tops of stumps. It causes white rot. *Gibbosus* is Latin for 'humped'.

Oak bracket *Pseudoinonotus dryadeus* (Hymenochaetaceae), common and widespread, especially in England, its felted surface exuding amber droplets, most noticeably towards the broad growing margin. It is found from late summer to early winter near the base of living and dead hardwoods, mainly oak but also beech, alder and birch, causing white rot. Greek *dryadeus* = 'associated with oak'.

Powderpuff bracket *Postia ptychogaster*, annual, scattered and widespread in Britain, on live or dead coniferous wood in mixed and coniferous woodland. Greek *ptyche + gaster* = 'fold' + 'stomach'.

Purplepore bracket *Trichaptum abietinum* (Polyporaceae), common and widespread, annual, found year-round, sporulating in late autumn, on dead wood of conifers, especially fir and spruce. Greek *trichaptum* = 'with clinging hairs', and Latin *abietinum* = 'inhabiting fir trees'.

Shaggy bracket *Inonotus hispidus* (Hymenochaetaceae), widespread and common in England, annual but evident throughout the year, sporulating in autumn, solitary or a limited number of tiers, saprobic on hardwoods, especially ash and apple, and causing white rot. *Hispidus* is Latin for 'hairy'.

Smoky bracket *Bjerkandera adusta*, widespread and common, annual, wood-rotting, evident throughout the year, sporulating in late summer and autumn, on dead wood of hardwoods. *Adusta* is Latin for 'scorched'.

Southern bracket *Ganoderma australe*, common and widespread, evident throughout the year on lower hardwood trunks. In the early stages of colonisation it is parasitic, but it causes white heart rot and as the host tree dies the fungus becomes saprobic. *Australe* is Latin for 'southern'.

Tinder bracket Alternative name for hoof fungus.

Willow bracket *Phellinus igniarius*, widespread but uncommon, perennial, sporulating in summer and autumn, parasitic and eventually saprobic, causing white rot on broadleaf trees, most commonly willows. *Igniarius* is Latin for 'relating to fire', from the charcoal appearance of its upper surface.

BRAIN FUNGUS 1) Gelatinous fungus-parasitic species of *Tremella* (Tremellales, Tremellaceae), Latin *tremella* = 'trembling', referring to the wobbly jelly-like structure.

Rarities include **brown brain** *T. steidleri* (from south England and south Wales), **conifer brain** *T. encephala* and **mulberry brain** *T. moriformis*.

Leafy brain *T. foliacea*, common and widespread, found throughout the year in woodland on dead wood of broadleaf trees and conifers, parasitic on the saprobic hairy curtain crust. In dry weather it shrivels to a hard blackish crust, but following rain the fruit body rehydrates and returns to translucence. *Foliacea* is Latin for 'leafy'.

Yellow brain *T. mesenterica*, widespread and common, found throughout the year, sporulating in late summer and autumn, on dead branches of hardwoods, especially birch and hazel, parasitic on saprobic *Peniophora* fungi, especially rosy crust. Greek *meso* + *entero* = 'middle' + 'intestine'.

2) Inedible gelatinous fungi, mostly in the genus *Exidia* (Latin for 'staining' or 'exuding') (Auriculariaceae), during dry weather shrivelling to leave a transparent rubbery patch on the host wood.

Crystal brain *E. nucleata*, widespread (especially in England) and fairly common, found throughout the year, particularly evident during the wetter autumn and winter, on rotting logs and branches of broadleaves, especially beech, ash and sycamore. Latin *nucleatus* here means 'little nut', describing the opaque white nodular calcium oxalate inclusions within the translucent fruit bodies.

Grey brain Alternative name for tripe fungus.

White brain *E. thuretiana*, widespread but infrequent, inedible, evident in autumn on rotting logs and branches of hardwoods, especially beech and ash. Gustave Thuret (1817–75) was a French botanist.

BRAMBLE From Old English *brœmbel*, deciduous or semi-evergreen shrubs, often spiny, of the genus *Rubus* (Rosaceae). Most taxa in the subgenus *Rubus* form a complex aggregate (around 320 microspecies) often known collectively as *Rubus fruticosus* (from Latin for 'bramble' and 'shrubby'). These common and widespread woody plants of woodland, scrub, banks, hedgerows, heathland and waste places can form dominant stands with a wide ecological tolerance, though they reach maximum vigour and diversity on acidic soils. The fruit (blackberry) is edible, and plants spread by bird-dispersed seeds and by tip-rooting stems.

Chinese bramble *R. tricolor*, a procumbent semi-evergreen neophyte shrub from China, producing trailing stems several metres long, first grown in British gardens in 1908, increasingly mass-planted for ground cover in parks and amenity areas, and becoming naturalised in hedges and woodland, and on roadsides and waste ground, with a widespread but scattered distribution. *Tricolor* is Latin for 'three-coloured'.

Stone bramble *R. saxatilis*, a perennial, stoloniferous deciduous herb, widespread and scattered in upland areas, usually on basic soils on mountain slopes, screes and in rocky woodland, occasionally in less acidic rocky heathland and on riverside shingle. *Saxatilis* is Latin for 'living among rocks'.

White-stemmed bramble *R. cockburnianus*, a deciduous neophyte shrub introduced from China in 1907, locally naturalised as a garden escape in Britain in woodland, and on roadsides, railway banks and waste ground. Henry Cockburn (1859–1927) was the British diplomat to China, 1880–1905.

BRAMBLING *Fringilla montifringilla* (Fringillidae), wingspan 26 cm, length 14 cm, weight 24 g, a winter visitor present from mid-September until March–April, numbers varying greatly each year (45,000 to 1.8 million birds) depending on weather conditions and abundance of its preferred food, beech mast, in Scandinavia and Siberia. It often flocks with the similar-looking chaffinch. The English name is probably derived from the Germanic *brāma* = 'bramble' or 'thorny bush'. Latin *fringilla* = 'finch', *mons* = 'mountain'.

BRANCHIOPOD Class of crustaceans of fresh, brackish and inland salt water in the orders Anostraca (fairy shrimp), Cladocera (*Daphnia* water fleas) and Notostraca (tadpole shrimp), characterised by having gills on many of the appendages, giving its name from Greek *branchia* + *pous* = 'gills' + 'foot'.

BRANCHIURA Subclass of the class Maxillopoda, containing the order Arguloidea, or parasitic fish lice.

BRANDY BOTTLE Alternative name for yellow water-lily, probably because its floral scent resembles wine dregs.

BRANT GOOSE See brent goose under goose.

BRASS (MOTH) Nocturnal species of *Diachrysia* (Noctuidae, Plusiinae) (Greek *diachrysos* = 'interwoven with gold').

Burnished brass *D. chrysitis*, wingspan 28–35 mm, widespread and common in habitats including gardens, hedgerows, riverbanks, fen, woodland edges and rough grassland. The species is double-brooded, with adults flying between June and September. Larvae feed on a number of herbaceous plants, including stinging nettle, white dead-nettle and spear thistle. *Chrysites* is Greek for 'like gold', from the metallic-coloured wing markings.

Scarce burnished brass *D. chryson*, wingspan 44–54 mm, scarce, local in river valleys, marsh, fen and damp woodland in southern England and south Wales. Adults fly in July–August. Larvae feed on hemp-agrimony. *Chryson* is Greek for 'gold'.

BRASSICACEAE (Formerly Cruciferae), the cabbage family (Latin *brassica* = 'cabbage'), including species of bitter-cress, charlock, garlic mustard, hedge mustard, mustard, rock-cress, scurvygrass, sea-kale, sea rocket, shepherd's-purse, thale cress and woad.

BREAM *Abramis brama* (Cyprinidae), length usually 30–55 cm, weight 2–4 kg, in bottom regions of nutrient-rich lowland ponds, lakes and canals, and in slow-flowing rivers, though spawning in shallow water in June–July. Diet is of insect larvae, worms and molluscs. It is fished for sport. Also known as **bronze bream**. 'Bream' dates from late Middle English, from Old French *bresme*. The genus name comes from the Greek for a kind of mullet.

Gilt-head bream *Sparus aurata* (Sparidae), length usually up to 35 cm, weight up to 7 kg, typically found at depths of 0–30 m over sandy substrates or near sea-grass, occasionally in estuaries. Fairly sedentary, it is particularly found in south-west Ireland, but is occasional off south England and west Wales. Diet is mainly small shellfish. The binomial is Latin for this species and for 'golden', from the gold bar marking between the eyes.

Silver bream *Blicca bjoerkna* (Cyprinidae), length up to 36 cm, weight up to 1.8 kg, in lowland lakes, slow-flowing rivers, streams and canals in east and south-east England and the Midlands, a bottom-feeding fish taking insect larvae, snails, crustaceans, worms and plant material. Spawning is in May–June in shallow, vegetated water.

BRIAR A prickly scrambling shrub, especially a wild rose, from Old English *brær* or *brer*.

BRICK, THE *Agrochola circellaris* (Noctuidae, Xyleninae), wingspan 33–8 mm, a nocturnal moth widespread and common in broadleaf woodland, scrub and gardens. Adults fly in August–October. Larvae feed on the flowers and leaves of wych elm, poplars, aspen, sallows and ash. The forewings range in colour from yellow to reddish-brown, the latter prompting the English name. Greek *agros* + *chole* = 'field' + 'bile' (suggesting the yellow colour), and Latin *circellaris* = 'small ring' (from wing markings).

BRIDEWORT 1) *Spiraea salicifolia* (Rosaceae), a deciduous neophyte shrub from central Europe, present in gardens by 1665, widespread and occasionally naturalised in woodland and on roadsides, riverbanks and waste places, flowering in June–September. Garden-escape naturalised hybrids include **intermediate bridewort** and **confused bridewort**. Its English name comes from being strewn in churches for festivals and weddings, and often made into bridal garlands. *Spiraea* is Latin for 'bridal wreath', from Greek *speiraira*, a plant used for garlands, in turn from *speira* = 'spiral'. *Salicifolia* is Latin for 'willow-leaved'.

2) Alternative name for meadowsweet, also used at weddings.

Pale bridewort *S. alba*, a strongly suckering, deciduous neophyte shrub from North America, cultivated by 1759, a naturalised garden escape, scattered mainly in the west and north, in woodland and hedgerows and on roadsides, railway banks, rough grassland and waste ground. Hybrids include **Billard's bridewort** and **intermediate bridewort**. *Alba* is Latin for 'white'.

BRIGHT (MOTH) 1) Members of the family Incurvariidae,

mostly *Incurvaria*, from Latin *incurvatus* = 'curved', referring to curved appendages on the mouth. Wingspan is usually 12–16 mm.

Common bright *I. oehlmanniella*, common on heathland and moorland, widespread but scattered, and found at altitude in the Cairngorms. Adults fly in June–July. Larvae feed on bilberry and cloudberry, initially mining the leaves then cutting out a portable case and dropping to the ground, where they feed on dead leaves. G. Oehlmann (d.1815) was a German insect dealer.

Feathered bright *I. masculella*, fairly common in woodland edge, scrub, hedgerow and garden, widespread though less common in Scotland. Adults fly during May. Larvae initially mine leaves of a number of deciduous trees and shrubs, favouring hawthorn; they later descend to the ground in a portable case and feed on dead leaves. *Masculus* is Latin for 'male', possibly referring to the distinctive comb-like antenna found only in the male.

Pale feathered bright *I. pectinea*, widespread but scattered in woodland and heathland in England, Wales and Ireland, commoner in south-west and north-east Scotland. Adults fly in April–May. Larvae mine leaves of birches and hazel, occasionally hornbeam and apple, living within a movable case, subsequently feeding in fallen leaves. *Pectinea* comes from Latin *pecten* = 'comb', from the distinctive antenna shape in the male.

Strawberry bright *I. praelatella*, widespread but scattered. Adults fly in sunshine in May–June. Larvae feed on rosaceous plants, including wild strawberry. *Praelatus* is among other things Latin for 'prelate', possibly referencing the purple forewing.

Striped bright *Phylloporia bistrigella*, wingspan 7–9 mm, widespread, found in heathland and scrub. Adults fly during the day in May–June. Larvae mine birch leaves, cutting out an oval case and dropping to the ground to feed on leaf litter. Greek *phyllon* + *poreia* = 'leaf' + 'a walking', from the larval mine, and Latin *bi* + *striga* = 'two' + 'furrow', from parallel wing markings.

2) Members of the family Prodoxidae in the genus *Lampronia* (Greek *lampros* = 'bright', referring to the wing colours). Wingspan is usually 12–15 mm. See also currant shoot-borer, and raspberry moth under moth.

Bramble bright *L. flavimitrella*, a rarity found in a few locations in woodland and hedgerows in south and south-east England, first recorded from Hampshire in 1974. Adults fly in sunshine in May–June. Larvae feed on fruit of bramble and raspberry. Latin *flavus* + *mitra* = 'yellow' + 'turban', from the adult's yellow head.

Northern bright *L. pubicornis*, a rarity found at a few limestone downlands and sandhill sites in north England and west Ireland. Adults fly in May–June. Larvae feed on young shoots of burnet rose. Latin *pubes* + *cornu* = 'pubic hair' + 'horn', describing the hairy male antenna.

Rose bright *L. morosa*, found in a few woodland, scrub and garden locations in Britain. Adults fly in May–June. Larvae feed inside rose buds. *Morosus* is Latin for 'morose', from the dingy wing colour.

Scarce bright *L. fuscatella*, wingspan 14–18 mm, a rarity found in a few widely scattered woodland and heathland locations in Britain. Adults fly in the afternoon in May–June. Larvae create a gall on a twig of birch, usually at a node, within which it feeds. *Fuscatus* is Latin for 'blackened', from the dark colour of both wings.

Wood bright *L. luzella*, scarce, local in woodland and bog in parts of southern Britain and in Ireland. Adults fly in May–July in afternoon sunshine. Larvae feed on the fruits of bramble, dewberry, cloudberry and raspberry. *Luzella* honours an Austrian army officer, J.F. Luz, who found the type specimen in the early nineteenth century.

BRIGHT-EYE Brown-line bright-eye *Mythimna conigera* (Noctuidae, Hadeninae), wingspan 30–5 mm, a nocturnal moth, widespread and common but perhaps in long-term decline, in grassland, woodland edges and rough meadows. Adults fly in June–July, visiting a variety of flowers. Larvae feed on a number of grasses. The English name describes a cone-shaped white spot on each of the forewings. Mythimna is a town on the Greek island of Lesbos. Latin *conus* + *gero* = 'cone' + 'wear'.

BRIMSTONE (BUTTERFLY) *Gonepteryx rhamni* (Pieridae, Coliadinae), wingspan 60–74 mm, found in England (having recently spread into the north), Wales and central-west Ireland, common in a range of habitats from scrubby grassland and woodland rides through chalk downland to gardens. Adults are most commonly seen in spring but can fly on warm days throughout the year. Larvae primarily feed on buckthorn on calcareous soils and alder buckthorn on moist acid soils, adults on a range of plants, including thistles, scabious, knapweeds and selfheal. Numbers after 2000 have been more or less static, and geographical range increased by 14% between 2005 and 2016. The word 'butterfly' may have originated from the yellow colour of the male of this species. Greek *gonia* + *pteryx* = 'angle' + 'wing', and Latin *rhamnus* = 'buckthorn'.

BRIMSTONE (MOTH) *Opisthograptis luteolata* (Geometridae, Ennominae), wingspan 32–7 mm, nocturnal, widespread and common in hedgerows, gardens, woodland, calcareous grassland and acid heathland. Adults appear in three overlapping generations in the south, graduating to one brood in the far north, and may fly at any time between April and October. Larvae feed on a variety of trees including blackthorn, hawthorn, rowan and wayfaring-tree. Greek *opisthen* + *graptos* = 'behind' + 'painted', with forewing markings on both upper and lower surfaces, and Latin *luteolata* = 'yellowish'.

BRINDLE (MOTH) 1) Nocturnal members of the family Noctuidae (Xyleninae).

Cloud-bordered brindle *Apamea crenata*, wingspan 36–44 mm, widespread and common in grassland, fen, moorland, woodland rides and gardens. Adults fly in June–July. Larvae feed on grasses. Apamea was a town in Asia Minor, having no intrinsic entomological meaning.

Clouded brindle *Apamea epomidion*, wingspan 40–6 mm, common in broadleaf woodland, scrub, parks and gardens throughout much of England, Wales and Ireland. Adults fly in June–July. Larvae feed on grasses. *Epomidios* is Greek for 'on the shoulder', from a wing marking.

Feathered brindle *Aporophyla australis*, wingspan 36–42 mm, scarce, on shingle beaches, dunes and sea-cliffs along the coasts of southern and south-east England from Cornwall to Suffolk and, locally, on the southern slopes of the South Downs and in south-west Wales. Adults fly in August–October. Larvae feed on sea campion, sorrel, wood sage, bramble and grasses. Greek *aporos* + *phyle* = 'difficult' + 'tribe', and Latin *australis* = 'southern'.

Goldenrod brindle *Xylena solidaginis*, wingspan 45–51 mm, predominantly nocturnal, on moorland, though in places also occurring in open woodland, found mainly in the north Midlands, northern England, Wales and central Scotland. Adults fly in August–September. Larvae feed on heather, bilberry, bog-myrtle, creeping willow and silver birch. Greek *xylinos* = 'wooden', and Latin *solidago* = 'goldenrod'.

Slender brindle *Apamea scolopacina*, wingspan 32–6 mm, local in woodland throughout much of England and Wales, very occasional in Scotland and Northern Ireland. Population increased by 137% between 1968 and 2007. Adults fly in July–August. Larvae feed on woodland grasses. *Scolopacina* is Latin for 'snipe-like' (*scolopax*), from a similar colouring to this bird.

Small clouded brindle *Apamea unanimis*, wingspan 30–8 mm, common in marsh, riverbanks, damp woodland and damp grassland throughout most of Britain. Adults fly in June–July. Larvae feed on grasses. *Unanimis* is Latin for 'harmonious'.

2) **Yellow-barred brindle** *Acasis viretata* (Geometridae, Larentiinae), wingspan 25–9 mm, nocturnal, widespread but local in broadleaf woodland, hedgerows, parks and gardens. Population increased by 131% between 1968 and 2007. Adults fly in May–June, and in the southern part of its range again

in August–September. Larvae feed on leaves and flowers of such woody plants as holly, ivy, dogwood, privet, hawthorn and guelder-rose. Greek *acasis* = 'brotherless' (the species originally believed to be monotypic), and Latin *viretum* = 'green spot'.

BRINE FLY More usually called shore fly in Britain, members of the Ephydridae.

BRISTLE-GRASS Species of *Setaria* (Latin *seta* = 'bristle', referring to the bristles on the spikelet), scattered and occasional neophytes found as casuals on waste ground, originating from wool shoddy, birdseed and soyabean waste, and as grain impurities.

BRISTLE-LEAF *Brachydontium trichodes* (Seligeriaceae), a widespread and scattered moss preferring shaded sandstone rock by streams in hilly districts, especially on vertical or overhanging surfaces, but is also found on pebbles. Greek *brachys* + *donti* = 'short' + 'tooth', and *trichoeidis* = 'hair-like'.

BRISTLE-MOSS Species of *Orthotrichum* (Orthotrichaceae), from Greek *orthos* + *trichos* = 'straight' + 'hair', acrocarpous mosses that form tufts 0.5–4 cm tall, all but anomalous, hooded and rock bristle-mosses being obligate epiphytes that grow on woody plants.

Rarities include **aspen bristle-moss** *O. gymnostomum* (on aspen in a few woods in the Highlands), **dwarf bristle-moss** *O. pumilum*, **Mitten's bristle-moss** *O. consimile* (a single site in Derbyshire found in 2006), and **showy bristle-moss** *O. speciosum* (remaining very locally frequent in eastern Scotland, and rediscovered in Cambridgeshire in 2008).

Anomalous bristle-moss *O. anomalum*, common and widespread on concrete, grave stones, wall tops and other man-made structures except in heavily polluted areas, and also common on exposed limestone. *Anomalum* is Latin for 'unusual'.

Blunt-leaved bristle-moss *O. obtusifolium*, rare, with most recent records from parkland trees in east Scotland. It was lost from England in the early twentieth century, but recent records from Norfolk, Cambridgeshire, Essex and Cardiganshire suggest that it may be recolonising. *Obtusifolium* is Latin for 'blunt-leaved'.

Elegant bristle-moss *O. pulchellum*, common and widespread except for the central Highlands, growing on willow and other trees. *Pulchellum* is Latin for 'pretty'.

Hooded bristle-moss *O. cupulatum*, common and widespread on limestone and walls. *Cupulatum* is Latin for cup-shaped.

Lyell's bristle-moss *O. lyellii*, widespread except for north Scotland and west Ireland, but uncommon, growing on a number of tree species, especially ash. Charles Lyell (1767–1849) was an English bryologist.

Pale bristle-moss *O. pallens*, almost entirely restricted to Weardale and eastern Scotland, but a recent record from near Watford suggests that it may be overlooked further south or be spreading. *Pallens* is Latin for 'pale'.

River bristle-moss *O. rivulare*, widespread in upland regions on trees, rocks and masonry by rivers. *Rivulare* is Latin for 'growing by rivers'.

Rock bristle-moss *O. rupestre*, widespread and scattered except in Scotland where it is much commoner on (especially igneous) boulders, particularly by water. In England and Wales it has been recorded on stones and trees. *Rupestre* is Latin for 'rock-loving'.

Shaw's bristle-moss Alternative name for smooth bristle-moss.

Slender bristle-moss *O. tenellum*, common and widespread, less so in Scotland. In some southern coastal areas it is abundant on willow and a variety of street trees. Further north and east it occurs on ash, and on old elder in hedgerows. Air pollution caused it to disappear from much of lowland England, but it is spreading back in the Midlands. *Tenellum* is Latin for 'tender'.

Smooth bristle-moss *O. striatum*, widespread and common

especially in west and north-east Britain, recovering in lowland areas previously subject to air pollution, growing on twigs of hazel and willow, and on trunks of ash and sycamore. *Striatum* is Latin for 'striped'.

Spruce's bristle-moss *O. sprucei*, scattered, absent from much of Scotland, restricted to, though often abundant in, silty zones of lowland rivers. Richard Spruce (1819–73) was an English botanist.

Straw bristle-moss *O. stramineum*, common and widespread though mostly local in east Ireland, the predominant bristle-moss of trees and shrubs in parts of mid-Wales and eastern Scotland, particularly on the trunks of ash and branches of hazel. *Stramineus* is Latin for 'straw'.

White-tipped bristle-moss *O. diaphanum*, common and widespread except for northern Scotland, very common on trees and shrubs in lowland regions, especially on elder and willow. It is particularly common in town centres, on concrete, rocks, brick walls and other inorganic structures, and is found on similar sites in farmyards. Each leaf ends in a white 'hair-point', giving its English name. Greek *diaphanis* = 'transparent'.

Wood bristle-moss *O. affine*, common and widespread on trees and shrubs in places with clean air, sometimes on rock or concrete. *Affine* is Latin for 'similar'.

BRISTLE WORM A term covering a number of species of polychaete.

BRISTLETAIL Small primitive wingless (apterogyte) insects in the orders Thysanura and Diplura.

1) **Thysanura: three-pronged bristletails**, with two families in Britain: Machilidae (seven species living under stones, organic seashore debris or low-growing vegetation) and Lepismatidae (living in houses and comprising silver-fish and firebrat).

Sea bristletail *Petrobius maritimus* (Machilidae), up to 1.5 cm long, widespread and scattered, found above high tide, usually under stones and on screes or in rock and wood crevices.

2) **Diplura: two-pronged bristletails**, small wingless, soil- or detritus-living apterygote insects (order Diplura) that feed on decaying matter, represented in Britain by 12 species of *Campodea* (Campodeidae).

BRITTLEGILL Generally inedible ectomycorrhizal fungi in the genus *Russula* with at least 239 recognised British and Irish species. *Russula* is Latin for 'reddish'; while many species do have red caps, many others do not.

Widespread and locally common species found in woodland include **bitter almond brittlegill** *R. grata* (crushing the gills produces an odour that explains the common name), **blackening brittlegill** *R. nigricans* (cut flesh goes reddish-brown before turning black), **bleached brittlegill** *R. exalbicans* (on birch), **blue band brittlegill** *R. chloroides* (oak), **Camembert brittlegill** *R. amoenolens*, **coral brittlegill** *R. velenovskyi*, **crab brittlegill** *R. xerampelina* (on beech, oak and pine, a scent of boiled crab prompting the common name), **crowded brittlegill** *R. densifolia*, **dawn brittlegill** *R. aurora*, **fragile brittlegill** *R. fragilis*, **geranium brittlegill** *R. fellea* (beech), **green brittlegill** *R. aeruginea* (birch), **humpback brittlegill** *R. caerulea* (pine), **milk white brittlegill** *R. delica*, **ochre brittlegill** *R. ochroleuca*, **oilslick brittlegill** *R. ionochlora* (the cap colours often merge as in oil spilt onto water), **powdery brittlegill** *R. parazurea* (mainly oak), **primrose brittlegill** *R. sardoni* (pine), **purple brittlegill** *R. atropurpurea* (oak and beech), **purple swamp brittlegill** *R. nitida* (birch), **sepia brittlegill** *R. sororia* (oak), **stinking brittlegill** *R. foetens* (young specimens give off an acrid smell which becomes fishy as the fruit bodies mature), **variable brittlegill** *R. versicolor* (birch), **winecork brittlegill** *R. adusta*, **yellow swamp brittlegill** *R. claroflava* (mainly birch), and **yellowing brittlegill** *R. puellaris* (birch, beech and conifers).

Rare species include **alpine brittlegill** *R. nana* (Lake District and Highlands), **beautiful brittlegill** *R. font-queri* (Highlands, plus three sites in Buckinghamshire), **burning brittlegill** *R. badia* (Highlands), **darkening brittlegill** *R. vinosa* (mostly

Highlands), **gilded brittlegill** *R. aurea*, **golden brittlegill** *R. risigallina*, **greasy green brittlegill** *R. heterophylla*, **intermediate brittlegill** *R. intermedia* (mostly Cumbria and Highlands), **lilac brittlegill** *R. lilacea* (England and south Wales), **minute brittlegill** *R. minutula*, **misfit brittlegill** *R. xenochlora* (Abernethy Forest and a few other sites), **naked brittlegill** *R. vinosobrunnea* (south-east England and Aberdeenshire), **pallid brittlegill** *R. raoultii*, **pelargonium brittlegill** *R. pelargonia*, **ruddy brittlegill** *R. rutila*, **russet brittlegill** *R. mustelina*, **Scottish brittlegill** *R. scotica* (Highlands plus Thornhill Park, Hampshire), **scurfy brittlegill** *R. melzeri*, **sunny brittlegill** *R. solaris* (mostly southern England and East Anglia), **tardy brittlegill** *R. cessans* (mostly Highlands), **velvet brittlegill** *R. violeipes*, **vinegar brittlegill** *R. acetolens* (southern England and south Wales), and **viscid brittlegill** *R. viscida*.

Bare-toothed brittlegill Alternative name for the flirt.

Birch brittlegill *R. betularum*, common and widespread, poisonous, found in summer on often damp soil, ectomycorrhizal on birch. Latin *betula* = 'birch'.

Bloody brittlegill *R. sanguinaria*, widespread but uncommon, found from July to October on soil and needle litter in coniferous woodland, mostly under pine. Latin *sanguis* = 'blood'.

Copper brittlegill *R. decolorans*, edible, scattered but mainly found in the Highlands in summer and autumn on coniferous woodland soil, ectomycorrhizal on Scots pine. *Decolorans* is Latin for 'staining'.

Freckled brittlegill *R. illota*, uncommon and scattered in Britain, found from August to November on often calcareous soil in woodland, often under oak and birch, occasionally conifers. *Illota* is Latin for 'unwashed', referring to the freckled appearance of the cap.

Fruity brittlegill *R. queletii*, widespread but uncommon, found from July to October, on soil in coniferous woodland, ectomycorrhizal on spruce. Lucien Quélet (1832–99) was a French mycologist.

Greencracked brittlegill *R. virescens*, widespread but uncommon, edible, found from August to October in deciduous woodland and parkland, ectomycorrhizal on beech, oak and especially sweet chestnut. *Virescens* is Latin for 'becoming green'.

Nutty brittlegill *R. illota*, uncommon, with a scattered distribution on soil in coniferous woodland, mycorrhizal on Scots pine. *Illota* is Latin for 'unwashed'.

Olive brittlegill *R. olivacea*, locally frequent in southern England and East Anglia, scattered elsewhere, ectomycorrhizal on beech.

Red swamp brittlegill *R. aquosa*, scattered in Britain, mostly in Scotland, found in summer and autumn on marshy ground with moss, often under birch. *Aquosa* is Latin for 'watery'.

Rosy brittlegill *R. rosea*, widespread and generally uncommon, found from August to October on woodland soil, mycorrhizal on beech and other hardwoods.

Scarlet brittlegill *R. pseudointegra*, mostly found in central and southern England, scattered elsewhere, in deciduous woodland, ectomycorrhizal mainly on oak. *Pseudointegra* is Latin for 'falsely entire'.

Slender brittlegill *R. gracillima*, widespread but uncommon, found from July to October on soil in damp deciduous woodland, ectomycorrhizal on birch. *Gracillima* is Latin for 'delicate'.

Willow brittlegill *R. laccata*, widespread but uncommon, mostly in the Highlands, often on sandy soils in woodland and heathland, ectomycorrhizal on willows.

BRITTLESTAR Marine echinoderms, members of the class Ophiuroidea, mostly in the order Ophiurida, with nine families in the British list. Individuals have a central 'disc' body from which extend five long, slender arms used to crawl along the sea floor, feeding as scavengers or detritivores. Many species are subtidal and not described here. See also serpent star.

Common brittlestar *Ophiothrix fragilis* (Ophiotrichidae),

the central disc reaching 2 cm, the five arms up to 10 cm, found from the low intertidal in crevices and under boulders to the sublittoral on hard substrates and coarse sediment on all British and Irish coasts, except for the east coast of Scotland, around the Humber estuary and north East Anglia and south Kent. Greek *ophios* + *thrix* = 'snake' + 'hair', and Latin for 'fragile'.

Crevice brittlestar *Ophiopholis aculeata* (Ophiactidae), the pentagonal disc 15 mm in diameter with five twisted arms 60 mm long, widespread around Britain and north-east Ireland, often in rock crevices among horse mussels or under rocks and shells in rock pools (though also reported from 2,000 m). It can form dense beds in tidal inlets in Scotland. Greek *ophios* + *pholis* = 'snake' + 'scale', and Latin *aculeata* = 'pointed'.

Daisy brittlestar, mottled brittlestar Alternative names for crevice brittlestar.

Dwarf brittlestar Alternative name for small brittlestar.

Sand burrowing brittlestar *Acrocnida brachiata* (Amphiuridae), with a flat central disc up to 5 mm diameter and five thin arms, a littoral and sublittoral benthic species usually buried in fine sand down to a depth of 40 m, scattered around all coastlines. Greek *acros* + *cnide* = 'tip' + 'nettle' or 'sting', and Latin *brachiata* = 'branched'.

Small bristlestar *Amphipholis squamata* (Amphiuridae), with a circular disc 3–5 mm in diameter and five spindly arms up to 20 mm long, found in the intertidal and shallow sublittoral under stones, among rock pool algae and occasionally on sandy bottoms, scattered around most coastlines. Greek *amphis* + *pholis* = 'both' + 'scale', and Latin *squamata* = 'scaled'.

BRITTLESTEM Inedible fungi, mostly saprobic, in the genera *Psathyrella* (Greek *psathyros* = 'friable'), with 105 species, and *Parasola* (13) (Psathyrellaceae). The hollow stems make them 'brittle'.

Common and widespread species on woodland soil, debris and litter include **chestnut brittlestem** *Ps. spadicea* (especially under beech), **common stump brittlestem** *Ps. piluliformis*, **conical brittlestem** *Pa. conopilus* (beech), **pale brittlestem** *Ps. candolleana*, **petticoat brittlestem** *Ps. artemisiae*, **red edge brittlestem** *Ps. corrugis*, **rootlet brittlestem** *Ps. microrhiza* (also in grassland), **slender stump brittlestem** *Ps. laevissima*, and **spring brittlestem** *Ps. spadiceogrisea* (also grassland).

Rarities include **medusa brittlestem** *Ps. caput-medusae*, **pygmy brittlestem** *Ps. pygmaea*, **solitary brittlestem** *Ps. solitaria* (Highlands), **spotted brittlestem** *Ps. maculata* and **yellowfoot brittlestem** *Ps. cotonea*. See also pleated inkcap under inkcap.

Clustered brittlestem *Ps. multipedata*, scattered mostly in England and south Wales, found from June to November, on dead wood in broadleaf woodland soil. Although *multipedata* is Latin for 'having many feet', groups of the 'many legs' within a cluster share a single 'foot', joining to form a common base.

Dune brittlestem *Ps. ammophila*, fairly common and widespread, found from June to November on coastal dunes, saprotrophic on decaying marram. Greek *ammos* + *philos* = 'sand' + 'loving'.

BROAD-BAR **Shaded broad-bar** *Scotopteryx chenopodiata* (Geometridae, Larentiinae), wingspan 25–30 mm, a widespread and common day- and night-flying (July–August) moth inhabiting open habitats, including calcareous grassland, hedgerows, heathland, dunes and woodland rides. Larvae feed on clovers and vetches. Greek *scotos* + *pteryx* = 'darkness' + 'wing'; *chenopodiata* is an incorrect reference to a food plant *Chenopodium*.

BROADLEAF **New Zealand broadleaf** *Griselinia littoralis* (Griseliniaceae), an evergreen neophyte shrub commonly planted near the sea in southern and south-western England and the Isle of Man, scattered elsewhere and in places self-sown and naturalised. Francesco Griselini (1717–83) was a Venetian naturalist. *Littoralis* is Latin for 'of the shore'.

BROADWINGS **Large broadwings** *Brachycercus harrisellus*

(Caenidae), a mayfly with a scattered, predominantly south and north-east England and Welsh Border distribution. Nymphs live in pools and stream margins. They are poor swimmers but are adapted for moving among mud and silt where they feed by collecting or gathering fine particulate organic detritus from the sediment. Adults are evident in July. Greek *brachys* + *cercos* = 'short' + 'tail', referring to one of a pair of small appendages at the end of the abdomen.

BROCADE Members of the nocturnal moth family Noctuidae, the English name referring to the rich patterns on the forewings.

Beautiful brocade *Lacanobia contigua* (Hadeninae), wingspan 36–42 mm, local on heathland and moorland, and in open woodland throughout Britain and the southern half of Ireland. Adults fly in June–July. Larvae feed on a variety of trees and bushes. *Lacanobia* is a typographic error for *Lachanobia*, from Greek *lachana* = 'vegetable' + *bios* = 'life'. *Contigua* is Latin for 'adjacent' (i.e. similar to a congeneric).

Dark brocade *Mniotype adusta* (Xylininae), wingspan 42–8 mm, previously widespread on heathland, chalk downland, fens, woodland, gardens, moorland, upland grassland and dunes, but decreasing, especially in southern coastal counties, and a species of conservation concern in the UK BAP. Adults fly from May to July. Larvae feed on a range of herbaceous and woody plants. Greek *mnion* + *typos* = 'moss' + 'mark of a blow' (here 'character'), the wing pattern resembling moss, and Latin *adustus* = 'scorched', from the dark wing colour.

Dusky brocade *Apamea remissa* (Xylininae), wingspan 36–42 mm, common in damp meadow, calcareous grassland, gardens, hedgerows and open woodland, but a species of conservation concern in the UK BAP. Adults fly in June–July. Larvae feed on the flowers and seeds of various grasses. Apamea was a town in Asia Minor, *remissus* is Latin for 'sent back': neither part of the binomial has entomological significance.

Flame brocade *Trigonophora flammea* (Xylininae), wingspan 44–52 mm, resident in the Channel Islands, elsewhere an immigrant from the Mediterranean, adults appearing along England's southern coast, with 2011 seeing highest numbers for 130 years, and possibly establishment of colonies. The flight period is October–November. Larvae feed on plants such as buttercup in early instars, later preferring ash and privet. Greek *trigonon* + *phoreo* = 'triangle' + 'to carry', and Latin *flammea* = 'flame-coloured', together describing forewing shape and colour.

Great brocade *Eurois occulta* (Noctuinae), wingspan 50–60 mm, scarce in bog, marsh, scrubby moorland and woodland in the Highlands. It also appears as an immigrant, mostly in north Scotland and the Scottish islands, northern England and down the east coast to Essex and Kent, with scattered records inland and as far west as Cornwall, Isle of Man, and the east coast of Ireland. Resident moths fly in July–August, immigrants from late August to September. Larvae mainly feed on bog-myrtle but also take birches and sallows. Greek *euroos* = 'fair-flowing' (from rivulet markings on the forewing), and Latin *occulta* = 'hidden', from the dark colour.

Light brocade *L. w-latinum* (Hadeninae), wingspan 37–42 mm, local on calcareous grassland, heathland and open woodland in the southern half of England (especially East Anglia) and parts of Cumbria and Wales. Adults fly in May–July. Larvae feed on a number of plants, including dyer's-greenweed, broom and bramble. The specific epithet derives from the w-shaped markings on the forewing.

Pale-shouldered brocade *L. thalassina* (Hadeninae), wingspan 35–8 mm, widespread and common in woodland, scrubland, moorland and fens. Adults fly in June–July, sometimes with a partial second brood in the south. Larvae feed on a variety of woody and herbaceous plants. Greek *thalassinos* = 'sea-like', from the wavy subterminal line on the forewing.

Sombre brocade *Dichonioxa tenebrosa* (Xylininae), wingspan 35–40 mm, first recorded in Guernsey in 2006, since when it

has appeared quite regularly. Recorded for the first time on mainland Britain at Durlston, Dorset, in 2008, with four at the same site in 2009, suggesting that it may be breeding. In Hampshire, recorded for the first time at Sandy Point, Hayling, in 2010. The binomial is in part from Greek *dichos* = 'double', and Latin *tenebrosa* = 'dark'.

Toadflax brocade *Calophasia lunula* (Oncocnemidinae), wingspan 26–32 mm, a rare RDB moth probably arriving in Britain in the 1950s, found on shingle beaches and grassy areas in parts of south-east England, mostly coastal. In Kent it occurs at Dungeness and along the coast to Folkestone. In Sussex, larvae can be common on shingle beaches and waste places within a mile of the seashore. Since 2002, it has been recorded from Greater London, Essex, Norfolk and Dorset, and may be expanding its range. In Hampshire it was first recorded in 1971 in Southampton, with one at Southsea in 1992 and one in Gosport in 1996, and likely to become established on shingle. It has two generations, sometimes overlapping, from May to August, and migrants sometimes appear away from the main stronghold in July–August. Larvae feed on purple and common toadflax. Greek *calon* + *phasis* = 'wood' + 'appearance' (the moth resembling dead wood), and Latin *lunula* = 'little moon', from the wing pattern.

BROCCOLI See cabbage.

BROME Grasses (Poaceae) in the genera *Brachypodium* (Greek *brachys* + *podi* = 'short' + 'foot', referring to short pedicels), *Bromus* (Greek *bromos* = 'oat', also giving the English 'brome') and *Anisantha* (Greek + *anis* + *anthos* = 'aniseed' + 'flower'); and generally perennial grasses in the genera *Bromopsis* (Greek meaning 'looking like brome') and *Ceratochloa* (Greek *cerato* + *chloe* = 'horn' + 'grass').

Scattered rare annuals include **foxtail brome** *A. rubens*, **lesser hairy brome** *Bromopsis benekenii*, **Patagonian brome** *C. brevis*, **Smith's brome** *Bromus pseudosecalinus*, **stiff brome** *Brachypodium distachyon*, **Thunberg's brome** *Bromus japonicus* and **western brome** *C. marginata*.

Barren brome *A. sterilis*, an annual archaeophyte, common and widespread in lowland England, Wales, eastern Scotland and south-east Ireland, on roadsides, railway banks, grassland, gardens and waste ground, sometimes a weed in winter cereal fields, often introduced with wool shoddy. *Sterilis* is Latin for 'sterile'.

California brome *C. carinata*, a neophyte, often short-lived or annual, 'escaping' from Kew in 1919, though not dispersing much until 1945, and only spreading rapidly since the 1960s, now scattered and widespread in England, scarcer elsewhere, favouring lighter soils, on verges, field margins and waste ground. *Carinata* is Latin for 'keeled'.

Compact brome *A. madritensis*, an annual European neophyte, scattered mainly in southern England and south Wales on waste and cultivated ground, roadside banks, walls, around docks, and on dunes and other sandy or rocky coastal habitats. Long-established (and possibly native) at a few sites in south-west Britain and the Channel Islands, it is otherwise seen as a casual, especially from wool shoddy. *Madritensis* is Latin for 'from Madrid'.

Drooping brome *A. tectorum*, an annual European neophyte introduced to Britain by 1776, on waste ground, roadsides and grassy habitats on sandy soils as a casual, often originating from grain, wool shoddy and grass seed. It can be naturalised on sandy banks, field margins and grass-heaths in Breckland, though it has recently declined there. Many recent records are from newly sown road verges. Latin *tectum* = 'roof'.

False brome *Brachypodium sylvaticum*, a common and widespread perennial of well-drained neutral to calcareous soils. In the lowlands, it favours shady habitats such as woodland and hedgerows, as well as railway banks and roadsides, and it has colonised downland following scrub invasion. Above 200 m, it can occur in more open sites, for example limestone grassland

and pavements, cliffs and scree. *Sylvaticum* is from Latin for 'forest' or 'wild'.

Field brome *Bromus arvensis*, an annual European neophyte, introduced by 1763, scattered and declining mostly in England, casual in arable and grass fields and on tracksides, docks and waste ground as a wool and grain alien, mostly on nutrient-poor soils. *Arvensis* is from Latin for 'farmland'.

Great brome *A. diandra*, an annual neophyte introduced from Morocco in 1804, scattered mostly as a grain, birdseed and wool alien, increasing in southern and eastern England since the mid-twentieth century, especially naturalised in East Anglia, in arable fields, waste ground and roadsides, and in open grassland and heathland on sandy soils. It is sometimes well-established on dunes. *Diandra* is Greek for 'two stamens'.

Hairy brome *Bromopsis ramosa*, a widespread and common perennial in lowland regions in shaded habitats, including woodlands and hedgerows, on moist, moderately base-rich soils. *Ramosa* is Latin for 'branched'.

Heath false brome *Brachypodium pinnatum*, a perennial, scattered, casual in Scotland, in grassland, scrub and open woodland, often on clay. *Pinnatum* is Latin for 'feather-shaped'.

Hungarian brome *Bromopsis inermis*, a rhizomatous perennial European neophyte cultivated in 1794 as a fodder grass, now naturalised or casual as a contaminant in verges, field margins and other rough grassland habitats, resistant to drought and persisting on sandy, well-drained soils. *Inermis* is Latin for 'without spines'.

Large-headed brome *Bromus lanceolatus*, an annual Mediterranean neophyte introduced by 1798, an uncommon casual in waste ground, originating from wool shoddy and birdseed, and as a garden escape. *Lanceolatus* is Latin for 'lance-shaped'.

Lesser soft brome *Bromus* x *pseudothominei* (*B. hordeaceus* x *B. lepidus*), a fertile annual grass hybrid, scattered and locally common, most frequent in sown grassland, but also found in a variety of ruderal habitats. Charles Thomine-Desmasures (1799–1824) was a French botanist.

Meadow brome *Bromus commutatus*, an annual locally common in central and southern England, scattered elsewhere, on unimproved damp meadow, and on verges and field margins. *Commutatus* is Latin for 'changing'.

Rescue brome *C. cathartica*, a Latin American neophyte, introduced as a fodder grass in 1788, scattered and probably increasing in England and Wales as a grain and wool alien on roadsides, waste ground and field margins, and in arable crops. Usually casual, but it has become naturalised in the Channel Islands and Scilly. *Cathartica* is Latin for 'purgative'.

Ripgut brome *A. rigida*, an annual Mediterranean neophyte first recorded in the wild in 1834, a wool, grain and agricultural seed alien, scattered mainly in England in grassland and disturbed or cultivated ground on light soils, usually casual but establishing on dunes and other sandy places near the sea in southern England and the Channel Islands. The name 'ripgut' (also applied to great brome) comes from the sclerotised nature of this grass which can harm feeding livestock. *Rigida* is Latin for 'rigid'.

Rye brome *Bromus secalinus*, an annual or biennial archaeophyte probably present since prehistoric times, initially an arable weed or possibly an alternative source of grain. It was common, especially in England, in the nineteenth and early twentieth centuries, but has undergone a dramatic decline, apart from few areas such as Norfolk and Worcestershire where it may be increasing as a grass-seed contaminant. Still found in cereal fields, it is also a casual on waste ground. *Secalinus* is Latin for 'resembling rye'.

Slender soft brome *Bromus lepidus*, annual, scattered, in improved grassland, and a contaminant of other grass-seed mixtures. It also occurs in arable fields and waste ground. *Lepidus* is Latin for 'graceful'.

Smooth brome *Bromus racemosus*, an annual, scattered mostly in England and Wales, on unimproved hay and water meadow, usually on damp, periodically flooded alluvial soils. While generally declining, it is often frequent on dry field margins. It is also found as a grass-seed casual in arable margins and on verges. *Racemosus* is Latin for 'having racemes' (flower clusters).

Soft brome *Bromus hordeaceus*, a winter-annual, common and widespread, on moderately fertile neutral soils, particularly in disturbed or open habitats, for example on coastal cliffs, sand and shingle (sspp. *ferronii* and *thominei*), on waste ground and waysides (ssp. *molliformis*), and in pasture and hay meadow (ssp. *hordeaceus*). *Hordeaceus* is Latin for 'like barley'.

Upright brome *Bromopsis erecta*, a winter-green perennial found on dry, relatively infertile calcareous soils in much of England, more scattered in Wales and Ireland, introduced in north England and Scotland, in ungrazed chalk and limestone grasslands, calcareous sand dunes, roadside banks, quarry spoil and waste ground.

BROOK-MOSS Species of *Hygrohypnum* (Amblystegiaceae) (Greek *hygros* + *hypnon* = 'wet' + 'lichen'). **Claw brook-moss** *H. ochraceum*, **drab brook-moss** *H. luridum* and **western brook-moss** *H. eugyrium* are locally common in upland Britain and Ireland, often on stream rocks.

Rarities include **arctic brook-moss** *H. smithii* (Highlands), **broad-leaved brook-moss** *H. duriusculum* (north Wales, north England and Scotland), and **soft brook-moss** *H. molle* (Highlands).

BROOKLIME *Veronica beccabunga* (Veronicaceae), a common and (apart from north-west Scotland) widespread herbaceous perennial growing on all but the most infertile substrates in a range of wetland habitats, for example in shallow water, by streams and ponds, in ditches, flushes and wet woodland rides. The English name is from Middle English *broklemok*, from 'brook' + *hleomoce*, the name of the plant in Old English. The genus name = for St Veronica; the specific name is from German *Bachbunge* = 'brook' + 'bunch'.

BROOKWEED *Samolus valerandi* (Primulaceae), a short-lived deciduous perennial with a generally coastal (except for north and east Scotland), occasionally inland distribution, found in small, often impermanent colonies on wet flushes, springs, ditches, lagoons and lake shores, on open mesic, often calcareous or somewhat saline soils. *Samolus* is the Latin name for this plant, possibly of Celtic origin.

BROOM *Cytisus scoparius* ssp. *scoparius* (Fabaceae), a common and widespread erect shrub of sandy acidic soils on heathland, open woodland, railway banks and, especially, on roadside banks and verges where it is often planted. This is the 'county' flower of Glasgow as voted by the public in a 2002 survey by Plantlife. The prostrate ssp. *maritimus* grows on western sea-cliffs exposed to wind, and also on shingle at Dungeness, Kent. 'Broom' comes from Old English *brom*. Greek *cytisos* is an old name for woody legumes, and also a kind of clover. *Scoparius* is from Latin *scopa* = 'broom'.

Black broom *C. nigricans*, a deciduous Central European neophyte shrub grown in gardens since 1730, first recorded in the wild in west Kent in 1970, now naturalised in a few places in southern England on waste ground and roadsides and in gravel-pits. *Nigricans* is Latin for 'black'.

Hairy-fruited broom *C. striatus*, a neophyte deciduous shrub introduced from the Iberian Peninsula by 1816, occasionally planted in parks and amenity areas. It reproduces freely from seed, sometimes forming extensive populations, naturalised on roadside banks, with a scattered distribution in Britain. *Striatus* is Latin for 'striped'.

Montpellier broom *Genista monspessulana* (Fabaceae), a deciduous Mediterranean neophyte shrub, cultivated by 1735, now a garden escape naturalised in parts of southern England on roadsides, banks, railway banks, gravel-pits and waste ground. *Genista* is Latin for 'broom', the species name for 'from Montpellier'.

Mount Etna broom *G. aetnensis*, a large neophyte deciduous shrub or small tree from Sicily and Sardinia, cultivated in 1823, naturalised in southern England since at least 1977 (found in Surrey) as a garden escape on waste ground. *Aetnensis* is the Latinised form of 'from Etna'.

Spanish broom *Spartium junceum* (Fabaceae), a neophyte Mediterranean shrub present by the mid-sixteenth century, planted in amenity areas and on roadsides. It sets seed readily, quickly becoming naturalised on light soils by roads, railways, and (in north Essex) along coastal cliffs. It is scattered in England and Wales, rare in Scotland. Greek *sparton*, a kind of grass, and Latin *junceum* = 'rush-like'.

White broom *C. multiflorus*, a neophyte deciduous shrub from the Iberian Peninsula, grown in gardens by 1752, recorded in the wild by 1957, widespread and scattered but increasing in abundance, naturalised on roadsides, railway banks and waste ground. *Multiflorus* is Latin for 'many flowers'.

BROOM-TIP *Chesias rufata* (Geometridae, Larentiinae), wingspan 28–32 mm, a scarce nocturnal moth found on heathland, open woodland, scrubby embankments and verges in parts of south-east England and discontinuously from Devon to Cumbria; also in Scotland (ssp. *scotica*) and Ireland (Co. Wexford and Co. Wicklow). With numbers declining by 90% over 1968–2007, this is a species of conservation concern in the UK BAP. Adults fly from April to July. Larvae feed on broom. The binomial derives from Chesium on the island of Samos, and Latin *rufus* = 'red'.

BROOMRAPE Chlorophyll-less root-parasitic herbs in the genus *Orobanche* (Orobanchaceae), from Greek *orobos* + *ancho* = 'vetch' + 'to strangle', referring to the plant's parasitic properties.

Bean broomrape *O. crenata*, a Mediterranean neophyte perennial herb introduced by 1845, occasionally casual on peas and beans in fields and gardens. *Crenata* is Latin for 'scalloped'.

Bedstraw broomrape *O. caryophyllacea*, found on hedge bedstraw, probably perennial, local in east Kent in stabilised dune grassland, and in scrub and hedges on chalk downland. *Caryophyllacea* is Latin for 'pinkish'.

Common broomrape *O. minor*, usually annual, found on a range of hosts, but mainly on species of Fabaceae and Asteraceae, scattered but mostly in southern England and East Anglia. Var. *minor*, the commonest variety, is probably an introduction, usually found on cultivated land and other disturbed ground. Var. *maritima* is native on dunes and cliffs on the south coast, with wild carrot, buck's-horn plantain and common restharrow its main hosts. *Minor* is Latin for 'smaller'.

Greater broomrape *O. rapum-genistae*, a perennial local in southern and south-west England, Wales and the Scottish Borders on gorse and broom, occasionally dyer's greenweed and other Fabaceae, mostly in scrub, hedges and tracksides. The specific name is Latin for 'turnip-broom'.

Ivy broomrape *O. hederae*, annual or perennial, scattered in (especially southern) England, Wales and Ireland, growing on the roots of ivy on coastal cliffs, rocky woodland, quarries and hedges. In south-east England it is increasing found, probably introduced, in artificial habitats including gardens. *Hedera* is Latin for 'ivy'.

Knapweed broomrape *O. elatior*, probably perennial, found on greater knapweed, mainly in chalk and limestone grassland in south and east England north to the Yorkshire Wolds. It can also form large populations in man-made habitats such as verges, railway banks and quarries. *Elatior* is Latin for 'taller'.

Oxtongue broomrape *O. picridis*, a rare annual or perennial found mainly on the ledges and calcareous debris of coastal chalk cliffs on the Isle of Wight, West Sussex and east Kent where it parasitises roots of hawkweed oxtongue and hawksbeard species. Latin *picris* = 'bitter'.

Thistle broomrape *O. reticulata*, perennial, found on thistles in rough grassland, on verges, and especially on river margins and flood plains where it is frequent, suggesting that seed is

dispersed by water. It is found on chalk and Magnesian limestone in Yorkshire. *Reticulata* is Latin for 'netted'.

Thyme broomrape *O. alba*, an annual possibly perennial parasitising wild thyme and perhaps other Lamiaceae, found locally in south-west England, the mid-Pennines, west Scotland, and north and west Ireland, mainly on base-rich rocky coastal slopes, but also inland on stabilised scree below limestone outcrops in the Pennines. *Alba* is Latin for 'white'.

Yarrow broomrape *O. purpurea*, annual, perhaps perennial, found on yarrow and possibly other Asteraceae on dry basic soils in clifftop grassland and on roadsides and grassy banks, usually near the sea, declining but currently recorded from Dorset, the Isle of Wight, Norfolk and Pembrokeshire. It also rarely occurs in disturbed artificial habitats. *Purpurea* is Latin for 'purple'.

Yorkshire broomrape Alternative name for thistle broomrape, reflecting the only county where it is found in Britain.

BROWN (BUTTERFLY) **Meadow brown** *Maniola jurtina* (Nymphalidae, Satyrinae), male wingspan 40–5 mm, female 42–60 mm, very common (though numbers declined by 8% over 2000–9), widespread except in mountainous areas, favouring grasslands but also found on woodland rides, field margins, hedgerows, road verges and gardens, and flying in June–September. Young larvae feed by day, more mature ones by night, on various grasses. Adults feed on the nectar of many species, especially bramble, knapweeds and thistles. Numbers and range appear to be stable, though there was a small decline in numbers (by 15%) between 2005 and 2014. *Maniola* is a diminutive of Manes, the souls of the beloved departed in Roman religion; *jurtina* is possibly a misprint for Juturna, a Roman fountain-nymph.

BROWN (MAYFLY) Members of the family Heptageniidae.

False March brown *Ecdyonurus venosus*, widespread in upland Britain in stream riffles on submerged plants and stones, scraping periphyton from the substrate or by gathering fine particulate organic detritus from the sediment. Adults emerge between April and July, in places to October, males swarming throughout the day into dusk. Greek *ecdyo* + *oura* = 'eclipse' + 'tail', and Latin *venosus* = 'venous'.

March brown 1) *Rhithrogena germanica*, similar in distribution and life history to false March brown, but adults emerge in March–May with notable synchronicity, hatching en masse typically around midday. Greek *rheithron* + *genos* = 'stream' + 'race'. 2) Angler's name for olive upright *Rhithrogena semicolorata*.

BROWN (MOTH) **Lunar marbled brown** *Drymonia ruficornis* (Notodontidae, Notodontinae), wingspan 35–40 mm, nocturnal, widespread but local in woodland, hedgerows, parks and gardens. Population levels increased by 117% between 1968 and 2007. Adults fly in April–May. Larvae feed on oak leaves. Greek *drymos* = 'oak coppice', and Latin *rufus* = 'red' + *cornus* = 'red' + 'horn'.

Marbled brown *D. dodonaea*, wingspan 33–8 mm, nocturnal, common in oak woodland in much of England and Wales, and parts of Scotland, with numbers declining in many places. Adults fly in May–June. Larvae feed on oak leaves. Dodona is a town in north-west Greece.

Pale shining brown *Polia bombycina* (Noctuidae, Hadeninae), wingspan 43–52 mm, nocturnal, scarce, in rough and scrubby grassland, downland, open woodland, and chalk and limestone habitats in the southern half of England and parts of Wales. Because of a reduction in range (for example recent attempts to find the species on Porton Down where it was once common, and where the habitat appears to be unchanged, have proved negative), this is a priority species in the UK BAP. Adults fly in June–July. Larval diet in Britain is uncertain. Greek *polios* = 'grey', and Latin *bombycina* = 'silken', from the glossy forewing.

BROWN ALGAE Members of the class Phaeophyceae, including wrack and kelp.

BROWN-EYE **Bright-line brown-eye** *Lacanobia oleracea* (Noctuidae, Hadeninae), wingspan 32–7 mm, a widespread nocturnal moth common in gardens, heathland, saltmarsh, woodland rides and a range of other habitats. Adults fly in May–June, with an occasional second brood in the south. Larvae feed on a range of plants, and can become a pest by eating tomatoes. *Lacanobia* is a typographic error for *Lachanobia*, from Greek *lachana* + *bios* = 'vegetable' + 'life', *oleracea* Latin also for 'vegetables'.

BROWN-TAIL *Euproctis chrysorrhoea* (Erebidae, Lymantriinae), wingspan 36–42 mm, a nocturnal moth local in scrub, hedgerows, parks and gardens, along the coasts on England from Yorkshire to the Isles of Scilly, and also inland. Adults fly in July–August. Larvae feed on bramble, hawthorn, blackthorn, dog-rose and sallows, living gregariously inside a silk web, causing sufficient damage to be a serious pest in some areas. The larval hairs are irritating, causing rashes on exposed human skin. The moth has white wings and body except for a tufted golden-brown abdominal tip. Greek *eu* + *proctos* = 'good' + 'anus', for the anal tuft, and *chrysos* + *rheo* = 'golden' + 'flow', for the golden-brown 'tail'.

BROWNIE Inedible fungi in the genera *Hypholoma* (Greek *hyphe* + *loma* = 'web' + 'fringe', probably referring to the thread-like partial veil that connects the cap rim to the stem of young fruit bodies, possibly to the thread-like rhizomorphs that radiate from the stem base) and *Psilocybe* (Greek *psilos* + *cybe* = 'smooth' + 'head', i.e. cap) (Strophariaceae).

    Blueleg brownie *P. cyanescens*, scattered in England, rare in south Wales and central Scotland, found in summer and autumn on buried wood and wood chip. Greek *cyaneos* = 'blue'.

    Mountain brownie *P. montana*, widespread and scattered in (often non-montane) Britain, found in summer and autumn on soil under grassland, dunes, scrub and pathways.

    Olive brownie *H. myosotis*, locally common with a scattered distribution found in summer and autumn in damp habitats including moorland, bog, wet heath and woodland, often with moss. *Myosotis* is Greek for 'mouse-ear'.

    Peat brownie *H. udum*, locally common with a scattered distribution found in late summer and autumn in plant debris among mosses on moorland and damp heathland. *Udum* is Latin for 'moist'.

    Rooting brownie *H. radicosum*, uncommon with a scattered distribution, found in summer and autumn on stumps and other dead wood in coniferous forest. *Radicosum* is Latin for 'rooting'.

    Snakeskin brownie *H. marginatum*, widespread and locally common, found in late summer and autumn on rotting wood or litter in coniferous woodland. A snakeskin pattern on the stem explains the common name. *Marginatum* is Latin for 'with a margin' referring to remnants of the partial veil that adheres to the cap margin.

    Sphagnum brownie *H. elongatum*, widespread and locally common on moorland, bog, wet heath and damp woodland, usually associated with mosses. *Elongatum* is Latin for 'elongated'.

BRUSSELS LACE *Cleorodes lichenaria* (Geometridae, Ennominae), wingspan 31–8 mm, a nocturnal moth found locally in scrub, woodland, plantations and rocky areas by the sea, in south-west England, a few places in central and southern England, west Wales, the west coast of Scotland, and over much of Ireland. Adults fly in June–August. Larvae feed on various lichens, giving the specific epithet. Cleora was the wife of Agesilaus, a fourth-century BCE Spartan king; *Cleorodes* uses Greek *eidos* = 'similar to' (i.e. to the genus *Cleora*).

BRUSSELS-SPROUT See cabbage.

BRYACEAE Moss family (Greek *bryo* = 'moss') that includes *Anomobryum* (slender silver-moss), *Plagiobryum* (hump-moss), *Rhodobryum* (rose-moss), and especially *Bryum*, with 55 recognised species in Britain (see thread-moss and bryum).

BRYONY Black bryony *Tamus communis* (Dioscoreaceae), a dioecious perennial twining climber common and widespread in England and Wales mostly on neutral or calcareous, well-drained soils, particularly those overlying chalk and limestone, though because it has a large tuber it avoids shallow or waterlogged soil. It grows well in hedgerows, woodland edges and along paths and in wasteland, but is often found in a depauperate, non-flowering state in woodland. Although bird-sown, it is not a good colonist. *Tamus* is from Latin *Tamnus*, which refers to another climbing plant; *communis* = 'common'.

    White bryony *Bryonia dioica* (Cucurbitaceae), a scrambling perennial herb widespread in England except for the south-west and north, occasional elsewhere, on well-drained, often base-rich, soils in hedgerows, scrub and woodland borders, and on rough ground. Being unpalatable to rabbits it can be locally abundant around warrens. *Bryonia* is the Ancient Greek name used by Dioscorides; *dioica* refers to male and female flowers being on separate plants.

BRYOPHYTA, BRYOPHYTE The phylum containing mosses, with the classes Andreaeopsida, Bryopsida, Oedipodiopsida, Polytrichopsida, Sphagnopsida and Tetraphidopsida. *Bryo* is Greek for 'moss' or 'lichen'.

BRYOPSIDA The largest class of mosses (Bryophyta).

BRYOZOA Small colonial aquatic 'moss animals', living in fresh or seawater in large numbers, filter feeders that use a retractable lophophore (a 'crown' of tentacles lined with cilia).

BRYUM Genus of moss, including thread-moss, in the family Bryaceae (Greek *bryo* = 'moss').

    Rarities include **blunt bryum** (or **matted bryum**) *B. calophyllum* (one site in Anglesey and scattered locations in Scotland), **blushing bryum** *B. elegans*, and **round-leaved bryum** *B. cyclophyllum*.

    Bicoloured bryum *B. bicolor*, scattered in Britain as far north as south-east Scotland, commonest in south-east England and Lancashire. Probably synonymous with *B. dichotomum*, which is common and widespread. It is/they are found on disturbed soil, especially base-rich clays, for example on waste ground and paths, in gardens, fields, quarries and dunes, and on roofs, walls and cliffs.

    Lesser potato bryum *B. subapiculatum*, widespread if scattered in north Scotland and Ireland, on often acid, generally ruderal habitats. *Subapiculatum* is Latin for 'nearly blunt'.

    Marsh bryum *B. pseudotriquetrum*, common and widespread in marsh, fen and flushes, and on damp soil by lakes, in dune slacks, and on wet rocks on cliffs. Greek *pseudotriquetrum* = 'falsely three-cornered'.

    Pea bryum *B. ruderale*, common and widespread in Wales and the southern half of England, scattered elsewhere, often on bare soil associated with human activity, including mine spoil. *Ruderale* is Latin for 'ruderal', growing in wasteland.

    Pill bryum *B. violaceum*, locally common, widespread in Wales and the southern half of England, more scattered elsewhere. *Violaceum* is Latin for 'violet-coloured'.

    Potato bryum *B. bornholmense*, scattered, commonest in south and south-west England, on fairly acidic soil in a variety of habitats, including waste ground. The specific name refers to the Danish island of Bornholm.

    Raspberry bryum *B. klinggraeffii*, widespread and especially common in England and Wales. Hugo von Klinggraeff (1820–1902) was a Prussian bryologist.

    Small-bud bryum *B. gemmiferum*, scattered, commoner in south-east England and East Anglia, on well-drained, often sandy, but periodically wet soil, for example riverbanks, by ponds and ditches, and in damp arable fields. *Gemmifera* is Latin for 'bearing buds'.

    Yellow-bud bryum *B. gemmilucens*, scattered in England, especially in the east, often on damp sites. *Gemmilucens* is Latin for 'having shiny buds'.

BUCCULATRICIDAE Family name for some bent-wing moths, from Latin *buccula* = 'visor', from the large eye caps on the head.

BUCKLER-FERN General family name of members of the Dryopteridaceae, but also some members of the genus *Dryopteris* (others being male-ferns), from Greek *drys* + *pteris* = 'oak' + 'fern'. A buckler is a round shield, here describing the shape of the indusium, the membranous 'shield' covering the sorus (spore-producing receptacles) on the frond underside.

Broad buckler-fern *D. dilatata*, common and widespread in woodland, hedges, heathland, ditches and montane habitats in shade on base-poor soils of intermediate fertility. *Dilatata* is Latin for 'expanded'.

Crested buckler-fern *D. cristata*, rare, local and declining in southern England and East Anglia on wet heath, marsh, fen and dune slacks. *Cristata* is Latin for 'crested'.

Hay-scented buckler-fern *D. aemula*, local and scattered in moist shady woodland and hedge banks, mostly in western regions but also acid habitats in Sussex and Kent. *Aemula* is Latin for 'imitating'.

Narrow buckler-fern *D. carthusiana*, common and widespread in open or semi-shade on moist, infertile, acid (often peaty) soil in grassland and heathland. Johann Cartheuser (1704–77) was a German physician and botanist.

Northern buckler-fern *D. expansa*, locally frequent in damp woodland, mountain crevices and scree in Cumbria, Wales and Scotland. *Expansa* is Latin for 'spread'.

Rigid buckler-fern *D. submontana*, scarce but local in north-west England, rare in north and south Wales, in limestone crevices and scree. *Submontana* is Latin for 'near montane'.

BUCK'S-BEARD *Aruncus dioicus* (Rosaceae), a large European neophyte perennial herb, grown in gardens since 1633, planted in estate woodlands, occasionally naturalised. *Aruncus* is the Latin name for goat's-beard, *dioicus* for having male and female flowers on separate plants.

BUCKTHORN *Rhamnus cathartica* (Rhamnaceae), a strongly calcicolous shrub or small tree, locally common in England, scattered in Wales and Ireland, in a variety of habitats, including as an undershrub in deciduous woodland, scrub and hedgerows, and in fen carr and damp alder woods. 'Buckthorn' dates from the sixteenth century, compounding 'buck' (male deer) + 'thorn', a translation of the herbalists' Latin name *spina cervina*. *Rhamnus* was the actual Latin name for the plant, from Greek *rhamnos* = 'branch', and Latin *catharticus* = 'purgative', from the former use of the berries as a purgative and laxative.

Alder buckthorn *Frangula alnus* (Rhamnaceae), a deciduous shrub, widespread and locally common in England and Wales, scattered in Scotland and Ireland, growing on a range of soils, but avoiding drought-prone and permanently waterlogged sites, found in scrub on fen peat, on the edges of raised mires, on heathland and in valley mires, and in scrub, hedgerows and woodland. The English and specific names may derive from the similar appearance to alder and because both trees are often found growing together. *Frangula* is the Latin name for this species, possibly from the Greek for a tree with 'fragile wood'; *alnus* is Latin for alder.

Mediterranean buckthorn *R. alaternus*, an evergreen neophyte shrub in gardens by 1629, naturalised in scrub and waste ground in a few places in south England, north Wales and west Ireland. Latin *alatus* = 'winged'.

BUCKWHEAT *Fagopyrum esculentum* (Polygonaceae), a non-persistent Asian neophyte annual present by the Tudor period, found in waste ground, field margins and woodland rides. Until the nineteenth century it was an important grain crop in Britain and parts of Ireland, and while now scattered and generally uncommon it is still occasionally grown for green manure, for pheasants and as a bee-plant. 'Buckwheat' dates from the sixteenth century, derived from Middle Dutch *boecweite* = 'beech wheat', the grains being shaped like beechmast. The binomial is from the Latin/Greek *fagus*/*phigos* = 'beech' + Greek *pyros* = 'fire', and Latin *esculentum* = 'edible'.

BUD MOTH 1) Members of the family Torticidae (Olethreutinae). Specifically, *Spilonota ocellana*, wingspan 12–17 mm, nocturnal, widespread and common in orchards, woodland, gardens and parks, more abundant in the south. Adults fly in July–August. Larvae mine the buds of a number of trees during spring, with a preference for rosaceous species including apple, pear and hawthorn. Greek *spilos* + *nota* = 'spot' + 'the back', from a forewing blotch that appears over the back when folded, and Latin *ocellus* = 'small eye', again from the forewing marking.

Larch bud moth *S. laricana*, wingspan 12–16 mm, local in woodland with larch throughout England, Wales and south Ireland. Adults fly in June–August. Larvae mine needles of larch and occasionally other conifer trees, overwinter, then mine the buds. *Laricana* is from Latin *larix* = 'larch'.

Nut bud moth *Epinotia tenerana*, wingspan 12–16 mm, nocturnal, widespread and common in woodland and freshwater margins. Adults fly in July–October. Larvae mine hazel and alder, feeding on catkins in spring then burrowing into leaf buds. Greek *epi* + *nota* = 'upon' + 'the back' (meaning obscure), and from Latin *tener* = 'soft'.

Pine bud moth *Pseudococcyx turionella*, wingspan 14–21 mm, with a scattered distribution, local in pine woodland throughout much of southern England and Scotland. Adults fly in May–June. Larvae mine the buds and shoots of Scots pine, occasionally other pine species. Greek *pseudis* + *coccyx* = 'false' + 'cuckoo' (meaning obscure), and Latin *turio* = 'leaf shoot'.

Spruce bud moth *Zeiraphera ratzeburgiana*, wingspan 12–15 mm, nocturnal, widespread, local in coniferous woodland. Adults fly in July–August. Larvae feed on young shoots of a number of coniferous tree species, including Norway spruce. Greek *zeira* + *phero* = 'a loose garment' + 'to carry' (perhaps describing the larval spinnings). Julius Ratzeburg (1801–71) was a German entomologist.

2) **Ash bud moth** *Prays fraxinella* (Praydidae), wingspan 14–18 mm, widespread and common. Adults fly in May–September. Larvae feed on leaves, buds and bark of ash in May–June and October. Greek *prays* = 'gentle', and Latin *fraxinus* = 'ash tree'.

BUDDLEIA See butterfly-bush and orange-ball-tree.

BUFF (MOTH) Clouded buff *Diacrisia sannio* (Erebidae, Arctiinae), wingspan 35–50 mm, widespread and locally common, in Scotland and Ireland particularly in western areas, on heathland, rough grassland and other open habitats. Males are both night- and day-flying from June to August, females are mainly nocturnal. Larvae feed on heather and a range of herbaceous plants. Greek *diacrisis* = 'separation', and Latin *sannio* = 'a mimic', probably from males and females resembling each other.

Reddish buff *Acosmetia caliginosa* (Noctuidae, Condicinae), male wingspan 27–30 mm, female 21–3 mm, an endangered RDB nocturnal moth confined to heathland where it requires successional stages created by periodic small-scale clearances and grazing. Currently probably restricted to a single locality on the Isle of Wight, though formerly it also occurred in Hampshire and Dorset. Protected in the UK WCA, and a priority species in the UK BAP. Adults fly in May–June. Larvae feed on saw-wort. Greek *a* + *cosmos* = 'without' + 'adornment', and Latin *caliginosa* = 'dark', both parts of the name describing the dull coloration.

Small dotted buff *Photedes minima* (Noctuidae, Xyleninae), wingspan 20–3 mm, nocturnal, widespread and common in damp grassland, woodland rides, marsh, riverbanks and other damp areas. Adults fly in June–August. Larvae feed on the stems of tufted hair-grass. Greek *photos* + *edos* = 'light' + 'delight', and Latin *minima* = 'smallest'.

BUFF ARCHES *Habrosyne pyritoides* (Drepanidae, Thyatirinae), wingspan 35–40 mm, a nocturnal moth found in open woodland, young plantations, hedgerows and gardens throughout England, Wales and Ireland. Numbers declined by 80% over 1968–2007. Adults fly in July–August. Larvae feed on bramble, and in places dewberry. Greek *habrosyne* = 'splendour', and *pyrites* + *eidos* = 'of fire' + 'form', from the brassy yellow forewing markings.

BUFF TIP *Phalera bucephala* (Notodontidae, Phalerinae), wingspan 55–68 mm, a nocturnal moth widespread and common in open woodland, scrub, hedgerows and gardens. Adults fly in June–July. When at rest, they resemble a broken birch twig. Caterpillars live gregariously and feed on a number of different deciduous trees, sometimes defoliating entire branches. The binomial is the Greek for having a white patch on the forewing, and from *boucephalos* = 'bull-headed' (probably with no entomological meaning).

BUFFALO-BUR *Solanum rostratum* (Solanaceae), an annual North American neophyte introduced in 1823, a frequent casual in Britain, mainly in England, on roadsides, arable fields and waste ground, originating from grain impurities, wool shoddy and birdseed. The English name references the spines on the stems, leaves, and flower heads. *Solanum* is Latin for 'quietude', referring to the narcotic properties of some species; *rostratum* = 'with a beak'.

Red buffalo-bur *S. sisymbriifolium*, an annual South American neophyte, a casual originating from wool shoddy in waste ground in southern and eastern England, and Co. Cork, field-planted in Norfolk as a trap crop for potato cyst nematode control. The specific name is Latin for having leaves like *Sisymbrium* species (e.g. hedge mustard).

BUG A member of the order Hemiptera (true bugs), which includes aphids, leafhoppers, greenfly, scale insects and shield bugs, and characterised by a mouthpart (proboscis) which can pierce plant tissue and suck out liquid such as sap. Over 1,650, and probably nearer 1,700 species are native to the British Isles.

BUGLE *Ajuga reptans* (Lamiaceae), a common and widespread rhizomatous perennial herb of damp deciduous woodland, shaded places and unimproved grassland on neutral or acidic soils, sometimes growing in flushed terrain. The English name is from the Latin *bugula* for this plant. The binomial is from the Latin for this plant, and *reptans* = 'creeping'.

Pyramidal bugle *A. pyramidalis*, a herbaceous perennial, local in north and west Scotland, west and central Ireland, and a few other upland sites, on free-draining slopes, rock crevices and shallow peat in open heathland and grassland usually overlying moderately acidic but sometimes neutral or basic soils. *Pyramidalis* is Latin for 'pyramid-shaped'.

BUGLOSS *Anchusa arvensis* (Boraginaceae), an annual archaeophyte arable weed, locally common in lowland regions on well-drained soils, and also near the sea on sandy heathland, disturbed dunes and waste ground. 'Bugloss' dates from late Middle English, from Latin *buglossus*, in turn from Greek *bouglossos*, viz. *bous* + *glossa* = 'ox' + 'tongue', from the leaf shape. *Anchusa* was a plant used as a cosmetic in classical times. *Arvensis* is from Latin for 'field'. See also viper's-bugloss.

BUGSEED *Corispermum intermedium* (Amaranthaceae), an annual European neophyte, introduced into cultivation in 1739, first recorded in the wild on railway sidings at Emscote (Warwickshire) in 1962, scattered as a rare casual in England and south Wales, usually on coastal sand. Greek *corios* + *sperma* = 'bug' + 'seed'.

BULB FLY Narcissus bulb fly *Merodon equestris* (Syrphidae), body length 15 mm, wingspan 22–4 mm, a bumblebee mimic common and widespread, especially in England and Wales, found from April to September. Larvae of this hoverfly attack bluebell bulbs in the countryside, but add daffodil bulbs to their repertoire in gardens. Greek *meros* + *donti* = 'part' + 'tooth', and Latin *equestris* = 'equestrian'.

Wheat bulb fly *Delia coarctata* (Anthomyiidae), a serious pest of winter cereal in parts of (especially eastern) England. Adults lay eggs on bare soil in autumn, which lay dormant

until January–March when they hatch; larvae invade crops of wheat, barley and rye. Plant tillers are attacked at the base of the stem, causing 'deadheart' symptoms as the tiller dies and can cause significant yield losses. Set-aside in the late twentieth century created ideal oviposition sites, but crop rotation as well as insecticides suppress infestation. *Delia* is Greek for 'dice', *coarctata* Latin for 'restricted'.

BULGAR Black bulgar *Bulgaria inquinans* (Leotiales, Bulgariaceae), an inedible ascomycete common and widespread, found in September–March on the dead wood of broadleaf trees, especially oak, but also sweet chestnut, beech and ash (and will almost certainly benefit from ash death as a result of ash dieback disease). *Inquinans* is Latin means 'soiling', referencing the dark brown stain that comes from handling this fungus.

BULIN Pulmonate snails in the family Enidae.

Lesser bulin *Merdigera obscura*, shell length 8.5–10.5 mm, width 3–4 mm, height 8.5–10.5 mm, common and widespread in England and Wales, more scattered in Scotland and Ireland, found, often covered by algae or hidden under its own excrement and soil, on trees in deciduous woodland, rocks, walls and other often vertical surfaces. Latin *merdigera* = 'excrement carrier', and *obscura* = 'hidden'.

Mountain bulin *Ena montana*, shell length 14–17 mm, width 6–7 mm, height 14–17 mm, a rarity found in southern and south-central England on well-drained calcareous soils, declining with changes in land management, and close to meeting the criteria for Vulnerable in the RDB.

BULL ROUT *Myoxocephalus scorpius* (Cottidae), length 15–30 cm, a widespread fish around the British coastline found among seaweed and stones on rocky bottoms with mud or sand, sometimes in rock pools, and at depths of 2–50 m. Greek *myos* + *cephale* = 'muscle' + 'head', and Latin *scorpius* = 'scorpion', this species belonging to the order of scorpionfish.

BULLACE *Prunus domestica* ssp. *insititia* (Rosaceae), the English name dating from Middle English, in turn from Old French *buloce* = 'sloe'. Better known as damson; see wild plum under plum.

BULLFINCH *Pyrrhula pyrrhula* (Fringillidae), wingspan 26 cm, length 16 cm, weight 21 g, a widespread resident mainly found in deciduous woodland, orchards and farmland, and a common garden visitor in winter (when numbers can be augmented with continental birds), with an estimated 190,000 territories in Britain in 2009, with the 39% decline in numbers between 1970 and 2015 apparently halted since numbers between 1995 and 2015 increased slightly by 9% (8% over 2010–15). It is Amber-listed in Britain, Green-listed in Ireland. Nests are in scrub or trees, usually with 4–5 eggs which hatch after 14–16 days, the altricial young fledging in 15–17 days. Diet is of seeds of fleshy fruits (e.g. cherries), buds and shoots (they can become pests in orchards); the young feed on invertebrates. The stocky shape of the head prompts the English name. The binomial comes from Greek *pyrrhoulas*, a bird mentioned by Aristotle that, unlike bullfinches though, ate worms.

BULLFROG American bullfrog *Lithobates catesbeianus* (Ranidae), up to 20 cm long, named for its deep croak, males the smaller sex, introduced from eastern North America, numbers increasing in the 1970s with both accidental and deliberate transport of tadpoles with aquaculture goldfish. Deliberate imports were banned in 1997. Records from garden ponds and wetlands in southern England increased during the 1980s, and breeding was confirmed in ponds on the East Sussex/Kent border in 1999 and 2000, and later elsewhere in south-east England Essex and Hampshire. All known populations have eventually, often with difficulty, been eradicated, but there is a strong possibility of unknown populations persisting locally. It has been classed by IUCN's Invasive Species Specialist Group as one of the 100 worst invasive species in the world. Greek *lithos* + *bates* = 'stone'

+ 'one who treads' (i.e. rock climber). Mark Catesby (1682/3–1749) was an English naturalist.

BULLHEAD 1) *Cottus gobio* (Cottidae), length 7–10 cm, a common and widespread fish in England and Wales, local in central Scotland, in shallow rivers, lakes and canals with stony or gravelly bottoms, feeding nocturnally on insects and crustaceans. Spawning is in March–April, the male creating a shallow hollow nest and guarding the eggs. The English name reflects the broad head. The binomial comes from Greek *cottos*, an unknown river fish, and Latin for 'gudgeon'. 2) Alternative name for bull rout.

BULLWORT *Ammi majus* (Apiaceae), an annual south European neophyte introduced by 1551, very scattered in parkland and neglected gardens, and as a casual on spoil tips and roadsides, originating from wool shoddy or birdseed. *Ammi* is the Latin for this plant, *majus* = 'greater'.

BULRUSH *Typha latifolia* (Typhaceae), an emergent rhizomatous herbaceous perennial, common and widespread except for the Highlands, in shallow water (up to 1 m) or on mud in reed swamp, at the edge of lakes, ponds, canals and ditches and, less frequently, by streams and rivers, favouring nutrient-rich sites. It spreads by wind-dispersed fruits, subsequently spreading by vegetative growth. Previously known as greater reedmace. The plant formerly known as bulrush is a sedge, *Schoenoplectus lacustris*, now called common club-rush. *Typha* is Ancient Greek for this species, *latifolia* Latin for 'broad-leaved'.

Lesser bulrush *T. angustifolia*, an emergent rhizomatous herbaceous perennial, common and widespread in England, scattered elsewhere and absent from the Highlands and much of Ireland, found in mesotrophic or eutrophic water in lakes, ponds and ditches. It tends to grow in deeper water than bulrush (up to 2 m), and tolerates mesotrophic as well as eutrophic conditions. In some places it grows as a floating raft. Formerly known as lesser reedmace. *Angustifolia* is Latin for 'narrow-leaved'.

BUMBLEBEE Members of the genus *Bombus* (Apidae). *Bombus* is Latin for 'buzz' or 'humming', and from such words 'bumblebee' itself.

Bilberry bumblebee *B. monticola*, queen body length 16 mm, wingspan 32 mm, worker length 12 mm, male 14 mm, a rare species associated with moorland and other upland habitats in western and northern Britain and eastern Ireland, but it has also been found at lowland grassland sites. Queens emerge from hibernation in April; workers are present from May, and males and new females from July to October. Nests are underground, started in old mammal nests. They can be parasitised by forest cuckoo bumblebee. For nectar and pollen, bilberries and sallow are preferred in spring, bird's-foot-trefoil, clovers and bramble in early to midsummer, and bell heather and bilberries in mid- to late summer. *Monticola* is Latin for 'mountain dweller'.

Broken-belted bumblebee *B. soroeensis*, queen body length 16 mm, wingspan 30 mm, worker length 12 mm, male 13 mm, very scattered in Britain, commoner in north Scotland, but also with a 2012 record from Dungeness, Kent, found in a range of habitats, from moorland in Scotland to calcareous grassland on Salisbury Plain. Nest-searching queens are found from June to August, according to latitude, the nests themselves being in old mouse and vole nests. Pollen and nectar are taken from a range of flowers.

Buff-tailed bumblebee *B. terrestris*, queen body length 20–2 mm, wingspan 30 mm, worker length 11–17 mm, male 14–16 mm, common and widespread, extending its range into northern Scotland in recent years, but with few records from Ireland. This probably has the widest diet of any British bumblebee. The species is partially bivoltine, with a winter generation under favourable circumstances of available forage in gardens and relatively warm winters, such as in 1992–2005. Normally queens are found between February and April, according to latitude, males between July and October. Records of workers in January–March are common

in southern England. Nests are underground in old mouse or vole nests, and are large, often with >500 individuals. Nests may be parasitised by vestal cuckoo bees. Captive nests, not of the British subspecies, are now used by commercial tomato and fruit growers, and some sexuals may escape and interbreed with wild bees. *Terrestris* is Latin for 'terrestrial'.

Cryptic bumblebee *B. cryptarum*, only recently distinguished from white-tailed bumblebee, with a scattered distribution, scarce outside Scotland, and with one record from Ireland.

Early bumblebee *B. pratorum*, queen body length 15–17 mm, workers 10–14 mm, males 11–13 mm, widespread in Britain, with a recent record also from Co. Sligo, strongly associated with gardens and woodlands, but also found in grassland, heathland and moorland. It is bivoltine in the south, with a smaller late-summer generation, univoltine towards the north. Nest-searching queens are among the first bee species to emerge throughout its range, being present from March to May, depending on latitude, and males are often seen by the end of May or June. Nests are underground in old mouse or vole nests, or in old bird nests, especially if these are in tree holes. Nests may be parasitised by forest cuckoo bumblebees. Visits are made to a variety of flowers for pollen and nectar. *Pratorum* is from Latin for 'meadow'.

Forest cuckoo bumblebee *B. sylvestris*, queen body length 15 mm, males 14 mm, a socially parasitic bumblebee, widespread but uncommon, taking nectar from a variety of flowers in a range of habitats. During spring the fertilised female enters a nest of the host, early bumblebee (possibly also heath and bilberry bumblebees in the Highlands), kills the host queen and lays her own eggs. All foraging and nest duties are carried out by the host workers. *Sylvestris* is from Latin for 'woodland' and 'wild'.

Great yellow bumblebee *B. distinguendus*, historically known from much of Britain and Ireland, though with a northern bias, but since 1999 only definitely recorded on some islands of the Inner and Outer Hebrides, Orkney, and the north coast of mainland Scotland, and a priority species in the UK BAP. Preferred habitat is unimproved grassland, with flower visits favouring bird's-foot-trefoil, red clover and common knapweed. Nests are usually underground in small mammal burrows or off rabbit burrows. *Distinguendus* is Latin for 'distinguished'.

Heath bumblebee *B. jonellus*, queen body length 16 mm, workers and males 12 mm, widely distributed in heathland, moorland and (less commonly) other habitats, especially in southern England and north Scotland, and scarce in eastern England. In southern lowland areas it is often bivoltine, with first generation queens searching for nest sites in March, and males and new females produced in May. These queens may found new nests in June, to produce sexuals in late August–September. In northern and upland areas nests are not founded until June, with males in late August–September. This species nests in a variety of situations, including roof spaces; old bird nests (usually in holes); moss and leaf litter on the surface of soil; and underground in old mouse or vole nests. Forest cuckoo bumblebees may parasitise the nests. Visits are made to a variety of flowers for pollen and nectar.

Ilfracombe bumblebee Alternative name for broken-belted bumblebee.

Large garden bumblebee *B. ruderatus*, queen body length up to 22 mm, workers 16 mm, males 15 mm, found in England, commoner in the southern half, but generally declining in status. Recent records focus on extensive river valley systems in south and central England, where it forages on plants such as comfrey, yellow iris and marsh woundwort, and in legume-rich habitats. Queens emerge from hibernation from April to June; workers are present from May, and males and new females from July to October. Nests may be only shallowly underground. An old small mammal nest is used by the queen as a starting point for her nest. Nests may be parasitised by Barbut's cuckoo bee. *Ruderatus* is Latin indicating a ruderal habitat.

Mountain bumblebee Alternative name for bilberry bumblebee.

Northern white-tailed bumblebee *B. magnus*, queen body length up to 22 mm, workers and males 15 mm, with a scattered distribution, commoner in north Scotland, in a variety of habitats taking pollen and nectar from a variety of flowers. *Magnus* is Latin for 'great'.

Red-tailed bumblebee *B. lapidarius*, queen body length 20–2 mm, workers 11–16 mm, males 14–16 mm common and widespread and continuing to spread into north Scotland, in a range of habitats, including gardens, with pollen and nectar taken from many flower species (possibly with a preference for yellow flowers). Queens emerge from hibernation in March, workers are present from April, and males and new females from July to October. The underground nests are started in old mammal nests. Populations are large, with 100–300 workers. The species uses 'traditional' hibernation sites – north-facing banks, usually within open woodland. Queens use these sites year after year. Nests may be parasitised by hill cuckoo bees. *Lapidarius* is from Latin for 'stone'.

Short-haired bumblebee *B. subterraneus*, formerly widespread across southern England, as far north as Humberside, but numbers and distribution rapidly declined from the 1950s as species-rich grassland habitat disappeared. The species was last seen at Dungeness, Kent, in 1988. It was declared extinct in the UK in 2000. Attempted reintroduction using New Zealand examples failed in 2009, but in 2012 some 100 specimens were brought from Sweden to repopulate areas where it previously thrived. By the summer of 2013, workers of the species were found within 5 km of the reintroduction site at Dungeness. Continued introduction of queens from Sweden has been effective. In spring, fertile queens seek out old mouse nests as the foundation for their nests. Once the new sexual forms have emerged, the nest disintegrates, the mated queens go into hibernation, and the workers and males can be found on flowers before eventually dying.

Small garden bumblebee *B. hortorum*, the queen being variable in size, body length 19–22 mm, wing span 35–8 mm, workers almost as large, common and widespread in a range of especially lowland habitats taking pollen and nectar from a range of flowers, favouring red clover nectar if available and, with a tongue length of 15 mm, also foxglove and honeysuckle. Queens emerge from hibernation from March to June; workers are present from late April, and males and new females from July to October. Nests may be only shallowly underground. An old small mammal nest is used by the queen as a starting point for her nest. Mature nests are medium-sized, with about 100 workers. *Hortorum* is from Latin for 'gardens'.

Tree bumblebee *B. hypnorum*, first recorded in Britain near Southampton in 2001, having colonised naturally from continental Europe, spreading rapidly into much of England and south Wales, reaching Scotland by 2013, and Ireland (Co. Antrim) in 2014. This expansion is probably encouraged by a warming climate, but is also because this is a species naturally associated with open clearings in woodlands, and suburban gardens are an analogous habitat. Queens, which vary in size but are usually around 18 mm long, emerge from hibernation in late February or March. This species has two generations a year: males (16 mm long) have been found in late May and again at the end of August into early September. Queens have been noted in November and early December. This bee prefers to nest above ground, often inhabiting bird boxes, but is also found nesting under eaves, behind soffit boards and in cavity walls. It also nests in compost bins. Pollen and nectar are taken from a range of flowers. Greek *hypnos* = 'sleep'.

White-tailed bumblebee *B. lucorum*, queen length 18–22 mm and wingspan 36 mm, workers 12–18 mm long, common and widespread in a variety of habitats taking pollen and nectar from a variety of flowers. Queens seek nests from March onwards; males are found in July–October. Nests are underground in old mouse or vole nests. Nests may be parasitised by gypsy cuckoo bees. *Lucorum* is from Greek *leucos* = 'white'.

BUNT A smut disease of grasses and cereals caused by the fungi *Tilletia tritici* and *T. laevis* which parasitise the host plant and produce masses of soot-like spores in the leaves, grains or ears.

BUNTING Birds in the family Emberizidae (see also yellow-hammer), with Lapland and snow buntings sometimes placed in a separate family, Calcariidae. 'Bunting' has an obscure etymology, but might come from Scots *buntin*, meaning 'short and thick'.

Cirl bunting *Emberiza cirlus*, wingspan 24 cm, length 16 cm, weight 25 g, a resident Red List seed-eating breeding bird restricted to south-west England, found in fields, hedges and scrub, especially near the coast around Kingsbridge, south Devon, with an estimated 860 breeding pairs in 2009, low-key survey identifying 42 territories at two sites in Devon and 16 territories in Cornwall. Farmland conservation practices may by 2017 have raised the population to nearer 1,000. Clutch size is 3–4, incubation lasts 13–14 days, and the altricial chicks fledge in 12–13 days. The binomial comes from Old German *Embritz* = 'bunting' and the Italian (Bolognese) name for a bunting.

Corn bunting *E. calandra*, wingspan 29 cm, length 18 cm, male weight 53 g, females 41 g, a seed-eating Red-listed resident, widespread in lowland (arable) regions in Britain, formerly widespread but now extinct in Ireland because of a decline in cereal farming, last breeding in Co. Mayo in the 1990s. There were an estimated 11,000 territories in Britain in 2006–10, but in Britain there was a 90% decline in numbers between 1970 and 2015 (and by 34% between 1995 and 2014, though an increase of 24% between 2014 and 2015). Males are often polygynous. The grass nests, usually on the ground, generally have 4–5 eggs which hatch after around 13 days, the altricial chicks fledging in 11–13 days. Greek *calandra* = 'calandra lark'.

Lapland bunting *Calcarius lapponicus*, wingspan 26 cm, length 16 cm, weight 24 g, an Amber List winter or passage visitor, averaging around 700 birds a year, mainly seen along the east coast of England (mainly Lincolnshire and north Norfolk) and south-east Scotland, feeding on seeds. Latin *calcaria* = 'spurs' (from the long claw on the hind toe of each foot).

Reed bunting *E. schoeniclus*, wingspan 24 cm, length 16 cm, weight 21 g, both a resident and winter visitor/bird of passage, historically associated with reedbeds and wet moorland, but from the 1960s extending into farmland (favouring oil-seed rape), and in winter into gardens, and is now widespread in Ireland and, in Britain, common inland of The Wash, absent only from parts of south-west England and the Highlands. Some 250,000 territories were recorded in Britain in 2009. Numbers dropped by 63% between 1975 and 2015 (mainly declining in 1975–85), and by 8% between 2010 and 2015. It remains Amber-listed in Britain (Green-listed in Ireland). It eats seeds, taking invertebrates in the breeding season. Nests, in a bush or tussock, typically have 4–5 eggs, incubation lasts around 13 days, and fledging of the altricial chicks takes 12–14 days. Greek *schoinos* = 'reed'.

Snow bunting *Plectrophenax nivalis*, wingspan 35 cm, length 16 cm, male weight 42 g, females 35 g, a scarce Amber List resident seed-eating breeder in the Highlands, with an estimated 60 pairs in 2007, more widespread in northern and eastern Britain and parts of coastal Ireland with wintering continental birds, with perhaps up to 15,000 birds. Nests, in rock crevices, contain 4–6 eggs, incubation lasts 12–13 days, and the altricial chicks fledge in 12–14 days. Greek *plectron* + *phenax* = 'cock's spur' + 'imposter' (from the long claw on the hind toe of each foot), and Latin *nivalis* = 'snowy'.

Yellow bunting Alternative name for yellowhammer.

BUPRESTIDAE Family name of jewel beetles, with 17 species on the British list, most being restricted to southern England. They are 2–12 mm in length, and are often of a metallic green sheen, giving them their common name. Eleven species rely on dead wood (saproxylic), and the endangered *Melanophila acuminata* (found in Surrey and Hampshire heathland) is the only British beetle to be strongly associated with fire-damaged wood. Six species in the genera *Trachys* and *Aphanisticus* have larvae which mine leaves or stems. Larval exit holes in wood have a characteristic D shape and are good indicators of the existence of a colony. *Boupestris* is Greek for a beetle that, when eaten by cattle, caused them to swell up and die (*bous + pretho* = 'cow' + 'to swell').

Golden buprestid *Buprestis aurulenta*, 13–22 mm long, occasionally found in softwoods imported from North America; adults may emerge from wood up to 50 years after processing. Latin *aureus* = 'golden'.

Two-spotted oak buprestid Alternative name for oak jewel beetle.

BUR-GRASS Species of *Tragus*, mainly **European bur-grass** (*T. racemosus*), an annual neophyte casual on waste ground originating from wool shoddy. *Tragos* is Greek for 'goat', *racemosus* Latin for 'having racemes' (flower clusters).

BUR-MARIGOLD Annual plants in the genus *Bidens* (Asteraceae), the English name reflecting the barbed fruit, the genus name coming from Latin *bis + dens* = 'twice' + 'tooth', the 'two-toothed' nature referring to the achene bristles.

London bur-marigold *B. connata*, a North American neophyte present by 1817, naturalised by 1977 along the Grand Union Canal and adjoining watercourses, on the Thames in west Kent, and the Trent and Mersey Canal in Cheshire in 1997. *Connata* is Latin for 'fused together'.

Nodding bur-marigold *B. cernua*, scattered, less common in Scotland, and generally declining, growing on the margins of slow-flowing rivers and streams, and by ponds and meres, often where there is winter flooding, and also in ditches and marshes. *Cernua* is Latin for 'nodding', the flower heads sometimes drooping.

Trifid bur-marigold *B. tripartita*, scattered, less common in Scotland, and especially declining in south-east England, growing on nutrient-rich mud or gravel by ponds, and by slow-flowing rivers and streams, often in places wet in winter but exposed in summer; also in ditches, peat workings and other damp habitats. *Tripartita* is Latin for 'having three parts', referring to the commonly three-lobed leaves.

BUR-REED Emergent or floating aquatic rhizomatous herbaceous perennials with prickly burs in the genus *Sparganium* (Typhaceae), from the Greek used by Dioscorides.

Branched bur-reed *S. erectum*, common and widespread, growing as a marginal in shallow water in lakes, rivers, streams, canals and ditches, sometimes found as large stands in swamps. It grows in mesotrophic or eutrophic water, and is tolerant of eutrophication. *Erectum* is Latin for 'upright'.

Floating bur-reed *S. angustifolium*, found in upland Britain and Ireland in clear, oligotrophic water, most frequent in lakes but also in pools, streams, canals and ditches, preferring water 0.3–1.5 m deep, away from the most exposed shallows. Many sites are exposed to strong winds. *Angustifolium* is Latin for 'narrow-leaved'.

Least bur-reed *S. natans*, scattered in upland Britain in shallow, sheltered, mesotrophic, highly calcareous to acidic waters at the edges of lakes, or in ponds, slowly flowing streams and drainage ditches. *Natans* is Latin for 'floating'.

Unbranched bur-reed *S. emersum*, common and widespread, tolerant of disturbance, in still or slowly flowing, mesotrophic or eutrophic waters in lakes, ponds, streams, canals and ditches. *Emersum* is Latin for 'emerged'.

BURBOT *Lota lota* (Lotidae), length 50 cm, the only freshwater gadiform fish, living in cold water and thought to be extinct in Britain, the last record being of a fish taken from the River Cam in 1969. However, two records in 2010 were possibly of this species, from the river Eden (Cumbria) and Great Ouse (Cambridgeshire). The binomial comes from Old French *lotte* for a fish also named *barbot* in Old French which, with 'burbot' itself, comes from Latin *barba* = 'beard', referring to the single barbel.

BURDOCK Monocarpic herbaceous perennials in the genus *Arctium* (Asteraceae), the flowers transforming into hooked burs, the genus name coming from Greek, possibly for mullein.

Greater burdock *A. lappa*, a common and widespread archaeophyte in England, scattered elsewhere, on stream and riverbanks, woodland clearings, verges, tracks and waysides, field borders, waste ground and other disturbed places. *Lappa* is Latin for 'bur'.

Lesser burdock *A. minus*, common and widespread in woodland, scrub, hedgerows, roadsides, railway banks, rough pastures, dunes and waste ground. *Minus* is Latin for 'smaller'.

Wood burdock *A. nemorosum*, common and widespread in woodland, rough ground and disturbed habitats, often on calcareous soil. *Nemorosum* is from Latin for 'woodland'.

BURNER (FUNGUS) See charcoal burner.

BURNET (MOTH) Members of the genus *Zygaena* (Zygaenidae, Zygaeninae). The genus name, used by Fabricius, is Greek for 'hammerhead shark', clearly with no entomological meaning. 'Burnet' is Middle English, from Anglo-French *burnete*, from *brun* = 'brown'.

Five-spot burnet *Z. trifolii*, wingspan 28–33 mm, day-flying (May–August), locally common in south England and south and west Wales, with scattered records in the Midlands and East Anglia, on calcareous grassland, heath, damp meadow, marsh and coastal cliffs. Larvae feed on bird's-foot-trefoil. *Trifolii* refers to clover (*Trifolium*), though this is not a food plant.

Mountain burnet Alternative name for Scotch burnet.

Narrow-bordered five-spot burnet *Z. lonicerae*, wingspan 30–46 mm, often common, day-flying in June–July, with different subspecies in England/south Wales/Scottish Borders, the Isle of Skye (rare), and Ireland. It is found on rough grassland, chalk downland, verges and sea-cliffs, larvae feeding on clover, trefoil, sainfoin and vetch. Adults take nectar from a variety of plants, including thistle, knapweed and scabious. *Lonicerae* refers to honeysuckle (*Lonicera*), though this is not a food plant.

New Forest burnet *Z. viciae*, wingspan 22–32 mm, once found in the New Forest, Hampshire, the subspecies *ytenensis* has not been seen since 1927 and is presumed extinct. Ssp. *argyllensis* was discovered at a single location in Argyll, Scotland, in 1963 – a steep, south-facing, herb-rich, grassy slope with rocky ledges – the only known site for this day-flying moth. Larvae feed on meadow vetchling and bird's-foot-trefoil. Adults are active in July, feeding on nectar from wild thyme, occasionally other species. This is an endangered species (RDB category 1) of high priority for conservation. *Viciae* refers to vetches (*Vicia*), a food plant on the Continent.

Scotch burnet *Z. exulans*, wingspan 25–33 mm, day-flying in June–July, restricted to parts of the Cairngorms near Braemar, where it can be common. Larvae feed near bare rocky mountain tops (700–850 m altitude), mainly eating crowberry but also cowberry, bilberry and heather; they are very sunshine-dependent, basking and feeding only when the sun is out. *Exulans* is Latin for 'exile', from the remote type locality in the Alps.

Six-spot burnet *Z. filipendulae*, wingspan 30–8 mm, the commonest of Britain's day-flying burnet moths, found throughout, though mainly coastal in Scotland in grasslands, verges, woodland clearings and sea-cliffs. Larvae mainly feed on bird's-foot-trefoil. Adults, which fly in June–August, are particularly attracted to thistles, knapweeds and scabious. *Filipendulae* refers to dropwort (*Filipendula*), incorrectly stated by Linnaeus to be a food plant.

Slender scotch burnet *Z. loti*, wingspan 25–35 mm, now restricted to the Hebridean islands of Mull and Ulva, where it inhabits steep coastal grassy slopes and hillocks by the sea.

Adults fly in sunshine from mid-June to early July. The male can be found on milkworts and thyme; the female usually feeds on nectar from flowers of the larval food plant, bird's-foot-trefoil. An RDB moth, this a priority species for conservation. *Lotis* refers to bird's-foot-trefoil.

Transparent burnet *Z. purpuralis*, wingspan 30–4 mm, a scarce sunshine-favouring moth flying in June–July. Subspecies *caledonensis* is found on the Inner Hebrides and in a few sites on the Scottish mainland in Kintyre and parts of west Argyllshire, usually on coastal grassy slopes. Ssp. *segontii* has not been seen from sea-cliffs on the Lleyn Peninsula since 1962, and is presumed extinct. Ssp. *sabulosa* is found on the Burren, west Ireland. Larvae and adults feed on thyme. *Purpura* is Latin for 'purple', referring to the forewing colour.

BURNET (PLANT) Herbaceous perennial species of *Sanguisorba* (Latin *sanguis* + *sorbeo* = 'blood' + 'staunch', the herb having once been used for staunching wounds) and *Poterium* (Latin for 'goblet' or 'drinking vessel'), both in the family Rosaceae. 'Burnet' is Middle English, from Anglo-French *burnete*, from *brun* = 'brown', referring to the dark crimson-brown colour of the flowers.

Fodder burnet *P. sanguisorba* ssp. *balearicum*, a south European neophyte herb introduced in 1803, a widespread relic of cultivation that occurs, especially in central and southern England, as a casual or naturalised plant on field edges, tracksides, banks, roadsides and railways.

Great burnet *S. officinalis*, common and widespread in south-west and northern England, the Midlands and Wales, scattered and occasional elsewhere, on alluvial or peaty soils in unimproved pasture, hay meadow and marshy meadow, on riverbanks and lake shores and in base-enriched flushes on grassy heathland. Latin *officinalis* = 'of pharmacological value'.

Salad burnet *P. sanguisorba* ssp. *sanguisorba*, found in dry, infertile chalk and limestone grassland, and also on boulder-clay. It is often abundant on downland, but also grows in rock crevices, scree, quarries and on roadside banks. It is locally common in England, Wales and central Ireland, absent from elsewhere in Ireland and most of Scotland.

White burnet *S. canadensis*, a rhizomatous North American neophyte perennial cultivated by 1633, occasionally naturalised in marshes and beside streams and lakes in (mainly) central Scotland.

BURNET-SAXIFRAGE *Pimpinella saxifraga* (Apiaceae), a herbaceous perennial, common and widespread except in north and south Scotland and north and west Ireland, on well-drained, especially base-rich soils on grazed and ungrazed chalk and limestone downland, in rough pasture and other grassland, in woodland edges and open rides, and occasionally on roadsides and rough ground. *Pimpinella* is Latin meaning 'bipinnate', *saxifraga* = 'rock-breaker'.

Greater burnet-saxifrage *P. major*, a herbaceous perennial, locally common in the south-west, south-east and Midlands of England, and in south and west Ireland, scattered elsewhere, mainly on chalk and limestone soils, but also on clay, usually growing on roadsides, hedge banks, railway banks and woodland edges. *Major* is Latin for 'greater'.

BURREN GREEN *Calamia tridens* (Noctuidae, Xyleninae), wingspan 37–42 mm, a nocturnal moth found on limestone pavements in the Burren, Co. Clare. Adults fly in July–August. Larvae feed on blue moor-grass. Greek *calamos* = 'reed', and Latin *tridens* = 'three-toothed' (from wing markings).

BUSH-CRICKET Long-horned 'grasshopper' (order Orthoptera, suborder Ensifera, family Tettigoniidae), with large hind legs and long antennae, including both macropterous and brachypterous species.

Bog bush-cricket *Metrioptera brachyptera*, male length 11–18 mm, female 13–21 mm, wing length 5–10 mm and 6–9 mm, with a scattered distribution in Wales and (predominantly southern) England, in lowland bog and heathland and clearings in damp heathy woodland. The binomial is from the Greek for 'moderate wing' and 'short wing'.

Dark bush-cricket *Pholidoptera griseoaptera*, length 13–20 mm, wing length 3–5 mm, a common omnivore associated with a range of habitats, for example gardens, waste ground, bramble, old hedges, woodland edges and rides, and scrub in much of the southern half of England and coastal Wales. Greek *pholido* + *ptero* = 'flakes' + 'wing', and Latin *griseo* = 'grey' + Greek *aptero* = 'wingless'.

Great green bush-cricket *Tettigonia viridissima*, Britain's largest bush-cricket, male length 40–50 mm, female 42–54 mm, wing length 33–40 mm and 34–44 mm, an omnivore favouring grassland, bramble, bracken, woodland and coastal scrub on light dry soil in southern England, East Anglia and south Wales. *Tettigonia* is Latin for 'leafhopper', in turn from Greek *tettigonion*, the diminutive of the onomatopoeic *tettix* = 'cicada'. *Viridissima* is Latin for 'very green'.

Grey bush-cricket *Platycleis albopunctata*, male length 20–5 mm, female 20–8 mm, wing length 15–19 mm and 14–21 mm, a scarce omnivore found in coarse grass along the coast from Pembroke to Suffolk, but particularly from Cornwall to Kent. Greek *platys* + *cleis* = 'broad' + 'key', and Latin *albopunctata* = 'white-spotted'.

Oak bush-cricket *Meconema thalassinum*, length 13–17 mm, wing length 11–13 mm, in woodland (especially in the canopy of old trees), hedgerows and gardens in southern Britain, feeding on small invertebrates. The male does not stridulate but signals to females by rapidly pattering his legs on a leaf. *Thalassinum* is Greek for 'of the sea'.

Roesel's bush-cricket *Metrioptera roeselii*, male length 13–26 mm, female 15–21 mm, wing length 7–10 mm and 4–8 mm, though f. *diluta* (becoming common in hot summers) is macropterous, with wing length 21–2 mm and 23–5 mm, increasingly common in south, south-east and eastern England in damp meadow and other grassland as well as saltmarsh and dunes in coarse vegetation; currently expanding its range westwards and northwards into urban wasteland and agricultural set-aside, with roadside rough grassland and scrub providing a habitat corridor. August Rösel von Rosenhof (1705–59) was a German entomologist.

Speckled bush-cricket *Leptophyes punctatissima*, male length 9–16 mm, female 11–18, common in open woodland, scrub, gardens and hedgerows in much of southern Britain including coastal Wales, mostly active at dusk and during the night. *Punctatissima* is Latin for 'very speckled'.

BUSH WEED **Straggly bush weed** *Rhodomela confervoides* (Rhodomelaceae), a widespread red seaweed, fronds up to 30 cm long, found on rocks and shells in intertidal pools. Greek *rhodon* + *melas* = 'rose-coloured' + 'black', and Latin for 'resembling the genus *Conferva*'.

BUSTARD Members of the family Otididae. **Little bustard** *Tetrax tetrax* and **MacQueen's bustard** *Chlamydotis macqueenii* are scarce accidentals.

Great bustard *Otis tarda*, wingspan 225 cm, length 90 cm, male weight 12 kg (with mute swan, the heaviest flying bird), females 4.2 kg, became extinct in Britain in 1832, but since 2004, >100 chicks have been imported from Russia, and since 2013, Spain (these birds being genetically closest to the original British population) for release on Salisbury Plain, the project run by the Great Bustard Group. In 2011 the group joined Natural England, RSPB and the University of Bath, after the RSPB had secured a €2.2 million (£1.8 million) EU LIFE+ grant on behalf of the partnership. The partnership ended in 2014 following 'differences in the preferred approach'. In 2009, one pair produced chicks, with four pairs doing so in 2010. Fifteen birds were living wild in 2011. At least 30 more young birds were released in 2015, adding to the then total of 26 adult and subadult birds, bringing the population closer to the point where it would become self-sustaining and thereby continue to grow through natural reproduction. Three wild chicks from the 2017 season joined the population. 'Bustard' probably derives from an Anglo-Norman French blend of Old French *bistarde* and *oustarde*. The genus name comes from the Greek for this bird; the specific name probably derives from Latin *tarda* = 'slow', reflecting the bird's walking pace, but it might be from a Hispanic name for the species.

BUTCHER BIRD Alternative name for red-backed shrike, and **great(er) butcher bird** for great grey shrike.

BUTCHER'S-BROOM *Ruscus aculeatus* (Asparagaceae), a dioecious, evergreen, rhizomatous shrub, native as far north as Suffolk and south Wales in dry woodland and hedgerows, and on cliffs and rocky ground near the sea. It is also naturalised further north and in Ireland in similar habitats, and in churchyards and near habitation. It reproduces using creeping rhizomes, and by (often bird-sown) seed. *Ruscum* is Latin for 'butcher's-broom', *aculeatus* = 'prickly' or 'thorny'.

Spineless butcher's-broom *R. hypoglossum*, a south-east European neophyte shrub, grown in Tudor gardens, naturalised on woodland edges and other shady habitats at a few places in south and central England. Greek *hypo* + *glossa* = 'under' ('slightly') + 'tongue'.

BUTOMACEAE Family name of flowering-rush, from Greek *bous* + *tomos* = 'ox' + 'cut', referring to the sharp leaf margins.

BUTTER (FUNGUS) (Black) witches' butter *Exidia glandulosa* (Auriculariaceae), a widespread but infrequently found, inedible fungus, evident throughout the year, mainly fruiting in late summer and autumn, on rotting logs and branches of broadleaf trees and on wound tissue on living wood. In wet weather it turns black and jelly-like, but during prolonged dry spells it shrinks to a series of cone-shaped olive-brown crusts. The English name reflects the butter-like consistency and greasy surface when wet together with its dark colour, though another suggestion is that it had the power to counteract witchcraft if thrown on to a fire. *Exidia* is Latin for 'exuding' or 'staining', *glandulosa* for 'provided with glands' (i.e. the papillae on the cap surface).

BUTTERBUR (MOTH) *Hydraecia petasitis* (Noctuidae, Xylininae), wingspan 44–50 mm, nocturnal, local in damp grassland, marsh and riverbanks throughout much of Britain. Adults fly in August–September. Larvae feed from April to July in the stem, then in the roots of butterbur. Greek *hydor* + *oiceo* = 'water' + 'to dwell' (for the damp habitat), and *petasitis* = 'butterbur'.

BUTTERBUR (PLANT) *Petasites hybridus* (Asteraceae), a dioecious rhizomatous perennial herb, common and widespread except for the Highlands, found on moist, fertile, often alluvial soils by streams and ditches, in wet meadow, marsh and flood plains, and on roadsides, spreading mostly from rhizome fragments. Female plants are frequent only in northern and central England. Male-only colonies are probably single clones, many perhaps deliberately planted as a source of pollen and nectar for hive bees. Greek *petasos* = 'like a broad-brimmed hat', describing the leaves.

Giant butterbur *P. japonicus*, a Japanese neophyte herb naturalised by streams and other damp habitats, scattered in Britain as far north as the Scottish Central Lowlands.

White butterbur *P. albus*, a European neophyte herb, common and widespread in Britain, especially north-west England and much of Scotland, and in Northern Ireland, in rough ground and woodland. *Albus* is Latin for 'white'.

BUTTERCUP Herbaceous plants in the genus *Ranunculus* (Ranunculaceae), from the diminutive of Latin *rana* = 'frog', referencing the plants growing in marsh and bog. The common name dates from the late eighteenth century, from the imagined

connection between the flower colour and the yellow of cream and butter.

Aconite-leaved buttercup *R. aconitifolius*, perennial, introduced from Europe by the Tudor period, naturalised in damp woodland, in ditches and by streams, and also occurring as a casual. The specific name is Latin for 'aconite-leaved'.

Bulbous buttercup *R. bulbosus*, perennial, common and widespread on well-drained, neutral or calcareous soils in dry grassland and dunes.

Celery-leaved buttercup *R. sceleratus*, annual, widespread in England, more scattered and mostly coastal elsewhere, found in shallow water or wet, disturbed, nutrient-rich mud, especially at the edges of ponds, ditches and streams poached by drinking livestock. It is salt-tolerant and frequent on grazed estuarine marshes. *Sceleratus* is Latin for 'hurtful', this being the most toxic buttercup (containing 2.5% protoanemonin): when the leaves are crushed, they bring out sores and blisters on human skin.

Corn buttercup *R. arvensis*, an annual archaeophyte of arable land on a range of soils, mostly in England. Because seeds are long-lived, plants may reappear on disturbed waste ground, in gardens, or new roadside verges, but population and range are declining with intensified agriculture and better seed screening. *Arvensis* is from Latin for 'field'.

Creeping buttercup *R. repens*, perennial, common and widespread, of very wide ecological tolerance though avoiding acidic soil, and most typical of disturbed habitats on damp or wet nutrient-rich soils, for example woodland rides, ditch sides, gardens and waste ground. It also occurs in damp grassland and dune slacks. *Repens* is Latin for 'creeping'.

Goldilocks buttercup *R. auricomus*, perennial, widespread and locally abundant, characteristic of deciduous woodland on chalk, limestone and other basic soils. It also grows in scrub, on roadsides and in churchyards, occasionally on moorland. Latin *auris* + *coma* = 'ear' + 'hair'.

Hairy buttercup *R. sardous*, annual, growing in damp coastal pasture, pond edges, wet hollows, moist road verges, farm tracks and gateways, generally restricted to thin turf or disturbed areas on damp, neutral, moderately fertile soils, mostly near the coast in southern and eastern England. *Sardous* is Latin for 'Sardinian'.

Jersey buttercup *R. paludosus*, perennial, growing in grassland which is damp in winter but dry in summer, found only on Jersey (discovered in 1872), thought to be an extreme rarity, but recently found at a number of sites on the west and south coast. *Paludosus* is from Latin for 'swamp'.

Meadow buttercup *R. acris*, perennial, common and widespread in damp meadow and pasture on a variety of soils, though avoiding very dry or acid conditions. It is characteristic of unimproved hay and water meadows and damp road verges, and also grows on dune grassland, in montane flushes and in tall-herb communities on rock ledges. Latin *acer* = 'bitter'.

Rough-footed buttercup *R. muricatus*, annual, introduced from south Europe, naturalised as a weed of cultivated ground in south-west England, particularly in bulb fields and gardens in the Scilly Isles, and as a grain and birdseed casual elsewhere. *Muricatus* is Latin for 'roughened'.

Small-flowered buttercup *R. parviflorus*, an annual of dry disturbed habitats on neutral and calcareous soils, especially near the coast in south-west England and south Wales, rare and declining elsewhere. Habitats include turf on cliff edges, slopes and banks, rabbit scrapes, tracks, poached gateways, building sites and gardens. *Parviflorus* is Latin for 'small-flowered'.

St Martin's buttercup *R. marginatus*, annual, introduced from the Mediterranean, found since 1950 as a naturalised weed in south-west England and in bulb fields in the Scilly Isles, and as a rare grain, birdseed and wild-flower mixture casual elsewhere. *Marginatus* is Latin for 'margined'.

BUTTERFISH *Pholis gunnellus* (Pholidae), eel-like, length up to 25 cm, probably found around most of the coastline from mid- to low-tide mark among seaweed, under rocks and in crevices, and subtidally to 40 m. It can remain above the waterline at low tide and breathe air. Diet is of crustaceans, molluscs and polychaetes. Spawning is in late winter and spring. The English name reflects the slimy skin. Greek *pholis* = 'scale'.

BUTTERFLY Day-flying insects of the order Lepidoptera with large patterned wings, and the club-tipped antennae, lack of a frenulum (a filament arising from the hind wing and matching up with barbs on the forewing) and enlargement of the humeral lobe of the hind wing that separates the superfamily Papilionoidea (in the suborder Rhopalocera, named from the Greek for 'club' and 'antennae') from other Lepidoptera (moths, though the Hesperiidae are sometimes distinguished as a separate superfamily). In this taxonomic scheme moths belong to the suborder Heterocera (Greek for 'varied' and 'antennae'). Many authorities argue that distinguishing butterflies from moths has no taxonomic merit. Of the 56 species for which there is robust data, 29 declined in numbers during 2005–14, and 25 species increased. Families are Papilionidae (swallowtail), Riodinidae (Duke of Burgundy), Hesperiidae (skippers), Pieridae (whites, yellows), Nymphalidae (browns, fritillaries) and Lycaenidae (coppers, hairstreaks, blues). 'Butterfly' may have originated from the yellow colour of male brimstones.

BUTTERFLY-BUSH *Buddleja davidii* (Scrophulariaceae), a common and widespread deciduous Chinese neophyte shrub introduced in the 1890s, now very well established in non-upland regions on waste ground, by railways, in quarries, on roadsides and generally in urban habitats, where it often grows on walls and neglected buildings. It prefers dry, disturbed sites where large populations can develop from its wind-dispersed seed. Its attractiveness to butterflies prompts its English name. Its genus name honours the English botanist Rev. Adam Buddle (1662–1715), the specific name the French missionary Père Armand David (1826–1900). See also orange-ball-tree.

Alternate-leaved butterfly-bush *B. alternifolia*, a deciduous Chinese neophyte shrub introduced by 1915, planted in woodland and amenity areas, and naturalised, often self-sown, as a garden escape in woodland and hedges, and on banks at a few sites in England. *Alterniflora* is Latin for 'alternating flowers'.

Weyer's butterfly-bush *Buddleja* x *weyeriana* (*B. davidii* x *B. globosa*), a deciduous shrub occasionally found as a naturalised garden escape in hedgerows and on roadsides and rough ground. It was raised in Dorset in 1914, and first recorded in the wild in Cornwall in 1976. The hybrid was developed by Major William van de Weyer of Dorset in the 1910s.

BUTTERFLY-ORCHID Herbaceous perennials in the genus *Platanthera* (Orchidaceae), from Greek *platys* + *anthos* = 'broad' + 'flower'.

Greater butterfly-orchid *P. chlorantha*, widespread and locally common, though overall declining, found in a variety of habitats, usually on well-drained calcareous soils, for example downland, rough pasture, hay meadow, scrub and woodland, and sometimes on dunes and railway embankments. *Chlorantha* is Greek for 'green-flowered'.

Lesser butterfly-orchid *P. bifolia*, widespread and locally common, though scarce in central and eastern England, on heathy pasture, grassland, open scrub, woodland edge and moorland, often among bracken, on a range of acidic and calcareous soils on sands, gravels and clays. It tolerates high soil moisture, also being found in acidic bog and calcareous fen. *Bifolia* is Latin for 'two-leaved'.

BUTTERWORT Insectivorous perennial herbs in the genus *Pinguicula* (Lentibulariaceae), the name dating from the late sixteenth century, from the plant's supposed ability to keep cows in milk, thereby maintaining a butter supply. *Pinguicula* is Latin for 'fat' or 'greasy'.

Common butterwort *P. vulgaris*, rosette-forming, locally common in northern England, Wales, Scotland and Ireland, scattered elsewhere in England, found in damp, nutrient-poor habitats, overwintering as a rootless bud, in bogs, in crevices of wet rocks, in base-poor as well as base-rich open flushes, and in open bryophyte-dominated fen communities. *Vulgaris* is Latin for 'common'.

Large-flowered butterwort *P. grandiflora*, rosette-forming, overwintering as a rootless bud which also functions as a vegetative propagule, found on wet rocks, flushed moorland and acidic bogs, locally common in south-west Ireland, and planted in damp places and occasionally naturalised in England and Wales. *Grandiflora* is Latin for 'large-flowered'.

Pale butterwort *P. lusitanica*, retaining its insect-trapping leaves throughout winter, locally frequent in south-west England, Dorset/Hampshire, south-west Wales, north and west Scotland and much of Ireland, growing on bare peat and at the base of tussocks by small water channels in moorland, acidic flushes and wet heathland, often where trampled by livestock or deer. *Lusitanica* is Latin for 'Portuguese'.

BUTTON (CADDIS FLY) See Welshman's button.

BUTTON (MOTH) Members of the family Tortricidae (Tortricinae), generally nocturnal, mostly species of *Acleris*, from Greek *acleros*, without entomological meaning.

Ashy button *A. sparsana*, wingspan 18–22 mm, widespread and common in woodland, gardens and hedgerows. Adults fly in August–October. Larvae feed on sycamore and beech leaves. *Sparsus* is Latin for 'speckled', from the forewing pattern.

Buff button *A. permutana*, wingspan 15–20 mm, scarce but found locally on limestone downland and sand dunes in parts of southern England, south Wales and the Burren, Co. Clare. Adults fly in August–September. Larvae feed on shoots and leaves of burnet rose and blackthorn. *Permutana* is Latin for 'change', here implying variation.

Caledonian button *A. caledoniana*, wingspan 13–15 mm, found locally on high moorland from 200 to 700 m and in bogs at lower elevations in northern parts of Britain, where widely distributed on hills throughout Scotland, northern England and Wales, southwards to Herefordshire and Exmoor, with unconfirmed records from Kent, Dorset and Devon. Adults fly during the day from July to September. Larvae mine leaves and shoots of bilberry and similar upland plants.

Dark-streaked button *A. umbrana*, wingspan 18–23 mm, a rarity found in woodland, fens and marshes at a few sites in England and Wales. Adults fly from July onwards, overwintering as adults. Larvae feed on leaves of a number of deciduous tree species. *Umbra* is Latin for 'shade', from the dark streak on the forewing.

Dark-triangle button *A. laterana*, wingspan 15–20 mm, widespread and common in woodland. Adults fly in August–September. Larvae feed on the leaves of a variety of trees and shrubs. Latin *later* = 'brick', from the brown and reticulate wing pattern suggesting a brick wall.

Elm button *A. kochiella*, wingspan 15–18 mm, local in hedgerows, gardens, orchards and parks throughout southern England, north to Yorkshire, and south Wales. With two generations, adults fly in June–July then from September, the latter generation overwintering to reappear in spring. Larvae feed on elm leaves. Gabriel Koch (1807–81) was a German entomologist.

Ginger button *A. aspersana*, wingspan 13–16 mm, widespread and locally common on grassland, downland, heaths and other open areas. Adults fly in July–August. Larvae feed on the leaves of wild strawberry, marsh cinquefoil, meadowsweet and other Rosaceae. *Aspersus* is Latin for 'sprinkled', from the forewing pattern.

Grey birch button *A. logiana*, wingspan 18–22 mm, found locally in birch woodland in parts of the Highlands, also since the 1990s appearing as a migrant in southern England. Adults emerge from September and overwinter, being found through to April. Larvae feed between spun leaves of birch. Latin *logium* = 'stage platform', from raised scales on the forewing.

Heath button *A. hyemana*, wingspan 12–19 mm, widespread if local on heathland and moorland. Adults emerge in September–October, overwinter and reappear in spring. Larvae feed on the shoots of heather, heaths and bilberry. *Hiems* is Latin for 'winter'.

Lichen button *A. literana*, wingspan 18–22 mm, found locally in oak woodland throughout much of Britain, rare in Ireland. Adults fly in August–September, overwinter, then reappear in April–May. Larvae feed on oak leaves. *Litera* is Latin for 'letter', from the black forewing markings.

Maple button *A. forsskaleana*, wingspan 12–17 mm, common in woodland and gardens throughout England, Wales and Ireland. Adults fly in July–August. Larvae feed on leaves of field and Norway maples and sycamore from September to June. Peter Forsskål (1732–63) was a Swedish entomologist.

Marbled button *A. maccana*, wingspan 19–25 mm, scarce, found in woodland and moorland in parts of the Highlands. Adults fly in August–October. Larvae feed on leaves of bilberry and bog-myrtle. *Maccus* is Latin for 'clown', possibly from contrasting colours on the wings.

Marsh button *A. lorquiniana*, wingspan 15–20 mm, a rarity restricted to the fens of East Anglia and Cambridgeshire, and a few sites in Hampshire and on the Isle of Wight. With two generations, adults fly in June–July and in September–October. Larvae mine the shoots, flowers and seeds of purple-loosestrife, living in a silk mesh. Pierre Lorquin (1797–1873) was a French entomologist.

Meadowsweet button *A. shepherdana*, wingspan 13–16 mm, scarce, found locally in fens, marshes, riverbanks and other damp areas in parts of southern England and East Anglia. Adults fly in August–September. Larvae feed on meadowsweet leaves. Edwin Shepherd (d.1883) was an English entomologist.

Northern button *A. lipsiana*, wingspan 17–24 mm, a rarity found on moorland and mountains in parts of Scotland and northern England. Adults fly between August and October, after which they overwinter to reappear in spring. Larvae feed in spun shoots or leaves of bilberry, cowberry and bog-myrtle. *Lipsiana* is Latin for 'from Leipzig', the German location of the type specimen.

Perth button *A. abietana*, wingspan 18–25 mm, scarce in coniferous plantations in parts of northern England and Scotland, first discovered in Perthshire in 1965. Adults fly from August, overwinter, and reappear in spring. Larvae feed on the needles of conifers. *Abies* refers to these food plants (Norway spruce *Picea abies*, firs *Abies* spp.).

Rusty birch button *A. notana*, wingspan 15–19 mm, widespread and common in woodland. With two generations (except in the far north), adults fly in July then October, the latter generation overwintering to reappear in spring. Larvae feed on birch leaves. *Nota* is Latin for 'spot', from the wing markings.

Rusty oak button *A. ferrugana*, wingspan 14–18 mm, widespread and common in woodland and heathland, though less common in Ireland. Adults fly in July and again in September–October, the second generation overwintering to reappear in spring. Larvae mainly feed on oak leaves, but have also been found on goat willow. *Ferrugo* is Latin for 'rust', from the forewing colour.

Sallow button *A. hastiana*, wingspan 17–24 mm, widespread and common in woodland, hedgerows, fens and freshwater margins. There are two generations over much of its range, with adults flying in June–July and again from August, after which they hibernate. In parts of Scotland there is just one brood, flying from September onwards. Larvae feed on willow leaves. Reinhart Hast was an eighteenth-century Finnish naturalist.

Small purple button *Spatalistis bifasciana*, wingspan 12–14 mm, scarce, found locally in woodland edges in parts of southern England and south Wales. Adults fly in May–June. Larvae may feed internally in berries or fruits of buckthorn, alder buckthorn

and dogwood, and they have also been found in dead leaves of oak and sweet chestnut. Greek *spatale* = 'luxury' (from the ornamental wings), and Latin *bifasciana* = 'two fasciae' (broad bands on the wing).

Sweet-gale button *A. rufana*, wingspan 17–24 mm, a rarity found locally on moorland and damp woodland scattered throughout Britain. Adults fly between August and October, after which they overwinter to reappear in spring. Larvae feed on bog-myrtle leaves, but have also been found on sallow, white poplar and meadowsweet. Latin *rufus* = 'red', from the forewing colour.

Tufted button *A. cristana*, wingspan 18–22 mm, commonly found in woodland throughout much of England (as far north as Lincolnshire) and Wales, plus one record from Co. Kerry. Adults fly from August to November, then again from March to May after hibernation. Larvae feed on leaves of a number of tree and shrub species, though mainly hawthorn and blackthorn. *Crista* is Latin for 'tuft', from the scale tuft on the forewing.

Viburnum button *A. schalleriana*, wingspan 15–20 mm, found locally in gardens and woodland throughout much of England and Wales, and a few sites in western Ireland. Adults fly from August to October, hibernate, then reappear in spring. Larvae feed on the leaves of guelder-rose and wayfaring-tree. Johann Schaller (1734–1813) was a German entomologist.

White-triangle button *A. holmiana*, wingspan 10–15 mm, widespread and common in gardens, hedgerows and woodland, though absent in north Scotland. Adults fly in July–August. Larvae feed on leaves of rosaceous trees and shrubs. *Holmiana* is Latin for 'of Stockholm', Sweden.

Yellow oak button *Aleimma loeflingiana*, wingspan 14–19 mm, commonly found in oak woodland, areas with scattered trees, parkland and gardens. Adults fly in July–August. Larvae feed mainly on oak leaves. Greek *aleimma* = 'oil', from the smudged greasy-looking forewing markins. Pehr Löfling (1729–56) was a Swedish botanist.

Yellow rose button *A. bergmanniana*, wingspan 10–14 mm, a widespread late-afternoon and nocturnal moth common in gardens, hedgerows and scrub. Adults fly in June–July. Larvae feed on leaves of roses. T. Bergmann (d.1784) was a Swedish entomologist.

BUTTON-GRASS *Dactyloctenium radulans* (Poaceae), an annual Australian neophyte grass, recorded in the wild in 1948, occasionally found in southern England as a casual on waste ground, originating from wool shoddy. Greek *dactylos* + *ctenion* = 'finger' + 'little comb', and Latin *radulans* = 'like a scraper'.

BUTTONWEED (ALGA) Alternative name for the brown seaweed thongweed.

BUTTONWEED (PLANT) *Cotula coronopifolia* (Asteraceae), an annual to perennial South African neophyte herb, widely grown in gardens since 1683, and also a wool alien, naturalised in wet habitats in England and Wales on tidal saline mud, and increasingly in inland sites, for example in areas of winter-flooded mining subsidence. The flower heads look like thick buttons, giving the English name. Greek *cotule* = 'small cup', from the cupped feature at the base of the leaves, and Latin *coronopifolia* = 'having leaves like *Coronopus*' (= swine-cress *Lepidium coronopus*), itself from Greek *corone* + *pous* = 'crown' + 'foot', referring to the cleft leaves.

Annual buttonweed *C. australis*, an Australian neophyte, very scattered in England and the Scottish Borders as a casual wool alien in arable fields and rough ground, though naturalised in south Devon since 1946. *Australis* is Latin for 'southern'.

BUXACEAE Family name for box and carpet box (Latin *buxus* = 'box tree').

BUZZARD *Buteo buteo* (Accipitridae), wingspan 120 cm, length 54 cm, male weight 780 g, females 1 kg, the commonest and most widespread British bird of prey, with 68,000 pairs

(57,000–79,000) estimated in 2009, greatest numbers being in Scotland, Wales, the Lake District and south-west England. There was an increase in numbers in Britain by 465% over 1970–2015, and by 80% between 1995 and 2015. It is found in a range of habitats, particularly woodland, moorland, scrub, pasture, arable, marsh, bog and villages. Nests are of twigs in trees, clutch size is 2–3, incubation takes 34 days, and the altricial young fledge in 44–52 days. Diet is of small mammals, birds and carrion, as well as earthworms and large insects when other prey is scarce. 'Buzzard' is late Middle English, from Old French *busard*, based on Latin *buteo* = 'falcon', which in turn provides the binomial. See also honey-buzzard.

Rough-legged buzzard *B. lagopus*, wingspan 135 cm, length 55 cm, male weight 900 g, females 1.3 kg, a winter visitor averaging around 70 records per annum, though rather more in influx years reflecting food shortage in Scandinavia, mainly seen in East Anglia and Cornwall/Scilly on coastal marsh and farmland feeding on small mammals. Greek *lagos* + *pous* = 'hare' + 'foot', a reference to the feathered feet.

BYRRHIDAE From Greek *byrrhos* = 'flame-coloured', the family name of pill beetles, with 13 species in 7 genera in Britain. These are small (1–10 mm) beetles which when disturbed fold their appendages into grooves on the underside of the body and play dead, giving the appearance of a seed or pill. Generally found in litter and moss or under stones, they are largely found in heathland and moorland. Adults and larvae both feed on mosses, probably the only beetle family to do so. See pill beetle.

BYTURIDAE From Greek *bytouros* = 'tusk', the family name of fruitworm beetles (3–5 mm length), with two species of *Byturus*. See raspberry beetle under beetle.

C

CABBAGE Members of the family Brassicaceae, specifically *Brassica oleracea*, a herbaceous biennial or perennial, native on some sea-cliffs, mostly on chalk and limestone in the south but also on other base-rich substrates. It also grows in coastal grassland and in quarries inland. Elsewhere it is a casual garden escape in waste places and on roadsides. 'Cabbage' is late Middle English, from Old French *caboche* = 'head'. *Brassica* is Latin for 'cabbage', *oleracea* = 'green/edible vegetable'.

Varieties are wild cabbage (var. *oleracea*), and a number of cultivated forms: **cabbage** (var. *capitata*, meaning 'having a head'), **Savoy cabbage** (var. *sabauda*, meaning 'of Savoy'), **kale**, a northern English form of 'cole' (var. *viridis* = 'green'), **cauliflower** (late sixteenth-century, from Old French *chou fleuri* = 'flowered cabbage', probably from Italian *cavolfiore* or modern Latin *cauliflora*; the original English form *colieflorie* or *coleflory* had its first element influenced by cole) and **broccoli** (mid-seventeenth-century, from the Italian plural of *broccolo* = 'cabbage sprout, head', a diminutive of *brocco* = 'shoot', based on Latin *broccus* = 'projecting', var. *botrytis*, from Greek *botrys* = 'cluster of grapes'), **Brussels-sprout** (from Old English *sprutan* = 'to sprout') (var. *gemmifera*) and **kohl-rabi** (dating from the early nineteenth century, ultimately from Latin *caulis + rapa* = 'cole' + 'turnip', var. *gongylodes*, from Greek *gongulos* = 'round').

Bastard cabbage *Rapistrum rugosum*, a widespread, locally common annual or short-lived perennial European neophyte, introduced by 1739, often brought in with grain and birdseed, mainly found as a casual of waste ground, but becoming naturalised in a variety of habitats, especially in England, where it is sometimes invasive, for example arable fields and open grassland. Latin *rapa* = 'turnip' (an alternative American name is turnipweed), and *rugosum* = 'wrinkled'.

Isle of Man cabbage *Coincya monensis* var. *monensis*, an annual or short-lived perennial, on the west coast of northern England and central Scotland, as well as the Isle of Man, mainly by the sea on open dunes and on the strand-line, rarely in fields and hedge banks near the sea. Seed might be dispersed by the sea. *Coincia* is Greek for 'stones', *monens* Latin for 'warning'.

Lundy cabbage *C. wrightii*, a short-lived perennial endemic to Lundy, found on sparsely vegetated rock on sea-cliffs, rapidly colonising bare soil after disturbance. Numbers fluctuate but overall are stable, though the species requires protection from grazing and shrub invasion, and is listed as a priority for conservation in the UK BAP.

Pale cabbage *B. tournefortii*, an annual Mediterranean neophyte, a rare and declining casual of waste ground in England, derived from wool shoddy and grain impurities. Joseph de Tournefort (1656–1708) was a French botanist.

Steppe cabbage *R. perenne*, a biennial or perennial European neophyte present since 1789, growing in arable fields, on waste ground and around docks where it is introduced with grain and birdseed. While usually casual, it is naturalised in a few places. *Perenne* is Latin for 'perennial'.

Wallflower cabbage *C. monensis* var. *cheiranthos*, an annual or biennial European neophyte naturalised especially in south Wales by docks, roadsides and railways, but mainly found as a casual in waste ground. Greek *cheiranthos* = 'resembling wallflower'.

CABBAGE-PALM *Cordyline australis* (Asparagaceae), an evergreen woody perennial New Zealand neophyte often planted by the sea, with seedlings produced in the Channel Islands, Cornwall, by the Thames in Surrey and Greater London, the Isle of Man, and east and south-east Ireland. Greek *cordyle* = 'club', referring to the root, and Latin *australis* = 'southern'.

CABBAGE ROOT FLY *Delia radicum* (Anthomyiidae), 6 mm length, common and widespread in England and Wales, less so elsewhere, a field and garden pest of brassicas, oriental greens, and radish, and probably the main pest of swedes and turnips. Adults can smell volatile chemicals produced by the leaves of the host plant from 20 m. Females land on the plant then move to the soil surface under which she lays her eggs. The 8 mm-long larvae feed on roots. There are three generations during summer, the first generation in late spring-early summer often the most damaging because roots are less developed. *Radicum* comes from Latin *radix* = 'root'.

CACAO MOTH *Ephestia elutella* (Pyralidae, Phycitinae), wingspan 15–20 mm, introduced, widespread, naturalised, nocturnal, found in dried goods in barns, warehouses and granaries, where it can be a pest. Larvae feed on cocoa, tobacco, hay, decaying vegetable matter, dried fruit, nuts, cereals and grain. Greek *ephestios* = 'beside the hearth' (i.e. 'domestic'), referring to its presence in buildings, and Latin *elutus* = 'insipid', from the wing colour.

False cacao moth *E. unicolorella*, wingspan 14–20 mm, local in gardens and farmland throughout much of southern England. Adults fly in June–September. Larvae feed on detritus and dead plant material, including leaves. *Unicolorella* is Latin for 'all one colour'.

CADDIS (FLY) Insect in the order Trichoptera whose larva (caddis worm) lives underwater, most in a protective case of sand, stone, twig, etc., some (see Hydropsychidae) using nets to catch detritus or prey, some (see Polycentropodidae) living in tubes or using trumpet-nets, others (see Rhyacophilidae) being free-living.

Land caddis *Enoicyla pusilla* (Limnephilidae), cased length 8–9 mm, diameter 1.5–2 mm, the case largely made up of grains of sand and pieces of oak leaf, scarce, found in the west Midlands (especially in the Wyre Forest, Worcestershire) in moist tree litter feeding on dead leaves. Females have vestigial wings. Greek *enoiceo* = 'to inhabit', and Latin *pusilla* = 'small'.

Square caddis Alternative name for small silver sedge, from the square cross section of the instar V case.

CADLINA **White Atlantic cadlina** *Cadlina laevis* (Cadlinidae), a nudibranch slug, up to 3.2 cm length, widespread and common among red algae-covered rocks from the shore to the sublittoral, feeding on sponges. The binomial may be from Greek *cados* = 'urn', and Latin *laevis* = 'smooth'.

CAENIDAE Mayfly family, with two genera: *Brachycercus* (see broadwings) and seven, perhaps eight, species of *Caenis* (see angler's curse). Nymphs are partly carnivorous, slow-moving, freshwater bottom-dwelling individuals; adults do not have hind wings. Caenis was a figure in Greek mythology.

CAGE (FUNGUS) **Red cage** *Clathrus ruber* (Phalales, Phallaceae), an inedible fungus local in the southern half of England, found in summer and autumn, with a latticed red sphere extending from a partially submerged 'ball', and foul-smelling, on compost in parks and gardens, occasionally on woodchip and bark mulch. The binomial is Latin for 'lattice' and 'red'.

CALAMINT Herbaceous perennials in the genus *Clinopodium* (Lamiaceae) (Greek *cline + podi* = 'bed' + 'foot'). The English name is from Greek *cala + menta* = 'good' + 'mint'. See also basil.

Common calamint *C. ascendens*, rhizomatous, local and scattered, commonest south of a line from the Severn estuary to the Wash, on hedge banks, verges, rough scrubby grassland and rocky outcrops, usually on dry calcareous soils. Latin *ascendens* = 'climbing up'.

Lesser calamint *C. calamintha*, mostly found in central-east and south-east England, scarce elsewhere, on banks and rough grassland on calcareous, sandy or gravelly soils, largely confined to roadsides, railway banks, churchyards and waste ground.

Wood calamint *C. menthifolium*, rhizomatous, in woodland edge and scrub, known since 1843 from a single dry chalk valley in the Isle of Wight. Cessation of coppicing in the 1940s led to a marked decline, now stemmed by a resumption of coppicing and clearance of invasive ground cover, and it was certainly still present in 2016. *Menthifolium* is Latin for 'mint-leaved'.

CALAMOPHYTES Subdivision comprising horsetails (Greek *calamos* + *phyton* = 'reed' + 'plant').

CALCARIIDAE Family name used by some taxonomists for Lapland and snow buntings which with other buntings, are more conventionally placed in the Emberizidae.

CALCEOLARIACEAE Family name of slipperwort (Latin *calceolaria* = 'slipper').

CALENDULA FLY *Napomyza lateralis* (Agromyzidae), scattered in England and Wales. Larvae mine leaves of plants such as groundsel. Greek *nape* + *myza* = 'glade' + 'suck', and from Latin *latus* = 'side'.

CALICIUM Crustose lichens in the genus *Chaenotheca* (Coniocybaceae), from Greek *chaeno* + *thece* = 'grasp' + 'case'. 'Calicium' derives from Latin *calyx* = 'cup'.

    Gold-headed calicium *C. chrysocephala*, local in England but not recorded in the Midlands or the southern half of Ireland, common in upland Wales and Scotland, on bark, especially of conifers. Greek *chrysos* + *cephale* = 'gold' + 'head'.

    Rusty calicium *C. ferruginea*, common and widespread in Britain, scattered in Ireland, on acid bark, unbarked stumps and fences, and pollution-tolerant. *Ferruginea* is Latin for 'rusty'.

    Verdigris(e) calicium *C. trichialis*, widespread if rather uncommon, scattered in Ireland, in dry, acid-barked crevices and on wood in moderately shaded and moist habitats. *Trichos* is from Greek for 'hair'.

CALLA-LILY Alternative name for altar-lily.

CALLIANASSIDAE Family name (Greek *calli* + *anassa* = 'beautiful' + 'lady') for a group of decapods, the deposit-feeding tunnelling ghost shrimp *Callianassa subterranea*, up to 40 mm long, in particular common around the coast from the lower shore to the shallow sublittoral in Wales and western Scotland.

CALLIPHORIDAE Family name (Greek *calli* + *phoros* = 'beautiful' + 'carrying') for blow flies, the name coming from an old English term for meat that had eggs laid on it, which was said to be fly-blown, with 38 species in 14 genera, including *Calliphora* (e.g. bluebottles), *Lucilia* (e.g. greenbottles) and *Pollenia* (cluster flies). Larvae generally scavenge on carrion or dung.

CALLITRICHACEAE Family name of water-starwort, from Greek *calli* + *thrix* = 'beautiful' + 'hair', referring to the stems.

CALOBRYALES Order of thalloid liverworts (Greek *calos* + *bryon* = 'beautiful' + 'moss'), including the family Haplomitriaceae (see Hooker's flapwort under flapwort).

CALOPLACA **Snow caloplaca** *Caloplaca nivalis* (Teloschistaceae), a crustose lichen found on soil, with three records from the Highlands. Greek *calos* + *plax* = 'beautiful' + 'flat plate', and from Latin *nivalis* = 'snow'.

CALOPTERYGIDAE Family name for damselflies and demoiselle flies (Zygoptera) (Greek *calos* + *pteryx* = 'beautiful' + 'wing').

CAMBERWELL BEAUTY *Nymphalis antiopa* (Nymphalidae, Nymphalinae), male wingspan 76–86 mm, female 78–88 mm, a rare migrant butterfly originating in Scandinavia and mainland Europe. In some years there is a relatively large influx, as occurred in 1995 and 2006, where individuals were reported in mid- and late summer from throughout the British Isles, though particularly in the east. It is associated with woodland but is found in other habitats, including gardens. Greek *nymphe* = 'nymph'. In Greek mythology, Antiope married Lycus, king of Thebes.

CAMPANULACEAE From Latin *campanula* = 'little bell', the family name of bellflower (including Canterbury-bells, harebell, lobelia and rampion).

CAMPHOR BEETLE Rove beetles in the genus *Stenus* (Staphylinidae) (Greek *stenos* = 'narrow').

CAMPION (MOTH) *Sideridis rivularis* (Noctuidae, Hadeninae), wingspan 27–30 mm, nocturnal, widespread, common in a range of habitats, including verges, damp meadow, moorland and sea-cliffs. Adults fly in May–June, with a second brood in August–September in the south. Larvae feed in the seed capsules of campions, ragged-robin and catchflies. Greek *sideros* + *eidos* = 'iron' + 'appearance', from the wing colour, and Latin *rivulus* = 'small stream', from a wavy line on the forewing.

CAMPION (PLANT) Perennial herbs in the genus *Silene* (Caryophyllaceae), the English name possibly derived from 'champion', *Silene* from the Greek for another plant (catchfly), possibly from Latin *silenus* = 'spittle' or 'foam'.

    Alpine campion *S. alpestris*, a neophyte with a slightly woody base, naturalised on a bank below Ben Lawers, mid-Perth, where it was deliberately planted. It was also planted at a mid-West Yorkshire site. *Alpestris* is Latin for 'alpine'.

    Bladder campion *S. vulgaris*, common (if declining) and widespread in open, grassy habitats, for instance arable fields, rough pasture, verges, quarries, railway banks, walls and waste ground. Able to tolerate partial shade, it can also grow in open woodland and on hedge banks. 'Bladder' describes the swollen calyx. *Vulgaris* is Latin for 'common'.

    Moss campion *S. acaulis*, cushion- or mat-forming, mainly on base-rich substrates in the Highlands, the Lake Distrct and north Wales, and rarely in north-west Ireland, growing on rock ledges, serpentine fellfields, cliff slopes and, in west Scotland, stabilised dunes. *Acaulis* is Latin for 'without stem'.

    Red campion *S. dioica*, common and widespread, though scattered in the Republic of Ireland, evident in lightly shaded habitats such as hedgerows and woodland clearings. It also grows as a non-flowering form in deep shade. It is also found in coastal habitats, including clifftop grassland and scrub, rock crevices and stabilised shingle, and also on montane screes and cliffs. *Dioica* is Latin for having male and female flowers on separate plants.

    Rose campion *S. coronaria*, a south-east European neophyte known in gardens by the mid-fourteenth century, a persistent garden escape found on light soils in a range of habitats, including heathland edge, dunes, railway banks, roadsides and waste ground, scattered in Britain, a rare casual in eastern Ireland. *Coronaria* (Latin *corona* = 'crown') refers to this plant's use for victors' garlands in classical times.

    Sea campion *S. uniflora*, found on much of the coast on rocky sea-cliffs, on shingle banks, drift lines and walls. It tolerates high levels of nutrient enrichment. It also occurs on montane lake shores, streamsides and river shingle, and in gullies, in parts of northern England, Wales and Scotland. Habitats also include metalliferous mine spoil, disused railway lines and ballast. *Uniflora* is Latin for 'single-flowered'.

    White campion *S. latifolia*, a short-lived, annual or biennial archaeophyte, common and widespread, though less so in western Scotland and western Ireland, on arable land, in hedge banks and waste ground, most abundantly on light soils. *Latifolia* is Latin for 'wide-leaved'.

CANARY FLY **Hairy canary fly** *Phaonia jaroschewskii* (Muscidae), scattered in England and scarce (a UK BAP species), found in wet bog (e.g. Thorne and Hatfield moors, Yorkshire), larvae feeding on sphagnum moss. The yellow colour gives the English name. Greek *phaios* = 'dark'.

CANARY-GRASS Species of *Phalaris* (Poaceae), the name probably from the Greek for a kind of grass, the English name denoting a Canary Islands origin, but also use as caged bird food. Specifically, *P. canariensis*, an annual neophyte from the Canaries via Europe, recorded in the wild by 1632, scattered and sometimes frequent on waste ground, walls, roadsides and pavement cracks as a casual originating from birdseed, grain and wool shoddy.

Awned canary-grass *P. paradoxa*, a casual annual Mediterranean neophyte cultivated by 1687, scattered and increasingly common in (mainly south-east) England, originating from birdseed, grain and shoddy on waste ground, and as a weed in arable fields and newly sown grass leys. *Paradoxa* is Latin here used to mean 'unusual'.

Bulbous canary-grass *P. aquatica*, a rhizomatous south European neophyte perennial introduced by 1778, sown as cover and food for pheasants. Scattered in England, it was recorded in the wild in 1912 and seems to be increasing, especially in East Anglia and south-east England, naturalised in woodland clearings, field borders, roadsides and waste ground. It also occurs as a casual originating from shoddy, and bird- and grass seed.

Confused canary-grass *P. brachystachys*, a casual annual Mediterranean neophyte recorded in 1908, occasionally found in England in docks and waste ground, originating from shoddy, birdseed and grain. Greek *brachys* + *stachys* = 'short' + 'spike'.

Lesser canary-grass *P. minor*, an annual scattered and usually casual on waste ground, but sometimes established among arable crops, including bulb fields in the Isles of Scilly and carrot fields in East Anglia, originating from grain, birdseed and shoddy.

Reed canary-grass *P. arundinacea*, a common and widespread rhizomatous perennial of ditches, riverbanks, carr and margins of canals, lakes and ponds, favouring sites where the water table fluctuates widely. It can also occur on rough ground and roadsides. *Arundinacea* is Latin for 'reedlike'.

CANCRIDAE Family name for edible crab (Latin *cancer* = 'crab').

CANDLESNUFF (FUNGUS) Inedible species of *Xylaria* (Xylariaceae) (Greek *xylon* = 'wood'). Specifically *X. hypoxylon*, widespread and common, antler-shaped, found throughout the year, though mainly in summer and autumn, on dead wood, usually of hardwoods. Greek *hypo* + *xylon* = 'beneath' + 'wood'.

Beechmast candlesnuff *X. carpophila*, common and widespread, found year-round, though mainly in summer and autumn, in beechwood on rotting beech mast, often buried in litter. Greek *carpo* + *philos* = 'a fruit' (here beech) + 'liking'.

Haw candlesnuff *X. oxyacanthae*, Vulnerable in the RDL (2006), found on buried hawthorn fruit. *Oxyacanthae* refers to hawthorn under a former binomial, *Crataegus oxyacantha*.

CANDYTUFT Species of *Iberis* (Brassicaceae), the genus name referring to Iberia. 'Candytuft' dates from the early seventeenth century, derived from Candy, an obsolete form of Candia, the former name of Crete.

Garden candytuft *I. umbellata*, an annual neophyte, cultivated by Tudor times, first recorded in the wild in 1858, widely distributed as a casual of waste ground and gardens. *Umbellata* is Latin for having umbels.

Perennial candytuft *I. sempervirens*, a perennial dwarf European neophyte shrub, introduced by 1731, occasionally escaping on to waste ground or walls, especially in southern England, scattered elsewhere. *Sempervirens* is Latin for 'evergreen'.

Wild candytuft *I. amara*, an annual, occasionally biennial, herb scattered and locally common on bare places on chalk grassland, including rabbit scrapes, and in quarries. It also occurs as an arable weed, and as a casual in ruderal habitats. *Amara* is Latin for 'bitter'.

CANIDAE Family name of red fox and dog (Latin *canis* = 'dog').

CANKER 1) A bacterial, fungal or viral disease of woody plants in which bark formation is locally prevented, and an area of dead tissue forms and grows. 2) The disease trichomonosis when in pigeons and doves.

CANNABACEAE Family name of hemp (Latin *cannabis*) and hop.

CANNON Dung cannon *Pilobolus crystallinus* (Mucorales, Pilobolaceae), a locally common and widespread fungus found

on the dung of deer, rabbit and livestock. Latin *pilus* + *bolus* = 'hair' + 'a throw', and *crystallinus* = 'crystal'.

CANTERBURY-BELLS *Campanula medium* (Campanulaceae), a herbaceous south European neophyte perennial, scattered in the UK and Isle of Man as a casual or naturalised on waste ground and other grassy habitats. The English name dates from the late sixteenth century, from the bells on the horses of Canterbury pilgrims. The binomial is from the Latin for 'little bells' and 'average'.

CANTHARIDAE Family name of the predatory soldier and sailor beetles of which there are 41 species in 7 genera in the British list. The name soldier beetle may come from the boldly patterned coloration in reds, blacks, yellows and oranges, recalling historical military uniforms. *Cantharos* is Greek for 'beetle'.

CAP (FUNGUS) Agaric fungi in a number of families. Scarce species, associated with fallen conifer cones, are **conifercone cap** *Baeospora myosura* (Marasmiaceae), and **pinecone cap** *Strobilurus tenacellus* and **sprucecone cap** *S. esculentus* (both Physalacriaceae).

Butter cap *Rhodocollybia butyracea* (Marasmiaceae), edible, common and widespread, found from June to December, generally in coniferous woodland, saprobic on conifer needles. Greek *rhodo* + Latin *collybia* = 'rose-red' (from the pinkish gill colour) + 'small coin', and Greek *bouterodys* = 'buttery'.

Crazed cap *Dermoloma cuneifolium* (Tricholomataceae), inedible, widespread and locally common, found in summer and autumn on unimproved grassland. Greek *derma* + *loma* = 'skin' + 'fringe', referring to how the pileipellis (topmost layer of hyphae that make up the pileus or cap skin) overhangs the margin of the cap. *Cuneifolium* is Latin for 'with leaves tapered to the base', referring to the gills reducing in width towards the stem.

Cucumber cap *Macrocystidia cucumis* (Marasmiaceae), inedible, widespread but uncommon, found from late summer to early winter often on disturbed soil associated with decomposing wood in broadleaf (especially beech) and mixed woodland; also on woodchip in parks and gardens. A cucumber odour gives it its common name. The binomial is from Greek meaning 'possessing large cystidia' (i.e. sterile cells that occur between the spore-bearing basidia), and Latin for 'relating to cucumber'.

Dark crazed cap *Dermoloma pseudocuneifolium* (Tricholomataceae), inedible, widespread but scarce, found in summer and autumn on unimproved or semi-improved grassland and parkland. Etymology as for crazed cap + the Greek *pseudo* = 'false'.

Liberty cap *Psilocybe semilanceata* (Strophariaceae), widespread, generally infrequent but locally common, hallucinogenic, found in late summer and autumn on pasture soil. Greek *psilos* + *cybe* = 'smooth' + 'head' (i.e. cap) and Latin *semilanceata* = 'half spear-shaped'.

CAPE-GOOSEBERRY *Physalis peruviana* (Solanaceae), a rhizomatous South American neophyte perennial, grown for its ornamental and sweet-tasting fruit, a casual occasionally found on tips in Britain (naturalised in Hertfordshire) and recorded in Co. Dublin. Its English name comes from its common cultivation in South Africa in the nineteenth century. Greek *physalis* = 'bladder'.

CAPE-PONDWEED *Aponogeton distachyos* (Aponogetonaceae), a tuberous South African neophyte perennial introduced in 1788, scattered in Britain, mainly in England, as an emergent aquatic in water up to 2 m deep, persisting in lakes and ponds as a relic of cultivation, or at sites where it has been introduced. It shows little sign of spreading to new sites without human help. The genus name references the healing springs at Aquae Aponi, Italy, plus *geiton* = 'neighbour', originally applied to a water plant found there. *Distachyos* is Greek for 'two spikes'.

CAPERCAILLIE *Tetrao urogallus* (Phasianidae), the largest member of the grouse family, wingspan 106 cm, length 74 cm, male weight 4.3 kg, females 2 kg, found in native and commercial pinewood in the Highlands feeding on pine needles and leaf buds, and on blueberries (with chicks eating insects). It had become extinct in the 1770s, then reintroduced to Perthshire in 1837, and numbers slowly increased. However, they declined greatly again from the 1960s because of deer fencing, predation and lack of suitable habitat, dropping from a high of 10,000 pairs in the 1960s to <1,000 birds in 1999. Slow recovery has followed conservation measures by the RSPB and other organisations, but wintering numbers in 2009–10 were estimated at 1,300 (800–1,900) individuals, a 34% decline since 2003–4. It remains a Red List species. Nests are on the ground, usually under low-growing branches. Clutch size is 7–11, incubation lasts 24–6 days, and the precocial young fledge in 14–21 days. Capercaillie have never been classified as game as they were extinct by the time the Game Act was passed in 1831. Gaelic *capull coille* means 'horse of the woods'. *Tetrao* is from the Greek for some kind of game bird; *urogallus* is a partial homophone of the German *Auerhahn* = 'mountain cock'.

CAPERER Angler's name for the caddis fly *Halesus radiatus* (Limnephilidae), common and widespread in streams, with omnivorous larvae. Adults fly from August to November. Halesus was the son of Agamemnon. *Radiata* is Latin for 'rayed'.

CAPNIIDAE Monogeneric stonefly family, sometimes known as winter stoneflies, including *Capnia atra bifrons*, the only fairly widespread species. *Capnia* may come from Greek *capnos* = 'smoke' or 'vapour'.

CAPRIFOLIACEAE Family name (Latin *capra* + *folia* = 'goat' + 'leaf') of honeysuckle, including elder, guelder-rose, snowberry and wayfaring-tree.

CAPRIMULGIDAE Family name of nightjar (Latin *capra* + *mulgeo* = 'goat' + 'suck'), 'goatsucker' being one of this bird's dialect names.

CAPSID BUG Members of the family Miridae (formerly Capsidae) (Heteroptera).

Apple capsid *Plesiocoris rugicollis*, common and widespread on apple, willow and poplar. Greek *plesios* + *coris* = 'near' + 'bug', and Latin *rugicollis* = 'red-necked'.

Black-kneed capsid *Blepharidopterus angulatus*, length 5–6 mm, widespread and very common throughout Britain on deciduous trees, predominantly preying on red spider mite. Greek *blepharis* + *ptero* = 'eyelash' + 'wing', and Latin *angulatus* = 'angular'.

Common green capsid *Lygocoris pabulinus*, length 5–6.5 mm, widespread and often very common, found on a range of both woody and herbaceous food plants, though especially nettle. Greek *lygos* + *coris* = 'twig' + 'bug', and Latin *pabulus* = 'food'.

Delicate apple capsid *Malacocoris chlorizans*, length 3–4 mm, widespread in England (particularly common in Kent), Wales and north-east Scotland, found from May to October in two broods on a number of deciduous tree species, particularly hazel, feeding on aphids, other small insects and mites. Greek *malacos* + *coris* = 'gentle' + 'bug', and *chloros* = 'green'.

Potato capsid *Calocoris norvegicus*, widespread and often abundant on a variety of plants. Greek *calo* + *coris* = 'beautiful' + 'bug', and Latin *norvegicus* = 'Norwegian'.

CARABIDAE Family name (Greek *carabis* = 'horned beetle') for ground beetles, with recent taxonomic thought placing the 362 species into 4 subfamilies. **Cicindelinae** with two genera of tiger beetle: Cicindela (four species) and Cylindera (one). *Cicindela* is actually Latin for 'glow-worm' (Lampyridae). **Brachininae** with two species of *Brachinus*; see bombardier beetle under beetle. *Brachion* is Greek for 'arm'. **Omophroninae** with one species, *Omophron limbatum*, semi-aquatic, burrowing in sand around flooded sandpits. *Omophron* is Greek for 'in agreement', *limbatum* Latin for 'rounded'. **Carabinae**, the 'traditional' ground beetles in 83 genera and 354 species. The most species-rich genus is *Bembidion* (Greek *bembex* = 'buzzing insect', 54 species).

CARAWAY *Carum carvi* (Apiaceae), a monocarpic perennial European archaeophyte herb present by 1375, sparsely scattered and naturalised in meadows, on dunes, roadsides and railway banks, and as a casual on waste ground and tips. *Caron* is Greek for this plant.

Whorled caraway *C. verticillatum*, a herbaceous perennial calcifuge found on the western edges of both Britain and Ireland, in marsh, streamsides, damp meadows and rushy pastures and on wet hillsides. This is the county flower of Carmarthenshire as voted by the public in a 2002 survey by Plantlife. *Verticillatum* is Latin for 'whorled'.

CARDER BEE 1) Species of bumblebee in the genus *Bombus* (Apiaceae), the English name reflecting the combing and plaiting (carding) of mosses, etc. to construct the nest.

Brown-banded carder bee *B. humilis*, queen length 16–18 mm, worker 10–15 mm, found intermittently along the south and west coasts of England and Wales, reaching furthest north on the Lleyn Peninsula and Anglesey. There are a few inland populations, for example on Salisbury Plain. It has undergone a major decline in distribution, closely linked to intensification of farming, with most remaining populations being on extensive areas of coastal grassland. Overwintered queens search for nesting sites during May and early June; workers fly between June and September; males during August–September. Queens establish nests on the surface in open grassland, often using an old mouse nest as a base. The nest is covered with dead grass and moss, gathered initially by the queen, later by workers. Field cuckoo bees can parasitise nests. Nectar is from a variety of flowers, with a strong preference for pollen from plants in the Fabaceae, Lamiaceae and Scrophulariaceae. *Humilis* is Latin for 'low'.

Common carder bee *B. pascuorum*, queen length 17–19 mm and wingspan 32–5 mm, workers 10–16 mm and wingspan 26–9 mm, drones 13–15 mm and wingspan 26–9 mm, widespread, possibly deceasing in abundance but increasing its range northwards, recently recorded from Orkney, and found in a variety of habitats, including gardens. Queens emerge from hibernation from March to June; workers are present from April onwards, and males and new females from July to October, the further north the later the date. Nests are made in tall open grassland, under hedges and in plant litter, generally gathering moss and dry grass to cover the nest; occasionally bird boxes or tree holes may be used. Pollen and nectar are taken from a variety of flowers, especially those with long corollae. *Pascuorum* is Latin for 'of pastures'.

Moss carder bee *B. muscorum*, queen length 17–19 mm and wingspan 32–5 mm, workers 10–16 mm and wingspan 26–9 mm, drones 13–15 mm and wingspan 26–9 mm, widespread but showing a severe decline since 1970, and now largely coastal with sizeable populations associated with some of the larger offshore islands. It is found in areas of tall flower-rich grasslands, especially those which have many flowers with long corollas. Queens emerge from hibernation in May, workers are present from June, and males and new females from July to early September. Nests are made in tall open grassland, with moss and dry grass used to make the nest covering. *Muscorum* is Latin for 'of moss'.

Red-shanked carder bee *B. ruderarius*, queen length 17 mm and wingspan 32 mm, workers 15 mm, drones 13 mm, widespread but very scattered following a drastic decline in the latter half of the last century, the strongest modern populations corresponding with larger areas of unimproved grasslands, for example on Tiree (Inner Hebrides). Queens leave their hibernation sites from mid- to late April. The queen constructs a nest of grass clippings and moss on the ground under cover of long vegetation or slightly underground, often based on an old mouse nest. Males and females are produced at the end of July and during August. Once the new sexual forms have hatched the nest disintegrates, the mated queens go into hibernation, and the workers and males can be found on flowers before eventually dying. Pollen and nectar are taken from a variety of flowers. *Ruderarius* is Latin for 'living in ruderal habitats'.

Shrill carder bee *B. sylvarum*, queen length 16–18 mm, workers 10–15 mm, a local lowland species which has shown a significant decline in range since the 1960s when it had been found in a variety of open, flower-rich situations, but in 1999 there were only four areas in England with known populations, three in south Wales, and a few records from the Burren, Ireland. Nests are usually built in a slight hollow on the ground among rough vegetation, or just underground, in early June. *Sylvarum* is Latin for 'living in woodland'.

2) **Wool carder bee** *Anthidium manicatum* (Megachilidae), widely distributed in southern England and Wales, rarer further north, and with recent records from south-west Scotland, flying from late May to early August, found in a range of habitats, including open woodland, chalk grassland, dunes and gardens, visiting a variety of flowers, nesting in the ground and in walls. Hairs are brought to the nest site in a ball and applied to the inner surface of the cavity by teasing them out with the mandibles, hence its common name. Greek *anthos* = 'flower', and Latin *manicatum* = 'sleeved'.

CARDIIDAE Family name for cockles (Latin *cardia* = 'heart', from the shape of the shell).

CARDINAL BEETLE Members of the family Pyrochroidae (Greek *pyro* + *chroa* = 'fire' + 'skin colour'). The English also name references their bright red colour.

Black-headed cardinal beetle *Pyrochroa coccinea*, 20 mm length, widespread in England and Wales, especially in southern England and the Welsh Borders. Larvae and adults are both carnivorous, the former living under loose bark or in rotting wood. Adults prey on insects visiting flowers and are found in May and June. *Coccinea* is Latin for 'scarlet'.

Common cardinal beetle Alternative name for red-headed cardinal beetle.

Red-headed cardinal beetle *P. serraticornis*, 12–18 mm length, widespread and locally common in England and Wales, found between May and July on flowers, trunks and stumps where it hunts other insects. Larvae are omnivorous feeding under tree bark feeding on bark beetle larvae, fungal hyphae and soft cambium tissue. *Serraticornus* is Latin for 'serrated horn'.

Scarce cardinal beetle *Schizotus pectinicornis*, 7–9 mm length, found in the Welsh uplands and Scottish Highlands, particularly associated with birch woodland, larvae hunting small invertebrates under loose bark. *Schizo* is Greek for 'split', *pectinocornis* Latin for 'comb-like horns'.

CARDUELIDAE Former family name for finches (Latin *carduus* = 'thistle', whose seeds are commonly a food plant), now placed in the Fringillidae.

CARL (MOTH) Mining moths in the genera *Tischeria* (from the German entomologist Karl von Tischer, 1777–1849) and *Coptotriche* (Greek *copto* + *tricha* = 'to cut up' + 'hair') in the family Tischeriidae.

Bordered carl *C. marginea*, wingspan 7–8 mm, common in woodland, hedgerow and heathland throughout the southern British Isles. With two generations a year, adults are in flight in May–June and August. Larvae mine the leaves of bramble, the second generation feeding through winter. *Marginea* is Latin for 'marginal', referencing the dark forewing edge.

Oak carl *T. ekebladella*, wingspan 8–11 mm, fairly common and widespread in deciduous woodland in England and Wales, with a more scattered distribution in Scotland and Ireland. Adults fly in May–July. Larvae mine leaves of oak and sweet chestnut in September–October, then overwinter inside the mine chambers. Count Claes Ekeblad was an eighteenth-century Swedish collector.

Rose carl *C. angusticollela*, wingspan 9 mm, found at a few scattered woodland and hedgerow sites in England, reaching as far north as Lancashire and Yorkshire. Adults fly in May–June.

Larvae mine leaves of dog-rose. Latin *angustus* + *collum* = 'narrow' + 'neck', from the incised larval intersegment divisions.

Small carl *T. dodonaea*, wingspan 6–7 mm, found locally in deciduous woodland in parts of England and Wales (and with two records from Ireland). Larvae mine leaves of oak and sweet chestnut, evident from September onwards. Larvae overwinter in mine chambers, adults emerging in June. Dodonaea was an Ancient Greek oracle.

CARNATION FLY *Delia cardui* (Anthomyiidae), scarce, with scattered records mostly from English gardens. Larvae mine leaves of sweet-william, though the specific epithet refers to *carduus*, Latin for 'thistle', not a known host plant.

CARNIVORA Order containing the families Canidae (fox), Felidae (cat), Mustelidae (mink, badger, ferret, otter, pine marten, polecat, stoat and weasel) and Phocidae (seals).

CARP Freshwater fish in the family Cyprinidae. The word is late Middle English from Old French *carpe*, in turn from Latin *carpa*.

Common carp *Cyprinus carpio*, up to 58 cm long and 9 kg in the wild (heavier where stocked), a central Asian fish introduced to England in the late fifteenth or early sixteenth century, though possibly stocked in monastery ponds before then, and to Ireland in the mid-seventeenth century. Particularly common and widespread in England, scattered or locally common elsewhere. This is an omnivorous bottom-feeder, growing well in still or slow-flowing, nutrient-rich warm water. Spawning is in early summer. Both parts of the binomial are Latin for 'carp'.

Crucian carp *Carassius carassius*, 15 cm long, weight up to 1.5 kg, probably native in south-east England, possibly a medieval introduction, and certainly introduced elsewhere in lowland Britain and Ireland. It is variable in shape and size, a deep-bodied form with a humped profile in lakes with abundant food, much slimmer in unfavourable conditions such as acid or low-oxygenated water, where it is more tolerant than many other fish species, for instance gulping oxygen at the water surface and using anaerobic respiration. Also tolerant of waters that freeze over in winter, it is often stocked for coarse fishing. Diet is of insect larvae and plants. Spawning is in May–July. The binomial comes from a Latinisation of *karass*, a common eastern European name for this species.

Grass carp *Ctenopharyngodon idella*, up to 120 cm long, weight 18 kg, native to East Asia, introduced to East Anglia in the 1960s to control water weeds in ponds and ditches, populations maintained by restocking, and now found in lakes and slow-moving waters, mainly in south-east and north-west England, and south and north Wales. Greek *cteis* + *pharyx* + *odon* = 'comb' + 'throat' + 'toothed', and *idios* = 'distinct'.

Mirror carp A form of common carp.

CARPENTER BEE Members of the genera *Ceratina* (Greek *cerato* = 'horn'), *Chelostoma* (*chele* + *stoma* = 'hoof' + 'mouth') and *Xylocopa* (*xylo* + *copa* = 'wood' + 'dagger') in the family Apidae, the name coming from their nesting habit of burrowing into hard plant material such as dead wood.

Blue carpenter bee *Ce. cyanea*, male length 5–7 mm, females 5–9 mm, considered a rarity until 1972 when it was found in abundance on a chalk downland site in east Hampshire. It subsequently proved to be widely distributed and locally common in south-facing chalk escarpments, heathland, disused sand quarries and open rides in deciduous woodland in south-east England, especially in the Weald. Females excavate nesting burrows in dead dry broken stems in which the pith has been exposed. Most nests have been found in bramble stems, a few in rose. From late summer, adults seek out hollow stems in which to pass the winter. Pollen and nectar are taken from a range of flowers. *Cyanea* is Latin for 'blue'.

Harebell carpenter bee *Ch. campanularum*, length 5–6 mm, abundant in gardens in southern and central England, flying in summer, taking pollen and nectar from members of the

Campanulaceae and Geraniaceae. Nests are in small-bore beetle burrows, often using woodworm-infested planks on garden sheds and fence-posts. English and specific names reference harebell *Campanula rotundifolia*.

Sleepy carpenter bee *Ch. florisomne*, length 7–10 mm, wingspan 0.3–0.6 mm, with a scattered distribution in England and Wales, mostly found in the interface between woodland and meadow where buttercups are present, flying from May to July, taking pollen and nectar from buttercups. Nests are in old beetle burrows in dead wood at sunny sites. Females also use thatch and similar, naturally occurring cavities, such as old reed stems. *Florisomne* comes from the Latin for 'flower' + 'sleepy', since this bee often sleeps in flowers.

Violet carpenter bee *X. violacea*, length 20–8 mm, one of the largest bees in Europe, but rare. It was recorded as breeding in west Wales in 2006, and present in England in gardens in Leicestershire, Cheshire and Kent in 2007, and Worcestershire and Northamptonshire in 2010. Gravid queens bore tunnels in dead wood. *Violacea* is Latin for 'violet'.

CARPET (MOTH) Mainly nocturnal members of the family Geometridae, most in the subfamily Larentiinae, and including the genera *Epirrhoe* (Greek *epirrhoe* = 'river', from the wavy lines on the forewings), *Thera* (an Aegean island), and *Xanthorhoe* (Greek *xanthos* + *rhoe* = 'yellow' + 'stream', from the yellowish wavy lines on the forewings of some species). Not to be confused with tapestry moth, sometimes referred to as carpet moth.

Autumn green carpet *Chloroclysta miata*, wingspan 34–40 mm, widespread (especially in upland areas) but local on moorland, and in broadleaf woodland, scrub, hedgerows and gardens. Adults are active in September–October, hibernate, then fly again in March–April. Larvae feed on leaves of woody species. Greek *chloros* + *clyzo* = 'green' + 'wash off', and Latin *miata* = 'urinate', from the yellowish-green wing colour.

Balsam carpet *X. biriviata*, wingspan 27–30 mm, uncommon in damp grassland, water meadow, marsh, riverbanks and damp areas in parts of south and south-east England, East Anglia and the east Midlands. The bivoltine adults fly in May–June and July–August. Larvae mainly feed on the introduced orange balsam in June and in August–September. *Biriviata* is from Latin *bi* + *rivus* = 'two' + 'stream', from the two wavy lines on the forewing.

Barberry carpet *Pareulype berberata*, wingspan 27–32 mm, formerly widespread across much of southern England, north to Yorkshire, but now an endangered RDB species, recorded from hedgerows and woodland edges at six sites in Wiltshire, one in Dorset and one in Gloucestershire. It is protected in the UK WCA (it is illegal to collect or disturb any of its stages), and a priority species in the UK BAP. The bivoltine adults fly in May–June and in August. Larvae feed on barberry, the Latin *berberis* giving the specific name. Greek *para* = 'beside' (affinity with) the genus *Eulype*.

Barred carpet *Martania taeniata*, wingspan 26–30 mm, scarce, in damp woodland and other sheltered, damp areas in northern England and Scotland, rare elsewhere in the British Isles. Adults fly in June–July. Larvae have not been recorded in the wild. *Taenia* is Latin for 'ribbon', referencing the forewing marking.

Beautiful carpet *Mesoleuca albicillata*, wingspan 34–8 mm, widespread and common in open broadleaf woodland and mature hedgerows. Adults fly in June–July. Larvae feed on bramble, raspberry, dewberry, wild strawberry and hazel. Greek *mesos* + *leucos* = 'middle' + 'white', the central part of the wing being white, and Latin *albus* + *cilla* = 'white' + 'tail', probably from the white abdomen.

Beech green carpet *Colostygia olivata*, wingspan 22–7 mm, local in woodland, moorland, downland and cliffs throughout much of Scotland and north-west England, spreading into Wales and southern England; one post-2000 record from Ireland (Co. Fermanagh). Adults fly in July–August. Larvae feed on bedstraw. Greek *colos* = 'stunted' + *stygios* = relating to the River Styx in the Underworld (from the black rivulet wing markings), and Latin *olivata* referencing the olive green wing colour.

Blue-bordered carpet *Plemyria rubiginata*, wingspan 22–8 mm, day-flying, widespread and moderately common especially in England and Wales, but also widespread elsewhere, in damp woodland, hedgerows, scrub, orchards and gardens. Population levels increased by 388% between 1968 and 2007. Adults fly in June–August. Larvae feed on a range of trees and bushes, especially alder and blackthorn. Greek *plemmyris* = 'flood tide', referencing the wavy wing pattern, and Latin *rubiginus* = 'rust', from reddish scales on the forewing.

Broken-barred carpet *Electrophaes corylata*, wingspan 22–30 mm, widespread and common in broadleaf woodland, hedgerows, scrub, parks and gardens. Adults fly in May–June. Larvae feed on leaves of broadleaf trees and shrubs, for example hazel, hawthorn, blackthorn, oak and birch. Greek *electron* + *phaos* = 'shining substance' + 'brightness', from the yellow-brown speckling on the forewing, and Latin *corylus* = 'hazel'.

Chalk carpet *Scotopteryx bipunctaria*, wingspan 32–8 mm, active in July–August, fairly common and widespread on chalk downland and limestone hills, favouring bare ground, in south England and Wales, becoming scarcer further north in England, and distribution declining since at least the 1970s. Larvae feed on bird's-foot-trefoil, vetches and clovers. This is a priority species in the UK BAP. Greek *scotos* + *pteryx* = 'dark' + 'wing', from the shaded forewing markings, and Latin *bipunctaria* = 'two-spotted', again for wing markings.

Chestnut-coloured carpet *T. cognata*, wingspan 26–30 mm, scarce, on moorland, lightly wooded hillsides, limestone downland and rocky areas, both inland and on the coast, in north England, north Wales and Scotland. Adults fly in July–August. Larvae feed on juniper. *Cognatus* is Latin for 'related' (i.e. to a congeneric).

Cloaked carpet *Euphyia biangulata*, wingspan 25–30 mm, local in woodland margins, hedgerows and sunken lanes in southern England, Wales and south Ireland. Adults fly in June–July. Larvae may feed on stitchwort. *Euphyia* is Greek for 'goodness of shape', *biangulata* Latin for 'two-angled', from wing markings.

Common carpet *Epirrhoe alternata*, wingspan 20–5 mm, widespread and fairly common in open habitats such as calcareous grassland, heathland, moorland, marsh, dune and garden, flying from dusk from May to October, with two generations in the south. Larvae feed on bedstraws and cleavers. Latin *alternata* refers to the alternating black and white wing bands.

Common marbled carpet *Dysstroma truncata*, wingspan 24–30 mm, widespread and common in such habitats as gardens, parks, broadleaf woodland, calcareous grassland, fens, heathland and moorland. There are two broods, flying in May–June, and from August to October, sometimes later. Larvae feed on a range of woody plants. Greek *dys* + *stroma* = 'bad' + 'bed', and Latin *truncata* = 'cut short', from the abbreviated wing markings.

Cypress carpet *T. cupressata*, wingspan 28–32 mm, first recorded in mainland Britain in 1984 in West Sussex, and now established across much of southern England, increasingly common in parks and gardens. The bivoltine adults fly in May–June and August–September. Larvae feed on Monterey and Leyland cypresses, referenced in the specific epithet (Latin *cupressus* = 'cypress').

Dark-barred twin-spot carpet *X. ferrugata*, wingspan 18–22 mm, widespread and common in gardens, woodland, hedgerows, downland, fen, moorland and dunes, but numbers declined by 91% over 1968–2007. In the south of its range adults fly in May–June and again in August. Further north there is usually only one brood in June–July. Larvae feed on plants such as bedstraws, docks and ground-ivy. *Ferrugo* is Latin for 'rust', from the wing colour.

Dark marbled carpet *Dysstroma citrata*, wingspan 25–30 mm, widespread in woodland, heathland, scrub, hedgerows and gardens. Adults fly in July–August. Larvae feed on birches, sallow, heather and bilberry. *Citrata* is Latin for 'citrus', from the orange wing markings.

Devon carpet *Lampropteryx otregiata*, wingspan 27–30 mm, scarce, in damp woodland mostly in parts of southern and south-west England, and south and west Wales, but with records as far north as Cumbria. Population levels increased by 1,279% between 1968 and 2007. The bivoltine adults fly in May–June and in August–September. Larvae feed on common marsh-bedstraw, possibly also fen bedstraw. Greek *lampros* + *pteryx* = 'bright' + 'wing', from the glossy wing. *Otregiata* is the Latin name given to Ottery St Mary, Devon, the type locality.

Dotted carpet *Alcis jubata*, wingspan 28–33 mm, local in parts of western Britain and north Scotland, with a few sightings from Ireland, one from Co. Antrim the only post-2000 record. Nevertheless, population levels increased by 1,009% over 1968–2007. Adults favour ancient woodland and scrub where lichen covers tree-trunks and branches on which they rest hidden using cryptic camouflage. They fly in June–July. Larvae feed on beard lichen and others. In Greek mythology, Alcis was the daughter of an Egyptian king. Latin *jubata* means having a crest.

Flame carpet *X. designata*, wingspan 25–8 mm, widespread and common in damp woodland, hedgerows and gardens. The bivoltine adults fly in May–June and in August. Larvae feed on cruciferous plants. *Designo* is Latin for 'mark out', from the sharply defined wing band.

Galium carpet *Epirrhoe galiata*, wingspan 28–32 mm, local on sea-cliffs, dunes and shingle beaches along the coasts of west Wales and southern England from Somerset to Kent, and inland, on calcareous grassland and grassy heathland in England and more rarely in Ireland and Scotland. With numbers declining by 79% over 1968–2007, this is a species of conservation concern in the UK BAP. In the south there are two generations; further north there is generally only one, with adults on the wing mainly between June and August. Larvae feed on bedstraws (*Galium* species, hence the English name).

Garden carpet *X. fluctuata*, wingspan 18–25 mm, adults seen in April–October, often evident resting during daytime on walls and fences, in a range of habitats including suburban gardens since larvae feed on a variety of cruciferous plants. Numbers, however, declined by 75% over 1968–2007. *Fluctus* is Latin for 'wave', from the wavy lines on the forewing.

Green carpet *Colostygia pectinataria*, wingspan 22–7 mm, common and widespread, flying just before dusk and into the night from May to September in habitats that include heathland, moorland, downland, open woodland and gardens, larvae mainly feeding on bedstraws. Population levels increased by 230% between 1968 and 2007. Greek *colos* = 'stunted' + *stygios* = relating to the River Styx in the Underworld (from the black rivulet wing markings), and Latin *pectinataria* = 'toothed like a comb', from the shape of the male antennae.

Grey carpet *Lithostege griseata*, wingspan 28–31 mm, in farmland, verges, waste ground and other open areas in Breckland (East Anglia), where it is currently known from five sites, a rare RDB species, and a priority in the UK BAP. Adults fly from May to July. Larvae feed on the developing seeds of flixweed and treacle-mustard. Greek *lithos* + *stege* = 'stone' + 'covering', and Latin *griseus* = 'grey', both referencing the wing colour.

Grey mountain carpet *Entephria caesiata*, wingspan 32–41 mm, in mountain and moorland habitats in Wales and from the Midlands north to northern Scotland, and widespread but scattered in Ireland. Numbers declined by 94% between 1968 and 2007. Larvae feed on plants such as heather and bilberry. Adults are often evident on rock from June to August. Greek *entephros* = 'ash-coloured', and Latin *caesius* = 'blue-grey', both describing the wing colour.

Grey pine carpet *T. obeliscata*, wingspan 28–36 mm, widespread and common in coniferous woodland, plantations, parks and gardens. Adults fly from May to July, and again in September–October. Larvae feed on conifer needles. Greek

obeliscos means something pointed, describing a tapering mark on the wing.

Juniper carpet *T. juniperata*, wingspan 26–9 mm, local on chalk downland and in gardens, in much of southern England, north Wales and Cumbria, less commonly on moorland in upland areas of Scotland and parts of Ireland. Population levels increased by 836% between 1968 and 2007. Adults fly in October–November. Larvae feed on juniper (giving the specific epithet) and are probably adapting to other garden conifers.

Large twin-spot carpet *X. quadrifasiata*, wingspan 29–32 mm, local in mature, often damp, woodland, fen and scrubby heathland throughout England, mainly in the southern counties. Adults fly in June–July. Larvae feed on low-growing plants. *Quadrifasciata* (*sic*) is Latin for 'four fasciae' or bands on the forewing.

Least carpet *Idaea rusticata* (Sterrhinae), wingspan 19–21 mm, local in gardens, verges, hedgerows, scrub, downland and embankments. From a low base, numbers increased by 74,684% between 1968 and 2007. The main increase occurred during the 1980s and 1990s. Since the 1970s, the moth has also dramatically increased its distribution, spreading from the London area into East Anglia, the Midlands and south-west England, and reaching Cheshire by 1999 and Lancashire in 2009. Adults fly in July–August. Larvae feed on ivy, traveller's-joy and other herbaceous plants. The genus name is the Greek *Idaios*, Mt Ida, from whence the gods watched the Trojan War. *Rusticata* is Latin for 'rural'.

Marsh carpet *Gagitodes sagittata*, wingspan 27–35 mm, scarce in fen and marsh in Cambridgeshire, Huntingdonshire, Norfolk and Suffolk, northwards to Yorkshire. Adults fly in June–July. Larvae feed on the ripening seeds of common meadow-rue from July to September. Latin *sagitta* = 'arrow', referencing a wing marking.

Netted carpet *Eustroma reticulata*, wingspan 20–5 mm, a rarity found in damp woodland around Lake Windermere and Coniston Water in the Lake District, a priority species in the UK BAP. Adults fly in July–August. Larvae feed on leaves of touch-me-not. Greek *eu* + *stroma* = 'well' + 'something spread out', and Latin *reticulata* = 'reticulated', both referring to the network wing pattern.

Oblique carpet *Orthonama vittata*, wingspan 23–6 mm, widespread but local in fen, water meadow, damp woodland, marsh and riverbanks, but with numbers declining by 85% over 1968–2007, a species of conservation concern in the UK BAP. The bivoltine adults fly in May–June and August–September. Larvae feed on bedstraw. Greek *orthos* + *nama* = 'straight' + 'stream', from the straight lines on the forewing, and Latin *vittata* = 'striped', referencing the diagonal wing stripe.

Pine carpet *Pennithera firmata*, wingspan 30–4 mm, widespread and common in pine woodland. Population levels increased by 336% between 1968 and 2007. Adults fly from late July to November. Larvae feed on needles of Scots and Corsican pine. Latin *penna* = 'wing' + the Aegean island Thera, and Latin *firmata* = 'confirmed', species in this genus being difficult to identify.

Pretty chalk carpet *Melanthia procellata*, wingspan 27–32 mm, common in woodland, scrub, hedgerows and gardens on calcareous soils throughout southern and central England and south Wales, but with numbers declining by 8% over 1968–2007, a species of conservation concern in the UK BAP. Adults fly in July–August. Larvae feed on traveller's-joy. Greek *melas* + *anthos* = 'black' + 'flower', from wing markings, and Latin *procella* = 'rainstorm', from the dark scales scattered on the light wing colour.

Red carpet *X. decoloraria*, wingspan 30–4 mm, locally common, occurring from Shropshire and Staffordshire northwards into Scotland (where a local subspecies *hethlandica* occurs on the Shetland Isles) and, less frequently, the northern half of Ireland. The favoured habitat is rocky moorland and grassy hillsides, where larvae feed on lady's-mantle. Adults fly at dusk from June to August. Numbers have declined by 88% over 1968–2007. *Decolor* is Latin for 'faded'.

Red-green carpet *Chloroclysta siterata*, wingspan 30–6 mm, widespread and common in broadleaf woodland, hedgerows, moorland and gardens. Population levels increased by 739% between 1968 and 2007. Adults fly in September–October, mated males dying, adult females hibernating to fly again in early spring. Larvae feed on deciduous trees. Greek *siteros* = 'relating to corn', referring to the fading of the wing colour from green to corn yellow.

Red twin-spot carpet *X. spadicearia*, wingspan 24–7 mm, fairly common, flying from dusk (April–August), widespread but with a more scattered distribution in (especially lowland) Scotland and Ireland, in habitats that include gardens, woodland, hedgerows, downland and marsh. Larvae feed on a number of herbaceous plants, favouring bedstraws, wild carrot and ground-ivy. *Spadix* is Latin for 'nut brown', from the colour of the median forewing band.

Ringed carpet *Cleora cinctaria* (Ennominae), wingspan 28–35 mm, on damp heathland in the New Forest, Hampshire, and adjoining counties, and in bogs and mosses in Scotland and northern England (ssp. *bowesi*). There are also scattered populations of the nominate race in Ireland. Adults fly in April–May. Larvae feed on birch, bog-myrtle, bilberry, bell heather and cross-leaved heath. Cleora was the wife of Agesilaus, king of Sparta in the fourth century BCE. *Cinctus* is Latin for 'girdled', referencing the white mark ringed with black on the forewing.

Ruddy carpet *Catarhoe rubidata*, wingspan 26–31 mm, on downland, embankments, scrub and sea-cliffs in England and Wales, especially in southern parts. Adults fly in June–July. Larvae feed on bedstraw. Greek *cata* + *rhoe* = 'downwards' + 'stream', referring to the wavy wing pattern, and Latin *rubidus* = 'reddish'.

Sandy carpet *Perizoma flavofasciata*, wingspan 26–32 mm, common in open woodland, woodland rides and edges, mature hedgerows, calcareous grassland and dunes, widespread, though absent from the southern half of Ireland. Adults fly in June–July. Larvae feed in campion flowers and seeds. The binomial is Greek for 'girdle' and Latin for 'yellow-banded', both referencing the wing markings.

Scorched carpet *Ligdia adustata* (Ennominae), wingspan 20–5 mm, local in broadleaf woodland, scrub, hedgerows and gardens throughout much of southern England, especially where there are calcareous soils, south Wales, and over much of Ireland. It is mainly double-brooded, flying in May–June and in August. Larvae feed on spindle. *Ligdia* seems to have no entomological meaning. *Adustus* is Latin for 'scorched', a reddish-brown colour being more pronounced on the underwing.

Sharp-angled carpet *Euphyia unangulata*, wingspan 25–8 mm, local in woodland and hedgerows throughout England and Wales. The species has undergone major increases in population levels, with high abundance in all five of the most recent years in a 1968–2007 study. This changed a steep decline over the initial 35-year period into a slight increase looking at the entire 40 years. Adults fly in June–July, sometimes August. Larvae feed on chickweed and stitchwort. *Euphyia* is Greek for 'goodness of shape', Latin *unangulata* = 'single-angled', from the wing marking.

Silver-ground carpet *X. montanata*, wingspan 24–8 mm, widespread and common in hedgerows, scrub, woodland, fens, chalk downland and gardens, adults mainly flying by night (May–July) but also by day when disturbed. Larvae eat such low-growing plants as bedstraws and cleavers. *Montana* is Latin for 'mountainous', its first collection being in Austria.

Sloe carpet *Aleucis distinctata* (Ennominae), wingspan 27–31 mm, scarce, in scrub, hedgerows, farmland, damp heathland and mosses in parts of south-east and southern England, a priority species in the UK BAP. Adults fly in April. Larvae feed on blackthorn. The binomial is from Greek for 'not white' and Latin for 'distinct'.

Spanish carpet *Scotopteryx peribolata*, wingspan 28–33 mm,

resident in the Channel Islands and a rare immigrant to the southern coast of England, mostly from Dorset to Cornwall, occasionally in Kent and Sussex. Adults fly in August–September. There are no records of breeding in mainland Britain. Greek *scotos* + *pteryx* = 'dark' + 'wing', from the shaded forewing markings, and *peribole* = 'encircling'.

Spruce carpet *T. britannica*, wingspan 18–25 mm, common and widespread in coniferous woodland, plantations, parks and gardens. Population levels increased by 1,731% between 1968 and 2007. With two broods, adults fly in May–July and September–October. Larvae feed on conifer needles.

Striped twin-spot carpet *Coenotephria salicata*, wingspan 29–31 mm, flying in May–July, associated with moorland, but also open woodland, dune and some grassland, fairly common in the northern and western British Isles. Larvae feed on bedstraws. The grey colour of the adult camouflages it when on its preferred resting substrate of tree-trunk or rock. Greek *coinos* + *tephra* = 'in common' + 'ashes' (i.e. sharing its ash colour with other genera), and *salicata* from Latin *salix* = 'willow', though this is not a food plant.

Twin-spot carpet *Mesotype didymata*, wingspan 24–9 mm, common on moorland and woodland in the uplands, and on grassland and verges in the lowlands throughout much of Britain and Ireland, but infrequent in south-east England. Adults fly from June to August. Larvae feed on a range of plants. Greek *mesos* + *typos* = 'middle' + 'marking', and *didymos* = 'twin', from the wing markings.

Water carpet *Lampropteryx suffumata*, wingspan 25–32 mm, widespread and common in damp woodland, scrub, hedgerows, chalk downland, moorland and gardens. Adults fly in April–May. Larvae feed on bedstraws and cleavers during early summer. Greek *lampros* + *pteryx* = 'bright' + 'wing', from the glossy wing, and Latin *sub* + *fumatus* = 'a little' + 'smoky', from the wing colour.

Waved carpet *Hydrelia sylvata*, wingspan 27–30 mm, scarce, in woodland, scrubby heathland and on sea-cliffs in southern England and Wales, and with three post-2000 records in Ireland. It has greatly decreased in numbers with the decline in coppicing, but also appears to be expanding its range in places, for example Sussex. Adults fly in June–July. Larvae feed on leaves of various hardwoods. Greek *hydrelos* = 'watery', from the wavy wing pattern, and Latin *sylva* = 'woodland', referencing the habitat.

White-banded carpet *Spargania luctuata*, wingspan 30–4 mm, scarce, found since the 1950s in woodland rides and clearings in south-east Kent, Thetford Forest, Norfolk, and a few places in Sussex. The bivoltine adults fly in May–June and July–August. Larvae feed on rosebay willowherb. *Sparganium* is the genus name for bur-reed, though this is not a food plant. *Luctuata* is from Latin *luscus* = 'mourning', from the black and white wing markings.

Wood carpet *Epirrhoe rivata*, wingspan 28–34 mm, local in woodland rides and edges, mature hedgerows, embankments and scrubby downland throughout much of England and Wales, scarcer as one moves northwards. Adults fly from June to August. Larvae feed on bedstraws. *Rivata* is Latin for 'riverlike', referencing the wavy wing pattern.

Yellow-ringed carpet *Entephria flavicinctata*, wingspan 34–9 mm, scarce, mainly in rocky upland limestone habitat in central and north-west Scotland but with local populations in parts of England (Yorkshire, Herefordshire), Wales (Breconshire) and the north coast of Ireland. An early brood can be on the wing in May, but more commonly this moth flies in July–August. Larvae feed on saxifrages and stonewort. Greek *entephros* = 'ash-coloured', and Latin *flavus* + *cinctatus* = 'yellow' and 'encircled', describing the yellow-brown speckling on the dark-coloured wing.

CARPET BEETLE Members of the genus *Anthrenus* (possibly if puzzlingly from Greek *anthrene* = 'wasp') in the family Dermestidae, adult length 1.5–3.5 mm. See also museum beetle under beetle.

Common carpet beetle *A. scrophulariae*, a common larval pest of upholstered furniture, carpets and textiles, but adults will feed on pollen and nectar. The specific name references *scrophularia* = 'figwort'.

Furniture carpet beetle *A. flavipes*, a common pest of upholstered furniture, larvae being able to digest keratin, the principal protein found in animal hair and feathers. *Flavipes* is Latin for 'yellow-footed'.

Varied carpet beetle *A. verbasci*: adults are pollen grazers, but larvae are common in houses, feeding on natural fibres and damaging carpets, furniture, clothing and insect collections. The specific name references *verbascum* = 'mullein'.

CARPET SHELL Filter-feeding bivalves in the genus *Tapes* (Veneridae), *tapes* being Greek for 'carpet'.

Banded carpet shell *T. rhomboides*, shell size up to 6.5 cm, widespread, burrowing in gravel, shell-gravel and coarse sand in the intertidal, sublittoral and shallow shelf. *Rhomboides* is Greek for 'diamond-shaped'.

Chequered carpet shell *T. decussatus*, shell size up to 7.5 cm, mainly found off the southern and western coasts of Britain and around Ireland burrowing in sand, muddy gravel and clay in the intertidal and shallow sublittoral. *Decussatus* is Latin for 'X-shaped'.

Golden carpet shell *T. aureus*, shell size up to 4.5 cm, found around all coasts except for most of the east coast north of Essex, and north-west England, as a shallow burrower in intertidal and sublittoral mud and gravel. *Aureus* is Latin for 'golden'.

Pullet carpet shell *T. corrugata*, shell size up to 5 cm, widespread in the intertidal and sublittoral burrowing in a variety of substrates or found in crevices or old piddock holes on rocky shores. *Corrugata* is Latin for 'wrinkled'.

CARRAGEEN *Chondrus crispus* (Gigartinaceae), 7–15 cm long, a common and widespread red seaweed found on rocks on the mid- to lower rocky shore and in rock pools, and down to 24 m. It can tolerate reduction in salinity so is present in some estuaries. A source of carrageenan (a sulphated polysaccharide), it has been used to make soups, jellies, etc., and in Ireland as a remedy for respiratory disorders. The name 'carrageen' was introduced in Ireland in about 1840 by Mr Todhunter from Donegal, probably taken from Carrigeen (or Carrigan) Head in Co. Donegal, itself from Irish *carraigín* = 'little rock'. *Chondrus* is Greek meaning something granular or 'gristle', *crispus* Latin for having wavy or curly margins.

Black carrageen Alternative name for clawed fork weed.

CARRION BEETLE 1) Members of the family Nitidulae.

Two-spotted carrion beetle *Nitidula bipunctata*, with a scattered distribution in Britain, feeding on carrion. *Nitidula* is the Latin diminutive of *nitidus* = 'shining' or 'bright', and *bipunctata* = 'two-spotted'.

2) Members of the family Silphidae (Silphinae).

Beet carrion beetle *Aclypea opaca*, 9–12 mm length, with a scattered distribution in England and Wales. Adults overwinter in litter in protected grassy sites, then in March–April leave their shelter to feed initially on young leaves of winter cereals and some wild plants before migrating to beet fields, feeding and laying eggs. Larvae are especially active in June–July, and cause more damage to beet than the adults. The binomial is Latin for 'without shield' and 'dark'.

Red-breasted carrion beetle *Oiceoptoma thoracinum*, 11–16 mm length, widespread in Britain, found in woodland, feeding on carrion and rotting fungus. Greek *oicos* + *ptoma* = 'home' + 'corpse', and from Latin *thorax* ('thorax').

CARROT *Daucus carota* ssp. *sativus* (Apiaceae), a biennial herb with a swollen root, much cultivated as a root crop, casual on waste places, and as a relic of cultivation. The binomial uses the Greek and Latin for 'carrot'; *sativus* is Latin for 'cultivated'.

Australian carrot *D. glochidiatus*, an annual Australian neophyte, casual in a few places in England and Scotland as a wool alien. Latin *glochidiatus* = 'barbed'.

Moon carrot *Seseli libanotis*, a usually biennial, though sometimes a short-lived monocarpic perennial, very local in East Sussex, Hertfordshire and Bedfordshire on chalk grassland, and in Cambridgeshire also found on chalky roadside banks and in a former chalk quarry. *Seselis* is Greek and Latin for meadow saxifrage, *libanotis* Latin for 'Lebanese'.

Sea carrot *D. carota* ssp. *gummifer*, a coastal biennial herb locally found from Anglesey to Kent, and in south-east Ireland, in cliff grassland and on stable dunes. *Gummifer* is Latin for 'gum-bearing'.

Wild carrot *D. carota* ssp. *carota*, a biennial herb common and widespread in lowland regions, on fairly infertile, well-drained often calcareous soils. Habitats include chalk downland, rough grassland on roadsides, waysides and railway banks, quarries, and chalk- and gravel-pits.

CARROT FLY *Chamaepsila rosae* (Psilidae), widespread and scattered, but can be a larval pest feeding on roots of mainly carrot, but also parsnip, parsley, celery and celeriac. Larvae can grow to a length of 10 mm. Greek *chamae* + *psilos* = 'dwarf' + 'smooth'.

CARYOPHYLLACEAE Family name for pinks, from Greek *carya* + *phyllo* = 'walnut' + 'leaf', referring to the aromatic scent, becoming used for cloves, and subsequently for clove pinks. Includes campion, chickweed, corncockle, mouse-ear, ragged-robin, soapwort and stitchwort.

CARYOPHYLLIIDAE Family name (Greek *carya* + *phyllo* = 'walnut' + 'leaf') of hard corals, with 31 species in 17 genera in British waters.

CASE-BEARER (MOTH) 1) Generally nocturnal members of the genus *Coleophora* (Coleophoridae), from Greek *coleos* + *phoreo* = 'sheath' + 'carry', from the portable cases made by the larvae, which also provide the English name.

Rarities include **basil-thyme case-bearer** *C. tricolor* (on unimproved grassland in Breckland, and a disused airfield in East Anglia), **bearberry case-bearer** *C. arctostaphyli* (heathland around Loch Garten on Speyside, and at Dinnet on Deeside), **betony case-bearer** *C. wockeella* (recently known from Chiddingfold Forest, Surrey, and Holland Wood, Petworth, West Sussex, last seen in 1997), **broom case-bearer** *C. saturatella* (parts of East Anglia and the Midlands), **buff blite case-bearer** *C. aestuariella* (coastal south-east England, discovered in Kent in 1981), **cliff case-bearer** *C. serpylletorum* (Great Orme, Caernarfonshire, where first recorded in Britain in 1964; the Lizard, west Cornwall; and east Kent), **daisy case-bearer** *C. ramosella* (Burren, Co. Clare, and Blean Wood, Kent), **drab case-bearer** *C. niveicostella* (southern downland), **Essex case-bearer** *C. vibicigerella* (in saltmarsh on the Thames estuary, rediscovered on the Isle of Sheppey in 1980), **estuarine case-bearer** *C. linosyridella* (Thames estuary in Kent and Essex, and at Berry Head, Devon in 2013), **fleabane case-bearer** *C. inulae* (declining, recently found in south-east Hampshire, West Sussex and Herefordshire), **grey blite case-bearer** *C. deviella* (on saltmarsh from Sussex to Lincolnshire), **Kent case-bearer** *C. galbulipennella* (shingle beaches between Dungeness and Hythe), **knapweed case-bearer** *C. conspicuella* (a few sites in Essex and Kent), **large gold case-bearer** *C. vibicella* (Hampshire and adjoining counties), **new sloe case-bearer** *C. prunifoliae* (recently added to the British list when discovered in Cornwall, now known from a number of English counties), **viviparous case-bearer** *C. albella* (found in 1985 in woodland at Inglestone Common, south Gloucestershire), and **water-dock case-bearer** *C. hydrolapathella* (in marsh in the Norfolk Broads and Suffolk; in 1989 also recorded from Co. Durham).

Agrimony case-bearer *C. follicularis*, wingspan 13–15 mm, local in damp meadow, marsh, riverbanks and woodland rides as far north as south Scotland, and including the southern half of Ireland and the Channel Islands. Adults fly in June–August. Larvae mine leaves of hemp-agrimony, common fleabane and ploughman's-spikenard. Latin *folliculus* = 'small bag', referring to the larval case.

Apple & plum case-bearer *C. spinella*, wingspan 10–12 mm, common in open woodland, scrub, orchards, gardens and hedgerows throughout much of England, Wales and south Scotland, more locally in Ireland. Adults fly in June–July. Larvae mine leaves of hawthorn, blackthorn and apple. *Spinella* references blackthorn *Prunus spinosa*.

Black-bindweed case-bearer *C. therinella*, wingspan 13–16 mm, local on waste ground and arable land in England south of a line from the Wash to the Severn estuary, and in Ireland from the Burren. Adults fly in June–July. Larvae mine seeds of black-bindweed in August. *Therinella* comes from Greek *theros* = 'summer', from when the adults are active.

Black-stigma case-bearer *C. hemerobiella*, wingspan 12–15 mm, local in hedgerows, gardens and orchards throughout much of south-east England and East Anglia. Adults fly in July. Larvae overwinter twice and mine leaves of hawthorn, apple, wild cherry, rowan, pear, plum and whitebeam, causing sufficient damage to be a pest in some areas. Greek *hemerobios* = 'living for a day', referencing a supposed similar appearance to lacewings *Hemerobius*.

Blackthorn case-bearer *C. coracipennella*, wingspan 8.5–11.5 mm, local in hedgerows, scrub and orchards scattered throughout much of Britain, though not in north Scotland, and rare in Wales. Adults emerge in June–July. Larvae mine leaves of blackthorn, apple, cherry and hawthorn. Latin *corax* + *penna* = 'raven' + 'wing', from the dark forewing.

Body-marked case-bearer *C. clypeiferella*, wingspan 12–15.5 mm, scarce, on waste ground, grassland and farmland in south-east England, East Anglia and Lincolnshire, but with a declining geographical range. Adults fly in July–August. Larvae mine the seedheads of fat-hen, creating their movable cases from seed fragments during September–October. Latin *clypeus* + *fero* = 'shield' + 'carry', from the ring of abdominal spines the adult uses to cut itself out of its cocoon.

Brown alder case-bearer *C. alnifoliae*, wingspan 12–13 mm, scarce, in woodland and freshwater margins in small, isolated colonies in south-east and central-southern England, as far north as Shropshire, and overall slowly increasing its range; rare in Wales. Adults mainly fly in July–August. Larvae mine leaves of alder. Latin *alnifoliae* = 'alder leaf'.

Buckthorn case-bearer *C. ahenella*, wingspan 10–13 mm, scarce on chalk downland and heathland in Hampshire and adjoining counties. In Hampshire, recorded from the New Forest, where the movable larval cases are found from July to October on leaves of alder buckthorn, and on the chalk, where found in small numbers in scrub containing buckthorn. Adults fly in May–June. Latin *aheneus* = 'copper' or 'bronze', from the forewing colour.

Buff birch case-bearer *C. milvipennis*, wingspan 11–13 mm, with a local, scattered distribution in England on heathland and in open woodland; rare in Wales, Scotland and Ireland. Adults fly in July. Larvae mine birch leaves in September–October. Latin *milvus* + *penna* = 'kite' (bird) + 'wing', from the forewing shape.

Buff rush case-bearer *C. caespititiella*, wingspan 8–10 mm, common in woodland clearings, damp meadow and marsh, both salt and fresh water, throughout much of England, Wales and south and east Ireland. Adults fly in May–June. In August larvae live without a case concealed in a rush seed. In September they feed on seeds of rushes from a case, often hidden in the seedhead. In late September or October, larvae either leave the food plant or hide in the seedhead for the winter. *Caespititiella* comes from Latin *caespis* = 'turf', probably referencing the habitat.

Bugloss case-bearer *C. pennella*, wingspan 15–20 mm, in dry sandy areas, calcareous grassland and shingle beaches in England, south-east of a line from Gibraltar Point to Portland Bill. Adults fly in June–July. Larvae mine viper's-bugloss, sometimes alkanet. Latin *penna* = 'wing'.

Burren case-bearer *C. pappiferella*, wingspan 11–13 mm, local in the central Highlands around Braemar, and on limestone pavement on the Burren, west Ireland. Adults fly in May–June. Larvae mine and make their cases using the hairy seeds of mountain everlasting. *Pappiferella* comes from Latin *pappus*, literally 'old man', here by analogy referring to the hairy seed, + *fero* = 'carry'.

Campion case-bearer *C. nutantella*, wingspan 14–19 mm, local on grassy slopes, verges and waste ground in parts of south-east England, extending as far into the Midlands as Oxfordshire, and East Anglia. Adults fly in June–August. Larvae feed inside seed capsules of Nottingham catchfly and bladder campion. Latin *nutans* = 'nodding'.

Clover case-bearer *C. alcyonipennella*, wingspan 11–15 mm, local on rough grassland, waste ground and verges in England, Wales and Ireland, but identification and distribution confused with small clover case-bearer. Adults have two generations, in May–June and in late July–August. Larvae feed in the seed heads of white clover, living in a portable case. Latin *alcyon* + *penna* = 'kingfisher' + 'wing', from the metallic colours of the forewing.

Common case-bearer *C. serratella*, wingspan 11–14 mm, widespread and locally common in woodland, hedgerows, heathland and fen. Adults fly in June–July. The larva feeds by inserting its head into small mines it creates on the leaves of birch, elm, alder or hazel. Latin *serratus* = 'serrated', describing the serrated leaf fragments used in building the portable larval case.

Common oak case-bearer *C. lutipennella*, wingspan 10–12 mm, in oak woodland and areas with scattered oak trees throughout much of England, Wales and south Scotland, and with one Irish record. Adults fly in July–August. Larvae mine oak leaves. Latin *luteus* + *penna* = 'yellow' + 'wing'.

Common rush case-bearer *C. alticolella*, wingspan 10–12 mm, widespread, the commonest and most widespread coleophorid, found on heathland, damp meadow, marsh and a range of other habitats. Adults fly at dawn, dusk and night in June–July. Larvae mine seeds of soft-rush and heath-rush, most evident in August–October. Latin *altus* + *colo* = 'high' + 'inhabit', from the type locality.

Dark elm case-bearer *C. limosipennella*, wingspan 10–13 mm, local in woodland margins and hedgerows throughout much of the southern half of England and much of Wales. Adults fly in June–July. Larvae mine elm leaves. *Limosipennella* comes from *Limosa limosa* (black-tailed godwit) + Latin *penna* = 'wing', from the supposed forewing shape.

Dark thistle case-bearer *C. paripennella*, wingspan 10–13 mm, locally common on verges, rough grassland and waste ground as far north as Aberdeenshire. Adults fly in July–August. Larvae mine leaves of thistles and knapweeds, mainly in April–June. *Paripennella* comes from *Parus* (tit) + *penna* = 'wing', though with no cogent resemblance of the moth's wing with the bird.

Downland case-bearer *C. lixella*, wingspan 16–21 mm, local on downland, calcareous grassland, dunes and rocky slopes in parts of the British Isles, predominantly coastal in the north of its range. Adults fly in July–August in the afternoon and at night. Larvae initially feed on the seeds of wild thyme, then after overwintering feed on grasses. *Lixella* comes from the weevil genus *Lixus*, from a supposed resemblance between the moth's antennae and the beetle's rostrum.

Dusted case-bearer *C. adspersella*, wingspan 13–16 mm, local in saltmarsh, waste ground, farmland and fen throughout much of England and Wales south of a line from the Wash to the Mersey estuary, and in south and east Ireland. Adults fly in June–August. Larvae mine seeds of grass-leaved orache and sea-purslane in autumn. Latin *adspersus* = 'sprinkled', from the dark scales on the forewing.

Eastern case-bearer *C. vestianella*, wingspan 12–14.5 mm, in waste ground, arable fields and sandhills with a scattered distribution that includes the Breckland district of East Anglia and the Lancashire sandhills, and recently found in the Scillies, Bedfordshire, north Lincolnshire, south-west Yorkshire, the Firth of Forth and Shetland. Adults fly in June–July. Larvae mine seeds of common orache. *Vestianella* comes from Latin *vestis* = 'garment', since the name was originally mistakenly given to a clothes moth.

Forest case-bearer *C. ibipennella*, wingspan 10–14 mm, local in oakwood in south-east England, extending in decreasing numbers northwards and westwards to Lincolnshire, Herefordshire, Cornwall and south Wales. Adults fly in July–August. Larvae mine oak leaves from April to October. *Ibipennella* comes from *Bubulcus ibis* (cattle egret) + Latin *penna* = 'wing', from the white forewing colour.

Glasswort case-bearer *C. salicorniae*, wingspan 12–14 mm, local in saltmarsh along the English coast from Cornwall to Yorkshire, and in parts of Wales, and with a single record in Ireland (Co. Wexford). Adults fly in late July–August. Larvae mine seeds of glasswort *Salicornia* (hence the specific epithet), then burrow into the mud to overwinter in a silk tube.

Goldenrod case-bearer *C. obscenella*, wingspan 10–13 mm, in woodland, grassland, cliffs, dune and saltmarsh, widespread but scattered and local. Adults fly in July–August. Larvae mine the seed heads of goldenrod and sea aster. *Obscenella* probably comes from Latin *ob* + *scenicus* = 'against' + (in one meaning) 'tent', referring to the larval case.

Gorse case-bearer *C. albicosta*, wingspan 12–15.5 mm, day-flying, widespread and common in rough grassland, downland and heathland. Adults fly in June–July. Larvae mine gorse seed pods. Latin *alba* + *costa* = 'white' + 'rib', from the forewing's white costal (leading margin) streak.

Grey alder case-bearer *C. binderella*, wingspan 8–12 mm, local in alder carr, woodland and heathland as far north as Inverness-shire, and in Ireland from Co. Kerry and Co. Leith. Adults fly in late June–July. Larvae mine leaves of alder, hazel, birch and hornbeam. C.F. Binder von Kriegelstein was a late eighteenth and early nineteenth-century Austrian entomologist.

Grey birch case-bearer *C. siccifolia*, wingspan 12–14 mm, scarce in hedgerows and heathland throughout much of Britain, but not north Scotland. Larvae mine leaves of birch, hawthorn, apple and rowan, overwintering twice. Latin *sicca* + *folia* = 'dry' + 'leaf', the larval case resembling a withered leaf.

Grey-rush case-bearer *C. glaucicolella*, wingspan 10–12 mm, widespread and common in marsh, fen, damp meadow, woodland rides and saltmarsh. Adults fly at dawn, dusk and night from June to August. Larvae mine seeds of rushes, the specific name derived from *Juncus glaucus*.

Hazel case-bearer *C. fuscocuprella*, wingspan 7.5–10 mm, local in woodland margins, woodland rides and heathland throughout much of southern England, with records north to Leicestershire; also recorded from the Burren. Adults fly in May–June. Larvae mine leaves of hazel and birch. Latin *fuscus* + *cupreus* = 'dusky' + 'coppery', from the forewing colour.

Hedge case-bearer *C. striatipennella*, wingspan 11–13 mm, widespread and common in rough grassland, waste ground, woodland rides, damp woodland, marsh, riverbanks and other damp areas. Adults fly from May to August in one or two generations. Larvae mine within seed capsules of lesser stitchwort, common mouse-ear and chickweeds. Latin *striatus* + *penna* = 'streaked' + 'wing', from the forewing markings.

Hereford case-bearer *C. sylvaticella*, wingspan 9–14 mm, scarce in woodland and moorland in parts of south-west England, the west Midlands, Wales, Scotland and Ireland. Larvae mine flowers and seeds of great wood-rush, overwintering twice. Latin *sylvaticus* = 'relating to woodland'.

Jointed-rush case-bearer *C. tamesis*, wingspan 11–14 mm, local in damp meadow, bog and woodland rides with a scattered distribution in Britain. Adults fly from mid-June to August.

Larvae mine the seeds of jointed rush. *Tamensis* comes from the Latin for Thames, its type locality.

Larch case-bearer *C. laricella*, wingspan 8–10 mm, widespread and common in coniferous plantations and mixed woodland. Adults fly in June–July. Larvae mine larch needles, in places causing sufficient damage to be a serious pest. *Laricella* comes from *larix* = 'larch'.

Large buff case-bearer *C. ochrea*, wingspan 15–19 mm, on calcareous grassland, limestone pavements and rock ledges on the North and South Downs, the Isle of Wight and the Cotswold Hills in England, and the Gower Peninsula, south Wales. Adults fly in July–August. Larvae mine leaves of common rock-rose. Greek *ochra* = 'ochre', from the forewing colour.

Large clover case-bearer *C. trifolii*, wingspan 15–20 mm, local on rough grassland, waste ground and verges, less numerous in Scotland, more numerous in the south, and rare in Ireland. Adults fly by day in June–July. Larvae feed inside the seed pods of ribbed and tall melilot, though *trifolii* refers to clover (*Trifolium*).

Least case-bearer *C. juncicolella*, wingspan 6–8 mm, local on heathland, moorland and gardens throughout much of Britain as far north as western Ross. Adults fly in late June–July in afternoon sun. Larvae mine leaves of heather and bell heather, fully developed in April–May. *Juncicolella* comes from *Juncus* = 'rush' + *colo* = 'inhabit', from the habitat of the type locality.

Ling case-bearer *C. pyrrhulipennella*, wingspan 11–14 mm, local on heathland, moorland and bogs throughout much of Britain, north to the Great Glen, and scattered in Ireland. Adults fly in May–July. Larvae mine leaves of heather and bell heather, mainly in April–May. The specific name comes from bullfinch *Pyrrhula* (if relating to forewing colour then from the dark bird's back, not its red front) + *penna* = 'wing'.

Lotus case-bearer *C. discordella*, wingspan 11–13 mm, widespread and common in rough grassland, downland, woodland clearings, and marsh and other damp places, rather coastal in Ireland. Adults mainly fly in July. Larvae mine leaves of bird's-foot-trefoil. *Discordella* comes from Latin *discors* = 'discordant' (i.e. different from congenerics).

Meadow case-bearer *C. mayrella*, wingspan 10–12 mm, widespread and common on grassland, verges and waste ground, if local in Scotland. Adults fly in June–July. Larvae mine flowers of white clover. *Mayrella* honours the Viennese U. Mayer, who had the type specimen by 1813.

Mugwort case-bearer *C. artimisicolella*, wingspan 11–13 mm, local on waste ground, rough grassland and verges throughout much of England (more numerous in the south and east) and north Wales. Adults fly in July–August. Larvae mine seeds and leaves of mugwort *Artemisia* (prompting the specific epithet) in September–October.

Northern case-bearer *C. vitisella*, wingspan 10–13 mm, in pine woodland, heathland and moorland in two disjunct areas: the Pennines, extending westwards into Lancashire and Cheshire and into the Clwyd Hills, Wales, and southwards to north Warwickshire; and in west-central Scotland and the Highlands. Adults fly from late May to early July. Larvae overwinter twice and mine leaves of cowberry, in April–May when in their second year. *Vitisella* comes from the Latin for cowberry, *Vaccinium vitis-idaea*.

Ochreous case-bearer *C. solitariella*, wingspan 10–12 mm, scarce, in hedgerows and open woodland in southern England, as far north as Lincolnshire, and east Wales. Adults fly in June–July. Larvae mine leaves of greater stitchwort, mainly in spring. Latin *solitarius* = 'solitary', a mistaken term for the actually gregarious larvae.

Orache case-bearer *C. saxicolella*, wingspan 13–16 mm, widespread, local on waste ground, farmland and rough ground. Adults fly in July–August. Larvae mine seeds of orache and goosefoot. Latin *saxum* + *colo* = 'rock' + 'inhabit', with rocky places one of the habitats on the Continent.

Osier case-bearer *C. lusciniaepennella*, wingspan 10–13.5

mm, widespread and common in freshwater margins, woodland, bog and fen. Adults fly in June–July. Larvae mine leaves of willows, sallows, osier, bog-myrtle and, in Scotland, occasionally birch. Latin *luscinia* + *penna* = 'nightingale' + 'wing', from the forewing shape.

Pale birch case-bearer *C. orbitella*, wingspan 10.5–14 mm, scattered and local in heathland and alder carr throughout Britain, with a couple of records in Ireland. Larvae mine leaves of birch and alder, possibly also hornbeam and hazel, from mid-August to October. Latin *orbus* = 'bereaved', from the dark forewing colour.

Pale elm case-bearer *C. badiipennella*, wingspan 9–11 mm, local in hedgerows, woodland margins and scrub throughout much of England. Adults fly in June–July. Larvae mine leaves of elm, preferring young trees. Latin *badius* + *penna* = 'chestnut brown' + 'wing'.

Pale orache case-bearer *C. versurella*, wingspan 11–15 mm, local on saltmarsh, waste ground and farmland in England as far north as Co. Durham, Wales, Isle of Man, and southern Ireland. Adults fly in June–September. Larvae mine seeds of orache and goosefoot. Latin *versura* = 'a turning', possibly because of the ringed pattern of the antennae.

Pale thistle case-bearer *C. peribenanderi*, wingspan 12.5–15 mm, common in rough grassland, waste ground, verges and woodland rides throughout much of England and Wales, rarer in south Scotland and Ireland. Adults fly in June–July. Larvae mine leaves of thistles, occasionally burdock, from July to September. *Peribenanderi* comes from Greek *peri* = 'about', and the Swedish entomologist Per Benander (1885–1976), with a pun, *per(i)* on his forename.

Penny-whin case-bearer, petty-whin case-bearer *C. genistae*, wingspan 11–13 mm, scarce in rough grassland, damp heathland, mosses and moorland in England north to Loch Ness, Scotland, and rare in north Wales. Adults fly in June–August. Larvae mine the leaves and sometimes flowers of petty whin, *Genista anglica*, which provides the specific name.

Pistol case-bearer *C. anatipennella*, wingspan 12–16 mm, local in scrub, hedgerows, parks and gardens in south England and Wales, occasional elsewhere in the British Isles. Adults fly in June–July. Larvae mine leaves of (mainly) blackthorn, apple, wild cherry, pear, hawthorn and rowan, living inside a movable case shaped like a flintlock pistol, prompting the English name, with peak activity in May and October. Latin *anas* + *penna* = 'duck' + 'wing', from the domestic duck colour and the white forewing.

Rannoch case-bearer *C. glitzella*, wingspan 10–14 mm, scarce in lowland heathland, moorland, pine woodland and sheltered rock ledges up to 600 m in the Highlands, from Perthshire to east Ross. Larvae feed on leaves of cowberry, overwintering twice. C.T. Glitz (1818–89) was a German entomologist.

Red-clover case-bearer *C. deauratella*, wingspan 10.5–12.5 mm, local on waste ground, grassland and verges in parts of southern England, the Midlands, East Anglia, Wales and Ireland. Adults fly in June–July. Larvae mine red clover seeds. *Deauratella* comes from Latin for 'gilt', from the glossy copper forewing colour.

Richardson's case-bearer See Dorset tineid moth under moth.

Rose case-bearer *C. gryphipennella*, wingspan 10.5–13 mm, widespread and common in open woodland, scrub, waste ground, downland and gardens. Adults fly at dusk and early morning in late June–July. Larvae feed on leaves of dog-rose and cultivated roses, building successively larger portable cases from cut-out leaf fragments, with peak activity in April–May and October. Latin *gryps* + *penna* = 'griffin' + 'wing'.

Saltern-rush case-bearer *C. adjunctella*, wingspan 8–11 mm, local in coastal saltmarsh from Cumbria round to Suffolk. Adults fly in May–June. Larvae mine the seed capsules of saltmarsh rush in July–September. Latin *adjunctus* = 'joined to', possibly with a pun on the food plant *Juncus*.

Saltmarsh case-bearer *C. atriplicis*, wingspan 12–14 mm, on saltmarsh throughout much of southern England, north Wales and east Ireland. Adults fly in June–August. Larvae mine seeds of glasswort, sea-purslane, annual sea-blite and grass-leaved orache (*Atriplex*, giving the specific epithet) in September–October.

Sandy case-bearer *C. lithargyrinella*, wingspan 11–13.5 mm, in open woodland and hedgerows (where larvae mine leaves of stitchwort) and cliffs and shingle beaches (where the food plant is sea campion), widespread but local, including a few sites in east Ireland. Adults fly in June–July. Larvae are particularly evident in March–April and October. Greek *lithargyros* = 'lead monoxide', from the forewing colour.

Scarce thorn case-bearer *C. trigeminella*, wingspan 9–10 mm, local in hedgerows and scrub in parts of south-east England, though with records north to Yorkshire. Adults fly in June. Larvae mine leaves and buds of hawthorn, rowan, wild cherry and apple. Latin *trigeminus* = 'one of a set of triplets', from a resemblance with two congenerics.

Scarce wood case-bearer *C. currucipennella*, wingspan 13–15 mm, local in woodland, most recently recorded in Kent, Essex, Oxfordshire, Monmouthshire, Derbyshire and Cheshire, and declining in range. Larvae mine leaves of hazel, hornbeam and oak. *Currucipennella* is from the Latin for an unidentified small bird + *penna* = 'wing'.

Scotch case-bearer *C. idaeella*, male wingspan 13–15 mm, females 11.5–13.5 mm, in pine woodland and heathland in east Highlands, from Kincardineshire to east Inverness-shire. Larvae mine leaves of cowberry, overwintering twice. *Idaeella* comes from the Latin for cowberry, *Vaccinium vitis-idaea*.

Sea-aster case-bearer *C. asteris*, wingspan 10–15 mm, local on coastal saltmarsh from Devon to south-east Yorkshire. Adults fly from June to September. Larvae mine seeds of sea aster, prompting the specific name, in September–October.

Sea-purslane case-bearer *C. salinella*, wingspan 8.5–13 mm, scarce in saltmarsh in parts of England, especially East Anglia. Adults fly in July–August. Larvae mine seeds of sea-purslane, grass-leaved orache and spear-leaved orache. Latin *salinus* = 'relating to salt'.

Sea-rush case-bearer *C. maritimella*, wingspan 9–11 mm, found on a few saltmarshes along the coast of England, Wales and east Ireland. Adults fly in July–August. Larvae mine the seed heads of sea rush *Juncus maritimus* (prompting the specific name) throughout much of the year.

Shaded case-bearer *C. potentillae*, wingspan 8–9.5 mm, local in rough grassland, damp woodland, fen, ditches and woodland rides in England south of a line from Lincolnshire to Herefordshire, central and west Scotland, and the Burren, Ireland. Larvae mine leaves of a number of herbaceous plants and grey willow and birch, particularly from September to May. The specific name comes from cinquefoil *Potentilla*.

Silver-streaked case-bearer *C. limoniella*, wingspan 10–11 mm, scarce, day-flying, in saltmarsh throughout much of southeast England and East Anglia, with records north to Lincolnshire. Adults fly in July–August. Larvae mine the flowers and seeds of common sea-lavender *Limonium* (giving the specific name) in April–May and, especially, September–October.

Small clover case-bearer *C. frischella*, wingspan 11.5–14.5 mm, scarce, local on rough grassland, waste ground and verges in England, Wales and Ireland, but identification and distribution confused with clover case-bearer. Adults fly by day in June–July. Larvae feed on clover seeds. J.L. Frisch (1666–1743) was a German entomologist.

Small rush case-bearer *C. taeniipennella*, wingspan 9–12 mm, widespread and locally common in fen, damp meadow and marsh. Adults fly in June–July. Larvae mine seeds of jointed and sharp-flowered rushes, most evident in September–October. Latin *taenia* + *penna* = 'band' + 'wing', from white forewing streaks.

Small streaked case-bearer *C. gardesanella*, wingspan 9.5–11 mm, scarce, in habitats that include saltmarsh, damp meadow, marsh, riverbanks, dry grassland, downland, waste ground and verges in England, from Hampshire to Essex. Adults fly in July–August. Larvae mine leaves of sea wormwood, mugwort, sneeze-wort, yarrow, tansy and oxeye daisy in May–June. *Gardesanella* comes from the type locality on the shore of Lake Garda, Italy.

Southern case-bearer *C. adjectella*, wingspan 9–10 mm, local in hedgerows and scrub in Bedfordshire, Cambridgeshire, Northamptonshire, Isle of Wight, south Devon and mid-Wales. Adults fly in July. Larvae mine blackthorn leaves. Latin *adjectus* = 'additional', for its addition to the genus in 1861.

Speckled case-bearer *C. sternipennella*, wingspan 10.5–15 mm, scarce in waste ground and farmland with a scattered distribution mostly in south and south-east England, but also north-east England, north Wales and the Isle of Man. Adults fly in July–August. Larvae mine seeds of orache and fat-hen from September to June. Latin *sterna* + *penna* = 'tern' + 'wing', possibly from similar wing shapes.

Spikenard case-bearer *C. conyzae*, wingspan 12–16 mm, on chalk downland and freshwater margins in parts of southern England, East Anglia and south Wales. Adults fly in June–July. Larvae mine leaves of ploughman's-spikenard and common fleabane, mostly in May–June. *Conyzae* references ploughman's-spikenard *Inula conyzae*.

Stitchwort case-bearer *C. lutarea*, wingspan 9–11 mm, in open woodland in southern England and south Wales. Adults fly in during daytime in May. Larvae mine seeds of greater stitchwort. Latin *lutarius* = 'pertaining to mud', from the fore-wing's earthy colour.

Surrey case-bearer *C. squamosella*, wingspan 11–13 mm, local on waste ground, verges and calcareous grassland in south and south-east England and the Breckland district of East Anglia. Adults fly in June–July. Larvae mine seeds of blue-fleabane. *Squamosella* comes from Latin *squamosus* = 'scaly', from the scattering of white scales on the forewing.

Tipped-oak case-bearer *C. flavipennella*, wingspan 11–13 mm, locally common in oakwood and areas with scattered oak trees throughout much of England and Wales; rare in Ireland and Scotland. Adults fly in July–August. Larvae mine oak leaves, predominantly in May and October. Latin *flavus* + *penna* = 'yellow' + 'wing'.

Toad-rush case-bearer *C. lassella*, wingspan 10–12 mm, scarce, in bare muddy habitats and dunes in south-west England, south-west Ireland and the Channel Islands. Adults fly from May to August. Larvae mine toad-rush seed capsules. *Lasella* comes from Latin *lassus* = 'weary' (i.e. a fainter colour than some congenerics).

Verge case-bearer *C. trochilella*, wingspan 11–14 mm, scarce, on downland, open scrub and grassy slopes in parts of England and Wales, north to Yorkshire, with a few records from Scotland and south Ireland. Adults fly in June–August. Larvae mine leaves of mugwort, oxeye daisy, tansy, wormwood and yarrow from April to July. Greek *trochilos*, a small bird (possibly willow warbler), relevance being obscure.

Violet case-bearer *C. violacea*, wingspan 9–11 mm, scarce, with a scattered distribution in open woodland, parks and scrub over much of Britain. Adults fly in the evening and at night in late May to June. Larvae feed on the leaves of a range of trees, shrubs and herbs, favouring Rosaceae. The fully developed cased larva is active in October and, after winter diapause, in April. *Violacea* supposedly describes the forewing colour, though this is more bronze than violet.

White-birch case-bearer *C. betulella*, wingspan 10–15 mm, local in heathland and birch woodland in parts of England and Wales. Adults fly in June–July. Larvae mine birch leaves, the specific name coming from Latin *betula* = 'birch'.

White-legged case-bearer *C. albitarsella*, wingspan 10–13 mm, local in hedgerows, open woodland, fen and downland throughout much of England and Wales, more numerous in the south. Adults mainly fly at night but also in sunlight from mid-June to August. Larvae preferably mine leaves of wild marjoram, but also wild basil, clary, ground-ivy, mints and selfheal. The larva starts feeding and making its movable case in August, overwinters half-grown, starts feeding again in early March and is full-grown in May. *Albitarsella* comes from Latin *albus* + *tarsus* = 'white' + '(part of) leg'.

White-oak case-bearer *C. kuehnella*, wingspan 14–17 mm, local in oak woodland and scrub throughout England, especially the southern half. Adults fly in July–August. Larvae mine oak leaves. *Kuehnella* honours a German, Prof Kühn, who sent specimens for determination in 1877.

White-sallow case-bearer *C. albidella*, wingspan 13–16 mm, local in fen, bog, damp woodland, marsh, riverbanks and damp areas throughout much of England and Wales, most frequently in the west; rare in Ireland. Adults fly in June–July. From May, larvae mine buds, later leaves, of willows, preferring young trees. Latin *albidus* = 'white', from the forewing colour.

Wood-rush case-bearer *C. otidipennella*, wingspan 10–13 mm, widespread but local in grassland, downland, heathland and woodland rides. Adults fly in May–June. Larvae mine seeds of wood-rush. Latin *otis* + *penna* = 'bustard' + 'wing', from the forewing shape.

Wormwood case-bearer *C. albicans*, wingspan 10–13 mm, local in saltmarsh on the east coast from Kent to south-east Scotland, and in Cheshire and Dorset. Adults fly in July–August. Larvae mine seeds and leaves of sea wormwood from August to June. *Albicans* comes from Latin *alba* = 'white'.

Woundwort case-bearer *C. lineolea*, wingspan 11–14.5 mm, local in hedgerows and waste ground throughout much of southern England and Wales, with records north to Yorkshire and Cheshire. Adults fly from late June–August. Larvae mine leaves of horehounds, betony and hedge woundwort from September to May, with a winter diapause. Latin *lineola* = 'small line', from streaks on the forewing.

Yarrow case-bearer *C. argentula*, wingspan 9.5–13 mm, common on waste ground and rough grassland, and by ditches and freshwater margins, throughout much of southern England, Wales and central Scotland. Adults fly in July–August. Larvae mine seeds and leaves of yarrow and sneezewort from August to November. *Argentula* comes from Latin *argentum* = 'silver', from the silvery streaks on the forewing.

2) Species of lichen case-bearers *Dahlica* (Psychidae, Nary-ciinae), the family name coming from Greek *psyche* = 'soul', using the analogy between metamorphosis and resurrection.

Lesser lichen case-bearer *D. inconspicuella*, male wingspan 9–13 mm, endemic, local, sometimes common in southern England north to Lancashire and Yorkshire, and nowhere else in the world. The rarely seen adults (winged male, wingless female) have been recorded in April. Larvae feed on lichens growing on rocks, tree-trunks, fence-posts and old walls. Latin *inconspicuella* = 'inconspicuous'.

Lichen case-bearer *D. lichenella*, a rarity recorded from parts of southern England, Cheshire, south Scotland and Perthshire, probably overlooked elsewhere. Females are parthenogenic and wingless, emerging in March–April; males have not been recorded in Britain. Larvae feed on lichens growing on tree-trunks, fence-posts and old walls, mosses and decaying vegetable matter.

Narrow lichen case-bearer *D. triquetrella*, a rarity known in Britain only as a parthenogenic wingless female form from a few sites in Hampshire, Kent, Essex and Cumbria. Larvae feed on lichen and algae growing on tree-trunks, fence-posts and old walls. Latin *triquetrus* = 'triangular', from the shape of the larval case.

CASTOR-OIL-PLANT *Ricinus communis* (Euphorbiaceae), an annual tropical neophyte herb, casual on waste ground as a garden throw-out in a few places mainly in England. *Ricinus* is Latin for this plant; *communis* = 'common'.

CASTORIDAE Family name for beaver (Latin *castor*); see beaver.

CAT *Felis silvestris* ssp. *catus* (Felidae), domesticated and feral carnivores, present in Britain since the Iron Age, in 2015 probably totalling 7.4 million with an estimated 20% being feral. Cat predation on wildlife is huge, but they take what would in any case be a 'doomed surplus'. Cat presence might lead to birds, in particular, selecting non-optimal habitat, thereby reducing productivity. Cats hybridise with and transmit disease to wildcats, threatening their survival as a pure subspecies. 'Cat' (Old English *catt*) comes from Latin *cattus* and Byzantine Greek *catta*. The binomial is Latin for 'wild cat'. See also wildcat (*Felis silvestris* ssp. *silvestris*).

CAT-MINT *Nepeta cataria* (Lamiaceae), an archaeophyte perennial herb, scattered and declining in England and Wales, on grassland, waysides, hedge banks, roadsides and rough ground on calcareous soils. Plants escaping from gardens also give rise to casual populations. **Garden cat-mint** *N.* x *faasenii* and **eastern cat-mint** *N. racemosa* are also occasionally found as garden throw-outs on waste ground. *Nepeta* may derive from Nepete, an Etruscan town which grew the plant; *cataria* is New Latin = 'relating to cats'.

CATAPYRENIUM **Tree catapyrenium** *Catapyrenium psoromoides* (Verrucariaceae), a Critically Endangered (RDB) lichen growing on bark, with a few records from southern England, and one each from north Wales and the eastern Highlands (currently known only from an ash tree in Milton Wood, Kinross). Greek *cata* + *pyren* = 'downward' + 'kernel', and *psora* + *homoios* = 'itch' + 'resembling'.

CATCHFLY Herbaceous plants in the genus *Silene* (Caryophyllaceae) (Latin *silenus* = 'spittle' or 'foam'). The English name reflects the sticky hairs on the stem that may prevent wingless insects climbing to eat the pollen.

Alpine catchfly *S. alpestris*, a rare montane rosette-forming perennial. In Angus, it grows on serpentine debris rich in magnesium and other metals (this is the county flower of Angus as voted by the public in a 2002 survey by Plantlife) and in Cumbria on copper-rich soil over Skiddaw slate.

Berry catchfly *S. baccifera*, a scarce perennial European neophyte, naturalised on a ditch bank on the Isle of Dogs (Surrey), and established in tall grass and woodland clearings at several sites in Norfolk. Elsewhere it is a casual on roadsides and hedge banks, originating from birdseed or, occasionally, as a garden escape. *Baccifera* is Latin for 'bearing berries'.

Italian catchfly *S. italica*, a perennial European neophyte cultivated in Britain by 1759, recorded in the wild in west Kent in 1863, naturalised on chalky roadside banks and chalk quarries between Dartford and Greenhithe. It is a scarce casual garden escape elsewhere in southern England.

Night-flowering catchfly *S. noctiflora*, an annual archaeophyte, scattered but declining, commonest in eastern England, on cultivated and waste ground, on dry, sandy and calcareous substrates, but also on heavier soils over oolitic limestone. *Noctiflora* is Latin for 'night-flowering'.

Nodding catchfly *S. pendula*, an annual south European neophyte in cultivation by 1731, known in the wild since 1898 (in Dorset), but having declined is now a scarce casual or naturalised garden escape on waste ground in southern England. *Pendula* is Latin for 'hanging down'.

Nottingham catchfly *S. nutans*, perennial, scattered and locally common in Britain, mainly on shallow, drought-prone, calcareous soils but also on acidic soil overlying shingle. By the coast it is found on grassy cliffs, dunes and shingle and, inland (where it is declining), on limestone rock and cliff ledges. It is also a casual at ports and on railway banks. This is the 'county' flower of the city of Nottingham as voted by the public in a 2002 survey by Plantlife. *Nutans* is Latin for 'hanging down'.

Sand catchfly *S. conica*, an annual of open habitats on free-draining sandy soils with a scattered distribution, commonest in East Anglia but overall declining. In coastal regions, it grows on stabilised dunes and sandy shingle, open pasture and waste ground; inland it also occurs at the edges of tracks across heathland and in abandoned arable fields. *Conica* is Latin for 'cone-shaped'.

Small-flowered catchfly *S. gallica*, a winter-annual archaeophyte, scattered and declining in numbers, on disturbed ground, mainly in arable fields on often acidic sandy or gravelly soils, and on old walls and waste ground. It also occurs in open, drought-prone coastal grassland, and on dunes in the Channel Islands. *Gallica* is Latin for 'French'.

Spanish catchfly *S. otites*, a perennial of shallow, well-drained, light calcareous soils. As a native it is confined to four sites in Norfolk, all Breckland grass-heaths, and to Suffolk where it is slightly more common, protected by the Suffolk Wildlife Trust's roadside verge scheme. It occurs elsewhere in England as a rare casual. *Otites* derives from Greek *otis* = 'bustard'.

Sticky catchfly *S. viscaria*, an evergreen perennial growing locally in Wales and Scotland (this is the 'county' flower of Edinburgh and Midlothian as voted by the public in a 2002 survey by Plantlife) on cliffs and steep rocky slopes, preferring mildly acidic to moderately basic soils, but many sites are on basic volcanic rocks such as basalt and dolerite. It is also a rare garden escape. *Viscaria* is Latin for 'sticky'.

Sweet-william catchfly *S. armeria*, an annual European neophyte, present by 1800, occasional in England and Wales as a garden escape or throw-out on waste ground, usually casual, sometimes persisting for a few years in disturbed, open sites. *Armeria* here refers to Deptford pink *Dianthus armeria*, reason unclear.

CATERPILLAR-PLANT *Scorpiurus muricatus* (Fabaceae), an annual introduced from south Europe found mainly in southern England as a casual on rough ground, as a wool alien or from birdseed. Greek *scorpios* + *oura* = 'scorpion' + 'tail', referring to the shape of the pods, and Latin *muricatus* = 'roughened'.

CATERPILLARCLUB **Scarlet caterpillarclub** *Cordyceps militaris* (Hypocreales, Cordycipitaceae), a rare parasitic fungus found mainly in grassland and on woodland edges in late summer and autumn. Growing on underground larvae and, more commonly, pupae of moths, this ascomycete parasitises its host, turning it into a mushy mess, then pushes up through the turf as a bright orange club. Greek *cordyle* + New Latin *ceps* = 'club' + 'head', and the Latin for 'military', possibly from the way the fungus attacks its host.

CATILLARIA **Laurer's catillaria** *Megalaria laureri* (Ramalinaceae), a corticolous lichen, Endangered in the RDB, restricted to a few sites in the New Forest and dependent on the presence of senescent beech with rain tracks. *Catillaria* was a previous genus name. Greek *megale* = 'great'. Johann Laurer (1798–1873) was a German lichenologist.

CAT'S-EAR *Hypochaeris radicata* (Asteraceae), a common and widespread herbaceous perennial of meadow, pasture, lawn, heathland, clifftops, dunes, shingle, roadsides, railway banks and waste ground, on slightly acidic, usually free-draining soils, very tolerant of drought. *Hypochaeris* was a name used by the Greek philosopher Theophrastus. *Radicata* is Latin for 'rooted'.

Smooth cat's-ear *H. glabra*, an annual locally common in southern England, Lincolnshire and especially East Anglia, scattered elsewhere, in open summer-dry grassland and heathy pasture, usually on acidic, nutrient-poor, sandy or gravelly soils, and also in dune grassland and on sandy shingle. It was formerly more widespread as an arable weed, and as a wool-shoddy alien. *Glabra* is Latin for 'smooth'.

Spotted cat's-ear *H. maculata*, a perennial herb on free-draining, usually base-rich substrates on chalk and limestone downland, coastal cliffs over limestone and serpentine, and wind-blown calcareous sand on the north Cornish coast. In Jersey, it grows on granite cliffs. *Maculata* is Latin for 'spotted'.

CAT'S-TAIL Grass species in the genus *Phleum* (Poaceae), from Greek *phleos*, a name given to a reed or grass. See also timothy.

Alpine cat's-tail *P. alpinum*, a scarce rhizomatous perennial on base-rich flushes and mires, more rarely with rocky habitats, occasionally with weakly acid substrates enriched by flushing with base-rich water, at two sites in the north Pennines and around 70 sites in the Highlands.

Purple-stem cat's-tail *P. phleoides*, a rare perennial, very local from Bedfordshire to East Anglia, on open habitats on free-draining sandy or chalky soils, especially in Breckland where it thrives on grazed grass-heath, verges and trackside banks, and around pits, rabbit warrens and other disturbed sites.

Sand cat's-tail *P. arenarium*, an uncommon annual of mobile or semi-fixed coastal dunes and sandy shingle around the British and Irish coastline except for north and west Scotland, and also found inland in Breckland on grass-heaths, sand banks and other disturbed sandy areas. *Arenarium* comes from Latin *arena* = 'sand'.

Smaller cat's-tail *P. bertolonii*, a widespread and generally common perennial of old meadow and pasture, downland, roadside banks and waste ground. It is occasionally found as a wool alien, on shoddy fields and waste tips. Antonio Bertoloni was a nineteenth-century Italian botanist.

CATTLE *Bos taurus* (Bovidae), descended from aurochs, a domesticated Neolithic introduction, the most important feral stock in Britain being a white breed (in practice semi-managed) at Chillingham Castle, Northumberland, numbering around 100 in 2015. There is a small reserve herd of about 20 animals near Fochabers, north-east Scotland. Other feral herds are in the New Forest, Vaynol (north Wales), Dynevor (south Wales), and Rum (Inner Hebrides). On Rum the cattle are a conservation tool, improving grazing for red deer in spring. 'Cow' comes from Old English *cu*; 'cattle' from Anglo-Norman French, *catel*. The binomial is Latin for 'ox' and 'bull'.

CATWORM **White catworm** *Nephtys cirrosa* (Nephtyidae), a widespread polychaete up to 10 cm long, living infaunally in sandy sediment in the intertidal and shallow sublittoral. The genus name may come from the Egyptian goddess Nephthys. *Cirrosa* is Latin for 'curled'.

CAUDATA Order of amphibians that includes newts (Salamandridae), from the Latin for having a tail (*cauda*).

CAULIFLOWER (PLANT) See cabbage.

CAULIFLOWER FUNGUS **Bonfire cauliflower** *Peziza proteana* f. *sparassoides* (Pezizales, Pezizaceae), a scarce inedible fungus mainly found scattered in England on burnt sites. *Pezis* is Greek for a kind of fungus, *proteana* means capable of changing, and *sparasso* = 'to tear'.

Wood cauliflower *Sparassis crispa* (Polyporales, Sparassidaceae), edible if collected when young and fresh, a widespread and locally common fungus forming a cauliflower-like cream mass, sometimes weighing several kilograms, from July to October. It is parasitic on the base of conifer trees, especially Scots pine, causing rot in infected timber. Greek *sparasso* = 'to tear' and Latin *crispa* = 'curly'.

CAVALIER (FUNGUS) Members of the genus *Melanoleuca* (Tricholomataceae) (Greek *melas* + *leucos* = 'black' and 'white').

Uncommon species include **clouded cavalier** *M. schumacheri* (mainly on Scottish dunes), **grooved cavalier** *M. grammopodia*, **smoky cavalier** *M. exscissa*, **spring cavalier** *M. cognata*, **stunted cavalier** *M. brevipes* and **warty cavalier** *M. verrucipes* (mainly on woodchip mulch).

Common cavalier *M. polioleuca*, common, widespread, edible but tasteless, mainly found in England, in summer and autumn on decomposing wood in broadleaf woodland, and near trees in parkland. Greek *polios* + *leucos* = 'grey' + 'white'.

Dune cavalier *M. cinereifolia*, found in summer and autumn in dunes at a number of coastal sites in Britain, often associated with marram-grass. *Cinereifolia* is Latin for 'ash-coloured leaves'.

CECIDOMYIIDAE Family name (Greek *cecis* = 'oak gall' + *myia* = 'a fly') of the very small (<1–3 mm length) gall midges, with 650 species in 148 genera.

CEDAR Evergreen trees in the genus *Cedrus* (Pinaceae). See also deodar.

Atlas cedar *C. atlantica*, introduced from Morocco in 1840, widely planted as a park ornamental, with records of self-sown trees in south-east England, growing best on soils of poor to medium nutrient status and of dry to medium soil moisture.

Cedar-of-Lebanon *C. libani*, introduced from the Near East in the 1630s, commonly planted in warm, low rainfall areas as a parkland ornamental, growing best on soils of poor to medium nutrient status and of dry to medium soil moisture.

CELANDINE **Greater celandine** *Chelidonium majus* (Papaveraceae), a perennial archaeophyte, common and widespread except in northern Scotland and western Ireland. Once grown as a cure for warts, it is now a garden weed or escape on waste ground, walls, hedgerows and other open habitats, on moist, base-rich soils. *Chelidonium* is Latin for this plant, derived from the Greek for the bird 'swallow'.

Lesser celandine *Ficaria verna* (Ranunculaceae), a common and widespread perennial herb growing in woodland, hedge banks, meadows, lawns, roadsides, coastal grassland, river and stream banks, and shaded waste ground, preferring damp, loamy or clay soils, and avoiding very dry, very acidic or permanently waterlogged sites. Subspecies *ficariiformis* occurs as a common naturalised garden escape. The binomial is Latin for 'fig-like' and 'spring'.

CELASTRACEAE Family name of spindle tree (Greek *celastros*).

CELERY **Wild celery** *Apium graveolens* (Apiaceae), a biennial or monocarpic perennial found on the coastline of Britain north to central Scotland, and in mainly south and east Ireland, on sea walls, by brackish ditches, on tidal riverbanks and drift lines, and upper saltmarsh. Inland it is occasionally found on disturbed ground in marshes, and by ponds and ditches. The binomial is Latin for 'celery' and 'strong-scented'.

CELERY FLY *Euleia heraclei* (Tephritidae), widespread and scattered in England and Wales, on plants such as wild angelica, hemlock, hogweed and alexanders, as well as celery and parsnip, adults active in May–August, the leaf-mining larval pests in a number of generations from spring to the end of autumn. Greek *eu* + *leios* = 'good' + 'smooth'. *Heraclei* relates to Hercules.

CELLAR FUNGUS *Coniophora puteana* (Boletales, Coniophoraceae), a causative agent of wet rot. Greek *conis* + *phora* = 'dust' + 'carrying'.

CENTAURY Herbaceous plants in the genus *Centaurium*, *Cicendia* (Latin, from Cicend, Albania) and *Exaculum* (Gentianaceae). The English name dates from late Middle English, from Latin *centaurea*, based on Greek *centauros* = 'centaur', because its medicinal properties were said to have been discovered by the centaur Chiron.

Common centaury *Cent. erythraea*, a common and widespread (though mainly coastal in Scotland) biennial, rarely annual, of mildly acidic to calcareous, well-drained, often disturbed, soils, found in chalk and limestone grassland, heathland, woodland rides and open scrub, dune grassland, quarries, spoil heaps and verges. *Erythros* is Greek for 'red'.

Guernsey centaury *E. pusillum*, an annual growing in moist, short turf in coastal dune slacks in Guernsey. Winter-flooding and rabbit grazing maintain open conditions, which this species requires. *Pusillum* is Latin for 'small'.

Lesser centaury *Cent. pulchellum*, an annual mainly found in the Channel Islands, southern England and south Wales, but local as far north as central Scotland, and also east Ireland. It grows on slightly acidic to calcareous soils, inland in dry grassland and heathland, woodland rides, marl-pits and other open, disturbed ground, and on the coast in open sandy and muddy grassy habitats, often by estuaries, dunes and upper saltmarsh. *Pulchellus* is Latin for 'pretty'.

Perennial centaury *Cent. scilloides*, a perennial on freely draining soils on the slopes of coastal cliffs in Pembrokeshire, and rediscovered in 2010 in grassland near Land's End. It has been found as a lawn weed in south-east England since 1974. *Scilloides* in Latin for resembling plants in the genus *Scilla* (squills).

Seaside centaury *Cent. littorale*, a biennial on coastal dunes, upper saltmarsh and calcareous, humus-rich turf near the sea, north of south Wales and north-east Yorkshire, and in Co. Londonderry, in places where competing vegetation is checked by grazing or trampling. *Littorale* is Latin for 'of the seashore'.

Slender centaury *Cent. tenuiflorum*, a rare annual of open, poorly drained sandy or clay soils on slumping coastal cliffs. It will not persist in closed vegetation, tending to appear a few years after the habitat is created, and disappear after about 10 years. It is now only found near Golden Cap, Dorset. A 1991 record from the Isle of Wight is probably a misidentification. *Tenuiflorum* is Latin for 'slender-flowered'.

Yellow centaury *Cic. filiformis*, an extremely rare annual in Devon and Sussex and rapidly declining in Cornwall and Dorset. In the Lizard Peninsula and New Forest it remains locally widespread. It is also found on St David's Peninsula, Wales, on grass-heath commons, and in south-west Ireland. It requires open heathland, growing on sandy and peaty soils of relatively high base status which are damp in winter and spring, and is found around seasonally flooded pools along rutted trackways, and an absence of competition. It is classified as Nationally Scarce and included as a species of principal importance for the purpose of conserving biodiversity in the Natural Environment and Rural Communities Act (2006). *Filiformis* is Latin for 'thread-like'.

CENTIPEDE Members of the class Chilopoda, subphylum Myriapoda, with elongated segmented bodies having one pair of legs per segment, the number of pairs varying but always with an odd number. They are generalist carnivores, with elongated mandibles and a pair of venom claws (forcipules) formed from a modified first appendage. Often nocturnal, they favour a range of moist habitats. The name is from the Latin for 'a hundred legs', though examples never have such a number.

CENTURION Soldier flies in the genera *Chloromyia* (Greek *chloros* + *myia* = 'green' + 'fly') and *Sargus* (Stratiomyidae).

Broad centurion *C. formosa*, 9 mm length, very common and widespread. Adults, flying from May to August, are usually found in damp habitats, though including woodland and gardens, and feed on (especially hogweed) nectar. Larvae feed on decomposing vegetation. *Formosa* is Latin for 'beautiful'.

Clouded centurion *S. cuprarius*, 9 mm length, with a putatively scattered distribution in England and south Wales (most records probably being iridescent centurion; a record from the Pevensey Levels, East Sussex, is one that has been confirmed). Latin *cupreus* = 'coppery'.

Iridescent centurion *S. iridatus*, 6–12 mm length, widespread and locally common in a variety of habitats from May to September. Larvae breed in (often cow) dung. *Iridatus* is Latin for 'iridescent'.

Twin-spot centurion *S. bipunctatus*, 12–13 mm length, widespread and locally common in England and Wales, plus a few records from Scotland and east Ireland. Adults fly in July–November in woodland edge, often sun-bathing on foliage in sheltered spots. Larvae are found in cow dung, compost,

rotting vegetation and decaying fungi. *Bipunctatus* is Latin for 'two-spotted'.

Yellow-legged centurion *S. flavipes*, 7–9 mm length, widespread and scattered, adults evident in May–October, often sun-bathing on foliage in sheltered locations along meadow and woodland edges. Larvae have been found in cow dung and pasture soil. *Flavipes* is Latin for 'yellow-legged'.

CEP *Boletus edulis* (Boletaceae), a common, widespread edible and tasty fungus containing high levels of the antioxidants ergothioneine and glutathione, found in summer and autumn on soil under broadleaf or coniferous trees. Also known as penny bun and king bolete. 'Cep' dates to the mid-nineteenth century, from French *cèpe*, and Gascon *cep* = 'mushroom', derived from Latin *cippus* = 'stake'. Greek *bolos* = 'lump of clay', and Latin *edulis* = 'edible'.

Summer cep Alternative name for summer bolete.

CEPHALOPOD(A) Class of Mollusca that includes squid and octopus (Greek *cephale* + *podi* = 'head' + 'foot').

CEPHIDAE Family name of stem borers (stem sawflies) (Greek *cephale* = 'head').

CERAMBYCIDAE Family name of longhorn beetles (with a wide range of body lengths, 2.5–30 mm, but with long antennae giving them their common name), named after Cerambos, a shepherd in Greek mythology who was transformed into a large beetle with horns, with 63 extant British species in 47 genera. Larvae are generally wood borers. Adults of non-native species often emerge from imported timber products, sometimes years after import. Adults are often found on flowers or on recently fallen timber.

CERATOPHYLLACEAE Family name of hornworts (Greek *ceratophyes* + *phyllon* = 'having horns' + 'leaf').

CERATOPOGONIDAE Family name (Greek *ceratophyes* + *pogon* = 'having horns' + 'beard') of biting midges, with 171 species in 20 genera, the most speciose being *Culicoides* (from Latin *culex* = 'midge') and *Forcipomyia* (Latin *forceps* = 'pincer' and Greek *myia* = 'fly'). Eggs are laid in damp soil, in which the larvae develop, feeding on nematodes, other insect larvae, fungi and parts of plants. As adults, most species feed on insect 'blood'. Males feed on nectar, if at all. Adult female *Culicoides*, however, with other members of the family, feed on blood, the bite causing great irritation in large mammals such as horses (in which they can transmit sweet itch) and deer, many species of bird, and people (the Highland midge *Culicoides impunctatus* in northern Scotland and north Wales). In Scotland adults begin to emerge in April and are on the wing until October. They are most active just before dawn and sunset, and are less active with wind speeds of over 6–8 mph, or humidity below 60–75%.

CEREAL Grasses whose starchy grain is used as human and animal food; see barley, maize, millet, oat, rye, triticale and wheat.

CERIANTHIDAE Family name (Greek *ceras* = 'horn') of two species of sea anemone in deep British waters.

CERTHIIDAE Family name of treecreeper, from Greek *certhios* for this bird.

CERVIDAE Family name for deer (Latin *cervus*).

CERYLONIDAE Beetle family (Greek *cerylos*, a seabird, reason obscure), with two uncommon *Murmidius* species found in stored products, and three widespread *Cerylon* species that occur in decaying ancient hardwood trees, probably feeding on fungal hyphae and slime moulds.

CESTODA Class of parasitic flatworms in the phylum Platyhelminthes, including tapeworms (Greek *cestos* = 'girdle').

CETACEA The order (or infraorder in the order Artiodactyla) that includes the suborders Mysticeti (baleen whales) and

Odontoceti (toothed whales, dolphins and porpoises), from Latin *cetus* = 'whale', in turn from Greek *cetos* = 'large fish' or 'sea monster'.

CHAFER Beetles in the family Scarabaeidae; see also cockchafer.

Bee chafer Alternative name for bee beetle.

Brown chafer *Serica brunnea* (Melolonthinae), 8–15 mm long, widespread, adults concealed at the base of grass tussocks, in moss or decaying tree stumps during the day, flying in the evening, usually above long grass. Larvae feed on grass roots, and can become a pest in lawns. Greek *sericon* = 'silk' and Latin *brunnea* = 'brown'.

Dune chafer *Anomala dubia* (Rutelinae), up to 12 mm long, widespread in coastal England and Wales. The binomial is Greek for 'uneven' and Latin for 'doubt'.

Garden chafer *Phyllopertha horticola* (Rutelinae), 10–12 mm long, widespread in Britain, though more local in Scotland, adults flying during sunny mornings in June–July, in woodland edges, hedgerows, gardens and parkland, feeding on various plants. Larvae feed on grass roots, often becoming a pest in lawns, especially during autumn. Greek *phyllo* + *pertha* = 'leaf' + 'plunder', and the Latin *hortus* = 'garden'.

Rose chafer *Cetonia aurata* (Cetoniinae), up to 20 mm long, with a scattered distribution in the southern half of England, found in grassland, scrub and gardens, adults feeding on dog rose and other flowers, larvae on decaying plant material, including compost. *Cetonia* is Latin for this kind of beetle, *aurata* = 'golden'.

Summer chafer *Amphimallon solstitialis* (Melolonthinae), 15–20 mm long, locally common in the southern half of England, declining but still locally abundant, crepuscular from May to August, habitats including meadow, hedgerows and gardens, as well as other urban sites, often flying at heights of 15–20 m around trees and buildings. Greek *amphis* + *mallon* = 'around' + 'ball of wool', and the Latin *solstitialis* = 'belonging to midsummer'.

Welsh chafer *Hoplia philanthus* (Rutelinae), 8–9 mm long, scarce with a scattered distribution in England and Wales in grassland, hedgerows and forest margins. Adults are day-active from June to August; larvae feed in often sandy soil on grass roots, and may damage lawns and sports turf. Greek *hople* = 'hoof', and *philos* + *anthos* = 'loving' + 'flower'.

CHAFFINCH *Fringilla coelebs* (Fringillidae), wingspan 26 cm, length 14 cm, weight 24 g, a resident breeder with an estimated 5.8 million territories in Britain in 2009, numbers having increased by 21% over 1970–2015, though this trend has slowed in the last couple of decades, and there was a weak decline (−12%) over 2010–15. Outbreaks of trichomonosis impacted numbers in 2006 and 2007. It is Britain's second commonest breeding bird, after the wren. It prefers open woodland, but is found in a range of habitats, often foraging on the ground for insects and seeds. The female builds a nest with a deep cup often in the fork of a tree, usually with 4–5 eggs, incubation lasting 12–13 days, and the altricial young fledging in 13–16 days. The English name, derived from Old English *ceaffinc*, comes from the bird's habit of foraging around barns, picking seeds out of the chaff. Latin *fringilla* = 'finch' and *coelebs* = 'unmarried' or 'single', the specific name given by Linnaeus, who saw only male chaffinches in his native Sweden, females from its northern breeding grounds wintering further south.

CHAFFWEED *Centunculus minimus* (Primulaceae), an annual with a scattered if mostly western and southern distribution, in open habitats on damp, sandy, usually acidic soils, often near the sea. Habitats include dunes, sandy cliffs, paths and tracks on heathland, and in forest rides. Latin *cento* + *unculus* = 'clothing' or 'patchwork garment' + 'little', and *minimus* = 'smallest'.

CHAGA *Inonotus obliquus* (Hymenochaetales, Hymenochaetaceae), a scarce inedible bracket fungus with a scattered distribution in England, locally common in Scotland, especially in the Highlands, saprobic on birch, and causing sterile conk trunk rot (referring to the fruiting bodies growing under the outer layers of wood surrounding the sterile conk once the tree is dead, thereby spreading the spores). The English name is a transliteration of the Russian word for this species. Greek *inodis* + *otos* = 'fibrous' + 'ear', and Latin *obliquus* = 'oblique'.

CHAIN WEED Creeping chain weed *Catenella caespitosa* (Caulacanthaceae), a common and widespread red seaweed found on the middle to upper levels of rocky shores, restricted to sheltered and shaded sites, sometimes found in estuaries, creating mats of interlaced fronds up to 20 mm long. The binomial is Latin for 'little chain' and 'growing in tufts'.

CHALARA DIEBACK Alternative name for ash dieback, caused by the fungus *Hymenoscyphus fraxineus*, previously *Chalara fraxinea*, hence the name.

CHALCIDIDAE Family name of parasitoid or hyperparasitoid wasps (Greek *chalcos* = 'copper', from the metallic colour), with six species in four genera, all scarce in England and Wales.

CHAMOMILE 1) *Chamaemelum nobile* (Asteraceae), a herbaceous perennial scattered, though mostly in south-west and southern England and south-west Ireland, in moderately acidic, seasonally damp short grassland, especially where mowing, grazing or trampling reduces competition; also in coastal grassland and on cliffs. The English and genus names are from Greek *chamae* + *mylo* = 'on the ground' + 'apple', from the apple scented flowers.

2) Species of *Anthemis* (Asteraceae) (Greek *anthos* = 'flower').

Austrian chamomile *A. austriaca*, an annual or biennial European neophyte scattered in England and Wales as a casual on tips and newly sown grass as a contaminant of grass and birdseed and wool shoddy.

Corn chamomile *A. arvensis*, a widespread but declining annual of calcareous or sandy soils, growing in arable fields, leys, field edges and waste ground, as well as on roadsides and disturbed ground near the sea. It is also a grass-seed alien. *Arvensis* is from Latin for 'field'.

Sicilian chamomile *A. punctata*, a herbaceous perennial neophyte grown in gardens since 1818, scattered in England, Wales, Ross-shire and south Ireland, naturalised on roadsides, railway banks, pathsides, cliffs and waste ground. It is also a casual from grain impurities. *Punctata* is Latin for 'spotted'.

Stinking chamomile *A. cotula*, an archaeophyte annual of arable crops, common and widespread in England, scattered elsewhere, and a grain-seed casual in Ireland. It generally favours heavy soils but can grow on light soils, including those over chalk. *Cotula* is Latin for 'small cup', referencing a cupped area at the base of the leaves.

Yellow chamomile *A. tinctoria*, a biennial or perennial European neophyte herb introduced by 1561, formerly cultivated for its yellow dye, now scattered as a casual, occasionally naturalised, on rough ground, usually on dry soils. *Tinctoria* is Latin indicating a plant used in dyeing.

CHAMPIGNON Fairy ring champignon, fairy ring mushroom or **Scotch bonnet** *Marasmius oreades* (Marasmiaceae), a widespread and common edible fungus found from spring to autumn, typically in rings on soil in short grass, including park grassland and garden lawns. Greek *marasmos* = 'drying out'. In Greek mythology the Oreiades were nymphs of mountains and valleys.

CHANTERELLE 1) Edible, often tasty fungi mostly in the genus *Cantharellus* (Cantharellales, Cantharellaceae), 'chanterelle' coming from French, ultimately from Greek *cantharos*, meaning a drinking vessel. Specifically, *C. cibarius*, a widespread and common though local, found in summer and autumn, on (generally broadleaf) woodland acid soil, often under oak, sweet chestnut, hazel and birch. *Cibarius* is from Latin *cibus* = 'food'.

Rare chanterelles include **ashen chanterelle** *C. cinereus*, blackening chanterelle *C. melanoxeros*, **golden chanterelle** *C. aurora*, **orange chanterelle** *C. friesii* and **pale chanterelle** *C. ferruginascens*.

Sinuous chanterelle *Pseudocraterellus undulatus*, edible, common and widespread in Britain and north and north-west Ireland, found in summer and autumn in litter in deciduous woodland, especially under beech. Greek *pseudis* + *crater* = 'false' + '(small) cup', and Latin *undulatus* = 'wavy'.

Trumpet chanterelle *C. tubaeformis*, edible, widespread but local, especially in spruce forests on acidic soil in west Wales, north-west England and Scotland, found in late summer and autumn often in large groups. *Tubaeformis* is Latin for 'trumpet-shaped'.

2) **False chanterelle** *Hygrophoropsis aurantiaca* (Boletales, Hygrophoropsidaceae), inedible, widespread and common in soil under conifers or on heathland in late summer and autumn. The genus name means resembling woodwaxes *Hygrophorus*. Latin *aurantiaca* = 'orange colour'.

CHAOBORIDAE Family name (Greek *chaoo* + *bora* = 'destroy' + 'flesh') of phantom midges, with six species in two genera. The aquatic larvae uniquely have antennae that are modified into grasping organs which impale or crush prey such as other insect larvae or small crustaceans, then carry it to the stylet (mouth).

CHAR(R) Freshwater fishes in the genus *Salvelinus* (Salmonidae), *salvelinus* being New Latin, possibly from German *Saibling* = 'charr', which itself may be a word of Celtic origin.

Arctic charr *S. alpina*, length 25–30 cm, populations in cold upland lakes having variable characteristics: historically, 15 separate species have been recognised but a 2007 review concludes that evidence for these being afforded full species status is poor (however, see haddy charr under charr). Deep-water fish spawn in late winter or spring, shallow-water examples in autumn. Diet is of small crustaceans and other small, often planktonic animals.

Brook charr More commonly referred to in Britain as brook trout.

Haddy charr *S. killinensis*, an endemic fish, possibly with full species status, found in the deeper parts of Loch Killin (Inverness-shire), which provides the specific name, Loch Doine, Trossachs, and perhaps Loch Builg, Cairngorms, moving to shallow waters for spawning.

CHARACTER (MOTH) **Chinese character** *Cilix glaucata* (Drepanidae, Drepaninae), wingspan 18–22 mm, nocturnal, common in hedgerows, scrub, woodland and gardens, widespread except for northern Scotland. Adults combine their wing pattern and resting posture to give the appearance of a bird dropping, a mimicry defence against predation. The bivoltine adults fly in May–June and August. Larvae feed on leaves of hawthorn, blackthorn and crab apple. The binomial comes from Cilix, a mythical king of Phoenicia, and the metallic glaucous blotch on the dorsal wings.

CHARADRIIDAE Family name for waders in the genera *Charadrius*, *Haematopus*, *Pluvialis* and *Vanellus*. Greek *charadrios* was a bird found in river valleys (*charadra* = 'ravine').

CHARADRIIFORMES Order that includes the families Charadriidae (plovers, etc.), Alcidae (auks), Burhinidae (stone curlew), Glareolidae (pratincoles and coursers), Haematopodidae (oystercatcher), Laridae (gulls), Recurvirostridae (avocet and stilt), Scolopacidae (woodcock, snipe, curlew, sandpiper, etc.), Sternidae (terns), Stercorariidae (skuas), and Phalaropiidae (phalaropes). Greek *charadrios* was a bird found in river valleys (*charadra* = 'ravine').

CHARALES, CHAROPHYCEAE, CHAROPHYTA Order, class and division of stoneworts (freshwater green algae), a sister group to the Chlorophyta. The prefix *charo* is Greek for 'graceful' or 'beautiful'.

CHARCOAL BURNER *Russula cyanoxantha* (Russulales,

Russulaceae), a common and widespread edible ectomycorrhizal fungus found from July to November, on deciduous woodland soil, especially under oak and beech. The English name reflects the stems being as brittle as a stick of charcoal. Latin *russula* = 'reddish', Greek *cyanos* + *xanthos* = 'blue' + 'yellow'.

CHARD See foliage beet under beet.

CHARLOCK *Sinapis arvensis* (Brassicaceae), a widespread and abundant annual archaeophyte arable weed, also found on roadsides, railways and waste ground. 'Charlock' is from Old English *cerlic* or *cyrlic*, of unknown origin. *Sinapi* is an Ancient Greek word for 'mustard', *arvensis* from Latin for 'field'.

CHASER Dragonflies in the genus *Libellula* (Libellulidae), genus and family names possibly from the Latin for 'small book', but more convincingly from *libella* = 'level' and by derivation 'dragonfly' from the level position of the wings.

Broad-bodied chaser *L. depressa*, body length 39–48 mm, wingspan 70 mm, widespread and common throughout England and Wales, associated with shallow still water (often the first coloniser of a new pond). Larvae take one to three years to develop, lying partially buried in the bottom mud. The main flight period is May–August. *Depressa* is Latin for 'depressed'.

Four-spotted chaser *L. quadrimaculata*, body length 39–48 mm, wingspan 75 mm, widespread and common, associated with well-vegetated sites on the margins of shallow ponds and lakes. Larvae take two years to develop. The main flight period is May–August. *Quadrimaculata* is Latin for 'four-spotted'.

Scarce chaser *L. fulva*, body length 40–9 mm, wingspan 74 mm, restricted to Norfolk/Suffolk, Sussex, Wiltshire/Somerset, Cambridgeshire, Kent and Dorset/Hampshire. Populations are stable and it may be expanding its range, but it remains listed as Scarce in the British RDB 3. It usually inhabits good quality, slow-flowing, meandering, well-vegetated rivers, though oviposition required open water. *Fulva* is Latin for a brown shade of yellow.

CHAT Small members of the family Turdidae, though some taxonomists have placed them in the Muscicapidae; see robin, redstart, stonechat, wheatear and whinchat. 'Chat' dates to the seventeenth century, probably imitative of the harsh call of some species.

CHECKERBERRY *Gaultheria procumbens* (Ericaceae), a low North American neophyte shrub introduced by 1762, widespread and locally naturalised in woodland and moorland. Originally planted as ground cover, now a scarce and scattered naturalised species. Jean-François Gaultier (1708–56) was a Canadian botanist. *Procumbens* is Latin for 'procumbent'.

CHELIFER Pseudoscorpions in the families Cheliferidae, Neobisiidae and Withiidae (Greek *chele* = 'claw'), length 2–3 mm.

Bog chelifer *Microbisium brevifemoratum* (Neobisiidae), found in sphagnum in Wales in 2005.

Cambridge's two-eyed chelifer *Roncocreagris cambridgei* (Neobisiidae), with a westerly distribution from north-west Scotland to Cornwall east to Hampshire, with records also from south-west Ireland and Northern Ireland, found in deciduous tree litter and grass and beneath stones.

Lazy chelifer *Withius piger* (Withiidae), associated with grain and other stored food products and warehouse debris, probably introduced via ships' cargo. *Piger* is Latin for 'lazy'.

Marram grass chelifer *Dactylochelifer latreillei* (Cheliferidae), local in south and south-east England, and around the Firth of Forth, associated with marram and sea couch-grass, in leaf bases and among the roots, under driftwood and in strand-line debris. Greek *dactylo* + *chele* = 'finger' + 'claw'.

Oak-tree chelifer *Larca lata* (Garypidae), length 1.6 mm, recorded from one locality in Windsor Great Park, from nest debris in a decaying oak tree.

Reddish two-eyed chelifer *Roncus lubricus* (Neobisiidae), locally common in the southern half of England and in Wales, scarce in Northern Ireland, found in dry leaf litter and moss, and under stones. *Lubricus* is Latin for 'slippery'.

CHENOPODIACEAE Formerly used name for the chenopod family now included with the Amaranthaceae (Greek *chen* + *podion* = 'goose' + 'little foot', referring to leaf shape).

CHERNES Pseudoscorpions in the genera *Allochernes* (Greek *allos* + *cherne* = 'other' + 'a needy thing'), *Dinocheirus* (*deinos* + *cheir* = 'terrible' + 'hand') and *Pselaphochernes* (*pselaphao* = 'to feel about') (Chernetidae).

Compost chernes *P. scorpioides*, length 1.5–2 mm, scattered in south and central England, with a recent record from Scotland, in compost, manure and straw debris, as well as leaf litter, dead wood and the nests of the red ant *Formica rufa*. It is often phoretic on flies.

Nest chernes Alternative name for terrible-clawed chernes.

Powell's chernes *Allochernes powelli*, length 2.3 mm, with scattered records in England and Wales, mainly from barn and stable refuse.

Small chernes *P. dubius*, length 1.5–1.7 mm, widespread and scattered in Britain, commoner in the south, associated with calcareous habitats in grassland and woodland soil and leaf litter, beneath stones and in decaying wood.

Terrible-clawed chernes *Dinocheirus panzeri*, length 2.1–2.6 mm, found throughout England, in south Scotland and Northern Ireland, mainly synanthropic, occurring in hay, straw and grain refuse in barns, inside chicken houses and pigeon lofts. It has also been found in bird, mammal and ant nests, and appears to be phoretic on birds.

CHERNETIDAE Family name of pseudoscorpions such as tree-chernes.

CHERRY Species of *Prunus* (Rosaceae). 'Cherry' is Middle English, from Old French *cherise*, from medieval Latin *ceresia*, in turn based on Greek *cerasos* = 'cherry tree'. Latin *prunus* = 'plum tree'.

Bird cherry *P. padus*, a widespread deciduous shrub or small tree, less common in southern England, and more scattered in the Republic of Ireland, found in damp woodland and scrub, streamsides and shaded rocky places; also in fen carr in East Anglia. It grows on a variety of soil types, but is most frequent on damp calcareous or base-rich substrates, and avoids very dry or very acidic conditions. *Padus* is Latin here meaning 'of the River Po'.

Dwarf cherry *P. cerasus*, a widespread Near Eastern archaeophyte shrub or small tree, long cultivated, and naturalised in hedgerows and woodland edge.

Fuji cherry *P. incisa*, a deciduous neophyte shrub or small tree in cultivation by 1913, found in a few woodlands in England as a garden escape. *Incisa* is Latin for 'incised' or 'deeply cut', describing the leaf.

Japanese cherry *P. serrulata*, a deciduous ornamental neophyte tree introduced from Japan and China by 1822, widely planted on roadsides and in parks. *Serrulata* is Latin for 'small-toothed', referring to the leaf margins.

Rum cherry *P. serotina*, a small or medium-sized North American neophyte tree, with a scattered distribution mostly in southern England, planted in gardens by 1629, naturalised in woodland and hedgerows, and on roadside verges, riversides and heathland. *Serotina* is Latin for 'late', referring to time of flowering.

St Lucie cherry *P. mahaleb*, a south European deciduous neophyte shrub or small tree planted in gardens by 1714, and naturalised in a few places in England in woodland, rough grassland and railway banks. *Mahaleb* is based on the Arabic word for this species.

Wild cherry *P. avium*, a common and widespread small to medium-sized tree found in woodland edge and hedges on fertile soils, and widely planted as an ornamental or fruit tree in parks and gardens. *Avium* is from Latin *avis* = 'bird'.

CHERVIL (BRYOZOAN) Sea chervil *Alcyonidium diaphanum* (Alcyonidiidae), common and widespread, forming an erect colony that can grow up to 50 cm long but more usually 15 cm, attached to rocks, shells or stones from the lower intertidal zone to shelly sands and coarse grounds offshore. Greek *alcyonion*, a kind of sponge, so called because it resembles a kingfisher (*alcyon*) nest, and *diapharos* = 'different'.

CHERVIL (PLANT) Herbaceous species of *Anthriscus* and *Chaerophyllum* (Apiaceae). The name is Old English, derived from Latin *chaerephylla*, in turn from Greek *chairephyllon*. *Anthriscus* is Latin for another, unidentified plant; *Chaerophyllum* is from Greek *chairo* + *phyllon* = 'to please' + 'leaf'.

Bur chervil *A. caucalis*, a widespread annual common near the sea on well-drained, often sandy or gravelly soils in open habitats such as dry grassland, hedge banks, roadsides, shingle, waste ground, gravel-pits and arable fields. *Caucalis* is Latin for an unspecified umbelliferous plant.

Garden chervil *A. cerefolium*, a perennial south-east European archaeophyte cultivated as a garden herb for at least a millennium, naturalised in open and ruderal habitats, including a sandy rock face near Bromsach, Herefordshire, since 1867, and as a casual on waste ground, roadsides and tips. *Cerefolium* is Latin for 'waxy leaves' (*cera* = 'wax').

Golden chervil *C. aureum*, a European neophyte perennial, naturalised or casual, very scattered in England and central Scotland in grassy habitats. *Aureum* is Latin for 'gold'.

Hairy chervil *C. hirsutum*, a European neophyte perennial cultivated by 1759, a naturalised garden escape on a verge in Cumbria since 1979, a riverbank in Lanarkshire since 1989, and possibly in north Yorkshire. *Hirsutum* is Latin for 'hairy'.

Rough chervil *C. temulum*, biennial, common and widespread in England, Wales, south-east and central Scotland, and south of the Moray Firth, and introduced, occasional and declining in Ireland. It favours rank grassland on verges and hedge banks, along woodland edges, and on railway banks and waste ground, tolerating light shade, rarely occurring on damp and acidic soils. The upper parts and the fruit contain toxins (including the volatile alkaloid chaerophylline) which cause gastro-intestinal inflammation in humans, but more commonly affects animals, for example a staggering gait and severe colic in cattle and pigs. *Temulentus* is Latin for 'drunken', from a putative similarity with these symptoms.

CHESTNUT (MOTH) 1) Members of the family Noctuidae, specifically *Conistra vaccinii* (Xyleninae), wingspan 28–36 mm, nocturnal, widespread and common in broadleaf woodland, scrub, hedgerows and gardens. Adults fly over winter from September to May. Larvae feed on leaves of oaks, birches, elm, blackthorn and other deciduous trees and shrubs. Greek *conistra* = 'somewhere covered in dust' (referencing speckles on the forewing). *Vaccinium* (bilberry) is a possible food plant.

Barred chestnut *Diarsia dahlii* (Noctuinae), wingspan 32–42 mm, widespread but local in broadleaf woodland, moorland and wooded heathland. Population levels increased by 288% between 1968 and 2007. Larvae feed on birches, sallows, bilberry and bramble. Greek *diarsis* = 'a raising up'. Georg Dahl was an early nineteenth-century Austrian collector.

Beaded chestnut *Agrochola lychnidis* (Xyleninae), wingspan 30–5 mm, common in broadleaf woodland, scrub, hedgerows, grassland, heathland and gardens throughout England, Wales, Ireland and southern Scotland, but with numbers declining by 93% over 1968–2007, a species of conservation concern in the UK BAP. Adults fly in September–October. Larvae initially feed on herbaceous plants then on leaves of trees and shrubs such as hawthorn. Greek *agros* + *chole* = 'field' + 'bile' (for the yellow wing colour). *Lychnis* is a possible food plant.

Dark chestnut *C. ligula* (Xyleninae), wingspan 30–8 mm, common in woodland, farmland and gardens, widespread but more local in Scotland. Adults fly in October–November, occasionally further into winter in the south. Larvae initially feed on deciduous trees and shrubs such as blackthorn, hawthorn, oaks and sallows, and when nearly full-grown on herbaceous plants, for example dandelion and docks. *Ligula* is Latin for 'little tongue' or 'strap', from the wing markings.

Dotted chestnut *C. rubiginea* (Xyleninae), wingspan 30–5 mm, in woodland, wooded heathland, farmland and hedgerows in parts of England, most commonly in the south-west. Adults fly in October–November, overwinter, and reappear the following spring. Larvae feed on deciduous trees, including apple. *Rubiginea* is from Latin *rubigo* = 'rust', from the forewing colour.

Flounced chestnut *A. helvola* (Xyleninae), wingspan 30–5 mm, widespread and common in broadleaf woodland, scrubby downland, heathland and moorland, but with numbers declining by 94% over 1968–2007 a species of conservation concern in the UK BAP. Adults fly in September–October. Larvae feed on a number of deciduous trees in the south, heather and bilberry on northern moors. *Helvola* is Latin for 'pale yellow', possibly describing the underwing colour.

Red chestnut *Cerastis rubricosa* (Noctuinae), wingspan 32–8 mm, widespread and common in broadleaf woodland, scrub and gardens in the south and moorland in the north. Adults fly in March–April. Larvae feed on herbaceous plants such as groundsel and bedstraws. Greek *cerastes* = 'horned', from the antennae, and Latin *rubricosa* meaning having a red ochre colour.

Southern chestnut *A. haematidea* (Xyleninae), wingspan 32–8 mm, discovered in Britain in 1990, known from just one heathland site in West Sussex until 1996 when it was also found on heathland in the New Forest, Hampshire. It was subsequently discovered elsewhere in the New Forest and on some east Dorset heaths. It is probably a long-overlooked resident due to its late season (October–November) and short flight period (half an hour at dusk). Larvae feed in April–July on flowers of bell heather and cross-leaved heath. *Haematidea* is Greek for 'blood red'.

2) **Horse chestnut** *Pachycnemia hippocastanaria* (Geometridae, Ennominae), wingspan 28–32 mm, a scarce nocturnal moth found in heathland in (especially southern) England and Wales. Adults fly in April–May, with a second brood in August. Larvae (despite both the specific and English names) feed on heather and cross-leaved heath. Greek *pachys* + *cneme* = 'thick' + 'shin', from the dilated male hind tibia, Latin *hippocastanaria* = 'horse chestnut'.

CHESTNUT (TREE) See horse-chestnut and sweet chestnut.

CHEVRON *Eulithis testata* (Geometridae, Larentiinae), wingspan 25–35 mm, a nocturnal moth widespread and common on heathland, moorland, fen, dunes and open woodland. Adults fly in July–August. Larvae feed on leaves of sallow, creeping willow, aspen and birch. Greek *eu* + *lithos* = 'good' + 'stone', with a stone-coloured wing, and Latin *testa* = 'burnt clay', from a yellow-brown wing colour.

CHEWING DISEASE Nigropallidal encephalomalacia, a fatal disease in horses caused by ingesting yellow star-thistle. In large quantities (for example where present in hay) it causes an untreatable neurological condition, with mouth lesions and ulcers, destroying the animal's ability to chew and swallow: death occurs through starvation or dehydration.

CHI See grey chi.

CHICK PEA *Cicer arietinum* (Fabaceae), a casual annual of rubbish tips, docks and waste ground, growing from grain, birdseed and food refuse. *Cicer* is Latin for 'chick pea', *arietinum* = 'ram's head'.

CHICKEN OF THE WOODS *Laetiporus sulphureus* (Polyporales, Fomitopsidaceae), an edible and tasty fungus, locally common, widespread (especially in England), with large fleshy cream or yellow, fan-shaped brackets said to resemble chicken's

feet, found from late spring to late summer on the trunks and stumps of especially oak, but also noted on beech, sweet chestnut, willow, apple and yew. Not the same as hen of the woods. The binomial is from the Latin for 'light pores' and 'sulphur-yellow'.

CHICKWEED Species of *Stellaria* (Latin *stella* = 'star', from the flower shape), *Moenchia* (honouring the German botanist Conrad Moench, 1744–1805) and *Myosoton* (like forget-me-not *Myosotis*) (Caryophyllaceae). 'Chickweed' derives from the late fourteenth century *chekwede*, in turn from Old English *cicene mete* = 'chicken food'.

Common chickweed *S. media*, a common and widespread annual, capable of two or three generations in a year, often overwintering, found in disturbed habitats, especially in nutrient-enhanced sites. It is a common weed in gardens and fields, also found on walls, and a characteristic plant of coastal strand-line. *Media* is Latin for 'middle'.

Greater chickweed *S. neglecta*, a herbaceous annual to short-lived perennial scattered throughout England and Wales, scarce in south and central Scotland, Isle of Man and Ireland, in damp, shaded habitats such as hedgerows, wood margins and streamsides, on a range of soils from poorly drained clays to damp sand and peaty alluvium. *Neglecta* is Latin for 'neglected' or 'overlooked'.

Lesser chickweed *S. pallida*, a locally abundant annual found in England, Wales, east Scotland and east Ireland, growing on light, well-drained soils. By the coast it is found on dunes and shingle; elsewhere it occurs on waste and cultivated ground, in gravel- and sandpits, on sandy tracks in conifer plantations, and occasionally in lawns and on walls. *Pallida* is Latin for 'pale'.

Upright chickweed *Moenchia erecta*, an annual found locally in southern England, the Midlands and Wales, growing on summer-dry soils in grazed grassland and heathland, on clifftops, pathsides, dunes and sandy shingle. It is also found in quarries, sandpits, and other disturbed habitats.

Water chickweed *Myosoton aquaticum*, a perennial herb locally common in much of England and Wales, naturalised and scarcer elsewhere, usually growing in moist habitats, including damp woods, alder and willow carr, the banks of watercourses and ditches, by ponds and in marshes.

CHICKWEED-WINTERGREEN *Trientalis europaea* (Primulaceae), a deciduous perennial herb locally common in parts of Scotland (this is the county flower of Nairn as voted by the public in a 2002 survey by Plantlife), the Scottish Borders and Yorkshire, occasional elsewhere south to Derbyshire, found on moist, acidic and humus-rich soils in birch, oak and especially pine woodland and on moorland, less commonly on heathland. It is highly localised, colonies often separated by apparently suitable ground. An ancient woodland indicator species, it is a good competitor but a poor colonist. *Trientalis* is Latin for a third of a foot, referring to the plant's height.

CHICORY *Cichorium intybus* (Asteraceae), an archaeophyte perennial herb, often used as a salad plant, locally common in England, scattered elsewhere, on roadsides, field margins and rough grassland on a range of soils, though with a preference for calcareous ones. 'Chicory' dates from late Middle English, from French *cicorée*, ultimately from Greek *cichorion*. *Intybus* is Latin for this plant, derived from Egyptian *tybi* (January), referring to the when the plant was usually eaten.

CHIFFCHAFF *Phylloscopus collybita* (Phylloscopidae), wingspan 18 cm, length 10 cm, weight 9 g, a common and widespread insectivorous woodland leaf warbler, a migrant summer breeder and, increasingly, Siberian birds as winter visitors to coastal southern England. There were 1.1 million territories in 2009, and an increase in numbers by 105% between 1970 and 2015, 96% between 1995 and 2014, and 20% between 2010 and 2015. The domed nest with a side entrance is built on or near the ground in brambles, nettles or other dense low vegetation. Clutch size is

5–6, incubation lasts 13–14 days, and the altricial young fledge in 14–16 days. The English name is an onomatopoeic representation of the repetitive 'chiff-chaff' song. Greek *phyllo* + *scopos* = 'leaf' + 'watcher'; the specific name is a corruption of Greek *collybistes* = 'money changer', the song apparently akin to the jingling of coins.

CHIGGER A biting larval form of mites in the family Trombiculidae, the most prevalent in Britain being **harvest mite**, which can cause severe itching in humans. The name dates from the eighteenth century, from the earlier *chigoe*, in turn from French *chique*.

CHILE PINE Alternative name for monkey-puzzle.

CHILEAN-IRIS *Libertia formosa* (Iridaceae), a rhizomatous herbaceous neophyte perennial naturalised as garden escapes on rough ground and rocky lakesides and coasts scattered and local from Dorset to Argyll, the Isle of Man and south-west Ireland. Marie Libert (1782–1863) was a Belgian botanist. *Formosa* is Latin for 'beautiful'.

CHILOPODA From Greek *chilos* + *podia* = 'thousand' + 'feet', the class (subphylum Myriapoda) of centipedes (not millipedes). The British list recognises 60 species, with 6 families in the orders Geophilomorpha, Scolopendromorpha, Lithobiomorpha and Scutigeromorpha.

CHIMABACHIDAE Family name for some tubic moths (Greek *cheima* + *bacche* = 'winter' + 'reveller').

CHIMNEY SWEEPER *Odezia atrata* (Geometridae, Larentiinae), wingspan 23–7 mm, a daytime moth flying in June–August, in wet grassland in the north, calcareous grassland and hay meadow further south, locally common in Scotland and northern England, becoming less so south of the Midlands and in Ireland. Larvae feed mainly on pignut seeds and flowers. It is completely black except for a white fringe on the forewing tip, prompting the English name. Greek *odos* + *exomai* = 'road' + 'to sit', describing a habit of resting on the ground, and Latin *ater* = 'black'.

CHINA-MARK Nocturnal moths in the family Crambidae, subfamily Acentropinae. Larvae of most species are entirely aquatic.

Beautiful china-mark *Nymphula nitidulata*, wingspan 18–22 mm, widespread and common in and around lakes, ponds and slow-moving waters. Adults fly in July–August. Larvae feed on bur-reeds, yellow water-lily and other aquatic plants. Greek *nymphe* = 'nymph', and Latin *nitida* = 'shining'.

Brown china-mark *Elophila nymphaeata*, wingspan 22–30 mm, widespread and common in and around lakes, ponds, canals and ditches. Adults fly in July–August. Larvae feed on plants such as water-lilies, bur-reeds and pondweeds. Greek *elos* + *philos* = 'water meadow' + 'loving', and *Nymphaea* = 'water-lily'.

Long-legged china-mark *Dolicharthria punctalis*, wingspan 22–7 mm, scarce, on cliffs and shingle beaches from Sussex to Cornwall and in the Scillies. Adults fly in July–August. Larvae may feed on dead and decaying plant material, though there is also evidence of living on flowers and leaves of bird's-foot-trefoil, red clover and bucks-horn plantain. Greek *dolicos* + *arthra* = 'long' + 'limbs', and Latin *punctum* = 'spot'.

Ringed china-mark *Parapoynx stratiotata*, wingspan 15–28 mm, common in and around ponds and slow-moving waters throughout England, Wales and Ireland. Adults fly in June–August. Larvae feed on pondweeds and other aquatic plants. Greek *para* + *pougx* = 'beside' + 'heron', and *Stratiotes* = water-soldier plant (i.e. living in the same habitat as these species).

Small china-mark *Cataclysta lemnata*, wingspan 18–24 mm, widespread and common in and around ponds and reedbeds. Adults fly in June–August. Larvae are semi-aquatic, though live mostly under water, feeding on duckweed, bulrush and unbranched bur-reed, and building floating cases made from fragments of these plants. Greek *catacluzo* = 'to flood', and Latin *lemna* = 'duckweed'.

CHINAMAN'S HAT *Calyptraea chinensis* (Calyptraeidae), a filter-feeding limpet-like shell, up to 15 mm across and 5 mm high, on shells or under stones at low water of spring tides on sheltered shores in south-west England, and in the sublittoral on stones and shell-gravel associated with soft substrata down to 20 m on the west coast of Britain and Ireland. Greek *calyptra* = 'veil'.

CHINESE GRASS CARP See grass carp under carp.

CHINK SHELL **Banded chink shell** *Lacuna vincta* (Littorinidae), shell length 12 mm, width 5 mm, widespread near the low tide level and in shallow water, feeding on detritus, and micro- and macroalgae. The binomial is Latin for among other things 'pool' or 'gap', and 'banded'.

CHIRONOMIDAE Family name (Greek *cheironomos* = 'a pantomime artist') of non-biting midges, with 619 species in 140 genera. The aquatic larval stages are known as bloodworms (a term also used for bristle worm larvae). Emerging adults tend to swarm above the water in mating 'dances'. Adults can feed on sucrose, though some may not feed at all.

CHIROPTERA From Greek *cheir* + *ptero* = 'hand' + 'wing', the order of bats, including the families Rhinolophidae (horseshoe bats) and Vespertilionidae.

CHITON A marine mollusc in the class Polyplacophora that has an oval flattened body with a shell of overlapping plates, living on hard surfaces in the intertidal or sublittoral, most being grazers on algae, bryozoans, etc. *Chiton* is Greek for 'tunic'.

CHIVES *Allium schoenoprasum* (Alliaceae), a bulbous herbaceous perennial widespread and scattered as a native in a range of habitats, usually on thin soils over limestone, serpentine and basic igneous rocks, sometimes growing in rank grass on deeper soils, and in rock crevices by streams. As an introduced garden throw-out it grows on roadsides and tips. 'Chives' derives from an Old French dialect variant of *cive*, from Latin *cepa* = 'onion'. The binomial is Latin for 'garlic' and Greek for 'rush-leek'.

CHLOROPERLIDAE Stonefly family, with *Chloroperla tripunctatum* and *Siphonoperla torrentium* found in parts of the western British Isles. This is the fisherman's small yellow sally. *Chloroperla* comes from Greek for 'yellowish-green round head'.

CHLOROPHYTA Division of green algae (Greek *chloros* = 'yellowish-green', *phyton* = 'plant'), including seaweeds (macroalgae) and the tiny algae seen for example on tree-trunks.

CHLOROPIDAE Family name (Greek *chloros* = 'yellowish-green') of frit flies and grass flies, generally 1–4 mm long, with 177 species in 43 genera. Larvae of most species feed on grasses, including cereal. See gout and swarming flies, and frit fly under fly.

CHOCOLATE-TIP Moths in the genus *Clostera* (Notodontidae, Pygaerinae), from Greek *closter* = 'spindle', from the shape of the abdomen. Specifically, *C. curtula*, wingspan 27–35 mm, nocturnal, local in woodland, scrub, plantations, hedgerows and gardens in the southern half of England, with a few records also from north-east Scotland. The English populations have two generations, with adults flying in April–May and August–September. In Scotland the species is single-brooded, flying in June–July. Larvae feed on aspen, poplars and willows. The English name comes from the dark brown tips of the forewing. *Curtula* is from Greek *curtos* = 'arched', from the larval shape.

Scarce chocolate-tip *C. anachoreta*, wingspan 37 mm, an endangered RDB nocturnal moth restricted to the shingle beaches at Dungeness, Kent. It is also an occasional migrant, with adults sometimes found elsewhere in the south. Adults fly in April and again in August. Larvae feed on sallow and poplars. Latin *anachoreta* = 'hermit' (anchorite), describing the larvae living in a 'cell' of spun leaves.

Small chocolate-tip *C. pigra*, wingspan 22–7 mm, an

occasionally day-flying moth found in fen, marsh, damp heath-land, mosses and moorland in southern England and, very sparsely, elsewhere in the British Isles. Adults fly from May to August in one or two generations. Larvae feed on creeping willow, eared willow and aspen. *Piger* is Latin for 'sluggish', from the sedentary larvae living in a spun case.

CHOKE *Epichloë typhina* (Hypocreales, Clavicipitaceae), a widespread and common pathogenic endophytic fungus that galls the stems of a number of grasses in grassland, woodland clearings and edges. The binomial is Greek for '(growing) on young grass', and Latin for 'smoky' or 'dull'.

CHOKEBERRY Deciduous North American neophyte shrubs. The English name reflects the astringency of the fruit, the genus name *Aronia* (Rosaceae) from Greek *aria* = 'whitebeam', because of the fruit's resemblance.

Black chokeberry *A. melanocarpa*, introduced in 1700, local as a naturalised garden escape in boggy habitat and on damp heathland in Caernarfonshire, Lancashire and Dorset. *Melanocarpa* is Greek for 'black fruit'.

Red chokeberry *A. arbutifolia*, cultivated by 1700, found as a naturalised garden escape in 1975 in woodland on acidic sandy soil in Surrey.

CHONDRICHTHYES Class containing cartilaginous fishes (sharks, skates and rays) (Greek *chondr* + *ichthys* = 'cartilage' + 'fish').

CHORDATA Phylum (from Greek *chorde* = 'string', for having a notochord or 'back-string'), consisting of the subphyla Leptocardii (see lancelet), Tunicata and Vertebrata.

CHOREUTIDAE Family name for nettle-tap, skeletoniser and twitcher moths. *Choreutes* is Greek for 'dancer' or the chorus in Greek drama, reflecting the jerky movement of these insects.

Diana's choreutis moth Alternative name for Inverness twitcher.

CHOUGH *Pyrrhocorax pyrrhocorax* (Corvidae), wingspan 82 cm, length 40 cm, weight 310 g, breeding along much of the rocky coastline of west and south Ireland, parts of west Scotland, Wales and the Isle of Man, and – having disappeared in 1947 – in 2001 returning to Cornwall (Lizard Peninsula), successfully breeding every year since 2002. Total number of pairs in Britain in 2014 was around 400, with a further 120–50 pairs on Man, and 2,400 individuals in Ireland. Stick nests are on sea-cliffs and caves, clutch size is 3–5, incubation takes 17–18 days, and the altricial young fledge in 31–41 days. Feeding is by probing the soil for invertebrates. 'Chough' was originally an onomatopoeic dialect name for another corvid, the jackdaw, based on its call. *Pyrrhocorax* was known originally as 'Cornish chough', then just 'chough'. Greek *pyrhos* + *corax* = 'flame-coloured' + 'raven', linking the bright red bill and legs to its membership of the crow family.

CHRYSALIS SNAIL Pulmonates in the families Chondrinidae, Lauriidae, Pupillidae and some Vertiginidae.

Common chrysalis snail *Lauria cylindracea* (Lauriidae), shell length 3–4 mm, width 1.8 mm, common and widespread in woodland, gardens and rocky habitat, often under ivy on walls.

English chrysalis snail *Leiostyla anglica* (Lauriidae), shell length 3.1–3.8 mm, width 1.7–2.1 mm, common and widespread, especially in western Britain and north-east England, and in Ireland where it is commoner than elsewhere in Europe, in wet, shaded habitats on neutral to base-rich soils usually attached to twigs or larger branches in leaf litter, and in the west coast of Ireland recorded from more open habitats such as acid coastal heath and rough pasture. Greek *leios* + *stylos* = 'smooth' + 'pillar', and Latin *anglica* = 'English'.

Large chrysalis snail *Abida secale* (Chondrinidae), shell length 6–11 mm, width 2.3–2.8, scattered in England locally common on chalk and limestone substrates on usually unshaded rocks and rubble. *Secale* is Latin for 'rye'.

Moss chrysalis snail *Pupilla muscorum* (Pupillidae), shell length 3–4 mm, width 1.6–1.8, common and widespread in coastal areas on grassland and dunes, and inland in much of England on sheep-grazed calcareous grassland. Declining in Ireland, it is now local and rare in northern coastal areas. The binomial is from the Latin for 'doll' or 'child', and 'moss-like'.

Toothless chrysalis snail *Columella edentula* (Vertiginidae), shell length 2.5–3.5 mm, widespread and locally common on damp woodland, pasture and marsh, and in neutral to base-rich fens and flushes. The binomial is Latin for 'small pillar' and 'toothless'.

CHRYSIDIDAE Family name for ruby-tailed wasps (Greek *chrysidion* = 'piece of gold').

CHRYSOMELIDAE Family name for seed and leaf beetles (Greek *chrysos* + *meli* = 'golden' + 'honey'), including 9 subfamilies and 62 genera, with over 283 British species; 35 of these species were recognised in 2015 as being endangered, critically endangered, vulnerable to extinction or extinct. Most are small to medium in length (1–18 mm); many are brightly coloured, often metallic. Adults and larvae feed on plant material, some becoming pests because of this, others because they carry and transmit disease. Subfamilies are: **Bruchinae**, 15 species of small (2–5 mm) seed beetles, pea weevils and bean weevils, the pollen-feeding adults found on white and yellow flowers. Larvae develop within the seeds of Fabaceae. **Cassidinae**, 13 species of tortoise beetles. **Chrysomelinae**, 43 species including bloody-nosed and leaf beetles, and Colorado and rosemary beetles. **Criocerinae**, 8 species including asparagus, cereal leaf and lily beetles. **Cryptocephalinae**, 22 species, including pot beetles. **Donaciinae**, 21 species of reed beetles. **Eumolphinae**, one species, the 5–6 mm *Bromius obscurus*, recorded on willowherbs in Cheshire. **Galerucinae**, 147 species, divided into 2 tribes: the Galerucini, 20 species with simple femora, and the Alticini, 127 species with the hind femora modified and enlarged for jumping (hence their common name, flea beetles). **Lamprosomatinae**, with one species, the 2–3 mm *Oomorphus concolor*, widespread in England and Wales in broadleaf woodland.

CHRYSOPHYCEAE, CHRYSOPHYTA Class and (old classification of) phylum of golden algae (Greek *chrysos* = 'golden').

CHTHONID Pseudoscorpions in the genus *Chthonius* (Chthoniidae) (Greek *chthon* = 'Earth'), 1.2–2.4 mm length.

Common chthonid *C. ischnocheles*, common and widespread, if less so in Scotland, found year-round beneath stones, bricks and other debris as well as woodland soil and litter, and occasionally bird and other animal nests. Greek *ischnos* + *chele* = 'slender' + 'claw'.

Dark-clawed chthonid *C. tenuis*, widespread and locally abundant in southern England and Wales, among dead leaves and humus, and under stones, with a preference for well-drained habitats on chalk and sand. *Tenuis* is Latin for 'slender'.

Dimple-clawed chthonid *C. (Epippiochthonius) tetrachelatus*, widespread, found under stones and rocks, in leaf litter and decaying vegetation, and in synanthropic habitats such as old gardens, quarries and waste ground. Greek *tetra* + *chele* = 'four' + 'claw'.

Halbert's chthonid *C. halberti*, with a few historic records from Devon and Dublin. In 2016, a specimen was found at Kimmeridge Bay, Dorset, beneath a large boulder. Names honour the Irish naturalist James Halbert (1872–1948).

Kew's chthonid *Chthonius (Ephippiochthonius) kewi*, scarce, endemic to Britain, most records coming from east England (Essex, Norfolk, Suffolk, Kent) in decaying drift-line debris and large pebbles on the upper shore. It has also been found on the Dorset coast, and inland from Lincolnshire, Norfolk, Sussex and Nottinghamshire.

Straight-fingered chthonid *C. orthodactylus*, locally abundant in the south-east Midlands, with two records from south Wales and one from central Ireland, found among dead leaves, humus and grass tussocks. Greek *orthodactylus* = 'straight-fingered'.

CHUB *Leuciscus cephalus* (Cyprinidae), up to 4.5 kg, in slow-flowing rivers in England, parts of Wales and in faster-flowing trout streams in south Scotland, as well as in some lakes and gravel-pits. Diet includes smaller fish, insect larvae and plants. Spawning is around May. Size and fighting qualities (but not taste) make it a prized angling fish. 'Chub' ('chubbe') was used in late Middle English. Greek *leyciscos* = 'white mullet', and *cephale* = 'head'.

CHYTRIDIOMYCOSIS A disease caused by the fungus *Batrachochytrium dendrobatidis* that infects the skin of adult amphibians and the mouthparts of tadpoles, widespread in England and evident in Wales and Scotland. It was first identified in the UK in 2004 and by 2011 had been found in all native amphibian species and two non-natives. The fungus probably disrupts movement of salts through the skin and can lead to cardiac arrest.

CICADA Homopteran bug, the only British species (*Cicadetta montana*, Cicadidae) being a rarity found in the New Forest. *Cicada* is Latin for this insect, *cicadetta* a diminutive, and *montana* = 'from mountains'.

CICADELLIDAE Family name for leafhopper bugs, from cicada, though this is taxonomically incorrect.

CICELY **Sweet cicely** *Myrrhis odorata* (Apiaceae), a herbaceous European neophyte perennial, smelling of aniseed when crushed, grown in gardens as a salad herb, or boiled as a relief for coughs and flatulence. It was recorded in the wild in 1777, and is now locally common in the northern two-thirds of Britain except for north-west Scotland, scattered elsewhere, in hedge banks, woodland edges, verges, pathsides, stream banks and other grassy habitats. The English name is late sixteenth-century, from Latin *seselis*. *Myrrhis* is from the Greek for 'perfume'.

CICONIIFORMES Order (Latin *ciconia* = 'stork') that includes the families Ardeidae (herons, egrets and bitterns), Ciconiidae (storks) and Threskiornithidae (spoonbills and ibis).

CIIDAE Family name of minute tree fungus beetles (1–3 mm length), with 22 British species in 7 genera. *Cis* here is Greek for 'woodworm'.

CINCH BUG **European cinch bug** *Ischnodemus sabuleti* (Lygaeidae), length 4–6 mm, with both micropterous and macropterous forms, expanding its range but remaining scattered in England as far north as south Yorkshire, both nymphs and adults found on grasses and reeds from summer into winter, often swarming over wetland habitats. Greek *ischnos* + *demos* = 'slender' + 'body', and Latin *sabulum* = 'sand'.

CINCLIDAE Family name of the dipper (Greek *cinclos* was a waterbird).

CINDER **Brittle cinder** *Kretzschmaria deusta* (Xylariales, Xylariaceae), a common, widespread pathogenic crust fungus that causes soft rot, parasitic on the roots and lower trunk of hardwoods, especially oak, beech and lime. New fruit bodies appear in spring, but old blackened specimens are evident year-round. *Kretzschmaria* may refer to the German botanist Horst Kretzschmar (b.1945). *Deusta* is Latin for 'burned', a reference to the cinder-like appearance and texture of mature fruit bodies.

CINNABAR *Tyria jacobaea* (Erebidae, Arctiinae), wingspan 32–42 mm, a generally nocturnal moth flying from May to August, the distinctively black and red adult flying by day if sunny or it is disturbed. Common throughout much of the British Isles, though more coastal in northern England and south Scotland, and absent from the Highlands, associated with open grassy habitats including waste ground, railway banks, gardens and woodland rides but most frequent on well-drained rabbit-grazed grassland, mature dunes and heathland. The equally distinctive

larvae, with their aposematic yellow and black hoops, generally feed gregariously on ragwort (*Senecio jacobaea*, giving the specific epithet) and related plants, groundsel, etc. *Tyrios* is Greek for Tyrian (of Tyre).

CINQUEFOIL Herbaceous plants in the genus *Potentilla* (Latin for 'small powerful one', referencing medicinal properties) and *Comarum* (Greek *comaros* = 'arbutus'), family Rosaceae. The name dates from Middle English, from Latin *quinque* + *folium* = 'five' + 'leaf'.

Alpine cinquefoil *P. cranzii*, a perennial of dry base-rich montane rock faces and ledges, close-grazed calcareous grassland and, occasionally, river shingle. It is very local in north-west Wales, northern England and the Highlands, reaching 1,065 m on Ben Lawers (mid-Perth), but descending to 250 m in Assynt (west Sutherland). Heinrich von Crantz (1722–99) was an Austrian botanist.

Brook cinquefoil *P. rivalis*, an annual or biennial North American neophyte naturalised on a pool edge near Bridgnorth, Shropshire since at least 1976. *Rivalis* is Latin for 'growing by streams'.

Creeping cinquefoil *P. reptans*, a common and widespread perennial of woodland rides, grassland, dunes, hedgerows, banks and roadsides, and waste and cultivated ground, generally on neutral to basic soils, scarce and probably introduced in much of Scotland. *Reptans* is Latin for 'creeping'.

Grey cinquefoil *P. inclinata*, a perennial European neophyte grown in gardens since 1806, a rare casual escape on waste ground, mostly in south-east England. *Inclinata* is Latin for 'bent over'.

Hoary cinquefoil *P. argentea*, perennial, scattered in Britain and generally declining in status though still common in south-east England and East Anglia, on dry, free-draining gravelly or sandy soils, in open grassy swards on commons, in pastures, on banks, and on tracks and waste ground. *Argentea* is Latin for 'silvery'.

Marsh cinquefoil *C. palustre*, a rhizomatous perennial, common and widespread (except for central and south-east England), found in permanently flooded swamps, and in mires and wet meadows where the summer water table lies below the soil surface. It prefers nutrient-poor but slightly or moderately base-rich water and grows on a wide range of soils. Habitats include lake margins, bog pools, peat cuttings and floating rafts of vegetation. *Palustre* is Latin for 'marsh'.

Rock cinquefoil *P. rupestris*, a rare perennial found in Montgomeryshire (Craig Breidden), Radnorshire (Wye flood zone at Boughrood) and east Sutherland (Strathfleet and Migdale). In three of its four native sites it grows in thin, dry mildly acidic soils in rocky or cliff habitats subject to summer drought. *Rupestris* is from Latin *rupes* = 'rock'.

Russian cinquefoil *P. intermedia*, a biennial or short-lived perennial neophyte introduced in 1786, occasionally found as a casual or naturalised grain alien in grassy places and waste ground in Britain.

Shrubby cinquefoil *P. fruticosa*, found in Lancashire and north-east England on basic, damp rock ledges and river flats subject to flooding. In Ireland it occurs in rocky places subject to flooding, usually around loughs and turloughs, especially in the west. It also occurs, especially in south-central England and central Scotland, as a garden escape or relic in waste ground. *Fruticosa* is Latin for 'shrubby'.

Spring cinquefoil *P. tabernaemontani*, a widespread but local perennial of dry basic grassland and rocky slopes. This is the county flower of Cromarty as voted by the public in a 2002 survey by Plantlife. *Tabernaemontanum* is the Latin form of the German *Bergzabern* = 'mountain cottage'.

Sulphur cinquefoil *P. recta*, a perennial European neophyte growing in gardens by 1648, scattered in Britain, rare in Ireland, found as a garden escape or as a contaminant of grass seed, naturalised on waste ground, roadside banks and other grassy habitats. *Recta* is Latin for 'erect'.

Ternate-leaved cinquefoil *P. norvegica*, a widespread but scattered annual to short-lived perennial European neophyte grown in gardens since 1680, casual, sometimes naturalised, in waste places, quarries and on old railway lines, often introduced with grain or birdseed. Records have generally declined, except in the Forth–Clyde valley where they may be increasing. *Norvegica* is Latin for 'Norwegian'.

CIRRIPEDIA The infraclass (class Maxillopoda) that comprises the barnacles (Greek *cirri* + *podi* = 'curls of hair' + 'foot').

CISTACEAE Family name for rock-roses, from Greek *cistos*, an evergreen shrub.

CLADOCERA Suborder (sometimes considered to have order status) of the Diplostraca that includes water fleas (Greek *clados* = 'branch').

CLADONIA Lichen genus in the Cladoniaceae (Greek *clados* = 'branch') that includes cup lichens.

CLAM Species of bivalve, the word dating from the early sixteenth century, from an earlier use meaning a clamp (reflecting the closed parts of the double shell), from Old English *clam(m)* = 'a bond'. Specifically, can be used as an alternative name for northern quahog, but this can lead to confusion because so many other bivalves incorporate this word.

American jack-knife clam *Ensis directus* (Pharidae), shell length up to 20 cm, an introduced filter-feeding bivalve from the west Atlantic first recorded in Britain in 1989, now spreading along the coast from the Humber estuary to Kent, and from Merseyside, and discovered in Milford Haven, south-west Wales, in 2002. It burrows in sand from the low intertidal to the nearshore shelf. The binomial is Latin for 'sword' and 'straight'.

Asiatic clam *Corbicula fluminea* (Corbiculidae), a filter-feeding bivalve of mainly Asian origin, shell length usually <30 mm, first recorded in England from the River Chet, Norfolk Broads, in 1998. By 2002, it had colonised all of the major rivers of the Broads, often being transported on boat hulls, and in places densities of >2,500 individuals per m² were recorded. It is now found in many rivers in this area, as well as in the Thames and the Great Ouse system, and is continuing to spread in south-east England. It is found in oligotrophic to eutrophic streams, rivers, lakes, and irrigation and drainage ditches, favouring oxygenated muddy and sandy sediments, but also occurring in gravel and stony substrates. It can alter ecosystems by increasing sedimentation through huge production of pseudofaeces, and by out-competing native species. The binomial is Latin for 'small basket' and *flumen* = 'river'.

Bamboo clam Alternative name for American jack-knife clam.

European clam Alternative name for basket shell; not to be confused with European spoon clam.

European fingernail clam Alternative name for horny orb mussel.

European spoon clam *Cochlodesma praetenue* (Periplomatidae), a filter-feeding bivalve, shell size up to 3.8 cm, widespread in the lower intertidal and across the shelf in mixed substrates from sand to muddy sand and gravel. Greek *cochlo* + *desma* = 'to turn' + 'chain' or 'bundle', and Latin *praetenuis* = 'very thin'.

Manila clam *Tapes philippinarum* (Veneridae), an Indo-Pacific filter-feeding bivalve, shell size up to 6.5 mm, introduced to Poole Harbour in 1988, via North America, as a possible commercial fishery and, naturalised since 1994, has spread to other locations on the south coast of England and to north Kent and Essex, in silty and muddy sand and gravels from the low intertidal down to 20 m. UK commercial production in 2012 was 5 tonnes. The binomial is Greek for 'carpet', and Latin for 'from the Philippines'.

Softshell clam *Mya arenaria* (Myidae), shell size to 1.5 cm,

a deep burrower in sand, sandy mud and muddy gravel, widespread in estuaries and sheltered shores from the low intertidal and nearshore to about 50 m. *Mya* is Latin for a marine mussel and *arenaria* from *arena* = 'sand'.

Truncate softshell clam *Mya truncata*, shell size to 80 mm, a deep burrower in sand and sandy mud, widespread from the low intertidal to shelf depths of about 70 m.

CLAMBIDAE Family name (Greek *clambos* = 'deficient') for fringe-winged beetles, 0.8–1.8 mm long, the ten British species having scattered distributions, mainly in decaying plant material.

CLARKIA *Clarkia unguiculata* (Onagraceae), an annual Californian neophyte commonly grown in gardens, escaping as a casual with a scattered distribution on tips and waste ground particularly in England and Wales. The name honours William Clark who made the first transcontinental expedition across America in 1804–6. *Unguiculata* is Latin for 'claw-like'.

CLARY *Salvia sclarea* (Lamiaceae), or **clary sage**, a herbaceous biennial or perennial south European neophyte, scattered in England as an occasional naturalised garden relic or throw-out on rough ground, walls and tips. 'Clary' dates to Late Middle English, from Old French *clarie*, from medieval Latin *sclarea*. The seeds have a mucilaginous coat, and some herbals recommended placing a seed, with its mucilaginous coat, on an eye with a foreign object in it so that it could make it easy to remove, hence the variant name 'clear eye'. The distilled essential oil from this plant is currently used in perfumes, as a flavouring for vermouths, wines, and liqueurs, and in aromatherapy, *Salvia* is Latin for this plant, associated with *salvis* = 'saving' or 'healing'; *sclarea* = 'clear'.

Annual clary *S. viridis*, an annual Mediterranean neophyte present by 1596, an occasional casual or naturalised garden escape or birdseed alien on roadsides, tips and waste ground in southern England and the Midlands. *Viridis* is Latin for 'green'.

Meadow clary *S. pratensis*, a herbaceous perennial, native in a dozen or so sites in southern England, naturalised elsewhere in England, on unimproved grassland, lane-sides, verges and disturbed ground on well-drained soil overlying chalk and limestone, occasionally as a casual on waste ground. *Pratensis* is from Latin for 'meadow'.

Sticky clary *S. glutinosa*, a herbaceous south European neophyte perennial present by 1596, a naturalised garden escape in woodland and hedgerows and on riverbanks and roadsides in a very few places in England and Scotland. *Glutinosa* is Latin for 'sticky'.

Whorled clary *S. verticillata*, a herbaceous south European perennial cultivated by the end of the sixteenth century, scarce and scattered in England and Wales, usually casual, sometimes naturalised, on verges, waste ground and railway banks, originating as a garden escape or grain contaminant. *Verticillata* is Latin for 'whorled'.

Wild clary *S. verbenaca*, a herbaceous perennial of grassland on banks, dunes and roadsides, locally common in south and central England, scarce elsewhere, usually on well-drained, base-rich soils, including calcareous clays that are wet in winter and dry in summer. It is protected under Schedule 8 of the WCA (1981, amended 1992). In south-east England, it is associated with at least eight churchyards because of the medieval practice of sowing it on graves in the belief it conferred immortality. In Ireland, it is almost exclusively coastal. *Verbenaca* is Latin for resembling the plant verbena.

CLAUSILIIDAE Family name for door snails, both the common and family names coming from the door-like clausilium which is unique to this family.

CLAW WEED Purple claw weed *Cystoclonium purpureum* (Cystocloniaceae), a common and widespread red seaweed up to 60 cm frond length, on rocks and stones in the intertidal in large pools, and in the shallow subtidal. Greek *cysti* + *clonion* = 'bladder' + 'small branch'.

CLAWS See eagle's claws.

CLAY (MOTH) Members of the family Noctuidae (subfamily Noctuinae), generally nocturnal. Specifically, *Mythimna ferrago* (Hadeninae), wingspan 35–40 mm, widespread and common in a range of habitats, including gardens, riverbanks, marsh and open woodland. Adults fly in July–August. Larvae feed on various grasses and on non-grasses such as dandelion and chickweed. Mithimna is a town on the Greek island of Lesbos. *Ferrago* is from Latin *ferrugo* = 'rust colour'.

Dotted clay *Xestia baja*, wingspan 35–40 mm, nocturnal, widespread and common in woodland, heathland, scrubby grassland and marsh. Adults fly in July–August. Larvae feed on herbaceous plants such as primrose, stinging nettle and dock in autumn, hibernate, then feed nocturnally on woody plants such as blackthorn, bog-myrtle, bramble and willows in spring. Greek *xestos* = 'polished' (from the glossy forewings), and *baja* from Latin *badius* = 'chestnut-coloured'.

Ingrailed clay *Diarsia mendica*, wingspan 28–35 mm, widespread and common in woodland, gardens and heathland. Adults fly in June–July. Larvae feed on a variety of woody and herbaceous plants. Greek *diarsis* = 'a raising up' and Latin *mendicus* = 'beggar', both etymologies obscure.

Plain clay *Eugnorisma depuncta*, wingspan 36–44 mm, scarce, in broadleaf woodland and rough grassland in parts of Scotland, north England and north Wales, and sporadically in southern England. Adults fly in July–September. Larvae feed on herbaceous plants. Greek *eu* + *gnorisma* = 'good' + 'recognition mark', and Latin *depuncta* = 'distinctly marked'.

Purple clay *D. brunnea*, wingspan 35–8 mm, widespread and common in broadleaf woodland and wooded heathland. Adults fly in June–July. Larvae feed on a range of woody and herbaceous plants. *Brunnea* is Latin for 'brown'.

Square-spotted clay *X. stigmatica*, wingspan 37–44 mm, scarce in ancient broadleaf woodland, woodland clearings and scrub. While found elsewhere in Britain, since 1980 the main areas have been the Chiltern beechwoods of Oxfordshire, Buckinghamshire and Berkshire, and acidic, thin-soiled areas around Guildford and Reading, the Brecklands of Norfolk and Suffolk, and the North York Moors. Adults fly in August. Larval diet is unclear. *Stigmatica* is from Greek *stigma* = 'pointed'.

Triple-spotted clay *X. ditrapezium*, wingspan 35–42 mm, local in damp, broadleaf woodland and fen, from Kent and East Anglia to Devon, in Wales, and in central and west Scotland. Adults mainly fly in July. In autumn, larvae feed on herbaceous plants, and in spring on buds of woody plants. Greek *di* + *trapezion* = 'two' + 'trapezium', from the forewing pattern.

CLEARWING Day-flying moths in the genera *Bembecia* (from a putative resemblance to a sand wasp, *Bembex*), *Pyropteron* (Greek *pyrosis* + *pteron* = 'burning' + 'wing') and *Synanthedon* (Greek *syn* + *anthedon* = 'close to' + 'flowery one'), in the family Sesiidae. In most species the wings are partly transparent, prompting their common name.

Currant clearwing *S. tipuliformis*, wingspan 17–20 mm, widespread in England and Wales, extending into parts of Scotland and Ireland, but nowhere particularly common. Larvae feed inside red and black currant shoots, occasionally gooseberry; adults, flying in June–July, are found in and around gardens, allotments and fruit fields, and on wild currants in damp woods and along stream banks. *Tipuliformis* is Latin for having the form of a cranefly (*Tipula*).

Fiery clearwing *P. chrysidiformis*, wingspan 15–23 mm, rare and endangered, flying in June–July, found beneath cliffs in Kent, larvae mining the roots of docks, particularly curled dock, and sorrel. A priority species in the UK BAP, and not to be disturbed without a licence. *Chrysidiformis* is Latin for having the form of a ruby-tailed wasp (*Chrysis*).

Large red-belted clearwing *S. culiciformis*, wingspan 23–7 mm, scarce in heathland and open woodland with a widespread but scattered distribution in Britain. Adults fly in May–June. Larvae burrow into the bark of birches and alder. *Culiciformis* is Latin for having the form of a mosquito (*Culex*).

Orange-tailed clearwing *S. andrenaeformis*, wingspan 18–22 mm, scarce in chalk downland, limestone grassland and woodland edges in southern England. Adults fly in May–June. Larvae mine the stems of wayfaring-tree and, less commonly, guelder-rose, creating galls and overwintering twice. *Andrenaeformis* is Latin for having the form of a mining bee (*Andrena*).

Red-belted clearwing *S. myopaeformis*, wingspan 18–26 mm, scarce in orchards, gardens, hedgerows, open woodland and scrub in southern England, northwards to Yorkshire, and parts of Wales. Adults fly in June–August. Larvae live under the bark of rowan and especially old fruit trees, particularly apple. *Myopaeformis* is Latin for having the form of a gadfly (*Myopa*).

Red-tipped clearwing *S. formicaeformis*, wingspan 17–19 mm, scarce in fen, marsh, riverbanks, gravel-pits and ponds in England, northwards to Dumfriesshire, and south Wales; widespread but rare in Ireland. Adults fly in May–July. Larvae mine the stems of willows, especially osier. *Formicaeformis* is Latin for having the form of an ant (*Formica*).

Sallow clearwing *S. flaviventris*, wingspan 17–20 mm, scarce in damp woodland and heathland in central-southern and south-east England. With a two-year life cycle, adults generally appear only in even years, flying in June–July. Larvae mine stems of sallow. *Flaviventris* is Latin for 'yellow-stomached'.

Six-belted clearwing *B. ichneumoniformis*, wingspan 15–21 mm, scarce on chalk and limestone grassland, downland, embankments and quarries in the southern half of England up to east Yorkshire, and south Wales. Adults fly in June–August. Larvae mine roots of bird's-foot-trefoil and kidney vetch. *Ichneumoniformis* is Latin for having the form of an ichneumon fly.

Thrift clearwing *P. muscaeformis*, wingspan 15–18 mm, relatively common in south-west England, local along the west British coastline generally, in south and west Ireland, and with a few sites on the north-east Scottish coast. Larvae feed on the roots and crown of thrift, adults also taking the nectar of this species as well as thyme. *Muscaeformis* is Latin for having the form of a fly (*Musca*).

Welsh clearwing *S. scoliaeformis*, wingspan 30–6 mm, flying in June–July, local in birchwood or birch-colonised heathland in parts of north and mid-Wales, central Scotland and south-west Ireland, and recently discovered in sizeable numbers on Cannock Chase, Staffordshire, and Sherwood Forest, Nottinghamshire. Larvae bore into mature birch trees to feed on the bark, overwintering three times. It is an RDB species and a priority species for conservation in Wales ('Section 42' list). *Scoliaeformis* is Latin for having the form of a dagger wasp (*Scolia*).

White-barred clearwing *S. spheciformis*, wingspan 26–31 mm, scarce in heathland, marsh, riverbanks and damp woodland in southern and central England, northwards to Yorkshire, and Wales. Adults fly in June. Larvae burrow into the wood of birches and alder, overwintering two or three times. *Spheciformis* is Latin for having the form of a sand wasp (*Sphex*).

Yellow-legged clearwing *S. vespiformis*, wingspan 18–20 mm, scarce in open woodland, parks and hedgerows in central-southern England, northwards to Yorkshire; occasionally locally common in gardens and on allotments. Adults fly in May–July. Larvae mainly live under the bark of oak, but also birches and elms, often becoming abundant in one-year-old stumps. *Vespiformis* is Latin for having the form of a wasp (*Vespa*).

CLEAVERS *Galium aparine* (Rubiaceae), a scrambling annual arable and garden weed, and in hedges, verges, soil heaps and waste ground, also growing in natural habitats such as riverbanks, scree and shingle. It grows in both tall-herb and ruderal communities, thriving on highly fertile soil. The globular fruits are burrs covered with hooked hairs which cling to mammal fur, facilitating seed dispersal. 'Cleavers' comes from Old English *clife*, related to 'cleave'. *Galium* comes from Greek *gala* = 'milk', referring to certain species used to curdle milk. *Aparine* is Greek for this plant.

Corn cleavers *G. tricornutum*, an archaeophyte annual of cereal fields and disturbed ground, mainly on dry calcareous soil, once locally common now greatly declined. *Tricornutum* is Latin for 'three-horned'.

False cleavers *G. spurium*, an annual European neophyte present by 1806, formerly an arable weed, but recently known only from a few allotments and nearby verges apart from a naturalised population in arable land around Saffron Walden, Essex. *Spurium* is Latin for 'false'.

CLEFTCLAM **Flexuose cleftclam** Alternative name for wavy hatchet shell.

CLEG Horseflies in the family Tabanidae, genus *Haematopota* (Greek *haima* + *potes* = 'blood' + 'drinker'), adult females feeding on mammalian blood, males on nectar. Body length is generally 9–13 mm. The name is late Middle English, from Old Norse *kleggi*.

CLEMATIS Deciduous climbing and scrambling plants in the family Ranunculaceae. See also traveller's-joy and virgin's-bower. *Clematis* is Ancient Greek for 'branch' or 'twig', coming to mean a climbing plant.

Himalayan clematis *C. montana*, introduced from Asia in 1831, very scattered in Britain, usually found close to habitation growing over hedges and walls and as a garden relic. Given its abundance in gardens and its vigour it is surprisingly uncommon in the wild (first record being 1928). *Montana* is Latin for 'montane'.

Orange-peel clematis *C. tangutica*, introduced from China in 1898, occasional and scattered in England, Wales and the Isle of Man on dunes, waste ground and quarries, and as a garden relic. *Tangutica* is Latin for 'from Tibet'.

Purple clematis *C. viticella*, introduced from south Europe by Tudor times, found locally as a garden escape in hedgerows and on wasteland in England as far north as Cheshire and Northumberland. *Viticella* is Latin for 'small vine'.

CLERIDAE Family name of chequered beetles, with eight species in the British list, associated with stored products, carrion and trees. *Necrobia* (ham beetles) scavenge dead flies and fly larvae while other species prey on wood borers.

CLICK BEETLE Members of the family Elateridae, the English name coming from the ability to launch themselves backwards (arching their back and snapping their head back) with a clicking noise, both an anti-predator defence and a self-righting mechanism. About half the species develop as larvae in the soil, and some (particularly *Agriotes*, the larvae of which are often called wireworms) can be serious crop pests. The other half spend their larval stage in dead wood and under bark, where they are predators on other invertebrates.

Chequered click beetle *Prostenon tessellatum*, 9–13 mm length, widespread and fairly common in England up to north Yorkshire and Cumbria, and in Wales, together with scattered records from south-west Scotland, the Cairngorms and around the Moray Firth, associated with (often damp) grassland. Larvae develop underground, feeding on decomposing plant material. Adults appear in May and rest on grass stems, fence-posts and flowers of various herbaceous species. *Tessellatum* is Latin for 'chequered'.

Chestnut-coloured click beetle *Anostirus castaneus*, 7–14 mm length, scarce with isolated records from Gloucestershire, south-east Wales, Norfolk, north Yorkshire and the Isle of Wight, in sandy soils, cliffs, quarries and open woodland. Larvae develop in soil and probably feed on roots of heathers, birches and willows.

Pupation occurs in August–September, adults emerging in mid-April. *Castaneus* is Latin for 'chestnut-coloured'.

Hairy click beetle *Synaptus filiformis*, 9–13 mm length, with a few records from around the Severn estuary and north Yorkshire, on riverbanks, fens and wet woodland. Adults emerge in May and are seen on tree foliage, especially that of willows, poplars and alder. Larvae may develop in waterlogged soil, but more likely live in dead wood. *Filiformis* is Latin for 'thread-like'.

Marsh click beetle *Actenicerus sjaelandicus*, 10–16 mm length, widespread, particularly common in west Wales, becoming scarcer into the Midlands, East Anglia and south and south-east England. Isolated populations occur in north-west England and north and west Scotland. Found in marsh, fen, bog and wet woodland. Adults emerge in May, feeding in the evening, on leaves and flowers of various herbaceous plants, sedges and trees, particularly willows. Larvae develop underground in soil and feed on plant roots. The specific epithet refers to the island of Zealand, Denmark.

Oak click beetle *Lacon quercus*, 9–12 mm length, recorded recently only from the ancient woodland of Windsor Forest, the predatory larvae found in dead standing wood or fallen branches of oak.

Sandwich click beetle *Melanotus punctolineatus*, 11–16 mm length, known from just three sites on dunes in Kent, and a priority species for conservation in the UK BAP. *Punctolineatus* is Latin for 'spot' and 'lined'.

Violet click beetle *Limoniscus violaceus*, 10–12 mm length, endangered, with only three records, all in ancient broadleaf woodland: Windsor Forest (Surrey/Berkshire), Bredon Hill (Worcestershire) and Dixton Wood (Gloucestershire). Adults are nocturnal and feed on hawthorn flowers. The predatory larvae feed in leaf litter and decaying wood.

CLINGFISH Largely detrivorous species in the genus *Lepadogaster* (Gobeisocidae), from Greek, *lepas* + *gaster* = 'limpet' + 'stomach', referencing the ventral sucker (adapted pelvic fins) allowing the fish to cling to rocks.

Connemara clingfish *L. candolii*, length up to 7.5 cm, a benthic species of the lower intertidal found off south-west Ireland, south-west England and west Scotland, living in cavities in the rock face and sea-grass meadows.

Shore clingfish *L. lepadogaster*, length 6.5 cm, found in south-west England, Wales, south-west Scotland and Ireland in rock pools and seaweed-covered shores attached to the underside of overhanging rock crevices.

CLITELLATA Class that includes oligochaetes and leeches, from Latin *clitella* = 'pack saddle', referencing the clitellum, or thickened section of the body near the head.

CLOSTRIDIUM See Bacillaceae.

CLOTHES (MOTH) Members of the family Tineidae, including the genus *Tinea*, Greek for a chewing 'worm', here used for moth larvae that eat clothes. Many species feed on animal waste and feathers, especially in bird nests, and hair, as well as clothes made of natural fibres, when they may become pests.

Rarities include **large Scotch clothes** *Archinemapogon yildizae* and **pied clothes** *Nemapogon picarella* (both central Highlands), and **white-blotched clothes** *Monopis monachella* (a rare visitor to England together with a small resident population on the Suffolk coastline, and possibly also in Norfolk).

Barred-white clothes *Nemapogon clematella* (Nemapogoninae), wingspan 12–15 mm, nocturnal, sometimes common in woodland over much of England and Wales, and with a few Scottish and Irish records. Adults fly in June–August. Larvae feed on fungus on decaying wood. Greek *nema* + *pogon* = 'thread' + 'beard', from the mouthpart bristles, and *clematella* a reference to traveller's-joy *Clematis vitalba*, though this is neither food plant nor habitat.

Brindled clothes *Niditinea striolella* (Tineinae), wingspan

10–14 mm, found throughout much of England, with records north to Yorkshire, but generally scarce. Adults fly in June–August. Larvae feed on detritus in bird nests, living within a silk tent. Latin *nidus* + *tinea* = 'nest' + 'moth', and *striolatus* = 'grooved'.

Brown-dotted clothes *Niditinea fuscella*, wingspan 11–17 mm, widespread and local indoors and in poultry houses and farm outbuildings, less numerous in the north and in Ireland. Adults mainly fly from late afternoon on in May–August. Larvae largely feed on feathers and other dry animal and vegetable matter. Latin *fuscus* = 'dusky', from the forewing colour.

Brown timber clothes *Monopis fenestratella* (Tineinae), wingspan 11–16 mm, recorded from bird nests in woodland in a few locations in England. Adults fly in May–August. Larvae feed on detritus in bird nests, decaying plant material, dead wood and fungi. *Monops* is Greek for 'one-eyed', *fenestratus* Latin for 'windowed', both describing the forewing spot.

Buff clothes *Tinea dubiella* (Tineinae), wingspan 9–15 mm, widespread but local, often common but in places declining rapidly. Almost entirely overlooked until 1979, when about half the case-bearing clothes moths in British collections were reidentified as buff clothes. Adults fly in late afternoon in May–September, probably in two generations, and are found inside houses, bird nests, barns, warehouses and granaries. Larvae feed on woollen materials, fur, feathers, hair, owl pellets, detritus in bird nests and stored goods, living within a silk case, causing sufficient damage to be a pest in some places. *Dubiella*, from Latin *dubius* = 'doubtful', reflects the difficulties of identification.

Case-bearing clothes *Tinea pellionella*, wingspan 9–16 mm, widespread but local in houses, bird nests, barns, warehouses and granaries, in many places having become more abundant than the common clothes moth. Adults generally fly in June–October, but may appear outside this period since they often live inside buildings. Larvae feed on woollen materials, fur, feathers, hair, owl pellets and stored goods, living within a silk case, and often causing sufficient damage to be a serious pest. *Pellio* is Latin for 'a furrier', from the larval feeding habit.

Cellar clothes *Dryadaula pactolia* (Dryadaulinae), wingspan 8–11 mm, a rarity found occasionally in the British Isles. Larvae feed on the cellar fungus *Rhacodium cellare*, found in wine cellars and distilleries. Adults appear in March–November, remaining indoors. Greek *dryas* + *daulos* = 'wood nymph' + 'shaggy', from the head texture. Pactolus is a river in Lydia (north-west Turkey) with golden sands, reflecting gold markings on the forewing.

Common clothes *Tineola bisselliella* (Tineinae), wingspan 9–16 mm, widespread but local in houses and other buildings. In many places once the most abundant clothes moth, but is now often outnumbered by case-bearing clothes moth. Larvae feed on woollen materials (including clothes and carpets), animal fur and feathers in bird nests, living within a silk tent. These moths are more or less continuously brooded except during winter. They may cause sufficient damage to be a serious pest, but numbers are declining because of the increased use of man-made fibres and the drier atmosphere created by central heating. The binomial is a diminutive of *tinea* = 'moth', and Latin *bisellium* = 'seat of honour'.

Felt clothes *Monopis imella*, wingspan 11–14 mm, scarce, with a wide and scattered distribution, often near the coast. Adults fly in June–September, perhaps in two generations. Larvae feed on sheep's wool, hair, animal carcasses and detritus in bird nests. *Imus* is Latin for 'lowest', etymological significance unknown.

Four-spotted clothes *Triaxomera fulvimitrella* (Nemapogoninae), wingspan 15–22 mm, locally distributed in woodland over much of mainland Britain, though especially in England. Adults fly in May–July. Larvae feed from September into winter in dead wood and on bracket fungus, especially those growing on birch, beech and oak. Latin *fulvus* + *mitra* = 'tawny brown' + 'mitre', from the red head.

Fulvous clothes *Tinea semifulvella*, wingspan 14–22 mm, nocturnal, widespread and common in bird nests. Adults fly from

late May to September. Larvae feed on detritus in bird nests and sheep's wool. Latin *semi* + *fulvus* = 'half' + 'tawny brown', from the colour of the distal half of the forewing.

Gold-sheen clothes *Nemapogon ruricolella*, wingspan 10–24 mm, scarce with a scattered distribution in woodland, occasionally heathland, in the southern half of England, with a few records in Wales and Ireland. Adults fly in June–August. Larvae feed on bracket fungus, especially razorstrop, and on dead wood, particularly birch. Latin *rus* + *colo* = 'countryside' and 'inhabit'.

Gold-speckled clothes *Nemaxera betulinella* (Nemapogoninae), wingspan 12–19 mm, local in woodland in England and Wales. Adults fly in May–August. Larvae feed in bracket fungus, especially *Piptoporus betulinus* (prompting the specific epithet), and in decaying birch wood.

Large brindled clothes *Triaxomera parasitella*, wingspan 16–21 mm, nocturnal, widespread in woodland in the southern third of mainland Britain, scarce further north. Adults fly in May–July. Larvae feed in dead wood and parasitic bracket fungi, especially *Coriolus*, prompting the specific name.

Large clothes *Morophaga choragella* (Scardiinae), wingspan 18–32 mm, local in south and south-east England, with scattered records as far north as Northumberland. Adults fly in June–August. Larvae feed in bracket fungus and dead wood. The binomial is a Latinisation of French *morille* = 'mushroom' + Greek *phago* = 'to eat', and *choragos* = 'chorus leader'.

Large pale clothes *Tinea pallescentella*, wingspan 12–15 mm, widespread but local inside houses and farm buildings, more numerous in the north, and becoming scarcer in the south during the twenty-first century. It may have been introduced from South America during the nineteenth century. Adults can be found on the wing at any time of the year. Larvae feed on hair, wool, fur and feathers. *Pallescens* is Latin for 'tending to be pale', from the forewing colour.

Pale-backed clothes *Monopis crocicapitella*, wingspan 10–16 mm, found in south Britain and south and east Ireland, local and sometimes abundant in places with flour, cereals, natural fibres and poultry. Adults fly in June–October. Larvae feed on detritus in bird nests, dry plant material and the like. Latin *croceus* + *caput* = 'yellow' + 'head'.

Pale corn clothes *Nemapogon variatella*, wingspan 10–14 mm, found in a few localities in southern England and Wales. Adults are evident from March to August. Larvae feed on bracket fungus, dead wood, and possibly dried grain and stored produce. Latin *variatus* = 'variegated'.

Silver-barred clothes *Infurcitinea argentimaculella* (Meessiinae), wingspan 8–9 mm, with a scattered distribution in southern England, though recorded from as far north as Cumbria, and south Wales. Adults fly in July–August. Larvae, active in April–June, feed on powdery lichens on rocks, walls and tree-trunks, constructing lichen-covered portable silk cases. Latin *in* + *furca* + *tinea* = 'no' + 'forked' + 'moth', reflecting the wing venation, and *argentum* + *maculum* = 'silver' + 'spot', from the forewing mark.

White-speckled clothes *Nemapogon wolffiella*, wingspan 10–14 mm, local and scarce in woodland in southern England and Wales, with records also from Cumbria and Northumberland. Adults fly in June–July from late afternoon onwards. Larvae feed on fungus and dead and decaying wood, often of birch. N.L. Wolff (1900–78) was a Danish entomologist.

Yellow-backed clothes *Monopis obviella*, wingspan 10–13 mm, nocturnal, local in barns, warehouses and granaries in much of southern Britain, though with records north to south Scotland. Adults fly in May–October, probably with two generations. Larvae feed on detritus in bird nests and woollen materials. *Obvius* is Latin for 'exposed', referencing the conspicuous dorsal streak.

CLOUD (MOTH) Silver cloud *Egira conspicillaris* (Noctuidae, Hadeninae), wingspan 36–42 mm, nocturnal, in rough grassland, hedgerows, scrub, open woodland and orchards in parts of the

Severn Valley, locally common in Somerset, Gloucestershire, Worcestershire and Herefordshire, with outlying records from Devon, Monmouth and Warwickshire. Individuals occasionally found in the eastern Home Counties may be of a near-extinct native population or migrant individuals. Adults fly in April–May. Larval diet is uncertain. Aegira was a city in the Peloponnese, Greece; *conspicilium* is Latin for 'a place to look out from', in this case the eyes.

CLOUDBERRY *Rubus chamaemorus* (Rosaceae), an annual herb of base-poor peats on moorland and blanket mire, spreading by extensively creeping rhizomes and by seed. It is locally common in north Wales and from the Pennines northward into Scotland (this is the county flower of Peeblesshire as voted by the public in a 2002 survey by Plantlife), usually found >600 m and reaching at least 1,160 m. The binomial is Latin for 'bramble' and 'ground mulberry'.

CLOVER (MOTH) Species of *Heliothis* (Noctuidae, Heliothinae), from Greek *heliotes* = 'of the sun', from their day-flying habit, wingspans 30–6 mm.

Marbled clover *H. viriplaca*, a rarity (RDB) evident in June–July, declining in numbers in recent years, associated with the East Anglian Breckland; other south and east England sightings on chalk downland may be relict populations but more probably represent migrants. Larvae feed on the flowers and unripe seeds of many herbaceous plants. Viriplaca was a Roman goddess.

Shoulder-striped clover *H. maritima*, a rarity flying in late June–July, local to a few usually early successional stage heaths in Hampshire, Dorset and Surrey. Larvae feed on heather and cross-leaved heath. Adults take nectar from various heathland flowers. An RDB species of declining distribution and priority conservation concern. *Maritima* reflects the saltmarsh habitat in the Vendée, France, where the type specimen was caught.

CLOVER (PLANT) Members of the genus *Trifolium* (Fabaceae), from the Latin here meaning leaves divided into three parts.

Alsike clover *T. hybridum*, a common, widespread annual European neophyte introduced as a forage crop in 1777, found on grassy banks, meadows, roadsides and waste ground. Alsike is a location in Sweden between Stokholm and Uppsala.

Bird's-foot clover *T. ornithopodioides*, a procumbent winter-annual locally common near the coast in England and Wales, on acidic sands, gravels and shingle, on bare ground in disturbed, often much trampled sites, preferring habitats that are moist in winter and parched in summer. Greek *ornithopodioides* = 'resembling a bird's foot'.

Bur clover *T. lappaceum*, an annual European neophyte herb, found locally in England, Wales and south Scotland as a casual on waste ground, mainly introduced with birdseed. *Lappaceum* is Latin for 'like a bur'.

Clustered clover *T. glomeratum*, a winter-annual generally found near the coast in southern England and East Anglia, rare elsewhere, in grassy habitats on light, drought-prone, often rather acidic sandy or stony soils such as pathside banks and cliff slopes; also in sandy pastures, arable land, and in the Scilly Isles as a weed of bulb fields. It is a rare casual inland. *Glomeratum* is Latin for 'clustered'.

Crimson clover *T. incarnatum* ssp. *incarnatum*, an annual European neophyte once much grown for fodder and a common casual, but now uncommon and only occasionally recorded. *Incarnatum* is Latin for 'flesh-coloured'.

Egyptian clover *T. alexandrinum*, an annual Mediterranean neophyte casual arriving in grass-seed mix in roadside and park sowings in Guernsey and a few places in south England. The specific name means from Alexandria, Egypt.

Hare's-foot clover *T. arvense*, a locally frequent annual found in much of Britain and east and north Ireland on open rocky or sandy habitats such as heathlands, sea-cliffs and dunes; also on railway ballast and waste ground inland, and in disturbed

grassland and set-aside on light, sandy soils. *Arvense* is from Latin for 'field'.

Hedgehog clover *T. echinatum*, an annual south-east European neophyte sporadically casual in England on tips and rough ground via birdseed. *Echinatus* is Latin here meaning 'prickly' (*echinus* = 'hedgehog').

Knotted clover *T. striatum*, a locally common, widespread winter-annual found in short, open communities around rock outcrops and on thin, relatively infertile drought-prone, often sandy soils. Habitats include well-drained pasture, grassy banks and verges, often near the coast. *Striatum* is Latin for 'striped'.

Long-headed clover *T. incarnatum* ssp. *molinerii*, a rare winter-annual found in short grassland that is severely droughted in summer on schist cliff slopes at five sites on the Lizard Peninsula, Cornwall, and on Jersey, found only within 200 m of the sea.

Narrow-leaved clover *T. angustifolium*, an annual European neophyte, found sporadically as a casual on waste ground, mainly introduced with wool shoddy. *Angustifolium* is Latin for 'narrow-leaved'.

Nodding clover *T. cernuum*, an annual European casual found on waste ground, derived from wool shoddy and granite ballast. *Cernuum* is Latin for 'nodding'.

Red clover *T. pratense*, a common and widespread perennial, found in a range of grasslands other than on very acidic soils, and common in waste ground. Large agriculturally selected variants (var. *sativum*) are extensively sown into stubble and as components of short-term leys. *Pratense* is from Latin for 'meadow'.

Reversed clover *T. resupinatum*, a south European annual neophyte present since 1713, found on waste ground and verges, and a contaminant of grain, birdseed and wool shoddy. Usually casual, but several persistent populations are known in grassland in southern England. Latin *resupinatum* = 'inverted'.

Rose clover *T. hirtum*, an annual European neophyte, local in England and south Scotland as a casual on waste ground, mainly introduced with wool shoddy. *Hirtum* is Latin for 'hairy'.

Rough clover *T. scabrum*, a winter-annual with a scattered, often near-coastal distribution in England, Wales, east Scotland and east Ireland, on thin, infertile, drought-prone soils over limestone, sand and gravel; also in summer-parched coastal clifftop grasslands. *Scabrum* is Latin for 'rough'.

Sea clover *T. squamosum*, a scarce annual found in dry upper saltmarsh, in brackish meadows and by tidal rivers and creeks as far north as Lincolnshire and south Wales. It is occasionally found inland in grassland on calcareous soils, and as a casual of waste ground and railways embankments. *Squamosum* is Latin for 'scaly'.

Starry clover *T. stellatum*, an annual Mediterranean neophyte naturalised on shingle at Shoreham, Sussex, since 1804, and recently found at a similar site at Browndown Ranges, Gosport, Hampshire. Elsewhere it has been declining as a casual on waste ground, possibly having arrived as a wool alien. *Stellatum* is Latin for 'starred'.

Strawberry clover *T. fragiferum*, a procumbent perennial, widespread but common in southern and central England. It is found behind saltmarsh and on earth sea walls, and inland in pasture or by tracks on damp clay soils, and sometimes in long-established amenity grassland. *Fragiferum* is Latin for 'strawberry-bearing'.

Subterranean clover *T. subterraneum*, a winter-annual found near the coast from Anglesea round to Yorkshire, and in Co. Wicklow, in grassland or heathland on thin, free-draining neutral to acidic sands, gravels and shingle; inland it occurs in summer-parched grasslands on chalk and limestone. Also introduced with wool shoddy.

Suffocated clover *T. suffocatum*, a rare procumbent winter-annual of thin, dry soils on rocky coasts or on acidic compacted sand and shingle, either in open turf or on bare ground, mostly in southern England and East Anglia.

Sulphur clover *T. ochroleucon*, a perennial found on chalky boulder-clays or, more rarely, chalk, from East Anglia into the south-east Midlands, declining and now rare in pastures, as grassland has been converted to arable or subjected to eutrophication, lack of grass cutting and scrub encroachment. Many former roadside sites have been destroyed. However, it does occur as a casual elsewhere in England. Greek *ochros* + *leucos* = 'pale' + 'white'.

Swedish clover Alternative name for Alsike clover.

Twin-headed clover *T. bocconei*, a rare winter-annual found on the Lizard Peninsula, Cornwall, and Jersey, on shallow soils over serpentine (rarely schist), favouring sheltered south-facing summer-droughted grasslands near the sea and requiring grazing to keep the turf short. Paolo Boccone (1633–1704) was an Italian botanist and monk.

Upright clover *T. strictum*, a winter-annual, local to Jersey and a very few sites in Cornwall, Hampshire and mid-Wales, on shallow soils over schist, basalt and serpentine, preferring rock outcrops and south-facing cliff slopes. *Strictum* is Latin for 'erect'.

Western clover *T. occidentale*, a stoloniferous perennial herb discovered in 1957, and described as a new species from Cornwall and the Channel Islands in 1961. It was found in Ireland in 1979 and in Wales in 1987. It is associated with dry, species-rich coastal grassland, often growing on cliff slopes or on stabilised sand. It is largely restricted to exposed sites liable to drenching by salt-laden winds, and rarely occurs >100 m from the sea. *Occidentale* is Latin for 'western'.

White clover *T. repens*, a common and widespread stoloniferous perennial found in grassland on all but the wettest or most acidic soils; also on waste ground and in other ruderal habitats. It is very tolerant of grazing, mowing and trampling, and is widely sown as a component of leys, and on roadsides. *Repens* is Latin for 'creeping'.

Zigzag clover *T. medium*, a rhizomatous perennial, mostly found in neutral grasslands on heavy soils, but also occurring in hedgerows and on woodland edges, and in ruderal habitats such as quarry spoil and railway banks. In upland areas it is also found on rocky streamsides and in tall-herb communities on rock ledges, and on heathland in Ireland. *Medium* is Latin for 'middle'.

CLOWN BEETLE See Histeridae.

CLUB (FUNGUS) Characterised by branched, club-shaped sporophores. Species of *Clavulinopsis* have tougher, less brittle fruit bodies that are solid rather than hollow compared with *Clavaria* and *Clavulina*. All species are inedible or tasteless. The etymological root of these genera is Latin *clava* = 'club'.

Rarities include **dark club** *Clavaria greletii*, **moss club** *Multiclavula vernalis*, **pendulous sedge club** *Pterula caricis-pendulae*, **skinny club** *Clavaria incarnata*, **straw club** *Clavaria straminea* and **yew club** *Clavicorona taxophila*.

Apricot club *Clavulinopsis luteoalba* (Agaricales, Clavariaceae), widespread, infrequent but locally common, emerging as a small yellow club-shaped fruiting body from late summer to autumn on soil in short grass. Latin *luteus* + *alba* = 'yellow' + 'white', referencing the white tips of the yellow clubs.

Bracken club *Typhula quisquiliaris* (Agaricales, Typhulaceae), common and widespread on dead bracken stems in autumn. Greek *typhos* = 'smoke' (here meaning 'slightly smoky'), and Latin *quisqualis* = 'of what kind'.

Giant club *Clavariadelphus pistillaris* (Gomphales, Clavariadelphaceae), widespread but uncommon in southern England and south-east Wales, very rare and local elsewhere, found in late summer and autumn emerging like clubs on calcareous soil and leaf litter under deciduous trees, especially beech. *Pistillaris* is Latin for 'pestle', which is club-shaped.

Handsome club *Clavulinopsis laeticolor*, widespread and fairly common, emerging as a golden yellow club-shaped fruiting body, from summer to late autumn on soil or leaf litter in open, often damp mixed woodland. *Laeticolor* is Latin for 'of a joyous colour'.

Moor club *Clavaria argillacea*, widespread but uncommon on sandy heathland soil in summer and early autumn. *Argillos* is Greek for 'potter's clay'.

Pipe club *Macrotyphula fistulopsa* (Typhulaceae), fairly common and widespread, found in autumn as tall, yellow spindles in leaf litter and on rotting fallen wood of beech, birch and other hardwoods. Latin *macro* + *typhula* = 'large' + 'slightly smoky', and *fistula* = 'pipe'.

Pointed club *Clavaria acuta*, widespread but uncommon, found in late summer and autumn on soil in grassy habitats, from gardens to unimproved pasture, and including woodland glades and rides. *Acuta* is Latin for 'pointed'.

Redleg club *T. erythropus*, widespread and locally common in England, uncommon elsewhere, on woody debris in often damp deciduous woodland in autumn. Greek *erythros* = 'red'.

Slender club *Macrotyphula juncea*, locally common and widespread, found in autumn as tall, whitish spindles in leaf litter, debris and fallen wood. *Juncea* is Latin for 'rush-like'.

Wrinkled club *Clavulina rugosa* (Cantharellales, Clavulinaceae), widespread and common fungus, found as a white antler-like branching fruiting body from late summer to early winter on woodland soils and leaf litter, among moss, often by paths. *Rugosa* is Latin for 'wrinkled'.

Yellow club *Clavulinopsis helvola* (Clavariaceae), widespread but only locally common, emerging as a yellow club-shaped fruiting body in late summer and autumn on soil in open mixed woodland or clearings. *Helvola* is Latin for 'honey-yellow'.

CLUB-RUSH Generally rhizomatous perennial species in the family Cyperaceae.

Bristle club-rush *Isolepis setacea*, common and widespread, in open, damp, generally acidic sites, especially with winter flooding, on sandy or gravelly tracks, the shores of lakes or ponds, in short grassland, on stream banks, and occasionally on the coast in dunes or in upper saltmarsh. Greek *isos* + *lepis* = 'equal' + 'scale', and Latin *setacea* = 'bristly'.

Common club-rush *Schoenoplectus lacustris*, common and widespread, in fresh water, from eutrophic and base-rich to oligotrophic and base-poor, and on silt, clay, peat or gravel. It is found in ponds, lakes, canals and slowly moving rivers, usually in water 0.3–1.5 m deep, but can be found in deeper water. Latin *schoenus* (from Greek *schoinos*) + *plecto* = 'reed' + 'to braid', and Latin *lacustris* = 'of lakes'.

Floating club-rush *Eleogiton fluitans*, stems usually floating, scattered and locally common especially in western regions, mainly on peaty, acidic substrates on the margins of slowly flowing streams, ditches and pools, and the sheltered shores of larger lakes, often in seasonally flooded sites. It also occurs in muddy hollows in grassland and heathland, and the wet floors of old quarries, and sand- and gravel-pits. Greek *eleos* + *geiton* = 'marsh' + 'neighbour', and Latin *fluitans* = 'floating'.

Grey club-rush *Sch. tabernaemontani*, most frequent in coastal sites growing in brackish water in rivers, tidal channels, lagoons and dune slacks, and in depressions in saltmarsh and in wet pasture. Inland, it has a more scattered distribution by lakes, ponds, slowly flowing rivers, streams and canals, and in flooded quarries and pits. *Tabernaemontanus* is the Latinised form of German *Bergzabern* = 'mountain cottage'.

Round-headed club-rush *Scirpoides holoschoenus*, rare and endangered, found at Braunton Burrows, north Devon, in damp dune slacks and on adjacent low dunes, and at Berrow Dunes, north Somerset, in a damp sandy hollow on a golf course. Elsewhere, it is an occasional introduction, especially in industrial areas, in southern England, south Wales and Northern Ireland. Latin *Scirpoides* = 'relating to rushes', and Greek *holos* + *schoinos* = 'whole' + 'reed'.

Sea club-rush *Bolboschoenus maritimus*, mainly found on saline ground or in shallow brackish water, usually rooted in mud but sometimes in gravel and shingle, in saltmarsh, tidal riverbanks, creeks, ditches and ponds, occasionally in inland freshwater habitats. Greek *bolbos* + *schoinos* = 'swelling' or 'bulb' + 'reed', from the difference from the genus *Schoenus* in having bulbous tubers.

Slender club-rush *I. cernua*, mainly coastal, locally common from the Hebrides round to Hampshire, in east Norfolk, and around Ireland, in wet coastal grassland, in open sites over damp sand, peat and mud, in short turf and sometimes in seepages on rocky cliffs. Also locally common in the New Forest in flushed acidic or base-rich turf and in old marl-pits. *Cernua* is Latin for 'nodding'.

Triangular club-rush *Sch. triqueter*, a rarity found on mud banks along the lower reaches of tidal rivers, where it may become submerged at the highest tides, with a small population on the River Tamar, south Devon, and on the Shannon estuary in Co. Limerick and Co. Clare. *Triquetra* is Latin for 'three-cornered'.

Wood club-rush *Scirpus sylvaticus*, locally common in much of England and Wales, lowland Scotland and north Ireland, often establishing extensive stands in swampy valley woodland and similar shady places, as well as in wet pasture, and on the margins of rivers, streams, lakes and ponds. It typically grows over iron-rich eutrophic silts. The binomial is Latin for 'rush' and 'of woodland'.

CLUB-TAIL Common club-tail *Gomphus vulgatissimus* (Gomphidae), body length 50 mm, wingspan 60–70 mm, a dragonfly associated with (despite its name) only a few unpolluted moderate- or slow-flowing rivers in south England and the Welsh borders. The main flight period is May–July. Greek *gomphos* = 'club' or 'bolt', and Latin *vulgatissimus* = 'commonest'.

CLUBIONIDAE Family name of night-hunting foliage spiders, with 25 British species, 3–10 mm body length, many being common and widespread.

CLUBLET Erect clublet *Cordylecladia erecta* (Rhodymeniaceae), a red seaweed with fronds up to 10 cm long, on sand-covered rocks in the low intertidal and subtidal, widespread on south and west coasts north to Shetland. Greek *cordyle* + *clados* = 'club' + 'branch'.

CLUBMOSS 1) Members of the family Lycopodiaceae (Greek *lycos* + *podi* = 'wolf' + 'foot').

Alpine clubmoss *Diphasiastrum alpinum*, in grass and heather on moorland and mountains, locally common in the British Isles except for the southern third of England. Greek *di* + *phasis* = 'double' + 'appearance'.

Fir clubmoss *Huperzia selago*, common and widespread in northern and western Britain and in Ireland on heathland, moorland and montane grassy and rocky habitats. Johann Huperz (1771–1816) was a German botanist. *Selago* is Latin for 'clubmoss'.

Hare's-foot clubmoss *Lycopodium lagopus*, first recorded in 2007, with one record from each of Westerness and Easterness on slopes >800 m. Greek *lycos* + *podi* = 'wolf' + 'foot', and *lagos* + *pous* = 'hare' + 'foot'.

Interrupted clubmoss *L. annotinum*, on moorland and mountain slopes on thin soil, local in central and north Scotland, plus one record from Cumbria. *Annotinus* is Latin for 'of the previous year'.

Issler's clubmoss *D. complanatum*, on heathland and lowland moor, declining in range, and recently recorded only from a few sites in Northumberland, Aberdeenshire and Sutherland. *Complanata* is Latin for 'flattened'.

Marsh clubmoss *Lycopodiella inundata*, formerly widespread, now very local on generally bare peaty soil on wet heathland. *Inundata* is Latin for 'flooded'.

Stag's-horn clubmoss *Lycopodium clavatum*, in heathland, moorland and montane habitats, usually in grassy sites. *Clavatum* is Latin for 'club-shaped'.

2) Species of *Selaginella* (Selaginellaceae), meaning resembling a small *Selago*, another moss-like plant.

Krauss's clubmoss *S. kraussiana*, a naturalised introduction in

damp shady habitats such as shrubberies, with a scattered distribution. Christian von Krauss (1812–90) was a German naturalist.

Lesser clubmoss *S. selaginoides*, in damp montane habitats among moss and short grass in the north and west of the British Isles, down to Yorkshire, mid-Wales and Co. Cork.

CLUBWORT *Eremonotus myriocarpus* (Jungermanniaceae), a scarce foliose liverwort found in upland Wales, Cumbria and the western Highlands on wet, base-rich rock and damp, gravelly soil on ledges. Greek *eremos* + *notos* = 'solitary' + 'back', and *myrias* + *carpos* = 'myriad' + 'fruit'.

CLUSIIDAE Family name (Latin *clusus* = 'closed') for druid flies (Bracycera), with ten species.

CLUSTER FLY Members of the blow fly genus *Pollenia* (Calliphoridae) (Latin *pollens* = 'powerful'), with eight British species, the English name reflecting their congregation, often in large numbers, to hibernate in attics and outhouses. Larvae parasitise earthworms.

CNIDARIA Phylum that includes the classes Anthozoa (sea anemones, corals, sea fans and sea pens), Hydrozoa (siphonophores and hydroids), Scyphozoa (jellyfish) and Staurozoa (stalked jellyfish), with a single orifice that enables digestion and respiration. They are distinguished by having cnidocytes (or nematocytes) that fire like harpoons and are used mainly to capture prey; these 'stinging cells' prompt the phylum name, from Greek *cnide* = 'nettle'.

COACH HORSE See devil's coach horse.

COALMAN *Tricholoma portentosum* (Agaricales, Tricholomataceae), an edible fungus locally common in Scotland, scattered and scarcer further south, in coniferous forest in late summer and autumn, often in small tufts joined at the stem bases, ectomycorrhizal on conifers, especially pines, on sandy soil. Greek *tricholoma* = 'hairy fringe', and Latin *portentosum* = 'portentous' or 'monstrous'.

COB Kentish cob Alternative name for the nuts of filbert.

COBITIDAE From Greek *cobitis* = '(a kind of) sardine', the family name of true loaches; see spined loach under loach.

COBWEB (FUNGUS) Yellow cobweb *Phlebiella sulphurea* (Polyporales), inedible, locally common in England, scattered in Wales and Scotland, on dead wood. Greek *phlebos* = 'blood', and Latin *sulphurea* = 'sulphur-yellow'.

COCCID(AE) Scale insects in the superfamily Coccoidea, suborder Sternorrhyncha.

COCCIDIOSIS A disease of mammals and birds which affects various internal organs, such as the liver, by a parasitic protozoan, for example *Eimeria* affecting poultry, cattle, etc.

COCCINELLIDAE From Latin *coccineus* = 'scarlet' (in turn from Greek *coccinos*), the family name for ladybird beetles, with 46 extant and non-occasional adventive species in 25 genera. The family comprises the ladybirds (subfamilies Chilocorinae, Coccinellinae and Epilachninae) (>3 mm length) and mostly smooth, brightly coloured and with spots or stripes, and the inconspicuous coccinellids (subfamily Coccidulinae, plus *Platynaspis luteorubra*, 1–3 mm long), generally hairy and usually without spots. Most species prey on smaller invertebrates, particularly aphids, scale insects and mites.

COCHLICOPIDAE Family name for pillar or moss snails (Greek *cochlos* + *copis* = 'spiral-shelled mollusc' + 'dagger').

COCKCHAFER *Melolontha melolontha* (Scarabaeidae, Melolonthidae), 30 mm length, widespread especially in England and Wales, associated with woodland edge, hedgerows and gardens in May–June (hence the alternative names May bug and May beetle), flying in the evening and at night, occasionally swarming in large numbers, and feeding on oak and hawthorn leaves. Larvae, which can grow to 40–6 mm in length, are fat white grubs (rook worms) that live in the soil feeding on plant roots. The English name, dating from the early eighteenth century, derives from 'cock' (expressing size or vigour) + Middle English 'chafer' from Old English *ceafor*, a scarabaeid beetle. *Milolonthi* is Greek for 'cockchafer'.

Northern cockchafer *M. hippocastani*, local and scarce in northern England and parts of Wales, Scotland and Ireland, listed in RDB 1, and found in woodland where larvae (40–50 mm long) feed on roots for 4–5 years before pupating. The specific name melds Greek *hippos* = 'horse' with Latin *castanum* = 'chestnut'.

COCKLE Members of the filter-feeding bivalve family Cardiidae (Greek *cardia* = 'heart' from the shell shape). Dog-cockles (Glycymerididae) live in deep marine waters.

Common cockle *Cerastoderma edule*, shell size up to 5.5 cm, common, widespread and abundant in the intertidal and shallow sublittoral in sand and muddy sand, mostly estuarine, and the basis of an often significant fishing industry. Greek *ceras* + *derma* = 'horn' + 'skin', and Latin *edulis* = 'edible'.

Lagoon cockle *C. glaucum*, shell size up to 4.5 cm, patchily distributed mainly on the south and west coasts but also East Anglia and Orkney, with a thin-shelled form found in brackish lagoons and a thicker shelled form in estuaries. *Glaucum* is Latin for 'glaucous'.

Little cockle *Parvicardium exiguum*, shell size up to 2 cm, widely distributed except down the east coast, burrowing into sand and mud from low in the intertidal to the upper shelf. Latin *parvus* + Greek *cardia* = 'small' + 'heart', and Latin *exiguum* = 'small'.

COCKLEBUR Annual South American neophytes in the genus *Xanthium* (Asteraceae), from Greek *xanthos* = 'yellow', producing 2.5 cm-long burs covered with stiff hooked spines.

Argentine cocklebur *X. ambrosioides*, present by 1910, naturalised in an arable field in Bedfordshire since 1973, and a scarce casual in fields, tips and waste ground elsewhere, originating from wool shoddy and soya bean waste. *Ambrosioides* means resembling *Ambrosia*, probably a comment on the scent.

Rough cocklebur *X. strumarium*, geographical origins obscure, present by Tudor times, occasionally found in England (recently declining), usually casual, but in places persisting for a few years, on estuarine shores, docks and waste ground, originating from grain, wool, oil-seed and soya bean waste. *Strumarium* is Latin for 'swelling', relating to the seed pods.

Spiny cocklebur *X. spinosum*, present by 1713, on waste ground, in sewage-works and railway sidings, scattered in England, declining and usually casual, sometimes persisting for a few years, originating from wool shoddy or birdseed.

COCKLESHELL Aniseed cockleshell (sometimes called **tawny cockleshell**) *Lentinellus cochleatus* (Russulales, Auriscalpiaceae), a widespread but uncommon edible fungus with a mild aniseed scent and taste found in late summer and autumn on the stumps of hardwood trees in autumn. *Lentinellus* references *Lentinus*, a similar fungus genus, via Latin *lentus* = 'pliant' or 'tough', and Latin *cochlea* = 'spiral'.

COCKROACH Members of the order Dictyoptera, with three native species, together with the non-native common cockroach, German cockroach and six other introduced species occasionally found indoors. The name dates from the early seventeenth century as cacaroch, derived from Spanish *cucaracha* = 'chafer', from *cuca*, a kind of caterpillar.

Common cockroach *Blatta orientalis* (Blattidae), male body length 18–29 mm, females 20–7 mm, widespread in houses and other warm places, including some landfill sites. *Blatta* is Latin for 'cockroach'.

Dusky cockroach *Ectobius lapponicus* (Blattellidae), 7–11 mm long, locally common in scrub and coarse vegetation on woodland margins and verges mostly in central-south England. Males are winged; females have vestigial wings and are flightless.

Adults are evident from May to September. Greek *ectos* + *bios* = 'outside' + 'life', *lapponicus* Latin for 'from Lapland'.

German cockroach *Blatella germanica* (Blattellidae), generally 1.1–1.6 cm long, widespread in houses, restaurants, bakeries and other warm places, including rubbish tips, often increasing to such numbers as to become a serious indoor pest.

Lesser cockroach *E. panzeri*, 5–8 mm long, mainly coastal in south England, on sea-cliffs, dunes and shingle beaches, with an increase in inland records (dry heathland, chalk grassland and occasionally woodland) since the end of the last century. Males are winged; females have vestigial wings and are flightless. Adults are evident in July–October.

Oriental cockroach Alternative name for common cockroach.

Tawny cockroach *E. pallidus*, 8–10 mm, widespread in southern England in woodland, heathland, chalk downland and meadows, usually among leaf litter but sometimes also in trees. Adults fly from late June to October. *Pallidus* is Latin for 'pale'.

COCK'S COMB *Plocamium cartilagineum* (Plocamiaceae), a widespread red seaweed, fronds 30 cm long, on rocks in the lower littoral and shallow subtidal, and as an epiphyte on other seaweeds. Greek *plocos* = 'braid', and Latin *cartilagineum* = 'gristly'.

COCK'S-EGGS *Salpichroa origanifolia* (Solanaceae), a rare perennial South American neophyte naturalised on rough ground on the south and south-east coasts and the Midlands, and recorded in Co. Kerry in 1999. Greek *salpinx* + *chroma* = 'trumpet' + 'colour', referring to the trumpet-shaped flowers, and Latin for having oregano-like leaves.

COCK'S-FOOT *Dactylis glomerata* (Poaceae), a common and widespread perennial grass of woodland, meadow, pasture and downland, maritime cliff grassland, fixed dunes, field margins, roadsides and waste ground on fertile soils. Greek *dactylos* = 'finger', referring to the finger-like shape of the inflorescence, and Latin *glomerata* = 'clustered'.

Slender cock's-foot *D. polygama*, a scarce perennial European neophyte recorded in the wild in 1934, naturalised in woodland in a few places in England. *Polygama* is Greek indicating the species has both single- and two-gender flowers.

COCKSPUR *Echinochloa crus-galli* (Poaceae), an annual grass known in Britain since at least 1690, but increasing in range after the Second World War when it was introduced with North American seed, and now scattered (mainly in England) as a casual of tips, and waste and cultivated ground, mainly from birdseed but also from wool shoddy, soya bean and other waste, and sometimes sown as food for game. Greek *echinos* + *chloa* = 'hedgehog' + 'young grass shoot', and Latin *crus* + *gallus* = 'foot' + 'cock'.

Yellow cockspur Synonym of yellow star-thistle.

COCKSPURTHORN Some species of North American neophyte hawthorns, specifically, *Crataegus crus-galli* (Rosaceae), a deciduous neophyte tree cultivated since 1691, scattered in England, planted in hedgerows and naturalised on roadsides, in scrub and in gravel-pits. *Crataegus* is Latin, possibly referring to the wood's hardness. Latin *crus* + *gallus* = 'foot' + 'cock'.

Broad-leaved cockspurthorn *C. persimilis*, in places self-sown from ornamental plantings in England and Wales. *Persimilis* is Latin for 'very similar'.

Hairy cockspurthorn *C. submollis*, **large-flowered cockspurthorn** *C. coccinioides* and **pear-fruited cockspurthorn** *C. coccinea* are all also occasionally naturalised on roadsides and rough ground, and in hedgerows and woodland in southern England.

Round-fruited cockspurthorn *C. succulenta*, occasionally planted but rarely naturalised in hedgerows and woodland in southern England, Worcestershire and Shropshire. *Succulenta* is Latin for 'fleshy'.

COELENTERATA Phylum name encompassing both

Ctenophora and Cnidaria, from Greek *coilos* = 'hollow-bellied', a reference to the hollow body cavity common to these two groups.

COENAGRIONIDAE Damselfly family (Greek *coinos* + *agrios* = 'shared' + 'wildness') comprising 12 British species.

COLCHICACEAE Family name of meadow saffron, from Latin *colchica* = 'from Colchis', a region in the eastern Black Sea, Turkey.

COLEOPHORIDAE Family name for case-bearer moths, from Greek *coleos* + *phoreo* = 'sheath' and 'carrying', the larvae constructing portable cases.

COLEOPTERA Order name of beetles (Greek *coleos* + *pteron* = 'sheath' + 'wing'), with the forewings hardened to form protective covers (elytra) for the underwings which they use for flying.

COLLAR-MOSS Species of *Splachnum* (Splachnaceae) (Greek *splachnon* for a type of moss).

    Cruet collar-moss *S. ampullaceum* and **round-fruited collar-moss** *S. sphaericum* are both widespread on herbivore dung in wet heathland and bog.

    Rugged collar-moss *S. vasculosum* is scarce on base-rich flushes in some uplands in northern England and Scotland. *Vasculum* is Latin for 'vessel'.

COLLEMA Jelly lichens (Collemataceae) (Greek *colla* = 'glue'). See also lichen.

    Many-branched collema *Collema multipartitum*, scattered, common only on the Burren, on hard limestone. *Partitum* is Latin for 'divided'.

    Plaited collema *Leptogium plicatile*, scattered, mainly in southern Britain and Ireland, usually on damp limestone and horizontal tombstones, occasionally on siliceous rock with calcareous seepage. Greek *leptos* = 'slender', and Latin *plicatile* = 'braided'.

    Schraderian collema *L. schraderi*, scarce, on calcareous grassland, limestone and mortar-rich walls. Heinrich Schrader (1767–1836) was a German botanist.

    Tenacious collema *C. tenax*, common and widespread, especially var. *tenax*, on soil, walls and dunes. *Tenax* is Latin for 'tenacious'.

COLLEMBOLA Order of springtails. *Collembola* is Latin for 'springtail', derived from Greek *colla* + *embolos* = 'glue' + 'plug', from the belief that the ventral tube has adhesive properties, though its function is actually for excretion and maintaining water balance.

COLLETES Mining bees in the genus *Colletes* (Colletidae) (Greek *colla* = 'glue'), sometimes called plasterer bees or polyester bees because of lining the brood cells with a transparent polyester-like membrane (waterproof, resistant to fungal attack and maintaining an appropriate level of humidity during larval development).

    Common colletes, heather colletes *C. succinctus*, three closely related species often considered together in the 'succinctus group', widespread on heathland and moorland. Adults fly in July–September. Nests, especially in the northern half of Britain, sometimes number many thousand in a small area, for example in north Yorkshire 60,000–80,000 nests have been reported from a 100 m length of riverbank. Nests are dug by females in south-facing (warm) bare or thinly vegetated earth banks. Females stock the burrow with pollen collected from heaths. Each egg is laid on a pollen mass, and the larva develops to emerge as an adult a year later. Latin *succinctus* = 'short'.

    Margined colletes *C. marginatus*, flying in summer, on dunes in southern England and south Wales, and on grassy heathland in Breckland. Pollen and nectar come from a range of plants. Nesting is in burrows, sometimes as aggregations, in light sandy soil.

    Northern colletes *C. floralis*, flying in summer, widely distributed in coastal Ireland, south-west Scotland and the Western

Isles, in dunes and machair. Pollen and nectar are taken from a number of flowers. The female excavates her nesting burrow in bare or very sparsely vegetated, firm sand, often as part of a large aggregation. *Floralis* is Latin for 'relating to Flora', goddess of flowers.

    Sea-aster colletes *C. halophilus*, flying in late summer and autumn, on dunes in coastal Lincolnshire and East Anglia, the Thames estuary and parts of the south coast. Pollen is mostly from sea aster, nectar from a range of plants. Nesting aggregations, sometimes very large, are burrows in bare soil and have been found in the sides of rabbit burrows. *Halophilus* is Greek for 'salt-loving'.

    Vernal colletes *C. cunicularius*, ground-nesting, flying in spring, in dune systems in north-west England and north-west and south Wales and in 2011 also recorded on sandy heathland and in disused sandpits in Shropshire, Worcestershire and Nottinghamshire. It can be locally abundant, for example at Kenfig dunes, West Glamorgan, there have been aggregations of up to 18,000 nests, but it remains listed as Rare in RDB 3. Pollen and nectar are taken from a few flowers, but mainly from creeping willow. *Cunicularius* is Latin for 'burrower'.

COLON (MOTH) **White colon** *Sideridis turbida* (Noctuidae, Hadeninae), wingspan 38–44 mm, nocturnal, on dunes and shingle beaches along the coasts of England, Wales, east Scotland and east Ireland; also inland on the heaths of Hampshire and Surrey, and at some Midlands localities. Adults fly in May–June, with a small second brood in August in the south. Larvae feed on such plants as sea rocket, restharrow, sandwort, spurrey and plantain. Greek *sideros* + *eidos* = 'iron' + 'appearance', from the rust-brown colour, and Latin *turbida* = 'muddy'.

COLONEL (FLY) Soldier flies in the genera *Odontomyia* (Greek *odontos* + *myia* = 'tooth' + 'fly') and *Oplodontha* (*oplon* + *donti* = 'armour' + 'tooth') (Stratiomyidae).

    Rarities include **barred green colonel** *Od. hydroleon* (in 1986 a colony was discovered in Dyfed; other records are from the North York Moors in 1988, Norfolk in 1998, and Somerset), **common green colonel** *Op. viridula* (despite its name relatively scarce, with records mainly from the north Midlands), **orange-horned green colonel** *Od. angulata* (recent records from two sites in Norfolk, and from an Oxfordshire fen), **ornate colonel** *Od. ornata* (locally common in the Somerset and Gwent Levels), and **silver colonel** *Od. argentata* (currently known only from fen and marsh at Thompson Common and East Walton Common, Norfolk, some chalk streams in south Hampshire, and a site in Berkshire).

    Black colonel *Od. tigrina*, widespread and locally common in the southern half of England, with a few records from Wales, in ponds and freshwater grazing marsh ditches with dense emergent vegetation. Larvae are aquatic, requiring a moderate depth of water all year. Adults fly from mid-May to mid-July. *Tigrina* is Latin for '(striped) like a tiger'.

COLT'S-FOOT *Tussilago farfara* (Asteraceae), a common and widespread rhizomatous perennial herb found, often as a pioneer, in a range of generally disturbed habitats, for example dunes and shingle, slumping cliff slopes, landslides, spoil heaps, seepage areas, rough grassland, riverbanks, waste ground and verges, and it can be a serious arable weed. The English name comes from the shape of the leaf. The binomial is Latin for 'cough dispeller' (*tussis* + *ago* = 'coughing' + 'act on'), referring to medicinal properties in soothing a coughs, though toxic pyrrolizidine alkaloids can result in liver problems, and *farfara* = 'like white poplar', from the leaf shape.

COLUBRIDAE Family name for grass and smooth snake (Latin *coluber* = 'snake').

COLUMBIDAE Family name for doves and pigeons (Latin *columba* = 'dove').

COLUMBIFORMES Order of pigeons and parrots/parakeets (Latin *columba* = 'dove').

COLUMBINE *Aquilegia vulgaris* (Ranunculaceae), a herbaceous perennial found locally on calcareous soil over limestone in England and Wales, typically growing in woodland glades and scrub, streamsides, in damp grassland and fen, and on scree slopes. Naturalised garden escapes are widespread. 'Columbine' comes from medieval Latin *columbina herba* = 'dovelike herb', from the resemblance of the spurred petals to doves. The binomial is Latin for 'eagle claw' and 'common'.

COLYDIIDAE Possibly from Greek *colon* + *idea* = 'colon' + 'appearance of', the family name of cylindrical bark beetles and narrow timber beetles, 1–6 mm length, with 10 British species, none common or widespread. Most live beneath bark, others in fungi, decaying vegetable matter and within the galleries of other wood-boring beetles. Some taxonomists consider this to be a subfamily, Colydiinae, in the family Zopheridae.

COMB-MOSS Species of *Ctenidium* (Hypnaceae) (Greek *ctenos* = 'comb').

    Alpine comb-moss *C. procerrimum*, a rarity found at two limestone rock sites in the Highlands. *Procerrimum* is Latin for 'very tall'.

    Chalk comb-moss *C. molluscum*, common and widespread in many calcareous habitats such as woods, banks, cliffs, flushes and grassland, on rocks or soil. Var. *sylvaticum* is fairly frequent on acidic soil and humus in woods in southern England. Var. *condensatum* is scarce on damp, shaded cliffs, in flushes, and in scree. Vars. *robustum* and *fastigiatum* are rare on base-rich rocks at high altitudes. *Molluscus* is here Latin for 'soft'.

COMB WEED Red comb weed Alternative name for the seaweed cock's comb.

COMFREY Herbaceous perennials in the genus *Symphytum* (Greek *symphis* + *phyton* = 'gluing together' + 'plant') (Boraginaceae). 'Comfrey' (fifteenth-century, *conforye*) derives from Latin *confervere* = 'to unite', as the roots and leaves contain allantoin, which helps new skin cells grow, along with other substances that reduce inflammation and keep skin healthy, and were once used to help bind wounds and mend fractures.

    Scarce naturalised neophytes are **bulbous comfrey** *S. bulbosum*, **Caucasian comfrey** *S. caucasicum* and **rough comfrey** *S. asperum*.

    Common comfrey *S. officinale*, locally frequent and widespread, on stream and riverbanks, in ditches, fens and marshes, and on damp rough grassland and verges, possibly introduced in much of north and west Britain and in Ireland. The potassium-rich leaves make this plant useful as a liquid fertiliser, mulch and compost activator. *Officinale* is Latin for having pharmacological properties.

    Creeping comfrey *S. grandiflorum*, a Caucasian neophyte, locally common in England and Wales, more scattered in Scotland, naturalised in woodland and hedges. *Grandiflorum* is Latin for 'large-flowered'.

    Norfolk comfrey *Symphytum* x *norvicense* (*S. asperum* x *S. orientale*), endemic, possibly of garden origin, recorded from five grassy verge sites in east Norfolk since 1999. *Norvicense* is New Latin meaning from Norfolk.

    Russian comfrey *Symphytum* x *uplandicum* (*S. officinale* x *S. asperum*), introduced for fodder in 1870, now widespread and locally common on rough and waste ground, railway banks and roadsides, and in hedge banks and woodland margins. *Uplandicum* is New Latin meaning from Uppland, Sweden.

    Tuberous comfrey *S. tuberosum*, in damp woodland, ditches, and stream banks. Common in lowland Scotland and northern England, but as an introduction in Wales, Ireland and much of England, it is more often found on verges, waste ground and other disturbed sites.

    White comfrey *S. orientale*, a south Russian neophyte

introduced by 1752, found in the southern half of England, scattered elsewhere, as an escape in shaded habitats such as hedgerows and copses, by roads and railways, and on waste ground. *Orientalis* is Latin for 'eastern'.

COMMA *Polygonia c-album* (Nymphalidae, Nymphalinae), wingspan 50–64 mm, following a decline starting in the mid-nineteenth century with a reduction in the favoured larval food plant, hops, has increased in numbers and range since the 1960s, possibly a consequence of global warming, and shifting larval preference to common nettle. Between 2000 and 2009 numbers increased by 34%, but there was a subsequent decline by 28% in 2005–14. It is nevertheless widespread in England and Wales, increasing in range by 57% in 1976–2014, and by 2017 it had reached as far north as Inverness; there are also a few records from east Ireland. With a second brood in some places, the flight period can be from March to October, with a peak in July. It is mainly a woodland species, but is often seen in gardens in late summer. Larvae favour nettle, adults the nectar of a variety of plants, including bramble, thistles and ivy. Older larvae resemble bird droppings, as defensive mimicry. The ragged outline of the adult wings, when closed, mimics a dead leaf, a bright white C (giving its specific epithet in Latin) in each underwing reinforcing this deceit since it looks like a small leaf tear, and resembles a comma, hence the common name. Greek *poly* + *gonia* = 'many' + 'angle'.

COMPANION (MOTH) Burnet companion *Euclidia glyphica* (Erebidae, Erebinae), wingspan 25–30 mm, day-flying (May–July), relatively common in the southern half of Britain and in Ireland, more scattered and scarcer further north, in open woodland, flower-rich hay meadow, pasture and downland. Larvae feed on clovers, medicks, lucerne and trefoils. The English name comes from the fact that it is often found in company with burnet moths. The binomial honours the Greek geometrician Euclid, and uses Greek *glyphe* = 'emblem', from the wing markings.

COMPOSITAE Former name of the family Asteraceae, from Latin *compositus* = 'compound', describing the flowers.

CONCH (MOTH) Members of the family Tortricidae (Tortricinae), including the genera *Aethes* (Greek *aethes* = 'unusual', perhaps because of the yellowish wing colour, unusual in this family), *Cochylis* and *Cochylidia* (*chogchyle* = 'conch'), and *Phtheochroa* (*phtheo* + *chroa* = 'to fade' + 'colour'). 'Conch' possibly refers to the general colour of these moths. Most species are nocturnal. Wingspans are usually 11–15 mm.

    Scarce species include **bank conch** *Gynnidomorpha luridana* (parts of England), **blue-fleabane conch** *Cochylidia heydeniana* (parts of southern England), **Breckland conch** *Falseuncaria degreyana* (Breckland heaths), **coast conch** *Gynnidomorpha permixtana* (a few places in southern England, Wales, Scotland and west Ireland), **dingy roseate conch** *Cochylidia subroseana* (occasional in south England, for example Cliffe, Kent, in 2014), **goldenrod conch** *Phalonidia curvistrigana* (a few sites in England and Wales), **juniper conch** *Aethes rutilana* (not recorded from southern England since 1961, probably extinct there, but in 1983 discovered in Wester Ross, and in 2011 again found in Scotland at Arkle, west Sutherland), and **silver carrot conch** *Aethes williana* (parts of southern England and the east Midlands).

    Birch conch *Cochylis nana*, wingspan 9–13 mm, widespread and common in birch woodland and heathland. Adults fly in May–June. Larvae feed inside birch catkins. *Nana* is Latin for 'dwarf'.

    Black-headed conch *Cochylis atricapitana*, widespread and locally common on chalk downland and rough ground, with a coastal preference particularly in Ireland. Adults fly in two generations in May–June and August. Larvae feed on common ragwort, early brood larvae (July) on the flowers, completing growth in the main stem. Later larvae feed in September–October in the stems and rootstock. Latin *ater* + *caput* = 'black' + 'head'.

Bluebell conch *Hysterophora maculosana*, widespread but local in woodland, including a few sites in Ireland. Adults fly in May–June. Larvae feed inside the seed capsules of bluebell. Greek *hysteros* + *phoreo* = 'behind' + 'to bear', and Latin *maculosus* = 'spotted'.

Buckthorn conch *Phtheochroa sodaliana*, wingspan 13–17 mm, local in dry pasture, heathland and fen in parts of England, rarely in Ireland. Adults fly from just before dusk onwards in June–July. Larvae feed inside buckthorn berries. Latin *sodalis* = 'companion'.

Burdock conch *Aethes rubigana*, wingspan 15–19 mm, widespread and locally common on waste ground and dry open areas. Adults fly in June–August. Larvae mine the seed heads of greater and lesser burdock. Latin *rubigo* = 'rust', from the brown forewing markings.

Chamomile conch *Cochylidia implicitana*, local on waste ground and verges throughout much of southern England, but found as far north as Lancashire. Adults fly from May to August. Larvae feed on flowers, seed and stems of stinking chamomile, mayweeds, goldenrods and other composites. Latin *implicitus* = 'embraced', from previous inclusion with a congeneric.

Common yellow conch *Agapeta hamana*, wingspan 15–24 mm, widespread and common in rough meadows and on waste ground. Adults fly from June to August. Larvae mine roots of various thistles. Greek *agapetos* = 'beloved', and Latin *hamus* = 'hook', from the shape of a forewing mark.

Devil's-bit conch *Aethes piercei*, wingspan 15–24 mm, local in damp habitats such as marsh and fen, but also recorded on dry pasture and downland; found in many parts of the British Isles, including a widespread distribution in Ireland. Adults fly in June–July. Larvae mine roots of devil's-bit scabious. The specific name honours the entomologist F.N. Pierce.

Downland conch *Aethes tesserana*, wingspan 12–19 mm, local in a range of habitats throughout much of southern England, with records north to Yorkshire, and a few from Wales. Adults fly in May–August. Larvae mine roots of ploughman's-spikenard, bristly oxtongue and hawkweed oxtongue. Latin *tessera* = 'a die', from the forewing pattern.

Fen conch *Gynnidomorpha minimana*, scarce on heathland and mosses in parts of southern England, southern Scotland and central Ireland. Adults fly in June–July, or in the south are double-brooded, May–June and July–August. Larvae feed on seeds of marsh lousewort. Latin *minimus* = 'smallest'.

Gold-fringed conch *Cochylis flaviciliana*, local in dry, calcareous grassland in parts of southern England, south Wales and on the Burren, Ireland. Adults fly in July–August. Larvae feed in seed heads of field scabious from July to October. Latin *flavus* + *cilium* = 'yellow' + 'cilium', from the forewing markings.

Hemlock yellow conch *Aethes beatricella*, wingspan 14–17 mm, local on waste ground and in woodland edges and hedgerows throughout much of England and Wales. Adults fly in June–July. Larvae mine hemlock stems. *Beatricella* honours the Hon. Mrs Beatrice Carpenter, who reared specimens in around 1880.

Hemp-agrimony conch *Cochylidia rupicola*, local in woodland, downland, marsh, riverbanks and ditches in England, south Wales and south and central Ireland, mainly coastal in the north of its range. Adults fly in June–July, mainly at night but also during sunny afternoons. Larvae feed on hemp-agrimony flowers and seeds in August–October. Latin *rupes* + *colo* = 'rocks' + 'to inhabit'.

Kentish conch *Cochylimorpha alternana*, wingspan 18–24 mm, scarce on chalk downland though locally common in parts of Kent. Adults fly at dusk in July–August. Larvae feed within the flower heads of greater knapweed. The binomial is from Greek for resembling *Cochylis* (= 'conch'), and Latin for 'alternating' colours on the forewing.

Knapweed conch *Agapeta zoegana*, wingspan 15–25 mm, widespread and common in rough meadow and on waste ground.

Adults fly in May–August. Larvae mine roots of common knapweed and small scabious. Johan Zoega (1742–88) was a Swiss entomologist.

Large saltmarsh conch *Phalonidia affinitana*, local in saltmarsh throughout much of England, Wales and Ireland. Adults fly in June–August. Larvae mine sea aster flower heads in July–August, boring into the stem in September to overwinter. *Affinis* is Latin for 'similar to', from the resemblance to a congeneric.

Little conch *Cochylis dubitana*, common on dry pasture and waste ground throughout southern England, but reaching as far north as Cumbria, south Wales and Ireland. Adults fly in June–August in two generations. Larvae feed inside the flowers and developing seed heads of Asteraceae in July and August–April. *Dubitana* is Latin for 'doubtful', from initial uncertainty that this was a distinct species.

Long-barred yellow conch *Aethes francillana*, widespread if local in dry pasture, rough grassland, downland and waste ground, mainly coastal in the north of its range. Adults fly in June–September. Larvae mine wild carrot seeds, living between flower heads spun together with silk. John Francillon (1744–1816) was a British entomologist.

Marbled conch *Eupoecilia angustana*, widespread and common on woodland edges, mountains and heathland. Adults fly in June–September. Larvae mine flowers of plantains, yarrow and heather. Greek *eupoicilos* = 'variegated', and Latin *angustus* = 'narrow', from the forewing pattern and shape.

Orange conch *Commophila aeneana*, wingspan 13–17 mm, local on rough meadow and waste ground in scattered localities in southern England, the east Midlands and East Anglia. Adults fly in May–July. Larvae mine common ragwort roots from September through winter. Greek *commos* + *philos* = 'decoration' + 'loving', and Latin *aeneus* = 'brazen', from the forewing pattern.

Oxtongue conch *Cochylis molliculana*, first recorded in Britain at Portland, Dorset, in 1993, since then rapidly colonising rough ground in southern England. On the Isle of Wight it appeared in Parkhurst Forest in 2006, and subsequently reported at several further sites on the island. Records from Suffolk began in 2011, Norfolk 2012. Adults fly from June to August. Larvae feed inside the seed heads of bristly oxtongue. *Molliculana* is derived from Latin *mollis* = 'soft'.

Plain conch *Phtheochroa inopiana*, wingspan 17–22 mm, local in dry pasture, fen and marsh, and on woodland edge, riverbanks and other damp areas throughout England, Wales and a few sites in south-east Ireland. Adults fly in June–August. Larvae mine roots of common fleabane and field wormwood, living inside a silk tent. Latin *inops* = 'poor', from the 'poverty' of the wing pattern.

Red-fringed conch *Falseuncaria ruficiliana*, widespread but local on downland, heathland and moorland. Adults fly in May–June. Larvae feed inside the seed capsules of primrose, cowslip, lousewort and goldenrod. Latin *rufus* + *cilia* = 'red' + 'eyelash', from the wing marking.

Rosy conch *Cochylis roseana*, wingspan 10–17 mm, common in southern England, but further north in Cheshire and Lancashire it is scarce and local. Adults fly from late May to August. Larvae feed inside the seed heads of teasel in August–May.

Rough-winged conch *Phtheochroa rugosana*, wingspan 16–20 mm, common in hedgerows and open woodland throughout England and the Welsh Borders. Adults fly around dusk in May–June. Larvae feed on berries and stems of white bryony. Latin *rugosus* = 'wrinkled'.

Scabious conch *Aethes hartmanniana*, wingspan 11–17 mm, local on chalk and limestone downland in parts of England and Wales. Adults fly in June–August. Larvae mine the roots of small and field scabious. Peter Hartmann (1727–91) was a German naturalist.

Scarce gold conch *Phtheochroa schreibersiana*, local in hedgerows, marsh, riverbanks and other damp areas in parts of southern England. Adults fly in May–June. Larvae feed within stems of elm, black poplar and bird cherry, creating galls. Karl Ritte von Schreibers (1775–1852) was a German entomologist.

Sheep's-bit conch *Cochylis pallidana*, local on dry pasture, downland and dunes in a few parts of the British Isles, including south-west Ireland. Adults fly in June. Larvae feed inside the seed heads of sheep's-bit. *Pallidus* is Latin for 'pale', from the dull white forewings.

Short-barred yellow conch *Aethes dilucidana*, local on downland and waste ground in southern England, with records north to Lancashire and Yorkshire. Adults fly in July–August. Larvae mine seeds of wild parsnip and hogweed, living between seeds spun together with silk. Latin *dilucidana* = 'clear' or 'bright'.

Silver coast conch *Aethes margaritana*, occasional in downland, waste ground and shingle beaches in parts of southern England. Adults fly in July–August. Larvae mine flowers and seed heads of yarrow, scented mayweed and tansy. Latin *margarita* = 'pearl', from the white forewing markings.

Small saltern conch *Gynnidomorpha vectisana*, wingspan 9–12 mm, local in saltmarsh, fen, wet heathland and freshwater marsh throughout much of England, Wales and Ireland. With two generations a year, adults fly in May–June and July–September, typically as dusk approaches, especially in warm weather. Larvae feed on arrowgrass, the first generation on flower heads, the later one on the shoots and rootstock. *Vectis* was the Latin name for the Isle of Wight where the species was first recorded.

Straw conch *Cochylimorpha straminea*, widespread and common in a range of habitats. The bivoltine adults fly in May–July and August–September. Larvae feed within the stems and developing seeds of common knapweed. *Straminea* is Latin for 'straw-coloured'.

Thistle conch *Aethes cnicana*, wingspan 14–17 mm, widespread and common in rough grassland, waste ground and damp woodland edges. Adults fly in June–July. Larvae mine seeds of (especially spear and marsh) thistles. *Cnicus* is a synonym of *Cirsium*, the larval food plant.

Wall-lettuce conch *Phalonidia gilvicomana*, uncommon in woodland, downland and waste ground in parts of southern England north into Glamorgan. Adults fly in June–July. Larvae mine seed heads of wall lettuce and nipplewort. Latin *gilvus* + *coma* = 'pale yellow' + 'mane'.

Water-mint conch *Phalonidia manniana*, local in marsh, riverbanks and other freshwater margins, predominantly in southern England, but also in north Wales and Lancashire eastwards to Yorkshire. Adults fly in June–July. Larvae mine stems of water mint and gypsywort. Josef Mann (1804–89) was an Austrian entomologist.

Water-plantain conch *Gynnidomorpha alismana*, in streams, small rivers and shallow pools in parts of England, and with one Irish record (Co. Cork). Adults fly in June–August. Larvae mine flowering shoots of water-plantain, *Alisma*, prompting the specific name.

White-bodied conch *Cochylis hybridella*, local on chalk downland and woodland on chalky soils and sand dunes throughout much of southern England and south Wales, with records north to Cheshire. Adults fly in July–August. Larvae feed inside the seed heads of oxtongues and hawk's-beard. *Hybridella* indicates an appearance like a cross between species.

Yarrow conch *Aethes smeathmanniana*, wingspan 12–19 mm, common and widespread in waste ground, shingle beaches and chalk downland. Adults fly in May–August. Larvae feed inside seed heads of yarrow, knapweed and other herbaceous plants. Henry Smeathman (1750–87) was a British entomologist.

CONCOLOROUS *Photedes extrema* (Noctuidae, Xyleninae), wingspan 26–8 mm, a nocturnal moth found in fen and ancient woodland in parts of south-east England and also appearing as a migrant in southern England. Previously thought to be confined to fenland in Cambridgeshire, it is more widespread, with isolated colonies also in Lincolnshire, Huntingdonshire

and Leicestershire. It is Endangered in the RDB and a priority species in the UK BAP. Adults fly in June–July. Larvae mine small-reed stems. Greek *photos* + *edos* = 'light' + 'delight', and Latin *extrema* = 'extreme', this being the smallest in its genus.

CONECAP Inedible homobasidiomycete fungi in the genus *Conocybe* (Agaricales, Bolbitiaceae) (Latin *conus* + *cybe* = 'cone' + 'lump').

Common conecap *C. tenera*, widespread, commonest in England, found in summer and autumn on short grassland, including lawns, golf courses and parks, and in dune slacks; it can also appear on leaf litter, sawdust and woodchip mulch, and on disturbed nutrient-rich soil in orchards and gardens. *Tener* is Latin for 'tender'.

Fool's conecap *C. filaris*, widespread if scattered, commonest though remaining local in England, on woodland soil, and in gardens on soil, compost and wood chip. *Filaris* is Latin for 'thread-like'.

Milky conecap *C. apala*, common and widespread, its appearance from June to October often triggered by heavy rainfall but which rarely survives beyond mid-afternoon on sunny days as the stems quickly collapse and the caps turn brown and rot. It favours close-cropped grassland and dune slacks but can also appear on leaf litter, sawdust and woodchip mulch. *Apala* is Latin for 'tender'.

Ringed conecap *C. arrhenii*, frequent and widespread in England, local elsewhere, appearing from spring to autumn on soil in parkland, open woodland and field margins, and often by paths. *Arrhenos* is Greek for 'male'.

Verdigris conecap *C. aeruginosa*, a rarity with a scattered distribution in England and Wales, with 23 records from 1987 to 2013, on soil under hardwoods such as beech and oak. *Aeruginosa* is Latin for 'rusty-coloured'.

CONEFLOWER *Rudbeckia laciniata* (Asteraceae), a herbaceous North American perennial neophyte cultivated by 1640, scattered in England and Scotland, naturalised on riverbanks, roadsides, waste ground, and in rough grassy habitats, originating as a garden escape or throw-out. *Rudbeckia* honours Olof Rudbeck and his son, also Olof, seventeenth-century Swedish botanists. Latin *laciniata* = 'divided into narrow lobes'.

CONEHEAD Species of bush-cricket in the genus *Conocephalus* (Tettigoniidae) (Greek for 'cone' + 'head').

Long-winged conehead *C. discolor*, male body length 16–21 mm, wing length of the macropterous form 16–18 mm, females 16–22 mm and 15–19 mm, in rough grassland and woodland rides, as well as damp habitats, first appearing in Britain in the 1940s and initially confined to the south coast. However, as a result of climate change there has been a dramatic population growth and its range has expanded more than 240 km since the 1980s, and it can now be found beyond the Thames and as far west as Wales. *Discolor* is Latin for 'variegated'.

Short-winged conehead *C. dorsalis*, male body length 11–15 mm, wing length 6–10 mm, females 12–18 mm and 5–8 mm, found in two distinct habitats: coastally on saltmarshes and dunes, particularly associated with rushes and grasses; and inland on lowland bog, fen, reedbeds and river floodplains, and by lakes and pools. Its range is expanding, probably due to climate change, and it is now found mainly south of a line from the Humber to the Severn estuaries, plus parts of coastal Wales and Lancashire. *Dorsalis* is Latin for 'dorsal'.

CONFUSED (MOTH) *Apamea furva* (Noctuidae, Xylinae), wingspan 34–42 mm, nocturnal, local on rocky places on coasts, and limestone hills inland, in Scotland, north England and Ireland; in southern England, known only from Folkestone, Kent, and parts of the south-west. Adults fly in July–August. Larvae feed on roots and stem bases of various grasses. The species is difficult to identify, hence 'confused'. Apamea was a town in Asia Minor. *Furvus* is Latin for 'dark', referencing the dark brown wing colours.

CONGER EEL See eel.

CONIFEROPSIDA In some classifications, a subdivision or class of the gymnosperms (Latin *conifer* + Greek *opsis* = 'cone-bearing' and 'appearance').

CONOPIDAE Family name (Greek *conops* = 'mosquito') of thick-headed flies (Bracycera), with 23 species in 7 genera.

CONVOLVULACEAE Family name (Latin *convolvere* = 'to twine around') of bindweed, together with dodder and morning-glory.

COOT *Fulica atra* (Rallidae), wingspan 75 cm, length 37 cm, weight 800 g, common and widespread, mainly on ponds, lakes, flooded gravel-pits, reservoirs, slow-flowing rivers and urban park lakes. There were an estimated 31,000 pairs in 2009, 180,000 individuals having overwintered in 2008–9. There was a 69% increase in numbers between 1975 and 2015 (19% in 1995–2015, though this figure includes an 18% decline in 2010–15). Nests are constructed near the water's edge or on small protrusions in the water, of dead reeds, grasses and sometimes paper or plastic litter. Clutch size is 5–7, incubation takes 21–4 days, and the precocial young fledge in 55–60 days; there are often two or three broods each year. Food is mostly aquatic plants, but also snails and insect larvae. 'Coot' comes from Middle English *cote*, probably related to Dutch *koet*. The binomial derives from Latin for 'coot' and 'black'. **American coot** *F. americana* is an occasional accidental.

COPEPOD(A) From Greek *cope* + *podi* = 'oar' + 'foot', small, mostly planktonic phytoplanktonic-feeding crustaceans commonly found in marine and freshwater habitats, a subclass in the class Maxillopoda, and major food items of many aquatic animals.

COPPER (BUTTERFLY) Species of *Lycaena* (Lycaenidae, Lycaeninae), possibly from Greek *lycaina* = 'she-wolf', which has no entomological significance.

Large copper *L. dispar*, with the male wingspan 44–8 mm, female 46–52 mm. Fenland drainage and over-collecting led to the extinction of this butterfly, which feeds primarily on water dock, last recorded in Cambridgeshire in 1851. Several attempts at reintroduction have failed. *Dispar* is Latin for 'unlike', from the visual intersexual disparity.

Small copper *L. phlaeas*, wingspan 26–36 mm, widespread, found throughout Britain apart from montane habitats and north-west Scotland (ssp. *eleus*) and Ireland (ssp. *hibernica*), remaining common despite recent declines in both numbers and distribution (by 37% and 16%, respectively over 1976–2014). It is found in open habitats such as grassland, wasteland, heathland and verges where larval food plants, mainly sorrels, are found. Adults fly in May–September, and have a broad diet. *Phleo* is Greek for 'to flourish', from the rich wing colours.

COPPER-MOSS Elongate copper-moss *Mielichhoferia elongata* (Mielichhoferiaceae), scarce, in cool, humid sites, often in shade, in crevices in extremely acidic, metal-rich rocks, but apparently absent from mine workings and metalliferous spoil, in Yorkshire and the east Highlands. Mathias Mielichhofer (1772–1847) was an Austrian botanist.

Tongue-leaf copper-moss *Scopelophila cataractae* (Pottiaceae), scarce in south-west England and Wales on toxic, zinc-rich substrates, usually where very damp, on seepages associated with abandoned metal-processing buildings or metal-rich spoil. Greek *scopelos* + *philos* = 'cliff' + 'loving', and Latin *cataracta* = 'waterfall'.

COPPERWORT Scarce foliose liverworts in the genus *Cephaloziella* (Cephaloziellaceae) (Greek *cephale* + *ziella* = 'head' + 'jelly') on copper-rich substrates in south-west England and north-west Wales.

COPSE-BINDWEED *Fallopia dumetorum* (Polygonaceae), a climbing annual, rare and local in south-central England, on hedges and woodland edge on well-drained soils, sometimes abundant following felling, thinning or coppicing. Gabriele Fallopia (1523–62) was an Italian anatomist. Greek *dumus* = 'thorn-bush'.

CORACIIFORMES From Greek *corax* = 'crow' or 'raven', the order of kingfishers, bee-eaters and rollers, though none of these birds is raven-like.

CORAL (ANIMAL) Marine invertebrates in the phylum Cnidaria, class Anthozoa, soft corals being in the order Alcyonacea, hard corals in the Scleractinia. Corals are typically colonial, consisting a many polyps each of which possess a set of tentacles surrounding a central mouth opening. An exoskeleton is excreted near the base, which can build up into blocks or reefs. Some corals catch small prey and plankton, but most get the majority of their energy and nutrients from photosynthetic unicellular dinoflagellates (zooxanthellae) that live inside their tissues. Latin *corallum* = 'coral'.

CORAL (FUNGUS) With fruit bodies resembling simple or branched coral, or clubs. The genera *Clavaria*, *Clavulina* and *Clavulinopsis* derive their names from Latin *clava* = 'club'. *Ramaria* and *Ramariopsis* come from Latin *ramus* + *aria* = 'branch' + 'possessing'.

Rarities include **ashen coral** *Tremellodendropsis tuberosa*, **beige coral** *Clavulinopsis umbrinella*, **lilac coral** *Ramariopsis pulchella*, **ochre coral** *Ramaria decurrens*, **orange coral** *Ramariopsis crocea* and **violet coral** *Clavaria zollingeri*.

Crested coral *Clavulina coralloides* (Cantharellales, Clavulinaceae), edible if tasteless, widespread and common, found as a grey, coral- or antler-like branching fruiting body from late summer to early winter on woodland soils and leaf litter, often by paths, occasionally grassland. *Coralloides* is Latin for 'resembling coral'.

Grey coral *Clavulina cinerea*, edible, widespread and common, found as a grey, coral- or antler-like branching fruiting body in summer and autumn on woodland soils and leaf litter, often by paths. *Cinerea* is Latin for 'grey'.

Ivory coral *Ramariopsis kunzei* (Agaricales, Clavariaceae), inedible, widespread and scattered, found in late summer and autumn on soil in grassy habitat, often under coniferous trees.

Meadow coral *Clavulinopsis corniculata*, edible, widespread but infrequent, found as a yellow antler-like branching fruiting body from early summer to autumn on soil in short unimproved grassland, occasionally in open grassy woodland. *Corniculata* is Latin for 'having small horns'.

Rosso coral *Ramaria botrytis*, edible, uncommon, with a cauliflower-like appearance, found from July to November, scattered in Britain in soil in broadleaf woodland, ectomycorrhizal on beech. *Botrys* is Latin for 'cluster of grapes'.

Salmon coral *Ramaria formosa*, poisonous, causing stomach pain and diarrhoea, uncommon and scattered in England and Wales, found from July to November in deciduous and mixed woodland, mycorrhizal on oak and other hardwoods. *Formosa* is Latin for 'beautiful'.

Upright coral *Ramaria stricta*, edible, widespread and, in England, locally common, found from July to November in woodland, mycorrhizal or, on dead wood, saprobic on beech and sometimes conifers, and also found on wood chippings. *Stricta* is Latin for 'upright'.

CORAL MOSS *Cladonia rangiformis* (Cladoniaceae), a widespread and locally common squamulose and fruticose lichen of dry neutral to basic grassland, coastal clifftops and dunes. Greek *clados* = 'branch', and Latin *rangifer* = 'reindeer'.

CORAL-NECKLACE *Illecebrum verticillatum* (Caryophyllaceae), a rare annual found in Cornwall and southern England, recorded in Hampshire in 1920 and spreading in the New Forest since 1950. It grows in periodically wet acidic or neutral soils on tracks, ditch margins, in grass-heath and grassland, and as a casual elsewhere in England, for example on railway clinker. Greek *illicium* = 'charming', and Latin *verticilla* = 'whorl'.

CORAL WEED *Corallina officinalis* (Corallinaceae), a common and widespread calcareous red seaweed up to 12 cm long, especially found on exposed coasts, typically forming a turf in rock pools from the mid-tidal level to the sublittoral fringe. Latin *corallum* = 'coral', and *officinalis* = 'of pharmacological use', since it was formerly used as a vermifuge, and is currently used in the cosmetics industry given its anti-inflammatory, slimming, firming and moisturising properties.

CORAL WORM *Salmacina dysteri* (Serpulidae), a polychaete 4–5 mm long, common if scattered, on rocks on the lower shore and, more often, sublittorally in dense aggregation of interwoven tubes. Salmacis was a fountain in Greek mythology, said to make weak those who drank its waters.

Filigreed coral worm *Filograna implexa*, a polychaete 4–5 mm long, encrusting bryozoans, corals, kelp holdfasts, and hard substrates in the lower intertidal and sublittoral, locally abundant on the coasts of south-west and west Britain, and of Ireland. Latin *filum* + *granum* = 'thread' + 'grain', and *implexa* = 'twisted'.

CORALBELLS *Heuchera sanguinea* (Saxifragaceae), a North American perennial herb, widespread but local in England and Scotland as a garden escape on waste ground. Johann von Heucher (1677–1747) was a German botanist. Latin *sanguinea* = 'blood-red'.

CORALBERRY *Symphoricarpos orbiculatus* (Caprifoliaceae), a deciduous North American neophyte shrub, introduced by 1730, scattered in England, naturalised in hedges and scrub and on waste ground, often in dense thickets. Greek *symphoricarpos* = 'fruit borne together', and Latin *orbiculatus* = 'round'.

Chenault's coralberry *Symphoricarpos* x *chenaultii* (*S. microphyllus* x *S. orbiculatus*), a hybrid deciduous shrub of garden origin, noted in 1910, widespread in Britain and probably increasing, naturalised in woodland, scrub and hedgerows and on waste ground, and planted for game cover. Léon Chenault (1853–1930) was a French nurseryman and botanist.

CORALROOT *Cardamine bulbifera* (Brassicaceae), a rhizomatous perennial herb mainly found on dry woodland slopes over chalk in the Chilterns, and in damp woodlands over clay in the Weald. Elsewhere it is an escape from cultivation by roads and in woodland and parkland, naturalised in places. *Cardamine* is from Greek *cardamon*, referring to the Indian spice, *bulbifera* Latin for 'bearing bulbs'.

Pinnate coralroot *C. heptaphylla*, a rhizomatous perennial neophyte herb from Europe, in gardens by 1683, found at a few sites in England and Scotland as a naturalised escape in woodland and waste ground. *Heptaphylla* is Greek for 'seven leaves'.

CORD-GRASS Rhizomatous perennial species of *Spartina* (Poaceae) (Greek *spartine*, a cord made from Spanish broom *Spartium*).

Common cord-grass *S. anglica*, a rhizomatous perennial originating in Southampton Water in about 1890 as an amphidiploid derivative of the hybrid Townsend's cord-grass, found in tidal mudflats and saltmarsh, much planted around the British Isles (and globally) to stabilise mud, forming extensive stands in many estuaries. *Anglica* is Latin for 'English'.

Prairie cord-grass *S. pectinata*, a North American neophyte found as a naturalised garden escape by lakes (Co. Galway since 1967, Northumberland since 1970) and a quarry (Hampshire since 1986). It may have potential for biomass production. *Pectinata* is Latin for 'like a comb'.

Small cord-grass *S. maritima*, scarce, found locally from Poole Harbour to the Wash on tidal mudflats, generally at relatively high elevations, such as in saltmarsh creeks and pans, and on bare ground behind sea walls.

Smooth cord-grass *S. alterniflora*, a North American neophyte naturalised on intertidal mudflats. It was present by 1816; it was planted in Southampton Water, and first recorded there in 1829, initially spreading (becoming one of the parents of Townsend's cord-grass, leading to the amphidiploid common cord-grass; see below and above respectively), then declining, and further reduced by dredging since the 1970s. It was planted in the Moray Firth in 1920, Essex in 1935 and Dorset in 1963. *Alterniflora* is Latin for 'alternating flowers'.

Townsend's cord-grass *Spartina* x *townsendii* (*S. maritima* x *S. alterniflora*), a spontaneous hybrid noted at Hythe, Southampton Water, in 1870, found in tidal mudflats and saltmarshes, spreading to Dorset and West Sussex, and scattered elsewhere. Attempts have been made to limit its spread. In southern England it has declined due to 'dieback' at many sites.

CORD-MOSS Species of *Entosthodon* (Greek *entosthi* + *donti* = 'within' + 'tooth', a reference to the position of 'teeth' inside the capsule) and *Funaria* (Latin *funis* = 'cord'), both Funariaceae.

Rarities are **Muhlenberg's cord-moss** *F. muhlenbergii* (found locally on limestone) and **pretty cord-moss** *F. pulchella* (mostly from south Wales).

Blunt cord-moss *E. obtusus*, common in western montane Britain and Ireland, and in south England, on wet peaty soil and in marshy areas. *Obtusus* is Latin for 'blunt'.

Common cord-moss *F. hygrometrica*, common and widespread as a colonist of bare, disturbed, nutrient-rich soils, characteristic of old bonfire sites. Greek *hygrometrica* = 'wet' + 'measure'.

Thin cord-moss *E. attenuatus*, common in western montane Britain and Ireland on wet, peaty soil, especially frequent near the coast. *Attenuatus* is Latin for 'thin'.

CORDULEGASTRIDAE Dragonfly family (Greek *cordule* + *gaster* = 'swelling' + 'stomach'), represented in the British Isles by golden-ringed dragonfly.

CORDULIIDAE Family name of emerald dragonflies (Greek *cordule* = 'swelling').

COREIDAE Family name of sap-sucking heteropteran bugs (Greek *coris* = 'bug').

CORIANDER *Coriandrum sativum* (Apiaceae), an annual Mediterranean archaeophyte cultivated for at least a millennium, scattered in Britain as a casual on disturbed ground, mostly from birdseed and garden escapes, but occasionally naturalised, for example on roadsides in north-west Essex. The English and genus names ultimately come from Greek *coriannon*, in turn perhaps from *coris* = 'bed bug', reflecting the scent and appearance of coriander seed. *Sativum* is Latin for 'cultivated'.

CORIXIDAE Family name (Greek *coris* = 'bug') of water boatmen, with 39 species in 9 genera in Britain.

CORKIR Also **corkie**, **korkie**, **korkir** and other spellings; applied loosely to any lichen used as a red dye, usually cudbear lichen *Ochrolechia tartarea*, but including limestone urceolaria *Aspicilia calcarea*.

CORKLET Latticed corklet *Cataphellia brodricii* (Hormathiidae), a sea anemone with a small oral disc with 100 or so short tentacles, found in south-west England and south Ireland under stones, in algal holdfasts, or attached to rock beneath sand or gravel, typically in the lower shore but also down to 20 m. Greek *catos* + *phellos* = 'down' + 'cork'.

CORMORANT *Phalacrocorax carbo* (Phalacrocoracidae), wingspan 145 cm, length 90 cm, male weight 2.5 kg, females 2.1 kg, widespread on rocky shores, coastal lagoons and estuaries, and increasingly seen inland at reservoirs, lakes and gravel-pits. In the UK the cormorant was almost exclusively a coastal breeder until 1981, when an inland tree-nesting colony became established at Abberton reservoir, Essex. By 2012 cormorants had bred at 89 inland sites in England. Britain had an estimated 8,400 breeding pairs, and (of international importance) 35,000 birds wintering in 2008–9. There was a 53% range expansion from 1981–4 to 2012 in Britain, and an 18% range expansion in Ireland. Population trends are unclear: SMP data suggest a 12% decline in numbers over 2000–14, while BBS data indicate an increase by 15% (though with an error range of −12% to +53%) between 1995 and 2014, and a decline by 9% in 2009–14. Coastal nests are of seaweed and twigs on rocky ledges. Clutch size is 3–4, incubation takes 28–31 days, and the altricial young fledge in 48–52 days. Diet is of fish, and there remains conflict with some fishermen. 'Cormorant' comes from Middle English, in turn from Old French *cormaran*, and medieval Latin *corvus marinus* = 'sea raven'. Greek *phalacros* + *corax* = 'bald' + 'raven', and Latin *carbo* = 'charcoal'.

CORN See maize.

CORN-BORER European corn-borer *Ostrinia nubilalis* (Crambidae, Pyraustinae), wingspan 26–30 mm, a rare migrant before the 1930s which began to colonise the area around London and south-east England, then becoming established in the Thames Valley and around Portsmouth, and now found throughout much of southern and south-west England and south Wales, with scattered records from north England and in Co. Cork. Adults fly at night in June–July. This is a pest on corn (maize) on the continent, but in Britain larvae mainly feed on mugwort. Greek *ostreion* = 'purple', and Latin *nubilum* = 'cloudy weather', from the dark wing markings.

CORNACEAE Family name of dogwood (Latin *cornus*).

CORNCOCKLE *Agrostemma githago* (Caryophyllaceae), an annual European neophyte, tolerant of a range of soil types, present in Britain as a weed of arable crops since the Iron Age. Widespread and common until the twentieth century, it dramatically declined with improved seed-cleaning, and is now extinct as an arable weed, an occasional casual, though a frequent component of wild-flower seed mixes. 'Corncockle' dates to the early eighteenth century, from corn + cockle, from Old English *coccul* = 'corncockle'. Greek *agros* + *stemma* = 'field' + 'garland'. *Githago* is Latin for this plant.

CORNCRAKE *Crex crex* (Rallidae), wingspan 50 cm, length 28 cm, male weight 170 g, females 140 g, a Red List migrant breeder arriving from Africa or south Europe in mid-April to leave in August–September, mostly confined to traditionally managed hayfields in the Outer Hebrides and Orkney in Scotland, and Co. Donegal and Co. Connemara in Ireland, numbering (five-year mean) 1,200 pairs in 2006–10, but since 2005 there has also been a reintroduced breeding population in the Nene Washes near Peterborough, with 22 singing males recorded in 2014. Nevertheless there has been a decline in numbers by 55% over 1970–15. Nests are in hollows in grassland or hayfield, clutch size is 8–12, incubation lasts 16–19 days, and the precocial young fledge in 34–8 days. Food is mostly of small invertebrates plus some seed. The Latin and to an extent the English name (from Old Norse *kráka*) come from the bird's disyllabic rasping vocalisation.

CORNEL Dwarf cornel *Cornus suecica* (Cornaceae), a rhizomatous herbaceous perennial, local in the Highlands and northern England, in wet, base-poor peats at moderate to high altitudes, mostly in montane dwarf-shrub heath communities, though extending into acid montane grassland. 'Cornel' dates from late Middle English, denoting the wood of the cornelian cherry, derived from Old French *corneille*, from Latin *cornus* = 'dogwood'. *Suecica* is Latin for 'Swedish'.

CORNELIAN-CHERRY *Cornus mas* (Cornaceae), a European neophyte shrub or small tree present by Tudor times, very scattered in Britain and recorded from Co. Antrim, in woodland, hedgerows, scrub and verges, but rarely naturalised. 'Cornel' dates from late Middle English, denoting the wood of the cornelian cherry, derived from Old French *corneille*, from Latin *cornus* = 'dogwood'. *Mas* is Latin for 'male'.

CORNFLOWER *Centaurea cyanus* (Asteraceae), a once common annual archaeophyte arable weed, known from the Iron Age, now a frequent casual garden escape or birdseed introduction on waste ground and roadsides. The binomial references the centaur Chiron, who in Greek myth discovered the medicinal uses of this plant, and the Latin name for this plant.

Perennial cornflower *C. montana*, a herbaceous perennial neophyte introduced from montane central and southern Europe by Tudor times, scattered and naturalised over much of Britain, scarce in Ireland, on roadsides, railway banks and waste ground. *Montana* is Latin for 'montane'.

CORNICULARIA Brown prickly cornicularia *Cetraria aculeata* (Parmeliaceae), a fruticose lichen, common and widespread on heathland, acidic coastal dunes and shingle. Latin *cetra* = 'leather shield', and *aculeata* = 'pointed (like a needle)'.

CORNSALAD Annuals in the genus *Valerianella* (Valerianaceae). The English name indicates its presence as a weed in cereal fields, with common cornsalad indeed being used in salad. The genus name is a diminutive of *Valeriana*, itself derived from Latin *valere* = 'to be healthy', from medicinal use in countering nervousness and hysteria.

Broad-fruited cornsalad *V. rimosa*, an archaeophyte known since the Iron Age, generally local in southern England and Ireland on rough ground and arable field margins on sand, calcareous clay and chalk soils, and declining. *Rimosa* is Latin for 'cracks'.

Common cornsalad *V. locusta*, a winter-annual common and widespread in lowland regions on thin soils around rock outcrops, and on dunes (often the dwarf var. *dunensis*) and coastal shingle. It also grows in disturbed habitats such as walls, railway tracks, paving, gardens and, rarely, on arable land. It is also grown as a winter salad crop. Other names for this plant are **lamb's lettuce**, and **rapunzel**: in most English translations of the fairy tale the witch names the girl after this plant. *Locusta* is Latin for 'locust', but here might mean 'growing in an enclosed space'.

Hairy-fruited cornsalad *V. eriocarpa*, a winter-annual, scattered and declining in southern and south-west England on drought-prone, stony substrates such as cliff edges, walls and quarries, probably native on bare limestone and chalk in Dorset and the Isle of Wight. *Eriocarpa* is Greek for 'woolly fruit'.

Keeled-fruited cornsalad *V. carinata*, an autumn-germinating annual archaeophyte, scattered and increasing though mainly in southern England and south-east Ireland, and scarce in Scotland, found mainly in sites associated with human activity such as walls, gravel paving, railway tracks and gardens. *Carinata* is Latin for 'keeled'.

Narrow-fruited cornsalad *V. dentata*, an archaeophyte known since the Bronze Age, scattered and declining, absent from most of Scotland, mostly on rough ground and the edges of arable fields, especially on chalky soils, but locally also on sand and calcareous clay. *Dentata* is Latin for 'toothed'.

CORONET Members of the nocturnal moth family Noctuidae, specifically *Craniophora ligustri* (Acronictinae), wingspan 30–5 mm, widespread but local in a variety of habitats, including hedgerows, woodland, calcareous grassland, fen and hillsides. Adults fly in June–July. Larvae mainly feed on ash, but also on hazel, alder and privet. Greek *cranion* + *phoreo* = 'skull' + 'to bear', from markings on the folded forewing, and Latin *ligustrum* = 'privet'.

Barrett's marbled coronet *Conisania andalusica* (Hadeninae), wingspan 35–42 mm, scarce, found on sea-cliffs and shingle beaches in parts of south-west England, Wales and southern Ireland. Adults fly in June–July. Larvae feed on coastal plants such as sea campion, sand spurrey and rock sea-spurrey. The English name honours the entomologist Charles Barrett (1836–1904).

Marbled coronet *Hadena confusa* (Hadeninae), wingspan 33–9 mm, local on calcareous grassland in southern England and along the coasts of Britain and north and west Ireland. Adults fly in May–June, sometimes with a partial second generation in south-east England. Larvae feed on seeds of bladder campion, sea campion and rock sea-spurrey. The binomial is from Hades, the Underworld in Greek mythology, and Latin for 'confused', from the mingled wing colours.

Varied coronet *H. compta*, wingspan 25–30 mm, an adventive colonist, relatively unknown in Britain until the late 1940s, when it began appearing in numbers in the south-east since when, having spread rapidly, it is now commonly found in calcareous grassland and gardens throughout much of the southern half of England. Adults fly in June–July. Larvae feed on seeds of bladder campion and, in gardens, sweet-william. *Comptus* is Latin for 'adored', from the attractive wing pattern.

CORTINARIACEAE Family name of webcap fungi in the genus *Cortinaria*, named from the partial veil or cortina (Latin for 'curtain') that covers the gills when the caps are immature. See gypsy (fungus) and saffron stainer under stainer.

CORVIDAE Crow family (Latin *corvus* = 'crow' or 'raven'). See also chough, jackdaw, jay, magpie, raven and rook.

CORYDALIS Plants in the family Papaveraceae (Greek *corydon* = 'lark', in turn from *corys* = 'crest').

Climbing corydalis *Ceratocapnos claviculata*, a widespread climbing or scrambling annual found in woodland and other shady habitats on base- and nutrient-poor, often peaty soils or over rocky outcrops. In Ireland it occurs on shaded boulder slopes. Greek *ceras* + *capnos* = 'horn' + 'smoke', and Latin *claviculata* = 'like a small club' or 'twig'.

Fern-leaved corydalis *Corydalis cheilanthifolia*, introduced from China, scattered and uncommon in southern and central England as a garden escape on walls. Latin *cheilanthifolia* = 'having wallflower-like leaves'.

Pale corydalis *Pseudofumaria alba*, a perennial neophyte introduced from Central Europe by Tudor times, a garden escape recently naturalised and probably expanding its range, found on walls and in churchyards. Greek *pseudofumaria* = 'false *Fumaria*', itself a genus name meaning 'earth smoke', and Latin *alba* = 'white'.

Yellow corydalis *P. lutea*, a perennial neophyte introduced from the Alps, commonly cultivated and since the eighteenth century widely naturalised in Britain, less so in Ireland, on walls and pavements, and canal and railway brickwork. It tolerates both shade and strong light, and roots in damp, base-rich, infertile substrates. *Lutea* is Latin for 'yellow'.

CORYLOPHIDAE Family name (Greek *corys* = 'helmet') for very small beetles (0.5–1.3 mm length), with 12 species in 5 genera. Feeding on mould, they are usually found under fungoid bark or in piles of rotting wood or vegetation, but *Orthoperus atomarus* is synanthropic, occurring in wine cellars on mouldy corks. Many species are rare, and all are found in the southern half of Britain.

COSMET Mostly nocturnal moths in the families Batrachedridae (Greek *batrachos* + *hedra* = 'frog' + 'seat', from the adult's posture), Cosmopterigidae (*cosmos* + *pteryx* = 'ornament' + 'wing'), Momphidae (with species of *Mompha*, from *momphe* = 'blame', entomological significance unknown) and Parametriotidae (*para* + *metriotes* = 'near' + 'moderation'). Adults often have raised scale tufts or head crests. Wings are narrow, and folded when at rest. Larvae of most species are leaf miners.

August cosmet *Sorhagenia rhamniella* (Cosmopterigidae, Chrysopeleiinae), wingspan 9–10 mm, local in woodland margins on chalky soils and fens in parts of south and south-east England and East Anglia, also with one Irish record. Adults fly in July–August. Larvae feed on buds, shoots and flowers of buckthorn and alder buckthorn. Ludwig Sorhagen (1836–1914) was a German entomologist. *Rhamniella* is derived from *Rhamnus* (buckthorn).

Brown cosmet *Mompha miscella*, wingspan 7–9 mm, local on chalk and limestone downland in parts of England; rare

in Wales and eastern Scotland. Adults fly in two generations, mainly May–June and July–August, but they can be found any time from April to October. Larvae mine leaves of rock-roses from October to April. *Miscella* is from Latin *misco* = 'mingle', from the blending of the forewing colours.

Buckthorn cosmet *S. lophyrella*, wingspan 8–11 mm, in woodland edges and hedgerows in parts of south and south-east England and East Anglia, occasionally further north, and with two records in Ireland. Adults fly in July. Larvae feed on buds and leaves of buckthorn and alder buckthorn, living between leaves spun together with silk. *Lophyrella* is from Greek *lophion* = 'small crest'.

Buff cosmet *M. ochraceella*, wingspan 14–16 mm, common in damp habitats throughout much of Britain south of the Moray Firth, and recorded in the east of Ireland. Adults fly in May–August. Larvae mine the stems and later the leaves of great willowherb. Latin *ochra* = 'ochre', from the forewing colour.

Bulrush cosmet *Limnaecia phragmitella* (Cosmopterigidae, Cosmopteriginae), wingspan 16–22 mm, widespread and common in fen, marsh and freshwater margins. Adults fly in July from dusk into the night. Larvae feed inside the seedheads of bulrushes throughout winter. Greek *limne* + *oiceo* = 'marshy lake' + 'to dwell'.

Clouded cosmet *M. langiella*, wingspan 9–11 mm, local in woodland in parts of England and Wales, more numerous in the west, rare elsewhere, including Ireland. Adults fly in July–August. Larvae mine leaves of enchanter's-nightshade and great willowherb, especially from June to August. H.G. Lang was an eighteenth-century German entomologist.

Common cosmet *M. epilobiella*, wingspan 10–13 mm, common at freshwater margins and on dry open areas and waste ground throughout England and Wales, less numerous in the north, rare in southern Scotland, and south and east Ireland. Adults hibernate and are found throughout the year, particularly in April–May and July–August. Larvae feed on great willowherb (*Epilobium*, prompting the specific name), especially in May–August.

Enchanter's cosmet *M. terminella*, wingspan 7–9 mm, scarce, local in shady woodland in much of England and Wales, with a few Irish records. Adults fly in July–August. Larvae mine leaves of enchanter's-nightshade in August–September. *Terminella* comes from Latin *termen* = 'tip'.

Fen cosmet *Cosmopterix lienigiella* (Cosmopterigidae, Cosmopteriginae), wingspan 10–13 mm, local in reedbeds in parts of south and south-east England and East Anglia (and with one Irish record). Larvae mine flowers and seed heads of lousewort. Greek *cosmos* + *pteryx* = 'ornament' + 'wing'. Mme Lienig (d.1855) was a Latvian entomologist.

Garden cosmet *M. subbistrigella*, wingspan 7–12 mm, common in freshwater margins, ditches, woodland clearings, waste ground and gardens across much of England, Wales and Ireland, occasional in Scotland. Adults emerge in late summer and may be seen on the wing any time until late spring. They may overwinter inside garden sheds, etc. Larvae mine seed pods of willowherbs. Latin *sub* + *bis* + *striga* = 'somewhat' + 'twice' + 'streak', from the two forewing markings.

Hawthorn cosmet *Blastodacna hellerella* (Parametriotidae), wingspan 10–11 mm, widespread and common in hedgerows, woodland edges, scrub, parks and gardens. Adults fly in June–July. Larvae mine hawthorn berries in September–October. Greek *blastos* + *dacno* = 'a shoot' + 'to bite'. J.F. Heller (b.1813) was an Austrian entomologist.

Hedge cosmet *Cosmopterix zieglerella*, wingspan 8.5–10.5 mm, scarce, in hedgerows in south-east England and East Anglia, from Kent northwards to Norfolk and westwards to Berkshire, with the first Staffordshire record in 2009. Adults fly in June–July. Larvae mine hop leaves. Dr Ziegler was an early nineteenth-century Austrian physician.

Kentish cosmet *M. sturnipennella*, wingspan 13–18 mm, scarce, confirmed in Surrey in 1950 and since spreading gradually, found in waste ground, heaths, embankments and urban areas in parts of England, commoner in the south but spreading northwards to the Scottish border. The bivoltine adults fly in July–August and September–May. Larvae mine rosebay willowherb. The first generation (May–June) can be identified by a gall within the stem. The second generation in July–August feeds either in a seed pod with a small hole, or in a stem gall high up on the plant. Latin *sturnus* + *pennella* = 'starling' and 'little wing'.

Large dark cosmet *M. conturbatella*, wingspan 14–18 mm, local in waste ground, woodland clearings, heathland, marsh and verges throughout much of Britain, but only recorded in the east of Ireland. Adults fly in June–July. Larvae feed on willowherbs, living between leaves spun together with silk. *Conturbatus* is Latin for 'confused', reflecting the mingled forewing colours.

Lime cosmet *Chrysoclista linneella* (Parametriotidae), wingspan 10–13 mm, scarce, in lime woodland, parkland, urban areas and on verges in southern England and south Wales, with records north to Yorkshire. There is one record from Co. Cork. Adults fly in May–September. Larvae feed under the bark of lime trees. Greek *chrysos* + *clystos* = 'gold' + 'washed', from the forewing markings, and *linneella* is in honour of Linnaeus (1707–78).

Little cosmet *M. raschkiella*, wingspan 14–18 mm, widespread and common on heathland, waste ground, verges and woodland clearings. With two generations, adults fly in May and August. Larvae mine leaves of rosebay willowherb from June to October. Johann Raschke (1763–1815) was a German entomologist.

Marbled cosmet *M. propinquella*, wingspan 11–12.5 mm, local in open woodland, waste ground and gardens throughout Britain. Adults fly in June–August. Larvae mine the willowherb leaves. *Propinquus* is Latin for 'close', from similarity to a congeneric.

Marsh cosmet *Cosmopterix orichalcea*, wingspan 8–10 mm, local in damp meadow, fen, marsh and by riverbanks in parts of southern England, East Anglia, south Wales, the Western Isles in Scotland, and Ireland. Adults fly in May–August. Larvae mine the blades of a number of grass species. *Orichalcum* is Latin for 'brass', from the colour of forewing markings.

Neat cosmet *M. divisella*, wingspan 10–13 mm, local in damp woodland, waste ground, open areas and damp shady habitats in parts of south Wales and southern England, though with records north to Cumbria. Adults fly from August, then hibernate until the following spring when they reappear on the wing. Larvae mine and create galls in willowherb stems. *Divisus* is Latin for 'divided', from the dorsal forewing which is paler than elsewhere.

New marsh cosmet *Cosmopterix scribaiella*, wingspan 10–11 mm, restricted to alder carr and reedbeds in Hampshire and neighbouring counties. Adults fly in June–July. Larvae mine leaf blades of common reed.

New neat cosmet *M. bradleyi*, wingspan 9–11 mm. In the early 1990s specimens from Herefordshire and Worcestershire previously thought to be neat cosmet were correctly identified; subsequent investigations revealed other overlooked records. It is now regularly recorded from damp meadows, ditches and marshes in the south Midlands and southern-central England. Adults emerge in July–September, hibernate, and reappear in spring. Larvae mine and create galls great willowherb stems. The specific name honours the British entomologist J.D. Bradley.

Oak cosmet *Dystebenna stephensi* (Parametriotidae), wingspan 8–10.5 mm, scarce in ancient oak woodland and parkland in parts of south-east England, Somerset, Herefordshire and Derbyshire. Adults fly during the day from late June to September, often resting on oak tree-trunks. Larvae feed in oak bark. Greek *dys* + *tebenna* = 'bad' + 'state robe'. James Stephens (1792–1852) was a British entomological writer.

Orange-blotch cosmet *Chrysoclista lathamella*, wingspan 11–13 mm, restricted to a few woodlands in England, plus one record from Co. Wicklow. Adults fly in June–August. Larvae feed within the bark of white and goat willows. John Latham (1740–1837) was a British naturalist.

Pine cosmet *Batrachedra pinicolella* (Batrachedridae), a small scarce nocturnal moth in coniferous woodland and plantations with a scattered local distribution in the southern half of England. Adults fly in June–July. Larvae mine young needles of Scots pine and Norway spruce. Greek *batrachos* + *hedra* = 'frog' + 'seat', from the adult's posture, and Latin *pinus* + *colo* = 'pine' + 'inhabiting'.

Poplar cosmet *B. praeangusta*, wingspan 14–15 mm, widespread and common in a range of habitats, less common in northern England and Scotland, but found as far north as Aberdeenshire, and with a scattered distribution in Ireland. Adults fly in July–August. Larvae feed in May–June on catkins, seeds and leaves of white poplar, aspen, white willow and sallow. Latin *prae* + *angusta* = 'very' + 'narrow', from the forewing shape.

Red cosmet *M. locupletella*, wingspan 9–12 mm, widespread but local in damp habitats. Adults fly in two generations, mainly May–July and August–September. Larvae mine willowherb leaves during April–May and in July–August. *Locuples* is Latin for 'opulent', from the rich forewing colours.

Rust-blotch cosmet *M. lacteella*, wingspan 9–13 mm, confined to woodland and waste ground in parts of England, Wales and Scotland. Adults fly in May–July. Larvae mine leaves of broad-leaved willowherb. *Lacteus* is Latin for 'milky', from the colour of the thorax and forewing blotch.

Scarce cosmet *M. propinquella*, wingspan 11–13 mm, local on waste ground and in dry open areas mainly in south-east England, but also as far north as Yorkshire. Adults fly from September, hibernating over winter and appearing again up to April. Larvae mine and create galls in great willowherb stems in July–August. *Propinquella* is Latin for 'resembling'.

Scarce violet cosmet *Pancalia schwarzella* (Cosmopterigidae, Antequerinae), wingspan 11–16 mm, on calcareous grassland in parts of south-east England, on mountainsides in north Wales and the Highlands, and on the Burren, Ireland. Adults fly in April–July. Larvae mine petioles (leaf stalks) of dog-violets, then move into the roots. Greek *pagcalos* = 'entirely beautiful'. C. Schwarz (d.1810) was a German entomologist.

Violet cosmet *Pancalia leuwenhoekella*, wingspan 10–12 mm, local on calcareous grassland, mainly in south England, but sparingly northwards to Easter Ross. Adults fly in April–June. Larvae mine petioles of dog violets, then move into the roots. Antony van Leeuwenhoek (1632–1723) was a Dutch scientist.

Wood cosmet *Sorhagenia janiszewskae*, wingspan 10–13 mm, scarce in damp heathland and woodland in parts of southern England and East Anglia. Adults fly in July–August. Larvae mine twigs of buckthorn and alder buckthorn, living between flowers and shoots spun together with silk. The specific name honours the Polish Professor J. Janiszewska.

Yellow-headed cosmet *Spuleria flavicaput* (Parametriotidae), wingspan 12–14 mm, widespread but local in hedgerows, woodland edge, scrub and gardens. Adults fly in the morning in May–June. Larvae feed from August to October by mining into hawthorn twigs. Arnold Spuler (1869–1937) was a German entomological writer. Latin *flavicaput* = 'yellow head'.

COSMOPOLITAN *Leucania loreyi* (Noctuidae, Hadeninae), wingspan 34–44 mm, a fairly common immigrant nocturnal moth appearing particularly in south-west England and south-east Ireland, in some years arriving in sufficient numbers to breed. Since 2000 it has extended its range in England north-eastwards, probably as a result of increasing temperatures. The main arrival period is from August to October, though adults have been recorded as early as May. The binomial is from Greek *leucanie* = 'throat', and in honour of a Dr Lorey of Dijon, France.

COSMOPTERIGIDAE Family name for cosmet moths and beautiful cosmopterix (Greek *cosmos* + *pteryx* = 'ornament' + 'wing').

COSMOPTERIX Beautiful cosmopterix *Cosmopterix pulchrimella* (Cosmopterigidae, Cosmopteriginae), wingspan 6.5–9 mm, a nocturnal moth arriving as an immigrant from southern Europe and first recorded in Britain at Walditch, Dorset, in 2001 (and in Guernsey in 2002), after which it has rapidly colonised southern England, for example noted at Portchester Castle, Hampshire, in 2007, often associated with farm buildings and old walls. Adults fly from late June to October. Larvae mine leaves of pellitory-of-the-wall. Greek *cosmos* + *pteryx* = 'ornament' + 'wing', and Latin *pulchrimella* = 'very beautiful'.

COSSIDAE Family name of goat, reed leopard, and leopard moths, from Latin *cossus*, a moth larva found under tree bark.

COTONEASTER From Latin for 'like a quince' (*cotoneum*), species of *Cotoneaster* (Rosaceae), a large genus of ornamental neophyte deciduous or evergreen shrubs, many from China or the Himalayas, increasingly becoming naturalised in rough and waste ground, scrub and roadsides via bird-sown seed from gardens and ornamental amenity plantings.

COTTIDAE Family name of sculpins, sea scorpion, bullhead and bull rout, *cottos* being Greek for an unknown freshwater fish. These fishes do not possess a swim bladder.

COTTONGRASS Rhizomatous perennial species of *Eriophorum* (Cyperaceae), from Greek *erion* + *phoreo* = 'wool' + 'to bear', reflecting, as does the English name, the fluffy white seed heads.

Broad-leaved cottongrass *E. latifolium*, scattered though scarce or extinct in central and eastern England, in wet, base-rich lowland meadows and mires, and in fens and calcareous flushes in the uplands. *Latifolium* is Latin for 'broad-leaved'.

Common cottongrass *E. angustifolium*, common and widespread if scarcer in central and eastern England in open, wet, peaty ground, often growing in standing water. Habitats range from upland blanket bogs and hillside flushes to wet heathland and (though declining) in marshy lowland meadow. This is the 'county' flower of Greater Manchester as voted by the public in a 2002 survey by Plantlife. *Angustifolium* is Latin for 'narrow-leaved'.

Hare's-tail cottongrass *E. vaginatum*, scattered in southern England, absent from south-east, central and eastern England, locally common elsewhere on wet heathland and mires, including blanket and raised bogs. It is typical of wet peaty moorlands, often (co-)dominant with heather, surviving or indeed increasing after burning. *Vaginatum* is Latin for 'with a sheath'.

Slender cottongrass *E. gracile*, declining in southern England to one site in Surrey and two in Hampshire. However, it appears to be stable in south Wales and the Lleyn Peninsula and in (mainly western) Ireland, in the wettest parts of bogs, transitional mires, poor fen and on the edge of alder carr, typically over liquid peats. Sites are calcareous or moderately acidic, and have some water movement. *Gracile* is Latin for 'slender'.

COUCH Grasses in the genera *Elymus* (Greek *elymos* = 'a cereal') and *Elytrigia* (*elytron* = 'sheath' or 'covering') (Poaceae).

Australian couch *Elymus scabrus*, an annual neophyte recorded in the wild in south Lancashire in 1960, and occasionally elsewhere since in arable fields and waste ground, originating from wool shoddy. *Scabrus* is Latin for 'rough'.

Bearded couch *Elymus caninus*, a widespread but scattered perennial of partially shaded sites in woodland, on riverbanks and roadsides on free-draining, mainly base-rich, soils, as well as on mountains up to 800 m on cliffs and ledges. *Caninus* is Latin for 'relating to dogs'.

Common couch *Elytrigia repens*, a common and widespread rhizomatous perennial found in fertile, disturbed habitats, for example waste ground, roadsides, railway banks, arable land and rough grassland, as well as in coastal areas (ssp. *arenosa*) on dunes, shingle and saltmarsh margins. It is a persistent weed of gardens and agricultural land, its dense network of rhizomes, which become entangled with roots of other plants, being difficult to remove, and regrowing from any fragments left behind. *Repens* is Latin for 'creeping'.

Neglected couch See common couch (ssp. *arenosa*) above.

Sand couch *Elytrigia juncea*, a common and widespread rhizomatous perennial growing on or just above the strand-line in loose sand, sometimes also on shingle. It often stabilises dune systems, forming low hummocky fore-dunes on the seaward side of the main marram dunes. *Juncea* is Latin for 'rush-like'.

Sea couch *Elytrigia atherica*, a rhizomatous perennial found around much of the British coastline south of south Scotland, and in Ireland, on the margins of brackish creeks, saltmarsh, and dunes, and on shingle banks and sea walls. On ungrazed sites it can form dense almost monospecific stands. *Ather* is Greek for 'awn'.

COUSIN GERMAN *Protolampra sobrina* (Noctuidae, Noctuinae), wingspan 34–9 mm, a rare RDB nocturnal moth found in pine woodland, birch woodland and scrub in parts of the Highlands, and a priority species in the UK BAP. Adults fly in July–August. Larvae feed at night on bilberry, heather, heaths, eared willow and birches. The English name dates from the fourteenth century, coming from medieval French *cousin germain* = 'related (cf. germane) cousin' or 'first cousin', for which the Latin is *sobrina*. Greek *proto* + *lampros* = 'first' + 'shining'.

COW From Old English *cu*; see cattle.

COW PARSLEY See parsley (2).

COW-WHEAT Annual hemiparasitic species of *Melampyrum* (Orobanchaceae), from Greek *melas* + *pyros* = 'black' + 'wheat', referring to the black-coloured bread that results from adding the seeds to flour made using other grains.

Common cow-wheat *M. pratense*, in woodland, scrub, heathland and upland moorland on well-drained, nutrient-poor acidic soils (ssp. *pratense*), more rarely, in southern England and south Wales, in scrub, hedgerows and deciduous woodland on chalk and limestone (ssp. *commutatum*). The seeds are distributed by ants, attracted by the elaiosomes. *Pratus* is Latin for 'meadow'.

Crested cow-wheat *M. cristatum*, mostly found on the margins and clearings of ancient pedunculate oak woodlands, and in associated hedge banks on chalky boulder-clay soils, very locally in East Anglia and the adjacent east Midlands. *Cristatum* is Latin for 'crested'.

Field cow-wheat *M. arvense*, probably native, mainly found on the roots of grasses, declining, found on slumping chalk cliff faces in the Isle of Wight, on a road verge at Honeydon, Bedfordshire, and a few verges and field margins in north-west Essex. Flowers are pollinated by bumblebees. The seeds may be dispersed by ants attracted by an elaiosome, and which carry them to their nests for food. *Arvum* is Latin for 'field'.

Small cow-wheat *M. sylvaticum*, in humid, lightly shaded situations on damp, nutrient-rich, acidic soils in wooded ravines and on upland cliff ledges, local and scarce, mostly in northern Scotland. *Sylva* is Latin for 'forest'.

COWBANE *Cicuta virosa* (Apiaceae), a herbaceous perennial very local in the north-west Midlands, East Anglia, and north and central Ireland, scattered and rare elsewhere, in shallow water on the margins of standing or slowly flowing water, or in deeper water on floating mats of vegetation. It is also found in tall-herb fen and marshy pasture. The toxin cicuotoxin is found in the sap, concentrated in the roots rather than stem and leaf, and cattle do graze on the plant without ill effect, despite its English name. This is the county flower of Co. Armagh as voted by the public in a 2002 survey by Plantlife. The binomial is Latin for 'hemlock' and 'stinking'.

COWBERRY *Vaccinium vitis-idaea* (Ericaceae), a common shrub widespread in British uplands on peaty heathland and moorland, in the understorey of oak, birch and pine woodland on acidic substrates, and on drier hummocks in blanket bogs. The binomial is Latin for 'bilberry' and 'vine of Mt Ida' (Mount of the Goddess in Ancient Greece).

COWHERB *Vaccaria hispanica* (Caryophyllaceae), an annual

European neophyte present in gardens by mid-Tudor times, now a garden escape, and also brought in as a grain contaminant, and in wool and birdseed. With a scattered distribution, it is found in hedges, waste ground and cultivated land, usually as a casual but occasionally naturalised. Latin *vacca* = 'cow' (from growing in cow pasture) and *hispanica* = 'Spanish'.

COWLWORT **Fingered cowlwort** *Colura calyptrifolia* (Lejeuniaceae), a foliose liverwort widespread in western Britain and in Ireland on shaded rocks by streams, and on a number of tree species, especially willow but also spruce and other conifers in forestry plantations. Greek *calyptra* = 'veil', and Latin *folium* = 'leaf'.

COWRIE Gastropods in the genus *Trivia* (Triviidae), from Trivia, a name for Diana in Greek mythology. 'Cowrie' dates from the mid-seventeenth century, from Hindi *kaurie*.

Arctic cowrie *T. arctica*, shell length 10 mm, width 8 mm, widespread except for Yorkshire and East Sussex, mainly in the sublittoral, on rocky coastlines, occasionally on the lower shore, feeding on ascidians.

Spotted cowrie *T. monacha*, shell length 12 mm, width 8 mm, widespread especially along the south and west coasts of Britain, and around Ireland, on the lower intertidal and the sublittoral on rocky coasts, associated with its prey, colonial sea squirts. *Monacha* is Greek for 'solitary'.

COWSLIP *Primula veris* (Primulaceae), a widespread and locally common herbaceous perennial on well-drained, species-rich grassland, increasingly on verges, on light, base-rich soils, less commonly on seasonally flooded soils, in scrub or woodland rides, and on calcareous cliffs. 'Cowslip' comes from Old English *cuslyppe*, from *cu* + *slipa* or *slyppe* = 'cow' + 'slime' (i.e. cow slobber or dung). The binomial is from a diminutive of Latin *primus* and *ver* = 'first' and 'spring'. This is the county flower of Northamptonshire, Surrey and Worcestershire as voted by the public in a 2002 survey by Plantlife.

Japanese cowslip *P. japonica*, a rhizomatous perennial neophyte planted by 1871, an occasional garden escape in wet meadows and damp woodland, and on riverbanks.

Mealy cowslip Alternative name for red cowslip.

Red cowslip *P. pulverulenta*, a herbaceous perennial neophyte from China, scattered in southern England and Scotland by streams and other damp habitats. *Pulverulenta* is Latin for 'powdery'.

Tibetan cowslip *P. florindae*, a rhizomatous perennial neophyte introduced in 1924, naturalised along stream banks, by ponds, and in marsh and damp meadow as a garden escape or relic, scattered in Britain, mostly in northern England and Scotland. The specific name was given by Frank Kingdom Ward, who introduced the species, after Mrs Florinda Annesley.

COYPU *Myocastor coypus* (Myocastoridae), head-body length 40–60 cm, tail length 30–45 cm, weight 4–9 kg, a large, semi-aquatic herbivorous rodent from South America introduced to fur farms in England in 1929. During the 1930s many animals escaped, most were recaptured, but a breeding colony became established in the Norfolk Broads. With two litters a year, each of 2–9 young, numbers rapidly increased to the extent that damage to waterway banks by its 6 m long burrows and consumption of waterside vegetation and crops gave them pest status. By 1960 there were around 200,000 coypu in East Anglia, and an eradication campaign began: culling plus the harsh winter of 1962–3 brought numbers down to 2,000 by 1970, but numbers once again rose. An intensive trapping programme began in 1981, bringing numbers down from perhaps 5,000 to just 40 by 1986. No sightings have been made since 1989. 'Coypu' comes from Spanish *coipú*, in turn from the Chilean Mapuche tribal language *koypu*. The genus name is from Greek *mys* = 'muscle' and Latin *castor* = 'beaver'.

CRAB Member of the order Decapoda, generally covered with a thick exoskeleton, composed primarily of calcium carbonate, and armed with a single pair of chelae (claws). See also hermit, spider and swimming crabs.

Angular crab *Goneplax rhomboides* (Goneplacidae), building branching burrows in mud and sand in the shallow sublittoral down to 100 m on the west coast of Scotland, but scattered southwards as far as Kent, and in north-east Ireland. The English and specific names reflect the angular/rhomboidal shape of the carapace, which is up to 4 cm long. Greek *gonia* = 'corner' + *plax* = 'plate'.

Bristly crab Alternative name for hairy crab.

Broad-clawed porcelain crab *Pisidia platycheles* (Porcellanidae), carapace up to 15 mm long, with large, flattened hairy claws, common and widespread under boulders on the middle and lower intertidal especially among mud and gravel, and sometimes found in the shallow subtidal. Greek *pisos* = 'pea', and *platys* + *chele* = 'wide' + 'claw'.

Brown crab Alternative name for edible crab.

Bryer's nut crab *Ebalia tumefacta* (Leucosiidae), carapace length up to 12 mm, on muddy sand and gravel common around the British coastline at depths of 2–15 m. The species was collected at Weymouth by a Mr Bryer before 1805. The binomial is probably from Greek *ebaios* = 'small' or 'poor', and Latin *tumidus* + *facta* = 'swollen(-backed)' + 'behaviour'.

Chinese mitten crab *Eriocheir sinensis* (Varunidae), carapace length up to 5.6 cm, found in rivers and lakes, burrowing into sediment banks. Adults migrate to estuary mouths to breed and females may overwinter in deep, fully saline water. Probably introduced to the Thames at Chelsea in 1935 from China via The Netherlands, it is now common in the Thames and Medway, with records from the Tyne (Newcastle), Tamar (near Plymouth) and recently Dungeness, Kent, and also from Southfields Reservoir near Castleford, Yorkshire. When population densities are high it can damage soft sediment banks by burrowing, increasing erosion and affecting flood defences. With the dense mat of hair on the claws, the binomial is from Greek *erion* + *cheir* = 'wool' + 'hand', also reflected in its common name, and the Latin for 'Chinese'.

Circular crab *Atelecyclus rotundatus* (Atelecyclidae), up to 5 cm diameter, common and widespread from the shallow sublittoral to >300 m depth on sand or gravel. Greek *ateles* + *cyclos* = 'imperfect' + 'circle', and Latin *rotundatus* = 'rounded'.

Common shore crab Alternative name for green shore crab.

Edible crab *Cancer pagurus* (Cancridae), carapace width up to 25 cm (typically 15 cm), common and widespread on the lower shore, shallow sublittoral and to 100 m depth, feeding on a variety of molluscs and crustaceans as well as carrion. It is an important commercial species in the UK with landings of 28,778 tonnes in 2013, worth £33.5 million. The binomial comprises the Latin and Greek for 'crab'.

Flying crab *Liocarcinus holsatus* (Polybiidae), carapace up to 4 cm long, widespread but patchily distributed in rock pools, in the shallow sublittoral and offshore, on a variety of substrates. Its speed through the water prompts the English name. Greek *leios* + *carcinos* = 'smooth' + 'crab'.

Furrowed crab *Xantho hydrophilus* (Xanthidae), carapace up to 22 mm long, 70 mm wide, living under stones on sandy and stony beaches, occasionally in rock pools but usually below the intertidal zone to a depth of 40 m along the west and south coasts of the British Isles, feeding at night on algae. Greek *xanthos* = 'yellow', and *hydrophilus* means 'water-loving'.

Green shore crab *Carcinus maenas* (Portunidae), with a carapace much broader than long (up to 8 cm across), common and widespread from high water to depths of 60 m, but predominantly a shore and shallow-water species. The binomial is from Greek *carcinos* = 'crab', and named for a Roman priestess.

Hairy crab *Pilumnus hirtellus* (Pilumnidae), carapace broader than it is long (up to 3 cm across), found on various substrates from muddy sand to rocks, under stones and sometimes among seaweed holdfasts, from the lower shore to depths of 80 m on all British coasts though most frequent in the south and west. In Roman mythology, Pilumnus was a deity of naure. *Hirtellus* is Latin for 'hairy'.

Harbour crab 1) *Liocarcinus depurator* (Polybiidae), carapace 51 mm wide and 40 mm long, widespread on the lower shore and sublittoral on fine, muddy sand and gravel. Latin *depuratus* = 'purified'. 2) Alternative name for green shore crab.

Helmet crab Alternative name for, among other species, masked crab.

Lesser hairy crab Alternative name for Risso's crab.

Long-clawed porcelain crab *Pisidia longicornis* (Porcellanidae), carapace width <10 mm, widespread if patchily distributed under boulders in the intertidal, and common in the circalittoral especially in bryozoan turf. Latin *longus* + *cornu* = 'long' + 'horn'.

Masked crab *Corystes cassivelaunus* (Corystidae), up to 4 cm long, 3 cm wide, widespread though less common in north Scotland, in burrows in the sand from the lower shore and shallow sublittoral to about 100 m. When buried, the antennae are brought together, the setae interlocking to form a respiratory tube. Greek *corys* = 'helmet', and Latin *cassis* + *velum* = 'helmet' + 'veil'.

Montagu's crab Alternative name for furrowed crab.

Pennant's nut crab *Ebalia tuberosa* (Leucoiidae), carapace 13–17 mm wide, common and widespread in the sublittoral down to 190 m, rarely intertidal, on muddy gravel and stone. *Tuberosa* is Latin for 'tuberous'.

Risso's crab *X. pilipes*, carapace up to 3 cm long, on the lower shore of sandy and stony beaches to 110 m depth, from around Ireland, the west coast of Britain, Shetland and on the south-west coast, and around the Thames Estuary. Antoine Risso (1777–1845) was a French naturalist. Latin *pilus* + *pes* = 'hair' + 'leg', an alternative name being lesser hairy crab.

Sand crab Alternative name for, among other species, masked crab.

Sponge crab *Dromia personata* (Dromiidae), carapace length up to 5.3 cm, local from north Wales to north Devon, from the lower shore down to 8 m. The last two pairs of legs, positioned dorsally, are used by the young to hold a sponge as camouflage. Greek *dromos* = 'running', and Latin *personata* = 'masked'.

Thumbnail crab *Thia scutellata* (Thiidae), carapace up to 20 mm long, burrowing in sand and mud from low water to 45 m, recorded from Moray Firth, Plymouth, the Channel Isles, the Scilly Isles, Carmarthen Bay, Cardigan Bay, Anglesey, Liverpool Bay, the Isle of Man and Galway Bay. Greek *theio* = 'to run', and the Latin *scutellata* = 'lozenge-shaped'.

Toad crab *Hyas coarctatus* (Oreogoniidae), carapace length up to 61 mm, a spider crab found on both hard and sandy bottoms from the intertidal down to a depth of 50 m, generally widespread but absent from the west coast of Ireland. In Greek mythology Hyas was a daughter of Atlas. *Coarctatus* is Latin for 'confined'.

CRAB SPIDER 1) Members of the family Thomisidae, body lengths 2–10 mm. Adults lie in wait for their prey on the ground, in vegetation or in flowers, often cryptically coloured and some, for example *Misumena vatia*, changing colour over a few days to match the flower colour. Prey is seized with large, often spiny front legs. The common name may reflect the ability to move sideways and/or because of 'crab-like' front legs.

Pale crab spider *Misumena vatia*, common and widespread in the southern half of England and Wales in grassy scrub and woodland edge. *Vatia* is Latin for 'bow-legged'.

2) **Running crab spider** Members of the family Philodromidae, with 15 British species in 3 genera, body lengths 4–7 mm, rapidly moving hunters on the ground or on leaves, sometimes in buildings. See also turf-running spider under spider.

CRABRONIDAE Family name (Latin *crabro* = 'hornet') that includes bee wolf, black wasps, digger wasps, mournful wasp, mimic wasps, Shuckard's wasp and wood borer wasps.

CRAKE Waterbirds in the family Rallidae, including corncrake.

Other species are **little crake** *Zapornia parva* and **Baillon's crake** *Z. pusilla*, both scarce accidentals. 'Crake' is imitative of the call, coming from Old Norse *kráka*, but in Middle English it denoted a crow or raven.

Spotted crake *Porzana porzana*, wingspan 40 cm, length 23 cm, weight 90 g, rare (Amber-listed) and scattered in Britain in shallow and densely vegetated freshwater wetlands, with 80 pairs estimated in 1999, and a five-year average of 33 pairs in 2010. Nests are in reed swamp, clutch size is 10–12, incubation lasts 18–19 days, and fledging of the precocial young takes 43–56 days. Diet is of insects, snails, worms, small fish and plant material. *Porzana* is a Venetian word for 'small rail', 'rail' itself deriving from Old French *raale*, related to *râler* (= 'to rattle'), perhaps imitative of its vocalisation.

CRAMBIDAE Family name of china-mark, grey, pearl, purple & gold, sable and veneer moths. Greek *crambos* = 'dry' or 'parched', from the yellow-brown colour of the forewing of many of the species, perhaps echoing the dry grassland habitat used by many of these moths.

CRAMP BALLS Alternative name for the fungus King Alfred's cakes, from the belief that carrying the fungus cured cramp attacks.

CRANBERRY *Vaccinium oxycoccus* (Ericaceae), a dwarf shrub found in the upland British Isles in bogs and (in south and south-west England) on very wet heaths, usually growing in moss. Latin *vaccinium* = 'bilberry', and Greek *oxy* + *coccus* = 'bitter' + 'berry'.

American cranberry *V. macrocarpon*, a neophyte shrub from North America where it is widely grown as a fruit crop, present in England by 1760, very occasionally found as a naturalised garden escape on acidic, boggy ground. Greek *macrocarpon* = 'large fruit'.

Small cranberry *V. microcarpum*, a dwarf shrub found in central and north Scotland, the Cheviot Hills and Northumberland, in sphagnum moss mires, generally forming colonies in drier microhabitats such as hummock tops. Greek *microcarpum* = 'small fruit'.

CRANE *Grus grus* (Gruidae), wingspan 2.3 m, length 1.2 m, weight 5.6 kg. Birds made a natural return to the UK in the late 1970s with the first fledged chick for 400 years produced in the Norfolk Broads in 1982. Small numbers pass through mostly south and east Britain in spring and autumn, and there are now small breeding population in the Norfolk Broads, the East Anglian Fens, Yorkshire and north-east Scotland. In 2016 there were 48 pairs across the UK, with a total population of 160 birds. Two eggs are usually laid, incubation lasts 30 days, and the precocial chicks fledge in 65–70 days. The Great Crane Project was a collaborative effort involving the WWT, RSPB and the Pensthorpe Conservation Trust: over 2010–15, eggs were imported from Germany, hatched, and 93 birds were hand-reared for release to the Somerset Levels and Moors. The total population of 'known alive' released birds at the end of 2015 plus known surviving juveniles from the year – the 'flock population' – was 78. Diet is mainly of plants, with some insects and worms. 'Crane' derives from Old English *cran*, and *grus* is Latin for this bird. The North American **sandhill crane** *G. canadensis* is a rare accidental.

CRANEFLY Mostly members of the family Tipulidae (Diptera), but also **hairy-eyed craneflies** (Pediciidae), **limoniid craneflies** (Limoniidae), **long-bodied craneflies** (Cylindrotomidae), **phantom craneflies** (Ptychopteridae) and **winter craneflies** (Trichoceridae). There are 94 British species. Adults have long, stilt-like legs and slender wings, and are also called daddy-long-legs. The soil-dwelling larvae (leatherjackets) are often garden and agricultural pests, feeding on roots of grass and crops, some species also feeding on the leaves and crown of crop plants.

European cranefly *Tipula paludosa*, body length 16 cm, leg length 50 mm, very common and widespread, flying in

April–October, particularly in autumn, in grassland, fields, parks and gardens, the soil-dwelling larvae often viewed as pests of grass and field crops. *Tipula* is Latin for a kind of water spider, *paludosa* = 'swampy'.

Marsh cranefly *T. oleracea*, body length 25–30 mm, wing length 18–28 mm, very common and widespread, flying in April–October, particularly in early summer, in damp grassland, parks and gardens, and often found in houses, attracted by lights. The soil-dwelling larvae are often viewed as root-eating pests of turf grass and root crops. *Oleracea* is Latin for 'resembling vegetables'.

Spotted cranefly *Nephrotoma appendiculata*, 13–15 mm body length, common and widespread in hedgerows and woodland from May to September. Adults feed on umbellifers such as cow parsley. The soil-dwelling larvae feed on grass roots. Greek *nephros* + *tomos* = 'kidney' + 'a cut', and Latin *appendiculata* = 'like an appendage'.

Tiger cranefly *N. flavescens*, 18 mm body length, common and widespread in lowland areas, especially in hedgerows, verges and other grassy habitats from April to August. Adults take nectar and pollen from umbelliferous flowers such as hogweed. The soil-dwelling larvae feed on grass roots. *Flavescens* is Latin for 'yellowish'.

CRANE'S-BILL Herbaceous species of *Geranium* (Geraniaceae), from Greek *geranos* = 'crane', this and the English name coming from the appearance of the fruit capsule of some of the species, with their long pointed beak shape. See also herb-robert and little-robin.

Armenian crane's-bill *G. psilostemon*, a perennial Near Eastern neophyte introduced in 1874, with a very scattered distribution in England and Scotland as a naturalised garden escape on grassy habitats, roadsides and waste ground. Greek *psilos* + *stema* = 'smooth' + 'stamen'.

Bloody crane's-bill *G. sanguineum*, a rhizomatous, perennial of base-rich grassland and scrub, rocky woodland, coastal cliffs and stabilised dunes, mainly on the coast but also inland on limestone pavement and cliff ledges, and in chalk and limestone grassland. It has a widespread but patchy distribution, often restricted to localised substrates such as dolerite and serpentine. It also occurs as a garden escape or throw-out on grassy banks, verges and waste ground. This is the county flower of Northumberland as voted by the public in a 2002 survey by Plantlife. *Sanguineum* is Latin for 'blood-red'.

Cut-leaved crane's-bill *G. dissectum*, a common and widespread annual archaeophyte of grassy and stony habitats, hedge banks, waysides and waste ground, and a common weed of flower borders, allotments and arable fields. Its ability to thrive in disturbed, nutrient-enriched habitats makes it probable that it is becoming more abundant in many areas. *Dissectum* is Latin for 'dissected'.

Dove's-foot crane's-bill *G. molle*, a common and widespread annual found in open habitats such as dry grassland, cultivated land, lawns, verges and waste ground. *Mollis* is Latin for 'soft'.

Dusky crane's-bill *G. phaeum*, a widespread if scattered perennial central European neophyte, in gardens since 1724, naturalised on roadsides and railway banks, and in churchyards and woodland edge. It often grows near habitation as a garden escape or throw-out. It favours shaded situations and moist, fertile soil. Greek *phaios* = 'dusky'.

French crane's-bill *G. endressii*, a widespread if scattered rhizomatous perennial Pyrenean neophyte introduced in 1812, now a garden escape on grassy or wooded banks and roadsides around habitation, and as a garden throw-out on waste ground. Philipp Endress (1806–31) was a German plant collector.

Hedgerow crane's-bill *G. pyrenaicum*, a perennial, possibly native of hedgerows, roadsides, field margins, rough grassy banks and waste places, often growing close to habitation, and also found as a garden escape or throw-out, widespread but especially common in England. The specific name is Latin for 'Pyrenean'.

Himalayan crane's-bill *G. himalayense*, a rhizomatous perennial neophyte introduced in the 1880s, found as a garden escape on grassy banks, verges and waste ground, mainly in urban areas in the Thames Valley, in south-west and north Scotland, occasionally elsewhere.

Knotted crane's-bill *G. nodosum*, a rare and rather scattered rhizomatous perennial European neophyte growing in gardens by 1633, found as a garden escape or throw-out in woodland edge, hedgerows, churchyards and in grassland on railway banks and roadsides. *Nodosum* can be translated from Latin as 'knotty'.

Long-stalked crane's-bill *G. columbinum*, an annual of dry grasslands and grassland-scrub mosaics: habitats include dunes, scrubby cliff slopes, hedge banks, field margins, chalk and limestone downland, railway banks and old quarries. It is usually found on calcareous soils, and is often a pioneer on disturbed sites, mainly in the southern half of England, Wales and south-east Ireland. *Columbinum* is Latin for 'dovelike'.

Meadow crane's-bill *G. pratense*, a perennial of damp hay meadows and lightly grazed pastures, mainly on calcareous soils, but increasingly restricted to verges, railway banks and streamsides. It remains widespread in England, Wales and much of Scotland. *Pratense* is from Latin for 'meadow'.

Pencilled crane's-bill *G. versicolor*, a rhizomatous perennial Mediterranean neophyte, in gardens by 1629, scattered but especially frequent in south-west and southern England as a garden escape or throw-out, in grassy roadsides and railway banks, in hedge banks and woodland edge. *Versicolor* is Latin for 'variously coloured'.

Purple crane's-bill *Geranium* x *magnificum* (*G. ibericum* x *G. platypetalum*), a sterile neophyte of garden origin, with a scattered distribution frequently naturalised on roadsides and waste ground.

Rock crane's-bill *G. macrorrhizum*, a rhizomatous perennial south European neophyte, grown in gardens since the sixteenth century, scattered and occasional in Britain as an escape or throw-out in hedge banks, open woodland and on roadsides and walls, naturalised in a few places, for instance south Devon. *Macrorrhizum* is Greek for 'large-rooted'.

Round-leaved crane's-bill *G. rotundifolium*, an annual of hedgerows, dry roadside banks and walls, spreading to verges, railway ballast and waste ground. Widespread in England, Wales, south Scotland and south Ireland, and increasing in range northwards with naturalised populations. *Rotundifolium* is Latin for 'round-leaved'.

Shining crane's-bill *G. lucidum*, a common and widespread annual, found on rock outcrops, favouring calcareous soils and characteristic of limestone districts. It is widespread in artificial habitats such as walls, churchyards, roadsides, waste ground and railway ballast, also as a garden escape, especially in southern England, and is expanding its range in such habitats. *Lucidum* is Latin for 'shining'.

Small-flowered crane's-bill *G. pusillum*, an annual of cultivated land, open summer-droughted grasslands, roadsides and waste places, favouring well-drained, sandy soils, widespread but particularly common in England. *Pusillum* is Latin for 'small'.

Wood crane's-bill *G. sylvaticum*, a rhizomatous perennial, widespread in northern England and Scotland except the far north, scarce, scattered and probably naturalised elsewhere, in hay meadows, hedge banks, verges, ungrazed damp woodlands, streamsides and mountain rock ledges. It has declined locally since the mid-twentieth century, especially on the edges of its range, having been lost from many meadows with increased use of fertilisers and a change to silage production. This is the 'county' flower of Sheffield as voted by the public in a 2002 survey by Plantlife. *Sylvaticum* is from Latin for 'forest'.

CRANGONIDAE A shrimp family (Greek *crangon* = 'shrimp'), the commonest species being brown shrimp.

CRASSULACEAE Family name (Latin *crassus* = 'thick') for stonecrops, together with pygmyweeds, navelwort, house-leek, orpine and pennywort.

CRAYFISH Edible freshwater decapod crustacea in the genera *Orconectes* (Greek *nectos* = 'swimmer') and *Procambarus* (Latin *cambarus* from Greek *cammarus* = 'sea crab') (Cambaridae) and *Astacus* (Greek *astacos* = 'crayfish' or 'lobster'), *Austropotamobius* (Latin *australis* + Greek *potamos* = 'southern' + 'river') and *Pacifastacus* (Latin *pacificus* = 'peaceful') (Astacidae), with one native and six introduced species.

Narrow-clawed crayfish Alternative name for Turkish crayfish.

Noble crayfish *Astacus astacus*, males 16 cm long, females 12 cm, from continental Europe, escaping from fish farms in the 1980s and found in a few sites in Gloucestershire and Somerset.

Red swamp crayfish *Procambarus clarkia*, up to 10 cm long, introduced from the southern USA in the 1980s. The first sightings were confirmed in a roadside marsh drain near Tilbury and the Grand Union Canal in 1990. Populations were found in Hampstead Heath ponds the following year.

Signal crayfish *Pacifastacus leniusculus*, males 16–18 cm long, females usually 12 cm, introduced from North America, with the first record in 1976, to be farmed for food, but escaping through watercourses and across land. It is now well-established in England (especially the south-east) and Wales, and since 1995 in a few places in south and central Scotland, found in flowing and still freshwater habitats, and in brackish waters. Mating takes place in autumn and females carry a dense clutch of 200–400 eggs on the underside of their tail over the winter. Once hatched, the young remain attached to the females until release in May–June. Adults shelter in interconnected riverbank burrows up to 2 m deep in winter and enter a state of torpor. They feed on fish and amphibian eggs, tadpoles, juvenile fish, aquatic invertebrates, detritus and aquatic vegetation. They grow faster, are more fecund, more aggressive and are tolerant of a wider range of conditions than white-clawed crayfish, and they also transmit disease, causing the native species to be driven to widespread local extinction. They 'signal' to other crayfish using a white-turquoise patch on top of the claws at the joint between the two 'fingers', which they open wide. *Leniusculus* is from Latin *lenis* = 'smooth'.

Spiny-cheek crayfish *Orconectes limosus*, up to 12 cm long, introduced from North America, probably in the 1990s, the first established population being found in a catfish pond in Warwickshire in 2001, and a population 5 km downstream in the River Arrow in 2002. There are also records from Nottinghamshire and Leicestershire, and the River Lee catchment, north London. They live in rivers, streams, ponds and lakes, preferring calm and turbid waters, and are tolerant of a range of environmental conditions, including polluted canals and organically enriched water bodies. *Limosus* is Latin for 'muddy'.

Turkish crayfish *A. leptodactylus*, usually 15 cm long, but reaching 30 cm, brought to Britain in the 1970s for sale in fish markets, escaping or deliberately introduced into the wild in and around London, and now with established populations in south-east England and other isolated populations in England and Wales. High mortality caused by crayfish plague has recently been reported in populations in the River Colne near Colchester, the Waveney in Norfolk and some populations around London. They are omnivorous, and found in lakes, rivers, streams and canals, including brackish water in the Waveney estuary. Greek *leptos* + *dactylos* = 'slender' + 'finger'.

Virile crayfish *O. virilise*, up to 10 cm long, introduced from North America, and recorded since 2004 from the River Lee and other river stretches in the London area. The species is parthenogenic: *virilise* is Latin for 'no husband'.

White-clawed crayfish *Austropotamobius pallipes*, up to 12 cm long, found in unpolluted, well-oxygenated and especially alkaline rivers and streams, with fragmented populations in

England and Wales having declined rapidly and by around 95% since the 1970s through habitat loss, pollution, crayfish plague (caused by the fungus *Aphanomyes astaci*) and competition from invasive species, especially the larger, more aggressive signal crayfish, though managing to persist in south Wales, Exmoor, Suffolk, the east Midlands, Dorset, Somerset, Gloucestershire and the North York Moors. With a British population of perhaps 20,000 in 2014, it is listed under the EU Habitat and Species Directive and in Schedule 5 of the WCA (1981), and classified as Endangered in the IUCN Red List. The species fares better in Northern Ireland where plague is absent, and in the Republic where it currently occurs at a low level. It emerges at night to feed on detritus, animal matter and plants. Mating takes place in autumn, eggs developing while attached to the mother's abdomen. After hatch, the juveniles remain attached to the mother before becoming independent at the beginning of summer. *Pallipes* is from Latin *pallidus* + *pes* = 'pale' + 'foot'.

CREEPER Blue-star creeper Alternative name for lawn lobelia.

CREEPHORN *Chondracanthus acicularis* (Gigartinaceae), a red seaweed, frond length up to 10 cm, generally scarce but locally common on sheltered, silty shores in the lower intertidal, south and west shores of England, Wales and Ireland, reaching its northern limit on the mid-west Irish coast. Greek *chondros* + *acantha* = 'granular' + 'thorn', and Latin *acus* = 'needle' (i.e. 'prickly').

CREEPING-JENNY *Lysimachia nummularia* (Primulaceae), a common and widespread evergreen perennial, probably naturalised in the north, found on damp, often clay-rich soil in shaded woodland and hedges, especially stream banks and damp grassland. The genus name is from King Lysimachus, who used the plant to calm his oxen. *Nummularia* is Latin for 'coin-shaped', another name for this plant being moneywort.

CRENELLA Mussels in the genus *Musculus* (Mytilidae), *musculus* being Latin for 'muscle', and so possibly a pun. 'Crenella' may come from the Latin for 'small notch'.

Green crenella *M. discors*, shell size up to 12 mm, widespread in the lower intertidal and shallow sublittoral, attached by the byssus to rocks and algae, sometimes forming rafts on gravel substrates. *Discors* is Latin for 'different'.

Marbled crenella *M. subpictus*, shell size up to 20 mm, widespread, mainly from the intertidal and sublittoral but also deeper, often embedded in the tests of ascidians (tunicates), sometimes attached by the byssus in kelp holdfasts, crevices and on coarse offshore substrates. *Subpictus* is Latin for 'almost painted'.

CRESCENT *Helotropha leucostigma* (Noctuidae, Xyleninae), a nocturnal moth, wingspan 37–44 mm, widespread but local in fen, reedbed, marsh, riverbanks and other damp habitats, of conservation concern under the UK BAP. Adults fly in August–September. Larvae feed on the stems of marshland plants such as yellow flag and great fen-sedge. Greek *helos* + *trophon* = 'marsh' + 'feeding', and *leucos* + *stigma* = 'white' + 'spot', from the forewing mark.

Green-brindled crescent *Allophyes oxyacanthae* (Noctuidae, Psaphidinae), a nocturnal moth, wingspan 35–45 mm, common in broadleaf woodland, scrub, hedgerows and gardens throughout Britain, but with numbers declining by 81% over 1968–2007, a species of conservation concern in the UK BAP. Adults fly in September–November. Larvae feed on hawthorn, blackthorn and other deciduous trees. The binomial is from Greek *allophyes* = 'changeable', from the moth's dimorphism, and the hawthorn food plant *Crataegus oxyacantha*.

Olive crescent *Trisateles emortualis* (Erebidae, Boletobiinae), a rare RDB nocturnal moth, wingspan 29–35 mm, found in a few broadleaf woodlands in southern England, a priority species under the UK BAP, having shown a 50% decline over the last three decades. Some records from the south and east coasts of

England are immigrants. Adults fly in June–July. Larvae feed on withered and fallen leaves of oak and beech. Greek *tris* + *ateles* = 'three times' + 'fruitless' (reflecting problems of taxonomic fit), and Latin *emortualis* = 'relating to death', possibly because the colour resembles dead leaves.

CRESS Herbaceous plants in the family Brassicaceae. See also bitter-, penny-, rock-, swine-, wart-, water-, winter- and yellow-cress.

Garden cress *Lepidium sativum*, a widespread but scattered annual casual of waste ground and ruderal habitats, mainly derived from birdseed and kitchen throw-out, it being the original 'cress' of 'mustard-and-cress' Latin *lepidus* = 'pretty', *sativum* = 'cultivated'.

Garlic cress *Peltaria alliacea*, a perennial neophyte persistent on open ground at the top of a beach at Rubha Phòile, Skye, planted in about 1995, and now producing seed. Greek *pelte* = 'small shield', and Latin *allium* = 'garlic'.

Hoary cress *L. draba*, a perennial rhizomatous European neophyte accidentally introduced to Swansea, Glamorgan, in 1802, and subsequently to other ports. It spread rapidly in the nineteenth century, especially in urban and industrial areas, and has continued to expand in Britain during the twentieth century, though not so in Ireland. It grows on roadsides, limestone and clinker railway ballast, waste ground, arable fields on light soils, dunes and other sandy ground, especially near the sea, and the upper fringe of saltmarsh. Greek *drabe* = 'acrid', from the taste of the leaves.

Mouse-ear cress Alternative name for thale cress.

Shepherd's cress *Teesdalia nudicaulis*, a scattered and local winter-annual of acidic, well-drained sandy soils on heathland, dunes, shingle and gravels, on sandy lake shores in Ireland (where it is rare), by railways and on coal and cinder tips, with a preference for bare or disturbed ground. Teesdale is the Co. Durham, east Pennine valley. *Nudicaulis* is Latin for 'naked stem'.

Thale cress *Arabidopsis thaliana*, a common and widespread winter-annual intolerant of competition, a pioneer species on rocky habitats, dunes and other open sandy or calcareous habitats. It is also very common as a garden weed and on waste ground. Because of its small genome, rapid growth and copious seed production, it is an important plant in research, in 2000 becoming the first plant to have its genome sequenced. *Arabidopsis* is from Greek *arabis* = 'rock-cress' (etymology obscure, but not meaning 'Arabian') + *opsis* = 'appearance'. *Thaliana* may be from Greek *thalos* = 'a shoot' or *thalia* = 'abundance', or after the German physician and botanist Johannes Thal (1542–83).

Tower cress *Pseudoturritis turrita*, a herbaceous biennial or perennial European neophyte first noted as naturalised on walls at St John's College, Cambridge, in 1722, and still present there. It has decreased elsewhere in England, and is now a rare naturalised or casual garden escape. *Turris* is Latin for 'tower'.

Trefoil cress *Cardamine trifolia*, a rhizomatous perennial European neophyte herb found in gardens by 1629, naturalised in woodland, quarries and churchyards, and on shaded roadsides, but very scattered in Britain. *Cardamine* was a species of cress mentioned by the Greek physician Dioscorides (d.90 CE).

Violet cress *Cochlearia acaulis*, an annual introduced from Portugal in 1824, now a weed in gardens and occasionally a garden escape on roadsides and waste ground, where in southern England and the Isle of Man it is sometimes naturalised. Latin *cochlear* = 'spoon', *acaulis* = 'unstemmed'.

CREST (MOTH) Members of the family Gelechiidae, nocturnal, some species having raised tufts (crests) on the forewing.

Rarities include **clay crest** *Helcystogramma lutatella* (grassy slopes on the Dorset coast between Portland and Kimmeridge), **Scotch crest** *Dichomeris juniperella* (central Highlands), and **Worcester crest** *D. ustalella* (Shrawley Wood, Worcestershire,

recorded as common in 1987, and in Monmouthshire in 1999 and 2014).

Fen crest *Brachmia inornatella* (Dichomeridinae), wingspan 11–15 mm, local in fen, marsh and riverbanks in East Anglia and south-east England, sometimes in high numbers. Adults fly in June–July. Larvae feed in stems of common reed. Nikolaus Brahm (1754–1821) was a German entomologist. *Inornatus* is Latin for 'plain'.

Gorse crest *B. blandella*, wingspan 9–12 mm, widespread and common in grassland and woodland over much of south and central England, east Wales and the Channel Islands, more local over the rest of Wales and northern England. The species seems to have expanded northward during the late 1990s and the first few years of the twenty-first century, since when this range expansion has faltered in north-west England with the moth now very local. Adults fly in early evening in July–August. Larvae feed on gorse, overwintering and reaching the final instar by June. Latin *blandus* = 'pleasant'.

Humped crest *Psoricoptera gibbosella* (Gelechiinae), wingspan 15–17 mm, local in mature oak woodland throughout much of southern England and south Wales, with records north to Cheshire. Adults fly in July–October. Larvae feed in a rolled oak leaf in spring. Greek *psoricos* + *pteron* = 'mangy' + 'wing', and Latin *gibbosus* = 'humped', both from the tufts of raised forewing scales.

Marjoram crest *Acompsia schmidtiellus* (Dichomeridinae), wingspan 15–17 mm, local on chalk and limestone downland and roadside verges in southern England, south Wales and Norfolk. Adults fly in July–August. Larvae feed in May–June on wild marjoram inside a rolled leaf. Greek *acompsos* = 'unadorned'. A. Schmidt (d.1899) was a German entomologist.

Orange crest *Helcystogramma rufescens* (Dichomeridinae), wingspan 14–18 mm, widespread and locally common in rough pasture, unimproved grassland, downland and woodland rides through much of England, lowland Wales, Ireland and the Channel Islands, local in north England, central Wales and Northern Ireland with just a few records from south-west and west Scotland. Adults fly in June–August. Larvae overwinter on grasses. *Rufescens* is Latin for 'reddish'.

Small crest *Anarsia spartiella* (Anacampsinae), wingspan 12–15 mm, widespread but generally local in heathland, downland and waste ground throughout most of England, Wales, Ireland (where it is mainly coastal) and the Channel Islands. Very local in north England and Scotland as far north as Inverness. Adults fly in June–July. Larvae feed in May–June on shoots of gorse, broom and dyer's greenweed, living inside a web. Greek *anarsios* = 'incongruous' (having a different labial palpus from all other gelechiids), and *Spartium* (Spanish broom) is a food plant.

Square-spot crest *Hypatima rhomboidella* (Anacampsinae), wingspan 15–19 mm, widespread and common in woodland, heathland and. Adults fly from late July to September. Larvae feed on birch and hazel living within a rolled leaf. Greek *hypatos* + *timeo* = 'best' + 'to honour', and *rhomboides* from the rhomboid-shaped brownish blotch on the forewing.

CRESTWORT Foliose liverworts in the genus *Lophocolea* (Lophocoleaceae) (Greek *lophos* + *coleos* = 'crest' + 'sheath').

Bifid crestwort *L. bidentata*, probably the commonest foliose liverwort in the British Isles, in a range of habitats on the ground in woodland, grassland and heathland, and on other plants and rocks. *Bidentata* is Latin for 'two-toothed' (bifid, or divided by a deep cleft).

Fragrant crestwort *L. fragrans*, common and widespread in western Britain and in Ireland on hard, shaded rocks in humid sites in oceanic woodland, especially by streams and waterfalls.

Great crestwort *L. bispinosa*, uncommon though probably spreading, probably introduced from New Zealand, well established on the Scilly Isles by the early 1970s, and recorded from a forestry track in south Wales in 2015. *Bispinosa* is Latin for 'two-spined'.

Southern crestwort *L. semiteres*, introduced from Australasia, first recorded in the Scilly Isles in 1955, with an increasing but still scattered distribution in England and south Scotland on earth and sandy banks in woodland and heathland. *Semiteres* is Latin for 'semi-smooth'.

Variable-leaved crestwort *L. heterophylla*, common and widespread, though scarce in north Scotland and scattered in Ireland, especially growing on the base of trees, but also in a range of other habitats, including decomposing logs and woodland litter. Further north and west, it becomes increasingly characteristic of ancient woodland. *Heterophylla* is Greek for 'differently leaved'.

CRIBELLATE ORB SPIDER Member of the family Uloboridae, possessing a cribellum (silk-spinning organ); see also triangle spider under spider.

CRICETIDAE Family name (Latin *cricetus* = 'hamster') for voles (muskrat, lemmings, etc.), sometimes retained as part of the Muridae.

CRICKET Long-horned 'grasshopper', in the order Orthoptera, suborder Ensifera, family Gryllidae (Latin *gryllus* = 'cricket'), together with mole cricket (Gryllotalpidae), with large hind legs and long antennae. See also water cricket (Veliidae). 'Cricket' is Middle English, derived from Old French *criquet*, from *criquer* = 'to crackle', but possibly of the same origin as the sport, the insect stridulation imitative of the sound made by a ball striking wood. Water crickets are species of bug.

Field cricket *Gryllus campestris*, male body length 19–23 mm, wing length 12–14 mm, female 17–22 mm and 10–14 mm, associated with south-facing short turf on sandy or chalk soils with patches of bare ground currently only at two sites in West Sussex. Endangered in the RDB. The binomial is Latin for 'cricket' and 'of the field'.

House cricket *Acheta domesticus*, body length 14–20 mm, wing length 9–12 mm, an introduction probably present since the Crusades, with few outdoor records and mostly confined to heated urban sites including domestic houses. The binomial is Latin for 'cicada' and 'household'.

Mole cricket *Gryllotalpa gryllotalpa*, male body length 35–41 mm, forewing length 14–17 mm, female 40–6 mm and 16–18 mm, a large rare omnivorous cricket that burrows into deep, loose damp soil at the edges of wetlands and seepages. Once commoner and more widespread it is now Vulnerable in the RDB, more or less limited to sites in the north New Forest, the Itchen Valley, East Sussex and Guernsey. Latin *gryllus* + *talpa* = 'cricket' + 'mole'.

Scaly cricket *Pseudomogoplistes vicentae*, male body length 8–11 mm, female 10–13 mm, wingless, restricted to the eastern half of Chesil Beach in Dorset, and perhaps also Branscombe in Devon, and Marloes in Wales. Endangered in the RDB. Greek *pseudo* + *mogos* = 'false' + 'trouble'.

Wood cricket *Nemobius sylvestris*, male body length 7–9 mm, wing length 3–4 mm, female 7–11 mm and 2–3 mm, associated with deep leaf litter in warm clearings in woodland edges and scrub in parts of the New Forest, Isle of Wight and south Devon. The binomial is from the Latin for 'woodland grove' and 'of woodland'.

CRIMSON & GOLD Scarce crimson & gold *Pyrausta sanguinalis* (Crambidae, Pyraustinae), a small scarce moth with a declining range (probably no longer found in England or Scotland), but local in dunes and on limestone pavements in parts of Ireland (mainly the Burren, but twenty-first-century records also from Co. Wexford and Co. Roscommon), north Wales and the Isle of Man, predominantly coastal, a priority species in the UK BAP. The bivoltine adults fly in sunshine and at night in June and August. Larvae feed on wild thyme. Greek *pyraustes* = 'a moth that gets singed in candle flame', from the black markings, and Latin *sanguinalis* = 'blood-red'.

CRINOIDEA Class that includes feather-stars and sea-lilies (Greek *crinon* + *eidos* = 'lily' + 'form').

CRISP-MOSS Widespread and generally common species in the genera *Pleurochaete* (Greek *pleura* + *chaete* = 'side' + 'mane'), *Tortella* (Latin *tortilis* = 'twisting') and *Trichostomum* (Greek *trichos* + *stoma* = 'hair' + 'mouth', alluding to the hair-like peristome teeth at the mouth of the capsule), family Pottiaceae.

Curly crisp-moss *Trichostomum crispulum*, on lime-rich substrates, often by streams. *Crispus* is Latin for 'curly'.

Frizzled crisp-moss *Tortella tortuosa*, commoner in north and west Britain and in Ireland, favouring calcareous dune sand and grassland. *Tortuosa* is Latin for 'twisted'.

Neat crisp-moss *Tortella nitida*, with a mainly western distribution on rocks and walls. *Nitida* is Latin for 'shiny' or 'neat'.

Side-fruited crisp-moss *P. squarrosa*, scattered in England and Wales on sandy or calcareous habitats, including dunes, and limestone grassland. *Squarrosa* is Latin here meaning 'overlapping leaves'.

Variable crisp-moss *Trichostomum brachydontium*, favouring damp upland habitats. Greek *brachys* + *donti* = 'short' + 'tooth'.

Yellow crisp-moss *Tortella flavovirens*, on lime-rich dune grassland. *Flavovirens* is Latin for 'yellow-green'.

CROCUS Cormous perennials in the genera *Crocus* (Latin for 'saffron yellow') and *Romulea* (named for Romulus, founder of Rome) (Iridaceae).

Eastern Mediterranean neophytes, scattered but often naturalised in grassy habitats, include **Bieberstein's crocus** *C. speciosus* (introduced in 1847), **early crocus** *C. tommasinianus* (1847), **golden crocus** *C. chrysanthus* (1874), and **Kotschy's crocus** *C. kotschyanus* (1896). A few other introduced species are occasionally recorded as garden escapes.

Autumn crocus 1) *C. nudiflorus*, from south-west Europe, cultivated in Tudor times, scattered in Britain, naturalised in meadows, pastures, amenity grasslands and on roadsides, spreading by rhizomes. This is the county flower of Nottinghamshire as voted by the public in a 2002 survey by Plantlife. *Nudiflorus* is Latin for 'naked flower' (the flower emerging before the leaf). 2) An alternative name for meadow saffron.

Sand crocus *Romulea columnae*, in short, open turf on freely draining sandy ground and coastal cliff slopes in the Channel Islands, as well as near Dawlish, south Devon, and in east Cornwall. *Columna* is Latin for 'column'.

Spring crocus *C. vernus*, the most commonly grown species, widespread in much of Britain and in south-east Ireland, naturalised in a grassy habitats, including churchyards and amenity grasslands, and on verges. *Vernus* is Latin for both 'vernal' and 'crocus'.

Yellow crocus *Crocus* x *stellaris*, of garden origin, scattered and naturalised in Britain in churchyards, roadsides, parks and other amenity grassland. *Stellaris* is Latin for 'star-like'.

CROSSBILL or COMMON CROSSBILL *Loxia curvirostra* (Fringillidae), wingspan 29 cm, length 16 cm, weight 43 g, breeding areas of residential birds in coniferous woodland including the Highlands, the north Norfolk coast, Breckland, the New Forest and the Forest of Dean, but as an irruptive species in some years there may also be migrant breeders over a wider area. There were an estimated 40,000 pairs (31,000–53,000 pairs) in 2009, with a 15% decline in numbers between 1995 and 2010, including a 46% decline in 2010–15. Clutch size is 3–4, incubation lasts 14–15 days, and fledging of the altricial young takes 20–5 days. Their mandibles, crossed at the tips, enable them to extract seeds from conifer cones and other fruits, and gives these birds both their English name and their binomial, from Greek *loxos* = 'crosswise' or 'crooked', and Latin *curvus* + *rostrum* = 'curved' + 'bill'.

Parrot crossbill *L. pytyopsittacus*, wingspan 32 cm, length 18 cm, weight 55 g, an Amber-listed resident breeder in Abernethy Forest and elsewhere in conifer forest or plantations in the

Highlands, and an occasional irruptive visitor when food sources on the Continent are poor, its main food being conifer seeds. Some 50–65 breeding pairs were estimated in 2008. Clutch size is 3–4, incubation lasts 14–16 days, and fledging of the altricial young takes 21–3 days. Greek *pituos* + *psittacus* = 'pine tree' + 'parrot'.

Scottish crossbill *L. scotica*, Britain's only endemic bird species, being viewed as having species status in 1978 (confirmed in 2006), until when it had been considered as a subspecies of *L. curvirostra*. It is confined to the Caledonian Scots pine forest and commercial pinewoods in the Highlands, with their diet of pine seeds extracted from the cone using their specialised bills where top and bottom overlap sideways. There are an estimated 6,800 pairs (4,100–11,400 pairs) in Britain in 2008, and it is Amber-listed. Clutch size is 3–4, incubation lasts 12–14 days, and fledging of the altricial young takes 20–2 days. *Scotica* is Latin for 'Scottish'.

CROSSWORT *Cruciata laevipes* (Rubiaceae), a herbaceous perennial, common and widespread in Britain as far north as central Scotland, scarce as an introduction in Ireland, on deep, well-drained neutral or calcareous soils, usually in ungrazed grassland, open scrub, hedge banks and woodland rides and edges, and on waysides. 'Crosswort' comes from the four-leaved whorls around the stem, plus wort, an old name for a medicinal or culinary herb. Latin *cruciata* = 'crossed', and *laevis* + *pes* = 'smooth' + 'foot'.

Caucasian crosswort *Phuopsis stylosa*, a procumbent annual Near Eastern neophyte, a garden escape scattered in England and Wales, occasionally naturalised on waste ground. Greek *phou* + *opsis* = 'a kind of valerian' + 'appearance', and Latin *stylosa* = 'having styles'.

CROTTLE *Parmelia saxatilis* (Parmeliaceae), a common and widespread foliose lichen found on usually base-poor (siliceous) rocks, and on acid-barked trees and other trees. It has been used in Scotland to make a golden- or reddish-brown dye for staining wool for making tweed, and was once considered a cure for the plague, and believed to be an effective treatment for epilepsy if found growing on an old skull, especially that of an executed criminal (giving the alternative name skull lichen). The name dates from the mid-eighteenth century, coming from Scottish Gaelic and Irish *crotal* and *crotan*. Latin *parma* = 'small shield' and *saxatilis* = 'growing among rocks'. Some species of *Ochrolechia*, including crab's-eye lichen are also known as crottle.

Black crottle *P. omphalodes*, foliose, locally common and widespread north and west of a line from the Severn estuary to the Tyne, found on base-poor rocks. Greek *omphalos* = 'navel'.

Dark crottle *Hypogymnia physodes*, foliose, widespread and very common on siliceous rocks, trees, heather stems and other acidic substrates. Greek *hypo* + *gymnos* = 'beneath' + 'naked', and *physis* + *eidos* = 'growth' + '(looking) like'. The English name has also been used for black crottle.

Powdered crottle Alternative name for netted shield lichen.

Stone crottle Alternative name for common greenshield lichen.

CROW Members of the genus *Corvus* (Corvidae), from Old English *crawe*, imitative of the bird's call. *Corvus* is Latin for 'crow' and 'raven'.

Carrion crow *C. corone*, wingspan 98 cm, length 46 cm, weight 510 g, common and widespread in a range of habitats in Britain but only occasional in Ireland. There were an estimated million territories in Britain in 2009, numbers having risen steadily since the 1960s, including by 98% between 1970 and 2015, 19% in 1995–2015, and 3% in 2010–15. Despite strong increases in south-east England there have been declines in south-west England, upland England and Wales. Nests are of twigs in trees or ledges, with clutches of 3–4 eggs, incubation lasting 18–20 days, the altricial chicks fledging in 29–30 days. The opportunistic omnivorous diet includes worms, insects, cereal grain, eggs, carrion and scraps. *Corone* comes from Greek for 'crow'.

Hooded crow *C. cornix*, wingspan 98 cm, length 46 cm, weight 510 g, common and widespread in Scotland and Ireland, geographically 'replacing' carrion crow, until 2002 thought to be a race of the latter but now given full species status. There were an estimated 260,000 territories in the UK in 2009, with a 19% increase in numbers between 1995 and 2014. Nests are of twigs in trees or ledges, with clutches of 3–6 eggs, incubation lasting 18–19 days, the altricial chicks fledging in 28–30 days. The opportunistic omnivorous diet includes worms, insects, cereal grain, eggs, carrion and scraps. *Cornix* is another Latin word for 'crow'.

CROWBERRY *Empetrum nigrum* (Ericaceae), a low-growing evergreen shrub common in upland Britain and Ireland on well-drained acidic soils on moorland and mountains, and on blanket bog where it can increase greatly after burning or where dry surfaces have been exposed by erosion. Greek *en* + *petros* = 'among' + 'rock', and Latin *nigrum* = 'black'.

CROWFOOT Annual or short-lived perennial species of *Ranunculus* (Ranunculaceae), the divided leaves suggesting a crow's foot. *Ranunculus* is the diminutive of Latin *rana* = 'frog', referencing the plants growing in marsh and bog. See also water-crowfoot.

Ivy-leaved crowfoot *R. hederaceus*, generally widespread but having declined in arable regions since the 1950s. It grows at the edge of small water bodies and river backwaters, and on cattle-poached edges of ponds, ditches and streams, in gateways, and on paths and tracks. It tolerates a broad range of pH and nutrient levels. *Hederaceus* is Latin for 'relating to ivy'.

Round-leaved crowfoot *R. omiophyllus*, common in south-west and north-west England, Wales, south-west Scotland and southern Ireland, in shallow water or on wet soil, for instance the margins of ponds and ditches, seepages, poached gateways and tracks in pastures and on heathland, and river backwaters, confined to acidic, mesotrophic or oligotrophic soils. *Omiophyllus* may mean 'same leaves' in Greek.

Three-lobed crowfoot *R. tripartitus*, often subaquatic, found in south-west and southern England and the Welsh coast in shallow water bodies over base- and nutrient-poor substrates in open sites which are winter-flooded but summer-dry, for example ditches and shallow ponds. It showed a decline in status and distribution in the twentieth century, but a number of new sites have recently been noted, for example in 2000 it was refound in Ireland. *Tripartitus* is Latin for 'having three parts'.

CROWN (FUNGUS) Rowan crown *Gymnosporangium cornutum* (Pucciniales, Pucciniaceae), a rust fungus with a distribution centring on north England and the Highlands where its host plant, rowan, is present. Greek *gymnos* + *spora* = 'naked' + 'seed', and Latin *cornutum* = 'horned'.

CROWNCUP Violet crowncup *Sarcosphaera coronaria* (Pezizales, Pezizaceae), a scarce fungus, scattered in England, poisonous when uncooked, found in late spring and summer on calcareous soils in broadleaf woodland, especially under beech. Greek *sarca* + *sphaera* = 'flesh' + 'sphere', and Latin *corona* = 'crown'.

CRUCIFERAE Former name for the family Brassicaceae (Latin *crux* = 'cross', *crucifer* = 'cross-bearer', reflecting the four-petalled flowers).

CRUET-MOSS Members of the genus *Tetraplodon* (Splachnaceae), found on dung, from the Greek *tetra* + *odonti* = 'four' + 'teeth', from the arrangement of the peristome teeth (projections round the capsule mouth). **Narrow cruet-moss** *T. angustatus* is limited to the Highlands. **Slender cruet-moss** *T. mnioides* is fairly widespread in upland Britain and Ireland.

CRUMBLECAP Fungi in the family Psathyrellaceae (Greek *psathyros* = 'friable') which tend to disintegrate when touched. **Common crumblecap** is an old name for pale brittlestem. **Trooping crumblecap** is an alternative name for fairy inkcap.

CRUST (FUNGUS) Inedible basidiomycetes characteristically encrusting or as small brackets with smooth fruit bodies, saprobic on the undersides of dead tree-trunks or branches.

Bleeding broadleaf crust *Stereum rugosum* (especially found on hazel and oak), **bleeding conifer crust** *S. sanguinolentum* (conifers) and **beeding oak crust** *S. gausapatum* (on oak), colour of these three species changing to blood-red when cut, **hairy curtain crust** *S. hirsutum* (oak and beech), **oak crescent crust** *Hymenochaete rubiginosa* (barkless oak), **waxy crust** *Vuilleminia comedens* (rolled-back hardwood bark), **wrinkled crust** *Phlebia radiata* (gelatinous, especially found on beech), and **yellowing curtain crust** *S. subtomentosum* (especially beech) are all common and widespread. See also netcrust, porecrust, tarcrust and toothcrust.

Blushing crust *Eichleriella deglubens* (Auriculariales, Auriculariaceae), widespread but scattered and infrequent, on dead branches of a variety of hardwoods, especially ash. August Eichler (1839–87) was a German botanist. Latin *deglubens* = 'peeling away'.

Cobalt crust *Terana caerulea* (Polyporales, Phanerochaetaceae), uncommon and scattered in England and Wales, rare in Ireland, found year-round but sporulating in autumn, on dead wood of hardwoods, especially ash. The meaning of *Terana* is unclear. Latin *caerulea* = 'dark blue', from the colour of the fruit body.

Glue crust *H. corrugata* (Hymenochaetales, Hymenochaetaceae), widespread and locally common, evident year-round, sporulating in summer and autumn in deciduous woodland on dead wood, favouring hazel and, to an extent, willows. It may be parasitised by hazel gloves (another fungus). Its common name comes from its ability to migrate between trees by gluing together twigs or small branches that are in contact. Greek *hymen* + *chaete* = 'membrane' + 'long hair', referring to the fine hairs (settae) on the upper surface, and Latin *corrugata* = 'wrinkled', again reflecting the nature of the surface.

Netted crust *Byssomerulius corium* (Polyporales, Phanerochaetaceae), widespread but infrequent, found from late summer to spring on dead branches of broadleaf trees. Greek *byssos* = 'fine thread' plus reference to the closely related genus, *Merulius*. *Corium* is Latin for 'leather', referring to the skin-like form taken by the fruit bodies.

Rosy crust *Peniophora incarnata* (Russulales, Peniophoraceae), widespread, on dead wood of gorse and broom, as well as some hardwood trees. *Peniophora* is Greek for 'a few moments', *incarnata* Latin for 'in the flesh'.

Toothed crust *Basidioradulum radula* (Hymenochaetales, Schizoporaceae), wood-rotting, forming irregular patches, sometimes covering the surface of a fallen branch of a hardwood tree, widespread but only occasionally found, fruiting from late summer to early winter. *Radula* is Latin for 'scraper' (this fungus having tooth-like structures).

CRUSTACEA(N) From Latin *crusta* = 'shell', a subphylum of the phylum Arthropoda, encompassing hard-shelled, usually aquatic animal with several pairs of legs, including crabs, lobsters, shrimps, prawns and sandhoppers, but also sessile animals such as barnacles.

CRYPHAEA Mosses in the genus *Cryphaea* (Cryphaeaceae) (Greek *cryphaios* = 'hidden').

Lateral cryphaea *C. heteromalla*, common and widespread except for upland Scotland, on the bark of trees and shrubs in woodland, scrub and orchards, more rarely on rocks, stonework and concrete. Greek *heteros* + *mallos* = 'different' + 'lock of wool'.

Multi-fruited cryphaea *C. lamyana*, on rocks and trees in the frequently flooded banks of large rivers in south-west England and south-west Wales.

CRYPTOPHAGIDAE Family name (Greek *cryptos* + *phago* = 'hidden' + 'to eat') of silken fungus beetles (1–11 mm length), with 103 British species in 11 genera, largely associated with fungi. *Cryptophagus micaceus* and several *Antherophagus* species are associated with the nests of tree-dwelling Hymenoptera

and bumblebees, respectively. Several *Cryptophagus* species are found in stored products. *Ootypus globosus* is mainly found in the dung of large herbivores.

CRYPTOSTIGMATA Former name of the order Oribatida (Greek *cryptos* + *stigma* = 'hidden' + 'mark').

CRYSTALWORT Thallose liverworts in the genus *Riccia* (Ricciaceae) (Marchantiales), named after an eighteenth-century Italian nobleman, P.F. Ricci.

Rarities include **black crystalwort** *R. nigrella* (Cornwall and Wales), **blue crystalwort** *R. crystallina* (Scilly Isles and Cornwall), **channelled crystalwort** *R. canaliculata*, **ciliate crystalwort** *R. crozalsii* (Scilly Isles, Cornwall, south Devon, parts of Wales), **Lizard crystalwort** *R. bifurca* (Lizard Peninsula, Cornwall), **pond crystalwort** *R. rhenana* (England and Ireland, in ponds and streams) and **Violet crystalwort** *R. huebeneriana*.

Cavernous crystalwort *R. cavernosa*, widespread and scattered on mud by reservoirs, lakes and ponds, sometimes in damp hollows in arable fields, gravel-pits and dunes.

Common crystalwort *R. sorocarpa*, common and widespread in arable fields, gardens, waste ground, footpaths, clifftops, rock ledges, and mud by lakes. Greek *soros* + *carpos* = 'vessel' + 'fruit'.

Floating crystalwort *R. fluitans*, popular with aquarium keepers and found as a throw-out, widespread and locally common in England, Wales and Ireland in stagnant or slow-moving water in ponds, pits, canals and ditches, often with duckweed. Terrestrial forms grow on mud, humus or stonework by water. *Fluitans* is Latin for 'floating'.

Glaucous crystalwort *R. glauca*, common and widespread on disturbed soil in coastal clifftops, mud by lakes, gardens, fields of vegetables and bulbs, and woodland tracks.

Least crystalwort *R. subbifurca*, scattered and widespread, in arable fields, gardens, rocky banks, footpaths and pond edges. *Subbifurca* is Latin for 'almost split in two'.

Purple crystalwort *R. beyrichiana*, scattered in mostly upland regions on acidic soils in sites subject to seasonal inundation. It also occurs on leached limestone soils in south Wales. Heinrich Beyrich (1796–1834) was a German botanist.

CUCKOO *Cuculus canorus* (Cuculidae), wingspan 58 cm, length 33 cm, male weight 130 g, females 110 g, a Red List summer visitor, wintering in Africa, arriving in late March–April, leaving in July–August, young birds doing so about a month later. The species is widespread (especially in southern and central England) but increasingly uncommon: 16,000 pairs (9,000–24,000 pairs) were estimated in 2009. There was a 57% decline in numbers between 1970 and 2015, and 43% between 1995 and 2015, though these figures incorporate a 15% increase in 2010–15. Declines have been most pronounced in Northern Ireland and south-east England. A brood parasite, the female lays her eggs in the nests of other birds, especially meadow pipits, dunnocks and reed warblers. A single female may visit 25–50 nests each year, selecting the species which had raised her. She visits the host's nest, pushes one egg out, lays her egg (which mimics the host's) and flies off. The single cuckoo egg in each nest hatches in 11–13 days. On hatching the cuckoo chick rolls out any of the host's chicks or unhatched eggs. The altricial cuckoo chick fledges in 17–21 days. Adults feed on insects, especially hairy caterpillars avoided by other birds, and beetles. The English and genus names are onomatopoeic representations of the male's call, *cuculus* also being Latin for 'cuckoo' (and *coccyx* Greek); *canorus* = 'melodious'. **Great spotted cuckoo** *Clamator glandarius*, **black-billed cuckoo** *Coccyzus erythropthalmus* and **yellow-billed cuckoo** *Coccyzus americanus* are all scarce accidentals.

CUCKOO BEE Some socially parasitic (nest-usurping) bumblebees in the genera *Bombus* (Latin for 'buzz'), *Nomada* (Greek *nomas* = 'nomad' or 'wanderer') and *Sphecodes* (Greek *sphex* + *odes* = 'wasp' + 'resembling'), all in the family Apidae. See also nomad bees, and common mourning bee under bee.

Armoured nomad bee See scabious cuckoo bee below.

Barbut's cuckoo bee *B. barbutellus*, male body length 15 mm, female 18 mm, widespread but uncommon in England and south Wales, occasional elsewhere. During spring the overwintered, fertilised female searches for a small nest of the host, small garden bumblebee (sometimes large garden bumblebee). She enters the nest to hide while acquiring the nest scent. She then dominates or kills the host queen and takes over egg-laying for the colony. These eggs produce only new *B. barbutellus* females and males, which take no part in the running of the colony, all foraging and nest duties being done by the host workers. James Barbut (1776–91 or 1799) was an English painter and naturalist.

Field cuckoo bee *B. campestris*, body length 10–20 mm (average male 15 mm, female 18 mm), common and generally widespread, but absent from north Scotland and Ireland. During spring the overwintered, fertilised female searches for a small nest of the host, common carder bee (sometimes brown-banded carder bee). She enters the nest then kills the host queen and takes over egg-laying for the colony. These eggs produce only new *B. campestris*, which take no part in the running of the colony, all work being done by the host workers. *Campestris* is from Latin *campus* = 'field'.

Forest cuckoo bumblebee See bumblebee.

Gypsy cuckoo bee *B. bohemicus* (Apidae), male body length 11–17 mm, female 15–20 mm, widespread though commoner in the north, and possibly declining in range in the south as a consequence of climate change. After emerging from hibernation during April the mated female seeks out a small *B. lucorum* nest with a few workers. She enters the nest to hides while acquiring the nest scent. She then dominates or kills the host queen and takes over egg-laying for the colony. These eggs produce only new *B. bohemicus* females and males, which take no part in the running of the colony, all work being done by the host workers. *Bohemicus* is Latin for 'Bohemian'.

Hill cuckoo bee *B. rupestris*, male body length no more than 16 mm, female 20–5 mm, widespread in unimproved grassland and gardens, particularly in southern England; it declined considerably during and after the 1940s but has shown a remarkable recovery in the twenty-first century. The host, red-tailed bumblebee, is itself widespread and is increasing its range northward. Adults take nectar mainly from Apiaceae, Lamiaceae and Asteraceae. In early summer, the female enters a nest of the host species and kills the resident queen. The cuckoo bee then establishes herself as the queen and lays eggs that will be reared by the host workers. Once egg-laying is completed, the cuckoo bee queen dies in the nest. *Rupestris* is from Latin for 'rock'.

Scabious cuckoo bee *N. armata*, very rare and declining, with the most recent records from near Oxford (1974) and a few sites on Salisbury Plain, Wiltshire (since 1991), listed as Endangered in RDB 1, associated with chalk grassland, and kleptoparasitic on *Andrena hattorfiana*. It is often found on flowers of small scabious, from which it obtains nectar. Its English name was the winning suggestion when, in 2010, *The Guardian* launched a competition for the public to suggest common names for previously unnamed species, though 'armoured nomad bee' had been in use, reflecting Latin *armata* = 'armed'.

Vestal cuckoo bee *B. vestalis*, male body length no more than 16 mm, female up to 21 mm, wingspan 37 mm, widespread in England and Wales, taking nectar from a variety of flowers in a range of habitats. After emerging from hibernation during April the mated female enters a buff-tailed bumblebee nest and hides while she acquires the nest scent. She then kills the host queen and lays her own eggs. Her offspring take no part in the running of the colony, all foraging and nest duties being undertaken by the host workers.

CUCKOO-PINT Alternative name for lords-and-ladies, possibly derived from cuckoo's pintle or pintel, meaning cuckoo's penis.

CUCKOO WASP 1) *Vespula austriaca* (Vespidae), male forewing length 10–11.5 mm, female 12–13 mm, an obligate social parasite of red wasp, fairly widespread in open habitats where nests of its host are found. The queen enters the host colony after the first red wasp workers have emerged, and kills the host queen or drives her away. The cuckoo wasp queen lays her eggs in the red wasp cells and they are then reared by the host workers. The binomial is the Latin diminutive of *vespa* = 'wasp', and 'Austrian'.

2) Solitary members of the genus *Chrysis* (Chrysididae) (Greek *chrysos* = 'gold'), with 14 species on the British list. They are socially parasitic (kleptoparasites), laying their eggs in the nests of unrelated host species, where their larvae consume the host egg or larva while it is still young, then food provided by the host. See ruby-tailed wasp under wasp (2).

CUCKOOFLOWER *Cardamine pratensis* (Brassicaceae), a common and widespread herbaceous perennial of wet grassy habitats on moderately fertile, seasonally waterlogged soils in woodland, wet meadows, fens and flushes. In upland areas it also grows in rush pasture. It is sometimes found in gardens and lawns. This is the county flower of Cheshire and Brecknockshire as voted by the public in a 2002 survey by Plantlife. The binomial is from Greek *cardamon*, for an Indian spice, and Latin for 'meadow'.

Greater cuckooflower *C. raphanifolia*, a rhizomatous herbaceous perennial south European neophyte, introduced in 1710, naturalised in damp, shaded habitats by rivers and lakes, very scattered but probably increasing in range in Britain. *Raphanifolia* is Latin for 'radish-like leaves'.

CUCUJIDAE Family name of some flat bark beetles (see also Silvanidae). *Cucujus* is derived from the Latin for a kind of beetle.

CUCUMBER (ECHINODERM) See sea cucumber.

CUCUMBER (PLANT) See Cucurbitaceae.

CUCURBITACEAE Name of the white bryony family (Latin *cucurbitus* = 'pumpkin') that includes rare neophyte casuals on tips and at sewerage works: **water melon** *Citrullus lanatus*, **melon** *Cucumis melo*, **cucumber** *C. sativus*, **squirting cucumber** *Ecballium elaterium*, **marrow** *Cucurbita pepo* and **pumpkin** *C. maxima*.

CUDWEED Herbaceous species of *Filago* (Latin *filum* = 'thread', from the cotton-like flower heads) and *Gnaphalium* (Greek *gnaphalion*, a plant whose soft white leaves were used as cushion stuffing) (Asteraceae). 'Cudweed', dating from the mid-sixteenth century, was given to cattle that had lost their cud.

Broad-leaved cudweed *F. pyramidata*, an archaeophyte annual of well-drained soils usually kept open through drought or disturbance. Historically an arable weed, the few current sites are in chalk quarries or on chalk spoil in southern England. *Pyramidata* is Latin for 'pyramid-like'.

Common cudweed *F. vulgaris*, an autumn- or spring-germinating annual, generally widespread in England (declining in the south-west), Wales, east Scotland and south-east Ireland, in dry, open, acidic to neutral (occasionally calcareous) habitats including grassland, quarries, dunes, sandy heathland, and arable and other cultivated ground. *Vulgaris* is Latin for 'common'.

Dwarf cudweed *G. supinum*, perennial, local in the Highlands and Hebrides on mountain-top fellfield communities, wet grassy slopes, cliffs, moraines and late snow-patches, growing in well-drained and stony sites that dry out in summer. *Supinum* is Latin for 'prostrate'.

Heath cudweed *G. sylvaticum*, a widespread and locally common (though overall, declining) perennial on dry, acidic, often sandy or gravelly soils on open habitats, for example heathland, heathy pasture, dunes, tracks and, especially, open woodland and forestry rides in former heathland. *Sylvaticum* is from Latin for 'woodland'.

Highland cudweed *G. norvegicum*, perennial, very local in the central Highlands on rock ledges gravel and scree on acidic, well-drained mineral soil. *Norvegicum* is Latin for 'Norwegian'.

Jersey cudweed *G. luteoalbum*, an annual or biennial of sandy fields, dune slacks and waste ground; now restricted as a native to the Channel Islands, the margins of two pools in Norfolk and an area of excavated shingle in east Kent discovered in 1996. Other recent records are casuals on waste ground, although it is persistent on heathland tracks at Holton Heath, Dorset. *Luteoalbum* is Latin for 'yellow-white'.

Marsh cudweed *G. uliginosum*, a common and widespread annual of acidic muddy ground subject to (winter) waterlogging, characteristic of trampled field entrances, compacted arable and cultivated land, reservoir margins, rutted tracks on heaths and wet rides in woodland. *Uliginosum* is from Latin for 'bog'.

Narrow-leaved cudweed *F. gallica*, an archaeophyte annual of well-drained, sandy and gravelly soils in open, disturbed sites. Changing agricultural practices and reduced rabbit disturbance led to its extinction in 1955. It was reintroduced to its last known site in Essex in 1994, and has been planted in Suffolk. It also survives on Sark since recorded in 1902. *Gallica* is Latin for 'French'.

Red-tipped cudweed *F. lutescens*, a winter- or spring-annual, declining in range and now local from Hampshire to Norfolk on dry, open, sandy acidic to neutral soils such as arable field margins, tracks, heathland and commons, and particularly characteristic of rabbit scrapes. *Lutescens* is Latin for 'yellowish'.

Small cudweed *F. minima*, a widespread, scattered and declining annual on dry, open, infertile, acidic to neutral soils in a range of habitats including arable fields, grassland, quarries, mine spoil, woodland tracks, sandy heaths, sandpits and dunes. *Minima* is Latin for 'smallest'.

CULICIDAE From Latin *culex* = 'gnat' or 'midge', the family name for species of mosquito, with 34 British species in 8 genera. Females are ectoparasites using piercing mouthparts to take blood from their host, and may serve as vectors for a number of diseases. Both sexes feed on nectar. Eggs are laid in stagnant water, and the larvae develop in this medium as filter feeders. Adults emerge from pupae as they float on the water. *Anopheles maculipennis* (from Greek for 'troublesome' and Latin for 'spotted wing') is one of the more common and widespread species, and *Aedes*, *Culex* and *Culiseta* are other important genera. Species and diseases (such as dengue fever and West Nile virus) new to Britain are anticipated with global warming. Increased numbers of garden water butts have already led to recent increased numbers of *Anopheles plumbeus* and *Culex pipiens* in towns.

CUMIN *Cuminum cyminum* (Apiaceae), an annual occasionally casual on waste ground, originating from birdseed and food waste. The Greek *cuminon* is this plant, *cyminon* its seed.

CUP (FUNGUS) Species of *Caloscypha* (Greek *calo* + *scyphos* = 'good' + 'cup'), *Disciotis* ('two-striped'), *Geopora* ('earth cup'), *Geopyxis* ('earth box'), *Helvella* (Latin for a kind of pot herb), *Peziza* (Greek for a stalkless fungus), *Plectania* ('plaited' or 'twisted'), *Pseudoplectania* ('false' + *Plectania*) and *Tarzetta* (a corruption of the Italian *Tazzeta* = 'little cup') (all in the order Pezizales), and *Dencoeliopsis* (Greek *coilos* = 'hollow') *Dumontinia* (after the American mycologist Kent Dumont, b.1941) and *Rutstroemia* (after the Danish mycologist Carl Rutström, 1758–1826) (all Heliotales).

Widespread but infrequent, or at best locally common, are **bleach cup** *Disciotis venosa* (poisonous if uncooked, edible and tasty if cooked), **blistered cup** *Peziza vesiculosa* (on dung, rotting straw, manured soil and woodchip mulch), **cedar cup** *Geopora sumneriana* (under Atlantic cedar), **cellar cup** *Peziza cerea* (concrete rubble, old mortar, rotting wood and other organic debris), **charcoal cup** *Peziza echinospora* (damp burnt sites), **glazed cup** *Helvella hemisphaerica* (rotting wood in deciduous woodland), **greater toothed cup** *Tarzetta catinus* (broadleaf woodland, mostly under beech, on rotten wood and woodchip), **oakleaf cup** *Rutstroemia sydowiana* (oak dead wood), **palamino cup** *Peziza repanda* (rotting stumps and woody debris), **sooty cup** *Helvella leucomelaena* (poisonous, on nutrient-poor sandy

woodland soils and heathland grass), **toothed cup** *Tarzetta cupularis* (woodland tracks, sawdust and woodchip), **vinegar cup** *Helvella acetabulum* (poisonous, in deciduous woodland), and **yellowing cup** *Peziza succosa* (woodland soil, especially under beech and hazel).

Scarce species include **anemone cup** *Dumontinia tuberosa*, **birch cup** *Dencoeliopsis johnstonii*, **corona cup** *Plectania melastoma*, **ebony cup** *Pseudoplectania nigrella*, **pouch cup** *Peziza saccardiana*, and **stalked bonfire cup** *Geopyxis* (on burnt ground).

Bay cup *Peziza badia* (Pezizaceae), common and widespread, poisonous if eaten raw or inadequately cooked, found from June to November on soil and paths in woodland. *Badia* is Latin for 'bay' or 'chestnut' brown.

Brown cup *Rutstroemia firma* (Rutstroemiaceae), common and widespread, inedible, found from late summer to winter on dead wood of oak. *Firma* is Latin for 'firm'.

Dune cup *Peziza ammophila*, inedible, scattered and very local around the British coastline in late summer and autumn on marram dunes. Dry sand blowing across the dunes falls into the cup, triggering discharge of the spores. Greek *ammophila* = 'sand lover'.

Golden cup 1) *Caloscypha fulgens* (Caloscyphaceae), rare, scattered in England, mostly from the southern counties, on woody litter and rotting wood under broadleaf trees such as oak, willow and birch. *Fulgens* is Latin for 'bright'. 2) Alternative name for golden bootleg fungus.

Layered cup *Peziza varia*, widespread and, in England, common, inedible, found from spring to autumn in deciduous or mixed woodland on buried rotting wood and other debris. *Varia* is Latin for 'variable'.

Mountain cup *Geopora arenosa* (Pyronemataceae), locally common, inedible, with most records in the Midlands, a few elsewhere in England and in Scotland, found in often sandy soil under grass or woodland leaf litter. *Arenosa* is Latin for 'sandy'.

CUP-LICHEN Species of *Cladonia*; see common cup lichen under lichen.

CUPLET **Crowded cuplet** *Merismodes fasciculata* (Agaricales, Niaceae), a local and scattered fungus in Britain, on dead branches and twigs of deciduous trees. Greek *merismos* = 'division', and Latin *fasciculus* = 'bunched'.

CUPRESSACEAE Juniper family (though Latin *cupressus* = 'cypress') which also includes coastal redwood, wellingtonia, red-cedar and cypresses.

CUPULA-MOSS **Lurid cupula-moss** *Cinclidium stygium* (Cinclidiaceae), in calcareous marsh, springs and fen in East Anglia, and in upland habitats in north-west England, the Southern Uplands and Highlands, and Ireland. Greek *cinclis* = 'lattice' or 'opening', and Latin *stygium* = 'stygian' or 'infernal'.

CURCULIONIDAE Family name of weevils (Latin *curculio*), containing 480 British species in 155 genera, divided into 14 subfamilies. They are characterised by their generally small size (1–6 mm length), elbowed (geniculate) antennae, and – apart from Scolytinae – elongated snout (rostrum). **Baridinae**, a small group in Britain. The two *Limnobaris* species are relatively widespread, though locally distributed, on sedges. **Bagoinae** are rare semi-aquatic weevils that feed on water weeds, particularly in drainage ditches. **Ceutorhynchinae**, a large and diverse subfamily, most species associated with one or a few species of herbaceous plant, including some of the 32 species of *Ceutorhynchus* which can be pests on brassica crops. Many species are (semi-)aquatic. **Cossoninae**, a group of wood-living species, several of which are partly synanthropic, occurring indoors in old floorboards, etc. **Cryptorhynchinae** contains three small species of *Acalles*, associated with dead twigs, and *Cryptorhynchus* (see withy weevil under weevil). **Curculioninae** contains the archetypal weevils *Curculio* species. Members include leaf miners, gall inquilines, and catkin-, seed- and leaf-feeders on a wide range of plants.

This also includes the flea weevils, which can leap to avoid disturbance. **Cyclominae** comprises the native *Gronops lunatus*, associated with spurreys, and *G. inaequalis*, an introduction first recorded from a landfill site in Kent in 1982. **Entiminae** are the broad-nosed weevils, with short broad rostrums. Several are pest species, including vine weevil. **Hyperinae** comprises 18 British species, including 16 species of *Hypera*, and **Lixinae** 11 species (several probably now extinct in Britain). **Mesoptiliinae** contains eight species of *Magdalis*, larvae found under bark. In **Molytinae** the 17 species include the abundant pine weevil. **Orobitinae** contains just violet weevil. **Scolytinae** comprises a large group of bark and ambrosia beetles, distinctive among the weevils by their lack of a rostrum.

CURLEW *Numenius arquata* (Scolopacidae), Britain's largest wader, wingspan 90 cm, length 55 cm, male weight 770 g, females 1 kg, found in winter around all coasts, with largest concentrations at Morecambe Bay, the Solway Firth, the Wash, and the estuaries of the Dee, Severn, Humber and Thames. Greatest breeding numbers, often on moorland, are found in north Wales, the Pennines, the Scottish Southern Uplands and east Highlands, Orkney and Shetland. Some 68,000 UK pairs bred in 2009, but there was a decline in numbers by 19% over 1980–2015, and by 48% over 1995–2015, with placement in the Red List. Wintering numbers in 2008–9 were estimated at 150,000 birds, but a decline in numbers by 24% was noted from 2002–3 to 2012–13. The decline in Irish numbers, with <100 pairs recorded in a partial survey in 2015, has made this species one of the country's highest conservation priorities. Nests are bare scrapes and usually have 4 eggs, incubation taking 27–9 days, the precocial chicks fledging in 32–8 days. Diet is omnivorous, but mainly of invertebrates. 'Curlew', derived from Old French *courlieu*, is supposed to be imitative of the call. Greek *neos* + *mene* = 'new' + 'moon' is a reference to the crescent-shaped bill. *Arquata* comes from Latin *arcuatus* = 'bow-shaped', again referring to bill shape.

Slender-billed curlew *N. tenuirostris* is a scarce accidental.
Small curlew Alternative name for black-tailed godwit.
Stone curlew *Burhinus oedicnemus* (Burhinidae), wingspan 81 cm, length 42 cm, weight 470 g, the only European member of the thick-knee family, a summer visitor (wintering in North Africa) with a UK population in 2010 of around 350–400 breeding pairs found on bare stony ground, heaths and chalk-soil arable fields in the Breckland area of East Anglia and Salisbury Plain, and at Minsmere, Suffolk. Diet is of worms and insects. Greek *bous* + *rhinos* = 'bull' + 'nose', and *oideo* + *cneme* = 'swelling' + 'knee', the specific name reflecting the alternative name, 'thick-knee'.

CURRANT Small shrubs in the genus *Ribes*, Latin for currant, derived from the Arabic name for a shrub with acidic fruit, in the family Grossulariaceae. 'Currant' comes from Middle English, translating Anglo-Norman French *raisins de Corauntz* = 'grapes of Corinth', which were seedless raisins (dried grapes). In the 1570s the word was (as now) applied to the unrelated northern European berry (genus *Ribes*), recently introduced in England, through its resemblance to the raisins.

Black currant *R. nigrum*, first cultivated for its edible fruit shortly after 1600 when plants were imported from Holland. It was first recorded in the wild in 1660 and occurs as a naturalised escape throughout the British Isles, although it may be native in fen carr and wet woodlands in East Anglia. *Nigrum* is Latin for 'black'.

Buffalo currant *R. odoratum*, introduced from North America in 1812, widespread but scattered, found as a garden escape on riverbanks and waste ground, and in hedgerows and chalk-pits. It has become naturalised in scrub, hedgerows and verges. *Odoratum* is Latin for 'fragrant'.

Downy currant *R. spicatum*, found locally from Lancashire and Yorkshire to Caithness in limestone woods, streamsides, and deep grikes in limestone pavement. *Spicatum* is Latin for 'pointed'.

Flowering currant *R. sanguineum*, with edible fruit,

introduced from North America in 1826, widely grown in gardens and for ornamental hedging. It was known from the wild by 1916, and has become naturalised throughout Britain in, among other habitats, woodland, verges, hedges and waste ground. *Sanguineum* is Latin for 'blood-red'.

Mountain currant *R. alpinum*, native in northern England, common only in the Peak District, found in limestone woods, rocky hedgerows and streamsides, often trailing over rocks in shaded places. It is also grown in gardens, and has become more widespread and naturalised in Britain on roadsides and wasteland, and as a relic of cultivation.

Red currant *R. rubrum*, edible, widespread and locally common. Small-fruited plants in fen carr and by streams in woods may be native, but large-fruited plants near old habitations are clearly relics or escapes from cultivation and it has become widely naturalised. *Rubrum* is Latin for 'red'.

CURSE (MAYFLY) See angler's curse.

CUSHION (FUNGUS) Ochre cushion *Hypocrea pulvinata* (Hypocreales, Hypocreaceae), widespread, associated with dead wood, often on the underside of birch polypore bracket fungus in woodland. Greek *hypo* + *creas* = 'under' + 'flesh', and Latin *pulvinum* = 'cushioned'.

CUSHION STAR Alternative name for starlet.

CUT-GRASS *Leersia oryzoides* (Poaceae), a rhizomatous perennial, very local in southern England on nutrient-rich mud around cattle-trampled edges of lakes and ponds, in wet meadow and ditches, and on canal and riverbanks. Daniel Leer was an eighteenth-century German botanist. Latin *oryzoides* = 'rice-like'.

CUTPURSE (WASP) *Aporus unicolor* (Pompilidae), female body length 10 mm, a spider wasp recorded, mainly from sunny coastal cliffs and landslips, but also inland on well-grazed downland and heathland, in the Channel Islands and south England, and at Bosherton, Dyfed. Adults visit umbellifer flowers. It preys on the purse web spider *Atypus affinis*, which it paralyses within the spider's burrow. The body shape of the wasp is adapted for gaining entry to the host's nest (head and thorax elongated, head flattened and forelegs powerfully developed). The wasp larva eats the paralysed spider and pupates among the remains of its prey. In 2010, *The Guardian* launched an annual competition for the public to suggest common names for previously unnamed species and this name won in 2012. The etymology of *Aporus* is unclear, but is possibly from Greek *apo* = 'separate from'; *unicolor* is Latin for 'single-coloured'.

CUTWORM Name given to the larvae of turnip moth, dark sword-grass and large yellow underwing from their habit of biting off or chewing the base of stems, roots, leaves and tubers of often cultivated plants. Larvae feed at night on the soil surface. They generally hatch in late June–July. A second generation hatches in August–September, overwinters in the soil, and comes to the surface to feed when environmental conditions are favourable.

Black cutworm Larvae of the dark sword-grass noctuid moth that feed underground on plant roots, increasingly being reported as a turf grass nuisance.

CUVIE, CUVY *Laminaria hyperborea* (Laminariaceae), a brown seaweed, or kelp, thallus up to 2 m long, common and widespread except between the Ouse and Thames estuaries because of turbidity and lack of hard substrates, growing on hard substrates from extreme low water to depths dependent on light penetration and sea urchin grazing (typically 8 m depth, but down to 30 m in clear water), forming dense 'forests' under suitable conditions. The binomial is Latin for 'thin flat leaf' and Greek relating to the Hyperboreans – mythical people who lived 'beyond the North Wind'.

CYANOBACTERIA Phylum of photosynthesising bacteria, formerly known as blue-green algae. This earlier name reflected their colour, from Greek *cyanos* = 'blue'. Cyanobacteria are prokaryotes, but the term 'algae' is generally reserved for eukaryotes. By producing oxygen as a by-product of photosynthesis, cyanobacteria probably converted Earth's early reducing atmosphere into an oxidising one, causing the 'Great Oxygenation Event', and radically changing the composition of the planet's life forms.

CYBAEIDAE Family name of the water spider, possibly from Greek *cybe* = 'head'.

CYCLOSTOMATA Superclass of jawless fishes with cartilaginous skeletons, including lampreys (Greek *cyclos* + *stomata* = 'circular' + 'mouths').

CYLINDER-MOSS Montagne's cylinder-moss *Entodon concinnus* (Entodontaceae), widespread and scattered in base-rich habitats. In southern England it is commonest in calcareous grassland, but also in dunes, quarries, and on limestone rock ledges and scree. It is occasionally found in mountain areas and in more acidic conditions. Jean Montagne (1784–1866) was a French botanist. Greek *entos* + *odous* = 'inside' + 'tooth', and Latin *concinnus* = 'skilfully joined'.

CYNIPIDAE Family name, from Greek *scnips* (*sic*), for an insect living under bark, containing most of the gall wasps, including oak gall wasps in genera such as *Andricus* (25 species) and *Neuroterus* (7).

CYPERACEAE Family name of sedges (Greek *cyperus*).

CYPHEL *Minuartia sedoides* (Caryophyllaceae), a mat- or cushion-forming herbaceous perennial of north and central Scotland, on base-rich rocks, flushed grassland, montane heath and mountain ledges. 'Cyphel' dates from Late Middle English, denoting house-leek, probably from Greek *cyphella* = 'hollows of the ears'. Jean Minuart was an eighteenth-century Spanish botanist and apothecary. Latin *sedoides* = 'similar to house-leek' (*Sedum*).

CYPRESS Introduced evergreen trees in the family Cupressaceae, the name dating to Middle English, from Old French *cipres*, from Late Latin *cypressus*, in turn from Greek *cyparissos*.

Lawson's cypress *Chamaecyparis lawsoniana*, introduced from North America in 1854 by collectors working for the Lawson & Son nursery in Edinburgh, widespread and commonly planted as a screening or windbreak tree, especially in towns, and for underplanting in some forestry plantations, and it not infrequently self-sown. *Chamae* here is Greek for 'false' (and certainly not 'dwarf').

Leyland cypress x *Cuprocyparis leylandii* (*Cupressus macrocarpus* x *Xanthocyparis nootkatensis*) first found at Welshpool in 1888 (though the parents have hybridised on at least 20 separate occasions), widespread and commonly planted as a fast-growing evergreen hedging and screening tree tolerant of soils of poor to medium fertility. *Cupro* comes from Latin *cupreus* = 'coppery'. Christopher Leyland was a nineteenth-century Liverpool banker.

Monterey cypress *Cupressus macrocarpa*, introduced from California, commonly planted in parks and gardens in southern and south-west England, self-sown in Jersey, the Scillies and southern Ireland. *Macrocarpa* is Greek for 'large fruit'.

Nootka cypress *Xanthocyparis nootkatensis*, introduced from North America in 1853, a common garden and parkland ornamental. Nootka Island is adjacent to Vancouver Island, British Columbia. *Xanthos* is Greek for 'yellow'.

Sawara cypress *Chamaecyparis pisifera*, introduced from Japan in 1843, commonly planted as a garden and park ornamental, and occasionally self-sown. *Pisifera* is Latin for 'bearing peas', from the size and shape of the cones. Sawara is a town in south Honshu, Japan.

CYPRINIDAE The carp family (Latin *cyprinus* = 'carp'), including barbel, bitterling, bleak, bream, chub, dace, goldfish, gudgeon, minnow, orfe, roach, rudd, sunbleak, tench and topmouth gudgeon.

CYPRINIFORMES Order of carp and its allies in the families Cyprinidae, Catostomidae, Cobitidae and Nemacheilidae.

D

DABBERLOCKS *Alaria esculenta* (Alariaceae), a brown seaweed with blades usually 30–50 cm, but up to 4 m, 25 cm wide, widespread except between south Yorkshire and Devon, found at low water and in the subtidal generally to 8 m depth on exposed rocky shores. The binomial is Latin for 'edible wings'.

DABCHICK Alternative name for little grebe.

DACE *Leuciscus leuciscus* (Cyprinidae), a fish usually up to 25 cm long and 70 g weight in the UK, widespread in England, scarce elsewhere, favouring rivers with moderate currents, though also found in gravel-pits, feeding on adult and larval insects, crustaceans and some plant material, and spawning in February–May. 'Dace' dates from late Middle English from Old French *dars*. Greek *leuciscos* = 'white mullet'.

Sea dace Alternative name for sea bass.

DADDY-LONG-LEGS Alternative name for craneflies.

DAFFODIL Specifically, *Narcissus pseudonarcissus* (Alliaceae), a bulbous perennial herb, scattered and locally common in England and Wales, less common and introduced in lowland Scotland and south-east Ireland, found in ash and oak woods, bracken stands, scrub and grassland. This is the county flower of Gloucestershire as voted by the public in a 2002 survey by Plantlife. There are a number of other species and hybrids occasionally found as naturalised or casual. *Narcissos* is Greek for daffodil and other narcissi, from *narce* = 'numbness', referencing their supposed narcotic properties.

Bunch-flowered daffodil *N. tazetta*, a Mediterranean neophyte present in Tudor times, mainly found as a naturalised relic of cultivation, especially in bulb fields in the Channel Islands. It is also planted on roadsides, and can be a casual garden throw-out on waste ground, scattered in southern and south-west England, south Wales and south-east Ireland. *Tazetta* is from Italian for 'little cup'.

Pheasant's-eye daffodil *N. poeticus*, a south European neophyte present by the early sixteenth century, scattered in Britain, naturalised in hedgerows, on roadsides, by tracks, and on waste ground from garden throw-outs. *Poeticus* is Latin for 'poetic'.

Spanish daffodil *N. hispanicus*, a Mediterranean neophyte in cultivation by 1629, scattered mainly in the southern half of Britain, naturalised from discarded or deliberately planted bulbs in woodlands, hedgerows, on roadsides, along tracks and paths and on waste ground. *Hispanicus* is Latin for 'Spanish'.

DAGGER (MOTH) Nocturnal moths in the genera *Acronicta* (Greek *acronyx* = 'nightfall') and *Simyra* (probably from the Phoenician town, Simyra) (Noctuidae, Acronictinae), the English name coming from black, dagger-like markings on the forewings.

Dark dagger *A. tridens*, wingspan 35–45 mm, common in woodland, fen, gardens and other habitats throughout much of England and Wales. Adults fly in June–July. Larvae feed in autumn on a range of trees and shrubs. Latin *tri* + *dens* = 'three' + 'tooth', from the three 'prongs' of the forewing markings.

Grey dagger *A. psi*, wingspan 30–40 mm, widespread and common in habitats such as woodland, hedgerows, gardens, scrubby heathland, calcareous grassland and fens, at low altitudes in the north. This is nevertheless a species of conservation concern in the UK BAP. Adults fly in June–August. Larvae feed on a variety of trees and shrubs. The specific name reflects a similarity between the forewing mark and the Greek letter *psi*.

Reed dagger *S. albovenosa*, wingspan 32–40 mm, scarce in reedbed, marsh and fen in East Anglia and parts of southern England. On the Isle of Wight there was a colony in Freshwater Marsh until 1989 and at Titchfield Haven from 2000. The bivoltine adults fly in May and again in July–August. Larvae feed on common reed and other reedbed plants. Latin *albus* + *venosus* = 'white' + 'veined'.

DAGGER FLY Members of the family Empididae (Bracycera), so called because of their sharp piercing mouthparts, with 208 species in 13 genera, the numerically largest being *Empis*, *Hilara* (see balloon fly) and *Rhamphomyia*. Adults are 1–15 mm in length and prey on other arthropods. Males of some species present the females with a prenuptial gift of food, usually a small fly. Larvae are also predaceous, generally found in moist soil, rotten wood or dung.

DAISY *Bellis perennis* (Asteraceae), a common and widespread rosette-forming, winter-green, stoloniferous perennial growing in mown, heavily grazed or trampled neutral and calcareous grassland, preferring those that are relatively wet for at least part of the year. It is a weed of lawns and recreational areas, verges and pasture; more natural habitats include stream banks, lake margins, dune slacks and the margins of upland flushes. The binomial is Latin for 'daisy' and 'perennial'.

Autumn oxeye (daisy) *Leucanthemella serotina*, a rhizomatous herbaceous perennial European neophyte, an occasional and scattered garden escape or throw-out in Britain, naturalised on rough ground and by ponds and ditches. Greek *leucos* + *anthemis* = 'white' + 'flower', and Latin *serotina* = '(flowering) late'.

Crown daisy *Glebionis coronaria*, a scarce, scattered annual European neophyte present by 1629, casual in arable fields, waste ground and roadsides, originating from wool shoddy and birdseed, as a grain contaminant and as a garden escape. Latin *gleba* + *ionis* = '(clod of) soil' + 'result of', and *corona* = 'crown'.

Oxeye daisy *L. vulgare*, a common and widespread herbaceous perennial, possibly declining in Scotland, found in grassy habitats, especially where cut or moderately grazed, preferring well-drained, nutrient-rich soils, and on coastal cliffs and stabilised dunes, and often common in urban waste ground, by railways and newly sown roadsides. *Vulgare* is Latin for 'common'.

Seaside daisy *Erigeron glaucus*, a herbaceous perennial North American neophyte naturalised, often as a garden escape, on rocky habitats and cliffs around the coastline of England, Wales and the Isle of Man, and very local in east Scotland and south Ireland. Greek *erion* + *geron* = 'wool' + 'old man', referring to the flowers occurring in spring turning grey like hair, and Latin *glaucus* = 'glaucous' (blue-green).

Shasta daisy *Leucanthemum* x *superbum* (*L. lacustre* x *L. maximum*), a rhizomatous perennial of American garden origin in 1890, common and widespread in Britain, scattered in Ireland, naturalised on rough ground and grassy habitats. Greek *leucos* + *anthemis* = 'white' + 'flower'. Mount Shasta is in the Cascade Range, California, giving its name because the plant's petals are snow-coloured.

DAISY-BUSH *Olearia* x *haastii* (*O. avicenniifolia* x *O. moschata*) (Asteraceae), a New Zealand neophyte shrub scattered in England and the Isle of Man on walls and waste ground. Johann Ölschläger, his name latinised to Olearius (1635–1711) was a German gardener. Sir Johann von Haast (1824–87) was a German-born geologist.

Mangrove-leaved daisybush (*O. avicenniifolia*), from New Zealand, naturalised on dunes in the Channel Islands, Devon and Midlothian. *Avicenniifolia* is Latin for having leaves like the mangrove *Avicennia*.

DALTONIA Irish daltonia *Daltonia splachnoides* (Daltoniaceae), a moss recorded from south Wales, west Scotland and, mainly, Ireland in high rainfall sites, often where it is splashed with or covered by running water. It grows on the branches, trunks and roots of willows, by streams in conifer plantations in south-west Ireland, and on humus and rotting logs, occasionally on damp, shaded rocks. *Splachnon* is Greek for 'moss'.

DAME'S-VIOLET *Hesperis matronalis* (Brassicaceae), a common and widespread perennial, sometimes biennial,

European neophyte herb grown since at least medieval times, with a preference for moist, shaded habitats, and found in hedgerows and woodland edges, on riverbanks, roadsides and waste ground, usually (as a garden escape) near habitation. It is naturalised where there is little competition, otherwise casual. Hesperis was a figure in Greek mythology, coming to mean 'of the evening'; *matronalis* is Latin for 'relating to 1 March', the Roman festival of married women (matrons).

DAMSEL BUG Members of the genera *Nabis* (possibly named after Nabis, a Spartan king) and *Himacerus* in the family Nabidae (Heteroptera), length usually 6–9 mm. A number of species are widespread and common on grassland: **broad damsel bug** *N. flavomarginatus*, **common damsel bug** *N. rugosus*, **field damsel bug** *N. ferus*, and **marsh damsel bug** *N. limbatus*.

Ant damsel bug *H. mirmicoides*, common in the southern half of England and the Welsh Borders, in low vegetation in dry open habitats. *Mirmicoides* is Greek for 'antlike'.

Heath damsel bug *N. ericetorum*, widespread but scattered as far north as Deeside, associated with heathland. *Ericetorum* is from Greek for 'heath'.

Reed damsel bug *N. lineatus*, 9.5–12 mm, widespread, scattered but locally common, associated with reeds, sedges and rushes, and some saltmarshes. *Lineatus* is Latin for 'lined'.

Tree damsel bug *H. apterus*, 8–11 mm long, the only arboreal nabid, preferring but not exclusive to deciduous trees, feeding on mites, aphids and other small insects, common in much of England and the Welsh Borders. *Apterus* is from Greek *apterygos* = 'wingless'.

DAMSELFLY Insects in the order Odonata, suborder Zygoptera, families Coenagrionidae, Lestidae (*Lestes*) and Platycnemididae (*Platycnemis*), also known as demoiselle flies and bluets, characterised by elongated bodies (more slender than those of dragonflies), large multifaceted eyes, and two pairs of independently manoeuvreable transparent wings that (unlike those of dragonflies) are folded at rest, along or above the abdomen. Larvae are aquatic, generally spending one to three years in fresh water.

Azure damselfly *Coenagrion puella*, 33 mm long, common and widespread in England, Wales and Ireland, expanding in Scotland; associated with ponds and lakes. The main flight period is May–July. Greek *coinos* + *agrios* = 'shared' (with the genus *Agrion*) + 'wild', and Latin *puella* = 'girl'.

Blue-tailed damselfly *Ischnura elegans*, 31 mm long, abundant and widespread throughout the lowland British Isles, nymphs found in a wide range of waters, including polluted or brackish. The main flight period is May–September. Greek *ischnos* + *oura* = 'slender' + 'tail', and Latin *elegans* = 'elegant' or 'graceful'.

Common blue damselfly or common bluet *Enallagma cyathigerum*, 32–5 mm long, in lakes and other waters, abundant and widespread. The main flight perids is May–August. Greek *enallax* + *agma* 'alternate' + 'splinter', possibly from the alternating abdominal stripes, and arguably *cyathos* + Latin *gerum* = 'cup' + 'carrier', from the shape of an abdominal mark.

Dainty damselfly *C. scitulum*, 32 mm long, formerly recorded as a scarce breeding species in Essex, locally extinct with the 1953 North Sea floods, rediscovered in Kent in 2010 in brackish borrow pits. *Scitulum* is Latin for 'neat'.

Emerald damselfly *Lestes sponsa*, locally common throughout Britain, scattered in Ireland, nymphs associated with still waters. The main flight period is May–October. The binomial is Greek for 'robber' and Latin for 'bride'.

Irish damselfly *C. lunulatum*, 31 mm long, scarce and local in central and north Ireland in sheltered mesotrophic lakes, large pools and cut-over bogs, not found in Britain. *Lunulatum* is Latin for 'crescent-shaped' (via *luna* = 'moon').

Large red damselfly *Pyrrhosoma nymphula*, 33–6 mm long, widespread and common in most freshwater habitats except where water flow is rapid. The main flight period is May–July.

Greek *pyrrhos* + *soma* = 'flame-coloured' + 'body', and Latin *nymphula* = 'like a small nymph'.

Northern damselfly *C. hastulatum*, 31–3 mm long, also known as spearhead bluet, a rare Red Book species found only in a few sedge-fringed lochans in the Highlands (Deeside, Speyside). *Hastulatum* is Latin for 'shaped like a little spear'.

Red-eyed damselfly *Erythromma najas*, 35 mm long, in lakes, canals and slow-flowing streams, locally common in southern England and the Welsh Borders. Greek *erythros* + *omma* = 'red' + 'eye', and *naias* = 'water nymph'.

Scarce blue-tailed damselfly *I. pumilio*, 26–31 mm long, rare, associated with south and south-west England and Wales, though with isolated colonies also in central England, nymphs found in shallow slow-flowing waters, especially where fed by seepages and flushes, with bare substrate and little vegetation. *Pumilio* is Latin for 'pygmy'.

Scarce emerald damselfly *L. dryas*, rare and local in Lincolnshire, East Anglia, particularly coastal Essex marshes and the Brecklands, and north Kent, Vulnerable in the RDB. *Dryas* = 'dryad', a woodland nymph in Greek mythology.

Small red damselfly *Ceriagrion tenellum*, 31 mm long, rare, larvae living in debris in acidic shallow pools and streams in heathland bog, mostly in south and south-west England and west Wales. *Tenellum* is Latin for 'delicate'.

Small red-eyed damselfly *E. viridulum*, 29 mm long, first recorded in Britain in 1999, initially increasing its range rapidly in south-east England, spread slowing down but having reached Devon and Yorkshire; breeding confirmed in 2002. Associated with ponds, ditches, lakes and occasionally brackish water. *Viridulum* is Latin for 'greenish'.

Southern damselfly *Coenagrion mercuriale*, 29–31 mm long, a rarity mainly of alkaline streams within acid heathland in the New Forest and water meadows on chalk-stream floodplains in Hampshire, and the Preseli mountains (Pembrokeshire), with smaller colonies in Devon, Dorset, Oxfordshire, Gower and Anglesey. Adults are on the wing from mid-May to August. *Mercuriale* is Latin for 'relating to the god Mercury'.

Variable damselfly or variable bluet *C. pulchellum*, 33 mm long, with a scattered but widespread distribution, common only in parts of Ireland, associated with cut-over bogs, vegetated ditches, canals and ponds. The main flight period is from May to the end of August. *Pulchellum* is Latin for 'beautiful'.

White-legged damselfly *Platycnemis pennipes*, 36 mm long, generally uncommon though locally abundant, found in the southern half of England and a few sites in Wales, nymphs associated with slow-flowing streams and canals. The flight period is May–August. Greek *platys* + *cneme* = 'broad' + 'knee', and Latin *penna* + *pes* = 'feather' + 'foot'.

DAMSON *Prunus domestica* ssp. *insititia* (Rosaceae), or **bullace**; see wild plum under plum. The word dates from late Middle English *damascene*, from Latin *damascenum prunum* = 'plum of Damascus'.

DANCE FLY Members of the families Hybotidae and Empididae.

DANDELION *Taraxacum officinale* agg. (Asteraceae), a taxonomically difficult group comprising 232 apomictic microspecies of which >40 are probably endemic and about 100 introductions. This tap-rooted herbaceous perennial is found in a wide range of habitats. Some microspecies are associated with natural and semi-natural habitats, for example dunes, chalk grassland, fens, flushes and cliffs, but most occur in disturbed habitats, including pasture, verges, lawns, tracks, paths and waste ground. 'Dandelion' is late Middle English from French *dent-de-lion*, a translation of the medieval Latin *dens leonis* = 'lion's tooth', because of the jagged shape of the leaves. Other common names are blowball, lion's-tooth, cankerwort, milk-witch, yellow-gowan, Irish daisy, monks-head, priest's-crown and puff-ball, faceclock, pee-a-bed, wet-a-bed and swine's snout. *Taraxacum* is Latin for this plant, possibly from Persian *tarashqun*. Latin *officinalis* means having pharmacological properties.

DAPHNIA, DAPHNIIDAE See water flea.

DAPPERLING Inedible fungi in the genera *Chamaemyces* (Greek *chamae* + *myces* = 'dwarf' + 'fungus'), *Cystolepiota* (*cistis* + *lepis* + *oti* = 'blister' or 'bladder' + 'scale' + 'ear'), *Lepiota*, *Leucoagaricus* (*leucos* + *agarico* = 'white' + 'agaric mushroom') and *Leucocoprinus* (*leucos* = 'white' + *Coprinus*, a genus that until recently contained all fungi commonly known as inkcaps) (Agaricales, Agaricaceae). 'Dapperling' means a little elegant fungus.

Widespread and common (often more so in the southern half of England) on soil and litter usually in deciduous woodland are **bearded dapperling** *Cystolepiota seminuda*, **cat dapperling** *Lepiota felina* (with a dappled leopard-like pattern on the cap, found in coniferous woodland), **chestnut dapperling** *Lepiota castanea*, **freckled dapperling** *Lepiota aspera* (causing alcohol intolerance, possibly toxic), **girdled dapperling** *Lepiota boudieri* (also often with currants in gardens), **lilac dapperling** *Cystolepiota bucknallii* (especially on nutrient-rich calcareous soils), and **stinking dapperling** *Lepiota cristata* (poisonous, also in damp shady gardens).

Rarities include **apricot dapperling** *Leucoagaricus sublittoralis*, **blushing dapperling** *Leucoagaricus badhamii*, **deadly dapperling** *Lepiota brunneoincarnata* (lethally poisonous, containing amatoxins), **dune dapperling** *Lepiota erminea*, **fatal dapperling** *Lepiota subincarnata*, toxic (containing amatoxins), **green dapperling** *Lepiota grangei*, **plantpot dapperling** *Leucocoprinus birnbaumii* and **smoky dapperling** *Leucoagaricus barssii*.

Dewdrop dapperling *Chamaemyces fracidus*, scattered and local in England, found in summer and autumn on soil in mixed woodland and pasture. *Fracidus* is Latin for 'mellow' or 'soft'.

Shield dapperling *Lepiota clypeolaria*, infrequent and scattered, found in late summer and autumn on nutrient-rich soil and leaf litter in woodland, especially under oak and beech. Latin *clypeus* = 'shield'.

Skullcap dapperling *Leucocoprinus brebissonii*, possibly toxic, local with scattered records in England, Wales and Northern Ireland, in soil, litter and forest floor dead wood in deciduous and coniferous woodland. Louis de Brébisson (1798–1888) was a French botanist.

White dapperling *Leucoagaricus leucothites*, locally common in England and Wales, found in summer and autumn usually in pasture and other grassland, often on verges or in gardens.

Yellowfoot dapperling *Lepiota magnispora*, uncommon but widespread and scattered, found in late summer and autumn on woodland soil, especially under oak and beech, occasionally in coniferous forest. *Magnispora* is Latin for 'large seeds'.

DARNEL *Lolium temulentum* (Poaceae), an annual Mediterranean archaeophyte, formerly a common and persistent weed of arable, now a rare casual on waste ground originating from grain, birdseed and wool shoddy. The name comes from Middle English, probably related to French (Walloon) *darnelle*. The binomial is Latin for rye-grass, specifically darnel, and *temulentum* = 'drunken'.

DART (MOTH) Members of the nocturnal moth family Noctuidae (Noctuinae and Xyleninae), including the genera *Agrotis* (from Greek *ager* = 'field') and *Euxoa* (*euxoos* = 'well-polished', describing the hindwing); see also heart & dart.

Archer's dart *A. vestigialis*, wingspan 30–5 mm, local on dunes around the coasts of the British Isles, and on heathland inland in southern England. Adults fly in July–September. Larvae feed on low plants such as bedstraws, greater stichwort and grasses. *Vestigium* is Latin for 'trace'.

Coast dart *E. cursoria*, wingspan 34–8 mm, scarce, on dunes in parts of east England, north-west England and Scotland, and very local in Ireland. Adults fly in July–September. Larvae feed on sea sandwort, couch and other herbaceous plants. *Cursoria* is

Latin for 'relating to a (sand) running track', perhaps an oblique reference to the habitat.

Crescent dart *A. trux*, wingspan 35–42 mm, local on coastal cliffs and rocky areas from the Isle of Wight westwards to north Wales, and in much of Ireland. Adults fly in July–August. Larvae feed on low-growing plants such as rock sea-spurret and thrift. *Trux* is Latin for 'rough' or 'wild'.

Deep-brown dart *Aporophyla lutulenta*, wingspan 36–44 mm, on calcareous grassland, rough meadows, downland, heathland, moorland, dunes and gardens throughout England, and with numbers declining by 81% over 1968–2007, a species of conservation concern in the UK BAP. Adults fly in August–October. Larvae feed on hawthorn, docks, sorrels and grasses, and heathland/moorland examples on heather. Greek *aporos* + *phyle* = 'difficult' + 'tribe', and Latin *lutulenta* = 'muddy', from the forewing colour.

Double dart *Graphiphora augur*, wingspan 35–42 mm, widespread in broadleaf woodland, scrub, hedgerows, parkland, marsh, riverbanks, fen and gardens, but with numbers declining by 98% over 1968–2007, a species of conservation concern in the UK BAP. Adults fly in June–July. Larvae feed on trees and shrubs such as sallows, birches, hawthorn and blackthorn. Greek *graphis* + *phoreo* = 'stylus' + 'to carry', from the forewing mark, and Latin *augur* = 'soothsayer', possibly because the broken outline of this mark looks like runic lettering.

Garden dart *E. nigricans*, wingspan 28–35 mm, widespread and local in gardens, allotments and farmland, and on downland and rough ground, but with numbers declining by 98% over 1968–2007, a species of conservation concern in the UK BAP. Adults fly in August–September. Larvae feed on low-growing herbaceous plants, including clovers, plantains and docks. *Nigricans* is Latin for 'blackish'.

Northern dart *Xestia alpicola*, wingspan 34–40 mm, scarce, on mountains and high moorland in the Highlands, Outer Hebrides, Orkneys, Pennines and the Lake District, and a priority species in the UK BAP. Adults fly in June–August. Larvae feed on crowberry, occasionally heather. Greek *xestos* = 'polished', from the glossy forewings, and Latin for a species living (*colo* = 'to inhabit') in the Alps.

Northern deep-brown dart *Aporophyla lueneburgensis*, wingspan 34–8 mm, on moorland, rough grassland and rocky areas throughout much of Scotland and north-west England. Adults fly in August–September. Larvae feed on heather, bilberry and bird's-foot trefoil. Latin *lueneburgensis* means from Lüneburg, Germany.

Sand dart *A. ripae*, wingspan 32–42 mm, scarce, on sandy beaches on the coasts of England and Wales, east Scotland, Jersey, Alderney, the Isle of Man, and south and east Ireland. Adults fly in June–July. Larvae feed at night on various dune plants, burrowing in the sand during the day. *Ripa* is Latin for 'riverbank', here used to mean shoreline.

Shuttle-shaped dart *A. puta*, wingspan 30–2 mm, common and often abundant in gardens, farmland, grassland, heathland and open woodland throughout England and Wales, with a few records from Scotland (as a stray or recent colonist) and Ireland (post-2000 in Co. Antrim and Co. Wicklow). Adults fly in July–August. Larvae feed on low-growing plants, including docks and dandelion. *Puta* has no entomological significance.

Square-spot dart *E. obelisca*, wingspan 35–40 mm, local on cliffs and rocky hillsides along the coasts of the southern and western British Isles, and the east coast of Scotland. Adults fly in August–October. Larvae feed on low-growing herbaceous plants. *Obelisca* is Latin for 'obelisk', from the black forewing streak.

Stout dart *Spaelotis ravida*, wingspan 42–50 mm, local in damp grassland, meadow and marsh, and also gardens, in much of England, with great variations in abundance and temporary extensions of range; on the east coast the species is possibly reinforced by immigration. Adults fly in June–September. Larvae feed on herbaceous plants such as sowthistle, dandelion and dock. *Spaelotis* is entomologically meaningless; *ravidus* is Latin for 'greyish'.

White-line dart *E. tritici*, wingspan 28–40 mm, widespread and common on dunes, sea-cliffs, heathland, moorland and downland, predominantly coastal in Ireland. With numbers declining by 94% over 1968–2007, this is a species of conservation concern in the UK BAP. Adults fly in July–August. Larvae feed on low-growing herbaceous plants. *Triticum* is Latin for 'wheat', mistakenly thought to be a food plant.

DARTER Dragonflies in the family Libellulidae, mainly *Sympetrum* (Greek *sym* + *petra* = 'together' + 'rock').

Black darter *S. danae*, length 29–34 mm, widespread, more local in central and eastern England and southern Ireland, breeding in peat moss and moorland pools and ditches. The main flight period is July–October. *Danaos* was a mythical king of Arabia.

Common darter *S. striolatum*, length 38–43 mm, widespread, in ponds (including small garden ponds) and other still, stagnant or even brackish waters; adults are frequently found away from water, resting on the tops of plants in woodland rides. The main flight period is June–October, into early November if the weather is warm. *Striolatum* is Latin for 'finely striped'.

Red-veined darter *S. fonscolombii*, length 38–40 mm, a frequent migrant found as far north as south Scotland, breeding nearly annually in shallow water bodies. The main flight period is May–October. *Fonscolombi* is from Latin *fons* = 'fountain' + 'of (St) Colombus'.

Ruddy darter *S. sanguineum*, length 34–6 mm, found in much of England, parts of Wales and central Ireland, increasing in range, associated with weedy ponds and ditches, adults favouring woodland. The main flight period is June–September. *Sanguineum* is Latin for 'blood-coloured'.

Scarlet darter *Crocothemis erythraea*, length 33–44 mm, a rare migrant from southern Europe, first recorded in the UK in 1994, most records being from southern England. In Greek mythology Themis was the guardian of justice and peace; *croco* may be from Greek *croce* = 'pebble' (given its colour, unlikely to be from *crocos* = 'saffron yellow'); *erythraia* is Greek for 'red'.

White-faced darter *Leucorrhinia dubia*, length 33–7 mm, in lowland peatbog, requiring relatively deep, oligotrophic, acidic bog pools with rafts of sphagnum at the edges in which to breed. Adults also require scrub or woodland as roosting and feeding sites. Found at isolated sites from the north Midlands northwards, including Foulshaw Moss in Cumbria where larvae were reintroduced in 2010. Major strongholds occur in the Highlands. Populations in Inverness-shire and Ross-shire are particularly important. English populations have declined in the last 35 years. It is the subject of BAPs in Cheshire and Cumbria. The main flight season is May–July. Greek *leucos* + *rhinos* = 'white' + 'nose', and Latin *dubia* = 'doubtful'.

Yellow-winged darter *S. flaveolum*, length 32–7 mm, an irregularly migrating dragonfly that can occur in large numbers (e.g. 1995, 2006) when it then often breeds, though colonies do not persist. Found in marginal vegetation in still water such as ponds, ditches and river backwaters. *Flaveolum* is Latin for 'yellowish'.

DASYTIDAE Family name (Greek *dasytes* = 'hairiness' or 'roughness') of soft-winged flower beetles, with nine species whose adults mainly occur in grassland, while larvae develop in wood, where they are predatory.

DAY-LILY Rhizomatous herbaceous perennials in the genus *Hemerocallis* (Xanthorrhoeaceae), from Greek for 'beautiful for a day'.

Orange day-lily *H. fulva*, of garden origin, scattered in Britain mainly in the south, in Lancashire and the Isle of Man, naturalised in rough ground and grassy habitats. *Fulva* is Latin for 'reddish-yellow' (i.e. 'orange').

Yellow day-lily *H. lilioasphodelus*, an East Asian neophyte, scattered in Britain, naturalised in rough ground and grassy habitats. The specific name combines lily and asphodel.

DEAD MAN'S BOOTLACES, DEAD MAN'S ROPE Alternative names for the brown seaweed mermaid's tresses.

DEAD MAN'S FINGERS (ANTHOZOAN) *Alcyonium digitatum* (Alcyoniidae), a common and widespread sea anemone, colonies forming thick, fleshy masses of irregular shape, typically of stout, finger-like lobes usually >20 mm in diameter. Height and breadth of colonies are up to 200 mm, attached to rocks, shells and stones where otherwise dominant macroalgae are inhibited by lack of light, occasionally on the lower shore but more common in the sublittoral down to 50 m. *Alcyonion* is Greek for a kind of sponge (*sic*), from its resemblance to the nest of a kingfisher (= *alcyon*). *Digitatum* is Latin for 'having fingers'.

DEAD MAN'S FINGERS (FUNGUS) *Xylaria polymorpha* (Sphaeriales, Xylariaceae), inedible, common throughout England and Wales, more scattered in Scotland and Ireland, evident throughout the year, though mainly in summer and autumn, on stumps of beech and other hardwood trees. Greek *xylon* = 'wood', *polymorpha* = 'many shapes'.

DEAD MOLL'S FINGERS *Xylaria longipes* (Sphaeriales, Xylariaceae), inedible, common and widespread, especially in England and Wales, found year-round, though mainly in summer and autumn, on stumps of beech and other hardwood trees. Greek *xylon* = 'wood', and Latin *longipes* = 'long foot'.

DEAD-NETTLE Species of *Lamium* (Lamiaceae), Latin for these plants, the English name coming from the superficial resemblance to true nettles.

Cut-leaved dead-nettle *L. hybridum*, an annual archaeophyte, common and widespread in lowland regions on cultivated, waste and disturbed ground on dry soils, often a weed of heavily fertilised, broad-leaved crops.

Henbit dead-nettle *L. amplexicaule*, a widespread annual archaeophyte, especially common in England, east Scotland and south-east Ireland, on open, cultivated and waste ground, usually on light, dry soils. It also occurs on walls, by railways and in pavement cracks. Its English name partly reflects its consumption by poultry (hen bite). *Amplexicaule* is Latin for 'clasping the stem'.

Northern dead-nettle *L. confertum*, an annual archaeophyte, locally frequent near the coast in Scotland and the Isle of Man, a scarce casual elsewhere, on cultivated and waste ground. *Confertum* is Latin for 'crowded'.

Red dead-nettle *L. purpureum*, a common and widespread annual archaeophyte frequently colonising fertile and disturbed soils, in cultivated and waste ground, gardens and hedgerows, on roadside verges, along railways, and in rough grassland.

Spotted dead-nettle *L. maculatum*, a rhizomatous or stoloniferous perennial neophyte herb introduced from Italy in 1683, naturalised on rough and waste ground, and roadsides, usually close to habitation, scattered but mostly in England and north-east Scotland. *Maculatum* is Latin for 'spotted'.

White dead-nettle *L. album*, a rhizomatous, sometimes stoloniferous perennial archaeophyte herb, common and widespread in lowland Britain except scarce in north and west Scotland, and local in Ireland, found in secondary woodland, and on hedge banks, waysides and rough ground, often growing on fertile soil close to habitation. *Album* is Latin for 'white'.

DEATHCAP *Amanita phalloides* (Agaricales, Amanitaceae), a generally infrequently found fungus but commoner in some localities during some years, more local in Scotland and Ireland. It is lethally poisonous: several toxins have been isolated but the constituent that damages the liver and kidneys is α-amanitin. Without treatment, coma and eventual death are almost inevitable. People hospitalised late into an amotoxin poisoning episode can only be saved by a liver transplant, and even then recovery is painful, protracted and not guaranteed. It is found in summer and autumn, on soil in broadleaf woodland, usually with oak. *Amanita* is Greek for this group of fungi; *phalloides*, from Latin *phallus*, reflects the phallic shape.

False deathcap *A. citrina*, a widespread, common inedible fungus of late summer and autumn, on soil in (especially beech) woodland. *Citrina* is Latin for 'yellow', from the cap colour.

DECAPODA Order in the class Malacostraca that includes crabs, shrimps, prawns, crayfish and lobsters, the name coming from the Greek for 'ten' + 'legs' (having five pairs of walking limbs).

DECEIVER *Laccaria laccata* (Agaricales, Hydnangiaceae), a widespread and very common edible fungus found from early summer to early winter on soil in mixed woodland and on heathland, ectomycorrhizal on a number of trees, including pines, beech, oak and birch. 'Deceiver' may derive from the very variable appearance. *Laccatus* is Latin for 'varnished'.

Amethyst deceiver *L. amethystina*, edible, very common and widespread, found from early summer to winter on woodland soil and leaf litter, especially under beech with which it is ectomycorrhizal.

Bicoloured deceiver *L. bicolor*, edible, widespread and locally common, found from early summer to winter on nutrient-poor acidic woodland and heathland soil, especially under pine and birch.

Gumtree deceiver *L. fraterna*, recorded from a few mostly southern English sites on soil under *Eucalyptus*, occasionally willow. *Fraterna* is Latin for 'brotherly'.

Sand deceiver *L. maritima*, with only three records since 1999 (all on sandy soils in north-east Scotland), Critically Endangered in the RDL (2006).

Scurfy deceiver *L. proxima*, edible, widespread and locally common, found in autumn on nutrient-poor acid soil under damp mixed woodland, boggy heathland and moorland. *Proxima* is Latin for 'next', from the close appearance to the type species.

Twisted deceiver *L. tortilis*, edible, widespread but uncommon, found from early summer to winter on bare, heavy soil in damp woodland, often under willow and alder. *Tortilis* is Latin for 'twisted'.

DEER Mammals in the family Cervidae. 'Deer' comes from Old English *deor*, also originally referring to any quadruped. See also muntjac and reindeer.

Barking deer Alternative name for Reeves's muntjac.

Chinese water deer *Hydropotes inermis*, shoulder height 42–65 cm, head-body length 82–106 cm, tail 2.5–9 cm, male weight 12–18.5 kg, females 14–17.5 kg, introduced from east China to Woburn Park, Bedfordshire, in 1896, and released into surrounding woodlands from 1901 onwards, but first reported in the wild in Buckinghamshire in 1940. It has a limited range in East Anglia and adjacent counties, and while continuing to spread had an estimated population in 2005 of <2,000. However, this is of international importance, because of increased threats to this species' existence (indeed, possible extinction) within its natural range. It prefers wetlands adjoining woodland and fen, though often forages on farmland, eating herbs, and some browse mainly at dawn and dusk. The rut is in December, when males fight with their tusks (downward-pointing elongated canine teeth). Gestation lasts 160–210 days. The female often has twins, sometimes up to four kids, in May–June, which, although weaned in three months, remain with their mother into winter. Greek *hydor* + *potes* = 'water' + 'drinker', and Latin *inermis* = 'weaponless', references to the lack of antlers and an association with wetland.

Fallow deer *Dama dama*, shoulder height 50–120 cm, head-body length 140–80 cm, tail 14–21 cm, male weight 40–63 kg, females 30–44 kg, a south-west Asian deer found in archaeological sites of the Roman period (e.g. Fishbourne, West Sussex), but probably permanently introduced by the Normans to both Britain and Ireland in around 1100 for hunting, now the most wide-spread deer in Britain (less common in Scotland) and Ireland,

mainly in deciduous woodland (especially where close to grassy habitat), farmland and parks. Estimated numbers in 2005 were 95,000 in England, <5,000 in Wales and <8,000 in Scotland. Some 60% of the diet is of grasses, with herbaceous plants and broadleaf browse also taken, plus acorns, beech mast and fruit in autumn. The rut is in October–November. Gestation is 230–5 days. Fawns (occasionally twins) are born in June–July, and are weaned by the time of the next rut. Females (does) usually first breed as yearlings. Outside the rut, males (bucks) usually live in small bachelor herds (5–10), away from the does and fawns. Fallow deer can be a pest in commercial forestry, eating leader shoots and stripping bark, and farmland, eating cereal. Culling, with sale of venison, is effective in limiting numbers, which are nevertheless generally increasing slightly, and spreading. 'Fallow' comes from Old English *f(e)alu* referring to the coat's pale brown or reddish-yellow colour. *Dama* is a Latin term covering roe deer, gazelle and antelope.

Red deer *Cervus elaphus*, up to 1.5 m at the shoulder, males nose-tail length 1.5–2 m, male weight 100–80 kg (up to 250 kg), females 1 m at the shoulder, 1–1.5 m length, 70–130 kg, Britain's largest land mammal, particularly associated with moorlands of the Scottish Highlands and Islands, with scattered populations found in north-west England, East Anglia, Exmoor and Ireland (where preferred habitats are woodland–grassland transition areas). Recent increases in numbers have been associated with the rise of deer stalking, under-culling of females, and colonisation of forestry plantations. In some parts of Scotland, density is so high that regeneration of native woodland has been prevented and higher culling rates are required. Culling of red deer for sport, meat (venison) or management is significant, with around 70,000 animals killed annually, though in many parts of the Highlands, annual culling rates (of 6–12% of hinds, 10–17% of stags) have not prevented population increases. Some 347,000 individuals were estimated in Scotland in 2005, 8,000 in England, <500 in Wales, and 4,000 in Ireland. Hybridisation with introduced sika deer poses a major threat to the genetic integrity of red deer, and in the Lake District and Ireland (where the Co. Kerry population may be the only original one remaining), populations are now essentially red-sika hybrids. About 80% of the Scottish population live in open-hill habitats year-round. Males (stags) and females (hinds) usually remain in separate groups for much of the year, females tending to monopolise relatively grass-rich habitats, males using heather-dominated areas. In winter red deer usually concentrate on sheltered lower ground, moving to higher feeding areas in summer. Summer diet is mainly of grasses, sedges and rushes; dwarf shrubs such as heather and bilberry become important in winter. Young trees are also browsed. Mating takes place from the end of September to November: during the rut, mature stags, 5–6 years old, compete, via roaring and fighting (using antlers) to defend groups of 10–15 hinds. Antlers, only grown by the males, are shed and regrown each year. Mature stags can develop up to 12 points (tines) on the antlers. Following the rut, stags and hinds again segregate. Calves (usually one, occasionally twins) are born from mid-May to early June, and are weaned by eight months old. Latin *cerva* and Greek *elaphos* both mean 'deer'.

Roe deer *Capreolus capreolus*, shoulder height 60–75 cm, weight 10–25 kg (males slightly heavier), now common and wide-spread in Scotland and parts of England. They became extinct in England, Wales and south Scotland during the eighteenth century but populations were reintroduced to south England (Dorset) and East Anglia in the nineteenth century, and there has been a substantial expansion in range, including into Wales, during the latter half of the twentieth century into the twenty-first. Estimated numbers in 2005 were 150,000 in England, 350,000 in Scotland and <1,000 in Wales. They are generally found in open woodland, especially at the edge between trees and open habitats (including arable), particularly active at dusk and dawn, browsing on buds, shoots and leaves. The rut is in

July–August. Mating is followed by delayed implantation (the only hoofed animal in which this occurs), gestation starting in early January. Females give birth, usually to twins, sometimes to single kids or triplets, between mid-May and mid-June. They have become a local pest of commercial forest and farmland and in places are exploited in game shoots and for meat. 'Roe' comes from Old English *raha*, possibly referring to a striped or spotted pattern. *Capreolus* in Latin for 'little goat'.

Sika deer *Cervus nippon*, shoulder height 50–120 cm, head-body length 140–80 cm, tail length 14–21 cm, male weight 40–63 kg, females 30–44 kg, native of Japan and China, introduced to Powerscourt Park, Co. Wicklow, from where many introductions were made to English parks from 1874 onwards, and at the same time this deer was brought to Scottish parks. Escapes and releases followed. In England current populations are around 200 in the Forest of Bowland (Lancashire), the Lake District, the New Forest, and south-east Dorset. Numbers in England were estimated in 2005 as 2,500, though it may now be double this figure, and 10,000 in Scotland. Figures everywhere, however, are confounded by high rates of hybridisation with red deer, and pure sika stock may actually be rare. Culling has reduced stock, especially in south England. Sika favour both woodland and grassland, browsing on and damaging trees, especially in winter (and are often seen as a forestry pest), and feeding on grasses and heather in summer. Calves are born in May–June, after a gestation of 220 days, and are weaned by the time of the next rut in October. Rutting includes antler-fighting between males. 'Sika' comes from Japanese *shika* = 'deer', and *nippon* = 'Japanese'. *Cervus* comes from Latin for 'deer'.

DEERFLY Members of the genus *Chrysops* (Tabanidae) (Greek *chrysos* = 'gold'), 8–10 mm body length, often found in damp environments. Larvae feed on small insects and pupate in mud at the water's edge. Adult females feed on mammalian blood in order for their eggs to mature. When they bite, they inject saliva with an anti-coagulant agent that prevents the blood clotting.

Black deerfly *C. sepulcralis*, mainly recorded from the New Forest in Hampshire and east Dorset. Larvae feed on organic matter. *Sepulcralis* is Latin for 'of graves'.

Splayed deerfly *C. caecutiens*, common and widespread in England and Wales, occasional in Scotland. Larvae feed on algae and organic matter in damp muddy soils. Adults are active from May to September; males and females feed on nectar and pollen. *Caecutiens* is Latin for 'blind'.

Square-spot deerfly *C. viduatus*, with a scattered distribution in England and Wales in wet meadows, valley mires and wet woodland. Larvae feed on organic matter in damp muddy soils. Adults are active from May to September. *Viduatus* is Latin for 'deprived'.

Twin-lobed deerfly *C. relictus*, widespread but scattered, preferring damp floodplain meadows. Larvae feed on organic matter and small invertebrates in damp muddy soils. Adults are active from May to September; males feed on pollen. *Relictus* is Latin for 'abandoned'.

DEERGRASS *Trichophorum germanicum* (Cyperaceae), widespread and locally common on peat on mires, moorland and wet heath. The binomial is Greek for 'bearing hair' and Latin for 'German'.

Northern deergrass *T. cespitosum*, locally common in south-west England, scattered elsewhere on wet peaty moors and bogs over acidic soils, persisting in burnt and deer-grazed areas. *Cespitosum* is Latin for 'tufted'.

DELICATE *Mythimna vitellina* (Noctuidae, Hadeninae), wingspan 36–43 mm, a migrant nocturnal moth commonest on the south and south-west coasts of England and south and east coasts of Ireland usually between August and October, sometimes earlier. Some of the earlier migrants may breed, giving rise to autumn adults. Mithimna is a town on the Greek

island of Lesbos; Latin *vitellus* = 'calf', possibly from the wings' calfskin colour.

DELPHINIDAE Family name of toothed whales, in particular dolphins (for which the Latin is *delphinus*).

DEMOISELLE FLY Members of the order Odonata, suborder Zygoptera, including species of *Calopteryx* (Calopterygidae) (Greek *calo* + *pteryx* = 'beautiful' + 'wing'). Also known as damsel flies.

Banded demoiselle *C. splendens*, length 45 mm, widespread but absent in Scotland. Nymphs are associated with large, slowly moving rivers and canals, especially those with muddy bottoms. The flight period is May–August. *Splendens* is Latin for 'bright'.

Beautiful demoiselle *C. virgo*, length 45 mm, widespread if scarcer towards the east and north, and found in Scotland only in the north, mainly found along streams and rivers, especially those with sand or gravel bottoms. The main flight period is May–August. *Virgo* is Latin for 'damsel'.

DENDROPHYLLIIDAE Family of hard corals with 12 species in 5 genera (Greek *dendros* + *phyllon* = 'stick' + 'leaf').

DEODAR *Cedrus deodara* (Pinaceae), introduced from western Asia in 1831, mostly planted in parks and gardens in England and east Scotland, with some records of self-sown trees. *Cedrus* is Latin for cedar; *deodara* comes from Sanskrit *devadaru* = 'divine'.

DEPRESSARIIDAE Family name for flat-body moths (Latin *depressus* = 'flat' or 'pressed down').

DERMAPTERA Order of earwigs (Greek *derma* + *pteron* = 'skin' + 'wing').

DERMESTIDAE From Greek *derma* + *esthio* = 'skin' + 'eat', a family of 40 mostly small (1.5–12 mm) beetles. Many are pests of stored products. The 'woolly bear' larvae may be common where protein-rich food sources are available, including carrion, leather, bones even dead insects in light fittings. See carpet, larder and hide beetles, and museum beetle under beetle.

DERODONTIDAE Beetle family (Greek *dere* + *odontos* = 'neck' + 'tooth') with one British species, the scarce *Laricobius erichsoni*, 3–4 mm, first recorded in 1971, found on conifers where both adults and larvae prey on adelgids.

DESMID A large group of unicellular freshwater green algae in the order Desmidiales (Charophyta), common on the surface of ponds and pools, often forming a green film, on mud and aquatic plants, and between sphagnum moss plants. Greek *desmos* = 'chain'.

DESPERATE DAN *Parmotrema crinitum* (Parmeliaceae), a widespread foliose lichen, especially common in oceanic western regions, on rocks, coastal cliffs and mossy trunks, particularly in ancient woodland for which it is an indicator. The English name comes from the black cilia on the surfaces and margins, echoing the chin stubble on the *Dandy* (comic) character Desperate Dan. Latin *parma* + Greek *trema* = 'small shield' + 'perforation', and Latin *crinis* = 'hair'.

DEUTZIA *Deutzia scabra* (Hydrangeaceae), a deciduous Far Eastern neophyte shrub possibly in gardens by 1822, scattered in Britain, planted in woodland and hedgerows, as a garden escape on waste ground, and as a relict of cultivation. Johan van der Deutz was an eighteenth-century Dutch patron of botany.

DEVIL'S COACH HORSE *Ocypus olens* (Staphylinidae, Staphylininae), up to 28 mm length, common and widespread in damp habitats in April–October. During the day it rests under stones and logs; at night it emerges to feed on slugs, worms, spiders, woodlice, other invertebrates and carrion. Its characteristic defensive pose is to raise its rear end, echoing a scorpion posture. If it continues to feel threatened it can emit a foul smell from its abdomen (Latin *olens* = 'smell'), can excrete an unpleasant fluid from its mouth and rear, and will also bite. Mating is in autumn. A female will lay a single egg at a time in a damp location such as leaf litter or moss. Larvae, which possess powerful jaws, live underground feeding on other invertebrates; they also can adopt the threatened display of raised rear and open jaws. Greek *ocys* + *pous* = 'swift' + 'foot'.

DEVIL'S FINGERS *Clathrus archeri* (Phalales, Phallaceae), an inedible fungus introduced from Australasia in the 1910s, local in southern and south-east England in summer and autumn, with arching red 'arms' extending from a partially submerged 'ball', and foul-smelling, on soil among leaf litter in grassy habitats close to trees. *Clathrus* is Latin for 'cage'. *Archeri* probably honours a Tasmanian amateur naturalist William Archer (1820–74).

DEW-MOSS Blue dew-moss *Saelania glaucescens* (Ditrichaceae), a rarity growing on mineral soil in sheltered sites on calcareous crags in the Highlands. Anders Saelan (1834–1921) was a Finnish botanist. Latin *glaucescens* = 'glaucous'.

DEWPLANT Species in the family Aizoaceae.

DIADUMENIDAE Family name of orange and orange-striped (sea) anemone.

DIAPENSIA *Diapensia lapponica* (Diapensiaceae), a long-lived, slow-growing, dense, cushion-forming evergreen dwarf shrub, restricted to a rock outcrop on an exposed mountain ridge at 760 m near Fort William, on acidic fellfield soil. The population (about 1,200 clumps) and the area it occupies have not changed markedly since discovery in 1951. *Diapensia* is a classical Greek name adopted by Linnaeus; *lapponica* is Latin for 'from Lapland'.

DIATOM Chrysophyta (Bacillariophyceae), a class of microscopic, single-celled, solitary or colonial freshwater or marine algae (also found in damp soil), with hard, siliceous cell walls, which occur in many forms. Some species form a brown scum on stones and other algae. Greek *diatomos* = 'cut in two'.

DICRANACEAE Family name of fork-mosses and sprig-moss (Greek *dicranon* = 'pitchfork').

DICTYNIDAE From Greek *dictyon* = 'net', the family name of mesh-webbed spiders (mesh-weavers), 1–4 mm body length, with 18 British species in 9 genera.

DICTYOPTERA Order of cockroaches (Greek *dictyon* + *pteron* = 'net' + 'wing').

DIGENEA From Greek *dis* + *genos* = 'double' + 'race', a subclass of the phylum Platyhelminthes, class Trematoda (flukes).

DIGGER WASP Members of the family Crabronidae (Latin *crabro* = 'hornet'), with over 110 British and Irish species. Females dig burrows in which to lay her eggs.

Blunt-tailed digger wasp *Crossocerus dimidiatus*, widespread, commoner in the north, flying from May to August, nesting in existing cavities such as burrows in rotten wood or soft mortar between stones or bricks, cells provisioned with snipe flies and other dipterans. Greek *crossos* + (possibly) *ceros* = 'fringe' + 'wax', and Latin *dimidiatus* = 'divided'.

Common spiny digger wasp *Oxybelus uniglumis*, widespread in England and Wales, rare in south Scotland and occasional, near the coast, in Ireland, usually found on open patches of bare, loose sand but also on heavier soils, such as in open woodland, flying from June to September. Burrows are dug in flat or sloping bare, sandy soil. The female closes her nest by scraping sand into the entrance, and begins hunting. Dipteran (mostly muscid) prey is captured in mid-air and on vegetation. Each cell is provisioned with 2–16 paralysed flies. After taking the last fly in, the female oviposits, then digs the next cell, with usually two or three cells in a nest. Greek *oxy* + *belos* = 'sharp' + 'sting', and Latin *uniglumis* = 'single husk'.

Field digger wasp 1) *Argogorytes mystaceus*, female body length 10–14 mm, males 8–11 mm, widespread but more scattered in Scotland and Ireland, in sunny sites especially in deciduous woodland edges and glades from late April to June. Both sexes visit a number of flowers; males are important pollinators of fly orchid. The nest is dug in soil in dry banks in moist woodland glades; females then prey on froghopper nymphs, breaking into the nymph's protective spittle, to provision her nest with up to 30 items for her developing young. The scarcer *A. fargeii* has a similar life history. Greek *argos* + *gorytos* = 'bright' + 'quiver', and *mystax* = 'moustache'. 2) *Mellinus arvensis*, body length 12–14 mm, widespread, favouring sandy, often coastal habitats, flying from June to September. Predation is mainly on Diptera which is sometimes hunted on mammalian droppings. Nests are constructed in sandy soil (sometimes under paving slabs), often in aggregations, excavated material left as small mounds. The burrow goes down 30–40 cm almost vertically; there may be one or several cells, each provisioned with 4–13 flies. Latin *mellinus* = 'like honey', and from *arvum* = 'field'.

Large-spurred digger wasp *Nysson spinosus*, female body length 8–12 mm, males 7–10 mm, widespread in England and Wales, locally common in Scotland and Ireland, kleptoparasitic on and found in the same habitats its host species *Argogorytes fargei* and, particularly, field digger wasp; the female finds the host nest by scent, enters the burrow and lays an egg on the prey at the bottom of the cell. After hatching, the *Nysoon* larva destroys the host egg and feeds on the stored prey. Greek *nysso* = 'to prick', and Latin *spinosus* = 'spiny'.

Ornate-tailed digger wasp *Cerceris rybyensis*, female body length 8–12 mm, males 6–10 mm, locally common and widely distributed in the southern half of England on sandy soils but also recorded on chalk grassland and heavier soils, flying in summer, visiting a variety of flowers. Nesting is in dense aggregations as burrows on level, exposed compacted soil: sites have included an unsurfaced road and an abandoned sand quarry in Dorset, and soil between paving stones in Greater London. Larvae are provisioned with small and medium-sized bees of various genera, which are paralysed by stinging. Greek *cercho* = 'to be harsh'.

Pale-jawed spiny digger wasp *Oxybelus mandibularis*, scarce in southern England, East Anglia and west Wales, on the coast and in heathland. Cells in nest burrows are provisioned with Diptera. *Mandibularis* is from Latin *mandibula* = 'jaw'.

Sand-tailed digger wasp *Cerceris arenaria*, locally common and widely distributed in England and Wales on sandy soils on coastal dunes and landslips and in heathland, flying in summer, preying on weevils and visiting flowers for water on hot days. Nests are deep burrows in the soil, from which side-tunnels radiate. Dense aggregations containing thousands of nests may form. Each cell is provisioned with 3–14 prey: a colony of 1,000 nests would cull around 100,000 weevils each year. *Arenaria* is Latin for 'relating to sand'.

Silver spiny digger wasp *Oxybelus argentatus*, found very locally on the coast (especially on dunes) and on inland heathland in southern England and west Wales, with records for Ireland from Co. Wexford. Cells in nest burrows are provisioned with Therevidae (Diptera). *Argentatus* is Latin for 'silvery'.

Slender-bodied digger wasp *Crabro cribrarius*, body length 12–14 mm, locally common and widespread in Britain, mainly on light, sandy soils, such as on lowland heathland and dunes, but also on heavier soils, such as open woodland and chalk grassland. Flight is in summer and early autumn, visiting mainly umbellifer flowers for nectar. Nest burrows are excavated in the soil. Cells are provisioned with paralysed flies. *Cribrarius* is Latin for 'sifted'. Other *Crabro* species, with similar biologies, are *C. peltarius* (locally common, especially in south England and Wales) and *C. scutellatus* (southern England).

Slender digger wasp *Crossocerus elongatulus*, found throughout Britain in a range of habitats, including scrub and woodland edge, probably bivoltine, flying from May (in the south) or June (north) to September. The female usually nests in the ground, but has also been found in the soft mortar of brickwork and stone walls, and in dead wood. Cells are provisioned with small dipterans. *Elongatulus* is Latin here meaning 'slender'.

Small-spurred digger wasp *Nysson dimidiatus*, scattered and locally common in England and Wales, a kleptoparasite of *Harpactus tumidus*, occurring in the same sparsely vegetated sandy areas as its host – heathland, coastal dunes, coastal land slips, open areas in woodland, embankments and occasionally gardens. Adults are in flight in June–November. Nectar is taken from umbellifers. *Dimidiatus* is Latin for 'divided'.

Spiny digger wasps Members of the genus *Oxybelus*.

Wesmael's digger wasp *Crossocerus wesmaeli*, locally common and widespread in England and Wales on sandy soils, including dunes, active in May–September. Nests are provisioned with small dipterans. Constantin Wesmael (1798–1872) was a Belgian entomologist.

White-mouthed digger wasp *Crossocerus leucostomus*, recorded mainly from pinewood flying in May–August, from Scotland and northern England south to Yorkshire, Rare in RDB 3 (1987), in 1991 considered to be Nationally Scarce. Nests are constructed in dead wood in warm, sunny situations, and cells are provisioned with small dipterans. *Leocostomus* is Greek for 'white-mouthed'.

DILL *Anethum graveolens* (Apiaceae), an aromatic annual archaeophyte grown as a culinary herb for at least a millennium, widespread but scattered as a casual in gardens and waste ground, as a garden escape or from birdseed or grain impurities, favouring light, well-drained soils. 'Dill' comes from Old English *dile* or *dyle*. *Anethum* is the Latin for anise or dill, *graveolens* for 'strong-scented'.

DINOFLAGELLATA Phylum of eukaryotic flagellate protists (Greek *dinos* = 'whirling', Latin *flagellum* = 'whip'). Most are marine plankton, but they are also common in freshwater habitats.

DIOSCOREACEAE Family name for black bryony, after the first-century CE Greek herbalist Dioscorides.

DIPLOPODA Class (subphylum Myriapoda) of millipedes, characterised by having two pairs of jointed legs on most body segments (Greek *diplo* + *podi* = 'double' + 'foot'). The British list recognises 72 species in 7 orders.

DIPLURA Order of two-pronged bristletails (Greek *diplo* + *oura* = 'two' + 'tail'). See bristletail (2).

DIPPER *Cinclus cinclus* (Cinclidae), wingspan 28 cm, length 18 cm, weight 64 g, a resident breeder, Amber-listed in Britain, Green in Ireland, found along fast-flowing rivers, mainly in upland areas but also on lowland rivers in south-west England. A broad estimate of breeding pairs in 2009 was 6,200–18,700. Numbers declined in Britain by 22% between 1975 and 2015, and 20% between 1995 and 2015, though these figures include a slight increase (7%) over 2010–15. Riverside nests, in holes, cracks and crevices, usually have 4–5 eggs which hatch after around 17 days, the altricial but downy birds fledging after 21–2 days, though if endangered pre-fledged chicks will escape into the water. Feeding is on large invertebrates, especially caddis flies, mayfly larvae and shrimps, caught from (often while walking along) the stream-bed. It characteristically bobs up and down when perched, giving it its English name. The Greek *cinclos* was used by Aristotle and others to describe small tail-wagging birds that live near water.

DIPRIONIDAE Family name of some sawflies (Greek *di* + *prion* = 'two' + 'saw').

DIPSACACEAE Family name of teasel, together with scabious, from Greek *dipsa* = 'thirst', referring to the attached leaf bases found in some species which can hold water.

DIPTERA Order of insects known as true flies (Greek *di* + *pteron* = 'two' + 'wing'), characterised by a pair of flight wings on the mesothorax and a pair of halteres, derived from the hind wings, on the metathorax. Members consume only liquid or fine granular foods (e.g. pollen), and their mouthparts and digestive tracts show various modifications for such diets. The Dipterists

Forum (2015) lists 7,094 British species in 107 families, with 3,386 species noted for Ireland (2013 update).

Suborders and families (in bold, see separate entries) (common names, number of genera/species) are: **Suborder Nematocera** (Lower Diptera) – **Anisopodidae** (wood gnats, 1/4), **Bibionidae** (2/18), Bolitophilidae (some fungus gnats, 1/16), **Cecidomyiidae** (gall midges, 148/650), **Ceratopogonidae** (biting midges, 20/171), **Chaoboridae** (phantom midges, 2/6), **Chironomidae** (non-biting midges, 140/619), **Culicidae** (mosquitoes, 8/34), Cylindrotomidae (long-bodied crane flies, 4/4), Diadocidiidae (1/3), Ditomyiidae (some fungus gnats, 2/3), Dixidae (meniscus midges, 2/15), Keroplatidae (some fungus gnats, 15/52), Limoniidae (limoniid craneflies, 50/217), Mycetophilidae (some fungus gnats, 61/482), Mycetobiidae (1/3), Pediciidae (hairy-eyed craneflies, 4/20), **Psychodidae** (moth flies, 21/99), Ptychopteridae (phantom craneflies, 1/7), **Scatopsidae** (black scavenger flies, 18/46), **Sciaridae** (dark-winged fungus gnats, 22/266), **Simuliidae** (blackflies, 3/35), Thaumaleidae (1/3), **Tipulidae** (craneflies, 8/87) and Trichoceridae (winter craneflies, 1/3).

Suborder Brachycera – Acartophthalmidae (1/2), Acroceridae (2/3), **Agromyzidae** (some leaf miners, 18/394), **Anthomyiidae** (root-maggots, etc. 29/245), **Anthomyzidae** (10/21), **Asilidae** (robber flies, 16/29), Asteiidae (3/8), Atelestidae (1/2), Athericidae (3/3), Aulacigastridae (1/1), **Bombyliidae** (bee-flies and villas, 4/9), Borboropsidae (1/1), Brachystomatidae (3/4), Braulidae (1/2), **Calliphoridae** (blow flies, bluebottles, greenbottles and cluster flies, 14/38), Camillidae (1/5), Campichoetidae (1/2), Canacidae (5/11), Carnidae (2/13), Chamaemyiidae (7/32), Chiropteromyzidae (1/1), **Chloropidae** (frit flies and grass flies, 43/177), Chyromyidae (3/11), **Clusiidae** (druid flies, 3/10), Cnemospathidae (1/1), Coelopidae (2/3), **Conopidae** (thick-headed flies, 7/23), Diastatidae (1/6), **Dolichopodidae** (long-legged flies, 46/301), **Drosophilidae** (some fruit flies, 13/63), Dryomyzidae (2/3), **Empididae** (dagger flies, 13/208), **Ephydridae** (shore flies, 39/151), Fanniidae (2/60), Helcomyzidae (1/1), **Heleomyzidae** (13/56), Heterocheilidae (1/1), **Hippoboscidae** (louse flies, 10/14), **Hybotidae** (dance flies, 20/180), **Lauxaniidae** (13/56), **Lonchaeidae** (lance flies, 5/46), Lonchopteridae (spear-winged flies, 1/7), Megamerinidae (1/1), **Micropezidae** (stilt-legged flies, 5/10), Milichiidae (6/19), **Muscidae** (house flies, 40/288), Nycteribiidae (3/3), Odiniidae (1/9), **Oestridae** (bot flies and warble flies, 5/11), Opetiidae (1/1), **Opomyzidae** (2/16), **Pallopteridae** (trembling-wing flies, 2/13), Periscelididae (1/3), Phaeomyiidae (1/2), **Phoridae** (scuttle flies, 13/344), **Piophilidae** (10/14), Pipunculidae (big-headed flies, 14/95), **Platypezidae** (flat-footed flies, 10/33), **Platystomatidae** (signal flies 2/2), Pseudopomyzidae (1/1), **Psilidae** (rust flies, 6/26), **Rhagionidae** (snipe flies, 5/15), **Rhinophoridae** (6/8), **Sarcophagidae** (flesh flies, 16/61), **Scathophagidae** (dung flies, 23/54), Scenopinidae (2/2), **Sciomyzidae** (marsh flies, 23/70), **Sepsidae** (ensign flies, 6/29), Sphaeroceridae (lesser dung flies, 38/138), Stenomicridae (2/2), **Stratiomyidae** (soldier flies, 16/48), Strongylophthalmyiidae (1/1), **Syrphidae** (hoverflies and drone flies, 68/282), **Tabanidae** (horseflies, deerflies and clegs, 5/30), **Tachinidae** (145/266), Tanypezidae (1/1), **Tephritidae** (some fruit flies, 33/76), **Therevidae** (stiletto flies, 6/14), Trixoscelididae (1/4), **Ulidiidae** (picture-winged flies, 11/20), Xylomyidae (2/3) and Xylophagidae (1/3).

DISCO Inedible fungi in a number of genera, families and orders, a name usually given to members of the Ascomycota that often form disc-shaped structures, but including some basidiomycetes.

Rarities include **heath sedge disco** *Lachnum callimorphum*, **juniper disco** *Pithya cupressina*, **midnight disco** *Pachyella violaceonigra* and **pink disco** *Aleurodiscus wakefieldiae*.

Common grey disco *Mollisia cinerea* (Heliotales, Dermateaceae), common and widespread, wood-rotting, found throughout the year as small discs on dead oak and beech wood. Latin *mollis* and *cinerea* = 'soft' and 'grey'.

Conifer disco *Lachnellula subtilissima* (Helotiales, Hyaloscyphaceae), widespread but scattered and infrequent, found from early spring to autumn as small cups on dead twigs of fir and, less frequently, pine and spruce. Greek *lachnellula* is a diminutive of *lachnos* = 'woolly' or 'downy', and Latin *subtilissima* = 'very delicate'.

Hairy nuts disco *Lanzia echinophila* (Helotiales, Rutstroemiaceae), a small cup fungus fairly common in the southern half of England, often on the cupules (nut cases) of oak and, especially, sweet chestnut. *Echinophila* reflects Greek *echinos* (= 'hedgehog') to mean 'prickly'.

Larch canker disco *Lachnellula willkommii*, widespread but scattered and uncommon, found from early spring to autumn as small cups on dead twigs of larch. Heinrich Willkomm (1821–95) was a German mycologist.

Larch disco *Lachnellula occidentalis*, widespread but infrequent, found from early spring to autumn as small cups on dead twigs of larch and spruce. *Occidentalis* is Latin for 'western'.

Lemon disco *Bisporella citrina* (Leotiales, Heliotiaceae), common and widespread, wood-rotting, found throughout the year, fruiting from late summer to early winter on the decaying trunks, branches and stumps of oaks and other hardwood trees. The binomial is Greek for 'two-spored' and Latin for 'lemon-coloured'.

Nut disco *Hymenoscyphus fructigenus* (Helotiales, Helotiaceae), common and widespread as small white saucers on fallen beech mast, occasionally on acorns and hazel nuts, from spring to winter but usually in summer and autumn. Greek *hymen* + *scyphos* = 'membrane' + 'cup', and Latin *fructus* + *genus* = 'fruit' + 'sort'.

Rush disco *Lachnum apalum*, widespread but uncommon, found as small cups on soft-rush and other rush species. *Apalos* is Greek for 'soft'.

Snowy disco *Lachnum virgineum*, common and widespread, found from early spring to summer as small cups on fallen beech mast, other plant debris, and occasionally bramble stems. *Virgineum* comes from Latin *virginens* = 'maidenly'.

Sulphur disco *Bisporella sulfurina* (Heliotiaceae), common, wood-rotting, particularly widespread in England, found throughout the year, fruiting from late summer to early winter, on old mycelia of other fungi which have fruited on the decaying trunks, branches and stumps of oaks and other hardwood trees.

DISCOMYCETE, DISCOMYCOTA Former names of, respectively, the class and division of fungi, the class including the cup, sponge, brain and some club-like fungi, now referred to as ascomycete and Ascomycota.

DISTEMPER See phocine distemper.

DISTICHIUM **Fine distichium** *Distichium capillaceum* (Ditrichaceae), a locally common moss on base-rich mountains in Wales, northern England, Scotland and coastal Ireland on damp rock ledges or in crevices. It also occurs on dunes and has been recorded from mortared walls. Greek *distichos* = 'in two rows', Latin *capillus* = 'hair'.

Inclined distichium *D. inclinatum*, scarce in bryophyte-rich turf on damp shell sand in the north and west of Britain, and rare on damp ledges of calcareous rocks, and occasional on old wall mortar in south Wales.

DITCH DUN Angler's name for the mayfly *Habrophlebia fusca* (Ephemerellidae).

DITRICHUM (MOSS) **Alpine ditrichum** *Ditrichum zonatum* (Ditrichaceae) on shallow, stony soils in high mountains in Wales, Cumbria and Scotland (descending lower in north-west Scotland), particularly on with disturbed soils with late snow-lie, or wind-scoured ridges. Greek *di* + *tricha* = 'two' + 'hair', Latin *zonatum* = 'banded'.

Awl-leaved ditrichum *D. subulatum*, on acidic banks under oak and above coastal creeks in Cornwall and Pembrokeshire. *Subula* is Latin for 'awl'.

Bendy ditrichum *D. flexicaule*, widespread and scattered, calcicolous, more common in England and Wales on upland limestone rocks, and on dunes and calcareous grassland. *Flexicaule* is Latin for 'bent stem'.

Brown ditrichum *D. pusillum*, uncommon on sandy or gravelly, acidic quarries and banks, and recorded from a few acidic arable fields. *Pusillum* is Latin for 'small'.

Curve-leaved ditrichum *D. heteromallum*, common and widespread except for central and eastern England and central Ireland, on disturbed mineral soil in such habitats as heathy banks, old quarries, streamsides and gravel tracks. In Scotland it is abundant by forest tracks. Greek *hetero* + *mallos* = 'different' + 'skein of wool'.

Cylindric ditrichum *D. cylindricum*, common and widespread in acidic habitats, in stubble, and sand- and gravel-pits, on stream banks, roadsides, and sometimes seasonally flooded ground at the edge of lakes.

Dark ditrichum *D. lineare*, in upland Cornwall, Wales, Cumbria and the Highlands on bare, often disturbed, acidic mineral soils on sites such as exposed, montane ridges, areas of late-lying snow, scree and lower down by forest tracks. *Lineare* is Latin for 'linear'.

Slender ditrichum *D. gracile*, calcicolous, widespread and locally abundant on crag ledges, calcareous grassland and dunes. *Gracile* is Latin for 'slender'.

DITTANDER *Lepidium latifolium* (Brassicaceae), a rhizomatous perennial herb native as far north as south Wales and East Anglia, scattered and naturalised or casual elsewhere, found on creek sides, ditches, sea walls, brackish grassland and upper saltmarsh. It is also naturalised in disturbed areas such as waste ground, dockland, railways and roadsides. Greek *lepis* = 'scale', referencing the shape of the seed pods, and Latin *latifolium* = 'broad-leaved'.

DITTANY **False dittany** *Ballota acetabulosa* (Lamiaceae), a herbaceous perennial European neophyte naturalised in a sod-hedge on the Isle of Man since the 1930s. 'Dittany' is late Middle English, from Old French *ditain* and Greek *dictanos*, from its presence on Mt Dicte, Crete. Latin *ballota* = 'little ball', and *acetabulum* = 'vinegar cup'.

DIVER Birds in the genus *Gavia* (Gaviidae), from Latin for 'sea mew' or 'common gull'. Legs are far back on their bodies to optimise swimming, and they are ungainly on land, only coming ashore to breed. **White-billed diver** *G. adamsii* is a scarce winter visitor averaging six birds a year. **Pacific diver** *G. pacifica* is an accidental first recorded in north Yorkshire in 2007, with <20 records since then.

Black-throated diver *G. arctica*, wingspan 120 cm, length 66 cm, male weight 3.4 kg, females 2.3 kg, breeding in north Scotland, mainly on the west, usually in lochans, and in Orkney and Shetland, with 220 pairs estimated in 2006. Winter migrants from Eurasia boosted numbers to 560 individuals in 2008–9, found much of the coastline except either side of the Irish Sea. Feeding is on marine and freshwater fish. Two eggs are usually laid, incubation lasts 28–30 days, and the precocial chicks fledge in 60–5 days. This is an Amber-listed species.

Great northern diver *G. immer*, wingspan 137 cm, length 80 cm, weight 4 kg, an Amber List fish- and crustacean-eating winter visitor from Greenland and Iceland, recorded from most coastlines where it is usually solitary, often farther out to sea than other divers (and occasionally seen inland on reservoirs), with 2,500 individuals estimated in 2008–9. It does very rarely breed, the first record of this being from Wester Ross (north-west Highlands) in 1970. *Immer* may be Germanic or Swedish, referring to fire ash and therefore to the dark colour of the plumage.

Red-throated diver *G. stellata*, wingspan 111 cm, length 61 cm, weight 1.6 kg. Evidence suggests a 30% decrease from 1,350 pairs in the 1980s to 945 pairs in 1994, but recovery means there are now around 1,300 pairs breeding in west and north Scotland,

Orkney and especially Shetland, as well as north-west Ireland, and it is a Green List species. Some 17,000 birds, including Eurasian migrants, were estimated in 2006 to be wintering on all coastlines. Feeding is on marine and freshwater fish. Breeding is usually in small water bodies in moorland. Two eggs are usually laid, incubation lasts 26–9 days, and the precocial chicks fledge in 47–54 days. *Stellata* is Latin for 'set with stars', a reference to the speckled back of the non-breeding plumage.

DIVING BEETLE Members of the genus *Dytiscus* (Dytiscidae, Dytiscinae) (Greek *dytes* = 'diver'). Both adults and larvae are predaceous. The larva passes digestive juices down hollow mandibles into the prey, turning the inside into a soup which the larva then sucks up, leaving the exoskeleton and hard parts. Adults have spurs on their tibia sharp enough to draw blood if mishandled, and larvae have mandibles strong enough to puncture human skin. Adults also secrete a steroid-based fluid from their thoracic glands as a defence.

Great diving beetle *D. marginalis*, length of around 30 mm and one of Britain's largest beetles, in still or slow-running water, preferably with vegetation, widespread and common, and evident throughout the year. Adults and larvae both predate anything they can tackle, including other water insects, tadpoles and very small fish. Larvae are lighter than water so must actively swim to stay submerged, or hold on to something. They breathe through two spiracles at the tip of the abdomen. They need damp soil by the edge of the water in order to pupate successfully. Adults periodically come to the surface, extruding the tip of the abdomen to replenish the air supply kept under the wings. They leave the water and fly to colonise new ponds. *Marginalis* is Latin for 'marginal'.

King diving beetle *D. dimidiatus*, the largest diving beetle with a length of 32–8 mm, with scattered records in England, having RDB 3 status (rare but not in immediate danger of extinction). Adults are actively predatory; larvae ambush prey and may prefer shaded waters. *Dimidiatus* is Latin for 'divided'.

DOCK Herbaceous plants in the genus *Rumex* (Polygonaceae), from Old English *docce*, and Latin *rumex* = 'sorrel'.

Argentine dock *R. frutescens*, a perennial South American neophyte naturalised on dunes in south-west England (e.g. Braunton Burrows, north Devon since 1929), and south Wales (e.g. Kenfig Dunes, Glamorgan, since 1934); locally in East Anglia and the Scottish Borders, and elsewhere, as a casual from wool shoddy and grain, especially around docks (e.g. Swansea). *Frutescens* is Latin for 'becoming shrubby'.

Broad-leaved dock *R. obtusifolius*, a very common and widespread perennial of field margins, hedge banks, roadsides, stream and riverbanks, ditches and former cultivated ground. Classified as injurious under the Weeds Act (1959). *Obtusifolius* is Latin for 'blunt-leaved'.

Clustered dock *R. conglomeratus*, perennial, common and widespread, less so in upland areas, and absent from north Scotland, in wet meadow, stream and riverbanks, ponds, ditches, and muddy pathsides, field margins and gateways where waterlogged in winter. *Conglomeratus* is Latin for 'clustered'.

Curled dock *R. crispus*, a common and widespread annual to short-lived perennial of waste ground, roadsides, disturbed pasture and arable, with many subspecies and hybrids, also in a range of coastal habitats including drift lines, shingle beaches, dunes and upper saltmarsh. Classified as injurious under the Weeds Act (1959). *Crispus* is Latin for 'curled'.

Fiddle dock *R. pulcher*, a biennial or short-lived perennial native in England south of a line from the Wash to the Severn estuary, and in south Wales, a local casual further north. It is an introduction in Ireland, generally found as a casual but in places also naturalised. It grows in dry coastal pastures and disturbed grassland on commons, churchyards and roadsides, mainly on lighter soils and often where the habitat is grazed or trampled. *Pulcher* is Latin for 'pretty'.

Golden dock *R. maritimus*, an uncommon annual to short-lived perennial, more evident in England, scattered elsewhere, though expanding its range in Ireland because of eutrophication, growing on the margins of pools, lakes, rivers and ditches, in clay-pits and wet hollows in marshy fields. Sites are usually waterlogged in winter. *Maritimus* is Latin for 'maritime'.

Greek dock *R. cristatus*, a perennial European neophyte, casual or naturalised on riverbanks, pathsides, dunes and waste ground. In around 1920 it was found by the River Rhymney, Monmouth, where it has persisted. It was recorded from the Thames at Kew Bridge in 1938 and from there spread rapidly in south-east England and the lower Thames Valley. In north Somerset it was first recorded in 1942 but it does not seem to be spreading. *Cristatus* is Latin for 'crested'.

Hooked dock *R. brownii*, a rhizomatous perennial Australian neophyte, first recorded in 1908, very scattered in England and Scotland as a casual in fields, and on tips and waste ground, usually introduced in wool shoddy. Robert Brown (1773–1858) was a Scottish botanist and explorer.

Marsh dock *R. palustris*, an annual, biennial or short-lived perennial, local in parts of southern and eastern England, typical of wet, nutrient-rich mud, most often in marsh and beside ponds and ditches, and in clay- and gravel-pits and on damp disturbed ground. It is also an occasional weed of dry open sites, and has been found as a ballast alien. *Palustris* is Latin for 'of marsh'.

Northern dock *R. longifolius*, a perennial, common and widespread in Scotland and northern England, recently having spread on roadsides as far south as Staffordshire; also on disturbed ground on streamsides and lake shores, in fields and around farms. *Longifolius* is Latin for 'long-leaved'.

Obovate-leaved dock *R. obovatus*, an uncommon annual South American neophyte first recorded in 1912, occurring as a casual on tips, on waste ground near mills and warehouses, and around docks, originating from wool shoddy, grain and birdseed, and declining as hygiene improves.

Patience dock *R. patientia*, a perennial European neophyte present by the Tudor period, uncommon and scattered in England, especially found in the London area, casual or naturalised on waste ground and riverbanks, especially near docks, breweries and other waste ground, originating as a garden escape and from grass and clover seed.

Russian dock *R. confertus*, a rhizomatous perennial European neophyte introduced in 1796, naturalised on a few roadside verges in Kent and Aberdeenshire, though previously found elsewhere. *Confertus* is Latin for 'crowded'.

Scottish dock *R. aquaticus*, an aquatic perennial growing around Loch Lomond and in Stirlingshire on seasonally flooded lake shores, beside ditches and streams, and in marshes and wet fields. This is the county flower of Stirlingshire as voted by the public in a 2002 survey by Plantlife.

Shore dock *R. rupestris*, a perennial found at about 40 sites in the Channel Island, south-west England east to Dorset, south and south-west Wales and Anglesey, on sand and shingle beaches, the base of sea-cliffs, among coastal rocks and in damp dune slacks, generally in places where fresh water feeds onto the shore. Threats include sea defence works, visitor pressure and winter storms, and reintroductions have recently been attempted at three sites in Devon and Cornwall as part of a recovery plan. *Rupestris* is Latin for 'relating to rocks'.

Water dock *R. hydrolapathum*, a perennial with a scattered distribution, locally common in southern England and recently extending its range in northern England, Scotland and Ireland, although some records may be escapes from ornamental plantings. It usually grows as an emergent on the margins of slow-flowing rivers and streams, by canals, lakes and ponds, and in ditches. It also colonises bare ground in marsh and fen. Greek *hydor* + Latin *lapathium* = 'water' + 'sorrel'.

Willow-leaved dock *R. salicifolius*, a perennial North American neophyte first recorded in the wild in 1900, usually casual on waste ground, around docks and by railways and canals. It derives from grain and birdseed, and has declined with improved seed-cleaning. *Salicifolia* is Latin for 'willow-leaved'.

Wood dock *R. sanguineus*, a perennial, widespread and common, less so in northern Scotland, in woodland edge, hedgerows, roadsides and waste ground, on a range of soils but favouring damp clay, and usually in relatively shaded habitats. *Sanguineus* is Latin for 'blood-red'.

DOCK BUG *Coreus marginatus* (Coreidae), length 13–15 mm, common and widespread in southern Britain and Ireland, feeding on docks, sorrels and other members of the Polygonaceae. Greek *coris* = 'bug', and Latin *marginatus* = 'margined'.

DODDER *Cuscuta epithymum* (Convolvulaceae), an uncommon annual, occasionally perennial, rootless twining herb, scattered, and locally abundant especially in southern England, East Anglia, Anglesey and western Ireland, parasitic on the stems of a variety of small shrubs and herbs on heathland, chalk downland and fixed dune grasslands, and also casual on field crops and in arable field borders in northern and western parts of its range. It is declining through loss of lowland heath, ploughing of chalk downlands, and an increase in scrub. 'Dodder' dates from the early seventeenth century as a variant of the dialect *dadder*, related to 'dither'. *Cuscuta* is Latin for dodder, probably of Arabic origin; the specific name is from Greek *epi* + *thymus* = 'upon' + 'an unnatural growth', referring to the parasitic nature of this plant.

Greater dodder *C. europaea*, a scarce annual, rarely perennial, rootless twining holoparasite of damp nitrophilous places, especially riverbanks, but also hedges and ditches, found locally in England as far north as Northamptonshire. Its primary host is common nettle, occasionally hop or other species, from which it can spread to a range of secondary hosts. It often grows close to flowing water, which may disperse the seeds.

Yellow dodder *C. campestris*, an annual, occasionally perennial, rootless twining North American neophyte herb parasitic on a range of cultivated plants, especially carrot, with a scattered distribution. *Campestris* is from Latin *campus* = 'field'.

DOG-HOBBLE *Leucothoë fontanesiana* (Ericaceae), an evergreen North American neophyte shrub naturalised where originally planted in Cardigan and Dumbarton. In its native USA the dense arching stems make terrain difficult for dogs, hence its name. In Greek mythology, Leucothoë was the daughter of King Orchamus of Persia. René Desfontaines (1750–1833) was a French botanist.

DOG LICHEN See under lichen.

DOG-ROSE *Rosa canina* (Rosaceae), an aggregate species of deciduous shrubs, generally common and widespread on well-drained calcareous to moderately acidic soils. Habitats include woodland, scrub, hedgerows, cliffs, riverbanks, rock outcrops, roadsides, railways and waste ground. The plant is rich in antioxidants. The fruit contains high levels of vitamin C, and is used to make rose-hip syrup and marmalade. This is the county flower of Hampshire as voted by the public in a 2002 survey by Plantlife. The binomial is the Latin for this plant (translated into English as its common name), from Greek *cynodoron* (*cyon* + *rhodon* = 'dog' + 'rose'), from the belief that the root was a cure for the bite of a mad dog.

Round-leaved dog-rose *R. obtusifolia*, a deciduous shrub, scattered in England (more frequent in the Midlands), Wales and south-east Scotland, found along woodland edges and in hedgerows, scrub, rough grassland and road verges, gravel-pits and wasteland, preferring well-drained calcareous to mildly acidic soils. *Obtusifolia* is Latin for 'blunt-leaved'.

DOG-VIOLET Herbaceous perennials in the genus *Viola* (Violaceae), Latin for 'violet'. See also violet and pansy.

Common dog-violet *V. riviniana*, common and widespread in a range of habitats, including deciduous woodland, hedge banks, road verges, meadows, heathland, moorland, montane grassland, rocky slopes and ledges, and often spreading in gardens. It shuns wet areas but thrives in most soil types, if avoiding very acidic habitats. This is the county flower of Lincolnshire as voted by the public in a 2002 survey by Plantlife. August Bachmann, the last name latinised to Rivinus (1652–1723) was a German botanist.

Early dog-violet *V. reichenbachiana*, common and widespread in England, Wales and Ireland, very local in Scotland, in deciduous woodland and hedge banks, usually in moderately shaded situations but also in the open where it can persist following woodland clearance. It is most frequent on calcareous soils, especially in woodland over limestone and chalk or base-rich clay. Heinrich Reichenbach (1823–89) was a German orchidist.

Heath dog-violet *V. canina* ssp. *canina* is found in acid habitats such as heathland, coastal dunes, stony riversides and lake shores, especially in Scotland. It also occurs on thin, heavily leached substrates overlying chalk and (as the rare ssp. *montana*) in fens in Cambridgeshire (remaining sites at Wicken Fen and Woodwalton Fen having statutory protection). Its pronounced decline since the mid-twentieth century has mainly been due to habitat loss, drainage and agricultural improvement, but also over- and undergrazing and possibly hybridisation with other *Viola* species. *Canina* is from Latin *canis* = 'dog'.

Pale dog-violet *V. lactea*, local in south and south-west England, and near-coastal parts of Wales and Ireland, on dry heathland. *Lactea* is Latin for 'milky'.

DOG WHELK See whelk.

DOGEND *Podostroma alutaceum* (Hypocreales, Hypocreaceae), an inedible fungus very scattered in England and Wales, found in late summer and autumn on damp woody debris, usually on wood of broadleaf trees, especially beech and hazel. Greek *podi* + *stroma* = 'foot' + 'bed', and possibly *aleuri* = 'flour'.

DOG'S-TAIL Grasses in the genera *Cynosurus* (Greek *cynos* + *oura* = 'dog' + 'tail', referring to the shape of the inflorescence) and *Lamarckia* (honouring the French naturalist Jean Lamarck, 1744–1829) (Poaceae).

Crested dog's-tail *C. cristatus*, a common and widespread perennial found in a variety of grasslands, particularly short and heavily grazed swards, growing in neutral to base-rich, fairly well-drained or damp soils. *Cristatus* is Latin for 'crested'.

Golden dog's-tail *L. aurea*, an annual Mediterranean neophyte introduced in 1770, very scattered and scarce on waste ground, originating from wool shoddy or as a garden escape. *Aurea* is Latin for 'golden'.

Rough dog's-tail *C. echinatus*, an annual neophyte naturalised on open sandy soils in the Channel Islands and the Scillies, sometimes a weed in bulb fields, and at a few sites in southern England. Elsewhere in Britain it is found as a grain and woolshoddy casual on waste ground and tips, occasionally in arable fields. *Echinatus* is Latin for 'prickly', from *echinus* = 'hedgehog'.

DOG-TOOTH Mosses in the genus *Cynodontium*, from Greek *cynos* + *donti* = 'dog' + 'tooth' (Rhabdoweisiaceae), from the appearance of a peristome tooth in some species.

Brunton's dog-tooth *C. bruntonii*, widespread, in uplands on dry, acidic to mildly base-rich rocks, usually in sheltered locations in ravines or shaded crags, occasionally on dry-stone walls.

Delicate dog-tooth *C. tenellum*, scarce, on sheltered, acidic rocks in the Highlands. *Tenellum* is Latin for 'tender'.

Jenner's dog-tooth *C. jenneri*, mainly found in upland Scotland on sheltered, generally acidic rocks in crevices, on dry ledges, and especially in the interstices of block scree, usually at fairly low altitudes.

Many-fruited dog-tooth *C. polycarpon*, scarce, in sheltered acidic rock crevices in the uplands. *Polycarpon* is Greek for 'many fruit'.

Strumose dog-tooth *C. strumiferum*, locally frequent, on dry acidic scree in the eastern Highlands. *Strumosus* is Latin for 'swollen'.

DOG'S TOOTH *Lacanobia suasa* (Noctuidae, Hadeninae), wingspan 32–7 mm, nocturnal, local in saltmarsh, grassland, moorland and farmland throughout England, Wales and coastal Ireland. It is bivoltine in the south, flying in May–June then August–September. In the north, the single generation appears in June–July. Larvae feed on such plants as docks, plantains, goosefoot and sea-lavender. The genus name is a typographical error for Lachanobia, from Greek *lachana* + *bioo* = 'vegetables' + 'to live', and Latin *suasum*, a smoky brown colour produced by dyeing, reflecting the forewing colour.

DOGWOOD *Cornus sanguinea* (Cornaceae), a deciduous shrub, widespread and locally common in lowland Britain, scattered in Ireland and peripheral upland Scotland, in woodland, scrub, hedgerows and shelter belts on limestone soils or base-rich clays, and sometimes dominant in hedges and scrub on chalk. The English name comes from the wood having formerly been used to make skewers, which were known as 'dogs'. The binomial is Latin for this species, and 'blood-red', from the red-tipped winter stems.

Red-osier dogwood *C. sericea*, a deciduous North American neophyte shrub in Britain by 1683, scattered in lowland regions, naturalised in woodland and along riversides, sometimes suckering to create thickets; also much planted in parkland, amenity plantings and on roadsides and sometimes occurring as an escape on waste ground. *Sericea* is Latin for 'silky'.

White dogwood *C. alba*, a deciduous, suckering east Asian neophyte shrub in Britain by 1741, scattered in lowland regions, especially in southern England, Lancashire and south Wales, in hedges, on roadsides, in parks and in amenity plantings. It is also a garden escape, and is planted for game cover. *Alba* is Latin for 'white'.

DOLICHOPODIDAE From Greek *dolichos* + *podi* = 'long' + 'leg', family of the small to medium-sized (1–9 mm length) long-legged flies, with 301 species in 46 genera. Adults and larvae are both generally predatory, found in a range of habitats.

DOLPHIN Marine cetacean mammals in the family Delphinidae.

Atlantic white-sided dolphin *Lagenorhynchus acutus*, male length 2.8 m, females 2.5 m, weight 200–30 kg, found off the coasts of Scotland, north-east and south-west England, and north-west Ireland, usually in large groups (pods), and occasionally reported from the North Sea, English Channel and all other British waters. Diet is of fish and squid. The English name comes from the white to pale yellow patch found behind the dorsal fin on each side. Latin *lagena* = 'bottle' + Greek *rhynchos* = 'nose', and *acutus* = 'sharp' in Latin.

Bottlenose dolphin Members of the genus *Tursiops*, formerly used specifically for common bottlenose dolphin.

Common bottlenose dolphin *Tursiops truncatus*, length 2–4 m, weight 150–650 kg, males usually larger than females, a generally inshore species, with resident populations known from Cardigan, the Moray Firth, the west coast of Ireland, and (confirmed in 2017) 28 individuals resident in shallow waters around St Ives, Cornwall, sometimes extending into Devon and Dorset waters. Transient groups, however, are not infrequent almost anywhere around the British coast except the southern North Sea and off south-east England. The total population in UK inshore waters is probably <300 individuals. Diet is of fish, squid and crustaceans. *Tursio* was a fish like a dolphin in Pliny's Natural History; *truncatus* is Latin for 'truncated'.

Risso's dolphin *Grampus griseus*, length typically 3 m, weight 300–500 kg, males slightly larger, particularly recorded, often in groups (pods) off the Hebrides, Scottish Minches and in the Irish Sea, especially near Bardsey Island off the north-west Welsh coast, as well as round Shetland and west coasts of the British

Isles and Ireland generally. Diet is of cephalopods, crustaceans and small fish. Antoine Risso (1777–1845) was a French naturalist. *Grampus* is derived from Latin *craspiscis* (*crassus piscis*) = 'fat fish', and *griseus* = 'grey'.

Striped dolphin *Stenella coeruleoalba*, male length 2.6 m, weight 160 kg, females 2.4 m and 150 kg, found in the western English Channel and off the south-west coasts of England, Wales and Ireland. Diet is of fish (mainly cod), cephalopods, and crustaceans. Greek *stenos* + *ella* = 'narrow' and 'little', and Latin *caeruleus* + *albus* = 'blue' + 'white', the specific name referring to the pattern of blue/dark grey and white stripes along the lateral and dorsal sides.

White-beaked dolphin *L. albirostris*, length 2.3–3.1 m, weight 180–355 kg, found all around the British Isles, particularly off the coasts of Scotland and in the North Sea but also south-west Britain and west Ireland. Diet is of fish and squid. *Albirostris* is Latin for 'white beak'.

DOMECAP Fungi in the genera *Lyophyllum* (Greek *lyo* + *phyllon* = 'loose' + 'leaf') and *Rugosomyces* (Latin *rugosus* = 'wrinkled' + Greek *myces* = 'fungus') (Agaricales, Lyophyllaceae).

Rarities include **clustered domecap** *L. decastes*, **lilac domecap** *R. onychinus* and **yellow domecap** *R. chrysenteron*.

Gristly domecap *L. loricatum*, edible, locally common with a scattered distribution, found in summer and autumn on soil in open woodland. *Loricatum* is Latin for 'armoured'.

Pink domecap *R. carneus*, inedible, saprobic, widespread but uncommon, found from July to October in grassy habitats, sometimes in woodland glades or edges. *Carneus* is Latin for 'flesh-coloured'.

Smoky domecap *L. gangraenosum*, uncommon, scattered, found in summer and autumn on soil in grassy areas in woodland and heathland. *Gangraenus* is Latin for 'gangrene'.

Violet domecap *R. ionides*, edible, saprobic, uncommon, scattered in Britain in late summer and autumn on soil and litter in deciduous woodland, and with creeping willow on dunes. *Ionides* is Greek for 'like a violet' (*ion*).

White domecap *L. connatum*, inedible, common and widespread, found in summer and autumn on soil in grass, in clearings in broadleaf woodland and near disturbed soil by woodland paths. *Connatum* is Latin for 'born together', reflecting the species being commonly found in groups.

DOOR SNAIL Keeled pulmonates in the family Clausiliidae, so called from the calcareous structure (clausilium) resembling a spoon- or tongue-shaped 'door' that can slide down to close the shell aperture. All are sinistral. See also tree snail under snail.

Cliveden door snail *Papillifera papillaris*, from south Europe, shell length 11–17 mm, width 2.5–4.5 mm. In 2004, examples were discovered on a marble balustrade at Cliveden, Buckinghamshire, probably feeding on lichens, and possibly having been transported with the stone from Italy in 1896. In 1993, however, the species had already been noted on stonework of Italian origin on Brownsea Island, Poole Harbour, Dorset. Latin *papilla* = 'nipple' or 'pimple'.

Common door snail Alternative name for two-toothed door snail.

Plaited door snail *Cochlodina laminata*, shell length 15–17 mm, width 4 mm, local and widespread in much of England, scattered and scarcer elsewhere, in places vulnerable, found in (ancient) woodland, climbing trees to feed on lichen and algae, especially after rain. Greek *cochlos* means a mollusc with a spiral shell (*cochlo* = 'to turn'), and Latin *laminata* = 'thin layer'.

Rolph's door snail *Macrogastra rolphii*, shell length 11–14 mm, width 3.4–4 mm, local in southern England, and scattered as far north as Lincolnshire, on litter, fallen timber and moss in deciduous woodland, usually on calcareous substrate. Greek *macrogastra* = 'large stomach'.

Thames door snail *Balea biplicata*, shell length 15–18 mm, width 3.8–4.5 mm, rare (RDB 3), having declined, all current known colonies being in the lower River Thames corridor near London, including a colony at Purfleet, Essex. It is locally abundant and populations are probably stable. It is usually found in litter beneath rough herbage and strand-line rubbish on upper riverbanks. *Balea* is Greek for 'knave', *biplicata* Latin for 'two-folded'.

Two-lipped door snail Alternative name for Thames door snail.

Two-toothed door snail *Clausilia bidentata*, shell length 10–11 mm, width 3 mm, widespread but having declined in the Midlands and around London through a combination of air pollution and a lack of calcareous substrate. It is generally nocturnal and commonly found in woodland and hedgerows under the bark of trees, which it climbs to feed on lichens, in rock crevices or on walls, being most active in wet weather. The binomial uses the Latin *clausilium* (defined above) and for 'two-toothed'.

DORIS Angled doris *Goniodoris nodosa* (Goniodorididae), a common and widespread sea slug up to 3 cm long, found from the intertidal to depths of 60 m, adults feeding on ascidians, juveniles on bryozoans. *Goniodoris* is from Greek *gonio* = 'angle' + Doris, a sea goddess in Greek mythology, and Latin *nodosa* = 'knotted'.

Rough-mantled doris *Onchidoris bilamellata* (Onchidorididae), a gregarious sea slug up to 4 cm long, widespread in the intertidal and shallow sublittoral to a depth of 20 m, feeding exclusively on barnacles. Greek *oncos* = 'swelling' + Doris, a sea goddess in Greek mythology, and Latin *bilamellata* = 'two-layered'.

DORMOUSE Members of the family Gliridae. 'Dormouse' comes from French *dormir* and Latin *dormire* = 'to sleep' + 'mouse'.

Common dormouse *Muscardinus avellanarius*, head-body length 60–90 mm, tail length 57–68 mm, weight 15–26 g, up to 43 g before hibernation, found mainly in south and south-west England (especially Devon, Somerset, Sussex and Kent), locally in the Midlands, and with scattered records from the Lake District and Wales, living in deciduous woodland, especially with hazel coppice. Numbers are declining: in 2016 a PTES report found the number of dormice counted at nestboxes in England and Wales since 2000 had fallen by 38%, and by 55% since the mid-1990s, though as part of Natural England's Species Recovery Programme, dormouse reintroductions began in 1993, and since then 26 reintroductions have taken place in 12 English counties. It was recorded in Ireland in Co. Kildare in 2010, and it appears to be spreading using hazel and bramble in hedgerows. This strictly nocturnal species spends much of its time climbing among tree branches in search of food (flowers, pollen, fruits, insects and nuts, hazel nuts favoured for building up pre-hibernation fat). During the day it sleeps in a nest of honeysuckle bark, leaves and grass in the undergrowth, or in a tree hollow or deserted bird nest, entering a state of torpor in cold or wet weather. Breeding season (and success) is weather-dependent. Females raise one, occasionally two litters a year, each usually of four young, which achieve independence after six to eight weeks. It hibernates from October until April or May in nests on the ground, in the base of coppiced trees, and under piles of logs or leaves. It is protected by the WCA (1981). The binomial derives in part from Latin *mus* = 'mouse' (but not 'dormouse', which is *glis*) and *avellanarius* = 'of hazel'.

Edible dormouse *Glis glis*, head-body length 140–90 mm, tail length 110–30 mm, weight 120–50 g (almost doubling just before hibernation), a European introduction, a few Hungarian animals being released from Walter (later Lord) Rothschild's collection at Tring Park, Hertfordshire, in 1902, and subsequently increasing in number and range. By the 1990s a population of around 30,000 was mainly concentrated in a 600 km² triangle between Aylesbury, High Wycombe and Whipsnade, but the species has also been found elsewhere, for example the New Forest, and numbers may now be around 1 million. They are nocturnal and arboreal, spending the day in holes or old nests in trees. Feeding is mainly on berries, apples and nuts, especially beech mast. Hibernation is from October to May. Gestation lasts 20–30 days, with a litter size usually of 4–7 (up to 11). Young leave the nest after 30 days, and are sexually mature after their second hibernation. They are viewed as a pest in orchards and when they enter attics, wall cavities and underfloor areas; >3,000 were killed in 2002–7. This dormouse was once eaten by the Romans, Etruscans and Gauls as a delicacy. *Glis* is Latin for dormouse.

Fat dormouse, grey dormouse, squirrel-tailed dormouse Alternative names for edible dormouse.

Hazel dormouse Alternative name for common dormouse.

DOT MOTH *Melanchra persicariae* (Noctuidae, Hadeninae), wingspan 37–40 mm, nocturnal, widespread and common in habitats such as gardens, waste ground, verges, hedgerows and woodland, throughout England, Wales, southern Scotland and the southern half of Ireland. With numbers declining by 91% over 1968–2007, this is a species of conservation concern in the UK BAP. Adults fly in July–August. Larvae feed on a variety of wild and garden plants. The English name references the white spot on the black forewing. Greek *melas* + *chros* = 'black' + 'colour', and Latin *persicaria* for the food plant redshank *Persicaria maculosa*, itself derived from 'peach tree'.

Straw dot *Rivula sericealis* (Erebidae, Rivulinae), wingspan 18–22 mm, nocturnal, widespread in damp grassland, fen, heathland, moorland, woodland and gardens, commoner in more southern parts of its range. The bivoltine adults fly in June–July and August–September. Larvae feed on grasses. Latin *rivulus* = 'rivulet', and *sericeus* = 'silky'.

DOTTEREL *Charadrius morinellus* (Charadriidae), wingspan 60 cm, length 21 cm, weight 110 g, a Red List migrant arriving from southern Europe at breeding grounds in the montane Highlands in April–May, with the population in 2011 estimated as 423 males, a 57% decline since the 980 males recorded in 1988. Birds leave in July–August. Autumn passage birds are seen in East Anglia in August–September. Nests are in bare scrapes, clutches are usually of 3 eggs, incubation (by the male, the polyandrous female seeking another male to lay another clutch) takes 24–8 days, and fledging of the precocial young 25–30 days. Food is of invertebrates, mainly insects. The English name, known by 1440, refers to the tolerance of the bird to human proximity, making it easy to catch, and also as a term for a simple person or a dotard (which use coming first being unclear), a meaning also reflected in the specific name, from Greek *moros*, meaning 'foolish'. *Charadrius* derives from Greek *charadrios*, a bird found in ravines and river valleys (*charadra* = 'ravine').

DOTTY Blackthorn dotty *Polystigma rubrum* (Phyllachorales, Phyllachoraceae), a scarce fungus, very scattered, mostly coastal, Vulnerable in the RDL (2006), producing red blister-like marks on leaves of blackthorn and related trees such as plum and damson. Greek *polystigma* = 'many points', and Latin *rubrum* = 'red'.

DOUBLE LOBED *Lateroligia ophiogramma* (Noctuidae, Xyleninae), wingspan 32–5 mm, a nocturnal moth widespread but local in fen, marsh, water meadow, damp woodland and gardens. Adults fly from June to August. Larvae mine grass stems. The binomial is from the Latin for 'side' or 'lateral' and Ligia, a nymph in mythology, and Greek *ophis* + *gramme* = 'snake' + 'mark'.

DOUGLAS-FIR *Pseudotsuga menziesii* (Pinaceae), introduced from North America in 1826, very widely planted for timber and as a parkland ornamental, with a few records of self-sown trees. It grows well on mineral soil of poor to medium fertility but requires a degree of moisture and good soil aeration. The tallest example in the UK is at Reelig Glen near Inverness, with a height of 66.4 m in 2013. The English name honours the Scots botanist David Douglas (1799–1834). *Tsuga* is the Japanese name for hemlock-spruce; *pseudo* is Greek for 'false'. The Scottish botanist Archibald Menzies (1754–1842) documented the tree on Vancouver Island in 1791.

DOVE Members of the family Columbidae, from Old English *dūfe* = 'dove' or 'pigeon'. The **oriental** or **rufous turtle dove** *Streptopelia orientalis* and the North and Central American **mourning dove** *Zenaida macroura* are both rare accidentals.

Collared dove *Streptopelia decaocto*, wingspan 51 cm, length 32 cm, weight 200 g, strongly dispersive, spreading rapidly across Europe from Turkey and the Middle East from the 1930s on, responding to changes in arable farming, first recorded breeding in Britain at Cromer (Norfolk) in 1955, subsequently spreading throughout the British Isles (in Ireland by 1959), especially in small towns and suburban habitats, feeding on seed and grain. There were 980,000 pairs (880,000–1,070,000) in Britain in 2009, with a 343% increase in numbers between 1970 and 2014 (BBS data) or even 799% (BTO data for 1970–2015), and 8% over 1995–2014 (BBS data), though a significant decline by 15% was noted for 2010–15. Nests are in trees, often close to habitation, usually with 2 eggs, hatching after 16–17 days, the altricial young fledging in 17–19 days. The English name comes from the black half-collar edged with white on the nape. Greek *streptos* + *peleia* = 'collar' + 'dove'. The song is *coo-COO-coo*, repeated 6 times, that is the sound *coo* for a total of 18 times, reflected in the specific name *decaocto*, Greek for 'eighteen'.

Rock dove *Columba livia*, wingspan 66 cm, length 32 cm, weight 300 g, ancestor of the domesticated and feral pigeon. The wild rock dove is now a rarity found only on sea-cliffs in the Hebrides, Orkney and Shetland, with perhaps 1,000 and no more than 5,000 birds of the 'original' species, and on west and north-west Ireland coasts. Nests are in holes in cliffs or on ledges in caves. Clutch size is usually two, incubation takes 16–19 days, and the altricial young fledge in 35–7 days. It mainly feeds on seed. *Columba* is Latin for 'dove'; *livia* for 'bluish'.

Stock dove *C. oenas*, wingspan 66 cm, length 33 cm, weight 300 g, widely distributed in the UK, except for parts of north Scotland and the northern half of Ireland, with particularly high densities in south-west England and the Midlands, most commonly seen in woodland (edge) in summer, stubble fields in winter. Some 260,000 territories were estimated in 2009. Numbers increased by 113% between 1970 and 2014, following the ban in the 1970s on toxic organochlorine pesticides (which had been used as a seed dressing), and there was an increase by 17% between 2010 and 2014. Over half of the European population is found in the UK, and it is Amber-listed. Nests are usually in tree holes (occasionally disused rabbit burrows, rock cracks and heavy ivy growth). There are usually 2 eggs that hatch after 21–3 days, the altricial young fledging in 28–9 days. Food is of cereal grain, and seeds and fruits of herbs and grasses, occasionally green plant parts and invertebrates. The word 'stock' in the English name comes from Old English *stocc* = 'stump, post or tree-trunk', so is referring to a dove that lives in hollow trees. *Oinas* is Greek for 'dove'.

Turtle dove *S. turtur*, wingspan 50 cm, length 27 cm, weight 140 g, a Red List migrant breeder, mainly found in southern and eastern England (scattered in Scotland and Ireland) in woodland edges, hedgerows and open land with scrub, arriving in late April/May, leaving between July and September. There were an estimated 14,000 territories in 2009, but numbers in Britain fell by 98% between 1970 and 2015 (and by 71% between 2010 and 2015), probably reflecting a decline in availability of its weed seed food and the impact of the disease trichomonosis (affecting >95% of tested birds in 2014). Clutch size is usually two, incubation lasts 15–16 days, and the altricial young fledge in 18–19 days. The deep, vibrating 'turrr, turrr' song gives both the English adjectival name and the specific name, which has nothing to do with the Latin *turtur* = 'turtle'.

DOWD (MOTH) Members of the family Blastobasidae, all species being recent introductions (how being unknown) from Madeira, and nocturnal. Greek *blastos* + *basis* = 'shoot' + 'step', from the shape of the antenna.

Furness dowd *B. adustella*, wingspan 15–20 mm, naturalised and widespread. Common in gardens, hedgerows and yew woodland. Adults fly in August–September. Larvae feed on fallen leaves and other decaying plant material and detritus. *Adustella* is Latin for 'swarthy', reflecting the wing colour.

Hampshire dowd *B. phycidella*, wingspan 17–19 mm, rare, found on a wooded south-facing cliff on Guernsey, first discovered in 1990. Before this it was known only from Southampton Docks, Hampshire, in 1930. Adults fly in June. Larvae feed on decaying wood and fallen pine needles. The specific name may describe the variable wing colour from *phycis*, Greek for a seasonally colour-changing littoral fish.

Marsh dowd *B. rebeli*, wingspan 12–14 mm, naturalised, widespread and expanding its range. Recorded in Hampshire (first in 1998) and West Sussex from meadow and damp woodland. Adults fly in July.

Wakely's dowd *B. laticolella*, wingspan 18–21 mm, naturalised, widespread and expanding its range (including east Ireland). Common in gardens, hedgerows and yew woodland. Adults fly in May–June and again in autumn. This species' success may reflect the broad larval diet which includes leaf litter, vegetation, stored products, and seed pods of tansy and tree lupin. Stanley Wakely (1892–1976) was an English entomologist. *Laticolella* is Latin for 'broad-necked'.

DOWITCHER North American wading birds, family Scolopacidae: **long-billed** *Limnodromus scolopaceus* and **short-billed dowitcher** *L. griseus*, both scarce accidentals. 'Dowitcher' is an Iroquoian word. Greek from *limne* + *dromos* = 'marsh' + '(running) race'.

DOWN-LOOKER FLY Some snipe flies in the genus *Rhagio*, so called because they perch head-down on tree-trunks.

DOWNY-ROSE Species of *Rosa* (Rosaceae), deciduous shrubs with soft hairy leaves.

Harsh downy-rose *R. tomentosa*, widespread in England and Wales, rare and scattered elsewhere, preferring calcareous to mildly acidic soils, on woodland edges and in hedgerows, where it thrives in relatively shaded conditions, and in more open habitats including scrub, rough grassland, disused quarries and less acidic heaths. *Tomentosa* is Latin for 'covered with fine hairs'.

Sherard's downy-rose *R. sherardii*, widespread and common (except in south-east England), in woodland edge, hedgerows, scrub, rough grassland, heathland, rock outcrops, cliffs and tracksides, growing on a variety of soils and able to tolerate moderately acid and wet conditions. William Sherard (1659–1728) was an English botanist.

Soft downy-rose *R. mollis*, suckering freely and sometimes forming dense thickets, found in Wales, the north Midlands northwards into much of Scotland, and northern Ireland in woodland, hedges, scrub and rough grassland, and on rocky streamsides, rock outcrops, cliffs, dunes, waste ground and roadsides. It grows in a variety of well-drained soils, but avoids very acidic conditions. *Mollis* is Latin for 'soft'.

DRAB (MOTH) Nocturnal species of *Orthosia* (Noctuidae, Hadeninae), from Greek *orthosis* = 'making straight', from the straight subterminal line on the forewing, wingspans 34–40 mm.

Clouded drab *O. incerta*, widespread and common in woodland, gardens and wherever there are oak trees. Adults fly from March to May. Larvae particularly feed on oaks, but also willows, birches, hawthorn, hazel and lime. *Incerta* is Latin for 'uncertain', from the variable wing colour.

Lead-coloured drab *O. populeti*, local in broadleaf woodland, parks and gardens throughout Britain, and with records in Co. Cavan and Co. Fermanagh. Adults fly in April. Larvae mainly feed on aspen and black poplar, initially on catkins then the leaves, prompting the specific name, Latin for 'poplar wood'.

Northern drab *O. opima*, local on saltmarsh, marsh, heathland, downland, golf courses and dunes throughout much of Britain, and with records from Co. Donegal and Co. Fermanagh. Adults fly in April–May, and feed on sallow blossom. Larvae feed in May–June on sallows, birches, ragworts, mugwort, sealavender and other plants. *Opima* is Latin for 'fat' or 'splendid'.

DRAGON-ARUM *Dracunculus vulgaris* (Araceae), a rhizomatous herbaceous Mediterranean neophyte perennial, scattered in southern and south-east England as a garden throw-out in hedges and rough ground. The English name reflects the foul scent produced during flowering. The binomial is Latin for 'little dragon' and 'common'.

DRAGONFLY Insects in the order Odonata, suborder Anisoptera, adults characterised by an elongated body, large multifaceted eyes and two pairs of independently manoeuvreable transparent wings that are usually held flat and away from the body. Larvae are aquatic, generally spending one to three years in fresh water. See also emerald dragonfly, and chaser, skimmer and darters.

Emperor dragonfly *Anax imperator* (Aeshnidae), body length 66–84 mm, hindwing 45–52 mm, widespread in south and central England and Wales, increasing its range northwards into south Scotland, and recently found in Ireland. Mostly associated with large, well-vegetated ponds and lakes and slow-moving waters. The main flight period is May–September. The binomial is Greek for 'chief' and Latin for 'emperor'.

Golden-ringed dragonfly *Cordulegaster boltonii* (Cordulegastridae), male body length 74–80 mm, female 80–5 mm (including its ovipositor, Britain's longest dragonfly), wingspan 100 mm, common on moors and heathland in southern England, Wales, the Lake District and much of Scotland, breeding in acidic running water. The main flight period is June–September. Greek *cordyle* + *gaster* = 'swelling' + 'stomach'.

Hairy dragonfly *Brachytron pratense* (Aeshnidae), body length 55 mm, Britain's smallest hawker, uncommon but increasing its range in England, Wales and Ireland, in unpolluted wellvegetated water bodies. The main flight period is May–July. Greek *brachys* = 'short', and Latin *pratum* = 'meadow'.

Lesser emperor dragonfly *A. parthenope*, body length 71 mm, a rare annual migrant associated with ponds and small lakes, first recorded in Gloucestershire in 1996, and having bred in Cornwall since at least 2011. Parthenope was a siren in Greek mythology.

DRAGON'S-TEETH *Tetragonolobus maritimus* (Fabaceae), a perennial herb introduced from central Europe by 1683, naturalised in calcareous grassland in southern England, having been introduced with grass seed. Elsewhere it is a casual on roadsides and waste ground. Greek *tetra* + *gonia* + *lobos* = 'four' + 'angle' + 'lobe', and Latin *maritimus* = 'maritime'.

DRAIN FLY Alternative name for some moth flies (Psychodidae).

DREISSENIDAE Family name of zebra mussel.

DREPANIDAE Family name for hook-tip and lutestring moths (Greek *drepane* = 'sickle', the forewings having hookshaped tips). Adults have poorly developed mouthparts and do not feed.

DRILIDAE Beetle family represented by *Drilus flavescens*, found mostly on the chalk downlands of south-east England. Larvae feed on snails, and overwinter in snail shells. The male is winged and 12 mm long, the female larviform, flattened, fleshy and rounded, and about five times the length of the male. Greek *drilos* = 'worm', and Latin *flavescens* = 'yellowish'.

DRILL (MOTH) Members of the genus *Dichrorampha* (Tortricidae, Olethreutinae) (Greek *dichroos* + ramphe = 'two-coloured' + 'hooked knife').

Broad-blotch drill *D. alpinana*, wingspan 13–15 mm, local in

gardens, meadows and rough grassland in the southern half of England, Wales and much of Ireland. Adults fly in June–August, often during the day. Larvae mine oxeye daisy roots. *Alpinana* is from Latin *alpinus* = 'alpine'.

Common drill *D. petiverella*, wingspan 10–13 mm, widespread and common in rough grassland, hedgerows and gardens. Adults fly in June–August during the late afternoon and early evening. Larvae mine the roots of tansy, yarrow and sneezewort. James Pettifer (1660–1718) was an English apothecary and naturalist.

Dark drill *D. sequana*, wingspan 9–11 mm, rare, local in scrub and cliffs in parts of England and Wales. Adults fly in June. Larvae mine the roots of oxeye daisy, tansy and feverfew. *Sequana* is from Latin *sequor* = 'to follow'.

Dingy drill *D. sedatana*, wingspan 12–16, scarce, in rough ground, verges, dry pasture, gardens and parks throughout Britain, predominantly coastal in the north of its range. Adults fly in May–June during the day. Larvae mine tansy roots. *Sedatus* is Latin for 'staid' or 'sedate'.

Downland drill *D. consortana*, wingspan 9–12 mm, widespread but found only locally on dry grassland and roadside verges, chiefly in southern England. Adults fly from mid-June to August. Larvae mine oxeye daisy roots. *Consortana* is from Latin *consors* = 'colleague'.

Gold-fringed drill *D. vancouverana*, wingspan 12–15 mm, widespread but local in dry grassland, rough meadow, downland and scrub, more numerous in the south, and also commonly recorded in Northern Ireland. Adults fly in June–July in the afternoon and towards sunset. Larvae mine roots of yarrow and tansy.

Lead-coloured drill *D. plumbana*, wingspan 11–14 mm, widespread but local in rough meadow, dry grassland and waste ground. Adults fly in May–June during the day into the evening. Larvae mine the roots of oxeye daisy and yarrow from August through to the following spring. *Plumbum* is Latin for 'lead', from the colour of the forewing stripes.

Narrow-blotch drill *D. flavidorsana*, wingspan 13–16 mm, local in dry pasture, gardens, orchards, parks and waste ground throughout much of southern Britain, predominantly coastal in the north of its range. Adults fly in June–July generally at night but also in the afternoon. Larvae mine tansy roots. Latin *flavus* + *dorsum* = 'yellow' + 'back', from the yellow dorsal blotch.

Obscure drill *D. aeratana*, wingspan 11–16 mm, widespread but local in waste ground, predominantly coastal in the north of its range. Adults fly in May–June. Larvae mine oxeye daisy roots. *Aeratus* is Latin for 'coppery', from the colour of the forewing stripes.

Round-winged drill *D. simpliciana*, wingspan 12–16 mm, local on waste ground and roadside verges throughout England and Wales, particularly in the south; probably under-recorded in Scotland; a few Irish records, though only one (in Co. Cork) post-dates 2000. Adults fly from July to September, becoming most active just before dark. Larvae mine mugwort roots. Latin *simplex* = 'plain', from the plain forewing pattern.

Sharp-winged drill *D. acuminatana*, wingspan 10–15 mm, widespread but local on meadows and waste ground. The bivoltine adults fly in April–May and August–September. Larvae mine roots of tansy and oxeye daisy. *Acuminatus* is Latin for 'sharpened', from the pointed forewing apex.

Silver-lined drill *D. plumbagana*, wingspan 10–15 mm, widespread but local on dry pasture, downland, waste ground and dunes, less common and mostly coastal in the north of its range. Adults fly in May–June during late afternoon and early evening. Larvae mine yarrow roots. *Plumbago* is Latin for 'black lead' (graphite), from the colour of the forewing stripes.

Sneezewort drill *D. sylvicolana*, wingspan 10–14 mm, local in marshy woodland, damp pasture, riverbanks and ditches in parts of east and south-east England. Adults fly in June–August. Larvae mine sneezewort roots. Latin *sylva* + *colo* = 'wood' + 'to live'.

Spike-marked drill *D. montanana*, wingspan 10–14 mm,

local on dry pasture in England north of Staffordshire, Scotland (common on the east coast), Wales and Ireland. Adults fly in June–August during late afternoon and early evening. Larvae mine the roots of yarrow and tansy. *Montanus* is Latin for 'montane'.

Square-spot drill *D. sequana*, wingspan 9–11 mm, local in verges, dry pasture, gardens, orchards, parks and waste ground throughout the southern half of England and Wales. Adults fly in the afternoon and towards dusk in May–June. Larvae mainly mine roots of yarrow but also of tansy. *Sequor* is Latin for 'to follow'.

DRILL (SNAIL) See oyster drill.

DRINKER *Euthrix potatoria* (Lasiocampidae, Pinarinae), wingspan 45–65 mm, a nocturnal moth widespread and common on damp grassland, fens, damp woodland, boggy heathland and gardens. It was identified as a declining species in a 35-year study (1968–2002), but its population has increased markedly in recent years, and its overall trend is now an increase. The four best years for this species since monitoring began in 1968 occurred between 2003 and 2006. Adults fly in July–August. Larvae feed on reeds and grasses. The English name comes from the caterpillar drinking drops of dew. Greek *eu* + *thrix* = 'well' + 'hair', from the hairy adult, and Latin *potatoria* = 'relating to drinking'.

DRONE FLY Name of some hoverflies (Syrphidae), especially in the genus *Eristalis*, from their resemblance to bee drones (males). *Eristalis* was used by Pliny for some kind of precious stone. The aquatic larvae, many living in drainage ditches, stagnant water, and water polluted with organic matter, including manure, are often known as rat-tailed maggots (q.v.).

Common drone fly *E. tenax*, body length and wingspan 15 mm, widespread and common in gardens and hedgerows. Females hibernate in buildings and crevices to emerge on warm days in late winter, leading to it being seen in virtually every month of the year. Adults feed on nectar. *Tenax* is Latin for 'tenacious'.

Tapered drone fly *E. pertinax*, wing length 8–13 mm, widespread and common from March to November in hedgerows and woodland rides. Adults feed on nectar. *Pertinax* is Latin for 'persistent'.

DROPWORT *Filipendula vulgaris* (Rosaceae), a perennial herb, mainly found in grassland on chalk and limestone downs, and in rough pasture, and also on coastal and inland heathland over basic rocks, including serpentine. It is often planted in churchyards in south-west Wales. With a scattered distribution, mostly in England and south-west Wales, it is declining in its southern range, but is stable elsewhere. Latin *filum* + *pendulus* = 'thread' + 'hanging', referring to the tubers which hang on the fibrous roots, and *vulgaris* = 'common'.

DROSERACEAE Family name for sundews (Greek *droseros* = 'dewy').

DROSOPHILIDAE Family name (Greek *drosos* + *philos* = 'dew' + 'loving') of some of the small to medium-sized (2.5–10 mm length) fruit flies (see also Tephritidae), with 63 species in 13 genera. Both adults and larvae feed on plant parts and/or products, and are often seen in and around compost. Larvae of a few species are leaf miners: the widespread and increasingly common *Scaptomyza flava*, for instance, has recently become a serious pest of brassica crops.

DRUID FLY Members of the family Clusiidae. Adults are generally 3–6 mm in length and tend to be found on tree-trunks. Larvae live under the bark feeding on small saprobic invertebrates.

DRUMSTICKS *Aulacomnium androgynum* (Aulacomniaceae), a common and widespread moss in Britain, rare and scattered in Ireland, growing on dead wood and living trees. The gemmae on the stem tip look like small green drumsticks. Greek *aula* + *mnium* = 'furrow' + 'moss', and Latin *androgynum* = 'androgynous'.

DRYOPHTHORIDAE A largely tropical weevil family

(Greek *drys* + *phthora* = 'tree' + 'destruction'), with four British species in two genera, including *Sitophilus*. Species are stored-product pests mostly found in granaries.

DRYOPIDAE Family name (Greek *drys* + *opsis* = 'tree' + 'appearance') of long-toed water beetles, 3–6 mm long, with nine British species. Most species cannot swim, and cling on to floating detritus. Larvae live in soil or decaying wood.

DUBLIN BAY PRAWN Alternative name for Norway lobster.

DUCK From Old English *duce*, members of the family Anatidae, different species associated with freshwater, estuarine and coastal habitats, and feeding by dabbling (e.g. *Anas*) and diving (e.g. *Aythya* and *Mergus*).

Ferruginous duck *Aythya nyroca*, wingspan 63–7 cm, length 38–42, male weight 470–730 g, female 465–727 g, a rare winter visitor from eastern Europe/west Asia, with an average of 13 records each year, mostly on water bodies in eastern England. Greek *aithyia* is an unidentified seabird mentioned by Aristotle, and Russian *nyrok* = 'duck'.

Long-tailed duck *Clangula hyemalis*, wingspan 76 cm, length 44 cm, weight 730 g, a winter visitor and passage migrant, most common from Northumberland northwards, but recorded from all coasts and with scattered inland sightings, with 11,000 individuals estimated in 2008–9. They dive for bivalves, crabs and small fish. With recent population and range declines this is a Red List species. Latin *clangere* = 'to resound' and *hyemalis* = 'of winter'.

Mandarin duck *Aix galericulata*, wingspan 71 cm, length 45 cm, male weight 630 g, females 520 g, a highly colourful ornamental species brought from China to Richmond Green, Surrey, just before 1745. Wild birds were increasingly noted from 1866, but free-flying colonies did not become well established until the 1950s since when the species has been noted over much of Britain, mainly in southern England, with an important colony in Windsor Great Park. Numbers in the 1970s were probably 300–400 pairs, rising to 2,300 pairs in 1988 (and perhaps 7,000 individuals). BBS data suggest an increase in numbers by 418% between 1995 and 2014. Nests are in tree cavities close to water. Clutch size is 9–12, incubation takes 28–30 days, and fledging of the precocial young takes 40–5 days. Diet is of insects, seed and plant material. Greek *aix* is an unidentified diving bird mentioned by Aristotle. Latin *galericulata* = 'wig'.

Ring-necked duck *Aythya collaris*, wingspan 61–75 cm, length 37–46 cm, male weight 540–910 g, female 490–890 g, a rare North American vagrant, though with a few birds recorded most years, and occasional small flocks occurring, for example four birds at Standlake, Oxfordshire, in April 2015. *Collaris* is from Latin *collum* = 'neck'.

Ruddy duck *Oxyura jamaicensis*, wingspan 58 cm, length 39 cm, male weight 670 g, females 480 g, a North American introduction spreading from post-1953 escapes from the WWTs collection at Slimbridge, Gloucestershire, the first record of breeding (other than the brief success of a Shropshire escaped pair in the 1930s) coming from the Chew Valley reservoir in Avon in 1960. This species particularly benefited from suitable habitat in west Midlands reservoirs, and subsequently spread elsewhere in Britain and in Europe, since 1973 endangering the Spanish white-headed duck *O. leucocephala* by mating with females and producing hybrid birds. As a consequence since 2005 this duck has been subject to a Government-led eradication programme with an estimated cost of £3.3–5.4 million. By early 2014, the cull had reduced the British population to 20–100, down from a peak of around 5,500 in 2000. Nests are on the ground, clutch size is 6–10, incubation lasts 25 days, and the precocial young fledge in 50–5 days. Diet is of seeds and roots of aquatic plants, aquatic insects and crustaceans. Greek *oxus* + *oura* = 'sharp' + 'tail'.

Sharp-tailed duck Alternative name for long-tailed duck.

Tufted duck *Aythya fuligula*, wingspan 70 cm, length 44 cm, weight 760 g, common and widespread, with 16,000–19,000 pairs in 2009. Numbers increased by 99% between 1975 and 2015, and by 38% between 1995 and 2015 (though with a downturn after 2010) and, augmented by wintering migrants from Iceland and northern Europe, there were 110,000 individuals in winter 2008–9, found on lakes, ponds, reservoirs, flooded gravel-pits and estuaries. The ground nests are close to water, clutch size is 8–11, incubation takes 25 days, and the precocial young fledge in 45–50 days. These diving ducks favour insects and molluscs, and also feed on plants. Latin *fuligo* + *gula* = 'soot' + 'throat'.

Winter duck Alternative name for pintail.

Wood duck *Aix sponsa*, wingspan 66–73 cm, length 47–54 cm, weight 500–700 g, introduced from eastern North America in the 1870s, kept in and released from captivity in many parts of Britain since then, but only sporadic breeding has been noted in the wild, and populations do not persist. *Sponsa* is Latin for 'bride', possibly referring to the male plumage being so striking that it looks like it is dressed for a wedding.

DUCKWEED Small aquatic species of *Lemna* (Greek for these plants), *Spirodela* (Greek *speira* + *delos* = 'spiral' + 'clear', referring to spiral vessels visible through the plant), and *Wolffia* (honouring the German physician and botanist Johann Friedrich Wolff, 1778–1806). Each plant mostly consists of a small 'thallus' or 'leaf' a few cells thick, often with air pockets that allow it to float on or just under the water surface. Reproduction is by vegetative budding. Originally placed in the family Lemnaceae these plants are now considered to be members of the Araceae, subfamily Lemnoideae.

Common duckweed *L. minor*, widespread and abundant, floating on still or slowly flowing, mesotrophic or eutrophic waters, as well terrestrially on exposed mud, or damp stonework and rocks. *Minor* is Latin for 'smaller'.

Fat duckweed *L. gibba*, common and widespread in England, scattered in Wales, Ireland and central Scotland (though frequent in the Forth & Clyde and Union Canals), found in still or slowly flowing, eutrophic ponds, canals, ditches and quiet river backwaters of rivers. In very eutrophic sites it can form dense masses which exclude other aquatics. *Gibba* is Latin for 'hunched'.

Greater duckweed *S. polyrhiza*, locally common in England, more scattered in Wales and Ireland, scarce in (central) Scotland, found in base-rich ponds, ditches, canals and slowly flowing rivers, and especially frequent in grazing marshes. *Polyrhiza* is Greek for 'many roots'.

Ivy-leaved duckweed *L. trisulca*, the only submerged duck-weed species in the British Isles, common in lowland regions in mesotrophic to eutrophic, still to slowly flowing waters where low nutrient levels or exposure prevent development of a dense blanket of floating duckweeds. *Trisulca* is Latin for 'three-furrowed'.

Least duckweed *L. minuta*, shade-tolerant, common and widespread in England, Wales and south-east Ireland, though first recorded in 1977, scarcer elsewhere, found on the surface of lakes, ponds, slowly flowing rivers, streams, canals and ditches. *Minuta* is Latin for 'small'.

Red duckweed *L. turionifera*, a North American neophyte recorded from a ditch in Stoborough, Dorset, and the South Forty Foot Drain in Lincolnshire (both in 2007), the very widely separated sites suggesting that it had been in Britain for some time but overlooked. It has since been found in the Gwent Levels, Monmouthshire (2010) and from a lake at Stourhead, Wiltshire (2013), spreading rapidly. *Turionifera* is Latin for 'bearing turions'.

Rootless duckweed *W. arrhiza*, local in southern England and Somerset, on ponds and ditches as small patches or scattered among other floating duckweeds. Flowers (only 0.3 mm across) have never been recorded in the British Isles, but it is the region's smallest 'flowering plant'. *Arrhiza* is Greek for 'without root'.

DUKE OF BURGUNDY BUTTERFLY *Hamearis lucina* (Riodinidae), male wingspan 29–32 mm, female 31–4 mm, frequenting scrubby grassland and woodland clearings, found only in England with a stronghold in central-southern parts, for example downland in the Cotswolds and from Wiltshire to West Sussex, and isolated colonies in the Lake District and the North York Moors. It has declined substantially in recent decades (in distribution and numbers by 84% and 42%, respectively, between 1976 and 2014, though – from a low base – with a slight recovery in 2005–14, by 3% and 67%, respectively), especially in woodlands where it is found in fewer than 20 sites. There was an increase in numbers by 26% from 2013 to 2014. It is listed as a Section 41 species of principal importance under the NERC Act. Key larval food plants are primrose and cowslip. Adults fly in May–June, feeding primarily on nectar of tormentil. Greek *hama* + *ear* = 'at the same time as' + 'spring', describing the flight period. Lucina was the goddess of childbirth (who brings light).

DULSE *Palmaria palmata* (Palmariaceae), a red seaweed, fronds generally 5–30 (up to 100) cm long, widespread but patchily distributed on the east coast, in the littoral and sublittoral down to 20 m, epilithic and epiphytic, especially on kelp stipes. It is collected for eating raw and dried, for baking, for example used in white soda bread in Ireland, and cooking as a flavouring. 'Dulse' comes from Irish/Scottish Gaelic *duileasc* or *duileasg*. Both parts of the binomial are from Latin *palma* for 'palm (of the hand)', here meaning 'broad'.

False dulse Alternative Irish name for red rags seaweed.

Pepper dulse *Osmundea pinnatifida* (Rhodomelaceae), a widespread red seaweed common on exposed to moderately sheltered rocky shores, often covering large areas with a turf-like growth in tidal pools and on rocks, in places subtidal. It is dried and used as a curry-flavoured spice in Scotland and the Channel Islands. Latin *pinnatifida* = 'pinnately cleft'.

DUN Angler's name for mayflies, especially those in the families Heptageniidae and Leptophlebiidae. Nymphs generally feed by scraping periphyton from the substrate or by gathering fine particulate organic detritus from the sediment. See also entry in Part I.

Autumn dun *Ecdyonurus dispar* (Heptageniidae), widespread, in riffle areas of rivers and larger streams, occasionally on wave-affected lake shores. Nymphs, with a flattened shape, are usually found on submerged plants and stones. There is one generation a year, adults emerging in June–October. Males swarm throughout the afternoon. Greek *ecdyo* + *oura* = 'eclipse' + 'tail', and Latin *dispar* = 'different'.

Brown May dun *Kageronia fuscogrisea* (Heptageniidae), scattered and local, for example in the Thames valley, south-west Scotland and Ireland. Nymphs live in the riffle sections of larger rivers, wave-affected shores of calcareous lakes and, atypically for Heptageniidae, among macrophytes in standing waters. There is one generation a year, with emergence in May–June. Males swarm throughout the day. *Fuscogrisea* is Latin for 'brown-grey'.

Claret dun *Leptophlebia vespertina* (Leptophlebiidae), common and widespread, preferring peaty or acidic waters and so tends to be less common in lowland waters. Larvae are found in the pools and margins of slow-flowing streams and in ponds and lakes where they climb upon the surface of leaves of aquatic plants or crawl in the surface layers of fine sediments. There is one generation a year. Emergence is between April and August during daylight at the surface of the water or, more typically, partially or entirely out of the water on a plant stem. Males swarm during the day and into dusk, hence Latin *vespertina* = 'of the evening'. Greek *leptos* + *phlebos* = 'thin' + 'of a vein'.

Ditch dun *Habrophlebia fusca* (Leptophlebiidae), scattered and local, though common in central England, and with few records north of the Central Belt of Scotland. Nymphs live in pools and stream margins on the leaves of aquatic plant or in the surface layers of fine sediments. There is one generation a year, with emergence partially or entirely out of the water on a plant stem during daylight from May to September. Greek *habros* + *phleps* = 'soft' or 'delicate' + 'vein', and Latin *fusca* = 'dark' or 'brown'.

Large brook dun *E. torrentis*, mainly an upland species but also found in parts of southern England. Life history is similar to large green dun. Overall emergence is from March to September: in upstream reaches the flight period can last for up to three months, but in lower reaches it may last one month. *Torrentis* is from Latin for 'torrent'.

Large green dun *E. insignis*, highly localised, with records from a few watercourses in upland Britain. Nymphs live in stream riffle areas, generally clinging to submerged plants and stones to feed by scraping periphyton from the substrate or by gathering fine particulate organic detritus from the sediment. Adults emerge between May and October, males swarming in the afternoon and at dusk. *Insignis* is Latin for 'notable'.

Purple dun *Paraleptophlebia cincta* (Leptophlebiidae), widespread but rarely abundant. Life history is similar to claret dun, with emergence from May to August. Greek *para* = 'near' + the genus *Leptophlebia* (*leptos* + *phlebos* = 'thin' + 'of a vein'), and Latin *cincta* = 'girdle'.

Scarce purple dun *P. werneri*, scarce at a few sites in southern England, burrowing in the sediment of calcareous streams. Flight is in May–June.

Sepia dun *L. marginata*, common and widespread, particularly tolerant of acidification, and found in waters with pH values of 4–5. Nymphs live in the pools and margins of slow-flowing streams and in ponds and lakes where they climb upon the surface of leaves of aquatic plants or crawl in the surface layers of fine sediments. There is one generation a year, which emerges between April and June during daylight at the surface of the water or, more typically, partially or entirely out of the water on a plant stem. Males swarm throughout the day.

Turkey dun Angler's name for *Paraleptophlebia submarginata* (Ephemerellidae).

Yellow (evening) dun See yellow hawk under hawk (mayfly).

Yellow May dun *Heptagenia sulphurea* (Heptageniidae), widespread and common. Nymphs live in the riffle sections of larger rivers and along wave-affected shores of calcareous lakes. There is one generation a year which often has a group of fast-growing individuals which emerge in May or June and a slower growing group that emerges in August–September. Emergence takes place by day, male swarming in the afternoon into dusk. Greek *hepta* + *genos* = 'seven' + 'type' and Latin *sulphurea* = 'sulphur-yellow'.

DUN-BAR *Cosmia trapezina* (Noctuidae, Xyleninae), wingspan 25–33 mm, a widespread nocturnal moth common in woodland, gardens and hedgerows. Adults fly in July–September. Larvae feed on a variety of deciduous trees and on the larvae of other moths, especially winter moth, and possibly also cannibalistically. Greek *cosmios* = 'well-ordered', and Latin for 'trapezium', from the wing pattern.

DUNG BEETLE See under beetle.

DUNG FLY Members of the family Scathophagidae.

Yellow dung fly *Scathophaga stercoraria*, 5–11 mm length, very common and widespread, especially in cattle-farming areas in more western parts of the British Isles. Adults are active from March to November. Females mostly forage in surrounding vegetation and only visit dung (preferably cow pats) to mate and lay eggs on the dung surface. Males spend most of their time on the dung pats, waiting for females and feeding on (often other dipteran) visiting insects that. Both sexes also take nectar. Eggs are laid in the dung, and larvae remain there feeding on other insect larvae. Greek *scata* + *phago* = 'dung' + 'to eat', and Latin *stercoraria* = 'dung'.

DUNLIN *Calidris alpina* (Scolopacidae), wingspan 40 cm, length 18 cm, weight 48 g, an Amber List wader, ssp. *schinzii* being a migrant breeder (wintering in West Africa), 8,600–10,600 pairs estimated in 2005–7, mostly in the uplands of Britain, greatest numbers in the Hebrides, Orkney, Shetland, the Flow

Country of Caithness and Sutherland, and the Pennines, as well as a few sites on machair in north-west Ireland. The nest is a plant-lined scrape, usually with 4 eggs, incubation taking 21–2 days, and fledging of the precocial young 19–21 days. Ssp. *alpina* from west Siberia winters round much of the coastline (less so in north-west Scotland and north-west Ireland), with 360,000 birds estimated in 2008–9, making this the commonest small coastal wader, though the RSPB has estimated that wintering numbers in Britain declined by 24% between 2002–3 and 2012–13. Wintering dunlin in Ireland are found at <10 sites, but numbers can be high, for instance the Shannon and Fergus estuary (Co. Clare) and Dundalk Bay (Co. Louth) both regularly support >10,000 birds. Small numbers of ssp. *arctica* from Greenland pass through in autumn. Coastal birds feed on invertebrates along the sea edge, roosting on fields, saltmarshes and the shore at high tide. Inland breeders have a mainly insect diet. The English name is a dialect form of 'dunling', derived from 'dun' meaning dull brown. Greek *calidris* was used by Aristotle for some grey-coloured waterside birds.

DUNNOCK *Prunella modularis* (Prunellidae), wingspan 20 cm, length 14 cm, weight 21 g, a resident breeder with 2.3 million territories in Britain in 2009, numbers declining by 29% between 1970 and 2015, but clearly recovering in the latter part of this period with a 22% increase between 1995 and 2015 (stabilising at 1% over 2010–15). It is Amber-listed in Britain, Green-listed in Ireland. It frequents a range of scrubby and wooded habitats, including gardens. Females are polyandrous, males polygamous: birds breed in groups of up to three males and three females, with two males and a female being the most common. Nests are lined cups of twigs and moss built in bushes, scrub and hedges. Clutch size is usually 4–5, incubation takes 14–15 days, and the altricial young fledge in 12–15 days. Diet is mainly insectivorous, with seed also taken in winter. 'Dunnock' either derives from Middle English *dunnākos* = 'little brown one', or *dunoke* and *donek*, from *dun* (from the brown and grey plumage). *Prunella* is from the German *Braunelle* for this bird, a diminutive of *braun* = 'brown' and *modulare* = 'to sing'.

DUST MITE See under house dust mite under mite.

DUTCH ELM DISEASE Causing the death of most elm trees in southern and central Britain during the 1970s (and continuing to affect re-suckering trees), the pathogenic fungus *Ophiostoma ulmi* (Ophiostomatales, Ophiostomataceae) (Greek *ophis* + *stoma* = 'serpent' + 'mouth', and Latin for 'elm') is a wilt disease: it blocks the vascular (water transport) system of the host tree, causing the branches to wilt and die. By 1980, >20 million elms had been killed. It is spread by bark beetles, mainly species of *Scolytus*.

DUTCHMAN'S-PIPE Alternative name for yellow bird's-nest.

DWARF (MOTH) Members of the family Elachistidae, the majority in the genus *Elasticha* (Greek *elastichos* = 'very small'), most being leaf miners of grasses, sedges and rushes.

Basil dwarf *Stephensia brunnichella*, wingspan 8–9 mm, scarce, in deciduous woodland and woodland margins on chalky soils in England and parts of Wales. Double-brooded, adults fly in May–June and August–September. Larvae mine leaves of wild basil in spring and autumn. The binomial honours the British entomologist James Stephens (1792–1852), and the Danish entomologist Morten Brünnich (1737–1827).

Black-headed dwarf *Elachista atricomella*, wingspan 11–13 mm, nocturnal, widespread and common on roadside verges, woodland edges and waste ground. Adults fly in May–September. Larvae mine leaves of cock's-foot in spring. Latin *ater* + *coma* = 'black' + 'hair'.

Bog dwarf *E. utonella*, wingspan 8–9 mm, scarce, in bogs and acid heaths local in England and east Ireland. Adults mainly fly in July–August. Larvae mine leaves of various sedges from

March to May. The specific name is Latin for Mount Uetilberg, near Zuruch, the type locality.

Broken-barred dwarf E. *freyerella*, wingspan 7–8 mm, in damp meadows and woodland rides, widespread and locally common. Adults fly from late April to June and again from late July to September. Larval mines on various grasses are evident from March to April and in July. Christian Freyer (1794–1885) was a German entomologist.

Brown-barred dwarf E. *subocellea*, wingspan 8–10 mm, widespread but scarce, in woodland edges and clearings, more numerous in the south. Adults fly in June–July. Larvae mine leaves of false-brome. Latin *sub* + *ocellus* = 'somewhat' + 'small eye', describes the forewing pattern.

Buff dwarf E. *subalbidella*, wingspan 10–13 mm, fairly widespread if local on moorland and heathland. Adults fly in May–June. Larvae mine leaves of purple moor-grass in autumn. Latin *sub* + *albidus* = 'somewhat' + 'whitish', from the forewing colour.

Cotton-grass dwarf E. *albidella*, wingspan 9–10 mm, widespread but local on bogs and acid heaths. Adults fly in June–August. Larvae mine leaves of deergrass, slender tufted-sedge, greater pond-sedge and common spike-rush. *Albidella* is Latin for 'whitish', from the forewing colour.

Dark dwarf E. *subnigrella*, wingspan 7–8 mm, widespread if local on dry grassland. Adults fly in May–June and again in August. Larvae mine grass leaves. Latin *sub* + *niger* = 'somewhat' + 'black'.

Dusky dwarf E. *serricornis*, wingspan 7–9 mm, widespread but scarce in bogs and damp woodland in widely separated localities. Adults fly in June–July. Larvae mine leaves of wood-sedge from autumn into spring. Latin *serra* + *cornu* = 'saw' + 'horn', from the serrate antennal shape.

Fen dwarf E. *pomerana*, wingspan 8–10 mm, local on fens at Stalham and Sutton in Norfolk, and Woodwalton Fen in Huntingdonshire. Adults are triple-brooded, flying from late April to September. Larvae mine leaves of wood small-reed in April–June, possibly also in August. Latin *pomerana* = 'Pomeranian'.

Field dwarf E. *consortella*, wingspan 7–8 mm, found in open grassland in widely scattered localities. The bivoltine adults fly from late April to September. Larvae mine leaves of annual meadow-grass. *Consortella* is from Latin *consors* = 'partner'.

Grey dwarf E. *bedellella*, wingspan 7–8 mm, scarce, in dry open areas scattered throughout England as far north as the Lake District, but most frequent in the south. Larvae mine leaves of meadow oat-grass in September–April and in June–July. George Bedell (1805–77) was a British entomologist.

Grey-spotted dwarf E. *cinereopunctella*, wingspan 7–9 mm, widespread but scarce, local on limestone grassland, fens and dune slacks. Larvae mine leaves of glaucous sedge. Latin *cinereus* + *punctum* = 'ash-coloured' + 'spot', from the forewing marking.

Hampshire dwarf E. *littoricola*, wingspan 7–8 mm, recently recorded only at one site on coastal grassland at Hurst Castle and Keyhaven in Hampshire, first in 1982, last in 1990. Adults fly in June–August. Larvae mine leaves of common couch. *Littoricola* is from Latin *litus* + *colo* = 'shore' + 'to inhabit', from the estuarine type locality (Garonne, south-west France).

Highland dwarf E. *eskoi*, wingspan 10–14 mm, scarce, on damp acid grassland in Scotland, where first recorded at Gordon Moss, Berwickshire and Auchencorth Moss, Midlothian in June 1956 and known from only six sites in total. Adults fly in early morning and late evening sunshine. Larvae are undescribed. Esko Suomalainen (1910–95) was a Finnish entomologist.

Honeysuckle dwarf *Perittia obscurepunctella*, wingspan 8–10 mm, in open woodland scattered throughout much of England, more numerous in the south, and scarce in Wales and Ireland. Adults fly during the day in April–May. Larvae mine leaves of honeysuckle in June–August. Greek *perittos* = 'strange', and Latin *obscurus* + *punctus* = 'obscure' + 'spot'.

Little dwarf E. *canapennella*, wingspan 8–10 mm, widespread and common in grassland. It has two generations in most of its range, flying from April to June and from July to September; in north Scotland it is single-brooded from June to August. Larvae mine leaves of grass species. Latin *canus* + *penna* = 'hoary' or 'grey' + 'wing', from the forewing colour.

Marsh dwarf E. *alpinella*, wingspan 10–13 mm, widespread but scarce in marsh, water meadows and other damp areas. Adults fly in June–September. Larvae mine the leaves of sedges from September, overwinter, and appear again in spring. Latin *alpinella* = 'alpine'.

Meadow dwarf E. *triatomea*, wingspan 8–9 mm, in open grassland, hillsides and clifftops, throughout the Midlands, East Anglia and south-east Scotland; also in central and west Scotland and (especially east) Ireland. Adults fly at night in June–July. Larvae mine leaves of fescues. Latin *tri* + *atomus* = 'three' + 'speck', from the forewing pattern (with in fact only two spots).

Moorland dwarf E. *kilmunella*, wingspan 9–11 mm, local on acid heathland, bogs and damp grassland in upland areas of Britain, as far south as Herefordshire, and very scarce in Ireland. Adults fly in the late afternoon in May–August. Larvae mine leaves of sedges. The specific name relates to Kilmun, Argyll, the type locality.

Narrow-barred dwarf E. *unifasciella*, wingspan 9–10, scarce, on grassland in England north to Worcestershire. Larvae mine leaves of cock's-foot and false-brome in spring. Latin *unus* + *fascia* = 'one' + 'band', from the forewing pattern.

Oblique-barred dwarf E. *adscitella*, wingspan 9–11, scarce, in woodland clearings in parts of England, Wales and south Ireland. Adults fly in May–July and again in August. Larvae mine leaves of tufted hair-grass and blue moor-grass. Latin *adscitella* = 'assumed'.

Obscure dwarf E. *humilis*, wingspan 9–10 mm, common in a range of grassy habitats throughout much of Britain, but very rare in Ireland. Adults fly from May to July in the south in two overlapping broods, and July in the north as a single brood. Larvae mine grass leaves, in the south mainly from March to June. Latin *humilis* = 'low'.

Pearled dwarf E. *apicipunctella*, wingspan 10–11 mm, local along woodland edges and in woodland clearings, except for south-west England. Adults emerge from late April to June in the south, in June–July in the north. There may occasionally be a second generation in the south. Larvae mine grass leaves over winter. Latin *apex* + *punctum* = 'tip' + 'spot', from the white spot on the forewing tip.

Pembroke dwarf E. *collitella*, wingspan 7–9 mm, in dry grassland very local along the south coast of England and Wales. Larvae mine leaves of sheep's-fescue, crested hair-grass and smooth meadow-grass. Latin *collitus* = 'smeared', from the wing pattern.

Red-brindled dwarf E. *rufocinerea*, wingspan 10–11 mm, widespread and common in grassland, bog and heathland. Adults fly in April–June. Larvae mine grass leaves. Latin *rufus* + *cinereus* = 'red' + 'ash-grey', from the forewing colour.

Saltern dwarf E. *scirpi*, wingspan 10–12 mm, scarce, in saltmarsh and brackish marsh local and scattered along the coast of England (for example Suffolk and Hampshire), Wales and south Ireland. Adults fly in June–July. Larvae mine leaves of sea club-rush and saltmarsh rush. *Scirpi* is from the genus name for club-rush, *Scirpus*.

Scarce dwarf E. *trapeziella*, wingspan 8–10 mm, local in deciduous woodland, ravines and gorges in three discrete regions: southern England, southern Ireland, and northern England and Scotland. Adults fly in May–July. Larvae mine wood-rush leaves in spring. Greek *trapezion* = 'trapezium', the shape produced if the forewing spots were to be joined.

Scotch dwarf E. *orstadii*, wingspan 7–9 mm, local on dry grassland in parts of Scotland, with the first records in England coming from four limestone grassland sites in the Cotswolds in 2013. Adults fly in May–July. Larvae mine grass leaves. E.T. Orstadius (1861–1939) was a Swedish entomologist.

Small bog dwarf *E. eleochariella*, wingspan 7–8 mm, scarce, in peat bog and marsh in parts of west Scotland and Ireland. Larvae mine leaves of common and glaucous sedge, spike-rush and common cottongrass. *Eleochariella* is from the genus name of spike-rush, *Eleocharis*.

Southern dwarf *E. stabilella*, wingspan 7–8 mm, in grassland in parts of southern England and south Wales, though with records as far north as Yorkshire. Adults fly in May–June and again in July–August. Larvae mine grass leaves. Latin *stabilis* = 'fixed', from a lack of variation.

Swan-feather dwarf *E. argentella*, wingspan 11–12 mm, widespread and common in grassland and saltmarsh. Adults fly at night in May–July. Larvae mine grass leaves. The English name is prompted by the feathery white wings. Latin *argenta* = 'silver'.

Sweet-grass dwarf *E. poae*, wingspan 7–9 mm, local on the fringes of ponds, lakes and slow-moving streams throughout much of England and Wales. There are two generations, with larvae mining reed sweet-grass leaves in April–May and July–early August, and adults flying in May–June and in August. *Poae* is from the former genus name of the food plant sweet-grass, *Poa*.

Triple-spot dwarf *E. maculicerusella*, wingspan 10–12 mm, common in fen and marsh and by rivers and canals, widespread and extending into Scotland as far as the south Highlands. Adults fly between May and August in one or two broods depending on latitude. Larvae mine leaves of reed canary-grass, common reed and occasionally other grasses. Latin *macula* + *cerussa* = 'spot' + 'white lead', from the forewing pattern.

Twin-barred dwarf *E. gleichenella*, wingspan 8–9 mm, widespread but local in shaded deciduous woodland. Adults fly between May and August. Larvae mine leaves of sedges and wood-rushes, mainly in spring. Wilhelm von Gleichen (1717–83) was a Viennese naturalist.

Two-spotted dwarf *E. biatomella*, wingspan 7–9 mm, local on grassland and heathland in England and Wales. Adults fly in May–June and July–August. Larvae mine sedge leaves. Latin *bi* + *atomus* = 'two' + 'speck', from the forewing spots.

Western dwarf *E. triseriatella*, wingspan 7 mm, scarce, on calcareous grassland in south-west England and Wales; also recorded in Dumfries and Galloway in 2001 and 2005. Adults fly in June–July. Larvae mine leaves of fescues and other grasses. Latin *tri* + *series* = 'three' + 'row', from the forewing pattern.

White dwarf *E. cahorsensis*, wingspan 8 mm, scarce, on dry pasture and downland in Wales and Cornwall. Adults fly in May–June. Larvae mine leaves of red fescue and other grasses. Cahors is a town in south-west France.

White-headed dwarf *E. albifrontella*, wingspan 8–9 mm, widespread and common in grassland, woodland and gardens. Adults fly in June–July. Larvae mine grass leaves from September to May. Latin *albus* + *frons* = 'white' + 'forehead'.

White-tipped dwarf *E. cingillella*, wingspan 8–10 mm, scarce, in damp woodland, previously with a scattered distribution in England but in 1987 reported only from Herefordshire. Larvae mine leaves of wood-millet from September to April. Latin *cingella* = 'small girdle', from a resemblance when the wings are folded.

Wood dwarf *E. obliquella*, wingspan 8–10 mm, local in woodland rides and edges scattered throughout much of England. Adults fly in April–July and again in August. Larvae mine leaves of false and upright bromes and other grasses as well as sedges in March–May. Latin *obliquus* = 'slanting', from the forewing marking.

Wood-rush dwarf *E. regificella*, wingspan 8–9 mm, scarce, with a scattered distribution in the British Isles, mostly in the south, in open woodland. Adults fly in July. Larvae mine wood-rush leaves. Latin *regificus* = 'royal', from the metallic gold forewing markings.

Yellow-barred dwarf *E. gangabella*, wingspan 9–10 mm, local in woodland in southern England and Wales, rare elsewhere, for example with two records from central Ireland. Adults fly in May–June. Larvae mine leaves of various grasses in August–October. Persian *gangaba* = 'porter', entomological significance unknown.

Yellow-headed dwarf *E. luticomella*, wingspan 10–11 mm, widespread but local on woodland edges and rides, roadside verges and waste ground. Adults fly in the evening from June to August. Larvae mine leaves of grasses such as cock's-foot and false-brome throughout winter. Latin *luteus* + *coma* = 'yellow' + 'hair'.

Yellow-tipped dwarf *E. bisulcella*, wingspan 8–10 mm, local in deciduous woodland and woodland edges throughout much of Britain and scattered localities in Ireland. Adults fly between May and September in the south, probably in two generations, and in July–August further north. Larvae mine leaves of tufted hair-grass and tall fescue from April to July. Latin *bi* + *sulcus* = 'two' + 'furrow', referencing the forewing line markings.

DYEBALL *Pisolithus arhizus* (Boletales, Sclerodermataceae), a rare inedible fungus with a few records especially in south and south-east England in summer and autumn, in well-drained soil, including sand- and gravel-pits, and coal tips, mycorrhizal on pine and other trees. Greek *pisos* + *lithos* = 'pea' + 'stone', and *arhiza* = 'without roots'.

DYER'S MADDER See madder.

DYSDERIDAE Family name (Greek *dysderis* = 'quarrelsome') of woodlouse spiders, a group with six rather than eight eyes.

DYSENTERY-HERB Short-fruited dysentery-herb *Monsonia brevirostrata* (Geraniaceae), an annual South African neophyte, introduced with wool shoddy, found as a casual in fields and waste ground in a few places in England. Lady Ann Monson (1714–76), was a plant collector in the Cape of Good Hope. *Brevirostris* is Latin for 'short-beaked'.

DYTISCIDAE From Greek *dytes* = 'diver' and *dyticos* = 'able to dive', the family name for diving beetles in the suborder Adephaga, with six subfamilies. **Copelatinae** has one British species, *Copelatus haemorrhoidalis*, found in well-vegetated ponds and ditches throughout much of southern Britain, local in Ireland. **Laccophilinae**, with three species of *Laccophilus*, recognisable by an ability to jump. **Hydroporinae**, with 64 species in 17 genera, often pioneer colonisers of new water bodies or found in shallow or ephemeral habitats. **Colymbetinae**, with seven species in two genera. **Agabinae**, with three genera, most found in vegetation and sediment around the edges of pools and rivers. **Dytiscinae**, with 14 species in 5 genera.

E

EAGLE Large birds of prey in the family Accipitridae. 'Eagle' comes from Middle English and Old French *aigle*, from Latin *aquila* = 'eagle'. **Greater spotted eagle** *Aquila clanga* and **short-toed eagle** *Circaetus gallicus* are rare accidentals.

Fish eagle, Fishing eagle See osprey.

Golden eagle *Aquila chrysaetos*, wingspan 1.8–2.3 m, length 66–102 cm, male weight 3.7 kg, females 5.3 kg, exterminated in England and Wales by 1850, and in Ireland by 1912, but surviving in small numbers on moorland and mountain habitat in Scotland. Numbers declined even more in the 1950s and 1960s because of eggshell thinning/lack of hatching via the effects of organochlorine pesticide bioaccumulation, then recovered following a ban on these chemicals. Birds remain mainly found in the Scottish Highlands and Islands, with records also from the south-west Scottish Uplands and the Lake District (never more than 2 pairs), with an estimated total of 442 breeding pairs in 2003, increasing by 15% to 508 pairs in 2016. It was reintroduced to Ireland (Glenveagh National Park, Co Donegal) in 2001, with only 12 chicks hatching in the period 2007–16 (with persecution and poor habitat), and just 4 pairs breeding in 2016. Nests are of twigs on rocky crags, 2 eggs are laid, incubation takes 43–5 days, fledging of the altricial young 65–70 days. Diet is of birds and mammals such as rabbits and hares, taken both alive and as carrion, the latter especially in the western Highlands. Irish golden eagles tend to eat crows and badger cubs, indicating their difficulty in finding appropriate food. Latin *aquila* = 'eagle', and Greek *chrysos* + *aetos* = 'gold' + 'eagle'.

Sea eagle Alternative name for white-tailed eagle, and a dialect name for osprey.

White-tailed eagle *Haliaeetus albicilla*, wingspan 1.8–2.5 m, length 66–94 cm, male weight 3.1–5.4 kg, female 4.6–6.9 kg. It became extinct in Britain during the early twentieth century mainly through illegal killing, last breeding in Skye in 1916 and Shetland in 1918. They were reintroduced to Rum in 1975 using Norwegian birds, and in the following 10 years, 82 young eagles from Norway were released there. The first successful breeding took place in 1985, and several pairs have nested successfully every year since. Further releases in the 1990s in Wester Ross ensured that the population became self-sustaining, and the species now breeds throughout the Hebrides and the mainland coast of north-west Scotland. In 2008, 15 chicks raised in Norway were released in Fife. By 2015 the breeding population exceeded 100 pairs. Reintroduction in Ireland began in 2007, with birds released each spring in Killarney National Park. In 2013, the first eaglets were born in Ireland since the reintroduction programme began – one in the Killarney National Park and two in Co. Clare. In 2015, five nests hatched chicks in counties Clare, Cork, Galway and Kerry. Nests are of twigs, usually in trees, with 1–3 eggs, incubation taking 38 days, fledging 11–12 weeks. Diet is varied, opportunistic and seasonal, including fish, birds and mammals. Greek *halos* + *aetos* = 'of the sea' + 'eagle', and Latin *albus* + *cilla* = 'white' + 'tail'.

EAGLE'S CLAWS *Anaptychia ciliaris* ssp. *ciliaris* (Physciaceae), a foliose to subfruticose lichen found on well-lit and nutrient-enriched tree-trunks. During the last century it suffered decline in range of 50–80% from sulphur dioxide pollution, excessive use of fertilisers, and especially the loss of elms through Dutch elm disease. It remains locally common in southern and eastern Britain, scattered elsewhere. In 2012 it was reported that there were only five sites in England where it was found on >10 trees, and it is vulnerable to local extinction. Cilia on the margins of the lobes are often pale and curved, giving the lobe tip the appearance of eagle's claws. Greek *ana* + *ptyche* = 'upon' + 'a fold', and Latin *ciliaris* = 'fringed with hair'.

EAR (MOTH) *Amphipoea oculea* (Noctuidae, Xyleninae), common and widespread in damp habitats. Adults fly in July–September. Larvae feed on the base of roots and stems of grasses and butterbur. Greek *amphi* + *poia* = 'around' + 'grass', and Latin *oculea* = 'eyed', from the ear-shaped forewing mark. Wingspan for all species 29–36 mm.

Crinan ear *A. crinanensis*, local on moorland, unimproved grassland and dunes, widespread in Scotland and northern England, mid- and western Wales, and much of the coast of Ireland. Adults fly in August–September. Larvae feed inside the stems of yellow iris. Crinan is a village in Argyll.

Large ear *A. lucens*, local in wet moorland, acid grassland, marsh and mosses in northern and western Britain. Adults fly in August–September. Larvae feed in the roots and stems of grasses, especially purple moor-grass. *Lucens* is Latin for 'shining'.

Saltern ear *A. fucosa*, local on coastal saltmarsh (hence 'saltern') and dunes on the coasts of England, Wales and east Scotland, extending inland at low density in some areas. Adults fly in August–September. Larvae feed in grass roots and stems. *Fucosa* is Latin for 'spurious' or possibly 'painted'.

EARPICK FUNGUS *Auriscalpium vulgare* (Russulales, Auriscalpiaceae), inedible, widespread and common, particularly in England, found in late summer and autumn, emerging from buried pine cones. The tiny spines on the cap prompt both its English and genus names, the latter from Latin *auris* + *scalpare* = 'ear' + 'to scratch'. *Vulgaris* is Latin for 'common'.

EARTH-MOSS Species in the genera *Aphanorrhegma* (Funariaceae), from Greek *aphanes* + *rhegma* = 'invisible' + 'break' or 'tear'; *Archidium* (Archidiaceae), possibly from *arche* = 'primitive'; *Ephemerum* (*ephemeros* = 'short-lived'), *Microbryum* (*micro* + *bryo* = 'small' + 'moss') and *Phascum* (*phascos* = a kind of tree moss) (Pottiaceae); and *Pleuridium* (*pleuron* = 'side') and *Pseudephemerum* (*pseudis* = 'false' + the genus *Ephemerum*) (Ditrichaceae).

Common and widespread species, often as pioneers on bare or disturbed soil, include **awl-leaved earth-moss** *Pleuridium subulatum*, **clay earth-moss** *Archidium alternifolium*, **cuspidate earth-moss** *Phascum cuspidatum*, **serrated earth-moss** *E. serratum*, **spreading earth-moss** *Aphanorrhegma patens*, and **taper-leaved earth-moss** *Pleuridium acuminatum*.

Scarce species are **clustered earth-moss** *E. cohaerens* and **sessile earth-moss** *E. sessile*.

Delicate earth-moss *Pseudephemerum nitidum*, common and widespread, less so in eastern England, the Highlands and central Ireland, on damp ground on lake margins, stream and ditch sides, muddy pasture, woodland rides, and arable fields. It is evident in summer and autumn but usually disappears by winter. *Nitidum* is Latin for 'bright'.

Strap-leaved earth-moss *E. recurvifolium*, locally common in the southern half of England. *Recurvifolium* is Latin for 'bending back'.

Swan-necked earth-moss *Microbryum curvicolle*, scattered in Britain, mainly England, on disturbed soil in calcareous grassland and quarries, as well as arable fields and woodland rides. *Curvicolle* is Latin for 'curved neck'.

EARTHBALL Round-capped inedible mycorrhizal fungi in the genus *Scleroderma* (Boletales, Sclerodermataceae) (Greek *scleros* + *derma* = 'hard' + 'skin').

Common earthball *S. citrinum*, very common and widespread, found from July to early December on bare, acid, often sandy soil or among moss on heathland and in deciduous woodland. *Citrinum* is Latin for 'citrus yellow'.

Leopard earthball *S. areolatum*, widespread and common, found from July to early December on soil, moss and other vegetation in damp deciduous woodland, usually under oak. *Areolatum* is from Latin *areola* = 'small open space', a reference to the pale annular region around each scale on the cap surface like a leopard's spots.

Potato earthball *S. bovista*, widespread and locally common, found in autumn on soil under deciduous woodland, scrub and heathland. *Bovis* is from Latin *bos* = 'of an ox'.

Scaly earthball *S. verrucosum*, poisonous, very common and widespread, found from July to early December on sandy or nutrient-rich dry soils, in grassland and woodland, especially under oak and beech. *Verrucosum* is Latin for 'wart-like', from scaly patches on the cap.

EARTHFAN Inedible fan-shaped fungi in the genera *Thelephora* (Thelephorales, Thelephoraceae), from Greek *thele* + *phoreos* = 'nipple' + 'bearer', and *Sistotrema* (Cantharellales, Hydnaceae), from Greek *sisto* + *trema* = 'point' + 'hole' (possibly 'torrent'). Specifically, *Thelephora terrestris*, scattered and locally common, found in July–November in coniferous woodland on sandy soil and litter, ectomycorrhizal on pine and spruce, but also found under hardwoods, on heathland and in mossy coastal dune slacks.

Carnation earthfan *T. caryophyllea* and **stinking earthfan** *T. palmata* are both uncommon.

Aromatic earthfan *S. confluens*, scattered in Britain found throughout the year in coniferous woodland encrusting dead wood, mainly Scots pine. *Confluens* is Latin for 'flowing together'.

Urchin earthfan *T. penicillata*, scattered but locally common, found from July to November on soil and litter mainly in wet acidic coniferous woodland, but also under deciduous trees. *Penicillata* is Latin for 'having finely divided tufts of hair'.

EARTHSTAR 1) Inedible fungi in the genus *Geastrum* (Geastrales, Geastraceae), from Greek *ge* + *aster* = 'earth' + 'star', which gives the English name, from the shape of the fruiting body, initially an onion shape (though known as an 'egg'), the outer layer of which splits open in a star-like manner. Many species are hygroscopic: in dry weather the 'petals' dry and curl up around the soft spore sac, protecting it, and the whole body can become detached from the ground. In damp weather, the 'petals' moisten and uncurl, lifting up the spore sac, allowing rain or animal movement to hit the sac which explosively releases spores during conditions that enhance germination prospects. A new and apparently endemic species, *Geastrum britannicus*, which resembles a human figure, was discovered at Cockley Cley, Norfolk, in 2000, and has since been identified from 15 other locations across southern England and Wales, for instance under yew at Cusop, Herefordshire.

Other uncommon or rare species include **beaked earthstar** (or **beret earthstar**) *G. pectinatum*, **Berkeley's earthstar** *G. berkeleyi*, **crowned earthstar** *G. coronatum*, **daisy earthstar** *G. floriforme*, **dwarf earthstar** *G. schmidelii*, **elegant earthstar** *G. elegans*, **field earthstar** *G. campestre*, **flask earthstar** *G. lageniforme*, **rayed earthstar** *G. quadrifidum*, **rosy earthstar** *G. rufescens*, **striate earthstar** *G. striatum*, **tiny earthstar** *G. minimum* and **weathered earthstar** *G. corollinum*. See also pepper pot.

Arched earthstar *G. fornicatum*, infrequent, with a scattered distribution in England and Wales, found in late summer and early autumn but persisting through winter in deciduous and especially coniferous woodland soils and leaf litter (including under yew), and often hollow rotting stumps. The common name comes from the extended rays standing on their tips, and *fornicatum*, from Latin, here means 'arched'.

Collared earthstar *G. triplex*, locally common and, especially in England, widespread, found in late summer and autumn, especially after rain, on deciduous woodland soil and leaf litter, often on slopes, and under wood chip in gardens. *Triplex* is Latin for 'threefold', here indicating having three layers.

Sessile earthstar *G. fimbriatum*, widespread but infrequent, found in late summer and early autumn, though often lasting throughout winter, on usually alkaline deciduous woodland soils and in leaf litter. That the fruiting body lies flat on the ground prompts 'sessile'. *Fimbriatum* is Latin for 'fringed', describing the ends of the rays.

2) **Barometer earthstar** (or **false earthstar**) *Astraeus hygrometricus* (Phallales, Diplocystidiaceae), found at a few locations in (mostly southern) England in late summer and early autumn on sandy soil in woods, especially under oak, or on dunes. The rays of this fungus open and close, reacting to the moisture content of the air and the soil on which it grows. *Hygrometricus* is Latin for 'water measurer', referring to this behaviour, similar to a barometer responding to fluctuations in atmospheric pressure which in turn gives the English name. In Greek mythology Astraeus, one of the Titans, was the god of dusk.

EARTHTONGUE Inedible tongue-shaped fungi in the genera *Geoglossum* (Greek *ge* + *glossa* = 'earth' + 'tongue'), *Microglossum* ('small' + 'tongue') and *Trichoglossum* (*trichos* = 'of hair') (Helotiales, Geoglossaceae).

Widespread but uncommon or only locally common species include **black earthtongue** *G. cookeanum* (on sandy unimproved grassland soil, often in dune slacks), **glutinous earthtongue** *G. glutinosum* (on grassland soil, occasionally lawns), and **hairy earthtongue** *T. hirsutum* (on usually acid soil or in short grass and moss).

Rarities include **dune earthtongue** *T. rasum* (north Devon and Cardiganshire in dunes and dune slacks), **elongate earthtongue** *G. elongatum* (dunes and calcareous grassland in Cornwall, Anglesey and Northern Ireland), **green earthtongue** *M. viride*, **olive earthtongue** *M. olivaceum*, **short spored earthtongue** *T. walteri*, **star earthtongue** *G. starbaekii* (Highlands), and **viscid earthtongue** *G. uliginosum* (grassland in Northern Ireland, Carmarthen and Argyllshire).

Dark-purple earthtongue *G. atropurpureum*, with a scattered distribution in western England, Wales, Scotland and Northern Ireland, a priority species in the UK BAP, found in late summer and autumn in short turfed, freely drained grassland in sites ranging from upland sheep walks through old pit tips to coastal cliff grasslands. *Atropurpureum* is Latin for 'dark purple'.

Plain earthtongue *G. umbratile*, widespread, on grassland soil, from under deciduous woodland to garden and parkland lawn. *Umbratile* is Latin for 'shading'.

EARTHWORM Species in the genera *Aporrectodea* (Latin *aporrectus* = 'not elongated'), *Lumbricus* (Latin for 'earthworm'), *Dendrobaena* (Greek *dendron* + *baino* = 'a stick' + 'to walk') and *Eisenia* (honouring the Swedish scientist Gustav Eisen, 1847–1940; see tiger worm), all in the family Lumbricidae. Epigeic species live on the surface of the soil in leaf litter, and tend not to make burrows. Endogeic species live in and feed on the soil, making horizontal burrows through the soil to move around and to feed, reusing these burrows to an extent. Anecic species make permanent vertical burrows in soil; they feed on leaves on the soil surface that they drag into their burrows. They also cast on the surface. Compost species tend to be found in places rich in rotting vegetation, preferring warm, moist environments.

Black-headed worm *A. longa*, 8–12 cm long, widespread, living in permanent burrows in alkaline soil, with casts up to 5 cm high, common in gardens.

Chestnut earthworm *L. castaneus*, 4 cm long, epigeic on woodland litter and under logs, widespread but scattered in Britain. *Castanea* is Latin for 'chestnut'.

Common earthworm 1) *L. terrestris*, up to 35 cm long, living in deep vertical burrows which aerate the soil, and surface 'worm casts' consisting of excreted soil. Darwin estimated that earthworms moved 100 tonnes of soil per hectare in a year. They can anchor themselves by broadening their tail to grip the sides of the burrow. They emerge at night to feed on fallen leaves and other decaying plant material, reaching densities of 20–40 worms per m² in an average garden lawn. *Terrestris* comes from Latin *terra* = 'earth'. 2) Sometimes used for *A. caliginosa*, but see grey worm below.

Grey worm *A. caliginosa*, 4–18 cm long, common and widespread, especially in woodland, pasture, gardens and other cultivated habitats, found in horizontal burrows in topsoil. *Caliginosa* is Latin for 'darkish'.

Mottled worm *A. icterica*, 5–14 cm long, with a few records from England, mainly from the Thames valley, in garden, orchard and meadow. *Ictericus* is Latin for 'jaundiced', meaning 'yellowish'.

Mucous worm *A. rosea*, 2–11 cm long, widespread and scattered in woodlands, pasture and gardens, preferring moist soil. *Rosea* is Latin for 'pink'.

Red earthworm *L. rubellus*, 2.5–10.5 cm long, common and widespread in soils with a pH of 5.5–8.7 (with a preference for neutral) and high in organic matter, preferably dung and faeces, feeding on highly decomposed organic material. *Rubellus* is Latin for 'reddish'.

Rosy-tip worm Alternative name for mucous worm.

Ruddy worm *L. festivus*, 5–11 cm long, widespread but scattered, living in leaf litter (epigeic). *Festivus* is Latin for 'festive' or 'with bright colours'.

EARWIG Elongate brownish members of the order Dermaptera, with cerci modified into (probably defensive) pincers, strongly curved in males. Females look after their eggs, laid in soil, over the winter. They are at the northern limits of their range (distribution controlled by winter temperatures). There are four native species and three introductions of very limited and local distribution (**bone-house, giant** and **ring-legged earwigs**). The name comes from Old English *ear wicga* (ear insect or beetle), reflecting a preference for dark places.

Common earwig *Forficula auricularia* (Forficulidae), 10–13 mm long, widespread, typically occurring in long vegetation and under litter in almost any habitat, including gardens. The binomial is Latin for 'little scissors' (for the pincers) and 'relating to the ear'.

Hop-garden earwig Alternative name for short-winged earwig.

Lesne's earwig *F. lesnei*, 6–8 mm long, uncommon and scarce but found in southern England, particularly on base-rich soil such as chalkland scrub and oakwoods, as well as nettles and rough vegetation.

Lesser earwig *Labia minor* (Labiidae), 6–7 mm long, native or long-established, associated with dung, compost and rubbish heaps, widespread but especially found in southern and eastern parts of Britain. The binomial is Latin for 'lips' and 'smaller'.

Short-winged earwig *Apterygida media* (Forficulidae), associated with woodland and scrub leaf litter and formerly (before insecticide use) Kentish hop gardens (hence the alternative name **hop-garden earwig**), found in south-east England and East Anglia. Greek *apterygida* = 'absence of' + 'little winged', and Latin *media* = 'middle'.

EARWORT Mainly upland foliose liverworts in the genera *Diplophyllum* (Greek *diplo* + *phyllon* = 'double' + 'leaf'), *Douinia* (honouring the French bryologist Charles Douin, 1858–1944) and, mostly, *Scapania* (Greek *scapos* = 'staff'), all Scapaniaceae.

Common and widespread species, associated with damp habitats, include **grove earwort** *S. nemorea* (woodland floor, decaying logs and heathland peat), **heath earwort** *S. irrigua* (by upland streams and in rush pasture), **northern earwort** *S. subalpina* (crevices in stream boulders and on gravelly lake shores), **water earwort** *S. undulata* (commonest of the earworts, on stream rocks, in flushes, and ditch edges), **waxy earwort** *Douinia ovata* (in base-poor habitats such as rock outcrops in woodland, upland crags, and on bark in Atlantic woodlands), and **white earwort** *Diplophyllum albicans* (on acidic peat, boulders and rock faces).

Rarities include **alpine earwort** *Diplophyllum taxifolium* (Highlands), **bird's-foot earwort** *S. ornithopodioides* (upland west Scotland), **bog earwort** *S. paludicola* (mainly Wales), **calcicolous earwort** *S. calcicola* (Highlands), **ciliate earwort** *S. praetervisa* (Highlands), **cloud earwort** *S. nimbosa* (west Scotland and Highlands), **Degen's earwort** *S. degenii* (Cumbria and, mainly, Highlands), **floppy earwort** *S. paludosa* (mainly Highlands), **least earwort** *S. curta* (mostly south-east and north England), **narrow-lobed earwort** *S. gymnostomophila* (Carmarthenshire

and Highlands), **obscure earwort** *S. parvifolia* (Highlands), **tongue earwort** *S. lingulata* and **untidy earwort** *S. cuspiduligera*.

Blunt-leaved earwort *Diplophyllum obtusifolium*, widespread if scattered, a pioneer of crumbling acidic soil in forestry plantations and disused quarries, and on pathsides. *Obtusifolium* is Latin for 'blunt-leaved'.

Lesser rough earwort *S. aequiloba*, on base-rich mountain rocks and calcareous turf. In north Scotland it also grows on calcareous coastal sand. *Aequiloba* is Latin for 'equal-lobed'.

Marsh earwort *S. uliginosa*, scattered in Britain, mainly in the Highlands, on base-poor, bryophyte-dominated, montane seepages and springs. *Uliginosa* is Latin for 'growing in marsh'.

Norwegian earwort *S. scandica*, common and widespread, often a pioneer on acidic soil banks in forests, disused quarries, cliffs and mountain tops. *Scandica* comes from Latin for 'climbing' (*scandens*).

Rough earwort *S. aspera*, widespread, on chalk, limestone, calcareous sandstone and shell sand, and in base-rich grassland. It is locally abundant in the Pennines on natural rock and dry-stone walls. It is also a typical member of the liverwort community on north-facing chalk grassland in southern England. *Aspera* is Latin for 'rough'.

Shady earwort *S. umbrosa*, widespread, locally common on decaying logs, soft acidic rock, humus-rich soil banks and occasionally bare peat on moorland and bog. *Umbrosa* is Latin for 'shady'.

Thick-set earwort *S. compacta*, widespread, in rocky habitats or on shallow soil, in disused quarries, and on dry-stone walls and gravelly heathland. *Compacta* is Latin for 'compact'.

Western earwort *S. gracilis*, common and widespread in western Britain and in Ireland on humid sites, especially wooded valleys, screes and acidic crags, usually on boulders or tree bases, sometimes on peat. It also occurs on wet heathland and bog in western regions, and on coastal cliffs and dry-stone walls. *Gracilis* is Latin for 'slender'.

ECHINODERM(ATA) From Greek *echinos* + *derma* = 'urchin' + 'skin', the phylum of marine invertebrates with an internal skeleton and a body that usually shows a five-point radial symmetry, such as starfish, sea cucumbers, brittlestars and sea urchins.

ECHINOIDEA Class of echinoderm that includes sea urchins and heart urchins.

ECHIURA From Greek *echis* = 'serpent', the phylum (though considered by some authorities to be a class in the phylum Annelida) that includes spoon worms, infaunal in the lower intertidal, sublittoral and deeper.

ECNOMIDAE Caddis fly family (Greek *ec* + *nomos* = 'out of' + 'custom') that includes *Ecnomus tenellus*, with scattered records from the southern half of England and Wales, found in still or slow-moving waters, the predatory tube-dwelling larvae associated with freshwater sponges, and adults in flight from June to September. *Tenellus* is Latin for 'delicate'.

EDWARDSIIDAE Family name of some sea anemones with 12 species in 5 genera.

EEL True eels, members of the order Anguilliformes. 'Eel' comes from Old English *æl*. See also sand eel.

Common eel Alternative name for European eel.

Conger eel *Conger conger* (Congridae), length generally 1.5–2 m, weight up to 25 kg, mainly found on the south and west coasts of England, Wales and Scotland, and around the Irish coast, in holes or crevices on rocky or sandy bottoms and in artificial structures, emerging at night to feed on fish, cephalopods and crustaceans. They migrate to deeper waters to spawn, then die. 'Conger' comes from Latin *conger* = 'sea eel', in turn from Greek *gongros*.

European eel *Anguilla anguilla* (Anguillidae), adults usually 60–80 cm long, common in rivers, ditches, canals and fens, and around the coast. Spawning is in the Sargasso Sea, from where the larvae, metamorphosing into a transparent stage ('glass eel'), drift across the Atlantic to enter the estuaries of their natal rivers, and start migrating upstream. After entering fresh water, glass eels metamorphose into elvers. As the eels grow, they change colour and become 'yellow eels'. After 5–20 years in fresh water, the eels become sexually mature, change colour again and become 'silver eels', and they migrate back to the Sargasso Sea. In freshwater habitats they tend to be bottom-dwelling and nocturnal, feeding on snails, crayfish, frogs, and fish eggs. Across Europe, elver numbers have crashed to 5% of the levels in the 1980s, possibly reflecting shifts in the pattern of the Gulf Stream which have affected migration. *Anguilla* is Latin for 'eel'.

EELGRASS *Zostera marina* (Zosteraceae), a perennial scattered around the coast, commoner in the west, in the subtidal zone down to a depth of around 4 m, on substrates of gravel, sand or sandy mud. It declined after an outbreak of wasting disease in the 1930s (the pathogen being the fungus *Labyrinthula macrocystis*), and has never fully recovered, and there was a further outbreak in southern England in 1987–92. Greek *zoster* = 'belt', from the ribbon-shaped leaves, and Latin *marina* = 'marine'.

Dwarf eelgrass *Z. noltei*, a perennial growing along much of the coastline in sheltered estuaries and harbours, found between half-tide and low tide on mixed substrates of sand and mud. Ernst Nolte (1791–1875) was a German botanist.

EELPOUT **Viviparous eelpout** *Zoarces viviparus* (Zoarcidae), length 30–50 cm, a fish found around the coasts of the Irish Sea, North Sea and English Channel (absent on the Atlantic coasts of Ireland) in the intertidal, including rock pools, down to 40 m, feeding on bottom-dwelling invertebrates. Adults mate during August–September using internal fertilisation, giving birth to 30–400 live young during winter. 'Eelpout' comes from Old English *æleputa* = 'eel' + 'pout'. The binomial is Greek for 'life-supporting' or 'refreshing', and Latin for 'living'.

EGGAR Moths in the family Lasiocampidae (Greek *lasios* + *campe* = 'hairy' + 'larva'), with an egg-shaped cocoons in some species.

Grass eggar *Lasiocampa trifolii* (Lasiocampinae), wingspan 40–55 mm, RDB Endangered, on dunes on the coasts of south-west and north-west England and south Wales, and inland on heaths in Dorset. In Kent, a distinctive pale subspecies occurs on Dungeness. Adults fly at night in August–September. Larvae feed on a number of herbaceous plants. *Trifolii* is from Latin for 'trefoil'.

Northern eggar Northern form of oak eggar.

Oak eggar *L. quercus*, wingspan 45–75 mm, widespread and common on heathland, moorland, scrub, hedgerows, dunes and coastal cliffs, with a larger northern form, males flying by day, females by night in July–August. The English name comes from the acorn-like shape of the cocoon. Larvae mainly feed on heather and bilberry, but other woody plants are also taken. *Quercus* is Latin for 'oak'.

Pale eggar *Trichiura crataegi* (Poecilocampinae), wingspan 25–30 mm, widespread and common in woodland edge, hedgerows, heathland, moorland, scrub and gardens, but a species of conservation concern in the UK BAP, with numbers having fallen by 86% over 1980–2007. Adults fly in August–September. Larvae feed on a range of woody plants. Greek *trichos* + *oura* = 'hair' + 'tail', from the adult's anal tuft, and Latin *crataegus* = 'hawthorn'.

Small eggar *Eriogaster lanestris* (Lasiocampinae), wingspan 30–40 mm, nocturnal, on heathland, scrub and downland, scarce and discontinuous throughout England and a few places in Wales and Ireland (especially in Co. Clare). The species' decline has been attributed to severe trimming of hedges during the larval season. Adults fly in February–March. Larvae feed gregariously in silk webs on hawthorn and blackthorn. Greek *erion* + *gaster* = 'wool' + 'belly', and Latin *lana* = 'wool', from the adult's anal tuft.

EGRET Members of the family Ardeidae. **Snowy egret** *Egretta*

thula is an occasional vagrant. The name is Middle English, from Old French *aigrette*.

Cattle egret *Bubulcus ibis*, wingspan 98 cm, length 50 cm, weight 350 g, a scarce visitor nevertheless increasing in frequency, and recorded as breeding in Somerset in 2008 following an influx in south-west England in 2007. Further breeding at Avalon Marshes was noted in 2017, with 5 nests producing 11 young. Sightings in Co. Kerry and elsewhere in Ireland were also noted in 2008. Latin *bubulcus* = 'herdsman', referring, like the English name, to this bird's association with cattle. *Ibis* originally but erroneously referred to the sacred ibis.

Great white egret *Ardea alba*, wingspan 155 cm, length 94 cm, weight 870 g, a winter visitor to England and Wales seen with increasing frequency, sometimes in large flocks, in marsh, reedbed and lakes, with most records in south-east England and East Anglia. Diet is of fish, insects and frogs. The first record of successful breeding was at Shapwick Heath NNR, Somerset, in 2012. By 2017, 7 nests producing 17 young were recorded in the Somerset Levels. The binomial is Latin for 'heron' and 'white'.

Little egret *Egretta garzetta*, wingspan 92 cm, length 60 cm, weight 450 g. During 1958–88, the average number in Britain was <15 p.a., mostly spring vagrants, but in 1989 there was an unprecedented early autumn influx involving at least 40 individuals and subsequently this became the norm, with a high proportion of individuals overwintering. Breeding was first noted on Brownsea Island, Poole Harbour, Dorset, in 1996. The species is now common along the south and east coasts of England, and in Wales. The estuaries of Devon and Cornwall, and Poole Harbour and Chichester Harbour, hold some of the largest concentrations; they are also common in East Anglia, and increasingly recorded inland, slowly increasing their range northwards, and also found in north-east Ireland. There were an estimated 700 breeding pairs in 2009, and 4,500 individuals, including migrants, wintering in 2008–9. The population increased by 33% over 2010–15, with over 1,000 pairs in 2015. Nests are in colonies in trees, with clutch size 4–5, incubation lasting 21–2 days, and the altricial young fledging in 40–5 days. Diet is mostly of small fish and amphibians. *Garzetta* is the Italian name for this bird.

EIDER *Somateria mollissima* (Anatidae), wingspan 94 cm, length 60 cm, weight 2.2 kg (Britain's heaviest duck), both a resident breeder (26,000 pairs in 2010, mainly from Northumberland northwards) and winter visitor (total numbers of 60,000 individuals in 2009–10, found in all coastal and estuarine waters). The nest, built close to the sea, is lined with small feathers plucked from the female's breast, collected as eiderdown for use in pillows, etc. since at least the fourteenth century, almost leading to the eider's extinction in the nineteenth century. Birds generally nest in colonies, the nest being a slight hollow in the ground in the shelter of rocks or vegetation. Clutch size is 4–6, incubation lasts 25–8 days, and the precocial chicks fledge in 65–75 days. The diet is mainly of shellfish, especially mussels. The European population (Red Listed) is currently declining overall at a rate of >40% (1988–2015), and British birds are on the UK Amber List. The English name, dating from the late seventeenth century, comes from Icelandic *æthur*, in turn from Old Norse *æthr* = 'duck'. Greek *somatos* + *erion* = 'body' + 'wool', and Latin *mollissimus* = 'very soft', both parts referring to the soft down feathers.

King eider *S. spectabilis* (Latin = 'remarkable') is a rare migrant (averaging three birds a year) recorded from north-east and west Scotland, Orkney and, especially, Shetland.

ELACHISTIDAE Family name (Greek *elastichos* = 'very small') for dwarf moths, mostly leaf miners of grasses, sedges and rushes.

ELAEAGNACEAE Family name of sea-buckthorn and oleaster (Greek *elaia* + *agnos* = 'olive' + 'pure').

ELASMOBRANCH Subclass of the class Chondrichthyes

(cartilaginous fish, with no swim bladders) that includes sharks, skates and rays (Greek *elasmos* + *branchia* = 'beaten metal' + 'gills').

ELATERIDAE Click beetle family (Latin *elatus* = 'elevated' or 'lifted off the ground'), members (3–20 mm in length) having an ability to launch themselves backwards. There are 73 species in 37 genera on the British list.

ELDER *Sambucus nigra* (Caprifoliaceae), a common and widespread deciduous shrub or small tree found on fertile soils in woodland, hedgerows, grassland, scrub, waste ground, roadsides and railway banks. Being resistant to rabbit grazing it often occurs around warrens. 'Elder' comes from Old English *ellærn*. Latin *sambuca*, a stringed instrument made of elder wood, and *nigra* = 'black'.

    American elder *S. canadensis*, a deciduous neophyte shrub present by 1761, scattered in Britain, commonest in north-east England, planted and naturalised along railway banks, hedges and roadsides, and also found in scrub and on waste ground.

    Dwarf elder *S. ebulus*, a rhizomatous perennial archaeophyte herb, scattered in hedgerows, on roadsides and waste ground. *Ebulus* is Latin for 'dwarf elder'.

    Red-berried elder *S. racemosa*, a deciduous neophyte shrub introduced from Europe in the sixteenth century, scattered and widespread but commonest in north-east England and Scotland, in woodland, hedges and waste ground, and planted as game cover in parts of its northern range. *Racemosa* is Latin for 'with racemes'.

ELECAMPANE *Inula helenium* (Asteraceae), a herbaceous perennial archaeophyte, widespread but scattered on roadsides and waste ground, and by woodland edges. The oligosaccharide inulin, derived from the roots, was traditionally used for chest infections, coughs, consumption and other pulmonary complaints. 'Elecampe' is late Middle English, from medieval Latin *enula* (in turn from Greek *helenion* for this plant) + *campana* = 'of the fields'. *Inula* is Latin for this plant, *helenium* from Helen of Troy (said to be collecting this plant when abducted).

ELEPHANT-EARS *Bergenia crassifolia* (Saxifragaceae), a rhizomatous perennial herb introduced from Siberia in 1765, a garden escape in woodland and hedgerows and on verges, railway banks, waste ground and chalk-pits, widespread if scattered, and spreading in the south, especially by the coast. The English name reflects the leaf shape. Karl von Bergen (1704–59) was a German physician and botanist. Latin *crassifolia* = 'thick leaves'.

ELEPHANT GRASS *Miscanthus* (Poaceae), an ornamental garden plant, increasingly planted as a biofuel crop, up to 2 m tall (hence its name), with a government target of 350,000 ha by 2020. It is harvested in April, which might adversely affect brown hare breeding, females using the crop as cover for the young. Greek *miscos* + *anthos* = 'stem' + 'flower', referring to the stalked spikelets.

ELF EARS *Normandina pulchella* (family uncertain), a crustose and squamulose lichen, common and widespread, less so in eastern Britain and the Midlands, usually on other lichens and mosses, and on bark and stone, in sites with high humidity. The genus name is obscure; Latin *pulchella* = 'beautiful'.

ELFCUP Fungi in the genera *Chlorociboria* (Heliotales) (Greek *chloros* + *ciboria* = 'green' + 'drinking cup') and *Sarcoscypha* (Pezizales, Sarcoscyphaceae) (Greek *sarcos* + *scyphos* = 'fleshy' + 'cup').

    Green elfcup *C. aeruginascens*, inedible, widespread and common on rotting bark-free dead wood of hardwoods, and fruiting in autumn and winter, but the fruiting bodies are uncommon, and green staining of the wood is more obvious and year-round evidence of its presence. Infected stained wood has been used in such decorative woodworking as Tunbridgeware. *Aeruginosa* is Latin for 'rust-coloured' or 'verdigris'.

    Ruby elfcup *S. coccinea*, edible, widespread but generally uncommon, scarce in Scotland and the southern half of Ireland, found throughout the year, mostly in the colder months, on damp decaying wood of hazel and other broadleaf trees, often under litter. *Coccinea* is Latin for 'scarlet'.

    Scarlet elfcup *S. austriaca*, edible, widespread and (especially in areas with high rainfall) often common, found throughout the year, mostly in the colder months, on damp decaying wood of both hardwood and conifer trees, often under litter or with moss. *Austriaca* is Latin for 'Austrian'.

    Turquoise elfcup *C. aeruginosa*, inedible, scattered and rare, on rotting bark-free dead wood of broadleaf trees, and fruiting in autumn and winter, but the fruiting bodies are uncommon, and green staining of the wood is more obvious and year-round evidence of its presence. Infected stained wood has been used in such decorative woodworking as Tunbridgeware. *Aeruginosa* is Latin for 'rust-coloured' or 'verdigris'.

ELM Deciduous trees in the genus *Ulmus* (Ulmaceae), taxonomically difficult, with a number of hybrids and subspecies sometimes given full species status. 'Elm' comes from Old English, related to the dialect German *Ilm*, and *ulmus* is Latin for these trees. Dutch elm disease killed most trees in southern and central Britain in the 1970s, and in many places elms persist only as hedgerow suckers, unidentifiable as to species. Named hybrids include **Cornish**, **Goodyer's**, **Huntington** and **Jersey elms**.

    Dutch elm *Ulmus* x *hollandica* (*U. glabra* x *U. minor* x *U. plotii*), a hybrid tree of hedgerows (widely planted) and field borders, rarely found in woodland.

    English elm *U. procera*, a large wide-topped tree, once viewed as characterising the English countryside (though also common in Wales, lowland Scotland and Ireland), found in hedgerows, rarely in woodland, preferring deep and moist soils. In most areas few mature trees remain, being very susceptible to Dutch elm disease which began in the 1960s. New sapling growth remains vulnerable, but the species often remains a major component of hedgerows. The natural distribution is confused by planting but probably does not extend much beyond England and Wales. Brighton and Hove is one of the few refuges for mature, full-size trees in Britain. *Procera* is Latin for 'tall'.

    European white-elm *U. laevis*, a neophyte formerly planted but now found in only a few woods and riverbanks. Reproduction is through vigorous suckering. It may be resistant to Dutch elm disease. It has been grown in Britain since around 1800. It was recorded from the wild for the first time in 1943 in Surrey, and in 1996 trees that had probably been planted 70–100 years previously were found in a woodland near Aberystwyth, Cardiganshire. *Laevis* is Latin for 'smooth'.

    Plot's elm *U. plotii*, found in hedgerows and field borders, once particularly common on neutral to base-rich soils in the east Midlands, mostly in moist, deep-soiled river valleys, but few mature trees remain, having succumbed to Dutch elm disease, new sapling growth also dying after a few years. *Plotii* honours the English naturalist Robert Plot (1640–96).

    Small-leaved elm *U. minor* ssp. *minor*, extremely variable, found in hedgerows, woodland edges and field borders, rarely in woodland but often forming small copses. Few mature trees remain, being vulnerable to Dutch elm disease, and new sapling growth still succumbs after a few years. It is, however, a major hedgerow constituent in its native area, especially in eastern England, and has been widely planted. *Minor* is Latin for 'smaller'.

    Wych elm *U. glabra*, widespread and (formerly more) common, largely non-suckering, of hedgerows, field borders and streamsides that also forms mixed or pure woodland, especially on limestone and other base-rich soils. It also colonises ungrazed grassland, rocky ground and waste and spoil heaps, and is widely planted. Although more resistant to Dutch elm disease than English elm, most mature trees outside Scotland have now been killed. 'Wych' was used in names of trees with pliant branches, from Old English *wic(e)*. *Glabra* is Latin for 'smooth'.

ELMIDAE Family name of riffle beetles, with 12 species in 8 genera.

ELVER Young eel, especially one migrating upriver, dating from the mid-seventeenth century as a variant of the dialect term 'eel-fare' (= eel passage or journey).

ELYSIA Green elysia *Elysia viridis* (Plakobranchidae), a sea slug up to 5 cm long, widespread if mainly recorded from the west coast of Britain, on the underside of macroalgal fronds in shallow water and rock pools. It uses solar energy via chloroplasts from its algal food. *Elysia* may come from Greek *elysios* = 'relating to Elysium'; *viridis* is Latin for 'green'.

EMBERIZIDAE Family name of buntings, including yellow-hammer, from New Latin *emberiza*, in turn from Swiss-German *Emmerilz* = 'bunting'. Lapland and snow bunting are sometimes placed in the Calcariidae.

EMERALD (DRAGONFLY) Members of the family Corduliidae.

Brilliant emerald *Somatochlora metallica*, length 50–5 mm, hindwing 34–8 mm, nationally rare and endangered, associated with slow-flowing water with soft substrate at a few Scottish and south-east English sites. Adults fly in June–August. Greek *somatos* + *chloros* = 'of the body' + 'green'.

Downy emerald *Cordulia aenea*, length 48–50 mm, hindwing 29–35 mm, with a breeding preference for ponds associated with deciduous woodland, with sparse emergent vegetation and organic-rich pond floors; found in south-east England and with a scattered western distribution elsewhere in Britain. The main flight period is May–July. Greek *cordyla* = 'club', and Latin *aenea* = 'bronze' or 'copper'.

Northern emerald *S. arctica*, length 40–55 mm, hindwing 34–6 mm, with breeding restricted to moorland bogs and pools in north-west Scotland and south-west Ireland. The main flight period is May–September.

Willow emerald *Chalcolestes viridis*, male length 42–7 mm, female 39–44 mm, with a peak in adult sightings in August–September, though individuals may be seen in November, usually near ponds, canals or other still water with overhanging trees, eggs being laid in the bark of willow or alder. A recent colonist, one specimen was found in 1978, another in 1992, then in 2009 there were c.400 records in south-east Suffolk. Since 2014 the rate of expansion has increased rapidly; in 2015 it was seen in Suffolk, Norfolk, Essex, Kent, Cambridgeshire, Hertfordshire, Surrey and West Sussex; and in 2016 it also colonised Bedfordshire, Lincolnshire, Northamptonshire and Buckinghamshire. Greek *chalcos* + *lestes* = 'copper' + 'robber', and Latin *viridis* = 'green'.

EMERALD (MOTH) Nocturnal members of the family Geometridae (Geometrinae).

Blotched emerald *Comibaena bajularia*, wingspan 23–7 mm, local, sometimes common, in oak woodland, hedgerows and gardens throughout much of England. Adults fly in June–July. Larvae feed on oak leaves, attaching pieces of leaf and other plant debris to themselves which act as camouflage. Greek *comus* + *baino* = 'bundle' + 'to go', and Latin *bajulus* = 'carrier', from this larval behaviour.

Common emerald *Hemithea aestivaria*, wingspan 24–7 mm, common in woodland, hedgerows, scrubby heathland, downland, gardens and parks throughout lowland England and Wales, and much of Ireland. Adults fly in June–July. Larvae feed on hawthorn, blackthorn, hazel and other woody species. Greek *hemitheos* = 'demigod', and Latin *aestivus* = 'relating to summer'.

Grass emerald *Pseudoterpna pruinata*, wingspan 30–5 mm, common on heathland, moorland, open woodland, scrub, shingle beaches, gravel-pits, verges and banks, widespread but scarcer north of south Scotland. Adults fly in June–July. Larvae feed on petty whin, gorse and broom. Greek *pseudos* + *terpnos* = 'false' + 'delightful', and Latin *pruinus* = 'frost', from the white lines on the wings.

Large emerald *Geometra papilionaria*, wingspan 40–50 mm, widespread and common in woodland, scrubby heathland, hedgerows, gardens and parks. Adults fly in June–July. Larvae mainly feed on birch, but also alder and hazel. Greek *geometreo* = 'to measure the earth', and Latin *papilio* = 'butterfly', from the resemblance.

Light emerald *Campaea margaritaria*, wingspan 30–40 mm, widespread and common in broadleaf woodland, areas with scattered trees, scrub, hedgerows, parks and gardens. Adults fly in June–August, often with a partial second generation in the south in August–September. Larvae feed from spring to autumn on a variety of deciduous trees. The binomial is from Greek *campe*, probably a pun, since with the stress on the first syllable this word means 'caterpillar', on the second 'a bending', and Latin for 'pearl', from the pale green colour of the adult.

Little emerald *Jodis lactearia*, wingspan 23–6 mm, common in open woodland, mature hedgerows, scrubby heathland, moorland and grassland throughout much of England, Wales, Ireland and west Scotland. Adults fly in May–June. Larvae feed on trees such as birch, hawthorn, blackthorn, hazel and oaks. Greek *iodes* = 'rust-like', and Latin *lacteus* = 'milky'.

Small emerald *Hemistola chrysoprasaria*, wingspan 28–32 mm, local in chalk and limestone downland, open woodland, scrubland and gardens throughout much of southern England, spreading into northern England, a species of conservation concern under the UK BAP. Adults fly in July–August. Larvae feed in September–October on traveller's-joy, overwinters, then feeds again in April–May. Autumn stages are brown, spring ones green, to match the colour of the food plant. They also eat cultivated clematis. Greek *hemi* + *stole* = 'half' + 'garment', from the thin wing scaling, and Latin *chrysoprasus* = 'green chalcedony'.

Small grass emerald *Chlorissa viridata*, wingspan 24–7 mm, a rarity occasionally found in damp heathland and mosses in parts of southern England, the Midlands, East Anglia and north-west England. Adults fly in June–July. Larvae mainly feed on heather and birch, but also other woody plants. Greek *chloros* = '(pale) green', and Latin *viridis* = 'green'.

Sussex emerald *Thalera fimbrialis*, wingspan 25–30 mm, formerly known from Sussex, hence its name, but now confined to Dungeness, Kent as a resident (though also an occasional migrant along the south coast from Dorset eastwards), found on shingle beaches. It is protected under the UK WCA, and a priority species under the UK BAP. Adults fly in July–August. Larvae feed on wild carrot. Greek *thaleros* = 'fresh', and Latin *fimbria* = 'fringe', from the fringed trailing wing edges.

EMPEROR (BUTTERFLY) See purple emperor.

EMPETRACEAE Former family name for crowberry (Latin *empetrus*), which since 2002 has been subsumed within the Ericaceae.

EMPIDIDAE Family name (Greek *empis* = 'gnat') for dagger and dance flies.

ENCAPHALOMALACIA Nigropallidal encephalomalacia See chewing disease.

ENCHANTER'S-NIGHTSHADE *Circaea lutetiana* (Onagraceae), a perennial rhizomatous herb, common and widespread (except for north Scotland) in damp, usually base-rich, shaded habitats such as ancient and secondary woodland, hedgerows, scrub, and stream and riverbanks, and as a garden weed. Circe was an enchantress in Homer's Odyssey. Lutetia is the Latin name for Paris.

Alpine enchanter's-nightshade *C. alpina*, a perennial herb very scattered in Wales, the Lake District and parts of Scotland, associated with seepages in rocky, moss-rich oak woodland, and among boulders and scree by the sides of streams and waterfalls.

Upland enchanter's-nightshade *Circaea* x *intermedia* (*C. lutetiana* x *C. alpina*), a perennial herb, locally frequent in upland north England, Wales, Scotland and northern Ireland, in damp wooded or shaded habitats, by streams and among wet rocks; also in gardens and on disturbed roadsides. It spreads by rhizomes, stolons and occasionally by seed, though it is usually sterile.

ENDIVE Alternative name for chicory, from medieval Latin *endivia*.

ENDOMYCHIDAE Family name (Greek *endon* + *mychos* = 'within' + 'inward') of handsome fungus beetles (2–6 mm length), with eight species in five genera. All species feed on fungi beneath bark and in leaf litter. See false ladybird under ladybird (2).

ENGRAILED *Ectropis crepuscularia* (Geometridae, Ennominae), wingspan 30–40 mm, a nocturnal moth, widespread and common in broadleaf woodland, hedgerows, parks and gardens. There are usually two generations, flying in March–April and July–August, but in the northern parts of the range it is single-brooded (April–May). Larvae feed on a variety of trees. 'Engrailed' means having semicircular indentations along the (wing) edge. Greek *ectropos* = 'turning out of the way', from the wavy band markings on the wings, and Latin *crepusculum* = 'twilight'.

ENSIGN FLY Members of the family Sepsidae, 2.5–4.5 mm length. The common name comes from the habit of 'waving' their wings, an activity accentuated visually by having dark patches near the wing tip. Adults are mainly found on animal (including livestock) and human excrement, less often on other decomposing organic matter, where eggs are laid and larvae develop.

EPERMENIIDAE Family name for lance-wing moths, from Greek *epi* + Armenia leading to *erminea* = 'ermine', referencing scale tufts on the dorsal forewing margins.

EPHEMERELLIDAE Mayfly family (Greek *ephemeros* = 'ephemeral') with two British species, yellow evening hawk and blue-winged olive.

EPHEMERIDAE Family of mayflies in the genus *Ephemera* (Greek *ephemeros* = 'ephemeral'), containing the largest mayfly species. Nymphs burrow in slow-moving or still waters.

EPHEMEROPTERA Order of mayflies (Greek *ephemeros* = 'ephemeral'), with 44 species in 7 families.

EPHYDRIDAE Family name (Greek *epi* + *hydor* = 'on' + 'water') of shore flies, with 151 species in 39 genera. These small flies are mainly associated with seashores, but are sometimes seen in inland ponds and pools.

EQUIDAE Family name for horses (Latin *equus*) and ponies.

EQUISETACEAE Family name for horsetails (Latin *equus* + *seta* = 'horse' + 'bristle').

EREBIDAE Family name (from Erebos, a place of darkness in Greek mythology) for ermine, fan-foot, footman, snout, tiger and underwing moths, and including the so-called tussock moths (Lymantriinae).

ERESIDAE Family name for ladybird spider (Greek *eresis* = 'discontinuation').

ERGOT Most commonly *Claviceps purpurea* (Hypocreales, Clavicipitaceae, from Latin *clava* + *ceps* = 'club' + 'head'), but a name more generally applied to a group of similar tiny ascomycetes fungi that occur on grasses, with rye-grass particularly susceptible, and in the past including some cereal crops, especially rye. In humans and other animals, when eaten with grain, toxic alkaloids cause the illness known as ergotism, and can be lethal. 'Ergot' dates from the late seventeenth century, derived from Old French *argot* = 'cock's spur' because of the appearance produced by the disease.

ERICACEAE Family name (Latin *erica* = 'heath') of heather, including azalea, bilberry, blueberry, bog-laurel, bog-rosemary, cowberry, cranberry, crowberry, heath, Labrador-tea, rhododendron and strawberry-tree.

ERINACEIDAE Family name for hedgehog (Latin *erinaceus*).

ERIOCAULACEAE Family name of pipewort (Greek *erion* + *caulos* = 'wool' + 'stem').

ERIOCRANIIDAE Family name (Greek *erion* + *cranos* = 'wool' + 'helmet') for small leaf-mining moths called 'purples'.

ERIOPHYIDAE Family name (Greek *erion* + *phye* = 'wool' + 'growth') of gall mites in the order Trombidiformes.

ERIRHINIDAE Family name (Greek *eri* + *rhine* = 'very' + 'rasp') for wetland weevils, with 13 British species in 7 genera. Most are found by the water's edge on floating or emergent vegetation or in litter.

ERMEL Moths, mostly nocturnal, in the families Ethmiidae (species of *Ethmia*, from Greek *ethmos* = 'sieve', black stigmata on the wings of some species possibly suggesting this), Roeslerstammiidae (*Roeslerstammia*, honouring the German entomologist Josef von Röslerstamm, 1787–1866), and Yponomeutidae (including *Swammerdamia*, *Paraswammerdamia* and *Pseudoswammerdamia*, honouring the Dutch entomologist Jan Swammerdam, 1637–80).

Birch ermel *Swammerdamia caesiella*, wingspan 9–13 mm, widespread and common in heathland and open woodland. Adults fly in May–June and again in August. Larvae feed on birch leaves in July and October, living gregariously in spun leaves. *Caesius* is Latin for 'bluish-grey', from the forewing colour.

Bordered ermel *E. bipunctella*, wingspan 19–28 mm, found on shingle beaches in south-east England, and scattered in parts of England south of a line from the Bristol Channel to the Wash, inland records perhaps the result of immigration. Adults fly in May–June, with a partial second generation in autumn. Larvae feed on flowers and leaves of viper's-bugloss, spun together with silk. Latin *bipunctella* = 'two-spotted'.

Brown ash ermel *Zelleria hepariella* (Yponomeutidae), wingspan 10–15 mm, widespread but local. Adults fly in July–August. Larvae feed on leaves of ash in May–June, living sometimes gregariously in spun leaves. Philipp Zeller (1808–83) was a German lepidopterist. Greek *hepar* = 'liver', from the forewing colour.

Brown pine ermel *Cedestis subfasciella* (Yponomeutidae), wingspan 9–11 mm, widespread but local in pine woodland, though rare in Ireland. Adults emerge during March–July. Larvae mine pine needles. Greek *cedestes* = 'a relation by marriage', indicating affinity with related species, and Latin *subfascia* = '(weakly developed) stripe', on the forewing.

Comfrey ermel *E. quadrillella*, wingspan 15–19 mm, with a scattered distribution in fen, marsh, damp woodland, waste ground and gardens, and on riverbanks in eastern and northern England. Adults generally fly at night, but sometimes in sunshine, in May–July. Larvae feed on leaves of comfrey, common gromwell and lungwort inside a silk web.

Copper ermel *Roeslerstammia erxlebella*, wingspan 12–14 mm, mainly found in open woodland in the southern half of England, with scattered records from Wales and Scotland. There are two generations, at least in the south of the range, with adults flying in May–June and August–September. Larvae mine young leaves of small-leaved lime and birch. Johann Erxleben (1744–77) was a German naturalist.

Copper-tipped ermel *Pseudoswammerdamia combinella*, wingspan 13–16 mm, with a scattered distribution but locally common in hedgerows and scrub. Adults fly in May–June. Larvae feed on blackthorn leaves in May–June, initially in galleries then often gregariously in spun leaves. *Combinella* is from Latin *combino* = 'unite', from the wings pressed together when at rest.

Dotted ermel *E. dodecea*, wingspan 17–21 mm, local in woodland rides, fen, scrub and chalk downland across parts of England, south and east of a line from the Wash to the Severn estuary. Adults fly from May to July. Larvae feed in August–September on common gromwell, living in a silk web. *Dodecim* is Latin for 'twelve', from the forewing spots.

Five-spot ermel *E. terminella*, wingspan 16–22 mm, an occasional immigrant, mostly in south-east England, establishing localised colonies on shingle beaches. Adults fly from May to July. Larvae feed on the flowers and developing seeds of viper's-bugloss. *Terminus* is Latin for 'end', for the series of dots on the forewing.

Gold pine ermel *C. gysseleniella*, wingspan 11–13 mm, local in pine woodland throughout much of England and Scotland. Adults fly in June–July. Larvae feed in spring internally in pine needles, working their way from the base to the tip, then feeding externally in a silk web. J.V.G. Gysselin was a nineteenth-century Austrian collector.

Grey pine ermel *Ocnerostoma friesei* (Yponomeutidae), wingspan 8–10 mm, local in pine woodland throughout much of England, south Wales and Scotland, but rare in Ireland. The species is bi- or trivoltine, flying in March–May, August and sometimes November. Larvae mine pine leaves, living between needles spun together with silk. Greek *ocneros* + *stoma* = 'unready' + 'mouth'. G. Friese was a German entomologist.

Hawthorn ermel *Paraswammerdamia nebulella*, wingspan 11–14 mm, widespread and common in hedgerows and open woodland. Adults fly in July. Larvae mainly feed on hawthorn, rowan and dog rose, initially mining the leaves then living, often communally, inside a silk web. *Nebula* is Latin for 'cloud', from the wing colour.

Little ermel *S. pyrella*, wingspan 10–13 mm, widespread and common in hedgerows, woodland, orchards and gardens. With two generations a year, adults emerge in May and August. Larvae feed on leaves of hawthorn, apple and pear (= *pyrus*, hence *pyrella*) beneath a silk web mainly in July and October.

Mountain ermel *E. pyrausta*, wingspan 17–23 mm, an adult rediscovered in 1996 at an altitude of about 1,000 m in the Cairngorms; two other records have been in 2001 (resting on a snow bed at 810 m in Aberdeenshire) and 2008 in east Ross-shire. Greek *pyraustes*, meaning a candle-singed moth, may refer to the black forewing markings.

Rowan ermel *S. compunctella*, wingspan 14–15 mm, uncommon and local in open woodland in England, south Wales and parts of Scotland. Adults fly in June–July. Larvae feed on leaves of rowan and hawthorn beneath a silk web, often communally. *Compunctus* is Latin for 'speckled', from the forewing markings.

Scotch ermel *S. passerella*, wingspan 9–11 mm, found only at 300–700 m in the Highlands. Adults emerge in late May. Larvae mine leaves of dwarf birch. *Passer* is Latin for 'sparrow'.

White-headed ermel *Paraswammerdamia albicapitella*, wingspan 10–13 mm, found in hedgerows and scrub, common and widespread as far north as the Central Lowlands of Scotland, where it is local and scarce. Adults fly in July. Larvae feed on blackthorn, initially mining the leaves then hibernating, often communally, in a silk enclosure. In spring they feed on the leaves from inside a silk web. *Albicapitella* is Latin for 'white-headed'.

White pine ermel *O. piniariella*, wingspan 8–10 mm, widespread in pine woodland throughout much of England and Scotland. Adults fly in June–July. Larvae mine Scots pine (hence *piniariella*, from *pinus* = 'pine'), living between needles spun together with silk.

ERMINE (MAMMAL) Alternative name for stoat, especially in northern England and Scotland, often used for the animal in its white winter coat, or for the fur itself. The word comes from Old English *hearma*, in turn from Old French *hermine*, probably from medieval Latin *mus Armenius* = 'Armenian mouse'.

ERMINE (MOTH) Nocturnal moths, mostly in the genus *Yponomeuta* (Yponomeutidae), from Greek *yponomeuo* = 'to make underground mines', larvae being leaf miners, together with some members in the Erebidae and Pyralidae.

Apple ermine *Y. malinellus*, wingspan 20–3 mm, local in orchards throughout much of Britain as far north as the Central Lowlands of Scotland, but less common than in the past, probably because of use of insecticides. Occasionally recorded from Ireland. Adults fly in July–August. Larvae feed on apple leaves in May–June, living gregariously within a silk web. *Malinellus* is Latin for 'relating to the apple tree' (*malus*).

Bird-cherry ermine *Y. evonymella*, wingspan 16–25 mm, widespread and common in woodland and scrub, more numerous in the north. Adults fly in July–August. Larvae feed on bird cherry, living gregariously inside a silk web, causing sufficient damage to be a serious pest in places. *Evonymella* is Latin for 'relating to the spindle tree' (*euonymus*), though this is not a food plant of this species.

Black-tipped ermine *Y. plumbella*, wingspan 16–20 mm, local in woodland on chalky soils throughout England, Wales and the southern half of Ireland. Often with two generations a year, adults fly in April–May and again in August. Larvae feed on leaves of spindle, living communally inside a silk web. Latin *plumbum* = 'lead', from the wing colour.

Buff ermine *Spilosoma lutea* (Erebidae, Arctiinae), wingspan 28–42 mm, widespread and common in a range of habitats, including gardens, hedgerows, parkland and woodland, less evident in Scotland, but a species of conservation concern under the UK BAP. Adults fly in May–July. Larvae feed in autumn on a variety of plants, ranging from stinging nettle and honeysuckle to hop and wild plum. Greek *spilos* + *soma* = 'spot' + 'body', and Latin *lutea* = 'yellow', from the wing colour.

Grey ermine *Y. sedella*, wingspan 15–18 mm, local in woodland, rocky areas and hedgerows in England and Wales, as far north as Lincolnshire and Worcestershire. Often with two generations a year, adults fly in April–May and August. Larvae feed on orpine or cultivated sedums (hence *sedella*, from sedum), living communally inside a silk web.

Orchard ermine *Y. padella*, wingspan 19–22 mm, common and widespread in hedgerows and scrubland. Adults fly in July–August. Larvae feed on blackthorn, hawthorn and cherry in a communal silk web. *Padella* is from *Prunus padus* = bird cherry.

Scarce ermine *Y. irrorella*, wingspan 19–25 mm, recorded only from woodland on chalky soils in parts of Hampshire, Kent and occasionally other parts of southern England. Adults fly in July–August. Larvae feed on spindle leaves inside a silk web. *Irroro* is Latin for 'to sprinkle with dew', from the forewing black spots.

Spindle ermine *Y. cagnagella*, wingspan 19–26 mm, locally common wherever spindle occurs, on chalk and limestone soils. Adults fly in July–August. Larvae feed on spindle leaves in May–June inside a communal silk web. The specific name is a misprint for *cognatella*, from Latin *cognatus* = 'related'.

Thistle ermine *Myelois circumvoluta* (Pyralidae, Phycitinae), wingspan 27–33 mm, commonly found on rough ground and chalk downland throughout southern Britain and eastern Ireland. Adults fly in June–July. Larvae feed on flowers and stems of thistles and greater burdock. Greek *myelos* = 'pith', and Latin *circumvoluta* = 'nearly rolling around'.

Water ermine *S. urticae*, wingspan 38–46 mm, found in fens, water meadows and marshes, south and east of a line from the Wash to the Severn estuary. Adults fly in June–July. Larvae feed on a number of marsh plants. *Urtica* is Latin for 'nettle', a food plant on the continent.

White ermine *S. lubricipeda*, wingspan 34–48 mm, widespread and common in gardens, hedgerows, grassland, heathland, moorland and woodland, but a species of conservation concern in the UK BAP. Adults fly in May–July. Larvae feed on a variety of herbaceous plants, especially stinging nettle. *Lubricipes* is Latin for 'swift-footed'.

Willow ermine *Y. rorrella*, wingspan 19–24 mm, local on riverbanks in much of England and Wales, particularly near the coast in the south and east. Adults fly in July–August. Larvae feed on willow leaves in May–July, living inside a communal silk web. *Irroro* is Latin for 'to sprinkle with dew', from the forewing black spots.

EROTYLIDAE Family name (Greek *erotylos* = 'a darling') of pleasing fungus beetles, with eight British species in four genera, 3–6 mm in length, generally widespread in Britain, mostly nocturnal often living beneath fungoid bark or in (usually bracken) fungi on trees.

ERYNGO Herbaceous perennials in the genus *Eryngium* (Apiaceae), from Greek for 'sea-holly'.

Blue eryngo *E. planum*, a European neophyte, scattered in Britain (mainly in England) as a garden escape on waste ground. *Planum* is Latin for 'flat'.

Field eryngo *E. campestre*, an archaeophyte found on well-drained neutral or calcareous soils in old pasture and coastal grassland in south-west England, with sites in Devon and Somerset having statutory protection, but elsewhere, locally, in southern England short-lived or casual populations are found on pasture, roadsides and rough ground. *Campestre* is Latin for 'of flat areas' (i.e. fields).

Tall eryngo *E. giganteum*, a biennial or perennial Caucasian neophyte introduced by 1820, scattered in England and probably increasing its range as a naturalised or, more frequently, casual garden escape in field edges and on waste ground.

ESCALLONIA *Escallonia macrantha* (Escalloniaceae), an evergreen neophyte shrub from Chile, planted for ornament and hedging, and a relic in coastal Britain from west Scotland to the Wash, the Isle of Man, and (especially south-west) Ireland. Antonio Escallon (1739–1819) was a Spanish botanist. Greek *macrantha* = 'large-flowered'.

ESOCIDAE Name of the pike family (Latin *esox* = 'pike', in turn from Greek *isox*, an unknown fish).

ETHMIIDAE Family name for some of the ermel moths, from Greek *ethmos* = 'sieve', black stigmata on the wings of some species possibly suggesting this.

EUCALYPTUS Species of gum (Greek *eu* + *calyptos* = 'good' + 'covered').

EUCINETIDAE Family name (Greek *eu* + *cinesis* = 'good' + 'movement') for plate-thigh beetles, with the one British species, *Eucinetus meridionalis*, having a few records from south and east England, found in detritus or tree bark, where adults and larvae eat fungi.

EUCNEMIDAE Family name (Greek *eu* + *cneme* = 'good' + 'knee') for false click beetles, with seven British species in six genera, larvae living in dead wood.

EULIPOTYPHLA Order (Greek *eu* + *lipos* + *typhlos* = 'truly' + 'fat' + 'blind') that includes shrews, mole and hedgehog, formerly in the Insectivora.

EUPHORBIACEAE Name of the spurge family, after Euphorbus, a first-century BCE Greek physician.

EURYTOMIDAE Family (Greek *eury* + *tomos* = 'wide' + 'cut') of tiny parasitoid wasps, with 93 species, including 8 species of *Tetramesa*, which form galls on grasses.

EVENING HAWK Yellow evening hawk *Ephemerella notata* (Ephemerellidae), a mayfly with a scattered British distribution, especially found in north-west England, and currently spreading northwards. Adults are active in May–June. Nymphs live in vegetation or in sand and gravel on stream-beds feeding on fine particulate organic debris. Greek *ephemeros* = 'short-lived', Latin *notatus* = 'marked'.

EVENING-PRIMROSE Biennial neophytes in the genus *Oenothera* (Onagraceae, from *onagros* = 'wild ass' or *onagra*, an unspecified plant), the genus name therefore possibly derived from Greek *oinos* + *thera* = 'wine' + 'booty', or possibly *onos* +

thera = 'ass' + 'hunting', though neither is etymologically relevant to these American plants.

Common evening-primrose *O. biennis*, from North America, both cultivated and naturalised by the seventeenth century, widespread and locally common in England, scarce elsewhere, on open ground on sandy soils, for instance dunes, riverbanks, waste ground, railway sidings and roadsides, in neglected cultivated fields, quarries, sandpits and rubbish tips. Large numbers often occur on disturbed ground but tend not to persist under competition. *Biennis* is Latin for 'biennial'.

Fragrant evening-primrose *O. stricta*, from Chile, naturalised on sandy, often near-coastal habitats, but also in the west Midlands and Thames Valley, locally common in the southern half of England and in Wales, scarce elsewhere. *Stricta* is Latin for 'erect'.

Intermediate evening-primrose *Oenothera* x *fallax* (*O. glazioviana* x *O. biennis*), a hybrid that commonly arises spontaneously, and often found as a garden escape, first recorded in the wild in 1978. It is mostly found in England and south Wales, and is scarce elsewhere, growing in open habitats on light soils, especially on dunes and sandy places, but also on waste ground, quarries, railway sidings and roadsides. *Fallax* is Latin for 'false' or 'deceptive'.

Large-flowered evening-primrose *O. glazioviana*, probably of European origin, present by the mid-nineteenth century, now widespread, locally common and expanding in England and Wales, scarce elsewhere, found especially near the sea but also inland, on open ground on sandy soils, for example dunes, waste ground, railway sidings, roadsides, former quarries and sandpits. Large numbers often occur on disturbed ground but tend not to persist under competition. Auguste Glaziou (1828–1906) was a French botanist.

EVERLASTING Perennial herbs in the family Asteraceae.

Mountain everlasting *Antennaria dioica*, stoloniferous, on thin, basic to mildly acidic soils in north England, Scotland, Wales and much of Ireland. Lowland habitats include calcareous grassland, heathland, coastal clifftops, dunes and machair. In upland areas, habitats include rock ledges, streamsides, scree, well-drained acidic grassland and moorland. *Antennaria* is from Latin *antenna*, because of the resemblance of the male flowers to insect antennae, and *dioica* means having male and female flowers on separate plants.

Pearly everlasting *Anaphalis margaritacea*, a rhizomatous North American neophyte in cultivation by the end of the sixteenth century, scattered in Britain (frequent in south Wales), naturalised on waste ground and short grassland, especially roadside verges, railway banks and coal mine slag-heaps, and in woodland clearings. *Anaphalis* is from the Greek name for this or a similar plant, and Latin *margaritacea* = 'pearl-like'.

EVERLASTING-PEA Scrambling perennial herbs in the genus *Lathyrus* (Fabaceae), Greek for 'pea' or 'pulse'.

Broad-leaved everlasting-pea *L. latifolius*, a neophyte from Europe, cultivated since the fifteenth century, a widespread if scattered naturalised garden escape on road and railway banks and waste ground. *Latifolius* is Latin for 'broad-leaved'.

Narrow-leaved everlasting-pea *L. sylvestris*, with a scattered distribution especially in England and Wales, in hedges, woodland edge and scrub, and as a garden escape on roadsides and railway banks. Away from the coast it prefers calcareous soils. *Sylvestris* is Latin for 'wild' (strictly 'of woods').

Norfolk everlasting-pea *L. heterophyllus*, a neophyte from Europe naturalised as a relic of cultivation on damp sand dune hollows at Burnham Overy Staithe (Norfolk), and by a flooded gravel-pit at Yateley (Hampshire). *Heterophyllus* is Greek for 'differently leaved'.

Two-flowered everlasting-pea *L. grandiflorus*, a European neophyte with a scattered distribution as a persistent garden escape in hedges and waste ground. *Grandiflorus* is Latin for 'large-flowered'.

EXILE Alternative name for the noctuid moth northern arches ssp. *marmorata*, found on the Shetlands.

EXTINGUISHER-MOSS Tuft-forming species of *Encalypta* (Encalyptaceae) (Greek *en* + *calypos* = 'in' + 'cover' or 'veil').

Rarities include **alpine extinguisher-moss** *E. alpina* (Highlands) and **ribbed extinguisher-moss** *E. rhaptocarpa* (mainly Cumbria and Highlands).

Common extinguisher-moss *E. vulgaris*, widespread but scattered, mainly in lowland regions on base-rich substrates. *Vulgaris* is Latin for 'common'.

Fringed extinguisher-moss *E. ciliata*, locally common in upland regions in base-rich rock crevices. *Ciliata* is Latin for 'fringed with hairs'.

Spiral extinguisher-moss *E. streptocarpa*, common and widespread on base-rich rock and stone (especially limestone) and mortar in walls, and in grassland on chalk and limestone. Greek *streptos* + *carpos* = 'twisted' + 'fruit'.

EYEBRIGHT Species of annual root hemiparasitic *Euphrasia* (Orobanchaceae), from the Greek goddess Euphrosyne, one of the Three Fates, her name meaning 'gladness' or 'delight'. Stace (2010) reports that 71 hybrids are known. 'Eyebright' comes from the plants' use in treating eye infections.

Arctic eyebright *E. arctica*, locally common in upland regions in damp, rough grassland, pasture and hay meadow, on riverbanks, and by roadsides.

Campbell's eyebright *E. campbelliae*, a rare endemic found in Lewis, Outer Hebrides, in damp, mossy, grazed turf in coastal sedge-rich communities or in grassy dwarf-shrub heath.

Chalk eyebright *E. pseudokerneri*, an endemic of herb-rich downland turf on chalk and soft limestone in southern and south-east England and East Anglia, occasionally found on harder limestones in Ireland or as forma *elongata* in damp fens in East Anglia, and recently discovered in calcareous flushes, a lead mine and coastal grassland in Cardiganshire (Dyfed). *Pseudo* is Greek for 'false'; *kerneri* honours the German botanist Anton Kerner von Marilaun (1831–98).

Common eyebright *E. nemorosa*, growing in much of the lowlands in short grassland, on heathland, downland and (in Scotland) dunes, in open scrub, woodland rides and upland moorland. *Nemorosa* is Latin for 'of woodland'.

Confused eyebright *E. confusa*, in grazed pasture, moorland and grassy heathland on free-draining acidic or calcareous soils, characteristic of hill pasture in north, west and south-west Britain, and elsewhere occasionally found in open vegetation on sandy soils. It is rare and mainly coastal in Ireland.

Cornish eyebright *E. vigursii*, an endemic local to Devon and Cornwall, characteristic of bent grass/dwarf gorse heaths. In coastal areas it occurs mainly on clifftops in short, species-rich turf around rock outcrops, and by tracks and paths, where scrub is suppressed. Inland it grows on lightly grazed damp heathland and open moorland. Chambre Corker Vigurs (1867–1940) was an English botanist.

Cumbrian eyebright *E. rivularis*, an endemic found in damp upland pasture, upland rocky flushes, seepage areas and wet rock ledges on montane cliffs, growing from 380 m at Langstrath (Cumbria) to 750 m on Snowdon, and also recorded from south Wales. *Rivularis* is Latin for 'growing by streams'.

English eyebright *E. officinalis* ssp. *anglica*, on damp grassland and heathland mainly in southern England and Wales, rather local elsewhere in England, south-west Scotland and Ireland. *Officinalis* is Latin for having pharmacological properties.

Foula eyebright *E. foulaensis*, scarce, in damp, open turf on coastal clifftops subject to salt spray, and at the upper fringe of saltmarshes, in north Scotland, the Outer Hebrides, Orkney and Shetland. Foula is an island in Shetland. Grazing by sheep or rabbits is essential for its survival.

Heslop-Harrison's eyebright *E. heslop-harrisoni*, a rare endemic found in west Scotland, Hebrides, Orkney and Shetland, largely restricted to turfy areas in saltmarshes immediately above the high water mark, associated with plantain species, named after the English botanist John Heslop-Harrison (1881–1967).

Irish eyebright *E. salisburgensis*, scarce, in calcareous grasslands, maritime dunes and montane limestone cliffs in west Ireland.

Marshall's eyebright *E. marshallii*, a rare endemic of coastal rocks and sea-cliff edges in north Scotland, including the Outer Hebrides, Orkney and Shetland, possibly parasitic on plantain species.

Montane eyebright *E. officinalis* ssp. *monticola*, very local in dry upland grassland in northern England, west Wales and southern Scotland.

Ostenfeld's eyebright *E. ostenfeldii*, a rarity found in north and north-west Scotland, Orkney, Shetland, the Lake District, and north-west and south Wales in sparsely vegetated areas in well-drained, exposed habitats such as limestone rock ledges, sea-cliffs, scree, serpentine debris and sandy coastal turf. Carl Ostenfeld (1873–1931) was a Danish botanist.

Pugsley's eyebright *E. rotundifolia*, a rare endemic of flushed basic turf on sea-cliffs on the extreme north of Scotland, parasitic on Scottish primrose. *Rotundifolia* is Latin for 'round-leaved'.

Rostkov's eyebright *E. officinalis* ssp. *pratensis*, found in often upland damp grassland, locally common in northern England, central and southern Scotland, Wales and Ireland.

Scottish eyebright *E. scottica*, associated with flushes and wet moorland in upland areas in Wales, northern England, Scotland and Ireland.

Slender eyebright *E. micrantha*, in heathland and moorland, widespread except in central and east England, usually associated with heather. It also grows on open clay and sandy substrates in disturbed habitats. *Micrantha* is Greek for 'small flower'.

Snowdon eyebright Alternative name for Cumbrian eyebright.

Sticky eyebright See English, montane and Rostkov's eyebright, all above.

Upland eyebright *E. frigida*, on damp, usually basic cliff ledges in the Lake District, the Southern Uplands and Highlands of Scotland, and west Ireland. *Frigida* is Latin for 'cold'.

Welsh eyebright *E. cambrica*, a rarity found in north-west Wales on well-drained, basic, sheep-grazed grassland on mountain slopes, and even more rarely in wetter, base-enriched flushes. *Cambrica* is Latin for 'Welsh'.

Western eyebright *E. tetraquetra*, on short turf on exposed coastal cliffs and on dunes around the British Isles except eastern England and north Scotland, and locally inland in southern England on chalk and limestone pasture. *Tetraquetra* is Greek for 'four-cornered'.

EYELASH (FUNGUS) Common eyelash *Scutellinia scutellata* (Pezizales, Pyronemataceae), inedible, common and widespread, found from June to November, in damp woodland on rotten wood or woody debris, and on livestock dung. A hair-like fringe around the edge of the cap prompts the English name. Latin *scutellus* = 'small shield'.

EYELASH WEED *Calliblepharis ciliata* (Cystocloniaceae), a red seaweed up to 30 cm long, mainly found from the south-east to the west coast of England, around the coast of Ireland and the west coast of Scotland, but also recorded from Orkney and Yorkshire. It is epilithic, occasionally epiphytic, growing to depths of 21 m, occasionally abundant on bedrock below the lower limit of kelp, and sometimes found in pools in the lower littoral. Greek *calliblepharis* = 'beautiful eyelids', and Latin *cilia* = 'eyelash'.

Lance-shaped eyelash weed *C. jubata*, found along west Britain and around Ireland, is epilithic and epiphytic and can grow on other seaweeds, on rocks in pools of the lower eulittoral zone, down into the *Laminaria* zone. *Jubata* is Latin for 'crested'.

F

FABACEAE Pea family (Latin *faba* = 'bean'), formerly known as, or including, Leguminosae Papilionaceae, Caesalpiniaceae and Mimosaceae.

FACELINA Boston facelina *Facelina bostoniensis* (Facelinidae), a widespread sea slug up to 5.5 cm long, found intertidally on the hydroid *Clava*.

FAGACEAE Family name for oak, beech and sweet chestnut (Latin *fagus* = 'beech').

FAIRY BEADS *Microlejeunea ulicina* (Lejeuneaceae), a common and widespread foliose liverwort in south and west Britain and in Ireland on sheltered tree-trunks and rocks in humid sites, often over other bryophytes. *Ulicina* is Latin for 'resembling gorse' (*ulex*).

FAIRY HELMET Bleeding fairy helmet Alternative name for the fungus burgundydrop bonnet.

FAIRY SHRIMP *Chirocephalus diaphanus* (Chirocephalidae), an anostracan crustacean, body length 25 mm, recently becoming restricted to Devon, Cornwall, Salisbury Plain, the New Forest and Cambridgeshire, found in and passively dispersed between small, temporary, often disturbed pools. Eggs survive when the habitat dries out. It feeds on microscopic animals and organic particles. It is Vulnerable in the RDB, fully protected by the WCA (1981), and a Species of Conservation Concern under the UK BAP. Greek *cheir* + *cephali* = 'hand' + 'head', and Latin *diaphanus* = 'transparent'.

FALCON Birds of prey in the family Falconidae; see hobby, kestrel, merlin and peregrine. **Amur falcon** *Falco amurensis* and **Eleanora's falcon** *F. eleanorae* are scarce accidentals, and **gyr falcon** *F. rusticolus* and **red-footed falcon** *F. vespertinus* rare visitors.

FALCONIFORMES Order that includes falcons (Falconidae), hawks (Accipitridae) and, in some schemes, osprey (Pandionidae), probably better placed in the Accipitridae).

FALSE-ACACIA *Robinia pseudoacacia* (Fabaceae), a widely planted deciduous tree introduced from North America before 1640, naturalised in southern Britain. It spreads by suckering (occasionally by seed), occurring on roadsides, in pavement cracks and on waste ground in urban areas. It is also found in woodland and on scrubby banks, vigorous only on light soils. *Robinia* honours the eighteenth-century French herbalist Jean Robin. Greek *pseudo* = 'false' and acacia (from *acis* = 'thorn').

FALSE-BROME See brome.

FALSE-BUCK'S-BEARD *Astilbe japonica* (Saxifragaceae), a perennial introduced from Japan, with a scattered distribution in (mainly northern) England and Scotland as a garden escape, and naturalised in damp habitat. Greek *a* + *stilbos* = 'without' + 'glittering', referring to the dullness of the leaves.

FALSE-FEATHER Bramble false-feather *Schreckensteinia festaliella* (Schreckensteiniidae), wingspan 10–12 mm, a widespread and common moth in open woodland and hedgerows. Adults are probably continuously brooded and can be found between March and September. Larvae feed on bramble and raspberry under a silk web. R. von Schreckenstein (d.1808) was a German entomologist. Latin *festum* = 'feast'.

FALSE-FLAX Alternative name for gold-of-pleasure.

FALSE MOREL *Gyromitra esculenta* (Pezizales, Discinaceae), an infrequently found fungus found near pine trees during spring and early summer, with a widespread but very scattered mostly upland British distribution. Greek *gyros* + *mitra* = 'round' +

'headband'. It is deadly poisonous, so *esculenta*, Latin for 'edible', is dangerously misleading.

Pouched false morel *G. infula*, inedible, infrequent in Kent, Surrey, Yorkshire and, especially, the Highlands, on soil or dead wood in, usually, coniferous forest. *Infula* is Latin for 'a band'.

FALSE SCORPION Alternative name for pseudoscorpion.

House false scorpion *Chelifer cancroides* (Cheliferidae), a pseudoscorpion, body length 2.5–4.5 mm, scattered in England, very common in stables, barns, grain stores, flour mills, factories and houses feeding on book lice and dust mites. Greek *chele* = 'claw', and Latin *cancroides* = 'crab-like'.

FALSE-SEDGE *Kobresia simpliciuscula* (Cyperaceae), a perennial found in stony flushes, base-rich mires and wet, grassy or sedge-rich turf on limestone or calcareous mica-schist. Generally montane, found locally in the Highlands, but growing as low as 360 m in Teesdale (north-west Yorkshire and Co Durham). Joseph von Cobres (1747–1823) was a German botanist. Latin *simpliciuscula* = 'unbranched'.

FALSE TRUFFLE *Elaphomyces granulatus* (Euriatales, Elaphomycetaceae), a widespread inedible fungus, infrequent (though parasitised by the more visible drumstick and snakestongue truffleclubs), present throughout the year but sporulating in summer and autumn, the ball-like semi-subterranean fruiting body found in coniferous woodland, nearly always under spruce. Greek *elaphos* + *myces* = 'deer' + 'fungus' (Americans call this 'deer truffle'), and Latin for 'granulated', a reference to the granular warts on the outer rind.

Carroty false truffle *Stephanospora caroticolor* (Russulales, Stephanosporaceae), inedible, scarce, a UK BAP species, with a few records mostly in English deciduous woodland on soil and litter. Greek *stephanos* + *spora* = 'poet' (itself from *stephos* = 'crown') + 'seed'.

Yellow false truffle *Rhizopogon luteolus* (Boletales, Rhizopogonaceae), inedible, found in late summer and autumn, widespread, with most records from the Highlands, found in sandy soil, ectomycorrhizal on pine. Greek *rhiza* + *pogon* = 'root' + 'beard', and the Latin *luteolus* = 'yellowish'.

FAN (CNIDARIAN) See sea fan under Anthozoa.

FAN (FUNGUS) Yellow fan *Spathularia flavida* (Rhytismatales, Cudoniaceae), inedible, widespread but uncommon, found in late summer and autumn on soil with moss and debris under coniferous trees. The binomial is from the Latin for 'spatula' and 'yellow'.

FAN-FOOT (MOTH) Nocturnal moths in the family Erebidae (Herminiinae). Specifically, *Herminia tarsipennalis*, wingspan 30–5 mm, widespread and common (less frequent in Scotland) in broadleaf woodland, hedgerows and gardens. Adults fly in June–July. Larvae feed on withered leaves of trees such as oak and beech. *Herminia* is the Latinised form of French *herminé* = 'covered in ermine', and Latin *tarsus* + *penna* = 'tarsus' (lower part of the leg) + 'feather', both parts referring to leg tufts which also prompts the English name.

Clay fan-foot *Paracolax tristalis*, scarce, in oak woodland in parts of southern England from Kent to Wiltshire, a priority species under the UK BAP. Probably in decline owing to changes in woodland management, especially cessation of coppicing. Adults fly in July–August. Larvae feed on pedunculate oak leaves, especially those fallen to the ground. Greek *para* + *colax* = 'alongside' + 'flatterer', meaning a similarity with related species, and Latin *tristis* = 'sombre', from the wing colour.

Common fan-foot *Pechipogo strigilata*, despite its English name now declining and uncommon, found in ancient deciduous woodland in parts of southern England. It is a priority species in the UK BAP. Adults fly in May–June. Larvae feed on withered oak leaves. Greek *pechys* + *pogon* = 'forearm' + 'beard', from the hairy foreleg, and Latin *striga* = 'streak', from the wing marking.

Dotted fan-foot *Macrochilo cribrumalis*, scarce, found in the

East Anglian fens and near the coast from Norfolk to Essex; there are also isolated colonies in east Kent, Sussex and Hampshire. Adults fly in June–August. Larvae feed on wood-rushes and sedges. Greek *macros* + *cheilos* = 'large' + 'lip', from the labial size, and Latin *cribrum* = 'sieve', from the forewing spots.

Shaded fan-foot *H. tarsicrinalis*, a rare RDB moth found in ancient woodland, thickets and copses in parts of south-east Suffolk, Norfolk and north Essex. In Norfolk, records from 1989 to 1991 were from just one woodland site, but all the larval food plant (bramble) was removed shortly after discovery of the species. It was rediscovered there in 2009 and was still present in 2015. Adults fly in June–July. Latin *tarsus* + *crinis* = 'tarsus' + 'hair'.

Small fan-foot *H. grisealis*, widespread and common in broadleaf woodland, scrub, hedgerows and gardens. Adults fly in June–August. Larvae feed on the leaves of a range of deciduous trees, often when withered and fallen. *Griseus* is Latin for 'grey', from the wing colour.

FAN WEED Red seaweeds/macroalgae (Rhodophyta).

Beautiful fan weed *Callophyllis laciniata* (Kallymeniaceae), up to 20 cm long, widespread if uncommon between the Hampshire and Lincoln coasts, attached to rock and kelp in the lower littoral and shallow sublittoral to at least 30 m. Greek *calos* + *phyllon* = 'beautiful' + 'leaf', and Latin *laciniata* = 'torn'.

Devonshire fan weed *Ahnfeltiopsis devoniensis* (Phyllophoraceae), mainly recorded from the lower littoral or subtidal rocky shore in south-west England, with occasional records elsewhere, for example Strangford Lough. Nils Ahnfelt (1801–37) was a Swedish botanist.

Norwegian fan weed *Gymnogongrus crenulatus* (Phyllophoraceae), with fronds 6–10 cm long, locally abundant on sand-covered rocks on the lower intertidal from Anglesey to Norfolk, parts of west Scotland and the southern half of Ireland. *Gymnos* is Greek for 'naked', but the etymology of the full genus name is obscure; *crenulatus* is Latin for 'notched', reflecting the fronds' small rounded serrations.

Papery fan weed *Stenogramma interruptum* (Phyllophoraceae), with fronds 7 cm long, subtidal, occasionally found on rocks and stones on the coasts of south-west England, south-west Wales and south-west and north-east Ireland. Greek *stenos* + *gramme* = 'narrow' + 'mark', and Latin for 'interrupted'.

Rosy fan weed *Rhodymenia pseudopalmata* (Rhodymeniaceae), with fronds up to 10 cm long, widespread on the south and west British coastline, and around Ireland, on rocks in shaded pools in the lower intertidal and subtidal, and epiphytic on kelp. Greek *rhodon* = 'rose', and Latin *pseudopalmata* = 'false' + 'palmated'.

FANNER (MOTH) Mining moths in the genus *Glyphipterix* (Glyphipterigidae, Glyphipteriginae), from Greek *glyphis* + *pteryx* = 'notch' + 'wing', from an indentation on the forewing.

Bog-rush fanner *G. schoenicolella*, wingspan 6–9 mm, day-flying, local in fen, bog and dune slacks from Dorset to Cornwall, in the fens of East Anglia, in west and north Scotland, and in Ireland. Adults fly from May to September. Larvae feed inside the seeds of black bog-rush *Schoenus nigricans*, which prompts the specific epithet.

Cotton-grass fanner *G. haworthana*, wingspan 11–15 mm, day-flying, widespread but local on damp moorland, heathland and mosses. Adults fly in May. Larvae feed within seed capsules of cotton-sedge, spinning the cotton heads together or to nearby vegetation, making them noticeable in late winter or early spring. Adrian Haworth (1767–1833) was an English naturalist.

Plain fanner *G. fuscoviridella*, wingspan 10–16 mm, nocturnal, common in dry grassland throughout much of England, Wales and south-east Scotland. Adults fly in May–June. Larvae feed inside the stems of wood-rush. Latin *fuscus* + *viridis* = 'brown' + 'green', from the forewing colour.

Sedge fanner *G. forsterella*, wingspan 8–11 mm, day-flying, local (though becoming more widespread and common) in damp grassland and woodland. Adults fly in May–June, feeding on

sedge flowers. Larvae feed from August to April within the ears of fox-sedge and other sedges. Johann Forster (1729–98) was a Polish/Prussian-born entomologist.

Speckled fanner *G. thrasonella*, wingspan 10–15 mm, nocturnal, widespread and common in a range of marginal aquatic habitats, including bog, damp heathland and mosses. Adults fly in May–August. Larvae probably feed inside the stems of rush species. Thraso was a character in a play by the Roman writer Terence, having no entomological significance.

Stonecrop fanner *G. equitella*, wingspan 9–10 mm, day-flying, on shingle beaches, rocky areas, old walls and dry sandy areas in a few parts of England and Wales. Larvae feed inside the stems of biting stonecrop. *Equitis* is Latin for 'cavalryman', having no entomological significance.

FANVAULT Species of rare (scattered and local) fungi in the genus *Camarophyllopsis* (Hygrophoraceae), generally on grassland, often in deciduous woodland. Greek *camara* = 'vaulted chamber', this and the English name reflecting the cap morphology.

FAT-HEN *Chenopodium album* (Amaranthaceae), a common and widespread annual of disturbed, nutrient-rich habitats, including cultivated fields and gardens, waste ground and soil heaps. Greek *chen* + *podion* = 'goose' + 'little foot', referring to the leaf shape, and Latin *album* = 'white'.

FATSIA *Fatsia japonica* (Araliaceae), an evergreen neophyte shrub introduced in 1838, though not recorded from the wild until 1989 (in south Devon), but scattered as a garden escape or relic of cultivation in woodland, and on roadsides and waste ground elsewhere, mainly in England. The genus name may be a misrendering of the Japanese *yatsude* = 'eight hands'.

FEATHER-GRASS Mexican feather-grass Alternative name for Argentine needle-grass.

FEATHER MITE See mite.

FEATHER-MOSS Mosses with feather-like foliage.

Rarities include **compact feather-moss** *Conardia compacta*, **elegant feather-moss** *Eurhynchium pulchellum* (east Highlands, Ulster and Suffolk Breckland), **Haller's feather-moss** *Campylophyllum halleri* (Highlands), **incurved feather-moss** *Homomallium incurvatum* (north England and southern Highlands), **Muhlenbeck's feather-moss** *Herzogiella striatella* (Highlands), **Portland feather-moss** *Eurhynchium meridionale* (mainly Portland, Dorset), **reflexed feather-moss** *Brachythecium reflexum* (Highlands), **rock feather-moss** *Lescuraea saxicola* (west Highlands), **round-leaved feather-moss** *Rhynchostegium rotundifolium* (known only from a roadside bank on limestone near Bisley, Gloucestershire, and ash coppice on chalk near Wilmington, East Sussex), **snow feather-moss** *Brachythecium glaciale* (Highlands), **swamp feather-moss** *Amblystegium radicale*, and **Yorkshire feather-moss** *Thamnobryum cataractarum* (one known site by the River Doe near Ingleton, North Yorkshire).

Beech feather-moss *Cirriphyllum crassinervium* (Brachytheciaceae), widespread, especially in England and Wales, locally frequent on base-rich rocks, old walls, and the base and roots of trees, for example in woodland on chalk in the south. The binomial may blend Latin *cirrus* = 'curl' with Greek *phyllon* = 'leaf', plus Latin *crassinervium* = 'large-veined'.

Blunt feather-moss *Homalia trichomanoides* (Neckeraceae), common and widespread on rocks, walls and the base of trees in shaded sites, including woodland, especially by ditches and streams. Greek *homalia* = 'evenness of surface' and *trichomanoides* = 'resembling trichomanes', the Greek name for a type of fern.

Brook-side feather-moss *Hygroamblystegium fluviatile* (Amblystegiaceae), locally common in south-west and northern England, Wales, and much of Scotland and Ireland, characteristic of fast-flowing streams on neutral to base-rich rocks, usually attached to rocks. Greek *hygros* + *amblys* + *stege* = 'wet' + 'blunt' + 'roof', and Latin *fluviatile* = 'riverine'.

Chalk feather-moss *Campylophyllum calcareum* (Hypnaceae), uncommon, scattered in Britain in low-lying, calcareous habitats, usually on stones and tree roots, in woodland and other shaded sites. Greek *campylos* + *phyllon* = 'curved' + 'leaf', and Latin *calcareum* = 'relating to chalk'.

Clustered feather-moss *Rhynchostegium confertum* (Brachytheciaceae), common and widespread on elder and other trees, stumps and logs, as well as rocks, walls and other man-made structures. It is found in light to moderate shade, in woodland, waysides and gardens. Greek *rhynchos* + *stege* = 'beak' + 'roof', and Latin *confertum* = 'crowded'.

Common feather-moss *Eurhynchium praelongum* (Brachytheciaceae), common and widespread on banks, in turf, on woodland floors, logs and tree-trunks, and in lawns. It also occurs in upland rush flushes and among purple moor-grass. Greek *eu* + *rhynchos* = 'well' + 'beaked', and Latin *praelongum* = 'very long'.

Common striated feather-moss *Eurhynchium striatum*, common and widespread on the ground in (especially long-established) woodland, favouring but not limited to base-rich soils. Latin *striatum* = 'striped'.

Constricted feather-moss *Hygroamblystegium humile*, widespread and scattered, mainly in England, in marsh and muddy sites on soil, stones and tree bases. *Humile* is Latin for 'low'.

Creeping feather-moss *Amblystegium serpens* (Amblystegiaceae), common and widespread in moist or sheltered places, on living and dead wood in woodland and hedges, and on soil and stones on banks, beside streams and rivers, at the base of walls, and on man-made habitats such as tarmac. Greek *amblys* + *stege* = 'blunt' + 'roof', and the Latin *serpens* = 'snake' (i.e. 'creeping').

Curve-stalk feather-moss *Rhynchostegiella curviseta* (Brachytheciaceae), locally common in southern England and the Midlands on at least mildly base-rich rocks, bridge supports and similar substrates by streams. Greek *rhynchos* + *stege* = 'beak' + 'roof', and the Latin *curviseta* = 'curved bristles'.

Curving feather-moss *Scorpiurium circinatum* (Brachytheciaceae), uncommon in south and south-west England, south Wales and the southern half of Ireland, especially on base-rich rocks and in turf. Greek *scorpios* + *oura* = 'scorpion' + 'tail', and Latin *circinatum* = 'rounded'.

Depressed feather-moss *Taxiphyllum wissgrillii* (Hypnaceae), scattered and widespread in lowland regions in calcareous habitats, mostly on rock, sometimes on soil and tree bases, usually in woodland. Greek *taxiphyllum* = 'yew-like leaves'.

Derbyshire feather-moss *Thamnobryum angustifolium* (Neckeraceae), a Critically Endangered endemic, with a colony known since 1883 from a Carboniferous Limestone rock-face site in Derbyshire of 3 m², with a few plants nearby, associated with a base-rich spring, and vulnerable to disturbance and desiccation. A second site was found in 2008 on the River Eden near Armathwaite, Cumbria. Fully protected by the WCA (1981), this was also the first bryophyte to have its own BAP. Greek *thamnos* + *bryon* = 'shrub' + 'moss', and Latin *angustifolium* = 'narrow-leaved'.

Dwarf feather-moss *Eurhynchium pumilum*, common and widespread, less soin mountain areas, on soil and stones in woodland, tolerating shade, and growing in crevices and on ledges in rocky valleys. It usually occurs on dry ground, but some sites may be seasonally wet. *Pumilum* is Latin for 'dwarf'.

Fertile feather-moss *Drepanocladus polygamus* (Amblystegiaceae), scattered and widespread, especially on coasts, on clifftops, in salt-sprayed turf, dune slacks and, in Scotland, upper saltmarsh. Inland it scarce in fen, marshy grassland, gravel-pits and base-rich, temporary wetlands. Greek *drepanon* + *clados* = 'sickle' + 'branch'.

Fine-leaved marsh feather-moss *Campyliadelphus elodes* (Amblystegiaceae), scattered and widespread in wet habitats, especially where periodically submerged, growing in grass, sometimes on wood, especially in base-rich marsh, on damp sand, and in dune slacks. It can be locally abundant on tussocks

in calcareous fen. Greek *campylos* + *adelphos* = 'curved' + 'brother', and *elodes* = 'marshy'.

Flagellate feather-moss *Hyocomium armoricum* (Hypnaceae), common and widespread, especially north of a line from Devon to north Yorkshire on wet rocks, stony banks and tree roots beside streams, usually in shade, in acidic or mildly base-rich habitats. Greek *hyo* + *come* = 'upsilon-shaped' + 'hair', and Latin *armoricum* = 'Armorican'.

Fountain feather-moss *Hygroamblystegium tenax*, widespread, less common in East Anglia, north Scotland and central Ireland, on stones and tree roots, and also on concrete, bricks, weirs and other man-made substrates. *Tenax* is Latin for 'tough'.

Foxtail feather-moss *Thamnobryum alopecurum*, common and widespread on wet rocks by streams. It also grows on the ground, coppice stools and tree bases in mildly base-rich woodland. Greek *alopex* + *ouros* = 'fox' + 'tail'.

Glasswort feather-moss *Scleropodium tourettii* (Brachytheciaceae), found on the coast from south-west Scotland to Norfolk, the Isle of Man and south and east Ireland, in open sites in turf, on stony ground, over rocks, and on clifftops, favouring warm, dry places. It is occasionally found inland in south England and the Welsh Borders. Greek *scleros* + *podi* = 'hard' + 'foot'.

Golden feather-moss *Campyliadelphus chrysophyllus* (Amblystegiaceae), locally common and widespread on stones, rocks and turf in chalk and limestone grassland, and in calcareous dunes. *Chrysophyllus* is Greek for 'golden-leaved'.

Hair-pointed feather-moss *Cirriphyllum piliferum*, common and widespread in woodland, at the base of old walls and among rocks, especially in base-rich habitats, and on clay, usually where there is shade or shelter. *Piliferum* is Latin for 'hair-bearing'.

Kneiff's feather-moss *Leptodictyum riparium* (Amblystegiaceae), widespread (but not in the Highlands) and locally common in wet places, usually on tree bases, debris in marsh, by ponds and slow-flowing streams, and in wet woodland. F.G. Kneiff (d.1832) was a German botanist and physician. Greek *leptos* + *dictyon* = 'slender' + 'net', and Latin *riparium* = 'riverside'.

Lesser striated feather-moss *Eurhynchium striatulum*, scattered in England, Wales and Ireland on calcareous rocks and walls, occasionally on tree roots. The habitat is usually lightly to moderately shaded, often in woodland, and in warm, dry environments. *Striatulum* is Latin for 'faintly striped'.

Long-beaked water feather-moss *Rhynchostegium riparioides*, common and widespread, semi-submerged for at least part of the year on stones, roots and wood by streams and rivers, growing best in running water, but also found in ditches and by canals and ponds. Latin *riparioides* = 'of riverbanks'.

Matted feather-moss *Sciuro-hypnum populeum* (Brachytheciaceae), common and widespread on stones, boulders, old walls and compacted soil, frequent in woodland and hedge banks, preferring light to moderate shade. It favours base-rich substrates, but also grows on slightly acidic, siliceous rocks. It also grows on trees with a base-rich bark. Greek *sciouros* + *hypnos* = 'squirrel' + 'sleep', and Latin *populus* = 'poplar'.

Megapolitan feather-moss *Rhynchostegium megapolitanum*, scattered and widespread, especially in the southern half of England and south Wales, particularly on dunes and chalk, occasionally on light, well-drained soil on banks, clifftops and walls. Greek *mega* + *polis* = 'large' + 'city'.

Neat feather-moss *Scleropodium purum*, common and widespread, especially in unimproved, acidic grassland and heathland, but also in chalk and limestone grassland, and on banks and rocks. It also grows in open woodland. *Purum* is Latin for 'pure'.

Ostrich-plume feather-moss *Ptilium crista-castrensis* (Hypnaceae), locally common in upland oak, birch and pine woodland (including ancient woodland) throughout the Highlands, and in upland oak in the Lake District and Snowdonia. It also grows on open, heathy hillsides in the Highlands, typically on shaded north-east-facing slopes. *Ptilium* is from Greek *ptilon* = 'wing'.

Portuguese feather-moss *Rhynchostegium alopecuroides*, scattered in upland areas, especially in south-west England and Wales, on base-poor rocks in clear, fast-flowing streams in woodland and in the open. Latin *alopecuroides* = 'like foxtail' (*Alopecurus*).

Prince-of-Wales feather-moss *Leptodon smithii* (Leptodontaceae), common across southern England, and south and north Wales, on mature trees in woodland, parks and hedgerows, occasionally on walls and calcareous rocks. Greek *leptos* + *donti* = 'slender' + 'tooth'.

Red-stemmed feather-moss *Pleurozium schreberi* (Hylocomiaceae), common and widespread, a calcifuge often found in heathland and in heathy woodland. The binomial is from Greek *pleura* = 'side', and the German naturalist Johann von Schreber (1739–1810).

River feather-moss *Brachythecium rivulare*, common and widespread in springs and wet ledges in the uplands, and on streamside boulders. It is also frequent in low-lying woodland and carr. *Rivulare* is Latin for 'growing by streams'.

Rough-stalked feather-moss *Brachythecium rutabulum*, common and widespread on trees, stumps and logs, and on soil and gravelly ground, on stones, rubble and rocks, walls, and in grassland and marsh. It is found in both shade, for instance woodland and hedge banks, and in the open, for instance gardens, waste ground and stream banks. *Rutabulum* is Latin for 'shovel'.

Rusty feather-moss *Sciuro-hypnum plumosum*, common and widespread except in the south and east Midlands and East Anglia, on rock, and the base and roots of trees, often in light shade, by fast-flowing, upland streams and larger rivers in the lowlands. *Plumosum* is Latin for 'feathery'.

Sand feather-moss *Brachythecium mildeanum*, widespread and common, especially in the southern half of Britain, in open habitats, moist grassland, marsh and dunes. Carl Milde (1824–71) was a German bryologist.

Scabrous feather-moss *Rhynchostegiella litorea*, locally common, mainly on chalk in southern England. *Litorea* is Latin for 'coastal'.

Showy feather-moss *Oxyrrhynchium speciosum* (Brachytheciaceae), widespread in lowland regions in wet woodland and carr, marsh and seepages, and on stream banks. Greek *oxys* + *rhynchos* = 'sharp' + 'beak', and Latin *speciosum* = 'showy'.

Silesian feather-moss *Herzogiella seligeri* (Plagiotheciaceae), mainly found in south and south-east England and Yorkshire on logs and stumps in broadleaf woodland, particularly coppiced sweet chestnut.

Silky wall feather-moss *Homalothecium sericeum* (Brachytheciaceae), common and widespread on base-rich rocks and locally abundant on limestone walls and crags. It is also common on trees with base-rich bark, such as ash and elder, and on brick walls, concrete and other man-made structures. Greek *homalos* + *thece* = 'smooth' + 'box', and Latin *sericeum* = 'silky'.

Smooth-stalk feather-moss *Brachythecium salebrosum*, scattered in Britain, mostly in England, on logs, tree-trunks, and on stones and rubble, in woodland and sheltered habitats. Contradicting the English name, *salebrosum* is Latin for 'rough'.

Streaky feather-moss *Brachythecium glareosum*, common and widespread in open, usually base-rich ground, particularly on chalk and limestone, on bare soil, rubble and rocks, in turf and on banks. *Glareosum* is Latin for 'gravelly'.

Swartz's feather-moss *Oxyrrhynchium hians*, common and widespread on soil on stream banks, woodland, hedge banks, grassland on chalk, clay and other base-rich soils, arable fields, parks and gardens. It also grows in seepages and on wet rock ledges. *Hians* is Latin for 'opening'.

Teesdale feather-moss *Rhynchostegiella teneriffae*, widespread and locally common on sandstone and base-rich rocks, preferring sites that are splashed by water rather than submerged.

Tender feather-moss *Rhynchostegiella tenella*, common and widespread except for upland Scotland, on base-rich rocks, usually where there is shade or shelter, including woodland and coastal cliffs. It also occurs on bridges, mortared walls and brickwork. *Tenella* is Latin for 'tender'.

Tiny feather-moss *Amblystegium confervoides*, widespread, scattered but rare on shaded, calcareous rocks, especially in limestone woodland and on hillside outcrops. *Conferva* is Latin for an unspecified healing water plant.

Tufted feather-moss *Scleropodium cespitans*, common and widespread in England and Wales, scarce and scattered in Scotland and Ireland, usually by lowland streams and rivers where there is occasional flooding, on the roots and trunks of trees, and on rocks and boulders. It also found on tarmac. *Cespitans* is Latin for 'tufted'.

Twist-tip feather-moss *Oxyrrhynchium schleicheri*, locally common in southern and northern England and the Welsh Borders, on soil in woodland and on sheltered banks, especially on light or sandy soil. Johann Schleicher (1768–1834) was a Swiss botanist.

Velvet feather-moss *Brachytheciastrum velutinum*, common and widespread in (especially south and east) Britain, though declining in places, and scarce in Ireland, on trees and dead wood, as well as stones and compacted soil. Greek *brachys* + *thece* = 'short' + 'box', and Latin *velutinum* = 'velvety'.

Wall feather-moss *Rhynchostegium murale*, common and widespread, more scattered in upland Scotland and in Ireland, in base-rich habitats, especially chalk and limestone, nearly always on rocks and stones in woodland in light shade. It is also widespread on the old walls. *Murale* is Latin for 'of walls'.

Whitish feather-moss *Brachythecium albicans*, common and widespread, though coastal (mainly on dunes) in Ireland, on well-drained, base-poor soils, especially sand. It occurs in short turf in pasture and over rocks, less often on waste ground. Mainly found on acidic substrates, it also occurs on leached soil over base-rich rocks. *Albicans* is Latin for 'whitish'.

Willow feather-moss *Hygroamblystegium varium*, common and widespread in wet ground by streams and ponds, on decomposing vegetation, wood, stones and soil. *Varium* is Latin for 'varied'.

Woolly feather-moss *Tomentypnum nitens* (Amblystegiaceae), found in calcareous fens. It is rare in southern Britain, but locally common in upland mires in the Highlands, and in the Borders and northern England, where it tends to grow in flushes. Latin *tomentus* = 'wool stuffing' and Greek *hypnos* = 'sleep', and Latin *nitens* = 'shiny'.

Wrinkle-leaved feather-moss *Rhytidium rugosum* (Rhytidiaceae), scattered in Britain on well-drained, calcareous grassland, and on south- and south-west-facing ledges on (especially montane) limestone cliffs, as well as limestone quarries, and occasionally in calcareous dunes. Greek *rhytis* = 'wrinkle', and Latin *rugosum* = 'wrinkled'.

Yellow feather-moss *Homalothecium lutescens*, common and widespread in lowland regions in unimproved calcareous grassland, and in other open sites, quarries, and on dunes. A prostrate form is locally frequent on rocks in limestone valleys. *Lutescens* is Latin for 'yellowish'.

Yellow starry feather-moss *Campylium stellatum* (Amblystegiaceae), common and widespread on (slightly) base-rich mires, flushes and dune slacks, sometimes over wet rocks. Greek *campylos* = 'curved', and Latin *stellatum* = 'star-like'.

FEATHER-STAR Marine echinoderms, members of the class Crinoidea, mostly in the order Comatulida, with five families in the British list. Individuals have a central 'disc' body from which extend five or more long, slender arms used to crawl along the sea floor, feeding as scavengers or detritivores in the shallow sublittoral.

FEATHER WEED Red seaweeds.

Bushy feather weed *Pterothamnion plumula* (Ceramiaceae), 3–15 cm long, common and widespread, less so on the east coast, on rocks in the lower intertidal and shallow subtidal down to

20 m. Greek *pteron* + *thamnion* = 'wing' + 'little bush' (diminutive of *thamnos*), and Latin *plumula* = 'feathery'.

Siphoned feather weed *Heterosiphonia plumosa* (Dasyaceae), with fronds up to 20 cm long, especially common in south-west Britain, scarce in north-west Scotland, widespread in Ireland, on rocks and as an epiphyte on seaweeds in the lower littoral and sublittoral. Greek *heteros* + *siphon* = 'different' + 'tube', and Latin *plumosa* = 'feathery'.

Soft feather weed *Plumaria plumosa* (Wrangeliaceae), common and widespread, with fronds 10 cm long, on vertical parts of the lower littoral rocky shore. *Plumaria* is, strictly, Latin for 'needlework', but here, with *plumosa*, refers to a feathery appearance.

FEATHERWORT Foliose liverworts in the genera *Adelanthus* (Adelanthaceae), from Greek *adelos* + *anthos* = 'concealed' + 'flower'; *Pedinophyllum* (Jamesoniellaceae), from *pedinos* + *phyllon* = 'flat' + 'leaf'; and *Plagiochila* (Plagiochilaceae), from *plagios* + *cheilos* = 'oblique' + 'margin'.

Rare and uncommon species include **British featherwort** *Pl. britannica*, **Carrington's featherwort** *Pl. carringtonii* (Highlands), **deceptive featherwort** *A. decipiens*, **petty featherwort** *Pl. exigua*, and **western featherwort** *Pl. atlantica* (west Wales, Cumbria and, mainly, western Scotland).

Craven featherwort *Pe. interruptum*, locally common, mainly found in the mid-Pennines and north-west Ireland on Carboniferous or metamorphic limestone, on the rock itself or on soil, in wooded gorges. *Interruptum* is Latin for 'interrupted'.

Greater featherwort *Pl. asplenioides*, common and widespread in damp turf in sheltered woodland and hedgerows, and in more open habitats such as chalky slopes, fen, stream banks, rocks and decomposing wood. *Asplenioides* is Latin for 'resembling spleenwort' (= *Asplenium*).

Killarney featherwort *Pl. killarniensis*, found in western Britain and western Ireland, typically on base-rich rocks and trees in Atlantic woodlands. In south-west England, it grows on rocky lane banks, and in sea-cliff turf.

Lesser featherwort *Pl. porelloides*, common and widespread in sheltered habitats, from woodland banks to streamsides and upland slopes. *Porelloides* is from Latin for 'small pore'.

Lindenberg's featherwort *A. lindenbergianus*, on rocky slopes in west and north-west Ireland and one site on Islay. Johann Lindenberg (1781–1851) was a German bryologist.

Prickly featherwort *Pl. spinulosa*, often found in upland western Britain and western Ireland in Atlantic woodland forming cushions on rocks, banks and trees. *Spinulosa* is Latin for 'spiny'.

Spotty featherwort *Pl. punctata*, found in western Britain and western Ireland forming clumps on trees and rocks by streams in Atlantic woodland. *Punctata* is Latin for 'spotted'.

FEBRUARY RED Fisherman's name for taenyopterygid stoneflies.

FELIDAE Cat family (Latin *felis* = 'cat').

FEN-SEDGE Great fen-sedge *Cladium mariscus* (Cyperaceae), a rhizomatous perennial, widespread, scattered and locally common especially in parts of the southern English coast, East Anglia, western Scotland and Ireland, in oligotrophic and mesotrophic habitats, usually on peat in swamps at lake and pond edges and along streams, and in tall-herb fens and fen carr. In England and Wales it is mainly found on calcareous sites; in the Hebrides and western Ireland it also grows in acidic habitats. Greek *clados* = 'branch', and Latin *mariscus* = 'marsh'.

FENNEL *Foeniculum vulgare* (Apiaceae), a herbaceous perennial archaeophyte used in cooking since Roman times, naturalised in the southern half of Britain, scattered, rare and casual further north and in Ireland, in marsh, and on sea walls, roadsides and waste ground. Its deep tap root allows it to survive in droughted habitats, making it very persistent. *Foeniculum* is the diminutive

of Latin *foenum* = 'hay', from the smell, and Latin *vulgare* = 'common'.

False fennel *Ridolfia segetum*, a scarce annual Mediterranean neophyte found in the south Midlands and north-east Lincolnshire as a casual on waste ground, originating from birdseed. *Segetum* is Latin for 'corn'.

Giant fennel *Ferula communis*, a herbaceous Mediterranean neophyte perennial naturalised on verges in west Suffolk since 1988, Northamptonshire (1996) and south Essex (2004), an occasional casual elsewhere. The binomial is Latin for 'fennel' and 'common'.

Hog's fennel *Peucedanum officinale*, a scarce herbaceous perennial of coastal grassland and scrubby habitats adjoining saltmarsh or brackish grazing marsh, on creek sides and on sea walls, and on waste ground, very local in east Kent, north Essex and east Suffolk. Greek *peucedanon* = 'hog's fennel', and Latin *officinale* for having pharmacological properties.

Sow fennel Alternative name for hog's fennel.

FENUGREEK *Trigonella foenum-graecum* (Fabaceae), an annual neophyte from the Mediterranean with a scattered distribution as a casual via birdseed on waste ground and tips. *Trigonella* is from Greek for 'three small angles' (*gonia* = 'angle'). The English and specific names are from Latin *foenum* + *graecum* = 'hay' + 'Greek', the Romans using the hay as fodder. **Blue fenugreek** (*T. caerulea*), **sickle-fruited fenugreek** (*T. corniculata*) and other fenugreeks are similarly found.

FERN (PLANT) A group of flowerless vascular plants that reproduce via spores.

Beech fern *Phegopteris connectilis* (Thelypteridaceae), on acid soils in damp shaded woodland, common in northern and western Britain, scattered in Ireland. Greek *phegos* + *pteris* = 'oak' or 'beech' + 'fern', and Latin *connectilis* = 'joined'.

Jersey fern *Anogramma leptophylla* (Pteridaceae), in damp shady hedge banks, common in Jersey and from one site in Guernsey. Greek *ano* + *gramme* = 'upward' + 'lined', because the sori mature first near the tips, and *leptos* + *phyllon* = 'slender' + 'leaf'.

Killarney fern *Trichomanes speciosum* (Hymenophyllaceae), one of the filmy-ferns (q.v.), scattered, with most records from south-west and northern England, west Scotland and south-west Ireland on sheltered damp rock faces, often at cave entrances or near a waterfall. *Trichomanes* was the name used for this plant by Dioscorides and Theophrastus. Latin *speciosum* = 'showy'.

Lemon-scented fern *Oreopteris limbosperma* (Thelypteridaceae), found on streamsides and ditch banks on nutrient-poor soils, particularly in upland parts of England and Wales, though also in south-east England. Greek *oreos* + *pteris* = 'of a mountain' + 'fern', and *limbos* + *sperma* = 'marginal' + 'seed'.

Limestone fern *Gymnocarpium robertianum* (Woodsiaceae), local in England and Wales, rare in Scotland and western Ireland on limestone rocks and scree, and on walls especially in eastern England. Greek *gymnos* + *carpos* = 'naked' + 'seed'.

Maidenhair fern *Adiantum capillus-veneris* (Pteridaceae), on limestone cliffs and crevices, but also grown in gardens and escaped onto moist parts of walls and bridges, thereby becoming widespread in England, Wales and parts of Ireland. Greek *adiantos* = 'unwetted', referring to the plant's water-repellancy, and Latin *capillus-veneris* = 'hair of Venus', a term used by Roman herbalists promoting this plant in alleviating respiratory problems.

Marsh fern Members of the family Thelypteridaceae, specifically, *Thelypteris palustris*, a rarity associated with shaded sites in marsh, fen and fen woodland, with a scattered distribution, but common only in East Anglia. Greek *thelys* + *pteris* = 'female' + 'fern', and Latin *palustris* = 'marshy'.

Oak fern *G. dryopteris*, in humid woodland or rocky sites on humus-rich but nutrient-poor acidic soils, mainly in north-west Britain. Greek *drys* + *pteris* = 'oak' + 'fern', reflecting the habitat.

Parsley fern *Cryptogramma crispa* (Pteridaceae), locally common in Wales, Scotland, northern and south-west England, and north-east Ireland on acid soil in rocky upland habitat. Greek *cryptos* + *gramma* = 'hidden' + 'mark', and Latin *crispa* = 'curled', describing the leaf.

Ribbon fern Members of the family Pteridaceae. Specifically, *Pteris cretica*, introduced, a garden escape naturalised on sheltered walls, old buildings and rock faces, especially in England. *Pteris* is Greek for this fern (*pteron* = 'wing'), *cretica* Latin for 'Cretan'.

Royal fern *Osmunda regalis* (Osmundaceae), widespread though commoner in western parts of Britain and Ireland, in open or semi-shaded woodland, fen, bog and heathland, and grown in gardens, on wet, moderately or acid nutrient-poor soil. *Osme* is Greek for 'scent', *regalis* Latin for 'royal'.

Water fern *Azolla filiculoides* (Salviniaceae), a small floating fern introduced from tropical America, first recorded at Pinner (north-west London) in 1883, now found in Britain north to central Scotland, east Ireland, and the Channel Islands, in fertile pools, canals and lakes with fairly high pH, in well-lit sites or partial shade, potentially becoming a dominant weed. Greek *azo* + *ollyo* = 'to dry' + 'to kill', the plants being killed by drought, and Latin *filiculoides* = 'fern-like'.

FERN (MOTH) *Horisme tersata* (Geometridae, Larentiinae), wingspan 31–6 mm, nocturnal, common in open woodland and hedgerows, and on downland and calcareous grassland throughout much of southern England and south Wales. Adults fly in June–July. Larvae feed on traveller's-joy and cultivated clematis. Greek *horisma* = 'boundary', from the black line across the wings, and Latin *tersus* = 'wiped clean'.

FERN BUG *Bryocoris pteridis* (Miridae), 2–4 mm long, widespread, feeding on the sporangia of ferns; this and bracken bug are the only bug species with this diet. Greek *bryo* + *coris* = 'full of' + 'bug', and *pteris* = 'fern'.

FERN-GRASS *Catapodium rigidum* (Poaceae), a locally common annual, widespread but scarce in Scotland, on sandy banks, stabilised shingle and rock outcrops, usually on calcareous substrates, and also in artificial habitats such as quarries, walls, pavements and railway ballast. Greek *cata* + *podion* = 'below' + 'little foot', the spikelets having short pedicels, and Latin *rigidum* = 'rigid'.

Sea fern-grass *C. marinum*, an annual of dry bare coastal habitats, rock crevices, grassy banks, clifftops, dunes and stabilised shingle, and on artificial habitats such as walls and pavements, and in the south increasingly inland by salt-treated roads.

FERN WEED Grateloup's fern weed *Grateloupia filicina* (Halymeniaceae), a red seaweed with fronds up to 12 cm long, often locally common in south-west England and Wales, west Scotland, and south and west Ireland, on rock in pools from the midlittoral to shallow subtidal. Jean-Pierre de Grateloup (1782–1862) was a French physician and naturalist. *Filicina* means resembling a fern (Latin *filix*).

Rounded fern weed *Osmundea hybrida* (Rhodomelaceae), with fronds 15 cm long, widespread and locally abundant on stones, limpets and mussels in rock pools, sometimes epiphytic. *Osme* is Greek for 'scent'.

FERRET *Mustela putorius furo* (Mustelidae), male (hob) head-tail length 38 cm, weight 1–2 kg, females (jills) 35 cm, 600–900 g, a domesticated form of the polecat kept as a pet and for rabbiting, introduced to England in the late eleventh or early twelfth centuries, with viable feral populations in north Yorkshire, Renfrewshire, Argyll, possibly Caithness, and some offshore islands, with a population in 2005 estimated at 2,500 (England 200, Scotland 2,250, Wales 50). 'Ferret' is late Middle English, in turn from Old French *fuiret*, based on Latin *furo*. The trinomial comes from Latin for 'weasel', *putor* = 'stench' (referring to the musky odour), and *furo* from *furonem* = 'thief'.

FESCUE Grasses (Poaceae) in the genera *Festuca* which,

with 'fescue', comes from Latin for 'stalk' or 'stem'; *Schenodorus*, meaning similar to (the sedge) *Schoenus*; and *Vulpia*, named after the German botanist Johann Vulpius (1760–1864).

Bearded fescue *V. ciliata*, a scarce annual mainly found in southern and eastern England, in disturbed sandy places. The native ssp. *ambigua* is found on tracks and paths through coastal dunes, and inland on sandy heathland, along roadsides and in patches of open grassland, spreading in some areas, for example the New Forest. The introduced ssp. *ciliata* is a rare casual from grain and wool shoddy. *Ciliata* is Latin for 'fringed with hairs'.

Blue fescue *F. longifolia*, a scarce, densely tufted perennial found on dry, rabbit-grazed heathland, sandy roadside banks and, in south Devon and the Channel Islands, on coastal clifftops and ledges. Several colonies in Breckland have been lost since the 1980s, and remaining populations there, and in Nottinghamshire and Lincolnshire, are small and at risk of disappearing because of shading. *Longifolia* is Latin for 'long-leaved'.

Confused fescue *F. lemanii*, a tufted perennial, probably native, local and scattered in Britain on grassy habitats on well-drained soils, for instance characteristic of limestone cliff rock crevices in Derbyshire. Dominique Léman (1781–1829) was a French botanist.

Dune fescue *V. fasciculata*, a locally frequent annual of dunes, particularly open, disturbed parts of fixed dunes, and sandy shingle, mainly from Cumbria and the Isle of Man round to the Wash, eastern Ireland, and the Channel Islands. *Faciculata* is Latin for 'bundled'.

Fine-leaved sheep's-fescue *F. filiformis*, a densely tufted perennial, scattered and locally common except in central and southern Ireland in heathland, moorland, open woodland and other grassy habitats, usually on acidic, sandy, well-drained soil. *Filiformis* is Latin for 'thread-like'.

Giant fescue *S. giganteus*, a perennial, common and widespread except for north Scotland and the Highlands, in damp woodland on neutral and base-rich soils, particularly by streams, in disturbed areas in clearings and on woodland margins, and as a colonist of secondary woodland. Dispersal is by fruits that stick to fur and clothing.

Hard fescue *F. brevipila*, a tufted perennial European neophyte, naturalised on roadsides, railway banks, golf courses and other amenity grasslands, especially on well-drained, acidic soils. *Brevipila* is Latin for 'short hair'.

Mat-grass fescue *V. unilateralis*, a scarce annual, probably native, scattered in the southern half of England on stony ground, dry banks and grassy tracks on chalk and limestone, as well as on railway ballast, walls and tips. *Unilateralis* is Latin for 'one-sided': stalks are directed to one side.

Meadow fescue *S. pratensis*, a common and widespread perennial of neutral grassland, usually on fertile soils, including pasture, hay and water meadow. Often sown for fodder, it has become naturalised on roadsides, railway banks and waste ground. *Pratensis* is from Latin for 'meadow'.

Rat's-tail fescue *V. myuros*, an annual archaeophyte, common and widespread, especially in England and Wales, growing by (and continuing to increase its range via) railways, on walls and waysides, in pavement cracks and on waste ground in built-up areas. Occasionally a weed of cultivation and introduced via wool shoddy, grain and grass seed. Greek *mys* + *ouros* = 'mouse' + 'tail'.

Red fescue *F. rubra*, an extremely variable tufted or rhizomatous perennial, with a number of subspecies recognised, common and widespread in grassy habitats, including lowland meadow and pasture, verges, saltmarsh, sea-cliffs, dunes, upland grassland, mountain slopes and rock ledges. *Rubra* is Latin for 'red'.

Rush-leaved fescue *F. arenaria*, a rhizomatous perennial, scattered and locally common round the coastline of Britain, though absent from west and north Scotland, on dunes (typically semi-mobile fore-dunes) and sandy shingle. *Arenaria* is Latin for 'related to sand'.

Sheep's-fescue *F. ovina*, a common and widespread,

morphologically variable, densely tufted perennial found in a range of usually well-drained grassy habitats, including lowland calcareous grassland, upland heathland and moorland, mountain slopes and sea-cliffs, and it is the main grass species in many upland pastures. Subspecies *ovina* and *hirtula* are usually found on acid soils, ssp. *ophioliticola* on calcareous or serpentine soils. *Ovina* is Latin for 'of sheep'.

Squirreltail fescue *V. bromoides*, a common and widespread annual of grassy habitats, including rough ground, on well-drained soil. *Bromoides* is Latin for 'resembling brome' (*Bromus*).

Tall fescue *S. arundinaceus*, perennial, common and widespread except in the Highlands, on neutral and basic soil in scrub, woodland edge, hedgerows, pasture, meadow, roadsides, railway banks and waste ground. *Arundinaceus* is Latin for 'reedlike'.

Various-leaved fescue *F. heterophylla*, a perennial European neophyte introduced by 1812, originally planted for ornament or ground cover, possibly for fodder, recorded in the wild in 1874. It has become a contaminant of grass-seed mixtures, and is scattered and naturalised in woodland on light soils. *Heterophylla* is Greek for 'differently leaved'.

Viviparous sheep's-fescue *F. vivipara*, a tufted perennial, common on both basic and acidic substrates in most upland regions on heathy pasture, open birch and oak woodland, and rocky habitats, and in a range of mountain communities including sites with late snow-lie. *Vivipara* is Latin for 'live-bearing'.

Wood fescue *F. altissima*, perennial, scattered in mainly western and northern Britain and north and east Ireland, in moist, wooded valleys, on rocky slopes, deciduous woodland margins and streamsides, on soils of a moderate base status. *Altissima* is Latin for 'tallest'.

FESTOON *Apoda limacodes* (Limacodidae), a scarce moth, wingspan 24–8 mm, in broadleaf woodland and hedgerows from Dorset to Kent into parts of the south Midlands and East Anglia. Adults fly in June–July, sometimes in sunlight but mostly at night. Larvae feed on oak and beech. Greek *apodi* = 'without foot', describing the almost 'footless' larva, and *leimax* + *eidos* = 'slug' (*Limax*) + 'form', again from the larval morphology.

FEVER FLY *Dilophus febrilis* (Bibionidae), 3.5–8 mm length, common and widespread, adults evident from spring to late summer, larvae usually feeding on organic matter around the base of grasses and on grass roots themselves, sometimes damaging amenity grass. It was formerly incorrectly believed that this fly was associated with fever-affected houses. Greek *di* + *lophos* = 'two' + 'crest', and Latin *febris* = 'fever'.

FEVERFEW *Tanacetum parthenium* (Asteraceae), an aromatic archaeophyte perennial herb from the Balkans, commonly cultivated as an ornamental, widely naturalised on walls, waysides, tips and waste ground. It seeds freely, but only disperses for short distances, so is most frequent near habitation. 'Feverfew' is a corruption of 'febrifuge', from its tonic and fever-dispelling properties. *Tanacetum* comes from medieval Latin *tanazita* (= 'tansy'), from Greek *athanasia* (= 'immortality'), having been used in winding sheets to discourage worms. *Parthenium* is from Greek *parthenos* = 'virgin', possibly from use in 'female medicine'.

FIBRECAP Inedible fungi in the genus *Inocybe* (Agaricales, Inocybaceae) (Greek *inodes* + *cybe* = 'fibrous' + 'head').

Scarce species include **blushing fibrecap** *I. whitei*, **collared fibrecap** *I. cincinnata*, **fleecy fibrecap** *I. flocculosa*, **foxy fibrecap** *I. vulpinella* (Wales, but also Dungeness, Kent, north Devon, Cheviot Hills, and the Inch Peninsula, Co. Kerry), **frosty fibrecap** *I. maculata*, **greenflush fibrecap** *I. corydalina*, **greenfoot fibrecap** *I. calamistrata*, **sand fibrecap** *I. arenicola*, **silky fibrecap** *I. duriuscula*, and **torn fibrecap** *I. lacera*.

Also infrequent, but widespread on often calcareous soil, commonly under beech, are **fruity fibrecap** *I. bongardii* (poisonous), **scaly fibrecap** *I. hystrix* (poisonous), **scurfy fibrecap** *I. petiginosa*, **straw fibrecap** *I. cookei* (poisonous, containing

muscarine, under beech, oak and sweet chestnut), and **woolly fibrecap** *I. stellatospora* (on soil and debris in mixed or coniferous woodland).

Beige fibrecap *I. sindonia*, frequent and widespread, poisonous, found in summer and autumn on often clay soil mostly in coniferous woodland. Latin *sindon* = 'fine linen'.

Bulbous fibrecap *I. napipes*, widespread and common, poisonous, found in summer and autumn on soil and litter in mixed woodland and parkland. Latin *napus* + *pes* = 'turnip' + 'foot'.

Deadly fibrecap *I. erubescens*, rare, lethally poisonous (containing muscarine), with cardiac failure without the antidote atropine, scattered in England, particularly recorded from the New Forest, in late summer and autumn on calcareous or neutral soil in deciduous woodland, especially under beech and hornbeam. *Erubescens* is Latin for 'becoming red'.

Lilac fibrecap See white fibrecap below.

Lilac leg fibrecap *I. griseolilacina*, locally common, widespread but mostly in England, found in summer and autumn on soil in deciduous woodland. Latin *griseolilacina* = 'grey-lilac'.

Pear fibrecap *I. fraudans*, locally common, found in autumn on deciduous woodland soil, often under beech or hazel. The odour of mature specimens is like overripe pear, hence the common name. *Fraudans* is Latin for 'cheating', probably referencing the scent which is misleading in an inedible fungus.

Split fibrecap *I. rimosa*, common and widespread, poisonous (containing muscarine), found in summer and autumn on soil in deciduous woodland, especially under beech. *Rimosa* is Latin for 'cracked'.

Star fibrecap *I. asterospora*, common and widespread, found in late summer and autumn, usually on calcareous soil in deciduous woodland, especially under oak. Greek *aster* + *spora* = 'star' + 'spore'.

White fibrecap *I. geophylla*, very common and widespread, very poisonous (containing muscarine), found in summer and autumn on soil in deciduous, occasionally coniferous woodland, often by tracksides. Var. *lilacina* (**lilac fibrecap**) has similar characteristics and distribution, and is also found on verges. *Geophylla* is Greek for 'earth' + 'leaf'.

FIDDLENECK Annual North American neophytes in the genus *Amsinckia* (Boraginaceae). The flower stems carry small flowers that curl over at the top like the head of a violin. Wilhelm Amsinck (1752–1831) was a patron of the Hamburg Botanical Gardens.

Common fiddleneck *A. micrantha*, present since 1836, a widespread and locally common weed of arable land and waste ground on light, sandy soils, originating as a contaminant of grain and from wool shoddy. *Micrantha* is Greek for 'small flower'.

Scarce fiddleneck *A. lycopsoides*, naturalised on rough ground on the Farne Islands since 1922, with a few scattered but dubious records elsewhere in England.

FIELD-SPEEDWELL See speedwell.

FIELDCAP Fungi mostly in the genus *Agrocybe* (Agaricales, Strophariaceae) (Latin *ager* + Greek *cybe* = 'field' + 'head').

Scarce if sometimes scattered species include **bearded fieldcap** *A. molesta* (grassy habitats), **common fieldcap** *A. pediades* (grassland and dune slacks), **dark fieldcap** *A. erebia* (woodland and hedgerow litter, and woodchip mulch), **netted fieldcap** *Bolbitius reticulatus* (hardwood stumps, woody debris and litter), **poplar fieldcap** *A. cylindracea*, edible (base of trunks, and soil near to broadleaf trees), **spring fieldcap** *A. praecox* (grassy habitats and woodland edges, and woodchip).

Mulch fieldcap *A. putaminum*, inedible, on woodchip mulch in parks and gardens recorded from a very few locations in England. *Putaminum* is from Latin *putamen* = 'husk'.

Wrinkled fieldcap *A. rivulosa*, inedible, added to the British list in 2004, since becoming fairly common in southern England (and it is also found in Wales, Scotland and Ireland) due to mulching flower beds using woodchip. Found in summer and autumn. *Rivulosa* is Latin for 'of streams'.

Yellow fieldcap *B. titubans*, inedible, widespread and common, found in summer and autumn on soil in nutrient-rich verges, manured grassland, straw, woodchip mulch and old cow pats (the genus name means 'of cow dung'). The fruiting body might last for less than a day. *Titubans* is Latin for 'staggering'.

FIELDFARE *Turdus pilaris* (Turdidae), wingspan 40 cm, length 26 cm, weight 100 g, a thrush which is a scarce breeder (seven pairs found in 2004 and 2008, one in 2005, 2006, 2007 and 2010), but common and widespread as a winter visitor, with 700,000 an average figure, depending on weather conditions on the continent, flocking in fields and (in bad weather) gardens, feeding on berries, insects and earthworms. It has been Red-listed in Britain since January 2013, but Green-listed in Ireland. The English name comes from Old English *feldefare*, perhaps from *feld* = 'field' and the base of *faran* = 'to travel'. Latin *turdus* = 'thrush', and *pilus* = 'hair' (a misreading of Greek *trichas* = 'thrush' and *trichos* = 'hair').

FIG *Ficus carica* (Moraceae), a spreading deciduous shrub or small tree introduced from the east Mediterranean at least a millennium ago, growing from discarded fruit or sewage waste, and first recorded in the wild in 1918. It has become naturalised on waste ground, in churchyards, on railway banks, cliffs and walls, and especially on the banks of urban rivers, some colonies being associated with warm-water industrial discharge, for example along a 16 km stretch of the River Don in Sheffield where 60- to 70-year-old trees were recorded in 1990. The binomial is Latin for 'fig tree' and '(dried) fig', the latter from Greek *carice* = 'fig'.

FIGURE OF EIGHT *Diloba caeruleocephala* (Noctuidae, Dilobinae), wingspan 30–40 mm, a fairly common nocturnal moth found in woodland, hedgerows and gardens over much of England, scarcer in Wales, Scotland and Ireland. Nationally there was an overall decline by 97% over 1968–2007. Adults fly in October–November. Larvae feed on leaves of hawthorn, blackthorn, wild roses and fruit trees such as apple. The English name comes from the creamy markings on the forewing, one or both of which can resemble the figure 8. Greek *dilobos* = 'two lobes', and Latin *caeruleus* = 'sky blue' + Greek *cephale* = 'head', from the putative larval head colour.

FIGURE OF EIGHTY *Tethea ocularis* (Drepanidae, Thyatirinae), wingspan 32–8 mm, a nocturnal moth commonly found in broadleaf woodland, plantations, parks, hedgerows and gardens throughout much of England and parts of Wales. Adults fly in May–July. Larvae feed on leaves of aspen and poplars. A white '80' mark on the forewing, from which the moth gets its name, varies in shape and intensity. The binomial comes from Tethys, mother of the river gods in Greek mythology, and Latin *ocularis* = 'relating to the eye', from the wing markings.

FIGWORT Generally herbaceous perennials in the genus *Scrophularia*, with the belief that the plant cured scrofula (the resemblance of the bulbous shape to swollen glands being explained by the 'doctrine of signatures'), together with *Phygelius* (Greek *phyge* + *helios* = 'flight' + 'sun'), both Scrophulariaceae. 'Figwort' dates to the sixteenth century, from the obsolete *fig* = 'piles' (haemorrhoids) + wort, reputedly being a cure.

Balm-leaved figwort *S. scorodonia*, native and common in the Channel Island and south-west England, naturalised in Hampshire and south Wales, casual in a few other places, on hedge banks, scrubby field borders, rough ground, disused quarries and old walls. *Scorodonia* is from Greek *scordo* = 'garlic'.

Cape figwort *P. capensis*, an evergreen neophyte South African shrub found naturalised by rivers in Co. Wicklow since at least 1970, and with a few records in Britain as a garden escape.

Common figwort *S. nodosa*, common and widespread in damp open or shaded habitats, preferring fertile soils, in woodland, hedge banks, ditches and riversides, sometimes in drier sites on waste ground. *Nodosa* is Latin for 'knotty'.

Green figwort *S. umbrosa*, locally scattered, rhizomatous, in fertile soils by streams and rivers, and in damp woodland. *Umbrosa* is Latin for 'shady'.

Water figwort *S. auriculata*, common and widespread in England and Wales, scattered and locally common in Ireland, scarce in Scotland, in wet habitats on lake, stream and canal margins, and in ditches, marsh and wet woodland. *Auriculata* is Latin for 'ear-shaped'.

Yellow figwort *S. vernalis*, a biennial, sometimes perennial European neophyte present by 1633, scattered and locally naturalised in Britain in woodland clearings, hedge banks and rough and waste ground, usually in shade. *Vernalis* is Latin for 'spring-flowering'.

FILBERT *Corylus maxima* (Betulaceae), a multi-stemmed deciduous neophyte shrub or small tree, introduced from south-east Europe by 1759, grown for its edible nuts in orchards (especially in Kent, the nuts known as Kentish cobs), now with a scattered distribution mostly in England gardens and parks, and as a relic of cultivation. 'Filbert' comes from Middle English *fylberd* and dialect French *noix de filbert* (because it is ripe around 20 August, the feast day of St Philibert). Greek *corylos* = 'hazelnut', Latin *maxima* = 'largest'.

FILICALES The order containing the typical ferns (Latin *filix* = 'fern').

FILICOPHYTINA One of the subdivisions of ferns (Latin *filix* = 'fern' and Greek *phyton* = 'plant').

FILMY-FERN Members of the family Hymenophyllaceae (Greek *hymen* + *phyllon* = 'membrane' + 'leaf', referencing the membranous fronds which are only one cell thick). **Tunbridge filmy-fern** *Hymenophyllum tunbrigense* and **Wilson's filmy-fern** *H. wilsonii* are both local in Ireland and western Britain, the former with records also from East Sussex, on damp rock and tree-trunks. See also Killarney fern under fern.

FINCH Derived from Old English *finc*, passerine birds in the family Fringillidae.

FINGER-GRASS Casual or naturalised species of *Digitaria* and *Paspalum* (Poaceae) occasionally found on waste ground and tips originating from wool shoddy and birdseed.

Water finger-grass *Paspalum distichum*, a stoloniferous perennial warm-temperate neophyte cultivated since 1776, recorded in the wild in 1924, and naturalised on damp, sandy ground below the walls of Mousehole Harbour, Cornwall, since 1971, by a canal in East London since 1984, and in a few other places in England and south Wales. Greek *paspalos* = 'millet', and *di* + *stichos* = 'two' + 'row'.

FINGERNAIL CLAM (or MUSSEL) See orb and pea mussels (Sphaeriidae).

FINGERS 1) See **dead man's fingers** (fungus and sea anemone). 2) **Devil's fingers** (fungus).

FINGERWORT Foliose liverworts in the genera *Kurzia*, honouring the German botanist Wilhelm Kurz (1834–78) and *Lepidozia*, from Latin *lepidus* = 'pretty' (Lepidoziaceae).

Bristly fingerwort *K. pauciflora*, locally common, scattered and widespread in bog, wet heathland, flushes, fen and wet woodland. *Pauciflora* is Latin for 'few flowers'.

Creeping fingerwort *L. reptans*, common and widespread on acidic substrates such as decomposing logs, bark of oak, birch and conifers, and peaty banks. It is locally abundant in scree and on rocky, heathery hill slopes. *Reptans* is Latin for 'creeping'.

Heath fingerwort *K. trichoclados*, scattered in upland regions, mainly in Scotland, in drier parts of bogs, wet heath, flushes, fen and wet woodland. Greek *tricha* + *clados* = 'hair' + 'branch'.

Pearson's fingerwort *L. pearsonii*, locally common in north-west Wales, Cumbria and west Scotland growing through mosses and liverworts in humid woodland, on rocky mountain slopes and on scree. William Pearson (1849–1923) was an English bryologist.

Rock fingerwort *L. cupressina*, locally abundant, upland and oceanic, on acidic rocks, peat, decaying logs and tree bases in humid, rocky woodland. South-easterly British examples are in Millstone Grit scree, while some noth-eastern colonies are on north-facing sandstone crags. *Cupressinus* is Latin for 'resembling cypress'.

Wood fingerwort *K. sylvatica*, scattered in upland regions in drier parts of bogs, wet heath, flushes, fen and wet woodland. *Sylvatica* is from Latin for 'woodland'.

FIR Members of the genus *Abies* (Pinaceae), Latin for 'fir', itself late Middle English, probably from Old Norse *fyri*. See also silver-fir.

Caucasian fir *A. nordmanniana*, introduced in 1848 from the Caucasus as an amenity species, best suited to areas of at least moderately nutrient-rich soil with >900 mm rainfall in western Britain; also self-sown in Surrey and west Wales. It has recently become more frequently planted as an alternative to Norway spruce for Christmas trees because it does not lose its needles indoors so readily. Alexander von Nordmann (1803–66) was a Finnish botanist.

Fraser fir *A. fraseri*, from North America, planted for the Christmas tree trade, mainly in England. John Fraser (1750–1811), who introduced it to Britain in 1811, was a Scottish botanist and collector.

Giant fir *A. grandis*, introduced from North America in 1831 as a parkland ornamental, increasingly used in plantation forestry, especially in the lowlands receiving >1,000 mm rainfall, preferring sheltered, fertile, well-drained soils (though also tolerant of less fertile conditions), and occasionally self-sown.

Grand fir Alternative name for giant fir.

Greek fir *A. cephalonica*, introduced from Greece by General Sir Charles Napier in 1824, quite widely planted as a parkland ornamental, with a few records of self-sown trees. Cephalonia is the largest of the Ionian islands.

Noble fir *A. procera*, introduced from North America in 1830 as a parkland ornamental and minor forestry species, growing best on moist mineral soils of poor nutrient status, mainly in Scotland, and occasionally self-sown. *Procera* is Latin for 'tall'.

Nordmann fir Alternative name for Caucasian fir.

FIREBRAT *Thermobia domestica* (Lepismatidae), 10–15 mm long, a bristletail found in warm indoor locations such as kitchens and bakeries, feeding on carbohydrate material. *Thermobia* incorporates the Greek *thermos* = 'heat'.

FIRECREST *Regulus ignicapilla* (Regulidae), wingspan 13–16 cm, length 8.5–9.5 cm, weight 4–7 g, with the congeneric goldcrest Europe's smallest bird, with breeding populations in south and east England, augmented with continental birds in winter. Information in 2010 indicated 1,000 territories in England, but BBS data for 2016 suggest the New Forest alone may contain 700 territories, with a further 500 elsewhere in Hampshire, so countrywide estimates may be far too low. BTO figures indicate a strong increase by 93% over 2010–15. The nest is a closed cup built in three layers with a small entrance hole near the top: the outer layer is made from moss, twigs, cobwebs and lichen, cobwebs being used to attach the nest to thin branches; the middle layer is moss, in turn lined with feathers and hair. Clutch size is 7–10, incubation takes 15–17 days, and the altricial young fledge in 19–20 days. Feeding is on aphids, springtails and other small insects, caterpillars and spiders (eggs, cocoons and adults). Goldcrests are members of the kinglet family: Latin *regulus* = 'little king' or 'prince', a diminutive of Latin *rex* = 'king'; Latin *ignis* and *capillus* = 'fire' + 'hair'.

FIREFUNGUS Pine firefungus *Rhizina undulata* (Pezizales, Rhizinaceae), inedible, scattered, widespread and uncommon, found from early summer to early autumn, on conifer debris, especially on old fire sites, and parasitic on conifer seedlings, causing 'group dying'. Greek *rhiza* = 'root', and Latin *undulata* = 'undulated'.

FIRETHORN *Pyracantha coccinea* (Rosaceae), an evergreen neophyte European shrub cultivated by 1629, widespread and common in the southern half of Britain, scarcer and scattered elsewhere, as a bird-sown garden escape or relic of cultivation in hedgerows and amenity areas, on roadsides, banks, railways and waste ground, and in walls, pavement cracks and quarries. Greek *pyr* + *acantha* = 'fire' + 'thorn', and Latin *coccinea* = 'scarlet', the berries generally being red.

Asian firethorn *P. rogersiana*, a deciduous neophyte Chinese shrub grown since 1911 in gardens, naturalised as bird-sown, scattered in England on road and riverbanks, on woodland edges, at the base of walls, in rough grassland and in gravel-pits.

FISH A limbless, cold-blooded vertebrate with gills and fins living in water, including lampreys, and cartilaginous and bony fish. Taxonomists do not considered these animals a formal taxonomic grouping in systematic biology: they are paraphyletic, that is, any clade containing all fish also contains the tetrapods, which are not fish. For this reason, the class Pisces, used in older references, is no longer used.

FISH EAGLE In the UK context another term for osprey.

FISSIDENTACEAE Family name for pocket-mosses (Latin *fissus* + *dens* = 'split' + 'tooth').

FLAG-MOSS *Discelium nudum* (Disceliaceae), scattered but locally common in lowland northern England on silty, often vertical banks, usually by rivers, lakes and other water bodies.

FLAGELLATE An organism with one or more whip-like organelles used for motion.

FLAME *Axylia putris* (Noctuidae, Noctuinae), wingspan 27–32 mm, a nocturnal moth widespread and common on farmland, downland and heathland, and in hedgerows, woodland edges and gardens. Adults fly in June–July, sometimes with a partial second generation in the autumn in the south. Larvae feed on low-growing plants. Greek *xylon* = 'cut wood', and Latin *puter* = 'rotten', from the forewing pattern's resemblance to decaying plankwood.

FLAME SHELL *Limaria hians* (Limidae), shell size up to 3 cm, a clam found in coastal sea beds of the north-east Atlantic, particularly in the sublittoral (5–30 m depth) off western Scotland, creating high-biodiversity reefs (important nursery sites for crustaceans, scallops and fish), of which Loch Alsh, between Skye and the mainland, is probably the world's largest. Loch Carron, just north of Loch Alsh, was heavily dredged in 2017, devastating the reef which will take decades to recover. Latin *limus* = 'mud', and *hians* = 'opening'.

FLAPWORT Foliose liverworts, including species of *Jungermannia* (Jungermanniaceae), named to honour the German botanist Ludwig Jungermann (1572–1653).

Rarities include **arctic flapwort** *Jungermannia polaris* (Highlands), **autumn flapwort** *Jamesoniella autumnalis*, **book flapwort** *Nardia breidleri* (Highlands), **brown flapwort** *Odontoschisma elongatum* (Highlands and Islands), **delicate flapwort** *Ju. caespiticia* (mainly Pennines), **horned flapwort** *Lophozia longidens* (Highlands), **kidney flapwort** *Ju. Confertissima* (mainly Highlands), **long-leaved flapwort** *Ju. leiantha* (north Yorkshire, Pennines and Cumbria), **Macoun's flapwort** *O. macounii* (Ben Heasgarnich and elsewhere in the western Highlands), **northern flapwort** *Ju. Borealis* (Highlands, and a few Welsh records), and **two-lipped flapwort** *Ju. subelliptica* (mainly Highlands).

Anomalous flapwort *Mylia anomala* (Myliaceae), locally common and widespread, except in central and eastern England, on bog and wet heathland growing in sphagnum hummocks and eroded peat surfaces and cuttings. Greek *myle* = 'molar' and Latin *anomala* = 'irregular'.

Bog-moss flapwort *Odontoschisma sphagni* (Cephaloziaceae), one of the commonest foliose liverworts on sphagnum hummocks bogs. It also grows on wet peat, and is sometimes found on heathland and in upland woodland. Greek *odontos* + *schisma* = 'tooth' + 'split', and Latin for 'of sphagnum' (moss).

Compressed flapwort *N. compressa*, in wet acidic rocks in upland streams, occasionally in sand and gravel.

Cordate flapwort *Ju. exsertifolia* ssp. *cordifolia*, widespread, locally common on upland stony flushes and stream rocks. Latin *exsertus* + *folium* = 'protruding' + 'leaf'.

Crenulated flapwort *Ju. gracillima*, common in northern and western Britain and in Ireland, and widespread in acidic habitats in south-east England, often as a pioneer of base-poor soil, benefiting from disturbance by trampling or weathering, and characteristic of soil on banks, paths, ditches and cliffs. *Gracillima* is Latin for 'delicate'.

Dark-green flapwort *Ju. atrovirens*, widespread in upland calcareous habitats, on rock, tufa or soil, often abundant on damp limestone or sandstone cliffs, or in damp, calcareous grassland. *Atrovirens* is Latin for 'dark green'.

Dwarf flapwort *Ju. pumila*, common and widespread, mainly in the uplands on damp rock, sometimes on soil, and on boulders and stones in flushes. *Pumila* is Latin for 'dwarf'.

Earth-cup flapwort *N. geoscyphus*, widespread and scattered, mostly in Britain, on mildly base-rich calcareous cliffs, and on damp, gravelly soil in disused quarries. *Geoscyphos* is Greek for 'earth cup'.

Egg flapwort *Ju. obovata*, common and widespread in upland valleys on stream rocks. *Obovata* is Latin for 'egg-shaped'.

Great mountain flapwort *Harpanthus flotovianus* (Geocalycaceae), mainly found in Highland bogs. Greek *harpe* + *anthos* = 'hook' + 'flower'.

Hooker's flapwort *Haplomitrium hookeri* (Haplomitriaceae), thallose, on damp, gravelly ground in the uplands, often where shallow water lies in winter. It is also sometimes found in dune slacks, in corries and below late-lying snow. Greek *haplo* + *mitrion* = 'simple' + 'small cap'. William Jackson Hooker (1785–1865) was an English botanist.

Ladder flapwort *N. scalaris*, common and widespread, usually on disturbed damp, acidic, mineral soil, for example on banks by gravel tracks, and forming dense cushions under late-lying snow. *Scalaris* is Latin for 'of a ladder'.

Large-celled flapwort *L. capitata*, uncommon, scattered in England, first recorded from a disused brick pit in Chawley, Berkshire in 1948. *Capitata* is Latin for 'headed'.

Marsh flapwort *Ja. undulifolia*, scattered in western Britain. Sites in Scotland are in sphagnum hummocks on the margins of base-rich flushes near the coast. Elsewhere it is found on raised bogs or valley mire. *Undulifolia* is Latin for 'wavy-leaved'.

Matchstick flapwort *O. denudatum*, widespread but generally uncommon in upland Britain and southern English heaths, especially on decomposing logs and on tree-trunks in humid woodlands, on peaty soil on heathland and moorland, and in dried-out parts of bogs. *Denudatum* is Latin for 'naked', here meaning 'without leaves'.

Round-fruited flapwort *Ju. sphaerocarpa*, locally common and widespread, mainly upland, on damp ledges by streams and on rock ledges, favouring sandstone and shale. *Sphaerocarpa* is Latin for 'round-fruited'.

Shining flapwort *Ju. paroica*, common and widespread on wet upland rock ledges and stream rocks.

Stipular flapwort *Harpanthus scutatus* (Geocalycaceae), with a preference for damp rocks, especially sandstone, but also on rotten logs and peat. It is most frequent in the upland west, but is also recorded from southern England. *Scutatus* is Latin for 'with shield'.

Taylor's flapwort *Mylia taylorii* (Myliaceae), common and widespread in humid and sheltered upland sites on rocks and tree bases, heathy banks in ravines, bogs and scree. *Mylia* is from

Greek *myle* = 'mill' or 'grinder'. Thomas Taylor (1775–1848) was an English botanist.

Transparent flapwort *Ju. hyalina*, widespread and scattered on rock ledges and stream rocks, especially in western and northern Britain. *Hyalina* is Greek for 'glass' (i.e. 'transparent').

Wedge flapwort *Leptoscyphus cuneifolius* (Lophocoleaceae), growing in Scotland and Ireland in humid woodlands, usually on birch trunks. Greek *leptos* + *scyphos* = 'slender' + 'cup', and Latin *cuneifolius* = 'wedge-shaped leaves'.

FLASHER *Phallus duplicatus* (Phallales, Phallaceae), an inedible fungus, one of the stinkhorns, found on soil in a mixed plantation at Moor Piece, mid-West Yorkshire, new to Britain in 2014. Greek *phallos* = 'penis', from the shape, and Latin *duplicatus* = 'doubled'.

FLASK-FUNGUS Pyromycete fungus with small, flask-shaped fruit bodies sometimes enclosed within a protective stroma, for example candlesnuff fungi.

FLAT-BODY (MOTH) Nocturnal moths in the family Depressariidae, mainly in the genera *Depressaria* (Latin *depressus* = 'flat' or 'pressed down', from the flat abdomen and flattened wings when at rest) and *Agonopterix* (Greek *agonios* + *pteryx* = 'non-angled' + 'wing'). Wingspans are generally 17–22 mm.

Locally common or scarce species on grassland and waste ground include **black-spot flat-body** *A. propinquella*, **carline flat-body** *A. nanatella* (calcareous grassland), **carrot flat-body** *D. douglasella*, **chervil flat-body** *D. sordidatella*, **coastal flat-body** *A. yeatiana*, **Cornish flat-body** *A. kuznetzovi* (Cornish clifftops), **fuscous flat-body** *A. capreolella* (calcareous grassland in the Isle of Wight, Kent, Co. Clare and Co. Galway), **Highland flat-body** *D. silesiaca* (Aberdeenshire, Inverness-shire and Morayshire), **lemon flat-body** *Hypercallia citrinalis* (now extinct in Britain, last recorded in Kent in 1979, but present in the Burren), **mugwort flat-body** *Exaeretia allisella* (west and north Midlands, northern England, north Wales, scattered in south-east England, Norfolk, Lincolnshire, and parts of Scotland), **pale flat-body** *A. pallorella* (calcareous grassland in south and south-east England, south Wales and western Ireland), **small purple flat-body** *A. purpurea* (downland and dry grassland north to Lancashire, occasional in Ireland), **straw flat-body** *A. kaekeritziana* (downland), and **thistle flat-body** *A. carduella*.

Angelica flat-body *A. angelicella*, often common in fen, marsh and damp woodland, and by riverbanks throughout much of England, Wales, south Scotland, and south-west and north-east Ireland. Adults fly in August–September. Larvae feed, often gregariously, among silk-spun leaves or shoots of wild angelica and hogweed in May–June.

Brindled flat-body *A. arenella*, widespread and common in habitats that include rough grassland, waste ground and gardens. Adults fly in April–May and August–October, the latter generation then overwintering. Larvae feed on thistles, knapweed, burdock and saw-wort, inside a rolled leaf. *Arenella* is from Latin *arena* = 'sand', from the forewing colour.

Broom flat-body *A. scopariella*, local in heathland and open woodland throughout much of Britain, and with one record from Ireland. Adults fly and overwinter from August to April. Larvae feed on broom from May to July between shoots spun together with silk. *Scoparius* is the specific name of broom.

Brown flat-body *D. badiella*, widespread but local in dry habitats such as sandhills and quarries. Adults fly in July–September. Larvae occur in May–July, feeding on cat's-ear, perennial sowthistle and dandelion, initially between spun leaves, later among the roots. *Badius* is Latin for 'chestnut brown', from the forewing colour.

Brown-spot flat-body *A. alstromeriana*, widespread and common on verges, waste ground, marsh and riverbanks, except for the far north, and scarce in Ireland. Adults are active from August to April. Larvae feed on hemlock flowers and leaves in May–July inside a spun or rolled leaf. Baron Clas Alströmer (1736–94) was a Swedish naturalist.

Common flat-body *A. heracliana*, wingspan 15–21 mm, widespread and common in gardens, waste ground, woodland edges and other habitats. Adults fly in February–April (after overwintering) and again in July–August. Larvae feed from May to July on cow parsley, rough chervil and hogweed (*Heracleum*, giving *heracliana*), living inside a rolled leaf.

Dark-fringed flat-body *A. nervosa*, widespread and common on heathland and grassland, and in open woodland. Adults are active in July–September, a few overwintering. Larvae feed in May–June on shoots of broom, gorse, etc. inside a silk web. *Nervus* is Latin for 'vein', from forewing markings.

Dawn flat-body *Semioscopis steinkellneriana*, local in woodlands, hedgerows, scrub and occasionally gardens across much of Britain. Males fly on sunny mornings in April. Larvae feed inside a folded leaf of blackthorn, hawthorn and rowan. Greek *semeioscopos* = 'diviner', from runelike markings on the forewing. Professor Steinkellner was an eighteenth-century Austrian priest and entomologist.

Dingy flat-body *D. daucella*, wingspan 21–4 mm, widespread and locally common in ditches, marsh, riverbanks, water meadow and damp moorland. Adults fly in September–April. Larvae initially mine the stems of water-dropwort and related plants, then move externally to the flowers and seed heads. *Daucella* comes from wild carrot *Daucus*, though this is not a food plant.

Dusted flat-body *A. assimilella*, local on heathland and in open woodland throughout much of Britain, scattered and rare in Ireland. Adults fly in April–June. Larvae feed on broom, initially inside stems then between stems, linked with silk. *Assimilis* is Latin for 'similar' (i.e. to other species).

Early flat-body *S. avellanella*, wingspan 20–6 mm, local in ancient small-leaved lime woodland in the southern half of England, and in mature birch woodland in Scotland and the northern half of England. Adults fly in March–April. Larvae feed inside a folded leaf of small-leaved lime, birch or hornbeam, though in Britain not hazel *Corylus avellana* which gives the specific name.

Estuarine flat-body *A. putridella*, a rarity found on estuarine saltmarsh in Kent and Essex. Adults fly in June–August. Larvae feed on hog's fennel in May–June inside a rolled leaf. *Putridus* is Latin for 'decaying', referring to the wing colour of light brown dead wood.

Fen flat-body *D. ultimella*, local in ditches, marsh and water meadow, and on riverbanks in southern through central England as far as Lancashire and Yorkshire; also in Wales and southern and eastern Ireland. Adults fly in August, overwinter, and reappear in spring. Larvae feed on stems and leaves of fool's-water-cress, greater water-parsnip and water-dropworts. *Ultimus* is Latin for 'furthest', from the dark terminal area on the forewing.

Gorse flat-body *A. umbellana*, wingspan 19–24 mm, widespread but local on heathland and moorland. Adults are active from August, hibernating, then reappearing in early spring. Larvae feed in May–August on gorse and hairy greenweed, inside a silk tent. *Umbella* is Latin for 'sunshade'.

Greenweed flat-body *A. atomella*, on rough grassland in southern England, from Dorset as far north as Worcestershire and Cambridgeshire, then again from Cheshire northwards to Northumberland; in Wales it has been found in Pembrokeshire and Monmouthshire. Adults fly in July–August. Larvae feed on dyer's greenweed inside a rolled leaf. This scarce and declining species has recently been added to the UK BAP. *Atomus* is Latin for 'speck', from the forewing markings.

Large carrot flat-body *A. ciliella*, wingspan 19–24 mm, widespread and common in rough grassland, hedgerows, waste ground and woodland edges. Adults fly in August–May. Larvae feed in June–September on umbelliferous plants inside a rolled leaf. *Cilium* is Latin for 'small hair', referencing the wing fringes.

Large purple flat-body *A. liturosa*, local in England, Wales and south Scotland, and only rarely recorded on the Irish west and south-east coasts. Adults fly in July–August. Larvae feed on St John's-wort between leaves spun together with silk. *Litura* is Latin for 'smear', from the smudged markings on the forewing.

Long-horned flat-body *Carcina quercana*, widespread and common in woodland, hedgerows, scrub and gardens. Adults fly in July–August. Larvae feed on leaves of a variety of deciduous trees, living between leaves spun together with silk. Greek *carcinos* = 'crab' (reason obscure), and Latin *quercus* = 'oak'.

Mountain flat-body *Levipalpus hepatariella*, wingspan 20–5 mm, found in a few scattered localities in the Highlands and the Isle of Coll, Inner Hebrides. Males fly in August–September. Larvae feed on mountain everlasting in silk tubes that extend into the soil, providing a defensive retreat following disturbance. Latin *levis + palpus* = '(s)light' + '(labial) palpus', and *hepatarius* = 'relating to the liver', from the forewing colour.

Pignut flat-body *D. pulcherrimella*, wingspan 15–19 mm, local in flower meadows and chalk grassland in Britain, from Kent to Cornwall along the south coast and less frequently in the Midlands, Wales and northern England; more frequent throughout much of Scotland; and in western Ireland. Adults fly in June–September. Later instars of larvae feed on flowers and seed heads of pignut, wild carrot and burnet-saxifrage. *Pulcherrimus* is Latin for 'very beautiful'.

Pimpinel flat-body *D. pimpinellae*, scarce, from Kent to Cornwall, and from the west Midlands to Cheshire, Lancashire and Yorkshire. Adults begin to hatch in September and overwinter, appearing the following spring. Larvae feed on young leaves of burnet-saxifrage *Pimpinella*, hence the specific name, on chalk and limestone grassland, and greater burnet-saxifrage along roadsides, hedgerows and in woodland, subsequently feeding on the flowerheads.

Powdered flat-body *A. curvipunctosa*, wingspan 15–17 mm, a rarity found in dunes, hedgerows and verges. Mainly recorded from Suffolk, but also from Kent, Somerset, Derbyshire and Hampshire. Adults fly from August, hibernate, then reappear in April–May. Larvae feed on chervils and cow parsley, living between flowers and leaves spun together with silk. Latin *curvus + punctum* = 'curved' + 'spot' describes the forewing pattern.

Red-letter flat-body *A. ocellana*, widespread and common in a range of habitats, including woodland, hedgerows, parks, gardens and at freshwater margins. Overwintering as adults, these are often seen in February–May and July–September. Larvae feed on willows between leaves spun together with silk. *Ocellus* is Latin for 'small eye', from the forewing marking.

Rolling carrot flat-body *A. rotundella*, wingspan 14–17 mm, a rarity local on downland and coastal cliffs in England and Wales, from Cornwall to Kent, northwards along the west coast to Cheshire; the Isle of Man, Yorkshire and Northumberland; scarce in Wales and Scotland. Adults live in September–May, overwintering in low vegetation. Larvae feed June–August on wild carrot in a folded leaf sewn with white silk. *Rotundus* is Latin for 'round', from the wing edge.

Ruddy flat-body *A. subpropinquella*, widespread and common on waste ground, hedgerows and verges, scarcer in Scotland and Ireland. Adults emerge from August, hibernate over winter, then reappear up to May. Larvae feed on knapweed and thistles inside a rolled leaf. Latin *sub + propinquus* = 'near' + 'close to' (i.e. to another species).

Sallow flat-body *A. conterminella*, local in damp woodland and scrub, fen, and marsh and by riverbanks in England, Wales, south Scotland and south-west Ireland. Adults fly in August–September. Larvae feed on the terminal shoots of willows in May–June, living in a silk web. *Conterminus* is Latin for 'bordering on' (i.e. similar to another species).

Sanicle flat-body *A. astrantiae*, scarce, in ancient woodland on calcareous soils at scattered sites in Britain, from Gloucestershire to Sussex and from Yorkshire to north Wales, with one record from east Ireland. Adults fly in August–September. Larvae feed on sanicle between shoots spun together with silk in May–June. *Astrantia* is another larval food plant.

Scotch flat-body *Exaeretia ciniflonella*, wingspan 20–3 mm, rare and endangered, in birchwood at Rannoch (Perthshire), Glen Affric (Inverness-shire) and Deeside (Aberdeenshire). Males fly in July–August. Larvae feed on birch between leaves spun together with silk. Latin *cinis* + *flo* = 'ash' + 'blow', probably from the grey wing colour.

Sea-holly flat-body *A. cnicella*, scarce, local in south and east coastal parts of England northwards to Yorkshire, favouring shingle and dune where the larval food plant, sea-holly, is found. Adults fly in June–July. Latin *cnicus* = 'safflower', which yields a red dye, the forewing colour.

Sloe flat-body *Luquetia lobella*, local in hedgerows and scrubland throughout much of southern England and East Anglia. Males fly in June. Larvae feed on blackthorn during autumn, between leaves spun together with silk. Latin *lobos* = 'lobe', from the ovate forewing.

Streaked flat-body *D. chaerophylli*, scarce, local in mature hedgerows, woodland edges and sunken lanes in southern England, the Midlands and East Anglia, with records north to Durham, and also in south Wales. Adults fly in July–August, hibernate, then reappear in spring up to April. Larvae feed on rough chervil (*Chaerophyllum*, hence the specific name) between flowers spun together with silk.

Twin-spot flat-body *A. bipunctosa*, on downland, heathland and damp grassland. Originally thought to be restricted to Dorset, Hampshire and the Isle of Wight, but after 1978 found in scattered localities in southern and eastern England as far north as Norfolk, and in Derbyshire, and in west and north Wales. Adults fly in July–September. Larvae feed on saw-wort inside a rolled leaf. *Bipunctosa* is Latin for 'two-spotted', one spot on each forewing.

White-spot flat-body *D. albipunctella*, scarce, in grassland and scrub in southern England, parts of the south Midlands and Suffolk, Lincolnshire, Yorkshire and Lancashire, and south Wales. Larvae feed on wild carrot, upright hedge-parsley, cow parsley and rough chervil, between leaves spun together with silk. *Albipunctella* is Latin for 'white-spotted'.

Yarrow flat-body *D. olerella*, scarce in dry grassland, heathland and dry sandy areas, with a disjunct distribution: at Woolmer Forest in Hampshire, in the Breckland district of Norfolk, and along the north-west edge of the Cairngorms at Rothiemurchus. Larvae feed on yarrow between leaves spun together with silk. *Oleris* is Latin for 'edible plants'.

FLAT-BROCADE MOSS *Platygyrium repens* (Pylaisiadelphaceae), locally common in the southern half of England and south-east Scotland on bark, especially in damp woodland, favouring willow and alder. Greek *platys* + *gyros* = 'flat' + 'round', and Latin *repens* = 'creeping'.

FLAT-FOOTED FLY Yellow flat-footed fly *Agathomyia wankowiczii* (Platypezidae), discovered in Britain in 1990, recorded from six sites in eastern England, seen on the surface of tree leaves, especially at woodland edges and on sunny spots in dappled light. It is the only fly species that can create a gall on a fungus, artist's bracket. Greek *agathos* + *myia* = 'excellent' + 'fly'.

FLAT-SEDGE *Blysmus compressus* (Cyperaceae), a rhizomatous perennial, locally common but declining in England, in marsh and fen, and in sedge-rich, damp grassland, calcareous seepages and stream borders. Greek *blysma* = 'a bubbling up', and Latin *compressus* = 'compressed'.

Saltmarsh flat-sedge *B. rufus*, common in west Scotland, locally frequent further south to Glamorgan and on the east coast south to Lincolnshire, as well as much of Ireland, in creeks and depressions in saltmarsh, brackish ditches and dune slacks. It also occurs on rocky shores and in freshwater seepages. *Rufus* is Latin for 'red'.

FLATBUG Or barkbug; members of the family Aradidae (Heteroptera), extremely flattened and suited to living under

bark, often associated with dead trees and rotting wood, generally feeding on fungi. There are seven species in two genera in the UK.

Pine flatbug *Aradus cinnamomeus*, length 3.5–4.5 mm, local in heathland in southern England where, untypically for the usually fungus-eating flatbugs, it feeds on young Scots pine tree sap. Greek *arados* = 'rattling', and Latin *cinnamomeus* = 'cinnamon-coloured'.

FLATWORM Members of the phylum Platyhelminthes.

Candy stripe flatworm *Prostheceraeus vittatus* (Euryleptidae), widespread, under stones on mud in the low intertidal and shallow subtidal. Greek *prosthe* + *ceraia* = 'in front of' + 'projection' or 'horn', and Latin *vittatus* = 'striped.'

New Zealand flatworm *Arthurdendyus triangulatus* (Geoplanidae), probably introduced from New Zealand with plant material in the 1960s, mainly found in cooler and wetter parts – Scotland, northern England and Ireland – where since the 1990s it has become locally abundant and widespread, but it has also been reported from central and southern England. At rest it is 1 cm wide and 6 cm long but when extended can be 20 cm long and proportionally narrower. It potentially poses a threat to its prey, native earthworms, and further spread could impact wildlife dependent on earthworms (e.g. badgers and moles), and even have a local deleterious effect on soil structure. Arthur Dendy (1865–1925) was an English zoologist.

FLAX 1) *Linum usitatissimum* (Linaceae), or linseed, an annual neophyte cultivated since at least the thirteenth century for its fibre (producing linen), increasingly grown since the 1990s for linseed oil, widespread in lowland regions, naturalised or a casual on verges, tips and waste ground where it is sometimes derived from birdseed. The binomial is Latin for 'flax' and 'most useful'.

Fairy flax *L. catharticum*, a locally common and widespread annual or biennial of dry, infertile calcareous or base-rich substrates, but also found in seepage sites. It occurs in calcareous grassland, moorland, mires and flushes, in fen-meadows, on outcrops of basic rock, road cuttings, quarry spoil and lead-mine debris, occasionally on dry heathland. *Catharticum* is Latin for 'purgative'.

Pale flax *L. bienne*, an annual, biennial or short-lived perennial, scattered but locally common in south and south-west England, south Wales and south-east Ireland, on dry grassy habitats and grassland-scrub mosaics, often near the sea, including field margins, roadsides and railway banks. It favours warm, sheltered, south-facing slopes and relatively infertile, drought-prone soils.

Perennial flax *L. perenne*, perennial, with a local, scattered distribution, mainly in eastern England, in open, well-drained locations on base-rich substrates, including lightly grazed calcareous grassland, dry banks and roadsides.

2) **New Zealand flax** *Phormium tenax* (Xanthorrhoeaceae), an evergreen perennial neophyte, persistent on coastal rocky sites, scattered in England (especially the south-west and in Lancashire), Isle of Man, and south-west and west Ireland, sometimes planted as a windbreak. Greek *phormion* = 'basketwork' or 'mat', and the Latin *tenax* = 'strong'.

FLEA Wingless insects, 1–6 mm in length, in the order Siphonaptera with flattened bodies (lateral compression) facilitating movement through fur or feathers using strong claws, the adults being ectoparasitic on mammals and birds, and are haemophagic (feeding on blood), with mouthparts having piercing stylets adapted for bloodsucking. Long hind legs facilitate jumping ability, human fleas able to jump a distance of >30 cm, over 200 times their body length. There are around 60 species on the British list, including introductions. Families are Ceratophyllidae (mostly on birds), Hystrichopsyllidae (mole flea), Ischnopsyllidae (on bats) and Pulicidae. Some species have general English appellations rather than truly specific names, for example *Typhloceras poppei* (Histrichopsyllidae), known as 'a wood mouse flea'. Most flea species use a number of host species; any specificity is more to do with host habitat and nesting habits. New hosts are generally found by detecting their body warmth, but vibration and raised carbon dioxide (breath) levels are also used. Larvae feed on detritus. See also **snow flea** (Mecoptera) and **water flea** (Cladocera). For **lucerne flea** see clover springtail under springtail.

Badger flea *Paraceras melis* (Ceratophyllidae), common and widespread, parasitic on fox, dog, cat, polecat, mole and fallow deer, though primarily on badger. It is the vector of *Trypanosoma pestanai*, the causal agent of a protozoan disease of badgers. *Melis* is Latin for 'badger'.

Cat flea *Ctenocephalides felis* (Pulicidae), adults 3 mm long, larvae 2 mm, also parasitic on dogs. These fleas can transmit diseases such as the bacterium *Bartonella*, murine typhus and apedermatitis; the tapeworm *Dipylidium caninum* can be transmitted when an immature flea is swallowed by pets or humans. Greek *ctenos* + *cephale* = 'comb' + 'head', and Latin *felis* = 'cat'.

Crow's nest flea *Ceratophyllus rossittensis* (Ceratophyllidae), scarce in England and Ireland, first recorded in Britain in the late 1940s. Greek *ceras* + *phyllon* = 'horn' + 'leaf'. Rossiten is a bird observatory near Klaliningrad, Russia.

Dog flea *Ct. canis*, adult length 1–4 mm, larvae up to 5 mm. The tapeworm *Dipylidium caninum* can be transmitted when an immature flea is swallowed by pets or humans. Latin *canis* = 'dog'.

Feral pigeon nest flea *Ce. columbae*, common and widespread, and can affect humans. Latin *columba* = 'pigeon'.

Grey squirrel flea *Orchopeas howardi* (Ceratophyllidae), common and widespread in England and Wales, scarce in Scotland. Greek *orchis* + *peos* = 'testicle' + 'penis'.

Hedgehog flea *Archaeopsylla erinacei* (Pulicidae), length 2–3.5 mm, common and widespread, occasionally found on foxes and dogs as well as hedgehogs, which can be heavily infested with up to a thousand individuals. Greek *archaeos* + *psylla* = 'ancient' + 'flea', and Latin *erinaceus* = 'hedgehog'.

Hen flea *Ce. gallinae*, 1–8 mm long, common and widespread. Latin *gallina* = 'hen'.

House martin flea *Ce. hirundinis* and *Ce. rusticus*, common and widespread. *Hirundo* is Latin for 'swallow', *rusticus* for 'countryside'.

House sparrow nest flea *Ce. fringillae*, scattered in Britain. Latin *fringilla* = 'finch'.

Human flea *Pulex irritans* (Pulicidae), 1–4 mm long, common and widespread, primarily a parasite of foxes, badgers and other hole-dwelling mammals, using humans when they began using caves. They are a vector of *Yersinia pestis* (plague). The binomial is Latin for 'flea' and 'irritant'.

Long-eared bat flea *Ischnopsyllus hexactenus* (Ischnopsyllidae), scattered and widespread. Greek *ischnos* + *psylla* = 'slender' + 'flea', and *hexi* + *ctenos* = 'six' + 'comb'.

Manx shearwater flea *Ce. fionnus*, very rare, recorded from the Hebrides.

Mole flea *Hystrichopsylla talpae* (Hystrichopsyllidae), up to 6 mm long, common and widespread. Greek *hystrix* + *psylla* = 'porcupine' + 'flea', and Latin *talpa* = 'mole'.

Moorhen flea *Dasypsyllus gallinulae* (Ceratophyllidae), common and widespread. Greek *dasys* + *psylla* = 'hairy' + 'flea', and Latin *gallinula* = 'hen'.

Noctule bat flea *I. elongatus*, scarce and scattered in England and Wales.

Rabbit flea *Spilopsyllus cuniculi* (Pulicidae), common and widespread, the main vector of the myxoma virus, adult females more or less permanently attaching themselves to the rabbit host, generally on the ear. Greek *spilos* + *psylla* = 'blemish' + 'flea', and Latin *cuniculus* = 'rabbit'.

Rat flea *Xenopsylla cheopis* (Pulicidae), scarce with records from a few English ports, vector of the bubonic plague and murine virus. Greek *xenos* + *psylla* = 'stranger' + 'flea'. Cheops was the pharaoh who built the Great Pyramid.

Red squirrel flea *Tarsopsylla octodecimdentata* (Ceratophyllidae), with occasional records from the Highlands and Northumberland, from nests rather than the squirrel itself. Greek *tarsos* + *psylla* = 'flat surface' + 'flea', and Latin *octodecimdentata* = 'eighteen-toothed'.

Serotine bat flea *I. intermedius*, scarce and scattered in England and Ireland.

Starling nest flea Alternative name for house sparrow nest flea.

FLEA BEETLE Members of the family Chrysomelidae, subfamily Galerucinae, with the hind femora modified and enlarged for jumping, and generally 1.3–2.5 mm in length.

Barley flea beetle *Phyllotreta vittula*, widespread but local in southern England, scarce in Wales, feeding on brassicas and grasses, adults on leaves, larvae in roots. Adults overwinter in moss, leaf litter and decomposing wood. Greek *phyllon* + *tretos* = 'leaf' + 'pierced', and Latin *vittula* = 'little ribbon'.

Beet flea beetle Alternative name for mangold flea beetle.

Belladonna flea beetle *Epitrix atropae*, scarce, with a scattered distribution in southern and eastern England, adults feeding on leaves, larvae on roots of various Solanaceae, especially deadly nightshade. Adults overwinter in grassy borders of ditches and hedgerows. Greek *epi* + *thrix* = 'upon' + 'hair', referring to the setae on the elytra. Atropos was one of the Fates in Greek mythology.

Bladderwort flea beetle *Longitarsus nigerrimus*, found in the New Forest and the Isle of Wight, with a few records elsewhere in England and south-west Ireland, Endangered in RDB 1. It is found in shallow boggy pools and peat bogs where (especially lesser) bladderwort occurs, adults feeding on leaves and stems above the water level, larvae on partially or wholly submerged plant parts. *Longitarsus* is Latin for 'long tarsus' and 'very black'.

Cabbage-stem flea beetle *Psylliodes chrysocephala*, 3–4.5 mm length, common and widespread, especially in England. Adults feed on leaves and possibly pollen of wild and cultivated brassicas. Larvae are stem miners and are also found in seed pods and leaf stalks. Greek *psylla* + *odes* = 'flea' + 'similar to', and *chrysos* + *cephale* = 'gold' + 'head'.

Chrysanthemum flea beetle *L. succineus*, common and widespread, mostly coastal in Ireland, adults feeding on leaves of Asteraceae, larvae in soil at the roots of common ragwort. *Succineus* is Latin for 'amber-coloured'.

Clover flea beetle Alternative name for corn flea beetle.

Corn flea beetle *Chaetocnema hortensis*, common and widespread, adults feeding on the leaves, larvae in the stems of various grasses (including cereals, sometimes in such numbers as to become a pest). Greek *chaeta* + *cneme* = 'bristle' + 'knee', and Latin *hortus* = 'garden'.

Flax flea beetle *L. parvulus*, common and widespread in much of England, more scattered in the north and in Wales. Adults feed on leaves, larvae on roots of flax, often causing economically significant damage to seedlings of commercial crops. *Parvulus* is Latin for 'small'.

Heath speedwell flea beetle *L. longiseta*, with a very few records from south-east England, found in woodland clearings, shady grassland and fallow fields, vulnerable to scrub encroachment, and limited by the distribution of its food plant, heath speedwell, adults feeding on leaves, larvae probably on roots. *Longiseta* is Latin for 'long bristle'.

Hop flea beetle *Ps. attenuata*, 2–2.8 mm length, scarce, with a scattered distribution, mainly recorded from south-east England, Endangered in RDB 1. Adults feed on hop leaves, flowers and cones, and on nettles; larvae feed on roots. *Attenuata* is Latin for 'thinned'.

Horsetail flea beetle *Hippuriphila modeeri*, widespread but scattered in damp habitats, adults feeding on leaves and stems, larvae on young stems of horsetails. Adults overwinter in moss and grass tussocks. Greek *hippouris* + *philos* = 'horsetail' + 'loving'. Adolph Modéer (1738–99) was a Swedish naturalist.

Iris flea beetle *Aphthona nonstriata*, 2.6–2.8 mm length, common and widespread. Adults eat strips from yellow iris leaves; larvae are leaf miners of iris. Greek *aphthonos* = 'plentiful', and Latin *nonstriata* = 'non-striped'.

Large flax flea beetle *A. euphorbiae*, widespread and locally common in England, except for the north, and in Wales, occasional in Ireland, a pest of flax and linseed, adults being active in spring and summer. Adults are also found on leaves of apples, strawberry and other plants. Larvae feed on roots. *Euphorbia* is one of the putative food plants.

Large striped flea beetle *Ph. nemorum*, 2.4–3.5 mm length, locally common and widespread in England and Wales, less common in Scotland, with one record from south Ireland. It is a pest of brassicas, including turnips (hence an alternative name, turnip flea beetle), adults and larvae (as miners) feeding from spring to late autumn. Latin *nemus* = 'glade'.

Lundy cabbage flea beetle *Psylliodes luridipennis*, 3 mm length, larvae mining the leaves of Lundy cabbage, a priority species in the UK BAP. Latin *luridus* + *penna* = 'pale yellow' + 'wing'.

Mallow flea beetle *Podagrica fuscipes*, 3–6 mm length, scarce, with a few records from southern England and the east Midlands, most from around the Thames Estuary, adults feeding on leaves of mallow, larvae on roots. Greek *podagrica* = 'swollen foot' and Latin *fuscipes* = 'brown'.

Mangold or **mangel flea beetle** *C. concinna*, common and widespread, especially in England and Wales, in a range of habitats, adults feeding on leaves of non-woody hosts and sometimes pollen, larvae on roots, and sometimes a pest of sugar beet. *Concinna* is Latin for 'neat'.

Mint flea beetle *L. ferrugineus*, with recent records only from Suffolk and Essex, and Endangered in RDB 1, in damp habitats feeding on mints, sometimes gypsyworts and germanders, adults on leaves, larvae on roots. *Ferrugineus* is Latin for 'rust-coloured'.

Moss flea beetle *Mniophila muscorum*, 1–1.5 mm length, widespread but scarce and scattered, the flightless adults and larvae (probably) feeding on mosses. Greek *mnion* + *philos* = 'moss' + 'loving', and Latin *muscus* = 'moss'.

Potato flea beetle *Ps. affinis*, 2.2–2.9 mm length, common and widespread in England and Wales, local and scattered in Ireland. Adults feed on leaves and stems of potato and wild Solanaceae; early instar larvae are root miners, final instar larvae feed externally on roots. *Affinis* is Latin for 'related'.

Ragwort flea beetle *L. jacobaeae*, 2.5–3.3 mm length, widespread and locally common, if mainly coastal in Ireland, adults feeding on leaves of ragwort (*Senecio jacobaea*, giving the specific name), larvae on root crowns and petioles of lower leaves.

Small striped flea beetle *Ph. undulata*, 2–2.8 mm length, widespread and common, though more scattered in Scotland and Ireland, feeding on wild and cultivated brassicas, adults on leaves and possibly pollen, larvae in roots. Adults overwinter in grass tussocks, under bark, in moss and in rotting logs. *Undulata* is Latin for 'waved'.

Striped flea beetle *Ph. striolata*, scarce and scattered in England and Wales, feeding on wild and cultivated brassicas, adults on leaves, larvae in roots and below-ground stems. *Striolatus* is Latin for 'finely lined'.

Turnip flea beetle Alternative name for large striped flea beetle.

Willow flea beetle *Crepidodera aurata*, 2.5–3.3 mm, common and widespread over much of England, more local and scattered elsewhere, adults feeding on the leaves of willows, and hibernating under bark and woody debris. Larvae feed in the roots. Greek *crepis* + *eidos* = 'slipper' + 'similar to', and Latin *aurata* = 'golden'.

Wood sage flea beetle *L. membranaceus*, with a scattered distribution in England and Wales, and a few records from Scotland and eastern Ireland, mainly found in woodland clearings, but also chalk grassland, fen, moorland and grassy sea-cliffs,

adults feeding on wood sage leaves, larvae on roots. *Membrana* is Latin for 'membrane'.

Yellow-striped flea beetle Alternative name for large striped flea beetle.

FLEABANE Herbaceous plants in the genera *Conyza* (Greek *conops* = 'flea'), *Dittrichia* (honouring the German botanist Manfred Dittrich, b.1934), *Erigeron* (Greek *eri* + *geron* = 'early' + 'old man', from the flowers occurring in spring which turn grey), *Inula* (Latin for elecampane, q.v.), and *Pulicaria* (Latin *pulex* = 'flea'), all in the Asteraceae. The name comes from the belief that, when burnt, the smoke repels fleas and other insects.

Alpine fleabane *E. borealis*, a rare rhizomatous perennial found in the central Highlands on unstable, basic, mostly south-facing cliff ledges of mica-schist. *Borealis* is Latin for 'northern'.

Argentine fleabane *C. bonariensis*, an annual neophyte, scattered but locally common in England as a casual on waste and cultivated ground, naturalised in Middlesex since 1993. *Bonariensis* is latinised 'from Buenos Aires'.

Bilbao fleabane *C. floribunda*, an annual South American neophyte first noted in 1992 at Southampton, now scattered and increasing in the southern half of England and eastern Ireland on waste ground. *Floribunda* is Latin for 'abundant flowers'.

Blue fleabane *E. acris*, an annual or perennial, widespread and locally common in England and Wales, less so in Scotland and Ireland, on open, well-drained, neutral or calcareous soils. Habitats include dunes, spoil heaps, railway ballast and industrial waste. It also grows on rock outcrops, especially of chalk and limestone, and on mortared walls. *Acris* is here Latin for 'bitter'.

Canadian fleabane *C. canadensis*, an annual neophyte present by 1690, common in much of England and Wales, scattered in northern England, Scotland and Ireland, in well-drained, open habitats such as pavements, waste ground, walls and railway ballast, and as a weed of cultivated ground. It is particularly characteristic of urban areas of southern England and East Anglia, and occasionally recorded from dunes, and sandy ground inland.

Common fleabane *P. dysenterica*, a rhizomatous perennial, common and widespread in lowland England, Wales and Ireland, scattered in Scotland, in damp, open habitats including marsh, water- and fen-meadow and tall-herb fen, by streams, canals and ditches, in dune slacks, sea-cliff seepages, damp woodland rides, hedge banks and verges. *Dysenterica* is from Greek, indicating the plant's efficacy in treating dysentery.

Guernsey fleabane *C. sumatrensis*, an annual South American neophyte first noted in Guernsey in 1961, naturalised in south Essex by 1974, spreading rapidly and well established in the London area by 1984, and now locally common over much of southern and central England and Wales. It was first recorded in Ireland in 1990. It grows in well-drained, open and disturbed ground such as wasteland, railway sides and docks, mainly in towns.

Hairy fleabane *I. oculus-christi*, a perennial east European neophyte present by 1759, naturalised in abandoned gardens and on rough ground in the Isle of Man and a few other locations. The specific name is Latin for 'eye of Christ'.

Hooker's fleabane *I. hookeri*, a Himalayan neophyte perennial naturalised on roadsides, quarries and open woodland in mid-west Yorkshire (since 1986), Lanarkshire (1993), Cumbria (1995) and Shropshire (2008). Sir Joseph Hooker (1817–1911) was an English botanist.

Irish fleabane *I. salicina*, a perennial known only from the northern half of Lough Derg, Co. Tipperary, along the limestone shoreline and on the islands in stony habitat between the flood line and surrounding scrub, declining because of increased lake eutrophication. Latin *salicina* = 'willow-like'.

Mexican fleabane *E. karvinskianus*, a perennial neophyte cultivated since 1836, widespread and especially common in the Channel Island, Scillies and southern England, naturalised on walls, rock outcrops and cliffs, in pavement cracks and on stony banks, usually spread by seed from nearby gardens. Wilhelm Karwinski von Karwin (1780–1855) was a Hungarian-born German explorer and naturalist.

Small fleabane *P. vulgaris*, an annual of damp, winter-flooded hollows in acidic, unimproved grassland on New Forest 'lawns', on commons and village greens, and on rutted tracks. Pony grazing often provides the disturbed open conditions needed for seedling survival. It is also found locally in Surrey. Latin *vulgaris* = 'common'.

Stinking fleabane *D. graveolens*, an annual Mediterranean neophyte, historically a common wool alien, now naturalised by some Hampshire roadsides, and a widespread and scattered casual elsewhere in England. *Graveolens* is Latin for 'strongly scented'.

Tall fleabane *E. annuus*, an annual or biennial North American neophyte introduced in 1816, scattered in Britain as a garden escape, usually casual but sometimes naturalised on rough ground and dunes and in newly sown grass. Latin *annuus* = 'annual'.

Woody fleabane *D. viscosa*, a perennial south European neophyte naturalised on rough ground in east Suffolk and by a harbour in East Sussex, casual in a few other places. Latin *viscosa* = 'sticky'.

FLEA'S EAR *Chlorencoelia versiformis* (Heliotales, Hemiphacidiaceae), a rare and critically endangered fungus with a few records from the southern third of England, found on the dead wood of broadleaf trees, especially ash (itself now threatened with ash dieback disease), and a priority species for conservation in the UK BAP.

FLEAWORT **Field fleawort** *Tephroseris integrifolia* ssp. *integrifolia* (Asteraceae), a biennial or short-lived perennial herb, found in England on shallow soils on chalk and, rarely, oolitic limestone, in short grassland, and on ancient earthworks and tracks, favouring warm, dry, south-facing sites. Greek *tephra* = 'ash' (i.e. 'grey'), and Latin *integrifolia* = 'complete (i.e. undivided) leaves'. The endemic ssp. *maritima* (**spathulate fleawort**) is found on South Stack, Anglesey, on soil derived from glacial drift on grassy coastal cliff slopes and crevices.

FLESH FLY Members of the family Sarcophagidae.

FLIRT *Russula vesca* (Russulales, Russulaceae), a common and widespread edible brittlegill fungus, found in July–October on deciduous woodland soil, especially mycorrhizal on beech. The binomial is from Latin for 'reddish' and 'small' or 'thin'.

FLIXWEED *Descurainia sophia* (Brassicaceae), an annual, rarely biennial, archaeophyte herb, locally abundant as a weed in arable fields in light soils in East Anglia. Elsewhere, it is local as a casual on roadsides and waste ground. 'Flixweed' dates from the late sixteenth century, derived from the obsolete *flix* (variant of *flux*) + weed. Francois Descourain (1658–1740) was a French botanist and physician. Greek *sophia* = 'wisdom'.

FLOOD-MOSS *Myrinia pulvinata* (Myriniaceae), widespread and scattered in Britain, confined to the boles and roots of trees, especially willows, growing in the flood zone of lowland rivers. *Myrinia* may come from Greek *myrios* = 'myriad', *pulvinata* certainly from Latin for 'little cushion'.

FLOSSFLOWER *Ageratum houstonianum* (Asteraceae), an annual Mexican neophyte with 'fluffy' flowers, hence the English name, a garden escape casual on waste ground, mainly found in the London area but also elsewhere in England, the Isle of Man, and central Scotland. *Ageratum* is from Greek for 'not ageing' (*geros* = 'old age'). Dr William Houston (1695–1733) was a Scottish-born surgeon, botanist and plant collector.

FLOUNDER *Platichthys flesus* (Pleuronectidae), length up to 50 cm, flattened laterally, usually right-eyed, a widespread nocturnal demersal fish present in inshore waters to depths of 50 m, often found in estuaries and sometimes fresh water (it has been found as far inland as Montgomeryshire on the River Severn). Diet is of small molluscs, crustaceans and polychaetes.

'Flounder' dates from Middle English, coming from Old French *flondre*, probably of Scandinavian origin (e.g. Old Norse *flythra*). Greek *platys* + *ichthys* = 'flat' + 'fish'.

FLOUR BEETLE Stored-product pests in the family Tenebrionidae, particularly in the genus *Tribolium* (Greek *tribolos* = 'three-pointed').

Broad-horned flour beetle *Gnatocerus cornutus*, 4 mm length, occasionally found outside with a scattered distribution in England, and indoors as a pest of stored cereal products. Greek *gnatos* + *ceras* = 'born' + 'horn', and Latin *cornutus* = 'horned'.

Confused flour beetle *T. confusum*, a widespread and common indoors pest of stored cereal products, found locally outdoors, mainly in the Midlands. Both adults and larvae feed on grain dust and broken kernels, but not undamaged whole grain. They often hitch-hike into the home in infested flour and rapidly multiply into large populations. The common name comes from difficulty in separating this species from rust-red flour beetle (i.e. with which it is 'confused').

Dark flour beetle *T. destructor*, a widespread and common indoors pest of stored cereal products, both adults and larvae feeding on grain dust and broken kernels, though not whole grain.

Depressed flour beetle *Palorus subdepressus*, 2.5–3 mm length, an uncommon pest of stored products, especially in damp locations, originally from Africa, found occasionally in the Midlands and south-east England. *Palorus* may be from Greek *palaius* = 'ancient'.

Red flour beetle Alternative name for rust-red flour beetle.

Rust-red flour beetle *T. castaneum*, a widespread and common indoors pest of stored cereal products, found locally outdoors. Both adults and larvae feed on grain dust and broken kernels, but not undamaged whole grain. They often hitch-hike into the home in infested flour and rapidly multiply into large populations. *Castaneum* is Latin for 'chestnut brown'.

FLOWER BEE Member of the genus *Anthophora* (Apidae) (Greek *anthos* + *phor* = 'flower' + 'thief'), 10–14 mm in length, feeding and collecting pollen and nectar, prompting the common name.

Fork-tailed flower bee *A. furcata*, found in much of England and Wales, flying in May–August or September in habitats including gardens, open woodland and moorland, visiting a range of flowers, though favouring Lamiaceae, and unusually for this genus excavating its nest burrows in rotten wood rather than in the soil. *Furcata* is Latin for 'forked'.

Four-banded flower bee *A. quadrimaculata*, found in southern (especially south-east) England, flying from June to August, with most records from gardens, the bees visiting flowers of various lamiates, particularly cat-mint and lavender, and nesting in walls and sandy banks. *Quadrimaculata* is Latin for 'four-spotted'.

Hairy-footed flower bee *A. plumipes*, found throughout much of England and south Wales, flying in March–May in a range of habitats, including gardens, open woodland and coastal sites, visiting a range of flowers, though with a preference for Lamiaceae (especially lungwort), and nesting in the ground and in walls. *Plumipes* is Latin for 'feather-footed'.

Potter flower bee *A. retusa*, was widespread in southern England until the end of the Second World War since when it has declined greatly, probably because of agricultural intensification, and it has been recorded from just four areas since 1990 (Farnborough, Hampshire; Isle of Wight; Isle of Purbeck, Dorset; Seaford Head, Sussex). It is listed as RDB 1 Endangered, and is included on the BAP Priority Species List. *Retusa* is Latin for 'blunted'.

FLOWER BEETLE **Ant-like flower beetle** Member of the family Anthicidae (Greek *anthicos* = 'relating to flowers').

Soft-winged flower beetle Member of the families Dasytidae (Greek *dasytes* = 'roughness' or 'hairiness') and Malachiidae (*malache* = 'mallow').

FLOWER BUG Common flower bug *Anthocorus nemorum* (Anthocoridae), 3–4 mm long, a predatory bug with a preference for lower vegetation, common and widespread. Greek *anthos* + *coris* = 'flower' + 'bug', and Latin *nemus* = 'glade'.

FLOWERING-RUSH *Butomus umbellatus* (Butomaceae), a submerged or emergent rhizomatous perennial, scattered but locally common in England, in calcareous, often eutrophic, water at the edges of rivers, lakes, canals and ditches, and in swamps, generally reproducing by lateral buds on the rhizome. Greek *bous* + *tomos* = 'ox' + 'cut', referring to sharp leaf margins, and Latin *umbellatus* = 'with umbels'.

FLUELLEN Annual archaeophytes in the genus *Kickxia* (Veronicaceae), a mid-sixteenth-century alteration of the Welsh *llysiau Llywelyn* = 'Llewelyn's herbs'. *Kickxia* probably honours the Belgian botanist Jean Kickx (1803–64).

Round-leaved fluellen *K. spuria*, on arable land and waste ground, usually on calcareous soils, widespread but probably declining in England south of a line from the Severn to the Humber estuaries. *Spuria* is Latin for 'false'.

Sharp-leaved fluellen *K. elatine*, mainly found in the southern half of England, Wales and south-east Ireland, on basic soils, including light soils over chalk and calcareous boulder-clay, on arable field margins as well as tracks, waste ground and in gardens. It has also been recorded on sandy soil, and on peat in Co. Cork. *Elatine* was the Ancient Greek name for toadflax.

FLUKE See Trematoda. The name comes from Old English *floc* = 'flat', describing the flattened, rhomboidal shape of these flatworms.

Liver fluke, common liver fluke or **sheep liver fluke** *Fasciola hepatica* (Fasciolidae), dwarf pond snails being their commonest vector in Britain, parasitising the liver of cattle and sheep, causing the disease fasciolosis, symptoms including diarrhoea, weight loss, anemia and reduced milk production, and occasionally infecting humans.

FLUTE Tree flute or **flute lichen** *Menegazzia terebrata* (Parmeliaceae), foliose, local in oceanic parts of western Britain and south-west Ireland in damp shady woods on acid bark of deciduous trees or mossy rock, an indicator of ancient woodland. The English names, and another one of **perforated lichen**, come from the hollow or inflated perforated thallus. *Menegazzia* dates to 1854, from the Veronese lichenologist Abramo Massalongo who named it after his friend Luigi Menegazzi. *Terebrata* is Latin for 'perforated'.

FLY See also Diptera.

Face fly *Musca autumnalis* (Muscidae), 7–8 mm length, common and widespread in England and Wales, getting its common name from its habit of landing on the faces of cattle where they feed on tears, sweat and blood (from the bites of other flies). Adults also feed on nectar. They are also evident in March–October sunning themselves on posts and fences. Eggs, hatching within hours, are laid in cattle dung in which the larvae (maggots) feed on the microbial flora and fauna. The binomial is Latin for 'fly' and 'autumnal'.

Frit fly *Oscinella frit* (Chloropidae), with a scattered distribution, the larvae boring into the stems of grasses, including cereals, and locally becoming a pest of oats and rye in particular, and of turf grasses. *Oscinella* is Latin for a small songbird. 'Frit' dates from the late nineteenth century, from the Latin *frit* = 'particle on an ear of corn'.

Head fly *Hydrotaea irritans* (Muscidae), 5–7 mm length, a common and widespread vector of bacterial pathogens that cause summer mastitis (a swelling of the teats) in non-lactating cows and heifers, and also affecting sheep, horses and other livestock. The flies live in trees and can only fly during mild, humid conditions and low wind speeds. They feed on body secretions around the nose, eyes and mouth, on organic debris at the base of the horns, and on mammalian dung. They are also attracted by blood of open wounds. Females deposit eggs in soil or in plant litter, often in pastures close to cow pats. Larvae prey on other insect larvae; they pupate to overwinter in the soil, and adults hatch late in the next spring, with a seasonal peak in activity in late summer.

Horn fly *Haematobia irritans* (Muscidae), length 4 mm, with a scattered distribution in England and Wales. Eggs are laid in fresh cow manure in which the developing larvae remain feeding on microbial organisms and their waste products. On emergence, adults can fly long distances to find a host. Having done so the fly remains on it and others in the same herd, the female leaving only to oviposit. The fly moves to different areas on the same animal to regulate its temperature and minimise exposure to the wind. Flies feed on cow blood, piercing the skin with sharp mouthparts, leading to irritation and sometimes to anaemia and secondary infections. Greek *haematobia* references bloodsucking; *irritans* is Latin for 'irritant'.

House fly Member of the family Muscidae, specifically *Musca domestica*, body length 6–7 mm, ubiquitous, widespread and synanthropic. Adults have a sponging/sucking mouthpart and feed on liquid or semi-liquid substances beside solid material which has been softened by saliva or vomit. Because of their large intake of food, they constantly deposit faeces, one of the features that make them vectors of bacterial, viral and parasitic pathogens. Larvae (maggots) feed on carrion, manure and kitchen waste. The binomial is Latin for 'fly' and 'of the house'.

Noon fly *Mesembrina meridiana* (Muscidae), length 15 mm, common and widespread in Britain, found in summer and autumn. Adults feed on nectar, but often sunbathe in grassland on (livestock) dung, in which eggs are laid and the consequent larvae develop, feeding on other fly larvae. They are also sometimes found in sunny spots in woodland, on leaves, trunks, and the ground itself.

Semaphore fly *Poecilobothrus nobilitatus* (Dolichopodidae), length 5–7 mm, common and widespread in England and Wales, the adult found in May–August in damp habitats such as muddy pond edges, often seen resting on the water surface. In 2010, *The Guardian* launched an annual competition for the public to suggest common names for previously unnamed species; this name was selected in 2012, given the male's behaviour of waving his white wing tips to attract mates.

Signal fly Members of the family Platystomatidae, superfamily Tephritoidea, the two British species (*Platystoma seminationis*, *Rivellia syngenesiae*) having scattered distributions. Adults, 5–7 mm long, have wings that are in almost constant motion as though giving signals. They feed on nectar and pollen in May–October, larvae on decomposing plant material.

Stable fly *Stomoxys calcitrans* (Muscidae), 6–8 mm length, common and widespread, particularly in England and Wales, as a synanthropic pest using its 'bayonet' mouthparts to feed on the blood of livestock, especially cattle and horses (which can become anaemic). Larvae (maggots) develop in rotting straw and manure. Greek *stoma* + *oxys* = 'mouth' + 'sharp', and Latin *calcitrans* = 'kicking'.

FLYCATCHER Birds in the family Muscicapidae. Other flycatcher species are scarce vagrants.

Pied flycatcher *Ficedula hypoleuca*, wingspan 22 cm, length 13 cm, weight 13 g, a summer migrant in Britain, wintering in West Africa, Red-listed in 2015, rare but Amber-listed in Ireland, found from April to September in mature deciduous woodlands mainly in western Britain, with 17,000–20,000 pairs in 2009. Numbers in Britain declined by 40% between 1995 and 2015, but increased by 20% in 2010–15, and indeed by 89% in 2014–15. Nests are usually in (especially oak) tree holes, with 6–7 eggs, incubation lasting 13–15 days, and fledging of the altricial young taking 16–17 days. Males practise successive or sequential polygyny. Diet is insectivorous, hunting from a perch. *Ficedula* is Latin for a small fig-eating bird described by Pliny. Greek *hypo* + *leucos* = 'below' + 'white'.

Red-breasted flycatcher *F. parva*, wingspan 20 cm, length 12 cm, weight 10 g, a scarce passage migrant averaging 80–90 records a year. Reports for 2016 include birds from Outer Hebrides to Hampshire. *Parva* is Latin for 'small'.

Spotted flycatcher *Muscicapa striata*, wingspan 24 cm, length 14 cm, weight 17 g, a widespread migrant breeder, wintering in southern and western Africa, found in deciduous woodland, parks and gardens, with an estimated 36,000 territories in Britain in 2009. There was a decline by 85% between 1970 and 2015, and by 44% between 1995 and 2015, making this a Red List species, though a strong increase (by 25%) was noted in 2010–15. It is Amber-listed in Ireland. The nest is built in a recess, often against a wall (or in an open-fronted nest box), with a clutch size of 4–5, incubation lasting 13–15 days, and the altricial young fledging in 13–16 days. They fly from, and return to, a perch to feed on flying insects. Latin *musca* + *capere* = 'a fly' + 'to seize', and *striata* = 'lined'.

FOOL'S-WATER-CRESS *Apium nodiflorum* (Apiaceae), a herbaceous perennial, locally common in Britain as far north as central Scotland (though also found in the Outer Hebrides), and in Ireland, in shallow water in streams, ditches, swamp and marsh, and on mud at the edges of ponds, lakes, rivers and canals, characteristic of nutrient-enriched sites. Latin *apium* = 'parsley' or 'celery', and *nodiflorum* = 'with nodal flowers'.

FOOT **Club foot** *Ampulloclitocybe clavipes* (Agaricales, Tricholomataceae), a widespread and common inedible fungus, found in summer and autumn on soil in broadleaf woodland, especially under beech. Latin *ampulla* = 'flask' + Greek *clytocibe* = 'sloping head', and Latin *clava* + *pes* = 'club-shaped' + 'foot'.

FOOTBALL JERSEY WORM *Tubulanus annulatus* (Tubulanidae), a nemertean (ribbon worm) reaching a length of 75 cm, usually 3–4 mm wide, on a variety of sublittoral (occasionally intertidal) substrates under stones or in macroalgae holdfasts. Latin *tubulus* = 'tube' and *annulatus* = 'ringed'.

FOOTBALLER *Helophilus pendulus* (Syrphidae), wing length 8.5–11.5 mm, a widespread hoverfly, common in April–November associated with muddy puddles, wet ditches and ponds, and sunny places along hedgerows, roadsides and field margins. Adults visit flowers to feed on nectar, but often rest on leaves, characteristically emitting a buzzing sound. Larvae feed on detritus in farmyard drains, wet manure and wet sawdust. Its common name comes from the yellow and black striped thorax said to resemble some footballers' strip, but it is also known as the sunfly. Greek *helos* = 'marsh' and Latin *pendulus* = 'suspended'.

FOOTMAN Nocturnal members of the moth family Erebidae, subfamily Arctiinae. Larvae generally feed on lichens.

Buff footman *Eilema depressa*, wingspan 28–36 mm, local in broadleaf and mixed woodland, scrubby downland and heathland, and fen throughout much of southern England and Wales, occasional in south-west Ireland. From a low base, population levels increased by 3,884% between 1968 and 2007. Adults fly in July–August. Larvae feed on lichen and algae growing on yew and other trees. Greek *eilema* = 'veil', and Latin *depressa* = 'low'.

Common footman *E. lurideola*, wingspan 28–35 mm, common in gardens, farmland, marsh and woodland throughout England, Wales, southern Scotland and parts of Ireland. Adults fly in July–August. *Luridus* is Latin for 'pale yellow', from the colour of the wing streaks.

Dingy footman *E. griseola*, wingspan 32–40 mm, common in fen, water meadow, damp grassland and rural gardens throughout southern England and Wales, since 2000 colonising north to Yorkshire and Lancashire. From a low base, population levels increased by 1,851% between 1968 and 2007. Adults fly in July–August. *Griseus* is Latin for 'grey', from the forewing colour.

Dotted footman *Pelosia muscerda*, wingspan 24–8 mm, a rare RDB moth, possibly now only breeding in Britain in fen and damp woodland in the Norfolk Broads, but recorded in a number of other sites where it may be a migrant. Adults fly in July–August. Larvae feed on algae and lichens growing on shrubs. Greek *pelos* = 'clay', from the brownish-grey wing colour, and Latin *muscerda* = 'mouse dung', from the forewing spots.

Four-dotted footman *Cybosia mesomella*, wingspan 25–33 mm, local on heathland, moorland, damp grassland, fen and open woodland throughout Britain (plus a single Irish record). Adults fly in June–August. Greek *cybos* = 'cube' or, here, 'die', from the forewing spots, and *mesos* + *melas* = 'middle' + 'black'.

Four-spotted footman *Lithosia quadra*, male wingspan 35–40 mm, female 40–55 mm, local in broadleaf woodland in southern England and west Wales, sporadically as far north as Ross-shire and the Isle of Man, and occasional in Ireland. Adults fly in July–September. Greek *lithos* = 'stone', from the colour or because larvae were thought to feed on lichen growing on rocks, and Latin *quadra* = 'square', the figure produced were the forewing spots to be joined.

Hoary footman *E. caniola*, wingspan 28–35 mm, scarce, local on sea-cliffs and shingle beaches, and inland on quarries, from Kent to Anglesey. Adults fly in July–September. *Caniola* derives from Latin *canus* = 'grey' (of hair, i.e. hoary).

Muslin footman *Nudaria mundana*, wingspan 19–23 mm, widespread but local on dry-stone walls and rocky areas. Population levels increased by 113% between 1968 and 2007. Adults fly in June–August. Latin *nudus* = 'naked', from the thin wing scales, and *mundus* = 'neat'.

Orange footman *E. sororcula*, wingspan 27–30 mm, found in England, rarely reaching further north than Cumbria, and Lincolnshire in the east, and scarce in Wales. Adults fly in May–June. *Sororcula* is Latin for 'little sister', from the affinity with a congeneric.

Pygmy footman *E. pygmaeola*, wingspan 24–8 mm, a rare RDB moth found on dunes and shingle beaches in parts of East Anglia and Kent, and a rare immigrant on southern coasts. Adults fly in July–August.

Red-necked footman *Atolmis rubricollis*, wingspan 25–35 mm, day- and night-flying (June–July), local in broadleaf woodland and coniferous plantations in south and west England and Wales, and parts of Scotland and Ireland. Larvae, which live in autumn, feed on green algae and lichens on tree-trunks. Greek *atolmia* = 'lack of courage', possibly because larvae hide under bark during the day, and Latin *ruber* + *collum* = 'red' + 'neck'.

Rosy footman *Miltochrista miniata*, wingspan 23–7 mm, local in broadleaf woodland, heathland and mature hedgerows throughout southern England, rare elsewhere. Nevertheless, population levels increased by 488% between 1968 and 2007. Adults fly in July–August. Larvae feed on tree-growing lichens, including dog lichen. Greek *miltos* + *christos* = 'red earth' + 'anointed', and Latin *minium* = 'vermilion'.

Scarce footman *E. complana*, wingspan 28–35 mm, local in heathland, moorland, woodland, gardens and dunes throughout the southern half of England and in coastal parts of Wales and Ireland. From a low base, population levels increased by 3,590% between 1968 and 2007. Adults fly in July–August. Larvae feed on lichens, occasionally mosses and leaves of low-growing plants. *Complano* is Latin for 'to make level', from the flat adult posture.

Small dotted footman *Pelosia obtusa*, wingspan 25–30 mm, Vulnerable in the RDB, found in reedbeds in East Anglia. It was first recorded in the Norfolk Broads in 1961 with the existence of a colony at Hickling confirmed by a number of sightings in the 1980s. Adults fly in June–July. Larvae feed on algae and lichens on fallen dead branches. *Obtusa* is Latin for 'blunt'.

Speckled footman *Coscinia cribraria*, wingspan 30–5 mm, Endangered in the RDB, associated with heathland, and a priority species in the UK BAP. It declined greatly in the early part of the twentieth century and no confirmed breeding localities are currently known. Recent records (2010) of probable resident moths are confined to a few heathland sites in the Wareham area of south-east Dorset. There are occasional continental migrants found on the south coast. Adults fly in July–August. Larvae feed on a number of heathland plants. Greek *coscinon* = 'sieve', and Latin *cribraria* = 'sifted', both describing the speckled forewing.

FORAMINIFERA A phylum or class of very small (usually <1 mm), often benthic marine amoeboid protists characterised by an external shell (test) made of calcareous and other materials, with various morphologies, extruding granular ectoplasm for catching food. The name is Latin for 'hole bearers' (*foramen* = 'hole').

FORESTER *Adscita statices* (Zygaenidae, Procridinae), wingspan 25–8 mm, a day-flying (May–July) moth with a widespread but scattered distribution in England, Wales, Ireland and west Scotland, on damp neutral grassland, calcareous grassland, heathland, dunes and woodland rides. Larvae feed on sorrel, adults on the nectar of flowers such as scabious and marsh thistle. Numbers and distribution are declining and this has become a priority species for conservation. The binomial is from Latin for 'enrolled' and an early genus name for thrift, *Statice*, erroneously believed to be a food plant.

Cistus forester *A. geryon*, wingspan 20–5 mm, local on flower-rich chalk and limestone grassland in southern and northern England and north Wales, favouring south-facing slopes, larvae feeding on rock-rose. Day-flying adults (May–July) feed on nectar of rock-rose, kidney vetch, thyme and bird's-foot trefoil. In Greek mythology Geryon was a three-headed monster slain by Hercules.

Scarce forester *Jordanita globulariae*, a rarity, mostly day-flying (June–July), of chalk downland, with only a few populations ranging from Gloucestershire to Kent. Larvae feed on knapweeds, adults on the nectar of knapweeds and salad burnet. *Jordanita* probably honours the German entomologist Heinrich Jordan (1861–1959). *Globularia* is a genus of putative food plants.

FORGET-ME-NOT Mainly species of *Myosotis* (Greek *mys* + *ous* = 'mouse' + 'ear') (Boraginaceae). The English name, dating from the late fourteenth century, is a translation of German *Vergissmeinnicht*: in German legend, God had named all the plants but one, which called out, 'Forget-me-not, O Lord,' to which God replied, 'That shall be your name.' There was a similar Greek legend. In medieval times the plant was often worn by ladies as a sign of faithfulness and enduring love.

Alpine forget-me-not *M. alpestris*, a herbaceous perennial found in heavily grazed limestone grassland on base-rich well-drained soils in the Pennines and on ungrazed cliffs in Perthshire.

Bur forget-me-not *Lappula squarrosa*, an annual European neophyte introduced by 1683, originating in bird- and grass seed, grain and wool shoddy, occasionally found as a casual on waste ground. *Lappula* is the diminutive of Latin *lappa* = 'bur', and *squarrosa* = 'having scales'.

Changing forget-me-not *M. discolor*, a common and widespread annual of open grassland and disturbed ground found in meadow, pasture, moorland edge, marsh, dune slacks, arable field margins, verges, railway tracks, chalk- and gravel-pits, and walls. *Discolor* is Latin for 'two-coloured'.

Creeping forget-me-not *M. secunda*, a stoloniferous annual to perennial herb, somewhat calcifugous, mainly upland but only scarce in the Midlands and eastern England, by streams and pools, and in wet pasture, moorland flushes and springs, with a preference for acid peaty soils. *Secunda* is Latin for 'second', here meaning having leaves down one side.

Early forget-me-not *M. ramosissima*, an annual of open habitats or bare ground in lowland regions, on dry, generally infertile soils, including chalk and limestone grassland, sandy heaths, stabilised dunes, edges of sandy arable fields, railway tracks, rocks, walls, quarry spoil and waste ground. *Ramosissima* is Latin for 'many-branched'.

Field forget-me-not *M. arvensis*, a common and widespread annual or biennial archaeophyte of well-drained open or disturbed ground, especially arable fields, but also woodland edges, grassland, hedges, scrub, roadsides, walls and quarries. *Arvensis* is from Latin for 'field'.

Great forget-me-not *Brunnera macrophylla*, a herbaceous Caucasian neophyte perennial, scattered in the UK, mainly in England, naturalised in woodland, rough ground and tips. Samuel Brunner (1790–1844) was a Swiss botanist. Greek *macrophylla* = 'large-leaved'.

Jersey forget-me-not *M. sicula*, an annual known from two sites on Jersey: in damp habitat on Ouaisné Common (now probably disappeared) and by a small pool at Noirmont. *Sicula* is Latin for 'Sicilian'.

Pale forget-me-not *M. stolonifera*, a mainly upland herbaceous perennial growing in northern England and southern Scotland by streams and along base-rich spring lines and flushes. *Stolonifera* is Latin for 'bearing stolons'.

Tufted forget-me-not *M. laxa*, a common and widespread annual or biennial herb of marsh, fen-meadow and rush pasture, and by lakes, ponds, rivers and streams, as well as disturbed wet places such as where trampled by livestock. *Laxa* is Latin for 'loose'.

Water forget-me-not *M. scorpioides*, a common and widespread herbaceous stoloniferous or rhizomatous perennial found in damp habitats, usually in fertile, calcareous to mildly acidic soils. Generally terrestrial, growing by lakes, ponds, rivers and streams, in marsh and fen, it can also form submerged patches or floating rafts. *Scorpioides* is Latin for 'scorpion-like', indicating a coiled shape.

White forget-me-not *Plagiobothrys scouleri*, an annual North American neophyte, a contaminant of grass seed found in a few places on bare, wet, sandy or gravelly hollows and in newly sown roadside grass. Greek *plagios* + *bothros* = 'oblique' + 'scar'. Dr John Scouler (1804–71) was a Scottish botanist.

Wood forget-me-not *M. sylvatica*, a widespread biennial or perennial, more scattered in Ireland, growing as a native, at least in England, on damp, fertile soils in woodland, rocky grassland, scree and rock ledges. It is more widespread in a wider range of habitats as a garden escape. *Sylvatica* is from Latin for 'woodland'.

FORK-MOSS Often growing in upland habitats, generally in tightly packed clumps.

Highland rarities include **arctic fork-moss** *Arctoa fulvella*, **bendy fork-moss** *Dicranum flexicaule*, **Blytt's fork-moss** *Kiaeria blyttii*, **dense fork-moss** *D. elongatum* and **long-leaved fork-moss** *Paraleucobryum longifolium* (all Dicranaceae), and **sickle-leaved fork-moss** *Kiaeria falcata*, **snow fork-moss** *K. glacialis* and **Starke's fork-moss** *K. starkei* (also Cumbria) (Rhabdoweisiaceae).

Broom fork-moss *D. scoparium*, common and widespread in a range of habitats, especially on the ground in woodland, but also on trees and logs, and on heathland, mires, dunes, acidic rocks and montane short turf. *Scoparium* is Latin for 'like (the plant) broom' (*Cytisus scoparius*).

Crisped fork-moss *D. bonjeanii*, common and widespread on damp grassland, and in bog and fen, at lake margins, occasionally on drier sites, for example chalk grassland and sea-cliff turf. Jean Bonjean (1780–1846) was a French botanist.

Dusky fork-moss *D. fuscescens*, widespread except for the Midlands and central Ireland, on hard substrates, especially tree bases, and on boulders. Higher in the mountains it also occurs on scree, loch margins and crags, and forms an association with stiff sedge on soil with late snow-lie. *Fuscescens* is Latin for 'becoming brown'.

Fragile fork-moss *D. tauricum*, common and widespread in Britain, having spread greatly during the twentieth century, mainly on decaying wood, stumps and old fence-posts. *Tauricum* is Latin for 'from the Crimea'.

Fuzzy fork-moss *D. leioneuron*, new to Britain in 1965, with a few records from suboceanic mountain habitats in Wales,

Cumbria and the Highlands. Greek *leios* + *neuron* = 'smooth' + 'nerve'.

Greater fork-moss *D. majus*, widespread on acidic soils, and on rock where there is a build-up of organic material, common in woodland and ravines in the north and west where it also grows on scree, crags and heathy banks. *Majus* is Latin for 'larger'.

Mountain fork-moss *D. montanum*, a lowland moss, despite its name, widespread in Britain, commonly on the base and trunk of trees and exposed roots, occasionally on branches.

Rugose fork-moss *D. polysetum*, scarce and scattered but slowly spreading in lowland England and Scotland in coniferous woodland, heathy birch woodland and mires. Greek *poly* = 'many', Latin *seta* = 'bristle'.

Rusty fork-moss *D. spurium*, scattered and rare in heathland in England (declining) and Scotland, favoured by shelter under leggy heather. *Spurium* is Latin for 'false'.

Scott's fork-moss *D. scottianum*, with a strongly western distribution, usually found on dry vertical acidic crags in humid sites, often near the coast, named after the bryologist Robert Scott (1757–1808).

Transparent fork-moss *Dichodontium pellucidum* (Rhabdoweisiaceae), common and widespread except for the lowland Midlands and eastern England, in wet habitats, particularly on gravel and silted rocks by streams and lake margins. A small form is found in exposed montane sites, usually on base-rich rock. Greek *dicho* + *donti* = 'in two' + 'tooth', and Latin *pellucidum* = 'transparent'.

Waved fork-moss *Dicranum bergeri*, rare if widespread, on raised bogs in Scots pine woodland, and in mires next to base-rich flushes. It has declined in England and Wales, and sites have been lost to afforestation in Scotland. *Bergeri* possibly refers to the German botanist Ernst Berger (1814–53).

Whip fork-moss *Dicranum flagellare*, scattered in England and Wales on decaying logs and stumps in woodland. *Flagellum* is Latin for 'whip'.

Yellowish fork-moss *Dichodontium flavescens*, scattered in upland regions on gravel and silted rocks by streams, and in other base-rich habitats. *Flavescens* is Latin for 'yellowish'.

FORK WEED Clawed fork weed *Furcellaria lumbricalis* (Furcellariaceae), a red seaweed with fronds 30 cm long, common and widespread on rocks in the lower intertidal and shallow subtidal often on sandy or muddy shores, tolerating low salinities. Latin *furca* = 'fork', and *lumbricus* = 'earthworm'.

Discoid fork weed *Polyides rotundus* (Polyidaceae), a common and widespread red seaweed with fronds up to 20 cm long, attached to rock and stones in sandy pools in the lower littoral and subtidally to 20 m.

FORKLET-MOSS Marsh forklet-moss *Dichodontium palustre* (Dicranaceae), widespread and common in wet upland habitats such as ditches, flushed grassland and the edges of streams and stony flushes. Greek *dicho* + *donti* = 'in two' + 'tooth', and Latin *palustre* = 'marshy'.

FORMICIDAE Family name for ants in the genera *Formica*, Latin for 'ant' (with 11 British species), *Lasius* (13), *Myrmica* (12), *Ponera* (1), *Temnothorax* (4) and *Tetramorium* (1).

FORSYTHIA *Forsythia* x *intermedia* (Oleaceae), a deciduous shrub of garden origin, naturalised as a relic or throw-out on rough ground and roadsides, scattered but mainly in England. William Forsyth (1737–1804) was a Scottish horticulturalist.

FOUR-SPOTTED *Tyta luctuosa* (Noctuidae, Aediinae), wingspan 22–5 mm, a scarce, vulnerable day- and night-flying moth (May–September) once widely distributed in south England, the east Midlands and East Anglia but now found only in a few scattered locations on dry, well-drained, often south-facing sandy or chalky sites. Larvae feed on field bindweed. Numbers and distribution are declining and this is a priority species for conservation. *Tyta* has no entomological meaning; *luctuosa* is

Latin for 'mournful', the black and white colouring suggesting mourning.

FOUR-TOOTH MOSS Pellucid four-tooth moss *Tetraphis pellucida* (Tetraphidaceae), common and widespread in Britain, scattered in Ireland, on decomposing stumps and logs. The English name refers to the large peristome teeth on the sporophyte capsule. *Tetra* is Greek for 'four'.

FOX Red fox *Vulpes vulpes* (Canidae), male head-body length 67–72 cm, tail 40 cm, weight 6–7 kg, females 62–7 cm, tail 40 cm, weight 5–6 kg, common and widespread (absent from the Isle of Man and the Scottish islands except Skye), in a wide range of habitats including urban areas. Pre-breeding numbers in Britain in 2008 were estimated at 258,000, with 33,000 (14%) living in towns. These largely solitary nocturnal hunters spend most of the day in sheltered lie-up nests. As opportunistic omnivores, foxes have a varied diet. In lowland rural areas small mammals, especially field voles and rabbits, are the main source of food, with earthworms, beetles, fruit (particularly blackberries) and small birds also eaten. In upland regions carrion is important, particularly during winter. Urban foxes prey on rodents and birds such as pigeons; scavenge from bins, edible litter, bird tables, and rubbish and compost heaps, and are often deliberately fed by the public; excess food is cached. Dens (earths) may be dug by the foxes, or they enlarge holes made by other animals. In urban areas, resting and birth dens are dug in waste ground or little-used parts of large gardens, and under garden sheds, etc. Courtship and mating vocalisation (screaming) is common in late winter and spring. The female (vixen) produces (usually) four or five cubs in spring. At about four weeks old, usually in late April/early May, cubs begin to emerge into the open. Foxes are susceptible to sarcoptic mange which causes fur loss and skin lesions. They are considered to be pests, for example preying on poultry and on ground-nesting game birds. There were about 190 fox hunts in England and Wales in 2002: these probably killed a small proportion of foxes compared with those captured in snares or shot. The Hunting Act (2005) banned hunting of wild animals with dogs in England and Wales; a ban in Scotland had been passed in 2002; hunting in Northern Ireland and the Republic remains legal. Road casualties probably make up 50% of fox mortality, especially in towns. 'Fox' comes from Old English, related to West Frisian *foks*, Dutch *vos* and German *Fuchs*. *Vulpes* is Latin for 'fox'.

FOX-AND-CUBS *Pilosella aurantiaca* (Asteraceae), a stoloniferous or rhizomatous perennial European neophyte herb, cultivated by 1629, common and widespread (less so in Ireland), naturalised as a garden escape on railway and roadside banks, on walls and in churchyards and other grassy and waste places. The name refers to the way that many as yet unopened flower heads hide beneath those that have opened. The binomial is Latin for having small hairs, and 'golden-coloured'.

Yellow fox-and-cubs *P. caespitosa*, a perennial European neophyte herb introduced in 1819, scattered and naturalised as a garden escape on grassy habitats, riverbanks, dunes, roadsides, walls, railway banks and rough ground. *Caespitosa* is Latin for 'tufted'.

FOX-SEDGE Perennial sedges (Latin *carex*) (Cyperaceae).

American fox-sedge *Carex vulpinoidea*, a neophyte wool alien occasional in England and the Glasgow area, naturalised on rough ground. *Vulpinoidea* is Latin for 'resembling a fox'.

False fox-sedge *C. otrubae*, locally common and widespread, though mainly coastal in north England and Scotland, usually on heavy soils, on stream and pond margins, in ditches, swamps, wet meadow and pasture, and upper saltmarsh, occasionally on damp roadsides and hedge banks.

True fox-sedge *C. vulpina*, found in a few places in (mainly south-east) England, in wet habitats, usually on heavy clay soils that are flooded in winter and dry in summer. It occurs by ditches and rivers, in meadows and sometimes in standing water, and can tolerate shade. *Vulpina* is Latin referring to a fox.

FOXGLOVE *Digitalis purpurea* (Veronicaceae), a common and widespread herbaceous biennial or short-lived perennial, common on acidic soils in hedge banks and woodland clearings, on heathland and moorland margins, riverbanks, montane rocky slopes, sea-cliffs and waste ground. It is often abundant in disturbed or burnt areas. As a common garden plant it can occur as an escape outside its native range and also on calcareous soils. This is the 'county' flower of Birmingham, Monmouthshire and Argyll as voted by the public in a 2002 survey by Plantlife. The thimble-like flowers provide an explanation for the 'glove' part of the name, but why 'fox' is obscure. (The suggestion that it is a corruption of 'folk's' is without foundation.) Old English *foxes glofa* was initially used for deadly nightshade. *Digitalis* is Latin for this plant (from *digitus* = 'finger'), and *purpurea* = 'purple'.

Fairy foxglove *Erinus alpinus*, a scattered and widespread semi-evergreen south-west European neophyte perennial herb in cultivation by 1739, found in the crevices of old walls, and on limestone or bricks with lime mortar. *Erinus* was used by Dioscorides for a plant similar to basil.

Straw foxglove *D. lutea*, a herbaceous European neophyte biennial cultivated by 1629, scattered in England, locally naturalised and increasing, on verges and banks, quarries, walls and waste ground. *Lutea* is Latin for 'yellow'.

FOXGLOVE-TREE *Paulownia tomentosa* (Paulowniaceae), a deciduous neophyte tree introduced from China in the 1830s, planted in parks, amenity areas and by roads, naturalised in a few places in the southern half of England and south Wales. Anna Paulovna was the daughter of Tsar Paul I. *Tomentosa* is Latin for 'covered with hairs'.

FOXTAIL Generally perennial grasses with dense silky or bristly brush-like flowering spikes in the genus *Alopecurus* (Poaceae) (Greek *alopex* + *ouros* = 'fox' + 'tail').

Alpine foxtail *A. magellanicus*, rhizomatous, scarce, found from north-west Yorkshire to the Highlands, in or by oligotrophic springs and flushes, often associated with late snowlie. It occurs on a range of acidic or slightly basic rocks up to 1,220 m on Braeriach, Aberdeenshire. *Magellanicus* means from the Magellan Straits.

Bulbous foxtail *A. bulbosus*, scarce, local from Pembrokeshire to Lincoln, on brackish grassland in unimproved coastal grazing marsh, at the edges of ditches and in trampled ground at the base of sea walls, and locally in upper saltmarsh.

Marsh foxtail *A. geniculatus*, common and widespread, frequent in fertile sites that are flooded in winter such as ditch sides, wet arable fields, pond margins and grazing marshes. *Geniculatus* is Latin for 'bent sharply'.

Meadow foxtail *A. pratensis*, common and widespread in lowland regions in grassland, particularly with damp, fertile soils, and on roadsides and woodland margins. *Pratensis* is from Latin for 'meadow'.

Orange foxtail *A. aequalis*, an uncommon annual, scattered but mostly found in England, in habitats associated with fresh water, including the margins of ponds, ditches, reservoirs, turloughs and flooded gravel-pits. *Aequalis* is Latin for 'equal' or 'similar'.

FRAGRANT-ORCHID Tuberous herbaceous perennials in the genus *Gymnadenia* (Orchidaceae). Greek *gymnos* + *adenas* = 'naked' + 'gland'.

Chalk fragrant-orchid *G. conopsea*, probably widespread and locally common on chalk and limestone grasslands, limestone pavement, less acidic heaths, and wetter habitats such as base-rich fens. It also grows in artificial habitats such as quarries and railway embankments. *Conopsea* is Latin for 'gnat-like'.

Heath fragrant-orchid *G. borealis*, in hilly grassland in south-west and north England, west Wales and Scotland, and

in mire in the New Forest and East Sussex. *Borealis* is Latin for 'northern'.

Marsh fragrant-orchid *G. densiflora*, scattered in base-rich chalk grassland, wet meadow, fen and ditches. *Densiflora* is Latin for 'densely flowered'.

FRANKENIACEAE Family name for sea-heath. Johan Franke (1590–1661) was a German-born Swedish botanist.

FRILLWORT Thallose liverworts in the genus *Fossombronia* (Fossombroniaceae), tentatively from Latin *fossa* = 'ditch' and Greek *ombros* = 'rain (storm)'.

Rarities include **fragile frillwort** *F. fimbriata*, **pitted frillwort** *F. foveolata*, and **sea frillwort** *F. maritima* (Cornwall and the Scillies).

Acid frillwort *F. wondraczekii*, common and widespread on acidic often otherwise bare ground, named after M.C. Wondraczek, who discovered it in Prague.

Common frillwort *F. pusilla*, common and widespread on acidic often otherwise bare ground. *Pusilla* is Latin for 'very small'.

Greater frillwort *F. angulosa*, near the sea in Cornwall, south-west and north-west Wales, and west Ireland in sheltered, shaded sites, for example hedge banks, streamsides, rocky slopes, and horizontal crevices at the foot of rock outcrops and sea-cliffs. *Angulosa* is Latin for 'angled'.

Spanish frillwort *F. caespitiformis*, in south and south-west England on acidic often otherwise bare ground. *Caespitosa* is Latin for 'tufted'.

Weedy frillwort *F. incurva*, widespread and scattered, found on lake and stream margins, in dune slacks, on moist, often sandy or gravelly soil on roadsides, paths and quarry floors, and on china-clay spoil and fly ash. *Incurva* is Latin for 'incurved'.

FRINGE-MOSS Species of *Racomitrium* (Grimmiaceae) (Greek *racos* + *mitrion* = 'tatter' + 'small cap').

Widespread and locally common upland species of rocky habitats include **bristly fringe-moss** *R. heterostichum*, **green mountain fringe-moss** *R. fasciculare* and **lesser fringe-moss** *R. affine* (for these three species, lowland colonies have also been found on slate and sandstone roofs and gravestones, and in railway cuttings), **dense fringe-moss** *R. ericoides*, **Himalayan fringe-moss** *R. himalayanum*, **long fringe-moss** *R. elongatum*, **narrow-leaved fringe-moss** *R. aquaticum* (especially where periodically flushed), **oval-fruited fringe-moss** *R. ellipticum*, and **slender fringe-moss** *R. sudeticum*.

Hoary fringe-moss *R. canescens*, widespread and scattered, favouring dunes and upland calcareous grassland. *Canescens* is Latin for 'grey'.

Woolly fringe-moss *R. lanuginosum*, common and widespread north of a line from Devon to north Yorkshire on open, stony, exposed ridges and plateaux, where it dominates *Racomitrium* heathland, and scree. It also grows on dry-stone walls, dry acidic turf, lowland shingle heath, and blanket and raised bogs. It is a rarity in old chalk grassland in south-east and eastern England. *Lanuginosum* is Latin for 'woolly'.

Yellow fringe-moss *R. aciculare*, common and widespread north of a line from Devon to north Yorkshire, scattered elsewhere, on wet, rocky sites, including the top of rocks in fast-flowing, base-poor rivers in the west and north. It grows in mountains, but it is commoner at relatively low altitudes. *Aciculare* is Latin for 'needle-shaped'.

FRINGECUPS *Tellima grandiflora* (Saxifragaceae), rhizomatous perennial introduced from North America in 1826, a widely distributed garden escape naturalised in nitrogen-rich soils in damp woods and hedgerows. *Tellima* is an anagram of the related genus *Mitella*; *grandiflora* is Latin for 'large-flowered'.

FRINGEWORT Foliose liverworts in the genus *Ptilidium* (Ptilidiaceae) (Greek *ptilon* = 'feather').

Ciliated fringewort *P. ciliare*, widespread and common in Scotland and Wales, with patchier distributions in England

and Ireland, on acidic grassland, rocky slopes, wall tops, dwarf-shrub heath, bog, dunes and heathy woodland. *Ciliare* is Latin for 'fringed with hair'.

Tree fringewort *P. pulcherrimum*, scattered and widespread in Britain on trunks and branches of birch, willow, juniper and other trees, rarely on logs and fallen branches, or on boulders. *Pulcherrimum* is Latin for 'very beautiful'.

FRINGILLIDAE Members of the finch family (Latin *fringilla* = 'finch').

FRITILLARY (BUTTERFLY) Members of the Nymphalidae, subfamilies Heliconiinae and Nymphalinae. 'Fritillary' comes from Latin *fritillus* = 'dice box', referring to the wing markings.

Dark green fritillary *Argynnis aglaja* (Heliconiinae), male wingspan 58–68 mm, locally abundant, widespread if less common in central and eastern England and in central Ireland, on calcareous and other grassland, and in places on moorland, dunes and in woodland clearings. Larvae primarily feed on violets, adults on thistle and knapweed nectar. Adults fly in June–August. Although this butterfly has declined considerably in range since the 1970s, especially in eastern England, it remains Britain's most widespread fritillary: between 1976 and 2014 figures show an overall decline in distribution by 33%, but this includes a recovery between 2005 and 2014 with a 44% increase, and within this (overall) contracted range numbers have actually increased, with an 186% increase over 1976–2014. In Greek mythology Argynnis was beloved by Agamemnon, and Aglaia was one of the three Graces.

Glanville fritillary *Melitaea cinxia* (Nymphalinae), wingspan 41–7 mm, named after Lady Eleanor Glanville (1654–1709), a lepidopterist who discovered this species in Lincolnshire. It flies from May to August, and currently found mainly on cliff grasslands on the south coast of the Isle of Wight with the occasional colony on the south Hampshire coast. It is also found on Guernsey and Alderney. The chief larval food plant is ribwort plantain; adults favour bird's-foot trefoil, dandelion, hawkweeds and thrift. After hatching, the gregarious larvae spin a silk web over the food plant in which they live and feed. After the fourth moult, caterpillars build a tent in which to hibernate, usually low down in vegetation. Despite being at the limit of its northern range and of such limited distribution, colonies of this butterfly appear to be stable (though, from a low base, witnessed a 66% decline in distribution and a 42% decline in numbers over 1976–2014), but are vulnerable to habitat change, so the species is of conservation concern. The genus name may be from Melitaea, a town in Thessaly, Greek *melitaios* = 'from Malta', or *melitoeis* = 'honeyed', from the nectarivorous adults. *Cinxia* comes from Latin *cinctus* = 'girdled'.

Heath fritillary *M. athalia*, male wingspan 39–44 mm, female 42–7 mm, confined to a small number of heathland and grassland sites in Somerset, Devon and Cornwall, and in woodland clearings in parts of Kent. It has also been reintroduced into sites in Essex. This is one of Britain's rarest butterflies, was on the brink of extinction in the late 1970s, and had a range that declined by 68% between 1976 and 2014. While it has responded well to management, distribution and numbers have continued to decline (by 12% and 79%, respectively, from 2005 to 2014). Larvae in woodland colonies favour common cow-wheat and foxglove; those on other sites germander speedwell and ribwort plantain. Adults, flying from May to August, feed on nectar of bramble, buttercups, knapweeds, heather, etc. Athalia was the daughter of Omri, King of Israel.

High brown fritillary *A. adippe*, wingspan 55–70 mm, once common in woodland across England and Wales, flying in June–September, but numbers recently crashing owing to climate change, cessation of coppicing and loss of traditional grazing allowing scrub invasion to shade out violets on which the caterpillars depend. Adults feed on nectar from bramble, thistles and knapweeds. This butterfly, once at high risk of extinction in Britain, declined in range and numbers by 96% and 62%, respectively over 1976–2014, even taking into account the series of warm springs and conservation effort which led to a 180% increase in numbers from 2013 to 2014 in its 30 remaining sites, especially in Exmoor, Dartmoor, parts of south Wales, and on limestone pavement at Morecambe Bay. The original specific name, *cydippe*, was suppressed for this species in favour of earlier use for a different species, and the invented name *adippe* allocated instead.

Marsh fritillary *Euphydryas aurinia* (Nymphalinae), male wingspan 30–42 mm, female 40–50 mm, found primarily in south-west England with small populations in north-west England, west Scotland, and north-west and south-west Wales. It is locally widespread in Ireland. It is found, flying in May–July, in open sunny habitats such as chalk hillsides, heathland, moorland and damp meadow. Larvae favour scabious species; adults find nectar from a number of plants, particularly buttercups and thistles. Caterpillars spin a silk web by binding together leaves of the food plant. The species responds badly to adverse weather and suffers from larval parasitism by a wasp, *Apanteles*. This has become one of Britain's most threatened butterflies, and is a priority species for conservation. In the twentieth century numbers generally declined by >10% each decade, but colony size does fluctuate: while there was an overall decline by 10% over 1976–2014, numbers actually increased by 71% from 1995 to 2005. Range declined greatly (by 79%) over 1976–2014. Greek *euphy* + *dryas* = 'good shape' + 'wood nymph'. Aurinia was a prophetess mentioned by Tacitus.

Pearl-bordered fritillary *Boloria euphrosyne* (Heliconiidae), male wingspan 38–46 mm, female 43–7 mm, with a scattered distribution of small colonies in woodland clearings south of a line from north Wales to Kent, in west Lancashire and Cumbria, and the southern Highlands, and on limestone pavement on the Burren, west Ireland. Colonies tend to die out with woodland regrowth. Adults fly from April to early June. Larvae feed by day on common dog (preference) and other violets; adult foods include nectar from bugle, bluebell, buttercups, hawkweeds, dandelion and thistles. Once common and widespread this is now one of Britain's most threatened butterflies because of the decline in coppicing, with a reduction in distribution by 95% between 1976 and 2014, and a drop in numbers by 71% over the same period, though a recovery with an increase by 45% was recorded between 2005 and 2014. However, it remains a priority species for conservation effort. Greek *bolos* = 'fishing net', from the reticulate wing pattern. Euphrosine was one of the three Graces.

Silver-washed fritillary *A. paphia*, male wingspan 69–76 mm, female 73–80 mm, found in (especially oak) woodland in southern Wales and England (absent from East Anglia), south Cumbria, and much of Ireland. Larvae feed during the day on common dog-violet, adults (flying in June–September) on honeydew and also nectar from bramble, thistles and other species. Despite an overall decline in the twentieth century, range and numbers increased by 56% and 141%, respectively, over 1976–2014, with these figures being 55% and 6% between 2005 and 2014. Aphrodite is sometimes called Paphia after an association with the town of Paphos on Cypress.

Small pearl-bordered fritillary *B. selene*, male wingspan 35–41 mm, female 38–44 mm, associated with (preferably damp) woodland clearings in western Britain (and marsh and moor in the north). Larvae feed on common dog and marsh violets; adults (flying in May–July) on the nectar of a variety of plants. While remaining locally abundant in Scotland, cessation of coppicing has led to a long-term decline in both distribution and numbers in England: between 1976 and 2014 there was a reduction in range by 76% and a decline in numbers by 58% (though with a 3% increase recorded over 2005–14), and it is a priority species for conservation effort. *Selene* is Greek for 'moon'.

FRITILLARY (PLANT) Snake's-head fritillary *Fritillaria* *meleagris* (Liliaceae), a possibly native bulbous herbaceous perennial scattered in England, now locally common only in the Thames Valley and Suffolk, growing in damp neutral grasslands, usually those managed for hay followed by grazing. It is often planted in other grassland habitats elsewhere in Britain, sometimes becoming naturalised. This is the county flower of Oxfordshire as voted by the public in a 2002 survey by Plantlife. The English and genus names come from Latin *fritillus* = 'dice box', from the markings on the flowers; *meleagris* means having spots like a guinea fowl.

FROG From Old English *frogga*, short-bodied, tail-less members in the family Ranidae, order Anura, with two native and five introduced species. See also bullfrog.

Common frog *Rana temporaria*, snout-vent length 6–9 cm, males being smaller. Common and widespread, they emerge from hibernation in late February, and spawning (often in garden ponds) usually takes place in early March, though sometimes as soon as January. Spawn is laid in clumps typically of 300–400 gelatinous eggs. Newly hatched tadpoles are mainly herbivorous, feeding on algae, detritus and plants but also small invertebrates. During metamorphosis the tadpole grows legs, with the rear legs first to grow. The tail is absorbed into the body, and it loses its gills and grows lungs. Tadpoles become fully carnivorous once their back legs develop, feeding on small water animals, including other tadpoles. Juvenile frogs (froglets) feed on invertebrates both on land and in water. Mature frogs eat only on land, feeding on insects (especially flies), worms, slugs and snails, catching the prey on long, sticky tongues. Adults range far from their breeding ponds and may be found almost anywhere in damp places from late February to October. They hibernate in running water, muddy burrows, or in decaying leaves and mud at the bottom of ponds. Oxygen is absorbed through the skin. Populations are generally stable, but have recently become susceptible to the diseases red-leg and chytridiomycosis. The binomial is Latin for 'frog' and 'temporary'.

Edible frog *Pelophylax esculentus*, male snout-vent length 6–11 cm, females 5–9 cm, a natural hybrid between marsh frog and pool frog, with a patchy range in southern and eastern England established through introductions and escapes. Greek *pelos* + *phylax* = 'mud' + 'guard', and Latin *esculentus* = 'edible'.

Italian pool frog *P. bergeri*, snout-vent length 8 cm, an introduction difficult to identify unambiguously, but occasionally found in south-east England. Leszek Berger (1925–2012) was a Polish herpetologist.

Marsh frog *P. ridibundus*, Europe's largest native frog, female snout-vent length up to 13 cm, males smaller, first introduced to Britain near the Romney, Walland and Denge marshes, Kent, in 1935, and now found in several areas of Kent and East Sussex. Other introductions include colonies in south-west and west London, with a number of further, generally accidental introductions in south-east England and near Norwich, including adult escapes and tadpoles released into ponds with shipments of ornamental fish. It is the most successful of introduced frogs, choosing breeding sites such as dykes and ditches not generally used by native amphibians, and it is a voracious predator. *Ridibundus* is Latin for 'laughing'.

Pool frog *P. lessonae*, female snout-vent length up to 9 cm, males much smaller, a rare native in Norfolk, the last population dying out in 1993. The species had been subject to impermanent introductions elsewhere in eastern and southern England from Europe since the nineteenth century. The species was reintroduced to Thompson Common, Norfolk, with a series of releases from 2005 on, with a good prognosis for success reported in 2015.

Southern marsh frog *P. perezi*, an introduction difficult to identify unambiguously, but occasionally found in south-east England.

FROGBIT *Hydrocharis morsus-ranae* (Hydrocharitaceae), a perennial floating plant locally frequent in England, scattered in Wales and Ireland, introduced in the Central Lowlands of Scotland, generally decreasing in Britain, its rosettes found in shallow, calcareous, mesotrophic or meso-eutrophic water in sheltered parts of lakes or in ponds, canals and ditches. Reproduction is mainly vegetative. Greek *hydor* + *charis* = 'water' + 'grace', and Latin *morsus* + *rana* = 'biting' + 'frog'.

FROGHOPPER Hemipteran insects (Auchenorrhyncha), with nine British species.

Common froghopper *Philaenus spumarius* (Aphrophoridae), length 5–7 mm, very variable in colour, widespread and very common. Nymphs and adults are found on a variety of plant species, the former protected from desiccation in spring by a frothy mass of bubbles secreted by the juvenile animal, hence the alternative names **spittlebug** and **cuckoo-spit insect**. Adults can jump a hundred times their own length and up to 70 cm in the air (more per body size than a flea), accelerating at around 4,000 m/s². Greek *philein* = 'love', and *spumarius*, from Latin *spuma* = 'froth', referred to the foam nests.

Red-and-black froghopper *Cercopis vulnerata* (Cercopidae), length 8–10.5 mm, present in much of England and Wales, nymphs feeding underground on roots but adults seen in a number of open and woodland habitats. Greek *cercis* = 'shuttle', and Latin *vulnerata* = 'wounded'.

FROSTWORT Upland foliose liverworts in the genus *Gymnomitrion* (Gymnomitriaceae) (Greek *gymnos* + *mitrion* = 'naked' + 'small cap').

Braided frostwort *G. concinnatum*, abundant in the hepatic-rich crust with dwarf willow that forms at the margin of very late-lying snow, on gravel terraces in fellfields, and lower down on crags and scree, and in gullies. *Concinnatum* is Latin for 'well made', literally 'skilfully joined'.

Coral frostwort *G. corallioides*, scarce, on friable rock at high altitudes in the Highlands.

Western frostwort *G. crenulatum*, in upland western Britain and (largely western) Ireland on a range of rock types. *Crenulatum* is Latin for 'with small rounded teeth'.

White frostwort *G. obtusum*, widespread and locally common on sheltered rock faces in scree, on crags and on large boulders. While most frequent on acidic rocks, it can tolerate quite base-rich conditions. *Obtusum* is Latin for 'blunt'.

FROUNCE The disease trichomonosis when in birds of prey.

FRUIT FLY See Drosophilidae and Tephritidae.

FRUIT MOTH Small moths whose larvae feed inside various fruits.

Apple fruit moth *Argyresthia spinosella* (Argyresthiidae), wingspan 9–11 mm, widespread and common in orchards and woodland, more numerous in the north. Adults fly (day and night) in May–July. Larvae feed inside apples, in some places becoming a serious orchard pest, and in rowan berries. Greek *argyros* + *esthes* = 'silver' + 'dress', from the metallic gloss on the wings. *Spinosella* refers to the food plant blackthorn *Prunus spinosa*.

Cherry fruit moth *A. pruniella*, wingspan 10–13 mm, nocturnal, widespread and common in gardens, woodland and orchards, more scattered in Scotland and Ireland. Adults fly in July from the evening on. Larvae feed inside the fruit of wild, commercial and ornamental cherries, often reaching pest numbers. *Pruniella* is from Latin *prunus*, which includes the cherries.

Plum fruit moth *Grapholita funebrana* (Tortricidae, Olethreutinae), wingspan 10–15 mm, local in gardens, orchards and hedgerows throughout much of the southern half of England, Wales, Ireland and occasionally in Scotland. Adults fly in June–August. Larvae feed inside the fruit of plum, sloe, wild cherry and other prunus species in August–September. Greek *graphe* + *litos* = 'drawing' + 'frugal', and Latin *funebrana* = 'funereal', from the dark forewing.

FUCACEAE, FUCALES Family and order of brown seaweeds, including wrack.

FUCHSIA *Fuchsia magellanica* (Onagraceae), a South American shrub, introduced in 1788, mainly found as a planted hedge or in abandoned rural gardens but also naturalised in hedgerows and scrub, by streams, among rocks and on walls. Nearly all the fuchsia hedges in western Britain and Ireland (commonly naturalised in the south and west) are the sterile cultivar 'Riccartonii' (or var. *macrostema*), which originated in a Scottish nursery before 1850. This is the county flower of the Isle of Man as voted by the public in a 2002 survey by Plantlife. Leonhart Fuchs (1501–66) was a German botanist.

FULMAR *Fulmarus glacialis* (Procellariidae), wingspan 107 cm, length 48 cm, male weight 880 g, females 730 g, most abundant on the coasts of north Scotland, Orkney and Shetland, with 500,000 breeding pairs in 2010, but recorded around all coastlines, including as winter visitors and birds of passage. Numbers declined by 22% in 1986–2014, and by 12% in 2009–14. Nests are on cliffs, a single egg being laid, incubation taking 52 days, fledging of the altricial young 46–50 days. Feeding is on fish and crustaceans at sea. They vomit a foul-smelling oil as a defence: the word 'fulmar' derives from Old Norse *fúll* = 'stinking, foul' + *már* = 'gull'. Latin *glacialis* = 'glacial', reflecting the bird's northern range.

FUMITORY Scrambling annuals in the genus *Fumaria* (Papaveraceae) (Latin *fumus terrae* = 'earth smoke'). See also ramping-fumitory.

    Common fumitory *F. officinalis*, a common and widespread archaeophyte of well-lit disturbed ground including arable, gardens, allotments and wasteland, usually on well-drained, fairly base- and nutrient-rich soils. Subspecies *officinalis* occurs throughout the range of the species; ssp. *wirtgenii* is most frequent on light soils in the east. *Officinalis* is Latin meaning having pharmacological value.

    Dense-flowered fumitory *F. densiflora*, an archaeophyte of arable (commonly in spring-sown cereals and root crops), mainly found in southern and south-east England and central-east Scotland, most frequently on chalk but also occurring on other free-draining soils.

    Few-flowered fumitory *F. vaillantii*, an archaeophyte of chalk and limestone soils in England. Never very abundant, it has declined since 1945 as a result of agricultural intensification, and is increasingly restricted to fields with spring-sown crops. Sébastien Vaillant (1669–1722) was a French botanist.

    Fine-leaved fumitory *F. parviflora*, an archaeophyte of chalky soils in England. Never very abundant, it has declined since 1945 as a result of agricultural intensification, and is increasingly restricted to field margins associated with spring-sown crops. *Parviflora* is Latin for 'small-flowered'.

FUNARIACEAE Family name of bladder-mosses and cord-mosses (Latin *funis* = 'cord').

FUNGI One of the five Kingdoms of living organisms – heterotrophic organisms that acquire nourishment as parasites or saprophytes. Many have a mutually beneficial and often quite specific relationship with a host plant, usually by the fungal mycelia, the hyphae serving as mycorrhiza interacting with the plant root. Many fungi are microscopic, for example moulds, mildews, smuts, rusts and yeasts, many of which are pathogenic. Larger fungi (commonly, but non-botanically, called mushrooms and fungi) take a variety of forms in their fruiting bodies, including cup and stem forms, brackets and gelatinous bodies. In 2016, the British Mycological Society listed 14,091 species recorded from the British Isles.

FUNGUS BEETLE Pests of stored grain in the families Mycetophagidae and Tenebrionidae, some larvae also feeding on fungi. Also, polypore fungus beetle (see Tetratomidae) and round fungus beetle (see Leiodidae).

    Black fungus beetle *Alphitobius laevigatus* (Tenebrionidae), 6 mm length, occasional outdoors in the Midlands, East Anglia and southern England, but also a pest of stored products. Greek *alphiton* = 'barley meal', and Latin *laevigatus* = 'burnished'.

    Hairy fungus beetle Members of the family Mycetophagidae, specifically, *Typhaea stercorea*, length 3 mm, widespread in England and Wales. The elytra have fine hairs. Greek *typhos* = 'smoke', and Latin *stercus* = 'dung'.

    Two-banded fungus beetle *Alphitophagus bifasciatus* (Tenebrionidae), 2.5 mm length, scarce, found locally outdoors in southern England and East Anglia feeding on fungi (including moulds), but more significantly is a pest of stored grain products. Greek *alphiton* + *phago* = 'barley meal' + 'eat', and Latin *bifasciatus* = 'two-striped'.

FUNGUS GNAT Members of the families Bolitophilidae, Ditomyiidae, Keroplatidae, Mycetophilidae and (dark-winged fungus gnats) Sciaridae. The larvae of most species feed on fungi or decaying plant matter, and some sciarids can become pests of house plants.

FUNNEL, FUNNEL-CAP (FUNGUS) Mainly members of the genus *Clitocybe* (Greek *clitos* + *cybe* = 'slope' + 'head'), together with *Lepista* and *Leucopaxillus* (Agaricales, Tricholomataceae), and *Faerberia* (Polyporales, Polyporaceae).

    Rarities include **aniseed funnel** *C. odora* (especially under beech), **bitter funnel** *Leucopaxillus gentianeus* and **mealy funnel** *C. vibecina*.

    Chicken run funnel *C. phaeophthalma*, inedible, smelling like the inside of a chicken house, widespread in (especially southern) England, found in summer and autumn on soil and leaf litter in broadleaf woodland, especially under beech. Greek *phaios* + *ophthalmos* = 'dark' + 'eye', the cap having a dark centre.

    Clouded funnel *C. nebularis*, inedible, common and widespread, found from late summer to early winter on woodland soil. *Nebula* is Latin for 'cloud', from the cap colour.

    Common funnel *C. gibba*, edible, widespread and common, found in summer and autumn on soil in deciduous woodland, hedgerows, rough grassland and heathland. *Gibba* is Latin for 'humped'.

    Firesite funnel *Faerberia carbonaria*, locally common with a scattered distribution in England, on soil at burnt sites in woodland.

    Fool's funnel *C. rivulosa*, poisonous, widespread and common in England, less common elsewhere, found from summer to early winter on often sandy soil on unimproved grasslands and lawns, and on dunes. *Rivulosa* is Latin for 'stream', referencing the ridges that form on mature caps.

    Fragrant funnel *C. fragrans*, poisonous, widespread and locally frequent, found from late summer to early winter on soil and leaf litter in grass or moss beneath deciduous trees.

    Frosty funnel *C. phyllophila*, inedible perhaps poisonous, widespread in England, less common elsewhere, found in late summer and autumn on soil and leaf litter in broadleaf woodland, especially under beech. *Phyllophila* is Greek for 'liking leaves'.

    Giant funnel *Leucopaxillus giganteus*, edible, widespread but uncommon, found in late summer and autumn in pasture, roadside hedges, parkland and woodland clearings.

    Mealy frosted funnel *C. ditopa*, inedible, widespread but rare, found in late summer and autumn on soil in coniferous woodland.

    Rickstone funnel-cap Alternative name for trooping funnel.

    Tawny funnel *Lepista flaccida*, edible but not tasty, very common and widespread, found in summer and autumn on soil in (mainly broadleaf) woodland. The binomial is Latin for 'goblet' and 'flaccid'.

    Trooping funnel *C. geotropa*, edible and tasty, widespread, especially in England, but local, found from late summer and to early winter on soil in broadleaf and mixed woodland, often in grass in clearings, in parkland, and on verges. Greek *ge* + *tropos* = 'earth' + 'turn', a reference to the downward-turning cap margin.

Twotone funnel *C. metachroa*, inedible, widespread, especially in England, but not common, found in late summer and autumn on soil in deciduous woodland. *Metachroa* is Greek for 'changing appearance'.

FURBELOW *Saccorhiza polyschides* (Phyllariaceae), a brown seaweed, the largest macroalga in Europe, thallus up to 3 m long, widespread except absent from the Northumberland coast to the Solent, growing from extreme low water to a depth of 35 m, usually attached to rocks. It may form dense stands in sheltered locations but can also tolerate strong currents. The flattened stipe has a frilly margin: 'furbelow' dates from the late seventeenth century, from French *falbala* = 'showy ornaments or trimmings'. Greek *saccos* + *rhiza* = 'sack' + 'root', and *polys* + *schidion* = 'many' + 'split off'.

FURROW BEE Species in the genus *Halictus* (Halictidae), from New Latin *halictus* for these bees, also known as sweat bees.

Bronze furrow bee *H. tumulorum*, widespread, found in open habitats, usually on sandy or calcareous soils. Primitively eusocial, with nests excavated in horizontal ground with a (near-)vertical burrow. Females are found from March to October, with males appearing in late June or early July. A wide range of flowers are visited for nectar. *Tumulus* is Latin for 'mound'.

Common furrow bee Alternative name for slender mining bee.

Downland furrow bee *H. eurygnathus*, thought to have become extinct, but in 2012 individuals were recorded at seven sites in chalk grassland on the South Downs, and in East Anglia. It is listed as Endangered (RDB 1). Females are active from early June to September, males from August to September, feeding on knapweed nectar. Nests are aggregated. Greek *eury* + *gnathos* = 'broad' + 'jaw'.

Southern bronze furrow bee *H. confusus*, with a few records (Rare in RDB 3) from Dorset to Kent and in Norfolk, on sandy heaths and other sandy habitats, adults flying in May–September, nests being excavated in sparsely vegetated ground in warm, sunny situations. Nectar and pollen are taken from a variety of flowers. *Confusus* is Latin for 'mixed'.

Yellow-legged furrow bee *H. rubicundus*, body length 10 mm, widespread in Britain, occasional in Ireland, rarely abundant, though found in dense nesting aggregations in northern Britain. Nests are in the ground. The species is eusocial, queens emerging from hibernation in April, workers found from May onwards and males and new females from July to October. A range of habitats are used, and a number of flowers, though mostly Asteraceae. *Rubicundus* is Latin for 'reddish'.

FURROW SHELL Filter-feeding bivalves in the family Semelidae.

Peppery furrow shell *Scrobicularia plana*, shell size up to 6.5 cm, widespread and common in detritus-rich intertidal mud in estuaries and sheltered bays, but rare or absent from northern Scotland. The binomial is Latin for 'small trench' and 'flat'.

White furrow shell *Abra alba*, shell size up to 2.5 cm, widespread and common from the lower shore to the shelf edge in a range of substrates but most often in fine silty sand. *Abra* may come from Greek *habros* = 'delicate' or 'splendid'; *alba* is Latin for 'white'.

FURUNCULOSIS A bacterial disease of salmon and trout, characterised by 'furuncles' or 'boils'.

FURZE Alternative name for gorse.

G

GADFLY A fly that bites livestock, especially a horsefly, warble fly, or bot fly, 'gad' derived from the Old Norse *gaddr* = 'spike' or 'nail'.

GADWALL *Anas strepera* (Anatidae), wingspan 90 cm, length 51 cm, male weight 830 g, females 700 g, resident in much of England, south Wales, east-central Scotland and eastern Ireland, and a winter visitor in south-west England and much of Ireland, found on lakes, reservoirs and gravel-pits, with rivers and estuaries in winter. There were 690–1,730 pairs (best estimate 1,200) in 2009. Breeding numbers increased by 126% between 1970 and 2015, by 105% between 1995 and 2015, by 32% between 2010 and 2015, and by 43% in 2015–16. Winter numbers were 25,000 in 2008–9. Nesting is on the ground, often some distance from open water, clutch size is 9–11, incubation lasts 24–6 days, and fledging of the precocial young takes 45–50 days. This dabbling duck eats underwater plant material, but chicks and adults take insects and adults sometimes molluscs in the breeding season. The origin of 'gadwall' (first recorded in 1676) is obscure: the American *Webster's Dictionary* suggests *gad well* = 'go about well', but it may come from the syllables *quedul*, of Latin *querquedula* = 'teal'. The binomial is Latin for 'duck' + 'noisy'.

GALATHEIDAE Family name of some squat lobsters. Galathea was a sea nymph in Greek mythology.

GALINGALE *Cyperus longus* (Cyperaceae), an uncommon rhizomatous herbaceous perennial, local in the Channel Islands and in south and south-west England, frequently introduced elsewhere in England, Wales and central Scotland, in marsh and wet pasture near the coast, and in base-rich sea-cliff seepages. Inland it also occurs, generally planted, on pond margins and in ditches. The name comes from Middle English *galingal*, ultimately traceable to the Persian name for this plant, *khulanjan*. *Cyperus* is Ancient Greek for 'sedge'.

Brown galingale *C. fuscus*, a rare herbaceous annual found locally in north Somerset, south Hampshire, Berkshire and Middlesex on moist disturbed ground around humus-rich margins of ponds and by ditches, often where there is winter flooding. *Fuscus* is Latin for 'dark brown'.

Pale galingale *C. eragrostis*, a rhizomatous herbaceous tropical American neophyte perennial, widespread and scattered in England and Wales, a garden escape naturalised on rough ground and roadsides, and by water, also originating from wool shoddy and birdseed. Greek *eros* + *agrostis* = 'love' + 'grass'.

GALL An abnormal growth produced by a plant prompted by a chemical stimulus by another organism, involving enlargement or proliferation of host cells, and providing shelter and nutrition for the invader. Most galls are caused by fungi (especially rusts and smuts) or insects such as aphids, mites, psyllids, gall midges (Cecidomyiidae), gall flies (Tephritidae), gall wasps (mainly Cynipidae but also Eurytomidae) and sawflies (Tenthredinidae). Galls can also be caused by viruses, bacteria and phytoplasmas. 'Gall' comes from Latin *galla* = 'oak apple'.

Cigar gall See reed cigar gall below.

Currant gall See common spangle gall wasp under gall wasp.

Hairy beech gall Created by the widespread gall midge *Hartigiola annulipes* (Cecidomyiidae) on beech leaves, the galls maturing in August and September. The gall then falls to the ground. Pupation takes place and adults emerge the following spring to lay eggs on the new growth of leaves. Latin *annulipes* = 'ring-footed'.

Hawthorn button top gall Caused by the gall midge *Dasineura crataegi* (Cecidomyiidae). Adults emerge from pupae in the ground beneath hawthorn, and terminal bud infestations start in March–April. Fully grown larvae fall to the ground in September–October to pupate after feeding and sheltering within the leaf rosette. Greek *dasys* + *neuron* = 'hairy' + 'nerve', and Latin *crataegus* = 'hawthorn'.

Little black pudding gall Caused by the gall midge D. *pteridis*, scattered, creating galls on bracken (*Pteridium*, giving the specific name) during autumn.

Reed cigar gall Caused by the fly *Lipara lucens* (Chloropidae), creating a cigar-shaped gall on reed stems, common and conspicuous in late autumn. Adults lay eggs in young reed shoots in spring; larvae feed on the stalks. Greek *liparos* = 'shiny', and Latin *lucens* = 'shiny'.

Robin's pincushion gall Also known as a bedeguar; see rose bedeguar gall wasp under gall wasp.

GALL FLY Members of the family Tephritidae.

GALL MIDGE Members of the family Cecidomyiidae.

GALL WASP Mostly members of the family Cynipidae. Many species can be generally called **oak gall wasps**, with >30 species in Britain. Most alternate between generations that are either asexual (all females) or sexual (males and females). The generation emerging as adults in summer has both sexes; the generation that develops as adults in winter-spring is all female. The alternating generations develop as larvae inside galls that often look very different and found on different parts of the oak tree. Females insert eggs into the appropriate part of the oak tree. On hatching, the legless grubs begin secreting chemicals that reorganise the oak's normal growth processes. Instead of producing normal oak tree tissues, the gall structures are created by the plant around the developing grubs. Most oak galls contain a single larva but some, for example oak apples, contain a number of larvae. Pupation takes place inside the galls. See also Eurytomidae.

Acorn gall wasp See knopper gall wasp below.

Apple gall wasp See oak apple gall wasp below.

Artichoke gall wasp or **oak artichoke gall wasp** *Andricus fecundator*, a widespread and fairly common oak gall species in Britain, laying eggs in buds at the shoot tip, which become enlarged during summer. The next (spring) generation develops small hairy pale green or brown galls on the male catkins. *Andricus* is New Latin, dating from 1861. Latin *fecundus* = 'fruitful'.

Bedeguar gall wasp See rose bedeguar gall wasp below.

Cherry gall wasp See oak cherry gall wasp below.

Cola nut gall wasp *A. lignicola*, widespread and often common in Britain, with eggs laid into the leaf axil buds and terminal buds of both pedunculate and sessile oak. Old galls persist for years, often in clusters of two to five. Exit holes are always close to the point of attachment. Latin *Lignicola* = 'living on wood'.

Common spangle gall wasp *Neuroterus quercusbaccarum*, widespread in Britain, causing disc-like galls on the underside of oak leaves in late summer–early autumn. The galls drop to the ground in autumn. Females emerge in spring to lay eggs on the male catkins. The next generation causes spherical fleshy galls on the leaves, known as currant galls. Greek *neuron* + *teras* = 'nerve' + 'monster', and Latin *quercus* + *bacca* = 'oak' + 'berry'.

Cottonwool gall wasp *A. quercusramuli*, scattered and scarce in England and Wales, the sexual generation creating galls on oak catkins; the asexual generation is a bud gall on oak. Latin *quercus* + *ramulus* = 'oak' + 'twig'.

Knopper gall wasp *A. quercuscalicis* became established in Britain during the 1970s and is now widespread. Eggs are laid during early summer in the developing acorns of pedunculate oak: instead of the usual cup and nut, the acorn is converted into a ridged woody structure in which the wasp larva develops. The next generation forms inconspicuous galls on the male catkins of Turkey oak. Latin *quercus* + *calix* = 'oak' + 'cup'.

Marble gall wasp or **oak marble gall wasp** *A. kollari*, widespread, forming hard, woody, spherical galls on oak, often persisting for several years. The alternate generation causes small, insignificant galls in buds of Turkey oak.

Oak apple gall wasp *Biorhiza pallida*, widespread in England and Wales and locally common in north-east Scotland, causing flattened rounded galls on oak twigs in spring. Males and females emerge in midsummer and eggs are laid on oak roots. The next (asexual) generation produces marble-like galls on the roots, from which females emerge in late winter to lay eggs in twig buds. Greek *bios* + *rhiza* = 'life' + 'root', and Latin *pallida* = 'pale'.

Oak cherry gall wasp *Cynips quercusfolii*, with a scattered, local distribution in Britain, forming spherical, pithy galls on the underside of oak leaves in late summer–autumn, often remaining attached to fallen leaves. The spring generation forms inconspicuous galls in oak buds. *Cynips* is from Greek *scnips* (*sic*) for an insect living under bark. Latin *quercus* + *folium* = 'oak' + 'leaf'.

Oriental chestnut gall wasp *Diplolepis kuriphilus*, discovered in a woodland in Kent in June 2015, and subsequently in Hertfordshire, creating galls on the buds, leaves and petioles of sweet chestnut, potentially weakening the tree and becoming a pest. Greek *diploos* + *lepis* = 'double' + 'scale'.

Oyster gall wasp *N. anthracina*, with a scattered but locally common distribution in England, usually found on the underside of the oak leaf. Latin *anthracina* = 'coal black'.

Rose bedeguar gall wasp *D. rosae*, widespread, producing a cluster of wiry red growths replacing a dog rose leaf. The gall is the product of a group of larvae, each living in their own chamber. Larvae overwinter in the gall, emerging as adults in spring.

Silk button spangle gall wasp *N. numismalis*, widespread but scattered in England and locally common in Scotland, creating disc-shaped galls on the underside of oak leaves in late summer–early autumn. The next (spring) generation forms small oval galls on the male catkins and leaf margins. Latin *numisma* = 'coin'.

Smooth spangle gall wasp *N. albipes*, locally common in England, producing saucer-shaped galls on oak, mainly on the underside of the leaf with each gall containing a single larva. Pupation occurs during winter while the galls are on the ground, and females emerge in spring. They lay eggs which give rise to small, oval, green galls which are attached to the leaf margins or the catkins. Males and females emerge in May–June. Latin *albipes* = 'white foot'.

GALL WEEVIL See weevil.

GALLANT-SOLDIER *Galinsoga parviflora* (Asteraceae), an annual South American neophyte introduced to Kew Gardens by 1796, from where it had escaped by 1860. It has since spread steadily, mainly in England, and particularly in London and other large urban areas, also occurring as a wool alien, becoming a weed on disturbed light soils, for example in cultivated fields, allotments and gardens, and in waste ground and paving in urban areas. Ignacio de Galinsoga (1766–97) was director of the Real Jardín Botánico de Madrid. Latin *parviflora* = 'small flower'. The English name is a corruption of the genus name.

GALLIFORMES Order of birds containing the family Phasianidae (Latin *gallus* = 'cockerel').

GANGLY LANCER *Nymphon gracile* (Nymphonidae), a sea spider (pycnogonid), body length 1 cm, with eight legs three to four times body length, widespread but patchily distributed, generally migrating into the sublittoral to breed during winter; males then carry the eggs back to the intertidal in March or April. In 2010, *The Guardian* launched an annual competition for the public to suggest common names for previously unnamed species, the judges commenting that this name 'captured both this sea spider's gangly nature and its large fang-like structures (chelifores) at the front of its head'. *Nymphon* may refer to a Greek temple of the nymphs; *gracile* is Latin for 'slender'.

GANNET Northern gannet *Morus bassanus* (Sulidae), wingspan 172 cm, length 94 cm, weight 3 kg, a migrant as well as resident breeder and bird of passage. Gannets arrive at their colonies from January onwards and leave between August and October. Non-breeding birds can be seen at any time around

the coast. The main migration period offshore is during autumn. The 21 gannetries around the British Isles, with 218,500 occupied nests in 2004, contained 56% of the global population. It is Amber-listed, though the 75,000 pairs on Bass Rock, Firth of Forth, in 2014 represents a 26% rise since 2009; and there were 60,300 pairs on St Kilda (64 km west-north-west of North Uist), and 36,000–39,000 pairs on Grassholm, Pembrokeshire. (The RSPB estimates there are 18 tons of plastic detritus on Grassholm, which gannets use to build nests, confusing it with seaweed, entangling and strangling themselves; many birds die, though some 50 a year are rescued.) Nests on the cliffs contain one egg, incubation takes 42–6 days, and fledging of the altricial young takes 84–97 days. Gannets dive for fish from a height of 25 m and can reach 100 km/h when they plunge into the water to a depth of up to 35 m. 'Gannet' comes from Old English *ganot*, meaning 'strong' or 'masculine' Greek *moros* = 'foolish', referencing the lack of fear shown by breeding gannets. Latin *bassanus* refers to Bass Rock.

GARDEN STAR-OF-BETHLEHEM See star-of-bethlehem.

GARGANEY *Anas querquedula* (Anatidae), wingspan 62 cm, length 39 cm, weight 380 g. Breeding birds arrive from wintering in central Africa from March and return from July, with a scattered distribution, mostly in England, and no more than 15–100 pairs. The favoured habitat is well-vegetated shallow wetland, including flooded meadow, these dabbling ducks feeding underwater on plants and insects. Clutch size is 8–9, incubation 21–3 days, and the precocial young fledging in 35–40 days. Small numbers of non-breeding birds also visit on passage migration in spring and autumn. 'Garganey' dates from the seventeenth century and comes from Lombard *gargenei*, ultimately from Latin *gargala* = 'tracheal artery', since the bony enlargement of the male's trachea differs from that in other duck species, being placed in the median line of the windpipe instead of on one side. Latin *anas* = 'duck'; *querquedula* is an onomatopoeic representation of the male's call.

GARLIC Bulbous herbaceous perennials in the genera *Allium* and *Nectaroscordum* (Alliaceae), including garlic *A. sativum*, of unknown origin, much cultivated, and a scarce casual as a throw-out on tips and waste ground. 'Garlic' comes from Old English *gar* + *leac* = 'spear head' + 'leek' (the leek that has spearhead-shaped cloves). *Allium* is Latin for garlic, *sativum* = 'cultivated'. Also, **broad-leaved garlic**, **wild garlic** and **wood garlic**: alternative names for ramsons.

Mediterranean neophytes with scattered distributions, naturalised on waste ground and roadsides include **few-flowered garlic** *A. paradoxum* (introduced from the Caucasus in 1823), **hairy garlic** *A. subhirsutum*, **honey garlic** *Nectaroscordum siculum* (1832, also found in woodland, under bracken), **Italian garlic** *A. pendulinum* (Essex and Norfolk), **keeled garlic** *A. carinatum* (1789, spreading by seed – var. *pulchellum* – or, more frequently, by bulbils – var. *carinatum*), **Neapolitan garlic** *A. neapolitanum* (southern and south-west England and East Anglia), **rosy garlic** *A. roseum* (introduced by 1752), **three-cornered garlic** *A. triquetrum*, and **yellow garlic** *A. moly*.

Field garlic *A. oleraceum*, scattered in England, very scattered in Wales and Scotland, an occasional introduction in Ireland, on dry, usually steeply sloping, calcareous grasslands, and on sunny banks in river floodplains. Latin *oleraceum* means relating to kitchen gardens.

GARRYACEAE Family name of spotted-laurel, honouring Nicholas Garry (1782–1856), Secretary to the Hudson Bay Company.

GASTEROSTEIDAE, GASTEROSTEIFORMES Family and order names of sticklebacks (Greek *gaster* = 'abdomen' or 'stomach' and *osteon* = 'bone').

GASTRODONTIDAE Family name for some of the glass snails (Greek *gaster* = 'abdomen' or 'stomach' and *odontos* = 'tooth').

GASTROPOD(A) Class in the phylum Mollusca of slugs and snails, characterised by a single flattened, muscular, ventral 'foot' (Greek *gaster* + *podi* = 'stomach' + 'foot').

GATEKEEPER *Pyronia tithonus* (Nymphalidae, Satyrinae), a common and widespread moth, male wingspan 37–43 mm, female 42–8 mm, found in England south of a line from Cumbria to south-east Yorkshire, but absent from the Isle of Man and Scotland. Also found in south and south-east coastal Ireland. Habitat involves tall grass close to scrub, so includes scrubby grassland, woodland rides, field margins, hedgerows and gardens. Larvae feed at night on grasses, adults (flying June–September) on honeydew and the nectar of a range of plants including bramble, ragwort and thistles. Range has seen a 15% increase since 1976, mostly from a northward movement, but numbers declined by 41% between 1976 and 2014. *Pyronia* is Greek for 'purchase of wheat'; however, the next word in most Greek lexicons is *pyropos* = 'fiery-eyed' (which is descriptive of the wing eyespot), and this moth may have been misnamed with an inappropriate word having been transcribed. Tithonus in Greek mythology was the beloved of Eos (Aurora), granted immortality but not eternal youth.

GAVIIDAE, GAVIIFORMES Family and order names of divers. Latin *gavia* = 'sea mew' or 'common gull'.

GEAN Alternative name for wild cherry, from Old French *guine*.

GELATINOUS LICHEN Foliose species in the genus *Collema* and *Leptogium* (Collemataceae), respectively from Greek for 'that which is glued' and *leptos* = 'slender'. See under lichen.

GELECHIIDAE Family name for crest, groundling, neb and sober moths. Greek *geleches* = 'sleeping (or resting) on the ground', many species feeding on low-growing plants as larvae, and flying close to the ground as adults.

GEM (FLY) Common and widespread soldier flies, 3–5 mm length, in the genus *Microchrysa* (Greek 'small' + 'gold'), family Stratiomyidae.

Black gem *M. cyaneiventris*, more frequent in the north and west, usually in deciduous woodland or by shaded water edges, adults flying from May to September. Larvae live in decaying vegetation and moss. Latin *cyano* + *venter* = 'blue' + 'stomach'.

Black-horned gem *M. polita*, in woodland, hedgerows and gardens, flying in March–September, feeding on nectar. Larvae live in dung, rotting vegetation and compost heaps. Latin *polita* = 'polished'.

Green gem *M. flavicornis*, adults found in a variety of habitats, including woods, gardens, hedgerows, fens and heathland, flying in May–September. Latin *flavicornis* = 'yellow horn'.

GEM (MOTH) *Nycterosea obstipata* (Geometridae, Larentiinae), wingspan 18–21 mm, a nocturnal migrant from southern Europe, appearing in late summer and early autumn mainly in southern Britain, with occasional records north to the Orkneys, in some years arriving in sufficient numbers to breed. Larvae feed on the leaves of herbaceous plants. Greek *nycteros* = 'nocturnal', and Latin *obstipata* = 'bent forward', referring to the diagonal mark on the forewing.

GENERAL A soldier fly in the genus *Stratiomys* (Stratiomyidae) (Greek *stratiotys* + *myia* = 'soldier' + 'fly').

Banded general *S. potamida*, common and widespread in England and Wales, becoming more so since the 1980s. The carnivorous larvae are amphibious, feeding in ponds, ditches and stream margins. Adults are slow clumsy fliers, feeding on umbellifers and bramble in wet and marshy areas in June–August. Greek *potamos* = 'river'.

Clubbed general *S. chamaeleon*, Endangered in RDB 1. Its only recent record in England has been Cothill Fen, Oxfordshire. There are two known sites in Wales (on Anglesey, larvae being found in base-rich seepages in fen and adults on hogweed flowers), and one in Scotland. Flight period is June–August. Greek *chamae* + *leon* = 'low' + 'lion'.

Flecked general *S. singulario*, locally common, with a scattered distribution in the southern half of England, south Wales and Ireland, mostly in estuarine and coastal grazing marshes, more rarely in inland fen. The detritus-feeding amphibious larvae live in mud and shallow water with emergent vegetation, or temporary pools, favouring slightly brackish conditions. Adults fly in May–September, feeding on nectar. Latin *singularis* = 'solitary'.

Long-horned general *S. longicornis*, Vulnerable in RDB 2, confined to grazing marsh and saltmarsh along the southern coast from the Solent to Suffolk, with concentration round the Thames Estuary. Larvae live in mud in strongly brackish pools and ditches. Inland records, such as from Wiltshire and Wicken Fen (Cambridgeshire), are unlikely to represent breeding populations. Adult flight period is May–July. Latin *longicornis* = 'long horn'.

GENTIAN Herbaceous plants in the genera *Gentiana* and *Gentianella* (Gentianaceae) (Latin *gentiana*, itself from Greek *gentiane*).

Alpine gentian *Gentiana nivalis*, an annual or biennial found in a very few places in the Highlands above 730 m on rock ledges and vegetated scree. This is the county flower of Perthshire as voted by the public in a 2002 survey by Plantlife. Latin *nivalis* = 'snowy'.

Autumn gentian *Gentianella amarella*, an annual or biennial of well-drained basic soils, in grazed chalk and limestone grassland, on calcareous dunes and machair, on spoil tips, and in cuttings and quarries. In Ireland only the endemic ssp. *hibernica* has been recorded; ssp. *amarella* is widespread in Britain north to central Scotland; and ssp. *septentrionalis*, which grows on machair and dunes, and in the Grampians on grassy mica-schist slopes and ledges and in limestone grassland, is local in the Highlands, the Cheviot Hills and parts of upland Yorkshire. Latin *amarella* = 'slightly bitter'.

Chiltern gentian *Gentianella germanica*, an annual or biennial, declining in range, local in north Hampshire and on the Chilterns, on shallow chalk soil, in chalk grassland and chalk-pits, sometimes in open scrub and woodland margins. This is the county flower of Buckinghamshire as voted by the public in a 2002 survey by Plantlife.

Dune gentian *Gentianella uliginosa*, an annual of coastal dunes, dune slacks and machair, usually in open ground or short vegetation maintained by grazing, disturbance or winter flooding, found locally in north Devon (rediscovered at Braunton Burrows in 1998), south Wales, and Colonsay, Inner Hebrides. Latin *uliginosa* = 'marshy'.

Early gentian *Gentianella anglica*, an endemic annual or biennial of shallow calcareous soils, found in England and south Wales on closely grazed chalk and limestone grassland and quarries, and on clifftops and dunes. Latin *anglica* = 'English'.

Field gentian *Gentianella campestris*, a biennial, occasionally annual, with a scattered if mainly northern distribution, on mildly acidic to neutral soils in open habitats such as pasture, hill grassland, grassy heathland, dunes, machair and verges. Sites are being lost through overgrazing in the uplands and the neglect of lowland pastures. *Campestris* comes from Latin *campus* = 'field'.

Marsh gentian *Gentiana pneumonanthe*, a scarce and declining perennial of damp acidic grassland and wet heathland, usually on relatively enriched soils, found locally from Dorset and East Sussex to north-east Yorkshire and Cumbria, as well as Anglesey. Greek *pneumon* + *anthos* = 'wind' + 'flower'.

Spring gentian *Gentiana verna*, a herbaceous perennial, local in northern England, naturalised in a few other places, on limestone grassland and calcareous glacial drift; also on hummocks in calcareous flush communities, and in western Ireland on limestone pavement and fixed dunes. This is the county flower of Co. Durham as voted by the public in a 2002 survey by Plantlife. Latin *verna* = 'spring'.

Willow gentian *Gentiana asclepiadea*, a herbaceous central European neophyte perennial, in gardens by 1629, a naturalised garden escape by streams and in woodland in a very few places in England and Scotland. Asclepius was the Greek god of healing.

GENTIANACEAE Gentian family, including centaury and yellow-wort.

GEOMETRIDAE Family name for thin-bodied moths with relatively large wing surfaces. Caterpillars have no prolegs, so move using a looping motion (and are known as loopers), 'measuring the Earth', hence Geometridae, from the Greek for 'Earth' and 'measure'. Subfamilies are: Archiearinae (orange and light orange underwings), Ennominae (including beauties, thorns, most umbers, and some carpets and waves), Larentiinae (including most carpets, pugs, treble-bars and some umbers and waves), and Sterrhinae (including mochas and most of the waves).

GEORISSIDAE Family name of beetles, represented in the British list by *Georissus crenulatus* (1.5–2 mm length), local though widespread, usually caked in the drying mud and silt of rivers. Larvae live a centimetre or so beneath the surface and, like the adults, feed on mud containing algae and other microorganisms. Adults, active in May–June, move very slowly and feign death when disturbed. Latin *crenulatus* = 'crenulate' (having a scalloped or notched outline or edge).

GEOTRUPIDAE Family name (Greek *geos* + *trypetes* = 'earth' + 'borer') of dor beetles, with eight species in the British list, seven in the subfamily Geotrupinae comprising large black heavily built, strongly flying beetles (10–26 mm), including minotaur beetle, Britain's only member of the 'roller' guild of dung beetles. The other species excavate short burrows beneath or beside dung.

GERANIACEAE Family name (Greek *geranos* = 'crane') of crane's-bills and stork's-bills, so named from the appearance of the fruit capsule of some of the species, which have a long pointed beak shape.

GERMAN-IVY *Delairea odorata* (Asteraceae), a perennial South African neophyte with a woody base, naturalised and trailing over hedges and walls in the Channel Islands and Scilly, Cornwall, Hampshire and southern Ireland. Eugene Delaire (1810–56) was head gardener at the botanical gardens in Orleans. Latin *odorata* = 'fragrant'.

GERMANDER Species of *Teucrium* (Lamiaceae), the English name coming from the medieval apothecary's Latin *gamandrea*, derived from Greek *chamae* + *drys* = 'low' + 'oak', because the leaves of some species were thought to resemble those of oak. *Teucrium* may reference a plant used medicinally by Teucer, a king of Troy, but Linnaeus may have named the genus after a Dr Teucer, a medical botanist.

Cut-leaved germander *T. botrys*, a biennial herb, possibly native, otherwise a European neophyte cultivated by 1633, found on bare ground within open grassland, arable field margins, and open fallow overlying chalk and limestone in southern England. Having generally declined through agricultural intensification, scrub encroachment and lack of grazing, it benefits from disturbance and at some sites thousands of plants have been recorded following cultivation or conservation management such as harrowing and turf cutting. *Botrys* is Greek for 'trusses'.

Wall germander *T. chamaedrys*, a herbaceous perennial, usually viewed as a European neophyte, recorded in the wild since 1710, scattered in England on walls, rocks and dry banks. A population known since 1945 from clifftop chalk grassland at Cuckmere Haven, East Sussex, may be native.

Water germander *T. scordium*, a stoloniferous perennial herb scattered in England in wetland habitats with fluctuating water levels, for instance margins of dune-slack pools, reed-fen and the banks of rivers, ponds and ditches. In places it is declining

as a result of drainage, reclamation and eutrophication; elsewhere populations are threatened by lack of management, scrub encroachment and shading. In Ireland, where distribution is stable, it is often recorded from turloughs. Greek *scordon* = 'garlic'.

GERMANDERWORT Thallose liverworts in the genus *Riccardia* (Aneuraceae). 'Germander' comes from the medieval apothecary's Latin *gamandrea*, derived from Greek *chamae* + *drys* = 'low' + 'oak'. 'Wort' is Middle English for 'plant' or 'herb'. **Delicate germanderwort** *R. multifida* and **jagged germanderwort** *R. chamedryfolia* are both common and widespread in damp rock, bog and peat habitats.

GHOST SHRIMP See Callianassidae.

GHOSTWORT *Cryptothallus mirabilis* (Aneuraceae), a thalloid liverwort, widespread but scattered in Britain, found in carpets of sphagnum around the base of birch trees in wet woodland, usually at a depth of about 15 cm, where leaf litter forms a compact layer. Greek *cryptothallus* = 'hidden thallus' and Latin *mirabilis* = 'wonderful'.

GIANT-RHUBARB *Gunnera tinctoria* (Gunneraceae), a perennial herb with large rhubarb-shaped leaves, introduced from South America in 1908, forming dense thickets in damp places and in woodland near lakes and rivers, often self-sown and naturalised in much of lowland Britain, and considered to be invasive in western Ireland. Johan Gunnerus (1718–83) was a Norwegian botanist. Latin *tinctoria* = 'used in dyeing'.

GIANT SEQUOIA Alternative name for wellingtonia.

GIGARTINALES Order of red algae in the class Florideophyceae.

GILL (FUNGUS) **Crimped gill** *Plicatura crispa* (Agaricales), inedible, with a scattered distribution in northern England and Scotland usually on dead wood of deciduous trees in woodland, northerly records on rowan and hazel. Latin *plicatus* = 'braided', and *crispa* = 'curly'.

GILLAROO A variety of brown trout (though viewed by some taxonomists as a distinct species, *Salmo stomachicus*) which eats snails and other benthic invertebrates, and only known from Lough Melvin on the border between Co. Leitrim and Co. Fermanagh. Irish *giolla rua* = 'red fellow', reflecting the fish's distinctive colouring.

GILLYGOBBLER **Green gillygobbler** *Hypomyces viridis* (Hypocreales, Hypocreaceae), a scarce mould fungus, with a few scattered Scottish and English records, expressing itself as dense clusters of 'pinheads' on old woodland polypore fungi. Greek *hypo* + *myces* = 'below' + 'fungus', and Latin *viridis* = 'green'.

GILTHEAD See gilt-head bream under bream.

GINGERTAIL Uncommon saprobic fungi in the genus *Xeromphalina* (Agaricales, Mycenaceae), found in autumn on dead wood of Scots pine. **Pinelitter gingertail** *X. cauticinalis* is mainly recorded from the Highlands, with one record from Brecon. **Pinewood gingertail** *X. campanella* is mainly recorded from northern England and the Highlands. Greek *xeromphalina* = 'little dry navel'.

GLADIOLUS Cormous perennial herbs in the genus *Gladiolus* (Latin for 'little sword') (Iridaceae).

Eastern gladiolus *G. communis*, a Mediterranean neophyte cultivated in Tudor times, scattered in lowland (and especially south-west) England, Wales and south-east Ireland, naturalised in field margins, roadsides and rough ground. *Communis* is Latin for 'common'.

Wild gladiolus *G. illyricus*, on acidic, brown earth soils on grass-heaths in the New Forest, usually associated with bracken in scrub. In classical antiquity, Illyria was a region in the Balkan Peninsula.

GLAND-MOSS **Slender gland-moss** *Tayloria tenuis* and

tongue-leaved gland-moss *T. lingulata* (Splachnaceae), both found at a few sites each in the Highlands on herbivore dung. Thomas Taylor (1786–1848) was an English botanist. *Tenuis* is Latin for 'slender', *lingulata* for 'tongue-shaped'.

GLASS SNAIL Pulmonate species in the genera *Zonitoides* (Greek *zone* = 'girdle') (Gastrodontidae) and *Aegopinella* (*aiga* + *pinella* = 'goat' + 'brush'), *Oxychilus* (*oxy* + *cheilos* = 'sharp' + 'margin') and *Nesovitrea* (*nesos* = 'island' and Latin *vitreus* = 'like glass') (Oxychilidae).

Clear glass snail *A. pura*, shell diameter 3.5–5 mm, height 2–2.7 mm, common and widespread in litter in deciduous (especially beech) woodland, usually on a calcareous substrate. *Pura* can mean 'clear' in Latin.

Draparnaud's glass snail *O. draparnaudi*, a European introduction, shell diameter 12–15 mm, height 6–7 mm, common and widespread in England, Wales, the Central Lowlands of Scotland, and the southern half of Ireland, in shady habitats, including gardens and waste ground, in leaf litter and under plants and stones, preying on invertebrates, including other snails. Jacques Draparnaud (1772–1804) was considered to be the father of malacology in France.

Glossy glass snail *O. navarricus*, shell diameter 8–10 mm, height 4.5–6 mm, a widespread post-Roman introduction to Britain from western Europe, still spreading, and introduced in Ireland probably in the 1970s, on rubble, roadsides and gardens. Latin *navarricus* = 'of Navarre' (north Spain).

Hollowed glass snail *Z. excavatus*, shell diameter 5.3–6 mm, height 3.5–4 mm, in the litter of acid woodland, widespread and locally common in southern England and western Britain, and the only obligate calcifuge land mollusc in Ireland, absent from the central plain and from eastern counties. Latin *excavatus* = 'hollowed'.

Pyrenean glass snail Alternative name for Pyrenean semi-slug.

Rayed glass snail *N. hammonis*, shell diameter 3.6–4.1 mm, height 1.9–2.1 mm, common and widespread in the litter of damp deciduous (often beech) woodland, unimproved pasture, heathland and marsh, and, tolerant of acid conditions, on some Irish blanket peat.

Shiny glass snail *Z. nitidus*, shell diameter 6–7 mm, height 3.5–4 mm, common and widespread but scarce in north-east England and upland Scotland, in damp habitats such as marsh and the margins of rivers, lakes and ponds, found under wood, rock and plants, feeding on decomposing plant material and fungi. Latin *nitidus* = 'bright'.

Smooth glass snail *A. nitidula*, shell diameter 6–10 mm, height 4–6 mm, common and widespread in litter in gardens, hedgerows, roadsides and fields, feeding on plant material.

GLASSCUP Common glasscup *Orbilia xanthostigma* (Orbiliales, Orbiliaceae), a common and widespread disc fungus found on rotting wood under many hardwood species. *Orbilia* comes from the circular shape (Latin *orbis*) and Greek *xanthos* + *stigma* = 'yellow' + 'pointed mark'.

GLASSHOUSE RED SPIDER MITE See red spider mite under spider mite.

GLASSWORT Species of *Salicornia* and *Sarcocornia* (Amaranthaceae). These plants were reduced to ash to provide glass-makers with alkali (carbonate of soda), hence the name. *Salicornia* comes from Latin *sal* + *cornu* = 'salt' + 'horn', referring to the horn-like branched fronds. *Sarcocornia* uses a prefix from Greek *sarcodis* = 'fleshy'.

Common glasswort *Sal. europaea*, an annual found around much of the coastline of the British Isles at all levels of sandy or muddy saltmarsh, in saltmarsh-sand dune transitions and wet, tidally submerged dune slacks.

Glaucous glasswort *Sal. obscura*, an annual found in mud in saltmarsh pans and creeks, local in England north to Lancashire and Lincolnshire. Latin *obscura* = 'indistinct' or 'dark'.

Long-spiked glasswort *Sal. dolichostachya*, an annual found along much of the coastline of Britain, except north-west Scotland, and of south-east Ireland, growing on open mud and muddy sand on intertidal flats and in the lowest parts of saltmarsh, and occasionally mid-marsh along the banks of creeks and runnels. Greek *dolichos* + *stachys* = 'long' + 'flower spike'.

One-flowered glasswort *Sal. pusilla*, an annual found from the south Devon coast round to Lincolnshire, and in south Wales and parts of Ireland, in the uppermost saltmarsh, on firm mud or sand in salt pans and on the drift line. *Pusillus* is Latin for 'very small'.

Perennial glasswort *Sar. perennis*, a woody perennial sub-shrub found along the south and south-east English and East Anglian coasts, and in Co. Wexford, on bare or sparsely vegetated eroding lower parts of saltmarsh, at higher elevations on saltmarsh drift lines, and on shell and shingle banks.

Purple glasswort *Sal. ramosissima*, a morphologically variable annual, widespread but scarcer in Scotland and Ireland, usually in the middle and upper saltmarsh, in closed common saltmarsh-grass swards, salt pans, creeks and drift lines. It also grows on firm sand and muddy shingle, and in brackish grazing marsh. *Ramosissima* is Latin for 'much-branched'.

Shiny glasswort *Sal. emerici*, a local annual scattered in England, Wales and Ireland, mainly found in the middle and upper saltmarsh, especially in mud associated with salt pans and runnels.

Yellow glasswort *Sal. fragilis*, an annual found in mud and muddy sand on intertidal flats and in the lowest parts of saltmarsh, absent from much of Scotland and Ireland. *Fragilis* is Latin for 'fragile'.

GLIRIDAE Family name of common and edible dormouse (Latin *glis* = 'dormouse').

GLOBE-THISTLE *Echinops exaltatus* (Asteraceae), a biennial to perennial southern European neophyte herb widely grown in gardens since 1822, scattered in Britain on roadsides and waste ground, usually casual, sometimes naturalised. Latin *echinus* = 'hedgehog' + Greek *opsis* = 'appearance', referring to the flower heads. Latin *exaltatus* = 'raised up'.

Blue globe-thistle *E. bannaticus*, a biennial to perennial south-eastern European neophyte herb, grown in gardens since 1832, scattered in Britain, locally common and naturalised on roadsides, railway banks and waste ground. *Bannaticus* refers to the Banat, a region now in Romania, Hungary and Serbia, where the plant was found.

Glandular globe-thistle *E. sphaerocephalus*, a biennial to perennial southern European neophyte herb grown in gardens since the late sixteenth century, scattered in Britain on roadsides, railway banks and waste ground, and in disused quarries, usually casual but sometimes persisting. Latin *sphaerocephalus* = 'spherical head'.

GLOBEFLOWER *Trollius europaeus* (Ranunculaceae), a perennial herb, local usually in upland parts of Wales, northern England, Scotland and north-west Ireland, in damp grassland, woodland, stream banks, lake margins and rock ledges, preferring basic soils, and often associated with limestone. This is the county flower of Co. Fermanagh as voted by the public in a 2002 survey by Plantlife. *Trollius* may deriv from Latin *trulleum* = 'basin', from a Swiss-German name for the flower, *trollblume*, or from German *trol* = 'globe' or 'round object'.

GLORY (MOTH) Kentish glory *Endromis versicolora* (Endromidae), wingspan 50–65 mm, the only British species in its family, with a declining distribution, now restricted to the Highlands (with possibly an isolated population in Worcestershire) in birch scrub and lightly wooded moorland, males flying by day and night in March–May, females by night. Larvae feed on silver birch, occasionally downy birch and alder. Greek *en* + *dromos* = 'in' + 'running', the dense abdominal hair resembling a tracksuit, and Latin *versicolora* = 'changing or reversing colour', from the wing pattern.

GLORY-OF-THE-SNOW *Scilla forbesii* (Asparagaceae), a bulbous herbaceous neophyte perennial from Turkey, scattered and often naturalised in Britain as a garden throw-out. The English name refers to the habit of flowering in alpine zones when the snow melts in spring. *Scilla* is the Greek name for a plant in a different genus, *Drimia maritima*. James Forbes (1773–1861) was a British botanist and gardener.

Lesser glory-of-the-snow *S. sardensis*, a bulbous herbaceous neophyte perennial from Turkey, scattered and naturalised in parts of England and Scotland as garden throw-outs. Latin *sardensis* = 'Sardinian'.

GLOSSOSOMATIDAE Family name (Greek *glossa* + *soma* = 'tongue' + 'body') of caddis flies whose domed, tortoise-shaped larval case is made of sand grains.

GLOVES (FUNGUS) Rare species of *Hypocreopsis* (Hypocreaceae), from Greek *hypo* + *creopsis* = 'under' + (probably) 'similar to snakes', a reference to the contorted 'fingers', a growth form that also gives the common name. **Hazel gloves** *H. rhododendri* and **willow gloves** *H. lichenoides* parasitise glue crust fungus on, respectively hazel (occasionally blackthorn) and willow. See also glue (fungus).

GLOW-WORM *Lampyris noctiluca* (Lampyridae), a widespread beetle, probably declining in numbers and range. Larvae paralyse then feed on snails. Adults do not feed, and live for no more than a fortnight. Larviform (wingless) females (15–25 mm length) are strongly bioluminescent so to attract the flying males (15–18 mm). She remains on one spot with a curled posture so as to turn the glow upwards, and often moving her tail segments from side to side. Once mated she stops emitting light, lays eggs, then dies. Greek *lampo* = 'to shine', and Latin *noctis* + *lux* = 'night' + 'light'.

Lesser glow-worm *Phosphaenus hemipterus*, for which only one extant colony is known, near Burlesdon in Hampshire. Both sexes are flightless and only weakly bioluminescent. Greek *phos* = 'light' + either *phaeno* = 'to show' or *phaenolos* = 'light-giving', and *hemi* + *pteryx* = 'half' + 'wing'.

GLUE (FUNGUS) Willow glue *Hymenochaete tabacina* (Hymenochaetaceae), a scarce encrusting fungus, scattered in Britain, sporulating in autumn, in deciduous woodland on dead wood, favouring willow but also recorded from ash and oak. Greek *hymen* + *chaete* = 'membrane' + 'long hair', referring to the fine hairs on the upper cap surface, and New Latin *tabacina* = 'relating to tobacco', from its scent.

GLYPHIPTERIGIDAE Family name (Greek *glyphis* + *pteryx* = 'notch' + 'wing', referring to an indentation on the forewing) for fanner moths, some of the smudge moths, and leek moth.

GNAPHOSIDAE Family name of ground spiders, with 33 British species in 11 genera. Greek *gnaphos* = 'wool comber's card'.

GNAT Small flies, often swarming as adults, members of the suborder Nematocera. 'Gnat' comes from Old English *gnætt*. See also fungus gnat.

Winter gnat *Trichocera annulata* (Trichoceridae), 8–10 mm long, a kind of cranefly, widespread and common in woodland, rough ground, parks and gardens, swarms (mating dances) gathering in sunlight throughout the year. Larvae feed on decaying vegetation. Greek *trichos* + *ceras* = 'hair' + 'horn', and Latin *annulata* = 'ringed'.

Wood gnat Member of the family Anisopodidae.

GNATHOSTOMATA A superclass comprising the jawed vertebrates (Greek *gnathos* + *stoma* = 'jaw' and 'mouth').

GOAT *Capra hircus* (Bovidae), a domesticated Neolithic

introduction with a number of feral populations, for example at Bagot's Park, near Uttoxeter (Staffordshire) and the Cheviot Hills (Northumberland); in Merioneth, Snowdonia and elsewhere in Wales; various parts of especially west and south-west montane Scotland; and in Ireland, again mainly in the west, but also the Bilberry goats in the town of Waterford (21 animals in 2005). There are perhaps 5,000–10,000 feral goats in Britain. 'Goat' comes from Old English *gat* = 'nanny goat'. The binomial is Latin for 'she-goat' and 'he-goat'.

GOAT TANG Alternative name for the seaweed discoid fork weed, tang being an old Norse word for washed-ashore seaweed.

GOAT'S-BEARD *Tragopogon pratensis* (Asteraceae), a scattered but widespread annual to perennial herb, absent from upland Scotland and much of Ireland, in tall grassland in meadow and pasture, on field margins, dunes, roadsides, railway banks and waste ground. The English name is a translation of the herbalists' Latin *barba hirci*, in turn a translation of Greek *tragopogon* (*tragos* + *pogon* = 'goat' + 'beard'), the long white pappus resembling a goat's beard. Latin *pratum* = 'meadow'.

GOAT'S-RUE *Galega officinalis* (Fabaceae), a perennial herb from Europe introduced into cultivation by 1568, but its spread is recent: it was not mapped in the 1962 *Atlas of the British Flora*, yet in the twenty-first century it has been found in parts of London in 75% of the tetrads in each 10 km square. It has mostly been an urban plant of waste ground, gravel-pits, roadsides, railway banks and tips, but is now spreading into the countryside, mainly in England. *Galega* comes from the Greek *galacto* + *ago* = 'milk' + 'to lead', and as with the English name refers to its use for stimulating lactation in goats and cattle. *Officinalis* is Latin for having pharmacological value.

GOBIIDAE Family name of species of goby.

GOBLET *Pseudoclitocybe cyathiformis* (Agaricales, Tricholomataceae), an edible fungus found from September to December, occasionally into spring on soil on woody debris in mixed woodland. The English name reflects the shape of the cap and the long slender stem. *Pseudoclitocybe* indicates that this species look very much like *Clitocybe* (funnel) mushrooms. *Cyathiformis* uses Greek and Latin to mean 'cup-shaped'.

Alder goblet *Ciboria caucus* (Heliotales, Sclerotiniaceae), inedible, infrequent in the southern half of Britain, from spring to early summer on old and rotting (previous season's) catkins of alder, less commonly hazel and willow, often submerged at water's edge. Both parts of the binomial are Latin for a drinking vessel.

Beechwood goblet *Tatraea dumbirensis* (Heliotaceae), inedible, Vulnerable in the RDL (2006), scattered in England and Wales, on dead wood of beech and oak.

Haw goblet *Monilinia johnsonii* (Sclerotiniaceae), associated with hawthorn leaves, scattered in England and Wales.

GOBLIN LIGHTS *Catolechia wahlenbergii* (Rhizocarpaceae), a lichen found in damp crevices on Ben Nevis, and on limestone in the central Highlands on Ben Alder and Aonach Beag, near Dalwhinnie. Being protected in the WCA (1981), the appropriate committee gave it a required English name. Göran Wahlenberg (1780–1851) was a Swedish botanist.

GOBY Fish in the family Gobiidae. 'Goby' dates from the mid-eighteenth century, from Latin *gobius* (= 'gudgeon'), in turn from Greek *cobios*, denoting some kind of small fish. Genera include *Gobius*, *Gobiusculus* (diminutive of *Gobius*), *Pomatoschistus* (Greek *poma*, -*atos* + *schistos* = 'cover' or 'operculum' + 'divided') and *Thorogobius* (incorporating Greek *thoros* = 'semen').

Black goby *Gobius niger*, length up to 18 cm, probably widespread, in shallow coastal or estuarine waters among *Zostera* or algae on sandy or muddy bottoms, at depths of 1–50 m, feeding on invertebrates and small fish. *Niger* is Latin for 'black'.

Common goby *P. microps*, length up to 9 cm, common and widespread in tide pools, estuaries, saltmarsh and brackish landlocked lagoons. While preferring open water over bare muddy or sandy sediment down to 10 m, it is often found among dense vegetation. There may be several spawnings between April and September, the female laying eggs under shells, stones or on aquatic plants, the male then guarding the eggs. Diet is mainly of small crustaceans. Greek *micros* + *ops* = 'small' + 'eye'.

Couch's goby *Gobius couchi*, length up to 9 cm, only recorded from four locations – Helford (south Cornwall), Portland Bill (Dorset), Lough Hyne (Co. Cork), Mulroy Bay (Co. Donegal) – in the lower intertidal and inshore waters, under stones or algae on sheltered muddy sand. In 1998 it was added to Schedule 5 of the WCA (1981). Jonathan Couch (1789–1870) was a Cornish physician and naturalist.

Giant goby *Gobius cobitis*, length up to 27 cm, in intertidal rock pools on sheltered shores down to 10 m in the south-west coast of England from Wembury to the Scillies, feeding on invertebrates and green macroalgae. In 1998 it was added to Schedule 5 of the WCA (1981). Greek *cobitis* is a type of sardine.

Leopard-spotted goby *T. ephippiatus*, length up to 13 cm, widespread except for the east coast, living in fissures of steep rock faces inshore or a short distance offshore, as well as in sheltered estuaries and sea lochs, and occasionally deep rock pools. Their depth distribution ranges from the low water of spring tides to 40 m, usually no more than 12 m, feeding on small crustaceans and polychaetes. Latin *ephippiatus* = 'mounted on a saddled horse', reason obscure.

Rock goby *Gobius paganellus*, length up to 12 cm, probably widespread, in rocky habitats from the intertidal down to 15 m, on shores with good seaweed cover in rock pools and under stones, feeding on small crustaceans and other invertebrates. Spawning is in spring, usually for kelp forest. *Paganus* is Latin for 'rustic'.

Sand goby *P. minutus*, length usually 4–5 cm, common and widespread on sandy or muddy substrates usually to a depth of 20 m, and present in estuaries, lagoons, saltmarsh and coastal waters, feeding mainly on amphipods. It breeds in summer, the male building a nest usually under a shell. *Minutus* is Latin for 'minute'.

Two-spotted goby *Gobiusculus flavescens*, length up to 6 cm, common and widespread in small shoals among macroalgae and eelgrass beds and over seaweed-covered rocks, in intertidal pools, and the sublittoral down to 20 m, feeding on zooplankton, including larval crustaceans. In summer the male becomes territorial and scoops out a nest in the sand, usually under a stone, in which the female lays her eggs. The male guards these until hatch. Juveniles move into deeper water to spend the winter. *Flavescens* is Latin for 'yellowish'.

GODETIA *Clarkia amoena* (Onagraceae), an annual North American neophyte, a garden escape found as a casual with a scattered distribution particularly in England and Wales. Charles Godet (1797–1889) was a French naturalist. Captain William Clark (1770–1838) was an American explorer. The specific name is Latin for 'delightful'.

GODWIT Wading shorebirds in the genus *Limosa* (Latin *limus* = 'mud') in the family Scolopacidae. **Hudsonian godwit** *L. haemastica* is a scarce accidental. 'Godwit' purports to imitate the bird's call.

Bar-tailed godwit *L. lapponica*, wingspan 75 cm, length 38 cm, male weight 300 g, females 370, an Amber List winter visiting shorebird, widespread but especially found on large estuaries, numbers starting to build in July–August and falling off in March–April, totalling 43,000 in Britain in 2008–9, and 2,500–3,500 in Ireland. They feed along the tidal edge or in shallow water, beginning on an ebbing tide, and do so continuously for up to six hours. Polychaetes, particularly lugworms, are the main food. *Lapponica* refers to Lapland.

Black-tailed godwit *L. limosa*, a Red List migrant shorebird, wingspan 70–82 cm (females larger), length 42 cm, male weight 280 g, females 340 g, with a small migrant breeding population

(wintering in West Africa) nesting on marsh and wet meadow: there were around 65 pairs in 2009, with 90% of these found in the Nene Washes. Some 43,000 birds, mostly from Iceland, were recorded overwintering in 2008–9, mostly in England and south Wales, with more birds (1,000–3,000 annually) along the coasts of south and east Ireland. Nesting is in scrapes, in loose colonies, with clutches usually of 4 eggs, incubation taking 22–4 days, the precocial chicks fledging in 25–30 days. Diet is of insects, worms and snails, but also plants, beetles, grasshoppers and other small insects during the breeding season. The female has a bill 12–15% longer than that of the male, reducing intersexual competition for food items.

GOERIDAE Family name (Greek *goeros* = 'mournful') of caddis flies, larvae constructing tubular cases of sand particles with small stones attached to the sides as ballast.

GOLD (MOTH) 1) Small day-flying moths in the family Micropterygidae, genus *Micropterix* (Greek *micros* + *pteryx* = 'small' + 'wing'). Adults, with wingspans of 7–11 mm, are among moths with working mandibles, and they feed on the pollen of a variety of flowers.

Black-headed gold *M. mansuetella*, local in much of England into parts of Scotland in damp habitats. Larval diet is not known. Latin *mansuetus* = 'tame', from the 'quiet' behaviour of adult females when feeding on pollen.

Plain gold *M. calthella* found throughout most of Britain, flying in daytime in May–June. Larvae may feed on sedges, adults on marsh-marigold (*Caltha*, giving the specific name).

Red-barred gold *M. tunbergella*, adults flying during the day in June–July, recorded from Wales and most English counties in woodland glades, but scarce or unknown in Scotland and Ireland. Adults feed on the pollen of trees such as oak, sycamore and hawthorn. Larval diet is not known. Karl von Thunberg (1743–1828) was a Swedish botanist.

White-barred gold *M. aruncella*, widespread, flying in daytime between May and August. Larval diet is not known. *Aruncella* comes from adults feeding on pollen of *Aruncus*.

Yellow-barred gold *M. aureatella*, day-flying, widespread in Britain except in parts of the east, in woodland and heathland. Larval diet is not known but probably includes bilberry. Latin *aureatella* = 'golden', from metallic markings on the forewings.

2) **Purple-bordered gold** *Idaea muricata* (Geometridae, Sterrhinae), wingspan 18–20 mm, local in the southern half of England and Wales, occasional in south-west and central Ireland, on damp heathland, fen, marsh and wet grassland, adults flying from dusk onwards in June–July. Larvae feed on marsh cinquefoil. Idaios refers to Mt Ida, from where the deities watched the battles of the Trojan War; *Murex* is the mollusc that yields the Tyrian purple dye, referencing the purple wing markings.

GOLD-OF-PLEASURE *Camelina sativa* (Brassicaceae), an annual archaeophyte, once a common arable weed, now mainly occasional as a casual derived from birdseed on waste ground, and as a garden weed. The English name refers to the yellow flower; although a weed, the plant was also used for oil derived from the seeds. Greek *chamae* + *linon* = 'dwarf' + 'flax', alluding to its being a weed that suppressed flax. Latin *sativa* = 'cultivated'.

Lesser gold-of-pleasure *C. microcarpa*, a rare casual annual of arable fields and waste ground, probably introduced with grain. Greek *microcarpa* = 'small fruit'.

GOLD SPOT *Plusia festucae* (Noctuidae, Plusiinae), wingspan 34–46 mm, a widespread and common moth in marsh, fen, woodland rides and upland grassland, and on heathland, moorland and riverbanks. It has two generations in the south, sometimes only one in the north, flying at night in June–July and August–September. Larvae feed on sedges and other waterside plants. Greek *plousios* = 'rich', referring to the silver or gold markins on the forewing. *Festucae* may come from Latin for 'stalk', or refer to fescue grass (as a food plant).

Lempke's gold spot *P. putnami*, wingspan 32–42 mm, nocturnal, local in fen, marsh, riverbanks and upland grassland throughout much of Norfolk, northern England and southern Scotland. Adults fly in July–August. Larvae feed on small-reeds and Yorkshire-fog.

GOLDCREST *Regulus regulus* (Regulidae), wingspan 13.5–15.5 cm, length 8.5–9.5 cm, weight 4.5–7.0 g, with the congeneric firecrest Europe's smallest bird, common and widespread, especially in coniferous woodland, with 610,000 territories estimated in 2009, and numbers augmented in autumn with migrants from Scandinavia. Although there was a decline in numbers by 15% over 1970–2015, the period 2010–15 saw a strong increase of 18%. The nest, often suspended on cobweb thread from a hanging branch, has an outer layer of moss, twigs, cobweb and lichen; a middle layer is of moss, this in turn lined by a layer of feathers and hair. Clutch size is 6–8, incubation takes 16–19 days, and fledging of the altricial young 17–18 days. Feeding is on aphids, springtails and other small insects, caterpillars and spiders (eggs, cocoons and adults). Goldcrests are members of the kinglet family: the binomial is Latin for 'little king', a diminutive of Latin *rex* = 'king'.

GOLDEN-HEAD MOSS *Breutelia chrysocoma* (Bartramiaceae), on mountain and moorland on unshaded, acidic ground, for example flushes, streamsides, wet heathland, grassy slopes, damp rock ledges, and open hazel and birch woodland in Scotland. It also grows in bogs in the more oceanic parts of Britain, and on limestone pavements in west Ireland. Johann Breutel (1788–1875) was a Moravian (Czech) botanist. Greek *chrysa* and Latin *coma* = 'golden' + 'hair'.

GOLDEN ORIOLE *Oriolus oriolus* (Oriolidae), wingspan 46 cm, length 24 cm, weight 68 g, a Red List insectivorous migrant breeder occasionally seen in deciduous woodland, especially in East Anglia, arriving in May, returning to the Continent in August. No breeding has been recorded since 2009, though singing males have been heard. The English name and the binomial come from Latin *aureolus* = 'golden', reflecting the yellow plumage.

GOLDEN-SAXIFRAGE Members of the genus *Chrysosplenium* (Saxifragaceae), perennial stoloniferous herbs. 'Saxifrage' comes from Latin *saxum + frangere* = 'rock' + 'to break'. *Chrysosplenium* is from Greek for 'golden band'.

Alternate-leaved golden-saxifrage *C. alternifolium*, local though widespread except in east England and north Scotland, in boggy ground in woods, by stream banks, and on mountain ledges, cliffs and gullies, often in shade and usually with a supply of alkaline water. It is also found in montane, bryophyte-dominated flushes. Latin *alternifolium* = 'alternate-leaved'.

Opposite-leaved golden-saxifrage *C. oppositifolium*, widespread except in east England, in boggy ground and seepages in woodland, and on stream banks and mountain ledges, usually in shade; also in grikes and sink holes in limestone in the Pennines. This is the county flower of Clackmannanshire as voted by the public in a 2002 survey by Plantlife. Latin *oppositifolium* = 'opposite-leaved'.

GOLDENEYE *Bucephala clangula* (Anatidae), wingspan 72 cm, length 46 cm, male weight 1 kg, females 750 g, a winter visitor from northern Europe with 27,000 birds estimated present in 2015–16 (though BBS data has indicated a 32% decline between 2002–3 and 2012–13), found on lakes, large rivers and sheltered coasts, particularly in northern and western Britain. Breeding was first recorded in Inverness-shire in 1970, since when birds have been attracted to nest in specially designed boxes constructed on trees close to water in Speyside, and around 200 pairs have been noted since 2009. Clutch size is 8–11, incubation lasts 28–32 days (the male abandoning his mate after 1–2 weeks), and the precocial chicks remain in the nest for 24–36 hours, to fledge in 57–66 days. These diving ducks feed on crustaceans, mussels,

insect larvae, small fish and plants. It is on the Amber List. Greek *boucephalos* = 'bull-headed', a reference to the bulbous head shape of the North American congeneric, bufflehead *B. albeola*. *Clangula* comes from Latin *clangere* = 'to resound'.

GOLDENRING Common goldenring Alternative name for golden-ringed dragonfly.

GOLDENROD *Solidago virgaurea* (Asteraceae), a herbaceous perennial, common and widespread, though absent from the east Midlands and central Ireland, on free-draining, usually acidic substrates in a range of habitats: in the lowlands, including woodland, hedge banks, heathland and coastal clifftops; in the uplands, cliff ledges, rocky streamsides, tall-herb communities, montane grass-heath and fellfield. The English and specific names derive from Latin *virga + aurea* = 'rod' or 'twig' + 'golden'. *Solidago* may come from Latin meaning 'heal' or 'make whole'.

Canadian goldenrod *S. canadensis*, a rhizomatous perennial North American herb cultivated by 1648, common and widespread, especially in England, Wales and central Scotland, naturalised on roadsides, by railways and on riverbanks, (urban) waste ground and spoil heaps.

Early goldenrod *S. gigantea*, a rhizomatous perennial North American herb cultivated by 1758, common and widespread, especially in England, south Wales, and south and central Scotland, naturalised on roadsides, by railways and on riverbanks, (urban) waste ground and spoil heaps. *Gigantea* is Latin for 'giant'.

Grass-leaved goldenrod *S. graminifolia*, a rhizomatous perennial North American herb, scattered in England, south Wales, and south and central Scotland, naturalised on roadsides, by railways, (urban) waste ground and spoil heaps. Latin *graminifolia* = 'grass-like leaf'.

Rough-stemmed goldenrod *S. rugosa*, a herbaceous perennial North American neophyte, scattered in England and west-central Scotland, naturalised on waste ground and rough grassland. Latin *rugosa* = 'wrinkled' (i.e. rough).

GOLDFINCH *Carduelis carduelis* (Fringillidae), wingspan 24 cm, length 12 cm, weight 17 g, a common and widespread resident breeder found in a variety of habitats where there are scattered shrubs and trees, rough ground with thistles and other seeding plants, with 1.2 million pairs in Britain in 2009, numbers increasing by 117% over 1995–2015, including by 15% in 2010–15. Nests, built by the female, of moss and lichen, attached to branches using spider thread, usually contain 4–6 eggs which hatch after 13–15 days, the altricial young fledging in 14–17 days. Diet is of small seeds, nestlings eating insects. *Carduelis* is Latin for 'goldfinch'.

GOLDFISH *Carassius auratus* (Cyprinidae), an ornamental introduction established in the wild from escapes and deliberate release, found in ponds, lakes and slow-moving rivers, mainly in England and Wales, feeding on insect larvae, crustaceans, copepods and plant material. Breeding is in June–July when conditions are exceptionally warm. It readily hybridises with crucian carp. *Carassius* comes from a Latinisation of *karass*, a common East European name for crucian carp, and *auratus* = 'golden'.

GOLDSINNY *Ctenolabrus rupestris* (Labridae), length 12–18 cm, a widespread species of wrasse, though rare in the North Sea and eastern English Channel, found among rocks and seaweed (particularly eelgrass) at depths of 1–50 m. Adults live in deeper waters, but young can be found inshore, including in rock pools. Greek *cteno + labros* = 'comb' + 'furious', and Latin *rupes* = 'rocks'.

GOLF-CLUB MOSS *Catoscopium nigritum* (Catoscopiaceae), found in Anglesey, the Sefton Coast, Co. Durham and north Northumberland in dune slacks, and also base-rich montane flushes in parts of Scotland. Greek *catoscopium* = 'looking down', referring to the pendulous capsules, and Latin *nigritum* = 'blackness'.

GONACTINIIDAE Family name of sea anemones (Greek

gonia + *actinia* = 'angle' + 'ray'), with two species in British waters. See sealoch anemone under sea anemone.

GOOD-KING-HENRY *Chenopodium bonus-henricus* (Amaranthaceae), a perennial archaeophyte, present in Roman times and once grown for its edible leaves, scattered and locally common except in north and west Scotland and Ireland, growing on disturbed, nitrogen-rich soil around farm buildings, and on roadsides and waste ground. It also sometimes occurs in limestone grassland. The English and specific names are a sixteenth-century translation of the German for this plant, *Guter Heinrich*, with 'king' interpolated: this was a herb of 'good' medicinal value. Greek *chen* + *pous* = 'goose' + 'foot', referencing the leaf shape.

GOOSANDER *Mergus merganser* (Anatudae), wingspan 90 cm, length 62 cm, male weight 1.7 kg, females 1.3 kg, first recorded as breeding in Britain in Perthshire in 1871, increasing numbers in Scotland, then since 1970 spreading across England into Wales, and now reaching south-west England, totalling around 3,500 pairs in 2009. Numbers increased by 122% over 1981–2015, though actually declined by 12% in 1995–2014, and fluctuate annually, for example increasing by 82% between 2014 and 2015, but declining by 41% between 2015 and 2016. They are mostly found in rivers in summer; birds are shot under licence on several, mostly Scottish, rivers to protect angling interests, but this probably has at most a local impact on population size. A few pairs nest in Co. Wicklow. In winter goosanders move to lakes, gravel-pits and reservoirs, occasionally to sheltered estuaries. Migrant Eurasian winter populations add to numbers in England and parts of east Scotland: an estimated total of 12,000 birds were present in 2008–9. Nests are in tree cavities (and nest boxes), clutch size is usually 8–11 eggs, incubation lasts 30–2 days, and the precocial young are taken by their mother in her bill to rivers or lakes immediately after hatching, and fledge in 60–70 days. This diving sawbill duck feeds mainly on fish, but takes other prey, for example molluscs and crustaceans. The English name, dating from the seventeenth century, probably comes from goose + *ander* as in the dialect word *bergander* = 'shelduck'. Latin *mergus* + *anser* = 'diver' + 'goose'.

Red-breaster goosander Alternative name for red-breasted merganser.

GOOSE Large birds in the family Anatidae, genera *Alopochen* (Greek *alopos* + *chen* = 'goose' + 'fox-like'), *Anser* (Latin for 'goose') and *Branta* (Latinised form of Old Norse *brandgás* = 'burnt (black) goose').

Barnacle goose *B. leucopsis*, wingspan 138 cm, length 64 cm, weight 1.7 kg, a winter visitor from Greenland and Spitzbergen, though there are also resident populations of feral birds that have escaped from captive flocks, widespread in Britain, mostly coastal in Ireland. Numbers in winter 2009–10 were 58,000 from Greenland, 33,000 from Spitzbergen, and 3,000 from the feral UK populations, with 900 breeding pairs in the last group in spring. Diet is of shoots and leaves. The species is Amber-listed because of declining numbers. The English name reflects the old belief that the bird grew from barnacles: its breeding grounds were not known, and barnacles looked a little like miniature geese. Greek *leucos* + *opsis* = 'white' + 'faced'.

Bean goose *Anser fabalis*, with sspp. *fabalis* (**taiga bean goose**) and *rossicus* (**tundra bean goose**), wingspan 158 cm, length 75 cm, male weight 3.4 kg, females 2.8 kg, a winter migrant mostly from Scandinavia, scattered in England but mostly near the coast in East Anglia. In 2009–10 there were records of 410 birds from the 'taiga' population and 320 from the 'tundra', feeding on grass, in cereal fields, and on root crops. The English name comes from the bird's habit in the past of grazing in bean field stubbles in winter. Latin *faba* = 'broad bean'.

Brent goose *B. bernicla* sspp. *bernicla* (**dark-bellied**), *nigricans* (**black**) and *hrota* (**light-bellied**), wingspan 115 cm, length 58 cm, weight 1.5 kg, a winter visitor from Canada, Greenland, Spitzbergen and Siberia. The black subspecies is mainly found on the coast of East Anglia; dark-bellied are found on the English, Welsh and east Scottish coasts (the main concentrations being on the Wash, north Norfolk coastal marshes, Essex estuaries, the Thames Estuary, and Chichester and Langstone Harbours); and most light-bellied birds are found at Strangford Lough and Lough Foyle, northern Ireland and at Lindisfarne, Northumberland. Total numbers were 95,000 in 2009–10. Estuaries and saltmarsh are favoured sites, with birds feeding of plants such as eelgrass. The species is Amber-listed because of declining populations. The specific name is medieval Latin for 'barnacle', this species once thought to be the same as barnacle goose.

Canada goose *B. canadensis* ssp. *canadensis*, wingspan 168 cm, length 95 cm, weight 4.6 kg, introduced to St James's Park, London, from North America by 1672 (possibly 1665), now common and widespread in Britain, scattered in Ireland, numbering 62,000 breeding pairs, and 190,000 wintering birds in 2008–9. Breeding in the wild has been recorded since 1731, but only became common and widespread in the 1930s, with particularly rapid expansion since the 1980s. BBS data suggest an 84% increase in numbers between 1995 and 2014. They are common in ponds, lakes, reservoirs and flooded gravel-pits, feeding on surrounding amenity grassland in urban parkland. They are viewed as pests because of their slimy faeces, trampling, eutrophication of water bodies, and aggression towards people during breeding. In the countryside they have also increasingly become a pest, eating newly sown grass, cereals and roots since the 1950s. Pairs are monogamous. The female lays 2–9 eggs (average 5) on an elevated ground nest, incubating for 24–8 days. The precocial young fledge in 6–9 weeks.

Egyptian goose *Alopochen aegyptiacus*, wingspan 144 cm, length 68 cm, male weight 2.1 kg, females 1.7 kg, a native of sub-Saharan Africa and the Nile Valley, introduced to London in the late seventeenth century, probably from the Cape, with records in England steadily increasing thereafter. By the mid-nineteenth century a number of free-flying colonies were established on estate parkland lakes, mainly in south and east England. Most birds are now found in East Anglia and the lower Thames Valley, failure to spread much beyond probably due to their nesting in the winter months, resulting in low productivity. There were around 1,100 breeding pairs in 2009, with a winter population of 3,400 individuals. Nests are usually a mound of grass or reeds. Clutch size is 5–8, and incubation 28 days. Diet is of seed and grass. Hybridisation with Canada geese has been recorded in the UK, and it might also hybridise with native species. Greek *alopex* + *chen* = 'fox' + 'goose' (referring to the ruddy colour of its back), and Latin *Aegyptius* = 'Egyptian'.

Greylag goose *Anser anser*, wingspan 164 cm, length 82 cm, male weight 3.6 kg, females 3 kg, widespread and generally common, either a resident breeder, truly wild in north Scotland (especially the Hebrides) with restocking or reintroduction elsewhere, including feral or habituated birds in suburban parks, or – in parts of Scotland, east Ireland and around the Shannon estuary – a winter visitor. In 2010 there were an estimated 46,000 breeding pairs, with winter numbers around 140,000 British-bred together with 88,000 migrants from Iceland. Numbers increased by 298% between 1970 and 2015, by 211% between 1995 and 2015, and by 9% between 2010 and 2015, but the species is Amber-listed. These geese normally pair for life, nesting on the ground among rushes, reeds or dwarf shrubs, or on a raft of floating vegetation. Clutch size is typically 4–6 eggs, incubation lasts around 28 days, and the precocial chicks fledge in 8–9 weeks. Feeding is especially on grass and cereals/stubble, but also roots and tubers, and water plants. This is the ancestor of most domestic geese. 'Lag' is dialect for 'goose'.

Lesser white-fronted goose *Anser erythropus*, an endangered vagrant winter visitor from Scandinavia and Siberia, most likely to be seen from December to March, with 122 records over 1950–2007, scattered mostly in England and Wales. A sighting of Britain's second lesser whitefront on the Severn led Sir Peter

Scott to set up the Wildfowl Trust at Slimbridge, Gloucestershire. Greek *erythros* + *pous* = 'red' + 'foot'.

Pink-footed goose *Anser brachyrhynchus*, wingspan 152 cm, length 68 cm, weight 2.5 kg, a winter migrant from Greenland, Iceland and Spitzbergen, widespread and scattered in Britain. Total numbers in 2009–10 were about 360,000, with numbers in England increasing, especially in Norfolk, probably due to improved protection at winter roosts. Feeding is on grass, roots and tubers, mostly on farmland. Greek *brachys* + *rhynchos* = 'short' + 'bill'.

Snow goose *Anser caerulescens* sspp. *atlanticus* (**greater**) and *caerulescens* (**lesser**), wingspan 148 cm, length 72 cm, weight 2.3 kg, a scarce visitor usually associated with migrating flocks of other goose species that breed in the high arctic. A population on Mull and Coll has been present following escape from a collection in the 1960s, and number 60–70 birds. Other escaped birds have been noted in the UK, occasionally breeding in England, for example on the Babingley river (Norfolk), Radwell (Bedfordshire) and Stratfield Saye (Hampshire), but none persistent. Latin *caerulescens* = 'bluish' (from *caeruleus* = 'dark blue'), this species also being called blue goose in North America.

White-fronted goose *Anser albifrons*, wingspan 148 cm, length 72 cm, weight 2.5 kg, a winter visitor, with ssp. *albifrons* migrating from Siberia, ssp. *flavirostris* from Greenland. Siberian birds are found around the Severn estuary and near the coast from Dorset to Lincolnshire, Greenland birds in Ireland and west and north-east Scotland. In 2009–10 there were around 2,400 birds from Siberia and 13,000 from Greenland. On the IUCN Red List because of declining breeding and therefore wintering populations. Feeding is on grass, clover, grain, winter wheat and potatoes. Latin *albifrons* = 'white-fronted'.

GOOSE BARNACLE See barnacle.

GOOSEBERRY *Ribes uva-crispa* (Grossulariaceae), an introduced spiny shrub with edible fruit in cultivation since the thirteenth century. Readily dispersed by birds, but not recorded in the wild until 1763, now widespread as a garden escape and throw-out in woodland, hedges and scrub, and as a relic of cultivation. 'Gooseberry' may be a version of French *groseille*, or its Latinisation as *grossulus* (= an unripe fig), or a corruption of either Dutch *kruisbes* or German *Krausbeere*. *Ribes* is the name of a plant with acidic fruit mentioned by Arabian physicians. Latin *uva* + *crispa* = 'grape' + 'curved'.

GOOSEFOOT Annual species of *Dysphania* and *Chenopodium* (Amaranthaceae). The name reflects the leaf shape, translated from Greek *chen* + *pous* (or *podion* = 'little foot'), which also gives *Chenopodium*. *Dysphania* derives from Greek *dysphanis* = 'obscure', probably referring to the inconspicuous flowers.

Scattered European archaeophytes, some dating to Iron Age or Roman times, casual on disturbed, often nutrient-rich soils, include **fig-leaved goosefoot** *C. ficifolium*, **many-seeded goosefoot** *C. polyspermum*, **maple-leaved goosefoot** *C. hybridum*, **nettle-leaved goosefoot** *C. murale*, **oak-leaved goosefoot** *C. glaucum* and **upright goosefoot** *C. urbicum*.

Rare and declining neophyte casuals on tips and in fields, all having arrived in wool shoddy and other waste material, include **clammy goosefoot** *D. pumilio*, **crested goosefoot** *D. cristata*, **keeled goosefoot** *D. carinata* and **nitre goosefoot** *C. nitrariaceum* from Australia; **pitseed goosefoot** *C. berlandieri*, **Probst's goosefoot** *C. probstii*, **slimleaf goosefoot** *C. pratericola* and **soyabean goosefoot** *C. bushianum* from North America; **foetid goosefoot** *C. hircinum* and **scented goosefoot** *D. multifida* from South America; and **grey goosefoot** *C. opulifolium*, **striped goosefoot** *C. strictum* and **Swedish goosefoot** *C. suecicum* from Europe.

Red goosefoot *C. rubrum*, a common and widespread in England, more scattered and only locally common elsewhere, on nutrient-rich mud around the margins of freshwater or brackish ponds and ditches trampled by livestock; also in cultivated and waste ground. *Rubrum* is Latin for 'red'.

Saltmarsh goosefoot *C. chenopodioides*, locally common in parts of Kent and the Thames estuary, growing in dry, brackish mud of ditches and salt pans in the upper saltmarsh and in coastal grazing marsh.

Stinking goosefoot *C. vulvaria*, an often prostrate archaeophyte of disturbed, nutrient-rich soil on sandy shingle beaches, dunes and coastal cliffs where the soil is enriched by seabird droppings, and as an occasional inland casual on waste ground in England. *Vulvaria* is Latin for 'foul smell'.

GOOSEGRASS Alternative name for cleavers, it being eaten by geese, and formerly used to feed goslings.

GORSE *Ulex europaeus* (Fabaceae), a common and widespread shrub of acidic soils in heathland (often where there is disturbance), undergrazed pasture, woodland rides, and on sea-cliffs, dunes, waste ground and railway embankments. This is the 'county' flower of Belfast as voted by the public in a 2002 survey by Plantlife. 'Gorse' comes from Old English *gors* or *gorst*, from an Indo-European root meaning 'rough' or 'prickly'. Latin *ulex* = 'gorse'.

Dwarf gorse *U. minor*, a small, sometimes procumbent shrub common in southern England, scattered elsewhere, in heathland on free-draining acidic, nutrient-poor soils over podzolised sands and gravels, and occasionally over superficial deposits overlying chalk. It can also be found in heathland scrub, secondary woodland and undergrazed heathy pasture. Latin *minor* = 'smaller' or 'lesser'.

Spanish gorse *Genista hispanica*, a densely spiny, small neophyte shrub, a broom rather than a gorse, from south-west Europe, cultivated by 1759, occasionally planted on roadsides and in amenity areas, from where it has sometimes established itself by seed onto nearby sandy or rocky banks. *Genista* is Latin for the plant 'broom', *hispanica* for 'Spanish'.

Western gorse *U. gallii*, a shrub of heathland common and widespread in the western half of England, Wales, south-west Scotland and Ireland, and occasional in Kent and East Anglia, on infertile acidic soils, including leached soils overlying chalk and limestone; also found on sea-cliffs, in under grazed or abandoned pastures, and on scrubby banks and waste ground. *Gallii* is Latin for 'of Gaul'.

GOSHAWK Northern goshawk *Accipiter gentilis* (Accipitridae), wingspan 150 cm, length 55 cm, male weight 850 g, females 1.5 kg. Deforestation and persecution led to a rapid decline in numbers during the nineteenth century, last recorded breeding in both England and Scotland being in the 1880s. Re-establishment came in the late 1960s and early 1970s when falconers brought in birds from central Europe and Scandinavia, and escapes and deliberately released birds began to breed and spread. The principal breeding range is Wales, the Forest of Dean, south Pennines, the Scottish Borders and north-east Scotland. In 2010, a five-year mean of 432 breeding pairs was estimated, and it is Green-listed. Nests are of sticks in large trees, clutch size is 3–4, incubation takes 35–8 days, and the altricial young fledge in 35–42 days. Diet is of birds and mammals. The name is a contraction of Old English *gōsheafoc*, meaning 'goose-hawk', perhaps because of its size and its barred grey plumage. Latin *accipiter* = 'hawk' (from *accipere* = 'to grasp'), and *gentilis* = 'noble' or 'gentle', because in the Middle Ages only the nobility had permission to fly goshawks for falconry.

GOTHIC *Naenia typica* (Noctuidae, Noctuinae), a nocturnal moth, wingspan 33–40 mm, widespread but local in gardens, marsh, hedgerows and damp woodland. Numbers declined by 76% over 1968–2007. Adults fly in June–August, nectaring at various flowers. The larval stage lasts from July to April, feeding on a variety of herbaceous and woody plants continuing over winter during mild weather. The forewings are blackish with a network of fine white lines, a pattern echoing elements of Gothic architecture, hence the English name. Naenia was the Roman goddess of funerals. *Typica* may come from Latin for 'symbol', or more likely Greek *typos* = 'pattern', from the distinctive wing pattern.

Beautiful gothic *Leucochlaena oditis* (Xyleninae), wingspan 28–36 mm, a rare RDB moth found on grassy slopes and sea-cliffs on the Isle of Wight, and on the south coasts of Cornwall, Devon and Dorset. Adults fly in September–October. Larvae feed nocturnally on grasses. Greek *leucos* + *chlaina* = 'white' + 'garment' (from the pale colour). The meaning of *oditis* is obscure, possibly from Greek *odites* = 'traveller'.

Bordered gothic *Sideridis reticulata* (Hadeninae), wingspan 32–7 mm, on chalk downland, heathland, sea-cliffs, quarries and embankments, a priority species in the UK BAP. Formerly widely distributed, but always local, in England and Wales from Yorkshire southwards, its range has declined substantially since the 1960s. The Brecklands of Norfolk, Suffolk and Cambridge are the main national stronghold. Subspecies *hibernica* occurs on the coast of southern Ireland. Adults fly in June–July. Larvae feed on such plants as soapwort, bladder campion and knotgrass. Greek *sideros* + *eidos* = 'iron' + 'form' (from the brown colour), and Latin *reticulata* = 'netlike' (from the vein pattern).

Feathered gothic *Tholera decimalis* (Hadeninae), wingspan 32–45 mm, widespread on rough grassland, downland, woodland edges, parks and gardens, though scattered and less common in Scotland and Ireland. With numbers having declined by 89% over 1968–2007, it is of conservation concern in the UK BAP. Adults fly in August–September. Larvae feed on grasses, initially on the leaves, moving to the stems at ground level. Greek *tholeros* = 'muddy' (from the colour), and Latin *decimus* = 'tenth' (possibly the tenth in a series described by the species' authority, Poda).

GOUT FLY *Chlorops pumilionis* (Chloropidae), 3–4 mm length, scattered and locally common in England and south Wales. Larvae eat into and overwinter in grasses, including cereals, becoming a pest mainly of barley (causing barley gout) – it is sometimes called barley gout fly – but also wheat and rye. Adults emerge from mid-May, followed by a second generation that can persist until October. *Chlorops* is similar to Greek *chloros* = 'green'. The specific name comes from Latin *pumilus* = 'dwarf'.

GOUTY-MOSS *Oedipodium griffithianum* (Oedipodiaceae), rare to occasional on peaty and acidic soil in rock crevices and in boulder-scree in parts of upland Britain. Greek *oedipodium* = 'swollen foot', referring to the swollen hypophysis (a strongly differentiated neck between the seta and spore-bearing part of the capsule), which also explains the English name. Griffith Hooper Griffiths (1823–72) was a British bryologist.

GRACILLARIIDAE Family name for some bent-wing, leaf-miner, midget and slender moths (Latin *gracilis* = 'slender', from the abdomen shape).

GRAMINEAE Former family name of grasses, now termed Poaceae.

GRAMMOPTERA Common grammoptera *Grammoptera ruficornis* (Cerambycidae), a longhorn beetle 3–7 mm length, widespread and common in deciduous woodland in England and Wales, more scattered in Scotland and Ireland. Adults are found on flowers, especially of hawthorn. Larvae feed in fungus-decaying rotting dead wood. Greek *gramme* + *ptero* = 'mark' + 'wing', and Latin *rufus* + *cornu* = 'red' + 'horn'.

GRANNOM *Brachycentrus subnubilus* (Brachycentridae), a sedge caddis fly much used by anglers, the male forewing 7–9 mm long, female 10–14 mm, locally common and widespread, especially in northern and western regions, favouring fast-flowing streams and rivers. Larvae filter-feed. Adults are evident from March to July, usually emerging in the morning. Greek *brachys* + *centron* = 'short' + 'point', and Latin *subnubilus* = 'somewhat cloudy'.

GRAPE-HYACINTH *Muscari neglectum* (Asparagaceae), a bulbous herbaceous perennial, scattered in Britain, mainly in England, on free-draining soil, native in East Anglia and naturalised elsewhere in grassland, hedgerows, pine plantations and rough ground, and on roadsides on a range of nutrient-poor soils. It is also a garden escape or throw-out near buildings, and on roadsides, allotments and waste ground. *Muscari* is from Greek for 'musk'.

Compact grape-hyacinth *M. botryoides*, a south European neophyte perennial, scattered and naturalised in central-south England and south-east Scotland as a garden throw-out. *Botryoides* is Greek for 'like a bunch (of grapes)'.

Garden grape-hyacinth *M. armeniacum*, a Balkan neophyte perennial introduced in 1878, scattered, often as a garden throw-out on free-draining soils in grassland and hedgerows and on dunes, roadsides, walls and waste ground. *Armeniacum* is Latin for 'Armenian'.

Tassel grape-hyacinth *M. comosum*, a European neophyte perennial, local, scattered and naturalised in England and Wales as a garden throw-out, on rough and cultivated ground (sometimes a persistent weed), grassy habitats and dunes. *Comosus* is Latin for 'with many leaves'.

GRAPE-VINE *Vitis vinifera* (Vitaceae), a perennial scrambling woody climber from southern Europe cultivated since Roman times, naturalised in hedges and scrub and along riverbanks, often as a garden escape or relic of cultivation, widespread but mainly in southern England. 'Grape' comes from Old French *grape*, which means both a bunch of grapes and a small bill-hook used for cutting the bunch off the vine. The binomial is Latin for 'vine' and 'wine-bearing'.

GRAPES (TUNICATE) See sea grapes.

GRASS Annual or herbaceous perennial monocotyledonous plants (though bamboos are woody) in the family Poaceae, generally with rhizomes or stolons, and with hollow, cylindrical stems, and thin, alternate leaves (blades). 'Grass' comes from Old English *gærs*, *græs* or *gres*, originally meaning all green herbage eaten by cattle.

GRASS-OF-PARNASSUS *Parnassia palustris* (Parnassiaceae), a perennial, rhizomatous herb widespread and locally common in England (except for southern regions), north Wales, Scotland and Ireland (except for the south), found in base-rich flushes in short grassland, mires, fens, dune slacks and machair. The last two habitats support the coastal ecotype, var. *condensata*. A marked decline in the southern part of its range has mainly been due to land drainage. This is the county flower of Cumbria and Sutherland as voted by the public in a 2002 survey by Plantlife. Mount Parnassus was the home of Apollo and the Muses, associated by Dioscorides with this species. Latin *palustris* = 'marshy'.

GRASS-POLY *Lythrum hyssopifolia* (Lythraceae), a herbaceous annual occasionally found in southern England and the Channel Islands on disturbed ground flooded or wet in winter but drying out in spring, and as a casual (from birdseed) elsewhere. Greek *lythron* = 'blood' (from the pink colour of the flowers), and Latin for 'hyssop-leaved'.

False grass-poly *L. junceum*, an annual Mediterranean neophyte herb found sporadically as far north as south Scotland as a casual, often of birdseed origin, in parks and waste ground. *Junceum* is Latin for 'rush-like'.

GRASS SNAIL Terrestrial pulmonates in the genus *Vallonia* (Valloniidae) (Latin *vallum* = 'rampart' or 'wall'). Shell diameters are 2–2.7 mm, height 1–1.5 mm.

Eccentric grass snail *V. excentrica*, widespread and locally common in England, Wales, Ireland and coastal Scotland in dry habitats, meadow, rock rubble and dunes.

Ribbed grass snail *V. costata*, shell diameter, common and widespread in England, central Ireland, north and south Wales, and many of the coastal regions of Ireland and Scotland, in dry, open habitats on calcareous substrates, grassy and sunny slopes, rubble and stone walls, and in short-grassed meadow and dunes. *Costata* is Latin for 'ribbed'.

Smooth grass snail *V. pulchella*, widespread and locally common in much of England and central Ireland, scattered in coastal areas elsewhere, at grass roots in damp meadow and wetland margins on calcareous substrates. *Pulchella* is Latin for 'beautiful'.

GRASS SNAKE See under snake.

GRASS-VENEER MOTH Members of the nocturnal moth family Crambidae (Crambinae), many adults adopting folded postures on grass stems. The family name, and that of the genus *Crambus*, comes from Greek *crambos* = 'dry' or 'parched', from the yellow-brown colour of the forewing of many of the species, perhaps echoing the dry grassland habitat used by many of these moths.

Banded grass-veneer *Pediasia fascelinella*, wingspan 24–30 mm, rare, on coastal dunes in south-east England and East Anglia. Adults fly in June–July. Larvae feed on dune grasses, living in a silk tent at the base of the stem. Greek *pediasios* (*pediacos*) = 'plain', from the habitat, and Latin *fascelinus* = 'small band', from the forewing pattern.

Barred grass-veneer *Agriphila inquinatella*, wingspan 22 mm, widespread and common in dry grassland and rough meadow, less numerous in the north. Adults fly from July to September. Larvae feed on the base stems of low-growing grasses, living in a silk gallery. Greek *agros* + *phileo* = 'field' + 'love', from the habitat, and Latin *inquinatus* = 'stained', from speckling on the wing.

Chequered grass-veneer *Catoptria falsella*, wingspan 18–24 mm, common on old walls and buildings throughout much of Britain, more numerous in the south, together with two post-2000 records from Co. Wicklow. Adults fly in July–August. Larvae feed on mosses growing on walls, especially *Tortula muralis*, in a silk tube or net. Greek *catoptron* = 'mirror', from the glossy markings, and Latin *falsus* = 'false'.

Common grass-veneer *A. tristella*, wingspan 22–30 mm, widespread and common in grassland and rough meadows. Adults fly from June to September. Larvae feed on the base stems of grasses, living in a silk gallery. Latin *tristis* = 'sad', from the dark shading sometimes seen on the forewings.

Dark grass-veneer *Crambus hamella*, wingspan 18–25 mm, scarce, on dry heathland in parts of England and south Scotland. Adults fly in July–August. Larvae probably feed on wavy hair-grass. Latin *hamus* = 'hook', from the shape of a forewing white streak.

Elbow-stripe grass-veneer *A. geniculea*, wingspan 20–5 mm, common in dry pasture, grassland, dunes and gardens, except in north Scotland. Adults fly from July to October. From September to May larvae feed on the base stems of low-growing grasses, especially sheep's-fescue, living in a silk gallery. *Geniculum* is Latin for 'little knee', from a pattern on the forewing.

Garden grass-veneer *Chrysoteuchia culmella*, wingspan 18–24 mm, widespread and common in grassland, rough meadows and gardens. Adults fly in June–July. Larvae feed on the stems of grasses, usually at the base, sometimes becoming a pest. Greek *chrysos* + *teucho* = 'gold' + 'to furnish with', from gold colour on the forewings, and Latin *culmus* = 'stalk', from the adult behaviour of resting head-down on grass stems.

Heath grass-veneer *Cr. ericella*, wingspan 22–4 mm, scarce, on moorland in parts of northern England and Scotland. Adults fly in late afternoon and evening in July–August. Larval stages in Britain are undescribed. *Ericella* is from Latin *erica* = 'heath', reflecting the habitat.

Hook-streak grass-veneer *Cr. lathionellus*, wingspan 18–22 mm, widespread and common in grassland and rough meadows. Adults fly in May–August. Larvae feed on the roots and stem bases of grasses. *Lathionellus* comes from Latona, mother of Artemis and Apollo.

Hook-tipped grass-veneer *Platytes alpinella*, nocturnal, wingspan 18–22 mm, rare, on dunes and shingle beaches from Devon to Lincolnshire and Yorkshire, plus some records from south-east Ireland, especially Co. Wexford. Adults fly in July–August. Larvae probably feed on mosses. Greek *platytes* = 'breadth', the genus having broader forewings than other Crambinae, and Latin *alpinella* = 'montane', this being a habitat used on the continent.

Inlaid grass-veneer *Cr. pascuella*, wingspan 21–6 mm, widespread and common in grassland, rough meadows and gardens. Adults fly in June–August. Larvae feed in the roots of grasses, living in a silk tent. *Pascuella* comes from Latin *pascuum* = 'pasture'.

Little grass-veneer *Platytes cerussella*, wingspan 10–15 mm, local on coastal shingle and sandy beaches throughout much of south England, and inland on dry sandy heaths in Breckland. Adults fly in June–July. Larvae feed on the roots of grasses and sedge. *Cerussa* is Latin for 'white lead', from the colour of the female's wings (the male's is brown).

Marsh grass-veneer *Cr. uliginosellus*, wingspan 18–23 mm, in bog in parts of England, Wales and Ireland. Adults fly in June–July. Larvae feed on grasses and sedges, living in a silk tent. *Uliginosus* is Latin for 'marshy'.

Northern grass-veneer *Ca. furcatellus*, wingspan 20–4 mm, in parts of the Highlands, the Lake District and north Wales at altitudes of 400–1,000 m. Adults fly in June–August. Larvae probably feed on clubmosses, possibly grasses. Latin *furcus* = 'forked', from a branched white stripe on the forewing.

Pale-streak grass-veneer *A. selasella*, wingspan 26 mm, widespread (not north Scotland) but local in saltmarsh. Adults fly in July–August. Larvae feed on the base stems of grasses, living in a silk gallery. Greek *selas* = 'bright', from the white streak on the forewing.

Pearl-band grass-veneer Alternative name for silver-stripe grass-veneer.

Pearl grass-veneer *Ca. pinella*, wingspan 18–24 mm, widespread and locally common on damp heathland, fen, marsh, riverbanks and other damp areas in Britain and in west and south-west Ireland. Adults fly in July–August. Larvae feed on the roots of grasses and sedges. *Pinella* is Latin, from *pinus*, suggesting a pinewood habitat.

Powdered grass-veneer *Thisanotia chrysonuchella*, wingspan 24–34 mm, local in calcareous grassland, dry sandy areas, dunes and sea-cliffs in parts of southern England and Breckland. Adults fly in May–June. Larvae feed on grasses in a silk tent at the base of the stem. Greek *this* + *ano* = 'beach' + 'upper part', and *chrysos* + *nychios* = 'gold' + 'of the night', from the gold marks on the forewing.

Saltmarsh grass-veneer *Pediasia aridella*, wingspan 20–6 mm, on the drier edges of coastal saltmarsh from Hampshire round to east Yorkshire, north-west England, south Wales and south-east Ireland. Adults fly in July–August. Larvae feed on the stems of common saltmarsh-grass, living in a silk tent. *Aridella* derives from Latin *aridus* = 'dry'.

Satin grass-veneer *Cr. perlella*, wingspan 21–8 mm, common in grassland and rough meadow. Adults fly in July–August. Larvae feed on the base stems of grasses, living in a silk gallery. *Perlella* derives from Latin *perla* = 'pearl', from the glossy pale forewing.

Scarce grass-veneer *Cr. pratella*, wingspan 22–5 mm, widespread but local and scattered on dry grassland in Britain. Adults are mainly nocturnal but sometimes also fly towards evening from late May to August. Larvae feed on the roots and stem bases of grasses. Latin *pratus* = 'meadow'.

Scotch grass-veneer *Ca. permutatellus*, wingspan 22–9 mm, in pine plantations and areas with scattered trees in parts of Aberdeenshire and Perthshire, rare elsewhere. Adults fly in July–August. Larvae probably feed on mosses. Latin *permutatus* = 'altered', from initial confusion with another species.

Silver-stripe grass-veneer *Ca. margaritella*, wingspan 20–4 mm, widespread but local in bog, damp heathland and mosses throughout much of Scotland, north England, south-west England, Wales and Ireland. Adults fly in July–August. Larvae probably feed on various grasses and/or mosses. Latin *margarita* = 'pearl', from the colour of the forewing stripe.

Straw grass-veneer *A. straminella*, wingspan 16–20 mm, widespread and common in grassland and rough meadow. Adults fly in July–August. Larvae feed on the base stems of low-growing grasses, especially sheep's-fescue, in a silk gallery. Latin *stramen* = 'straw', from the dominant forewing colour.

Waste grass-veneer *Pediasia contaminella*, wingspan 17–19 mm, a rarity on dry grassland and golf courses, mostly coastal from Hampshire to Norfolk. Adults fly in July–August. Larvae feed in a silk tent on the leaves and stems of grasses. *Contaminatus* is Latin for 'contaminated', from the dirty-coloured forewing speckling.

White-streak grass-veneer *A. latistria*, wingspan 22–7 mm, local on dry heathland, dunes and other dry sandy areas throughout much of England and Wales, but rare in Scotland, commoner near the coast. Adults fly in July–August. Larvae feed in a silk tent on the base stems of low-growing grasses. Latin *latus* + *stria* = 'broad' + 'furrow', from the broad white forewing stripe.

Wood grass-veneer *Cr. silvella*, wingspan 22–6 mm, on damp heathland, mosses and bog in Surrey, Hampshire, Dorset, Norfolk and south-west Ireland. Adults fly in July–August. Larvae feed on sedges. *Silvella* is from Latin *silva* = 'woodland', though this is not a particularly used habitat.

GRASS-WRACK Alternative name for eelgrass.

GRASSHOPPER Member of the order Orthoptera (q.v. for general description), suborder Caelifera, in the family Acrididae, with large hind legs and short antennae.

Blue-winged grasshopper *Oedipoda caerulescens*, male body length 17–20 mm, female 25–6 mm, in coastal dunes and south-facing cliffs, disused quarries and stony fields in the Channel Islands. Greek *oidos* + *podos* = 'swelling' + 'foot', and Latin *caerulescens* = 'becoming blue'.

Common green grasshopper *Omocestus viridulus*, male body length 15–19 mm, female 17–22 mm, common and widespread, particularly associated with long damp grass such as in unimproved, lightly grazed pasture. Greek *omo* + *cestos* = 'similar to' + 'girdle' or 'variegated'; *viridulus* is Latin for 'greenish'.

Egyptian grasshopper *Anacridium aegyptium*, body length 50–80 mm, is an occasional casual introduction.

Field grasshopper *Chorthippus brunneus*, male body length 15–19 mm, female 19–25 m, common and widespread apart from the Highlands, associated with short vegetation in dry, sunny situations, particularly in downland and coastal grassland. Greek *chortos* + *hippos* = 'grass' + 'horse', and Latin *brunneus* = 'brown'.

Green grasshopper See common green grasshopper above.

Heath grasshopper *C. vagans*, male body length 13–18 mm, female 16–21 mm, only in dry heathland, often in areas of pure heather in the New Forest and east Dorset. Latin *vagans* = 'wandering'.

Jersey grasshopper *Euchorthippus elegantulus*, male body length 10–13 mm, female 16–22 mm, found only in Jersey near the sea on sunny dune systems, verges and unimproved pasture. Greek *eu* + *chortos* + *hippos* = 'good' + 'grass' + 'horse'.

Large marsh grasshopper *Stethophyma grossum*, Britain's largest grasshopper, male body length 22–9 mm, female 29–36 mm, limited to acid bog and tussocky grassland in parts of Hampshire–Dorset and a few other south English and East Anglian sites. Greek *stethos* + *phyma* = 'breast' + 'swelling', and Latin *grossum* = 'coarse'.

Lesser marsh grasshopper *C. albomarginatus*, male body length 14–17 mm, female 17–21 mm, common in coarse grassy habitats (together with dunes and saltmarsh) in south and east England, east Midlands, south Wales and south-west Ireland. Latin *albomarginatus* = 'white-edged'.

Lesser mottled grasshopper *Stenobothrus stigmaticus*,

Britain's smallest grasshopper, male body length 10–12 mm, female 12–15 mm, found on dry grassy slopes and vegetated dunes on the Isle of Man. Greek *stenos* + *bothros* = 'narrow' + 'hole', and Latin *stigmaticus* = 'marked'.

Meadow grasshopper *C. parallelus*, male body length 10–16 mm, female 16–22 mm, common and generally widespread (especially in south Britain) in coarse grasses in a range of habitats.

Mottled grasshopper *Myrmeleotettix maculatus*, male body length 12–15 mm, female 13–19 mm, widespread but not common, in dry sunny situations with short turf and bare ground, usually on free-draining soils. Habitats include the steep sides of disused quarries, road and rail cuttings, heaths and dunes. Present on several small, offshore islands. Greek *myrmos* + *leon* + *tettix* = 'ant' + 'lion' + 'grasshopper', and Latin *maculatus* = 'spotted'.

Rufous grasshopper *Gomphocerippus rufus*, male body length 14–18 mm, female 16–22 mm, local to rough dry chalk and limestone grassland often close to scrub in southern England. Greek *gomphos* + *ceras* + *pous* = 'fastening' + 'horn' + 'foot', and Latin *rufus* = 'reddish'.

Stripe-winged grasshopper *Stenobothrus lineatus*, male body length 15–19 mm, female 17–23 mm, on chalk and limestone grassland (also some heaths and dunes) in southern England and East Anglia. *Lineatus* is Latin for 'lined' or 'striped'.

Woodland grasshopper *Omocestus rufipes*, male body length 12–17 mm, female 18–20 mm, in woodland rides and clearings and on grassland and heath near woodland and scrub in south and south-west England. *Rufipes* may refer to the reddish legs of the male.

GRAYLING (BUTTERFLY) *Hipparchia semele* (Nymphalidae, Satyrinae), male wingspan 51–6 mm, females 54–62 mm, a 'brown' butterfly with six geographically isolated subspecies with a generally near-coastal distribution in Britain and Ireland, though extending inland into heathland in the New Forest and East Anglia. It is associated with sunny dry habitats where there are bare patches, such as heathland, dunes, coastal grassland, chalklands and old quarries, flying from July to September. Larvae feed on grasses, especially fescues, hair-grass and bents; adults favour the nectar of bramble, ling, bell heather, thistles and bird's-foot trefoil. A severe population decline during the last century has continued, particularly inland, range declining by 62% and numbers by 58% over 1976–2014, though there was a recovery in numbers (an increase by 10%) between 2005 and 2014, and it remains a priority species for conservation in the UK BAP. Hipparchia was a Greek astronomer of the second century BCE; in Greek mythology, Semele was the mortal mother of Dionysus by Zeus.

GRAYLING (FISH) *Thymallus thymallus* (Salmonidae), length up to 55 cm, weight 2.3 kg, with a patchy distribution as far north as the south Highlands (and absent in Ireland), often stocked into streams and lakes, favouring cold water, and sensitive to pollution. Its omnivorous diet includes small crustaceans and insects. Spawning is from March to May in redds excavated in gravelly shallows. This is an important species for coarse fishing in Britain, with no closed season in Scotland, where it has been introduced. Greek *thymallos* = 'thyme-smelling', from the fragrance of freshly caught fish.

GREBE Members of the family Podicipedidae, in the genera *Podiceps* and *Tachybaptus*, together with **pied-billed grebe** *Podilymbus podiceps*, which is an occasional accidental. Chicks are precocial. Grebes do not walk well: the names Podicipedidae and *Podiceps* come from Latin *podicis* + *pes* = 'vent' + 'foot', referring to the placement of the legs towards the rear of the body, more suitable for swimming. 'Grebe' itself, dating from 1766, comes from French *grèbe*, possibly from Breton *krib* = 'a comb', since some species are crested.

Black-necked grebe *Podiceps nigricollis*, wingspan 58 cm, length 31 cm, male weight 360 g, females 260 g, a rare breeding bird, though increasing in number (from <20 pairs in 1968–72 to

40 pairs in 2009), and a winter visitor in southern England, East Anglia and south Wales (130 individuals in 2008–9). Clutch size is 3–4, with incubation taking 20–2 days. It is found on lakes, reservoirs and flooded gravel-pits, and feeds on fish, insects and crustaceans. Latin *niger* + *collis* = 'black' + 'neck'.

Great crested grebe *P. cristatus*, wingspan 88 cm, length 48 cm, male weight 1.1 kg, females 910 g, common and widespread in lakes, reservoirs, flooded gravel-pits and rivers, except in the Highlands and interior south-east Ireland. There were 5,300 breeding pairs in 2009, and winter numbers, augmented by migrants, were 19,000 birds. Numbers weakly declined by 8% between 1995 and 2015, and by 11% in 2010–15. They have an elaborate courtship display, rising out of the water as a synchronised pair and shaking their heads which are often holding plant material. Nests are near the water's edge or on platforms of vegetation in the water; when leaving the nest the adult usually hides the eggs, and keeps them warm, by covering them with often rotting plant material. Clutch size is 3–4, incubation takes 27–9 days, and fledging takes 71–9 days. Diet is mainly of fish. *Cristatus* is Latin for 'crested'.

Little grebe *Tachybaptus ruficollis*, wingspan 42 cm, length 23–9 cm, weight 140 g, common and widespread on lakes, gravel-pits, canals, slow rivers and estuaries, except in upland areas, with an estimated 5,300 pairs (3,900–7,800) in 2009, and with 16,000 individuals wintering in 2008–9. Numbers in Britain decreased by 27% over 1975–2015, though this included a 10% increase in 2010–15. It nests near the water's edge; when leaving the nest the adult usually hides the eggs by covering them with plant material. Nests on ponds and lakes were more successful than those on rivers and streams. Clutch size is 4–6, incubation lasts 20–1 days, and fledging takes 44–8 days. Diet is of adult and larval insects and small fish. Greek *tachys* + *bapto* = 'fast' + 'to sink under', and Latin *rufus* + *collis* = 'red' + 'necked'.

Red-necked grebe *P. grisegena*, wingspan 81 cm, length 45 cm, weight 820 g, a scarce Red List breeder in lakes and large ponds (first records in Cambridgeshire and Scotland in 1988), with <20 individuals present in summer, and wintering numbers in 2008–9 only 55, mostly found on the coast. Diet is of insects, molluscs, crustaceans and fish. Latin *griseus* + *gena* = 'grey' + 'cheek'.

Slavonian grebe *P. auritus*, wingspan 62 cm, length 34 cm, weight 410 g, a Red List bird breeding in north-east Scotland, with 30 pairs in 2009. In winter they number around 1,100 individuals, found around the British coastline, with the Moray Firth, Firth of Forth, Clyde Estuary and Islay in Scotland, and Pagham Harbour, Sussex, being important sites, and also found on the north, east and south-west Irish coasts. Nests are near the water's edge or on platforms of vegetation in the water. Clutch size is 4–5, incubation takes 22–5 days, and fledging takes 55–60 days. Diet is of fish and insect larvae. *Auritus* is Latin for 'eared'.

GREEN (MOTH) Brindled green *Dryobotodes eremita* (Noctuidae, Xyleninae), wingspan 32–9 mm, nocturnal, widespread and common where there are oak trees in broadleaf woodland, parks and gardens, more locally in Scotland and Ireland. Population levels increased by 287% between 1968 and 2007. Adults fly in August–September. Larvae feed on buds then leaves of oaks. Greek *drys* + *bosco* + *eidos* = 'oak' + 'to feed on' + 'form', and Latin *eremita* = 'hermit', from larval spinning between leaves.

Frosted green *Polyploca ridens* (Drepanidae, Thyatirinae), wingspan 30–5 mm, nocturnal, local in broadleaf woodland, hedgerows and gardens throughout the southern half of England, and in Wales. Adults fly in April–May. Larvae feed on oak leaves. Greek *polys* + *ploce* = 'many' = 'twisting', from the wing patern, and Latin *ridens* = 'laughing'.

Marbled green *Nyctobrya muralis* (Noctuidae, Bryophilinae), wingspan 27–34 mm, local in rocky areas and on dry-stone walls throughout much of southern England, south Wales and south and east Ireland, predominantly coastal. Adults fly at night in July–August. Larvae feed on lichens, often on walls. Greek *nyctos* + *bryo* = 'of night' + 'to be full of', and from Latin *murus* = 'wall'.

GREEN ALGAE Members of the division Chlorophyta.

GREENBOTTLE Large blow flies of the genus *Lucilia* (from Latin *lux* = 'light') (Calliphoridae) with a dark metallic green body. Adults feed on nectar, detritus or animals, larvae on carrion. The commonest and most widespread is *Lucilia sericata*, the specific name being Latin for 'silken' (known, more in Australia, as sheep maggot fly), followed by *L. caesar*, which lay their eggs in sheep wool. Larvae migrate down the wool to feed on the skin surface which can cause massive lesions and secondary bacterial infections. Blow fly strike (myiasis) affects 1 million sheep as well as 80% of UK sheep farms each year.

GREENFINCH *Chloris chloris* (Fringillidae), wingspan 26 cm, length 15 cm, weight 28 g, a common and widespread resident breeder found in a variety of habitats with trees, including gardens. Populations declined during the late 1970s and early 1980s, but increased dramatically during the 1990s, with 1.7 million pairs estimated in Britain in 2009. Numbers, however, had declined by 46% between 1995 and 2015, with outbreaks of trichomonosis, especially in 2006 and 2007, heavily impacting this species. Between 2010 and 2015 the decline was by 40%, an annual change of –9.8%. Nests are in trees and bushes, usually with 4–5 eggs which hatch after 14–15 days, the altricial young fledging in 14–16 days. Adult diet is of large seeds. Nestlings take insects. The binomial comes from the Greek *chloros* = 'green'.

GREENFLY See aphid.

GREENGAGE *Prunus domestica* ssp. *italica* (Rosaceae); see wild plum under plum. Sir William Gage (1657–1727) introduced this plum from France in 1724.

GREENLEAF WORM *Eulalia viridis* (Phyllodocidae), a widespread polychaete 5–15 cm long, in crevices, barnacle and mussel beds and on kelp holdfasts from the intertidal to the shallow sublittoral. Greek *eulalia* = 'well-spoken', Latin *viridis* = 'green'.

GREENSHANK *Tringa nebularia* (Scolopacidae), wingspan 69 cm, length 32 cm, weight 190 g, a resident and summer Amber List migrant breeder in north and west Scotland in scrapes in wet moorland and by peatland pools, averaging 1,000–1,100 pairs. Clutch size is 3–4, incubation takes 25–7 days, and the precocial chicks fledge in 29 days. As a passage migrant it can be found across the UK, inland around lakes and freshwater marsh, and at coastal wetlands and estuaries, with largest numbers in autumn (approaching 5,000 birds) close to the coast. It is also a winter visitor to estuaries in south-west England, Wales, the Solway Firth and much of Ireland, averaging 600–770 birds. It feeds on invertebrates and small fish. 'Greenshank' describes the green legs. *Tringa* comes from Greek *tryngas*, a wading bird mentioned by Aristotle, and Latin *nebularia* = 'mist', reflecting the favoured damp marshy habitat.

GREENTAIL Alternative name (especially by anglers) for grannom, a caddis fly which carries its greenish egg-mass at the tip of the abdomen.

GREENWEED Deciduous shrubs in the genus *Genista* (Fabaceae), *genista* being Latin for the plant 'broom'.

Dyer's greenweed *G. tinctoria* ssp. *tinctoria*, in rough pasture, old meadow and grassy heathland, and on cliffs, verges and field edges on heavy soils, usually calcareous to slightly acidic clays, widespread in England and Wales, but having declined in the last half-century with the loss of old grassland. Subspecies *littoralis* is a rare small procumbent shrub of clifftop grassland and maritime heaths on the Lizard Peninsula (Cornwall), north Devon and south-west Wales. *Tinctoria* is Latin indicating a plant used in dyeing, in this case producing a yellow colour.

Hairy greenweed *G. pilosa*, low, usually prostrate, growing on rocky clifftops in coastal grassland or heathland in south-west England and south-west Wales, but declining in range. It also grows in limestone grassland in Brecon and on rock ledges in the Cadair Idris range. *Pilosa* is Latin for covered with long hair.

GREY (MOTH) Moths in the families 1) Noctuidae, 2) Geometridae and 3) Crambidae.

1) Specifically, the **grey** *Hadena caesia* (Noctuidae, Hadeninae), wingspan 32–42 mm, a rare RDB nocturnal moth found on rocky areas and shingle beaches, on the south and west coasts of Ireland, the Isle of Man and the Inner Hebrides. Adults fly in May–August. Larvae feed on sea campion. *Hadena* is Greek for Hades, *caesia* Latin for 'bluish-grey'.

Early grey *Xylocampa areola* (Psaphidinae), wingspan 32–40 mm, nocturnal, widespread in broadleaf woodland, scrub, hedgerows and gardens, commoner in the south. Adults fly in March–May. Larvae feed on honeysuckle. Greek *xylon* + *campe* = 'wood' + 'caterpillar' (from the larval resemblance to a twig), and Latin for a space demarcated by a line (from dark-ringed markings).

Poplar grey *Subacronicta megacephala* (Acronictinae), wingspan 38–45 mm, nocturnal, widespread and common, especially in the south, in woodland, fen, parks and gardens. Adults fly in May–August. Larvae feed on poplars and aspen, resting by day with the head curled back. Greek *sub* + *acronyx* (i.e. the genus *Acronicta*) = 'near' + 'nightfall', and *mega* + *cephale* = 'large' + 'head'.

2) Members of the Geometridae.

Bordered grey *Selidosema brunnearia* (Ennominae), wingspan 37–43 mm, in heathland, moorland and dunes in the New Forest, Hampshire and adjoining counties, with isolated populations in north-west England and west Scotland, and scattered but widespread in Ireland. Adults fly in July–August, mostly at night, though males are also evident in daylight when warm. Larvae feed on heather and bird's-foot-trefoil. Greek *selidos* = 'row of benches' (referring to a broad band on the forewing), and Latin *brunus* = 'brown'.

Mottled grey *Colostygia multistrigaria* (Larentiinae), wingspan 26–31 mm, widespread and common on heathland, moorland, downland and woodland margins. Adults fly at night in March–April. Larvae feed on bedstraws and cleavers during early summer. Greek *colos* + *Stygios* = 'stunted' + 'relating to the River Styx', from the dark wing markings, and Latin *multus* + *striga* = 'many' + 'furrow'.

3) Members of the family Crambidae (subfamily Scopariinae), nocturnal moths in the genera *Eudonia* (Greek *eudo* = 'to rest', from the adult's resting behaviour) and *Scoparia* (Latin *scopae* = 'twigs', from the shape of the wing tufts).

Base-lined grey *S. basistrigalis*, wingspan 21 mm, local in deciduous woodland throughout England, less numerous in the north, and in south Wales. Adults fly in July. Larvae feed on the moss *Mnium hornum* growing on soil which is not deeply shaded. Latin *basistrigalis* = 'base-lined', from the black line on the forewing.

Common grey *S. ambigualis*, wingspan 15–22 mm, widespread and common in deciduous woodland and on moorland. Adults fly in May–July. Larvae feed on various mosses. *Ambigualis* is from Latin for 'uncertain'.

Ground-moss grey *E. truncicolella*, wingspan 18–22 mm, widespread and common in woodland, moorland and gardens. Adults fly in July–August. Larvae feed on ground mosses. Latin *truncus* + *colo* = 'trunk' + 'to inhabit', from the adult behaviour of resting on tree-trunks.

Highland grey *E. alpina*, wingspan 20–5 mm, scarce on mountains in parts of the Highlands, and at lower altitudes in the Orkneys and Shetlands. Adults fly in June–July, mostly at night but sometimes in afternoon sunshine. Larvae probably feed on mosses and lichens.

Large grey *S. subfusca*, wingspan 21–30 mm, widespread and common in woodland edges, gardens and rough ground. Adults fly in June–August. Larvae mine the roots of oxtongue, colt's-foot and butterbur. *Subfusca* is Latin for 'brownish'.

Little grey *E. lacustrata*, wingspan 15–20 mm, common in a range of habitats throughout most of Britain. Adults fly in July–August. Larvae feed on mosses, usually on tree-trunks or walls. *Lacustrata* derives from Latin *lacus* = 'lake'.

Marsh grey *E. pallida*, wingspan 16–19 mm, fairly widespread, local in marsh, riverbanks and damp areas. Adults fly in June–July. Larvae feed on mosses and lichens on the ground. *Pallida* is Latin for 'pale'.

Meadow grey *S. pyralella*, wingspan 17–22 mm, common on chalk downland and waste ground throughout much of Britain, less numerous in Scotland. Adults fly in June–July. Larvae may feed on dead leaves of ribwort plantain.

Moorland grey *E. murana*, wingspan 20 mm, local in moorland and mountains throughout much of south-west England, Wales, northern England and Scotland. Adults fly in June–August, possibly in two generations. Larvae feed on mosses growing on rocks and walls, *murana* coming from Latin *murus* = 'wall'.

Narrow-winged grey *E. angustea*, wingspan 17–22 mm, common on coastal dunes and rough ground by the sea, more local inland. Adults fly from July into late autumn. Larvae feed on mosses on walls and dunes. *Angustea* comes from Latin *angustus* = 'narrow'.

Northern grey *S. ancipitella*, wingspan 18–21 mm, found in woodland mainly in parts of west England, less numerous in the south, and locally in Wales and Scotland. Adults fly in July–August. Larvae feed on lichens and mosses. *Ancipitella* comes from Latin *ancipitius* = 'narrow', or from *anceps* = 'doubtful' (from difficulty in identification).

Pied grey *E. delunella*, wingspan 17–18 mm, widespread but local on heathland and especially in woodland. Adults fly in July–August. Larvae feed on lichens and mosses, often on tree-trunks. Latin *de* + *luna* = 'lacking' + 'moon', from a dark blotch that obscures a crescent mark in related species.

Small grey *E. mercurella*, wingspan 16–19 mm, widespread and common in a variety of habitats, including woodland and gardens. Adults fly in June–September. Larvae feed on mosses, often on tree-trunks and walls. *Mercurella* relates to the symbol of Mercury on the forewing.

White-line grey *E. lineola*, wingspan 19 mm, widespread but local in woodland and heathland, favouring the coastal counties of east and south Britain, Guernsey and the Burren. Adults fly in July–August. Larvae feed on lichens, especially *Parmelia* species. *Lineola* is Latin for 'line'.

GREY CHI *Antitype chi* (Noctuidae, Xylfeatinae), a nocturnal moth, wingspan 32–7 mm, on moorland and upland grassland throughout much of Scotland, northern England, the west Midlands, parts of Wales and the northern half of Ireland. Numbers declined by 80% over 1968–2007. Adults fly in August–September. Larvae feed on docks and sorrels. Each grey forewing usually has a cross-shaped black mark in the centre, similar to the Greek letter *chi*, giving it its English and specific names. An 'antitype' is something that is represented by a symbol, here the black mark on the forewing that resembles *chi*, hence the binomial.

GREYHEN Female black grouse.

GREYLING Fungi in the genus *Tephrocybe* (Agaricales, Lyophyllaceae) (Greek *tephros* + *cybe* = 'ash-grey' + 'head').

Rancid greyling *T. rancida*, scattered but uncommon on soil and litter in deciduous and mixed woodland.

Sphagnum greyling *T. palustris*, scattered but uncommon on sphagnum moss in moorland or damp coniferous woodland. *Palustris* is Latin for 'of swamp'.

Sweet greyling *T. osmophora*, a Critically Endangered (RDL, 2006) and UK BAP fungus, with records from Surrey and

Herefordshire on soil and litter under beech and conifers. Greek *osmophora* = 'bearing a scent'.

GRILSE Young Atlantic salmon that has returned to its natal river to spawn, having spent one winter at sea, from Welsh *gleisiad*, in turn from *glas* = 'blue'.

GRIMMIA Mosses in the genera *Grimmia* and *Schistidium* (Grimmiaceae), the first name honouring the German physician and botanist Johann Grimm (1737–1821). *Schistidium* comes from Greek *schistos* = 'divided', probably referring to the split calyptra (which protects the capsule containing the embryonic sporophyte).

Many are upland species, uncommon or at best locally common in parts of the Highlands, Cumbria or Wales, for example **black grimmia** *G. incurva*, **brown grimmia** *G. elongata*, **dapple-mouthed grimmia** *G. tergestina*, **gravel grimmia** *G. retracta*, **Hartman's grimmia** *G. hartmanii* (also scattered in Ireland), **large grimmia** *G. elatior*, **north grimmia** *G. longirostris*, **round-fruited grimmia** *G. orbicularis* (usually coastal), **sand grimmia** *G. arenaria*, **spreading-leaved grimmia** *G. curvata*, **string grimmia** *G. funalis*, **sun grimmia** *G. montana* and **twisted grimmia** *G. torquata*, and **black mountain grimmia** *S. atrofuscum*, **compact grimmia** *S. confertum*, **Dupret's grimmia** *S. dupretii*, **frigid grimmia** *S. frigidum*, **mealy grimmia** *S. pruinosum*, **robust grimmia** *S. robustum*, **rough grimmia** *S. papillosum*, **stook grimmia** *S. trichodon* and **upright brown grimmia** *S. strictum*.

Broadleaf grimmia *S. platyphyllum*, mainly in upland south Wales and from the Pennines to the south-east Highlands, on rocks in rivers, most abundant on base-rich examples. *Platyphyllum* is Greek for 'broad-leaved'.

Copper grimmia *G. atrata*, in north-west Wales, Cumbria, the Southern Uplands and the Highlands, favouring acidic rock rich in heavy metals on cliffs, often associated with flushed sites. *Atrata* is Latin for 'in black'.

Donn's grimmia *G. donniana*, the commonest grimmia in upland Britain, occasionally in south and east Ireland, typical of acidic rock outcrops and scree, frequently on rocks rich in heavy metals. At lower altitudes it may occur on dry-stone walls and on mine waste.

Elegant grimmia *S. elegantulum*, scattered with a mainly western distribution, typically in shaded limestone in open woodland, on the sides of roadside walls and bridges, and in churchyards.

Flat-rock grimmia *G. ovalis*, scattered in Britain, mainly in the south Welsh Borders, Cumbria and the Highlands on base-rich to neutral rocks such as basalt and dolerite, and it can form deep cushions on roofing tiles. *Ovalis* is Latin for 'oval'.

Great grimmia *G. decipiens*, scattered and uncommon on igneous outcrops, sarsen stones, Old Red Sandstone and Millstone Grit, mostly in lowland regions, preferring sunny sites. Several colonies in the Welsh Borders and in southern England are on stone-tiled roofs. *Decipiens* is Latin for 'deceiving'.

Grey-cushioned grimmia *G. pulvinata*, common and widespread, predominantly lowland, on usually base-rich rocks, tolerating moderate pollution, so a characteristic urban and suburban species, growing on wall tops, mortar, tombstones, asbestos roofs and concrete. In more natural habitats, it is found on cliffs or boulders of limestone or calcareous sandstone, sometimes on serpentine. *Pulvinata* is Latin for 'cushion-shaped'.

Hair-pointed grimmia *G. trichophylla*, common and widespread except for much of the south Midlands, eastern England and central Ireland, on siliceous rocks, growing in scree, on cliffs, sarsen stones, roofs, walls and other acidic stonework. *Trichophylla* is Greek for 'hairy-leaved'.

Hedgehog grimmia *G. crinita*, recorded from a few sites in south-west, south and central England, declining and Critically Endangered in the RDB. *Crinita* here is Latin for 'long-haired'.

Hoary grimmia *G. laevigata*, scattered in England, Wales and the central Scottish lowlands, on exposed, acidic or slightly base-rich rocks, including sandstone roof tiles particularly near the Welsh Borders. *Laevigata* is Latin for 'smooth'.

River grimmia *S. rivulare*, common and widespread in Britain north of a line from Devon to north Yorkshire, scarce in southern England, and scattered in Ireland, on rocks, wall bases and tree roots in fast-flowing rivers submerged for part of the year. *Rivulare* is Latin for 'relating to streams'.

Seaside grimmia *S. maritimum*, common and widespread except between Dorset and south Yorkshire, in coastal rock crevices, especially where sheltered or with trickles of fresh water.

Sessile grimmia *S. apocarpum*, widespread and increasingly frequent in humid parts of the north and west, on rocks and bridge piles by rivers, shaded masonry and flat gravestones. In the uplands it favours base-rich, siliceous rocks. *Apocarpum* is Latin for having carpels free of each other.

Thickpoint grimmia *S. crassipilum*, common and widespread on limestone and base-rich sandstone, and especially man-made substrates such as calcareous walls, and on tarmac. Latin *crassipilum* = 'thick point'.

Toothless grimmia *G. anodon*, Britain's rarest moss, recorded on Arthur's Seat, Edinburgh, in 1864, disappeared by 1912, but rediscovered there in 2005, still present in 2015. *Anodon* is Greek for 'toothless'.

Water grimmia *S. agassizii*, recorded from acidic upland streams in north Wales, calcareous streams in Teesdale, rocks in acidic, lowland rivers in Scotland, flushed igneous outcrops in the Galloway Hills, and seasonally flooded hollows in mica-schist rocks on Ben Lawers.

GRISETTE *Amanita vaginata* (Agaricales, Amanitaceae), an edible, widespread but locally common fungus, found in summer and autumn on soil under broadleaf woodland, occasionally heathland. 'Grisette' comes from French *gris* (grey), a reference to the colour of stem and cap. *Amanita* is Greek for a kind of fungus; *vaginata* is Latin for 'sheathed'.

Rare species include **mountain grisette** *A. nivalis* (Highlands) and **pale grisette** *A. lividopallescens* (seven records from England, two from Wales, one each from Scotland and Ireland, associated with beech and oak).

Orange grisette *A. crocea*, edible, widespread (but unrecorded from much of the south-central and east Midlands and East Anglia), generally infrequent though locally common, evident in summer and autumn on soil under broadleaf trees, especially beech and birch, often in open, grassy or mossy sites. *Crocea* is Latin for 'saffron yellow'.

Snakeskin grisette *A. ceciliae*, inedible, widespread but infrequent, found in late summer and autumn on soil in broadleaf or mixed woodland. Cecilia Berkeley was wife of the English mycologist Miles Joseph Berkeley, who co-authored the first description of this species in 1854.

Tawny grisette *A. fulva*, edible, common and widespread, evident in summer and autumn on acidic soil in woodland, especially under birch. It is often the most abundant *Amanita* in the conifer forests of west Wales, north-west England and Scotland. *Fulva* is Latin for 'yellow'.

GROMWELL Species of *Lithospermum* (Boraginaceae). 'Gromwell' comes from Old French *gromil*, the second syllable derived from Latin *milium* = 'millet'. *Lithospermum* comes from Greek *lithos* = 'stone' and Latin *spermum* = 'seeded', the plants having been used to break up kidney or gall bladder stones.

Common gromwell *L. officinale*, a rhizomatous herbaceous perennial, locally common in England, very local elsewhere, in grassland, hedgerows and wood margins, mostly on base-rich soils. *Officinalis* is Latin for having pharmacological properties.

Field gromwell *L. arvense*, a declining archaeophyte annual present as an arable weed since the Bronze Age, locally frequent in England, elsewhere occasionally found as a casual on waste ground and other disturbed habitats, favouring light, dry, calcareous soils. Seed is short-lived and populations depend upon regular disturbance for survival. *Arvensis* is from Latin for 'field'.

Purple gromwell *L. purpureocaeruleum*, a herbaceous perennial with creeping woody stems found over chalk and limestone, scattered in the southern half of England and in north and south Wales. Inland it grows in woodland edges and rides, and on banks in partial shade, while on the coast it grows in scrub on slumped cliffs and slopes. It also occurs as a garden escape on roadsides and waste ground. Latin *purpureocaeruleum* = 'purply blue'.

GROOVE MOSS Alternative name for species of *Aulacomnium*, **bog groove-moss** for bog bead-moss, **bud-headed groove-moss** for drumsticks, and **mountain groove-moss** for swollen thread-moss.

GROSSULARIACEAE Family name for species of *Ribes* (Latin *grossulus* = an unripe fig). See gooseberry and currant.

GROUND BEETLE Members of the family Carabidae (Greek *carabis* = 'horned beetle'), with subfamilies Omophroninae, Brachininae (bombardier beetles), Cincindelinae (tiger beetles), and Carabinae – which includes both the largest and most frequently encountered species, such as the two violet ground beetles, *Nebria brevicollis* and black cock. Most are ground-dwelling predators or scavengers, but others burrow in the soil, forage on tree branches or live in the intertidal zone. Most are largely nocturnal, but several large-eyed species are diurnal.

Blue ground beetle *Carabus intricatus* (Carabinae), length 24–35 mm and Britain's largest ground beetle, with small populations in nine or ten sites in Devon (Dartmoor) and Cornwall, and one recently discovered at Coed Maesmelin, south Wales. It lives in moist deciduous oak and occasionally beech woodland. Most sites are ancient pasture woodlands with sparse ground vegetation, high humidity and an abundance of mosses. Both adults and larvae are nocturnal carnivores, feeding on slugs. With habitat loss one reason for its fragile status (it was considered extinct in Britain until a specimen was found in 1994), this is a priority species for conservation in the UK BAP.

Crucifix ground beetle *Panagaeus cruxmajor*, 8–10 mm length, with large red spots on the wing cases giving the appearance of a red background behind a black cross. In 2008 it was recorded at Wicken Fen, Cambridgeshire, after an absence of over 50 years. Before this discovery it was thought to survive at only three places in the UK (and at one of these it had not been seen for 10 years). Endangered in the RDB, it is a priority for conservation in the UK BAP. It shelters under wood during the day, and is a nocturnal predator probably mainly feeding on semi-aquatic snails. *Panagaeus* is from Greek *panages* = 'hallowed'. Latin *cruxmajor* = 'greater cross'.

Necklace ground beetle *C. monilis*, a large (22–6 mm long), widespread but very scarce and declining in England and Wales (35% between 1985 and 2010, a priority species in the UK BAP), and extinct in Scotland and Ireland. Reasons for the decline probably include use of pesticides, a shift from spring to autumn cultivation, loss of hedgerows and field margins, and habitat fragmentation. The predatory and scavenging adults are active in open habitats, mostly fields and field margins, from April to September, with peak activity midsummer. Larvae live on the soil surface, in the litter layer or in the soil, preying on other invertebrates. The pattern on the wing cases resembles a beaded necklace: *monilis* is Latin for 'necklace'.

Violet ground beetle Usually refers to *C. violaceus* but also applicable to *C. problematicus*, both 20–30 mm long, both widespread in Britain, but the former absent from Ireland while the latter is present in north and west Ireland. Both species are abundant in woodland, rough grassland, heathland and moorland, and *C. violaceus* is common in gardens. Both are found throughout the year although *C. problematicus* is a spring breeder (and commoner then), *C. violaceus* an autumn breeder. Adults and larvae are nocturnal predators, the former often hiding under stones, leaf litter and logs during the day. Diet includes

garden invertebrates such as slugs and pest insects. *Violaceus* is Latin for 'violet', *problematicus* for 'problematic'.

GROUND BUG Nettle ground bug *Heterogaster urticae* (Lygaeidae), length 6–7 mm, in open habitats, associated with nettles, throughout southern Britain. Greek *heteros* + *gaster* = 'different' + 'stomach', and Latin *urtica* = 'nettle'.

GROUND-ELDER *Aegopodium podagraria* (Apiaceae), a common and widespread rhizomatous herbaceous perennial archaeophyte known since Roman times, found in disturbed habitats, especially hedgerows, verges, churchyards, neglected gardens and waste ground, typically occurring near habitation. Its persistence makes it a pernicious garden weed. The English name comes from the superficial similarity of the leaves and flowers to those of elder. *Aegopodium* is a corruption of Greek *aix* or *aigos* + *pous* = 'goat' + 'foot', from a fancied resemblance in the leaf shape to the foot of a goat. *Podagraria* is derived from Latin (and Greek) *podagra* = 'gout', for which at one time it was considered a cure.

GROUND-IVY *Glechoma hederacea* (Lamiaceae), a stoloniferous perennial herb common and widespread except for north-west Scotland, in woodland, grassland, hedgerows and waste ground, usually on fertile, often damp soil. Its common and specific names are misleading: the leaves are kidney- or fan- rather than ivy-shaped. *Glechoma* was the Greek name for pennyroyal, another lamiate plant.

GROUND-PINE *Ajuga chamaepitys* (Lamiaceae), an annual or biennial herb local in England on arable field margins and bare tracks on calcareous soils, and on chalk downland, generally declining but in places responding to conservation measures. *Ajuga* comes from Latin for 'without a collar'. *Chamaepitys* comes from Greek for 'dwarf pine', prompted by the resinous scent when crushed.

GROUNDHOPPER Flightless grasshoppers in the genus *Tetrix* (Tetrigidae), possibly from Greek for 'tertiary', with reduced forewings. All species can swim.

Cepero's groundhopper *T. ceperoi*, male body length 8–10 mm, female 10–13 mm, found along the south English and south Wales coast in open sunny situations such as dunes, dune slacks and shingle banks where there is food (mosses and lichens).

Common groundhopper *T. undulata*, male body length 8–9 mm, female 9–11 mm, more associated with but not restricted to southern Britain, in open habitat with bare ground and short vegetation in the presence of its main food, mosses and lichens.

Slender groundhopper *T. subulata*, male body length 9–12 mm, female 11–14 mm, in bare mud and short vegetation in damp, unshaded mossy locations, particularly with calcareous soils, in south and east-central England, south Wales and central Ireland. *Subulatus* is Latin for 'tapering'.

GROUNDLING (MOTH) Nocturnal members of the family Gelechiidae (mostly subfamily Gelechiinae), wingspans generally 10–17 mm, genera including *Gelechia* (Greek *geleches* = 'sleeping or resting on the ground', many species feeding on low-growing plants as larvae, and flying close to the ground as adults); *Stenolechia* (*stenos* = 'narrow', reflecting the narrow forewings, plus *echia* as a curtailed form of *Gelechia*); *Bryotropha* (Greek for 'moss-eating'); *Caryocolum* (Greek *caryon* + Latin *colo* = 'nut' + 'to inhabit', the larval food plants coming from the Caryophyllaceae); *Chionodes* (Greek for 'snowy', from white scales surrounding forewing markings); *Mirificarma* (Latin *mirificus* + *arma* = 'wonderful' + 'weapons', from the length of the male reproductive organ); *Pseudotelphusa* (Latin *pseudo* = 'false' and the genus *Telphus*, named for an Arcadian nymph, in which this moth was originally placed); *Recurvaria* (Latin *recurvus* = 'bent back', from the shape of the labial palps); *Scrobipalpa* (Latin *scrobis* + *palpus* = 'ditch' + 'palp', from a furrow under the forewings); and *Teleiodes* (Greek *teleios* + *eidos* = 'perfect' + 'form').

Apple groundling *Gelechia rhombella*, local in orchards, gardens, hedgerows, parkland and open deciduous woodland in central and south England, very local elsewhere in England and Wales. Adults fly in July–August. Larvae feed in a folded leaf or between spun leaves of (crab) apple and pear in May–June. *Rhombella* is Greek for 'rhombus-shaped' (if the wing spots were 'joined').

Atlantic groundling *Scrobipalpa clintoni*, very local, restricted to shingle and sandy beaches on the west and south-west coast of Scotland. It can be abundant as a larva but adults have only occasionally been recorded, flying in April–June. Larvae mine stems of curled dock in July. E.C. Pelham-Clinton, later Duke of Newcastle (1920–88), discovered this moth.

Barred groundling *Teleiodes sequax*, local on chalk downland and limestone pavements in England and Wales, and on acid pasture and heath in Scotland as far north as Banffshire. Adults fly in May–July. Larvae feed on rock-rose in May–June living between leaves spun together with silk. *Sequax* is Latin for 'following'.

Beautiful groundling *Caryocolum marmorea*, widespread but scarce on sandy coasts, occasionally inland. Adults fly in May–August. Larvae feed on mouse-ears, initially as a leaf miner from late January, then in spring entering the plant stem, and finally feeding externally on the basal leaves from a sand-covered silk tube on the ground. Latin *marmor* = 'marble', referencing the variegated forewing.

Black-dotted groundling *Stenolechia gemmella*, local in oak woodland and on mature oak trees in hedgerows and parkland throughout much of England and Wales; one Irish record. Adults fly in July–September. Larvae feed within young shoots of oak in April–June. *Gemella* is Latin for 'little gem'.

Black groundling *G. nigra*, wingspan 13–18 mm, scarce, in woodland, riverbanks and suburban habitats in parts of southern England and the Midlands. There has been a small but significant increase in records since 2000, extending the range into north-west England and north Wales. Adults fly in June–August. Larvae feed in May–June on poplar and aspen leaves. Latin *nigra* = 'black', from the dark forewing.

Black Isle groundling *Caryocolum blandelloides*, rare, on dunes at a few Scottish localities, including Loch Fleet in east Sutherland and Black Isle in Easter Ross. Adults fly in July–August. Larvae feed on common mouse-ear. Latin *blandellus* = 'attractive'.

Black-speckled groundling *Carpatolechia proximella*, widespread and common in open woodland and scrub. Adults fly in May–July. Larvae feed on leaves of birch and alder in August–September. *Proximella* is Latin for 'very similar' (i.e. to other congerics).

Black-spotted groundling *P. scalella*, scarce in oak woodland in parts of southern and central England, north to Yorkshire. Adults fly in May–June. Larvae probably feed on oak leaves and use moss for pupation. *Scalella* is from Latin *scala* = 'ladder', reflecting a rung-like wing pattern.

Brindled groundling *R. nanella*, local in orchards, gardens, woodland and hedgerows, mainly in southern England, occasionally north to Cheshire and Lincolnshire, and in north Wales, though overall distribution appears to be contracting. Adults fly in July–August. Larvae are found in orchards where they can be a pest, feeding on apple, pear, plum and blackthorn; they initially mine leaves and later feed inside a bud after hibernation. Latin *nanus* = 'dwarf'.

Bucks-horn groundling *Scrobipalpa samadensis*, very local to rare on saltmarsh, occasionally vegetated shingle, from Lincolnshire to Sussex, becoming locally common in most of the remaining coastal areas around the British Isles, including Ireland. Adults fly in June–September. Larvae mine roots and leaves of bucks-horn plantain and sea plantain in April–August. Samaden, south-west Switzerland, is the species' type locality.

Buff groundling *Prolita solutella*, wingspan 16–21 mm, scarce on dry pasture and dry heathland with recent records only from the central Highlands on dry, herb-rich, often cattle-grazed pasture, and the Lizard Peninsula in Cornwall on dry coastal grassland heath. Adults fly in sunshine in May–June. Larvae feed on the prostrate form of petty whin (Scotland) and hairy greenweed (Cornwall), within a silk tent in July–September. The binomial reflects Greek *litos* = 'plain' and Latin *solutus* = 'unrestrained', both parts reflecting the simple and non-distinctive wing pattern.

Campion groundling *Caryocolum viscariella*, locally common in dry grassland, hedgerows and open woodland in England (absent from East Anglia and adjacent counties), previously thought to be restricted to the north but recently recorded in many southern counties; there are scattered records from Wales, in Scotland mainly in the south, and in west Ireland. Adults fly in June–August. Larvae feed from late March to June on campions and sticky catchfly (the specific name referring to this food plant, *Silene viscaria*), initially on shoots and leaves then burrowing into and feeding inside the stem.

Cinerous groundling *Bryotropha terrella* (Anomologinae), widespread and common in grassland and mossy habitats. Adults fly in June–August. Larvae feed from October to March on common bent and other grass stems, at the base, and on mosses, living inside a silk tent. *Terrella* derives from Latin *terra* = 'earth', from its prevalence in low grassland.

Coast groundling *Caryocolum vicinella*, locally common on sea-cliffs and coastal shingle in coastal areas of southern England, Wales and east Scotland, less common in the Channel Islands, Ireland, the Isle of Man, south-west Scotland and the Hebrides. Adults fly in July–September. Larvae feed on sea campion from late March to June, spinning young shoots and leaves together. Latin *vicinus* = 'neighbouring' (i.e. similar to congenerics).

Colt's-foot groundling *Scrobipalpa tussilaginis*, forewing 6 mm, on sparsely vegetated open ground on the south Devon coast (first found at Axmouth in 1983), Dorset (from 1987, 26 records from 24 sites), and Hampshire (recorded at Milford-on-Sea in 1987, but not since). Adults fly in June–July and August–September. Larvae mine leaves of colt's-foot (= Latin *tussilago*).

Common groundling *Teleiodes vulgella*, common in gardens, scrub, open woodland and hedgerows throughout England and Wales, rare in Ireland. Adults fly in June–July. Larvae feed in April–May on a number of trees and shrubs, including hawthorn and blackthorn, living between leaves spun together with silk. Latin *vulgus* = 'people', reflecting this moth's abundance.

Common sea groundling *Scrobipalpa nitentella*, widely scattered and local, but often common, on saltmarsh and sandy beaches in coastal parts of the British Isles. Adults fly in June–August. Larvae mine leaves of annual sea-blite, goosefoot, orache, sea beet and sea-purslane, in a silk tent, in September–October. Latin *nitens* = 'shining'.

Confluent groundling *Caryocolum junctella*, wingspan 9–10 mm, in woodland and hedgerows in parts of north-west England, Wales and north Scotland. Adults fly from July onwards then hibernate, flying again in spring. Larvae feed in May–June on mouse-ear and lesser stichwort, initially mining the leaf then moving to shoot tips. Latin *junctus* = 'joined', referencing a forewing pattern.

Cornish groundling *Nothris congressariella* (Anacampsinae), wingspan 15–20 mm, on waste ground and dunes locally, occasionally common, in south-west England, the Isles of Scilly and the Channel Islands, possibly expanding its range, with a record from Dorset in 2003. Adults fly in May–July and September–October. Larvae feed between spun leaves and shoots of balm-leaved figwort, itself very localised and dictating the moth's distribution. Greek *nothros* = 'sluggish'. It was first announced as a species at a congress in 1857, hence its specific name.

Crescent groundling *Teleiodes luculella*, wingspan 9–13 mm, common in oak woodland throughout England and Wales, occasional in Ireland. Adults fly in May–July. Larvae feed on oak leaves in September–October. *Luculella* (from *lucus*) is Latin for 'a little grove', probably reflecting a key habitat.

Cypress groundling *G. senticetella*, first found in 1988 in Essex as an adventive introduced with plant material, since spreading into London and surrounding counties, parts of the Midlands, Hampshire and Wiltshire, plus one record from Devon in 2008, found in parks and gardens. Adults fly in July–August. Larvae feed in and on leaves of cypress species, juniper and red-cedar from autumn to spring. Latin *senticetella* = 'briar bush', a putative habitat.

Dark groundling *B. affinis*, found around old buildings and in urban areas, common and widespread over much of England, but local in the Channel Islands, Wales, parts of north England, east Ireland and south Scotland, and very local in the rest of Scotland, west Ireland and Northern Ireland. Adults fly in June–July. Larvae feed in March–May on mosses, often on old walls or thatch, living within a silk gallery. *Affinis* is Latin for 'akin to', from its similarity to another species.

Dark-striped groundling *G. sororculella*, widespread and common in woodland, scrub, wet heathland, fen, vegetated dunes and marsh, more scattered and local in Scotland and Ireland. Adults fly in July–August. Larvae feed on leaves and catkins of sallow and grey and other willows in May–June. Latin *sororculella* = 'little sister' (from *soror*), reflecting similarity with a congeneric.

Desert groundling *B. desertella*, widespread but local in sandy coastal areas. In England, the rare inland records are associated with dry sandy areas such as Breckland. Adults fly in May–August. Larvae feed in October–February on moss, living inside a silk tent. Latin *desertus* = 'waste' or 'desert', from its habitat preference.

Dotted grey groundling *Athrips mouffetella*, common in woodland and gardens throughout much of England and Wales, rare in south-east Scotland and west Ireland. Adults fly in July–September. Larvae mainly feed on honeysuckle in May–June, spinning together terminal shoots. *Athrips* comes from Greek *athroos* + *ips* = 'in masses' + 'larva', referencing the gregarious larvae of some species in this genus. Thomas Mouffet (1553–1604) was an English physician and naturalist.

Downland groundling *Chionodes fumatella*, local on sandy beaches and dunes, and inland in dry sandy areas over much of England and north-west Wales; very local in north England, east Scotland as far north as the Shetland Islands, and south Ireland. Adults fly by day and night in June–August. Larval diet is probably mosses as on the continent. Latin *fumatus* = 'smoked', from the dark brown wings.

Dull red groundling *B. senectella*, wingspan 9–13 mm, in urban areas, old buildings, rocky areas, dunes and dune slacks, locally common over much of England and Wales, becoming more scattered in north England, west Wales, Scotland and Ireland. Adults fly in July–August. Larvae feed in April–May on mosses, living inside a silk tent. *Senectella* comes from Latin *senex* = 'old age', suggested by the grey speckling on the wing.

Dusky groundling *Aroga velocella*, wingspan 14–19 mm, local on heathland and moorland and in woodland clearings across most of south and east England, less common in the Midlands and north England, and scarce, probably extinct, in Scotland. With two generations a year, adults fly in May and August in afternoon sunshine and at night. Larvae are found throughout much of the year on sheep's sorrel, feeding in a silk tent around the stem base. *Aroga* is probably a meaningless term. *Velocella* is from Latin *velox* = 'swift', from rapid movements made between resting.

Eastern groundling *Chionodes distinctella*, with a scattered distribution on downland, coastal shingle, breckland, dunes and other sandy habitats, at best local but occasionally common, inland and on the coast in central and south England, but restricted to coastal locations in northern England, Wales, Scotland and Ireland. Adults fly in June–August. Larvae feed from a sand-covered, silk tube vertically attached to moss that

is touching the upper leaves of bird's-foot-trefoil, and possibly sheep's sorrel and cat's-ear. Latin *distinctus* = 'distinct' (i.e. from similar species).

Elm groundling *Carpatolechia fugitivella*, local in woodland, parks, gardens and hedgerows throughout much of Britain; rare in Ireland. Adults fly in June–August. Larvae feed in April–May on elm, living between leaves spun together with silk. Latin *fugitivus* = 'fugitive'.

Goosefoot groundling *Scrobipalpa atriplicella*, widely distributed but local on dry open grassland, wasteland, and disturbed and cultivated habitats throughout much of England, rare in Scotland, Wales and Ireland. With two (maybe three) generations a year, adults are evident from April to September. Larvae mine species of orache and goosefoot, feeding in spun leaves then flowers and seeds, from May to October. *Atriplicella* refers to orache (*Atriplex*).

Gorse groundling *Mirificarma mulinella*, widespread and locally common on heathland and waste ground. Adults, recorded nectaring on ragwort, fly in July–September. Larvae feed on flower buds of gorse and broom in April–May. *Mulinella* is Latin for 'mule-coloured'.

Grand groundling *G. turpella*, wingspan 16–22 mm, found in woodland and parkland. Since 1990, it has been located in only Kent, Middlesex, Surrey and Wiltshire. Adults fly in July. Larvae feed on black and Lombardy poplars in May, living inside a rolled leaf. *Turpella* derives from Latin *turpis* = 'base', from its dingy colour.

Greenweed groundling *M. lentiginosella*, local on grassy heathland, downland and old meadow in parts of southern England, with records north to Lancashire, though probably declining in range. Adults fly in July–August. Larvae feed in May–June on dyer's greenweed between terminal shoots spun with silk. Latin *lentiginosus* = 'freckled'.

Grey sallow groundling *G. muscosella*, scarce in marsh, fen, riverbanks, shingle beaches and other damp areas. In recent years it has been recorded only from Wicken Fen, Cambridgeshire; Dungeness and Whitstable, east Kent; and at Blacktofts Sands Nature Reserve, south-west Yorkshire in 1998. Adults fly in June–August. Larval diet may be catkins and shoots of willows and poplars, but *muscosella* comes from Latin *muscosus* = 'mossy', suggesting a moss larval diet.

Hazel groundling *Teleiodes wagae*, wingspan 9–13 mm, in mixed deciduous woodland, hedgerows and scrub in Hampshire (first UK record in 1976), Sussex, Surrey and Kent; first recorded in the British Isles from the Burren, Co. Clare in 1961. Adults fly in May–June. Larvae feed during August–September on hazel, birch and sweet chestnut, within a rolled leaf. Antoni Waga (1799–1890) was a Polish entomologist.

Heather groundling *Neofaculta ericetella* (Anacampsinae), wingspan 13–18 mm, widespread and common on heathland and moorland, occasionally in gardens where heather is grown. Adults fly in April–June. Larvae feed on flowers moving on to shoots of heathers and cross-leaved heath. *Ereticella* is Latin for relating to 'heath'.

Horseshoe groundling *Altenia scriptella*, scarce in hedgerows and woodland margins in southern England from Somerset to East Anglia, and in Herefordshire and Worcestershire. Adults fly in June–July. Larvae feed in folded leaves of field maple in August–September. Latin *scriptus* = 'marked', from markings on the forewing.

House groundling *B. domestica*, common near buildings (including urban areas) and around dry-stone walls over much of England, local in northern England, Wales, Ireland, Isle of Man, and south and central Scotland. Recorded from 2009 onwards in a few locations in north-east Scotland: it may be extending its range or has previously been overlooked there. Adults fly in May–August. Larvae feed in February–April on moss, often on walls, inside a silk gallery. *Domesticus* is from Latin for 'house'.

Large groundling *Teleiopsis diffinis*, wingspan 13–18 mm, common on grassland and heathland throughout much of Britain, rare in Ireland. Adults have two or more probably overlapping broods and fly any time between May and October. Larvae feed on sheep's sorrel roots inside a silk tent. *Diffinis* is Latin for 'distinct'.

Long-winged groundling *G. cuneatella*, in marsh, riverbanks and other damp areas, only having been found in England at 15 sites from Kent, north to Lancashire and Yorkshire and mainly as singletons; six individuals were found at a site in Warwickshire in 2007. Adults fly in August–September. Larvae feed within rolled willow leaves in May–July. Latin *cuneus* = 'wedge', the wings having several wedge-shaped streaks.

Mallow groundling *Platyedra subcinerea* (Apatetrinae), wingspan 16–18 mm, local in rough ground and gardens in south and east England and the Channel Islands, and declining in part of its range. Adults fly from August, and overwinter, sometimes hibernating in thatch, to reappear in spring. Larvae feed on flowers and seeds of mallow and hollyhock. Greek *platus* + *edra* = 'wide' + 'foundation', referencing the broad hindwing, and Latin *subcinerea* = 'slightly ash-grey'.

Meadow groundling *Caryocolum proxima*, in open grassland, hedgerows and parks in parts of England and declining in range (only three sites noted since 2000). Adults fly in July–August. Larvae feed in April–July on flowers and seeds of mouse-ear and chickweed. *Proxima* is Latin for 'close', reflecting a similar appearance to congenerics.

Moorland groundling *Xenolechia aethiops*, wingspan 18 mm, scarce on heathland and moorland, especially where recently burned, scattered over parts of Britain (plus two records from central Ireland), but with a recent serious decline in numbers and geographical range. Adults fly in April–June. Larvae feed on bell heather in June–July, initially leaf-mining then in a silk gallery among the leaves. Greek *xenos* = 'foreign' + *echia* as a curtailed form of *Gelechia*. Latin *aethiops* = 'Ethiopian' ('African'), from the dark-coloured forewing.

Mottled groundling *Neofriseria singula*, local on heathland in south and south-east England, from Hampshire and Kent northwards to Suffolk, and the Brecks of East Anglia. Adults fly in June–July. Larvae feed on sheep's sorrel in a silk tent in May. *Singula* is Latin for 'particular' (i.e. distinct from a congener).

Mouse-ear groundling *Caryocolum fraternella*, local in dry grassland, heathland and damp open areas in Britain as far north as East Lothian, and in Ireland. Adults fly in July–August. Larvae feed on terminal shoots of mouse-ears and stichworts from April–May. Latin *fraternus* = 'fraternal', referencing a similar appearance to congenerics.

Narrow groundling *Caryocolum alsinella*, wingspan 8–10 mm, scarce in sandy, coastal areas mostly south of a line from the Humber to Morecambe Bay, including several sites in Wales and a few in the Scilly Isles and south-east and west Ireland. Further north it is rare, with records from the Isle of Man and the Outer Hebrides (on machair). Adults fly in July–August. Larvae initially mine leaves of mouse-ears in May, then move to flowers and seeds in June. Spring sandwort is a food plant on the Burren. *Alsinella* references the Alsineae, a former subfamily name for some food plants.

Northern groundling *Athrips tetrapunctella*, wingspan 9–10 mm, in grass and grass-heath in the Highlands and limestone pavement in the Burren, west Ireland. Adults fly at dawn and dusk in May–June. Larvae feed on bitter and tufted vetch inside a rolled leaf in June–August. Greek *tetra* = 'four' and Latin *punctella* = 'spotted', referencing markings on the forewings.

Obscure groundling *B. similis*, widespread but local in urban areas with old walls, and open country with dry-stone walls and rocky outcrops, less common in Ireland. Adults fly in June–August. Larvae feed in April–May on mosses on rocks and old walls. *Similis* is Latin for 'similar to' (i.e. congenerics).

Perth groundling *B. galbanella*, mainly found in Scots pine and larch woodland in the Highlands but with records also from coastal sites in north-east Scotland and central lowland Scotland. There is also a record from upland limestone in Yorkshire. Adults fly in May–August. Larvae feed on moss, living in a silk tube or net. Latin *galbanus* = 'yellowish', from the colour of the forewing speckling.

Pine groundling *Exoteleia dodecella*, widespread and locally common in conifer plantations. Adults fly in June–July. Larvae initially feed internally on pine needles (with some records on larch) and subsequently externally on the shoots, spinning the needles together. Greek *dodeca* = 'twelve', referring to each anterior wing having six spots.

Pointed groundling *Scrobipalpa acuminatella*, widespread and common in rough meadow, verges, gardens, parkland, field margins and chalk downland, more scattered in Ireland. Adults fly in April–June and July–September. Larvae mine leaves of thistles and maybe coltsfoot in July–October. Latin *acuminatus* = 'pointed', from the pointed forewings.

Polished groundling *B. politella*, local and sometimes common on dry unimproved grassland and moorland in Scotland and upland parts of Wales and northern England, very local in the Midlands and south England, and in Ireland. Adults fly in June–July. Larvae feed in April–May on moss, living inside a silk tent. Latin *politus* = 'polished', referencing the forewing gloss.

Rock groundling *Scrobipalpa murinella*, scarce on grass-heath and rocky areas in south Aberdeenshire and parts of the inner Western Isles of Scotland, on limestone pavement on the Burren, west Ireland, and at single sites in west Sutherland, east Perthshire and Co. Fermanagh. Adults fly in April–June. Larvae mine leaves and flowers of mountain everlasting in July–September. Latin *murinus* = 'mouse-like', this being the smallest of its genus.

Sallow-leaf groundling *Carpatolechia notatella*, widespread in woodland, heathland, scrub and fen. Adults fly in May–July. Larvae feed in August–October on sallow leaves. Latin *notatus* = 'marked'.

Saltern groundling *Scrobipalpa instabilella*, scarce on salt-marsh, mudflats and estuaries along parts of the coast of England and Wales, uncommonly in south Scotland and east Ireland. Adults fly in June–August. Larvae mine leaves of sea-purslane, grass-leaved orache, sea aster and glasswort in February–May. Latin *instabilis* = 'variable', referencing variable colour and pattern of the forewing.

Sandhill groundling *B. umbrosella*, wingspan 9–11 mm, on dunes over much of the coastal British Isles and, rarely, dry sandy inland areas. Adults fly in May–August. Larvae feed in April–May on mosses and grasses in a silk tube or web. Latin *umbrosus* = 'shady', describing the dark forewing.

Scarce groundling *Caryocolum kroesmanniella*, rare, in open woodland with a scattered distribution in England. Adults fly in July–August. Larvae feed on stichwort. D.W. Krössmann was a mid-nineteenth-century German schoolmaster.

Sea-aster groundling *Scrobipalpa salicorniae*, scarce in salt-marsh from Cornwall to Lincolnshire, and in Co. Durham; Lancashire, and Pembrokeshire, and East Lothian and Nairn in Scotland. Adults fly at dusk in June–September. Larvae mine leaves of glasswort (species of *Salicornia*, giving the specific name), sea aster and annual sea-blite in April–June.

Sea-blite groundling *Scrobipalpa suaedella*, scarce, local on saltmarsh from Dorset to north-west Norfolk. Adults fly in June–August. Larvae mine flowers and leaves of shrubby sea-blite (*Suaeda*, giving the specific name) in May–June.

Seathorn groundling *G. hippophaella*, wingspan 17–21 mm, scarce, recorded from coastal sandhills in England, intermittently, from Camber in East Sussex to Spurn Head in south-east Yorkshire. Adults fly in September. Larvae feed within rolled or spun leaves of sea-buckthorn (*Hippophae*, giving the specific name) in June–July.

Short-barred groundling *Caryocolum blandella*, scarce, in hedgerows and woodland throughout much of England and

Wales, less common in south Scotland and south and east Ireland. Adults fly in July–August. Larvae feed on greater stitchwort, initially (April) in a mine, then from May in spun shoots, and finally seed capsules. Latin *blanda* = 'attractive'.

Six-spot groundling *Prolita sexpunctella*, day-flying, local on heathland, moorland, damp heaths and mosses in parts of Wales, northern England and Scotland. Adults fly in sunshine in May–June. Larvae feed on heather between leaves spun together with silk from August into spring. Latin *sexpunctella* = 'six-pointed'.

Southern groundling *Caryocolum blandulella*, wingspan 8–11 mm, found on dunes at Sandwich, Kent (last in 1983) and Carmarthenshire (last in 2006). Adults fly in July–August. Larvae feed in May–June on little mouse-ear. Latin *blanda* = 'attractive'.

Suffused groundling *Carpatolechia alburnella*, local in woodland, scrub and heathland throughout England, north Wales and the Highlands. Adults fly in June–August. Larvae feed between spun or folded birch leaves in May–June. Latin *alburnus* = 'whitish', describing the forewing colour.

Summer groundling *Scrobipalpa obsoletella*, on saltmarsh and sandy beaches in coastal south nd east England and south Wales, though also recorded inland, becoming more local northwards; recorded as far north as Aberdeenshire; and two geographically well-separated Irish records. With two generations a year, adults have been noted from May to September. Larvae mine orache stems from May to October. Latin *obsoletus* here means 'ordinary' (with no spots near the base of the forewing, unlike other congenerics).

Tawny groundling *Pseudotelphusa paripunctella*, wingspan 9–16 mm, widespread but scarce, found in two distinct ecotypes: on woodland edges, heaths and hedgerows (where larvae feed on oak, more so in the south of the range); and on moorland, fen, bog, damp heathland and mosses (bog-myrtle, especially in the north). Adults fly in May–June. Latin *par* + *punctum* = 'pair' + 'point', from the black spots on the forewing which are arranged in spots.

Thatch groundling *B. basaltinella*, in urban habitats and old buildings, local in south and east England, plus a small number of records from widely scattered parts of northern England and a single record from east Scotland. Adults fly in May–August. Larvae feed on moss, living within a silk tent. Latin *basaltes* = 'basalt', from the dark grey wing colour.

Thicket groundling *G. scotinella*, scarce, very local with scattered records in mature hedgerows, scrub and gardens across the southern half of England. Adults fly in July–August. Larvae feed within blackthorn stems and flowers in April–May. *Scotinella* comes from Greek *scoteinos* = 'dark-coloured', from the speckling on the forewing.

Three-colour groundling *Caryocolum tricolorella*, local in open woodland and hedgerows throughout parts of England and Wales, and rare in Ireland. Adults fly in June–August. Larvae feed on greater stitchwort, initially (December–January) mining the leaves, later (into May) in buds and between spun shoots. Latin *tricolorella* refers to the three wing colours.

Western groundling *B. dryadella*, found on coastal limestone cliffs, dunes and chalk quarries in Cornwall and Devon, and two sites bordering the Thames estuary in Essex and Kent. Adults fly in May–August. Larvae feed on moss within a silk tent. *Dryadella* relates to *Dryas* (mountain avens), though this is not a food plant nor found in the same habitats.

White-barred groundling *R. leucatella*, local in hedgerows, and in gardens and orchards with scrubby growth or isolated larger trees throughout much of England and north Wales, and slowly extending its range with increasing numbers and wider distribution into Yorkshire and Lancashire since 2000. Adults fly in June–July. Larvae feed between two spun leaves of hawthorn, apple and (less commonly) rowan in May–June. *Leucatella* comes from Greek *leucos* = 'white', from the white forewing mark.

White-spot groundling *Neofriseria peliella*, on shingle beaches in parts of south-east England, notably in Kent and East Sussex, plus records (2009, 2013) from Kempton Nature Reserve, Middlesex. Adults fly in July. Larvae feed on sheep's sorrel in May–June in a silk tube. *Peliella* comes from Greek *pelos* = 'mud', from the forewing colour.

Winter groundling *Scrobipalpa costella*, wingspan 9–16 mm, common in a range of habitats, from hedgerows, damp woodland and gardens to shingle beaches, throughout Britain, though less numerous in the north, and mostly coastal in Ireland. Adults are probably double-brooded, flying from April to November. Larvae mine leaves of bittersweet and possibly deadly nightshade in May and August–October. *Costella* comes from Latin *costa*, the anterior margin of the wing on which there is a distinctive blotch.

Winter oak groundling *Carpatolechia decorella*, in deciduous woodland and scrub, widespread but local across much of England, Wales and Scotland, and very local in Ireland. Adults emerge in July and overwinter, sometimes found again the following spring. Larvae feed inside silk-folded leaves of oak and dogwood in May–June. Latin *decorus* = 'attractive'.

Wood groundling *Parachronistis albiceps*, widespread but local, occasionally common, in England from the Midlands southwards; very local in south-west England, Nottinghamshire, Yorkshire, Wales and the Channel Islands. Adults fly in June–July. Larvae feed inside young hazel shoots and buds in spring. Greek *para* + *chronos* = 'beside' in the sense of 'contrary to' + 'time', since this is a replacement name (i.e. mistimed), and Latin *albus* + *caput* = 'white' + 'head'.

GROUNDSEL *Senecio vulgaris* (Asteraceae), a common and widespread annual of open and disturbed ground, a weed in waste ground, gardens, arable fields and other open habitats, and (ssp. *denticulatus*) on dunes and coastal cliffs. The name comes from Old English *grundeswylige*, meaning 'ground swallower' (i.e. a weed). There is also a seventh-century word *gundesuilge*, meaning 'swallower of eye discharge', which could refer to the early use of groundsel as an eye drop and a poultice. *Senecio* may be derived from Latin *senex* = 'old man', in reference to the downy seed head of seeds; Latin *vulgaris* = 'common'.

Eastern groundsel *S. vernalis*, an annual European neophyte introduced in 1803, occasionally found in England and south-east Ireland on verges, newly grassed areas and disturbed land, originating as a grass-seed contaminant. Latin *vernalis* = 'of spring'.

Heath groundsel *S. sylvaticus*, a widespread and locally common annual found on heathland, in cleared and burnt woodland, and on banks and sea-cliffs, usually on sandy, non-calcareous soils. *Sylvaticus* is from Latin for 'forest'.

Sticky groundsel *S. viscosus*, an annual European neophyte introduced in 1660, though possibly native, widespread in Britain, more scattered in Ireland, on free-draining disturbed substrates, for example sands, gravels and cinders, on roadsides, banks, walls, pavements, railway ballast, coastal shingle (where it might be native) and dunes, in gravel-pits, and on open rough and waste ground. Latin *viscosus* = 'sticky'.

Tree groundsel *Baccharis halimifolia*, a deciduous North American shrub introduced by 1683, naturalised by the sea in east Dorset and west Hampshire. The binomial is from Greek *baccaris*, a plant with a fragrant root, and Latin for having leaves like sea orache *Atriplex halimus*.

Welsh groundsel *S. cambrensis*, an annual not identified as this species (probably a hybrid between groundsel and Oxford ragwort) until 1957, now found only in Wales, with 19 known sites. The first confirmed record came from Flintshire (now Clwyd) in 1948, and it is still found in the Wrexham area. It was also found in Leith, near Edinburgh in 1982, though now extinct there. It grows in open or disturbed sites, mostly verges, but also rough ground, and footpaths, and in cracks in walls. *Cambrensis* is Latin for 'Welsh'.

GROUSE Members of the family Phasianidae. 'Grouse' is an early sixteenth-century word, possibly related to medieval Latin *gruta* or Old French *grue* = 'crane'.

Black grouse *Tetrao tetrix*, wingspan 72 cm, length 48 cm, male (blackcock) weight 1.2 kg, female (greyhen) 930 g, a game bird found in upland Britain with 5,100 males estimated in 2005, but habitat loss and overgrazing have resulted in a severe population decline, making this a Red List species. Males have an elaborate courtship display in arenas (leks) in open vegetation. Clutch size is 6–11, incubation lasts 25–7 days, and the precocial young fledge in 10–14 days. Diet comprises buds, shoots, catkins and berries. Both binomial names come from Greek, referring to some kind of game bird.

Red grouse *Lagopus lagopus*, possibly an endemic subspecies of willow grouse, as ssp. *scotica*, or a distinct species; birds in Ireland are sometimes thought to belong to ssp. *hibernica*. Wingspan is 60 cm, length 40 cm, and weight 600 g. It is common and widespread in upland regions, with 230,000 pairs estimated in 2009, but the population declined in the recent past, perhaps linked to disease and loss of moorland which sustains the food plant of heather, and it is on the Amber List. However, numbers increased by 14% over 1970–2015, by 19% over 1995–2015, and by 15% over 2010–15. Nests are in the heather, clutch size is 6–9, incubation takes 19–25 days, and the precocial young fledge in 12–13 days. Heather moorland management is often driven by grouse shooting (which can be traced back to 1853 and the invention of the breech loading gun) which brought in more revenue than sheep grazing. Greek *lagos* + *pous* = 'hare' + 'foot', a reference to the feathered feet and toes typical of this cold-adapted bird.

GROUSE WING *Mystacides longicornis* (Leptoceridae), a common and widespread caddis fly found in slow-moving rivers, streams, canals and ponds. Nymphs, found on mud, sand, plants and woody debris, are mainly gatherers, but are also shredders, grazers and predators. The case is made of sand and small stones, with some plant material. Adults are evident from May to September. Greek *mystacos* = 'upper lip', and Latin *longicornis* = 'long-horned'.

GRUIFORMES Order of coots, moorhen and rails (Rallidae) and cranes (Gruidae) (Latin *grus* = 'crane').

GRYLLIDAE Cricket family (from Gryllos, a figure in Ancient Greek comedy, in turn named after an Egyptian dance), comprising three rare native species, together with house cricket.

GUDGEON *Gobio gobio* (Cyprinidae), 9–13 cm long, common in England, less so elsewhere, in rivers, streams, lakes and gravel-pits, especially those with sandy or gravelly bottoms. An omnivorous bottom-feeder favouring small animals, it spawns in and around May. 'Gudgeon' dates from late Middle English, from Old French *goujon*, in turn from Latin (*gobio* and *gobius*) for this species.

Topmouth gudgeon *Pseudorasbora parva*, 2–7.5 cm long, a Far Eastern fish accidentally imported to, and escaping from, a fish farm near Romsey, Hampshire, in the late 1980s, and entering part of the River Test catchment. There have been a few other escapes elsewhere in England. The binomial comes from *Rasbora*, a tropical fish genus, derived from a native word in the East Indies, prefixed by Latin *pseudo* = 'false', and *parva* = 'small'.

GUELDER-ROSE *Viburnum opulus* (Caprifoliaceae), a common and widespread deciduous shrub of neutral or calcareous soils, much planted in parks and gardens, found in woodland, scrub and hedgerows, in fen carr and alder and willow thickets, and on stream banks, preferring damp places, but also found in dry habitats. American and Asian guelder-rose probably have subspecies rather than species status. The name comes from Gueldersland, a Dutch province, where the tree was first cultivated, and the Dutch *geldersche roos*, meaning the tree with a 'rose-like flower'. *Viburnum* is the Latin name for the related wayfaring-tree, and *opulus* for a kind of maple.

GUILLEMOT *Uria aalge* (Alcidae), wingspan 67 cm, length 40 cm, weight 690 g, an Amber List auk spending most of its life ay sea, but nesting on rocky cliffs from March to the end of July, except between the Humber estuary and the Isle of Wight, with 880,000 breeding pairs in Britain in 2002 (950,000 in the UK), making it the most abundant seabird. Numbers increased by 57% between 1986 and 2014, by 6% in 2000–14, and (strongly) by 17% in 2009–14. Numbers of breeding pair at Skomer, for example, increased from 2,000 in 1972 to 25,000 in 2013 (though were about 100,000 in the 1930s). The single egg takes 28–37 days to hatch, the altricial chick fledging in 18–25 days. Diet is mainly of small fish, with some invertebrates, caught by surface diving. The name guillemot probably derives from French Guillaume, or 'William'. The binomial comes from Greek *ouriaa*, a waterbird mentioned by the rhetorician Athenaeus, and the Danish *aalge* = 'auk' (from Old Norse *alka*). The arctic **Brunnich's guillemot** *U. lomvia* is a scarce accidental.

Black guillemot *Cepphus grylle*, wingspan 55 cm, length 31 cm, weight 420 g, an Amber List auk found year-round along much of the Scottish coast, especially the larger western sea lochs and the Hebrides, Orkney and Shetland, as well as the Isle of Man, much of the Irish coastline, and at a few coastal sites in Cumbria and Anglesey. Some 19,000 breeding pairs were estimated for Britain in 2002, while wintering numbers for the British Isles may be 58,000–80,000. Nests are among boulders on cliffs and in rock crevices, with 1–2 eggs, incubation lasting 23–40 days, and fledging of the altricial young taking 31–51 days. Food is mostly fish. Greek *cepphos* was a waterbird described by Aristotle, and Swedish *grissla* = 'guillemot'.

GULL Members of the family Laridae, in the genera *Larus* (Greek *laros*, then Latin *larus* = 'seabird'), *Chroicocephalus* (Greek *chroia* + *cephale* = 'colour of' + 'head'), *Hydrocoloeus*, *Pagophila*, *Rhodostethia*, *Rissa* and *Xema*, associated with coastal habitats, but many species now commonly found feeding and breeding inland. See also kittiwake.

Caspian gull *L. cachinnans* and **ring-billed gull** *L. delawarensis* are scarce visitors; and **American herring gull** *L. smithsonianus*, **Audouan's gull** *L. audouinii*, **Bonaparte's gull** *C. philadelphia*, **Franklin's gull** *L. pipixcan*, **glaucous-winged gull** *L. glaucescens*, **great black-headed gull** *L. ichthyaetus*, **ivory gull** *P. eburnea*, **laughing gull** *L. atricilla*, **Ross's gull** *Rh. rosea* and **slender-billed gull** *C. genei* are all accidentals.

Black-headed gull *C. ridibundus*, wingspan 105 cm, length 36 cm, male weight 330 g, common and widespread, with 130,000 breeding pairs in Britain in 2002, and 14,000 in Ireland, and an increase by 102% estimated between 2000 and 2014, with similar numbers breeding inland (often in towns) as on the coast. Most of the breeding population is resident, with numbers greatly supplemented during winter by birds from Scandinavia and eastern Europe, especially in east and south-east England, to total 2.1 million birds. Even so, this species remains on the Amber List. It breeds in colonies in reedbeds, marsh, and islands in lakes, reservoirs and flooded gravel-pits, nesting on the ground. Clutch size is 2–3, incubation takes 23–6 days, and the semi-precocial young fledge in 34–6 days. It opportunistically feeds on insects, earthworms, plant material and scraps in towns, and invertebrates in ploughed fields. 'Black-headed' is misleading: rather, the summer adult has a chocolate-brown head, but even this colour disappears in winter, leaving two dark spots on the side. Latin *ridibundus* = 'laughing' (*ridere* = 'to laugh').

Common gull *L. canus*, wingspan 120 cm, length 41 cm, weight 400 g, an Amber List resident breeder with an estimated 48,700 pairs in 2002 (an increase of 36% since 1988), and 1,600 in Ireland (an increase of 222% since 1988), and with around 700,000 wintering birds, found inland (mainly near marsh and lakes and on farmland, but also in towns) but more abundant by the coast. Nesting, in scrapes or on cliffs, is colonial, with clutches usually of 3 eggs, incubation taking 22–8 days, and fledging of the semi-precocial young 34–6 days. Feeding is on invertebrates, with some fish and carrion, and a preference for foraging on the ground. The English name was coined by Thomas Pennant in 1768 because he believed it to be the most numerous gull in Britain. Latin *canus* = 'grey'.

Glaucous gull *L. hyperboreus*, wingspan 158 cm, length 65 cm, male weight 1.8 kg, females 1.4 kg, an Amber List winter visitor with 150 birds noted in Britain in 2009, found around the coast, and inland where gulls gather at rubbish tips to scavenge and roost on reservoirs. 'Glaucous' refers to the blue-green or grey colour. Latin *hyperboreus* = 'far north'.

Great black-backed gull *L. marinus*, wingspan 158 cm, length 71 cm, weight 1.7 kg, an Amber List bird found around the coast in the breeding season, but especially in the Hebrides, Orkney and Shetland, which offer extensive areas of the preferred breeding habitat of well-vegetated rocky coastline with stacks and cliffs. The twentieth century saw widespread expansion of both breeding range and numbers. An estimated 17,000 pairs bred in Britain in 2002. Numbers increased by 7% between 1986 and 2014, and by 28% between 2000 and 2014. Irish breeding numbers were 2,300 pairs in 2002, a 28% decline over 1988 figures. At other times of the year this species also congregate inland, roosting on reservoirs, and scavenging on landfill sites. Winter numbers in Britain in 2006 were 76,000 (71,000–81,000) birds. This species nests almost exclusively in coastal habitats, but is occasionally found inland at freshwater sites and on roofs of buildings. Nests are lined with grass, seaweed or moss, or objects such as rope or plastic. Clutch size is usually 3, incubation takes 27–8 days, and the semi-precocial chicks fledge in 50–5 days. It is omnivorous, mostly feeding on animals including other seabirds, and it also takes carrion, scavenges, and kleptoparasitises other seabirds. Latin *marinus* = 'marine'.

Herring gull *L. argentatus*, wingspan 144 cm, length 60 cm, male weight 1.2 kg, females 950 g, widespread, found all year around the coast and inland around rubbish tips, fields, reservoirs and lakes, especially during winter (numbers in Britain in 2006 estimated at 730,000). Breeding population in Britain in 2002 was an estimated 132,000 pairs (declining by 13% over 1988 numbers), with SMP suggesting a decline of 38% between 2000 and 2014, but BTO has reported a strong increase by 47% in 2009–14. Over half of the UK breeding population is confined to <10 sites. There were 6,200 pairs in Ireland in 2002 (decreasing by 81% since 1988). These trends have placed this species on the Red List. It prefers to nest on cliffs and offshore islands, though other habitats used include dunes, shingle banks and, increasingly, rooftops of buildings in towns. Nests are on cliff ledges or roof tops, with 2–4 (usually 3) eggs laid, incubation taking 28–30 days, the semi-precocial chicks fledging in 35–40 days. While primarily a coastal feeder, it takes advantage of food supplies available indirectly from human activity, especially waste from the fishing industry and on landfill. A decline in the marine fish stock and increases in the number of landfill sites has pushed distribution inland, but the former is probably a major cause for the overall decline in numbers. Latin *argentatus* = 'decorated with silver'.

Iceland gull *L. glaucoides*, wingspan 145 cm, length 56 cm, weight 750 g, an Amber List winter visitor with 210 birds noted in Britain in 2009, recorded from around the coast, occasionally inland at large gull roosts on reservoirs and at rubbish tips. Despite its name, it does not breed in Iceland: the main breeding grounds are in Greenland and eastern Arctic Canada. *Glaucoides* is Latin for a colour similar to *glaucus* = 'blue-green' or 'grey'.

Lesser black-backed gull *L. fuscus*, wingspan 142 cm, length 58 cm, weight 830 g, an Amber List with an estimated 112,000 breeding pairs in Britain in 2002 (an increase by 40% since 1988, and representing 40% of the European population), but data suggest a subsequent decline by 48% between 2002 and 2013. There were 4,800 pairs in Ireland in 2002 (increasing by 86% since 1988). The largest colony is on Walney Island, Cumbria,

with a third of the UK population. More than half the UK population is found in <10 sites. They are increasingly common in urban habitats, including inland locations such as the west Midlands. In winter they are mainly found from south Scotland southwards, with a total of 130,000 birds in Britain in 2006, augmented by Scandinavian migrants arriving from October. Nests are usually on the ground on offshore and lake islands, dunes and coastal cliffs, but small numbers also nest on roof tops, for example in Dublin. Clutch size is usually 3, incubation takes 24–7 days, and fledging of the semi-precocial young 30–40 days. The diet is omnivorous and includes scavenging on dumps. *Fuscus* is Latin for 'dark'.

Little gull *Hydrocoloeus minutus*, wingspan 78 cm, length 26 cm, weight 120 g, a winter visitor with 400–800 birds p.a., mostly seen around the coasts of England, Wales, east Scotland and east Ireland, and a passage migrant with 200–700 birds in spring and late summer/autumn. Greek *hydor* + *coloios* = 'water' + 'a kind of web-footed bird', and Latin *minutus* = 'small'.

Mediterranean gull *L. melanocephalus*, wingspan 96 cm, length 37 cm, weight 320 g, very rare until the 1950s, now an Amber List bird breeding in increasing numbers, with 543–92 pairs at 37 sites in the UK in 2008, and 1,016–34 pairs at 34 sites in 2010. A decline by 15% was noted in 2010–15. It arrived in Ireland in 1995 and first bred in 1996 in Co. Wexford. It is also widespread in winter in much of England, Wales, Ireland and east Scotland, with 1,800 birds recorded in Britain in 2008–9. It often nests in black-headed gull colonies. Clutch size is usually 3, incubation taking 23–5 days, the semi-precocial chicks fledging in 35–40 days. In summer birds often feed on insects, in winter marine fish and molluscs. Greek *melas* + *cephale* = 'black' + 'head'.

Sabine's gull *Xema sabini*, a scarce passage visitor migrating between Greenland and Spitzbergen and the tropical Atlantic, with an average of 140 records p.a., mostly on the English, Welsh and south and west Irish coasts. *Xema* is a made-up word, while the English and specific names honour Gen. Sir Edward Sabine (1788–1883), an English scientist and explorer.

Yellow-legged gull *L. michahellis*, wingspan 150 cm, length 63, weight 1.2 kg, Amber-listed, until recently considered to be a race of herring gull, a northward spread from the Mediterranean having made it a regular winter visitor in parts of England, Wales and north-east Ireland on reservoirs and rubbish tips, in fields, on coastal marshes and in large evening gull roosts, often with lesser black-backed gulls. A winter population of 1,100 was estimated for Britain in 2009. The first recorded of breeding was in 1995; by 2010 there were six known breeding sites, but only one or two pairs bred true (non-mixed pairs). These are omnivorous scavengers. Karl Michahelles (1807–34) was a German zoologist and physician.

GUM (TREE) Species in the genus *Eucalyptus* (Myrtaceae), species listed here all being fast-growing evergreen Tasmanian neophyte trees. Greek *eu* + *calyptos* = 'good' + 'covered', referring to the operculum of the flower bud which protects the developing parts as the flower develops.

Cider gum *E. gunnii*, hardy, widely planted in gardens, parks, amenity areas and, occasionally, for small-scale forestry since 1840, particularly in England and Wales. It has become naturalised in woodland and on roadsides in south-east England and Essex.

Ribbon gum *E. viminalis*, planted in gardens since 1885, used in small-scale forestry in Ireland, and recorded as naturalised on the Isle of Man. *Viminalis* is Latin for 'bearing wickerwork (osier) shoots'.

Southern blue-gum *E. globulus*, planted in gardens since 1829, found on roadsides and, rarely, in small-scale forestry. It is self-sown in the Scilly Isles, the Isle of Man, south-west Ireland and the Channel Islands. *Globulus* is Latin for 'ball'.

Urn-fruited gum *E. urnigera*, grown in gardens since 1860, planted for small-scale forestry in Ireland, and naturalised on the Isle of Man. *Urnigera* is Latin for 'urn-bearing'.

GUNNEL **Rock gunnel** Alternative name for butterfish.

GUNNERACEAE Family name of giant-rhubarb. Johan Gunnerus (1718–83) was a Norwegian botanist.

GUTWEED *Ulva intestinalis* (Ulvaceae), a common and widespread green seaweed with fronds 10–30 cm in length, 6–18 mm in diameter. It grows on rocks, mud and sand, and in rock pools. It is also a common epiphyte on other algae and shells. It can be abundant in brackish water and saltmarsh. The inflated, hollow fronds resemble large mammalian intestines, hence its common and specific names. *Ulva* is classical Latin for 'sedge', but in New Latin = 'sea lettuce'.

GWYNIAD *Coregonus pennantii* (Salmonidae), an endemic glacial relict whitefish found native only in Lake Bala (Llyn Tegid), north Wales, where the population is threatened by eutrophication and by ruffe (introduced to the lake in the 1980s and eating gwyniad eggs and fry). As a conservation measure, gwyniad eggs were transferred to the nearby Llyn Arenig Fawr between 2003 and 2007; a 2012 survey indicated at least one successful subsequent generation. The name comes from Welsh *gwyn* = 'white'. Greek *core* + *gonia* = 'pupil of the eye' + 'angle', leading to 'angle-eyed'. Thomas Pennant (1726–98) was a Welsh naturalist.

GYALECTA **Elm gyalecta** *Gyalecta ulmi* (Gyalectaceae), a crustose lichen with scattered records in Britain, growing on moss and the shaded northern side of calcareous rocks, formerly commoner on elm tree bases but having declined following Dutch elm disease, and now a priority species in the UK BAP. Greek *gyalon* = 'hollow vessel', and Latin *ulmus* = 'elm'.

GYMNOSPERM A plant, usually a conifer, whose seeds develop either on the surface of scales or leaves, which are modified to form cones, or are solitary as seen in yew, in contrast to angiosperms, from Greek *gymnos* + *sperma* = 'naked' + 'seed'.

GYPSY, THE (FUNGUS) *Cortinarius caperatus* (Agaricales, Cortinariaceae), rare, inedible, found in late summer and autumn in the Highlands on acid soil in coniferous woodland and heathland, it being ectomycorrhizal with conifers and possibly beech, as well as with ericaceous plants such as heather, heath and bilberry. *Cortinarius* refers to the partial veil or *cortina* (Latin for 'curtain') that covers the gills when caps are immature. Latin *caperatus* = 'wrinkled', a comment on the furrowed surface of most mature caps.

GYPSYWORT *Lycopus europaeus* (Lamiaceae), a rhizomatous herbaceous perennial, common and widespread in England, Wales and west Scotland, scattered elsewhere, in wet habitats on organic and mineral soils, including stream, lake and ditch banks, fen, fen carr, the top of beaches and dune slacks. It tolerates temporary flooding, often a colonist of newly exposed mud and shallow water. 'Gypsywort' derives from 'Egyptian herb', because it was reputed to have been used by gypsies to stain their skin brown. Greek *lycos* + *pous* = 'wolf' + 'foot'.

GYRINIDAE Family name (Greek *gyrinos* = 'tadpole', in turn from *gyros* = 'round') for whirligig beetles, in the suborder Adephaga, with 12 British species in Britain, some of them widespread (11 species of *Gyrinus*, plus hairy whirligig beetle *Orectochilus villosus*).

GYROPHORA From Greek *gyros* + *phoros* = 'circle' + 'carrying', 'gyrose' meaning marked with wavy lines, a name used in the dye industry for some foliose lichens in the genus *Umbilicaria*; see rock tripe.

H

HAHNIIDAE Family name for lesser cobweb spiders (named after the German zoologist, Carl Hahn, 1786–1835), with seven British species in two genera.

HAIR-GRASS Species of *Deschampsia* (French for 'of the fields', but actually named after the French naturalist Louis Deschamps, 1765–1842), *Aira* (Greek for a kind of rye-grass), *Corynephorus* (Greek *coryne* + *phoros* = 'club' + 'bearing'), *Koeleria* (honouring the German botanist George Koeler, 1764–1807) and *Rostraria* (Latin *rostrata* = 'beaked' or 'curved'), all Poaceae.

Bog hair-grass *D. setacea*, a scarce perennial, local in the New Forest, north Scotland and the Hebrides, and western Ireland, in heathland depressions, peaty or stony margins of lochs and shallow pools, and on acid bogs, favouring bare areas flooded in winter but dry in summer, and perhaps where there is lateral water movement. *Setacea* is Latin for 'bristly'.

Crested hair-grass *K. macrantha*, a perennial of grasslands on infertile, mainly calcareous substrates locally in southern Britain; also on scree, quarry heaps and old lead workings. In northern regions and in Ireland it is more coastal, often in dry, sandy, base-rich grassland on clifftops and dunes. *Macrantha* is Latin for large-flowered.

Early hair-grass *A. praecox*, a common and widespread annual, if declining in England and central Ireland, in sandy, gravelly and rocky habitats, generally on thin acidic soils around rock outcrops, on walls, clifftops, heathland and dunes. *Praecox* is Latin for 'early'.

Grey hair-grass *C. canescens*, a rare perennial local in Jersey, Suffolk, Norfolk and Lancashire, probably introduced in parts of north Scotland, on dunes, sandy shingle and open sand, and also on sandy heathland on acidic soils in east Suffolk, and naturalised at sites in Worcestershire and Staffordshire. It benefits from mobile sand, mature tufts being reinvigorated by partial burial. *Canescens* is Latin for 'greyish'.

Mediterranean hair-grass *R. cristata*, an annual neophyte recorded in the wild in 1902, originating from wool shoddy and birdseed, declining in frequency, and now a rare casual of tips, docks and waste ground. *Cristata* is Latin for 'crested'.

Silver hair-grass *A. caryophyllea*, a widespread and generally common annual, though declining especially in south-east England, on well-drained sandy and rocky habitats, clifftops, heathland, summer-dry grassland, ant hills and stabilised dunes, as well as on stone walls and railway ballast. Greek *carya* + *phyllon* = 'walnut' + 'leaf', referring to the aromatic scent.

Somerset hair-grass *K. vallesiana*, a rare perennial found on sheep-grazed turf around rock outcrops on dry, sunny south-facing Carboniferous limestone slopes on seven sites in the Mendip Hills. *Vallesiana* may come from Latin for 'furrowed'.

Tufted hair-grass *D. cespitosa*, a common and widespread perennial, with ssp. *cespitosa* growing on poorly drained, mildly acidic, neutral or basic soils in rough and marshy grassland, fen-meadow and grass-heath. It rapidly colonises bare ground and tolerates some disturbance. Subspecies *parviflora* grows in woodland and hedgerows, while ssp. *alpina* is found in the west and central Highlands, and locally in western Ireland, on open montane habitats. *Cespitosa* is Latin for 'tufted'.

Wavy hair-grass *D. flexuosa*, a perennial of acid heathland, dry parts of moorland, hill pasture and open woodland, common and widespread on appropriate habitat, growing on a range of freely draining base-poor substrates. *Flexuosa* is Latin for 'winding'.

HAIRCAP Mosses in the family Polytrichaceae, and the genera *Oligotrichum* (Greek for 'a few hairs'), *Pogonatum* (*pogon* = 'beard') and *Polytrichum* ('many hairs').

Aloe haircap *Pog. aloides*, a fairly common colonist of bare, loose, acidic soils, relatively widespread in the uplands, but declining in lowland regions, on the roots of fallen woodland trees.

Alpine haircap *Pol. alpinum*, widespread and common in upland regions on grassy or heathy upland slopes, on stony banks, among scree, on cliff ledges, and occasionally on dry moorland peat.

Bank haircap *Pol. formosum*, common and widespread, typically in deciduous woodlands in the lowlands, on soils from strongly acidic to nearly neutral, and in upland woodlands on leached soils and well-drained moorlands. It also grows on lowland heathland. *Formosum* is Latin for 'beautiful'.

Bristly haircap *Pol. piliferum*, a common and widespread colonist and pioneer of dry, acidic substrates, especially sands and gravels and on bare patches in heathy grassland in the lowlands. It is also common in upland areas on disturbed, stony substrates. *Piliferum* is Latin for 'having hair'.

Common haircap *Pol. uliginosum*, common and widespread in damp, acidic habitats, tolerating shade and moderate amounts of pollution and nutrient enrichment. It can be abundant on moorland in the uplands, and is also frequent in western lowland Britain in wet woodland, bog and ditches, by lake margins, and on heathland. In south-east England it is frequent in former gravel and sandpits by pools under scrub. *Uliginosum* is from Latin for 'marsh'.

Dwarf haircap *Pog. nanum*, scattered but declining, largely restricted to coastal areas, though with a few inland locations, as well as growing on mountain tops in the Inner Hebrides. *Nanum* is Latin for 'dwarf'.

Hercynian haircap *O. hercynicum*, on loose or disturbed, acidic mineral soils in upland regions, abundant in stony detritus on mountain tops and scree. The Hercynian Mountains are in central Europe.

Juniper haircap *Pol. juniperinum*, common and widespread on dry, exposed, acidic habitats, frequently as a pioneer on recently disturbed or burnt soil. Favourite habitats include dry grassy heathland, grassland in fixed dunes, quarry and colliery spoil, and forestry tracks.

Northern haircap *Pol. sexangulare*, a Highland rarity on moist siliceous soils in late-lying snowbeds, damp screes and rock faces, and sheltered gullies. *Sexangulare* is Latin for 'six-angled'.

Slender haircap *Pol. longisetum*, widespread and scattered, often on disturbed, acidic ground, for example burnt heathland and woodland, growing on decomposing wood as well as soil. *Longisetum* is Latin for 'long bristles'.

Strict haircap *Pol. strictum*, widespread in upland regions, but local and declining, typical of raised and blanket bogs, valley mires and very wet heathland. *Strictum* is Latin for 'erect'.

Urn haircap *Pog. urnigerum*, widespread on acidic mineral soils, most frequent in upland areas on gravelly tracks, quarries and disturbed ground. In lowland regions it tends to grow in disturbed anthropogenic habitats, including burnt ground. *Urnigerum* is Latin for 'urn-bearing'.

HAIRSTREAK Butterflies in the family Lycaenidae, subfamily Theclinae.

Black hairstreak *Satyrium pruni*, wingspan 34–40 mm, with a restricted distribution following a line of heavy clay soils between Oxford and Peterborough, in woodland edge and hedgerows where the larval food plant, blackthorn (prompting the specific name) is present, with wild plum an occasional alternative food. Adults fly in June–July and feed on aphid honeydew in the tree canopy or along the top of dense scrub, but will also come down to feed on various nectar sources, privet and bramble flowers being favourites. Distribution declined by 61% between 1976 and 2014, numbers by 54% (with equivalent figures of 11% and 87% in 2005–14), but status is considered to be stable. The genus name is from the Greek Satyros, a satyr who danced with nymphs, referencing the dancing flight of these butterflies.

Brown hairstreak *Thecla betulae*, the largest British hairstreak, male wingspan 36–41 mm, female 39–45 mm, locally distributed in southern England, south-west Wales and the Burren, west Ireland, in woodland edge, scrub and hedgerows where the main larval food plant, blackthorn, is present. Adults fly from late July, peaking in August, and can remain on the wing until November. They congregate in the canopy (especially where this is of ash) to mate and feed on aphid honeydew. Adults may also feed lower down on nectar from bramble, hemp-agrimony and common fleabane. Exacerbated by hedgerow removal and inappropriately timed flailing, range and numbers declined by 49% and 15%, respectively, between 1976 and 2014, numbers continuing to fall by 58% in 2005–14, and it is a priority species for conservation. Thecla was a martyr in the Greek Orthodox Church. *Betulae* relates to birch, though not a food plant.

Green hairstreak *Callophrys rubi*, wingspan 27–34 mm, widespread but local, partly a consequence of the variety of food plants it uses and the range of open habitats (from chalk grassland to heath and moorland) it frequents. Associated with scrubby plants and hedgerows this butterfly is in flight in April–July, and has the widest range of larval food plants of any British species, including bramble, bilberry, gorse, broom, bird's-foot-trefoil and dogwood. Larvae are cannibalistic, and adults also feed on honeydew. Both numbers and distribution show an overall long-term decline, by 41% and 30% between 1976 and 2014 (and by 34% and 14% in 2005–14), and conservation status remains under review. Greek *callos* + *ophrys* = 'beautiful' + 'eyebrow', probably from the scales between the eyes, and Latin *rubus* = 'bramble' as a larval food.

Purple hairstreak *Favonius quercus*, male wingspan 33–40 mm, female 31–8 mm, Britain's commonest hairstreak, associated with oak (*Quercus*), the larval food plant (giving the specific name), particularly in southern England and Wales, and with a scattered distribution in Ireland. Overall distribution and numbers declined by 30% and 54% between 1976 and 2014 (and by 15% and 10% over 2005–14). However, paradoxically, there has been a recent increase in numbers in some places, and an extension of this butterfly's range especially in the Midlands and northern England, and south-west and central Scotland, as well as in urban areas possibly related to improvements in atmospheric quality. Adults, flying in June–September, largely remain in the canopy feeding mainly on honeydew; but they are driven down to seek fluid and nectar, for example from bramble, during periods of drought. Favonius was the Roman god of the western wind, the herald of spring.

White-letter hairstreak *Satyrium w-album*, wingspan 25–35 mm, widespread but uncommon in England and Wales; absent from Scotland and Ireland. Flower buds of elm (preferentially wych elm) are the only larval food plant. Adults fly, in July–September, around the tree canopy feeding on honeydew, but occasionally come down to ground level to take nectar from plants such as privet, thistles and bramble. This butterfly suffered as a result of Dutch elm disease in the 1970s and early 1980s, with a 45% decline in distribution and a 96% decline in numbers over 1976–2014 (and by 41% and 77%, respectively, in 2005–14), and this remains a priority species for conservation. The specific name refers to the white (Latin *album*) W shape on the underside of the hindwing.

HAIRY SNAIL Name for the keeled pulmonate family Hygromiidae, though not all species possess hairs, and in many only the juvenile form. Hairs may help adhesion to wet leaves. Species in other families may also have hairs, for example cheese snail. Specifically, *Trochulus hispidus* (Latin for 'small wheel' and 'bristly' or 'hairy'), shell diameter 5–11 mm, height 3–6 mm, common and widespread in a variety of open, often wet and disturbed habitats.

German hairy snail *Pseudotrichia rubiginosa*, shell diameter 6–8 mm, height 4.5–5 mm, first recorded in 1981 from the tidal Thames frontage of Syon Park, Hounslow, Greater London; it

has since been found elsewhere within the Thames corridor in the London area and as far west as Oxfordshire, and in the River Medway, Kent, typically among strand-line debris in bare mud often beneath willows. Vulnerable in the RDB. Greek *pseudo* + *trixa* = 'false' + 'hair', and Latin for 'rust-coloured'.

HALICTIDAE Family name for furrow bees (or sweat bees, so named from their habit of landing on people and licking sweat from the skin to obtain the salt) and some mining bees, from New Latin *halictus* for these bees.

HALIPLIDAE Family (Greek *haliploos* = 'submerged') of crawling water beetles (suborder Adephaga), with 19 British species in 3 genera, length 2.5–4.5 mm, found in slow-moving or static lowland water bodies. Adults are partly carnivorous and swim clumsily using the legs on each side alternately. Larvae feed on algae and duckweeds.

HALOCLAVIDAE Family (Greek *halos* + Latin *clava* = 'salty' + 'club') of sea anemones with four species in three genera in British waters.

HALORAGACEAE Family name for water-milfoil (possibly Greek *halos* + *ragos* = 'salty' + 'snakes').

HAM BEETLE Red-shouldered ham beetle *Necrobia ruficollis* (body length 5–7 mm) and **red-legged ham beetle** *N. rufipes* (3.5–7.0 mm), both with scattered distributions in the southern half of England, feeding on dead flies and fly larvae, and dead meat (including stored meat, hence the common name) and skin. Greek *necros* = 'dead', and Latin for 'red shoulder' and 'red foot', respectively.

HAMPSHIRE-PURSLANE *Ludwigia palustris* (Onagraceae), a scarce perennial herb found in acid, seasonally flooded pits pools overlying base-rich clays, in runnels within valley mires, and in poached areas in pasture or woodland clearings. Its range continues to expand in the New Forest, where it benefits from livestock puddling the ground. A colony was found in Dorset in 1996. Christian Ludwig (1709–73) was a German physician and botanist. Latin *palustris* = 'marshy'.

HARD FERN Members of the genus *Blechnum* (Blechnaceae), specifically, *B. spicant*, generally common in heathland, moorland, rocky and grassy slopes and in semi-shade in woodland on infertile acid soils. *Blechnon* is Greek for this fern; *spicant* is Latin for 'tufted' or 'having spikes'.

Greater hard-fern *B. cordatum*, a South American garden escape naturalised in shady habitats and by streams in south-west England, west Scotland, and south-west and north-east Ireland. *Cordatum* is Latin for 'heart-shaped'.

Little hard-fern *B. penna-marina*, a South American introduction recorded as naturalised in Shropshire in 2007. *Penna-marina* is Latin for 'sea-feather'.

HARD-GRASS *Parapholis strigosa* (Poaceae), an annual of damp sparsely vegetated sites around the British and Irish coasts, absent north of central Scotland, characteristic of the upper parts of grazed saltmarsh, but also on mud banks, shingle ridges, saltmarsh-sand dune transitions and sea walls. It occasionally and increasingly grows inland by salt-treated roads. Greek *para* + *pholis* = 'near' + 'scale', and Latin *strigosa* = 'with bristles'.

Curved hard-grass *P. incurva*, a scarce annual of sparsely vegetated coastal habitats, mostly from Pembrokeshire to Lincolnshire, and the Channel Islands, including gravelly mud banks, shingle ridges, rock ledges, clifftops and upper saltmarsh, as well as on sea walls, with a few records around docks, and inland as a wool and ballast alien.

One-glumed hard-grass *Hainardia cylindrica*, a casual annual European neophyte cultivated in 1806 and recorded in the wild in 1940, found in waste ground and sewage-works, mostly originating from wool shoddy, grain, ballast and birdseed. *Hainardia* honours the twenty-first-century Swiss botanist and ecologist Pierre Hainard.

HARE Lagomorph animals in the genus *Lepus* (Latin for 'hare'), family Leporidae. Their large eyes are set in the sides of the head allowing for a field of vision close to 360°. When running they can reach 70 kph, and can change direction sharply. They occupy ground surface dens (forms) – sheltered areas of flattened vegetation – using both a daytime resting area and a separate foraging area used at night. The English name comes from Old English *hara*.

Blue hare Alternative name for mountain hare.

Brown hare *L. europaeus*, head-body length 52–65 cm, tail length 8.5–12 cm, weight 3–4 kg, locally abundant and widespread in lowland regions (especially mixed arable farmland) in Britain, introduced in the Iron Age, on the Isle of Man, and in Ireland where it was introduced for greyhound coursing and as a game animal in the nineteenth century. Expansion of brown hares in Northern Ireland has been at the expense of the Irish hare. A general decline in numbers in Britain followed the Ground Game Act (1880); there were increases in some central and eastern regions pre-1940, but declines continued post-war. The estimated British population in the winters of 1991–2 and 1992–3 was 817,500 (60% in arable, 23% in pastoral, 11% in marginal upland, 6% in upland areas). Mating takes place from February to September. Females 'box' to fight off unwanted male attention ('mad March hares'). Gestation lasts 41–2 days. Females produce 3–4 litters a year, each with generally 2–4 precocial offspring (leverets) which are left alone during the day, fed by the mother for about 5 minutes at sunset. Feeding, usually in groups, is mainly on grass, including cereals, and root crops in winter. Some 390,000 are shot annually in Britain, yet this is a priority species in the UK BAP. Coursing has been illegal in the UK since the Hunting Act (2002), but is still legal in Ireland.

European hare Alternative name for brown hare.

Irish hare *L. timidus* ssp. *hibernicus*, a subspecies of mountain hare, though some taxonomists give it species status – it is larger (head-body length 60 cm, including a 7 cm tail, male weight 3 kg, females 3–3.6 kg), has a preference for lowland habitat, and a coat that does not turn white in winter. Numbers have been declining since the 1970s because of intensification of agriculture, possibly to be exacerbated by competition, disease and hybridisation with introduced European (brown) hare if this spreads out of Northern Ireland into the Republic. Nevertheless, the population in 2011 was estimated at 59,700–86,900 (more than three times the number believed present in 2002), and some estimates for 2016 gave 500,000. They use a variety of habitats, from coastal grassland and saltmarsh to moorland, but are most abundant on lowland pasture and areas that provide short grass, herbs and heather. In upland areas they feed on young heather, herbs, sedges and grasses, with willow, gorse and bilberry in winter. In the lowlands, grasses account for up to 90% of the diet. They reingest their droppings to allow them to break down and digest cellulose. Breeding is in January–August. Gestation lasts 50 days. Females can produce up to three litters a year, each litter containing an average of four young (leverets), which are independent after three weeks. This is a legally protected species under the Irish Wildlife Act, the European Habitats Directive and the International Berne Convention. The trinomial is Latin for 'hare', 'timid' and 'Irish'.

Mountain hare *L. timidus*, head-body length 45–55 cm, including a 4–8 cm tail, weight 2–5.3 kg, females being slightly the heavier, native to the Highlands, generally >500 m altitude, and introduced to the Southern Uplands, the Peak District and on some Scottish islands including Hoy (Orkney), Mainland (Shetland), Mull and Skye. Irish hare is a subspecies. Population densities fluctuate periodically, possibly a response to parasite burden, varying at least tenfold, and reaching a peak about every ten years. A BTO survey in 2013 indicated a decline in Scottish numbers of 43% since 1995. Mating takes place from January. Gestation lasts 50 days. Females produce 1–4 litters a year, starting in March, each with 1–3 offspring, occasionally more. It is found

on moorland, feeding on grasses (preferred when available in summer), heather, sedges and rushes. From October to January the coat changes from brown to white or grey and back to brown during February–May. It is a priority species in the UK BAP, and listed in Annex 5 of the EC Habitats Directive (1992) as a species 'of community interest whose taking in the wild and exploitation may be subject to management measures', but it is subject to licensed shooting as small game, and it is shot by some gamekeepers to prevent spread of a tick-borne virus to grouse chicks.

HARE-TAIL MOSS *Myurium hochstetteri* (Myuriaceae), found in west Scotland and the Hebrides usually growing near the sea, on rocky or grassy banks, cliff ledges and in rock crevices. *Myurium* may be from Greek *mys* + *oura* = 'mouse' + 'tail'. Ferdinand von Hochstetter (1829–84) was a German geologist and naturalist.

HAREBELL *Campanula rotundifolia* (Campanulaceae), a rhizomatous perennial herb, common and widespread except for south-west England and central and southern Ireland, on dry, open, infertile habitats such as grassland, dunes, rock ledges, roadsides and railway banks, tolerating a wide range of soil pH, and heavy-metal tolerant races have been recorded. This is the county flower of Yorkshire, Dumfriesshire and Co. Antrim as voted by the public in a 2002 survey by Plantlife. The English name possibly comes from its presence in places frequented by hares. The binomial is Latin for 'little bell' and 'round leaf'.

HARE'S-EAR Species of *Bupleurum* (Apiaceae), the English name coming from the leaf shape, the genus name from Greek *bous* + *pleuron* = 'ox' + 'side', from the supposed effect of cattle swelling on eating this plant.

Shrubby hare's-ear *B. fruticosum*, an evergreen European neophyte shrub occasionally naturalised on waste ground and railway banks as a garden escape in southern England, the Midlands and East Anglia. *Fruticosum* is Latin for 'shrubby'.

Sickle-leaved hare's-ear *B. falcatum*, a biennial or short-lived perennial herb, possibly neophyte, occasionally found in hedge banks and field borders, on ditch banks and on verges. *Falcatum* is Latin for 'sickle-shaped'.

Slender hare's-ear *B. tenuissimum*, an annual colonist of thinly vegetated or disturbed and often brackish coastal sites from south Wales to Lincolnshire. It also grows on commons near Malvern, Worcestershire. *Tenuissimum* is Latin for 'very slender'.

Small hare's-ear *B. baldense*, a rare annual found in rabbit-grazed coastal grassland over calcareous substrates. Near Beachy Head, East Sussex, it grows where turf gives way to bare chalk by the cliff edge, and in Devon also in clifftop turf. In the Channel Islands it grows in open turf on consolidated dunes, occasionally on cliffs. *Baldense* means from Monte Baldo, north Italy.

Smallest hare's-ear Alternative name for slender hare's-ear.

HARE'S-TAIL *Lagurus ovatus* (Poaceae), an annual grass naturalised on dunes in the Channel Islands and southern England, elsewhere scattered as a casual garden escape or introduction from wool shoddy or grain, on walls, pavements, roadsides and tips. Greek *lagos* + *oura* = 'hare' + 'tail', and Latin *ovatus* = 'oval'.

HARPOON WEED *Asparagopsis armata* (Bonnemaisoniaceae), a red seaweed originating in the Pacific and Indian oceans, first recorded in Ireland in Galway Bay in 1939 and Britain in 1949 at Lundy, and gametophytes (sexual plants) now well established in open sandy pools of the lower intertidal and subtidal, on rock or epiphytic in the Channel Islands, southern England (Swanage to the Scillies), and south and west Ireland, and tetrasporophytes in west Scotland up to Shetland. The English name refers to its harpoon-like hooks. The binomial is Greek for 'resembling asparagus' and Latin for 'spiny'.

HARRIER Birds of prey in the genus *Circus* (Accipitridae). The English name comes from the verb 'to harry'. The genus name comes from Greek *circos* = 'circle', referring to a bird of prey named for its circling flight. **Northern harrier** *C. hudsonius* and **pallid harrier** *C. macrourus* are scarce accidentals.

Hen harrier *C. cyaneus*, wingspan 110 cm, length 48 cm, male weight 350 g, females 500 g, found in the breeding season on upland heather moorlands of Wales, northern England, the Isle of Man, north and south Ireland and Scotland. Feeding on birds and small mammals, its impact on the number of grouse available to shoot is a cause of conflict and threatens its survival, and it is on the Red List. In 2016 the recorded number of breeding pairs in the UK/Isle of Man was 575 (4 in England, 35 in Wales, 30 in the Isle of Man, 46 in Northern Ireland, 460 in Scotland). This represents a decline of nearly 14% since 2010, when there were 633 pairs in the UK, and of 27% since 2004, when there were 749, mostly a result of illegal killings. In winter these birds move to lowland farmland, heathland, coastal marshes, fen and river valleys, those recorded from in southern England and East Anglia being visitors from mainland Europe. The ground nest is built of sticks, lined with grass and leaves, clutch size if 4–5, incubation lasts 34 days, and the precocial chicks fledge in 37–42 days. Latin *cyaneus* = 'dark blue'.

Marsh harrier *C. aeruginosus*, wingspan 122 cm, length 52 cm, male weight 540 g, females 670 g, almost died out as a British breeding species, but has recovered since the 1960s, partially changing its behaviour to nest in farmland, and many individuals now overwinter. It is mainly found in south-east and eastern England, with some birds in south-west and north-west England, and Scotland, breeding in and hunting for small mammals and birds over reedbeds and marshes, as well as farmland near wetlands. In 2010 there were up to 327 known breeding pairs, with a suggested actual total of around 400 pairs, though it remains on the Amber List. Clutch size is 4–5, incubation takes 31–8 days, and the altricial young fledge in 35–40 days. Latin *aeruginosus* = 'rust-coloured'.

Montagu's harrier *C. pygargus*, wingspan 112 cm, length 45 cm, male weight 270 g, females 380 g, a rare Amber List summer visitor from Africa, found south of a line from the Severn estuary to the Wash, with a five-year mean of 16 breeding pairs noted in 2010. Five pairs successfully raised 13 chicks in 2016. Since the 1960s most nests have been in cereal crops. Clutch size is 4–5, incubation lasts 28–9 days, and fledging of the altricial young takes 35–40 days. The English name honours the British soldier and naturalist George Montagu (1753–1815). Greek *pyge* + *argos* = 'rump' + 'shining white'.

HART'S-TONGUE *Asplenium scolopendrium* (Aspleniaceae), a common and widespread fern found in shady places on moist soils in woodland and stream banks, and in urban areas on old walls and rail and canal bridges. The English name refers to the leaf shape. *Asplenium* is Latin for 'without a spleen', referring to the plant's putative medicinal properties in removing obstructions from the liver and spleen. *Scolopendrium* is from Greek meaning 'like a millipede'.

HARTWORT *Tordylium maximum* (Apiaceae), an annual or biennial European neophyte (possibly native) present by 1670, very occasionally found in south-east England in neutral grassland and in grassy thorn scrub. *Tordylion* is Greek for this plant, *maximum* Latin for 'largest'.

HARVESTMAN Members of the order Opiliones, arachnids with small bodies and long, thin legs. Harvestmen differ structurally from spiders in that the connection between the cephalothorax and abdomen is fused so that the body appears to be a single oval structure. They possess a single pair of eyes in the middle of the head, oriented sideways. In general, the legs are longer than those of most spiders. They are often omnivorous, and catch insect prey using hooks on the ends of the legs, and unlike spiders can eat solid food.

Two-spotted harvestman *Nemastoma bimaculatum* (Nemostomatidae), common and widespread, found year-round in woodland litter. Greek *nema* + *stoma* = 'thread' + 'mouth', and Latin *bimaculatum* = 'two-spotted'.

HAT (FUNGUS) Witch's hat See blackening waxcap under waxcap.

HAT (GASTROPOD) See Chinaman's hat.

HATCHET SHELL Members of the bivalve family Thyasiridae, all bacterial chemosynthesisers.

Northern hatchet shell *Thyasira gouldi*, shell size up to 10 mm, in a few sheltered sea lochs and inlets around Shetland and mainland Scotland, in mud and muddy sand.

Wavy hatchet shell *T. flexuosa*, shell size up to 12 mm, widespread from the intertidal and shelf to over 150 m in mud and muddy sand, chemosymbiotic, with a dependence on bacteria, though it can also suspension-feed. *Flexuosa* is Latin for 'wavy'.

HAWFINCH *Coccothraustes coccothraustes* (Fringillidae), wingspan 31 cm, length 18 cm, weight 58 g, Britain's largest finch, a rather uncommon Red List resident breeder, mainly found in deciduous woodland in southern England, but also elsewhere in lowland England (e.g. Forest of Dean), north and south Wales, and south-east Scotland, with an estimated 500–1,000 pairs in 2011, there having been a decline in numbers by 37–45% between 1990 and 1999. Large numbers in winter 2017–18 reflected an irruption of continental birds. Nests, in bushes and trees, contain 2–7 (usually 4–5) eggs, incubation lasting 11–13 days, and fledging of the altricial young taking 12–13 days. It eats large hard seeds (e.g. cherries), buds and shoots, and (in summer) invertebrates, especially caterpillars. The 'haw' of its English name is the hawthorn berry. Greek *coccos* + *thraustes* = 'kernel' or 'seed' + 'broken'.

HAWK (BIRD) Medium-sized diurnal birds of prey in the family Accipitridae. See buzzard, eagle, goshawk, harrier, honey-buzzard, kite, osprey and sparrowhawk.

HAWK (MAYFLY) Yellow hawk or **yellow evening hawk** *Ephemerella notata* (Ephemerellidae), widespread, having spread northwards since the 1990s. Nymphs live in streams on submerged plants and stones, feeding on fine particulate organic detritus from the sediment. Adults emerge in May–June, males swarming at dusk. Greek *ephemeros* = 'ephemeral', and Latin *notata* = 'distinguished'.

HAWK-MOTH Members of the family Sphingidae. The reared-up posture of some larvae has been considered to resemble that of the Sphinx, hence the family and, for some, the genus name. Rare vagrant species are: **bedstraw** (*Hyles gallii*), **spurge** (*H. euphorbii*), **striped** (*H. livornica*) **oleander** (*Daphnis nerii*), **silver-striped** (*Hippotion celerio*), and **willowherb hawk-moth** (*Proserpinus proserpina*).

Broad-bordered bee hawk-moth *Hemaris fuciformis* (Macroglossinae), wingspan 38–48 mm, local in open woodland and clearings, over much of southern England. Larvae feed on honeysuckle and bedstraw, adults on honeysuckle and other plants such as bugle, ragged-robin and rhododendron. Adults are active in late-morning and early afternoon sunshine in May–August. Greek *hemera* = 'day' (from its diurnal habit), and Latin *fucus* + *forma* = 'drone' + 'shape' (from its resemblance to a bee).

Convolvulus hawk-moth *Agrius convolvuli* (Sphinginae), wingspan 80–120 mm, nocturnal, a fairly common migrant from southern Europe, appearing in late summer and autumn, with occasional major influxes (as in 2003), when it breeds in small numbers, but it usually cannot overwinter. Adults feed on garden flowers, especially tobacco plant. Larvae feed on bindweed (*convolvulus*, giving the specific name) and morning-glory. In Greek mythology Agrios was a giant who fought the gods, the genus name reflecting this moth's large size.

Death's-head hawk-moth *Acherontia atropos* (Sphinginae), a rare immigrant from southern Europe, appearing in late summer and autumn mostly in southern and eastern England, in some years in sufficient numbers to breed in small numbers. With a wingspan of 80–120 mm, this is the largest moth to occur in Britain. Adults enter beehives in search of honey (the image of a skull on the thorax may resemble a bee to a bee, allied with production of bee-like pheromones), and if handled they emit a loud squeak. Larvae feed on woody and deadly nightshade, and on potato where it can locally become a pest. In Greek mythology, Acheron was the River of Pain in the Underworld, and Atropos, one of the three Fates, cut the thread of life.

Elephant hawk-moth *Deilephila elpenor* (Macroglossinae), wingspan 45–60 mm, widespread and common in hedgerows, gardens, woodland edges and heathland, and on rough grassland and sand dunes, increasing its range in Scotland in recent years. Adults fly in May–July, visiting flowers such as honeysuckle for nectar. The English name comes from the caterpillar's resemblance to an elephant's trunk. Larvae mainly feed on willowherbs, but are also found on other plants, including bedstraws. Greek *deile* + *phileo* = 'evening' + 'to like', from the moth's crepuscular habit. Elpenor was one of Odysseus' companions turned by Circe into swine: the larval shape resembles a pig's snout.

Eyed hawk-moth *Smerinthus ocellata* (Smerinthinae), wingspan 70–80 mm, nocturnal, common in gardens, orchards, parks, fens, scrub and woodland throughout England, Wales and Ireland. Adults fly in May–July. When disturbed, it opens its forewings to reveal eyespots to deter would-be predators (allied with body shaking). Larvae feed on leaves of sallows, apples and other trees. Greek *merinthos* = 'thread', probably referring to the thread-like haustellum, and Latin *ocellata* = 'eyed'.

Hummingbird hawk-moth *Macroglossum stellatarum* (Macroglossinae), wingspan 40–50 mm, an immigrant day-flyer which sometimes occurs in large numbers, for instance the large influx in 2000 when these moths were noted in parks and gardens all over Britain, especially in the southern half, and again in 2006. Caterpillars feed on bedstraws and wild madder; adults hover to feed on nectar of a number of flowers. Overwintering may occur in mild winters. Greek *macros* + *glossa* = 'large' and 'tongue'. *Stellatarum* relates to Latin *stella* = 'star', and Stellatae is an old name for Rubiaceae, the family that contains the larval food plants bedstraw and madder.

Lime hawk-moth *Mimas tiliae* (Smerinthinae), wingspan 55–70, nocturnal, common in broadleaf woodland, parks and gardens throughout much of England and south Wales; the first confirmed Irish record was from Dublin in 2010. Adults fly in May–June. Larvae feed on the leaves of lime, together with some other trees. In Greek mythology Mimas was a giant who fought the gods. *Tiliae* refers to lime (*Tilia*).

Narrow-bordered bee hawk-moth *H. tityus*, wingspan 37–42 mm, a bumblebee mimic with a wide but scattered distribution, inhabiting marshy woodland, damp pasture, moorland and chalk downland. Larvae feed on devil's-bit scabious and field scabious. Adults fly from mid-April to July and take nectar from such plants as bugle, louseworts and marsh thistle. Generally scarce, though with a probably stable distribution, and a priority species for conservation. The Greek god Tityos was punished for attempting to seduce the goddess Leto by having vultures eat his self-renewing liver.

Pine hawk-moth *Sphinx pinastri* (Sphinginae), wingspan 65–80 mm, nocturnal, local in coniferous woodland, plantations, heathland and gardens throughout much of eastern and south-east England, spreading to reach Yorkshire by the 1990s. Adults fly in May–June. Larvae mainly feed on Scots pine needles, but are also known from maritime pine (giving the specific name) and other conifers.

Poplar hawk-moth *Laothoe populi* (Smerinthinae), wingspan 65–90 mm, nocturnal, widespread and common in parkland, gardens, fen, woodland, heathland and moorland. Adults fly in May–July. When at rest the hindwings are held forward of the forewings, and the abdomen curves upwards at the rear. Larvae

feed on the leaves of aspen, poplars and sallows. Laothoe was a mistress of Priam, King of Troy. *Populus* = poplar and aspen.

Privet hawk-moth *S. ligustri*, wingspan 90–120 mm, nocturnal, common on downland and in open woodland, hedgerows and gardens mostly in the southern half of Britain. Adults fly in June–July. Larvae feed on leaves of privet (*Ligustrum*), lilac, guelder-rose and ash.

Small elephant hawk-moth *D. porcellus*, wingspan 40–5 mm, widespread if local in calcareous grassland, chalk downland, heathland, dunes, shingle beaches and damp grassland, less common in Scotland and Ireland. Adults fly in May–July. Caterpillars mainly feed on bedstraws, but also rosebay willowherb and purple-loosestrife. The larval shape resembles a pig's snout, hence Latin *porcellus* = 'piglet'.

HAWKBIT Herbaceous plants in the genera *Leontodon* (Latin *leo* = 'lion' and Greek *odonti* = 'tooth') and *Scorzoneroides* (Italian *scorzone*, an alteration of medieval Latin *curtion* = 'venomous snake', plants once thought to be a cure for snake-bite) in the Asteraceae. The English name comes from the medieval belief that hawks ate the plant to improve their eyesight (c.f. hawkweed).

Autumn hawkbit *S. autumnalis*, a common and widespread perennial of meadow and pasture, open scrub, heathland, moorland, saltmarsh, fixed dunes and roadsides in the lowlands, and also of screes, flushes and lake margins in the uplands.

Lesser hawkbit *L. saxatilis*, perennial or biennial, common and widespread in England, Wales and Ireland, more scattered in Scotland (commoner in the west), on often heavily grazed and trampled grassland, dry moorland, limestone and other basic rock outcrops, fixed dunes, tracks, and sand- and gravel-pits, preferring well-drained, calcareous to mildly acidic soils, but also in periodically wet habitats such as dune slacks and pond margins. *Saxatilis* is Latin for 'growing among rocks'.

Rough hawkbit *L. hispidus*, perennial, common and widespread in Britain as far north as the Central Lowlands of Scotland, locally common in Ireland, on dry, neutral or calcareous soils, in hay meadow, pasture and other grasslands, on verges, railway banks and rock ledges. *Hispidus* is Latin for having bristly hairs.

HAWKER Dragonfly in the genus *Aeshna* (Aeshnidae), possibly as erroneous New Latin from Greek *aischros* = 'ugly'. The English name reflects a zigzagging patrolling behaviour when hunting.

Azure hawker *A. caerulea*, body length 57–67 mm, hindwing 38–41 mm, associated with boggy pools, widespread but rarely abundant in the Highlands. The main flight period is from late May to August. *Caerulea* is Latin for 'blue'.

Brown hawker *A. grandis*, body length 70–7 mm, hindwing 41–9 mm, common and widespread, though rare in Scotland and south-west England, breeding in standing or slow-flowing water, but adults hunting away from such habitat. The main flight period is July–September. *Grandis* is Latin for 'great'.

Common hawker *A. juncea*, body length 74 mm, common and widespread in a variety of habitats but favouring acidic uplands pools and so less common in south-east and eastern England. The main flight period is June–November. *Juncea* is Latin for 'rush-like'.

Migrant hawker *A. mixta*, body length 63 mm, found throughout England and Wales, extending its range into Scotland and Ireland, breeding in standing or slow-flowing water, but adults found in a variety of habitats, especially resting on low vegetation. Increasingly common since the 1940s when it began migrating from continental Europe (and continues doing so), giving it its common name. *Mixta* is Latin for 'mixed'.

Norfolk hawker *A. isosceles*, body length 67 mm, a Red Book dragonfly restricted to unpolluted fens and grazing marshes of the Broads of Norfolk and north-east Suffolk.

Southern hawker *A. cyanea*, body length 67–76 mm, hindwing 43–53 mm, common in England, Wales and upland Scotland, breeding in small vegetated ponds, including garden ponds, though often hunting away from such habitats. The main flight period is July–October. *Cyanea* is Latin for 'dark blue'.

Southern migrant hawker *A. affinis*, body length 60 mm, a rare migrant becoming commoner in the UK and a potential colonist, for example many individuals noted in parts of Kent and Essex in 2010, with breeding and oviposition noted. *Affinis* is Latin for 'related to'.

HAWK'S-BEARD Members of the genus *Crepis* (Asteraceae) (Greek *crepis* = 'shoe').

Beaked hawk's-beard *C. vesicaria*, a biennial, sometimes annual or perennial Mediterranean neophyte first recorded in 1713 in Kent. It spread rapidly, reaching the west coast of Ireland in 1896. It has not spread much into northern England or Scotland. It is found in lightly mown or grazed grassland on roadsides, lawns, railway banks and walls, and in waste places. *Vesicaria* is Latin for 'like a bladder'.

Bristly hawk's-beard *C. setosa*, a generally casual annual first recorded in Essex in 1843, very scattered, usually in newly sown grass-clover leys, but also in arable fields and on waste ground. *Setosa* is Latin for 'bristly'.

French hawk's-beard *C. nicaeensis*, an annual Mediterranean neophyte, casual as a grain and grass-seed alien on roadsides and rough ground mainly in southern England. *Nicaeensis* refers to Nice, south France.

Marsh hawk's-beard *C. paludosa*, perennial, common in Britain north of a line from the Severn estuary to that of the Humber, locally common in Ireland, declining at the southern edges of its range. In upland regions, it grows on rocky, wooded streamsides, and in sheltered gullies and flushed banks. At lower altitudes it occurs in fen, damp meadow and ditches, and on verges. *Paludosa* is from Latin for 'swamp'.

Narrow-leaved hawk's-beard *C. tectorum*, an annual European neophyte first recorded in the wild in 1874, scattered, and now well established around the mouth of the Shannon. It grows on roadsides, soil banks verges and waste ground, usually as a contaminant of grain and grass seed. *Tectorum* is from Latin *tectum* = 'roof'.

Northern hawk's-beard *C. mollis*, a winter-green perennial, very local and declining in the northern Pennines, Scottish Borders and southern Highlands, on species-rich grassland and wood pasture on shallow base-rich soils. *Mollis* is Latin for 'soft'.

Rough hawk's-beard *C. biennis*, a biennial with a scattered distribution in Britain north to central Scotland, and in Ireland, locally common in rough grassland and woodland margins on chalk soils in south-east England, elsewhere introduced, often with grass seed, persisting locally in pasture, arable fields and on field margins, roadsides, dry banks and waste ground. *Biennis* is Latin for 'biennial'.

Smooth hawk's-beard *C. capillaris*, a common and widespread winter-green annual, an early colonist of waste ground, verges, lawns, spoil heaps and rocky bank, locally increasing in man-made habitats. *Capillaris* is Latin for 'hair-like'.

Stinking hawk's-beard *C. foetida*, an annual or biennial archaeophyte, now found only at Dungeness, Kent, becoming extinct in 1980, but following its reintroduction in 1992 a new population has been established in a shingle-heath community, with reseeding reported in 2008. *Foetida* is Latin for 'stinking'.

HAWKWEED 1) Microspecies (411 currently recognised) of *Hieracium*, a group of tap-rooted herbaceous perennials growing in a range of habitats, though with a preference for infertile, rocky substrates. The English name and that of the genus (Greek *hieracion*, in turn from *hierax* = 'hawk') refers to the belief that hawks ate the plant to strengthen their eyesight (c.f. hawkbit).

2) **Mouse-ear-hawkweeds** See under this name; also **orange hawkweed**, another name for fox-and-cubs. These are all species of *Pilosella*.

HAWTHORN Members of the genus *Crataegus* (Rosaceae). Specifically, *C. monogyna*, a common and widespread deciduous

shrub or tree of hedgerows (planted as a spiny stock-proof quickset), scrub and woodland edge, and as an understorey in open woodland on a wide range of soils. 'Hawthorn' derives from Old English *haga-* or *haguthorn*, meaning 'hedge thorn'. *Crataegus* is from Latin, possibly referring to the wood's hardness. Greek *monogyna* = 'one ovary'.

Large-sepalled hawthorn *C. rhipidophylla*, a European neophyte, mostly planted in hedgerows, but also self-sown, with a scattered distribution. Greek *rhipos* + *phyllon* = 'fan' + 'leaf'.

Midland hawthorn *C. laevigata*, native to the Midlands and east and south-east England in ancient woodland, woodland edge, old hedgerows and boundary banks on clay soils. It is widely planted and naturalised both inside and outside its core area, though remaining local in Scotland and Ireland. *Laevigata* is Latin for 'smooth'.

Oriental hawthorn *C. orientalis*, a neophyte from the Near East introduced in 1810, planted in hedgerows, gardens and parks, and naturalised on roadsides, banks, quarries and sandpits, and in rough grassland, with scattered records in England and north-east Ireland.

Various-leaved hawthorn *C. heterophylla*, a neophyte from the Caucasus, planted in parks, occasionally self-sown, naturalised in urban woodland in parts of south-east England, west Wales, and central Scotland. *Heterophylla* is Greek for 'differently leaved'.

HAZEL (MOTH) **Scalloped hazel** *Odontopera bidentata* (Geometridae, Ennominae), wingspan 32–40 mm, widespread and common in woodland, scrub, hedgerows, plantations, parks and gardens. Adults fly at night in May–June. Larvae feed on a variety of trees. Greek *odontos* + *peras* = 'tooth' + 'end', and Latin *bidentata* = 'two-toothed', both parts (and the English name) describing the dentate wing margins.

HAZEL (TREE) *Corylus avellana* (Betulaceae), a common and widespread deciduous shrub with edible nuts found in dry or damp, calcareous to mildly acidic soils, though favouring moist, base-rich conditions. It is native in the understorey of many woods, in scrub, hedgerows, on riverbanks, and limestone pavement, but it is also widely planted in copses and can be an indicator of age in species-rich hedgerows. 'Hazel' comes from Old English *hæsel*. Greek *corylos* = 'hazelnut'. *Avellana* means coming from Avella, Italy. See also filbert.

Turkish hazel *Corylus colurna*, a neophyte tree introduced from the Near East by 1664, occasionally cultivated in parks, gardens, plantations (for example on Lancashire dunes) and, increasingly, as a street tree. *Colurna* may be the Latinised form of the vernacular name.

HAZELCUP Inedible fungi in the genus *Encoelia* (Helotiales, Leotiomycetes) (Greek *en* + *coelia* = 'on' + 'stomach'), growing on hazel. **Green hazelcup** *E. glauca*, Vulnerable in the RDL (2006), is found in south Devon, Inner Hebrides and Northern Ireland. **Spring hazelcup** *E. furfuracea* is infrequent but widespread in much of England, occasional elsewhere.

HEART & CLUB *Agrotis clavis* (Noctuidae, Noctuinae), wingspan 35–40 mm, a nocturnal moth common on chalk downland, dry open areas and gardens, and in coastal dunes throughout much of England and Wales, with recent records also from Ireland (Co. Galway, Co. Clare and Co. Wexford). Adults fly in June–July. Larvae feed on low-growing plants. The English name refers to the shapes of the dark stigmata on the generally pale forewings. Greek *agrotes* = 'of the field'. Latin *clavis* = 'key', referring to the 'club' part of the wing pattern.

HEART & DART *Agrotis exclamationis* (Noctuidae, Noctuinae), wingspan 30–40 mm, a nocturnal moth widespread, common and often abundant in a range of habitats, yet overall numbers have declined by 76% over 1968–2007. Adults fly in May–July. Larvae feed on a number of wild and cultivated plants. The English name refers to the shapes of the dark stigmata on the generally pale forewings. Greek *agrotes* = 'of the field'. Latin *exclamationis* refers to the exclamation-mark shape on the wing.

HEART URCHIN An echinoid marine invertebrate in the order Spatangoidea, in which the body is usually oval or heart-shaped. See also sea potato.

HEARTEASE Alternative name for wild pansy, formerly used as a love charm, the arrangement of the petals symbolising love.

HEATH (BUTTERFLY) Species of *Coenonympha* (Nymphalidae, Satyrinae), from Greek *coinos* + *nymphe* = 'shared' + 'nymph' (i.e. is a nymphalid).

Large heath *C. tullia*, wingspan 35–40 mm, confined to boggy habitats below 500 m mostly in north Wales, north-west England and Scotland, plus a scattered distribution in Ireland. Populations in the north have almost no spots at all; those in the south have very distinctive spots. Adults fly in June–July. Larvae favour cottongrass, adults the nectar of ling, cross-leaved heath, hawkweeds, tormentil and white clover. Drainage of wetland habitat has led to a major decline in numbers and range in England and Wales, with a national decline in distribution over 1976–2014 by 58%, and of numbers over 2005–14 by 49%. In 2016 a new stable colony was confirmed at Heysham Moss, from a captive breeding programme at Chester Zoo, to add to the two other sites in Lancashire, but in 2017 fire damaged most of the site. Tullia was a Roman name.

Small heath *C. pamphilus*, male wingspan 33 mm, female 37 mm, smallest of the 'browns', common and widespread, though a number of colonies have disappeared in recent years, and there was a reduction in range by 57% over 1976–2014. Nevertheless, despite an ongoing loss of range over 2005–14 of 7%, numbers over this period did increase by 18%. It is found in a variety of open habitats. Larvae feed at night on a variety of fine grasses, adults (flying in May–October) on the nectar of plants such as bramble, buttercups, scabious and yarrow. Pamphilos was a son of Aegyptus in Greek mythology.

HEATH (MOTH) Members of the family Geometridae, subfamily Ennominae.

Common heath *Ematurga atomaria*, wingspan 22–30 mm, a day-flying moth (May–August) widespread and common on heathland, moorland and some grasslands (including calcareous downland) and in woodland rides and glades. Larvae feed on heathers, cross-leaved heath, trefoils, vetches and clovers. Greek *ematos* + *ergon* = 'day' (poetic) + 'work', and Latin *atomus* = 'atom', from the speckles on the wings.

Latticed heath *Chiasmia clathrata*, wingspan 20–5 mm, day- and night-flying, widespread and common, though not in mountainous regions, in open woodland, moorland, heathland, chalk grassland, gardens and waste ground. There are usually two generations, especially in the south, flying in May–June and August–September. Larvae feed on clovers, trefoils and lucerne. Numbers have declined by 85% over 1968–2007. Greek *chiasma* = 'the mark of chi', and Latin *clathrata* = 'furnished with a lattice' (i.e. with a barred wing pattern).

HEATH (PLANT) Low-growing evergreen shrubs, mainly in the genus *Erica* (Ericaceae), from Old English *hæth*, and *Erica* from Latin for 'heath'. See also bell heather under heather.

Blue heath *Phyllodoce caerulea*, very local in the mid-Highlands on acidic, free-draining sites on steep rocky slopes with prolonged snow, usually in dwarf shrub communities, though sometimes in herb-rich grassland. Pyllodoce was a sea nymph mentioned by Virgil. Latin *caerulea* = 'dark blue'.

Cornish heath *E. vagans*, the county flower of Cornwall as voted by the public in a 2002 survey by Plantlife, locally abundant in heathland over serpentine and gabbro in the Lizard Peninsula, and also on moist gley soils, naturalised as a garden escape in a few places elsewhere in the British Isles, though possibly native in Co. Fermanagh. Latin *vagans* = 'wandering'.

Corsican heath *E. terminalis*, a Mediterranean neophyte introduced by 1765, naturalised on dunes in Co. Londonderry and on roadsides and waste ground in a few places in Britain. *Terminalis* is Latin for 'terminal'.

Cross-leaved heath *E. tetralix*, common and widespread (less so in the Midlands and eastern England) in mire and wet heathland, though also on drier heath in south-west England, usually on nutrient-poor organic soils, occasionally in mesotrophic or eutrophic habitats. Greek *tetralix* means having leaves in groups of four.

Dorset heath *E. ciliaris*, the county flower of Dorset as voted by the public in a 2002 survey by Plantlife, on moist heathland, extending into relatively dry heath, and also into wet valley bogs, mainly on drier hummocks, locally common in Dorset, south Devon and west Cornwall, and naturalised in south Hampshire and western Co. Galway. *Ciliaris* is Latin for 'fringed with hair'.

Irish heath *E. erigena*, in damp or boggy but well-drained base-rich moorland, usually on slopes, often near streams or on lake shores in western parts of Co. Mayo and Co. Galway. *Erigena* comes from Greek for 'born in Ireland'.

Mackay's heath *E. mackayana*, scarce, on peat bogs in Counties Kerry, Donegal, Mayo and Galway. James Mackay (1775–1862) was an English botanist who lived in Ireland.

Portuguese heath *E. lusitanica*, a south-west European neophyte in cultivation by about 1800, naturalised as a garden escape in (near) coastal south and south-west England on heathland, and on roadsides and railway banks. *Lusitanica* is Latin for 'Portuguese'.

Prickly heath *Gaultheria mucronata* (Ericaceae), a dwarf South American neophyte shrub introduced by 1828, scattered and occasionally naturalised in open woodland and scrub and on roadsides on acidic, sandy soils. Jean-François Gaultier (1708–56) was a Canadian botanist. Latin *mucronata* = 'pointed'.

St Dabeoc's heath *Daboecia cantabrica*, on heathland and moorland, often on rocky sites, on thin acidic soils over quartzites or mica-schists, avoiding peat, locally common in Co. Galway, and Co. Mayo, planted and naturalised in a few sites in southern England. St Dabeoc was an Irish saint. *Cantabrica* is Latin for 'Cantabrian'.

Tree heath *E. arborea*, a Mediterranean neophyte in gardens by 1658, an occasionally naturalised escape in southern England and the Channel Islands in woodland and hedgerows, on roadsides and on grassy waste ground. *Arborea* is Latin for 'tree-like'.

HEATH-GRASS *Danthonia decumbens* (Poaceae), a common and widespread perennial of pasture, heathy grassland and moorland, favouring mildly acidic soils. It is also found in chalk and limestone grassland where rooted into acidic, superficial or leached horizons. It is also locally common in damp montane grassland. Étienne Danthoine (1739–94) was a French botanist. *Decumbens* is Latin for 'prostrate'.

HEATHER *Calluna vulgaris* (Ericaceae), the county flower of Staffordshire as voted by the public in a 2002 survey by Plantlife, a common and widespread low shrub, on heathland, moorland and nutrient-poor grasslands, and in open woodland on acidic soils, ranging from dry exposed habitats to wet peat bogs. It may be declining, particularly in much of England, through loss of heathland to forestry, agriculture, mineral workings and scrub colonisation. 'Heather' comes from Old English *hadre* or *hedre*, of unknown origin. Greek *calluno* = 'to brush', referring to its use as a broom, Latin *vulgaris* = 'common'.

Bell heather *Erica cinerea*, the county flower of Flintshire as voted by the public in a 2002 survey by Plantlife, a low-growing shrub, common and widespread (less so in the Midlands and eastern England), on thin, acidic, peaty or mineral soils in well-drained sites, on dry heathland, and as an occasional undershrub in open-canopy pine or oak woodland. It is also recorded, though declining, from some calcareous grasslands that are leached and acidic at the surface. Latin *erica* = 'heath', *cinerea* = 'ash-coloured' or 'grey'.

HEBE Evergreen shrubs of the genus *Veronica*, a few species and hybrids found locally in mostly southern Britain as garden escapes on waste ground, walls, etc. The name is that of the Greek goddess of youth, thereby coming to mean 'pubescent' or 'hairy'.

HEBREW CHARACTER *Orthosia gothica* (Noctuidae, Hadeninae), wingspan 30–5 mm, a widespread, often abundant nocturnal moth common in a variety of habitats. Adults fly in March–April, feeding on sallow blossom. Larvae feed on woody plants such as oak, birch and hawthorn, and herbaceous plants such as meadowsweet and stinging nettle. The English name comes from the dark mark on the forewing which resembles a letter in the Hebrew alphabet, or alternatively a Gothic arch (though in effect it is Norman), which prompted the specific name. The genus name is from Greek *orthosis* = 'making straight' (referring to the subterminal line towards the apex of the forewing).

Setaceous Hebrew character *Xestia c-nigrum* (Noctuinae), wingspan 35–42 mm, nocturnal, widespread, common and often abundant, especially in autumn when numbers may be reinforced by immigration, in a variety of habitats, including gardens, woodland, heathland and marsh. In the southern half of its range, there are two broods, flying in small numbers in May–June, but far more commonly in August–September. In the north there is one generation, flying in July–August. Larvae especially feed on stinging nettle, but other plants are used. Greek *xestos* = 'polished', from the glossy forewings, and Latin *niger* = 'black', from the colour of the C-shaped forewing mark.

HEDGE ACCENTOR Alternative name for dunnock.

HEDGE BROWN Alternative name for gatekeeper (butterfly).

HEDGE-PARSLEY Species of *Torilis* (Apiaceae), a genus name of obscure origin.

Knotted hedge-parsley *T. nodosa*, an annual with a scattered, often coastal distribution, especially in eastern England, in dry, sparsely vegetated habitats, for example grassland, clifftops, arable fields and waste ground, occasionally in disused sand- and gravel-pits. Latin *nodosa* = 'knotted'.

Spreading hedge-parsley *T. arvensis*, an annual, rarely biennial, archaeophyte, rare and declining, on a range of soils in southern England in arable fields in autumn-sown cereals, and on waste and disturbed ground. *Arvensis* is from Latin for 'field'.

Upright hedge-parsley *T. japonica*, an annual or rarely biennial herb, common and widespread except for the Highlands, on dry neutral and basic soil, in woodland rides, hedgerows and rough grassland, and on verges.

HEDGE SPARROW Not a true sparrow but an alternative name for dunnock.

HEDGEHOG (FUNGUS) Species of *Hydnum* (Cantharellales, Hydnaceae) (Greek *hydnon* = 'truffle'). Terracotta (or rufous) hedgehog *H. rufescens* and wood hedgehog *H. repandum* are both widespread and common, edible and very tasty, found from late summer to early winter on soil and leaf litter in both coniferous and broadleaf woodland. Latin *rufescens* = 'reddishbrown', *repandum* = 'bent back', since part of the cap's edge is often upturned.

HEDGEHOG (MAMMAL) *Erinaceus europaeus* (Erinaceidae), head-body length 15–30 cm, weight generally 1.0–1.5 kg, widespread, favouring deciduous woodland, scrub, hedgerows, open grassland and suburban gardens, avoiding very wet habitats and conifer plantations. A pre-breeding population of 1,555,000 was estimated for Britain in 1995: 1,100,000 in England, 310,000 in Scotland and 145,000 in Wales. Irish numbers are probably around 1 million. Distribution has become patchy, with numbers declining: in 2017 the People's Trust for Endangered Species found urban populations had probably fallen by up to a third since 2000 and rural populations had declined by at least a half, mainly because of habitat loss and fragmentation, and a decline

in prey abundance. They are solitary, mostly sleeping by day, active at night, feeding on earthworms, slugs, snails, beetles, fungi, fruit and carrion. Females usually have two litters a year, with 4–5 young per litter. Gestation lasts up to five weeks with the young (hoglets) suckled for four weeks. Leaf and wood piles are used for hibernation, which usually takes place between November and March–April, depending on weather. Adults can curl into a ball and have around 5,000 spines that provide protection against predators. 'Hedgehog' dates from around 1450, from Middle English *heyghoge*, from *heyg, hegge* = 'hedge', because it frequents hedgerows, and *hoge, hogge* = 'hog', from its piglike snout and its habit of rooting around in the undergrowth for food. The binomial is Latin for 'European hedgehog'.

HELEOMYZIDAE Family (Greek *heleos* + *myzo* = 'pity' or 'mercy' + 'to suck') of small to medium-sized flies, with 56 species in 13 genera, found in a variety of habitats. Larvae feed on fungi, and decaying plant and animal matter.

HELICIDAE Family name for some keeled pulmonate snails (Greek *helix*, meaning something spiral or twisted).

HELICODONTIDAE Family name for cheese snail (Greek *helix*, meaning something spiral or twisted + *donti* = 'tooth').

HELIOTROPE Winter heliotrope *Petasites fragrans* (Asteraceae), a dioecious rhizomatous perennial North African neophyte herb, widespread and naturalised on rough ground, though only the lowlands in northern England and Scotland. 'Heliotrope' comes from Old English *eliotropus*, originally applied to a number of plants whose flowers turn towards the sun, via Latin, in turn from Greek *heliotropion*, from *helios* + *tropos* = 'sun' + 'a turn'. Greek *petasos* + *eidos* = 'broad-brimmed hat' + 'appearance', and Latin for 'fragrant'.

HELIOZELIDAE Family of leaf-mining lift moths (Greek *helios* + *zelos* = 'sun' + 'eagerness', from the behaviour of flying in bright sunlight).

HELLEBORE Herbaceous perennials in the genus *Helleborus* (Ranunculaceae). The English and genus names come from Old French *ellebre* or *elebore*, derived from the Greek for this plant, *helleboros*, in turn from *elein* + *bora* = 'to injure' + 'food'.

Corsican hellebore *H. argutifolius*, introduced from Corsica and Sardinia in 1710, a garden escape with a scattered naturalised distribution in England north to Lancashire on rocks and in rocky grassland, and as a casual on hedge banks and waste ground. *Argutifolius* is Latin for 'sharply pointed leaves'.

Green hellebore *H. viridis*, local in shady habitats, usually on chalk or limestone, in woodland glades, scrub and hedge banks, but also common as a naturalised garden escape. *Viridis* is Latin for 'green'.

Stinking hellebore *H. foetidus*, intolerant of deep shade so usually found in small colonies in woodland glades or open scrub, on scree slopes, rock ledges and hedge banks on shallow, nutrient-rich, moist calcareous soils, but it is also common and widespread in gardens and as an adventive in urban habitats. *Foetidus* is Latin for 'stinking'.

HELLEBORINE Species of rhizomatous herbaceous perennial orchids in the genera *Cephalanthera* (Greek *cephale* + *anthos* = 'head' + 'flower') and *Epipactis* (Greek *epipactis* or *epipegnyo*, referring to a milk-curdling property claimed for some species), both Orchidaceae. 'Helleborine' dates from the late sixteenth century, ultimately derived from the Greek for this plant, *helleboros*, in turn from *elein* + *bora* = 'to injure' + 'food'.

Broad-leaved helleborine *E. helleborine*, locally common and widespread though absent from upland Scotland, on calcareous to slightly acidic soils in coniferous and deciduous woodland, hedgerows, banks, streamsides, roadsides, alder carr, dune slacks, limestone pavement and scree. It can invade secondary woodland and also occurs in urban habitats: it is for unknown reasons actually more common in and around Glasgow than anywhere else in Britain, with 75% of recorded populations in parks, gardens, cemeteries, golf courses, railway embankments, roadsides and coal-spoil tips.

Dark-red helleborine *E. atrorubens*, local in north Wales, Derbyshire, northern England, the Highlands, the Hebrides and western Ireland, mostly on rock or well-drained skeletal soils overlying limestone. This is the county flower of Banffshire as voted by the public in a 2002 survey by Plantlife. *Atrorubens* is Latin for 'dark red'.

Dune helleborine *E. dunensis*, an endemic found locally from north Lincolnshire to Northumberland, in Lancashire and Cumbria, Anglesey, and central Scotland, on dunes and in woodland on mine spoil and soils polluted with heavy metals. This is the county flower of Lanarkshire as voted by the public in a 2002 survey by Plantlife.

Green-flowered helleborine *E. phyllanthes*, an endemic scattered in England, Wales and Ireland, growing in sparsely vegetated, shaded places on dry, acidic, humus-poor substrates, for example pine and birch scrub on the Bagshot Sands, beechwood, hazel coppice on sandy alluvium, and on dunes. Greek *phyllon* + *anthos* = 'leaf' + 'flower'.

Lindisfarne helleborine *E. sancta*, endemic, in dune slacks on Holy Island, Northumberland. *Sancta* is Latin for 'sacred'.

Marsh helleborine *E. palustris*, locally common but overall declining in England, Wales and Ireland, very scarce in Scotland, in neutral to calcareous fen, marsh, damp pasture, meadow and dune slacks, preferring flushed or seasonally flooded sites where competition is reduced. *Palustris* is from Latin for 'marsh'.

Narrow-leaved helleborine *C. longifolia*, scattered and declining in woodlands on calcareous soils, usually on chalk and hard limestone, favouring permanent patches of light and steep, rocky slopes with an open tree canopy; also found along woodland edges and rides, and in scrub. *Longifolia* is Latin for 'long-leaved'.

Narrow-lipped helleborine *E. leptochila*, scattered in south England, Shropshire and south Wales in deep beechwood shade on calcareous substrates; under birch on stony soils and spoil, often polluted with lead and zinc; and on the edges of dune slacks growing among creeping willow. Greek *leptos* + *cheilos* = 'slender' + 'lip'.

Red helleborine *C. rubra*, on well-drained sloping sites in deciduous (especially beech) woodland on calcareous soils, declining and now (RDB Critically Endangered) only found at three sites (in north Hampshire, Buckinghamshire, and the Cotswolds, Gloucestershire). *Rubra* is Latin for 'red'.

Violet helleborine *E. purpurata*, uncommon, in southern and central England and the Welsh Marches in densely shaded beechwood, particularly those on clay-with-flints, and on calcareous and occasionally acidic sands and clays in mixed woodland and hazel and hornbeam coppice. *Purpurata* is Latin for 'purplish'.

White helleborine *C. damasonium*, shade-loving, locally common in south and south-east England on well-drained soils on chalk and oolitic limestone, usually in woodland with little ground cover, especially beechwood, but also extending into chalk scrub. Damasonium is a classical Greek name.

HELMINTH A general, commonly prefixed term for parasitic worms that live inside their host, usually in the intestines or blood, for example platyhelminth (flat worms such as flukes and tapeworms) and nemathelminth (round worm or nematode).

HELOPHORIDAE Water scavenger beetles, with 20 species in one genus, *Helophorus* (Greek *helos* + *phoresis* = 'marsh' + 'carrying'). The 2–7 cm elongate beetles are generally found in stagnant water and impermanent ponds, often ephemeral pools on grassland, though three species are associated with dry ground, and are known as mud beetles. The (aquatic) larvae but not the adults are predaceous.

HELVELLA Black helvella, another name for the fungus elfin saddle.

HEMIPTERA Order (Greek *hemi* + *pteryx* = 'half' + 'wing') to which the true bugs belong, divided into three suborders: Heteroptera, in which the front wing is divided into two 'regions', and the formerly recognised Homoptera, in which the front wing is not divided, now taxonomically separated into Auchenorrhyncha and Sternorrhyncha. Over 1,650, and probably nearer 1,700 hemipteran species are native to the British Isles. See bug.

HEMLOCK *Conium maculatum* (Apiaceae), a biennial archaeophyte, common and widespread except for central Wales, the Pennines and upland Scotland, in damp habitats such as ditches and riverbanks, and in drier habitats, including rough grassland, waste ground and roadsides. It often colonises disturbed areas such as exposed mud. It is highly toxic, being emetic and convulsive, leading to paralysis of the central and peripheral nervous system, with death caused by respiratory failure. 'Hemlock' comes from Old English *hymlice* or *hemlic*. *Conium* comes from Greek *coneion*, the name for hemlock, both the plant and the poison derived from it. *Maculatum* is Latin for 'spotted'.

HEMLOCK-SPRUCE Or simply **hemlock**, evergreen conifers in the genus *Tsuga* (from the Japanese for these trees) (Pinaceae), introduced from North America, the branches said to resemble giant hemlock leaves.

Eastern hemlock-spruce *T. canadensis*, introduced in 1736, mainly grown in parks.

Western hemlock-spruce *T. heterophylla*, introduced in 1852, very shade-tolerant but drought-sensitive, widely planted for timber, often mixed with hardwoods, and in parks. Greek *heterophylla* = 'different leaf'.

HEMP *Cannabis sativa* (Cannabaceae), an annual herb occurring as a non-persistent casual with a scattered (especially urban) distribution. It has been grown in gardens since at least the early fourteenth century, and was cultivated up to the eighteenth century as a fibre crop, and there are present-day trials of it for fibre and oil-seed. It is also a constituent of birdseed mixtures, is used as bait by anglers, and (as cannabis) is cultivated illegally as a drug plant. 'Hemp' comes from Old English *henep* or *hænep*, of Germanic origin, and also from Greek *cannabis*. *Sativa* is Latin for 'cultivated'.

HEMP-AGRIMONY *Eupatorium cannabinum* (Asteraceae), a common and widespread herbaceous perennial, mostly coastal in Scotland, growing on base-enriched soils in damp habitats, including pond and stream sides, sometimes in rough ground and grassy habitats. 'Hemp' comes from Old English *henep* or *hænep*, of Germanic origin; 'agrimony' is late Middle English, ultimately from Greek *argemone* = 'poppy'. *Eupatorium* comes from King Mithridates VI of Pontus, also known as Eupator Dionysius. Latin *cannabinum* = 'hemp-like'.

HEMP-NETTLE Annuals in the genus *Galeopsis* (Lamiaceae) (Greek *gale* + *opsis* = 'weasel' + 'appearance', the flower supposedly resembling a weasel's face). The English names come from Old English *henep* (or *hænep*) and *netle* (or *netele*), both parts being of Germanic origin.

Bifid hemp-nettle *G. bifida*, a common and generally widespread weed of arable, waste and cultivated ground, less often woodland clearings and ditch sides. *Bifida* is Latin for 'split in two'.

Common hemp-nettle *G. tetrahit*, common and widespread in woodland clearings, ditches, fens, riverbanks, verges and wet heathland, but most frequent in arable, waste and cultivated ground. *Tetrahit* is Greek for 'in four parts'.

Downy hemp-nettle *G. segetum*, a weed of arable and waste ground in England and Wales, a declining casual mostly in root crops. *Segetum* is Latin for 'pertaining to corn fields'.

Large-flowered hemp-nettle *G. speciosa*, an archaeophyte, locally common in north England, Scotland and north Ireland, scarce elsewhere, on cultivated, marginal and waste ground, often a weed of root crops (especially potatoes) on peaty soils. *Speciosa* is Latin for 'splendid'.

Red hemp-nettle *G. angustifolia*, an archaeophyte of arable land, waste places and open ground on calcareous substrates, and also found on coastal sand and shingle. Scattered and generally declining in England and Wales, but increasing on ground disturbed by gravel extraction in central Ireland. *Angustifolia* is Latin for 'narrow-leaved'.

HEN OF THE WOODS *Grifola frondosa* (Polyporales, Meripilaceae), a widespread but local fungus, edible if picked young, resembling the texture and meatiness of chicken breast, giving it its English name. It is evident from late summer to late autumn emerging from a common, repeatedly branching stem to form a rosette shape, parasitic on the wood of broadleaf trees, especially beech and oak, creating a white rot. Not the same species as chicken of the woods.

HENBANE *Hyoscyamus niger* (Solanaceae), a biennial archaeophyte herb known since the Bronze Age, scattered though mainly in England on dry, calcareous, preferably disturbed soils, particularly on chalk, and on coastal sandhills, sandy open areas and waste ground, especially associated with animal manure. 'Henbane' comes from Old English *hennebane* and Old French *hanebane*. Although toxic its leaves, combining the therapeutic actions of the alkaloids hyoscyamine and hyoscine, have been used as an antispasmodic, hypnotic, and mild diuretic. Greek *choiros* + *cyamos* = 'pig' and 'bean', and Latin *niger* = 'black'.

HENBIT See under dead-nettle.

HEPATICOPSIDA Class name for liverworts. See also hornwort (Anthocerotopsida) and moss (Bryopsida). *Hepaticos* is Greek for relating to the liver.

HEPIALIDAE Family of swift moths (Greek *epialos* = 'fever', from the fitful flight behaviour). See also ghost moth under moth.

HEPTAGENIIDAE A mayfly family (Greek *hepta* + *genos* = 'seven' + 'type'), comprising ten species in five genera, generally uncommon in Britain, favouring hilly parts where there are fast streams and clear lakes. The only widespread species are yellow May dun and (very important to anglers) olive upright (= March brown). Stream-dwelling nymphs of all species are flattened, adapted to clinging to stones.

HERALD, THE *Scoliopteryx libatrix* (Erebidae, Scoliopteryginae), wingspan 40–5 mm, a nocturnal moth common in broadleaf woodland, fen, heathland, parks and gardens. It overwinters as an adult (which feeds on ivy flowers) so can be one of the last species to be seen in one year and one of the first in the next, possibly giving its English name, though this might rather reflect the moth looking like a herald's clothing. It is also sometimes found in large groups hibernating inside barns, cellars and outbuildings. Larvae feed on grey and creeping willow, aspen, and poplars. Greek *scolios* + *pteryx* = 'crooked' + 'wing', and Latin *libatrix* = one who makes a libation to the gods.

HERALD OF WINTER *Hygrophorus hypothejus* (Agaricales, Hygrophoraceae), a widespread, common edible waxcap fungus found in autumn and winter on pinewood soil, often on the edges of grassy rides. This is the only member of its family to thrive after the first frosts of winter, hence its English name. Greek *hygro* + *phoros* = 'water' + 'bearer', it being moist and sticky, and *hypo* + *theio* = 'under' + 'like sulphur', the gills and stem having a sulphur-yellow colour.

HERALD SNAIL Members of the genus *Carychium* (Ellobiidae), possibly from Greek *caryon* = 'nut', shell lengths 1.6–2.3 mm, widths 0.9–1.1 mm.

Long-toothed herald snail *C. tridentatum*, common and widespread, in deep leaf litter in mature woods (an ancient woodland indicator) with persistent leaf litter such as beech and holly, as well as on wetland margins. *Tridentatum* is Latin for 'three-toothed'.

Short-toothed herald snail *C. minimum*, common and widespread, terrestrial but preferring wet habitats (not peat), possibly an indicator of ancient woodland.

HERB BENNET(T) Alternative name for wood avens, from Old English *herbe beneit*, from medieval Latin *herba benedicta* = 'blessed herb', because of its clove-scented root.

HERB CHRISTOPHER Alternative name for baneberry, possibly so called because it was thought to be efficacious against the plague (for which St Christopher was invoked), or because the flowers are found in a spike above the leaves like the infant Christ on St Christopher's shoulder.

HERB-PARIS *Paris quadrifolia* (Melanthiaceae), a rhizomatous herbaceous perennial scattered in England, local in parts of Wales and Scotland, in moist, calcareous, usually ancient, woodland, but it can spread into secondary woods which are adjacent to primary woodland. It flowers and fruits most freely in the open stages of the coppice cycle, but persists in deep shade. It is also occasionally found in grikes on limestone pavement. The English name derives from the apothecaries' *herba paris* = 'pair herb', since it has twice two leaves, twice four stamens, twice two outer and twice two inner parts of the perianth, twice two styles, and twice two cells to the ovary. The binomial comes from Latin for 'equal' (*par*) and 'four-leaved'.

HERB-ROBERT *Geranium robertianum* (Geraniaceae), very common and widespread, annual or biennial, shade-tolerant, found on a range of soil types, except strongly acidic ones. Habitats include woodland, hedgerows, walls, shaded banks, scree, limestone pavements (ssp. *celticum* in south Wales and central-west Ireland) and coastal shingle (ssp. *maritimum*), and also in disturbed artificial habitats, including gardens. The English name comes from medieval Latin *herba Roberti*, possibly because of the red flowers and an association with the eighth-century St Rupert of Salzburg, who was invoked against bleeding wounds (for which herb-robert was a treatment). 'Geranium' comes from Greek *geranos* = 'crane', referring to the beak-like fruit capsule.

HERMIT CRAB *Pagurus bernhardus* (Paguridae), carapace length up to 3.5 cm, with a large right-hand pincer, related more closely to lobsters than to crabs. It has a 'naked' hind body, only the front end being covered with shell, and it adopts the empty shells of gastropod molluscs (especially dog whelks), carrying them around (hence 'hermit') and swapping them for a larger shell as it grows. It is common and widespread on rocky and sandy substrates from mean tide level to 140 m. Greek *pagouros*, from *pagos* = 'rock'.

Hairy hermit crab *P. cuanensis*, carapace length up to 1.5 cm, with a larger right pincer, 7 mm in length, all visible parts of the crab being very hairy, found in the intertidal but more typically in the shallow sublittoral to 100 m depth, around much of the coastline, though not recorded east Britain except between the Firth of Forth and Northumberland.

Small hermit crab *Diogenes pugilator* (Diogenidae), carapace length up to 1.1 cm, with a large left-hand claw, its soft abdomen protected by using an empty snail shell which it changes as it grows larger. It is found in shallow intertidal waters down to 10 m off the south-west coasts of Britain as far north as Anglesey and the south and west coasts of Ireland. Diogenes of Sinope was a Greek philosopher (412 or 404 BCE to 323 BCE). *Pugilator* is Latin for 'boxer'.

St Piran's hermit crab *Clibanarius erythropus* (Diogenidae), up to 1.5 cm carapace length, in rock pools and the sublittoral in the Channel Island and Cornwall, feeding on organic debris, and macroalgae with associated fauna and epiphytic algal flora. Always uncommon, it had not been recorded in Cornwall since 1985 (probably having been affected by the 1967 Torrey Canyon oil spill) until rediscovery near Falmouth in March 2016, and given its English name following a vote on BBC's *Springwatch* in June 2016, in honour of St Piran, patron saint of Cornwall.

Latin *clibanarius* = 'soldier clad in armour', and Greek *erythros* + *pous* = 'red' + 'foot'.

HERON Generally long-necked, water- or marsh-based prey-stalking birds in the family Ardeidae (Latin *ardea* = 'heron'). **Great blue heron** (*Ardea herodias*), **green heron** (*Butorides virescens*) and **squacco heron** (*Ardeola ralloides*) are scarce vagrants. 'Heron' comes from Middle English *heroun* and *heiron*, from Anglo-Norman *heiron*.

Black-crowned night heron *Nycticorax nycticorax*, wingspan 115–18 cm, length 58–65 cm, weight 800 g, a scarce visitor, with an average of <20 records p.a., usually as single birds, from late March to mid-May by birds 'overshooting' on their return to Europe from wintering in Africa, or in autumn by juveniles, scattered around Britain. Breeding was recorded for the first time in Britain at Westhay Moor, Somerset, in 2017, with two successfully fledged young. The English name reflects their habit of catching fish and frogs at night. Greek *nyctos* + *corax* = 'night' + 'raven'.

Grey heron *Ardea cinerea*, wingspan 155–95 cm, length 84–102 cm, weight 1–2 kg, common and widespread, with 13,000 pairs estimated in 2007–11, and 11,000 occupied nests in 2016. Numbers fluctuate dramatically depending on the harshness of winter, but overall declined in Britain by 17% between 1975 and 2015, by 25% between 1995 and 2015, and by 10% over 2010–15. Some 61,000 individuals, augmented by migrants, wintered in 2008–9. Nests are colonial in trees, with clutches of 3–4 eggs, incubation taking 27 days, and the altricial chicks fledging in 50–5 days. Diet is fish, as well as small birds (e.g. ducklings), small mammals and amphibians. Latin *ardea* + *cinerea* = 'heron' + 'ash-grey'.

Purple heron *A. purpurea*, wingspan 135 cm, length 84 cm, weight 870 g, generally a passage visitor with scattered records on wetlands in England and Wales but, following a breeding attempt at Minsmere, Suffolk, in 2007 which was curtailed when flooding led to the departure of the birds, the successful breeding of a pair at Dungeness in 2010 becomes the first confirmed breeding of this species in Britain. Diet is mainly fish. Latin *purpurea* = 'purple'.

HESPERIIDAE Family name of skipper butterflies, named after their rapid, darting flight. Hesperia was one of the Hesperides, the nymphs who, in Greek mythology, guarded the golden apples of Hera.

HETEROCERIDAE Family (Greek *hetero* + *ceras* = 'different' + 'horn') of mud beetles, with eight British species in two genera, 2–5 mm in length, most associated with saltmarsh, some with riverbanks and pond sides. Adults live in shallow tunnels that they dig in bankside mud, but they also fly well.

HETEROPTERA Suborder of true bugs (Hemiptera), with a generally flattened body and wings folded flat over the resting body. Front wings are divided into two 'regions', hence Heteroptera from Greek *hetero* + *pteron* = 'different' + 'wing'. Major groups are shieldbugs and allies in the families **Acanthosomatidae** (4 species found in the British Isles), **Alydidae** (1), **Coreidae** (11), **Cydnidae** (8), **Pentatomidae** (20), **Rhopalidae** (11), **Scutelleridae** (4) and **Thyreocoridae** (1); plant bugs and allies in the families **Anthocoridae** (33), **Aradidae** (7), **Berytidae** (9), **Ceratocombidae** (1), **Cimicidae** (4), **Lygaeidae** (86), **Microphysidae** (7), **Miridae** (q.v.) (231), **Nabidae** (12), **Piesmatidae** (3), **Reduviidae** (8) and **Tingidae** (25); and water bugs and allies in the families **Aepophilidae** (2), **Corixidae** (39), **Dipsocoridae** (2), **Gerridae** (10), **Hebridae** (2), **Hydrometridae** (2), **Mesoveliidae** (1), **Naucoridae** (2), **Nepidae** (2), **Notonectidae** (4), **Pleidae** (1), **Saldidae** (22) and **Veliidae** (5).

HIDE BEETLE 1) Members of the family Trogidae, with two species of *Trox* (Greek *trogo* = 'to gnaw') recorded recently, scattered and scarce in England and Wales, mainly found in association with dry carcasses but also in large bird nests, particularly those of hole-nesting species.

2) *Dermestes maculatus* (Dermestidae). Both adults and larvae feed on a variety of animal-based foods, particularly raw skins, rawhide, and carcasses, producing a keratin-digesting enzyme. Larvae, in particular, are voracious eaters. *Dermestes* is Greek for 'skin eater', *maculatus* Latin for 'spotted'.

HIGHFLYER Nocturnal moths in the genus *Hydriomena* (Geometridae, Larentiinae), possibly from Greek *hydria* + *meno* = 'water pot' + 'to remain'.

July highflyer *H. furcata*, wingspan 26–39 mm, widespread and common in a range of habitats, including coppiced woodland, woodland margins, hedgerows, fen, heathland, scrub, parks, gardens and moorland. Adults fly in July–August. Larvae feed on a variety of plants, for example sallows, hazel, bilberry and heather. *Furcata* is Latin for 'fork', from the forewing markings.

May highflyer *H. impluviata*, wingspan 30–4 mm, widespread in damp woodland, marsh, riverbanks and other damp habitats. Adults fly in May–July. Larvae feed on alder leaves in autumn. Larval spinnings might get waterlogged after rain, and the *impluvium* was the basin in the central court of a Roman house into which rainwater was directed.

Ruddy highflyer *H. ruberata*, wingspan 30–4 mm, widespread but local on moorland, heathland, bogs and damp woodland, more rarely in the southern half of England. Adults fly in May–June. Larvae feed on sallows, especially eared and grey willows. Latin *ruberata* = 'ruddy'.

HINTAPINK *Russula paludosa* (Russulales, Russulaceae), an uncommon edible brittlegill fungus, most records coming from the Highlands, found from August to October on acid soil in damp, mossy coniferous woodland. The binomial is Latin for 'reddish' and 'marsh'.

HIPPOBOSCIDAE Family (Greek *hippos* + *bosco* = 'horse' + 'to feed') of louse flies and keds, with 14 species in 10 genera, many brachypterous (wingless), and obligate parasites of (often hoofed) mammals and birds.

HIPPOCASTANACEAE Former family name (and Latin) for horse-chestnuts, now (together with maples) in the Sapindaceae.

HIPPURIDACEAE Family name of mare's-tail (Greek *hippos* + *oura* = 'horse' + 'tail').

HIRUDINEA Class, or subclass in the class Clitellata, that includes leeches (Latin *hirudo* = 'leech').

HIRUNDINIDAE Family name of swallows and martins (Latin *hirundo* = 'swallow').

HISTERIDAE Family name for hister beetles and clown beetles, with 47 species in 21 genera. *Histeros* is Greek for 'coming after' or 'behind'. Generally strong-flying species, most are carnivorous and are often found on carrion, where they usually arrive late and feed on other scavenging species. When disturbed, many species feign death.

HOAR-MOSS Species of *Hedwigia* (Hedwigiaceae). Johann Hedwig (1730–99) was a German botanist.

Fringed hoar-moss *H. ciliata* scattered and rare, favouring upland parts of Wales and Cumbria, especially on igneous rocks, but more frequent on sandstone roof tiles. *Ciliata* is Latin for 'fringed with hair'.

Green hoar-moss *H. integrifolia*, local in upland west Britain and south-west and east Ireland on unshaded boulders and scree, and below cliffs. *Integrifolia* is Latin for 'entire (undivided) leaves'.

Starry hoar-moss *H. stellata*, common and widespread in upland regions on hard, unshaded, south- or west-facing, acidic, siliceous or igneous rocks on moors and mountains, a rarity on trees in the Midlands and eastern England. *Stellata* is Latin for 'starry'.

HOBBY *Falco subbuteo* (Falconidae), wingspan 74–84 cm, length 29–36 cm, male weight 180 g, females 240, a migrant breeder arriving from April and leaving in September–October, once rare but now widespread (and Green-listed) in farmland in lowland England, Wales and south Scotland, more scattered further north in these regions, with 2,800 pairs estimated in 2009, and at least 3,000 pairs in 2016. Nests tend to be in abandoned corvid nests in trees, clutch size being 2–3, incubation taking 29 days, and fledging of the altricial young 30–4 days. Diet is of small birds and large insects taken in flight. 'Hobby' comes from Old French *hobet*, a small bird of prey. Latin *subbuteo* = 'close to being a buzzard'.

HOGWEED *Heracleum sphondylium* (Apiaceae), a common and widespread herbaceous perennial on neutral to calcareous soils in rough grassland, especially on roadsides, woodland rides, scrub, riverbanks, stabilised dunes, coastal cliffs, montane tall-herb vegetation and waste ground. 'Hogweed' may reflect a piglike smell or its use in feeding pigs. *Heracleum* is named for Hercules, who was supposed to have used it for its healing properties: the roots are rich in carbohydrates and the plant yields essential oils with sedative, expectorant and tonic properties. Greek *spondylos* = 'vertebra', referring to the segmented stem.

Giant hogweed *H. mantegazzianum*, a widespread and locally abundant, tall (up to 5.5 m), herbaceous biennial or perennial Caucasian neophyte introduced by 1817, found in derelict gardens, neglected urban sites and waste ground, on tips, roadsides and by streams and rivers, forming large colonies if permitted. It spreads by prolifically produced seed. Chemicals (furanocoumarins) in the sap can cause photodermatitis or photosensitivity, where the skin becomes very sensitive to sunlight and may suffer blistering, pigmentation and long-lasting scars. The WCA (1981) lists it on Schedule 9, Section 14, meaning it is an offence to cause giant hogweed to grow in the wild in England and Wales (with similar legislation effective in Scotland and Northern Ireland). It can also be the subject of Anti-Social Behaviour Orders where occupiers of giant hogweed infested ground can be required to remove the weed or face penalties. Paolo Mantegazzi (1831–1910) was an Italian physiologist.

HOLLOWROOT *Corydalis cava* (Papaveraceae), a rhizomatous perennial herb introduced from Europe, scattered in England and Wales, a garden escape naturalised since the seventeenth century in woodland and hedgerows, and a casual on waste ground. Latin *corydalis* = 'crested lark' (in turn from Greek *corys* = 'helmet', referring to the shape of the flower), and *cava* = 'hollow'.

HOLLY 1) *Ilex aquifolium* (Aquifoliaceae), a common and widespread evergreen shrub of deciduous woodland, especially on acidic soils where beech and oak dominate, and often a locally dominant undershrub. Leaf spines become smoother with height as risk of grazing by large mammals declines. It is also found in wood pasture, scrub and hedgerows, and is commonly planted in amenity areas and parkland. 'Holly' comes from Old English *holegn*. The binomial is Latin for 'holly' and 'holly-like leaves'.

Highclere holly *Ilex* x *altaclerensis* (*I. aquifoluma* x *I. perado*), an evergreen neophyte shrub or small tree, in cultivation just before 1800, scattered in the UK as a garden escape or planted in woodland, hedges, parks and amenity areas, and naturalised on roadsides and waste ground. Bird- and self-sown plants have commonly been recorded. The common name comes from its early cultivation at Highclere Castle, Berkshire.

2) **New Zealand holly** *Olearia macrodonta* (Asteraceae), an evergreen New Zealand neophyte shrub introduced in 1886, scattered but mainly in south-west England, Ireland and the Isle of Man, naturalised in hedges and scrub, and on roadsides, sea-cliffs, dunes and waste ground. Johann Ölschläger (1635–1711) was a German horticulturalist, his name latinised to Olearius. Greek *macros* + *odontos* = 'large' + 'teeth', from the leaf shape.

HOLLY-FERN *Polystichum lonchitis* (Dryopteridaceae), uncommon, in rock crevices, scree and gullies in northern England, Scotland and western Ireland. Greek *polystichum* = 'many rows of spores' and *lonchitis* = 'spear-shaped'.

Fortune's holly-fern *Cyrtomium fortunei*, a neophyte introduced from east Asia, naturalised on old walls and banks, especially in towns in southern England and the Midlands. *Cyrtomium* comes from Greek *cyrtos* = 'arched', referring to the plant's habit. Robert Fortune (1812–80) was a Scottish horticulturist and collector in China.

House holly-fern *C. falcatum*, introduced from east Asia, a garden escape naturalised on walls and rocks in a few places in west and south-east England, and west Scotland. *Falcatum* is Latin for 'sickle-shaped'.

HOLLYHOCK *Alcea rosea* (Malvaceae), a herbaceous biennial or perennial neophyte, possibly introduced from China by the sixteenth century, or originating in gardens, certainly commonly grown there, casual as an escape or throw-out on waste ground, mainly in central and southern England, scattered elsewhere. 'Hollyhock' comes from 'holy' + Old English *hocc*, which is a mallow. *Alcea* is Latin for this plant, *rosea* = 'rosy'.

HOLLYWORT Hutchins' hollywort *Jubula hutchinsiae* (Jubulaceae), an upland foliose liverwort found in western Britain and south and west Ireland on wet rocks in wooded, oceanic ravines, usually at the side of streams just above water level, often near waterfalls. Ellen Hutchins (1785–1815) was an Irish botanist.

HOLOTHUROIDEA Class of marine echinoderms that includes sea cucumbers (Greek *holothourion*) and sea gherkin.

HOLY-GRASS *Hierochloe odorata* (Poaceae), a rare rhizomatous perennial found at a few sites in Scotland in reedbeds, sedge swamp, willow carr, riverbanks and wet meadows, as well in south-west Scotland at stream-flushed bases of coastal cliffs and along the upper saltmarsh. This is the county flower of Kinross as voted by the public in a 2002 survey by Plantlife. It is abundant on Lough Neagh at its only site in Ireland. Its English name comes from its being strewn before church doors on saints' days because of its scent (from coumarin, a fragrant organic chemical compound) when bruised by trampling. Greek *hieros* + *chloe* = 'sacred' + 'grass'.

HOMOBASIDIOMYCETES Class of fungus (Greek *homos* = 'same', Latin *basidium* = 'small pedestal', and Greek *myces* = 'fungus') more or less equivalent to the Agaricomycetes in other classification systems.

HOMOPTERA Formerly recognised suborder of bugs (Hemiptera), now divided into Auchenorrhyncha and Sternorrhyncha, from Greek *homos* + *ptero* = 'equal' or 'same' + 'wing', since wings are uniform in texture and do not overlap when folded.

HONESTY *Lunaria annua* (Brassicaceae), a herbaceous European neophyte, cultivated in Tudor times, a common and widespread garden escape in England and Wales, more scattered elsewhere, naturalised and casual near habitation on scrubby banks and waste ground, beside paths, and in hedgerows and secondary woodland. *Lunaria* is Latin for 'moonlike', reflecting the shape of the seed pods; *annua* (= 'annual') is misleading, given its biennial nature.

HONEWORT *Trinia glauca* (Apiaceae), a monocarpic, dioecious biennial restricted to dry limestone habitats in south Devon, north Somerset and west Gloucestershire, forr example short-grazed, open, species-rich turf. In heavily grazed grassland it can be perennial until the opportunity arises for it to flower. 'Honewort' dates to the mid-seventeenth century, from the obsolete 'hone' (meaning a swelling, for which the plant was believed to be a remedy) + 'wort' (or herb). The binomial is Latin referring to 'triple' and 'glaucous'.

HONEY (MOTH) Twin-spot honey *Aphomia zelleri* (Pyralidae, Galleriinae), wingspan 20–37 mm, scarce, nocturnal, found on dunes and other dry sandy areas usually in coastal parts of Norfolk, Suffolk and east Kent. Female adults fly in June–August, the smaller males doing little more than flutter their wings, and both sexes often running on the sand. Larvae feed on moss found in such locations, using a vertical tube in the sand. The larva is full-fed by May then pupates in the tube. Greek *aphomoios* = 'unlike' (other species). Phillip Zeller (1808–83) was a German entomologist.

HONEY-BUZZARD *Pernis apivorus* (Accipitridae), wingspan 142 cm, length 56 cm, weight 730 g, an Amber List migrant breeder, nesting in woodland, with 29–47 pairs (mean 41) reported for recent years in 2010, and passage visitor, mainly found in a band between south-west England and East Anglia from May to August. Two eggs are usually laid, incubation taking 30–5 days, and fledging of the altricial young 40–4 days. It is neither a buzzard nor does it feed on honey, but excavates bee and (especially) wasp nests to feed on the grubs. 'Buzzard' is late Middle English, from Old French *busard*, based on Latin *buteo* = 'falcon'. *Pernes* was used by Aristotle for a bird of prey. Latin *apis* + *vorus* = 'bee' + 'eating'.

HONEY FUNGUS *Armillaria mellea* (Agaricales, Tricho-lomataceae), widespread, very common and edible, found in summer and autumn, parasitic on and around both broadleaf and coniferous trees. 'Bootlace' growth under stripped-off bark is bioluminescent. The fungus is honey-yellow in colour. In 2017 it had been top of the Royal Horticultural Society's list of most serious diseases for 22 years, being found on at least 70 host genera. Greek *meli* = 'honey'. *Armillaria* derives from Latin *armilla* = 'bracelet', from the bracelet-like frill on the fruiting body.

HONEYBEE *Apis mellifera* (Apidae), common and widespread especially in lowland England and Wales. Each colony has one queen (though possibly several queen grubs) who lays up to 1,500 eggs, can sting many times, and lives 2–4 years; up to 80,000 workers (infertile females) who build the comb, regulate nest temperature, forage, guard, nurse and clean, can sting once (then die), living 4–6 weeks in summer, up to 6 months during the winter huddle; and 100–500 drones (males) whose role is to mate with the queen, stingless, living for a few days or weeks. They survive periods of cold and drought by storing honey. A new, sexually active generation appears at the end of the year when the colony declines and the old occupants die off. Importation of Italian honeybees took place from 1920, following the 'Isle of Wight disease' epidemic (a mite infection), but it was neglect and a loss of bee-keeping experience resulting from World War I that contributed most to the loss of colonies. The native form remains important over large areas of the British Isles though pure examples are comparatively rare. The Italian bee performs well in warm summers, particularly in southern England. Honeybee populations have recently declined, including colony collapse, because of habitat changes, a reduction in flowering plant diversity, neonicotinoid pesticides and parasitic varroa mites. The binomial is Latin for 'bee' and 'honey-bearing'.

HONEYBELLS *Nothoscordum borbonicum* (Alliaceae), a bulbous herbaceous South American neophyte perennial introduced in 1770, occasional in southern and south-west England, naturalised as a garden escape on roadsides, in arable fields and on waste ground. Greek *nothos* + *scordo* = 'false' + 'garlic', lacking as it does the taste and smell of garlic. *Borbonicum* means 'from Réunion', an island (once called Ile Bourbon) in the Indian Ocean.

HONEYCOMB WORM *Sabellaria alveolata* (Sabellariidae), a frequently gregarious segmented annelid that builds tubes from sand or shell fragments, found intertidally (occasionally subtidally) in exposed areas. Tubes are often densely aggregated forming a honeycomb pattern, and can form reefs several metres across and a metre deep. It is abundant on the south and west coasts of Britain with isolated records from the south-east and east coasts, and has scattered records from around Ireland. *Sabellaria* is the Latin diminutive of *sabulum* = 'sand' or 'gravel'; *alveo* is Latin for 'channel'.

HONEYSUCKLE *Lonicera periclymenum* (Caprifoliaceae), a common and widespread perennial deciduous twining shrub or woody climber, found in woodland, scrub and hedgerows, preferring freely drained, moderately basic to acidic soils, but also growing on poorly drained base-rich clays. This is the county flower of Warwickshire as voted by the public in a 2002 survey by Plantlife. 'Honeysuckle' comes from Old English *honisuge* (though the thirteenth-century *hunisuccle* may refer to red clover). *Lonicera* honours the German botanist and herbalist Adam Lonitzer (1528–86). Periclymenon was an argonaut from Greek mythology who had the ability to change his shape, the honeysuckle also changing in shape (and colour). Other species are also *Lonicera* except for Himalayan honeysuckle (*Leycesteria*).

Box-leaved honeysuckle *L. pileata*, an evergreen Chinese neophyte shrub, introduced in 1900, scattered as a garden escape or self-sown on roadsides and in parks, woodland, hedges and amenity areas. *Pileata* is Latin for 'cap-shaped'.

Californian honeysuckle *L. involucrata*, a deciduous shrub introduced in 1824, scattered and possibly increasing in range, naturalised in rough grassland and waste ground. Involucrate here means having a ring of bracts around the base of a flower.

Fly honeysuckle *L. xylosteum*, a deciduous shrub of basic or neutral soils in woodland, hedgerows and scrub, possibly native on the South Downs chalk in West Sussex, otherwise a European neophyte present by the late seventeenth century, bird-sown and widely naturalised elsewhere. Greek *xylo* + *osto* = 'wood' + 'bone'.

Garden honeysuckle *L. x italica* (*L. caprifolium* x *L. etrusca*), a deciduous climbing shrub of garden origin since 1730, scattered in Britain, naturalised in scrub and hedges, on roadsides and on waste ground.

Henry's honeysuckle *L. henryi*, an evergreen Chinese neophyte shrub, naturalised from garden throw-outs or bird-sown along a line from the London area to Merseyside/west Yorkshire. Augustine Henry (1857–1930) was an Irish plant collector in China.

Himalayan honeysuckle *Leycesteria formosa*, a deciduous neophyte shrub, introduced in 1824, scattered and probably increasing as a garden escape in woodland and hedgerows, and on waste ground, and planted as cover for pheasants. *Formosa* is Latin for 'beautiful'.

Japanese honeysuckle *L. japonica*, a semi-evergreen twining neophyte shrub introduced in 1806, local in woodland, scrub, hedgerows and on waste ground, sometimes forming extensive thickets in southern England.

Perfoliate honeysuckle *L. caprifolium*, a twining deciduous south European neophyte shrub, present by the sixteenth century, scattered and naturalised in woodland, scrub and hedgerows, perhaps mainly bird-sown. *Caprifolium* is Latin for 'goat-leafed'.

Tartarian honeysuckle *L. tatarica*, a deciduous Asian neophyte shrub introduced in 1752, locally naturalised in Britain in woodland, hedgerows, roadsides and rough ground.

Wilson's honeysuckle *L. nitida*, an evergreen Chinese neophyte shrub, one clone introduced in 1908 never flowering, but the second, available since 1939, flowering freely, widespread except in upland areas, locally common and increasing as a garden escape and self-sown into woodland, scrub, hedgerows and waste ground. Ernest Henry Wilson (1876–1930) was an English plant collector in China. *Nitida* is Latin for 'shiny'.

HOOF FUNGUS *Fomes fomentarius* (Polyporales, Poly-poraceae), a common inedible bracket fungus, commoner in northern England and Scotland, evident throughout the year but sporulating in spring and early summer, on trunks of birch, less so beech and sycamore. The flesh was formerly used for lighting fires (it burns very slowly), hence its alternative name of tinder bracket. *Fomes* is Latin for 'tinder' or 'fuel', *fomentarius* for 'used for tinder'.

HOOK (MOTH) **Silver hook** *Deltote uncula* (Noctuidae,

Eustrotiinae), wingspan 20–2 mm, day- and night-flying (May–July), widespread but local, and largely absent from most of central England, in damp habitats such as marsh, bog, damp heathland and fen, larvae feeding on sedges and grasses. Greek *deltotos* means shaped like the capital Greek letter *delta*, from the triangular forewing, and Latin *uncula* = 'little hook', from the shape of a forewing mark.

HOOK-MOSS Species of *Cratoneuron* (Greek *cratis* + *neuro* = 'rib' + 'nerve'), *Drepanocladus* (*drepano* + *clados* = 'sickle' + 'branch'), *Palustriella* (Latin *palustris* = 'marshy'), *Pseudocalliergon* (Greek *pseudo* + *calli* + *ergon* = 'false' + 'beautiful' + 'work') and *Sanionia* (all Amblystegiaceae), and *Hamatocaulis* (Latin *hamus* + *caulis* = 'hook' + 'stem') and *Warnstorfia* (named after Carl Warnstorf, 1837–1921) (both Calliergonaceae).

Widespread and locally common species include **chalk hook-moss** *D. sendtneri*, **intermediate hook-moss** *D. cossonii* (especially in upland calcareous flushes, springs and fens), **Kneiff's hook-moss** *D. aduncus* (lowland pools, ditches and fens, with large plants in mineral-rich dune slacks), **large hook-moss** *Ps. lycopodioides* (calcareous dune slacks), **lesser curled hook-moss** *Pa. decipiens* (Highlands, in calcareous mires and springs), and **varnished hook-moss** *H. vernicosus* (mainly Wales and Cumbria, in neutral flushes and fens).

Curled hook-moss *Pa. commutata*, common and widespread in wet, base-rich habitats. *Commutata* is Latin for 'changing'.

Fern-leaved hook-moss *C. filicinum*, common and widespread in wet, base-rich habitats, and frequent in gravelly gateways and tracksides, and on old tarmac. *Filicinum* is Latin for 'fern-like'.

Floating hook-moss *W. fluitans*, common and widespread in nutrient- and base-poor, still water, especially in the uplands. *Fluitans* is Latin for 'floating'.

Ringless hook-moss *W. exannulata*, common and widespread especially in upland areas, often abundant in flushes with neutral water, and in fens. *Exannulata* is Latin meaning 'without rings'.

Rusty hook-moss *D. revolvens*, common and widespread, mainly in upland regions, requiring moderately base-rich water, and forming red patches in flushed habitats.

Sickle-leaved hook-moss *S. uncinata*, common and widespread, especially in upland regions, in gravelly stream edges, montane ledges, boulder tops and mossy turf. Latin *uncinata* refers to a hook.

St Kilda hook-moss *S. orthothecioides*, local in north Scotland, Orkney, Shetland and St Kilda, growing within 100 m of the sea or a sea-cliff edge, in lightly or ungrazed herb-rich grassland and heathland. Greek *orthos* + *thece* = 'straight' + 'small box'.

HOOK-TIP Moths in the family Drepanidae, subfamily Drepaninae (Greek *drepanon* = 'a reaping hook', referring to the shape of the tips of the forewings). Adult drepanids have poorly developed mouthparts and do not feed. Also, beautiful hook-tip (Erebidae); in Greek mythology Erebia was a region of darkness between Earth and Hades.

Barred hook-tip *Watsonalla cultraria*, wingspan 20–8 mm, found on beech (the larval food) and in beechwood where present in the southern half of Britain. First recorded from Northern Ireland (Co. Down) in 1999 and again in 2004, and from the Republic in 2007 (Co. Waterford) and 2008 (Co. Clare). This moth is nocturnal, but males also fly in sunshine (two broods, May–September) when they can be seen among higher tree branches. *Cultraria* derives from Latin *culter* = 'knife' or 'ploughshare', from the curved shape of the forewing.

Beautiful hook-tip *Laspeyria flexula* (Boletobiinae), wingspan 23–7 mm, nocturnal, local in woodland, parks, scrub, hedgerows, orchards and gardens throughout much of southern England and south Wales. Adults fly in June–August. Larvae feed on lichens growing on tree bark. J.H. Laspeyres (1769–1809) was a German entomologist. *Flexula* is the diminutive of Latin *flexa* = 'bent', from the forewing shape.

Oak hook-tip *W. binaria*, wingspan 18–30 mm, nocturnal, and although still common in oak woodland, hedgerows, parks and gardens throughout England (and also found in parts of Wales and in Co. Wicklow) is nevertheless a species of conservation concern in the UK BAP because numbers have declined by 81% over the last 35 years. The bivoltine adults fly in May–June and in August. Larvae feed on oak leaves. *Binaria* refers to the two black forewing spots.

Pebble hook-tip *Drepana falcataria*, wingspan 27–35 mm, widespread and common in (especially birch) woodland, heathland and gardens. Adults fly at night in May–June and again in August. Larvae feed on leaves of birch, sometimes alder. *Falcataria* refers to the falcate (sickle) shape of the forewing.

Scalloped hook-tip *Falcaria lacertinaria*, wingspan 27–35 mm, widespread and common in deciduous, especially birch, woodland, scrub, heathland, gardens and hedgerows. The bivoltine adults fly in May–June and in August, and while at rest have an unusual posture with the wings held in an arch over the back, resembling a dried leaf. Larvae feed on birch in June–July and August–September. The binomial is Latin for 'sickle-shaped' and 'lizard-shaped', the latter probably from the shape of the larva.

Scarce hook-tip *Sabra harpagula*, wingspan 25–35 mm, a rare RDB moth found in deciduous woodland in parts of the Wye Valley on the Monmouthshire and Gloucestershire border. Adults fly in June–July. Larvae feed on leaves of small-leaved lime. *Harpagula* is a diminutive of Latin *harpago* = 'grappling hook', from the forewing shape.

HOOK WEED Bonnemaison's hook weed *Bonnemaisonia hamifera* (Bonnemaisoniaceae), a Pacific Ocean red seaweed, the gametophyte (up to 20 cm long) introduced by 1893 (Isle of Wight) and found on rocks and other algae in the lowest intertidal and subtidal, rare on south and west coasts, the tetrasporophyte first recorded in Falmouth (Cornwall) and Studland (Dorset) in 1893, epiphytic on *Corallina* in lower intertidal pools and the subtidal, now common and widely distributed on the south and west coasts up to Shetland. The first record in Northern Ireland was in 1972 from Sandeel Bay. The genus, family and English names honour the nineteenth-century French algologist M. Bonnemaison. The specific name comes from Latin *hamus* = 'hook'.

HOOPOE *Upupa epops* (Upupidae), wingspan 44 cm, length 27 cm, weight 68 g, a passage visitor with as many as 100 birds arriving in spring (mostly seen as single birds) as they migrate to Europe from Africa, overshoot and land on the south England and East Anglian coasts (though individuals have been seen in Shetland); they are less commonly seen in autumn. There are around 40 known cases of breeding, usually in years with warm, dry springs or early summers, witness four pairs in 1977 when nests were discovered in Avon, Somerset, Surrey and Sussex. Nests are in tree holes, clutch size is 7–8, incubation takes 15–16 days, and the altricial young fledge in July or early August after 26–30 days. They largely feed on the ground taking larvae, pupae and adults of large insects. The binomial comes from Latin and Greek for 'hoopoe', like the English name, onomatopoeic versions of the bird's vocalisation.

HOP *Humulus lupulus* (Cannabaceae), a scrambling, perennial climber, widespread and common, probably native in moist, open woodland, fen carr and hedgerows, but in many places a naturalised escape from cultivation (grown for flavouring beer, especially in Kent and the south-west Midlands) or as an ornamental. This is the county flower of Kent as voted by the public in a 2002 survey by Plantlife. The Middle English name was *hoppe*, the Anglo-Saxon *hoppan* = 'to climb'. *Humulus* is Latin for this plant, possibly derived from *humus* = 'ground' or 'soil', because of its trailing nature. *Lupulus*, from *lupus*, means 'little wolf', since according to Pliny when growing among osiers it strangles them by its 'embrace', as a wolf does a sheep.

HOPPER See sand hopper.

HOREHOUND Herbaceous perennials in the genera *Ballota* (Latin for 'little ball') and *Marrubium* (Hebrew meaning 'bitter juice'), family Lamiaceae. 'Horehound' comes from Old English *har* = 'hoar' (i.e. grey) + *hune*, the plant's name.

Black horehound *B. nigra*, an archaeophyte present since the Bronze Age, common in England and north Wales, local elsewhere, in hedgerows, field borders, walls, waysides and waste ground, often on disturbed nutrient-rich soils near habitation. *Nigra* is Latin for 'black'.

Common horehound Alternative name for white horehound.

White horehound *M. vulgare*, scattered, mainly in England and Wales, probably native near the sea on open, exposed clifftop grasslands and slopes overlying limestone and chalk, and on sandy banks and verges in Breckland. It has been cultivated for tea and as a remedy for coughs and colds, and is naturalised in rough and waste places; it has also been introduced in wool shoddy. *Vulgare* is Latin for 'common'.

HORMATHIIDAE Family name (Greek *hormos* = 'cord' or 'chain') of sea anemones with 12 species in 8 genera in British waters.

HORN (SEAWEED) See velvet horn.

HORN OF PLENTY *Craterellus cornucopioides* (Cantharellales, Cantharellaceae), a widespread and locally common edible and tasty chanterelle fungus found in summer and autumn or early winter among leaf litter in broadleaf woodland, especially under beech. Cornucopia in Greek mythology was the horn of plenty, providing endless supplies of food.

HORNBEAM *Carpinus betulus* (Betulaceae), a deciduous tree, found as native in southern and south-east England in both pure and mixed woodland on base-poor sandy or loamy clays, or clay-with-flints; coppiced plants are often the chief member of the shrub layer in oakwood. It is also extensively planted in woodland, on roadsides, in amenity areas and for hedging, both within and outside its native range. 'Hornbeam' comes from Old English *horn* (from the wood's hardness) + *beam* = 'tree'. The binomial is Latin for hornbeam and birch.

HORNET *Vespa crabro* (Vespidae), body length 25–35 mm, common and widespread in England and Wales, becoming more common in the north Midlands and into northern England since the 1970s, associated with ancient woodlands and buildings in rural (and increasingly urban) areas. Queens emerge from overwintering sites from early April. Workers emerge in June–July. Males and females mainly emerge during September. Prey includes other species of social wasp, honeybees, flies, butterflies, moths (hornets can forage in moonlight) and spiders. Exudates are collected from the damaged roots and branches of oak trees, and ash and lilac twigs are ring-barked to encourage sap flow. Honeydew is also taken. Nests, built from pulp collected from decayed wood, are usually in aerial situations, particularly inside hollow trees, but are also in attics and outhouses. Some nests are constructed at the base of tufts of grass and in cavities in ant nests. 'Hornet' comes from Old English *hyrnet*. *Vespa* is Latin for 'wasp', *crabro* for 'hornet'.

Asian (giant) hornet *V. mandarinia*, the world's largest hornet, with the queen's body length up to 3 cm and workers up to 2.5 cm, with a wingspan of 75 mm, and a stinger of 6 mm, feeding on large insects including honeybees (and honey). Common in Europe after being accidentally introduced to France in 2004 in a shipment of pottery from China, they were first noted in Jersey and Alderney in summer 2016, and recorded near Tetbury (Gloucestershire) in September, prompting DEFRA to set up a 5 km surveillance zone with the aim of finding and exterminating this invasive species, which it did in October when a nest was destroyed. In 2017 a nest found at Woolacombe, Devon, was destroyed, and in 2018 an adult was identified in Bury, Lancashire,

in a cauliflower grown in Boston, Lincolnshire, with surveillance initiated in both locales.

HORNTAIL WASP Sawflies in the family Siricidae. See wasp.

HORNWORT 1) Members of the non-vascular phylum Anthocerotophyta, family Anthocerotaceae, the common name referring to the elongated horn-like structure, which is the sporophyte. The thin, flattened, green plant body of a hornwort is the gametophyte. They are found in damp stubble and fallow fields, poached field corners, ditch sides and woodland tracks. 'Hornwort' comes from the spores being produced in slender horn-shaped structures. **Dotted hornwort** *Anthoceros punctatus* is common and widespread, more so in the west, **field hornwort** *A. agrestis*, also common and widespread, mainly in the east. Greek *anthos* + *ceros* = 'flower' + 'horn'.

2) Species of submerged rootless aquatic perennials in the genus *Ceratophyllum* (Ceratophyllaceae), Greek for 'horned leaf'.

Rigid hornwort *C. demersum*, scattered in England and Wales, rare and local in Scotland and Ireland, associated with base-rich, still or slowly flowing waters such as eutrophic ponds, canals and lakes, up to 1 m depth. Latin *demersum* = 'submerged'.

Soft hornwort *C. submersum*, in ponds and ditches in grazing (salt)marshes, local and mainly near the sea in England and south Wales, but extending into the Midlands, and very rare in Ireland (first recorded in Co. Down in 1989). *Submersum* is Latin for 'submerged'.

HORNWRACK *Flustra foliacea* (Flustridae), a bryozoan, widespread and common on rocky shores usually forming a clump 6–10 cm high in the shallow sublittoral. (Not to be confused with horned wrack.) *Flustra* may be from an Anglo-Saxon word for 'braiding'; *foliacea* is Latin for 'leafy'.

HORSE *Equus caballus* (Equidae), a single-toed ungulate that includes the domestic horse, present since the Middle Pleistocene, and feral ponies; see pony. The earliest remains found in Britain are from Pakefield, Suffolk (700,000 BCE) and Boxgrove, West Sussex (500,000 BCE). Konic horses (= 'little horse' in Polish) have been introduced from the Netherlands since 2006 to trample and poach damp areas to maintain this habitat for marsh plants and wildlife, for example at Stodmarsh and Ham Fen (Kent) and Minsmere (Suffolk). Height is generally 1.4–1.8 m at the withers, but can be taller or, often, smaller depending on breed. Adult mass is 380–1,000 kg. Gestation is 11–12 months. 'Horse' comes from Old English *hors*, itself of Germanic origin. The binomial comprises two Latin words for 'horse'.

HORSE-CHESTNUT *Aesculus hippocastanum* (Sapindaceae), a deciduous tree introduced from the Balkans in the early seventeenth century, common and widespread in lowland Britain, much planted in parkland, large gardens and estates, churchyards, urban streets and village greens, and also found in some woodlands. It sometimes self-sows in scrub, waste ground and rough grassland, and can regenerate in woodland, but is rarely fully naturalised. 'Horse' has the connotation of being 'coarse'. The genus name is Latin for this species; the specific name melds Greek *hippos* = 'horse' with Latin *castanum* = 'chestnut'.

Indian horse-chestnut *A. indica*, a deciduous tree introduced from the Himalayas in 1851, scattered, planted in woodland and parks and on roadsides, occasionally naturalised.

Red horse-chestnut *A. carnea*, a deciduous neophyte of garden origin, planted in parkland and large gardens, and as a street tree, over much of Britain. It is probably self-sown at sites in west Kent, Surrey and north Hampshire. *Carnea* is Latin for 'flesh-coloured'.

HORSE MUSSEL *Modiolus modiolus* (Mytilidae), shell size up to 22 cm, widespread from the low intertidal, sublittoral and across the shelf, often forming large beds especially where the current is strong. *Modiolus* is Latin for a 'small measure'.

Bean horse mussel *M. phaseolina*, shell size up to 2 cm, patchily widespread, attached by the byssus in crevices or among epifauna from the low intertidal to moderate shelf depths. *Phaseolina* is Latin for a 'small bean'.

Bearded horse mussel *M. barbatus*, shell size up to 6.5 cm, in the low intertidal and sublittoral in the English Channel, Celtic and Irish Seas and west Ireland. Hairs on the broad end of the shell give the English name. *Barbatus* is Latin for 'bearded'.

HORSE-NETTLE *Solanum carolinense* (Solanaceae), a rare casual North American neophyte perennial found in south-east England on waste ground from soyabean waste. *Solanum* is Latin for this family (nightshade).

HORSE-RADISH *Armoracia rusticana* (Brassicaceae), a long-lived herbaceous perennial archaeophyte, widespread and locally common in England and Wales, scattered elsewhere, persisting in old gardens and allotments and (being sterile, with seed-set unknown in Britain) spreading by root fragments to roadsides, waste ground, railways, sandy seashores and riverbanks. It was introduced before 1500, initially as a medicinal herb, but by 1650 it had replaced dittander as a vegetable cultivated for hot relishes. *Armoracia* is Greek for this plant, *rusticana* Latin for 'countryfied'.

HORSEFLY Members of the family Tabanidae in the genera *Tabanus* (Latin for 'horsefly'), *Atylotus* (Greek for without a phallus) and *Hybomitra* (Greek *hybos* + *mitra* = 'humpbacked' + 'headband' or 'mitre'). Adult females feed on mammalian blood, especially deer, cattle and horses. Males often feed on nectar.

Scarce species are: **black-legged horsefly** *H. micans* (with records from southern England, the Midlands, Cumbria and Wales), **broad-headed horsefly** *H. lurida* (mainly Highlands), **broadland horsefly** *H. muehlfeldi*, **Cheshire horsefly** *A. plebeius*, **downland horsefly** *T. glaucopis*, **four-lined horsefly** *A. rusticus*, **golden horsefly** *A. fulvus*, **Levels yellow-horned horsefly** *H. ciureai* (grazing marsh from Hampshire to Norfolk), **narrow-winged horsefly** *T. maculicornis*, **pale giant horsefly** *T. bovinus*, **plain-eyed brown horsefly** *T. miki*, **plain-eyed grey horsefly** *T. cordiger*, **saltmarsh horsefly** *A. latistriatus*, **scarce forest horsefly** *H. solstitialis*, **slender-horned horsefly** *H. montana* and **striped horsefly** *H. expollicata*.

Band-eyed brown horsefly *T. bromius*, 14–15 mm length, locally abundant in England and Wales, becoming less so further north, favouring wet grassland, especially where cattle or horses are grazing. In Ancient Greece Bromius, from *bremein* = 'boisterous', was an alternative name for Bacchus.

Bright horsefly *H. distinguenda*, 15–18 mm length, widespread and scattered in Britain, less common in Scotland, in wet heath, bog, wet woodland edge and damp meadow. The flight period is June–August. Both sexes feed on nectar. Latin *distinguenda* = 'distinguished'.

Dark giant horsefly *T. sudeticus*, length up to 25 mm and at up to 1.5 g the heaviest fly in Europe, with a scattered distribution in Britain, more common in boggy habitats in northern and western parts, and in the New Forest, as well as in south and south-west Ireland, especially where there is grazing livestock, with peak female feeding activity in July–August. *Sudeticus* is Latin for 'sweating'.

Hairy-legged horsefly *H. bimaculata*, 13–16.5 mm length, locally common with scattered records in England, Wales and the Highlands, in woodland edge, heath woodland, fen and marsh. Larvae prey on invertebrates in wet soil and wood detritus. Females are active in May–August. *Bimaculata* is Latin for 'two-spotted'.

Large marsh horsefly *T. autumnalis*, 13–16 mm length, with a scattered distribution in England and Wales, probably extending its range, favouring wet grassland and marsh, especially where cattle are grazing, with peak female activity in June–July.

HORSEHAIR WORM See Nematomorpha.

HORSESHOE Red-spotted horseshoe *Protula tubularia* (Serpulidae), a polychaete bristleworm 3–5 mm long that creates calcareous tubes on hard substrates in the lower shore and

sublittoral on the south and west coasts of Britain, and at a few sites in Yorkshire and Ireland.

HORSESHOE WORM Member of the phylum Phoronida, marine animals usually 2 cm long and 1.5 mm wide that filter-feed with a horseshoe-shaped lophophore (crown of tentacles), building erect tubes of chitin to support and protect their bodies, usually living in the intertidal. There are five species in the genus *Phoronis* in British waters, *P. hippocrepia* and *P. muelleri* being common and widespread. Phoronis was the surname of Io who, in Greek mythology, changed into a heifer to wander the Earth before being restored to her original form. Greek *hippos* + *crepis* = 'horse' + 'shoe'.

HORSETAIL Members of the genus *Equisetum* (Equisetaceae) (Latin *equus* + *seta* = 'horse' + 'hair'), subdivision Calamophytes. 'Horsetail' is descriptive of the stems and branches.

Field horsetail *E. arvense*, common and widespread on grassland, marsh, dune slacks, wasteland, roadsides, parks and gardens on soils of moderate or high pH and nutrient status. *Arvensis* comes from Latin for 'field'.

Great horsetail *E. telmateia*, common and widespread (less so in Scotland) in moist soils in (semi-)shaded woodland or woodland edge, and hedgerows. *Telmateia* is Greek for 'of marshland'.

Mackay's horsetail *E. trachyodon*, on sandy lakesides and damp dunes, scattered in around 25 sites in Ireland, five in Scotland, in Anglesey and three sites in Cheshire. James Mackay (1775–1862) was an English botanist who lived in Ireland. *Trachyodon* is Greek for 'rough-toothed'.

Marsh horsetail *E. palustre*, widespread and very common at pond and stream sides, and in ditches in pasture and marsh. *Palustre* is from Latin for 'marsh'.

Moore's horsetail *E. moorei*, on dunes in Co. Wexford and Co. Wicklow. David Moore (1808–79) was a Scottish botanist.

Rough horsetail *E. hyemale*, found by streams and in ditches, damp woodland and moorland flushes, scattered but mostly found in the north, though decreasing in range. *Hyemale* is from Latin for 'winter'.

Shady horsetail *E. pratense*, on often shaded stream banks and grassland flushes, mostly in upland areas in Scotland, northern England and north Ireland. *Pratense* is from Latin for 'meadow'.

Shore horsetail *E. litorale*, local and scattered in ditches, dune slacks and river- and lakesides. *Litorale* is Latin for 'shore'.

Variegated horsetail *E. variegatum*, scattered in northern England, coastal Wales, Scotland and Ireland on dune slacks, river and lake sides, and mountain slopes.

Water horsetail *E. fluviatile*, common and widespread, semi-aquatic, in still or very slow-moving water up to 1.5 m deep, often in monospecific stands. Latin *fluvius* = 'river'.

Wood horsetail *E. sylvaticum*, common and widespread, especially in the north and west in damp woods, by streams and in marshy grassland, requiring shade, shelter and damp ground in lowland Britain, and in upland moorland. *Sylvaticum* is Latin for 'of the forest'.

HOTLIPS *Octospora humosa* (Pezizales, Pyronemataceae), a scattered, uncommon discomycete fungus associated with moss and grass on heathland and acid grassland. In 2010, *The Guardian* launched a competition for the public to suggest common names for previously unnamed species: the species looks a bit like lips and bright orange in colour. The binomial is Greek for 'eight seeds', and Latin for 'full of earth'.

HOTTENTOT-FIG *Carpobrotus edulis* (Aizoaceae), a naturalised creeping mat-forming succulent perennial from South Africa, found on sea-cliffs, rock and sand in the Channel and Scilly Isles, and locally from Cornwall to north Wales and Suffolk, and occasionally in Lancashire, Isle of Man, and south and east Ireland, where it can form monospecific stands, out-competing the native flora. Greek *carpos* + *broteos* = 'fruit' + 'edible', and Latin *edulis* = 'edible', since the leaves and fruit can be eaten.

HOUND'S-TONGUE *Cynoglossum officinale* (Boraginaceae), a herbaceous biennial of mainly lowland regions, declining, on disturbed ground, favouring dry, often base-rich soils such as on dunes, shingle, grassland, woodland margins, field edges and gravelly waste. Being unpalatable to grazing animals (toxic components are pyrrolizidine alkaloids) it is often found close to rabbit burrows. Its English and genus names (Greek *cyon* + *glossa* = 'dog' + 'tongue') may come from the texture of the leaves or the unpleasant smell. *Officinale* is Latin for having pharmacological use.

Green hound's-tongue *C. germanicum*, a herbaceous biennial or short-lived perennial found (though declining) in southern England, mainly on calcareous, freely draining, loamy soils in clearings in or margins of deciduous woodland, sometimes also in hedge banks. Latin *germanicum* = 'German'.

HOUSE-LEEK *Sempervivum tectorum* (Crassulaceae), an evergreen perennial grown in gardens since at least 1200, widespread and more or less naturalised on tiled roofs and porches (planted as supposed protection against fire, lightning and thunderbolts), old walls and in churchyards. It is also occasionally found on stabilised dunes. The binomial is Latin for 'everlasting' and 'relating to the roof'.

HOUSE-MOTH Members of the family Oecophoridae.
Brown house-moth *Hofmannophila pseudospretella*, wingspan 15–26 mm, an Asian species introduced in the mid-nineteenth century, widespread and common outdoors in woodland, hedgerows and grassland, and indoors in houses and barns. Adults can be seen at any time in the year, but especially in summer. Larvae feed on detritus and refuse. Ottmar Hofmann (1835–1900) was a German entomologist. *Pseudospretella* adds Greek *pseudos* = 'false' to a clothes moth specific name *spretella*.

White-shouldered house-moth *Endrosis sarcitrella*, wingspan 15–21 mm, widespread and common in houses, and around sheds, etc. Adults fly at any time of year. Larvae live on dried plant and animal debris and grain. Greek *endrosos* = 'bedewed', and *sarcos* = 'flesh'.

HOVERFLY Members of the family Syrphidae.
Bog hoverfly *Sericomyia silentis*, length 16 mm, wingspan 19–28 mm, common and widespread, though less recorded in the south and east Midlands, found in May–November with a preference for wet heathland and upland moor. Larvae are especially found in peat ditches and pools. Latin and Greek *serico* + *myga* = 'silk' + 'fly'. Latin *silentis* = 'silent'.

Chequered hoverfly *Melanostoma scalare*, length 6–9 mm, common and widespread from April to November in flower-rich habitats, including gardens. Greek *melano* + *stoma* = 'black' + 'opening', and Latin *scala* = 'ladder'.

Long hoverfly *Sphaerophoria scripta*, length 7–12 mm, wingspan 5–7 mm, widespread and common (less so in Scotland) evident from April to November on flowers in grassland and wasteland, adults feeding on nectar, larvae on ground-dwelling aphids. Latin and Greek *sphaero* + *phoria* = 'round' + 'bring', and Latin *scripta* = 'written' or 'marked'.

Marmalade hoverfly *Episyrphus balteatus*, length 9–12 mm, wingspan 14–18 mm, probably the commonest and most widespread hoverfly in Britain, mainly feeding on nectar but also pollen in gardens, parks, hedgerows and sunny parts in woodland throughout the year (hibernating as adults but emerging during warm days), peaking in August, and with coastal numbers sometimes boosted by migrants. Larvae feed on aphids. *Balteatus* is Latin for 'girdled'.

Migrant hoverfly *Eupeodes corollae*, length 6–11 mm, wingspan 10–12 mm, common and widespread, coastal numbers in particular often boosted by migration, found in March–November on

flowers in gardens, fields, verges and hedgerows. Larvae feed on aphids. *Corolla* is Latin for 'crown'.

Pellucid hoverfly *Volucella pellucens*, length 15–16 mm, wingspan 20–30 mm, common and widespread, found in May–October in hedgerows and woodland taking nectar from, especially, bramble and umbellifers. Larvae live in the nests of social wasps and bumblebees, eating waste products and host larvae. *Vollucella* is derived from Latin *volucris* = 'winged animal', and *pelluscens* = 'transparent'.

Pied hoverfly *Scaeva pyrastri*, length 11–15 mm, wingspan 19–25 mm, widespread and common (less so in Scotland) evident in May–November, numbers boosted by migration but fluctuating year to year, in gardens, wasteland and meadows, adults feeding on nectar, larvae on aphids. *Scaeva* is Latin for 'unfortunate', *pyrastri* probably relating to pear.

Thick-legged hoverfly *Syritta pipiens*, common and widespread, though less so in northern Scotland and Ireland, found in April–November in gardens, meadows and hedgerows. Larvae live and feed in compost, manure and other rotting organic matter. *Pipiens* comes from Latin *pipio* = 'chirp'.

Vagrant hoverfly Alternative name for migrant hoverfly.
Wannabee hoverfly *Pocota personata*, wingspan 28 mm, arguably the finest hoverfly bumblebee mimic, scarce but widespread in England, adults evident in April–June associated with very old trees in woodland and semi-natural parkland, especially beech, but also around old sycamore and ash, and they also favour hawthorn flowers. Larvae develop in tree cavities filled with water or wet detritus, often high up. In 2010, *The Guardian* launched a competition for the public to suggest common names for previously unnamed species. In 2012, this name was chosen. Greek *pocos* = 'fleece', and Latin *personata* = 'masked'.

White-footed hoverfly *Platycheirus albimanus*, common and widespread in March–November in hedgerows and gardens, adults feeding on nectar, larvae on aphids. Greek *platys* + *cheir* = 'broad' + 'short', and Latin *albus* + *manus* = 'white' + 'hand'.

HUCKLEBERRY Garden huckleberry *Solanum scabrum* (Solanaceae), a scrambling annual neophyte, possibly African, a rare casual on tips, sewage farms and (spread with sludge fertiliser) fields at a few sites in eastern England. 'Huckleberry' is possibly a variant of 'hurtleberry' (c.f. whortleberry). *Solanum* is Latin for this family (nightshade), *scabrum* Latin for 'rough'.

HUMP-BACKED FLY Alternative name for scuttle fly (see Phoridae).

HUMP-MOSS Species in the genera *Amblyodon* (Greek for 'blunt tooth') and *Meesia* (both Meesiaceae), and *Plagiobryum* (Greek for 'slanting moss') (Bryaceae).

Alpine hump-moss *P. demissum*, local in the Highlands in damp crevices on base-rich or calcareous rock on mountains >750 m. *Demissum* is Latin for 'low-lying'.

Broad-nerved hump-moss *M. uliginosa*, found on base-rich montane flushes in Cumbria and the Highlands, and more rarely in base-rich coastal dune slacks in north-west England and parts of Scotland. *Uliginosa* is Latin for 'growing swamp'.

Short-tooth hump-moss *A. dealbatus*, scattered, in base-rich montane flushes, and short, damp turf in coastal dune slacks. *Dealbatus* is Latin for 'whitish'.

Zierian hump-moss *P. zieri*, in montane Britain, rare in Ireland, in crevices of damp, base-rich, shaded rock outcrops and cliffs, and in gullies. J. Zier (d.1796) was a Polish botanist.

HUMPBACK (FUNGUS) *Cantharellula umbonata* (Agaricales, Tricholomataceae), inedible, rare, with a scattered distribution in Britain, commoner in Scotland, on soil under unimproved grassland or coniferous forest. *Cantharos* is Greek for a small drinking cup; *umbona* is Greek for 'nipple'.

HUNTSMAN Green huntsman Alternative name for green spider.

HUTCHINSIA *Hornungia petraea* (Brassicaceae), a

winter-annual very local in the Channel Islands, south-west and north England, and Wales, on open habitats on calcareous soils and rocks subject to summer drought, especially on Carboniferous limestone and on fixed, open dunes. It also occurs on garden walls and in chalk-pits. Ellen Hutchins (1785–1815) was an Irish botanist, Ernst Hornung (1795–1862) a German naturalist. *Petraea* is Latin for 'growing among rocks'.

HYACINTH *Hyacinthus orientalis* (Asparagaceae), a bulbous herbaceous south-west Asian neophyte perennial, scattered in Britain and the Isle of Man, mostly in southern England, as persistent garden throw-outs. 'Hyacinth' was a Spartan prince in Greek mythology. Latin *orientalis* = 'eastern'.

Starry hyacinth, winter hyacinth Alternative names for autumn squill.

HYACINTH-OF-PERU Alternative name for Portuguese squill.

HYBOTIDAE Family name for dance flies (Bracycera), with 180 species in 20 genera, the most numerous being *Platypalpus*. Adults are generally 2.5–5.5 mm long. They show fast erratic movement in flight, and hunt other small insects on tree bark, moving rapidly in complex patterns, giving them their common name. Larvae live in soil and plant litter, sometimes water, and also prey on insects. *Hybo* is a Greek prefix for 'humped'.

HYDRACARINA See Hydrachnellae.

HYDRACHNELLAE Or Hydracarina, a class or 'informal' group of superfamilies comprising mainly freshwater 'water mites', with around 300 British species. *Hydra* comes from the Greek prefix *hydor* = 'water'.

HYDRAENIDAE Family of very small beetles (1–3 mm long), represented by 31 species in 5 genera. Although all species crawl rather than swim they are found in and near water bodies. Many of the aquatic species occur in silt and vegetation at the edges of water bodies, although several occur in fast-flowing water. See moss beetle.

HYDRANGEA *Hydrangea macrophylla* and other species (Hydrangeaceae), a deciduous Japanese neophyte shrub cultivated by 1788, widespread but scattered as a sometimes persistent garden escape in woodland, hedgerows, stream banks and waste ground, and as a relic of cultivation. Greek *hydor* + *aggos* = 'water' + 'jar', referring to the cup-shaped fruit. *Macrophylla* is Greek for 'large leaf'.

HYDROBATIDAE Family name of petrels (Greek *hydor* + *bates* = 'water' + 'tread').

HYDROBIIDAE Family name of the fresh- and brackish water spire snails (Greek *hydor* = 'water').

HYDROCHARITACEAE Family name (Greek *hydor* + *charis* = 'water' + 'delight') for frogbit, including naiad, tapegrass, water-soldier and waterweed.

HYDROCHIDAE Family of beetles represented by one British genus, *Hydrochus*, containing seven species of elongate crawling aquatic beetles sometimes referred to as water scavenger beetles (q.v.). They are generally found in plant litter in well-vegetated ponds and bogs. Judging from their mouthparts the larvae are predators, living on wet ground and seizing small animals from the water. Adults survive winter in underwater debris, feeding on algae. Generally sluggish, they feign death when disturbed.

HYDROCOTYLACEAE Family name (Greek *hydor* + *cotyle* = 'water' + 'cup-shaped') of pennywort.

HYDROID Animals in the class Hydrozoa, with a small free-swimming ciliated planula larva about 1 mm long, which settles and metamorphoses into a sessile (attached), usually colonial polyp stage, which in turn frees a gamete-producing male or female medusa. Many British examples are sublittoral, and are in the orders Anthoathecata and Leptothecata.

Club-headed hydroid *Clava multicornis* (Hydractiniidae), colonial, on the middle to lower intertidal, usually as an epiphyte on fucoid algae, but also on rocks or shells and in rock pools, widespread but scattered, if absent from the east coast of Britain. *Clava* is Latin for 'club', *multus* + *cornus* = 'many' + 'horn'.

Double-toothed hydroid *Obelia bidentata* (Camanulariidae), colonial, stems reaching 15 cm, on substrates such as wood, shells, wrecks and on sandy bottoms, sometimes algae, sublittoral, rarely intertidal in pools. Greek *obelias* = 'round cake', and Latin *bidentata* = 'two-toothed'.

Oaten pipes hydroid *Tubularia indivisa* (Tubulariidae), widespread, solitary, tentacled, 10–15 cm high on hard substrates in the intertidal or the shallow sublittoral. Oaten pipes are a type of reed pipe made from dried oat stalks, echoing the long yellow 'stem' of this hydroid. The binomial is Latin for 'undivided tube'.

Sea-thread hydroid *O. dichotoma*, widespread, colonial, up to 35 cm tall usually found on floats, pilings, rocks, shells and other solid objects. Greek *diche* + *tomos* = 'split in two' + 'a cut'.

Snail fur hydroid *Hydractinia echinata* (Hydractiniidae), forming a horny mat 3 mm thick on gastropod shells occupied by hermit crabs in the intertidal and subtidal, widespread though less common on the east coast. Greek *hydor* + *actis* = 'water' + 'ray', and Latin *echinata* = 'like an urchin'.

HYDROPHILIDAE Family (Greek *hydor* + *philos* = 'water' + 'loving') of water scavenger beetles, with 70 species in 18 genera. Members of the subfamily Hydrophilinae are aquatic, but the subfamily Sphaeridiinae are terrestrial. Most *Cercyon* species and all *Sphaeridium* are associated with dung. Other *Cercyon* species occur in wet debris such as rotting seaweed. The aquatic species are generally poor swimmers, moving more by walking underwater than diving or swimming.

HYDROPSYCHIDAE Family (Greek *hydor* = 'water' + Psyche, a Greek nymph) of net-spinning caddis flies. Found in flowing waters, larvae construct 'retreats' fixed to rocks or roots using a silk thread, equipped with trapping nets, positioned perpendicular to the current, in order to capture drifting organic particles and small animals. The retreats are sealed off to allow pupation.

HYDROPTILIDAE Family (Greek *hydor* + *ptilon* = 'water' + 'wing') of micro-caddis flies, larvae rarely >5 mm long, these not building a protective case until the final instar when they construct a typically purse-shaped case, either portable or stuck to the substrate, in which they finish growth and pupate.

HYDROZOA Class in the phylum Cnidaria that includes siphonophores, sea firs and sea beard (Greek *hydor* + *zoon* = 'water' + 'animal'). The eight orders found in British waters include Anthoathecata and Leptothecata.

HYGROBIIDAE Family name (Greek *hygros* + *bios* = 'wet' or 'moist' + 'life') for screech beetle, in the suborder Adephaga.

HYGROMIIDAE Family name for hairy snails (Greek *hygros* = 'wet' or 'moist').

HYMENOPTERA From Greek *hymen* + *pteron* = 'membrane' + 'wing', the order incorporating bees, wasps, ichneumonids, ants and sawflies, with 60 families in the British list.

HYPERICACEAE Family name for St John's-worts, tutsans and rose-of-Sharon (Greek *hyper* + *eicon* = 'above' + 'picture', these plants traditionally hung above pictures, the scent warding off evil spirits).

HYSSOP (PLANT) *Hyssopus officinalis* (Lamiaceae), an evergreen south European neophyte shrub cultivated as a culinary herb by at least 1200, a naturalised garden escape in limestone quarries and on old walls at sites in Dorset, Gloucestershire and Berkshire, and as a scarce casual on waste ground in a few other places in England. 'Hyssop' comes from Old English *ysope*, from Greek *hyssopos* and Hebrew *esob*, though the plant in classical times was probably marjoram. *Officinalis* is Latin for having pharmacological properties.

HYSSOP (MOSS) Hasselquist's hyssop *Entosthodon fascicularis* (Funariaceae), widespread and scattered, mostly found in England and Wales on arable fields and other recently disturbed soil, occasionally on thin soil overlying limestone. Fredrik Hasselquist (1722–52) was a Swedish traveller and naturalist who considered this moss to be the hyssop alluded to by Solomon at Jerusalem. Greek *entos* + *odon* = 'within' + 'tooth', a reference to the position of 'teeth' inside the capsule, and Latin *fascis* = 'bundle'.

I

IBIS Glossy ibis *Plegadis falcinellus* (Threskiornithidae), wingspan 88 cm, length 60 cm, weight 630 g, previously a scarce visitor, but in 2014 a pair unsuccessfully attempted to breed at Frampton Marsh RSPB reserve, Lincolnshire. 'Ibis' derives from Latin, in turn Greek (and possibly Egyptian *hab*) for this bird. Greek *plegados* and Latin *falcis* both mean 'sickle', a reference to the bill shape.

ICELAND MOSS *Cetraria islandica* (Parmeliaceae), a foliose lichen whose leaf-like habit gives it the appearance of a moss, locally common and widespread on rock in upland regions in Dartmoor, north Wales, the northern half of England, Scotland and western Ireland. Latin *cetra* = 'leather shield', and Latin *islandica* = 'Icelandic'.

ICELAND-PURSLANE *Koenigia islandica* (Polygonaceae), an annual of bare, damp, basaltic gravel pans and screes, recorded only from two locations, at altitudes of up to 715 m (near the summit of Storr, Skye) and at 385–523 m on Mull. 'Purslane' dates from late Middle English, from Old French *porcelaine*, probably from Latin *porcil(l)aca*, variant of *portulaca*, Latin for 'purslane'. Johann Koenig (1728–85) was a Polish/Latvian-born botanist.

ICHNEUMONIDAE Family name (Greek *ichneumon* = 'tracker') for parasitoid ichneumon wasps that lay their eggs in the larvae or pupae of other Hymenoptera, Coleoptera, Lepidoptera and other insects.

IDE Alternative name for orfe, from *id*, Swedish for this fish species, originally referring to its bright colour.

INCURVARIIDAE Family name for some of the bright moths. Latin *incurvatus* = 'curved' refers to curved appendages on the mouth. Larvae mine leaves, create a portable protective case, then drop to the ground to feed on leaf litter.

INDIAN-RHUBARB *Darmera peltata* (Saxifragaceae), a rhizomatous perennial herb introduced from North America in 1873, with a scattered distribution, found in damp habitats as a garden escape. Karl Darmer (1843–1918) was a German horticulturalist. Latin *peltata* = 'round shield', referring to the leaf shape.

INFECTIOUS BOVINE KERATOCONJUNCTIVITIS See New Forest eye.

INFUSORIA(N) An outdated collective term for minute freshwater organisms such as ciliates, euglenoids, protozoa, unicellular algae and small invertebrates, from Latin *infusorium* = 'pouring vessel'.

INKCAP Fungi in the genera *Coprinus* (Greek *copros* = 'dung'), *Coprinellus* (i.e. similar to *Coprinus*), *Coprinopsis* (*opsis* = 'appearance' of *Coprinus*) and *Parasola* (Greek for 'parasol') (Agaricales). The gills beneath the cap are initially white, then pink, then turn black, secreting a black spore-filled liquid, prompting the English name.

Widespread but infrequent or only locally common species include **bonfire inkcap** *Coprinopsis jonesii* (on burnt soil), **hare's-foot inkcap** *Coprinopsis lagopus* (on humus-rich soil and leaf litter in woodland, and on woodchip mulch; the young fruiting body is densely covered in 'fur', resembling a hare's paw, hence its common name), **humpback inkcap** *Coprinopsis acuminata* (on herbivore dung, soil and decomposing wood), and **snowy inkcap** *Coprinopsis nivea* (on old cow or horse dung and manured straw).

Scarce species include **distinguished inkcap** *Coprinus alopecius*, **dune inkcap** *Coprinopsis ammophilae*, **firerug inkcap** *Coprinellus domesticus*, **magpie inkcap** *Coprinopsis picacea*, **midden inkcap** *Coprinus sterquilinus* (associated with weathered dung, especially that of horse and rabbit), **pied inkcap** *Coprinopsis stangliana*, and **sawdust inkcap** *Coprinopsis scobicola*.

Common inkcap *Coprinopsis atramentarius*, common and widespread, edible (with caution, since there is an adverse reaction including palpitations and nausea when taken with or a day or two before or after alcohol), found from spring to early winter in fields, waste ground and gardens, near stumps of broadleaf trees or from buried wood, less frequently near the base of living trees, and able to push through tarmac and paving because of the strength of the vertically aligned hyphae in the stems. *Atramentum* is Latin for 'black ink'.

Fairy inkcap *Coprinellus disseminatus* (Psathyrellaceae), widespread and common, edible but tasteless, found from spring to early winter on rotting stumps of broadleaf trees, spreading onto adjacent soil. Latin *disseminatus* = 'disseminated'.

Glistening (or **mica**) **inkcap** *Coprinellus micaceus*, edible, widespread and common, appearing from early summer to winter on rotting stumps of broadleaf trees. Latin *mica* = 'grain', referring to the small granules (veil fragments) that glisten like mica flecks on the surface of the immature cap.

Pleated inkcap *Parasola plicatilis* (Psathyrellaceae), common and widespread, inedible, saprobic, found from May to November in grassy habitats and on woodchip. *Plicatus* is Latin for 'folded'.

Shaggy inkcap *Coprinus comatus* (Agaricaceae), widespread and common, edible and tasty, found in late summer and autumn on soil in short grassy habitats. Also known as lawyer's wig. *Comatus* is Latin for 'hairy', from the shaggy scales on the cap.

INOCYBACEAE Family name for fibrecap and oysterling fungi (Greek *inos* + *cybe* = 'fibre' + 'head').

INSECT(A) Class of arthropod invertebrates with a chitinous exoskeleton, six legs (or secondarily legless), a segmented body (head, thorax, abdomen), compound eyes, a pair of antennae, and two pairs of wings, except Diptera (one pair), or secondarily wingless (apterous). 'Insect' derives from Latin *in* + *secare* = 'into' + 'to cut'.

INSECTIVORA The now generally disused order name for the so-called 'waste basket' taxon that included shrews, hedgehog and mole, now placed in the Eulipotyphla.

IRIDACEAE From Greek *iris* = 'rainbow', family name of iris and related species.

IRIS Rhizomatous herbaceous perennials in the genera *Iris* (Greek for 'rainbow') and *Hermodactylus* (from the Greek god Hermes + *dactylos* = 'finger', referring to the tubers), both in the family Iridaceae.

Neophyte species, generally garden throw-outs, casual if occasionally naturalised include **Algerian iris** *I. unguicularis* (Mediterranean, found in west Cornwall, south Somerset and the Isle of Man), **beaked iris** *I. ensata* (east Asian, persisting in reed swamp in Surrey and west Kent), **blue iris** *I. spuria* (European, in grassy habitats), **English iris** *I. latifolia* (despite its common name a Pyrenean species mostly recorded from grassy habitats in Shetland), **purple iris** *I. versicolor* (North American, introduced in 1732, naturalised in reed swamp and by ponds and streams), **Siberian iris** *I. sibirica* (central European, cultivated in Tudor times, found in damp rough grassland), **smooth-leaved iris** *I. laevigata* (east Asian, naturalised in ponds and ditches in the New Forest), **snake's-head iris** *H. tuberosus* (Mediterranean, naturalised in hedgerow banks and grassy habitats), and **Turkish iris** *I. orientalis* (east Mediterranean, in scrub and rough ground).

Bearded iris *I. germanica*, an archaeophyte of garden origin, scattered in Britain, mainly in England, with records also from Co. Waterford, in waste ground and former gardens, and on roadsides and railway banks, generally spreading vegetatively. *Germanica* is Latin for 'German'.

Stinking iris *I. foetidissima*, widespread and locally common in England and Wales, scattered elsewhere, and introduced in Ireland, highly tolerant of drought and shade, in hedge banks and woodland, and on sheltered sea-cliffs, mostly in scrub on calcareous soils. *Foetidissima* is Latin for 'stinking'.

Yellow iris *I. pseudacorus*, common and widespread, generally in lowland regions, in wet meadow, damp woodland, fen, margins of lakes, ponds and streams, wet dune slacks, and in northern and western Britain also on shingle, upper saltmarsh and raised beaches. This is the county flower of Wigtownshire as voted by the public in a 2002 survey by Plantlife. *Pseudacorus* is Greek for 'false sweet-flag'.

IRISH MOSS Alternative name for carrageen.

False Irish moss *Mastocarpus stellatus* (Phyllophoraceae), a red seaweed up to 17 cm long, widespread but especially common on western coastlines, on rocky shores, in exposed areas growing among barnacles and mussels, often in large continuous mats, and on less exposed shores under fucoids. It is mainly associated with the lower shore and rock pools, but is also found in the shallow sublittoral and even deeper waters. In Ireland it is collected more or less interchangeably with carrageen for culinary and pharmaceutical purposes. Greek *mastos* + *carpos* = 'breast' + 'fruit', and Latin *stellatus* = 'starry'.

ISOPOD(A) Order of small crustaceans (class Malacostraca) with flattened bodies, for example woodlice and sea skaters. Greek *isos* + *podi* = 'same' + 'foot', referring to the single thoracic leg. There are 62 native and introduced species in 14 families, some having only been found in greenhouse environments such as Kew Gardens and the Eden Project. Aquatic species mostly live on the seabed or bottom of freshwater bodies, but a few can swim for a short distance. Terrestrial species are generally found in cool, moist places. Some species can roll into a ball to conserve moisture or as a defence mechanism.

ITCH **Sweet itch** See pruritus.

IVY (COMMON IVY) *Hedera helix* (Araliaceae), a common and widespread evergreen woody climber characteristic of woodland, scrub and hedgerows, and also frequent on trees. Walls and rock outcrops. It often carpets the ground in secondary woodland. It prefers basic to moderately acidic soils. *Hedera* is Latin for 'ivy', *helix* Greek for 'ivy', and also 'winding around'.

Atlantic ivy *H. hibernica*, an evergreen perennial climber with a scattered distribution, commoner to the west, in woodland, scrub and hedgerows, and on walls, rock outcrops and cliffs. It can also carpet the ground in woodland. It avoids only very acidic soils. *Hibernica* is Latin for 'Irish'.

Persian ivy *H. colchica*, an evergreen neophyte perennial climber from the Caucasus cultivated since 1851, scattered in Britain, particularly England, as a garden escape in woodland, hedges and scrub and on roadsides, railway banks, walls and waste ground. *Colchica* is Latin for 'from Colchis', a Black Sea region of Turkey.

IXODIDA Or Metastigmata, the order name of ticks, with three families in Britain: Amblyommidae and Argasidae (both being soft ticks) and Ixodidae (hard ticks). Greek *ixos* = 'mistletoe' or 'birdlime' is used here in the sense of something sticking, as ticks do on their host.

IXODIDAE Family name of 15 British ticks in the genus *Ixodes*.

J

JACK 1) A fungus: see slippery jack. 2) A small northern pike (under 8 lb).

JACK BY THE HEDGE Alternative name for garlic mustard.

JACK O'LANTERN *Omphalotus illudens* (Agaricales, Omphalotaceae), a fungus found at a few sites in south and south-east England, Near Threatened in the RDL (2006), on dead wood and woody debris in mostly oak woodland. The gills are bioluminescent in low light, giving this species its common name. It is poisonous (containing illudin) but not lethal to humans, ingestion leading to vomiting, diarrhoea and cramping. Greek *omphalos* = 'navel', from the central depression in the cap, and Latin *illudens* = 'deceiving'.

JACKDAW *Corvus monedula* (Corvidae), wingspan 70 cm, length 34 cm, weight 220 g, common and widespread in woodland, grassland (including clifftop) and gardens, with 1.3 million pairs estimated in Britain in 2009, and other data suggesting an increase in numbers by 148% between 1970 and 2015. Roosting of these social birds is often communal. Nests are in cavities, with clutches of 4–5, incubation taking 20 days, and the altricial young fledging in 32–3 days. Diet is of insects, young birds and eggs, fruit and seeds, as well as material found by scavenging. The early English name for this bird was 'daw'; 'jackdaw' appears in the sixteenth century, possibly a compound of Jack, used in animal names to signify a small form. *Corvus* is Latin for 'raven'; the specific name comes from Latin *moneta* = 'money', reflecting the bird's supposed fondness for picking up coins.

JACOB'S-LADDER *Polemonium caeruleum* (Polemoniaceae), a perennial herb, largely restricted as a native to steep limestone scree, usually in partial shade, locally common in the Yorkshire Dales and the Peak District, but also on andesite debris and river cliffs in Northumberland. It is confined to sites where the soil remains moist. Introduced populations are more widespread, growing in hedgerows, on riverbanks and in other places near habitation. The English name dates from the mid-eighteenth century, coming from the resemblance to a ladder of the rows of slender leaves, and referencing Jacob's dream of a ladder reaching to heaven (Genesis 28:12). The binomial refers to Polemon, a second-century Greek philosopher, with *caeruleus* Latin for '(dark) blue'. This is the county flower of Derbyshire as voted by the public in a 2002 survey by Plantlife.

JAPAN WEED **Siphoned japan weed** *Dasysiphonia japonica* (Dasyaceae), a red seaweed with fronds 60 cm long, probably entering Britain and Ireland in the 1990s with imported oyster spat from Japan, now found along the west coast of Britain as far north as Orkney, and Co. Clare and Co. Galway in Ireland, on sand-covered rocks in the lower intertidal. Greek *dasysiphonia* = 'hairy (or dense) siphon'.

JAPANESE-LANTERN *Physalis alkekengi* (Solanaceae), a rhizomatous perennial European neophyte herb grown for its ornamental fruit by the mid-Tudor period, an occasional but persistent garden escape in England and Wales on railway banks and waste ground. *Physalis* is from Greek *physa* = 'bladder'; *alkekengi* is from Arabic *al-kakanj* = 'winter cherry'.

JAPWEED Alternative name for wireweed, an invasive brown seaweed.

JASMINE Scrambling deciduous shrubs in the genus *Jasminum* (Oleaceae), 'jasmine' used since the mid-sixteenth century, from French *jasmin* and obsolete French *jessemin*, in turn from Arabic *yasamin*, itself from Persian *yasamin*, with *jasminum* the Latin form.

Red jasmine *J. beesianum*, a Chinese neophyte introduced in

1906, naturalised on walls and rough ground in a few places in Britain. The specific name comes from Bee's Nursery, Chester.

Summer jasmine *J. officinale*, an Asian neophyte present by the mid-sixteenth century, scattered and naturalised, mainly in the Thames Valley, on walls, pavements, roadsides, banks and waste ground, and as a throw-out and relic of cultivation. *Officinale* is Latin for having pharmacological value.

Winter jasmine *J. nudiflorum*, a Chinese neophyte cultivated by 1844, a garden escape in woodland, scrub and hedgerows, and on walls, roadsides and waste ground. It also grows as a throw-out on roadsides and as a relic of cultivation. The winter-green stems bear flowers in winter and early spring. *Nudiflorum* is Latin for 'naked flowers', the plant flowering before leafing.

JAY *Garrulus glandarius* (Corvidae), wingspan 55 cm, length 34 cm, weight 170 g, common and widespread, if scarce in north Scotland and western Ireland, associated with deciduous woodland but increasingly seen in gardens, with 170,000 territories in Britain in 2009, and a 25% increase in numbers between 1995 and 2015, though the 2% increase over 2010–15 statistically indicates no change. Nests are of twigs in trees, the 4–6 eggs hatching after 16–19 days, fledging of the altricial young taking 21–3 days. Diet is of invertebrates, fruit and seed (especially acorns which are cached for overwinter retrieval – a single jay can store as many as 5,000 acorns), and will also take nestling birds and small mammals. Latin *garrulus* = 'chattering' or 'noisy', reflecting the jay's harsh vocalisation; *glandarius* = 'producing acorns' (*glans* = 'acorn').

JELLY (FUNGUS) Several inedible species in the orders Tremellales and Auriculariales.

Amber jelly *Exidia recisa* (Auriculariales, Auriculariaceae), widespread and fairly common, evident throughout the year but mainly in winter on dead and rotting logs and branches, mainly of willows. *Exidia* is Latin for 'staining', *recisa* for 'cut off', a reference to the truncated shape of the fruit body.

Birch jelly *E. repanda*, widespread, scattered and uncommon, evident throughout the year but mainly in winter on dead and rotting logs and branches of birch. *Repandus* is Latin for 'bent backwards' or 'turned up'.

Pine jelly *E. saccharina*, widespread, scattered and uncommon, most records coming from the Highlands, on dead and rotting logs and branches of pine. *Sacchar* is Greek for 'sugar'.

Toothed jelly See jelly tooth.

Willow jelly Alternative name for amber jelly.

JELLY EAR *Auricularia auricula-judae* (Auriculariales, Auriculariaceae), a widespread, common edible fungus, found throughout the year but mainly fruiting in late summer and autumn on dying branches and trunks of broadleaf trees, especially elder. *Auricula* is Latin for 'earlobe', *judae* a reference to Judas (not Jew, *pace* an alternative name of Jew's-ear).

JELLY FUNGUS, JELLY HEDGEHOG Alternative names for jelly tooth.

JELLY LICHEN Species in the genus *Collema* (Collemataceae), from Greek for 'that which is glued'. See river jelly lichen under lichen.

JELLY ROT *Phlebia tremellosa* (Polyporales, Meruliaceae), a common and widespread encrusting gelatinous fungus found in autumn and winter on dead wood of a number of deciduous trees. Greek *phlebos* = 'vein', and Latin *tremellosa* = 'trembling'.

JELLY SKIN **Frilly fruited jelly skin** *Leptogium burgessii* (Collemataceae), a jelly lichen, found on mossy, shaded, nutrient-rich trees and rocks, especially in old woodland in the oceanic west. Greek *leptos* = 'slender'. *Burgessii* honours the lichenologist J.M. Burgess.

JELLY SPOT Species of inedible fungi in the genus *Dacrymyces* (Dacrymycetales, Dacrymycetaceae) (Greek *dacry* + *myces* =

'tear' (weeping) + 'fungus'). **Orange jelly spot** *D. chrysospermus* and **pine jelly spot** *D. ovisporus* have only a few records each.

Common jelly spot *D. stillatus*, common and widespread, found throughout the year, especially during wet weather, though the often merging gelatinous fruiting bodies mainly appear in late summer and autumn on rotting wood, including structural timber such as fence-posts. *Stilla* is Latin for 'a drop'.

JELLY TONGUE Alternative name for jelly tooth fungus.

JELLY TOOTH *Pseudohydnum gelatinosum* (Auriculariales), a widespread but infrequent inedible gelatinous fungus, found in late summer and autumn on dead and rotting logs and stumps of coniferous trees, especially spruce. The binomial is Greek meaning 'similar to fungi of the genus *Hydnum*', and Latin for 'gelatinous'.

JELLY BABY 1) *Leotia lubrica* (Leotiales, Leotiaceae), a widespread but infrequently found inedible gelatinous fungus found in late summer and autumn on soil in damp woodland, often under other vegetation. Greek *leiotes* = 'smooth', and Latin *lubrica* = 'slippery'.

2) Fungi in the genus *Cudonia* (Rhytismatales, Cudoniaceae): **cinnamon jelly baby** *C. confusa* and **redleg jelly baby** *C. circinans*, both Highland rarities Near Threatened in the RDL (2006).

JELLY DISC Species of gelatinous fungi.

Beech jellydisc *Neobulgaria pura* (Heliotales, Heliotaceae), widespread but infrequent, inedible, found in summer and autumn on dead wood of beech and other hardwoods. *Neobulgaria* may be a reference to the leathery skin, *bulga* being Latin for a leather pouch for carrying wine. *Pura* is Latin for 'pure' or 'clean'.

Bog jellydisc *Sarcoleotia turficola* (Geoglossales, Geoglossaceae), with a few records from the Highlands, north England and Ireland, on bog and wet heath, on soil with sphagnum. Greek *sarca* + *leiotes* = 'flesh' + 'smooth'.

Purple jellydisc *Ascocoryne sarcoides* (Heliotaceae), small, very common and widespread, wood-rotting, found from late summer to early winter on dead tree-trunks and branches, especially of beech. Greek *ascos* + *coryne* = 'sac' + 'club', and *sarca* = 'flesh'.

Violet jellydisc *Ombrophila violacea* (Heliotaceae), inedible, widespread but infrequent in litter, moss and soil under deciduous trees. Greek *ombros* + *philos* = 'rain' + 'loving', and Latin *violacea* = 'violet'.

JELLYFISH Marine cnidarians in the class Scyphozoa (true jellyfish) and Staurozoa (stalked jellyfish). They have a gelatinous umbrella-shaped bell which pulsates to provide locomotion, and trailing tentacles that are used to capture prey or defend against predators by emitting toxins in a sting. 'Jellyfish' was first used in 1796.

Barrel jellyfish *Rhizostoma pulmo* (Rhizostomatidae), common in the north-east Atlantic and Irish Sea, numbers recently increasing with raised sea temperatures (e.g. blooms off the Cornwall, Devon and Dorset coast in 2014 and 2015), typically up to 40 cm in diameter, but exceptionally reaching 90 cm, 150 cm in length, and up to 35 kg in weight, making it the largest jellyfish in British waters. Eight thick arms end in 'paddles' covered in frilly, cauliflower-like tissue where they meet the body which are small, dense tentacles around hundreds of tiny mouths used to catch and eat plankton. The sting is weak. Greek *rhizo* + *stoma* = 'root' + 'mouth', and Latin *pulmo* = 'lung'.

Blue jellyfish *Cyanea lamarckii* (Cyaneidae), the bell growing up to 30 cm in diameter, recorded all around the coastal British Isles. Greek *cyaneos* = 'dark blue'. Jean-Baptiste de Monet, Chevalier de Lamarck (1744–1829) was a French naturalist.

Compass jellyfish *Chrysaora hysoscella* (Pelagiidae), with a bell growing up to 30 cm in diameter, found round all coasts, especially in the south-west and off Wales. Greek *chrysos* + *aoros* = 'gold' + 'ugly'.

Dustbin-lid jellyfish, frilly-mouthed jellyfish Alternative names for barrel jellyfish.

Kaleidoscope jellyfish *Haliclystus auricula* (Lucernariidae),

a stalked funnel-shaped jellyfish up to 2.5 cm high with eight arms radiating from the mouth, fixed to the substrate by a stalk that is the same length as the bell, found on algae and seagrass in the low intertidal and shallow sublittoral, widespread though absent from Yorkshire to Dorset. Its English name was the winning suggestion when, in 2010, *The Guardian* launched a competition for the public to suggest common names for previously unnamed species. Greek *hals* + *clysta* = 'sea' + 'pipe', and Latin *auricula* = 'earlobe'.

Lion's mane jellyfish *Cyanea capillata* (Cyaneidae), the largest known jellyfish species, with its bell sometimes measuring >2 m in diameter, and tentacles up to 37 m long, living for a year, found in the North Atlantic, extending to the Irish Sea, North Sea and occasionally the English Channel. Its sticky stinging tentacles, grouped into eight clusters, each cluster containing over 100 tentacles arranged in a series of rows, are used to capture and eat a variety of prey. Being stung causes pain (rarely death) in humans. They tend to be found near the surface, most often seen during late summer and autumn when they have grown to a large size and currents sweep them towards shore. *Capillata* is Latin for 'hairy'.

Moon jellyfish *Aurelia aurita* (Ulmaridae), common and widespread, with a smooth, flattened saucer-shaped bell with eight simple marginal lobes, usually growing to 25 cm in diameter but sometimes 40 cm. It is sporadic in appearance, forming massive local populations in some areas. Although a pelagic species it may be found washed up on the shore. It occurs up estuaries and in harbours, and is especially common in Scottish sea lochs. Latin *aurelia* = 'golden', *aurita* = 'ear-like'.

St John's jellyfish *Lucernariopsis cruxmelitensis* (Kishinouyeidae), a stalked jellyfish with a broad funnel-shaped bell up to 1.2 cm in diameter and 0.8 cm in height, found in southwest England and west Ireland on moderately exposed rocky shores in the low intertidal and shallow sublittoral zones attached to macroalgae. Its English name derives from white spots on the surface of the bell that form the shape of a Maltese cross, and was the winning suggestion when, in 2010, *The Guardian* launched a competition for the public to suggest common names for previously unnamed species. Latin *lucerna* = 'lamp' + Greek *opsis* = 'appearance', and Latin for 'Maltese cross'.

JEWEL BEETLE Member of the family Buprestidae.

Devil's-bit jewel beetle *Trachys troglodytes*, with a scattered distribution in (mainly southern) England and Wales, a leaf miner on devil's-bit scabious in summer, present on sphagnum in winter. Greek *trachys* = 'rough', and *trogle* + *dytes* = 'hole' + 'diver'.

Ground-ivy jewel beetle *T. scrobiculatus*, with a scattered distribution south of a line from the Severn estuary to the Humber estuary, a leaf miner on ground-ivy and henbane. *Scrobiculus* is Latin for 'small trench'.

Hawthorn jewel beetle *Agrilus sinuatus*, with a scattered distribution in much of southern and central England, a leaf miner of hawthorn. Greek *agrios* = 'living in the countryside', and Latin *sinuatus* = 'curved'.

Oak jewel beetle *A. biguttatus*, with a scattered distribution in much of England, previously thought of as a rare, vulnerable species but it appears to be increasingly common, and is causing concern because of links with acute oak decline. *Biguttatus* is Latin for 'two-spotted'.

JEW'S EAR Alternative name for jelly ear fungus.

JOHNSON-GRASS *Sorghum halepense* (Poaceae), a rhizomatous North African neophyte perennial cultivated by the late seventeenth century, scattered as a casual of waste ground in the Channel Isles, southern England, and south and west Wales, introduced via birdseed, wool shoddy and grain impurities. Italian *sorgo* = 'tall cereal grass', and Latin *halepense* = 'from Aleppo' (Syria).

JONQUIL *Narcissus jonquilla* (Alliaceae), a bulbous perennial herb occasionally found as a garden relic in south-east England. The English and specific names come from Spanish *junquillo*, referring to *Juncus* (rush), because the leaves are rush-like. *Narcissos* is Greek for daffodil and other narcissi, from *narce* = 'numbness' (i.e. for their supposed narcotic properties).

Campernelle jonquil *Narcissus* x *odorus*, a bulbous perennial herb grown in gardens and naturalised in woodland and hedge banks, and on roadsides and waste ground. It is also found as a relic of cultivation, especially in bulb fields in the Channel Islands. *Odorus* is Latin for 'odorous'.

JORUNNA White jorunna *Jorunna tomentosa* (Discodorididae), a sea slug 6 cm long, widespread but scattered on the lower shore and shallow sublittoral on rocky coasts, feeding on encrusting siliceous sponges. *Tomentosa* is Latin for 'densely hairy'.

JUGLANDACEAE Family name for walnuts and wingnuts (Latin *juglans* = 'walnut tree').

JUNCAGINACEAE Family name for arrowgrass (Latin *juncago* = 'rush-like').

JUNEBERRY *Amelanchier lamarckii* (Rosaceae), a widespread but scattered neophyte North American shrub or small tree introduced in 1746, found on acidic, usually sandy soils and naturalised in open woodland and scrub, and on dry heathland and roadsides. On the New Forest and Surrey heaths it was initially planted but is now spreading from bird-sown fruits. *Amelanchier* is from Old French *amelancier*. Jean-Baptiste de Monet, Chevalier de Lamarck (1744–1829) was a French naturalist.

JUNGERMANNIALES Order of liverworts with thin leaf-like flaps on either side of the stem, including the family Pleuroziaceae, named after the German botanist Ludwig Jungermann (1572–1653).

JUNIPER Common juniper *Juniperus communis communis* (Cupressaceae), a widespread but very local evergreen tree or shrub found in southern and western regions on limestone and acid soils. Subspecies *hemisphaerica* is confined to coastal cliffs in Cornwall. Subspecies *nana* is found on moorland and rocks in western and northern regions of the British Isles. The binomial is Latin for 'juniper' and 'common'; *nana* = 'dwarf'.

K

KALE Type of cabbage with crinkled leaves, and sea-kale.

KANGAROO-APPLE *Solanum laciniatum* (Solanaceae), a casual, occasionally naturalised Australian neophyte shrub found on rough ground and coastal sand, mainly in south-west and southern England. The binomial is Latin for 'nightshade', and 'torn' (divided into lobes).

KATERETIDAE Family name (Greek *cateres* = 'fitted out') of short-winged flower beetles (length 1.5–4.2 mm), with nine British species in three genera. *Kateretes* species are common and widespread in wetlands where they feed on sedge and rush flowers. *Brachypterus* (Greek for 'short-winged') species, including **nettle pollen beetle** *B. urticae*, are also widespread and found on nettles. *Brachypterolus linariae* and **antirrhinum beetle** *B. pulicarius* are found on toadflax flowers.

KED **Sheep ked** *Melophagus ovinus* (Hippoboscidae), 4–6 mm length, a wingless parasite that bites into sheep to feed on their blood, leading to irritation and possibly anaemia. In 2010 it was reported to have been largely eradicated in the UK, though persisting in the feral Soay sheep of St Kilda, Outer Hebrides. Greek *mele* + *phago* = 'probe' + 'to eat', and Latin *ovis* = 'sheep'. See also louse fly.

KELCH-GRASS *Schismus barbatus* (Poaceae), an annual Mediterranean neophyte, a rare casual in fields and waste ground, originating from wool shoddy. *Schisma* is Greek for 'a slitting', *barbatus* Latin for 'bearded' or 'barbed'.

KELLYCLAM **Suborbicular kellyclam** *Kellia suborbicularis* (Kelliidae), a filter-feeding bivalve, shell size up to 1 cm, in the intertidal and sublittoral around all coasts.

KELP Large brown seaweeds in the genera *Laminaria* and *Saccharina* (Laminariaceae), cuvie (q.v.) in particular often forming 'forests' in the sublittoral. See also oarweed.

Sugar kelp *Saccharina latissima*, common and widespread, up to 4 m long, occasionally found in rock pools, but more usually from the sublittoral fringe down to a depth of 30 m, preferring sheltered conditions. The English and genus names refer to a whitish, sweet-tasting powder which forms on the dried frond. Latin *latissima* = 'very broad'.

KELP FLY Species of fly in a number of families, mainly Anthomyiidae and Coelopidae, that feed on stranded, rotting seaweed in the intertidal zone. Specifically, *Coelopa frigida*, common and widespread on British shorelines. Adults (10 mm length) fly short distances when disturbed. Greek *coilos* = 'hollow', and Latin *frigida* = 'cold'.

KELT A salmon or sea trout after spawning and before returning (as some do) to the sea.

KERATOCONJUNCTIVITIS See New Forest eye.

KERRIA *Kerria japonica* (Rosaceae), a deciduous neophyte shrub from China and Japan, in gardens since 1804, widespread and locally common as a throw-out in woodland and hedgerows, and on riverbanks, roadsides and rubbish tips. William Kerr (d.1814) was an English plant collector.

KESTREL *Falco tinnunculus* (Falconidae), wingspan 76 cm, length 34 cm, male weight 190 g, females 220 g, common and widespread (though on the Amber List) in grassland, scrub, heathland, moorland, farmland and, increasingly, in towns, often seen hovering/hunting over motorway verges. There was an estimated 45,000 pairs in 2009, but numbers are generally declining, for example by 50% over 1970–2015, including >65% in Scotland since 1994. Nests are on ledges (including those on buildings) and in hollows (including nest boxes), with a clutch of 4–5 eggs, incubation taking 28–9 days, and the altricial young

fledging in 32–7 days. Diet is of small mammals and birds. **American kestrel** *F. sparverius* and **lesser kestrel** *F. naumanni* are rare accidentals. 'Kestrel' possibly reflects the bird's call, from the French *crécerelle* a diminutive of *crécelle* (a musical rattle and a bell used by lepers); similarly, *tinnunculus* comes from Latin *tinnio* = 'to ring'.

KETTLEWORT **Common kettlewort** *Blasia pusilla* (Blasiaceae), a common and widespread mainly but not exclusively upland thalloid liverwort, on damp usually non-calcareous substrates along ditches, river and roadside banks, old quarries, forestry tracks and arable fields. *Pusilla* is Latin for 'small'.

KIDNEY (MOTH) **Double kidney** *Ipimorpha retusa* (Noctuidae, Xyleninae), wingspan 26–32 mm, local in damp woodland, fen, marsh and along riverbanks throughout much of England and Wales. Adults fly at night in July–September. Larvae feed on willows and black poplar. Greek *ipos* + *morphe* = 'mouse trap' + 'shape', and Latin *retusa* = 'blunt', from the forewing shape.

KILLER WHALE Misnomer for orca (q.v.); it is a dolphin (family Delphinidae), not a whale but rather a whale killer.

KING ALFRED'S CAKES *Daldinia concentrica* (Xylariales, Xylariaceae), an inedible fungus, widespread in England and Wales, scattered in Scotland and Ireland, found throughout the year, sporulating in spring, encrusting dead wood, especially though not exclusively ash and beech. The common name comes from their burnt appearance. Also called cramp balls. *Concentrica*, Latin for 'concentric', describes how the spores are ejected to create a spore print extending outwards from the cap edge.

KINGCUP Alternative name for marsh-marigold.

KINGFISHER *Alcedo atthis* (Alcedinidae), wingspan 25 cm, length 16 cm, weight 40 g, a brightly coloured Amber-listed bird, widespread, especially in central and southern England, becoming less common further north but, following some declines last century, it is currently increasing its range in Scotland and Ireland. The British population was 3,800–6,400 pairs in 2009. There was a non-significant 19% decline in numbers between 1975 and 2015, a 17–21% decline between 1995 and 2015, and a 3% decline in 2010–15. Nests are in tunnels dug in vertical banks along streams and rivers. There can be up to three broods in a season. Clutch size is 5–7, incubation lasts 20–1 days, and the altricial young fledge in 23–6 days. Food is of small fish and large aquatic insects caught by plunge-diving from a perch or while hovering. The binomial comes from Latin *alcedo* = 'kingfisher' (itself from Greek *halcyon*) and Atthis, a beauty mentioned in the poetry of Sappho.

KITE Birds of prey in the genus *Milvus*, Latin for 'red kite', family Accipitridae. 'Kite' comes from Middle English *kite* or *kete*, meaning 'kite', 'bittern' or 'bird of prey'.

Black kite *M. migrans*, a scarce, usually spring visitor averaging seven records a year, most coming from East Anglia and Cornwall. Its only breeding record in Britain was in 2011 at a secret location in the Inverness area. *Migrans* is Latin for 'migrating'.

Red kite *M. milvus*, wingspan 185 cm, length 63 cm, male weight 1 kg, female weight 1.2 kg, once common scavengers in towns as well as rural habitats, but persecuted until, by the mid-twentieth century, there were only a very few birds in old oakwood in mid-Wales. With protective measures and, starting in 1989, reintroductions from Wales, Sweden, Spain and Germany into England and Scotland, numbers and range greatly increased. By 2004 there were 375 known occupied territories, with at least 200 pairs producing fledged young. In 2007, 30 birds were released in Co. Wicklow. In 2011 the RSPB reported that non-breeding birds were regularly seen in all parts of Britain, and recently had become regular visitors to Northern Ireland, and there were probably 1,800 breeding pairs in Britain (7% of the world population), about half being in Wales. A 2012 estimate by RBBP suggested numbers approaching 2,500 pairs. BBS data indicate an increase in numbers by >1,000% between 1995 and 2015, and BTO data suggest an 85% increase in 2010–15, and the species was Green-listed in 2015. Nests are built in trees, 2 eggs are usually laid, incubation takes 31–2 days, and fledging of the altricial young takes 50–60 days. While mainly feeding on carrion and worms, they may opportunistically take small mammals.

KITTEN (MOTH) Noctural moths in the genus *Furcula* (Latin for 'little fork', from the larva's two anal appendages) (Notodontidae, Cerurinae). Ear-like projections on the larval head prompts the name 'kitten'.

Alder kitten *F. bicuspis*, wingspan 30–5 mm, local in woodland, thickets, copses and gardens, from Cornwall northwards through Wales and the west Midlands to south Lancashire, and in south-east England, Norfolk and Co. Durham. The one record from Ireland is from Legatillida, Co. Fermanagh, in 2001. Adults fly in May–July. Larvae feed on leaves of birches and alder. Latin *bi* + *cuspis* = 'two' + 'point', from the two larval appendages.

Poplar kitten *F. bifida*, wingspan 35–48 mm, local in woodland, plantations and gardens throughout England and the Welsh Borders. Adults fly in May–August. Larvae feed on leaves of poplars and aspen, occasionally sallows. *Bifida* is Latin for 'cleft in two'.

Sallow kitten *F. furcula*, wingspan 27–35 mm, widespread and common in woodland rides and clearings, hedgerows, parks, moorland, heathland and gardens. In the southern half of its range there are two generations, flying in May–June then again in August. In the north it is single-brooded, flying in June–July. Larvae feed on leaves of willows, aspen and poplars.

KITTIWAKE Black-legged kittiwake *Rissa tridactyla* (Laridae), wingspan 108 cm, length 39 cm, weight 410 g, a migrant and resident breeder, with 370,000 pairs in Britain and 50,000 in Ireland in 2002. The decline in numbers by 62% over 1986–2014 (and by 13% in 2006–14) was because of a dearth of the main prey, sandeels, through overfishing and climate change, and it is Red-listed. After breeding from February to August, birds move into the Atlantic where they spend winter. Nests are on sea-cliff ledges, with the 2 eggs hatching after 25–32 days, and fledging of the semi-precocial chicks takes 33–54 days. The English name comes from the sound of its call. The genus name is from the Icelandic name *Rita*, and Latin *tridactyla* = 'three-fingered'.

KNAPWEED Herbaceous perennial species of *Centaurea* (Asteraceae), the common name coming from late Middle English as knopweed, from 'knop' (because of the hard rounded flower 'head'), the genus name referencing the centaur Chiron who discovered the medicinal uses of these plants. See also star-thistle.

Scattered and uncommon European neophytes include **brown knapweed** *C. jacea* (in southern England in well-drained grassland), **giant knapweed** *C. macrocephala* (on waste ground), and **silver** (or **ragwort**) **knapweed** *C. cineraria* (on coastal cliffs and walls in east Dorset and south Hampshire).

Chalk knapweed *C. debeauxii*, in rough ground and other grassy habitats on light, often calcareous soil, locally common in southern, central and eastern England, rare or absent further north. Jean Debeaux (1826–1910) was a French botanist.

Common knapweed *C. nigra*, common and widespread in meadow and pasture, sea-cliffs, roadsides, railway banks, scrub, woodland edge, field borders and waste ground, on often damp and heavy soils. *Nigra* is Latin for 'black'.

Greater knapweed *C. scabiosa*, winter-green, widespread and locally common in England and south Wales, more scattered elsewhere, on dry, usually calcareous soils, in grassland, scrub and woodland edge, on cliffs, roadsides, railway banks, quarries and waste ground. *Scabiosa* refers to the rough (Latin *scaber*) leaves being a supposed cure for scabies.

Slender knapweed Alternative name for chalk knapweed.

KNAWEL Species of *Scleranthus* (Caryophyllaceae), from German *Knauel* = 'knotgrass'. Greek *scleros* + *anthos* = 'hard' + 'flower'.

Annual knawel *S. annuus*, a widespread but scattered annual or biennial found on disturbed dry sandy soil on heathland, waste ground and arable fields.

Perennial knawel *S. perennis* ssp. *perennis*, biennial, restricted to dry, basic igneous (doleritic) rocks in pockets of shallow soil at Stanner Rocks, Radnor. Subspecies *prostratus* is an endemic biennial or short-lived perennial found in Breckland on acidic sandy soils in heathland, lost from other habitats since around 1960.

KNIGHT (FUNGUS) Members of the genera *Tricholoma* (Greek *trichos* + *loma* = 'hair' + 'fringe', though this does not describe the cap margins of most of these species) and *Melanoleuca* (Greek *melas* + *leucos* = 'black' + 'white') (Agaricales, Tricholomataceae).

Widespread but uncommon species include **beech knight** *T. sciodes* (possibly toxic, on calcareous soil in broadleaf woodland, ectomycorrhizal on beech and oak), **deceiving knight** *T. sejunctum* (most records from the southern half of England and the Highlands, on oak and birches), **gassy knight** *T. inamoenum* (on soil in often calcareous grassland and deciduous and woodland), **larch knight** *T. psammopus* (coniferous woodland, on larch), and **scaly knight** *T. vaccinum* (coniferous forest, mainly on spruce).

Rarities (Vulnerable or Critically Endangered in the RDL, 2006) include **acrid knight** *T. aestuans*, **bitter knight** *T. acerbum*, **booted knight** *T. focale* (Highlands), **chemical knight** *T. stiparophyllum*, **giant knight** *T. colossus*, **poplar knight** *T. populinum*, **robust knight** *T. robustum*, **scented knight** *T. apium*, **stinky knight** *T. bufonium*, **tacked knight** *T. pessundatum* (Highlands), **upright knight** *T. stans* (Highlands), **white knight** *T. album*, and **yellow staining knight** *T. sulphurescens*.

Aromatic knight *T. lascivum*, widespread and locally common, inedible, ectomycorrhizal, found in autumn on soil and litter in deciduous woodland. *Lascivum* is Latin for 'lasciviousness'.

Ashen knight *T. virgatum*, widespread and locally common, inedible, ectomycorrhizal, found in summer and autumn on usually acid soil and litter of broadleaf and coniferous woodland. *Virgatus* is Latin for 'striped', from the radial streaks on the cap.

Bald knight *M. melaleuca*, common and widespread, inedible, mainly found in England, in summer and autumn on soil generally in grassy habitats, including broadleaf woodland and parkland.

Birch knight *T. fulvum*, common and widespread, tasteless, found from June to October on soil in damp deciduous woodland, often at the edges, ectomycorrhizal on hardwoods, mainly birch. *Fulvum* is Latin for 'tawny brown'.

Blue spot knight *T. columbetta*, widespread and locally common, edible, mycorrhizal, in summer and autumn on soil in broadleaf woodland. *Columbetta* is Latin for 'small dove'.

Burnt knight *T. ustale*, widespread and locally common, inedible (possibly poisonous), in late summer and autumn on soil in broadleaf woodland, ectomycorrhizal on beech and other hardwoods. *Ustale* is Latin for 'roasted' or 'burnt', from the cap colour.

Dark scaled knight *T. atrosquamosum*, widespread, edible, mainly in deciduous woodland on alkaline soil, locally common in southern England, scarce elsewhere, in late summer and autumn, ectomycorrhizal mainly on beech. Latin *ater* + *squamosus* = 'black' + 'scaly'.

Girdled knight *T. cingulatum*, tasteless, locally common in England and Wales, scarce elsewhere, from June to October in woodland, scrub and dune slacks, mycorrhizal on willow. *Cingulus* is Latin for 'girdle'.

Grey knight *T. terreum*, widespread and locally common (e.g. in the Highlands) inedible, ectomycorrhizal, found from August to October on often calcareous soil on the edge of coniferous forest. *Terreum* is Latin for 'earthy', referring to its colour.

Matt knight *T. imbricatum*, widespread and locally common, inedible, in late summer and autumn on sandy soil in coniferous forest, mycorrhizal on Scots pine. *Imbricatum* is Latin for 'covered in scales'.

Soapy knight *T. saponaceum*, common and widespread, inedible, from June to October on woodland soil, with a soapy smell when the gills are crushed. *Saponaceum* is Latin for 'soapy'.

Sulphur knight *T. sulphureum*, widespread and locally common, poisonous, in late summer and autumn on soil in broadleaf woodland, mainly ectomycorrhizal on oak and beech.

Yellow knight *T. equestre*, poisonous, locally common in the Highlands, occasional in England and Wales, from July to October, mainly ectomycorrhizal on Scots pine, but also on birch and oak. *Equestre* is Latin for 'cavalry', possibly from the saddle shape of many mature specimens.

Yellowing knight *T. scalpturatum*, widespread and locally common, dubiously edible (and reported by some as poisonous) from late June to early November in deciduous woodland, ectomycorrhizal on beech, oak and lime. *Scalpturatum* is Latin for 'scratched' or 'engraved', a reference to the scales on the cap surface.

KNOT *Calidris canutus* (Scolopacidae), wingspan 59 cm, length 24 cm, weight 140 g, an Amber List (Red List in Ireland) winter visiting shorebird from North America and Greenland found on large muddy estuaries, with 330,000 birds estimated in the UK in 2008–9, and high numbers also in Ireland, with Dundalk Bay (Co. Louth) and Strangford Lough (Co. Down) supporting most birds (7,500 and 10,000, respectively). They feed in large flocks, mainly on mussels and crustaceans. The English and specific name, recalling King Knut, may come from the fact that they generally feed at the tidal margin. Greek *calidris* was used by Aristotle for some grey-coloured waterside birds.

KNOT GRASS (MOTH) *Acronicta rumicis* (Noctuidae, Acronictinae), wingspan 30–43 mm, nocturnal, widespread on grassland, heathland, marsh, gardens and a range of other habitats, but with numbers declining by 75% over 1968–2007, a species of conservation concern in the UK BAP. Flying in May–July, there is a second brood in southern England in August–September. Larvae feed on a range of woody and herbaceous plants. Greek *acronyx* = 'nightfall', and Latin *rumex* = 'sorrel'.

Light knot grass *A. menyanthidis*, wingspan 33–41 mm, nocturnal, local on acid moorland in northern England, Wales, Scotland and Ireland. Adults fly in May–July. Larvae feed on bog-myrtle, heather, bilberry, bogbean (*Menyanthes*, giving the specific name) and other moorland plants.

KNOT-HORN (MOTH) Nocturnal moths in the family Pyralidae, subfamily Phycitinae.

Scarce species include **agate knot-horn** *Nyctegretis lineana* (on dunes and beaches from Norfolk to Kent, larvae favouring restharrow), **birch knot-horn** *Ortholepis betulae* (heathland, larvae feeding on birch), **dark spruce knot-horn** *Assara terebrella* (coniferous woodlands in southern England, larvae feeding in spruce cones), **gorse knot-horn** *Pempelia genistella* (heathland, from Sussex to Dorset, mainly coastal), **hoary knot-horn** *Gymnancyla canella* (sandy beaches from Sussex to Dorset, larvae feeding on prickly saltwort), **Kent knot-horn** *Moitrelia obductella* (chalk downland and quarries in Kent and Sussex, especially the North Downs, larvae feeding on marjoram), **large clouded knot-horn** *Homoeosoma nebulella* (chalky or sandy rough ground, especially in East Anglia north to Yorkshire, and eastern Ireland, larvae feeding on spear thistle), **lime knot-horn** *Salebriopsis albicilla* (woodland in the Wye Valley, discovered in 1964, and a few woods in Somerset and Warwickshire, larvae feeding on small-leaved lime), **new pine knot-horn** *D. sylvestrella* (recorded in Dorset in 2001, subsequently found elsewhere in southern and south-east England), **ornate knot-horn** *Pempeliella ornatella* (downland

in south England and on the Burren, larvae feeding in thyme roots), **pine blossom knot-horn** *Vitula biviella* (first record in 1997 at Lydd, Kent, now with small breeding populations in Kent and Suffolk), **rosy-striped knot-horn** *Oncocera semirubella* (downland and limestone cliffs from Kent to Somerset, and in Norfolk, larvae feeding on white clover, restharrow and bird's-foot-trefoil), **scarce aspen knot-horn** *Sciota hostilis* (one wood in Warwickshire), **silver-edged knot-horn** *Pima boisduvaliella* (dunes and shingle beaches from Norfolk to Kent, and in Lancashire), **spruce knot-horn** *Dioryctria schuetzeella* (first British record in 1980 in Kent, since extending its range in southern and south-east England), **white-barred knot-horn** *Elegia similella* (mature oak trees, mainly in southern and south-east England), **willow knot-horn** *Sciota adelphella* (a rare continental immigrant found on the coast of south and south-east England, sometimes arriving in sufficient numbers to breed, and established since 1992 in south-east Kent), and **wormwood knot-horn** *Euzophera cinerosella* (waste ground in parts of England north to Yorkshire, and south Wales).

Ash-bark knot-horn *Euzophera pinguis*, wingspan 23–8 mm, common in woodland and hedgerows throughout much of southern Britain. Adults fly in July–August. Larvae burrow into the bark of ash trees, and can eventually kill the tree if present in great numbers. Greek *eu* + *zopheros* = 'very' + 'dusky', and Latin *pinguis* = 'fat'.

Beautiful knot-horn *Rhodophaea formosa*, wingspan 20–3 mm, local in hedgerows in much of southern England and East Anglia. Adults fly in June–August. Larvae feed inside rolled leaves of elm. Greek *rhodon* + *phaios* = 'rose' + 'dark', and Latin *formosa* = 'beautiful'.

Broad-barred knot-horn *Acrobasis consociella*, wingspan 19–22 mm, widespread but local in oak and other deciduous woodland and areas with scattered oak trees. Adults fly in July–August. Larvae feed on oak leaves. Greek *acron* + *basis* = 'point' + 'step', from the 'tooth' at the base of the antenna, and Latin *consocius* = 'companion', from the gregariously feeding larvae.

Brown knot-horn *Matilella fusca*, wingspan 25–8 mm, widespread but local on heathland and moorland. Adults fly in June–July. Larvae feed on heather and bilberry, and adapting to cultivated heathers so becoming more widespread in gardens on acid soils. The genus name is 'obscure'; *fusca* is Latin for 'dusky'.

Brown pine knot-horn *Dioryctria simplicella*, wingspan 21–30 mm, local in pine woodland throughout much of England, especially East Anglia, parts of Wales and Scotland, and with one record from Co. Down. Adults fly in July–September. Larvae feed on pine needles. Greek *dioryctes* = 'digger', from the larvae of some species burrowing into cones, and Latin *simplex* = 'simple'.

Chalk knot-horn *Phycitodes maritima*, wingspan 18–22 mm, local in rough meadow and waste ground throughout much of England, especially East Anglia, and with records from south and east Ireland. Adults fly in May–August, possibly in two generations. Larvae mine flowerheads of yarrow, tansy, groundsel and common ragwort. Greek *Phycita* + *eidos* = 'in the shape of the genus *Phycita*', and Latin *maritima* here meaning 'coastal', a common location.

Dark pine knot-horn *D. abietella*, wingspan 26–33 mm, local in hedgerows throughout much of southern England and East Anglia, and scattered in Wales, Scotland and Ireland. Adults fly in July–August. Larvae feed on conifer needles, buds and cones. *Abies* is Latin for 'fir'.

Dingy knot-horn *Hypochalcia ahenella*, wingspan 22–32 mm, on chalk downland, in quarries, embankments and dry open areas throughout much of southern England, rare elsewhere in Britain. Adults fly in June–August. Larvae feed on common rock-rose. Greek *hypochalcos* = 'containing copper', and Latin *ahenius* = 'bronze', from the forewing colour.

Dotted oak knot-horn *Phycita roborella*, wingspan 24–9 mm, common in oak woodland and areas with scattered trees throughout much of England and Wales, scattered in Ireland.

Adults fly in July–August. Greek *phycitis* = 'a red stone', and *roborella* from the larval food plant pedunculate oak *Quercus robur*.

Double-striped knot-horn *Cryptoblabes bistriga*, wingspan 18–20 mm, local in oak woodland throughout much of England, Wales and Ireland. Adults fly in June–July. Larvae feed on a variety of tree and shrub species, usually inside a folded leaf. Greek *cryptos* + *blabe* = 'hidden' + 'damage', from the larval damage to the food plant, and Latin *bi* + *striga* = 'two' + 'striations', from the forewing pattern.

Ermine knot-horn *Phycitodes binaevella*, wingspan 22–7 mm, local in rough ground and meadows throughout England, Wales and Ireland. Adults fly in July–August. Larvae mine flowerheads of thistles, tansy and mugwort. Latin *bi* + *naevus* = 'two' + '(body) mole', from the forewing spots.

Grey knot-horn *Ac. advenella*, wingspan 18–24 mm, common in (especially old uncut hawthorn) hedgerows and gardens throughout Britain. Adults fly in July–August. Larvae feed on leaves of hawthorn, blackthorn and sometimes rowan. *Advena* is Latin for 'stranger'.

Heath knot-horn *Apomyelois bistriatella*, wingspan 18–25 mm, on burnt areas in heathland throughout southern England into parts of the Midlands and south Wales, and in Co. Wicklow, but erratically and in mobile colonies. Adults fly in June–July. Larvae feed on the fungus *Daldinia vernicosa/concentrica* which grows on burnt gorse and dead birch. Greek *apo* = 'away from' the genus *Myelois*, and Latin *bi* + *stria* = 'two-striated', from the forewing marking.

Heather knot-horn *Pempelia palumbella*, wingspan 21–7 mm, local on heathland throughout much of Britain together with eastern Ireland. Adults fly in July–August. Larvae feed on heather. *Palumbus* is Latin for 'woodpigeon', from the forewing colour.

Marbled knot-horn *A. marmorea*, wingspan 18–23 mm, local in hedgerows and scrub throughout England and Wales, more numerous in the south from Dorset to Kent. Adults fly in June–August. In May larvae spin a silk web that looks like a small piece of sheep's wool on the underside of a twig of blackthorn, hawthorn or rowan, leaving the web to feed. *Marmorea* is Latin for 'marbled'.

Powdered knot-horn *Delplanqueia dilutella*, wingspan 18–23 mm, local on limestone cliffs and grassy slopes on the coasts of much of the British Isles, and inland on chalk downland in southern England and south Wales. Adults fly in May–June. Larvae feed on wild thyme, often near a nest of yellow ant, creating a silk gallery often including debris from the ant nest. *Delplanqueia* may be in honour of the French entomologist A. Delplanque. Latin *diluta* = 'pale'.

Saltmarsh knot-horn *Ancylosis oblitella*, wingspan 18–22 mm, local on saltmarsh and waste ground in parts of south-east England where first recorded in 1956. In 1976, there was a population explosion or large-scale immigration, and examples were reported along the south coast and inland as far as Hertfordshire, and since spreading into other parts of the Midlands. Adults are bivoltine and fly in May and July–August. Larvae feed on goosefoot and orache. Greek *agcylos* = 'curved', describing the palps, and Latin *oblitus* = 'smeared', from the forewing colour.

Samphire knot-horn *Epischnia asteris*, wingspan 27–30 mm, found on sea-cliffs in Dorset, south-west England and Wales. Adults fly in August. Larvae feed on golden-samphire in August–September and again in April–May after overwintering. Greek *epischnaino* = 'to dry', and Latin *asteris* refers to the food plant *Aster*.

Sandhill knot-horn *Anerastia lotella*, wingspan 19–27 mm, local on coastal dunes and slacks in England, Wales, south Scotland and east Ireland, and inland on the Breckland sands of Norfolk and Suffolk. Adults fly in July. Larvae feed on the roots of dune grasses. Greek *anerastos* = 'unloved', and possibly Latin *lotus* = 'bath' (i.e. 'washed'), from the pale wing colour.

Scarce clouded knot-horn *Homoeosoma nimbella*, wingspan 16–21 mm, uncommon in rough meadow and waste ground in parts of East Anglia, Cornwall, the Scilly Isles and south Wales. Adults fly in June–July. Larvae mine flower heads of yarrow, common ragwort and yellow chamomile. *Nimbus* is Latin for 'rain-cloud', from the forewing shading.

Scarce oak knot-horn *A. tumidana*, wingspan 19–24 mm, uncommon and local, probably often migrant, mostly recorded from southern and south-east England, with a few records elsewhere north to Lancashire. Adults fly in July–September. Larvae probably feed on oak leaves. *Tumidus* is Latin for 'swelling', from the ridge of raised forewing scales.

Small clouded knot-horn *Phycitodes saxicola*, wingspan 14–20 mm, widespread but local in dunes, shingle and other dry open coastal habitats. Adults fly in June–August. Larvae mine flower heads of chamomile, groundsel and ragwort. *Saxicola* is Latin for 'rock-dwelling'.

Spindle knot-horn *Nephopterix angustella*, wingspan 20–5 mm, local though increasing in range in hedgerows, scrub and gardens throughout southern England and East Anglia. With two generations, adults fly in June–July and September–October. Larvae feed on berries of spindle from September to November. Greek *nephos* + *pteryx* = 'cloud' + 'wing', and Latin *angustus* = 'narrow', from the forewing shape.

Thicket knot-horn *Ac. suavella*, wingspan 22–4 mm, local in hedgerows and scrub throughout southern England and East Anglia. Adults fly in July–August. Larvae feed on leaves of hawthorn and, preferably, blackthorn in May–June. *Suavis* is Latin for 'agreeable'.

Twin-barred knot-horn *Homoeosoma sinuella*, wingspan 18–23 mm, common in dry open areas, calcareous grassland, dunes and waste ground throughout southern Britain, together with records from Co. Wicklow. Adults fly in June–August. Larvae feed on plantain roots. *Sinosus* is Latin for 'curved', from the forewing markings.

Warted knot-horn *A. repandana*, wingspan 20–5 mm, common in oak and other deciduous woodland and areas with scattered oak trees throughout England and Wales. Adults fly in July–August. Larvae feed on oak leaves. *Repandus* is Latin for 'bent backwards', possibly from the labial morphology.

KNOTGRASS *Polygonum aviculare* (Polygonaceae), a common and widespread annual of open and disturbed ground including arable land (where it is a serious weed), gardens, waste ground and seashore. Greek *poly* + *gonia* = 'many' + 'angle', and Latin *aviculus* = 'small bird'.

Cornfield knotgrass *P. rurivagum*, an annual archaeophyte, rare, scattered and decreasing in south and south-east England, in arable fields and more rarely of ruderal habitats, especially on light chalky soils and calcareous clays. *Rurivagum* is from Latin *rus* + *vago* = 'countryside' + 'wander'.

Equal-leaved knotgrass *P. arenastrum*, a common and widespread annual archaeophyte of open and trampled ground, especially on well-drained soils, on tracks, roadsides and pavements, around ponds and field gateways. Latin *arena* + *aster* = 'sand' + 'star'.

Lesser red knotgrass *P. arenarium*, an annual Mediterranean neophyte cultivated in 1807, scattered as a casual from grain or birdseed, on waste ground, especially around docks. *Arenarium* is Latin for 'sandy'.

Northern knotgrass *P. boreale*, a rare annual of open, disturbed ground on roadsides and paths, cultivated land, sandy beaches, dunes and coastal shingle, scattered and mainly coastal in Britain, especially Scotland. *Boreale* is Latin for 'northern'.

Ray's knotgrass *P. oxyspermum*, an annual, biennial or short-lived perennial of coastal sand, shingle or shell beaches, with a scattered and declining range. John Ray (1627-1705) was an English naturalist. Greek *oxys* + *sperma* = 'sharp' + 'seed'.

Red knotgrass *P. patulum*, an annual European neophyte occasionally found, and possibly declining, as a casual grain and wool-shoddy alien on tips and waste ground. *Patulum* is Latin for 'spreading'.

Sea knotgrass *P. maritimum*, a perennial herb of sand and shingle beaches, growing above the spring tide limit. In the 1960s it had been recorded from just four post-1930 sites (west Cornwall, north Devon and the Channel Islands) but fieldwork during the 1990s considerably extended its known range in south and south-west England, its recent spread correlating with a run of mild winters and hot summers. In Ireland it persists at the Co. Waterford site where it was discovered in 1973.

KNOTWEED Perennial species of *Persicaria* (referring to the similarity of the leaves to those of a peach tree = *persicum*) and *Fallopia* (after the Italian anatomist Gabriele Fallopia, 1523–62, superintendent of the botanical garden at Padua) (Polygonaceae).

Alpine knotweed *P. alpina*, a rhizomatous European neophyte herb introduced in 1816, naturalised on shingle by the River Dee, verges and waste ground in a few localities in Scotland and the Midlands.

Chinese knotweed *P. weyrichii*, a herbaceous rhizomatous east Asian neophyte introduced by 1917, naturalised on a roadside verge, grassland and woodland at Wasdale Hall (Cumbria) and locally in rough ground in Herefordshire and Yorkshire. Dr Heinrich Weyrich was a nineteenth-century surgeon on a Russian naval expedition to the Far East.

Giant knotweed *F. sachalinensis*, a rhizomatous neophyte herb from Sakhalin and Japan, introduced in 1869, naturalised in Ireland by 1896 and in Britain by 1903, now scattered and locally common. The shoots emerge in spring and grow rapidly up to 4 m tall. Roots can extend to a depth of 2 m. In autumn, when shoots are killed by frost, food reserves are translocated to the rhizomes which form a deep mat and can be >2 m deep and 15–20 m long. The plant forms extensive thickets on waste ground, roadsides, riverbanks and lake and sea loch shores, and it has become a serious weed (see also Japanese knotweed below).

Himalayan knotweed *P. wallichii*, a rhizomatous neophyte herb introduced before 1900, a garden escape found on streamsides, hedge banks, roadsides, railway banks and waste ground, growing in dense stands. Seed is only occasionally set, but it establishes readily from broken rhizomes. Nathaniel Wallich (1786–1854) was a Danish-born surgeon and botanist.

Japanese knotweed *F. japonica*, a common and widespread rhizomatous Far Eastern neophyte grown in British gardens since 1825, and first recorded in the wild in 1886. Almost all plants are female. It forms dense thickets on waste ground, roadsides and railway banks, along canal, stream and riverbanks, and on sea loch shores. Rhizome fragments are dispersed in garden and other rubbish, and by river floods. In winter the plant dies back beneath ground but by early summer the bamboo-like stems shoot to >2 m, suppressing growth of other plants. With its deeply penetrating and spreading rhizomes it is very hard to remove by hand or with chemicals. Under provisions made in the WCA (1981), it is an offence to cause Japanese knotweed to grow in the wild. An amendment to the Anti-Social Behaviour, Crime and Policing Act (2014) includes Japanese knotweed. If excavation and physical removal is used to effect control, the volume of excavated soil can extend to 3 m vertically and 7 m horizontally from the above-ground growth. Disposal costs range from £25–50 per tonne (not including landfill tax), so excavation of even a relatively small infestation can cost several thousand pounds in waste charges alone. Off-site disposal can result in total treatment costs >£10,000.

Lesser knotweed *P. campanulata*, a stoloniferous neophyte herb introduced from the Himalayas in 1909, sparsely scattered but expanding, found on damp roadsides, hedge banks and streamsides, establishing from rhizome fragments. *Campanulata* is Latin for 'bell-shaped'.

Soft knotweed *P. mollis*, a scarce herbaceous Himalayan neophyte introduced in 1840, naturalised in rough ground and roadside verges in a few English and Scottish locations. *Mollis* is Latin for 'soft'.

KOHL-RABI See cabbage.

KOHUHU *Pittosporum tenuifolium* (Pittosporaceae), an evergreen New Zealand neophyte shrub or tree in places planted near the sea as an ornamental, self-sown in Cornwall, a relic elsewhere mainly in southern England. Kohuhu is the Maori name for this plant. Greek *pitta* + *spora* = 'resin' + 'seed', and Latin *tenuifolium* = 'narrow-leaved'.

KOROMIKO *Veronica salicifolia* (Veronicaceae), an evergreen neophyte shrub from both Chile and New Zealand, introduced in 1843, a garden escape on roadsides, hedgerows, sea-cliffs, quarries, pavement cracks and walls, widespread but mainly naturalised in south-west England and the Isle of Man. Koromiko is a Maori word for a hebe plant. Veronica is named after the saint; *salicifolia* is Latin for 'willow-leaved'.

KRILL Small marine crustaceans in the order Euphausiacea, family Euphausiidae. *Krill* is Norwegian meaning 'small fry of fish'.

L

LABIATAE Former name for Lamiaceae.

LABRADOR-TEA *Rhododendron groenlandicum* (Ericaceae), an evergreen North American neophyte shrub naturalised in a few places in Britain on bogs and other wet peaty habitats. Greek *rhodon* + *dendron* = 'rose' + 'tree', and Latin *groenlandicum* = 'from Greenland'.

LABRIDAE Family name of wrasses (Greek *labros* = 'furious').

LABURNUM *Laburnum anagyroides* (Fabaceae), a common and widespread small neophyte tree from montane Europe introduced by the Tudor period, much planted in parks, gardens and on waysides, thriving mainly on acid soils and frequently self-seeding in waste ground and on roadsides and railway banks. In western Britain it was formerly planted for hedging, and persists locally. *Laburnum* is the Latin name for this tree, *anagyroides* means like the genus *Anagyris*.

LACE-BUG Members of the family Tingidae (Heteroptera), so called because of the delicate wing pattern and pronotum, with 14 genera in Britain.

Gorse lacebug *Dictyonota strichnocera*, length 3 mm, locally common in south-east England and the Midlands, associated with gorse and also broom. Greek *dictyon* = 'net', and *strychnos* + *ceras* = 'nightshade' + 'horn'.

Spear thistle lacebug *Tingis cardui*, length 3–4 mm, common throughout most of Britain, mainly on spear thistle. *Tingis* is New Latin for a type of insect, *cardui* = 'of thistles'.

LACE-WEAVER SPIDER *Amaurobius similis* (Amaurobiidae), widespread and common, females (found all year) with a body length up 12 mm, males (June–November) to 8 mm, in woodlands in tree bark crevices (where webs are common), logs and leaf litter, and often found around buildings in holes in walls and fences. *Amaurobius* is Greek for 'living in the dark', *similis* Latin for 'similar'.

Black lace-weaver spider *A. ferox*, widely distributed in England, mainly near the coast in Wales, with a few records for Scotland, females (found all year) with a body length to 16 mm, males (spring and autumn) to 12 mm, often in cellars, houses and gardens, and under logs in woodland or hedgerows, building an irregular, lace-like web around a crude funnel. *Ferox* is Latin for 'wild' or 'fierce'.

Window lace-weaver spider *A. fenestralis*, female length up to 12 mm, males (June–November) to 8 mm common and widespread, found in woodlands with webs common on tree-trunks in crevices and under loose bark, sometimes in masonry cracks, and often around windows and sills, hence its name. The web usually forms an open funnel leading to the spider's retreat in the deepest part of the crevice, and its strands are 'fluffed' to tangle the insect (and other spider) prey. *Fenestra* is Latin for 'window'.

LACERTIDAE Family name of lizards (Latin *lacerta*).

LACEWING Predatory insects in the order Neuroptera, with 14 species of **green lacewing** (Chrysopidae), 29 species of **brown lacewing** (Hemerobiidae). Females lay their eggs on a thread of hardened mucus attached to a leaf, suspending them in the air.

Bordered brown lacewing *Megalomus hirtus* (Hemerobiidae), found at Holyrood Park, Edinburgh in 2015, previously recorded there in 1982. It flies in June–August, feeding on wood sage. It is a UK BAP priority species. Latin *hirtus* = 'hairy'.

Common green lacewing *Chrysoperla carnea*, wingspan 10–20 mm, common and widespread, peak activity in May–August, adults and larvae preying on aphids in various habitats, often entering houses in preparation for hibernation (unique behaviour in this insect group). Greek *chrysos* = 'gold' with New Latin *perla* = an unspecified insect, and Latin *carnea* = 'flesh-coloured'.

Giant lacewing *Osmylus fulvicephalus* (Osmylidae), wingspan 50 mm, adults found in streamside vegetation in April–August, larvae hunting in damp moss and debris. It is widespread if local in England and Wales, scarce in Scotland and Ireland. Greek *osme* = 'scent', and Latin *fulvus* + Greek *cephale* = 'yellow' + 'head'.

LACKEY *Malacosoma neustria* (Lasiocampidae, Malacosomatinae), wingspan 25–35 mm, a moth found commonly on heathland and in gardens and open woodland in the southern half of England, Wales and west and south-east Ireland, a species of conservation concern in the UK BAP, numbers having declined by 90% since 1980. Adults fly in July–August. Larvae feed gregariously in a silk web of silk on hawthorn, blackthorn, plum, apple and other trees and shrubs. Greek *malacos* + *soma* = 'soft' + 'body', from the larva; *neustria* has a problematic interpretation.

Ground lackey *M. castrensis*, a nocturnal moth, wingspan 30–40 mm, in saltmarsh and on shingle beaches in England from Kent to Suffolk. Adults fly in July–August. Larvae feed gregariously on sea wormwood, sea-lavender and other saltmarsh plants in spun tents. *Castrensis* is Latin for 'relating to a tent'.

LADY For the butterfly, see painted lady; also a subtidal sea slug, called scarlet lady.

LADY-FERN Members of the family Woodsiaceae. Specifically, *Athyrium filix-femina*, found in semi-shade in scree slopes and other rocky places, woodland, hedgerows and streamsides on damp but well-drained soils to an altitude of 1,000 m. Greek *a* + *thyrium* = 'without' + 'shield', referring to the enclosed sori, and Latin *filix-femina* = 'lady-fern'.

Alpine lady-fern *A. distentifolium*, local in central Scotland in damp acid rocky habitat, from 455 m in the Breadalbanes (mid-Perth) to 1,220 m (Ben Macdui, Aberdeenshire). Latin *distentus* + *folium* = 'distended' + 'leaf'.

LADYBIRD 1) Members of the beetle family Coccinellidae, generally 3–7 mm length, commonly yellow, orange, red or black with small black spots (or red spots on black) on the elytra (wing covers). 'Ladybird' derives from 'Our Lady's bird' or 'Lady beetle' from Mary (Our Lady), who in early iconography was often depicted wearing a red cloak, and the spots of the seven-spot ladybird symbolised her seven joys and seven sorrows.

Adonis' ladybird *Hippodamia variegata*, widespread but local in England and Wales, recently increasing in numbers, and abundant in suitable habitat + ruderal, weedy plants on sandy, open soils, leaf litter, wasteland and post-industrial sites + feeding on aphids during the summer. In Greek mythology Hippodamia was the wife of Pirithous, king of the Lapiths in Thessaly, at whose wedding the Battle of the Lapiths and Centaurs took place. *Variegata* is Latin for 'variegated'.

Bryony ladybird *Henosepilachna argus*, first recorded in 1997 at Molesey, Surrey, subsequently spreading over Surrey, crossing the Thames in 2000 to Hampton Court and westward to Staines in 2001. Its spread northwards in Europe is possibly a result of global warming. Adults feed on white bryony leaves. Greek *enas* + *epi* + *lachnos* = 'one' + 'on' + 'woolly hair', and *argos* = 'bright'.

Cream spot ladybird *Calvia quattuordecimguttata*, common and widespread, especially in England and Wales. Adults are active in April–October, feeding on aphids and psyllids, often found on leaves and shrubs along hedgerows. They overwinter in plant litter, bark crevices and beech nuts. The binomial is from Latin *calvus* = 'smooth', and 'fourteen-spotted'.

Cream-streaked ladybird *Harmonia quadripunctata*, common and widespread in England and Wales, adults active in April–October, often in pine woodland (though also on nettle), feeding on aphids. The binomial is Latin for 'harmony' and 'four-spotted'.

Eighteen-spot ladybird *Myrrha octodecimguttata*, with a widespread but scattered distribution, rather local, adults evident during most months, living in trees with a preference for the crowns of Scots pine. In Greek mythology Myrrha was the mother of Adonis. Latin *octodecimguttata* = '18-spotted'.

Eleven-spot ladybird *Coccinella undecimpunctata*, common and widespread, preferring warm, moist places, and populations have greatly declined after hard winters. It is found in a variety of habitats, feeding on aphids, and overwintering in leaf litter and buildings. The binomial is from Latin *coccineus* = 'scarlet', and for '11-spotted'.

Eyed ladybird *Anatis ocellata*, at 8–9 mm long Britain's largest ladybird, widespread but locally common, restricted by the need for pine. Adults are active from spring to summer, feeding on pine aphids. The binomial is Latin for 'duck-like' and 'like a small eye'.

Five-spot ladybird *Coccinella quinquepunctata*, found on the shingle edges of some Welsh and Scottish river estuaries, feeding on aphids. Latin *quinquepunctata* = 'five-spotted'.

Fourteen-spot ladybird *Propylea quattuordecimpunctata*, common and widespread, especially in England and Wales, adults active in May–September in a variety of habitats, particularly where there are shrubs. Adults and larvae both feed on aphids. If handled adults, as a defence, exude pungent orange liquid from their joints, a form of controlled bleeding. The binomial is Greek for 'an entrance' and Latin for '14-spotted'.

Harlequin ladybird *Harmonia axyridis*, 5–8 mm length, a variably coloured and patterned, highly invasive east Asian ladybird, most commonly found on deciduous trees and on low-growing plants such as nettles. It preys on (often pest) aphids, coccids and other Hemiptera, eggs and larvae of butterflies and moths, and other insects, as well as pollen and sugars from honeydew, fruit and nectar. It also preys on and out-competes other ladybirds (helped by its large size, broad diet, tolerance of a wide range of habitats and climate conditions, long reproductive period and high reproductive rate), and has been largely responsible for the decline of at least seven native ladybird species, including the two-spot which decreased in numbers by 44% over 2004–9. Most eggs are laid in June–July, adults emerging from pupae in August, and moving to overwintering sites in September–November. They sometimes manage two generations in one year. Overwintering is usually in large aggregations. They live for about a year and are reproductively active for three months. First recorded in Essex in 2004, harlequins can disperse easily, hence their rapid spread over England and parts of Wales and Scotland, reaching the north coast of mainland Scotland and the Shetlands (though probably not overwintering there) by 2014 – the most rapid spread of an invasive animal on record. There are a few records from south and east Ireland. Greek *axyres* = 'uncut'.

Heather ladybird *Chilocorus bipustulatus*, with a scattered distribution, mostly found on heathland, feeding on coccids, but also recorded from coastal scrub, dunes and marsh. Overwintering is in litter or bark crevices. Greek *cheilos* = 'margin', and Latin *bipustulatus* = 'two-blistered'.

Hieroglyphic ladybird *Coccinella hieroglyphica*, widespread but uncommon, largely confined to heathland and some acid grassland. Feeding is on heather aphid, eggs and larvae of heather leaf beetle, and chrysomelids in the genera *Altica* and *Galerucella*. Overwintering is in litter under heather, pine and gorse.

Kidney-spot ladybird *Chilocorus renipustulatus*, common and widespread in England and Wales, adults active in April–October usually in woodland, often on tree bark, feeding on scale insects. Its two spots are sometimes indented to give the appearance of a kidney, hence its common name. Latin *renipustulatus* = 'kidney-blistered'.

Larch ladybird *Aphidecta obliterata*, widespread and generally common. Adults are active in spring and summer, found on coniferous trees, especially larch, feeding on aphids and scale insects, and overwintering in bark crevices. Greek *aphis* + *dectes* = 'aphid' + 'biter', and Latin *obliterata* = 'obliterated'.

Nineteen-spot ladybird Alternative name for water ladybird, although the number of spots actually varies from 15 to 21.

Orange ladybird *Halyzia sedecimguttata*, formerly closely associated with ancient woodland but now increasing in range because it has adapted to feed on sycamore and ash, and it has become more widespread and locally common. There was a population explosion in Epping Forest during 1998–9, and it is now recorded from many woods throughout London, on Hampstead Heath and even appears on garden trees and the central London parks. Adults are active in April–October, feeding on mildew, then overwintering in leaf litter or sheltered tree parts. The binomial is from Greek *halysis* = 'chain', and Latin for '16-spotted'.

Pine ladybird *Exochomus quadripustulatus*, widespread and common, especially in England and Wales, and frequently recorded in urban locations. Adults are active in April–October, mainly found on pine but also on other trees and shrubs, such as hawthorn, when pine is absent, feeding on aphids and scale insects. It overwinters in leaf litter, foliage and bark crevices of evergreen trees and shrubs. The binomial is from Greek *exo* + *choma* = 'out of' + 'mound', and Latin for 'four-spotted'.

Scarce seven-spot ladybird *Coccinella magnifica*, 7–8 mm length, with records from southern England and a few Midlands sites, found close to (but not in) wood ant nests in woodland and heathland, feeding on aphids. *Magnifica* is Latin for 'magnificent'.

Seven-spot ladybird *Coccinella septempunctata*, 6–8 mm length, Britain's commonest ladybird (though there was a massive reduction in numbers in 1999–2000, and these have since fluctuated) with a widespread distribution, usually red but sometimes yellow. Both larvae and adults feed on aphids. Adults are active in March–October, and hibernate in crevices, often in outbuildings and around window frames. In warm years, sizeable numbers may migrate from continental Europe. Latin *septempunctata* = 'seven-spotted'.

Sixteen-spot ladybird *Tytthaspis sedecimpunctata*, common and widespread, mainly in (especially southern) England. Adults are active in summer, particularly associated with grassland and hedgerows, feeding on pollen, nectar and fungi. The binomial is from Greek *tytthos* + *aspis* = 'small' + 'shield', and Latin for 'sixteen-spotted'.

Small brown ladybird *Rhyzobius litura*, 2–3 mm length, common and widespread, especially in England and lowland Wales, from spring to late autumn in dense patches of vegetation. Greek *rhyzo* = 'growl', and Latin *lituro* = 'to erase'.

Striped ladybird *Myzia oblongoguttata*, 6–8 mm length, with a scattered distribution but locally abundant where there is Scots pine where it feeds on aphids, being active in April–August. Greek *myzo* = 'to suck in', and Latin *oblongoguttata* = 'oblong-spotted'.

Ten-spot ladybird *Adalia decempunctata*, variable in colour, and not always with ten spots, very common (less so in the north) and widespread in woodland and gardens with trees. Adults and larvae both feed on aphids. Adults are active in March–October, and often hibernate in large groups in sheds or under loose tree bark. Greek *adales* = 'unhurt', and Latin *decempunctata* = 'ten-spotted'.

Thirteen-spot ladybird *Hippodamia tredecimguttata*, occasional in Ireland, and not seen in Britain since the 1950s, indeed may never have been resident, but it has recently been reported from marsh in the West Country. Latin *tredecimguttata* = 'thirteen-spotted'.

Twenty-four-spot ladybird *Subcoccinella viginiquattuorpunctata*, widespread, especially in England and Wales, though not common, the generally flightless adults being most evident in summer and autumn as a plant-eater (as are larvae) in grassland. Adults overwinter in grass tussocks or moss. The binomial is Latin for 'almost scarlet' and '24-spotted'.

Twenty-two-spot ladybird *Psyllobora vigintiduopunctata*, widespread and common, especially in England and Wales.

Adults are active in April–August, and feed on mildew by grazing on the soil surface or on low vegetation. The binomial is from Greek *psylla* + *bora* = 'psyllid' + 'food', and Latin for '22-spotted'.

Two-spot ladybird *Adalia bipunctata*, very common (less so in the north) and widespread. As an example of polymprphism, most are red with two black spots but some are black with red spots, the latter more common further north where the dark colour helps it to absorb heat from the sun. Both larvae and adults feed on aphids and similar small invertebrates (and as such are often used in pest control), and are found in any habitat with appropriate prey, including gardens. Adults are active in March–November, and overwinter by hibernating, often in groups. *Bipunctata* is Latin for 'two-spotted'.

Water ladybird *Anisosticta novemdecimpunctata*, widespread in England and Wales, adults active from April to late summer usually by water, feeding on aphids on waterside rushes. It overwinters between leaves and in stems of reeds, grass tussocks and other dense vegetation. It is also known as the 19-spot ladybird although the number of spots on the elytra varies between 15 and 21. The binomial is Greek *anison* + *stictos* = 'anise' + 'dotted', and Latin for '19-spotted'.

2) **False ladybird** *Endomychus coccineus* (Endomychidae), a fungus beetle 4–6 mm long, living under the bark of dead or dying trees especially beech and birch, widespread especially in southern England and Wales. Greek *endon* + *mychos* = 'within' + 'inward', and Latin *coccineus* = 'scarlet'.

LADY'S-MANTLE

Perennial herbs in the genus *Alchemilla* (Rosaceae), including the aggregate *A. vulgaris*. Leaves shaped like a mantle (loose cloak) may have prompted its English name. *Alchemilla* may come from the Arabic *al-kimiya*, a plant valued in alchemy.

Alpine lady's-mantle *A. alpina*, found in montane grassland and grass-heath, scree, cliffs, rocky streamsides, in well-drained habitats in areas of late snow-lie, from near sea level in north-west Scotland to 1,270 m on Ben Macdui (Aberdeenshire), and in the Lake District, Co. Kerry and Co. Wicklow. Soils range from acidic to strongly calcareous.

Clustered lady's-mantle *A. glomerulans*, typically found in ungrazed or lightly grazed base-poor grassy habitats in the central and northern Scottish mountains; a depauperate form grows in heavily grazed seepages and screes below cliffs. In Teesdale and Craven it occurs on roadsides and in species-rich hay meadows. *Glomeros* is Greek for 'cluster'.

Crimean lady's-mantle *A. tytthantha*, a neophyte from the Crimea, very occasionally naturalised in central and southern Scotland, probably as escapes from botanical gardens. Greek *tytthos* + *anthos* = 'small' + 'flower'.

Hairy lady's-mantle *A. filicaulis* ssp. *vestita*, widespread in rough pasture, grassy hill slopes, banks and mountain flushes, woodland edges and verges; it is also local on superficial clay-with-flints on chalk downs. Latin *filum* + *caulis* = 'thread' + 'stem', and *vestita* = 'clothed'.

Large-toothed lady's-mantle *A. subcrenata*, recorded in species-rich hay meadows, along the margins of those which are more intensively managed, and from unimproved pasture. First found in 1951, it has only ever been known from a very restricted area of Teesdale (north-west Yorkshire), and one site in Weardale (Co. Durham). Since 1990 it has been recorded in only two sites, and a search of three others in 1996 failed to detect it. *Subcrenata* is Latin for 'partly scalloped'.

Pale lady's-mantle *A. xanthochlora*, widespread and common in north England, Wales, much of Scotland and the northern half of Ireland, on neutral or slightly calcareous grassland, lowland pastures, hay meadows, streamsides, woodland edges and verges. Greek *xanthos* + *chloros* = 'yellow' + 'green'.

Rock lady's-mantle *A. wichurae*, in grazed base-rich grassland and herb-rich rock ledges on outcrops and cliffs in north England and Scotland, sometimes colonising damp scree and cracks

in limestone and basalt. It prefers moist soils. Max Wichura (1817–66) was a German botanist.

Shining lady's-mantle *A. micans*, not recognised in Britain until 1976, and currently known only from Northumberland. The largest population occurs in species-rich grazed pasture on shallow soil overlying Carboniferous limestone. It has also been recorded from a rough pasture, from tall herbage in an ungrazed hay meadow and from a roadside verge. Casual plants were recorded in Lanarkshire in 1986 and 1992. *Micans* is Latin for 'glittering'.

Silky lady's-mantle *A. glaucescens*, local on limestone grassland and grassy banks by roads and rivers in northern England, and on limestone grassland in north-west and south-east Scotland and north-west Ireland. *Glaucescens* is Latin for 'becoming glaucous' (greenish grey).

Silver lady's-mantle *A. conjuncta*, a neophyte from the Alps, grown as a rockery plant since around 1800, now naturalised in a few places in Britain in montane grassland and streamsides, and in the lowlands on roadsides, riverbanks and in rough grassland. *Conjuncta* is Latin for 'joined'.

Slender lady's-mantle *A. filicaulis* ssp. *filicaulis*, with a scattered distribution from south Wales and the Midlands to north Scotland, and Co. Sligo, in calcareous or neutral grassland, grass-heath, herb-rich banks and seepages; also on rock outcrops, and mountain ledges of basic rock. Latin *filix* + *caulis* = 'fern' + 'stem'.

Smooth lady's-mantle *A. glabra*, found in northern England, Wales, Scotland and the northern third of Ireland on damp soils in pasture, hay meadows, grass-heath, roadsides and herb-rich banks kept moist by seeping water; also among tall-herb vegetation on mountain ledges. *Glabra* is Latin for 'smooth'.

Soft lady's-mantle *A. mollis*, common and widespread, introduced from the Carpathians by 1874, frequently grown in gardens and naturalised on roadsides, riverbanks, rough ground and garden refuse dumps. *Mollis* is Latin for 'soft'.

Starry lady's-mantle *A. acutiloba*, found in Co. Durham and Northumberland in unimproved, species-rich hay meadows, or on the margins of meadows which have been reseeded or are more intensively managed. It is also locally frequent on herb-rich verges, and occasional on railway banks. It has been naturalised in Lanarkshire since 1992, and in Angus. *Acutiloba* is Latin for 'pointed lobes'.

Velvet lady's-mantle *A. monticola*, very local in north-west Yorkshire and Co. Durham on unimproved, species-rich hay meadows, and locally frequent on verges. *Monticola* is Latin for 'mountain dweller'.

LADY'S-SLIPPER

Cypripedium calceolus (Orchidaceae), a rhizomatous herbaceous perennial orchid once fairly widespread in grassy woodland clearings on limestone in northern England on well-drained soil, but suffering from collecting, especially in the nineteenth century, and since the 1970s found as a native at a single site in the Yorkshire Dales. Since 2003 it has been planted in a few places (using plants raised at Kew) as part of a conservation recovery programme. One non-secret site is Gait Barrows, Lancashire, where by 2012 there were 80 flowering plants on limestone pavement. Greek *Cypris* + *podilon* = an alternative name for Aphrodite + 'slipper', and Latin *calceolus* = 'small shoe'.

LADY'S SMOCK

Alternative name for cuckooflower.

LADY'S-TRESSES

Perennial orchids in the genera *Goodyera* (honouring the English botanist John Goodyer, 1592–1664) and *Spiranthes* (Greek *speira* + *anthos* = 'coiled' + 'flower', from the spiral shape of the inflorescence) (Orchidaceae).

Autumn lady's-tresses *S. spiralis*, rhizomatous, locally frequent in (especially southern) England, Wales and Ireland, on unimproved, grassland on dry calcareous soils, and on clifftops and dunes; also on lawns and, rarely, on less acidic heathland. It is often prompted to flowering when grazing or mowing ceases. *Spiralis* is Latin for 'spiralled'.

Creeping lady's-tresses *G. repens*, evergreen, found mainly in the Highlands, but also Cumbria and planted in Norfolk, in semi-natural and planted coniferous woodland, usually of Scots pine, growing in slight to moderate shade in moist layers of moss and pine needles. It is also found under pine on old dunes. *Repens* is Latin for 'creeping'.

Irish lady's-tresses *S. romanzoffiana*, rhizomatous, found in north and west Ireland and west Scotland in nutrient-poor, periodically flooded or flushed vegetation, often on peaty soils by streams and lake margins, and common among purple moor-grass in pastures grazed by cattle or ponies. Count Nicholas Romanzoff was a nineteenth-century Russian who financed collecting expeditions.

LAEMOPHLOEIDAE Family name of lined flat bark beetles (Greek *laimos* + *phloeos* = 'throat' + 'tree bark'), with 11 species in 4 genera. Several are predators, hunting underneath bark. Others are synanthropic and can be serious pests of stored grains (especially *Cryptolestes*). See flat grain beetle under beetle.

LAGOMORPHA The order name of rabbit and hares (Greek *lagos* + *morphe* = 'hare' + 'shape').

LAGOON WORM Tentacled lagoon worm *Alkmaria romijni* (Ampharetidae), a polychaete (bristleworm) up to 5 mm long, with eight thread-like and slimy tentacles, recorded from only 27 sites in Britain, in lagoons and sheltered estuaries living in a mud tube in muddy sediments from the Humber Estuary to south Cornwall, and in Pembrokeshire. It is protected under Schedule 5 of the WCA (1981). Successful translocations of populations threatened by redevelopment have been made on the Medway estuary and on the Isle of Wight.

LAMB'S-EAR *Stachys byzantina* (Lamiaceae), a stoloniferous perennial south-west Asian neophyte herb in cultivation by 1782, scattered in Britain and south-east Ireland as a garden escape or throw-out. Sometimes self-sown, it often persists on roadsides, waste ground and in quarries. The English name reflects the curved shape and white, soft, fur-like coating of the leaves. *Stachys* is Greek for 'flower spike', *byzantina* Latin for from Byzantium or Constantinople (i.e. Istanbul).

LAMB'S-QUARTERS In Britain, an alternative name for fat-hen (not in the USA where it is another name for pitseed goosefoot).

LAMIACEAE Family name of dead-nettle (Latin *lamium*) and mint, together with a large number of other herbaceous species.

LAMINARIALES Order of brown seaweeds, including the families Chordaceae (mermaid's tresses) and Laminariaceae, the kelps. *Lamina* is Latin for 'thin plate' or 'layer'.

LAMNIFORMES Order of sharks that includes basking shark and porbeagle (Greek *lamna*).

LAMPREY Eel-shaped ectoparasitic jawless cartilaginous fish with a round sucking mouth in the family Petromyzontidae.

Brook lamprey *Lampetra planeri*, length 12–14 cm, common and widespread in streams, rivers and lakes. Adults spawn in spring in gravel, then die. Eggs hatch within a few days. The blind larvae bury themselves in soft sediment with the mouth protruding, and filter-feed on detritus and other organic matter for 3–5 years before maturing. Adults do not feed, but develop teeth which are used for gripping stones in order to build nests. It is declining in parts of its range, and is an RDB species. The English and genus names are from Latin *lambere* + *petra* = 'to lick' + 'stone', from the attachment to stones. Latin *planus* = 'flat'.

River lamprey *L. fluviatilis*, length 25–40 cm for sea-going forms, up to 28 cm for the lake forms, a demersal and anadromous fish found close to the coast, migrating upstream in many British and Irish rivers in August to spawn, after which they die. Larvae (ammocoetes) spend several years in soft sediment before migrating to the sea as adults. They then spend 2–3 years in marine habitats before making the return trip. Its mouth is a circular sucking disk with sharp teeth, and it feeds as an

ectoparasite by clinging on to and rasping the flanks or gills of a fish. *Fluviatilis* is Latin for 'riverine'.

Sea lamprey *Petromyzon marinus*, length up to 120 cm, weight up to 2.3 kg, a common and widespread demersal and anadromous fish found offshore, migrating upstream in many British and Irish rivers in August to spawn, then dying. Larvae burrow in sand and silt in quiet waters downstream from spawning areas, filter-feeding on plankton and detritus. After a few years, larvae metamorphose and as adults return to the sea, to feed as ectoparasites on fish, using a suction-like toothed mouth to grasp and rasp, feeding on the host's blood. Greek *petra* + *myzo* = 'stone' + 'to suck'.

LAMPYRIDAE Family name (Greek *lampas* = 'lamp') for the bioluminescent glow-worms, which are strongly sexually dimorphic: a small, winged male and a much larger larviform female.

LANCE FLY Members of the family Lonchaeidae (Diptera), adults small and robust, mainly found in woodland, larvae generally feeding under bark and in decomposing wood.

LANCE-WING Narrow-winged moths in the family Epermeniidae, in the genera *Epermenia* (Greek *epi* + Armenia leading to *erminea* = 'ermine', referencing scale tufts on the dorsal forewing margins) and *Phaulernis* (*phaulos* + *ernos* = 'paltry' + 'a shoot', from the species that has a small-scale 'tooth' on the forewing). Wingspan is generally 9–12 mm.

Carrot lance-wing *P. aequidentellus*, scarce, found in southern and south-west England, predominantly in coastal sites. Adults fly in June–July and in September–October in two generations. Larvae mine wild carrot and burnet-saxifrage. Latin *aequus* + *dens* = 'equal' + 'tooth', from the forewing tufts.

Chalk-hill lance-wing *P. insecurella*, scarce on limestone grassland and chalk downland, and on some sandy soils in England, south of a line from the Severn estuary to the Wash, and a priority species under the UK BAP. Adults fly in late-afternoon sunshine in May–August in two generations. Larvae feed on shoots then the leaves of bastard-toadflax. *Insecurus* is Latin for 'insecure', from uncertainty of differentiation from congenerics.

Garden lance-wing *E. chaerophyllella*, wingspan 12–14 mm, widespread and common in rough ground, hedgerows and damp grassland. With two or three generations a year, adults fly in March–May and from late June to September. Larvae mine leaves of umbellifers, including rough chervil (*Chaerophyllum*, giving the specific name), feeding gregariously in a web.

Large lance-wing *E. falciformis*, local in damp woodland, marsh and freshwater margins throughout much of southern England, rarely in Wales, and recorded from one site in Ireland. With two generations a year, adults fly in June–July and August–September. Larvae mine stems and leaves of wild angelica and ground-elder. *Falciformis* is Latin for 'scythe-shaped'.

Little lance-wing *E. profugella*, a rarity of downland, rough ground and damp meadow scattered in parts of England. Larvae mine seedheads of wild carrot, burnet-saxifrage, wild angelica and ground-elder. *Profugus* is Latin for 'fugitive'.

Scale-tooth lance-wing *P. dentella*, scarce, day-flying, in hedgerows, waste ground and dry grassland in southern England, from East Anglia westwards to Herefordshire and Cornwall. Adults fly in June. Larvae feed on umbellifers. Latin *dens* = 'tooth', from the small-scale 'tooth' on the forewing.

Scarce lance-wing *E. farreni*, a rarity in open areas and waste ground in parts of East Anglia, the south Midlands and Kincardineshire. Larvae mine seed capsules of wild parsnip. William Farren (1836–87) was an English entomologist.

Yellow-spotted lance-wing *P. fulviguttella*, widespread but local in grassland, open woodland and damp meadows. Adults fly in July–August. Larvae feed in seed heads of hogweed and wild angelica in September–October. Latin *fulviguttella* = 'yellow-spotted'.

LANCELET *Branchiostoma lanceolatum* (Branchiostomatidae), a filter-feeding fish-like chordate in the class Leptocardii, lacking jaw and obvious sense organs, narrow and up to 8 cm long (like a lance), found off the south coasts of Britain and Ireland as far north as Anglesea and Liverpool Bay, and along the southern North Sea, with scattered records around Scotland, and north-east England, in the sandy sublittoral at depths down to 30 m partially buried. Greek *branchia* + *stoma* = 'gills' + 'mouth', and Latin *lanceolatum* = 'lance-shaped'.

LANDHOPPER See woodhopper.

LANGOUSTINE Alternative name for Norway lobster.

LANTERN BEETLE See Mab's lantern beetle under beetle.

LAPPET *Gastropacha quercifolia* (Lasiocampidae, Pinarinae), a nocturnal moth, wingspan 50–90 mm (females being larger), common in hedgerows, scrub, downland, open woodland and gardens throughout southern England and south Wales, north to Cumbria. Adults fly in June–July. Larvae feed on the leaves of woody plants, particularly hawthorn and blackthorn. The English name describes the caterpillar, which has fleshy 'lappets' or skirts along the sides. Greek *gaster* + *pachys* = 'stomach' + 'thick', from the large abdomen. *Quercifolia* refers not to oak leaves as the food plant but to the adult's resemblance to a cluster of dried oak leaves when at rest.

Pine-tree lappet *Dendrolimus pini*, wingspan 45–70 mm, until recently a rare vagrant to the south coast and Channel Islands, but in 2007 it found resident in small numbers in pine woodland in Inverness-shire, Scotland. Larvae feed gregariously on Scots pine. Greek *dendron* + *limos* = 'tree' + 'hunger', from the potential of the larvae to defoliate.

LAPWING *Vanellus vanellus* (Charadriidae), wingspan 84 cm, length 30 cm, weight 230 g, found on farmland throughout Britain, particularly in lowland areas of northern England, the Scottish Borders and eastern Scotland. In the breeding season they favour spring-sown cereals, root crops, unimproved pasture, meadows and fallow fields, but are also found on wetlands with short vegetation. In 2009 there were an estimated 140,000 breeding pairs in Britain, but over 1980–2015, there was a decline in numbers by 52% (starting in the mid-1980s), including by 11% over 2010–15, and it was placed on the Red List in 2009. In winter they flock on pasture and ploughed fields, mainly in south-west England and south-west Ireland, with an estimated 620,000 birds in Britain 2006–7. The ground nests usually contain 4 eggs, incubation takes 25–34 days, and the precocial young fledge in 35–40 days. Food is mainly insects and worms. The English name refers to the bird's wavering flight, as does the binomial which derives from Latin *vannus*, a winnowing fan.

LARCH Deciduous conifers in the genus *Larix* (Pinaceae), from the Middle High German *Larche*, based on Latin *larix* = 'larch'.

European larch *L. decidua*, introduced from central Europe, widely planted in forestry plantations and as an ornamental in parks. It grows in open situations on soils of poor to medium nutrient status, light, moist but free-draining, and does not tolerate either very dry or waterlogged soils. It regenerates from seed freely, especially onto disturbed soil or rocky ground.

Hybrid larch *Larix* x *marschlinsii*, a hybrid between European and Japanese larch first noticed at Dunkeld (east Perth) in 1904, now widely planted in forest plantations, best suited (but not limited) to mineral soils of low nutrient status. Carl von Salis-Marschlins (1762–1818) was a Swiss naturalist.

Japanese larch *Larix kaempferi*, introduced from Japan in 1861 and widely used in forestry and parks, as well as land reclamation. It is light-demanding and requires >1000 mm of rainfall for good growth. It is cold hardy though vulnerable to spring and autumn frosts, and best suited to mineral soils of poor nutrient status. Engelbert Kaempfer (1651–1716) was a German naturalist and collector.

LARDER BEETLE Members of the family Dermestidae (see

also hide beetle), specifically *Dermestes lardarius*, with adults 7–10 mm length. Larvae (10–12 mm) are often found in corners at the base of skirting boards, and behind cookers where they feed on grease deposits and food spillage. They can sometimes be found in bathrooms where they feed on human skin scales. They are also found in bird nests and sometimes on dead animals. Greek *derma* = 'skin'.

Black larder beetle *D. haemorrhoidalis*, with adults 7–10 mm length. Larvae (10–12 mm) are often found in corners at the base of skirting boards, and behind cookers where they feed on grease deposits and food spillage. They may bury into wood or plaster to pupate. They are also found in bird nests and on animal carcasses. Greek *haima* + *rheo* = 'blood' + 'flow'.

Peruvian larder beetle *D. peruvianus*, introduced from the Neotropics, having a few records from the Midlands and south-east England.

LARIDAE Family name of gulls and some terns (Latin *larus* = 'gull').

LARK Songbirds in the family Alaudidae, with elongated hind claws and a song delivered on the wing, and typically crested. 'Lark' comes from Old English *laferce* and *læwerce*, but of unknown ultimate origin. See also skylark and woodlark.

Short-toed lark *Calandrella brachydactyla* is a scarce visitor; **lesser short-toed lark** *C. rufescens*, **crested lark** *Galerida cristata*, **bimaculated lark** *Melanocorypha bimaculata*, **black lark** *M. yeltoniensis*, **calandra lark** *M. calandra* and **white-winged lark** *M. leucoptera* are all accidentals.

Shore lark *Eremophila alpestris*, wingspan 32 cm, length 16 cm, weight 37 g, a wintering visitor from Scandinavia found on sand and shingle coasts, mainly in eastern England; in a good year, a few hundred may arrive, in others it can be scarce, and it is Amber-listed. Exceptionally, a mated pair were recorded at Windmill End Nature Reserve, Dudley (west Midlands) in 2013. Greek *eremos* + *philos* = 'desert' + 'loving', and Latin *alpestris* = 'alpine' (or mountains generally).

LARKSPUR *Consolida ajacis* (Ranunculaceae), an annual garden escape introduced from Mediterranean Europe by the Tudor period, found on waste ground and as an arable weed, usually on dry soils in chalky or sandy areas, particularly in central and southern England. 'Larkspur' comes from the resemblance of the calyx and petals to the lark's long straight hind claw (spur). The binomial is Latin for 'solid' (and 'comfrey'), and after the Greek Trojan hero Ajax.

LASIOCAMPIDAE Family name (Greek *lasios* + *campe* = 'hairy' + 'larvae') for eggar and other moths.

LATRIDIIDAE Family name (Greek *latris* = 'handmaiden'), of minute brown scavenger beetles (1–3 mm length), with 56 species in 8 genera, found in compost and leaf litter, fungoid bark, rotting fungi, mature Myxomycetes, and – indoors – in mouldy areas of plaster, wallpaper and wood.

LATTICE FUNGUS Alternative name for red cage.

LATTICE-MOSS Species of *Cinclidotus* (Greek *cinclis* = 'lattice') and *Dialytrichia* (*dialyo* + *tricha* = 'to separate' + 'hair') in the family Pottiaceae. 'Lattice' refers to the peristome around the capsule mouth.

Fountain lattice-moss *C. riparius*, rare at a few English sites in river flood zones (*riparius* being Latin for 'of riverbanks').

Pointed lattice-moss *D. mucronata*, mainly found in the southern half of England and south Wales on tree boles and rocks in the flood zone of rivers and streams, sometimes on the base of trees, and on stones and walls away from running water. It is fairly frequent on base-rich sandstone in south-west England, and is increasing on tarmac. *Mucronata* is Latin for 'having a short tip'.

Smaller lattice-moss *C. fontinaloides*, widespread and locally common on intermittently but frequently submerged rocks, tree roots and stonework, particularly in upland regions, on limestone and on siliceous rocks, but avoiding very acidic water. Latin *fontinaloides* = 'of springs'.

LAURACEAE Family name for bay (= Latin *laurus*).

LAUREL 1) Alternative name for bay.

2) Some species of *Prunus* (Rosaceae).

Cherry laurel *P. laurocerasus*, a common and widespread evergreen neophyte shrub or small tree from south-east Europe, in gardens by 1629, naturalised in woods and scrub. The binomial is Latin for 'plum tree' and 'cherry laurel'.

Portuguese laurel *P. lusitanica*, an evergreen neophyte shrub from south-west Europe present in parks and gardens by 1648, scattered and locally common in woodland and scrub, and on waste ground. *Lusitanica* is Latin for 'Portuguese'.

LAURUSTINUS *Viburnum tinus* (Caprifoliaceae), an evergreen south European neophyte shrub cultivated since the sixteenth century, commonest in southern England, Lancashire and north Wales, scattered elsewhere and rare in Scotland and Ireland, a garden escape or planted on sea-cliffs, banks, woodland, rough grassland, roadsides, railway banks, amenity plantings and waste ground. Populations can become well-established, especially on soils overlying chalk and limestone in coastal localities. The binomial is Latin for wayfaring-tree and laurustinus.

LAUXANIIDAE Family name for flies (2–7 mm length) in the superfamily Lauxanioides, with 56 species in 13 genera. Some adults feed on nectar. Larvae are mostly saprophages, feeding in leaf litter.

LAVENDER Garden lavender *Lavandula angustifolia* (Lamiaceae), a small evergreen Mediterranean neophyte shrub grown as an ornamental and culinary herb by at least the seventeenth century, found as a garden escape, throw-out or relic of cultivation, especially in southern England, almost always in urban habitats. It is commercially cultivated in fields in East Anglia. The Anglo-Norman French *lavendre* and the genus name are medieval Latin for 'lavender', from *lavare* = 'to bathe', from its use as a bath scent; *angustifolia* is Latin for 'narrow-leaved'. (See also sea-lavender, *Limonium* spp.)

LAVENDER-COTTON *Santolina chamaecyparissus* (Asteraceae), an evergreen Mediterranean neophyte shrub cultivated by the mid-sixteenth century, scattered in Britain, persistent on rough ground, and as a relic of cultivation, and recently established on some sandy shores. Latin *sanctus* + *linum* = 'holy' + 'flax', and Greek *chamaecyparissus* = 'dwarf cypress'.

LAVER Purple laver *Porphyra umbilicalis* (Bangiaceae), up to 20 cm long, a common and widespread red alga abundant singly or in dense colonies on rocky shores, especially in the upper intertidal, tolerant of both air exposure and wave action, and often growing on mussels. It is boiled and eaten as a jelly and used to make laver bread in Wales, especially in Pembrokeshire and Carmarthenshire, and eaten cold with vinegar in Cornwall. 'Laver' is late Old English, referencing a water plant mentioned by Pliny. The binomial comes from Greek for 'purple', and Latin possibly here meaning 'central'. Other laver species include *P. dioica* and *P. linearis*.

LEAD-MOSS *Ditrichum plumbicola* (Ditrichaceae), known from about 30 sites, most in Wales, restricted to the most toxic parts of abandoned lead mines, growing on fine, preferably damp silt-like spoil. Greek *di* + *tricha* = 'two' + 'hair', and Latin *plumbum* + *colo* = 'lead' + 'inhabit'.

LEAD POISONER Alternative name for the fungus livid pinkgill.

LEAF BEARER Stalked leaf bearer *Phyllophora pseudoceranoides* (Phyllophoraceae), a widespread red seaweed with fronds up to 10 cm long, on shaded rock pool sides in the lower intertidal and on subtidal rocks. Greek *phyllon* + *phora* = 'leaf' + 'carrying', and *pseudo* = 'false' + (*Fucus*) *ceranoides* (horned wrack).

LEAF BEETLE Members of the family Chrysomelidae.

Alder leaf beetle *Agelastica alni* (Galerucinae), 6–7 mm length, previously considered extinct in the UK but, since being found in Manchester in 2004, it has increased its range in northern England from the Mersey across to the Humber estuary, feeding in damp open areas, including alder carr, on the leaves of alders, though also found on hazel, birches and hornbeam if alder is absent. Greek *agelasticos* = 'living in groups', and Latin *alnus* = 'alder'.

Brooklime leaf beetle *Prasocuris junci* (Chrysomelinae), 4–5 mm length, widespread and locally common, especially in England, feeding on leaves of brooklime and water-speedwells. *Prasocuris* is Greek for a larva that feeds on leek, and Latin *juncus* = 'rush'.

Broom leaf beetle *Gonioctena olivacea* (Chrysomelinae), 3.7–7.5 mm length, locally common and widespread in Britain, feeding on leaves of broom, sometimes dyer's greenweed, laburnum, gorse or lupin. Greek *gonia* + *ctenos* = 'angle' + 'of a comb', and Latin *olivacea* = 'olive green'.

Celery leaf beetle *Phaedon tumidulus* (Chrysomelinae), 3.5–4 mm length, common and widespread in England, Wales and southern Scotland, more scattered north of Scotland's Central Lowlands, and in Ireland. Adults are found throughout the year in a range of habitats feeding on leaves (occasionally pollen) of umbellifers, and can become pests of celery. They overwinter in grass tussock and under bark. Greek *phaios* = 'dark' or 'grey', and Latin *tumidus* = 'swollen'.

Cereal leaf beetle *Oulema rufocyanea* and *O. melanopus* (Criocerinae), difficult to separate without detailed examination, 4–5 mm length, widespread and locally common especially in England and Wales. Adults and larvae feed on cereals (often becoming pests) and grasses; oats and barley are preferred over wheat, which is in turn preferred over rye, maize and wild grasses. Greek *oulon* = 'gums', and Latin *rufocyanea* = 'red-blue' and Greek *melas* = 'black'.

Hawthorn leaf beetle *Lochmaea crataegi* (Galerucinae), 3.7–5.5 mm length, common and widespread in England and Wales, scattered in Ireland, in woodland and hedgerows, feeding on hawthorn: adults on leaves, overwintered adults mainly on pollen, and larvae mining fruits which causes them to fail to set seed. Greek *lochme* = 'bush', and Latin *crataegus* = 'hawthorn'.

Imported willow leaf beetle *Plagiodera versicolora* (Chrysomelinae), body length 2.5–4.8 mm, fairly common in southern England and the Midlands, with a few records further north in England and in Wales and Ireland. There are usually three generations over spring and summer. Both adults and larvae feed on leaves of (especially crack) willow. Greek *plagios* + *dere* = 'slanting' + 'throat', and Latin *versicolora* = 'variously coloured'.

Iridescent green tansy beetle *Chrysolina graminis* (Chrysomelinae), 7.7–10.5 mm length, once widespread but now scarce, a UK BAP species of conservation concern, with a scattered distribution in England, most records coming from south Yorkshire and East Anglia on fens and the banks of rivers with wide floodplains, feeding on leaves of tansy and water mint. Conservation efforts include habitat management near York, and at Woodwalton Fen and Wicken Fen, East Anglia. *Graminis* is from Latin *gramen* = 'grass'.

Mint leaf beetle *Chrysolina herbacea* (Chrysomelinae), 7–11 mm length, with a scattered distribution in Britain, common in central-southern parts, in wet habitats, feeding on leaves of mints and other Lamiaceae. Greek *chrysos* = 'gold', and Latin *herbacea* = 'herbaceous'.

Rainbow leaf beetle *Chrysolina cerealis* (Chrysomelinae), 7–8 mm length, only known from Snowdon and nearby montane grassland. Endangered in RDB 1, and a species of conservation concern in the UK BAP, feeding on leaves and, especially, flowers of wild thyme. *Cerealis* is Latin for 'relating to grain'.

Red poplar leaf beetle *Chrysomela populi* (Chrysomelinae), 10–12 mm length, widespread in England and Wales, feeding on leaves of willows and sapling poplars. Greek *chrysos* + *melas* = 'gold' + 'black', and Latin *populus* = 'poplar'.

Skullcap leaf beetle *Phyllobrotica quadrimaculata* (Galerucinae), 5–7 mm length, widespread though not common, adults feeding on leaves of skullcaps, larvae the roots, in damp, open habitats. Greek *phyllon* + *broticos* = 'leaf' + 'voracious', and Latin *quadrimaculata* = 'four-spotted'.

Strawberry leaf beetle *Galerucella tenella* (Galerucinae), 3–4 mm length, widespread though scattered, with Scottish records mostly in the west, favouring but not restricted to damper habitats, feeding on rosaceous plants (becoming a pest on cultivated plants), and often hibernating in grass tussocks and moss. Latin *galerum* + *eruca* = 'helmet' + 'caterpillar', and *tenella* = 'tender'.

Tansy (leaf) beetle See iridescent green tansy beetle above.

Willow leaf beetle *Lochmaea caprea* (Galerucinae), 4–6 mm length, widespread and locally common, adults feeding on willow and sometimes birch leaves; larvae on willow leaves. They have also been found feeding on tips of crowberry, causing dieback. *Caprea* is Latin for 'of willow'.

LEAF-CUTTER BEE
Species of *Megachile* (Megachilidae) (Greek *mega* + *cheilos* = 'large' + 'lip') that use cut leaves to line the cells of their nests, the leaf discs probably helping to prevent desiccation of the larval food supply. Males are usually 8–10 mm long, females 10–12 mm.

Black-headed leaf-cutter bee *M. circumcincta*, with a scattered distribution, on coastal dunes and heathland, declining in southern England. Latin *circum* + *cincta* = 'around' + 'banded'.

Brown-footed leaf-cutter bee *M. versicolor*, common in England, especially the south-east and south, and Wales on a variety of habitats from heathland to gardens and brownfield sites. *Versicolor* is Latin for 'variously coloured'.

Coast(al) leaf-cutter bee *M. maritima*, widely distributed and locally common in southern England, extending north to Caernarfonshire, Lancashire and Norfolk, and to the Isle of Man. In Ireland it has been recorded from Co. Wicklow and Co. Wexford. While associated with coastal habitats, especially where there is light, sandy soil, it has also been recorded on lowland heathland and, rarely, on chalk grassland. Nectar and probably pollen is taken from a variety of flowes, flight taking place in summer. Nest burrows are excavated in the soil. Cells are constructed from sections of green leaves from various plants, round pieces used for the end walls, oval ones for side walls.

Patchwork leaf-cutter bee *M. centuncularis*, widespread in (especially southern) England and Wales, with a few records from Scotland and eastern Ireland, visiting a range of flowers, especially thistles, in gardens in June–August, constructing nests in holes in deadwood and walls using cut green leaf fragments. Latin *centuncularis* = 'a hundred little hooks'.

Silvery leaf-cutter bee *M. leachella*, found on coastal dunes in south England and south Wales.

Willughby's leaf-cutter bee *M. willughbiella*, common and widespread, less so in Scotland and Ireland, nesting in a variety of habitats, including gardens and brownfield sites, especially in conurbation centres in England. Nests are in wood or soil, cells being constructed with pieces of cut leaves. Flying in summer, pollen and nectar are taken from a variety of flowers. Francis Willughby (1635–72) was an English naturalist.

Wood-carving leaf-cutter bee *M. ligniseca*, found in the southern half of England and south Wales, though uncommon, flying in summer, associated with ruderal-dominated and post-industrial sites, recorded feeding on Himalayan balsam, bramble and thistles. Nests are most frequently found in old trees and fence-posts. Nesting holes have a large diameter and cells

constructed from cut leaves of trees such as sycamore. Latin *lignum* + *secare* = 'wood' + 'to carve'.

LEAF MINER (BEETLE)
Beech leaf miner *Orchestes fagi* (Curculionidae, Curculioninae), 2–3 mm length, common and widespread in broadleaf woodland with a varied shrub layer, wooded pasture, parkland and hedgerows where the host is present. Adults become active on evergreens during February but feed on hawthorn leaves as they open, then move to beech as soon as its buds burst. Greek *orchestes* = 'dancer', and Latin *fagus* = 'beech'.

LEAF MINER (FLY)
Flies in the families Agromyzidae and Anthomyiidae, adults usually 2–3 mm long, whose larvae burrow between the two surfaces of a leaf.

Scarce species include **bean leaf miner** *Liriomyza congesta*, **black wheat leaf miner** *Agromyza albipennis*, **chick pea leaf miner** *L. cicerina* (mining leaves of melilots and restharrows), **geranium leaf miner** *A. nigrescens* (herb-robert), **poplar leaf miner** *Aulagromyza populi* and **tomato leaf miner** *L. bryoniae*. See also blotch miner.

Beet leaf miner *Pegomya hyoscyami* (Anthomyiidae), a leaf-mining pest of beetroot, spinach beet, Swiss chard and related plants, found locally in England. Greek *pegos* + *myia* = 'well put together' + 'fly', and Latin *Hyoscyamus* = 'henbane'.

Cabbage leaf miner *Phytomyza rufipes* (Agromyzidae), widespread if scattered in England and south and south-west Wales. Larvae mine leaves of cabbage and rape. Greek *phyton* + *myzo* = 'plant' + 'to suck', and Latin *rufus* + *pes* = 'red' + 'foot'.

Cereal leaf miner 1) *A. nigrella*, scarce, with a scattered distribution in England and south Wales. Larvae mine leaves of wheat, rye and grasses, mainly in July. *Nigrella* is from Latin *niger* = 'black'. 2) *Chromatomyia nigra* (Agromyzidae), with a scattered distribution in Britain, possibly in Ireland. Larvae mine leaves of wheat, rye and grasses, mainly in June–August. Greek *chroma* + *myia* = 'colour' + 'fly', and Latin *nigra* = 'black'.

Chrysanthemum leaf miner *Ch. syngenesiae*, locally common but scattered in England and Wales. In July–August, larvae mine leaves of oxeye daisy, common ragwort, tansy and sowthistle, as well as chrysanthemums. Greek *syn* + *genesis* = 'together' + 'origin'.

Columbine leaf miner 1) *Ph. aquilegiae*, scattered in the southern two-thirds of England, locally common in gardens. Larvae have usually been recorded mining leaves of aquilegia, giving the specific name. 2) *Ph. miniscula*, common and widespread, if scattered, throughout Britain, especially in gardens, larvae mining leaves of aquilegia and meadow-rue.

Iris leaf miner *Cerodontha iraeos* (Agromyzidae), with a scattered distribution in England, south Wales and west Scotland. Larvae mine leaves of yellow iris in July–August. Greek *ceras* + *odontos* = 'horn' + 'tooth', and from Latin for 'iris'.

Pea leaf miner 1) *Ch. horticola*, with scattered records from England, Wales and north-east Scotland. Larvae mine the leaves of a range of plants, not only garden peas, in both late spring and early autumn. *Horticola* is from Latin for 'little garden'. 2) *L. pisivora*, found in a few sites in southern England. Larvae mine leaves of garden peas and vetches in July–August. *Pisivora* is Latin for 'pea-eating'.

LEAF MINER (MOTH)
Members of the Gracillariidae, Lyonetiidae and Neptulicidae. See also pine leaf-mining moth (Tortricidae) under helm.

Apple leaf miner *Lyonetia clerkella* (Lyonetiidae, Lyonetiinae), wingspan 7–9 mm, nocturnal, widespread and common in gardens, orchards and hedgerows. There are two or more generations a year, later ones overwintering to reappear in spring. Larvae, active in May–November, peaking in August–October, mine leaves of apple, wild cherry, bird cherry, hawthorn, birch, cotoneaster and rowan, becoming a serious pest in some areas. The binomial honours the French (later naturalised Dutch) naturalist Pierre Lyonet (1706–89), and the Swedish entomologist Karl Clerk (1710–65).

Azalea leaf miner *Caloptilia azaleella* (Gracillariidae, Gracillariinae), wingspan 10–11 mm, a naturalised nocturnal moth found locally in parks, gardens and woodland throughout much of southern England, spreading into northern England, Wales and Ireland. An accidental introduction in azaleas and rhododendrons, this native of east Asia occurs in the wild among these ornamental species, and more widely in greenhouses. With two or three generations a year, adults are active from May to October. Greek *calos* + *ptilon* = 'good' + 'feather'.

Firethorn leaf miner *Phyllonorycter leucographella* (Gracillariidae, Lithocolletinae), wingspan 7–9 mm, nocturnal, accidentally introduced with garden plants, the first British record being in Essex in 1989. Since then it has spread rapidly throughout England and Wales northwards to south Scotland, and also Ireland, and it has become naturalised in gardens, parks and woodland. The multivoltine adults can be found in April–October. Larvae mine the leaves of rosaceous trees and bushes, mainly firethorn, but also common in hawthorn, apple, rowan and whitebeam, evident from late summer through autumn. Greek *phyllon* + *oryctes* = 'leaf' + 'digger' (i.e. a leaf miner), and *leucos* + *graphe* = 'white' + 'marking'.

Horse chestnut leaf-miner *Cameraria ohridella* (Lithocolletinae), wingspan 8 mm, an introduced small nocturnal leaf-mining moth now common in parks and urban areas throughout much of England and Wales, first recorded in Wimbledon, London, in 2002, and reaching pest numbers around Greater London and along the Thames Valley by 2006. There may be three generations each year. Larvae mine leaves of horse-chestnut, often leading to complete defoliation. Greek *camara* = 'vaulted chamber'.

Laburnum leaf-miner *Leucoptera laburnella* (Lyonetiidae, Cemiostominae), wingspan 7–9 mm, common in gardens and other suburban areas north to the Highlands, and along the south and east Irish coast. Adults mainly fly in July–August. Larvae mine leaves of laburnum and occasionally dyer's greenweed. Greek *leucos* + *pteron* = 'white' + 'wing'.

Onion leaf miner Alternative name for leek moth.

Rose leaf miner *Stigmella anomalella* (Nepticulidae), wingspan 5–6 mm, widespread and common in England, Wales, Ireland and north-east Scotland, adults flying in both May and August. Larvae mine leaves of cultivated and wild roses in July and October–November. Greek *stigma* = 'small mark' and *anomalos* = 'anomalous'.

LEAF MINER (SAWFLY) See sawfly.

LEAF-ROLLER Weevils in which the female lays an egg within the leaf epidermis or on the surface, then rolls the leaf around it into a case tied with 'silk' in which the larva feeds.

Birch leaf roller *Deporaus betulae* (Rhynchitidae), 5 mm length, a common and widespread tooth-nosed snout weevil found on birch in spring and summer. The female cuts most of the way through a leaf, producing an inverted cone, in which she lays an egg. Pear leaf roller also uses birch leaves in this way.

Hazel leaf-roller *Apoderus coryli* (Attelabidae), 6–8 mm length, widespread in England and Wales, locally common in woodland and hedgerows in May to July, feeding on hazel. Greek *apo* + *dere* = 'away from' + 'neck', and the Latin *corylus* = 'hazel'.

Oak leaf-roller *Attelabus nitens* (Attelabidae), 4–6 mm length, widespread in England and Wales, locally common in woodland, feeding on oak. *Attelabos* is Greek for a kind of wingless locust, *nitens* Latin for 'bright'.

Pear leaf roller (or **pear leaf-rolling weevil**) *Byctiscus betulae* (Rhynchitidae), a tooth-nosed snout weevil, 4.5–7 mm length, with a scattered distribution in England and Wales, feeding on young birch leaves, and a potential pest of young pear trees in the south of its range. Greek *byctes* = 'swelling', and Latin *betula* = 'birch'.

LEAF SCORCH Strawberry leaf scorch *Diplocarpon earlianum* (Dermateaceae), a fungal pathogen affecting the leaves, sometimes the whole strawberry plant, overwintering in plant debris. Greek *diploos* + *carpos* = 'double fruit'.

LEAF SKELETONISER Apple leaf skeletoniser *Choreutis pariana* (Choreutidae), wingspan 11–15 mm, a day-flying moth found locally in orchards, gardens and hedgerows throughout Britain. Adults fly in July and September in two generations, and may overwinter to reappear the following spring. Larvae feed on leaves of apple, crab apple and other fruit trees, causing enough damage to be a serious pest in some areas. Greek *choreutes* = 'dancer', and Latin *par* = 'pair', from the two bands on the forewing.

LEAF WARBLER Members of the family Phylloscopidae: see warbler and chiffchaff. **Arctic warbler** *Phylloscopus borealis*, **dusky warbler** *P. fuscatus*, **greenish warbler** *P. trochiloides* and **Radde's warbler** *P. schwarzi* are scarce visitors; **eastern Bonelli's warbler** *P. orientalis*, **eastern crowned warbler** *P. coronatus*, **green warbler** *P. nitidus*, **Hume's warbler** *P. humei* and **western Bonelli's warbler** *P. bonelli* are all vagrants, as is **Iberian chiffchaff** *P. ibericus*, though four males of this last species held territory in 2010, these being new records in Kent, Yorkshire and Gwent.

LEAF WEED Broad leaf weed Alternative name for the brown seaweed sea petals.

LEAF WEEVIL See weevil.

LEAFHOPPER Members of the sap-sucking bug family Cicadellidae (Auchenorrhyncha), with 201 species recognised in the British list. Many species are common and widespread, but few have common English names. Some are associated with particular habitats, for example *Cicadella viridis*, common in damp grassland and marsh. Many are limited to one or a few host plants, for example *Ulopa reticulata*, widespread and locally common on heathland, living on heather and heaths.

LEATHERBUG Member of the squash bug family Coreidae (Heteroptera).

Breckland leatherbug *Arenocoris waltlii*, length 7–8 mm, an RDB bug local to Breckland, feeding on common storksbill, believed extinct but rediscovered in 2011. Greek *arena* + *coris* = 'sand' + 'bug'. Joseph Waltl (1805–88) was a German physician and naturalist.

Cryptic leatherbug *Bathysolen nubilus*, length 5.5–7 mm, an historically scarce ground-dwelling bug now more widely distributed in southern England and East Anglia in dry, scarcely vegetated habitats feeding on members of the pea family, especially black medick. Greek *bathys* + *solen* = 'deep' + 'channel' or 'pipe', and Latin *nubilus* = 'cloudy'.

Dalman's leatherbug *Spathocera dalmanii*, length 5–6.5 mm, scarce but increasing its range from Dorset to Norfolk, in heathland, acid grassland and dunes, feeding on sheep's sorrel. Greek *spathe* + *ceras* = 'blade' + 'horn'. Johan Dalman (1787–1828) was a Swedish physician and naturalist.

Denticulate leatherbug *Coriomeris denticulatus*, length 7–9 mm, common in England as far north as Yorkshire on dry open habitats on sandy and chalky soil, feeding on black medic and other members of the pea family. Latin *coreum* and Greek *meros* = 'leather' + 'a part'.

Fallen's leatherbug *Arenocoris falleni*, length 6–7 mm, local in coastal dunes between south Wales and Norfolk, and in the Brecklands, feeding on storksbill, increasing inland in gravel-pits. Carl Fallén (1764–1830) was a Swedish naturalist.

Rhombic leatherbug *Syromastus rhombeus*, length 9.5–10.5 mm, until recently confined to a few coastal sites between south Wales and Suffolk but currently expanding inland into dry grassland, feeding on sandworts, spurreys and other members of the Caryophyllaceae. Greek *syra* + *mastos* = 'skin' + 'breast'.

Slender-horned leatherbug *Ceraleptus lividus*, length 9–11 mm, local to sites in south and central England on dry open (often grassland) habitat, larvae feeding on clovers and trefoil. Greek *ceras* + *leptos* = 'horn' + 'slender', and Latin *lividus* = 'blue-grey'.

LEATHERJACKET Soil-dwelling larvae of craneflies (Tipulidae), often pests of grass and some crops by eating roots and stem bases.

LECANORA **Tarn lecanora** *Lecanora achariana* (Lecanoraceae), a crustose lichen found in a few upland sites in Wales, Cumbria and northern Scotland. Greek *lecane* = 'dish'. The Swedish Erik Acharius (1757–1819) is sometimes called the 'father of lichenology'.

LEECH (ANIMAL) Haematophagous members of the class Hirudinea, or subclass in the class Clitellata. A posterior sucker is used mainly for leverage, an anterior sucker, consisting of the jaw and teeth, for feeding. Large adults can consume up to 10 times their body weight in a single meal, with 5–15 ml being an average volume. Specifically, *Erpobdella octoculata* (Erpobdellidae), common and widespread, a predator of aquatic invertebrates. Greek *herpo* + *bdella* = 'to creep' + 'leech'.

Medicinal leech Species of *Hirudo* (Latin for 'leech'), in particular *H. medicinalis* (Hirudinidae), up to 20 cm long, scarce, scattered, but following a decline populations are now generally stable, usually found in small water bodies with a muddy substrate and fringing vegetation. It requires relatively warm water (19–23°C) in which to feed and breed. Egg cocoons are laid on marginal plants. Mammalian and possibly avian blood is required to enable successful breeding. Anti-coagulant properties have led to the animal's use for bloodletting (phlebotomy) since ancient times, a practice reaching its peak in the nineteenth century, and recently showing some revival. Collection from the wild is illegal in the UK without a licence, being listed in Schedule 5 of the WCA (1981).

LEECH (FUNGUS) **Elm leech** *Hypsizygus ulmarius* (Agaricales, Lyophyllaceae), an uncommon tasteless and tough fungus with a scattered distribution mostly in England, found in autumn as tufts on living oak, poplar, horse-chestnut and elm. Greek *hypsos* + *zygos* = 'high' + 'a joining', and Latin *ulmus* = 'elm'.

LEEK *Allium porrum* (Alliaceae), a bulbous herbaceous perennial archaeophyte, widely grown in gardens and fields, occasionally found as a casual throw-out on tips and waste ground. *Allium* is the Latin for 'garlic' as well as 'leek', as is *porrum*.

Broad-leaved leek *A. nigrum*, a European neophyte, scattered in rough ground in a few places in southern England and the Midlands. *Nigrum* is Latin for 'black'.

Sand leek *A. scorodoprasum*, mainly found between Derbyshire and Aberdeenshire, occasional elsewhere, spreading mainly by bulbils in dry grassland, scrub and waste ground, on verges and tracksides and by railways, on sandy riverbanks, open woodland on well-drained soil, and a number of coastal habitats. Greek *scorodon* + *prason* = 'garlic' + 'leek'.

Wild leek *A. ampeloprasum*, very local in south-west England and west and south-east Ireland, occasional elsewhere, in rank vegetation in sandy and rocky habitats near the sea, especially in old fields and hedge banks, on cliff slopes, by paths, and in ditches and other disturbed habitats. This is the 'county' flower of Cardiff as voted by the public in a 2002 survey by Plantlife. Greek *ampelos* + *prason* = 'vine' + 'leek'.

LEGIONNAIRE Soldier flies in the genera *Beris* and *Chorisops* (Greek *choris* + *opsis* = 'apart' + 'appearance') (Stratiomyidae).

Bright four-spined legionnaire *B. nagatomii*, 3–4 mm length, first described from a suburban garden in 1979, widespread but scattered in a variety of habitats in England and Wales. Adults mainly fly in June–August. Akira Nagatomi (1928–2005) was a Japanese expert on flies.

Common orange legionnaire *B. vallata*, 5–6 mm length, very common and widespread in a wide variety of damp habitats, including wet woodland, but also in dry grassland. Adults fly in May–September.

Dull four-spined legionnaire *C. tibialis*, 3–4 mm length, locally common and widespread in England and Wales, with a few records from Scotland and Ireland, in moist, shaded habitats, including damp woodland, marsh and fen, flying in June–September. Larvae live in wood debris and tree rot holes.

Long-horned black legionnaire *B. geniculata*, common and widespread, though scarce in Ireland. Adults, found in damp habitats (as are larvae) fly in May–September. *Geniculata* is Latin for having a knee-like joint.

Murky-legged black legionnaire *B. chalybata*, common and widespread, especially in England and Wales, recorded in Shetland (for the first time) in 2012. Larvae have been found in rotting vegetation. Adults are found in damp, usually shaded, habitats, in flight in April–August. *Chalybatus* is Latin for 'steel-coloured'.

Scarce orange legionnaire *B. clavipes*, uncommon but widely distributed throughout England and Wales, rare in Scotland and Ireland. Adults are associated with damp habitats, flying in May–July. Larvae have been recorded from damp moss. Latin *clava* + *pes* = 'club' + 'foot'.

Short-horned black legionnaire *B. fuscipes*, generally uncommon, scattered in England, Wales, south-west Scotland and Ireland. Adults are mainly found in fen, marsh and wet woodland, flying in May–July. Latin *fuscus* + *pes* = 'dark' + 'foot'.

Yellow-legged black legionnaire *B. morrisii*, widespread and locally common in England and Wales, becoming less so north of the Midlands, with only occasional records from Scotland, associated with damp woodland margins and other shaded habitats, adults flying in May–September. Larvae are found in decomposing vegetation and in roots of wild angelica.

LEGUMINOSAE Former name (Latin *legumen* = 'legume') of the family Fabaceae.

LEIODIDAE Family of round fungus beetles, with 93 species in 20 genera, most associated with fungi, generally widespread though more local in Scotland.

LEJEUNIACEAE Family name for pounceworts, cowlwort and fairy beads (foliose liverworts). Alexandre Lejeune (1779–1858) was a French physician and botanist.

LEMNACEAE Former family name of duckweed, now in the Araceae.

LEMON (GASTROPOD) See sea lemon.

LEMON-BALM See balm.

LEMON PEEL FUNGUS Alternative name for hare's-ear fungus.

LENTEN-ROSE *Helleborus orientalis* (Ranunculaceae), a rhizomatous perennial introduced from Turkey by at least 1839, sometimes planted, and recently naturalised in parks and woodland, and also as a relic of cultivation, occasionally as a casual on waste tips. The binomial is from Greek for this plant, *helleboros*, in turn from *elein* + *bora* = 'to injure' + 'food', and Latin for 'eastern'.

LENTIBULARIACEAE Family name of bladderwort, including butterwort. *Lentibula* (altered from Latin *lenticula* by the Swiss botanist Conrad Gesner) = 'lenticulate'.

LENTIL *Lens culinaris* (Fabaceae), a casual annual found at a few locations on rubbish tips and waste ground, originating from grain and food refuse. 'Lentil' is from Old French *lentille*, in turn from *lens*, Latin for 'lentil'. *Culina* is Latin for 'kitchen' or 'food'.

LEOPARD (MOTH) *Zeuzera pyrina* (Cossidae, Zeuzerinae), wingspan 35–60 mm, nocturnal, common in open woodland, scrub, gardens, orchards and parks throughout England (less so in the north) and Wales. Adults fly in June–July. Larvae feed on the wood of a number of tree species, overwintering two or three times. The English name reflects the black spots on the white wings. *Zeuzera* is a misprint for *zenzera*, Italian for 'gnat', in turn from *zenzero* = 'ginger', for its pungency. *Pyrina* references *Pyrus* (pear), one of the trees mined by the larva.

Reed leopard *Phragmataecia castaneae*, wingspan 15–23 mm, of conservation concern in the UK BAP, confined to reedbeds at Chippenham Fen and Wicken Fen, Cambridgeshire, and at Bure Marsh and Hickling Broad in the Norfolk Broads, with a small and declining population in Dorset. Adults fly in June–July. Larvae mine reed stems, overwintering twice. The binomial is from *phragmites* (reed) + Greek *oiceo* = 'to dwell', and Latin *castanea* = 'sweet chestnut', though this is neither a food plant nor describing a chestnut coloration.

LEOPARDPLANT *Ligularia dentata* (Asteraceae), a herbaceous perennial neophyte from the Far East very scattered in England and Wales, naturalised in damp shady habitats. *Ligularia* is Latin for 'small tongue', *dentata* = 'toothed'.

Przewalski's leopardplant *L. przewalskii*, from China, found in the River Tyne valley, Northumberland. Nikolai Przewalski (1839–88) was a Russian geographer and explorer.

LEOPARD'S-BANE *Doronicum pardalianches* (Asteraceae), a rhizomatous perennial European neophyte herb present since the sixteenth century, common and widespread in Britain, scarce in Ireland, naturalised in woodland and plantations, and on roadsides and shaded habitats. *Doronicum* is from Arabic *darawnaj*; the specific name incorporates *pardus* = 'leopard'. Other species, with similar habitats and distributions, are: **eastern** (*D. columnae*), **Harpur-Crewe's** (*D.* x *excelsum*), **plantain-leaved** (*D. plantagineum*) and **Willdenow's** (*D.* x *willdenowii*) **leopard's-banes**.

LEPIDOPTERA From Greek *lepis* + *pteron* = 'scale' and 'wing', the order that comprises butterflies and moths, which have two pairs of wings covered with fine gossamer scales.

LEPIDOSTOMATIDAE From Greek *lepis* + *stoma* = 'scale' + 'mouth', a caddisfly family with cased larvae. See sedge.

LEPORIDAE Family name of rabbit and hares (Latin *lepus* = 'hare').

LEPRARIA Brimstone-coloured lepraria *Chrysothrix chlorina* (Chrysothricaceae), a powdery or crustose lichen, rare, scattered in upland areas, especially Wales and the Highlands on sheltered siliceous rock. Greek *chrysos* + *thrix* = 'gold' + 'hair', and *chloros* = 'green'.

LEPTINELLA *Cotula squalida* (Asteraceae), a herbaceous New Zealand neophyte scarce in lawns and roadsides in England and Scotland, a similar story for **hairless leptinella** (*C. dioica*) in the London area. *Leptinella* is a now superseded genus name, from Greek *leptos* = 'slender', referring to the ovary. *Cotula* is from Greek *cotyle* = 'small cup', reflecting the cupped area at the base of the leaves. *Squalida* is Latin for 'rough' or 'dirty', *dioica* = 'dioecious'.

LEPTOCERIDAE Family name of longhorn caddisflies (Greek *leptos* + *ceros* = 'slender' + 'horn'). See sedge and silverhorns.

LEPTON Adanson lepton *Lasaea adansoni* (Lasaeidae), an intertidal filter-feeding bivalve, shell size up to 3 mm, widespread on rocky shores, but absent from the south-east and east English coast. *Lepton* is Greek for 'small'. The genus name may be from Greek *las* = 'stone'. Michel Adanson (1727–1806) was a French naturalist.

LEPTOPHLEBIIDAE From Greek *leptos* + *phlebos* = 'thin' + 'of a vein', a mayfly family with three genera and six species in Britain, many being what anglers call 'dun'.

LEPTOTHECATA An order of Hydrozoa (Greek *leptos* + *thece* = 'thin' + 'case'); see hydroid.

LESKEA Pleurocarpous moss species. Nathanael Leske (1751–86) was a German natural scientist.

Scarce Highlands species include **brown mountain leskea** *Pseudoleskea incurvata*, **nerved leskea** *Pseudoleskeella nervosa*, **patent leskea** *Pseudoleskea patens* and **plaited leskea** *Ptychodium plicatum*.

Chained leskea *Pseudoleskeella catenulata* (Leskeaceae), on unshaded, exposed, calcareous, sedimentary and metamorphic mountain rocks in the Highlands, Lake District, north Pennines and north Wales. *Catenulata* is Latin for 'small chain'.

Fine-leaved leskea *Orthothecium intricatum* (Plagiotheciaceae), locally common in upland areas on moist, base-rich rock in overhangs and crevices, on limestone and on base-rich siliceous rocks. Greek *orthos* + *thece* = 'straight' + 'case', and Latin *intricatum* = 'entangled'.

Many-flowered leskea *Pylaisia polyantha* (Hypnaceae), widespread and scattered in Britain on the bark of several tree species, most frequently ash and elder, in hedgerows and open woodland. It was lost from some areas following sulphur dioxide air pollution, which acidified bark, but is now increasing again. Greek *pylaios* = 'at the entrance', and *poly* + *anthos* = 'many' + 'flowers'.

Many-fruited leskea *Leskea polycarpa* (Leskeaceae), common and widespread on trees in the flood zones of lowland streams, and ditches, and beside standing water. It also grows on stones, woodwork and brickwork, occasionally away from water, for instance on willows in moderately humid sites or in hedge tops. *Polycarpa* is Greek for 'many-fruited'.

Red leskea *O. rufescens*, in upland regions, mainly in Scotland, on wet, base-rich limestone and siliceous rocks. *Rufescens* is Latin for 'becoming red'.

Spruce's leskea *Platydictya jungermannioides* (Plagiotheciaceae), widespread but scattered in damp, shaded rock crevices, usually limestone, in woodland, often in but not confined to upland areas. Richard Spruce (1819–73) was an English botanist. Greek *platys* + *dictyon* = 'wide' + 'net'. Ludwig Jungermann (1572–1653) was a German botanist.

Wispy leskea *Pseudoleskeella rupestris*, mainly recorded from the Highlands, for example at Ballochbuie (Aberdeenshire), but also Cumbria and Snowdonia. *Rupestris* is from Latin *rupes* = 'rock'.

LESTIDAE Emerald damselfly family (Greek *lestes* = 'robber').

LETTUCE (LICHEN) Frilly lettuce *Platismatia glauca* (Parmeliaceae), foliose, common and widespread on acidic bark and twigs, occasionally rock. It has been used as a source of brown dye, especially for wool. Greek *platys* = 'flat' and Latin *glauca* = 'blue-green'.

LETTUCE (PLANT) 1) Species of *Lactuca* (Latin *lac* = 'milk', from the plants' milky 'juice') and *Mycelis* (derivation obscure) in the family Asteraceae.

Blue lettuce *L. tatarica*, a herbaceous perennial European neophyte introduced in 1784, with a scattered distribution on disturbed sandy coastal habitats, including dunes and waste ground. It also grows inland on rough ground, railway sidings and tips. *Tatarica* is Latin for from the Tatar Mountains, Russia.

Garden lettuce *L. sativa*, an annual or biennial archaeophyte, scattered in England, cultivated since at least 1200 as a salad plant, and often escaping to waste ground or persisting on abandoned arable land; it is also a birdseed alien occurring as a casual on tips. *Sativa* is Latin for 'cultivated'.

Great lettuce *L. virosa*, an annual or biennial herb, sensitive to grazing, native on coastal cliffs, inland rock outcrops and perhaps dunes, but more widespread as a plant of rank calcareous grassland, woodland edge, quarries, tracks, rough ground and road banks (road development has greatly helped its spread since 1980), locally common from England north to central Scotland, favouring the eastern half of these regions. *Virosa* is Latin for 'stinking'.

Least lettuce *L. saligna*, an autumn- or spring-germinating annual local on disturbed ground, saltmarsh, sandy shingle and old sea walls in East Sussex, west Kent (Isle of Grain, refound in 1999) and south Essex (benefiting from cattle grazing). *Saligna* is Latin for 'willow-like'.

Prickly lettuce *L. serriola*, an annual or biennial archaeophyte common and widespread in England and coastal Wales, scattered elsewhere, first recorded in Ireland in 1996. It is found on roadsides, waste ground, gravel-pits and sea walls, occasionally shingle banks and dunes, often rapidly colonising freshly disturbed soil, recently spreading northwards and westwards. *Serriola* is Latin for 'serrated'.

Wall lettuce *Mycelis muralis*, a winter-green perennial common and widespread in England and Wales, scattered and probably introduced in Scotland and Ireland, on shaded walls, rock outcrops and hedge banks, and in woodland, woodland edge and scrub, especially on chalk and limestone but also on acidic rocks in some areas. In the Burren it grows on limestone pavement. *Muralis* is Latin for 'of walls'.

2) **Lamb's lettuce** Alternative name for common cornsalad.

LETTUCE (SEAWEED) See sea lettuce.

LEUCOBRYACEAE Family name for swan-neck mosses (Greek *leucos* + *bryon* = 'white' + 'moss').

LEUCOIIDAE Family name (Greek *leucos* = 'white') of crabs including Bryer's and Pennant's nut crab.

LEUCTRIDAE Stonefly family (Greek *leucos* = 'white') comprising six widespread species of *Leuctra* favouring fast-water streams and some lake margins. Fisherman's needle fly.

LIBELLULIDAE Dragonfly family, also known as chasers, skimmers, gliders and darters, represented by 14 species. *Libellula* is Latin for 'small book', but the origin is more convincingly from *libella* = 'level' and by derivation 'dragonfly' from the level position of the wings.

LICE 1) Small wingless insects in the order Phthiraptera (Greek *phtheir* + *aptera* = 'lice' + 'wingless'), with two suborders, **Mallophaga** (biting or chewing lice, found on birds) and **Anoplura** (sucking lice, found on mammals) that live as generally host-specific ectoparasites. Bodies are flattened and often possess strong claws, two features that allow them to cling to their host's body. A louse will tend to spend all its life on its host, but it may transfer from one host individual to another. Food is blood, scraped skin or feather, and feeding often leads to irritation in the host, especially if there is an infestation of lice. **Body lice** (*Pediculus humanus humanus*) and **head lice** (*P. h. capitis*) parasitise humans. 'Louse' is from Old English *lus*, plural *lys*, of Germanic origin. *Pediculus* is Latin for 'louse'.

2) **Fish lice** Ectoparasitic brachiurans (order Arguloidea, family Argulidae), with one common British species, *Argulus foliaceus*, with scattered records, mostly in England, from marine, brackish and especially fresh waters. They cause the severe disease state argulosis in many (especially freshwater) fish species, including those of aquacultural value. *Argulus* is New Latin for 'lazy one'; *foliaceus* = 'leafy'.

LICHEN Or lichenised fungi: fungi (mycobionts) with a symbiotic relationship sustained by photosynthesising photobionts, viz. green algae (Chlorophyta) (phycobionts) and/or blue-green algae (Cyanobacteria) (cyanobionts) that form identifiable growth forms (thallus, thalli). They are often pioneers of unvegetated surfaces such as rock (saxicolous lichen) or tree bark (corticolous lichen). Crustose lichen grow completely attached to the substrate. Squamulose lichen have scale-like outgrowths. Foliose lichen are leaf-like with distinctly different upper and lower surfaces. Fruticose lichen are shrubby, branched, beard-like or strap-shaped. Lichenised fungi belong to the Ascomycetes. Some 1,873 species were listed by Smith *et al.* (2009), and while a few are now no longer present (so the actual figure may be around 1,800) new examples are being discovered each year. 'Lichen' dates from the early seventeenth century, coming via Latin from Greek *leichen*.

Rarities include **bat's wing lichen** *Collema nigrescens* (a foliose jelly lichen), **beaded rim lichen** *Lecanora cinereofusca* (first recorded in Britain from Loch Sunart in 1983 on willow and rowan, and since on hazel, ash, holly, oak and willow in Glen Creran, Glen Nant and Glen Stockdale in Argyll, and

at Coille Tokovaig on Skye), **bean-spored rim lichen** *Lecania dubitans* (crustose, on aspen in Speyside), **bird-perch gristle lichen** *Ramalina polymorpha* (fruticose), **black and grey shrubby lichen** *Bryoria bicolor* (fruticose), **black tufted lichen** *Pseudephebe pubescens* (fruticose), **crowded sunk-shield lichen** *Immersaria athroocarpa* (crustose), **ear-lobed dog lichen** *Peltigera lepidophora*, **flesh-coloured alpine lichen** *Placopsis gelida* (crustose), **forked hair lichen** *Bryoria furcellata* (caespitose, in Caledonian pinewood), **fringed shield lichen** *Parmelina carporrhizans* (foliose), **Gold-eye lichen** *Teloschistes chrysophthalmus* (fruticose), **golden pine lichen** *Vulpicida pinastri* (mainly eastern Highlands), **Lilliput ink lichen** *Placynthium asperellum*, **matt felt lichen** *Peltigera malacea* (north-east Scotland), **mealy-cracked leafy lichen** *Ramalina pollinaria* (fruticose and epiphytic), **pitted stone lichen** *Verrucaria calciseda* (crustose, in Co. Clare and Co. Sligo), **spiky starburst lichen** *Imshaugia aleurites* (foliose), **stump lichen** *Cladonia botrytes* (Highlands), **two-toned rock foam lichen** *Stereocaulon symphycheilum* (Skye, Cumbria and north-west Wales), and **yellow rail lichen** *Cyphelium tigillare* (crustose, mainly Highlands).

Abraded camouflage lichen See camouflage lichen below.

Alternating dog lichen *Peltigera didactyla* (Peltigeraceae), foliose, widespread, if scattered in Ireland, on moss or recently disturbed soil, for example roadsides, banks, quarries and urban waste ground. Greek *pelte* = 'small shield', and *di* + *dactylos* = 'two' + 'finger'.

Antler lichen *Pseudevernia furfuracea* (Parmeliaceae), fruticose, widespread, especially in the north, on acid bark, wood, less often on siliceous rock. It is sensitive to nitrogen pollution, so a good indicator of clean air. The flattened, strap-like lobes branch like antlers, prompting the English name. Greek *pseudes* = 'false' + the genus *Evernia*, itself from *evernes* = 'sprouting well' (loosely, 'branched'), and Latin *furfur* = 'bran'.

Aromatic lichen *Toninia aromatica* (Bacidiaceae), crustose, scattered but locally abundant in England and Wales, common in Ireland, in base-rich rock crevices and lime mortar in walls. Despite its English and specific name it is not especially aromatic. *Toninia* is probably from Greek *tonos* = 'something stretched'.

Asterisk lichen *Arthonia radiata* (Arthoniaceae), crustose, common and widespread on nutrient-rich smooth bark. The black branched apothecia (lirellae) are roughly star-shaped, prompting the English name. The etymology of *Arthonia* is unknown; *radiata* is Latin for 'rayed'.

Barnacle lichen *Thelotrema lepadinum* (Graphidaceae), fairly widespread, commoner in western regions, on deciduous trees (on rough bark, resembling barnacles), an indicator of ancient woodland, preferring shade, and sensitive to air pollution. Greek *thele* + *trema* = 'nipple' + 'hole', and *lepas* = 'limpet'.

Beaded rosette lichen *Physcia tribacia* (Physciaceae), foliose, scattered, commonest in southern Britain, on sheltered, nutrient-rich bark and on vertical walls and rock. Greek *physce* = 'blister' and Latin *tribax* = 'worn down'.

Beard lichen Members of the genus *Usnea* (Parmeliaceae), fruticose and generally draping over branches. Very sensitive to sulphur dioxide pollution. *Usnea* comes from Arabic *oshnah* = 'moss'.

Beret lichen Species of the fruticose genus *Baeomyces* (Baeomycetaceae) (Greek *baios* + *myces* = 'slender' + 'fungus').

Bitter wart lichen *Pertusaria amara* f. *amara* (Pertusariaceae), warty and crustose, widespread and common on deciduous tree bark, occasionally on moss on rock. Latin *pertusus* = 'perforated', and *amara* = 'bitter'.

Black and blue lichen *Toninia sedifolia*, crustose, widespread and locally common, on basic rocks, lime mortar, soil, calcareous turf and dunes. *Sedifolia* is Latin for 'sedum-like leaves'.

Black-eye lichen *Tephromela atra* (Mycoblastaceae), crustose, common and widespread on siliceous or mildly basic rocks, often dominant in coastal sites. Greek *tephros* + *melas* = 'ash-coloured' + 'black', and Latin *atra* = 'black'.

Black hair lichen Alternative name for ciliate strap lichen.

Black saddle lichen *Peltigera neckeri*, widespread but scattered on soil among moss. Noël Necker (1730–93) was a Belgian botanist.

Black-shielded lichen Alternative name for black-eye lichen.

Black tar lichen *Verrucaria maura* (Verrucariaceae), widespread, very common on acid rocks in the upper littoral zone, probably excluded from lower tidal levels by grazing. The dark, often black colour resembles an oil spill. Latin *verruca* = 'wart', and Greek *mauros* = 'dark'.

Black woolly lichen *Ephebe lanata* (Lichinaceae), common in much of western Britain, especially Scotland, and in Ireland, on wet, acidic, often siliceous rocks in upland areas. Greek *ephebos* = 'a youth', and Latin *lanata* = 'woolly'.

Blistered camouflage lichen *Xanthoparmelia loxodes* (Parmeliaceae), foliose, widespread, mainly lowland and coastal on well-lit rocks, in some inland sites typical of basalt outcrops. Greek *xanthos* = 'yellow' + the genus *Parmelia*, and *loxos* = 'slanting'.

Blistered lichen *Lasallia pustulata* (Umbilicariaceae), foliose, scattered, commoner in upland western Britain and south-west Ireland, often abundant on nutrient-rich, acidic, coastal or upland rocks. Greek *las* = 'stone', and Latin *pustula* = 'blister'.

Bloody comma lichen *Arthonia cinnabarina*, crustose, widespread and locally common, especially in western Britain and southern England, on nutrient-rich, smooth bark. Greek *cinnebari* here means 'red-coloured'.

Bloody heart lichen *Mycoblastus sanguinarius* (Mycoblastaceae), crustose, widespread in upland regions on bark and branches of deciduous trees, on siliceous rocks, and growing over moss. Black discs break to reveal a red pigment resembling flecks of blood, giving its English and specific (*sanguinarius* = 'bloody') names. Greek *myces* + *blastos* = 'fungus' + 'branch'.

Blue blister lichen Alternative name for black and blue lichen.

Blue-grey rosette lichen *Physcia caesia*, foliose, pollution-tolerant, common and widespread on urban walls, tarmac and similar substrates as well as in upland areas. Latin *caesia* = 'blue-grey'.

Board lichen *Trapeliopsis flexuosa* (Trapeliaceae), crustose, widespread and locally common on fences and stumps, and recorded from acid bark and siliceous rock. Greek *trapelos* + *opsis* = 'changeable' + 'appearance', and Latin *flexuosa* = 'bent'.

Borrerian lichen *Punctelia borreri* (Parmeliaceae), foliose, mainly in southern England and south Wales, scattered and expanding its range, on well-lit nutrient-rich tree bark, less often rocks. *Punctelia* is from Latin for 'spotted' (punctured). William Borrer (1781–1862) is the so-called father of British lichenology.

Brittle lichen or **brittle globe lichen** *Sphaerophorus fragilis* (Sphaerophoraceae), fruticose, common and widespread on rock and scree in upland regions. Greek *sphaera* + *phoros* = 'ball' + 'carrying'.

Broom beard lichen *Usnea wasmuthii*, common and widespread in western and northern Britain and in north-west Ireland, on trunks and branches of deciduous trees.

Brown cobblestone lichen *Acarospora fuscata* (Acarosporaceae), crustose, calcifuge, widespread and common on well-lit, nutrient-rich siliceous rocks and walls, and also found in south-east England and Suffolk. Greek *acares* + *spora* = 'tiny' + 'seed', and Latin *fuscata* = 'somewhat brown'.

Brown mushroom lichen *Baeomyces rufus*, fruticose, widespread on moist, often shaded, disturbed soil or litter. Latin *rufus* = 'red'.

Brown shingle lichen *Fuscopannaria sampaiana*, on basic bark, probably extinct in England and scarce in Wales (a few records around Snowdon), but present in the central Highlands and Southern Uplands and becoming more frequent in woodlands of sheltered valleys in the western Highlands. It is rare in south-west Ireland, in moss and bark in old woodland. Latin *fuscus* + *pannus* = 'dark' or 'brown' + 'rag'.

Buck's-horn lichen *Cladonia cervicornis* ssp. *cervicornis* (Cladoniaceae), squamulose and fruticose, common and widespread on acidic heathland and moorland soil and on dunes. Latin *cervus* + *cornu* = 'deer' + 'horn'.

Button lichen Crustose species in the genus *Buellia* (Caliciaceae).

Camouflage lichen *Melanelixia subaurifera* (Parmeliaceae), fruticose, widespread and abundant on smooth-barked branches and twigs of deciduous trees in well-lit sites, the dull green or brown colour prompting the English name. Greek *melas* = 'black' (from the thallus underside) + *lix* = 'lye' or 'ashes', and Latin *subaurifera* = 'somewhat ear-like'.

Chewing gum lichen Alternative name for radiated wall lichen, from the appearance of the discarded gum often found flattened on pavements.

Ciliate strap lichen *Heterodermia leucomela* (Physciaceae), fruticose, Endangered in RDB, restricted to the extreme west of England, Wales and south-west Ireland, with the largest populations in Cornwall and the Scilly Isles, and outlying populations in Pembrokeshire, Caernarfonshire (Lleyn Peninsula) and Anglesey, in coastal heaths on mosses and short turf, sometimes in rock crevices. Greek *heteros* + *derma* = 'different' + 'skin', and *leucos* + *melos* = 'white' + 'black', referring to the contrasting colours of the upper and lower surfaces.

Cinder lichen *Aspicilia cinerea* (Megasporaceae), crustose, on siliceous rocks, scattered in upland regions, but distribution uncertain owing to confusion with *A. intermutans*. Greek *aspis* + *cilium* = 'shield' + 'hair', and Latin *cinerea* = 'grey'.

City dot lichen *Scoliciosporum chlorococcum* (Scoliciosporaceae), crustose, very pollution-tolerant, common and widespread on damp, shaded, nutrient-rich bark. Greek *scolex* + *spora* = 'worm' + 'seed', and *chloros* + *coccos* = 'green' + 'grain'.

Clam lichen or **common clam lichen** See olive and black imbricated lichen below.

Clementine lichen *Physcia clementei*, a foliose rosette lichen, scattered mainly in the southern half of England, in Ireland, and parts of Wales and western Scotland on nutrient-rich bark, rocks and walls.

Clustered water lichen *Dermatocarpon luridum* (Verrucariaceae), foliose, locally abundant on wet rocks in northern and western Britain, usually on the margins of lochs and rivers, often periodically submerged. Greek *derma* + *carpos* = 'skin' + 'fruit', and Latin *luridum* = 'pale yellow'.

Cobblestone lichen See brown and rimmed cobblestone lichens, above and below respectively.

Comma lichen See bloody comma lichen above.

Common beard lichen *Usnea subfloridana*, fruticose, common and widespread, draping on trunks and, especially, branches. *Floridana* is Latin for 'having many flowers'.

Common coal dust lichen *Polysporina simplex* (Acarosporaceae), widespread on siliceous rocks, granite memorials, slate roofs and walls, as well as on weakly calcareous rocks. Greek *poly* + *spora* = 'many' + 'seed', and Latin *simplex* = 'simple'.

Common cup lichen *Cladonia pyxidata*, squamulose and fruticose, widespread and generally abundant on tree-trunks, rotting wood, well-drained humus-rich soil and heathland, and in degraded peaty habitat. *Pyxidata* is Latin for 'boxed'.

Common greenshield lichen *Flavoparmelia caperata* (Parmeliaceae), foliose, widespread and locally common, more scattered in eastern England, absent from the Highlands and northern Scotland, on trunks of deciduous trees and on other base-poor substrates. Latin *flavus* + *parma* = 'yellow' + 'small shield', and *caperata* = 'wrinkled'.

Common ink lichen *Placynthium nigrum* (Placynthiaceae), common and widespread on natural and manufactured calcareous substrates, including mortar. See ink lichen below.

Concentric boulder lichen *Porpidia crustulata* (Porpidiaceae), crustose, widespread and often common on upland siliceous rock and scree on exposed slopes and along sheltered streams.

Concentric lichen *Rhizocarpon petraeum*, crustose, widespread and common in northern and western regions on base-poor rocks, often on stone buildings and dry-stone walls. Greek *rhiza* + *carpos* = 'root' + 'fruit', and Latin *etraeum* = 'growing among rocks'.

Confluent-shielded lichen *Lecidea confluens* (Lecideaceae), scattered and widespread on non-basic rock. Greek *lecis* = 'small dish', and Latin *confluens* = 'confluence'.

Confused brown-headed cup lichen *Cladonia ramulosa*, squamulose and fruticose, widespread and often abundant on wet acidic heathland and rotting wood. *Ramulosa* is Latin for 'small branched'.

Coralloid rosette lichen *Heterodermia japonica* (Physciaceae), foliose, locally common in southern and south-west England, western Wales, western Scotland and Ireland, on tree-trunks or over mosses on rocks. Greek *heteros* + *derma* = 'different' + 'skin'.

Covered lichen *Pertusaria hymenea*, warty and crustose, widespread and common on tree-trunks, occasionally on rocks. Greek *hymen* = 'membrane'.

Crab's eye lichen *Ochrolechia parella* (Pertusariaceae), crustose, common and widespread including on rocks in the upper littoral zone. The large apothecia are said to resemble crab eyes. Greek *ochros* + lechia = 'pale' + 'words'.

Cracked ruffle lichen *Parmotrema reticulatum* (Parmeliaceae), foliose, locally common in southern England and southern Ireland, scattered up the west coast, on rocks and well-lit tree-trunks. Latin *parma* + Greek *trema* = 'small shield' + 'perforation', and Latin *reticulatum* = 'netlike'.

Crescent frost lichen *Physconia perisidiosa* (Physciaceae), foliose, widespread but local on nutrient-rich bark, occasional on walls.

Crescent map lichen *Rhizocarpon lecanorinum* (Rhizocarpaceae), crustose, locally abundant in upland Britain on base-poor though sometimes metal-rich rocks. Greek *lecane* = 'dish' or 'pot'.

Crisped gelatinous lichen *Collema crispum*, foliose, scattered in England and Wales on damp, often shaded calcareous substrates, for example walls, crumbling mortar and rocks. *Crispum* is Latin for 'wrinkled'.

Crowned lichen Alternative name for frilly fruited jelly skin.

Cudbear lichen *Ochrolechia tartarea*, crustose, mainly found in upland areas especially in the north and west, on base-poor, especially siliceous rocks, old walls and nutrient-poor bark. Cudbear is a purple dye extracted from this and other lichens used to dye wool and silk, a process patented in 1758 by Dr Cuthbert (Cudbear) Gordon, a Scottish chemist. Greek *tartaron* = 'crust'. See also corkir.

Daisy-flowered lichen *Cladonia bellidiflora*, widespread but rarely abundant in upland Britain, scattered in Ireland, mainly in moorland and upland heath, on soil, rocks and scree. Latin *bellidiflora* = 'daisy flower'.

Dark shadow lichen *Phaeophyscia sciastra* (Physciaceae), foliose, scattered and local in upland western Britain on rock, occasionally bark. Greek *phaios* + *physce* = 'dark' + 'blister', and *scia* = 'shadow'.

Delicate spurious cup lichen *Cladonia parasitica*, squamulose, widespread in Britain, scattered and coastal in Ireland, on decomposing wood.

Dog lichen Used in general for *Peltigera* (Greek *pelte* = 'small shield'), but specifically for *P. canina*, foliose, widespread but local and scattered on dune grassland and stabilised shingle and, inland, on grassland and heathland over gravelly and sandy soils. The fruiting bodies resemble dogs' teeth, prompting the English name, and dog lichens were in past times used to treat rabies. *Canina* is Latin for 'canine'.

Dot lichen Species in the generally widespread but scattered, mainly upland crustose genus *Bacidia* (Ramalinaceae), from Latin for 'little rod'.

Dust lichen *Lepraria incana* agg. (Stereocaulaceae), crustose but thin and powdery, scattered but mainly in Dorset, south-east

and north-west England, East Anglia, Wales and the Isle of Man on shaded acid rocks, walls and trunks. Greek *lepras* = 'rough' or 'scaly', and Latin *incana* = 'hoary'.

Earthy marsh lichen *Placynthiella uliginosa* (Trapeliaceae), crustose, fairly widespread on peaty and sandy heathland soils, and on decomposing wood. The genus name may be from Greek *plax* + the diminutive of *cynthos* = 'plate' + the mountain Cynthos (birthplace of Diana). Latin *uliginosus* = 'marshy'.

Elegant starburst lichen *Xanthoria elegans*, foliose, common and widespread on nutrient-enriched rocks, walls and concrete. Greek *xanthos* = 'yellow'.

Even grey lichen Alternative name for smooth loop lichen.

Fence lichen *Cetraria sepincola* (Parmeliaceae), fruticose, common on trees in central Wales, Cumbria, the Southern Uplands and the Highlands. Latin *cetra* = 'leather shield'.

Field dog lichen *Peltigera rufescens*, widespread among mosses on dunes and well-drained basic and ultrabasic soils. Latin *rufescens* = 'becoming red'.

Finger-scale foam lichen *Stereocaulon dactylophyllum* var. *dactylophyllum* (Stereocaulaceae), initially squamulose then tufted, locally common in upland areas on siliceous and metalliferous rocks. Greek *stereos* + *caulos* = 'hard' + 'stem', and *dactylos* + *phyllon* = 'finger' + 'leaf'.

Fingered cup lichen *Cladonia digitata*, squamose, widespread but uncommon, on damp trunks and decaying wood. *Digitata* is Latin for 'having fingers'.

Flaccid gelatinous lichen *Collema flaccidum*, in Ireland mainly in the north and west on acidic seepage rocks by streams, waterfalls and lough edges, and on damp mossy woodland rocks.

Floury dog lichen *Peltigera collina*, foliose, widespread but local generally in old woodland in upland regions on mossy trunks and rock in moist habitats. *Collina* is Latin for 'hill-loving'.

Flowery lichen See witches' whiskers lichen below.

Flute lichen See flute.

Foam lichen Species of *Stereocaulon* (Stereocaulaceae), from Greek *stereos* + *caulos* = 'hard' + 'stem'.

Fragrant gelatinous lichen *Collema fragrans*, mainly found in old-growth beechwood in the New Forest, around wound tracks on veteran beech trees. Elsewhere it is known from two trees in Devon, one in Dorset and one in Savernake Forest, Wiltshire.

Fringed button lichen Alternative name for netted rock tripe.

Fringed coastal rosette lichen *Physcia tenella* ssp. *marina*: see little ciliated lichen (ssp. *tenella*) below.

Fringed cup lichen *Cladonia fimbriata*, shade-loving, common and widespread on soil and decomposing wood, often on raised bog in Ireland. *Fimbriata* is Latin for 'fringed'.

Fringed rosette lichen *Physcia leptalea*, foliose, widespread, but somewhat scattered, local and declining, on slightly basic bark and branches. *Leptaleios* is Greek for 'slender'.

Frost lichen Foliose species of *Physconia* (Physciaceae), from Greek *physce* = 'blister' (i.e. inflated).

Gelatinous moss lichen *Polychidium muscicola* (Placynthiaceae), widespread in upland areas, especially the Highlands, on damp moss or soil, commonly at the base of trees, and over acidic rocks, often in and near streams. Greek *poly* + *chidion* = 'many' + 'hide', and Latin *muscicola* = 'moss-bearing'.

Glaucous leafy lichen Alternative name for frilly lettuce.

Globe lichen *Sphaerophorus globosus*, fruticose, locally common in upland regions in Britain and western Ireland on mossy, siliceous rocks, trees and peaty soil.

Golden dot lichen *Arthrorhaphis citrinella* (Arthrorhaphidaceae), widespread and scattered on moss and soil in crevices on siliceous upland rock faces. Greek *arthron* + *rhaphis* = 'joint' + 'needle', and Latin *citreus* = 'of the citrus tree' (i.e. 'yellow').

Golden-hair lichen *Teloschistes flavicans*, fruticose, mostly on short coastal grassland anchored to other lichens or bryophytes, or on rock, occasionally epiphytic on trees. It is very sensitive to pollution, is contracting its range, and is currently RDB Vulnerable in south-west England, south-west and north-west Wales, and south-west Ireland. The English name reflects its bright orange colour. *Flavicans* is Latin for 'yellowish'.

Golden shield lichen *Xanthoria parietina* (Teloschistaceae), foliose, widespread and common on nutrient-enriched bark (especially of elder), rock, and stone and concrete walls, especially near the coast. It is tolerant of nitrogen pollution. From its yellow colour it was once used to treat jaundice. Latin *paries* = 'wall'.

Granulated coral-crusted lichen *Pertusaria coccodes*, warty and crustose, on well-lit, nutrient-rich bark of deciduous trees, occasionally on smooth siliceous rocks, particularly in south and south-east England. Greek *coccos* + *eidos* = 'grain' + 'similar to'.

Great ciliated lichen Alternative name for eagle's claws.

Green loop lichen *Hypotrachyna sinuosa* (Parmeliaceae), foliose, widespread and locally abundant in western Britain and western Ireland on well-lit acidic rocks and trees. Greek *hypo* + *trachys* = 'beneath' + 'rough', and Latin *sinuosa* = 'curved'.

Green powdery stellated lichen *Physconia distorta* (Physciaceae), foliose, widespread and scattered, on nutrient-rich bark, occasionally on moss and rock. Greek *physce* = 'blister' (i.e. inflated) and Latin for 'distorted'.

Green satin lichen *Lobaria virens*, foliose and epiphytic, widespread, especially in upland areas, especially western Scotland, shade-tolerant on often acid bark in old deciduous woodland (an ancient woodland indicator). Greek *lobos* = 'lobe' and Latin *virens* = 'green'.

Green starburst lichen *Parmeliopsis ambigua* (Parmeliaceae), foliose, on conifer bark in central and eastern Scotland, having become widespread on acidified substrates as a result of sulphur dioxide pollution, now contracting in range as pollution levels decline. Latin *parma* + *ops* = 'small shield' + 'appearance', and *ambiguus* = 'doubtful'.

Green turfy lichen *Cladonia caespiticia*, widespread, commoner in southern England and western Britain on mossy banks, trees and stumps in sheltered old woodlands (an ancient woodland indicator). *Caespiticius* is Latin for '(made of) turf'.

Green veiny lichen See pixie gowns lichen below.

Grey cloudy lichen *Dermatocarpon miniatum* (Verrucariaceae), foliose, occasional in western Scotland, common in western Ireland, especially on calcareous rocks in the Burren, and on rocky seashores and around lakes on siliceous rocks subjected to periodic wetting. *Miniatum* is Latin for 'vermilion-coloured'.

Grey starburst lichen *Parmeliopsis hyperopta* (Parmeliaceae), foliose, on acid, especially conifer bark in central and eastern Scotland, having extended its range as a result of sulphur dioxide pollution, now rather rare as pollution levels decline. *Hyperopta* is Greek for 'above sight'.

Grey stone lichen Alternative name for crottle.

Grey tree lichen *Diploicia canescens* (Caliciaceae), crustose, common and widespread in England, Wales and Ireland, rather coastal in Scotland, on generally shaded rocks, walls and nutrient-enriched bark. Greek *diploos* + *icos* = 'double' + 'relating to', referencing the two-celled ascospores, and Latin *canescens* = 'becoming grey'.

Hairless-spined shield lichen Alternative name for New Forest parmelia.

Hammered shield lichen See netted shield lichen below.

Handwriting lichen *Graphis scripta* (Graphidaceae), crustose, widespread and common on smooth-barked broadleaf trees. Its growth pattern is of long, narrow, curved, often forked apothecia (or lirellae), such squiggles prompting its English name and the binomial, from Greek *graphe* = 'writing' and Latin *scripta* = 'written'. There are two forms: *serpentina* (commoner in Scotland) with curved lirellae and *recta* with straight lirellae, these growing in similar places but never on the same branch.

Heath lichen *Icmadophila ericetorum* (Icmadophilaceae), crustose, widespread, particularly common in the Highlands, on peat and decomposing wood. Greek *icmados* + *philos* = 'moisture' + 'loving', and Latin *ericetorum* = 'of heath'.

Hoary rosette lichen *Physcia aipolia*, foliose, common and

widespread on nutrient-rich bark of deciduous trees. Greek *aei* + *polios* = 'always' + 'grey'.

Hollowed lichen *Diploschistes scruposus* (Graphidaceae), crustose, common and widespread on siliceous or slightly basic nutrient-rich rocks. Greek *diploos* + *schistos* = 'double' + 'divided', and Latin *scruposus* = 'like a small sharp stone'.

Hooded rosette lichen *Physcia adscendens*, foliose, an indicator of nitrogen enrichment, common and widespread on bark of deciduous trees, and on high-nutrient rocks, walls and concrete. *Adscendens* is Latin for 'ascending'.

Horny-cupped lichen *Pachyphiale carneola* (Gyalectaceae), crustose, widespread but local and largely absent from central and eastern England, generally in old broadleaf woodland on mature trees, and an ancient woodland indicator. Greek *pachys* + *phiale* = 'thick' + 'broad flat bowl', and Latin *carneola* = 'fleshy'.

Horsehair lichen *Bryoria fuscescens*, fruticose, common and widespread in upland regions, scattered in the lowlands, on bark (favouring conifers), and mosses, old walls, and exposed silica-rich rocks. It is fairly sensitive to sulphur dioxide air pollution, and intolerant of ammonia. *Fuscescens* comes from Latin *fuscus* = 'dark'.

Imbricated lichen See olive and black and sinuous imbricated lichens below. (Imbricated means overlapping like roof tiles.)

Incurved yellow-warted lichen *Arctoparmelia incurva* (Parmeliaceae), foliose, on base-poor rocks in upland parts of north and west Britain, with a few modern records from Northern Ireland. Greek *arctos* = 'northern' (literally 'bear') + the genus *Parmelia* (from Latin *parma* = 'small shield').

Inelegant hollow-shielded lichen *Phlyctis agelaea* (Phlyctidaceae), crustose, widespread, scattered and rare in damp, sheltered woodland, on smooth bark of deciduous trees, occasionally on rock. Greek *phlyctis* = 'blister', and *agelaios* = 'gregarious'.

Inflated lichen Alternative name for dark crottle.

Ink lichen Species of *Placynthium* (Placynthiaceae), possibly from Greek *plax* + *cynthos* = 'plate' + the mountain Cynthos (birthplace of Diana).

Ivory rock lichen See sea ivory.

Jelly lichen Species of *Collema* (e.g. see river jelly lichen below).

Jointed lichen *Usnea articulata*, once widespread in southern England and Wales, but highly pollution-sensitive and, while recovering, remains rare (RDB Near Threatened, a UK BAP priority species) except from the New Forest west to central Cornwall, generally draped over branches. It is rare in Ireland, found in the south and west.

Kidney lichen Species of *Nephroma* (Greek *nephros* = 'kidney'), including the rare **arctic kidney lichen** *N. arctica* (with a very few records in the Highlands) and **reversed lichen**. Kidney-shaped apothecia are produced on the lower surface of the lobe tips.

Lakezone lichen *Staurothele fissa* (Verrucariaceae), crustose, scattered in upland regions on permanently submerged siliceous rocks in unpolluted rivers and streams. Greek *stauros* + *thele* = 'cross' + 'nipple', and Latin *fissa* = 'split'.

Leafy ash lichen *Ramalina fraxinea*, fruticose, common and widespread on preferably well-lit basic bark and branches, very sensitive to sulphur dioxide pollution, now recovering in the cleaner air. *Fraxinus* is Latin for 'ash tree'.

Lemon-coloured rock lichen *Psilolechia lucida* (Pilocarpaceae), crustose but powdery, common and widespread often under shade, on base-poor rock, especially crevices in dry-stone walls and urban brick and stonework. Greek *psilos* + *lechia* = 'naked' or 'smooth' + 'words', and Latin *lucida* = 'bright'.

Lemon-coloured wall lichen *Caloplaca citrina* agg. (Teloschistaceae), crustose common and widespread on rocks, and in towns on walls and other artificial substrates. True *Caloplaca citrina* does not occur in the British Isles, the name having been wrongly applied to a complex of species. Greek *calos* + *plax* = 'beautiful' + 'flat plate', and Latin *citrina* = 'lemon yellow'.

Lightfootian lichen *Fuscidea lightfootii* (Fuscideaceae), widespread if commoner in western regions, on bark and twigs, especially in damp woodland. Latin *fuscus* = 'dark' or 'brown'. John Lightfoot (1735–88) was a Scottish lichenologist.

Little ciliated lichen *Physcia tenella* ssp. *tenella*, foliose, widespread, often abundant on twigs, bark, walls and rocks, an indicator of nitrogen enrichment, and often the dominant lichen on tree-trunks in urban areas. Ssp. *marina* is found near the coast. *Tenellus* is Latin for 'tender'.

Little dot lichen *Mycoporum antecellens* (Mycoporaceae), crustose, widespread and scattered, locally frequent, mainly in upland regions, in old woodlands on the smooth bark of deciduous trees such as hazel, holly and rowan. Greek *myces* + *spora* = 'fungus' + 'seed', and the Latin *antecellens* = 'antechamber'.

Little emerald lichen *Acarospora smaragdula*, crustose, local on sheltered, slightly base-rich siliceous rocks, sites rich in heavy metals and on stones below metal run-off. *Smaragdos* is Greek for an emerald.

Little fleshy-shielded lichen *Massalongia carnosa* (Massalongiaceae), foliose, uncommon, widespread and scattered on wet streamside upland acidic mossy rocks. The binomial is Latin for 'mass' + 'long', and for 'fleshy'.

Loop lichen Species of *Hypotrachyna* (Greek *hypo* + *trachys* = 'beneath' + 'rough').

Lurid lichen *Romjularia lurida* (Lecideaceae), squamulose, scattered though mostly in western Britain and western Ireland, on basic rock. *Romjul* is Norwegian for the period between Christmas and New Year, the diagnosis and most of the description of the species having been undertaken in this period in 1995. Latin *lurida* = 'pale yellow'.

Many-dotted lichen *Pertusaria multipuncta*, warty and crustose, mainly in southern and western England, Wales and western Scotland, on deciduous trees with smooth, acidic bark in old woodland (an ancient woodland indicator). *Multipuncta* is Latin for 'many-spotted'.

Map lichen *Rhizocarpon geographicum*, crustose, widespread and common, especially in western and northern Britain, especially on exposed, base-poor rocks. Lichen patches grow next to each other, bordered by a black line of spores, giving the appearance of a map.

Mealy-rimmed shingle lichen *Pannaria conoplea* (Pannariaceae), foliose, epiphytic, scattered, local and declining in upland Britain, locally abundant in western Ireland, on basic bark and mossy rock in humid old woodland (an ancient woodland indicator). Latin *pannus* = 'tattered garment', *conoplea* possibly from *canopus* = 'canopy'.

Mealy shadow lichen *Phaeophyscia orbicularis*, foliose, widespread and common, on bark, occasionally rock. *Orbicularis* is Latin for 'circular'.

Mealy spreading lichen Alternative name for spiky starburst lichen.

Membranous dog lichen *Peltigera membranacea*, common and widespread among moss over soil and rock, on tree-trunks, and on damp turf on dunes and lawns.

Monk's hood lichen See dark crottle under crottle.

Moss lichen *Diploschistes muscorum*, crustose, widespread, especially common in the southern half of England and north-west England, initially parasitic on *Cladonia* species on the top of walls and on dunes and heaths, developing into independent thalli which spread over moss. Latin *muscorum* = 'mossy'.

Moss-shingle lichen *Protopannaria pezizoides* (Pannariaceae), crustose, local in generally upland northern and western regions on peat, damp rocks and mossy trunks. Greek *protos* = 'first' + the genus *Pannaria*, itself possibly from Latin *pannus* = 'tattered garment', and *pezis*, Greek for a sessile fungus.

Mottled-disc lichen *Trapeliopsis granulosa*, crustose, widespread, common on heathland and moorland on damp, peat banks, decaying tree stumps, plant debris and soil. *Granulosa* (Latin for 'granular') reflects the texture.

Mottled ink lichen *Placynthium tantaleum*, scarce and scattered, on intermittently submerged rock. The habitat is reflected in the specific name: Tantalus was a king of Phrygia, punished in Hades by standing up to his neck in water which flowed away when he tried to drink, and with overhead fruit moved away by wind when he tried to feed.

Netted shield lichen *Parmelia sulcata* (Parmeliaceae), foliose, common and widespread on tree branches and trunks, occasionally rocks, tolerant of pollution. Latin *parma* = 'small shield' and *sulcatus* = 'furrowed'.

Nipple lichen Alternative name for papillary lichen.

Oak lichen *Pyrrhospora quernea* (Lecanoraceae), crustose, common and widespread on acidic to intermediate, moderately nutrient-rich rough bark, not only of oak. Greek *pyrrhos* + *spora* = 'flame-coloured' + 'seed', and Latin *quernea* = 'oak-like'.

Obscure black-shielded lichen *Rinodina sophodes* (Physciaceae), crustose, widespread, commoner in upland regions, on smooth, nutrient-rich twigs, favouring ash. Greek *rhinos* + *dinos* = 'nose' + 'rolled around', and *sophodes* = 'wise'.

Oederian lichen *Rhizocarpon oederi*, crustose, locally abundant in upland regions on base-poor but iron-rich rock. *Oidos* is Greek for 'swelling'.

Old growth rag lichen *Platismatia norvegica* (Parmeliaceae), foliose, found in upland pinewoods, mainly the Highlands. Greek *platys* = 'flat', and Latin *norvegica* = 'Norwegian'.

Olive and black imbricated lichen *Hypocenomyce scalaris* (Ophioparmaceae), crustose, common and widespread in Britain, local and scattered in Ireland, on acid-barked trees, fences and sheltered, siliceous rocks. Greek *hypo* + *cenos* + *myces* = 'beneath' + 'empty' + 'fungus', and Latin *scala* = 'ladder'.

Orange-fruited elm lichen *Caloplaca luteoalba*, crustose, greatly declined since the 1970s with the loss of elms from Dutch elm disease. It is listed in the UK BAP, classified as Endangered in the RDB, and included in EN's Species Recovery Programme. It is currently only known in England from three trees in Norfolk, Suffolk and Oxfordshire where it grows on black poplar or horse-chestnut. It still occurs on elm in eastern Scotland. *Luteoalba* is Latin for 'yellow-white'.

Orange sea lichen *Caloplaca marina*, crustose, widespread, on rock and walls in and just above the supralittoral zone.

Pale bark-cup lichen *Ramonia dictyospora* (Gyalectaceae), possibly endemic but local, widespread and scattered, on nutrient-rich, usually slightly decomposed bark, mostly on the trunks of mature broadleaf trees in open woodland, especially by streams. Latin *ramus* = 'branch', and Greek *dictyon* + *spora* = 'net' + 'seed'.

Papillary lichen *Pycnothelia papillaria* (Cladoniaceae), crustose and granular, on peat mainly in the Highlands, local elsewhere, including western Ireland. Greek *pycnos* + *thele* = 'compact' + 'nipple', and Latin *papilla* = 'nipple'.

Parchment lichen *Lobaria amplissima* (Lobariaceae), foliose and epiphytic, widespread if rare and declining in upland Britain, especially western Scotland, southern England, and scarce in Ireland, in old deciduous woodland on bark and branches. *Amplissima* is Latin for 'largest'.

Pearly lichen *Parmotrema perlatum*, foliose, widespread, common in much of western and southern Britain, recently mostly absent from central and eastern England because of sulphur dioxide pollution, but now recolonising as pollution levels drop. It is found on acid bark and siliceous rock in well-lit sites. *Perlatum* is Latin for 'completed'.

Pepper moon lichen Alternative name for shiny brown-shield lichen.

Peppered moon lichen *Sticta fuliginosa* (Lobariaceae), foliose, common and widespread in upland regions on mossy trees and rocks, generally in old moist forest. Greek *stictos* = 'dotted' and Latin *fuliginosa* = 'sooty'.

Peppered rock-shield lichen *Xanthoparmelia conspersa*, foliose, widespread and locally common, especially in upland regions, on exposed, base-poor rocks and walls. *Conspersa* is Latin for 'sprinkled'.

Pepperpot lichen *Pertusaria pertusa* (Pertusariaceae), crustose, widespread and common on trees, rock and walls. The apothecia have narrow openings like the top of a pepper pot, and the binomial is from Latin *pertusus* = 'perforated'.

Perforated lichen See flute.

Pincushion starburst lichen *Xanthoria polycarpa* (Teloschistaceae), foliose, common and widespread on nutrient-enriched trees, an indicator of nitrogenous air pollution. Greek *xanthos* = 'yellow', and *poly* + *carpos* = 'many' + 'fruit'.

Pink bullseye lichen *Placopsis lambii* (Agyriaceae), crustose, on damp, siliceous, upland rocks, frequently by streams or lakes, often on metal-rich sites.

Pitted lichen *Lobaria scrobiculata*, foliose and epiphytic, scattered, especially in upland areas, generally declining but most abundant in the western Highlands, on shaded trees in old woodland. Latin *scrobiculata* = 'small trench'.

Pixie foam lichen *Stereocaulon pileatum*, widespread but uncommon, on metal-rich rock, including quarries and tips, in shaded, moist conditions. *Pileatum* is Latin for 'covered with a cap'.

Pixie gowns lichen *Peltigera venosa*, foliose, uncommon, most records coming from northern England and Scotland, especially the Highlands, on sparsely vegetated open upland sites, mine spoil and forest trails. Its English name was the winning suggestion when, in 2010, *The Guardian* launched a competition for the public to suggest common names for previously unnamed species, though **green veiny lichen** had been used in the nineteenth century. *Venosa* is Latin for 'veined'.

Plum-fruited felt lichen *Degelia plumbea* (Pannariaceae), foliose, in well-lit woodland and scrub on basic mossy bark, typically on hazel and ash in western Ireland, and sometimes found near the sea on silica-poor rocks. *Plumbea* is Latin for 'lead-coloured'.

Pollution lichen *Lecanora conizaeoides*, crustose, common and widespread though scattered in northern Scotland and Ireland, on tree-trunks and branches, fences, walls, etc. It is the most tolerant of all lichens to sulphur dioxide, a useful indicator of acidic pollution, and particularly common in cities (often the only species surviving in polluted cities in the mid- to late twentieth century), but now increasingly being replaced by less pollution-tolerant species as air has become cleaner.

Powdered moon lichen Alternative name for floury sticta.

Powdered speckled shield lichen *Punctelia subrudecta*, foliose, widespread in the southern half of Britain and the Isle of Man (scarcer further north) and in Ireland, locally common on trees and siliceous rock, spreading into urban areas with the reduction in sulphur dioxide pollution.

Powdery starburst lichen *Xanthoria ulophyllodes*, foliose, found in eastern Britain on nutrient-enriched bark. Greek *oulos* + *phyllodes* = 'curly' + 'leaves'.

Prickly lichen *Cetraria muricata*, fruticose, common and widespread, especially in upland areas, forming mats on acid heathland, fallen tree-trunks and mossy rocks. *Muricata* is Latin for 'pointed'.

Puffed shield lichen Alternative name for dark crottle.

Purple rock lichen Alternative name for black crottle.

Radiated wall lichen *Lecanora muralis* (Lecanoraceae), crustose, common and widespread on stone substrates as well as concrete, paving slabs and tarmac. It is resistant to air pollution. Greek *lecane* = 'dish', and Latin *murus* = 'wall'.

Ragged hoary lichen Alternative name for oak moss.

Red beard lichen *Usnea rubicunda*, widespread but mainly in southern and western Britain and western Ireland, draping over trunks and branches. *Rubicunda* is Latin for 'red'.

Red-eyed shingle lichen *Pannaria rubiginosa*, on mossy, nutrient-rich bark in sheltered, humid, old woodland in western Britain (especially upland Scotland) and western Ireland, where it is also found in hazel and willow coppice. *Rubiginosa* is Latin for 'rusty-coloured'.

Reindeer lichen *Cladonia rangiferina* (Cladoniaceae), fruticose, found in the Highlands and, locally in other mainly upland sites, on branches and on the ground. *Rangifer* is Latin for 'reindeer'. It is an important food for reindeer in Scandinavia and North America.

Reversed lichen *Nephroma laevigatum* (Nephromataceae), foliose, common and widespread in upland western regions in mature oceanic woodlands on mossy tree-trunks, damp rocks and walls, and an ancient woodland indicator. One of the kidney lichens, Greek *nephros* = 'kidney', and Latin *laevigatum* = 'smooth'.

Rimmed cobblestone lichen *Acarospora glaucocarpa*, crustose, scattered and locally common on calcareous rock, rarely on granite, often where seasonally flushed with water. Not recently recorded in Ireland. Greek *glaucocarpa* = 'glaucous-fruited'.

Rimmed wart lichen *Pertusaria velata*, crustose, on the bark of mature and veteran tree-trunks, especially of ash, beech and oak, in sheltered but well-lit sites in ancient woodland and parkland. There is a viable population in the New Forest, but elsewhere there has been a major decline in range, with just two recent records in Scotland and one in Wales, and RDB considers it to be Vulnerable. *Velata* is Latin for 'veiled'.

Ring lichen Species of Arctoparmelia; see incurved yellow-warted lichen above.

River jelly lichen or **river gelatinous lichen** *Collema dichotomum*, on boulders and rocks in fast-flowing, mostly upland streams where it is submerged for at least three-quarters of the year, avoiding shade and acidic waters. In Wales it is Critically Endangered, and in England it is found in the upper reaches of the River Coquet, Northumberland, but with a significant population in Scotland it is classified as Vulnerable in the RDB. Its one known site in Northern Ireland is on basalt rock in the Glenarm River, Co. Antrim. Greek *dicha* + *tomos* = 'in two' + 'cut'.

Rock dimple lichen *Gyalecta jenensis* var. *jenensis* (Gyalectaceae), crustose, common and widespread on damp and often shaded calcareous rocks. Var. *macrospora* is rare, growing on siliceous rocks, especially granite. Once thought to be confined to the Channel Islands, since 2000 it has been found in west Cornwall and the Scillies, on granite boulders and old walls. Greek *gyalon* = 'hollow vessel'.

Rock foam lichen *Stereocaulon saxatile*, scattered in the Highlands in high gullies, favouring gravelly soil. *Saxatile* is Latin for 'growing among rocks'.

Rock hair lichen Alternative name for horsehair lichen.

Rock shingle lichen *Vahliella leucophaea*, squamulose, local, mostly in western Britain, and in western and northern Ireland, in damp crevices in often basic coastal rocks, sometimes on tree bases. Martin Vahl (1749–1804) was a Danish/Norwegian botanist. The paradoxical specific name is from Greek *leucos* + *phaios* = 'dusky' + 'bright'.

Rosette lichen Species of *Physcia* (Physciaceae) (Greek *physce* = 'blister' or 'inflated').

Rusty-rock lichen *Tremolecia atrata* (Hymeneliaceae), crustose, widespread in upland regions, locally common on siliceous, iron-rich rocks. The binomial is Greek for 'trampling', and Latin for 'clothed in black'.

Rusty-shielded lichen Name given to both *Caloplaca crenularia* and *C. ferruginea*, crustose, the former found on base-poor rocks, typically near the coast and locally common inland on damp rocks and old walls, the latter recently split off from the former, with a scattered distribution including more inland sites, commoner and increasing in Ireland, and found on tree bark. *Crenularia* is New Latin for 'notched', *ferruginea* for 'rusty'.

Sanguineous lichen Alternative name for bloody heart lichen.

Saturnine gelatinous lichen *Leptodium saturninum* (Collemataceae), a jelly lichen locally abundant in northern Britain, especially the Highlands, on conifer bark, occasionally on acidic rocks among mosses, at moderately high altitude. Greek *leptos* = 'slender'.

Scaly dog lichen *Peltigera praetextata*, widespread and locally common on soil, mossy boulders, tree bases and bark, and on moist, heathy banks. *Praetextata* is Latin for 'protected'.

Scaly ink lichen *Placynthium flabellosum*, with scattered records mainly from the Highlands. *Flabellum* is Latin for 'small fan'.

Scaly mottled-disc lichen *Trapeliopsis wallrothii*, crustose, mainly on thin soil and rocks on the west coast of Britain. Karl Wallroth (1792–1857) was a German botanist.

Scaly stippled lichen *Endocarpon pusillum* (Verrucariaceae), squamulose, Near Threatened in the RDB, in small populations on the Dorset coast from Durlston to Portland on slumping cliffs and old stabilised spoil heaps from quarries. Other records have been from south Wales and Yorkshire. Greek *endon + carpos* = 'within' + 'fruit', and Latin *pusillum* = 'very small'.

Scarlet cup lichen *Cladonia coccifera*, common and widespread on humus-rich soils, but also on heathland, occasionally on wood. *Coccos* is Greek for the cochineal insect, which provides a red dye, hence *coccifera* = 'scarlet'.

Scarlet lichen *Haematomma ochroleucum* (Haematommataceae), powdery and crustose, common and widespread in Britain, scattered in Ireland, on dry surfaces of siliceous or slightly calcareous rocks and walls. Greek *haima* = 'blood', and *ochros + leucos* = 'pale' + 'white'.

Scrambled egg lichen *Fulgensia fulgens* (Teloschistaceae), known from six sites in England (Isle of Wight westwards), and in south-west Wales in the Castlemartin Cliffs and Dunes SSSI, and at Brownslade and Linney Burrows, all on blown sand over limestone, and on the cliff edge at Pen-y-holt Down and Flimston Down. It usually grows directly on the acrocarpous moss *Trichostomum crispulum*. The English name reflects the overlapping lemon-yellow scales which have white lobed margins. Latin *fulgens* = 'shining'.

Script lichen See white script lichen and handwriting lichen, below and above respectively.

Seaweed lichen *Lichina confinis* (Lichinaceae), fruticose, found from the middle to upper intertidal, especially in northern and western regions where sheltered and exposed to direct sunlight. *Confinis* is Latin for 'related'.

Shadow lichen Foliose species in the genus *Phaeophyscia* (Physciaceae) (Greek *phaios + physce* = 'dark' + 'blister').

Shaggy strap lichen *Ramalina farinacea*, fruticose, widespread and common on nutrient-rich bark and branches. *Farina* is Latin for 'flour'.

Shiny brown-shield lichen *Melanelixia fuliginosa* (Parmeliaceae), foliose, common and widespread on siliceous rocks and deciduous tree bark and fences. Greek *melas + helix* = 'black' + 'twisted', and Latin *fuliginosa* = 'sooty'.

Short perforated lichen *Cladonia uncialis* ssp. *biuncialis*, squamulose and fruticose, widespread on damp heathland, moorland and bog edges, mainly coastal in England though also inland in northern England. *Uncialis* is Latin for the twelfth part of something.

Shrubby starburst lichen *Xanthoria candelaria*, foliose, widespread on nutrient-enriched bark, wooden posts and rocks. *Candela* is Latin for candle.

Sinoper lichen *Acarospora sinopica*, crustose, on nutrient-poor rocks in upland western Britain, and north-west and east Ireland. The English and specific names refer to Sinop in Turkey, the source of a reddish-brown earth pigment, the colour of this lichen.

Sinuous imbricated lichen Alternative name for smooth loop lichen.

Skull lichen Alternative name for crottle.

Slender cup lichen *Cladonia gracilis*, squamulose, common and widespread on well-drained acid woodland floors and decomposing wood, dry parts of heathland and moorland, and (in Ireland) on dunes. *Gracilis* is Latin for 'slender'.

Smooth loop lichen *Hypotrachyna laevigata*, foliose, common and widespread in upland western Britain and Ireland on hardwood bark in damp woodland, occasionally on rocks and overgrowing moss. *Laevigata* is Latin for 'smooth'.

Snow lichen *Flavocetraria nivalis* (Parmeliaceae), on montane heath in the Highlands, scarce and declining further with a warming climate and drier summers. Latin *flavus + cetra* = 'yellow' + 'leather shield', and *nivalis* = 'of snow'.

Socket lichen *Solorina saccata* (Peltigeraceae), foliose, widespread in upland and some coastal regions on base-rich rock and calcareous grassland. Latin *saccata* = 'sack-like'.

Sooty-knobbed lichen *Cyphelium inquinans* (Caliciaceae), crustose, common in the southern half of England and central Wales, scattered elsewhere in Britain, two records in southern Ireland, on dead wood and acid bark. Greek *cyphos* = 'curved', and Latin *inquinans* = 'staining'.

Speckled lichen *Lecanographa lyncea* (Roccellaceae), crustose, scattered in England (mostly in the south and in East Anglia) and Wales on dry, generally well-lit bark of mature oak and other hardwoods (an ancient woodland indicator). Greek *lecane + graphe* = 'dish' + 'drawing'. Lynceus was a sharp-sighted Argonaut in Greek mythology.

Speckled sea-storm lichen *Cetrelia olivetorum* (Parmeliaceae), foliose, scattered and local on mossy well-lit acid-barked trees in moist, often mature woodland. Latin *cetra* = 'leather shield', and *olivetorum* = 'olive-coloured'.

Spongy gelatinous lichen *Solorina spongiosa*, foliose, scarce, scattered in Britain, recent Irish records only from Co. Sligo and Co. Wicklow, in moist, often grazed or compacted calcareous grassland.

Stag lichen Alternative name for oak moss.

Star rosette lichen *Physcia stellaris*, foliose, widespread but scattered and local, especially found in northern and western parts, possibly extending its range, growing on deciduous trees.

Starburst lichen Species of *Xanthoria* (Greek *xanthos* = 'yellow').

Starry breck lichen *Buellia asterella* (Caliciaceae), crustose, growing in the Suffolk Breckland but probably becoming extinct during 1991–2002 because open grassland closed up to deny it the required light and calcareous mineral soil. *Asterella* is from Greek for 'little star'.

Stellated star lichen *Psora decipiens* (Psoraceae), crustose, widespread but local on coastal and montane calcareous sand or mineral-rich soil. Greek *psora* = 'itch', and Latin *decipiens* = 'deceiving'.

Straggling lichen *Psoroma hypnorum* (Pannariaceae), mainly in northern and western Britain, rare and coastal elsewhere, on soil and mossy trunks and in dune grass. Greek *psora + oma* = 'itch' + 'morbid', and *hypnos* = 'sleep'.

String-of-sausage lichen *Usnea articulata* (Parmeliaceae), fruticose, locally common in southern England from the New Forest to Cornwall, and in Pembrokeshire and the Brecon Beacons, and rare elsewhere in Wales and in south and south-west Ireland. It has disappeared from other regions owing to a sensitivity to sulphur dioxide air pollution. It is most frequently found hanging over branches and trunks of trees, favouring open sites. When well grown, its straggly, hair-like stems that can be up to a metre long, swell at intervals along the main branches to create sausage-like strings 3 mm in diameter. *Osnah* is Arabic for 'moss' or 'lichen', *articulata* Latin for 'jointed'.

Stygian lichen *Melanelia stygia* (Parmeliaceae), foliose, local in the Cairngorms and scattered in the western Highlands. Greek *melas* = 'black', and *stygia* (relating to the River Styx in Greek mythology) meaning 'very dark'.

Sulphurous mountain lichen *Lecanora orosthea*, crustose, common and widespread in Britain, scattered in Ireland, on siliceous rock and walls, and on bark, for example of beech and birch. Greek *oros + thea* = 'mountain' + 'aspect'.

Sulphurous pin-headed lichen *Chaenotheca furfuracea*

(Coniocybaceae), crustose, widespread, commonest in upland Britain, on soil, in shaded humid crevices, and on roots, unbarked stumps, mosses and rock. Greek *chaeno* + *thece* = 'grasp' + 'case', and Latin *furfur* = 'bran'.

Tailed loop lichen *Hypotrachyna taylorensis*, foliose, widespread and locally abundant in western Britain and south-west Ireland on mossy trees, especially oak, and rocks in old woodland. Greek *hypo* + *trachys* = 'beneath' + 'rough'. Thomas Taylor (1775–1848) was an English botanist.

Tar lichen See black tar lichen above.

Thrush lichen *Peltigera leucophlebia*, foliose, widespread in upland regions among short grasses or moss in moist, often calcareous habitats. Greek *leucos* + *phleps* = 'white' + 'vein'.

Tiny button lichen *Amandinia punctata* (Caliciaceae), crustose, common and widespread on preferably acidic bark, fencing and rock, and tolerant of air pollution. Latin *amanda* = 'deserving of love', and *punctata* = 'spotted'.

Tubular lichen *Hypogymnia tubulosa* (Parmeliaceae), foliose, common and widespread favouring acid substrates, such as siliceous rocks, acid-barked trees and heather. It is an indicator of old-growth woodland, and rather sensitive to air pollution. Greek *hypo* + *gymnos* = 'beneath' + 'naked'.

Unpolished lichen *Arthonia pruinata*, crustose, widespread and locally common, especially in western Britain and southern England, on old, well-lit, smooth-barked trees. *Pruinata* is Latin for 'frosted'.

Variable-shielded lichen *Lecanora varia*, crustose, common and widespread, less so in western Ireland, on wood though rarely on bark and usually on horizontal surfaces.

Veiled black-shielded lichen *Porpidia speirea*, crustose, in upland areas of northern and western Britain, rare and scattered in Ireland, on base-rich rocks such as limestones and schists. *Speira* is Greek for 'something wrapped around'.

Vesicle-shielded lichen *Pertusaria albescens*, common and widespread on bark of deciduous trees. *Albescens* is Latin for 'becoming white'.

Wall lichen *Xanthoria parietina*, foliose, widespread, very common on nutrient-enriched bark (especially on elder), stonework and cement, often abundant on coastal rocks, and having responded positively to nitrate/ammonia deposition from air pollution and bird droppings. *Paries* is Latin for 'wall'.

Wart lichen Crustose lichens in the genus *Pertusaria* (Latin *pertusus* = 'perforated').

White coral-crusted lichen *Pertusaria corallina*, locally common in upland and coastal northern and western Britain and in Ireland on exposed, base-poor rock. *Corallina* is Latin for 'coral-red'.

White paint lichen *Phlyctis argena* (Phlyctidaceae), crustose, common and widespread, on bark of deciduous trees and basic walls, the smoothness and white colour giving its English name. Greek *phlyctis* = 'blister', and *arges* = 'bright' or 'white'.

White script lichen *Graphis alboscripta* (Graphidaceae), a rare endemic epiphyte found in hazel woodland on the west coast of Scotland. Its growth pattern is of long, narrow, curved, often forked apothecia (or lirellae), such squiggles prompting its English name and the binomial, from Greek *graphe* = 'writing' and Latin *alba* + *scripta* = 'white' + 'written'.

Whitish radiating lichen *Solenopsora candicans* (Catillariaceae), rosette-forming, widespread in England and Wales, rare in Scotland and Ireland, on hard limestone, gravestones and monuments. The thallus has radiating, irregular marginal lobes, prompting the English name. Greek *solen* + *opsis* = 'pipe' + 'appearance', and Latin *candicans* = 'becoming white'.

Witches' whiskers lichen *U. florida*, mainly found in south and south-west England and Wales, though declining (it is sensitive to ammonia pollution), considered as Near Threatened in the RDB, and a priority species in the UK BAP. It is scattered elsewhere, and found in broadleaf tree canopies. It contains usnic acid, an effective antibiotic and antifungal agent which can halt infection and is effective against tuberculosis. This lichen lent itself to treating surface wounds before sterile gauze and antibiotics. Its English name was the winning suggestion when, in 2010, *The Guardian* launched a competition for the public to suggest common names for previously unnamed species, though in fact it had previously been called **flowery lichen**. *Florida* is Latin, here meaning 'having many flowers'.

Wood blood-clot lichen *Strangospora microhaema* (order Lecanorales), confined to mossy, acid-barked trees in boggy ancient woodland and wood pasture in high rainfall areas in west Scotland, and five trees in three known sites in mid- and west Wales. It is almost certainly under-recorded because of its tiny size (the red apothecia being 0.05–0.3 mm diameter). Greek *strangos* + *spora* = 'twisted' + 'seed', and *micros* + *haema* = 'small' + 'blood'.

Wrinkled shield lichen Alternative name for common greenshield lichen.

Yellow candle lichen Alternative name for shrubby starburst lichen.

Yellow-edged frost lichen *Physconia enteroxantha*, foliose, widespread, commoner in Wales, northern Britain and northern and eastern Ireland on well-lit nutrient-rich bark, occasionally on walls. Greek *enteron* + *xanthos* = 'intestine' + 'yellow'.

Yellow ground lichen Alternative name for scrambled egg lichen.

Yellow pine dust lichen *Chrysothrix flavovirens* (Chrysothricaceae), crustose, widespread and probably increasing in range, on dry, well-lit, usually nutrient-poor wood and bark. Greek *chrysos* + *thrix* = 'gold' + 'hair', and Latin *flavovirens* = 'yellow-green'.

Yolk of egg lichen *Candelariella vitellina* f. *vitellina* (Candelariaceae), crustose, common and widespread on nutrient-rich siliceous rocks and walls, and on wood including bark. *Vitellus* is Latin for 'yolk'.

LIFT (MOTH) Leaf-mining moths in the genera *Antispila* (Greek *anti* + *spilos* = 'opposite' + 'spot', from forewing markings) and *Heliozela* (*helios* + *zelos* = 'sun' + 'eagerness', from the behaviour of flying in bright sunlight) (Heliozelidae). Wingspans are generally 5–7 mm.

Alder lift *H. resplendella*, found in many parts of southern Britain and throughout Ireland. Adults fly from late May to July, often in sunshine around branches of alder, the leaves of which are mined by the larvae. *Resplendeo* is Latin for 'to shine', from the forewing gloss.

Birch lift *H. hammoniella*, widespread, adults flying in May–June. Larvae mine leaves of (preferentially young) birch in July–August. *Hammonia* is the latinised form of Hamburg.

Four-spot lift *A. metallella*, wingspan 8–9 mm, a locally common dogwood-mining moth with a scattered distribution in England and the Welsh Marches. Adults fly in May, preferring sunshine. *Metallum* is Latin for 'metal', from the metallic gloss on the forewing.

Oak satin lift *H. sericiella*, with a scattered distribution in the southern half of Great Britain and throughout Ireland. Adults emerge in May and are found on or near oak. Larvae initially mine oak twigs, later entering the leaf. *Sericus* is Latin for 'silky', from the forewing gloss.

Yellow-spot lift *A. treitschkiella*, with a scattered distribution in southern England. Adults fly in June–July. Larval mines in dogwood are evident in August–October. Georg Treitschke (1776–1842) was a German entomologist.

LILAC *Syringa vulgaris* (Oleaceae), a strongly suckering deciduous south-east European neophyte shrub or small tree introduced by 1597, scattered throughout, though scarce in north-west Scotland and western Ireland, as a relic of cultivation in sites of former habitation, or naturalised in hedges, on roadsides, railway banks and waste ground. 'Lilac' dates from the early seventeenth century, from obsolete French, via Spanish and Arabic from Persian *lilak*, a variant of *nilak* = 'bluish'. Greek *syrinx* = 'pipe' (i.e. hollow stem) and Latin *vulgaris* = 'common'.

LILIACEAE Family name for lily (Latin *lilium*), including fritillary and tulip.

LILY 1) Bulbous herbaceous perennials in the genera *Lilium* (Latin for 'lily') and *Gagea* (honouring the English botanist Sir Thomas Gage, 1719–87), in the family Liliaceae, 'lily' itself coming from Old English *lilie*.

Martagon lily *L. martagon*, a European neophyte present by Tudor times, though its presence in some ancient woods far from habitation, for example near Tintern, Lower Wye Valley, suggests a few populations may be native. It is scattered in lowland regions usually as small clumps near woodland edges or in coppiced woodland. Martagon means 'child of Mars'.

Pyrenean lily *L. pyrenaicum*, a neophyte present by Tudor times found in woodland, woodland edge, hedgerows and road-sides, often as isolated clumps originating as garden throw-outs. With a generally scattered distribution, sizeable, long-established and naturalised populations are known from south-west England, south Wales and east Scotland.

Snowdon lily *G. serotina*, confined to damp, often shaded rock ledges and crevices at five sites in Snowdonia. This is the county flower of Caernarfonshire as voted by the public in a 2002 survey by Plantlife. *Serotina* is Latin for 'late-flowering'.

2) **Cuban lily** Alternative name for Portuguese squill.

3) **Jersey lily** *Amaryllis belladonna* (Alliaceae), a bulbous herbaceous South African neophyte perennial cultivated in the Channel Islands, and a common relic in old fields and rough ground there, and as a casual in Scilly and Cornwall. Amaryllis was a shepherdess in Greek mythology, coming to mean 'sparkling'. *Belladonna* is Italian for 'beautiful lady'.

4) **May lily** *Maianthemum bifolium* (Asparagaceae), a rare rhizomatous herbaceous perennial of free-draining acidic soils in oak-birch woodland, very local in Co. Durham, Yorkshire and north Lincolnshire, introduced in a few other places. The binomial is from Greek for 'May blossom', and Latin for 'two-leaved'.

5) **Peruvian lily** *Alstroemeria aurea* (Alstroemeriaceae), a herbaceous neophyte perennial from Chile, very scattered in Britain, naturalised as a garden escape on rough ground. Baron Clas von Alstroemer (1736–94) was a Swedish botanist. Latin *aurea* = 'gold'.

6) **Radnor lily** Alternative name for early star-of-bethlehem.

LILY BEETLE, SCARLET LILY BEETLE *Lilioceris lilii* (Chrysomelidae, Criocerinae), a pest of garden lilies and fritil-laries native to Eurasia. In 1939 a colony was found in a garden at Chobham, Surrey. The species initially spread slowly, expanding into much of southern England by 1990, but over the next two decades it spread more rapidly and by 2009 had been found in every English county. It was first reported from Scotland and Northern Ireland in 2002 where it has become widespread. The first report from the Republic of Ireland was in 2010 and it is now established here. Both adults (8–10 mm long, evident in April–September) and larvae (6–8 mm long) feed on and can defoliate lily leaves, adults also feeding on flowers and seed pods. Larvae disguise themselves as bird droppings by covering them-selves in excrement. Listed as fourth in the Royal Horticultural Society's list of worst garden pests of 2014, sixth in 2016. Latin *lilium* = 'lily' and Greek *ceras* = 'horn'.

LILY-OF-THE-VALLEY *Convallaria majalis* (Asparagaceae), a rhizomatous herbaceous perennial of free-draining, nutrient-poor soils, most frequent in ash woods on limestone, but also found in hedge banks, limestone pavement grikes, acidic woods in south-east England, scrub, and in a fen in Cumbria. It is also commonly naturalised as a garden escape. Latin *convallaria* = 'of the valley', *majalis* = '(month of) May'.

False lily-of-the-valley *Maianthemum kamtschaticum*, a North American neophyte, naturalised and spreading in

woodland in West Porlock Wood, Somerset, since 1983. Also recorded from mid-west Yorkshire and south Lancashire. The binomial is Greek for 'May flowers', and Latin for 'from Kamchatka'.

LIMACIDAE Family name for keelback slugs, from Latin *limax* = 'slug' (and 'snail').

LIMACODIDAE Family name for the festoon and triangle moths (Greek *leimax* + *eidos* = 'garden' + 'form'), but referring to the Latin for 'slug' (*limax*) and the shape and movement of the larva.

LIME *Tilia* x *europaea* (*T. cordata* x *T. platyphyllos*) (Malvaceae), a deciduous tree, found in a few woods with both parents, but widely planted, readily propagated from suckers and therefore common in woodland, scrub, shelter belts, avenues, parkland, roadsides, and as an urban street tree. 'Lime' dates from the mid-seventeenth century, from the Provençal *limo* and Arabic *lima*. *Tilia* is Latin for this tree.

Large-leaved lime *T. platyphyllos*, local and scattered in England and Wales in old deciduous woodland on generally calcareous soils, typically as a large tree or coppice stool, less common in Scotland and Ireland. It is often planted on roadsides, and in gardens, parkland and plantations. Seedlings are frequent, but saplings rare. Vegetative reproduction is by new shoots from the tree base. *Platyphyllos* is Greek for 'broad-leaved'.

Small-leaved lime *T. cordata*, mostly found in England and Wales in mixed deciduous oak or ash woodland on a range of soil types, and often on steep slopes and cliffs. Regeneration by seed occurs, mainly in southern England, but is rare. Vegeta-tive reproduction is by shoots growing from fallen trees, or by layering, and it often occurs as an ancient coppiced tree. It has been much planted in parks and as a street tree. *Cordata* is Latin for 'heart-shaped', describing the leaves.

LIME BEETLE *Stenostola dubia* (Cerambycidae), a longhorn beetle up to 15 mm long with a scattered distribution in England and Wales, Nationally Scarce in the RDB. Larvae feed in lime. Greek *stenos* + *stole* = 'narrow' + 'garment', and Latin *dubia* = 'uncertain'.

LIMICOLAE Alternative name for Charadriiformes as (migra-tory) shore birds.

LIMNANTHACEAE Family name of meadow-foam (Greek *limne* + *anthos* = 'marsh' + 'flower').

LIMNEPHILIDAE Caddis fly family, including land caddis (Greek *limne* + *philos* = 'pond' + 'loving').

LIMNICHIDAE From Greek *limne* = 'pond', a beetle family with a single British species, *Limnichus pygmaeus* (1.8 mm length), scarce and scattered in southern and eastern England, on wet rocks in the splash zone of running water. It can withdraw its appendages tightly against the underside of the body to appear dead.

LIMPET Gastropod molluscs, from Old English *lempedu*, from medieval Latin *lampreda* = both 'limpet' and 'lamprey'.

American slipper limpet *Crepidula fornicata* (Caliptaeidae), shell length up to 5 cm, introduced from North America to Essex between 1887 and 1890, now recorded from Cardigan Bay round to Spurn Head, with introductions to Ireland not persisting. They are typically attached to shells (especially mussels and oysters) and stones on soft substrates around the low water mark and the shallow sublittoral, commonly in curved chains of up to 12 animals, large shells at the bottom of the chain. Greek *crepis* = 'slipper', and Latin *fornicata* = 'arched over'.

Atlantic plate limpet Alternative name for common tortoise limpet.

Black-footed limpet *Patella depressa* (Patellidae), shell length up to 3 cm, mainly found on the south and west coasts of Britain on exposed rocky shores from the middle to the lower intertidal, grazing on microalgae. *Patella* is Greek for 'pan' or 'plate', *depressus* Latin for 'low'.

Blue-rayed limpet *P. pellucida*, shell length up to 2 cm, common and widespread except around the Wash, on macroalgae in the lower eulittoral to a depth of 27 m. Latin *pellucida* = 'transparent'.

China limpet *P. ulyssiponensis*, shell width up to 6 cm, widespread around Britain except between the Humber estuary and the Isle of Wight, and found in north-east Ireland, grazing on microalgae on exposed rocky shores (at densities of up to 1,000 per m²) from the lower intertidal to the shallow sublittoral, and in shallow rock pools on the middle shore and on overhanging rocks.

Common keyhole limpet *Diodora graeca* (Fissurellidae), with a conical shell up to 4 cm width, widespread on the south and west coasts of Britain and of Ireland on rocks and under stones in the low water zone, feeding on sponges. *Diodos* is Greek for 'passageway', *graeca* Latin for Greek.

Common limpet *P. vulgata*, shell up to 6 cm wide, common and widespread on substrates firm enough for its attachment from the high intertidal to the sublittoral fringe. It is abundant on all rocky shores of all degrees of wave exposure although highest densities coincide with high exposed conditions. Although it grazes on microalgae, it is generally not abundant where there is dense growth of macroalgae. *Vulgata* is Latin for 'common'.

Common slipper limpet See American slipper limpet above.

Common tortoise limpet *Testudinalia testudinalis* (Lottiidae), shell up to 2.5 cm wide, mostly from north England, Scotland and north-east Ireland, from the midshore to depths of 50 m on boulders and small smooth stones, especially those that have red algae. Latin *testudo* = 'tortoise'.

Keyhole limpet See common keyhole limpet above.

Lake limpet *Acroloxus lacustris* (Acroloxidae), 3 mm shell diameter, 2 mm width, widespread and locally common in often stagnant ponds, lakes and slowly flowing vegetation-rich waters. Greek *acron* + *loxos* = 'summit' + 'slanting', and Latin *lacuster* = 'lake'.

River limpet *Ancylus fluviatilis* (Planorbidae), shell length 5–8 mm, a common and widespread bottom-living pulmonate snail found in unpolluted fast-flowing water on the sides of stones in moderate flows. Greek *ancylos* = 'curved', Latin *fluvius* = 'river'.

Rough limpet Alternative name for China limpet.

Slit limpet *Emarginula fissura* (Fissurellidae), shell up to 10 mm long and 8 mm high, on rocks and boulders among sponges (on which it feeds) from the low intertidal to greater depth around much of the coastline but absent from the east English Channel and southern North Sea. The binomial is Latin for 'unbordered' and 'fissure'.

Wautier's limpet *Ferrisia wautieri* (Planorbidae), shell length 4 mm, a European introduction first recorded in Britain from hothouse tanks in Glasgow in 1931 and discovered in the wild in 1976. Now occasional and scattered in England and south Wales in still or slowly flowing waters attached to aquatic plants.

LINACEAE Family name of flax (Latin *linum*) and allseed.

LINDEN Old name for lime tree.

LINE (MOTH) Double line *Mythimna turca* (Noctuidae, Hadeninae), wingspan 37–45 mm, nocturnal, scarce, contracting in distribution, found in parks, open woodland, damp grassland, woodland edges and scrub in parts of Cornwall, Devon and Somerset, south-west and central Wales, Cheshire and Co. Wexford (one record in 2010). Adults fly in June–July. Larvae feed at night on grasses. Mithimna is a town on Lesbos, Greece. *Turca* is Latin for the dye Turkey red (obtained from madder), for the wing colour.

LINES (MOTH) Treble lines *Charanyca trigrammica* (Noctuidae, Xyleninae), wingspan 35–40 mm, nocturnal, common in grassland, heathland, gardens and open woodland, and on dunes and verges throughout England and Ireland, more local in Wales. Adults fly in May–July. Larvae generally live below the surface and chew through the stems of plants such as plantains, knapweeds and dandelion at ground level. Greek *chara* + *nyctos* = 'delight' + 'night', and Latin *trigrammica* = 'three-lined', describing the forewing pattern. Not to be confused with clay triple-lines.

LING (PLANT) Alternative name for heather, from Old Norse *lyng*.

LINNET *Linaria cannabina* (Fringillidae), wingspan 24 cm, length 14 cm, weight 19 g, a resident breeding finch, widespread (commoner along the east coast) on heathland, rough ground, hedges and saltmarsh, and in parks and gardens, with 410,000 territories in Britain in 2009. Numbers declined by 55% from 1970 to 2015, but while populations in England and Wales continue to decline, if only slightly, those in Scotland and Northern Ireland are increasing. It is Red-listed in Britain, Amber-listed in Ireland. Nests are in bushes, usually with 4–5 eggs which hatch after 13–14 days, the altricial young fledging in 13–14 days. Diet is mainly of seeds, nestlings also taking insects. 'Linnet' dates from the early sixteenth century, from Old French *linette*, in turn from *lin* = 'flax', the bird feeding on seeds of flax (hemp) from which linen is made. Latin *linarius* + *cannabis* = 'linen weaver' + and 'hemp'.

LINSEED Old English *lin* + *sæd* = 'seed' + 'flax'; see flax.

LINYPHIIDAE Family name of money spiders (Greek *linyphos* = 'linen clothes': another name for this group is sheet weavers, from the web shape), with 280 British species in 122 genera, 1.5–3.5 mm long (Erigoninae) or 1.2–7.2 mm long (Linyphiinae). See also horse-head and invisible spiders under spider, and horrid ground weaver under weaver.

LIOCRANIDAE Family name (Greek *leios* + *cranos* = 'smooth' + 'helmet') of running foliage spiders, with 12 British species in 5 genera, 2.5–6 mm body length.

LIQUORICE Wild liquorice *Astragalus glycyphyllos* (Fabaceae), a straggling perennial milk-vetch, widespread but scattered, commonest in England, in woodland edges and scrubby grassland on railway banks and verges, mainly on calcareous soils. It has sometimes been used as a tea. Greek *astragalos* = 'ankle bone', from the seed shape, and *glycis* + *phyllon* = 'sweet' + 'leaf'.

LITTLE-ROBIN *Geranium purpureum* (Geraniaceae), local in the Channel Islands, southern and south-west England, south Wales and southern Ireland, with ssp. *purpureum* an upright annual in stony or rocky habitats near the sea, on sheltered cliffs, disused railway lines, and particularly by roads and fields on Cornish hedge banks where numbers are increasing. Ssp. *forsteri* is a prostrate plant of stabilised tops of shingle beaches. Greek *geranos* = 'crane', from the beak-like fruit.

LIVERWORT Members of the phylum Marchantiophyta (named in honour of the French botanist Nicolas Marchant, d.1678), non-vascular plants with single-celled rhizoids, some species growing as a flattened leafless thallus (thallose), but most being 'leafy' (foliose) and looking like a flattened moss. See the orders Calobryales, Marchantiales and Metzgeriales. The name comes from the old belief that this was a plant that cured diseases of the liver, probably derived from the appearance of some thalloid liverworts, which resemble a liver in outline.

Rarities include **Dumortier's liverwort** *Dumortiera hirsuta* (Glenarm Glen, Co. Antrim, and south-west Scotland), and **hyaline liverwort** *Athalamia hyalina* (Highlands).

Common liverwort *Marchantia polymorpha* (Marchantiaceae), common and widespread on soil or rock, in wet, shaded areas, possibly symbiotic with the fungus liverwort navel. *Polymorpha* is Greek for 'many-shaped'.

Crescent-cup liverwort *Lunularia cruciata* (Lunulariaceae), common and widespread, except for the Highlands and central Ireland, in damp, shady habitats such as the bases of walls and path edges. It can be a weed of gardens and greenhouses. *Lunularia* is Latin for 'little moon', *cruciata* for 'crossed'.

Great scented liverwort *Conocephalum conicum* (Conocephalaceae), thallose, common and fairly widespread, usually on damp, shady, mildly base-rich to neutral substrates, for example rocks by rivers, streams and waterfalls. The strong-smelling thalli can grow to 17 mm wide, large for a liverwort. It occasionally produces capsules which have cone-shaped heads, hence the binomial from Greek *conos* + *cephale* = 'cone' + 'head', and Latin for 'conical'.

Hemisphaeric liverwort *Reboulia hemisphaerica* (Aytoniaceae), thallose, common and widespread, often on calcareous soil on boulders, and in crevices in cliffs, rocks, limestone pavement and walls. Eugène de Reboul (1781–1851) was an Italian botanist.

Narrow mushroom-headed liverwort *Preissia quadrata* (Marchantiaceae), thallose, commoner in upland regions, favouring moist, montane and subalpine limestone, schist and other calcareous rock ledges. It also grows in mossy, calcareous turf, in flushes, fen and dune slacks down to sea level. Johann Preiss (1811–83) was a German-born British naturalist. *Quadrata* is Latin for 'in fours'.

Orobus-seed liverwort *Targionia hypophylla* (Targioniaceae), thallose, with a scattered distribution, favouring seasonally dry sites, especially on limestone, thin soil on rocky banks, and wall crevices. *Hypophylla* is Greek for 'under leaves'.

Pale liverwort *Chiloscyphus polyanthos* (Lophocoleaceae), foliose, locally common and widespread in fen, mixed with mosses, and on leaf litter and wood. Greek *cheilos* + *cyphos* = 'edge' + 'curved', and *poly* + *anthos* = 'many' + 'flowers'.

Scented liverwort See great scented liverwort above.

Umbrella liverwort Alternative name for common liverwort, from its umbrella-like gametophores (reproductive structures).

LIZARD Reptiles in the family Lacertidae, dating from late Middle English, from Old French *lesard*, from Latin *lacertus* = 'lizard'. See also slow worm.

Common lizard *Zootoca vivipara*, common and widespread in Britain, more local and scattered in Ireland, length 15–16 cm, in heathland, woodland clearings, disused railway tracks, open meadows and hedgerows. Adults emerge from hibernation in March. Mating is in April–May. In spring both sexes bask in the open to absorb the heat from the sun; in summer it is usually the pregnant females seen basking. In July, the young are born in an egg-sac that breaks either during birth or soon afterwards (ovi-vivipary). Adults feed on small insects, spiders, and other invertebrates, including centipedes, small snails and earthworms. The species is protected under the WCA (1981). Greek *zoon* + *tocos* = 'animal' + 'offspring' (i.e. bringing forth young), and Latin *vivus* + *pareo* = 'alive' + 'bring forth' (viviparous).

Sand lizard *Lacerta agilis*, length 16–20 cm, a rarity only naturally found on sandy heathland in Surrey, Dorset and Hampshire and coastal sand dune systems in Merseyside. A captive breeding programme has reintroduced the species to further sites in these areas and restored its range with releases in north Wales, Devon, Cornwall and West Sussex. They emerge from hibernation in March or April. Mating occurs in May–June. This is the only native lizard to lay eggs (oviparity), laid in burrows dug by the female in loose sand during June–July. The eggs hatch in one or two months depending on weather. The binomial is Latin for 'lizard' and 'agile'.

Viviparous lizard Alternative name for common lizard.

Wall lizard *Podarcis muralis*, length up to 20 cm, a European introduction with a scattered number of small colonies in southern England; 31 viable populations were reported in 2012. At Boscombe, Bournemouth, for example, 50–60 individuals were released onto sandstone cliffs in 1992; numbers in 2017 exceeded 1,300. They require very particular microhabitats – south-facing and often near-vertical, with multiple refuges (crevices or vegetation), and close to sandy substrate or rocks for egg-laying. Greek *podargos* = 'swift-footed', and Latin *murus* = 'walls'.

LOACH Freshwater fishes. 'Loach' is a Middle English word, from Old French *loche*.

Spined loach *Cobitis taenia* (Cobitidae), length 8–10 cm, weight 20–50 g, a rarity found in slow-flowing rivers, lakes, canals, (stagnant) ponds, ditches and reservoirs, mainly in the east Midlands and East Anglia. Preferring oxygen-rich waters, it nevertheless tolerates low oxygen, gulping air from the surface. Active at night, riverbed sand containing small animals and other organic material is ingested, then, stripped of nutrients, ejected through the gills. Spawning is in April–June. Greek *cobitis* = '(a kind of) sardine', and Latin *taenia* = 'band'.

Stone loach *Barbatula barbatula* (Balitoridae), length 14 cm, common and widespread except for the northern half of Scotland, and introduced to Ireland, in clean, fast-flowing waters, searching at night for invertebrate food in stream-bed gravel and stones. Latin *barbatula* reflects the presence of barbels.

LOB WORM, LOBWORM Alternative name for common earthworm, or (blow) lugworm.

LOBE SHELL *Philine aperta* (Philinidae), a sea slug up to 7 cm long, the shell being internal, with a scattered distribution, hunting small infaunal prey under sand or mud in the sublittoral. Secretion of sulphuric acid from the skin protects it from predators. *Aperta* is Latin for 'open'.

LOBELIA Species of *Lobelia* (named in honour of the Flemish botanist Mathias de L'Obel, 1538–1616), *Downingia* (after the American gardener Andrew Downing, 1815–52) and *Pratia* (after a French naval officer, Ch.L. Prat-Bernon, d.1817), in the family Campanulaceae.

Californian lobelia *D. elegans*, an annual neophyte, probably introduced with grass seed, noted in 1978, recorded from a few English locations (e.g. in Sussex and Buckinghamshire) and apparently increasing in range as a casual and sometimes naturalised in grassy habitats. *Elegans* is Latin for 'elegant'.

Garden lobelia *L. erinus*, an annual South African neophyte commonly escaping from gardens, naturalised over much of Britain, especially in England, on rough ground, and in pavement cracks. *Erinus* is a name used by Dioscorides for a plant similar to basil.

Heath lobelia *L. urens*, a rare rhizomatous perennial herb found locally in southern England on rough pasture and grassy heathland on infertile acidic soil that is often seasonally waterlogged, germination apparently stimulated by disturbance. *Urens* is Latin for 'stinging'.

Lawn lobelia *P. angulata*, an invasive herbaceous perennial New Zealand neophyte, occasional and scattered in Kent, Surrey and Scotland (e.g. Edinburgh since the 1930s) as a weed on damp lawns. *Angulata* is Latin for 'angular'.

Water lobelia *L. dortmanna*, a rosette-forming perennial herb of oligotrophic lakes with acidic substrates, locally common in Cumbria, Wales, upland Scotland and western Ireland. It is slow-growing, with little ability to withstand shade or competition, and is confined to water <2 m deep. Dortmann was a sixteenth-century apothecary in Groeningen.

LOBSTER Decapod crustaceans in the family Nephropidae, having long bodies with muscular tails, which live in crevices or burrows on the sea floor. Three of their five pairs of legs have claws, including the first pair, which are usually larger than the others. Not to be confused with squat lobsters.

Common lobster *Homarus gammarus*, up to 1 m in length, though 50 cm is more usual, common and widespread on rocky substrates, living in holes and excavated tunnels from the lower intertidal to 60 m depth. They emerge to feed at night when they scour the seabed for marine worms, starfish, other crustaceans and carrion. It is widely caught using lobster pots, with a minimum landing size of 8.7 cm carapace length, and a closed season from October to December. In 2013, the UK exported

7,400 tonnes worth £75 million. *Homar* is Old French and *gammarus* Latin for 'lobster'.

Montagu's plated lobster Alternative name for black squat lobster, the name honouring the British soldier and naturalist George Montagu (1753–1815).

LOGANBERRY *Rubus loganobaccus* (Rosaceae), a hybrid (*R. idaeus* x *R. vitifolius*) spiny deciduous shrub accidentally created in 1881 in Santa Cruz, California, by horticulturist James Logan (1841–1928), grown for its fruit, and brought over to Europe in 1897. It has become naturalised as a garden escape in hedges and on roadsides, railway banks and waste ground. *Rubus* is Latin for 'bramble', *bacca* = 'berry'.

LONCHAEIDAE Family name of lance flies (Greek *lonche* = 'spear').

LONDONPRIDE *Saxifraga* x *urbium* (*S. spathularis* x *S. umbrosa*) (Saxifragaceae), a perennial stoloniferous herb of garden origin, widespread and naturalised in shaded or damp places in woods, by streams, on banks and walls, and among rocks. *Saxifraga* is Latin for 'rock-breaker'.

False Londonpride *Saxifraga* x *polita* (*S. spathularis* x *S. hirsuta*), locally naturalised in shaded or damp habitats in woods, by streams, on banks and among rocks, in north-west England, south-west and west Ireland, and scattered in Scotland. *Polita* is Latin for 'polished'.

Lesser Londonpride *S. cuneifolia*, introduced from southern Europe in 1768, found in Somerset, northern England and central Scotland as a local naturalised garden escape on rocks and old walls. *Cuneifolia* is Latin for leaves tapered at the base (*cuneus* = 'wedge').

Scarce Londonpride *Saxifraga* x *geum* (*S. umbrosa* x *S. hirsuta*), introduced from the Pyrenees by 1754, a local garden escape in damp, shaded, often rocky habitats, for example woodland, roadside banks and streamsides, in Cornwall, northern England and central Scotland.

LONG-HORN (MOTH) Day-flying moths in the family Adelidae. Antennae, especially in males, are long (often more than twice the length of the forewing). Larvae of most species build portable protective cases, and feed entirely or partially on leaf litter.

Brassy long-horn *Nemophora metallica* (Adelinae), wingspan 15–20 mm, scarce in southern England, East Anglia and a few sites in Wales, in dry grassland, chalk downland and some heathland. Adults fly in June–July. Larvae initially feed on seeds then leaves, mainly of scabious, and litter. Greek *nema* + *phoreo* = 'thread' + 'carry', from the long antennae.

Buff long-horn *Nematopogon metaxella* (Nematopogoninae), wingspan 15–17 mm, in south England and south Wales, and scattered in south Ireland, in damp woodland, marsh and fen. Adults fly in June–July in the afternoon and at dusk. Larvae feed on leaf litter. Greek *nema* + *pogon* = 'thread' + 'beard', from the long antennae and the hairy labial palpus. L. and T. Metaxa were nineteenth-century Italian naturalists.

Coppery long-horn *Nemophora cupriacella*, wingspan 12–16 mm, scarce and local, mostly in southern and eastern England, in dry grassy habitats, including downland. Adults are mainly evident in July. Larvae feed on scabious, initially on seeds, later on fallen and lower leaves. *Cupriacella* comes from Latin *cupreus* = 'copper', from the forewing colour.

Early long-horn *Adela cuprella* (Adelinae), wingspan 14–16 mm, local in the British Isles, particularly in south-west England and coastal parts of south Wales and East Anglia, in heathland, fen and some woodlands. Adults often 'dance' in groups above sallow, and feed from the flowers. Eggs are laid in sallow catkins but on hatching the larvae drop to the ground and feed in leaf litter. Greek *adelos* = 'unseen', from the larvae being 'unseen' in their case, and from Latin *cupreus* = 'coppery', from the metallic forewing colour.

Green long-horn *A. reaumurella*, wingspan 14–18 mm, common and widespread in England, Wales and southern Scotland, more local in east and south-east Ireland. Adults fly in May–June. Larvae feed on leaf litter. René de Réaumur (1683–1757), the French scientist who introduced the Réaumur temperature scale, was also a noted entomologist.

Horehound long-horn *Nemophora fasciella*, wingspan 13–16 mm, local mainly in East Anglia and the south-east Midlands. Adults fly in July. Larvae feed on black horehound, at first eating the seeds. *Fasciella* is Latin for 'small band', from the forewing markings.

Large long-horn *Nemophora swammerdamella*, largest of the long-horns, wingspan 17–21 mm, widespread and generally common in deciduous woodland. Adults fly in May–June. Larvae feed on leaf litter and overwinter twice. Jan Swammerdam (1637–80) was a Dutch entomologist.

Little long-horn *Cauchas fibulella* (Adelinae), wingspan 8–11 mm, scattered but often common in Britain. Adults fly in June, especially in sunshine, visiting flowers of the food plant, germander speedwell. Larvae feed initially on seeds, later on leaves close to the ground. *Fibulella* is Latin for 'small clasp', from the merged pattern on the folded forewings.

Meadow long-horn *C. rufimitrella*, wingspan 9–12 mm, fairly common in southern Britain and in Ireland on damp pasture and damp heathland. Male antennae are twice, females' one and a half times as long as the forewing. Adults fly in May–June, visiting flowers of their food plants, cuckooflower and garlic mustard. Larvae feed within seed pods of these plants, then on the leaves. Latin *rufus* + *mitra* = 'red' + 'turban', from the adult's red head.

Pale-brown long-horn *Nematopogon pilella*, wingspan 15–19 mm, scarce, in open woodland and moorland in parts of north-west England, Scotland and south Wales, and on the Burren. Larvae feed on dead leaves of bilberry. *Pilella* is Latin for 'small hair', from the long antennae.

Sandy long-horn *Nematopogon schwarziellus*, wingspan 14–17 mm, widespread and fairly common throughout England, Wales, Ireland and north-east Scotland, in woodland, hedgerow, acid grassland, heathland and moorland. Adults fly early in the morning and from late afternoon to dusk in May–June. Larvae feed on dead leaves throughout winter and all the following year until the spring after that. C. Schwarz (d.1810) was a German entomologist.

Scarce long-horn *Nematopogon magna*, wingspan 17–18 mm, scarce (proposed as a future RDB species) on heathland and moorland in parts of north Scotland and southern Ireland, a vulnerable priority species in the UK BAP. Larvae feed on dead leaves. *Magna* is Latin for 'large'.

Small-barred long-horn *A. croesella*, wingspan 11–14 mm, widespread but local in England and Wales in woodland edge, scrub, downland and fen. Adults fly in May–July. Larvae initially feed on flowers and seed of sea-buckthorn and wild privet, later on fallen leaves. Croesus was king of Lydia, here referencing the purple forewing with its golden marks.

Small long-horn *Nemophora minimella*, wingspan 10–14 mm, widespread but local and scarce on dry grassland, including pasture and downland. Adults fly in July. Larvae feed on scabious, initially on seeds then on the lower leaves. *Minimus* is Latin for 'smallest'.

Yellow-barred long-horn *Nemophora degeerella*, wingspan 16–23 mm, fairly widespread and common in England and Wales, south-west Scotland and Ireland, in deciduous, often damp, woodland and in mature hedgerows. Males are active in May–June, often 'dancing' in groups in sunshine, flying into the air about 60 cm then dropping back down. Larvae feed on leaf litter. Baron Karl DeGeer (1720–78) was a Swedish naturalist.

LONG-LEGGED FLY Members of the family Dolichopodidae.

LONGHORN BEETLE Members of the family Cerambycidae, their long antennae (sometimes longer than their bodies) giving them their common name. Specifically, *Pyrrhidium sanguineum*, 6–16 mm length, Vulnerable in the RDB, with a few (often old) records from England, and more recently recorded as locally common in Wales (an alternative name is **Welsh oak longhorn beetle**), found on deciduous trees, with a preference for oak. Larvae feed under the bark of dead branches and trunks. Greek *pyrrhos* = 'flame-coloured', and Latin *sanguineum* = 'blood-coloured'.

Other scarce species include **eyed** (or **twin-spot**) **longhorn** *Oberea oculata* (parts of southern England and East Anglia), **six-spotted longhorn** *Anoplodera sexguttata* (particularly eastern England), **small black longhorn** *Stenurella nigra* (southern England, with a few records from the Midlands, larvae favouring decaying oak), and **white-clouded longhorn beetle** *Mesosa nebulosa* (southern third of England, larvae feeding on dead wood of deciduous trees, especially oak).

Black-and-yellow longhorn *Rutpela maculata*, a wasp mimic, 13–20 mm length, widespread and common in England, Wales and south-west Scotland. Adults frequent flowers in summer, especially those of bramble and umbellifers. Larvae feed in dead wood of a variety of tree species. *Rutpela* is an anagram of *Leptura*, *maculata* Latin for 'spotted'.

Black-clouded longhorn *Leiopus nebulosus*, 5–10 mm length, common and widespread, found in summer and autumn, larvae and adults both associated with broadleaf trees, including oak, alder and lime. Greek *leios* + *pous* = 'smooth' + 'foot', and Latin *nebulosus* = 'cloudy'.

Black-spotted longhorn *Rhagium mordax*, one of Britain's largest longhorns (12–23 mm), widespread in deciduous woodland and hedgerows, often on hawthorn. Larvae feed in dead wood of a number of tree species, favouring oak. Greek *rhaga* = 'vigour', and Latin *mordax* = 'biting'.

Black spruce longhorn *Tetropium castaneum*, introduced, uncommon and widely scattered, larvae reported as feeding in various conifer species. *Castaneum* is Latin for 'chestnut'.

Black-striped longhorn *Stenurella melanura*, 6–10 mm length, with a scattered distribution in England and Wales, commoner in the southern half of England. Adults visit a variety of flowers. Larvae feed on dead wood and in the roots of a number of tree species. Greek *stenos* + *oura* = 'narrow' + 'tail', and *melas* + *oura* = 'black' + 'tail'.

Four-banded longhorn *Leptura quadrifasciata*, introduced, 10–20 mm length, common and widespread in Britain, larvae feeding in dead wood of deciduous trees, favouring birches. Greek *leptos* + *oura* = 'slender' + 'tail', and Latin *quatuor* + *fasciata* = 'four' + 'bundled'.

Golden-bloomed grey longhorn *Agapanthia villosoviridescens*, up to 20 mm length, a stem borer that breeds in thistles and other herbaceous plants, widespread over much of England, mainly found in May–June in moist meadow and hedgerows. Adults feed on umbellifers and nettles, and produce an attractive scent when disturbed. Greek *agape* + *anthos* = 'love' + 'flower', and Latin *villosus* + *viridescens* = 'hairy' or 'rough' + 'green'.

Greater thorn-tipped longhorn *Pogonocherus hispidulus*, 5–8 mm length, widespread in Britain and locally common, adults found in spring and autumn on deciduous trees, especially oak and lime. Larvae feed on small deadwood. Greek *pogon* = 'beard', and Latin *hispidus* = 'shaggy'.

House longhorn Alternative name for old house borer.

Larch longhorn *Tetropium gabrieli*, 8–18 mm length, introduced from central Europe, with a scattered distribution in Britain, larvae feeding in the bark of diseased or dying larch. Adults, active in May–August, are crepuscular and do not visit flowers.

Lesser thorn-tipped longhorn *P. hispidus*, 4–6 mm length, with a scattered distribution in England and Wales. Adults are found from spring to autumn on deciduous trees. Larvae feed in dead wood.

Rufous-shouldered longhorn *Anaglyptus mysticus*, 8–15 mm length, widespread and locally common in England, rarer in Wales, in woodland and hedgerows in spring and summer, adults especially evident on hawthorn flowers. Larvae feed in dead branches and stumps. Greek *ana* + *glyptos* = 'throughout' + 'carved', and Latin *mysticus* = 'mysterious'.

Speckled longhorn *Pachytodes cerambyciformis*, 7–12 mm length, common in southern and south-west England and in Wales, less so in northern England and Scotland. Adults are found on a variety of flowers, especially bramble, in May–August. Larvae live in both deciduous and coniferous trees, especially in roots. Greek *pachys* + *eidos* = 'thick' + 'form'. Cerambos was a shepherd in Greek mythology who was transformed into a large beetle with horns.

Tawny longhorn *Paracorymbia fulva*, 9–14 mm length, in coastal areas from Pembrokeshire to Norwich, with inland records in Hampshire and a few from the west Midlands, but scarce (e.g. first recorded in Sussex in 2012). Adults are found in June–July in broadleaf woodland; larvae have not been found. Greek *para* + *corymbos* = 'near' + 'head', and Latin *fulva* = 'tawny'.

Tobacco-coloured longhorn *Alosterna tabacicolor*, 6–10 mm length, widespread and locally common in England and Wales, with a few records from south Scotland, larvae feeding in small dead wood and decaying stumps of both deciduous and coniferous trees.

Two-banded longhorn *Rhagium bifasciatum*, one of Britain's largest longhorns (12–22 mm), widespread but local, adults found in spring and summer, often in woodland, feeding on nectar of hawthorn, bramble and hogweed. Larvae bore holes and feed in dead wood. Latin *bifasciatum* = 'two-banded'.

Variable longhorn *Stenocorus meridianus*, 15–27 mm length, widespread and common in England and Wales. Adults are found in May–June on woodland edges feeding on pollen from a variety of flowers. Larvae feed on dead wood, including roots, of deciduous and, sometimes, coniferous trees. Greek *stenos* + *coris* = 'narrow' + 'bug', and Latin *meridianus* = 'midday'.

LONGLEAF *Falcaria vulgaris* (Apiaceae), a rhizomatous herbaceous European neophyte perennial introduced by 1726, scattered and local in southern and central England in pasture, arable fields, roadsides, clifftop grassland, scrub, chalk quarries, gravel-pits, railway ballast, riverbanks and waste ground. Latin *falcatus* = 'sickle-shaped' (describing the leaves) and *vulgaris* = 'common'.

LONGSPUR Lapland longspur Alternative name for Lapland bunting, named after the long claw on the hind toe of each foot.

LOOPER Larvae of geometrid moths have no prolegs (i.e. have only true legs at the front end and claspers at the rear) and so move forward by bending in a loop, hence 'loopers' and the family name, Geometridae (earth measurers).

Drab looper *Minoa murinata* (Larentiinae), wingspan 14–18 mm, day-flying in April–July in southern England and Wales, including the Welsh Borders, in wooded rides and coppices. Larvae feed on wood spurge. Nationally scarce and with a declining distribution, this is a priority species for conservation. *Minous* is Greek for Minoan (Cretan), with no entomological significance; *murinus* is Latin for 'mouse-coloured'.

Rannoch looper *Macaria brunneata* (Ennominae), wingspan 25–30 mm, scarce (a priority species for conservation), day-flying in June–July, restricted to clearings in mature pine and birch woodlands in central Scotland, where caterpillars feed on bilberry, possibly cowberry. Also occasionally found as a migrant, particularly in eastern England. Since 2009, there has been an increase in the number of records in Sussex, where there might therefore be a local breeding population. *Macaria* is Greek for 'happiness', *brunneata* from Latin for 'brown'.

LOOSESTRIFE Perennial herbs in the genus *Lysimachia* (Primulaceae), named after King Lysimachus (360–281 BCE), ruler of Thrace, Asia Minor and Macedon, who used the plant to calm his oxen. 'Loosestrife' dates from the mid-sixteenth century, erroneously translating via Latin the Greek *lysimacheion* as from *lyein* + *mache* = 'undo' + 'strife' rather than it being the king's name. See also purple-loosestrife (Lythraceae).

    Dotted loosestrife *L. punctata*, a widespread and increasingly common evergreen European neophyte cultivated since 1658, naturalised in rough grassland and on woodland edges, roadsides and waste ground, spreading by rhizomes. *Punctata* is Latin for 'spotted'.

    Fringed loosestrife *L. ciliata*, an uncommon semi-evergreen North American neophyte introduced by 1732, found as a garden throw-out, spreading by rhizomes to form sparse colonies, usually in moist shaded sites. *Ciliata* is Latin for 'fringed with hairs'.

    Lake loosestrife *L. terrestris*, a rhizomatous North American neophyte introduced by 1781, naturalised in reedbeds and lakeside meadows in a few places in the Lake District and Lancashire. *Terrestris* is Latin for 'growing on the ground'.

    Tufted loosestrife *L. thyrsiflora*, locally common in central Scotland, naturalised, local and declining in England, colonies growing in shallow water in permanently wet habitats such as fens on river floodplains, lake margins, ditches and canal sides. Latin *thyrsus* + *flora* = 'stalk' + 'flower'.

    Yellow loosestrife *L. vulgaris*, a semi-evergreen perennial, scattered and common except in north Scotland, spreading vegetatively using rhizomes to form colonies in permanently wet places usually on organic soils. Habitats include riverbanks and streamsides, marsh, tall-herb fen, fen carr, ponds and ditches. It withstands shade in wet open woodland. *Vulgaris* is Latin for 'common'.

LOPHIIDAE Family name of angler fish (monkfish) (Greek *lophos* = 'crested').

LORDS-AND-LADIES *Arum maculatum* (Araceae), a rhizomatous herbaceous perennial, common and widespread (except in upland Scotland) in woodland, hedgerows and other shaded habitats on moist, well-drained and base-rich soils. The name refers to the plant's likeness to male and female genitalia, differentiated by the stamen colour. The poker-shaped inflorescence (spadix) is shrouded by a hooded spathe which at its base bulges out and wraps around with about 30° of overlap. Above the male flowers is a ring of hairs forming a trap for insects which are attracted to the spadix by its faecal smell and a temperature up to 15°C warmer than ambient. The insects are trapped beneath the hairs and are dusted with pollen by the male flowers before escaping and carrying the pollen to the spadices of other plants, where they pollinate the female flowers. In autumn extremely poisonous red berries remain after the spathe and other leaves have withered away; they contain oxalates of saponins which have needle-shaped crystals that irritate the skin and mouth, resulting in swelling of the throat, difficulty in breathing, burning pain, and an upset stomach. The sap is also toxic. For culinary uses, see starch-root. *Aron* is the Greek name for this plant; Latin *maculatum* = 'spotted'.

    Italian lords-and-ladies *A. italicum*, in woodland, hedge banks, scrub and field borders, preferring shaded, moist environments and a deep, well-drained nutrient-rich soil. Subspecies *neglectum* is native in southern and south-west England, the Channel Islands and Scilly; ssp. *italicum* is an introduced, naturalised garden throw-out with a scattered distribution throughout Britain and Ireland.

LOTIDAE Family name of ling (gadiform fishes); see also burbot and rockling.

LOUSE See lice and sea louse.

LOUSE FLY Members of the family Hippoboscidae.

    Swift louse fly *Crataerina pallida*, a common obligate blood parasite of adult and nestling swifts.

LOUSEWORT *Pedicularis sylvatica* (Orobanchaceae), a perennial, rarely biennial, root-hemiparasite, common and widespread except in the east Midlands, declining in southern England, on acidic soils on damp grassy heaths, moorland, upland flushed grassland and drier parts of bogs and marshes. The name comes from the belief that these plants, when eaten, were responsible for lice infestations in stock. The binomial is from Latin 'relating to lice' and 'of the forest'.

    Marsh lousewort *P. palustris*, an annual to biennial root-hemiparasitic herb, common and widespread except in south-east England and the Midlands, on a range of base-rich to acidic, moist habitats, including wet heathland, valley bog, wet meadow, ditches, fen and hillside flushes, but generally declining with loss of habitat. *Palustris* is from Latin for 'marsh'.

LOVAGE *Levisticum officinale* (Apiaceae), a herbaceous perennial Near Estern archaeophyte present for at least a millennium, widely grown as a culinary herb, local and scattered in Britain, naturalised on rough ground and by walls and paths. 'Lovage' derives from the Middle English *loveache* (i.e. 'love' + 'ache', the latter being an old name for parsley), from Old French *levesche* and late Latin *levisticum*, the genus name (ultimately derived from *ligusticus* = 'of Liguria', where the herb was much grown). *Officinale* is Latin meaning having pharmacological properties, the roots having diuretic-like properties. Leaves are often used in salads and soups.

    Scots lovage *Ligusticum scoticum*, a herbaceous perennial found round the coastline of Scotland and Northern Ireland on rock crevices and free-draining skeletal soils, including cliffs, rocky shores, spray-covered shingle, stabilised dunes and stone walls. *Ligusticos* is Greek for 'Ligurian', *scoticum* Latin for 'Scottish'.

LOVE-GRASS Species of *Eragrostis* (Poaceae), annual Mediterranean, African and Australian neophytes, with around 50 species introduced with wool shoddy and as grain impurities, occasionally found as casuals on waste ground. Greek *eros* + *agros* = 'love' + 'grass'.

LOVE-IN-A-MIST *Nigella damascena* (Ranunculaceae), an annual garden escape flowering in June–July, found in the southern half of England and in Wales on waste ground, old walls and in pavement cracks. It readily regenerates from seed and populations can be persistent in disturbed sites where there is little competition. The thread-like green bracts give a hazy appearance (as in 'mist') to the flowers. Latin *niger* = 'black', referring to the seeds, and *damascena* = 'Damascan'.

LOVE-LIES-BLEEDING *Amaranthus caudatus* (Amaranthaceae), an annual neophyte of garden origin, an occasional escape on waste ground in England and south Wales. *Amarantos* is Greek for 'unfading flower', *caudatus* Latin for 'having a tail', referencing the pendant inflorescence. The plant formerly symbolised hopeless love.

LUCANIDAE Family name of stag and rhinoceros beetles. *Luca* is Latin for stag beetle.

LUCERNE *Medicago sativa* ssp. *sativa* (Fabaceae), a perennial Mediterranean neophyte herb cultivated as a fodder and green manure crop in Britain in the seventeenth century, now widespread, common (especially in England) and naturalised on field margins, tracksides, rough grassland and waste ground. See also sickle medick under medick. 'Lucerne' dates from the mid-seventeenth century, coming from French *luzerne*, in turn from Provençal *luzerno* = 'glow-worm', a reference to the shiny seeds. *Medicago* is from Greek *medice*, a kind of clover from Media; *sativa* is Latin for 'cultivated'.

LUCERNE BUG *Adelphocoris lineolatus* (Miridae), 8–10 mm long, locally common in England and parts of Wales, in dry and damp grassland, feeding on lucerne, clovers, restharrow and other legumes. Greek *adelphos* + *coris* = 'brother' + 'bug', and Latin *lineolatus* = 'lined'.

LUCINA Boreal lucina *Lucinoma borealis* (Lucinidae), a filter-feeding bivalve, shell size to 4 cm, widespread from the low intertidal to the shelf edge in silty sand. Lucina was the Roman goddess of childbirth, but also meaning 'light-bringing'; *borealis* is Latin for 'northern'.

LUGWORM Blow lugworm *Arenicola marina* (Arenicolidae), a common and widespread polychaete, much prized by anglers as bait, 120–200 mm in length, found from high water neap tidal level to the middle or lower shore in sand and muddy sand, digging a U or J-shaped burrow (20–40 cm deep) with characteristic depressions at the head end and a cast of defecated sediment at the tail end. It feeds on detritus and microorganisms in ingested sediment. Latin *arena* + *colo* = 'sand' + 'inhabit'.

LUMBRICIDAE Family name of earthworms, with three genera that include common and widespread species in Britain and Ireland.

LUMINOUS MOSS *Schistostega pennata* (Schistostegaceae), with refractive, lens-like cells causing it to shine like cats' eyes, scattered and local in Britain, especially in south-west England, Wales and the Pennines in dry, crumbling earth just inside rabbits burrows, in dark sandstone crevices, caves and old mines. Greek *schistos* + *stege* = 'divided' + 'roof', and Latin *pennata* = 'feathered'.

LUMPSUCKER *Cyclopterus lumpus* (Cyclopteridae), male length 30–40 cm, females generally 50 cm, found round much of the coast in deep water (50–300 m), but moving to the intertidal and subtidal in March–May to lay eggs in seaweed, which the male guards and aerates. Diet is of crustaceans, polychaetes and small fishes. Its ventral sucker (modified pelvic fins) allows it to live on the sea floor attached to rocks and boulders. Greek *cyclos* + *pteryx* = 'circle' + 'wing', from the circle-shaped pectoral fins.

Cornish lumpsucker Alternative name for shore clingfish.

LUNGWORT (LICHEN) Tree lungwort *Lobaria pulmonaria* (Lobariaceae), foliose, epiphytic, growing on the bark of mature deciduous trees, most commonly in central and western Scotland. It is sensitive to atmospheric pollution and loss of old-growth habitat. Greek *lobos* = 'lobe', and Latin *pulmonaria* = 'relating to the lung'.

LUNGWORT (PLANT) *Pulmonaria officinalis* (Boraginaceae), a herbaceous perennial European neophyte in gardens by 1597, naturalised in woodlands and scrub, on banks, and rough and waste ground. The English and the genus name (Latin *pulmo* = 'lung') come from the phlegm-like markings on the leaves, and through sympathetic magic were thought to cure pulmonary infections. *Officinalis* is Latin for having pharmacological properties.

Narrow-leaved lungwort *P. longifolia*, found in Dorset, Hampshire and the Isle of Wight, mostly on base-rich clay soils, in coppiced woodland, wood pasture and bracken heathland, and a rare garden escape elsewhere. *Longifolia* is Latin for 'long-leaved'.

Red lungwort *P. rubra*, a European neophyte, naturalised in shaded habitats, very scattered in England, commoner in Scotland. *Rubra* is Latin for 'red'.

Suffolk lungwort *P. obscura*, found on poorly drained chalky boulder-clay in rides and clearings in three ancient woods in east Suffolk (Burgate Wood, Stubbing's Wood, Gittin Wood), all with a long history of coppice management; cessation of coppicing has reduced its already very local distribution. *Obscura* is Latin for 'dark'.

Unspotted lungwort Alternative name for Suffolk lungwort.

LUPIN Species of *Lupinus* (Fabaceae), both the English and genus names (Latin *lupus* = 'wolf'), coming from the belief that these plants robbed the soil of nutrients.

Garden lupin *L. polyphyllus*, a biennial or short-lived perennial neophyte herb, naturalised on roadside and railway banks, river shingle and waste places. Introduced to cultivation in 1826, it is now rarely grown, its hybrid, Russell lupin being more common and most modern records, at least south of Scotland, are actually probably the hybrid. *Polyphyllos* is Greek for 'many-leaved'.

Narrow-leaved lupin *L. angustifolius*, a Mediterranean neophyte herb grown as a seed or silage crop in southern England and south Wales, a scarce casual in waste ground. *Angustifolius* is Latin for 'narrow-leaved'.

Nootka lupin *L. nootkatensis*, a perennial North American neophyte herb introduced to gardens in 1794 and recorded in the wild by the River Dee in 1862, having escaped from the grounds of Balmoral Castle, now local on river shingle, riverbanks and moorland, mostly in upland Scotland. Nootka Sound is on the west coast of Vancouver Island, British Columbia.

Russell lupin *Lupinus* x *regalis* (*L. arboreus* x *L. polyphyllus*), a short-lived perennial herb first cultivated in Britain in 1937 by George Russell of York, and recorded in the wild since 1955, now naturalised and quite frequent in England and Scotland, rare in Wales and Ireland, on rough ground, motorway and railway banks, and riverside shingle.

Tree lupin *L. arboreus*, a short-lived, semi-evergreen Californian neophyte shrub first cultivated in 1793, a widespread, scattered garden escape and also planted on dunes and (in Cornwall) china-clay tips. It also occurs on coastal shingle, roadsides, railway banks and waste ground. *Arbor* is Latin for 'tree'.

White lupin *L. albus*, a Balkan neophyte herb grown as a seed or silage crop in southern England, a scarce casual in waste ground. *Albus* is Latin for 'white'.

Yellow lupin *L. luteus*, a Mediterranean neophyte herb used as a seed crop in southern England, a scarce casual on acid soils in waste ground. *Luteus* is Latin for 'yellow'.

LUTESTRING (MOTH) Members of the family Drepanidae, subfamily Thyatirinae. The name may have nothing to do with lute strings, though wings do tend to have wavy lines, but rather deriving from 'lustrine', a shiny fabric.

Common lutestring *Ochropacha duplaris*, wingspan 27–32 mm, widespread and common in woodland, scrub, heathland and gardens. Numbers increased by 203% between 1968 and 2007. Adults fly at night in June–August. Larvae mainly feed on birch leaves. Greek *ochros* + *pachys* = 'pale yellow' + 'thick', and Latin *duplaris* = 'containing double', from the two black forewing spots.

Oak lutestring *Cymatophorina diluta*, wingspan 33–6 mm, local in mature oak woodland throughout England, and also in some rural gardens. Of conservation concern in the UK BAP since numbers have declined by 82% since 1980. Adults fly at night in August–September. Larvae feed on oak leaves. Greek *cyma* + *phoreo* = 'a wave' + 'to carry', from the wavy lines on the forewings, and Latin *diluta* = 'diluted', from the pale wing colour.

Poplar lutestring *Tethea or*, wingspan 38–43 mm, local in broadleaf woodland throughout much of southern England and Wales (ssp. *or*); also (ssp. *hibernica*) in the northern part of Ireland (plus Co. Kerry), and Scotland (ssp. *scotica*). Adults fly in June–July. Larvae mainly feed on aspen leaves but also those of other poplars. The binomial comes from Tethys, the wife of the god Oceanus, and the forewing marks that resemble the letters O and R.

Satin lutestring *Tetheella fluctuosa*, wingspan 35–8 mm, local and scattered in broadleaf, especially birch, woodland and heathland throughout England (especially the south-east), Wales and parts of Ireland. Adults fly at night in June–August. Larvae mainly feed on birch leaves but also on alder. The binomial is the diminutive of Tethys, and the Latin *fluctuosa* = 'wavy', from the forewing pattern.

LYCAENIDAE Family name for copper (Lycaeninae), hairstreak (Theclinae) and blue and argus (Polyommatinae) butterflies. The origin of the name is obscure, possibly from Greek *lycaina* = 'she-wolf', but it may have no explicit entomological meaning. The blues have an association with ants which care for and milk the caterpillars; in exchange, the caterpillars eat the ant larvae.

LYCHNIS *Hadena bicruris* (Noctuidae, Hadeninae), wingspan 30–40 mm, a nocturnal moth common and widespread in gardens, farmland, hedgerows and woodland, and on verges and dunes. Adults fly in June–July, sometimes again in August–September in the south. Larvae feed on the seed capsules of campion and sweet-william (and species of Caryophyllaceae formerly known as *Lychnis*). The binomial is from Hades, the Underworld in Greek mythology, and Latin *bicruris* = 'two legs', here referring to marks on the forewing.

Striped lychnis *Cucullia lychnitis* (Cucullinae), wingspan 42–7 mm, scarce, on calcareous grassland and verges, mainly in southern England, occasionally in East Anglia and Lincolnshire. It is a priority species in the UK BAP. Adults fly at night in June–July. Larvae feed on the flowers of mullein (*Verbascum lychnitis*, giving the specific name). Latin *cucullus* = 'cowl', describing the thoracic crest.

LYCIDAE Family name for net-winged beetles, 5–9 mm length, all scarce, found in association with ancient woodland where larvae live in white rot areas of decay in dead wood.

LYCOPHYTE, LYCOPHYTINA Subdivision of clubmosses and quillworts.

LYCOSIDAE Family name for wolf spiders (Greek *lycos* = 'wolf'), with 38 species in 8 genera.

LYMANTRIIDAE Old family name (Greek *lymanter* = 'destroyer') for so-called tussock moths, now considered to be a subfamily (Lymantriinae) in the family Erebidae.

LYME DISEASE, LYME BORRELIOSIS A bacterial infection spread to humans by being bitten (usually when in heathland or woodland) by ticks such as sheep tick infected with *Borrelia burgdorferi*. A rash is followed by symptoms that include fatigue, fever and muscle and joint pain, with more severe problems including inflammation of the heart muscles and of the membranes surrounding the brain and spinal cord. Treatment is by antibiotics.

LYME-GRASS (PLANT) *Leymus arenarius* (Poaceae), a common and widespread rhizomatous perennial growing on coastal dunes, sometimes also on fine shingle; it is important in stabilisation of mobile dunes and is widely planted to bind sand. Inland it is a scarce casual or naturalised garden escape. *Leymus* is an anagram of the former genus name *Elymus*; *arenarius* comes from Latin *arena* = 'sand'.

LYME GRASS (MOTH) *Longalatedes elymi* (Noctuidae, Xyleninae), wingspan 34–8 mm, scarce, on dunes along the east coast of Britain from Suffolk to Angus; there is also a small colony at Camber, Sussex, probably introduced with the food plant. Adults fly at night in June–August. Larvae mine the lower stem of lyme-grass (formerly *Elymus*).

LYMEXYLIDAE Family name (Greek *lyme* + *xylon* = 'ruin' + 'wood') for timberworm beetles, with two British species, wood borers associated with wood pasture and woodland.

LYMNAEIDAE Family (Greek *limne* = 'marsh' or 'pond') of pond snails, freshwater pulmonate gastropods.

LYONETIIDAE Family name for some bent-wing moths. Pierre Lyonet (1706–89) was a French naturalist.

LYPUSIDAE Family name for some of the tubic moths.

LYTHRACEAE Purple-loosestrife family (Greek *lythron* = 'blood', here in the sense of 'red').

M

MACROLEPIDOPTERA An artificial, non-taxonomic grouping of 'large moths' and butterflies with a wing span generally >20 mm, from Greek *macros* = 'large' and Lepidoptera (q.v.). See Microlepidoptera. The largest micros actually have a greater wing span than the smallest macros.

MACTRIDAE Family name for otter shell and trough shell (Greek *mactra* = 'trough').

MADDER *Rubia tinctorum* (Rubiaceae), a herbaceous evergreen Asian neophyte, formerly cultivated for the production of a red dye, locally naturalised or casual, especially in Somerset and Lincolnshire. Latin *ruber* = 'red', referring to the reddish dye obtained from the roots, and Latin *tinctorum* = 'related to dyeing'.

Field madder *Sherardia arvensis*, a widespread annual of open, droughted grasslands, dunes, arable fields, waste ground, waysides, verges and lawns, common in some coastal localities and locally inland, but a rare casual in places. It may be native only in some western coastal habitats, an archaeophyte elsewhere. William Sherard (1659–1728) was an English botanist. Latin *arvensis* = 'field'.

Wild madder *R. peregrina*, a scrambling, evergreen perennial of hedge banks, scrub, walls, cliffs and other rocky places near the coast from north Wales to Kent, and in the southern half of Ireland, and locally on calcareous soils inland. *Peregrina* is Latin for 'foreign'.

MADWORT *Asperugo procumbens* (Boraginaceae), a now rare herbaceous European neophyte annual arable weed, introduced with grain or wool shoddy, also casual on rough and waste ground. The binomial is from Latin for 'rough' (i.e. with prickly leaves) and 'procumbent'.

MAERL Three species of coralline red algae growing loose in beds of fragmented nodules in the sublittoral. *Lithothamnion glaciale* (Hapalidiaceae) is most abundant in the sea lochs of west Scotland, Orkney and Shetland, recorded along the east coast south to Flamborough, occasional elsewhere, found both as a thin encrusting species on rock and as a loose-lying algal gravel. *L. corallioides* is patchily distributed along west and south-west Ireland, south-west Wales and a few sites off the south coast of England. *Phymatolithon calcareum* has been recorded along Shetland, Orkney and along the east coast of Scotland, and south coast of England, but is particularly abundant around the west coasts of Ireland and Scotland in the sublittoral or lower littoral. *Lithothamnion* comes from Greek *lithos* + *thamnion* = 'stone' + 'small shrub'; *Phymatolithon* incororates *phymatos* = 'of a swelling'.

MAGGOT Rat-tailed maggot Larvae of some droneflies in the tribe Eristalini, for example in the genus *Eristalis*, that live in water with low amounts of oxygen: drainage ditches, stagnant water, water polluted with organic matter such as manure, and garden water butts, rain- and leaf-filled buckets. The common name comes from the overall shape of the larva with its tube-like, three-segmented, telescoping breathing siphon located at its posterior end which serves as a snorkel, allowing it to breathe while submerged. The siphon is usually about as long again as the maggot's body (20 mm when mature), but can be extended as long as 150 mm.

MAGIC MUSHROOM Alternative name for liberty cap.

MAGPIE (BIRD) *Pica pica* (Corvidae), wingspan 56 cm, length 45 cm, male weight 240 g, females 200 g, common and widespread. The population trebled between 1970 and 1990, especially with increases in urban and suburban numbers, then stabilised (a decline of 1% between 1995 and 2011, an increase by 6% between 2013 and 2014, and a non-significant decrease of 2% between 2014 and 2015). There were an estimated 650,000

breeding pairs in Britain in 2014. There were also high numbers in Ireland (a density of 16 pairs per km² in Dublin the highest recorded), though the population did decline slightly (by 2%) between 1998 and 2010. Nests are of twigs, usually in trees, with clutches of 5–6, incubation taking 20 days, and the altricial chicks fledging in 26–31 days. The opportunistic omnivorous diet includes the eggs and young of other birds, but magpies are not responsible for a decline in songbird numbers. 'Pie' is derived from French, itself from Latin *pica* = 'black and white' or 'pied'. Magpie is derived from 'magot pie', which first appeared in *Macbeth*, the name becoming established from about 1600 onwards in the Midlands and southern England.

MAGPIE (MOTH) 1) *Abraxas grossulariata* (Geometridae, Ennominae), wingspan 35–40 mm, widespread and common on moorland, in gardens and allotments. Adults fly at night in July–August. Larvae feed on currants and gooseberry, where they can become a pest, as well as spindle, blackthorn, hawthorn, hazel and heather. *Abraxas* is a Coptic word to express the number of days in a year; it has no entomological significance. (*Ribes*) *grossularia* is the food plant gooseberry.

Clouded magpie *A. sylvata*, wingspan 35–44 mm, local in woodland, in England, Wales and (especially north-east) Ireland. Adults fly at night in May–July. Larvae feed on elm. *Sylva* is Latin for 'woodland'.

2) **Small magpie** *Anania hortulata* (Crambidae, Pyraustinae), wingspan 24–8 mm, widespread and generally common, though less frequent in the north. Adults fly at night in June–July. Larvae feed from a rolled or spun leaf in August–September before hibernating in a silk cocoon in a hollow stem or under bark. The main food plants are nettles, but woundwort, mint, horehound and bindweed are also taken. Greek *anania* = 'without pain' (a litotes indicating the moth's beauty), and Latin *hortulus* = 'small garden'.

Spotted magpie *A. coronata*, also known as **elder pearl**, wingspan 22–6 mm, nocturnal, common on rough ground and in gardens throughout England, Wales and Ireland. Adults fly in June–July. Larvae feed on young elder leaves. *Coronata* is Latin for 'crowned'.

MAIDEN'S BLUSH *Cyclophora punctaria* (Geometridae, Sterrhinae), wingspan 18–25 mm, local in oak woodland throughout much of England, Wales and Ireland. Numbers increased by 240% between 1968 and 2007. The bivoltine moths fly at night in May–June and in August. Larvae feed on oak leaves. Greek *cyclos* + *phoreo* = 'ring' + 'to carry', and Latin *punctaria* = 'dotted', describing marks on the forewing.

MAIZE *Zea mays* (Poaceae), a Central American neophyte introduced in Tudor times, much cultivated since the 1970s, mainly for livestock feed, but also scattered (mainly in England and Wales) as a relic or escape from cultivation found on field edges and waste ground, and also as a casual of birdseed and kitchen waste. 'Maize' dates from the mid-sixteenth century, coming from Spanish *maíz* and Taíno (indigenous Caribbean) *mahiz*. *Zea* is Greek for a kind of grain.

MAJIDAE Family name for common spider crab (Greek *maia*, a large crab).

MAJOR (FLY) Soldier fly in the genus *Oxycera* (Stratiomyidae) (Greek *oxys* + *ceras* = 'pointed' + 'horn').

Four-barred major *O. rara*, body length 7 mm, wing length 6 mm, widespread if local in England as far north as Yorkshire, and in Wales, adults found in June–August in damp meadow and marsh, and around pools and streams, often on tree leaves. Larvae have been found in damp moss in seepages. *Rara* is Latin for 'rare'.

Round-spotted major *O. dives*, wing length 6 mm, Vulnerable in RDB 2. Recent records are from Tayside, the Galashiels area of Borders, moorland in Northumberland, Lower Teesdale (Durham), and North Yorkshire. Breeding sites are wet mossy flushes in meadows, moorland or woodland. Adults fly in June–September. *Dives* is Latin for 'rich'.

Twin-spotted major *O. leonina*, Vulnerable in RDB 2, with a few records from Norfolk and Suffolk.

MALACHIIDAE Family name (Greek *malache* = 'mallow'), for some of the soft-winged flower beetles, with 17 British species in 11 genera. Larvae are predatory in dead wood; adults are usually found in grassland, among tree leaves or under bark.

MALACHITE BEETLE Species of *Malachius* (Malachiidae), from Greek *malache* = 'mallow' and the green mineral malachite (which has the same etymology), from the body colour.

Common malachite beetle *M. bipustulatus*, length 5.5–6 mm, common and widespread in England and Wales. Adults are found in April–August on open-structured flowers feeding on pollen and nectar, as well as on other insects. Larvae eat small invertebrates, particularly other insect larvae, finding them under bark, on grass roots and in litter. *Bipustulatus* is Latin for 'two-spotted' ('blistered').

Scarlet malachite beetle *M. aeneus*, length 5–7 mm, once widespread in southern England but in 2006 known from just eight sites in Essex, Hertfordshire, Bedfordshire, Surrey, Hampshire and Wiltshire, a decline due to habitat loss and intensification of farming. It was listed as Near Threatened by Natural England in 2014. Adults are active from May to mid-June, feeding on flowers in meadows and hedgerows. Larvae are predators of other invertebrates. *Aeneas* was the son of Aphrodite.

MALACOSTRACA Class of crustaceans that includes the orders Amphipoda, Cumacea, Decapoda, Isopoda, Mysida and Stomatopoda. Greek *malacos* + *ostracon* = 'soft' + 'shell' (a state that is accurate immediately after moult, before the shell hardens).

MALE-FERN Some members of the genus *Dryopteris* (Dryopteridaceae), others being buckler-ferns. Specifically, *Dryopteris filix-mas*, common and widespread in semi-shade on moderately moist, fairly fertile substrates in a range of habitats, for example woodland, hedgerows, roadsides, gardens and churchyards. Greek *drys* + *pteris* = 'oak' + 'fern', and Latin *filix-mas* = 'male fern'.

Borrer's male-fern *D. borreri*, widespread in semi-shade on moderately moist, fairly fertile substrates in a range of habitats. William Borrer (1781–1862) was an English botanist.

Golden-scaled male-fern Alternative name for scaly male-fern.

Mountain male-fern *D. oreades*, uncommon, on rocky and scree slopes >240 m in northern England, Wales, Scotland, Co. Down and Co. Kerry. *Oreades* is from Greek *oros* = 'mountain'.

Narrow male-fern *D. cambrensis*, widespread (though absent from much of southern and eastern England) in semi-shade on moderately moist, fairly fertile substrates in a range of habitats. *Cambrensis* is Latin for 'Welsh' (Cambria).

Scaly male-fern *D. affinis*, common and widespread in well-lit, often moist woodland and scrub on base- and nutrient-poor soils. *Affinis* is Latin for 'similar'.

MALLARD *Anas platyrhynchos* (Anatidae), a common and widespread dabbling duck found in a range of freshwater habitats, and common in urban rivers and ornamental lakes, wingspan 90 cm, length 58 cm, male weight 1.2 kg, females 980 g, with 100,000 pairs (61,000–146,000 pairs) in 2009, and 680,000 individuals in winter 2008–9, resident numbers being augmented by Icelandic and Scandinavian birds. Numbers increased by 213% between 1975 and 2015, including levelling off after around 2000. Slight declines in northern and western Scotland and in parts of eastern England, and a decline in the wintering population, have led to it being Amber-listed. Nests are in vegetation near water, made of grass and leaves, lined with down, clutch size 11–14, incubation 27–8 days, fledging of the precocial young in 50–60 days. Diet is opportunistically omnivorous. This is the ancestor of most domestic ducks. 'Mallard' originally referred to any wild drake

and is derived from Old French *malart* or *mallart*. Latin *anas* = 'duck', and Greek *platys* + *rhynchos* = 'broad' + 'bill'.

MALLOPHAGA Suborder of biting or chewing lice, external parasites of birds (Greek *mallos* + *phagos* = 'lock of wool' + 'eating').

MALLOW (MOTH) *Larentia clavaria* (Geometridae, Larentiinae), wingspan 36–40 mm, nocturnal, common on verges, rough ground, gardens, marsh and riverbanks throughout England and Wales, local in Scotland, and a few non-recent records from eastern Ireland. Adults fly in September–October. Larvae mainly feed on mallows and also use garden hollyhocks. Larentia was foster mother to Romulus and Remus; this has no entomological significance. *Clavaria* is either from Latin *clavus*, a stripe on a toga, or *clava* = 'club' (i.e. a band), for the forewing pattern.

MALLOW (PLANT) Species in the genus *Malva* (Malvaceae), from Old English *meal(u)we*, from Latin *malva* for this plant. Occasionally found in waste ground via wool shoddy and other waste material are the annual neophytes **Chinese mallow** *M. verticillata*, **French mallow** *M. nicaeensis*, **least mallow** *M. parviflora* and **small mallow** *M. pusilla*. See also musk-mallow and tree-mallow.

Common mallow *M. sylvestris*, a drought-tolerant herbaceous perennial archaeophyte, common in lowland England and Wales, scattered and local elsewhere on well-drained, often nutrient-rich soils in unshaded habitats, on roadsides, railway banks, waste ground and field edges, often near habitation. *Sylvestris* is from Latin for 'woodland' or 'wild'.

Dwarf mallow *M. neglecta*, an annual herbaceous archaeophyte (present in Roman times) common in much of England, but often found as a casual in north England, Wales, Scotland and Ireland, growing in waste ground, gateways, paths and rough ground, and on roadsides, often near habitation, with a preference for shallow, dry soils, tolerant of grazing and mowing but not of competition with more vigorous plants. *Neglecta* is Latin for 'overlooked'.

Rough mallow *M. setigera*, a herbaceous annual European neophyte present by the eighteenth century, found in field and woodland edges, naturalised in Somerset, Oxfordshire and Kent, casual elsewhere in southern England and south Wales. *Setigera* is Latin for 'bearing bristles'.

MALTESE-CROSS *Silene chalcedonica* (Caryophyllaceae), a herbaceous perennial Russian neophyte in gardens by the late Tudor period, scattered in Britain as a casual or, rarely, a naturalised escape on waste ground and roadsides. This is the 'county' flower of Bristol as voted by the public in a 2002 survey by Plantlife. The flower is in the shape of a Maltese cross. The binomial refers to Silenus, foster father of the Greek god Bacchus, and Latin for 'from Chalcedon' (north-west Turkey).

MALVACEAE Family name for lime (formerly Tiliaceae), mallow (Latin *malva*) and hollyhock, together with other occasional casuals introduced via birdseed, wool, etc.

MAMMAL Warm-blooded animal in the class Mammalia with, among other distinguishing characteristics, hair and mammary glands. The Mammal Society lists 30 native terrestrial mammals (+ 4 terrestrial island species), 17 native bats, 2 native marine mammals, 11 introduced and naturalised species, 22 cetacean species found in and around British waters, 9 vagrant species (4 bats, 5 pinnipeds), 4 feral species, and 2 domesticated species. Orders are Diprotodontia (infraclass Marsupialia), Rodentia, Eulipotyphla (formerly Insectivora), Lagomorpha, Carnivora, Chiroptera, Artiodactyla, Perissodactyla, Pinnipedia and Cetacea.

MANGE See sarcoptic mange.

MANGEL-WURZEL, MANGOLD-WURZEL See root beet under beet.

MANGOLD FLY Alternative name for beet miner fly.

MANTIS SHRIMP A marine crustacean in the order Stomatopoda. Species have powerful claws used to attack and kill prey by spearing, stunning or dismemberment. The only commonly recorded species in British waters is *Rissoides desmaresti* (Squillidae), 10 cm long, which creates a simple burrow system in sandy and gravelly mud sediments from the lower shore down to 50 m depth.

MANTLE Royal mantle *Catarhoe cuculata* (Geometridae, Larentiinae), wingspan 22–7 mm, a nocturnal moth found locally in calcareous grassland, quarries, embankments, hedgerows, woodland rides and sea-cliffs throughout much of southern England, rarely northwards to central Scotland. Adults fly in June–July. Larvae feed on bedstraws. Greek *cata* + *rhoe* = 'downward' + 'stream', and a typographical error for Latin *cucullata* = 'cowl', from the dark thorax and wing base when resting.

MAP Bracken map *Rhopographus filicinus* (Pleosporales), a common and widespread inedible fungus found in autumn forming a series of irregular black lines on dead bracken stems. Greek *rhops* + *graphe* = 'understorey' + 'drawing', and Latin *filix* = 'fern'.

MAPLE Deciduous trees in the genus *Acer* (Sapindaceae), *acer* being Latin for 'maple' but also meaning 'sharp', possibly a reference to the hardness of maple wood which was used by the Romans for spears. See also sycamore. Stace (2010) notes that there are >12 other introduced species recorded in the wild, as well as planted as ornamentals in Britain.

Ashleaf maple *A. negundo*, a dioecious North American neophyte introduced by 1688, in England and outh Wales commonly planted in town parks, gardens, amenity sites and streets, sometimes found on railway banks. Establishment from seed is rare, but regeneration has been reported in south-east England where both sexes are present. *Negundo* comes from a Sanskrit name for *Vitex negundo* from the similarity of the maple's leaves to this species.

Cappadocian maple *A. cappadocicum*, a Near Eastern neophyte introduced by 1838, widely planted, especially in England, in parks and gardens, and on roadsides, and in south-east England self-sown or spreading by suckering in grassland, hedgerows and waste ground. Cappadocia is a region of Anatolia, Turkey.

Field maple *A. campestre*, common and widespread in England, Wales and southern Scotland, more local elsewhere, in woodland, scrub and hedgerows on a range of moist, usually base-rich calcareous and clay soils. Continental and Asian forms are also often planted in amenity areas, along roads and in hedgerows. *Campestre* is from Latin *campus* = 'field'.

Norway maple *A. platanoides*, a European neophyte, cultivated by 1683, widespread and commonly planted in woodland, hedgerows, urban amenity sites and gardens, and along roads. It tolerates a wide range of soil types and is frequently self-sown, becoming naturalised in secondary woodland, rough grassland, scrub and urban waste ground. *Platanoides* means 'resembling the plane tree' (*Platanus*).

Silver maple *A. saccharinum*, a North American neophyte introduced in 1725, commonly planted in town parks, large gardens and roadsides. While rarely setting seed in Britain, self-sown examples have been reported from the London area. *Saccharinum* is Latin for 'sugary'.

MARASMIACEAE From Greek *marasmos* = 'drying out', parachute fungi that recover when wetted after dehydration.

MARBLE (MOTH) Members of the family Tortricidae, subfamily Olethreutinae.

Apple marble *Eudemis porphyrana*, wingspan 17–21 mm, found in (often old) woodland in parts of southern England. Adults fly at night in July–August. Larvae mostly feed on apple leaves. *Eudemis* is probably a combination of Greek *eu* = 'well'

and a contraction of the genus *Pandemis*. *Porphyra* was a purple dye, from the reddish-brown forewing markings.

Arched marble *Olethreutes arcuella*, wingspan 14–18 mm, day-flying, with a scattered distribution in woodland and on heaths, rare in Scotland and Ireland. Adults fly in May–August. Larvae feed on dead leaves, including litter. Greek *olethros* = 'destruction', from the damage to plants, and Latin *arcus* = 'bow', from the shape of the forewing band.

Barred marble *Celypha striana*, wingspan 16–22 mm, common on dry pasture and downland throughout England and Wales, scarcer in the north of its range. Adults fly at night in June–August. Larvae mine dandelion roots. Greek *celyphos* = 'husk' or 'pod', and Latin *stria* = 'streak', from the forewing pattern.

Bearberry marble *Argyroploce arbutella*, wingspan 13–16 mm, scarce on moorland and in mountains throughout much of the Highlands, down to sea level in the Hebrides and Co. Clare, Ireland. Adults fly in May–June on sunny afternoons. Larvae mine bearberry shoots. Greek *argyros* + *plocos* = 'silver' + 'braid', and *arbutella* after *Arbutus unedo* (strawberry-tree), from similarity in colour.

Bilberry marble *Apotomis sauciana*, wingspan 13–16 mm, local in woodland and on moorland throughout much of England and Wales, and on moorland in north Scotland. Adults fly in the afternoon and evening in June–August. Larvae feed on shoots of bilberry and, in Scotland, bearberry. Greek *apotome* = 'a cutting off', from the distinct separation of dark and pale colour on the forewing, and Latin *saucius* = 'wounded', from a pinkish colour that penetrates the white ground colour of the wings.

Birch marble *Ap. betuletana*, wingspan 16–20 mm, nocturnal, widespread and common in heathland, woodland edges and birch woodland. Adults fly in July–September. Larvae feed on birch leaves in April–May. *Betuletum* is New Latin for 'birch thicket'.

Black-edged marble *Endothenia nigricostana*, wingspan 11–15 mm, local in woodland margins and dry pastures throughout much of England and Wales, scarcer in the north. Adults fly in the late afternoon and after dark in June. Larvae feed on woundworts, eating down from the flower into the stem and roots during spring. Greek *endothen* = 'inside', from the larval feeding behaviour, and Latin *niger* + *costa* = 'black' + 'rib', from the forewing vein pattern.

Blotched marble *En. quadrimaculana*, wingspan 18–22 mm, widespread but local in water meadows and marsh and on riverbanks. Adults fly in the evening and at night in June–September. Larvae feed inside the roots and lower stems of marsh woundwort, occasionally spear-mint. Latin *quadrimaculana* = 'four-spotted', from the forewing pattern.

Bordered marble *En. marginana*, wingspan 11–16 mm, widespread and common in rough ground, damp meadow, heathland and fen. Adults fly in May–August. Larvae feed in September–June in the seed heads of betony, teasel, hemp-nettles, lousewort, yellow-rattle and ribwort plantain. *Margo* is Latin for 'margin', from the male hindwing pattern.

Buff-tipped marble *Hedya ochroleucana*, wingspan 16–21 mm, common in gardens, orchards and hedgerows throughout much of England and Wales, more numerous in the south, rare in Scotland, and two records from south-east Ireland. Adults fly at night in June–July. Larvae feed on leaves of roses and apple. Greek *hedys* = 'pleasing', and *ochros* + *leucos* = 'pale yellow' + 'white'.

Bugle marble *En. ustulana*, wingspan 9–13 mm, scarce in damp woodland and waste ground in parts of (especially south and south-east) England and Wales, rare in Scotland, and one record from Ireland (Co. Waterford). Adults fly in afternoon sunshine during June–July. Larvae mine bugle roots, overwinter, then feed in leaf stalks. *Ustilo* is Latin for 'scorch', from the blackened forewing.

Common marble *C. lacunana*, wingspan 16–18 mm, nocturnal, widespread and common on woodland edges and verges and in rough grassland, hedgerows and gardens. Adults fly in May–August. Larvae feed on the leaves of a variety of herbaceous and some woody plants. *Lacuna* is Latin for 'hole', from the wing pattern.

Cowberry marble *Stictea mygindiana*, wingspan 15–20 mm, scarce, local on heathland and moorland throughout much of northern England, and parts of Scotland and western Ireland. Adults fly in May–June in the afternoon and early evening. Larvae feed on cowberry, bearberry and bog-myrtle, living between shoots spun together with silk. The binomial is from Greek *stictos* = 'dotted', and after a Danish student, Mr Mygind (d.1787).

Diamond-back marble *Eu. profundana*, wingspan 14–20 mm, widespread and common in oak woodland, in England, Wales and south-east Ireland. Adults fly at night in July–August. Larvae feed on oak leaves. Latin *profundus* = 'deep', for the breadth of the forewing.

Downland marble *En. oblongana*, wingspan 13–15 mm, scattered and local on field edges, wasteland, chalk downland and flower meadows in parts of England and Wales. Adults fly in the late afternoon and evening in June–July. Larvae feed on common knapweed. *Oblongus* is Latin for 'long', from the elongated forewing.

Great marble *Pseudosciaphila branderiana*, wingspan 21–7 mm, scarce, local to stands of aspen, in woodland, gardens and parkland across much of southern England, occasionally north to Yorkshire. Adults fly at night in June–July. Larvae feed on aspen leaves. Greek *pseudo* = 'false' and the genus *Sciaphila*, and (probably) in honour of E. Brander, an eighteenth-century collector.

Hawkweed marble *C. rurestrana*, wingspan 12–16 mm, first recorded at Tintern, Monmouthshire, in 1962, and at Ifracombe, north Devon, in 1985, and also recorded at Merthyr Tydfil, Glamorgan. Larvae have not yet been found in Britain. *Rurestris* is Latin for 'rural' (*rus* = 'countryside').

Heath marble *En. ericetana*, wingspan 14–20 mm, nocturnal, widespread but local in rough pasture, flower meadows and farmland. Adults fly in July–August. Larvae feed on stems and roots of woundworts and corn-mint. *Ericetum* is New Latin for 'heathland', though this is not a habitat used in Britain.

Highland marble *Phiaris metallicana*, a rarity on moorland and in mountains in the Highlands. Adults fly in June–July. Larvae feed on bilberry leaves in May–June. Greek *phiaros* = 'bright', and Latin *metallicus* = 'metallic', from the silvery metallic wing margins.

Lakes marble *C. rufana*, wingspan 16–19 mm, rare on rough ground in parts of Wales and north-west England. Adults fly in May–July in the afternoon and evening. Larvae feed on tansy and mugwort roots. *Rufus* is Latin for 'red'.

Large marble *Ph. schulziana*, wingspan 14–25 mm, widespread but local on moorland, heathland and mountains, rare in southern England. Adults fly in the late afternoon and evening in June–August. Larvae probably feed on leaves of crowberry and heather, inside a silk tube. Dr J.D. Schulz was an eighteenth-century German entomologist.

Mistletoe marble *C. woodiana*, wingspan 16–18 mm, found only in orchards in Herefordshire, Worcestershire, Gloucestershire and Monmouthshire, a priority species in the UK BAP. Adults fly at night in July–August. Larvae mine leaves of mistletoe. Dr J.H. Wood (1841–1914) discovered the species in Herefordshire in 1878.

Moss marble *C. aurofasciana*, wingspan 12–14 mm, local in a few woodlands in southern and central England and south Wales. Adults fly in June–July in the late afternoon and evening. Larvae feed on mosses and liverworts, often on tree-trunks. Latin *aurum* + *fascia* = 'gold' + 'band', from the forewing colour and pattern.

Mottled marble *Bactra furfurana*, wingspan 13–19 mm, widespread if local in marsh, riverbanks and ditches. Adults fly at night in June–July. Larvae feed inside rush stems. Greek *bactron* = 'a club', describing the maxillary palpus, and Latin *furfur* = 'bran', from the brown forewing markings.

Mountain marble *Ph. obsoletana*, wingspan 15–19 mm, local on moorland in the Highlands and the Shetlands. Adults fly in

early morning during June–July. Larvae probably feed on bilberry, bearberry and cowberry. *Obsoletus* is Latin for 'worn-out', from the rather drab colour.

Narrow-winged marble *A. sororculana*, wingspan 17–20 mm, nocturnal, widespread in woodland and heathland, less numerous in the south. Adults fly in May–July. Larvae feed on birch leaves. *Sororcula* is Latin for 'little sister', from a similarity with some congenerics.

Northern marble *Ph. palustrana*, wingspan 14–16 mm, scarce, local on heathland and lightly wooded slopes in parts of the Highlands and north Wales, rare in north England. Adults fly in May–August, sometimes on sunny afternoons but more typically towards dusk. Larvae feed on mosses. *Paluster* is Latin for 'marshy', habitat used on the Continent.

Oak marble *Lobesia reliquana*, wingspan 12–14 mm, day-flying, widespread if local in woodland, gardens, orchards and parks, though scarce in Scotland, and recorded from south-west and eastern Ireland. Adults emerge in May–June. Larvae mainly feed on leaves and new shoots of oaks, but also on blackthorn and birch. Greek *lobesis* = 'ruin', possibly from destructive feeding habits, and Latin *reliquus* = 'remaining'.

Olive marble *Ph. micana*, wingspan 13–18 mm, widespread but local in marsh and on riverbanks, more numerous in the north. Adults fly in July–August. Larvae feed on mosses.

Pine marble *Piniphila bifasciana*, wingspan 12–16 mm, nocturnal, local in pine woodland throughout much of Britain and eastern Ireland. Adults fly in June–July. Larvae feed on young pine shoots. The binomial is from Greek for 'pine-loving' and Latin for 'two-banded'.

Rannoch marble *Ap. infida*, wingspan 18–19 mm, found in parts of north-east Scotland. Adults fly at night in July–August. Larvae feed on sallow leaves. *Infidus* is Latin for 'untrustworthy', from its variable appearance.

Roseate marble *C. rosaceana*, wingspan 15–19 mm, nocturnal, local in rough pasture and grassland in parts of southern England and south Wales. Adults fly in June–July. Larvae mine dandelion and sowthistle roots. *Rosaceus* is Latin for 'rosy'.

Rush marble *B. lancealana*, wingspan 11–20 mm, nocturnal, widespread and common in damp heathland, marshes, riverbanks and moorland. Adults have two generations, flying in July and October. Larvae feed on rushes and deergrass. *Lanceola* is Latin for a 'small lance', from the narrow forewings.

Sallow marble *Ap. capreana*, wingspan 17–22 mm, local in woodland, marsh, riverbanks and ditches in the southern half of England and Wales, south-west Scotland and south-west Ireland. Adults fly at night in June–August. Larvae feed on leaves of goat willow *Salix caprea* (prompting *capreana*) during spring.

Saltern marble *B. robustana*, wingspan 16–23 mm, scarce in saltmarsh mainly on the south and east coasts of England, but with records west to Devon and south Wales, and north to Yorkshire. Adults fly at night in June–July. Larvae mine stems of sea club-rush. *Robustus* is Latin for 'robust'.

Scarce sedge marble *B. lacteana*, wingspan 10–17 mm, in damp meadow and marsh and on riverbanks in parts of southern England, first discovered in 1996. Adults fly in June–August. Larvae feed on sedges and rushes. *Lac* is Latin for 'milk', from the wing colour.

Shore marble *Lobesia littoralis*, wingspan 12–14 mm, widespread but local on sea-cliffs, saltmarsh and shingle beaches, and inland in gardens. The bivoltine adults fly in the late afternoon and at night in June–July and September. Larvae mainly feed on the flower heads and seed of thrift, but also of bird's-foot-trefoil. *Littoralis* is Latin for 'relating to the shore'.

Short-barred marble *Ap. semifasciana*, wingspan 18–19 mm, nocturnal, widespread but scattered in woodland and along freshwater margins. Adults fly in July–August. Larvae feed on sallow catkins and leaves. Latin *semifasciana* = 'semi-banded', from the forewing marking.

Shoulder-spot marble *H. atropunctana*, wingspan 14–17

mm, nocturnal, local in woodland, heathland and dry pasture throughout much of Britain, though known from just a few sites in the south. Adults fly in May–June or slightly later further north, and sometimes have a second generation in August in the south. Larvae feed on the terminal shoots of bog-myrtle, birches and willows. *Atropunctata* is Latin for 'black-spotted', from the forewing marks.

Silver-striped marble *C. rivulana*, wingspan 15–19 mm, local in open woodland, damp meadow, chalk downland, heathland and sandhills, with a scattered distribution throughout Britain. Adults fly in July–August by day and at night. Larvae feed on the flowers and terminal shoots of a range of herbaceous and woody plants. *Rivulus* is Latin for 'small stream', from the wavy forewing pattern.

Smoky-barred marble *Lobesia abscisana*, wingspan 10–13 mm, nocturnal, common on waste ground and in flower meadows in England north to Yorkshire and west to Devon, and in Wales and eastern Ireland. Adults are bivoltine, flying in May and in July–August. Larvae mine shoots of creeping thistle. *Abscisus* is Latin for 'abrupt', probably from the sharply defined edges of the wing markings.

Spurge marble *Lobesia occidentis*, wingspan 12–14 mm, nocturnal, local in open ancient woodland, saltmarsh and sandy beaches in parts of south-east England and the Midlands. Adults are bivoltine, flying in May–June and July–August, managing a third brood in exceptional summers. Larvae feed on the leaves and mine the stems of wood and sea spurge. *Occidens* is Latin for 'western'.

Teasel marble *En. gentianaeana*, wingspan 15–19 mm, common on rough ground and dry pasture throughout much of England and Wales, less numerous in the north. Adults fly in June–July. Larvae mine the seed heads of teasel. *Gentiana* (gentian) is not a food plant in Britain.

Thyme marble *C. cespitana*, wingspan 12–16 mm, day-flying, local on chalk downland, dry pasture, sandhills and sea-cliffs throughout Britain, predominantly coastal in the north of its range. Adults fly in June–August. Larvae feed on leaves of wild thyme, common sea-lavender and thrift. *Caespitis* is Latin for 'of turf'.

White-backed marble *H. salicella*, wingspan 19–24 mm, common in woodland, gardens, orchards, parks and marsh throughout England and Wales. Adults fly at night in June–August. Larvae feed on leaves of willow (= *salix*, hence *salicella*), aspen and black poplar.

White-shouldered marble *Ap. turbidana*, wingspan 19–23 mm, widespread and common in (especially birch) woodland and heathland. Adults fly at night in June–July. Larvae feed on birch leaves. *Turbidus* is Latin for 'confused', from the forewing pattern.

Willow marble *Ap. lineana*, wingspan 18–22 mm, nocturnal, in damp pasture and marsh, by riverbanks and ditches in parts of England and Wales. Adults mainly fly in July–August. Larvae feed on willow leaves. *Lineus* is Latin for 'flaxen', from the whitish forewing colour.

Woodland marble *Orthotaenia undulana*, wingspan 15–20 mm, nocturnal, widespread and common in woodland edges, hedgerows, scrub, dry pasture, dunes and moorland, less numerous in the north. Adults fly in May–July. Larvae feed on the leaves of a variety of plants. Greek *orthos* + *tainia* = 'straight' + 'band', and Latin *undulana* = 'wavy', both from aspects of the forewing pattern.

Woundwort marble *En. pullana*, wingspan 10–14 mm, scarce, by marsh, riverbanks and other damp areas in a few parts of the British Isles. Adults fly in May–June. Larvae feed on marsh woundwort. *Pullus* is Latin for 'dark'.

MARBLED (MOTH) **Rosy marbled** *Elaphria venustula* (Noctuidae, Xyleninae), wingspan 19–23 mm, scarce, nocturnal, in deciduous woodland, heathland, verges and rough ground in parts of southern and south-east England and East Anglia.

Adults fly in May–June. Larvae probably feed on tormentil and creeping cinquefoil. Greek *elaphria* = 'lightness' or 'agility', from the moth's behaviour, and Latin *venustula* = 'charming'.

MARCHANTIALES Order of thalloid liverworts. Nicolas Marchant (d.1678) was a French botanist.

MARCHANTIOPHYTA Phylum name for liverworts, with the classes Marchantiopsida (orders Blasiales, Lunulariales, Marchantiales and Sphaerocarpales), Haplomitriopsida (order Calobryales) and Jungermanniopsida (orders Jungermanniales, Fossombroniales, Metzgeriales, Pallaviciniales, Pelliales, Pleuroziales, Porellales and Ptilidiales). Nicolas Marchant (d.1678) was a French botanist.

MARE'S-TAIL (PLANT) *Hippuris vulgaris* (Hippuridaceae), a widespread, locally common herbaceous perennial with two growth forms: plants with long, flaccid stems are submerged aquatics, sometimes abundant in clear calcareous water; more rigid, stiffly erect plants grow as emergents at the edge of lakes and ponds, in swamps or in upland flushes. Greek *hippos* + *ouros* = 'horse' + 'tail', and Latin *vulgaris* = 'common'.

MARE'S-TAIL (SEAWEED) See sea mare's-tail.

MARGARITIFERIDAE Family name for freshwater pearl mussel (Greek *margarites* = 'pearl', Latin *fero* = 'to carry').

MARIGOLD 1) Both the **annual field marigold** (*Calendula arvensis*) and the **perennial pot marigold** (*C. officinalis*) (Asteraceae) are generally casual garden escapes or throw-outs in fields, waste ground and tips, pot marigolds being particularly widespread.

Corn marigold *Glebionis segetum*, a mainly spring-germinating annual archaeophyte known since the Iron Age, locally common and widespread (though declining), casual or naturalised on light, sandy or loamy soils deficient in calcium, in arable fields and other disturbed habitats, on roadsides and waste ground. 'Marigold', dating from the late fourteenth century, comes from Mary (a reference to the Virgin) and 'golde' for the colour of the flower. Formerly placed in the genus *Chrysanthemum*, the binomial is from Latin *gleba* = 'clod' and *segetum* = 'of cornfields'.

2) **African marigold** (*Tagetes erecta*) and **French marigold** (*T. patula*) are casual Mexican neophytes, not infrequent garden escapes and throw-outs; southern marigold (*T. minuta*) is a wool alien casual in a few places in England. Tagetes was an Etruscan god who sprang up from ploughed earth. *Patulus* is Latin for 'spreading'.

3) **Dwarf marigold** (*Schkuhria pinnata*) is a casual Central American neophyte wool alien occasionally found in England. Christian Schkuhr (1741–1811) was a German botanist. Latin *pinnata* = 'feather-shaped'.

MARJORAM **Wild marjoram** *Origanum vulgare* (Lamiaceae), a herbaceous perennial, common and widespread (decreasing northwards) on dry, infertile, calcareous soils naturalised in grassland, hedge banks, and scrub, and a colonist of sparsely vegetated ground. 'Marjoram' is late Middle English, from Old French *majorane*, in turn from medieval Latin *majorana*. Greek *oros* + *ganos* = 'mountain' + 'beauty', and Latin *vulgaris* = 'common'.

MARRAM *Ammophila arenaria* (Poaceae), a rhizomatous perennial grass of coastal dunes, important in the stabilisation of mobile dunes and blow-outs, and widely planted to bind sand. Inland, it is a rare casual, with several attempts recently made to establish it on golf courses. 'Marram' dates from the mid-seventeenth century, from Old Norse *marálmr* (*marr* + *hálmr* = 'sea' + 'haulm'). Greek *ammos* + *philos* = 'sand' + 'loving', and from Latin *arena* = 'sand'.

Purple marram *Calamagrostis epigejos* x *Ammophila arenaria*, a vigorous rhizomatous perennial hybrid growing on coastal dunes in Suffolk, Norfolk, Northumberland and Sutherland, and planted to stabilise them in East Anglia and Hampshire.

MARSH-ELDER *Iva xanthiifolia* (Asteraceae), an annual

North American neophyte present by 1905, scattered and occasional in England in damp habitats, and as a casual in cultivated and waste ground, originating from birdseed, grain impurities and wool shoddy. The genus name refers to *Ajuga iva*, an American species of bugle which has a similar scent. Greek *xanthiifolia* = 'yellow-flowered'.

MARSH FLY Members of the family Sciomyzidae.

MARSH-MALLOW *Althaea officinalis* (Malvaceae), a herbaceous perennial locally common in coastal habitats in England, Wales and southern Ireland, on the banks of ditches containing brackish water, in brackish pastures, and in the transition zone between upper saltmarsh and freshwater habitats; also as a garden escape. It is intolerant of grazing, and declined during the twentieth century because of drainage and development. Greek *althaino* = 'to heal', and Latin *officinalis* for a plant with pharmacological properties, having especially been used to alleviate sore throats.

MARSH-MARIGOLD *Caltha palustris* (Ranunculaceae), a common and widespread perennial herb, of wet pasture, marsh, ditches, water margins and wet woodland, in muddy, fairly base-rich soils of intermediate fertility. A small form, var. *radicans*, is found in mountain flushes and lake shores in northern and western regions. *Caltha* is Greek for this plant, from *calathos* = 'goblet', from the flower shape; *palustris* is from Latin for 'marsh'.

MARSH-ORCHID Tuberous herbaceous perennials in the genus *Dactylorhiza* (Orchidaceae) (Greek *dactylos* + *rhiza* = 'finger' + 'root'). See also spotted-orchid.

Early marsh-orchid *D. incarnata*, widespread and locally common (though overall declining), on damp calcareous soils in meadow, marsh, ditches, fen, flushes and dune slacks, and also on more acidic soils in bog and damp heathland. Ssp. *coccinea* (Latin for 'scarlet') is found in machair grassland and damp dune slacks, and occasionally on the wet terraces of slumped sea-cliffs, and inland in calcareous fens, flushes and on highly saline fly-ash waste from power stations. Ssp. *ochroleuca* (Greek for 'yellow-white') is restricted to moist, periodically inundated calcareous fens in East Anglia. Ssp. *pulchella* (Latin for 'pretty') is more widespread, in acidic valley bogs, marsh and damp heathland, often growing with sphagnum. It also occurs in marsh on more neutral substrates. *Incarnata* is Latin for 'flesh-coloured'.

Irish marsh-orchid *D. kerryensis*, scattered in western and southern Ireland (e.g. Co. Kerry) in marsh, fen, wet meadow and dune slacks, occasionally on peat overlying limestone.

Narrow-leaved marsh-orchid *D. traunsteinerioides*, scattered and local especially in East Anglia, Yorkshire, north Wales, western Scotland and western Ireland, in damp, neutral to base-rich often grassy habitats such as marsh, water meadow, flushes and fen. Traunstein is a town in Bavaria, Germany.

Northern marsh-orchid *D. purpurella*, common and widespread in Wales, northern England south to the north Midlands, Scotland and much of Ireland (though declining in the south), on neutral to base-rich soils in dune slacks, fen, marsh, wet meadow, flushes and ditches, and on verges. It is also common in old quarries and urban waste ground, where it colonises drier sites such as rubble. *Purpurella* is Latin for 'pale purple'.

Southern marsh-orchid *D. praetermissa*, common and widespread in England and Wales on calcareous marsh, fen, damp meadow, roadsides, dune slacks, less acidic bogs and wet heathland. It also colonises artificial habitats such as quarries and industrial waste tips. *Praetermissa* is Latin for 'overlooked' or 'neglected'.

Western marsh-orchid Alternative name for Irish marsh-orchid.

MARSHWORT Herbaceous perennials in the genus *Apium* (Apiaceae) (Latin *apium* = 'parsley' or 'celery').

Creeping marshwort *A. repens*, known from Port Meadow, Oxfordshire, in neutral grassland and adjacent open soil where trampling by horses and cattle keeps the habitat open, and possibly also in Berkshire and Essex on damp meadows and shallow water in ditches and ponds. *Repens* is Latin for 'creeping'.

Lesser marshwort *A. inundatum*, widespread but scattered in base-poor oligotrophic or mesotrophic shallow streams, ditches, ponds, canals and backwaters, and in bare mud subject to periodic desiccation. *Inundatum* is Latin for 'flooded'.

MARSUPIAL Members of the infraclass Marsupialia, the young carried in a protective pouch (the marsupium), represented in the British fauna by the introduced red-necked wallaby.

MARTEN Pine marten *Martes martes* (Mustelidae), male head-body length 51–4 cm, tail 26–7 cm, weight 1.5–2.2 kg, females 46–54 cm, 18–24 cm and 0.9–1.5 kg, a generally arboreal generalist predator which, until the nineteenth century, was found throughout much of mainland Britain, the Isle of Wight and some Scottish islands. Habitat fragmentation, persecution by gamekeepers and the animals being killed for their fur drastically reduced this distribution. By the 1920s, the main population was restricted to north-west Scotland, with small numbers in north Wales and the Lake District. In the last 50 years or so numbers and range have expanded in Scotland, with a population of at least 4,000. In England, recent records from upland areas have all been misidentifications, escapes from wildlife parks or reintroductions. In 2015, however, a marten may have been photographed near Bude, Cornwall, and another more definitely in the Shropshire hills near the Welsh border – the first definitely wild example in England for a century. Photographic evidence also confirmed its presence in the New Forest in 2016. In Wales the main strongholds are the Cambrian Mountains, Snowdonia and the uplands of south Wales. In 2015 the Vincent Wildlife Trust began transferring Scottish pine martins to woodland east of Aberystwyth, with 40 adults to be released over two years. It remains one of the rarest native mammals in Britain, with a total population of 3,000–4,000. Martens and their dens are fully protected by the WCA (1981). They are much commoner and more widespread in Ireland: current distribution is largely concentrated in western counties and the midlands, and they are present in around 50% of their historical range. The total population of pine marten in Ireland is about 2,700 individuals, and is expanding. Pine martens prefer well-wooded habitats, and feed on small rodents, birds, beetles, carrion, eggs and fungi, together with berries in autumn. They den in old badger setts and squirrel dreys, tree hollows, rock crevices, and openings under large tree roots. The mating season is July–September. Following delayed implantation, gestation lasts a month. Young (kits) are born in early spring in litters of 1–5, are weaned in 6–8 weeks, and become fully independent after 6 months. 'Marten' derives from Middle English, in turn from Old French *martrine*. Latin *martes* = 'marten'.

MARTIN Birds in the family Hirundinidae. **Crag martin** *Ptyonoprogne rupestris* and **purple martin** *Progne subis* are vagrants.

House martin *Delichon urbicum*, wingspan 28 cm, length 12 cm, weight 19 g, common and widespread from April to October, though Amber-listed in 2002, with 510,000 pairs (360,000–660,000) estimated in 2009. There was a decline in numbers by 10% over 1995–2015, and by 7% over 2010–15. The nest is constructed using >1,000 beak-sized pellets of mud, originally on cliffs, but now more commonly having adopted the underside of eaves on buildings in villages and suburbia. Clutch size is 4–5, incubation takes 13–19 days, and fledging of the altricial young takes 19–25 days. Feeding is on flying insects. *Delichon* is an anagram of the Greek *chelidon* = 'swallow', and *urbicum* comes from Latin *urbs* = 'town'.

Sand martin *Riparia riparia*, wingspan 28 cm, length 12 cm, weight 14 g, a common and widespread summer visitor, breeding colonially in nests in burrows up to 1 m long in sandy banks (e.g. riverbanks and by gravel-pits), with 114,000 nests (54,000–174,000 nests) estimated in 2009. There was a 22% increase in numbers between 1995 and 2014, with a weak decline (−10%) in 2010–15, but there have been two major crashes in numbers in the last 50 years following droughts in its sub-Saharan wintering quarters. Green-listed in Britain but Amber-listed in Ireland. Clutch size is 4–5, incubation lasts 14–15 days, fledging of the altricial young takes 19–21 days. Feeding is on aerial invertebrates. The binomial comes from Latin *ripa* = 'bank'.

MASON (POLYCHAETE) See sand mason.

MASON BEE Members of the genus *Osmia* (Megachilidae) (Greek *osme* = 'scent'). Their name comes from their use of mud and other 'masonry products' in constructing their nests, which are made in gaps or holes in wood, rock, etc.

Blue mason bee *O. caerulescens*, 8–10 mm length, locally common and widely distributed in England and Wales, bivoltine, flying in April–July and in August, habitats including woodland and gardens, collecting pollen and nectar from a variety of flowers. Nests are built inside cavities, including insect exit burrows in dead wood and masonry crevices. *Caeruleus* is Latin for '(dark) blue'.

Fringe-horned mason bee *O. pilicornis*, 10 mm length, local in southern England as far north as Herefordshire, and in Mid Glamorgan, in open broadleaf woodland, especially on chalky soils, where management has adversely affected numbers and range. In spring it is found flying over banks, paths and in coppiced clearings. A number of flower species are visited for pollen and nectar. Nests are built in existing burrows in dead wood such as stumps and fallen branches. *Pilicornis* is Latin for 'hairy-horned'.

Gold-fringed mason bee *O. aurulenta*, 10 mm length, locally common in southern England, and exclusively coastal in Wales, north-west England, south-west Scotland and Ireland, in coastal dunes, shingle ridges, grassland, landslips and former quarries, and inland mostly on calcareous grassland. Flying in April–August, it visits a range of flowers for pollen and nectar. Females often build nests in empty snail shells, but also in burrows. *Aurulenta* is Latin for 'full of gold'.

Hairy-horned mason bee Alternative name for fringe-horned mason bee.

Large mason bee *O. xanthomelana*, 12–13 mm length, formerly widespread but drastically contracting its distribution because of habitat change and/or climatic change so that it is now near extinction, its only known sites currently being at Sandown Bay, Isle of Wight, where the population consists of a few tens of individuals, and at Porth Ceiriad and Porth Neigwl on the Lleyn Peninsula, north Wales, on eroded soft-rock cliffs where the food plant bird's-foot-trefoil is found. Nectar is also taken from horseshoe vetch, bramble and bugle. Sites also need a supply of fresh water from seepages for nest cell construction. The nest consists of five or six pitcher-shaped cells, standing as a cluster with their bases inserted into the soil, or inside a burrow. *Xanthomelana* is Greek for 'yellow-black'.

Red mason bee *O. bicornis*, males 8–10 mm length, females 10–12 mm, common and widespread in England, Wales and lowland Scotland, and a recent arrival in Ireland, active in March–June. As well as in burrows in soil and dead wood, nests are in holes in walls and timber. Adults visit a large variety of flowers for pollen and nectar. *Bicornis* is Latin for 'two-horned', describing the female's head.

Two-coloured mason bee *O. bicolor*, 9–11 mm length, found in southern England and south Wales, closely correlated with grassland and open deciduous woodland on chalk and limestone soils. Flight is in April–June, pollen and nectar taken from a range of flowers. Females establish their nests in empty snail shells; when the nest is completed she covers the shell with a mound of dead grass stems, beech scales or leaf fragments. *Bicolor* is Latin for 'two-coloured'.

Wall mason bee *O. parietina*, 10 mm length, found in north Wales, north-west England and a few locations in Scotland,

listed as Rare in RDB 3, flying in May–July in unimproved grassland where its pollen source, common bird's-foot-trefoil, grows; it takes nectar from other flowers. Females make their nests in dry-stone walls, dead wood and other cavities. *Paries* is Latin for 'wall'.

MASON WASP Members of the family Vespidae, nesting in existing burrows or constructing their own. See also potter wasp.

Black-headed mason wasp *Odynerus melanocephalus*, found in southern England and the south Midlands on grassland, heathland and other open habitats on light clayey soils, flying in June–July. Nests are in burrows usually on level, exposed soil, provisioned with weevil larvae and small caterpillars. Greek *odyne* = 'pain', and *melanocephalus* = 'black-headed'.

Fen mason wasp *O. simillimus*, very rare, having been recorded from marshy areas in East Anglia, but until its rediscovery in 1986 it had been considered extinct in Britain. *Simillimus* is Latin for 'like'.

Purbeck mason wasp *Pseudepipona herrichii*, found on heathland in the Poole Basin, Dorset, Endangered in RDB 1. It requires exposed ground with a clay content (as a nesting site), open water (to assist nest building) and heathland with bell heather (for nectar and prey foraging). The shallow nests, excavated in dense aggregations in bare ground, contain up to three cells provisioned with 12–20 paralysed caterpillars of the tortrix moth heath button.

Spiny mason wasp *O. spinipes*, scattered in much of England and Wales, flying in June–July, visiting flowers with a short corolla and accessible nectaries. Extra-floral nectaries, and aphid honeydew are also taken. Nests are in vertical banks of hard earth. The digging site is wetted with water and a cluster of 5–6 cells excavated just behind the vertical face. Excavated material is used to build a 'chimney' up to 30 mm long to prevent rain entering the burrows or to reduce the risk of kleptoparasitism. Several females may nest close together in small aggregations. The female hunts for weevil (*Hypera*) larvae: up to 30 beetle larvae have been found in a cell. The egg is laid before the prey is collected and hatches in a few days. The prey is eaten by the larva over a few weeks. *Spinipes* is Latin for 'spiny feet'.

Wall mason wasp *Ancistrocerus parietum*, widespread in England and Wales, local in Scotland and Ireland, in a variety of habitats, including urban. Adults emerge from overwintering sites and mate in spring, and females search for nest sites. They are tube-dwellers generally using hollow cavities in plants, but a range of nesting sites includes thatched roof straw, holes in walls and disused burrows of wood-boring insects. The female clears away debris and pith from the tube or hole and plugs the inner end with clay softened with water. When a cell is completed an egg is laid together with larval provision of paralysed caterpillars. Greek *ancistron* + *ceras* = 'fish hook' + 'horn', and Latin *paries* = 'wall'.

MASTERWORT *Imperatoria ostruthium* (Apiaceae), a rare herbaceous perennial archaeophyte, scattered in northern England, Scotland and north Ireland, naturalised in moist or damp grassy habitats, including marshy pasture, on hillsides and by streams and rivers. The binomial is Latin for 'imperial' and probably *struthio* = 'wort' (i.e. herb).

MASTIGOPHORA Division of single-celled freshwater and marine protozoans, from Greek for 'bearing a whip'. Most species are capable of self-propelled movement using one or several flagella.

MAT-GRASS *Nardus stricta* (Poaceae), a common, widespread and often dominant tufted rhizomatous perennial of base-poor, infertile, peaty soils, on moorland and mountain, as well as lowland mire, heathland and acidic grassland. The binomial is from Greek *nardos*, the name for spikenard, although the reason for this is obscure, and Latin for 'upright'.

MAXILLOPODA Class of crustacean that includes barnacles and copepods (Latin *maxilla* + *podos* = 'jaw' + 'foot').

MAY BEETLE, MAY BUG Alternative names for common cockchafer.

MAY FISH Alternative name for allis and twaite shad, anadromous fish that enter rivers to spawn in April–May.

MAY TREE Alternative (folk) name for hawthorn.

MAYFLY Members of the order Ephemeroptera, with 50 species in 9 families: Ameletidae, Baetidae, Caenidae, Ephemerellidae, Ephemeridae, Heptageniidae, Leptophlebiidae, Potamanthidae and Siphlonuridae. The aquatic nymphs have three 'tails', prefer clear moving or still fresh water, and generally feed on algae and plant debris. Species are widely distributed, but with a variety of forms (e.g. cylindrical and burrowing, streamlined and flattened) associated with different habitats. On emergence, adults have a final moult (see subimago). Many adults live less than a day, a few species up to a week. Mating swarms are characteristic. Eggs are laid directly in the water or on submerged vegetation.

Drake mackerel mayfly *Ephemera vulgata* (Ephemeridae), fairly common in south-east England and up the east coast as far as Humberside. Nymphs live in pools and margins of muddy rivers, digging into the substrate to form a tubular burrow, and using their gills to force the water through this burrow to filter fine particulate organic detritus from the water. The species generally has a two-year life cycle. Adults fly in May–August. Emergence takes place during daylight on the water surface, occasionally on a stick or plant stem partially out of the water. Males swarm throughout the day, often continuing until dusk. Greek *ephemeros* = 'ephemeral', and Latin *vulgata* = 'common'.

Green drake mayfly *E. danica*, commonest and widespread. Nymphs live in unpolluted lakes and fast-flowing streams with a sand or gravel bed, creating a tubular burrow and using their gills to force the water through this, filtering or collecting fine particulate organic detritus. It usually has a two-year life cycle, but in the warmer waters of southern England it can complete its life cycle in one year. The main flight period is towards the end of May, but adults are often present between April and November. Emergence takes place during daylight on the water surface, occasionally on a stick or plant stem partially or entirely out of the water. Males swarm throughout the day, often continuing until dusk. Latin *danica* = 'Danish'.

Northern iron blue mayfly *Baetis muticus* (Baetidae), common and widespread though less so in south-east England, the nymphs found in the riffle sections of streams, living in gravel, sand or mud, feeding by scraping algae from stones or by collecting fine particulate organic detritus from the sediment. There is a slow-growing winter generation and a faster summer generation, emerging during April–October, males swarming in daylight. The binomial is from the Spanish river Baetis, and Latin *muticus* = 'curtailed'.

Pale evening mayfly *Procloeon bifidum* (Baetidae), widespread but local, nymphs living in stream margins and pools, feeding by scraping algae from stones or by collecting fine particulate organic detritus from the sediment. Adults are found in April–October. *Bifidum* is Latin for 'bifid' (i.e. divided in two by a deep notch).

Southern iron blue mayfly *B. niger*, widespread but absent from Ireland, and whose abundance has declined in some areas by as much as 80% in recent decades. Life history is similar to northern iron blue. *Niger* is Latin for 'black'.

Striped mayfly *E. lineata*, a rarity currently known only from the mid-Thames valley and the river Wye. Nymphs live in pools and river margins, creating a tubular burrows in the substrate, using their gills to force the water through this and filtering or collecting fine particulate organic detritus from the water column. Adults are present in July, emerging at dusk or dawn. The species has a two-year life cycle. *Lineata* is Latin for 'striped'.

Summer mayfly *Siphlonurus lacustris* (Siphlonuridae), widespread in localised pockets. Nymphs typically live in the pools and margins of rivers and streams, and lakes at high altitude. Adults emerge in May–September, although the main flight period is in June–August. Males swarm throughout the day, including dawn and dusk. Greek *siphlos* + *ouros* = 'maimed' + 'tail', and from Latin for 'lake'. There are few records for **northern summer mayfly** *S. alternatus* and **scarce summer mayfly** *S. armatus*.

Turkey brown mayfly *Paraleptophlebia submarginata* (Leptophlebiidae), widespread in Britain, particularly in the Midlands and Wales. Nymphs live in pools and stream margins, burrowing into gravel, sand or mud on the bed of the watercourse, filtering fine particulate organic detritus from the sediment. There is one generation a year. Emergence is in April–July. Males swarm during the day and into dusk. Greek *para* = 'near' + the genus *Leptophlebia* (*leptos* + *phlebos* 'thin' + 'of a vein'), and Latin *submarginata* = 'somewhat bordered'.

Upland summer mayfly *Ameletus inopinatus* (Ameletidae), scarce in northern England and Scotland. Nymphs favour streams >300 m, but are also found in some Highland lochs, feeding on fine particulate organic detritus in the sediment. Adults emerge in May–October, males swarming in the afternoon. *Ameletes* is Greek for 'not worthy of attention', *inopinatus* Latin for 'unimaginable'.

Yellow mayfly *Potamanthus luteus* (Potamanthidae), very rare, recently lost from the river Usk and suffering a population crash on the Wye. Nymphs mainly live in the pools and margins of larger rivers, living among stones and sand feeding on fine particulate organic detritus. Adults fly in May–July. Greek *potamos* + *anthos* = 'river' + 'flower', and Latin *luteus* = 'yellow'.

MAYWEED (ALGA) Alternative name for the kelp cuvie.

MAYWEED (PLANT) Species of *Matricaria* (Latin *matrix* = 'womb', referring to its use in midwifery) and *Tripleurospermum* (Greek *treis* + *pleura* + *sperma* = 'three' + 'rib' + 'seed', referring to the achenes) (Asteraceae). See also pineappleweed.

Scented mayweed *M. chamomilla*, an annual archaeophyte of arable fields and waste ground, locally common and widespread in England and Wales, a scattered casual in Scotland and Ireland, usually on light soils, but sometimes on loams and heavy clays. *Chamomilla* is from Greek *chamaimelon* = 'earth-apple', because of the apple-like smell of the flowers.

Scentless mayweed *T. inodorum*, a common and widespread annual archaeophyte of arable fields, farm tracks and waste ground on disturbed, fertile soils; also on roadsides, railway ballast and spoil heaps. *Inodorum* is Latin for 'scentless'.

Sea mayweed *T. maritimum*, a perennial, sometimes biennial, herb widespread and locally common in a range of coastal habitats, including open sand, shingle, cliffs, walls and waste ground, and also rarely inland on verges. *Maritimum* is Latin for 'of the sea'.

MAZEGILL Inedible bracket fungi in various genera and families. As the fruit body matures some of the pore walls break down to form slits with a mazelike appearance.

Rarities include **anise mazegill** *Gloeophyllum odoratum* (Gloeophyllales, Gloeophyllaceae), **spongy mazegill** *Spongipellis delectans* (Polyporales, Polyporaceae), and **timber mazegill** *G. trabeum* (recent records from Devon, Surrey and Norfolk).

Birch mazegill *Lenzites betulina* (Polyporales, Polyporaceae), annual, widespread and local in England, rare elsewhere, found throughout the year, sporulating in summer and autumn, as a bracket on the underside of dead wood, especially of birch and willow. Harald Lenz (1798–1870) was a German mycologist. Latin *betula* = 'birch'.

Common mazegill *Datronia mollis* (Polyporaceae), common and widespread, found throughout the year, sporulating from spring to late autumn, brackets found in tiers on dead wood, especially beech. *Mollis* is Latin for 'soft'.

Conifer mazegill *G. sepiarum*, widespread but uncommon, sporulating from late summer to late autumn on coniferous deadwood, causing a virulent brown rot.

Dyer's mazegill *Phaeolus schweinitzii* (Polyporales, Fomitopsidaceae), widespread and locally common, annual, found in summer and autumn often as overlapping tiers on pine and spruce, often parasitic on the roots. It can kill its host, turning saprobic and feeding on the dead wood. This fungal infection, Schweinitzii butt rot, can cause significant economic forestry losses. Its English name comes from its use in dyeing various shades of yellow, orange and brown, depending on fruitbody age and the metal used as a mordant. Greek *phaios* + Latin *olus* = 'dark' + 'somewhat'. Lewis von Schweinitz (1780–1834) was an American mycologist.

Oak mazegill *Daedalea quercina* (Fomitopsidaceae), widespread (including an east Irish distribution), sporulating in late spring and late autumn, usually solitary but occasionally in tiers on the trunks of oak trees and, less commonly, other broadleaf trees. Daedalus constructed a labyrinth at Knossos, Crete, here reflecting the labyrinth gill-like pores. Latin *quercus* = 'oak'.

MEADOW CAP **Pink meadow cap** Alternative name for pink waxcap.

MEADOW-FOAM *Limnanthes douglasii* (Limnanthaceae), an annual Californian neophyte, a frequent casual as a garden escape on waste ground, roadsides and sea and lake shores, persisting in a few locations, for example in west Cornwall, Cumbria and Co. Waterford. Greek *limne* + *anthos* = 'marsh' + 'flower'. David Douglas (1799–1834) was a Scottish botanist.

MEADOW-GRASS Species of *Poa* (Poaceae), *poa* being Greek for these grasses, commonly used as fodder.

Alpine meadow-grass *P. alpina*, a scarce perennial local mainly in the Highlands and Hebrides, but also the Lake District, Snowdonia, and Co. Kerry and Co. Sligo, on damp mountain rock faces and slopes on calcareous substrates.

Annual meadow-grass *P. annua*, very common and widespread in disturbed and man-made habitats such as overgrazed and trampled grassland, lawns, arable fields, waste ground, paths, waysides and wall tops. Perennial variants occur in montane and coastal grassland. It is also a wool and birdseed alien, and a common garden weed.

Broad-leaved meadow-grass *P. chaixii*, a perennial European neophyte cultivated since 1802, scattered in Britain, rare in eastern Ireland, naturalised in woodland. Dominique Chaix (1730–99) was a French botanist.

Bulbous meadow-grass *P. bulbosa*, a scarce perennial, mostly scattered in southern England and East Anglia, in grassland and sandy or rocky sites near the sea, mainly on dunes and stabilised shingle, but also on bare chalk and limestone. The few inland records may come from recent natural range extensions or introductions with sand and ballast.

Early meadow-grass *P. infirma*, a scarce annual found in the Channel Islands and with a scattered distribution in southern England, mostly growing near the sea in open, trampled grassland, on clifftop paths, lawns, rough ground and in stabilised dunes and other sandy habitats. *Infirma* is Latin for 'weak'.

Flattened meadow-grass *P. compressa*, a rhizomatous perennial found on well-drained soils on rough or stony ground, cinders, dry grassy banks and walls scattered in Britain, introduced in Ireland. Some populations on waste ground are probably wool aliens. *Compressa* is Latin for 'flattened'.

Glaucous meadow-grass *P. glauca*, a scarce perennial, very local in Snowdonia, the Lake Distrct and the Highlands on damp mountain rock faces, ledges and scree on calcareous substrates.

Narrow-leaved meadow-grass *P. angustifolia*, a rhizomatous perennial, common and widespread in much of England, scattered in Wales and Scotland, on dry grassland, walls, rough ground and railway embankments, typically on well-drained, relatively infertile soils. *Angustifolia* is Latin for 'narrow-leaved'.

Rough meadow-grass *P. trivialis*, a common and widespread stoloniferous perennial of open woodland, meadow, pasture, walls, and waste and cultivated ground; it also grows in marsh and beside ponds, ditches and streams. It is often used in amenity grassland, and is a common wool alien. *Trivialis* is Latin for 'ordinary'.

Smooth meadow-grass *P. pratensis*, a common and widespread rhizomatous perennial of meadow, pasture, and rough and waste ground, commonly used in amenity grassland. It prefers well-drained, neutral soils of moderate to high fertility, and is tolerant of grazing and trampling. *Pratensis* is from Latin for 'meadow'.

Spreading meadow-grass *P. humilis*, a common and widespread rhizomatous perennial found in a variety of grassland habitats, often on sandy soil, for example neutral meadow, dunes, roadsides, wall tops and riverbanks. *Humilis* is Latin for 'low'.

Swamp meadow-grass *P. palustris*, a perennial European neophyte, scattered, sporadic and naturalised following introduction in the nineteenth century as fodder, and as a grain and wool alien, in marsh, fen, ditches, willow carr and on water margins, as well as around docks, by railways and on waste ground. *Palustris* is from Latin for 'marsh'.

Wavy meadow-grass *P. flexuosa*, a rare perennial, local in the Highlands on acidic rock ledges, scree and mountain plateaux. *Flexuosa* is Latin for 'wavy'.

Wood meadow-grass *P. nemoralis*, a common and widespread perennial, probably introduced in Ireland, in woodland clearings, hedgerows and other shaded places, local on walls and, in mountains, on rock ledges. It has been sown in woodlands and parks as an ornamental, and in places may have been introduced with wool shoddy, grass seed or soil. *Nemoralis* is from Latin for 'woodland grove'.

MEADOW-RUE Rhizomatous perennial herbs in the genus *Thalictrum* (Greek for this plant) (Ranunculaceae).

Alpine meadow-rue *T. alpinum*, of mountain habitats, especially in Scotland, on damp rock ledges, at the edges of stony streams and flushes, and in thin grassland.

Common meadow-rue *T. flavum*, with a scattered distribution in England, Wales and Ireland, naturalised in Scotland, declining because of drainage and agricultural intensification, found in fen, ditches and stream banks, and tall vegetation in wet meadow and fen carr, where the substrate or water is baserich. *Flavum* is Latin for 'yellow'.

French meadow-rue *T. aquilegiifolium*, a neophyte introduced from Europe by 1629, with a scattered distribution as a garden escape, naturalised on roadsides and railway banks, and as a casual on waste tips. Latin *aquilegiifolium* = 'with aquilegia-like leaves'.

Lesser meadow-rue *T. minus*, widespread, scattered, but locally common, in calcareous or other base-rich habitats where competition is low, including dunes, scrubby banks, rocky lake and river edges, limestone grassland and pavement and montane rock ledges. It also occurs in habitats such as hedge banks and roadsides as a garden escape. *Minus* is Latin for 'small' or 'less'.

MEADOWCAP Inedible fungi in the genus *Porpoloma* (Agaricales, Tricholomataceae) (Greek *porpe* + *loma* = 'buckle' or 'brooch' + 'fringe').

Aromatic meadowcap *P. spinulosum*, mostly recorded from south-east England, Vulnerable in the RDL (2006), on soil under, especially, ash and beech. *Spinulosum* is Latin for 'spiny'.

Mealy meadowcap *P. metapodium*, scarce, widespread and scattered, on soil in grassland, often associated with coniferous woodland. Greek *metapodium* means 'in common with a foot'.

MEADOWSWEET *Filipendula ulmaria* (Rosaceae), a common, widespread perennial herb of damp or wet habitats on moderately fertile, neutral or calcareous soils, characteristically where water levels fluctuate, and absent from permanently waterlogged ground. Typical habitats are wet woodland, damp meadow, swamp and tall-herb fen, damp roadsides, ditches and

railway banks, and upland tall-herb communities. It occasionally grows in drier conditions, for example north-facing chalk grassland. As well as culinary and medicinal uses, the sweet scent of the crushed plant made it a valuable plant to strew on floors, giving its English name. Alternatively it might be a corruption of 'mead sweet', being used to flavour mead. Latin *filum* + *pendulus* = 'thread' + 'hanging', probably describing the root tubers that hang below the fibrous roots, and *ulmaria* = 'elm-like', (poorly) describing the leaf shape.

Giant meadowsweet *F. camtschatica*, a neophyte perennial herb from east Asia, planted in damp sites, occasionally naturalised with a scattered distribution, mainly in Scotland. Kamchatka is a region in the Russian Far East.

MEAL MOTH *Pyralis farinalis* (Pyralidae, Pyralinae), wingspan 18–30 mm, nocturnal, common in farm buildings, barns, warehouses and granaries, also sometimes in houses, across much of Britain, and also the south and east coast of Ireland. Adults fly in June–August. Larvae feed on stored grain and other dry plant matter. *Pyralis* was an unknown insect (or bird) which according to Pliny lived in fire (Greek *pyr*). *Farinalis* is from Latin *farina* = 'flour', the habitat.

Indian meal moth *Plodia interpunctella* (Phycitinae), wingspan 14–20 mm, nocturnal, introduced, frequent in food warehouses, larvae becoming a pest by feeding on cereals, grains, dried fruits, nuts and dried insect remains. *Plodia* may be an invented name. Latin *inter* + *punctus* = 'between' + 'dotted', from the wing markings.

MEALWORM Larvae of some of the darkling beetles (Tenebrionidae) used for feeding insectivorous birds in the wild and in captivity, but also often used imprecisely for larvae of yellow mealworm beetle.

Dark mealworm beetle *Tenebrio obscurus*, adults 12–18 mm in length, occasionally recorded from outdoors in the east Midlands and East Anglia, but a common pest of stored grain products. The protein-rich larvae, sometimes referred to as mini mealworms, are commonly used as food for pet reptiles. Latin *tenebrio* = 'lover of darkness', *obscurus* = 'dark'.

Lesser mealworm beetle *Alphitobius diaperinus*, adults 6 mm long, with a scattered outdoor distribution in England, but a common pest of stored grain products. Late instar larvae and adults do eat nuisance house flies in poultry houses, but are generally viewed as pests themselves, eating chicken feed and causing irritation to the birds, as well as being the vector of a number of animal pathogens. Greek *alphiton* = '(grain) meal' and *diapeiro* = 'to perforate'.

Yellow mealworm beetle *T. molitor*, adults 12.5–18 mm in length, with a scattered distribution outdoors in England and Wales, but a common pest of stored grain products. The protein-rich larvae are commonly used as food for wild birds and pets (especially birds and reptiles). *Molitor* is Latin for 'miller'.

MEALYBUG Member of the family Pseudococcidae (Sternorrhyncha), with 59 species in 26 genera found in Britain. Many have become pests of commercial and ornamental plants.

MEASLES A leaf spot fungus; see paint measles.

MECOPTERA Order of scorpion-flies (Greek *mecos* + *pteron* = 'long' + 'wing'), with three species in the Panorpidae, one (snow flea) in the Boreidae.

MEDICK Species of *Medicago* (Fabaceae), dating from the late Middle English (Latin *medica*, in turn from Greek *medice poa* = 'median grass'), mostly annual.

Black medick *M. lupulina*, a common and widespread annual or short-lived perennial of dry grassland and disturbed places on relatively infertile neutral or calcareous soils, often in sunny pasture and on roadside banks, waste ground and walls. *Lupulina* is Latin for 'resembling hops'.

Bur medick *M. minima*, a procumbent winter-annual found locally but declining in East Anglia and south-east England on

heathland, dunes and shingle, occasionally as a casual elsewhere in England, introduced with wool shoddy. *Minima* is Latin for 'smallest'.

Early medick *M. praecox*, a widespread but scattered procumbent Mediterranean neophyte found as a casual via wool shoddy on waste ground and tips. *Praecox* is Latin for 'early flowering'.

Shore medick *M. littoralis*, a procumbent or scrambling Mediterranean neophyte naturalised on beaches at Whitstable, Kent, since 2001. *Littoralis* is Latin for 'of the shore'.

Sickle medick *M. sativa* ssp. *falcata*, perennial, local in East Anglia, casual elsewhere, but declining in status near the coast, on grassy heaths, sea walls, roadsides (often confined to the rear of verges, being sensitive to mowing), mainly on calcareous soils and sands. *Sativa* is Latin for 'cultivated', *falcata* for 'sickle-shaped'. See also lucerne.

Spotted medick *M. arabica*, a procumbent winter-annual, common and increasing in the southern half of England, scattered elsewhere, in grassy habitats, often on light, sandy and gravelly soils. It grows as a weed in lawns and frequently occurs as a casual in fields and waste ground via wool shoddy. Latin *arabica* = 'Arabian'.

Strong-spined medick *M. truncatula*, a procumbent or scrambling Mediterranean neophyte, scattered and occasional as a casual on waste ground via seed or wool shoddy, though found naturalised in West Sussex in 1997. *Truncatulus* is Latin for 'almost cut off'.

Tattered medick *M. laciniata*, a procumbent North African neophyte, an occasional casual via wool shoddy, etc. on waste ground. *Laciniata* is Latin for 'torn' (i.e. divided into lobes).

Toothed medick *M. polymorpha*, procumbent and scrambling, recorded from open sandy and gravelly habitats by the coast in southern England, and from grassland on summer-parched banks, particularly in south-west England. Inland, it occurs as a casual, especially from wool shoddy. *Polymorpha* is Greek for 'many-shaped'.

MEDLAR *Mespilus germanica* (Rosaceae), an archaeophyte European shrub or small tree long cultivated for its fruit, occasionally found in hedges or woods, and as a relic of cultivation, mainly in England and Wales. 'Medlar' is late Middle English, from Old French *medler*, Latin *mespila*, and Greek *mespilon*. *Germanica* is Latin for 'German'.

MEGACHILIDAE Family name (Greek *mega* + *cheilos* = 'large' + 'lip') for leaf-cutter bees *Megachile*, mason bees *Osmia*, and wool carder bee.

MEGALOPODIDAE Family name (Greek *megalos* + *podos* = 'large' + 'foot') for beetles, with three species in a single genus, *Zeugophora*, which feed on young poplar and aspen where they skeletonise the leaves by eating small round windows between the veins. The leaf-mining larvae feed in a black blotch in May–July.

MEGALOPTERA Order (Greek *megalos* + *pteron* = 'large' + 'wing') that includes alderflies and snake flies, Alder fly larvae are aquatic and predaceous, snake flies terrestrial.

MELANDRYIDAE Family name (Greek *melas* + *drys* = 'black' + 'tree') of false darkling beetles (3–16 mm length), with 17 British species in 9 genera, all generally widespread but rare and of conservation concern, found in association with fungi and fungoid bark, particularly in ancient woodland.

MELANTHIACEAE Family name (Greek *melas* + *anthos* = 'black' + 'flower') for herb-paris.

MELICK Rhizomatous perennial grasses in the genus *Melica* (Poaceae), from Latin *melica* for a grass, possibly sorghum.

Mountain melick *M. nutans*, uncommon, scattered on basic soil over limestone and other base-rich rocks, growing in shade in deciduous woodland and in limestone pavement. *Nutans* is Latin for 'nodding' or 'pendant'.

Wood melick *M. uniflora*, scattered and locally common

(absent from north Scotland) in woodland clearings and edges, shady hedge banks and rock ledges, mainly on free-draining, base-rich soil. *Uniflora* is Latin for 'single-flowered'.

MELILOT Species of *Melilotus* (Fabaceae) from Greek *mel* + *lotos* = 'honey' + 'lotus', referring to the fragrant foliage (from the presence of coumarin in the tissues), enjoyed by bees.

Furrowed melilot *M. sulcatus*, an annual neophyte from the Mediterranean with a scattered distribution as a casual via birdseed on waste ground and tips. *Sulcatus* is Latin for 'furrowed'.

Ribbed melilot *M. officinalis*, a biennial neophyte herb native to Europe but introduced as seeds with clover from North America, first recorded in 1835, widely naturalised on roadsides, field borders and grassy banks, and found as a casual on waste ground. *Officinalis* is Latin for having pharmacological properties.

Small melilot *M. indicus*, an annual neophyte from south Europe with a scattered distribution as a casual on waste ground, sometimes on cultivated ground. It has usually been introduced via wool shoddy and birdseed. *Indicus* is Latin for 'Indian'.

Tall melilot *M. altissimus*, a biennial or short-lived perennial European archaeophyte, widespread but commonest in southern England and the Midlands, present since at least Tudor times, in disturbed grassland and on roadsides, and as a casual in field borders and waste ground. *Altissimus* is Latin for 'tallest'.

White melilot *M. albus*, an annual or biennial European neophyte mainly occurring as a casual on waste ground, in railway sidings and by roadsides. *Albus* is Latin for 'white'.

MELITTIDAE Family of burrowing bees (Greek *melissa/ melitta* = 'honey bee') in the genera *Dasypoda*, *Macropis* and *Melitta*. See also mining bee.

MELOIDAE From Greek *mele* = 'probe', the family name of oil beetles (10–35 mm length), with ten British species in three genera. They are all nest parasites of solitary bees, found in habitats such as grasslands, woodlands and gardens. Adults are flightless. Female *Meloe* and *Lytta* dig short burrows in the soil and lay batches of eggs underground. *Sitaris* eggs are laid within the host's nest hole. Newly hatched larvae climb on vegetation or sit in the host's burrow, the former (with three-clawed feet, and called triungulins) attaching themselves to passing insects. Those surviving are taken to the nests of the host bees. Once in the nest they feed on pollen collected by the bee for her own larvae and change to a grub-like form. Pupation takes place in the host burrow. *Sitaris* adults do not feed. Also known as blister beetles: see Spanish fly.

MENYANTHACEAE Family name of bogbean and fringed water-lily (Greek *menyo* + *anthos* = 'to disclose' + 'flower').

MERCURY Species of *Mercurialis* (Euphorbiaceae). 'Discovered' by the Roman god Mercury, annual mercury was medicinal, while dog's mercury was *mercurialis canina*, the useless (indeed toxic) mercury.

Annual mercury *M. annua*, an annual archaeophyte (known from deposits from Viking York) mainly present in the southern half of England, scattered elsewhere, in disturbed waste and cultivated ground, particularly in allotments and gardens, walls, and roadsides, thriving on light, nutrient-rich soils.

Dog's mercury *M. perennis*, a rhizomatous perennial herb, common and widespread except in north Scotland, scattered in Ireland where probably native only in the Burren. It is usually found on damp but free-draining base-rich soils. In the lowlands it is largely restricted to shaded sites, an indicator of ancient woodland, and also found in older secondary woodland, hedgerows and shaded banks. In the uplands it occurs on unshaded basic crags, scree, cliff ledges and in ravines, particularly on moist north-facing slopes, and also grows in limestone pavement grikes. Latin *perennis* = 'perennial'.

MERGANSER Red-breasted merganser *Mergus serrator* (Anatidae), wingspan 78 cm, length 55 cm, weight 1.1 kg, a widespread and locally common sawbill duck in especially western Britain and Ireland, in both freshwater and coastal habitats, with 2,200 breeding pairs. Wintering migrants from Eurasia, found around much of the British and south-east Irish coasts, bring numbers up to 9,000 birds, though BBS data indicate a decline by 20% between 2002–3 and 2012–13, and BTO suggest an 18% decline over 2010–15. Nests are on the ground, clutch size is 8–10, incubation takes 31–2 days, and the precocial young fledge in 60–5 days. Their diet of fish can bring conflict with fishermen. The species is Green-listed. 'Merganser' has been used since the seventeenth century, coming from Latin *mergus* + *anser* = 'diver' + 'goose'. The specific name *serrator* = 'one who saws'.

MERLIN *Falco columbarius* (Falconidae), Britain's smallest bird of prey, once popular in falconry (especially for ladies on account of its size), with a wingspan of 50–73 cm, length 24–33 cm, male weight 180 g, females 230 g. It has a widespread distribution, but is still recovering from a population crash in the late twentieth century (remaining on the Amber List), a 2008 survey suggesting 900–1,500 (mean 1,160) breeding pairs. Nests are in upland areas in abandoned corvid or hawk nests in trees, or in scrapes in heather moorland, with a clutch size of 4–5, incubation taking 30 days, and fledging of the altricial young 28–31 days. Birds move to lowland regions in winter, numbers augmented by migrating birds from Iceland Feeding is mainly on small birds. 'Merlin' comes from Old French *esmerillon* via Anglo-Norman *merilun*.

MERMAID'S TRESSES *Chorda filum* (Chordaceae), a widespread annual brown seaweed with thallus 2–6 m (up to 10 m) long, found on small stones and gravel in rock pools in the low intertidal and sublittoral down to 5 m. The binomial is Latin for 'string' + 'thread'.

MERVEILLE DU JOUR *Griposia aprilina* (Noctuidae, Xyleninae), wingspan 42–52 mm, a widespread if scattered nocturnal moth found in broadleaf woodland, especially mature oak woods, parks, hedgerows and gardens. Adults fly in September–October. Larvae feed on buds, flowers and later leaves of oak. *Gripos* is Greek for 'aquiline', *aprilis* Latin for April, here referring to the colour of spring leaves.

Scarce merveille du jour *Moma alpium* (Acronictinae), wingspan 30–5 mm, a rare RDB nocturnal moth almost entirely restricted to large areas of mature oak forest, once found from Cornwall to Kent, but now restricted to the New Forest, West Sussex and the Kentish Weald. Adults fly in June–July. Larvae feed on oak. The binomial is from Momus, the Roman god of mockery, and Latin *alpium* = 'alpine'.

MESH-WEAVER Small mesh-weaver *Dictyna pusilla* (Dictynidae), male body length 2.5 mm, female 1.8–3 mm, a scarce, probably declining UK BAP priority spider with recent records from the Highlands (e.g. Abernethy Forest) plus a few scattered ones from England and Wales, found from spring to late summer on low, dry or dead vegetation spinning an irregular retreat. Greek *dictyon* = 'net', and Latin *pusilla* = 'very small'.

MESOSTIGMATA Order of mites (Greek *mesos* + *stigmata* = 'middle' + 'marks'), body size 0.2–2 mm, free-living in soil or decaying organic matter. Many are parasites of vertebrates (except amphibians and fishes) and invertebrates (e.g. see varroa mite under mite), and some are economically important (e.g. as pests of stored products).

METASTIGMATA Order of ticks (or Ixodida) (Greek *meta* + *stigmata* = 'next to' + 'marks').

METRIDIIDAE Sea anemone family represented in Britain by plumose anemone, the Greek *metros* here being used in the sense of 'surrounded', referring to the tentacles.

METZGERIALES Order of thalloid liverworts, named after the German botanist Johann Metzger (1789–1852), including the families Aneuraceae (ghostwort and germanderworts), Fossombroniaceae (frillworts), Metzgeriaceae (veilworts), Moerckiaceae (ruffworts), Pallaviciniaceae (veilworts), Pelliaceae (pellias) and Petalophyllaceae (petalwort).

MEXICAN-TEA *Dysphania ambrosioides* (Amaranthaceae), an annual Central American neophyte, cultivated by 1640, occasional in Britain as far north as south Scotland as a casual on waste ground, originating from wool shoddy, soybean waste and birdseed. Greek *dys* + *phaino* = 'bad' + 'to show', referring to the inconspicuous flowers, and 'resembling *Ambrosia*', from the scent.

MEZEREON *Daphne mezereum* (Thymelaeaceae), a deciduous shrub of calcareous woodland, chalk-pits and fen. Local in England and Wales, native populations have suffered from habitat loss, but it is common as a garden escape, and is often bird-sown. Daphne was a Greek nymph who was transformed into a laurel (Greek *daphne*) to escape Apollo. The English and specific names are late fifteenth-century, ultimately from Arabic *mazaryun*.

MICHAELMAS-DAISY Perennial herbaceous plants in the genus *Aster* (Asteraceae), from Greek, then Latin, *astro* = 'star', the English name coming from these plants flowering around Michaelmas (29 September).

Common Michaelmas-daisy *Aster* x *salignus* (*A. novi-belgii* x *A. lanceolatus*), a widespread and locally common, vigorous hybrid neophyte of garden origin (by 1815), naturalised in river and lakeside habitats and in fen, as well as along roadsides and railways, and on waste ground. *Salignus* is Latin meaning 'like willow' (*salix*), from the leaf shape.

Confused Michaelmas-daisy *A. novi-belgii*, a widespread and scattered North American neophyte introduced by 1710, naturalised on hedge banks, rail- and roadsides, and waste ground. New Latin *novi-belgii* = 'from New York'.

Glaucous Michaelmas-daisy *A. laevis*, a North American neophyte cultivated by 1758, rare and scattered in England and recorded from Co. Tyrone, naturalised in grassy habitats and waste ground. *Laevis* is Latin for 'smooth'.

Hairy Michaelmas-daisy *A. novae-angliae*, a North American neophyte, scattered in Britain as far north as central Scotland, naturalised on rough ground. New Latin *novae-angliae* = 'from New England'.

Late Michaelmas-daisy *Aster* x *versicolor* (*A. laevis* x *A. novi-belgii*), established in English gardens by 1790, scattered in Britain and eastern Ireland in waste places, by railways and on roadsides. *Versicolor* is Latin for 'variously coloured'.

Narrow-leaved Michaelmas-daisy *A. lanceolatus*, a North American neophyte introduced by 1811, widespread and locally common along railways and riverbanks, on roadsides and on waste ground. *Lanceolatus* is Latin for 'lance-shaped'.

MICROLEPIDOPTERA An artificial, non-taxonomic grouping of 'small moths' (Greek *micros* = 'small') with a wing span generally of <20 mm, though the largest micros actually have a greater wing span than the smallest macros. See Macrolepidoptera.

MICROPEZIDAE Family name (Greek *micros* + *pezos* = 'small' + 'walking on foot') for stilt-legged flies, named after their long middle and rear legs, with fore legs smaller than the other pairs, in the superfamily Nerioidea, with ten species in five genera. Adults are generally predators on smaller insects such as aphids; larvae are probably phytophages or saprophages in decomposing ground vegetation.

MICROPTERIGIDAE Family name (Greek *micros* + *pteryx* = 'small' + 'wing') for small day-flying moths called golds.

MIDGE Small mosquito-like insects, members of the families Cecidomyiidae, Ceratopogonidae (biting midges), Chaoboridae (phantom midges), Chironomidae (non-biting midges), Dixidae and Psychodidae (owl midges, etc.). See also under gall.

Blackberry leaf midge *Dasineura plicatrix* (Cecidomyiidae), with a scattered distribution in England and Wales, adults evident in late spring and summer, larvae found in galls of blackberry. Greek *dasys* + *neuron* = 'hairy' + 'nerve', and Latin *plicatrix* = 'folded'.

Highland midge *Culicoides impunctatus* (Ceratopogonidae), 2 mm length, the commonest biting midge in the Highlands, a nuisance to warm-blooded animals, including deer and horses, and responsible for about 90% of midge bites on humans. *Culicoides* is Latin for a small midge (*culex*) and *impunctatus* = 'unspotted'.

Owl midge Alternative name for some moth flies (Psychodidae).

MIDGET Nocturnal leaf-mining moths in the genus *Phyllonorycter* (Gracillariidae, Lithocolletinae), from Greek *phyllon* + *orycter* = 'leaf' + 'a digging' (i.e. a leaf miner), often with two generations a year. Wingspan is generally 7–9 mm.

Scarce species include **clover midget** *P. insignitella* (a few scattered localities in northern Scotland, western Ireland, Yorkshire, Durham and Kent), **gold-bent midget** *P. roboris* (mature oakwood in a few places in England, east Wales and east Ireland), **osier midget** *P. viminetorum* (osier beds and other marshy habitats in Cambridgeshire, east Suffolk, north Hampshire, south Wiltshire and west Suffolk), **sandhill midget** *P. quinqueguttella* (dunes and heathland in a few coastal locations), **scarce midget** *P. distentella* (oakwoods in Gloucestershire, Monmouthshire and Herefordshire, and the Blean Wood complex near Canterbury, Kent), **scarce aspen midget** *P. sagitella* (woodland in the Midlands and Wales), **southern midget** *P. dubitella* (woodland and scrub and on riverbanks in southern and central England, and parts of Wales), and **winter poplar midget** *P. comparella* (damp woodlands in England).

Beech midget *P. maestingella*, common in woodland containing beech trees. Adults fly in May–June and August. Larvae mine beech leaves in spring. The specific name probably honours an eighteenth-century German entomologist of this name.

Broad-barred midget *P. froelichiella*, wingspan 9–10 mm, common in damp woodland throughout much of Britain, less so in Scotland, and rare in Ireland. Adults fly in July–August. Larvae mine alder leaves in autumn. F.A.G. Frölig was a nineteenth-century German entomologist.

Broom midget *P. scopariella*, widespread but scarce in heathland and open woodland. Adults fly in May–July. Larvae mine twigs of broom (*Cytisus scoparius*, giving the specific name) in spring.

Brown apple midget *P. blancardella*, common in woodland, orchards and gardens in southern England and Wales, less common and scattered in south Scotland and Ireland. Adults fly in April–May and August–September. Larvae mine leaves of apples in July–November. Steven Blankaart (1650–1704) was a Dutch entomologist.

Cherry midget *P. cerasicolella*, local in woodland and orchards throughout England north to Yorkshire, and parts of Wales. Adults fly in May and August. Larvae mine leaves of wild and cultivated cherries in August and September–November. *Cerasicolella* is from the food plant dwarf chery *Prunus cerasus* and Latin *colo* = 'to inhabit'.

Common alder midget *P. rajella*, widespread and common in damp woodland. Adults fly in April–May and July–August. Larvae mine alder leaves in June–November. John Ray (1627–1705) was an English naturalist.

Common oak midget *P. quercifoliella*, widespread and common in oak woodland and areas with scattered oaks. Adults fly in April–May and August–September. Larvae mine oak leaves, hence the specific name from *quercus* + *folium* = 'oak' + 'leaf'.

Common thorn midget *P. oxyacanthae*, widespread and common in hedgerows and woodland. Adults fly in May and July–August. Larvae mine hawthorn leaves, giving the specific name from the formerly termed *Crataegus oxyacantha*, occasionally those of pear and quince, from August to November.

Dark alder midget *P. kle(e)mannella*, common in damp woodland throughout Britain, rarer in Scotland; there have been a couple of records from south-west Ireland. Adults fly in May and (mostly) August. Larvae mine alder leaves in July and August–November, peaking in October. Christian Kleeman (1735–89) was a German entomologist.

Dark hornbeam midget *P. esperella*, local in woodland in England, especially the Midlands and East Anglia, with isolated records elsewhere from the south-west and the Welsh borders. Adults fly in May and August. Larvae mine hornbeam leaves from July to November.

Elm midget *P. tristrigella*, common in hedgerows in England, Wales and south-west Scotland. Adults fly in May and August. Larvae mine elm leaves in July–November. Latin *tri* + *striga* = 'three' + 'streak', from the forewing markings.

Fiery oak midget *P. lautella*, wingspan 6–7 mm, widespread and locally common in woodland and areas with scattered oaks, less so in Scotland. Adults fly in May and August. Larvae mine leaves of oak, preferring saplings from July to November. *Lautella* is Latin for 'well washed', here meaning brightly coloured.

Garden midget *P. messaniella*, widespread and common in woodland, particularly in the south. Adults fly any time between April and November. Larvae mine leaves of oaks, beech, sweet chestnut and hornbeam. The specific name refers to Messina, Sicily.

Gold birch midget *P. cavella*, local in birch woodland and heathland in England, Wales and the Highlands. Adults fly in June–July. Larvae mine birch leaves in August–October. *Cavus* is Latin for 'hollow'.

Gorse midget *P. ulicicolella*, wingspan 6–7 mm, scarce on heathland, downland and grassland in the coastal counties of south England, from Norfolk to Cornwall, and north-west England. Adults fly in June–August. Larvae mine the shoots and spines of gorse (*Ulex*), which with Latin *colo* = 'inhabit' gives the specific name.

Grey alder midget *P. strigulatella*, local in damp woodland and field edges, in local, isolated colonies in south and south-west England, East Anglia, the Midlands, Welsh borders, and parts of Scotland. Adults fly in May and late July–August. Larvae mine grey alder leaves in June and September–October. *Strigula* is Latin for a small linear mark, here found on the forewing.

Hawthorn midget *P. corylifoliella*, widespread but scattered in hedgerows, scrubland and open woodland, less common in northern England, Scotland and Ireland. Adults fly in May and August. Larvae mine leaves of hawthorn, apple, rowan and birch in July and September–October. The specific name is from Latin *corylus* + *folium* = 'hazel' + 'leaf', though this is not a food item.

Honeysuckle midget *P. trifasciella*, common in woodland and hedgerows throughout England and Wales, more local in Scotland, mostly coastal in Ireland. Adults have three generations, flying in May, August and November. Larvae mainly mine leaves of honeysuckle in April and July–November. *Trifasciella* is Latin for 'three-banded', from the forewing markings.

Hornbeam midget *P. tenerella*, local in woodland in parts of the south-east Midlands, Kent and East Anglia. Adults fly in May and July–August. Larvae mine hornbeam leaves in June and August–November. Latin *tenella* = 'delicate'.

Large midget *P. emberizaepenella*, wingspan 9–10 mm, widespread but local and rarely common in woodland, scrub and hedgerows, and rarely recorded in Ireland. Adults fly in May and August. Larvae mine leaves of honeysuckle and snowberry from July and more commonly in September–October. The specific name is from *Emberiza* (bunting) + the Latin *pennae* = 'wings', from a putative resemblance to this bird's coloration.

London midget *P. platani*, wingspan 8–10 mm, discovered new to Britain in London in 1990, since when it has been recorded on London plane trees in (sub)urban parks and streets throughout the southern counties, East Anglia and the Midlands.

Adults fly in late April–May and August. Larvae mine leaves of London plane (*Platanus*, giving the specific name) in late autumn.

Long-streak midget *P. salicicolella*, widespread and common in (especially damp) woodland and scrub, especially in the south, more localised in Scotland and less common but widespread in Ireland Adults fly in May and July–August. Larvae mine leaves of sallow (*Salix* + Latin *colo* = 'inhabit', giving the specific name), particularly in autumn.

Maple midget *P. acerifoliella*, common in woodland and hedgerows in south Wales and England as far north as Derbyshire, with isolated records in Northumberland and Cumbria. Adults fly in May and August. Larvae mine leaves of field maple (*Acer campestre* + *folium* = 'leaf', giving the specific name) in July and September–October.

Pale oak midget *P. heegeriella*, common in oak woodland throughout England (especially in the south), Wales and parts of Scotland, except in the far north, and with a few records in east Ireland and Co. Kerry. Adults fly in May and August. Larvae mine leaves of oak in July and September–October. Ernst Heeger (d.1866) was an Austrian entomologist.

Red birch midget *P. ulmifoliella*, widespread and common in birch woodland. Adults fly in May and August. Larvae mine leaves of birch in July and more commonly September–October. The specific name is from *Ulmus* (elm) + *folium* = 'leaf', though this is not a food plant.

Red hazel midget *P. nicellii*, common in woodland, widespread but more numerous in the south. Adults fly in May and August. Larvae mine hazel leaves in July and September–October. Graf von Nicelli was a nineteenth-century German entomologist.

Rowan midget *P. sorbi*, widespread and common in woodland, more numerous in the north. Adults fly in April–May and August. Larvae mine leaves of rowan and sometimes other species of *Sorbus* (hence the specific name) in July–November.

Sallow midget *P. hilarella*, widespread and common in woodland edges and woodland rides. Adults fly in late May–June and August. Larvae mine leaves of sallow in July and September–November. *Hilaris* is Latin for 'cheerful'.

Scarce brown midget *P. mespilella*, occasional in woodland in parts of England, south Wales, Scotland and south and east Ireland. Adults fly in May and August. Larvae mine leaves of wild service-trees, sometimes those of other rosaceous trees such as pear and medlar (*Mespilus*, giving the specific name), in July–August and October.

Scarce oak midget *P. kuhlweiniella*, previously known from a few scattered records and last recorded in 1949, but in the early 1980s it was discovered in a Norfolk woodland and has since shown to be fairly common but localised in the region. An individual was also found in Bedfordshire in 2014. Adults fly in May and July–August. Larvae mine oak leaves.

Sloe midget *P. spinicolella*, widespread and common in hedgerows and scrub. Adults fly in May and August–September. Larvae mine leaves of blackthorn (*Prunus spinosa*, giving the specific name) in August–November.

Small alder midget *P. stettinensis*, common in damp woodland throughout much of England, Wales and parts of western Scotland. Adults fly in May and August. Larvae mine alder leaves in July–November. Stettin (now Szczecin) is a city in Poland.

Small birch midget *P. anderidae*, scarce on heathland, moorland and mosses in south and south-east England, north East Anglia, north Welsh and adjacent English counties to Cumbria, and west Perthshire. Adults fly in May and August. Larvae mine leaves of (preferably seedling) birch. *Anderida* is Latin for the type locality, Abbot's Wood, East Sussex.

Small elm midget *P. schreberella*, often common in hedgerows throughout much of England south of the Humber estuary and east Wales. Adults fly in May and August. Larvae mine leaves of elm (though rarely wych elm) from July to November. Johann von Schreber (1739–1810) was a German entomologist.

Sycamore midget *P. geniculella*, common in woodland and hedgerows in Britain, and recorded from east Ireland. Adults fly in May and August. Larvae mine sycamore leaves in summer and autumn. *Geniculum* is Latin for 'little knee', from forewing markings.

Upland midget *P. junoniella*, found in moorland in Wales and from the Midlands northwards to much of Scotland, especially the Highlands. Adults fly in June–July, in places with another brood in autumn. Larvae mine cowberry leaves. *Junoniella* refers to the Greek goddess Juno, with no entomological significance.

Viburnum midget *P. lantanella*, in hedgerows and open woodland on calcareous soils in the southern half of England and parts of Wales. Adults fly in May and August. Larvae mine leaves of wayfaring-tree (*Viburnum lantana*, giving the specific name) and guelder-rose in July and September–November.

Western midget *P. muelleriella*, found in ancient oak woodland, mainly in the south Midlands and Welsh Border, into Cumbria and parts of Scotland. Adults fly in May and August. Larvae mine oak leaves in July and September–October. The specific name refers to one of the (unhelpfully) many German entomologists called Müller.

White-bodied midget *P. joannisi*, common in woodland and hedgerows in south Wales and England as far north as Yorkshire. Adults fly in May and August. Larvae mine Norway maple leaves, mostly in autumn.

White oak midget *P. harrisella*, common in oak woodland and areas with scattered oaks, widespread but less so in Scotland and Ireland. Adults fly in May–June and July–August. Larvae mine oak leaves in July–November. The specific name probably honours the English entomologist Moses Harris (1731–88).

Willow midget *P. viminiella*, common in marsh, riversides and other wetland habitats in England and Wales, with scattered records in Scotland, and a couple of records in Ireland. Adults fly in May–June and August. Larvae mine leaves of (particularly narrow-leaved) willows such as crack willow and osier, and also recorded from aspen, in June–July and August–September. *Vimen* is Latin for an osier twig.

MIGNONETTE Herbaceous plants in the genus *Reseda* (Resedaceae) (Latin *resedo* = 'to calm'). 'Mignonette' dates from the early eighteenth century, from French *mignonnette*, a diminutive of *mignon* = 'small and sweet', describing the fragrant flowers.

Corn mignonette *R. phyteuma*, an annual or biennial European neophyte present by the mid-eighteenth century, naturalised in arable fields, and a casual on waste ground, introduced with wool shoddy and grain, in a few locations in England and south Wales. *Phyteuma* is Greek for a kind of scented plant, probably this genus.

Garden mignonette *R. odorata*, an annual or biennial Mediterranean neophyte present by the mid-eighteenth century, a casual garden escape at a few sites in England and the Isle of Man on waste ground and roadsides. *Odorata* is Latin for 'scented'.

White mignonette *R. alba*, an annual or perennial European neophyte, present by Tudor times, usually a grain, wool or birdseed alien, but sometimes a garden escape, found in a range of man-made habitats including roadsides, railway banks, arable land and waste ground, often near the sea, occasionally naturalised. *Alba* is Latin for 'white'.

Wild mignonette *R. lutea*, a biennial or perennial, common in England, rather scattered or absent elsewhere, on well-drained calcareous soils in open habitats such as waste ground and verges, in marginal grassland, disused railway land, quarries and arable land, in disturbed chalk and limestone grassland and on fixed dunes. *Lutea* is Latin for 'yellow'.

MILACIDAE Family name for keeled slugs and smooth jet slug.

MILDEW A thin, superficial, powdery, usually whitish ascomycete fungal growth consisting of minute hyphae that penetrate living plants (often being host-specific) and organic matter. Powdery mildews are generally in the order Erysiphales, family Erysiphaceae (common species including oak mildew *Erysiphe alphitoide*, and mint mildew *Neoerysiphe galeopsidis* found on hedge woundwort and other Lamiaceae) or are fungus-like organisms or 'water moulds' or 'downy mildews' in the family Peronosporaceae. More generally, mildew is sometimes taken to be synonymous with mould (q.v.).

MILK-PARSLEY *Thyselium palustre* (Apiaceae), a biennial or short-lived perennial herb, mainly found on damp peat, often in sites flooded in winter. Characteristic of tall-herb fen, it can survive in fen scrub and alder carr, mainly in East Anglia and Somerset, with a few other English locations. *Palustre* is from Latin for 'marsh'.

Cambridge milk-parsley *Selinum carvifolia*, a herbaceous perennial of fen, damp meadow and rough grazed marshy pasture on calcareous peaty soils. It was last recorded in Lincolnshire in 1931 and in Nottinghamshire by 1952, and is currently confined to three statutorily protected sites in Cambridgeshire (Chippenham Fen, Sawston Hall Fen and Snailwell Meadows), where it is thriving. It is Vulnerable in the RDL (2005), and listed on Schedule 8 of the WCA (1981). Greek *selinon* = 'parsley', and Latin *carvifolia* = having foliage like caraway (*Carum carvi*).

MILK-VETCH Perennial herbaceous plants in the genus *Astragalus* (Fabaceae) (Greek *astragalos* = 'ankle bone', referring to the seed shape).

Alpine milk-vetch *A. alpinus*, known from four sites in the Highlands in flushed calcareous grassland, and on base-rich ledges and rocky outcrops.

Chick-pea milk-vetch *A. cicer*, introduced from Europe, an occasional casual in England, and naturalised on a hedgebank in Midlothian. *Cicer* is Latin for 'chickpea'.

Lesser milk-vetch *A. odoratus*, introduced from the Mediterranean, with a scattered distribution in grassy habitats in southern and central England. *Odoratus* is Latin for 'fragrant'.

Purple milk-vetch *A. danicus*, in short unimproved turf on well-drained calcareous soils, mostly on chalk and limestone, but also on dunes and machair. In Scotland, it also grows on Old Red Sandstone sea-cliffs and on mica-schist. Status is mostly stable but during the twentieth century it declined substantially on the chalk in southern England and limestone in north-east England, largely due to agricultural improvement or lack of grazing. *Danicus* is Latin for 'Danish'.

MILKCAP Fungi in the genus *Lactarius* (Russulales, Russulaceae), from Latin for 'milk-producing', referencing the milky latex that exudes from the gills when cut or torn.

Widespread but uncommon species include **alder milkcap** *L. obscuratus* (in damp woodland, carr and swamp), **bearded milkcap** *L. pubescens* (poisonous, in grassland, heathland, scrub or woodland under birch, with which it is ectomycorrhizal), **birch milkcap** *L. tabidus*, **lilac milkcap** *L. lilacinus* (damp deciduous woodland and carr, often under alder), **lilacscale milkcap** *L. spinulosus* (under birch), **orange milkcap** *L. aurantiacus* (acid woodland soil, and under dwarf willow in dune slacks), **pale milkcap** *L. pallidus* (usually under beech), **sooty milkcap** *L. fuliginosus* (beech), **two-spored milkcap** *L. acerrimus* (oak), **watery milkcap** *L. serifluus* (oak), **willow milkcap** *L. aspideus*, and **woolly milkcap** *L. torminosus* (poisonous, in grassy woodland or heathland under birch).

Rarities, mostly from the Highlands, include **fishy milkcap** *L. volemus*, **hoary milkcap** *L. scoticus*, **larch milkcap** *L. porninsis*, **lurid milkcap** *L. luridus*, **pine milkcap** *L. musteus*, **rollrim milkcap** *L. resimus*, **whiskery milkcap** *L. mairei* and **yellow bearded milkcap** *L. repraesentaneus*.

Beech milkcap *L. blennius*, inedible, common and widespread, found in summer and autumn on deciduous woodland soil, especially under beech with which it is ectomycorrhizal. *Blennos* is Greek for 'slimy'.

Coconut milkcap *L. glyciosmus*, edible, common and widespread, found in late summer and autumn on deciduous woodland soil, nearly always under birch with which it is mycorrhizal. Greek *glycys* + *osme* = 'sweet' + 'smell'.

Curry milkcap *L. camphoratus*, edible, common and widespread, found in late summer and autumn on coniferous woodland soil, generally under pine, occasionally in broadleaf woodland under birch. Caps have a curry odour when dried. *Camphoratus* is Latin for 'like camphor'.

Delicious milkcap Alternative name for saffron milkcap.

False saffron milkcap *L. deterrimus*, inedible, widespread and common, found in late summer and autumn on acid soil in coniferous woodland, often under spruce. *Deterrimus* is Latin for 'worst', a comment on the taste.

Fenugreek milkcap *L. helvus*, inedible (possibly poisonous), with a scattered distribution, commoner in Scotland, found in late summer and autumn on soil either under conifers in upland sites, or under birch in heathland and scrub in southern England. The cap, when cut, gives off an odour of fenugreek. *Helvus* is Latin for 'creamy' or 'amber'.

Fiery milkcap *L. pyrogalus*, inedible (possibly poisonous), widespread and locally common, found in late summer and autumn on soil in often coppiced deciduous woodland, under hazel with which it is ectomycorrhizal. Greek *pyr* + *gala* = 'fire' + 'milk'.

Fleecy milkcap *L. vellereus*, inedible, common and widespread, found in late summer and autumn in damp acid woodland soil, especially under birch, oak and beech. The caps are covered in fine fleece-like fibres (Latin *vellus* = 'fleece').

Grey milkcap *L. vietus*, tasteless, common and widespread, found in late summer and autumn in soil under birch with which it is ectomycorrhizal. *Vietus* is Latin for 'shrivelled'.

Liver milkcap *L. hepaticus*, inedible, with a scattered distribution, commoner in southern England, found in late summer and autumn on soil under pine. *Hepaticos* is Greek for 'relating to the liver'.

Mild milkcap *L. subdulcis*, edible, very common and widespread, found in late summer and autumn in soil in broadleaf woodland, under beech, with which it is ectomycorrhizal, and also birch and oak. Latin *sub* + *dulcis* = 'under' + 'sweet', describing the initially mild then sweet taste followed by a slight bitterness.

Oakbug milkcap *L. quietus*, inedible, very common and widespread, found in late summer and autumn on soil in deciduous woodland, under oak with which it is ectomycorrhizal. A rather unpleasant oily smell, of bedbugs according to some, contributes to its common name. *Quietus* is Latin for 'tranquil'.

Peppery milkcap *L. piperatus*, inedible, locally common and widespread, found in summer and autumn in soil and leaf litter in deciduous woodland, commonly under beech. *Piperatus* is Latin for 'peppery'.

Rufous milkcap *L. rufus*, inedible, common and widespread, found from early summer to autumn on acid soil in coniferous woodland, especially under pine, occasionally birch. *Rufus* is Latin for 'red'.

Saffron milkcap *L. deliciosus*, edible and tasty, relatively common in Scotland, less so elsewhere, found in summer and autumn on coniferous woodland soil, especially under pine and spruce, occasionally in oakwoods. *Deliciosus* is Latin for 'delicious'.

Slimy milkcap Alternative name for beech milkcap.

Spruce milkcap Alternative name for false saffron milkcap.

Tawny milkcap *L. fulvissimus*, inedible, widespread and locally common, found in late summer and autumn on soil in deciduous woodland, often under oak, lime, hornbeam or beech on base-rich soil. *Fulvissimus* is Latin for 'very reddish-brown'.

Ugly milkcap *L. turpis*, inedible and possibly carcinogenic, common and widespread, found in late summer and autumn in damp soil in broadleaf woodland, under birch with which it is ectomycorrhizal. *Turpis* is Latin for 'ugly'.

Yellowdrop milkcap *L. chrysorrheus*, poisonous, widespread, common and often abundant, found in late summer and autumn on acid soil in deciduous woodland under oak, with which it is ectomycorrhizal. Greek *chrysos* + *rheos* = 'gold' + 'stream'.

MILKWORT Perennial members of the genus *Polygala* (Polygalaceae), once used as an infusion to increase the flow of a nursing mother's milk, hence its common name, and Greek *polys* + *gala* = 'much' + 'milk' also from the belief that cattle grazing on this plant produced more milk.

Chalk milkwort *P. calcarea*, found in grazed chalk and limestone grassland, in southern England north to south Lincolnshire, usually on warm south-facing slopes. As a poor competitor it disappears if insufficient grazing allows coarser grasses to become dominant. *Calcarea* is Latin for 'of chalky land'.

Common milkwort *P. vulgaris*, common and widespread in short, moderately infertile, often acidic grassland, dunes, heathland and fen-meadow. Latin *vulgaris* = 'common'.

Dwarf milkwort *P. amarella*, a rarity of base-rich substrates. On the Kent North Downs it is usually found in well-grazed chalk grassland, and in northern England in limestone grassland, on rock ledges and in fissures in limestone scars, sometimes on damp stream banks. *Amarella* is Latin for 'somewhat bitter'.

Heath milkwort *P. serpyllifolia*, widespread, often common, found on acidic grassland, moorland, heathland and mire. *Serpyllifolia* is Latin for 'with leaves like thyme'.

MILLER (FUNGUS) *Clitopilus prunulus* (Agaricales, Entolomataceae), edible, with a 'mealy' smell and taste (hence its name), widespread and common, found in summer and autumn on soil in grass, usually near trees, for example in woodland clearings and under hedgerows. Greek *clitos* + *pilos* = 'slope' + 'hair', and Latin *prunulus* = 'small plum tree'.

MILLER (MOTH) *Acronicta leporina* (Noctuidae, Acronictinae), wingspan 35–45 mm, nocturnal, widespread and common in broadleaf woodland, heathland, moorland, fen and gardens. Adults fly in June–August. Larvae feed on birch, alder and grey willow. Usually a grey colour, the whitest forms of the adult, giving the common name, occur in parts of Scotland, with darker ones in industrial parts of northern England and the Midlands. Greek *acronyx* = 'nightfall', and Latin *leporina* = 'like a hare', for the hare's white winter coat and the moth's wing colour.

MILLER'S THUMB Alternative name for bullhead, so named from its stocky, thumb-like shape, and referring to millers who developed broad thumbs from rubbing grain between their fingers.

MILLET Species of grass (Poaceae) in the genera *Sorghum* (from Italian *sorgo*, a tall cereal grass), *Milium* (Latin for millet), and *Panicum* (Latin for wild millet). Other *Panicum* (and *Echinochloa*) species are occasional casual wool, birdseed and grain aliens on waste ground.

Common millet *P. miliaceum*, a common and widespread Asian neophyte annual introduced by 1596 and recorded in the wild in 1872, a bird- and oil-seed casual on waste ground and in woodland around pheasant feeding areas, and also a grain contaminant in arable fields. *Miliaceum* is Latin for 'millet-like'.

Early millet *M. vernale* ssp. *sarniense*, an annual found at two sites in Guernsey in short turf on fixed dunes. *Vernale* is Latin for 'flowering in spring'. Sarnia was the Latin name of Guernsey.

Great millet *S. bicolor*, an annual African neophyte cultivated by 1596, recorded from the wild in 1890, and probably increasing in abundance on cultivated and waste ground, originating from birdseed and wool shoddy. *Bicolor* is Latin for 'two-coloured'.

Wood millet *M. effusum*, a widespread and locally common perennial of damp, deciduous woodland and shaded banks, growing on calcareous to mildly acidic clay and loam, and in western Scotland over rocks. It is absent from north Scotland, and scattered in Ireland. It is an indicator of ancient woodland in parts of eastern England, but can colonise sites that are disturbed by felling or fire, and it has spread to more recent woodland in some upland areas. *Effusum* is Latin for 'loose'.

MILLIMETRE MOSS *Micromitrium tenerum* (Pottiaceae), Britain's smallest moss, seen recently (2004 and subsequently) only on the edge of Roadford Reservoir (east of Launceston), Devon. Greek *micros* + *mitrion* = 'small' + 'small cap', and Latin *tenerum* = 'soft' or 'tender'.

MILLIPEDE From Latin *mille* + *pes* = 'thousand' + 'leg', members of the class Diplopoda, subphylum Myriapoda, characterised by having two pairs of jointed legs on most body segments. Most millipedes have elongated cylindrical or flattened bodies with more than 20 segments, though pill millipedes are shorter and can roll into a ball. These detritivores favour leaf litter, dead wood and soil, with a preference for humid conditions.

Bristly millipede *Polyxenus lagurus* (Polyxenidae), widespread south of a line from the Severn to the Humber estuaries, increasingly restricted to coastal sites further north, found on old walls and trees and in maritime habitats, feeding on lichens and algae. Greek *poly* + *xenos* = 'many' + 'stranger', and Latin *lagurus* = 'like a hare'.

Pill millipede *Glomeris marginata* (Glomeridae), common and widespread, but not north of the Central Lowlands of Scotland, typically in deciduous woodland on calcareous loam, and having an association with bare rock, including screes. It can roll into an egg shape when disturbed. The binomial is Latin for 'of a ball' and 'bordered'.

Spotted snake millipede *Blaniulus guttulatus* (Blaniulidae), males 8–12 mm long, females 12–15 mm, widespread and locally common, including in gardens, feeding on roots and tubers, and sometimes becoming a pest. Greek *blanos* = 'blind', and Latin *guttatus* = 'spotted'.

MIMETIDAE Family name for pirate spiders (Greek *mimos* = 'imitator'), with four British species in the genus *Ero*.

MIND-YOUR-OWN-BUSINESS *Soleirolia soleirolii* (Urticaceae), an evergreen, procumbent, carpet-forming perennial neophyte herb from Corsica and Sardinia, introduced into gardens in 1905, now scattered and locally common, on damp paths, shaded banks and roadside walls, and in sheltered places in churchyards and gardens, usually close to habitation. It can become a serious weed of lawns. The binomial honours Captain Joseph Soleirol, who collected many Corsican plants in the early nineteenth century.

MINING BEE 1) Solitary, ground-nesting bees in the family Andrenidae, which includes 68 species of *Andrena* in the British list. New Latin *andrena* derives from Greek *anthrene* = 'wasp'.

Ashy mining bee *A. cineraria*, length 10–14 mm, forewing length 9–11 mm, widespread in England and Wales, occasional in Ireland, and probably increasing its range generally on open sunny habitats including gardens, in southern England mainly on calcareous grassland, flying in March–April (later further north), visiting a range of flowers for pollen and nectar. The burrow entrance is left open during foraging trips, but at the end of these flights, during rain and when disturbed they are closed. *Cineraria* is Latin for 'ashy'.

Early mining bee *A. haemorrhoa*, common and widespread, generally on open habitats, flying in March–June, visiting a range of flowers. Greek *haima* + *rheo* = 'blood' + 'flow'.

Girdled mining bee *A. labiata*, with a scattered distribution in England and Wales, mainly in the south, very local on sandy soils in grassland and woodland edge, improving in status following a decline over 1950–90, flying in May–June, visiting a range of flowers for pollen and nectar. The burrow entrance is left open during foraging trips, but at the end of these flights, during rain and when disturbed they are closed. *Labiata* is Latin for 'lipped'.

Grey mining bee Alternative name for ashy mining bee.

Gwynne's mining bee *A. bicolor*, body length 8–10 mm, forewing length 6–8 mm, common and widespread, though mainly coastal in Scotland and Ireland, found in a range of habitats, bivoltine (flying in March–June and June–August) visiting a range of flowers. *Bicolor* is Latin for 'two-coloured'.

Tawny mining bee *A. fulva*, male body length 9–11 mm, female 12–14 mm, male forewing length 6.5–9 mm, female 10–11 mm, common in much of lowland England and Wales, flying in March–June, visiting a range of flowers, and nesting in open grassland including lawns, flower beds and mown banks, nest entrances surrounded by volcano-shaped mounds. *Fulva* is Latin for 'tawny'.

2) Some members of the family Halictidae, including 32 species in the genus *Lasioglossum* (Greek *lasios* + *glossa* = 'hairy' + 'tongue').

Brassy mining bee *L. morio*, body length 5–7 mm, forewing length 4 mm, common and widespread in (especially the southern half of) England and Wales. Females are active in March–October, males in late June–October, taking pollen and nectar from a range of flowers. Being primitively eusocial, the females usually nest in burrows in large aggregations in exposed soil. *Morio* is Latin for 'fool'.

Least mining bee *L. minutissimum*, body length 4.5–5.5 mm, forewing length 3–4 mm, locally common in southern England, the Midlands, East Anglia, and south Wales, often in grassland and post-industrial habitats, creating nest burrows in bare ground. *Minutissimum* is Latin for 'very small'.

Neat mining bee *L. nitidiusculum*, mainly found in southern England, but also recorded elsewhere in England and in south Wales, in sandy habitats, including dunes, evident from late spring to early autumn, visiting a range of plants for pollen (though favouring yellow flowers) and nectar. The female nests solitarily in steeply sloping, sandy soil. Latin *nitidiusculum* = 'slightly bright'.

Shaggy mining bee *L. villosulum*, body length 6–7 mm, forewing length 4–5.5 mm, widely distributed in England and Wales, north into south Scotland, and also in Ireland. Found in a range of habitats though in Wales and north-west England it is largely coastal. The species is bivoltine but there is a peak of activity in July and August, with a number of plants visited for pollen (those with yellow flowers being favoured) and nectar. This is a solitary mining species, although nests may be found in aggregations as burrows in bare soil. *Villosulum* is Latin for 'covered with soft hair'.

Slender mining bee *L. calceatum*, body length 7.5–10 mm, forewing length 5.5–7 mm, common and widespread, found from late spring to early autumn in a range of habitats, including grassland and post-industrial sites, taking pollen and nectar from a variety of flowers. Nests are excavated in the ground. *Calceus* is Latin for 'slipper'.

Yellow-footed mining bee *L. xanthopus*, locally common in the southern third of England, mainly on calcareous grassland, coastal landslips and cliffs, females flying from April to at least August. Nest burrows are rarely found, suggesting they occur singly and are obscured by low vegetation, but an aggregation of about a thousand burrows has been reported from Cambridgeshire. Greek *xanthos* + *pous* = 'yellow' + 'foot'.

3) Members of the family Melittidae (see also under bee).

Hairy-legged mining bee *Dasypoda hirtipes* (Melittidae), found in southern England, East Anglia and south Wales, on sandy soils (especially on heathland and dunes) in summer, pollen and nectar collected from (especially yellow) Asteraceae. Females dig nests in sandy, sparsely vegetated, level soil at an oblique angle, resulting in a 'fan' of spoil to one side of the entrance. Some sites contain sizeable nest aggregations. Greek *dasys* + *podos* and Latin *hirtus* + *pes* both mean 'hairy' + 'foot'.

MINING FLY Carrot mining fly *Napomyza caroti* (Agromyzidae), a local pest of carrot in Norfolk. Latin *napus* + Greek *myzao* = 'turnip' + 'to suck'.

MINK American mink *Neovison vison* (Mustelidae), head-body length 56 cm, tail 29 cm, male weight 5 kg, females 4 kg, a

North American semi-aquatic carnivore introduced to fur farms from 1929 through to the 1980s, with escapes (and misguided releases) noted from the 1930s, though only becoming widespread and abundant from the 1950s. Estimated numbers in 2005 were: England 16,500, Scotland 19,450 and Wales 1,000, but by 2017, a UK figure totalling 110,000 was suggested. The first known escape from a fur farm in Ireland was in Co. Tyrone in 1961, and again escapes and releases have been common. These opportunistic predators eat a range of mammals, birds and fish (fewer fish being taken when in competition with otters). Water voles are especially vulnerable prey. Gestation takes a month. Females have one litter a year, usually in April–May, with 4–6 young (kits), which become weaned after 8 weeks. 'Mink' is a late Middle English word denoting the fur, coming from Swedish. Greek *neos* = 'new' + the Swedish *vison* = 'weasel'.

MINNOW *Phoxinus phoxinus* (Cyprinidae), length 8–10 cm, common and widespread, often in shoals, in cool well-oxygenated streams with sandy or gravelly beds and ponds and pools, feeding on insect larvae, crustaceans and plants, including algae, spawning in spring. 'Minnow' dates from late Middle English, from Anglo-Norman French *menu*. Greek *phoxinos* = 'minnow'.

MINOR (MOTH) Members of the family Noctuidae (Xyleninae), larvae mostly feeding on grasses.

Cloaked minor *Mesoligia furuncula*, wingspan 22–8 mm, day-flying in July–September, widespread, with a more scattered distribution in Scotland and Ireland, commonest on dunes, grassy slopes and other coastal habitats but also found inland in England, for instance on chalk downland. Greek *me* = 'not' + the genus *Oligia*, and *furuncula* the diminutive of Latin *fur* = 'thief'.

Haworth's minor *Celaena haworthii*, wingspan 25–32 mm, day-flying (August–September), common on moorland in northern England, Ireland, Wales and Scotland, but also on southern heaths and the East Anglian fens. Larvae mine stems of cottongrass. Numbers have declined by 94% over 1968–2007. Greek *celainos* = 'dark'. Adrian Haworth (1767–1833) was an English entomologist.

Least minor *Photedes captiuncula*, wingspan 15–18 mm, a rarity local to northern England and western Ireland (including the Burren), favouring limestone grassland and grassy clifftops. Males fly by day and night, females at night, in June–August. Larvae feed inside stems of glaucous sedge. Greek *photos* + *edos* = 'light' + 'delight', from the day-flying habit, and Latin *captio* = 'deception'.

Marbled minor *Oligia strigilis*, wingspan 22–5 mm, nocturnal, widespread and common in grassland, woodland rides, gardens and other habitats, more local in Scotland and Ireland. Adults fly in May–June. Greek *oligos* = 'small', and Latin *striga* = 'line', from the wing pattern.

Middle-barred minor *O. fasciuncula*, wingspan 22–6 mm, nocturnal, widespread and common in damp grassland, marsh, fen, woodland and gardens. Adults fly in June–July. *Fascia* is Latin for 'band', from the forewing marking.

Rosy minor *Litoligia literosa*, wingspan 25–30 mm, nocturnal, fairly widespread and common in calcareous grassland, fen, scrub, gardens, sea-cliffs and dunes, but with numbers declining by 93% over 1968–2007, a species of conservation concern in the UK BAP. Adults fly in July–August. Greek *litos* = 'smooth' + the genus *Oligia*, and Latin *literosa* = 'learned', from 'writing' marks on the forewing.

Rufous minor *O. versicolor*, wingspan 22–8 mm, nocturnal, local in woodland, heathland, sea-cliffs and gardens throughout much of England, more scattered in Wales, Scotland and Ireland. Adults fly in June–July. *Versicolor* is Latin for 'changeable colour'.

Tawny marbled minor *O. latruncula*, wingspan 24–7 mm, nocturnal, widespread and common in grassland, woodland rides, gardens and other habitats. Adults fly in May–July. Latin *latrus* = 'petty thief'.

MINT Herbaceous perennial rhizomatous species of *Mentha*

(Lamiaceae), the Latin name of a Greek nymph Minthe, who was turned into a mint. See also pennyroyal and peppermint. There are also a number of often common and widespread hybrids: **apple**, **bushy**, **false apple**, **sharp-toothed**, **tall** and **whorled mint**. Some are commonly grown as culinary and medicinal herbs.

Corn mint *M. arvensis*, rarely annual, common and widespread, though decreasing northwards, in arable fields, woodland rides, marshy pasture and waste ground. Commonly hybridises, with bushy, tall and whorled mint, having locally common and scattered distributions. *Arvensis* is from Latin for 'field'.

Corsican mint *M. requienii*, a neophyte introduced in 1829, scattered, mainly in southern England, as a weed in cultivated ground, occasionally in damp grassy and rocky habitats, in woodland, along tracks, paths and pavements, and on tips. Esprit Requien (1788–1851) was a French botanist.

Round-leaved mint *M. suaveolens*, widespread if scattered, in damp habitats, probably native only in south-west England and Wales, elsewhere occurring as a garden escape, often forming extensive colonies on roadsides and waste ground. *Suaveolens* is Latin for 'fragrant'.

Spear mint *M. spicata*, an archaeophyte, widespread if more local in north Scotland and Ireland, naturalised in damp habitats, and on rough ground, usually close to habitation. *Spicata* is Latin for 'pointed'.

Water mint *M. aquatica*, common and widespread, associated with permanently wet habitats, often partially or wholly submerged, and in marsh, wet pasture, dune slacks, fen and wet woodland. It spreads clonally by rhizomes, and by detached rhizome fragments, often dispersed by water. Commonly hybridises.

MINTWEED *Salvia reflexa* (Lamiaceae), a scarce annual North American neophyte, scattered in England in fields and on waste ground, originating from bird- and grass seed, grain and wool shoddy. The binomial is Latin for 'sage' and 'bent back'.

MIRIDAE Family name for capsid bugs. See bracken and fern bugs.

MISCANTHUS From Greek *mischos* + *anthos* = 'stem' + 'flower', species of silver-grass grown as biomass crops, occasionally casual.

MISTLETOE *Viscum album* (Santalaceae), a hemiparasite of trees in orchards, hedgerows, parklands and gardens, locally common and widespread in the southern half of England, local and often introduced elsewhere. Its most frequent hosts are apple trees, followed by lime and hawthorn (the commonest native hosts), poplars, maples, willows and false-acacia. This is the county flower of Herefordshire as voted by the public in a 2002 survey by Plantlife. *Viscum* is Latin for 'bird lime', made from mistletoe berries, and *album* = 'white'.

MITE Small arachnids, with four pairs of legs: see Acari, Astigmata, Mesostigmata, Oribatida and Trombidiformes.

Cotswold mite Local name for harvest mite.

Feather mite A group of mites (Astigmata, Psoroptidia) ectoparasitic (blood-feeding) on and tending to reduce condition in birds, becoming a pest when affecting poultry.

Flour mite *Acarus siro* (Acaridae), unfed female length 0.35–0.65 m, male 0.32–0.40 mm, a serious pest of stored food including flour and cheese, and also paper, tobacco and birds nests. Latin *acarus* = 'mite' (from Greek *acari*), and Greek *siros* = 'pit for keeping cereal'.

Fuchsia gall mite *Aculops fuchsiae* (Eriophyidae), native of South America, first recorded in England in 2007. It distorts and discolours young fuchsia shoot tips, and was second in the Royal Horticultural Society's list of worst garden pests of 2017. Parts of southern England are most affected, especially in May–September.

Gall mite Members of the family Eriophyidae, with a body size around 0.5 mm, small enough to pierce and feed on individual plant cells, causing the surrounding cells to enlarge and multiply to form a gall. Some are host-specific, for example walnut leaf gall mite *Aceria erinea*, and cauliflower gall mite *A. fraxinivora* on ash keys. See fuchsia gall mite above.

Harvest mite *Trombicula autumnalis* (Trombiculidae), the commonest 'chigger' in Britain, adult body size 1 mm, the larvae (0.5 mm) living parasitically on humans (causing severe itching), mammals and ground-nesting birds. Eggs are laid in damp soil. After hatching, larvae climb blades of grass and wait for a potential host. They attach themselves to the host and feed on its tissues for several days, then fall off and develop over three stages of nymph to adults. Greek *trombodes* = 'timid'.

House dust mite *Dermatophagoides pteronyssinus* and *D. farinae* (Pyroglyphidae), body size 0.2–0.3 mm, living off human skin scales that have been partially digested by mould, and thriving in humid environments. They are found in bedding, carpets, soft furnishings and clothing. In people allergic to dust mites it is often proteins in their droppings that cause the allergy. Greek *derma* + *phago* = 'skin' + 'to eat', and *pteron* + *nysso* = 'feather' + 'afflict'.

Varroa mite *Varroa jacobsoni* and especially *V. destructor* (Varroidae), parasites on honeybees, discovered in England in 1992, that weaken the host by feeding on their haemolymph (c.f. blood), reducing the honey crop and spreading disease (varoosis). The mites can be controlled by monitoring the infestation in colonies and the use of control methods to keep mite numbers below levels that are harmful. Control includes management methods and medicinal controls (using varroacides, though recently these mites have shown resistance to the active compounds, pyrethroids), most effective when combining methods at different times of the year depending on level of infestation (integrated pest management). *Varroa* derives from modern Latin, from Varro, Marcus Terentius, referencing his work on bee-keeping. *Destructor* is Latin for 'destroyer'.

Velvet mite *Trombidium holosericeum* (Trombididae), 3–5 mm body length, common and widespread in loose soil, feeding on small invertebrates and their eggs, found from March to October but most evident in spring when they emerge from hibernation. The binomial is from Greek *trombodes* = 'timid', and *holos* + *sericon* = 'entire' + 'silk' (i.e. silky to the touch).

MITTEN CRAB Chinese mitten crab See crab.

MNIACEAE Some species of thyme-moss (Greek *mnion* = 'moss').

MOCHA (MOTH) Nocturnal members of the family Geometridae (Sterrhinae). Specifically, *Cyclophora annularia*, wingspan 18–22 mm, scarce, in scrubland, woodland and hedgerows, mainly in southern England, scattered in the Midlands and East Anglia. The bivoltine adults fly in May–June and July–August. Larvae feed on field maple leaves. Greek *cyclos* + *phoreo* = 'ring' + 'to carry', from the forewing marking, and Latin *annulus* = 'small ring'.

Birch mocha *C. albipunctata*, wingspan 20–5 mm, widespread but local in woodland and heathland. Adults fly in May–June and, in the south, also in August. Larvae feed on birch leaves. *Albipunctata* Is Latin for 'white-spotted'.

Blair's mocha *C. puppillaria*, wingspan 28–36 mm, an increasingly common immigrant from mainland Europe, appearing in southern England and south-east Ireland, especially in late summer and early autumn. There is as yet no evidence of breeding in Britain. *Pupilla* is Latin for 'eye pupil'.

Dingy mocha *C. pendularia*, wingspan 26–9 mm, rare in scrubby heathland and damp grassland in parts of Dorset, Hampshire and Wiltshire, scattered elsewhere in England, a priority species in the UK BAP. The bivoltine adults fly in May–June and July–August. Larvae feed on willow leaves. *Pendularia* may refer to birch *Betula pendula*, the food plant of a related species.

False mocha *C. porata*, wingspan 25–30 mm, local in coppiced woodland, heathland and woodland clearings throughout much of England and Wales, but dramatically declining in numbers and distribution since the mid-twentieth century, and a priority

species in the UK BAP. The bivoltine adults fly in May–June and August–September. Larvae feed on oak leaves. *Porus* is Latin for 'tufa', from the colour of this porous limestone.

Jersey mocha *C. ruficiliaria*, wingspan 25–32 mm, local in woodland on Jersey, present since at least 1917, but not recorded on the British mainland until 2003, when a female was trapped at Portland, Dorset, and on the Isle of Wight, recorded for the first time at Bonchurch in 2008, with further records at the same site in 2010 and 2011. Adults fly in August. Latin *rufus* + *cilia* = 'red' + 'fringed with hair'.

MOCK-ORANGE *Philadelphus coronarius* (Hydrangeaceae), a deciduous European neophyte shrub present by Tudor times, with a scattered distribution, occasionally planted in woodland, but more frequent as a garden throw-out or relic of cultivation in hedges, and on roadsides and waste ground. Philadelphus was a Greek king of Egypt. Latin *corona* = 'crown'.

MOLANNIDAE From Greek *molannos* = 'white', a caddisfly family, including *Molanna angustata*, fairly widespread in England, occasional in Wales, cased larvae found in still or slow-moving water, adults in flight in May–September. *Angustata* is Latin for 'narrowed'.

MOLE *Talpa europaea* (Talpidae), head-body length 12 cm (males larger), the small eyes hidden behind fur, and ears small ridges in the skin. It is common and widespread, though absent from most islands except Anglesey, the Isle of Wight and a few of the Inner Hebrides, and is not present in Ireland. Originally found in deciduous woodland, it is now common in agricultural habitats, especially pasture. Pre-breeding population estimates in 1995 were 31 million: 19,750,000 in England, 8 million in Scotland and 3,250,000 in Wales. By 2010, a further estimate was 33–40 million, with numbers increasing rapidly, aided by a ban on strychnine poisoning. In 2014, a wet spring and a long summer led to suggestions that post-breeding numbers had doubled. It lives in a continually extended underground tunnel system, displaced earth pushed to the surface, resulting in molehills. It mainly feeds on earthworms which are paralysed by toxins in its saliva, consuming about half its body weight each day. Mating occurs in March–April. Gestation lasts 4–5 weeks. Litter size is 2–7. The young suckle for 4–5 weeks before forced to disperse. *Talpa* is Latin for 'mole'.

MOLLUSC(A) An invertebrate phylum, with members in land, freshwater and marine environments. Universal features are a mantle with a cavity used for breathing and excretion; a rasping feeding organ, or radula, except for bivalves; and a characteristic nervous system. Classes include Gastropoda, Cephalopoda and Bivalvia. Depending on the authority there are 213 or 229 non-marine mollusc species native to the British Isles, with a further 36–52 naturalised introductions (reflecting uncertainty in level of naturalisation). *Molluscus* is Latin for 'soft', referring to the soft parts of these animals.

MOMPHIDAE Family name (Greek *momphos* = 'blame') for some of the cosmet moths, most of which are miners of species of willowherb.

MONARCH *Danaus plexippus* (Nymphalidae, Danainae), wingspan 95–100 mm, the largest butterfly seen in the British Isles, and one of the rarest migrants when blown across the Atlantic during their mass migration across North America. The total number of records for the British Isles is <500, almost all from south and south-west England and southern Ireland. The most recent major migration was in 1999 with 300 reports. Five sightings were recorded from Cornwall and the Scilly Isles in October 2014. Larvae are not found in Britain, with an absence of their food plant, milkweed, from which they would absorb toxic steroids (cardenolides) that trigger heart failure in vertebrates. In Greek mythology Danaus was king of Argos, Plexippus a Greek hero.

MONERA One of the five kingdoms in taxonomy, comprising unicellular organisms with a prokaryotic cell organisation (having no nuclear membrane), such as bacteria. *Moneres* is Greek for 'single'.

MONEYWORT Alternative name for creeping-jenny.

Cornish moneywort *Sibthorpia europaea* (Veronicaceae), a herbaceous perennial found locally in south-west England, the Channel Islands, East Sussex, south Wales, Cheshire and Co Kerry, on acidic soils in damp, shady places, including woodland, banks by small streams and ditches, wet heathland and shaded lawns. John Sibthorp (1758–96) was an English botanist.

MONKEY-PUZZLE *Araucaria araucana* (Araucariaceae), introduced from Chile in 1795 and widely planted in parks and gardens, named after the Araucarian Indians of Patagonia/the Arauco Indians of Chile.

MONKEYFLOWER *Mimulus guttatus* (Phrymaceae), a widespread and locally common herbaceous North American neophyte perennial introduced in 1812, naturalised in wet habitats by streams and ponds, and in damp meadows and open damp woodland. Hybrids include coppery, hybrid and Scottish monkeyflower. This is the 'county' flower of Tyne and Wear as voted by the public in a 2002 survey by Plantlife. The binomial comprises the diminutive form of Greek *mimos* = 'imitator', and Latin for 'spotted'.

MONK'S HEAD Alternative name for the fungus trooping funnel.

MONK'S-HOOD *Aconitum napellus* (Ranunculaceae), a tuberous perennial herb of calcareous to slightly acidic soil along stream banks, often in shade, in damp open woodland and sometimes in damp meadow in south-west England and south Wales, and common and widespread as garden escapes on roadsides and waste ground. *Aconiton* is Greek for this plant, *napellus* Latin for 'like a small turnip', from the shape of the roots.

Hybrid monk's-hood *Aconitum x stoerkianum* (*A. napellus* x *A. variegatum*), a perennial herb with annually renewed tuberous rhizomes, mainly a naturalised garden escape in northern and western parts, usually in shade or in tall vegetation. Habitats include damp roadsides, pasture and woodland, and waste ground.

MONK'S-RHUBARB *Rumex alpinus* (Polygonaceae), a rhizomatous perennial European neophyte herb, scattered in Scotland and England south to Staffordshire, East Anglia, Surrey and south-west England, naturalised in verges, and by streams and old buildings. Latin *rumex* = 'sorrel'.

MONOGENEA Class of Platyhelminthes, small parasitic flatworms (flukes) mainly found on the skin or gills of fish, generally <20 mm long.

MONOTOMIDAE Family name (Greek *monas* + *tomos* = 'single' + 'a cut') of root-eating beetles (which, despite their name, are predatory), with 23 British species. Habitats include compost heaps, dung, bird nests and under bark. Graveyard beetle *Rhizophagus* spp. are subterranean and indeed have been found in coffins.

MONTAGU SHELL **Two-toothed montagu shell** *Kurtiella bidentata* (Montacutidae), bivalve, shell size up to 6 mm, common and widespread, intertidal in mud-filled crevices, and in mud and muddy sand in the sublittoral and shelf. Commensal with brittlestars, sipunculans and polychaetes. George Montagu (1753–1815) was a British soldier and naturalist. Greek *curtos* = 'curved', and Latin *bidentata* = 'two-toothed'.

MONTBRETIA *Crocosmia x crocosmiiflora* (Iridaceae), a cormous rhizomatous herbaceous perennial of French garden origin, reaching England in 1880, widespread and locally common, especially in Ireland, naturalised in hedgerow banks, verges, waste ground and woodland. Antoine Coquebert de Montbret

(1780–1801) was a French botanist. *Crocosmia* is Greek for 'saffron smell'.

Giant montbretia *C. masoniorum*, from South Africa, occasionally naturalised in waste ground in Britain, locally common in north-east Ireland.

Potts' montbretia *C. pottsii*, from South Africa, locally naturalised on roadsides in south-west England, north Wales, Lancashire and western Scotland.

MONTIACEAE Family name for blinks, spring beauty and pink purslane.

MOONWORT *Botrychium lunaria* (Ophioglossaceae), a fern of open, mostly upland habitats, including dry grassland and heathland, on (lead) mine spoil and quarries, especially in the north and west. Growing from an underground caudex it sends one fleshy green leaf above the surface of the ground. The sterile part of the leaf has fan-shaped leaflets. The fertile part has rounded, grape-like clusters of sporangia, prompting the binomial from Greek *botrys* = 'cluster of grapes', and Latin *luna* = 'moon', from the moon-shaped spore vessels.

MOOR-GRASS Perennial grasses (Poaceae).

Blue moor-grass *Sesleria caerulea*, uncommon, rhizomatous, in well-drained, mainly open habitats on limestone, including grassland and heathland, scree and cliffs, and on limestone pavement. It extends locally into open woodland in western Ireland and northern England, and is found on sandy loams over micaceous schists in Perthshire. Leonardo Sesler (d.1785) was an Italian botanist. Latin *caerulea* = 'blue'.

Purple moor-grass *Molinia caerulea*, widespread and common, especially in heathland, moorland, bog and fen, but also in open birchwood, mountain grassland and cliffs, on mildly basic to strongly acidic peats and mineral soils that are permanently or seasonally wet. Fr. Juan Ignacio Molina (1740–1829) was a Spanish/Chilean naturalist. Latin *caerulea* = 'blue'.

MOORHEN *Gallinula chloropus* (Rallidae), wingspan 52 cm, length 34 cm, weight 320 g, common and widespread in small lakes, ponds and slow-flowing rivers, tending to avoid uplands, but including towns. Some 270,000 territories were estimated in Britain in 2009, with 320,000 individuals having wintered in 2008–9. Numbers declined in Britain by 30% between 1975 and 2015 (wintering numbers crashing in 2009–12), by 12% between 1995 and 2015, and by 10% between 2010 and 2015. The basket-shaped ground nests are constructed in dense vegetation, clutch size is 5–7, incubation lasts 20–2 days, and the precocial young fledge in 53–6 days. Feeding is from water and land, the diet is of water plants, seeds, fruit, grasses, insects, snails, worms and small fish. 'Moor' is used in the old sense of 'marsh'. Latin *gallinula* = 'little hen' and Greek *chloros* + *pous* = 'green' + 'foot'.

MORACEAE Family name for (black) mulberry and fig, *morus* being Latin for 'mulberry tree'.

MORDELLIDAE Family name (Latin *mordax* = 'biting') for tumbling flower beetles, with 17 species in 5 genera, all with a southern British distribution.

MOREL Or **yellow morel** *Morchella esculenta* (Pezizales, Morchellaceae), an edible fungus, widespread but uncommon in Britain and eastern Ireland, fruiting from March to June, the honeycombed cap evident on (preferably calcareous) soil in woodland, scrub, grassland and gardens. 'Morel' dates from the late seventeenth century, from French *morille*. *Morchella* may derive from German *Morchel* = 'fungus'; *esculenta* is Latin for 'edible'. See also false morel.

Black morel *M. elata*, edible, widespread but generally uncommon, though increasing in southern England and the Midlands associated with bark mulch used in parks and gardens, fruiting from March to June, the honeycombed cap evident on well-drained soil in grassland in or near coniferous woodland. *Elatus* is Latin for 'tall'.

Semifree morel *Mitrophora semilibera*, inedible, locally common and widespread in damp deciduous woodland in England, occasional elsewhere, found in spring. Greek *mitra* + *phoros* = 'cap' + 'bearing', and Latin *semilibera* = 'half free'.

Thimble morel *Verpa conica*, edible, widespread and locally common in England, found in spring under hawthorn, especially on damp soil. The binomial is Latin for 'penis' and 'conical'.

MORNING GLORY North American neophytes in the genus *Ipomoea* (Convolvulaceae), from Greek *ips*, a grub that eats vines, describing the coiled flower bud. **Common morning-glory** *I. purpurea* (Latin for 'purple'), **ivy-leaved morning-glory** *I. hederacea* (Latin for 'ivy-like') and **white morning-glory** *I. lacunosa* (Latin for 'with holes') are all casual on waste ground in southern England, originating from soyabean waste, birdseed, etc.

MORTAR BEE **Red mortar bee** Alternative name for red mason bee.

MOSCHATEL *Adoxa moschatellina* (Adoxaceae), a rhizomatous herbaceous perennial common and widespread in Britain, except for the Fens and north-west Scotland, and also recorded from Co. Antrim and (as an introduction) Co. Dublin. It favours damp, humus-rich soil, and grows in woodland, hedges and rocky montane habitats. 'Moschatel' dates from the mid-eighteenth century, from French *moscatelle*, Italian *moscatella*, in turn from *moscato* = 'musk'; similarly *moschatellina* is from Latin meaning 'musky', referring to the scent when the plant is wet. *Adoxa* is from Greek *a* + *doxa* = 'without' + 'glory', a comment on the plant's 'humble growth'.

MOSQUITO From Spanish for 'little fly', members of the family Culicidae. Larvae are aquatic. Adult females are haematophagous.

Banded mosquito *Culiseta annulata*, scattered and widespread, especially in south-east England. Adults are evident from spring to autumn, often resting in buildings, caves and hollow trees, where they also hibernate. Both sexes feed on nectar, but females take blood from mammals, including livestock and humans, often causing the skin to blister. As an adult it freely attacks humans as well as cattle to take blood, often causing blisters. Larvae favour water with a high nitrogen content. *Culiseta* is from Latin *culex* = 'gnat' or 'midge', and Latin *annulata* = 'banded'.

MOSS Members of the phylum Bryophyta, non-vascular flowerless plants with a 'stem and leaf' morphology, growing as acrocarps (forming short cushions) or pleurocarps (free-branching), often favouring shaded and/or damp conditions. There are some 760 species in the British Isles.

MOSS (LICHEN) See reindeer lichen under lichen, and coral, Iceland, oak and tree mosses.

MOSS ANIMAL See Bryozoa.

MOSS BEETLE Members of the family Hydraenidae, 1–3 mm in length.

Black moss beetle *Hydraena nigrita*, widespread but scattered, rare in Ireland, on the margins of fast streams in wooded areas, often living in moss on rocks along the margins or in flood refuse. Greek *hydor* = 'water', and Latin *nigrita* = 'blackness'.

Red-legged moss beetle *H. rufipes*, widespread but scattered, rare in Ireland, in clean water in rivers and lakes, local declines related to reductions in water quality. Latin *rufipes* = 'red-legged'.

Sculptured moss beetle *Enicocerus exsculptus*, widespread but scattered, most abundant in northern England and southern Scotland, on the edges of swift freshwater streams, often living in moss on stones either in the middle of the flow or on margins. Greek *henicos* + *ceras* = 'single' + 'horn', and Latin *exsculptus* = 'sculptured'.

MOSS PIGLET See tardigrade.

MOSS PIXY CAP Alternative name for the fungus hairy leg bell.

MOSSCAP Inedible fungi in the genus *Rickenella* (Hymenochaetales, Rickenellaceae). Adalbert Ricken (1851–1921) was a German mycologist. **Collared mosscap** *R. swartzii* (Olof Swartz, 1760–1818, was a Swedish botanist) and **orange mosscap** *R. fibula* (Latin for 'fastener') are both common and widespread, found from summer to winter on grassland, including lawns, often with moss.

MOSSEAR Inedible fungi in the genus *Rimbachia* (Agaricaceae, Tricholomataceae), named after Rimbach, Germany. **Spidery mossear** *R. arachnoidea* (New Latin for 'spider-like') and **veined mossear** *R. bryophila* (Greek *bryon* + *philos* = 'moss' + 'loving') are both very scattered, Near Threatened in the RDL (2006), and associated with moss, often in woodland.

MOTACILLIDAE Family name for wagtails and pipits (Greek *muttex*, a bird described by the lexicographer Hesychius of Alexandria).

MOTH Members of the insect order Lepidoptera in the suborder Heterocera (from Greek for 'varied' and 'antennae') with often feathery antennae (not clubbed, as with butterflies), a frenulum (a filament arising from the hind wing and matching up with barbs on the forewing). A number of suborders have been distinguished, but authorities differ on taxonomic status, and many indeed argue that distinguishing butterflies from moths has no taxonomic merit. 'Moth' comes from Old English *moththe*, of Germanic origin (*Motte*) though, cited in 950, may derive from Scandinavian *mott* = 'maggot'. Moths may be day- and/or night-flying. Over 2,400 species have been recorded in the British Isles, including around 1,400 in the Irish list. However, 62 species became extinct in Britain during the twentieth century and several more are thought to have been lost. Forty-year (1968–2007) national population trends for the larger moths (macrolepidoptera) show that 227 species decreased in abundance, 67% of the larger moths assessed, while the remaining 110 species (33%) became more abundant.

Families (common names, number of genera/species in Britain excluding rare migrants and occasional adventives) are: **Adelidae** (long-horns, 4/13), **Argyresthiidae** (argents, 1/22), **Autostichidae** (obscures, 2/4), **Bedelliidae** (bindweed bent-wing, 1/1), **Blastobasidae** (dowds, 1/5), **Bucculatricidae** (some bent-wings, 1/10), **Chimabachidae** (some tubics, 2/3), **Choreutidae** (twitchers and nettle-taps, 4/6), **Coleophoridae** (case-bearers, 1/101), **Cosmopterigidae** (cosmets, 4/11), **Cossidae** (goat and leopard moths, 3/3), **Crambidae** (china-marks, greys, pearls, purple & golds, sables and veneers, including grass- and water-veneers, 32/92), **Depressariidae** (flat-body moths, 8/49), **Douglasiidae** (spear-wing, 1/1), **Drepanidae** (hook-tips and lutestrings, 13/15), **Elacistidae** (dwarfs, 3/46), **Endromidae** (Kentish glory, 1/1), **Epermeniidae** (lance-wings, 2/8), **Erebidae** (ermines, fan-foots, footmen, snouts, tigers and underwings, 47/63), **Eriocraniidae** (purples, 3/8), **Ethmiidae** (some ermels, 1/5), **Gelechiidae** (crests, groundlings, nebs and sobers, 46/135), **Geometridae** (beauties, carpets, emeralds, mochas, pugs, rivulets, thorns, treble-bars, umbers and waves, 135/285), **Glyphipterigidae** (fanners and smudges, 5/14), **Gracillariidae** (14/84), **Hepialidae** (swifts, 4/5), **Heliozelidae** (lifts, 2/5), **Incurvariidae** (brights, 2/5), **Lasiocampidae** (eggars and lappets, 9/11), **Limacodidae** (festoons and triangles, 2/2), **Lyonetiidae** (some bent-wings, 2/7), **Lypusidae** (some tubics, 2/4), **Micropterigidae** (golds, 1/5), **Momphidae** (some cosmets, 1/15), **Nepticulidae** (leaf miners, 5/57), **Noctuidae** (arches, brindles, brocades, chestnuts, clays, coronets, daggers, darts, drabs, ears, gothics, minors, pinions, quakers, ranunculus, rustics, sallows, shoulder-knots, square-spots, underwings and wainscots, 153/290), **Nolidae** (black arches and silver-lines, 6/8), **Notodontidae** (chocolate-tips, kittens and prominents, 13/23), **Oecophoridae** (some tubics and streaks, 15/20), **Opostegidae** (some bent-wings 2/4), **Parametriotidae** (some cosmets, 4/6), **Peleopodidae** (long-horned flat-body, 1/1), **Plutellidae** (smudges, 3/7), **Praydidae** (ash bud moth, 1/2),

Prodoxidae (brights, etc., 1/7), **Psychidae** (some case-bearers, smokes and sweeps, 14/16), **Pterophoridae** (plumes, 20/38), **Pyralidae** (knot-horns and tabby moths, 44/62), **Roeslerstammiidae** (copper ermel, 1/1), **Saturniidae** (emperor moth, 1/1), **Schreckensteiniidae** (bramble false-feather, 1/1), **Scythrididae** (owlets, 1/10), **Scythropiidae** (hawthorn moth, 1/1), **Sesiidae** (clearwings, 5/15), **Sphingidae** (hawk-moths, 9/12), **Stathmopodidae** (alder signal, 1/1), **Tineidae** (clothes moths, 15/34), **Tischeriidae** (carl moths, 2/4), **Tortricidae** (bells, buttons, conchs, drills, marbles, piercers, rollers, shades, shoots, tortrix and twists, 88/347), **Yponomentidae** (ermines and some ermels, 7/20), **Ypsolophidae** (smudges and stem-moths, 2/15) and **Zygaenidae** (burnets and foresters, 3/10).

As well as the species given below, see separate entries including: arches, argent, argent & sable, bar, barred, beauty, belle, bent-wing, birch, black arches, border, brass, bright, brimstone, brindle, brocade, brown, bud, buff, burnet, button, carl, carpet, case-bearer, chestnut, clay, clearwing, clothes, coronet, cosmet, crest, dagger, dart, dwarf, ear, flat-body, fruit, gold, green, grey, house-, kitten, leopard, lift, long-horn, marble, midget, mountain, neb, nettle-tap, obscure, piercer, pinion, plume, prominent, pug, purple, purple & gold, quaker, smoke, smudge, smut, snout, sober, spinach, spot, square-spot, stem-, straw, streak, striped, sweep, thorn, tiger, tissue, triangle, tussock, twist, twitcher, umber, veneer, wave, wax, winter and Y moths.

Antler moth *Cerapteryx graminis* (Noctuidae, Hadeninae), wingspan 27–39 mm, mainly day-flying (July–September) but also found at night, widespread and common on downland and moorland. Larvae feed on grasses and can devastate large areas of grassland on high ground. Adults feed on various plants, including ragwort and thistles. Greek *ceras* + *pteryx* = 'horn' + 'wing' (from a streak on the forewing shaped like an antler), and Latin *graminis* relating to grass.

Apple fruit moth See fruit moth.

Apple moth See light brown apple moth below.

Apple pith moth *Blastodacna atra* (Parametriotidae), wingspan 11–13 mm, nocturnal, in orchards and gardens throughout much of England, south Wales and south Scotland, but rare in Ireland. Adults fly in May–September. Larvae mine twigs of apple, often causing die-off and making this moth a pest. Greek *blastos* + *dacno* = 'shoot' + 'bite', and Latin *ater* = 'black' or 'dark'.

Autumnal moth *Epirrita autumnata* (Geometridae, Larentiinae), wingspan 40–4 mm, nocturnal, widespread and common in open woodland, woodland rides, heathland, gardens and scrub, but problems in identifying individuals to specific status mean that all *Epirrita* species are under-recorded. Adults fly in September–October. Larvae feed on birch, alder and heather. Greek *epirrheo* = 'to flow'. See also small autumnal moth below.

Bee moth *Aphomia sociella* (Pyralidae, Galleriinae), wingspan 18–44 mm, nocturnal, commonly found around bee and wasp nests and in beehives throughout Britain. Adults fly from June to August. Larvae feed on debris in and contents of the comb inside bee and wasp nests. Greek *aphomoios* = 'unlike' (other genera) and Latin *socius* = 'associated with' (possibly from the social behaviour of the larvae).

Beet moth *Scrobipalpa ocellatella* (Gelechiidae, Gelechiinae), wingspan 13 mm, nocturnal, local on vegetated shingle and saltmarsh along parts of the coast of southern England from Cornwall to Suffolk; rare in south-west Wales and with single records from north Somerset and southern Ireland. The bivoltine adults fly in May–July and August–October. Larvae mine leaves, buds and stems of sea beet in August–September and January–June. Latin *scrobis* + *palpus* = 'trench' + 'palp' (from the wing furrow), and *ocellatum* = 'small with eye-like marks' (from forewing markings).

Black-veined moth *Siona lineata* (Geometridae, Ennominae), wingspan 35–40 mm, a daytime and dusk flier (May–July) restricted to a few locations in south-east Kent, though formerly also found in other southern counties, feeding on wild herb species, bird's-foot-trefoil and black knapweed on rough chalk downland and similar grasslands. A priority RDB species for conservation. The binomial comes from the Swiss town Sion, and the black lines on the underwings.

Box tree moth *Cydalima perspectalis* (Crambidae), wingspan 40–5 mm, an introduced moth from India and the Far East, adults first recorded in the UK in south-west London in 2008 (possibly in Kent in 2007), and larvae from private gardens in 2011. By the end of 2014 the moth, which may have two to three generations a year, had become established in parts of London and surrounding counties (Sussex, Essex, Buckinghamshire). The larvae, which after four weeks can have a length of 35–40 mm, feed on box leaves under a silk web to cause severe defoliation. Reports of the caterpillar to the Royal Horticultural Society increased by an order of magnitude in 2015 compared with 2014 (>150 vs 20), making it the 'top' pest species, and in 2017 it was seventh in the Royal Horticultural Society's list of worst garden pests.

Broom moth *Ceramica pisi* (Noctuidae, Hadeninae), wingspan 32–7 mm, nocturnal, widespread and common in heathland, moorland, gardens and a range of other habitats, but with numbers declining by 84% over 1968–2007, a species of conservation concern in the UK BAP. Adults fly from May to July. Larvae feed on broom, bracken, heather, bramble, sea-buckthorn and sallow. Greek *ceramicos* = 'relating to pottery' (from the earthenware colour) and Latin *pisum* = 'pea' (though not a food plant).

Cabbage moth *Mamestra brassicae* (Noctuidae, Hadeninae), wingspan 37–45 mm, nocturnal, common in gardens, allotments, open woodland and other habitats. It has two or three overlapping generations, adults flying in May–September. Larvae feed on cabbage and other brassicas, and can become a serious pest. Mamestra is the capital of Lesser Armenia; *brassicae* is Latin referring to various brassicas as food plants.

Carrion moth *Monopis weaverella* (Tineidae, Tineinae), wingspan 13–18 mm, generally uncommon but widespread. Adults fly in May–August. Larvae scavenge on dead animals and faeces, often in bird nests. Greek *monopos* = 'one-eyed', from the spot on the forewing. R. Weaver was a nineteenth-century British entomologist.

Cocksfoot moth *Glyphipterix simpliciella* (Glyphipterigidae, Glyphipteriginae), body length 3–4 mm, wingspan 6–9 mm, day-flying, common and widespread on grassland. Adults fly from May to July, feeding on buttercups and other flowers where they are often found in large numbers on the same flower. Larvae feed inside the seeds of cock's-foot and tall fescue. Greek *glyphis* + *pteryx* = 'notch' + 'wing', from an indentation on the forewing, and Latin *simplex* = 'simple'. See also fanner moth.

Codling moth *Cydia pomonella* (Tortricidae, Olethreutinae), wingspan 14–22 mm, widespread and common in gardens, orchards and hedgerows. Adults fly in July–August, sometimes with a second generation in September–October. Larvae feed on the fruit of apple, quince, pear, plum, walnut and whitebeam and other cultivated and wild fruit, often becoming a pest. To 'coddle' is to cook gently: 'codlin' or 'codling' was a name given to apples that need very little cooking to become soft, derived from Anglo-Norman French *quer de lion* = 'lionheart'. Greek *cydos* = 'glory', and Pomona, goddess of fruit trees, in turn from *pomum* = 'fruit', Linnaeus referring to a pear rather than an apple.

Cork moth *Nemapogon cloacella* (Tineidae, Nemapogoninae), wingspan 10–18 mm, widespread and locally common, generally in woodland where larvae feed mainly on bracket fungi. They also feed on cereals, grain and dried fruit, so may be found in barns and warehouses. Adults mainly fly at dawn and dusk. Greek *nema* + *pogon* = 'thread' + 'beard', from the head bristles, and Latin *cloaca* = 'sewer', from the belief that this moth lives in drains.

Corn moth *Nemapogon granella*, wingspan 9–16 mm, local and scattered in the southern half of Britain. Adults fly in March–September. Larvae feed on grain and other stored vegetable products in barns, granaries, etc. (so are a commonly transported pest), as well as in various bracket fungi in woodland. Latin *granum* = 'grain'.

Cypress tip moth *Argyresthia cupressella* (Argyresthiidae), wingspan 8–9 mm, nocturnal, introduced from North America with its ornamental coniferous food plants. First found in the British Isles in Suffolk in 1997, but expanding its range rapidly, especially into East Anglia, and north and west England. Adults fly in June–July. Larvae feed on cypress and juniper, feeding inside the leaves and young shoots. Greek *argyros* + *esthes* = 'silver' + 'dress', from the metallic-looking forewing, and Latin *cupressella* = 'relating to cypress'.

December moth *Poecilocampa populi* (Lasiocampidae, Poecilocampinae), wingspan 30–45 mm, common and widespread in woodland, scrub, hedgerows and gardens. Adults fly in October–December. Larvae feed in spring on deciduous trees. Greek *poicilos* + *campe* = 'varied' + 'larva', from the variability of the larva's appearance, and Latin *populus* = 'poplar', a putative larval food plant.

Dew moth *Setina irrorella* (Erebidae, Arctiinae), wingspan 26–32 mm, scarce, day-flying (June–July) along the south coast and sporadically from south Wales to the Hebrides. It also occurs on calcareous grassland inland on the North Downs in Surrey, and the Cotswolds, and in the Burren, west Ireland. Larvae feed on lichens growing on rocky shores, shingle and downland. Its name may come from its habit of hanging from a blade of grass or leaf when at rest, giving it a transparent appearance. Greek *ses* = 'moth' and Latin *irroro* = 'to sprinkle with dew', probably from the black wing spots, possibly (also) from dew-covered wings as adults rest on grass stems at dawn.

Diamond-back moth *Plutella xylostella* (Plutellidae), wingspan 13–15 mm, a common immigrant from mainland Europe, sometimes appearing in great numbers (e.g. in June 2016) in all types of habitat throughout the British Isles, becoming established under favourable conditions, and occasionally reaching pest proportions on farms growing cabbages and sprouts. Two or more generations occur between May and September. Larvae feed on the leaves of cruciferous vegetables and a range of other plants. Greek *plutos* = 'washed', from the smudged pattern on the wings, and *xylostella* from fly honeysuckle *Lonicera xylosteum*, mistaken to be a food plant.

Dorset tineid moth *Eudarcia richardsoni* (Tineidae, Meessiinae), wingspan 7–9 mm, found only at two sites on the Dorset coast (Portland and Isle of Purbeck), and not known outside of Britain and therefore believed to be endemic. Adults emerge in June–July. Larvae inhabit sea-cliffs and scree, living inside a portable case made from grains of rock and fragments of lichen attached to the rock surface. They feed on algae and lichen. A BAP Priority Species. *Eudarcia* may come from Greek *eudarces* = 'bright-eyed', but is probably a meaningless neologism. N.M. Richardson (1855–1925) was an entomologist who discovered it on Portland.

Dried currant moth *Cadra cautella* (Pyralidae, Phycitinae), wingspan 16–20 mm, introduced with and feeding as a larva on dried fruit, stored nuts, seeds and other foodstuffs, widespread as a pest inside warehouses. It flies at night and rests on walls and windows during the day. Adults are evident from May to October, in two generations. *Cadra* appears to be a meaningless neologism; *cautella* may be from Latin *cautus* = 'cunning'.

Early moth *Theria primaria* (Geometridae, Ennominae), wingspan 32–7 mm, nocturnal, common in open woodland, woodland edges, scrub and mature hedgerows throughout England, Wales and Ireland, more local in Scotland. Adult males fly in January–February; females are almost wingless. Larvae feed on blackthorn and hawthorn. Greek *thereios* = 'wild animal' (the original specific name was *rupicapraria* = 'chamois', from a similarity in colour), and Latin *primaria* = 'of the first' (i.e. first part of the year).

Emperor moth *Saturnia pavonia* (Saturniidae, Saturniinae), wingspan 40–60 mm, widespread and common on heathland, moorland bog, fen, field margins, hedgerows and other scrub habitats, males flying by day, females at night in April–May. Larvae eat a variety of plants, depending on habitat. The binomial is from Saturnia or Juno (a daughter of Saturn), to whom the peacock (Latin *pavo*) was sacred.

Fisher's estuarine moth *Gortyna borelii* (Noctuidae, Xyleninae), wingspan 42–58 mm, nocturnal, Vulnerable in the RDB, protected in the UK WCA, and a priority species in the UK BAP. The only current natural British site is an area of sea walls and grassland on the Hamford Water estuary, north-east Essex. The total population has nevertheless been estimated at 1,000–5,000 adults. A colony in north Kent is probably derived from an unauthorised introduction. Adults fly in September–October. Larvae feed in the roots and stems of hog's fennel. *Gortyna* is a town in Crete. M. Borel was an early nineteenth-century French collector.

Fox moth *Macrothylacia rubi* (Lasiocampidae, Lasiocampinae), male wingspan 52–60 mm, female 58–78 mm, a generally widespread, often common moth, males flying in the afternoon, females at night in May–June. Larvae feed on heather, bilberry and creeping willow on moorland and heathland, bramble and meadowsweet in wet habitats, and salad burnet on downland. Greek *macros* + *thylacos* = 'large' + 'pouch', from the large cocoon, and *Rubus* (bramble), the food plant.

Ghost moth *Hepialus humuli* (Hepialidae), wingspan 50 mm though females can reach 80 mm, common and fairly widespread in lowland areas on disturbed weedy habitats in both town and countryside. Adults, with non-functional mouthparts, do not feed, and fly in June–July. Larvae feed underground on the roots of grasses, nettles and docks. The English name comes from the white males which can be seen at dusk hovering over grassy areas. Greek *hepialos* = 'fever', from the moth's fitful flight. This moth has no affinity with hop (*humulus*).

Goat moth *Cossus cossus* (Cossidae, Cossinae), wingspan 68–96 mm, locally widespread in fen, marsh, riverbanks, damp areas, heathland and woodland edges, less numerous in the north, predominantly coastal. It has declined in numbers and range since at least the 1960s and is a priority species in the UK BAP. Adults fly in June–July. Larvae burrow into the trunks of deciduous trees and feed on the wood. Willows and poplars are favoured. Because of the long digestion period required for their food material, larvae often live for four or five years before pupating. This moth gets its English name from the strong 'goaty' odour of the caterpillar. *Cossus* was a larva found under tree bark, eaten by the Romans.

Gypsy moth *Lymantria dispar* (Erebidae, Lymantriinae), wingspan 32–55 mm, common in the East Anglian and southern fens in the early part of the nineteenth century, but by 1900 clearance and drainage led to its extinction as a breeding species. Migrant males occasionally still appear in coastal south England, and since 1995 the species has been resident in small numbers in parts of London, where it has been subject to an eradication campaign owing to potential pest status, the larvae voraciously feeding on a number of deciduous trees. Sightings should be reported to DEFRA. Adults fly in July–August. Greek *lymanter* = 'destroyer', referencing its activities as a pest; *dispar* is Latin for 'unalike', referencing different intersexual colouring (male brown, female white).

Hawthorn moth *Scythropia crataegella* (Scythropiidae), wingspan 11–15 mm, nocturnal, common in hedgerows and scrub throughout England north to Yorkshire, and in south Wales. Adults fly in June–July, sometimes again in September. Larvae mine leaves of hawthorn, blackthorn and cotoneaster, feeding gregariously within a silk web, from May to October. Greek *scythropos* obscurely means looking sad or angry. *Crataegella* refers to the larval food plant, hawthorn.

Heart moth *Dicycla oo* (Noctuidae, Xyleninae), wingspan 32–8 mm, nocturnal, rare, found in parts of south England, and a priority species in the UK BAP. Very erratic in abundance, occurring commonly in one of its locations for a few years, then declining to very low numbers. Only seen in numbers in parts of Surrey, though it survives in low density in a few other counties of south England. Adults fly in June–July. Larvae feed during the night on pedunculate oak leaves, hiding during the day in a tent made of leaves spun together with silk. Greek *di* + *cyclos* = 'two' + 'ring', in turn reflecting the specific name which describes the roundish marks on the forewing, though the outer one is usually more heart-shaped, giving the English name.

Honeysuckle moth *Ypsolopha dentella* (Ypsolophidae), wingspan 18–23 mm, nocturnal, widespread and common in woodland, hedgerows and gardens. Adults fly in July–August. Larvae feed on honeysuckle underneath a silk web. The binomial comes from a malformation of Greek *ypsilophos* = 'high-crested', and Latin *dens* = 'tooth', from the apex of the forewing which projects upwards when the wings are folded.

Hornet moth *Sesia apiformis* (Sesiidae), wingspan 33–8 mm, a day-flying clearwing moth found in parks, hedgerows, fens and areas with scattered trees in southern England, East Anglia and the Midlands, though also recorded from Durham, Wales and south-east Scotland. Adults fly in June–July. Larvae burrow into the wood of black poplar and congenerics, overwintering 2–3 times. *Apiformis* is Latin for 'shaped like a bee'. See also lunar hornet moth below.

Leek moth *Acrolepiopsis assectella* (Glyphipterigidae, Acrolepiinae), wingspan 11–14 mm, a day-flying mining moth local in gardens and allotments throughout much of south-east England, with records west to Devon. The bivoltine adults fly in June–July, and from October through winter to early spring. Larvae mine stems and roots of alliums such as leek, chives and onions (another name for this moth being onion leaf miner), occasionally reaching pest proportions. Greek *opsis* = 'appearance' + *Acrolepia*, a related genus, and Latin *assector* = 'to accompany'.

Light brown apple moth *Epiphyas postvittana* (Tortricidae, Tortricinae), wingspan 16–25 mm, an Australian moth first reported in Britain from Cornwall in 1936. It has since expanded its range across most of the British Isles, and is commonly found in gardens. The bivoltine adults fly in May and October. Larvae feed on a range of plants (though not especially fond of apple they can be a pest on fruit trees). Greek *epi* + *phyas* = 'upon' + 'shoot', and Latin *post* + *vitta* = 'behind' + 'band', from the forewing pattern.

Lobster moth *Stauropus fagi* (Notodontidae, Dicranurinae), wingspan 45–60 mm, nocturnal, common in mature woodland throughout much of the southern half of England, parts of Wales and south-west Ireland. Adults fly in May–July. The lobster-like appearance of the caterpillar prompts the English name, the forelegs resembling outstretched claws and the body swelling and cantilevering back like a lobster's tail. Larvae feed in July–September on leaves of trees such as oak, birch, hazel and beech. Greek *stauros* + *pous* = 'cross' + 'foot', and Latin *fagus* = 'beech'.

Lunar hornet moth *Sesia bembeciformis* (Sesiidae), wingspan 32–42 mm, a day-flying moth widespread and common in fen, woodland, heath, moorland and hedgerows, with most records in the southern half of Britain. Adults fly in spring and summer. Larvae burrow into the wood of willows and poplars. Greek *ses* = 'moth' (or its larva), and Latin *bembeciformis* = having the form of a sand wasp (*Bembix*).

Many-plumed moth Alternative name for twenty-plume moth.

March moth *Alsophila aescularia* (Geometridae, Ennominae), wingspan 25–35 mm, widespread and common in open woodland, parks and gardens. Adult males fly at night in March–April, when the wingless females may be found crawling on tree-trunks. Larvae feed on various trees. Greek *alsos* + *philos* = 'grove' + 'loving', and Latin *aesculus*, here meaning 'oak' (a food plant) rather than 'horse-chestnut'.

Marsh mallow moth *Hydraecia osseola* (Noctuidae, Xyleninae), wingspan 40–50 mm, nocturnal, RDB Endangered, a priority species in the UK BAP, found in brackish marsh. The two known breeding areas are in the Romney Marsh/Rye area on the border of East Sussex and Kent, and on the banks of the Medway between Maidstone and Rochester in Kent. Adults fly in September. Larvae feed in stems and roots of marsh-mallow. Greek *hydor* + *oiceo* = 'water' + 'to live in', and Latin *osseus* = 'bone', from the colour.

Marsh moth *Athetis pallustris* (*sic*) (Noctuidae, Xyleninae), male wingspan 26–34 mm, female 18–22 mm, a rare RDB nocturnal species found in damp meadows, and a priority species in the UK BAP. Formerly much more widespread it is now confined to the coastal belt of Lincolnshire where it occurs on two nature reserves and possibly just two other sites. Adults fly in May–June. Larvae feed on plantains and meadowsweet. Greek *athetos* = 'without position' (as a taxonomically anomalous genus), and a misspelling of Latin *palustris* = 'marsh'. See also rosy marsh moth below.

Mediterranean flour moth *Ephestia kuehniella* (Pyralidae, Phycitinae), wingspan 22 mm, an introduced nocturnal moth, larvae frequently found on flour, although they also feed on other foodstuffs in warehouses, granaries, etc. It can be a pest in flour mills and bakeries, where adults may rest on walls and windows by day. Greek *ephestios* = 'beside the hearth' (i.e. 'domestic'), and in honour of the late nineteenth-century German, Professor Kühn.

Mint moth *Pyrausta aurata* (Crambidae, Pyraustinae), wingspan 18–20 mm, day- and night-flying (March–October), locally common in England, Wales and south Scotland in open woodland, marsh, calcareous grassland and gardens. Larvae mainly feed on species of Lamiaceae. Greek *pyr* + *auo* = 'fire' + 'to burn' (i.e. getting burned when drawn to flame), and Latin *aurata* = 'golden'.

Mouse moth *Amphipyra tragopoginis* (Noctuidae, Amphipyrinae), wingspan 33–8 mm, widespread and common in gardens, woodland, dunes, moorland and fen, but with numbers declining by 85% over 1968–2007, a species of conservation concern in the UK BAP. Adults fly at night in July–September. If disturbed, they will scuttle away mouse-like rather than take flight, hence the English name. Larvae feed on a number of herbaceous and woody plants. Greek *amphi* + *pyr* = 'around' + 'fire', and *tragopoginis* from goat's-beard *Tragopogon*, a food plant.

Ni moth *Trichoplusia ni* (Noctuidae, Plusiinae), wingspan 30–40 mm, nocturnal, an immigrant north to Cumbria, increasingly annual in occurrence, sometimes in high numbers, and breeding in favourable years. Adults are evident in July–September. Larvae feed on a variety of herbaceous plants. Greek *trichos* = 'hair' + the genus *Plusia*, and Greek letters *nu* + *iota* = 'n' + 'i', from wing markings that resemble these.

November moth *Epirrita dilutata* (Geometridae, Larentiinae), wingspan 38–44 mm, widespread and common in broadleaf woodland, scrub, hedgerows and gardens, but problems in identifying individuals to specific status mean that all *Epirrita* species are under-recorded. Adults fly at night in September–November. Larvae feed in late spring on a number of tree and shrub species. Greek *epirrheo* = 'to flow', from the rivulet pattern on the wings, and Latin *dilutus* = 'washed out' or 'pale'. See also pale November moth below.

Nut leaf blister moth *Phyllonorycter coryli* (Gracillariidae, Lithocolletinae), wingspan 7–9 mm, nocturnal, widespread and common in woodland, scrub and hedgerows. Adults have two broods, flying in May and August. Larvae mine leaves of hazel (= *Corylus*, giving the specific name) in summer and autumn. Greek *phyllon* + *orycter* = 'leaf' + 'a digging'.

Orange moth *Angerona prunaria* (Geometridae, Ennominae), wingspan 35–45 mm, nocturnal, local in broadleaf woodland, scrubby heathland and mature hedgerows throughout the southern half of England and parts of Ireland. Adults fly in June–July. Larvae feed on woody plants such as blackthorn (*Prunus spinosa*, giving the specific name), hawthorn, birch and heather. In Greek mythology Angerona was the goddess of silence. Not the same species as orange (moth) (q.v.).

Pale November moth *Epirrita christyi dilutata* (Geometridae, Larentiinae), wingspan 38–42 mm, common in beech woodland throughout much of England and southern Scotland, but problems in identifying individuals to specific status mean that all *Epirrita* species are under-recorded. Adults fly at night in September–November. Larvae feed in late spring on a number of tree and shrub species. William Christy (1863–1939) was a British entomologist.

Pear leaf blister moth *Leucoptera malifoliella* (Lyonetiidae, Cemiostominae), wingspan 7–8 mm, in gardens, orchards and hedgerows throughout England, Wales and south Scotland, sometimes uncommon, but in hot summers found in huge numbers. Adults fly in June–July. Larvae mine leaves of apple, pear and hawthorn from August to October, causing sufficient damage to be a serious pest in high-population years. Greek *leucos* + *pteron* = 'white' + 'wing', and Latin *malus* + *folium* = 'apple' + 'leaf'.

Peppered moth *Biston betularia* (Geometridae, Ennominae), wingspan 35–60 mm, nocturnal, widespread and common in woodland, scrub, hedgerows, parks and gardens. A favourite of genetic studies, this species has demonstrated industrial melanism, where dark individuals became the dominant form in parts of northern England when air pollution was pronounced, but this melanic form (*carbonaria*) declining in these areas as air quality has improved. Adult males fly in May–August. Larvae feed on a number of tree species, as well as herbaceous species such as mugwort and goldenrod. In Greek mythology, Biston was a son of Mars. Latin *betulus* = 'birch', one of the food plants.

Pine leaf-mining moth *Clavigesta purdeyi* (Tortricidae, Olethreutinae), wingspan 10–12 mm, local in pine woodland and plantations throughout much of England and Wales, spreading into south Scotland and south-east Ireland, probably continuing to expand its range. Adults fly at night in July–September. Larvae feed on pine needles. Latin *clavus* + *gestus* = 'purple stripe on the Roman tunic' + 'bearing', from the forewing colouring. Capt. W. Purdey (1844–1922) was a British entomologist.

Pine resin-gall moth *Retinia resinella* (Tortricidae, Olethreutinae), wingspan 16–22 mm, with a scattered distribution locally common in pine woodland throughout much of Britain, especially in the north and east. Larvae have a two-year development, beginning by boring into a small twig of (usually Scots) pine, causing the host plant to exude resin which then forms a gall inside which the larva lives. In the second year the gall is enlarged. Pupating inside the gall, the adults then emerge in June. Because of the two-year cycle, moths are commonest in odd years. Greek *rhetine* and Latin *resina* = 'resin'.

Portland moth *Actebia praecox* (Noctuidae, Noctuinae), wingspan 35–40 mm, nocturnal, on sandy beaches of England, Wales and north-west Ireland. Adults fly in August–September, and if disturbed while resting during the day will fall to the ground and feign death. Larvae feed on creeping willow, wormwood, bird's-foot-trefoil and tree lupin. Greek *acte* + *bioo* = 'coast' + 'to live', and Latin *praecox* = 'early', the wing colour echoing the fresh green of spring.

Raspberry moth *Lampronia corticella* (Prodoxidae), wingspan 9–12 mm, widespread, scattered and generally scarce in gardens and hedgerows in Britain, with a few records in Ireland. Adults fly in the afternoon in May–June. Larvae mainly feed on raspberry in March–April, initially in the flower receptacle then in the buds and shoots, making this a serious pest in a few areas. Greek *lampros* = 'bright', referring to the wing colours, and *corticella* from Latin *cortex* = 'bark'.

Rosy marsh moth *Coenophila subrosea* (Noctuidae, Noctuinae), wingspan 35–42 mm, a Vulnerable RDB nocturnal moth found in bogs in the Tregaron and Borth bogs in mid-Wales, and (discovered in 2005) at Roudsea Wood in Cumbria. Adults fly in late July–August. Larvae mainly feed on bog-myrtle and crowberry. Greek *coinos* + *philos* = 'community' + 'loving', and Latin *subrosea* = 'almost rosy-coloured'.

Short-cloaked moth *Nola cucullatella* (Nolidae, Nolinae), wingspan 16–18 mm, common in hedgerows, gardens, scrub and woodland throughout much of England and parts of Wales. Adults fly at night in June–July. Larvae feed on blackthorn, hawthorn, apple, pear and plum. Nola is a town in Campania, Italy. Latin *cucullus* = 'cowl', describing the dark basal area on the forewing that resembles a cloak when the moth is at rest.

Skin moth *Monopsis laevigella* (Tineidae, Tineinae), wingspan 13–20 mm, widespread and common in and around bird nests and farm buildings. Adults fly from dusk onwards from May to September, with two generations. Larvae feed in December–March on dead animals, animal waste, owl pellets, bird nest detritus and droppings, living inside a silk tent. Greek *monops* = 'one-eyed', from the forewing spot, and Latin *laevigo* = 'to pulverise', adults described as being gold-dusted.

Small autumnal moth *Epirrita filigrammaria*, wingspan 30–8 mm, nocturnal, common on moorland throughout much of Scotland, Wales and northern England, rare elsewhere in England, and scattered in Ireland. Numbers have declined by 81% over 1968–2007. Adults fly in August–September. Larvae feed on heather, bilberry and sallow. Latin *filum* = 'thread' and Greek *gramma* = 'line', from the thinly banded wing pattern.

Sweet gale moth *Acronicta cinerea* (Noctuidae, Acronictinae), wingspan 32–40 mm, nocturnal, scarce on moorland and farmland in Scotland, from Galloway to Sutherland, and near the coast in Aberdeenshire and Morayshire. In west Ireland there is a second brood in July–August. Larvae feed on a range of plants. Greek *acronyx* = 'nightfall', and Latin *cinerea* = 'grey'.

Tapestry moth *Trichophaga tapetzella* (Tineidae, Tineinae), wingspan 15–22 mm, rare, found in bird nests and around old buildings, previously distributed throughout much of the British Isles but now absent from most of its original range. Adults fly in May–August. Larvae feed on natural and manufactured-product fur, hair and feathers, and on owl pellets, and were formerly a pest in unheated buildings. Greek *trichos* + *phagein* = 'hair' + 'to eat', and Latin *tapete* = 'carpet'.

Thyme moth *Scrobipalpa artemisiella* (Gelechiidae, Gelechiinae), wingspan 10–12 mm, scarce, on sandhills, dry limestone slopes, grass-heath and (in west Scotland) machair widespread but very local. Adults fly at dusk in June–August. Larvae feed on wild thyme in May–June, initially inside a leaf, later inside a silk web on the underside of a stem. Latin *scrobis* + *palpus* = 'ditch' + 'palp', from a furrow on the labial palp, and *Artemisia* = 'wormwood', though this is not a food plant.

Turnip moth *Agrotis segetum* (Noctuidae, Noctuinae), wingspan 27–40 mm, nocturnal, widespread and common in gardens, farmland, parkland, oak woodland and dunes. The bivoltine adults fly in May–June and August–September. This is one of the cutworms: larvae live underground and feed on the roots of root vegetables, herbaceous plants and other cultivated crops, becoming a serious pest, their habit of biting off the shoots of small seedlings giving rise to the name. Greek *agrotes* = 'field', and Latin *seges* = 'cornfield' (loosely any field with a crop).

Vine moth *Eupoecilia ambiguella* (Tortricidae, Tortricinae), wingspan 12–15 mm, scarce, local on heathland and woodland in parts of southern England and East Anglia. Adults fly at night in May–June. Larvae mine the flowers and berries of alder buckthorn, honeysuckle and ivy, overwintering in a case made of leaf fragments. It is a pest of grape-vine on the continent, hence its English name. Greek *eupoicilos* = 'variegated', from the forewing pattern, and Latin *ambiguus* = 'doubtful'.

White satin moth *Leucoma salicis* (Erebidae, Lymantriinae), wingspan 37–60 mm, nocturnal, local in plantations, hedgerows, scrub, parks and gardens throughout England, the Welsh Borders, and the west and east coasts of Ireland. Adults fly in July–August. Larvae feed on sallow (*salix*, giving the specific name) and poplar species. Greek *leucos* + *come* = 'white' + 'hair'.

MOTH FLY Members of the family Psychodidae.

MOTHER OF PEARL *Pleuroptya ruralis* (Crambidae, Spilomelinae), wingspan 26–40 mm, the largest of the so-called micromoths, and indeed larger than many macro-moths, widespread and common in rural gardens, waste ground and rough pastures. Adults fly from dusk onwards in July–August. Larvae mainly feed on leaves of stinging nettle. Greek *pleuron + ptyon* = 'rib' + 'winnowing fan', and Latin for 'rural'.

MOTHER SHIPTON *Callistege mi* (Erebidae, Erebinae), wingspan 25–30 mm, a day-flying moth (May–July) on wasteland, hay meadow, downland, heathland, moorland and other open habitats, fairly widespread but scarcer in Scotland and Ireland. Larvae feed mainly on grasses, medicks, lucerne, bird's-foot-trefoil and clovers. The pattern on the forewing vaguely resembles a witch's face, hence its English name, that of a sixteenth-century Yorkshire witch. Greek *callos + stege* = 'beautiful' + 'roof', and the Latinised form of Greek letter *mu*, from the hindwing underside M pattern.

MOTHERWORT *Leonurus cardiaca* (Lamiaceae), a neophyte rhizomatous perennial herb introduced from continental Europe in the Middle Ages as a medicinal herb and later as an impurity of imported grain, recorded in the wild in 1597, scattered in Britain, naturalised on waysides and in waste places, often near habitation. Its common name comes from its traditional use in midwifery of preventing uterine infection. It contains the alkaloid leonurine, a mild vasodilator which has a relaxing effect on muscles, so it has also been used as a cardiac tonic. Greek *leon + ouros* = 'lion' + 'tail', and *cardia* = 'heart', from its medicinal use.

MOTTLEGILL Common and widespread inedible fungi in the genera *Panaeolina* and *Panaeolus* (Agaricales), both genera named from Greek *pan + aiolos* = 'all' + 'variegated', reflecting the variegated or mottled gill colouring. **Banded mottlegill** *Panaeolus cinctulus*, **brown mottlegill** *Panaeolina foenisecii*, **dewdrop mottlegill** *Panaeolus acuminatus* and **turf mottlegill** *Panaeolus fimicola* are all found in grassy habitats, including lawns and amenity grassland. **Egghead mottlegill** *Panaeolus semiovatus* and **Petticoat mottlegill** *Panaeolus papilionaceus* are both associated with (mainly horse) dung and upland manured pasture.

MOULD Tiny fungi in the genus *Hypomyces* (Hypocreales, Hypocreaceae) (Greek *hypo + myces* = 'under' + 'fungus'). See also mildew and slime mould.
     Bolete mould *H. chrysospermus*, widespread and common, expressing itself as dense clusters of 'pinheads' on old woodland bolete fungi. Greek *chrysos + sperma* = 'gold' + 'seed'.
     Orange polypore mould *H. aurantius*, widespread but uncommon, expressing itself as dense clusters of 'pinheads' on old woodland polypore fungi. *Aurantius* is Latin for 'golden'.
     Orchid mould Parasitic, forming extensive and often circular dark brown or black blotches on living leaves of common marsh-orchid.
     Pink polypore mould *H. rosellus*, uncommon, with a scattered distribution expressing itself as dense clusters of 'pinheads' on old woodland polypore fungi. *Rosellus* is the Latin diminutive of *roseus* = 'rose-red'.

MOUNTAIN-ASH Alternative name for rowan.

MOUNTAIN-LAUREL *Kalmia latifolia* (Ericaceae), an evergreen North American neophyte shrub present in gardens by 1734, naturalised in a few wet, acidic places in south-east England. Pehr Kalm (1716–79) was a Swedish/Finnish traveller. Latin *latifolia* = 'broad-leaved'. See also sheep-laurel.

MOUNTAIN MOTH Members of the family Geometridae, subfamily Ennominae.
     Black mountain moth *Glacies coracina*, wingspan 38–48 mm, scarce but widespread in the central Highlands on mountains and moorlands from 600 m and higher, larvae feeding on crowberry. Males fly in sunshine in June–July. *Coracina* is Latin for 'raven-black'.
     Netted mountain moth *Macaria carbonaria*, wingspan 23–5 mm, day-flying in April–June, local on moorlands and mountains of the Highlands. Caterpillars feed at night, only on bearberry. Adults take nectar from a few upland plant species. Rare (RDB) and declining, this is a priority species for conservation. Greek *macaria* = 'happiness' and Latin *carbonaria* = 'pertaining to charcoal' (i.e. black), from the wing pattern.

MOUSE Small rodents in the family Muridae.
     Field mouse Alternative name for wood mouse.
     Harvest mouse *Micromys minutus*, Europe's smallest rodent, head-body length 50–80 mm, length of the prehensile tail 50–70 mm, weight 4–6 g, mainly found in the southern half of England, with scattered populations in Cheshire, Yorkshire and Wales, in cereal crops, long grass, reedbeds, bramble patches and grassy hedges. The few reports from Scotland are probably reintroductions. The pre-breeding population in 2005 was estimated at 1,415,000 in England, and 10,000 in Wales. Numbers and range are declining: in the 1990s a Mammal Society survey found that they were absent in 30% of the sample sites they were recorded in during the 1970s. They are active by day (especially in winter) and night (especially in summer), though particularly at dusk and dawn. Diet is mainly of seeds, cereal, fruits and berries, and on insects, flowers and green shoots in spring when other food is scarce. They build small shelter nests close to the ground, and breeding nests of woven grass in the stems of plants well above the ground. Breeding is generally in May–October. Gestation lasts 19 days. Females produce 3–7 litters a year, with 1–8 young, which are weaned by 15 days. They do not hibernate, but spend winter mostly underground. A captive breeding programme was set up at Chester Zoo in the late 1980s, and a number of populations have been successfully reintroduced or restocked in Cheshire and elsewhere. The binomial comes from Greek and Latin for 'small (mouse)'.
     House mouse *Mus musculus*, head-body and tail lengths each 75–90 mm, male weight 25 g, females 30 g. They are common and widespread in a range of urban buildings, and in the countryside will occupy farm buildings, refuse tips and hedgerows. Urban populations generally shelter in wall crevices and beneath floors, building nests of paper and cloth. They are nocturnal generalist, opportunistic omnivores, though with a preference for seed. Rural populations construct shallow grass-lined burrows and, nearby, storage chambers to cache seed, grass, fruit, insect larvae, earthworms, etc. Urban populations may breed throughout the year if the food supply is sufficient, producing 5–10 litters a year, each litter having 4–9 young. Gestation is 20 days. The young are weaned in two weeks. Females can become sexually mature after six weeks, but males take up to 18 months before they can reproduce. For rural populations the breeding season runs from spring to autumn with females producing a litter every five weeks. They are generally viewed as pests, transmitting diseases, contaminating food and damaging packaging. The complete reference genome was sequenced in 2002. The binomial is Latin for 'mouse' and 'little mouse'.
     Wood mouse *Apodemus sylvaticus*, head-body length 80–100 mm, tail length 70–95 mm, average male weight 27 g, females 23 g, Britain and Ireland's most common and widespread rodent, found in woodland, hedgerows, farm land, scrub, grassland, dunes and gardens (often commensal). Nests, lined with chopped grass, are usually underground but in colder winters wood mice use buildings that are not occupied by house mice. Underground burrows contain a number of tunnels and chambers; some chambers act as food stores, with the nest room located deeper in the earth. They are primarily nocturnal seed-eaters, but take other plant material, and (especially in woodland) invertebrates. Seed crop size is reflected in mouse numbers in the same autumn and the following summer, but pre-breeding populations of >38 million

have been estimated as common. Breeding takes place from February to October, females normally producing a number of litters in a year. Gestation lasts 25–6 days. Litter size is 4–7, the young becoming independent after 18–22 days and sexually mature afer 2 months, so young born in spring and early summer can breed in the same year. Greek *apodemos* = 'abroad' and Latin *sylvaticus* = 'relating to woodland'.

Yellow-necked mouse *A. flavicollis*, head-body length and tail length both 90–135 mm, weight 15–30 g, found mainly in the Welsh borders, the Cotswolds, south and south-east England and East Anglia, with a pre-breeding population estimated to be up to 750,000. They prefer mature woodland, hedgerows and wooded gardens, the underground nests consisting of a ball of dry grass, moss and leaves. Food, mainly sought at night, comprises seedlings, buds, fruit, nuts, adult and larval insects, and spiders, which can be stored in burrows. Breeding takes place from February to October, peaking in July–August. Gestation lasts 23 days. Three litters of 3–10 young (which are weaned after 3 weeks) are produced each year. *Flavicollis* is Latin for 'yellow-necked'.

MOUSE-EAR Species of *Cerastium* (Caryophyllaceae), from Greek *ceras* = 'horn', from the shape of the seed capsule.

Alpine mouse-ear *C. alpinum*, a herbaceous mat-forming montane perennial, found mainly in the Highlands, but also occasional in the Lake District and Snowdonia, favouring basic rocks, especially abundant on mica-schists though also found on limestone and occasionally serpentine, in species-rich dwarf-herb communities.

Arctic mouse-ear *C. nigrescens*, a herbaceous montane, tufted perennial mainly found in highland Scotland, but also north Wales, on acidic and hard basic rocks, for example serpentine on Unst (Shetland), usually in wet, thinly vegetated crevices and on ledges in north-facing corries, but also on a montane fellfield on Skye. It is rarely found <700 m and reaches 1,200 m on Ben Nevis. This is the county flower of Shetland as voted by the public in a 2002 survey by Plantlife. *Nigrescens* is Latin for 'becoming black'.

Common mouse-ear *C. fontanum*, a common and wide-spread perennial herb found in a range of usually fertile habitats, including neutral pasture and meadow, calcareous and acidic grassland, rush pasture, heathland and mires, springs and flushes, montane grassland, dunes and shingle, cultivated ground, waste ground and walls. *Fontanum* is Latin for 'growing in running water'.

Dwarf mouse-ear *C. pumilum*, a winter-annual scattered in the southern half of England and in Wales on chalk and limestone substrates, mainly in open patches within grazed grassland; also found as an introduction in quarries, on spoil heaps and on ballast along railway lines. *Pumilum* is Latin for 'dwarf'.

Field mouse-ear *C. arvense*, a widespread, often common herbaceous perennial, though less common and declining in the west of its range, on dry calcareous to slightly acid sandy soils, in grassland, and on dry roadsides, wayside banks, arable field margins, sandy or gravelly waste ground, and dunes. *Arvense* is from Latin for 'field'.

Grey mouse-ear *C. brachypetalum*, an annual of dry, calcareous, open grassland and railway banks, first recorded in Britain in Bedfordshire in 1947, so abundant on an embankment at Sharnbrook Summit in 1954 that it was considered most likely to be native. It was also found on the Bedfordshire–Northamptonshire border in 1973. Its occurrence in old grassland on chalk in west Kent in 1978 might indicate that it is a relic of a more widespread distribution in this area. Stace (2010), however, considers it to be probably introduced and naturalised. *Brachypetallum* is Greek for 'short-petalled'.

Little mouse-ear *C. semidecandrum*, an annual or overwintering herb scattered and locally common in England and Wales, less so in Scotland, and coastal in Ireland, on well-drained sandy or calcareous soils, on dry banks and grassy and heathy habitats,

fixed dunes, disturbed sandy areas near the sea and walls. It also grows on bare places on limestone. Latin *semidecandrum* = 'half of ten (i.e. five) flowers'.

Sea mouse-ear *C. diffusum*, an annual of light, dry, sandy or gravelly soil, widespread and common on coastal habitats that include grassland, fixed dunes and sandy banks. Inland, it occurs locally in open habitats such as dry grassland, by paths, on wall tops, verges and railway ballast. It has declined in many inland railway sites, but this may have been offset by recent records from salt-treated roadsides. *Diffusum* is Latin for 'loosely spreading'.

Starwort mouse-ear *C. cerastoides*, a herbaceous mat-forming montane perennial found in the Highlands on wet acidic rocks, often where there is late snow-lie. *Cerastoides* is from Greek *ceras* = 'horn'.

Sticky mouse-ear *C. glomeratum*, a common and widespread annual growing in disturbed, often nutrient-enriched habitats, tolerant of trampling and particularly common around farms, in gateways, on field edges, in bare patches in improved grassland, beside tracks and in waste ground. It is also frequent on dunes and shingle. *Glomeratum* is Latin for 'clustered'.

MOUSE-EAR-HAWKWEED *Pilosella officinarum* (Asteraceae), a widespread and locally common stoloniferous, perennial herb of dry habitats such as short grassland, heathland, dunes, scree, rock outcrops, quarries and cliffs, on both base-rich and acidic substrates. The binomial is from Latin for 'small hairs', and for having pharmacological properties.

Shaggy mouse-ear-hawkweed *P. peleteriana*, a herbaceous stoloniferous perennial found in a few places in England and Wales on steep, well-drained slopes in chalk and limestone grassland, dry dolerite rock shelves and quarry waste, shallow soils overlying granite, and dunes.

Shetland mouse-ear-hawkweed *P. flagellaris* ssp. *bicapitata*, an endemic perennial of grassy limestone rocky outcrops, heathy granulitic gneiss and feldspar sea-banks in three localities in Shetland, first discovered in 1962. *Flagellaris* is Latin for 'with whips', *bicapitata* for 'two-headed'.

Spreading mouse-ear-hawkweed *P. flagellaris* ssp. *flagellaris*, a perennial, stoloniferous European neophyte introduced in 1816, naturalised on railway banks (since 1869) and roadsides, scattered in central and central-south England and central Scotland.

Tall mouse-ear-hawkweed *P. praealta*, a perennial European neophyte, scattered in Britain, naturalised on verges, walls, railway banks and rough ground. *Praealta* is Latin for 'very tall'.

MOUSE-TAIL MOSS Species of *Isothecium* (Lembophyllaceae) (Greek *isos* + *thece* = 'equal' + 'box'), and *Myurella* (Plagiotheciaceae) (Greek *mys* + *oura* = 'mouse' + 'tail').

Dwarf mouse-tail moss *M. tenerrima*, scarce in the Highlands on base-rich soil among rocks and crevices. *Tenerus* is Latin for 'tender'.

Holt's mouse-tail moss *I. holtii*, locally common in montane western Britain and parts of Ireland on rocky woodland stream banks on the sides of boulders, sometimes on tree bases and roots, usually close to water, and mainly on siliceous and base-poor rocks. George Holt (1852–1921) was a Manx-born bryologist.

Larger mouse-tail moss *I. alopecuroides*, common and widespread in woodland, on stream banks and other sheltered habitats, often on the lower part of tree-trunks and on the roots. It also grows on base-rich rocks. *Alopecuroides* means like the grass foxtail (Greek *alopex* + *oura* = 'fox' + 'tail').

Slender mouse-tail moss *I. myosuroides*, common and widespread on boulders and tree-trunks in woodland, but may occur in the open on siliceous and non-calcareous substrates. Greek *mys* + *oura* = 'mouse' + 'tail'.

Small mouse-tail moss *M. julacea*, scarce in the Highlands, north-west Yorkshire and west Galway on base-rich soil among rocks and crevices. *Julacea* may be from Greek *iulus* = 'plant down'.

MOUSETAIL *Myosurus minimus* (Ranunculaceae), a rare annual declining in status found in (especially south-east)

England on seasonally flooded, nutrient-rich soils in farmland disturbed by machinery or animals, locally persisting in habitats such as coastal grazing marshes. Greek *mys* + *oura* = 'mouse' + 'tail', and Latin *minimus* = 'smallest' or 'least'.

MOUSETAIL PLANT *Arisarum proboscideum* (Araceae), a rhizomatous herbaceous Mediterranean neophyte perennial, very scattered in southern and south-west England as a garden throw-out on rough ground and in hedges. *Arisarum* is a plant named by Dioscorides, and Latin *proboscideum* = 'elephant trunk'.

MRS GRIFFITHS'S LITTLE FLOWER *Halurus flosculosus* (Wrangeliaceae), a widespread red seaweed of the mid- and lower intertidal. Amelia Griffiths (1768–1858) was an amateur phycologist from Torquay. Latin *flosculosus* = 'woolly'.

MUD BEETLE 1) or **minute bog beetle** *Sphaerius acaroides* (Sphaeriusidae), length 0.6–0.8 mm, the only representative of its family, in moss adjoining small bodies of water, mainly in Dorset, feeding on microscopic algae. Greek *sphaera* = 'ball', Latin *acaroides* = 'like a mite'.

2) Some terrestrial water scavenger beetles (Helophoridae) in the genus *Helophorus* (Greek *helos* + *phoros* = 'marsh' + 'carry').

New Forest mud beetle *H. laticollis*, recently recorded from five areas in the New Forest in shallow, exposed, grassy pools on peaty heathland. Latin *latus* + *collum* = 'broad' + 'neck'.

Turnip mud beetle *H. porculus* (with a more northerly distribution) and *H. rufipes*, widespread but scattered, occasional larval pests of white turnips, swedes, beans, kale, cabbage and lettuce. Greek *helios* + *phora* or *phoros* = 'sun' + 'motion', and New Latin *porculus* = 'little pig'.

Wheat mud beetle *H. nubilus*, widespread but scattered, adults being active in May–September. Larvae are active from November, often feeding on winter wheat during cold weather when most other pests are inactive, occasionally becoming a serious pest, for example in East Anglia in 2007. *Nubilus* is Latin for 'cloudy'.

3) Members of the Heteroceridae (q.v.).

MUDWORT *Limosella aquatica* (Scrophulariaceae), a scattered, declining annual found on muddy edges, which may dry out in summer, of rivers, lakes, pools, ditches, rutted tracks and roadsides. In the Burren it grows in limestone solution hollows, but generally may prefer mildly acidic, nutrient-enriched soils. Latin *limosa* = 'full of mud', and *aqua* = 'water'.

Welsh mudwort *L. australis*, a scarce annual to short-lived perennial found in a few sites in Wales in mudflats and saltmarsh pools. *Australis* is Latin for 'southern'.

MUGILIDAE Family name of (grey) mullets, from Latin *mugil* for these coastal fishes.

MUGWORT *Artemisia vulgaris* (Asteraceae), a common and (in the lowlands) widespread aromatic archaeophyte perennial herb of waste ground and verges, usually on fertile soils. The seeds are often distributed by human activities, especially in urban areas and along roads and railways. Artemis was the Greek goddess of chastity. Latin *vulgaris* = 'common'.

Annual mugwort *A. annua*, a south-east European neophyte introduced in 1741, recorded in the wild in Middlesex in 1923. It is very scattered in England on waste ground, originating in wool shoddy and birdseed.

Breckland mugwort Alternative name for field wormwood.

Chinese mugwort *A. verlotiorum*, a rhizomatous herbaceous neophyte perennial, naturalised and locally abundant on rough ground. As it flowers in October and November, it rarely sets seed, and its spread is often by detached pieces of rhizome. It was first recorded in the wild in Middlesex in 1908. By 1950 it was well-established in Surrey and Middlesex, but while remaining frequent in the London area, spread through England, reaching the north-east, and Moray has been relatively slow. Bernard Verlot (1836–97) was a French collector.

Hoary mugwort *A. stelleriana*, a rhizomatous perennial

north-east Asian neophyte herb, introduced by 1865, naturalised on dunes in west-central Scotland and the Solway Firth, and on waste ground in a few inland sites elsewhere. Georg Steller (1709–46) was a German naturalist.

Slender mugwort *A. biennis*, an annual to biennial North American neophyte herb present by 1804, on waste ground and roadsides, and in pavement cracks, originating from wool shoddy, oil- and birdseed, and as a grain impurity, sometimes naturalised, more usually casual, in southern England and south Wales. *Biennis* is Latin for 'biennial'.

MULBERRY (FUNGUS) Wood mulberry *Bertia moriformis* (Coronophorales, Bertiaceae), a decay fungus found on the dead wood of trees such as oak, beech, sycamore and larch throughout the Britain, less common in Ireland. *Moriformis* is Latin for 'mulberry-shaped'.

MULBERRY (PLANT) Black mulberry *Morus nigra* (Moraceae), a small long-cultivated Near Eastern neophyte tree, frequently planted in parks and large gardens, naturalised on waste ground and walls, and as a relic of cultivation, scattered in England and Wales. 'Mulberry' comes from Old English *morberie*, from Latin *morus* (mulberry tree) + berry. The Latin *nigra* = 'black'.

MULLEIN (MOTH) *Cucullia verbasci* (Noctuidae, Cucullinae), wingspan 45–50 mm, in calcareous grassland, woodland clearings, woodland rides, scrub, verges, shingle beaches, parks and gardens throughout much of England and Wales, mostly in the south. Adults fly at night in April–May. Larvae feed on the flowers of mullein, especially great mullein, giving their common name, but in gardens are also found on buddleia. Latin *cucullus* = 'cowl', from a crest on the thorax, and the food plant *Verbascum* (mullein).

MULLEIN (PLANT) Herbaceous species of *Verbascum* (a corruption of Latin *barbascum* = 'with a beard', from the hairy leaves) in the family Scrophulariaceae. 'Mullein' dates from late Middle English, from Old French *moleine*, and Latin *mollis* = 'soft'.

Casual, sometimes naturalised European neophytes scattered in England and Wales on waste ground are **Broussa mullein** *V. bombyciferum* (from Turkey, found in Guernsey, southern England and the Isle of Man), **Hungarian mullein** *V. speciosum*, **moth mullein** *V. blattaria*, **nettle-leaved mullein** *V. chaixii*, **orange mullein** *V. phlomoides* (from Turkey, found in the Channel Islands, southern and central England and the Isle of Man), and **purple mullein** *V. phoeniceum*.

Caucasian mullein *V. pyramidatum*, perennial, casual sometimes naturalised on rough ground in southern England. Latin *pyramidatum* = 'pyramid-like'.

Dark mullein *V. nigrum*, biennial or short-lived perennial, locally common especially on English limestone on verges and embankments, in hedge banks and other grassy habitats, on walls and in cultivated ground, with a preference for well-drained calcareous soil. *Nigrum* is Latin for 'black'.

Dense-flowered mullein *V. densiflorum*, biennial, scattered in England and south-east Ireland as a casual, occasionally naturalised garden escape on waste ground, roadsides and rubbish tips.

Great mullein *V. thapsus*, a common and widespread biennial, often a garden escape, of scrub and hedge banks, waysides, railway banks and sidings, rough grassy habitats, waste ground and quarries, preferring well-drained soils, especially those over a sand, gravel or chalk substrate. The ancient Greek settlement of Thapsos was near modern Syracuse, Sicily.

Hoary mullein *V. pulverulentum*, perennial, found in East Anglia over chalky soil on verges and railway banks, in old quarries and gravel-pits, in hedge banks and rough ground, and locally on coastal shingle; elsewhere in England and in Wales it is scattered as a casual on waste ground. *Pulverulentum* is Latin for 'powdery'.

Twiggy mullein *V. virgatum*, a possibly native biennial, found on dry banks, walls, field margins, rough grassland, pastures and sheltered sea-cliffs in south-west England; elsewhere a casual of waste ground, verges, tracks and disturbed dunes. *Virgatum* is Latin for 'twiggy'.

White mullein *V. lychnitis*, a biennial, occasionally short-lived perennial locally common in southern England, a scattered casual or naturalised escape elsewhere in Britain, on dry, usually calcareous soil, in rough pasture, recently cleared woodland and waste ground, and on railway embankments, tracksides and verges. *Lychnitis* is Latin for 'like (the plant) *Lychnis*'.

MULLET **Thick-lipped (grey) mullet** *Chelon labrosus* (Mugilidae), length 30–70 cm, weight up to 4.5 kg, widespread though especially common in south Scotland and the English Channel, a demersal catadromous species usually found in shallow inshore waters or entering brackish lagoons and fresh water, including around power stations and sewage outfalls. It swallows mud to extract diatoms, epiphytic algae, small invertebrates and detritus. Spawning is in spring. It is a sport and seine fishery species. 'Mullet' dates from late Middle English, from Old French *mulet*, ultimately from Greek *myllos* = 'red mullet'. Greek *chelon* = 'tortoise', and Latin *labrosus* = '(thick)-lipped'.

MUNG-BEAN *Vigna radiata* (Fabaceae), an introduced Asian vegetable, casual on waste ground, especially near docks and factories in southern England. Dominico Vigna (d.1647) was an Italian botanist. *Radiata* is Latin for 'spreading rays'.

MUNTJAC **Reeves's muntjac** *Muntiacus reevesi* (Cervidae), shoulder height 45–52 cm, head-body length 77–90 cm, tail 13–18 cm, male weight 12–17 kg, females 10–16 kg, a deer with characteristic barking vocalisation, from China, first successfully introduced to Woburn, Bedfordshire in 1894, deliberately released into surrounding woodlands from 1901 onwards. Releases, translocations and escapes from other localities from the 1930s onwards resulted in wide establishment in south-east England, and it is still spreading, now reaching the West Country and south Yorkshire. The estimated population in England in 2005 was 40,000, with a few in Scotland (<50) and Wales (<250). In 2008 they were found Northern Ireland and reported from Co. Wicklow. They favour deciduous woodland with a developed understorey, and unthinned young conifer plantations, but have also become adapted to some suburban habitats, for example gardens (viewed as a pest), hedgerows and railway embankments, feeding on shrubs and herbaceous plants, with bramble and raspberry important. Mating and birth take place year-round. Gestation lasts 210 days. Females average three kids every two years. 'Muntjac' comes from Malay *menjangan* and Javanese *mindjangan* = '(small) deer'. It is named after the naturalist John Reeves (1774–1856).

Chinese muntjac Alternative name for Reeves's muntjac.

MUREX **Hedgehog murex** Alternative name for oyster drill.

MURIDAE Family name for mice (Latin *mus* = 'mouse') and rats, some taxonomists including Cricetidae.

MUSCICAPIDAE Family name for Old World flycatchers (Latin *musca* + *capere* = 'a fly' + 'to seize').

MUSCIDAE Family name of house flies (Latin *musca* = 'fly'), with 288 species in 40 genera. Habitats and feeding habits vary. Larvae mainly develop in decaying plant material or manure. Some are vectors of disease, particularly of livestock. See fly and Diptera.

MUSHROOM As with 'toadstool', a botanically non-meaningful term for the fleshy, spore-bearing fruiting body of a fungus, typically produced above ground on soil or on its food source. Many species with 'mushroom' as part of their common name belong to the genus *Agaricus*, from Greek *agaricon* = 'mushroom' or 'tree fungus' (Agaricales, Agaricaceae). See also oyster (fungus).

Blushing wood mushroom *Agaricus silvaticus*, edible and tasty but infrequently found, though widespread, evident in late summer and autumn on coniferous woodland soil, especially under spruce. *Silvaticus* is from Latin for 'woodland'.

Button mushroom See cultivated mushroom below.

Clustered mushroom *A. cappellianus*, with a couple of records from Avon. *Capellianus* is Latin for 'relating to goats'.

Cultivated mushroom *A. bisporus*, in the wild an uncommon, edible and tasty fungus with a scattered distribution in England, Wales and Ireland, evident from early summer to autumn on compost, manure and manured soil. The commercially cultivated variety is *A. bisporus* var. *bisporus*. *Bisporus* is Latin for 'two-spored', there being two spores on each basidium, other *Agaricus* having four-spored basidia.

Fairy cake mushroom Alternative name for poisonpie.

Fairy ring mushroom See fairy ring champignon under champignon.

Field mushroom *A. campestris*, widespread, common, edible and tasty, found in summer and autumn, trooping on soil in grassy habitats such as pasture. Latin *campus* = 'field'.

Horse mushroom *A. arvensis*, widespread, common, edible and tasty, found in late summer and autumn on soil in grassy habitats such as pasture. The common name may come from the affinity with horse manure, and its common occurrence near stables or fields in which horses graze; another suggestion is that 'horse' was used to mean 'coarse', or less palatable than field mushroom. *Arvensis* is from Latin for 'field'.

Inky mushroom *A. moelleri*, inedible, with a locally frequent but scattered distribution in England, much less frequent in Wales and Ireland, appearing in late summer on soil in deciduous woodland, with a preference for calcareous conditions. Alfred Möller (1860–1921) was a German mycologist.

Lilac mushroom *A. porphyrizon*, inedible (possibly poisonous), with a locally frequent but scattered distribution in England and south Wales, on woodland soil. *Porphyros* is Greek for 'russet' or 'purple'.

Macro mushroom *A. urinascens*, edible and tasty, widespread but local and rare, found from early summer to autumn in permanent pasture, verges, woodland edges and grassy woodland clearings. *Urinascens* is Latin for 'with urine'.

Magic mushroom Alternative name for liberty cap.

Pavement mushroom *A. bitorquis*, uncommon, edible and tasty, with a scattered distribution in England, very rare elsewhere, evident from early summer to autumn favouring sandy soils, often on roadsides, but also on manured soil, and noted growing through asphalt. *Bitorquis* is Latin for 'two-collared', a reference to the double ring that results when the partial veil covering the young gills tears from the cap rim.

Rosy wood mushroom *A. dulcidulus*, uncommon but edible with a scattered distribution, appearing in late summer and early autumn, on humid soil with a rich layer of decomposing oak leaves. *Dulcis* is Latin for 'sweet'.

Salty mushroom *A. bernardii*, uncommon though edible, with a scattered distribution in England, found coastally and in fields and roadsides, its tolerance of (roadside) salt probably favouring this last habitat.

Scaly wood mushroom *A. langei*, edible, with a locally frequent but scattered distribution in England, much less frequent elsewhere in the British Isles, appearing from early summer to early autumn on soil in coniferous or mixed woodland.

Soft slipper mushroom Alternative name for peeling oysterling.

St George's mushroom *Calocybe gambosa* (Lyophyllaceae), widespread but local, edible and tasty, found in spring (traditionally fruiting on St George's Day) in pastures or, less commonly, in mixed woodland or in verges close to hedgerows. Greek *calos* + *cybe* = 'beautiful' + 'head', and Latin *gambosa* = 'club-footed'.

Strawberry mushroom Alternative name for plums and custard.

Tufted wood mushroom *A. impudicus*, edible, with a locally frequent but scattered distribution, appearing in late aummer and autumn on soil in coniferous woodland. *Impudicus* is Latin for 'immodest'.

Wood mushroom *A. silvicola*, widespread but infrequent, edible and tasty, found in late summer and autumn on woodland soil. *Silvicola* is from Latin *sylva* + *colo* = 'tree' + 'to inhabit'.

MUSK *Mimulus moschatus* (Phrymaceae), a herbaceous North American neophyte perennial introduced in 1826, scattered but scarce in the Midlands, East Anglia and Ireland, casual or naturalised in damp, often shaded places, including muddy edges of ditches, damp woodland rides, by ponds and in damp pasture. The binomial comprises the diminutive of Greek *mimos* = 'imitator', and Latin for 'musk-scented'.

MUSK-MALLOW *Malva moschata* (Malvaceae), a perennial herb, native to England and Wales, probably introduced in (most of) Scotland and Ireland, in verges, hedge banks, woodland edges, pasture, field borders, riverbanks and grassy waste ground, preferring well-drained soils in unshaded or lightly shaded situations. It is tolerant of moderate levels of grazing or mowing. The binomial is Latin for 'mallow' and 'musk-scented'.

Greater musk-mallow *Malva alcea*, a herbaceous perennial European neophyte present by Tudor times, scattered in England and south Wales, local as a naturalised garden escape on woodland edges, hedgerows, roadsides and chalk-pits. *Alcea* is Greek for 'mallow'.

MUSKRAT *Ondatra zibethicus* (Cricetidae), head-body and tail lengths both 20–35 cm, weight 0.6–2 kg, a North American semi-aquatic rodent introduced to and escaped or released from fur farms during the 1920s. Most were soon recaptured, but some persisted, damaging riverbanks, including on the rivers Wey (Surrey) and Arun (West Sussex), and a group near Shrewsbury (Shropshire) that escaped in 1930 had, by 1931, spread along 30 km of the Severn, later extending to 60 km from Welshpool (Montgomeryshire/Powys) to Bewdley (Worcestershire). Intensive trapping programmes in 1932–34 managed to eradicate the species from England. Similar stories of escape and recapture come from east-central Scotland and Co. Tipperary. The English name may have an Algonquian origin, *muscascus* = 'it is red', or Abenaki *moskwas*, evidenced in the Old English name for the animal, 'musquash'. Also because of its musky odour and a resemblance to rats, it later became 'muskrat'. *Ondatra* comes from the Latinised Huron for the animal, *ondathra*, and Latin *zibethicus* = 'musky'.

MUSLIN (MOTH) *Diaphora mendica* (Erebidae, Arctiinae), wingspan 28–38 mm, common in England, Wales, parts of Ireland and south-east Scotland, males flying by night, females by day, in May–June in open woodland, hedgerow, downland and suburban habitats, larvae feeding on herbaceous plants. The binomial comprises Greek for 'distinction' and Latin for 'beggar', from the sexual dimorphism – the drab brown male in contrast to the white female which prompts the English name.

Round-winged muslin *Thumatha senex*, wingspan 15–20 mm, widespread but local in fen, bog, moorland, damp grassland and marsh, especially in East Anglia. Adults fly at night in July–August. Larvae feed on lichens, especially dog lichen, and mosses. *Thumatha* appears to be a meaningless neologism; *senex* is Latin for 'old', possibly from the thin wing scaling.

MUSSEL Filter-feeding bivalve molluscs in various families; see Dreissenidae, Margaritiferidae, Mytilidae (see also horse mussel), Sphaeriidae (see orb and pea mussels) and Unionidae.

Blue mussel *Mytilus edulis* (Mytilidae), shell size up to 15 cm, common and widespread on all coasts in the intertidal and sublittoral down to 5 m on the rocky shores of open coasts attached to the rock surface by the byssal thread, and on rocks and piers in sheltered harbours and estuaries, often found as dense masses. There are large commercial beds in the Wash, Morecambe Bay,

Conway Bay and the estuaries of south-west England, north Wales, and west Scotland. Commercial production in the UK in 2012 was 26,000 tonnes, worth £27.3 million. Greek *mytilos* = '(sea) mussel', and Latin *edulis* = 'edible'.

Depressed river mussel *Pseudanodonta complanata* (Unionidae), shell length 5–8 cm, scattered east of a line from Somerset through the Welsh Borders to south Yorkshire, Near Threatened in the RDB and a priority species in the UK BAP. Some of the largest populations in Europe are in East Anglia. It is found in lentic habitats or slowly flowing rivers, on fine sand or muddy substrates up to 11 m depth, and is very sensitive to pollution and eutrophication. It uses a variety of fish species as larval hosts. Greek *pseudo* = 'false' + *Anodonta* (see duck mussel below), and Latin *complanata* = 'flattened'.

Duck mussel *Anodonta anatina* (Unionidae), shell length 8–15 cm, widespread in England, more scattered elsewhere (in Ireland along the Shannon–Erne system and in Lough Neagh), found in slowly flowing rivers, lakes and creeks in a muddy or sandy substrate usually at 2–3 m depth. There are a number of host fish for larval development. Greek *ano* + *odontos* = 'upward' + 'tooth', and Latin *anatina* = 'relating to a duck'.

Freshwater pearl mussel *Margaritifera margaritifera* (Margaritiferidae), occasionally producing a pearl, shell length up to 15 cm, locally common on coarse sand or fine gravel in clean fast-flowing rivers in Scotland and non-central Ireland. A large, thick-shelled form (var. *durrovensis*) is unique to Ireland, confined to the River Suir and now close to extinction. It has been reduced to single populations in England and Wales. Larvae (glochidia) attach to the gills of juvenile salmonids during summer, encysting in the gills of the fish and carried around until the following spring when they hatch and fall as spat to the bottom of the river to start life as a mussel. Over half the world's population exists in Scotland, with populations in over 50 rivers, mainly in the Highlands, though 75% of sites surveyed in 2010 had suffered 'significant and lasting criminal damage'. It has been protected since 1998. Greek *margarites* = 'pearl' and Latin *fero* = 'to bear'.

Mediterranean mussel *Myt. galloprovincialis*, shell size up to 15 cm, found in the intertidal and sublittoral in south-west England, south-west Ireland and south Wales.

Northern bay mussel *Myt. trossulus*, shell size up to 15 cm, found in intertidal Loch Etive, west Scotland, probably a postglacial relict restricted to the low salinity area of the loch.

Painter's mussel *Unio pictorum* (Unionidae), shell length 10 cm, occasionally up to 14 cm, widespread and locally common in England, edging into east Wales, in rivers, sometimes in lakes, reservoirs and canals, mainly in lowland regions, inside a sandy or silty substrate, up to 6 m depth, and avoiding muddy and stony bottoms. In the past the shell was used as a conveniently sized and shaped receptacle for holding artist's paint, giving the common name. Latin *unio* 'single pearl', and *pictor* = 'painter'.

Scottish pearl mussel See freshwater pearl mussel above.

Swan mussel *A. cygnea*, shell length 10–20 cm, widespread and locally common in England, scattered elsewhere, mostly in lentic habitats with muddy substrates and little or no vegetation. Red-listed in Ireland as Vulnerable and as a Priority Species in Northern Ireland, having been displaced by drainage schemes, excess siltation, eutrophication and, recently, by the invasive zebra mussel. When bitterling, a fish, comes close to this mussel it becomes host to its minute, parasitic, glochidium larvae; this is used in larval dispersal. *Cygnus* is Latin for 'swan'.

Swollen river mussel *U. tumidus*, shell length 8 cm, occasionally up to 12 cm, widespread but declining in England, edging into north-east Wales, in clean, well-oxygenated, slowly flowing rivers and lakes, favouring sandy substrates up to 9 m deep. *Tumidus* is Latin for 'swollen'.

Zebra mussel *Dreissena polymorpha* (Dreissenidae), native of the Black and Caspian seas, shells (typically 20 mm long, maximum 40 mm), often striped, giving them their English name, first recorded in England at Wisbech in 1825, probably entering via imported Baltic timber, and in Scotland at Edinburgh in 1833. By the 1850s it had spread throughout the canal system of England and Wales. It was first recorded in Ireland in 1994. British numbers declined during the 1980s, but since 1999 there has been an increase in abundance throughout much of southern and eastern England. Adults can attach to boats and are then transported to new sites. It has been found in slow rivers, canals, docks, reservoirs and pipelines, attaching in dense colonies to solid surfaces such as stones, native mussels and canal walls using byssal threads. It can colonise soft sediments once dead zebra mussel shells have accumulated to create a suitable substrate. It can have an adverse impact on native species, and has become a major pest in water treatment works, power station cooling system intakes and aquaculture facilities by clogging pipes, filters and turbines. *Polymorpha* is Greek for 'many-shaped'.

MUSTARD Species of *Sinapis*, *Sisymbrium* and other genera (Brassicaceae). See also charlock and treacle-mustard.

Ball mustard *Neslia paniculata*, an annual neophyte introduced from Europe by the seventeenth century, casual, mostly in England, waste ground as an impurity of grain and birdseed, now rare following improved seed-cleaning techniques. *Paniculata* is Latin for 'with panicles'.

Black mustard *Brassica nigra*, common and widespread, especially in England, Wales and south-east Ireland, an often persistent annual by rivers as a member of the tall-herb community in the flood zone, and on sea-cliffs and shingle. It is also widespread as a casual on roadsides and waste ground, and on arable field margins. The binomial is Latin for 'cabbage' and 'black'.

Chinese mustard *B. juncea*, an annual east Asian neophyte herb, present by 1710, a rare and declining casual of waste ground, derived from wool shoddy and birdseed. *Juncea* is Latin for 'rush-like'.

Garlic mustard *Alliaria petiolata*, a common and widespread herbaceous biennial found in a range of generally shaded habitats, for example disturbed woodland, woodland edge and clearings, hedge banks, verges, waste ground and gardens, with a preference for fertile, moist soils, but only avoiding very acidic conditions. *Alliaria* is Latin for 'of the garlic family', *petiolata* for having petioles.

Hare's-ear mustard *Conringia orientalis*, a casual annual sporadically found in arable fields and waste ground, usually derived from birdseed. *Conringia* honours the German intellectual Hermann Conring (1606–81). Latin *orientalis* = 'eastern'.

Hedge mustard *Sisymbrium officinale*, a common and widespread herbaceous annual or biennial archaeophyte of dry, neutral or base-rich soils, preferring open habitats and frequent in cultivated ground, and on hedge banks, roadsides and waste ground. *Sisymbrium* is an Ancient Greek name for several plants; *officinale* means possessing pharmacological properties.

Hoary mustard *Hirschfeldia incana*, an annual or short-lived perennial European neophyte herb, cultivated by 1771 and also introduced with wool shoddy, increasingly naturalised, especially in England and Wales, in places such as railways, roadsides and waste ground. It is often associated around docks with grain imports and birdseed, and frequently occurs as a casual. C. Hirschfeldt was an eighteenth-century horticulturalist. Latin *incana* = 'hoary'.

Horned mustard *S. polyceriatum*, an annual European neophyte, initially rare as a grain impurity, now increasing in range in parts of England. Greek *poly* + *ceras* = 'many' + 'horn'.

Russian mustard *S. volgense*, a rhizomatous herbaceous perennial neophyte first recorded in the wild via grain contamination in 1896 at Bristol, naturalised on roadsides, docks and waste ground, declining but still found in a few places in England. Latin *volgense* means from the Volga region.

Tower mustard *Turritis glabra*, a biennial, rarely perennial, herb, recorded from around 30 sites, local and declining

in England (but thriving in Breckland), occasional in Wales and casual in Scotland, on grassy habitats such as verges and on disturbed (including waste) ground on free-draining sandy soils over chalk and limestone. Latin *turris* = 'tower', and *glabra* = 'smooth'.

White mustard *Sinapis alba*, an annual archaeophyte grown for fodder and green manure, for mustard seeds, or as a salad plant, naturalised or casual in arable fields and on waste ground and roadsides, especially on calcareous soil, widespread and locally common except in the north. The binomial is Latin for 'mustard' and 'white'.

MUSTELIDAE Weasel family (Latin *mustelus* = 'weasel' or 'stoat', in turn possibly from *mus* = 'mouse', viz. 'mouse-taker'). See also badger, ferret, marten, mink, otter, polecat and stoat.

MUTILLIDAE Probably from Latin *mutilus* = 'maimed', the family name of velvet ants.

MYCETOPHAGIDAE Family name of hairy fungus beetles (Greek *mycetos* + *phago* = 'of fungi' + 'to eat'), generally 1–6 mm length, with 15 British species in 6 genera, usually associated with fungi and found in rotting wood and decaying plant material. These beetles are mostly distributed in England and Wales, some with very few records.

MYRIAPODA The subphylum comprising the Chilopoda (centipedes) and Diplopoda (millipedes) (Greek *myrios* + *podos* = 'myriad' + 'of feet').

MYRICACEAE Family name for bog-myrtle and bayberry (Greek *myrice* = 'tamarisk').

MYRTACEAE Family name for myrtles (Latin *myrtus*): species of gum (eucalyptus) and Chilean myrtle.

MYRTLE Chilean myrtle *Luma apiculata* (Myrtaceae), an evergreen neophyte shrub or small tree, introduced in 1844, grown in gardens in mild areas and becoming naturalised in woodland, hedgerows and scrub and on roadsides in south-west England, south-west Ireland and the Isle of Man. *Apiculata* is Latin for 'with a short point'.

MYSIDA From Greek *mysis* = a closing of the eyes or lips, an order of marine crustaceans in the class Malacostraca; see opossum shrimp.

MYSTICETI From Greek *mystis* + *cetos* = 'mystic' + 'whale', the suborder of baleen whales (rorquals), families Balaenidae and Balaenopteridae.

MYTILIDAE Family name for bivalves; see mussel, horse mussel and crenella.

MYXOMA VIRUS, MYXOMATOSIS From Greek *myxa* = 'mucus', a tumour of connective tissue containing mucus or gelatinous material. The myxoma virus causes myxomatosis, the disease affecting rabbits which develop skin tumours, and in some cases blindness, followed by fatigue and fever; they usually die within 14 days of contracting the disease. The disease is spread by contact with an affected animal or by being bitten by fleas that have fed on an infected rabbit. The virus was introduced from South America to France in 1952 to control rabbits, and accidentally reached England in 1953. Spreading rapidly, >99% of British rabbits died within a few years. The few resistant animals, however, survived and bred. The rabbit population has now somewhat recovered, but the disease resurges from time to time as different strains evolve. See also rabbit.

N

NAIAD Aquatic herbaceous species of *Najas* (Hydrocharitaceae) (Greek *naias*, a river nymph).

Holly-leaved naiad *N. marina*, annual, growing in mesoeutrophic water over deep substrates of peat or silty mud in the Norfolk Broads, a decline due to eutrophication now reversed as measures to reduce excessive nutrient input take effect.

Slender naiad *N. flexilis*, annual, usually found in deep, clear, mesotrophic lakes where the water receives base enrichment from nearby basalt, limestone or calcareous dune sand. It has apparently been lost from its only English site, Esthwaite Water (Lake District), and some lakes in east Scotland because of eutrophication, but is still found in the Hebrides and western Ireland. *Flexilis* is Latin for 'flexible'.

NAIL FUNGUS *Poronia punctata* (Xylariales, Xylariaceae), a disc fungus scattered in England and Wales in rough pasture on horse dung, evident over July–November, Near Threatened in the RDL (2006). Looking a bit like a broad-headed nail gives its common name. *Poronia* is probably from Latin *porus* = 'pore', from holes on the top of the cap, and *punctatus* = 'punctured'.

NANOPHYIDAE Family (Greek *nano* + *phye* = 'small' + 'growth') that includes two species of weevil: *Nanophyes marmoratus*, widespread on purple-loosestrife in wet areas; *Dieckmanniellus gracilis*, with a few scattered records in southern England, associated with water purslane.

NARCISSUS FLY **Large narcissus fly** Alternative name for the hoverfly narcissus bulb fly.

NARTHECIACEAE Family name for bog asphodel (Greek *narthex* = 'hollow stem').

NASTURTIUM 1) *Tropaeolum majus* (Tropaeolaceae), a common herbaceous annual neophyte of garden origin, a frequent casual on waste ground, and naturalised on Sark. **Chilean flame nasturtium** *T. speciosum* is an occasional casual, scrambling through bushes and hedgerows. Greek *tropaion* = 'trophy', referring to the shape of the flowers.

2) Genus name of water-cresses (Brassicaceae). *Nasturtium* comes from Latin *nasus tortus* = 'twisted nose', because of its taste.

NAVEL (FUNGUS) Examples of Agaricales, so called from the central depression in mature caps suggesting a navel.

Rarities include **golden navel** *Chrysomphalina chrysophylla* (Hygrophoraceae), **liverwort navel** *Loreleia marchantiae* (Agaricomycetes), with mosses and liverworts, penetrating the rhizoids of common liverwort possibly as a symbiosis, and **orange navel** *Haasiella venustissima*, **slender navel** *Fayodia bisphaerigera* and **verdigris navel** *Arrhenia chlorocyanea* (all Tricholomataceae).

Cinnamon navel *Omphalina pyxidata* (Tricholomataceae), inedible, widespread and scattered on grassland and semifixed dunes. Greek *omphalos* = 'navel', and the Latin *pyxidata* = 'box-like'.

Heath navel *Lichenomphalia umbellifera* (Hygrophoraceae), common in upland parts of Wales, Cumbria and the Highlands, on acid unimproved grassland, heathland and moorland, and on stumps in mixed woodland. Greek *leichen* + *omphalos* = 'lichen' + 'navel', and Latin *umbellifera* = 'bearing umbels'.

NAVELWORT *Umbilicus rupestris* (Crassulaceae), a perennial stonecrop frequent in south-west England, Wales, central-west Scotland and Ireland, naturalised elsewhere, on walls, in rock crevices and on stony hedge banks, mainly on acidic substrates. In Cornwall it is sometimes epiphytic on tree branches. The English and genus names reflect the umbilicate (navel) shaped round leaves with their central depression. *Rupestris* is from Latin for 'rock'.

NEB (MOTH) Generally nocturnal members of the family Gelechiidae, mostly in the subfamily Anomologinae. 'Neb' is a Scottish or northern English word for 'nose'.

Black neb *Monochroa lutulentella*, wingspan 14–16 mm, uncommon, found in fen and marsh and on riverbanks in England south and east of a line from Devon to Herefordshire and the Wash; also in Derbyshire, Monmouthshire and on the Burren. Adults fly in June–August. Larvae probably feed on meadowsweet roots. Greek *monos* + *chros* = 'single' + 'surface (colour)', and Latin *lutulentus* = 'muddy', from the brown forewing.

Bracken neb *Mo. cytisella*, wingspan 10–12 mm, widespread but local in woodland and on heathland, moorland and hillsides in southern England, the Midlands and Wales, becoming more localised in Ireland and northern England, and rare in south Scotland. Adults fly in June–August. Larvae feed in May–June in bracken stems, though the specific *cytisella* is erroneously from *Cytisus* (broom) as a food plant.

Bright neb *Argolamprotes micella*, wingspan 10–14 mm, local in hedgerows, open woodland and gardens in south-west England and coastal south Wales with a few scattered records in Hampshire, Wiltshire and East Sussex, slowly extending its range eastwards and northwards from its first British record in Devon in 1963. Adults fly in June–July. Larvae feed on raspberry and blackberry in April–May. Greek *argos* = 'shining' + the genus *Lampros*, and Latin *mico* = 'to shine', from the glossy forewing.

Brilliant neb *Metzneria aprilella*, wingspan 15–18 mm, local on mainly calcareous grassland, verges and waste ground in central-southern England, very local in parts of eastern and south-east England, the Midlands and south Yorkshire, and rare in south Wales. Adults fly in May–August. Larvae feed from September to March inside greater knapweed seedheads. *Metzneria* honours the German entomologist Metzner (d.1861). *Aprilella* refers to April, reflecting the moth's spring-fresh beauty, not season of emergence.

Brown-veined neb *Me. neuropterella*, wingspan 14–24 mm, local on chalk downland in south-east England. Adults fly in June–August. Larvae feed from October to May in the seedheads of dwarf thistle and common knapweed. Greek *neuron* + *pteron* = 'vein' + 'wing'.

Buff-marked neb *Mo. lucidella*, wingspan 12–14 mm, local, occasionally common, in marsh, water meadow and fen and by riverbanks in England and Wales, very local in Ireland, and rare in south-west Scotland. Adults fly in late afternoon in June–August. Larvae feed in May–June inside common spike-rush stems. *Lucidus* is Latin for 'bright', from the forewing markings.

Burdock neb *Me. lappella*, wingspan 16–20 mm, local in field borders, hedgerows, verges, rough pasture, downland, dunes, scrub, waste ground, gardens and woodland edge throughout England and north Wales; rare in Scotland. Adults fly in May–August. Larvae feed in September–April on seedheads of lesser and greater burdock (*Arctium lappa*), the latter prompting the specific name.

Carline neb *Me. aestivella*, wingspan 13–16 mm, in calcareous grassland, and some dunes and quarries, throughout much of southern and eastern England and south Wales; rare in northern England, north Wales and south-east Ireland. Adults fly in June–July. Larvae feed in October–April on carline thistle seedheads, overwintering in a silk chamber. *Aestivus* is Latin for 'relating to summer'.

Coast neb *Mo. moyses*, wingspan 8–9 mm, scarce, local on the edges of saltmarsh and ditches on the coast of south East Anglia and south-east England, from Suffolk to Hampshire. Adults fly in June–July. Larvae mine sea club-rush leaves in August–September. Moyses is an alternative spelling of (Old Testament) Moses, the baby hidden in bulrushes, from larvae found mining vegetation in borrow-dykes.

Common plain neb *Mo. tenebrella*, wingspan 10–12 mm, day-flying, widespread but local in scrub, unimproved grassland and heathland. Adults fly in June–July. Larvae feed on the roots and lower stems of sheep's sorrel. *Tenebra* is Latin for 'darkness'.

Dark fleabane neb *Apodia bifractella*, wingspan 9–12 mm, locally common in damp meadow, open downland and salt-marsh in southern and eastern England; local in the Midlands, northern England and Wales. Adults fly in late afternoon in July–September. Larvae feed in October–April inside seed heads of common fleabane, ploughman's-spikenard and sea aster. Greek *a + podos* = 'without' + 'foot', larvae being almost apodal, and Latin *bi + fractus* = 'two' + 'broken', from discontinuous wing markings.

Dingy neb *Mo. conspersella*, wingspan 11–12 mm, scarce in fen and marsh in eastern and south-east England, but only recorded with any regularity at Wicken, Woodwalton and Catfield Fens, Adults fly in July. Larvae feed within stems and leaves of yellow loosestrife in September–May. *Conspersus* is Latin for 'sprinkled', from forewing markings.

Flame neb *Chrysoesthia drurella*, wingspan 8–9 mm, locally common on waste ground, farmland and other open areas over much of southern England and the Midlands, very local in south-west and northern England, parts of Wales and the Channel Islands. The bivoltine adults fly in May–June and August–September. Larvae mine leaves of goosefoot and orache. Greek *chrysos + esthes* = 'gold' + 'clothing', from yellow forewing markings. Dru Drury (1775–1803) was a British entomologist.

Heather neb *Aristotelia ericinella*, wingspan 12–13 mm, locally common on heaths, mosses and moorland in England and Wales, very local in central Ireland, rare in Scotland. Adults fly in July–August from the afternoon onwards. Larvae feed in June–July on heather, living between shoots spun together with silk. *Aristotelia* is from the Greek philosopher Aristotle; Latin *erica* = 'heath' (rather than heather).

Isle of Wight neb *Me. littorella*, wingspan 9–14 mm, confined to open ground on the south coast of the Isle of Wight, where it has persisted for over 150 years on sparsely vegetated rock under-cliff and landslips where vegetation has started to recolonise; also recorded from a site each in East Sussex and Guernsey. Adults fly in May–June. Larvae feed in July–March inside buck's-horn plantain seedheads. *Littoris* is Latin for 'of the seashore'.

Kentish neb *Mo. niphognatha*, wingspan 13–15 mm, with regular records only from a single freshwater reedbed in Kent (first recorded in 1984), with single records from Southsea, Hampshire (2009) and south Devon (2002). Adults fly in June–July. Larvae mine amphibious bistort stems. Greek *niphas + gnathos* = 'snow' + 'jaw', from the white labial palpus.

Knotweed neb *Mo. hornigi*, wingspan 9–14 mm, rare, in a variety of habitats, for example gardens, parks and damp open areas, in England from a line from Wiltshire to Bedfordshire and in south Yorkshire; probably more widespread and possibly fairly common, but elusive. Adults fly in June–August. Larvae mine pale persicaria stems in September–April. Johann von Hornig (1819–86) was an Austrian entomologist.

Light fleabane neb *Ptocheuusa paupella*, wingspan 10–12 mm, locally common in damp grassland, ditches, woodland rides and saltmarsh in England south of a line from the Wash to the Severn estuary, rare further north and west with a few records from Nottinghamshire, Yorkshire, north Wales and south-east Ireland. The bivoltine adults fly in June and August–September. Larvae feed on the seedheads of a number of plants, especially common fleabane and golden-samphire, from September to April, and again in summer. Greek *ptocheuo* = 'to be a beggar', and Latin *pauper* = 'poor', possibly from the wan forewing colour.

Meadow neb *Me. metznieriella*, wingspan 14–19 mm, common on grassland, downland, waste ground and verges throughout England, Wales and central lowland Scotland, and locally in Ireland. Adults fly in May–August. Larvae are evident from October to April on seeds of common knapweed and saw-wort.

Northern neb *Xystophora pulveratella* (Gelechiinae), wingspan 10–11 mm, local in upland grassland and grass-heath in parts of the Highlands. Adults fly in May–June. Larvae feed on bird's-foot-trefoil and bitter-vetch among spun leaves in August–September. Greek *xyston + phoreo* = 'something scraped'

+ 'to carry', and Latin *pulveratus* = 'dusted', from forewing sprinkle markings.

Notch-wing neb *Mo. suffusella*, wingspan 10–12 mm, scarce in bog, fen, marsh, riverbanks, damp areas and saltmarsh in parts of north Wales, southern England and East Anglia, as far north as Yorkshire, and a few Irish records. Adults fly from late May to July. Larvae generally feed inside cottongrass. *Suffusus* is Latin for 'suffused'.

Painted neb *Eulamprotes wilkella*, wingspan 8–10 mm, local on shingle beaches on the coast, occasionally inland on dry sandy areas, in parts of England, Wales and east Scotland; rare in south-east Ireland. The bivoltine adults fly in June and August. Larvae feed in March–May below ground in silk tubes on moss. Greek *eu* = 'well' ('affinity with') + the genus *Lamprotes*. B. Wilkes (d.1749) was an English entomological artist.

Pembroke neb *Mo. elongella*, wingspan 12–15 mm, scarce and endangered, on dunes in north Devon, south-east England, Norfolk, Pembrokeshire and Anglesey; also a few records on chalk downland. Adults fly in June–August. Larvae feed inside the stems of silverweed in April–May. *Elongella* is Latin for 'elongated', from forewing markings.

Saltern neb *Mo. tetragonella*, wingspan 9–11 mm, scarce and endangered, in drier parts of saltmarsh in Dorset, Essex, Norfolk, Lincolnshire, Co. Durham and Glamorgan. Adults fly in the evening in June–July. Larvae feed in April–May in stems and leaves of sea-milkwort. Greek *tetra + gonia* = 'four' + 'angle', from the pattern of the four forewing spots.

Scarce marsh neb *Mo. divisella*, wingspan 15–16 mm, in fen, marsh, riverbanks and other damp areas in East Anglia. Adults fly in June–July. Larvae have not been recorded in Britain. *Divisus* is Latin for 'divided', from contrasting colours on the forewing.

Sedge neb *Mo. arundinetella*, wingspan 9–10 mm, scarce in reed swamp, fen and marsh and by riverbanks in England, south of a line from the Severn estuary to the Wash. Adults fly in June–July. Larvae mine the leaves of pond-sedges in March–May. *Arundinetum* is New Latin for 'reedbed', from the habitat.

Silvery neb *Psamathocrita argentella*, wingspan 10–11 mm, endemic, known from a few saltmarshes in Dorset, Hampshire, Isle of Wight and West Sussex where it can be locally common. Adults fly in June–July. Larvae feed on flowers and seeds of sea couch from July to September. Greek *psamathos + critos* = 'sea sand' + 'chosen', and Latin *argentum* = 'silver', from the forewing colour.

Six-spot neb *Chrysoesthia sexguttella* (Apatetrinae), wingspan 8–10 mm, locally common on waste ground, farmland and other open areas in England and Wales, more local in Ireland and central and east Scotland; scarce in west and north Scotland. It has a coastal bias in the western and northern parts of the British Isles and within Ireland. The bivoltine adults fly in May–June and August. Larvae mine leaves of goosefoot and orache. Latin *sex + gutta* = 'six' + 'spot', there being usually three yellow spots on each forewing.

Thrift neb *Aristotelia brizella*, wingspan 9–10 mm, in salt-marsh on the south and east coasts of England, from Cornwall to Durham, and from Lancashire on the west coast. The bivoltine adults fly in May–June and July–August. Larvae feed in June–July and September–October on seedheads of thrift, occasionally sea-lavender. *Brizella* comes from *Briza* (quaking-grass), though the moth has no connection with the plant.

Twilight neb *Eulamprotes immaculatella*, wingspan 8–13 mm, in habitats including sea-cliffs, damp meadow, limestone pavement and grassland: widely distributed but with an extremely patchy distribution in England and Wales; more evenly recorded in the northern half of Scotland; and also found in the Burren, Ireland. Adults have been noted in sunshine from midday to early evening in June–September. Larval food plant is possibly slender St John's-wort.

Two-spotted neb *E. atrella*, wingspan 11–13 mm, local on downland, rough ground, dry pasture and in woodland clearings throughout much of England and Wales, rare in Ireland. Adults fly in July–August. Larvae feed in March–May in the stems of St John's-worts. *Atrella* is from Latin *ater* = 'black', from the dark forewing colour.

Unmarked neb *E. unicolorella*, wingspan 10–13 mm, a rarity on chalk downland, limestone grassland, dry herb-rich fields, waste ground and dry open areas throughout much of southern England, south-east Wales and western Ireland, occasional elsewhere. Adults fly in May–July. Larvae may feed within stems of perforate St John's-wort. *Unicolorella* is Latin for 'one-coloured'.

Wainscot neb *Mo. palustrellus*, wingspan 15–19 mm, on waste ground, dry pasture, verges and dunes, south-east of a line from Somerset to Norfolk. A few recent scattered records elsewhere in south-west England, south Wales and single records from the Midlands and Ireland may indicate a range expansion. Adults fly in June–August. Larvae feed on (especially curled) dock, feeding internally on roots and stems in May–July. *Palustris* is from Latin for 'marsh'.

White-border neb *Isophrictis striatella*, wingspan 11–13 mm, local on waste ground, verges, gardens, damp grassland and fen in central, eastern and southern England and the Channel Islands, very local in the rest of England, rare in Wales, and with one record in Scotland (near Aberdeen). Adults fly in July–August. Larvae feed in October–May inside the seedheads of tansy (drier habitats) and sneezewort (moister habitats). Greek *isos* + *phrisso* = 'equal' + 'to bristle', and Latin *stria* = 'streak', from the forewing markings.

NECKERA Species of moss in this genus (Neckeraceae), named after the Belgian botanist Noel de Necker (1730–93).

Crisped neckera *N. crispa*, common and widespread, less so in eastern Britain and the Midlands, on limestone and other base-rich rocks and walls, usually in lightly shaded places. It also grows on bark, and is occasionally noted on calcareous grassland.

Dwarf neckera *N. pumila*, common in south and south-west England, Wales, central-west Scotland and Ireland, scattered elsewhere, on trees and shrubs in sheltered woods and scrub, especially on ash, hazel, sycamore and willow. *Pumila* is Latin for 'dwarf'.

Flat neckera *N. complanata*, common and widespread on shaded, base-rich rocks, walls, and masonry, and on bark at the base of trees and on coppice stools in woodland in eastern England. It also colonises shrubs in the west of its range. *Complanata* is Latin for 'flattened'.

NEEDLE FLY Fisherman's name for leuctrid stoneflies.

NEEDLE GRASS The perennial **American needle-grass** *Nasella neesiana* (from South America), the perennial **Argentine needle-grass** *N. tenuissima* and the annual **Mediterranean needle-grass** *S. capensis* (Poaceae), all casual or naturalised in a few places in England, originating in wool shoddy.

NEMATOCERA From Greek *nematos* + *ceras* = 'thread' + 'horn', a suborder of Diptera with thin, segmented antennae and mostly aquatic larvae, including mosquitoes, craneflies, gnats, blackflies and midges.

NEMATODA, NEMATODE Roundworms (Greek *nematos* = 'thread'), a phylum with more species than insects themselves, parasitic on animals including humans. They are found on mountain tops, the arctic ice cap and on the ocean floor (and even in beer mats).

NEMATOMORPHA Phylum (Greek *nematos* + *morphe* = 'thread' + 'shape') superficially similar to Nematoda, commonly known as horsehair worms. Found in damp or wet habitats, adults are free-living, but larvae are parasitic on arthropods.

NEMERTEA Phylum containing ribbon worms. Nemertea was a nereid (sea nymph) in Greek mythology.

NEMONYCHIDAE Primitive weevil family (Greek *nema* + *onychos* = 'thread' + 'of a claw') represented in the British list by the

3–5 mm long *Cimberis attelaboides*, with a scattered distribution. Larvae feed on pollen within Scots pine catkins.

NEMOURIDAE A generally common and widespread stonefly family (Greek *nema* + *oura* = 'thread' + 'tail') associated with ponds and streams.

NEPTICULIDAE Family name for the often monophagous leaf-, seed- and bark-mining pygmy moths (Greek diminutive of *neptis* = 'granddaughter', from their small size), with 57 species.

NERITE Freshwater nerite or **river nerite** *Theodoxus fluviatilis* (Neritidae), shell width 5–9 mm (can reach 11–13 mm), height 4–6.5 mm, a common and widespread snail found in rivers, lakes and brackish water, especially in England and the Republic of Ireland, on stony substrates, grazing on diatoms and biofilm. *Nerites* is Greek for a kind of shellfish. Greek *theos* + *doxa* = 'god' + 'glory' or 'dignity', and Latin *fluvius* = 'river'.

NETCRUST Orange netcrust *Pseudomerulius aureus* (Boletales, Tapinellaceae), a scarce inedible fungus, with most records in Surrey, on dead pinewood in coniferous woodland or heathland. Greek *pseudes* = 'false' + the genus *Merulius*, and Latin *aureus* = 'golden'.

NETTLE Members of the genus *Urtica* (Urticaceae), *urtica* being Latin for these plants. 'Nettle' comes from Old English *netle* or *netele*, of Germanic origin (*Nessel*).

Common nettle *U. dioica*, a common and widespread rhizomatous and stoloniferous perennial herb found in a range of habitats, including woodland, scrub, unmanaged grassland, fen, riverbanks, hedgerows, roadsides, and cultivated and waste ground, preferring nutrient-rich soils. Subspecies *dioica* has conspicuous stinging leaf hairs (commonly known as **stinging nettle**); ssp. *galeopsifolia*, found in damp habitats, is stingless. *Dioica* is Latin, meaning having male and female flowers on separate plants.

Small nettle *U. urens*, a spring-germinating annual archaeophyte with stinging leaf hairs, common and widespread, especially in England, on well-tilled arable land, especially fields of broad-leaved crops, and allotments, gardens, farmyards and waste ground, preferring light, often sandy, soils of high fertility. *Urens* is Latin for 'stinging'.

NETTLE RASH *Leptosphaeria acuta* (Pleosporales, Leptosphaeriaceae), a common, widespread pathogenic fungus found at the base of dead stems of nettle, evident year-round but mainly maturing in late winter and spring. Greek *leptos* + *sphaera* = 'slender' + 'sphere', and Latin *acuta* = 'sharp' or 'pointed'.

NETTLE-TAP Moths in the family Choreutidae.

Common nettle-tap *Anthophila fabriciana*, wingspan 10–15 mm, widespread and common on waste ground, hedgerows and gardens. Adults fly during the day from May through summer. Larvae feed on nettle and pellitory-of-the-wall. Greek *anthos* + *phileo* = 'flower' + 'to love'. Johann Fabricius (1745–1808) was a Danish entomologist.

Miller's nettle-tap See small twitcher under twitcher.

NEUROPTERA Order (Greek *neuron* + *pteron* = 'nerve' + 'wing') that includes lacewings, in the families Chrysopidae (green lacewings, 23 British species), Coniopterygidae (7), Hemerobiidae (29), Osmylidae (see giant lacewing under lacewing), Sisyridae (sponge flies, 3 species, the aquatic larvae feeding on freshwater sponge).

NEW FOREST EYE Infectious bovine keratoconjunctivitis, a highly contagious eye infection of, especially, young cattle caused by the bacterium *Moraxella bovis*, with flies transmitting infection between individuals. It is the commonest eye disease in cattle in the UK.

NEWT Tailed amphibians in the genera *Lissotriton* (Greek *lissos* = 'smooth' + Triton, a sea demigod), *Mesotriton* (Greek *mesos* = 'middle') and *Triturus* (from Triton) all Salamandridae.

Alpine newt *M. alpestris*, length up to 12 cm, introduced from Europe by 1903 to Newdigate, Surrey, with colonies since reported from Kent, south-east London, east Surrey, Birmingham, Shropshire, Sunderland, Co. Durham and near Edinburgh, but these all remain at risk.

Common newt Alternative name for smooth newt.

Great crested newt *T. cristatus*, males (which develop a jagged crest in the breeding season) up to 15 cm long, females to 18 cm, widespread in England and Wales, local in lowland Scotland, breeding in well-vegetated ponds. Though usually only aquatic from March to August, some populations remain in ponds all year. More commonly adults leave their ponds in July–August and live on land until hibernation in September. The aquatic juveniles feed on protozoa and unicellular algae, taking small arthropods and worms as they develop. Adults are active at night, mostly bottom-feeding in the water, taking invertebrates, occasionally larger prey such as adult smooth newts and large dragonflies, and continuing to prey on invertebrates on land. These newts range widely from their breeding ponds, and can be found in a variety of habitats in late summer. Numbers declined during the twentieth century as a consequence of habitat change, though there an estimated 75,000 populations in the UK. In England and Wales the species is protected under the Conservation of Habitats and Species Regulations (2010) and the WCA (1981). In Scotland, it is protected under the Conservation (Natural Habitats, &c.) Regulations (1994). It is a priority species under the UK BAP and is a Species of Principal Importance in England in the NERC Act (2006) (section 42 in Wales), and in Scotland under the Nature Conservation (Scotland) Act (2004). It is an offence to capture, kill or injure this species, disturb it in a place used for shelter, or damage a breeding site or resting place. A licence is required for survey work. Relocation and mitigation have been tried with varying levels of success where permission has been given for habitat alteration or destruction. *Cristatus* is Latin for 'crested'.

Italian crested newt *T. carnifex*, an introduction with a few records of temporary escapes, with two extant populations: at Newdigate, Surrey, since the 1920s, and in the Birmingham area centred round several garden ponds, having spread just 600 m from the original introduction site. *Carnifex* is Latin for 'executioner'.

Palmate newt *L. helveticus*, length 7–11 cm, males smaller than females, widespread in Britain, breeding in shallow pools (including garden ponds) and larger bodies of water, more tolerant of acidic waters than smooth newt. They emerge from hibernation in early March. The breeding season continues until late May. In July the adults leave the water, and are fully terrestrial during August–September, preparing for hibernation by feeding on worms and other small invertebrates. They typically hibernate in deep leaf litter from late September. *Helveticus* is Latin for 'Swiss'.

Smooth newt *L. vulgaris*, common and widespread in England, Wales and the central Scottish lowlands, scattered elsewhere in Scotland and in Ireland, length 10–11 cm, males smaller than females. Having emerged from hibernation in March, they breed in ponds through to May, the males developing a wavy crest, later absorbed during the terrestrial phase. The female deposits her eggs individually on aquatic plants, wrapping each egg in a leaf. Adults leave the water in July to spend the rest of the summer and winter close to the breeding pond, hiding in leaf litter and long grass and under stones. They feed after dark on small invertebrates, preparing for hibernation in late September. Latin *vulgaris* = 'common'.

Warty newt Alternative name for great crested newt.

NIGER (or NYJER) *Guizotia abyssinica* (Asteraceae), an annual East African neophyte introduced in 1806, scattered in Britain, mainly England, as a casual on sewage farms, roadsides and waste ground, often originating from birdseed, but also grain, oil-seed and wool shoddy. Francois Guizot (1787–1874) was a French statesman.

NIGHT CRAWLER European night crawler *Eisenia hortensis* or *Dendrobaena veneta*, a lumbricid earthworm.

NIGHT HERON See black-crowned night heron under heron.

NIGHTINGALE *Luscinia megarhynchos* (Turdidae), wingspan 24 cm, length 16 cm, weight 21 g, a Red-listed bird found locally in deciduous woodland. BTO surveys in 2012 and 2013 recorded 5,500 males, most in Kent, Essex, Suffolk, and East and West Sussex. There was a 48% decline in 1995–2015, but no change was noted in 2010–15. Nests are on or near the ground in dense vegetation, with 4–5 eggs, incubation lasting around 14 days, and fledging of the altricial chicks taking 11–13 days. 'Nightingale' has been in use for over 1,000 years, with its Old English form *nihtgale* and *nihtegala* meaning 'night songstress' (though only the male sings). *Luscinia* is Latin for nightingale, and Greek *megas* + *rhynchos* = 'large' + 'bill'.

NIGHTJAR *Caprimulgus europaeus* (Caprimulgidae), wingspan 60 cm, length 27 cm, weight 83 g, an Amber List crepuscular and nocturnal bird arriving from Africa between late April and mid-May, leaving in August–September, found on heathland, moorland, in open woodland with clearings, and in recently felled conifer plantations, particularly numerous in southern England with relatively high numbers in the New Forest, Dorset and Surrey heathlands, and Thetford Forest in Suffolk. Also found in the Midlands (e.g. Cannock Chase, Staffordshire), parts of Wales, northern England and south-west Scotland, and occasional elsewhere. BBS estimated 4,600 (3,700–5,500) vocalising males in 2004. The (usually) 2 eggs are laid directly onto the ground, taking 18 days to hatch, the semi-precocial young fledging in 18–19 days. Feeding is on flying insects, especially moths and beetles. The English name refers to the bird's nocturnal activity and the churring vocalisation of the male. *Caprimulgus* is derived from Latin *capra* = 'goat' and *mulgere* = 'to milk', a reference to the myth that nightjars suck milk from goats (giving it the dialect name 'goatsucker'). **Egyptian nightjar** *C. aegyptius*, **red-necked nightjar** *C. ruficollis* and **common nighthawk** *Chordeiles minor* are all rare vagrants.

NIGHTSHADE Species of *Solanum* (Latin for 'nightshade') and *Atropa* (after Atropos, one of the three Fates, who cut the thread of life) (Solanaceae).

Casual, occasionally naturalised neophytes found on cultivated and waste ground are **garden nightshade** *S. scabrum* (probably from Africa), **green nightshade** *S. physalifolium* (South America, scattered as far north as central Scotland, but mainly eastern England), **leafy-fruited nightshade** *S. sarachoides* (South America, introduced in 1897), **red nightshade** *S. villosum* (Europe), **small nightshade** *S. triflorum* (North America, naturalised in Norfolk and the Cheviots), and **tall nightshade** *S. chenopodioides* (Argentina, naturalised on pavements and in rough grassland in the Channel Islands and, since 1989, the London area).

Black nightshade *S. nigrum*, a common and widespread annual weed, particularly in England, Wales and south-east Ireland, on cultivated and waste ground, especially on nutrient-rich soils. *Nigrum* is Latin for 'black'.

Deadly nightshade *A. belladonna*, a rhizomatous perennial herb, scattered if mainly in England, on dry disturbed ground, field margins, hedgerows and open woodland, native only on calcareous soils, but occurring on a wider range of soils as an introduction, often as a relic of cultivation as a medicinal herb. Every part of the plant is poisonous to humans (the root mostly so, berries least) because of the alkaloid atropine, though wildlife seems to be unaffected. This is a valuable plant in the treatment of eye diseases, atropine serving to dilate the pupil. Also, preparations, locally applied, lessen irritability and pain, and have been used as a lotion to counter neuralgia, gout, rheumatism and sciatica. *Belladonna* is Italian for 'beautiful lady'.

Sticky nightshade Alternative name for red buffalo-bur.

Woody nightshade Alternative name for bittersweet.

NIPPLEWORT *Lapsana communis* (Asteraceae), a common and widespread annual or perennial herb, growing in disturbed and shaded places, and thriving over a range of soil acidity and moisture, in open woodland, scrub, hedgerows, waste ground, gardens, railway banks, roadsides and old walls. 'Nipplewort' comes from the closed flower buds which were said to resemble nipples. *Lapsane* is Greek for this plant, *communis* Latin for 'ordinary'.

NIT-GRASS *Gastridium ventricosum* (Poaceae), an annual of well-drained grassland on calcareous substrates, possibly native in grassland in south-west Britain and in the Channel Islands. It is also a wool-shoddy and grain casual in southern and south-east England and south Wales. Greek *gaster* and Latin *venter* both mean 'stomach', describing the swollen base of the spikelets.

Eastern nit-grass *G. phleoides*, an infrequent south-west Asian neophyte annual of fields, docks and tips in England, originating from wool shoddy. Greek *phleos* means a kind of rush or reed.

NITIDULIDAE From the diminutive of Latin *nitidus* = 'bright', the family name of sap beetles and pollen beetles, 1–8 mm in length, with 91 British species in 16 genera in 5 subfamilies. **Carpophilinae** contains 31 species in 3 genera. *Carpophilus* are largely found in association with fungi, particularly on mouldy fruit, and several species have become stored-product pests. Several *Epuraea* species are found in association with wood-boring beetles in conifers; others develop in subterranean nests, particularly of bumblebees. **Cryptarchinae**, with six species in three genera, are mostly found on sap runs. **Cybocephalinae** consists of *Cybocephalus fodori*, a predator of scale insects, recorded from birch trees on Putney Heath, London. **Meligethinae** has 39 species in 2 genera that attack a wide range of plant families, but each beetle species is host-specific. Two exceptions are pollen beetle *Meligethes aeneus* and *M. viridescens*, have become pests on oil-seed rape and other brassicas. **Nitidulinae** has 14 species in 7 genera, with several (particularly *Nitidula* and *Omosita*) associated with dried carrion and bones, others found in fungi. See also pollen beetle.

NOCTUIDAE From Latin *noctus* = 'night', from the time of flight of most species, though some authorities argue the name comes from *noctua*, the Latin name of an owl sacred to Minerva, and this group has sometimes been referred to as 'owlets'. This is the family name for (among others) arches, brindle, brocade, chestnut, clays, coronet, dagger, dart, drab, ear, gothic, minor, pinion, quaker, ranunculus, rustic, sallow, shoulder-knot, square-spot, sword-grass, underwing and wainscot moths.

NOCTULE *Nyctalus noctula* (Vespertilionidae), head-body length 37–48 mm, wingspan 320–400 mm, weight 18–40 g, a relatively widespread bat in England, Wales and south-west Scotland, though it has become scarce in areas of intensive agriculture. The population was cautiously estimated in 1995 as 50,000 (45,000 in England, 4,750 in Wales, 250 in Scotland). Surveys in Britain up to 2015 indicate a 16.3% increase on a 1999 base year value, equivalent to an annual increase of 1%. They primarily frequent deciduous woodland, roosting in rot holes and woodpecker holes. They forage mainly at dusk for up to two hours and for about half an hour at dawn, flying at up to 50 kph, often above tree level (though they also hunt moths around street lamps), with steep dives when chasing insect prey. Young are born in maternal colonies in late June–July. Females usually have one young, independent within six weeks. The maternity colonies frequently change location, mothers carrying the young between roosts during lactation. The young are left in crèches while the mothers forage. During summer, males

are solitary or form small groups. Hibernation is mainly in tree holes or rock crevices. Echolocation calls in the range 20–45 kHz, peaking at 25 kHz. The English (and specific) name is late eighteenth-century, ultimately from Italian *nottola* = 'small night creature'. *Nyctalus* is from Greek *nyx* = 'night', Nyx also being the goddess of the night.

NODDING-MOSS Some species of *Pohlia* (Mielichhoferiaceae), named after the Austrian botanist Johann Pohl (1782–1834). **Crookneck nodding-moss** *P. camptotrachela* (Greek *camptos* + *trachelos* = 'bent' + 'neck') is a widespread calcifuge in upland areas, **pretty nodding-moss** *P. lescuriana* widespread and scattered in Britain, rare in south Ireland.

NOLIDAE Family name for black arches and related moths, named after Nola, a town in Campania, Italy.

NOMAD BEE Members of the genus *Nomada* (Apidae) (Greek *nomas* = 'wandering'), wasp-mimicking cuckoo bees that, as social parasites, lay their eggs in the nests of *Andrena* solitary bees, the larvae feeding on pollen stored by the host. Of the 28 species found in the British Isles, only 11 are abundant or widespread.

Armed nomad See scabious cuckoo bee under cuckoo bee.

Dark nomad bee *N. sheppardana*, length 4–6 mm, Britain's smallest nomad bee, locally common in southern England, East Anglia and the Welsh Borders, a small kleptoparasite of mining bees (especially *Lasioglossum parvulum*), habitats including post-industrial sites.

Fabricius' nomad bee *N. fabriciana*, length 7–12 mm, widespread in England and Wales, especially southern parts, sporadic in Scotland and Ireland, bivoltine (flying in March–June, and June–August), visiting a variety of flowers for nectar. It is a kleptoparasite, laying its eggs in the nests of *Andrena* solitary bees. Johan Fabricius (1745–1808) was a Danish entomologist.

Goldenrod nomad bee *N. rufipes*, length 10 mm, widespread in England and Wales in a number of habitats, especially heathland, flying in July–September, visiting a variety of flowers, particularly heathers and ragworts, for nectar. It is a kleptoparasite, laying its eggs in the underground nests of *Andrena* solitary bees. *Rufipes* is Latin for 'red-footed'.

Gooden's nomad bee *N. goodeniana*, length 10–13 mm, laying its eggs in the nests of *Andrena* solitary bees, larvae feeding on pollen stored by the host, widespread in England, Wales and parts of Scotland, active in April–June.

Marsham's nomad bee *N. marshamella*, length 9 mm, widespread, especially in England, in a variety of habitats, including gardens, visiting a number of flowers for nectar. It is both univoltine and bivoltine, depending on which *Andrena* host is attacked, though it is mainly a kleptoparasite in the burrows of *A. carantonica* nests. Thomas Marsham (1747 or 1748–1819) was an English entomologist.

Red-horned nomad bee *N. ruficornis*, length 6.5–8 mm, widespread, common in England, flying in April–June, found in grassland and hedgerows, kleptoparasitic on *Andrena* bees, especially early mining bee, larvae feeding on the host larvae in the nest burrow. *Ruficornis* is Latin for 'red-horned'.

NOSTOC Blue-green algae, capable of living on carbon dioxide, water and nitrogen.

NOTCHWORT Foliose liverworts. Rarities include **alpine notchwort** *Anastrophyllum alpinum* (north-west Scotland), **alpine jagged notchwort** *Lophozia opacifolia* (Highlands), **chalk notchwort** *Lo. perssonii* (in England on chalk or Magnesian limestone), **curled notchwort** *A. saxicola* (Cairngorms and north Scotland), **Donn's notchwort** *A. donnianum* (Highlands), **fen notchwort** *Leiocolea rutheana* (Scarning Fen, Norfolk, and – probably no longer – Orton Moor, Cumbria), **Fitzgerald's notchwort** *Le. fitzgeraldiae* (north-west Ireland, north-west Wales and western Highlands), **flush notchwort** *Tritomaria polita* (Highlands), **Gillman's notchwort** *Le. gillmanii* (Highlands and northern England), **Joergensen's notchwort** *A. joergensenii* (north-west

Scotland and Cairngorms), and **Welsh notchwort** *G. acutiloba* (north-west Wales, Merseyside and Highlands).

Bantry notchwort *Leiocolea bantriensis* (Mesoptychiaceae), common and widespread in calcareous flushes in the uplands, and fens in the lowlands. Bantry is in Co. Cork. Greek *leios* + *coleos* = 'smooth' + 'sheath'.

Bog notchwort *Cladopodiella fluitans* (Cephaloziaceae), locally common and widespread in bog and wet heathland pools on the surface of sphagnum carpets, sometimes fully submerged. Greek *cladion* + *podos* = 'branch' + 'of a foot', and Latin *fluitans* = 'floating'.

Capitate notchwort *Lophozia excisa* (Lophoziaceae), a common and widespread calcifuge, on damp soil, typically in depressions in heathland, dunes and spoil tips. *Excisa* is Latin for 'cut away'.

Comb notchwort *A. minutum*, common and widespread in upland regions, mainly on rocky or heathery slopes.

Heller's notchwort *A. hellerianum*, scarce, on decaying logs in sessile oakwoods in parts of Wales, Cumbria and north Ireland, and pine logs in the Caledonian forest in the Highlands.

Hill notchwort *Lo. sudetica*, common and widespread in well-drained, acidic, upland sites such as boulder tops, scree and dry-stone walls.

Holt notchwort *C. francisci*, locally common on heathland in south and south-west England. George Holt (1852–1921) was a Manx-born bryologist.

Inflated notchwort *Gymnocolea inflata* (Scapaniaceae), common and widespread in damp, acidic habitats such as wet heathland, bogs and the edge of peaty pools. It also grows on non-base-rich shale, gravel and sand, and in scree. Greek *gymnos* + *coleos* = 'naked' + 'sheath', and Latin *inflata* = 'swollen'.

Jagged notchwort *Lo. incisa*, common and widespread on upland bogs. *Incisa* is Latin for 'incised'.

Large cut notchwort *T. exsectiformis*, locally common and widespread, often upland, mostly on decaying logs, occasionally on peat and organic matter on boulders. Latin *ex* + *sectus* + *forma* = 'from' + 'cut' + 'shape'.

Lax notchwort *Hygrobiella laxifolia* (Cephaloziaceae), common and widespread on permanently moist upland rock faces in stream gorges. *Laxifolia* is Latin for 'loose-leaved'.

Lesser notchwort *Lo. bicrenata*, widespread if scattered on bare acidic soil, especially on heathy banks in conifer plantations, and also on stabilised dunes, soil on wall tops and boulders, and mine spoil. *Bicrenata* is Latin for 'bi-scalloped'.

Lyon's notchwort *T. quinquedentata*, common and widespread, favouring humid, sheltered base-rich upland sites, frequent among other bryophytes on steep, mossy slopes in Atlantic woodlands. It also grows on base-rich rock faces, and hummocks in calcareous seepages. Latin *quinquedentata* = 'five-toothed'.

Mountain notchwort *Le. alpestris*, locally common and widespread, favouring calcareous conditions on wet rock outcrops and on flush margins, in drier calcareous turf, and in dune slacks.

Orkney notchwort *Anastrepta orcadensis* (Scapaniaceae), locally common on heathery slopes in the Highlands and Islands, and elsewhere in upland Britain and Ireland on woodland floors and in scree crevices. Greek *ana* + *streptos* = 'similar to' + 'twisted', and Latin *orcadensis* = 'Orcadian'.

Ragged notchwort *Le. heterocolpos*, scattered but mainly found in the Highlands in steep valleys on base-rich sandstone and calcareous cliffs, usually growing over cushions of moss. Greek *heteros* + *colpos* = 'different' + 'breast', here meaning 'curved'.

Scarce notchwort *Le. badensis*, scattered and widespread, favouring calcareous conditions, typical of disused quarries, on flush margins, tufa, rocks by streams and damp limestone in woodland, and in base-rich dune slacks.

Top notchwort *Le. turbinata*, scattered and widespread, on calcareous substrates, with chalk and limestone grasslands its most typical habitat, but locally abundant in woodlands, and in dune slacks. *Turbinata* is Latin for 'like a (spinning) top'.

Tumid notchwort *Lo. ventricosa*, common and widespread on acidic substrates (rock, peat, sphagnum and decomposing wood), especially in upland habitats, including bogs, wet heathland and woodland. 'Tumid' means swollen, *ventricosa* here meaning the same, from Latin *venter* = 'stomach'.

NOTERIDAE From Greek *noteros* = 'wet' or 'damp', aquatic beetles in the suborder Adephaga, formerly considered part of the Dytiscidae, with the two representatives in the British Isles both predatory, and found mainly in stagnant water in association with floating rafts of vegetation, often burrowing through mud at the bottom of the pond. The larger *Noterus clavicornis* is commoner than *N. crassicornis*, but both are quite widespread.

NOTHOFAGACEAE Family name for southern beeches, from Greek *nothos* = 'spurious' prefixing Fagaceae, the (true) beech family name.

NOTODONTIDAE From Greek *notos* + *odontos* = 'back' + 'teeth', from the dorsal scale forewing tufts which point upwards when the wings are folded, the family name of chocolate-tip, kitten and prominent moths, including buff tip, puss moth and oak processionary.

NOTOSTRACA From Greek *notos* + *ostracon* = 'back' + 'shell', the order of Crustacea that comprises the one family, Tropsidae, tadpole shrimps.

NUDIBRANCH From Latin *nudus* = 'naked' and Greek *branchia* = 'gills', marine gastropods, including many sea slugs or shell-less molluscs.

Red-gilled nudibranch *Flabellina pellucida* (Flabellinidae) an uncommon sea slug up to 4 cm long, recorded from west and north-east Scotland, Anglesey and Pembrokeshire, the Dart estuary (south Devon), and Lough Hyne, Co. Cork. Latin *flabellum* = 'small fan', and *pellucida* = 'transparent'.

NULLIPORE Small red seaweeds (Rhodophyceae) forming calcareous encrustations. 'Nullipore' dates from the nineteenth century (Latin *nullus* = 'no' + pore).

NUT CRAB Species of *Ebalia*; see crab.

NUT-MOSS *Diphyscium foliosum* (Diphysciaceae), common and widespread in upland regions on acidic soil in sheltered rock crevices and on rocky banks, often on shaded sides of streams. Greek *di* + *physce* = 'two' + 'blister', and Latin *foliosum* = 'leafy'.

NUTCRACKER *Nucifraga caryocatactes* (Corvidae), wingspan 55 cm, length 32 cm, weight 160 g, an omnivorous bird, accidental, usually seen in August–September, with about 7 records p.a. Latin *nux* + *frangere* and Greek *caryon* + *cataseio*, both = 'nut' + 'to shatter'.

NUTHATCH *Sitta europaea* (Sittidae), wingspan 24 cm, length 14 cm, weight 24 g, a resident breeding bird common and widespread in deciduous woodland in Britain, with an estimated 220,000 territories in Britain in 2009, and with numbers increasing by 250% between 1970 and 2014, and by 94% in 1995–2014, though stabilising at 5% in 2010–15. These increases have been accompanied by range expansion into northern England and central Scotland. The tree-cavity nest usually has 6–8 eggs which hatch after 16–17 days, the altricial young fledging in 24–5 days. Feeding is mainly of invertebrates found on tree-trunks and branches, often sought by moving downwards; nuts and seeds are also eaten in winter, as well as fat from bird feeders. Greek *sitte* = 'nuthatch'. **Red-breasted nuthatch** *S. canadensis* is a vagrant.

NUTMEG *Anarta trifolii* (Noctuidae, Hadeninae), wingspan 30–5 mm, a nocturnal moth common in gardens, waste ground, downland, fen and open woodland throughout England, more local in Wales, Scotland and south and east Ireland. The bivoltine adults fly between May and September. Larvae feed on orache, goosefoot and other herbaceous plants. *Anarta* is a cockle shell mentioned by Pliny, with no entomological significance. *Trifolium* is clover, though not a food plant.

Large nutmeg *Apamea anceps* (Xyleninae), wingspan 35–40 mm, nocturnal, local in calcareous grassland, farmland, gardens and woodland edges in England mainly in the south-east and East Anglia, and with numbers declining by 93% over 1968–2007, a species of conservation concern in the UK BAP. Adults fly in June–July. Larvae feed on grasses. Apamia was a town in Asia Minor, with no entomological significance. *Anceps* is Latin for 'doubtful', from the original uncertainty that this was a distinct species.

NUTRIA Alternative name for coypu, both the animal and its fur.

NYCTEOLINE Oak nycteoline *Nycteola revayana* (Nolidae, Chloephorinae), wingspan 20–5 mm, a nocturnal moth, widespread but local in broadleaf woodland, parks and gardens. Adults appear in late autumn, overwinter, and reappear in early spring. Larvae feed on oak. Greek *nyx* = 'night', possibly + *eos* = 'dawn'. M. Revay was an eighteenth-century French entomologist.

NYMPHAEACEAE Water-lily family, with one species of *Nymphaea* (Latin for water-lily, named after the water goddess in Greek mythology) and three of *Nuphar*.

NYMPHALIDAE From Greek *nymphe* = 'nymph', the largest butterfly family in the British Isles, members being of medium to large size and often with brightly coloured upper wings (underwings are usually dull).

O

OAK (HYDROID), OAK (SEAWEED) See sea oak.

OAK (MOTH) Scalloped oak *Crocallis elinguaria* (Geometridae, Ennominae), wingspan 32–41 mm, widespread and common in woodland, scrub, hedgerows, parks and gardens. Adults fly at night in July–August. Larvae feed on a variety of woody plants. Greek *crocos* + *callos* = 'crocus' + 'beautiful', from the yellow colour, and Latin *elinguaria* = 'tongueless'.

OAK (TREE) Broadleaf trees in the genus *Quercus* (Latin for 'oak') (Fagaceae). A single jay can store as many as 5,000 acorns, those not retrieved and subsequently germinating being a crucial factor in oak dispersal.

Evergreen oak *Q. ilex*, an evergreen neophyte present in Tudor times, planted in parks, large gardens, churchyards and cemeteries, and becoming well-established in copses and woodland, and on dunes. It prefers light, warm soils, and is frequently planted near the coast. It is commonly self-sown, especially in southern and eastern England. *Ilex* is Latin for 'holly', reflecting the shape and evergreen nature of the leaves.

Holm oak Alternative name for evergreen oak.

Pedunculate oak *Q. robur*, common and widespread, long-lived, deciduous, of mature forest, coppice woodland and ancient wood pasture, growing on a range of soils, typically where heavy and fertile, but not thriving on thin soils over limestone or acidic peat, nor much found above 300 m. It is fairly tolerant of waterlogging. It is widely planted in hedges and woodland. The peduncle is a short stalk on which the acorn grows. *Robur* is Latin for 'strength' (of the wood, as well as the tree itself).

Red oak *Q. rubra*, a deciduous North American neophyte, introduced by 1724, common and widespread in England and Wales, more scarce and scattered in Scotland and Ireland, widely planted for ornament in parks, gardens and roadsides, and occasionally for forestry, hedging and screening, especially on light, sandy soils. It is often self-sown and has become naturalised in a few places. Latin *rubra* = 'red'.

Sessile oak *Q. petraea*, common and widespread, long-lived, deciduous, forming mature forest or coppice woodland, especially on well-drained, shallow, moderately to strongly acidic often sandy soils. It is the characteristic species of upland oakwoods. Acorns grow directly on the outer twigs (i.e. are sessile not pedunculate). *Petraea* comes from Latin *petra* = 'rock'.

Turkey oak *Q. cerris*, a deciduous neophyte from southern Europe, present by 1735, widespread and especially common in England, north and south Wales, central Scotland and south-east Ireland, planted in woodlands, town parks, estates, large gardens and along roads, especially on acidic, sandy soils. It seeds freely, and has become naturalised on free-draining soils in other habitats including railway embankments and waste ground, spreading into calcareous grassland and heathland. *Cerris* is the Latin name for this species.

OAK MOSS (OAKMOSS) *Evernia prunastri* (Parmeliaceae), fruticose, common and widespread on trunks, small branches and twigs of oak and other trees. The distinct, complex scent has made this a useful fixative in the French perfume industry, and as the base note in a number of fragrances; the English name is a translation of the French *mousse de chêne*. Greek *evernes* = 'sprouting well' (loosely, 'branched'), and Latin *prunus* + *aster* = 'plum tree' + 'star'.

OAK RAG Alternative name for both lungwort and pitted lichen.

OARWEED *Laminaria digitata* (Laminariaceae), a brown kelp seaweed with a thallus 1–2 m long, common and widespread except between the Ouse and Thames estuaries because of turbidity and lack of hard substrates, attached to bedrock in the lower intertidal and sublittoral fringe, down to a depth of 20 m in clear water. In exposed locations with strong wave action it may extend upwards into the lower eulittoral. It is also found in rock pools in the midlittoral and higher on wave-exposed coasts. Latin *lamina* = 'thin leaf', and *digitata* = 'with fingers', referring to the finger-like segments of the thallus blade.

OAT *Avena sativa* (Poaceae), a widespread annual archaeophyte known since the Iron Age, but now much less frequently grown as a crop than before mechanical transport, when there was great demand for it as horse food. It is often found as a relic of arable crops, and an occasional casual on field edges, roadsides and waste ground. It does not persist and is never naturalised. 'Oat' derives from Old English *ate* (plural *atan*). The binomial is Latin for 'oats' and 'cultivated'.

Bristle oat *A. strigosa*, an annual south-west European neophyte, for many centuries the main cereal cultivated in the north and west where conditions were unfavourable for oat *A. sativa*. It was first recorded in the wild in 1790, and is now a scattered, occasional arable weed, or casual as a grain impurity. *Strigosa* comes from Latin *stria* = 'bristle'.

Slender oat *A. barbata*, an annual Mediterranean neophyte scattered in Britain as far north as central Scotland as a casual on waste ground, originating as a grain contaminant, and naturalised on Guernsey since 1970. *Barbata* is Latin for 'bearded'.

Wild oat *A. fatua*, an annual archaeophyte, a common weed on arable land, and also found on roadsides and waste ground, especially in England, where it is increasing, being more scattered (though expanding its range) elsewhere. *Fatua* is Latin for 'simple'.

Winter wild oat *A. sterilis*, an annual Mediterranean neophyte present by 1640, found as a weed of winter cereal crops, predominately on heavier clay soils. It was first recorded in the wild in 1910 at Port Meadow, Oxfordshire, and has spread to neighbouring counties, though overall declining in frequency. It is also an uncommon wool or grain alien on waste ground.

OAT-GRASS Generally perennial grasses (Poaceae).

Downy oat-grass *Avenula pubescens*, widespread and especially common on chalk and limestone in England, in moist or dry, neutral and calcareous grassland, on roadsides and railway banks, in open woodland, coastal cliffs, fixed dunes, and some less acidic heaths. It tolerates low levels of mowing, grazing and manuring, but not artificial fertilisers or competition. *Avenula* is a diminutive of Latin *avena* = 'oat', and *pubescens* = 'downy'.

False oat-grass *Arrhenatherum elatius*, common and widespread in neutral to base-rich habitats, frequent in grassland, but especially common on verges, hedge banks, and river sides. It stabilises limestone scree, and colonises muddy, calcareous cliffs and maritime shingle. Var. *bulbosum* is often found on roadside banks and as an arable weed on light soils. Greek *arrhen* + *anthos* = 'male' + 'flower', the male flowers having awns or bristles, and Latin *elatius* = 'raised up' (i.e. 'tall').

French oat-grass *Gaudinia fragilis*, an annual or short-lived perennial of meadow, pasture and wayside on calcareous clay soils, and an occasional casual around docks and on tips. It was first recorded in the wild in 1903 and has increased in range since 1980, though many new records were of well-established but overlooked populations. It has a very similar distribution to corky-fruited water-dropwort, and this, together with its preference for old meadows, suggests that it might be native in its core areas. Alternatively, it may have been introduced in the late nineteenth and early twentieth centuries with grass seed imported from southern Europe. Jean Gaudin (1766–1833) was a Swiss botanist.

Meadow oat-grass *Av. pratensis*, on calcareous rendzina and brown earth soils, over chalk and limestone, as well as glacial deposits and basic igneous rocks, locally common and scattered in Britain on well-grazed downland, scree, cliffs and limestone pavement, occasionally in open ash woods, on dunes and on montane ledges. *Pratensis* is from Latin for 'meadow'.

Yellow oat-grass *Trisetum flavescens*, common in England, Wales and south-east Scotland, scattered elsewhere. It is an introduction in Shetland and its native status is doubtful in other areas of north and west Scotland, and in north Wales and south-west England. The introduced ssp. *purpurascens* is widely sown in seed mixtures in well-drained neutral and calcareous grasslands such as pasture, hay meadow, downland, banks and roadsides. It is highly palatable to stock and susceptible to damage by heavy trampling. The binomial is Latin for 'three-bristled' and 'yellow'.

OBLIQUE-BARRED Marsh oblique-barred *Hypenodes humidalis* (Erebidae, Hypenodinae), wingspan 14–15 mm, a scarce nocturnal moth, one of the smallest of the macrolepidoptera (smaller than the larger microleps), found in bog, moorland, damp heathland, fen, water meadow and marsh, from west England through west Wales to the Moray Firth in Scotland, and in north and west Ireland. Adults fly in June–August. Larval diet may consist of sedges and rushes, marsh cinquefoil or cross-leaved heath, depending on habitat. The binomial comprises the genus *Hypena* + Greek *eidos* = 'appearance of', and Latin for 'damp', from the habitat.

OBSCURE (MOTH) Nocturnal species of *Oegoconia* (Autostichidae), probably a misprint of the Greek *oecos* + *gonia* = 'house' + 'corner' (i.e. found in houses), wingspans 11–17 mm.

Four-spotted obscure *O. quadripuncta*, common in hedgerows and woodland throughout much of southern England and south Wales, though with records as far north as Lancashire, and scattered, local and coastal in Ireland. Adults fly in July–August. Larvae feed on leaf litter and decaying plant material. *Quadripuncta* is Latin for 'four-spotted', from the forewing markings.

Scarce obscure *O. deauratella*, local in hedgerows and woodland throughout much of the south-east half of England. Adults fly in June–July. Larvae probably feed on leaf litter and decaying plant material. *Deauratus* is Latin for 'gilded', from the yellow forewing.

Straw obscure *O. caradjai*, first recorded in Britain in 1981, from around haystacks, thatched roofs, farm buildings, barns and granaries in south-west and south-east England, and the south-east Midlands. Adults fly in June–August. Larvae feed on leaf litter and decaying plant material. Aristide von Caradja (1861–1955) was a German entomologist.

OCEANSPRAY *Holodiscus discolor* (Rosaceae), a deciduous neophyte North American shrub grown in gardens since 1827, scattered, naturalised in hedgerows and scrub and on cliffs, roadsides, railway banks and walls. The cascading white flowers prompt the English name. Greek *holos* + *discos* = 'entire' + 'disc' (i.e. unlobed), and Latin *discolor* = 'variously coloured'.

OCHRE (MOTH) Brindled ochre *Dasypolia templi* (Noctuidae, Xyleninae), wingspan 35–40 mm, local on grassland, sea-cliffs, upland grassland, moorland and dunes throughout much of Scotland, northern and south-west England, and north and east Ireland, predominantly coastal. With numbers declining by 94% over 1968–2007, this is a species of conservation concern in the UK BAP. Adults fly at night in September–October. Larvae feed internally on roots of hogweed, wild angelica and other umbellifers. Greek *dasys* + *polios* = 'shaggy' + 'grey', from the grey scales on the body, and Latin *templum* = 'temple'.

OCTOPUS Mostly in the sublittoral or deeper, cephalopod molluscs (family Octopodidae) (Greek *octo* + *pous* = 'eight' + 'foot').

Curled octopus or **horned octopus** *Eledone cirrhosa*, mantle length up to 50 cm, on rocky coasts in the intertidal and sublittoral, down to 500 m. *Eledone* is Greek for a kind of octopus, *cirrhos* = 'yellow'.

ODONATA Order of dragonflies and damselflies, with 62 species recognised by the British Dragonfly Society, in the suborders Zygoptera (damselflies) and Anisoptera (true dragonflies).

Nymphs, which have three 'tails', live in freshwater habitats. Adults have long, slender bodies and large compound eyes. Mouthparts of nymphs and adults are adapted for a carnivorous diet; adult mandibles are toothed (Odonata coming from Greek *odontos* = 'of teeth'). Nymphal mouthparts (mask) have a modified lower lip (labium) which extends rapidly to catch prey. Final instar nymphs emerge from the water on a plant stem; the skin splits and the adult emerges, wings and body expanding; the adult flies off, leaving behind the cast nymphal skin. The resting adult damselfly holds its wings vertically above its body, the dragonfly holds its wings horizontally.

ODONTOCERIDAE A caddis fly family (Greek *odontos* + *ceros* = 'teeth' + 'horn') represented in Britain by *Odontocerum albicorne*, the omnivorous cased larvae found in fast-flowing streams with rocky substrate. *Albicornis* is Latin for 'white-horned'.

ODONTOCETI Suborder of toothed whales (Greek *odontos* = 'teeth').

OECOPHORIDAE Moth family (Greek *oicos* + *phoreo* = 'house' + 'to carry', referencing the portable larval cases) including house-moths, tubics and streaks.

OEDEMERIDAE Family (Greek *oedema* = 'swelling') of false blister beetles, 5–17 mm length, with ten British species in four genera.

OESTRIDAE Family name of bot flies and warble flies, whose larvae penetrate the skin of mammals, including deer and livestock. Greek *oistros* = a sting that drives the recipient mad.

OIL BEETLE Species of *Meloe* (Meloidae) (Greek *mele* = 'probe'). Their common name comes from their ability to produce a bitter oil from their knee-joints, containing the toxin cantharadin. They are nest parasites of solitary bees. Adults are flightless. See also Spanish fly. **Mediterranean oil beetle** *M. mediterraneus*, **rugged oil beetle** *M. rugosus* and **short-necked oil beetle** *M. brevicollis* are all very rare.

Black oil beetle *M. proscarabaeus*, up to 30 mm length, with a widespread but scattered distribution in Britain, less common further north. Found in March–June in meadow and coastal grassland.

False oil beetle Alternative name for swollen-thighed beetle.

Violet oil beetle *M. violaceus*, up to 30 mm length, found in March–June in meadow and woodland in west and north Britain.

OIL-SEED RAPE See rape.

OLD LADY *Mormo maura* (Noctuidae, Xyleninae), wingspan 55–65 mm, a nocturnal moth, widespread but local in marsh, gardens, hedgerows, scrub and woodland. Adults fly in July–August. Larvae feed on blackthorn, hawthorn, ivy, dock and chickweed. Mormo, a she-monster in Greek mythology, has no entomological significance. *Maurus* is Latin for an inhabitant of Mauritania, the type locality.

OLD MAN OF THE WOODS *Strobilomyces strobilaceus* (Boletales, Boletaceae), a rare fungus of little culinary value, scattered in Britain, found in July–November on soil and litter in deciduous woodland. Greek *strobilos* + *myces* = 'something twisted' + 'fungus'.

OLD MAN'S BEARD (LICHEN) Alternative name for beard lichen, especially *Usnea subfloridana*; see lichen.

OLD-MAN'S-BEARD (PLANT) Alternative name for traveller's-joy. When the plant has finished flowering, the developing seeds (achenes) retain part of the flower which has long, silky hairs (the old man's beard), which help in seed dispersal.

OLEACEAE Family name (Latin *olea* = 'olive') for ash, including jasmine, lilac and privet.

OLEASTER Shrubs in the genus *Elaeagnus* (Elaeagnaceae) (Latin *olea* = 'olive' and Greek *elaia* + *agnos* = 'olive' + 'pure', possibly referring to the fruit).

Broad-leaved oleaster *E. macrophylla*, an evergreen neophyte Japanese shrub planted in gardens in 1879; also used in hedging in Guernsey and Sark. It has been recorded from waste ground in southern England. *Macrophylla* is Greek for 'large-leaved'.

Spreading oleaster *E. umbellata*, a deciduous or semi-evergreen neophyte from east Asia cultivated since 1829, with a scattered distribution in England and south Wales, bird-sown and naturalised in hedgerows and quarries, and on roadsides and waste ground. *Umbellata* is Latin for 'with umbels'.

OLIGOCHAETA, OLIGOCHAETE From Greek *oligos* + *chaeta* = 'a few' + 'bristle', a class (or subclass of annelid worm in the class Clitellata), with few bristles (setae). Almost all British taxa are in the order Crassiclitellata, with Lumbricidae (earthworms) the only significant family, containing 12 genera. Many species have a clitellum, a thickened region that secretes cocoons for enclosing eggs. See earthworm and Polychaeta.

OLIVE (MAYFLY) Angler's name for mayflies (Ephemeroptera), especially species of *Baetis* (from the Roman name for the Spanish river Guadalquivir), *Serratella* (Latin for 'saw-shaped') and *Rhithrogena*.

Blue-winged olive *Serratella ignita* (Ephemerellidae), widespread and one of the commonest mayflies, but numbers have greatly fallen in recent decades and it has almost vanished from some English rivers, a consequence of fine sediment run-off from farmland (preventing oxygen transferring into the egg, and encouraging fungal growth) and phosphate pollution (affecting egg development). Nymphs live in fast-lowing streams and rivers, especially where aquatic vegetation is present. Occasionally reported from stony shores of upland lakes. They are usually found clinging to, or crawling among submerged plants and stones. *Ignita* is Latin for 'glowing'.

Dark olive *Baetis atrebatinus* (Baetidae), nymphs recorded from a few sites in Hampshire, Dorset, the river Teifi in south Wales, and Ireland in the riffle area of calcareous streams where they feed by scraping algae from submerged stones or by collecting fine particulate organic detritus from the sediment. Adults are found during the day into the evening in May–October, suggesting two generations a year. *Atrebatinus* incorporates Latin *ater* = 'black'.

Lake olive *Cloeon simile* (Baetidae), common and widespread. Nymphs live in pools and stream margins or in deeper water in larger lakes, feeding by scraping algae from stones or by collecting fine particulate organic detritus from the sediment. There are usually two generations a year, one of which overwinters as nymphs and emerges in the spring. Adults are seen between March and November, males swarming throughout the day. Greek *cloios* = 'collar', and Latin *similis* = 'like'.

Large dark olive *B. rhodani*, common and widespread. Nymphs live in in-stream vegetation in riffle areas, feeding by scraping algae from submerged stones or by collecting fine particulate organic detritus from the sediment. There is a slow-growing winter generation and a much faster summer generation emerging from May to September. *Rhodanus* is Latin referring to the Rhone in France.

Medium olive *B. vernus*, common in England and Wales, local elsewhere, with a similar life history to large dark olive, adults evident in April–October. *Vernus* is Latin for 'vernal'.

Pond olive *C. dipterum*, with a similar life history to lake olive. Adults are found in May–October. *Diptera* is Latin for 'fly' (insect).

Scarce iron blue olive *B. digitatus*, recorded from a few streams in Wales and south England, and a single site in Scotland. Nymphs crawl among in-stream vegetation in riffle areas, scraping algae from submerged stones or collecting fine particulate organic detritus from the sediment. There is a slow-growing winter generation and a much faster summer generation emerging in May–September. *Digitatus* is Latin for 'having digits'.

Scarce olive *B. buceratus*, widespread if local in Wales, the Midlands and southern England, nymphs living in calcareous stream riffles scraping algae from submerged stones or collecting fine particulate organic detritus from the sediment. With two generations a year, adults are diurnal in April–October. *Buceratus* is Latin for 'horned'.

Small dark olive *B. scambus*, with a similar life history to large dark olive, adults evident in February–November. *Scambos* is Greek for 'curved'.

OLIVE (MOTH) *Ipimorpha subtusa* (Noctuidae, Xyleninae), wingspan 27–30 mm, local in broadleaf woodland, marsh, riverbanks, gravel-pits, gardens and parks throughout England, Wales, southern Scotland and the eastern half of Ireland. Population levels increased by 698% between 1968 and 2007. Adults fly at night in July–August. Larvae feed on aspen and poplar leaves. Greek *ipos* + *morphe* = 'mouse trap' + 'shape', the forewing shaped like a piece of wood in such a trap, and Latin *sub* + *(re)tusus* = 'a bit' + 'blunt', again from the forewing shape.

ONAGRACEAE Family name of willowherbs, as well as evening-primrose, fuchsia and enchanter's-nightshades, from Greek *onagros* = 'wild ass' or *onagra*, an unspecified plant, etymological significance unclear.

ONCHIDORIS Fuzzy onchidoris *Onchidoris muricata* (Onchidorididae), a sea slug up to 1.4 cm long, widespread but commoner on the west coast of Britain and north-east England, on the lower intertidal to depths to 15 m, mainly feeding on encrusting bryozoans. Greek *oncos* = 'swelling' + Doris, a sea goddess in Greek mythology, plus Latin *muricata* = 'pointed'.

ONION *Allium cepa* (Alliaceae), a bulbous herbaceous perennial (including **shallot** and **spring onion**) cultivated in gardens and fields with a scattered distribution as a relic of cultivation or a casual as a throw-out on tips and waste ground. 'Onion' dates to Middle English, from Old French *oignon*, based on Latin *unio*. *Allium* is Latin for 'garlic' or 'leek', *cepa* for 'onion'.

Wild onion *A. vineale*, common in the southern half of England, scattered elsewhere, on dry, neutral or calcareous soils, generally in summer-dry grasslands, hedgerows, roadsides and cultivated ground, and once a serious weed of cereal crops in south-east England. It also occurs on coastal cliff ledges in west Scotland. *Vineale* is Latin for 'vine-like'.

ONION FLY *Delia antiqua* (Anthomyiidae), 5–7 mm length, scarce, with scattered records in England and Wales. There are several generations a year, the first usually emerging in May–June to attack onion, shallot and leek seedlings. Second and third generations attack mature plants. Damage is caused by maggot-like larvae (10 mm length) feeding on roots, moving from one plant to another in the soil. *Delia* is either named from the Greek island of Delos, or from one of the names of the Greek moon goddess Artemis; *antiqua* is Latin for 'old'.

OONOPIDAE Family (Greek *oon* = 'egg') of spiders, with three British species, including pink prowler.

OPHIDIA(N) (Reptile) of the order including snakes (Greek *ophis* = 'snake').

OPHIURIDA, OPHIUROIDEA Respectively, the class and order of Ophiuroidea to which most British brittlestar species belong, from Greek *ophis* = 'snake', describing the long sinuous arms.

OPILIONES Order of harvestmen, arachnids with small bodies and long, thin legs, the scientific name coming from Latin *opilio* = 'shepherd', because shepherds in parts of Europe sometimes used stilts to monitor their flocks from a distance.

OPISTHOBRANCH(IA) A large group of snails with the gill located behind the heart (hence the name, from Greek *opisthen* + *branchia* = 'behind' + 'gills'), in contrast with prosobranchs and pulmonate snails, where the respective respiratory organs are located in front of the heart. See also Prosobranch and Pulmonata.

OPOMYZIDAE Family (Greek *opos* + *myzao* = 'vegetable juice' + 'to suck') of a group of small flies, with 16 species. Adults are found in grassland and cereal fields. Larvae feed on grass stems, including those of cereals, and some species can become agricultural pests, for example *Opomyza florum* and *Geomyza tripunctata* in early sown wheat in England and Wales.

OPOSSUM SHRIMP A marine crustacean in the order Mysida, family Mysidae, recorded from estuarine and subtidal waters, often found at the edge of an incoming tide. The name comes from the female's brood pouch (marsupium), in which larvae spend several weeks.

OPOSTEGIDAE Family name for some bent-wing moths (Greek *ops* + *stege* = 'face' + 'a covering').

ORACHE Mostly annual species of *Atriplex* (Amaranthaceae), from Middle English *orage*, from Anglo-Norman French *arasche*, and from Latin *atriplex*. A number of hybrids are naturalised.

Babington's orache *A. glabriuscula*, rare but widespread around the coast close to the strand-line on sand and shingle beaches, and in waste places near the sea. Charles Babington (1808–95) was an English botanist. Latin *glabriuscula* = 'slightly smooth', referring to the leaves.

Common orache *A. patula*, common and widespread on cultivated ground, roadsides and waste ground in towns, and on fertile soils in disturbed semi-natural habitats such as riverbanks and pond margins. *Patula* is Latin for 'spreading'.

Early orache *A. praecox*, recognised as a British species in 1975, found in west and north Scotland and north-east England on sand and shingle beaches around the margins of sea lochs and other sheltered inlets and bays, growing in the lowest part of sparsely vegetated strandlines. *Praecox* is Latin for 'early'.

Frosted orache *A. laciniata*, on the lower parts of sand and shingle beaches around most of the British and Irish coastline, occasionally on saltmarsh-dune transitions and saltmarsh drift lines. *Laciniata* is Latin for 'aciniate' (i.e. divided into deep narrow lobes).

Garden orache *A. hortensis*, a neophyte, possibly Asian, cultivated since the mid-sixteenth century as a food crop until the eighteenth century, now grown as an ornamental, first recorded from the wild in 1824 (Surrey), and a casual in Britain (mainly England) on waste ground and roadsides. *Hortus* is Latin for 'garden'.

Grass-leaved orache *A. littoralis*, in open, usually sandy or silty habitats near the sea, commoner on the east coastline, often along saltmarsh drift lines, estuarine banks and sea walls, and on waste ground around docks. Inland it also grows in saline habitats, and since the 1980s as a colonist by salt-treated roads, especially in eastern England. *Littoralis* is Latin for 'of the shore'.

Long-stalked orache *A. longipes*, confirmed as a British species in 1977, local on coastlines of Britain north to south Scotland, and in Co. Waterford, found in silty estuarine salt-marsh, usually in tall, ungrazed vegetation covered by brackish water during high spring tides. *Longipes* is Latin for 'long leg', here referring to the stalk.

Purple orache Alternative name for shining orache.

Shining orache *A. sagittata*, a European neophyte found on salted roadsides on the A13, south Essex, since 2006, and at Queenborough, Sheppey, Kent since 2013. *Sagittata* is Latin for 'arrow-shaped'.

Shrubby orache *A. halimus*, a European neophyte shrub, introduced into cultivation by 1640 and recorded from the wild in south Devon by 1900. It has been naturalised on rough ground at Hengistbury Head, Hampshire, since 1923, and planted as a wind break in many coastal areas, especially in southern England. *Halimos* is Greek for 'of the sea'.

Spear-leaved orache *A. prostrata*, common and widespread except in upland areas, on beaches, saltmarsh and other open saline habitats near the sea, and inland in disturbed areas on moist, fertile, neutral soils, for instance ditch and pond margins, cultivated land and waste ground. It also grows in inland salt-marsh and is increasingly found along salt-treated roadsides. *Prostrata* is Latin for 'prostrate' or 'spreading'.

ORANGE (MOTH) **Frosted orange** *Gortyna flavago* (Noctuidae, Xyleninae), wingspan 35–40 mm, widespread and common in rough grassland, woodland, waste ground, fen, marsh and gardens. Adults fly at night in August–October. Larvae feed inside the stems of thistles, burdocks and foxglove. Gortyna is a town in Crete, with no entomological significance. *Flavago* derives from Latin *flavus* = 'yellow'. Not the same species as orange moth (q.v.).

ORANGE (SPONGE) See sea orange.

ORANGE-BALL-TREE *Buddleja globosa* (Scrophulariaceae), a deciduous South American neophyte shrub introduced in 1774, scattered as a garden escape or throw-out on roadsides and waste ground, and as a relic of cultivation. The Rev. Adam Buddle (1662–1715) was an English botanist.

ORANGE PEEL FUNGUS *Aleuria aurantia* (Pezizales, Otidiaceae), widespread, common and edible, on bare soil in woodland or low grassland, often favouring gravelly ground, found from late summer to late autumn. Greek *aleuron* = 'wheat flour', and Latin *aurantia* = 'golden'.

ORANGE TIP *Anthocharis cardamines* (Pieridae, Pierinae), wingspan 40–52 mm, a common and widespread butterfly in Britain (ssp. *britannica*), though more local in Scotland (where its range is nevertheless spreading possibly with climate warming), and Ireland (ssp. *hibernica*), the orange tip found only on the male. It lives in a wide range of habitats, for example hedgerows, woodland margins and rides, damp meadows and gardens, adults flying in April–July. Larvae, which are also cannibalistic, feed on the developing seed pods, flowers and leaves of a range of plants, including lady's smock, hedge garlic and hedge mustard. Adults also have a wide diet, including bramble, dandelion, stitchwort, hawkweeds, vetches and bitter-cress (*Cardamine*, giving the specific name). Numbers increased by 59% over 2005–14, and range is also generally increasing. Greek *anthos* + *charis* = 'flower' + 'grace'.

ORB MUSSEL Freshwater filter-feeding fingernail mussels in the family Sphaeriidae.

Horny orb mussel *Sphaerium corneum*, common and widespread, shell width 9–11 mm, height 7–13 mm, in shallow, unpolluted freshwater habitats such as lakes, ponds and slowly moving rivers. It climbs plants to improve feeding location. Greek *sphaeria* = 'orb', and Latin *cornu* = 'horn'.

Lake orb mussel *Musculium lacustre*, shell width 7–15 mm, height 8–10 mm, common and widespread in England, scattered elsewhere, in small and temporary water bodies and swamps, not very deep and inside fine sediments, tolerating anaerobic substrates, and able to crawl on plants. Latin *musculus* = 'mussel' and *lacustris* 'of lakes'.

Nut orb mussel *S. rivicola*, shell width 15–25 mm, height 17–20 mm, locally common in south-east England and the Midlands, in well-oxygenated hard-water rivers and canals, almost always associated with river snail. *Rivicola* is Latin for 'living in a stream'.

Oblong orb mussel *M. transversum*, shell width 7–15 mm, height 8–10 mm, introduced from North America to the English canal system in the 1850s, spread until widely distributed, then declined after 1900 for unknown reasons to become a rarity, though surviving in scattered localities in the Midlands, often in urban sites, in muddy anaerobic substrates. *Transversum* is Latin for 'transverse', here implying 'oblong'.

ORB SPIDER See spider and cribellate orb spider.

ORB WEAVER Members of the Araneidae. See also weaver (Linyphiidae).

Four-spotted orb weaver *Araneus quadratus*, female body length up to 17 mm, males about half this, and Britain's heaviest spider, one found at Juniper Hall, Surrey, in 1979 (exceptionally) weighing 2.25 g. It is widespread but patchily distributed, on vegetation such as tall grassland, heather and gorse that has the height and strength to support the large orb web which is built close to the ground to catch jumping insects. The female builds the more elaborate web, with a funnel-shaped retreat to the side. Adults are found in late summer and autumn, males peaking in August, females between August and October. Greek *aranea* = 'spider' and Latin *quadratus* = 'square'.

Missing sector orb weaver *Zygiella x-notata*, a spider up to 7 mm body length, common and widespread, found year-round though mainly in late summer and autumn. It conceals itself within a silk tube above an orb web with a characteristic missing sector between '11 and 12 o'clock'. Running through the middle of this section is a single silk thread which leads to the spider's retreat. It is typically found on buildings and street furniture, often in the outside corners of windows. *Zygiella* is a diminutive of Greek *zygos* = 'yoke', and Latin *x-notata* = 'marked with an x'.

Walnut orb weaver Alternative name for toad spider.

ORCA *Orcinus orca* (Delphinidae), male length 6–9 m and weight >6 tonnes, females 5–7 m and 3–4 tonnes, the largest dolphin species, migratory groups recorded from Shetland, Orkney, north and west Scotland, the Irish coastline and the south and west coasts of England and Wales, an apex predator feeding on cephalopods, fish, seals and smaller dolphins. Pollution, especially by polychlorinated biphenyls (PCBs), has greatly impacted this and other dolphin species: orcas were present in the North Sea until the 1960s, when PCB pollution peaked. The UK's last resident pod off north-west Scotland was reduced to eight members in January 2016 with the death of a female which had not produced a calf in 19 years. A pod of perhaps six individuals was observed between Dunoon and Gourock on the lower Clyde estuary in April 2018. The binomial comes from Latin for '(like a) whale'.

ORCHAL Flat-leaved orchal *Roccella fuciformis* (Roccellaceae), a fruticose lichen, Near Threatened in the RDB, found in south-west England and south-west Wales on sheltered vertical rock faces on sea-cliffs.

ORCHID Often tuberous herbaceous perennials, often not flowering every year, in a number of genera, but mainly species of *Orchis* and *Ophrys* (Orchidaceae). See also butterfly-, fragrant-, marsh- and spotted-orchids. There are many hybrids. Orchids are obligate mycorrhizal symbionts, the fungi being crucial during orchid germination as the seed has virtually no energy reserve and obtains its carbon from the fungus. *Orchis* is Latin for 'testicle', prompting the common and genus name for these plants from the shape of the roots. *Ophrys* is Greek for 'eyebrow', in Latin coming to mean a plant with two leaves.

Bee orchid *Ophrys apifera*, locally frequent in much of England, Wales, south-west Scotland and Ireland, on calcareous, well-drained soils, in grassland, scrub, railway banks, roadsides, lawns, dunes and limestone pavement, as well as disturbed sites such as quarries, gravel-pits and industrial waste ground. The flowers have a bee shape, but attract these insects for pollination using scent rather than visual mimicry. This is the county flower of Bedfordshire as voted by the public in a 2002 survey by Plantlife. Latin *apis* + *fero* = 'bee' + 'carry' (i.e. is attractive to bees).

Bird's-nest orchid *Neottia nidus-avis*, a widespread and locally common chlorophyll-less mycorrhizal saprophyte found in deep leaf litter of densely shaded beechwood on chalky soils, less commonly in mixed deciduous woodland and mature hazel coppice on soils derived from limestones and base-rich clays and sands. The short, underground stem and the mass of roots resemble a bird's nest. The binomial names are, respectively, Greek and Latin for 'bird's nest'.

Bog orchid *Hammarbya paludosa*, a pseudobulbous herb uncommon in the Highlands and the New Forest, in boggy habitats with acidic water subject to lateral movement, typically growing in saturated sphagnum, but also on peaty mud and among grasses on the edges of flushes, and vulnerable to drainage (lowlands) and overgrazing (uplands). The binomial is in part from Greek *hamma* = 'knot', and Latin *palus* = 'bog'.

Burnt orchid *Neotinea ustulata*, with a much reduced distribution, remaining local in southern England, requiring warm, dry conditions and often found in grazed chalk and limestone grassland on south-facing slopes, as well as on sandy and gravelly soils in river meadows and on dunes. This is the county flower of Wiltshire as voted by the public in a 2002 survey by Plantlife. *Ustulata* is Latin for 'burnt'.

Coralroot orchid *Corallorhiza trifida*, a scarce chlorophyll-less saprophyte local in northern England and Scotland usually in damp, shaded alder and willow carr on raised bogs and lake margins, but also in dune slacks with creeping willow. More rarely, it grows in tall-herb fen, in sphagnum in birch and pine woods, and on moorland. This is the county flower of Fife as voted by the public in a 2002 survey by Plantlife. The binomial is Greek for 'coral' + 'root', and Latin for 'divided into three'.

Dense-flowered orchid *Neotinea maculata*, local in western Ireland on base-rich rocky or gravelly substrates, for example in limestone pavement, old pasture, hill grassland, dunes and verges. *Maculata* is Latin for 'spotted'.

Early-purple orchid *Orchis mascula*, widespread and generally common on neutral and calcareous soils, most frequent in woodland, coppice and calcareous grassland, but also in hedgerows and scrub, on roadsides and railway banks, and on limestone pavement and moist cliff ledges. It prefers shade in the south, more open habitats further north. Latin *mascula* = 'masculine' or 'vigorous'.

Early spider orchid *Ophrys sphegodes*, winter-green, scarce and scattered in southern England in old, species-rich, heavily grazed grassland on chalk and Purbeck limestone. It has also been found on disturbed ground in limestone quarries and old spoil tips.

Fen orchid *Liparis loeselii*, pseudobulbous, rare, restricted in East Anglia to species-rich fen on infertile soils, and to old peat cuttings. In north Devon and south Wales it grows in young dune slacks, but everywhere is declining. Greek *liparos* = 'oily' or 'smooth', referring to the glossy leaf surface. Johannes Loesel (1607–55) was a German botanist and physician.

Fly orchid *Ophrys insectifera*, scattered in (mainly southern) England, north Wales and Ireland, shade-tolerant, usually on chalk and limestone soil in open deciduous woodland and scrub, but also in grassland, chalk-pits, limestone pavement, disused railways and spoil tips. In Ireland and Anglesey it is found only in open calcareous flushes and fens.

Frog orchid *Coeloglossum viride*, restricted in southern England to dry, well-grazed, base-rich grassland on chalk downland and dunes, and in chalk-pits. Elsewhere it has been found in a wider range of calcareous grasslands, flushes, limestone pavement, rocky ledges, roadsides and quarries, but has become locally extinct at many sites. Greek *coilos* + *glossum* = 'hollow' + 'tongue', referring to the hollow spur on the tongue-like labellum, and Latin *viride* = 'green'.

Ghost orchid *Epipogium aphyllum*, a chlorophyll-less saprophyte growing in leaf litter or rotting stumps in beechwood on chalk. Always rare, it was seen regularly at a few Chiltern sites in 1953–87, then disappeared and was thought to be extinct until one plant was discovered in 2009. *Aphyllum* is Greek for 'without leaves'.

Green-winged orchid *Anacamptis morio*, widespread in England, Wales and Ireland, once frequent but now local, on base-rich to mildly acidic soils, most commonly found in hay meadow and pasture, but also on dunes, heathland and roadsides, and in quarries, gravel-pits and lawns. Although scarce in south-west Scotland this is the county flower of Ayrshire as voted by the public in a 2002 survey by Plantlife. Greek *anacampto* = 'bend back', referring to the flower spur, and Latin *morio* = 'fool'.

Lady orchid *Orchis purpurea*, on thin calcareous soils, typically over chalk on the North Downs in Kent, possibly locally on clay, ragstone and Carboniferous limestone elsewhere in southern England, but probably extinct at most of these sites, growing in open hazel, beech or ash woodland and scrub, occasionally in grassland. It was recorded in 2006 as having been established at Hartslock, south Oxfordshire in 1998.

Late spider-orchid *Ophrys fuciflora*, found at a very few sites in east Kent on well-drained chalky soil in species-rich, closely grazed grassland. *Fuciflora* comes from Latin *fucus* + *flora* = 'wrack' (or lichen) + 'flower'.

Lesser tongue-orchid *Serapias parviflora*, found in rabbit-grazed grassland in gorse and bramble scrub on coastal cliffs in east Cornwall since 1986, augmented by planting. Serapis was an Egyptian god; *parviflora* is Latin for 'small-flowered'.

Lizard orchid *Himantoglossum hircinum*, winter-green, scarce with a scattered distribution on chalk and limestone grassland, on roadsides and in quarries, occasionally on calcareous dunes and heathland, restricted to around 20 sites in Kent and Sussex, Somerset, Devon, Gloucestershire, Norfolk (recorded in 2015 for the first time since 1956) and the North and South Downs, with at least six sites being on golf courses (including the Open course, Royal St George's, Sandwich, Kent). Greek *himantos* + *glossa* = 'of a leather strap' + 'tongue', and Latin *hircinum* = 'goat-scented'.

Loose-flowered orchid *Anacamptis laxiflora*, locally common in Jersey and Guernsey in base-rich or calcareous marshy meadow. Following a decline in the early twentieth century, sites are now protected and populations have been stable since 1970. *Laxiflora* is Latin for a loose flowering habit.

Man orchid *Orchis anthropophora*, local in south-east England (mainly in Kent), declining but still found locally west to Dorset and north to south Lincolnshire, in old chalk and limestone quarries, calcareous grassland and verges. It tolerates considerable shade and is often found in grassland at the edge of scrub. *Anthropos* is Greek for 'man'.

Military orchid *Orchis militaris*, habitat destruction and collecting leading to it being considered extinct in the 1920s, but it was refound in chalk grassland in the Chilterns, Buckinghamshire, in 1947, and in a disused chalk quarry in west Suffolk in 1954, where populations remain stable, and it appears sporadically in two Oxfordshire sites.

Monkey orchid *Orchis simia*, believed to have become extinct until a native Kent population was found in 1955. Seed from this site was used to establish a second population in Kent. In 1974, a few plants appeared in south-east Yorkshire, only persisting until 1983. Oxfordshire (including at Hartslock) and Kent (including Park Gate Down) currently have two populations each, all of which are increasing in number, aided by management of grazing on chalk grassland. *Simia* is Latin for 'monkey'.

Musk orchid *Herminium monorchis*, scattered and decreasing in southern England on short turf on soils overlying chalk or oolitic limestone, and on quarry floors. *Herminium* comes from the Greek god Hermes.

Pyramidal orchid *Anacamptis pyramidalis*, locally common in England, coastal Wales, Ireland and Hebridean Scotland on well-drained calcareous soils in grazed downland and dune slacks, and on clifftops, and also in semi-stable dunes, scrub, verges and churchyards. It has colonised disturbed ground in former quarries, industrial wasteland and railway embankments. This is the county flower of the Isle of Wight as voted by the public in a 2002 survey by Plantlife.

Small-flowered tongue-orchid Alternative name for lesser tongue-orchid.

Small-white orchid *Pseudorchis albida*, scattered in upland regions, on well-drained hill pasture and mountain grassland, stream banks and cliff ledges on dry, acidic or calcareous soils; also on recently burnt moorland, though not persisting when heather regrows. *Albida* is Latin for 'white'.

ORCHIDACEAE Family name for orchids, including helleborine, lady's-slipper, lady's-tresses and twayblade. *Orchis* is Latin for 'testicle' as well as these plants, from the shape of the roots.

OREGANO See marjoram. The name is late eighteenth-century from the Spanish, ultimately from Greek *oreiganon*, from *oros* + *ganos* = 'mountain' + 'brightness'.

OREGON-GRAPE *Mahonia aquifolium* (Berberidaceae), an evergreen neophyte shrub introduced from North America in 1823, commonly planted as game cover, and a garden escape, often bird-dispersed, in hedgerows, woodland, wasteland and urban habitats north to central Scotland, but rare in northern Ireland. Bernard MacMahon was a nineteenth-century American horticulturalist.

ORFE *Leuciscus idus* (Cyprinidae), length 40–60 cm, weight 4 kg, a Scandinavian and central European fish, introduced from the late nineteenth century to a number of scattered locations in Britain for coarse fishing and as an ornamental, favouring ponds, lakes and clean slow-flowing rivers. They eat insects, crustaceans, molluscs and small fish. In spring they spawn over gravel or vegetation. 'Orfe' perhaps ultimately derives from Greek *orphos* = 'sea perch'. Greek *leyciscos* = 'white mullet', and *id* = the Swedish for this species.

ORIBATIDA From Greek *oribates* = 'mountain ranging', the order (or suborder in the order Sarcoptiformes) of oribatid mites (formerly known as Cryptostigmata), ranging in body size from 0.2–1.4 mm, most species found in soil and litter. Sometimes known as beetle mites.

ORIOLE, ORIOLIDAE See golden oriole.

ORMER **Green ormer** *Haliotis tuberculata* (Haliotidae), a sea slug 9 cm long, 6.5 cm wide, recorded from rocky shores down to depths of 40 m, at the northern edge of its range in the Channel Islands and south Devon. It grazes on algae, especially sea lettuce. Considered to be a great delicacy, numbers have recently declined, and harvesting is restricted to 'ormering tides', from 1 January to 30 April, which occur on the full or new moon and two days following that. Greek *halos* = 'of the sea', and Latin *tuberculata* = 'humped'.

ORNITHOSIS (Replacing **psittacosis**), a zoonotic infectious disease caused by the bacterium *Chlamydia psittaci*, found in parrots, pigeons and other birds, when it is known as **avian chlamydiosis**, and causing pneumonia in humans.

OROBANCHACEAE Family name of broomrape, from Greek *orobos* + *ancho* = 'vetch' + 'to strangle', referring to the plant's parasitic properties.

ORPINE *Sedum telephium* (Crassulaceae), a perennial herb, widespread but local in woodland edge, hedge banks, roadsides, rocky banks and limestone pavement. It is an ancient woodland indicator. Many populations have become naturalised as garden escapes, and the native range is now obscured. 'Orpine' dates from Middle English, from Old French *orpine*, probably an alteration of *orpiment*, applied to a yellow-flowered sedum. *Sedum* is Latin for house-leek, *telephium* for a kind of succulent plant.

ORSODACNIDAE Beetle family comprising the widespread *Orsodacne cerasi*, and *O. humeralis*, a scarce species of south-east England, both found on flowers, particularly of rowan or hawthorns, in grassland and scrub surrounding broadleaf woodland. *Orsodacne* is Greek for a bud-eating insect.

ORTHOPTERA From Greek *orthos* + *pteron* = 'straight' + 'wing', the order of grasshoppers (suborder Caelifera) and crickets and bush-crickets (suborder Ensifera). There are 27 native British species and a number of non-native, naturalised species. The head has a 'vertical face' and lies forward of a large pronotum. Antennae can be short (grasshoppers sometimes referred to as short-horned grasshoppers) or long (crickets sometimes referred to as long-horned grasshoppers). The symptomatic enlarged hind legs are used for jumping (especially in grasshoppers) and for stridulation which produces the animal's 'song'. Metamorphosis is slight in this exopterygote group.

ORTHOTRICHACEAE Family name for some yolk-mosses (Greek *orthos* + *trichos* = 'straight' + 'of hair').

OSIER *Salix viminalis* (Salicaceae), a common and widespread archaeophyte shrub or small tree, frequently coppiced and pollarded, which grows by streams and ponds, in marsh, fen, osier beds and landscaped areas, much planted for basketry and amenity, sometimes as distinct cultivars. It is also now being planted for biomass production. 'Osier' is late Middle English, from Old French *(h)osier* = 'willow twig', cognate with medieval Latin *auseria* = 'osier bed'. *Salix* is Latin for 'willow', *viminalis* for 'bearing wickerwork (osier) shoots'.

Broad-leaved osier *Salix x smithiana* (*S. viminalis x S. caprea*), a common and widespread shrub or small tree often found with its parents in hedgerows and waste ground, and in osier beds, once of value in basketry, now planted for biomass production.

Eared osier *Salix x stipularis* (*S. viminalis x S. caprea x S. aurita*), a shrub or small tree of damp habitats, hedgerows and scrub, probably much planted, scattered in northern England, south and central Scotland, and northern Ireland.

Fine osier *Salix x forbyana* (*S. purpurea x S. viminalis x S. cinerea*), a widespread but scattered deciduous shrub of wet habitats, especially moist thickets and riverbanks, and often found as a relic of osier cultivation having been much used in basketry.

Shrubby osier *Salix x fruticosa* (*S. viminalis x S. aurita*), a shrub or small tree found in hedgerows and thickets, with a scattered distribution, commoner in the north and west of Britain and Ireland. It sometimes occurs naturally, but it has also been planted for basketry and, recently, for biomass production. Only female plants are known. *Fruticosa* is Latin for 'shrubby'.

Silky-leaved osier *Salix x holosericea* (*S. viminalis x S. cinerea*), a common and widespread shrub or small tree found in damp habitats. *Holosericea* is Latin for 'silky'.

OSMERIDAE Family name of smelt (Greek *osmeros* = 'odorous').

OSOBERRY *Oemleria cerasiformis* (Rosaceae), a deciduous west coast North American neophyte shrub cultivated by 1848, recorded as a naturalised garden escape in rough grassland in Middlesex (1977), and subsequently Surrey, East Sussex and Kent. 'Osoberry' comes from Spanish *oso* = 'bear' + berry, from its use as a food for grizzly bears. Greek *oemleria* = 'pathway', and Latin *cerasiformis* = 'cherry-like'.

OSPREY *Pandion haliaetus* (Accipitridae), wingspan 158 cm, length 56 cm, weight 1.5 kg, a breeding migrant, arriving from wintering in West Africa in late March and April, leaving in August–September, mainly found in Scotland, for example Loch Garten (Cairngorms), Wigtown (Dumfries and Galloway) and Loch of the Lowes (Perthshire), with some sites in north-east England. In 2001 it began breeding in England at Bassenthwaite (Cumbria), and at Rutland Water (where it had been introduced in the same year), with 117 ospreys having fledged in the latter area by 2017. Eight chicks were translocated from Scotland to Poole Harbour, Dorset, in 2017, and 14 in 2018, with the hope that they will return there to breed after migration and maturation, perhaps by 2021. There are also two pairs in north Wales, at Dyfi and Glaslyn. In 2010 a BTO report considered it likely that the number of breeding pairs in Britain each year was then >200, perhaps as high as 250, and it is on the Amber List. Nests are of twigs in tall trees (and artificial nesting platforms), with 2–3 eggs laid, incubation taking 37 days, and fledging of the altricial young 53 days. Diet is of fish. 'Osprey' was first recorded around 1460, derived via Anglo-French *ospriet* and medieval Latin *avis prede* = 'bird of prey'. *Pandion* comes from the mythical Greek king of Athens, Pandion II, *haliaetus* from Greek *hali* + *aetos* = 'sea' + 'eagle'.

OSTEICHTHYES Name of the superclass of bony (in contrast with cartilaginous) fish (Greek *osteon* + *ichthys* = 'bone' + 'fish').

OSTREIDAE Family name for oysters (Greek *ostrea*).

OTTER *Lutra lutra* (Mustelidae), head-body length 60–80 cm, tail 32–56 cm, male weight 8.2 kg, females 6 kg, a once widespread semi-aquatic mammal which, while remaining common in much of Ireland, by the 1970s had become restricted in Britain mainly to Scotland (particularly the islands and north-west coast), west Wales, parts of East Anglia and south-west England, a decline largely caused by habitat destruction, general water pollution and the presence of organochlorine pesticides during the 1950s and 1960s. With the chemicals withdrawn from use in the 1960s, and improvements in water quality, otters have gradually returned to many areas: a sighting in Kent in 2011 then made all counties in England having at least one recent record. The population in Britain in 2005 (estimated at England 977, Scotland 7,948, Wales 540) has continued to increase. From virtual extinction in most of England during the early 1970s, EA site monitoring indicated population increases in England were 5.8% over 1977–9, 9.6% in 1984–6, 23.4% in 1991–4, 36.3% in 2000–2 and 58.8% in 2009–10. Otters live by streams or on the coast, the latter nevertheless requiring fresh water to cleanse the fur of the salt which can affect its insulating properties. In some places, especially in Ireland, otters have spread into urban areas via canals and other artificial water bodies. River otters mainly hunt at night. Diet is of fish, especially eel, salmon, trout and perch. Coastal otters hunt by day and night, eating bottom-living marine fish, crabs, sea urchins, etc. Deposition of musky faeces (spraints) in prominent places identifies an otter's territory and facilitates social contact. Gestation lasts just over two months. The two or three young (cubs) in a litter are usually born in summer, in dens (holts) in a riverbank hole, tree root system, or under a pile of rocks, to emerge after about ten weeks, though are often nursed for 15 weeks. 'Otter' comes from Old English *otr*, of Germanic origin. *Lutra* is Latin for 'otter'.

OTTER SHELL Bivalves in the genus *Lutraria* (Mactridae). The intended genus name was *Lutaria* (Latin *lutum* = 'silt') but was accidentally written as *Lutraria* (from *Lutra* = otter). These are deep burrowers in muddy sand and gravel from the low intertidal and nearshore shelf.

Common otter shell *L. lutraria*, shell size to 13 cm, widespread, becoming commoner since 2000.

Narrow otter shell *L. angustior*, shell size to 11.5 cm, recent records mostly off the south and south-west coasts but extending along the west to Orkney and Shetland. *Angustior* is Latin for 'narrower'.

OUZEL See ring ouzel.

OWL Hunting birds, some nocturnal, in the families Tytonidae and Strigidae. **Hawk owl** *Surnia ulula*, **scops owl** *Otus scops* and **Tengmalm's owl** *Aegolius funereus* are all rare vagrants.

Barn owl *Tyto alba* (Tytonidae), wingspan 90 cm, length 34 cm, weight 300 g, widespread in Britain but scarce in the Highlands and now relatively common only in central and southern Ireland, with 4,000 pairs (3,000–5,000 pairs) estimated in 1995–7, recovering from a decline in the 1950s and 1960s as organochlorine pesticides were banned, and according to BBS increasing by 227% over 1995–2014. It is Green-listed. A cavity nester, the species gets its English name for adopting outbuildings, including barns, as equivalents, with a clutch size of 4–6, incubation taking 32 days, and the altricial young fledging in 53–61 days. Chick survival is greatly dependent on availability of food: small mammals and small birds hunted by day (especially dusk) and night along field edges and verges. Greek *tyto* = 'owl', and Latin *alba* = 'white'.

Eagle owl *Bubo bubo* (Strigidae), wingspan 188 cm, length 75 cm, females slightly larger than males, known in captivity in Britain since at least the seventeenth century. Of the 440 captive eagle owls registered between 1994 and 2007, 73 were reported as having escaped and not recovered. If the same proportion is applied to a conservative estimate of the British captive population over the same period, around 65 birds p.a. would have escaped. Some breeding has been recorded in northern England and Scotland, but the population size, while small, is unknown, though 12–40 pairs were suggested by GWCT in 2016. A female in north Yorkshire who had successfully reared 23 young, was shot dead in January 2006. Eagle owls may naturally colonise Britain in the future as the population recovers on the European mainland, and isotope analysis of a bird found in Norfolk in 2006 indicated overseas origin. Nests are on rocky ground, with 1–2 eggs usual, incubation takes 31–6 days, and chicks leave the nest after 22–5 days. Chicks can walk at five weeks old and by seven weeks can take short flights. Little is known about diet in Britain, but rodents, rabbit and pheasant have been recorded. The binomial is Latin for 'eagle owl'.

Little owl *Athene noctua* (Strigidae), wingspan 56 cm, length 22 cm, weight 180 g, introduced from Europe, released/breeding in the wild since 1889, and from the 1930s widespread in England, scattered in Wales and southern Scotland. There were an estimated 5,700 breeding pairs (3,700–7,700 pairs) in Britain in 2009, BBS data suggesting numbers falling by 58% between 1995 and 2014. It is found in lowland farmland with hedges and copses, parkland and orchards. It nests in tree holes or wall cavities, and is found in old orchards, hunting by day though mostly at night in lowland farmland and other open landscapes, including grassland and moorland, feeding on small mammals, small birds, beetles and earthworms. Clutch size is 3–4, incubation lasts 29–31 days, and the altricial young fledge in 37–40 days. The genus name is that of the Greek goddess, who was a goddess of the night; *noctua* is the Latin name of an owl sacred to Minerva, Athena's Roman equivalent.

Long-eared owl *Asio otus* (Strigidae), wingspan 95 cm, length 36 cm, weight 290 g, widespread but thinly scattered, especially in south-west England and Wales, though the owl most likely to be noted in Ireland. Northern birds migrate southwards, including birds from Europe wintering in Britain. Its nocturnal habit makes it difficult to monitor numbers: there was a wide estimate of 1,600–5,300 pairs in Britain in 2007–11, with similar numbers in Ireland. Nests are in (often coniferous) trees using old stick nests of corvids or hawks. Clutch size is 3–4, incubation takes 28 days, and the altricial young fledge in 29–34 days. Diet is of small mammals and birds. *Asio* is Latin for a type of eared owl; *otus* refers to a small-eared owl.

Screech owl Alternative name for barn owl, from its call.

Short-eared owl *Asio flammeus*, wingspan 102 cm, length 38 cm, weight 330 g, scattered if widespread as a breeding bird in Britain, mostly in northern England and Scotland, with a winter influx of continental birds to England, especially around the coast. It is a rare breeder in Ireland, and wintering birds are especially evident along the east coast. A wide estimate of 610–2,140 pairs is given by BBS for Britain in 2007–11, and it is Amber-listed. Nests are on the ground with clutches of 4–7, incubation taking 27 days, the altricial young fledging in 26–32 days. Hunting of small mammals, especially voles, is by day and night, requiring extensive areas of open ground (for example moorland, dune or young forest plantation). Latin *flammeus* = 'flame-coloured'.

Snowy owl *B. scandiacus*, wingspan 154 cm, length 60 cm, weight 2.1 kg, a rare former migrant breeder: a pair bred for eight years on Fetlar, Shetland, until 1976, when the male failed to return; one or two females then summered every year until 1993 but no partner ever appeared. It is an occasional visitor to north Scotland (an adult visited the Hebrides each year in 2003–11, and an adult was recorded in the Cairngorms in February 2013), and west Ireland (a bird was recorded from the Burren in June 2014, possibly having arrived the previous autumn, and a female in the Gleninagh Mountain, Co. Clare in April 2015). *Scandiacus* is New Latin for 'Scandinavia'.

Tawny owl *Strix aluco* (Strigidae), wingspan 100 cm, length 38 cm, male weight 420 g, females 520 g, common and widespread in Britain (absent from Ireland), an Amber List bird found in woodland and, increasingly in suburbs, with 50,000 pairs in 2005, numbers declining by 37% between 1970 and 2015, slowing to 20% between 1995 and 2015, and 5% over 2010–15. These owls usually nest in tree holes, but they also use old magpie nests and squirrel dreys. Clutch size is 2–3, incubation takes 30 days, and the altricial young fledge in 35–9 days. Feeding is on small mammals, small birds (especially in towns), frogs, beetles and worms, hunting at night helped by the asymmetrically placed ears which give excellent directional hearing, and the large eyes at the front of the head with a (stereoscopic) visual overlap of 50–70%. Greek *strix* = 'owl' and Italian *allocco* = 'tawny owl', in turn from Latin *ulucus* = 'screech-owl'.

OWLET 1) Generally day-flying moths in the genus *Scythris* (Scythrididae), the long usually brownish wings prompting the common name, and reflected by the genus and family names from Greek *scythros* = 'sullen'. Larvae feed inside a silk tent.

Scarce species include **bronze owlet** *S. fallacella* (on limestone slopes in Lancashire, Yorkshire and Cumbria, larvae feeding on rock-rose), **copper owlet** *S. crassiuscula* (chalk and limestone grassland as far north as Cumbria, larvae feeding on rock-rose leaves), **goosefoot owlet** *S. limbella* (pastures, field edges and waste ground in Kent and East Anglia north to Yorkshire, larvae feeding on goosefoot, orache and good-King-Henry), **least owlet** *S. siccella* (sandy areas at Chesil Beach, Dorset, larvae feeding on thyme, bird's-foot-trefoil, kidney vetch, thrift and restharrow), **ling owlet** *S. empetrella* (heathland in Dorset and Hampshire, adults hopping around rather than flying, larvae feeding on heathers and cross-leaved heath), **Norfolk owlet** *S. inspersella* (dry open areas in north-west Norfolk, first discovered in 1977, with later records at Alethorpe Wood and Bawtry Forest, Yorkshire, larvae feeding on willowherbs), **sand owlet** *S. cicadella* (in parts of south-east England and Suffolk, but with no recent records), and **sorrel owlet** *S. potentillella* (dry sandy areas near Ipswich, Suffolk, recorded in 2004).

Black owlet *S. grandipennis*, wingspan 12–20 mm, local on heathland, downland, waste ground and dry open areas in southern Britain and Ireland. Adults fly in June, with a possible second generation in August. Larvae feed on shoots of gorse, living gregariously. Latin *grandis* + *penna* = 'large' + 'wing'.

White-dusted owlet *S. picaepennis*, wingspan 9–12 mm, widespread but scattered in grassy areas, dry sandy habitats and calcareous grassland. Adults fly in July. Larvae feed on bird's-foot-trefoil and wild thyme. Latin *pica* + *penna* = 'magpie' + 'wing', from the white markings on the dark wing.

2) Name sometimes given to some moths in the family Noctuidae, *noctua* being the Latin name of an owl sacred to Minerva.

OXALIDACEAE Family name for wood-sorrel, from the oxalic acid in leaves and roots.

OXEYE Oxeye daisy See daisy.

Yellow oxeye *Telekia speciosa* (Asteraceae), herbaceous perennial Balkan neophyte in cultivation by 1739, scattered and naturalised in Britain as a garden escape in rough grassland and by lakes and rivers, preferring damp soil. Samuel Teleki de Szek was a patron of the nineteenth-century German botanist J.C. Baumgarten. *Speciosa* is Latin for 'showy'.

OXLIP *Primula elatior* (Primulaceae), a perennial herb, found in East Anglia in deciduous woodland on damp chalky boulder-clay soils, especially where seasonal flooding occurs; rarely in wet alder woodland, damp meadows and ancient hedgerows. Elsewhere in England it is an occasional garden escape. This is the county flower of Suffolk as voted by the public in a 2002 survey by Plantlife. 'Oxlip' derives from Old English *oxanslyppe*, from *oxa* + *slyppe* = 'ox' + 'slime'. *Primula* is the female diminutive of Latin *primus* = 'first' (i.e. first or early flowering); *elatior* = 'taller'.

OXTONGUE (FUNGUS) Alternative name for beefsteak fungus.

OXTONGUE (PLANT) Species in the family Asteraceae.

Bristly oxtongue *Helminthotheca echioides*, an annual or biennial archaeophyte locally common in England, Wales and south-east Ireland on grassland, roadsides, field margins, riverbanks and waste ground, especially on lime-rich clay soils. Greek *helminthos* + *thece* = 'of a worm' + 'ovary', and Latin for resembling *Echium* (viper's-bugloss).

Hawkweed oxtongue *Picris hieracioides*, a biennial or perennial herb, mainly found in England and south Wales on calcareous soils in less heavily grazed chalk and limestone grassland, on roadsides and railway banks, and in quarries and lime-pits, and scattered elsewhere. Greek *picros* = 'bitter' or 'pungent', and Latin for resembling *Hieracium* (hawkweed).

OXYCHILIDAE Family name (Greek *oxys* + *cheilos* = 'sharp' + 'lip') for some of the glass snails (others being in the Vitrinidae). See also cellar and garlic snails, both under snail.

OXYOPIDAE From Greek *oxys* + *opos* = 'sharp' + 'of the eye' (i.e. sharp-sighted), the family name for lynx spider.

OXYTROPIS Perennial herbs in the genus *Oxytropis* (Fabaceae) (Greek *oxys* + *trope* = 'sharp' + 'turning', from the bitter taste).

Purple oxytropis *O. halleri*, a rarity in south-west and (mainly) highland Scotland on mountain rock ledges and grassy slopes on limestone and schists, and on base-rich sandstone sea-cliffs and calcareous sand dunes. Albrecht von Haller (1708–77) was a Swiss naturalist.

Yellow oxytropis *O. campestris*, confined to ledges on limestone and calcareous schists. Its two inland sites lie in the range 500–640 m at Clova (Angus), but its third site is on sea-cliffs in the range 25–180 m. *Campestris* is from Latin *campus* = 'field'.

OYSTER (BIVALVE) Generally members of the family Ostreidae, filter-feeding, 'oyster' coming from Middle English, in turn from Old French *oistre*, via Latin *ostrea* from the Greek *ostreon* = 'oyster'. See also saddle oyster.

American oyster *Crassostrea virginica*, up to 18 cm long, introduced to the south-east and south-west of England and north Wales on the lower intertidal and shallow sublittoral. Latin *crassus* = 'thick' + *Ostrea*.

Common oyster *Ostrea edulis*, shell size up to 10 cm, cemented to hard substrates and forming dense beds from low in the intertidal to 50 m, found on all coasts, severely depleted in the wild, especially in the North Sea, but extensively farmed on the south and west coasts; the only managed Scottish fishery is in Loch Ryan, which has a large self-sustaining population. Total UK commercial production in 2012 was 110 tonnes. *Edulis* is Latin for 'edible'.

Eastern oyster Alternative name for American oyster.

Edible oyster, European flat oyster Alternative names for common oyster.

Frond oyster *Dendrostrea frons*, shell size up to 5 cm, an intertidal introduction from the west Atlantic, with records from the Mullet Peninsula, Co. Mayo (in 1974), St Ives, Cornwall (1986), and Waterville, Co. Kerry (2013). Greek *dendron* = 'a stick' + *Ostrea*.

Pacific oyster *Crassostrea gigas*, shell size up to 30 cm, an introduction native to Japan and north-east Asia, recorded in 1926 on the River Blackwater, Essex, then deliberately introduced from Canada during the 1960s for commercial purposes, the first subsequent record from the wild being in 1965. They attach to hard substrates in intertidal and shallow subtidal zones of estuaries and coastal waters. In muddy or sandy substrates they can settle on small rocks, shells or other oysters and can create reefs by cementing their shells to each other, forming dense layers. Farms are widespread except for the north-east coast of England and east coast of Scotland. Escapes have established populations in parts of south-west and south-east England, and in Wales. In 2012, 1,200 tonnes were produced in the UK, but where these oysters establish wild populations economic losses may occur through loss of mussel and other bivalve fisheries. *Gigas* is Latin for 'giant'.

Portuguese oyster Alternative name for Pacific oyster.

Small oyster *Heteranomia squamula* (Anomiidae), bivalve, shell width up to 1.5 cm, found around Britain and much of Ireland, often on exposed shores, from the low intertidal to a depth of 1,000 m, attached by its byssus to many substrates from kelp in the sublittoral to deepwater corals. Greek *heteros* + *anomos* = 'different' + 'irregular', and Latin *squamula* = 'small scale'.

OYSTER (FUNGUS) A member of one of a number of genera, all Agaricales, including **oyster mushroom** *Pleurotus ostreatus* (Pleurotaceae), widespread, common (more local in Ireland), edible and tasty, found throughout the year, though fruiting mostly in summer and autumn, saprophytic on stumps, trunks and felled timber of hardwoods, especially beech. Greek *pleura* + *otos* = 'side' + 'ear', from the lateral attachment of the stem, and Latin *ostreatus* = 'oyster-like'.

Rarities include **marram oyster** *Hohenbuehelia culmicola* (most records from Kent, on dead marram stems), **orange mock oyster** *Phyllotopsis nidulans*, and **spatula oyster** *H. auriscalpium* and **woolly oyster** (both on beech deadwood).

Branching oyster *Pl. cornucopiae*, edible, widespread but commonest in England, found in summer and early autumn on stumps and other dead wood of broadleaf trees, especially oak and beech. *Cornucopia* is Latin for 'horn of plenty'.

Mealy oyster *Ossicaulis lignatilis* (Lyophyllaceae), inedible, scattered in England and the Welsh Border, on dead wood of beech, ash and other hardwoods. Latin *ossis* + *caulis* = 'of bone' + 'stem', and *lignum* = 'wood'.

Pale oyster *Pl. pulmonarius*, inedible, scattered and widespread, on dead wood in deciduous woodland. Latin *pulmonis* = 'lungs'.

Veiled oyster *P. dryinus*, inedible, widespread though generally infrequent, found in late summer and autumn on stumps and other dead wood of deciduous trees, mainly beech, oak and horse-chestnut. *Dryinus* is from Greek *dryos* = 'of a tree (especially oak)'.

OYSTER DRILL *Ocenebra erinaceus* (Muricidae), shell 5 cm long, 2.5 cm wide, widespread but mainly along the south and west coasts of Britain and around Ireland, mainly in the sublittoral on rocks and under stones to depths of 150 m, but often found on the lower intertidal of sheltered rocky shores in summer. It feeds on bivalves by boring through the shell, and also takes barnacles and polychaetes. *Erinaceus* is Latin for 'hedgehog'.

American oyster drill or **Atlantic oyster drill** *Urosalpinx cinerea*, an accidental introduction with American oyster from North America probably in the early twentieth century, though first recorded in 1927 from Essex, shell length 4 mm, width 2 mm, found on the Essex and Kent coasts, in the lower shore and sublittoral to a depth of about 12 m, especially in estuaries, feeding on oysters. Greek *oura* + *salpinx* = 'tail' + 'trumpet', and Latin *cinerea* = 'grey'.

OYSTER THIEF *Colpomenia peregrina* (Scytosiphonaceae), a balloon-like brown seaweed 1–7 cm in diameter, up to 25 cm in height when young, collapsing when old, native of the Pacific, introduced from North America to Cornwall and Dorset via France in 1907, and first recorded in Ireland in 1934, now widespread (reaching Orkney in 1940), commoner on western shores, usually epiphytic on various seaweeds, though also growing on rock and shells, in midlittoral rock pools and into the sublittoral, especially on sheltered coastlines. When attached to oysters it floats away with the oyster when the air-filled thalli grow large enough, hence the English name, though this does not occur in Britain. Latin *peregrina* = 'wandering'.

OYSTERCATCHER *Haematopus ostralegus* (Charadriidae), wingspan 80–5 cm, length 40–5 cm (the bill accounting for 8–9 cm), weight 540 g, common and widespread, breeding on almost all coasts, and over the last 50 years more birds have started breeding inland, totalling 110,000 pairs in 2009. Numbers in Britain increased overall by 66% between 1975 and 2015, though peaking in 1999, and with a decline since then (–9% in 2010–15). Most birds spend winter on the coast, joined on the east coast by birds from Norway, to total 320,000 individuals (2009 estimate). The nest is a bare scrape on pebbles or, inland, on gravelly islands in flooded gravel-pits, reservoirs, etc. Clutch size is 2–3, incubation lasts 24–7 days, and the precocial young fledge in 34–7 days. 'Oystercatcher' was coined for the oyster-eating North American species *H. palliatus*, but British birds feed more on mussels and cockles. Individuals specialise either by hammering the bivalve prey through the shell (and have short blunt bills), or by prising the two shells apart (and have longer pointed bills). Greek *haema* + *pous* = 'blood' + 'foot', and Latin *ostrea* oyster + *legere* to collect or pick.

OYSTERLING Inedible saprobic fungi in a number of genera and families, generally in the order Agaricales, but also Polyporales.

Widespread but infrequent species include **elastic oysterling** *Panellus mitis*, **flat oysterling** *Crepidotus applanatus*, **grounded oysterling** *Cr. autochthonus*, **Miller's oysterling** *Clitopilus hobsonii*, **moss oysterling** *Arrhenia acerosa*, **olive oysterling** *Sarcomyxa serotina*, **pale oysterling** *Cr. caspari* and **small moss oysterling** *A. retiruga*.

Rarities include **cinnabar oysterling** *Cr. cinnabarinus* (records from ash wood in Whippendell Wood, Watford, Hertfordshire, over 1995–7 were the first in Britain), and **marram oysterling** *Campanella caesia* (parts of south and south-east England and west Wales, on dead marram).

Bitter oysterling *Panellus stipticus* (Mycenaceae), widespread and locally common, possibly emetic, found throughout the year on dead wood of oak and other hardwoods. *Stipticus* is Latin for 'binding' (i.e. having styptic properties, constricting damaged blood vessels to stem bleeding).

Grass oysterling *Cr. epibryus*, widespread but scattered, commoner in England, found in late summer and autumn on deciduous hardwood twigs, and the dead leaves and stems of bracken and other herbaceous plants in woodlands and on heathland. Greek *epi* + *bryo* = 'on' + 'to swell'.

Lilac oysterling *Panus conchatus* (Polyporales, Polyporaceae), widespread and locally common in deciduous woodland on dead wood. *Conchatus* is Latin for 'spiral' or 'snail-like'.

Peeling oysterling *Cr. mollis*, widespread and common in late summer and autumn on trunks, large branches and stumps of dead broadleaf trees. *Mollis* is Latin for 'soft'.

Round-spored oysterling *Cr. cesatii*, widespread, especially in England, on the ground on dead branches of a variety of hardwoods.

Smoked oysterling *Resupinatus applicatus* (Tricholomataceae), widespread and, in England, locally common, found in summer and autumn on dead wood in deciduous woodland. The binomial is from Latin for 'overturned' and 'laid on'.

Variable oysterling *Cr. variabilis*, common and widespread, especially in England, found from early summer to early winter on woody debris in broadleaf woodland, and at the base of hedgerows.

Yellowing oysterling *Cr. luteolus*, scattered, commoner in England, on the ground on dead branches of a variety of hardwoods, as well as dead stems of bracken, bramble and stinging nettle. *Luteolus* is Latin for 'yellowish'.

OYSTERPLANT *Mertensia maritima* (Boraginaceae), a herbaceous perennial local but scattered on gravelly beaches and shingle, sometimes on sand. Seeds can survive prolonged immersion in seawater, and dispersion in sea currents enables colonisation of new though often transient sites. Range expansion in the north has been balanced by contraction in the south, where it has nearly disappeared. In Ireland it declined during the last century, but is now increasing. Losses result from storms, recreational activities, shingle removal and grazing. Franz Mertens (1764–1831) was a German botanist.

P

PAEONIACEAE Family name of garden peony, after Paeon, physician of the gods in Greek mythology.

PAGURIDAE Family name for most hermit crabs (Greek *pagouros* = 'crab').

PAINT MEASLES *Phoma violacea* (Pleosporales), a rare usually leaf spot or soil fungus, with most of the few records from emulsion paint. Greek *phois* = 'blister' and Latin for 'violet-coloured'.

PAINTED LADY *Vanessa cardui* (Nymphalidae, Nymphalinae), male wingspan 58–70 mm, female 62–74 mm, a migrant butterfly travelling from the desert fringes of North Africa, reaching and becoming widespread in Britain and Ireland. The round trip to and from Africa incorporates a series of steps involving up to six generations. In some years it is abundant, with around 11 million estimated in the 2009 invasion, and high numbers also in 2015, flying in April–October, becoming particularly abundant in late summer, and occupying open habitats, including gardens. Larvae and nectar-feeding adults are found on a variety of plants but favour thistles and common nettle. Long term distribution and population trends both show an increase, though year-to-year variation results from differences in migration strength. *Vanessa* comes from Jonathan Swift's poem 'Cadenus and Vanessa', *cardui* from *Carduus* (thistle).

PALAEMONIDAE Family name for common and grass prawn, from Palaemon, a sea goddess in Greek mythology.

PALLOPTERIDAE Family name (Greek *pallo* + *pteron* = 'quiver' + 'wing') of trembling-wing flies, with 13 species, 12 being *Palloptera*.

PAMPAS-GRASS *Cortaderia selloana* (Poaceae), a tussock-forming, dioecious perennial South American neophyte cultivated since 1848, recorded in the wild in 1925, now naturalised especially in the Channel Islands and much of the southern half of England on roadsides, railway banks and rough grassland on sea-cliffs and dunes. *Cortadera* is Spanish for 'cutter'. Friedrich Sello (1789–1831) was a German botanist and explorer.

Early pampas-grass *C. richardii*, a tussock-forming, dioecious perennial New Zealand neophyte naturalised on cliffs and on waste ground in a few places in Britain and the Isle of Man.

PANDIONIDAE Family name of osprey, recently often subsumed into the Accipitridae, from the mythical Greek king of Athens, Pandion II.

PANSY Members of the genus *Viola* (Violaceae), Latin for 'violet'. 'Pansy', regarded as symbolic of thought or remembrance, derives from late Middle English, from French *pensée* = 'thought', from Latin *pensare* = 'to think'. See also violet and dog-violet.

Dwarf pansy *V. kitaibeliana*, an annual herb found in the Scilly and Channel Islands on short coastal turf, open disturbed areas on sandy soils and eroding coastal dunes, around rabbit burrows, and in arable fields. It also grows under the shelter of coastal bracken in the Scillies, and in thin soil on granite sea-cliffs in Guernsey. Paul Kitaibel (1757–1817) was a Hungarian botanist.

Field pansy *V. arvensis*, an annual archaeophyte, common and widespread in lowland regions on cultivated and waste ground. *Arvensis* is from Latin for 'field'.

Garden pansy *Viola* x *wittrockiana* (*V. tricolor* x *V. arvensis*), an annual or perennial neophyte, grown in gardens since at least 1816, scattered and widespread as a garden escape, often found on waste ground and verges. It is usually casual, but can persist for a few years. The hybrid specific honours the Swedish botanist Veit Wittrock (1839–1914).

Horned pansy *V. cornuta*, a neophyte annual or short-lived perennial introduced from the Pyrenees in 1776, widely grown in gardens, sometimes escaping to become established on roadsides, woodland edges and railway embankments, and in hedgerows and rough grassland. It also occurs as a casual on waste ground, locally common in north-east Scotland but overall remaining scattered and very local in Britain. *Cornuta* is Latin for 'bearing horns'.

Mountain pansy *V. lutea*, a perennial of grazed grassland on hill slopes and rock ledges in upland northern England, Wales and Scotland (this is the county flower of Selkirkshire as voted by the public in a 2002 survey by Plantlife), though it was also rediscovered in Somerset in 1990. Usually found on calcareous rocks, it is actually a mild calcifuge, preferring leached soil; it also grows on metalliferous soils. It is also found on dunes in west Ireland. *Lutea* is Latin for 'yellow'.

Wild pansy *V. tricolor*, a widespread and locally common annual or perennial, on dunes and other sandy areas, on acidic grassland on heaths and hills, and in cultivated ground, gardens and waste ground. *Tricolor* is Latin for 'three-coloured'.

PANTHERCAP *Amanita pantherina* (Amanitaceae), a widespread but infrequently found (perhaps lethally) poisonous and certainly hallucinogenic fungus that contains the psychoactive chemical compounds ibotenic acid and muscimol as well as muscazone and muscarine, found in late summer and autumn on alkaline soils, especially under beech and oak. *Amanita* is Greek for a kind of fungus; *panthera* Latin for 'panther', this and the common name referencing the brown-and-white spotted cap.

False panthercap Alternative name for grey spotted amanita.

PANURIDAE Family represented by bearded tit (Greek *panu* + *oura* = 'exceedingly' + 'tail').

PAPAVERACEAE Family name of poppies (Latin *papaver*), greater celandine and fumitories.

PAPILIONACEAE Former name of the family Fabaceae (Latin *papilio* = 'butterfly').

PAPILIONIDAE Family name for the swallowtail butterfly (Latin *papilio* = 'butterfly').

PARACHUTE (FUNGUS) Inedible, generally minute Agaricales, mostly in the Marasmiaceae (Greek *marasmos* = 'drying out', these species recovering if wetted after dehydration), and found on dead wood or leaves.

Box parachute *Marasmius buxi*, found in Surrey, Berkshire and Gloucestershire, Vulnerable in the RDL (2006), evident from spring to autumn on deadwood on fallen box leaves. *Buxus* is Latin for 'box tree'.

Cabbage parachute *Gymnopus brassicolens*, uncommon, mostly recorded from deciduous woodland in southern England on soil, litter and deadwood, often under beech. Greek *gymnos* + *pous* = 'naked' + 'foot', and Latin *brassica* + *colens* = 'cabbage' + 'inhabiting'.

Collared parachute *Marasmius rotula*, common and widespread, found in late summer and autumn on surface roots and small dead wood in broadleaf woodland. *Rotula* is Latin for 'wheel', from the collar, gills and the outer rim of the cap resembling the hub, spokes and rim.

Foetid parachute *G. foetidus*, widespread but infrequent in the southern half of England, occasional elsewhere in Britain, poorly edible, on woodland soil, especially under beech and ash.

Garlic parachute *Mycetinis alliaceus*, uncommon and scattered in Britain, on leaf litter of hardwoods, mostly beech. Greek *myces* = 'fungus', and Latin *allium* = 'garlic'.

Goblet parachute *Marasmiellus vaillantii*, locally common and widespread, found in summer and autumn on deadwood in broadleaf woodland and parkland. Sébastien Vaillant (1669–1722) was a French botanist.

Grass parachute *Marasmius curreyi*, with a scattered distribution in England on dead grass stems in grassland, heathland, woodland and gardens. Frederick Currey (1819–81) was an English botanist.

Hairy parachute *Crinipellis scabella*, widespread but scarce outside England, found on the decaying stems of grasses on grassland and waste ground. Latin *crinis* + *pellis* = 'hair' + 'skin', and *scabella* = 'footstool'.

Holly parachute *Marasmius hudsonii*, uncommon, with a scattered distribution, found in late summer and autumn on leaf blades and fallen twigs in woodland. William Hudson (1730–93) was an English naturalist.

Horsehair parachute *G. androsaceus* (Tricholomataceae), widespread and common, found from spring to late autumn often carpeting large areas on conifer (especially pine) deadwood and litter, and on dead heather. Greek *andros* = 'masculine'.

Leaf parachute *Marasmius epiphyllus*, uncommon, with a scattered distribution, found in late summer and autumn on leaf blades and fallen twigs in woodland. Greek *epiphyllos* = 'on the leaf'.

Pearly parachute *Marasmius wynneae*, common in England, Wales and Ireland in summer and autumn on surface roots and small dead wood in broadleaf woodland.

Stinking parachute *G. perforans*, with a scattered distribution in Britain, records focusing around the Scottish Borders, in soil and litter in coniferous plantations, especially under Norway spruce. *Perforans* is Latin for 'perforating'.

Twig parachute *Marasmiellus ramealis*, very common and widespread, found in summer and autumn on dead twigs and on dead bramble stems. *Ramus* is Latin for 'branch'; *ramealis* comes from the diminutive (i.e. 'twig').

PARAKEET Escaped and naturalised members of the parrot family (Psittacidae).

Monk parakeet or **Quaker parakeet** *Myiopsitta monachus*, wingspan 48 cm, length 29 cm, weight 100 g, a small South American parrot, with escapes from cage and aviary, early colonies in England in the 1980s dying out, but a colony at Borehamwood, Hertfordshire, first noted in 1993, grew to 40 birds by 2005. In 2011 the total population in Hertfordshire and the London area (especially the Isle of Dogs) was around 150. Nests are built of sticks in trees and pylons, and are communal, each female laying 5–12 eggs which hatch in 24 days. In 2011, DEFRA announced plans to control this species, because of threats to crops and native wildlife, by trapping and rehoming, removing nests, and shooting. By 2014 the cull, costing £260,000, had brought numbers down to around 50, and in 2016 the Rare Breeding Birds Panel reported that these birds were maintaining a population of 20–30 pairs in the London area but control measures were preventing further increases. Greek *myia* + *psittacos* = 'a fly' + 'parrot', and Latin *monachus* = 'monk'.

Ring-necked parakeet or **rose-ringed parakeet** *Psittacula krameri*, wingspan 45 cm, length 40 cm, male weight 130 g, females 110 g, an Afro-Asian introduction, the first naturalised populations, probably from escaped aviary birds, recorded in 1969 when groups were noted at a number of sites in Kent. Expansion of numbers and range during the 1970s were mostly in south England, but breeding was noted in Greater Manchester in 1974, in Merseyside in 1980, then increasingly throughout much of England and south Wales, though most numbers remaining in the south-east, where sizeable flocks are found. A breeding population of 8,600 was estimated by BBS in 2012, with an increase in total numbers by >1,300% between 1995 and 2014. Nests are in tree holes (where native hole-nesters might be displaced), with a clutch of 3–4, incubation lasting 22–4 days, the altricial young fledging in 40–50 days. Birds feed on seed, fruit and flowers, with a negative impact on orchards and vineyards. *Psittacula* is a diminutive of Latin *psittacus* (from Greek *psittacos*) = 'parrot'; *krameri* commemorates the Austrian physician and naturalist Wilhelm Kramer (d.1765).

PARAMETRIOTIDAE Family name (Greek *para* + *metriotes* = 'near' + 'moderation') for some cosmet moths, some taxonomists retaining them in the Elachistidae.

PARASOL Or **field parasol** *Macrolepiota procera* (Agaricales, Agaricaceae), a widespread tasty fungus, especially common in England, rare in Scotland and Ireland, found in late summer and autumn on soil in woodland clearings, on verges, in neglected pasture and on grassy seaside cliffs. Variety *pseudo-olivascens* is found under conifers. Greek *macros* + *lepis* = 'large' + 'scale', and Latin *procera* = 'tall'.

Shaggy parasol *Chlorophyllum rhacodes*, edible (though can cause an allergenic reaction), common and widespread, found from early summer to autumn on soil in woodland, especially with conifers. Greek *chlorophyllum* = 'green leaves' (i.e. gills) and *rhacos* = 'piece of rag', from the shaggy appearance of the cap.

Slender parasol *M. mastoidea*, edible, fairly widespread, locally common in England and Wales, found in summer and autumn on alkaline soil in grassy areas in woodland clearings and edges, and calcareous dunes. Greek *mastos* = 'breast'.

PARCHMENT WORM *Chaetopterus variopedatus* (Chaetopteridae), a filter-feeding polychaete up to 25 cm long, that lives in a tough, flexible tube of a parchment-like material in sand and stone or shell-gravel from low water and deeper on rock, widespread though not recorded between Yorkshire and Sussex. Greek *chaeta* + *pteron* = 'bristle' + 'wing', and Latin *varius* + *pedatus* = 'varied' and 'having feet'.

PARENT BUG *Elasmucha grisea* (Acanthosomatidae), length 7–9 mm, common and widespread, overwintering as adults. New adults are found from August onwards, especially on birch and alder. Females guard the eggs, giving the English name. Metathoracic and abdominal glands discharge a foul-smelling secretion as a defensive behaviour. Greek *elasma* = 'thin plate', and Latin *grisea* = 'grey'.

PARIDAE Family name of most tits (Latin *parus* = 'tit').

PARMELIA **New Forest parmelia** *Parmelinopsis minarum* (Parmeliaceae), a rare foliose lichen with scattered records from the New Forest westwards to the Scillies, and in north-west Wales, on bark on woodland trees. Latin *parma* = 'small shield' and *minus* = 'less'.

PARMELIACEAE A genus-rich lichen family (Latin *parma* = 'small shield'). See lichen, crottle and oak moss.

PARNASSIACEAE Family name for grass-of-Parnassus, named after Mount Parnassus.

PARR One- to three-year-old salmon.

PARROT'S-FEATHER *Myriophyllum aquaticum* (Haloragaceae), an ornamental subaquatic and emergent perennial introduced from South America by 1878, becoming naturalised where thrown out, now widespread, though first recorded in the wild only in 1960 in Surrey. In Ireland it was first seen in Co. Down in 1990. It grows in small sheltered, eutrophic water bodies (e.g. ponds, reservoirs, canals and ditches), often causing flooding by blocking drainage channels, and it can rapidly displace native species. It is listed under Schedule 9 to the WCA (1981), whereby it is an offence to plant or allow it to grow in the wild. Only female plants are known in Britain, spreading clonally by vegetative fragmentation. Greek *myrios* + *phyllon* = 'numberless' + 'leaves'.

PARSLEY 1) Biennial species of *Petroselinum* (Greek *petra* + *selinon* = 'rock' + 'parsley') and *Sison* (Greek for the related species, honewort), both Apiaceae. 'Parsley' dates to Old English *petersilie*, via late Latin based on Greek *petroselinon*.

Corn parsley *P. segetum*, locally common south of a line from the Severn estuary to Yorkshire, and in Pembrokeshire, on well-drained calcareous soils on arable field margins, grassy banks, roadsides, railway and riverbanks, by sea walls, in drained estuarine marshes, on rough waste ground and occasionally as a garden weed. *Segetum* is from Latin *seges* = 'corn'.

Garden parsley *P. crispum*, a Mediterranean archaeophyte cultivated for at least a millennium, with a scattered distribution as small but persistent colonies on cliffs, banks and waste ground in coastal areas. The crisped form grown in gardens as a culinary herb occurs as a casual close to habitation. *Crispum* is Latin for 'curled', describing the leaf margins.

Stone parsley *S. amomum*, also known as **bastard stone parsley**, found in the southern half of England, Cheshire and north Wales, on clay and drained neutral to calcareous soils, mainly in hedgerows, and on banks, scrubby grassland and waste ground. *Amomon* is Greek for a spice plant from India.

2) **Cow parsley** *Anthriscus sylvestris* (Apiaceae), a common and widespread herbaceous perennial, characteristic of roadsides and hedgerows, but also found in abandoned pasture and woodland edges, and on railway banks, waste and cultivated ground, avoiding very wet or dry conditions. *Anthriscos* is Greek for some kind of parsley-like plant; Latin *sylvestris* = 'wild'.

3) **Fool's parsley** *Aethusa cynapium* (Apiaceae), an annual of hedge banks, waste ground, arable fields and other cultivated ground, common and widespread except for upland Scotland, and introduced in Ireland. *Aethus* is from Greek *aithos* = 'burnt', referencing the plant's pungency and toxicity; *apium* is Latin for 'parsley'.

PARSLEY-PIERT *Aphanes arvensis* (Rosaceae), a common and widespread winter- or, less frequently, spring-germinating annual of well-drained soils in arable fields, bare patches in grassland and lawns, heaths, woodland rides, waste ground, gravel-pits and along railways. 'Piert' is English dialect, an alteration of *pert*. Greek *aphanes* = 'inconspicuous', and from Latin for 'field'.

Slender parsley-piert *A. australis*, a common and widespread winter- or, less frequently, spring-germinating annual of acidic, sandy or gravelly soils in woodland rides, on dunes, in sand- and gravel-pits, on tracks and verges, and in dry rocky habitats. *Australis* is Latin for 'southern'.

PARSNIP Wild parsnip *Pastinaca sativa* ssp. *sylvestris* (Apiaceae), a biennial locally common in England and Wales in neutral and calcareous grassland, and on roadsides, railway banks, and rough ground. The cultivated ssp. *sativa* is local and generally scattered. *Pastinaca* comes from Latin *pastus* = 'food', referring to the edible root. *Sylvestris* and *sativa* are Latin for 'wild' and 'cultivated'. 'Parsnip' is late Middle English from Old French *pasnaie*, from Latin *pastinaca*, with a change in the ending via *neep*, Scottish dialect for 'turnip'.

PARSNIP MOTH *Depressaria radiella* (Depressariidae), wingspan 23–8 mm, nocturnal, widespread and common on waste ground and dry grassland. Adults begin to hatch in September and overwinter, often in buildings, appearing until the following May. Larvae feed on wild parsnip and hogweed, living gregariously within rolled leaves. Latin *depressus* = 'pressed down', from both the flat abdomen and the flattened wings when at rest, and *radiella* is a diminutive of *radius* = 'ray'.

PARTRIDGE Game birds in the family Phasianidae. 'Partridge' comes from Middle English *partrich*, from Latin *perdix* for these birds.

French partridge Alternative name for red-legged partridge.

Grey partridge *Perdix perdix*, wingspan 46 cm, length 30 cm, weight 390 g, a fairly widespread game bird in lowland arable land, but it has undergone a serious decline throughout most of its range and is a Red List species. Some 43,000 territories were estimated in 2009, but a decline in numbers by 92% was noted between 1970 and 2015, including by 15% over 2010–15. Nests are on the ground, clutch size (the largest of any bird species) is 13–16, incubation lasts 23–5 days, and the precocial young fledge in 14–16 days. Diet is of grass, cereal and clover leaves, weed seeds and cereal grain, chicks feeding on insects, especially in the first two weeks. The Greek word for this species may derive from *perdesthai* = 'to break wind', referring to the whirring noise of the bird's wings in flight.

Red-legged partridge *Alectoris rufa*, wingspan 48 cm, length 33 cm, weight 490 kg, probably arriving naturally in the Channel Islands, introduced from Europe to England in 1673, but not becoming common until releases in Suffolk from about 1770 into the early nineteenth century. From there and with further introductions this (not highly regarded) game bird is now found in much of arable England and parts of Scotland and Wales. Introductions in Ireland have not persisted. There was an estimated 82,000 territories in 2009, and a 13% increase in numbers between 1995 and 2014 (BBS data). Nests are on the ground, clutch size is 10–16, incubation lasts 23–4 days, and the precocial chicks fledge in 10 days. Food comprises seed and roots, with some damage to sugarbeet. Greek *alectoris* = 'chicken', and Latin *rufa* = 'red'.

PASQUEFLOWER *Pulsatilla vulgaris* (Ranunculaceae), a perennial rhizomatous herb of species-rich turf on the slopes of chalk or oolite escarpments, lost from many sites as a result of agricultural changes: there were only 33 known 'natural' sites in the UK in 2015, from Gloucestershire across to Lincolnshire. It also occurs as a garden escape and may have been planted on the Wiltshire–Dorset border. Called 'passeflower' in the late sixteenth century, from French *passe-fleur*, it transformed into its present name from flowering around Easter (French *Pâques*). This is the county flower of Cambridgeshire and Hertfordshire (which has one colony of up to 10,000 plants) as voted by the public in a 2002 survey by Plantlife. Latin *pulsus* = 'a push' or *pulsator* = 'fighter', reason unknown, and *vulgaris* = 'common'.

PASSERIDAE Family name for sparrows (Latin *passer*).

PASSERINE, PASSERIFORMES Order of perching birds (Latin *passer* = 'sparrow'). Nearly half of the bird species of Britain and Ireland are passerines.

PATH-MOSS Cornish path-moss *Ditrichum cornubicum* (Ditrichaceae), on bare, acidic soil contaminated with copper in and near copper mines in Cornwall. The binomial is Greek for 'two-haired' and Latin for 'Cornish'.

PAULOWNIACEAE Family name of foxglove-tree. Anna Paulownia was the daughter of Tsar Paul I.

PAUROPODA Class of tiny miriapods (Greek *pauros* + *podia* = 'small' + 'feet').

PAWWORT Upland foliose liverworts in the genera *Barbilophozia* and *Tetralophozia* (Scapaniaceae).

Scarce species include **bog pawwort** *B. kunzeana*, **four-fingered pawwort** *B. quadriloba*, **greater pawwort** *B. lycopodioides*, **Hatcher's pawwort** *B. hatcheri*, and **monster pawwort** *T. setiformi*.

Atlantic pawwort *B. atlantica*, in north Wales, northern England and Scotland locally abundant on mossy boulders, in turf on north-facing slopes, and on dry-stone walls.

Bearded pawwort *B. barbata*, among other liverworts on boulder tops, dry-stone walls, rock ledges and heathy grass. *Barbata* is Latin for 'bearded'.

Common pawwort *B. floerkei*, locally abundant on mossy boulders, turf on north-facing slopes, and on dry-stone walls. Heinrich Flörke (1764–1835) was a German botanist.

Trunk pawwort *B. attenuata*, widespread, on decomposing stumps and trees with acidic bark as well as on sandstone and igneous boulders. *Attenuata* is Latin for 'shortened'.

PEA Members of the genera *Pisum* (Latin for 'pea') and *Lathyrus* (Greek *lathyros* = 'pea'), family Fabaceae. **Fodder pea** *L. annuus*, **Indian pea** *L. sativus* and (as a garden escape) **sweet pea** *L. odoratus* are climbing annual neophytes, scattered and casual on waste ground.

Black pea *L. niger*, a perennial European neophyte herb cultivated by Tudor times, occurring as a naturalised garden escape in a few grassy, scrubby and rocky places in southern England. *Niger* is Latin for 'black'.

Garden pea *P. sativum*, a widespread if scattered climbing or sprawling annual, cultivated for at least two millennia, a casual on field margins, roadsides waste ground. Latin *sativum* = 'cultivated'.

Marsh pea *L. palustris*, a generally declining perennial herb of base-rich fen, reedbed and fen-meadow in eastern England and central Ireland; also, rarely, on marshy ground by rivers. *Palustris* is from Latin for 'marsh'.

Sea pea *L. japonicus*, a perennial herb, scattered around the coast, commoner in south-east England and Suffolk/Essex, forming large and conspicuous patches on shingle beaches.

Tuberous pea *L. tuberosus*, a scrambling perennial herb introduced from Europe by at least Tudor times, widespread but scattered in Britain, casual in hedgerows, rough grassland, waste places and occasionally arable field margins.

PEA BEETLE Members of the family Chrysomelidae, subfamily Bruchinae, also known as pea weevils. Specifically, *Bruchus pisorum*, 3.4–4.5 mm length, probably originating from the eastern Mediterranean, found in stored peas or, where escaped, on wild peas, larvae living in pea seeds, adults on pollen. Latin *bruchus* is derived from Greek *brouchos*, a wingless locust, and *pisorum* is from *pisum* = 'pea'.

PEA MOTH *Cydia nigricana* (Tortricidae, Olethreutinae), wingspan 12–16 mm, common in allotments, farmland and rough ground throughout much of Britain, more numerous in the south, widespread but less common in Ireland. Adults fly in May–August during late afternoon and early evening. Larvae feed in the pods of pea (becoming a serious pest in places), sweet pea and common vetch. Greek *cydos* = 'glory', and Latin *nigricana* = 'blackish'.

Cream-bordered green pea *Earias clorana* (Nolidae, Chloephorinae), wingspan 16–20 mm, in fen, damp woodland, heathland and mosses throughout much of southern England, East Anglia and Lincolnshire. Adults fly at night in May–June, sometimes with another generation in autumn. Larvae feed on willows. Greek *ear* = 'spring' and *chloris* = 'green', both reflecting the wing colour.

PEA MUSSEL, PEA SHELL Freshwater filter-feeding bivalves in the genus *Pisidium* (Sphaeriidae), from a diminutive of Latin *pisum* = 'pea'.

Arctic-alpine pea mussel *P. conventus*, shell width 2.2–2.8 mm, height 2–2.4 mm, scarce and local in north Wales, the Lake District, Highlands and north and east Ireland, in cold standing often deeply shaded montane waters, up to 750 m altitude and down to 300 m depth. *Conventus* is Latin for 'assembly'.

Caserta pea mussel (or clam) *P. casertanum*, shell width 4–5 mm, height 3–4 mm, common and widespread (possibly the world's most widely distributed non-marine mollusc), in small temporary water bodies, springs, swamp, and streams with little water current, preferring clear water, and tolerant of a range of trophic conditions. Caserta is a town and province in Campania, Italy.

False orb pea mussel *P. pseudosphaerium*, shell width 2.5–3.2 mm, height 2–2.6 mm, local in grazing marsh, drains and ponds in eastern and southern England, and Somerset. It was added to the Irish list in 1969 from sites along the Royal Canal in Co. Kildare and Co. Westmeath (still present in 2016), and from the Lagan Canal near Moira, Co. Antrim (not seen recently). Latin *pseudo* + *sphaera* = 'false' + 'sphere'.

Fine-lined pea mussel *P. tenuilineatum*, shell width 1.5–2.4 mm, height 0.8–1.4 mm, rare, scattered and declining in England (mainly in the south) and Welsh Borders, in rivers, canals and lake shores in muddy substrates in clean, base-rich water. Latin *tenuis* + *linea* = 'thin' + 'line'.

Giant pea shell (or mussel) *P. amnicum*, shell width 7–11 mm, height 4–8 mm, common and widespread in England, more scattered and local elsewhere, in rivers, canals and the littoral zones of large lakes, generally in clear, unpolluted water in a gravelly or sandy substrate near plants, and usually at no more than 10 m depth. Latin *amnis* = 'stream'.

Globular pea mussel *P. hibernicum*, shell width 2.5–3.5 mm, height 2.5–3.5 mm, widespread and scattered in lakes, pools and slowly flowing rivers, usually in clean clear water not below 10 m

depth, and preferring substrates with rich organic compounds. *Hibernicum* is Latin for 'Irish'.

Henslow's pea mussel *P. henslowanum*, shell width 3–5 mm, height 2–3.5 mm, widespread and scattered, common in England and central Scotland, in small running waters, marsh, canals and lakes, between roots of water plant, preferring eutrophic conditions with a sandy or rocky substrate rich in organic compounds on a calcareous base, and usually up to 25 m depth. John Henslow (1796–1861) was an English naturalist.

Humpbacked pea mussel *P. supinum*, shell width 3–5 mm, height 3–5 mm, extinct in Ireland, widespread but scattered in England in lowland rivers and canals, with clean, slowly moving well-oxygenated hard water, usually in mud or silt, sometimes on sandy substrates. *Supinum* is Latin for 'supine' rather than 'humpbacked'.

Iridescent pea mussel *P. pulchellum*, shell width 1.5–2.5 mm, height 2–3 mm, widespread, scattered and local in slowly moving, clean, well-oxygenated, often base-rich waters to a depth of 10 m. Red-listed as Vulnerable in Ireland, and a Priority Species in Northern Ireland. *Pulchellum* is Latin for 'pretty'.

Lilljeborg's pea mussel *P. lilljeborgii*, shell width 3–4.5 mm, height 2.5–3.5 mm, locally common and scattered in upland, usually soft-water, well-oxygenated, unpolluted lakes to a depth of 10 m. Red-listed as Vulnerable in Ireland, and a Priority Species in Northern Ireland. Wilhelm Lilljeborg (1816–1908) was a Swedish zoologist.

Porous pea mussel *P. obtusale*, shell width 2–3.5 mm, height 2–3.5 mm, widespread but scattered and declining, in shallow lentic water, swamp and marsh, requiring emergent plants. Latin *obtusus* = 'blunt'.

Pygmy pea mussel *P. moitessierianum*, shell width 1.5–2 mm, the smallest freshwater bivalve in the world, scattered in England and Wales and extinct in Ireland since 1924, in slowly moving lowland rivers and canals, in well-oxygenated hard water over unpolluted substrates varying from fine mud to sand. Prosper-Antoine Moitessier (1807–67) was a French collector.

Quadrangular pea mussel Alternative name for rosy pea mussel.

Red-crusted pea mussel *P. personatum*, shell width 3.5–4 mm, height 2–2.5 mm, common and widespread in ponds, lakes (often at considerable depth) and ditches, under damp litter in wet woodland and alder carr, and artificial habitats can be rapidly colonised. Latin *personatum* = 'masked'.

River pea shell Alternative name for giant pea shell.

Rosy pea mussel *P. milium*, shell width 2.5–3 mm, height 2–2.5 mm, common and widespread in both lentic and slow-moving lotic water with a muddy substrate to a depth of 10 m. *Milium* is Latin for 'thousands'.

Shining pea mussel *P. nitidum*, shell width 2.5–3.5 mm, height 2–3 mm, common and widespread in lakes (to a depth of 20 m) and lowland rivers with water plants, in sandy and muddy substrates, preferring clean, well-oxygenated water. *Nitidum* is Latin for 'shining'.

Short-ended pea mussel *P. subtruncatum*, shell width 3–4 mm, height 2–2.5 mm, common and widespread in slow-flowing water or lake littorals on muddy substrates to depths usually of 5–20 m, rarely above 300 m altitude. *Subtruncatum* is Latin for 'somewhat cut off' (i.e. 'short-ended').

PEA URCHIN Alternative name for green sea urchin.

PEA WEEVIL See pea beetle.

PEACH (FUNGUS) Wrinkled peach *Rhodotus palmatus* (Agaricales, Physalacriaceae), inedible, found July–November on stumps and trunks of elm. Initially benefiting from an increase in decomposing trees following Dutch elm disease in the 1970s, abundance has since declined so that it is now uncommon. Greek *rhodon* + *otos* = 'rose' (i.e. 'red') + 'ear', and Latin *palmatus* = 'palm-like', from the resemblance of the wrinkled cap to the lines on a hand.

PEACH (TREE) *Prunus persica* (Rosaceae), a deciduous tree long-introduced from China via Europe, scattered in England and Wales on waste ground, introduced with discarded fruit, rarely reaching maturity. It is also occasionally planted on roadsides and occurs as a relic of cultivation. *Prunus* is Latin for 'plum tree'. 'Peach' dates from late Middle English, from Old French *pesche*, from medieval Latin *persica*, from Latin *persicum malum* = 'Persian apple'.

PEACH BLOSSOM *Thyatira batis* (Drepanidae, Thyatirinae), wingspan 32–8 mm, a nocturnal moth widespread and common in scrub, hedgerows and gardens. Adults fly in June–July. Larvae feed on bramble in late summer. Thyatira is a town in Asia Minor; *batos* is Greek for 'bramble'.

PEACOCK (BUTTERFLY) *Aglais io* (Nymphalidae, Nymphalinae), male wingspan 63–8 mm, female 67–75 mm, a highly mobile and widespread butterfly. Numbers fluctuate, but it saw an overall increase by 17% from 1976 to 2014, and 21% from 2005 to 2014. It is absent from parts of north Scotland, though 'infilling' seems to be taking place, and range increased by 16% from 1976 to 2014 (3% over 2005–14). It is generally single-brooded but in 'good' years may produce a small second brood. Individuals hibernate, often in buildings, wood piles and tree hollows. Adults can therefore be seen in flight from January (with a mild winter) through to October, though peaks tend to be in April and, particularly, August. The primary larval food plant is common nettle; the nectarivorous adults favour thistles and buddleias but use a wide variety of plants. On hatching, larvae build a communal web near the top of the plant from which they emerge to bask and feed. 'Peacock' comes from the eyespot pattern on the upper wing which startles or confuses a would-be predator; a hissing sound created by rubbing the wings together also serves as a warning. The dark underwing provides camouflage. Greek *aglaios* = 'beautiful'. In Greek mythology Io was loved by Zeus, metamorphosed into a heifer by his jealous wife Hera and placed in the charge of Argus, who had a hundred eyes; on Argos' death, Hera placed his eyes on the peacock's tail.

PEACOCK MOTH *Macaria notata* (Geometridae, Ennominae), wingspan 28–32 mm, local in broadleaf woodland and scrubby heathland in southern England, East Anglia, north-west England, west Scotland and south-west Ireland. Population levels increased (from a low base) by 2,409% between 1968 and 2007. Adults fly at night in May–June, sometimes again in August in the south. Larvae mainly feed on birch and sallow. Greek *macaria* = 'happiness', and Latin *nota* = 'mark'.

Sharp-angled peacock moth *M. alternata*, wingspan 27–32 mm, local in broadleaf woodland, scrubby heathland and dunes throughout much of England, more commonly in the south, south-west Wales, and south-east Ireland. Adults fly at night in May–June. Larvae feed on sallows, alder, blackthorn and sea-buckthorn.

PEACOCK WORM *Sabella pavonina* (Sabellidae), up to 30 cm long, living in a muddy tube that projects up to 10 cm above sand or mud from which the head (with a crown of feathery tentacles) projects during feeding, widespread in the low intertidal and shallow sublittoral. *Sabella* is a diminutive of Latin *sabulum* = 'sand' or 'gravel', *pavonina* a diminutive of *pavo* = 'peacock'.

PEACOCK'S TAIL *Padina pavonica* (Dictyotaceae), a brown seaweed with a fan-shaped thallus 12 cm long and 1–10 cm wide, in rock pools and on stones on the mid- to lower littoral along parts of the south English, Pembrokeshire and south Irish coasts. Greek *pados* = 'tree', and Latin *pavonica* = 'peacock-like'.

PEAR *Pyrus communis* (Rosaceae), a deciduous archaeophyte tree or shrub of garden origin, widespread, locally common in England and Wales, naturalised in hedges, woodland margins and old gardens, and on railway banks and waste ground. The binomial is Latin for 'pear' and 'common'.

Plymouth pear *P. cordata*, a small deciduous tree, probably native, possibly an archaeophyte, restricted to two hedgerows near Plymouth (known since 1870), and three sites near Truro (found in 1989). *Cordata* is Latin for 'heart-shaped', referring to the leaves.

Wild pear *P. pyraster*, probably spiny stock of formerly cultivated (non-spiny) pear *Pyrus communis*.

PEARL (MOTH) Species of nocturnal moth in the family Crambidae.

Scarce species include **beautiful pearl** *Agrotera nemoralis* (restricted as a breeding species to hornbeam woodlands in Kent – Blean and, in 1992 only, Orlestone Forest – until 2014 when another colony was found in Torpoint, Cornwall), **fulvous pearl** *Udea fulvalis* (a continental immigrant found in south England in some years in sufficient numbers to breed, especially in and around Dorset), **mountain pearl** *U. uliginosalis* (Highlands, and on islands such as Rhum and the Orkneys), **Scotch pearl** *U. decrepitalis* (Highlands), **translucent pearl** *Paratalanta hyalinalis* (downland and woodland clearings in parts of southern and south-central England), and **yellow pearl** *Mecyna flavalis* (downland and grassy slopes in parts of coastal south England, East Anglia and south Wales).

Bordered pearl *Paratalanta pandalis* (Pyraustinae), wingspan 25–9 mm, in open woodland on chalky soils in parts of England and Wales, more numerous in the south, and in west Ireland. Adults fly in May–August. Larvae feed on a number of herbaceous plants, later stages feeding from a portable case constructed from leaves. Greek *para* = 'close to' + Atalanta, a huntress in Greek mythology, and Panda, Roman goddess of agriculture, from the corn-coloured forewing.

Chequered pearl *Evergestis pallidata* (Glaphyriinae), wingspan 24–9 mm, common in damp woodland throughout southern England, less abundantly but still widespread in south-west Scotland and Ireland. Adults fly in June–August. Larvae feed on winter-cress and other cruciferous plants. Greek *euerges* + *esthes* = 'well-fashioned' + 'garment', and Latin *pallidus* = 'pale', from the pale yellow wings.

Cinerous pearl *Anania fuscalis* (Pyraustinae), wingspan 20–6 mm, widespread but local in open woodland, moorland and damp meadows. Adults fly in June. Larvae feed on yellow-rattle and common cow-wheat. Greek *ananios* = 'without pain' (i.e. pleasurable to see), and Latin *fuscus* = 'dark' or 'brown'.

Coastal pearl *Mecyna asinalis* (Spilomelinae), wingspan 25–9 mm, local on limestone downland and coastal cliffs from Sussex to south Wales, and in south and west Ireland. Adults fly in May–October, and are probably bivoltine. Larvae feed on wild madder leaves. Greek *mecyno* = 'to lengthen', from the long forewing and abdomen, and Latin *asinus* = 'ass', from the grey wing colour.

Dark-bordered pearl *E. limbata*, wingspan 20–3 mm, uncommon, on rough ground in southern England. Since its arrival on the Isle of Wight in 1994, it has persisted at low density and has recently established small resident populations in Sussex and elsewhere. Adults fly in June–September. Larvae feed on garlic mustard, hedge mustard and other cruciferous plants. *Limbata* is from Latin *limbus* = 'border', from the dark wing edges.

Dusky pearl *Udea prunalis* (Spilomelinae), wingspan 20–4 mm, widespread, often very common in thickets, copses and hedgerows. Adults fly in June–July. Larvae feed on a number of herbaceous and woody plants. Greek *oudeos* = 'surface of the Earth', because larvae and adults feed near the ground, and *prunalis* comes from *Prunus spinosus* (blackthorn), a larval food plant.

Elder pearl See spotted magpie under magpie (moth).

Fenland pearl *An. perlucidalis*, wingspan 21–3 mm, first recorded in 1957 in a Huntingdonshire fen, and since locally found in fen, reedbeds and damp meadows, mostly in East Anglia and the east Midlands, occasionally further south and west. Adults fly in June–July. Larvae feed on thistles. Latin *per* + *lucidus* = 'very' + 'bright'.

Golden pearl *A. verbascalis*, wingspan 22–6 mm, found in open areas where the larval food plant, wood sage, grows such as heaths and shingle beaches, in parts of southern England, East Anglia and Lincolnshire. Adults fly in June–July. *Verbascalis* refers to *Verbascum* (mullein), though this is not a food plant.

Goldenrod pearl *A. terrealis*, wingspan 24–8 mm, local in rocky areas mostly in coastal parts from south-west England to south-west Scotland. Adults fly in June–July. Larvae feed on leaves and (more commonly) flowers of goldenrod. Latin *terreus* = 'earthen', from the wing colour.

Lesser pearl *Sitochroa verticalis* (Pyraustinae), wingspan 30–4 mm, local in farmland and rough grassland throughout southern England, in some years also appearing as a migrant from mainland Europe, reinforcing the resident population. Adults fly in June–July. Larvae feed on a variety of herbaceous plants. Greek *sitochroos* = 'colour of ripe wheat', and Latin *vertex* = 'highest point', reason unknown.

Long-winged pearl *A. lancealis*, wingspan 26–34 mm, common in damp woodland and fen in southern England, East Anglia and Wales. Adults fly in June–July. Larvae feed on hemp-agrimony, hedge woundwort and other herbaceous plants. Latin *lancea* = 'light spear', from the forewing shape.

Madder pearl Alternative name for coastal pearl.

Marbled yellow pearl *E. extimalis*, wingspan 25–31 mm, local on chalk downland and calcareous soils in parts of East Anglia and the Thames Estuary, and a migrant on the south coast of England, in some years arriving in sufficient numbers to breed. There is one post-2000 record from Co. Wicklow. Adults fly in June–July, though migrants may appear into autumn. Larvae feed on seed heads of cruciferous species. *Extimalis* is the superlative of Latin *exter* = 'outermost'.

Ochreous pearl *A. crocealis*, wingspan 22–5 mm, in water meadows, marsh, riverbanks and damp areas throughout much of England and Wales, more local further north. Adults fly in July–August. Larvae feed on leaves of common fleabane and ploughman's-spikenard. Latin *croceus* = 'saffron yellow-coloured'.

Olive pearl *U. olivalis*, wingspan 24–8 mm, widespread and common in woodland, gardens and waste ground. Adults fly in June–July. Larvae feed on a variety of herbaceous plants. *Oliva* is Latin for 'olive tree', possibly a food plant on the Continent.

Orange-rayed pearl *Nascia cilialis* (Pyraustinae), wingspan 24–7 mm, locally common in marsh and fen and along freshwater margins in the Cambridgeshire and East Anglian fens and, very locally, in Hampshire. Adults fly in June–July. Larvae feed on greater pond-sedge, occasionally other sedges. The binomial is possibly from Nascia, the Greek goddess of birth, and Latin *cilia* = 'eyelashes', from the fringes on the wing.

Pale straw pearl *U. lutealis*, wingspan 23–6 mm, widespread and often very common on waste ground, gardens and woodland edges. Adults fly in July–August. Larvae feed on bramble, thistle, wormwood and scabious. *Luteus* is Latin for 'clay-coloured', from the wing colour.

Rusty-dot pearl *U. ferrugalis*, wingspan 18–22 mm, an immigrant arriving from mainland Europe from June to November, generally peaking in August. It is commonest on waste ground, mostly by the coast but also inland, around much of the British Isles, and is sometimes abundant. *Ferrugo* is Latin for 'rust-coloured'.

Starry pearl *Cynaeda dentalis* (Odontiinae), wingspan 22–8 mm, on calcareous grassland and shingle beaches in southern and south-east England. Adults fly in July. Larvae mine stems and leaves of viper's-bugloss. Greek *cyon* = 'dog' and Latin *dens* = 'tooth', from the forewing pattern.

Straw-barred pearl *Pyrausta despicata* (Pyraustinae), wingspan 14–19 mm, with a widespread but scattered distribution on dunes, dry heathland, downland and open areas on chalk or limestone. The bivoltine adults fly in sunshine and at night in May–June and July–August. Larvae feed on plantain, often gregariously. Greek *pyraustes* = 'singed in a flame', perhaps from the black marks on the forewings, and Latin *despicata* = 'contemptible'.

Sulphur pearl *S. palealis*, wingspan 26–32 mm, local on heathland, downland, waste ground and clifftops throughout much of southern England and East Anglia. Adults fly in June–July. Larvae feed on umbellifers. *Palea* is Latin for 'chaff', reflecting the yellowish wing colour.

Woundwort pearl *A. stachydalis*, wingspan 23–5 mm, in farmland and rough ground in parts of southern England and south Wales. Adults fly in June–August. Larvae feed on leaves of hedge woundwort *Stachys*, prompting the specific name.

PEARLWORT Species of *Sagina* (Caryophyllaceae), from Latin *sagina* = 'fattening', sheep grazing on these plants reputedly fattened up.

Alpine pearlwort *S. saginoides*, a perennial herb of the Highlands and Hebrides, usually on base-rich, well-drained soils. It is a poor competitor, found on steep ground where there is late snow-lie and severe wind-scour which provide open conditions.

Annual pearlwort *S. apetala*, a widespread and scattered annual of open, often (semi-)artificial habitats such as heathland and paths but also on gravelly and sandy sites, and sea-cliffs. *Apetala* is Latin for 'without petals'.

Heath pearlwort *S. subulata*, a perennial, mat-forming herb scattered mainly in southern and south-west England, Wales, Scotland and west Ireland, usually in dry, open, sandy or gravelly habitats, including heathland, pasture, grassy slopes, and trackways through heaths and moors. It also occurs on basalt gravel terraces on the Trotternish Mountains, Skye. *Subulata* is Latin for 'awl-shaped'.

Knotted pearlwort *S. nodosa*, a herbaceous perennial, mainly in mires and springs with base-rich water, but also in open, calcareous, sandy habitats, especially dunes and dune slacks and sometimes in drier calcareous grassland. Widespread but only locally common, its decline in southern England since 1950 reflects loss of open calcareous habitats. Latin *nodosa* = 'knotted'.

Procumbent pearlwort *S. procumbens*, a common and widespread mat-forming perennial found on rocks, cliffs and riverbanks, but also growing in many artificial, disturbed and fertile habitats, for example spoil heaps, mining waste, paths, verges and urban pavements. It is a common garden weed, especially of lawns. It tolerates a range of soils and heavy trampling.

Sea pearlwort *S. maritima*, a widespread annual around the coast in rock crevices, and on clifftops, stabilised shingle, dune slacks and disturbed areas in the upper saltmarsh; also on walls and tracks, in pavements, on sandy coastal roadsides, and occasionally by salted inland roads.

Slender pearlwort *S. filicaulis*, a widespread and scattered annual of often artificial habitats such as paths, walls and bare cultivated soil. Latin *filicaulis* = 'with thread-like stamens'.

Snow pearlwort *S. nivalis*, a very rare cushion-forming, perennial herb of the central Highlands, usually found in open, unstable habitats with little plant cover, generally growing on soft calcareous schist. It grows in both dry and damp conditions, often through a bryophyte mat on a rock face or on gravelly ground. *Nivalis* is from Latin for 'snow'.

PEBBLE (MOTH) Garden pebble *Evergestis forficalis* (Crambidae, Glaphyriinae), wingspan 25–8 mm, widespread and common in gardens, allotments and farmland, especially where brassicas are grown, the caterpillars often becoming pests in September–October. The bivoltine adults fly at night in May–June and July–September. Greek *euerges* + *esthes* = 'well-fashioned' + 'garment', and Latin *forfex* = 'scissors' or 'shears', from the appearance of the folded wings.

PECTINIDAE Family name for scallop (Latin *pecten*).

PEEWIT Alternative name for the lapwing in imitation of its display call.

PELECANIFORMES Order of cormorants, gannet and shag, from Greek *pelecan* (leading to the Latin *pelecanus*) = 'pelican'.

PELLIA Common and widespread thallose liverworts in the genus *Pellia* (Pelliaceae) (Greek *pellos* = 'dark-coloured'), found in damp habitats such as springs, flushes, fens, dune slacks and wet rocks. **Endive pellia** *P. endiviifolia* is found on base-rich sites, **Nees' pellia** *P. neesiana* on acid sites, and **overleaf pellia** *P. epiphylla* on neutral or acid sites.

PELLITORY Bastard pellitory, wild pellitory Alternative names for sneezewort.

PELLITORY-OF-THE-WALL *Parietaria judaica* (Urticaceae), a perennial herb widespread in England and Wales, east and central Scotland, and the southern half of Ireland, growing from the cracks and mortar crevices of brick and stone walls, on building rubble, rocks, cliffs and steep-sided hedge banks, preferring dry, sunny, sheltered locations, and often found in built-up areas. Presence on the walls of abbeys and priories may be connected to its use by medieval herbalists as a remedy for urinary disorders. 'Pellitory' dates from late Middle English as an alteration of *parietary*, from Old French *paritaire*, based on Latin *paries* = 'wall', the root of the genus name. *Judaica* is Latin for 'Jewish'.
Eastern pellitory-of-the-wall *P. officinalis*, a rhizomatous European neophyte perennial, naturalised in two woodland/hedgerow sites (Middlesex and Essex). *Officinalis* is Latin for having pharmacological value.

PELT Foliose lichens in the genus *Peltigera* (Peltigeraceae) (Greek *pelte* = 'small shield').
Concentric pelt *P. elisabethae*, found at a few sites in the Highlands among moss on soil.
Flat-fruited pelt *P. horizontalis*, common and widespread, especially in upland areas, on moss on soil, trunks, stumps and rocks, especially in old woodland.
Many-fruited pelt *P. polydactylon*, scattered but rare among mosses on soil at moderate to high altitudes. *Polydactylon* is Greek for 'many fingers'.
Scaly pelt Alternative name for both flat-footed pelt and ear-lobed dog lichen.
Veinless pelt Alternative name for matt felt lichen.

PENNY BUN Alternative name for cep or king bolete.

PENNY-CRESS Herbaceous plants in the family Brassicaceae.
Alpine penny-cress *Noccaea caerulescens*, a herbaceous perennial with disjunct populations from Somerset, through west Wales and the Pennines to the southern Highlands, virtually confined to rocks or soils containing lead or zinc, found on spoil heaps and mine waste and on metalliferous river gravels. It has also been recorded as a rarity on outcrops and scree of limestone and other base-rich rocks, particularly in Scotland. Domenico Nocca (1758–1841) was an Italian botanist. Latin *caerulescens* = 'becoming blue'.
Caucasian penny-cress *Pachyphragma macrophyllum*, a rare rhizomatous perennial Caucasian neophyte, locally naturalised in woodland in Somerset, Dorset, Hertfordshire and Shropshire. Greek *pachys* + *phragma* = 'thick' + 'wall', and *macros* + *phyllon* = 'large' + 'leaf'.
Field penny-cress *Thlaspi arvense*, a locally common and widespread annual archaeophyte found as an arable weed, especially with broad-leaved crops and mainly on heavier soils. It is also a common weed in verges, waste ground and gardens. Greek *thlaeo* = 'to crush', referring to the flattened seed capsule, and from Latin for 'field'.
Garlic penny-cress *T. alliaceum*, an annual European neophyte in cultivation by 1714, a naturalised or casual weed of arable fields, field margins and other cultivated land in a few places in south-east England. *Alliaceum* is Latin for 'like garlic'.
Perfoliate penny-cress *Microthlaspi perfoliatum*, an annual of bare or sparsely vegetated habitats on oolitic limestone, local on scree, old quarries, stony banks and open pasture in Wiltshire, Oxfordshire, Gloucestershire and Worcestershire. Elsewhere it is a casual of waste ground, though populations have persisted on railway embankments: since it has poor seed dispersal, many colonies away from the Cotswolds may have resulted from seed spread in the slipstream of trains.

PENNY WEED *Zanardinia typus* (Cutleriaceae), a scarce brown seaweed with horizontal flat fronds up to 20 cm wide, patchily distributed from south-west Wales round to the Isle of Wight, and in south-west Ireland, mainly found on silty boulders or bedrock from shallow water to 20 m depth.

PENNYROYAL *Mentha pulegium* (Lamiaceae), a herbaceous perennial of seasonally inundated grassland overlying silt and clay, scattered and declining, mainly in England and Wales. Most populations are confined to pools and poached areas on heavily grazed commons and village greens, but habitats also include damp heathy pasture, lake shores and coastal grassland. 'Pennyroyal' dates from the mid-sixteenth century, from Anglo-Norman French *puliol* (based on Latin *pulegium* = 'thyme') + *real* = 'royal'. *Mentha* comes from the Latin name of a Greek nymph, Minthe, who was turned into a mint, and Latin *pulegium* = 'relating to fleas' (*pulex* = 'flea'), since this plant was believed to repel these insects.

PENNYWORT Herbaceous perennial species of *Hydrocotyle* (Hydrocotylidae) (Greek *hydor* + *cotyle* = 'water' + 'cup'). The neophyte **New Zealand pennywort** *H. novae-zelandiae* and **hairy pennywort** *H. moschata* are found in lawns and turf in a few places.
Floating pennywort *H. ranunculoides*, a North American neophyte either rooted in mud or growing as free-floating or emergent colonies on the surface of still or slowly moving water. It can form dense colonies that exclude native species. Rooted plants flower and set seed; floating plants spread by stem fragments. It is sold in aquatic plant nurseries, first recorded in the wild in 1990 in south Essex, now locally common in England, Wales and north-east Ireland. Its high growth rate (up to 20 cm per day), competitive ability and rapid rate of spread are cause for concern. It is listed under Schedule 9 of the WCA (1981), and eradication measures currently total £250,000–£300,000 in herbicide per annum (in 2010 estimated at £1,800–£2,000 per km for removal). Latin *ranunculoides* = 'resembling ranunculus' (buttercup).
Marsh pennywort *H. vulgaris*, locally common and widespread, mat-forming in lake sides, carr, mire, fen, fen-meadow, swamp, marsh, dune slacks and wet hollows in stabilised shingle. In very oceanic areas it grows in drier habitats. *Vulgaris* is Latin for 'common'.
Wall pennywort Alternative name for navelwort.

PENTATOMOIDEA Heteropteran superfamily that contains the shieldbugs, from Greek *pente* + *tomos* = 'five' + 'a cut', usually having antennae with five segments.

PEONY Garden peony *Paeonia officinalis* (Paeoniaceae), a perennial herb from southern Europe, introduced by Tudor times, a garden escape or throw-out in woodland, on verges and waste ground, and as a relic of cultivation. Paeon was physician of the gods in Greek mythology. *Officinalis* is Latin for having pharmacological properties.

PEPPER Sweet pepper *Capsicum annuum* (Solanaceae), an annual tropical American neophyte, occasionally casual on tips and at sewerage works in southern England. The binomial is from the Latin for this plant and for 'annual'.

PEPPER POT *Myriostoma coliforme* (Agaricomyces, Geastraceae), a rare inedible earthstar fungus that for over a century was thought to have become extinct in Britain, but rediscovered in 1996 on sandy soil in the Channel Islands, and in 2006 on the edge of oak woodland at one of its previously known sites in east Suffolk, and close by on a verge in 2010 and 2014. It fruits in autumn but can persist over winter. Greek *myrias* + *stoma* = 'myriad' + 'mouth', referring to the openings through which the spores are ejected, and for the same reason Latin *coliforme* = 'strainer'.

PEPPERED-SAXIFRAGE *Silaum silaus* (Apiaceae), a herbaceous perennial, locally common in England, scarce in Wales and south-east Scotland, in damp, unimproved neutral grassland, usually on clay. Habitats include hay and water meadow, species-rich pasture and roadsides, occasionally on chalk downland, railway banks and vegetated shingle. *Silaus* is Latin for a kind of parsley.

PEPPERMINT *Mentha* x *piperita* (*M. aquatica* x *M. spicata*) (Lamiaceae) a rhizomatous perennial herb, widespread but scattered, on damp ground and waste places. Glabrous (smooth) plants are probably garden escapes or throw-outs; pubescent (hairy) plants have probably arisen spontaneously. *Mentha* comes from the Latin name of a Greek nymph Minthe, who was turned into a mint; *piperita* is Latin for 'pepper-like'.

PEPPERWORT Annuals, occasionally biennials, in the genus *Lepidium* (Brassicaceae) (Greek *lepis* = 'scale', referring to the seed pod shape).

Field pepperwort *L. campestre*, scattered, an archaeophyte of open grassland and arable fields, especially on sandy or gravelly soils; also found on roadsides and walls, in gardens and waste places. It can be persistent, but frequently occurs as a casual. *Campus* is Latin for 'field'.

Least pepperwort *L. virginicum*, a North American neophyte in cultivation by 1713, scattered and declining in range, on railways and waste ground, and in arable fields, derived from grain impurities, and in wool shoddy and birdseed. Usually casual, it sometimes persists. *Virginicum* is Latin for 'Virginian'.

Narrow-leaved pepperwort *L. ruderale*, an archaeophyte common in the southern half of England, scattered and scarce elsewhere, naturalised on banks and bare wasteland near the sea, where in places it might be native, and (increasingly) on salted road verges. It is also common as a casual of gardens and waste ground. *Ruderale*, strictly from Latin *rudera* = 'rubbish', refers to a plant growing in a disturbed habitat.

Perfoliate pepperwort *L. perfoliatum*, a European neophyte present by 1640, a rare casual on roadsides, and by docks as a seed and grain impurity.

Smith's pepperwort *L. heterophyllum*, a widespread and locally common perennial of acidic soils in dry heathy and gravelly habitats, frequent on shingle, railway ballast and embankments, and occasional in arable fields. Greek *heterophyllum* = 'differently leaved'.

PERCH *Perca fluviatilis* (Percidae), average length 20 cm, weight 0.25 kg, a common and widespread predatory fish that feeds on invertebrates and smaller fish in slow-flowing lowland rivers, lakes and ponds. Spawning is in March–June. In the 1960s and 1970s an ulcer disease (epizootic haematopoietic necrosis) seriously affected the population in Britain, with some waters losing almost all its perch; the disease died down in the 1980s but its effects are still felt. It is fished for food and in game angling. Greek *perce* = 'perch' and Latin *fluvius* = 'river'.

PERCIDAE Perch family, including ruffe and zander.

PERCIFORMES Order of fishes that includes Percidae, Moronidae and Sparidae.

PEREGRINE *Falco peregrinus* (Falconidae), wingspan 74–120 cm, length 34–58 cm, male weight 670 g, females 1.1 kg, widespread, but on the edge of extinction in Britain in the 1960s because of illegal killing by gamekeepers, the impact of egg collectors, and mainly because of the effects of pesticide bioconcentration to lethal and sublethal (eggshell thinning) levels. Numbers began to recover following a ban on certain organochlorine pesticides. A BTO survey in 2014 estimated the breeding population at 1,769 pairs, 22% greater than in 2002 (1,437), largely because of increases in lowland areas, particularly in England, where many peregrines were breeding on man-made structures. The number of breeding peregrines in England was almost twice that reported in 2002 (828 vs 469). In Wales the population was stable at 280 pairs. Numbers fell in the Isle of Man (22 vs 31) and Scotland (523 vs 573). There was a slight increase in Northern Ireland (103 vs 81). In both Scotland and England, upland populations showed greatest declines. In East Anglia, the population increased from no pairs in 2002 to 44 pairs in 2014. Sixteen pairs were recorded in the Channel Islands. Nests are on cliff edges and, increasingly, their urban equivalent on buildings. Clutch size is 3–4, incubation takes 31–3 days, and fledging of the altricial young 39–40 days. Diet is of medium-sized birds (in urban areas, particularly feral pigeons), usually caught in a steep dive (stoop) that is often 240 km/h and not uncommonly 300 km/h (390 km/h has been recorded outside Britain), the fastest speed of any bird species. 'Peregrine' means 'pilgrim falcon', because falconers' birds were caught fully grown on migration, not taken from the nest. The binomial is Latin for 'falcon' and 'pilgrim' or 'traveller' (*per* + *ager* = 'through' + 'field').

PERISSODACTYLA From Greek *perisso* + *dactylos* = 'uneven' + 'toe', order of odd-toed ungulates (hoofed animals with their weight borne mostly or entirely by the third toe) that includes the Equidae (horses and ponies).

PERIWINKLE (PLANT) Herbaceous perennials in the genus *Vinca* (Apocynaceae), Latin for these plants. 'Periwinkle' comes from late Old English *peruince*, in turn from late Latin *pervinca*.

Greater periwinkle *V. major*, a Mediterranean neophyte in gardens by 1597, widespread, commoner in the south, more scattered in the north, on waste ground, verges, shaded banks and woodland. *Major* is Latin for 'greater'.

Intermediate periwinkle *V. difformis*, a spreading European neophyte in cultivation by 1890, found in southern and south-west England, south Wales and Co. Waterford as a naturalised garden escape on grassy banks and verges. *Difformis* is Latin for 'differing shape'.

Lesser periwinkle *V. minor*, a common and widespread archaeophyte naturalised in woodland, roadside banks and verges, and waste ground. *Minor* is Latin for 'lesser'.

PERIWINKLE (SNAIL) Intertidal species of *Littorina* (Latin *littus* = 'shore') and the monotypic *Melarhaphe* (Greek *melas* + *raphe* = 'black' + 'seam'), in the Littorinidae.

Common periwinkle *L. littorea*, shell height 5.2 cm, common and widespread on rocky coasts from the upper shore into the sublittoral, grazing on algae or suspension-feeding. In sheltered conditions they can also be found in sandy or muddy habitats such as estuaries and mudflats. They tend to form clusters on sites that are more favourable, such as rock pools. The species has some commercial value and is gathered by hand at a number of localities, particularly in Scotland and in Ireland where the industry is valued at around £5 million per year.

Flat periwinkle *L. obtusata*, shell height 1.5 cm, widespread and common on shores feeding on brown seaweeds from mid- to lower tidal levels on rocky shores and occasionally into the sublittoral. *Obtusata* is Latin for 'blunt'.

Rough periwinkle *L. saxatilis*, shell height 1.8 cm, width 1.4 cm, widespread and common from the upper eulittoral zone to the littoral fringe of the intertidal, typically in crevices, empty barnacle shells and under stones. It also occurs on saltmarsh on the base of cord-grass, on firm mud banks, and submerged in sheltered, brackish lagoons attached to macrophytes. *Saxatilis* is Latin for 'growing among rocks'.

Small periwinkle *M. neritoides*, shell height 0.9 cm, common and widespread, high on the rocky shore in cracks and crevices around most of the coastline, feeding on black lichens and detritus. *Nerites* is Greek for a kind of shellfish.

PERLIDAE Possibly from Latin *perula* via *perla* = 'pearl', describing the round head, the stonefly family comprising the large *Dinocras cephalotes* and *Perla bipunctata*, both common in stony upland streams, particularly in western and northern regions.

PERLODIDAE Possibly from Latin *perula* via *perla* = 'pearl', describing the round head, the stonefly family comprising the monospecific genera *Diura* (uncommon), *Isogenus* (catastrophic recent decline, now only known from the Welsh River Dee) and *Perlodes microcephalus* (common), plus *Isoperla* (*I. obscura* possibly extinct, *I. grammaticus* being common), all with a patchy western and northern British distribution. These are the fisherman's yellow sally.

PERSICARIA Annuals in the genus *Persicaria* (Polygonaceae), the name reflecting the similarity of the leaves to those of peach *Prunus persicaria*.

Nepal persicaria *P. nepalensis*, a Himalayan neophyte introduced by 1917, naturalised in cultivated and rough ground in several nurseries in Somerset, Dorset and Hampshire, and a casual from birdseed on waste ground.

Pale persicaria *P. lapathifolia*, widespread and generally common in cultivated fields, and on waste ground and edges of lakes, ponds and streams. *Lapathifolia* is Latin for having leaves like sorrel (*lapathus*).

PERTUSARIA **Alpine moss pertusaria** *Pertusaria bryontha* (Pertusariaceae), an endangered crustose lichen, with a few records from the Highlands and one from Cumbria. Latin *pertusus* = 'perforated', and Greek *bryon* = 'lichen'.

PETALWORT *Petalophyllum ralfsii* (Petalophyllaceae), a scarce thallose liverwort with a scattered distribution on base-rich, damp ground in dune slacks and in short, sandy coastal grass. Greek *petalos* + *phyllon* = 'outspread' + 'leaf'.

PETER (CADDIS FLY) Species of *Agrypnia* (Phryganeidae) (Greek *agrypnos* = 'watchful').

Dark peter *A. obsoleta*, found in Scotland and Ireland, the mainly predatory larvae favouring soft substrates, including plants and wood, in the littoral of especially acid standing waters. Adults fly in June–October.

Speckled peter *A. varia*, locally common and widespread, larvae of ponds and lakes, and tolerant of brackish conditions. Life history is similar to dark peter. *Varia* can mean 'speckled'.

PETREL 1) Oceanic birds in the family Hydrobatidae. The name comes from a corruption of *pitteral*, referring to the bird's pitter-pattering across the water as it feeds from the ocean surface.

Leach's petrel *Oceanodroma leucorhoa* (Procellariidae), wingspan 46 cm, length 20 cm, weight 45 g, an Amber-listed migrant breeding on cliffs in the Hebrides, Orkney and Shetland, and at one site in Ireland, The Stags of Broadhaven, Co. Mayo. It is more widespread as a passage migrant. There were an estimated 48,000 pairs in 2000. They feed at sea on fish and crustaceans, taking these from the surface, during the day, returning to their burrow or crevice nests at night to avoid predation. They lay one egg, incubating this for 41–2 days, the altricial young fledging in 63–70 days. Most birds migrate in winter to the tropics, although a few remain in the North Atlantic. William Leach (1791–1836) was a British zoologist. Greek *oceanos* + *dromos* = 'ocean' + 'runner', and *leucos* + *orrhos* = 'white' + 'rump'.

Storm petrel *Hydrobates pelagicus* (Hydrobatidae), wingspan 38 cm, length 16 cm, weight 27 g, an Amber-listed migrant breeding largely on islands on the west coasts of both Britain and Ireland, Orkney and Shetland, with 26,000 nests in the early 2000s. They feed at sea on fish and crustaceans, taking these from the surface, during the day, returning to their burrow or crevice nests at night to avoid predation. They lay one egg, incubating this for 38–50 days, the altricial young fledging in 56–73 days. They migrate in September–October to waters off South Africa. 'Storm' comes from the association of this bird with bad weather by sailors. Greek *hydor* + *bates* = 'water' + 'walker', and *pelagicos* = 'of the sea'.

Wilson's petrel *Oceanites oceanicus* (Procellariidae), a scarce visitor from the Southern Ocean with an average of nine records p.a. off the coasts of Cornwall and the Scilly Isles. Alexander Wilson (1766–1813) was a Scottish-American ornithologist. The binomial refers to the mythical Oceanids, the 3,000 daughters of Tethys, and the Latin *oceanus* = 'ocean'.

2) **Manx petrel** Alternative name for Manx shearwater.

PETROMYZONTIDAE, PETROMYZONIFORMES Family and order of lampreys, from Greek *petra* + *myzo* = 'stone' + 'to suck', from the habit of gripping stones in order to build nests.

PETUNIA *Petunia* x *hybrida* (Solanaceae), of garden origin, a frequent escape on rough ground, scattered mostly in England and Wales, the name being the latinised form of the Brazilian *petun* = 'tobacco' (another member of the Solanaceae).

PHACELIA *Phacelia tanacetifolia* (Boraginaceae), a Californian neophyte grown since 1832 in gardens as an ornamental and for its nectar to encourage bees and hoverflies, as well as in fields as green manure, and a contaminant of grass seed, increasingly found as a casual on waste ground, and among crops and newly sown grass. Greek *phacelos* = 'bundle', and Latin *tanacetifolia* = 'leaves like tansy' (*Tanacetum*).

PHAEOPHYCEAE Class of brown algae, including the order Laminariales (wrack and kelp) (Greek *phaios* + *phycos* = 'dark' + 'seaweed').

PHALACRIDAE From Greek *phalacros* = 'bald', a beetle family with 14 species (1.5–3 mm length).

PHALACROCORACIDAE Family name of cormorant and shag (Greek *phalacros* + *corax* = 'bald' + 'raven').

PHALAROPE Migrant shorebirds in the genus *Phalaropus* (Scolopacidae), from Greek *phalaris* + *pous* = 'coot' + 'foot', since phalaropes and coots both have lobed toes.

Grey phalarope *P. fulicarius*, wingspan 42 cm, length 21 cm, male weight 50 g, females 63 g, an arctic breeding passage visitor normally wintering in the Atlantic but sometimes seen on British and especially Irish coasts, usually after storms, between October and January, with variable numbers but in general 160–200 birds p.a. *Fulica* is Latin for 'coot'.

Red-necked phalarope *P. lobatus*, wingspan 36 g, length 18 cm, weight 36 g, a Red List migrant breeder in the Outer Hebrides, Orkney and Shetland, with 15 sites and 20–7 males recorded over 2006–10. There have been few breeding records in Ireland since the 1970s, but breeding was reported from Co. Mayo in 2015. The nest is a grass-lined depression on top of a small mound, clutch size is usually four, incubation by the male lasts 17–21 days, and the precocial chicks fledge in 19–21 days. Females in this polyandrous species leave the nest after laying to seek other males, sequentially, to initiate further broods. Small numbers (around 30 birds) are also seen on passage in autumn, usually along the east coast, especially in East Anglia. Diet is of small invertebrates, eaten while swimming (often spinning in a small, rapid circle), wading or walking on land. *Lobatus* is Latin for 'lobed', referring to the toes.

PHANEROGAM From Greek *phaneros* + *gamos* = 'visible' + 'marriage', the subdivision of all flowering (seed-producing) plants; also known as spermatophytes.

PHANTOM MIDGE Members of the family Chaoboridae.

PHARIDAE Family name of razor shells (Greek *pharos* = 'mantle').

PHASCUM Floerke's phascum *Microbryum floerkeanum* (Pottiaceae), a cleistocarpous moss associated with chalk and limestone habitats in England. Greek *phascon* = 'moss tuft', and *micros* + *bryon* = 'small' + 'moss'.

PHASIANIDAE Family name of pheasant (Latin *phasianus*), capercaillie, partridge, ptarmigan and quail.

PHASMIDA Order of stick insects (Greek *phasma* = 'monster' or 'vision').

PHEASANT *Phasianus colchicus* (Phasianidae), wingspan 80 cm, length 71 cm, male weight 1.4 kg, females 980 g, a game bird introduced from Europe, with wild populations recorded by the eleventh century. Recent introductions have brought in a number of races and breeds for sport shooting. It is common and widespread, with around 2.2 million birds present in summer, and some 38 million birds released each year for the shooting season (1 October–1 February). It is commonly seen in and around arable fields, but also in broadleaf woodland and reedbeds, feeding on plant material and insects. Nests are on the ground, clutch size is 10–14, incubation lasts 23–8 days, and the precocial young fledge in 11–13 days. 'Pheasant' comes from Middle English and Old French *fesan*, from Greek *phasianos* = '(bird) of Phasis', the name of a Caucasian river from which the bird is said to have spread westwards. The binomial is Latin for 'pheasant' and for 'of Colchis', a country on the Black Sea where this species became known to Europeans.

Golden pheasant *Chrysolophus pictus*, wingspan 70 cm, length 88 cm, weight 630 g, scattered in Britain, mainly in south-central England, East Anglia, north-west Wales and south-west Scotland, introduced from China, with the first records of breeding in Norfolk and Suffolk, as a game bird as well as an ornamental, in the 1870s. There were perhaps 1,000–2,000 birds in the 1980s and 1990s, but only around 100 pairs in 2009. In 2016 the Rare Breeding Birds Panel concluded that populations were no longer self-sustainable, though illegal releases were still being made. The bird favours dense, usually coniferous forest, roosting in trees at night but nesting on the ground, and feeding on leaves and buds. Clutch size is 5–12, incubation takes 22–3 days, and the precocial young fledge in 12–14 days. Its rarity in China makes even a low British population significant – the IUCN lists the species as Near Threatened. Hybridisation with Lady Amherst's pheasant provides a further threat. Greek *chrysolophos* = 'with golden crest', and Latin *pictus* = 'painted'.

Lady Amherst's pheasant *C. amherstiae*, wingspan 78 cm, length 90 cm, weight 740 g, scarce and scattered in England, named after Sarah, Countess Amherst, wife of the Governor General of Bengal, who was responsible for sending the first specimen of this species (native of China) to London in 1828. It was breeding in captivity by 1871, and had a few further discrete introductions afterwards. In the late 1960s numbers (mainly in Bedfordshire) were around 250, but numbers were only 5–10 pairs in 2009, a decline caused by predation, hybridisation with golden pheasant and loss of appropriate woodland habitat. In 2016 the Rare Breeding Birds Panel concluded that populations were no longer self-sustainable, though illegal releases were still being made.

PHEASANT SHELL *Tricolia pullus* (Phasianellidae), shell length 9 mm, width 5 mm, widespread along the south and west coasts of Britain, and around Ireland, on rocky coasts from the lower shore to about 35 m, usually among red algae. *Pullus* is here Latin for 'dark'.

PHEASANT'S-EYE *Adonis annua* (Ranunculaceae), an annual known from Iron Age deposits, an arable weed of dry soils on chalk and limestone, and also recorded from tracks and other disturbed habitats. Range is declining because of improved seed-cleaning and increased use of agrochemicals, and it is now a rare

casual in southern England, with small populations restricted to field edges. In Greek mythology, Adonis was not only the god of beauty and desire, but also of plants.

PHEASANT'S-TAIL *Anemanthele lessoniana* (Poaceae), a perennial New Zealand neophyte, local in southern England and east Ireland as a garden escape in rough ground and pavement cracks. Greek *anemos* = 'wind'. René Lesson (1794–1849) was a French physician and naturalist.

PHILODROMIDAE Family name of running crab spiders (Greek *philos* + *dromos* = 'fondness of' + 'running'), with 15 British species in 3 genera. See turf-running spider under spider.

PHILOPOTAMIDAE Family name (Greek *philos* + *potamos* = 'liking' + 'river') of so-called fingernet caddisflies whose larvae favour fast-flowing water, using mesh bags to capture food items.

PHILOSCIIDAE Family name (Greek *philos* + *scia* = 'fondness of' + 'shadow') for common striped woodlouse, the only species in this isopod family recently recorded in Britain.

PHLOIOPHILIDAE Beetle family (Greek *phloios* + *philos* = 'bark' + 'loving'), the single British representative, *Phloiophilus edwardsi*, being scattered in England, Wales and the Highlands, associated with the fungus *Peniophora quercina*.

PHOCIDAE Family name for earless seals (Greek *phoce* = 'seal').

PHOCINE DISTEMPER A virus of the genus *Morbillivirus* that is pathogenic for pinnipeds, particularly seals, first recognised in 1988 following a massive epidemic in harbour and grey seals in north-west Europe.

PHOCOENIDAE Family name for harbour porpoise (Greek *phocaina* = 'porpoise').

PHOENIX *Eulithis prunata* (Geometridae, Larentiinae), wingspan 30–5 mm, a widespread nocturnal moth found in gardens, allotments and woodland. Adults fly in July–August. Larvae feed on black currant, red currant, and gooseberry. Greek *eu* + *lithos* = 'good' + 'stone', the colour of some species resembling sandstone, and Latin *prunata* from a supposed food plant *Prunus* (rather than the more accurate *Ribes*).

Small phoenix *Ecliptopera silaceata*, wingspan 23–7 mm, nocturnal, widespread and common in woodland rides, gardens, hedgerows, calcareous grassland, heathland and fen, and on allotments and verges, but a species of conservation concern in the UK BAP because numbers have declined by 77% over the last 35 years. The bivoltine adults fly in May–July and August–September, except in the north where there is only one generation. Larvae feed on willowherbs and enchanter's-nightshade. Greek *ecleipo* + *ops* = 'to fail' + 'face', and Latin *sil* for a yellowish earth, describing the wing colour.

PHOLADIDAE Family name of boring bivalve piddocks (Greek *pholas* = 'hole-living').

PHOLCIDAE Family name (Greek *pholcos* = 'squint-eyed') for daddy-long-legs spiders or cellar spiders, with three British species.

PHORIDAE Family name (Greek *phora* = 'movement') of the <1–8 mm length scuttle flies in the superfamily Platypezoidea, with 344 species in 13 genera, by far the most speciose being *Megaselia*.

PHORONIDA Phylum of horseshoe worms. In Greek mythology, Phoronis was the surname of Io, a mortal princess loved by Zeus.

PHRYGANEIDAE From Greek *phryganon* = 'dry stick', describing the larval case, the family name of some caddis flies.

PHRYMACEAE Family name of monkeyflowers, the North American *Phryma* being the type genus (not found in Britain).

PHTHIRAPTERA Order of lice, *phtheir* being Greek for 'louse', *apteryx* = 'wingless'.

PHYLLOPHORACEAE From Greek *phyllon* + *phoros* = 'leaf' + 'carrying', the family of red algae in the order Gigartinales.

PHYLLOSCOPIDAE Family name of leaf warblers (Greek *phyllon* + *scopa* = 'leaf' + 'twigs').

PHYSCIA **Southern grey physcia** *Physcia tribacioides* (Physciaceae), a foliose lichen, local in southern England and west Wales, scattered in (especially southern) Ireland, on well-lit nutrient-rich bark, especially close to the coast, having become rare in England following the loss of mature elm trees, though also growing on other tree species. Vulnerable in the RDB, protected by the WCA (1981). Greek *physce* = 'blister' (i.e. inflated), and *tribax* = 'worn down'.

PHYSIDAE Family name of the freshwater pulmonate bladder snails (Greek *physa* = 'bladder').

PHYTOLACCACEAE Family name of pokeweeds (Greek *phyton* + *laccos* = 'plant' + 'pit').

PICIDAE, PICIFORMES Family and order names of woodpeckers and wryneck (Latin *picus* = 'woodpecker').

PICK-A-BACK-PLANT *Tolmiea menziesii* (Saxifragaceae), a rhizomatous perennial introduced from North America in 1812, widespread and locally naturalised in nitrogen-rich soils in damp woodland, by streams and other moist, shady places, occasionally on waste ground. The common name reflects the many small flowers that are carried in a loose raceme. William Tolmie (1812–66) was a Scottish botanist, and Archibald Menzies (1754–1842) a Scottish plant collector.

PICKERELWEED *Pontederia cordata* (Pontederiaceae), an aquatic rhizomatous herbaceous North American neophyte perennial, scattered in England north to Lancashire, and in Co. Cork, naturalised as a marginal on pond edges. Pickerel is the North American name for some pike species. Guilio Pontedera (1688–1754) was an Italian botanist. Latin *cordata* = 'heart-shaped', describing the leaves.

PICTURE-WINGED FLY Members of the family Ulidiidae, named after the brightly coloured wing patterning. Adults feed on plant material, take nectar and pollen or prey on small insects; larvae feed on decaying plant material and in roots.

PIDDOCK Members of the boring bivalve family Pholadidae, and the filter-feeding Petricolidae.
　　American piddock *Petricolaria pholadiformis* (Petricolidae), shell size up to 6.4 cm, introduced from North America not later than 1890 to the River Crouch, Essex, now established from Lyme Regis, Dorset, to the Humber Estuary in the intertidal on clay, peat or soft-rock shores. Latin *petra* + *colo* = 'rock' + 'inhabit', and Greek *pholas* = 'hole-dwelling'.
　　Atlantic great piddock or **great piddock** Alternative name for oval piddock.
　　Common piddock *Pholas dactylus* (Pholadidae), shell size up to 15 cm, a borer into soft rock and submerged wood from the low intertidal, sublittoral and shallow shelf along the English Channel, south-west England, the Irish Sea and south-west Ireland Greek *pholas* = 'hole-dwelling' and *dactylos* = 'finger'.
　　Little piddock *Barnea parva* (Pholadidae), shell size up to 4 cm, a borer into soft rocks such as chalk, red sandstone and cement stone, in the intertidal and sublittoral from Pembrokeshire to Essex. *Parva* is Latin for 'small'.
　　Oval piddock *Zirfaea crispata* (Pholadidae), shell size up to 9 cm, a widespread but patchily distributed borer in soft sedimentary rock, from the low intertidal to depths of 10 m. *Crispa* is Latin for 'curly'.
　　White piddock *Barnea candida* (Pholadidae), shell size up to 6.5 cm, a widespread borer into soft substrates such as clay,

mudstone, shale and chalk, found in the intertidal and sublittoral to about 10 m. *Candida* is Latin for 'white'.

PIERCER (MOTH) Members of the family Tortricidae, subfamily Olethreutinae, predominantly in the genera *Cydia* (Greek *cydos* = 'glory', from the species' beauty), *Grapholita* (*graphe* + *litos* = 'design' + 'simple') and *Pammene* (*pan* + *mene* = 'all' + 'moon', from the circular wing markings of many species, suggesting a full moon). Wingspan is generally 10–14 mm.
　　Rare species include **ash-bark piercer** *Pammene suspectana* (East Anglia), **Breckland piercer** *Cydia millenniana* (sandy heathland in Breckland and elsewhere in East Anglia), **cabbage piercer** *Selania leplastriana* (on sea-cliffs in parts of Kent, Dorset and south Devon), **crescent piercer** *Grapholita orobana* (woodland, fen and marsh, and on riverbanks in parts of south and east England), **Devon piercer** *P. ignorata* (first recorded at Axminster, Devon, in 1986; in 2006, two specimens were found in Wetmoor, Gloucestershire; a further Gloucestershire record was at Lineover Wood, Cheltenham, in 2009; two specimens were taken in Northumberland in 2011; a possible record was from Wolverhampton in 2015), **early oak piercer** *P. giganteana* (woodland and heaths in parts of Britain, plus one post-2000 record from Co. Cork), **Guildford piercer** *P. agnotana* (known from a male in woodland at Newlands Corner, Surrey in 1961, with further specimens taken in Dorset in 2007 and Norfolk in 2011), **Inverness piercer** *P. luedersiana* (damp moorland in the Highlands), **Isle of Wight piercer** *G. gemmiferana* (on cliffs at Luccombe Chine on the Isle of Wight, and Sidmouth in south Devon), **Kent fruit piercer** *G. lobarzewskii* (in orchards at a few sites in Kent, and also from Kenilworth, Warwickshire, in 2009), **larch piercer** *C. illutana* (first reported in Britain in Berkshire in 1984, though the first known British specimen was later shown to have been taken at Southsea, Hampshire, in 1975), **liquorice piercer** *G. pallifrontana* (downland and scrub in parts of southern England, declining in range and numbers), **maple piercer** *P. trauniana* (woodland and scrub in parts of south and south-east England), **sainfoin piercer** *G. caecana* (chalk downland in southern England, especially Kent and Wiltshire), **sallow-shoot piercer** *C. servillana* (damp woodland margins and hedgerows in parts of southern England and south Wales), and **scarce spruce piercer** *C. pactolana* (coniferous woodland in parts of south-east England).
　　Acorn piercer *Pammene fasciana*, local in deciduous, especially oak, woodland, and parkland throughout much of England, more widely distributed in the south, Wales and parts of Ireland. Adults fly in June–July, and are partially diurnal, males in particular flying at sunrise and in the afternoon. Larvae feed in oak acorns and nuts of sweet chestnut, creating spongy galls. *Fascia* is Latin for 'a band', from the forewing marking.
　　Beech-mast piercer *P. herrichiana*, in woodland in parts of southern England, but extent of true distribution unknown owing to confusion with acorn piercer. Larvae mine beech mast. Gottlieb Herrich-Schäffer (1799–1874) was a German entomologist.
　　Black-bordered piercer *P. argyrana*, widespread but scattered and local in oak woodland. Adults fly at night in April–May. Larvae feed inside an oak gall, often that of a hymenopterous insect. *Argyros* is Greek for 'silver', from forewing markings.
　　Black-patch piercer *P. ochsenheimeriana*, wingspan 8–11 mm, scarce in coniferous woodland and plantations in parts of England, central Wales, south Scotland and Co. Louth. Adults fly in May–June during the afternoon. Larvae feed in needles of Scots pine and other conifers. Ferdinand Ochsenheimer (1767–1822) was a German actor and entomologist.
　　Black piercer *P. germana*, local in hedgerows and open woodland throughout much of England, in parts of Wales, and post-2000 in Co. Kerry and Co. Wicklow. Adults fly in May–June. Larvae feed in fruit of plum and shoots of oak, hawthorn and blackthorn. *Germanus* is Latin for 'related', from an affinity with a congeneric.

Blotched piercer *P. albuginana*, local in oak woodland and orchards, mainly in southern England, but recorded from as far north as Lancashire. Adults fly in May–June. Larvae feed inside oak galls formed by usually hymenopterous insects. *Albugines* is Latin for 'head scurf', from the ochreous sprinkled scales on the forewing.

Dark gorse piercer *G. internana*, wingspan 9–10 mm, day-flying, local on heathland and moorland throughout much of Britain, more frequently in the south. Adults fly in April–June, males being more active. Larvae feed on gorse seeds. *Internus* is Latin for 'inner', from the location of wing markings.

Deep-brown piercer *G. funebrana*, local in woodland, gardens, orchards, parks and scrub throughout much of Britain. Adults fly in June–July in sunshine and at night. Larvae feed in August–September inside the hips of dog-rose and other roses, occasionally on rowan berries. *Funebris* is Latin for 'funereal', from the dark brown forewing.

Drab oak piercer *P. splendidulana*, widespread but scarce in oak woodland, commoner in the south, and with a few records from Ireland. Adults fly in April–June from late afternoon into the early evening. Larvae feed on oak leaves. Latin *splendidulus* (diminutive of *splendidus*) = 'bright', from the forewing's metallic gloss.

Greenweed piercer *G. lathyrana*, wingspan 9–11 mm, local in dry pasture in a few places in England, particularly in Sussex and Surrey. Adults fly in sunshine in April–May. Larvae feed on dyer's greenweed. *Lathyrus* is the genus name for peas and vetches rather than the true food plant.

Large beech piercer *C. fagiglandana*, wingspan 13–19 mm, common in beech and other deciduous woodland in much of the southern half of England and south Wales, local in Ireland. Adults fly in June–July. Larvae feed inside beech nuts. Latin *fagus + glans* = 'beech' + 'nut'.

Little beech piercer *Strophedra weirana*, local in beechwood in much of the southern half of England and Wales. Adults fly at sunrise and in the late afternoon in June. Larvae mine beech leaves. Greek *strophos + edra* = 'twisted cord' + 'seat', from the larval habit of fastening food plant leaves using silk. John Weir (1822–94) was a British entomologist.

Little oak piercer *St. nitidana*, wingspan 8–10 mm, local in oakwood in much of England and Wales. Adults fly in the afternoon in May–June. Larvae mine oak leaves. *Nitidus* is Latin for 'shining', from the glossy forewing.

Marbled piercer *C. splendana*, wingspan 12–16 mm, widespread and locally common in oak and other deciduous woodland. Adults fly in July–August from evening into the night. Larvae feed inside oak acorns and sweet chestnut nuts. *Splendens* is Latin for 'magnificent'.

Northern crescent piercer *G. lunulana*, wingspan 11–17 mm, locally associated with hedgerows, rough pasture and south-facing hillsides in parts of northern England, Scotland, north Wales and, in Ireland, on the Burren and Co. Wexford. Adults fly by day in May–June. Larvae feed in the seed pods of a variety of herbaceous plants, including vetches. *Lunula* is Latin for a moon-shaped ornament, from the forewing marking.

Obscure birch piercer *P. obscurana*, on damp heathland and mosses in parts of England, south and central Scotland, and central Ireland. Adults fly in the afternoon and early evening in May–June. Larvae feed on birch catkins on the continent. *Obscurus* is Latin for 'obscure', from the indistinct wing pattern.

Obscure silver-striped piercer *C. cosmophorana*, wingspan 8–14 mm, scarce in pine woodland and heathland in parts of eastern England and south Scotland. Adults fly during the day in sunshine in May–June, and sometimes again in August as a small second brood. Larvae feed in one-year-old galls of pine resin-gall moth, old galls of dark pine knot-horn moths and in resinous nodules on pine bark. Greek *cosmos + phoreo* = 'ornament' + 'to wear', from the forewing markings.

Orange-spot piercer *P. aurana*, wingspan 9–13 mm,

widespread and common in hedgerows, rough ground and woodland edges. Adults fly in June–July during the afternoon. Larvae feed on hogweed seeds. *Aurum* is Latin for 'gold'.

Pale-bordered piercer *G. janthinana*, wingspan 9–11 mm, common in hedgerows, gardens and woodland edges throughout much of England (especially the southern half), Wales and Ireland. Adults fly in July–August in the afternoon and early evening. Larvae feed inside hawthorn berries. *Ianthinus* is Latin for 'violet-coloured', from the forewing colour.

Pine-bark piercer *C. coniferana*, uncommon in pine woodland in parts of Britain. Adults fly in May–August, with two generations in the south. Larvae feed inside damaged bark of Scots pine. *Conifer* is Latin for 'cone-bearing'.

Pine-cone piercer *C. conicolana*, occasional in pine woodland in southern England, East Anglia and south Wales. Adults fly in May–June during daylight and dusk. Larvae feed inside cones of, especially, Scots pine. *Conus + colo* is Latin for 'cone' and 'inhabit'.

Purple-shaded piercer *P. gallicana*, widespread on dry pasture, downland and dunes. Adults fly in July–August during the afternoon and from dusk onwards. Larvae feed on umbellifer seeds. Latin *gallicus* = 'French'.

Pygmy piercer *P. populana*, in woodland, marsh, riverbanks, fen and dunes in parts of England, Wales and Ireland. Adults fly in July–August. Larvae feed on leaves of goat and creeping willow, and osier. *Populus* is Latin for 'poplar', though this is not a food plant.

Red piercer *Lathronympha strigana*, wingspan 14–18 mm, widespread and common in open woodland and dry pasture, most numerous in southern England, least so in Scotland. Adults fly at sunrise and from dusk in June–July, sometimes with a second brood in August–September. Larvae feed on St John's-wort leaves. Greek *lathre + nymphe* = 'secretly' + 'nymph', from the larvae feeding in a spun case, and Latin *striga* = 'streak'.

Regal piercer *P. regiana*, wingspan 13–16 mm, widespread and common in parks, gardens and woodland. Adults fly in May–July. Larvae mine seeds of sycamore, occasionally Norway maple. *Regius* is Latin for 'royal' or 'magnificent', from the rich forewing markings.

Restharrow piercer *C. microgrammana*, scarce on dry pasture and waste ground in coastal parts of southern England, especially East Anglia, south Wales and east Ireland. Adults fly in June–July. Larvae mine restharrow seed heads. Greek *micros + gramma* = 'small' + 'something drawn', from the forewing line markings.

Sycamore piercer *P. aurita*, an adventive first discovered in Britain in 1943 in Kent. It has since become widely established, local in parks, gardens and woodland throughout much of England and Wales, spreading into south Scotland, and with post-2000 records from Co. Wicklow and Co. Wexford. Adults fly in July–August, from late afternoon into the night. Larvae mine sycamore seeds. *Aurita* is Latin for 'eared'.

Triangle-marked piercer *P. spiniana*, of woodland margin, hedgerow and scrub throughout much of England, and parts of Wales and Ireland. Adults fly in August–September. Larvae feed on flowers and leaves of blackthorn (*Prunus spinosa*, prompting the specific name) and hawthorn.

Triple-stripe piercer *G. compositella*, wingspan 9–10 mm, day-flying, widespread and common in dry grassland and flower meadows, especially in the south. Adults fly in May–June and August. Larvae feed on white and red clovers.

Vagrant piercer *C. amplana*, wingspan 13–20 mm, a rare immigrant appearing in increasing numbers in the West Country, notably at Portland, Dorset. The first authenticated record was from south Devon in 1990, and it is increasingly recorded along much of the south coast, together with one record from Spurn, North Humberside. Adults fly in August. Larvae feed on oak acorns, and nuts including those of hazel, walnut, sweet chestnut and beech. *Amplus* is Latin for 'large'.

Vetch piercer *G. jungiella*, widespread and common in woodland, downland and scrub. Adults fly in sunshine in April–May in the north; in the south a second generation may occur. Larvae feed on the leaves of bitter-vetch and bush vetch. *Jungiella* possibly refers to the eighteenth-century German collector R.C. Jung.

PIERIDAE Butterfly family that includes the whites and yellows. In mythology, Pieria in Greece was the home of the Muses. Subfamilies are **Dismorphiinae** (wood whites), **Pierinae** (orange tip and the whites) and **Coliadinae** (clouded yellow and brimstone).

PIGEON From Middle English *pejon* = 'young dove', members of the family Columbidae; see also woodpigeon.

Feral pigeon *Columba livia* f. *domestica*, wingspan 66 cm, length 32 cm, weight 300 g, domesticated from the rock dove, some then becoming feral, in particular adopting urban habitats, and becoming common and widespread, with 540,000 (440,000–640,000) breeding pairs estimated in 2009. Even so, a decline in numbers by 18% between 1995 and 2014 has been estimated by BBS. Flocks are of up to 500 birds, depending on food supply. The crude nests are 'constructed' on building ledges, in roof spaces, and similar locations. Birds breed at any time of year if there is sufficient food, with up to six clutches a year. Clutch size is usually 2, incubation takes 16–19 days, and the altricial young fledge in 35–7 days. The favoured diet is of grain, as seed, bread, etc., but birds will scavenge on most edible material deliberately or otherwise left by humans. *Columba* is Latin for 'dove', *livia* for 'bluish'.

PIGGYBACK Very small inedible fungi in the genus *Asterophora* (Agaricales, Lyophyllaceae) (Greek *aster* + *phoros* = 'star' + 'carrying'). **Powdery piggyback** *A. lycoperdoides* and **silky piggyback** *A. sericeum* are both widespread but uncommon, parasitic on old, decaying brittlegill and milkcap fungi.

PIGMY See under 'pygmy' spellings.

PIGNUT *Conopodium majus* (Apiaceae), a common and widespread (though locally declining) herbaceous perennial, in damp meadow and pasture, hedgerows, verges and woodlands, and characteristic of some northern hay meadows. The underground tuber, resembling a chestnut, has in the past been eaten as a root vegetable as well as for feeding pigs. Greek *conos* + *podion* = 'cone' + 'little foot', and Latin *majus* = 'larger'.

Great pignut *Bunium bulbocastanum*, a tuberous perennial herb of dry chalk soils, in (especially former) arable fields, and in broken turf on chalk downs, field edges, in hedgerows and scrub and on verges, locally in Buckinghamshire, Hertfordshire, Bedfordshire and Cambridgeshire. *Bounion* is Greek for a kind of earth nut, *bulbocastanum* Latin for having chestnut-like bulbs (tubers).

PIGWEED Species of *Amaranthus* (Amaranthaceae), occasional casuals introduced from the Americas, found as garden escapes and originating from wool shoddy and birdseed.

PIKE *Esox lucius* (Esocidae), length 70–100+ cm, weight up to 20 kg, common and widespread, favouring still or slow-flowing streams and shallow weedy places in lakes, an ambush predator feeding on invertebrates, fish (including small pike), frogs, waterfowl and rodents. Eggs are laid in vegetated shallows in February–May. 'Pike' is a Middle English word meaning 'pointed', referring to the pointed jaw akin to the infantry weapon. *Esox* comes from Greek *isox*, used for an unknown large fish, then in turn the Latin *esox* for this fish. An early common name, *luci* (now used in heraldry as *lucy*) is the basis of *lucius*.

PIKE-PERCH Alternative name for zander.

PIKE'S WEED *Pikea californica* (Dumontiaceae), a red seaweed with fronds 10–14 cm long, on rock and in pools in wave-exposed sites from the low intertidal down to 14 m, possibly introduced on boats from the US Pacific seaboard during World War II, first recorded from the Scilly Isles in 1983, and found near Falmouth (Cornwall) in 2015.

PILL BEETLE Member of the family Byrrhidae. Specifically, *Byrrhus pilula*, 8–10 mm length, widespread, found in damp habitats, adults and larvae probably feeding on mosses. Greek *byrrhos* = 'flame-coloured', and Latin *pilula* = 'globule'.

Banded pill beetle *B. fasciatus*, with a scattered distribution in Britain, including upland habitats. Adults and larvae feed on moss. *Fasciatus* is Latin for 'bundled'.

Mire pill beetle *Curimopsis nigrita*, 2 mm length, first discovered in England in 1977, and recorded from three sites in Yorkshire and Lincolnshire: Thorne and Hatfield Moors are the largest, both supporting many small and fragmented populations, and a small isolated population at Haxey Grange Fen, all associated with damp open peat and mire litter. Listed as Endangered. Greek *courimos* = 'cut off', and Latin *nigrita* = 'blackish'.

Northern pill beetle *B. arietinus*, scarce, with a scattered distribution, mostly in northern England. Adults and larvae feed on moss. *Arietinus* is Latin for 'ram-like'.

PILL BUG Alternative name for pill woodlouse.

PILLWORT *Pilularia globulifera* (Marsileaceae), a scarce grass-like fern, submerged for at least part of the year on infertile mud on pond and lake margins, with a scattered and declining distribution, common only in western and southern Britain and west Ireland. The binomial is from Latin for 'globule' and 'globule-bearing'.

PIMPERNEL Low-growing herbaceous members of the genera *Anagallis* (Greek for this plant, from *ana* + *agallein* = 'again' + 'to delight in', referring to the opening and closing of the flowers in response to sunlight) and *Lysimachia* (after King Lysimachus, who used loosestrife, *Lysimachia*, to calm his oxen), both Primulaceae.

Bog pimpernel *A. tenella*, a creeping, evergreen perennial of wet open sites. In south and east Britain it is very local on bare soil or bryophyte mats in calcareous dune slacks and short-sedge fens, sometimes on acidic bogs, and many sites have been lost to grassland improvement, eutrophication and drainage. In the west it is more widespread, growing in peaty mires, hillside flushes and rush pastures, where livestock or periodic flooding keeps sites open. *Tenella* is Latin for 'delicate'.

Scarlet pimpernel *A. arvensis* ssp. *arvensis*, a common and widespread winter-annual, occasionally a short-lived perennial, absent from much of Scotland, common in open habitats such as waste ground, and as an arable or garden weed. It also grows in rocky and bare sites including around rabbit burrows, and on coastal cliffs, chalk downland, heaths and dunes. *Arvensis* is from Latin for 'field'. **Blue pimpernel** (ssp. *foemina*) is a rarer archaeophyte with a similar distribution on arable land.

Yellow pimpernel *L. nemorum*, a common and widespread evergreen perennial of herb-rich deciduous woodland, though declining in southern England with loss of woodland; also found in old hedges, woodland glades, damp grassland, fen and marsh, as well as shaded gullies and cliffs in upland areas. It is typical of mesic brown earth soils kept relatively open by slope, shade, vernal herbs and disturbance, and avoids places where litter accumulates. *Nemorum* is from Latin for 'woodland glade'.

PIMPLET **Glaucous pimplet** *Anthopleura thallia* (Actiniidae), a sea anemone with a column up to 5 cm, with up to 100 tentacles, recorded from south-west England, Skokholm and the Outer Hebrides on rocky shores exposed to strong wave action, in pools, crevices, or among dense aggregations of mussels, preferably burrowing in any gravel. Greek *anthos* + *pleura* = 'flower' + 'side'. *Thallia* refers to 'thallus'.

PIN (FUNGUS) Inedible members of the genus *Cudoniella* (Helotiales, Helotiaceae).

Oak pin *C. acicularis*, widespread, more common but still infrequent in England and Wales, found from late summer to winter on rotting wood and stumps of oak and other hardwoods. *Acicularis* is Latin for 'needle-like'.

Spring pin *C. clavus*, infrequent though with a widespread distribution, less common in Ireland, found in spring and summer on often submerged woody and leafy debris in damp woodland habitats. *Clavus* is Latin for 'club'.

PIN-PALP Ground beetles in the genus *Bembidion* (Carabidae, Carabinae), all three species being of conservation concern in the UK BAP.

Pale pin-palp *B. testaceum*, length 4.5–5.5 mm, scarce and scattered, affected by changes in river habitat.

Scarce four-spot pin-palp *B. quadripustulatum*, length 4 mm, often associated with wetland, river and pond margins where litter accumulates, but currently much associated with brownfield sites. There has been a 50% decline in numbers in the last three decades, and while it appears to have expanded into the Ouse and Nene catchments of East Anglia in the last 15 years, there is no indication of a change in status elsewhere.

Thorne pin-palp *B. humerale*, found only in Thorne and Hatfield Moors, Yorkshire, endangered by habitat changes associated with peat cutting and changes in drainage.

PINACEAE Family name for pines (Latin *pinus*), including spruces, hemlock-spruces, firs, Douglas-fir, larches and cedars.

PINCERWORT Foliose liverworts in the genus *Cephalozia* (Cephaloziaceae) (Greek *cephale* = 'head').

Widespread but scattered on saturated peat are **blunt pincerwort** *C. pleniceps*, **bog pincerwort** *C. macrostachya*, **pale pincerwort** *C. leucantha* and **scissors pincerwort** *C. loitlesbergeri*.

Scarce species include **Irish pincerwort** *C. hibernica* (a few acidic bogs in west and south-west Ireland) and **snow pincerwort** *C. ambigua* (Highlands).

Chain pincerwort *C. catenulata*, widespread, mainly upland but also on heathland in southern England, on sphagnum and on damp, peaty soil. *Catenula* is Latin for 'small chain'.

Forcipated pincerwort *C. connivens*, common and widespread on bogs, acidic tussocks in fen and damp, peaty soil in heathland in lowland and submontane regions, occasionally on decomposing wood and moist sandstone rock. *Connivens* is Latin for 'converging'.

Moon-leaved pincerwort *C. lunulifolia*, common and widespread, especially in woodland (characteristic of decomposing logs and stumps), but also in open habitats such as rocky hill slopes, peatland and heathy habitat, occasionally on acidic rock faces. *Lunulifolia* is Latin for 'moon-shaped leaf'.

Two-horned pincerwort *C. bicuspidata*, common and widespread in acidic habitats on organic substrates such as peat and decomposing wood, and damp inorganic mineral soils and rocks. *Bicuspidata* is Latin for 'two-horned'.

PINCUSHION Species of moss, mainly in the genus *Ulota* (Orthotrichaceae).

Scattered species in upland regions include **balding pincushion** *U. calvescens* (mainly west Scotland and the Burren, especially on birch, hazel and rowan), **club pincushion** *U. coarctata* (Scotland, on birch, elder, oak and willow), **Drummond's pincushion** *U. drummondii* (especially Scotland), **Hutchins' pincushion** *U. hutchinsiae* (west Scotland and west Ireland, on igneous rocks), and **mountain pincushion** *D. crispula* (especially the Highlands, on scree and rocks).

Bruch's pincushion *U. bruchii*, common and widespread on twigs and branches of trees, rarely on rocks. Philipp Bruch (1781–1847) was a German pharmacist and bryologist.

Common pincushion *Dicranoweisia cirrata* (Rhabdoweisiaceae), common and widespread in Britain, scattered in Ireland, on trees, thatch and other organic substrates, and on exposed rocks and old stone walls in hilly districts. *Dicranos* is Greek for 'two-headed', *cirratus* Latin for 'curly'.

Crisped pincushion *U. crispa*, common and widespread, often upland, on twigs and branches, occasionally on rocks.

Frizzled pincushion *U. phyllantha*, common and widespread,

formerly mainly coastal but now much commoner inland, on tree branches and trunks, occasionally on rocks. Greek *phyllon* + *anthos* = 'leaf' + 'flower'.

Greater pincushion or **long-shanked pincushion** *Ptychomitrium polyphyllum* (Ptychomitriaceae), common and widespread, particularly in coastal districts, though absent from much of lowland England, on exposed or slightly shaded, acidic or weakly base-rich rocks. Greek *ptyche* + *mitrion* = 'fold' or 'leaf' + 'small cap', and *poly* + *phyllon* = 'many' + 'leaf'.

PINE Coniferous trees in the genus *Pinus* (Pinaceae) (Latin *pinus* = 'pine tree'). **Chile pine** is another name for monkey-puzzle.

Austrian pine *P. nigra nigra*, introduced from central Europe in 1835, widely planted in shelter belts and as an ornamental, preferring light soils, commonly self-sown and naturalised, for example on heathland. *Nigra* is Latin for 'black'.

Bhutan pine *P. wallichiana*, introduced from the Himalayas in 1823, grown as a parkland ornamental on non-calcareous soils, and recorded as self-sown in Surrey. Nathaniel Wallich (1786–1854) was a Danish-born physician, botanist and plant collector.

Corsican pine *P. nigra laricio*, an evergreen Mediterranean neophyte tree introduced in 1814, widely planted in shelter belts and as an ornamental, especially near the sea, and as a forestry tree on sandy soil in Scotland, though tolerating a wide range of soils, commonly self-sown and naturalised. *Laricio* means resembling larch (*Larix*).

Dwarf mountain pine *P. mugo*, introduced from Europe in 1774, very hardy and tolerant of a range of soil conditions. Often planted in upland areas, for example as wind breaks for young forestry plantations. *Mugho* is Italian for this species.

Lodgepole pine *P. contorta*, introduced from North America in 1851, widely planted and occasionally self-sown. It is a pioneer and light-demanding species which grows well on a range of nutrient-poor soils. Tolerance of acid peat soils, and resistance to winter cold, spring frost, exposure, air pollution and salt-laden winds all help explain its extensive use in upland forestry and in Ireland. *Contorta* is Latin for 'twisted'.

Macedonian pine *P. peuce*, introduced from the Balkans in 1864, mainly grown as a parkland ornamental on a wide range of soil moisture and from poor to medium soil nutrient regimes. *Peuce* is Greek for 'pine'.

Maritime pine *P. pinaster*, introduced from Mediterranean Europe by at least Tudor times, planted for timber or in shelter belts. It is a light-demanding pioneer species adapted to acid, poor or medium fertility soils and of moderately dry to moist soil moisture status, naturalised and invasive in Dorset and Hampshire heathland, and often planted near the sea in southern and south-west England. *Pinaster* is Latin for 'wild pine'.

Monterey pine *P. radiata*, introduced from California in 1832, commonly planted in south-west England and coastal Wales in parks and small forest plantations, and occasionally self-sown. *Radiata* is Latin for 'rayed'.

Scots pine *P. sylvestris*, native in the Highlands, widely planted elsewhere as an ornamental in parks and gardens and (less frequently in recent years) in forestry plantations, with self-set trees as light-demanding pioneer colonisers in heathlands, typically on base- and nutrient-poor soils. *Sylvestris* is Latin for 'wild', literally 'of woodland'.

Western yellow pine *P. ponderosa*, introduced from North America in 1827, common as a park ornamental. *Ponderosa* is Latin for 'heavy'.

Weymouth pine *P. strobus*, introduced from North America in 1605, planted as an ornamental in parks on poor to rich sandy or sandy loam soils of moderately dry to medium soil moisture. *Strobus* is Latin for 'pine (cone)', from Greek *strobos* = 'a whirling around'.

PINE CONE BUG *Gastrodes grossipes* (Lygaeidae), length 6–7 mm, scattered but locally common across Britain, associated with

Scots pine. Adults overwinter and mate in spring, and nymphs are found in May–July. Greek *gaster* + *eidos* = 'stomach' + 'form', and Latin *grossus* + *pes* = 'thick' + 'foot'.

PINEAPPLEWEED *Matricaria discoidea* (Asteraceae), an annual neophyte of disturbed, usually fertile ground, on roadsides, waste ground, tracks and field gateways, and in arable crops. In cultivation by 1781, it was first recorded in the wild as an escape from Kew Gardens in 1871, and in Ireland in 1894, but it became one of the fastest spreading plants in the twentieth century, dispersal aided by the transport of seeds on tyres and footwear. It produces a pineapple aroma when crushed. Latin *matrix* = 'womb', from the plant's former use in midwifery, and *discoidea* = 'disc-like'.

PINION (MOTH) 1) Nocturnal members of the Noctuidae (Xyleninae) in the genera *Agrochola* (Greek *agros* + *chole* = 'field' + 'bile'), *Cosmia* (*cosmios* = 'well-ordered'), *Lithophane* (*lithos* + *phaino* = 'stone' + 'to appear to be') and *Perizoma* (*perizoma* = 'girdle', from the mark like a belt when the wings are folded).

Brown-spot pinion *A. lychnidis*, wingspan 30–5 mm, common in broadleaf woodland, parkland, heathland, fen, scrub, hedgerows and gardens throughout Britain, but with numbers declining by 82% over 1968–2007, a species of conservation concern in the UK BAP. There are two records from Ireland, both in Co. Down. Adults fly in September–October. Larvae initially feed on herbaceous plants such as meadowsweet, sorrel and bladder campion, then on the leaves of deciduous trees. *Lychnis* (now *Silene*) is a genus of food plants.

Lesser-spotted pinion *C. affinis*, wingspan 28–35 mm, local in woodland, hedgerows and gardens throughout much of southern England and Wales. Adults fly in July–August. Larvae feed on elms, and the species underwent a rapid decline as a result of Dutch elm disease during the 1970s. *Affinis* is Latin for 'related' (i.e. to a congeneric).

Lunar-spotted pinion *C. pyralina*, wingspan 29–34 mm, local in hedgerows, scrub, woodland, parks and gardens throughout much of the southern half of England. Adults fly in July–August. Larvae feed on the leaves of a number of deciduous trees, favouring elms. *Pyralina* in an invented word meaning resembling a *Pyralis* species.

Pale pinion *L. socia*, wingspan 38–42 mm, local in broadleaf woodland and parks throughout much of England, Wales and Ireland. Adults fly in October–November, overwinter, and reappear the following spring when breeding occurs. Larvae feed on deciduous trees and later on herbaceous plants, such as docks. *Socia* is the Latin here meaning 'allied' (i.e. to a similar species).

Stone pinion Alternative name for Blair's shoulder-knot.

Tawny pinion *L. semibrunnea*, wingspan 40–4 mm, local in broadleaf woodland, parks and gardens in much of the southern half of England, rare in Wales, and with one record in Ireland (Co. Wicklow). Adults fly in October–November, overwinter, and reappear the following April–May. Larvae mainly feed on ash. *Semibrunnea* is Latin for 'half brown', from the darker dorsal (lower) part of the forewing.

White-spotted pinion *C. diffinis*, wingspan 29–36 mm, in woodland and thickets in parts of southern England, extending sparsely as far as the Lake District. It underwent a rapid decline as a result of Dutch elm disease during the 1970s, and is now a priority species in the UK BAP. Adults fly in July–September. Larvae feed on elms. *Diffinis* is an invented word conflating *differens* (= 'differing') and *affinis* (= 'related').

2) **Pretty pinion** *Perizoma blandiata* (Geometridae, Larentiinae), a moth found on moorland, upland pasture limestone hills and verges in the northern half of Scotland, Ireland (especially the Burren), and locally in Cumbria and a few other scattered sites in England. Larvae feed on eyebright flowers and seeds. Adults fly in the afternoon and at night in May–August. *Blandus* is Latin for 'smooth'.

PINK Species of *Dianthus* (Greek *Di* = 'of Zeus' + *anthos*

= 'flower', i.e. 'divine flower') (see also sweet-william) and *Petrorhagia* (*petra* + *rhagos* = 'rock' + 'break'), both Caryophyllaceae. 'Pink' comes from the verb meaning to decorate with a perforated or zigzag pattern, describing the frilled edge of the flowers. Specifically, *Dianthus plumarius*, a herbaceous perennial south-east European neophyte in gardens by 1629, scattered in England and Wales, naturalised on roadside banks and railway cuttings, especially on well-drained soils over chalk and limestone, and on old mortared walls. *Plumarius* is Latin for 'feather-like'.

Cheddar pink *D. gratianopolitanus*, perennial, mainly confined to crevices and ledges on Carboniferous limestone cliffs at Cheddar Gorge, Somerset, where it was once abundant before the lower slopes were stripped by plant collectors. Populations have recently increased following statutory protection, scrub clearance and reinstatement of grazing. It persists at some nearby sites where it has been planted, and is a scarce casual elsewhere in England. This is the county flower of Somerset as voted by the public in a 2002 survey by Plantlife. The specific name is New Latin for 'from Grenoble'.

Childing pink *P. nanteuilii*, an annual recorded from stabilised shingle in West Sussex and stabilised dunes in Jersey. It is also recorded as a rare casual around dockyards, in dry or reseeded grassland, and as a garden weed. Nanteuil is in western France.

Clove pink *D. caryophyllus*, a perennial European neophyte, an occasional casual garden throw-out on old walls, tips and verges. *Caryophyllus* is Latin for 'pink-coloured' or 'clove-scented'.

Deptford pink *D. armeria*, an annual or short-lived perennial scattered and declining in Britain (and probably never found in Deptford, south-east London), on basic, unimproved, drought-prone soils that experience occasional disturbance, for example short grassland, woodland edge, scrub, tracksides, railway sidings, dune slacks and waste ground. It was recorded, new to Ireland, at Horse Island, Co. Cork in 1992 on sub-maritime grassland, and in 2012 a possibly native site was found at Inis Meáin, Co. Clare, in species-rich calcareous grassland. *Armeria* is Latinised Old French *armoires*, meaning a cluster-headed dianthus.

Maiden pink *D. deltoides*, a perennial of dry, usually base-rich, soils overlying chalk, limestone, mica-schist or basalt (sometimes on metal-rich mining spoil), or sandy soils and dunes. It grows in dry grassland, and is also a garden escape. While widespread, if scattered, it is generally declining because of overgrazing and nutrient enrichment, or undergrazing and scrub encroachment. This is the county flower of Roxburghshire as voted by the public in a 2002 survey by Plantlife. *Deltoides* is Latin for 'triangle-shaped'.

Proliferous pink *P. prolifera*, an autumn-germinating annual of freely draining substrates. In Norfolk, it has been (re-)recorded since 1985 from the open edge of sandy grass-heath, and in Bedfordshire from a sandpit. It is a scarce casual at a few other sites in England. *Prolifera* is Latin for 'bearing offspring'.

Sea pink Alternative name for thrift.

PINK EYE See New Forest eye.

PINK-SORREL Species of *Oxalis* (Oxalidaceae), Greek for 'sorrel', from *oxys* = 'acid', referring to oxalic acid in the leaves and roots. Specifically, *O. articulata*, a rhizomatous perennial North American neophyte herb usually reproducing through semi-woody rhizome fragments, occasionally naturalised in disturbed areas, on stony and sandy waste ground, roadsides and seashores, always associated with human habitation, being widespread, commonest in southern England and south-east Ireland, scattered elsewhere. *Articulata* is Latin for 'jointed'.

Annual pink-sorrel *O. rosea*, an annual Chilean neophyte garden weed, naturalised in Jersey and west Cornwall, locally casual in eastern England. *Rosea* is Latin for 'rose-red'.

Garden pink-sorrel *O. latifolia*, a perennial Central and South American neophyte once grown as an ornamental, now a garden weed from which it spreads to rubbish tips It spreads vigorously by easily detached bulblets that are resistant to most herbicides. Mainly found in southern England, it is uncommon and scattered elsewhere. *Latifolia* is Latin for 'broad-leaved'.

Large-flowered pink-sorrel *O. debilis*, a perennial South American neophyte, introduced in 1826, once grown as an ornamental, becoming naturalised in gardens and on waste ground. It spreads rapidly by easily detached bulblets that are resistant to most herbicides, and in places it has become an almost ineradicable weed, mainly found in southern England, scattered elsewhere. *Debilis* is Latin for 'weak'.

Pale pink-sorrel *O. incarnata*, a perennial South African neophyte grown in gardens since 1739, occasionally escaping to nearby disturbed, shaded sites, hedge banks, stone walls and pavement cracks, mainly found in southern England, scattered elsewhere. *Incarnata* is Latin for 'flesh-coloured'.

PINKGILL Inedible fungi in the genus *Entoloma* (Greek *entos* + *loma* = 'inside' + 'fringe') plus tan pinkgill *Rhodocybe* (*rhodon* + *cybe* = 'rose-red' + 'head') (Agaricales, Entolomataceae), common in unimproved grassland or open woodland.

Scarce species include **April pinkgill** *E. aprile*, **big blue pinkgill** *E. bloxamii*, **black pinkgill** *E. aethiops*, **bluefoot pinkgill** *E. catalaunicum*, **drab pinkgill** *E. indutoides*, **hairy pinkgill** *E. tjallingiorum*, **oysterling pinkgill** *E. byssisedum*, **sepia pinkgill** *E. jubatum*, and **stump pinkgill** *E. euchroum*.

Also uncommon are **bicoloured pinkgill** *E. dichroum* (southern half of England, Wales, very rare in Scotland and Ireland), **excentric pinkgill** *E. excentricum* (in soil on especially calcareous grassland, occasionally dune grassland), **livid pinkgill** *E. sinuatum* (poisonous, commoner in England), **pine pinkgill** *E. nitidum* (acidic coniferous woodland soil, especially under pine), **rosy pinkgill** *E. roseum* (under often calcareous grassland), **tan pinkgill** *R. gemina* (in grassy habitat, often in deciduous woodland), and **yellow-foot pinkgill** *E. turbidum* (on soil in unimproved grassland and open coniferous woodland).

Aromatic pinkgill *E. pleopodium*, widespread and locally common on grassland soil, sometimes under trees. Greek *pleos* + *podion* = 'full' + 'small foot' (i.e. stem).

Blue edge pinkgill *E. serratulum*, widespread and common, found from early summer to late autumn on soil in both acid and calcareous grassland, unimproved pasture and woodland clearings. *Serratulum* is Latin for 'finely serrated', describing the gill edges.

Cream pinkgill *E. sericellum*, widespread but only locally common, found in summer and autumn in pasture, in grassy and mossy areas under broadleaf trees in parkland, and on the edges of open deciduous woodland. *Sericon* is Greek for 'silk'.

Crow pinkgill *E. corvinum*, locally common and widespread, found in summer and autumn on soil among pasture grass. *Corvus* is Latin for 'crow'.

Felted pinkgill *E. griseocyaneum*, widespread but scattered and locally common, found in summer and autumn on upland acid soils under often sheep-grazed pastures. Latin *griseocyaneum* = 'grey-blue'.

Honey pinkgill *E. cetratum*, widespread and common in Britain, scarce in Ireland, found in summer and autumn on coniferous woodland soil, often among moss, and particularly under spruce. *Cetratum* is Latin for 'armoured'.

Indigo pinkgill *E. chalybaeum*, poisonous, widespread, commoner in west Scotland, north-west England and Wales, found in summer and autumn on soil in moist upland forest, occasionally lowland woodland. Greek *chalyps* = 'steel', from the grey colour.

Lilac pinkgill *E. porphyrophaeum*, locally common in unimproved moist grassland in northern Britain and an occasional find in southern Britain and Ireland, found from spring to autumn. Greek *porphyros* + *phaeos* = 'purple' + 'dusky'.

Mealy pinkgill *E. prunuloides*, widespread but uncommon, fruiting from late summer to late autumn on unimproved, often sheep-grazed pasture. *Prunulus* is Latin for 'pruinose' (i.e. as if covered in a fine white powder).

Mousepee pinkgill *E. incanum*, poisonous, widespread but scattered and uncommon, with a distinctive odour said to resemble mouse urine (from acetamide, derived from acetic acid), found in late summer and autumn on calcareous soil under unimproved grassland, occasionally woodland clearings. *Incanum* is Latin for 'grey' or 'hoary'.

Papillate pinkgill *E. papillatum*, widespread and locally common, found from spring to autumn on soil in short, often calcareous grassland, including lawns. Latin *papillatum* = 'papillate' (having a small nipple-like projection on the cap).

Pimple pinkgill *E. hebes*, widespread but scattered and locally common (especially in England), on woodland or parkland soil, usually under broadleaf trees, often in grassy areas. *Hebes* is Latin for 'blunt'.

Shield pinkgill *E. clypeatum*, widespread, commoner in England, found in spring and early summer on soil under rosaceous bushes and trees such as hawthorn, wild cherry and wild roses, with which it is probably mycorrhizal. Latin *clypeus* = 'shield', from the round shape of the cap.

Silky pinkgill *E. sericeum*, widespread and common, found in summer and autumn on soil in unimproved grassland, verges, moorland and woodland clearings. *Sericum* is Latin for 'silk'.

Star pinkgill *E. conferendum*, common and widespread, found from spring to late autumn on soil in grassland, parks and grassy woodland clearings. *Conferendus* is Latin for 'gathered together'.

Wood pinkgill *E. rhodopolium*, widespread and common, poisonous (toxins including muscarine), occurring in summer and autumn on soil in broadleaf woodland and woodland edges. Greek *rhodon* + *polos* = 'rose-red' + 'pole' ('stem').

PINKWEED *Persicaria pensylvanica* (Polygonaceae), an annual North American neophyte introduced by 1800, an occasional casual on waste ground, often associated with soya bean waste. *Persicaria* refers to the similarity of the leaves to those of the peach tree *Prunus persica*.

PINMOULD Species of microscopic fungi in the genus *Mucor* (Latin for 'mould'), including common pinmould *M. mucedo* (*mucedus* = 'mouldy'), the mould found on bread.

PINNIPEDIA From Latin *pinna* + *pedis* = 'wing' (here 'fin') + 'foot', the superfamily in the order Carnivora that includes seals.

PINNOTHERIDAE From Greek *pinne* + *ther* = 'bivalve' (*sic*) + 'wild beast', the family name of pea crabs.

PINTAIL *Anas acuta* (Anatidae), wingspan 88 cm, length 58 cm, male weight 900 g, females 700 g, the breeding population (only around 20, maximum 37 pairs) found in parts of eastern England and south-west Scotland, wintering birds (29,000 individuals mostly from Europe) widespread, and including Ireland. Wintering numbers in Britain seem to have declined by 43% between 2002–3 and 2012–13 (BBS data). The small breeding population and significant winter numbers make this an Amber List species. Its rarity as a breeding species is because of its specific habitat requirements (shallow pools in open grassland). The nest is a scrape lined with down and plant material, clutch size 7–9, incubation 22–4 days, fledging of the precocial young 40–5 days. This omnivorous dabbler feeds on the mud bottom at depths of 10–30 cm. The binomial comes from Latin *anas* = 'duck' and *acuta* which comes from *acuere* = 'to sharpen', referring, like the English name, to the pointed tail of the male in breeding plumage.

PIOPHILIDAE Family name (Greek *pion* + *philos* = 'milk fat' + 'loving') for a group of scavenging flies, with 14 species in 10 genera, including **cheese fly** *Piophila casei* (the larvae used to putrefy pecarino cheese in Italy, eaten with the live maggots) and (nest) skipper fly.

PIPEFISH Species of *Nerophis* (Greek *neros* = 'water' or, here, 'swimmer') and *Syngnathus* (*syn* + *gnathos* = 'together', meaning fused + 'jaw'), family Syngnathidae. These fishes have a distinctive shape with the head set at an angle to the body. The trunk is short and fat, the tail tapering, curled and prehensile. They pair for life, and the female transfers her eggs to the male which he self-fertilises in his pouch and carries until hatch.

Broad-nosed pipefish *S. typhle*, length 15–20 cm, in seaweed and sea-grass in shallow coastal waters in parts of southern England and Wales, feeding on small crustaceans. *Typhle* possibly comes from Greek *typhlos* = 'blind'.

Deep-snouted pipefish Alternative name for broad-nosed pipefish.

Greater pipefish *S. acus*, length generally 33–5 cm, widespread except for the east coast of Britain and west Ireland in shallow water in seaweed or rock pools and to depths of 90 m, as well as on sand and mud in estuarine mouths, feeding on small crustaceans. Latin *acus* = 'needle'.

Lesser pipefish *S. rostellatus*, length 10–17 cm, locally common mainly in parts of the English and Welsh coastline in shallow water, feeding on small crustaceans. *Rostellatus* is Latin for 'with beak'.

Nilsson's pipefish Alternative name for lesser pipefish, named after the Swedish zoologist Sven Nilsson (1787–1883).

Straight-nosed pipefish *N. ophidion*, length 15–20 cm, found around much of the coast of England, Wales and Ireland at 2–25 m in sea-grass and seaweed, feeding on copepods and other zooplankton. *Ophidion* is diminutive of Greek *ophis* = 'snake'.

Worm pipefish *N. lumbriciformis*, length up to 15 cm, mostly found under boulders or in seaweed on the lower shore and sublittoral down to 30 m on the south and west coasts of Britain and Ireland, feeding on small crustaceans. Latin *lumbriciformis* = 'worm-like'.

PIPEWORT *Eriocaulon aquaticum* (Eriocaulaceae), a scarce aquatic herbaceous perennial, very local in west Scotland and west Ireland, whose rosettes grow on peat or inorganic substrates at the edge of oligotrophic lakes and pools. Greek *erion* + *caulos* = 'wool' + 'stem', and Latin *aquaticum* = 'aquatic'.

PIPISTRELLE Bats in the genus *Pipistrellus* (Vespertilionidae). The name is late eighteenth-century, from French, in turn from Italian *pipistrello*, a variant of *vipistrello*, from Latin *vespertilio* = 'bat', from *vesper* = 'evening'.

Bandit pipistrelle Alternative name for common pipistrelle.

Brown pipistrelle Alternative name for soprano pipistrelle.

Common pipistrelle *P. pipistrellus*, head-body length 35–45 mm, wingspan 200–35 mm, weight 3–8 g, with soprano pipistrelle Europe's smallest bat, widespread and scattered. Common and soprano pipistrelles were identified as separate species in 1999, initially based on the frequency of their echolocation calls (common at 45–70 kHz, soprano at 55–80 kHz). The UK population was estimated in 2005 as 2,430,000. A 1999 estimate for Northern Ireland was 1,150,000. Roost surveys up to 2015 indicate a 81% increase on a 1999 base year value, equivalent to an annual increase of 3.8%. Summer roosting is largely in building cavities and tree holes. They generally emerge around 20 minutes after sunset to fly 2–10 m above ground level searching for insect prey (aerial hawking) over a range of habitats, for example woodland, hedgerows, grassland, farmland and suburban areas, an individual capturing around 3,000 small insects a night. Mating is from July to September. Warm summers allow a gestation of around 44 days, but in colder conditions pregnancy can last for 80 days. Females form maternity colonies where they give birth to a single young in June–July. Young are fed solely on their mother's milk for three to four weeks. After four weeks the young can fly and at six weeks they become independent. In winter the bats hibernate in crevices of buildings and trees, and in bat boxes.

Kuhl's pipistrelle *P. kuhlii*, head-body length 40–50 mm, wingspan 210–50 mm, weight 5–10 g, a Mediterranean bat first reported from the British mainland in 1991, since when there

have been over ten records, and there is a maternity colony on Jersey. Heinrich Kuhl (1797–1821) was a German naturalist.

Nathusius' pipistrelle *P. nathusii*, head-body length 46–55 mm, wingspan 228–50 mm, weight 6–16 g, widespread but scarce: individuals have been recorded from Shetland (where it was first recorded in Britain), Orkney, Jersey and Guernsey, and the Isle of Wight as well as on mainland Britain and Ireland. It was initially considered to be a vagrant, status subsequently upgraded to winter visiting migrant as records accumulated, and since the 1990s it has been known to breed in the UK, with a small number of maternity colonies found in Northern Ireland and England. The Bat Conservation Trust gave an estimate of 16,000 individuals in the UK in 2005 (though a web site dedicated to this species refers to a total of only 1,200 records in 2013). Also, the summer breeding population is supplemented by migratory individuals during winter. They roost in buildings, rock crevices and tree holes, and forage at 3–15 m for medium-sized insects near rivers, canals, lakes, as well as along woodland rides and edges. Females form large maternity colonies of up to 350 bats, each giving birth to a single young in June–July. Lactation, flight and independence of young are similar to common pipistrelle. Hibernation roosts are in cliff crevices, caves and tree hollows. Echolocation while hunting is at 36–40 kHz, with social calling at 20–30 kHz. It is uncertain after whom this species is named.

Soprano pipistrelle *P. pygmaeus*, head-body length 35–45 mm, wingspan 190–230 mm, weight 3–8 g, (with common pipistrelle) Europe's smallest bat, widespread and scattered. Soprano and common pipistrelles were identified as separate species in 1999, initially based on the frequency of their echolocation calls (soprano at 55–80 kHz, common at 45–70 kHz). The UK population was estimated in 2005 as 1.3 million. A 1999 estimate for Northern Ireland was 580,000. Field surveys up to 2015 indicate a 52.4% increase on a 1999 base year value, equivalent to an annual increase of 2.7%. Summer roosting is largely in building cavities and tree holes. They generally emerge around 20 minutes after sunset to fly 2–10 m above ground level searching for insect prey (aerial hawking), more selective in habitat selection than common pipistrelle, often using wetland habitats, including over lakes and rivers, as well as woodland edge and hedgerows, and in suburban gardens and parks. Mating is from July to September. Warm summers allow a gestation of around 44 days, but in colder conditions pregnancy can last for 80 days. Females form maternity colonies where they give birth to a single young in June–July. Lactation, flight and independence of young are similar to common pipistrelle. In winter the bats hibernate in crevices of buildings and trees, and in bat boxes. *Pygmaeus* is Latin for 'pygmy'.

PIPIT Passerine birds in the genus *Anthus* (Motacillidae), from Greek *anthos*, a small grassland bird described by Aristotle. 'Pipit' is probably imitative of the bird's call. Other pipits are scarce visitors or vagrants.

Meadow pipit *A. pratensis*, wingspan 24 cm, length 14 cm, weight 15–22 g, a widespread insectivorous resident breeder, probably the commonest passerine in upland areas, though birds tend to move southwards to lowland regions in winter. Wintering migrants from the Continent are common in south-central England and much of Ireland. It favours open habitat such as pasture, bog and moorland, but also occurs in arable fields. In winter, it also uses saltmarsh and open woodland. An estimated 2 million pairs (1.8–2.3 million) bred in Britain in 2009. Numbers peaked in 1976, but there was an overall 34% decline between 1970 and 2015, though only a 9% decline over 1995–2015, and a strong increase by 19% over 2010–15. It is Amber-listed in Britain, but Green-listed in Ireland where there are 500,000 to 1 million pairs. The ground nests usually contain 3–5 eggs, incubation lasting 13–15 days, and the altricial young fledging in 12–14 days. *Pratensis* is from Latin for 'meadow'.

Rock pipit *A. petrosus*, wingspan 25 cm, length 16 cm, weight

24 g, a resident breeder found round much of the rocky coastline, feeding at the tidal edge, often in seaweed, on insects, marine shrimps and small molluscs. There were an estimated 34,000 pairs in Britain in 1988–91, numbers augmented in winter by Scandinavian birds. Clutch size is 4–5, incubation takes 14–15 days, and fledging of the altricial young takes 15–16 days. *Petrosus* is from Greek *petros* = 'rock'.

Tree pipit *A. trivialis*, wingspan 26 cm, length 15 cm, weight 24 g, an insectivorous migrant breeder, wintering in sub-Sahara, widespread in Britain in April–September, especially in the western uplands, favouring open birchwood, heathland scrub and young conifer plantations, with an estimated 88,000 pairs (55,000–121,000) in 2009. There was a decline in numbers by 69% in Britain between 1970 and 2015, but a strong increase (15%) in 2010–15. However, numbers do fluctuate, for example an increase by 16% over 1995–2015, but a decline by 21% between 2014 and 2015. It was Red-listed in 2009. The ground nest usually contains 4–5 eggs, incubation lasting around 13 days, and the altricial young fledging in 13–14 days. *Trivialis* is Latin for 'common'.

Water pipit *A. spinoletta*, wingspan 26 cm, length 18 cm, weight 23 g, an insectivorous Amber List winter visitor particularly found in southern England, East Anglia and the east Midlands, averaging 190 birds p.a. The Florentine *spinoletta* means 'little pipit'.

PIRIMELA Toothed pirimela *Pirimela denticulata* (Pirimelidae), a crab, 12 mm carapace length, 15 mm width, locally common off the south and east coasts in burrows in sandy sediments, or on underwater vegetation, at depths down to 250 m.

PIRRI-PIRRI-BUR *Acaena novae-zelandiae* (Rosaceae), a prostrate dwarf neophyte perennial herb introduced to Britain from New Zealand as a wool contaminant, with a scattered distribution on free-draining soil, naturalised in moderately disturbed, sparsely vegetated sites, for example dunes, cliffs, heathland, conifer plantations on sandy soils, old gravel workings, roadsides and disused railways, in places becoming a nuisance because of rapid dispersal via sticky seed heads and by overgrowing native plants. *Acaena* is Greek for 'thorn'.

PISAURIDAE Family name of nursery-web, raft and fen raft spiders, from Pisaurum, the Latin name for Pesaro in the Marche region of Italy.

PITCHERPLANT *Sarracenia purpurea* (Sarraceniaceae), an insectivorous herbaceous perennial North American neophyte planted and naturalised in peat bogs in central Ireland, with a few less permanent sites in England and Scotland. Michel Sarrazin (1659–1734) was a Canadian physician and botanist.

PLAICE *Pleuronectes platessa* (Pleuronectidae), length up to 60 cm, flattened laterally, usually right-eyed, widespread, mainly living on sandy bottoms, though also on gravel and mud, usually at depths of 10–50 m but occurring in the range 0–200 m, and fish in their first year found in very shallow water, often in sandy tidal pools. Diet is of small benthic animals. UK landings in 2013 totalled 4,100 tonnes, valued at £4 million. 'Plaice' dates from Middle English, from Old French *plaiz*, from Latin *platessa* (giving the specific name), and Greek *platys* = 'flat'. *Pleuronectes* comes from Greek for 'side'.

PLAIT-MOSS Species of *Hypnum* (Hypnaceae) (Greek *hypnon* = '(tree) moss').

Rarities are **golden plait-moss** *H. bambergeri*, **revolute plait-moss** *H. revolutum*, and **Vaucher's plait-moss** *H. vaucheri* (all Highlands).

Cypress-leaved plait-moss *H. cupressiforme*, very common and widespread on acidic to slightly base-rich bark and siliceous rock. Latin *cupressiforme* = 'cypress-shaped'.

Downy plait-moss *H. callichroum*, found in montane Britain and in west Ireland in sheltered sites, on acidic or base-rich rock (not limestone), in short grass, in woodland, on stream banks and in ravines, and scree. Greek *cala* + *chroa* = 'beautiful' + 'colour'.

Great plait-moss *H. lacunosum* var. *lacunosum*, common and widespread in calcareous sites, on well-drained soil or over siliceous rock and limestone, often in sunny locations. *Lacunosum* is Latin for 'pitted'.

Heath plait-moss *H. jutlandicum*, common and widespread in acidic heathland, upland grassland, woodland, conifer plantations, tracksides, and between boulders in scree.

Hook-leaved plait-moss *H. hamulosum*, locally common in upland Britain on calcareous rock or shallow soil where the ground faces north or east. It can be locally abundant in species-rich grass on hill ridges in north-west Scotland. *Hamulus* is Latin for 'small hook'.

Lindberg's plait-moss *H. lindbergii*, widespread and scattered on damp, often slightly calcareous, sandy or stony ground beside tracks, in old quarries, beside streams and on dunes. Sextus Lindberg (1835–89) was a Swedish physician and bryologist.

Mamillate plait-moss *H. andoi*, widespread, locally abundant and generally the commonest *Hypnum* on acidic bark and rock, dominant in upland woodland. 'Mamillate' means having nipple-like protrusions.

Pellucid plait-moss *H. imponens*, scattered in Britain, with most records from southern and northern England and south Scotland in wet peaty ground on heathland and bog. *Imponens* is Latin for 'deceiving'.

Roof plait-moss *H. lacunosum* var. *tectorum*, scattered and widespread in calcareous sites, on well-drained soil or over siliceous rock and limestone, often in sunny locations. *Tectum* is Latin for a 'covering'.

PLANARIAN Flatworms in the class Turbellaria (Latin *planus* = 'flat'), found in damp and fresh- and saltwater habitats.

PLANE (TREE) London plane *Platanus* x *hispanica* (Platanaceae), a hybrid (*P. occidentalis* x *P. orientalis*) long-lived tree, possibly originating in Oxford Botanical Gardens in the seventeenth century, extensively planted in streets and parks. It is vigorous even in polluted air (shedding its bark) and where root space is restricted, and can be repeatedly pruned. It has been recorded in the wild since at least 1939. *Platanos* was the Greek name for oriental plane.

PLANIPENNIA Suborder of green, brown and giant lacewings (Latin *planus* + *penna* = 'flat' + 'wing').

PLANORBIDAE Family name (Latin *planus* + *orbis* = 'flat' + 'circle') for the freshwater pulmonate ramshorn snails and some limpets.

PLANT BUG Specifically can refer to members of the family Miridae (Heteroptera).

Meadow plant bug *Leptoptema dolabrata*, common and widespread, preferring damp habitats and feeding on grasses. Greek *leptos* + *ptenos* = 'slender' + 'wing', and Latin *dolabrata* = 'axe-shaped'.

Tarnished plant bug *Lygus rugulipennis*, common and widespread in a variety of habitats, feeding on many plant species. Greek *lygos* = 'flexible' and Latin *rugula* + *penna* = 'wrinkle' + 'wing'.

PLANTAGINACEAE Family name for plantain (Latin *plantago*), including shoreweed.

PLANTAIN Species of *Plantago* (Plantaginaceae), the English and genus name coming from Latin *plantago* = 'plantain', in turn from *planta* = 'sole of the foot' because of the broad prostrate leaves of most species.

Branched plantain *P. arenaria*, an annual south European neophyte introduced in 1804, declining though occasionally naturalised, usually casual in England and Wales on sandy waste ground, and in docklands. *Arenaria* is from Latin *arena* = 'sand'.

Buck's-horn plantain *P. coronopus*, a herbaceous perennial common round the coast, scattered inland mostly in England increasingly occurring by salt-treated roads, more generally growing near the sea on dry, open, often trampled habitats on acidic to basic stony or sandy soils, on grassland, heathland, dunes and shingle, sea-cliffs and walls, waste ground and paths. Greek *corone* + *pous* = 'crown' + 'foot', from the leaf shape.

Glandular plantain *P. afra*, an annual south European neophyte, a scattered casual in England and south Wales on waste ground, originating from birdseed and grain impurities. *Afer* is Latin for 'African'.

Greater plantain *P. major*, a common and widespread herbaceous perennial of open habitats, most frequent on trampled paths and tracks, disturbed field edges and roadsides, and in gardens, growing in a range of soils, avoiding only very acidic sites. Subspecies *intermedia* is more scattered on damp, open habitats, including upper saltmarsh, stream banks, and mud by ponds and reservoirs, as well as on salt-treated road verges. *Major* is Latin for 'greater'.

Hoary plantain *P. media*, perennial, locally common in England, scattered elsewhere, generally on calcareous but also heavy clay soils, the main habitats being downland grass, calcareous pasture and mown grassland, less commonly in hay meadows and on fixed dunes. Latin *media* = 'middle'.

Ribwort plantain *P. lanceolata*, a common and widespread perennial, found over all but very acidic soils, in meadow and pasture, in upland grassland, on rock crevices, dunes, cliffs, roadsides, riverbanks, cultivated and waste ground, lawns and walls. Latin *lanceolata* = 'lance-shaped'.

Sea plantain *P. maritima*, perennial, growing in middle and upper saltmarsh, coastal grass and heathland, rocks and cliffs, occasionally shingle beaches and inland saltmarsh. In upland Britain, especially Scotland, it is found in species-rich pasture, on stream banks, scree, and in stony flushes. It is increasingly colonising salted road verges.

PLANTHOPPER Members of the sap-sucking bug families Cixiidae (12 British species), Issidae (2) and Delphacidae (76) (Auchenorrhyncha). Many species are common and widespread, mostly in grassland.

PLANTLICE Jumping plantlice Members of the family Psyllidae (Sternorrhyncha), with 42 species in 13 genera.

PLATANACEAE Family name for plane trees.

PLATES Pink plates *Mesophyllum lichenoides* (Hapalidiaceae), a red seaweed consisting of thin, brittle, leafy calcified fronds epiphytic on coral weed in tidal pools, local in south England west of the Solent, along the west coast north to Shetland, and around Ireland. The binomial is Greek for 'intermediate leaved' and Latin for 'lichen-like'.

PLATYCNEMIDIDAE Damselfly family (Greek *platys* + *cneme* = 'flat' + 'knee'). See white-legged damselfly under damselfly.

PLATYHELMINTH(ES) Phylum of flatworms (Greek *platys* + *helminthos* = 'flat' + 'of flatworms'), consisting of unsegmented, soft-bodied bilaterally symmetrical invertebrates, with no specialised circulatory or respiratory organs, their flattened shape maximising the diffusion of oxygen and nutrients through the body. The phylum includes the non-parasitic turbellarians, and the parasitic cestodes (tapeworms), trematodes (flukes) and monogeneans.

PLATYPEZIDAE Family name (Greek *platys* + *pezos* = 'flat' + 'walking') of 1.5–6 mm long flat-footed flies, with 33 species in 10 genera, mainly found in woodland. Larvae are fungus feeders, living in decaying wood or under bark.

PLATYPODIDAE Family name (Greek *platys* + *pous* = 'flat' + 'foot') of weevils, with a single British representative, pinhole borer.

PLATYSTOMATIDAE Family name of signal flies (Greek *platys* + *stoma* = 'flat' + 'mouth').

PLECOPTERA Order of stoneflies (Greek *plecos* + *pteron* = 'wickerwork' + 'wing'), with 34 species in 7 families.

PLEURONECTIDAE, PLEURONECTIFORMES Family and order name of flatfishes (Greek *pleura* + *nectos* = 'side' + 'swimming').

PLOUGHMAN'S-SPIKENARD *Inula conyzae* (Asteraceae), a biennial or perennial herb locally common in England and Wales on dry sites, mainly on chalk or limestone, less frequently on sand and gravel, typically in areas of open soil or stony ground. It grows in dry grassland, on banks, woodland edges, rides and scrub, in quarries, scree, on the more vegetated parts of dunes, and on roadsides and rough ground. 'Spikenard' dates from Middle English, from medieval Latin *spica nardi*, in turn from Greek *nardostachys*, an aromatic substance from an Indian plant. *Inula* is Latin for elecampane, *conyzae* coming from Greek *conops* = 'flea', as in fleabane.

PLOVER Wading birds in the family Charadriidae; see also lapwing. 'Plover' ultimately derives from Latin *pluvia* = 'rain', the reason obscure though suggestions include because the migration arrival coincides with the start of the rainy season, and from a supposed restlessness when rain approaches.

American golden plover *Pluvialis dominica* is a scarce visitor; **Pacific golden plover** *Pluvialis fulva*, **semipalmated plover** *Charadrius semipalmatus*, **killdeer** *C. vociferus*, **sociable plover** *Vanellus gregarius*, **white-tailed plover** *V. leucurus*, **Caspian plover** *Anarhynchus asiaticus*, **greater sand plover** *A. leschenaultii* and **lesser sand plover** *A. mongolus* are all scarce vagrants.

Golden plover *Pluvialis apricaria*, wingspan 72 cm, length 28 cm, weight 220 g, found in May–September on their breeding grounds on moorland in the Southern Uplands and Highlands, the Hebrides, Orkney and Shetland, the Peak District, north Yorkshire, Wales and Devon. In winter they move to lowland fields, forming large flocks. There were up to 59,000 pairs p.a. in 1980–2000, but there was a reduction in breeding numbers by 16% over 1995–2014, and while one report suggests 420,000 birds wintering in 2006–7, RSPB data indicate a decline in numbers of 25% between 2002–3 and 2012–13. Nests are on the ground, clutch size is usually four, incubation takes 28–31 days, and the precocial young fledge in 25–33 days. Feeding is at night on invertebrates, mainly beetles and worms. Latin *pluvia* = 'rain', and *apricaria* = 'basking in sun'.

Grey plover *P. squatarola*, wingspan 77 cm, length 28 cm, weight 240 g, an Amber-listed winter visitor, adults starting to arrive from the Continent in July and the young in August–September. Peak numbers are between November and March, and birds leave in April–May. Some 43,000 birds were estimated to overwinter in 2008–9, found along many coasts, preferring large muddy and sandy estuaries, scarce or absent in north Scotland and west Ireland. Largest numbers are generally found on the Wash, Thames, Blackwater, Medway, Dee, Ribble and Humber estuaries, and Chichester and Langstone Harbours. Diet is of shellfish and worms. *Squatarola* comes from Venetian *sgatarola* for a type of plover.

Kentish plover *Charadrius alexandrinus*, no longer breeding in Britain, but a scarce European visitor or passage migrant, with an average of 34 records p.a. Greek *charadrios*, a bird found in ravines and river valleys (*charadra* = 'ravine'), and Latin *alexandrinus* named after Alexandria.

Little ringed plover *C. dubius*, wingspan 45 cm, length 14 cm, weight 40 g, a migrant arriving in March and leaving in late June–July, previously a rare vagrant but noted as breeding in Britain in 1938, subsequently colonising a large part of England and Wales, taking advantage of man-made habitats such as gravel-pits and reservoir shores, with 1,200 pairs estimated in 2007, an 80% increase on 1984 numbers. Nests are in scrapes, clutch size is usually four, incubation takes 24–5 days, and fledging of the precocial young 25–7 days. Diet is of invertebrates, mainly insects. *Dubius* is Latin for 'doubtful', because it was once thought that this species might be a variant of ringed plover.

Ringed plover *C. hiaticula*, wingspan 52 cm, length 19 cm, weight 64 g, a Red List bird widespread on beaches and, inland, by flooded gravel-pits, with 5,300 pairs estimated in 2007. Ringed plovers from Europe winter in Britain, with 36,000 resident and wintering birds in 2008–9, though RSPB data indicate a 42% decline between 2002–3 and 2012–13. Also, birds from Greenland and Canada pass through on migration. Nests are in scrapes with four eggs usually laid, incubation taking 23–5 days, and fledging of the precocial young also 23–5 days. Diet in summer is of invertebrates, in winter mainly marine worms, crustaceans and molluscs. Latin *hiatus* + *colo* = 'gap' + 'to inhabit'.

PLUM Small trees or shrubs in the genus *Prunus* (Rosaceae), with edible fruit, originating from the Near East, the name deriving from Old English *plume*, in turn from Latin *prunus* for these fruit trees.

Cherry plum *P. cerasifera*, cultivated by at least Tudor times, common and widespread especially in England and Wales as a street tree, and naturalised on roadsides and in hedgerows and woodland. The binomial is Latin for 'plum tree' and 'cherry-bearing'.

Wild plum *P. domestica*, an archaeophyte long in cultivation, common and widespread, naturalised in hedges, woodland edge, scrub and waste ground. Three subspecies, ssp. *domestica* (**plum**), ssp. *insititia* (**bullace, damson**) and ssp. *italica* (**greengage**) are recognised, all important for their edible fruit. Many plants are relics of cultivation, but ssp. *domestica* is still being introduced from discarded plum stones.

PLUMBAGINACEAE Family name for thrift and sea-lavender (Latin *plumbago* = 'lead-coloured').

PLUME (MOTH) 1) Nocturnal moths in the family Pterophoridae (mostly Pterophorinae), with narrow-lobed or divided wings. The usual resting posture is T-shaped, with wings extended at right angles to the body and narrowly rolled up, often resembling a piece of dried grass, and so unnoticed by predators even when resting in exposed situations in daylight. The usually hairy or bristly larvae are generally stem- or root-borers; others are leaf-feeders. Most specific names include Greek *dactylos* = 'finger'.

Scarce species include **Breckland plume** *Crombrugghia distans* (East Anglia and southern England west to Devon in dry, sandy habitats, larvae feeding on hawkweed), **citron plume** *Hellinsia carphodactyla* (chalky southern and eastern England, recently expanding into the Midlands and south Yorkshire, larvae feeding inside flowers of ploughman's-spikenard), **cliff plume** *Agdistis meridionalis* (grassy slopes and cliffs in south-west England, larvae feeding on rock sea-lavender), **dingy white plume** *Merrifieldia baliodactylus* (chalk downland and waste ground in southern England, larvae feeding on marjoram), **goldenrod plume** *Platyptilia calodactyla* (woodland in Norfolk, Kent and Sussex), **horehound plume** *Wheeleria spilodactylus* (chalky ground in southern England, especially the Isle of Wight, and south Wales), **Irish plume** *Platyptilia tesseradactyla* (the Burren, larvae mining mountain everlasting), **mountain plume** *Stenoptilia islandicus* (Highlands, larvae mining saxifrage flowers and seeds), **reedbed plume** *Emmelina argoteles* (discovered at Wicken Fen in Cambridgeshire in 2005, larvae feeding on hedge bindweed), **saxifrage plume** *Stenoptilia millieridactyla* (limestone pavements on the Burren, increasing in gardens in northern England, probably introduced as larvae on imported plants), **scarce goldenrod plume** *H. chrysocomae* (woodland in south-east England), **short-winged plume** *Pselnophorus heterodactyla* (woodland in Suffolk, Gloucestershire, Cumbria and Scotland, larvae feeding on wall lettuce, nipplewort and marsh hawk's-beard), **small goldenrod plume** *H. osteodactylus* (oakwood in northern England and southern Scotland, rare in north Wales), **small scabious plume** *Stenoptilia annadactyla* (first recognised in Britain in 2005 when larvae were reared having been found in the Brecklands of East

Anglia and from Settle, Yorkshire), **spotted plume** *Porrittia galactodactyla* (open woodland and dry pasture in the Brecklands, and parts of southern England and south Wales, larvae feeding on burdock), **sundew plume** or **marsh plume** *Buckleria paludum* (damp heathland and moorland in parts of England and Wales), **tamarisk plume** *Agdistis tamaricis* (Channel Islands, first recorded on Jersey in 2007), **western thyme plume** *Merrifieldia tridactyla* (chalk downland in parts of south-west England, and on the Burren and elsewhere in western Ireland), and **wood sage plume** *Capperia britanniodactylus* (southern Britain, where the larval food plant, wood sage, grows in the open).

Beautiful plume *Amblyptilia acanthadactyla*, wingspan 17–23 mm, widespread in woodland, heaths, mountains and gardens, having become increasingly common since the 1990s. The bivoltine adults are on the wing in July and from September onwards, flying after hibernation until May. Larvae feed in summer, with peaks in June–August, on flowers and young leaves of a range of herbaceous plants. Greek *amblys* + *ptilon* = 'blunt' + 'insect wing', from blunt forewing tips, and *acantha* = 'thorn'.

Brindled plume *Amblyptilia punctidactyla*, wingspan 18–23 mm, widespread but local in woodland, heaths and riverbanks. The bivoltine adults fly in July and then from September, hibernating until the following spring. Larvae feed on flowers and unripe seeds a number of herbaceous plants. *Punctum* is Latin for 'dot'.

Brown plume *Stenoptilia pterodactyla*, wingspan 20–6 mm, widespread and common in downland, open woodland and dry pastures, more numerous in the south. Adults fly in June–August. Larvae mine shoots and stems of germander speedwell in August–March, moving to flowers in April–May. Greek *steno* + *ptilon* = 'narrow' + 'insect wing', and *pteron* = 'feather'.

Common plume *Emmelina monodactyla*, wingspan 18–27 mm, widespread and common in gardens, rough ground, hedgerows and woodland edges. Adults are found throughout the year. Larvae feed on the leaves and flowers of a range of plants, in two overlapping generations from late May to September. *Monos* is Greek for 'single'.

Crescent plume *Marasmarcha lunaedactyla*, wingspan 18–22 mm, local on downland, shingle beaches and dunes in southern England, south Wales and south-east Ireland. Adults fly in July. Larvae feed on common restharrow. Greek *marasmos* + *arche* = 'decay' + 'beginning', and Latin *luna* = 'moon', from a crescent mark on the forewing.

Dowdy plume *Stenoptilia zophodactylus*, wingspan 16–23 mm, local on sea-cliffs, dry grassland and dunes throughout much of England and Wales. Adults are found from July to September in two or more overlapping broods. Larvae feed in two or more generations in June–October or later, on centaury, yellow-wort and gentians. *Zophos* is Greek for the gloom of the Underworld.

Dusky plume *Oidaematophorus lithodactyla*, wingspan 26–9 mm, local on grassland, verges and embankments throughout much of England, Wales and south and west Ireland. Adults fly in July–August. Larvae feed on the stems then leaves and flowers of common fleabane and ploughman's-spikenard. Greek *oidema* + *phoreo* = 'swelling' + 'carry', from scale tufts on the legs, and *lithos* = 'stone'.

Hemp-agrimony plume *Adaina microdactyla*, 13–17 mm, common on downland throughout England, more numerous in the south and East Anglia, Wales and Ireland. Double-brooded, adults fly in May–June and August. Larvae mine hemp-agrimony stems, creating a gall.

Hoary plume *Platyptilia isodactylus*, wingspan 19–29 mm, widespread but scattered, often common in fen, swamp and marsh. The bivoltine adults fly in June and August–September. Larvae mine marsh ragwort. *Isos* is Greek for 'equal'.

Mugwort plume *H. lienigianus*, wingspan 17–21 mm, scarce on dry grassland and waste ground throughout much of England, less numerous in the north. Adults mainly fly in July. Larvae feed on mugwort. The specific name honours the Latvian entomologist Mme Lienig (d.1855).

Plain plume *H. tephradactyla*, wingspan 18–23 mm, widespread but local in woodland in Britain and west and south Ireland. Adults fly in July. Larvae feed on goldenrod flowers and seed heads. *Tephra* is Latin for 'ashes', from the grey forewing.

Rose plume *Cnaemidophorus rhododactyla*, wingspan 18–26 mm, in hedgerows and woodland in parts of south-east England. Adults fly in July–August. Larvae feed on roses, first on developing leaves then on flower buds and the flowers themselves. Eggs hatch in autumn and larvae overwinter when small, probably inside a stem. They restart feeding in May–June. *Cnemidophoros* is Greek for wearing greaves (leg armour), from the leg tufts; *rhodon* is Greek for 'rose-red'.

Saltmarsh plume *Agdistis bennetii* (Agdistinae), wingspan 24–30 mm, local in coastal saltmarsh throughout much of southern and eastern England. Adults fly in July–August. Larvae feed on sea-lavender. E. Bennet was a nineteenth-century English entomologist.

Small plume *Oxyptilus parvidactyla*, wingspan 13–18 mm, widespread but scattered and scarce on heathland, dry pasture and chalk downland. Adults fly in the afternoon in June–July. Larvae feed on mouse-ear-hawkweed. Greek *oxys* + *ptilon* = 'sharp' + 'insect wing', and Latin *parvus* = 'small'.

Tansy plume *Gillmeria ochrodactyla*, wingspan 23–7 mm, in flower meadows and gardens, locally common with a scattered distribution in England and Wales. Adults fly in July. Larvae feed on tansy, mining down a shoot, hibernating in the roots, and burrowing up another shoot in spring. Max Gillmer (1857–1923) was a German entomologist. Greek *ochros* = 'pale'.

Thyme plume *Merrifieldia leucodactyla*, wingspan 18–25 mm, widespread but local on chalk downland, sea-cliffs and sandhills. Adults fly in June–August. Larvae feed in July–August on wild thyme leaves. Greek *leucos* = 'white'.

Triangle plume *Platyptilia gonodactyla*, wingspan 20–30 mm, widespread and fairly common in open, grassy habitats, woodland margins and waste ground. The bivoltine adults fly in May–June and in autumn. Larvae feed on coltsfoot. Greek *gonia* = 'angle'.

Twin-spot plume *Stenoptilia bipunctidactyla*, wingspan 17–25 mm, widespread and common in damp heathland, downland and flower meadows. The bivoltine adults fly in May–October. Larvae mine flowers and seeds of scabious. Latin *bipunctis* = 'two-spotted'.

White plume moth *Pterophorus pentadactyla*, wingspan 26–34 mm, the only completely all-white plume moth, widespread and common in dry grassland, waste ground and gardens. Adults fly at dusk and at night in June–July, sometimes with a second generation in September. Larvae feed on bindweed leaves in May–June and August–September. Greek *pteron* + *phoreo* = 'feather' + 'carry'. *Pente* is Greek for 'five', from the number of wing lobes.

Yarrow plume *Gillmeria pallidactyla*, wingspan 23–7 mm, widespread and common in a range of habitats, including waste ground and gardens. Adults fly in June–July. Larvae mine stems and leaves of yarrow, sneezewort and possibly tansy. Latin *pallidus* = 'pale'.

2) **Twenty-plume moth** *Alucita hexadactyla* (Alucitidae), wingspan 14–16 mm, nocturnal, widespread and common in gardens and open woodland. Adults can be found at almost any time of year, peaking in April–May and July–August. Larvae feed on honeysuckle leaves and buds. Latin *alucita* = 'gnat', and Greek *hex* + *dactylos* = 'six' + 'finger', from the wing division.

PLUMS AND CUSTARD *Tricholomopsis rutilans* (Agaricales, Tricholomataceae), a widespread, common inedible saprobic fungus found in July–October on or close to decomposing conifer stumps. *Tricholomopsis* means similar to the genus *Tricholoma*. Latin *rutilans* = 'glowing' or 'reddening'. See also prunes and custard.

PLUSIA Golden plusia *Polychrysia moneta* (Noctuidae, Plusiinae), a nocturnal moth, wingspan 32–7 mm, common in gardens and parks throughout much of England and Wales. Adults fly in June–August. Larvae feed on larkspur and cultivated delphiniums. Greek *polys* + *chrysos* = 'much' + 'gold', from the forewing colour. Moneta, the temple of Juno where the Roman mint was located, here describes the metallic wing markings.

PLUTELLIDAE Family containing some of the smudge moths, together with diamond-back moth (Greek *ploutos* = 'wealth').

POACEAE The grass family (Greek *poa* for a grass used for fodder).

POCHARD *Aythya ferina* (Anseridae), wingspan 77 cm, length 46 cm, weight 930 g, a common and widespread duck in winter when the few residents, averaging 330–700 pairs in 2006–10, and mainly breeding in lakes and flooded gravel-pits in eastern England, lowland Scotland and Northern Ireland, have numbers augmented with east European and Russian migrants, together totalling 38,000 individuals. Winter numbers have nevertheless declined by 41% between 2002–3 and 2012–13. Clutch size is 8–10, incubation lasts 25 days, and the precocial young fledge in 50–5 days. They feed by diving and sometimes dabbling, often at night, on plants, snails, small fish and insects. With numbers declining, it was moved from being an Amber List species in 2001 to the Red List in 2010. Greek *aithuia* is an unidentified seabird mentioned by Aristotle, and Latin *ferina* = 'wild game'.

POCKET-MOSS Mostly species of *Fissidens* together with *Octodiceras* (Fissidentaceae), from Latin *fissus* = 'split'.

Scarce species include **fountain pocket-moss** *O. fontanum*, **large Atlantic pocket-moss** *F. serrulatus* and **many-leaved pocket-moss** *F. polyphyllus* (all in water or on wet rock), and **Portuguese pocket-moss** *F. curvatus* (on soil).

Atlantic pocket-moss *F. monguillonii*, on submerged rocks and on riverbanks in south-west England, and parts of Wales and Ireland. Eugène Monguillon (1865–1940) was a French botanist.

Beck pocket-moss *F. rufulus*, locally frequent in the Pennines and the Welsh uplands, on limestone or siliceous submerged rocks by rivers and lakes. Latin *rufus* = 'reddish'.

Common pocket-moss *F. taxifolius* var. *taxifolius*, widespread and common on soil and in rock crevices, in woodland, on shaded banks, in arable fields and undisturbed garden borders; var. *pallidicaulis* grows in rock crevices, occasionally on soil on rocky stream banks, mostly in west Scotland. Latin *taxifolius* = 'yew-leaved'.

Curnow's pocket-moss *F. curnovii*, common in west Britain, scattered in coastal Ireland, by streams, especially in steep-sided valleys, on acidic soil banks, in crevices and on rock. William Curnow (1809–87) was a market gardener and amateur bryologist in Devon.

Fatfoot pocket-moss *F. crassipes*, common and widespread, especially in England, on limestone or siliceous submerged rocks by rivers and lakes. Latin *crassus* + *pes* = 'thick' + 'foot'.

Green pocket-moss *F. viridulus*, common and widespread, especially in England and Wales, on calcareous to slightly acidic soil banks in woodland and next to streams. Latin *viridulus* = 'greenish'.

Herzog's pocket-moss *F. limbatus*, scattered in England and Wales on calcareous to slightly acidic soil banks and rocks in woodland and next to streams. Theodor Herzog (1880–1961) was a German bryologist. Latin *limbatus* = 'bordered'.

Lesser pocket-moss *F. bryoides*, common and widespread on neutral or mildly acidic soil in mainly lowland woodland, arable fields and gardens, and next to streams. *Bryoides* is Latinised Greek for 'moss-like'.

Maidenhair pocket-moss *F. adianthoides*, common and widespread in flushes and fens, on wet rock faces, and calcareous grassland. *Adiantos* is Greek for 'maidenhair fern'.

Narrow-leaved pocket-moss *F. gracilifolius*, scattered on limestone and siliceous rocks. Latin *gracilifolius* = 'slender-leaved'.

Petty pocket-moss *F. pusillis*, widespread but scattered on limestone and siliceous rocks. Latin *pusillus* = 'very small'.

Purple-stalked pocket-moss *F. osmundoides*, common and widespread in upland regions in damp, usually base-rich habitats, including rock faces, by streams and in flushes and wet, calcareous grassland. *Osmundoides* means like *Osmunda* fern.

River pocket-moss *F. rivularis*, calcifuge, mostly in south-west England and Wales on shaded, moist or submerged, neutral to acidic rocks in lowland streams and by lakes.

Rock pocket-moss *F. dubius*, common and widespread on calcareous grassland and dry-stone walls, occasionally on moist siliceous rocks. Latin *dubius* = 'uncertain'.

Short-leaved pocket-moss *F. incurvus*, widespread, mainly lowland (more scattered in Scotland and Ireland) on calcareous to slightly acidic soil in woodland and a number of open habitats.

Slender pocket-moss *F. exilis*, common and widespread in lowland regions, more scattered in Scotland and Ireland, on neutral to acidic loam and clay in deciduous woodland, stream-sides and damp grassland. Latin *exilis* = 'slender'.

Welsh pocket-moss *F. celticus*, common in west Britain and the Weald, scattered in Ireland, on shaded soil banks, especially by woodland streams and ditches, growing on acidic soil, though avoiding peaty or very acidic substrates.

POCKET PLUM *Taphrina pruni* (Taphrinales, Taphrinaceae), a widespread and locally common fungal pathogen on blackthorn fruit and leaves. Greek *taphros* = 'trench', and Latin *prunus* = 'plum'.

PODICIPEDIFORMES Order of grebes, from Latin *podicis* + *pes* = 'vent' + 'foot', referring to the placement of the legs towards the rear of the body, more suitable for swimming.

POISONER **Lead poisoner** Alternative name for the fungus livid pinkgill.

POISONPIE Poisonous fungi in the genus *Hebeloma* (Agaricales, Strophariaceae), from Greek *hebe* + *loma* = 'youth' + 'fringe', with a 'veil' covering the gills in the young cap. Specifically *H. crustuliniforme*, widespread and common, found in late summer and autumn on soil in open woodland. Latin *crustuliniforme* means in the form of a thin crust of bread.

Widespread but uncommon species include **bitter poisonpie** *H. sinapizans* (on calcareous soil in broadleaf woodland, often under beech), **dwarf poisonpie** *H. pusillum* (in deciduous woodland or parkland, usually under willow), **rooting poisonpie** *H. radicosum* (on deciduous woodland soil under oak or beech, an 'ammonia fungus' associated with mole and rodent latrines), **sweet poisonpie** *H. sacchariolens*, **veiled poisonpie** *H. mesophaeum* (on sandy acidic soil favouring pine and, sometimes, birch woodland), and **willow poisonpie** *H. vaccinum* (mainly on dunes under willow in parts of coastal Wales).

Pale poisonpie *H. leucosarx*, widespread and common, found in late summer and autumn on soil in usually deciduous woodland. Greek *leucosarx* = 'white flesh'.

POKEWEED Species of *Phytolacca* (Greek *phyton* + *laccos* = 'plant' + 'pit'), including **Indian pokeweed** *P. acinosa* and **Chinese pokeweed** *P. polyandra*, herbaceous perennial neophytes occasionally found in the Channel Islands and the southern mainland as garden escapes.

POLECAT *Mustela putorius* (Mustelidae), male head-body length 33–45 cm, tail 12–19 cm, weight 0.8–1.9 kg, females 32–9 cm, 12–19 cm, 0.5–1.1 kg, probably native but arguably (post)-Norman, once widespread in Britain, now scattered in woodland and farmland with hedges and copses. The estimated population in 2008 was 47,00 (England 28,000, Wales 18,500, Scotland 500), but by 2017 one total estimate was 63,000. A Vincent Wildlife Trust-funded survey reported in 2016 that polecats were more widespread in Britain than at any time in the previous 100 years,

maintaining their range in Wales and the west Midlands, and considerably expanding in south-west England and East Anglia. A Cumbrian population (probably initiated by releases) was expanding into parts of western Northumberland, and there was a presence in the Yorkshire Dales. Polecats were recolonising Dumfriesshire. Reintroduced populations in Argyll and in Perthshire–Angus remained established. Rabbits are a major food item in summer, rats in winter. Birds are also taken, and frogs may be important in spring when gathering to spawn. Dens are often in rabbit burrows, but polecats frequently move into farmyards in winter, denning in hay bales, under sheds and in rubbish tips. Gestation lasts 40–3 days. Females give birth (litter size 5–10) in late May–early June, the young beginning to take meat after 3 weeks, though they take 2–3 months to become independent. They reach adult size by autumn, and breed at one year old. It is a priority species in the UK BAP. Polecats have long been domesticated as ferrets for hunting vermin, and polecat-ferret 'hybrids' are often recorded. The first part of its name possibly comes from the French *pole* (*poule*) = 'chicken', referring to the its fondness for poultry, or it could be a variant of the Old English *ful* = 'foul'. In Middle English, it was called *foulmart*, meaning 'foul marten', because of its strong odour. Latin *mustela* = 'weasel', *putor* = 'stench'.

POLEMONIACEAE Family name of Jacob's-ladder, from Polemon, a second-century Greek philosopher.

POLICEMAN'S HELMET Alternative name for Indian balsam.

POLLAN *Coregonus autumnalis* (Salmonidae), a rare glacial relic, endemic to Ireland, known only from Lough Neagh (remaining abundant and commercially fished), Lower Lough Erne (severely declined), Lough Ree (maintained by restocking), Lough Derg (also restocked), and Lough Allen (confirmed in 2006). Feeding is mostly of insect larvae and small crustaceans, and spawning is in December–January in shallow rocky areas where wave action oxygenates the developing eggs. It is Endangered in the Irish RDB. 'Pollan' dates from the early eighteenth century, from Irish *pollan*, perhaps based on *poll* = 'pool'. Greek *core* + *gónia* = 'pupil of the eye' + 'angle', leading to 'angle-eyed'.

POLLARD Alternative name for chub.

POLLEN BEETLE Members, with sap beetles, of the family Nitidulidae. Specifically, *Meligethes aeneus*, 2–3 mm length, widespread in Britain, and a pest of oil-seed rape and other brassicas, both adults and larvae feeding on pollen and nectar, but larvae also feeding on flower buds thus causing damage. Greek *melos* + *getheo* = 'song' + 'to rejoice', and Latin *aeneus* = of 'bronze' or 'copper'.

Nettle pollen beetle *Brachypterus urticae*, length <2 mm, common and widespread in England, Wales and north Scotland in a range of habitat, including woodland and urban sites, adults and larvae mainly feeding on nettles but also umbellifers. Greek *brachys* + *pteron* = 'broad' + 'wing', and Latin *urtica* = 'nettle'.

POLYBIIDAE Family name of some crabs, including velvet swimming crab *Necora puber* and five widespread species of *Liocarcinus*.

POLYCENTROPODIDAE Family name of trumpet-net and tube-making caddisflies (Greek *poly* + *centros* + *podos* = 'many' + 'centre' + 'of a foot'). The flattened larvae live in rock hollows and other objects, stretching silk threads across nearby surfaces to capture prey and organic debris. *Polycentropus* species make slender, tubular structures surrounded by silk threads among the stems of aquatic plants, responding to prey that are caught in these.

POLYCHAETA, POLYCHAETE Class of annelid worm, the body covered with a number of chitinous bristles (chaeta), hence many being called bristle worms. Many are tubeworms. See also Oligochaeta.

POLYGALACEAE Family name for milkworts, from Greek *polys* + *gala* = 'much' + 'milk', from the belief that cattle grazing on these plants produced more milk.

POLYGONACEAE Family name for knotweeds, docks, knotgrass and sorrel (Greek *polys* + *gonia* = 'many' + 'angles').

POLYPLACOPHORA Class name for chitons (Greek *polys* + *plax* + *phoreo* = 'many' + 'plate' + 'to carry').

POLYPODY Ferns in the genus *Polypodium* (Polypodiaceae) (Greek *polys* + *podos* = 'many' + 'feet'). Specifically, *P. vulgare*, common and widespread on base-poor substrates, for example rock outcrops and walls, and on tree branches and trunks as an epiphyte. Latin *vulgare* = 'common'.

Intermediate polypody *P. interjectum*, common (especially further south) and widespread on generally shaded and calcicolous habitats such as rocks and walls. Latin *interjectum* = 'intermediate'.

Southern polypody *P. cambricum*, scattered but locally common on base-rich rocks and walls, and epiphytic on treetrunks, favouring moist conditions. *Cambricum* is Latin for 'Welsh'.

POLYPORE A group of inedible saprobic basidiomycete bracket fungi, mostly annual, in the order Polyporales that form fruiting bodies with pores or tubes on the underside. Genera include *Polyporus* (Greek *polys* + *poros* = 'many' + 'pore') with 27 species, *Meripilus* (*meros* + *pileos* = 'a part' + 'cap'), both Polyporaceae, and *Piptoporus* (*pipto* + *poros* = 'to fall' i.e. easily detachable + 'pore'), in the Fomitopsidaceae.

Bay polypore *Po. durus*, widespread and, in England, locally common, found from spring to early winter, though sporulating in summer and autumn, on living and decaying broadleaf trees, especially beech. Latin *durus* = 'hard'.

Birch polypore Alternative name for razorstrop fungus.

Blackfoot polypore *Po. leptocephalus*, common and widespread on dead wood of beech and other hardwoods. Greek *leptos* + *cephale* = 'slender' + 'head'.

Fringed polypore *Po. ciliatus*, locally common and widespread in Britain, found from spring to late summer on dead wood, particularly of beech and oak, on the floor of deciduous woodland. Latin *cilia* = 'hair' or 'eyelashes'.

Giant polypore *M. giganteus*, common and widespread, found from summer to late autumn, as large compound rosettes on stumps and the base of hardwoods, favouring beech.

Oak polypore *Piptoporus quercinus*, occasional, scattered, mainly in England, a UK BAP species, and listed in Schedule 8 of the WCA (1981), legal protection covering picking and destruction, found on oak trunks and stumps. *Quercinus* is Latin for 'of oak'.

Sulphur polypore Alternative name for the fungus chicken of the woods.

Tuberous polypore *Po. tuberaster*, widespread but scattered and uncommon, found in spring and summer on dead wood of beech and other hardwoods.

Umbrella polypore *Po. umbellatus*, scattered and scarce, found in summer and autumn parasitic on the base of oak and other hardwoods, or saprobic on dead wood. *Umbellus* is Latin for 'umbrella', from the preference for shady sites.

Winter polypore *Po. brumalis*, a widespread but infrequently found (and very rare in Ireland), found from late autumn to spring, sporulating in winter, found on the dead wood of broadleaf trees. Latin *brumalis* = 'relating to winter'.

POLYTRICHACEAE Family name (Greek *polys* + *trichos* = 'much' + 'hair') of haircaps and smoothcaps, species of moss.

POMPILIDAE Family name for spider wasps (Latin *pompilus* = 'pilot fish'), reason obscure. See wasp.

POND-SEDGE Rhizomatous perennial species of *Carex* (Cyperaceae), Latin *carex* = 'sedge'.

Greater pond-sedge *C. riparia*, common in much of England, scattered and locally common elsewhere, in reed swamp, on pool and lake margins, marsh and wet woodland, and along the banks of slow-flowing rivers and canals. It favours base-rich, mesotrophic or eutrophic waters. *Riparia* is from Latin for 'riverbank'.

Lesser pond-sedge *C. acutiformis*, common in lowland regions, shade-tolerant, in base-rich, mesotrophic and eutrophic water, on pond and lake margins, marsh, fen-meadow, and tallherb fen, and carr. *Acutiformis* is Latin for 'sharply pointed'.

POND SKATER Water bugs of the family Gerridae, with seven British species of *Gerris* and two of *Aquarius* (see river skater). Pond skaters are attracted to movement in the surface film such as the struggles of a non-aquatic insect that has fallen into the water, and will quickly 'row' across the water to secure its prey. **Common pond skater** *G. lacustris* (Latin for 'of a lake') and **toothed pondskater** *G. odontogaster* (Greek *odontos* + *gaster* = 'tooth' + 'stomach') are common and widespread in pools and ponds, including garden ponds. Other species are widespread but local.

POND SNAIL Freshwater pulmonate gastropods in the family Lymnaeidae.

Dwarf pond snail *Galba truncatula*, shell length 5–12 mm, width 2.5–6 mm, widespread and locally common, feeding on algae and plant detritus in shallow well-aerated neutral or alkaline water. It is amphibious by small streams and at the margins of wetlands and bodies of fresh water and can be very common in damp field margins where it may crawl considerable distances from open water. This is the main vector of the liver fluke. Latin *galba* = 'fat belly', *truncatula* = 'of a small trunk'.

Ear pond snail *Radix auricularia*, shell length 14–24 mm, width 12–18 mm, the ear-shaped aperture being around five times greater than the actual spire and the lowest whorl representing 90% of the shell volume, common and widespread, less so further north and west, in base-rich lakes, ponds and slowly moving rivers with mud bottoms feeding on algae and detritus. Latin *radix* = 'root' or 'base', *auricularia* = 'relating to the ear'.

Great pond snail *Lymnaea stagnalis*, shell length 45–60 mm, width 20–30 mm, common and widespread in England, Wales and Ireland in well-vegetated lentic (including stagnant) and slow-flowing waters, periodically floating on the surface to replenish its air supply. Greek *limne* and Latin *stagnum*, both = 'marsh' or 'pond'.

Marsh pond snail *L. fusca*, shell length 10–25 mm, width 6–12 mm, common and widespread in the margins of medium to large water bodies, and in shallow, sometimes temporary habitats, and in Ireland also in inter-drumlin fens and neutral to alkaline flushes in coastal areas. Latin *fusca* = 'dark'.

Mud pond snail *Omphiscola glabra*, shell length 9–12 mm, width 3–4 mm, Vulnerable in the RDB, in small standing lownutrient waters with poor aquatic flora. Believed to be extinct in Ireland but rediscovered in Co. Waterford in an acidic mire near Carrickavrantry in 2009. Latin *glabra* = 'smooth'.

Wandering (pond) snail *Radix balthica*, shell length 20 mm or more, common and widespread in Britain, though since the 1980s declining in much of Ireland (Vulnerable, if stable, in Northern Ireland's RDB, 2009), in streams and stagnant waters, tolerant of different pH levels, salinity concentrations and temperatures, preferring calcareous water, feeding on algae, bacterial biofilms and detritus. The complete mitochondrial genome has been obtained (released in 2010): length of the mitochondrial DNA is 13,993 nucleotides in 37 genes. *Balthica* is Latin for 'Baltic'.

PONDWEED Aquatic rhizomatous herbaceous perennials, mainly species of *Potamogeton* (Greek *potamos* + *geiton* = 'river' + 'neighbour'), together with *Groenlandia* (Latin = 'of Greenland') and *Zannichellia* (named after the Italian botanist Giovanni

Zannichelli, 1662–1729), all Potamogetonaceae. See also Cape-pondweed.

American pondweed *P. epihydrus*, despite its name a native species discovered in 1943 in three peaty lochans in South Uist, Outer Hebrides, in oligotrophic and base-poor water <1 m deep. It has also been established from North American plants in the mesotrophic Rochdale Canal and Calder & Hebble Navigation in north England since 1907. Greek *epi* + *hydor* = 'upon' + 'water'.

Blunt-leaved pondweed *P. obtusifolius*, widespread, locally frequent, characteristic of mesotrophic or meso-eutrophic, acidic or neutral standing waters in lakes, ponds and flooded mineral workings, or in canals and river backwaters. Latin *obtusifolius* = 'blunt-leaved'.

Bog pondweed *P. polygonifolius*, generally widespread and locally common, but scarce in central and eastern England, and declining in the south-east, in shallow water in lakes, pools, river backwaters, streams and ditches, or in a dwarf, subterrestrial state in wet moss lawns. It is usually restricted to acidic water, but is occasionally found in calcareous but nutrient-poor sites. Greek *polys* + *gonia* = 'many' + 'angles' + Latin *folia* = 'leaf'.

Broad-leaved pondweed *P. natans*, common and widespread, generally as a floating-leaved aquatic in still or slowly flowing water such as ponds and ditches, but occasionally also with submerged phyllodes in more rapid streams and rivers. With a wide ecological tolerance, it grows in oligotrophic to eutrophic and base-poor to base-rich water over a range of substrates. It can grow in shallow swamp or in water >5 m deep, but is most frequent at depths of 1–2 m. Latin *natans* = 'swimming'. Hybrids, with limited distribution, include **ribbon-leaved pondweed** and **Schreber's pondweed**.

Curled pondweed *P. crispus*, common and widespread, though absent from montane Wales and Scotland, in mesotrophic and eutrophic waters such as lakes, ponds, streams, canals, ditches and disused mineral workings. It is more tolerant of eutrophication than most British pondweeds. Latin *crispus* = 'curled'.

Fen pondweed *P. coloratus*, scattered (absent in north and east Scotland), and generally declining because of drainage and eutrophication. The largest populations are possibly in abandoned brickfields at Orton Pit, Peterborough. It is still relatively common in central Ireland. It favours shallow, calcium-rich but nutrient-poor waters in lakes, pools, clay-pits, shallow streams and ditches, growing over a range of substrates, but especially base-rich peat. Latin *coloratus* = 'coloured'.

Fennel pondweed *P. pectinatus*, common and widespread, especially in England, in eutrophic or brackish waters, forming often dense stands in lakes, streams, ditches, ponds and flooded mineral workings. It is tolerant of disturbance, for example in canals. Latin *pectinatus* = 'like a comb'.

Flat-stalked pondweed *P. friesii*, locally common but scattered in lowland regions, in calcareous and often rather eutrophic, still or very slowly flowing waters, including lakes, sluggish rivers and streams, canals, and flooded mineral workings. It expanded through the canal network, then declined as this became disused or dominated by pleasure boat traffic. *Friesii* means from Friesland, or may honour the Dutch botanist Willem Hendrik de Vriese (1806–62).

Grass-wrack pondweed *P. compressus*, in still or slowly flowing, mesotrophic and slightly to moderately base-rich waters, for instance sluggish rivers, ditches, canals and flooded mineral workings. It has been in gradual decline for over 150 years, has all but disappeared from lakes and rivers, and is declining in grazing marsh ditches. Remaining locally common in central England and east Wales, however, some of the most vigorous surviving populations are in canals, especially the Montgomery branch of the Shropshire Union Canal. Latin *compressus* = 'compressed'.

Hairlike pondweed *P. trichoides*, scattered in Britain, mainly in England, in still or slowly flowing, mesotrophic or eutrophic waters including lakes, ponds, rivers and flooded mineral workings, and often colonising disturbed sites, for example recently cleared canals and ditches. Greek *trichoides* = 'resembling hair'.

Horned pondweed *Z. palustris*, widespread and locally common, in shallow-water habitats such as clear chalk streams, eutrophic lakes and ponds, and brackish lagoons, ponds and ditches. It often colonises disused mineral workings. Latin *palustris* = 'marsh'.

Lesser pondweed *P. pusillus*, widespread and locally common, absent from upland Scotland, in standing or slowly flowing water in sheltered lakes and reservoirs, ponds, rivers, canals, ditches and flooded mineral workings, favouring mesotrophic to base-rich conditions, and tolerating slightly brackish water. Latin *pusillus* = 'small'.

Loddon pondweed *P. nodosus*, confined to a few calcareous and moderately eutrophic rivers (Bristol Avon, Dorset Stour and the river Loddon in Berkshire) in slow-flowing water, preferring gravelly substrates and avoiding soft clays, reproducing vegetatively. Latin *nodosus* = 'with nodes'.

Long-stalked pondweed *P. praelongus*, scattered but absent from southern England, usually growing at depths >1 m in clear, mesotrophic water in lakes, rivers and canals. *Praelongus* is Latin here meaning 'long and slender'.

Opposite-leaved pondweed *G. densa*, scattered in England, scarce in Wales and Ireland, in shallow, clear, base-rich water, occasionally in lakes and rivers, but more usually in smaller waters such as streams (often fast-flowing), canals, ditches and ponds. *Densa* is Latin for 'compact'.

Perfoliate pondweed *P. perfoliatus*, widespread and locally common, generally in mesotrophic or eutrophic water, but occasionally in oligotrophic sites, most vigorous at depths of 1 m or more, but it can grow in shallower water.

Red pondweed *P. alpinus*, scattered, less common in southern regions, in still or slow-flowing water in lakes, rivers, canals, ditches and flooded mineral workings, often where silt has accumulated. It characteristically grows in mesotrophic, often neutral or mildly acidic water.

Sharp-leaved pondweed *P. acutifolius*, confined to shallow, species-rich drainage ditches in lowland grazing marsh, typically in calcareous, mesotrophic or meso-eutrophic water. Found locally in southern and eastern England, but since 1960 it has decreased in Norfolk, and become extinct in the London area and critically endangered in Dorset. Several vigorous populations survive in Sussex. Latin *acutifolius* = 'sharp-leaved'.

Shetland pondweed *P. rutilus*, with submerged leaves, in the Outer Hebrides and Shetland in unpolluted, mesotrophic or eutrophic lochs and adjoining streams, usually where there is some base enrichment. Latin *rutilus* = 'ruddy'.

Shining pondweed *P. lucens*, with submerged leaves, locally common in England and Ireland, scarce elsewhere, in relatively deep, calcareous water in lakes, larger rivers, canals, flooded chalk- and gravel-pits and major fenland drains, in clear, nutrient-poor, unpolluted waters as well as more eutrophic and turbid sites. Latin *lucens* = 'shining'.

Slender-leaved pondweed *P. filiformis*, scattered and local in Northumberland, Scotland and Ireland, in the shallow edges of lakes, typically where the water is base-rich, eutrophic or slightly brackish. It is occasionally found in rivers, streams and ditches. Latin *filiformis* = 'thread-like'.

Small pondweed *P. berchtoldii*, common and widespread in a range of still or slowly flowing waters. Friedrich von Berchtold (1781–1876) was a Czech botanist.

Various-leaved pondweed *P. gramineus*, scattered though rare in the southern half of England, in relatively shallow water in lakes, reservoirs, streams, canals and ditches, tolerating a wide range of water quality, but absent both from very acidic and oligotrophic sites and from the most eutrophic. Hybrids include the less common **bright-leaved pondweed** and **long-leaved pondweed**. *Gramineus* is Latin for 'grass-like'.

PONTEDERIACEAE Family name of pickerelweed, honouring the Italian botanist Guilio Pontedera (1688–1754).

PONY *Equus caballus* (Equidae), small forms of the domestic horse, with a number of semi-feral regional breeds: New Forest, varying in size; Dartmoor, up to 125–30 cm at the withers; Exmoor, 125 cm; Welsh (cobs), varying in size; Fell (Lake District and Northumberland), up to 125–30 cm; and Western Isles, up to 135 cm. These are all free-ranging generally in rough grassland and moorland, but usually subject to annual round-ups, with some foals taken from the herd for taming. 'Pony' dates from the mid-seventeenth century, probably from French *poulenet* = 'small foal', ultimately from Latin *pullus* = 'young animal'. The binomial comprises two Latin words for 'horse'.

POPLAR Trees in the genus *Populus* (Salicaceae), the name dating from Middle English, from Old French *poplier*, from Latin *populus* = 'poplar'. See also aspen.

Black poplar *P. nigra*, a common and widespread, taxonomically complex species with many varieties (e.g. var. *betulifolia*, found especially in England and Wales, by ponds and in hedgerows, especially on lowland flood plains, and as an amenity urban tree), named forms (e.g. fastigiate = Lombardy poplar, much planted in parks and as a windbreak) and hybrids. Latin *nigra* = 'black'.

Grey poplar *Populus* x *canescens* (*P. alba* x *P. tremula*), widespread if scattered, increasing in range in Scotland and Ireland, rarely naturalised, often growing as a solitary, usually male tree, planted in windbreaks and as an amenity tree, especially in damp woods and by streams. *Canescens* is Latin for 'greyish'.

Lombardy poplar A fastigiate form of black poplar.

White poplar *P. alba*, a common and widespread neophyte introduced during the sixteenth century, most frequent as a female in amenity plantings along roadsides and in parks; also in windbreaks and on coastal dunes. It suckers freely and sometimes forms dense thickets. Latin *alba* = 'white'.

POPPY Herbaceous plants named from Old English *popig* or *papæg*, from Latin *papaver* = 'poppy', mainly in the genus *Papaver*, together with *Argemone* (Greek *argemon* = 'cataract', for which argemone was thought to be a cure), *Eschscholzia* (after the Estonian physician and botanist Johann von Eschscholtz, 1793–1831), *Glaucium* (Greek *glaucos* = 'glaucous', from the leaf colour) and *Meconopsis* (Greek *mecon* + *opsis* = 'poppy' + 'appearance'), all in the Papaveraceae.

Atlas poppy *P. atlanticum*, a perennial neophyte introduced from Morocco, a garden escape found on pavements, walls, verges, churchyards and disturbed ground, with a scattered but locally common distribution in the southern half of England, uncommon elsewhere in Britain.

Californian poppy *E. californica*, a perennial neophyte grown since 1826, now a garden escape in disturbed ground on tips and roadsides, sometimes naturalised in quarries and on railway tracks and waste ground. It has long been naturalised on dunes in Guernsey.

Common poppy *P. rhoeas*, a common and widespread annual archaeophyte associated with disturbed ground in towns and arable fields, favouring well-lit, moist (but well-drained), base- and nutrient-rich soils. This is the county flower of Essex and Norfolk as voted by the public in a 2002 survey by Plantlife. *Rhoeas* is Greek for 'poppy'.

Long-headed poppy *P. dubium*, a common and widespread annual archaeophyte of railways, canalsides, wasteland, gardens and arable, in well-lit sites on moist (but well-drained), base- and nutrient-rich soils. *Dubium* is Latin for 'uncertain'.

Mexican poppy *A. mexicana*, an annual introduced from Central America, scattered in southern and central England as a casual on waste ground, a wool and grain alien, and a garden escape.

Opium poppy *P. somniferum*, an annual archaeophyte, a common and widespread weed of gardens, allotments and disturbed ground, especially in well-lit sites on dry to moist, base-rich well-drained soil. *Somniferum* is Latin for 'sleep-bringing', from the narcotic properties.

Oriental poppy *P. pseudoorientale*, introduced from south-east Asia, a naturalised garden escape or discard on generally well-drained soils throughout Britain.

Prickly poppy *P. argemone*, an annual archaeophyte found in arable, especially field edges, most frequent on light sandy, gravelly and chalky soils, and occasionally wasteland, mainly in southern and central England, including East Anglia.

Rough poppy *P. hybridum*, an annual archaeophyte of waste ground and arable, mainly found on calcareous soil in southern England and East Anglia.

Welsh poppy *M. cambrica*, a long-lived perennial native of Wales and Ireland, introduced elsewhere, but overall nationally scarce, growing naturally in rocky woodlands and on cliffs, but commoner in the built environment as a garden escape, favouring cool shady places on fairly fertile soil. This is the county flower of Merioneth as voted by the public in a 2002 survey by Plantlife. *Cambrica* is Latin for 'Welsh'.

Yellow horned-poppy *G. flavum*, a biennial or perennial mostly found widespread on coastal shingle, occasionally in loose rock and on eroding cliffs, and on bare chalk clifftops. The few inland records are of casual occurrences. Latin *flavum* = 'yellow'.

Yellow-juiced poppy *P. lecoqii*, a casual annual archaeophyte found in disturbed sites, including town pavements, scattered in England (common only on southern chalkland), Wales, Ireland and south-east Scotland on heavy soils. Henri Lecoq (1802–71) was a French botanist.

PORCELAIN CRAB See crab.

PORCELAIN FUNGUS *Oudemansiella mucida* (Agaricales, Physalacriaceae), common and widespread, edible, saprobic, found from summer to winter, in clumps on trunks and branches of beech. Cornelius Oudemans (1825–1906) was a Dutch mycologist. Latin *mucida* refers to the mucus layer that covers the cap.

PORCELLANIDAE Family name for porcelain crabs, from Latin *porcellus* = 'piglet', the curved shape of the upper surface of the shell resembling the raised back of a pig.

PORCUPINE Large rodents (**crested porcupine** and **Himalayan porcupines**) in the genus *Hystrix* (Greek for 'porcupine') (Histricidae), occasionally escaping into the wild, but not surviving for more than a few years. 'Porcupine' comes from late Middle English, from Old French *porc espin*, in turn from Latin *porcus* + *spina* = 'pig' + 'thorn'.

PORECRUST Inedible fungi, mostly Polyporales, in which small 'cushions' fuse to form a crust.

Scarce species include the bracket **frothy porecrust** *Oxyporus latemarginatus* (usually on dead wood of beech) and the crustose **hazel porecrust** *Dichomitus campestris* (causing white rot of beech and oak).

Bleeding porecrust *Physisporinus sanguinolentus* (Meripilaceae), a common and widespread annual sporulating in summer and autumn, found on damp wood. Greek *physis* + *spora* = 'nature' + 'spore', and Latin *sanguinus* + *lentus* = 'bloody' + 'tough' or 'sticky'.

Cinnamon porecrust *Fuscoporia ferrea* (Hymenochaetaceae), widespread, more common in England, evident throughout the year but sporulating in autumn, on dead branches of broadleaf trees, especially hazel, causing a white rot. The genus *Fuscoporia* was established in 2007, the name coming from Latin *fusca* + *porus* = 'dark brown' + 'pore'. *Ferrea* is Latin for 'hard', but here probably refers to the ferrous (rusty) colour.

Common porecrust Alternative name for splitgill.

Green porecrust *Ceriporiopsis pannocincta* (Phanerochaetaceae), in southern and south-east England and East Anglia on fallen branches, logs and stumps of a variety of hardwoods. Greek *ceras* + *poros* = 'horn' + 'callus', and Latin *pannus* + *cinctus* = 'cloth' + 'belt'.

Pink porecrust *C. gilvescens*, widespread but at best locally common in England, on fallen branches, logs and stumps of hardwoods. Latin *gilvus* = 'pale yellow'.

Rusty porecrust *F. ferruginosa*, widespread, commoner in England, evident throughout the year but sporulating in autumn, on dead branches of hardwoods, especially beech. Latin *ferruginosa* = 'rusty'.

Split porecrust *Schizopora paradoxa* (Hymenochaetales, Schizoporaceae), a common and widespread annual found throughout the year though sporulating in late summer and autumn on decaying wood. Greek *schizo* + *poros* = 'split' + 'callus', and Latin *paradoxa* = 'paradox'.

PORIFERA Phylum including sponges (Latin *porus* + *fero* = 'pore' + 'to bear').

PORPOISE Harbour porpoise or **common porpoise** *Phocoena phocoena* (Phocoenidae), body length 1.3–1.5 m, weight 50–60 kg, the smallest, commonest and most widely distributed cetacean in northern Europe, favouring shallow, cold waters. They may be residential off south-west and west England, west Wales, west Scotland, the Northern Isles and east Scotland, with greatest numbers between July and October. They were once regular summer visitors to the south coast of England and the southern part of the North Sea, but are now rarely seen there. Diet is of fish, cephalopods and crustaceans. The main mating season is summer, and birth takes place 10–11 months later. Calves are suckled for 4–8 months, and females usually reproduce every 1–2 years. 'Porpoise' comes from French *pourpois*, Old French *porpais*, in turn from medieval Latin *porcopiscus*, a compound of *porcus* = 'pig' and *piscus* = 'fish', from a resemblance of the snout to that of a pig, or the sound of a porpoise breathing resembling a pig snort. Greek *phocaina* = 'big seal' as described by Aristotle, in turn from *phoce* = 'seal'.

PORTUGUESE MAN O' WAR *Physalia physalis* (Physaliidae), a pelagic warm-water species occasionally driven into to shore, especially on south-west coasts, by winds and currents. Each 'individual' is composed of a group of polyps specialised for movement, catching prey, feeding and breeding. A large gas-filled float (pneumatophore) reaching up to 30 cm in height allows it to float on the surface. The crest running along the top of the pneumatophore acts as a sail when raised. The jellyfish has many digestive polyps which hang down and secrete digestive juices onto the prey that has been caught and immobilised by the sting of the long, contractile tentacles. Tentacles may hang down several metres and have a bead-like appearance, each 'bead' containing specialised stinging cells (nematocysts). *Physalis* is Greek for 'bladder'.

PORTULACACEAE Family name of common purslane (Latin *portulaca* = 'carrying milk').

PORTUNIDAE Family name of the swimming, harbour, shore and sand crabs. Portunus was the Roman god of the harbour.

POT BEETLE Species of *Cryptocephalus* (Chrysomelidae, Cryocephalinae) (Greek *cryptos* + *cephale* = 'hidden' + 'head').

Hazel pot beetle *C. coryli*, 5.8–7.5 mm length, with a scattered distribution in the east Midlands and southern England in clearings and rides in broadleaf woodland on south-facing slopes and calcareous grassland, Endangered in RDB 1, and a priority species in the UK BAP. Adults and larvae feed on birch leaves, larvae also taking catkins and fallen leaves. *Corylus* is Latin for 'hazel', though this is not a food plant.

Pashford pot beetle *C. exiguus*, 2–3.2 mm length, feeding on sorrel, with recent records (to 2000) only from Pashford Poors Fen, Suffolk. However, drainage of surrounding fields has led to drying of the site and this species may now be extinct. *Exiguus* is Latin for 'short'.

Rock-rose pot beetle *C. primarius*, 4.5–8 mm length, Endangered in RDB 1, and a priority species in the UK BAP. Widely scattered with recent records only from a few locations in south England, mostly calcareous grassland, especially warm, sheltered, dry south-facing slopes where adults feed on petals, anthers and pollen of rock-rose, larvae on stems and leaves. *Primarius* is Latin for 'distinguished'.

Six-spotted pot beetle *C. sexpunctatus*, 4.5–6.5 mm length, scarce, with a very scattered distribution in England and Scotland, on calcareous grassland and broadleaf woodland, Vulnerable in RDB 2, and in the UK BAP. Adults feed on tree leaves and possibly also pollen; larvae feed on hazel catkins and leaves, and leaves of low-growing plants. *Sexpunctatus* is Latin for 'six-spotted'.

Ten-spotted pot beetle *C. decemmaculatus*, 3.5–5 mm length, with a few records in the north-west Midlands and the Highlands, Vulnerable in RDB 2, and in the UK BAP. Found in damp broadleaf woodland, especially on wet hillsides or in areas with quaking bogs, feeding on leaves of willows, alder and downy birch. *Decemmaculatus* is Latin for 'ten-spotted'.

POTAMOGETONACEAE Family name of pondweed (Greek *potamos* + *geiton* = 'river' + 'neighbour').

POTATO (ECHINODERM) See sea potato.

POTATO (PLANT) *Solanum tuberosum* (Solanaceae), a herbaceous rhizomatous South American neophyte very commonly grown in fields, gardens and allotments as a vegetable, introduced in the late Tudor period, a widespread perennial of cultivated and wasteland, rubbish tips and on coastal sand and shingle associated with dumped domestic waste. It is usually casual in disturbed habitats, but production of tubers allows some populations in stable sites to become established. 'Potato' comes from Spanish *patata*, a variant of Taíno (indigenous Caribbean) *batata* = 'sweet potato', the English word originally denoting sweet potato and acquiring its current sense in the late sixteenth century. *Solanum* is Latin for the congeneric 'nightshade', *tuberosum* = 'tuberous'.

POTTER WASP Member of the family Vespidae (Eumeninae), with 23 species in 8 genera in the British list. Most species prey on caterpillars, and build pot-shaped mud nests. See also mason wasp.

Heath potter wasp *Eumenes coarctatus*, scarce, found from south Devon to East Sussex, north to Buckinghamshire, flying in April–October, visiting flowers of bramble, gorse and heather for nectar, in heathland with bare clay and nearby water used to construct the nests (pots) which are built on heather and gorse, and provisioned with caterpillars, mainly geometrids and tortricids, for the wasp larvae. Greek *eu* + *menes* = 'well' + 'force' or 'vigour', and Latin *coarctatus* = 'confined'.

POTTIA Species of moss in the family Pottiaceae.

Blunt-fruited pottia *Tortula modica*, common and widespread on disturbed field and garden soil, on waste ground, banks, anthills and molehills, in quarries, by tracks and paths, and on walls. Latin *tortula* = 'small twist', referring to the peristomes, *modica* = 'moderate'.

Common pottia *T. truncata*, common and widespread, absent from the Highlands. Latin *truncata* = 'truncated'.

Heim's pottia *Hennediella heimii*, common and widespread around most coasts on bare, salty and brackish soil, and recently found inland on salted road verges. Roger Hennedy (1809–77) was a Scottish botanist.

Lance-leaved pottia *T. lanceola*, widespread, favouring calcareous grassland. Latin *lanceola* = 'small lance'.

Oval-leaved pottia *Pterygoneurum ovatum*, widespread but scattered and, having declined greatly during the twentieth century, now a rarity found on recently disturbed calcareous soil in pits and quarries, among rocks (mainly chalk) and on sea-cliffs. Greek *pteryx* + *neuron* = 'wing' + 'nerve', and Latin *ovatum* = 'oval'.

Round-fruited pottia *Pottiopsis caespitosa*, scattered in southern England and the Midlands on bare soil on chalk grassland, limestone banks, disused chalk-pits, and calcareous coastal sand. Latin *caespitosa* = 'tufted'.

Short pottia *H. macrophylla*, scattered in England and Scotland on shaded, bare, compacted soil especially by streams. Greek *macrophylla* = 'large-leaved'.

Smallest pottia *Microbryum davalliana*, common and widespread in lowland regions on disturbed, lime-rich soil, for example in arable fields, calcareous grassland and woodland rides. Greek *micros* + *bryon* = 'small' + 'moss'. Edmund Davall (1763–98) was a Swiss-English botanist.

Starke's pottia *M. starkeana*, a scarce calcicolous moss found in England and Wales on disturbed, lime-rich soil usually in coastal areas or on metal mine sites.

Upright pottia *M. rectum*, widespread but scattered in England, Wales and Ireland on disturbed soil in calcareous grassland, quarries, cliffs and arable fields. Latin *rectum* = 'upright'.

POUCHWORT Foliose liverworts in a number of families. Rarities include **heath pouchwort** *Gongylanthus ericetorum* (with a few records from the Scilly Isles and south Cornwall), **Nees' pouchwort** *Calypogeia neesiana* (in upland areas on eroding peat on blanket bogs), and **turps pouchwort** *Geocalyx graveolens* (Geocalycaceae) (smalling of turpentine, on boulders in western Scotland and Ireland).

Blue pouchwort *Calypogeia azurea* (Calypogeiaceae), a common and widespread upland calcifuge found on peaty and sandy tracksides, rocky slopes, and heathland, moorland and blanket bog. Greek *cala* + *geios* = 'beautiful' + 'of the earth'.

Bog pouchwort *C. spagnicola*, locally common and widespread, growing almost exclusively on sphagnum hummocks and in bog pools.

Common pouchwort *C. fissa*, common and widespread in woodland, heathland and bog on acidic soil, peat, decomposing logs, and sphagnum or purple moor-grass tussocks. Latin *fissa* = 'split'.

Meylan's pouchwort *C. integristipula*, in woodland on sandstone outcrops in the Weald, south-east England, and on sandstone, gritstone and peaty banks in northern England, Scotland and north-west Ireland. Charles Meylan (1868–1941) was a Swiss botanist. Latin *integrus* + *stipula* = 'whole' + 'stalk'.

Mueller's pouchwort *C. muelleriana*, a common and widespread calcifuge found on peat, occasionally mineral soil or clay, in woodlands, especially conifer plantations, and in the mountains on bogs and heathland. Walther Müller (1833–87) was a German botanist.

Notched pouchwort *C. arguta*, common and widespread on acidic soil and rock faces, in woodland and lane banks. *Argutus* is Latin for 'distinct'.

Straggling pouchwort *Saccogyna viticulosa* (Geocalycaceae), in western Britain and in Ireland straggling through other bryophytes in shaded, moist habitat in oceanic conditions, particularly on woodland banks and damp rock faces. Greek *saccos* + *gyne* = 'sack' + 'woman', and Latin *viticula* = 'vine'.

Swedish pouchwort *C. suecica*, in upland north and west Scotland and north Ireland on decomposing logs, especially of conifers, in humid wooded valleys. *Suecica* is Latin for 'Swedish'.

POUNCEWORT Foliose liverworts in the family Lejeuniaceae, many in the genus *Lejeunea* and variant names, after the French physician and botanist Lejeune (1779–1858).

Rarities include **Atlantic pouncewort** *Lejeunea mandonii* (Lizard Peninsula, west Scotland and south-west Ireland), **Rossetti's pouncewort** *C. rossettiana* (scattered on base-rich upland rocks and coastal cliffs), and **Holt's pouncewort** *L. holtii*, **Irish pouncewort** *L. hibernica*, and **yellow pouncewort** *L. flava* (all found in Co. Kerry on boulders and tree-trunks).

Long-leaved pouncewort *Aphanolejeunea microscopica*, in shaded, humid sites on rocks, especially by streams and waterfalls. Greek *aphanes* = 'secret'.

Mackay's pouncewort *Marchesinia mackaii*, in western Britain and mainly western Ireland on shaded, dry, base-rich rocks, especially in woodland over limestone, occasionally on trees, especially yew. James Mackay (1775–1862) was an English botanist who lived in Ireland.

Micheli's least pouncewort *L. cavifolia*, common and widespread on rocks and trees in shaded and humid locations, especially in woodland. Pier Micheli (1679–1737) was an Italian priest and botanist. *Cavifolia* is Latin for 'hollow-leaved'.

Minute pouncewort *Cololejeunea minutissima*, on tree-trunks in humid sites, occasionally on rock, with a mainly coastal and southern distribution in Britain, but with an increasing number of inland records, and it has now spread northwards along the west coast as far as the Outer Hebrides. It is widespread and scattered in Ireland.

Pearl pouncewort *L. patens*, common and widespread, characteristic of base-rick upland rocky outcrops. *Patens* is Latin for 'spreading'.

Pointed pouncewort *Harpalejeunea molleri*, in upland western Britain and western Ireland in humid sites in oceanic districts, on rocks and trees. Greek *harpe* = 'hook' or 'sickle'. F.H. Møller (b.1887) was a Danish botanist.

Rock pouncewort *C. calcarea*, a common and widespread upland calciphile found on shaded limestone and other base-rich rocks.

Toothed pouncewort *Drepanolejeunea hamatifolia*, locally common in oceanic upland areas in western Britain and western Ireland on trees and, especially, rocks. Greek *drepane* = 'sickle', and Latin *hamus* + *folia* = 'hook' + 'leaf'.

Western pouncewort *L. lamacerina*, locally common in western Britain and in Ireland on rock faces by valley streams, occasionally on trees and rocks in upland areas.

POWAN *Coregonus clupeoides* (Salmonidae), a rare glacial whitefish endemic to two Scottish lochs (Loch Lomond, and Loch Eck in Argyll) and introduced to Loch Sloy and the Carron Valley Reservoir. In March 2010 around 400 hatchery-reared pollan were introduced into Loch Lomond, where threats include powan egg-eating ruffe introduced into these waters in the 1990s. 'Powan' is a mid-seventeenth-century Scottish variant of 'pollan', another whitefish. Greek *core* + *gonia* = 'pupil of the eye' + 'angle', leading to 'angle-eyed', and Latin *clupeoides* = 'herring-like'.

POWDERCAP Fungi with granular cap surfaces in the genera *Cystoderma* and *Cystodermella* (Agaricales, Agaricaceae) (Greek *cystis* + *derma* = 'blister' + 'skin').

Cinnabar powdercap *Cystodermella cinnabarina*, probably inedible, rare, very scattered, most records coming from Scotland on soil in deciduous and, more usually, coniferous woodland under pine.

Earthy powdercap *Cystoderma amianthinum*, edible, widespread and common, found in summer and autumn on acid soil in short grass and moss in coniferous woodland and on moorland and heathland, as well as on lawns sometimes forming fairy rings. *Amiantos* is Greek for 'pure' or 'unspotted'.

Pearly powdercap *Cystoderma carcharias*, inedible, widespread but uncommon, found in late summer and autumn on acid soil in short grass in coniferous woodland and on moorland and heathland. *Carcharos* is Greek for 'sharp' or 'jagged'.

Pine powdercap *Cystoderma jasonis*, inedible, widespread but scattered and uncommon, in upland areas on soil in coniferous plantations, especially under pine and spruce.

POX Nettle pox *Calloria neglecta* (Heliotales), the conidial stage of *Cylindrocolla urticae* (its anamorph), widespread and common in Britain, fruiting in spring and forming orange patches of up to 1 mm diameter on dead standing stems of nettle.

PRAWN Decapod crustaceans with long narrow muscular abdomens, long slender legs, and long antennae. 'Prawn' is generally used for shrimp species that have commercial value, but a morphological distinction is that in prawns the first three of the five pairs of legs have small pincers, while in shrimps only two pairs are claw-like. **Dublin Bay prawns** are actually Norway lobsters.

Chamaeleon prawn *Hippolyte varians* (Hippolytidae), up to 3.2 cm long, widespread if patchy around the coasts of Britain and north-east Ireland in rock pools, and among macroalgae and eelgrass beds from the lower intertidal to a depth of 150 m. Hippolyte was Queen of the Amazons. *Varians* is Latin for 'various'.

Common prawn *Palaemon serratus* (Palaemonidae), common and widespread, usually in groups, in crevices and under stones from intertidal pools to the shallow subtidal. They are also found in estuaries and sea-grass beds. Palaemon was a sea goddess in Greek mythology. Latin *serratus* = 'serrated'.

Grass prawn *P. elegans*, widespread but scattered, on rocky shores in rock pools around mid-tide level, sometimes higher. *Elegans* is Latin for 'graceful'.

PRAYDIDAE Family name (Greek *prays* = 'gentle') for ash bud moth. Also *Prays peregrina*, wingspan 13–15 mm, a naturalised accidental introduction new to science found in parks and gardens in south-east England, first recorded on Parliament Hill, north London in 2003. *Peregrinus* is Latin for 'stranger'.

PRETENDER *Boletus pseudoregius* (Boletales, Boletaceae), a rare fungus, recorded from just 14 locations between 1963 and 2012, in woodlands of central and southern England favouring calcareous sites, found under single oak trees in open mature woodland, often in bolete 'hot spots' such as parts of the New Forest and Wye Valley. Greek *bolos* = 'lump', and Latin *pseudoregius* = 'false' + 'royal'.

PRIDE-OF-INDIA *Koelreuteria paniculata* (Sapindaceae), a deciduous neophyte tree introduced from China in 1763, a garden escape in south-central and south-east England and south Wales, first found naturalised in Kent in 1908. Joseph Kölreuter (1733–1806) was a German botanist. *Paniculata* is Latin for 'with flower clusters'.

PRIMROSE *Primula vulgaris* (Primulaceae), a common and widespread evergreen herbaceous perennial, found often on heavy soils in woodland, on north-facing banks (avoiding direct sunlight), in hedgerows, damp grassland, coastal slopes and shaded montane cliffs. In East Anglia woodland populations have declined since the 1970s in response to a series of hot, dry summers. Seeds are usually dispersed by ants. This is the county flower of Devon as voted by the public in a 2002 survey by Plantlife. *Primula* is a diminutive of Latin *primus* = 'first', meaning early flowering, and *vulgaris* = 'common'.

Bird's-eye primrose *P. farinosa*, a short-lived herbaceous perennial locally common in northern England, on damp grassy or peaty habitat on limestone and on wet calcareous flushes. *Farinosa* is Latin for 'mealy' or 'powdery'.

Scottish primrose *P. scotica*, an endemic herbaceous perennial found in the extreme north of mainland Scotland and Orkney in moist, well-drained, usually heavily grazed, open grassland habitats, often on calcareous substrates. Sites include dunes, clifftops, the transition zone between grassland and maritime heath, mosaics of heath and machair, and around rock outcrops. This is the county flower of Caithness as voted by the public in a 2002 survey by Plantlife.

PRIMULACEAE From a diminutive of Latin *primus* = 'first' (early flowering), the family name of primrose that includes cowslip, creeping-jenny, loosestrife, oxlip and pimpernel.

PRINCE *Agaricus augustus* (Agaricales, Agaricaceae), a widespread but infrequent, edible and tasty fungus, found in late summer and autumn on soil under broadleaf and, especially, coniferous trees. Greek *agaricon* = 'mushroom'; *augustus* may refer to the month when it is commonly found.

PRIVET Shrubs in the genus *Ligustrum* (Oleaceae), Latin for these plants. 'Privet' dates from the mid-sixteenth century, etymology unknown.

Garden privet *L. ovalifolium*, a common and widespread semi-evergreen or evergreen Japanese neophyte, introduced in 1842, much planted for hedging and a persistent relic in old gardens, or a garden throw-out in hedges, and waste ground and railway banks. *Ovalifolium* is Latin for 'oval-leaved'.

Wild privet *L. vulgare*, common and widespread (absent from north Scotland), deciduous or semi-evergreen, a native in hedgerows, woodland and scrub, preferring well-drained, calcareous or base-rich soils, and often planted, particularly in hedges and woodland, occurring as a garden escape. It is probably native in coastal parts of Ireland, but is mostly introduced there. *Vulgare* is Latin for 'common'.

PROCELLARIIDAE Family of fulmars and shearwaters (Latin *procella* = 'storm').

PROCELLARIIFORMES Order of albatrosses, fulmars and shearwaters, and petrels (Latin *procella* = 'storm').

PROCESSIONARY (MOTH) Caterpillars form often large nose-to-rear processions when moving, though in oak processionaries they often actually form arrow-headed processions, with one leader and subsequent rows containing several caterpillars.

Oak processionary *Thaumetopoea processionea* (Notodontidae, Thaumetopoeinae) wingspan 25–35 mm, nocturnal, populations of this adventive having been found since 2006 in several boroughs in west and south-west London and the Elmbridge and Spelthorne districts of Surrey; in Pangbourne, west Berkshire (2010); and in the boroughs of Bromley and Croydon in south London (2012). Vagrant populations occur along the south coast from time to time. It is resident in the Channel Islands. Adults appear from July to September. Larvae feed on oak leaves, living gregariously in a silk web, and can defoliate the trees. Tiny severely irritating hairs on larger larvae contain a defensive toxin (thaumetopoein) that can cause persistent or severe (occasionally life-threatening) allergic reactions in people and pets, and are easily dispersed on air currents. This species is covered by UK Plant Health legislation under which it is illegal to knowingly keep, store or sell it, and occurrences should be reported to DEFRA. In 2015 the Forestry Commission and partners started deploying about 900 pheromone traps around the outskirts of the known outbreak area in London to trap male adult moths as they emerge from pupation. Results will help update knowledge of the pest's distribution. Nests are removed and destroyed. The genus name is a typographic error from Greek *thaumatopoios* = 'wonder-working', from the processionary behaviour of the larvae.

Pine processionary *T. pityocampa* is a rare immigrant.

PRODOXIDAE Family name for some of the bright moths, Greek *prodoxos* meaning judging something before experiencing it.

PROMINENT (MOTH) Nocturnal members of the family Notodontidae (Notodontinae). Adults do not feed. The English name comes from the tuft of hair on the trailing edge of the forewing which protrudes upwards at rest. Larvae often have 'humps' on their back (not so in great prominent), and may hold their tail segments in the air.

Coxcomb prominent *Ptilodon capucina*, wingspan 35–40 mm, widespread and common in woodland, scrub, parks and gardens. Adults fly in May–June and, in the south, again in August. Larvae feed on a number of deciduous trees. Adults feign death when handled. Larvae raise their head over their back when alarmed. Greek *ptilon* + *odontos* = 'feather' + 'tooth', and *capucina* from Late Latin *cappa* = 'cap', here a pointed hood, for the thoracic crest.

Great prominent *Peridea anceps*, wingspan 50–65 mm, local in old oak woodland, hedgerows and gardens throughout much of England and Wales. Adults fly in April–June. Larvae feed on oak leaves in late summer. Greek *perideies* = 'very timid', and Latin *anceps* = 'doubtful'.

Iron prominent *Notodonta dromedarius*, wingspan 35–40 mm, widespread and common in broadleaf woodland, heathland, marsh and gardens. Except in the north, there are two broods, adults flying in May–June and August. The single brood appears in June–July. Larvae feed on birch and alder leaves. Greek *notos* + *odontos* = 'back' + 'teeth', from the dorsal scale forewing tufts which point upwards when the wings are folded. *Dromedarius*, Latin for 'dromedary', references either the larval 'humps' or forewing tufts.

Lesser swallow prominent *Pheosia gnoma*, wingspan 45–50 mm, common in woodland, heathland, moorland, downland, parks and gardens. The bivoltine adults fly in May–June and August. Larvae feed on birch. Greek *pheos* = 'a spine', and *gnoma* = 'mark'.

Maple prominent *Ptilodon cucullina*, wingspan 35–40 mm, local in broadleaf woodland, scrub, chalk downland, hedgerows and gardens in much of south-east England and East Anglia. Adults fly in May–July. Larvae feed on (mainly young) field maple, occasionally sycamore. *Cucullus* is Latin for 'hood', from the thoracic crest.

Pale prominent *Pterostoma palpina*, wingspan 35–55 mm, widespread and common in woodland, scrub and gardens. Adults fly in May–June, and in the south again in August. Larvae feed on poplars, aspen and sallows. Greek *pteron* + *stoma* = 'feather' + 'mouth', *palpina* from the long labial palpus.

Pebble prominent *N. ziczac*, wingspan 40–5 mm, widespread and common in damp woodland, hedgerows and gardens. Adults fly between May and August in either one or two broods depending on latitude. Larvae feed on sallow and poplar leaves. *Ziczac* is Latinised 'zigzag', from the humped larval profile.

Plumed prominent *Ptilophora plumigera*, wingspan 33–44 mm, scarce on chalk downland in parts of southern England, from Dorset to Sussex, and East Anglia. Adults fly in November–December. Larvae feed on field maple, occasionally sycamore. Greek *ptilon* + *phoreo*, and Latin *pluma* + *gero*, both = 'feather' + 'to carry'.

Scarce prominent *Odontosia carmelita*, wingspan 38–45 mm, local in long-established birch woodland, from southern England discontinuously to the Highlands, and in Ireland, but absent from Wales. Adults fly in April–May. Larvae feed on birch leaves. Greek *odontos* + *osis* = 'tooth' + 'thrust', from the dorsal forewing 'tooth'. *Carmelita* is a Latinised form of Carmelite, an order of monks who wear cowls, referencing the thoracic crest.

Swallow prominent *Pheosia tremula*, wingspan 45–65 mm, widespread and common in woodland, scrub, gardens and parks. Adults fly in May–July, with two generations in the south. Larvae feed on poplars, aspen and sallows. *Tremula* is Latin for 'shaking'.

White prominent *Leucodonta bicoloria*, wingspan 38–42 mm, rediscovered in Co. Kerry in 2008. Prior to that it had been unrecorded since 1938, and thought to be extinct in the British list. Adults fly in June–July. Larvae feed on birch. Greek *leucos* + *odontos* = 'white' + 'tooth', from the white forewing, and Latin *bicoloria* = 'two-coloured', from the black and orange wing markings.

PRONGWORT Foliose liverworts in the genus *Herbertus* (Herbertaceae).

Juniper prongwort *H. aduncus*, on acidic substrates in oceanic-montane heathland in north-west Wales, Cumbria, western Scotland and western Ireland on rocky slopes and ledges, sometimes in wooded ravines. *Aduncus* is Latin for 'bent inwards'.

Northern prongwort *H. borealis*, a rare endemic found in a single locality in the Beinn Eighe nature reserve, north-west Scotland, in rocky, acidic, montane dwarf shrub heath. *Borealis* is Latin for 'northern'.

Straw prongwort *H. stramineus*, a usually montane species found in north-west Wales, the Lake District and the western

Highlands and Islands, generally on calcareous substrates, for example rock ledges, turf banks and boulders. *Stramineus* is Latin for 'straw-coloured'.

PROSOBRANCH(IA) From Greek *proso* + *branchia* = 'forward' + 'gills', a large group (subclass) of snails with the lung located in front of the heart. See also Opisthobranch and Pulmonata.

PROTIST A single-celled organism that has a cell nucleus. Protists (Greek *protos* = 'first') make up one of the five kingdoms of nature.

PROTOZOA From Greek *protos* + *zoon* = 'first' + 'animal', single-celled organisms found in every aquatic environment, including animal bodies.

PROTURA Order of tiny soil insects (Greek *protos* + *oura* = 'first' + 'tail').

PROWLER Pink prowler *Oonops domesticus* (Oonopidae), male body length 1.5 mm, females 2 mm, a scarce and scattered synanthropic, nocturnal spider found year-round, confined to houses and other warm buildings, feeding on such prey as booklice. Greek *oon* + *opsis* = 'egg' + 'appearance'.

PRUNELLIDAE Family name for accentors, including dunnock, from the Latin diminutive of *prunus*, here meaning 'brown'.

PRUNES AND CUSTARD *Tricholomopsis decora* (Agaricales, Tricholomataceae), an inedible saprobic fungus, locally common in the Highlands, scarce elsewhere, found in June–October on stumps and dead wood of pine. The genus name means similar to the genus *Tricholoma*. *Decora* is Latin for 'decorous'. See also plums and custard.

PRURITUS Or **sweet itch** (from Latin *prurigo* = 'itching'), a medical condition in horses caused by an allergic response to the bites of *Culicoides* midges.

PSEPHENIDAE Family name (Greek *psephos* = 'pebble') for water-penny beetles, with one British representative, *Eubria palustris*.

PSEUDOCYPHELLARIA See specklebelly.

PSEUDOSCORPION Members of the order Pseudoscorpiones (Arachnida), having a flat, pear-shaped body protected by armoured plates, but unlike scorpions their abdomen is short and rounded at the rear. They possess two pincers (pedipalps) equipped with cutting teeth used to catch and inject paralysing venom into prey such as springtails or mites. They are found in a variety of habitats including houses, where they feed on (and may be phoretic on) moth and carpet beetle larvae, ants, mites, flies and booklice.

PSILIDAE Family of rust flies (Greek *psilos* = 'naked' or 'smooth'), with 26 species in 6 genera. Larvae live and feed in stems and roots of herbaceous plants, and can become pests of crops, for example species of *Psila*, and carrot fly.

PSOCOPTERA Order (Greek *psocho* + *pteron* = 'rubbed' + 'wing') of barkfly species (winged, found outdoors) and booklice (not winged, indoors), very small soft-bodied insects. Some 98 species have been recorded in Britain, of which 68 are barkflies.

PSORA Rusty alpine psora *Psora rubiformis* (Psoraceae), a crustose lichen, on calcareous mica-schist, known in Britain only from Ben Lawers, southern Highlands. Greek *psora* = 'itch' and Latin *rubiformis* = 'red appearance'.

PSYCHIDAE From Greek *psyche* = 'soul' (and by analogy, 'butterfly' or 'moth', from the state between metamorphosis and resurrection), the family name for case-bearer, smoke and sweep moths.

PSYCHODIDAE From Greek *psyche* + *eidos* = 'moth' (via 'soul') + 'form', the family name for the small (<2 mm length) moth flies, with 99 species in 21 genera, their common name coming from their short hairy bodies and wings. Some species are also called drain flies and owl midges.

PSYCHOMYIIDAE From the Greek nymph Psyche + *myia* = 'fly', the family name of caseless, trumpet-net and tube-making caddis flies, larvae living in silk galleries disguised by silt and other material.

PSYLLID(AE) From Greek *psylla* = 'flea', see plantlice.

PTARMIGAN *Lagopus muta* (Phasianidae), wingspan 57 cm, length 35 cm, male weight 500 g, females 400 g, a Green-listed game bird breeding in the highest mountains of the Highlands, with 8,500 pairs estimated in 2007. Clutch size is 5–8, incubation lasts 21–3 days, and the precocial young fledge in 10–15 days. Rubbish left by tourists is causing a decline in numbers by attracting crows which eat ptarmigan chicks. Diet is of shoots, leaves, leaf buds, berries and insects. 'Ptarmigan' dates from the late sixteenth century, from Scottish Gaelic *tarmachan*; the spelling with 'p-' was introduced later, inappropriately prompted by Greek words starting with 'pt-'. Greek *lagos* + *pous* = 'hare' + 'foot', a reference to the feathered feet and toes typical of this cold-adapted bird, and Latin *muta* = 'mute'.

PTERIDOPHYTA Major grouping of vascular plants containing ferns, horsetails, mosses, and quillworts (Greek *pteris* + *phyton* = 'fern' + 'plant').

PTEROPHORIDAE Family name (Greek *pteron* + *phoros* = 'wing' + 'carry') of plume moths, with subfamilies Agdistinae (3 species, with forewings that are not cleft or separated into 'plumes') and Pterophorinae (35 species, with forewings usually consisting of two curved spars with more or less bedraggled bristles trailing behind).

PTERYGOTA Subclass (Greek *pterygotos* = 'winged') containing insects other than the primitive wingless bristletails, springtails and proturans.

PTILIIDAE Family of featherwing beetles (Greek *ptilon* = 'wing' or 'feather'), with 75 species in 18 genera in the British list. The family includes (at least some of) the smallest insects in Britain, all 1 mm or less in length, including the smallest British beetle, the ironically named *Nephanes titan*.

PTILODACTYLIDAE From Greek *ptilon* + *dactylos* = 'wing' or 'feather' + 'finger', a beetle family, with a single British species, *Ptilodactyla exotica*, found in moist soil in glasshouses at Kew and the Cambridge University Botanic Gardens, feeding on spores and hyphae of microfungi on vegetation.

PTINIDAE Family name (Greek *ptenos* = 'feathered') for (some of the) wood-borer beetles, with 56 British species in 25 genera. Most species are small (1–7 mm) wood borers, such as furniture and death-watch beetles; a few are found in stored products.

PUFFBALL Members of the genera *Bovista* (Old German *Vohenvist* = 'fox flatulence', referring to the smell of the spore dust releases), *Calvatia* (Latin = 'bald skull') and *Lycoperdon* (Greek *lycoperdon* = 'wolf's flatulence', again from the scent), all Agaricales (Agaricaceae).

Scarce species include **dwarf puffball** *B. dermoxantha*, **fen puffball** *B. paludosa*, **flaky** (or **Venus**) **puffball** *L. mammiforme*, **least puffball** *B. limosa* and **spiny** (or **spring**) **puffball** *L. echinatum*.

Widespread but nevertheless uncommon are **dusky** (or **blackening**) **puffball** *L. nigrescens* (possibly mildly toxic, on acid soil in coniferous woodland, and in grass on heathland and dunes), **grey puffball** *B. plumbea* (short grass), **mosaic puffball** *L. utriforme* (unimproved grassland), and **soft puffball** *L. molle* (woodland soil).

Brown puffball *B. nigrescens*, inedible, widespread and locally frequent, found in late summer and autumn, on acid soil associated with grasslands. *Nigrescens* is Latin for 'becoming black'.

Common puffball *L. perlatum*, edible when young, common and widespread, found in summer and autumn on soil and leaf litter in mixed woodland, occasionally pasture and grassed dunes. *Perlatum* is Latin for 'completed'.

Giant puffball *C. gigantea*, widespread, locally common, edible and tasty, growing up to 80 cm diameter and weighing several kilograms, found in summer and autumn on nutrient-rich soil adjacent to wooded areas, hedgerows, parks and gardens.

Grassland puffball *L. lividum*, inedible, locally common and widespread, found in summer and autumn on sandy soil in grassland, heathland and dunes. *Lividum* is Latin for 'lead-coloured'.

Meadow puffball *L. pratense*, edible when young, common and widespread, saprotrophic, found in summer and autumn on grassland soil. *Pratense* is from Latin for 'meadow'.

Pestle puffball *L. excipuliforme*, edible when young, common and widespread in late summer and autumn on soil in open woodland and under hedgerows. *Excipula* is Latin for 'receptacle'.

Stump puffball *L. pyriforme*, edible when fresh, very widespread and common, found from summer to early winter, the only British puffball to grow on hardwood stumps and rotting wood. *Pyriforme* is Latin for 'pear-shaped'.

PUFFIN Or **Atlantic puffin** *Fratercula arctica* (Alcidae), wingspan 55 cm, length 28 cm, weight 400 g, an auk with 580,000 pairs estimated in 2002 but, with numbers declining (in concert with its main prey species, sand eel) and 80–90% of the UK population in the 'best' ten sites, it has become an endangered Red List species. Important colonies are at Bempton Cliffs (Yorkshire), the Farne Islands and Coquet Island (Northumberland), South Stack (Anglesey), the Isle of May (Fife coast), Orkney, Shetland and west Ireland (also Great Saltee, Co. Wexford). Having wintered out at sea, summer migrants arrive at the breeding colony in March–April and leave in mid-August, but some birds remain in the North Sea area over winter. Nests are in burrows, dug by the puffin or using abandoned rabbit burrows, containing a single egg, incubation lasting 36–43 days, and fledging of the altricial chick (puffling) 34–44 days. Diet is of fish caught by diving to a depth of up to 60 m. The serrated beak can hold up to 80 (small) fish. Puffin is a Middle English word (*pophyn* or *poffin*) originally used for the salted meat of Manx shearwater *Puffinus puffinus*. Medieval Latin *fratercula* = 'friar', the black and white plumage echoing monastic robes. *Arctica* refers to the bird's northerly distribution.

PUG (MOTH) Nocturnal members of the family Geometridae (Larentiinae), many in the genus *Eupithecia* (Greek *eu* + *pithecos* = 'good' + 'dwarf', from the attractive appearance and small size).

Scarce species include **bilberry pug** *Pasiphila debiliata* (heathland, moorland and open woodland mainly in southern England), **bleached pug** *E. expallidata* (coppiced woodland, woodland rides and clearings in England and Wales, with a few coastal records in Ireland, larvae feeding on goldenrod), **Channel Islands pug** *E. ultimaria* (first recorded in Guernsey in 1984, now established on other Channel Islands and along parts of the southern English coast, for example the first Hampshire record in 1995, Dorset in 2003, larvae feeding on tamarisk), **dentated pug** *Anticollix sparsata* (damp woodland, fen and marsh in isolated colonies from Somerset to Yorkshire, and in south Wales, larvae feeding on yellow loosestrife), **lead-coloured pug** *E. plumbeolata* (open woodland and dunes in parts of England and Wales, with scattered Scottish and, more frequently, Irish records, larvae feeding on cow-wheat), **pauper pug** *E. egenaria* (until recently restricted to mature lime woodland in the Wye Valley and parts of East Anglia, but in parts of east and south-east England the population is recovering or possibly being reinforced by continental migrants), **scarce pug** *E. extensaria* (saltmarsh in east England from Essex to Yorkshire, but mainly around the Wash, larvae feeding on sea wormwood), and **valerian pug** *E. valerianata* (fen, marsh and damp woodland).

Angle-barred pug *E. innotata*, wingspan 18–24 mm, placed in a complex taxonomic group consisting of a number of forms previously thought to be separate species but now considered to be different forms of the same species using different food plants. The nominate form feeds on sea-buckthorn and occurs coastally from Yorkshire to Essex. **Ash pug**, originally given specific status *E. fraxinata*, is now known as *E. innotata* f. *fraxinata*, and feeds on ash. It is widely distributed throughout much of Britain. **Tamarisk pug** is the form *tamariscata*, and is found on tamarisk in a number of southern counties. Adults fly in two generations, in May–June and August. *Innotata* is Latin for 'unmarked'.

Bordered pug *E. succenturiata*, wingspan 20–3 mm, widespread in open areas, waste ground, verges and gardens, most frequent in East Anglia and the west Midlands, local in Scotland and Ireland. Adults fly in July–August. Larvae feed on mugwort. Latin *succenturiata* literally refers to a replacement centurion, here meaning 'substitute'.

Brindled pug *E. abbreviata*, wingspan 19–22 mm, widespread and common in deciduous woodland, especially of oak, but also in hedgerows and gardens. Adults fly in April–May. Larvae feed on leaves of oak and hawthorn. *Abbreviata* is Latin for 'shortened', the hindwing being shorter than the forewing.

Campanula pug *E. denotata*, wingspan 19–22 mm, on woodland margins, hedgerows, scrubby grassland and gardens in southern England; also on dry grassland, heaths and clifftops along the coasts of south-west England, west Wales and west Ireland (ssp. *jasioneata*, also known as **jasione pug**). Adults fly in July. Larvae feed in the seed capsules of bellflower; jasione pug feeds on sheep's-bit. Denotata is Latin for 'distinct' (i.e. from other species).

Cloaked pug *E. abietaria*, wingspan 21–3 mm, uncommon, in coniferous woodland and plantations in northern England and mid-Wales, with a scattering of records elsewhere in England, Scotland and Ireland. Adults fly in June–July. Larvae feed in cones of Norway spruce (*Picea abies*, giving the specific name) and other conifers.

Common pug *E. vulgata*, wingspan 15–18 mm, widespread and common in gardens, hedgerows, scrub, fen, heathland and woodland. Subspecies *clarensis* is restricted to the limestone habitat of the Burren, Co. Clare, but the species is itself found throughout Ireland. Adults fly in May–August in two generations. Larvae feed on a variety of trees and herbaceous plants. Latin *vulgus* = 'common'.

Currant pug *E. assimilata*, wingspan 17–22 mm, widespread and locally common in gardens, allotments, hedgerows and open woodland. Adults fly in May–June, with a second brood in the south of its range in August. Larvae feed on hop and currants. *Assimilis* is Latin for 'similar'.

Cypress pug *E. phoeniceata*, wingspan 18–22 mm, first recorded in Cornwall in 1959, since spreading along the south coast (e.g. Isle of Wight in 1965, Hampshire mainland from 1973), Somerset and northwards to the south Midlands, East Anglia (Norfolk 1988) and Lincolnshire. There was a single record from Co. Cork in 1993. Adults fly in August–September. Larvae feed on cypress species, and on the Continent *Juniperus phoenicea*, prompting the specific name.

Double-striped pug *Gymnoscelis rufifasciata*, wingspan 15–19 mm, widespread and common in parks, gardens, hedgerows, verges, heathland, moorland and woodland. It generally has two broods, flying in April–May and August, though it can be found on the wing as early as January in mild winters. Larvae feed on a range of plants. Greek *gymnos* + *scelos* = 'naked' + 'leg', and Latin *rufifasciata* = 'red-banded'.

Dwarf pug *E. tantillaria*, wingspan 16–19 mm, widespread and common in coniferous plantations, gardens and parks. Adults fly in June–July. Larvae feed on conifer needles. *Tantillus* is Latin for 'so small'.

Edinburgh pug Local name for the *millieraria* subspecies of Freyer's pug in Scotland.

Fletcher's pug Alternative name for pauper pug.

Foxglove pug *E. pulchellata*, wingspan 18–22 mm, widespread and common in woodland, calcareous grassland, embankments and waste ground. Adults fly in May–July. Larvae feed on the flowers and developing seeds of foxglove. *Pulchellus* is the Latin diminutive of *pulcher* = 'pretty'.

Freyer's pug *E. intricata*, wingspan 20–4 mm, common in gardens, parks and plantations throughout much of England. In Scotland subspecies *millieraria* (**Edinburgh pug**) occurs on moorland. In Ireland subspecies *hibernica* (**Mere's pug**) occurs on the limestone of the Burren and a site in Northern Ireland. Adults fly in May–June. Larvae feed on juniper and cypresses. *Intricata* is Latin for 'intricate', from the large number of forewing markings.

Goldenrod pug *E. virgaureata*, wingspan 17–23 mm, widespread but local on verges, rough ground, woodland edges and clearings, commoner in the north and west. There are two generations in the south, adults flying in May–June and July–August. Larvae of the second brood feed on flowers of goldenrod (*Solidago virgaurea*, giving the specific name) and ragwort.

Green pug *Pasiphila rectangulata*, wingspan 15–20 mm, widespread and common in parks, gardens, hedgerows, scrub and woodland. Adults fly in June–July. Larvae feed on the flowers of apple, pear, hawthorn, blackthorn and cherry.

Grey pug *E. subfuscata*, wingspan 17–21 mm, widespread and common in gardens, heathland and woodland, downland and marsh. Adults fly in May–June, sometimes again in August. Larvae feed on the leaves and flowers of a number of herbaceous and woody plants. *Subfuscus* is Latin for 'somewhat dark'.

Haworth's pug *E. haworthiata*, wingspan 12–14 mm, local in open woodland, scrub, hedgerows and gardens on alkaline soils throughout England, Wales and Ireland. Adults fly in June–July. Larvae feed on the buds and flowers of traveller's-joy. Adrian Haworth (1767–1833) was an English entomologist.

Jasione pug Refers to campanula pug ssp. *jasioneata*.

Juniper pug *E. pusillata*, wingspan 17–21 mm, widespread and common on moorland and calcareous grassland on native forms of the larval food plant, juniper, and gardens and parks on cultivated forms. Adults fly in July–September. *Pusilla* is Latin for 'very small'.

Larch pug *E. lariciata*, wingspan 19–22 mm, widespread and common in coniferous plantations, gardens and parks. Adults fly in May–June. Larvae feed on needles of European larch (*Larix*, giving the specific name).

Lime-speck pug *E. centaureata*, wingspan 16–20 mm, widespread and common in gardens, hedgerows, scrub and saltmarsh, and on verges, particularly south-west England, the Welsh Borders, East Anglia and coastal parts of Ireland. Adults, which resemble bird droppings, fly in April–September, sometimes as two broods. Larvae feed on a range of herbaceous plants, including knapweed (*Centaurea*, giving the specific name).

Ling pug A heathland race of wormwood pug.

Maple pug *E. inturbata*, wingspan 13–18 mm, local in woodland, hedgerows and scrub, particularly on calcareous ground throughout much of (especially south-east) England. Adults fly in July–August. Larvae feed on field maple flowers. *Inturbata* is Latin for 'not confused'.

Marbled pug *E. irriguata*, wingspan 18–20 mm, in oak woodland and mature hedgerows in central-southern England, best known from Hampshire where it is a notable New Forest speciality and often common there during May. More generally, adults fly in April–June. Larvae feed on oak leaves. *Irriguata* is Latin for 'well-watered', from 'rivulet' forewing markings.

Marsh pug *E. pygmaeata*, wingspan 14–18 mm, local in fen, marsh, damp meadow, dunes and waste ground throughout much of central and northern England, Scotland and Ireland. Adults fly in May–June. Larvae feed on the flowers and developing seed of field mouse-ear. *Pygmaeus* is Latin for 'pygmy', this being one of the smallest pugs.

Mere's pug Name for the *hibernica* subspecies of Freyer's pug in Ireland.

Mottled pug *E. exiguata*, wingspan 20–2 mm, common in woodland, scrub, hedgerows and gardens throughout England, Wales and Ireland, occasional in Scotland. Adults fly in May–June. Larvae feed on hawthorn, blackthorn, dogwood and other trees. *Exiguus* is Latin for 'very small' (for a moth generally rather than a pug moth).

Narrow-winged pug *E. nanata*, wingspan 13–17 mm, widespread and common on heathland, moorland and downland and in gardens. Adults fly in April–June, sometimes with a second generation in August. Larvae feed on heather flowers in autumn. *Nana* is Latin for 'dwarf'.

Netted pug *E. venosata*, wingspan 17–22 mm, local in hedgerows, woodland rides, sea-cliffs and gardens, mainly in chalk and limestone districts throughout much of England, Wales and Ireland, and also the Hebrides. Adults fly in May–June. Larvae feed in the seed capsules of bladder and sea campion near the coast. *Venosa* is Latin for 'veined'.

Oak-tree pug *E. dodoneata*, wingspan 19–22 mm, common in oak woodland, mature hedgerows, scrub and suburban habitats throughout much of England and Ireland, less common in Wales and Scotland. Adults fly in May–June. Larvae feed on leaves of pedunculate oak and hawthorn. Dodona is a town in north-west Greece where oracular pronouncements were made by wind blowing through oak trees, *dodoneus* coming to mean 'belonging to oaks'.

Ochreous pug *E. indigata*, wingspan 15–18 mm, widespread and common in coniferous woodland, plantations and gardens. Adults fly in April–May. Larvae feed on young shoots or buds of Scots pine and European larch. *Indiges* is Latin for 'in want of', from the forewings having very few markings.

Pimpinel pug *E. pimpinellata*, wingspan 20–4 mm, local on calcareous grassland, downland, quarries, verges and woodland rides throughout much of England and Ireland, with some Welsh records. Adults fly in July–August. Larvae feed in September on the seed capsules of burnet-saxifrage (*Pimpinella*, giving the English and specific names).

Pinion-spotted pug *E. insigniata*, wingspan 18–22 mm, found along hedgerows, woodland edges and sunken lanes in central and southern England, with records north to Yorkshire. Adults fly in April–May. Larvae feed on hawthorn and apple. *Insignis* is Latin for 'marked'.

Plain pug *E. simpliciata*, wingspan 21–3 mm, local on sea-cliffs, saltmarsh, waste ground and other open areas throughout much of England, Wales and Ireland, mostly coastal. Adults fly in June–August. Larvae feed on the seeds of orache and goosefoot. *Simplex* is Latin for 'simple', from having few wing markings.

Satyr pug *E. satyrata*, wingspan 18–24 mm, common on chalk downland, heathland, fen, open woodland and rough ground throughout much of southern England; also on moorland in northern England, Scotland and Ireland (ssp. *callunaria*) and Shetland (ssp. *curzoni*). Adults fly in May–June. Larvae feed on a range of plants, depending on habitat. In Greek mythology satyrs were woodland deities associated with Bacchus.

Shaded pug *E. subumbrata*, wingspan 16–20 mm, local on chalk downland, verges, woodland rides, saltmarsh and sea-cliffs throughout much of south-east England, with isolated colonies north to Yorkshire and north Wales, and widespread but local in Ireland. Adults fly in June–July. Larvae feed on the flowers and seeds of a number of herbaceous plants, including scabious and hawk's-beards. *Subumbrata* is Latin for 'somewhat shaded'.

Slender pug *E. tenuiata*, wingspan 14–16 mm, widespread and common in damp woodland, scrub, fen, dunes, marsh, gravel-pits and gardens. Adults fly in June–July. Larvae feed in early spring on willow catkins. *Tenuis* is Latin for 'slender'.

Sloe pug *Pasiphila chloerata*, wingspan 17–19 mm, common in mature hedgerows, scrub and woodland edges throughout much of (especially south and east) England. Adults fly in

May–June. Larvae feed in early spring on blackthorn flowers. *Chloe* is Greek for the young green of spring vegetation, from the forewing colour.

Tawny speckled pug *E. icterata*, wingspan 20–3 mm, widespread and common in gardens, hedgerows, verges, grassland, fen and open woodland. Adults fly in July–August. Larvae feed on the flowers and leaves of yarrow and sneezewort. *Icteros* is Greek for a yellow bird which cured jaundice, so coming to mean jaundice itself, here reflecting the yellowish (tawny) wing colour.

Thyme pug *E. distinctaria*, wingspan 16–18 mm, found on south-facing chalk and limestone hillsides, sea-cliffs, quarries and dunes along the coasts of England and Wales from Dorset to Cornwall northwards to the Inner Hebrides, including the coasts of Ireland and Scotland; also inland in a few sites in southern and northern England. Adults fly in June–July. Larvae feed on flowers of wild thyme. *Distinctus* is Latin for 'distinct'.

Toadflax pug *E. linariata*, wingspan 11–16 mm, common in calcareous grassland, rough ground and gardens throughout England and Wales, rare in Scotland. Adults fly in June–August. Larvae feed in the flowers and seed pods of common toadflax (*Linaria*, giving the specific name).

Triple-spotted pug *E. trisignaria*, wingspan 15–20 mm, local in fen, marsh, damp woodland and verges in England, Wales, north-east Scotland and (particularly Northern) Ireland. Adults fly in June–July. Larvae feed on the flowers and developing seeds of wild angelica, occasionally hogweed. Latin *tri* + *signum* = 'three' + 'mark' (on the forewing).

V-pug *Chloroclystis v-ata*, wingspan 14–19 mm, widespread and common in parks, gardens, hedgerows, scrub and woodland, scarcer in north-east England and Scotland. Population increased by 103% between 1968 and 2007. There are two generations in the south, flying in May–June and August, but further north there is just one in June–July. Larvae feed on the flowers of a variety of plants. Greek *chloros* + *clyst* = 'green' + 'wash away'; *v-ata* comes from the black V-shaped mark on the forewing.

White-spotted pug *E. tripunctaria*, wingspan 17–21 mm, local in woodland, hedgerows, marsh and fen, and on verges, throughout much of England and Ireland, with records also from Wales and Scotland. The bivoltine adults fly in May–June and July–August. Larvae feed on flowers and seeds of umbellifers, the spring generation also on elder. Latin *tripunctaria* = 'three-spotted', from the forewing pattern.

Wormwood pug *E. absinthiata*, wingspan 21–3 mm, widespread and common in gardens, saltmarsh, woodland and grassland. Adults fly in June–July. Larvae feed on a number of low-growing plants, e.g. wormwood (*Artemisia absinthium*, giving the specific name). **Ling pug** (f. *goossensiata*), once considered to be a separate species, is now considered a heathland race which feeds on heather and heaths.

Yarrow pug *E. millefoliata*, wingspan 21–3 mm, on waste ground, dunes, shingle beaches, verges and rough grassland in southern and south-east England and East Anglia. Adults fly in June–July. Larvae feed in the flowers and developing seeds of yarrow (*Achillea millefolium*), giving the specific name.

PULMONATA An 'informal' group of slugs and snails (no longer considered to be an order or subclass) that have the ability to breathe air by virtue of a simple lung (Latin *pulmo*) rather than through gills. See also Opisthobranch and Prosobranch.

PUMPKINSEED *Lepomis gibbosus* (Centrarchidae), typical length 10 cm, a North American fish established from a number of twentieth-century introductions in a few south England waters feeding on invertebrates, breeding in late spring. Greek *lepos* = 'scale', and Latin *gibbosus* = 'humped'.

PURPLE (MOTH) Small leaf-miners in the family Eriocraniidae, mostly species of *Eriocrania* (Greek *erion* + *cranos* = 'wool' + 'helmet'), wingspans generally 9–14 mm.

Common oak purple *Dyseriocrania subpurpurella*, common throughout much of England, Wales, Ireland, and south-west and north-east Scotland. Adults fly by day in April–May. Larvae mine oak leaves. Latin *subpurpurella* = 'somewhat purple'.

Early purple *E. semipurpurella*, the commonest of its genus that feed on birch, widespread but scattered in England, Wales and north-east Scotland. Adults fly in March–April, larvae feed in March–May. Latin *semipurpurella* = 'half purple'.

Large birch purple *E. sangii*, locally common but scattered in England, Wales and north-east Scotland. Adults fly in March–April. Leaf mines in the larval food plant birch can be found in March–May. John Sang (1828–87) was an English entomologist.

Mottled purple *E. sparrmannella*, local and scattered in England, Wales, and south-west and north-east Scotland, and recorded in Northern Ireland. Adults fly in April–May. Larvae feed in June–August. Both stages are found on birch. Anders Sparrmann (1748–1820) was a Swedish entomologist.

Small birch purple *E. salopiella*, widespread but local in Britain. Larvae feed on birch in May–June. The type locality was in Shropshire (Salop), giving the specific name.

Small hazel purple *Paracrania chrysolepidella*, scarce in a few places in England (especially Cornwall) and Ireland. Adults fly in April. Larvae feed on hazel and hornbeam in April–May. Greek *para* + *cranos* = 'near' + 'helmet', and *chrysos* + *lepis* = 'gold' + 'scale', from the wing colour.

Washed purple *E. cicatricella*, local and widespread in England and Wales, April-flying, larvae mining birch during May. *Cicatrix* is Latin for 'scar', from the wing marking.

White-spot purple *E. unimaculella*, widespread and relatively common over much of Britain, adults flying in sunshine in March–April. Larvae feed on birch, mines particularly evident in April–May. Latin *unimaculella* = 'one-spotted'.

PURPLE & GOLD (MOTH)

Members of the genus *Pyrausta* (Crambidae, Pyraustinae), *pyraustes* being Greek for a moth that gets singed in candle flame. Species, wingspans 15–22 mm, are found locally on chalk downland and dry grassy areas throughout much of Britain, favouring very short turf, the bivoltine adults flying in sunshine and at night in May–June and July–August.

Common purple & gold *P. purpuralis*, whose larvae feed on wild thyme, occasionally corn mint.

Scarce purple & gold *P. ostrinalis*, whose larvae feed on wild thyme. *Ostreon* is Greek for 'purple'.

Small purple & gold *P. aurata*, found throughout much of England, Wales and south Scotland. Larvae feed on many species in the Lamiaceae. *Aurata* is Latin for 'golden'.

PURPLE-BARRED

Small purple-barred *Phytometra viridaria* (Erebidae, Boletobiinae), wingspan 19–23 mm, a mostly day- but also night-flying moth (May–July) found on acid heathland, chalky downland and open woodland, most common in southern England, and local elsewhere to north Scotland and Ireland. Larvae are loopers in the early instars, semi-loopers in later ones, and feed on milkworts and lousewort. Greek *phyton* + *metron* = 'plant' + 'measure', and Latin *viridis* = 'green'.

PURPLE EMPEROR

Apatura iris (Nymphalidae, Apaturinae), wingspan 70–92 mm, a butterfly of great beauty, sunlight creating a refracted purple sheen in the male (the female is a deep brown), confined to deciduous woodland in central-southern England, flying in June–August. It spends most of its time in the tree canopy, feeding on aphid honeydew, but sometimes descends to feed on sap or (males only) animal droppings, carrion or moist soil that provide salts and minerals. Larvae feed on goat willow and other willow species. Following declines during the twentieth century, its status now appears to be more stable, but liable to fluctuations: between 1976 and 2014, distribution declined by 47% yet numbers increased by 69%, but in the shorter term (2005–14) these figures were, respectively, an increase of 135% and a decrease of 34%, and it remains a species of conservation concern. Greek *apatao* + *oura* = 'to deceive' + 'tail', from the 'deceptive' colour of the male upper wing and the slightly elongated hindwing. Iris

was the personification of the rainbow in Greek mythology, here reflecting the iridescent upper side of the male's wing.

PURPLE-LOOSESTRIFE

Lythrum salicaria (Lythraceae), a herbaceous perennial, common and widespread except in north Scotland, growing on the margins of slow-flowing rivers, canals, lakes and flooded gravel-pits, and in tall-herb fen and willow carr. It grows best in permanently wet, or periodically flooded, non-acidic fertile soils. Greek *lythron* = 'gore', for the red flower colour, and Latin *salicaria* = 'willow-like' (*salix* = 'willow').

PURSLANE

Common purslane *Portulaca oleracea* (Portulacaceae), a succulent annual Mediterranean neophyte grown in gardens since at least 1200, sometimes found as a persistent arable weed, especially in the Channel Islands and Scilly. Elsewhere in Britain it is a rare casual on waste ground, introduced in birdseed and wool shoddy. 'Purslane' dates from late Middle English, from Old French *porcelaine*, probably from Latin *porcil(l)aca*, variant of *portulaca*, the Latin for 'purslane'. See also Iceland-purslane (Polygonaceae), Hampshire-purslane (Onagraceae) and sea-purslane (Amaranthaceae). *Portulaca* is Latin for 'purslane', from 'milk-carrying', *oleracea* for resembling an edible vegetable.

Pink purslane *Claytonia sibirica* (Montiaceae), an annual to perennial North American neophyte present by 1768, scattered (commoner in the north and west), naturalised in damp, bare sites, often colonising and rapidly spreading in open woodlands or hedgerows, or by shaded streams where it can be washed downstream to new sites. John Clayton (1694–1773) was a Virginian plant collector.

PUSS MOTH

Cerura vinula (Notodontidae, Thaumetopoeainae), a nocturnal moth named because the adult has a thick 'fur' and the thick pointed antennae are said to resemble cat ears, widespread and common in gardens, hedgerows, open woodland, moorland and scrub. Adults fly in May–July. Larvae feed on leaves of aspen, poplars, and goat willow. The caterpillar can camouflage itself effectively among the leaves, but if disturbed it will rear up, raising its head with false eyespots resembling a face and wave two long extendable whip-like appendages from the tail. If further deterrent is needed, it can also squirt formic acid from its thorax. Greek *ceras* + *oura* = 'horn' + 'tail', from the larval morphology, and Latin *vinula* the diminutive of *vinum* = 'wine', from the colour of the larva's dorsal 'saddle'.

PYCNOGONIDA

Class of marine arthropods (Greek *pycnos* + *gonia* = 'compact' or 'solid' + 'joint' or 'angle'), usually with small bodies and eight long legs, known as sea spiders, some species found in the littoral. See gangly lancer.

PYGMY (MOTH)

Members of the family Nepticulidae, mostly *Ectoedemia* (Greek *ectos* + *oidema* = 'outside' + 'swelling', from the larval mining behaviour) and *Stigmella* (diminutive of *stigma* = 'brand' or 'small dot', probably from the metallic fascia on the forewing of many species). Wingspans are generally 4–7 mm.

Scattered, often widespread but only locally common species include **barred sycamore pygmy** *S. speciosa* (England as far north as Yorkshire, and parts of Wales, larvae mining sycamore leaves), **bent-barred pygmy** *E. angulifasciella* (parts of England and Wales, and three records from Ireland, larvae mining roses), **bilberry pygmy** *S. myrtillella*, **black-headed pygmy** *S. atricapitella* (mainly southern Britain, larvae mining oak), **black-spot sallow pygmy** *E. intimella* (southern half of England and Wales, and two sites in Ireland), **broken-barred pygmy** *E. minimella* (larvae mining birch and hazel), **broom pygmy** *Trifurcula immundella* (heathland and some grassland), **chestnut pygmy** *S. samiatella* (eastern England, Hampshire and the Welsh Borders, larvae mining oak and sweet chestnut), **common rowan pygmy** *S. nylandriella*, **common birch pygmy** *S. betulicola* (heathland with birch trees in a few places in England, Wales, Ireland and north-east Scotland), **dewberry pygmy** *E. rubivora* (parts of southern and eastern England, and on the Burren, larvae mining bramble and dewberry), **double-barred pygmy** *S. continuella* (southern

Britain and Ireland. Larvae mining birch), **dusty apple pygmy** *Bohemannia pulverosella* (mainly in the southern half of Britain), **five-spot pygmy** *E. quinquella* (East Anglia and a few other sites in the southern half of England, larvae mining oak), **four-spot pygmy** *B. quadrimaculella* (alder woods in the Midlands and Welsh Borders, with two records in Ireland), **least thorn pygmy** *S. perpygmaeella* (south Britain, larvae mining hawthorn), **new holm-oak pygmy** *E. heringella* (first noted in Britain on holm oak in the grounds of the Natural History Museum in London in 1996, subsequently spreading and now found across a large part of south-east England, especially Essex and Suffolk, and as far north as Cambridgeshire), **Norway-maple pygmy** *E. sericopeza* (mostly south-east Midlands), **oak-bark pygmy** *E. atrifrontella* (southern England and south Wales), **orange-headed pygmy** *S. svenssoni* (occasional in Britain, and with two records from Co. Cork, larvae mining oak), **pale birch pygmy** *S. confusella* (parts of England, Wales, Ireland and north-east Scotland), **short-barred pygmy** *S. luteella* (parts of England, Wales and north-east Scotland, occasional in Ireland, larvae mining birch), **sycamore-seed pygmy** *E. decentella* (woodland, parks and gardens in the southern half of England), **virgin pygmy** *E. argyropeza* (commoner in the south, larvae mining aspen), and **willow pygmy** *S. obliquella* (parts of southern Britain, rare in Ireland).

Barred rowan pygmy *S. sorbi*, commonest in Wales, northern England and parts of Scotland and Ireland. Adults fly in May. Larvae mine leaves of rowan *Sorbus aucuparia*, giving the specific name, in June–July.

Beech pygmy *S. hemargyrella*, fairly common and widespread, less so in Scotland. The bivoltine adults fly in April–May and July–August. Larvae feed on beech leaves in July–November. Greek *hemi* + *argyros* = 'half' + 'silver', from the colour of the forewing banding.

Black-poplar pygmy *S. trimaculella*, with a scattered distribution in (especially eastern) England, occasionally found in Wales, Scotland and Ireland. The bivoltine adults fly in May and August. Larvae mine poplar leaves in June–July and September–October. Latin *tri* + *maculus* = 'three' + 'spot', from the forewing marks.

Brassy pygmy *S. aeneofasciella*, local to parts of southern England, and from Lancashire to Co. Durham. With two generations, adults in May lead to July larvae, and adults in August to larvae in September–November. Larvae mine leaves of agrimony and cinquefoils. Latin *aeneus* + *fascia* = 'bronze' + 'band', from the forewing marking.

Coarse hazel pygmy *S. floslactella*, widespread and common. The bivoltine adults fly in May and August. Larval mines on hazel, occasionally hornbeam, can be found in June–July, but more often in September–October. Latin *flos* + *lac* = 'river' + 'milk', from the creamy forewing.

Coast bramble pygmy *E. erythrogenella*, mostly found, sometimes abundantly, on the Suffolk and Hampshire–Sussex coasts. Adults fly in July. Larval mines on bramble are active in October–December. Greek *erythros* + *gen-* = 'red' + 'begetting', from the red discoloration in bramble leaves caused by the mine.

Common fruit-tree pygmy *S. oxyacanthella*, common throughout much of the southern British Isles, including Ireland. Adults fly in June. Larvae mine leaves of hawthorn, apple and occasionally other tree species in September–October. *Oxyacanthella* comes from the former name of Midland hawthorn *Crataegus oxyacantha*, a food plant.

Common oak pygmy *S. roborella*, with a scattered, widespread but probably under-recorded distribution in southern England, Wales and north-east Scotland. The bivoltine adults fly in May–June and August–September. Larvae mine leaves of oak (including *Quercus robur*, giving the specific name) in June–July and October–November.

Common thorn pygmy *S. crataegella*, common and widespread throughout England, Wales and Ireland. Adults fly in May–June, and larvae mine leaves of hawthorn (*Crataegus*, giving the specific name) in June–August.

Cowberry pygmy *E. weaveri*, a submontane species of high moorland and mountains in Wales, part of the Peak District and the Cairngorms, where it can be fairly common if cowberry is present. Adults emerge at any time between April and August. Larvae tend to hatch in August and mine cowberry leaves. R. Weaver was a nineteenth-century English collector.

Drab birch pygmy *S. lapponica*, fairly common but with a scattered distribution. Adults fly in May. Larvae are evident in birch leaves in June–July. *Lapponica* refers to Lapland, the type locality being in north Norway.

Golden pygmy *S. aurella*, the commonest and most widespread in the British Isles, though less evident in Scotland. Adults fly in May and again in late summer. Larvae mine bramble leaves. *Aureus* is Latin for 'golden', from the metallic forewing band.

Greenish thorn pygmy *S. hybnerella*, widespread and common, especially in the southern British Isles, including Ireland. The bivoltine adults fly in April–May and July–August. Larvae mine hawthorn leaves in May–June and in July–September. Jacob Hübner (1761–1826) was a German entomologist.

Grey-poplar pygmy *E. turbidella*, often common where it occurs in Essex, Cambridgeshire and Middlesex. Adults fly in May–June. Larvae mine leaves of grey poplar. *Turbidus* is Latin for 'confused', from the poorly defined forewing pattern.

Hypericum pygmy *E. septembrella*, fairly common in England, Wales and Ireland. The bivoltine adults fly in May–June and August. Larvae mine St John's-wort in July and October–December. *Septembrella* refers to September, when the first specimens were reared – not the usual month of natural emergence.

Large birch pygmy *E. occultella*, common over much of England and Wales, parts of Scotland, and widespread but local in Ireland. Adults fly in May–July. Larval mines in birch leaves are evident in August–October. *Occultus* is Latin for 'concealed', from the larval behaviour.

Maple-seed pygmy *E. louisella*, found in East Anglia, south-east Midlands and southern Welsh Marches. Adults fly in April–May. There may be further generations in late summer and autumn. Larvae mine the samara (seeds) of field maple.

Narrow-barred pygmy *S. centifoliella*, local in England, generally in the south and east. The bivoltine adults fly in April–May and July–August. Larvae mine leaves of roses (including *Rosa centifolia*, giving the specific name) in June–July and from September sometimes to December.

Nut-tree pygmy *S. microtheriella*, wingspan 3–4 mm, widespread though patchily distributed but often common. Adults fly in May and August. Larvae generally mine leaves of hazel, occasionally hornbeam, in July and in October–November. Greek *micros* + *therion* = 'small' + 'animal'.

Pinch-barred pygmy *E. atricollis*, widespread in hedgerows and orchards through much of England and Wales, and with two records from Ireland. Adults emerge in June. Larvae mine leaves of hawthorn, apple, pear and wild cherry. Latin *ater* + *collum* = 'black' + 'neck'.

Red elm pygmy *S. lemniscella*, widespread, becoming scarcer further north and in Ireland. The bivoltine adults fly in May and August, and larval mines on elm are evident in June–July and September–October. *Lemniscus* is Latin for 'ribbon'.

Red-headed pygmy *S. ruficapitella*, common and widespread in England and Wales, less so in Scotland and Ireland. The bivoltine adults fly in May–June and July–August. Larvae feed on oak leaves in June–July and September–November. Latin *rufus* + *caput* = 'red' + 'head'.

Sallow pygmy *S. salicis*, widespread and fairly common over much of the (especially southern) British Isles. The bivoltine adults fly in April–May and July–August. Larval mines on especially rough-leaved willows and sallow are evident in June–July and September–November. *Salicis* is from Latin *salix* = 'willow'.

Scrubland pygmy *S. plagicolella*, fairly common throughout England, Wales and Ireland. The bivoltine adults fly in May–June and August. Larvae mine leaves of blackthorn and congeners, evident in July and September–October. Latin *plaga* + *colo* = 'flat open ground' + 'inhabit'.

Small beech pygmy *S. tityrella*, widespread and fairly common over much of the (especially southern) British Isles. Adults fly in April–May and July–August. Larvae mine beech leaves. In Greek mythology Tityrus was a shepherd who sang under a beech tree.

Sorrel pygmy *Enteucha acetosae*, one of the world's smallest moths, some specimens having a wingspan of 3–4 mm. Scarce in eastern England and eastern Ireland, there are two or three generations, with adults flying in April and July. Larvae feed in May–October by mining leaves of sheep's and common sorrel (*Rumex acetosa*, giving the specific name). Greek *en* + *teucho* = 'in' + 'to make' (i.e. well-made).

Spotted black pygmy *Ectoedemia subbimaculella*, found in much of England (especially East Anglia) and Wales, and parts of Ireland. Larvae mine oak in October–November, and may also overwinter in fallen leaves. Latin *sub* + *bi* + *macula* = 'somewhat' + 'two' + 'spot', from the forewing pattern.

White-banded pygmy *E. albifasciella*, found, sometimes commonly, in the southern half of Britain, north-east Scotland, and Ireland. Adults fly in June. Larval mines in oak appear in August–September. Latin *albus* + *fascia* = 'white' + 'band', from the forewing pattern.

White-spot pygmy *E. heringi*, found in the southern half of England and Wales, and with two records from Ireland. Adults fly in July. Larval mines can be found in oak leaves from October to December. Erich Hering (1893–1967) was a German entomologist.

PYGMY-MOSS Rounded pygmy-moss *Acaulon muticum* (Pottiaceae), scattered in Britain, rare in Ireland, on bare, base-poor, well-drained soil in arable fields and gravel-pits, by tracks, and on banks and ant hills. Greek *acaulon* = 'stemless' and Latin *muticum* = 'curtailed'.

PYGMYWEED Members of the family Crassulaceae, specifically *Crassula aquatica*, an annual currently only known from shallow water or on wet mud on the side of the River Shiel (west Inverness-shire). Latin *crassus* = 'thick', here meaning 'fleshy'.

New Zealand pygmyweed *C. helmsii*, a perennial neophyte herb introduced from New Zealand in 1927 as a pond ornamental, first discovered in the wild in 1956 at Greensted, Essex. Since the late 1970s it has spread rapidly throughout Britain. The first Irish record was in 1994 at Gosford Castle, Co. Armagh. It is found in sheltered waters up to 3 m deep or as an emergent on damp ground, growing on soft substrates in a variety of habitats, including ponds, lakes, reservoirs, canals and ditches, and can tolerate a wide range of water chemistry. It can form dense, virtually pure stands, a nuisance species out-competing other plants. Richard Helms (1842–1914) was an Australian naturalist.

PYRALIDAE Family name for knot-horn and tabby moths, *pyralis* being an unknown insect (or bird) which according to Pliny lived in fire (Greek *pyr*).

PYROCHROIDAE Family name (Greek *pyr* + *chroos* = 'fire' + 'appearance') of cardinal beetles.

PYTHIDAE Beetle family (Greek *pytho* = 'to cause decay'), with a single British representative, *Pytho depressus*, 14–15 mm long, found in the Highlands, larvae spending at least two years under pine bark. *Depressus* is Latin for 'low'.

Q

QUAHOG Edible, filter-feeding burrowing clams, the word dating from the mid-eighteenth century, from the Narragansett *poquauhock* = 'horse fish'.

Northern quahog *Mercenaria mercenaria* (Veneridae), shell size up to 13 cm, a shallow burrower in coarse but muddy intertidal sediments, first successfully introduced from eastern North America to Southampton Water in 1925, now found in scattered localities in south-east England, the English Channel, south-west and north-west Wales, and west Scotland. UK commercial production in 2012 was 8.6 tonnes. *Mercenarius* is Latin for 'a mercenary' or 'wages', given because of the Native American use of the purple inner shell ('wampum') as money and jewellery.

Ocean quahog *Arctica islandica* (Arcticidae), shell width up to 13 cm, found around the British and north-east Irish coasts from just below low tide to depths of 500 m, buried in sand, often with the shell hidden and a small tube (siphon) extending to the seabed surface to allow this bivalve to breathe, capture food and expel waste. It is on the OSPAR List of threatened species (Region II – Greater North Sea).

QUAIL *Coturnix coturnix* (Phasianidae), wingspan 34 cm, length 17 cm, weight 100 g, widespread as Britain and Ireland's only migrant game bird (from North Africa), arriving in late April–May, reaching the northern fringes of its breeding range here in summer. Traditional strongholds are in Wiltshire and Dorset, and in good years the Welsh Marches, East Anglia, low-lying parts of northern England and parts of south Scotland. It is an Amber List species because of a partial recovery from its historical decline as a UK breeding bird: in 2010 there were only around 540 males, and a population estimate of 840 was made in 2013 (RBBP). Numbers vary greatly between years, however, depending on the scale of irruptions, so for example >30% fewer quail were reported in 2010 than in 2009. But while in 2009 almost a third of birds were in Scotland, in 2010 Scotland held only 16% of the UK total (i.e. the distribution was more southerly). Nests are on the ground in arable or grass fields or field margins, clutch size is 8–13, incubation lasts 17–20 days, and the precocial chicks fledge in 18–20 days. Diet is of seeds and insects. 'Quail' is Middle English, from Old French *quaille*, in turn from medieval Latin *coacula*, probably imitative of its call. *Coturnix* is Latin for this bird.

QUAKER Nocturnal moths in the family Noctuidae (Hadeninae), the name probably reflecting the 'neat' appearance of both this group and quakers.

Common quaker *Orthosia cerasi*, wingspan 34–40 mm, widespread, often abundant, common in a range of lowland habitats, including woodland and gardens. Adults fly in March–April. Larvae feed on oak, sallow, hawthorn, hazel and birch. Greek *orthosis* = 'making straight', from the straight subterminal line on the forewing, and *Prunus cerasi* (dwarf cherry), another possible food plant.

Powdered quaker *O. gracilis*, wingspan 35–40 mm, widespread and common in marsh, damp woodland and gardens, but a species of conservation concern in the UK BAP. Adults fly in April–May, feeding on sallow blossom. Larvae feed on various woody and herbaceous plants, especially bog-myrtle in Scotland. *Gracilis* is Latin for 'slender'.

Red-line quaker *Agrochola lota*, wingspan 33–40 mm, widespread and common in woodland rides, scrub, hedgerows, fen, marsh, heathland and gardens. Adults fly in September–October. Larvae initially feed at night on the catkins of poplars and willows, later moving to the leaves. Greek *agros* + *chole* = 'field' + 'bile' (i.e. a yellow colour), and Latin *lotus* = 'washed', from merged wing colours.

Small quaker *O. cruda*, wingspan 25–30 mm, widespread and common in broadleaf woodland, areas with scattered trees and heathland. Adults fly in March–April, feeding on sallow blossom. Larvae feed on oak, sallow, field maple, hazel and birch. *Cruda* is Latin for 'premature', from the early appearance of the adults.

Twin-spotted quaker *Anorthoa munda*, wingspan 38–44 mm, widespread and common in broadleaf woodland and gardens, though more scattered in Scotland. Adults fly in March–April, and feed on sallow blossom. Larvae feed on a number of woody plants. Greek *anorthoa* = 'not straight' and Latin *munda* = 'neat'.

Yellow-line quaker *Agrochola macilenta*, wingspan 32–6 mm, widespread and common in broadleaf woodland, scrub, heathland, moorland, hedgerows and garden. Adults fly in September–November. In the north of their range larvae feed on heather, but further south they generally feed on leaves of oak, beech and other deciduous trees. Latin *macilenta* = 'meagre', from the weak forewing markings.

QUAKING-GRASS *Briza media* (Poaceae), a rhizomatous perennial, common and widespread except for north and north-west Scotland, usually in unimproved, well-grazed grassland on infertile, well-drained calcareous soils, but also in old meadow and pasture on neutral and sometimes acidic soils, in drier parts of fens. *Brizo* is Greek for 'to nod', from the plant's behaviour, *media* Latin for 'average'.

Greater quaking-grass *B. maxima*, an annual Mediterranean neophyte present by 1633, scattered mainly in England and Wales, naturalised or casual on dry open habitats such as field margins, cultivated ground, dunes, sea-cliffs, waste ground and wall tops, and in pavement cracks.

Lesser quaking-grass *B. minor*, an annual Mediterranean archaeophyte, scattered in England and Wales, locally common in the south and south-west, mainly on light, base-poor soils in arable habitats, including bulb fields. It is also a casual by roadsides, on walls and waste ground.

QUEEN ANNE'S LACE Alternative name for a number of roadside umbellifers (Apiaceae), in particular wild carrot and cow parsley.

QUEEN'S EXECUTIONER *Megapenthes lugens* (Elateridae), an uncommon click beetle, adults feeding on nectar from a range of flowers and growing to a length of 7–10 mm, with a few records from south and south-east England but recently only from Windsor Great Park. Larvae feed on weevil and other beetle larvae in decaying wood of beech and elm. In 2010, *The Guardian* launched a competition for the public to suggest common names for previously unnamed species. The rationale for this choice was 'the link to Windsor and the royals. The executioner is to represent that it kills (and eats) the larvae of others and links to its black colour (the hood of an executioner is traditionally black).' Greek *megas* + *penthos* = 'great' + 'sorrow', and Latin *lugens* = 'mourning'.

QUILLWORT Submerged freshwater plants with quill-like leaves in the family Isoetaceae. Specifically, *Isoetes lacustris*, locally frequent on stony substrates in clear mountain lakes to a depth of 6 m in the north and west British Isles and south Devon. Greek *isos* + *etas* = 'equal' + 'year', meaning evergreen, and Latin *lacus* = 'lake'.

Land quillwort *I. histrix*, submerged in winter on clifftop hollows, on Guernsey, Alderney and the Lizard Peninsula. *Hystrix* is Greek for 'porcupine'.

Spring quillwort *I. echinospora*, with a scattered and generally western distribution, in clear waters on silty substrates. Greek *echinos* + *spora* = 'hedgehog' (i.e. spiny) + 'seed'.

QUINCE *Cydonia oblonga* (Rosaceae), a long-cultivated deciduous neophyte Asian garden shrub or small tree occasionally planted in woodland and hedgerows, and on roadsides. 'Quince' comes from Middle English, from Old French *cooin*. Cydonia is the name of a city in ancient Crete on the site of present-day Chania.

Chinese quince *Chaenomeles speciosa*, a small deciduous neophyte shrub, in gardens by 1796, occasionally found as a bird-sown escape in woodland, hedgerows, scrub, rough grassland and waste ground. Greek *chaino* + *melon* = 'to gape' + 'apple', and the Latin *speciosa* = 'showy'.

Japanese quince *Ch. japonica*, a spiny deciduous Japanese neophyte shrub, in gardens by 1869, scattered in England being planted in hedgerows and woods.

QUINOA *Chenopodium quinoa* (Amaranthaceae), an annual South American neophyte, widely grown for game (sometimes human) food, and an occasional casual on waste ground in England and east Scotland. 'Quinoa' comes from the Spanish spelling of Quinua, Peru, where the grain originated. Greek *chen* + *podion* = 'goose' + 'little foot', describing the leaf shape.

R

RABBIT *Oryctolagus cuniculus* (Leporidae), head-body length 34–40 cm, tail length 4–8 cm, weight 1–2 kg, males (bucks) slightly heavier than females (does), a widespread Mediterranean animal probably brought to England by the Romans for food but not then found in the wild, for which the first references are in the late twelfth century when it was kept as food and for their fur in warrens or cunicularia. Early references to warrens in Scotland, Wales and Ireland were thirteenth-century. Escapes into the wild followed, but numbers did not increase greatly until the later eighteenth century, possibly as a tipping point was reached in population growth, probably responding to changes in game husbandry (killing of predators), agriculture (greater provision of cereals and winter crops as food) and the rural landscape (Parliamentary enclosures providing hedgerows as shelter). In 1953–4, however, myxomatosis reduced rabbit populations in Britain and Ireland by 99%. Subsequent population growth from genetically resistant rabbits or from those acquiring immunity means that numbers largely recovered, though outbreaks still occur. In 2005, a population of 40–45 million rabbits was likely, populations increasing threefold every two years. However, rabbit haemorrhagic disease virus, which arrived in England in 1992, and reached Scotland in 1995, has now greatly impacted numbers, a new strain reducing UK numbers in 2016 to perhaps <4 million. Rabbits are found where good burrowing conditions exist, especially in short grass, dunes and railway banks, feeding (generally at night) on a range of plants including grasses, cereal crops, root vegetables and young shoots of herbaceous plants. Social groups vary from a single pair to up to 30 rabbits. Breeding in chambers in the warren tunnel complex mainly takes place in January–August, producing a litter of 3–10 young (kittens), which are weaned by 18–25 days. Males are able to mate at 4 months, females at 3.5 months. Gestation is 30 days, and there are 3–7 litters p.a. Rabbit damage is estimated at over £100 million p.a., half of this through crop losses. 'Rabbit' is a late Middle English word, from Old French *rabotte* = 'young rabbit', in turn perhaps from Flemish *robbe*. Greek *oryco* + *lagos* = 'to dig' + 'hare', and Latin *cuniculus* = 'tunnel'.

RABBIT MOSS *Cheilothela chloropus* (Dichitraceae), local to Somerset and south Devon in short rabbit-grazed turf on cliffs by the sea and limestone hills, often on summer-dry rocky ground. Greek *cheilos* + *thele* = 'margin' + 'nipple', and *chloros* + *pous* = 'green' + 'foot'.

RADISH Species in the genus *Raphanus* (Brassicaceae), from Old English *rædic*, from Latin *radix* = 'root'. *Raphanus* is from Greek *raphane* for this plant, in turn from *ra* + *phainomai* = 'quickly' + 'appear'.

Garden radish *R. sativus*, an annual probably Mediterranean archaeophyte, scattered as a casual on waste places as a garden escape. *Sativus* is Latin for 'cultivated'.

Sea radish *R. raphanistrum* ssp. *maritimus*, a herbaceous biennial or perennial of coastal grassland, dunes, shingle, cliffs and disturbed ground by the sea, absent from the east coast north of Lincolnshire, and in north Scotland.

Wild radish *R. raphanistrum* ssp. *raphanistrum*, a common and widespread annual archaeophyte found as a casual or weed in cultivated fields and on roadsides and waste ground.

RADIX Big-ear radix Alternative name for ear pond snail.

RAGGED-ROBIN *Silene flos-cuculi* (Caryophyllaceae), a common and widespread herbaceous perennial found in wet grassland, rush pasture, fen-meadow, tall-herb fen, damp woodland margins and other moist habitats. *Silene* is from the Greek for another plant (catchfly), and Latin *flos-cuculi* = 'cuckoo-flower'.

RAGWEED *Ambrosia artemisiifolia* (Asteraceae), an annual wind-pollinated North American neophyte cultivated in Britain by 1759 and recorded as a casual since 1836, introduced with oilseed, grain and other agricultural seed, and with animal feed, and spreading rapidly since the 1960s. It prefers full sun and nutrient-rich, slightly acidic soils, tolerating dry conditions. Favouring disturbed habitats, it can invade cereal fields, pasture and other grassland. It is currently scattered and rare in Britain, but by natural seed dispersal and, more significantly, with a warming climate, its distribution and abundance will increase. Its pollen (one plant can generate up to a billion grains per season) is a major allergen in humans (causing allergic rhinitis), and will extend the hay fever season into autumn, posing a serious public health problem. Once established, ragweed is difficult to eradicate because of its long-lived seed, its capacity to re-sprout after cutting, and its ability to evolve resistance to herbicides. *Ambrosia*, Greek for 'immortality', was the food of the gods. Latin *artemisiifolia* = 'with Artemisia-like leaves'.

Giant ragweed *A. trifida*, an annual North American neophyte present by 1700, on waste ground and docks as a casual arriving with birdseed, soya beans and grain, local and declining as hygiene improves. *Trifida* is Latin for 'divided into three'.

Perennial ragweed *A. psilostachys*, a North American neophyte naturalised or casual in grassland, on waste ground and on dunes, first recorded in 1903 in Lancashire, persisting there but disappearing from its few other recorded sites. Greek *psilos* + *stachys* = 'naked' + 'ear of grain'.

RAGWORM *Hediste diversicolor* (Nereididae), a common and widespread intertidal polychaete 6–12 cm long, living in muddy substrates in a U or J-shaped burrow up to 20 cm depth, and also under stones on mud where the burrow is adjacent to the stone. Greek *hedos* = 'dwelling-place', and Latin *diversicolor* = 'variously coloured'.

RAGWORT 1) Species of *Senecio* (Asteraceae), from Latin *senex* = 'old man', referring to the 'hairy' pappus (modified calyx). 'Ragwort' probably comes from the ragged appearance of the leaves of this 'wort' (plant).

Rare neophytes include **chamois ragwort** *S. doronicum* (mid-Perthshire), **golden ragwort** *S. doria* (south-west Scotland and Hebrides), **Magellan ragwort** *S. smithii* (north Scotland, Orkney and Shetland), **purple ragwort** *S. grandiflorus* (Guernsey), **shoddy ragwort** *S. pterophorus* (ten records in England), **woad-leaved ragwort** *S. glastifolius* (Scillies), and **wood ragwort** *S. ovatus* (Lancashire and Yorkshire).

Broad-leaved ragwort *S. sarracenicus*, a herbaceous perennial European neophyte, in gardens by 1600, scattered and naturalised by streams and rivers, and in fen, swamp and marshy grassland.

Common ragwort *S. jacobaea*, a widespread herbaceous perennial, common in grassland, especially in neglected or overgrazed pasture; it also grows in scrub, open woodland, waste ground, verges and waysides, and on dunes, rocks, screes and walls. It is particularly toxic to horses and cattle, less so to pigs and chickens, and is classified as injurious under the Weeds Act (1959). *Jacobaea* refers to St James (Jacobus).

Fen ragwort *S. palustris*, thought to have become extinct in Victorian times but with a site discovered in 1972 in a ditch near Ely, Cambridgeshire. Seeds are infertile, but Natural England has propagated more plants from rhizomes and stems, and reintroduced the species to Woodwalton Fen and Wicken Fen. *Palustris* is from Latin for 'marsh'.

Hoary ragwort *S. erucifolius*, a herbaceous perennial common and widespread in England and Wales, local in eastern Ireland, in hay meadow and pasture, downland, and disturbed habitats such as field edges, railway banks, roadsides, waste places, shingle banks and fixed dunes, usually on neutral or calcareous soils, especially clays that are wet in winter but dry in summer. *Erucifolius* is Latin for having leaves like a kind of crucifer (*eruca*).

Marsh ragwort *S. aquaticus*, a common and widespread herbaceous biennial, sometimes perennial, in marsh, wet meadow and rush pasture, and by streams, ponds and ditches. Drainage and agricultural intensification have led to a decline in southern and eastern England.

Narrow-leaved ragwort *S. inaequidens*, a perennial herb with a woody base, naturalised at a few sites, and it may be spreading in south-east England. Elsewhere, it is a common and widespread casual on arable and waste ground, originating from wool shoddy. Latin *inaequidens* = 'unequal teeth', describing the leaf margin.

Oxford ragwort *S. squalidus*, a herbaceous perennial neophyte (sometimes with a woody base) common and widespread in England, Wales and the Central Lowlands of Scotland, scattered elsewhere. In 1794 it was abundant on many walls in and about Oxford as an escape from Oxford Botanic Garden, then found at scattered localities until the 1850s, when it began to spread rapidly along the expanding railway system, the seeds carried by the trains' slipstream wind, and germination favoured by the clinker that echoed the volcanic substrate of the plant's native Sicily (Mt. Etna). Its range continued to expand and it has spread into other habitats, for instance urban waste ground, walls, roadsides and gardens. In Ireland it was naturalised in Cork city by 1845 but subsequent spread has been relatively slow. *Squalidus* is Latin for 'neglected'.

Silver ragwort *S. cineraria*, a herbaceous but woody-based evergreen Mediterranean neophyte present by 1633, scattered but especially found in southern England, naturalised on cliffs and rough ground near the sea, and as a casual inland on verges and waste ground. *Cineraria* is Latin for 'ash-grey'.

2) Species of the evergreen shrubs *Brachyglottis* (Asteraceae) (Greek *brachys* + *glottis* = 'short' + 'tongue'), including **Monro's ragwort** (*B. monroi*) and **hedge ragwort** (*B. repanda*), both New Zealand neophytes, found on the Isle of Man and Llandudno, and in the Scillies, respectively.

Shrub ragwort *Brachyglottis* x *jubar*, of garden origin, scattered and naturalised on mostly coastal rough ground and dunes, and planted on urban roadsides.

3) **Chinese ragwort** *Sinacalia tangutica* (Asteraceae), a rhizomatous herbaceous perennial very scattered in the UK, naturalised in some damp shady habitats. *Tangutica* is a Latinised form of 'Tibetan'.

RAIL Water rail *Rallus aquaticus* (Rallidae), wingspan 42 cm, length 26 cm, male weight 140 g, females 110 g, widespread but scattered, avoiding upland regions, with 1,100 territories estimated for 2006–10, apparently declining (in places predation by mink is a problem), but retaining Green List status. Nests are in reedbeds or marsh, a little above the water level, clutch size is 6–11, incubation 19–22 days, and the precocial young fledge in 20–30 days. Two clutches in a year is common. The omnivorous diet is mainly of small fish, snails and insects. The binomial is Latin for 'water rail'. 'Rail' itself derives from Old French *raale*, related to *râler* (= 'to rattle'), perhaps imitative of the vocalisation.

RALLIDAE Family name of rails (Latin *rallus*), waterbirds that include coot, moorhen, crakes, gallinules and rail, as well as the terrestrial corncrake.

RAMALINA Dotted ramalina *Ramalina farinacea* (Ramalinaceae), an epiphytic fruticose lichen, widespread on trees, and tolerant of acid conditions and nitrogen pollution. Latin *ramalina* = 'small branch' and *farinacea* = 'floury'.

RAMPING-FUMITORY Scrambling annuals in the genus *Fumaria* (Papaveraceae) (Latin *fumus terrae* = 'earth smoke') See also fumitory.

Rarities are **Martin's ramping-fumitory** *F. reuteri* (since 1980 reported only from the Isle of Wight and Cornwall) and **western ramping-fumitory** *F. occidentalis* (endemic in Cornwall and the Scilly Isles on arable, especially field margins, hedge banks, verges and waste ground).

Common ramping-fumitory *F. muralis*, common and widespread (less so in central-eastern England), associated with arable (declining in recent years), waste ground, hedge banks and other disturbed habitats in moist, fairly base- and nutrient-rich soils. *Muralis* is from Latin for 'wall'.

Purple ramping-fumitory *F. purpurea*, an endemic of hedge banks, arable, waste ground and gardens on acidic, freely draining soils, most abundant in western parts of England and Wales, south-east Scotland, Orkney, and eastern Ireland. Most records are from spring-sown crops, and in the Scilly Isles in bulb fields.

Tall ramping-fumitory *F. bastardii*, found in arable (generally spring-germinating, typically found in spring-sown crops), waste ground and hedge banks, with a scattered distribution mostly on the western edges of Britain and in Ireland, usually on freely draining, acidic soils.

White ramping-fumitory *F. capreolata*, of scrub, hedge banks and cliffs, occasionally found in arable and wasteland and gardens. Although widespread, it has declined in its inland sites, where it may only have been casual, but the distribution is stable on or near the coast. *Capreolus* is Latin for 'tendril'.

RAMPION Herbaceous perennial species of *Phyteuma* (Campanulaceae), from *Reseda phyteuma* (corn mignonette). 'Rampion' dates from the late sixteenth century, a variant of medieval Latin *rapuncium*.

Oxford rampion *P. scheuchzeri* (in limestone crevices at Inchnadamph, west Sutherland) and **spiked rampion** *P. spicatum* (in just nine sites in East Sussex) are especially rare.

Round-headed rampion *P. orbiculare*, scarce in southern England on chalk grassland, open scrub, earthworks and verges. This is the county flower of Sussex as voted by the public in a 2002 survey by Plantlife.

RAMSHORN Members of the freshwater pulmonate snail family Planorbidae, the English name echoing the shell's shape. Specifically, *Anisus vorticulus*, sometimes called **lesser ramshorn** or **little whirlpool ramshorn**, 5 mm diameter shell, local and declining in southern England and East Anglia in unpolluted calcareous water with a preference for ditches >3 m wide and >1 m deep, with a moderate emergent vegetative cover, often occurring in ditches in fields that flood in winter (enabling young snails to colonise new ditches).

Flat ramshorn *Hippeutis complanatus*, shell diameter 3–5 mm, height 0.7–1.2 mm, common and widespread in lowland regions in vegetated calcium-rich ponds and ditches. *Complanatus* is Latin for 'flattened'.

Great ramshorn *Planorbarius corneus*, shell diameter 22–35 mm, height 10–17 mm, common and widespread in England, scattered in lowland regions elsewhere, probably introduced in Ireland, grazing on algae in well-vegetated, base-rich, lentic or slowly moving water. Latin *planus* + *orbis* = 'flat' + 'circle', and *corneus* = 'horny'.

Keeled ramshorn *Planorbis carinatus*, shell diameter 9–15 mm, with a ridge (keel) at the centre of the periphery of the body whorl, common and widespread in lowland regions, in ponds, lakes and slowly moving streams with muddy bottoms. Latin *carina* = 'keel'.

Margined ramshorn *Planorbis planorbis*, shell diameter 9–18 mm, height 2.5–4 mm, common and widespread in much of England and central Ireland, more scattered elsewhere, in shallow standing and slowly flowing water on a mud substrate, as well as ponds, ditches and temporarily drying flood water, up to 1 m depth.

Nautilus ramshorn *Gyraulus crista*, shell diameter 2–3 mm, height 0.7–1.1 mm, widespread and locally common in medium to very small vegetated habitats including slow streams, drains and marsh and fen pools, from mildly acidic to very base-rich. Greek *gyros* = 'circle', and Latin *crista* = 'crest'.

Shining ramshorn *Segmentina nitida*, shell diameter 4–5 mm, height 1–1.5 mm, declining (Endangered in the RDL, a priority species in the UK BAP), now confined to the Norfolk Broads, the Lewes and Pevensey Levels (East Sussex), Somerset

Levels and one site in East Yorkshire, in well-vegetated base-rich ponds and grazing marsh ditches. Latin *segmentum* = 'segment', and *nitida* = 'shiny'.

Smooth ramshorn *G. laevis*, shell diameter 3–4.5 mm, height 1–2.5 mm, widespread but scattered in nutrient- and oxygen-rich, unpolluted shallow water with moderately rich vegetation. *Laevis* is Latin for 'smooth'.

Thames ramshorn *G. acronicus*, shell diameter 5–7 mm, height 1–1.5 mm, restricted to short sections of rivers near their confluence with the Thames, characterised by clean, slowly flowing calcareous water and densely vegetated margins. Latin *acronus* = 'extremity'.

Trumpet ramshorn *Menetus dilatatus*, a North American introduction present since the 1860s, shell diameter 2–3 mm, height 0.8–1 mm, scattered in England and Wales in canals and lakes, often associated with discharges of heated water. *Menetus* is probably from Greek *mene* = 'moon', and Latin *dilatatus* = 'dilated'.

Twisted ramshorn *Bathyomphalus contortus*, shell diameter 3–6 mm, height 1–2 mm, common and widespread in lowland regions, in small nutrient-poor water bodies and marshes. In Ireland it is often the only snail present in pools or runnels in and around raised bogs. Greek *bathys* + *omphalos* = 'deep' + 'navel', and Latin *contortus* = 'twisted'.

Whirlpool ramshorn *Anisus vortex*, shell diameter 7–10 mm, height 0.8–1.1 mm, common in England and central Ireland, more scattered elsewhere, in clear, unpolluted vegetated ponds and streams. Greek *anisos* = 'unequal', and Latin *vortex* = 'whirlpool'.

White-lipped ramshorn *A. leucostoma*, shell diameter 5–8 mm, height 1.5 mm, common and widespread in lowland regions in streams, ponds and ditches which dry out in summer, and in marsh. Greek *leucos* + *stoma* = 'white' + 'mouth'.

White ramshorn *G. albus*, shell diameter 5–8 mm, height 1.3–1.8 mm, common and widespread, especially in lowland regions, in vegetation and bottom mud of lakes and large ponds, preferring but not requiring base-rich water, and often found at considerable depth, for example to 20 m in Lough Neagh. Latin *albus* = 'white'.

RAMSONS *Allium ursinum* (Alliaceae), a common and widespread bulbous herbaceous perennial mostly found in moist woodlands (an ancient woodland indicator), sometimes growing in more open situations, for example river and hedge banks, in scree and on coastal cliff ledges. 'Ramsons' comes from Old English *hramsan*, plural of *hramsa* = 'wild garlic'. *Allium* is Latin for 'garlic', *ursinum* = 'of a bear'.

RANIDAE Family name of frogs, including bullfrog (Latin *rana* = 'frog').

RANNOCH-RUSH *Scheuchzeria palustris* (Scheuchzeriaceae), a rhizomatous herbaceous perennial of base-poor, wet habitats, found in acid runnels, pools or semi-submerged pool edges. Current sites are all on Rannoch Moor. Johann Scheuchzer (1672–1733) was a Swiss naturalist. *Palustris* is from Latin for 'marsh'.

RANUNCULACEAE Buttercup family, from Latin for 'little frog' (diminutive of *rana*), referencing the (near-)aquatic habitat of many of the species.

RANUNCULUS (MOTH) Nocturnal moths in the family Noctuidae.

Feathered ranunculus *Polymixis lichenea* (Xyleninae), wingspan 35–40 mm, local on sea-cliffs, dunes, shingle beaches and grassy slopes in England, Wales and east Ireland. Adults fly in August–October. Larvae feed on various herbaceous plants. Greek *polys* + *mixis* = 'much' + 'mixing', from the mingled forewing colours, and *lichenea* from the lichen coloration.

Large ranunculus *P. flavicincta*, wingspan 40–50 mm, local in gardens, waste ground and damp meadow in the southern half of England and Wales, with scattered records from northern England. Adults fly in September–October. Larvae feed on a number of herbaceous plants, and they can be a pest on garden flowers. Latin *flavus* + *cinctus* = 'yellow' + 'belt', from the forewing pattern.

Small ranunculus *Hecatera dysodea* (Hadeninae), wingspan 32–4 mm, an RDB moth found on verges, chalk-pits, gardens, allotments and rough ground in parts of south and south-east England. The species declined rapidly in the first half of the twentieth century, disappearing from the British Isles completely by 1941. It subsequently recolonised and is becoming increasingly common again: following two examples recorded in Gravesend, Kent, in 1997, a resident population became established in north-west Kent from 1998 onwards. There were also records in Essex and Suffolk in 1999. It seems to be expanding quite rapidly north and westwards. The first modern records in Hampshire were in 2009. Adults fly in June–July. Larvae feed on lettuce. In Greek mythology Hecate was goddess of the Underworld. *Dysodes* is Greek for 'evil-smelling'.

RAPE *Brassica napus* (Brassicaceae), an annual or biennial herb of disturbed ground on roadsides, waste and cultivated ground, and docks. The acreage of ssp. *oleifera* (**oil-seed rape**) has dramatically increased since 1980, grown for its oil-bearing seed, less commonly for fodder and green manure, and it is now widespread as a casual on field margins and roadsides. Ssp. *rapifera* (**swede**), cultivated as a root crop, is occasionally found as a casual. *Brassica* is Latin for 'cabbage', *napus* and *rapa* (giving the English name) for 'turnip'.

RAPHIDIIDAE Family name of snake flies (Greek *raphis* = 'needle').

RASPBERRY *Rubus idaeus* (Rosaceae), a common and widespread shrub of open woodland, downland scrub, heathland, and sometimes hedgerows. It also occurs on waste ground as an escape from cultivation. In upland regions it grows on base-rich rocky ledges and cliffs. It spreads by bird-dispersed fruit and by suckering, often forming thickets. *Rubus* is Latin for 'blackberry' ('bramble'), *idaeus* = 'from Mt Ida' (Crete).

Purple-flowered raspberry *R. odoratus*, a spiny, erect, deciduous North American shrub cultivated by 1739, occasionally naturalised as a garden escape in rough grassland, and on waste ground, reproducing by suckering. *Odoratus* is Latin for 'fragrant'.

RASPWORT **Creeping raspwort** *Haloragis micrantha* (Haloragaceae), an introduced Australian perennial herb discovered in 1988 as naturalised on cut-over wet peaty heath near Lough Bola, Galway. Greek *halos* + *rhagos* = 'sea' + 'grapes', from the coastal habitat and bunching habit of the fruit in some species, and *micrantha* = 'small-flowered'.

RAT 1) Rodents in the genus *Rattus* (Muridae).

Alexandrine rat, house rat, Old English rat, roof rat, ship rat Alternative names for black rat.

Black rat *R. rattus*, head-body length 120–40 mm, tail length 140–260 mm, weight 110–340 g, native to Asia, present in England by Roman times, initially quite widespread, but greatly reduced in number and range in competition with brown rat from the eighteenth century, and now having low numbers and a very restricted distribution in disused buildings and warehouses in ports such as Bristol and Tilbury (London). One recent population has been on the Shiant Isles, near Harris, Outer Hebrides: in winter 2015–16 the Shiant Isles Recovery Project (RSPB and Scottish Natural Heritage) began a poisoning campaign to eradicate this predator of seabird eggs and young, where in April 2012 there was a pre-breeding population of 3,600 rats. In Ireland there remains one population on Lambay Island off Co. Dublin. They are primarily nocturnal, and good climbers and swimmers. They are opportunistic generalist omnivores, though their main food is seed (grain) and fruit. Breeding is between April and November. Females produce up to five litters of 5–8 pups. Young are sexually mature at five weeks. They were a pest when numbers were high, causing damage, spoiling food and as a vector of disease, fleas on black rats debatably bringing bubonic plague (Black Death) to England, in 1348–9, killing 3 million people (over half the population), and again in 1666.

Brown rat *R. norvegicus*, head-body length 150–270 mm, tail length 105–240 mm, weight 200–300 (up to 600) g, an Asian species accidentally introduced from Russia via ships in the late 1720s, common and widespread, with an estimated minimum British pre-breeding population is 6.5 million, not including urban habitats. They thrive in a range of lowland habitats, often associated with human activity, for example farms, industrial sites, rubbish tips, allotments, smallholdings, sewage farms, and sewers, selecting sites that provide abundant food, especially cereals and waste human food. They are good climbers and swimmers. They dig burrows and use a network of runs, living in family groups within loose colonies. While generalist omnivores, they do require cereal in some form (e.g. grain or bread) and water. Breeding takes place all year in the built environment. Females can breed at 3–4 months old, and if food is abundant can breed continuously, though typically have five litters a year. Litter size increases from around six in young females of 150 g, to 11 in females of 500 g, the maximum recorded being 22. Young are weaned at three weeks. They are a serious pest, causing damage, spoiling food and transmitting diseases such as leptospirosis and Weil's disease. They also pose problems for ground-nesting birds, especially on islands. *Norvegicus* is Latin for 'Norwegian'.

Common rat, Norway rat, sewer rat Alternative names for brown rat.

2) **Water rat** Alternative, taxonomically misleading name for water vole.

RAULE See southern beech.

RAVEN *Corvus corax* (Corvidae), wingspan 135 cm, length 64 cm, male weight 1.3 kg, females 1.1 kg, breeding in upland south-west England, Wales, north Pennines, Lake District and much of Scotland and Ireland. There were an estimated 7,400 pairs in Britain in 2009. Numbers increased by 38% over 1970–2015, by 45% over 1995–2015, and 10% over 2010–15. The nest is a deep stick bowl with an inner layer of roots, mud, and bark, and lined with soft material, usually in a large tree or on a cliff ledge. Clutch size is 4–6, incubation lasts 21–2 days, and the altricial young fledge in 39–41 days. The opportunistic omnivorous diet includes small vertebrates, carrion, and grain, berries and fruit. 'Raven' is from Old English *hræfn*. The binomial comprises the Latin and Greek for 'raven'.

RAY Thornback ray *Raja clavata* (Rajidae), length 85–100 cm, weight 2–4 kg, a cartilaginous fish common around all coasts and the most abundant ray in inshore waters (particularly in spring). The empty egg-cases (mermaid's purses) are commonly washed up on the shore. The English name reflects the row of 30–50 thorns found from the back of the head to the first dorsal fin. The binomial is Latin for 'ray' and 'studded'.

RAZOR SHELL Long, narrow filter-feeding bivalves in the family Pharidae, in the genera *Ensis* (Latin for 'sword'), together with American jack-knife clam.

Common razor shell *E. ensis*, shell length up to 12 cm, common and widespread, burrowing in fine sand in the low intertidal to the nearshore shelf.

Pod razor shell *E. siliqua*, shell length up to 20 cm, common and widespread, burrowing in sand in the low intertidal to the nearshore shelf. *Siliqua* is Latin for 'pod' or 'husk'.

RAZORBILL *Alca torda* (Alcidae), wingspan 66 cm, length 38 cm, weight 710 g, an Amber List auk breeding in colonies from March to the end of July around all coasts except between the Humber estuary and the Isle of Wight, with the largest colonies in north Scotland. There were an estimated 110,000 pairs in Britain in 2002, and 51,500 individuals in Ireland. The British population increased by 49% over 1986–2014 (SMP data; BTO gives 58%), by 6% over 2000–14 (SMP), and by 15% over 2009–14 (BTO). About 20% of the global population nests in the British Isles. Birds only come to shore to breed, and winter in the North Atlantic. Nests are on sea-cliffs, contain a single egg that is incubated for 32–9 days, the altricial chick fledging in 14–24 days. Fish are caught by diving to depths of up to 140 m. *Alca* is from Norwegian *alke*, *torda* from Swedish *törd*, both names for this species.

RAZORSTROP FUNGUS *Piptoporus betulinus* (Polyporales, Fomitopsidaceae), a very common inedible annual bracket fungus found throughout the year, sporulating in late summer and autumn on trunks of dead birch trees. Barbers used to 'strop' or sharpen their cut-throat razors on tough, leathery strips cut from the surfaces of these fungi, hence the name. Greek *pipto + poros* = 'to fall' (i.e. easily detachable) + 'pore', and Latin *betula* = 'birch'.

RECURVIROSTRIDAE Family name for avocet and black-winged stilt (Latin *recurvus + rostrum* = 'curving back' + 'beak').

RED (MOTH) Barred red *Hylaea fasciaria* (Geometridae, Ennominae), wingspan 27–44 mm, widespread and common where there are coniferous trees (the larval food). Adults fly at night in June–August, often with a partial second generation in the south in August–September. Greek *hylaios* = 'belonging to woodland', and Latin *fascia* = 'band', from the forewing marking.

RED ALGAE Division of Rhodophyta, though some taxonomists now place red algae into the class Rhodophyceae.

RED-CEDAR Introduced evergreen trees in the family Cupressaceae.

Japanese red-cedar *Cryptomeria japonica*, introduced from China in 1842 and from Japan in 1846, many cultivars commonly grown as park and garden ornamentals in lowland regions, and in small forestry plots in Wales, preferring areas with >1,200 mm annual rainfall. Best growth is in south-west England and Wales on slightly dry to moist soils. Greek *cryptos + meros* = 'hidden' + 'a part', from the cryptic flower parts.

Western red-cedar *Thuja plicata*, introduced from North America in 1853, common in parks as an ornamental, as a wind-break species, and for underplanting larch or hardwoods in some plantations. It is shade-tolerant, and grows best on medium to very nutrient-rich moist soils. Greek *thya*, a tree with fragrant wood, and Latin *plicata* = 'braided'.

RED-EYE Woodlice in the genus *Trichoniscoides* (Trichonis-cidae), the name incorporating the Greek *trichos* = 'hair'.

Coastal red-eye *T. saeroeensis*, 2 mm long, widespread, typically found in the supralittoral zone associated with coastal erosion banks, thinly vegetated shingle or saltmarsh strand-line, usually several centimetres below the surface in damp peaty soil. It has also been recorded up to 12 km inland deep inside limestone caves and mines, and on the summit of limestone mountains.

Sar's red-eye *T. sarsi*, records forming a band across south-east, east and central England into eastern Ireland. In 2010 specimens were collected from a coastal site in east Scotland. It is characteristically associated with synanthropic sites, for instance old gardens and churchyards, though some coastal sites in Kent appear to be semi-natural, and usually found beneath paving slabs, stones and rubble.

RED FOX See fox.

RED-HOT-POKER *Kniphofia uvaria* (Xanthorrhoeaceae), a rhizomatous herbaceous South African neophyte perennial, scattered in Britain, naturalised on dunes and waste ground, often near the sea, as garden throw-outs. Johann Kniphof (1704–63) was a German botanist. Latin *uvaria* = 'bunch of grapes'.

RED-LEG A recently occurring disease of frogs and other amphibians recognised by a redness on the underside of the legs and abdomen, often probably involving a virus followed by secondary infection by the bacterial pathogen *Aeromonas hydrophila*, but red-leg is a name for a generalised set of haem-orrhagic symptoms rather than a specific disease or pathogen.

Mortality can be high. Transfer of spawn between ponds is discouraged, since this facilitates spread of disease organisms.

RED RAGS *Dilsea carnosa* (Dumontiaceae), a widespread red seaweed, up to 50 cm long, 25 cm wide, in the lower littoral to shallow sublittoral attached to rocks. *Carnosa* is Latin for 'fleshy'.

REDLEAF One of the pathogenic fungi in the genus *Exobasidium* (Exobasidiales, Exobasidiaceae) whose hosts are generally ericaceous species. Particularly widespread are blaeberry and cowberry redleafs.

REDPOLL Finches in the genus *Acanthis* (Fringillidae), from Greek *acanthis*, an unidentified small bird (possibly a finch) mentioned by Aristotle. 'Poll' is an old name for 'head'; the English name reflects the red patch on this bird's head.

Lesser redpoll *A. cabaret*, wingspan 22 cm, length 12 cm, weight 11 g, widespread in deciduous and coniferous woodland, and in scrub, as a resident breeding finch in Scotland, north and east England and Wales. It is less common elsewhere in England, but does occur in winter when numbers are augmented by continental birds. There were an estimated 220,000 pairs in Britain in 2009, numbers declining by 87% from 1970 to 2015, though up by 38% over 1995–2015, and stabilising (–4%) in 2010–15. It is Red-listed in Britain, Green-listed in Ireland. The cup nest is built in a shrub or tree, usually with 4–5 eggs which hatch after around 12 days, the altricial young fledging in 14–15 days. *Cabaret* comes from Old Dutch *cambret* = 'entertainment offered at a restaurant'.

Mealy redpoll (or **common redpoll**) *A. flammea*, wingspan 23 cm, length 14 cm, weight 14 g, a winter and passage Amber-listed seed-eating migrant finch found mainly from October to April along the east coastline, moving westwards in search of food, so later in winter turning up in suitable habitat inland. Total numbers are usually around 300 p.a. It also occasionally breeds, mainly in the Outer Hebrides, with 2–6 pairs at five sites noted in Scotland in 2010. Latin *flamma* = 'flame'.

REDSHANK (BIRD) *Tringa totanus* (Scolopacidae), wingspan 62 cm, length 28 cm, male weight 110 g, females 130 g, breeding around all coasts but particularly in north England and Scotland, with 25,000 pairs in the UK in 2009 (Amber-listed). Numbers in Britain declined by 55% over 1975–2015, and by 39% over 1995–2015, though stabilising after 2000. It is declining as a breeding bird (Red-listed) in the Irish midlands (especially found in the Shannon Callows) and further north. Wintering birds are especially common in south-west England and around much of the Irish coastline, but in Britain numbers (130,000 estimated in 2008–9) have declined by 26% over 2002–3 to 2012–13. They breed in damp habitats such as saltmarsh and wet meadow and around lakes, but during winter most birds are found on estuaries and coastal lagoons, as many as half having arrived from Iceland. The ground nests usually contain 4 eggs, incubation takes 24 days, and the precocial young fledge in 25–35 days. Inland, the diet is mainly of invertebrates, especially earthworms and leatherjackets, from estuarine mud they take crustaceans, molluscs and marine worms. 'Redshank' describes the red legs. *Tringa* was a wading bird mentioned by Aristotle. *Totanus* comes from *totano*, the Italian name for this species.

Spotted redshank or **dusky redshank** *T. erythropus*, wingspan 64 cm, length 30 cm, weight 170 g, an Amber List wintering wader with over half the population found at <10 coastal wetland sites, especially in north Kent and Essex, Hampshire, west Wales and south Ireland, with an overwintering population averaging 100 birds. Wintering birds remain until April–May. Migration to Africa, totalling an average of 420 birds, is mostly in September. Diet is of insect larvae, shrimps and worms taken from water and muddy shores. Greek *erythros* + *pous* = 'red' + 'foot'.

REDSHANK (FLOWERING PLANT) *Persicaria maculosa* (Polygonaceae), a common and widespread annual of open ground preferring nutrient-rich soils, by ponds, lakes, streams and ditches, in waste places, on roadsides and railways, and in places a weed of cultivated land. The binomial is from medieval Latin likening the leaves to those of a peach tree (*Prunus persica*), and Latin *maculosa* = 'spotted'.

REDSHANK (MOSS) *Ceratodon purpureus* (Dichitraceae), common and widespread on acidic, well-drained substrates, particularly peaty or sandy soil. It is frequent in heathland and acidic grassland, on walls and roofs, and can be abundant on fire sites. Greek *ceras* + *odontos* = 'horn' + 'tooth', and Latin *purpureus* = 'purple'.

Scarce redshank *C. conicus*, scarce and declining in southern England and the Midlands, on shallow, calcareous soil over limestone, on walls and paths, and in quarries. Latin *conicus* = 'conical'.

REDSTART *Phoenicurus phoenicurus* (Turdidae), wingspan 22 cm, length 14 cm, weight 15 g, an Amber List migrant breeding chat, wintering in tropical Africa, mainly found between April and September in the north and west of Britain, and in Co. Wicklow, with greatest concentrations in Wales, particularly in oakwood. There were 100,000 pairs (70,000–130,000) in Britain in 2009, with an increase in numbers by 66% in 1970–2015, 47% in 1995–2015, and 20% (strong increase) in 2010–15. It is also a passage migrant in spring and autumn, for example in south-east Ireland, in coastal scrub, thickets and woodland. Nests are in cavities, with 6–7 eggs, incubation lasting 13–14 days, the altricial young fledging in 16–17 days. There can be two broods in a year. Diet is mainly insectivorous. 'Start' means 'tail', from Old English *steort*. Greek *phoinix* + *oura* = 'crimson' + 'tail'.

Black redstart *P. ochruros*, wingspan 24 cm, length 14 cm, weight 16 g, a Red List resident and migrant breeding chat in Britain. This originally cliff-dwelling montane bird spread throughout Europe after adopting buildings as nesting sites, first nesting in England in 1926 on the Wembley Empire Exhibition building, and 20–65 pairs are now found each year on other structures and a variety of industrial and brownfield sites (54 being recorded in 2010) in London (e.g. Battersea Power Station and Kings Cross station), with up to 20 pairs in 2010, and Birmingham and its surroundings (especially by railway stations and canal-side brownfield sites), with records from Liverpool, Manchester, Nottingham and Ipswich, though described following post-war regeneration of bomb sites. In London, Enfield, Tower Hamlets, Newham and Greenwich hold greatest numbers of breeding pairs. The species is also a passage migrant, found on the south and east coasts, and in winter on the coasts of Wales and south and south-west England, with a few at inland sites, arriving from the Continent in October–November, returning in March–April. They also winter on the south and east coasts of Ireland, where it is Green-listed. The cavity nest, of grass, leaves and moss lined with hair and feathers, contains 4–6 eggs, incubation takes 12–16 days, and the altricial young fledge in 12–19 days. Food is mainly of insects, with some seed and fruit. The male plumage is mainly black or dark grey with a reddish-brown lower rump and tail (= 'start'). Greek *ochros* = 'pale yellow' + 'tail'.

REDUVIIDAE Family name of bugs (Heteroptera), including assassin bugs, from Latin *reduvia* = 'hangnail' or 'remnant', possibly prompted by the lateral flanges on the abdomen of many species.

REDWING *Turdus iliacus* (Turdidae), wingspan 34 cm, length 21 cm, weight 63 g, a thrush which is a scarce breeder (5–10 pairs were likely in the Highlands in 2010), but common and widespread as a winter visitor, with 650,000 an average figure, depending on weather conditions on the Continent, flocking in fields and (in bad weather) gardens, feeding on berries, insects and earthworms. It has been Red-listed in Britain since January 2013, but Green-listed in Ireland. The English name comes from its red flanks and underwing. Latin *turdus* = 'thrush', and *iliacus* = 'of the flank' (from *ile* = 'flank').

REDWOOD Coastal redwood *Sequoia sempervirens* (Cupressaceae), introduced from California via Crimea in 1844, commonly planted in lowland regions as an ornamental in parks. Well-suited to south-west England and Wales in areas with >1,250 mm rainfall but it will grow elsewhere on moist soils of poor to medium nutrient status. Sequoia was a famous Native American of the Cherokee tribe. An etymological study published in 2012, however, suggested that the name may have originated from the Latin *sequi* (= 'follow') since the number of seeds per cone in the newly classified genus *Sequoia* showed a mathematical sequence with the other genera in the suborder. *Sempervirens* is Latin for 'evergreen'.

Giant redwood See wellingtonia.

REED Common reed or **Norfolk reed** *Phragmites australis* (Poaceae), a common and widespread rhizomatous and stoloniferous grass of swamp and fen, forming large stands in mud and shallow water in ditches, rivers, lakes and ponds. Salinity as low as 1 ppt causes stress. Reedbeds provide material for traditional thatching, especially in the Norfolk Broads. Dieback has been noted in the Broads since the mid-twentieth century as a consequence of eutrophication from agricultural run-off and increased boat-based recreation, leading to rapid growth and stem-weakening. Reedbeds have been planted beside artificial water bodies, and have been created for breeding birds. 'Reed' comes from Old English *hreod*. Greek *phragmites* = 'growing in hedges' (*phragmos* = 'barricade'), and Latin *australis* = 'southern'.

REED BEETLE Members of the genus *Donacia* (Chrysomelidae, Donaciinae), *donax* being Latin for a kind of reed. Adults feed on leaves of water plants, larvae feed underwater, taking oxygen via plant vessels. **Water-lily reed beetle** *D. crassipes* and **Zircon reed beetle** *D. aquatica* are both scarce.

REEDLING Bearded reedling Alternative name for bearded tit.

REEDMACE Greater reedmace Former name of bulrush; **lesser reedmace**, former name of lesser bulrush.

REEVE Female ruff.

REINDEER *Rangifer tarandus* (Cervidae), common and widespread in the Late Glacial, but becoming extinct as the climate warmed. In 1952, eight domesticated reindeer from Arctic Sweden were introduced into a 120 ha enclosure near Loch Morlich on Glen More, Aviemore, Inverness-shire, and in 1953 a further 17 animals were placed in a larger enclosure at nearby Airgiod Meall. Further animals were subsequently introduced from elsewhere in Sweden, Norway and Russia. Reindeer were restricted to these upland heather moorland sites (totalling 2,400 ha) until 1990, when another herd was established in a 1,200 ha area near Tomintoul. All these free-ranging animals are owned by the Reindeer Co. Ltd, the aim being to provide meat and hides, though they have also become a tourist attraction. There are currently the aimed maximum of around 150 animals. They graze on dwarf shrubs, grasses, lichen and fungi, and artificial feed in provided year-round. 'Reindeer' comes from late Middle English, in turn from Old Norse *hreindýri*, from *hreinn* + *dyr* = 'reindeer' + 'deer'. *Rangifer* may go back to the Saami (Lapp) *raingo* for this animal; *tarandus* probably originated from the ox-sized deer species (*tarandos*) mentioned by Aristotle and Theophrastus.

REINDEER MOSS See reindeer lichen under lichen.

REPTILE Cold-blooded air-breathing vertebrates in the class Reptilia, with horny scales or plates, including snakes, lizards and turtles.

RESEDACEAE Family name for mignonette, including weld (Latin *resedo* = 'to calm').

REST HARROW *Aplasta ononaria* (Geometridae, Geometrinae), wingspan 26–31 mm, a rare nocturnal moth found on calcareous grassland and dunes in parts of Kent, flying in June–July, occasionally elsewhere in southern England as a migrant. Larvae feed on common restharrow (*Ononis*, giving the specific name). Greek *aplastos* = 'unmoulded', from the weak wing markings.

RESTHARROW Species of *Ononis* (Fabaceae), Greek *ononis* meaning this plant. 'Restharrow' dates from the mid-sixteenth century, from the obsolete *rest* = 'arrest' + harrow, because the tough stems impeded a harrow's progress.

Common restharrow *O. repens*, a rhizomatous perennial subshrub, widespread in lowland Britain in grasslands on base-rich, well-drained, light soils, and on calcareous boulder-clay. On the coast it occurs on dunes and shingle, and inland it often colonises sandy or gravelly road verges. *Repens* is Latin for 'creeping'.

Small restharrow *O. reclinata*, a rare annual of thin, dry, calcareous soils with a low organic content, especially on coastal cliffs of limestone and, in Scotland, greywacke. In the Channel Islands it occurs in dune turf. *Reclinata* is Latin for 'leaning'.

Spiny restharrow *O. spinosa*, a woody perennial of infertile calcareous grasslands on well-drained chalk, limestone and heavy calcareous clay soils, locally common especially in England (though declining in the south-east). It also occurs on coastal grazing marsh and earth sea walls.

Yellow restharrow *O. natrix*, a dwarf shrub neophyte introduced from Europe by the seventeenth century, naturalised on waste ground in Berkshire since at least 1947, a rare casual elsewhere, sometimes having been introduced with ballast. *Natrix* is Latin for a water snake, reason obscure.

RHABDOWEISIACEAE From Greek for 'fluted column' (*rhabdos* = 'rod'), describing the striate capsules, the family name of some fork-, streak-, dog-tooth and yolk-mosses.

RHAGIONIDAE Family name for snipe flies (Greek *rhagion* for a kind of spider), with 15 British species in 5 genera.

RHAMNACEAE Family name for buckthorn (Greek *rhamnos*) and alder buckthorn.

RHINOLOPHIDAE Family name for horseshoe bats in the genus *Rhinolophus* (Greek *rhinos* + *lophos* = 'nose' + 'crest').

RHINOPHORIDAE Family name (Greek *rhinos* + *phoros* = 'nose' + 'carrying') of some flies, with eight species in six genera, parasitoids of woodlice and other arthropods.

RHIZOPODA Phylum of the Protoctista that contains amoebas and cellular slime moulds (Greek *rhiza* + *podos* = 'root' + 'foot').

RHODODENDRON *Rhododendron ponticum* (Ericaceae), a common and widespread evergreen neophyte shrub in parks and gardens by 1763, naturalised on heathy and rocky hillsides, rocky stream banks and ravines, and as an understorey in woodland on acid soils. It regenerates from seed freely and can form dense thickets as an invasive weed. Greek *rhodon* + *dendron* = 'rose-red' + 'tree'. *Ponticum* means 'from Pontus' (south Black Sea coast).

RHODOPHYTA Division commonly referred to as red algae (Greek *rhodon* + *phyton* = 'rose' + 'plant'), including coralline algae, mostly multicellular, macroscopic and marine, and using sexual reproduction. Some taxonomists place all red algae into the class Rhodophyceae.

RHUBARB *Rheum* x *rhabarbarum* (Polygonaceae), a rhizomatous herbaceous perennial neophyte of garden origin in at least Tudor times, commonly cultivated, but also found on waste ground, roadsides and stream banks as relics of cultivation. (Not to be confused with giant-rhubarb, q.v.) *Rheum* is medieval Latin for this plant, possibly from Greek *rheo* = 'to flow', from its purgative properties.

Ornamental rhubarb *R. palmatum*, a north-east Asian neophyte cultivated by 1763, scattered and occasionally naturalised in rough grassland and woodland, and on verges and waste ground. *Palmatum* is Latin meaning shaped like the palm of the hand.

RHYACOPHILIDAE Family name (Greek *rhyacos* + *philos* = 'stream' + 'loving') of predatory uncased caddis flies, found in the fastest sections of unpolluted streams, catching food material in mesh nets.

RHYNCHITES Strawberry rhynchites *Neocoenorrhinus germanicus* (Rhynchitidae), a tooth-nosed snout weevil, 2–3 mm length, widespread in England and Wales, and present if uncommon in south-west Scotland and Northern Ireland. Adults feed on strawberry leaves in early spring and can also become a pest of raspberry, occasionally blackberry. Greek *neo* + *coinos* + *rhynchos* = 'new' + 'in common' + 'beak', and Latin *germanicus* = 'German'.

RHYNCHITIDAE Family name (Greek *rhynchos* = 'beak') of tooth-nosed snout weevils, with 18 British species in 9 genera. Most species are relatively hostplant-specific. The larva can often be found developing within leafbuds and young shoots. In a few species the female laying an egg on the leaf surface before rolling the leaf around it for protection.

RIB-LEAF MOSS *Tortula atrovirens* (Pottiaceae), locally common on coastal habitats from Dorset to north Wales, Isle of Man, east Ireland and east Scotland. *Tortula* is Latin for 'small twist', *atrovirens* = 'dark green'.

RIBBONWORT Alternative name for veilwort.

RICCIACEAE Family name of crystalworts (thallose liverworts), named after an eighteenth-century Italian nobleman, P.F. Ricci.

RIDGEBACK Woodlice, up to 4 mm long, in the genus *Haplophthalmus* (Trichoniscidae) (Greek *haploos* + *ophthalmos* = 'single' or 'simple' + 'eye').

Common ridgeback *H. mengii*, widespread but scattered in England and Wales, scarce in Scotland and Ireland, in a range of semi-natural and synanthropic habitats, typically on the underside of stones and dead wood. It also occurs in humus-rich soil, compost and, by the coast, in peaty soil that accumulates below shingle or boulders.

Southern ridgeback *H. montivagus*, scattered in southern England across chalk and limestone, typically in ancient woodland on the underside of stones and dead wood, or among humus-rich soil and limestone rubble, often in damp sites. Latin *mons* + *vago* = 'mountain' + 'wander'.

Spurred ridgeback *H. danicus*, common south of a line between the Wash and the Severn estuary and in south Wales, scarcer elsewhere in England and Wales, rare in Scotland and Ireland, in damp habitats in rotten wood and compost. Latin *danicus* = 'Danish'.

RIFFLE BEETLE Members of the family Elmidae, generally found among stones and gravel in the riffles of fast-flowing rivers and streams. Measuring 1–5 mm in length, they walk on the substrate rather than swim. They are also found among stones on lake shorelines, slow-flowing sections of rivers, and on submerged logs. The adult holds a small bubble of air under the wing cases which act as gills. They are collector-gatherers or scrapers, feeding on algae and detritus.

RING OUZEL *Turdus torquatus* (Turdidae), wingspan 40 cm, length 24 cm, weight 110 g, a Red-listed migrant breeder, wintering in North Africa and the Mediterranean, found mainly in steep-sided valleys in upland areas of Ireland, Scotland, northern England, north-west Wales and Dartmoor. In Ireland it breeds on scree slopes in just a handful of sites. When on spring and autumn migration they are often seen on the south and east coasts of England, and in south-east Ireland, where they favour short grassy habitat. Breeding occurs from mid-April to mid-July, commonly with two broods. There were 6,200–7,500 pairs

in Britain in 1999, with range contractions and a 58% decline in numbers since 1988–91, a trend shown in 2010 to have continued, even more pronounced in Ireland. The cup-shaped nests are in bushes or among rocks, with an average of 4 eggs, incubation lasting 13 days, and fledging of the altricial young 14 days. Diet is of insects, earthworms and berries. 'Ouzel' (or 'ousel') is an old name for blackbird, from Old English *osle*. The adult is all black except for a white crescent on its upper breast/lower throat, hence 'ring'. Latin *turdus* = 'thrush', *torque* = 'torc' (neck ring).

RINGLET *Aphantopus hyperantus* (Nymphalidae, Satyrinae), male wingspan 42–8 mm, female 46–52 mm, a common and widespread 'brown' butterfly, absent from north-west England and north and north-west Scotland, the Isle of Man and the Channel Islands. Found in a variety of habitats though favouring sheltered and damp places. Larvae feed at night on grasses; adults, flying in June–August, feed primarily on the nectar of bramble, privet, vetches, ragwort and thistles. This species increased its range by 63% from 1976 to 2014, and between 2005 and 2014 numbers increased by 72%. The English name comes from the rings (eyespots) on the underwings. Var. *lanceolata* has rings elongated to form teardrops. Sometimes the rings are greatly reduced or absent. A form in Co. Kerry is of normal size until 200 m, when it starts to be replaced by a dwarf form that, at 350 m, takes over completely. Greek *aphantos* + *pous* = 'made invisible' + 'foot'. Hyperanthus was a figure in Greek mythology.

Mountain ringlet *Erebia epiphron*, male wingspan 32–40 mm, female 32–42 mm, Britain's only truly mountain butterfly, found particularly in the Lake District (at 500–700 m altitude) and west-central Highlands (350–900 m altitude), flying in July. It prefers moist or boggy ground in sheltered depressions where the primary larval food plant, mat grass, is abundant. Adults mainly feed on hawkweeds, thyme and tormentil. It is a priority species for conservation, with an overall population decline of 12% since the 1970s, though numbers did increase by 24% between 1995–9 and 2005–9. There was a decline in range by 63% over 1976–2014. In Greek mythology Erebos was the region of darkness between Earth and Hades, reflected by the dusky wing colour. *Epiphron* is Greek for 'thoughtful'.

RIPIPHORIDAE Family name (Greek *rhipis* + *phoros* = 'fan' + 'bearing') for wedge-shaped beetles, with one British species, wasp-nest beetle.

RIVER-MOSS Alternative name for cryphaea.

RIVER SKATER *Aquarius najas* (Gerridae), scattered and local in England, Wales and south-west Scotland, on the stony margins of rivers. Latin *aquarius* = 'of water' and New Latin *najas* = 'snake'. See also pond skater.

RIVULET *Perizoma affinitata* (Geometridae, Larentiinae), wingspan 24–30 mm, a nocturnal moth common in open woodland, hedgerows, verges and chalk downland throughout much of Britain and north-east Ireland. Adults fly in May–June. Larvae feed in red campion seed capsules. Greek *perizoma* = 'girdle', from the belt pattern on the forewings when folded, and Latin *affinitata* = 'related', from similarity to a congeneric.

Barred rivulet *P. bifaciata*, wingspan 20–6 mm, widespread but local on chalk downland and other habitats on calcareous soils, including woodland rides and waste ground, scarce in Scotland. Adults fly at night in July–August. Pupation sometimes lasts two winters. Larvae feed on red bartsia seeds. *Bifaciata* is a typographic error for *bifasciata* (Latin for 'two-banded').

Blomer's rivulet *Venusia blomeri*, wingspan 24–8 mm, nocturnal, in deciduous woodland and hedgerows in (mostly western) England and Wales, and recorded at Lochaline, west Scotland in 2013. Adults fly in June–July. Larvae feed on wych elm leaves during autumn. Venus was the Roman goddess of love. Charles Blomer was an eighteenth-century naturalist.

Grass rivulet *P. albulata*, wingspan 20–6 mm, flying in the late afternoon and early evening in May–July, widespread, more thinly in Ireland, on dry grassland, limestone hills, dune slacks and shingle beaches, larvae feeding on yellow-rattle seeds. Numbers have declined by 94% over 1968–2007. *Albulus* is Latin for 'whitish'.

Heath rivulet *P. minorata*, wingspan 18–20 mm, scarce in parts of Scotland with scattered records from Cumbria and the Pennines in England, and from the Burren and Mountains of Mourne in Ireland, living on moorland, limestone grassland and upland pasture. Larvae feed on eyebright seeds. Adults fly in the afternoon and into the night in July–August. *Minor* is Latin for 'smaller'.

Small rivulet *P. alchemillata*, wingspan 14–18 mm, widespread and common in open woodland, hedgerows, chalk downland, gardens and waste ground. Adults fly at night in June–July. Larvae feed in the flowers and seed capsules of common hemp-nettle and hedge woundwort. *Alchemilla* is lady's-mantle, though not a food plant.

ROACH (FISH) *Rutilus rutilus* (Cyprinidae), length generally 35 cm, a common and widespread coarse fish in lowland Britain in the deeper parts of slow to medium-flowing rivers, canals, reservoirs and gravel-pits, including nutrient-enriched, oxygen-poor waters. In Ireland it is found in the Shannon, Corrib, Boyne, Dee, Liffey and Cork Blackwater systems as well as many small lakes. Its diet includes plankton, algae, water plants, crustaceans, snails and insect larvae. Spawning is in April–June. 'Roach' dates from Middle English, from Old French *roche*. Latin *rutilus* = '(golden) red'.

ROBBER FLY Members of the family Asilidae. The powerfully built adults (body length generally 10–15 mm), with piercing mouthparts, feed mainly (some exclusively) on other insects during daylight in open, sunny habitats, generally waiting in ambush then catching their prey in flight. Larvae usually live in soil, sometimes leaf litter; later instars are predaceous.

Rarities include **bumblebee robber fly** *Laphria flava* (Caledonian pine forest), **downland robber fly** *Machismus rusticus* (chalk grassland), **false slender-footed robber fly** *Leptarthrus vitripennis* (chalk grassland), **golden-tabbed robber fly** *Eutolmus rufibarbis* (Breckland and Surrey and West Sussex heaths), **Manx (or Irish) robber fly** *M. cowini* (Isle of Man. south-east Ireland, Cumbria and Worcestershire, totalling 15 records), **orange-legged robber fly** *Dioctria oelandica* (old oak woodland), and **scarce red-legged robber fly** *Di. cothurnata* (five recent records from Gwent, Dorset, Somerset and Yorkshire).

Brown heath robber fly *Machismus cingulatus*, scattered and widespread in England and Wales, in dry sandy habitats such as dunes and heathland from June to October. Adults capture prey during short flights, mainly taking smaller Diptera. Greek *mache* = 'combat', and Latin *cingulum* = 'girdle'.

Common awl robber fly *Neoitamus cyanurus*, 12–19 mm length, widespread and locally common in ancient (particularly oak) woodland, especially evident along rides in May–October. Greek *neos* + *itamos* = 'new' + 'hasty', and Latin *cyaneus* = 'dark blue'.

Common red-legged robber fly *Dioctria rufipes*, locally common and widespread in scrub and deciduous woodland in May–July, preying on insects, including parasitic wasps. Greek *dioctes* = 'pursuer', and Latin *rufus* + *pes* = 'red' + 'foot'.

Dashed striped slender robber fly *Leptogaster guttiventris*, scattered and widespread in Britain in habitats ranging from dry grassland and dunes to open woodland in sandy heathland. Adults are active in May–August. Greek *leptos* + *gaster* = 'slender' + 'stomach', and Latin *gutta* + *venter* = 'spotted' + 'stomach'.

Dune robber fly *Philonicus albiceps*, 12–20 mm length, locally abundant on dunes, including east Ireland but absent from northern Scotland. Adults are active in June–September. Greek *philos* + *neicos* = 'loving' + 'strife', and Latin *albus* + *ceps* = 'white' + 'head'.

Fan-bristled robber fly *Dysmachus trigonus*, locally common in England and Wales, evident in May–August, especially in sandy coastal areas. Greek *dys* + *mache* = 'bad' + 'combat', and Latin *trigonus* = 'triangular'.

Golden-haired robber fly *Choerades marginatus*, associated with ancient deciduous woodland in the southern half of England.

Hornet robber fly *Asilus crabroniformis*, locally common and widespread in southern England and Wales, becoming scarcer because of habitat loss. It breeds in dung on heathland and downland, larvae feeding in June–October on beetle grubs in the soil. Adults find a perch for smaller insects to fly past, which they catch on the wing, preferring dung beetles, but also eating bees and grasshoppers. *Asilus* is Latin for 'gadfly', *crabronis* = 'hornet'.

Kite-tailed robber fly *M. atricapillus*, widespread in England (common in the south and south-east) and Wales, rare in Scotland, found during summer especially in scrubby grassland, heathland, dunes and woodland edge, adults active in June–October. Latin *ater* + *capillus* = 'black' + 'hair'.

Northern robber fly *Rhadiurgus variabilis*, mostly recorded from northern Scotland, and on the Scottish Biodiversity List of species of principal importance for biodiversity conservation (2005), found on riverbanks. Most records are from the Spey valley. Flight period is June–August. *Rhadiurgus* may incorporate Greek *rhadinos* = 'slender'.

Pied-winged robber fly *Pamponerus germanicus*, generally scarce though locally common along the western coast of Britain, especially Wales, plus a few sites on the east coast of Scotland, on dunes, adults evident in May–July. Larvae are found in decomposing plant litter. Greek *pan* + *poneros* = 'all' + 'vicious', and Latin *germanicus* = 'German'.

Slender-footed robber fly *Leptarthrus brevirostris*, 7–10 mm length, locally common and widespread in Britain, found in May–August mainly on chalk grassland, though with a broader range of habitats, including woodland, in the Midlands. Latin *brevirostris* = 'short-beaked'.

Small yellow-legged robber fly *Di. lineatus*, 8–11 mm length, locally common and widespread in England and Wales in low vegetation and patches of light in woodland from May to August. *Linearis* is Latin for 'lineal'.

Spring heath robber fly *Lasiopogon cinctus*, scarce, scattered through England and Wales as far north as Cumbria, more frequently found in the south, especially in Hampshire, West Sussex and Surrey, favouring heathlands and dunes, where adults fly in May–August, becoming scarcer after June. Greek *lasios* + *pogon* = 'hairy' + 'beard', and Latin *cinctus* = 'girdle'.

Stripe-legged robber fly *Di. baumhaueri*, 8–13 mm length, common and widespread in England and Wales, adults found in hedgerows, woodland edge and grassland in May–August. Mathias Baumhauer (1759–1818) was a German entomologist.

Striped slender robber fly *Leptogaster cylindrica*, common and widespread in England and Wales, though scarce north of the Midlands, found in long grassland in May–August, preying on leaf-dwelling aphids.

Violet black-legged robber fly *Di. atricapilla*, 9–12 mm length, common and widespread in England, mostly recorded in grassland and heathland from May to July, adults perching on lower parts of the grass stem to attack then feed on smaller flies and predatory hymenopterans. Larvae often live in dung or decaying organic matter, but are also found in soil. The species is protected at 24 sites in the UK, including heathland on the Isle of Purbeck (Dorset), New Forest, Windsor Forest, Essex estuaries, Severn estuary and Thorne Moor (Yorkshire). Latin *ater* + *capillus* = 'black' + 'hair'.

ROBIN *Erithacus rubecula* (Turdidae), wingspan 21 cm, length 14 cm, weight 18 g, common and widespread in woodland, parks and gardens, with 6.7 million territories in 2009, a 55% increase in numbers between 1970 and 2015, a 17% increase between 1995 and 2015, and 12% between 2010 and 2015. Numbers did drop sharply between 2008 and 2012 when three severe winters occurred. Nests are in holes and crevices, including those in human artefacts, with 4–5 eggs, incubation lasting 14–16 days, and fledging of the altricial young taking 13–16 days. Food is earthworms, insects, seed and fruit. Robins are more readily habituated to people than on the Continent (where they are hunted), for instance following gardeners who might be revealing prey by digging, etc. The robin topped a BBC *Springwatch* 2015 poll of >200,000 people in choosing the UK's first national bird, receiving 34% of the votes. Greek *erithacos* = 'robin' and the Latin diminutive form of *ruber* = 'red'.

ROBIN'S-PLANTAIN *Erigeron philadelphicus* (Asteraceae), a herbaceous North American neophyte perennial present by 1778, casual, occasionally naturalised, on walls, banks, verges and waste ground as a garden escape in (mainly southern) England. Greek *eri* + *geron* = 'early' + 'old man', from the spring flowers turning grey.

ROBLE See southern beech.

ROCK-BRISTLE Species of moss in the genus *Seligeria* (Seligeriaceae).

Widespread and scattered species include **Donn's rock-bristle** *S. donniana* and **dwarf rock-bristle** *S. pusilla* (both mainly on shaded limestone rock), **recurved rock-bristle** *S. recurvata* (particularly on sandstone), and **sharp rock-bristle** *S. acutifolia* (calcareous rock in upland areas).

Rarities include **bentfoot rock-bristle** *S. campylopoda* (south-east Wales and south England), **Irish rock-bristle** *S. oelandica* (Benbulbin range, Co. Sligo, Cliffs of Magho, Co. Fermanagh, and Craig y Cilau NNR, Breconshire), **long rock-bristle** *S. diversifolia* (one site each in the Wolds and Inverness-shire), **short rock-bristle** *S. brevifolia* (north Wales, south Pennines and Highlands), **triangular rock-bristle** *S. patula* (south Wales, Cumbria and Co. Sligo), **trifid rock-bristle** *S. trifaria* (south Wales and Cumbria), and **water rock-bristle** *S. carniolica* (a single population in Northumberland).

Chalk rock-bristle *S. calcarea*, common in England on chalk faces and stones, occasionally also found on oolitic limestone.

English rock-bristle *S. calycina*, common in England on chalk on small, shaded stones in woodland and in chalk-pits. *Calycina* is Latin for 'cup-like'.

ROCK COOK *Centrolabrus exoletus* (Labridae), length 15 cm, a fish found among seaweed and eelgrass around the shores of Scotland and south-west England, feeding on small invertebrates. Females lay eggs in 'nests' of fine algae in rock crevices in summer. Greek *centron* + *labros* = 'sting' + 'furious', and Latin *exoletus* = 'faded'.

ROCK-CRESS Species of *Arabis* and *Arabidopsis* (Brassicaceae). *Arabis* is medieval Latin, from Greek (feminine of *araps*), and *opsis* = 'appearance'.

Alpine rock-cress *Arabis alpina*, a perennial, mat-forming montane herb of shaded ledges on basic cliffs growing at 820–50 m, discovered in 1887. The populations at the original sites on Skye have generally remained stable. A Somerset record is of a garden escape on an old wall, possibly persisting since 1900. In mid-west Yorkshire it survives on a limestone scar.

Bristol rock-cress *A. scabra*, a short-lived perennial growing in shallow soils, on scree, and on rock ledges, known from the Avon Gorge (Carboniferous limestone) since 1686. While scrub invasion and wire-netting to prevent rockfalls have eliminated some sites and reduced the open ground needed for germination, populations remain healthy. *Scabra* is Latin for 'rough'.

Hairy rock-cress *A. hirsuta*, a widespread and locally common biennial or perennial of dry rock outcrops and in grassland on base-rich substrates; also on dunes, and on bridges and walls. *Hirsuta* is Latin for 'hairy'.

Northern rock-cress *Arabidopsis petraea*, a perennial herb of open montane sites in north-west Wales and the Highlands,

on acidic and basic rocks, cliff faces, screes and sea-cliffs. It is also found on river shingle and serpentine slopes in Shetland. *Petraea* comes from Latin *petra* = 'rock'.

Sand rock-cress *Arabidopsis arenosa*, an annual European neophyte, scattered in England as a casual on waste ground. *Arenosa* is Latin for 'sandy'.

ROCK HAIR Alternative name for horsehair lichen.

ROCK-MOSS Species in the genus *Andreaea* (Andreaeaceae), the name honouring the German apothecary Johann Andreae (1724–93), associated with upland regions, mostly on wetted rocks in the Highlands, often where there is late snow-lie.

ROCK-ROSE Species in the genera *Helianthemum* (Greek *helios* + *anthemis* = 'sun' + 'flower') and *Tuberaria* (from Latin for 'humped', from the thickened rootstock) in the Cistaceae.

Common rock-rose *H. nummularium*, a prostrate subshrub mostly confined to chalk and limestone grassland in England, but extending into mildly acid pastures and heathland on well-drained soils in east Scotland, and on base-rich soils over basalt in north-east England and east Scotland. This is the county flower of Berwickshire as voted by the public in a 2002 survey by Plantlife. *Nummularium* is Latin for 'coin-shaped'.

Hoary rock-rose *H. oelandicum*, a small, prostrate, perennial subshrub often abundant in short, open, rocky Carboniferous limestone grassland, on rock and cliff edges. Subspecies *incanum* occurs in Wales and north-west England, the endemic ssp. *levigatum* occurs on Cronkley Fell, north-west Yorkshire, and ssp. *piloselloides* grows in western Ireland. *Oelandicaum* refers to Öland, Sweden.

Spotted rock-rose *T. guttata*, an autumn- and spring-germinating annual found in the Channel Islands, north-west Wales (this is the county flower of Anglesey as voted by the public in a 2002 survey by Plantlife), west and south-west Ireland, and the Scottish island of Coll, in bare patches of thin, dry soil overlying hard igneous rock in open areas within heathland near the sea. In Ireland it sometimes grows on sites burned the previous year. It is typically found in lichen-rich communities. *Guttata* is Latin for 'spotted'.

White rock-rose *H. apenninum*, a perennial subshrub of dry rocky limestone grassland, generally on south-facing slopes in south Devon and Somerset. Latin *apenninum* = 'from the Apennines'.

ROCK TRIPE Foliose lichens in the genus *Umbilicaria* (Umbilicariaceae) (Latin *umbilicus* = 'navel'), often common on exposed siliceous rocks and boulders, particularly in the Highlands.

ROCKET Mainly annual species in the family Brassicaceae, particularly in the genus *Sisymbrium* (*sisymbrion* being Greek for several plants) but also in other genera. See also wall-rocket.

Uncommon species introduced with birdseed or wool shoddy, casual and sometimes naturalised on waste ground, include **eastern rocket** *S. orientale* (a European neophyte present by 1739), **false London rocket** *S. loeselii* (European, introduced by 1787, found in and around London and in the north-east Midlands), **French rocket** *S. erysimoides*, and **garden rocket** *Eruca vesicaria* (an archaeophyte).

Hairy rocket *Erucastrum gallicum*, a European neophyte, widespread and scattered, occasionally persisting along tracks on chalk soils, but usually a casual of roadsides and waste ground, though having spread on Salisbury Plain as a result of disturbance from army activity. The binomial is Latin for 'resembling *Eruca*' (see garden rocket above), and for 'French'.

London rocket *S. irio*, a European neophyte present since at least the mid-seventeenth century, frequent in London after the Great Fire of 1666 but absent between the early nineteenth and mid-twentieth centuries, now occasionally naturalised in waste ground, in pavement cracks and on roadsides, banks and walls, but more frequently found elsewhere, mostly in England, as a casual, sometimes with grain imports and historically as a wool alien. *Irio* is Latin for a kind of cress.

Perennial rocket *S. strictissimum*, a herbaceous perennial European neophyte first recorded in 1658, a garden escape in churchyards and on walls and waste ground, where it is naturalised, sometimes casual, in a few places in England, north-east Scotland and north-east Ireland. Latin *strictissimum* = 'very erect'.

Sea rocket *Cakile maritima*, common and widespread on sandy seashores and on fore-dunes, often frequent along the winter storm tideline where there is a rich source of nutrients. It is occasional on shingle beaches, and is sometimes found as a casual elsewhere. Seeds are dispersed by tides. *Cakile* is an old Arabic name for this plant.

Tall rocket *S. altissimum*, a neophyte present by 1768 but remaining relatively rare until populations were reinforced by plants accidentally brought in by troops returning from continental First World War battlefields, where it was abundant. It is now widespread, though generally common only in England, naturalised or casual on waste ground, and by roads and railways. It is a contaminant of bird- and grass seed. *Altissimum* is Latin for 'very tall'.

Wild rocket As a salad plant, another name for perennial wall-rocket.

ROCKLING Species of ling fish, family Lotidae. **Northern rockling** *Ciliata septentrionalis*, **four-bearded rockling** *Enchelyopus cimbrius* and **three-bearded rockling** *Gaidropsarus vulgaris* all have a few records from around the British coastline.

Five-bearded rockling *Ciliata mustela*, length up to 25 cm, found around the coasts of Britain and south-east Ireland in shallow water down to 20 m, usually over sand and under intertidal rocks. Diet is mainly of small crustaceans. Spawning in winter and spring is offshore, hatched young moving inshore in early summer. 'Five-bearded' refers to the number of barbels. Latin *cilium* = 'eyelash' (= hair-like, probably a reference to the barbels), and Latin *mustela* = 'weasel'.

Shore rockling *Gaidropsarus mediterraneus*, length up to 25 cm, recorded from south-west and north-east England, north Wales and west Ireland, but probably distributed all around Britain and Ireland, in pools on rocky shores and on sublittoral rocky ground, favouring sites with algal or sea-grass cover, feeding on small crustaceans and worms.

RODENT(IA) From Latin *rodere* = 'to gnaw', the order of mammals with teeth adapted for gnawing using two pairs of continuously growing incisors.

RODGERSIA *Rodgersia podophylla* (Saxifragaceae), a rhizomatous perennial, planted in damp habitat by ponds and streams, sometimes spreading vegetatively, with a scattered distribution, especially in England. Admiral John Rodgers was a nineteenth-century US Naval officer who led an expedition to Japan where the plant was discovered.

ROESLERSTAMMIIDAE Family name for the copper ermel, honouring the German entomologist Josef von Röslerstamm (1787–1866).

ROLLER (BIRD) *Coracias garrulus* (Coraciidae), an accidental from Europe, with recent records including Pembrokeshire (2005), Cleveland (2006), south Wales (2007), Barra, Outer Hebrides (2013), and Kent, Cambridgeshire and Co. Cork (2014). Greek *coracias* = a type of crow, and Latin *garrulus* = 'chattering' or 'noisy'.

ROLLER (MOTH) Members of the genus *Ancylis* (Tortricidae, Olethreutinae) (Greek *agcylis* = 'hook', from the sickle-shaped forewing), wingspan generally 12–16 mm.

Scarce species include **dark roller** *A. upupana* (birch woodland and heathland in parts of the southern half of England), **fen roller** *A. paludana* (parts of East Anglia), **festooned roller** *A. geminana* (fen, marsh, damp heathland and mosses), **Rannoch roller** *A. tineana* (scrub and birch woodland in north-east

Scotland, in 2006 discovered at Whixall Moss, Shropshire, and recently at Fenn's moss, Denbighshire).

Aspen roller *A. laetana*, local in woodland, gardens, orchards and parks throughout much of Britain, though mainly in south and south-east England, and recorded only at one site in Ireland (Co. Kerry). Adults fly at night in May–June. Larvae feed on leaves of aspen, occasionally black poplar. *Laetus* is Latin for 'pleasing'.

Bilberry roller *A. myrtillana*, local on moorland, damp heathland and mosses throughout much of Wales, the north Midlands, northern England, Scotland and Ireland. Adults fly in afternoon sunshine and at night in May–July. Larvae feed on leaves of bilberry (*Vaccinium myrtillus*, giving the specific name) and bog-bilberry.

Bridge roller *A. uncella*, wingspan 15–20 mm, widespread but local on heathland, moorland and mosses. Adults fly in May–June. Larvae feed on bell heather, cross-leaved heath and birch. *Uncus* is Latin for 'hook', echoing the genus etymology.

Broken-barred roller *A. unguicella*, wingspan 12–18 mm, widespread but local on especially upland moorland and heathland. Adults fly in May–July. Larvae feed on heather and bell heather. *Unguis* is Latin for 'claw', from the forewing shape.

Buckthorn roller *A. unculana*, scattered in woodland, downland and scrub throughout much of southern England, East Anglia and the Midlands. Adults fly in June–July. Larvae feed on buckthorn and alder buckthorn leaves. *Unculus* is Latin for 'small hook', from the forewing shape.

Common roller *A. badiana*, widespread and common in woodland, and on moorland, downland and waste ground. With two generations, adults fly in the late afternoon into the evening in April–May and July–August. Larvae feed on leaves of vetch, vetchling and clover. *Badius* is Latin for 'chestnut-coloured', from the forewing markings.

Creeping willow roller *A. subarcuana*, local in dry pasture and dunes at a few sites in the British Isles. Adults fly, possibly in two generations, from May to August. Larvae feed on creeping willow leaves. *Subarcuana* is Latin for 'somewhat bowed'.

Hook-tipped roller *A. apicella*, local in woodland and fen in parts of southern England and south Wales, rare in northern England. The bivoltine adults fly in May–June and July–August. Larvae feed on buckthorn and alder buckthorn leaves. *Apicis* is Latin for 'of a tip' (apex), from the forewing shape.

Little roller *A. comptana*, wingspan 11 mm, local on chalk downland and limestone cliffs in parts of England and Wales. The bivoltine adults fly in April–June and July–September. Larvae feed on leaves of wild strawberry, cinquefoil, salad burnet and related plants. *Comptus* is Latin for 'adorned'.

Red roller *A. mitterbacheriana*, wingspan 11–14 mm, widespread and common in deciduous woodland, less common in the north and in Ireland. Adults fly in the evening into the night in May–June. Larvae feed on leaves of oak and beech. Ludwig Mitterpacher (1734–1814) was a Hungarian naturalist.

Small buckthorn roller *A. obtusana*, local in woodland, hedgerows and heathland throughout much of the southern half of England, and in Co. Clare, Ireland. Adults fly in May–July, often during the afternoon. Larvae feed on buckthorn and alder buckthorn leaves in September–October. *Obtusus* is Latin for 'blunt', the forewing being less sickle-shaped than congenerics.

Small festooned roller *A. diminutana*, wingspan 12–18 mm, with a scattered distribution in some woodlands in England and Wales, plus one record from Ireland (Co. Roscommon). Adults fly, possibly in two generations, from May to August. Larvae feed on leaves of eared and grey willow. *Diminutana*, Latin for 'diminutive', reflects its small size (for moths generally, not for the genus).

Triangle-marked roller *A. achatana*, wingspan 14–18 mm, widespread and common in hedgerows, thickets and copses expanding its range throughout much of the southern half of England, with records north to Yorkshire, in Wales and less frequently in Ireland. Adults fly at night in June–July. Larvae feed on hawthorn and blackthorn leaves. *Achates* is Latin for 'agate', from the colour of forewing markings.

ROLLRIM Fungi (Boletales) in the genera *Paxillus*, Latin for 'peg' (Paxillaceae), and *Tapinella* (*tapis* = 'carpet') (Tapinellaceae), with inrolled cap margins, especially in young specimens.

Alder rollrim *P. rubicundulus*, widespread but scarce, found on soil under alder.

Brown rollrim *P. involutus*, poisonous, widespread and common, found in late summer and autumn on soil under birch and other broadleaf trees and on heathland. *Involutus* is Latin for 'inrolled'.

Oyster rollrim *T. panuoides*, inedible, saprobic, scattered and uncommon in Britain, scarce in Ireland, found in autumn on decaying conifer wood. The specific name means similar to a species of *Panus*, this in turn being Latin for 'swelling' or 'tumour', the morphology of the young fruiting body as it emerges.

Velvet rollrim *T. atrotomentosa*, inedible, saprobic, widespread but uncommon in England and Wales, scarce in Ireland, locally common in the Highlands, found in June–November on stumps and other dead wood of pines. Latin *ater* + *tomentosa* = 'black' + 'hairy'.

ROOK *Corvus frugilegus* (Corvidae), wingspan 90 cm, length 45 cm, weight 310 g, common and widespread, with 990,000 pairs (860,000–1,130,000 pairs) estimated in Britain in 2009. Data suggesting a decline in numbers by 5% between 2010 and 2014 is statistically non-significant. Roosting and nesting is communal in trees, with flocks in pasture and other open habitats. The stick nests contain 3–4 eggs, incubation taking 15–17 days, the altricial young fledging in 32–4 days. Diet is of invertebrates, especially beetles and earthworms, cereal grain, small vertebrates and carrion. 'Rook' has evolved from Middle English *rok* or *roke*, in turn from Old English *hrōc*. *Corvus* is Latin for 'raven', *frugilegus* for 'food-gathering'.

ROOK WORM Alternative name for the larva of common cockchafer.

ROOT-MAGGOT FLY Members of the family Anthomyiidae.

ROPE Dead man's rope Alternative name for the seaweed mermaid's tresses.

RORQUAL Baleen whales (Balaenopteridae) with 25–100 parallel, pleated throat grooves. When these whales eat, the grooves expand, allowing them to take large gulps of water, forcing it through their baleen to filter out tiny organisms. The French word 'rorqual' derives from Norwegian *røyrkval* = 'furrow whale'.

ROSACEAE The rose (Latin *rosa*) family, which includes agrimonies, lady's-mantles, hawthorns, meadowsweets, strawberries, apples, cinquefoils, salad burnet, almond, blackthorn, cherries, plums, pears, brambles, rowan, service-trees and whitebeams.

ROSE Deciduous or occasionally evergreen, spiny shrubs in the complex genus *Rosa* (Rosaceae), from Latin *rosa* for these plants, much hybridised and selected in cultivation as ornamental garden plants, many naturalised or found as casuals. See also dog-rose, downy-rose and sweet-briar.

Often casual or naturalised east Asian neophyte garden escapes include **Japanese rose** *R. rugosa* (introduced in 1796, but not successfully grown until reintroduction in 1845, now found in hedgerows and on dunes, sea-cliffs, verges and waste ground), **many-flowered rose** *R. multiflora* (cultivated since 1804, in woodland, hedges, copses and scrub and on railway embankments and waste ground), **memorial rose** *R. luciae* (cultivated since 1880, in coastal scrub, and on cliffs, grassy banks and beaches), and North American neophytes include **prairie rose** *R. setigera* (introduced in 1800, in scrub in Jersey, Guernsey and west Kent) and **Virginian rose** *R. virginiana*.

Burnet rose *R. spinosissima*, low, suckering and deciduous, widespread on coastal dunes and sea-cliffs, and also locally inland on sandy, less acidic heaths, in scrub and hedgerows on chalk and limestone, and on basic cliff ledges in upland areas. 'Burnet' is Middle English, from Anglo-French *burnete*, from *brun* = 'brown'. Latin *spinosissima* = 'very spiny'.

Field rose *R. arvensis*, deciduous, common and widespread except in north England, Scotland and west Wales where it is more scattered and local. It often climbs over other plants, growing on a variety of soils, though avoiding very acidic sites, and found on woodland edges, in clearings and along rides, on roadsides and railway embankments and in scrub and hedgerows. *Arvensis* is from Latin for 'field'.

Red-leaved rose *R. ferruginea*, a deciduous European neophyte, suckering freely to form thickets, seeds dispersed by birds, widespread and naturalised as a garden escape in woodland, hedgerows and scrub, and on roadsides, railway embankments, dunes and waste ground. Latin *ferruginea* = 'rust-coloured'.

Red rose (of Lancaster) *R. gallica*, a long-cultivated spiny deciduous European neophyte, scarce and very scattered in England and Wales as a naturalised garden escape in scrub and hedgerows, on roadsides and railway banks, and in rough grassland. This is the county flower of Lancashire as voted by the public in a 2002 survey by Plantlife. Latin *gallica* = 'French'.

Short-styled rose *R. stylosa*, deciduous, mainly found in the southern third of England (frequent in the south-west), south Wales and southern Ireland on well-drained calcareous soils including those overlying clay and sand. Tolerating slightly shaded habitats, it is found in open woodland, hedgerows, disused quarries and scrub. Latin *stylosa* = 'having styles'.

White rose (of York) *Rosa* x *alba*, a deciduous neophyte of garden origin, scarce and scattered in England as a persistent or naturalised escape in hedgerows, and on roadsides and waste ground. Latin *alba* = 'white'.

ROSE-MOSS *Rhodobryum roseum* (Bryaceae), widespread but scattered in short grassland in open woodland, heathland, dunes and chalk grassland, especially on and around yellow meadow ant nests. It is also occurs on grassy rock ledges in hilly sites, and occasionally forms almost pure swards on the sandy soils of the Breckland. Greek *rhodon* + *bryon* = 'rose' + 'moss', and Latin *roseum* = 'rosy'.

ROSE-OF-HEAVEN *Silene coeli-rosa* (Caryophyllaceae), an annual south-west European neophyte in cultivation by 1713, a scarce, scattered casual garden escape on waste ground in England. *Silene* is from the Greek for another plant (catchfly), Latin *coeli-rosa* = 'rose of heaven'.

ROSE-OF-SHARON *Hypericum calycinum* (Hypericaceae), a low-growing, rhizomatous Near Eastern neophyte shrub, widely cultivated since its introduction in 1676, and naturalised in hedgerows and on roadsides and railway banks throughout Britain, most commonly in the south. Greek *hyper* + *eicon* = 'above' + 'picture', plants in this genus having traditionally been hung above pictures to ward off evil spirits. Latin *calycinum* means having a conspicuous calyx.

ROSE WEED Holmes's rose weed *Rhodymenia holmesii* (Rhodymeniaceae), a scarce red seaweed with fronds up to 8 cm long, mainly found in south and south-west England, Wales and south-east and north-east Ireland, epilithic on sponges and soft rocks from the lower intertidal down to 25 m. Edward Holmes (1842–1930) was a British phycologist.

ROSEGILL Fungi in the genus *Volvariella* (Agaricales, Pluteaceae), the genus name referring to the volva, a cup-like structure that forms around the stem base by the remnants of the membranous tissue (universal veil) which covers the embrionic fruit body.

Angel rosegill *V. reidii* (Aberdeenshire) and **piggyback rosegill** *V. surrecta* (scattered in England) are both rare.

Silky rosegill *V. bombycina*, edible, scattered and uncommon, found from July to November on dead deciduous trunks and large branches, particularly of beech, maple and elm, often emerging from knot holes and other damaged areas high up on standing trees. Latin *bombycina* = 'silky', from the silky cap.

Stubble rosegill *V. gloiocephala*, edible, common and widespread, found from July to November, saprotrophic in nutrient-rich damp pasture, verges and old brassica and grain stubble fields, and on dung heaps, rotted straw and manured ground. Greek *gloios* + *cephale* = 'sticky' + 'head'.

ROSEMARY *Rosmarinus officinalis* (Lamiaceae), an evergreen Mediterranean neophyte shrub grown as a culinary herb since at least the fourteenth century, a throw-out or relic of cultivation, especially in southern England, close to habitation. Self- and bird-sown plants have been recorded from pavement cracks, walls and waste ground. *Rosmarinus* is Latin for this plant, derived from *ros* + *marinus* = 'dew' + 'maritime', from it growing on sea-cliffs on the Mediterranean coast.

ROSEROOT *Sedum rosea* (Crassulaceae), a rhizomatous perennial herb of sea-cliffs and moist rock ledges in at least slightly base-rich mountainous parts: while found at sea level in western Scotland and Ireland, it is usually found >300 m. In western Ireland, it occasionally occurs on coastal limestone pavement. *Sedum* is Latin for 'house-leek'.

ROSETTE Inedible fungi, mostly Polyporales.

Blushing rosette *Abortiporus biennis* (Coriolaceae), annual, widespread in England, scattered elsewhere, but generally infrequent, sporulating from spring to late summer on decomposing wood.

Roothole rosette *Stereopsis vitellina* (Stereopsidaceae), Vulnerable in the RDL (2006), in the Highlands under Scots pine. Greek *stereos* + *opsis* = 'hard' + 'appearance', and Latin *vitellina* = 'little egg yolk'.

Woolly rosette *C. pannosa*, scattered, Vulnerable in the RDL (2006) and Critically Endangered as a species of priority interest in the UK BAP (2010), numbers having greatly declined since the 1960s, found on often mossy soil in deciduous woodland, mainly under beech and oak. *Pannosus* is Latin for 'ragged'.

Zoned rosette *Podoscypha multizonata* (Meruliaceae), scattered in England and south Wales, found in late summer and autumn on woodland soil, usually under oak, and a UK BAP species of concern. Greek *podos* + *scypha* = 'foot' + 'cup'.

ROSETTE LICHEN **Coralloid rosette lichen** See under lichen.

ROSS WORM *Sabellaria spinulosa* (Sabellariidae), a widespread polychaete found on hard substrates on exposed coasts where sand is available for tube building, usually solitary but it can form thin crusts or large reefs up to several metres across and 60 cm high, generally subtidal but occasionally in the low intertidal. Latin *sabulum* = 'sand' and *spinula* = 'thorn'.

ROT Fungal diseases.

Butt rot Causing decay of the lower part of a tree. **Schweinitzii butt rot** is a disease of conifers caused by the bracket fungus Dyer's mazegill.

Dry rot *Serpula lacrimans* (Boletales, Coniophoraceae), scattered in the countryside, most records coming from England, relatively common in buildings, manifesting as a rust-brown encrusting wrinkled patch with a cream margin on dead wood, including that used in construction and sometimes spreading over walls and fabrics. *Serpula* is Latin for 'little snake', *lacrimans* for 'weeping'.

Fomes rot Alternative name, often used by forestry managers, for root rot. *Fomes* is Latin for 'tinder' or 'fuel'.

Root rot *Heterobasidion annosum* (Russulales, Bondarzewiaceae), a widespread and very common root bracket fungus found at the base of and on the roots of coniferous trees, causing acute decay. Trees become infected via wounds to the bark and by entry of spores into untreated stumps of freshly felled trees. Sporulation is generally in late summer and autumn. Greek *heteros* + *basidium* = 'different' + 'small pedestal', and Latin *annosum* = 'old' (full of years).

Wet rot Usually caused by *Coniophora puteana* (Boletales, Coniophoraceae), sometimes known as **cellar fungus**, widespread and common, especially in England, mainly found in late summer and autumn as a crust on decaying trunks and dead wood, including structural timbers, and one of the main causes of wet rot in buildings. Greek *conis* + *phoreo* = 'dust' + 'to bear'.

ROTIFER(A) From Latin for 'wheel bearer', a phylum of microscopic animals of ponds that feed on bacteria, detritus and other rotifers. The corona around the mouth moves in a way that resembles a wheel.

ROUNDHEAD Mostly inedible fungi in the genera *Leratiomyces* (honouring the French botanist Auguste-Joseph Le Rat, 1872–1910, + Greek *myces* = 'fungus') and *Stropharia* (Greek *strophos* = 'twisted') (Agaricales, Strophariaceae).

Rarities include **conifer roundhead** *S. hornemannii* (Caledonian Forest), **dune roundhead** *S. halophila* (one record on dunes in each of Norfolk, Devon, Merseyside and Co. Cork), **redlead roundhead** *L. ceres* (introduced from Australia, first recorded in Britain in 1957), and **slender roundhead** *L. squamosus* var. *thraustus*, and **wine roundhead** *S. rugosoannulata* (both on wood chip).

Blue roundhead *S. caerulea*, widespread and locally common, saprobic, mainly in alkaline areas of humus-rich beech woodland, inedible (possibly containing the hallucinogens psilobin and psilocybin), found in July–October in grass and leaf litter. *Caerulea* is Latin for 'blue'.

Dung roundhead *S. semiglobata*, common and widespread, saprobic, found in June–November on dung on commons, grazed heaths and pasture. *Semiglobata* is Latin for 'half-rounded'.

Garland roundhead *S. coronilla*, widespread and (especially in England) locally common, saprobic, found in June–November in lowland pasture and lawns. *Coronilla* is Latin for 'crown', from the crown-like pattern on the upper surface of the stem ring.

Peppery roundhead *S. pseudocyanea*, widespread but uncommon, found in July–October, saprobic in unimproved grassland, old lawns, churchyards and parkland, with a strong smell of freshly ground pepper. *Pseudocyanea* is Latin here meaning 'nearly blue'.

Smoky roundhead *S. inuncta*, scattered in England and Wales, scarce elsewhere, found in summer and autumn on grassland soil, sometimes under trees. *Inunctus* is Latin for 'oil-covered'.

Verdigris roundhead *S. aeruginosa*, common and widespread, possibly containing the hallucinogens psilobin and psilocybin, found in July–October, saprobic, mainly in grassland or leaf litter in alkaline humus-rich beech woodland and parkland. *Aeruginosus* is Latin for 'verdigris' (dark blue-green).

ROWAN *Sorbus aucuparia* (Rosaceae), a common and widespread small to medium-sized deciduous tree of woodland, cliffs, rock outcrops and rocky riversides. It can also be bird-sown from planted trees on waste ground and by railways. It avoids calcareous and heavy soils and dense shade. Also known as mountain-ash. *Sorbus* is Latin for this tree, *aucupor* = 'bird catcher', from *avis* + *capere*, describing the use of the fruit as bait for fowling.

Chinese rowan *S. glabriuscula*, a deciduous archaeophyte from China, scattered and naturalised in southern England on rough ground and cliffs, and in woodland.

False rowan *S. pseudomeinichii*, an endemic deciduous shrub or small tree known only from three specimens at Glen Catacol, Arran, on stream banks. Hans Meinich (1817–78) was a Norwegian plant collector and botanist.

RUBIACEAE From Latin *rubra* = 'red', for the red dye obtained from the roots of some plants, the name of the bedstraw family, including cleavers, madder, squinancywort and woodruff.

RUDD *Scardinius erythrophthalmus* (Cyprinidae), length 20–30 cm, found in lowland England and Ireland, scarce in Wales and Scotland, favouring clear, vegetated slow-flowing or still waters, spawning in spring, and feeding on larval and small adult insects, crustaceans and plant material. 'Rudd' reflects the archaic *rud* = 'red colour'. *Scardinius* derives from *scarus*, an unidentifiable species of fish in ancient Greece, and Greek *erythros* + *ophthalmos* = 'red' + 'eye'.

RUFF *Philomachus pugnax* (Scolopacidae), wingspan 53 cm, length 25 cm, male weight 180 g, females 110 g, a Red List (Green List in Ireland) bird of passage in spring and autumn seen while moving from Scandinavia and Central Europe to winter in Africa, particularly on the east and south coasts of Britain, and in south-east Ireland. About 800 birds overwinter, generally near the coast, and it occasionally breeds. They feed on mudflat invertebrates. The original English name (from at least 1465) is 'ree', possibly from a dialect word meaning 'frenzied'. A later name *reeve*, still used for the female, may come from the shire-reeve, a feudal officer, equating the male bird's ornamental breeding plumage to the official's robes. The name *ruff*, first recorded in 1634, comes from the exaggerated collar fashionable from the mid-sixteenth to the mid-seventeenth century, the male bird's neck ornamental feathers resembling this neckwear. Both parts of the binomial refer to the aggressive behaviour of the bird at its mating arenas: from Greek *philo* + *mache* = 'loving' + 'battle', and Latin *pugnax* = 'combative'.

RUFFE *Gymnocephalus cernua* (Percidae), length usually 12 cm, but can reach 25 cm, widespread but locally common only in the Midlands and East Anglia in slow-flowing streams, canals and lakes, feeding on insect larvae, crustaceans and small fish. Spawning is in April–May. 'Ruffe' is probably a variant on 'rough', reflecting its rough-textured scales. Greek *gymnos* + *cephale* = 'naked' + 'head', and Latin *cernua* = 'drooping'.

RUFFWORT Thallose liverworts in the genus *Moerckia* (Moerckiaceae), honouring the Danish botanist Axel Moerck (Mørch) (1797–1876).

Alpine ruffwort *M. blyttii*, found in the Highlands above 750 m in gravelly wet acidic sites, especially where there is late-lying snow. Axel Blytt (1843–98) was a Norwegian botanist.

Irish ruffwort *M. hibernica*, scarce in gravelly, calcareous sites in upland parts of Britain as well as Ireland, especially in flushes on limestone and calcareous streamsides and on tufa faces by waterfalls, and in fen and dune slacks in lowland sites. *Hibernica* is Latin for 'Irish'.

RUFOUS (MOTH) Slender-striped rufous *Coenocalpe lapidata* (Geometridae, Larentiinae), wingspan 28–34 mm, scarce and local in north and west Scotland, parts of Northern Ireland and possibly northern England. Its preferred habitat is damp heathland, upland pasture and open moorland, especially where there are rushes. Adults fly in September–October, sometimes in the afternoon, more usually after dark. Larval diet probably includes meadow buttercup. Greek *coinos* + *calpis* = 'in common' + 'a vessel for drawing water', and Latin *lapis* = 'stone', from the predominant wing colour or montane habitat.

Small rufous *Coenobia rufa* (Noctuidae, Xyleninae), wingspan 22–5 mm, local in fen, bog and damp grassland throughout much of England, Wales, south Scotland, and the southern half of Ireland. Adults fly at night in July–August. Larvae mine rush stems. Greek *coinobios* = 'living in a community', and Latin *rufa* = 'red'.

RUPPIACEAE Family name of tasselweed. Heinrich Rupp (1688–1719) was a German botanist.

RUPTUREWORT Species of *Herniaria* (Caryophyllaceae). The Latin and English names both reflect the earlier belief that the plant could cure hernias; it was also used as a diuretic in the treatment of chronic cystitis and urethritis, activated by component saponins and flavonoids.

Fringed rupturewort *H. ciliolata*, an endemic mat-forming perennial growing in the Channel Islands and Cornwall on coastal cliff slopes, dune grassland, rock outcrops, heathland, path edges and stone-faced banks. *Ciliolata* is Latin for 'having small hairs'.

Hairy rupturewort *H. hirsuta*, a scarce annual European neophyte, naturalised but now rare on waste ground and railway sidings at Burton-on-Trent, Staffordshire (introduced as a grain alien). Elsewhere in England, it is a casual originating from wool shoddy on waste ground. *Hirsuta* is Latin for 'hairy'.

Smooth rupturewort *H. glabra*, an annual or short-lived perennial found as a native in eastern England on compacted sandy or gravelly soils, with habitats often kept open by seasonal standing water or other disturbance. It is a casual in other sites, possibly as a garden escape. *Glabra* is Latin for 'smooth'.

RUSH Rhizomatous monocotyledonous members of the family Juncaceae (*Juncus* and wood-rushes *Luzula*). Species of *Juncus* (Latin for 'rush'), mostly perennials, are generally associated with wet or damp habitats. A further 30 or so introduced species have been recorded, a few persisting in a few places for a few years. See also soft-rush.

Alpine rush *J. alpinoarticulatus*, found locally in the north Pennines, and the Scottish Borders and Highlands, in open wet turf in marsh and flushes, and by lakes and streams, usually on base-rich soil, often over limestone. *Alpinoarticulatus* is from Latin for 'alpine' and 'jointed'.

Baltic rush *J. balticus*, common and widespread, in north Scotland in dune slacks and other damp areas in maritime sand, mud or peat, commonly beside estuaries. It also occurs inland in north-east Scotland on river terraces, flood plains and marsh.

Blunt-flowered rush *J. subnodulosus*, locally common in England, Wales and Ireland, and south-west and central-west Scotland, in dense stands in fen, marsh, wet meadow, dune slacks and ditches and by water, usually in peaty and base-rich conditions; it also tolerates brackish water. *Subnodulosus* is Latin for 'somewhat with nodes'.

Broad-leaved rush *J. planifolius*, an Australasian or South American neophyte, found in Ireland on the shore of Lough Truscan in 1971, new to the Northern Hemisphere, and currently found in at least 40 km^2 in Connemara, Co. Galway, on base-poor peaty soil by damp pathsides, in wet meadow and on lake shores. *Planifolius* is Latin for 'flat-leaved'.

Bulbous rush *J. bulbosus*, common and widespread, ranging from terrestrial plants to submerged and floating aquatics, found in or by water and in often seasonally wet habitats, in acidic to neutral soils. It also grows in some calcareous turloughs in the Burren.

Chestnut rush *J. castaneus*, scarce on montane bog and flushes in the Highlands. *Castaneus* is Latin for 'chestnut'.

Compact rush *J. conglomeratus*, a common and widespread rhizomatous herbaceous perennial, characteristic of fairly acid conditions in damp fields, ditches, open woodland and margins of still or running water. *Conglomeratus* is Latin for 'clustered together'.

Dwarf rush *J. capitatus*, a scarce autumn-germinating annual, very local in the Channel Islands, Cornwall and Anglesey, on bare ground with standing water in winter and droughted in summer, growing in heathland, around serpentine outcrops, on granite sea-cliff ledges and in dune slacks. *Capitatus* is Latin for 'having a head', referring to the growth habit.

Frog rush *J. ranarius*, a common and widespread annual growing in damp coastal brackish sites and sometimes inland, for example coastal mud- and sand-flats above high-water mark and the margins of saline and brackish lakes. It is also found on bare mud and waste ground associated with inland salt flashes and on highly basic lime-waste tips. *Ranarius* is Latin for 'frog-like'.

Hard rush *J. inflexus*, common and widespread in England and Wales, and in lowland Scotland and Ireland, in wet habitats by rivers, ponds and lakes, and in marsh, wet fields, ditches, dune slacks and fen, on neutral or base-rich soils, frequently on heavy clay. *Inflexus* is Latin for 'bent inwards'.

Heath rush *J. squarrosus*, common and widespread on wet peaty heath and moorland, raised and valley mires, and upland flushes on acidic substrates. *Squarrosus* is Latin for 'having scales'.

Jointed rush *J. articulatus*, common and widespread in a range of freshwater and brackish wet habitats such as damp fields, wet heathland, marsh, ditches, flushes, dune slacks, and margins of ponds, lakes and streams, avoiding the most acid soil. *Articulatus* is Latin for 'jointed'.

Leafy rush *J. foliosus*, scattered in western Britain and in Ireland, in wet fields, marsh and on the muddy edges of lakes and ponds. *Foliosus* is Latin for 'leafy'.

Pygmy rush *J. pygmaeus*, a rare annual found in seasonally wet, compacted open ground such as gateways and wheel tracks, less often in natural areas of erosion and in quarries, on serpentine heathland on the Lizard Peninsula, Cornwall.

Round-fruited rush *J. compressus*, scattered in Britain, mainly in England, rare in central Ireland, in marsh, wet grassland, often near the sea and sometimes in brackish conditions. *Compressus* is Latin for 'compressed'.

Saltmarsh rush *J. gerardii*, common and widespread, in saline habitats, mostly in upper coastal saltmarsh, but also around coastal rock pools, in spray-drenched clifftop turf and at inland saline sites. John Gerard (1545–1612) was an English horticulturalist and herbal author.

Sea rush *J. maritimus*, common and widespread on saltmarsh and saline dune slacks, except for north Scotland. It also occurs on sites subject to freshwater seepage on low rocky clifftops and stony sea loch shores. It tolerates a range of salinities and soil moisture.

Sharp-flowered rush *J. acutiflorus*, common and widespread in wet habitats on acidic soils, particularly damp meadow and pasture, marsh, bog and wet heathland, and by ditches and ponds. *Acutiflorus* is Latin for 'sharp-flowered'.

Sharp rush *J. acutus*, scarce, scattered on the coasts of the Channel Islands, north Cornwall, north Devon, Kent, Norfolk, parts of Wales, and south-east Ireland, in saline or brackish dune slacks, in upper saltmarsh, and on sandy shores and shingle banks. *Acutus* is Latin for 'sharp'.

Slender rush *J. tenuis*, a North and South American neophyte present since the 1790s, scattered and locally abundant in damp open ground by roads and lakes, on paths and in woodland and forest rides. *Tenuis* is Latin for 'slender'.

Thread rush *J. filiformis*, local and scattered but spreading, on stony, silty edges of lakes and reservoirs, usually in a fringing zone of periodically flooded wet marshy pasture. *Filiformis* is Latin for 'thread-like'.

Three-flowered rush *J. triglumis*, uncommon, local in north Wales and northern England, and in the Highlands and Hebrides, on base-rich damp rocky or gravelly montane sites, flushes and small marshes. *Triglumis* is Latin for 'three-glumed'.

Three-leaved rush *J. trifidus*, uncommon in the Highlands and Shetland on bryophyte- or lichen-rich montane sites on shallow soil or in rock crevices, one of the main angiosperms found on windswept, often almost snow-free plateau edges over 1,000 m, but it also occupies sites that are snow-covered for several months, and reaches 1,310 m in the Cairngorms. *Trifidus* is Latin for 'divided into three'.

Toad rush *J. bufonius*, common and widespread where the water table is seasonally high, including margins of ponds, lakes and streams, marsh and dune slacks. It also grows around brackish lakes and on estuarine mud- and sand-flats, and is often a weed of disturbed ground, including tracks and roadsides. *Bufonius* is Latin for 'toad-like'.

Two-flowered rush *J. biglumis*, in damp rocky or gravelly sites, flushes and marsh on base-rich but relatively competition-free habitats in species-rich localities in the Highlands and the Hebrides. *Biglumis* is Latin for 'two-glumed'.

RUSSIAN-VINE *Fallopia baldschuanica* (Polygonaceae), a climbing perennial central Asian neophyte introduced into gardens in 1894, first recorded in the wild in 1936, common in England, scattered elsewhere, increasing in abundance and range, though rarely naturalised. Its vine-like stems scramble over trees, scrub, hedges and neglected outbuildings. The Italian anatomist Gabriele Fallopia (1523–62) was superintendent of the botanical garden at Padua. *Baldschuanica* refers to Baljuan, Turkistan.

RUSSULA Common yellow russula Alternative name for ochre brittlegill.

RUST A number of plant diseases caused by pathogenic fungi mostly in the order Pucciniales, commonly manifesting as a coloured powder. Rust usually affects healthy plants and infection is limited to plant parts such as leaves, petioles, shoots, stem and fruits. Systemic infection may cause deformities such as retarded growth, witches broom, stem canker, or creation of galls. Plants with severe rust infection may be stunted or discoloured. Many rusts belong to the genera *Puccinia* and *Uromyces*.

RUST FLY Member of the family Psilidae.

RUSTGILL Inedible fungi in the genus *Gymnopilus* (Agaricales, Strophariaceae) (Greek *gymnos* + *pileos* = 'naked' + 'cap').

Common rustgill *G. penetrans*, actually uncommon, scattered, found from summer to late autumn on stumps or buried dead wood in coniferous woodland. Latin *penetrans* = 'penetrating'.

Magenta rustgill *G. dilepis*, possibly poisonous, with a few records from East Anglia and some other sites in southern England, growing in autumn on stumps or on dead wood of conifer trees, especially pines.

Scaly rustgill *G. sapineus*, possibly poisonous, infrequent, on dead conifer wood in forest floor litter in June–November. *Sapineus* is Latin for 'of pine or fir trees'.

Spectacular rustgill *G. junonius*, widespread and common, found from summer to early winter on stumps or on dead wood of usually broad-leaved, occasionally conifer trees.

RUSTIC Members of the nocturnal moth family Noctuidae. Specifically, *Hoplodrina blanda* (Xyleninae), wingspan 31–5 mm, widespread and common in a range of habitats, mainly in low-lying areas, including woodland, rough meadow and gardens, but with numbers declining by 78% over 1968–2007, a species of conservation concern in the UK BAP. Adults fly in June–August, sometimes with a second brood in October. Larvae feed on dock, plantains and chickweed. Greek *hoplon* = 'weapon' + *(Carad)rina* (i.e. differing from *Caradrina*), and the Latin for 'smooth'.

Ashworth's rustic *Xestia ashworthii* (Noctuinae), wingspan 35–40 mm, flying from late June to August, found only in mountainous north Wales on slate moorland and limestone hills, and a priority species for conservation. Larval food includes heather, bilberry and heath bedstraw. Greek *xestos* = 'polished', from the glossy forewings of some species in the genus. Joseph Ashworth discovered this species at Llangollen, Denbighshire in 1853.

Autumnal rustic *Eugnorisma glareosa* (Noctuinae), wingspan 32–8 mm, widespread and often common in heathland, moorland, rough grassland and downland, but with numbers declining by 94% over 1968–2007, a species of conservation concern in the UK BAP. Adults fly in August–September. Larvae feed on heather, bedstraws, bluebell and birch. Greek *eu* + *gnorisma* = 'well' + 'recognition mark', and Latin *glareosa* = 'gravelly', possibly from a habitat.

Black rustic *Aporophyla nigra* (Xyleninae), wingspan 40–6 mm, widespread on heathland, moorland, calcareous grassland, verges, gardens and woodland rides, but numbers declined by 75% over 1968–2007. Adults fly in September–October. Larvae feed on heather, docks, clovers and grasses. Greek *aporos* + *phyle* = 'difficult' + 'tribe', and Latin *nigra* = 'black'.

Brown rustic *Rusina ferruginea* (Xyleninae), wingspan 32–40 mm, widespread and common in broadleaf woodland, gardens, parks, calcareous grassland, heathland and moorland. Adults fly in June–July. Larvae feed on dock, plantain and vetches. Latin *russus* = 'reddish', and *ferruginea* = 'rust-coloured'.

Clancy's rustic *Caradrina kadenii* (Xyleninae), wingspan 28–34 mm, appearing as a migrant in southern England for the first time at New Romney, Kent, in 2002, recorded by Sean Clancy, hence its English name. It now regularly arrives in numbers along the south coast, mostly in September–October, occasionally as early as May, but there is as yet no evidence of breeding. Caradrina refers to a river in Albania. *Kadenii* may refer to a collector, C.G. Kaden.

Common rustic *Mesapamea secalis* (Xyleninae), wingspan 27–30 mm, widespread and common in calcareous grassland, gardens, farmland, heathland and woodland. Only separated from lesser common rustic and Remm's rustic in the 1980s, individuals are mostly recorded as an aggregate of the three species. Adults fly in July–August. Larvae mine the stems of grasses and some cereals. Greek *me* = 'not' + the genus *Apamea*, and Latin *secale* = 'rye', a food plant.

Dotted rustic *Rhyacia simulans* (Noctuinae), wingspan 45–60 mm, local in gardens, hedgerows, woodland, moorland and mountains, discontinuously across much of Britain, more scattered in Wales and Scotland. Adults fly in June–July, then aestivate in cool places such as tunnels and outbuildings, reappearing in autumn. Larval diet is unknown. Greek *rhyax* = 'stream', from the wavy forewing pattern, and Latin *simulans* = 'pretending', probably from feigning death during aestivation.

Flounced rustic *Luperina testacea* (Xyleninae), wingspan 30–5 mm, widespread and common on calcareous grassland, dunes, farmland and open woodland, more local in Scotland and Ireland. Adults fly in August–September. Larvae feed inside the roots and lower stems of grasses and some cereals. *Testacea* is Latin for 'brick-coloured'.

Heath rustic *X. agathina*, wingspan 28–36 mm, widespread if local on heathland and moorland, but with numbers declining by 95% over 1968–2007, a species of conservation concern in the UK BAP. Adults fly in September. Larvae feed on heather. Agatha is the Latinised form of the type locality Agde, a town in Provence, France.

Hedge rustic *Tholera cespitis* (Hadeninae), wingspan 34–40 mm, widespread and common in rough grassland and gardens (though despite the English name not hedgerows), but with numbers declining by 97% over 1968–2007, a species of conservation concern in the UK BAP. Adults fly in August–September. Greek *tholeros* = 'muddy', and Latin *cespitis* = 'turf', referencing the larval grass diet.

Lesser common rustic *M. didyma*, wingspan 22–30 mm, distinguished from common rustic in the 1980s, with much of its range, habitats and phenology still undifferentiated. *Didimos* is Greek for 'double', from a forewing mark.

Light-feathered rustic *Agrotis cinerea*, wingspan 33–40 mm, scarce in parts of England on calcareous grassland, quarries, sea-cliffs and shingle beaches, and in the Welsh hills. Adults fly in May–June. Larvae feed on wild thyme. Greek *agrotes* = 'of the field', and Latin *cinerea* = 'ashy', from the grey wing colour.

Mottled rustic *C. morpheus*, wingspan 32–8 mm, widespread and common in lowland habitats, including grassland, farmland, heathland, scrub, woodland and gardens, less common in Scotland and Ireland, and with numbers declining by 84% over 1968–2007, a species of conservation concern in the UK BAP. Adults fly in June–August, with another smaller brood in October in the south. Larvae feed on a number of herbaceous plants, especially stinging nettle and dandelion. Morpheus was the Greek god of dreams, used here with no entomological significance.

Neglected rustic *X. castanea*, wingspan 36–42 mm, widespread but local on heathland, moorland and bog, and with numbers declining by 76% over 1968–2007, a species of conservation

concern in the UK BAP. Adults fly in August–September. Larvae feed on heathers and cross-leaved heath. *Castanea* is Latin for 'chestnut-coloured'.

Northern rustic *Standfussiana lucernea* (Noctuinae), wingspan 36–46 mm, afternoon- and night-flying (June–September) with a scattered but declining distribution mainly around the west coast of England and Wales, and north Scotland, but also locally inland, on cliffs, quarries, scree slopes and other rocky habitats. Larvae feed on saxifrages, harebell and biting stonecrop, as well as some grasses. Maximilian Standfuss (1854–1917) was a German entomologist. Latin *lucerna* = 'lamp', from light attracting the moth.

Oak rustic *Dryobota labecula* (Xyleninae), wingspan 27–32 mm, first recorded in Jersey in 1991 and now established in the Channel Islands. First recorded in England at Freshwater, Isle of Wight, in 1999, and now established at Totland and Ventnor. It appeared in mainland Hampshire in 2005, and has been recorded on a number of subsequent occasions. Adults fly in October–December. Larvae feed on evergreen oak. Greek *drys* + *bosco* = 'oak' + 'to feed', and Latin *labecula* = 'stain'.

Remm's rustic *M. remmi*, wingspan 28–34 mm, uncommon in calcareous grassland, gardens, farmland, heathland and woodland in much of England and Wales, but only distinguished from common rustic in the 1980s, with much of its range, habitats and phenology still undifferentiated. H. Remm is an Estonian entomologist.

Rosy rustic *Hydraecia micacea* (Xyleninae), wingspan 28–45 mm, widespread and common in gardens, hedgerows, fen and woodland rides, and on rough ground and at freshwater margins, but with numbers declining by 86% over 1968–2007, a species of conservation concern in the UK BAP. Adults fly in August–October. Larvae feed in the tap roots of plants such as dock, plantain, woundwort and burdock. Greek *hydor* + *oiceo* = 'water' + 'dwell', and Latin *micacea* = 'glittering'.

Sandhill rustic *L. nickerlii*, wingspan 32–42 mm, represented by four allopatric subspecies, all restricted to sandy and rocky coasts. Subspecies *guennei* is a vulnerable RDB species known from Anglesey and other places along the coast of north Wales, and from Lancashire. Subspecies *knilli* is an endangered RDB species found along the cliffs of the Dingle Peninsula, Co. Kerry. Subspecies *leechi* is an endangered RDB species known from a single strip of shingly sand 400 m by 200 m in south-west Cornwall. Subspecies *demuthi* occurs along north Kent from Faversham to the Isle of Sheppey, and on the Essex coast north to Aldeburgh, Suffolk. Larvae feed on sand couch, saltmarsh-grass, bulbous meadow-grass and red fescue. Franz Nickerl (1813–71) was a Czech entomologist.

Six-striped rustic *X. sexstrigata*, wingspan 36–8 mm, widespread and common in water meadow, marsh, fen, damp woodland, hedgerows, gardens and downland. Adults fly in July–August. Larvae feed on a range of herbaceous plants. *Sexstrigata* is Latin for 'six-lined'.

Square-spot rustic *X. xanthographa*, wingspan 32–5 mm, widespread and common in grassland, gardens, waste ground and woodland rides. Adults fly in August–September. Larvae feed by night during winter, mainly on grasses, but also other low-growing plants. Greek *xanthos* + *graphe* = 'yellow' + 'marking', from the forewing.

Vine's rustic *Hoplodrina ambigua* (Xyleninae), wingspan 32–4 mm, following an expansion in range and numbers in recent decades now common in grassland, heathland, woodland rides and gardens throughout the southern half of England, and local in parts of Wales, south Scotland and south-west Ireland. Population increased by 433% over 1968–2007. There are two generations in May–October, with the second brood generally more numerous. Larvae feed on plantain, dock and dandelion. *Ambigua* is Latin here meaning 'doubtful', from an initial taxonomic uncertainty.

RUSTWORT 1) *Nowellia curvifolia* (Cephaloziaceae), a foliose liverwort, the commonest member of the flora that develops on decomposing logs in woodland in western Britain and Ireland, occasional in eastern Britain. John Nowell (1802–67) was an English botanist. Latin *curvifolia* = 'curved leaf'.

2) Foliose liverworts in the genus *Marsupella* (Gymnomitriaceae) (Greek *marsypos* = 'bag').

Rarities include **arctic rustwort** *M. arctica* (new to Britain in 2013), **Boeck's rustwort** *M. boeckii*, **compact rustwort** *M. condensata*, **rounded rustwort** *M. sparsifolia*, **snow rustwort** *M. brevissima* (all in the Highlands), and **western rustwort** *M. profunda* (in china-clay pits in Cornwall).

Alpine rustwort *M. alpina*, found in north-west Wales, the Lake District and the Highlands on wet rocks in sheltered montane habitats that include humid scree and snowbeds.

Funck's rustwort *M. funckii*, mainly upland, found in moist, acidic, sandy or gravelly soil, usually in open habitats, including mine spoil and on the tops of rocks. Heinrich Funck (1771–1839) was a German bryologist.

Notched rustwort *M. emarginata*, common and widespread on rocks or gravel in and by streams and rivers in upland areas, and by snowbeds. *Emarginata* is here Latin for 'with a notched margin'.

Scorched rustwort *M. adusta*, found in north-west Wales, the Lake District, the Highlands and north Ireland on the upper surface of acidic rocks, frequent in scree or where there is late snow-lie. *Adusta* is Latin for 'burnt'.

Speckled rustwort *M. sphacelata*, in upland areas in wet acidic rocks and gravels in burns and seepages, particularly by snowbeds. *Sphacelata* is Latin for 'appearing to be dead'.

Spruce's rustwort *M. sprucei*, on the upper surface of acidic rocks in upland areas, frequent in scree and sites with late snow-lie. Richard Spruce (1819–73) was an English botanist.

Stabler's rustwort *M. stableri*, found in north-west Wales, Cumbria and the Highlands on sheltered, steep, wet rock surfaces. George Stabler (1839–1910) was an English bryologist.

RUSTYBACK *Asplenium ceterach* (Aspleniaceae), a small fern growing in well-lit base-rich dry habitats such as rock crevices and old walls, with a general south-west British distribution. *Asplenium* is Latin for 'without a spleen', referring to the plant's putative medicinal properties in removing obstructions from the liver and spleen. *Ceterach* may be from Persian *chetrak* = 'fern'.

RUTACEAE Family name (Latin *ruta* = 'rue') for the garden herb rue *Ruta graveolens* and citrus trees such as orange and lemon.

RYE *Secale cereale* (Poaceae), an annual now only occasionally cultivated but sometimes growing in wheat and barley fields as a grain alien or relic of cultivation, and also found, usually as a casual, from bird- or grass seed, on roadsides, waste ground and pavements. 'Rye' derives from Old English *ryge*, of Germanic origin. *Secale* is Latin for 'rye', *cereale* = 'relating to agriculture', Ceres being the Greek goddess of farming.

RYE-GRASS Species of *Lolium* (Poaceae), the Latin for rye-grass, specifically darnel.

Flaxfield rye-grass *L. remotum*, an annual of arable land, once a weed of flax fields, now scattered and rare on waste ground and tips, introduced with wool shoddy and grain.

Italian rye-grass *L. multiflorum*, a common and widespread annual or short-lived perennial Mediterranean neophyte found in leys, on field margins, in gateways, along farm tracks, and on roadsides and rough ground. *Multiflorum* is Latin for 'many-flowered'.

Mediterranean rye-grass *L. rigidum*, a rare and declining casual annual of waste ground, docks and tips, introduced with grain, shoddy, birdseed and soya bean waste. *Rigidum* is Latin for 'rigid'.

Perennial rye-grass *L. perenne*, a common and widespread species of improved lowland pasture, ley and hay meadow, but found in many other habitats, for example downland, lawns, amenity grassland, verges and ruderal habitats, favouring fertile, heavy, neutral soils, but also growing under mildly acidic or basic conditions.

S

SABELLIDAE From the diminutive of Latin *sabulum* = 'sand', the family name of some polychaete worms. See peacock worm.

SABLE (MOTH) Members of the family Crambidae (Pyraustinae).

Diamond-spot sable *Loxostege sticticalis*, wingspan 24–9 mm, mainly a common immigrant from southern Europe, but it may also be resident of the Breckland region of East Anglia, and elsewhere in Norfolk was, for example, recorded at Cley in 2013 and Cromer in 2014. In recent years, the frequency and number of arrivals have increased, and records have occurred northwards to Shetland, with arrivals typically in late summer or early autumn. In Hampshire and on the Isle of Wight the species is most frequent on the coast, but very irregular in appearance, with a notable influx in 1995 when at least 12 were recorded, but often absent for many years. Larvae feed on mugwort and field wormwood. Greek *loxos* + *stege* = 'slanting' + 'covering', and *stictos* = 'spotted', from the forewing pattern.

Silver-barred sable *Pyrausta cingulata*, wingspan 14–18 mm, scarce, day-flying, local on chalky ground and dunes, mainly on the coast, throughout much of England, Wales and south Scotland, and at a few sites in Ireland. The bivoltine adults fly in May–June and July–August. Larvae feed on wild thyme. Greek *pyr* + *auo* = 'fire' + 'burn' (i.e. a moth that gets burned in a candle flame), and Latin *cingula* = 'girdle', from white marks on the forewing that make a continuous band when the wings are folded.

Wavy-barred sable *P. nigrata*, wingspan 14–17 mm, day-flying, local on chalk downland and limestone pavements throughout much of southern England and in the Lake District. The bivoltine adults fly in overlapping generations from April to October. Larvae feed on thyme, marjoram and other wild herbs. *Nigrata* is Latin for 'blackened'.

White-spotted sable *Anania funebris*, wingspan 20–3 mm, nocturnal, along woodland rides and edges and on limestone pavements and cliffs in parts of England and Wales, rare in Scotland and west and south-west Ireland, a priority species in the UK BAP. Adults fly using a spinning motion from late May to July. Larvae feed on leaves and flowers of goldenrod, sometimes dyer's greenweed. Greek *ananios* = 'without pain' (litotes for a pleasing appearance), and Latin *funebris* = 'funereal', from the black ground colour of the forewings with their white markings.

SADDLE FUNGUS 1) Members of the genus *Helvolla*, Latin for a kind of pot herb (Pezizales, Helvellaceae). They are probably all saprobic, but some may be mycorrhizal. The saddle-shaped cap may have a number of major undulations and many minor curled contortions.

Elastic saddle *H. elastica*, widespread but infrequent, edible but tasteless, found from early summer to autumn on soil and leaf litter in coniferous and mixed woodland, often by paths. *Elastica* is Latin for 'flexible'.

Elfin saddle *H. lacunosa*, edible but tasteless, widespread and locally common, found in autumn on nutrient-rich, often burnt soil in broadleaf and coniferous woodland. *Lacunosa* is Latin for 'having holes', a reference to elongated oval troughs in the stem surface.

Felt saddle *H. macropus*, common and widespread, inedible, found from early summer to early autumn on soil in broadleaf woodland (often under birch), occasionally coniferous woodland. Greek *macros* + *pous* = 'large' + 'foot', here meaning a long stem.

Palefoot saddle *H. leucopus*, with six records from Wales, mostly on sandy dune soil, Vulnerable in the RDL (2006). Greek *leucos* + *pous* = 'white' + 'foot'.

White saddle *H. crispa*, poisonous, widespread, especially frequent in England, found in spring and early summer on woodland soil under hardwoods, frequently in grass. *Crispa* is Latin for 'curled', from the contorted cap.

2) **Dryad's saddle** *Polyporus squamosus* (Polyporales, Polyporaceae), common and widespread, edible, annual, found from spring to early autumn, sporulating in summer, parasitic on broadleaf trees and their stumps, especially beech, sycamore and elm. When growing on trunks it forms saddle-shaped brackets. Dryads are wood nymphs in Greek mythology. Greek *poly* = *poros* = 'many' + 'pores', and Latin *squamosus* = 'scaly', from the pattern on the cap.

SADDLE OYSTER *Anomia ephippium* (Anomiidae), bivalve, shell width up to 6 cm, found on hard substrates in sheltered conditions in the low intertidal and sublittoral on the south and west coasts of Britain, north to Shetland, and on the north-east and south-west coasts of Ireland. Greek *anomos* = 'irregular', and *ephippos* = 'mounted on a horse'.

Ribbed saddle oyster *Monia patelliformis*, shell width up to 4 cm, found around Britain and in north-east Ireland attached by its byssus to rocks and hard substrates in the intertidal and sublittoral. Greek *monias* = 'solitary', and Latin for 'dish-shaped'.

SAFFLOWER *Carthamus tinctorius* (Asteraceae), an annual or biennial cultivated since at least 1551 for the oil extracted from its achenes and the red and saffron dyes from its flowers, now locally common in southern England, scarce elsewhere, derived from birdseed and found as a casual on waste places. The contraction of 'saffron flower' comes from Old French *safran*, based on Arabic *za'faran*. *Carthamus* comes from an Arabic word meaning to dye or paint; *tinctorius* is Latin for 'used in dyeing'.

Downy safflower *C. lanatus*, a southern European annual neophyte present in Tudor times, scattered in England and Wales on waste ground, derived from wool shoddy and birdseed. *Lanatus* is Latin for 'woolly'.

SAFFRON (LICHEN) **Mountain saffron** *Solorina crocea* (Peltigeraceae), foliose, widespread if local in the Highlands on rock in areas of late snow-melt. Latin *crocea* = 'saffron-coloured'.

SAFFRON (PLANT) **Meadow saffron** *Colchicum autumnale* (Colchicaceae), a cormous herbaceous perennial, local in southern and central England, scattered and scarce elsewhere, naturalised in Scotland, on damp grassland, but most frequently found in clearings and rides within woodland, as it is toxic to livestock and often destroyed when found in grazed pasture. Popular in gardens, it often becomes naturalised when discarded into suitable habitat. 'Saffron' dates to Middle English, from Old French *safran*, based on the Arabic *za'faran*, ultimately (perhaps contentiously) from Persian *zarparan* = 'yellow stigmas'. The binomial is Latin for 'from Colchis', a region in Turkey bordering the Black Sea, and 'autumnal'.

SAGARTIIDAE Family name of sea anemones, with eight species in five genera in British waters.

SAGE 1) *Salvia officinalis* (Lamiaceae), a woody south European perennial neophyte, much cultivated as a culinary herb, scattered in England and Wales as a casual garden escape on rough ground. 'Sage' derives from Old French *sauge*, from the Latin for this plant, *Salvia*, derived from *salvis* = 'saving' or 'healing', from assumed medicinal properties, also reflected in *officinalis*, meaning having beneficial pharmacological properties.

2) Also in the Lamiaceae, species of *Phlomis* (Greek for mullein, perhaps due to the similarity of the leaves) and *Teucrium* (named after Teucer, a Trojan king who used the plant as a medicine).

Jerusalem sage *P. fruticosa*, an evergreen Mediterranean neophyte shrub, naturalised in a few places on rough ground and sea-cliffs in England and south Ireland. Latin *fruticosa* = 'shrubby'.

Turkish sage *P. russeliana*, a herbaceous perennial neophyte naturalised in a few places in England on roadsides railway banks and other rough ground.

Wood sage *T. scorodonia*, a rhizomatous perennial herb, common and widespread except in central-east England and central Ireland, on well-drained, acidic to mildly calcareous mineral soils in habitats that include woodland, hedgerows, scrub, heathland, limestone grassland and pavement, mountain ledges, dunes and shingle, and among bracken. *Scorodon* is Greek for 'garlic'.

SAILOR BEETLE So called because of their bright colours, species of *Cantharis* (Cantharidae), Latin for 'Spanish fly'. See also soldier beetle.

Grey sailor beetle *C. nigricans*, 7–11 mm long, common and widespread, adults preying on other insects on flowers. Larvae prey on ground-dwelling invertebrates. The common name comes from the greyish down on the wing cases (elytra). *Nigricans* is Latin for 'black'.

Rustic sailor beetle *C. rustica*, 9–13 mm long, common and widespread in England and Wales, scattered in Scotland, among tall grasses and in open woodland in May–June hunting other insects on flowers. Larvae prey on ground-dwelling invertebrates.

SAINFOIN *Onobrychis viciifolia* (Fabaceae), a perennial herb locally frequent in a dwarf form in unimproved chalk grassland in southern and eastern England. Variants introduced in the seventeenth century and widely cultivated for fodder until the nineteenth century have been naturalised on grassy banks, roadsides and tracks on calcareous soils throughout Britain. The species is increasing as a constituent of wild-flower mixtures and as a contaminant of grass seed. 'Sainfoin' dates from the mid-seventeenth century, coming from Old French *saintfoin*, from modern Latin *sanum foenum* = 'wholesome hay', a reference to supposed medicinal properties. Greek *onos* + *brycho* = 'ass' + 'to eat (noisily)', referring to its use as a forage plant, and Latin *viciifolia* meaning having leaves like vetch (*vicia*).

SALAD Salmon salad *Guepinia helvelloides* (Auriculariales), an uncommon fungus scattered in England and south Wales, on woodland soil often in grass, and in parkland.

SALAMANDRIDAE Family name for newts (Greek *salamandra* = 'salamander').

SALICACEAE Family name for willow (Latin *salix*), including sallows and poplars.

SALLOW (MOTH) Some members of the Noctuidae, mostly in the subfamily Xyleninae, all nocturnal. Specifically, *Cirrhia icteritia*, wingspan 27–35 mm, widespread and common in broadleaf woodland, fen, heathland, moorland and gardens, but with numbers declining by 85% over 1968–2007, a species of conservation concern in the UK BAP. Adults fly in September–October. Larvae initially feed on sallow and poplar catkins, then move to a variety of herbaceous plants. 'Sallow' comes from Old English *salo* = 'dusky', of Germanic origin. Greek *cirrus* = 'curl', and *icteros* = 'jaundice', from the yellow forewing.

Angle-striped sallow *Enargia paleacea*, wingspan 40–60 mm, scarce, found in birch woodland and heathland in the Midlands, northern England and Scotland, with one (post-2000) record from Ireland in Co. Fermanagh. Also a migrant on the southern coast of England. Adults fly in July–August. Larvae feed on birches and aspen. Greek *enargeia* = 'bright appearance', and Latin *palea* = 'chaff', from the yellow forewing.

Barred sallow *Tiliacea aurago*, wingspan 27–32 mm, common in broadleaf woodland, mature hedgerows, scrub, downland and gardens throughout much of England, plus a few records from Wales. Adults fly in September–October. Larvae feed on the buds then flowers and leaves of beech and field maple. *Aurago* is Latin for 'affinity with gold', from the forewing colour.

Bordered sallow *Pyrrhia umbra* (Heliothinae), wingspan 27–35 mm, local on dunes and shingle beaches along the coasts of England, Wales, east Scotland and east Ireland, and inland on calcareous grassland in southern England and on the Burren. Adults fly in June–July. Larvae feed on flowers and seeds of restharrows, henbane and sea sandwort. Greek *pyrrhos*

= 'flame-coloured' and Latin *umbra* = 'shadow', from the shaded part of the forewing.

Centre-barred sallow *Atethmia centrago*, wingspan 32–6 mm, widespread and common in broadleaf woodland, hedgerows and gardens. Adults fly in August–September. Larvae feed on ash, initially mining the buds then moving to the flowers and leaves. Greek *a* + *ethmos* = 'not' + 'sieve' (i.e. 'unspotted'), and Latin *centrum* = 'centre'.

Dusky-lemon sallow *C. gilvago*, wingspan 32–8 mm, local in broadleaf woodland, hedgerows and parks throughout much of England, the Welsh Borders, and northwards to central Scotland, but with numbers declining by 94% over 1968–2007, a species of conservation concern in the UK BAP. Adults fly in August–October. Larvae feed on the seeds, flowers and later the leaves of wych and English elm from April to June. *Gilvago* is from Latin *gilvus* = 'pale yellow'.

Dusky sallow *Eremobia ochroleuca*, wingspan 34–7 mm, common, mainly in south-east England and East Anglia but extending into the Midlands and up to Yorkshire, with scattered records from north-west England. Adults fly by day and night, in July–September. Habitat preferences are dry calcareous grassland, dunes and shingle beaches, where larvae feed on a number of grasses, and adults on nectar, particularly of knapweeds and ragwort. Greek *eremos* + *bioo* = 'solitary' + 'to live', and *ochros* + *leucos* = 'yellow' + 'white'.

Orange sallow *T. citrago*, wingspan 28–33 mm, common in broadleaf woodland, parks and gardens throughout much of England, more local in Wales and Scotland, and rare in Ireland (four sites on the east coast). Adults fly in August–September. Larvae feed on the leaf buds then later the leaves of species of lime. *Citrago* is from Latin *citrus* = 'yellow'.

Pale-lemon sallow *C. ocellaris*, wingspan 32–6 mm, local in plantations, verges and parks throughout much of south-east England and East Anglia. Adults fly in September–October. Larvae initially feed on the catkins of black poplar then move to a variety of herbaceous plants. *Ocellus* is Latin for 'little eye', from the forewing marking.

Pink-barred sallow *Xanthia togata*, wingspan 27–30 mm, widespread and common in broadleaf, often damp, woodland, marsh, fen, heathland and gardens. Adults fly in September–October. Larvae initially feed on catkins of sallows and poplars, later stages then taking a variety of herbaceous plants. Greek *xanthos* = 'yellow' and Latin *togata* = 'toga-wearing', from the purple stripe on both a senatorial toga and the moth's forewing.

SALLOW (TREE) The common name for broad-leaved species of willow, which in the British Isles are **great sallow** or goat willow (*Salix caprea*) and **common sallow** or grey willow (*Salix cinerea*). 'Sallow' comes from Old English *salh* = 'willow', of Germanic origin.

SALMON Atlantic salmon *Salmo salar* (Salmonidae), a large anadromous fish reaching a length of 120 cm and a weight of 30 kg, common and widespread, increasingly so as rivers become less polluted. Adults aged 1–4 years (smolts) migrate to the Atlantic, then – using the Earth's magnetic field and a 'chemical memory' of the natal river's odour – return to its headwaters (leaping up to 3 m over obstacles such as rapids, weirs and small waterfalls), not feeding, and hollow out redds in riverbed gravel in which to lay eggs in late autumn and early winter. They then generally die, though some return to the sea, ready to spawn again after one or two years. Eggs hatch into alevins, then into fry and, at around 10 cm, parr which eventually smoltify, becoming tolerant of salinity and therefore able to migrate to the sea. There they feed and grow for a further year (the grilse phase) or up to four years, before returning to their river of birth. Diet is largely of invertebrates when young, other fish in the adult stages. Salmon are important sports fish for fly angling during their annual upstream runs. Genetically distinct salmon in southern chalk rivers (Frome, Piddle, Avon, Test, Itchen) may be a subspecies. 'Salmon' comes from Middle English *samoun*, from Anglo-Norman French *saumoun*, from Latin (which also gives the genus name) *salmo* = 'salmon'. *Salar* could mean 'leaper' but more probably 'resident of salt water'.

SALMONBERRY *Rubus spectabilis* (Rosaceae), a deciduous, suckering neophyte North American shrub introduced in 1827, scattered but commoner in north-west England, Scotland and Northern Ireland, planted in gardens for ornament and as game cover in estate woodland, naturalised in woodland and hedges, often forming large thickets. *Rubus* is Latin for 'bramble', *spectabilis* for 'spectacular'.

SALMONIDAE, SALMONIFORMES Salmon (Latin *salmo*) and trout family and order, in the genera *Coregonus* (whitefish, vendace), *Oncorhynchus* (rainbow trout), *Salmo* (salmon, trout), *Salvelinus* (charr, brook trout) and *Thymallus* (grayling).

SALPINGIDAE Family name (Greek *salpinx* = 'war trumpet') of 2–4 mm long beetles, with 11 British species in 6 genera, generally found in dead wood and under bark, preying on other insects, particularly bark beetle larvae. The blind, flightless *Aglenus brunneus* is a pest of (especially mouldy) stored products.

SALSIFY *Tragopogon porrifolius* (Asteraceae), an annual or biennial Mediterranean neophyte herb, scattered in (especially south-east) England and Wales escaping from cultivation (the root being used in cooking) and naturalised or casual on sea walls, cliffs, rough grassland and verges. Greek *tragos* + *pogon* = 'goat' + 'beard', and Latin *porrifolius* = 'leek-like leaves'.

Slender salsify *Geropogon glaber*, a scarce annual Mediterranean neophyte cultivated in 1704, found in gardens, parks and waste ground, originating from birdseed and grain impurities. Greek *geron* + *pogon* = 'old man' + 'beard', and Latin *glaber* = 'smooth'.

SALTICIDAE Family name of jumping spiders (Latin *saltare* = 'to jump'), with 39 British species in 17 genera. See zebra spider under spider.

SALTMARSH-GRASS Species of *Puccinellia* (Poaceae), named after the Italian botanist Benedetto Puccinelli (1808–50).

Borrer's saltmarsh-grass *P. fasciculata*, a perennial, locally common in south-west and southern England, East Anglia and south-east Ireland, on muddy coastal sites, in saltmarsh around cattle-trampled depressions, on earth sea walls and vehicle tracks, and in Kent by salt-treated roads. William Borrer (1781–1862) was an English botanist. *Fasciculata* is Latin for 'bundled'.

Common saltmarsh-grass *P. maritima*, a common and widespread stoloniferous perennial of saltmarsh, often dominant in the lower and middle parts, and in pans in the upper marsh; also locally on bare saline soils above the tidal limit, on sea walls and by ditches, as well recently as an occasional colonist by salt-treated roads.

Reflexed saltmarsh-grass *P. distans*, a perennial locally common and widespread on the eastern half of England and Scotland, more scattered elsewhere, on muddy ground near the sea, the upper saltmarsh, sea walls and coastal rocks, as well as in saline areas inland, and as a colonist by salt-treated roads, favouring compacted, poorly drained heavy soils. *Distans* is Latin for 'separated'.

Stiff saltmarsh-grass *P. rupestris*, annual or biennial, mainly found, though declining, from the Severn estuary round to the Wash on bare saline soils above the tidal limit, on tracks and in grazing marshes, sometimes on muddy shingle and in rock crevices. It also occurs inland by saline springs in Cheshire and salt-treated roads in Kent. *Rupestris* is from Latin for 'rock'.

SALTWORT Subspecies of *Salsola* (Amaranthaceae), from Latin *salsa* = 'salted'.

Prickly saltwort *S. kali* ssp. *kali*, a woody annual of sand and shingle beaches, usually on the drift line, around the British and Irish coastline, declining in parts of south and south-west England because of disturbance caused by recreation.

Spineless saltwort *S. kali* ssp. *tragus*, a bushy annual European neophyte first recorded in 1875, scarce and scattered, mostly in England, on waste ground and arable fields, introduced with wool shoddy, birdseed and as a contaminant of grain. Usually casual, it may persist at some sites, for example on ash tips in south Essex. *Tragus* is Latin for 'rank smell' (from Greek *tragos* = 'goat').

SALVER Olive salver *Catinella olivacea* (Leotiomycetes), a fungus found on rotten wood, especially wet logs, from spring to autumn, with a scattered distribution in England and Wales. Latin *catinella* = 'small bowl', *olivacea* = 'olive-coloured'.

SAMPHIRE Herbaceous perennials, the name dating from the sixteenth century, earlier as *sampiere* from French *(herbe de) Saint Pierre* = 'St Peter('s herb)', referring to the patron saint of fishermen because the plants grow in salt-sprayed habitats.

Golden-samphire *Inula crithmoides* (Asteraceae), on sea-cliff ledges and crevices and in open turf on calcareous or base-rich rocks from Anglesey round to Suffolk, in south-west Scotland, and in south and east Ireland. In south-east England it also occurs in the lower saltmarsh, and above this on moderately organic soils, often where drift-line litter accumulates. *Inula* is Latin for elecampane (*I. helenium*). *Crithmoides* means 'like *Crithmum*' (i.e. rock samphire).

Rock samphire *Crithmum maritimum* (Apiaceae), found round most of the British coastline from the Outer Hebrides round to Suffolk, and around Ireland, on salt-sprayed crevices and ledges on sea-cliffs, coastal rocks and stabilised shingle, as well as in maritime grassland and artificial habitats such as harbour walls. Greek *crithme* = 'barley' here alludes to the seed.

SAND EEL Fishes in the family Ammodytidae (Greek *ammos* + *dytis* = 'sand' + 'diver'). Excessive fishing of sand eels in the North Sea is linked to a decline in the breeding success of puffins, kittiwakes, terns, fulmars and shags.

Greater sand eel *Hyperoplus lanceolatus*, length up to 20 cm, found around much of the coastline over sandy substrates from the shore down to 150 m, feeding on plankton, fish larvae and crustaceans. Spawning is in spring and summer. Greek *hyper* + *hoplon* = 'beyond' + 'armour', and Latin *lanceolatus* = 'lance-shaped'.

Lesser sand eel *Ammodytes tobianus*, length usually 15 cm, found from mid-tide levels over sandy shores to the shallow sublittoral to depths of 30 m, feeding nocturnally on small invertebrates. They bury themselves 20–50 cm deep in the sand during winter.

SAND-GRASS Early sand-grass *Mibora minima* (Poaceae), a rare winter-annual grass of coastal dunes in Wales, on open nutrient-poor substrates, and in the Channel Islands on gravelly cliff slopes. A Lancashire population found in 1996 probably reflects spread from north Wales. It has also been recorded from a few other dune sites in England and Scotland.

SAND HOPPER Mainly refers to *Talitrus saltator* (Talitridae), a common and widespread supralittoral amphipod up to 2 cm long, usually found beneath or among debris and decaying algae deposited at the high water mark or, during the day, buried at depths of 10–30 cm in the substrate. Latin *talitrus* = 'a rap with the finger', *saltator* = 'jumper'.

SAND MASON *Lanice conchilega* (Terebellidae), a widespread polychaete up to 30 cm long found in intertidal and subtidal sediments. Latin *concha* + *lego* = 'shell' + 'collect'.

SAND STAR *Astropecten irregularis* (Astropectinidae), a starfish up to 20 cm diameter, partly buried on sand or sandy mud from the sublittoral to 1000 m, widespread but not recorded between Yorkshire and Hampshire. Latin *astra* + *pecten* = 'star' + 'comb'.

SANDERLING *Calidris alba* (Scolopacidae), wingspan 42 cm, length 20 cm, weight 60 g (40–100) g, a passage migrant in spring and autumn and an Amber List winter visitor to the (especially sandy) coastlines of England (scarce in the south-west), Wales, south-east Scotland and Ireland, with 17,000 birds in the UK in 2008–9. A 2015–16 survey estimated 12,900 sanderlings were using non-estuarine coastal sites, an increase of 79% compared with 2006–7. They feed on marine worms, crustaceans and molluscs, rushing along the tidal margin with a distinctive action because, uniquely among British waders, they have no hind toe. 'Sanderling' comes from Old English *sand-yrðling* meaning 'sand-ploughman'. Greek *calidris*, used by Aristotle for some grey-coloured waterside birds, and Latin *alba* = 'white'.

SANDPIPER Shorebirds in the family Scolopacidae.

Baird's sandpiper *Calidris bairdii*, **broad-billed sandpiper** *C. falcinellus*, **buff-breasted sandpiper** *C. subruficollis* and **white-rumped sandpiper** *C. fuscicollis* are all scarce visitors; **least sandpiper** *C. minutilla*, **marsh sandpiper** *Tringa stagnatilis*, **sharp-tailed sandpiper** *C. acuminata*, **solitary sandpiper** *T. solitaria*, **spotted sandpiper** *Actitis macularius*, **stilt sandpiper** *C. himantopus* and **Tereks sandpiper** *Xenus cinereus* are accidentals.

Common sandpiper *Actitis hypoleucos*, wingspan 40 cm, length 20 cm, weight 50 g, an Amber List migrant breeder wintering in West Africa, found by rivers, lakes, lochs and reservoirs in Scotland, Wales, Northern Ireland and north and south-west England, with around 15,000 pairs in 2009. Numbers declined in Britain by 46% over 1975–2015, by 15% over 1995–2015, and by 7% over 2009–15. The nests, in low vegetation or among stones close to water, usually have 4 eggs, incubation taking 21–2 days, the precocial chicks fledging in 26–8 days. It is also a scarce winter visitor found along the south coast, and a passage migrant (average 100 birds) in spring and autumn found elsewhere in the UK, near freshwater sites and on some estuaries. Diet is mainly of insects caught on the ground or in shallow water. Greek *actites* = 'coast-dweller', and *hypo* + *leucos* = 'beneath' + 'white'.

Curlew sandpiper *Calidris ferruginea*, a passage visitor mostly in August and September migrating from Siberia to Africa, seen along most English coasts, the east coast of Scotland and the south and east coasts of Ireland, feeding on shore invertebrates, with around 670–740 birds recorded p.a. *Calidris* was used by Aristotle for some grey-coloured waterside birds. Latin *ferruginea* = 'rust-coloured'.

Green sandpiper *Tringa ochropus*, wingspan 59 cm, length 22 cm, weight 75 g, an Amber List wader with one breeding site in north-west Scotland (with three pairs) known in 2004–9, more commonly a winter visitor to south Britain and south Ireland (900–1,000 birds in 2008–9), and an occasional passage migrant, feeding on mudflat invertebrates. *Tringas* was a wading bird mentioned by Aristotle. Greek *ochropus* + *pous* = 'ochre' + 'foot'.

Pectoral sandpiper *C. melanotos*, wingspan 46 cm, length 21 cm, male weight 94 g, females 68 g, a scarce passage migrant from America and Siberia, the majority in late summer and autumn. Breeding was confirmed at Loch of Strathbeg, north-east Scotland, with three possibles in the Outer Hebrides, in 2004. Greek *melas* + *notos* = 'black-backed'.

Purple sandpiper *C. maritima*, wingspan 44 cm, length 21 cm, weight 65 g, a winter visitor found on most rocky coasts, commoner from Shetland to Yorkshire, the Clyde Estuary to north-west Wales, Cornwall to Dorset, and north and west Ireland, with 13,000 birds estimated in Britain in 2008–9. It is a very occasional breeder, with one secret site (two pairs) recorded in Scotland in 2010. It is Amber-listed in the UK, Green-listed in Ireland. It feeds on invertebrates, with some plant material, often on seaweed-covered rocks near the tidal margin.

Wood sandpiper *T. glareola*, wingspan 56 cm, length 20 cm, weight 65 g, an Amber List (Green List in Ireland) summer migrant, with maximum breeding numbers (27 pairs) recorded in 2007 and probably 2010, with at least 10 sites in the Highlands and one in Caithness. The ground nest usually contains

4 eggs, incubation taking 22–3 days, fledging of the precocial young 29–31 days. It is also a passage migrant in spring and autumn, breeding in northern Europe and wintering in Africa, and recorded in south and east England and around the Irish coast. Feeding is mostly on invertebrates in shallow water and mudflats. Latin *glarea* = 'gravel'.

SANDWORM **Lagoon sandworm** *Armandia cirrhosa* (Opheliidae), a scarce polychaete <8 mm long, recorded from the Keyhaven-Lymington lagoons, Hampshire, and Portland Harbour and Fleet Lagoon, Dorset. *Cirrhosa* comes from Latin *cirrus* = 'curl'.

SANDWORT Members of the family Caryophyllaceae.

Rarities include **English sandwort** *A. norvegica* ssp. *anglica* (endemic in mid-West Yorkshire), **fringed sandwort** *A. ciliata* ssp. *hibernica* (endemic in the Ben Bulben range, Co. Sligo), **recurved sandwort** *M. recurva* (Caha Mountains on the Cork-Kerry border, and the Comeragh Mountains, Co. Waterford), **Teesdale sandwort** *M. stricta* (Widdybank Fell, Upper Teesdale, Co. Durham).

Arctic sandwort *Arenaria norvegica* ssp. *norvegica*, a scarce perennial herb growing in west and north-west Scotland and west Ireland over limestone, serpentine and other basic rocks on rocks, scree, river gravel and fellfields. Latin *arenaria* = 'sandy', *norvegica* = 'Norwegian'.

Fine-leaved sandwort *Minuartia hybrida*, an annual locally common but declining in central-south England and East Anglia, scattered elsewhere in England, introduced in Ireland, on light soils in dry rocky, grassland on chalk and limestone, but more frequent in artificial habitats such as abandoned arable fields, quarries, old walls, tracks, and railway banks and sidings. Jean Minuart was an eighteenth-century Spanish apothecary and botanist.

Mossy sandwort *A. balearica*, a widespread but scattered perennial Mediterranean neophyte grown in gardens since 1787, now a well-naturalised escape on walls, paths and stony banks, often in damp, shaded sites.

Mountain sandwort *M. rubella*, a cushion-forming montane herbaceous perennial, very local in central and northern Scotland on base-rich rocks, including limestone and soft calcareous schists. *Rubella* is Latin for 'reddish'.

Sea sandwort *Honckenya peploides*, a succulent perennial with creeping stolons that make it well adapted to growing in mobile substrates, common around the British shoreline on shingle beaches and shifting sand in foreshore communities, and one of the pioneer colonists of open fore-dunes. Gerhard Honckeny (1724–1805) was a German botanist. *Peploides* means resembling petty spurge *Euphorbia peplus*.

Slender sandwort *A. leptoclados*, a winter- or rarely summer-annual, widespread and scattered, commonest in the southern half of England and south-east Ireland, on dry, shallow, neutral to basic soils in a range of open habitats such as cultivated and waste ground, old walls and quarries, and in bare places in calcareous grassland, on roadsides and railway tracks. Greek *leptos* + *clados* = 'slender' + 'branch'.

Spring sandwort *M. verna*, a perennial, basicolous, cushion-forming herb, characteristic of Carboniferous limestone, found in short grassland, on pavement and scree. It also grows on base-rich volcanic rock in north Wales and basalt in north Ireland, on metal-rich soils (including those derived from serpentine), and on lead mine spoil, for example in the Pennines. It prefers open sites with reduced competition. *Verna* is Latin for 'spring'.

Three-nerved sandwort *Moehringia trinervia*, a common and widespread annual of open, often moist ground, generally in woodland but also in shaded hedge banks, occasionally in unshaded places, for example on walls. It prefers slightly acidic substrates. Paul Möhring (1710–92) was a German physician and naturalist.

Thyme-leaved sandwort *A. serpyllifolia*, a common and

widespread winter- or rarely summer-annual of dry, shallow, neutral to basic soils, in open habitats such as rock outcrops, screes, walls, mine spoil heaps, quarries, railway ballast, waysides, arable field margins and (ssp. *lloydii*, mostly in south-west England) dunes. *Serpyllifolia* means having leaves like Breckland thyme *Thymus serpyllum*.

SANICLE *Sanicula europaea* (Apiaceae), a widespread and locally common herbaceous perennial growing on leaf mould and moist soil in deciduous woodland, and locally in hedge banks and shaded roadsides. The substrate is usually calcareous. *Sanicula* is the diminutive of Latin *sanus* = 'healthy'.

SANTALACEAE Family name for bastard-toadflax and mistletoe, from Greek, in turn derived from Sanskrit *chandana* = 'fragrant'.

SAPINDACEAE Family name for maples (including sycamore) and horse-chestnuts, from Latin *sapo* + *indicus* = 'soap' + 'from India', after soap-berry (*Sapindus*), the pulp being used as a soap.

SARCOPHAGIDAE Family name (Greek *sarcos* + *phago* = 'flesh' and 'to eat') for flesh flies, with 61 species in 16 genera, 34 species being in the genus *Sarcophaga*. They are ovoviviparous, depositing hatched or hatching maggots instead of eggs on dead insects and snails or small vertebrate carrion, dung, decaying material, or open wounds of mammals, hence their common name. Some flesh fly larvae are internal parasites of other insects. Some species, for example *Amobia signata*, are kleptoparasites, laying eggs in the nests of solitary bees and wasps: when the larvae hatch, they eat food provided by host for its own larvae.

SARCOPTIC MANGE Or **canine scabies**, a highly contagious skin disease in domestic and farm mammals and, in wildlife, foxes caused by the mite *Sarcoptes scabiei*, which burrows through the skin. It is transmissible to humans. An allergic reaction causes intense irritation, heavy scratching by the infected animal and hair loss, together leading to poor condition, suppression of the immune system, hypothermia, starvation and, often, death.

SARRACENIACEAE Family name of pitcherplant and trumpets, honouring Dr Michel Sarrazin (1659–1734), a physician and botanist in Quebec.

SATELLITE *Eupsilia transversa* (Noctuidae, Xylennae), wingspan 32–42 mm, a nocturnal moth widespread and common in broadleaf woodland, scrub, gardens and moorland. Population levels increased by 116% between 1968 and 2007. Adults emerge in September, and can be seen throughout winter in mild conditions. Larvae feed on a number of deciduous trees, and when approaching maturity can be carnivorous on larvae of other moth species, aphids, etc. Greek *eu* + *psilos* = 'well' + 'bare', and Latin *transversa* = 'lying across', from the whitish forewing band.

SATURNIIDAE Family name for emperor moth.

SAUCER BUG *Ilyocoris cimicoides* (Naucoridae), widespread and common throughout (especially south) England and Wales in weedy muddy pond shallows hunting invertebrates, tadpoles and small fish. Greek *ilys* + *coris* = 'mud' + 'bug', and Latin *cimex* = 'bug'.

SAVORY **Winter savory** *Satureja montana* (Lamiaceae), a small evergreen south European neophyte shrub used in cooking, present by the mid-sixteenth century, a naturalised garden escape on walls in south Hampshire and north Somerset, and a casual elsewhere, mainly in southern England and south Wales. The binomial is Latin for 'savory' and 'montane'.

SAVOY CABBAGE See cabbage.

SAW-WORT *Serratula tinctoria* (Asteraceae), a widespread but local herbaceous perennial in England and Wales, on calcareous grassland, hay and fen-meadow, wet heathland and heathy mire, scrub, open woodland, clifftops, verges and railway banks. Latin

serratula = 'little saw', from the serrated leaves, and *tinctoria* for a plant used as a source of a dye (in this case, yellow).

Alpine saw-wort *Saussurea alpina*, a herbaceous perennial locally common in the Highlands, occasional in north England and north Wales, on damp, base-rich cliffs, scree and other open ground, occasionally in flushed areas and riverside shingle. Nicolas-Theodore de Saussure (1767–1845) was a Swiss botanist and chemist.

SAWBILL Specialist group of ducks which have saw-tooth edges to slender hooked bills that help them grasp fish. See goosander and merganser.

SAWFLY Including members of the Tenthredinidae and Siricidae (Hymenoptera, suborder Symphyta), distinguished from most other hymenopterans by a broad connection between abdomen and thorax, and by their caterpillar-like larvae. The name comes from the saw-like appearance of the ovipositor which the females use to cut into plants to lay eggs. There are around 107 genera and about 500 species on the British list. Most are leaf miners, but some species create galls on roses and willows. Body length is often 9–10 mm. See also wasp.

Alder sawfly *Eriocampa ovata* (Tenthredinidae), mostly from southern England, larvae camouflaged as bird droppings feeding on alder leaves, adults found in May–August. Greek *erion* + *campe* = 'wool' + 'caterpillar', and Latin *ovata* = 'oval'.

Berberis sawfly *Arge berberidis* (Argidae), the first British record being in Essex in 2002, since recorded in most southern counties, and now reaching Yorkshire. Larvae feed on leaves of berberis and mahonia. Adults fly in June–August. Greek *arges* = 'bright'.

Birch sawfly *Cimbex femoratus* (Cimbicidae), the largest British sawfly, adults up to 25 mm long, birch leaf-eating larvae 45 mm, widespread but not common. Adults are evident over May–August, larvae June–September. *Cimbex* is Greek for a sawfly, *femor* Latin for 'thigh'.

Bramble sawfly *Arge cyanocrocea*, adults 8 mm long, common and widespread especially in southern England, larvae feeding on bramble, adults on nectar and pollen of umbellifers in May–July. Latin *cyaneus* + *croceus* = 'blue' + '(saffron) yellow'.

Cherry slug sawfly Alternative name for pear slug sawfly.

Curled rose sawfly *Allantus cinctus* (Tenthredinidae), 6–9 mm, locally abundant mainly in England. Larvae feed on leaves of roses and strawberry. Adults feed on umbellifers in May–August. Greek *allas* = 'sausage', and Latin *cinctus* = 'belt'.

Geranium sawfly *Ametastegia carpini* (Tenthredinidae), common and widespread, larvae feeding on geranium leaves, adults evident in summer. Greek *ameta* + *stege* = 'immediate' + 'covering', and Latin *carpinus* = 'hornbeam'.

Gooseberry sawfly *Nematus ribesii* (Tenthredinidae), 6–9 mm long, common and widespread, adults evident in April–September in gardens and allotments where larvae feed on gooseberry and currant leaves. Greek *nema* = 'thread', and Latin *ribes* = 'currant'.

Hawthorn sawfly *Trichiosoma tibiale* (Cimbicidae), occasional in England and Wales, larvae feeding on hawthorn and other species. Adults fly in May–July. Greek *trichos* + *soma* = 'hairs' + 'body', and Latin *tibiale* = 'pipe-like'.

Large rose sawfly *Arge pagana*, with a similar distribution, abundance and life history to rose sawfly. *Pagana* is Latin for 'rustic'.

Oak slug sawfly *Caliroa annulipes* (Tenthredinidae), scattered in England and Wales. Larvae resemble slugs, and feed on the leaves of oak and lime, mainly in autumn. Adults are evident in May–August. Greek *cala* + *rhoe* = 'beautiful' + 'to flow', and Latin *annulus* + *pes* = 'ring' + 'leg'.

Pear slug sawfly *Caliroa cerasi*, 4–6 mm long, widespread. Larvae resemble slugs, and feed on the leaves of various deciduous trees, including pear, cherry and hawthorn, often becoming pests.

They cover themselves in slime, making them unpalatable to predators. Adults fly in June–July. *Cerasus* is Latin for 'cherry'.

Pine sawfly *Diprion pini* (Diprionidae), widespread and locally common, often a pest in pinewoods, larvae eating pine needles. Greek *di* + *prion* = 'two' + 'saw'.

Plum sawfly *Hoplocampa flava* (Tenthredinidae), 4–6 mm long, widespread. Larvae mine leaves of plum and blackthorn. Adult appearance peaks in April–May. Greek *hoplon* + *campe* = 'tool' or 'weapon' + 'caterpillar', and Latin *flava* = 'yellow'.

Poplar sawfly *Cladius grandis* (Tenthredinidae), widespread but local, with two generations between May and August. Larvae feed on poplar leaves. Greek *cladion* = 'branch' and Latin *grandis* = 'great'.

Rose sawfly *Arge ochropus* (Argidae), widespread if mostly found in south-east England, larvae feeding on rose leaves. Adults feed on nectar and pollen of plants such as hogweed and tansy in May–September. *Ochropus* is Latin for 'yellow-legged'.

Solomon's-seal sawfly *Phymatocera aterrima* (Tenthredinidae), widespread, scattered but expanding in range, mainly in England, adults evident in May–June, larvae feeding on leaves of Solomon's-seal in summer. Greek *phyma* + *ceras* = 'swelling' + 'horn', and Latin *aterrima* = 'very black'.

Turnip sawfly *Athalia rosae* (Tenthredinidae), 7–8 mm, fairly common with most records from southern England, larvae feeding on cruciferous crops, often becoming pests, adults evident in May–June, with a further generation possible. *Athales* is Greek for 'withered', referring to the larval damage.

SAWGILL Scaly sawgill *Neolentinus lepideus* (Polyporales, Polyporaceae), a scarce fungus scattered in Britain, on fences and dead pine wood. *Neolentinus* is Greek for 'new' + *Lentinus* (see tiger sawgill below).

Tiger sawgill *Lentinus tigrinus*, inedible, scarce, with a scattered distribution in England, found in summer in damp and wet habitats on decomposing wood of willow, sometimes poplar. *Lentus* is Latin for 'pliable' or 'tough'.

SAXIFRAGE Herbaceous plants, mostly perennial, in the genus *Saxifraga* (Saxifragaceae), from Latin *saxum* + *frangere* = 'rock' + 'to break', believed to indicate a medicinal use for treatment of kidney stones rather than breaking rocks apart by the roots.

Rarities include **drooping saxifrage** *S. cernua* and **Highland saxifrage** *S. rivularis* (both Highlands), **livelong saxifrage** *S. paniculata* (planted on limestone in west Yorkshire since at least 1988, and noted as a relic of cultivation in Lanarkshire in 1992), **marsh saxifrage** *S. hirculus* (upland northern England, Scotland and Co Mayo). (Not related to peppered-saxifrage. **Mountain meadow saxifrage** is an alternative name for moon carrot. Both of these species are in the Apiaceae.)

Alpine saxifrage *S. nivalis*, rhizomatous, local on damp, shady, base-rich rocks and cliffs, mainly in the Highlands, but also Cumbria, north-west Wales and north-west Ireland. *Nivalis* is from Latin for 'snow'.

Celandine saxifrage *S. cymbalaria*, usually annual, introduced from the Mediterranean in 1880, in damp shaded habitats, with a scattered distribution as a garden weed and garden escape. *Cymbalaria* is Latin for 'cymbal-like'.

Irish saxifrage *S. rosacea*, stoloniferous, locally common in west and south-west Ireland, especially common among rocks in and by mountain streams, but also on cliff ledges, scree slopes and sea-cliffs, and in gullies. *Rosacea* is Latin for 'rosy'.

Kidney saxifrage *S. hirsuta*, stoloniferous, as a native found in (mainly south-west) Ireland in damp, shaded habitats such as woods, north-facing cliffs and banks, and by streams and on siliceous rocks in the mountains. Naturalised populations in Britain, often derived from garden escapes, are often found on limestone. *Hirsuta* is Latin for 'hairy'.

Meadow saxifrage *S. granulata*, rhizomatous, common and widespread in moist but well-drained, often lightly grazed, base-rich and neutral grassland; in unimproved pasture and

hay meadow; on grassy banks; rarely, on shaded riverbanks and in damp woodland; and locally naturalised as a garden escape. *Granulata* is Latin for 'granulated'.

Mossy saxifrage *S. hypnoides*, stoloniferous, widespread and locally common on damp rocks, scree and cliffs and by mountain streams. Substrates are often base-rich, but it also grows on acidic rocks. It is also a garden escape. *Hypnoides* means resembling the moss *Hypnum*.

Purple saxifrage *S. oppositifolia*, prostrate to caespitose, locally common in the west Highlands, northern England, Wales and west and north-west Ireland on open, moist but well-drained, base-rich rocks and stony ground, mainly on cliff faces, ledges, stony flushes and scree slopes. This is the county flower of Co. Londonderry/Derry as voted by the public in a 2002 survey by Plantlife. *Oppositifolia* is Latin for 'opposite-leaved'.

Pyrenean saxifrage *S. umbrosa*, introduced from the Pyrenees by 1640, scattered but widespread as a naturalised garden escape in shady or damp places by streams, on banks and among limestone rocks. *Umbrosa* is Latin for 'shady'.

Round-leaved saxifrage *S. rotundifolia*, rosette-forming, introduced from Europe by Tudor times, an occasional naturalised garden escape in woodland and by shady streams in northern England and central Scotland. *Rotundifolia* is Latin for 'round-leaved'.

Rue-leaved saxifrage *S. tridactylites*, a winter-annual, locally common (more scattered in Scotland) in dry, open habitats, for example sandy grassland, limestone pavement, cliffs and scree, and on man-made structures such as mortared walls, pavements and railway tracks. It prefers base-rich substrates, often on skeletal soils or bare rock. *Tridactylites* is Greek for 'three-fingered'.

Starry saxifrage *S. stellaris*, stoloniferous, found in wet flushes, by mountain streams or on wet rock and cliff faces, usually in base-poor soil. It is generally found between 200 and 1,000 m in northern England, north Wales, Scotland and (locally) in Ireland, but it has been recorded at 1,340 m on the top of Ben Nevis. *Stellaris* is Latin for 'starry'.

Strawberry saxifrage *S. stolonifera*, stoloniferous, introduced from the Far East in 1771, occasionally naturalised on walls and in churchyards in Cornwall, Gloucestershire and the Isle of Man. *Stolonifera* is Latin for 'bearing stolons'.

Tufted saxifrage *S. cespitosa*, cushion-forming, rare on well-drained base-rich rocks in a few sites in the Highlands and at Cwm Idwal, north Wales. It is found >600 m on mossy ledges, in crevices and on boulder-scree slopes, and is vulnerable to drought. A restocking programme begun at the Welsh site in 1978 has been successful, though population size has fluctuated. *Cespitosa* is Latin for 'tufted'.

Yellow saxifrage *S. aizoides*, found in the Highlands, less commonly in northern England and north-west Ireland, by the side of mountain streams, in stony flushes and over wet rocks, occasionally on wet scree and sand dunes. *Aizoides* means like the genus *Aizoon* (iceplant).

SAXON *Hyppa rectilinea* (Noctuidae, Xyleninae), wingspan 38–44 mm, a scarce nocturnal moth of moorland, open woodland, plantations and scrub in Scotland, the Lake District, and Co. Kerry, Ireland. Adults fly in June–July. Larvae feed on woody plants such as sallows, bramble and bilberry. The genus name is probably invented; *rectilinea* (Latin for 'straight line') describes a line in the centre of each forewing.

SCABIES Canine scabies See sarcoptic mange.

SCABIOUS Herbaceous perennials in the family Dipsacaceae. In medieval times the rough (Latin *scaber*) leaves of these plants were believed to relieve the itch of scabies and other afflictions of the skin including sores caused by bubonic plague, hence the common name. *Scabiosa* is Latin for 'infected', *scabies* = 'itch'.

Devil's-bit scabious *Succisa pratensis*, common and widespread in moist to moderately free-draining habitats, and favouring mildly acidic soils, in woodland rides, heathland,

SCABIOUS 414

grassland and mires, and on cliff ledges. The common name comes from the shape of the rhizomes which appear to have been bitten off just under the ground, traditionally by the devil who was angry that the plant could cure so many ills. Latin *succisa* = 'cut off' or 'truncated', and *pratum* = 'meadow'.

Field scabious *Knautia arvensis*, generally widespread but absent from western and northern Scotland, on light, well-drained, especially basic soils, in chalk and limestone grassland, rough pasture, hedgerows and woodland edge, and as a colonist on verges, railway embankments and waste ground, and also a weed of cultivation, especially in field edges on chalk. Christoph Knaut (1638–94) was a German botanist. *Arvensis* is from Latin for 'field'.

Giant scabious *Cephalaria gigantea*, a Caucasian neophyte scattered in England and Scotland, naturalised on rough grassy habitats. Greek *cephale* = 'head', from the clustered flower heads.

Small scabious *Scabiosa columbaria*, locally common in Britain north to central Scotland, usually on dry, relatively infertile soils. Habitats include chalk downland, hill slopes, and occasionally on cliffs and rock outcrops and in chalk and limestone quarries. *Columbaria* is Latin for 'dovelike'.

Sweet scabious *Sc. atropurpurea*, a European neophyte locally naturalised or casual in Cornwall, south-east and eastern England, and south-east Scotland on rough ground, often near the sea. *Atropurpurea* is Latin for 'deep purple'.

SCALE INSECT Members of the superfamily Coccoidea (Sternorrhyncha). Females have elongated oval bodies. **Armoured scales** are members of the family Diaspididae, with 78 species in 36 genera in Britain. **Soft scales** are members of the Coccidae, with 27 species in 12 genera. Both sexes are often wingless, legless and with greatly reduced antennae. Many are widespread plant pests.

SCALEWORT Foliose liverworts in the genera *Frullania* (Frullaniaceae), *Porella* (Porellaceae) and *Radula* (Radulaceae). Locally common species include **Carrington's scalewort** *R. carringtonii* (upland, west Scotland and the Scottish islands, and west and south-west Ireland, on wet, base-rich coastal rocks), **cliff scalewort** *P. cordaeana* (silty tree bases and rocks by lowland rivers, occasionally on base-rich, siliceous rocks), **Holt's scalewort** *R. holtii* (south-west and west Ireland and west Scotland on shaded wet rocks in caves and by waterfalls), **lesser scalewort** *F. microphylla* (western edges of Britain and Ireland usually on rock, generally on the coast), **Lindenberg's scalewort** *R. lindenbergiana* (upland rocks), and **pale scalewort** *R. voluta* (damp rock, especially by waterfalls).

Bitter scalewort *P. arboris-vitae*, widespread except for the east Midlands and East Anglia, on base-rich rock in upland areas, sheltered crags, boulders, and occasionally on ash and hazel in humid western woodlands. Lowland records are from limestone woodland and north-facing chalk grassland. Arborvitae is a species of *Thuja*, to which the specific name refers.

Broad scalewort *P. obtusata*, on base-rich igneous rock outcrops or base-rich sandstone near the west coast of Scotland, and on igneous rocks in east Scotland, sometimes just above the high-tide mark. It is also found near the coast elsewhere except east England. *Obtusus* is Latin for 'blunt'.

Brown scalewort *R. aquilegia*, in upland west Wales, the Lake District, western Scotland, the Hebrides, Orkney, Shetland, and west and south-west Ireland in shady sites, usually on wet rocks. It also occurs on coastal clifftop grassland in Shetland. *Aquilegus* is Latin for a drawer of water.

Dilated scalewort *F. dilatata*, common and widespread on a number of trees, especially ash, willow and poplar, and on rock and in turf, especially in coastal areas.

Even scalewort *R. complanata*, common and widespread on trees with a preference for sheltered, moist conditions. It also grows on rocks by streams and lake margins, and on sea-cliffs. *Complanata* is Latin for 'flattened'.

Pinnate scalewort *P. pinnata*, occasional in south-west England, Wales and Ireland in non-calcareous rivers on rocks and tree roots, submerged when water levels are high.

Sea scalewort *F. teneriffae*, common on the west coast of Britain and Ireland on rocks and trees.

Spotty scalewort *F. fragilifolia*, common in western Britain and Ireland, and in southern England, usually on base-enriched rocks, and in mountainous areas, occasionally on bark.

Tamarisk scalewort *F. tamarisci*, abundant in western Britain and in Ireland, more scattered in eastern England, on rocks and trees, and also in (especially coastal) turf.

Wall scalewort *P. platyphylla*, common and widespread on calcareous cliffs and boulders, and on the tops of old walls, including in the middle of towns. In lowland England it is often found on the base of ash trees in ancient woodland and on the base of ash and beech on chalky banks. *Platyphylla* is Greek for 'broad-leaved'.

SCALLOP (BIVALVE) Edible filter-feeding shellfish in the family Pectinidae. 'Scallop' dates to Middle English, a shortening of the Old French *escalope*.

Great scallop *Pecten maximus*, shell size up to 15 cm, common and widespread, free-living on the surface of sand and fine gravel from sublittoral to mid-shelf depths. In 2010, the UK fleet landed 43,000 tonnes of scallops, worth £54 million, mostly by dredging. *Pecten* is Greek for 'scallop', *maximus* Latin for 'greatest'.

Humpback scallop *Talochlamys pusio*, shell size up to 4 cm, low in the intertidal, sublittoral and nearshore shelf, first byssally attached to rocks and on rough ground, later cemented to rocks, cobbles or dead shells. Greek *talas* + *chlamys* = 'poor' + 'cloak', and Latin *pusio* = 'small child'.

King scallop Alternative name for great scallop.

Tiger scallop *Palliolum tigerinum*, shell size up to 3 cm, widespread on coarse sediments from the low intertidal to moderate shelf depths. *Pallium* is Latin for 'mantle'.

Variegated scallop *Mimachlamys varia*, shell size up to 6.5 cm, widespread from the low intertidal to moderate shelf depths, generally free-living. Greek *mimos* + *chlamys* = 'mimic' + 'cloak'.

SCALLOP (MOTH) **Brown scallop** *Philereme vetulata* (Geometridae, Larentiinae), wingspan 24–30 mm, locally associated with habitats on calcareous soils (woodland, scrub, hedgerows and fen) in southern England and south Wales, north to Cumbria, and a few recent records from Ireland. Adults fly at night in June–July. Larvae feed on the buckthorn leaves. Greek *phileremos* = 'fond of solitude', possibly because the larva secludes itself within spun leaves, and Latin *vetulata* = 'oldish', perhaps because lines on the forewing suggest wrinkles.

SCALYCAP Inedible saprobic fungi, mostly in the genus *Pholiota* (Agaricales, Strophariaceae) (Greek *pholis* = 'scale', as found on the cap).

Widespread but locally common or uncommon species include **alder scalycap** *P. alnicola* (damp shaded deciduous woodland and carr, usually on hardwood deadwood, especially alder, birch and willow), **bonfire scalycap** *P. highlandensis* (burnt ground), **flaming scalycap** *P. flammans* (decaying conifer stumps and logs), **golden scalycap** *P. aurivella* (hardwood deadwood, especially beech), **shaggy scalycap** *P. squarrosa* (base of broadleaf trees, mostly beech), **sticky scalycap** *P. gummosa* (stumps and buried decaying wood), and **hedgehog scalycap** *Phaeomarasmius erinaceus* (woody debris in damp deciduous woodland, especially under willow).

Conifer scalycap *P. astragalina*, very scattered, from the New Forest to the Highlands, Critically Endangered in the RDL (2006) and a UK BAP species, found on decaying coniferous wood. *Astragalinos* is Greek for 'goldfinch'.

SCAPANIACEAE Family name of some foliose liverworts (Greek *scapane* = 'small digging tool').

SCARAB **Corrugated scarab** *Brindalus porcicollis* (Scarabaeidae, Aphodiinae), 3.3–4.3 mm long, possibly extinct, recently known only from a sandy beach at Whitesand Bay, Cornwall (RDB 1 status, Endangered). Latin *porcus* + *collum* = 'pig' + 'neck'.

SCARABAEIDAE Family name (Latin *scarabaeus* = 'beetle') for chafers and the majority of dung beetles, with at least 83 species on the British list in 7 subfamilies. **Aegialiinae:** three species, 4–5 mm, associated with decaying vegetation in dry sandy habitats. **Aphodiinae:** 53–5 species, 49 of which are dung beetles in the genus *Aphodius*. They rarely excavate tunnels, and larvae are found in surface dung or decaying vegetable matter. **Cetoniinae:** six species, none common. **Eupariinae:** *Saprosites mendax*, introduced accidentally from Australia, and the South African *S. natalensis*, both now rare, in south-east England, under bark. **Melolonthinae:** eight species of chafer that feed on plant roots. **Rutelinae:** three species of chafer. **Scarabaeinae:** seven species of *Onthophagus*, 4–11 mm, mainly found in dung, but also in carrion and rotting fungi.

SCATHOPHAGIDAE Family name of dung flies (Greek *scatos* + *phago* = 'dung' + 'to eat'), with 54 species in 23 genera. The richest fauna is in central Scotland. Adults prey on other small insects. The larvae of only five or six *Scathophaga* species actually live in dung, other larvae being leaf miners, stem borers or (sometimes aquatic) predators.

SCATOPSIDAE Family name of scavenger flies (Greek *scatos* + *opsis* = 'dung' + 'appearance').

SCAUP *Aythya marila* (Anatidae), wingspan 78 cm, length 46 cm, weight 1 kg, Britain's rarest breeding duck (a handful each year), most birds (5,200 in 2008–9) evident around the coast (mainly estuaries), with a few inland, as winter migrants from Eurasia. Over 30,000 scaup used to winter in the Firth of Forth, feeding on waste grain from breweries and distilleries until tighter sewage regulations stopped these discharges. Wintering numbers declined by 47% between 2002–3 and 2012–13. With a vulnerable European population, scaup were moved from the Amber List to Red in 2014. The diet is mainly of molluscs and crustaceans caught by diving; indeed the common name, dating from the late seventeenth century, derives from Scots and northern English *scalp* = 'mussel-bed'. *Aithuia* was an unidentified seabird mentioned by Aristotle, and Greek *marile* = 'charcoal embers'. An average of two vagrant North American **lesser scaup** *A. affinis* is recorded p.a.

SCAVENGER BEETLE **Minute brown scavenger beetle** Members of the family Latridiidae. See also water scavenger beetle under beetle.

SCAVENGER FLY **Black scavenger fly,** members of the family Scatopsidae, with 46 species in 18 genera.

SCHELLY *Coregonus stigmaticus* (Salmonidae), a glacial relict whitefish endemic to four Lake District water bodies – Brothers Water, Red Tarn and Ullswater (all with stable populations), and Haweswater (declining), with the major threats being water abstraction and predation by cormorants (now being culled). 'Schelly' is a local variant of 'skelly' for the same fish. Greek *core* + *gonia* = 'pupil of the eye' + 'angle', leading to 'angle-eyed', and *stigma* = 'mark'.

SCHEUCHZERIACEAE Family name for Rannoch-rush. Johann Scheuchzer (1672–1733) was a Swiss naturalist.

SCHRECKENSTEINIIDAE Family name for the bramble false-feather moth. R. von Schreckenstein (d.1808) was a German entomologist.

SCIARIDAE Family name (Greek *scia* = 'shadow') for dark-winged fungus gnats, with 266 species in 22 genera, commonly found in moist environments, including woodland and moorland. Adults feed on nectar of umbellifers and other flowers. Larvae feed on fungi, but also eat dead leaves and compost. Some species have become a pest in mushroom farms, and they are often found in plant pots in houses.

SCIOMYZIDAE Family name (Greek *scia* + *myzo* = 'shadow' + 'to suck') for marsh flies, with 70 species in 23 genera. Adults (2–12 mm length) are common along the edges of ponds and rivers, and in marshy areas. They drink dew, take nectar, and some species feed on gastropods. Larvae also prey on or parasitise snails and slugs.

SCIRTIDAE Family name (Greek *scirtao* = 'to leap') for aquatic marsh beetles, with 20 species of adult lengths of 1.5–6 mm. *Scirtes orbicularis* (coastal sites in the Severn estuary, south-east England and East Anglia) and *S. hemisphaericus* (widespread in England and Wales, scattered in Ireland), with enlarged hind femora, can jump in a manner similar to flea beetles.

SCIURIDAE Family name for squirrels (Latin *sciurus*), from Greek *sciouros*, from *scia* + *oura* = 'shadow' + 'tail', alluding to the animal sitting in the shadow of its tail.

SCLERACTINIA Order of hard corals, containing calcium carbonate, from Greek *scleros* + *actinos* = 'hard' + 'ray', with 13 families in British waters.

SCOLOPACIDAE Family name of wading birds (Greek *scolopax* = 'woodcock').

SCORCHED WING *Plagodis dolabraria* (Geometridae, Ennominae), wingspan 28–32 mm, a nocturnal moth widespread if local in broadleaf woodland, scrub, parks and gardens. Adults fly in May–June. Larvae feed on a variety of deciduous trees. Greek *plagios* + *eidos* = 'slanting' + 'shape', and Latin *dolabra* = 'pickaxe', from the shape of the wing markings.

SCORPION 1) **Yellow-tailed scorpion** *Euscorpius flavicaudis* (Euscorpiidae), 35–45 mm long, native to south-west Europe, probably introduced in shipments of Italian masonry, and found in dockyards at Sheerness, north Kent, with a population possibly of 10,000–15,000, dating from the 1860s. Other possible sightings are from docks at Harwich, Tilbury, Portsmouth and Southamptom. They live in cracks in old walls and old railway sleepers, feeding nocturnally on spiders, woodlice and fellow scorpions. An ambush predator, it does sting (mild for humans), but generally captures prey using the claws. Females have live births. Activity is low or absent during winter. Greek *eu* + *scorpion* = 'good' + 'scorpion', and Latin *flavus* + *cauda* = 'yellow' + 'tail'.

2) **Book scorpion** *Cheiridium museorum* (Cheiridiidae), actually a pseudoscorpion, 1.4 mm body length, an abundant and widespread synanthrope found in houses, outbuildings and bird nests close to buildings, feeding on book lice and other very small invertebrates. Greek *cheir* = 'hand'.

3) **Sea scorpion** A fish; see sea scorpion.

4) **Water scorpion** A bug; see water scorpion.

SCORPION-FLY Members of the order Mecoptera. The tip of the male's abdomen is generally turned up, giving a scorpion-like appearance. The head faces downwards like a beak with biting jaws, both larvae and adults preying on other insects. See also snow flea.

SCORPION-MOSS **Hooked scorpion-moss** *Scorpidium scorpioides* (Calliergonaceae), common and widespread, dominant in upland regions in mineral-rich flushes, and locally abundant beneath sedges and bog-rush in fen, dune slacks and coastal flushes. Less commonly, it grows on flushed upland rocks and cliff ledges.

SCORPION-VETCH Plants in the genus *Coronilla* (Fabaceae), Latin for 'small crown'.

Annual scorpion-vetch *C. scorpioides*, an introduced annual from south Europe, casual as far north as central Scotland, often from birdseed.

Scrubby scorpion-vetch *C. valentina*, an introduced Mediterranean shrub with a scattered distribution in southern and central England, especially south Devon on cliffs and rough ground. *Valentina* means from Valencia, Spain.

SCOTER Sea ducks in the genus *Melanitta* (Anatidae) (Greek *melas* + *netta* = 'black' + 'duck'). Other species on the British list are **surf scoter** *M. perspicillata* (a scarce visitor) and, not seen for many years, **black scoter** *M. americana* and **white-winged scoter** *M. deglandi* (vagrants). 'Scoter' dates from the late seventeenth century, meaning obscure but possibly an error for 'sooter', a reference to dark plumage.

Common scoter *M. nigra*, wingspan 84 cm, length 49 cm, weight 1 kg, a diving seaduck mainly breeding in small lochs in north Scotland and north-west Ireland, but whose numbers have substantially declined (52 pairs in 2007), and it is a Red List species. Migrants from Eurasia in winter, however, are found around the British and east Ireland coastline, and number around 100,000 individuals. Clutch size is 6–8, incubation lasts 30 days, and the precocial young fledge in 45–50 days. Scoters dive down to 30 m to feed on molluscs. *Nigra* is Latin for 'black'.

Velvet scoter *M. fusca*, wingspan 94 cm, length 54 cm, male weight 1.8 kg, females 1.4 kg, a winter visitor from Eurasia to the east coast, especially in Scotland, north-east England and Norfolk, numbering 2,500 individuals in 2008–9, and Red-listed as a vulnerable species. They dive to feed on molluscs as well as crabs, sea urchins, small fish and plants. *Fusca* is Latin for 'dusky brown'.

SCOTTISH WILDCAT See wildcat.

SCOUR WEED **Black scour weed** Alternative name for the seaweed landlady's wig.

SCRAPTIIDAE Family name for false flower beetles (2.5–5 mm length), with 13 extant British species. Adults are found on flowers, particularly of hawthorn, from June to August. Larvae develop in dead wood or leaf litter. Latin *scrapta* = 'prostitute'.

SCREW-MOSS Species in the genera *Stegonia* (Greek *stege* = 'roof' or 'covering', referencing strongly concave leaves), *Syntrichia* (*syn* + *trichos* = 'together' + 'hair'), *Hennediella* (honouring the Scottish algologist Robert Hennedy, 1809–77), and *Tortula* (Latin for 'small twist', referring to the peristomes), in the family Pottiaceae.

Scarce and scattered species include **clay screw-moss** *Syn. amplexa* (in England, probably introduced from North America), **dog screw-moss** *T. canescens* (declining in south-west England and Wales on ledges of disused quarries and similar places), **marble screw-moss** *Syn. papillosa* (usually on mature, particularly urban street trees, more rarely on walls, stones and tarmac), and **Stanford screw-moss** *H. stanfordensis* (shaded, bare, compacted soil, particularly by streams).

Rarities include **brown screw-moss** *Syn. princeps* (Midlands, north Wales and mainly Scotland), **hood-leaved screw-moss** *Stegonia latifolia* (Highlands) and **Norway screw-moss** *Syn. norvegica* (Highlands).

Great hairy screw-moss *Syn. ruralis*, common and widespread on calcareous substrates on walls, rocks and sandy ground, and also typically on old thatch.

Intermediate screw-moss *Syn. montana*, common and widespread in lime-rich sites such as calcareous rocks and walls and sunny, exposed, stony ground. *Montana* is Latin for 'montane'.

Sand-hill screw-moss *Syn. ruraliformis*, common around most coasts forming extensive mats on loose sand in unstable dunes, sandy coastal banks and cliffs, and sometimes on sandy roadsides and heathland. It is increasingly frequent inland in south-east England and East Anglia on concrete and on corrugated asbestos roofs.

Small hairy screw-moss *Syn. laevipila*, common and widespread on a variety of trees, occasionally on walls and rocks. *Laevipila* is Latin for 'smooth hair'.

Wall screw-moss *T. muralis*, widespread, the commonest moss on many mortared or base-rich walls. It also grows on concrete, roof tiles and other man-made structures, as well as outcrops of natural, base-rich rock, occasionally on wood. *Muralis* is from Latin for 'wall'.

Water screw-moss *Syn. latifolia*, widespread, most abundant in southern, lowland regions, and common on shaded tarmac roads in parts of western Britain. Despite its English name it typically grows on trees, rocks and walls, and also in river flood zones. *Latifolia* is Latin for 'broad-leaved'.

SCROPHULARIACEAE Family name of figwort, including butterfly-bush and mullein. Latin *Scrophularia* reflects the idea that some of these plants cured scrofula.

SCULPIN **Shorthorn sculpin** Alternative name for bull rout.

SCURVYGRASS Species of *Cochlearia* (Brassicaceae) (Greek *cochlarion* = 'spoon', referring to the shape of the basal leaves).

Common scurvygrass *C. officinalis*, biennial or perennial, common and widespread, generally in damp coastal habitats, including saltmarsh. It also grows under hedges, and is a recent occasional colonist of salted roadsides. Subspecies *scotica* grows in north Scotland and west Ireland, also on stony coastal shores, shingle spits, dunes and short, grazed grassland on clifftops and saltmarsh. *Officinalis* is Latin for having pharmacological properties.

Danish scurvygrass *C. danica*, a common and widespread winter-annual of clifftops, dunes and walls and pavements in coastal towns, preferring open ground on well-drained sandy soils or bare rock. It used to occur on railway ballast, and since the early 1980s has spread rapidly along salt-treated roads in Britain and Northern Ireland (not in the Republic, where grit is used). A preference for the central reservations of motorways and dual carriageways is becoming less apparent, and it is now common on many single-carriageway roads in England and Wales. Latin *danica* = 'Danish'.

English scurvygrass *C. anglica*, biennial or perennial, in saltmarsh on soft, silty substrates, and in firmer areas of mud near the high water mark of estuaries, widespread but scarce in Scotland.

Mountain scurvygrass *C. micacea*, a possibly endemic perennial of micaceous soils, local in flushes, by springs and on streamsides, also on ledges and cliffs and in ravines in highland Scotland.

Pyrenean scurvygrass *C. pyrenaica*, biennial or perennial, in damp, open habitats, including cliffs, wet gullies, mossy flushes and spoil heaps by old lead and zinc mines. Subspecies *pyrenaica* is only known from northern England and Skye, usually on more basic soils than ssp. *alpina* which is more widespread in northern England, western Ireland and the Highlands.

Tall scurvygrass *C. megalosperma*, an annual or short-lived perennial south-west European neophyte present by 1648, noted in the wild in Surrey in 1948; naturalised in cultivated and waste ground in Nottinghamshire since 1974; recorded by the roadside at Humbie, East Lothian, in 1994; and by the River Tweed at Tweedmill, Berwickshire, on wet rocks in 2006. *Megalosperma* is Greek for 'large-seeded'.

SCUTTLE FLY Members of the family Phoridae (Greek *phora* = 'movement'). Larvae range from parasites, through parasitoids, specialised predators and fungus feeders, to polyphagous saprophages. The adult escape behaviour of running rapidly across a surface prompts the name scuttle fly. A pronounced hump to the thorax, evident when viewed from the side, gives an alternative name for this group, hump-backed flies.

SCYPHOZOA Greek *scypha* + *zoon* = 'cup' + 'animal', class in the phylum Cnidaria that includes the true jellyfishes, with eight families in three orders in British waters.

SCYTHRIDIDAE Family name for owlet moths (Greek *scythros* = 'sullen').

SCYTHROPIIDAE Family name for the hawthorn moth (Greek *scythropos* = 'sad countenance').

SCYTODIDAE Family name of spitting spider (Greek *scytos* = 'leather').

SEA ANEMONE Members of the class Anthozoa, characterised by a sessile polyp attached at the bottom to the surface beneath it by an adhesive foot (basal disc), with a column-shaped body ending in an oral disc. The mouth, which is also the anus, is in the middle of the oral disc surrounded by tentacles armed with cnidocytes, stinging cells that are both defensive and used to capture prey. A number of other species are found in the deeper subtidal.

Beadlet anemone *Actinia equina* (Actiniidae), common and widespread in the intertidal zone, with a base up to 5 cm diameter base, moderately or firmly adhesive, with a smooth column, and up to 192 tentacles (which retract on being disturbed) arranged into six circles. Greek *actis* = 'beam', and Latin *equina* = 'equine'.

Chocolate tiny anemone Alternative name for ginger tiny anemone.

Cloak anemone *Adamsia carciniopados* (Hormathiidae), generally widespread in the sublittoral, not on the east coast, attached to hermit crabs: the base and lower part of the column form two lobes which envelop the crab and its gastropod shell. The disc and tentacles always occur beneath the belly of the crab, the span of the tentacles being up to 5 cm. The anemone gets somewhere to live and feeds on scraps of the crab's food in exchange for help in the crab's defence. As the anemone grows, it secretes a membrane (carcinoecium), which overlies the crab's original snail shell and expands the crab's living space so the anemone does not have to change substrate and the crab does not have to seek a larger shell as they both grow. Greek *carcinos* + *pados* = 'crab' + 'child'.

Dahlia anemone *Urticina felina* (Actiniidae), column base up to 15 cm diameter with up to 160 tentacles arranged in multiples of 10, found on the lower shore and subtidally, attaching to rock, typically in crevices and gullies, sometimes forming dense carpets. It also occurs in estuaries with hard substrates. *Urticina* is a diminutive of the Latin *urtica* = 'nettle', and *felina* = 'gut'.

Daisy anemone *Cereus pedunculatus* (Sagartiidae), the trumpet-shaped column up to 12 cm high, with short tentacles numbering between 500 and 1,000, arranged in multiples of 6 around the oral disc which is usually 3–7 cm but can be as wide as 15 cm, found in pools, holes and crevices, or attached to stones beneath the surface of sediments from the midshore to 50 m depth, widespread but rare on the east coast of Britain. Latin *cereus* = 'wax' and *pedunculatus* = 'with a stem'.

Elegant anemone *Sagartia elegans* (Sagartiidae), with a base up to 30 mm, with around 200 tentacles spanning 4 cm, common or locally abundant in the intertidal in pools, under stones and beneath overhangs, and in the sublittoral and deeper.

Gem anemone *Aulactinia verrucosa* (Actiniidae), base up to 25 mm diameter and broader than the column which is up to 50 mm tall; up to 48 tentacles are present in cycles of 6, reaching 15 mm in length. It is common on south and west shores in Britain, and is found on Shetland, and all Irish coasts, typically in crevices in shallow water and rock pools. Greek *aulos* + *actis* = 'tube' + 'beam', and Latin for 'rough'.

Ginger tiny anemone *Isozoanthus sulcatus* (Parazoanthidae), colonial, polyps up to 1 cm high, 0.2–0.3 cm in diameter, and numbering up to 17 per cm², each with 16–30 short blunt tentacles, found on vertical rocks, stones and shells in the intertidal and shallow sublittoral, and in rock pools, in south and south-west England, Wales, and south and west Ireland. Greek *isos* + *zoon* + *anthos* = 'equal' + 'animal' + 'flower', and Latin for 'furrowed'.

Jewel anemone *Corynactis viridis* (Corallimorphidae), brilliantly coloured, short and squat with a smooth column and up to 100 tentacles, widespread except the east coast of Britain, from the lower shore into the subtidal on rocks and beneath

overhangs, often in dense aggregations. Greek *coryne* + *actis* = 'club' + 'beam', and Latin *viridis* = 'green'.

Olive green wart anemone *Phellia gausapata* (Sagartiidae), column height up to 6 cm, with up to 120 irregularly arranged tentacles, on rocky coasts to 15 m depth, attached to rock faces or in pools and crevices, mainly in Shetland, the Outer Hebrides, and north and west Ireland. Algae or encrusting animals such as bryozoans or serpulids are often found growing on it. Greek *phellia* = 'stony ground', and *gausos* + *apate* = 'crooked' + 'illusion'.

Orange anemone *Diadumene cincta* (Diadumenidae), up to 35 mm in height with up to 200 tentacles, widespread but scattered in Britain on hard substrates, especially bivalve shells in pools in the intertidal and sublittoral.

Orange-striped anemone *D. lineata*, column diameter and height both 1–2 cm, with up to 100 tentacles, scattered around the British coastline on hard substrates, often intertidally, in harbours and brackish inshore waters, associated with mussel and oyster shells and stones, sometimes fouling boats and piers.

Parasitic anemone *Calliactis parasitica* (Hormathiidae), up to 8 cm high by 5 cm wide, with up to 700 slender tentacles, typically found in a mutualistic association with a hermit crab, but occasionally solitary on rock or empty shells. It is occasionally intertidal, recorded from south-west England, Pembrokeshire, Co. Cork and Tralee Bay. Greek *calli* + *actis* = 'good' + 'beam'.

Plumose anemone *Metridium senile* (Metridiidae), when expanded the tentacles forming a 'plume' above a conspicuous parapet at the top of the smooth column, which can be 30 cm high, common and widespread, attached to a hard substrate in overhangs and caves and beneath boulders on the lower shore, and on pier piles and rock faces to at least 100 m. Greek *metridios* = 'fruitful', and Latin *senile* = 'aged'.

Red-speckled anemone *Anthopleura ballii* (Actiniidae), with a base up to 5 cm in diameter, and a trumpet-shaped column up to 10 cm tall. There are up to 96 tentacles arranged in five circles, 3 cm in length with a span of 12 cm. It is found along the south and south-west coasts of England and west Ireland in crevices under boulders and can be buried in sand attached to stones or shells, in midshore rock pools and subtidally to 25 m. Greek *anthos* + *pleuron* = 'flower' + 'side'.

Sandalled anemone *Actinothoe sphyrodeta* (Sagartiidae), diameter of the column base 2 cm, up to 120 tentacles, common and widespread except for the east coast, on hard substrates, usually in the sublittoral. Greek *actis* + *thoe* = 'beam' + 'quick', and *sphyra* = 'hammer'.

Sealoch anemone *Protanthea simplex* (Gonactiniidae), the column reaching lengths of up to 2 cm, with numerous tentacles up to 1.5 cm long, which can span 7 cm, from the Firth of Clyde along the west coast of Scotland, particularly in sea lochs, and in 2006 it was found in Killary Harbour, Connemara. Greek *protos* + *anthos* = 'first' + 'flower', and Latin for 'simple'.

Small snakelocks anemone *Sagartiogeton undatus* (Sagartiidae), column up to 12 cm tall, 6 cm across the base, with long tentacles arranged in multiples of six, found on all but the east coast of Britain, buried in sand or gravel, attached to a stone or shell 10–15 cm into the sediment or in rock crevices, on the lower shore and sublittoral. *Undatus* is Latin for 'wavy'.

Snakeshead anemone *Anemonia viridis* (Actiniidae), up to 7 cm across the base with around 200 tentacles having an overall span of 18 cm, widespread along the south and west coasts of both Britain and Ireland, mainly in the intertidal, especially in rock pools from mid-tide downwards but not extending far into the sublittoral, and frequently present, often in large numbers, on the leaves of eelgrass. *Viridis* is Latin for 'green'.

Starlet sea anemone *Nematostella vectensis* (Edwardsiidae), rarely >15 mm in length, with 9–18 tentacles, found in Norfolk, Suffolk, the Blackwater Estuary and Hamford Water in Essex, Hampshire and Dorset in brackish lagoons at or above high water, typically in mud, muddy sand and muddy shingle. Greek *nematos* = 'thread' + Latin *stella* = 'star', and Latin *vectus* = 'carried'.

Strawberry anemone *Actinia fragacea* (Actiniidae), with a base up to 10 cm, scattered though mainly on south and west coasts, on the lower shore attached to rocks. Greek *aulos* + *actis* = 'tube' + 'beam', and Latin *fragum* = 'strawberry'.

Trumpet anemone *Aiptasia mutabilis* (Aiptasiidae), up to 12 cm in height, with around 100 tentacles, local in north Wales, Pembrokeshire, south-west and south England and the Scilly Isles on the lower shore in pools, under stones or beneath overhangs, often in macroalgal holdfasts. Greek *aeptos* = 'one unable to fly', and Latin *mutabilis* = 'variable'.

Worm anemone *Scolanthus callimorphus* (Edwardsiidae), the column measuring up to 14 cm length and 12 mm diameter, and with up to 16 tentacles, found in Weymouth Bay and a few sites in west Ireland and west Scotland, burrowing in sand or gravel from the low intertidal to the shallow sublittoral. Greek *scolos* + *anthos* = 'thorn' + 'flower', and *callos* + *morphe* = 'beautiful' + 'shape'.

SEA BASS *Dicentrarchus labrax* (Moronidae), length up to 100 cm, a widespread demersal fish found in the littoral zone usually to 10 m in depth, and in summer entering estuaries often penetrating some distance upriver. Spawning is in inshore waters on early summer. It feeds on small fish, polychaetes, cephalopods, and crustaceans. It is a good sports fishing species. UK commercial landings in 2013 totalled 800 tonnes, valued at £5.6 million. 'Bass' is a late Middle English alteration of the dialect *barse*, of Germanic origin. Greek *di* + *centron* + *archos* = 'two' + 'sting' + 'anus', referencing the presence of two anal spines. *Labrax* is Greek for 'bass'.

SEA BEECH *Delesseria sanguinea* (Delesseriaceae), a membranous red seaweed up to 30 cm long found around much of the coast, though scarcer in the east, on rocks in deep shaded pools in the lower eulittoral and subtidally to at least 30 m. It is characteristic of the understorey flora in kelp forests, and is occasionally epiphytic on kelp stipes. Jules Delessert (1773–1847) was a French naturalist. *Sanguinea* is Latin for 'blood-coloured'.

SEA BELT Alternative name for sugar kelp.

SEA-BLITE Species of *Suaeda* (Amaranthaceae), from the Arabic name of these plants.

Annual sea-blite *S. maritima*, a common and widespread annual found in the middle and lower saltmarsh, an early colonist of intertidal mud- and sand-flats, sometimes occurring higher up in salt pans and drift lines, and on shell and shingle banks. It is also occasionally found by salted roads.

Shrubby sea-blite *S. vera*, an evergreen shrub found from Dorset to Lincolnshire on shingle drift lines and the upper saltmarsh, sometimes beside brackish creeks and ditches in coastal grazing marsh. Latin *vera* = 'true'.

SEA-BUCKTHORN *Hippophae rhamnoides* (Elaeagnaceae), a thorny deciduous shrub or small tree locally common and widespread around the coast on stabilised dunes and coastal banks, and often planted as an amenity ornamental inland, spreading by rhizomes and layering and often forming dense thickets. It is wind-pollinated, flowering in winter or early spring on bare wood and fruiting in autumn. The genus name is Greek for another species, probably a prickly spurge. *Rhamnoides* means resembling *Rhamnus* (buckthorn).

SEA CUCUMBER Echinoderm invertebrates in the class Holothuroidea. See also sea gherkin.

Brown sea cucumber *Aslia lefevri* (Cucumariidae), with 10 tentacles around the mouth which can account for up to 10 cm of the 15 cm body length, found under stones and in rock crevices in areas of moderate water movement and clean water conditions from the lower shore to 50 m predominantly on the west coasts of Britain and Ireland.

SEA DACE Alternative name for sea bass.

SEA FINGER (SEAWEED) **Green sea finger** *Codium fragile* ssp. *fragile* (Codiaceae), a green seaweed introduced from Japan via mainland Europe to the River Yealm (Devon) in 1939, and recorded from Orkney, west and south-east Scotland, south and south-west England, west Wales and Ireland, often dominating in infralittoral rocky reef communities. *Codium* may be from Greek *codion* = 'fleece'. See also spongeweed.

SEA FINGERS (SEA ANEMONE) Sea anemones in the genus *Alcyonium* (Alcyoniidae), in the shallow sublittoral and deeper. *Alcyonion* is Greek for a kind of sponge (possibly sea anemone), so called from its resemblance to the nest of the kingfisher (= *alcyon*), even though this bird uses a burrow.

SEA FIR **Seagrass sea fir** *Laomedea angulata* (Campanulariidae), a colonial hydroid found on eel grasses to 8 m depth, from south England, north Wales, Isle of Man and a few locations in Ireland. Laomedon was king of Troy, *angulata* is Latin for 'angular'.

SEA GHERKIN *Pawsonia saxicola* (Cucumariidae), a sea anemone up to 15 cm in length, with 10 cm long tentacles round the mouth, found in rock crevices and under boulders in the lower shore and deeper along the south and west coasts of the British and Ireland. *Saxicola* is Latin for 'rock dweller'.

SEA GIRDLE Alternative name for the kelp oarweed.

SEA GRAPES *Molgula manhattensis* (Molgulidae), a sea squirt 1–3 cm across, solitary but often occurring in dense clusters attached to bedrock, boulders, stones and shells, often in harbours, in the littoral and sublittoral to depths of 90 m, widespread if scattered around Britain and east Ireland. Greek *molgos* = 'skin'. Found across the North Atlantic, *manhattensis* reflects its presence on the western shores.

Orange sea grapes *Stolonica socialis* (Styelidae), a colonial sea squirt in which the 2 cm high individual zooids rise singly from a stoloniferous base, resulting in dense aggregations, found on subtidal rocks from the western English Channel round to north Wales, and in north-west and south-east Ireland. *Stolonica* is from the Latin for 'branch'.

SEA GULL See gull.

SEA HARE **Spotted sea hare** *Aplysia punctata* (Aplysiidae), an opisthobranch sea slug usually 7 cm long but growing to 20 cm, widespread in shallow water, occasionally on the lower shore or in rock pools, feeding on macroalgae. Greek *aplysia* = 'dirty (colour)' and Latin *punctata* = 'spotted'.

SEA-HEATH *Frankenia laevis* (Frankeniaceae), a mat-forming evergreen semi-woody perennial of upper saltmarsh and saltmarsh-dune transitional habitats, occasionally on shingle beaches and chalk sea-cliffs. It is mainly found from Hampshire to Lincolnshire, and is sporadic and probably introduced elsewhere in England and Wales. A twentieth-century decline in eastern England has been due to disturbance and destruction of saltmarsh. Johan Franke (1590–1661) was a German-born Swedish botanist. Latin *laevis* = 'smooth'.

SEA-HOLLY *Eryngium maritimum* (Apiaceae), a spiny perennial herb mainly found on embryo and mobile dunes, occasionally on shingle, on most British and Irish coastlines but having recently disappeared from north and east Scotland. This is the 'county' flower of Merseyside as voted by the public in a 2002 survey by Plantlife. *Eryngium* is from Greek for 'sea-holly'. See also eryngo.

SEA HORSETAIL Alternative name for sea mare's-tail.

SEA IVORY *Ramalina siliquosa* (Ramalinaceae), a fruticose lichen, widespread and often abundant on siliceous rocks and stone walls above high tide and on sea-cliffs, tolerant of salt spray, and occasionally (in Ireland) up to 40 km inland. Latin *ramalia* = 'twigs', *siliqua* = 'pod'.

SEA-KALE *Crambe maritima* (Brassicaceae), a herbaceous perennial locally abundant around the British coastline on shingle beaches, occasionally on dunes where these overlie shingle, and on cliffs. Greek *crambe* means a kind of cabbage.

Abyssinian sea-kale *C. hispanica*, an annual East African neophyte cultivated as an industrial oil-seed crop, a very scattered casual on riverbanks and tracksides. *Hispanica*, Latin for 'Spanish', is misleading.

Greater sea-kale *C. cordifolia*, a herbaceous perennial Caucasian neophyte introduced in 1822, occasionally naturalised on roadside and waste ground in England, originating as a garden throw-out. Latin *cordifolia* = 'heart-shaped', referring to the leaves.

SEA-LAVENDER
Herbaceous perennials in the genus *Limonium* (Plumbaginaceae), from Greek *leimon* = 'meadow', and *leimonion* for this plant. 'Lavender' comes from Anglo-Norman French *lavendre* and medieval Latin *lavendula*. Rare local endemics include **broad-leaved, Giltar, Irish, Logan's, Purbeck, small** and **St David's sea-lavenders**.

Common sea-lavender *L. vulgare*, on ungrazed or lightly grazed muddy saltmarsh, occasionally also growing among nearby rocks and on sea walls, around the coastline of Britain, except from Ayrshire northwards and round to the Firth of Forth. A decline has occurred where grazing has intensified or where common cord-grass has expanded. *Vulgare* is Latin for 'common'.

Florist's sea-lavender *L. platyphyllum*, a European neophyte found in rough grassland, usually close to the sea, in a few places in Kent and Essex. *Platyphyllum* is Greek for 'broad-leaved'.

Lax-flowered sea-lavender *L. humile*, on ungrazed or lightly grazed muddy estuarine saltmarsh, patchily on coastlines from south-west Scotland to Norfolk, and frequent in Ireland. *Humile* is Latin for 'low'.

Matted sea-lavender *L. bellidifolium*, mat-forming on upper saltmarsh and the saltmarsh-dune transition zone. Scarce (RDL), recorded from Norfolk (e.g. at Blakeney Point), with a few local losses in the west of this county resulting from human trampling or burial under shifting dunes. *Bellidifolium* is Latin for having leaves like daisy (*bellis*).

Rock sea-lavender *L. binervosum* agg., a group of apomictic plants comprising nine species and many infraspecific taxa, many of which are British and Irish endemics. They occur in England, Wales and Ireland in coastal habitats including sea-cliffs, dock walls, shingle banks and saltmarsh. *Binervosum* is Latin for 'two-veined'.

Rottingdean sea-lavender *L. hyblaeum*, deciduous, found as a naturalised garden escape on and around chalk cliffs at Rottingdean and Seaford (both East Sussex) and as a garden escape on the sea front at West Bay, Dorset, apparently spread by gulls. Hybla was an ancient city in Sicily.

Tall sea-lavender *L. procerum* ssp. *procerum*, found in south and south-west England and Wales on rocks, sea-cliffs, stabilised shingle, upper saltmarsh and the saltmarsh-dune transition zone, and locally on the stonework of sea defences and harbour walls; ssp. *cambrense*, on cliff ledges and steep south- to south-west-facing rocky cliff slopes of Carboniferous limestone in Pembrokeshire; ssp. *devonianum*, on rocks and sea-cliffs near Torquay, south Devon. *Procerum* is Latin for 'tall', *cambrense* = 'Welsh'.

Western sea-lavender *L. britannicum*, on coastal rocks and sea-cliffs. Subspecies *britannicum* and *coombense* are local in Devon and Cornwall, respectively; ssp. *transcanalis* is found on saltmarsh and pebble beaches in south-west Wales; and ssp. *celticum* on rocky shores, sea-cliffs, upper saltmarsh, sea walls and coastal railway banks from north-west Wales to Cumbria.

SEA LEMON
Doris pseudoargus (Dorididae), a common and widespread sea slug up to 12 cm long, on the lower shore beneath large boulders, and offshore down to 300 m, mainly feeding on sponges. Doris was a sea goddess in Greek mythology, Argos a giant with a hundred eyes.

SEA LETTUCE
Ulva lactuca (Ulvaceae), a common and widespread green seaweed with fronds up to 25 cm found throughout the intertidal, and in more northerly latitudes and in brackish

habitats also in the shallow sublittoral and in estuaries. Rotting sea lettuce can produce toxic hydrogen sulphide which might endanger people. *Ulva* is classical Latin for 'sedge', but in New Latin = 'sea lettuce'. Latin *lactuca* = 'lettuce'. *Ulva rigida* is also known as 'sea lettuce'.

SEA LOUSE
Speckled sea louse *Eurydice pulchra* (Cirolanidae), a predatory intertidal isopod, widespread on open coast and estuarine sandy beaches, less common in south-east England, distribution shifting upshore on spring tides and downshore on neap tides. In Greek mythology Eurydice was the wife of Orpheus, who tried to return her from Hades after her death. Latin *pulchra* = 'beautiful'.

SEA MARE'S-TAIL
Halurus equisetifolius (Wrangeliaceae), a red seaweed up to 22 cm long, widespread but not common, especially from Anglesey to south-east England in pools in the littoral and on rocks in the shallow subtidal. Latin *equus* + *seta* + *folium* = 'horse' + 'bristle' + 'leaf'.

SEA MAT
Membranipora membranacea (Membraniporidae), a common and widespread bryozoan that forms encrusting lacy mat-like colonies, usually on kelp in the lower intertidal and shallow sublittoral.

SEA-MILKWORT
Glaux maritima (Primulaceae), a herbaceous perennial found on most of the British coastline, typically forming dense monospecific colonies on moist saline soils, for example on saltmarsh, strandlines, damp shingle, wet sand, brackish dune slacks, aerobic mud and spray-drenched rock crevices. Although a poor competitor, it can grow in open communities with other halophytes. *Glaux* was used by Dioscorides, though for a different plant.

SEA OAK (HYDROID)
Dynamena pumila (Sertulariidae), a common and widespread erect intertidal hydroid, up to 7 cm tall, found mainly on kelp and fucoid species where it can form dense populations, but also on rock. Greek *dynamis* = 'strength', and Latin *pumila* = 'dwarf'.

SEA OAK (SEAWEED)
Halidrys siliquosa (Sargassaceae), a widespread and locally common brown seaweed, thallus length of 30–130 cm, in the mid- to lower littoral (and in the upper eulittoral when in rock pools), also forming a zone in the sublittoral down to 10 m. Greek *hals* + *drys* = 'sea' + 'oak', and Latin *siliqua* = 'pod', a reference to the air bladders on the fronds.

SEA ONION
Alternative name for spring squill.

SEA ORANGE
Suberites ficus (Suberitidae), typically a massively lobed sponge but it can be a cushion, globular, elongated or encrusting, 10–40 cm in diameter, widespread though less so on the east coast, on rock and other hard substrates from the shallow sublittoral to greater depths. *Suber* is Latin for 'cork', from having a corky texture, *ficus* for 'fig'.

SEA PEN
Marine cnidarians in the order Pennatulacea, with a feather-like appearance (like quill pens), found in the sublittoral and deeper.

SEA PETALS
Petalonia fascia (Scytosiphonaceae), a widespread brown seaweed with fronds up to 30 cm long, generally growing on stones and shells in shallow pools in sheltered locations, particularly in harbours. Greek *petalon* = 'leaf', and Latin *fascia* = 'bundle'.

SEA POTATO
Echinocardium cordatum (Loveniidae), a common and widespread heart urchin, usually 6 cm long although it can grow up to 9 cm, living in a permanent burrow 8–15 cm deep in sandy sediments in the intertidal and sublittoral to 200 m. Greek *echinos* + *cardia* = 'sea urchin' + 'heart', and Latin *cor* = 'heart'.

SEA-PURSLANE
Atriplex portulacoides (Amaranthaceae), a common and widespread shrub of muddy or sandy upper saltmarsh, often fringing intertidal pools and creeks, and forming extensive stands on ungrazed saltings. Absent in Scotland except

for the south-west coastline. In western Britain and Ireland it is also local on coastal rocks and cliffs. *Atriplex* and *portulaca* are both Latin for this plant.

Pedunculate sea-purslane *A. pedunculata*, an annual associated with upper saltmarsh, presumed extinct since the late 1930s until rediscovered near Shoeburyness, Essex, in 1987, where it is threatened by encroaching sea couch, but further colonies and sites have been established nearby using plants and seeds derived from the original colony.

SEA SCORPION Long-spined sea scorpion *Taurulus bubalis* (Cottidae), length up to 20 cm, a common and widespread fish found over rocky ground among algae from the intertidal (including rock pools) to depths of 30 m, feeding on small molluscs and crustaceans. Some related species have venomous spines, hence the English name. The binomial comes from the Latin for 'little bull' and 'oxen', a reference to the head shape.

SEA SLUG See slug.

SEA SPIDER Member of the class Pycnogonida.

SEA-SPURREY Species of *Spergularia* (Caryophyllaceae) (Latin *spergulinus* = 'scattering'). 'Spurrey' dates to the late sixteenth century, from Dutch *spurrie*.

Greater sea-spurrey *S. media*, a common and widespread perennial found in saltmarsh and on muddy beaches and dune slacks. Inland, it occasionally colonises salt-treated roadsides. *Media* is Latin for 'middle'.

Greek sea-spurrey *S. bocconei*, an annual Mediterranean neophyte, first recorded in Britain in 1901, frequent in the Channel Islands, rare in Cornwall, and casual elsewhere in southern England, south Wales and Co. Wexford, on free-draining sandy soils on waste ground by the sea. Paolo Boccone (1633–1704) was an Italian botanist.

Lesser sea-spurrey *S. marina*, a common and widespread annual of saltmarsh, muddy shingle and brackish grazing pastures. Inland, it is increasingly frequent beside salt-treated roads. *Marina* is Latin for 'marine'.

Rock sea-spurrey *S. rupicola*, a herbaceous perennial locally common on the coastline from south-west Scotland to Hampshire, and around Ireland, on maritime rocks and cliffs, in crevices and on rock surfaces, sometimes in guano-enriched sites. It also grows in clifftop grassland and on sea walls. *Rupicola* is Latin for 'rock-dwelling'.

SEA SQUIRT Sac-like marine invertebrate filter feeders, class Ascidiacea, subphylum Tunicata. See also sea grapes and orange sheath tunicate under Tunicata.

Dirty sea squirt *Ascidiella aspersa* (Ascidiidae), up to 130 mm long and usually attached to the substratum by the left side, solitary but often found in dense unfused aggregations, common on all British coasts but more so in the south and west, in the lower intertidal and sublittorally, and in some estuaries. Greek *ascidion*, diminutive of *ascos* = 'wineskin', and Latin *aspersa* = 'sprinkled' (possibly from *asper* = 'rough').

Hairy sea squirt *A. scabra*, solitary, <4 cm long, widespread, on hard substrates and macroalgae from the intertidal down to 300 m. *Scabra* is Latin for 'rough'.

Leathery sea squirt *Styela clava* (Styelidae), solitary, up to 12 cm in height, native to the north-west Pacific, first recorded in Plymouth, south Devon, in 1953 possibly introduced on the hulls of war ships following the end of the Korean War in 1951, now found on the south and west coasts of England as far north as Cumbria; in Loch Ryan and Androssan marina, Scotland; in Cork and Fenit Harbours, Ireland, and the Channel Isles in shallow water on hard surfaces, especially in sheltered warm-water docks and harbours. *Clava* is Latin for 'club'.

Lesser gooseberry sea squirt *Distomus variolosus* (Styelidae), colonial, on the coasts of south England, Wales, Isle of Man, south-west and south-east Ireland, and the Outer Hebrides, on rocks, around the bases of hydroids and on macroalgal holdfasts.

Greek *di* + *stoma* = 'two' + 'mouths', and Latin *variolosus* = 'variegated'.

Neptune's heart sea squirt *Phallusia mammillata* (Ascidiidae), Britain's largest sea squirt, the overall shape being conical, with a broad base, up to 12 cm tall and 8 cm broad, in warm sheltered sites along the south and south-west English coast, and in south-west Ireland. In 2010, *The Guardian* launched an annual competition for the public to suggest common names for previously unnamed species. In 2011, this name was a winner: Neptune was the Roman god of the sea, and the animal is heart-shaped. Greek *phallos* = 'penis', and Latin *mammillata* = 'breast-like'.

Orange-tipped sea squirt *Corella eumyota* (Corellidae), solitary, 2–4 cm long, often attaching (along their right side) to conspecifics to form aggregations that foul underlying substrates, which include boulders, macroalgae and ship hulls. A Southern Hemisphere species, it was probably introduced with imported bivalves, first recorded at Brighton (Sussex), Gosport (Hampshire) and Weymouth (Dorset) in 2004, and rapidly spreading (probably on the hulls of leisure craft) around the south coast of England, northwards to Lowestoft (Suffolk) by 2009, and to Oban, west Scotland.

Yellow sea squirt *Ciona intestinalis* (Cionidae), solitary, up to 15 cm long, widespread on bedrock, boulders and artificial surfaces in harbours, from the lower shore down to 500 m. The binomial is from Greek demigoddess Chione, and Latin *intestinalis* = 'internal'.

SEA STAR Orange sea star *Caloplaca verruculifera* (Teloschistaceae), a crustose lichen widespread on both silica-rich and calcareous rocks, especially in nutrient-rich coastal sites such as near bird perches. Greek *calos* + *plax* = 'beautiful' + 'flat plate', and Latin *verruca* + *fero* = 'wart' + 'carry'.

SEA URCHIN Echinoderm marine invertebrates in the class Echinoidea, with fivefold symmetry. They have a hard round shaped body (test) with long spines which are used for protection, to move about, and to trap food particles, though most species feed on algae or are omnivorous. The mouth (Aristotle's lantern), is in the middle on the underside of the body. See also heart urchin.

Edible sea urchin *Echinus esculentus* (Echinidae), up to 16 cm in diameter at 7–8 years of age (the largest diameter recorded was 17.6 cm), common and widespread except between mid-Yorkshire to Dorset, on rocky substrates in the sublittoral fringe and deeper, feeding on macroalgae, bryozoans, barnacles and other encrusting invertebrates. *Echinos* is Greek for both 'hedgehog' and 'sea urchin'. The common and scientific names suggest that it is edible to humans (*esculentus* being Latin for 'edible'), but only the reproductive organs (roe) can be eaten, and demand in Britain is minimal.

Green sea urchin 1) *Echinocyamus pusillus* (Echinocyamidae), 1 cm diameter, widespread but scattered, buried in coarse sand or gravel below extreme low water, feeding on detritus, algae and foraminiferans. Greek *echinos* + *cyamos* = 'sea urchin' + 'pebble'.

2) *Psammechinus miliaris* (Parechinidae), up to 57 mm diameter (more typically 35 mm), generally widespread, intertidally on rocky shores under stones and macroalgae, and aso subtidally in sea-grass beds or on mixed coarse substrates. Greek *psammos* + *echinus* = 'sand' + 'sea urchin', and Latin *miliaris* = 'like millet', probably from body surface lesions that resemble millet seed.

3) Alternative name for northern sea urchin.

Northern sea urchin *Strongylocentrotus droebachiensis* (Strongylocentrotidae), up to 8 cm diameter, in the lower infralittoral and upper circalittoral, with recent records only from Shetland. Greek *strongylos* + *centron* = 'round' or 'compact' + 'spine'.

Purple sea urchin *Paracentrotus lividus* (Parechinidae), up to 7 cm diameter, particularly common on the west coast of Ireland, but also found in the Channel Islands, and a few places in west Scotland, south-west England and the Scillies, on the lower rocky shore in rock pools and in the shallow sublittoral down to depths of 3 m. It uses its spines and teeth to bore into soft rocks, its burrow providing protection from both wave action and desiccation at low tide. Greek *para* + *centron* = 'near' + 'spine', and Latin *lividus* = a blue-lead colour.

SEAHORSE Species of *Hippocampus* (Syngnathidae), from Greek *hippos* + *campos* = 'horse' + 'sea animal' (or *campe* = 'curvature'). These fishes have the head set at an angle to the body. The trunk is short and fat, the tail tapering, curled and prehensile. They pair for life, and the female transfers her eggs to the male which he self-fertilises in his pouch and carries until hatch. Feeding is on small crustaceans.

Long-snouted seahorse *H. guttulatus*, length up to 15 cm, along the south coast of England and south-west Wales at depths of 1–20 m, especially in eelgrass meadows, clinging by the tail or swimming upright. *Guttulatus* is Latin for 'snouted'.

Short-snouted seahorse *H. hippocampus*, length up to 15 cm, along the south coast of England, with large populations around the Channel Islands and Ireland. In 2007, colonies were found in the Thames estuary at Dagenham, Tilbury and Southend. They are found in shallow muddy waters, in estuaries, or inshore among seaweed and sea-grasses, clinging by the tail or swimming upright.

Spiny seahorse Alternative name for long-snouted seahorse.

SEAL Earless or true seals are members of the family Phocidae. The Conservation of Seals Act (1970) protects common and grey seals during the breeding season, although seals causing damage to fishing gear or taking fish from nets may be killed under licence. The Act also allows seals to be given year-round protection when appropriate, for instance following the 1988 outbreak of phocine distemper, common and grey seals were thus protected in England, Wales and Scotland. **Bearded seal** *Erignathus barbatus*, **hooded seal** *Cystophora cristatus* and **harp seal** *Pagophilus groenlandicus* are rare vagrants. 'Seal' comes from Old English *seolh*.

Common seal *Phoca vitulina*, male head-body length 140–85 cm, weight up to 150 kg, females up to 180 cm and 130 kg, with a British population of around 55,000 (with some estimates of 100,000), and an Irish population of 5,000. They feed on a variety of fish, travelling up to 50 km from haul-out sites and remaining at sea for several days, diving for up to 10 minutes, reaching depths of 70 m. They haul out on rocky shores or intertidal sandbanks to rest, to give birth and suckle their pups, and to mate. The most important haul-out areas are around Scotland, particularly the Hebrides, Orkney and Shetland; the Wash; and around Strangford Lough and in south-west Ireland. Gestation lasts 11 months. Females give birth to a single pup in June or July. Pups can swim and dive when just a few hours old. They are fed on extremely fatty milk, doubling their birth weight in 3–4 weeks before weaning. Greek *phoce* = 'seal', and *vitulina* = 'calf-like'.

Grey seal *Halichoerus grypus*, male head-body length 200–50 cm, weight 230–300 kg, female length 180–200 cm, weight 150–80 kg, found in Britain mainly along exposed rocky north and west coasts, but also in the south-west and off the east coast, around the Isle of May, the Farne Islands (Northumberland), and along the west coast of Ireland. At the start of the 2000 breeding season, Britain held around 124,000 grey seals, with a further 300–400 around the Isle of Man and Northern Ireland. Ireland as a whole has around 4,000 animals. One of the rarest seals in the world, the UK population represents 40% of the world population and 95% of the European population. They usually rest by day at low tide and at sunset at a haul-out site, while hunting at night and at high tide in the coastal zone up to 80 m deep although they can dive >200 m. They feed opportunistically, with sand eel and cod the most important prey. At haul-out sites they sometimes form groups of several hundred animals, especially when they are moulting in spring. In autumn they congregate at traditional sites on land (rookeries) to breed. Pregnancy last 8½ months. The timing of births varies around the coast, beginning in September in west Wales, in October in west Scotland, and November in the Farne Islands. Pups remain on land suckling for 18–21 days; the milk contains up to 60% fat, so pups can gain 2 kg each day. Greek *halichoerus* = 'sea pig', and Latin *grypus* = 'hook-nosed'.

Harbour seal Alternative name for common seal.

Ringed seal *P. hispida*, head-body length 135–40 cm, weight 50–90 kg, the smallest true seal, associated with Arctic waters but vagrant round north Scottish shores, for example Shetland and (in 2007) the Kyle of Sutherland. Its English name comes the ring like markings on its fur. *Hispidus* is Latin for 'bristly'.

SEAMINE Bearded seamine *Ripartites tricholoma* (Agaricales, Tricholomataceae), an inedible fungus, widespread especially in England, found in summer and autumn on woodland soil and litter. *Ripartes* comes from Latin *ripa* = 'stream bank', Greek *trichos* + *loma* = 'hair' + 'fringe'.

SEAWEED Marine macroalgae which are either, green, red or brown, which often zone themselves down the shore.

Bearded red seaweed *Anotrichium barbatum* (Wrangeliaceae), thought to be extinct in Britain but found in 1998 on small stones in an area of mixed muddy gravel at Tremadog Bay near Pwllheli, north-west Wales. The binomial translates as 'bearded twig weed'.

SEDGE (CADDIS FLY) Names particularly used by anglers.

Bicolor sedge *Triaenodes bicolor* (Leptoceridae), common and widespread in ponds, lakes, canals, slow rivers and wet fens, on living plant material, occurring only in standing water. Larvae are both grazers and shredders. The case is made of a spiral of plant material. Adults fly in June–September. Greek *triaina* + *eidos* = 'trident' + 'appearance of', and Latin *bicolor* = 'two-coloured'.

Black sedge *Silo nigricornis* (Goeridae), in cold-water rivers and streams with substrates of coarse gravel, boulders or bedrock. The tubular larval case is made of sand grains with small pebbles at the sides providing ballast. Larvae are mainly grazers and scrapers. Adults fly in May–September. Greek *silos* = 'snub-nosed', and Latin *niger* + *cornu* = 'black' + 'horn'.

Brown sedge *Anabolia nervosa* (Limnephilidae), common and widespread in streams. Nymphs feed on algae and small insects. Cases use sand grains with small sticks attached. Adults fly from late July to October. Greek *anabole* = 'mound' and Latin *nervosa* = 'sinewy'.

Cinnamon sedge *Limnephilus lunatus* and *L. marmoratus* (Limnephilidae), common and widespread in streams, ponds, lakes and marsh. The larval case uses leaf fragments, sand grains and other debris. Adults are evident, respectively, in May–November and June–October. Greek *limne* + *philos* = 'pond' + 'loving', Latin *lunatus* = 'moon-shaped', from the dark-edged pale 'half-moon' on the rear margin of the wing and *marmoratus* = 'marbled'.

Dark spotted sedge *Philopotamus montanus* (Philopotamidae), in neutral to alkaline upland, stony streams on coarse gravel and boulders. Larvae are filter-feeding, building long, tubular nets attached to rocks. Adults fly from February to October. Greek *philos* + *potamos* = 'liking' + 'river'.

Great red sedge *Phryganea bipunctata* and *P. grandis* (Phryganeidae), Britain's largest caddis flies (forewing length 18–28 mm), widespread, in ponds, lakes, canals and slow rivers, flying from May to August. Greek *phryganon* = 'dry stick', describing the larval case, and Latin for, respectively, 'two-spotted' and 'large'.

Grey sedge Alternative name for silver sedge.

Large cinnamon sedge *Potamophylax latipennis* (Limnephilidae), common and widespread, in streams, rivers and lakes. Adults fly in May–November. Greek *potamos* + *phylax* = 'river' + 'a guard', and Latin *latipennis* = 'broad-winged'.

Marbled sedge *Hydropsyche contubernalis* (Hydropsychidae), common and widespread in rivers and streams on gravel and boulders. The larva spins a net in gravel in slow moving water and passively filter-feeds fine particulate matter. Adults fly in

May–September. Greek *hydor* = 'water' + the Greek nymph Psyche, and Latin *contubernalis* = 'friend'.

Medium sedge *Goera pilosa* (Goeridae), common and widespread, especially in chalk streams of south-east England, in stony streams and lakes. The tubular larval case is made of sand grains with small pebbles at the sides providing ballast. Larvae are mainly grazers and scrapers. Adults fly in May–September. Greek *goeros* = 'mournful', and Latin *pilosa* = 'hairy'.

Mottled sedge *Glyphotaelius pellucidus* (Limnephilidae), frequent and widespread, less so in Scotland, in rivers, streams, lakes, ponds and marsh, especially in or near deciduous woodland. Nymphs mainly feed by shredding, but also predation and grazing. Larval cases use dead leaves. Adults fly in late April–June and August–October with a summer diapause. Greek *glypho* + *tylos* = 'to carve' + 'knot', and Latin *pellucidus* = 'transparent'.

Sandfly sedge, sand sedge *Rhyacophila dorsalis* (Rhyacophilidae), widespread and common in acid stony rivers and streams. Adults fly in May–October. Larvae are uncased predators. Greek *rhyacos* + *philos* = 'of streams' + 'loving', and Latin *dorsalis* = 'dorsal'.

Scarce brown sedge *Ironoquia dubia* (Limnephilidae), recorded in Suffolk, Windsor Forest and south-east Berkshire/north-east Hampshire, but with no recent records, and a priority species in the UK BAP, the English name 'manufactured' for this latter. Adults fly in September–October. They lay eggs in damp dead leaves at the bottom of a stream which will have dried out over summer. Larvae leave the water in spring to pass the summer in damp leaves at the margins, pupating there in September, behaviour unique to UK caddis species.

Silver sedge *Odontocerum albicorne* (Odontoceridae), common and widespread in neutral stony rivers and streams, preferably on woody debris. Larval cases are of sand grains. Feeding is by predation, grazing and shredding. Adults fly in May–September. Greek *odontos* + *ceras* = 'tooth' + 'horn', and Latin *albicorne* = 'white-horned'.

Small grey sedge *Glossosoma intermedium* (Glossosomatidae), recorded only from four Lake District streams rising from base-rich rocks and with a moderately fast-flowing current. The species was added to the UK BAP in 2007, the English name 'manufactured' for this purpose. The tortoise-shaped larval case is made of large sand grains. Rapid larval growth occurs from April to July, then larvae change into a pupa in small stone structures that are attached to larger stones. The mothlike adults emerge in April–May, flying in early evening and at night. Greek *glossa* + *soma* = 'tongue' + 'body'.

Small red sedge *Tinodes waeneri* (Psychomyiidae), common and widespread in flowing water on gravel or bedrock. Larvae are primarily grazers. Adults fly in April–November. Greek *teino* = 'to stretch'.

Small silver sedge *Lepidostoma hirtum* (Lepidostomatidae), common and widespread, especially in upland areas, in slow-flowing rivers, streams and stony lakes. Larvae are shredders and scrapers living in woody debris and other substrates in cases made of sand grains until instar V when plant fragments are used. Adults fly in May–September. Greek *lepidos* + *stoma* = 'scale' + 'mouth', and Latin *hirtum* = 'bristly'.

Small yellow sedge *Psychomyia pusilla* (Psychomyiidae), widespread and quite frequent in flowing water, larvae building galleries on rock surfaces in standing or slow-moving water, with a preference for neutral to alkaline water. Larvae are predominantly grazers. Adults fly in May–September. The binomial is from the Greek nymph Psyche + *myia* = 'fly', and Latin for 'small'.

Window-winged sedge *Hagenella clathrata* (Phryganeidae), declining, recorded from two sites near Aviemore, one in Galloway, two on the Shropshire Welsh border, three in Staffordshire, one in Greater London, and three sites in Surrey, Endangered in RDB 1, and a priority species in the UK BAP, the English name 'manufactured' for this latter. Sites are on edge of lowland raised and quaking bogs, and in wet heaths, with small pools drying to soggy by midsummer, when the adults fly,

doing so in daylight. Birch provides some food and case-making material. Larvae are predaceous. *Clathrata* is Latin for 'latticed'.

SEDGE (PLANT) From Old English *secg*, species of *Carex* (Latin for these plants), rhizomatous perennial grass-like plants with triangular cross-sectioned stems typically found in wet habitats.

Rarities include **bird's-foot sedge** *C. ornithopoda* (Derbyshire, Yorkshire and Cumbria), **bristle sedge** *C. microglochin* (Ben Lawers, mid-Perth), **club sedge** *C. buxbaumii* (four sites scattered in Scotland), **elongated sedge** *C. elongata*, **estuarine sedge** *C. recta* (Wick river, Caithness, and the River Beauly and Kyle of Sutherland, both Easter Ross), **hare's-foot sedge** *C. lachenalii* (central Highlands), **saltmarsh sedge** *C. salina* (Loch Duich, Wester Ross), and **string sedge** *C. chordorrhiza* (possibly a glacial relic at a few sites in the Highlands).

Bottle sedge *C. rostrata*, generally common and widespread but absent from parts of central and eastern England, in emergent stands on lake, pond, stream and ditch margins, swamp and bog pools, wet meadow, flushes, wet dune slacks, and alder and willow carr, usually in nutrient-poor acidic water. *Rostrata* is Latin for 'beaked'.

Brown sedge *C. disticha*, common and widespread, generally in lowland areas (though also common in the south-east Scottish Uplands) in damp meadow, marsh and fen, on lake and pond edges, occasionally in dune slacks. It favours non-shaded base-enriched substrates, especially with a fluctuating water table. *Disticha* is Greek for 'two-ranked'.

Carnation sedge *C. panicea*, common and widespread in damp habitats on neutral to moderately acidic soils, including wet grassland, hay and water meadow, mires, heath and flushes in montane grassland. In north and west Britain and in Ireland it is also recorded from upper saltmarsh. *Panicea* is Latin for 'made from bread'.

Common sedge *C. nigra*, common and widespread: tussock forms may occur in stagnant, acidic sites, and rhizomatous tufted examples from calcareous mires. It is found in marsh, fen, mire, flushes, wet grassland and dune slacks. *Nigra* is Latin for 'black'.

Curved sedge *C. maritima*, scarce, found along the north and east coasts of Scotland, and the Outer Hebrides, in damp dune slacks and open sand, often by freshwater seepage.

Cyperous sedge *C. pseudocyperus*, shade-tolerant, locally common in England, scattered elsewhere, a pioneer found along the edges of lakes, ponds, reservoirs, rivers, ditches and canals, as well as in reed swamp and tall-herb fen. It frequently colonises extraction pits, and is found in alder woodland and willow carr.

Dioecious sedge *C. dioica*, locally common in northern England, Wales, Scotland and Ireland, scattered and scarce elsewhere in England, on wet, neutral to base-rich mires, calcareous mud and on the margins of base-rich flushes and springs. *Dioica* is Latin for dioecious (having male and female flowers on separate plants).

Distant sedge *C. distans*, common and widespread on rocky shores and sea-cliffs, and in coastal grassland and upper saltmarsh. Inland, in south and east England, it grows in damp grassland, marsh and fen.

Divided sedge *C. divisa*, locally common in coastal south and east England, and Co. Wexford, in brackish ditches, marsh, dune slacks and damp grassland.

Dotted sedge *C. punctata*, local from south-west Scotland to Hampshire, and in south and west Ireland, on wet coastal rock ledges and cliffs, and sandy patches in saltmarsh. *Punctata* is Latin for 'spotted'.

Downy-fruited sedge *C. filiformis*, local in Wiltshire, Gloucestershire, Oxfordshire and Surrey, on calcium-rich soils, in (especially damp) grassland, woodland rides and along roadsides. *Filiformis* is Latin for 'thread-like'.

Dwarf sedge *C. humilis*, local from Dorset to Herefordshire, on grazed chalk and limestone grassland, field margins and rock outcrops. *Humilis* is Latin for 'dwarf'.

Few-flowered sedge *C. pauciflora*, mainly found in the Highlands, scattered elsewhere in upland Scotland, northern England, north Wales and north-east Ireland, on raised and blanket bog, often growing on hummocks. *Pauciflora* is Latin for 'few flowers'.

Fingered sedge *C. digitata*, local from north Somerset, Gloucestershire and Monmouthshire to Cumbria and north Yorkshire, in deciduous woodland over limestone, and on scree, rock outcrops and limestone pavement. *Digitata* is Latin for 'fingered'.

Flea sedge *C. pulicaris*, rhizomatous or tufted, common and widespread though absent from much of central and eastern England, on wet often calcareous soils, and more acidic soils where flushed by mineral-enriched groundwater. Habitats include bog, damp grassland, wet heathland and montane rock ledges. *Pulex* is Latin for 'flea'.

Glaucous sedge *C. flacca*, common and widespread in neutral and calcareous grassland, upper saltmarsh and base-rich mountain flushes, and a pioneer of disturbed bare sites. *Flacca* is Latin for 'flabby'.

Green-ribbed sedge *C. binervis*, generally common and widespread, though scarce in central and eastern England, on acidic soils in lowland heathland, moorland, rocky mountainsides, favouring open habitats. *Binervis* is Latin for 'two-nerved' ('veined').

Grey sedge *C. divulsa* ssp. *divulsa*, common and widespread in the southern half of England (scattered elsewhere), much of Wales, and south and east Ireland, tolerating a range of soils, except very acidic, in hedge banks, scrub, along woodland edges, verges and rough grassland. Subspecies *leersii* is found on chalk and limestone in woodland rides and grassland, commoner in southern England and East Anglia. *Divulsa* is Latin for 'torn (apart)'.

Hair sedge *C. capillaris*, in base-rich upland grasslands in northern England and, especially, Scotland, especially where flushed by calcareous springs, and damp limestone and mica-schist slopes, reaching 1,035 m on Ben Lawers (mid-Perth), but descending to the coast in north Scotland. *Capillaris* is Latin for 'relating to hair'.

Hairy sedge *C. hirta*, common and widespread, though absent from north Scotland, in damp grassy habitats, particularly those influenced by human activity, for example waste ground, verges and hedge banks. It is also common in rough grassland, occasionally dunes, marsh and damp woodland. *Hirta* is Latin for 'hairy'.

Long-bracted sedge *C. extensa*, common and widespread, more scattered in the east coast, on estuarine flats, upper salt-marsh and moist coastal rocks. *Extensa* is Latin for 'extended'.

Oval sedge *C. leporina*, common and widespread, especially on acidic, often upland grassland, and moorland. It is also common in lowland heathland, damp meadow, woodland edges and ruderal habitats. *Leporina* is Latin for 'like a hare'.

Pale sedge *C. pallescens*, common and widespread in Britain, scattered in (mostly northern) Ireland, usually on mildly acidic to neutral soils, in damp grassland, woodland clearings and stream banks. *Pallescens* is Latin for 'becoming pale'.

Pendulous sedge *C. pendula*, widespread and locally common, absent from much of upland Scotland and central Ireland, on damp base-rich, heavy soil in shaded habitats in deciduous woodland, by ponds, streams and tracksides, and in hedgerows.

Pill sedge *C. pilulifera*, common and widespread in dry habitats on base-poor, usually acidic sandy or peaty soils, for example heathland, dune and upland grassland and moorland. *Pilulifera* is Latin for 'bearing small balls'.

Prickly sedge *C. muricata* ssp. *pairae*, common in much of England, Wales, lower upland Scotland and south-east Ireland, on well-drained, light, sometimes sandy and acid soils, shade-intolerant on hedge banks and verges, in rough meadow and

heathland and on rocky slopes. Subspecies *muricata* is rare (six extant sites from Shropshire to north-west Yorkshire) on limestone pavement, grassy slopes and scree. *Muricata* is Latin for 'roughened'.

Remote sedge *C. remota*, common and widespread, though absent from much of north Scotland, in often seasonally flooded woodland, banks and ditches, very shade-tolerant.

Rock sedge *C. rupestris*, local in the Highlands and north Scotland on base-rich substrates on cliff ledges and crevices, and on rocky and grassy slopes. *Rupestris* comes from Latin *rupes* = 'rock'.

Russet sedge *C. saxatilis*, in the Highlands on base-rich substrates on flat mountain tops and gentle slopes, and in flushes and hollows with late snow-lie. *Saxatilis* is Latin for 'growing among rocks'.

Sand sedge *C. arenaria*, common and widespread in sandy habitats where it can be a dominant plant of fixed dunes and dune slacks, and in ruderal habitats. It is also locally common on inland dunes and heaths, especially on the Lincolnshire coversands and in Breckland, and occasionally found on railway clinker. *Arena* is Latin for 'sand'.

Sheathed sedge *C. vaginata*, mainly recorded from the Highlands, but also from the Borders and the Lake District, on flushed mountain grassland, bog and rock ledges. It is usually found >700 m. *Vaginata* is Latin for 'sheathed'.

Slender sedge *C. lasiocarpa*, scattered in upland regions, and also in parts of Dorset/Hampshire, Somerset and East Anglia, in reed swamp and other vegetation on the margins of lakes, pools and slow-flowing streams, and in flushes in fen, generally in nutrient-poor water. *Lasiocarpa* is Greek for 'woolly fruit'.

Smooth-stalked sedge *C. laevigata*, scattered if absent from much of central and eastern England, north Scotland and central Ireland, in damp woodland on heavy clay soil, often where there is flushing with base-rich water. Commonest in shaded sites, but sometimes found in more open situations. *Laevigata* is Latin for 'smooth'.

Soft-leaved sedge *C. montana*, local in south-west England, the New Forest, south Wales and a few sites in the Midlands north to Derbyshire, in rough grassland, heathland and woodland rides, often in partial shade. *Montana* is Latin for 'montane'.

Spiked sedge *C. spicata*, common and widespread in England, scarce and scattered elsewhere, on rough grassland, verges, railway embankments, woodland clearings, scrub and waste ground, favouring moist, neutral or slightly base-rich, heavy soils. *Spicata* is Latin for 'spiked'.

Star sedge *C. echinata*, common and widespread, though more scattered in central and eastern England, in seasonally or permanently waterlogged habitats on acidic to base-rich habitats such as bog, wet heathland and upland flushes, as well as in wet meadow. *Echinata* is Latin for 'prickly'.

Stiff sedge *C. bigelowii*, found in upland north Wales, northern England and Scotland, scattered in Ireland, on well-drained montane grassland, sedge-heath and stony ground, in corries with late snow-lie, and wet gullies. John Bigelow (1804–78) was an American botanist.

Tawny sedge *C. hostiana*, common and widespread in northern and western regions, more scattered in central and south-east England, in base-rich fen, flushed valley bog, marsh and wet meadow. In lowland Ireland it tolerates more acidic sites. Latin *hostia* = 'host' or 'victim'.

Water sedge *C. aquatilis*, growing in the central Scottish lowlands on riverbanks and the margins of lakes, bog and reed swamp, and throughout the British Isles on upland sites it often grows on gently sloping peat. *Aquatilis* is Latin for 'growing in water'.

White sedge *C. canescens*, absent from much of lowland Britain, but local in some lowland bog, floating sphagnum rafts in lowland mires, and in upland regions in boggy woodland, nutrient-poor upland bog wet heathland, and mountain slopes. *Canescens* is Latin for 'greyish'.

SEED BEETLE Some members of the family Chrysomelidae, subfamily Bruchinae, also known as bean weevils. Also some leaf beetles (Galerucinae).

Adzuki bean seed beetle *Calosobruchus chinensis*, 2–2.3 mm length, a major introduced pest of stored pulses, occasionally found in the wild on umbellifers, but unable to breed outside in this country. Greek *calos* = 'beautiful' and *brucos* (see bean seed beetle below).

Bean seed beetle *Bruchus rufimanus*, 3.1–5.3 mm length, common and widespread in England, less so in Wales, as a pest of both field and stored beans. Confusingly, *broucos* is Greek for 'brush' or 'wingless locust' (c.f. Latin *bruchus* = 'locust' or 'caterpillar'). Latin *rufus* + *manus* = 'red' + 'hand'.

Cowpea seed beetle *C. maculatus*, 2.7–3.8 mm length, a major pest of stored legumes and pulses, especially chickpeas. *Maculatus* is Latin for 'spotted'.

SEED BUG **Western conifer seed bug** *Leptoglossus occidentalis* (Coreidae), length 20 mm, a North American pest squash bug first recorded in Britain (following migration from continental Europe) in Dorset in 2007, soon spreading into southern England and further inland as far north as Merseyside and Yorkshire, feeding on conifers, especially pines, and entering buildings in a search for hibernation sites. Greek *leptos* + *glossa* = 'slender' + 'tongue', and Latin *occidentalis* = 'western'.

SEED-EATER Ground beetles in the genera *Amara* and *Harpalus* (Carabidae, Carabinae), though many other carabids also eat seeds as at least part of their diet.

Brush-thighed seed-eater *Harpalus froelichii*, on open vegetation on sandy soil, showing a 54% decline over 1985–2010. It has disappeared from the coasts of Suffolk and Norfolk, and it is now restricted to the East Anglian Breckland. Recent survey work under the Scarce Ground Beetle Project has discovered populations at five sites in Breckland. Greek *harpaleos* = 'greedy'. Josef Frölich (1766–1841) was a German physician and entomologist.

SEED FLY **Bean seed fly** *Delia platura* (Anthomyiidae), 6 mm length, common and widespread in England and Wales, less so elsewhere. A garden pest of French and runner beans, the larvae (8 mm length) feed on roots and seeds. There are 3–5 generations during spring and summer. The first generation in May is usually the most damaging.

SEED MOTH **Hollyhock seed moth** *Pexicopia malvella* (Gelechiidae, Apatetrinae), wingspan 17–20 mm, locally common in grassland, saltmarsh and gardens in south and east England, rare in Wales, contracting its range as the larval food plants (marsh-mallow and hollyhock) become less common. Adults fly at night in June–August. Latin *pexus* + *copia* = 'woolly' + 'abundance', describing the labial palpus or antenna, or both, and *malva* = 'mallow'.

Spruce seed moth *Cydia strobilella* (Tortricidae, Olethreutinae), wingspan 10–15 mm, local in coniferous woodland in the southern half of England, and in parts of Scotland and Wales, but scarce in between. Adults fly in sunshine in May, and also at night. Larvae feed in the cones of Norway spruce, in places becoming a pest. Greek *cydos* = 'glory', and Latin *strobilus* = 'pine cone'.

SEGESTRIDAE Family name for tube-weaving spiders (genus *Segestria*) (Latin *segestria* = 'covering') See snake-back spider under spider.

SELFHEAL *Prunella vulgaris* (Lamiaceae), a common and widespread herbaceous perennial of woodland clearings, meadow, pasture, lawns, roadsides and waste ground, typically on neutral and calcareous moist, fertile soil. Infusions have been used herbally as an astringent, styptic and tonic, prompting its common name. In the fifteenth and sixteenth centuries German herbalists

used *Prunella* and *Brunella* interchangeably: the name could be from the south German *braun* (and from the Latin *prunus*) meaning 'purple' (not north German *braun* = 'brown'), from the flower colour, or more likely from *Bräune* (quinsy), which this plant was believed to cure. *Vulgaris* is Latin for 'common'.

Cut-leaved selfheal *P. laciniata*, a rare, scattered rhizomatous perennial herb, possibly native, otherwise a neophyte naturalised in dry calcareous soils in grassland and on roadsides, waste ground and woodland rides. *Laciniata* is Latin for 'torn into narrow divisions'.

Large-flowered selfheal *P. grandiflora*, a herbaceous perennial neophyte found as an occasional casual in grassy habitats. *Grandiflora* is Latin for 'large-flowered'.

SEMELIDAE Family name for the bivalve furrow shells, after Semele, who was the mother of Dionysus (fathered by Zeus) in Greek mythology.

SEMI-SLUG Keeled pulmonates, members of the glass snail family Vitrinidae, shell width 5–7 mm, height 3.5–6 mm.

Greater semi-slug *Vitrina major*, scattered in the southern third of England and in south Wales, declining, an indicator of ancient woodland, and found in other damp shaded habitats. Latin *vitrum* = 'glass', *major* = 'greater'.

Pyrenean semi-slug *Semilimax pyrenaicus*, scattered in Ireland in moist litter in floodplain alder woodland and willow carr, and in more open, often disturbed (but ungrazed) habitats. *Limax* is Latin for 'slug'.

Winter semi-slug *V. pellucida*, common and widespread in moist, shaded woodland, grassland, waste ground and rubble, most commonly seen in autumn, feeding on decomposing leaves and liverworts. *Pellucida* is Latin for 'transparent'.

SENNA **Scorpion senna** *Hippocrepis emerus* (Fabaceae), an introduced European shrub, local in southern and central England naturalised on roadsides and banks. 'Senna' dates from the sixteenth century, ultimately from Arabic *sana*, seeds of various species being used as a laxative. Greek *hippos* + *crepis* = 'horse' + 'shoe', from the shape of the pod.

SENTINEL **Dun sentinel** *Assiminea grayana* (Assimineidae), an estuarine and saltmarsh snail, shell height 4–6 mm, width 3 mm, found at or just above the high tide level from Kent to the Humber estuary, Dee estuary, north Lancashire and the Shannon estuary.

SEPSIDAE Family name of ensign flies, with 29 species in 6 genera, the most speciose being *Sepsis* (Greek for 'putrefaction') and *Themira*.

SERAPHIM *Lobophora halterata* (Geometridae, Larentiinae), wingspan 20–5 mm, a nocturnal moth, widespread but local in broadleaf woodland. Adults fly in May–June. Larvae feed on aspen and black poplar. Greek *lobos* + *phoreo* = 'lobe' + 'to bear', and Latin *halter* = 'dumb-bell', both describing the lobe on the male's hindwing.

Small seraphim *Pterapherapteryx sexalata*, wingspan 22–5 mm, nocturnal, widespread but local in damp woodland, fen and marsh, and on riverbanks. In southern England there are usually two generations, flying in May–June then July–August; otherwise one brood appears in June–July. Larvae feed in autumn on sallows. Greek *pteron* + *phero* + *pteryx* = 'wing' + 'to carry' + 'wing', describing the male's hindwing where a lobe resembles an ancillary wing, and Latin *sex* + *ala* = 'six' + 'wing', for the same reason.

SERICOSTOMATIDAE A caddis fly family (Greek *sericon* + *stoma* = 'silk' + 'mouth'). See Welshman's button.

SERIN *Serinus serinus* (Fringillidae), wingspan 22 cm, length 12 cm, weight 13 g, a finch that has been recorded annually in Britain in small numbers since the 1960s and which has bred occasionally since the 1970s (in Devon, Dorset, Sussex, East Anglia and Jersey) but no more than one or two pairs a year. It is mainly a passage migrant seen in southern England, especially in April–May. 'Serin' is French for 'canary'.

SEROTINE *Eptesicus serotinus* (Vespertilionidae), head-body length 58–80 mm, wingspan 320–80, weight 15–35 g, a bat found in England largely south of the Wash and in Wales. The English population was estimated in 1995 as 15,000. In Wales, records are rare and numbers unestimatable. Roost surveys up to 2015 indicate a 22.1% decrease on a 1999 base year value, equivalent to an annual decrease of 1.6%. Roosting and hibernation are almost always in buildings. They hunt insects (mainly flies and moths in spring, cockchafers and dung beetles in summer) by flying at tree-top height (up to 10 m) and around street lamps, occasionally taking prey from the ground. Mating is in autumn. Maternity roosts are established from May. Females give birth to a single young in July. At three weeks the young start flying and are independent at six weeks. Echolocation calls in the range 15–65 kHz, peaking at 25–30 kHz. 'Serotine' is late eighteenth-century from French *sérotine*, in turn from Latin *serotinus* = 'belated', here meaning 'of the evening'. *Eptesicus* comes from Greek *epten* + *oicos* = 'I fly' + 'house'.

SERPENT STAR *Ophiura ophiura* (Ophiuridae), starfish with a disc up to 35 mm diameter with five arms up to 140 mm, widespread and common over sand and muddy sand from the lower intertidal to depths of 200 m. Greek *ophis* = 'snake'.

Black serpent star Alternative name for black brittlestar.

Little serpent star Alternative name for serpent's table brittlestar.

SERPENTES Suborder of reptiles containing snakes (Latin *serpens* = 'snake').

SERPULIDAE Family name (Latin *serpula* = 'little snake') of some polychaete bristleworms and tubeworms. See also horseshoe worm.

SERRADELLA *Ornithopus sativus* (Fabaceae), an annual introduced herb from Europe, naturalised on china-clay waste in Cornwall since 1978, and a rare casual elsewhere in southern England. 'Serradella' comes from Portuguese *serradela*, from Latin *serratus* = 'serrate'. Greek *ornis* + *pous* = 'bird' + 'foot', and Latin *sativus* = 'cultivated', it being used as forage and green manure on the continent.

Yellow serradella *O. compressus*, an annual introduced from Europe, naturalised on a sandy bank in west Kent since 1957, and a rare casual elsewhere in southern England and the Channel Islands. *Compressus* is Latin for 'compressed'.

SERVICE-TREE *Sorbus domestica* (Rosaceae), a shrub or small tree archaeophyte found on coastal cliff ledges and in cliff scrub and gorge woodland, mostly on limestone, but occasionally on mudstone or shale. It was originally known as a single tree in the Wyre Forest, Worcestershire, first described in 1678, destroyed in 1862; the five trees now existing near there are probably cuttings from the original. Its coastal cliff habitat was discovered in 1973: 22 trees are now known in Glamorgan, and eight in England, the largest English population being in the Horseshoe Bend SSSI at Shirehampton, near Bristol. A further population was found on a cliff in the Camel Estuary, Cornwall. It is also occasionally grown in gardens and parks. *Sorbus* is Latin for 'rowan'.

Arran service-tree *S. pseudofennica*, an endemic deciduous shrub or small tree, the 400–500 specimens known only from the sides of stream ravines and granite crags, and on rocky moorland, at Glen Catacol, Arran. *Pseudofennica* is Latin for 'falsely Finnish'.

Swedish service-tree *S. hybrida*, a deciduous neophyte from Scandinavia, frequently grown in gardens and parks since 1779, along streets and on roadsides, naturalised in semi-natural woodland, rough grassland and chalk-pits in Kent and Aberdeen.

Wild service-tree *S. torminalis*, locally common in England and Wales on a variety of soils, often clayey, or over limestone, an indicator of ancient woodland, found in hedgerows and scrub,

and planted in parks and plantations. The plant was supposedly effective against colic (Latin *tormina*).

SESIIDAE Family name of clearwing moths (Greek *ses* = 'moth'). See also hornet moth under moth.

SEWAGE WORM Alternative name for sludge worm.

SEXTON BEETLE Or burying beetles, members of the genus *Nicrophorus* (Silphidae, Nicrophorinae) (Greek *necros* + *phero* = 'corpse' + 'to carry').

Black sexton beetle or **black burying beetle**, *N. humator*, 20–30 mm length, widespread but more common in central and south-east England and East Anglia, adults active in April–September, flying to carrion, attracted by the smell. If the dead animal is small they crawl underneath and excavate the soil so that the corpse sinks into the ground. They may skin the corpse and amputate limbs to make burying easier. The female digs a passage off from the carcass and lays her eggs, then returns to feed on the corpse. She feeds the larvae by regurgitating liquid food until they can feed off the carcass themselves. During this time the male may remain or may leave. *Humator* is from Latin *humus* = 'soil' or 'ground'.

Common sexton beetle or **common burying beetle**, *N. vespillo*, 12–22 mm length, widespread and common in central and southern England and Wales, more local and scattered further north and in Ireland. Adults are active in April–October, flying to carrion, attracted by the smell. If the dead animal is small both the male and female bury the corpse and cover it with soil. The female digs a passage off from the carcass and lays her eggs, then returns to feed on the corpse. *Vespillo* is Latin for 'undertaker'.

SHAD Coastal planktivorous anadromous fish in the herring family (Clupeidae), genus *Alosa* (Latin *alausa* = 'shad'), that return to fresh water in April–May to spawn above gravel substrates. Juveniles stay in rivers for up to two years, then have an estuarine phase before moving to the sea. A recent decline in numbers comes from overfishing (as by-catch) and, mostly, because of disrupted migration coming from dams, sluices, weirs and pumping stations. 'Shad' comes from Old English *sceadd*, possibly from Norse *skadd* = 'small whitefish'.

Allis shad *A. alosa*, 30–60 cm, found for example in the rivers Tamar, Wye and Usk.

Twaite shad *A. fallax*, 20–40 cm, found for example in rivers flowing into the Severn estuary, and some in south-west England and the Solway Firth. 'Twaite' is Middle English, from Old Norse *thveit* = 'paddock', literally 'cut piece'. Latin *fallax* = 'deceptive' though another suggestion uses *halec* = 'pickle' derived from Greek *alas* = 'salt'.

SHADE (MOTH) Members of the family Tortricidae, mostly Tortricinae. (Not to be confused with the noctuid shades.)

Widespread but local species include **bluebell shade** *Eana incanana* (in woodland throughout much of Britain, rare in Ireland, larvae feeding on bluebell and oxeye daisy flowers), **deep brown shade** *Neosphaleroptera nubilana* (hedgerows, scrub and gardens in parts of England and Wales, larvae feeding on hawthorn and blackthorn leaves), **dotted shade** *E. osseana* (widespread but local on grassland, downland and moorland), **hedge shade** *Isotrias rectifasciana* (hedgerows and woodland edge in England, Wales and southern Scotland, larvae feeding on hawthorn leaves), **May shade** *C. communana* (in England and Wales on meadow, rough grassland, dry pasture, downland and fen), **meadow shade** *C. pasiuana* (rough pasture, marsh, gardens, woodland edge and other open areas throughout much of England, larvae feeding on flowers of a number of Asteraceae and buttercups).

Autumnal shade *Exapate congelatella*, wingspan 18–22 mm, day-flying, widespread but local in hedgerows (more so in the south) and moorland (in the north). Adults fly in October–December. Larvae feed on the leaves of various woody plants, including hawthorn, privet and bramble. Greek *exapate* = 'gross deceit' (a tortricid looking like a tineid), and Latin *congelatus* = 'frozen', adults appearing with early frosts.

Coast shade *Cnephasia conspersana*, wingspan 15–20 mm, nocturnal, on chalk downland and heathland, mainly on the coasts of England, Wales and Ireland, but also inland on the chalk in south-east England and on the limestone of the Burren, west Ireland. Adults fly in July. Larvae feed on Rosaceae and Compositae, spinning flowerheads together and living inside. Greek *cnephas* = 'darkness', and Latin *conspersus* = 'sprinkled', from grey speckling on the whitish forewing.

Dover shade *C. genitalana*, wingspan 15–22 mm, scarce in gardens, woodland edge, rough grassland and chalk downland in parts of southern and eastern England and south Wales. Adults fly at night in July–August. Larvae feed on flowers and leaves of a range of herbaceous plants, for example buttercups and ragworts. *Genitalana* reflects separation of this as a distinct species using its genitalia.

Large mottled shade *Eana penziana*, wingspan 20–5 mm, scarce on some cliffs along the northern coasts of Britain and Ireland (ssp. *colquhounana*) and on hills and mountains in northern regions (ssp. *bellana*). The latter occurs from Lancashire northwards, flying in June–July; the coastal subspecies is found from Wales northwards, and in west Ireland. It has two generations, flying in May and in August–September. Larvae of *bellana* feed on roots of sheep's-fescue, those of *colquhounana* (named for the nineteenth-century Glaswegian physician Dr H. Colquhoun) on sea plantain and thrift, both subspecies living in silk tubes. The eighteenth-century Consiliar D. Pentz collected the type specimen in Sweden.

Long-winged shade *C. longana*, wingspan 15–22 mm, nocturnal, common on downland and rough ground throughout much of England and Wales. Adults fly in July–August. Larvae feed on a number of herbaceous plants, spinning flowerheads together and living inside. *Longana* (from Latin *longus*) reflects the elongated forewing.

Silver shade *Eana argentana*, wingspan 20–7 mm, recorded from a couple of grassy mountainous sites in the Highlands. Adults fly in June–July. Larvae feed inside a silk tent on grasses, herbs and mosses. Latin *argentum* = 'silver', from the wing colour.

Winter shade *Tortricodes alternella*, wingspan 19–23 mm, common in deciduous woodland, especially of oak, throughout England, Wales, southern Scotland and the southern half of Ireland. Adults fly in February–April. Larvae feed on the leaves of a variety of trees. Greek *eidos* = 'resembling' + *Tortrix*, and Latin *alternus* = 'alternate', from the alternation of grey and brown on the forewing.

SHADES (MOTH) Angle shades *Phlogophora meticulosa* (Noctuidae, Xyleninae), wingspan 45–50 mm, nocturnal, widespread and common in a range of habitats, including gardens, hedgerows, fen and woodland. As a common migrant it is sometimes also found in large numbers at coastal sites. Adults fly in May–October in two generations. Larvae feed on a variety of herbaceous and woody plants. Greek *phlox* + *phoreo* = 'flame' + 'to carry', and Latin *meticulosa* = 'fearful'.

Small angle shades *Euplexia lucipara*, wingspan 27–32 mm, nocturnal, widespread and common in gardens, parks, woodland, heathland and moorland. Adults fly in June–July, sometimes with a second generation in August–September. Larvae feed on bracken and other ferns, and on birches and sallows. Greek *eu* + *plexis* = 'well' + 'weaving', from the resting position, and Latin *luciparens* = 'light-bearing', from a yellow spot on a black band on the forewing resembling a lamp shining out of darkness.

SHADOW (FUNGUS) Bog shadow *Arrhenia umbratilis* (Tricholomataceae), inedible, with two recent records from South Uist, Outer Hebrides, in 2012, in acidic grassland. Greek *arrhen* = 'male', and Latin *umbratus* = 'shading'.

SHAG *Phalacrocorax aristotelis* (Phalacrocoracidae), wingspan 98 cm, length 72 cm, weight 1.9 kg, breeding on coastal sites, mainly

in Orkney, Shetland, the Inner Hebrides and the Firth of Forth, with over half the population found at <10 sites, making this a Red List species. SMP data for 2004 suggest 26,600 occupied nests, but BBS data indicate a decline in numbers by 48% over 1986–2014, by 38% in 2000–14. With migrant influx, wintering numbers in 2002 were 110,000, found around all coastlines, with some also inland especially in east England. Nests are of seaweed and twigs on rocky ledges. Clutch size is usually 3, incubation takes 30–1 days, and the altricial young fledge in 48–58 days. Shag feed on fish by diving to a depth of up to 45 m. The English name may be a reference to the 'shaggy' crest. Greek *phalacros* + *corax* = 'bald' + 'raven', and Latin *aristotelis* = 'of Aristotle'.

SHAGGY-MOSS Species of *Rhytidiadelphus* (Hylocomiaceae) (Greek *rhytis* + *di* + *adelphos* = 'wrinkle' + 'two' + 'brother').

Big shaggy-moss *R. triquetrus*, common and widespread in calcareous woodland, on acidic ground in pinewood, chalk grassland, dunes and churchyards. *Triquetrus* is Latin for 'triangular'.

Little shaggy-moss *R. loreus*, common and widespread, less so in the Midlands and eastern England, in acidic woodland in upland areas, and in acidic grassland and on heathy slopes in montane western Britain. *Loreus* is Latin for 'made of thongs'.

SHAGGY-SOLDIER *Galinsoga quadriradiata* (Asteraceae), an annual South American neophyte first recorded in the wild in 1909 in Middlesex, perhaps introduced with garden plants, though it is also a birdseed alien. It grows in arable fields, waste ground, roadsides, derelict urban sites and pavement cracks, mostly found in conurbations in England. The Spanish physician Ignacio de Galinsoga (1766–97) was director of the Real Jardín Botánico de Madrid. Latin *quadriradiata* means spreading 'four rays'.

SHALLON *Gaultheria shallon* (Ericaceae), a low North American neophyte shrub introduced by 1826, widespread and locally naturalised in woodland and open heathland on acidic, sandy and peaty soils. Originally planted as cover for game birds, it has in places become a serious weed, especially on lowland heath where it regenerates rapidly after clearance. In the New Forest, pigs penned in infested areas have proved effective in uprooting it. Jean-François Gaultier (1708–56) was a Canadian physician and botanist. 'Shallon' is from Chinook *kikwu-salu* for this plant.

SHALLOT See onion.

SHANK Fungi in the order Agaricales.

Rooting shank *Xerula radicata* (Physalacriaceae), common and widespread, poorly edible, found in summer and autumn on decomposing hardwood and coniferous wood, often deep beneath leaf litter. Greek *xeros* = 'dry', and Latin *radicata* = 'rooted'.

Velvet shank *Flammulina velutipes* (Physalacriaceae), common and widespread, edible, found in autumn and winter, one of the few agarics that survive heavy frost, on trunks, stumps and branches of dead and diseased hardwoods, especially ash, beech and oak. Latin *flammulina* = 'little flame', from the orange caps, and *velutinus* + *pes* = 'velvety' + 'leg' (i.e. stem).

Whitelaced shank *Megacollybia platyphylla* (Marasmiaceae), inedible, widespread and locally common, especially in England, found from late spring to early autumn on decomposing stumps and buried wood. Greek *megas* + *collybos* = 'large' + 'small coin', from the flattish cap, and *platys* + *phyllon* = 'broad' + 'leaf' (here meaning 'gills').

SHANKLET Members of the fungus genera *Collybia* (Greek *collybos* = 'small coin', from the often round flattish caps) and *Dendrocollybia* (*dendron* = 'tree') (Tricholomataceae), the former generally living on the hardened remains of old fungi in other genera.

Branched shanklet *D. racemosa* and **lentil shanklet** *C. tuberosa* are both rare.

Piggyback shanklet *C. cirrhata*, widespread on soil and litter, especially in coniferous woodland. *Cirrhata* is from Greek for 'having tendrils' (possibly from *cirrhos* = 'yellow').

Splitpea shanklet *C. cookei*, widespread on soil, litter and old fungi in broadleaf woodland, especially under beech and oak.

SHANNY *Lipophrys pholis* (Blenniidae), length 16 cm, a common and widespread intertidal fish found in rock pools, or under stones and seaweed (and often found around man-made structures) emerging at high tide to feed on barnacles, and other invertebrates. In winter they move from rock pools to the shallow subtidal. Spawning is in April–August. Greek *lipo* + *ophrys* = 'fatty' + 'eyebrow', and *pholis* = 'scale'.

SHARK (FISH) Cartilaginous fishes, some species occasionally found inshore.

Basking shark *Cetorhinus maximus* (Cetorhinidae), the largest fish in British waters, up to 9.8 m long, generally living in open water but migrating towards the shore in summer, seen 'basking' – swimming slowly at the surface with the wide mouth open to filter-feed on plankton. Greek *cetos* + *rhine* = 'whale' + 'shark', and Latin *maximus* = 'greatest'.

SHARK (MOTH) *Cucullia umbratica* (Noctuidae, Cucullinae), wingspan 42–52 mm, common in rough ground, chalk and limestone grassland, dunes, shingle beaches and gardens, widespread though less common in Scotland. Adults fly at night in June–July. Larvae mainly feed on the flowers and leaves of perennial sowthistle, hawk's-beard and mouse-ear-hawkweed. Latin *cucullus* = 'hood', from the shape of the thoracic crest, and *umbratica* = 'belonging to the shade', from the larva hiding under leaves during daytime.

Chamomile shark *C. chamomillae*, wingspan 40–2 mm, widespread but local on calcareous grassland, verges, waste ground, farmland and gardens, showing a preference for sandy or light calcareous soils, and probably most common near the coasts in England, Wales, Scotland to Perthshire, and Ireland (especially the east coast). Adults fly at night in April–June. Larvae feed on composites such as chamomiles, mayweed and feverfew.

SHEARS (MOTH) *Hada plebeja* (Noctuidae, Hadeninae), wingspan 30–5 mm, widespread and common on heathland, dunes, downland and open woodland. Adults fly at night in May–July, with a partial second generation in the south. Larvae feed on dandelion, hawkweeds, lucerne and other herbaceous plants. *Hada* is from Hades, the Underworld in Greek mythology, and Latin *plebeja* = 'plebeian'.

Dingy shears *Apterogenum ypsillon* (Xyleninae), wingspan 32–42 mm, nocturnal, local in damp woodland, marsh and fen throughout much of England, scarce in Wales, Scotland and north-east and east Ireland. Adults fly in June–July. Larvae feed at night on sallows and poplars, hiding in loose bark during the day. Greek *apteros* + *genos* = 'wingless' + 'kind', and the Greek letter *ypsilon*, from the wing markings considered by British entomologists to resemble shears, giving the common name.

Glaucous shears *Papestra biren* (Hadeninae), wingspan 32–40 mm, local on moorland, particularly in much of north-west England, the north Midlands, parts of Wales, Scotland and Ireland. Adults fly at night in May–June. Larvae feed on heather, bilberry, bog-myrtle, creeping willow and meadowsweet. *Papestra* is an entomologically meaningless neologism; *biren* is from Latin *bi* + *renes* = 'two' + 'kidneys', from the wing markings.

Tawny shears *Hadena perplexa*, wingspan 27–36 mm, nocturnal, common in dry open areas, farmland and gardens throughout much of England (especially East Anglia and Cornwall), Wales, south-west Scotland and parts of Northern Ireland. Adults fly in May–June, sometimes again in autumn in the south. Larvae feed campions, rock sea-spurrey and Nottingham catchfly. *Perplexa* is Latin for 'intricate', describing the complexity and variety of the wing pattern.

SHEARWATER Members of the petrel family (Procellariidae). The birds fly close to the water and cut or 'shear' the tips of waves to move across wave fronts with minimal need for active flight.

Balearic shearwater *Puffinus mauretanicus*, **Cory's shearwater** *Calonectris borealis*, **great shearwater** *P. gravis*, **Scopoli's shearwater** *C. diomedea* and **sooty shearwater** *P. griseus* are all passage visitors.

Manx shearwater *Puffinus puffinus*, wingspan 82 cm, length 34 cm, weight 420 g, a migrant breeder, with 300,000 pairs estimated in 2000, nesting in burrows in colonies on offshore islands, notably Skomer and Skokholm in south Wales, Rum in the Inner Hebrides, and the Blasket Islands, Co. Kerry, where it is safe from rats and other ground predators. Birds leave their nest sites in July, to migrate to the coast of South America, where they spend the winter, returning in late February and March. A single egg is laid, incubation takes 47–55 days, and fledging of the altricial young 62–76 days. With legs set far back these birds shuffle or drag themselves, rather than walk. Small fish, plankton, molluscs and crustaceans are taken at sea during the day, the (non-incubating/brooding) birds returning to their burrows at night as a defence against predators such as gulls. This is the longest-lived bird in Britain, with one ringed at Copeland Island, Co. Down, being in 2003 at least 55 years old. It is Amber-listed, with 90% of the global population nesting in the British Isles. They are named after the Isle of Man but became extinct there in the eighteenth century when rats were introduced. A small breeding colony has recently become established on the Calf of Man. The binomial records a name shift: Manx shearwaters were called Manks puffins in the seventeenth century; *puffin* is an Anglo-Norman word for the cured carcass of a nestling shearwater.

SHEEP *Ovis aries* (Bovidae), probably descended from mouflon, a domesticated Neolithic introduction, with two feral forms from islands in the St Kilda archipelago, >60 km west-north-west of North Uist, Outer Hebrides. Those from Soay (Old Norse for 'island of sheep'), standing only 50 cm at the shoulder, were known in the Bronze Age. Some sheep were translocated to Hirta in 1930, when the St Kildans left the islands, and the sheep in Soay and Hirta have become feral. The breed had previously been introduced elsewhere, for example Holy Island (Firth of Clyde), Lundy (Bristol Channel), Woburn (Bedfordshire) and Cardigan Island. Boreray sheep are also named after a St Kilda island, and again are now feral there, with 204 ewes and a few rams recorded in 2012. Total numbers of feral sheep in the UK in 2005 were estimated at 150 in England, 1,850 in Scotland and 100 in Wales. Soay sheep are listed as 'Category 4: At Risk' by the Rare Breeds Survival Trust, and Broreray as 'Category 2: Endangered'. 'Sheep' comes from Old English *scep* or *sceap*. The binomial comprises two Latin words for 'sheep'.

SHEEP-LAUREL *Kalmia angustifolia* (Ericaceae), an evergreen North American neophyte shrub cultivated by 1736, found as a garden relic, and naturalised in Britain in a few places on acidic, boggy areas on moorland and waste ground. Pehr Kalm (1716–79) was a Swedish/Finnish traveller. Latin *angustifolia* = 'narrow-leaved'. See also mountain-laurel.

SHEEP'S-BIT *Jasione montana* (Campanulaceae), a herbaceous biennial widespread and locally common, especially in the west, by the coast on acidic, shallow, well-drained soils on cliffs, in maritime grasslands and heathland and on stabilised dunes, as well as inland on heathland, walls, hedge banks and railway cuttings. Disturbed, open habitats and recently burnt ground are often colonised. *Jasione* is Greek for this plant, *montana* Latin for 'montane'.

SHELDUCK *Tadorna tadorna* (Anatidae), wingspan 112 cm, length 62 cm, male weight 1.2 kg, females 1 kg, a resident breeder, with 15,000 pairs in 2009, numbers in winter rising to 61,000, with migrant visitors adding to native stock. Numbers increased by 127% between 1970 and 2014, mostly in the 1970s and 1980s since they declined by 6% between 1995 and 2014. They are widespread, mainly in coastal areas, but also around inland waters such as reservoirs and flooded gravel-pits. Nests are in rabbit burrows or tree holes. Clutch size is 8–10, incubation lasts 29–31 days, and fledging of the precocial young takes 45–50 days. Diet is small shellfish, aquatic snails and insects. This is an Amber-listed species. 'Shelduck' (c.f. 'sheldrake') comes from Middle English, probably from dialect *sheld* = 'pied'. *Tadorna* derives from *tadorne*, French for this species.

SHELL (MOTH) Members of the family Geometridae (Larentiinae).

Dingy shell *Euchoeca nebulata*, wingspan 23–5 mm, nocturnal, local in damp woodland and marsh and on riverbanks throughout much of England and Wales, rare in Scotland and eastern Ireland. Population levels increased by 214% between 1968 and 2007. Adults fly in June–July. Larvae feed on alder leaves. Greek *eu* + *choichos* = 'good' + 'earth' or 'clay', and Latin *nebula* = 'cloud', both describing colour.

Scallop shell *Hydria undulata*, wingspan 25–30 mm, common in damp woodland and scrub throughout much of England, Wales, south Scotland and Ireland. Adults fly in June–July. Larvae feed on shoot tips of bilberry, aspen and sallows. Greek *hydor* = 'water', and Latin *undulata* = 'wavy', from the wing markings.

Yellow shell *Camptogramma bilineata*, wingspan 20–5 mm, widespread and generally common in a range of habitats, including gardens, with a preference for damper sites, and less common in upland areas. Population levels increased by 101% between 1968 and 2007. Adults fly (May–August) from dusk onwards but also during the day if disturbed. Larvae feed on low-growing plants, including cleavers, chickweed, docks and sorrel. Greek *campos* + *gramma* = 'bent' + 'line', and Latin *bilineata* = 'two-lined', from the wing pattern.

SHEPHERD'S-NEEDLE *Scandix pecten-veneris* (Apiaceae), an annual archaeophyte of arable, especially on calcareous clay, declining, and scattered in southern and eastern England. The binomial is Greek for 'chervil' and 'comb of Venus', venus comb being another name for this plant. The English name come from the long fruit, up to 1.5 cm long with a 'beak' which can measure up to 7 cm and is lined with comb-like bristles.

SHEPHERD'S-PURSE *Capsella bursa-pastoris* (Brassicaceae), an annual archaeophyte that can germinate throughout the year in suitable conditions, common in disturbed and nutrient-rich habitats. It is abundant on waste ground and in gardens, and frequent in cultivated fields. It avoids very wet and acidic soils. *Capsella* is Latin for 'little case'. The English name and (in Latin) the specific name come from the triangular flat seed pods which are purse-like.

Pink shepherd's-purse *C. rubella*, a casual annual or biennial European neophyte, occasionally found in southern England and the south Midlands on cultivated and waste ground, possibly as a grain impurity. *Rubella* is Latin for 'pale red'.

SHIELD (FUNGUS) Saprobic, generally inedible members of the genus *Pluteus* (Pluteaceae), from the Latin for 'armoured breastwork', with 40 British species.

Blackedged shield *P. atromarginatus* (on conifers), **deer shield** *P. cervinus*, **dwarf shield** *P. nanus*, **fleecy shield** *P. hispidulus*, **ghost shield** *P. pellitus*, **goldleaf shield** *P. romellii*, **lion shield** *P. leoninus*, **satin shield** *P. plautus*, **scaly shield** *P. petasatus*, **veined shield** *P. thomsonii*, **velvet shield** *P. umbrosus*, **willow shield** *P. salicinus*, **wrinkled shield** *P. phlebophorus*, and **yellow shield** *P. chrysophaeus* (especially on elm) are all widespread, scattered but often uncommon on stumps, deadwood and woody debris.

Ashen shield *P. cinereofuscus*, scattered mainly in England, on soil and litter in deciduous woodland. *Cinereofuscus* fuses the Latin for 'grey' and 'dark'.

SHIELD-FERN Members of the genus *Polystichum* (Dryopteridaceae) (Greek for 'many rows of spores'). See also holly-fern and sword-fern.

Hard shield-fern *P. aculeatum*, in (semi-)shady woodland on moist, fairly infertile, base-rich soils, more common in upland areas. *Aculeatum* is Latin for 'pointed'.

Soft shield-fern *P. setiferum*, generally common and widespread on moist, moderately base-rich, fertile soils in woodland and on hedge banks. *Setiferum* is Latin for 'bristle-bearing'.

SHIELD-MOSS Species of *Buxbaumia* (Buxbaumiaceae). Johann Buxbaum (1693–1730) was a German physician and botanist.

Brown shield-moss *B. aphylla*, widespread and scattered in Britain in shady habitats, especially coniferous woodland, favouring decomposing wood, or humic well-drained sandy soils, and also recorded from colliery spoil. *Aphylla* is Greek for 'leafless'.

Green shield-moss *B. viridis*, locally common in the Highlands mainly on fallen dead wood, mostly spruce, pine, alder and birch. It also grows on bare patches of ground and on wood ant nests. *Viridis* is Latin for 'green'.

SHIELDBUG Species in the superfamily Pentatomoidea (Heteroptera), characterised by a triangular or semi-elliptical scutellum. Nymphs are similar to adults except smaller and without wings, and also have stink glands.

Birch shieldbug *Elasmostethus interstinctus* (Acanthosomatidae), length 8–11.5 mm, widespread and common, particularly on birch; larvae are also found on hazel and aspen. Greek *elasmos* + *stethos* = 'thin plate' + 'breast'.

Bishop's mitre shieldbug *Aelia acuminata* (Pentatomidae), length 8–9 mm, widespread and locally common in Wales and across England as far north as Lincolnshire, feeding on grasses in tall dry grassland and dunes. *Acuminata* is Latin for 'pointed'.

Blue shieldbug *Zicrona caerulea* (Pentatomidae), length 5–7 mm, widespread on heathland, grassland and woodland rides throughout Britain, only occasionally in Scotland and Ireland, feeding on beetle larvae. *Caerulea* is Latin for 'blue'.

Bordered shieldbug *Legnotus limbosus* (Cydnidae), length 3.5–4.5 mm, common and widespread in south England, scarcer further north, and absent from Scotland and Ireland. Found on bedstraws on dry sunny sites. *Legnotos* is Greek for 'with a coloured border', *limbosus* Latin for 'bordered'.

Brassica shieldbug *Eurydema oleracea* (Pentatomidae), length 6–7 mm, widespread but local in south and much of central England, feeding on brassicas. Greek *eurys* + *demas* = 'broad' + 'body', and Latin *oleracea* = 'resembling herbs'.

Bronze shieldbug *Troilus luridus* (Pentatomidae), length 10–11 mm, in woodland in England, Wales and parts of Ireland, though not common, feeding on caterpillars and plants. Troilus was a Trojan prince. *Luridus* is Latin for 'wan' or 'pale yellow'.

Common green shieldbug *Palomena prasina* (Pentatomidae), length 12–13.5 mm, common and widespread throughout England, Wales and Ireland, essentially absent from Scotland, in many habitats and often seen in gardens. *Prasina* is Latin for 'leek green'.

Crucifer shieldbug Alternative name for brassica shieldbug.

Forget-me-not shieldbug *Sehirus luctuosus* (Cydnidae), length 7–9 mm, widespread but local in England, especially on warm open chalky sites, favouring seeds of forget-me-not. *Luctuosus* is Latin for 'mournful'.

Gorse shieldbug *Piezodorus lituratus* (Pentatomidae), length 10–13 mm, widespread, common where gorse is present, though larvae also feed on broom, dyer's greenwood, and other plants. *Lituratus* is Latin for 'erased'.

Green shieldbug See common green shieldbug above.

Hairy shieldbug *Dolycoris baccarum* (Pentatomidae), length 11–12 mm, common and widespread in a variety of habitats (though particularly woodland edge and hedgerows), and favouring coastal areas further north. Feeds on a variety of plants. Greek *dolos* + *coris* = 'deceit' + 'bug', and Latin *baccarum* = 'berry (shaped)'.

Hawthorn shieldbug *Acanthosoma haemorrhoidale* (Acanthosomatidae), length 13–15 mm, common and widespread on hawthorn in woodland and hedgerows, particularly in southern

England, becoming scarce in Scotland. Greek *acantha* + *soma* = 'thorn' + 'body', and *haema* + *rein* = 'blood-red' + 'to flow'.

Heath shieldbug *L. picipes* (Cydnidae), length 3–4 mm, scarce in southern and eastern England on dune and heathland, feeding on bedstraws. Latin *picipes* = 'woodpecker-like feet'.

Heather shieldbug *Rhacognathus punctatus* (Pentatomidae), length 7–9 mm, widespread, but scarce, feeding mainly on heather beetle larvae and so confined at low densities to some heathlands and moorlands. Greek *rhacos* + *gnathos* = 'rags' + 'jaw', and Latin for 'spotted'.

Juniper shieldbug *Cyphostethus tristriatus* (Acanthosomatidae), length 9–10.5 mm, formerly found only in juniper woodland in south England, now more widespread and common across southern and central England owing to widespread planting of juniper and particularly Lawson's cypress in gardens. Greek *cyphos* + *stethos* = 'crooked' + 'breast', and Latin *tristriatus* = 'three-striped'.

Mottled shieldbug *Rhaphigaster nebulosa* (Pentatomidae), length 14–16 mm, first recorded in Britain from the London area in 2010 (following range expansion on the continent), well-established in south London the following year, and continuing to spread, feeding on a number of deciduous tree species. Greek *rhaphe* + *gaster* = 'seam' + 'stomach', and Latin for 'cloudy'.

Pied shieldbug *Tritomegas bicolor* (Cydnidae), length 5.5–7.5 mm, widespread and common in England and Wales in hedgerows and woodland edge, mainly on white dead-nettle and black horehound. Greek *tritos* + *megas* = 'the third' + 'great', and Latin *bicolor* = 'two-coloured'.

Red-legged shieldbug *Pentatoma rufipes* (Pentatomidae), length 11–14 mm, widespread and common, particularly associated with oak, hazel and alder in woodland, and trees in gardens and (especially cherry and apple) orchards. Adults feed on caterpillars and other insects as well as fruit. Greek *pente* + *tomos* = 'five' + 'section', and Latin *rufipes* = 'red-footed'.

Scarab shieldbug *Thyreocoris scarabaeoides* (Thyreocoridae), length 3–4 mm, found in parts of southern England, occasional in East Anglia and Lincolnshire, in moss and litter and associated with violets on dry sandy or chalky soils. Greek *thyreos* + *coris* = 'door-shaped shield' + 'bug', and Latin *scarabaeus* = '(scarab) beetle'.

Sloe shieldbug Alternative name for hairy shieldbug.

Small grass shieldbug *Neottiglossa pusilla* (Pentatomidae), in parts of south, south-east and central England, nymphs feeding on meadow-grass. Greek *neottia* + *glossa* = 'bird's nest' + 'tongue', and Latin *pusilla* = 'very small'.

Spiked shieldbug *Picromerus bidens* (Pentatomidae), widespread, Scottish records being scarcer and more recent, particularly associated with heathland but also recorded from flower-rich meadows and chalk downland, feeding on insect larvae and plant sap. Greek *picros* + *meros* = 'bitter' + 'part', and Latin *bis* + *dens* 'twice' + 'tooth'.

Tortoise shieldbug *Eurygaster testudinaria* (Scutelleridae), length 9–11 mm, widespread in south Britain and Ireland, expanding northwards, in grassland. Nymphs feed on grasses. Greek *eurys* + *gaster* = 'broad' + 'stomach', and Latin *testudinaria* = 'tortoise-like'.

Turtle shieldbug *Podops inuncta* (Pentatomidae), length 5–6 mm, widespread and common in much of south, south-east and central England and East Anglia, ground-dwelling in grassland. Greek *podos* + *opsis* = 'foot' + 'appearance', and Latin *inuncta* = 'smeared with oil'.

Woundwort shieldbug *Eysarcoris venustissimus* (Pentatomidae), length 5–7 mm, common and widespread in the Welsh Borderland and England as far north as Yorkshire in woodland edges and hedgerows, larvae feeding on hedge woundwort and other members of the Lamiaceae. Latin *venustissimus* = 'most elegant'.

SHINER *Lactarius uvidus* (Russulales, Russulaceae), an inedible milkcap fungus, locally common especially in Scotland, found in late summer and autumn in damp acid soil in broadleaf woodland under birch and willow, with which it is ectomycorrhizal. Latin *lactarius* = 'milk-producing', from the milky latex that exudes from the gills when torn, and *uvidus* = 'damp', from the habitat.

SHINING CLAW Pseudoscorpions (length 1.5–2.2 mm) in the genus *Lamprochernes* (Greek *lampros* + *cherne* = 'shining' + 'needy thing') (Chernetidae).

Chyzer's shining claw *L. chyzeri*, scattered and scarce in England, found beneath the bark of beech and birch trees, and in compost. It is phoretic on flies. Kornél Chyzer (1836–1909) was a Hungarian naturalist.

Knotty shining claw *L. nodosus*, widespread in England and Wales, scarce elsewhere, especially found in summer in compost, dung heaps and rotting wood of dead trees, feeding on small larvae, mites and tiny flies. *Nodosus* is Latin for 'knotty'.

Savigny's shining claw *L. savignyi*, scarce and scattered in compost and manure, and phoretic on flies. Marie de Savigny (1777–1851) was a French zoologist.

SHIPWORM Wood-boring marine bivalves in the family Teredinidae that drill tunnels using a pair of small valves at their anterior end.

Big-ear shipworm *Psiloteredo megotara*, bores into floating timber in the intertidal and below, widely recorded from around the coast. Greek *psilos* + *teredon* = 'naked' + 'woodworm'.

Great shipworm *Teredo navalis*, bores into wood, with scattered records from around Britain, mostly from the English Channel, south-west England and the Irish Sea. *Teredon* is Greek for 'woodworm', *navalis* Latin for 'ship'.

Norway shipworm *Nototeredo norvagica*, bores into wooden structures in a number of locations round the British and Irish coastline. Greek *notos* + *teredon* = 'back' + 'woodworm', and misprinted Latin *norvegica* = 'Norwegian'.

SHOO-FLY PLANT Alternative name for apple-of-Peru.

SHOOT (MOTH) Generally nocturnal members of the family Tortricidae (Olethreutinae), most with a wingspan of 11–15 mm.

Bramble shoot *Notocelia uddmanniana*, wingspan 15–20 mm, widespread and common in gardens, hedgerows and woodland. Adults fly in June–July. Larvae mostly mine leaves of bramble. Isaac Uddman (1733–81) was a Finnish entomologist.

Brindled shoot *Gypsonoma minutana*, scarce, in woodland edges and rides, parks and gardens throughout much of south-east England and East Anglia. Adults fly in July. Larvae mine leaves of aspen and poplars in May–June. Greek *gypsos* + *nomao* = 'chalk' + 'to distribute', from the extensive white areas on the wing of most moths in this genus, and Latin *minutus* = 'very small' (though the wingspan is average for the genus).

Common cloaked shoot *G. dealbana*, widespread and common in woodland. Adults fly in July–August. Larvae mine buds, young shoots and leaves of hazel, hawthorn, poplar and sallow. *Dealbo* is Latin for 'to whitewash', from the mainly white wings.

Dark pine shoot *Pseudococcyx posticana*, widespread but scattered and local in pine woodland and heathland. Adults become active towards the evening in May–June. Larvae mine buds and shoots of Scots pine. Greek *pseudo* + *coccyx* = 'false' + 'cuckoo', and Latin *posticus* = 'posterior'.

Elgin shoot *Rhyacionia logaea*, wingspan 14–18 mm, in pine woodland at a few sites in north-east Scotland. Adults fly in April–May from mid-afternoon into the night. Larvae feed on needles and young shoots of pines and Sitka spruce. Greek *rhyacion* = 'small stream', from the wavy rivulet forewing markings, and *logaios* = 'picked out'.

Orange-spotted shoot *R. pinicolana*, wingspan 16–23 mm, common in areas with pine throughout England and parts of Wales. Adults fly in July–August. Larvae feed on needles and young shoots of Scots pine (*pinus*, which with *colo* = 'to inhabit' gives the specific name).

Poplar shoot *G. oppressana*, in woodland edges, parks and gardens in parts of the southern half of England and Wales. Adults fly in June–July. Larvae mine poplar leaves, overwinter, then attack the buds. *Oppressus* is Latin for 'pressed down'.

Rosy cloaked shoot *G. aceriana*, local in woodland, gardens, orchards and parks throughout much of southern Britain, rare in Ireland. Adults fly in July. Larvae mine poplar shoots and buds, later moving into the stems. *Acer* is Latin for 'maple', though this is not a food plant.

Small pine shoot *Clavigesta sylvestrana*, scarce, local in pine woodland in parts of southern England. Adults fly in June–July. Larvae feed inside buds, shoots and flowers of pines, silver-fir and Norway spruce. Latin *clavus* (the purple stripe on the Roman tunic) + *gero* = 'to bear', from the forewing pattern, and a reference to Scots pine *Pinus sylvestris*.

Spotted shoot moth *R. pinivorana*, wingspan 15–19 mm, widespread and common in areas with pine, more numerous in Scotland. Adults fly from July, usually at night but sometimes in daylight. Larvae feed on buds and young shoots of Scots pine (*pinus*, which with *voro* = 'to devour' gives the specific name).

White cloaked shoot *G. sociana*, common in woodland, gardens, orchards and parks and on freshwater margins throughout much of Britain, rare in north Ireland. Adults fly in July–August. Larvae mine catkins of black poplar, aspen and sallow, later burrowing into and feeding on leaf buds. *Socius* is Latin for 'associate', for a similar species.

SHOOTING STAR *Sphaerobolus stellatus* (Geastrales, Geastraceae), a widespread and locally common inedible fungus, found on dead wood in deciduous woodland and on livestock dung in grassland. Greek *sphaera* + *bolos* = 'sphere' + 'lump', and Latin *stellatus* = 'starry'.

SHORE BUG Members of the family Saldidae (Heteroptera), common and widespread predatory species, including **common shore bug** *Saldula saltatoria*, found in many ponds, waterways and ditches.

SHORE CRAB See crab.

SHORE FLY Members of the family Ephydridae.

SHOREWEED *Littorella uniflora* (Plantaginaceae), a herbaceous perennial, mostly of upland regions, in oligotrophic or mesotrophic lakes, reservoirs, streams, ponds and winter-flooded dune slacks, on stones, gravel, sand, peat and mud to a depth of 4 m. Latin *litus* = 'seashore', *uniflora* = 'single-flowered'.

SHOULDER Flame shoulder *Ochropleura plecta* (Noctuidae, Noctuinae), wingspan 25–30 mm, a nocturnal moth widespread and common in a variety of habitats, including gardens, woodland edge, meadow and wetlands. The bivoltine adults fly in May–June and August–September. Larvae feed on plants such as groundsel, ribwort plantain and bedstraws. Greek *ochros* + *pleura* = 'pale' + 'rib', and *plecte* = 'twisted rope', both describing the pattern on the leading wing edge.

SHOULDER-KNOT Members of the nocturnal moth family Noctuidae (Xyleninae).

Blair's shoulder-knot *Lithophane leautieri*, wingspan 39–44 mm, which, as a colonist first recorded in Britain on the Isle of Wight in 1951, has rapidly spread northwards, with records from Cumbria. Population levels increased by 7,878% (from a low base) between 1968 and 2007. It is now common in coniferous woodland and especially suburban parks and gardens throughout much of England and Wales, and along the east coast of Ireland. Adults fly in October–November. Larvae feed on flowers and leaves of often ornamental conifers such as cypresses. Greek *lithos* + *phaino* = 'stone' + 'to appear', from the cryptic forewing. M. Leautier discovered the species in France in the early nineteenth century.

Grey shoulder-knot *L. ornitopus*, wingspan 32–8 mm,

common in oak woodland and parks throughout much of the southern half of England, and much of Wales and Ireland. Population levels increased by 1,269% (from a low base) between 1968 and 2007. Adults fly in September–October, overwinter, and reappear in spring. Larvae feed on oaks, and may be cannibalistic. Greek *ornis* + *pous* = 'bird' + 'foot', from the similarity of a forewing mark to a claw.

Minor shoulder-knot *Brachylomia viminalis*, wingspan 29–34 mm, widespread in damp woodland, marsh, fen and gardens, more scattered in Ireland. With numbers declining by 82% over 1968–2007, it is a species of conservation concern in the UK BAP. Adults fly in July–August. Larvae feed in spring on sallows, including *Salix viminalis* (giving the specific name). Greek *brachys* + *loma* = 'short' + 'fringe'.

Rustic shoulder-knot *Apamea sordens*, wingspan 34–42 mm, widespread and common in grassland, farmland, gardens and woodland rides. Adults fly in May–June. Larvae feed on grasses and some cereal crops. Apamea was a town in Asia Minor. *Sordens* is Latin for 'dirty'.

SHOVELER Northern shoveler *Anas clypeata* (Anatidae), wingspan 77 cm, length 48 cm, weight 630 g. These ducks particularly breed around the Ouse Washes, the Humber and the north Kent marshes, and in smaller numbers in south-west England, Scotland and Ireland, averaging around 1,000 pairs in 2006–10. In winter, breeding birds move south, to be replaced by an influx of continental birds from further north. Britain is home to >20% of the north-west European population (18,000 individuals overwintering in 2008–9), making it an Amber List species. The nest is a shallow depression in grassland, lined with plant material and down, clutch size is 9–11, incubation 22–3 days, and the precocial young fledge in 40–5 days. Using the large spatulate bills that give them their English name, shovelers sift plant material and insects from the water surface. The binomial is Latin for 'duck' + 'shield-bearing'.

SHREW Members of the family Soricidae, in the genera *Sorex* (Latin for 'shrew'), *Neomys* (Greek *neo* + *mys* = 'new' + 'mouse') and *Crocidura* (Greek *crocis* + *oura* = 'cloth pile' + 'tail', from the bristly tail). Young shrews may travel by forming a caravan behind the mother, each carrying the tail of the sibling in front in its mouth. 'Shrew' comes from Old English *screawa*.

Common shrew *Sorex araneus*, head-body length 55–82 mm, tail length 24–44 mm, weight 5–12 g, common and widespread in Britain, absent from Ireland, in deciduous woodland, hedgerows and grassland. There was an estimated pre-breeding population in 1995 of 41.7 million: 26 million in England, 11.5 million in Scotland and 4.2 million in Wales. It feeds on insects, slugs, spiders, worms and amphibians, occasionally small rodents, and needs to eat 200–300% of its body weight each day to survive, feeding every 2–3 hours. Breeding takes place in April–September. Gestation lasts 24–5 days, with 2–4 litters a year, each of 5–7 young, which are independent in 22–5 days. Latin *arena* = 'sand'.

Greater white-toothed shrew *Crocidura russula*, head-body length 60–90 mm, tail length 3–4.3 cm, weight 11–14 g, in woodland, hedgerows and farmland in the Channel Islands. It has been introduced to Ireland, initially discovered in 2007 in owl pellets in Co. Limerick and Co. Tipperary, but spreading into Counties Cork, Waterford, Kilkenny and Offaly by 2014, and out-competing populations of the native pygmy shrew, largely by more effective predation on insects. Breeding is in March–September, producing up to four litters which contain 2–10 young. *Russula* is Latin for 'reddish'.

Lesser white-toothed shrew *C. suaveolens*, head-body length 50–75 mm, tail length 24–44 mm, weight 3–7 g, in scrub on the Scilly Isles, Jersey and Sark, probably introduced from the continent in the Bronze Age. Population size in 1995 was around 14,000. It feeds on small crustaceans (e.g. sandhoppers) on the beach at low tide, as well as on soil invertebrates, tunnelling in leaf litter and active under logs and brushwood. Breeding is in March–September. Gestation lasts 24–32 days. There are 2–4 litters each year, with up to five pups in each. *Suaveolens* is Latin for 'fragrant'.

Pygmy shrew *S. minutus*, head-body length 40–60 mm, tail length 32–46 mm, weight 2.4–6.0 g, common and widespread (and the only shrew present on Orkney and the Outer Hebrides) in deciduous woodland and, especially, grassland, as well as peatland in Ireland. There was an estimated pre-breeding population in Britain in 1995 of 8.6 million: 4.8 million in England, 2.3 million in Scotland and 1.5 million in Wales. The recent introduction of greater white-toothed shrew to Ireland may lead to declining numbers of pygmy shrew. Feeding is on insects (especially beetles) and other invertebrates (especially woodlice), and with one of the highest metabolic rates of any animal, it must eat every two to three hours. Breeding lasts from April to August. Gestation is 22–5 days. Females have up to five litters a year, with usually 4–6 young per litter. Lactation lasts 22 days, then the young disperse. *Minutus* is Latin for 'minute'.

Water shrew *Neomys fodiens*, head-body length up to 10 cm, tail length 8 cm, weight 12–18 g, widespread in Britain, though rather local in north Scotland, and absent from Ireland. The estimated pre-breeding population in 1995 was about 1.9 million: 1.2 million in England, 400,000 in Scotland and 300,000 in Wales. It uses burrows, living close to fresh water, hunting small fish, crayfish, water snails, insects and aquatic larvae, amphibians (especially newts) and occasionally small rodents, taking advantage of its paralytic bite using venom-producing glands under its jaw. Breeding takes place in April–September. Gestation lasts 24 days. Litters, two or three a year, are each of 4–8 young. *Fodiens* comes from Latin *fodire* = 'to dig'.

SHRIKE Members of the genus *Lanius* (Laniidae). 'Shrike' comes from Old English *scríc* = 'shriek', describing the bird's shrill call.

Isabelline shrike *L. isabellinus* and **woodchat shrike** *L. senator* are scarce visitors. **Brown shrike** *L. cristatus*, **lesser grey shrike** *L. minor*, **long-tailed shrike** *L. schach*, **masked shrike** *L. nubicus* and **southern grey shrike** *L. meridionalis* are all accidentals.

Great grey shrike *L. excubitor*, wingspan 32 cm, length 24 cm, weight 68 g, a scarce winter visitor found in open habitats throughout Britain, if mainly in eastern England, usually with just over 60 birds p.a., though numbers sometimes approach 200. It mainly hunts insects and rodents, but also other small vertebrates, hunting from a perch, and often storing prey by impaling it using thorns or barbed wire as a larder, giving it an alternative name of 'greater butcher bird', the genus name being Latin for 'butcher' or 'executioner'. Latin *excubitor* = 'watchman'.

Red-backed shrike *L. collurio*, wingspan 26 cm, length 17 cm, weight 30 g. These shrikes had not bred in Britain since the last examples of the Breckland (East Anglia) population became extinct in 1992, but two sites were found in 2010 (with two pairs in Dartmoor, one pair in the Highlands). The Dartmoor pair also bred in 2011 and 2013, and a male was noted in 2017. It is a scarce spring and autumn passage migrant seen along the south and east coasts, with a total overall number of about 250 birds. The cup nest is built in scrub from plant material and lined with moss. Clutch size is 4–6, incubation lasts 14 days, and the altricial young fledge in 14–15 days. It mainly hunts insects (especially beetles), but also small vertebrates, hunting from a perch, and often storing the prey by impaling it using thorns or barbed wire as a larder, giving it an alternative name of 'butcher bird'. Greek *collurion* was a bird mentioned by Aristotle.

SHRIMP 1) Decapod crustaceans with long narrow muscular abdomens, long slender legs, and long antennae, that feed near the seafloor on most coasts and estuaries, as well as in rivers and lakes. (In prawns the first three of the five pairs of legs have small pincers, while in shrimps only two pairs are claw-like.) See also fairy, mantis, opossum and tadpole shrimps.

Brown shrimp *Crangon crangon* (Crangonidae), up to 8.5 cm long, common and widespread on sandy and muddy substrates often buried with only the eyes and antennae above the sediment surface. *Crangon* is Greek for 'shrimp'.

Burrowing mud shrimp *Callianassa subterranea* (Callianassidae), up to 4 cm long, creating multi-branched burrow systems up to 80 cm deep in sandy mud sediments from the lower shore to the shallow sublittoral, most records coming from south England and west Scotland. Greek *cala* + *anassa* = 'beautiful' + 'mistress' or 'queen'.

Common shrimp Alternative name for brown shrimp.

Hooded shrimp *Athanas nitescens* (Alpheidae), up to 2 cm long, found from the lower intertidal to depths of 60 m, usually beneath stones where the substrate is gravelly, and also under macroalgae, common and widespread around the south and west coasts of Britain, and the east and west coasts of Ireland. Greek *athanatos* = 'immortal'.

Pink shrimp *Pandalus montagui* (Pandalidae), 4–5 cm long, found around Britain and the south and east coasts of Ireland to depths of 100 m, but occasionally in rock pools on the lower shore.

2) Amphipod crustaceans.

Freshwater shrimp *Gammarus pulex* (Gammaridae), up to 2 cm long, common and widespread in flowing water. *Gammarus* is Latin for a kind of lobster, *pulex* = 'flea'.

Japanese skeleton shrimp *Caprella mutica* (Caprellidae), males up to 5 cm long, females 1.5 cm, a north-west Pacific amphipod recorded from the west coast of Scotland (first recorded at Oban, in 2000), south and south-west England, and the west coast of Ireland, on a range of natural substrates such as hydroids and algae, and artificial ones including buoys, mooring ropes and boat hulls. The binomial is Latin for 'little goat' and 'curtailed'.

Lagoon sand shrimp *G. insensibilis*, up to 2 cm long, found on the central south coast, with outliers in, Essex, north Norfolk and the Humber estuary, in sheltered, shallow, brackish water with sediments ranging from organic muds to shingle, often associated with the green alga *Chaetomorpha linum*.

Sand digger shrimp *Bathyporeia pelagica* (Bathyporeiidae), 6–8 mm long, common and widespread, with a patchier distribution in Ireland, in wet, fine to medium sand, from slightly above mean tide level into the shallow sublittoral. Greek *bathys* + *poreia* = 'broad' + 'gait', and Latin *pelagica* = 'oceanic'.

SIALIDAE Family name of alderflies (Greek *sialis* = a kind of bird).

SIBBALDIA *Sibbaldia procumbens* (Rosaceae), a locally common montane perennial herb found in the Highlands, particularly on rocky grassy slopes with late snow-lie, in corries and on stony substrates. Sir Robert Sibbald (1641-1722) was professor of medicine at the University of Edinburgh. *Procumbens* is Latin for 'procumbent'.

SICKENER *Russula emetica* (Russulales, Russulaceae), a widespread, generally infrequent but locally common, poisonous fungus, found in August–October on soil under pine, occasionally damp mossy heathland. *Russula* is Latin for 'reddish'.

Beechwood sickener *R. nobilis*, a common and widespread poisonous fungus, causing nausea, vomiting, stomach pain and diarrhoea if eaten, found in August–October, ectomycorrhizal on beech. *Nobilis* is Latin for 'noble' or 'notorious'.

SIEVE-TOOTHED MOSS *Coscinodon cribrosus* (Grimmiaceae), on friable, shaly rock on coastal clifftops in west Britain, in crevices, occasionally on igneous rocks on the coast, slate in inland mountains, and (rarely) on slate walls. Greek *coscinon* = 'sieve', and Latin *cribro* = 'to sieve'.

SIGNAL (MOTH) **Alder signal** *Stathmopoda pedella* (Stathmopodidae), wingspan 10–14 mm, widespread but local in damp woodland in southern and eastern England, north to Yorkshire. Adults fly in July. Larvae feed on alder. Greek *stathmos* + *podos*

= 'balance' + 'foot', and Latin *pedis* = 'of the foot', both being comments on the adult resting posture where the hindleg projects horizontally between the other two pairs.

SIGNAL-GRASS Species of *Brachiaria* (Greek *brachia* = 'forearm', the inflorescence branches resembling signal arms) and *Urochloa* (Greek *oura* + *chloa* = 'tail' + 'grass', referring to the awns), both Poaceae, casual birdseed aliens, occasionally found on waste ground.

SIGNAL-MOSS Species in the genus *Sematophyllum* (Sematophyllaceae) (Greek *sematos* + *phyllos* = 'sign' + 'leaf').

Bark signal-moss *S. substrumulosum*, a rarity which, at most sites, for example in Cornwall, southern Ireland and south Wales, grows on acidic bark and humus in woodland. On the Isles of Scilly, it is also found in deep depressions on coastal heathland. At its only known site in south-east England (Kingley Vale, West Sussex), it grows on yew in ancient woodland.

Prostrate signal-moss *S. demissum*, a rarity in north-west Wales and south-west Ireland on moist, acidic or mildly base-rich rocks and boulders by woodland streams. *Demissus* is Latin for 'unassuming'.

Sparkling signal-moss *S. micans*, local in upland regions, mainly in west Scotland and south-west Ireland, but also north Wales and the Lake District, close to water, on damp, acidic to slightly base-rich rocks in woodland. *Micans* is Latin for 'glittering'.

SILAGE EYE Bovine iritis, an eye infection in cattle caused by the bacterium *Listeria monocytogenes* when the animals eat silage which has not fermented sufficiently to kill the microorganism.

SILK-MOSS Species of *Isopterygiopsis* (Greek *isos* + *pteryx* + *opsis* = 'equal' + 'wing' + 'appearance'), *Plagiothecium* (*plagios* + *thecion* = 'slanting' + 'box') and *Pseudotaxiphyllum* (*pseudes* + *taxo* + *phyllon* = 'false' + 'arrange' + 'leaf') (Plagiotheciaceae).

Widespread and locally common species include **alder silk-moss** *Plagiothecium latebricola* (especially southern England and Wales, on decaying plant material in damp habitats), **bright silk-moss** *Pl. laetum* (soil, litter, logs, tree bases, rocks and boulders, often in woodland), **dentated silk-moss** *Pl. denticulatum* var. *denticulatum* (soil, rocks, logs and tree bases, mostly in woodland, occasionally upland scree and rock crevices), **neat silk-moss** *I. pulchella* (crevices and under overhangs on moist, base-rich cliffs), and **swamp silk-moss** *Pl. ruthei* (England and Wales).

Highlands rarities include **alpine silk-moss** *Pl. platyphyllum*, **Donnian silk-moss** *Pl. denticulatum* var. *obtusifolium*, **Mueller's silk-moss** *I. muelleriana* and **round silk-moss** *Pl. cavifolium* (also found other montane areas).

Curved silk-moss *Pl. curvifolium*, common and widespread in Britain, rare in Ireland, on soil, litter, logs, tree bases and rocks, mainly in woodland (including coniferous plantations). Latin *curvifolium* = 'curved-leaved'.

Elegant silk-moss *Pseudotaxiphyllum elegans*, a common and widespread calcifuge growing on shaded, acidic soil, rock, logs and tree roots in woodland or in rock crevices, often abundant in species-poor habitats such as conifer plantations.

Juicy silk-moss *Pl. succulentum*, common and widespread on soil, siliceous rock and limestone, and on tree bases, in both acidic and calcareous habitats, usually in woodland.

Waved silk-moss *Pl. undulatum*, common and widespread on acidic soil, rocks, wood and grass in woodland, heathland, blanket bog, and scree. Latin *undulatum* = 'wavy'.

Woodsy silk-moss *Pl. nemorale*, common and widespread on soil, siliceous rock and limestone, and on tree bases; growing in both acidic and calcareous habitats, usually in woodland. Latin *nemorale* = 'woodland glade'.

SILKY-BENT Annual grasses in the genus *Apera* (Poaceae), Greek for 'unpouched'.

Dense silky-bent *A. interrupta*, a neophyte mainly an arable weed in sandy fields in East Anglia, but also scattered in England on verges and trackways, and grazed grassy or grass-heath habitats. It is also a casual on waste ground originating from wool shoddy and as a seed impurity.

Loose silky-bent *A. spica-venti*, an archaeophyte of open sandy habitats, mostly as a weed in arable fields, especially in East Anglia, and in waste ground and roadsides elsewhere, often as a grain alien. Latin *spica-venti* = 'ear-wind'.

SILPHIDAE From Greek *silphe*, meaning an insect with an unpleasant odour, the family of large (9–30 mm length) sexton, burying and carrion beetles, with 20 species in 7 genera on the British list. **Silphinae** are largely associated with carrion, although some (e.g. *Silpha atrata*) prey on snails, the two species of *Aclypea* are herbivorous and *Dendroxena quadrimaculata* is an arboreal predator of caterpillars, hunting by scent. See carrion beetle, and black snail beetle under beetle. **Nicrophorinae** contain six species of *Nicrophorus*, strong-flying burying beetles. See sexton beetle.

SILURIAN *Eriopygodes imbecilla* (Noctuidae, Hadeninae), wingspan 24–7 mm, a rare moth (Vulnerable, RDB category 2), discovered in Britain in 1972, known from small upland areas of north-west Monmouthshire and on the Monmouthshire–Herefordshire border, on moorland where larvae feed on bilberry and heath bedstraw. Adults are mainly nocturnal but will fly on sunny afternoons, in June–July. Greek *erion* + *pyge* + *eidos* = 'wool' + 'rump' + 'form' (i.e. resembling the genus *Eriopyga*), from the anal tuft of the male, and Latin *imbecilla* = 'feeble', because the fat-bodied, short-winged female was formerly thought to be flightless.

SILURIDAE Family name of Eurasian catfish that includes the introduced wels, from Greek *silouros* for a kind of large river fish, perhaps catfish.

SILVANIDAE Family that includes some flat bark beetles (see also Cucujidae) and flat grain beetles (2–7 mm long), with 12 species in 10 genera. Largely predatory, several genera are primarily found beneath tree bark; the remaining species are mostly found in vegetable refuse and stored produce, many of the latter frequently imported on foodstuffs. Silvanus was a Roman deity of the woods and fields.

SILVER (MOTH) Clouded silver *Lomographa temerata* (Geometridae, Ennominae), wingspan 22–6 mm, common in woodland, scrub, hedgerows, parks and gardens throughout England, Wales and Ireland, more local in Scotland. Adults fly at night in May–June, with autumn sightings in the south almost certainly migrants. Larvae feed on hawthorn, blackthorn, plum, wild cherry and crab apple. Greek *loma* + *graphe* = 'border' + 'drawing', from the forewing markings, and *temero* = 'to stain', from the dark mark that 'stains' the white wings.

SILVER-FIR European silver-fir *Abies alba* (Pinaceae), introduced from Central Europe, common and widespread, shade- but not drought-tolerant, and suited to soils of poor to medium nutrient status with good soil moisture. It was formerly much planted for timber and as a parkland ornamental, now less so because it is more susceptible to rust fungus and woolly aphids than congeners. It is often self-sown in mixed woodland. *Abies* is Latin for 'fir', itself late Middle English, probably from Old Norse *fyri*. *Alba* is Latin for 'white'.

SILVER-FISH *Lepisma saccharina* (Lepismatidae), a bristletail, 12–25 mm long, a wingless insect common in food cupboards, feeding on carbohydrate material such as flour, paper and glue. Greek *lepisma* = 'scale' and Latin *saccharina* = 'sugary'.

Heath silverfish *Dilta littoralis* (Machilidae), local in south England in heathland leaf litter.

SILVER-GRASS Rhizomatous perennial east Asian neophytes (**Chinese silver-grass** *Miscanthus sinensis* and **giant silver-grass** *M. giganteus*) (Greek *mischos* + *anthos* = 'stem' + 'flower'), cultivated as biomass crops for biofuel and fibre, and sometimes found on waste ground.

SILVER-LINE Brown silver-line *Petrophora chlorosata* (Geometridae, Ennominae), wingspan 31–7 mm, a common and widespread moth of woodland, heathland and moorland, adults flying from dusk in April–June, larvae feeding on bracken. Greek *petros* + *phoreo* = 'rock' + 'to bear', from stone-coloured wing markings, and *chloros* = 'pale green'. (NB not silver-lines.)

SILVER-LINES Nocturnal moths in the family Nolidae, subfamily Chloephorinae.

Green silver-lines *Pseudoips prasinana*, wingspan 30–5 mm, widespread and common in deciduous woodland and rural gardens. Population levels increased by 144% between 1968 and 2007. Adults fly in June–July. Larvae feed on oak, birch, hazel and beech leaves. Greek *pseudos* + *ips* = 'false' + 'a vine bud-eating grub', and Latin *prasinus* = 'leek green'.

Scarce silver-lines *Bena bicolorana*, wingspan 40–5 mm, local in broadleaf woodland and parks throughout much of England, south-east of a line from the Severn to the Wash, and in Cumbria, Lincolnshire and south Yorkshire, as well as in south and central Wales. Adults fly in June–August. Larvae feed on oak and birch leaves. Bena is a meaningless neologism; *bicolorana* is Latin for 'two-coloured'.

SILVER-MOSS *Bryum argenteum* (Bryaceae), widespread and often abundant in disturbed habitats which may be very dry and rich in nutrients, for example soil on and by roads (including cracks between paving slabs), in arable fields, on waste ground and railway lines. It may also be found on stone, for example on walls, buildings, roofs, concrete and tarmac. Semi-natural habitats include dunes, banks of streams and rivers, and unstable soil on lowland cliffs. Greek *bryon* = 'moss', and Latin *argenteum* = 'silver'.

Slender silver-moss *Anomobryum julaceum*, common and widespread in upland regions on moist, often slightly base-rich soil by streams, waterfalls and lakes, in flushes, and on wet rock; also in disused quarries and gravel-pits. Var. *concinnatum* favours more base-rich, drier montane sites. Greek *anomos* + *bryon* = 'irregular' + 'moss', and Latin *iulus* = 'plant down'.

SILVER-STILETTO See stiletto (fly).

SILVER Y See Y.

SILVERHORNS Caddis flies in the family Leptoceridae.

Black silverhorns *Mystacides nigra* and *M. azurea*, widespread but local, *M. nigra* in slow-moving rivers, streams, canals and lakes, not found in Ireland. Larvae are found in cases built of sand grains with some plant material on sand, mud, plants and woody debris, and are omnivorous. Adults are on the wing in May–September. Greek *mystacos* = 'upper lip'. The specific names are Latin for 'black' and 'azure'.

Brown silverhorns *Athripsodes cinereus*, widespread and fairly frequent, with a similar life history to black silverhorns. Adults fly in June–November. Greek *athrips* + *eides* = 'amorous' + 'appearance of', and Latin *cinereus* = 'ash-grey'.

SILVERLEAF FUNGUS *Chondrostereum purpureum* (Agaricales, Cyphellaceae), inedible, widespread and common, found throughout the year, sporulating in summer and autumn, sometimes encrusting but more often as small brackets on broadleaf trees, entering wounds on fruit trees to cause silver leaf disease. Greek *chondros* + *stereos* = 'grain' + 'firm', and Latin *purpureum* = 'purple'.

SILVERWEED *Potentilla anserina* (Rosaceae), a common and widespread stoloniferous perennial herb of open grassland or on bare ground. Habitats include land subject to seasonal flooding, upper saltmarsh, dunes, rough ground and roadsides. Latin *potentilla* = 'small powerful one', from the medicinal properties of some members of the genus (= cinquefoils), and *anserina* = 'goose-like'.

SILVERWORT Foliose liverworts in the genus *Anthelia* (Antheliaceae) (Greek *anthelion*, meaning the downy plume of

a reed), found in damp rock or peat habitats in upland regions such as the Highlands and north Wales.

SIMAROUBACEAE Family name for tree-of-heaven, from the Guianan vernacular *simarouba*.

SIMULIIDAE From Latin *simulo* = 'imitate', the family name for blackflies, with 35 species in 3 genera, 31 species belonging to *Simulium*. Larvae are aquatic. Adult males feed on nectar, females on blood.

SIPHLONURIDAE From Greek *siphlos* = 'maimed', a mono-generic mayfly family (three species of *Siphlonurus*), nymphs favouring pools, lakes and rivers in hilly parts of northern and western Britain. The only common species is **summer mayfly**.

SIPHONAPTERA Order name of fleas (Greek *siphon* + *aptera* = 'siphon' + 'wingless').

SIPHONOPHORA, SIPHONOPHORE From Greek *siphon* + *phoreo* = 'siphon' + 'to bear', an order in the Hydrozoa, with 11 families recorded in British waters. 'Individuals' in each species are actually a colony composed of highly specialised individual animals (zooids), as exemplified by Portuguese man o' war.

SIPHUNCULATA Order of sucking lice (diminutive of Latin *sipho* = 'siphon').

SIRICIDAE Sawfly family containing woodwasps or horntail wasps (Greek *seiren*, a kind of wasp).

SISKIN *Spinus spinus* (Fringillidae), wingspan 22 cm, length 12 cm, weight 15 g, a common and widespread resident breeding and (in much of England) winter visiting finch (common in Irish gardens), with an estimated 420,000 pairs in Britain in 2009, numbers fluctuating but overall increasing by 54% between 1995 and 2015. It favours coniferous woodland, the bowl-shaped nest usually built of twigs, grass and moss at the end of a high branch, with 4–5 eggs, incubation lasting 12–13 days, and fledging of the altricial young taking 13–15 days. Diet is mainly of conifer seed. 'Siskin' dates from the sixteenth century, derived from the Middle Dutch *siseken*. *Spinos* was a bird described by Aristophanes.

SISYRIDAE Family name of sponge flies (Neuroptera) (Greek *sisyra* = 'garment of skin').

SKATER See pond and river skaters.

SKIMMER Dragonflies in the genus *Orthetrum* (Libellulidae) (Greek *orthos* + *etron* = 'straight' + 'abdomen').

Black-tailed skimmer *O. cancellatum*, fairly common in southern England, parts of Wales and Ireland, increasing its range northwards, found at open water sites with bare shoreline patches where patrolling males bask in sunshine. The main flight period is May–August. *Cancellatum* is Latin for 'lattice-worked'.

Keeled skimmer *O. coerulescens*, with a patchy distribution but locally common in western Britain and Ireland, associated with pools and streams in wet heathland. The main flight period is April–September. *Coerulescens* is Latin for 'blue'.

SKIPPER The most 'primitive' of British butterflies, family Hesperiidae, with short wings and stout bodies, named after their rapid, darting flight.

Chequered skipper *Carterocephalus palaemon* (Heteropterinae), male wingspan 29 mm, female 31 mm, extinct in England since 1976, now confined in apparently stable populations to west-central Scotland, centred on damp tussocky grassland in the Fort William area, where its main larval food is purple moorgrass, on the blades of which caterpillars create a protective tube. Adults fly in May–July, and take the nectar of bramble, bluebell and ground-ivy. Greek *carteros* + *cephale* = 'strong' + 'head', from the broad head. Palaemon was a sea god in Greek mythology.

Dingy skipper *Erynnis tages* (Pyrginae), wingspan 27–34 mm, Britain's most widely distributed skipper, despite a decline in range (by 48% since the 1970s) due to changes in farming practice, making it a priority species for conservation. Nevertheless, numbers increased by 69% between 2005 and 2014. Strongholds are in central and southern England, but colonies are found throughout the British Isles, including north Scotland and Ireland where, while scarce, it is found on limestone outcrops. It flies in April–August, favouring warm open habitats such as chalk and limestone grassland, dunes, railway embankments and disused quarries, basking on bare ground. Larvae, which create a protective tent, favour bird's-foot-trefoil, as do adults which also feed on the nectar of (especially) yellow flowers such as buttercups, horseshoe vetch and hawkweeds. The binomial comes from the Erynnes, or Furies, in Greek mythology, from this species' restless movements, and Tages, a boy with the wisdom of an old man who instructed the Etruscans in divination.

Essex skipper *Thymelicus lineola* (Hesperiinae), wingspan 26–30 mm, widespread and common south and east of a line from the Humber to the Severn estuaries, with a few records also from south Wales and south-east Ireland. A butterfly of especially dry grassland, its recent increase in range (more than doubling in the last few decades), particularly northwards, is partly explained by the habitat corridors of grass-covered motorway and trunk road embankments. Numbers, however, declined by 66% over 2005–14. Feeding on a range of grasses (especially cocksfoot) a few days after hatch the caterpillar creates a tube by spinning together the edges of a leaf blade from which it emerges to feed. Adults fly from June to September, feeding primarily on thistles, but nectar of other species is also taken. *Thymelicos* is Greek for a member of the chorus in Greek drama, dancers whose movements are echoed by this species, and Latin for 'small line', a mark on the male wing.

Golden skipper Name given to some of the skippers, for example small skipper.

Grizzled skipper *Pyrgus malvae* (Pyrginae), wingspan 23–9 mm, found mostly in southern and central England, with scattered colonies elsewhere in England and in Wales. Adults fly in April–June. Habitats are characterised by warmth, shelter and sparse vegetation, for example chalk downland, woodland edges, rides and clearings, unimproved grassland and occasionally heathland, where adults can bask on bare earth. Larvae feed on a number of plants, especially Rosaceae. Adults prefer the nectar of bird's-foot-trefoil and buttercups. Long-term this species is in decline, with range contracting by 53% between 1976 and 2014 (though showing a 7% increase over 2005–14), and numbers by 37% in 1976–2014 (though with zero overall change in abundance in 2005–14), and it remains a priority species for conservation. *Pyrgos* is Greek for 'battlement', from a chequered wing pattern, and Latin *malva* = 'mallow', incorrectly believed to be a food plant.

Large skipper *Ochlodes sylvanus* (Hesperiinae), male wingspan 29–34 mm, female 31–6 mm, common and widespread in England, Wales and south-west Scotland, but absent from Ireland. In 2005–14, its range extended by 12%, and numbers increased by 23%. It is particularly associated with tall grassland but is also found in urban parks and graveyards. The primary larval food plant is cocksfoot, but other grasses are taken. Caterpillars create protective tubes on their host plant. Adults fly in May–August and have a variety of nectar sources. Greek *ochlodes* = 'turbulent', from the flight behaviour, and Latin *sylva* = 'woodland'.

Lulworth skipper *T. acteon* (Hesperiinae), wingspan 24–8 mm, flying in June–September, on calcareous grassland where the larval food plant tor-grass is found, between Weymouth and the Isle of Purbeck in south Dorset, rarely 8 km from the coast. While overall long-term abundance (1976–2014) declined by 76%, recent trends (2005–14), admittedly from a low base level, show an upturn in numbers by 39%. Larvae overwinter in a cocoon and on emerging create a protective tube on the host plant. Actaeon was a hunter in Greek mythology who was turned into a stag after viewing a bathing Artemis, then eaten by his own hounds.

Silver-spotted skipper *Hesperia comma* (Hesperiinae), male wingspan 29–34 mm, female 32–7 mm, restricted to grazed chalk downland in south and south-east England, with sheep's-fescue the larval food plant. Its range contracted in the twentieth century with a reduction in grazing stock and onset of myxomatosis which greatly reduced the (grazing) rabbit populations. Caterpillars form a protective tent by spinning several leaf blades together. Adults, flying in July–September, mainly feed on thistle nectar. While remaining a rare species, slight range increases have been noted since 1980, and numbers increased by 12% in 2005–14. In Greek mythology Hesperia was one of the Hesperides, the nymphs who guarded the golden apples of Hera. The Greek *comma* describes the comma mark on the male wing.

Small skipper *T. sylvestris*, wingspan 27–34 mm, widespread and common throughout England and Wales, its range expanding northwards in recent years. While overall long-term abundance (1976–2014) declined by 75%, recent trends (2005–14) show an upturn in numbers by 27%. It favours rough grassland with Yorkshire-fog the main larval food plant, though other grass species are also eaten. Larvae overwinter in a cocoon and on emergence create a protective tube on the host plant. Adults (flying in June–September) take nectar from a variety of species, including thistles and red clover. *Sylvestris* is Latin for 'woodland' or 'wild'.

SKIPPER FLY Nest skipper fly *Neottiophilum praeustum* (Piophilidae), scarce and scattered in (mostly southern) England. Larvae are parasitic in bird nests, sucking nestling blood. 'Skipper' refers to the ability of the larvae to jump or skip several centimetres by bending then suddenly straightening. Greek *neottia* + *philos* = 'bird's nest' + 'loving', and Latin *praeustum* = 'scorched' (brown at the edge).

SKUA Seabirds in the genus *Stercorarius* (Stercorariidae), from Latin *stercus* = 'dung' since the food disgorged by other birds when pursued by skuas was once thought to be excrement. 'Skua' comes from Faroese *skúgvur*, which is the great skua.

Arctic skua *S. parasiticus*, wingspan 118 cm, length 44 cm, weight 450 g, a Red List summer migrant breeding in Shetland, Orkney the Outer Hebrides, and on some coastal moorlands of north and west Scotland, with 2,100 pairs reported in the Seabird 2000 survey, but with only 557 apparently occupied territories in 2010 (RBBP report), meaning this species had declined greatly in one decade, and the BTO has similarly calculated an 80% decline over 1986–2014, 71% over 2000–14 and 56% over 2009–14. The ground nest usually has two eggs, incubation taking 25–8 days, and the altricial young fledging in 25–30 days. It is also a passage migrant. The summer diet is mostly of birds, small mammals and insects, but in winter, it changes to fish, mostly by kleptoparasitism, hence the specific *parasiticus*.

Great skua *S. skua*, wingspan 136 cm, length 56 cm, male weight 1.3 kg, females 1.5 kg, migrating in April to the northernmost islands of Scotland from their wintering grounds off the coasts of Spain and Africa, to leave in July. Some 9,600 pairs were estimated in 2002, the BTO suggesting a 19% increase in number from then until 2014, and that 60% of the global population nests in the British Isles, where it is Amber-listed. In Ireland it is essentially a passage migrant, but a few pairs breed on isolated islands off the west coast. It breeds on coastal moorland and rocky islands, usually laying two eggs in grass-lined nests, incubation lasting 26–32 days, and fledging of the altricial young 40–51 days. Diet is mostly fish, obtained from the sea, by scavenging or by kleptoparasitism.

Long-tailed skua *S. longicaudus*, wingspan 111 cm. length 50 cm, weight 300 g, a scarce passage visitor occasional off the coasts of north Scotland, west Ireland and south England. Latin *longus* + *cauda* = 'long' + 'tail'.

Pomarine skua *S. pomarinus*, a scarce passage migrant found on the coastline in spring and autumn, wintering off the coast of West Africa. 'Pomarine' derives from the thin plates (cere) that overlay the base of the bill in all skuas (Greek *pomato* + *rhinos* = 'lid' + 'nose').

SKULLCAP *Scutellaria galericulata* (Lamiaceae), a widespread, locally common herbaceous perennial associated with ponds, rivers, canals, marsh, fen, fen-meadows, wet woodland and dune slacks, as well as coastal boulder beaches in Scotland. 'Skullcap' reflects the helmet-shaped calyx. The binomial is Latin for 'shield-shaped' and 'small helmet'.

Lesser skullcap *S. minor*, a perennial of wet heathland, bog, marsh and damp woodland on acidic, oligotrophic soils, widespread but locally common only in southern and south-west England, Wales, west Scotland and southern Ireland. *Minor* is Latin for 'lesser'.

Somerset skullcap *S. altissima*, a perennial neophyte naturalised in woodland edge and hedgerows at a few sites in England and the Scottish Borders, for example in the Wadbury Valley, Somerset, recorded on a wall and wooded slope and persisting since 1929, and in woodland edge at Holmbury St Mary, Surrey, since 1972 and apparently increasing. *Altissima* is Latin for 'tallest'.

SKUNK-CABBAGE American skunk-cabbage *Lysichiton americanus* (Araceae), a widespread if scattered, herbaceous North American neophyte perennial introduced in 1901, often planted beside ponds and streams in parks and gardens, which can rapidly become established as a persistent invasive escape in swampy ground. Greek *lysis* + *chiton* = 'loosening' + 'cloak': as the fruit ripens the spathe is removed from the spadix. **Asian skunk-cabbage** is probably a subspecies.

SKYLARK *Alauda arvensis* (Alaudidae), wingspan 33 cm, length 18 cm, male weight 42 g, females 35 g, widespread and locally common in open countryside, from farmland to moorland, easy to see and hear when in its distinctive song flight. There were an estimated 1.4 million territories in 2009, but numbers declined by 59% between 1970 and 2015, and by 24% between 1995 and 2014, according to BBS (mainly because of a shift from spring- to autumn-sown cereal), and is Red-listed in Britain, Amber-listed in Ireland. The grass nest on the ground usually contains 3–4 eggs, incubation taking 13–14 days, the altricial chicks fledging in 11–16 days. Summer diet is mostly insectivorous, with seed and grain also eaten, especially in winter. 'Lark' derives from Old English *laferce* or *lawerce*. The binomial comes from Latin for 'lark' and 'field'.

SLATER Isopod crustacean in the family Ligiidae. In Greek mythology Ligeia was a water nymph.

Carr slater *Ligidium hypnorum*, up to 9 mm long, found mainly south of a line between the Wash and the Severn estuary, most abundant in south-east England and East Anglia, in waterlogged habitats among moss and leaf litter, in sedge and grass tussocks, or beneath dead wood and stones. Further north and west it is local in habitats such as ancient woodland and fen carr. *Hypnos* is Greek for 'sleep'.

Common sea slater *Ligia oceanica*, up to 30 mm long, widespread on rocky coasts within the littoral fringe, especially in crevices, and rock pools, and under stones.

SLENDER Leaf-mining moths with slender abdomens, wingspan usually 9–12 mm, in the family Gracillariidae (Gracillariinae).

Blackthorn slender *Parornix torquillella*, common in scrub, woodland, hedgerows and gardens throughout much of Britain, especially the southern half. Adults fly at night in May–August. From July to November larvae mine leaves of blackthorn and wild plum, subsequently feeding in the fallen leaves. Greek *para* = 'alongside' + the genus *Ornix* (*ornis* = 'bird', being 'feathery-winged'), and Latin *torquis* = 'collar', from the black spots round the larval 'head'.

Brown birch slender *Parornix betulae*, common and widespread on heathland and open woodland, especially the southern half of Britain. Adults fly at night in May and August. Larvae

mine leaves of birch then create and feed in a leaf fold. *Betula* is Latin for 'birch'.

Brown oak slender *Acrocercops brongniardella*, scarce, nocturnal, local in open woodland, mainly south of a line from Lincolnshire to Lancashire, and in south and east Ireland. Adults fly in July–August; they can overwinter and be found again in April–May. Larvae mine leaves of oak, and have been found throughout the year. Alexandre Brongniart (1770–1847) was a French zoologist.

Clouded slender *Caloptilia populetorum*, wingspan 11–14 mm, scarce in moorland, heathland and open woodland, with a discontinuous distribution in south and central England from Kent to Glamorgan north to Lincolnshire, in Cumbria and in west and north Scotland; rarely recorded in Ireland. Adults fly from July–August onwards, overwintering, then again in March–May. Larvae initially mine birch then feed inside a leaf fold. Greek *calos* + *ptilon* = 'beautiful' + 'wing', and Latin *populetum* = 'poplar thicket', though poplar is not the food plant.

Clover slender *Parectopa ononidis*, wingspan 7–9 mm, scarce, local on chalk and limestone downland and rough meadows in parts of south and central England. Adults fly in two, possibly three generations between May and August. Larvae mine clovers and restharrow (*Ononis*, giving the specific name), found in April–May, July and from September onwards, overwintering. Greek *parectopos* = 'somewhat out of the way', possibly describing the forked wing venation.

Common slender *Gracillaria syringella*, widespread and common in gardens and woodland. Adults fly in May and July. Larvae initially mine leaves of privet, ash and lilac (*Syringa*, giving the specific name), then live gregariously within a leaf roll, at some sites causing sufficient damage to be a serious garden pest. *Gracillaria* is from Latin *gracilis* = 'slender'.

Feathered slender *Caloptilia cuculipennella*, scarce, scattered and widespread in England, with a few records elsewhere, on sea-cliffs and rough ground near the sea and woodland inland. Adults fly in September, perhaps earlier. Larvae initially mine ash and wild privet, then live in a leaf fold. Latin *cuculus* + *penna* = 'cuckoo' + 'feather', from the wing colour.

Garden apple slender *Callisto denticulella*, common in gardens, orchards, hedgerows and woodland in Britain north to Perthshire, uncommon in Ireland. Adults fly in May–June. Larvae initially mine leaves of apple, then create and feed in a leaf fold, and are active from June to November. Callisto in Greek mythology was a nymph of Lycaon loved by Zeus. Latin *denticulella* = 'little teeth'.

Gold-dot slender *Euspilapteryx auroguttella*, widespread and common in coppiced woodland and rough grassland, and on verges. The bivoltine adults fly in May and August from dusk into night. Larvae initially mine leaves of St John's-wort then feed in a leaf roll. Greek *eu* + *spilos* + *pteryx* = 'well' + 'spot' + 'wing', and Latin *aurum* + *gutta* = 'gold' + 'spot', from the forewing markings.

Hawthorn slender *Parornix anglicella*, widespread and common in woodland, hedgerows and gardens, more scattered and local in Scotland. The bivoltine adults fly in late afternoon and nocturnally in April–May and August. Larvae initially mine then feed in leaf folds of hawthorn, occasionally wild service-tree. Latin *anglicus* = 'English'.

Hazel slender *Parornix devoniella*, widespread and common in woodland and hedgerows, rarer in Scotland. The bivoltine adults fly during the evening in May and July–August. Larvae initially mine then feed in leaf folds of hazel. Dawlish, south Devon, is the type locality, giving the specific name.

Little slender *Calybites phasianipennella*, local in damp woodland, fen and cereal field margins, with a scattered distribution in England, rare in Scotland and Ireland. Adults fly in August–September and overwinter as adults, after which they can be found until April or May. Larval food plants include docks, knotgrass, sorrels and yellow loosestrife. Initially, a mine is created, and larvae later create and feed in two successive conical leaf rolls. Greek *calybites* = 'living in a hut', from the larval behaviour, and Latin *phasianus* + *penna* = 'pheasant' + 'feather', from the wing coloration.

Maple slender *Caloptilia semifascia*, local in woodland edges and hedgerows in southern England and Wales, widespread, but tending to occur in isolated if sometimes well-populated colonies; also present in Yorkshire and Cumbria. Adults fly from late July to October and again, after hibernating, until May. Larvae feed on field maple (sometimes sycamore) in June–July, occasionally in late August–September, initially in a leaf mine, then a leaf cone and finally a sequence of two leaf rolls. Latin *semi* + *fascia* = 'half' + 'band', from the forewing pattern.

Mugwort slender *Leucospilapteryx omissella*, wingspan 7–8 mm, scarce in waste ground, downland and verges throughout much of south-east England, with records north to Yorkshire. The bivoltine adults fly in May and August, usually in the late afternoon. Larvae mine leaves of mugwort in June–July and August–October. Greek *leucos* + *spilos* + *pteryx* = 'white' + 'spot' + 'wing', from wing markings, and Latin *omissella* = 'omitted', this species initially being omitted from texts.

New oak slender *Caloptilia robustella*, common in oak woodland and areas with scattered trees throughout Britain, as far north as Aberdeenshire; rarer in the west and north of its range. Adults haven a number of generations in a year between April and November. Larvae initially mine the leaves of oak, beech and sweet chestnut, then create and feed in a leaf fold. Latin *robustus* = 'oaken', from the food plant.

Northern slender *Parornix loganella*, scarce, nocturnal, local in woodland, hillsides and moorland in parts of northern England and Scotland, especially the Highlands, and south-west Ireland. Adults fly in May–June. Larvae present in July–August, feed in leaves of birch first in a gallery then in a leaf fold. R.F. Logan (1827–87) was a Scottish entomologist.

Pale red slender *Caloptilia elongella*, wingspan 14–16 mm, common and widespread. The bivoltine adults fly in June and September, after which they overwinter and reappear in spring. From July to October, larvae feed on alder, initially mining the leaves and feeding on sap, later on the parenchyma; they subsequently spin part of a leaf into a fold inside which they feed. Latin *elongus* = 'very long', from the long forewing.

Pointed slender *Parornix finitimella*, in the southern half of Britain as far north as Yorkshire, common in woodland, gardens, hedgerows and scrub in the south. Adults are bivoltine in the south, flying during the evening in May and August. Larvae mine leaves on blackthorn, then feed in fallen leaves. *Finitimus* is Latin for 'very close' (i.e. to a congeneric).

Red birch slender *Caloptilia betulicola*, wingspan 14–16 mm, common and widespread. The bivoltine adult flights peak in June–July, and September–October, the moths overwintering as adults to reappear in early spring. Larvae initially mine birch, especially in August–October, then feed inside a rolled leaf. *Betulicola* is Latin for 'birch-inhabiting'.

Ribwort slender *Aspilapteryx tringipennella*, common and widespread on downland, rough grassland, waste ground and roadside verges as far north as Shetland. Adults fly in afternoon and evening sunshine in May and with another generation in August. Larvae initially mine leaves of ribwort plantain, then create and feed within a leaf roll. Greek *a* + *spilos* + *pteryx* = 'without' + 'spot' + 'wing', and Latin *tringa* + *penna* = 'sandpiper' + 'wing', from the coloration.

Rowan slender *Parornix scoticella*, widespread if less so in Ireland, in woodland, orchards, hedgerows and moorland in the south, rarer and more associated with upland in Scotland. Adults have two generations in the south, flying during the evening in May and August. Further north there is one generation in August. Larvae mine leaves of rowan, whitebeam and apple, then create and feed in a leaf fold. The type locality is in Scotland, giving the specific name.

Scarce alder slender *Caloptilia falconipennella*, wingspan

12–14 mm, rare, with a local and scattered distribution in the southern half of England and Wales in marsh, riverbanks, ditches and other damp areas. Adults emerge in September, overwinter to reappear in spring, with flight peaks in April and July. Larvae initially mine alder leaves, then create and feed in a leaf fold. Latin *falco* + *penna* = 'falcon' + 'wing', from the wing colour.

Small red slender *Caloptilia rufipennella*, a recent colonist reported for the first time in 1970 but now common in woodland, gardens, parks and fens throughout much of Britain. Adults emerge in July–August and overwinter, appearing again in spring. Particularly from August to October larvae mine leaves of sycamore, occasionally other acers, then create and feed inside a leaf fold. Latin *rufus* + *penna* = 'red' + 'wing'.

Sulphur slender *Povolnya leucapennella*, wingspan 12–14 mm, widespread but scarce, with a local, scattered distribution in oak woodland. Adults are active at night from July to October. Larvae mine oak leaves, then create a fold at the edge, followed later by two or more 'cones' formed by rolling the leaf tip. Dalibor Povolný was a twentieth-century Czech entomologist. Greek *leucos* = 'white' + Latin *penna* = 'wing'.

Sycamore slender *Caloptilia hemidactylella*, wingspan 12–14 mm, very rare and endangered, recorded from a few sites in woodland edge habitat and hedgerows in parts of the Midlands, notably in Whittlebury Forest, Northamptonshire. Larvae mine sycamore leaves, later moving into leaf folds to feed. Adults fly from September and overwinter. Greek *hemi* + *dactylos* = 'half' + 'finger', from the wing shape.

White-triangle slender *Caloptilia stigmatella*, wingspan 12–14 mm, widespread and common in hedgerows, woodland and marshes, more scattered in Scotland. Adults have two generations, flying in the south between March and May, and from June to October. Larvae feed on sallow and other willows, and aspen and other poplars. Latin *stigmatus* = 'branded', from the forewing mark.

Yellow-triangle slender *Caloptilia alchimiella*, widespread and common in oak woodland and areas with scattered trees. Adults rest by day on trunks and fly in the evening from May to September, with two generations, at least in the south. Larvae mine oak leaves, then create and feed inside a leaf fold. *Alchimiella* is from Latinised Arabic *al-kimia* = 'alchemy', from the golden forewing markings.

SLIDER See terrapin.

SLIME MOULD A generic name for several types of unrelated eukaryotic protist organism. When food is abundant, they live as single-celled organisms, but when food is scarce they aggregate to resemble a gelatinous slime and start moving as a single body. In this state they are sensitive to airborne chemicals and can detect food sources. They can change the shape and function of parts of the overall mass and may form stalks that produce fruiting bodies, releasing spores light enough to be carried on the wind. They help in the decomposition of plant material, feeding on bacteria, yeasts, and fungi. Slime moulds are usually found on the forest floor, on soil or on decaying logs of hardwood trees.

SLIMECAP Scarce woodland soil fungi, Vulnerable in the RDL (2006), in the genus *Limacella* (Amanitaceae) (Latin *limus* = 'slime').

SLIPPER LIMPET See limpet.

SLIPPERWORT **Annual slipperwort** *Calceolaria chelidonioides* (Calceolariaceae), an annual Central and South American neophyte, an occasional casual, sometimes naturalised, on cultivated waste ground in Britain. *Calceolaria* is from the diminutive Latin *calceolus* = 'slipper', and Greek *chelidon* = 'swallow' (bird).

SLIPPERY JACK *Suillus luteus* (Boletales, Suillaceae), a widespread, common and edible (if tasteless) fungus, found from August to November on damp, shaded soil under coniferous trees, especially ectomycorrhizal on Scots pine. The English name describes the cap which is slimy when wet. *Suillus* is from

Latin *sus* = 'pig', from the greasy nature of the cap, and *luteus* = 'yellow', though here possibly 'grimy'.

SLOE Alternative name for blackthorn, derived from Old English for 'plum'.

SLOUGH-GRASS American slough-grass *Beckmannia syzigachne* (Poaceae), an annual or short-lived perennial North American neophyte, occasionally casual on waste ground and around docks (e.g. Avonmouth), from birdseed and grain. Johann Beckmann (1739–1811) was a German botanist. Greek *syzygos* + *achne* = 'jointed' + 'chaff' or 'achene'.

SLOW WORM *Anguis fragilis* (Anguidae), a 40–5 cm long legless lizard, widespread and locally common in Britain. They spend much of their time in burrows. They can be seen in waste ground, allotments and gardens (using compost heaps for their burrows and feeding on slugs and snails), along railway embankments, and in woodland and churchyards. They hibernate from October to March. Breeding takes place in May. The young (70–100 mm long) are born in an egg membrane that breaks soon after birth (ovo-vivipary), and take 4–5 months to develop, being produced in late August–September. The species is protected under the WCA (1981). The binomial is Latin for 'snake' and 'fragile'.

SLUDGE WORM *Tubifex tubifex* (Tubificidae), a common and widespread clitellate, up to 20 cm long living in freshwater and estuarine muds in a variety of habitats, being especially abundant in polluted sediments and marginal habitats not inhabited by many other species. It often appears red owing to the possession of the respiratory pigment haemoglobin which facilitates oxygen uptake in oxygen-poor environments.

SLUG Hermaphroditic terrestrial and marine pulmonate gastropod molluscs. Their shells are very small on the body or are contained internally. Terrestrial slugs produce three forms of mucus that deter predators, facilitate locomotion and prevent death from desiccation. They use two pairs of tentacles to sense their environment, the upper pair (optical tentacles) used to sense light, the lower pair (oral tentacles) providing the sense of smell. Both pairs can retract and extend themselves to avoid hazards, and can be regrown if damaged, or lost accidentally or by predation. They are generally decomposers or omnivores, but a few are predators (e.g. leopard slugs hunting other slugs). There are 32 recognised British land species, though in 2014 a scientific article suggested that undetected and undescribed species might raise numbers by 20%. **Round-backed slugs** are in the family Arionidae, **keelback slugs** Limacidae, **keeled slugs** Milacidae, and **shelled slugs**, with a small shell perched on the mantle at the rear end, Testacellidae.

Arctic field slug *Deroceras agreste* (Agriolimacidae), length 5 cm, scattered in Britain mostly in northern England and Scotland, mostly in upland parts, but also in East Anglia (and at one site in Ireland at Brittas Bay, Co. Wicklow), in damp grassy habitats and marsh, feeding on live and decomposing plant material. *Agreste* is from Latin for 'field'.

Ash-back slug *Limax cinereoniger* (Limacidae), with a maximum length >20 cm the largest slug in the world, widespread but scattered, locally rare and in many places declining, in often old deciduous woodland, feeding on fungi. Latin *limax* = 'slug' and *cinereoniger* = 'ashy-black'.

Atlantic shelled slug *Testacella maugei* (Testacellidae), probably introduced, animal length 6–12 cm, shell length 13–17 mm, width 7–11 mm and height 3.5 mm, scattered and generally declining in south and south-west England and south Wales, living 10–30 cm below the soil surface, nocturnal, feeding on earthworms. Often found in gardens and nurseries, in Devon and Cornwall it has also been recorded among rocks and on sea-cliffs. Latin *testa* + *cella* = 'shell' + 'room'.

Balkan three-band slug *Ambigolimax nyctelius* (Limacidae), introduced, length up to 5 cm, in a few woodland sites mostly in

southern England. Latin *ambigo* + *limax* = 'go around' + 'slug', and Greek *nyctos* = 'nocturnal'.

Black slug *Arion ater* agg. (Arionidae), a species complex with three species: *A. ater*, *A. rufus* and *A. vulgaris*, ranging in length from 7.5–18 cm at maturity, common and widespread, especially seen after rain and at night from spring to autumn. They feed on carrion and dung as well as vegetable matter, preferring decomposing vegetation to living plants and, while characteristically found in woodland, consequently rarely do much damage when present in the garden. Arion was a Greek poet and musician. *Ater* is Latin for 'black'.

Blue-black soil slug Alternative name for garden slug.

Blue spot slug *Polycera elegans* (Polyceridae), length usually up to 4 cm, on rocky coasts feeding on erect bryozoans, recorded only from Kilkieran Bay (Co. Galway), Sheep Haven (Co. Donegal), Skomer Island, Lundy, and south Cornwall. Greek *poly* + *ceras* = 'many' + 'horn', and Latin *elegans* = 'elegant'.

Brown soil slug Alternative name for common garden slug.

Budapest (keeled) slug *Tandonia budapestensis*, length 5–7 cm, common and widespread, though scarce in upland Scotland, a south-east European synanthrope introduced in 1921, continuing to spread, found in greenhouses, parks, gardens and ploughed fields (a pest of root crops, especially potato, and ornamental plants). It can burrow into the soil to a depth of 37 cm. *Tandonia* may be from Greek *tanaos* + *odonti* = 'stretched' + 'teeth'.

Celtic sea slug *Onchidella celtica* (Onchidiidae), body length 13 mm, width 6 mm, an intertidal pulmonate found on Upper Loch Fyne, the Farne Islands, south-west England and Jersey, foraging downshore as the tide retreats, feeding on small algae and returning up shore with the rising tide. *Onchidella* is a diminutive from Greek *oncos* = 'swelling'.

Common garden slug *A. distinctus*, length up to 5 cm, common and widespread, often in moist, sheltered, disturbed habitats, for example gardens (where they are a common pest) and roadsides, feeding on live plant material. Not the same species as garden slug *A. hortensis*.

Durham slug Alternative name for green-soled slug.

Dusky slug *Arion subfuscus*, length 5–7 cm, common and widespread in deciduous woodland, grassland and gardens, under stones and dead wood, eating usually decaying plant parts close to the ground, and fungi. *Subfuscus* is Latin for 'dusky' or 'brownish'.

Ear shelled slug *Testacella haliotidea*, probably introduced, animal length 6–12 cm, shell length 6–10 mm, width 4–7 mm and height 2 mm, uncommon and scattered, mainly in England, in parks and gardens, hunting earthworms in the soil at night. Greek *halas* + *otos* = 'salt' + 'ear'.

Field slug Keeled pulmonates in the genus *Deroceras* (Agriolimacidae) (Greek *dere* + *ceras* = 'neck' + 'horn').

Garden slug *Arion hortensis*, length 2.5–3.5 cm, common and widespread in England and Wales, scattered elsewhere, in deciduous woodland litter as well as in gardens, feeding on living plants and becoming a horticultural pest, and also synanthropic in cellars. *Hortensis* is from Latin for 'garden'. Not the same species as common garden slug *A. distinctus*.

Ghost slug *Selenochlamys ysbryda* (Trigonochlamydidae), an introduced, possibly Crimean blind keeled pulmonate, length 6.4 cm, first scientifically described in 2008 from south Wales, with a number of records now from this region together with a few from England, mainly from gardens, predominantly burrowing, living up to a metre underground and rarely, at night, coming to the surface, feeding on earthworms. Its white colour and nocturnal habits give it both its common and specific name, 'ysbryd' being Welsh for ghost or spirit. Greek *selene* + *chlamys* = 'moon' + 'cloak'.

Golden shelled slug *Testacella scutulum*, probably introduced, body length 8–12 cm, shell length 6–7 mm, width 3.5–4.5 mm, uncommon and scattered, mainly in England and especially from the London area, hunting earthworms in the soil of parks and gardens. *Scutulus* is Latin for 'small shield'.

Great grey slug Alternative name for leopard slug.

Green cellar slug *L. maculatus*, introduced from the Black Sea area, length 8–14 cm, common and widespread in Ireland (an alternative name being **Irish yellow slug**), scattered in Britain, in buildings and crevices of stone walls, and on roadsides, feeding on fresh and decomposing plant material. *Maculatus* is Latin for 'spotted'.

Green-soled slug *Arion flagellus*, length 6–10 cm, an Iberian introduction first recorded in Britain in 1945–6, widespread if patchily distributed in disturbed habitats, waste ground, acidic grassland, roadsides and gardens. In Ireland it was localised until the late 1980s when an expansion in numbers and range began. First noticed in gardens in 1995 the subsequent increase in numbers has been huge, in some areas reaching plague proportions. In the British Isles generally its rapid rate of reproduction may overwhelm and make it a threat to native slugs. *Flagellum* is Latin for 'whip'.

Greenhouse slug Alternative name for Iberian three-band slug.

Grey field slug Alternative name for netted (field) slug.

Grey sea slug *Aeolidia papillosa* (Aeolidiidae), up to 12 cm length, widespread in the intertidal and sublittoral feeding on sea anemones. *Aiolos* is Greek for 'variegated', *papillosa* Latin for 'nipple-like', referring to structures on the body that aid respiration.

Hedgehog slug *Arion intermedius*, length 1.5–2 cm, common and widespread in leaf litter of damp deciduous woodland and in grassland, feeding on plants and fungi.

Iberian three-band slug *Ambigolimax valentianus*, introduced, length up to 5 cm, scattered and spreading in lowland Britain and north-east Ireland, near greenhouses, plant nurseries and allotments, under dead wood and in compost bins. Latin *ambigo* + *limax* = 'to go around' + 'slug', and *valens* = 'strong'.

Irish yellow slug Alternative name for green cellar slug.

Jet slug See smooth jet slug below.

Kerry slug *Geomalacus maculosus* (Arionidae), length up to 9 cm, locally common in south-west Ireland in heather moorland and rough pasture up to 300 m, rarely in oak woodland, mostly on acid sandstone substrate, feeding on lichen, liverworts, algae, moss and fungi. Greek *ge* + *malacos* = 'earth' + 'soft', and Latin *maculosus* = 'spotted'.

Lagoon sea slug *Tenellia adspersa* (Tergipedidae), 8 mm long, recorded from the Firth of Forth, near St Osyth (Essex), the Fleet (Dorset), the Bristol Channel, off Pembrokeshire and Liverpool Bay, in the intertidal and shallow sublittoral, including estuaries and harbours, feeding on hydroids. Latin *tenellus* = 'tender', *adspersa* = 'scattered'.

Large red slug *Arion rufus*, length 7–14 cm, widespread but scattered, absent from north Scotland and south Ireland, omnivorous, in woodland, gardens (not as a pest) and some dry coastal habitats such as dunes. *Rufus* is Latin for 'red'.

Lemon slug Alternative name for slender slug.

Leopard slug *L. maximus*, so called because of its dark blotches, 10–20 cm long, common and widespread in woodland (especially in Ireland) but mostly in synanthropic habitats such as gardens, cellars and damp outbuildings, able to climb trees, walls, etc. It feeds on living and decomposing plants, detritus, fungi, and other slugs (hunting these at up to 15 cm a minute). A pair mates while suspended from a mucous string. *Maximus* is Latin for 'largest'.

Marsh slug *Deroceras laeve* (Agriolimacidae), length up to 2.5 cm, common and widespread in a range of habitats, including gardens, and it can become a greenhouse pest. It is omnivorous but prefers live and decaying plant material. Greek *dere* + *ceras* = 'neck' + 'horn', and Latin *laeve* = 'smooth'.

Netted (field) slug *D. reticulatum*, length 4–6 cm, the commonest slug in Britain, and widespread in most lowland habitats, particularly on cultivated or disturbed ground such as arable fields and gardens where, although it is omnivorous,

it prefers fresh plant material and can become a major pest. It produces milky mucus when irritated. *Reticulatum* is Latin for 'netted'.

Northern field slug Alternative name for arctic field slug.

Orange-clubbed sea slug *Limacia clavigera* (Polyceridae), 2 cm long, widespread though absent from south-east England, on encrusting bryozoans in the shallow sublittoral down to depths of 80 m. Latin *limax* = 'slug', and *clavus* + *gero* = 'club' + 'to carry'.

Pepper-and-salt slug *Caloria elegans* (Facelinidae), length up to 1.5 cm, recorded from Torbay, Lundy, Bristol Channel, Cardigan Bay, off the Llyn Peninsula, Lough Swilly (Co. Donegal) and off Larne, Northern Ireland, feeding among hydroids on rocks. *Calor* is Latin for 'heat'.

Portuguese slug Alternative name for vulgar slug.

Round-backed slug Members of the family Arionidae. Unlike keelback and field slugs they have no vestigial shell remaining; the respiratory opening is located in the middle or the anterior (frontal) half of the mantle shield; and they can roll into a ball for protection, contracting the body to form a hunchback then pulling together the foot sole.

Royal flush sea slug *Akera bullata* (Akeridae), length up to 6 cm, with an external shell, recorded from scattered locations, especially in Northern Ireland and the Isle of Man in fine mud to muddy sand in sheltered bays, from the lower shore to 370 m, herbivorous during summer but feeding on sediment the rest of the year. In 2010, *The Guardian* launched a competition for the public to suggest common names for previously unnamed species. This was a winner, describing the escape behaviour by flapping and exuding purple ink. *Bullatus* is Latin for 'blistered'.

Rusty false-keeled slug *Arion fasciatus*, length 6 cm, common and widespread from the Midlands and south Wales north, scattered elsewhere in England and in Ireland, in deciduous woodland litter. *Fasciatus* is Latin for 'bundled'.

Silver false-keeled slug *Arion circumscriptus*, length 3–4 cm, common and widespread in cool, humid woodland, roadsides and gardens, feeding on plants and fungi. Latin *circumscriptus* = 'written around'.

Silver slug *Arion silvaticus*, length up to 4 cm, widespread and locally common in moist habitats in herbage and litter, tolerant of non-calcareous habitats and found in acid woodland and heathland, and on sea-cliffs and mountains. *Silvaticus* is Latin for 'woodland' or 'wild'.

Slender slug *Malacolimax tenellus* (Limacidae), length 5 cm, scattered, local and declining in Britain, an indicator of ancient woodland, feeding on fungi. Greek *malacos* = 'soft' + Latin *limax* = 'slug', and Latin for 'tender'.

Smooth jet slug *Milax gagates* (Milacidae), length 4.5–5 cm, scattered in lowland regions, especially in coastal grassland and rocks, under stones and in moist ground litter, feeding on fresh plants, in places a pest of root crops such as carrot and potato. *Gagates* is Greek for 'jet black'.

Solar-powered sea slug Alternative name for green elysia.

Sowerbury's (keeled) slug *T. sowerbyi*, introduced from the Balkans, length 6 cm, widespread, scattered and locally common, often a pest in gardens and other synanthropic habitats in moist litter and at the base of plants. James Sowerby (1787–1871) was an English naturalist.

Spanish slug Alternative name for both Iberian three-band slug and vulgar slug.

Spanish stealth slug Alternative name for green-soled slug.

Spotted keeled slug *T. rustica* (Milacidae), length 7–10 cm, with a few records from south-east England, Gloucestershire and Co. Cork, in deciduous woodland and rocky grassland on calcareous substrates. *Rustica* is Latin for 'rural'.

Tawny soil slug *Arion owenii*, length up to 3.5 cm, common in south-west England, west Scotland and north-west Ireland, scattered elsewhere, in a variety of moist sheltered habitats in ground litter and soil.

Tramp slug *Deroceras invadens* (Agriolimacidae), length 2–3.5

cm, common, widespread and increasing, probably introduced from southern Europe, with the first record in Britain in 1930, mostly found in disturbed sites (it is the commonest slug in Manchester gardens). It demonstrated the highest crawling speed (4.9 mm/s) when compared to 27 other terrestrial slug and snail species.

Tree slug *Lehmannia marginata* (Limacidae), length 12 cm, common and widespread in woodland and rocky habitats, climbing trees, rocks and walls to graze algae, fungi and lichen.

Vulgar slug *Arion vulgaris*, introduced from the Iberian Peninsula, first recognised in Britain in 2012, length up to 8 cm, scattered with a concentration of records in Devon, omnivorous, and with a rate of population growth that could threaten to out-compete native slug species. *Vulgaris* is Latin for 'common'.

Worm slug *Boettgerilla pallens* (Boettgerillidae), Caucasian, first recorded in Britain in 1972, in Ireland in 1973, increasingly if patchily widespread, length up to 60 mm, with a small (1.5–3 mm) shell, found under rubble, bricks, etc. in gardens, and in grassland and woodland. Caesar Böttger (1888–1976) was a German zoologist.

Yellow (cellar) slug *L. flavus*, introduced from southern Europe by 1700, length 8–10 cm, widespread but occasional in Britain, and scarce in Ireland, strongly synanthropic, usually found in damp areas such as cellars, kitchens, outhouses and gardens, able to climb walls, feeding on fresh and decomposing plant material, fungi and detritus. *Flavus* is Latin for 'yellow'.

Yellow-plumed sea slug *Berthella plumula* (Pleurobranchidae), length up to 6 cm, widely distributed, an opisthobranch often found under stones, in clear rock pools on the lower shore and in shallow water up to 10 m depth, common on colonies of star ascidian.

Yellow skirt slug *Okenia elegans* (Goniodorididae), length up to 8 cm, scattered in south-west England, and on Skomer, the Saltees (Co. Wexford) and the Skerries, Portrush (Co. Antrim), often burrowed inside the test of the sea squirt *Polycarpa scuba*, on which it feeds. Lorenz Oken (1779–1851) was a German naturalist.

SMALL-REED Rhizomatous perennial grasses in the genus *Calamagrostis* (Poaceae), from Greek for these plants (*calamos* + *agrostis* = 'reed' + 'grass').

Narrow small-reed *C. stricta*, rare in near-neutral bog, marsh, fens and lake margins, very scattered from Suffolk to north Scotland. *Stricta* is Latin for 'upright'.

Purple small-reed *C. canescens*, scattered in Britain, mostly in central-east England, in lakeside marsh, tall-herb fen, damp wetland, and alder and willow carr. *Canescens* is Latin for 'greyish'.

Scandinavian small-reed *C. purpurea*, apomictic, with six known sites in central Scotland and two in Cumbria, in marsh and willow carr, especially where winter-flooded.

Scottish small-reed *C. scotica*, restricted to one site in Caithness in rush-dominated pasture.

Wood small-reed *C. epigejos*, widespread, common in parts of central, southern and eastern England, uncommon elsewhere, usually on light sands or heavy clays, in damp woodland, fen, ungrazed or lightly grazed grassland, on sea-cliffs and dunes, and as a colonist of artificial habitats, such as roadsides and railway banks. Greek *epi* + *geios* = 'on' + 'earth'.

SMEARWORT *Aristolochia rotunda* (Aristolochiaceae), a climbing perennial neophyte introduced from Mediterranean Europe by at least the sixteenth century, naturalised on chalk slopes at Woldingham, Surrey, since at least 1918. The common name comes from its use as ointment. Poultices derived from the leaves were used to heal chronic sores. Roots were used to remedy sheep cough. Greek *aristos* + *locheia* = 'best' + 'childbirth', referring to the value of the plant in helping childbirth, and Latin *rotunda* = 'round'.

SMELT *Osmerus eperlanus* (Osmeridae), length generally 15–18 cm but up to 30 cm, a migratory inshore fish especially found from the Tay to the Thames, from the Clyde to the Conwy, and off south-west Ireland, feeding on small crustaceans and small fish, moving into the lower stretches of rivers to spawn above sand or gravel in March–May. 'Smelt' dates from the sixteenth century, from Middle Dutch/Middle Low German *smelten* (to smelt). Greek *osmiris* (*osmeros*) = 'odorous', the fish said to smell of cucumber.

SMEW *Mergellus albellus* (Anatidae), wingspan 62 cm, length 41 cm, male weight 700 g, female 580 g, an Amber List winter visitor from Scandinavia and Russia found from December to March on lakes and reservoirs in small numbers (180 birds in 2008–9) south of a line from the Severn estuary to the Wash, feeding on fish and insect larvae. 'Smew' is at least seventeenth-century, probably related to Dutch *smient* = 'wigeon' and German *Schmeiente* = 'small wild duck'. *Mergellus* is a diminutive of *Mergus*, used by Pliny to refer to an unspecified waterbird; *albellus* comes from Latin *albus* = 'white'.

SMILO-GRASS *Oryzopsis miliacea* (Poaceae), a perennial south European neophyte established in Jersey, where it may be a garden escape, and in west Kent. Elsewhere in England it is an occasional casual of waste ground and docks, originating with wool shoddy and birdseed. Greek *oryza* + *ops* = 'rice' + 'appearance'.

SMOCK Lady's smock Alternative name for cuckooflower.

SMOKE (MOTH) Members of the family Psychidae.

Brown smoke *Taleporia tubulosa* (Taleporiinae), wingspan 15–19 mm, local, sometimes common in woodland and rough meadows, and on rocks and old walls, throughout much of the southern British Isles. Adults appear in May–June. Females are wingless. Larvae feed on lichens, dead insects and decaying vegetable matter, inside a movable case constructed of silk coated with lichen, fine sand and bark, and overwintering once or twice. Greek *talaiporia* = 'hard work', describing the larva's effort in dragging its case, and Latin *tubulosa* = 'small tube', from the elongated case.

Dotted-margin smoke *Diplodoma laichartingella* (Naryciinae), wingspan 10–15 mm, with a scattered and local distribution in woodland. Adults fly in May–June. Larvae build a portable case of fragments of plant matter and other particles, and feed on lichens, decaying plant matter, dead insects and detritus. The larval period usually lasts for two years. Greek *diplous* + *doma* = 'double' + 'house', from the double case built by the larva. Johann von Laicharting (1754–97) was an Austrian entomologist.

Shining smoke *Bacotia claustrella* (Psychinae), male wingspan 13–15 mm, female wingless, a rarity found in woodland in the southern counties of England as far north as Worcestershire. Adults are occasionally seen in June–July. Larvae feed on lichen growing on tree-trunks, living within a movable case, but resting with the case projecting at right angles from the bark, resembling a bud or stunted side shoot. They feed from autumn, overwintering and feeding again in spring up to May. Arthur Bacot (1866–1922) was a British entomologist. *Claustrum* is Latin for 'that which closes'.

Virgin smoke *Luffia ferchaultella* (Psychinae), locally common in damp shady places, widespread in (especially southern) England and Wales. The male is unknown, the female wingless and parthenogenic. Larva feed, often gregariously, on lichen growing on tree-trunks and fence-posts, within a movable case. The binomial honours the Channel Islands collector William Luff (1851–1910), and the French naturalist René Ferchault de Réaumur (1683–1757).

White-speckled smoke *Narycia duplicella* (Naryciinae), wingspan 7–12 mm, fairly common, with a scattered distribution in England and Wales and parts of Scotland, in woodland, gardens, parks and orchards. Adults fly in May–July often resting on trunks of such trees as oak, hawthorn and yew. Larvae construct portable cases and attach themselves to trunks, fences or other surfaces to feed on lichen, and are particularly active from March to June. *Duplicella* is Latin for 'double chamber', describing the larval case. Narycia was a Greek city.

SMOKE-TREE *Cotinus coggygria* (Anacardiaceae), a deciduous European neophyte shrub, commonly grown in gardens and parks, naturalised in places on roadsides and railway banks. The plume-like seed clusters give a smoky haze to the branch tips. The binomial comprises the Latin and Greek for this tree.

SMOOTHCAP Mosses in the genus *Atrichum* (Polytrichaceae).

Common smoothcap *A. undulatum*, common and widespread in shaded, well-drained habitats, especially in lowland woodlands, but also on heathland and in unimproved and semi-improved grasslands. *Undulatum* is Latin for 'wavy'.

Fountain smoothcap *A. crispum*, on rocky stream banks, gravelly and peaty substrates on lake shores, reservoirs and by rivers, and in ditches, restricted to acidic substrates and is frequent on upland moors in south-west and north-west England and Wales, local in eastern Ireland. *Crispum* is Latin for 'curly'.

Lesser smoothcap *A. angustatum*, declining, mainly found in south-east England on bare, moderately acid, disturbed, damp, shady, loamy or sandy soils. *Angustatum* is Latin for 'narrow'.

Slender smoothcap *A. tenellum*, scattered and uncommon in disturbed, acidic, open habitats on moist, sandy humus and loams, including streamsides, lake edges, woodland rides, ditch banks and heathy grassland. *Tenellum* is Latin for 'tender'.

SMUDGE (MOTH) So called from the smudged or 'smeared' colours on the wings, members of the genera *Acrolepia* (Greek *acron* + *lepis* = 'top' + 'scale', from the scaly head), *Acrolepiopsis* (as previous + *opsis* = 'appearance', i.e. similar to), *Digitivalva* (Latin *digitus* + *valva* = 'finger' + 'valve', here meaning clasper of the male genitalia) and *Orthotelia* (Greek *orthos* + *telos* = 'straight' + 'end', from the termen, or outer edge, of the forewing) in the family Glyphipterigidae; *Eidophasia* (Greek *eidos* + *phasis* = 'beautiful' + 'appearance'), *Plutella* (Greek *plutos* = 'washed', i.e. smudged appearance of the wings) and *Rhigognostis* (Greek *rhigos* + *gnostes* = 'cold' + 'one who knows', from the adults overwintering) (Plutellidae); and *Ypsolopha* (Greek *hypsilophos* = 'high-crested') (Ypsolophidae).

Rarities include **arctic smudge** *Plutella haasi* (one specimen found in Wester Ross in 1954, but searching in 2009 and 2010 found larvae elsewhere in north-west Scotland), **dark smudge** *Y. horridella* (in parts of southern and eastern England, and north Wales), **elusive smudge** *Acrolepiopsis marcidella* (found in Hampshire in 1986, with more records from 1995 onwards, and in 1997 also discovered in Dorset), and **Oban smudge** *Acrolepiopsis betulella* (a few records from Scotland, and one trapped in Co. Durham in 2012).

Barred smudge *Ypsolopha alpella*, wingspan 14–18 mm, local in oak woodland throughout much of southern England and south Wales, north to Yorkshire. Adults fly at night in August. Larvae feed on oak leaves underneath a silk web. *Alpella* is from Latin *Alpes* = 'Alps', the type locality.

Bitter-cress smudge *Eidophasia messingiella*, wingspan 14–16 mm, local on waste ground throughout much of southern England and Wales, north to Lancashire. Adults fly in June–July. Larvae feed on hoary-cress under a silk web. Herr Messing (*c*.1800–70) was the German amateur entomologist who discovered the species.

Bittersweet smudge *Acrolepia autumnitella* (Acrolepiinae), wingspan 11–13 mm, common in hedgerows and waste ground in southern England, with records north to Lancashire. Adults generally have two generations, emerging in July then October, the second brood overwintering. Larvae mine leaves of bittersweet, occasionally deadly nightshade from May to October, with peaks in June and August–September. *Autumnus* is Latin for 'autumn', from the phenology of the second generation.

Coast smudge *Rhigognostis annulatella*, wingspan 16–21 mm, mainly on rocky coastlines from the West Country northwards to north-east England, Scotland and Ireland. Adults fly

from July onwards and overwinter, occurring on the wing until April. Larvae feed on scurvygrass under a silk web. *Annulatus* is Latin for 'ringed', from forewing streaks that appear as two rings when folded.

Elm smudge *Y. vittella*, wingspan 16–20 mm, nocturnal, widespread but local in woodland and areas with scattered trees. Adults fly in July–August. Larvae feed on elm and beech leaves under a silk web. *Vitta* is Latin for 'band', from the black forewing streak.

Fleabane smudge *D. pulicariae* (Acrolepiinae), wingspan 12–14 mm, scarce, day-flying, local in rough grassland, woodland rides and waste ground in England as far north as Northumberland, Wales and the southern half of Ireland. Adults fly from August and hibernate over winter, reappearing the following spring. Larvae mine leaves of common fleabane (*Pulicaria*, giving the specific name) and hemp-agrimony.

Grey-streaked smudge *P. porrectella*, wingspan 14–17 mm, widespread and locally common. The bivoltine adults fly at night in May and July–August. Larvae feed on dame's-violet, often gregariously, in April–May and June–July in a silk web. *Porrectus* is Latin for 'outstretched', from the antennae extending when the moth is at rest.

Hooked smudge *Y. nemorella*, wingspan 21–4 mm, nocturnal, widespread if local in woodland, less common in Scotland and Ireland. Adults fly in July–August. Larvae feed on honeysuckle underneath a silk web. *Nemoris* is Latin for 'woodland grove', from the habitat.

Pied smudge *Y. sequella*, wingspan 18–20 mm, common in woodland and urban areas throughout England and Wales. Adults fly at night in July–August. Larvae feed on field maple leaves under a silk web. *Sequens* is Latin for 'following'.

Plain smudge *Y. lucella*, wingspan 16–18 mm, scarce, nocturnal, in oakwoods in England north to Lincolnshire. Adults fly in July–August. Larvae feed oak leaves underneath a silk web. *Lucus* is Latin for 'woodland grove', from the habitat.

Reed smudge *O. sparganella* (Orthoteliinae), wingspan 18–28 mm, nocturnal, widespread (absent from north Scotland) but local in ponds, canals, marsh, riverbanks and other damp habitats. Adults fly in July–August. Larvae feed inside the stems of branched bur-reed (*Sparganium*, prompting the specific name) and reed sweet-grass, with other reeds and rushes also used.

Rock-cress smudge *R. senilella*, wingspan 19–23 mm, distributed from north England northwards through Scotland, where it is widespread and common in rocky coastal and montane habitats. Rare in west Wales and north-west Ireland. Adults fly from August onwards, overwintering until spring. Larvae feed on rock-cress, flixweed and dame's-violet, under a silk web. *Senilis* is Latin for 'elderly', from the whitish parts of the head.

Scotch smudge *R. incarnatella*, wingspan 17–21 mm, scarce, in scattered locations on embankments, verges and rough ground from north England to the Highlands and parts of Ireland. Adults emerge in September and overwinter in this stage until around April, finding protection in thick cover. Larvae feed on dame's-violet under a silk web. Latin *incarnatella* = 'very flesh-coloured', from the forewing colour and markings.

Spikenard smudge *D. perlepidella* (Acrolepiinae), wingspan 10–12 mm, scarce, day-flying, recorded from chalk downs of south-east England, and on the limestone hills of Herefordshire, Gloucestershire and Somerset. Adults fly in May–June. Larvae mine leaves of ploughman's-spikenard. *Perlepidus* is Latin for 'very pretty'.

Spindle smudge *Y. mucronella*, wingspan 26–33 mm, nocturnal, local in woodland on chalky soils throughout southern England and the Midlands. Rare in Ireland. Adults fly in August–April, overwintering in dense cover. Larvae feed on spindle underneath a silk web. *Mucro* is Latin for 'sharp point', from the forewing tip.

Variable smudge *Y. ustella*, wingspan 15–20 mm, nocturnal, widespread and common in oakwood and areas with scattered oaks, overwintering as an adult. Larvae feed on oak leaves under a silk web. *Ustus* is Latin for 'burnt', from the scorched appearance of forewings in some morphs.

Wainscot smudge *Y. scabrella*, wingspan 20–2 mm, common in gardens and open woodland throughout England, Wales and Ireland. Adults fly at night in July–August. Larvae feed on leaves of apple, hawthorn and cotoneaster underneath a silk web. *Scaber* is Latin for 'rough', from the raised forewing scale tufts.

White-shouldered smudge *Y. parenthesella*, wingspan 16–22 mm, widespread and common in woodland. Adults fly at night in August–September. Larvae feed on the leaves of a number of tree species, including oaks and hazel. *Parenthesis* is Greek for 'insertion', from a blotch that interrupts the forewing ground colour.

Wood smudge *Y. sylvella*, wingspan 18–20 mm, local in oakwood throughout much of southern England and south Wales, north to Yorkshire. Adults fly at night in August–September. Larvae feed on oak leaves under a silk web. *Sylva* is Latin for 'woodland', from the habitat.

SMUT (FUNGUS) Plant diseases caused by pathogenic fungi mostly in the order Ustilaginales, often attacking grasses, including cereals.

SMUT (MOTH) Members of the genus *Psychoides* (Tineidae, Teichobiinae), from Greek *psyche* + *eidos* = 'soul' + 'appearance', from a resemblance to members of the family Psychidae.

Fern smut *P. filicivora*, wingspan 10–12 mm, local in parts of south and south-west England and south Wales, mainly on the coast, and widespread in Ireland, perhaps originating from ferns imported from the Far East and spreading through trade. Adults fly during daytime in an extended generation throughout summer. Larvae feed on a number of fern species such as male-fern *Dryopteris filix-mas*, prompting the specific name, and including indoor ferns.

Hart's-tongue smut *P. verhuella*, wingspan 9–12 mm, local in damp, shady woodland throughout much of England and Wales, more numerous in the west. Adults fly in June–July. Larvae, prominent in April–May, mine fronds of hart's-tongue, creating and living within a movable case, leaving it when feeding on the fern frond. Q.M.R. Verhuell was a nineteenth-century Dutch entomologist.

SNAIL Terrestrial, freshwater or marine pulmonate gastropod molluscs with a coiled shell large enough for the animal to retract completely into. See also amber, bladder, chrysalis, door, glass, grass, hairy, herald, pond and whorl snails, and slug.

Banded snail Alternative name for brown-lipped snail.

Blind snail *Cecilioides acicula* (Ferussaciidae), lacking eyes, shell length 4–5.5 mm, width 1.2 mm, widespread and locally common but scarce in Scotland and the northern half of Ireland, found in (preferably calcareous) soil, usually at a depth of 20–40 cm, occasionally 2 m, in organic litter of rock crevices, among roots and in ant hills. *Acicula* is Latin for 'small needle'.

Brown-lipped snail *Cepaea nemoralis* (Helicidae), highly polymorphic, with shell colour and stripe pattern very variable, diameter 18–25 mm, height 12–22 mm, common and widespread, though a rare introduction in north Scotland, in a variety of habitats, including gardens where, feeding on dead plant material, it is not a pest. *Nemoralis* is Latin for 'of woodland'.

Brown snail *Zenobiella subrufescens* (Hygromiidae), shell diameter 6–10 mm, height 4–6 mm, generally common and widespread, though less so in the eastern half of England, and declining in Ireland (Vulnerable), an indicator of ancient broadleaf woodland, often associated with great wood-rush, and commonly climbing tree-trunks. *Subrufescens* is Latin for 'nearly reddish'.

Carthusian snail *Monacha cartusiana* (Hygromiidae), shell diameter 9–18 mm, height 6–10 mm, introduced from south Europe in prehistoric times, mainly found, though declining, in southern England and East Anglia in grassy habitats, including roadsides, waste ground and gardens. *Monachos* is Greek for

'solitary'. Carthusians are a monastic order with a white habit, its often whitish shell prompting this snail's name.

Cellar snail *Oxychilus alliarius* (Oxychilidae), shell diameter 9–14 mm, height 4.5–6 mm, common and widespread in moist shaded places including gardens, damp woodland, hedgerows, scrub and old buildings. Greek *oxys* + *cheilos* = 'sharp' + 'margin', and Latin *allium* = 'garlic'.

Cheese snail *Helicodonta obvoluta* (Helicodontidae), shell diameter 11–15 mm, height 5–7 mm, the common name coming from the shell's shape resembling a wheel of cheese, local on chalk and limestone in south and south-east England, in leaf litter and with stones in deciduous woodland. Greek *helix* + *donti* = 'helical' + 'tooth'.

Copse snail *Arianta arbustorum* (Helicidae), shell diameter 18–25 mm, height 12–22 mm, common and widespread in Britain (in both woodland and open habitats up to 1,200 m altitude), rare in Ireland (in ancient woodland), moving a distance of only 7–12 m in a year, usually using water currents, feeding on plants, dead animals and faeces.

Craven door snail *Clausilia dubia* (Clausiliidae), sinistral, shell length 11–14 mm, width 2.7–3.2 mm, mostly recorded from northern England on rocks, trees and old walls. It hides in crevices and litter, climbs vertical surfaces in moist weather, and feeds on epiphytic lichens and algae. *Clausum* is Latin for 'enclosed space', from which comes clauciliium, a calcareous 'closing' device that effectively seals the shell aperture in such animals.

Crystal snail *Vitrea crystallina* (Pristilomatidae), shell diameter 3–4 cm, height 1.4–2.1 cm, common and widespread under litter and stones in woodland and grassland, including verges. See also milky crystal snail below. The Latin *vitrum* and Greek *crystallinos* both mean 'glass'.

Dwarf snail *Punctum pygmaeum* (Punctidae), shell diameter 1.2–1.6 mm, height 0.6–0.8 mm, common and widespread in wet unimproved pasture, wetland margins and litter in damp deciduous woodland. The binomial is Latin for 'point' and 'dwarf'.

Edible snail Alternative name for Roman snail.

Flat valve snail *Valvata cristata* (Valvatidae), widespread and common in lowland regions, 2–4 mm in length, in stagnant and slow-flowing fresh water. See also valve snail below. The binomial is Latin for 'valved' and 'crested'.

Garden snail *Cornu aspersum* (Helicidae), shell diameter 25–40 mm, height 25–35 mm, a Mediterranean introduction dating from Roman times, common and widespread in lowland regions, mainly nocturnal in a range of habitats, including gardens, feeding on living and decaying plant matter, sometimes scavenging on dead animal material. The binomial is Latin for 'horn' and 'speckled'.

Garlic snail *Oxychilus alliarius* (Oxychilidae), shell diameter 6–8 mm, height 3.5–4 mm, common and widespread in damp habitats, including gardens, feeding on plant material and detritus. Its common name comes from its production of a garlic-scented secretion when disturbed. Greek *oxys* + *cheilos* = 'sharp' + 'margin', and Latin *allium* = 'garlic'.

Girdled snail *Hygromia cinctella* (Hygromiidae), shell diameter 10–12 mm, height 6–7 mm, introduced from southern Europe, first noted in Devon in 1950, local and scattered but spreading rapidly in England and Wales in the 1970s, and recorded in central Scotland in 2008, found in tall-herb vegetation, waste ground, roadsides, gardens and nurseries. Greek *hygros* = 'wet', and Latin *cinctus* = 'girdle'.

Glossy pillar snail Alternative name for slippery moss snail.

Glutinous snail *Myxas glutinosa* (Lymnaeidae), shell length 12–15 mm, scarce, scattered and declining in central and west Ireland, Endangered in the RDB, favouring hard substrates in slow-moving waters of deep rivers, lakes and canals, and thought to have become extinct in Britain before rediscovery in Llyn Tegid, north Wales, in 1998 where it is restricted to the littoral, beneath boulders and large stones in silt-free areas. Greek *myxa* = 'slime' and Latin *glutinosa* = 'gluey'.

Green snail *Ponentina subvirescens* (Hygromiidae), shell diameter 5–9 mm, height 4–6.5 mm, found in the Scilly Isles, south-west England and south-west Wales in grassy and rocky habitats on well-drained non-calcareous soils. Latin *subvirescens* = 'a bit green'.

Heath snail *Helicella itala* (Hygromiidae), shell diameter 9–25 mm, height 5–12 mm, widespread but patchy in England (declining in the south and east since 1900), near the coast in Wales and west and north Scotland, and coastal and central Ireland, on calcareous grassland, dunes and sea-cliffs. *Helicella* is from Greek *helicos* = 'twisted', and *itala* Latin for 'Italian'.

Hedge snail *Hygromia limbata* (Hygromiidae), shell diameter 12–17 mm, height 8–14 mm, found in south Devon (introduced from Europe in the 1910s), extending to a few other sites north to the Midlands and south Wales in damp grassy habitats in low vegetation and litter in gardens and other disturbed places. *Limbata* is Latin for 'rounded'.

Kentish snail *Monacha cantiana* (Hygromiidae), shell length 16–20 mm, height 11–14 mm, introduced from Europe in the Late Roman period, spreading since medieval times, now common and widespread in England and south Wales, local elsewhere in Britain, in hedges, gardens and waste ground, and on roadsides and railway banks. Greek *monachos* = 'solitary', and Latin *cantiana* = 'Kentish'.

Lagoon snail *Paludinella littorina* (Assimineidae), an uncommon intertidal snail, shell height 2 cm, found sporadically from Pembrokeshire to Hampshire in crevices, caves, under rocks and in lagoonal shingle, at or just above the water line. Latin *paludis* = 'marsh', *littoralis* = 'shore'.

Lapidary snail *Helicigona lapicida* (Helicidae), shell diameter 14–18 mm, height 7.5–8.5 mm, widespread and locally common in England and Wales, but threatened by the destruction of old hedgerows and by air pollution. In Ireland it was probably introduced to its one known site, limestone rocks by the River Blackwater near Fermoy, Co. Cork. It is also found in rocks and walls (especially when covered with ivy) and tree-trunks, especially of beech, hornbeam and sycamore. Greek *helix* + *gonia* = 'twisted' (helical) + 'angle', and Latin *lapicida* = 'stonecutter'.

Large-mouthed valve snail *Valvata macrostoma*, found south of a line from the Wash to the Severn estuary in marsh and calcium-rich canals. Endangered in RDB 1: of the six areas it was recorded in since the 1970s, it has disappeared in two, and the other populations have decreased. Latin *valvata* 'having sliding doors', and Greek *macrostoma* = 'large-mouthed'.

Least slippery snail *Cochlicopa lubricella* (Cochlicopidae), shell length 4.5–6.7 mm, width 1.8–2.3 mm, common and widespread in dry, often base-rich grassland and on calcareous dunes. Greek *cochlos* + *cope* = 'snail' + 'handle', and Latin *lubricus* = 'slippery'.

Lister's river snail *Viviparus contectus* (Viviparidae), widespread and locally common in central and eastern England, shell height 25–55 mm, width 20–40 mm, in slow-flowing, unpolluted, vegetated, hard-water rivers and canals. Its viviparous habit is rare in snails. *Contectus* is Latin for 'connected'.

Looping snail *Truncatella subcylindrica* (Truncatellidae), a scarce RDB species, shell length 5 mm, recorded from Pagham Harbour, West Sussex, Isle of Wight, The Fleet, Dorset, and St Mawes Bay, Cornwall, in shingle with rotting vegetation and fine sediment to 15 cm depth, at high water mark, occasionally in mud under stones, moving with a looping movement. The binomial is from Latin for 'a bit truncated' and 'near-cylindrical'.

Milky crystal snail *Vitrea contracta*, shell diameter 2–3 mm, height 1–1.4 mm, common and widespread in soil and litter in dry habitats on calcareous substrate, for example rocky grassland, rubble and woodland with rocks. *Contracta* is Latin for 'shortened'.

Mouse-eared snail *Myosotella myosotis* (Ellobiidae), 6.5–8 mm shell length, common and widespread in saltmarsh around the British and Irish coastline. Greek *mys* + *ous* = 'mouse' + 'ear'.

Plated snail *Spermodea lamellata* (Valloniidae), shell diameter and height around 2 x 2 mm, widespread, scattered and declining, found in deciduous woodland litter. Ireland may have 50% of the current world population. Greek *sperma* + *eidos* = 'seed' + 'appearance', and Latin *lamellata* = 'plated'.

Point snail *Acicula fusca* (Aciculidae), shell width 2–3 mm, common and widespread in England, Wales and west Scotland, but showing a 55% decline in records since 1980 in Ireland, found in damp moss and leaf litter in woodland on calcareous soil. The binomial is Latin for 'pointed' and 'dark'. Not to be confused with pointed snail.

Pointed snail *Cochlicella acuta* (Cochlicellidae), shell length 9–15 mm, width 4–7 mm, found in most coastal areas except the east coast of Britain, on dunes and grassland on sandy calcareous soil, and in some inland sites in England and, especially, Ireland in limestone quarries and dry pasture. Introduced to Britain probably in the Late Prehistoric, it has recently extended its range eastwards along the south English coast in response to rising sea temperatures in the English Channel. Greek *cochlis* = 'snail' + Latin *cella* = 'room', and Latin *acuta* = 'sharp'. Not to be confused with point snail.

Prickly snail *Acanthinula aculeata* (Valloniidae), shell diameter and height about 2 x 2 mm, common and widespread in woodland and hedgerow, though perhaps declining. Transverse ridges forming long spines at the periphery gives a 'prickly' appearance. Greek *acantha* = 'thorn' and Latin *aculeata* = 'pointed'.

River snail *Viviparus viviparus*, widespread and locally common in central and eastern England, scarce elsewhere, shell height 25–35 mm, width 20–6 mm, in slow-moving, lowland rivers, canals and lakes, preferring well-oxygenated, base-rich water, sometimes in dense clusters with thousands of individuals on submerged branches and on man-made objects, occasionally scattered in bottom mud, filter-feeding on plankton and organic detritus. Its viviparous habit is rare in snails.

Rock snail *Pyramidula pusilla* (Pyramidulidae), shell width 2.5–3 mm, common and patchily widespread in western England, Wales and the Republic of Ireland, scarce elsewhere, on walls and (usually) limestone or chalk rocks. *Pusilla* is Latin for 'very small'.

Roman snail *Helix pomatia* (Helicidae), highly edible (the escargot of French cuisine), shell diameter 30–50 mm, height 30–45 mm, a Mediterranean introduction dating from the Romans, scattered mainly in southern England in undisturbed grassy or scrubby waste ground. *Pomatia* derives from Greek *poma* = 'cover'.

Round-mouthed snail *Pomatias elegans* (Pomatiidae), length 12.5–16 mm, width 7–11.5 mm, locally common and widespread, though declining, in the southern third of England and in north and south Wales, on chalk and limestone soils, and calcareous dunes. In 1976, a small colony was found on Finavarra, Co. Clare; the species is probably a relict native in Ireland. Greek *poma* = 'cover', and Latin *elegans* = 'elegant'.

Rounded snail *Discus rotundatus* (Patulidae), shell diameter 5.7–7 mm, height 2.4–6 mm, common and widespread in woodland, hedgerows and gardens, in litter and under logs and stones.

Sandbowl snail or **sand ambersnail** *Quickella arenaria* (Succineidae), limpet-like with a shell length 5–8 mm, scattered and uncommon in Britain (Endangered in the RDB, NERC Act (2006) Species of Principal Importance in England, UK BAP priority species) but widespread in central and west Ireland, in grazed or sparsely vegetated dune slacks and on shallow flooded sandy or muddy and calcareous soils on the stems of tall marginal vegetation or submerged macrophytes. Latin *arena* = 'sand'.

Silky snail *Ashfordia granulata* (Hygromiidae), with a hairy shell, diameter 7–9 mm. height 5–7 mm, widespread but patchily distributed in damp shaded woodland, hedgerows and marsh, locally common in southern Ireland in scrubby pasture, along the west Scottish coast, and in a swathe from south-east England to East Anglia (often near water), rare in central and north Ireland, the Midlands and upland Scotland.

Slender-toothed herald snail Alternative name for long-toothed herald snail.

Slippery moss snail *Cochlicopa lubrica*, shell length 5–7.5 mm, width 2.4–2.9 mm, common and widespread in damp grassland and woodland litter, feeding on dead plant material, microbial fungi and detritus. *Lubricus* is Latin for 'slippery'.

Strawberry snail *Trochulus striolatus* (Hygromiidae), shell diameter 10–13.5 mm, height 6–8 mm, common and widespread in a variety of damp habitats, mainly nocturnal but seen in daylight especially after rain, feeding on low-growing vegetation, including strawberries, lettuce and garden flowers. Greek *trochos* = 'wheel', and Latin *striolatus* = 'striped'.

Striped snail *Cernuella virgata* (Hygromiidae), shell diameter 12–23 mm, height 8–15 mm, introduced from the Mediterranean probably in Roman times, locally common and spreading round much of the coastline, and inland south of a line from Yorkshire to the Severn estuary, and in central Ireland, in dry habitats, for example dunes and calcareous grassland, and also on roadsides and in cultivated fields. It climbs to the top of tall plants, including cereals, to escape heat, often in groups, and it also climbs walls. It has partly spread by colonising the ballast and dry banks of railway lines. Latin *cernuus* = 'acrobat', here meaning 'turned towards the earth', and *virgata* = 'striped'.

Taylor's spire snail *Marstoniopsis insubrica* (Hydrobiidae), an endangered rarity, 3 mm shell height, in lentic or very slow-running waters (including canals) in a few places in England, south Wales and central Scotland.

Thin pillar snail Alternative name for least slippery snail.

Three-toothed moss snail *Azeca goodalli* (Azecidae), shell length 5–8 mm, width 2.4–3 mm, widespread and locally common in England and Wales, local in central-east Scotland, in moist, moderately open woodland in moss and ground litter, usually on calcareous soil.

Top snail *Trochoidea elegans* (Hygromiidae), shell diameter 6–12 mm, height 5–11 mm, with a few records from south-east England (where it was introduced from Europe in the 1890s) and Norfolk, apparently stable, in grassy habitats on base-rich substrates. *Trochos* is Greek for 'wheel'.

Tree snail *Balea perversa* (Clausiliidae), a sinistral door snail, shell length 7–10 mm, width 2.5–2.7 mm, common and widespread on rough tree bark, walls and rocky slopes, occasionally in ground litter, feeding on moss, lichens, algae and cyanobacteria. Latin *perversa* can mean 'turned the wrong way'.

Valve snail *Valvata piscinalis*, widespread and common in lowland regions, shell height 3.5–5.0 mm, width 4–6 mm, feeding on algae and detritus in clear lentic or, preferably, slow-flowing fresh water. *Piscina* is Latin for 'fish pool'.

Vertigo snail Members of the family Vertiginidae, more commonly known as whorl snails (q.v.). *Vertigo* is Latin for 'turning around'.

Wall snail Alternative name for tree snail.

White-lipped snail *Cepaea hortensis* (Helicidae), shell diameter 16–22 mm, height 15–16 mm, common and widespread in Britain and east-central Ireland in habitats such as waste ground, woodland, hedgerows and grassland (often in dense vegetation and damp conditions); in Scotland, it inhabits dunes and cliffs. Food includes nettle, ragwort and hogweed. *Hortensis* is from Latin *hortus* = 'garden'.

White snail *Theba pisana* (Helicidae), shell diameter 12–25 mm, height 9–20 mm, local in coastal dunes in south-west England and south Wales. The binomial refers to Thebes and Pisa.

Wrinkled snail *Candidula intersecta* (Hygromiidae), shell diameter 7–12 mm, height 4–7 mm, introduced from southern Europe, possibly post-medieval, common and widespread in England, Wales and the southern half of Ireland, and near the coast of both Scotland and the northern half of Ireland, in dry open habitats, under stones, and in low vegetation, often in dunes. The binomial is Latin for 'shining white' and 'intersected'.

SNAKE Elongated, legless, scaled, carnivorous reptiles of the suborder Serpentes (distinguished from legless lizards by a lack of eyelids and external ears), represented in Britain by adder, grass snake, the recently distinguished barred grass snake and smooth snake. The skin is periodically shed whole (sloughing). Legend has it that, in the fifth century, St Patrick exterminated Ireland's snakes by driving them into the sea, but their absence is because Ireland split off from the European land mass before Great Britain, which was recolonised while there was still a land-bridge with the rest of Europe.

Barred grass snake *Natrix helvetica* (Colubridae), distinguished as a separate species, rather than a subspecies of grass snake, in 2017. Clearly the natural history of this species is very similar to grass snake, differences evidentially genetic rather than morphological or behavioural, and hybridisation is apparent. *Natrix* is Latin for a water snake, from *natare* = 'to swim'; *helvetica* is Latin for 'Swiss'.

Grass snake *N. natrix*, non-venomous, males typically reaching 100 cm, females 130 cm, though larger individuals have been recorded, locally common in England, Wales and central Scotland, but protected under the WCA (1981). Eggs are laid in June–July, hatching in autumn. The species hibernates. It favours rough land and pastures, usually close to a standing body of water (it is a strong swimmer), and does visit gardens. The diet is almost entirely on amphibians, but some individuals may also take small fish.

Smooth snake *Coronella austriaca* (Colubridae), non-venomous, males reaching a maximum of 60 cm, females 68 cm (though adults are smaller in Surrey), a rarity protected under the WCA (1981), confined to south-east Dorset, south-west Hampshire and a small area of east Hampshire and west Surrey, on dry heath slopes with mature heather and dwarf gorse usually south-facing, but some near Poole are north-facing slopes, with damp heath at the bottom. Much of their time is spent underground. After emerging from hibernation, mating occurs from mid-March to May, possibly also in autumn, usually producing young every other year when conditions are unfavourable. Young are born in a membrane that breaks soon after birth. Adults mainly feed on lizards, but will take small mammals; young feed on other reptiles almost their own size. The binomial is Latin for 'small crown' and 'Austrian'.

SNAKE FLY Members of the order Megaloptera (sometimes considered in the separate Raphidioptera), in the family Raphidiidae, with four species of *Raphidia* scarce and scattered in England. Adults are usually found on oak. Larvae mainly live under tree bark.

SNAKE STAR Brooding snake star Alternative name for small brittlestar.

SNAKEROOT, SNAKEWEED Alternative names for bistort.

SNAKE'S-HEAD FRITILLARY See fritillary (plant).

SNAKEWORT *Conocephalum salebrosum* (Conocephalaceae), a thalloid liverwort favouring relatively dry calcareous substrates in upland regions. Greek *conos* + *cephale* = 'cone' + 'head', and Latin *salebrosum* = 'rough'.

SNAPDRAGON *Antirrhinum majus* (Veronicaceae), an annual or short-lived perennial south-west European neophyte herb, cultivated since Tudor times, widely naturalised on old walls, waysides, pavement cracks, waste ground and rubbish tips. Greek *antirrhinum* = 'snout-like', and Latin *majus* = 'larger'.

Lesser snapdragon Alternative name for weasel's-snout.

Trailing snapdragon *Asarina procumbens*, a stoloniferous perennial south European neophyte scattered in Britain, naturalised on dry banks, cliffs and walls. *Asaron* is Greek for a kind of low stemless shrub, *procumbens* Latin for 'procumbent'.

SNEEZEWEED *Helenium autumnale* (Asteraceae), a herbaceous perennial North American neophyte, a casual garden throw-out, especially in the London area. The name comes from the practice of using the dried leaves to make snuff. The genus name refers to Helen of Troy, who was said to be collecting this plant when she was abducted.

SNEEZEWORT *Achillea ptarmica* (Asteraceae), a widespread and locally common (though declining) herbaceous perennial of damp habitats on a range of soils, including fen- and water meadows, marsh, wet heath, hill slope flushes and occasionally in wet woodland. It is also found in churchyards, and on roadsides and waste ground. The plant is poisonous to farm animals, including cattle and horses, but its roots have in the past been used to induce (not prevent) sneezing, hence its name, and as a cure for toothache. Achilles used plants in this genus to staunch his soldiers' wounds at the siege of Troy. Greek *ptarmicos* = 'inducing sneezing'.

SNIPE *Gallinago gallinago* (Scolopacidae), wingspan 46 cm, length 26 cm, weight 110 g, a common and widespread resident and summer breeding Amber List wader, with 76,000 pairs estimated in Britain (80,000 in the UK) in 2009, but numbers decreasing by 81% over 1975–2015, though with a weak increase by 8% in 2010–15. Wintering numbers, in Britain, augmented by Faeroe Islands, Icelandic and Scandinavian birds, were 1 million in 2004–5. A drumming sound is produced as part of the courtship display flight, created by a vibration of modified outer tail feathers held out at a wide angle in the slipstream of a dive. Breeding is commonest in marsh, moorland and lakeside habitats. The ground nests usually contain four eggs, incubation taking 18–20 days, fledging of the precocial young 19–20 days. Feeding is on invertebrates on wet ground or in shallow water, with some seeds. 'Snipe', dating to Middle English, is probably of Scandinavian origin (c.f. the Icelandic *myrisnipa*). Latin *gallina* = 'hen' and the suffix *-ago*, 'resembling'.

Great snipe *G. media* and **Wilson's snipe** *G. media delicata* are scarce vagrants from northern Europe and North America, respectively.

Jack snipe *Lymnocryptes minimus*, wingspan 40 cm, length 18 cm, weight 55 g, a winter visitor arriving September–November, leaving in February–March, with 100,000 of these waders estimated in Britain in 2004–5, also numerous in Ireland, widespread in lowland wetland habitats. Feeding is in wet grass or mud, and on insect larvae and adults, molluscs, worms and plant material, especially seeds. 'Jack' may reflect the human male name, an early belief being that this species was the small male of the snipe. Greek *limne* + *cryptos* = 'marsh' + 'hidden', and Latin for 'smallest'.

SNIPE FLY Members of the family Rhagionidae, medium to large flies with slender bodies and stilt-like legs. Mouthparts are adapted for piercing: some species feed on blood, some prey on other insects. Larvae are also predatory, most terrestrial, some aquatic. *Rhagio* species are sometimes called 'downlooker' flies because they perch head-down on tree-trunks.

Scarce species include **black-fringed moss snipe fly** *Ptiolina obscura* (scattered from Grampian to Dorset), **large fleck-winged snipe fly** *R. notatus* (mainly in Scotland), **limestone snipe fly** *Symphoromyia immaculata* (on chalk and limestone grassland in England), **liverwort snipe fly** *Spania nigra* (scattered in northern Scotland, northern England, Norfolk, Kent and Hampshire), **pale-fringed moss snipe fly** *P. nigra*, **silver-banded snipe fly** *C. erythrophthalmus* (North Yorkshire: Swaledale in 1979 and Dalby Forest, North York Moors in 2013), **tree snipe fly** *C. laetus* (Windsor Forest and Windsor Great Park, Berkshire, plus Cambridgeshire in 1988, Great Kimble, Buckinghamshire in 2014, and probably Epping Forest, Essex, in 2015), **wood snipe fly** *R. annulatus* (Dorset, Surrey, Oxfordshire, Co. Westmeath), and **yellow downlooker snipe fly** *R. strigosus* (four sites in the Thames Valley in Oxfordshire and Berkshire from 1957 to 1980, and Box Hill and Headley Heath, Surrey, up to 1999).

Black snipe fly *Chrysopilus cristatus*, common and widespread in damp habitats, especially on tall vegetation (catching small insects flying past a perching spot) in marsh and fen, in May–September. The predaceous larvae are found in rotting wood and leaf mould, and in damp and wet soil, the latter probably being the more usual habitat. Greek *chrysos* + *pilos* = 'gold' + 'hair', and Latin *cristatus* = 'crested'.

Downlooker snipe fly *Rhagio scolopaceus*, 8–16 mm length, common and widespread in May–July on leaves in moist vegetation, especially in hedgerows and woodland edges. Larvae are soil-dwellers. Greek *rhagion*, a kind of poisonous spider, and *scolopos* = 'pointed'.

Little snipe fly *C. asiliformis*, widespread in England and Wales in habitats including scrub, woodland edge, wetlands, and gardens where it is usually encountered on foliage. Flight period is from June to August, peaking in July. Larvae are probably soil-dwellers. *Asiliformis* is Latin for 'fly-shaped'.

Marsh snipe fly *R. tringarius*, 8–14 mm length, common and widespread in Britain, with some Irish records, in marsh and wet meadow and on tree leaves in May–October. Larvae live in the soil. *Tringa* is Latin for 'sandpiper'.

Moorland snipe fly *Symphoromyia crassicornis*, upland, common in Scotland, less so in northern England and Wales. Adults fly in June–September in hillside seepages, stream and riverbanks, deciduous woodland and wet meadow. *Crassicornis* is Latin for 'thick-horned'.

Small fleck-winged snipe fly *R. lineola*, 6–8 mm length, common and widespread in woodland and hedgerows in May–September. Adults prey on smaller flies and other insects. Larvae feed on earthworms and beetle larvae. *Lineola* is Latin for 'lined'.

SNOUT (FLY) Soldier flies in the genus *Nemotelus* (Stratiomyidae) (Greek *nemos* + *tele* = 'pasture' or 'glade' + 'distant'), body lengths 4–7 mm, wing lengths 3.5–6.2 mm.

All-black snout *N. nigrinus*, common and widespread in England and Wales, with a few records from Ireland, in flower-rich grassland, adults taking nectar from a range of flowers in May–August. *Nigrum* is Latin for 'black'.

Barred snout *N. uliginosus*, common and widespread in estuaries and saltmarsh, with a few inland records from saline habitats in Cheshire and Cambridgeshire, and from waste ground and unimproved grassland in the Midlands. Adults fly in June–September. *Uliginosus* is Latin for 'damp'.

Fen snout *N. pantherinus*, scattered, widespread in England and Wales, with a few Irish records. Adults fly in June–August on freshwater coastal marsh, inland fen and wet meadow, sometimes locally common. Larvae are amphibious. *Pantherinus* is Latin for 'panther-like'.

Flecked snout *N. notatus*, common and widespread in saltmarsh, estuaries and brackish ditches and at inland saline sites, for example Cheshire brine pits. Larvae have been found in saline pools. The flight period is from late June to August. *Notatus* is Latin for 'notable'.

SNOUT (MOTH) Nocturnal members of the family Erebidae in the genera *Hypena* (Greek *hypene* = 'moustache' or 'beard', from the labial palps, in the Hypeninae) and *Schrankia* (after the German entomologist Franz Schrank, 1747–1835, Hypenodinae). Specifically, the **snout** *Hypena proboscidalis*, wingspan 25–38 mm, widespread and common in a range of habitats, including woodland, scrub, hedgerows, gardens, rough meadow and marsh. The bivoltine adults fly in June–August and in late autumn. Larvae feed on stinging nettle. *Proboscidalis* (from Greek *proboscis*) describes the long labial palps.

Beautiful snout *H. crassalis*, wingspan 25–30 mm, local in open woodland, moorland, bog and heathland throughout England and Wales, less common in the north, and over much of Ireland. Adults fly in June–July. Larvae mainly feed on bilberry. *Crassus* is Latin for 'thick', from the hairy labial palps.

Bloxworth snout *H. obsitalis*, wingspan 28–36 mm, a rare RDB moth found on sea-cliffs, grassy slopes and under-cliffs in parts of south-west England and on the Channel Islands. There are two generations, flying in July–August and

September–October. The second brood overwinters as an adult and reappears in spring. Larvae mainly feed on pellitory-of-the-wall. *Obsitus* is Latin for 'filled', describing forewings covered with scales or patterned.

Buttoned snout *H. rostralis*, wingspan 27–32 mm, scarce, in hedgerows, woodland edges and gardens throughout much of south-east England, formerly more widespread, a priority species in the UK BAP. Adults fly in August–October, overwinter, and reappear in spring. Larvae feed on hop leaves. *Rostrum* is Latin for 'beak', from the long labial palps.

Pinion-streaked snout *S. costaestrigalis*, wingspan 16–22 mm, local in damp woodland and meadow, fen and boggy heathland throughout England and Wales, and in the Highlands Adults fly in July–October, possibly in two generations. Larvae feed on heather, wild marjoram and wild thyme. Latin *costa* + *striga* = 'lower wing margin' + 'streak'.

White-line snout *S. taenialis*, wingspan 18–24 mm, local in moorland, plantations and damp woodland in parts of southern England and Wales. Adults fly in July–August. Larval diet is unknown, but may involve heathers. *Taenia* is Latin for 'band', from the wing marking.

SNOW FLEA *Boreus hyemalis* (Boreidae), length 3–5 mm, a wingless scorpionfly that can jump up to 5 cm, living in and feeding on mosses. Adults are evident in autumn and winter, including when snow is on the ground. Larvae, active in late spring and summer, live in a channel in the soil, and pupate at the end of the burrow. It seems to prefer sandy soils, and is widespread in heathland, except in the mild south-west England. The binomial is from Greek for 'north' and 'of winter'.

SNOW-IN-SUMMER *Cerastium tomentosum* (Caryophyllaceae), a mat-forming perennial Italian neophyte herb, in gardens by 1648, records from the wild dating from 1915 being garden escapes or discards, widely naturalised in dry habitats such as roadsides, railway banks, waste ground, dunes and coastal shingle. Greek *ceras* = 'horn', from the shape of the seed capsule, and Latin *tomentosum* = 'covered with dense hair'.

SNOWBERRY *Symphoricarpos albus* ssp. *laevigatus* (Caprifoliaceae), a common and widespread rhizomatous North American neophyte shrub introduced in 1817, naturalised in woodland, scrub and hedgerows and on waste ground, once planted as cover for game. It reproduces by suckering and spreads very slowly; dense thickets are usually a consequence of close initial planting. The trinomial is from Greek *syn* + *phoreo* + *carpos* = 'together' + 'to carry' + 'fruit', referring to the fruit clusters, and Latin for 'white' and 'smooth'.

SNOWDROP Winter- or early spring-flowering bulbous herbaceous perennial European neophytes. Specifically, *Galanthus nivalis* (Alliaceae), cultivated by Tudor times, common and widespread (except in Highland Scotland and western Ireland) in damp woodland and other shaded places, especially common in parks, gardens and churchyards, and also on verges, by watercourses and in damp grassland. Greek *gala* + *anthos* = 'milk' + 'flower', referring to the flower colour, and Latin *nivalis* = 'snow'.

Greater snowdrop *G. elwesii*, introduced in 1875, scattered in Britain, mainly in the south, in open woodland, roadsides, cemeteries, parkland and churchyards. Henry Elwes (1846–1922) was an English botanist.

Green snowdrop *G. woronowii*, scattered in England in grassy habitats and rough ground. Georg Woronow (1874–1931) was a Russian botanist.

Pleated snowdrop *G. plicatus*, introduced in 1818, scattered in Britain (mainly in the south) in deciduous woodland, hedgerows, roadsides, churchyards, cemeteries and parkland, and as a relic of cultivation. *Plicatus* is Latin for 'folded'.

SNOWFLAKE Bulbous herbaceous perennials in the genus *Leucojum* (Alliaceae) (Greek *leucos* + *ion* = 'white' + 'violet').

Spring snowflake *L. vernum*, possibly native at two sites in Somerset and Dorset, but most populations, scattered in Britain, are almost certainly neophyte relics of cultivation or garden throw-outs, naturalised in woodland and scrub, by watercourses and in grassy places. Latin *vernum* = 'spring'.

Summer snowflake *L. aestivum*, scattered, commoner in southern England, in winter-flooded carr, occasionally in other damp habitats such as meadows and woodland rides, as well as a garden escape near habitation. This is the county flower of Berkshire as voted by the public in a 2002 survey by Plantlife. *Aestivum* is Latin for 'summer'.

SOAPWORT *Saponaria officinalis* (Caryophyllaceae), a rhizomatous herbaceous perennial archaeophyte, scattered, widespread, sometimes invasive in a range of man-made and marginal habitats, often near habitation, including hedge banks, roadsides, railway banks and waste ground. It is naturalised by streams and in damp woods, especially in south-west England and north Wales. Leaves and roots are rich in saponins which produce a lather in water and can be used for washing. *Saponaria* is Latin for 'soapy', *officinalis* for having pharmacological value.

Rock soapwort *S. ocymoides*, a herbaceous perennial south European neophyte grown in gardens since 1768, widespread but scattered, on walls, churchyards, tips and beaches, usually as a casual, sometimes naturalised. *Ocymoides* is Latin for 'resembling basil' (*ocimum*).

SOAY SHEEP See sheep.

SOBER (MOTH) Nocturnal moths of generally soft coloration in the family Gelechiidae.

Ash-coloured sober *Acompsia cinerella* (Dichomeridinae), wingspan 16–19 mm, widespread and local in a range of habitats, with more scattered records from Ireland. Adults fly in June–July. Larvae feed on moss, often at the base of a tree. Greek *acompsos* = 'unadorned', and Latin *cinereus* = 'ash-coloured'.

Birch sober *Anacampsis blattariella* (Anacampsinae), wingspan 16–19 mm, locally common in heathland and woodland in England, but absent from the south-west and much of northern England. Very local in all but east Wales and only one record from Scotland (Berwickshire). Adults fly in July–September, resting on birch trunks during the day. Larvae feed on birch in May–June, inside a rolled leaf. Greek *anacampsis* = 'a bending back', from an indentation on the forewing, and from *Verbascum blattaria* (moth mullein), though this is not a food plant.

Black sober *An. temerella*, wingspan 12 mm, rarely encountered as an adult, recorded from south and north Wales, north-west England, one location in the Western Isles of Scotland and a few scattered sites in Ireland. Since 1970 there have been only 14 records from seven localities in the British Isles. Adults fly in July–August. Larvae feed in the terminal shoots of creeping willow. *Temere* is Latin for 'by chance', probably reflecting how the larva was originally obtained.

Eyelet sober *Thiotricha subocellea* (Thiotrichinae), wingspan 10–11 mm, local in dry grassland in south and east England, very local in Wales, scarce in north England and east Scotland. Adults fly in July–August. Larvae feed on seeds of wild marjoram, inside a movable case. Greek *theion* + *trichos* = 'sulphur' + 'hair', and Latin *sub* + *ocella* = 'somewhat' + 'little eye', from a black spot on the forewing.

Lichen sober *Dichomeris alacella* (Dichomeridinae), wingspan 13–14 mm, formerly scarce but increasing distribution and numbers during the first decade of the twenty-first century, occurring locally in a range of tree habitats in southern and eastern England as far north as Worcestershire and Monmouthshire, and East Anglia. Adults fly in July–August. Larvae feed on lichens on tree-trunks and wooden fences. Greek *dicha* + *meros* = 'in two' + 'a part', from the forewing division, and Latin *alacer* = 'lively', from the adult behaviour.

Poplar sober *An. populella*, wingspan 14–19 mm, locally common over much of England and parts of Wales. Absent from large parts of northern England, Ireland and south Scotland, but found locally in the Highlands, associated with aspen. Habitats include roadsides, parks, dunes, damp woods, stream edges and amenity plantings. Adults fly in July–September. Larvae feed in April–June in rolled leaves of poplars (*Populus*, giving the specific name) and willows, inside a spun or rolled leaf (in Scotland, on aspen, from May to July).

Silver-barred sober *Syncopacma taeniolella* (Anacampsinae), wingspan 11–14 mm, locally common on rough ground in England south of a line from Shropshire to the Wash, becoming less common northwards; also in small numbers in parts of Wales, Scotland and Ireland. Adults fly in July. Larvae feed in May–June on bird's-foot-trefoil, sometimes clovers and medicks, living between leaves spun together with silk. Greek *sygcope* + *acme* = 'cutting short' + 'point', from the short maxillary palps, and Latin *taeniola* = 'small band', from the white forewing bands that appear as a girdle when the wings are folded at rest.

Slate sober *Sy. albipalpella*, wingspan 8–10 mm, on damp heathland and mosses in south-east England, a priority species in the UK BAP, possibly now extinct owing to habitat loss. Adults fly in July. Larvae feed on petty whin, between leaves spun together with silk. Latin *albus* + *palpus* = 'white' + 'palp', describes the labial palp.

Vetch sober *Aproaerema anthyllidella* (Anacampsinae), wingspan 10–12 mm, widespread but local on rough grassland, dry pasture and dunes, less common in Scotland. The bivoltine adults fly in May–June and August–September. First generation larvae mine leaves of kidney vetch (*Anthyllis*, the main food plant, giving the specific name), clovers, restharrow, sainfoin and lucerne, the second generation feeding on flowers and seeds. Greek *aproaireo* = 'not what was chosen before' (there having been a precedent for first choice of name).

Western sober *Sy. suecicella*, wingspan 8–9 mm, confined to dry heathland on the Lizard, Cornwall, first discovered in 1984, a priority species in the UK BAP. Larvae feed on hairy greenwood, between leaves spun together with silk. Latin *suecica* = 'Swedish'.

White-shouldered sober *Sophronia semicostella* (Gelechiinae), wingspan 18–19 mm, occasionally common in chalk downland, dry heathland, vegetated shingle, disused gravel-pits, dry coastal and acidic grasslands, and open woodland in southern and central England and East Anglia, local to very local in northern England and Wales. Adults fly in the afternoon and at dusk and beyond in June–July. Larval diet is unknown in Britain but may be sweet vernal-grass, as on the continent. Greek *sophron* = 'sober', and Latin *semi* + *costa* = 'half' + 'rib', from the white streak on the forewing.

SOFT-BROME See brome.

SOFT-GRASS Creeping soft-grass *Holcus mollis* (Poaceae), a common and widespread rhizomatous perennial of well-drained acidic, sometimes neutral soils, in oak and birch woodland, open conifer plantations, hedge banks and heathland, and under bracken. It is locally an arable weed. *Holcos* is Greek for a kind of grass, *mollis* Latin for 'soft'.

SOFT-RUSH *Juncus effusus* (Juncaceae), a common and widespread rhizomatous herbaceous perennial, found by rivers, ponds and lakes, and in marsh, ditches, wet woodland, open heathland and moorland on acid soil, usually with a sandy or peaty substrate. *Juncus* is Latin for 'rush', *effusus* for 'loose' or 'spread out'.

Great soft-rush *J. pallidus*, a rhizomatous herbaceous Australasian neophyte perennial found as a casual at a few sites in England, originating from wool shoddy. *Pallidus* is Latin for 'pale'.

SOLANACEAE Family name for nightshade (Latin *solanum*), including henbane, potato, tobacco and tomato.

SOLDIER BEETLE So called because of their bright colours, members of the family Cantharidae. See also sailor beetle.

Blue soldier beetle *Ancistronycha abdominalis*, a scarce predator with a scattered distribution from the Midlands, north into Scotland, and a record from near Builth Wells, Powys, in 2014 being the first in Wales for 14 years. Greek *ancistron* + *onychos* = 'fish hook' + 'of a claw', and Latin *abdominalis* = 'abdominal'.

Common red soldier beetle *Rhagonycha fulva*, 7–11 mm length, common and widespread, adults seen in grassland, woodland, along hedgerows and in parks and gardens in June–July preying on aphids and other insects visiting the (often umbellifer) flowers they are resting on but also eating nectar and pollen. Larvae prey on ground-dwelling invertebrates such as slugs and snails, and live at the base of long grasses. Greek *rhagos* + *onychos* = 'a break' + 'claw', and Latin *fulva* = 'tawny' or 'yellow-brown'.

SOLDIER (FLY) Members of the family Stratiomyidae, mostly in the genus *Oxycera* (Greek *oxys* + *ceras* = 'sharp' + 'horn').

Some of the rarities are **dark-winged soldier** *O. analis* (scattered in England as far north as Leicestershire, mostly in fen or marsh), **hill soldier** *O. pardalina* (usually in upland areas, often near calcareous streams in scrub or woodland edge), **long-horned soldier** *Vanoyia tenuicornis* (as far north as a line from Anglesey to the Humber estuary, in coastal marsh and landslip, fen and wet meadow), **pygmy soldier** *O. pygmaea* (often near the coast, usually in base-rich seepages and springs in fen or grassland, but also in woodland on limestone where larvae live in saturated moss in seepages and trickles).

Delicate soldier *O. nigricornis*, locally common, widespread in England and Wales, and also found in east Ireland, in fen, carr, pond and river margins. Larvae live in shallowly submerged vegetation or wet litter. Adults fly in June–August. *Nigricornis* is Latin for 'black-horned'.

Three-lined soldier *O. trilineata*, locally common, scattered in England and Wales on coastal grazing marsh, and in fen, wet meadow, and seepages and springs on coastal landslips. The amphibious larvae tolerate mildly brackish conditions. Adults fly in June–August. *Trilineata* is Latin for 'three-lined'.

White-barred soldier *O. morrisii*, Vulnerable in RDB 2, scattered in England. About half of recent records relate to coastal cliffs, landslips and grazing marshes, from Cornwall to the Isle of Wight, Somerset, north Wales, Anglesey, Lancashire, Dumfries and Galloway, Suffolk and Norfolk, and the remainder inland in England as far north as Lancashire. Adults frequent seepages associated with landslips and springs, occasionally in fens, in June–August.

Yellow-tipped soldier *O. terminata*, Vulnerable in RDB 2, its range shrinking: in 2002 it was known only from Rockingham Forest (Northamptonshire), Somerset, and the Monnow valley (Hereford–Monmouth). Habitats include shaded sandy riverbanks (Monnow) and a limestone stream. Adults fly in June–July. *Terminata* is Latin for 'ended'.

SOLOMON'S-SEAL *Polygonatum multiflorum* (Asparagaceae), a rhizomatous herbaceous perennial, scattered, most common in ash-field maple woods over chalk and limestone in central-south England, less frequently over non-acidic substrates. The English name may come from depressions in the roots resembling royal seals, or that the cut roots resemble Hebrew characters. Greek *polys* + *gonia* = 'many' + 'angles', and Latin *multiflorum* = 'many-flowered'.

Angular solomon's-seal *P. odoratum*, very local around the Severn estuary, and in the Peak District and north-west England, in ancient ash woods, often in crevices and on outcrops of limestone. *Odoratum* is Latin for 'scented'.

Garden solomon's-seal *Polygonatum x hybridum* (*P. multiflorum x P. odoratum*), of garden origin, common and widespread in much of Britain, less so in Ireland, in woodland edges, scrub, hedgerows, roadsides, rough ground and churchyards.

Whorled solomon's-seal *P. verticillatum*, a rarity usually found on moist, nutrient-rich, often basic, soils in wooded montane gorges and on wooded riverbanks in the Highlands (Perth and Angus) and south Northumberland. *Verticillatum* is Latin for 'whorled'.

SORGHUM See great millet under millet.

SORICIDAE Family name for shrews (Latin *sorex*).

SORREL Herbaceous perennials in the genus *Rumex* (Latin for 'sorrel') and *Oxyria* (Greek *oxys* = 'sharp', from the bitter taste) (Polygonaceae). See also wood-sorrel and yellow-sorrel.

    Common sorrel *R. acetosa*, very common and widespread, on neutral to slightly acidic soil in meadow, pasture, woodland clearings, mountain ledges and shingle beaches. *Acetosa* is Latin for 'acidic' or 'sour'.

    French sorrel *R. scutatus*, a woody-based European neophyte cultivated in Tudor times, naturalised as a garden escape on road-sides, railway banks, old walls and waste ground, very scattered in (mainly northern) England. *Scutatus* is from Latin for 'shield'.

    Mountain sorrel *O. digyna*, locally common mainly in the Highlands, also Cumbria and north Wales, and scarce in western Ireland, on damp mountain ledges, shaded gullies, and streamsides. *Digyna* is Greek for 'two women', here meaning 'two carpels'.

    Sheep's sorrel *R. acetosella*, common and widespread, on dry heathland, non-calcareous dunes, shingle beaches and short grassland on acidic, impoverished, sandy or stony soils. *Acetosella* is Latin for 'acidic' or 'sour'.

SORREL MOTH Pygmy sorrel moth See sorrel pygmy under pygmy (moth).

SOUTHERN BEECH Neophyte hardwood trees in the genus *Nothofagus* (Greek *nothos* = 'false' + *Fagus* = 'beech') from Chile and western Argentina. Two species in particular have been planted for forestry and in parkland, in places becoming self-sown and naturalised: **rauli** (*N. alpina*), introduced in 1910, which grows on a range of soils, growing well in areas of high rainfall; and **roble** (*N. obliqua*), possibly introduced in 1849, planted in parks, arboreta, amenity areas and large gardens, sometimes used as a hedging plant, growing well on a range of soils, except on chalk.

SOUTHERNWOOD *Artemisia abrotanum* (Asteraceae), a neophyte aromatic shrub, scattered, on roadsides and waste ground, usually casual, occasionally naturalised. Artemis was the Greek goddess of the hunt and the natural environment. *Abrotos* is from Greek meaning either 'divine' or 'unfit to be eaten'.

SOWBREAD *Cyclamen hederifolium* (Primulaceae), a long-lived aestivating perennial European neophyte present in the Tudor period, common in parks and gardens, seeding in dry, shaded locations, and increasingly found naturalised in woodland and hedges, widespread but mainly in the southern half of England. Seeds can be carried hundreds of metres, chiefly by ants. Greek *cyclos* = 'circle', from twisted flower-stalks, and Latin *hederifolium* = 'ivy-leaved'.

    Eastern sowbread *C. coum* and **spring sowbread** *C. repandum* are Mediterranean neophytes, scattered and naturalised in southern England on verges and other grassy habitats. *Coum* is Latin for from Kos, Greece, *repandum* for 'bent backwards'.

SOWTHISTLE Species of *Sonchus* (Asteraceae) (Greek *sonchos* for these plants). Pigs are said to be particularly fond of the succulent leaves and stems, giving the English name. See also blue-sowthistle.

    Marsh sowthistle *S. palustris*, a herbaceous perennial of tall vegetation beside rivers on nitrogen-rich soils, also tolerant of moderately saline conditions. Urbanisation has caused a decline in the Thames Valley and in Kent. It may be increasing in the Broads and east Suffolk. It became extinct in Cambridgeshire through drainage, but has recently spread from plants introduced to Woodwalton Fen. A Hampshire population, found in 1959, is probably native. *Palustris* is from Latin for 'marsh'.

    Perennial sowthistle *S. arvensis*, a common and widespread creeping perennial with a preference for disturbed, nutrient-enriched soils, growing on verges, ditch and riverbanks, sea walls and the upper parts of beaches and saltmarsh; it is also a common arable weed, and is found on waste ground. *Arvensis* is from Latin for 'field'.

    Prickly sowthistle *S. asper*, a common and widespread overwintering annual of rough grassland, scrub, verges, railway banks, arable fields, gardens and waste ground, preferring dry, disturbed, sandy soils. *Asper* is Latin for 'rough'.

    Smooth sowthistle *S. oleraceus*, a common and widespread overwintering annual of disturbed or trampled grassland, coastal cliffs, verges, arable fields, walls, pavement cracks, gardens and waste ground, intolerant of grazing but invasive on bare ground. *Oleraceus* is Latin for 'vegetable'.

SOYABEAN *Glycine max* (Fabaceae), an introduced vegetable, casual on tips and waste ground, especially near docks and factories in lowland regions. Greek *glycys* = 'sweet', from the sweet leaves and roots of some species, and an abbreviated Latin *maximum* = 'largest'.

SPAGHETTI ALGAE *Chaetomorpha linum* (Cladophoraceae), a widespread green seaweed with filaments 5–30 cm long, in the intertidal and supralittoral, often free-floating or in groups of hundreds, even thousands of individuals in sandy sites, on rocks and around tide pools as a loosely entangled mass. Greek *chaite* + *morphe* = 'hair' + 'shape', and Latin *linum* = 'flax'.

    Sea spaghetti Alternative name for the brown seaweed thongweed.

SPANGLE (MOTH) Gold spangle *Autographa bractea* (Noctuidae, Plusiinae), wingspan 37–42 mm, common throughout much of northern Britain and Ireland and from the Midlands down to Gloucestershire. Further south, sightings are probably of migrants. Preferred habitats are sheltered meadows by streams and woodland clearings in the hills, and it may also be found in gardens. Adults fly at night in July–August. Larvae feed on herbaceous plants. Greek *autographos* = 'written in one's own hands' (suggesting that character marks on the wings are self-produced), and Latin *bractea* for a thin plate of metal (gold leaf), referencing the importance of this colour on the wings.

SPANISH-DAGGER *Yucca gloriosa* (Asparagaceae), an evergreen woody perennial south-east North American neophyte naturalised on a few dunes and coastal gravel-pits in south-west England and south Wales. Yucca is a Caribbean name, though for cassava, not this plant.

SPANISH FLY *Lytta vesicatoria* (Meloidae), an occasional vagrant blister beetle, 15–45 mm long, though with recent records from the Isle of Wight and Kent probably reflecting residency. Adults feed on leaves of ash, lilac, privet and white willow; larvae are parasitic on the brood of ground-nesting bees. The adult secretes an irritating substance, cantharidin, used medically to remove warts, and also putatively as an aphrodisiac. *Lytta* is Greek for a form of madness in dogs, *vesica* is Latin for 'blister'.

SPARASSIDAE Family name of green spider (Greek *sparasso* = 'to tear').

SPARK (FUNGUS) See wood spark.

SPARROW From Old English *spearwa*, birds in the genus *Passer* (Latin for 'sparrow'), in the family Passeridae.

    Hedge sparrow Alternative name for dunnock.

    House sparrow *P. domesticus*, wingspan 24 cm, length 14 cm, weight 34 g, a widespread resident breeder, largely associated with buildings, with 5.3 million pairs (4.8–5.8 million) in Britain in 2009. Nevertheless there has been a 72% decline in numbers over 1970–2015 (levelling off after 2000, with a decline of 3% in 2010–15). There may have been some recovery recently in Wales, Scotland and Northern Ireland, but the distribution has become patchy, with high numbers in places and none elsewhere in the space of a few kilometres in similar habitat, especially in towns. Greater London lost 70% of its sparrows between 1994 and 2001. The recent decline has probably been because of reductions in nest site availability and in insect food for nestlings. (A crash in numbers in the early twentieth century was triggered by the rise of the motor vehicle at the expense of horse-drawn transport, sparrows having benefited from seed found in horse droppings and spillage from nosebags and stables.) It has been Red-listed in Britain since 2002, but is Amber-listed in Ireland where numbers have recently been stable. Nests are usually in cavities, with clutches of 4–5, eggs hatching in 13–15 days, the altricial chicks fledging in 15–17 days. Diet is largely insectivorous, with nestlings taking insects. *Domesticus* is Latin for 'domestic' or 'of the house'.

Tree sparrow *P. montanus*, wingspan 21 cm, length 14 cm, weight 24 g, a resident breeder, Red-listed in Britain, Amber-listed in Ireland (where the population is stable). It is now mainly found across the Midlands and in southern and eastern England, though distribution extends into south Scotland and, patchily, eastern Ireland. There were 180,000 territories in Britain in 2009, with numbers declining from the 1970s to the 1990s: overall by 90% from 1970 to 2015, though offset by an increase (from a low point) by 125% from 1995 to 2015 (including by 8% in 2010–15). The decline may have been due to changes in farming practice, for example increased use of herbicides and loss of winter stubble fields. The nest, often built in a tree cavity, usually contains 5–6 eggs, incubation lasting 12–13 days, and fledging of the altricial young taking 15–18 days. Diet is largely of seed or grain, with nestlings fed on insects. *Montanus* is Latin for 'mountaineer' (from *mons* = 'mountain').

SPARROWHAWK *Accipiter nisus* (Accipitridae), wingspan 62 cm, length 33 cm, male weight 150 g, females 260 g, common and widespread in a range of habitats, though most often in woodland, with 35,000 pairs in 2009. There was an increase in numbers by 83% over 1970–2015, though this incorporates a weak decline (−11%) in 2010–15. Nests are of twigs built in trees, clutch size is 4–5, incubation takes 33 days, and the altricial young fledge in 27–31 days. Diet is mainly of birds: males can catch birds up to thrush size, but females, being up to 25% larger (the greatest intersexual size difference in any bird species), can catch birds up to pigeon size. 'Sparrowhawk' comes from Middle English *sperhauk* and Old English *spearheafoc*. Latin *accipiter* = 'hawk' (from *accipere* = 'to grasp') and *nisus* = 'sparrowhawk'.

SPATANGOIDEA From Greek *spatanges* = 'sea urchin', the order of echinoid marine invertebrate (Echinodermata) in which the body is usually oval or heart-shaped; see heart urchin.

SPATTER-DOCK *Nuphar advena* (Nymphaeaceae), a perennial water-lily with erect leaves held above the water, introduced from North America in 1772, planted in lakes and ponds where it becomes naturalised through rhizomatous growth, and now with a very scattered distribution. *Nuphar* is Arabic (Persian) for water-lily, *advena* Latin for 'newly arrived'.

SPEAR-MOSS Members of the genera *Calliergon* and *Calliergonella* (Greek *calos* + *ergon* = 'beautiful' + 'work') and *Straminergon* (incorporating Latin *stramen* = 'straw') (Calliergonaceae).

Giant spear-moss *Calliergon giganteum*, scattered and uncommon in wet, calcareous fens and flushes.

Heart-leaved spear-moss *Calliergon cordifolium*, common and widespread in marsh and wet woodland, especially in lowland regions. *Cordifolium* is Latin for 'heart-shaped leaves'.

Pointed spear-moss *Calliergonella cuspidata*, common and widespread in moist, base-rich habitats, including marsh, bog and flushes, in grassland, and among rocks. On soils such as clay it often occurs in lawns, but it also grows in relatively dry places in calcareous habitats such as chalk and limestone grassland. *Cuspidata* is Latin for 'pointed'.

Straw spear-moss *Straminergon stramineum*, common and widespread except for the south Midlands and eastern England in wet, acidic habitats often associated with sphagnum. *Stramen* is Latin for 'straw'.

Three-ranked spear-moss *Calliergon trifarium*, found in the Highlands and west Scotland, and locally in west Ireland in base-rich bogs. *Trifarium* is Latin for 'three-rowed'.

Twiggy spear-moss *Calliergon sarmentosum*, common in mountain regions in base- or mineral-rich flushes and springs and on wet, gravelly substrates. *Sarmentosum* is Latin for 'twiggy'.

SPEAR-WING **Bugloss spear-wing** *Tinagma ocnerostomella* (Douglasiidae), wingspan 8–9 mm, a small day-flying stem-feeding moth found in dry open areas in southern and south-east England, from Devon to Lincolnshire, distribution limited by that of its larval food plant, viper's-bugloss. Adults fly in June–July. The binomial is from Greek *tinagma* = 'a jerk', from the zigzag flight, and from a similarity to the genus *Ocnerostoma*.

SPEAR-WINGED FLY Members of the family Lonchopteridae, generally 2–5 mm long, with long pointed wings.

Yellow spear-winged fly *Lonchoptera lutea*, widespread and often common, especially in unshaded sites. Larvae live in leaf litter. Latin *lutea* = 'yellow'.

SPEARWORT With spear-shaped leaves, species of *Ranunculus* (Ranunculaceae), from the diminutive of Latin *rana* = 'frog', referencing the plants growing in marsh and bog. Spearwort comes from Anglo-Saxon *sperewyrt*, and describes the leaf shape.

Adder's-tongue spearwort *R. ophioglossifolius*, a very rare annual herb of marshland requiring winter flooding, bare wet mud for seedling establishment, reduced summer water levels, and low competition. The substrate at the two existing sites, both in Gloucestershire, is base-rich Lias clay: the larger population, at Badgeworth, has been dependent on conservation management since 1962; Inglestone Common, managed by livestock grazing, went down to a single plant and was subject to reintroductions in 2016. Greek *ophis* + *glossa* = 'snake' + 'tongue'.

Greater spearwort *R. lingua*, a stoloniferous perennial found in base-rich, still or slowly flowing water in fen and marsh, on ditch, canal and pond edges, around reservoirs and in flooded gravel-pits and quarries. It is frequently introduced to ponds and other wetlands and the distinction between native and introduced populations is compromised. In Ireland the native plant remains locally frequent. *Lingua* is Latin for 'tongue'.

Lesser spearwort *R. flammula*, a common and widespread perennial usually growing in oligotrophic or mesotrophic water over neutral to acid substrates, especially where there are seasonal fluctuations in water level, such as on lake shores, streamsides, dune slacks, marsh, water meadow, bog and ditches. *Flammula* is Latin for 'small flame'.

SPECKLE **Holly speckle** *Trochila ilicina* (Heliotales, Dermateaceae), a common and widespread speckling fungus on the upper parts of dead holly leaves. Greek *trochila* = 'wheel' and Latin *ilicina* = 'like holly' (*ilex*).

SPECKLEBELLY Large leafy lichens in the genus *Pseudocyphellaria* (Lobariaceae), on mossy trees and rocks in the Highlands and other upland areas, including **Norwegian specklebelly** *P. norvegica*, new to science in 1979, found (together with **ragged specklebelly** *P. lacerata*) in ash-hazel woodland (arguably the wettest woodland in Wales) in the Nant Gwynant Valley facing Snowdon. Some species were formerly used as a brown dye for wool.

SPECKLED WOOD *Pararge aegeria* (Nymphalidae, Satyrinae), male wingspan 46–52 mm, female 48–56 mm, a widespread and common butterfly having increased its distribution by 71% from 1976 to 2014, and numbers by 84%, associated with woodland clearings, scrub and hedgerows, and also frequent in gardens. Its appearance changes from north to south, forming a cline: individuals in the north are dark brown with white spots, those in more southerly locations dark brown with orange spots. Larvae favour a number of grass species; adults feed on honeydew and sap as well as nectar. Adults fly in April–October. Greek *para*

= 'close to' + the genus *Arge*. Aegeria was one of the Camenae, prophetic nymphs in Greek mythology.

SPECTACLE *Abrostola tripartita* (Noctuidae, Plusiinae), wingspan 27–30 mm, a nocturnal moth its English name coming from the raised tufts of thoracic scales which resembles a pair of spectacles when viewed from the front. It is widespread and common in gardens, heathland, rough grassland, fen, woodland edge and waste ground. Population levels increased by 108% between 1968 and 2007. Adults fly between May and September in one or two generations. Larvae feed on stinging nettle. Greek *abros* + *stole* = 'graceful' + 'robe', and Latin *tripartita* = 'triply divided', for the forewing divisions.

Dark spectacle *A. triplasia*, wingspan 28–32 mm, nocturnal, widespread and common in a range of habitats, including riverbanks, hedgerows and gardens, though rare in Scotland. Adults fly in June–July, sometimes with another generation in autumn. Larvae feed on stinging nettle and hop. *Triplasius* is Latin for 'triple', for the forewing divisions.

SPEEDWELL Species of *Veronica* (Veronicaceae). 'Speedwell' comes from the archaic 'speed well' (farewell), possibly from the fact that the flowers wilt and fall off soon after being picked. *Veronica* probably comes from Latin *vera* + *nica* = 'true' + 'image': when St Veronica wiped Christ's forehead on the way to the crucifixion, an image of his face was left on the cloth she had used. See also water-speedwell.

Widespread but scattered neophytes, mostly annual and casual, are **American speedwell** *V. peregrina* (North American, found in gardens and parks), **common field-speedwell** *V. persica* (Near Eastern, present by 1825, in cultivated areas and waste ground), **crested field-speedwell** *V. crista-galli* (Caucasian, present by 1813, naturalised in Somerset and south-west Ireland, elsewhere casual on roadsides, cultivated ground and waste ground), **French speedwell** *V. acinifolia* (European, in gardens and parks in southern England, East Lothian and Co. Down), **garden speedwell** *V. longifolia* (European, a woody-based perennial present by 1731, in rough grassland and hedgerows and on roadsides and waste ground), **large speedwell** *V. austriaca* (European, naturalised in rough ground and dunes as far north as central Scotland), and **slender speedwell** *V. filiformis* (Turkey and the Caucasus, introduced by 1808, naturalised on lawns, roadsides, grassy banks and streamsides, and in churchyards).

Alpine speedwell *V. alpina*, a herbaceous perennial of open, often rocky sites, local and >760 m in the central Highlands, on slightly moist ground with late snow. Most sites are subject to base enrichment from flushing.

Breckland speedwell *V. praecox*, an annual European neophyte introduced by 1775, naturalised on free-draining, regularly disturbed sandy soils on arable field edges, sandy banks, and open rough grassland, local and declining in west Norfolk and west Suffolk. *Praecox* is Latin for 'early'.

Fingered speedwell *V. triphyllos*, an annual archaeophyte, declining, very local in Norfolk and Suffolk on sandy arable field margins. Regular disturbance is required to maintain open ground for germination. *Triphyllos* is Greek for 'three-leaved'.

Germander speedwell *V. chamaedrys*, a common and widespread stoloniferous perennial herb of woods, hedge banks, damp grassland and verges, rock outcrops, scree, railway banks and waste ground. It also occurs on ant hills on chalk downland. 'Germander' comes from the medieval apothecary's *gamandrea*, in turn from Greek *chamae* + *drys* = 'on the ground' + 'oak' because the leaves of some species putatively resemble those of the oak.

Green field-speedwell *V. agrestis*, a widespread but declining annual archaeophyte of cultivated and rough land, including gardens and allotments, preferring well-drained and acidic soils, though occurring on calcareous substrates where there is surface leaching. *Agrestis* is Latin for 'field'.

Grey field-speedwell *V. polita*, an annual of cultivated fields and gardens, usually on light, sandy if often calcareous soils,

common and widespread, except in upland Scotland and in Ireland apart from the south-east. *Polita* is Latin for 'smoothed' or 'polished'.

Heath speedwell *V. officinalis*, a common and widespread herbaceous perennial in woodland clearings and rides, on banks, in grassland and on heathland, on well-drained, often moderately acidic or leached soils. In some grassland it is confined to raised ground or anthills. *Officinalis* is Latin for having pharmacological properties.

Ivy-leaved speedwell *V. hederifolia*, an archaeophyte annual common and widespread except in upland Scotland, found on cultivated and waste ground, woodland rides, hedge banks, walls, banks and gardens, on sandy, loam or clay soils. *Hederifolia* is Latin for 'ivy-leaved'.

Marsh speedwell *V. scutellata*, a common and widespread herbaceous perennial found in a range of wetland habitats, for example pond and lake margins, marsh, fen, wet grassland, hillside flushes, bog and wet heathland, usually on acidic soils. *Scutellata* is Latin for 'covered with small shields (plates)'.

Rock speedwell *V. fruticans*, a somewhat woody perennial, local on calcareous, dry open slopes and rock ledges >540 m in the Highlands. *Fruticans* is Latin for 'becoming shrubby'.

Spiked speedwell *V. spicata*, a herbaceous perennial scattered in England and Wales on well-drained, nutrient-poor soils. This is the county flower of Montgomeryshire as voted by the public in a 2002 survey by Plantlife. In East Anglia, ssp. *spicata* usually grows on acidic to base-rich sandy soils in open, short, grazed grassland. Elsewhere, ssp. *hybrida* grows in thin soils on base-rich cliffs, grassland and rocks. *Spicata* is Latin for 'spiked'.

Spring speedwell *V. verna*, a very rare annual of infertile sandy soils, in short grassland and uncultivated habitats depending on sheep or rabbit grazing to keep the habitat open. It has declined considerably, currently being found in only 12 sites in two 10 km squares in west Suffolk, but is flourishing where it has been introduced to two reserves. *Verna* is Latin for 'spring'.

Thyme-leaved speedwell *V. serpyllifolia*, a herbaceous perennial, common and widespread in woodland rides, grassland, heathland, flushes, damp rock ledges, cultivated and waste ground, lawns and damp paths. *Serpyllifolia* is Latin for having leaves like *Thymus serpyllum* (Breckland thyme).

Wall speedwell *V. arvensis*, a common and widespread annual of arable fields, grassland (often restricted to ant hills), heathland, dunes, gravelled paths and tracks, waste ground, banks, walls and pavements, usually on dry soils. *Arvensis* comes from Latin for 'field'.

Wood speedwell *V. montana*, a widespread if scattered, locally frequent herbaceous perennial of damp basic to mildly acidic soils in long-established, mixed deciduous woodland, scrub and shaded hedge banks, on loamy, sandy and heavy clay soils. *Montana* is Latin for 'montane'.

SPERMATOPHYTA The subdivision of all flowering (therefore seed-producing) plants (Greek *sperma* + *phyton* = 'seed' + 'plant'); also known as phanerogams.

SPHAERIIDAE Family name (Greek *sphaeria* = 'sphere') for pea and orb mussels, also known as fingernail clams.

SPHAERITIDAE Beetle family (Greek *sphaeria* = 'sphere') represented by *Sphaerites glabratus*, local in Scotland and north England in coniferous forest in decaying fungi, dung and carrion.

SPHAERIUSIDAE Family of mud or minute bog beetles (Greek *sphaeria* = 'sphere'), with a single, rare species.

SPHAGNUM Bog-mosses in the family Sphagnaceae (Greek *sphagnos*, a type of moss), with 36 species in Britain.

SPHECIDAE Old family name (Greek *sphex* = 'wasp') for some parasitoid wasps, most now moved to the family Crabronidae.

SPHENOPHYTINA Subdivision of horsetails (Greek *sphen* + *phyton* = 'wedge' + 'plant').

SPHINDIDAE From Greek *sphindos* = 'speed', a family of small (2–2.5 mm) beetles, with two British species, *Sphindus dubius* and *Aspidiphorus orbiculatus*, both with a scattered distribution in England and Wales, and associated with slime moulds on trees.

SPHINGIDAE Family name of hawk-moths. The reared-up posture of some larvae has been considered to resemble that of the Sphinx, prompting the Latin name.

SPIDER Members of the class Arachnida, order Araneae, with eight legs and jaws (chelicerae) that inject venom. Anatomically, spiders differ from other arthropods in that the usual body segments are fused into the cephalothorax and abdomen, joined by a small cylindrical pedicel. The abdomens have appendages modified into spinnerets that extrude silk to create lines and webs. Webs vary widely in size, shape and the amount of sticky thread used. Some taxa do not build webs but are active hunters. There are 649 species in 254 genera and 34 families in the British list, excluding vagrants, those only found in heated buildings, and those recorded only from the Channel Islands.

Families (see separate entry for those in bold) (common names, number of genera/species) are: **Agelenidae** (funnel-web and cardinal spiders, 3/11), **Amaurobiidae** (tangled nest spider, 2/5), Anyphaenidae (buzzing spider, 1/2), **Araneidae** (some orb weavers, 16/32), **Atypidae** (purse web spider, 1/1), **Clubionidae** (foliage spiders, 2/25), Corinnidae (ant-mimicking spiders, 1/2), **Cybaeidae** (water spider, 1/1), Dictynidae (mesh-webbed spiders, mesh-weavers, 9/18), **Dysderidae** (woodlouse spider, 2/4), **Eresidae** (ladybird spider, 1/1), **Gnaphosidae** (ground spiders, 11/33), **Hahniidae** (lesser cobweb spiders, 2/7), **Linyphiidae** (money spiders, 122/280), **Liocranidae** (running foliage spiders, 5/12), **Lycosidae** (wolf spiders, 8/38), **Mimetidae** (pirate spiders, 1/4), Nesticidae (comb-footed cellar spider, 1/1), **Oonopidae** (2/3), **Oxyopidae** (lynx spider, 1/1), **Philodromidae** (3/15), **Pholcidae** (daddy-long-legs or cellar spiders, 2/3), **Pisauridae** (nursery-web and raft spiders, 2/3), **Salticidae** (jumping spiders, 17/39), **Scytodidae** (spitting spider, 1/1), Segestriidae (tube-weaving spiders, 1/3), **Sparassidae** (green spider, 1/1), **Tetragnathidae** (long-jawed, or orb-weaver, and cave spiders, 4/14), **Theridiidae** (comb-footed and false widow spiders, 20/57), **Theridiosomatidae** (ray spider, 1/1), **Thomisidae** (crab spiders, 6/26), **Uloboridae** (cribellate orb spiders, 2/2), **Zodariidae** (ant-eating spider, 1/4) and **Zoridae** (ghost spiders, 1/4).

Ant-eating spider *Zodarion italicum* (Zodariidae), 2–3 mm body length, mainly recorded from the East Thames Corridor in south Essex and north Kent, with recent reports also from Newhaven, West Sussex (1998), Dibden Bay, south Hampshire (1999), and Padworth, Berkshire (2005). It favours dry, warm, sunny open habitats containing a proportion of bare ground, for example some grassland, disused railway lines and brownfield sites, preying on ants, especially small black ant. *Zodarion* is Greek for 'bovine', reason unclear.

Buzzing spider *Anyphaena accentua* (Anyphaenidae), 4.5–7 mm body length, widespread in most of England and Wales, more scattered in Scotland, rare in Ireland, in woodland, hunting and mating on tree and shrub leaves. During courtship the male emits a high-pitched buzzing sound by vibrating its abdomen on a leaf, hence its common name. Adults are mostly found in early to midsummer, females sometimes surviving into autumn. Immatures live in ground vegetation, leaf litter and under bark in autumn and winter. Greek *anyphaina* = 'unafraid'.

Cardinal spider *Tegenaria parietina* (Agelenidae), with a scattered distribution in south, central and eastern England, in houses and, especially, old buildings and walls in late summer and autumn, using a sheet web to catch insect prey. It shows extreme sexual dimorphism for leg length, with mature males having legs almost twice as long as those of equivalent sized females (with a body length up to 20 mm). Spiders living in Hampton Court allegedly used to terrify Cardinal Wolseley, hence the common name, or conversely because he regarded

them as lucky and forbade anyone to harm them. *Tegenaria* may translate as 'carpet weaver'. *Parietina* is Latin for 'old walls'.

Cave spider *Meta menardi* (Tetragnathidae), body length 10–15 mm, scattered but local in permanently dark, damp habitats such as caves, mines, sewers, houses, cellars, limestone pavement, hollow trees and railway tunnels, sometimes building an orb web but usually living in wall crevices, preying on beetles, flies, hibernating Lepidoptera, slugs, isopods and millipedes. Emerging spiderlings are attracted to light (unlike the adults), helping them disperse to new sites. *Meta* is Latin for 'goal' or 'destination'.

Cellar spider Alternative name for daddy-long-legs spider. See also comb-footed cellar spider below.

Cobweb spider Alternative name for daddy-long-legs spider and giant house spider. See also lesser cobweb spider below.

Comb-footed cellar spider *Nesticus cellulanus* (Nesticidae), male body length 3.7–4.5 mm, female 4–5.5 mm, local and widespread in Britain, scarce in north Ireland, in permanently dark, damp habitats such as mines, caves, cellars, sewers and hollow trees where it builds a small horizontal tangle web.

Comb-footed spider Members of the family Theridiidae. See also cupboard, false widow, mothercare and rabbit hutch spiders below. Specifically, *Enoplognatha ovata* (Greek *enoplos* + *gnathos* = 'armed' + 'jaw', and Latin for 'oval'), body length 6 mm, widespread and abundant, typical of open habitats such as verges, gardens and woodland clearings, feeding on insects caught in a tangled web. Males generally mature in June, females in July. After mating, males die and females establish themselves in rolled leaves to produce and guard their egg-sac. Once the young emerge in September, the female dies, and the spiderlings descend to ground level to overwinter.

Common house spider See house spider below.

Cotton's Amazon spider *Glyphesis cottonae* (Linyphiidae), tiny, scarce, with records from south-central England, Shropshire and Cheshire, in sphagnum bog in September–May, a UK BAP priority species.

Cross spider Alternative name for garden spider.

Cucumber (green) orb spider *Araniella cucurbitina* (Araneidae), female body length up to 8 mm, males up to 5 mm, common and widespread in Britain, more scattered in the west and north, and recorded from Co. Wicklow, spinning a small web among leaves in hedgerows or woodland edge, and in gardens, mainly in May–July. *Araniella* is a diminutive of Greek *aranea* = 'spider', and Latin *cucurbitina* = 'small gourd'.

Cupboard spider *Steatoda grossa* (Theridiidae), female length 6–10.5 mm, one of the so-called false widow spiders, scattered in Britain, commonest in south-west England, though expanding northwards, found in and close to buildings, using a scaffold web to catch prey, including pill woodlice. Its bite can be painful to humans. Greek *steatos* = 'fat' and Latin *grossa* 'thick'.

Daddy long-legs spider *Pholcus phalangioides* (Pholcidae), a common and widespread synanthropic species with a female body length of 7–10 mm, males 6–9 mm, with notably long legs, found almost exclusively in buildings where the average temperature throughout the year is >10 °C, favouring undisturbed parts of houses such as cellars (hence an alternative name, cellar spider) and bathrooms where it can establish large colonies, creating untidy webs. They hang upside down on the web and if disturbed will vibrate rapidly. They are effective predators of household pests including other spiders. Adults are found year-round, with peaks in males often in late spring and autumn. Greek *pholcos* = 'bandy-legged' or 'squint-eyed' (the former more apposite), and *phalangion* + *oides* = 'spider' = 'similar to'.

Diadem spider Alternative name for garden spider.

Diamond spider *Thanatus formicinus* (Philodromidae), length 7 mm, with only three previous records and considered extinct (last seen in the UK in 1969), the warm summer of 2017 saw a specimen in Clumber Park, Nottinghamshire. Its habitat is damp heathland, appearing in March–April. A black diamond shape on the abdomen prompts the English name.

Diving bell spider Alternative name for water spider.

Evening spider Alternative name for toad spider.

False widow spider Members of the genus *Steatoda* (Theridiidae), from Greek *stear* + *eidos* = 'fat' or 'tallow' + 'similar to', mainly noble false widow spider, but see also cupboard and rabbit hutch spiders, above and below respectively.

Fen raft spider *Dolomedes plantarius*, one of Britain's largest arachnids: females can reach 70–80 mm across (including legs). It is semi-aquatic, in neutral to alkaline freshwater habitats of fens and grazing marsh, hunting prey such as pond skaters, water beetles, tadpoles and efts across the water surface. It was first discovered in the UK in 1956 at Redgrave and Lopham Fen on the Norfolk–Suffolk border. In 1988 another population was identified in marsh ditches on the Pevensey Levels, East Sussex. In 2003, a third population was discovered on a disused canal near Swansea, south Wales, and strongholds now include the wetland reserves Red Jacket Fen and Pant-y-Sais Fen, between Swansea and Neath. This is a priority UK BAP species, and IUCN-listed. Populations have declined as a result of the loss of its lowland aquatic habitat. Between 2010 and 2015, 6,000 hand-reared, three-month old spiderlings and 56 adult females with their nurseries containing hundreds of week-old spiderlings were released at new sites in the Norfolk and Suffolk Broads, and numbers have been increasing. Adult males are rarely found after July, but females can survive until autumn. Nursery web building is concentrated in July–August, but the breeding season is protracted, adult females generally making two breeding attempts during summer. After the first brood disperses, females usually move back to the water surface to resume hunting; after their second brood has dispersed they often remain at the nursery until they die. Greek *dolomedes* = 'deceitful'. See also raft spider below.

Foliage spider General name for members of the family Clubionidae.

Fox-spider or **common fox-spider** *Alopecosa pulverulenta* (Lycosidae), a common and widespread wolf spider, body length 5–10 mm, found from spring to midsummer in open habitats such as heathland, grassland, moorland and dunes, actively hunting prey on the ground. Greek *alopex* = 'fox', and Latin *pulverulenta* 'dusty'.

Funnel-web spider Includes members of the families Agelenidae and Amaurobiidae.

Garden spider *Araneus diadematus* (Araneidae), 13 mm body length, common and widespread, found wherever the habitat can provide supports for its large orb web, for example woodland, scrub, hedgerows and verges, buildings (sometimes inside) and gardens, spinning large webs up to 40 cm across at heights of 1.5–2.5 m. Both sexes are mature in late summer and autumn, females surviving through to late autumn. The binomial is Latin for 'spider' and 'crowned'. See also lesser garden spider below.

Ghost spider Member of the family Zoridae, with four species of *Zora*, only *Z. spinimana* being common and widespread in the southern half of Britain and central-eastern Scotland, scattered elsewhere, an active hunter mainly in grassland.

Giant house spider *Tegenaria gigantea* (Agelenidae), female body length up to 18 mm, males, 12–15 mm, widespread across eastern, central and northern England, more sporadic in southwest England, Wales and Scotland, often found in houses, garages and sheds, but also in rock and tree crevices, rabbit holes and dense vegetation. Within houses it is most evident in autumn when males wander in search of the more sedentary females. A large leg span (females typically 45 mm, males highly variable at 25–75 mm) allows rapid movement – a distance equivalent to 330 times an adult female body length in 10 seconds. *Tegenaria* is Latinised Greek = 'carpet weaver'.

Green spider *Micrommata virescens* (Sparassidae), female body length up to 15 mm, male 10 mm, rare, scattered and widespread but declining in England and Wales, a priority species for conservation. It shows a preference for damp sheltered woodland,

often found on the lower branches of young oak, on tall grass or sedge tussocks, sitting head-down waiting to ambush passing insects. In midsummer the female stitches several leaves together to create a space to enclose the egg-sac. Females are mature from May to September; males have a short season and are rare. A bite on a human may cause local swelling and pain, though this usually disappears within 48 hours. Greek *micro* + *omma* = 'small' + 'appearance', and Latin *virescens* = 'green'.

Ground spider Member of the family Gnaphosidae, with body lengths 2–18 mm and short legs, that actively hunt prey (mainly other spiders) on the ground.

Hobo spider *Eratigena agrestis* (Agelenidae), male body length 8–11 mm, female 11–15 mm, a European funnel-web species first recorded in Britain from Wilverley Plain, Hampshire in 1949, since then colonising northwards and now with a wide if patchy distribution throughout much of England and Wales, and is colonising Scotland (first record for Glasgow in 2013), usually in sparse grassy vegetation and under stones, particularly in city brownfield sites and alongside railway tracks. *Eratigena* is an anagram of *Tegenaria*, in which genus it had been placed until 2013. *Agrestis* comes from Latin for 'field'.

Horse-head spider *Stemonyphantes lineatus* (Linyphiidae), a money spider, male body length 4 mm, female: 6–7.5 mm, widespread in much of England, though patchy in some areas, and scattered in Wales, Scotland and Ireland, found mainly from autumn and winter to midsummer, usually at or near ground level in a variety of habitats, including chalk downland. Greek *stemon* + *nyphantes* = 'thread' + 'wreaths', and Latin *lineatus* = 'lined'.

House spider or **common house spider** *Tegenaria domestica* (Agelenidae), female body length 9–10 mm, male 6–9 mm, common and widespread, almost entirely found within buildings year-round, but evident mostly from spring to midsummer and autumn, trapping prey in sheet webs. A large leg span allows rapid movement – a distance equivalent to 330 times an adult female body length in 10 seconds. Latin *domestica* = 'of the house'. See also giant house spider above.

Invisible spider *Drapetisca socialis* (Linyphiidae), body length up to 4 mm, widespread and locally common in Britain, scarce in north Ireland, usually found on the lower trunks of trees, particularly beech, spinning a very fine, almost invisible web close to the trunk to trap insects. It also hunts by running over the bark and among leaf litter. Adults occur in late summer and autumn, peaking in September–October. Greek *drapetes* = 'fugitive', and Latin *socialis* = 'sociable'.

Jumping spider Members of the family Salticidae, with body lengths of 2–10 mm, actively hunting on foliage, ground vegetation and walls, leaping onto their prey. See zebra spider below.

Labyrinth spider *Agelena labyrinthica* (Agelenidae), widespread in much of southern England and coastal Wales, scattered in northern England, found in rough grassland, field edges and low gorse and heather bushes, spinning a large sheet web with a funnel retreat on or above the ground. It waits at the end of the funnel for prey, mainly grasshoppers, to become entangled. In late summer the female builds a large elaborate chamber in the vegetation to enclose her eggs, giving this species its common name. Adults are found mainly in July and August. Greek *agele* = 'herd' (i.e. living gregariously).

Lace-weaver spider See main entry for lace-weaver spider.

Ladybird spider *Eresus sandaliatus* (Eresidae), males 6–9 mm long, females 10–16 mm, previously recorded from heathland sites in the Poole–Bournemouth area and the Isle of Wight, known since 1979 from seven individuals at one site in Dorset. Following successful short-distance translocations, it has been introduced to other sites in the area. A priority species in the UK BAP, which aims to establish it at up to ten new sites across its natural range. It builds a silk-lined burrow 10 cm deep covered by a silk and debris 'roof'. Greek *eresis* = 'discontinuation'.

Leopard spider Alternative name for snake-back spider.

Lesser cobweb spider Member of the family Hahniidae.

Lesser garden spider *Metellina segmentata* (Tetragnathidae), common and widespread, found year-round in a variety of habitats including gardens, waste ground, grassland, woodland and hedgerows.

Long-jawed spider or **long-jawed orb weaver** Members of the family Tetragnathidae, with body lengths 2.5–11 mm, adults spinning an orb web to catch prey. They are usually found in vegetation close to ground level and frequently close to water. When perturbed they lie flat out with their legs stretched in a straight line, hiding behind a plant, hence an alternative name 'stretch spiders'.

Lynx spider *Oxyopes heterophthalmus* (Oxyopidae), abundant on a few small heathland areas in west Surrey, hunting prey on the ground. Greek *oxys* + *pes* = 'quick' + 'foot', and *heteros* + *ophthalmos* = 'different' + 'eye'.

Money spider Members of the family Linyphiidae, making up almost a third of all the spider species in the UK, found in woodlands, hedgerows, meadows, parks and gardens. Adults, especially members of the subfamily Linyphiinae, make hammock-shaped sheet webs, then position themselves underneath to wait for small insects to be trapped. Many Erigoninae, however, are ground-running predators. Money spiders often land on people in summer (said to bring good fortune, hence the common name), ballooning through the air on strands of silk, picked up by the wind. *Tenuiphantes tenuis* is probably the commonest spider in the UK; at 2–3 mm long it is usually found in low-growing vegetation and leaf litter.

Mothercare spider *Phyloneta sisyphia* (Theridiidae), body length 3–4 mm, common and widespread in England and Wales, less so in Scotland, scarce in Ireland, found in early to midsummer, with most males in May–June, most females in June. It builds an inverted cup-shaped retreat covered with plant material, below which it spins a tangle web, typically in hedgerows or on gorse and heather, but also on nettles and thistles. The common name comes from the mother's behaviour towards its young: she guards the egg-sac then after hatch regurgitates food for the spiderlings. She dies before the young leave the nest and they eat her body. Greek *phylon* + *neta* = 'tribe' + 'spun'. Sisyphus, king of Ephyra, was punished by the gods for his deceitfulness by having to roll a rock to the top of a mountain, after which, repeatedly, the rock would fall back of its own weight.

Mouse spider *Scotophaeus blackwalli* (Gnaphosidae), body length 9–12 mm, widespread and scattered, though locally common, in England and Wales, scarce in Scotland and Ireland, found year-round though mostly in summer, in houses, sheds and gardens, and can be fairly common on fences and under bark of dead trees in urban sites. They move quickly hunting for insects on the ground at night. The abdomen is covered in silver-grey hair that resembles mouse fur, hence their name. Greek *scotos* + *phaios* = 'darkness' + 'dusky'. John Blackwall (1790–1881) was an English naturalist.

Noble false widow spider *Steatoda nobilis*, native of Madeira and the Canary Islands, first recorded in 1879 from Torquay, accidentally introduced in a cargo of bananas. With a scattered distribution in southern England its range is extending northwards as winters become warmer, and by 2015 there were two records from Scotland. It constructs a cobweb which is an irregular tangle of very strong sticky silk. Males stridulate during courtship by scraping 10–12 teeth on the abdomen against a file on the rear of the carapace. While venomous, its bite has a generally mild effect on humans, though there are one or two reports each year of serious effects.

Nursery-web spider *Pisaura mirabilis* (Pisauridae), female body length 12–15 mm, male 10–13 mm, locally common and widespread in England and Wales, scarce in Scotland and north-east Ireland, mainly found in May–July in habitats with tall vegetation. Adults stretch out on stems and leaves moving rapidly to capture passing insects. After mating, the female lays her eggs in a silk cocoon which she transports in her fangs. Just

before hatching, she spins a silk tent (the nursery web of the English name) and releases the spiderlings inside it. Pisaurum is the Roman name for the north Adriatic Italian town of Pesaro. Latin *mirabilis* = 'amazing'.

Pirate spider Four species of *Ero* (Mimetidae), 3–4 mm body length, found in summer and autumn in low vegetation in hedgerows, heathland and gardens, getting its name from a habit of invading the webs of other spiders, tweaking the strands to lure the owners out, then attacking and eating them. *Ero aphana* is an RDB 2 species.

Purse web spider *Atypus affinis* (Atypidae), female body length up to 15 mm, males 9 mm, widely distributed in England and Wales, with scattered coastal records in south-west Scotland, but scarce, very local and declining. Most of its tubular web, up to 80 mm in length, is underground, and a friable substrate facilitates burrowing. Beneath the tube the spider has a silk-lined burrow which can extend down a further 500 mm. In heathland it prefers loose sand with young heather hanging loosely over, and gravel banks covered with tufts of heather. In chalk and other grassland it makes its tubes in undisturbed areas around scrub edge and at the base of anthills. It is long-lived, possibly up to eight years, spending almost all this time inside its purse web, spearing insect and woodlice prey from the tube entrance, or stabbing its fangs through the purse into the prey, dragging the victim inside, then repairing the hole. The only British member of its genus and family, the binomial is Latin for 'atypical' and 'related'.

Rabbit hutch spider *Steatoda bipunctata* (Theridiidae), one of the so-called false widow spiders, female body length 7 mm, male 5 mm, widespread, locally common in England, scarce elsewhere. Both sexes are mature throughout the year and females may survive for several years. The species is mainly found in and around buildings (including animal pens) but also on old and dead trees living under bark and in cavities. The scaffold web includes sticky lines attached to the substrate under tension. The struggles of a captured prey breaks the attachment of these lines, lifting the captive into the air to leave it suspended and readily available to the spider. Greek *stear* = 'fat' ('tallow') and Latin *bipunctata* = 'two-spotted'.

Raft spider *Dolomedes fimbriatus* (Pisauridae), female body length up to 20 mm, males 9 mm, mostly found in spring and early summer, very local in pools, sphagnum bog and ditch margins in southern English heathlands, though populations appear to be stable, very scattered elsewhere in Britain. Adults assume a hunting position on emergent water plants with the front legs held on the water surface to detect vibrations from trapped invertebrates; then, using the surface tension of the water it runs across the surface to catch its prey. Greek *dolomedes* = 'deceitful', and Latin *fimbriatus* = 'fringed with hair'. See also fen raft spider above.

Ray spider *Theridiosoma gemmosum* (Theridiosomatidae), 1.5–2.5 mm body length, found in May–September, locally abundant south of a line from Anglesey to the Wash, in damp habitats spinning a horizontal, umbrella-shaped orb web low down in vegetation. Greek *theridion* + *soma* = 'small animal' + 'body', and Latin *gemma* = 'gem'.

Running foliage spider Member of the family Liocranidae.

Shepherd spider Alternative name for harvestmen (class Opiliones), this name coming from the Latin *opilio* = 'shepherd', because shepherds in parts of Europe sometimes used stilts to monitor their flocks from a distance.

Six-eyed spider Alternative name for the woodlouse spider family, Dysderidae.

Snake-back spider *Segestria senoculata* (Segestriidae), body length 7–10 mm, common and widespread especially in the north and west, living in a tube constructed in a hole in a wall, under bark and stones or in rubble. Threads act as trip wires, radiating from the tube. The spider waits in the entrance of the retreat for insect prey to touch the threads before rushing out to seize it, taking it back to the retreat for consumption. Adults are found throughout the year, numbers peaking in spring and early summer. The common name comes from the row of dark-coloured patches on the elongated abdomen which resemble the markings on some snakes. Another common name is **leopard spider**, for the same reason. The binomial is from Latin for 'covering' and 'six-eyed'.

Spitting spider *Scytodes thoracica* (Scytodidae), 3–8 mm body length, possibly introduced and naturalised, a slow-moving, nocturnal, synanthropic, six-eyed species found on walls and ceilings of heated buildings, widespread in the southern half of England, scarce in north England and south Wales. An adhesive, venomous fluid is squirted from the chelicerae onto the prey from a distance of 10 mm or more at 30 m/second. Squirting is also used as a defence against other spiders. Greek *scytodes* = 'leather-like', and Latin *thoracica* = 'of the thorax'.

Stretch spider Alternative name for long-jawed spiders from their behaviour when perturbed, lying flat out with their legs stretched in a straight line, hiding behind a plant.

Tangled nest spider Member of the family Amaurobiidae.

Toad spider *Nuctenea umbricata* (Araneidae), female body length up to 15 mm, males 11 mm, common and widespread though scattered in Scotland and Ireland, with its flat body concealed in daytime in dead trees under bark and in fissures in posts and fences, including in gardens, emerging to construct a new web each evening (hence an alternative name, evening spider), between its retreat and surrounding plants, waiting in the web centre for trapped night-flying insects. Adults are mainly found between late spring and autumn. The walnut colour of the abdomen gives it another common name, **walnut orb weaver**. Latin *nux* + *teneo* = 'nut' + 'to hold', and Latin *umbricata* = 'living in shadow'.

Triangle spider *Hyptiotes paradoxus* (Uloboridae), male length 3–4 mm, females 5–6 mm, scattered and rare (RDB 3) in England and Wales. It spins a triangular segment of orb web (containing two radial threads) within usually yew or box foliage, keeping it taut by holding on to one corner of the triangle. When an insect is caught in the web the spider slackens and tightens the tension of the web until its prey is entangled before advancing to wrap it. Both sexes reach maturity in late summer, mating occurring in the following spring. Greek *hyptios* = 'laid back', and Latin *paradoxus* = 'strange'.

Turf-running spider *Philodromus cespitum* (Philodromidae), common and widespread in much of the southern half of Britain in lowland areas, but rare in Scotland and Ireland, in heathland, hedgerows, scrub and woodland, mainly in summer, rapidly hunting for small insects on the ground and on leaves. Greek *philos* + *dromos* = 'loving' + 'running', and Latin *cespitum* 'tufted'.

Wasp spider *Argiope bruennichi* (Araneidae), female body length up to 20 mm, males 5 mm, first recorded in Britain in 1922 at Rye, East Sussex, and for many years restricted to a few sites close to the coast in Sussex, Kent, Hampshire and Dorset. Since the 1970s it has increased its range, spreading along the whole south coast and into East Anglia, with scattered records from the Midlands. Adults mature in August–September when the large orb webs, up to 300 mm across, can be found, especially in coastal chalk grassland, rough grassland, waste ground and verges, trapping insects such as grasshoppers. Black and yellow horizontal bands on the abdomen prompt its common name. Argiope was a naiad (water nymph) in Greek mythology. Morten Brünnich (1737–1827) was a Danish zoologist.

Water spider *Argyroneta aquatica* (Cybaeidae), 8–15 mm body length, widespread and scattered, locally common in the south, in clean, vegetated fresh water with little current, for example ponds, lakes, dykes and canals, living under water for most of its life. It creates an underwater silk cell that is filled with air by dragging small bubbles down from the water surface. Once filled, oxygen levels in this 'diving bell' remain stable as oxygen produced by green plants diffuses into it from the surrounding water. The spider can stay inside the bell for up to a day. Aquatic insect prey is taken back to the bell for eating, and it is also used for mating, egg-laying and overwintering. The spider occasionally feeds at the surface, and moulting usually takes place out of water. Adults are found year-round but are most active from spring to late summer. Spiderlings can be found at the surface walking among water plants. Adults can give a painful bite to humans. Greek *argyros* + *netos* = 'silver' + 'spun'.

Widow spider See false widow spider above.

Woodlouse spider Members of the family Dysderidae (a group with six rather than eight eyes). Specifically *Dysdera crocata*, female body length 11–15 mm, male 9–10 mm, widespread but scattered though locally abundant in the south, often synanthropic, in damp cellars and kitchens and under stones, logs and debris in gardens, churchyards and waste ground, mainly preying on woodlice at night. Adults are found year-round, peaking in spring and summer. Their strong jaws (chelicerae), capable of penetrating woodlice, can give a painful bite to humans. Greek *dysderis* = 'quarrelsome' and Latin *crocata* = 'saffron-coloured'.

Zebra spider One of three species of *Salticus* (Salticidae) (Latin *saltare* = 'to jump'), 6–7 mm body length with black and white abdominal stripes on the female, brown and white on the male, *S. scenicus* being the commonest and most widespread, *S. cingulatus* less so, and *S. zebraneus* local in southern England. They move in short, jerky bursts, and as jumping spiders catch their prey by leaping on it. Even though they are just a few millimetres long, they can leap a distance of up to 100 mm.

SPIDER BEETLE Members of the family Ptinidae, resembling spiders, having round bodies with long, slender legs, and lacking wings. Both larvae and adults are scavengers of plant and animal debris, and are often pests of stored products. See also biscuit and cigarette beetles, both under beetle.

Species found locally in the Midlands are **brown spider beetle** *Ptinus clavipes* (length 3 mm), **globular spider beetle** *Trigonogenius globulus* (2–3 mm), **golden spider beetle** *Niptus hololeucus* (3.4–5 mm), also in south-east England, and **shiny spider beetle** *Mezium affine* (1.5–3 mm).

Australian spider beetle *Ptinus tectus*, length 2.5–4 mm, introduced from Australia in around 1900, locally common with a scattered distribution in England and Wales. *Ptenos* is Greek for 'feathered', *tectus* Latin for 'hidden'.

White-marked spider beetle *P. fur*, length 2–4.3 mm, scattered in England and Wales. *Fur* is Latin for 'thief'.

SPIDER CRAB Marine decapods covered with a thick exoskeleton, composed primarily of calcium carbonate, and with long legs. See also toad crab under crab.

Common spider crab *Maja brachydactyla* (Majidae), carapace up to 20 cm long, common and widespread in south and west Britain and west Ireland from extreme low water to 50 m on rocky or coarse sandy substrates. *Maia* is Greek for a kind of large crab. Greek *brachys* + *dactylos* = 'short' + 'fingers'.

Great spider crab *Hyas araneus* (Oregoniidae), up to 10 cm long and 8 cm wide, common and widespread among rocks and seaweed on the lower shore to a depth of 50 m. In Greek mythology, Hyas was a daughter of Atlas. Latin *araneus* = 'spider'.

Slender spider crab *Macropodia tenuirostris* (Inachidae), carapace length 3.2 cm and breadth 1.1 cm, widespread and locally common, including in estuaries. Greek *macros* + *podos* = 'large' + 'foot', and Latin *tenuis* + *rostrum* = 'narrow' + 'beak'.

SPIDER MITE Members of the family Tetranychidae, <1 mm in size, generally living on the underside of leaves of different plant species, and commonly pests of crops. Many species spin a protective (anti-predator) silk webbing, hence the 'spider' part of the name.

Red spider mite or **two-spotted spider mite** *Tetranychus urticae* (Tetranychidae), a common and widespread sap-sucking mite that attacks foliage, causing a mottled appearance, and in severe cases, leaf loss and plant death, a pest of ornamental and edible greenhouse plants, houseplants, and garden plants during summer. Greek *tetra* + *onychos* = 'four' + 'claws', and Latin *urtica* = 'nettle'. The most commonly used biological control is a predatory mite (*Phytoseiulus persimilis*), originally from Chile, now widely used in preference to pesticides.

SPIDER WASP Members of the family Pompilidae, with 21 species in 3 subfamilies and 12 genera. All species are solitary. Adults feed on nectar. Females capture and paralyse prey (generally spiders), placing these in a tunnel, and laying a single egg in the host, which serves as the larval food source throughout its development.

Bloody spider-hunting wasp *Homonotus sanguinolentus*, local to heathland in Dorset and the New Forest, Endangered in RDB 1, flying in June–August, parasitising spiders in their webs. Hosts die after 10–11 days after the wasp larva hatches, having been sucked dry. Greek *homos* + *notos* = 'alike' + 'back', and Latin *sanguis* + *lentus* = 'blood' + 'fullness'.

Leaden spider wasp *Pompilus cinereus*, with a scattered, largely coastal distribution, scarce in Scotland and Ireland, on coastal dunes and inland in sandpits and sandy heaths with bare sand exposure, flying in May–September. Nests are excavated in loose sand and provisioned with a single large spider. Latin *cinereus* = 'grey'.

Red-legged spider wasp *Episyron rufipes*, locally common and widespread in lowland England on sandy soils, including dunes, grasslands and gardens. Away from the south-east heathlands and coastal systems, inland records are scattered north to Humberside. In Wales, north-west England and Ireland it is mostly confined to the coast. Flight is in May–September. Nectar is taken from umbellifers and other flowers. This is a specialist hunter of orb-web spiders: Araneidae (*Araneus*) and Metidae (*Meta*) are the usual prey, with a few records for Lycosidae. The nest is excavated in loose sand after the prey has been captured. The prey is stored during this process above ground in a nearby plant. Latin *rufipes* = 'red-legged'.

SPIGNEL *Meum athamanticum* (Apiaceae), an aromatic herbaceous perennial of neutral or mildly acidic soils in dry, unimproved upland pasture, hay meadow and roadsides, from Cumbria scattered north to the Highlands, and in north Wales. Sometimes discouraged by dairy farmers as the taste is transmitted to milk and butter. 'Spignel' dates from the early sixteenth century, possibly from Anglo-Norman French *spigurnelle*, the name of an unidentified plant. Greek *meon* means this plant, and *athamanticum* = 'like plants in the genus *Athamanta*'.

SPIKE (FUNGUS) Members of the Gomphidiaceae (Boletales), from Greek *gomphos* = 'club', more specifically a large conical bolt used mainly in shipbuilding, which the fruit body resembles. All are inedible, widespread and scattered in coniferous woodland soil, including **rosy spike** *Gomphidius roseus* and **slimy spike** *G. glutinosus*.

Copper spike *Chroogomphus rutilus*, on grass, favouring sites under pine. *Chroa* is Greek for 'body surface', *rutilus* Latin for 'reddish'.

Larch spike *G. maculatus*, under larch. *Maculatus* is Latin for 'spotted'.

SPIKE-RUSH Rhizomatous perennial species of *Eleocharis* (Cyperaceae) (Greek *elos* + *charis* = 'marsh' + 'grace' or 'beauty'), associated with wet or damp habitats.

Common spike-rush *E. palustris*, a common and widespread emergent found on the margins of ponds, lakes and slow-flowing streams, in fen, marsh, swamp, wet meadow, ditches, dune slacks and saltmarsh, on a range of organic and mineral soils. *Palustris* is from Latin for 'marsh'.

Dwarf spike-rush *E. parvula*, very local in Dorset and west Hampshire, Merioneth–Caernarfonshire (Gwynedd) and Co. Londonderry, on estuarine mud and in tidal pans in brackish grazing marshes, avoiding strongly saline areas. *Parvula* is Latin for 'very small'.

Few-flowered spike-rush *E. quinqueflora*, scarce and scattered in central and eastern England, locally common elsewhere, particularly in north England and Scotland, on base-rich marsh and fen, calcareous flushes on peaty soils and wet paths, and in coastal cliff seepages, dune slacks and upper saltmarsh. It is often dependent on grazing, cutting or other disturbance. *Quinqueflora* is Latin for 'five-flowered'.

Many-stalked spike-rush *E. multicaulis*, locally common and particularly found in southern England, Norfolk, east Yorkshire, western Britain, and Ireland, mainly on acid bogs, wet heathland, valley mires, pools and wet hollows over peat, at the edge of acidic lakes, and locally in dune slacks. *Multicaulis* is Latin for 'many-stalked'.

Needle spike-rush *E. acicularis*, widespread but uncommon and scattered, on the margins of lakes, ponds, reservoirs and rivers where there is winter flooding. It also grows submerged in shallow, still or slow-moving mesotrophic to eutrophic water. *Acicularis* is Latin for 'needle-like'.

Northern spike-rush *E. mamillata* ssp. *austriaca*, local in northern England and south Scotland in the middle stretches of upland rivers, usually in relatively slack water and on gravel with some silt deposition. It also grows in ditches, pools, runnels and springs. Latin *mamillata* = 'mamillate'.

Slender spike-rush *E. uniglumis*, mainly found in coastal habitats in damp dune slacks, saltmarsh, and short brackish grassland. It also occurs inland in base-rich wet meadow and calcareous marsh, and locally, for example in Oxfordshire, by springs with higher than normal sodium content. *Uniglumis* is Latin for 'single-glumed'.

SPINACH (MOTH) *Eulithis mellinata* (Geometridae, Larentiinae), wingspan 27–30 mm, nocturnal, common in woodland, gardens and allotments throughout lowland Britain, but with numbers declining by 96% over 1968–2007, a species of conservation concern in the UK BAP. Adults fly in June–August. Larvae feed on black and red currant. Greek *eu* + *lithos* = 'good' + 'stone', and Latin *mel* = 'honey', both from the wing colour.

Dark spinach *Pelurga comitata*, wingspan 25–30 mm, widespread and common in rough ground, gardens and allotments, though with numbers declining by 96% over 1968–2007, a species of conservation concern in the UK BAP. Adults fly in July–August. Larvae feed on goosefoot and orache. Greek *pelourgos* = 'worker in clay', and Latin *comitor* = 'to accompany'.

Northern spinach *E. populata*, nocturnal, widespread and common on moorland, heathland and woodland, generally on high ground so less common in southern and eastern England. Adults fly in July–August. Larvae feed on bilberry. Latin *populata* = 'waste'.

SPINACH (PLANT) *Spinacia oleracea* (Amaranthaceae), an annual or short-lived perennial in cultivation as an edible vegetable since at least 1400, an occasional casual garden escape on waste ground, a relic of cultivation, and a birdseed alien. 'Spinach' is from the Latinised *isbinakh*, Arabic for spinach. *Oleracea* is Latin for 'resembling a vegetable'.

Tree spinach *Chenopodium giganteum* a very large annual, first recorded in the wild in Britain in 1926 (Roxburghshire), scattered in England and Scotland as a casual on rubbish tips and waste ground, originating from wool shoddy and birdseed. It also occurs as a garden escape. Greek *chen* + *podion* = 'goose' + 'little foot', from the leaf shape.

SPINDLE *Euonymus europaeus* (Celastraceae), a deciduous shrub or small tree, common and widespread in Britain (as far north as central Scotland) and Ireland, in hedges, scrub and open deciduous woodland on free-draining base-rich soils, particularly where overlying chalk and limestone. Also planted in woodlands, hedgerows and gardens from where it can become naturalised.

The hard timber was once used for making spindles, giving the English name. Greek *eu* + *onyma* = 'good' + 'name'.

Evergreen spindle *E. japonicus*, an evergreen neophyte shrub or small tree introduced from Japan by 1804, widely planted for hedging, especially in coastal areas in south and south-west England, Lancashire and south-east Ireland; also found as a garden throw-out and relic of cultivation in woodland and on sea-cliffs and roadsides, naturalised along the south coast.

Large-leaved spindle *E. latifolius*, a deciduous neophyte European shrub planted in gardens by 1730, and in hedges and naturalised by bird-sown seed. It has a scattered distribution in woodland, in hedgerows and on roadsides. *Latifolius* is Latin for 'broad-leaved'.

SPINDLES Unbranched club fungi, often growing in clusters, in the genera *Clavaria* and *Clavulinopsis* (Agaricales, Clavariaceae) (Latin *clava* = 'club').

Golden spindles *Clavulinopsis fusiformis*, inedible, widespread but only locally common, emerging as a yellow club-shaped fruiting body in dense tufts in late summer and autumn on soil in short unimproved grassland, occasionally in open grassy woodland. *Fusiformis* is Latin for 'spindle-shaped'.

Rose spindles *Clavaria rosea*, a rarity found in a few locations in Britain in summer and autumn on soil in grassland habitats and on broad-leaved woodland litter.

Smoky spindles *Clavaria fumosa*, fairly common and widespread, found in summer and autumn on soil under usually unimproved pasture, but also on amenity grassland. Latin *fumosa* = 'smoky'.

White spindles *Clavaria fragilis*, edible, fairly common and widespread, found in summer and autumn, growing as white worm-like spindles, usually in clusters, on grassland soil or woodland litter. *Fragilis* is Latin for 'fragile'.

SPINDLESHANK Alternative name for the fungus spindle toughshank.

SPINE (FUNGUS) **Swarming spine** *Mucronella calva* (Agaricales, Clavariaceae), saprotrophic on decaying logs and trunks, the fruit bodies resembling hanging spines, scattered and uncommon, mainly in southern and eastern England. *Mucro* is Greek for a point, *calva* Latin for 'smooth'.

SPIONID **Bee spionid** *Spiophanes bombyx* (Spionidae), a polychaete bristleworm 5–6 cm long living in a sandy tube which protrudes above the surface, found from the low intertidal to depths of 60 m round much of the British and north-east Irish coastline. In Greek mythology Spio was a sea nymph. *Bombyx* is Latin for 'silkworm'.

SPIRAEA Deciduous shrubs in the genus *Spiraea* (Rosaceae), this being Latin for 'bridal wreath', from Greek *speiraira*, a plant used for garlands, in turn from *speira* = 'spiral'.

Himalayan spiraea *S. canescens*, a Himalayan neophyte in cultivation by 1837, found in a few places in England and Wales as a garden escape naturalised in scrub and rough grassland. *Canescens* is Latin for 'greyish'.

Japanese spiraea *S. japonica*, a Japanese neophyte in cultivation by around 1850, widespread, scattered and occasionally naturalised as a garden escape or throw-out in scrub, hedgerows and quarries, and on heaths and roadsides, waste ground and rubbish tips.

Russian spiraea *S. media*, a Russian neophyte cultivated by 1789, naturalised in a disused limestone quarry and on a lane-side in Cumbria since 1989. *Media* is Latin for 'middle'.

SPIRE SHELL Small marine snails.

Common spire shell *Rissoa parva* (Rissoidae), shell up to 5 mm high, 3 mm wide, common and widespread on rock from mean tide level to 15 m depth, but most common around mean low water spring tides. Also found on algae, in kelp holdfasts and under rocks, grazing on microalgae. *Parva* is Latin for 'small'.

Laver spire shell *Peringia ulvae* (Hydrobiidae), shell height 4.5–5.0 mm, width 2.5–3.0 mm, common around virtually the entire British and Irish coastline, in coastal brackish waters, estuaries and saltmarsh in the upper half of the tidal zone (though it has been recorded from a depth of 100 m), feeding on detritus and grazing on seaweed (including laver). In Ireland, especially, this snail is an important food for wading birds. *Ulva* is Latin for 'sea lettuce'.

SPIRE SNAIL *Hydrobia ventrosa* (Hydrobiidae), shell height 3–4 mm, width 1.5–2.0 mm, in non-tidal brackish water such as lagoons, scattered from south Wales to the Humber, especially round the Severn estuary, and widespread (though declining) around Ireland. Vulnerable in the RDL. Greek *hydor* = 'water', and Latin *venter* = 'belly'.

Jenkins' spire snail *Potamopyrgus antipodarum* (Hydrobiidae), a common, widespread invasive snail introduced from New Zealand to Ireland in 1837 and England in 1859, shell length 5 mm, found in especially nutrient-rich lentic or slow-flowing fresh and brackish water as a nocturnal grazer-scraper on algae and detritus. Greek *potamos* + *pyrgos* = 'river' + 'tower'.

Swollen spire snail *Mercuria confusa* (Hydrobiidae), an Endangered (RDB 1) species of conservation concern in the UK BAP, shell height 3–4 mm, with small populations on the River Alde, Suffolk, and in the Thames estuary at Barking Creek, and strongholds on the Suffolk-Norfolk border including the lower reaches of the rivers Waveney and Yare, and in West Sussex on the lower River Arun. It is found in Ireland on the estuaries of the Shannon, Suir, Barrow and Nore. It is typically found on bare mud exposed at low tide beneath emergent vegetation. *Mercuria* refers to Mercury, the Roman messenger of the gods.

Taylor's spire snail *Marstoniopsis insubrica* (Amnicolidae), 3 mm wide, 2 mm tall, a rare riverine species (Endangered in RDB 1), scattered in England, and the Scottish Central Lowlands.

SPITTLEBUG Alternative name for some members of the bug family Aphrophoridae (Auchenorrhyncha), including common froghopper; also known as cuckoo-spit insect.

Alder spittlebug *Aphrophora alni*, a common and widespread sap-feeding bug, adults found on a range of trees (especially alder, willow, poplar and birch) and shrubs, nymphs preferring herbaceous plants. Greek *aphros* + *phoreo* = 'foam' + 'to bear', and Latin *alnus* = 'alder'.

Willow spittlebug *A. pectoralis* and *A. salicina*, the name usually applied to the former but both species live in willows, the nymph sometimes damaging growth tops. The former is very local in south and central England, the latter scattered in England north to Yorkshire and in parts of Ireland. *Pectoralis* is Latin for 'pectoral', *salicina* from *salix* = 'willow'.

SPLASH (FUNGUS) **Scarlet splash** *Cytidia salicina* (Polyporales, Corticiaceae), a rare fungus with a scattered distribution from Yorkshire (one 2004 record) north to the Highlands, Endangered in the RDL (2006), found on dead branches of willow in damp woodland. Greek *cytis* = 'small trunk', and Latin *salicina* from *salix* = 'willow'.

SPLEENWORT Ferns in the genus *Asplenium* (Aspleniaceae), from Greek for 'without spleen', referring to the value of curing disorders of the spleen.

Black spleenwort *A. adiantum-nigrum*, requiring bright light and growing on dry substrates, including rocks, urban walls and bridges, and in churchyards and gardens. Latin *adiantum* + *nigrum* = 'unwetted', describing the water-repelling leaves, + 'black'.

Forked spleenwort *A. septentrionale*, local in rock crevices in acid, generally upland parts of the western and northern British Isles. *Septentrionale* is Latin for 'northern'.

Green spleenwort *A. viride*, local in base-rich (mainly limestone) rock crevices in upland parts of the British Isles. *Viridis* is Latin for 'green'.

Irish spleenwort *A. onopteris*, on dry banks and mostly coastal and limestone rocks, very local in south-west Ireland. Greek *onos* + *pteron* = 'ass' + 'wing'.

Lanceolate spleenwort *A. obovatum*, on rocks, hedge banks and walls, common in south-west England and west Wales, scattered elsewhere. *Obovatum* is Latin for 'inverse'.

Maidenhair spleenwort *A. trichomanes*, common and widespread, shade-tolerant, on rocks and old walls and bridges. *Trichomanes* is Greek for 'slave hair'.

Sea spleenwort *A. marinum*, in often sea-sprayed rock crevices and on cliffs and walls around the British coastline except for a stretch between Hampshire and south-east Yorkshire.

SPLIT (FUNGUS) Host-specific leaf pathogens in the genus *Lophodermium* (Rhytismatales, Rhytismataceae) (Greek *lophos* + *derma* = 'crest' + 'skin').

Holly leaf split *L. neesii*, Vulnerable in the RDL (2006), with a scattered distribution as a pathogen on fallen holly leaves.

Ivy leaf split *L. hedericola*, Vulnerable in the RDL (2006), recorded from various sites in Wales on mortared walls on Irish ivy. *Hedericola* is Latin for 'ivy-inhabiting'.

Juniper split *L. juniperinum*, scarce, with a scattered distribution, most records from the Highlands, pathogenic on juniper needles.

Pine needle split *L. pinastri*, common and widespread, the spores dispersing in late summer and infecting pine needles, which turn brown by the following spring and fall off (called needlecast).

SPLITGILL *Schizophyllum commune* (Agaricales, Scizophyllaceae), an inedible saprobic bracket fungus, locally common and widespread in England and south Wales, scarce elsewhere, found throughout the year on the bark of diseased broadleaf trees and on stumps and other dead wood. It has also recently been found on silage, growing through perforations in plastic bales. Greek *schizo* + *phyllon* = 'split' + 'leaf' (i.e. gills), and Latin *commune* = 'common'.

SPONGE Animal in the phylum Porifera. Bodies contain pores and channels allowing water to circulate, enabling them to obtain food and oxygen and to remove wastes.

Boring sponge See yellow boring sponge below.

Breadcrumb sponge *Halichondria panicea* (Halichondriidae), a suspension-feeder widespread on rocky substrates from the intertidal and sublittoral down to 550 m; in wave-sheltered areas, it can grow up to 20 cm thick, and in tidal rapids may be several metres across. As a protective device it produces a smell similar to exploded gunpowder. Greek *halos* + *chondros* = 'sea' + 'something granular' (the latter prompting the common and specific names), and Latin *panicea* = 'made of bread'.

Carrot sponge *Amphilectus fucorum* (Esperiopsidae), 2–15 cm high, a fairly widespread encrusting or creeping branching soft sponge found in the shallow sublittoral, often overgrowing the holdfasts of *Laminaria* or sessile invertebrates. Greek *amphis* + *lectos* = 'both sides' + 'selected one', and from Latin *fucus* = 'seaweed'.

Chalice sponge *Phakellia ventilabrum* (Axinellidae), cup- or fan-shaped, up to 45 mm in height, found on the west coast of Scotland, south-west Wales and west Ireland in sheltered areas on lower intertidal rock. Greek *phacelos* = 'cluster', and Latin *venter* + *labrum* = 'stomach' + 'lip'.

Crumb-of-bread sponge *Hymeniacidon perlevis* (Halichondriidae), recorded from Shetland, Orkney and a few locations on the east coast, but more abundant on the south, south-west and west coasts of Britain, and from the Ilen estuary, Co. Cork. It is generally intertidal, colonising a variety of surfaces from exposed rocks and seaweed holdfasts to muddy gravel. Growth form ranges from thin sheets on the lower to midshore (growing along crevices) to massive flanged or turreted forms on the lower shore in wave-sheltered sites and in the sublittoral where conditions are silty. Greek *hymen* + *acis* = 'membrane' + 'point', and Latin *per* + *laevis* = 'through' + 'smooth'.

Fig sponge Alternative name for sea orange.

Golfball sponge *Tethya aurantium* (Tethyidae), a spherical sponge up to 10 cm in diameter, widespread along the south and west coastline of Britain, and around Ireland, on rocks and stones in the shallow sublittoral to depths of 130 m, often in kelp forests. Tethys was a sea goddess in Greek mythology. *Aurantium* is Latin for 'orange'.

Goosebump sponge *Dysidea fragilis* (Dysideidae), widespread, up to 50 cm in its encrusting form, 30 cm in its massive form, the surface covered in conical projections 1.5 mm in height. Larger specimens can be found in tidal, rocky estuarine areas, but it generally occurs on the lower shore and sublittoral, in rock crevices, and on stones, shells and gravel. Greek *dys* + *idea* = 'bad' + 'appearance'.

Orange pipe sponge *Leucosolenia botryoides* (Leucosoleniidae), a delicate, soft, white, tubular calcareous sponge growing up to 2 cm wide and 1 cm thick, generally widespread and common from the mid-intertidal to the shallow sublittoral, in tide pools and the underside of stones, attaching itself to rocks and macroalgae. Greek *leucos* + *solen* = 'white' + 'pipe', and *botrys* = 'cluster'.

Vase sponge *Sycon ciliatum* (Sycettidae), a common and widespread tubular or vase-shaped calcareous sponge up to 5 cm tall and 7.5 mm wide, mainly found on the shore under overhangs or attached to rocks, shells and macroalgae on the lower shore. It is common in the shallow sublittoral down to 100 m. Greek *sycon* = 'fig', and Latin *ciliatum* = 'fringed with hairs'.

White-tit sponge *Polymastia mamillaris* (Polymastiidae), common and widespread, a low, spreading cushion up to 2 cm thick with a base of up to 12 cm diameter, and erect projections up to 12 cm in height, often found at the sediment-bedrock interface on upward-facing rock, or growing out of crevices, preferring silty conditions in the littoral, but usually occurring at a depth in the range 5–15 m. Greek *poly* + *mastos* = 'many' + 'breast', and Latin *mammillia* = 'breast'.

Yellow boring sponge *Cliona celata* (Clionaidae), widespread, occurring in a boring form, with yellow papillae (rounded protuberances) sticking out of limestone, and a large massive wall-shaped form covered with flattened papillae. It can form a thick plate-like structure standing on its edge, large examples growing up to 1 m across and 50 cm high. The massive form is only common around the south-west coasts reaching its easterly limit in Dorset. In Greek mythology Clio was a sea nymph. *Celata* is Latin for 'concealed'.

Yellow staghorn sponge *Ax. dissimilis*, a branching, finger-like sponge, usually 15 cm high, with branches 1.5 cm long, at a few sites around Mull, south-west England, west Wales, and much of Ireland, in exposed open coasts, on upward-facing bedrock and other hard surfaces in the intertidal and sublittoral. *Dissimilis* is Latin for 'different'.

SPONGE FLY Species in the order Neuroptera, family Sisyridae, including *Sisyra fuscata*, common and widespread, adults found in streamside vegetation, the aquatic larvae feeding on sponges.

SPONGEWEED *Codium fragile* ssp. *tomentosoides* and ssp. *atlanticum* (Codiaceae), a widespread but uncommon green seaweed now believed to be native, mainly found in pools in the upper and midlittoral. Greek *codion* = 'fleece', Latin *tomentosus* = 'hairy' or 'woolly'.

SPOONWORM Infaunal intertidal and sublittoral members of the phylum Echiura.

SPOONBILL *Platalea leucorodia* (Threskiornithidae), wingspan 122 cm, length 85 cm, a scarce visitor, averaging 75 records p.a. with scattered records in England, Wales and east Scotland. The first confirmed breeding record since 1668 was in Suffolk in 1998 (though the two eggs were predated), and in 1999 in Lancashire where two of three eggs hatched and subsequently fledged. The first Scottish record was in 2008 at Kirkcudbright, Dumfries and Galloway, and it has bred successfully at Holkham, north Norfolk, since at least 2010 when six pairs nested in trees and fledged ten young. Non-breeding birds turn up elsewhere in substantial numbers, including 47 on Brownsea Island, Poole Harbour, Dorset, in October 2014. Latin *platalea* = 'broad' (referring to the shape of the bill), and Greek *leucos* + *erodios* = 'white' + 'heron'.

SPOONWORT Purple spoonwort *Pleurozia purpurea* (Pluroziaceae), a thallose liverwort common in the Highlands and western Ireland on bogs and steep, peaty or rocky slopes. Greek *pleura* + *ozos* = 'side' + 'branch'.

SPOT (FUNGUS) Fungal plant pathogens, usually affecting leaves.

Black spot *Diplocarpon rosae* (Dermateaceae), genetically diverse, the most serious disease of roses, infecting leaves and reducing plant vigour. Greek *diploos* + *carpos* = 'double fruit'.

Coral spot *Nectria cinnabarina* (Nectriaceae), a common and widespread pathenogenic 'pinhead' fungus found throughout the year, mainly in summer and autumn, parasitic on dying or saprobic on dead twigs and branches of beech and other hardwoods. Greek *necros* = 'dead' and *cinnabari* = 'cinnamon' (coloured).

Tar spot A common and widespread fungal disease of attached and fallen leaves of sycamore and maple (by *Rhytisma acerinum* and other species) and willow (by *R. salicina*).

SPOT (LICHEN) See blood spot.

SPOT (MOTH) Nocturnal members of the Geometridae and Noctuidae.

Brindled white-spot *Parectropis similaria* (Geometridae, Ennominae), wingspan 33–9 mm, local in broadleaf woodland, favouring ancient woodland in rural areas, occasionally in gardens throughout much of the southern half of England. Adults fly in May–June. Larvae feed on leaves of pedunculate oak, birch, hazel and hawthorn. Greek *para* + *ectropos* = 'near' + 'turning out of the way' (or 'farmer'), from the sinuous wing markings, and Latin *similis* = 'similar'.

Marbled white spot *Deltote pygarga* (Noctuidae, Eustrotiinae), wingspan 20–2 mm, common in woodland, heathland and moorland throughout much of the southern half of England, and in Wales and Ireland. Population levels increased by 195% over 1968–2007. Adults fly in May–July. Larvae feed in autumn on a variety of grasses. Greek *deltotos* means 'shaped like the capital Greek letter *delta*', and *pyge* + *argos* = 'rump' + 'bright', from the white spot covering the anus when the wings are folded.

Treble brown spot *Idaea trigeminata* (Geometridae, Sterrhinae), wingspan 23–5 mm, local in woodland rides, gardens and hedgerows, and on chalk downland throughout the southern half of England and parts of Wales. Numbers increased by 4,312% (from a low base) over 1968–2007. Adults fly in May–July. Larvae feed on often withered leaves of ivy, knotgrass and other low-growing plants. Greek *Idaios* means from Mt Ida, where the gods and goddesses watched the Trojan War, and Latin *trigeminata* = 'threefold', from the three-spotted forewing markings.

White spot *Hadena albimacula* (Noctuidae, Hadeninae), wingspan 30–8 mm, Vulnerable in the RDB, on shingle beaches, chalk sea-cliffs and limestone cliffs in southern England, a priority species in the UK BAP. Populations are known at Dungeness (Kent), near Gosport (Hampshire), and at least one site between Sidmouth and Seaton, south Devon. There may be other populations on the coasts of Dorset, Kent and the Isle of Wight. Adults fly in June–July. Larvae feed on seeds of Nottingham catchfly at night, hiding in the shingle during the day. *Hadena* is from the Greek for Hades, the Underworld in Greek mythology, and Latin *albimacula* = 'white-spotted'.

SPOTTED (MOTH) White-pinion spotted *Lomographa bimaculata* (Geometridae, Ennominae), wingspan 22–6 mm, common in woodland, scrub, hedgerows and gardens throughout England, Wales and the southern half of Ireland. Adults fly at night in May–June. Larvae feed on hawthorn and blackthorn. Greek *loma* + *graphe* = 'border' + 'drawing', from markings at the edge of the forewings, and Latin *bimaculata* = 'two-spotted'.

SPOTTED-LAUREL *Aucuba japonica* (Garryaceae), an evergreen Japanese neophyte shrub in cultivation by 1783, rather scattered as far north as central Scotland, in woodland, hedges and parks, and on roadsides, usually deliberately planted or as a relic of cultivation, occasionally naturalised. *Aucuba* is the Latin form of the Japanese name for the genus.

SPOTTED-ORCHID Common and widespread tuberous herbaceous perennial in the genus *Dactylorhiza* (Orchidaceae) (Greek *dactylos* + *rhiza* = 'finger' + 'root'). See also marsh-orchid.

Common spotted-orchid *D. fuchsii*, on neutral or base-rich soils in deciduous woodland, scrub, roadsides, chalk grassland, meadow, marshes, dune slacks, fen and mildly acidic heathland. It can be abundant in artificial habitats such as waste ground, former gravel-pits, quarries and railway embankments. This is the county flower of West Lothian as voted by the public in a 2002 survey by Plantlife, while ssp. *hebridensis* is that of the Western Isles. Leonhart Fuchs (1501–66) was a German botanist.

Heath spotted-orchid *D. maculata*, on acidic soils in grassland, moorland, heathland, flushes and mires. *Maculata* is Latin for 'spotted'.

SPRAWLER *Asteroscopus sphinx* (Noctuidae, Psaphidinae), wingspan 39–49 mm, common in broadleaf woodland, gardens, and areas with scattered trees throughout England and parts of Wales and Ireland. With numbers declining by 87% over 1968–2007, this is a moth of conservation concern in the UK BAP. Adults fly at night in October–December. Larvae feed on the leaves of various deciduous trees. Greek *aster* + *scopos* = 'star' + 'view'. Reference to the moth genus *Sphinx* describes the dorsal hump on the larva.

Rannoch sprawler *Brachionycha nubeculosa*, wingspan 48–60 mm, a rare RDB moth found in birch woodland in the Highlands, virtually confined to the Rannoch area, the Spey Valley around Aviemore being a traditional stronghold. It has also been confirmed recently, after a 40-year interval, at sites in Glen Affric and Glen Moriston. Adults fly at night in March–April. Larvae feed on birch leaves. Greek *brachys* + *onychos* = 'short' + 'claw', and Latin *nubecula* = 'small cloud', from the wing shading.

SPRIG-MOSS *Aongstroemia longipes* (Dicranaceae), a rarity in the northern half of Scotland, in moderately base-rich, damp, disturbed sites by hydro-electric dams and in upland gravel-pits. Greek *angeion* + *stroma* = 'vessel' + 'bed', and Latin *longipes* = 'long-footed'.

SPRING BEAUTY *Claytonia perfoliata* (Montiaceae), a widespread, scattered but locally common annual North American neophyte introduced by 1794, on open, sandy, disturbed ground, as a garden and farmland weed, and under light shade, such as Breckland pine-belts. John Clayton (1694–1773) was an American plant collector. Latin *perfoliata* = 'perfoliate'.

SPRING ONION See onion.

SPRING-SEDGE *Carex caryophyllea* (Cyperaceae), a rhizomatous perennial, common and widespread except in north Scotland, in meadow and pasture on dry, base-rich soils; also in grassland and heathland on mildly acidic soil, and in upland pasture and seepages. 'Sedge' comes from Old English *secg*. *Carex* is Latin for sedge. Greek *carya* + *phyllo* = 'walnut' + 'leaf', referring to the aromatic scent.

Rare spring-sedge *C. ericetorum*, a perennial intolerant of grazing or competition, restricted to dry grasslands on infertile calcareous soils from Cambridgeshire and Suffolk north-west to Cumbria and Co. Durham. Greek *ereice* = 'heathland', though not found in this habitat.

SPRINGTAIL Order of Collembola, with around 300 British species, small wingless insects generally living in soil or high-humidity leaf litter, though the common *Podura aquatica* is found on pond surfaces. Their name comes from a forked springing organ on the rear end that enables the animals to leap into the air on disturbance. With six abdominal segments rather than the 11 that all other insects have, their taxonomic status as insects has been questioned. There are 6–8 instars, but metamorphosis is essentially absent.

Clover springtail or **lucerne flea** *Sminthurus viridis* (Sminthuridae), widespread and often abundant on clover and other legumes in summer, then moving to grasses and mosses. Greek *sminthos* + *ouros* = 'field mouse' + 'tail', and Latin *viridis* = 'green'.

Seashore springtail *Anurida maritima* (Neanuridae), wingless, up to 3 mm length, common and widespread though not recorded on the north-east Scottish or south-east Irish coasts, primarily limited to the upper intertidal, found in large clusters moving over rocks in search of food or floating on the surface film of upper shore rock pools. Its body is covered with hydrophobic hairs which allow it to stay above the water surface. It retreats into rocky crevices, or shelters under weeds during high tide, retreating one hour before the tide begins to rise.

SPRUCE Introduced evergreen conifers in the genus *Picea* (Pinaceae), from the Latin for these trees. 'Spruce' dates from the late seventeenth century, from late Middle English *pruce* = 'Prussian'.

Blue spruce Alternative name for Colorado spruce.

Colorado spruce *P. pungens*, a North American introduction mainly grown in parts of England for the Christmas tree trade. *Pungens* is Latin for 'prickly'.

Engelmann spruce *P. engelmannii*, introduced from North America in 1862, grown in small-scale forestry and for shelter belts, mainly in western regions. George Engelmann (1809–84) was a German-born American physician and botanist.

Norway spruce *P. abies*, originally from Scandinavia, possibly introduced over a millennium ago and certainly growing in England by the sixteenth century, common in plantations for the Christmas tree market and for timber products. It is an early successional, moderately shade-tolerant tree. Best growth is on moist, well-aerated soils of poor to medium fertility such as sandy loams. *Abies* is Latin for 'fir'.

Serbian spruce *P. omorika*, introduced from the Balkans in 1889, used on a small scale for shelter belts and in plantations, and often planted in parks. It requires >650 mm rainfall, is tolerant of atmospheric pollution, and grows on a range of soils from shallow limestones to deep acidic peats. *Omorika* is the Serbian name for this tree.

Sitka spruce *P. sitchensis*, introduced from North America in 1832, especially grown in western high rainfall areas, tolerant of exposed sites, favouring moist soils of poor or medium nutrient status. It is widely used in forestry (the most frequently planted conifer, often on former heathland and moorland), as an ornamental in parks, and often self-sown. *Sitchensis* is Latin for coming from Sitka, Alaska.

White spruce *P. glauca*, introduced from North America pre-1700, grown on a small scale for timber, as shelter belts, and as a park ornamental. *Glauca* is Latin for 'glaucous' or 'blue-grey'.

SPUR-MOSS Species of *Oncophorus* (Rhabdoweisiaceae), from Greek *oncos* + *phoreo* = 'swelling' + 'to bear', referencing the swelling at the base of the capsule, scarce in the Highlands in damp grassland by base-rich flushes or below calcareous crags.

SPURGE Species of *Euphorbia* (Euphorbiaceae), named after Euphorbus, Greek physician to Juba II, King of Mauretania. 'Spurge' comes from late Middle English as an abbreviation of Old French *espurge*, itself from Latin *expurgare* = 'cleanse', from the purgative properties of the plant's milky latex.

Balkan spurge *E. oblongata*, a rhizomatous perennial neophyte herb, naturalised on grassy banks in Hampshire since at least 1993, occasional elsewhere in southern England, west Wales, and Lanarkshire. *Oblongata* is Latin for 'oblong-shaped'.

Broad-leaved spurge *E. platyphyllos*, an annual archaeophyte, declining in range because of agricultural intensification, found locally in southern and south-central England on cultivated and waste ground, usually on calcareous clays but sometimes on lighter chalk or limestone soils, and generally found in arable field margins, occasionally on roadsides. *Platyphyllos* is Greek for 'broad-leaved'.

Caper spurge *E. lathyris*, a biennial archaeophyte of disturbed ground and waste places, often near human habitation, also persistent in a few woods in southern England, and casual or naturalised in other areas, especially elsewhere in England, originating as an escape from cultivation or from birdseed. It is often grown in gardens, and has a reputation for deterring moles, with mole spurge an alternative name. *Lathyris* is Greek for 'spurge'.

Coral spurge *E. corallioides*, a rhizomatous perennial Mediterranean herb grown in gardens since 1752, naturalised in woodland, hedgerows and pavement cracks and on roadsides, lane-sides and rubbish tips, scattered mainly in southern England. *Corallioides* is Latin for 'coral-like'.

Cypress spurge *E. cyparissias*, a rhizomatous perennial neophyte herb, introduced by 1640 (though possibly native on chalk grassland in Kent), widely naturalised in Britain as a garden escape in waste places such as tracksides, roadsides, walls and sandy banks, and in rough grassland and scrub. It colonises arable margins, and can become established on dunes.

Dwarf spurge *E. exigua*, an annual archaeophyte of arable land, common in southern and eastern England, scattered elsewhere and declining in status with agricultural intensification, less frequently occurring in other areas of disturbed ground such as gardens, waste ground and bare patches in dry grassland. It favours dry, light, base-rich soils in sunny situations. *Exigua* is Latin for 'meagre'.

Irish spurge *E. hyberna*, a rhizomatous perennial of woodland glades, hedgerows and shaded stream banks, growing best when in dappled sun, mainly found in south-west Ireland, scattered elsewhere in Ireland, local in south-west England and the mid-Thames Valley. *Hyberna* (c.f. *hiberna*) is Latin for 'relating to winter' (not *hibernica* = 'Irish').

Leafy spurge *E. esula*, a perennial European neophyte herb, part of a complex taxonomic group, naturalised on trackways, hedgerows, waste ground and verges, scattered in Britain as far north as central Scotland. Its DNA has been analysed to provide a model to study weed characteristics. *Esula* is the latinised form of the Celtic for 'sharp', referring to the acrid latex.

Mole spurge Alternative name for caper spurge.

Petty spurge *E. peplus*, an annual archaeophyte, common and widespread apart from upland Scotland, on cultivated, disturbed and waste ground, often growing close to habitation, favouring well-drained, nutrient-rich soils in sun-warmed situations. *Peplos* was a long garment worn by women in Ancient Greece.

Portland spurge *E. portlandica*, a biennial or short-lived perennial, local in a range of habitats round the south and west coastline of Britain and around Ireland, for example on cliffs, coastal grasslands, shingle and semi-fixed dunes. Portland refers to Portland Bill, Dorset.

Sea spurge *E. paralias*, a perennial herb local to the coastline from west Scotland south and round to Lincolnshire, and around Ireland, favouring free-draining mobile or semi-stable dunes, and along the drift line of sandy foreshores and, less frequently, on shingle. *Paralios* is Greek for 'sea littoral'.

Sun spurge *E. helioscopia*, an annual archaeophyte, common and widespread especially in the lowlands, on cultivated and disturbed ground in gardens, waste ground and arable fields, especially with root and leaf crops. It prefers dry, well-drained, neutral to base-rich soils in sun-warmed locations. Seeds may be dispersed by ants. Greek *helios* + *scopos* = 'sun' + 'view', from the habit of the flower turning its head with the sun.

Sweet spurge *E. dulcis*, a rhizomatous perennial European neophyte herb, in gardens by 1634, with a scattered distribution, scarce in Northern Ireland, naturalised in woodland and scrub, and on shaded roadsides and riverbanks. *Dulcis* is Latin for 'sweet'.

Upright spurge *E. stricta*, a rare annual or biennial known as native from about ten sites in Gloucestershire, Monmouthshire and Glamorgan, in open deciduous woodland, tracks and hedge banks on limestone soils, but introduced in a few other places in England and Wales. Populations tend to decline as shade and competition increase, but soil disturbance stimulates germination of even long-buried seed. *Stricta* is Latin for 'upright'.

Wood spurge *E. amygdaloides*, a rhizomatous perennial herb of neutral or acidic soils in mature woodland and shaded hedge banks, occasionally found in scrub and around rock outcrops. In woodland it is light-demanding and may reappear from buried seed after coppicing. It is also cultivated as a garden plant, where it can be invasive. *Amygdaloides* is Latin for 'almond-like'.

SPURGE BUG Members of the family Stenocephalidae, represented by two uncommon species of *Dicranocephalus* (Greek *dicranos* + *cephale* = 'two-headed' + 'head'), associated with spurges locally on dunes (*D. agilis*, 12–14 mm long) and woodland (*S. medius*, 8–11 mm long) in southern England.

SPURGE-LAUREL *Daphne laureola* (Thymelaeaceae), an evergreen shrub of heavy, neutral to basic soils in deciduous woodland, often in shade. It reproduces by self-layering and by seed, but as it flowers in early spring and requires cross-pollination by flies or moths, seed-set is often poor. Locally common in England and Wales, it is also a relic of cultivation or introduction in many sites, and across much of its range, including Scotland and Ireland, is more often associated with pheasant-rearing estates and parkland than woodland. In Greek mythology Daphne was a naiad (water nymph) transformed into a laurel to escape Apollo.

SPURREY Species of *Spergula* and *Spergularia* (Caryophyllaceae) (Latin *spargere* = 'to scatter'). 'Spurrey' dates from the late sixteenth century, derived from Dutch *spurrie*, probably *spergula*. See also sea-spurrey.

Corn spurrey *Spergula arvensis*, the only native populations of this annual calcifuge being those of var. *nana* on granite cliff ledges in the Channel Islands. Elsewhere it is a common and widespread archaeophyte of disturbed habitats on light, often sandy soils, commonly in arable fields but also on seashores, roadsides and waste ground. *Arvensis* is from Latin for 'field'.

Pearlwort spurrey *S. morisonii*, an annual European neophyte naturalised at a few sites, for example in East Sussex and Co. Kildare, on sandy or peaty cultivated land, including arable fields. Robert Morison (1620–83) was a Scottish botanist.

Sand spurrey *Spergularia rubra*, an annual or biennial, found in open habitats on free-draining acidic sands and gravels, for example on heathland, gravel- and sandpits, railway yards and waste ground. It occasionally grows on stabilised shingle and dunes, and is tolerant of trampling. *Rubra* is Latin for 'red'.

SPURWING Mayflies in the family Baetidae.

Large spurwing *Procloeon pennulatum*, known from England and Wales, rare in Scotland. Nymphs live in pools and stream margins feeding by scraping algae from submerged stones and other structures, or by gathering or collecting fine particulate organic detritus from the sediment. Adults are evident in May–October. Greek *pro* + *cloios* = 'in front of' + 'collar', and Latin *pennulatum* = 'winged'.

Small spurwing *Centroptilum luteolum*, common and widespread, less common in Scotland where it is associated with standing water. There are usually two generations a year, with adults present from April to November. Nymphs live in pools and margins of rivers and streams or on lake shores. They swim in short, darting bursts among the substrate, or climb among the vegetation, feeding by scraping algae from submerged stones and by taking fine particulate organic detritus from the sediment.

Greek *centron* + *ptilon* = 'centre' + 'wing', and Latin *luteolum* = 'yellowish'.

SQUAMATA From Latin *squamatus* = 'scaly', the order that includes lizards and snakes. Squamates periodically shed their skin, snakes (sloughing) doing so in one piece, and have uniquely jointed skulls and jaws that enable them to open their mouths wide to consume large prey.

SQUARE-SPOT (MOTH) 1) Specifically, *Paradarisa consonaria* (Geometridae, Ennominae), wingspan 40–5 mm, local in broadleaf woodland, scrub, parks and gardens throughout England, the main population centres being from Hampshire and the Isle of Wight to Kent, the Wye Valley and south Cumbria, and also in south Wales and south Ireland. Adults fly at night in April–June. Larvae feed on a number of tree species, including oaks and birches. Greek *para* = 'alongside' + the genus *Darisa*, and Latin *consonus* = 'fitting'.

2). Some members of the Noctuidae (Noctuinae), all nocturnal.

Double square-spot *Xestia triangulum*, wingspan 36–46 mm, widespread and common in deciduous woodland, hedgerows and gardens. Adults fly in June–July. After hibernation, larvae feed on such plants as meadow buttercup, primrose, cow parsley and wood spurge. Greek *xestos* = 'polished', from the glossy forewings, and Latin *triangulum* = 'triangle', from the wing markings.

Fen square-spot *Diarsia florida*, wingspan 30–8 mm, possibly an ecotype of small square-spot, local in bog and fen in East Anglia, northern England and Scotland. Adults fly in June–July. Larvae feed on a variety herbaceous plants. Greek *diarsis* = 'a raising up', reason obscure, and Latin *florida* = 'bright'.

Small square-spot *D. rubi*, wingspan 28–33 mm, widespread and common in damp woodland and grassland, gardens and rough meadow, but with numbers declining by 87% over 1968–2007 a species of conservation concern in the UK BAP. South of Scotland it is double-brooded, flying in May–June and August–September. In Scotland a single generation flies in July–August. Larvae feed on a variety of herbaceous plants, and also bramble (*rubus*, giving the specific name).

SQUASH BUG Member of the bug family Coreidae, so called because of the damage done to squash plants by American examples.

SQUAT LOBSTER Dorsoventrally flattened marine decapod crustaceans with long tails held curled beneath the thorax, similar morphologically to, but taxonomically separate from lobsters, in the genera *Galathea* (Galatheidae), named after the sea nymph Galatea in Greek mythology, and *Munida* (Munididae), possibly from Latin *munio* = 'to protect'.

Common squat lobster *G. strigosa* and **rugose squat lobster** *M. rugosa* are common in the subtidal and deeper on rock ledges and in sand under boulders.

Black squat lobster *G. squamifera*, reaching lengths of 6.5 cm, with a carapace length of up to 3.5 cm, common and widespread under stones and rocks on the lower intertidal and in crevices and fissures in the subtidal to depths of 70 m. Latin *squamus* + *fero* = 'scale' + 'to bear'.

SQUID Members of the class Cephalopoda, order Teuthida, and around British shores mainly in the family Loliginidae, mostly in deep water. They have a head, mantle and 10 arms including a pair of long tentacles used to capture small fish and crustacean prey. **Common squid** *Loligo vulgaris*, up to 54 cm body length, is commercially valuable off the coasts of south-west England and north-west Scotland.

SQUILL Bulbous herbaceous perennials in the genus *Scilla* (Asparagaceae). *Scilla* is the Greek name for a plant in a different genus, *Drimia maritima*.

Scattered casual or naturalised European neophytes found in grassy habitats include **alpine** (or **early spring**) **squill** *S. bifolia* (present in Tudor gardens, in England and Scotland), **Greek** squill *S. messeniaca* (in England), **Portuguese squill** *S. peruviana* (also in woodland, mainly in southern and south-west England; *peruviana* mistakingly reflects the country, but in fact it was that some bulbs came from a ship called 'Peru'), **Pyrenean squill** *S. liliohyacinthus* (England and Scotland, often spreading in open woodland), **Siberian squill** *S. siberica* (introduced in 1796, found on free-draining, often sandy soil), and **Turkish squill** *S. bithynica* (introduced in 1920, mainly found in southern England).

Autumn squill *S. autumnalis*, in open, drought-prone grasslands and heathy vegetation in rocky or sandy habitats near the sea in the Channel Islands and south-west England; also on river terrace gravels in Surrey and south Essex.

Spring squill *S. verna*, locally common on the west coast of Britain, the east coast south to Co. Durham, and the east coast of Ireland, on short turf and coastal heath on clifftops and rocky slopes. Where there is a pronounced oceanic climate (e.g. Anglesey) it can occur on inland heathland. This is the county flower of Co. Down as voted by the public in a 2002 survey by Plantlife. *Verna* is Latin for 'spring'.

SQUINANCYWORT *Asperula cynanchica* (Rubiaceae), a rhizomatous herbaceous perennial of dry, calcareous grassland and dunes, locally common in chalk and limestone in England, south Wales (mainly dunes) and western Ireland. 'Squinancywort' dates to the early eighteenth century, from medieval Latin *squinantia* (formed by a confusion of Greek *synanche* with *cynanche* both denoting throat diseases) + wort. Latin *asper* = 'rough', and Greek *cyon* + *ancho* = 'dog' + 'to strangle', referring to the plant's toxicity.

SQUIRREL Arboreal rodents in the genus *Sciurus* (Sciuridae), the English name coming from Middle English, in turn a shortening of the Old French *esquireul*, from a diminutive of Latin *sciurus*, itself from Greek *sciouros*, from *scia* + *oura* = 'shadow' + 'tail' (alluding to the squirrel sitting in the shadow of its tail).

Grey squirrel *S. carolinensis*, head-body length 23–30 cm, tail length 19.5–24 cm, male weight 440–650 g, females 400–720 g, with a number of introductions from North America beginning in England (Cheshire) in 1876, Ireland (Co. Longford) in the early twentieth century, common and widespread in England, Wales, central Scotland and the eastern half of Ireland, and still expanding its range. Estimated population size in 1995 was 2,520,000 grey squirrels in Britain (2 million in England, 320,000 in Wales, 200,000 in Scotland). There are colonies of white grey squirrels in Kent, Sussex and Surrey, and of black grey squirrels especially around Bedfordshire and Cambridgeshire. It favours deciduous woodland, but is common in suburban parks and gardens, often habituated to people. Diet is of tree seeds such as of oak, beech and hazel, but also flowers, buds, shoots, fungi, peanuts from bird feeders, and bird eggs and young. Seeds are buried for retrieval in winter. Domed nests (dreys), 25–30 cm in diameter, used for shelter and breeding, are built of sticks in trees, often in branch forks. Tree holes are also used. Mating occurs in late winter and again in June–July, depending on weather and availability of food. Gestation lasts 44 days. Females can have two litters a year, in early spring and summer, each with 1–7 (usually 3–4) young (kittens). Weaning lasts up to ten weeks. They can breed at 10–12 months old. They are serious pests of forestry, stripping the bark of thin-barked species such as beech and sycamore, and also of gardens and horticultural crops. They are implicated in the decline in numbers and range of red squirrels not because of aggression but probably by having a more generalist diet, for example (unlike reds) eating unripe acorns and hazel nuts, and therefore faring better (out-competing reds) in deciduous and mixed woodland. Greater ecological fitness also comes from producing more young per litter, and being able to store four times as much fat in winter than the red. They also transmit parapox virus, fatal to reds but not to greys. Research in Ireland and Scotland suggests that pine martens might suppress grey squirrel populations, enhancing survival of red squirrels. It

has been illegal to keep grey squirrels without a licence since 1937, and it is illegal to release them into the wild. *Carolinensis* is modern Latin referring to the Carolinas in the USA.

Red squirrel *S. vulgaris*, head-body length 18–24 cm, tail length 15–20 cm, weight 250–350 g, once common and widespread, with a preference for coniferous forest, but also living in deciduous woodland, parks and treed gardens, in these last habitats, however, generally out-competed by grey squirrels. Numbers and range since the 1920s have greatly contracted because of the grey squirrel, as well as parapox (squirrelpox) virus carried by the grey. Populations remain on Jersey; Brownsea Island, Poole Harbour (Dorset); the Isle of Wight; Thetford Chase (East Anglia), Cannock Chase (Staffordshire), Hope Forest (Derbyshire), around Formby (Merseyside), and at sites in Co. Durham, Northumberland and Cumbria; in a few places in Wales, including Anglesey; in (especially Caledonian) pine forest in Scotland; and in many parts of Ireland (where it died out in the late seventeenth century, but with a number of reintroductions during the nineteenth century). There are an estimated 138,000 red squirrels in the UK (120,000 in Scotland, 3,000 in Wales, 15,000 in England). In a 2013 study across northern woodlands, the number of red squirrels had risen by 7% compared to the previous spring. In 2016, the Isle of Man government sought the public's views on licensed introduction of red squirrels to the wild; the Mammal Society, however, argued that there were neither scientific grounds nor conservation value in the proposal. They are predominantly seed-eaters, favouring pine, but also feeding on buds, fungi and bird eggs. Seeds are buried for retrieval in winter. Domed nests (dreys), 25–30 cm in diameter, used for shelter and breeding, are built of sticks in trees, often in branch forks. Mating can occur during February–March and again in June–July, depending on weather and availability of food. Gestation last 45–8 days. Females have 1–2 litters a year, usually of 2–3 young (kittens). Juveniles are weaned after 8–12 weeks, but do not breed until they are one year old. *Vulgaris* is Latin for 'common'.

SQUIRREL-TAIL MOSS *Leucodon sciuroides* (Leucodontaceae), widespread but generally uncommon, more scattered in Ireland, on trees with base-rich bark such as ash and sycamore on base-rich rock outcrops, stone monuments or gravestones. Greek *leucos* + *odontos* = 'white' + 'teeth', and Latin *sciuroides* = 'squirrel-like'.

Lesser squirrel-tail moss *Habrodon perpusillus* (Pterigynandraceae), a scattered rarity in Britain, commonest in Cumbria, growing on the bark of isolated mature trees in grassland and parkland or at the edges of woodland, in well-lit situations. Greek *abros* + *odontos* = 'delicate', and Latin *perpusillus* = 'very small'.

SQUIRT See sea squirt.

ST JOHN'S-WORT Perennial herbs in the genus *Hypericum* (Hypericaceae), from Greek *hyper* + *eicon* = 'above' + 'picture', plants in this genus having traditionally been hung above pictures for the scent to ward off evil spirits. Plants flower around St John's Day, 24 June.

Hairy St John's-wort *H. hirsutum*, widespread and common in the eastern half of the British Isles, scarce in the west, on well-drained, neutral to basic soils in open or partially shaded habitats such as rough grassland, woodland clearings, riverbanks, roadside banks and verges. *Hirsutum* is Latin for 'hairy'.

Imperforate St John's-wort *H. maculatum*, rhizomatous, locally common especially in southern England, Wales, and west and south-east Ireland, in damp shaded places and in ruderal habitats such as quarries and rough, grassy places. *Maculatum* is Latin for 'spotted'.

Marsh St John's-wort *H. elodes*, stoloniferous, locally common (but with drainage schemes, generally declining) in south-west and southern England, Wales, western Scotland and Ireland, on peaty soils in damp, acidic, nutrient-poor, sometimes terrestrial habitats, but usually in shallow water, or forming floating mats in deeper water. It grows in heathland pools, on pond and slow-flowing stream margins, and in bog seepages. *Elodes* is Greek for 'marshy'.

Pale St John's-wort *H. montanum*, locally common, virtually confined to chalk and limestone habitats in England and Wales, preferring well-drained soils, and growing by hedges and in thickets, scrub, rough grassland, open woodland, and limestone pavement. Populations are generally small and vulnerable to habitat loss. *Montanum* is Latin for 'montane'.

Perforate St John's-wort *H. perforatum*, rhizomatous, common and widespread in open woodland, meadow, and along roadsides and railways, extending into Scotland using these corridors during the twentieth century.

Slender St John's-wort *H. pulchrum*, widespread and locally common in heathland, hedge banks and woodland glades on non-calcareous, fairly dry, sandy, peaty, or leached soils. A dwarf form with prostrate to procumbent stems (var. *procumbens*) occurs in exposed habitats in north and west Scotland and in islands off west Ireland. *Pulchrum* is Latin for 'beautiful'.

Square-stalked St John's-wort *H. tetrapterum*, rhizomatous, common and widespread in eutrophic moist or wet habitats including damp meadow, marsh and streamsides. Greek *tetra* + *pteron* = 'four' + 'wing'.

Toadflax-leaved St John's-wort *H. linariifolium*, a short-lived perennial local in south-west England and the Channel Islands, occasionally in south England and Wales, on rocky slopes and banks with a southerly aspect, on well-drained acidic soils. Latin *linariifolium* = 'linear-leaved'.

Trailing St John's-wort *H. humifusum*, a widespread but local short-lived perennial of open well-drained habitats such as heathland, moorland, woodland and roadside banks, on light acidic soils. *Humifusum* is Latin for 'trailing'.

Wavy St John's-wort *H. undulatum*, rhizomatous, in non-calcareous wet fields and streamsides, and mildly acidic bog, with a preference for waterlogged areas, found in south-west England and south-west and west Wales, declining in status because of agricultural intensification. *Undulatum* is Latin for 'wavy'.

ST MARK'S FLY *Bibio marci* (Bibionidae), males 12 mm, females 14 mm in length, common and widespread in hedgerows and woodland edge, adults evident from spring to late summer, generally emerging around St Mark's Day (25 April), hence the common name. Larvae feed on decaying plant material and on grass roots, sometimes damaging amenity turf grass. *Bibio* is Latin for a small insect reportedly generated by wine.

ST PATRICK'S CABBAGE *Saxifraga spathularis* (Saxifragaceae), a perennial, stoloniferous herb locally common in acid conditions in high rainfall parts of west and south-west Ireland, in humid, rocky woods, on shady mountain cliffs and relatively unshaded south-facing slopes. Rare elsewhere in Ireland, and naturalised in a few places in Britain. St Patrick is the patron saint of Ireland. Latin *saxum* + *frangere* = 'rock' + 'to break', and for 'with a spathe'.

ST PAUL'S-WORT Annual species of *Sigesbeckia* (Asteraceae), named by Linnaeus after the German botanist Johann Siegesbeck (1686–1755), a critic of Linnaeus, probably because the plant is small-flowered, unpleasant-smelling, weedy, and grows in mud.

Eastern St Paul's-wort *S. orientalis*, an Asian neophyte cultivated in 1730, found in 1938 as a casual in fields and waste ground in south-east Yorkshire derived from wool shoddy, and in a few other places in England since.

Western St Paul's-wort *S. serrata*, a tropical American neophyte naturalised in south Lancashire on waste and cultivated ground, and a scattered casual in a few other places in England. *Serrata* is Latin for 'serrated', referring to the leaves.

ST WINIFRED'S (OTHER) MOSS *Chiloscyphus pallescens* (Lophocoleaceae), despite its name a foliose liverwort, common and widespread mainly on rocks, aquatic in streams, semi-aquatic on flushes and wet ground. St Winifred was a seventh-century nun, decapitated by her suitor Caradoc when she took holy orders. Greek *cheilos* + *scyphos* = 'margin' + 'cup', and Latin *pallescens* = 'becoming pale'.

STAFF-VINE *Celastrus orbiculatus* (Celastraceae), a woody east Asian neophyte climber grown in gardens since 1870, naturalised on a wooded roadside at Shottermill (Surrey) since 1985, in woodland near West Porlock (Somerset) since 1993, and also in Hampshire. *Celastros* is the Greek name for another tree species, possibly privet, *orbiculatus* Latin for 'round'.

STAG BEETLE *Lucanus cervus* (Lucanidae), the largest British beetle (males, 25–75 mm, females 25–50 mm), widespread but declining in the southern third of England, commonest in the south-east, scattered elsewhere in England and in Wales. The preferred habitat is oak woodland, but it is also found in gardens, hedgerows and parkland. Larvae live in and feed on old trees and rotting wood, creating a series of tunnels, taking up to six years to develop before they pupate. Adults emerge in May–June; they do not feed, relying on fat stored during the larval stage, and they die in August once the eggs have been laid in decaying wood. Males fly in search of mates during summer evenings. Once found, the male displays his massive, antler-like mandibles to her, and uses them to fight off rival males. Latin *lucus* = 'grove' (*lucanus* being used by Pliny to mean this beetle), and *cervus* = 'deer' or 'stag'.

Lesser stag beetle *Dorcus parallelipipedus*, length 18–32 mm, widespread in the southern half of England, local in Wales. Adults are found in woodland, orchards and parkland and along hedgerows during summer, mating, then females laying eggs in decaying wood (ash, beech and apple are commonly used) in which the larvae live in and feed on. The binomial is Latin for 'gazelle' and 'parallel-footed'.

STAG'S HORN (LICHEN) Alternative name for oak moss.

STAGSHORN Inedible wood-rotting fungi in the genus *Calocera* (Dacrymycetales, Dacrymycetaceae) (Greek *calos* + *ceras* = 'beautiful' + 'horn').

Pale stagshorn *C. pallidospathulata*, with a gelatinous fruiting body, widespread in England, first discovered in 1969 (under larch in Wykeham Forest, north Yorkshire), but only locally common, scattered in Wales and Scotland, rare in west and south-east Ireland. Found in autumn on dead and rotting branches and stumps of conifers. *Pallidospathulata* is Latin for 'pale-sheathed'.

Small stagshorn *C. cornea*, widespread and common, with a finger-like gelatinous fruiting body, found throughout the year though mainly in autumn on dead and rotting branches and twigs of broadleaf trees. *Cornea* is Latin for 'made of horn'.

Yellow stagshorn *C. viscosa*, with an antler-like gelatinous fruiting body, widespread and common, found throughout the year but mainly in autumn on stumps and roots of conifers. *Viscosa* is Latin for 'sticky'.

STAINER Saffron stainer *Cortinarius olearioides* (Agaricales, Cortinariaceae), a rare inedible fungus, with a scattered distribution in England, Vulnerable on the RDL (2006), in autumn on soil in deciduous woodland and parkland mainly under beech. Latin *cortina* = 'curtain' and *olea* = 'olive'.

Yellow stainer *Agaricus xanthodermus* (Agaricaceae), a poisonous fungus that turns bright yellow when handled or cut, locally found, especially in the southern half of England, rare elsewhere in the British Isles, found in late summer and autumn on soil among grass in pasture and woodland. Greek *agaricon* = 'mushroom', and *xanthos* + *derma* = 'yellow' + 'skin'.

STALKBALL Fungi in the genera *Onygena* (Onygenales, Onygenaceae), *Phleogena* (Atractiellales, Phleogenaceae) and *Tulostoma* (Agaricales, Agaricaceae), from Greek *tylos* + *tomos* = 'swelling' + 'cut'.

Rarities include **feather stalkball** *O. corvina* (on old feathers and bird dung), **horn stalkball** *O. equina* (on old horn and hooves, mostly of sheep), **scaly stalkball** *T. melanocyclum* (dunes in parts of Wales, Lancashire and Norfolk), and **white stalkball** *T. niveum* (on limestone rocks and walls at Inchnadamph, Sutherland, and Craig Leek, Aberdeenshire).

Fenugreek stalkball *P. faginea*, an inedible jelly fungus scattered and infrequent in the southern half of England in deciduous woodland on the bark and dead wood of various hardwoods, smelling of fenugreek when scraped. *Faginea* is Latin for 'beechwood'.

Winter stalkball *T. brumale*, inedible, saprobic, common on the coast of England and Wales, occasional inland, and scarce in Scotland and Ireland, trooping from September to January in moss or short grass mainly on sandy, alkaline soil, for example in dune slacks. Latin *bruma* = 'winter'.

STAPHYLEACEAE Family name for bladdernut (Greek *staphyle* = 'cluster of grapes').

STAPHYLINIDAE Family name of rove beetles (Greek *staphyle* = 'cluster of grapes'), containing 1,120 species (excluding probable recent extinctions) – around a quarter of Britain's beetle species – in 278 genera.

Divided into 19 subfamilies. **Omaliinae** contains 71 British species, many widespread and common in a variety of habitats, often be found on dung or fungi. **Proteininae**, 11 species, common in decaying vegetable matter, rotting fungi and carrion. **Micropeplinae**, five species found mainly in warm decaying vegetable matter such as compost heaps. **Pselaphinae**, 52 small, often rare species, mainly feeding on orobatid mites. Many resemble ants in general appearance, and some are myrmecophilous. **Phloeocharinae**, represented by *Phloeocharis subtilissima*, widespread in moss or beneath tree bark. **Tachyporinae**, 66 species especially common in decaying vegetation and fungi. **Trichophyinae**, represented by *Trichophya pilicornis*, widespread but local in decaying vegetation and moss, particularly associated with sawdust. **Habrocerinae**, *Habrocerus capillaricornis*, in leaf litter and other decaying vegetation. **Aleocharinae**, 454 similar species, large numbers found in decaying vegetation, dung, carrion and rotten fungi. **Scaphidiinae**, probably three species. **Piestinae**, represented by *Siagonium quadricorne*, found under bark. **Oxytelinae**, 94 species, mostly found near water or in marshy places. The fossorial legs of the 27 species of *Bledius* reflect their habit of digging shallow galleries through the clay or sand of waterside areas, where they can be detected by the spoil heaps left behind. **Oxyporinae**, represented by *Oxyporus rufus*, associated with the gills of fungi. **Scydmaeninae**, 32 small species of ant-resembling stone beetles, associated with leaf litter, rotten wood and other decaying organic matter. **Steninae**, 75 species (74 being in the genus *Stenus*), found near water, for example in wet moss. **Euaesthetinae**, four small species, widespread but local in a range of habitats, including grass tussocks, marsh vegetation and flood debris. **Pseudopsinae**, represented by *Pseudopsis sulcata*, local but widespread mainly in decaying vegetable matter. **Paederinae**, 62 species, widespread in decaying organic matter, particularly in marsh and riparian habitats. **Staphylininae**, 185 species, including the largest staphylinids found in Britain. Unusual species include *Emus hirtus*, a dung-feeding bee mimic now restricted to Kent; *Velleius dilatatus*, found in hornet nests; and *Remus sericeus*, a specialist of decaying seaweed. See devil's coach horse.

STAPHYLINOIDEA Superfamily (Greek *staphyle* = 'cluster of grapes') of burying, carrion and rove beetles.

STAR See starfish, starlet, sand star and serpent star, and the fungus shooting star.

STAR-MOSS **Heath star-moss** *Campylopus introflexus* (Leucobryaceae), South American, introduced by 1941, common and widespread by the 1960s, a pioneer species of blanket bog, bare peat, burning, ploughing for forestry and ditching. It also grows on rotting logs and old fence-posts, on thin soil at the edge

of tracks, mine waste, shingle, and occasionally on roof tiles and thatch. Greek *campylos* + *pous* = 'curved' + 'foot', and Latin *introflexus* = 'curved inside'.

STAR-OF-BETHLEHEM 1) Spring-flowering bulbous herbaceous perennial in the genus *Gagea* (Liliaceae), honouring the English botanist Sir Thomas Gage (1719–87). See also Snowdon lily under lily.

Early star-of-bethlehem *G. bohemica*, in crevices, on ledges and in grazed turf on dolerite slopes at Stanner Rocks, Radnorshire (Powys). With an alternative name of **Radnor lily**, this is the county flower of Radnorshire as voted by the public in a 2002 survey by Plantlife. *Bohemica* is Latin for 'from Bohemia', Czech Republic.

Yellow star-of-bethlehem *G. lutea*, scattered in Britain, local in moist, base-rich, shady habitats including woodland, hedgerows, limestone pavement, pasture, riverbanks and stream banks. *Lutea* is Latin for 'yellow'.

2) **(Common) star-of-bethlehem** *Ornithogalum umbellatum* (Asparagaceae), a bulbous perennial herb, with white star-shaped flowers, possibly native on sandy soils in Breckland, otherwise introduced, scattered in Britain in rough pasture, dry, grassy banks and open woods. Greek *ornithos* + *gala* = 'birds' + 'milk', probably from the white colour of the flowers, and Latin *umbellatum* = 'shaded' or 'with umbels'.

Drooping star-of-bethlehem *O. nutans*, a Mediterranean neophyte cultivated by 1648, scattered in Britain, mainly in southern England and East Anglia, a garden escape in scrub, woodland edges, hedgerows, on roadsides and in rough grassy habitats. Populations are usually small and short-lived but some are persistent. *Nutans* is Latin for 'drooping'.

Spiked star-of-bethlehem *O. pyrenaicum*, in ash-elm woodland, scrub, hedgerows and verges on calcareous soils, scattered and local in southern England (a feature of green lanes on the borders of Somerset and Wiltshire) and the south Midlands, once harvested as a substitute for asparagus, and occasionally grown in gardens, sometimes escaping and becoming naturalised. *Pyrenaicum* is Latin for 'Pyrenean'.

STAR-THISTLE Species of *Centaurea* (Asteraceae), named after a centaur (Chiron) who discovered the plant's medicinal uses. See also knapweed.

Golden star-thistle Synonym of yellow star-thistle.

Lesser star-thistle *C. diluta*, an annual Mediterranean neophyte scattered as a casual on waste ground in England, Isle of Man and central Scotland, brought in with birdseed. Latin *diluta* here means 'weak'.

Maltese star-thistle *C. melitensis*, an annual neophyte scattered as a casual on waste ground in England and south-east Scotland, brought in with birdseed and wool shoddy. *Melitensis* is Latin for 'Maltese'.

Red star-thistle *C. calcitrapa*, a biennial Roman archaeophyte scattered in England, naturalised on waste ground and dry grassland, and on well-drained soils. It may be native in Sussex, and in Kent it has been recorded from two sites since the nineteenth century. Elsewhere it is declining as a casual from shoddy and birdseed. *Calcitrapa* comes from medieval Latin *calcatrippa* = 'thistle', itself a compound of either *calyx* = 'heel' or *calcare* = 'to tread', and *trapa*, a Germanic word meaning 'trap'.

Rough star-thistle *C. aspera*, a herbaceous perennial European neophyte cultivated by 1722, naturalised on dunes and in sandy fields in Guernsey since 1788, and in Jersey since 1839. In southern England, it also occurs as a casual on light soils in coastal waste ground. *Aspera* is Latin for 'rough'.

Yellow star-thistle *C. solstitialis*, a winter-annual (occasionally biennial) southern European neophyte introduced with grain, lucerne and sainfoin seed, and shoddy, scattered and occasional in England in arable and waste ground. In large quantities it is toxic to horses, causing the untreatable neurological condition 'chewing disease' (q.v.). *Solstitialis* is Latin for 'pertaining to the (summer) solstice', referencing the ability to flower very late into the summer.

STAR-WORT *Cucullia asteris* (Noctuidae, Cucullinae), wingspan 43–50 mm, a scarce nocturnal moth of saltmarsh along the coasts of England (mainly from Hampshire round to Lincolnshire) and south Wales, occasionally inland in woodland clearings and rides, and gardens in southern England. Adults fly in June–August, visiting flowers. Larvae feed on the flowers of sea aster (prompting the specific name), sea wormwood and goldenrod. Latin *cucullus* = 'cowl', from the thoracic crest.

STARCH-ROOT Synonym for lords-and-ladies. The starchy tubers were used in Elizabethan times to starch cuffs and ruffs. The roasted root is edible and when ground was once sold as 'Portland sago' as a substitute for arrowroot. It was also used to make 'saloop', a drink popular before the introduction of tea or coffee.

STARFISH Members of the class Asteroidea, usually with five arms (though some species have more), usually generalist predators from the intertidal (including rock pools) to the deep sublittoral. They can regenerate damaged or lost arms.

Bloody Henry starfish *Henricia oculata* (Echinasteridae), 12 cm across, widespread on all coasts except east Britain on a variety of substrates from the low intertidal to 100 m. It does prey on sponges, hydroids and ectoprocts, but is less predatory than other British starfishes, mainly suspension-feeding by filtering plankton and organic detritus. *Oculatus* comes from Latin for 'eye'.

Common starfish *Asterias rubens* (Asteriidae), up to 52 cm in diameter, more commonly 10–30 cm, found intertidally and subtidally on all coasts, especially among beds of mussels and barnacles, preying upon bivalves by forcing the shell open with its tube-feet. Greek *asterias* = 'starry' and Latin *rubens* = 'red'.

Common sun star *Crossaster papposus* (Solasteridae), up to 34 cm diameter with 10–12 arms (rarely 8–16), on sand, stones, mussel and oyster beds from the infralittoral fringe to 50 m around all coasts, if patchily around England. Greek *crossoi* + *aster* = 'fringe' + 'star', and *pappos* = 'downy hair'.

Purple sun star *Solaster endeca* (Solasteridae), up to 40 cm in diameter, although usually 20 cm, with 9–10 tapering arms (occasionally 7–13), frequently recorded on the north and west coasts of Britain and Ireland, and also from Shetland to Northumberland, on muddy gravel or silty rock from the infralittoral to 500 m. Latin *sol* + Greek *aster* = 'sun' + 'star', and Greek *endeca* = 'eleven'.

Seven-armed starfish *Luidia ciliaris* (Luidiidae), typically 40 cm diameter, occasionally up to 60 cm, on sand, gravel, mixed sediments and rock from the lower shore to 400 m around the north, west and south-west coasts of Britain and all around Ireland. *Ciliaris* is Latin for 'with fine hairs'.

Snowflake star Synonym for common sun star.

Spiny starfish *Marthasterias glacialis* (Asteriidae), up to 70 cm diameter, commonly 25–30 cm, from low water to 200 m in a variety of habitats from sheltered muddy sites to wave-exposed rock faces, recorded from the south-west and west coasts of England to the west coast of Scotland, Orkney and Shetland, and around Ireland. *Glacialis* Is Latin for 'frozen'.

STARFISH FUNGUS *Aseroe rubra* (Phallales, Phallaceae), introduced from Australia, probably via The Netherlands in 1828, evident in summer and autumn in woodland leaf litter, now a rarity in southern England (last recorded at Oxshott Heath in 2008 where it had grown for at least ten years). *Rubra* is Latin for 'red'.

STARFLOWER Spring starflower *Tristagma uniflorum* (Alliaceae), a bulbous herbaceous South American neophyte perennial introduced in 1832, a weed of arable fields, and growing on roadsides and waste ground, in churchyards, and as a relic of cultivation. It is naturalised on sandy soil in mild regions such as in west Cornwall, Scilly and the Channel Islands, a casual elsewhere in England and Wales. Greek *tri* + *stagmos* = 'three' + 'a drop', and Latin *uniflorum* = 'single-flowered'.

STARLET *Asterina gibbosa* (Asterinidae), a starfish up to 5 cm diameter with five (rarely four or six) very short, broad-based arms, common on most British coasts but scarce in the north-east and not recorded from Lincolnshire to Hampshire, under boulders and stones on the lower shore on sheltered and semi-exposed rocky coasts and also in rock pools, extending into the sublittoral to depths of 100 m. Greek *asterina* = 'small star', and Latin *gibbosa* = 'humpbacked'.

STARLING *Sturnus vulgaris* (Sturnidae), wingspan 40 cm, length 22 cm, weight 78 g, common and widespread in both countryside and towns, with 1.8 million pairs estimated in Britain in 2009, supplemented by large numbers arriving in autumn to overwinter. Flocks of half a million birds have been reported, especially at night-time roosts. BBS data, however, suggest that numbers declined by >80% from 1970 to 2015 (by 5% between 2010 and 2015), so it is Red-listed in Britain, Amber-listed in Ireland. The cavity nest contains 4–5 eggs which hatch after 12–15 days, the altricial chicks fledging in 19–22 days. Feeding is usually on the ground on insects (favouring leatherjackets), though seed and fruit are also taken, and scraps in towns. 'Starling' has evolved from Old English *stærlinc*. The binomial comes from Latin for 'starling' and 'common'.

STATHMOPODIDAE Family name for (alder) signal moth (Greek *stathme* + *podos* = 'carpenter's rule' + 'foot').

STAUROZOA From Greek *stauros* + *zoon* = 'a cross' + 'animal', the class in the phylum Cnidaria that includes the stalked jelly-fishes, with four families in the order Stauromedusae found in British waters. See jellyfish.

STEEPLE-BUSH *Spiraea douglasii* (Rosaceae), a deciduous neophyte shrub from North America, cultivated since 1827, naturalised by roadsides, and in hedges and waste places. Hybrids include **confused bridewort** and **Lange's spiraea**. *Spiraea* is Latin for 'bridal wreath', from Greek *speiraira*, a plant used for garlands, in turn from *speira* = 'spiral'. Davis Douglas (1799–1834) was a Scottish plant collector.

STELLATE BARNACLE See barnacle.

STEM BORER Sawflies in the family Cephidae.

Reed stem borer *Calameuta filiformis*, widespread in England and south Wales, larvae feeding in the stems of reeds and grasses. Adults found from late spring to autumn. *Filiformis* is Latin for 'thread-shaped'.

Wheat stem borer *Cephus pygmeus*, widespread and locally common in England and south Wales, larvae boring into the stems of wheat, other cereals and grasses. Peak adult activity is June–July. Greek *cephos* = 'head', and Latin *pygmeus* = 'pygmy'.

STEM FLY Spinach stem fly *Delia echinata* (Anthomyiidae), scarce, with scattered records in Britain. Larvae have been recorded as leaf miners in common mouse-ear, red campion, corncockle and slender sandwort, as well as garden pinks and spinach. *Echinatus* is Latin for 'prickly'.

STEM GALLER Sallow stem galler *Hexomyza schineri* (Agromyzidae), very rare, with records from Surrey, Hertfordshire and Warwickshire. Each larva lives in an oval gall up to 10 mm long on a twig of black poplar or aspen in May–June; sometimes several are arranged spirally around a twig. Greek *hex* + *myzo* = 'six' + 'to suck'. Ignaz Schiner (1813–73) was an Austrian entomologist.

STEM-MOTH Species of *Ochsenheimeria* (Ypsolophidae), named after the German playwright and entomologist Ferdinand Ochsenheimer (1767–1822).

Cereal stem-moth *O. vacculella*, wingspan 12–14 mm, scarce, in open woodland, grassland and around arable fields in parts of England and Wales. Adults fly in sunshine in July–August. Larvae feed on grasses and cereal crops such as wheat and rye, and can cause stem wilt. Latin *vaccula* = 'heifer', from horn-like antennae.

Feathered stem-moth *O. taurella*, wingspan 11–12 mm, scarce in grassland in parts of England and Wales, rare in Ireland. Adults fly in early afternoon sunshine in July–September. Larvae mine leaves and later stems of cock's-foot, meadow-grass and other coarse grasses. Latin *taurus* = 'bull', from horn-like antennae.

Variable stem-moth *O. urella*, wingspan 9–12 mm, scarce but widespread. Adults fly in morning sunshine in July–August. Larvae feed on couch-grass, brome and other grasses, mining leaves then later feeding inside stems. *Urella* is from Greek *ouros* = 'aurochs', from bull-like antennae.

STEM SAWFLY See stem borer.

STENOCEPHALIDAE Family name of spurge bugs (Greek *stenos* + *cephale* = 'narrow' + 'head').

STERCORARIIDAE Family name of skuas, from *stercorarius*, Latin for 'relating to dung': the food disgorged by other birds when pursued by skuas was once thought to be excrement.

STERNIDAE Family name of most terns, from *stearn*, used in Old English for these birds.

STERNORRHYNCHA Suborder of true bugs (Hemiptera), from Greek *sternon* + *rhynchos* = 'chest' + 'nose' or 'snout', referring to the position of the mouthparts, which lie between the front legs and near the chest. Members usually have soft bodies and long, thread-like antennae. Most species are wingless. The group includes the superfamily Adelgoidea which contains the family **Adelgidae**; superfamily Aleyrodoidea with the family **Aleyrodidae** (22 species in 17 genera in Britain, see whitefly); superfamily Aphidoidea with the family **Aphididae** (around 630 species in 160 genera, see aphid); superfamily Coccoidea with the main British families **Coccidae** (27 species in 12 genera), **Diaspididae** (78 species in 36 genera) and **Pseudococcidae** (around 59 species in 26 genera, see scale insect and mealybug); and superfamily Psylloidea with the family Psyllidae (42 species in 13 genera, see plantlice).

STEW FUNGUS Brown stew fungus Alternative name for sheathed woodtuft.

STICK INSECT Introduced members of the order Phasmida, long slender insects that mimic twigs, some of which have escaped to form small, localised colonies: **prickly, unarmed** and **smooth stick insects**, originally from New Zealand, and **Corsican stick insect**, are all known from locations in Devon, Cornwall and the Scillies.

Water stick insect *Ranatra linearis* (Nepidae), a common underwater predatory bug, long and thin like a stick insect, found throughout England and Wales, camouflaged and hiding in vegetation to capture passing invertebrates, tadpoles and small fish. Its long thin tail extends above water to act as a breathing tube. *Ranatra* is of unknown etymology (not Latin *rana* = 'frog'). *Linearis* is Latin for 'linear'.

STICKLEBACK Fishes in the family Gasterosteidae, from Old English *sticel* + *bæc* = 'thorn' + 'back'.

Fifteen-spined stickleback *Spinachia spinachia*, with 14–17 spines, length usually 8–15 cm but up to 22 cm, found in all coastal waters, though especially in south and west Britain, in eel grass and bladderwrack, feeding on fish fry and small crustaceans. Spawning is in May–June: the male defends a territory, displays to females, builds a nest of seaweed in the bladderwrack, and cares for the eggs and fry. *Spinachia* comes from Latin *spina* = 'spine'.

Nine-spined stickleback *Pungitius pungitius*, number of spines 7–12, length 5 cm, locally common, especially in England, in streams, lakes, ponds and rivers, favouring thick submerged vegetation. In spring the male defends a territory, displays to females, builds a nest, and cares for the eggs and fry. It is generally planktivorous. *Pungitius* is Latin for 'needle'.

Sea stickleback Synonym of fifteen-spined stickleback.

Three-spined stickleback *Gasterosteus aculeatus*, length usually 3–4 cm, occasionally up to 8 cm, the smallest British freshwater fish, common and widespread in a range of freshwater habitats and other waters. There are three forms: *leiurus*, the commonest; the anadromous *trachurus*, spawning in rivers, wintering in the sea; and *semi-armatus*, rare, living in brackish water on the east coast. In spring the male defends a territory, displays to females, builds a nest, and cares for the eggs and fry. It is generally planktivorous. Greek *gaster* = 'abdomen' or 'stomach' and *ostoun* = 'bone', and Latin *aculeatus* = 'spiny'.

STICTA Foliose lichens in the genus *Sticta* (Lobariaceae) (Greek *stictos* = 'dotted'). See also peppered moon lichen under lichen.

Floury sticta *S. limbata*, on mossy trees and rocks, mostly in western Britain and western and south-west Ireland, generally in old woodland and parkland. There is a fishy smell when wet. *Limbatus* is Latin for 'bordered'.

Stinking sticta *S. sylvatica*, widespread and locally common in generally upland, western parts of both Britain and Ireland, among moss and detritus and on damp, acid rocks and cliff faces in shaded, humid mature woodland. The English name reflects a fishy smell when the lichen is damp. *Sylvatica* is Latin for 'growing in woodland'.

STILETTO (FLY) Members of the family Therevidae, mostly in the genus *Thereva* (Greek *thereuo* = 'to hunt').

Rarities include **cliff stiletto** *T. strigata* (records only from Torquay, Devon; Isle of Wight; and Shakespeare Cliff, Kent), **dark northern stiletto** *T. valida* (two recent records from Tayside), **forest silver-stiletto** *Pandivirilia melaleuca* (all records from Windsor Forest or nearby, plus a recently reared larva from Greenwich Park, Greater London), **golden Scottish stiletto** *T. handlirschi* (Spey valley and the Moray Firth area), **large plain stiletto** *T. cinifera* (Cornwall, south Wales, and Pett Levels, near Hastings, East Sussex), **light Scottish stiletto** *T. inornata* (Spey and Dee vallies), **northern silver stiletto** *Spiriverpa lunulata* (Spey Valley and Easter Ross in the Highlands, the Tywi, Dyfed, plus unconfirmed recent records from Anglesey, Northamptonshire and Essex).

Coastal silver stiletto *Acrosanthe annulata*, locally common, of coastal dune systems, occasionally on sandy inland localities such as Breckland. Adults fly in April–August, possibly bivoltine. Greek *acros* + *anthos* = 'at the end of' + 'flower', and Latin *annulata* = 'ringed'.

Common stiletto *T. nobilitata*, widespread but only locally common. Adults are usually found in May–September in scrubby habitats, hedgerows, tall grassland and dry heaths. Larvae live in leaf litter, feeding on insects, worms and decomposing vegetation. *Nobilitata* is Latin for 'ennobled'.

Crochet-hook stiletto *T. plebeja* scattered in England, especially found in the south and in East Anglia, but generally declining in status. Adults fly in May–August. *Plebeia* is Latin for 'common'.

Small plain stiletto *T. fulva*, recorded recently from coastal sites in Mid and West Glamorgan, Pembrey Burrows (Dyfed), Sandwich (Kent) and inland sites in Surrey and Hampshire. Adults are most evident in June–July. *Fulva* is Latin for 'yellow-brown'.

Southern silver stiletto *Cliorismia rustica*, At Risk in RDB 3 and a priority species for conservation in the UK BAP, with a local distribution on sandy rivers in the Welsh Borders and in Surrey and Sussex, extending northwards with single records from the 1980s for Cumbria and Yorkshire. Recently, it has also been recorded from rivers in Cumbria, Yorkshire, Northumberland and Perthshire. In 2008, larvae were collected from sand banks from five rivers in Cheshire. The flight period is May–August.

Swollen silver-stiletto *Dialineura anilis*, restricted to a few coastal dune systems, mainly in Wales (Mid and West Glamorgan, Dyfed, Gwynedd, Anglesey) where it is locally frequent, and in Merseyside and the north-west Highlands. Adults fly in May–July. Greek *dialis* + *neuron* = 'aerial' + 'nerve', and *anileos* = 'cruel'.

Twin-spot stiletto *T. bipunctata*, locally common on the coastline of Britain, mostly recorded from coastal dune systems of the Moray Firth, Cumbria, Dyfed, Mid Glamorgan, Somerset and Dorset to East Sussex, and Norfolk, together with a few recent inland records from heathland in Breckland and north Lincolnshire. Adults fly in May–September. *Bipunctata* is Latin for 'two-spotted'.

STILT **Black-winged stilt** *Himantopus himantopus* (Recurvirostridae), an occasional visitor from the continent, with occasional records of successful breeding, for instance in Norfolk in 1987, and two examples in 2014, the RSPB reserves at Cliffe Pools, Kent, and Medmerry near Chichester, West Sussex. A changing climate may bring these birds to Britain more regularly in future. At 23 cm, the legs represent 60% of this bird's height (giving it its name), allowing it to feed on aquatic invertebrates in deeper water than other waders. Greek *himas* + *pous* = 'strap' + 'foot'.

STILT BUG Members of the family Berytidae.

STILT-LEGGED FLY Members of the family Micropezidae.

STILTBALL **Sandy stiltball** *Battarrea phalloides* (Agaricaceae), a very rare inedible homobasidiomycete fungus in the southern half of England, Near threatened in the 2006 RDL, a priority species in the UK BAP, found in late summer and autumn, generally associated with rotting wood in sandy soil. Giovanni Battara (1714–89) was an Italian mycologist.

STING WINKLE Alternative name for oyster drill; similarly, **American sting winkle** is a synonym for American oyster drill.

STINGING NETTLE Popular name for common nettle *Urtica dioica* ssp. *dioica*.

STINK BUG Alternative name for shieldbugs, from the pungent odour produced by the nymphs of some species.

STINKHORN Fungi in the genera *Phallus* and *Mutinus* (Phallales, Phallaceae). The cap is initially covered with a smelly substance that attracts insects, which then distribute spores via their feet. Specifically, *Phallus impudicus*, (poorly) edible, widespread but infrequent, if locally abundant, evident in June–October associated with rotted and often buried wood, more commonly coniferous, in parks and woodland. Greek *phallus* = 'penis', from the shape, and Latin *impudicus* = 'immodest'.

Dog stinkhorn *Mutinus caninus*, inedible, widespread but infrequent, if locally abundant, found in summer and autumn on soil among leaf litter in mixed woodland. The stems tend to collapse within an hour or so following extrusion from the soil. In Greek mythology Mutinus was another name for Priapus, and by extension 'penis', from the shape.

Latticed stinkhorn Synonym of red cage fungus.

Sand stinkhorn *P. hadriani*, inedible, scarce, local on marram dunes around the British coast. Hadrianus Junius (1512–75) was a Dutch botanist.

STINT Shorebirds in the genus *Calidris* (Scolopacidae), from Greek *calidris*, used by Aristotle for some grey-coloured waterside birds. **Long-toed stint** *C. subminuta* and **red-necked stint** *C. ruficollis* are accidentals. 'Stint' dates to Middle English, of unknown etymology.

Little stint *C. minuta*, wingspan 36 cm, length 13 cm, weight 24 g, a passage visitor migrating between Scandinavia/Russia and Africa, mostly juveniles in autumn, totalling around 460 birds, and fewer numbers of adults in spring. Diet is of tidal-margin and mudflat invertebrates. *Minuta* is Latin for 'small'.

Temminck's stint *C. temminckii*, mostly a passage visitor migrating between arctic Eurasia and southern Europe/Africa, seen in south-east, east and north-west England between May and June and late July to October, mainly on freshwater marsh, pools and lakes; they also visit estuaries. Diet is of insects, worms,

crustaceans and molluscs. The species has also bred at a few secret sites in Scotland, last confirmed in 1993; a pair was present at a breeding site in 2006 and 2007 but there was no evidence of success. Coenraad Temminck (1778–1858) was a Dutch zoologist.

STIPPLE-SCALE **Limy soil stipple-scale** *Placidium squamulosum* (Verrucariaceae), a squamulose lichen, widespread but local on calcareous soil, including dunes, wall tops and humus-filled rock crevices. Greek *plax* + *idion* = 'plate' (+ diminutive), and Latin for 'squamulose' (scaly).

STITCHWORT Species of *Stellaria* (Caryophyllaceae), from Latin for 'star-like', perennial herbs once thought to be a remedy for stitches in the side.

Bog stitchwort *S. alsine*, locally common and widespread, mat-forming, winter-green, in wet and often acid habitats including bog, wet grassland, riverbanks and ditches, and characteristic of cattle-poached areas. It has declined in southern and eastern England since the mid-twentieth century because of drainage and the reseeding or conversion to arable of wet grassland. *Alsine* is Greek for an undetermined plant, but probably chickweed (other species of *Stellaria*).

Greater stitchwort *S. holostea*, common and widespread, winter-green, in hedgerows, woodland margins, and verges. It tolerates a range of soils, but does best on moist, mildly acid and infertile ones, avoiding permanently wet and particularly freely drained substrates. Greek *holosteon* = 'entire bone' is facetiously applied to this 'weak' plant.

Lesser stitchwort *S. graminea*, common and widespread, on often dry neutral or acidic soil in woodland clearings, old pasture, hay meadow, grass-heath, hedge banks and verges. It tolerates moderate nutrient enrichment. *Graminea* is Latin for 'grass-like'.

Marsh stitchwort *S. palustris*, rhizomatous, widespread but scattered, absent from north Scotland and south-west Ireland, in base-rich wet habitats such as damp pasture, grassy fen and marsh, especially in places with winter standing water. *Palustris* is from Latin for 'marsh'.

Wood stitchwort *S. nemorum*, stoloniferous, in northern England, Wales and non-highland Scotland on fertile soils, mostly in damp, shaded sites, generally by streamsides and ditches and in wet woodland and damp hedge banks. *Nemorum* is from Latin for 'woodland glade'.

STOAT *Mustela erminea* (Mustelidae), male head-body length 275–312 mm, tail 95–140 mm, weight 200–445 g, females 242–92 mm, 95–140 mm, 140–280 g, common and widespread in a range of habitats. Numbers crashed in the mid-1950s following the impact of myxomatosis on rabbits, a major prey species (though other small mammals and birds are also taken), but populations recovered, and the pre-breeding population in Britain in 2005 was an estimated 462,000 (England 245,00, Scotland 180,000, Wales 37,000). Irish stoat *M. erminea* ssp. *hibernica* (sometimes called 'weasel' as the true weasel does not occur in Ireland) is a near-endemic subspecies, with >90% of the global population found in Ireland; it is also present in the Isle of Man. Females (including the current year's young) mate in early summer, but with delayed implantation do not give birth until the following spring. Actual gestation is four weeks. Litter size is 6–12 young (kits) which are weaned after 5 weeks and independent at 12 weeks. Some stoats (ermine) have a white winter coat. Stoats are legally protected in Ireland. 'Stoat' may have its roots in the Belgic and Dutch word *stout* = 'naughty', or Gothic *stautan* = 'to push'. *Mustela* is Latin for 'weasel', *erminea* New Latin for 'ermine' or 'stoat'.

STOCK Species of *Matthiola* (from Pierandrea Mattioli, 1501–77, an Italian physician and botanist) and *Malcolmia* (from the eighteenth-century English plantsman William Malcolm) (Brassicaceae).

Hoary stock *Matthiola incana*, a short-lived perennial, possibly native and certainly naturalised on sea-cliffs, shingle and other habitats by the sea in southern England, introduced elsewhere, occasionally inland where it is a garden escape on walls and banks. *Incana* is Latin for 'hairy'.

Night-scented stock *M. longipetala*, an annual neophyte from the Balkans introduced in the late eighteenth or the early nineteenth century, scattered as a casual on waste ground and in proximity to gardens. *Longipetala* is Latin for 'long petals'.

Sea stock *M. sinuata*, a biennial or short-lived perennial, local and declining in southern England and south Wales, on dunes and sea-cliffs. Most colonies are on young mobile dunes, probably spreading by seeds floating in seawater. *Sinuata* is Latin for 'wavy'.

Virginia stock *Malcolmia maritima*, an annual neophyte which, despite its common name, was introduced from the Balkans by 1713, a garden escape found as a casual on waste ground, rubbish tips and roadsides, mostly in England, scarce elsewhere.

STOMATOPODA From Greek *stoma* + *podos* = 'mouth' + 'foot', an order of marine crustaceans in the class Malacostraca (mantis shrimp).

STONECHAT *Saxicola rubicola* (Turdidae), wingspan 20 cm, length 12 cm, weight 15 g, a resident and migrant insectivorous breeder in south and west Britain and Ireland on heathland, in scrub and conifer woodland, and on grassy and heathy coastal sites, with 59,000 pairs (39,000–79,000) estimated in Britain in 2009. Numbers increased by 50% between 1970 and 2015 and by 29% between 1995 and 2015, declined after 2008, but recovered a little over 2012–15. It is also a winter visitor along Britain's east coast. The cup-shaped nest made of grass is built in scrub or bramble, with 5–6 eggs which hatch after around 15 days, the altricial chicks fledging in 14–15 days. 'Chat' is an onomatopoeic representation of the bird's call which sounds like two small stones being knocked together. Latin *saxum* + *colere* = 'rock' + 'to dwell', and *rubicola* = 'bramble-dweller'.

STONECROP Succulents in the genera *Sedum* (Latin for house-leek, from *sedere* = 'to sit') and *Crassula* (Latin *crassus* = 'thick' or 'fleshy') (Crassulaceae). See also orpine and roseroot.

Biting stonecrop *S. acre*, a common and widespread perennial of dry, undisturbed, open habitats on thin acidic or basic soils, for example shingle, dunes and cliffs. It is also common on walls, roofs, gravel tracks, pavements and road verges. *Acre* is Latin for having a bitter taste.

Butterfly stonecrop *S. spectabile*, a perennial introduction from the Far East with a scattered distribution as a garden escape or throw-out, and naturalised in a Wiltshire woodland. *Spectabile* is Latin for 'spectacular'.

English stonecrop *S. anglicum*, a creeping perennial, widespread, more abundant in western Britain and in Ireland, associated with rocks, dunes, shingle, and dry grassland. It also grows on old walls, rocky hedge banks, quarries and mine spoil. Latin *anglicum* = 'English'.

Hairy stonecrop *S. villosum*, biennial or perennial, locally abundant in northern England and Scotland, on base-enriched, stony ground, by streamsides in hilly areas, and in montane, often bryophyte-rich flushes. *Villosum* is Latin for 'hairy'.

Mossy stonecrop *C. tillaea*, an annual found in southern England and East Anglia. Since the mid-twentieth century it has also become naturalised in south-west England (in Cornwall, not recorded until 1988, it has since been found in car- and caravan parks), and it has colonised forestry tracks in north-east Scotland. It grows on bare, often compacted, sandy or gravelly soil kept open by disturbance. Michaelangelo Tilli (1655–1740) was an Italian physician and botanist.

Pale stonecrop *S. nicaeense*, an evergreen, mat-forming perennial introduced from the Mediterranean in 1769, naturalised on dry sunny banks and roadsides in Kent and a few other locations in England. *Nicaeense* is Latin for 'from Nice', southern France.

Reflexed stonecrop *S. rupestre*, a perennial from Europe

cultivated for its salad leaves by the seventeenth century, widespread and locally common, naturalised on old walls, rock outcrops, roadsides and waste ground. *Rupestre* is Latin for 'growing among rocks'.

Rock stonecrop *S. forsterianum*, a mat-forming perennial of open, dry, well-drained habitats, including rocks, screes and wooded cliffs. Naturalised populations are particularly found in south-west England and Wales in churchyards, and on walls, waste ground, mine waste and railway land.

Spanish stonecrop *S. hispanicum*, an annual or short-lived perennial introduced from south-east Europe by 1732, scarce and scattered, naturalised on walls, gravel-pits and stony waste ground. *Hispanicum* is Latin for 'Spanish'.

Tasteless stonecrop *S. sexangulare*, a perennial introduced from Europe, scattered in England, Wales and Ireland as a garden escape, naturalised on walls, banks, cliffs and other rocky habitats, and in churchyards. *Sexangulare* is Latin for 'six-angled'.

Thick-leaved stonecrop *S. dasyphyllum*, a perennial introduced from Europe, scattered in England, Wales and Ireland on old walls and limestone rocks, and in quarries and churchyards. *Dasyphyllum* is Greek for 'hairy-leaved'.

White stonecrop *S. album*, a locally common and widespread creeping perennial archaeophyte that may occur as a native on limestone rocks at the eastern end of the Mendips and in south Devon, growing on open, dry sites such as walls, roofs, shingle and footpaths. Latin *album* = 'white'.

STONEFLY Insects in the order Plecoptera, with 34 species in 7 families. Nymphs resemble non-winged adults, with two 'tails', and flattened bodies adapted to clinging to stones in running water, the preferred habitat of most species. Adults live for two to three weeks, crawling among waterside stones (prompting their common name). Mouthparts have weak biting structures, and adults of many species do not feed.

Northern February red stonefly *Brachyptera putata* (Taeniopterygidae), endemic and scarce, found in Scotland, particularly the north-east and Highlands, favouring highly oxygenated rivers with a shallow or moderate gradient on open heaths or upland pastures. Outside of Scotland, it is now probably extinct. Larvae are generally found during winter among loose stones and cobbles, usually below riffles where water flow is moderate, feeding on filamentous algae. Adults emerge in February–April. Greek *brachys* + *pteron* = 'short' + 'wing', and Latin *puta* = 'clean'.

Rare medium stonefly *Isogenus nubecula* (Perlodidae), internationally Critically Endangered, recently recorded only from the River Dee, a few specimens having been found in 2017 for the first time in 22 years. Larvae favour fast-flowing streams, 25–30 cm in depth, hunting mayfly and midge larvae above unstable cobbles and gravel. Adults are on the wing in March–April, drinking but not feeding. Greek *isos* + *genos* = 'similar' + 'type', and Latin *nubecula* = 'small cloud'.

Winter stonefly Members of the genus *Capnia*.

STONEWORT Freshwater multicellular green algae in the order Charales, family Characeae, the thallus generally attached to soft substrates, using chlorophyll to photosynthesise. 'Stonewort' comes from plants becoming encrusted in calcium carbonate. Genera include *Chara* (Greek *charis* = 'grace'), *Nitella* (Latin for 'splendour'), and *Lamprothamnium* (Greek *lampros* + *thamnion* = 'shiny' + 'small shrub').

Rarities include **coral stonewort** *C. tomentosa* (in ten or so low-nutrient lakes on limestone in central and western Ireland), **least stonewort** *N. confervacea* (in nutrient-poor lakes to a depth of 4 m in west and north Scotland, and north-west Ireland), **opposite stonewort** *C. contraria* (scattered and widespread), and **rugged stonewort** *C. rudis* (central and western Ireland in limestone lakes).

Bearded stonewort *C. canescens*, thallus up to 30 cm long, in mildly brackish, shallow sandy alkaline lakes and lagoons, recorded from East Anglia, Cambridgeshire, the Outer Hebrides, and south and west Ireland. Latin *canescens* = 'grey'.

Common stonewort *C. vulgaris*, thallus up to 50 cm long, common and widespread, especially in southern and eastern England and in Ireland in habitats from puddles and ditches to lakes, favouring coastal and calcareous sites in Wales and Scotland. Latin *vulgaris* = 'common'.

Dark stonewort *N. opaca*, thallus up to 10 cm long, widespread and locally common especially in upland areas, in a range of habitats from moderately fast-flowing streams to pools (including those drying out seasonally), favouring oligo- and mesotrophic acid waters. *Opaca* is Latin for 'shaded'.

Delicate stonewort *C. virgata*, thallus up to 50 cm long, common and widespread in a range of freshwater habitats, favouring acidic water, but also found in nutrient-poor alkaline sites. *Virgata* is Latin for 'twiggy'.

Foxtail stonewort *L. papulosum*, up to 40 cm long, in fairly still, brackish, nutrient-poor water <2 m deep on sand, gravel, small pebbles and silty substrates in The Fleet (Dorset), Fort Gilkicker Moat (Hampshire), Bembridge (Isle of Wight), Great Deep (West Sussex), and in nine saline lagoons in the Outer Hebrides, and first recorded in Ireland in 1973. Vulnerable in the RDB. *Papulosum* is Latin for 'pimpled'.

Hedgehog stonewort *C. aculeolata*, thallus up to 50 cm long, frequent if less so in central Ireland, rare in England and Wales, in low-nutrient pools and lakes <1 m deep, often associated with fen peat. *Aculeolata* is Latin for 'with small prickles'.

Pointed stonewort *N. mucronata*, thallus up to 40 cm long. The species itself is rare in moderately nutrient-enriched canals and lakes, but var. *gracillima* is increasing in canals and ornamental ponds. Latin *mucronata* = 'pointed'.

Slender stonewort *N. gracilis*, thallus up to 20 cm long, recorded from Sutherland, Ayrshire, parts of Wales, and Co. Wicklow, in small nutrient-poor, shallow, peaty lakes. Latin *gracilis* = 'slender'.

Smooth stonewort *N. flexilis*, thallus up to 10 cm long, found in southern and eastern England and central Ireland, widespread but scarce in Scotland, in moderately nutrient-rich pools and lakes. Latin *flexilis* = 'flexible'.

Strawberry stonewort *C. fragifera*, locally common in the northern third of Ireland, and recorded from a few sites in the Lizard Peninsula, Cornwall. *Fragifera* is Latin for 'strawberry-bearing'.

STORED NUT MOTH *Paralipsa gularis* (Pyralidae, Gallerinae), wingspan 21–32 mm, a south-east Asian moth naturalised following accidental introduction in dried fruit in warehouses in London since about 1921. Larvae feed on stored seeds and nuts, and mixed dried food, in places causing sufficient damage to become a pest. Greek *para* = 'near' + the genus *Alipsa*, and Latin *gula* = 'throat', referring to the larva's voracious appetite.

STORK Large, long-legged wading birds in the genus *Ciconia* (Latin for 'stork'). Both **black stork** *C. nigra* and **white stork** *C. ciconia* are occasional visitors.

STORK'S-BILL Annuals in the genus *Erodium* (Geraniaceae), from Greek *erodios* = 'heron', referring to the long pointed 'beak' shape of the fruit capsule, which also gives the English name.

Many species are neophyte wool aliens, introduced with shoddy, scattered in fields as casuals (e.g. **eastern stork's-bill** *E. crinitum* and **western stork's-bill** *E. cygnorum*, both from Australia, and **Mediterranean stork's-bill** *E. botrys*, **soft stork's-bill** *E. malacoides* and **three-lobed stork's-bill** *E. chium*, all Mediterranean).

Common stork's-bill *E. cicutarium*, widespread and locally common in well-drained sandy and rocky places, dunes, dry grasslands and heathland, on roadsides, stone walls and railway ballast. *Cicutarium* comes from a resemblance of the leaves to the genus *Cicuta*.

Musk stork's-bill *E. moschatum*, an archaeophyte, widespread

on dunes, roadsides, walls, field margins and waste ground. In the Scilly Isles it is a common weed of bulb fields. *Moschatum* is Latin for 'musk-scented'.

Sea stork's-bill *E. maritimum*, found around the coast from south-west Scotland to Kent, and in south and east Ireland, on trampled and closely grazed clifftop grassland, disturbed dunes and sea-cliffs, and around seaside settlements on walls and pavements. Inland, it has been found in limestone grassland (Somerset), in heathland by sandy tracks and gravel workings (with increasing numbers in Dorset), and occasionally as an introduction on railway ballast. *Maritimum* is Latin for 'maritime'.

Sticky stork's-bill *E. lebelii*, on bare ground in stabilised dunes, scattered round the coast, mainly in the west in Britain, and the east in Ireland.

STRAGGLE WEED Spiny straggle weed *Gelidium spinosum* (Gelidiaceae), a red seaweed with fronds 20–60 mm long, widespread if less so on the east coastline, in pools in the lower intertidal and subtidal. Latin *gelidus* = 'stiff', and *spinosum* = 'spiny'.

STRANGLER Inedible fungi in the genus *Squamanita* (Tricholomataceae), from Latin *squama* = 'scale', describing the 'squamules' on the cap.

Powdercap strangler *S. paradoxa*, Near Threatened in the RDL (2006), with a few records in England and south Wales, found in summer and autumn in grassland parasitic on earthy powdercap. *Paradoxa* reflects the initially puzzling taxonomic status.

Strathy strangler *S. pearsonii*, Vulnerable in the RDL (2006), found in grassland at two sites in the Highlands, and one near Bangor, Gwynedd, discovered in 2007. Strathy is a settlement in Sutherland. Arthur Pearson (1874–1954) was an English mycologist.

STRAPWORT *Corrigiola litoralis* (Caryophyllaceae), an annual confined as a native to periodically flooded muddy shingle around the margins of Slapton Ley, south Devon, but casuals have occasionally been recorded on railway ballast and waste ground in northern England.

STRATIOMYIDAE Family name for soldier flies (Greek *stratiotys* + *myia* = 'soldier' + 'fly'), with 48 species in 16 genera. Adults are 3–20 mm in length. Larvae may be terrestrial or aquatic, and can be scavengers (mainly), predators or algal feeders. Adults feed on nectar or not at all. See centurion, colonel, gem (fly), general, legionnaire, major (fly), snout (fly) and soldier (fly).

STRAW (MOTH) Barred straw *Gandaritis pyraliata* (Geometridae, Larentiinae), wingspan 28–33 mm, nocturnal, widespread and common in gardens, hedgerows, verges, woodland edges and scrubby grassland. Adults fly in June–July. The unique resting posture has the forewings extended to completely cover the hindwings. Larvae feed on bedstraws and cleavers. *Pyraliata* means resembling the genus *Pyralis*.

Bordered straw *Heliothis peltigera* (Noctuidae, Heliothinae), wingspan 34–42 mm, generally migrant (occasionally found year-round) mostly seen in June–August, sometimes arriving from North Africa in large numbers in south coastal England, larvae feeding on garden plants such as marigolds, and also restharrow, ploughman's-spikenard, mayweed, and sticky groundsel. Greek *heliotes* = 'of the sun', from its daytime flight behaviour, and Latin *pelta* + *gero* = 'small crescent-shaped shield' + 'to carry', from the wing marking.

STRAWBERRY Stoloniferous perennials in the genus *Fragaria* (Latin for 'strawberry') and also barren strawberry (*Potentilla*, from Latin for 'little powerful one', from the medicinal properties of some species), all Rosaceae.

Alpine strawberry Variant of wild strawberry.

Barren strawberry *P. sterilis*, common and widespread in relatively infertile, dry soils in open woodland, woodland edges, scrub, grassy hedge banks and rock crevices, also occasionally in meadows and on walls. In the lowlands it is usually found in partially shaded sites but it extends into open habitats in upland areas.

Garden strawberry *F. ananassa*, of garden origin much cultivated for its fruit, widespread, scattered and naturalised in waste ground and field margins. The specific name alludes to the fruit smelling like pineapple (*Ananas*).

Hautbois strawberry *F. moschata*, a European neophyte in cultivation by 1629, naturalised in woodland, scrub and hedgerows, widespread but now rare. *Moschata* is Latin for 'musk-scented'.

Wild strawberry *F. vesca*, common and widespread on dry soils in woodland and scrub, and on hedge and railway banks, roadsides, and basic rock outcrops and screes in upland areas. It colonises open ground in quarries and chalk-pits, and grows on walls. *Vesca* is Latin for 'thin' or 'meagre'.

STRAWBERRY-BLITE *Chenopodium capitatum* (Amaranthaceae), a scarce and declining annual introduced with wool shoddy, birdseed and esparto, probably from North America via Europe, naturalised on cultivated land and as a casual on waste ground in a few places in England and Ireland. Greek *chen* + *podion* = 'goose' + 'little foot', from the leaf shape, and Latin *capitatum* = 'having a head'.

STRAWBERRY-TREE *Arbutus unedo* (Ericaceae), a small tree native in Co. Kerry, Co. Cork and Co. Sligo in heathy scrub and young woodland on rocky slopes and lake shores, generally on shallow soil or rooted into rock. It also occurs as an escape from cultivation, in places invasive. It was formerly more abundant in Ireland, but by the sixteenth century became extinct or rare except in Co. Cork and Co. Kerry, probably because of its use for charcoal. It is naturalised, often on calcareous slopes, in eastern Ireland, England and Wales. *Arbutus* is Latin for this species; *unedo* is Latin for 'I eat (only) one', referring to the fruit's marginally palatable qualities.

STREAK (MOTH) 1) *Chesias legatella* (Geometridae, Larentiinae), wingspan 30–5 mm, nocturnal, common on heathland, moorland, open woodland, hedgerows and gardens throughout much of Britain, confined to the north in Ireland, but a species of conservation concern in the UK BAP. Adults fly in September–October. Larvae feed on broom. Chesias is another name for the goddess Diana. *Legatus* is Latin for 'ambassador', who would wear a stripe on his toga, echoing the forewing stripe or streak.

2) Members of the family Oecophoridae.

Light streak *Pleurota bicostella*, wingspan 23–5 mm, widespread but local on moorland and heathland, particularly in Ireland. Adults fly in June–July at dawn and just before dusk. Larvae feed on bell heather and cross-leaved heath inside a silk web. Greek *pleura* = 'side', and Latin *bi* + *costa* = 'twice' + 'costa' (anterior wing margin), from the wing streaks.

Ruddy streak *Tachystola acroxantha*, wingspan 13–15 mm, introduced from Australia, first noted in Devon in 1908, naturalised and increasing its range throughout much of England, north to Lancashire, locally common. Adults fly at night from May to October. Larvae feed on leaf litter. Greek *tachys* + *stola* = 'swift' + 'a white band worn by priests', and *acron* + *xanthos* = 'tip' + 'yellow', from the wing markings.

Southern streak *P. aristella*, wingspan 20–2 mm, on dry grassland and hillsides on Jersey. Adults fly in July–August. Larvae feed on low-growing herbaceous plants. *Aristos* is Greek for 'best'.

STREAK-MOSS Members of the genus *Rhabdoweisia* (Rhabdoweisiaceae), from Greek for 'fluted column' (*rhabdos* = 'rod'), describing the striate capsules, found in sheltered acidic montane habitats including woodland and ravines.

STREAMER *Anticlea derivata* (Geometridae, Larentiinae), wingspan 30–4 mm, a nocturnal moth widespread and common in open woodland, hedgerows, gardens and scrub. Adults fly in April–May. Larvae feed on leaves and flowers of wild roses, especially dog rose. Anticlea was the name of Ulysses' mother.

Derivo is Latin for diverting a stream (*rivus*), from rivulet-like wing markings.

STREPSIPTERA Order of stylopids (Greek *strepsis* + *pteron* = 'a twisting' + 'wing'), with 18 British species, minute insects whose early stages are usually endoparasites of other insects. Adult males are free-living, with front wings reduced to tiny club-like structures, and are mouthless. The grub-like females generally remain inside the host.

STRIGIDAE Family name for some owls (Greek *strix* = 'owl').

STRIGIFORMES Order that includes owls and nightjars (Greek *strix* = 'owl').

STRIPE (MOTH) **Shoulder stripe** *Earophila badiata* (Geometridae, Larentiinae), wingspan 25–30 mm, nocturnal, widespread and common in woodland, hedgerows, scrub and gardens. Adults fly in March–April. Larvae feed on wild roses, especially dog rose. Greek *earos* + *philos* = 'of spring' + 'loving', and Latin *badiata* = 'chestnut-coloured'.

STRIPED (MOTH) **Crescent striped** *Apamea oblonga* (Noctuidae, Xyleninae), wingspan 42–50 mm, scarce, on saltmarsh, mudflats, estuaries and brackish marshes along the coasts of England and Wales, and in Co. Louth and Co. Wexford. Adults fly at night in June–August. Larvae feed on roots and lower stems of saltmarsh-grass and meadow-grass. Apamea was a town in Asia Minor. *Oblonga* (Latin for oblong) refers to the black band on the forewing.

Oblique striped *Phibalapteryx virgata* (Geometridae, Larentiinae), wingspan 22–5 mm, scarce, local on sandy heathland in Breckland, chalk downland in southern England, and dunes at various sites along the coasts of England and Wales. The bivoltine adults fly at night in May–June and August. Larvae mainly feed on lady's bedstraw. Greek *phibalos* + *pteryx* = 'slender' + 'wing', and Latin *virgata* = 'twig', referencing a stripe.

STRIPPER **Birch bark stripper** *Xenotypa aterrima* (Diaporthales, Gnomoniaceae), an encrusting fungus, Vulnerable in the RDL (2006), with a few records from Surrey and Co. Londonderry, on dead branches of birch. Greek *xenos* + *type* = 'stranger' + 'type', and Latin *aterrima* = 'very black'.

STROPHARIACEAE Family name (Greek *strophos* = 'twisted') of fungi in the genera *Leratiomyces*, *Naucoria* and *Stropharia*; see aldercap and roundhead.

STUBBLE-MOSS Species in the genera *Gyroweisia* and *Weissia* (Pottiaceae).

Slender stubble-moss *G. tenuis*, common and widespread on damp shaded stone or brickwork, usually calcareous or enriched by mortar or percolating moisture.

STURNIDAE Family name of starling (Latin *sturnus*).

STYELIDAE Family of sea squirts in the class Ascidiacea.

STYLOPID From Greek *stylos* = 'pillar', a member of the insect order Strepsiptera.

SUCCINEIDAE Family name (Latin *succinum* = 'amber') for the terrestrial pulmonate amber snails. See also sandbowl snail under snail.

SUCKER **Stone sucker** Alternative name for lamprey, reflecting its ability to use its mouth to move stones on the riverbed to create a spawning depression (redd).

SUCKERS **Octopus suckers** *Collema fasciculare* (Collemataceae), a rare jelly lichen, swelling when wet, found in moist, shady habitats on elm and ash in three wet woodlands in Wales; at a number of sites in the Highlands, and two records from hazel woodland in coastal south-west Scotland; and in Ireland among moss on trunks of mature trees, usually ash and hazel, in western oceanic woodlands. Greek *collema* = 'that which is glued', and Latin *fasciculus* = 'bundle'.

SUIDAE Pig (Latin *sus*) family; see wild boar.

SULPHUR BEETLE *Cteniopus sulphureus* (Tenebrionidae), 7–8 mm length, commonly with a coastal distribution in England and Wales, and inland in the east Midlands and East Anglia, habitat including dune and heathland. Adults live in flowers, especially members of the Apiaceae, feeding on pollen and nectar. It belongs to the 'comb-clawed beetles', and the binomial is from Greek *ctenos* + *pous* = 'comb' + 'foot', and Latin for 'sulphurous', from the yellow colour.

SULPHUR-TRESSES Alpine sulphur tresses Alternative name for the lichen green witch's hair.

SUMACH Deciduous shrubs or small trees in the genus *Rhus* (Greek for 'sumac'), family Anacardiaceae. 'Sumach' comes from Arabic *summaq*.

Shining sumach *R. copallina*, a North American neophyte naturalised in a few places on roadsides, mainly by suckering, for instance in west Kent since 2006. *Copallina* is Latin for 'resinous'.

Stag's-horn sumach *R. typhina*, a North American neophyte, introduced by 1629, mostly found in England and Wales, local elsewhere, as a garden escape or throw-out on roadsides, railway banks and waste ground. It is naturalised in places, reproducing by suckers. *Typhina* is Latin for 'resembling *Typha*' (bulrush).

Tanners' sumach *R. coriaria*, a European neophyte naturalised in a few places on rough grassland, mainly by suckering. *Corium* is Latin for 'leather' (i.e. used in tanning).

SUMMER-CYPRESS *Bassia scoparia* (Amaranthaceae), a bushy annual Asian neophyte, in cultivation by 1629, recorded from the wild in 1866 (Surrey), scattered in England (increasing, on roadsides around the Humber estuary), scarce elsewhere, and first recorded in Ireland (Co. Limerick) in 1988. It is found in hedgerows and on roadsides and waste ground, especially around ports, usually casual, originating with grain, oil-seed and wool shoddy. Ferdinando Bassi (1710–74) was an Italian botanist. *Scoparia* is Latin for 'like a broom'.

SUMMER-MOSS *Anoectangium aestivum* (Pottiaceae), locally common in mountain regions in damp, slightly calcareous crevices and overhangs on siliceous rock faces. Greek *anoixis* + *angeion* = 'an opening' + 'vessel', and Latin *aestivum* = 'summery'.

SUN STAR See common and purple sun star under starfish.

SUNBLEAK *Leucaspius delineatus* (Cyprinidae), length 4–6 cm, a continental fish accidentally introduced to, and escaped from, a fish farm near Romsey, Hampshire in the late 1980s, and entering a part of the River Test catchment. It has now also been found in the Somerset Levels. Greek *leucaspius* = 'white aspius' and Latin *delineatus* = 'without lateral line'.

SUNBURST LICHEN *Lichenomphalia alpina* (Hygrophoraceae), locally common, recorded from upland parts of Wales, Cumbria and the Highlands, on peat in heathland and moorland, and in acid unimproved grassland. In 2010, *The Guardian* launched an annual competition for the public to suggest common names for previously unnamed species, and this was a winner in 2011 prompted by its bright yellow colour. *Omphalos* is Greek for 'navel'.

SUNDEW Insectivorous, rosette-forming perennial herbs in the genus *Drosera* (Droceraceae) (Greek *droseros* = 'dewy'), associated with damp, nutrient-poor habitats. The leaves are covered with sticky hairs that trap insects, digested to supplement otherwise poor mineral nutrition.

Great sundew *D. anglica*, growing in the wetter parts of raised and blanket bogs, in wet valley bogs, on stony lake shores and, rarely, in calcareous mires. Scattered, but mostly found in the Highlands, Ireland and the New Forest, but it has been declining in Ireland and England since the nineteenth century due to drainage, eutrophication and peat extraction. Latin *anglica* = 'English'.

Oblong-leaved sundew *D. intermedia*, locally common in mostly western and southern parts of the British Isles, its presence decreasing because of drainage, afforestation, peat extraction and loss of habitat. It grows in wet heathland, valley and raised bogs, and on the edge of oligotrophic lochs, usually on wet acidic peat. It is only found in sphagnum when the moss forms a fringe around bog pools.

Round-leaved sundew *D. rotundifolia*, locally common and widespread, except in southern-central and eastern England, on damp acid heathland and moorland, bogs and upland flushes, growing in sphagnum moss or on bare peat. This is the county flower of Shropshire as voted by the public in a 2002 survey by Plantlife. *Rotundifolia* is Latin for 'round-leaved'.

SUNFLOWER *Helianthus annuus* (Asteraceae), a tall annual North American neophyte cultivated by Tudor times, widely grown in gardens and on allotments, sometimes cultivated on a field scale for its oil-bearing seeds or for game-bird cover. It also derives from birdseed, and from wool shoddy. It is widespread as a casual, sometimes persisting, on waste ground and roadsides. Greek *helios* + *anthos* = 'sun' + 'flower', and Latin *annuus* = 'annual'.

Other occasionally found casual garden escapes or species derived from waste products are **lesser sunflower** *H. petiolaris*, **perennial sunflower** *H.* x *laetiflorus* and **thin-leaved sunflower** *H.* x *multiflorus*.

SUNFLY Alternative name for the footballer *Helophilus pendulus*, a hoverfly that shows a preference for bright sunny days, but this common name may derive from a misreading of its genus *Helo-* (*helos* being Greek for marsh) as *Helio-* (from *helios*, Greek for sun) plus *philos* = 'loving'.

SUNSET SHELL Faroe sunset shell *Gari fervensis* (Psammobiidae), a filter-feeding bivalve, shell size up to 5 cm, common and widespread, burrowing in fine silty sand from the low intertidal down to 100 m. *Fervens* is Latin for 'glowing'.

SUSPECTED, THE *Parastichtis suspecta* (Noctuidae, Xyleninae), wingspan 29–34 mm, a nocturnal moth local in fen, damp woodland, scrubby heathland and moorland throughout Britain, though rare in Wales. Adults fly in July–August. Larvae feed on birch and sallow. Greek *para* + *stictos* = 'beside' + 'spotted', from forewing markings, and Latin *suspecta* = 'mistrusted' (i.e. whether this was a distinct species, it being rather variable in its features).

SWALLOW *Hirundo rustica* (Hirundinidae), wingspan 34 cm, length 18 cm, weight 19 g, common and widespread from March to October, wintering in sub-Sahara, with 860,000 pairs estimated in 2009. Numbers increased by 26% over 1995–2015, but declined by 13% in 2010–15, though these trends hide fluctuations and show strong regional variation. It was moved from the Amber to the Green list in Britain in 2015, but remains Amber-listed in Ireland. It constructs a bowl-shaped nest out of mud in barns and other buildings. Clutch size is 4–5, incubation lasts 17–19 days, and the altricial chicks fledge in 20–2 days. Feeding is on flying insects. 'Swallow' is from Old English *swealwe*. It is also known as **barn swallow**. The binomial comes from Latin for 'swallow' and 'rural'.

Red-rumped swallow *Cecropis daurica* is a scarce visitor; **cliff swallow** *Petrochelidon pyrrhonota* and **tree swallow** *Tachycineta bicolor* are vagrants.

Sea swallow Alternative name for Arctic tern, so called because of the long tail streamers and general shape that echo the morphology of the swallow.

SWALLOW-TAILED MOTH *Ourapteryx sambucaria* (Geometridae, Ennominae), wingspan 40–50 mm, nocturnal, widespread and common in woodland, scrub, hedgerows, parks and gardens, except for northern Scotland. Adults fly in July. Larvae have a preference for ivy but do feed on other woody plants, including elder (*Sambucus*, giving the specific name). Greek *oura* + *pteryx* = 'tail' + 'wing', from the tail-like projection on the hindwing.

SWALLOWTAIL *Papilio machaon* (Papilionidae), Britain's

largest butterfly (male wingspan 76–83 mm, female 86–93 mm) and one of its rarest. The subspecies *britannicus* is confined to the fens of the Norfolk Broads, partly due to the distribution of the sole larval food plant, milk-parsley. Recent (2014) breeding of subspecies *gorganus* may have taken place at grassland sites in Kent, Sussex and Hampshire, larvae feeding on a variety of umbellifer species. Although starting from a low base, numbers of residential swallowtails increased by 30% between 2005 and 2014. Early instars resemble bird droppings as a form of mimicry (mimesis). Adults, flying in May–August (peaking in June) take nectar from a variety of (often pink or mauve) flowers, for example thistles and ragged-robin. Latin *papilio* = 'butterfly'. Machaon was a doctor serving the Greeks in the Trojan War.

SWAN Large, long-necked birds in the family Anatidae, genus *Cygnus* (Latin for 'swan'). 'Swan' dates back to Old English.

Black-necked swan *C. melancorypha* and **trumpeter swan** *C. buccinator* are occasional non-persistent escapes.

Bewick's swan *C. columbianus* ssp. *bewickii*, adult male wingspan 47–55 cm (average 52 cm), average female wingspan 50.4 cm; length 1.2–1.4 m; male (cob) weight 3.4–7.8 kg (6.4 kg average), females (pens) 5.7 kg. Arriving from Siberia in mid-October, around 7,000 birds (2005 data) overwinter until March, mainly in eastern England (especially in the Ouse and Nene Washes, Cambridgeshire), around the Severn estuary and in Lancashire, with records also from south-west England, south Wales and Yorkshire, feeding in fields on stubble, grain and root crops. Cygnets stay with their parents all winter and the family group is sometimes joined by offspring of previous years: family groups of up to 15 have been noted. The species is named after the engraver Thomas Bewick (1753–1828). *Columbianus* is New Latin for coming from the Columbia River, the type locality in the Pacific Northwest.

Black swan *C. atratus*, adult wingspan 1.6–2 m, length 1.1–1.4 m, weight 3.7–9 kg, introduced from Australia around 1791, recorded breeding in the wild in Surrey in 1851, and nesting from time to time on the Thames near London since the early twentieth century. By the mid-1980s the species had spread to a number of locations in the Home Counties, and since the 1990s has been recorded scattered throughout the UK, though not always breeding or persisting. From the end of the 1990s there has been a particularly well-established, self-sustaining breeding population in the Norfolk Broads. The monogamous pair may nest singly or in colonies, building a nest of reeds and grasses close to the water's edge, often on a small island. Clutch size is 4–8. Breeding is often early in the year and mortality of cygnets is high. *Atratus* is Latin for 'clothed in black'.

Mute swan *C. olor*, adult wingspan 2–2.4 m; length 1.2–1.7 m; male (cob) weight 11–12 kg (with great bustard, the heaviest flying bird), females (pens) 8.5–9 kg, a common and widespread resident, though absent from most upland regions, found in shallow water bodies and slow-flowing rivers, including in urban areas and parks. Food includes submerged and emergent water plants, land plants, insects, molluscs, small fish, frogs and worms. These long-necked monogamous birds mate for life, nesting on large mounds built with waterside vegetation in shallow water on islands in the middle or at the edge of a water body. Usually territorial, a few colonies have >100 pairs, notably at Abbotsbury Swannery, Dorset, and a flock of non-breeding birds has been reported on the Tweed estuary at Berwick-upon-Tweed, with a maximum count of 787 birds. The UK population, which had been fairly stable since the 1960s, increased by 86% from 1975 to 2015, though with a 'plateau' after around 2000. In 2009, breeding numbers were estimated as 6,400 pairs (5,800–7,000 pairs), with 79,000 birds overwintering, the increase attributed to the banning of lead weights for fishing: these had commonly been ingested, leading to death. Milder winters have also increased overwinter survival, with knock-on effects on breeding success. Females lay up to 7 eggs (average 4) and brood for 35–41 days. The

precocial young (cygnets) cannot fly before an age of 120–50 days. Having been on the Green List during 2009–15 the species is once again Amber-listed because of the international importance of its UK wintering population. The British monarch has the right to ownership of all unmarked mute swans in open water, but Queen Elizabeth II has only exercised her ownership on parts of the Thames and surrounding tributaries. This ownership is shared with the Vintners' and Dyers' Companies, who were granted rights of ownership by the Crown in the fifteenth century. See also swan-upping in Part I. The name 'mute' comes from it being less vocal than other swan species, not because it is silent. Both *cygnus* and *olor* mean 'swan' in Latin.

Polish swan A young mute swan that is white instead of the usual grey.

Tundra swan *C. columbianus*, with two subspecies; see Bewick's swan, ssp. *bewickii*, above. The North American ssp. *columbianus* only has two accepted British records.

Whooper swan *C. cygnus*, adult wingspan 2–2.8 m; length 1.4–1.6 m; male (cob) weight 9.8–11 kg, females (pens) 8.2–9.2 kg. Essentially a winter visitor from Iceland found in Ireland, much of Scotland, northern England and parts of East Anglia, its winter population (15,000 birds) and small breeding numbers (up to five pairs have recently bred in Orkney) make it an Amber List species. Cygnets stay with their parents all winter, sometimes joined by offspring from previous years. They feed in fields during the day, eating crops and stubble, then roost at night on open water. The English name reflects the bird's 'whooping' call.

SWAN-NECK MOSS Species of *Campylopus* (Leucobryaceae) (Greek *campylos* + *pous* = 'curved' + 'foot').

Awl-leaved swan-neck moss *C. subulatus*, in gravelly or sandy habitats, including footpath edges, gravel parking sites and old tarmac. *Subulatus* is Latin for 'awl-shaped'.

Bristly swan-neck moss *C. atrovirens*, widespread and locally abundant in upland western Britain and Ireland in wet, peaty habitats and on acidic rocks, often forming large patches in wet montane heath, and on boggy hillsides. *Atrovirens* is Latin for 'black (i.e. very) green'.

Brittle swan-neck moss *C. fragilis*, widespread and locally common on well-drained, grassy or heathy slopes, coastal heath and fen carr. It is more tolerant of base-rich conditions than most congenerics and is found on limestone grassland, shell sand and calcareous sandstone rock faces. *Fragilis* is Latin for 'brittle'.

Compact swan-neck moss *C. brevipilus*, widespread but scattered as cushions on boggy, often undisturbed sites, on shallow peat in wet heathland and at the edge of pools. *Brevipilus* is Latin for 'short hair'.

Dwarf swan-neck moss *C. pyriformis*, common and widespread on bare, acidic peaty or sandy soils in woodland or moorland, often following disturbance on banks, ditches or burnt ground. *Pyriformis* is Latin for 'pear-shaped'.

Rusty swan-neck moss *C. flexuosus*, common and widespread in acidic habitats such as peat, rotting stumps and logs, wet acidic montane rocks, and on scree. *Flexuosus* is Latin for 'winding'.

Schimper's swan-neck moss *C. schimperi*, a rare montane species. Wilhelm Schimper (1808–80) was a French botanist.

Schwarz's swan-neck moss *C. gracilis*, in upland western Britain and Ireland usually in flushed sites on organic soil on rocky banks and ledges, and wet rock. *Gracilis* is Latin for 'slender'.

Shaw's swan-neck moss *C. shawii*, hyperoceanic, locally frequent in the Outer Hebrides, but with only a small number of sites in mainland north-west Scotland and south-west Ireland, in bog and on flushed heathy slopes and rock ledges.

Silky swan-neck moss *C. setifolius*, local on the west coasts of Wales, Cumbria and Scotland, and coastal Ireland, a hyperoceanic endemic to Britain and Ireland growing in humid sites, most abundant in leggy heather on north- and north-east-facing rocky slopes, and in turf at the base of wet crags and on rocky ledges in ravines. *Setifolius* is Latin for 'bristly leaved'.

Stiff swan-neck moss *C. pilifer*, scarce, mostly on the west coasts of Britain and Ireland in crevices in dry, usually unshaded rocks or on thin, stony soil overlying rocks. *Pilifer* is Latin for 'having hairs'.

SWARMING FLY Yellow swarming fly *Thaumatomyia notata* (Chloropidae), 3 mm length, widespread and scattered, found during spring and summer in flower meadows feeding on nectar, sometimes entering houses, especially roof voids, in large numbers to overwinter. Larvae prey on aphids in grass roots. Greek *thauma* + *myia* = 'a wonder' + 'fly', and Latin *notata* = 'marked'.

SWEDE See rape.

SWEEP (MOTH) Members of the family Psychidae (Psychinae), females being apterous. Rarities include **birch sweep** *Proutia betulina* (south-east and eastern England), **black sweep** *Pachythelia villosella* (heathland in Dorset and Hampshire), and **netted sweep** *Whittleia retiella* (saltmarsh from Hampshire to Suffolk).

Brown sweep *Sterrhopterix fusca*, male wingspan 18–25 mm, in a few scattered locations in England and Wales. Males, emerging in June–July, are attracted by scent to the wingless, grub-like females, which remain partly within the pupal case. Larvae feed on grasses, oak, hawthorn, birch and heather inside a portable silken case, overwintering twice. Greek *sterrhos* + *pteryx* = 'stiff' + 'wing', and Latin *fusca* = 'dusky'.

Common sweep *Psyche casta*, male wingspan 12–15 mm, common in the southern half of Britain and Ireland. Larvae feed on grasses, dead leaves, lichen and decaying vegetable matter throughout the year, particularly in May–June, inside a portable case often attached to leaves and tree-trunks. The female is wingless and grub-like. Greek *psyche* = 'soul' (by analogy, a butterfly or moth from its metamorphosis and 'resurrection'), and Latin *casta* = 'chaste', from parthenogenesis in some conspecifics.

Dusky sweep *Acanthopsyche atra*, male wingspan 16–22 mm, uncommon on heathland and moorland in Berkshire, Hampshire and Dorset, north Wales, the Pennines and the Lake District; and the Highlands. Adults emerge in May–June, the male flying in the afternoon and evening. Larvae live on or near the ground feeding on grasses, heather and sallow, inside a portable silken case covered with grass and/or heather, overwintering once or twice. Greek *acantha* = 'prickle' + the genus *Psyche*, from a spur on the foreleg, and Latin *atra* = 'black', for the colour of the head and body scales.

Round-winged sweep *Epichnopterix plumella*, male wingspan 10–14 mm, local in grassland in south-east England, especially around the East Anglian coast, and in a few moorland sites in Derbyshire, Lancashire and Wales. Males fly in sunshine in April–June. Larvae feed on grasses inside a portable case. Greek *epichnoos* + *pteryx* = 'wool-like covering' + 'wing', and Latin *pluma* = 'feather down', both describing the hair-like forewing scales.

SWEEPER (MOTH) See chimney sweeper.

SWEET-BRIAR *Rosa rubiginosa* (Rosaceae), a widespread deciduous shrub characteristic of scrub and hedgerows on chalk and limestone (commonest in south-east England), but also in quarries, on railway embankments and on waste ground. It is often frequent as a colonist of undergrazed chalk grassland. *Rubiginosa* is Latin for 'rusty'.

Small-flowered sweet-briar *R. micrantha*, a tall, climbing, deciduous shrub found in southern and central England, Wales and south Ireland in woodland, scrub and hedgerows, and also more open habitats such as chalk grassland, heathland, sea-cliffs, disused quarries and railway embankments. It grows on a range of well-drained soils, but is most frequent on calcareous substrates, and avoids acidic sites. *Micrantha* is Greek for 'small-flowered'.

Small-leaved sweet-briar *R. agrestis*, a deciduous shrub of open scrub on grassland overlying chalk or limestone, generally uncommon and scattered in England, Wales and Ireland. *Agrestis* is from Latin for 'field'.

SWEET CHESTNUT *Castanea sativa* (Fagaceae), an archaeophyte deciduous tree, probably introduced in Roman times from southern Europe, a major constituent of coppiced woodland in south-east England. It is also planted in hedgerows, woodland edge, parkland and amenity areas, and in large gardens, tolerating a range of soils, but thriving on moist, sandy ones. While widespread it is only naturalised in southern England. *Castanea* is Latin for 'chestnut', *sativa* for 'cultivated' (for its wood and edible seed or 'nuts').

SWEET-FLAG *Acorus calamus* (Acoraceae), a rhizomatous herbaceous perennial neophyte introduced during the sixteenth century, scattered but locally common only in parts of England, marginal in streams, canals, ponds and lakes in shallow, nutrient-rich calcareous water. Acorus is a name used by Theophrastus for a plant with an aromatic rhizome; *calamos* is Greek for 'reed'.

SWEET-GRASS Perennials in the genus *Glyceria* (Poaceae) (Greek *glycys* = 'sweet', for the sweet-tasting grains).

Floating sweet-grass *G. fluitans*, common and widespread in marsh, swamp and muddy pond margins, and forming floating rafts in shallow water by ditches, streams, ponds and lakes, tolerant of high levels of disturbance and nutrient enrichment. *Fluitans* is Latin for 'floating'.

Hybrid sweet-grass *Glyceria* x *pedicellata* (*G. fluitans* x *G. notata*), common and widespread, stoloniferous, in ponds, ditches, streams and swampy depressions in pastures, growing with one or both parents, or with neither. Although sterile it spreads vegetatively, and detached ramets are carried by water to new sites. *Pedicellata* is Latin for 'small pedicels'.

Plicate sweet-grass *G. notata*, stoloniferous, widespread except in north Scotland, and generally common, especially in England, in ditches, streams and muddy pond margins, occurring on more calcareous substrates than other sweet-grasses. *Notata* is Latin for 'marked'.

Reed sweet-grass *G. maxima*, rhizomatous, common in most of England (scattered in the north), in Wales, much of Ireland and Scotland (absent from the Southern Uplands and Highlands), in ditches, canals, lakes and ponds, rooted on the bank or in the water, and often forming floating rafts. It is also occasionally found in seasonally flooded grassland. *Maxima* is Latin for 'greatest'.

Small sweet-grass *G. declinata*, common and widespread in muddy habitats such as pond margins, cattle-trampled places and wet fields. *Declinata* is Latin for 'turned away'.

SWEET-WILLIAM *Dianthus barbatus* (Caryophyllaceae), a mat-forming biennial or short-lived perennial southern European neophyte herb, cultivated by the mid-Tudor period and scattered in Britain, usually as a casual, as a garden escape or throw-out on roadside banks, railways, dunes and waste ground. Greek *di* + *anthos* = 'two' (double) + 'flower', and Latin *barbatus* = 'bearded'.

SWIFT (BIRD) *Apus apus* (Apodidae), wingspan 45 cm, length 16 cm, weight 44 g, a common migrant evident between April and August, wintering in Africa, widespread especially in southern and eastern Britain. The British population was estimated at 87,00 (64,000–111,000) breeding pairs in 2009. Numbers declined by 51% between 1970 and 2015, by 47% between 1995 and 2015, and by 24% over 2010–15, and the species was Amber-listed in 2009. They live a continually aerial life, catching flying insects and airborne spiders for food and moisture; mating; and sleeping (with half their brain at a time) all on the wing, coming down only for a short period each year to breed. They use their very short legs primarily for clinging to vertical surfaces that include the nests made of airborne material mixed with saliva constructed in holes and under overhangs (including on buildings). Clutch size is 2–3, incubation lasts 19–25 days, and the altricial chicks fledge in 37–56 days. *Apus* is Latin for 'swift', in turn from Greek *a* + *pous* = 'without foot'.

Alpine swift *A. melba* is a rare visitor; **little swift** *A. affinis*,

Pacific swift *A. pacificus*, **pallid swift** *A. pallidus*, **chimney swift** *Chaetura pelagica* and **needle-tailed swift** *Hirundapus caudacutus* are all vagrants.

SWIFT (MOTH) Members of the family Hepialidae.

Common swift *Korscheltellus lupulina*, wingspan 25–40 mm, the commonest swift moth, distributed throughout Britain, though more scattered in Scotland. Adults, which do not feed, fly in May–June. Larvae live underground, feeding on plant (especially grass) roots. *Lupulina* refers to hop *Humulus lupulus*, a putative food plant.

Gold swift *Phymatopus hecta*, wingspan 22–33 mm, widespread and sometimes common in rough grassland, heathland and open woodland. Adults, which do not feed, fly at dusk in June–July. Larvae feed on the roots of bracken and some grasses. Greek *phyma* = 'swelling', and *hecticos* = 'hectic' or 'feverish', from the flight behaviour (echoing the family name, from *hepialos* = 'fever').

Map-winged swift *K. fusconebulosa*, wingspan 30–5 mm, local but widespread in Britain, less so in southern and eastern England, often associated with moorland, heathland, pasture and open woodland. Adults, which do not feed, fly at dusk in June–July. Larvae live underground, feeding on bracken roots. Latin *fuscus* + *nebulosa* = 'dark' + 'cloudy'.

Orange swift *Triodia sylvina*, wingspan 32–48 mm, common and widespread in England and Wales, scattered in Scotland, in moorland, woodland, gardens and waste ground. Population levels increased by 150% between 1968 and 2007. Adults mainly fly at dusk in July–September. Larvae feed on the roots of various plants, including bracken, dock and dandelion. *Sylvina* is Latin for 'belonging to woodland'.

SWIMMING CRAB Members of the family Polybiidae.

Arch-fronted swimming crab *Liocarcinus navigator*, carapace 3 cm long and broad, found on the south and west coasts of both Britain and Ireland, and at Flamborough Head, on mixed sandy substrates from the shallow sublittoral to >100 m depth. Greek *leios* + *carcinos* = 'smooth' + 'crab', and Latin *navigator* = 'navigator' or 'sailor'.

Marbled swimming crab *L. marmoreus*, carapace up to 3.5 cm long, widespread but scattered on fine sands and gravel on the lower shore and sublittoral. *Marmoreus* is Latin for 'marbled'.

Velvet swimming crab *Necora puber*, carapace up to 8 cm long, widespread though absent from eastern Britain except between the Firth of Forth and Northumberland, on stony and rock substrates intertidally and in shallow water. Greek *necora* = 'dead' and Latin *puber* = 'downy haired'.

Wrinkled swimming crab *L. corrugatus*, carapace up to 4.3 cm long and 4.1 cm wide, on coarse sand and gravel from the shallow sublittoral down to 100 m, scattered on the west coasts of Britain and Ireland, and common on Orkney. *Corrugatus* is Latin for 'wrinkled'.

SWINE-CRESS *Lepidium coronopus*, an annual, rarely biennial, archaeophyte, common and widespread in England as far north as Yorkshire, Wales and south-east Ireland, scarce and scattered elsewhere, on nutrient-rich, often compacted soils in open, dry or winter-wet habitats, for example farmyards, waste ground, paths and, especially, puddled gateways. Greek *lepis* = 'scale', referring to the shape of the seed pods, and *corona* + *pous* = 'crown' + 'foot', from the divided leaves.

Lesser swine-cress *L. didymum*, an annual or biennial neophyte, possibly native to South America, introduced via Europe by the early eighteenth century, common in urban and industrial areas, and spreading into rural areas where it is widespread but scattered. It is found in damp, often winter-wet soils, on cultivated and waste ground, and often in gardens and lawns, by paths and roadsides and on tips. *Didymos* is Greek for 'in pairs' or 'twofold'.

SWISS CHARD See foliage beet under beet.

SWORD-FERN **Western sword-fern** *Polystichum munitum*

(Dryopteridaceae), a North American garden escape with records from Surrey (since 1980) and Cornwall (2001). Greek *poly* + *stichos* = 'many' + 'rows', and Latin *munitum* = 'armed'.

SWORD-GRASS *Xylena exsoleta* (Noctuidae, Xyleninae), wingspan 58–68 mm, a nocturnal moth found on moorland, rough grassland and upland woodland, generally restricted to uplands in parts of Scotland and northern England, rare elsewhere, and a priority species in the UK BAP. Adults fly from September onwards, overwinter, and reappear in March–April. Larvae feed on a variety of herbaceous plants. Greek *xylon* = 'wood' and Latin *ex(s)oletus* = 'mature' or 'old', the adult being the colour of old, rotting wood.

Dark sword-grass *Agrotis ipsilon* (Noctuinae), wingspan 35–50 mm, a widespread migrant nocturnal moth, in some years arriving in sufficient numbers to breed in small numbers. Adults can arrive at any time from March to November, though most commonly from August to October. Larvae (black cutworm) feed on or below the ground and at night on herbaceous plants and their roots, becoming an increasingly reported nuisance, for example as a turf grass pest. The binomial is from Greek *agrotes* = 'field', and the letter *ypsilon* (Y), from the forewing marking.

Red sword-grass *X. vetusta*, wingspan 50–7 mm, widespread but local on moorland, upland grassland, boggy heathland and damp woodland. Adults fly at night in September–October, hibernate, and reappear in March–April. Larvae feed on a variety of herbaceous and woody plants. *Vetusta* is Latin for 'old', suggesting, with the genus name, the adult's colour of old, rotting wood.

SYCAMORE (MOTH) *Acronicta aceris* (Noctuidae, Acronictinae), wingspan 35–45 mm, local in parks, gardens, woodland and scrub throughout the southern half of England. Adults fly at night in June–August. Larvae feed on sycamore and field maple (both *acer*, giving the specific name), and horse-chestnut. Greek *acronyx* = 'nightfall'.

SYCAMORE (TREE) *Acer pseudoplatanus* (Sapindaceae), a very common, widespread and fully naturalised deciduous European neophyte found in Scotland by the sixteenth century and widely planted in Britain from the late eighteenth century onwards in plantations, woodland, parkland, large gardens and roadsides, prolifically self-sowing in a range of natural, semi-natural and man-made habitats, though avoiding particularly acidic and waterlogged soils. In upland areas it is often restricted to sites associated with habitation. The timber – hard and strong, pale cream, with a fine grain – is used for furniture, and also kitchenware as the wood does not taint or stain food. In Wales, sycamore was used in the traditional craft of making 'love spoons'. 'Sycamore' dates to Middle English, from Old French *sic(h)amor*, via Latin from Greek *sycomoros*, from *sycon* = 'fig' + *moron* = 'mulberry'. *Acer* is Latin for 'maple' but also means 'sharp', possibly a reference to the hardness of the wood which was used by the Romans for spears. *Pseudoplatanus* is Greek and Latin for 'false plane tree'.

SYNGNATHIDAE, SYNGNATHIFORMES Family and order (Greek *syn* + *gnathos* = 'together' or 'fused' + 'jaws') of pipefishes and seahorses.

SYRPHIDAE Family name (Greek *syrphos* = 'a small flying insect') for hoverflies and drone flies, with 282 species in 68 genera. Many adults have black and yellow stripes, mimics of bees and wasps. They are commonly seen hovering in front of food plants from which they take nectar and pollen. Larvae can be soil- or water-dwelling, and saprotrophic, eating decaying matter, or insectivorous, feeding for example on garden pests such as aphids and thrips.

T

T-HEADED WORM *Scalibregma inflatum* (Scalibregmatidae), a polychaete 5–10 cm long, widespread round Britain and north-east Ireland in the low intertidal and shallow sublittoral buried deep in sand or mud. Greek *scala* + *bregma* = 'ladder' + 'front of the head', and Latin *inflatum* = 'swollen'.

TABANIDAE Family name of horseflies, deerflies and clegs (Brachcera, Tabanoidea), with 30 species in 5 genera. *Tabanus* is Latin for 'gadfly'. Females feed on mammalian blood, including that of livestock (and humans), and possess serrated mandibles that rip or slice flesh.

TABBY Nocturnal moths in the family Pyralidae (Pyralinae). 'Tabby' dates to the late sixteenth century, denoting a kind of silk taffeta, originally striped and later with a watered finish; tabby moths generally have dark wavy patterns on the forewing.

Double-striped tabby *Hypsopygia glaucinalis*, wingspan 23–30 mm, common around sheds and farm buildings, in gardens, haystacks and thatched roofs throughout southern Britain, though numbers and distribution are declining. Adults fly in July–August. Larvae feed on dry plant material such as thatch and hay, and detritus in bird nests. Greek *hypsos* + *pygaios* = 'height' + 'of the rump', from the habit of flexing the abdomen upwards when at rest, and Latin *glaucus* = 'blue-grey', from the wing colour.

Large tabby *Aglossa pinguinalis*, wingspan 29–38 mm, widespread but declining and local around farm buildings, barns, warehouses and granaries. Adults fly in June–August. Larvae feed on detritus, straw, chaff and animal waste, for example sheep droppings. Greek *a* + *glossa* = 'without' + 'tongue', and Latin *pinguis* = 'fat', putatively a part of the larval diet.

Long-legged tabby *Synaphe punctalis*, wingspan 22–7 mm, local on shingle beaches, dunes, chalky ground and dry open areas in mostly coastal parts of southern England and East Anglia. Adults fly at night (males also sometimes in daylight) in June–August. Larvae feed on mosses. Greek *synaphe* = 'union', and Latin *punctalis* = 'spotted', from the forewing markings.

Rosy tabby *Endotricha flammealis*, wingspan 18–23 mm, common in woodland, heathland, gardens and waste ground throughout southern Britain. Adults fly in July–August. Larvae feed on bird's-foot-trefoil, and deciduous tree such as oak and sallow, later instars favouring leaf litter. Greek *endon* + *trichos* = 'within' + 'hair', from the morphology of the forewing base, and Latin *flammeus* = 'flame-coloured', from the wing colour.

Scarce tabby *Pyralis lienigialis*, wingspan 22–6 mm, found in small numbers (and probably declining) in farm buildings and inside houses in parts of the south Midlands. Adults fly in June–September. Larvae feed on dry plant material such as hay, barley and other stored cereals. *Pyralis* refers to an animal that can supposedly live in fire (Greek *pyr*). Mme Lienig (d.1855) was a Latvian entomologist.

Small tabby *A. caprealis*, wingspan 23–7 mm, in lowland England, associated with old farm buildings, barns and granaries, but declining in numbers and distribution. Adults fly in July–August. Larvae feed on decaying plant material, including hay. *Caprealis* is probably a misprint for *cuprealis* (Latin *cupreus* = 'coppery'), from the sprinkling of this colour on the wings.

TACHINIDAE Family name (Greek *tachys* = 'swift') for some flies (Bracycera, Oestroidea), with 266 species in 145 genera. Larvae are parasitoids or parasites, typically of caterpillars.

TADPOLE SHRIMP *Triops cancriformis* (Triopsidae) has existed virtually unchanged for over 200 million years, formerly more widespread in southern Britain in temporary pools (eggs survive drying out), recently known from just one site in the New Forest, until 2004 when another population was found at Caerlaverock Wetland Centre, Dumfries and Galloway. It is Endangered in the RDB, and protected under the WCA (1981). *Cancriformis* is Latin for 'crab-shaped'.

TAENIOPTERYGIDAE Stonefly family (Greek *taenio* + *pteryx* = 'band' + 'wing'). *Brachyptera risi* is very common, with a generally western British distribution; *B. putata*, an endemic species, therefore of international importance, is mostly found in highly oxygenated streams in north Scotland. *Taeniopteryx nebulosa*, preferring muddy streams, is the February red of anglers.

TAIL-MOSS Members of the genus *Anomodon* (Anomodontaceae) (Greek *anomados* = 'not smooth').

Long-leaved tail-moss *A. longifolius* (Welsh Borders, mid-Pennines and central Highlands) and **slender tail-moss** *A. attenuatus* (south-east Highlands and the River Eden, Cumbria) are both rare.

Rambling tail-moss *A. viticulosus*, widespread and common on well-drained, lightly shaded, base-rich rocks and dry-stone walls in the lowlands. It also occurs on chalk, on hedge banks and in grassland, as well as at the base of old trees, especially ash. Other habitats include masonry by rivers, and shaded concrete and brickwork. *Viticulosus* is Latin for 'like the chaste tree' (*Vitex*).

TAIL WEED Straggly tail weed *Rhodomela lycopodioides* (Rhodomelaceae), an intertidal red seaweed found in rock pools in north-east England, Scotland and north-east Ireland. Greek *rhodon* + *melas* = 'rose-coloured' + 'black', and Latin *lycopodioides* = 'resembling the genus *Lycopodium*'.

TALPIDAE Family name for mole (Latin *talpa*).

TAMARISK (HYDROID) Sea tamarisk *Tamarisca tamarisca* (Sertulariidae), a colonial hydroid growing up to 15 cm in and around the littoral in moderate to strong tides on rocks or shells, occasional in north-east England, north Wales, west Scotland and around Ireland.

TAMARISK (PLANT) *Tamarix gallica* (Tamaricaceae), a deciduous south-west European neophyte shrub, rarely a small tree, introduced by Tudor times, much planted in coastal sites in England and Wales, rarely in Ireland, as it is resistant to wind. It can spread by suckering. It is also found inland as a garden escape on waste ground and tips. *Tamarix* is Latin for this plant, after the River Tamaris on the border of the Pyrenees where it grows; *gallica* = 'French'.

TAMARISK-MOSS Members of the genera *Abietinella* (Latin for 'associated with fir'), *Heterocladium* (Greek *heteros* + *clados* = 'different' + 'branch') and *Thuidium* (associated with *Thuja*).

Rarities include the calicolous **dimorphous tamarisk-moss** *H. dimorphum* (Highlands), the calcicolous **fir tamarisk-moss** *Abietinella abietina* var. *abietina*, **lesser tamarisk-moss** *T. recognitum* (on limestone pavement), and the acidophilous **Wulfsberg's tamarisk-moss** *H. wulfsbergii* (western montane regions).

Common tamarisk-moss *T. tamariscinum* (Thuidiaceae), common and widespread in damp woodland, in hedge banks and grass.

Delicate tamarisk-moss *T. delicatulum*, mainly found in western Britain and in Ireland in damp grassland and flushes, on mountain ledges, and banks and soil in woodland, quarries and roadsides.

Philibert's tamarisk-moss *T. assimile*, scattered and widespread on grazed, grass over chalk and limestone, in base-rich dunes, including machair, less frequently on damp, base-rich rocks in woodland, and on mountain rock ledges. *Assimile* is Latin for 'similar'.

Prickly tamarisk-moss *A. abietina* var. *hystricosa*, locally common in southern and south-east England on shallow soils in ancient, unimproved, closely grazed grassland overlying chalk or limestone, and on banks, in quarries and in calcareous dunes. *Hystricosa* derives from the Greek *hystrix* = 'porcupine' (i.e. 'prickly').

Slender tamarisk-moss *H. flaccidum*, locally common in upland areas on base-rich rock, especially calcareous sandstone, particularly in steep-sided woodlands, sometimes growing on flint in southern England. *Flaccidum* is Latin for 'flaccid' or 'weak'.

Wry-leaved tamarisk-moss *H. heteropterum*, common in mountain regions on acidic or slightly base-rich rocks in humid woodlands and gorges. Greek *heteros* + *pteron* = 'different' + 'wing' (i.e. feathered).

TANGLE Alternative name for the brown seaweeds cuvie and oarweed.

TANSY *Tanacetum vulgare* (Asteraceae), an aromatic, rhizomatous perennial herb, probably an archaeophyte, grown in medieval gardens as a medicinal (vermicide in children and abortifacient) or culinary herb, escapes from cultivation being widely naturalised. It is common and widespread in grassy habitats by rivers, roads and railways, and on waste ground. 'Tansy' dates from Middle English, from Old French *tanesie*, probably from medieval Latin *athanasia* = 'immortality' (morphing into the genus name), in turn from Greek. *Vulgare* is Latin for 'common'.

Rayed tansy *T. macrophyllum*, a European neophyte perennial, introduced in 1803, scattered (possibly increasing in central Scotland), naturalised since the early twentieth century in grassy habitats, including roadsides. *Macrophyllum* is Greek for 'large-leaved'.

White tansy Alternative name for sneezewort.

TAPEGRASS *Vallisneria spiralis* (Hydrocharitaceae), a submerged herbaceous perennial neophyte introduced in 1818 as an aquarium plant. In the River Lea Navigation canal, Middlesex, it is found in shallow water on mud and gravel. Antonio Vallisneri (1661–1730) was a Spanish naturalist. *Spiralis* is Latin for 'spiralled'.

TAPEWORM Parasitic flatworms (class Cestoda) that infect the digestive tract in humans, pets, livestock (e.g. *Moniezia* in cattle and sheep) and wildlife (e.g. *Hymenolepis erinacei* in hedgehogs).

Dog tapeworm *Dipylidium caninum*, transmitted by dog flea, and can also affect humans.

TARCRUST General name for encrusting or globular fungi in the orders Boliniales and Xylariales, found on dead wood.

Beech tarcrust *Biscogniauxia nummularia* (Xylariales, Xylariaceae), pathogenic, widespread and locally common in England and Wales, uncommon in Scotland and Ireland, found on fallen branches of beech. *Nummularia* is Latin for 'coin-like', from the shape of the encrustations.

Common tarcrust *Diatrype stigma* (Xylariales, Diatrypaceae), common, widespread, inedible, found throughout the year in deciduous woodland on dead branches, especially beech and oak. Greek *dia* + *trypa* = 'through' + 'hole', and *stigma* = 'mark'.

Spiral tarcrust *Eutypa spinosa* (Diatrypaceae), widespread and locally common, encrusting, most associated with beech. Greek *eu* + *type* = 'good' + 'wound', and Latin *spinosa* = 'spiny'.

Thick tarcrust *Camarops polysperma* (Boliniales, Boliniaceae), rare, scattered in the southern half of England, on trunks and logs of alder and beech. Greek *camara* + *opsis* = 'vaulted chamber' + 'appearance', and *polysperma* = 'many-seeded'.

TARDIGRADE Water bears (or moss piglets), members of the phylum Tardigrada (meaning 'slow stepper'), eight-legged segmented herbivorous or bacteriophage animals, generally 0.5 mm in size, that live in freshwater and marine or very damp habitats (e.g. wet moss), able to demonstrate cryptobiosis, surviving extremes of temperature (a few minutes at 151°C, 30 years at −20°C, a few days at −200°C, and a few minutes at −272°C), pressure (some examples live at great ocean depths), dehydration and radiation.

TARE Scrambling annual species of *Vicia* (Fabaceae) (Latin for vetch). 'Tare' comes from Old English *taru*.

Hairy tare *V. hirsuta*, common and widespread on rough and disturbed ground, including road and railway banks, scrubby grassland, hedgerows, sheltered sea-cliffs and shingle beaches; also along the edges of arable fields, and on waste ground. Latin *hirsuta* = 'hairy'.

Slender tare *V. parviflora*, on calcareous clays in the southern half of England that are wet in winter but dry in summer, found in hedgerows, on tracks and verges, grassy banks, coastal cliffs and (declining) arable field edges; also, less frequently, on urban waste ground. *Parviflora* is Latin for 'small-flowered'.

Smooth tare *V. tetrasperma*, common and widespread in England and Wales, scattered and mostly introduced in Scotland and Ireland, in hedgerows, scrub, woodland edge, and rough grassland on roadsides, railway banks and coastal cliffs; also in disturbed habitats such as urban waste ground and arable field margins. *Tetrasperma* is Greek for 'four-seeded'.

TARRAGON *Artemisia dracunculus* (Asteraceae), a woody-based neophyte perennial grown as an aromatic herb by Tudor times, scattered in England as a casual garden escape on waste ground, and also originating from wool shoddy. 'Tarragon' dates from the mid-sixteenth century, from medieval Latin *tragonia* and *tarchon*, Greek *tarchon*, and perhaps from an Arabic alteration of Greek *dracon* = 'dragon'. The binomial is from Artemis, the Greek goddess of the hunt, and Latin *dracunculus* = 'little dragon', possibly describing the twisted root.

TASSELWEED Submerged plants in the genus *Ruppia* (Ruppiaceae). Heinrich Ruppius (1688–1719) was a German botanist.

Beaked tasselweed *R. maritima*, an annual or perennial local around much of the coastline in brackish shallow water in coastal lakes, saltmarsh pans, rock pools, creeks and ditches near the sea. It is also found as a dwarf variant on tidal mudflats, especially in north-east Scotland. A recent inland record from near Nantwich and Sandbach, Cheshire, is in an area of natural salt deposits.

Spiral tasselweed *R. cirrhosa*, a perennial of relatively deep tidal waters, scattered and local around the coast, mostly from Dorset to Lincolnshire, and in the Outer Hebrides, in coastal lakes, tidal inlets, creeks and brackish ditches. *Cirrhosa* is Greek for 'curly'.

TAXACEAE Family name of yew (Latin *taxus*).

TEAL *Anas crecca* (Anatidae), wingspan 61 cm, length 36 cm, weight 330 g, a common and widespread resident duck, in 2009 numbering 1,600–2,800 (most probably 2,100) pairs. Numbers increased by 75% over 1995–2015, including by 45% in 2010–15. In winter teal congregate in low-lying wetlands, particularly in south-west England, the Midlands, Yorkshire, parts of Wales, and south-east Ireland, of which many are continental birds from around the Baltic and Siberia, with total numbers being 21,000 individuals. Britain is home to a significant percentage of the north-west European wintering population, making it an Amber List species. The nest is a deep hollow lined with dry leaves and down, built in dense vegetation near water, clutch size is 8–11, incubation takes 21–3 days, and fledging of the precocial young takes 25–30 days. This dabbling duck feeds on insects, worms and molluscs during the day in the breeding season, and mainly seeds in winter, feeding at night. 'Teal' derives from Middle English, possibly related to *teling*, the Dutch name for this bird. Latin *anas* = 'duck', and *kricka*, the Swedish name for this species (probably onomatopoeic of the male's call). An average of 4 vagrant **blue-winged teal** *A. discors* and 22 **green-winged teal** *A. carolinensis* are recorded each year.

TEAPLANT Deciduous neophyte shrubs from China in the genus *Lycium* (Solanaceae), from the Greek name of a medicinal tree from Lycia, Asia Minor, rare, scattered and naturalised in rough ground and on walls.

TEARDROPPER *Cylindrobasidium laeve* (Physalacriaceae), a widespread inedible fungus, common in England, evident throughout the year, sporulating in late summer and autumn, the white encrusting fruiting body attached to dead wood, favouring cut surfaces on logs. Latin *cylindrus* + *basidium* = 'cylinder' + 'small pedestral', and *laevis* = 'smooth'.

TEASEL Herbaceous plants in the genus *Dipsacus* (Dipsacaceae), from Greek *dipsa* = 'thirst', referring to the cup-like formation made where sessile leaves merge at the stem. Rainwater collects in this receptacle, preventing sap-sucking insects from climbing the stem. The flowers are often visited by insects, seed heads by birds. 'Teasel' comes from Old English *tæsl* or *tæsel* for this plant.

Biennial European neophytes, casual on rough ground, include **cut-leaved teasel** *D. laciniatus*, and **yellow-flowered teasel** *D. strigosus*, the latter introduced from Russia in 1820, sites also including chalk-pits in Cambridgeshire.

Fuller's teasel *D. sativus*, perennial, the prickly seed heads formerly used in fulling (cleansing of wool to eliminate oils and other impurities) – still so in Somerset – now scattered, mostly in England, as a casual from birdseed, sometimes naturalised on waste ground. *Sativus* is Latin for 'cultivated'.

Small teasel *D. pilosus*, biennial, local and scattered in England and Wales, introduced on the Isle of Man, on woodland edges and in clearings, in scrub and hedgerows, on ditch sides and stream and riverbanks, as well as in quarries and on waste ground, preferring damp, calcareous soils. Needing disturbance for germination, agricultural improvement or lack of woodland management may account for its decline in many places. *Pilosus* is Latin for 'hairy'.

Wild teasel *D. fullonum*, biennial, common and widespread in Britain north to central Scotland, local further north and in Ireland, often colonising bare ground after disturbance, and found in rough grassland, woodland edge and hedgerows, by railways, and on roadsides and waste ground on a range of soil types. *Fullonum* is New Latin for 'relating to fullers', from their use in wool carding (see Fuller's teasel above).

TECTIBRANCH Sea slugs, which may have an external shell (Greek *tecton* = 'roof').

TELLIN, TELLINIDAE Cockle-like filter-feeding burrowing bivalves (Greek *telline* for these shellfish).

Baltic tellin *Macoma balthica*, shell size up to 3 cm, common and widespread intertidally on all coasts in muddy sands, especially in estuaries and sheltered bays. *Macoma* is New Latin for this tellin.

Bean-like tellin *Tellina fabula*, shell size up to 2 cm, common and widespread in fine silty and muddy sand from the low intertidal to 30 m. *Fabula* is Latin for 'small bean'.

Blunt tellin *Arcopagia crassa*, shell size up to 6.5 cm, widespread on all coasts from the low intertidal and across the shelf in coarse sand and shell-gravel. Greek *acros* + *pagios* = 'at the edge' + 'solid', and Latin *crassa* = 'thick'.

Thin tellin *T. tenuis*, shell size up to 2 cm, common and widespread in (silty) sand from the mid-intertidal to 10 m depth. *Tenuis* is Latin for 'thin'.

TELOSCHISTACEAE Family name (Greek *telos* + *schistos* = 'end' + 'divided') of a number of lichens (Ascomycota), including species of *Teloschistes*, *Caloplaca*, *Fulgensia* and *Xanthoria*.

TENCH *Tinca tinca* (Cyprinidae), length usually 60 cm, weight 1.8 kg, common and widespread in England, more scattered elsewhere, in vegetated slow-flowing rivers, shallow lakes and gravel-pits with muddy beds, spawning in July–August when temperatures have been at least 18°C for two weeks, so they do not breed every year in shaded waters. Feeding is at night on small animals. In winter they may lie dormant on bottom mud. Known as 'doctor fish' in medieval times, tench slime was applied to cure ailments such as toothache, headache and jaundice. 'Tench' dates from Middle English, from Old French *tenche*, itself from *tinca*, Latin for this species.

TENEBRIONIDAE Family name (Latin *tenebrio* = 'lover of darkness') of darkling beetles, so named from the dark elytra of most species, of varying size (1.5–25 mm length), with 45 British species in 32 genera, mostly recorded from England and Wales. Many species are synanthropic, including pests of stored products, especially species of *Tribolium* in flour and grain. *Tenebrio molitor* is the mealworm sold as food for birds and reptiles. Members of the subfamily Diaperinae are found in fungi and decaying wood. *Myrmechixenus* species are associated with ants. See also flour beetle, fungus beetle, sulphur beetle and mealworm, and churchyard beetle under beetle.

TENTHREDINIDAE From Greek *tenthredon*, meaning a kind of wasp, the family name for sawflies (suborder Symphyta), most of which are leaf miners, others gall-creators.

TEPHRITIDAE Family name for some of the small to medium-sized (2.5–10 mm length) fruit flies (see also Drosophilidae), with 76 species in 33 genera. The name comes from Greek *tephritis* = 'ash-grey', despite the bright colours of the bodies and wings of many species. Both adults and larvae feed on plant parts and/or products (often Asteraceae), some larvae being leaf miners or found in galls. See celery fly.

TEREDINIDAE Family name of shipworms, a group of wood-boring bivalves (Greek *teredon* = 'woodworm').

TERN Seabirds in the family Sternidae. 'Tern' and *Sterna* come from *stearn*, used in Old English.

Caspian tern *Hydroprogne caspia*, **gull-billed tern** *Gelochelidon nilotica* and **white-winged black tern** *Chlidonias leucopterus* are scarce visitors. **Aleutian tern** *Onychoprion aleuticus*, **bridled tern** *O. anaethetus*, **Cabot's tern** *Sterna acuflavida*, **Forster's tern** *S. forsteri*, **lesser-crested tern** *S. bengalensis*, **royal tern** *S. maxima*, **sooty tern** *O. fuscatus* and **whiskered tern** *C. hybrida* are all accidentals.

Arctic tern *Sterna paradisaea*, wingspan 80 cm, length 34 cm, weight 110 g, Amber-listed, though Britain's most abundant breeding tern, with 53,000 pairs in 2004, with a 39% increase over 1986–2014, and a 34% increase in numbers between 2000 and 2014 (SMP data; BTO gives 15%). Arrival is in May–June, many birds recorded inland during their migration north in late April/early May. Migration south begins in late July–August. This species has the greatest known bird migration distance, with one individual breeding on the Farne Islands (Northumberland) recorded in June 2016 as having travelled 96,000 km to Antarctica and back, having spent much time in the Southern Ocean and Weddell Sea. It is a coastal breeding bird in much of Scotland, north-east England and north Wales, but in Ireland it breeds inland on the freshwater Lough Corrib (Co. Galway) and Lough Conn (Co. Mayo) as well as at many sites on the coast. Nests are in scrapes, the 1–2 eggs hatching in 20–4 days, the semi-precocial chicks fledging in 21–4 days. Feeding is on surface fish, with a maximum dive of 50 cm. Latin *paradisus* = 'paradise'.

Black tern *Chlidonias niger*, wingspan 66 cm, length 23 cm, weight 73 g, a visitor migrating between Scandinavia and southwest Europe, on spring passage in May and autumn passage from July to September on lakes, gravel-pits and reservoirs. Greek *chelidonios* = 'swallow-like' and Latin *niger* = 'black'.

Common tern *Sterna hirundo*, wingspan 88 cm, length 33 cm, weight 130 g, a widespread Amber List summer migrant seabird, arriving in April and leaving in August–September, breeding on shingle beaches, except for south-west England, the southern half of Wales and south-west Scotland, and it now also breeds at inland sites in parts of England on gravel-pits and reservoirs, feeding along rivers and over fresh water. There were 10,000 pairs estimated in 2000, but a 30% decline occurred between 1986 and 2014, and a 20% decline between 2000 and 2014, with a crash of 55% between 2014 and 2015. Birds nest colonially on the ground, with clutch size 2–3, incubation taking 21–2 days. Chicks are semi-precocial (fed at the nest for 5 days, then accompanying the adults), fledging taking 22–8 days. Food is mainly fish, caught by plunge-diving. Latin *hirundo* = 'swallow', a reference to the similarity to this bird with its light build and long forked tail.

Little tern *Sternula albifrons*, wingspan 52 cm, length 23 cm, weight 56 g, an Amber List summer migrant arriving in April–May, leaving in August–September, around the British coastline at suitable breeding beaches, the largest colonies found along the east coast of Scotland, and parts of eastern and southern England, for example Blakeney Point and Great Yarmouth (Norfolk), Minsmere (Suffolk) and Langstone Harbour (Hampshire). It is uncommon on the west and east coasts of Ireland, mainly breeding on shingle beaches in Co. Louth, Co. Wicklow and Co. Wexford. There were around 1,900 pairs in Britain in 2000, with a minimum of 1,391 pairs at the 54 colonies sampled, though declines of either 27% (SMP) or 12% (BTO) were estimated between 1986 and 2014, and of 9% in 2000–14 (SMP) or an increase by 4% (BTO). Clutch size is 2–3, incubation takes 18–22 days, and the precocial chicks fledge in 19–20 days. Feeding is at sea on small fish and invertebrates, the bird often hovering before plunge-diving. *Sternula* is a diminutive of *Sterna* (tern), Latin *albus* + *frons* = 'white' + 'forehead'.

Roseate tern *Sterna dougallii*, wingspan 76 cm, length 36 cm, weight 110 g, a Red List migrant breeder, present from May to August, wintering in West Africa. The British population was around 100 breeding pairs in 2015, compared to 24 in 1999 and 86 in 2010; >90% are found on Coquet Island, Northumberland. Anglesey and the Firth of Forth are other important sites. It is also a rare breeder in Ireland, mostly at Lady's Island, Rosslare (Co. Wexford) and at Rockabill (Skerries) and Maiden Rock, Dalkey (Co. Dublin). Passage birds can be seen along the south and east coasts, with Dungeness (Kent) a regular site. Birds nest colonially in scrapes, with 1–2 eggs, incubated for 21–6 days, the semi-precocial chicks fledging in 22–30 days. Food is mainly fish, caught by plunge-diving. In summer adults have a pink tinge to their underparts, prompting the English name. Peter Dougall (1777–1814) was a Scottish physician and collector.

Sandwich tern *Sterna sandvicensis*, wingspan 100 cm, length 38 cm, weight 250 g, an Amber List migrant breeder with colonies scattered around the British and Irish coasts, including the north Norfolk coast, Minsmere (Suffolk) and Dungeness (Kent) from March to September. There were an estimated 11,000 pairs in Britain in 2000, with a 10% decline over 1986–2014, 3% in 2000–14, and 18% in 2009–14. Birds nest colonially in scrapes, mainly on the coast but with some colonies inland, clutch size 1–2, incubation taking 21–9 days, the semi-precocial young fledging in 28–30 days. Feeding is mainly of fish, caught by plunge-diving. *Sandvicensis* is from Latin for 'of Sandwich', referring to this town in Kent.

TERRAPIN Red-eared terrapin or **red-eared slider** *Trachemys scripta* (Emydidae), shell length up to 30 cm, native to the Mississippi Basin, with perhaps 33,000 individuals imported as pets during the cartoon mutant ninja turtle craze of the late 1970s and early 1980s, many young animals subsequently being illegally released as they grew too large for tanks. They overwinter in pond bottoms. Summers are too cold for eggs to hatch, and no successful breeding has been reported, but animals can live for >40 years, and populations have built up in a number of urban ponds, especially in and around London, and colonies have been reported in Jersey, Bournemouth–Poole, Swindon, Coventry and Cardiff. Greek *trachys* + *emys* = 'rough' + 'tortoise', and Latin *scripta* = 'written'.

TESTACELLIDAE Family name for shelled slugs (Latin *testa* + *cella* = 'shell' + 'chamber').

TETRAGNATHIDAE Family name (Greek *tetra* + *gnathos* = 'four' + 'jaw') of long-jawed spiders or long-jawed orb weavers, with 14 British species in 4 genera. See also cave and lesser garden spiders, both under spider.

TETRANYCHIDAE Family name of spider mites (Greek *tetra* + *onychos* = 'four' + 'claws').

TETRATOMIDAE Family name (Greek *tetra* + *tomos* = 'four' + 'cut') of polypore fungus beetles (3–4 mm length), with four species in two genera, feeding on decaying fungi, widespread especially in England and Wales.

TETRIGIDAE Possibly from Latin *tetricus* or *taetricus* = 'harsh' or 'severe', a grasshopper family comprising three species of groundhopper.

TETTIGONIIDAE Bush-cricket family comprising ten native species; see also conehead and wartbiter. *Tettigonia* is Latin for 'leafhopper', in turn from Greek *tettigonion*, the diminutive of the onomatopoeic *tettix* = 'cicada'.

THALLOBIONTA, THALLOPHYTA A polyphyletic group of simple plant or plantlike organisms, including algae and lichens.

THATCH MOSS *Leptodontium gemmascens* (Pottiaceae), in southern England and East Anglia in rough, acidic grasslands on decaying leaf bases of grasses and rushes, particularly characteristic of wheat or reed thatch typically ten years or more old. Greek *leptos* + *odontos* = 'thin' + 'teeth', and Latin *gemmascens* = 'with buds'.

THEREVIDAE Family name (Greek *thereuo* = 'to hunt') for stiletto flies, with 14 species in 6 genera, 2.5–18 mm long, with long slender abdomens. Adults mainly feed on nectar, honeydew and pollen. Larvae are predatory, often found on sandy or shingle riverbanks and generally feeding on insect larvae.

THERIDIIDAE Family name (Greek *thereuo* = 'to hunt') of comb-footed spiders, with 57 British species in 20 genera, including false widow spiders.

THERIDIOSOMATIDAE Family name of ray spider (Greek *thereuo* + *soma* = 'hunt' + 'body').

THICK-HEADED FLY Member of the family Conopidae. Adults generally have a broad head (hence the common name), and length is usually 7–15 mm. Adults, commonly bee or wasp mimics, feed on nectar using long probosces. Larvae are internal parasites of bees: the adult female intercepts the host in flight; her abdomen is modified to form a 'can opener' to pry open the segments of the host's abdomen as the egg is inserted. The larva first feeds on the bee's haemolymph, then consumes gut tissue. When the larva is ready to pupate it manipulates its host into digging itself into the ground. It kills the host and pupates inside the body. The adult emerges the following year.

THICK-KNEE Synonym of stone curlew.

THIEF For the seaweed, see oyster thief.

THIMBLEBERRY *Rubus parviflorus* (Rosaceae), an erect deciduous neophyte North American shrub introduced in 1818, occasionally naturalised in Britain in rough grassland and waste ground, and on railway banks. The binomial is Latin for 'bramble' and 'small-flowered'.

THISTLE Herbaceous members of the Asteraceae characterised by prickles on stem and leaves in the genera *Carduus* (Greek *cardos* = 'thistle'), *Carlina* (derived from Carolus, from the legend that Charlemagne used this plant to cure his army of the plague), *Cirsium* (*cirsion* = 'thistle'), *Echinops* (*echinos* = 'hedgehog', i.e. 'prickly'; see globe-thistle), *Onopordum* (Greek for this thistle, derived from *onos* + *pordon* = 'donkey' + 'body', or *porde* = 'flatulence', from the putative effect of consumption), *Scolymus* (Greek for 'scaly' or 'artichoke') and *Silybum* (Greek for thistles with edible stems).

Rarities include **golden thistle** *Scolymus hispanicus*, **Plymouth thistle** *Carduus pycnocephalus*, **reticulate thistle** *Onopordum nervosum* and **yellow thistle** *Cirsium erisithales* (Leigh Woods, Somerset).

Broad-winged thistle *Carduus acanthoides*, a European neophyte biennial, scattered and casual in Britain on rough and waste ground. *Acanthoides* means 'resembling *Acanthus*' (Greek *ace* + *anthos* = 'thorn' + 'flower').

Bull thistle Synonym of spear thistle.

Cabbage thistle *Cirsium oleraceum*, a perennial European neophyte present by the mid-sixteenth century. While generally rare and casual in drier habitats, it has been known from marshland by the River Tay in Perthshire since 1912; by a tidal river in Co. Wexford since 1958; on a south Lancashire stream bank since 1978; and on road- and streamsides in Co. Fermanagh since 1988. *Oleraceum* is Latin for 'of the vegetable garden'.

Carline thistle *Carlina vulgaris*, a monocarpic perennial, common and widespread (coastal in Scotland), in well-grazed grassland on dry, infertile calcareous or base-rich soils, as well as in more open habitats, including rock ledges, scree, coastal cliffs and dunes. *Vulgaris* is Latin for 'common'.

Cotton thistle *Onopordum acanthium*, an archaeophyte biennial known in the Iron Age, locally common in southern and central England and East Anglia, scattered elsewhere, in fields, hedgerows and waste ground, often near market gardens and farm buildings, perhaps dispersed with manure or contaminated straw. Greek *ace* + *anthos* = 'thorn' + 'flower'.

Creeping thistle *Cirsium arvense*, a perennial growing in overgrazed pasture, hay meadow, rough grassy habitats, hedgerows, roadsides, arable fields and other cultivated land, and in urban habitats and waste ground. Plants regenerate from rhizome fragments broken up by ploughing or other disturbance, and it is classified as injurious under the Weeds Act (1959). *Arvense* is from Latin for 'field'.

Dwarf thistle *Cirsium acaule*, a rosette-forming perennial of short grassland on base-rich soils, particularly on chalk and limestone, south of a line from north Yorkshire to Glamorgan, occasional elsewhere. Northern and western limits appear to be determined by summer warmth and in areas such as the Yorkshire Wolds and Derbyshire it is virtually confined to south-west-facing slopes. It benefits from the sward being grazed, but cannot tolerate heavy trampling. *Acaule* is Latin for 'without stem'.

Field thistle Synonym of creeping thistle.

Marsh thistle *Cirsium palustre*, a common and widespread monocarpic perennial of bog, fen, marsh, wet grassland, damp woodland, mountain springs and flushes, and tall-herb vegetation on mountain ledges. *Palustre* is from Latin for 'marsh'.

Meadow thistle *Cirsium dissectum*, a stoloniferous perennial local in England and Wales as far north as Yorkshire, and in Ireland, found in fen, fen-meadow, bog margins and poorly drained meadow on acid to neutral, usually peaty, soils, often in sites subject to vertical or lateral water movement. *Dissectum* is Latin for 'dissected'.

Melancholy thistle *Cirsium heterophyllum*, a perennial found in northern England and Scotland, and scattered in other upland places (rare in Ireland), on stream banks, hay meadow, moist verges and damp woodland edge, often declining due to a shift from traditionally managed hay meadows to silage-making, and from unsympathetic management of roadside verges. *Heterophyllum* is Greek for 'different-leaved'.

Milk thistle *Silybum marianum*, an annual or biennial archaeophyte, scattered as a casual, or naturalised in rough pasture, on grassy banks, in hedgerows and on waste ground as an introduction with wool shoddy, as a seed contaminant, and as a garden escape. *Marianum* is Latin for 'of St Mary': the white veins of the leaves were believed to have originated from the milk of the Virgin Mary, who once fell upon this plant. Because of its potassium nitrate content, the plant is toxic to cattle and sheep.

Musk thistle *Carduus nutans*, biennial, sometimes perennial, common and widespread in England and Wales, scattered elsewhere, and an introduction as a seed contaminant in north Scotland, Isle of Man and Ireland. It is mainly found on calcareous

soils, but also on sandy or shingly ground, in rough pasture, roadsides and disturbed habitats. *Nutans* is Latin for 'nodding'.

Scotch thistle Synonym of milk thistle.

Slender thistle *Carduus tenuiflorus*, a widespread annual or biennial of dry, coastal grasslands, seabird colonies, sea walls, upper parts of beaches, sandy waste ground and roadsides. Inland, it occurs on well-drained soils, often as an introduction with wool shoddy. *Tenuiflorus* is Latin for 'slender-flowered'.

Spear thistle *Cirsium vulgare*, a common and widespread monocarpic perennial growing in habitats as varied as overgrazed pasture and rough grassland, sea-cliffs, dunes, drift lines and fertile, disturbed habitats such as arable fields and waste ground. It is listed as an 'injurious weed' in the Weeds Act (1959) and subject to control by landowners. *Vulgare* is Latin for 'common'.

St Mary's thistle Alternative name for milk thistle.

Tuberous thistle *Cirsium tuberosum*, a rare herbaceous perennial (Near Threatened in England, Vulnerable in Wales) declining in its two main areas in Wiltshire (in tall, unimproved, often ungrazed calcareous grassland) and Glamorgan (on clifftop grassland over Jurassic limestone) because of habitat destruction and, particularly in Wiltshire, hybridisation with dwarf thistle, exacerbated by grazing. It became extinct as a native in its single Cambridgeshire locality in 1973, and plants reintroduced in 1987 failed to survive. *Tuberosum* is Latin for 'tuberous'.

Welted thistle *Carduus crispus*, a herbaceous biennial common and widespread in England and south-east and central Scotland, scattered elsewhere, and introduced in north Scotland, in woodland edges, hedge banks, rough grassland, verges, railway banks and waste ground on basic soils, especially on clay with a high nutrient status. In Ireland, it is a native on dry banks and waste ground. *Crispus* is Latin for 'curly'.

Woolly thistle *Cirsium eriophorum*, a monocarpic perennial locally common in central-southern England, scattered as far north as Co. Durham, in dry, often ungrazed, grasslands, scrub and woodland on limestone, chalk and lime-rich clay. *Eriophorum* is Greek for 'bearing wool'.

THOMISIDAE Family name (Greek *thomisso* = 'to bind') of crab spiders, with 26 British species in 6 genera.

THONGWEED *Himanthalia elongata* (Himanthaliaceae), a common and widespread brown seaweed, though scarce on the coasts of south-east England and East Anglia, with a thallus up to 30 mm wide and 25 mm high, plus a long strap-like reproductive receptacle up to 2 m long, attached to gently sloping hard substrates on the lower littoral of moderately exposed shores. Greek *himantos* + *thalia* = 'strap' + 'abundance', and Latin *elongata* = 'elongated'.

THORN (MOTH) Nocturnal moths in the family Geometridae (Ennominae), including the genera *Ennomos* (Greek for 'lawful') and *Selenia* (Greek *selene* = 'moon', from moon-shaped wing spots). Caterpillars generally resemble (mimic) twigs.

August thorn *E. quercinaria*, wingspan 42–50 mm, local in woodland, parks, hedgerows, scrubby downland and gardens throughout England, Wales and Ireland, but with numbers declining by 85% over 1968–2007, a species of conservation concern in the UK BAP. Adults fly in August–September. Larvae feed on trees such as oaks (*Quercus*, prompting the specific name) and beech.

Canary-shouldered thorn *E. alniaria*, wingspan 38–42 mm, widespread and common in woodland, scrub, parks and gardens. Adults fly in July–October. Larvae feed on trees such as birches, goat willow and alder (*Alnus*, prompting the specific name).

Dusky thorn *E. fuscantaria*, wingspan 35–40 mm, common in woodland, hedgerows and gardens throughout much of England, less so in the south-east and north, and with one (post-2000) record from Co. Cork. Following a population decline by 98% over 1968–2007, it is now actually found at higher numbers, though remaining a species of conservation concern in the UK

BAP. Adults fly in August–October. Larvae feed on ash. Latin *fuscans* = 'becoming dusky', from part of the forewing colour.

Early thorn *S. dentaria*, wingspan 28–40 mm, widespread and common in woodland, scrub, hedgerows, parks and gardens. There are two generations (except in the far north), in April–May and August–September. Larvae feed on a variety of woody plants. Latin *dens* = 'tooth', referencing the dentate edge of the forewing.

Feathered thorn *Colotois pennaria*, wingspan 35–45 mm, widespread and common in broadleaf woodland, parks and gardens. Adults fly in September–November. Larvae feed in spring on a number of broadleaf trees. Greek *colos + otoeis* = 'stunted' + 'with ears', from the reduced labial palp, and Latin *penna* = 'feather', from the antenna shape.

Large thorn *E. autumnaria*, wingspan 40–50 mm, scarce, in broadleaf woodland, scrub and gardens in parts of the southern half of England, particularly the south-east and East Anglia. Adults fly in September–October, prompting the specific name. Larvae feed on a number of deciduous trees and shrubs.

Little thorn *Cepphis advenaria*, wingspan 23–30 mm, day- and dusk-flying in May–June, in woodlands and mature scrub, particularly in southern England and south Wales, and also found in south-west Ireland. Larvae favour bilberry, but will eat bramble and other plants. Greek *cepphos* = 'small seabird', with no entomological meaning, and Latin *advena* = 'a stranger'.

Lunar thorn *S. lunularia*, wingspan 38–44 mm, widespread but local in open woodland and scrub, usually in small numbers, and as an occasional immigrant to the south coast of England. Adults fly in May–June. Larvae feed on a variety of woody plants. *Lunula* is Latin for 'little moon', like the genus name referring to the moon-shaped wing spots.

Purple thorn *S. tetralunaria*, wingspan 30–8 mm, common in woodland, scrub, gardens and other habitats throughout much of England, and parts of Wales and Scotland. Adults are bivoltine except in the north of their range, flying in April–May then July–August. Where there is one generation, moths are evident in May. Larvae feed on a variety of woody plants. Latin *tetra + luna* = 'four' + 'little moon', from the four spots, one on each wing.

September thorn *E. erosaria*, wingspan 30–42 mm, common in woodland, parks and gardens throughout much of England, spreading into Wales and Scotland. With numbers declining by 87% over 1968–2007, this is a species of conservation concern in the UK BAP. Adults fly in July–October. The caterpillar resembles a twig and feeds on oaks, birches, limes and beech. Latin *erosus* = 'eroded', from the appearance of bites out of the outer wing edge.

THORN-APPLE *Datura stramonium* (Solanaceae), an annual American neophyte in gardens by 1597, grown for alkaloids used to treat asthma, now scattered, mainly in England, on cultivated and disturbed ground, usually as a casual, but sometimes naturalised. It is generally a garden escape, or originates from soyabean waste, birdseed, oil-seed or grain. The binomial is from the Hindu vernacular *dhatura*, and Latin *stramonium* = 'spiked fruit'.

Longspine thorn-apple *D. ferox*, an annual Mexican neophyte mainly arriving in wool shoddy, an occasional casual in fields in England and Wales. *Ferox* is Latin for 'wild'.

THOROW-WAX *Bupleurum rotundifolium* (Apiaceae), an annual archaeophyte, once an arable weed of chalk and limestone soils, now a rare birdseed casual in England. 'Thorow-wax' dates from the sixteenth century, from 'through' + 'wax', because the stem appears to grow through the leaves. Greek *boupleuros* = 'ox-rib', a name given to another plant, and Latin *rotundifolium* = 'round-leaved'.

False thorow-wax *B. subovatum*, a scarce annual neophyte casual on waste ground, and in gardens. *Subovatum* is Latin for 'somewhat ovate'.

THRACIA Filter-feeding bivalves in the genus *Thracia* (Thraciidae), meaning 'from Thrace' (south-east Europe).

Distorted thracia *T. distorta*, shell size up to 2.7 cm, in rock crevices in the low intertidal in low numbers, with a patchy distribution, mostly in southern England, west Scotland, Orkney and Shetland.

Kidneybean thracia *T. phaseolina*, shell size up to 4 cm, widespread from the low intertidal across the shelf in sand and muddy gravel. *Phaseolos* is Greek for 'kidney bean'.

THREAD-MOSS Mosses in a number of genera, especially *Bryum* (Bryaceae), from Greek *bryon* = 'moss', and *Pohlia* (Mielichhoferiaceae), honouring the Austrian botanist Johann Emanuel Pohl (1782–1834).

Scattered and uncommon or locally common species include **bent-bud thread-moss** *P. proligera* (mostly upland Britain), **cernuous thread-moss** *B. uliginosum*, **Don's thread-moss** *B. donianum* (England, Wales and Ireland, generally in lowland areas on well-drained soil, and in crevices on walls and bridges), **dune thread-moss** *B. dunense* (mainly in East Anglia), **Duval's thread-moss** *B. weigelii* (Highlands, Cumbria and a few locations in Wales in base-poor springs and flushes, and by streams, pools, lakes and snowbeds), **fat-bud thread-moss** *P. filum* (upland regions, especially the Highlands), **gravel thread-moss** *P. andalusica* (mostly south-west England), **orange-bud thread-moss** *P. flexuosa* (upland regions, especially west Scotland), **river thread-moss** *B. riparium* (parts of upland Britain by stream sides and other damp habitats), **Scottish thread-moss** *P. scotica* (endemic, the largest populations in Argyll, with smaller populations in Dunbartonshire and Easter Ross, in silt, sand and gravel subject to flooding), **slender thread-moss** *Orthodontium gracile* (usually on rock in humid, base-poor habitats), **small-mouthed thread-moss** *B. imbricatum* (many records coming from Lancashire), **swollen thread-moss** *Aulacomnium turgidum* (Highlands, in short vegetation on heathy base-rich ledges and mountain tops, especially with dwarf shrubs and moss), **tight-tufted thread-moss** *B. crebrerrimum* (scattered in Britain), **tufted thread-moss** *B. caespiticium* (mainly parts of England and Wales, in waste ground, quarries and dune slacks, and on tops of walls), **Welsh thread-moss** *B. gemmiparum* (aquatic, in rock crevices in the beds of lowland streams in south-central Wales, north Devon and west Ireland), and **yellow-tuber thread-moss** *B. tenuisetum* (calcifuge, in open habitats on peaty soils or damp sands and clays).

Rarities include **arctic thread-moss** *B. arcticum* (Highlands), **blunt-leaved thread-moss** *P. obtusifolia* (Highlands), **Dixon's thread-moss** *B. dixonii* (Highlands and Hebrides), **Knowlton's thread-moss** *B. knowltonii* (five sites in north-west Wales, north Lincolnshire, west Norfolk, Northumberland and Scotland), **Milde's thread-moss** *B. mildeanum* (south Wales, Lake District and Highlands), **saltmarsh thread-moss** *B. salinum* (Scotland, with a record from each of Cumbria, Somerset and Co. Galway), **Schleicher's thread-moss** *B. schleicheri* (Stirlingshire), and **Warne's thread-moss** *B. warneum* (four sites in England, two in Wales and a few in Scotland).

Alpine thread-moss *B. alpinum*, common, especially in hilly regions in south-west England, Wales, the Lake District, Scotland and west Ireland, on acidic rocks subject to seepage and on peaty moorland.

Archangelic thread-moss *B. archangelicum*, scattered and widespread in a variety of habitats. Archangel is a town in north-west Russia.

Blunt-bud thread-moss *P. bulbifera*, widespread but scattered in upland western Britain. *Bulbifera* is Latin for 'bearing bulbs'.

Cape thread-moss *Orthodontium lineare* (Orthodontiaceae), common and widespread in Britain, scattered in the Highlands and north Scotland, and in Ireland, in base-poor habitats on sandy and peaty banks, and on siliceous cliffs, and tolerating pollution. Greek *orthos + odonti* = 'straight' + 'teeth', and Latin *lineare* = 'linear'.

Capillary thread-moss *B. capillare*, very common and widespread in base-rich to slightly acidic soils, grassland, woodland rides, soil banks and waste ground. It also grows on trees, logs, walls, roofs and rocks. *Capillare* is Latin for 'hair-like'.

Crimson-tuber thread-moss *B. rubens*, common and widespread on soil in arable and waste ground, among grass in fields and gardens, and beside paths. *Rubens* is Latin for 'red'.

Drooping thread-moss *B. algovicum*, widespread but scattered, on rock, waste ground, dune slacks and the tops of walls.

Drummond's thread-moss *P. drummondii*, widespread in upland regions in moist, ruderal habitats by streams, on tracks, in disused gravel-pits and in montane rock crevices. Thomas Drummond (1793–1835) was a Scottish bryologist.

Flabby thread-moss *B. subelegans*, common and widespread in Britain up to south-east Scotland, scattered further north. *Subelegans* is Latin for 'almost elegant'.

Golden thread-moss *Leptobryum pyriforme* (Meesiaceae), common and widespread especially in lowland regions in habitats that include arable fields, reservoir margins and woodland cliffs, and in deep shade on logs in damp woodland. Relatively tolerant of air pollution, it also occurs on the mortar of brick walls in towns. Greek *leptos* + *bryon* = 'slender' + 'moss', and Latin *pyriforme* = 'pear-shaped'.

Long-fruited thread-moss *P. elongata*, a calcifuge of upland regions on shallow, shaded, often peaty soil among boulders, in crevices, and on ledges and banks. *Elongata* is Latin for 'elongated'.

Ludwig's thread-moss *P. ludwigii*, in north-west Wales, Cumbria and the Highlands by streams and in montane flushes, often where snow lies late, and on wet soil overlying rock outcrops, and in scree. Friedrich Ludwig (1851–1918) was a German botanist.

Many-seasoned thread-moss *B. intermedium*, scattered and widespread, scarce in Ireland, in unshaded, basic soil on roadsides, stream banks, rock ledges and dunes. *Intermedium* is Latin for 'intermediate'.

Nodding thread-moss *P. nutans*, widespread and often abundant, though in places declining, on acidic, peaty, sandy or gravelly soil on heathland and moorland, in acidic grassland, bog, woodland and dunes, as well as on rock and dry-stone walls, and in gravel-pits, sandpits and abandoned industrial sites. *Nutans* is Latin for 'nodding'.

Opal thread-moss *P. cruda*, in upland regions on base-rich substrates, mainly in shaded rock crevices and ledges, but also in ravines and on coastal cliffs above the tidal zone in north Britain. *Cruda* is Latin for 'raw'.

Pale-fruited thread-moss *P. annotina*, common and widespread except in the south Midlands, eastern England and central Ireland, on well-drained but damp soil, such as by streams and pool margins, on waste ground and wet rocks, in gravel-pits and woodland rides. *Annotinus* is Latin for 'of the previous year'.

Pale glaucous thread-moss *P. wahlenbergii*, common and widespread in moist habitats. On lower ground it prefers roadsides, marsh, waste ground, wet rocks and damp woodland rides. Var. *glacialis* favours montane springs and flushes. Var. *calcarea* is a rarity of calcareous habitat in Cornwall, Isle of Wight and the North Downs. Göran Wahlenberg (1780–1851) was a Swedish naturalist.

Pale thread-moss *B. pallens*, common and widespread in moist soil in a variety of habitats, including in chalk grassland in southern England. *Pallens* is Latin for 'pale'.

Pink-fruited thread-moss *P. melanodon*, common and widespread on moist clay banks in the lowlands, especially by streams, by paths, and in fields and woodland rides. Greek *melas* + *odontos* = 'black' + 'teeth'.

Sauter's thread-moss *B. sauteri*, widespread and scattered, commonest in south-west and south-east England, and south Wales. Anton Sauter (1800–81) was an Austrian botanist.

Syed's thread-moss *B. laevifilum*, scattered in England, usually on tree-trunks or branches, sometimes on decaying wood, soil or rocks. Latin *laevis* + *filum* = 'smooth' + 'thread'.

Tall-clustered thread-moss *B. pallescens*, widespread but scattered, favouring metal-polluted sites, including abandoned metal mines, but also in places without high metal concentrations such as quarries and dunes, in crevices of rocks and walls, and on concrete. *Pallescens* is Latin for 'becoming pale'.

Tozer's thread-moss *Epipterygium tozeri* (Mielichhoferiaceae), common in south and south-west England and Wales, scattered in Ireland, on shaded, disturbed, non-calcareous banks beside lanes, ditches and rivers. Greek *epi* + *pterygion* = 'upon' + 'little feather'. The Rev. John Tozer (1790–1836) was an English plant collector.

Twisting thread-moss *B. torquescens*, widespread and scattered, scarce in Scotland, generally on unshaded, dry sites in low vegetation. *Torquescens* is Latin for 'twisting'.

Wall thread-moss *B. radiculosum*, common and widespread in Wales and the southern half of England, more scattered elsewhere, calcicolous and favouring mortared walls, limestone and calcareous soil. *Radiculum* is Latin for 'little root'.

Yellow thread-moss *P. lutescens*, widespread, common in south and south-west England and Wales on soil by paths, in hedge banks and ditches, and on streamsides. *Lutescens* is Latin for 'yellowish'.

THREADWORT Foliose liverworts in a number of genera, especially *Cephaloziella* (Cephaloziellaceae) (Greek *cephale* + *ozos* = 'head' + 'twig').

Rarities include **chalk threadwort** *C. baumgartneri*, **lobed threadwort** *C. integerrima* (mainly Cornwall), **snow threadwort** *Pleurocladula albescens* (Highlands), **spurred threadwort** *C. elachista* (on bogs, often growing through sphagnum), and **toothed threadwort** *C. dentata* (Lizard Peninsula).

Common threadwort *C. divaricata*, common and widespread in acidic habitats such as peaty, sandy and gravelly soil in heathland, coastal clifftops, acidic rock on mountain slopes and scree, and mine spoil. *Divaricata* is Latin for 'spreading'.

Entire threadwort *C. calyculata*, mainly found in Cornwall on coastal heathland, copper mine spoil and coastal creek banks; sites in Glamorgan and Somerset are on soil in limestone grassland, and two colonies have recently been found in Pembrokeshire on damp soil in heathland surrounded by limestone grassland. *Calyculata* is Latin for 'having a small calyx'.

Hairy threadwort *Blepharostoma trichophyllum* (Pseudolepicoleaceae), on upland crags and in gorges, growing through moss cushions or directly on calcareous rock. In sheltered woodland it is sometimes found on tree-trunks. It also grows in soil gaps in calcareous grassland, and stony edges of flushes and streams. Greek *blepharis* + *stoma* = 'eyelash' + 'mouth', and *trichos* + *phyllon* = 'hair' + 'leaf'.

Hampe's threadwort *C. hampeana*, common and widespread on peat, sphagnum, decomposing logs and acidic mine spoil. Georg Hampe (1795–1880) was a German botanist.

Heath threadwort *C. stellulifera*, widespread but scattered in Britain on peat, sphagnum, decomposing logs and mine spoil. *Stellulifera* is Latin for 'carrying a small star'.

Horsehair threadwort *Sphenolobopsis pearsonii* (Scapaniaceae), in upland sites on humid, shaded outcrops of neutral and acidic shale and sandstone in wooded gorges, in scree beds, and on granite and gneiss cliffs. The binomial is from the genus *Sphenolobus* + Greek *opsis* = 'appearance', and honouring the English bryologist William Pearson (1849–1923).

Irish threadwort *Telaranea nematodes* (Lepidoziaceae), on peaty banks or rocks in shaded, humid woodland near the coast in south-west Ireland, and at West Penwith, Cornwall. Greek *tele* + *aranea* = 'far off' + 'spider', and *nematos* = 'thread'.

Red threadwort *C. rubella*, widespread if scattered, substrates including soil, peat, sphagnum, mine spoil, and especially decomposing logs. *Rubella* is Latin for 'reddish'.

Turner's threadwort *C. turneri*, scattered in south and (especially) south-west England, south Wales and south Ireland, locally common on acidic soil among tree roots overhanging coastal creeks, and on lane banks and ditchsides, usually in light shade. Dawson Turner (1775–1858) was an English bryologist.

THRIFT *Armeria maritima* (Plumbaginaceae), a common and widespread perennial herb of sea-cliffs, stone walls, shingle and saltmarsh. Inland, it grows on montane rock ledges, stony flushes and windswept moss-heaths, around old lead workings and other metalliferous mine wastes. It also occurs inland beside salt-treated roads and as a garden escape. This is the county flower of the Isles of Scilly, Pembrokeshire and Bute as voted by the public in a 2002 survey by Plantlife. It is a cushion plant, and 'thrift' may refer to the closely packed leaves which conserve moisture in salt-laden sea air. *Armeria* is a Latinised form of Old French *armoires*, for a cluster-headed dianthus.

Jersey thrift *A. arenaria*, on fixed dunes and coastal headlands on Jersey, and naturalised on cliffs at Bournemouth, Dorset. *Arena* is Latin for 'sand'.

THRIPS Sap-sucking flies, also called thunderflies, in the order Thysanoptera, with fringes of long hairs on the four wings.

THROATWORT Or **blue throatwort** *Trachelium caeruleum* (Campanulaceae), a herbaceous perennial Mediterranean neophyte, naturalised on walls in Guernsey, Jersey and parts of Greater London. The English name derives from the old belief that the plant could cure a sore throat. Greek *trachelos* = 'throat', and Latin *caeruleum* = 'blue'.

THROSCIDAE Family name (Greek *throsco* = 'to leap') of beetles closely related to click beetles, though generally smaller (1.5–3.5 mm), with five British species in two genera.

THRUSH Birds in the genus *Turdus* (Turdidae) (Latin *turdus* = 'thrush'). 'Thrush' comes from Old English *thrysce*. A number of other thrushes are scarce vagrants.

Mistle thrush *T. viscivorus*, wingspan 47 cm, length 27 cm, weight 130 g, widespread in woodland, parks and gardens, with an estimated 170,000 territories in Britain in 2009, but with a 55% decline in numbers between 1970 and 2015 and a 28% decline between 1995 and 2015 (if with a weak 2% increase in 2010–15), it remains on the Red List. It is Green-listed in Ireland. The cup nest, built against a trunk or in a forked branch, usually contains 4 eggs which hatch after 15–16 days, the altricial young fledging in 14–17 days. It can begin breeding in February, and there are normally two broods a year. Feeding is on earthworms, slugs, insects and (especially in winter) berries. Mistletoe is favoured where available, reflected in the bird's English and specific names (from Latin *viscum* + *vorare* = 'mistletoe' + 'to devour').

Song thrush *T. philomelos*, wingspan 34 cm, length 23 cm, weight 83 g, common and widespread, especially in wooded habitats, including gardens, with 1.2 million territories in Britain in 2009, but with a 50% decline in numbers between 1970 and 2015 it remains on the Red List, though recovery is indicated an increase by 15% between 1995 and 2015 (9% over 2010–15). It is Green-listed in Ireland. Winter numbers are augmented by birds from the Continent. The mud-lined grass cup-shaped nest in a hedge or bush usually contains 4 eggs which hatch after 14–15 days, fledging of the altricial young taking 14–15 days. There are commonly two, sometimes three broods. Feeding is on earthworms and other invertebrates, fruit and, especially in dry weather, snails (the shells broken by hammering on a stone 'anvil'). The specific name is after Philomela, who in Greek myth was turned into a nightingale (*philos* + *melos* = 'lover' + 'song').

THYASIRIDAE Family name for hatchet shells, bacterial chemosynthesising bivalves.

THYME Species of *Thymus* (Greek *thymos* = 'thyme', in turn from *thyo* = 'to perfume' or 'sacrifice', possibly from being burned at altars) and *Clinopodium* (*clino* + *podos* = 'to slope' + 'foot'), both Lamiaceae.

Basil thyme *C. acinos*, an annual of open habitats in dry grassland, rocky ground and arable fields, in Britain (where it is locally common, especially in England) usually growing on calcareous soils, in Ireland (where it is introduced) on sandy and gravelly sites. It is also a rare casual of waste ground, quarries and road and railway banks. *Acinos* was a name used by Pliny referring to an aromatic herb, possibly wild basil or this species.

Breckland thyme *T. serpyllum*, a prostrate perennial herb confined to around 22 sites in west Suffolk and west Norfolk on dry sandy heathland and grassland, and on inland dunes, especially in areas grazed by rabbits or sheep. *Serpyllum* is another Latin word for 'thyme'.

Garden thyme *T. vulgaris*, a dwarf evergreen Mediterranean neophyte shrub cultivated as an aromatic herb by at least Tudor times, occasionally naturalised on old walls, stony banks, rough grassland and waste ground. *Vulgaris* is Latin for 'common'.

Large thyme *T. pulegioides*, a prostrate perennial herb, locally common in southern, central and eastern England, scarce elsewhere and perhaps introduced in Scotland and Ireland, on bare ground and grassland on chalk, occasionally on sands and gravels on heathland and dunes. *Pulex* is Latin for 'flea'; *pulegioides* means 'resembling *pulegium*' or pennyroyal, which repels fleas.

Wild thyme *Thymus polytrichus*, a widespread and locally common herbaceous perennial of free-draining, calcareous or base-rich substrates, growing in short grassland on heathland, downland, sea-cliffs and dunes, and around rock outcrops and hummocks in calcareous mires. It is also found in upland grassland and cliffs and ledges. Greek *polytrichus* = 'many hairs'.

THYME-MOSS Species in a number of genera, particularly *Mnium* (Greek *mnion* = 'moss'), *Plagiomnium* (*plagios* + *mnion* = 'slanting' + 'moss') and *Rhizomnium* (*rhiza* + *mnion* = 'root' + 'moss').

Rarities include **alpine thyme-moss** *P. medium* (Highlands, with >10% of all UK populations in the Cairngorms National Park), **ambiguous thyme-moss** *M. ambiguum* (Highlands), and **large-leaf thyme-moss** *R. magnifolium* (Highlands in wet, base-poor, montane habitats, and base-rich flushes down to 400 m altitude).

Bordered thyme-moss *M. marginatum*, widespread and scattered, most records coming from south Wales, north-west England, the Highlands and Skye, in damp, sheltered, base-rich soil, on ledges and in crevices. *Marginatum* is Latin for 'bordered'.

Dotted thyme-moss *Rhizomnium punctatum* (Cinclidiaceae), common and widespread on damp soil and rock, and rotting wood, in acidic to base-rich habitats. *Punctatum* is Latin for 'spotted'.

Felted thyme-moss *R. pseudopunctatum*, widespread and scattered in upland regions and in East Anglia in fen and base-rich marsh. *Pseudopunctatum* is Latin for 'false spotted'.

Hart's-tongue thyme-moss *P. undulatum*, common and widespread on base-rich or neutral soil in woodland, grassland and on rocky banks, sometimes in seepages. *Undulatum* is Latin for 'wavy'.

Long-beaked thyme-moss *P. rostratum*, common and widespread in damp, calcareous habitats on soil and rock. *Rostratum* is Latin for 'beaked'.

Many-fruited thyme-moss *P. affine*, widespread, common in England and Wales, scattered elsewhere, in damp base-rich or slightly acidic habitats in woodland and grassland. *Affine* is Latin for 'similar'.

Marsh thyme-moss *P. ellipticum*, widespread and scattered on wet ground in flushes, beside streams, and in damp grassland and woodland. *Ellipticum* is Latin for 'elliptical'.

River thyme-moss *Pseudobryum cinclidioides* (Plagiomniaceae), scattered in parts of upland Wales, Cumbria and Scotland, in marsh, springs and fen woodland, locally frequent on riverbanks. Greek *pseudes* + *bryon* = 'false' + 'moss', and *cinclis* = 'lattice'.

Short-beaked thyme-moss *M. thomsonii*, mainly recorded from north-west Wales, the north Pennines, Cumbria and the

Highlands in damp, sheltered, lightly shaded, base-rich soil, on ledges and in crevices.

Starry thyme-moss *M. stellare*, widespread, common in western Britain on rocks and soil in damp and shady, base-rich habitats. *Stellare* is Latin for 'starry'.

Swan's-neck thyme-moss *M. hornum*, common and widespread on acidic soil, logs, rocks and tree bases, often abundant in woodland; mainly lowland, but also in montane rock crevices. *Hornum* is Latin for 'of the present year'.

Tall thyme-moss *Plagiomnium elatum*, common and widespread in marsh, fen and flushes, and beside streams. *Elatum* is Latin for 'tall'.

Woodsy thyme-moss *P. cuspidatum*, widespread and scattered in lowland areas on soil, rock, stumps and lower tree-trunks in base-rich habitats. *Cuspidatum* is Latin for 'pointed'.

THYMELAEACEAE Family name for species of *Daphne* (Thymelaea being an alternative name for this Greek mythological figure): see mezereon and spurge-laurel.

THYSANOPTERA Order of thrips (Greek *thysanos* + *pteron* = 'fringe' + 'wing').

THYSANURA Order of bristletails (Greek *thysanos* + *oura* = 'fringe' + 'tail').

TICK Members of the order Ixodida, the largest of the Acari (unfed length usually 2–4 mm, engorged normally 8–12 mm), with three families in Britain: Amblyommidae and Argasidae (both soft ticks) and Ixodidae (hard ticks). They have modified mouthparts with teeth turned backward to maintain a grip on hosts: active stages are mostly external parasites, feeding primarily on blood of (non-fish) vertebrates. They have a three-host life cycle, for example using starling, hedgehog and sheep. *Ixodes* comes from Greek *ixos* = 'mistletoe' or 'birdlime', used here in the sense of something sticking as ticks do on their host. Common names generally reflect their (main) hosts, including **seabird tick** *Ixodes uriae* (scattered around the coast), **sand martin tick** *I. lividus* (scattered in England and Wales), **rabbit tick** *I. ventaloi* (on Lundy and the Scillies) and **long-legged bat tick** *I. vespertilionis* (a rarity scattered in England and Wales).

Caster bean tick, deer tick, meadow tick, ornate dog tick, wood tick Synonyms of marsh tick.

Dog tick *I. canisuga*, common and widespread on fox, badger, sheep, cat and horse as well as dog, in which large infestation can cause dermatitis pruritus, alopecia and anaemia. Latin *canis* + *sugo* = 'dog' + 'to suck'.

Hedgehog tick *I. hexagonus*, parasitic on dog and cat (the commonest tick found on cats, the second most common on dogs), fox, sheep and horse as well as hedgehog, common and widespread though especially in south-east England. It can transmit Lyme disease. Greek *hex* + *gonia* = 'six' + 'angles'.

Marsh tick *Dermacentor reticulatus* (Amblyommidae), locally common in southern England and west Wales near the coast, feeding on a number of wild mammals, livestock (causing redwater fever in cattle), and on humans. Greek *derma* + *centeo* = 'skin' + 'to prick', and Latin for 'made like a net'.

Sheep tick *I. ricinus*, common and widespread, mainly in rough grassland, moorland, woodland and areas when wild deer and rabbit are abundant, also feeding on sheep, cattle, dogs and humans. It is an agent of Lyme disease, louping-ill, rickettsia and pyaemia in sheep, and other livestock diseases. *Ricinus* is Latin for this tick.

TIGER BEETLE Members of the family Carabidae (Cicindelinae). These diurnal predators move at speed in open, dry situations primarily in spring and early summer. Larvae live at the bottom of vertical tunnels which act as pitfall traps for small invertebrates. *Cicindela* is Latin for 'glow-worm', used for these beetles because of their metallic appearance and sometimes flashing behaviour.

Cliff tiger beetle *Cylindera germanica*, the smallest tiger beetle (8–11 mm), since 1970 found only on some warm, south-facing eroding coastal cliffs, close to seepages of water, in Dorset and the Isle of Wight. Larvae and adults hunt invertebrates that are active on the surface, particularly ants. Adults breed in summer, and rarely fly, moving by running rapidly over the ground. Showing a 60% decline in the previous century, this is a species of priority conservation concern in the UK BAP. Greek *cylindros* = 'cylinder', and Latin *germanica* = 'German'.

Common tiger beetle, field tiger beetle Synonyms of green tiger beetle.

Dune tiger beetle *Cicindela maritima*, a scarce predatory beetle of dunes and sandy beaches along the drift line and intertidally, on both sides of the Bristol Channel, north-west Wales, Norfolk and Kent. Adults breed during spring and summer. Adults and larvae both feed on insects; adults are fast runners and fly well when hunting their prey.

Green tiger beetle *Ci. campestris*, the commonest of the tiger beetles, widespread, 10–15 mm in length, adults evident in April–September, favouring warm and sandy sites on heathland, moorland, grassland and dunes, and also often evident on brownfield sites. Adults have strong mandibles with several teeth, allowing them to feed on a range of small invertebrates, including spiders, caterpillars and ants. Although ground hunters, when disturbed they will fly short distances before running away. Long legs make this one of the country's fastest-moving ground-dwelling insects. Breeding takes place in summer. Larvae have a spine on their back that anchors them in vertical burrows that serve as pitfall traps. *Campestris* is from Latin for 'field'.

Heath tiger beetle *Ci. sylvatica*, the largest tiger beetle (15–19 mm length), found on dry, open sandy soils in heathland, sometimes in open coniferous woodland, historically much more widespread in southern England, but numbers have halved since 1970, and the species recorded only from Sussex, Surrey, Dorset and Hampshire. Adults are active in March–August, particularly in sunshine, running rapidly while hunting for prey, but also readily flying. The life cycle lasts two years, with overlapping generations. Both adults and larvae feed on invertebrates, particularly ants and caterpillars. *Sylvatica* is from Latin for 'woodland'.

Northern dune tiger beetle *Ci. hybrida*, 15 mm in length, Vulnerable in the RDB, confined to coastal sites in Lancashire and Cumbria. With a 46% decline since the 1980s even within its limited range, this is a species of priority conservation concern in the UK BAP. Adults emerge in midsummer and overwinter before breeding the following midsummer. Larvae generally overwinter in a half-grown state, and are predatory, like the adult. Adults can fly well.

Wood tiger beetle Synonym of heath tiger beetle.

TIGER MOTH Members of the family Erebidae (Arctiinae).

Cream-spot tiger *Arctia villica*, wingspan 45–60 mm, local in open woodlands and grassy habitats, including clifftops, saltmarsh and dunes in southern England and south Wales. Adults fly at night in May–June. Larvae feed on a range of herbaceous plants. Greek *arctos* = 'bear', from the hairy larva, and Latin *villica* means a female housekeeper of a villa, possibly from the colourful wings.

Garden tiger *A. caja*, wingspan 45–65 mm, widespread and fairly common in gardens, water meadow, marsh, fen, dunes and open woodland. With numbers declining by 92% over 1968–2007, however, this is a species of conservation concern in the UK BAP. Adults fly at night in July–August. Caterpillars – often known as woolly bears – feed on various herbaceous plants. Caia was a Roman lady's name, possibly describing the colourful wings.

Jersey tiger *Euplagia quadripunctaria*, wingspan 42–52 mm, day- and night-flying in July–September, until recently found only in the Channel Islands and parts of the south coast. On the mainland it is commonest in south Devon, but is expanding its range and colonies have recently appeared in Dorset, the Isle of Wight, and other southern counties. A population in parts of London is either a consequence of natural range expansion or accidental introduction. Habitats include gardens, rough and disturbed ground, hedgerows and coastal cliffs. Larvae feed on a range of plants. Adults take nectar from buddleia and other flowers. Greek *eu* + *plagios* = 'very' + 'oblique', and Latin *quadripunctaria* = 'four-spotted', terms referring to the patterns on the fore- and hindwings, respectively.

Ruby tiger *Phragmatobia fuliginosa*, wingspan 30–5 mm, widespread and often common, day- and night-flying from April to September, in habitats including downland, heathland, moorland, woodland clearings, dunes and gardens. Population levels increased by 296% between 1968 and 2007. Larvae feed on a range of herbaceous plants. Greek *phragmos* + *bioo* = 'fence' + 'to live'. and Latin *fuliginosa* = 'sooty', from the forewing colour.

Scarlet tiger *Callimorpha dominula*, wingspan 45–55 mm, day- and night-flying in June–July, locally common in south and south-west England, south Wales and parts of north-west England in damp habitats, as well as rocky coastal cliffs, occasionally woodland and gardens. Larvae feed on herbaceous plants, bramble and sallows. This is one of the few tiger moths with developed mouthparts, allowing adults to feed on nectar. *Calos* + *morphe* = 'beautiful' + 'shape', and the Latin diminutive of *domina* = 'household mistress', possibly from the colourful wings.

Wood tiger *Parasemia plantaginis*, wingspan 32–8 mm, widespread but local on heathland and downland, but it has disappeared from many sites, especially in southern England. Males often fly in sunshine, but females tend to be mainly nocturnal, both flying in May–July. Larvae feed on a variety of herbaceous plants. A neck secretion provides a chemical defence against birds, another from the abdomen deters ants. Greek *parasemon* = 'mark of distinction', as with the wing pattern, and Latin *plantago* = 'plantain'.

TIGER WORM *Eisenia fetida* (Lumbricidae), widespread and common epigean earthworms, rarely found in soil, favouring organic material such as rotting vegetation, compost and manure. Banding prompts the English name. It exudes a pungent liquid when handled (an anti-predator response), leading to the specific name, Latin for 'foul-smelling'.

TIGER'S EYE *Coltricia perennis* (Hymenochaetaceae), an inedible fairly widespread (less so in Ireland) and locally found fungus with a cup-like fruiting body, found throughout the year but sporulating in summer and autumn on humus-rich sandy soil on woodland edges and acidic heathland, often by the sides of paths. Latin *coltricia* = 'seat', and *perennis* = 'perennial'.

TILIACEAE Former family name for lime tree (Latin *tilia*), now Malvaceae.

TIMMIA **Norway timmia** *Timmia norvegica* and **sheathed timmia** *T. austriaca* (Timmiaceae), both scarce mosses recorded from the Highlands. Joachim Timm (1734–1805) was an eighteenth-century German botanist.

TIMOTHY *Phleum pratense* (Poaceae), a common and widespread perennial grass growing in meadow, pasture, rough grassland, field margins and waysides, sown in grasslands, and found as a casual from wool shoddy and birdseed. 'Timothy' is named after a Swedish immigrant Timothy Hanson, a farmer who introduced it to the Carolinas from Maryland in *c*.1720. From here (rather than its native Europe) it was brought over to become a major source of hay and cattle fodder to British farmers in the mid-eighteenth century. Greek *phleos*, the name given to a reed, and from Latin *pratense* = 'meadow'.

TINEIDAE Family name for clothes moths (Latin *tinea* = 'moth').

TINGIDAE From New Latin *tingis*, for certain insects, the family name for lace-bugs.

TIPHIA **Small tiphia** *Tiphia minuta* (Tiphiidae), a parasitic wasp, 4–7 mm long, widespread south of the Humber, in heathland, downland and other grasslands, open woodland, and coastal dunes, flying in summer. The female burrows into the soil to find larvae, usually of dung beetles; she breaks into the host's cell, stinging the larva to temporary paralysis. *Tiphe* is Greek for a kind of insect.

TIPULIDAE Family name for craneflies, with 87 species in 8 genera, the most speciose being *Tipula* and *Nephrotoma*. *Tipula* is actually Latin for a kind of water spider. Adults have long, stilt-like legs and slender wings, and are also called daddy-long-legs. The soil-dwelling larvae (leatherjackets) are commonly garden and agricultural pests, feeding on roots of grass and crops, some species also feeding on the leaves and crown of crop plants.

TISCHERIIDAE Family name for carl moths. Carl von Tischer (1777–1849) was a German entomologist.

TISSUE (MOTH) *Triphosa dubitata* (Geometridae, Larentiinae), wingspan 38–48 mm, widespread but local in woodland, scrub, hedgerows, calcareous grassland and acid heathland, rare in north Scotland. Adults fly in August–September, overwintering in this stage, reappearing in April–May. Larvae feed on buckthorn and alder buckthorn. Greek *tri* + *phos* = 'three' + 'light', from the gloss on the wing, and Latin *dubitata* = 'doubtful' (i.e. whether a distinct species).

Scarce tissue *Hydria cervinalis*, wingspan 40–8 mm, local in hedgerows, gardens, parks and verges throughout much of England, scattered in Wales and Scotland. Adults fly in April–June. Larvae feed on barberry. Greek *hydor* = 'water', and Latin *cervinus* = 'tawny' (*cervus* = 'deer').

TIT Small birds in the families Aegithalidae, Panuridae and Paridae. Also known as 'titmouse', dating from Middle English *tit* + obsolete *mose*: the change in the ending in the sixteenth century was due to the association with 'mouse', probably because of the bird's size and quick movements. **Penduline tit** *Remiz pendulinus* is an accidental.

Bearded tit *Panurus biarmicus* (Panuridae), wingspan 17 cm, length 12 cm, weight 15 g, in reedbeds in parts of east, south and north-west England (Leighton Moss). In Ireland a handful of pairs breed in Co. Wexford. The largest population has been found in reedbeds at the mouth of the River Tay in (Perth and Kinross, Scotland), where there may be >250 pairs. Overall in Britain, there was a minimum of 630 pairs at 72 sites indicated by BBS data in 2016, and a five-year mean of 630 pairs. Nests are in reedbeds, often colonial. Clutch size is 4–8, incubation lasts 10–14 days, and the altricial young fledge in 12–13 days. Diet is insectivorous, with reed seeds taken in winter. Males have a black 'moustache' rather than a 'beard'. Greek *panu* + *oura* = 'exceedingly' + 'tail'. *Biarmicus* is from 'Biarmia', a Latinised form of Bjarmaland, formerly part of Arkhangelsk Oblast, Russia.

Blue tit *Cyanistes caeruleus* (Paridae), wingspan 18 cm, length 12 cm, weight 11 g, common and widespread in woodland, hedgerows, parks and gardens, with 3.4 million territories estimated in 2009, population size increasing by 21% over 1970–2015, if with a weak decline (–6%) over 2010–15. Nests are in tree holes and nest boxes, with clutches usually of 8–10, incubation taking 13–15 days, the altricial young fledging in 18–21 days. Diet is of insects, including caterpillars, seeds and nuts. The binomial comes from the Greek for 'dark blue' and Latin for 'sky blue'.

Coal tit *Periparus ater* (Paridae), wingspan 19 cm, length 12 cm, weight 9 g, common and widespread in (especially coniferous) woodland, parks and gardens, with 760,000 territories in Britain in 2009. Numbers have been stable, with an increase by 15% over 1970–2015, though this included a decline by 16% in 2010–15. Nests are in tree cavities, rock crevices and occasionally old burrows and dreys. Clutch size is 9–10, incubation lasts 14–16 days, and the altricial young fledge in 16–19 days. Food is insects and spiders, and in winter seed and nuts. *Periparus* is from Greek prefix *peri* = 'about' or 'around' and Latin for 'tit'; *ater* is Latin for 'dusky-black'.

Crested tit *Lophophanes cristatus* (Paridae), wingspan 18 cm, length 12 cm, weight 12 g, largely confined to ancient Caledonian pine forests and Scots pine plantations in the Highlands, especially in Inverness-shire and Strathspey, with 1,000–2,000 pairs estimated in 2007. The tree-cavity nests usually have 5–6 eggs that hatch after 13–16 days, the altricial young fledging in 20–5 days. Diet is mostly of insects and spiders, with plant material especially conifer seeds, taken in winter. Greek *lophos* + *phanes* = 'crest' + 'light colour', and Latin *cristatus* = 'crested'.

Great tit *Parus major* (Paridae), wingspan 24 cm, length 14 cm, weight 18 g, common and widespread in woodland, parks and gardens, with 2.5 million territories estimated in Britain in 2009, and with a growth in numbers by 80% over 1970–2015 (40% over 1995–2015, though including a weak decline of –6% over 2010–15). Nests are in tree holes and nest boxes, with a clutch size usually of 7–9 (but up to 12), incubation taking 13–15 days, fledging of the altricial young 18–21 days. Feeding is on insects and spiders, especially in summer, and seeds and nuts. Size of the black stripe (badge) on the breast of a male, controlled by testosterone level, is an indicator of status in inter-male competition and female selection. Latin *parus* and *maior* = 'tit' and 'greater'.

Long-tailed tit *Aegithalos caudatus* (Aegithalidae), wingspan 18 cm, length 14 cm, weight 9 g, common and widespread in woodland, hedgerows, scrub, parks and gardens, with 330,000 territories in Britain in 2009, an increase in numbers by 97% over 1970–2015, though only by 12% between 1995 and 2015, and with a slight decline (–7%) over 2010–15. Nests are built of moss bound by spider webs and covered in lichens. Clutch size is 6–8, incubation lasts 15–18 days, and the altricial young fledge in 16–17 days. Older previous offspring often help adults raise their chicks. Diet is insectivorous, with seed also taken in winter. Greek *aigithalos* = 'tit' and Latin *cauda* = 'tail'.

Marsh tit *Poecile palustris* (Paridae), wingspan 19 cm, length 12 cm, weight 12 g, found in England, Wales and south Scotland (most abundant in south Wales and southern and eastern England), despite their name most commonly in broadleaf woodland, also parks and gardens. There were 41,000 territories in Britain in 2009, but declines in number of 75% between 1970 and 2015, 32% in 1995–2014, and 22% in 2010–15 justify it being Red-listed in 2002. It nests in pre-existing tree holes. Clutch size is 7–9, incubation taking 14–16 days, the altricial young fledging in 18–21 days. Food is insects and spiders, and in winter seeds, fruit and nuts, with some items cached. Greek *poicilos* = 'spotted' and Latin *palus* = 'marsh'.

Willow tit *Poecile montana*, wingspan 19 cm, length 12 cm, weight 12 g, found in England, Wales and south Scotland in willow scrub in damp habitats, with 3,400 pairs probably optimistically estimated in Britain in 2009. It has been Red-listed since 2002 owing to a decline by 93% over 1970–2015, 77% over 1995–2015, and 10% over 2010–15, one of the greatest overall declines in any bird species, with loss of damp scrub woodland an important reason. The nesting hole is often excavated by the birds themselves in decaying wood. Clutch size is usually 6–8, incubation lasts around 14 days, and the altricial young fledge in 17–20 days. Food is of invertebrates, with seed also taken in winter. Latin *montanus* = 'mountaineer' (from *mons* = 'mountain').

TOAD Short-bodied, tail-less members of the order Anura, with two native species, and a rare introduction. 'Toad' comes from Old English *tadde* or *tada*, an abbreviation of *tadige*, of unknown origin.

Common toad *Bufo bufo* (Bufonidae), males up to 8 cm long, females up to 13 cm, common and widespread in Britain, though in 2016 numbers were estimated to have fallen by two-thirds since the 1980s, especially in south-east England; in eastern England, numbers had recovered since 2005, but not enough to reverse previous losses. Toads favour woodlands and damp areas (including gardens) with plenty of cover. They emerge from hibernation in late February and return to the same pond

year after year. Breeding is in February–March. Spawn is laid in strings each of which contains a double row of eggs. Tadpoles metamorphose in June–July. Adults spend the day in hollows in the ground, coming out after dark to feed on ants, slugs and earthworms. *Bufo* is Latin for 'toad'.

Midwife toad *Alytes obstetricans* (Alytidae), adults up to 5 cm long, tadpoles, 6–7 cm, so named because after the female expels her eggs, the male fertilises them externally then wraps them in a string around his legs to protect them from predators. When they are ready to hatch, the male wades into shallow water, where he allows the tadpoles to emerge. Colonies are currently found in Bedfordshire (where first introduced, accidentally, having persisted since the early twentieth century), Devon, York, Northamptonshire, Worksop (Nottinghamshire) and, in Powys, Llandrindod Wells since 2004, and Newbridge-on-Wye since 2007. Greek *alytos* = 'continuous', from the connected egg-mass, Latin *obstetrix* = 'midwife'.

Natterjack toad *Epidalea calamita* (Bufonidae), males up to 7 cm long, females up to 8 cm, a rarity now confined to a few sites in coastal dunes and lowland heaths. In England, the dune habitat is protected by a number of NNRs, for example at Hoylake, Ainsdale Sand Dunes, North Walney and Sandscale Haws. In Scotland, the species is confined to the Solway Firth, with a reserve at Caerlaverock. In Wales the species became extinct in the late twentieth century, but in 2003 and 2004 tadpoles were successfully translocated from Sefton (Lancashire) to newly created ponds at Gronant Dunes LNR (Denbighshire) and Talacre Dunes (Flintshire). In 2015, the Republic of Ireland announced spending of €250,000 (€48,000) p.a. to farmers and landowners to manage new breeding sites around Castlemaine Harbour and at Castlegregory, Co Kerry. The preferred habitat consists of small, shallow temporary pools where invertebrate predator numbers are kept low due to seasonal desiccation of the ponds. Breeding is from March to May. Adults lay a string of eggs, each string containing a single row of eggs. Tadpoles metamorphose in June–July. The relatively short legs give them a distinctive gait. The species is protected under the WCA (1981), and classified as Endangered, with a number of reintroduction programmes being implemented. 'Natterjack' dates from the eighteenth century, possibly derived from its loud croak + Jack as a proper name. *Epidalea* may be from Greek *epidalia* = 'plumage', and Latin *calamus* = 'reed'.

TOADFLAX Herbaceous plants in the genera *Linaria* (Latin for 'flax-like'), *Chaenorhinum* (Greek *chaeno* + *rhinos* = 'gape' + 'nose', from the flower shape), and *Cymbalaria* (Latin for 'cymbal-like'), in the family Veronicaceae. See also bastard-toadflax.

Annual toadflax *L. maroccana*, **Balkan toadflax** *L. dalmatica*, **Italian toadflax** *Cy. pallida*, **pale toadflax** *L. repens* and **prostrate toadflax** *L. supina* are neophytes casual on waste ground, roadsides, walls and by railways.

Common toadflax *L. vulgaris*, a perennial generally common and widespread but absent from much of Ireland and highland Scotland, growing in waste ground, hedge banks, verges, railway embankments and cultivated land, especially on calcareous soils. *Vulgaris* is Latin for 'common'.

Ivy-leaved toadflax *Cy. muralis*, a common and widespread south European neophyte perennial introduced by 1602, found on walls and bridges, pavements, and in other well-drained stony places, often near habitation. It is also found on shingle beaches. *Muralis* is from Latin for 'wall'.

Purple toadflax *L. purpurea*, an Italian neophyte perennial in cultivation by 1648, common and widespread in lowland regions as a naturalised garden escape on waste ground, roadsides, railways, pavements, and walls.

Sand toadflax *L. arenaria*, an annual south-west European neophyte planted at Braunton Burrows, north Devon, and naturalised on semi-fixed dunes, casual in a few other locations. *Arena* is Latin for 'sand'.

Small toadflax *Ch. minus*, a spring-germinating annual archaeophyte, common and widespread except in north Scotland in open habitats on well-drained, often calcareous, soils, including arable fields, waste ground, walls, quarries, and especially along railways. *Minus* is Latin for 'less'.

TOADSTOOL With mushroom, a botanically non-meaningful term for the fleshy, spore-bearing fruiting body of a fungus, typically produced above ground on soil or on its food source.

TOBACCO *Nicotiana tabacum* (Solanaceae), an annual Central and South American neophyte, casual or relic on a few tips in south-east England and north-west Midlands. 'Tobacco' dates from the mid-sixteenth century, from Spanish *tabaco*. Jean Nicot was a sixteenth-century diplomat who introduced the plant to France. **Sweet tobacco** (*N. alata*), **red tobacco** (*N. forgetiana*) and the North American **wild tobacco** (*S. rustica*) have similar descriptions.

TOFIELDIACEAE Family name for Scottish asphodel, honouring the English botanist Thomas Tofield (1730–79).

TOMATILLO *Physalis ixocarpa* (Solanaceae), an annual American neophyte introduced with wool shoddy, an occasional casual in England and Scotland. 'Tomatillo' is a diminutive of 'tomato'. Greek *physalis* = 'bladder', and *ixos* = 'sticky' (and 'mistletoe berry') + *carpos* = 'fruit'.

TOMATO *Solanum lycopersicum* (Solanaceae), a scrambling annual neophyte from Central and South America in cultivation for its fruit by 1595, now found on waste ground, tips and sewage-works. Being frost-sensitive, most populations are killed each year, with new plants germinating from seeds discarded as fresh fruit or from human sewage. 'Tomato' dates to the early seventeenth century, from the French, Spanish or Portuguese *tomate*, from Nahuatl (Aztec) *tomatl*. Latin *solanum* = 'nightshade', and Greek *lycos* = 'wolf' + Latin *persica* = 'peach', from the belief that tomatoes were poisonous.

TONGUES OF FIRE *Gymnosporangium clavariiforme* (Pucciniaceae), a locally common rust fungus with a scattered distribution in Britain, producing orange fruit bodies on its initial host, juniper, subsequently producing yellow depressions on the leaves of its secondary host, hawthorn, whose fruits (haws) sprout small white tubes that produce spores of the rust (which must then reach a juniper plant). Greek *gymnos* = 'naked' + 'seed', and Latin *clavariiforme* = 'club-shaped'.

TOOTH (FUNGUS) Usually inedible species in the genera *Hericium* (Russulales, Hericiaceae), resembling a mass of fragile icicle-like spines, growing from (usually) beech (Latin *hericium* = 'hedgehog', a reference to the spiny fertile surfaces); *Pseudohydnum*, from Greek *pseudo* + *hydnon* = 'false' + a kind of edible fungus (Auriculariales); and *Bankera*, *Hydnellum* (*hydnon* = 'edible fungus'), *Phellodon* (Greek *phellos* + *odontos* = 'cork' + 'teeth') and *Sarcodon* (Greek *sarcos* + *odontos* = 'flesh' + 'teeth') (Thelephorales, Bankeraceae).

Rarities include **bitter tooth** *S. scabrosus* (south-east England and Highlands), **black tooth** *Ph. niger* (southern England, Wales and Highlands), **blue tooth** *Hy. caeruleum* (Highlands, ectomycorrhizal on Scots pine), **coral tooth** *He. coralloides* (England, on beech), **devil's tooth** *Hy. pecki* (Highlands), **drab tooth** *B. fuligineoalba* (Highlands), **greenfoot tooth** *S. glaucopus* (Highlands), **mealy tooth** *Hy. ferrugineum* and **orange tooth** *Hy. aurantiacum* (both Highlands and a few English sites, ectomycorrhizal on Scots pine), **scaly tooth** *S. squamosus*, rare (southern England and, mainly, Highlands), and **spruce tooth** *B. violascens* (12 records over 1996–2012 in Aberdeenshire/north-east Highlands).

Bearded tooth *He. erinaceus*, edible, uncommon and scattered in mostly southern England, growing from the trunks of often veteran beech. Near Threatened in the RDL (2006), listed on Schedule 8 of the WCA (1981), and a priority species in the updated UK BAP (2010). *Erinaceus* is Latin for 'hedgehog'.

Fused tooth *Ph. confluens*, uncommon, found in lowland regions (south and south-east England, Norfolk) and uplands (south Pennines, Lake District, Highlands) in late summer and autumn in deciduous (beech, sweet chestnut, oak) and coniferous woodland. *Confluens* is Latin for 'flowing together'.

Grey tooth *Ph. melaleucus*, uncommon, scattered in southern England, Wales and the Highlands, in late summer and autumn on deciduous and coniferous woodland soil, often with bilberry. Greek *melas* + *leucon* = 'black' + 'white'.

Ridged tooth *Hy. scrobiculatum*, scarce (a priority species in the UK BAP), slightly commoner in England than in Wales or Scotland, found in late summer and autumn on soil in woodland, tracksides, verges and banks. *Scrobiculatum* is Latin for 'small trench'.

Tiered tooth *He. cirrhatum*, edible, uncommon, found in especially southern England, growing from the trunks and stumps of mainly beech, but also oak and horse-chestnut in deciduous woodland. *Cirrhatum* is Greek for 'curled'.

Velvet tooth *Hy. spongiosipes*, scarce, with a scattered distribution in southern England, East Anglia and south Wales, in late summer and autumn on soil in mainly deciduous woodland, tracksides, verges and banks, under oak and (being ectomycorrhizal on) sweet chestnut in England, and a priority species in the UK BAP.

Woolly tooth *Ph. tomentosus*, uncommon, scattered in southern England, the Welsh Borders and the Highlands, in late summer and autumn on soil in coniferous and mixed woodland, often associated with bilberry. *Tomentosus* is Latin for 'matted' or 'hairy'.

Zoned tooth *Hy. concrescens*, scarce, with a scattered distribution in the northern and southern thirds of Britain, in late summer and autumn on soil in broadleaf woodland in England, especially under oak and sweet chestnut, and perhaps ectomycorrhizal on Scots pine, so also found on soil in native pine forest in the Highlands. A priority species in the UK BAP. *Concrescens* is Latin for 'congealed'.

TOOTH-STRIPED (MOTH) Nocturnal species of *Trichopteryx* (Geometridae, Larentiinae), from Greek *trichous* + *pteryx* = 'holding three' + 'wing', from a lobe on the male hindwing that resembles a third wing.

Barred tooth-striped *T. polycommata*, wingspan 33–6 mm, a rarity on chalk downland, in scrub, hedgerows and open woodland, discontinuously in Britain from Sussex to Cumbria, a priority species in the UK BAP. Adults fly in March–April. Larvae feed on wild privet and ash in late spring and summer. Greek *poly* + *comma* = 'many' + 'comma mark', from forewing markings.

Early tooth-striped *T. carpinata*, wingspan 30–4 mm, widespread and common in broad-leaved woodland, scrubby heathland and fen. Population levels increased by 220% between 1968 and 2007. Adults fly in April–May. Larvae feed on honeysuckle, alder, sallows and birches. Latin *carpinus* = 'hornbeam', not a normal food plant.

TOOTHCRUST Orchard toothcrust *Sarcodontia crocea* (Meruliaceae), scattered in England, Vulnerable in the RDL (2006), and a priority species in the UK BAP, found on living and dead wood of apple. Greek *sarcos* + *odontos* = 'flesh' + 'teeth', and Latin *crocea* = 'saffron yellow'.

Weeping toothcrust *Dacryobolus sudans* (Fomitopsidaceae), Endangered in the RDL (2006), found recently only in the Highlands, with four records between 1997 and 2013, in coniferous plantations on trunks and fallen branches of pine. Greek *dacry* + *obolos* = '(eye) tear' + 'small coin', and Latin *sudans* = 'sweating'.

TOOTHPICK-PLANT *Ammi visnaga* (Apiaceae), a rare annual Mediterranean neophyte, scattered in parkland and neglected gardens, and as a casual on spoil tips and roadsides, originating from wool shoddy or birdseed. *Ammi* is Greek for a North African plant, probably this species, *visnaga* American

Spanish for 'toothpick', from the oval leaves being dissected into small lance-shaped segments.

TOOTHWORT *Lathraea squamaria* (Orobanchaceae), a chlorophyll-less annual or perennial herb, parasitic on the roots of a range of woody plants, especially hazel, ash and elm in deciduous woodland, hedgerows, and river and stream banks, on moist fertile soil, locally abundant but absent from south-west England, East Anglia, west Wales and north Scotland. A resemblance of the flowering and fruiting stem to a row of teeth gives it its common name. Greek *lathraios* = 'hidden', from its underground existence, and Latin *squama* = 'scale'.

Purple toothwort *L. clandestina*, a chlorophyll-less European neophyte root parasite mainly found on alder, willows and poplars, introduced by 1888, scattered and locally common in damp, shaded places in woodland and along hedgerows, especially near stream margins where running water can facilitate dispersal. *Clandestina* is Latin for 'hidden'.

TOP SHELL Marine gastropods.

Flat top shell *Gibbula umbilicalis* (Trochidae), shell width 2.2 cm, height 1.6 cm, on the west coast of Britain and on suitable shores in Ireland, from the upper intertidal into the sublittoral on sheltered rocky sites, feeding on microalgal films and detritus. Latin *gibbus* = 'bent', *umbilicatus* = 'navel-shaped'.

Grey top shell *G. cineraria*, shell width 1.7 cm, height 1.5 cm, a common and widespread algal grazer and detritivore, sometimes found in rock pools, more generally on lower levels of rocky shores on algae and under stones to a depth of 130 m. *Cineraria* is Latin for 'ashy grey'.

Grooved top shell *Jujubinus striatus* (Trochidae), shell height 10 mm, width 8 mm, locally common in the Channel Islands, and with recent records from Bognor Regis, Falmouth, and the Isles of Scilly, at low water spring tides down to 300 m on macroalgae and eelgrass. Jujube is another name for red date *Ziziphus jujuba*. *Striatus* is Latin for 'grooved'.

Lined top shell Synonym of thick top shell.

Painted top shell *Calliostoma zizyphinum* (Calliostomatidae), shell length and width both 3 cm, common and widespread on macroalgae-covered rocky shores from extreme low water to 300 m depth feeding on sessile invertebrates, especially hydroids. Greek *cali* + *stoma* = 'beautiful' + 'mouth', and *zizyphon* = 'jujube' (red date *Ziziphus jujuba*).

Pearly top shell *Margarites helicinus* (Margaritidae), shell length 9 mm, widespread north of Dublin and Yorkshire, especially common on the coasts of west Scotland, Orkney and Shetland, from the lower shore to the sublittoral under stones, in pools and on macroalgae. Greek *margarites* = 'pearl', and *helix* = 'helical'.

Purple topshell Synonym of flat topshell.

Thick top shell *Phorcus lineatus* (Trochidae), shell reaching up to 3 cm tall, 3 cm basal diameter, abundant, grazing on microalgae in the midshore region of moderately exposed rocky shores in Britain reaching its northern limits on Anglesey and eastern limits in West Sussex, and found on all but the east coast in Ireland. Greek *phorcos* = 'white' or 'grey', and Latin *lineatus* = 'lined'.

Toothed top shell Synonym of thick top shell.

TOR-GRASS *Brachypodium rupestre* (Poaceae), a perennial of dry, relatively infertile calcareous soils, mainly in England. In chalk grassland it is often dominant over large areas, and is also found in scrub, quarries, and on railway banks and roadsides. Greek *brachys* + *podion* = 'short' + 'small foot', referring to the short pedicels, and Latin *rupestre* = 'growing among rocks'.

TORMENTIL *Potentilla erecta* (Rosaceae), a common and widespread perennial herb found on more or less acidic soils in grassland, moorland, heathland, blanket and raised bog, open woodland, woodland edge, roadsides and hedge banks. Subspecies *erecta* is common in lowland areas, ssp. *strictissima* more

in upland parts in acidic habitats on mineral soils. 'Tormentil' dates to late Middle English, from French *tormentille*, possibly from Latin *tormentilla* = 'a little torment', from the plant's use in relieving pain. Latin *potentilla* = 'small powerful one', referencing medicinal properties, and *erecta* = 'upright'.

Trailing tormentil *P. anglica*, a procumbent perennial, widespread (but absent from most of the northern half of Scotland), on heathland, dry banks, and woodland and field edges, usually on well-drained acidic soils, but avoiding podzols. It also occurs on waste ground and railway banks. Latin *anglica* = 'English'.

TORTOISE BEETLE Species of *Cassida* (Chrysomelidae, Cassidinae), from Latin *cassis* = 'helmet', the rounded shape resembling a tortoise. Specifically, *Cassida vibex*, 5.5–7.5 mm length, widespread in the southern half of England, more local elsewhere in England and in Wales and Ireland, adults found on roadside and field margins, often on thistles and yarrow in summer. *Vibex* is Latin for 'weal'.

Bordered tortoise beetle *C. vittata*, 5–6.5 mm length, scattered in England and Wales, common in the south, feeding on Chenopodiaceae and various plants in other families. *Vittata* is Latin here meaning 'striped', from the elytral stripe.

Fleabane tortoise beetle *C. murraea*, 6.5–9 mm length, common in southern and south-west England and south Wales, scattered in other parts of England, with a preference for damp habitat supporting its food plants common fleabane and marsh thistle. Johan Murray (1740–91) was a Swedish physician and botanist.

Green tortoise beetle *C. viridis*, 7–10 mm length, widespread, especially in England and Wales, feeding on plants in the Lamiaceae, especially white dead-nettle and mints. Larvae are spiny allowing them to hold a bundle of cast skins and droppings over their back, used to deter predators and parasites. Adults grip on to a leaf and pull themselves down, presenting no opportunity for predators to grip. *Viridis* is Latin for 'green'.

Pale tortoise beetle *C. flaveola*, 4–6 mm length, widespread in long grass and coarse vegetation, feeding on chickweed and other Caryophyllaceae. *Flaveola* is Latin for 'yellowish'.

Thistle tortoise beetle *C. rubiginosa*, 6–8 mm length, common and widespread in England and Wales, more scattered in Scotland and Ireland, in rough grassland containing the host plants, thistles and other Asteraceae. Larvae have twin tail-spikes which can be used to carry dead skins and droppings in a kind of parasol. Latin *rubiginosa* = 'rust-coloured', referring to this beetle's ability to produce a red liquid (haemolymph) from the head as a defensive behaviour.

TORTOISESHELL **Small tortoiseshell** *Aglais urticae* (Nymphalidae, Nymphalinae), male wingspan 45–55 mm, females 52–62 mm, a common and widespread butterfly, though numbers have recently been declining in the south, possibly a consequence of parasitism by a fly, *Sturmia bella*, thus causing some conservation concern. Between 1976 and 2014, for example, overall numbers declined by 73%, but this includes a resurgence by 146% in 2005–14. Similarly, there was an increase in range by 13% in 2005–14. This butterfly usually has two broods each year, and hibernates in buildings, wood piles and hollow trees. On hatching from common nettle, larvae build a communal web, usually at the top of a nettle plant, from which they emerge to bask and feed. Adults can be seen from March to October, peak flight periods being in April–May and August. Adults feed on the nectar of a range of plants, including thistles, bramble and Michaelmas-daisy. Greek *aglaia* = 'beauty', and Latin *urtica* = 'nettle'.

TORTRICIDAE A large moth family including tortrix (from Latin for 'twisted') and twist species (both so called because the larvae of many species live in 'spinnings' where a leaf is twisted or rolled around and attached together with silk). See also bell, button, conch, drill, marble, piercer, roller, shade and shoot moths.

TORTRIX (MOTH) Members of the family Tortricidae,

mostly in the subfamily Tortricinae (Latin *tortrix* = 'twisted'), most species having nocturnal adults. As a group, sixth in the Royal Horticultural Society's list of worst garden pests of 2017.

Barred fruit-tree tortrix *Pandemis cerasana*, wingspan 16–25 mm, widespread and common in woodland, orchards and gardens. Adults fly in June–September. Larvae feed on the leaves of a number of tree species, including fruit trees such as dwarf cherry *Prunus cerasus*, giving the specific name. Greek *pandemos* = 'belonging to the people', here meaning 'common'.

Bilberry tortrix *Aphelia viburnana*, wingspan 15–22 mm, widespread and common on moorland, bogs and in woodland. Adults fly in June–August. Larvae feed on the leaves of a variety of herbaceous plants, including bilberry. Greek *apheleia* = 'plainness', from the wing colour, and *viburnana* from wayfaring-tree *Viburnus lantana*, a food plant on the Continent.

Brown oak tortrix *Archips crataegana*, male wingspan 19–22 mm, female 23–8 mm, local in woodland in parts of England, Wales and Ireland. Adults fly in June–August. Larvae feed on the leaves of a number of trees, including hawthorn *Crataegus*, giving the specific name. Greek *archi* + *ips* = 'chief' + 'larva'.

Carnation tortrix *Cacoecimorpha pronubana*, wingspan 14–24 mm, adventive, naturalised following accidental introduction in garden plants in southern England in 1905, now common throughout much of Britain and south and east Ireland. Adults fly in May–June and again in August–September. Larvae feed on leaves of privet, sea-buckthorn, spindle and other deciduous trees and shrubs. Greek *morphe* = 'shape' and like the genus *Cacoecia*, in turn from *cacos* + *oicos* = 'bad' + 'house', describing the spun larval cases on (fruit) trees (i.e. as pests). *Pronubana* means having an affinity with *Noctua pronuba*, both having yellow hindwings.

Cereal tortrix *Cnephasia pumicana*, wingspan 15–20 mm, scarce, probably a recent arrival and distinguished in 1986, found in south-east England. Adults fly in June–August. Larvae feed on grasses, including cereals. Greek *cnephas* = 'darkness', and Latin *pumica* = 'pumice', both names describing the grey forewings.

Chequered fruit-tree tortrix *Pandemis corylana*, wingspan 18–21 mm, widespread and common in woodland and gardens. Adults fly in July–August. Larvae feed on leaves of a number of deciduous tree species, including hazel *Corylus*, giving the specific name.

Cherry bark tortrix *Enarmonia formosana* (Olethreutinae), wingspan 15–19 mm, widespread but local in gardens, parks and orchards. Adults fly in June–July. Larvae feed on the bark of cherry trees. Greek *enarmonios* = 'harmony', and Latin *formosus* = 'beautiful'.

Cyclamen tortrix *Clepsis spectrana*, wingspan 16–22 mm, widespread and common in damp woodland, fen, saltings, marsh and gardens, less recorded in northern Scotland. Adults fly in May–September. Larvae, most commonly found in May–June, feed on leaves of a variety of plants. Greek *clepto* = 'to conceal', from the larval cases, and Latin *spectrum* = 'image'.

Dark fruit-tree tortrix *Pandemis heparana*, wingspan 16–24 mm, widespread and common in woodland, orchards and gardens. Adults fly in June–August, later in the north. Larvae feed on leaves of a number of tree species, including fruit trees on which they can become pests, attacking the flowers of apple, pear, plum, currant, etc. Greek *hepar* = 'liver', from the forewing colour.

Flax tortrix *Cnephasia asseclana*, wingspan 15–18 mm, widespread and common in gardens, farmland and other open areas. Adults fly in June–August. Larvae feed on a variety of herbaceous plants, initially mining leaves, then spinning together leaves or flowers. Latin *assecla* = 'attendant' (i.e. there are similar species).

Fruitlet mining tortrix *Pammene rhediella* (Olethreutinae), wingspan 9–12 mm, widespread but local in hedgerows, orchards and scrub, less numerous in the north. Adults fly in May–June, especially in sunshine. Larvae mine flowers and unripe fruit of hawthorn, rowan, crab apple and occasionally cultivated apple and pear. Greek *pan* + *mene* = 'all' + 'moon', from the circular wing

markings of some species. Hendrik van Rheede tot Draakenstein (1636–91) was a Dutch colonial administrator and naturalist.

Garden rose tortrix *Acleris variegana*, wingspan 14–18 mm, widespread and common in gardens and woodland. Adults fly in July–September. Larvae favour feeding on rose leaves but also use other rosaceous plants, for example blackthorn. Greek *acleros* = 'unallotted', and Latin *variegana* = 'variegated', from the forewing patterns in some forms.

Green oak tortrix *Tortrix viridana*, wingspan 18–23 mm, widespread and common in oak woodland, parks, gardens and areas with scattered trees. Adults fly in May–June. Larvae feed on leaves of oak and other trees which they sometimes severely defoliate. Greek *tortrix* = 'twister', referencing the larval cases, and Latin *viridis* = 'green', from the forewing colour.

Grey tortrix *Cnephasia stephensiana*, wingspan 18–22 mm, widespread and common in a range of habitats. Adults fly in July–August. Larvae feed on a variety of herbaceous plants. James Stephens (1792–1852) was an English entomologist.

Holly tortrix *Rhopobota naevana*, wingspan 12–16 mm, widespread and common in woodland, heathland and gardens. Adults fly in July–September. Larvae mine leaves of a number of trees, including holly, apple and hawthorn. Greek *rhops* + *bosco* = 'shrub' + 'to eat', from the food plants, and Latin *naevus* = '(skin) mole', from the dark patch on the hindwing.

Larch tortrix *Zeiraphera griseana* (Olethreutinae), wingspan 16–22 mm, widespread but local in woodland, parks and gardens, though rare in Ireland. Adults fly in July. Larvae feed on needles of European larch and other conifers. Greek *zeira* + *phero* = 'loose garment' + 'to carry', referring to the larval cases, and Latin *griseus* = 'grey', from the wing colour.

Large fruit-tree tortrix *Archips podana*, wingspan 18–26 mm, common in a range of habitats, including gardens, orchards, hedgerows and woodland, throughout England, Wales, Ireland and south-west Scotland. Adults mainly fly in June–July, but can be found from May to September. Larvae feed on leaves, flowers and fruit of deciduous trees, including apple, plum, sloe and cherry (sometimes becoming a pest), occasionally conifers. Nikolaus Poda von Neuhaus (1723–98) was a German physicist and entomologist.

Light grey tortrix *Cnephasia incertana*, wingspan 14–18 mm, widespread and common in woodland edges and hedgerows. Adults fly in June–July. Larvae feed on herbaceous species including plantains and sorrels. Latin *incertus* = 'uncertain', from the variability of the forewing pattern or the uncertainty of its status as a distinct species.

Marbled orchard tortrix *Hedya nubiferana* (Olethreutinae), wingspan 15–21 mm, widespread and common in hedgerows, gardens and orchards. Adults fly in June–August. Larvae mimic bird droppings and feed on the leaves of deciduous tree species, including hawthorn, blackthorn and wild cherry. Greek *hedys* = 'pleasing', and Latin *nubifer* = 'cloud-bearing', from the grey colour on part of the forewing.

Notch-wing tortrix *Acleris emargana*, wingspan 18–22 mm, widespread and common in woodland, gardens and hedgerows, less numerous in the north. Adults fly in July–September. Larvae feed on the leaves and shoots of trees such as birches, sallow and poplars. *Emargana* is Latin for 'emarginate', from the forewing shape.

Pine cone tortrix *Gravitarmata margarotana* (Olethreutinae), wingspan 14–18 mm, an adult first recorded in Britain in May 2011 in Clowes Wood, Kent. Latin *gravis* + *armata* = 'heavy' + 'armed', and *margarita* = 'pearl', from the white forewing marking.

Plum tortrix *Hedya pruniana*, wingspan 15–18 mm, widespread and common in woodlands, hedgerows, gardens and orchards. Adults fly in May–July. Larvae feed on leaves of blackthorn, plum and wild cherry, species of *Prunus*, giving the specific name.

Red-barred tortrix *Ditula angustiorana* (Tortricinae), wingspan 12–18 mm, widespread and common in woodland, parks and gardens, less numerous in the north. Adults fly in June–July, both sexes at night, males sometimes in sunshine. Larvae feed on a variety of trees and shrubs, including yew and rhododendron which have few grazers. Greek *ditulos* = 'two-humped', from the thorax crest, and Latin *angustior* = 'narrower', referring to the forewing.

Rhomboid tortrix *Acleris rhombana*, wingspan 13–19 mm, widespread and common in woodland, gardens and hedgerows. Adults fly in August–October. Larvae feed on leaves of various deciduous trees and shrubs. *Rhombana* refers to the rhomboid mark on the forewing.

Rose tortrix *Archips rosana*, wingspan 15–24 mm, widespread but local in gardens and orchards. Adults fly in July–September. Larvae feed on leaves of various fruit trees and bushes as well as cultivated roses.

Strawberry tortrix *Acleris comariana*, wingspan 13–18 mm, widespread but local in gardens, market gardens, fen and woodland. The bivoltine adults fly in June–July and August–November. Larvae feed on strawberry and related plants such as marsh cinquefoil, and can become a pest in strawberry fields. *Comarium* is the former genus name of marsh cinquefoil.

Summer fruit tortrix *Adoxophyes orana*, wingspan 17–22 mm, naturalised, local in orchards, gardens and parks in parts of south-east England. Adults are bivoltine, flying in June and August–September. Larvae feed on the flowers of deciduous shrubs and trees, including fruit trees, especially apple, and can become a pest, prompting the genus name from Greek *adoxos* + *phye* = 'ignoble' + 'character'. Latin *orana* = 'coastal', reason unknown.

Timothy tortrix *Aphelia paleana*, wingspan 18–22 mm, widespread and common on rough grassland, waste ground, coastal sandhills and various damp habitats, scarcer in the north. Adults fly in June–August. Larvae feed on leaves of various herbaceous plants, including grasses. *Palea* is Latin for 'chaff', from the whitish-yellow forewing.

Variegated golden tortrix *Archips xylosteana*, wingspan 15–23 mm, widespread and common in woodland and gardens, less so in Scotland. Adults fly in July–August. Larvae feed on leaves of deciduous trees and shrubs. *Xylosteana* is from fly honeysuckle *Lonicera xylosteum*, an occasional food plant.

Willow tortrix *Epinotia cruciana* (Olethreutinae), wingspan 12–15 mm, widespread but local in a range of habitats. Adults fly in June–August. Larvae mine leaves of terminal shoots of creeping and eared willow and sallows. Greek *epi* + *noton* = 'upon' + 'the back', and Latin *crux* = 'cross', from the pattern evident when the wings are crossed at rest.

TOTTER GRASS Synonym of quaking-grass.

TOUGHSHANK Generally inedible fungi in the genera *Gymnopus* (Greek *gymnos* + *pous* = 'naked' + 'foot'), and *Rhodocollybia* (*rhodon* + *collybos* = 'red' + 'small coin') (Marasmiaceae).

Widespread but scattered and uncommon species include **conifer toughshank** *G. acervatus* (in England and Scotland, usually under pine), **pine toughshank** *G. putillus* (a few records from Yorkshire, the Highlands and Co. Wicklow, on soil, litter or dead branches in coniferous plantations), **redleg toughshank** *G. erythropus* (on decomposing wood in broadleaf woodland), and **toothed toughshank** *R. prolixa* (England and Scotland, on soil and leaf or needle litter in woodland).

Clustered toughshank *G. confluens*, widespread and common, found from late summer to early winter on buried decomposing wood in leaf litter in broad-leaved and mixed woodland. *Confluens* is Latin for 'flowing together' (i.e. 'clustered').

Greasy toughshank Synonym of butter cap.

Russet toughshank *G. dryophilus*, widespread and common, found from late spring to early winter on decomposing wood in leaf litter in broadleaf woodland, especially under oak, and coniferous woodland. Greek *drys* + *philos* = 'oak' + 'loving'.

Spindle toughshank *G. fusipes*, common and widespread in England, more scattered elsewhere, found from early summer to late autumn in usually dry soil and litter at the base of hardwood trees, especially oak and beech, in deciduous woodland. Latin *fusus* + *pes* = 'spindle' + 'foot'.

Spotted toughshank *R. maculata*, common and widespread, saprobic, found from June to November on needle litter, lignin-rich soil or well rotted buried wood, mainly under conifers, occasionally under hardwoods. *Maculata* is Latin for 'spotted'.

TOWER SHELL Common tower shell *Turritella communis* (Turritellidae), a sea snail up to 3 cm in length and 1 cm wide, found around most of the British and Irish coastline, abundant on muddy sediment in the sublittoral down to 200 m, on the seabed filtering seawater for food particles. It can be gregarious and occur in large numbers. The binomial is Latin for 'turret' and 'common'.

TRACHEOPHYTA From Greek *trachys* + *phyton* = here meaning 'windpipe' + 'plant', an old name for the division of vascular plants – land plants that have lignified tissues for conducting water and minerals throughout the plant.

TRACHINIDAE Family name of weever fish (Greek *trachys* = 'rough').

TRAVELLER'S-JOY *Clematis vitalba* (Ranunculaceae), a climbing herbaceous perennial growing on hedges, scrubland, fences and railway cuttings, in open and partially shaded conditions on dry, base-rich, often calcareous and moderately nutrient-rich soils in the southern half of England, naturalised further north to central Scotland. The French name *viorné* is shortened from the Latin *viburnum*; Latinised into *viorna*, this was interpreted by Gerard in his *Herbal* (1597) as *vi(am)-ornans*, a plant decking the road with its flowers, bringing joy to the traveller, and he thus gave the plant one of its English names (the other being old-man's-beard, q.v.). *Clematis* is Ancient Greek for 'branch' or 'twig', coming to mean a climbing plant; Latin *vitis* + *alba* = 'grape-vine' + 'white'.

TREACLE-MUSTARD *Erysimum cheiranthoides* (Brassicaceae), an annual European archaeophyte present in the Bronze Age/Roman period, in England locally frequent in arable fields (scattered elsewhere) but also common on waste ground, roadsides and railways, preferring sandy substrates. Greek *erysimon* = 'hedge mustard', in turn from *eryomai* = 'to help' or 'save', referring to medicinal properties, and *cheiranthoides* = 'resembling *Cheiranthus*'.

TREASUREFLOWER *Gazania rigens* (Asteraceae), a perennial South African neophyte, naturalised on coastal rocks and walls in the Channel Islands, Scilly Isles, south-west England, Isle of Wight, north Wales and Isle of Man. Theodorus of Gaza was a fifteenth-century Italian scholar and translator of the works of Theophrastus. *Rigens* is Latin for 'rigid' or 'stiff'.

Plain treasureflower *Arctotheca calendula* (Asteraceae), an annual South African neophyte occasional in England and south-east Scotland in arable fields and waste ground as a wool-shoddy alien. Greek *arctos* + *theca* = 'bear' + 'cup', and from the Latin *calendae*, the first day of the month, referring to the long flowering season.

TREBLE-BAR Moths in the family Geometridae (Larentiinae). Specifically, *Aplocera plagiata*, wingspan 37–43 mm, nocturnal but readily disturbed by day, common and widespread in gardens, field margins, downland, heathland and moorland, larvae feeding on St John's-wort. In the south there are two generations, flying in May–June and August–September. In northern England and Scotland, there is usually one brood (July–August). Greek *haplos* + *ceras* = 'simple' + 'horn', referring to the simple antennae, and Latin *plaga* = 'stripe', from the forewing mark.

Lesser treble-bar *A. efformata*, wingspan 35–41 mm, flying in two generations during dusk in May–September, common in southern England, becoming more scattered further north, and in Wales and around the Firth of Forth in Scotland. Habitats include gardens, waste ground, field margins and chalk grassland. Larvae mainly feed on St John's-wort. Latin *ex* + *formatus* = 'out of' + 'formed' (i.e. different from a congeneric).

Manchester treble-bar *Carsia sororiata*, wingspan 20–30 mm, day- but more often night-flying (July–September) locally common in northern England, Scotland and parts of Ireland (e.g. Co. Mayo) on damp moorland and peat bog, where larvae feed on bilberry, cranberry, cowberry and similar plants. Greek *carsios* = 'across', from a transverse stripe on the forewing, and Latin *soror* = 'sister' (i.e. akin to a similar species).

TREE-CHERNES Pseudoscorpions in the family Chernetidae.

Common tree-chernes *Chernes cimicoides*, length 2.2–2.7 mm, widespread throughout England, as far north as York and into mid-Wales, found under the bark of dry, dead and overmature trees. Greek *cherne* = 'a needy thing', and Latin *cimicoides* = 'like a bug'.

Large tree-chernes *Dendrochernes cyrneus*, length 3.5–4.2 mm, a rarity associated with ancient woodlands, for example Burnham Beeches, Windsor Park, St James's Park and Sherwood Forest, beneath dry bark and in decaying wood, mainly of oak. *Dendros* is Greek for 'tree'.

Wider's tree-chernes *Allochernes wideri*, length 2.2–2.5 mm, widespread in England as far north as Lincoln, associated with dead and overmature trees, especially oak, where it occurs beneath the bark and in wood during the early stages of decay.

TREE-MALLOW *Malva arborea* (Malvaceae), a herbaceous biennial, rarely native more than 100 m from the coast, usually in shallow, nutrient-enriched soils, often in vegetation in seabird roosts, and on waste ground enriched by garden waste. Plants are killed by severe frost so it is restricted to mild microclimates, though it has been recorded from as far north as Aberdeenshire. The binomial is Latin for 'mallow' + 'tree-like'.

Garden tree-mallow *Malva* x *clementii*, a semi-woody perennial European neophyte herb present by 1731, found as a garden escape or throw-out on roadsides and waste ground, and in other urban habitats, scattered in England and Wales and also recorded from Sutherland. It occasionally appears on new road verges. It can be persistent, but is often only casual.

Smaller tree-mallow *M. pseudolavatera*, a herbaceous annual or biennial neophyte, scarce and scattered in England and south Wales, present by 1723 (possibly native in south-west England and the Channel Islands), in rough ground near the sea, including bulb fields, old quarries and roadsides. *Pseudolavatera* is Greek for 'false' + *Lavatera*, the former genus name of these plants.

TREE MOSS (LICHEN) Synonym of antler lichen.

TREE-MOSS (MOSS) *Climacium dendroides* (Climaciaceae), widespread and most frequent in damp habitats, especially where water levels fluctuate during the year such as at the edges of lakes and reservoirs, in dune slacks and turloughs, flushes, wet grassland, damp scrub and woodland. It is sometimes present in dry, base-rich or sandy grassland. Greek *climax* = 'ladder' and *dendron* = 'tree'.

TREE-OF-HEAVEN *Ailanthus altissima* (Simaroubaceae), a deciduous neophyte tree from China, present by 1751, planted in urban streets and parks in England, but rarely found elsewhere. It is very tolerant of atmospheric pollution. It grows rapidly, and especially in south-east England spreads by suckering, rarely setting seed. Indonesian *ailantho* = 'tree-heaven', and Latin *altissima* = 'highest'.

TREECREEPER *Certhia familiaris* (Certhiidae), wingspan 19 cm, length 12 cm, weight 10 g, a common and widespread resident breeder found in deciduous and mixed woodland, with 200,000 territories in Britain in 2009. A non-significant decline in numbers by 10% was noted over 1970–2015, incorporating a 7%

increase in 1995–2015, and an 8% increase in 2010–15. Nests are in tree crevices, with clutches of 5–6 eggs which hatch in 13–17 days, the altricial young fledging in 15–17 days. Numbers and survival rates are reduced by wet winter weather. Feeding is on invertebrates caught by moving upwards on tree-trunks. *Certhios* is a small tree-dwelling bird described by Aristotle, and Latin *familiaris* = 'familiar'. **Short-toed treecreeper** *C. brachydactyla* is a vagrant from southern Europe.

TREEHOPPER Members of the bug family Membracidae (Auchenorrhyncha), with two British species: *Centrotus cornutus*, local and scattered in woodland in England and Wales; and *Gargara genistae*, locally associated with broom in south-east England and East Anglia.

TREFOIL Species of *Lotus*, for example bird's-foot-trefoil (q.v.), or any plant with a three-lobed leaf, for example clovers *Trifolium*.

Hop trefoil *Trifolium campestre* (Fabaceae), a common and widespread winter-annual, declining in the north and west of its range, on grassland on dry, relatively infertile neutral or base-rich soils; also on spoil heaps from slate and limestone quarries. *Campestre* is from Latin for 'field'.

Large trefoil *T. aureum*, an annual European neophyte introduced in 1815, casual or naturalised in rough grassland and on waste ground, via birdseed, wool shoddy or grain, declining in status and now local. *Aureum* is Latin for 'golden'.

Lesser trefoil *T. dubium*, a common and widespread winter-annual of hay meadows, waste ground and lawns. While most frequent in dry grasslands, it can be abundant in winter-flooded meadows and damp pastures, and can thrive in nutrient-enriched sites. It also occurs in open habitats such as rock outcrops, quarry spoil and railway ballast. *Dubium* is Latin for 'uncertain'.

Slender trefoil *T. micranthum*, a winter-annual of neutral or moderately acidic soils, widespread and common in England and Wales, rare in Scotland and Ireland, found on the coast in sandy or gravelly grassland, and inland in drought-prone pasture, on paths and verges, and as a weed in lawns. It is tolerant of grazing, mowing and heavy trampling. *Micranthum* is Greek for 'small-flowered'.

TREMATODA From Greek *trematodes* = 'pierced with holes', a class in the phylum Platyhelminthes that includes two subclasses (Aspidogastrea and the much commoner Digenea) of parasitic flatworms – flukes – which are internal parasites of tissue or blood. Most trematodes have a complex life cycle with at least two hosts, a vertebrate primary host, in which the flukes sexually reproduce, and an intermediate host, in which asexual reproduction occurs, which is usually a snail.

TREMBLING-WING FLY Members of the family Pallopteridae, small to medium-sized flies with wings (that vibrate, giving the common name) considerably longer than the abdomen. Adults feed on nectar; larvae of different species are phytophagous or predatory.

TREMELLALES Order of basidiomycete fungi (Greek *trema* = 'hole') that includes jelly fungi, containing both teleomorphic and anamorphic species, most of the latter being yeasts.

TRIANGLE (MOTH) *Heterogenea asella* (Limacodidae), wingspan 15–20 mm, nocturnal, a rarity (RDB) found in oak and beech woodland. There are post-1980 records from Essex, Kent, East Sussex, Hampshire, Wiltshire, Devon and Cornwall. Adults fly in June–July. Larvae feed on oak and beech. Greek *heteros* + *genos* = 'different' + 'kind', from the anomalous structure of the larva and pupa, and *asella* from the woodlouse *Oniscus asella*, reflecting the larval morphology (small ovate body, very short legs and a retracted head).

Gold triangle *Hypsopygia costalis* (Pyralidae, Pyralinae), wingspan 16–23 mm, nocturnal, commonly found in gardens, woodland and farmland throughout England, Wales and south-east Scotland, and (one record, post-2000) in Co. Wicklow.

Adults fly in July–August. Larvae feed on dry plant material such as thatch and hay. In one resting posture it adopts a 'triangular' shape, with the hindwings hidden by the forewings, giving it its English name; at full rest, all four wings are splayed out. Greek *hypsos* + *pugaios* = 'height' + 'relating to the rump', from the moth's behaviour in flexing its abdomen while at rest. *Costa* is Latin for 'rib', referring to the two gold spots on the anterior margin of the forewing.

TRICHOLOMA Streaky tricholoma Synonym of (the fungus) coalman.

TRICHOMONOSIS From Greek *trichos* + *monas* + *osis* = 'hair' + 'single' or 'unit' + a suffix indicating in disease an increase in production, a disease in birds caused by the protozoan parasite *Trichomonas gallinae*. Epidemics occurred in 2006 and 2007, with smaller scale mortality events in subsequent years. Greenfinch and chaffinch have been most affected, but the disease has also been found in other bird species, for example dunnock, house sparrow, great tit and siskin. The disease is also known as 'canker' when seen in pigeons and doves, and as 'frounce' when evident in birds of prey. The disease affects the back of the throat and the gullet, birds showing signs of general illness (e.g. lethargy) and having difficulty in swallowing or laboured breathing.

TRICHONISCIDAE Family name for some woodlouse species (Greek *trichos* = 'hair').

TRICHOPTERA Order of caddis flies (Greek *trichos* + *pteron* = 'hair' + 'wing'), with 199 British and Irish species, of which three have only been recorded in Ireland.

TRICLAD Free-living flatworms; planarians.

TRIPE FUNGUS *Auricularia mesenterica* (Auriculariaceae), inedible, widespread and common in England, less so in Wales, Scotland and Ireland, mainly fruiting in late summer and autumn on dead and rotting branches and trunks of broadleaf trees, especially beech and elm. Latin *auricula* = 'earlobe', from the shape, and Greek *mesenterion* = 'mid-intestine'.

TRIPE (LICHEN) See rock tripe.

TRIPLE-LINES Clay triple-lines *Cyclophora linearia* (Geometridae, Sterrhinae), wingspan 23–33 mm, a nocturnal moth found locally in beech woodland, hedgerows, parks and gardens throughout much of England, Wales and Ireland. Adults fly in May–July, sometimes again in autumn. Larvae feed on beech leaves. Greek *cyclos* + *phoreo* = 'ring' + 'to carry', and Latin *linea* = 'line', both describing forewing markings. Not to be confused with treble lines.

TRITICALE *Triticosecale*, derived from artificial hybridisation of wheat and rye, developed in the late nineteenth century and not yet widely planted, but recorded from the wild in mid-West Yorkshire in 1996, and increasingly found as a casual relic of cultivation and on roadsides and waste ground. The name blends *Triticum* = 'wheat' and *Secale* = 'rye'.

TROGIDAE Family name of hide beetles (Greek *trogein* = 'gnaw'), with two species of *Trox* recorded recently.

TROGOSSITIDAE Beetle family (Greek *trogein* + *sitos* = 'gnaw' + 'grain'), with five British species. *Lophocateres pusillus* is a pest of stored products. *Thymalus limbatus* and *Ostoma ferrugineum* are found beneath bark. *Tenebroides mauritanicus* is a pest of cereals. *Nemozoma elongatum* preys on the rare weevil *Acrantus vittatus*.

TROMBIDIFORMES Order of mites that includes the suborder Prostigmata, found in a range of terrestrial, freshwater and marine habitats. Many are plant parasitic, but this group includes gall mites, spider mites, velvet mite and chiggers. *Trombidium* is New Latin for 'small timid one', from Greek *tromeo* = 'to tremble'.

TROPAEOLACEAE Family name for nasturtium (Greek *tropaion* = 'trophy', referring to the shape of the flowers).

TROUGH SHELL Bivalves in the family Mactridae with two short siphons, each with a horny sheath, shallow burrowers in sand or fine gravel. **Cut trough shell** *Spisula subtruncata* (shell size up to 2.6 cm), **thick trough shell** *S. solida* (4.5 cm) and **rayed trough shell** *Mactra stultorum* (5 cm) are all widespread from the low intertidal to the nearshore shelf.

TROUT Fishes in the family Salmonidae. 'Trout' comes from late Old English *truht* (c.f. French *truite*), from Latin *tructa*, based on Greek *trogein* = 'to gnaw'.

Brook trout *Salvelinus fontinalis*, length 25–65 cm, weight 0.3–3 kg, a North American fish introduced to British fisheries from the late nineteenth century, but acclimatised in only a few places, favouring cool, clear and at least mildly acid waters. This is more properly called **brook charr**. *Salvelinus* is New Latin, possibly from German *Saibling* = 'charr'; *fontinalis* = 'of a spring'.

Brown trout *Salmo trutta*, common and widespread, length up to 100 cm, weight 20 kg, though in small rivers a mature weight of <1.0 kg is common. The species includes purely freshwater populations in rivers and lakes (respectively, *S. trutta* morpha *fario* and morpha *lacustris*), and an anadromous form (sea trout, morpha *trutta*) which migrates to the sea for much of its life, returning to fresh water to spawn. Spawning in fresh water is in early winter, when the female lays her eggs in a hollow (redd) in the gravel riverbed. Trout grow from egg-attached alevins, through fry and parr stages into adults. Aquatic invertebrates are the commonest prey, but piscivory is most frequent in large specimens. The binomial is simply the Latin for 'salmon' and 'trout'. See also gillaroo.

Ferox trout *Salmo trutta ferox*, a large (length up to 80 cm), piscivorous variety of brown trout found in oligotrophic lakes, particularly in Ireland, north Scotland, Cumbria and Wales. Some view it as a distinct species, being reproductively isolated from 'normal' brown trout even in the same lake, but it is unclear whether the ferox of different lakes are all of a single origin. 'Ferox', Latin for 'wild' or indeed 'fierce', was first used for this fish by the scientist Sir William Jardine in 1835.

Rainbow trout *Oncorhynchus mykiss*, up to 2.3 kg, a North American species introduced to numerous fish farms, with some escapes (e.g. into Loch Lomond in the 1960s), but very few becoming self-sustaining populations. Greek *oncos* + *rynchos* = 'hook' + 'nose', a reference to the hooked jaw of the male in the mating season. *Mykiss* is derived from the Kamchatkan name for this fish, *mykizha*.

Sea trout Anadromous form of brown trout.

TRUE BUGS See bug and Hemiptera.

TRUE FLIES See Diptera.

TRUE LOVER'S KNOT *Lycophotia porphyrea* (Noctuidae, Noctuinae), wingspan 25–30 mm, a nocturnal moth sometimes seen in sunshine, flying in June–August, common and widespread on heathland and moorland, larvae feeding on heathers and heaths. Greek *lycophos* = 'twilight', from the dusky white wing pattern said to resemble a true lover's knot (not from time of day of flying), and *porphyreos* = 'purple', from the brown wings tinged with purple in some forms.

TRUFFLE The fruiting body of some subterranean ascomycete fungi whose mycelia form symbiotic mycorrhizal relationships with the roots of several tree species. 'Truffle' comes from Latin *tuber* = 'swelling', 'lump' or indeed 'truffle', which became *tufer*. Confusing, cubed and Michael's **fold truffles** have been recorded from a very few places in England. See also false truffle.

Brain fold truffle *Hydnotrya cerebriformis* (Pezizales, Pezizaceae), a rare fungus with a few records on calcareous soil and leaf litter from Gloucestershire, Wiltshire, Oxfordshire, south Devon, Sussex and Derbyshire in woodland under beech, and

Glamorganshire under hazel. Greek *hydnon* + *tryma* = 'tuber' or 'truffle' + 'hole', and Latin *cerebrum* + *forma* = 'brain' + 'shape'.

Red fold truffle *H. tulasnei*, a rare fungus with a few scattered records in England on woodland soil mainly under beech and oak.

Summer truffle *Tuber aestivum* (Tuberales, Tuberaceae), a rare edible and very tasty subterranean fungus found in summer and autumn, typically ectomycorrhizal on beech but also on other broadleaf trees on calcareous soils in England. The binomial is from Latin for 'tuber' or 'truffle' and 'summer'.

TRUFFLECLUB Parasitic fungi (Hypocreales, Cordycipitaceae) found on false truffle.

Drumstick truffleclub *Cordyceps capitata*, widespread, scattered and uncommon, found in summer and autumn mainly in conifer woodland, particularly with spruce. Growing from underground false truffle, a parasitic ascomycete often hidden in mossy woodlands or areas deep in needle litter when its brown cap may only just protrude through the surface. Greek *cordyle* + New Latin *ceps* = 'head', and Latin *caput* = 'head'.

Snaketongue truffleclub *Elaphocordyceps ophioglossoides*, widespread and locally common in woodland. Greek *elaphos* = 'stag' + *Cordyceps* (see drumstick truffleclub above), and *opheos* + *glossa* = 'small snake' + 'tongue'.

TRUMPET Black trumpet, trumpet of death Synonyms of the fungus horn of plenty, which is actually edible.

TRUMPETS *Sarracenia flava* (Sarraceniaceae), an insectivorous herbaceous perennial North American neophyte planted in bogs in south Devon, Hampshire and Surrey. Michel Sarrazin was an eighteenth-century physician and botanist in Quebec. *Flava* is Latin for 'yellow'.

TUBELET White tubelet *Henningsomyces candidus* (Agaricales, Marasmiaceae), a widespread but scattered and infrequent fungus on dead wood on the floor of deciduous woodland. Paul Hennings (1841–1908) was a German mycologist. Greek *myces* = 'fungus'. Latin *flava* = 'white'.

TUBENOSE Petrels, storm-petrels and shearwaters, so-called because of the nostrils lying on the upper mandible, associated with excretion of excess salt or with a sense of smell which may be better developed than in most birds.

TUBEWORM Marine member of the annelid class Polychaeta which lives in a tube made from sand particles or in a self-secreted calcareous tube.

Calcareous tubeworm *Serpula vermicularis* (Serpulidae), 5–7 cm long, fairly widespread but commonest off the north-west coast of Scotland attached to hard substrates from low water to the sublittoral down to 250 m. In some sheltered areas the tubes aggregate to form small reefs. *Serpula* is Latin for 'little snake', *vermiculus* = 'worm'.

Sinistral spiral tubeworm *Spirorbis spirorbis* (Serpulidae), a few mm long, the tube being sinistral (coiling to the left), common on rocky shores around Britain, particularly the west coast of Scotland, and also in Ireland, on wrack and kelp in the low intertidal and shallow sublittoral Latin *spira* + *orbis* = 'coil' + 'ring'.

Spindle-shaped tubeworm *Owenia fusiformis* (Oweniidae) up to 10 cm long, widespread but scattered in sand or muddy sand, at or below low water, on sheltered beaches. *Fusiformis* is Latin for 'spindle-shaped'.

TUBIC (MOTH) Members of the families Chimabachidae, Lypusidae and Oecophoridae. The larvae of many species feed on bark or decaying leaves in a portable case.

Scarce species include **gold-base tubic** *Oecophora bractella* (in ancient woodland in parts of southern England, the Midlands and south Wales), **heath tubic** *Denisia subaquilea* (heathland and moorland throughout most of Britain, but absent from south and east of a line from the Severn estuary to that of the Humber, larvae feeding on bilberry), **Kent tubic** *Bisigna procerella* (first

recorded in Britain in 1976 in woodland at Hamstreet, Kent, and subsequently found in other woods in Kent), **ling tubic** *Amphisbatis incongruella* (moorland and heathland, larvae feeding on heather, sometimes mouse-ear-hawkweed and wild thyme), **new tawny tubic** *Batia internella* (a few sites in Herefordshire, Hertfordshire and Wiltshire in pine and larch woodland, larvae feeding on lichens growing on larch and Scots pine), **northern tubic** *D. similella* (woodland in central and north-west Scotland, reappearing in south Scotland and north England and the Midlands as far south as Herefordshire, larvae feeding on fungus under dead wood or bark), **orange-headed tubic** *Pseudatemelia josephinae* (widespread but scattered in woodland), **pied tubic** *D. albimaculea* (woodland, parkland and gardens in parts of southern England), **scarce forest tubic** *Esperia oliviella* (ancient woodland in parts of south-east England), **silver-streaked tubic** *Schiffermuellerina grandis* (now probably confined to Devon, Somerset and, recorded in 2005, Worcestershire), **straw-coloured tubic** *P. subochreella* (woodland in southern England, as far north as Worcestershire and Leicestershire, and in parts of Wales), and **yellow-headed tubic** *P. flavifrontella* (woodland in southern England, the Midlands and Wales).

Common tubic *Alabonia geoffrella* (Oecophoridae), wingspan 17–21 mm, day-flying, common in woodland, hedgerows, gardens and occasionally marsh throughout southern England, East Anglia, the Midlands and Wales. Adults fly in May–June. Larvae feed in the bark of decomposing wood. Latin *ala* + *bona* = 'wing' + 'good'. Etienne Geoffroy (1727–1810) was a French entomologist.

Golden-brown tubic *Crassa unitella* (Oecophoridae), wingspan 12–16 mm, locally common in deciduous woodland in south England and Wales, less so as far north as Lancashire and Yorkshire. Adults fly at night from June to August. Larvae feed on fungus on and under dead bark from September to May. Latin *crassa* = 'thick', and *unitas* = 'unity', from the single-coloured forewing.

Greater tawny tubic *Batia lambdella* (Oecophoridae), wingspan 11–17 mm, nocturnal, widespread but local on heathland, scrub and open woodland, more numerous in the south. Adults fly in July–August. Larvae feed in (particularly gorse) bark. Greek *batos* = 'thorn-bush', possibly from where it was discovered, and the Greek letter *lambda* (L) from the shape of a mark on the forewing.

Lesser tawny tubic *Batia lunaris*, wingspan 7–10 mm, nocturnal, common in woodland, plantations and wooded parkland throughout much of England and Wales. Adults fly from July to August. Larvae feed under tree bark on fungi, possibly lichens, and decaying wood. *Lunaris* is Latin for 'crescent-shaped', from the forewing marking.

March tubic *Diurnea fagella* (Chimabachidae), male wingspan 26–30 mm, the flightless females 15–20 mm, day-flying, widespread and common in oak woodland, birch woodland, scrub and hedgerows. Males are often evident resting on tree-trunks and fly from March to May. Larvae feed on deciduous trees and shrubs, mainly in August–October, usually between leaves spun together with silk. Latin *diurnus* = 'relating to day', from time of flight activity, and *fagella* referencing beech *Fagus*, a food plant.

November tubic *D. lipsiella*, male wingspan 22–5 mm, the flightless female 15–18 mm, widespread but local in oak woodland. Males fly at night and by day from March to May. Larvae feed on oak, living between leaves spun together with silk, and sometimes on bilberry. *Lipsia* is New Latin for Leipzig, Germany.

Small tubic *Borkhausenia fuscescens* (Oecophoridae), wingspan 7–12 mm, widespread and common in woodland, hedgerows, farmland and farm buildings. Adults fly in July–August. Larvae feed on dried plant matter, for example dead leaves and birds' nests, usually within a silk case. Moritz Borkhausen (1732–1807) was a German entomologist. Latin *fuscescens* = 'tending to be dusky', describing the forewings.

Spring tubic *Dasystoma salicella* (Chimabachidae), wingspan

17–20 mm, local in heathland, scrub and woodland in England and Wales, scarce in Scotland (Kirkcudbrightshire and Perthshire) and Ireland. The female is flightless; males fly around midday in April. Larvae feed from between two overlapping leaves sewn together on a variety of plants including sallow (*Salix*, giving the specific name), blackthorn, meadowsweet and bog-myrtle, from May to September. Greek *dasys* + *stoma* = 'thick' + 'mouth', from the densely scaled labial palp.

Sulphur tubic *Esperia sulphurella*, wingspan 12–16 mm, widespread and common in woodland, hedgerows and gardens, north to central Scotland. Adults fly at night in May–June. Larvae feed on dead wood and associated fungi in late winter and spring.

Tinted tubic *Crassa tinctella*, wingspan 14–16 mm, nocturnal, local in ancient woodland throughout much of southern Britain, from Kent to Devon north to Lincolnshire, Cheshire and north Wales. Adults fly in May–June. Larvae feed under decaying wood and on lichens. *Tinctus* is Latin for 'dyed', from the forewing colour, possibly suggesting saffron dye.

Treble-spot tubic *Telechrysis tripuncta* (Depressariidae), wingspan 11–14 mm, local in woodland and hedgerows in parts of England and Wales; there is one record from west Ireland. Adults fly at dawn and dusk in May–June. Larvae feed on dead and decaying wood, with a likely preference for hazel. Greek *telos* + *chrysos* = 'the end' + 'gold', and Latin *tripuncta* = 'three-spotted', both describing the forewing markings.

TUBULAR WEED Dumont's tubular weed *Dumontia contorta* (Dumontiaceae), a red seaweed that germinates in autumn and develops long thin fronds up to 50 cm long in late winter and early spring, common and widespread on rocks in shallow rock pools in the upper intertidal down to the shallow subtidal.

TUFA-MOSS Members of the family Pottiaceae. The lower parts of the tufts of these mosses often become hard and petrified with calcium salts, giving them their common name.

Blunt-leaf tufa-moss *Gymnostomum calcareum*, scattered, on damp, shaded, sheltered soft limestone and tufa. Greek *gymnos* + *stoma* = 'naked' + 'mouth', and Latin *calcareum* = 'calcareous'.

Hook-beak tufa-moss *Hymenostylium recurvirostrum*, locally common in upland limestone flushes, and in crevices on damp, base-rich rocks and the mortar of old walls. Greek *hymen* + *stylos* = 'membrane' + 'pillar', and Latin *recurvus* + *rostris* = 'bent back' + 'bill'.

Luisier's tufa-moss *G. viridulum*, mainly found in southwest England, south Wales and Ireland on shaded soft, highly calcareous substrates, including lime mortar, old walls and thin soil on limestone rock ledges. Alphonse Luisier (1872–1957) was a Portuguese bryologist. *Viridulum* is Latin for 'greenish'.

Robust tufa-moss *H. insigne*, scarce in crevices in the western Highlands. *Insignis* is Latin for 'notable'.

Verdigris tufa-moss *G. aeruginosum*, common and widespread on (mostly) upland north of a line from the Severn estuary to north Yorkshire, on rocks and in crevices kept moist by base-rich seepage water, and on old damp walls with lime mortar. *Aeruginosum* is Latin for 'verdigris'.

Whorled tufa-moss *Eucladium verticillatum*, common and widespread in wet, shaded sites where base-rich water trickles from calcareous rocks, and by streams and flushes, one of the most important tufa-forming mosses. Greek *eu* + *cladion* = 'good' + 'small branch' (i.e. repeatedly branched) and Latin *verticillatum* = 'whorled'.

TUFT (FUNGUS) Members of the genus *Hypholoma* (Agaricales, Strophariaceae). *Hypholoma* is from Greek for 'mushroom with threads', probably referring to the thread-like partial veil that connects the cap rim to the stem of young fruit bodies, possibly to the thread-like rhizomorphs that radiate from the stem base. Also, **beech tuft** is an alternative name for porcelain fungus.

Brick tuft *H. lateritium*, widespread, locally common,

inedible, found in summer and autumn on surface roots, stumps and dead wood, particularly oak. *Later* is Latin for 'brick'.

Conifer tuft *H. capnoides*, widespread and frequent, edible, found in late summer and autumn on stumps and other dead wood of conifers. *Capnos* is Greek for 'smoke', describing the smoky grey gills.

Sulphur tuft *H. fasciculare*, widespread and common, inedible, found throughout the year but mainly evident in summer and autumn on stumps and other dead wood usually of broadleaf trees. *Fasciculus* is Latin for 'small bundle'.

TUFTED MOSS Black-tufted moss *Glyphomitrium daviesii* (Rhabdoweisiaceae), on (near) coastal rocks, mainly in western Scotland, more rarely in Ireland and Wales. Greek *glyphe* + *mitrion* = 'carving' + 'small cap'.

TUFTED-SEDGE *Carex elata* (Cyperaceae), a tussock-forming perennial, locally common in East Anglia and Ireland, scattered elsewhere and uncommon in Scotland, in oligotrophic and mesotrophic (sometimes eutrophic) often calcareous marshy habitats such as fen, reed swamp, margins of lakes, ponds, rivers and canals, ditches with seasonal flooding, and wet alder and willow woodland. 'Sedge' comes from Old English *secg*. The binomial is Latin for 'sedge' and 'tall'.

Slender tufted-sedge *C. acuta*, a rhizomatous perennial scattered in Britain as far north as central Scotland, and in Ireland, in shallow water or wet ground at stream, canal, lake and pond margins, in swamp, ditches, flood meadow and marsh, usually in calcareous and mesotrophic or eutrophic conditions, at sites subject to frequent flooding. Being shade-tolerant, it can grow under riverside trees and in wet woodland. *Acuta* is Latin for 'sharp' or 'pointed'.

TULIP Bulbous southern European neophyte perennial herbs in the genus *Tulipa* (corruption of Persian *thoulyban*, *tulbend* or *tulipant* or the Turkish *dulband* = 'turban', a reference to the shape of the flower) (Liliaceae), cultivated by Tudor times, and the cause of tulipomania (leading to the high price of bulbs) in the seventeenth century.

Garden tulip *T. gesneriana*, first recorded in the wild in 1955, scattered in Britain and south-east Ireland on rough ground, roadsides, quarries, churchyards and amenity grasslands. It can persist when planted or thrown out in favourable habitats. Conrad Gesner (1516–65) was a Swiss horticulturalist.

Wild tulip *T. sylvestris*, recorded in the wild in 1790, widely naturalised by the late eighteenth and nineteenth centuries but declining during and after the twentieth century. It is found in open woodland, orchards, hedgerows, grassy banks and waste ground. Populations originate as discarded bulbs, deliberate planting or garden relics. *Sylvestris* is Latin here meaning 'wild'.

TUNICATA, TUNICATE Subphylum of the Chordata, including sea squirts (class Ascidiacea). See also ascidiacea and sea grapes.

Light bulb tunicate *Clavelina lepadiformis* (Clavelinidae), a colonial sea squirt that grows up to 20 mm high, groups of transparent zooids being joined at the base by short stolons, looking like a chandelier, found off most coasts attached to rocks, stones and macroalgae in the sublittoral to depths of 50 m. *Clavelina* is Latin for 'small club', *lepas* Greek for 'limpet'.

Orange sheath tunicate *Botrylloides violaceus* (Styelidae), a colonial sea squirt that forms lobed sheets 2–3 mm thick, on artificial surfaces in shallow water, especially in harbours and marinas, and also attached to macroalgae and other unitary sea squirts, recorded from a few sites on the south coast of England, East Anglia, and Milford Haven, Pembrokeshire. New Latin *botryllus* = 'cluster', and *violaceus* = 'violet'.

TUNICFLOWER *Petrorhagia saxifraga* (Caryophyllaceae), a mat-forming European neophyte perennial introduced by 1774, naturalised on waste ground at Tenby, Pembrokeshire, recorded on pavement at Llanidloes, Montgomeryshire (Powys), in 1999, and occurring elsewhere in England and Wales as a scarce casual on walls, railway banks and roadsides, and in chalk-pits. Greek *petra* + *rhaga* and Latin *saxum* + *frango*, both meaning 'rock' + 'break'.

TUNING FORK WEED Brown tuning fork weed *Bifurcaria bifurcata* (Sargassaceae), a brown seaweed with fronds 30–50 cm long, usually found in rock pools on the middle and lower shore, particularly on exposed beaches, along the south and south-west coasts of England, south-west and north-west Wales, and west Ireland. The English name and binomial reflect the thallus shape which is narrowly forked from about halfway up.

TURBAN FUNGUS Synonym of false morel.

TURBELLARIA Class of platyhelminths consisting of free-living flatworms, including triclads and planarians.

TURDIDAE From Latin *turdus* = 'thrush', the family also including blackbird, bluethroat, chats, fieldfare, nightingale, redstarts, redwing, ring ouzel, robin and wheatear.

TURF-MOSS Species of *Rhytidiadelphus* (Hylocomiaceae) (Greek *rhytis* + *di* + *adelphos* = 'wrinkle' + 'two' + 'brother').

Scarce turf-moss *R. subpinnatus*, a rarity from Wales, central Scotland, west Lancashire and Co. Waterford, among grasses and on stream banks in humid deciduous woodland. *Subpinnatus* is Latin for 'almost plumed'.

Springy turf-moss *R. squarrosus*, common and widespread on unimproved grassland: where the grass is short, for instance from mowing or grazing, it can form extensive swards. It also occasionally grows in heathland, and in flushes and other wetlands. *Squarrosus* is Latin for 'rough'.

TURKEYTAIL *Trametes versicolor* (Polyporaceae), a common and widespread inedible generally saprobic fungus, annual if evident year-round and sporulating in autumn and winter, in deciduous woodland on standing dead or fallen timber. *Trametes* is from Greek meaning 'one who is thin' (in cross section); *versicolor* Latin for 'several colours', referencing the coloured bands on the upper surface which also gives the English name.

TURNIP *Brassica rapa* (Brassicaceae), an annual or biennial herb, found on river and canal banks, and as a casual on roadsides, in arable fields and on tips. **Wild turnip** (ssp. *campestris*) is found in semi-natural habitats. Two other subspecies are widely cultivated: **turnip-rape** (*oleifera*) is a birdseed or oilseed species; ssp. *rapa* (**turnip**) is a frequent relic of cultivation. 'Turnip' probably comes from *turn* (from its shape, as though turned on a lathe) + Middle English *nepe* = 'turnip', from Old English *næp*, in turn from Latin *napus*. The binomial is Latin for 'cabbage' and 'turnip'.

TURNIP ROOT FLY *Delia floralis* (Anthomyiidae), with a few records from England and locally common in Scotland, a pest of brassica crops, the larvae burrowing into the roots and basal stems, particularly during July–August. *Delia* is Greek for 'dice', Latin *flora* = 'flower'.

TURNSTONE *Arenaria interpres* (Scolopacidae), wingspan 54 cm, length 23 cm, weight 120 g, an Amber List winter visitor found on most coastlines, especially rocky ones. Birds from northern Europe pass through in July–August and again in spring. Canadian and Greenland birds arrive in August–September and remain until April–May. A few birds may breed, and non-breeding birds may also stay through the summer. An estimated 51,000 birds overwintered in the UK in 2008–9, with sizeable numbers also in Ireland. Diet is of sandhoppers and other invertebrates. The binomial is Latin for 'inhabiting sand' and 'messenger'.

TURPSWORT See turps pouchwort under pouchwort.

TURTON Minute turton *Turtonia minuta* (Turtoniidae), a filter-feeding bivalve, shell size 4 mm, widespread, attached by a byssus in crevices, and among algae and epifauna in the intertidal and shallow sublittoral. William Turton (1762–1835) was an English naturalist.

TUSSOCK (MOTH) Named after the tufts of hair on the larvae.

Dark tussock *Dicallomera fascelina* (Erebidae, Lymantriinae), wingspan 35–45 mm, local on heathland, moorland, dunes and shingle beaches, discontinuously in southern and north-west England, east Scotland, and central and north-east Ireland. Adults fly at night in July–August. Larvae feed on heather, broom, creeping willow, bramble and hawthorn. Greek *di* + *calos* + *meros* = 'two' + 'beauty' + 'thigh', from ornamental scales on the legs, and Latin *fasciculus* = 'small bundle' (i.e. larval tussocks).

Nut-tree tussock *Colocasia coryli* (Noctuidae, Pantheinae), wingspan 27–35 mm, nocturnal, widespread and common in broadleaf woodland and gardens, more local as one moves north. Adults fly in April–June and July–September in the south (double-brooded); May–June in the north. Larvae feed on leaves of hazel (*Corylus*, giving the specific name), birch, hornbeam, oaks and field maple. *Colocasia* is a plant found in Egypt, with no entomological significance in its use.

Pale tussock *Calliteara pudibunda* (Erebidae, Lymantriinae), wingspan 40–60 mm, nocturnal, common in gardens, hedgerows, parks, woodland and scrub throughout England and Ireland, and parts of Wales. Adults fly in May–June. Larvae feed on various deciduous trees and shrubs. *Calliteara* may come from Greek *calos* + *ear* = 'beauty' + 'spring', from an early appearance by the adults, and Latin for 'immodest'.

TUSSOCK-SEDGE Perennials in the genus *Carex* (Cyperaceae), Latin for 'sedge'. 'Sedge' comes from Old English *secg*.

Fibrous tussock-sedge *C. appropinquata*, mainly found in fen and willow carr, having declined in East Anglia and lowland sites in Yorkshire as a result of drainage and a series of dry summers, but sites in the Scottish Borders and central Ireland appear to be stable. *Appropinquata* is Latin for 'closer'.

Greater tussock-sedge *C. paniculata*, generally common and widespread (absent from the Highlands) in a range of generally base-rich habitats, including swamp and fen, margins of lakes, ponds, canals and ditches, open fen carr and wet woodland, tolerating moderate shade. *Paniculata* makes reference to the panicles.

Lesser tussock-sedge *C. diandra*, widespread if less common in England, growing in wet peaty habitats, tolerating acidic soils and those with calcareous seepage, often found on the edges of pools and in swamp, but also on the margins of wet woodland and in fen carr. *Diandra* is Greek for 'two stamens'.

TUTSAN *Hypericum androsaemum*, a shrub of damp or shaded habitats including woodland and hedgerows. It is also widely cultivated and occurs outside its native range in semi-natural, often drier habitats, having been spread by birds. 'Tutsan' dates from late Middle English, from Anglo-Norman French *tutsaine* = 'all wholesome', the leaves being used in treating wounds; indeed Greek *andros* + *aima* = 'man' + 'blood', from these healing properties. Greek *hyper* + *eicon* = 'above' + 'picture', plants being hung above pictures to ward off evil spirits.

Stinking tutsan *H. hircinum*, a Mediterranean neophyte shrub cultivated in gardens by 1640, widespread but scattered, naturalised in shaded habitats, occasionally in more open places. *Hircinum* is Latin for 'smelling like a goat' (*hircus*).

Turkish tutsan *H. xylosteifolium*, a rhizomatous evergreen Caucasian neophyte shrub introduced by 1870, first recorded as naturalised in 1978 at Eaves Wood, Silverdale (Lancashire), and subsequently noted in 1979 at Monkton Moor (West Yorkshire), and a few other sites in England, and near Edinburgh. It has been found as a garden escape in woodland, on railway banks and in waste ground. Greek *xyle* + *osteon* = 'wood' + 'bone', and Latin *folium* = 'leaf'.

TWACHEL Angler's name for an earthworm, from the Middle English *angletwitche*, or *angletwatche*.

TWAYBLADE Species of *Neottia* (Orchidaceae), Greek for 'bird's nest'. 'Twayblade' dates from the late sixteenth century, from *tway* as a variant of *twain* = 'two' + 'blade'.

Common twayblade *N. ovata*, a common and widespread rhizomatous herbaceous perennial on a range of calcareous to mildly acidic soils, habitats including grassland, woodland, hedgerows, scrub, dunes, dune slacks, limestone pavement and heathland; in Anglesey and Ireland it also grows in fen. It is often found, sometimes in large colonies, on railway embankments and in disused quarries and sandpits. *Ovata* is Latin for 'oval'.

Lesser twayblade *N. cordata*, locally common in Scotland, south-west England and northern England south to Derbyshire, in Wales and (scattered) Ireland, in moorland and peat bogs in wet, acidic conditions, often in moss. *Cordata* is Latin for 'heart-shaped'.

TWIGLET Inedible fungi in the genera *Simocybe* (Greek *simo* + *cybe* = 'flat' + 'head') and *Tubaria* (probably from Latin *tuba* = 'pipe' or 'connection') (Agaricales, Inocybaceae).

Dingy twiglet *S. centunculus*, scattered and locally common in England, scarce elsewhere, in late summer and autumn in deciduous woodland on stumps and logs. *Centunculus* is Latin for 'little patchwork'.

Felted twiglet *T. conspersa*, widespread and locally common in late summer and autumn in damp litter beneath deciduous trees, on well-rotted hardwood, and on wood chip used as mulch in parks and gardens. *Conspersa* is Latin for 'scattered', from being found in groups.

Hawthorn twiglet *T. dispersa*, common and widespread in much of England, scarce elsewhere, found in late summer and autumn in soil on decaying hawthorn seeds. *Dispersa* is Latin for 'scattered'.

Scurfy twiglet *T. furfuracea*, common and widespread, found in autumn on woody debris in a variety of habitats. *Furfur* is Latin for 'bran'.

TWIN FAN WORM *Bispira volutacornis* (Sabellidae), a colonial sedentary tube polychaete with a crown of <200 tentacles, in shaded rocky overhangs, or in rock crevices in the sublittoral, sometimes in deep intertidal rock pools, widespread in south-west England, with scattered records from Flamborough Head, the Isle of Man, west Scotland and Ireland. Latin *bi* + *spira* = 'two' + 'coil', and *voluta* + *cornu* = 'spiral' + 'horn'.

TWINFLOWER *Linnaea borealis* (Caprifoliaceae), a creeping perennial, woody at the base, of Scots pine woodland, mostly in the east Highlands (this is the county flower of Inverness-shire as voted by the public in a 2002 survey by Plantlife), with local introduction in northern England, growing in slight to moderate shade, on fairly bare ground or leaf litter. Seedling establishment seems largely restricted to disturbed ground. Carl von Linné (Carolus Linnaeus), 1707–78, was a Swedish naturalist. *Borealis* is Latin for 'northern'.

TWIST (MOTH) Members of the family Tortricidae (Tortricinae).

Brassy twist *Eulia ministrana*, wingspan 18–25 mm, widespread and common in woodland, damp heathland and mosses. Adults fly at night in May–June. Larvae feed on leaves of various deciduous trees and shrubs, including hazel, ash, buckthorn, rowan, birch and bilberry. Greek *eu* + *leios* = 'well' + 'smooth', from the merged forewing pattern, and Latin *minister* = 'servant'.

Brindled twist *Ptycholoma lecheana*, wingspan 16–20 mm, widespread and common in woodland and orchards. Adults fly at night in June–July. Larvae feed on the leaves of many tree and shrub species. Greek *ptycos* + *loma* = 'fold' + 'border', from the forewing pattern. Johan Leche (1704–64) was a Finnish entomologist.

Brown-barred twist *Epagoge grotiana*, wingspan 13–17 mm, widespread and common in deciduous woodland and on dunes. Adults fly in June–July, especially at dawn and dusk. Larvae feed on leaves of trees and shrubs, including oaks, hawthorn and bramble. The binomial is from Greek for 'enticing', and probably in honour of Hugo Grotius (1583–1645), a Dutch national hero.

Cinquefoil twist *Philedone gerningana*, wingspan 13–16 mm, widespread but local on upland heathland; also at lower altitudes, but absent from much of southern England. Adults fly in June–September, males during the day, females from dusk onwards. Larvae feed on leaves of a range of plants, including cinquefoil. Greek *philedonos* = 'fond of pleasure', males revelling in sunshine. Johann Gerning (1745–1802) was a German entomologist.

Common twist *Capua vulgana*, wingspan 13–19 mm, widespread and common in woodland, downland and acid grassland. Adults fly at dusk in May–June. Larvae feed on the leaves of a range of plants, including alder, hornbeam and rowan. Capua is a town in Campania, Italy. *Vulgus* is Latin for 'the common people', suggesting a visually undistinguished species.

Dark-barred twist *Syndemis musculana*, wingspan 15–22 mm, nocturnal, widespread and common in open woodland, mature hedgerows and high moorland. Adults fly in May–June from late afternoon into the night. Larvae feed on bramble, birch, oak and many other plants in July–October. Greek *syn* = 'with' + the genus *Pandemis*, and Latin *musculus* = 'small mouse', from the greyish forewing.

Forest twist *Choristoneura diversana*, wingspan 19–23 mm, occasionally found in gardens, scrub and fens in England and Wales. Larvae feed on the leaves of various plants. Greek *choristos* + *neuron* = 'separated' + 'vein', and Latin *diversus* = 'different'.

Great twist *Ch. hebenstreitella*, wingspan 24–30 mm, local in woodland throughout much of England, Wales and southern Scotland. Adults fly at dusk in June–July. Larvae feed on the leaves of a range of deciduous trees and on lower-growing plants. Johann Hebenstreit (1703–51) was a German entomologist.

Heath twist *Philedonides lunana*, wingspan 12–16 mm, on moorland and heathland scattered throughout much of northern England, Scotland and Ireland. Adults fly in daylight in March–May. Larvae feed on the leaves of a number of plants, including heather, bilberry, bog-myrtle and cinquefoil. *Philedonides* is from Philedone + the Greek *eidos* = 'form', indicating affinity with this other genus, and Latin *luna* = 'moon', from the crescent forewing markings.

Heather twist *Argyrotaenia ljungiana*, wingspan 12–16 mm, common on heathland and moorland throughout much of Britain. In the north, adults fly in April–June, but usually have two generations in the south, with adults evident in April–May and June–July. Larvae feed on leaves of bog-myrtle, heather, bilberry and other moorland plants. Greek *argyros* + *tainia* = 'silver' + 'band', from the forewing mark. Sven Ljung (1757–1828) was a Swedish entomologist.

Larch twist *Ptycholomoides aeriferana*, wingspan 17–21 mm, nocturnal, common in coniferous plantations spreading throughout much of south-east England, having been first recorded in Kent in 1951. Adults fly in July–August. Larvae feed on larch needles in May–June. *Ptycholoma* and Greek *eidos* = 'form', indicating affinity with this other genus, and Latin *aerifer* = 'bearing brass', from the golden wing colour.

Large ivy twist *Lozotaenia forsterana*, wingspan 20–9 mm, nocturnal, widespread and common in wooded gardens, hedgerows and woodland edges, as well as suburban habitats such as parks and gardens. Adults fly in June–July. Larvae feed on a variety of woody plants, with a preference for ivy. Greek *loxos* + *tainia* = 'oblique' + 'band', from the forewing pattern. Johann Forster (1729–98) was a Polish Prussian naturalist.

Long-nosed twist *Sparganothis pilleriana*, wingspan 15–20 mm, scarce and local in saltmarsh and damp heathland throughout southern England and south Wales. Adults fly in July–August. Larvae feed on a variety of herbaceous plants. Greek *sparganotheis* = 'swathed', from raised scales on the forewings or the larval spun cases. Matthias Piller (1733–88) was an Austrian entomologist.

Northern grey twist *Aphelia unitana*, wingspan 17–24 mm, on acid grassland and downland in parts of Britain, not in the south-east, more numerous in the north, and southern Ireland. Adults fly at night in June–July. Larvae feed on leaves of a number of herbaceous species. Greek *apheleia* = 'plainness', and Latin *unitas* = 'uniformity', both describe the plain wings.

Obscure twist *Clepsis senecionana*, wingspan 13–16 mm, widespread but local on heathland, moorland and in woodland. Adults fly on sunny afternoons in May–June. Larvae feed on leaves of herbaceous plants, including bilberry and bog-myrtle. Greek *clepto* = 'to conceal', from the larval cases, and Latin *senecio* = 'ragwort', though this is not a food plant.

Orange pine twist *Lozotaeniodes formosana*, wingspan 20–6 mm, locally common in pine woodland in much of southern England and Wales. Adults fly at night in June–August. Larvae feed on Scots pine needles. *Formosa* is Latin for 'beautiful'.

Pale twist *Clepsis rurinana*, wingspan 16–20 mm, local in woodland in southern England; also in Lancashire, Yorkshire, Caernarfonshire, Argyll and the Burren, Co. Clare. Adults fly in June–July. Larvae feed on leaves of a number of shrubs and trees. Latin *rus* = 'countryside'.

Pine twist *Archips oporana*, wingspan 19–28 mm, scarce, in coniferous woodland and plantations in parts of south-east England, East Anglia and the Midlands. Adults fly in June–July from afternoon into dusk. Larvae feed among conifer needles. Greek *archi* + *ips* = 'chief' + 'larva', and *opora* = 'late summer'.

Privet twist *Clepsis consimilana*, wingspan 13–19 mm, nocturnal, widespread and common in gardens. Adults fly in June–September. Larvae feed on (preferably dead) leaves of privet, ash, lilac and ivy. *Consimilis* is Latin for 'entirely similar'.

Thicket twist *Pandemis dumetana*, wingspan 18–22 mm, nocturnal, local and scattered in woodland, fruit farms, orchards, gardens, fenland and chalk landscapes in eastern parts of England. Adults fly in July–August. Larvae feed on flowers and leaves of garden strawberry and other herbaceous species. *Dumetum* is Latin for 'thicket'.

White-barred twist *Olindia schumacherana* (Chlidanotinae), wingspan 11–16 mm, widespread but local in woodland, marsh, riverbanks and other damp areas. Adults fly in June–July, males in daylight, females at night. Larvae feed on leaves of a range of herbaceous plants. Olindia has no entomological significance. Christian Schumacher (1757–1830) was a German entomologist.

White-faced twist *Pandemis cinnamomeana*, wingspan 18–24 mm, locally common in woodland throughout much of England and Wales. Adults fly at night in June–July. Larvae feed on leaves of a number of deciduous trees. Greek *cinnamomon* = 'cinnamon', the forewing colour.

Yellow-spot twist *Pseudargyrotoza conwagana*, wingspan 11–15 mm, day-flying, widespread and common in a range of habitats, including gardens. Adults fly in May–July. Larvae mainly feed leaves and fruit of ash, but also privet berries. Greek *pseudos* = 'false' + the genus *Argyrotoza*. Mr Conway was an eighteenth-century British entomologist.

TWITCH Synonym of common couch.

TWITCHER (MOTH) Species of *Choreutis* and *Prochoreutis* (Choreutidae), *choreutes* being Greek for a dancer or the chorus in Greek drama, reflecting the jerky movement of these insects.

Inverness twitcher *C. diana*, wingspan 14–18 mm, in birch woodland only at Glen Affric, Inverness-shire. Adults fly in July–August, visiting thistle flowers. Larvae mine leaves of birch. *Diana* here comes from Latin *dius* = 'divine'.

Silver-dot twitcher *P. sehestediana*, wingspan 9–12 mm, scarce, day-flying, in damp grassland and moorland in parts of southern England and west Scotland. Adults fly in May and July–August in two generations. Larvae feed on leaves of

skullcap and lesser skullcap. Ove Sehestedt (1757–1838) was a Norwegian entomologist.

Small twitcher *P. myllerana*, wingspan 10–14 mm, scarce, in damp grassland and moorland in parts of Britain and south-west Ireland. Adults fly in May–September, probably with three overlapping generations. Larvae feed on leaves of skullcap and lesser skullcap. Otto Friedrich Müller (1730–84) was a Danish naturalist.

Vagrant twitcher *P. micalis*, day- and night-flying, an immigrant from southern Europe found on the south coast of England, usually scarce but sometimes in sufficient numbers to breed but not able to establish itself. Adults fly in June–August. Larvae mine leaves of common fleabane. Latin *mico* means 'to flicker' or 'quiver', describing the flight behaviour.

TWITE *Linaria flavirostris* (Fringillidae), wingspan 23 cm, length 14 cm, weight 16 g, a resident and migrant breeding finch on moorland in the Highlands, northern England and north Wales. In winter some remain in north and west Scotland near the coast; others, augmented by continental visitors, move to coastal fields and saltmarsh in eastern England. There were an estimated 10,000 pairs (6,000–15,000 pairs) in Britain in 1999. In Ireland it is a declining breeding species (around 100 pairs) mainly on the north and west coasts, and also a scarce winter visitor (650–1,000 birds) to the north-east coast, largely of Scottish birds. It is Red-listed in the UK and Ireland. Nests are in bushes, usually with 5–6 eggs which hatch after around 13 days, the altricial young fledging after 15–16 days. Diet is of seeds, nestlings taking insects. The English name is imitative of the call. Latin *linarius* = 'linen weaver' and *flavus* + *rostris* = 'yellow' + 'billed'.

TYPHACEAE Family name of bulrush, including bur-reed (Greek *typha* for these plants).

TYTONIDAE Family name of barn owl (Greek *tyto*).

U

ULIDIIDAE Family name for picture-winged flies, with 20 species in 11 genera.

ULMACEAE Family name for elm (Latin *ulmus*).

ULOBORIDAE Family name (Greek *ouloboros* = 'lethal') for cribellate orb spiders, with two British species, the introduced *Uloborus plumipes*, widespread in garden centre greenhouses since the 1990s, and *Hyptiotes* (see triangle spider under spider).

UMBELLIFERAE Former name for the umbel-bearing family Apiaceae (Latin *umbella* = 'umbrella' or 'sunshade').

UMBER (MOTH) Nocturnal members of the family Geometridae.

Barred umber *Plagodis pulveraria* (Ennominae), wingspan 28–33 mm, widespread but local in ancient broadleaf woodland and on limestone pavements. Population levels increased by 128% between 1968 and 2007. Adults fly in May–June. Larvae feed on a variety of deciduous trees, including oaks, birches, hazel, hawthorn and sallows. Greek *plagios* + *eidos* = 'slanting' + 'shape', referring to the forewing, and Latin *pulvus* = 'dust', from the powdering of the wing scales.

Dark umber *Philereme transversata* (Larentiinae), wingspan 29–37 mm, local, particularly in woodland, scrub, hedgerows and fen where associated with calcareous soils, throughout England, Wales and central-south Ireland. Adults fly in July. Larvae feed on leaves of buckthorn and alder buckthorn. Greek *phileremos* = 'fond of solitude', perhaps from the larval behaviour, and Latin *transversus* = 'going across', from the forewing banding.

Mottled umber *Erannis defoliaria* (Ennominae), wingspan 30–40 mm, widespread and common in woodland, scrub, hedgerows, grassland, heathland, moorland and gardens. Adult males fly in October–December, when the flightless females climb up tree-trunks. Larvae feed on a number of tree species as well as dog-rose. Greek *erannos* = 'lovely', and Latin describing larval defoliation.

Scarce umber *Agriopis aurantiaria* (Ennominae), wingspan 27–35 mm, widespread and fairly common, despite its name, in broadleaf woodland, scrub and gardens, more local and scattered in Wales, Scotland and Ireland. Adult males fly in October–November, when the wingless females climb up tree-trunks. Larvae mainly feed on a number of hardwood species. Greek *agrios* + *ops* = 'wild' + 'face', and Latin *aureum* = 'gold', from the rough scales and hair colour on the frons.

Small waved umber *Horisme vitalbata* (Larentiinae), wingspan 30–5 mm, common in open woodland, scrub, chalk downland and hedgerows throughout England and south Wales. Population levels increased by 167% between 1968 and 2007. Adults fly in May–June and again in August. Larvae feed on traveller's-joy (*Clematis vitalba*, giving the specific name). Greek *horisma* = 'boundary', from distinct lines on both sets of wings.

Waved umber *Menophra abruptaria* (Ennominae), wingspan 36–42 mm, nocturnal, common in broadleaf woodland, scrub, hedgerows, parks and gardens throughout England, less so in the north. Adults fly in April–June. Larvae feed on privet, lilac and ash. Greek *mene* + *ophrus* = 'moon' + 'eyebrow', from the crescent-shaped edge of the wings, and Latin *abruptus* = 'broken off', from the clear-cut margin of the wing pattern.

UNCERTAIN, THE *Hoplodrina octogenaria* (Noctuidae, Xyleninae), wingspan 28–34 mm, a widespread and common nocturnal moth found in a range of generally lowland habitats, including woodland, rough meadows and gardens, more local in Scotland and Ireland. Adults fly in June–August, sometimes with a later second brood. Larvae feed on docks, plantains and chickweed. Greek *hoplon* = 'weapon' + *(Carad)rina* (i.e. differing from *Caradrina* in the genital armature), and Latin here meaning

'smooth', as in the wing scales. Taxonomic difficulty prompts the English name.

UNDERWING Moths in the families Erebidae, Geometridae and Noctuidae, with brightly coloured hindwings (underwings) which contrast with the brownish (hence cryptic) forewings. When disturbed they flash their 'petticoats' to surprise an attacker such as a bird – a 'disguise and surprise' tactic.

1) Nocturnal moths in the genus *Catocala* (Erebidae, Erebinae) (Greek *cato* + *calos* = 'below' + 'beautiful', from the brightly coloured underwings). **Dark crimson underwing** *C. sponsa* and **light crimson underwing** *C. promissa* are both resident in ancient oak woodland in parts of the New Forest, the latter also in two large woodlands in south Wiltshire, and with some migrant records, Rare or Vulnerable in the RDB. Larvae feed on oak leaves.

Red underwing *C. nupta*, wingspan 65–75 mm, common in woodland, parks, scrub, marsh and gardens throughout much of England and Wales, slowly increasing its range northwards. Adults fly in August–September. Larvae feed on poplars, aspen, and crack and white willow. *Nupta* is Latin for 'bride', reflecting the bright underwing (c.f. petticoat).

2) Moths in the family Noctuidae. Rarities include **broad-bordered white underwing** *Anarta melanopa* (on moorland above 600 m in much of Scotland, especially the central Highlands, larvae feeding on crowberry, cowberry, bilberry and bearberry), **Guernsey underwing** *Polyphaenis sericata* (on a number of the Channel Islands, not recorded on the mainland, larvae feeding on honeysuckle and oak), **lunar yellow underwing** *Noctua orbona* (dry sandy areas, heaths and open woodland, discontinuously in parts of Britain (and especially in East Anglia), larvae feeding on grasses and herbaceous plants), and **small dark yellow underwing** *Coranarta cordigera* (declining on moorland in central Scotland, larvae feeding on bearberry).

Beautiful yellow underwing *Anarta myrtilli* (Hadeninae), wingspan 20–2 mm, day-flying in April–August, frequenting heathland and moorland, widespread but scattered. Larvae feed on heathers. Greek *anarta* = 'sea cockle', having no entomological significance, and *Vaccinium myrtillus* (bilberry), though not a food plant in Britain.

Blossom underwing *Orthosia miniosa* (Hadeninae), wingspan 31–6 mm, nocturnal, found locally in oakwood and mature hedgerows throughout much of southern England and Wales, though extending north to Cumbria, and also appearing as a migrant in the south. Adults fly in March–April, visiting sallow flowers. Larvae feed on oak, doing so gregariously when young, as well as some herbaceous plants. Greek *orthosis* = 'making straight', from a line on the forewing, and Latin *minium* = 'red lead', from a flushed colour on the forewing.

Broad-bordered yellow underwing *Noctua fimbriata* (Noctuinae), wingspan 45–55 mm, nocturnal, widespread and common in broadleaf woodland, parkland, heathland and gardens. Population levels increased by 984% between 1968 and 2007. Adults fly in July–September. Larvae feed on a range of mostly woody plants, including birches, sallows and sycamore. Latin *noctus* = 'night' and *fimbriatus* = 'fringed', from the orange hindwing fringe.

Copper underwing *Amphipyra pyramidea* (Amphipyrinae), wingspan 40–52 mm, nocturnal, common in woodland, scrub, hedgerows, gardens and parks throughout much of England, Wales, south Scotland and Ireland. Adults fly in August–October. Larvae mainly feed on oak, but also other trees. Greek *amphi* + *pyr* = 'around' + 'fire', from the behaviour around a flame, and Latin *pyramis* = 'pyramid', from a conical hump on the larva.

Large yellow underwing *N. pronuba*, wingspan 45–55 mm, probably the most abundant of the larger moths of the British Isles (numbers often enhanced in the south by immigration) found in a range of habitats. Population levels increased by 186% between 1968 and 2007. Adults fly at night in July–September. Larvae, as cutworms, feed on or below the ground on various plants, sometimes becoming a pest of early brassicas and lettuce. *Pronuba* is Latin for 'bridesmaid', reflecting the bright underwing (c.f. petticoat).

Least yellow underwing *N. interjecta*, wingspan 31–6 mm, nocturnal, common in hedgerows, gardens, fen, dunes and other open areas throughout England, Wales and Ireland, rare in Scotland. Adults fly in July–August. Larvae feed on grasses, hawthorn and herbaceous plants such as meadowsweet and mallow. *Interjecta* is Latin for 'placed between'.

Lesser broad-bordered yellow underwing *N. janthe*, wingspan 30–40 mm, nocturnal, widespread in woodland, hedgerows, gardens, heathland and moorland, commoner in the south. Adults fly in July–September. Larvae feed on herbaceous plants such as white dead-nettle, broad-leaved dock and scentless mayweed. *Janthinos* is the Greek for 'violet-coloured', describing the forewing.

Lesser yellow underwing *N. comes*, wingspan 37–45 mm, widespread and common in gardens, downland, heathland, woodland and moorland. Adults fly in July–September. Larvae feed on nettle, broad-leaved dock and foxglove. *Comes* is Latin for 'comrade'.

Lunar underwing *Omphaloscelis lunosa* (Xyleninae), wingspan 32–8 mm, widespread and common in habitats that include grassland, parks, downland and gardens, more local further north. Population levels increased by 137% between 1968 and 2007. Adults fly at night in August–October. Larvae feed on grasses. Greek *omphalos* + *celis* = 'navel' or 'middle point' + 'stain', and Latin *luna* = 'moon', from the hindwing spot.

Pearly underwing *Peridroma saucia* (Noctuinae), wingspan 45–56 mm, nocturnal, a migrant which can occur almost anywhere in the British Isles, generally more common in the south, and though most frequent in September and October has appeared in all months. In some years it arrives in large numbers, and can breed. Greek *peridromos* = 'running around', from the forewing circular mark, and Latin *saucia* = 'wounded', from a reddish tinge on the wing.

Small yellow underwing *Panemeria tenebrata* (Metoponiinae), wingspan 19–22 mm, day-flying in May–June, widely distributed though local, as far as south-east Scotland and south-west Ireland on flower-rich grassland such as downland, meadows and verges, larvae feeding on the flowers and seeds of mouse-ears. Adults take nectar from a variety of plants, including buttercups and dandelion. Greek *panamereios* = 'all day', and Latin *tenebrata* = 'shaded', from the dark areas on both sets of wings.

Straw underwing *Thalpophila matura* (Xyleninae), wingspan 38–46 mm, nocturnal, common on calcareous grassland, moorland, dunes, woodland rides, gardens and parks throughout England and Wales, but mainly coastal in Scotland and Ireland (where it is also found on the Burren). Adults fly in July–August. Larvae feed on grasses. Greek *thalpos* + *philos* = 'summer heat' + 'loving', and Latin *matura* = 'ripe', from the ripe-corn yellow hindwing.

Svensson's copper underwing *Amphipyra berbera*, wingspan 47–56 mm, nocturnal, common in woodland, scrub, hedgerows, gardens and parks throughout much of England, Wales and south Scotland. Adults fly in July–September. Larvae feed on oak and other deciduous trees. *Berbera* refers to Barbary, North Africa, the type locality.

3) Moths in the family Geometridae (Archiearinae).

Light orange underwing *Boudinotiana notha*, wingspan 33–6 mm, scarce, day-flying in March–April, local in open woodland in the southern half of England. Larvae feed on aspen. Brendon E. Boudinot is an American entomologist, who recently moved this species out of *Archiearis*. *Nothos* is Greek for 'bastard', indicating taxonomic affinity with orange underwing.

Orange underwing *Archiearis parthenias*, wingspan 35–9 mm, day-flying in March–April, common in scrub and heathland, especially near birch woods in England, Wales and upland Scotland. Larvae eat birch catkins then the leaves. Greek *arche* + *ear* = 'beginning' + 'spring', from the early emergence, and *parthenias* = 'son of a concubine', the latter indicated by the bright underwing colour.

UNGULATE Herbivorous mammals with hooves (Latin *ungula* = 'hoof'): horses (Perissodactyla), sheep, cattle and deer (Artiodactyla).

UNIONIDAE Family name for some freshwater mussels (Latin *unio* = 'single pearl').

UPRIGHT (MAYFLY) **Olive upright** *Rhithrogena semicolorata* (Heptageniidae), common and widespread, the angler's March brown, found in the riffle sections of rivers, feeding by scraping periphyton from the substrate or by gathering fine particulate organic detritus from the sediment. Because growth rates vary with water temperature, the period over which adults emerge is variable, and they are found between April and September. Emergence is from dawn to dusk. Greek *rheithron* + *genos* = 'stream' + 'race' or 'kind', and Latin for 'semicoloured'.

UPUPIDAE, UPUPIFORMES Family and order names for hoopoe (Latin *upupa*).

URCEOLARIA **Limestone urceolaria** *Aspicilia calcarea*, a common and widespread lichen on limestone, including walls (but not on mortar), scattered in Scotland. Greek *aspis* + *cilium* = 'shield' + 'hair', Latin *calcarea* = 'relating to lime'.

URCHIN Dialect name for hedgehog: see also sea urchin.

URODELA Alternative name for Caudata (Amphibia) (Greek *ouros* + *delos* = 'tail' + 'visible').

URTICACEAE Family name for nettle (Latin *urtica*).

USHER **Spring usher** *Agriopis leucophaearia* (Geometridae, Ennominae), wingspan 23–8 mm, a nocturnal moth common in deciduous, especially oak, woodland, mature hedgerows, parks and gardens throughout England, more locally elsewhere. Adult males fly in February–March. The wingless females climb up tree-trunks and the males fly weakly to them. Larvae mainly feed on oaks. Greek *agrios* + *ops* = 'wild' + 'face', from the rough scales on the frons, and *leucophaios* = 'whitish grey'.

USTILAGINALES Order of small parasitic fungi, including smut.

V

V-MOTH *Macaria wauaria* (Geometridae, Ennominae), wingspan 25–30 mm, nocturnal, widespread if local in gardens and allotments, but with numbers declining by 99% over 1968–2007, a species of conservation concern in the UK BAP. Adults fly in July–August. Larvae feed on currants and gooseberry. There are four dark brown stains, the second from the inside having a V-shaped angle extended from the leading forewing edge. *Macaria* is Greek for 'happiness'. *Vau* is Latin for the letter V or 'double V'.

Yellow V moth *Oinophila v-flava* (Tineidae, Hieroxestinae), wingspan 10 mm, only found commonly on the Isles of Scilly where it has become established in the open. Elsewhere it has been recorded only indoors, usually in warehouses and wine cellars where larvae feed on fungus and wine corks. In the open larvae feed on dry plant material, garden cuttings and the bark of shrubs. Adults fly in July–September during the evening. Greek *oinos* + *philos* = 'wine' + 'loving', and 'V' + *flava* = 'yellow'.

VALERIAN Herbaceous perennials in the genera *Valeriana* (Latin *valeo* = 'to be strong', from the medicinal properties of the root, forr example in treating epilepsy) and *Centranthus* (Greek *centron* + *anthos* = 'spur' + 'flower', from the flower having a spur-like base) (Valerianaceae).

Common valerian *V. officinalis*, common and widespread. Subspecies *sambucifolia* grows in damp grassland, marsh, fen, water margins and ditches, and wet woodland throughout the British Isles, ssp. *collina* in dry rough ground, calcareous grassland, hedge banks and woodland rides in southern and central regions. *Officinalis* is Latin for having pharmacological properties.

Marsh valerian *V. dioica*, locally common but declining in Britain north to south Scotland, in calcareous mires, marshy grassland, water meadows, willow fen carr and alder woodland. *Dioica* is Latin for having male and female flowers on separate plants.

Pyrenean valerian *V. pyrenaica*, a neophyte present by the late seventeenth century, scattered but mainly in central and southern Scotland, naturalised in damp woods and shady hedge banks.

Red valerian *C. ruber*, a neophyte present by Tudor times, widespread and common, though in Scotland mainly in the south-east, Central Lowlands and around the Moray Firth, naturalised on sea-cliffs, limestone rock outcrops and pavements, rocky waste ground, railway banks, old walls and buildings, and in other well-drained, disturbed and open habitats. *Ruber* is Latin for 'red'.

VALLONIIDAE Family name for the pulmonate grass snails, and plated snail and prickly snail (Latin *vallum* = 'rampart' or 'wall').

VALVATIDAE Family name for valve snail (Latin *valvatus* = 'having folding doors').

VAMPIRE'S BANE *Mycetinis scorodonius* (Agaricales, Marasmiaceae), a minute fungus, Near Threatened in the RDL (2006), on leaf litter and dead wood in deciduous, mainly beech woodland, at a few sites in England and south Wales. Greek *myces* = 'fungus', and *scordon* = 'garlic'.

VAPOURER *Orgyia antiqua* (Erebidae, Lymantriinae), male wingspan 25–30 mm, a widespread and common tussock moth (caterpillars having hair tufts), more so in the south, less so in the west, on heathland, moorland and scrubby habitats, and in parks and gardens. Males fly during the day in July–October. The functionally wingless females wait by their pupal cocoon for a male to find them and after mating lay their eggs on the cocoon and die. Caterpillars feed on a range of deciduous trees, and may become a pest in towns. Greek *orgyia* = 'fathom' (i.e. the length of outstretched arms), here relating to the posture of adult males with their forelegs extended forward, and Latin *antiqua* = 'ancient'.

Scarce vapourer *O. recens*, male wingspan 35–40 mm, once quite widespread but currently confined to a few mostly English locations, especially eastern East Anglia and north Lincolnshire. Larvae feed over winter on deciduous trees such as oak, hawthorn and sallow. Males fly in sunlight in June–October. The functionally wingless females wait on their pupal cocoon for a male to find them and after mating lay their eggs on the cocoon and die. *Recens* is Latin for 'recent'.

VARIOLARIA Warty and crustose lichens in the genus *Pertusaria* (Pertusariaceae) (Latin *pertusus* = 'perforated'). *Variolaria* is from Latin for 'marked with small indentations', itself from *variola* = 'smallpox'.

Milky white variolaria *P. lactea*, locally common in upland and coastal northern and western Britain on damp, well-lit, base-poor or slightly calcareous rocks. *Lactea* is Latin for 'milky'.

Whitewash variolaria *P. aspergilla*, widespread, locally common on base-poor rocks and walls in upland northern and western Britain, scattered in Ireland. Latin *aspergillo* = 'scatter'.

VASCULAR PLANTS Division of land plants that have lignified tissues for conducting water and minerals throughout the plant.

VEILWORT Thallose liverworts. Specifically, *Pallavicinia lyellii* (Pallaviciniaceae), occasionally found in upland north and west Britain and in Ireland among purple moor-grass in wet areas on the fringes of bogs; in south and south-east England it grows at a few scattered sites in woodlands, on sandstone, damp sandy soil or leaf litter. The binomial honours the Italian botanist Lazarus Pallavicini (1719–85) and, probably, the English scientist Sir Charles Lyell (1797–1875).

Blueish veilwort *Metzgeria fruticulosa* (Metzgeriaceae), common and widespread on twigs and branches of deciduous trees, especially willow, elder and elm. Johann Metzger (1789–1852) was a German botanist. *Fruticulosa* is Latin for 'shrubby'.

Downy veilwort *Apometzgeria pubescens*, on limestone in sheltered woodland, fairly common in the Pennines and parts of Scotland, uncommon in north Wales, and very rare in Ireland. Greek *apo* = 'separate (from)' + *Metzgeria*, and Latin *pubescens* = 'downy'.

Forked veilwort *M. furcata*, common and widespread on a number of trees and shrubs, especially ash, sycamore and willow, sometimes on rocks, walls and gravestones. *Furcata* is Latin for 'forked'.

Hooked veilwort *M. leptoneura*, found in north-west Wales, Cumbria, west Scotland, the Hebrides and west Ireland on damp faces of siliceous or igneous rock, especially in the spray zone of waterfalls. Greek *leptos* + *neuron* = 'slender' + 'nerve'.

Rock veilwort *M. conjugata*, relatively frequent, in humid gorges and woodlands in upland north and west Britain, favouring slightly calcareous, siliceous or base-rich igneous rock. *Conjugata* is Latin for 'joined'.

Whiskered veilwort *M. consanguinea*, common and widespread in south and west Britain and in Ireland mainly on twigs and branches of deciduous trees, especially willow and alder, sometimes spruce. Colonies are also found on scree and rocks. *Consanguinea* is Latin for 'kindred' (literally 'joined with blood').

VELIIDAE Family name of water crickets.

VELVET HORN *Codium tomentosum* (Codiaceae), a green seaweed up to 30 cm long, attached to exposed rocks in rock pools on the lower shore, mainly on the south coast west of the Isle of Wight, west Britain, and much of Ireland. Greek *codeia* = 'head', and Latin *tomentosum* = 'hairy', from the covering of the fronds.

VELVETLEAF *Abutilon theophrasti* (Malvaceae), an annual European neophyte herb introduced with wool shoddy, birdseed and oil-seed, an occasionally casual in waste places, fields and on rubbish tips in England and Wales. *Abutilon* is from Arabic for a mallow-like plant. Theophrastus (371–287 BCE) was a Greek philosopher.

VENDACE *Coregonus albula* (Salmonidae), length up to 30 cm, a rare glacial relict, Britain's rarest fish, recorded from the deep, clean, nutrient-poor lakes of Bassenthwaite (declared locally extinct in 2001, but a specimen found in 2013, and two in 2014) and Derwentwater, and (now extinct because of eutrophication) in Lochmaben (Dumfries and Galloway), together with refuge populations in Loch Skeen (Border region) using eggs from Bassenthwaite in the early 2000s (now established), and the Daer Reservoir using Derwentwater eggs (failed). It feeds on small crustaceans, molluscs and insect larvae, and spawns in winter over sand or gravel in shallow water. This is a priority species in the UK BAP. 'Vendace' dates from the eighteenth century, coming from old French *vendoise*. Greek *core* + *gonia* = 'pupil of the eye' + 'angle' (i.e. 'angle-eyed'), and Latin *albulus* = 'whitish'.

VENEER (MOTH) Nocturnal members of the family Crambidae.

　　Bulrush veneer *Calamotropha paludella* (Crambinae), wingspan 22–33 mm, local in fen, marsh, riverbanks and gravel-pits in parts of southern England (with some records from Cheshire and Lancashire) and south Wales, increasing in recent years and an occasional wanderer away from its breeding habitat. Also one record from Co. Waterford, post-2000. Adults fly in July–August. Larvae feed inside leaves, stems and upper rootstock of bulrush in September–May. Greek *calamos* + *trophe* = 'reed' + 'food', and Latin *paludis* = 'marsh'.

　　Reed veneer *Chilo phragmitella* (Crambinae), wingspan 30–40 mm, local in reedbed and fen throughout much of England, Wales and parts of south and east Ireland. Adults fly in June–July. Larvae feed inside the stems of common reed (*Phragmites*, giving the specific name) and reed sweet-grass. Greek *cheilos* = 'lip', from the elongated labial palp.

　　Rush veneer *Nomophila noctuella* (Spilomelinae), wingspan 26–32 mm, a common immigrant from mainland Europe, appearing throughout the British Isles, abundantly in some years, but most frequently and numerously in the south. Adults fly in May–September. Larvae feed on clovers, other leguminous plants, and some grasses. Greek *nomos* + *philos* = 'pasture' + 'loving', and Latin *noctu* = 'by night'.

VENERIDAE From Venus, the Roman goddess of love, the family name for filter-feeding marine bivalves, including carpet and venus shells, artemis and Manila clam.

VENUS SHELL Filter-feeding bivalves in the family Veneridae.

　　Banded venus *Clausinella fasciata*, shell size up to 2.6 cm, widespread, in the sublittoral and shallow shelf as a shallow burrower in coarse, shell, muddy and sandy gravel. Latin *clausum* = 'enclosed space', and *fasciata* = 'bundled'.

　　Rock venus *Irus irus*, shell size up to 2.6 cm, in crevices and empty borings in the intertidal and sublittoral in the English Channel, north into the Irish Sea and south and west of Ireland. Irus is a beggar in the Odyssey, who challenged Odysseus on his return to Ithaca.

　　Striped venus (clam) *Chamelea striatula*, shell size up to 4.5 cm, widespread, found from the low intertidal to the outer shelf in fine to coarse sand frequently with a silty component. Greek *chamae* = 'on the ground', and Latin *striatula* = 'striped'.

　　Warty venus *Venus verrucosa*, shell size up to 6.5 cm, widespread except for east Britain, burrowing in sand and gravel from the intertidal down to 100 m. Venus was the Roman goddess of love. *Verrucosa* is Latin for 'warty'.

VENUS'S-LOOKING-GLASS *Legousia hybrida* (Campanulaceae), an annual European archaeophyte scattered in southern and eastern England on arable fields usually on calcareous soils. It is also a casual in disturbed sites, including motorway embankments.

　　Large venus's-looking-glass *L. speculum-veneris*, an annual

European neophyte present by 1596, naturalised in arable fields in north Hampshire since 1916, occasionally a casual garden escape and grain impurity elsewhere in southern England. The specific name is Latin for 'Venus's looking-glass'.

VERBENACEAE Family name of vervains (Latin *verbena* for these plants).

VERNAL-GRASS Sweet vernal-grass *Anthoxanthum odoratum* (Poaceae), a common and widespread loosely tufted perennial in old pasture and meadow, hill grassland, heathland and the drier parts of bogs, and on dunes, most frequent on acidic soils, avoiding very dry or waterlogged sites. The leaves smell of freshly cut hay with a hint of vanilla, prompting the 'sweet' in its English name; 'vernal' (spring) refers to the early flowering (April–July). Greek *anthos* + *xanthos* = 'flower' + 'yellow', from the colour of the panicles, and Latin *odoratum* = 'fragrant'.

VERONICA Hedge veronica *Veronica* x *franciscana* (Veronicaceae), an evergreen shrub, widespread in coastal locations as a garden escape or planted in hedge banks and on cliffs, frequently naturalised, and on waste ground, walls and pavement cracks. The plant is named for St Veronica.

VERONICACEAE Family name of speedwell, including foxglove and toadflax, named for St Veronica.

VERTIGINIDAE Family name for whorl snails, sometimes called vertigo snails (Latin *vertigo* = 'turning around').

VERVAIN *Verbena officinalis* (Verbenaceae), an archaeophyte perennial herb present since the Neolithic, locally common south of a line from Pembrokeshire to the Wash, scattered elsewhere, usually in open habitats on freely draining, often calcareous soils, particularly in rough grassland and scrub, on roadsides, on sheltered coastal cliffs and rock outcrops, sometimes on stream banks. 'Vervain' dates from late Middle English, from Old French *verveine*, in turn from Latin *verbena*. *Officinalis* is Latin for having pharmacological properties, having been used as an astringent, diaphoretic and antispasmodic agent, and to counter fever. **Argentinian vervain** *V. bonariensis* and **slender vervain** *V. rigida* are scarce South American neophyte perennials, casual on waste ground.

VESPERTILIONIDAE Family name for all British bats excluding horseshoe bats (Latin *vespertilio* = 'bat', from *vesper* = 'evening').

VESPIDAE Family name of both solitary and (eu)social wasps (Latin *vespa* = 'wasp'), including mason and potter wasps, and hornet.

VESTAL, THE *Rhodometra sacraria* (Geometridae, Sterrhinae), wingspan 22–8 mm, a nocturnal immigrant moth from southern Europe, appearing most frequently on the south coast, but found throughout England and Wales, north to north Scotland, and scattered in Ireland, sometimes breeding. Greek *rhodon* + *metron* = 'rose' + 'a measure', from the colour of the wing band, and Latin *sacer* = 'holy', referencing a priestess, or vestal virgin (hence the English name), from the light yellow wings.

VETCH Species of Fabaceae, mainly in the genus *Vicia* (Latin for 'vetch'). See also bitter-vetch.

　　Scattered but uncommon scrambling casual neophytes on rough ground, often arriving as grain impurities or in wool shoddy, include **Bithynian vetch** *V. bithynica*, **fine-leaved vetch** *V. tenuifolia* (introduced by 1799, recorded from the wild in 1859), **fodder vetch** *V. villosa* (introduced in 1815), and **Hungarian vetch** *V. pannonica* (naturalised in west Kent since 1971).

　　Bush vetch *V. sepium*, a common and widespread climbing or scrambling perennial of hedge banks, woodland edge, lightly grazed grasslands and occasionally (north-west Scotland and north-west Ireland) dunes, most frequently on neutral or basic soils. Latin *sepe* = 'hedge'.

　　Common vetch *V. sativa*, a common and widespread annual of grassy and wayside places, particularly on dry and sandy sites. Having been grown as a fodder crop, it has widely escaped and become naturalised in many ruderal habitats. Subspecies *nigra* (Latin = 'black') is procumbent or climbing in grassy and waste ground, particularly on dry and sandy sites such as dunes, shingle, sea-cliffs and heathland. Subspecies *sativa* (= 'cultivated') and *segetalis* (= 'crop') are found on waste and cultivated ground and field edges.

Crown vetch *Securigera varia*, an introduced perennial from Europe, scattered in grassy habitats and rough ground. It is considered to be good forage when fed as hay to or grazed by ruminants, but is toxic to non-ruminant (farm) animals such as horses because of the presence of nitroglycosides. Latin *securis* + *gero* = 'axe' + 'bearing', from the fruit shape.

Horseshoe vetch *Hippocrepis comosa*, a herbaceous perennial, local in England and Wales on dry calcareous grassland and clifftops. Greek *hippos* + *crepis* = 'horse' + 'shoe', from the shape of the pod, and Latin for 'hairy'.

Kidney vetch *Anthyllis vulneraria*, a widespread perennial of rock outcrops and open turf on south-facing slopes, on free-draining neutral to base-rich, often calcareous, soils. On the coast it is found on sea-cliffs, shingle and dunes. It is increasing as an alien on roadsides. Ssp. *carpatica* has been introduced in grass-seed mixtures, and occurs on roadsides and in waste places in England and Scotland. Ssp. *lapponica* is found in Scotland and Ireland on dunes, machair, coastal and inland cliffs and rock ledges, and recorded from ruderal habitats, including railway banks and forest tracks. It grows over a range of base-rich substrates. Ssp. *polyphylla* has a scattered distribution, naturalised on grassy banks. Greek *anthos* + *ioulos* = 'flower' + 'downy', and Latin *vulneraria* = 'wound healer'.

Spring vetch *V. lathyroides*, a procumbent or weakly climbing annual, scattered around the coast (except north-west Scotland and west Ireland) on dunes and grassland on sandy soils, and locally inland on disturbed ground, old walls, and heathland. *Lathyroides* means 'resembling species of *Lathyrus*' (vetchlings).

Tufted vetch *V. cracca*, a common and widespread scrambling perennial of hedgerows, verges, woodland edge, scrubby grassland, pasture, hay meadow, riverbanks, marsh and tall-herb fen, avoiding permanently wet sites. *Cracca* is Latin for a kind of vetch.

Wood vetch *V. sylvatica*, a climbing or scrambling light-demanding perennial, local and scattered, in hedges, woodland edge and clearings, scrub, rough grassland on cliffs, shingle, scree and railway banks. *Sylvatica* is from Latin for 'woodland'.

Yellow vetch *V. lutea*, a widespread but scattered and uncommon annual found in a variety of coastal habitats, including scrubby grassland and cliffs, and on shingle. In south Scotland it is confined to sheltered sea-cliffs. Inland it is a casual on roadsides, quarries and railway banks. *Lutea* is Latin for 'yellow'.

VETCHLING Species in the genus *Lathyrus* (Fabaceae), from Greek for 'pea' or 'pulse'.

Grass vetchling *L. nissolia*, a widespread annual, more common in England and Wales, in open, often disturbed, habitats on chalk and heavy calcareous clay soils, for example grassy banks, verges, railway banks, and woodland rides, as well as coastal grassland and shingle.

Hairy vetchling *L. hirsutus*, a scrambling European annual neophyte, scattered in England, on grassy banks and waste ground as a garden escape or birdseed or grain contaminant. Populations can persist, as in a few areas around London, but are usually casual. *Hirsutus* is Latin for 'hairy'.

Meadow vetchling *L. pratensis*, a common and widespread rhizomatous perennial of moderately fertile soils on roadside and railway banks, hedges, unimproved pasture, hay meadow and other grassy habitats. *Pratensis* is from Latin for 'meadow'.

Yellow vetchling *L. aphaca*, persistent populations of this scrambling annual are in open grassy habitats on chalk, limestone and calcareous clay soils, especially near the coast. It is also a casual in waste ground, and as an arable weed, possibly introduced as a contaminant of legume crops. *Phace* is Greek for 'lentil'.

VIBURNUM Wrinkled viburnum *Viburnum rhytidophyllum* (Caprifoliaceae), an evergreen Chinese neophyte shrub, introduced in 1900, scattered in Britain, mostly in southern England, occasionally naturalised in woodland, hedgerows, parks and amenity areas, and on roadsides. *Viburnum* is Latin for 'wayfaring-tree', and Greek *rhytis* + *phyllon* = 'wrinkle' + 'leaf'.

VILLA Bee-flies (Bombyliidae) in the genus *Villa* (Latin for 'shaggy hair').

Downland villa *V. cingulata*, with a scattered southern England distribution, feared extinct, having not been recorded during the latter half of the twentieth century, but rediscovered in the Cotswolds in 2000. With further records from east Gloucestershire, Oxfordshire and Buckinghamshire in the first decade of the twenty-first century, reports since 2010 have extended the known range to include Wiltshire and Berkshire, and in 2013 to south Hampshire and Middlesex. Many records are from calcareous grassland, but the Middlesex one is from flower-rich neutral grassland. Adults feed on nectar, and flight is from June to August. *Cingulata* is Latin for 'collared'.

Dune villa *V. modesta*, a rarity of coastal dunes, reported from Cornwall, Hampshire, south Wales, Lancashire and East Anglia. Adults feed on nectar using short mouthparts. *Modesta* is Latin for 'moderate'.

Heath villa *V. venusta*, found on heathland in Dorset, believed to have become extinct in the 1950s, but possibly seen in Wiltshire in 2012, and confirmed as records from Moor Copse, Berkshire in 2014 and 2015. *Venusta* is Latin for 'pretty'.

VIOLACEAE Family name for violets, dog-violets and pansies (Latin *viola*).

VIOLET 1) Perennial herbs in the genus *Viola* (Violaceae). See also dog-violet and pansy.

Fen violet *V. persicifolia*, local on damp, peaty base-rich soils in seasonally wet fen, and in Ireland on the margins of turloughs. It survives at Wicken Fen and Woodwalton Fen (Cambridgeshire). It was believed to be extinct at Otmoor (Oxfordshire), but a small population was found in 1997. Its distribution in west and central-north Ireland is stable. It is a weak competitor, favouring areas subject to fluctuating water levels, cattle trampling or peat digging. *Persicifolia* is Latin for 'with peach-like leaves'.

Hairy violet *V. hirta*, locally common in lowland England and Wales, scattered and scarce elsewhere, mainly on calcareous soils in short grassland or open scrub on downland, rocky slopes, limestone pavement, woodland borders and rides, sometimes on base-flushed but more acidic riverside habitats; also on roadsides and railway banks. *Hirta* is Latin for 'hairy'.

Marsh violet *V. palustris*, widespread in upland areas and also parts of southern England and East Anglia, found in bog, wet heathland, marsh, carr and wet woodland, especially on acidic soils. It is commonly associated with *Sphagnum*, and is also found in non-calcareous dune slacks. *Palustris* is from Latin for 'marsh'.

Sweet violet *V. odorata*, widespread though often local in lowland regions. In the north and west of its range many plants are introductions, and it may not be native north of Cumbria and Co. Durham. In Ireland it is probably introduced in the north and west. It is usually found on calcareous or other base-rich soils. Habitats include open woodland, hedge banks and scrub, less frequently shaded road and railway banks and verges. Alien populations are naturalised in churchyards and elsewhere. *Odorata* is Latin for 'fragrant'.

Teesdale violet *V. rupestris*, found in mid-west Yorkshire, Co. Durham and south Cumbria in dry, open limestone grassland on bare or eroded slopes or hummock tops. Significant populations were discovered in the Craven Pennines in 1977, but colonies were lost on Widdybank Fell (Co. Durham) through erosion and when a reservoir was built in 1970, and have been reduced at Arnside through trampling and undergrazing. Some remaining populations, however, remain sizeable. *Rupestris* is from Latin for 'rock'.

2) **Dog's-tooth-violet** *Erythronium dens-canis* (Liliaceae), a bulbous herbaceous southern European neophyte perennial, scattered in England, Scotland and north-east Ireland, planted in woodland and naturalised in a few parkland sites. *Erythros* is Greek for 'red', *dens-canis* Latin for 'dog's tooth'.

VIPERIDAE Family name of the adder, from Latin *viper*, derived from *vivus* + *pario* = 'alive' + 'to produce', describing live birth.

VIPER'S-BUGLOSS *Echium vulgare* (Boraginaceae), a widespread and, especially in southern and eastern England and central Scotland, locally common biennial of grassy and disturbed habitats on well-drained soils, growing in bare places on chalk and limestone grassland, on heathland, in quarries and chalk-pits, in cultivated and wasteland, along railways and roadsides, and on coastal cliffs, dunes and shingle. This is the county flower of East Lothian as voted by the public in a 2002 survey by Plantlife. 'Bugloss' dates from late Middle English, from Latin *buglossus*, in turn from Greek *bouglossos*, viz. *bous* + *glossa* = 'ox' + 'tongue', from the leaf shape. Viper's-bugloss gets its name from the small nutlike seeds resembling a snake's head and the spotted stems thought to look like snakeskin. *Echis* is Greek for 'viper', *vulgare* Latin for 'common'.

Giant viper's-bugloss *E. pininana*, a herbaceous annual or biennial neophyte found as garden escapes on rough ground in the Channel Islands, Scillies, southern England, west Wales, Isle of Man, and Co. Waterford and Co. Dublin. Latin *pinus* + *nana* = 'pine' + 'dwarf'.

Purple viper's-bugloss *E. plantagineum*, a scarce herbaceous annual or biennial archaeophyte, common in Jersey, local in west Cornwall and the Scillies, as a weed in arable fields, on cliffs and in open sandy habitats by the coast, and elsewhere as a garden escape or casual. Latin *plantagineum* = 'plantain-like'.

VIPER'S-GRASS *Scorzonera humilis* (Asteraceae), a rare herbaceous perennial of damp, unimproved grasslands and fen-meadow on infertile, neutral or mildly acidic soil, first recorded in Wareham Meadows, Dorset, in 1914, at a site that still supports a large population. It was recorded at another site in Dorset (Corfe Common), in 1927. A small Warwickshire population, discovered in 1954, had disappeared by 1967. Two sites near Bridgend, Glamorgan, were discovered in 1996 and 1997. *Scorzonera* is Spanish for 'black root', once thought to be a cure for snake-bite, *humilis* Latin for 'low'.

VIRGINIA-CREEPER *Parthenocissus quinquefolia* (Vitaceae), a perennial neophyte climber introduced from North America by 1629, very widely grown for its autumn colour, now well-established, especially in England, on old walls, in hedgerows and scrub, and on railway banks and roadsides. Greek *parthena* = 'virgin' and Latin *cissos* = 'ivy', and Latin for 'five-leaved'.

False Virginia-creeper *P. vitacea*, a perennial neophyte climber introduced from North America, grown for its autumn colour, locally naturalised on old walls, and in hedgerows and scrub. *Vitacea* is Latin for 'of the grape *Vitis* family'.

VIRGIN'S-BOWER *Clematis flammula* (Ranunculaceae), a scrambling or weakly climbing perennial introduced from the Mediterranean by Tudor times, occasionally naturalised in southern England on coastal cliffs, shingle and dunes, rarely inland. *Clematis* is Greek for 'branch' or 'twig', coming to mean a climbing plant. *Flammula* is Latin for 'small flame'.

VIRUS A microscopic often pathogenic agent consisting of a strand of DNA or RNA, surrounded by a protective capsule of protein. It can reproduce only by invading a living cell. The word dates to the late Middle English referring to snake venom, from the Latin for 'poison'.

VITACEAE Family name (Latin *vitis* = 'vine') for grape-vine, Virginia-creeper and Boston-ivy.

VITRINIDAE Family name (Latin *vitrum* = 'glass') for glass snails (but see also Oxychilidae) and semi-slugs.

VIVIPARIDAE Family name of river snails (Latin *vivus* + *pareo* = 'alive' + 'to appear').

VOLE Blunt-nosed plant-eating rodents in the family Cricetidae. 'Vole' dates from the early nineteenth century, originally vole-mouse, from Norwegian *vollmus* = 'field mouse'.

Bank vole *Myodes glareolus*, head-body length 100–50 mm, tail length around 50 mm, weight 15–40 g, common and widespread in Britain in deciduous woodland, hedgerows and scrub. They may have been accidentally introduced to Ireland in the 1920s, though they were only discovered in Co. Kerry in 1964; since then they have steadily spread throughout Munster and Leinster at a rate of 1–4 km p.a. By 1990 they had 'crossed' the Shannon and are now found in Co. Galway and elsewhere in Connaught. They generally live in shallow underground burrows lined with dried grass, moss and feathers though where there is thick cover they construct ground-level nests of grass, etc. They eat berries, bulbs, seeds, fruits and fungi, and hunt for insect adults and larvae, snails and earthworms. They climb well, and in winter may strip the bark from young trees, including beech, maples and larch, up to several metres above the ground to eat the soft tissue underneath. They are active during both day and night but are more nocturnal in summer. In winter they create a seed cache near their burrow. Breeding usually starts in April and ends in September, but can occur year-round if there is enough food available. Gestation is 16–18 days. The female can become pregnant again before weaning (which lasts two weeks) is complete. Litter size is usually 3–5 pups, with females having 4–5 litters a year. Early born pups can themselves breed in the same season as their birth. **Skomer vole** (*M. glareolus skomerensis*), found on the island of Skomer (south-west Wales), is much larger than the mainland bank vole. There may be 20,000 individuals on the island in late summer. Greek *myodes* = 'mouse-like', *glareolus* the Latin diminutive of *glarea* = 'gravel'.

Common vole *Microtus arvalis*, not present on the mainland, but one subspecies has been introduced to Orkney, and another subspecies to Guernsey, though this is more likely a glacial relict; see Orkney and Guernsey voles below. Greek *micros* + *otos* = 'small' + 'ear', and Latin *arvalis* = 'field'.

Field vole *Microtus agrestis*, head-body length 95–135 mm, tail length 25–44 mm, weight 20–50 g, common and widespread in Britain in grassland, deciduous woodland, hedgerows and scrub. It feeds on grasses, herbs, root tubers, moss and other plants, and gnaws bark during winter. Breeding can take place at any time but peaks in spring and summer. The nest is made on or just under the surface, often in a clump of grass or sedge. Gestation last three weeks and up to a dozen young are born. These suckle for 12 days and leave the nest at 21 days, reaching sexual maturity soon afterwards. Females become fertile again soon after giving birth. *Agrestis* is Latin for 'field'.

Guernsey vole *Microtus arvalis* ssp. *sarnius*, head-body length 97–134 mm, weight 42–51 g, found only in Guernsey, possibly a prehistoric introduction, but more likely a glacial relict, feeding on grass in damp meadow. Mating can take place from February through to autumn. Average litter size is 2–5. *Sarnia* is Latin for 'Guernsey'.

Orkney vole *Microtus arvalis* ssp. *orcadensis*, introduced in the Neolithic and found on five islands; some authorities consider each has its own subspecies. Head-body length is 97–134 mm, weight 42–51 g. The current total population is estimated at >4 million. Habitats include wet heathland, marsh, blanket bog, waste ground, verges and gardens, in which runways are created. Diet is mainly of grasses and rushes. Breeding is from March to October, with 2–3 litters a year. Gestation lasts 16–24 days, with litter size of 3–8. Latin *orcadensis* = 'of Orkney'.

Short-tailed vole Synonym of field vole.

Skomer vole Subspecies of bank vole (above), found on this south Wales island.

Water vole *Arvicola amphibius*, head-body length 140–220 mm, weight 110–40 g, a widespread but rare and seriously declining semi-aquatic rodent, particularly infrequent in Scotland, and absent from Ireland. It showed the most catastrophic decline of any mammal in Britain during the twentieth century. Intensification of agriculture in the 1940s and 1950s caused loss of habitat, with water pollution a later additional problem, but the most rapid period of decline was during the 1980s and 1990s as the predatory American mink spread, and between 1989–90 and 1996–8 the population fell by almost 90%, becoming fragmented and locally extinct. In 2004 there was an estimated pre-breeding population size of 875,000. Survey data for 2007–11, indicate a continued decline of 22%. The species did not gain legal protection until 1998, under Schedule 5 of the WCA (1981), Section 9, which protects place of shelter, not the vole itself. It is a UK BAP priority species. There have been a number of reintroduction projects, for example in 2012 and 2013, Gwent Wildlife Trust released over 200 water voles to Magor Marsh. Not seen in Cornwall since 1989, the species was reintroduced to the Bude river catchment in 2014. In the South Downs National Park, 450 animals were released at Titchfield Haven in 2013, 600 further upstream in 2014, and 190 near Soberton in 2015. In August 2016 around 100 animals were released at Malham Tarn, Yorkshire Dales, where this vole had not been seen for 50 years. Water voles dig burrows, which often have underwater entrances, in grassy banks along slow moving rivers, ditches, streams, lakes, ponds and canals. They are active during daytime, and are good swimmers. One population in Glasgow lives in rough urban grassland with no nearby open water; they create a burrow system and forage above ground. They need to eat 80% of their body weight every day, feeding on grasses, reeds and sedges in spring and summer, and roots, bark and fruit in autumn and winter. Breeding lasts from March to October. Females produce 2–5 litters a year, each of 2–8 pups which become independent after 28 days. Those born by July can breed that autumn. Latin *arvum* + *cola* = 'field' + 'inhabitant'.

W

WAGTAIL Passerine birds in the genus *Motacilla* (Motacillidae), from Greek *muttex*, a bird described by Hesychius of Alexandria, and Latin *cilla*, used in the sense of 'tail'. **Citrine wagtail** *M. citreola* is a scarce visitor. The reason for tail-wagging is unclear, though it might be to indicate the bird's vigilance to potential predators.

Grey wagtail *M. cinerea*, wingspan 26 cm, length 18 cm, weight 18 g, an insectivorous resident breeder, summer migrant and winter visitor, favouring habitat by fast-flowing rivers in summer, with highest densities in the hills of England, Scotland and Wales. In winter they are often seen around farmyards and lowland streams, and in city centres. They are scarce in the Midlands and eastern England in summer, and in upland areas in winter. Some 38,000 pairs were found in Britain in 2009, though with a 39% decline in numbers between 1975 and 2015 and a 12% decline over 1995–2015), and despite a strong increase by 15% in 2009–15, it is Red-listed in Britain, though Green-listed in Ireland where it remains common. The nest, often built between riverside stones and roots, contains 3–6 eggs which hatch after around 13 days, the altricial chicks fledging in 14–15 days. Latin *cinereus* = 'ash-coloured'.

Pied wagtail *M. alba* ssp. *yarrellii*, wingspan 28 cm, length 18 cm, weight 21 g, a common and widespread resident breeder (and migrant breeder in parts of the Highlands and Islands), with an estimated 470,000 pairs (410,000–520,000) in Britain in 2009, recent numbers showing a slight overall increase (by 38% over 1970–2015, 13% in 2010–15). Britain and Ireland together hold almost the entire world population of this distinctive dark-backed subspecies. It is found in a variety of habitats, including urban areas, but tending to avoid uplands. Nests are in crevices, with 5–6 eggs which hatch after around 13 days, the altricial chicks fledging in 14–15 days. Feeding is on insects taken in flight or from the ground; in towns these birds often feed in paved areas, car parks, etc., and have become habituated to people, and at dusk they often gather in trees and on roofs to roost in large groups. Latin *alba* = 'white'.

White wagtail *M. alba* ssp. *alba* pass through Britain on their way to breed in Iceland and Scandinavia in spring, and on their return in autumn, though a few may overwinter.

Yellow wagtail *M. flava*, wingspan 25 cm, length 17 cm, weight 18 g, an insectivorous migrant breeder, wintering in Africa, found in England, the Welsh Marches, and south-east and central-west Scotland from March to September, in arable, wet pasture and upland hay meadow, and especially on cattle-grazed grassland, with an estimated 15,000 territories in Britain in 2009, declining in numbers by 97% between 1975 and 2015, including by 53% in 2010–15. Britain holds almost the entire world population of the distinctive race *flavissima*, Red-listed in 2009. Nests are often in tussocks, with 5–6 eggs, incubation lasting around 14 days, the altricial chicks fledging in 13–15 days. The Latin *flavus* = 'yellow'.

WAINSCOT Nocturnal moths in the family Noctuidae, many in the genus *Mythimna* (Hadeninae), after Mithimna, a town on the Greek island of Lesbos. Many species are associated with reedbeds.

Blair's wainscot *Sedina buettneri* (Xyleninae), wingspan 28–34 mm, Endangered in the RDB, having bred in the Isle of Wight, discovered in 1945, until the habitat was destroyed in 1951. In 1996 another population was discovered in Dorset, since when several colonies have been found nearby. It is also an occasional migrant in south-east England. Adults fly in September–October. Larvae feed in the stems and roots of lesser pond-sedge. Sedyn is an old name for Stettin (now Szczecin), a town in Poland. J.G. Büttner was a nineteenth-century Latvian entomologist.

Brighton wainscot *Oria musculosa* (Xyleninae), wingspan 28–34 mm, first discovered near Brighton in the late nineteenth century, giving its English name, subsequently found in many southern and Midland counties, but now having declined (with a population crash in the 1970s) and possibly restricted to Salisbury Plain. Adults fly in July–August. Larvae mine stems of grasses and cereal crops, subsequently eating the seeds. *Oria* possibly comes from Greek *horeion* = 'granary', from the larval diet. *Musculus* (Latin for 'little mouse') may refer to the female's behaviour of running down plant stalks.

Brown-veined wainscot *Archanara dissoluta* (Xyleninae), wingspan 27–33 mm, local in reedbed, marsh, fen, riverbanks and gardens throughout much of southern England west to Devon, discontinuously north to Lancashire and north-west Wales, and in central Ireland. Adults fly in July–September. Larvae mine stems of common reed. *Archanara* is entomologically meaningless; *dissoluta* is Latin for 'dissolved', from the absence of wing spots in this species compared to congenerics.

Bulrush wainscot *Nonagria typhae* (Xyleninae), wingspan 45–50 mm, widespread and common in fen, marsh and reedbed, less so in Wales and Scotland. Adults fly in July–September. Larvae feed in stems of reedmace and lesser reedmace (= *Typha*, prompting the specific name). Nonagria is an old name for Andros, a Greek island in the Aegean.

Common wainscot *M. pallens*, wingspan 30–5 mm, widespread and common in grassland, woodland rides and gardens. In the south there are two generations, adults flying from May to October; further north they fly in July–August. Larvae feed on grasses. *Pallens* is Latin for 'pale', after the forewing shade.

Devonshire wainscot *Leucania putrescens* (Hadeninae), wingspan 32–6 mm, scarce on grassy slopes and clifftops in parts of south-west England and south Wales, predominantly coastal. Adults fly in July–August. Larvae feed by night on grasses. Greek *leucos* = 'white', and Latin *putrescens* = 'decaying', the forewing being the colour of decaying wood.

Fen wainscot *Arenostola phragmitidis* (Xyleninae), wingspan 32–6 mm, local in and around reedbeds, from the south coast westwards to Dorset, and north to Yorkshire, Lancashire and Cumbria. Adults fly in July–August. Larvae mine stems of common reed (*Phragmites*, prompting the specific name). Greek *arena* + *stole* = 'sand' + 'garment', from the sandy colour.

Fenn's wainscot *Protarchanara brevilinea* (Xyleninae), wingspan 30–8 mm, in reedbed in Norfolk and Suffolk, often recorded where the reeds are regularly cut and most numerous two or three years after cutting. An RDB rarity, this is a priority species in the UK BAP. Adults fly in July–August. Larvae mine stems, later leaves of common reed. The genus name is entomologically meaningless. *Brevilinea* is Latin for 'short line', from the forewing marking.

Flame wainscot *Senta flammea* (Hadeninae), wingspan 32–40 mm, scarce in reedbed, fen and marsh in parts of East Anglia, where it is resident, and as a migrant in south-central England, including Sussex, Hampshire and the Isle of Wight. Adults fly in May–July. Larvae feed on common reed. *Senta* may derive from Latin *sentio* = 'to feel', from the labial palps being viewed as 'feelers'.

L-album wainscot *M. l-album*, wingspan 30–5 mm, an immigrant found in damp coastal habitats such as brackish ditches in southern England, in most years arriving in sufficient numbers to breed. Since 2000 it has extended its range northwards, probably as a result of an increase in temperatures. Adults fly in July and again in September–October. Overwintering larvae feed on grasses, especially marram. *L-album* is Latin describing the white L-shape forewing mark.

Large wainscot *Rhizedra lutosa* (Xyleninae), wingspan 42–50 mm, widespread and common in reedbed as far as southern Scotland, but with numbers declining by 83% over 1968–2007, a species of conservation concern in the UK BAP. Adults fly in August–October. Larvae feed in stems and roots of common reed, prompting the genus name from Greek *rhiza* + *hedra* =

'root' + 'seat'. *Lutosa* is Latin for 'muddy', describing the wing colour or the habitat.

Mathew's wainscot *M. favicolor*, wingspan 34–42 mm, scarce, possibly a race of common wainscot, very locally common in saltmarsh from south Suffolk to east Kent and from West Sussex to west Hampshire. There are isolated records from the Isle of Wight and inland from the New Forest, Hampshire, Surrey and East Anglia. Adults fly in June–July. Larvae feed on saltmarsh-grass and other grasses. *Favicolor* is Latin for the 'honeycomb colour' of the wings.

Mere wainscot *Photedes fluxa* (Xyleninae), wingspan 26–30 mm, scarce in fen and open woodland in south and central England, though currently more or less restricted to Cambridgeshire, Suffolk and Northamptonshire. Adults fly in July–August. Larvae mine stems of wood small-reed. Greek *phos* + *edos* = 'light' + 'delight', from daytime flying, and Latin *fluxa* = 'pale', from the indistinct forewing markings.

Morris's wainscot *Photedes morrisii*, wingspan 26–34 mm, an endangered RDB moth found on south-facing hillsides, undercliffs and rocky areas by the sea along a strip of coastline between Charmouth, Dorset, where in 1837 it was discovered by the Rev. F.O. Morris (1810–93), and Sidmouth, Devon. Adults fly in June–July. Larvae mine tall-fescue stems.

Obscure wainscot *Leucania obsoleta*, wingspan 36–40 mm, local in reedbed, fen and freshwater margins throughout much of southern and eastern England, and parts of Wales. Adults fly in May–July. Larvae feed by night on common reed. *Obsoletus* is Latin for 'obscure', describing the weak wing markings.

Powdered wainscot Synonym of reed dagger.

Rush wainscot *Globia algae* (Xyleninae), wingspan 32–45 mm, a rare RDB moth found in reedbed, marsh and gravel-pits in southern England, largely in Sussex, Surrey (a recent discovery), the Norfolk Broads and in Breckland wetlands (near Brandon) on the Suffolk-Norfolk border. Adults fly in August–September. Larvae mine stems of common club-rush, reedmace and yellow iris. *Globia* may come from Greek for 'gills' or Latin *globus* = 'ball'; *alga*, originally meaning 'seaweed', became applied to 'lichen', another larval food plant.

Shore wainscot *M. litoralis*, wingspan 36–42 mm, on dunes and sandy beaches in parts of England, Wales, southern Scotland, and Ireland, predominantly coastal (hence *litoralis*). Adults fly in June–August. Larvae feed by night on marram, burrowing in the sand during the day.

Shoulder-striped wainscot *Leucania comma*, wingspan 32–7 mm, widespread and common in fen, marsh, grassland, gardens and damp woodland, but a species of conservation concern in the UK BAP. Adults fly in June–July. Larvae feed by night on cock's-foot and other grasses. *Comma* describes a wing streak, though not actually like the punctuation mark.

Silky wainscot *Chilodes maritima* (Xyleninae), wingspan 29–36 mm, local in reedbed on the south and east coasts of England, with scattered records elsewhere in England, Wales and eastern Ireland. Adults fly in June–August. Larvae feed on insects, including pupae of other wainscots, inside reed stems. *Chilodes* is from another genus, *Chilo* + *eidos* = 'shape'. *Maritima* reflects the type locality rather than typical habitat.

Small wainscot *Denticucullus pygmina* (Xyleninae), wingspan 23–9 mm, widespread and common in fen, marsh, riverbanks, damp woodland and gardens. Adults fly in August–September. Larvae mine stems of lesser pond-sedge, glaucous sedge and some grasses. Latin *dens* + *cucullus* = 'tooth' + 'hood', and Latin *pygmina* = 'pygmy'.

Smoky wainscot *M. impura*, wingspan 31–8 mm, widespread and common in grassland, woodland rides, gardens and dunes. Adults fly in June–August. Larvae feed on grasses and reed. *Impura* is Latin for 'unclean', from the dark hindwing colour.

Southern wainscot *M. straminea*, wingspan 32–40 mm, local in reedbed and fen throughout much of England south of a line from the Bristol Channel to the Wash, less common

further north, but occurring in mid-Wales, Cheshire, Yorkshire and Cumbria, and scattered in Ireland. Adults fly in July. Larvae feed on reed and reed canary-grass. *Straminea* is Latin for 'straw-coloured', for the forewing.

Striped wainscot *M. pudorina* (Hadeninae), wingspan 35–43 mm, local in marsh, fen and boggy heathland in much of England, south Wales and the southern half of Ireland. Adults fly in June–July. Larvae feed on grasses, reeds and rushes. *Pudor* is Latin for 'shame', which might lead to pink blushing, as with the forewing colour.

Twin-spotted wainscot *Lenisa geminipuncta* (Xyleninae), wingspan 27–32 mm, local in reedbed, marsh, fen and gardens throughout much of southern and eastern England and south Wales, generally most frequent by the coast, and in Co. Wexford, Ireland. Adults fly in August–September. Larvae mine reed stems. Latin *geminipuncta* = 'twin-spotted'.

Webb's wainscot *Globia sparganii*, wingspan 32–40, scarce, local in reedbed, marsh and gravel-pits on the coasts of southern England from Scilly and southern England through East Anglia to Lincolnshire, and in south Wales. Adults fly in August–September. Larvae mine the stems of yellow iris, reedmace, branched bur-reed (*Sparganium*, prompting the specific name) and other aquatic plants.

White-mantled wainscot *Archanara neurica*, wingspan 26–9 mm, a rare RDB moth restricted to reedbed at Walberswick and Minsmere in Suffolk, and a priority species in the UK BAP. Adults fly in July–August. Larvae mine reed stems. *Neuron* is here Greek for 'vein'.

WAKAME *Undaria pinnatifida* (Alariaceae), an annual brown seaweed, thallus 1–2 m long, introduced from Japan via France on ships' hulls, found on pontoons in the Hamble estuary in the Solent, Hampshire, in 1994, subsequently recorded patchily from Pembrokeshire round to Norfolk, growing in the subtidal, especially on artificial structures. As a fouling organism that may out-compete native species it is undesirable. Latin *pinnatifida* = 'pinnately cleft'.

WALL (BROWN) *Lasiommata megera* (Nymphalidae, Satyrinae), wingspan 45–53 mm, a butterfly that gets its name from the characteristic behaviour of resting with wings two-thirds open on bare surfaces, including bare ground. Once found throughout England, Wales, Ireland and parts of Scotland, it has suffered severe declines over the last few decades, continuing with a 36% reduction in range from 2005 to 2014, in numbers by 25%. It is now confined to primarily coastal regions as far north as south-west Scotland and the Isle of Man, particularly on unimproved grassland, wasteland, cliff edges and hedgerows, and it is a priority species for conservation. Larvae feed at night on a range of grasses, including bents; adult food includes nectar from daisy, hawkweeds, knapweeds, ragwort, ragged-robin, this-tles and yarrow. Adults fly in May–October, with peaks in May and August. Greek *lasios* + *ommata* = 'hairy' + 'eyes'. Megaera was one of the Furies in Greek mythology, perhaps describing a 'restless' movement.

WALL-ROCKET Species of *Diplotaxis* (Brassicaceae) (Greek *diploos* + *taxis* = 'double' + 'row', from the double row of seeds in the pod).

Annual wall-rocket *D. muralis*, an annual or short-lived perennial European neophyte first recorded in 1778, widespread and locally common in England and Wales, scattered in Scotland and Ireland, favouring dry open habitats, most frequent in waste ground by railways and roads, but also found on rocks and walls, and as a garden weed. *Muralis* is from Latin for 'wall'.

Perennial wall-rocket *D. tenuifolia*, a perennial archaeophyte, widespread if commonest in England and Wales, and rare in Ireland, favouring warm dry habitats, in waste ground, on walls and banks, and in quarries and railway sidings. Grown as a salad plant (as wild rocket), it is increasingly found as a garden escape casual. *Tenuifolia* is Latin for 'slender-leaved'.

White wall-rocket *D. erucoides*, an annual European neophyte cultivated by 1736, an increasingly rare casual in England on waste ground and pathsides, introduced as grain impurity and in wool shoddy. *Erucoides* means resembling rocket-salad *Eruca*.

WALL-RUE *Asplenium ruta-muraria* (Aspleniaceae), a common and widespread fern favouring well-lit base-rich dry habitats such as rocks and old walls. *Asplenium* is from Greek for 'without spleen', referring to the value of curing disorders of the spleen, and Latin for 'rue' + 'wall'.

WALLABY **Red-necked wallaby** or **Bennett's wallaby** *Macropus rufogriseus* ssp. *rufogriseus* (Macropodidae), the Tasmanian subspecies, more properly known as Bennett's wallaby, head-body length up to 90 cm, males being larger, weight 13.8–18.6 kg. A colony in the Peak District originated from five animals that escaped from a collection at Leek, Staffordshire, in 1939 or 1940, numbers peaking at 40–50 in 1962, thereafter fluctuating but in general declining, especially following harsh winters. Only a single individual was observed in 2006. A Sussex colony originated from a captive group near Horsham; by 1940 there was a small breeding colony in the Ashdown Forest and St Leonard's Forest. These populations, and a few other escapees, have now become extinct. A population on the island of Inchconachan, Loch Lomond, started with two pairs from Whipsnade introduced in 1975, numbered 43 in 1999, and persists there eating oak and birch. Escapes on the Isle of Man at various times since the 1960s have led to an estimated 100+ animals (probably also including a few parma wallabies) in 2010. Escapes from private collections continue: one, for example, was photographed at Highgate Cemetery in 2013. 'Wallaby' comes from *wolaba*, an Australian Aboriginal word. Edward Bennett was Secretary of the Zoological Society of London 1831–36. Greek *macros* + *pous* = 'large' + 'foot', and Latin *rufogriseus* = 'red-grey'.

WALLABY-GRASS *Rytidosperma racemosum* (Poaceae), a neophyte that behaves as an annual, occasionally casual in fields and waste ground, originating from wool shoddy. Greek *rytidosperma* = 'red-seeded', and Latin *racemosum* = 'clustered'.

Swamp wallaby-grass *Amphibromus neesii*, an Australian neophyte perennial, scattered in England on waste ground and fields, originating from wool shoddy. Greek *amphis* = 'apart' + the genus *Bromus*. Christian Nees von Esenbeck (1776–1858) was a German botanist.

WALLFLOWER *Erysimum cheiri* (Brassicaceae), a herbaceous perennial Mediterranean archaeophyte garden escape widely naturalised on cliffs and old walls, particularly on warm calcareous substrates, tolerating nutrient-poor, thin, dry soils. Greek *eryo* = 'to cure', from the medicinal value of some species, and *cheiri* = 'red-flowered'.

WALNUT *Juglans regia* (Juglandaceae), a neophyte deciduous tree present since Roman times as a nut tree and ornamental, widespread if scarce in Scotland and Ireland, commonly planted in urban areas and sometimes also in the wild. It is often self-sown or buried by grey squirrels, especially in the south, with plants appearing in copses, rough grassland, gravel-pits and on waste ground and railway embankments. 'Walnut' comes from late Old English *walh-hnutu*, from a Germanic compound meaning 'foreign nut'. *Juglans* is Latin for 'walnut tree', *regia* for 'royal'.

Black walnut *J. nigra*, a neophyte North American deciduous tree introduced in the seventeenth century as an ornamental in parks and large gardens and occasionally self-sown. *Nigra* is Latin for 'black'.

WALRUS *Odobenus rosmarus* (Odobenidae), average male head-body length 365 cm, weight 1,270 kg, females 300 cm and 850 kg, an Arctic species recorded as an occasional vagrant, for example a young male hauled up on the shore at North Ronaldsay, Orkney, for two days in March 2013. 'Walrus' dates from the early eighteenth century via Dutch *walrus*, perhaps from Old Norse *hrosshvalr* = 'horse whale'. Greek *odous* + *baino*

= 'tooth' + 'to walk', from a belief that walruses used their tusks in this way, and Danish *rosmar* = 'walrus'.

WARBLE FLY Species of *Hypoderma* (Oestridae) (Greek *hypo* + *derma* = 'beneath' + 'skin'), whose larvae penetrate the skin of mammals, including cattle and deer. Eradicated in the UK in 1990. A warble is a skin lump.

WARBLER Passerine birds in a number of genera and families. See also leaf warbler (Phylloscopidae), blackcap and whitethroat.

Great reed warbler *Acrocephalus arundinaceus*, **lanceolated warbler** *Locustella lanceolata*, **melodious warbler** *Hippolais polyglotta* and **subalpine warbler** *Sylvia cantillans* are all scarce visitors. A number of other warblers are vagrants.

Aquatic warbler *Acrocephalus paludicola* (Acrocephalidae), wingspan 18 cm, length 13 cm, weight 12 g, a scarce insectivorous Red List autumn passage visitor migrating between breeding grounds in eastern Europe and its winter home in Senegal, seen in small numbers (averaging 10 p.a.) in reedbeds along the south England coast. It is the only internationally endangered passerine breeding in Europe. Greek *acros* + *cephale* = 'pointed' + 'head', and Latin *palus* + *colere* = 'marsh' + 'to inhabit'.

Cetti's warbler *Cettia cetti* (Cettidae), wingspan 17 cm, length 14 cm, male weight 15 g, females 12 g, first breeding in England (Kent) in 1973, now found in England (mainly in the south and in East Anglia) and (mainly south) Wales in damp habitats close to wetland. Numbers fell by over a third between 1984 and 1986, but in the absence of severe winters during 1986–2009, population growth and range expansion continued. The first breeding records north of the Humber were made in 2006. There were around 2,000 singing males (i.e. territory-holding, often attracting more than one female) recorded in 2010, and an increase by 44% was estimated for 2010–15. Clutches are of 4–5 eggs, incubation taking 16–17 days, the altricial young fledging in 14–16 days. Diet is insectivorous. Francesco Cetti (1726–78) was an Italian zoologist.

Dartford warbler *Sylvia undata* (Sylviidae), wingspan 16 cm, length 12 cm, weight 10 g, an insectivorous Amber-listed resident breeder, found in heathland and scrub at a few sites in southern England and East Anglia, with 3,200 pairs estimated in 2009, but the run of colder-than-average winters between 2008 and 2011 have had a severe impact on numbers, cutting them by 122% over 2010–15. (The breeding population had already collapsed from 450 pairs to 10 following the severe winters of 1961–2 and 1962–3. Also, the 1976 drought led to many heathland fires, impacting breeding success.) The cup-shaped nest in gorse or heather usually contains 3–5 eggs, incubation lasts 12–14 days, and the altricial young fledge in 10–14 days. Two broods are usual in one season. Latin *Sylvia* = a woodland sprite, in turn from *silva* = 'wood', and *undata* = 'wavy'.

Garden warbler *S. borin*, wingspan 22 cm, length 14 cm, weight 19 g, a common and widespread insectivorous migrant breeder, wintering in tropical Africa, favouring deciduous woodland and woodland edge, with coppice and rides, less common in gardens, arriving in late April–May, leaving in mid-July. Some 170,000 territories were found in Britain in 2009. There was a decline in numbers by 19% between 1995 and 2015, and by 15% in 2010–15. In August–September continental birds can be seen along the east and south coasts as passage migrants, often eating fruit. In Ireland it is a scarce summer visitor to woodlands in the Midlands and Northern Ireland from April to September, as well as an occasional passage migrant in spring and autumn to headlands on the south and west coasts. The cup-shaped nest near the ground usually contains 4–5 eggs, incubation lasts 11–13 days, as does fledging of the altricial young. *Borin* comes from a local name for the bird in the Genoa area of Italy, in turn derived from Latin *bos* = 'ox', because this bird was believed to accompany oxen.

Grasshopper warbler *Locustella naevia* (Locustellidae), wingspan 17 cm, length 13 cm, weight 14 g, a widespread if scattered insectivorous migrant breeder, found in scrub, dense grassland, reedbed borders, new forestry plantations and gravel-pits, wintering in North and West Africa, with an estimated 13,000 territories in Britain in 2009, but declining numbers placing it in the Red List: BBS data indicate an 18% decline between 1995 and 2014. It is Amber-listed in Ireland. Nests, close to the ground in dense vegetation, usually contain 5–6 eggs, incubation lasting 14 days, the altricial chicks fledging in 12–13 days. Latin *locustella* = 'little grasshopper' and *naevius* = 'spotted'.

Icterine warbler *Hippolais icterina* (Acrocephalidae), wingspan 22 cm, length 14 cm, weight 13 g, a scarce woodland insectivorous migrant breeder (one or two pairs a year, often none) and passage visitor from the Continent. Icterine refers to its yellowish colour, from Latin *icterus*. Greek *hypolais* refers to a small ground-nesting bird.

Marsh warbler *A. palustris*, wingspan 20 cm, length 13 cm, weight 13 g, a scarce insectivorous Red List migrant breeder in areas of dense vegetation, arriving from south-east Africa in late May or June. In the 1970s it bred only in Worcestershire, where 40–70 pairs were recorded each year, but this population became extinct by the end of the 1990s. There were up to nine pairs in 2010, at sites in the Isle of Wight, Sussex, Norfolk (four sites) and Northumberland. It is also a passage migrant, with up to 50 records p.a. The cup nest, in dense vegetation, has a clutch size of 4–5, incubation lasts 12–14 days, and the altricial chicks fledge in 10–11 days. *Palustris* is from Latin for 'marsh'.

Pallas's warbler *Phylloscopus proregulus* (Phylloscopidae), wingspan 14 cm, length 9 cm, weight 7 g, an insectivorous leaf warbler breeding in south Siberia and the Far East, small numbers wintering in Europe, including Britain, arriving in October–November after a 5,000 km migration, with up to 300 birds having been noted, though the average is 70–80. Peter Pallas (1741–1811) was a Russian zoologist. Greek *phyllion* + *scopos* = 'leaf' + 'watcher', and Latin *pro* + *regulus* = 'close to' and 'little king' or 'prince'.

Reed warbler *A. scirpaceus*, wingspan 19 cm, length 13 cm, weight 13 g, a locally common and widespread insectivorous migrant breeder, wintering in sub-Sahara, found from April to October, with greatest concentrations in East Anglia and along the south coast, with only a few breeding in southern Scotland and (where it is Amber-listed) in Ireland. There were an estimated 130,000 pairs (100,000–160,000) in Britain in 2009, numbers having doubled since 1970, and increasing by 83% between 1981 and 2015, and by 13% between 1995 and 2014. The basket nests built in reeds contains 3–5 eggs, incubation lasting around 12 days, and the altricial chicks fledging in 11–13 days. Latin *scirpus* = 'reed'.

Savi's warbler *L. luscinioides*, wingspan 20 cm, length 14 cm, weight 18 g, a scarce Red-listed insectivorous migrant breeder, wintering in Africa, found from April to August in marsh with reedbeds, mostly in East Anglia and south-east England. Over 2006–10, only 2–10 pairs at 7 sites were recorded. Nest average 5 eggs that take 10–12 days to hatch, the altricial young fledging in 11–15 days. Paolo Savi (1798–1871) was an Italian ornithologist. Latin *locustella* = 'little grasshopper', and *luscinia* = 'nightingale'.

Sedge warbler *A. schoenobaenus*, wingspan 19 cm, length 13 cm, weight 12 g, a common and widespread migrant breeder, wintering in sub-Sahara, found from April to October in reedbeds and similar wetlands, with an estimated 290,000 territories in Britain in 2009, the decline in numbers by 42% between 1975 and 2015 including 24% between 2010 and 2015. High interannual variation in population size reflects changes in adult survival rates which, in turn, are related to changes in rainfall on the wintering grounds. The cup-shaped nest of grass and stems is on or near the ground, with a clutch size of 3–4, incubation lasting 14 days, and the altricial chicks fledging in 13–14 days. Diet is of insects, with some fruit in autumn to boost fat reserves. Greek *schoiniclos* + *baino* = 'reed' + 'to walk'.

Willow warbler *P. trochilus*, wingspan 19 cm, length 11 cm, weight 10 g, a common and widespread insectivorous migrant breeder, wintering in sub-Sahara, present from April to September, favouring young woodland, with 2.4 million territories in Britain in 2009, though a decline in numbers by 44% between 1970 and 2015, by 8% between 1995 and 2015, and by 13% in 2010–15 has made them an Amber List species (though Green-listed in Ireland, where there are 1 million pairs). The nest, close to the ground, usually contains 5–7 eggs, incubation lasting 13–14 days, the altricial young fledging in 13–16 days. *Trochilos* was a bird mentioned by Aristotle, probably wren (indeed an earlier common name was 'willow wren').

Wood warbler *P. sibilatrix*, wingspan 22 cm, length 12 cm, weight 10 g, a Red-listed (in 2009) insectivorous migrant breeder mainly found from April to August in south and west Britain (with highest densities in the oakwoods of west Wales) and (rare) in eastern Ireland, with number of males estimated at 6,500 (5,900–7,000) in Britain in 2009, but with a decline in numbers by 56% between 1995 and 2015, slightly offset by an 11% increase in 2010–15. The dome-shaped nest is built in low scrub. Clutch size is 5–6, incubation lasts around 13 days, and fledging of the altricial young 13 days. Latin *sibilare* = 'to whistle'.

Yellow-browed warbler *P. inornatus*, wingspan 17 cm, length 10 cm, weight 7 g, an insectivorous Siberian breeding passage visitor, mostly seen on the east coast from September to November, with a few birds overwintering. *Inornatus* is Latin for 'unadorned'.

WARBURG'S MOSS *Molendoa warburgii* (Pottiaceae), found in the west Highlands and Islands on moist, calcareous rock faces, often by streams and waterfalls, usually with seepage of base-rich water. Ludwig Molendo (1833–1902) was a German bryologist, Edmund Warburg (1908–66) an English botanist.

WART-CRESS Berry wart cress *Sphaerococcus coronopifolius* (Sphaerococcaceae), a red seaweed with fronds up to 30 cm long, on subtidal rock down to 20 m on the south and west coastline of Britain and around Ireland. Greek *sphaera* + *coccos* = 'sphere' + 'grain', and *corone* + *pous* = 'crown' + 'foot', with Latin *folius* = 'leaf'.

WART WEED Slender wart weed *Gracilaria gracilis* (Gracilariaceae), a generally widespread red seaweed with fronds up to 50 cm long, on rocks in the intertidal and subtidal, especially on sandy shores. Latin *gracilis* = 'slender'.

WARTBITER *Decticus verrucivorus* (Tettigoniidae), body length 31–7 mm, wing length 22–7 mm, an omnivorous bush-cricket of heathland and chalk downland with bare ground in five sites in southern England (two resulting from reintroduction). The heavy body allows only rare, short flights (up to 4 m). Loss of habitat, especially grassland, has led to rapid population decline, and it is Vulnerable in the RDB. Its English name comes from the earlier practice of allowing the cricket to bite warts off the skin. *Decticos* is Greek for 'able to bite', *verruciverus* Latin for 'wart-eating'.

WARTY-CABBAGE *Bunias orientalis* (Brassicaceae), a perennial, occasionally biennial, east European neophyte herb cultivated by 1731, scattered in Britain, persistent on rough grassy waste ground, roadsides, docks and railways.

WASP Insects in the order Hymenoptera, suborder Apocrita, excluding bees and ants. They may be social (eusocial) or solitary, and predatory or parasitic (including kleptoparasitic and parasitoidal). See also digger and mason wasps.

1) Solitary or (eu)social member of the families Vespidae (Latin *vespa* = 'wasp'), characterised by a narrow waist and a sting and typically yellow with black stripes. Other families, here treated separately, are Chrysididae (ruby-tailed wasp, see 2 below), Crabronidae (q.v. and 3 below), Siricidae (horntail wasps, see 4 below) and Sphecidae (sand wasps, see 5 below). Some species construct a paper nest from wood pulp; others burrow or are kleptoparasitic in the burrows of other species. There are about 240 social wasps in the British list. See also cuckoo

wasp (Chrysididae and Vespidae), digger wasp (Crabronidae), mason wasp (Vespidae), potter wasp (Vespidae), spider wasp (Pompilidae, see also cutpurse), weevil-wasp (Crabronidae) and hornet (Vespidae).

Common wasp *Vespula vulgaris* (Vespidae), body length 20 mm, common and widespread, though slightly declining in status in the last few decades. Flight ranges from spring (queens) to November (workers). Nectar is taken from a variety of flowers, and sugars from decaying fruit. Nest sites are mostly underground with entrance tunnels sometimes reaching 45 cm in length. Often found in roof spaces, where the nests can be up to 1.2 m across. The nest envelope is variously coloured yellowish to brown due to the workers collecting both rotted and fairly sound wood fibres. Average sized mature colonies have about 7,500 small cells and 2,300 large cells. They rear about 10,000 workers, 1,000 queens and 1,000 males. *Vespula* is from the Latin diminutive of *vespa* = 'wasp', and *vulgaris* = 'common'.

German wasp *V. germanica*, worker body length 12–15 mm, queens up to 20 mm, common and widespread in England, Wales and Ireland, scarce in Scotland, but seriously declining in status in the last few decades. Flight ranges from spring (queens) to November (workers). Nectar is taken from a variety of flowers, and sugars from decaying fruit. Nest sites are mostly underground with entrance tunnels 3–20 cm in length. Open sites are preferred, often in heathland or grassland. Nest paper is grey, due to workers collecting well-weathered wood fibres.

Norwegian wasp *Dolichovespula norwegica*, worker body length 13 mm, queens 17 mm, a social wasp, widespread if not common, and markedly declining in status since the 1990s. It frequents a variety of habitats, with a preference for heathland and moorland. Spring queens fly from April to June, workers from July to October and newly emerged sexuals from September to October. Spring queens visit flowers of cotoneaster and berberis for their nectar. Queens build nests in spring, hanging them from the branches of low bushes. Each tends her brood until the new adults emerge in June – workers which enlarge the nest and forage for food. In August, new males and females are produced. Mating takes place outside the nest, and the fertilised queens feed on nectar from hogweed and other Apiaceae before seeking overwintering sites. The males feed on the same umbels, and on bramble and other late fruits. Workers may visit figwort which has shallow flowers enabling them to reach the nectaries. Greek *dolichos* = 'long' + *Vespula*.

Red wasp *V. austriaca*, widespread in a variety of open habitats. The queen constructs the initial 'queen nest' and rears the first workers which start to forage and carry out all the nest building and brood rearing activities. Larvae are fed small insects and spiders. The first cells, in which the workers and most of the males are reared, are small; later in the season larger cells are built, in which the queens and a few males are reared. Latin *austriaca* = 'Austrian'.

Tree wasp *D. sylvestris*, a social wasp 22 mm long, widespread if not common, and markedly declining in status since the 1990s. It frequents a variety of habitats, and is often seen in suburban gardens. It has a relatively short life cycle: spring queens fly from early May–June; workers from the end of May–September; autumn queens July–September; and males July–September. Individuals in late summer visit flowers of Apiaceae. They also visit so-called 'wasp-flowers', for example figwort, cotoneaster and snowberry. Despite its common name, nests are not always found in trees, but require at least partial cover. Many are found at ground level or in cavities in the ground. Other locations are in the eaves, roof spaces and cavity walls of buildings, and small bird nest boxes. The nest paper is made from mostly well-seasoned wood. *Sylvestris* is Latin for both 'woodland' and 'wild'.

2) **RUBY-TAILED WASP** Members of the family Chrysididae. See also cuckoo wasp.

Northern osmia ruby-tailed wasp *Chrysura hirsuta*, body length 7–11 mm, recorded from only three sites in Scotland

(Vulnerable in RDB 2) in habitats related to the species of mason bee host: upland base-rich grassland at Blair Atholl (Perthshire) on *Osmia inermis*; Caledonian pinewood at Loch Garten on *O. uncinata*; and upland pasture with rocky outcrops and stone walls Whithorn Wigtownshire on wall mason bee *O. parietina*. Greek *chrysos* = 'gold', and Latin *hirsuta* = 'hairy'.

Ruby-tailed wasp *Chrysis ignita*, length 10 mm, a rare parasitoid that lays its eggs in the nests of other solitary wasps or those of solitary bees, for example red mason bee. After hatching the larva eats the host's egg or grub. It is a priority species in the UK BAP. *Ignita* is Latin for 'fiery', from the tail colour.

Shimmering ruby-tail *Chrysis fulgida*, length 10 mm, an endangered parasitoid that lays its eggs in the nests of the eumenid wasp *Symmorphus crassicornis*, found in Surrey and north Hampshire on scrubby heathland and open woodland where its host is found near aspen and creeping willow. In 2010, *The Guardian* launched an annual competition for the public to suggest common names for previously unnamed species, and this was a winner in 2011, describing one of Britain's more brightly coloured wasps. *Fulgida* is Latin for 'shining'.

3) **BLACK WASP, WOOD BORER WASP** With others, in the family Crabronidae, body length 5–7 mm.

Club-horned wood borer wasp *Trypoxylon clavicerum*, widely distributed in England (as far north as Yorkshire) and Wales, flying in summer, recorded from a variety of habitats, for example woodland, parkland, gravel-pits, sandy sites, chalk grassland and heathland. Nests are constructed in old insect burrows or galleries in dry wood or hollow plant stems, and provisioned with small spiders. Greek *trypa* + *oxys* = 'hole' + 'sharp', and Latin *clava* + *cera* = 'club' + 'wax'.

Horned black wasp *Passaloecus corniger*, widespread in England and Wales in a variety of habitats, including gardens, from May to September. Nests have been found in wooden posts, old timber and occasionally in chloropid fly *Lipara lucens* galls on common reed, cells provisioned with aphids. *Passalos* is Greek for 'peg', *corniger* Latin for 'horn-bearing'.

Little black wasp *Pemphredon lethifera*, common and widespread in England, Wales, and also found in south Scotland and parts of Ireland, in a range of habitats from May to September. Nests are mainly in stems of *Rubus* species but females may also use small branches, old wood, or the cigar galls formed on common reed by the chloropid fly *Lipara lucens*. *Pemphredon* is Greek for this kind of wasp, *lethifera* Latin for 'deadly'.

Melancholy black wasp *Diodontus tristis*, sporadic in England north to Yorkshire and in Wales, on sandy soils (including heathland), flying from late spring to early autumn. The burrow may be multi-branched and have over 20 cells, each cell provisioned with 20–40 wingless aphids. Greek *diodos* + *odontos* = 'passageway' + 'tooth', and Latin *tristis* = 'sad'.

Minute black wasp *D. minutus*, common in southern England and East Anglia, sporadic elsewhere in England and in Wales, flying in summer. Nests are excavated in sandy slopes with up to 10–15 cells, each provisioned with about 30 winged aphids.

Mournful wasp *Pemphredon lugubris*, widespread, especially in England and Wales, found in May–September in habitats where dead and decaying wood are present as substrates for nesting burrows where the cells are provisioned with aphids. *Lugubris* is Latin for 'lugubrious'.

Pale-footed black wasp *Psenulus pallipes*, with a scattered distribution in England and south and east Wales in a range of habitats, flying in summer. Nests have been noted in stems, straw (including thatch) and old beetle holes in wood, each with 5–6 cells provisioned with up to 30 aphids. Greek *psene* for a kind of insect living inside a plant, and Latin *pallium* + *pes* = 'a cover' + 'foot'.

Shuckard's wasp *Pemphredon inornata*, locally common and widespread in England and Wales, occasional in Scotland, rare in Ireland, found in May–September in habitats such as reedbed, scrub, woodland and gardens, where nests are created in hollow stems, branches and old wood as well as the cigar galls formed on common reed by the chloropid fly *Lipara lucens*, the cells provisioned with aphids. William Shuckard (1803–68) was an English entomologist. *Inornata* is Latin for 'unadorned'.

Slender wood borer wasp *T. attenuatum*, body length 8 mm, widely distributed, flying in summer, and recorded from habitats such as woodland, parkland, gravel-pits, sandy habitats, chalk grassland and heathland. Nests are constructed in pre-existing cavities and are provisioned with small spiders. *Attenuatum* is Latin for 'shortened'.

Two-coloured mimic wasp *Mimesa bicolor*, scarce, with scattered records from the southern half of England, Vulnerable in RDB 2, in a range of open habitats with light soils, such as heathland, occasionally recorded from coastal dunes, in flight from June to September. Nests are dug in sandy soil in warm, sheltered situations, and provisioned with a range of small Homoptera, usually leafhoppers. Flower visits are mainly to Apiaceae.

4) HORNTAIL WASP, WOOD WASP Non-social sawflies (Siricidae).

Banded horntail wasp Synonym of greater horntail wasp.

Greater horntail wasp *Urocerus gigas*, up to 40 mm length, widespread and locally common in Britain, adults active in May–October, larvae feeding on coniferous wood.

Lesser horntail wasp *Sirex noctilio*, males 9–32 mm, females 15–36 mm length, occasional in southern England and the north-east Midlands. The female lays two eggs on a symbiotic host-tree pathenogenic fungus *Amylostereum areolatum* for the wood-boring larvae to feed on, and sometimes a mucoid substance that is toxic to the host pine tree, all of which leads the wasp to be viewed as a forestry pest. Greek *seiren* means this kind of wasp, and Latin *noctus* = 'of the night'.

5) SAND WASP Members of the family Sphecidae. Eggs are laid on caterpillars to provision the wasp larvae.

Hairy sand wasp *Podalonia hirsuta*, locally common around the coasts of southern and south-west England, Wales, Lancashire, and Norfolk, on dunes and heathland. Females fly from March to September, males in June–September, taking nectar from bramble, willow, thyme, etc. The angled burrows have one cell which is provisioned with a single large noctuid caterpillar. *Hirsuta* is Latin for 'hairy'.

Heath sand wasp *Ammophila pubescens*, body length up to 19 mm, on heathland in southern England and coastal East Anglia, adults feeding on bramble and heathers. Nests are short, unicellular burrows dug in the sand, often in aggregations. Each is provisioned with about six caterpillars or sometimes sawfly larvae. Greek *ammos* + *philos* = 'sand' + 'loving', and Latin *pubescens* = 'downy'.

Mud wasp *P. affinis*, very local in southern England and East Anglia, on dunes and heathland. Females fly in May–September, males in June–August, taking nectar from a number of plants. Each angled burrow has one cell which is provisioned with a single large noctuid caterpillar. *Affinis* is Latin for 'related'.

Red banded sand wasp *A. sabulosa*, body length 20–4 mm, found throughout much of England and Wales on heathland and dunes. Adults are active in June–September. The female digs a burrow in sand then hunts for caterpillars. About half of all cells are provisioned with one large caterpillar carried back on foot, it sometimes being over ten times as heavy as the wasp. Other cells are provisioned with 2–5 smaller caterpillars. An egg is laid on the first caterpillar provisioned; after the last caterpillar has been interred the wasp closes the burrow, camouflaging the entrance with debris. *Sabulosa* is Latin for 'sandy'.

6) ICHNEUMON WASP Members of the family Ichneumonidae, parasitoids that lay their eggs in the larvae or pupae of other Hymenoptera, Coleoptera, Lepidoptera and other insects. *Ichneumon* is Greek for 'tracker'.

7) GALL WASP (q.v.), belonging to Cynipidae or Eurytomidae (see both).

WASPCLUB **Yellow waspclub** *Cordyceps forquignonii* (Hypocreales, Cordycipitaceae), a parasitic fungus, widespread, scattered and uncommon in England, growing on dead insects, usually Diptera, in woodland soil. Greek *cordyle* + New Latin *ceps* = 'club' + 'head'.

WATCHMAN **Lousy watchman** Alternative name for dor beetle, from it often having a large infestation of mites.

WATER BEAR See tardigrade.

WATER BEETLE **Long-toed water beetle** Members of the family Dryopidae, found in wet mud.

WATER BOATMAN Members of the family Corixidae (New Latin *coris* = 'bug'), with 39 species found in Britain, in 9 genera, 5–15 mm long. Sometimes all members have been called lesser water boatmen, to distinguish them from backswimmers (Notonectidae), incorrectly referred to as **greater water boatmen**. **Common water boatman** or **lesser water boatman** is now generally used for *Corixa punctata*, which is very common throughout the British Isles. Water boatmen differ from other water bugs in their herbivorous feeding habit (plant debris, algae, etc.). Hind legs are used for swimming, but they spend much of the time on pond bottoms, surfacing to trap air beneath the wing cases. Most species fly well in warm weather. Males produce a courtship 'song' by rubbing their hairy front legs against a ridge on the side of the face, or in the case of *Micronecta scholtzi* by stridulation of his penis ridge leading to a sound level of up to 99.2 dB. Water boatmen can bite when handled, with a stinging sensation lasting around half an hour.

WATER-CRESS *Nasturtium officinale* (Brassicaceae), a perennial herb, common and widespread, except in highland Scotland, both in and beside clear shallow streams and in ditches, ponds and marsh. It is also cultivated in cress beds, though not as commonly as hybrid water-cress. *Nasturtium* is from Latin *nasus tortus* = 'twisted nose', referring to the taste, and Latin *officinale* means a plant with pharmacological properties. See also fool's-water-cress.

Hybrid water-cress *Nasturtium* x *sterile* (*N. officinale* x *N. microphyllum*), a widespread if scattered perennial growing in and beside water, especially in streams, ditches and ponds, either with or without its parents. It was first cultivated commercially in about 1808, near Gravesend, Kent, and is the most widely cultivated water-cress, it being usually frost-hardy, and many wild colonies will have arisen from this source.

Narrow-fruited water-cress *N. microphyllum*, a widespread and scattered perennial growing in and around streams, ditches, canals and ponds, and in marsh. *Microphyllum* is Greek for 'small-flowered'.

WATER CRICKET In general terms, members of the bug family Veliidae (Heteroptera), which in the British Isles comprises two species of *Velia* and three of *Microvelia*, found on the surface of ponds and slow-moving water, feeding on small insects that have fallen on the water. Water cricket more specifically refers to the widespread and common *V. caprai*.

WATER-CROWFOOT Species of *Ranunculus* (Ranunculaceae), the divided leaves suggesting a crow's foot. *Ranunculus* is the diminutive of Latin *rana* = 'frog', referencing the plants growing in marsh and bog.

Brackish water-crowfoot *R. baudotii*, a widespread subaquatic annual or perennial growing in coastal water bodies, including lagoons, ditches, pools and dune slacks, usually in water 0.5–1 m deep, but it can grow in shallower water or as a dwarf terrestrial form on wet mud. Inland sites include flooded mineral workings and canals, not all receiving saline drainage water. Herr Baudot was a nineteenth-century German botanist.

Common water-crowfoot *R. aquatilis*, a common, widespread (less so in north Scotland) subaquatic annual or short-lived perennial in shallow water in marsh, ponds and ditches, and at the edge of slow-flowing streams, canals and sheltered lakes, mainly in eutrophic and mildly base-rich water, preferring lightly disturbed habitat. *Aquatilis* is Latin for 'growing in water'.

Fan-leaved water-crowfoot *R. circinatus*, a widespread though declining perennial of clear, base-rich, standing or very slowly flowing water, most common in lakes, flooded gravel-pits, and sluggish streams, rivers, canals and ditches, usually growing at depths of 1–3 m in meso-eutrophic water, though present in shallower water if it does not dry up in summer. *Circinatus* is Latin for 'rounded'.

River water-crowfoot *R. fluitans*, a widespread perennial subaquatic herb of large, rapidly flowing rivers with a stable substrate, usually in base-rich and meso-eutrophic water. In Ireland, it is found only in the highly polluted Six Mile Water, Co. Antrim. *Fluitans* is Latin for 'floating'.

Stream water-crowfoot *R. penicillatus*, a widespread perennial subaquatic herb of moderately to rapidly flowing rivers and streams, in oligotrophic and mesotrophic water. Ssp. *pseudofluitans* is the dominant plant in many base-rich streams, most frequent where the water flow is broken by riffles. *Penicillatus* is Latin meaning 'ending in a tuft of hair'.

Thread-leaved water-crowfoot *R. trichophyllus*, a widespread, locally common subaquatic annual or perennial of shallow, still or very slowly flowing water, most frequently in ponds, dune slacks and drainage ditches, but also in larger sheltered sites. It tolerates a range of water chemistry but is most frequent in mesotrophic or eutrophic water. *Trichophyllus* is Greek for 'hair-like leaves'.

WATER-DROPWORT Herbaceous perennial species of *Oenanthe* (Apiaceae), from Greek *oenos* + *anthos* = 'wine' + 'flower', referring to the wine-like scent of the flowers. 'Water dropwort' derives from the resemblance of some of the smaller species to dropwort.

Corky-fruited water-dropwort *O. pimpinelloides*, tuberous, locally common in southern England, scattered elsewhere in England, and a scarce introduction in Co. Clare, in hay meadow and (especially horse-grazed) pasture and on roadsides, in both damp and dry conditions, being the only water-dropwort in Britain that can grow in dry habitats. Latin *pimpinelloides* = 'like pimpernel'.

Hemlock water-dropwort *O. crocata*, tuberous, locally common though largely absent from much of east England, upland Scotland and central Ireland, in ditches, on the banks of streams, rivers, canals, lakes and ponds, in marsh and wet woodland, and on wet boulders and sea-cliffs. Containing the convulsant oenanthotoxin, it is rapidly and highly poisonous to humans, horses and cattle, and has traditionally been used to poison rats and moles. *Crocata* is Latin for 'saffron yellow'.

Narrow-leaved water-dropwort *O. silaifolia*, scattered and decreasing in south-east and central England and south-east Yorkshire in damp grassland that receives calcareous winter flood water in winter, for example hay meadows and streamsides. *Silaifolia* is Latin for 'like the genus *Silaum*' (peppered-saxifrage) which has finely cut leaves.

Parsley water-dropwort *O. lachenalii*, in coastal areas around the British Isles apart from north and east Scotland, growing in upper saltmarsh, in rough grassland in drained estuarine marsh, by brackish dykes and the lower parts of tidal rivers, while inland in England (where it is scattered and declining), it grows in base-enriched habitats such as marsh, fen-meadow and tall-herb fen. Werner de la Chena (1736–1800) was a Swiss botanist.

River water-dropwort *O. fluviatilis*, scattered in southern and eastern England and in central Ireland, declining because of eutrophication and dredging, most abundant in clear, meso-eutrophic water of calcareous streams and rivers but also in canals and ditches. In flowing water, propagation is usually via rooting at nodes or by vegetative fragmentation, while flowering is more frequent in still water. *Fluviatilis* is Latin for 'of a river'.

Tubular water-dropwort *O. fistulosa*, stoloniferous, scattered but declining in Britain north to central Scotland, and in Ireland, commoner in eastern England, growing in wet habitats such as meadow and pasture in flood plains, marsh, fen, and by streams, lakes and ponds. *Fistulosa* is Latin for 'tube-like'.

WATER FINGERS *Siphula ceratites* (Icmadophilaceae), a rare tuft lichen (Near Threatened in the RDB), first recorded in 1955 in Wester Ross, since recorded from a few other sites in north-west Scotland, including Skye, on wet peatland soil. Latin *siphula* = 'small tube', and Greek *ceratites* = 'horn-like'.

WATER FLEA Species of small, filter-feeding cladoceran crustaceans with 15 species of *Daphnia* (Daphniidae) in the British Isles, found in a variety of freshwater habitats. Body size is usually 1–5 mm. The common name comes from the similarity of their swimming movement to that of jumping fleas.

WATER-LILY Bottom-rooted aquatic perennial herbs in the families Nymphaeaceae and Menyanthaceae.

Fringed water-lily *Nymphoides peltata* (Menyanthaceae), rhizomatous, in water 0.5–2 m deep in lakes, ponds, slowly flowing rivers, canals and large fenland ditches. As a native in the Thames Valley and East Anglia it is found in calcareous and eutrophic water, but elsewhere it is also much grown as an ornamental and has become naturalised from material planted or discarded in the wild. *Nymphoides* means resembling *Nymphaea*. *Peltata* is Latin for 'round shield', referring to the leaf shape and the position of the stalk in the leaf centre.

Least water-lily or **least yellow water-lily** *Nuphar pumila* (Nymphaceae), with both submerged and floating leaves, mostly in oligotrophic and mesotrophic ponds and lochs in upland Scotland, and in the eutrophic Cole Mere, north Shropshire. *Nuphar* is Arabic for 'water-lily', *pumila* Latin for 'dwarf'.

White water-lily *Nymphaea alba* (Nymphaeaceae), with floating leaves and flowers, common and widespread, in still or slowly moving water to about 2 m depth, with a preference for base-rich ponds, lakes and canals of average fertility in partially shaded or well-lit sites. *Nymphaea* was a water goddess in Greek mythology. *Alba* is Latin for 'white'.

Yellow water-lily *Nuphar lutea*, with both submerged and floating leaves, growing in water 1.5 m deep, showing preference for well-lit, base-rich and nutrient-rich ponds, pools and lakes. *Lutea* is Latin for 'yellow'.

WATER MEASURER *Hydrometra stagnorum* (Hydrometridae), a common and widespread water bug in England and Wales, but with a more scattered distribution in Scotland and Ireland, in ponds, ditches and slowly moving streams, hunting on the water surface for mosquito larvae, water fleas, etc. Greek *hydrometra* = 'water measurer', and Latin *stagnum* = 'swamp'. The endangered **lesser water measurer** *H. gracilenta* has only been found in parts of the Broads and Pevensey Levels.

WATER-MILFOIL Submerged freshwater perennials in the genus *Myriophyllum* (Greek *myrios* + *phyllon* = 'numberless' or 'ten thousand' + 'leaves') and, more generally, of the family Haloragaceae. See also parrot's-feather.

Alternate water-milfoil *M. alterniflorum*, common in upland areas in both standing and flowing waters, including rapidly flowing peaty streams. In Scotland and Ireland it occurs in a range of habitats, including highly calcareous sites, but in south-east England it is confined to acidic, mesotrophic or oligotrophic waters.

Spiked water-milfoil *M. spicatum*, common and widespread in meso-eutrophic or eutrophic and often calcareous waters.

Whorled water-milfoil *M. verticillatum*, with a scattered distribution, but particularly found in eastern England and central Ireland, in clear or slightly turbid, still or slowly flowing calcareous water in lakes, streams, canals and ditches. It occurs over both peaty and inorganic substrates. *Verticillatus* is Latin for 'whorled'.

WATER MITE See Hydrachnellae.

WATER-MOSS Species of *Fontinalis* (Fontinalaceae), from Latin for 'pertaining to a fountain or spring'.

Alpine water-moss *F. squamosa*, common and widespread in upland regions in rapidly flowing, acidic, nutrient-poor streams and rivers, attached to rocks or tree roots. *Squamosa* is Latin for 'scaly'.

Greater water-moss *F. antipyretica*, common and widespread in a range of still and flowing water bodies, usually attached to stones, rocks or waterside trees, but sometimes occurring in loose masses on the bottom of shallow, still waters. It tolerates a range of water conditions and periods of emersion. It is also frequent around turloughs in Ireland. *Antipyretica* is from Greek for 'anti-fire' or 'anti-fever'.

WATER-PARSNIP Herbaceous perennials in the family Apiaceae.

Greater water-parsnip *Sium latifolium*, with submerged leaves, very local and decreasing in south-east and eastern England, Somerset and central Ireland, generally in ditches, growing among other emergent species, or in reed swamp, preferring alkaline conditions. *Sium* is from Greek *sion* for this kind of marsh plant, and Latin *latifolium* = 'broad-leaved'.

Lesser water-parsnip *Berula erecta*, both a submerged perennial of rivers and streams, and an emergent species at the margins of lakes, ponds and ditches, and in marsh. It is also found on seasonally flooded ground. Spread is via stolons or rhizomes. It is locally common in much of England, Wales and Ireland, but scarce in Scotland. *Berula* is Latin for this kind of marsh plant. *Erecta* is Latin for 'upright'.

WATER-PEPPER *Persicaria hydropiper* (Polygonaceae), a widespread and generally common annual of mud on the margins of ponds and lakes, canals and streams, or in wet depressions such as stock-trampled field gateways and in damp meadow. It is characteristic of shaded habitats that are waterlogged in winter, often on base-poor soils. *Persicaria* is New Latin for having leaves like a peach tree, *hydropiper* Latin for 'water-pepper'.

Small water-pepper *P. minor*, a scattered and uncommon annual of wet marshy sites, winter-flooded ground beside ponds, lakes and ditches, and damp pasture trampled by stock. *Minor* is Latin for 'smaller'.

Tasteless water-pepper *P. mitis*, an annual scattered in England, Wales and Ireland, in wet habitats with nutrient-rich soil beside ponds, lakes and rivers and in shallow ditches, damp hollows in fields, and stock-trampled pasture. *Mitis* is Latin for 'mild'.

WATER-PLANTAIN Herbaceous aquatic perennials in the family Alismataceae. Specifically, *Alisma plantago-aquatica*, common and widespread except in upland Scotland, in meso-trophic or eutrophic conditions on mud at the shallow edge of still or slow-flowing waters, and in marsh and swamp. It frequently colonises newly cleaned ditches and recently flooded mineral workings. *Alisma* is Greek for this plant. The specific name is Latin for 'plantain' + 'water'.

Floating water-plantain *Luronium natans*, stoloniferous, local in Wales, the north Midlands and northern England, with a few records from central Scotland and Co. Galway, in mesotrophic or oligotrophic lakes, pools and slow-flowing rivers, and little-used or relict canals. In deep water it grows as rosettes of submerged leaves, but it produces floating leaves and flowers in shallow water and on exposed mud. *Luronium* seems to be a neologism; *natans* is Latin for 'swimming'.

Lesser water-plantain *Baldellia ranunculoides*, with a scattered distribution, declining in England, found at the water's edge of streams, ponds and ditches where potential competitors are restricted by fluctuating water levels or disturbance, usually in calcareous or brackish waters. This is the county flower of Dumbartonshire as voted by the public in a 2002 survey by Plantlife. Count Bartolomeo Bartolini-Baldelli (1804–68) was an Italian botanist. *Ranunculoides* is Latin for resembling *Ranunculus* (buttercups).

Narrow-leaved water-plantain *A. lanceolatum*, emergent, locally common in England, scarcer elsewhere, in shallow water or on mud at the edge of slow-flowing water, in many areas common in canals, and favouring eutrophic, calcareous water. *Lanceolatum* is Latin for 'lance-shaped'.

Ribbon-leaved water-plantain *A. gramineum*, an annual or short-lived perennial, probably introduced from North America, in shallow, eutrophic water at the margins of lakes, rivers and fenland drains. It has been found at Westwood Great Pool, Worcestershire, since 1920, and the River Glen, Lincolnshire in 1955, reappearing in 1991; at both sites populations vary annually but are usually small and sometimes fail to appear. A few other sites have shown recent local extinction. It is Critically Endangered (RDB); in 1991 English Nature's Species Recovery Programme included action to conserve this plant, which has been introduced to new sites in Norfolk, Cambridgeshire and Northumberland. *Gramineum* is Latin for 'grass-like'.

WATER-PURSLANE *Lythrum portula* (Lythraceae), a widespread but scattered annual of acidic or calcium-deficient silty soils at the muddy edges of pools and in temporarily flooded habitats. *Lythron* is Greek for 'blood', hence 'red', *portula* for 'purslane'.

WATER RAT Alternative, taxonomically misleading name for water vole.

WATER SCORPION *Nepa cinerea* (Nepidae), a widespread and common bug, especially in southern Britain, in unpolluted, well-vegetated ponds or slow-flowing waters, ambush-preying on invertebrates, tadpoles and small fish. Its long thin tail extends above water to act as a breathing tube. The binomial comprises Latin for 'scorpion' and 'grey'.

WATER-SHIELD Carolina water-shield *Cabomba caroliniana* (Cabombaceae), a submerged North American perennial herb, an aquarium throw-out in the Forth and Clyde Canal (growing in factory-heated water in 1969, no longer found) and in the Basingstoke Canal (in unheated, eutrophic, calcareous water since 1991). Reproduction is by rooting of stem fragments. *Cabomba* is Spanish for the fruit of this plant.

WATER-SOLDIER *Stratiotes aloides* (Hydrocharitaceae), a herbaceous perennial found in calcareous, meso-eutrophic lakes, ponds and ditches in East Anglia and Lincolnshire. Elsewhere, with a scattered distribution, especially in England, introduced colonies also occur in other habitats, including canals. Native populations have been in long-term decline, probably due to eutrophication; introduced populations are often short-lived. All plants are female, reproducing vegetatively. *Stratiodes* is Greek for 'soldier', *aloides* for 'like the genus *Aloe*'.

WATER-SPEEDWELL Aquatic annuals in the genus *Veronica* (Veronicaceae). 'Speedwell' comes from the archaic 'speed well' (farewell), possibly from the fact that the flowers wilt and fall off soon after being picked. *Veronica* probably comes from Latin *vera* + *nica* = 'true' + 'image': when St Veronica wiped Christ's forehead on the way to the crucifixion, an image of his face was left on the cloth she had used. See also speedwell.

Blue water-speedwell *V. anagallis-aquatica*, scattered, locally common on fertile substrates by rivers, streams and ponds, in ditches and in flooded pits in lowland regions. It grows as a vegetative plant submerged in shallow water, as a flowering emergent, and as a terrestrial plant in marsh and disturbed ground at the water's edge. Greek *ana* + *agallein* = 'again' + 'to delight in', referring to the opening and closing of the flowers in response to sunlight.

Pink water-speedwell *V. catenata*, scattered, locally common, less so outside Scotland, in shallow water and on the muddy edges of rivers, streams, ponds and lakes, in dune slacks, and in pits. *Catenata* is Latin for 'linked in a chain'.

WATER-STARWORT Herbaceous species of *Callitriche* (Callitrichaceae) (Greek *cali* + *thrix* = 'beautiful' + 'hair', referring to the stems).

Autumnal water-starwort *C. hermaphroditica*, a submerged annual in mesotrophic lakes, canals and gravel-pits, scattered if uncommon in southern and eastern England, probably increasing in the Midlands.

Blunt-fruited water-starwort *C. obtusangula*, a perennial of permanent, still or slow-flowing mesotrophic to eutrophic waters mostly in England, Wales and Ireland, extending into brackish water in coastal grazing marsh. It also grows on wet mud as water levels drop. *Obtusangula* is Latin for 'blunt-angled'.

Common water-starwort *C. stagnalis*, a common and widespread annual or perennial on rutted tracks, in ephemeral pools and at the margins of ditches and rivers, as well as in shallow permanent water. *Stagnalis* is from Latin for 'pool'.

Intermediate water-starwort *C. brutia* ssp. *hamulata*, a perennial found in both deep, still water and fast-flowing rivers, particularly in acidic, oligotrophic water. It may be abundant in temporary water bodies such as pools, ditches and reservoir edges. Brutia is the former name of Calabria, Italy. *Hamulata* is Latin for 'with small hooks'.

Narrow-fruited water-starwort *C. palustris*, annual, possibly perennial, partially submerged, discovered on clay in the dry bed of a turlough in Co. Galway in 1999, near Loch Lomond in 2000, and in Stirlingshire in 2005. *Palustris* is Latin for 'marsh'.

Pedunculate water-starwort *C. brutia* ssp. *brutia*, a common and widespread annual or perennial in ephemeral pools, ruts and poached muddy ground, although also in permanent water.

Short-leaved water-starwort *C. truncata*, annual, occasionally perennial, scattered, in rivers, canals, ditches, lakes and gravel-pits, usually in base-rich mesotrophic or eutrophic waters, occasionally in Anglesey as a terrestrial plant on wet mud. *Truncata* is Latin for 'cut off'.

Various-leaved water-starwort *C. platycarpa*, locally common and widespread, perennial, in most types of often flowing water and as a terrestrial form on wet mud, though most frequent in eutrophic water, particularly ditches and canals. Greek *platys* + *carpa* = 'broad' + 'fruit'.

WATER STRIDER Synonym of pond skater.

WATER VENEER *Acentria ephemerella* (Crambidae, Acentropinae), a small moth, wingspan 11–13 mm, widespread and common in and around lakes, ponds and slow rivers. There are two forms of the female, one wingless, which lives under the water, and one winged. The fully winged males mate with females usually on the water surface. The flight period is June–August. Larvae live entirely under water, feeding on plants such as pondweeds and Canadian waterweed. Greek *a* + *centron* = 'without' + 'spur', from the vestigial tibial spurs, and *ephemeros* = 'living for just one day' (average adult lifespan is in fact two days).

Giant water-veneer *Schoenobius gigantella* (Schoenobiinae), male wingspan 25–30 mm, female 41–6 mm, local in reedbeds in parts of southern England and East Anglia, with records north to Yorkshire, predominantly coastal. Adults fly at night in July, sometimes into August. Larvae mine stems of reed and reed sweet-grass, but cut part of a leaf to use as a raft to float to a new food plant. Greek *schoinos* + *bioo* = 'reed' + 'to live', and Latin *gigantea* = 'huge'.

Pale water-veneer *Donacaula forficella* (Schoenobiinae), wingspan 24–32 mm, local in marsh, riverbanks and other damp areas throughout much of southern England, with records north to Yorkshire, and in Wales and Ireland. Adults fly at night in June–July. Larvae mine stems of reed, reed sweet-grass and sedges, but cut part of a leaf to use as a raft to float to a new food plant. Greek *donax* + *aule* = 'reed' + 'dwelling-place', and Latin *forfex* = 'shears', from the way the forewings overlap when at rest, like scissor blades.

Scarce water-veneer *D. mucronella*, male wingspan 22–6 mm,

female 29–35 mm, nocturnal, in marsh, riverbanks and other damp areas in southern England, East Anglia, South Wales, central Scotland and Ireland. Adults fly in June–July, sometimes into August. Larvae mine stems of reed, reed sweet-grass and sedges. Latin *mucro* = 'sharp point', from the forewing tip.

WATER-VIOLET *Hottonia palustris* (Primulaceae), a stoloniferous perennial scattered in England (locally common in the east) and Wales, occasional as naturalised in Ireland and the Isle of Man, in still, shallow, base-rich, clear and non-eutrophicated water bodies such as ponds and ditches, many colonies having been lost to drainage, eutrophication, boat traffic and trampling by stock. This is the county flower of Huntingdonshire as voted by the public in a 2002 survey by Plantlife. Petrus Houttuyn (Hotton) (1648–1709) was a Dutch botanist. *Palustris* is from Latin for 'marsh'.

WATERLOUSE Freshwater malacostracan crustaceans in the family Asellidae (order Isopoda), resembling woodlouse.

One-spotted waterlouse *Proasellus meridianus*, widespread and locally common in lakes, canals, rivers and ditches, among water plants and under stones. *Meridianus* is Latin for 'midday'.

Two-spotted waterlouse *Asellus aquaticus*, common and widespread in lakes, canals, rivers, ditches and urban ponds, among water plants, under stones and submerged dead wood, and among roots of riparian trees. It is tolerant of organically polluted waters, high salinity, low pH and high metal concentrations. *Asellus* is Latin for 'little ass'.

WATERMARK DISEASE A dieback disease caused by the bacterium *Erwinia salicis*, first recognised as a serious disease of cricket-bat willow (see white willow under willow) in 1924, especially evident in Essex and Suffolk, but also found elsewhere. It is occasionally also found in other willows, especially goat, grey and white willows.

WATERWEED Herbaceous aquatic perennials in the family Hydrocharitaceae.

Canadian waterweed *Elodea canadensis*, a common and widespread oxygenating neophyte first recorded in Ireland in 1836 and in Britain in 1842, subsequently spreading rapidly, growing in mesotrophic to eutrophic conditions, favouring still or slowly flowing water at depths of 0–3+ m, where silt accumulates. All plants are female and reproduce vegetatively. *Elodea* is from Greek *helodes* = 'marshy'.

Curly waterweed *Lagarosiphon major*, an oxygenating South African neophyte, recorded as naturalised in 1944, locally abundant in England and Wales, less common elsewhere, spreading rapidly through release of discarded material from garden ponds or aquaria, growing in lakes, ponds, flooded mineral workings, and canals. Only female plants have been recorded, and reproduction is by vegetative fragmentation. Greek *lagaros* + *siphon* = 'empty' + 'tube'.

Esthwaite waterweed *Hydrilla verticillata*, found in Esthwaite Water, Lake District, in 1914 but not been seen there since 1941. It was recorded from two sites at and near Rusheenduff Lough, Co. Galway, in 1935 and 2004, and in Kirkcudbrightshire in 1999. *Hydrilla* is a diminutive of Greek *hydra* = 'water serpent', and Latin *verticillata* = 'whorled'.

Nuttall's waterweed *E. nuttallii*, a common and widespread North American neophyte herb, first recorded as naturalised in Britain in 1966 in Oxfordshire, since then spreading rapidly, and now widespread. It was first recorded in Ireland in 1984, at Lough Neagh. It grows in still or slowly flowing, shallow or deep, generally eutrophic water, and can be common even in highly disturbed canals and rivers. It effectively colonises new habitats, though all plants are female and it must spread vegetatively. Thomas Nuttall (1786–1859) was an English botanist.

WATERWORT Annuals in the genus *Elatine* (Elatinaceae), Greek for this plant.

Eight-stamened waterwort *E. hydropiper*, a rarity in the south-west Midlands, north-west Wales, central Scotland and north-east Ireland, in mesotrophic or eutrophic shallow water or on damp mud or silty sand exposed at the water's edge. *Hydropiper* is Greek for 'water-pepper'.

Six-stamened waterwort *E. hexandra*, an annual with a scattered local distribution, mostly in the west and south, on exposed mud on the margins of lakes, reservoirs, ponds and flooded gravel-pits, or submerged on open substrates in shallow, oligotrophic to eutrophic water. When submerged it occasionally persists as a short-lived perennial. *Hexandra* is Greek for 'with six stamens'.

WAVE (MOTH) Nocturnal members of the family Geometridae, many in the genera *Idaea* (Greek *Idaios* = 'pertaining to Mt Ida', from where the gods and goddesses watched the Trojan War) and *Scopula* (Latin for a small broom, from tufts on the tibia of some species), most species characterised by wavy lines (rivulets) on all wings.

Bright wave *I. ochrata* (Sterrhinae), wingspan 21–4 mm, Endangered in the RDB and a priority species in the UK BAP, probably resident on golf courses, shingle beaches and dunes in Kent (from Sandwich to Walmer), Suffolk and Essex, with records from other parts of southern England likely to be of immigrant origin. Adults fly in June–August. Larvae feed on smooth tare, hare's-foot clover and other coastal species. *Ochrata* is Latin referring to the ochre or yellowish wing colour.

Common wave *Cabera exanthemata* (Ennominae), wingspan 30–5 mm, widespread and common in often damp broadleaf woodland. Adults fly in May–August, with two broods in the south. Larvae feed on goat and grey willow, aspen and poplar. In Greek mythology Cabera was the daughter of the god Proteus. Greek *exanthemata* = 'a skin eruption', from the sprinkling of dots on the wings.

Common white wave *C. pusaria*, wingspan 25–8 mm, widespread and common in woodland, scrub and gardens. Adults fly in May–August, sometimes in two generations. Larvae especially feed on birches, sallows and alder. *Pusa* is Latin for 'little girl', from the delicate appearance.

Cream wave *S. floslactata* (Sterrhinae), wingspan 29–33 mm, local in broadleaf woodland, hedgerows, damp grassland and gardens throughout much of England, Wales and Ireland. Adults fly in May–June. Larvae feed on bedstraws, woodruff, bush vetch, dandelion, etc. Latin *flos* + *lactis* = 'flower' + 'milk', from the cream wing colour.

Dotted border wave *I. sylvestraria*, wingspan 20–3 mm, scarce and possibly endangered, in scrubby heathland, mostly in south-east England, with scattered colonies from Sussex to Yorkshire and Lancashire. Adults fly in July–August. Larvae feed on herbaceous plants, including dandelion and knotgrass. *Sylvestris* is from Latin for 'woodland', not its usual British habitat, so the name might be in honour of the late eighteenth-century French entomologist Israel Sylvestre.

Dwarf cream wave *I. fuscovenosa*, wingspan 19–22 mm, local in hedgerows, scrubby grassland, woodland edges and gardens throughout England and (especially north) Wales, scarcer further north. Numbers increased by 600% over 1968–2007. Adults fly in June–July. Larvae feed on herbaceous plants, including dandelion and knotgrass. Latin *fuscus* + *venosus* = 'dusky' + 'veined'.

Grass wave *Perconia strigillaria* (Ennominae), wingspan 36–41 mm, preferring lowland heathland but also found on moorland and in open woodland, fairly common in parts of southern England, local elsewhere in the British Isles, sometimes flying by day but generally from dusk in May–July. Larvae mainly feed on heathers, heaths, gorse and broom. Greek *peri* + *conios* = 'all around' + 'dusty', from the sprinkled markings, and the diminutive of Latin *striga* = 'furrow', from the forewing marking.

Lesser cream wave *S. immutata*, wingspan 24–7 mm, local in wet grassland, marsh and other damp areas throughout much of England, Wales and Ireland. Adults fly in June–August. Larvae feed on meadowsweet and common valerian. *Immutata* is Latin for 'unchanged', both upper and lower wings being white.

Mullein wave *S. marginepunctata*, wingspan 25–8 mm, local in grassland and saltmarsh along most of the Welsh, English and south and east Irish coastline, but uncommon inland, except in the London area where it is locally abundant. This is a species of conservation concern in the UK BAP. Adults fly in June–July, and again in August–September in the south. Larvae feed on mugwort, yarrow, marjoram, sage and horseshoe vetch. Latin *margo* + *punctum* = 'margin' + 'spot', from the wing markings.

Plain wave *I. straminata*, wingspan 28–33 mm, widespread but relatively uncommon in open woodland, scrubby heathland, hedgerows, rough grassland and gardens. Numbers increased by 634% over 1968–2007. Adults fly in July. Larvae feed on various herbaceous plants, including dandelion and knotgrass. Latin *stramen* = 'straw', from the colour.

Portland riband wave *I. degeneraria*, wingspan 26–31 mm, a rare RDB moth, breeding in grassland and scrub on coastal limestone in Portland, Dorset. Also a rare immigrant from mainland Europe, appearing on the southern coast of England, where it has become much more common since the start of the twenty-first century. Adults fly in June–July. Larvae feed on various herbaceous plants. *Degener* is Latin for 'degenerate', reason obscure.

Riband wave *I. aversata*, wingspan 23–30 mm, widespread and common in gardens, hedgerows, woodland, heathland, calcareous grassland and fen, more numerous in the south. Adults fly in June–August. Larvae feed on docks, dandelions and other herbaceous plants. *Aversata* is Latin meaning belonging to the underside, from a discal spot being more pronounced on the lower wing surface.

Rosy wave *S. emutaria*, wingspan 23–6 mm, scarce, on heathland, bog and saltmarsh in parts of southern and eastern England, south Wales and south-east Ireland, mostly coastal. Adults fly in June–July. Larvae feed on low-growing plants such as sea beet. *Emuto* is Latin for 'to change', as an initially pink wing colour fades with age.

Satin wave *I. subsericeata*, wingspan 22–5 mm, common in woodland, scrub, hedgerows, downland, rough grassland and gardens throughout much of southern England, less frequent in south-west and northern England, Wales and Ireland. Adults fly in June–July. Larvae feed on herbaceous plants, including knotgrass, chickweed and dandelion. Latin *sub* + *sericus* = 'somewhat' + 'silken', from the weak wing gloss.

Silky wave *I. dilutaria*, wingspan 20–2 mm, an RDB rarity flying from dusk in June–July, recorded from one or two places in Wales, and occasionally in England, for example Avon Gorge, on limestone slopes with low bushes, where larvae feed on common rock-rose. *Dilutus* is Latin for 'washed out', from the pale colour.

Single-dotted wave *I. dimidiata*, wingspan 13–18 mm, widespread and common in woodland, hedgerows, gardens, fen, marsh and riverbanks, absent from northern Scotland. Adults fly in June–August. Larvae feed from autumn through winter on cow parsley, hedge bedstraw and burnet-saxifrage. *Dimidio* is Latin for 'to divide into half', from the markings being on the dorsal half (lower part) of the forewing.

Small dusty wave *I. seriata*, wingspan 19–21 mm, common in gardens, hedgerows and rough ground, occasionally indoors, throughout Britain. Numbers increased by 155% over 1968–2007. In the south two generations, adults fly in June–July and August–September. Further north a single brood flies in July–August. Larvae feed on herbaceous plants, including ivy, and perhaps plant detritus. *Seriata* is from Latin *series* = 'a row', from the series of spots as wing markings.

Small fan-footed wave *I. biselata*, wingspan 15–20 mm, widespread and common in broadleaf woodland, heathland and gardens. Adults fly in June–August. Larvae feed from August through winter on withered leaves of a number of herbaceous plants. *Biselata* (an original misprint for *bisetata*) is from Latin *bis* + *seta* = 'twice' + 'bristle', from the tuft on the hind leg tibia.

Small scallop wave *I. emarginata*, wingspan 22–5 mm, local and declining in damp woodland, rough grassland, rural gardens, riverbanks and other damp areas throughout England and Wales. Adults fly in June–August. Larvae feed on bedstraw. *Emarginata* is Latin for 'deprived of the margin', from the scalloped wing edge.

Small white wave *Asthena albulata* (Larentiinae), wingspan 14–18 mm, widespread, if scarcer in Scotland, in broadleaf, particularly ancient woodland. Adults fly in May–July. Larvae feed on the leaves of such trees as hazel, hornbeam and birch, as well as wild rose. Greek *asthenes* = 'weak', from the indistinct wing pattern, and Latin *albulata* = 'whitish', from the wing colour.

Small yellow wave *Hydrelia flammeolaria* (Larentiinae), wingspan 14–20 mm, common in broadleaf woodland and hedgerows throughout England and Wales, rare in Scotland, and north-east Ireland. Adults fly in June–July. Larvae feed on leaves of field maple, sycamore and, in the north, more commonly alder. Greek *hydrelos* = 'watery', from the rivulet wing markings, and Latin *flammeolus* = 'flame-coloured', from lines on the wings.

Smoky wave *S. ternata*, wingspan 20–9 mm, local on heathland and moorland throughout much of Scotland, Wales and western England. Adults fly in June–July. Larvae feed on bilberry and heather. Latin *terni* = 'three each', from the three marking on all four wings.

Sub-angled wave *S. nigropunctata*, wingspan 29–34 mm, Vulnerable (RDB), breeding in undercliffs and grassland and along the edges of woodland in a few sites in Kent, in particular at Folkestone Warren and Hamstreet. Also increasingly common as an immigrant from mainland Europe on the southern coast of England. Adults fly in July–August. Larvae feed on various herbaceous plants. Latin *niger* + *punctum* = 'black' + 'spot', describing the forewing spots.

Tawny wave *S. ornata*, wingspan 21–4 mm, an RDB rarity, breeding on East Anglian heathland in the Breckland, and on dunes in the Aldeburgh-Thorpeness area, and also appearing as an immigrant from mainland Europe in southern England from the Wash to Cornwall. The bivoltine adults fly in June–July and August–September. Larvae feed on clovers, lucerne, bindweeds, vetches and thyme. *Ornata* is Latin for 'adorned'.

Weaver's wave *I. contiguaria*, wingspan 20–1 mm, found only on a few hillsides in north-west Wales in upland moorland, resting on rocks and walls during the day in July–August. Larvae mainly feed on heather, possibly also crowberry and navelwort. In terms of conservation a priority species in Wales ('Section 42' list). *Contiguus* is Latin for 'touching', from the wing markings.

Welsh wave *Venusia cambrica* (Larentiinae), wingspan 27–30 mm, local on moorland and in open woodland, in northern and western England, Wales and Scotland, and occasionally recorded in Ireland. Adults fly in July–August. Larvae feed on leaves of rowan and birches. *Venusia* refers to the Roman goddess Venus, for the attractiveness of this moth. *Cambrica* means 'Welsh', the type locality being Hafod, Cardiganshire.

WAX MOTH *Galleria mellonella* (Pyralidae, Galleriinae), wingspan 29–40 mm, found around beehives throughout much of England and Wales, with a recent record also from Co. Cork. Adults fly at night in June–October. Larvae live inside bee nests and beehives, feeding on the honeycomb. Use for biodegrading plastics (c.f. wax) was mooted in 2017. Greek *galeros* = 'cheerful', though here probably a homophonic pun actually referring to the galleries created by the larvae. Mellona was the Roman goddess of bee-keeping.

Lesser wax moth *Achroia grisella*, wingspan 16–24 mm, local throughout much of England, Wales and Ireland, less common than previously because of improvements in bee-keeping practices, though it can still be seen as a beehive pest in places. Adults fly at night from July into autumn. Larvae feed on the wax of old honeycombs, as well as dried fruit and dead insects. *Achroia* is Greek for 'lack of colour', *griseus* Latin for 'grey'.

WAXCAP Inedible, sometimes poisonous fungi in the genus *Hygrocybe* (Agaricales, Hygrophoraceae), from Greek *hygros* + *cybe* = 'wet' + 'head' (fruiting bodies are generally moist, though the cap is waxy), usually associated with short grass in which there appears to be a mutualistic relationship with moss.

Scattered and uncommon species, usually on unimproved grassland, old lawns and churchyards, include **bitter waxcap** *H. mucronella* (also in heathland, dunes and woodland clearings), **citrine waxcap** *H. citrinovirens* (also heathland and amenity grassland), **date waxcap** *H. spadicea* (Near Threatened in the RDL 2006 and a priority species in the UK BAP), **dingy waxcap** *H. ingrata*, **earthy waxcap** *H. fornicata*, **fibrous waxcap** *H. intermedia*, **glutinous waxcap** *H. glutinipes* (thick glutinous mucus clinging to the stems in wet weather together with slimy or sticky caps prompt the common name), **goblet waxcap** *H. cantharellus*, **grey waxcap** *H. lacmus*, **heath waxcap** *H. laeta* (also moorland and dunes), **limestone waxcap** *H. calciphila* (Near Threatened in the RDL 2006), **nitrous waxcap** *H. nitrata*, **pink waxcap** *H. calyptriformis* (poisonous), **slimy waxcap** *H. irrigata* (the slimy cap and stem give its common name), **splendid waxcap** *H. splendidissima* (upland), and **vermilion waxcap** *H. miniata*.

Rarities include **alpine waxcap** *H. xanthochroa* (mostly Highlands), **amethyst waxcap** *H. viola* (eight records over 1992–2014 in southern England and Pembrokeshire), **blushing waxcap** *H. ovina*, **lilac waxcap** *H. lilacina* (Highlands), **mountain waxcap** *H. salicis-herbaceae* (Cumbria and Highlands), and **slender waxcap** *H. radiata*.

Blackening waxcap *H. conica*, common and widespread, evident in late summer and autumn in both acid and calcareous soil under short grass, including upland pasture and verges. The fruiting body turning black with age prompts the English name. *Conica* is Latin for 'conical'.

Butter waxcap *H. ceracea*, common and widespread, found from late summer to early winter in soil under short non-fertilised grass, including upland pastures, churchyards, old lawns, parkland, dunes and occasionally woodland edges. *Cera* is Latin for 'wax'.

Cedarwood waxcap *H. russocoriacea*, fairly common, widespread, found in late summer and autumn on soil under short grass in unimproved pasture. Latin *russus* + *corium* = 'reddish' + 'skin'.

Crimson waxcap *H. punicea*, widespread and locally common, found in late summer and autumn often in upland sites on acidic soil under short grass in unimproved pasture and heathland. *Punicea* is Latin for 'pink' or 'purple'.

Dune waxcap *H. conicoides*, locally common around parts of the British coast on sandy soils, especially on dunes covered in short grass. *Conicoides* is Latin for 'similar to conica' (= 'conical') (i.e. to blackening waxcap).

Golden waxcap *H. chlorophana*, edible, locally common and widespread, from late summer to early winter in soil under short non-fertilised grass, including sheep-grazed upland pastures, churchyards and parkland. Greek *chloros* + *phana* = 'green' + 'pale', though the English name 'golden' is a more accurate description.

Honey waxcap *H. reidii*, widespread and fairly common, found in late summer and autumn on acidic soil under short grass in unimproved pasture. There is a honey smell if the base of the stem is cut. Derek Reid (1927–2006) was a British mycologist.

Meadow waxcap *H. pratensis*, edible, widespread and very common, found in late summer and autumn on acidic soil in short-grassed meadow and pasture. Var *pallida* is known as pale waxcap. *Pratensis* is from Latin for 'meadow'.

Oily waxcap *H. quieta*, widespread and locally common, found in late summer and autumn often on both acid and basic soil under short grass in unimproved pasture, heathland and woodland clearings. An oil-like odour comes from crushing the gills, hence its common name. *Quieta* is Latin for 'quiet' (i.e. 'inconspicuous').

Orange waxcap *H. aurantiosplendens*, widespread but scattered, in late summer and autumn in soil under various grassland habitats from unimproved pasture, through heathland grasses to woodland clearings. Latin *aurantiosplendens* = 'orange bright'.

Pale waxcap See meadow waxcap above.

Parrot waxcap *H. psittacina*, widespread and locally common, found in late summer and autumn on short unfertilised mown or cropped grass, including verges and churchyards. *Psiittace* is Greek for 'parrot', reflecting the colourful fruiting cap.

Persistent waxcap *H. acutoconica*, widespread and locally common, found in late summer and autumn in low-nutrient grassland usually in sandy but alkaline soil. *Acutoconica* is Latin for 'sharply conical'.

Scarlet waxcap *H. coccinea*, widespread and locally common, evident in late summer and autumn in often acid soil under short grass, including upland pastures, churchyards, old lawns, parkland and woodland clearings. *Coccinea* is Latin for 'scarlet'.

Snowy waxcap *H. virginea*, very common and widespread in late summer and autumn usually in acid soil under short grass in upland pasture, woodland clearings, parkland, churchyards and lawns, unlike other waxcaps tolerant of lightly fertilised conditions. *Virginea* is Latin for 'maiden'.

Spangle waxcap *H. insipida*, widespread and common in late summer and autumn in short unimproved grasses, including amenity grassland. *Insipida* is Latin for 'flavourless'.

Toasted waxcap *H. colemanniana*, widespread and locally common in late summer and autumn in both acid and calcareous soil under short mossy grass, including upland pasture. *Colemanniana* may refer to the English botanist William Coleman (d.1863).

Yellow foot waxcap *H. flavipes*, widespread and locally common, especially in western parts, in late summer and autumn in short grassland. *Flavipes* is Latin for 'yellow-footed'.

WAXWING *Bombycilla garrulus* (Bombycillidae), wingspan 34 cm, length 18 cm, weight 63 g, a winter visitor from the Continent, mostly to eastern regions, but in irruptive years spreading through much of the rest of Britain and occasionally urban north-east and eastern Ireland, feeding on berries, especially rowan and hawthorn, but also cotoneaster and rose, often in suburban ornamental plantings. An irruption in October 2010 brought in 2,500 birds from Scandinavia and Siberia, numbers echoed in November–December 2012, and again in winter 2017–18. Total wintering numbers are usually around 10,000 birds. The English name is prompted by the red wing markings that look like sealing wax. Greek *bombycos* = 'silk' and Latin *cilla* = 'tail', and *garrulus* = 'chattering'. **Cedar waxwing** *B. cedrorum* is a North American vagrant recorded in Shetland (1985) and Nottinghamshire (1996).

WAYFARING-TREE *Viburnum lantana* (Caprifoliaceae), a deciduous shrub, common and widespread south of a line from south Wales to Lincolnshire, more scattered and often introduced to the north, and scarce in north Scotland and Ireland. It is found in woodland, scrub and hedgerows, especially on base-rich soils, characteristic of chalk and limestone districts, and frequently planted on roadsides. John Gerard in his *Herball* (1597) thought that the French name, *viorne*, derived from the Latin *via* = 'road', and so a plant (with its white flowers) seen by wayfarers. The binomial is Latin and Late Latin for this species.

WEASEL *Mustela nivalis* (Mustelidae), Britain's smallest carnivore, male head-body length 195–215 mm, tail 42–52 mm, weight 105–30 g, females 173–83 mm, 34–43 mm and 55–70 g, common and widespread in a range of habitats, with an estimated British population in 2005 of 450,000 (England 308,000, Scotland 106,000, Wales 36,000). It is absent from Ireland, though here stoats are sometimes referred to as 'weasels'. Weasels specialise in hunting small rodents, but also take young rabbits, as well as birds and eggs, hence attempts at control by gamekeepers. They do not hibernate and can hunt under deep snow. Following delayed implantation, gestation lasts 34–7 days. There is usually one litter a year, with 4–6 young, which are weaned at 3–4 weeks, and independent by 8–9 weeks. 'Weasel' comes from Old English *wesle* or *wesule*. The binomial is Latin for 'weasel' and 'snowy'.

WEASEL'S-SNOUT *Misopates orontium* (Veronicaceae), an uncommon annual archaeophyte, scattered mostly in southern Britain on light soils in arable fields, wasteland and bare soil, with a distribution that has halved over the last 40 years. Flowers are followed by a hairy fruit which resembles a weasel's nose, prompting the common name. *Misopates* was a name given by Dioscorides to several plants. The Orontes region is in Syria.

Pale weasel's-snout *M. calycinum* a scarce casual on waste ground in England, originating from birdseed. *Calycinum* is Latin for having a conspicuous calyx.

WEAVER (SPIDER) Members of the Linyphiidae. See also orb weaver (Araneidae).

Horrid ground weaver *Nothophantes horridus*, a money spider, body length 2.5 mm, endemic in the UK, Critically Endangered in the RDB, a priority species in the UK BAP, and since 1989 recorded from only four sites in Plymouth within an area of 1 km². It is difficult to find because of its small size, nocturnal habit and living inside rock crevices. One site (Shapter's Field Quarry) was lost to development, but in 2015 Plymouth City Council threw out a planning application for housing that would have destroyed a site at Radford Quarry. The fourth site, on industrial land in the Cattedown area, was only recognised in February 2016. The common and specific names come from the body and legs being hairy, Latin for 'bristly' being *horridus*. Greek *nothos* + *phantos* = 'spurious' + 'visible'.

Sheet weaver Alternative name for members of the Linyphiidae.

WEBBER Juniper webber *Dichomeris marginella* (Gelechiidae, Dichomeridinae), wingspan 13–24 mm, locally common on chalk and limestone downland, and in parks and gardens on cultivated junipers in southern and eastern England and the Midlands, very local in Wales, a few parts of northern England and the Channel Islands, and recorded from the Burren in Ireland. Adults fly at night in June–August. Larvae feed on juniper needles inside a silk web. Greek *dicho* + *meros* = 'in two' + 'a part', and Latin *marginis* = 'of an edge', both parts describing the white streak 'division' on the forewing.

WEBCAP Generally inedible fungi in the genus *Cortinarius* (Agaricales, Cortinariaceae), the common and generic names referring to the partial veil or *cortina* (Latin for 'curtain') that covers the gills when caps are immature.

Widespread but uncommon species, generally found in woodland, include **birch webcap** *C. triumphans* (possibly toxic, especially found under birch), **blue-girdled webcap** *C. collinitus* (on boggy acid soil in coniferous woodland), **bruising webcap** *C. purpurascens* (especially under oak and beech), **cinnamon webcap** *C. cinnamomeus* (coniferous woodland, and under birch on heathland), **contrary webcap** *C. variicolor* (often under beech in England, under pine in coniferous forest in Scotland), **cream webcap** *C. ochroleucus* (especially in grass under oak and beech), **deadly webcap** *C. rubellus* (highly poisonous, containing the toxin orellanine which, if eaten, destroys the kidneys and liver, especially found under pine and spruce), **earthy webcap** *C. hinnuleus* (especially under oak, beech and birch), **frosty webcap** *C. hemitrichus* (especially under birch), **girdled webcap** *C. trivialis* (possibly toxic, especially found under alder, willow, poplar and birch), **marsh webcap** *C. uliginosus* (possibly toxic, often under willows, and in alder carr), **orange webcap** *C. mucosus* (mostly Highlands, especially under Scots pine and birch), **pearly webcap** *C. alboviolaceus* (under beech and other broadleaf trees), **pelargonium webcap** *C. flexipes* (on boggy soil, often among mosses, and under birch and spruce), **purple stocking webcap** *C. mucifluoides* (especially under birch), **sepia webcap** *C. decipiens* (often under birch and willow), **silky webcap** *C. evernius* (coniferous woodland and unimproved grassland), **slimy webcap** *C. mucifluus* (mostly Highlands, especially under Scots pine), **wrinkled webcap** *C. livido-ochraceus* (especially under beech, but also oak, birch and, in Caledonian forest, pine), and **yellow webcap** *C. delibutus* (under birch or beech).

Vulnerable in the RDL (2006) are **fool's webcap** *C. orellanus* (poisonous, orellanine causing often fatal liver and kidney failure, usually found under oak), **ruddy webcap** *C. rubicundulus* (including in ancient oak woodland in Cumbria and birchwood and coniferous forest in Scotland), and **scaly webcap** *C. pholideus* (especially under birch).

Near Threatened in the RDL (2006) are **splendid webcap** *C. splendens* (southern England and the south-west Midlands, under beech, occasionally hazel), and **violet webcap** *C. violaceus* (especially under birch and beech).

Other rarities include **bitter bigfoot webcap** *C. sodagnitus*, **bitter webcap** *C. infractus*, **blushing webcap** *C. cyanites* (Highlands), **Caledonian webcap** *C. caledoniensis* (Highlands, most Abernethy Forest), **foxy webcap** *C. vulpinus* (11 records between 1988 and 2010 from Kent, Buckinghamshire, Hertfordshire, Hampshire and Perthshire), **freckled webcap** *C. spilomeus* (local in England and Scotland), **golden webcap** *C. humicola* (four records over 1992–2008 from Hampshire, Gloucester and Shropshire), **goliath webcap** *C. praestans* (one record each from Gloucestershire, Somerset and Buckinghamshire), **greenfoot webcap** *C. glaucopus* (very scattered from Somerset to Kent, and Highlands), **hard webcap** *C. durus* (possibly Hampshire, and Abernethy Forest, Highlands), **honey webcap** *C. talus* (mainly Highlands), **hotfoot webcap** *C. bulliardii* (southern half of England), **incense webcap** *C. subtortus* (southern England, Cumbria and Highlands), **mealy webcap** *C. aleuriosmus* (six records from Surrey, Oxfordshire and Berkshire), **mealy bigfoot webcap** *C. caerulescens*, **spring webcap** *C. vernus* (southern England and central Scotland), **tawny webcap** *C. callisteus* (mostly Highlands), and **yellow capped webcap** *C. xanthocephalus* (18 records between 1957 and 2008 from south-east England, the Midlands and central Scotland).

Bloodred webcap *C. sanguineus*, widespread and locally common in late summer and autumn on soil in dark, damp, often mossy coniferous woodland. *Sanguineus* is Latin for 'blood-red'.

Blueleg webcap *C. amoenolens*, with a scattered distribution in England and parts of Wales in autumn under beechwood on chalk soil. *Amoenus* is Latin for 'pleasing'.

Dappled webcap *C. bolaris*, possibly poisonous, locally common from late summer to early winter on woodland soil under beech (though the soil is often acid), sometimes oak. *Bolaris* is from Latin, meaning either 'red brick-coloured' or more probably 'netted', from the scaly cap.

Gassy webcap *C. traganus*, locally common on soil in Caledonian pinewood in the Highlands. *Traganos* is Greek for 'sucking pig'.

Goatcheese webcap *C. camphoratus*, uncommon, found in late summer and autumn with records from the Scottish Highland and south England on acid soil in coniferous woodland where it is mycorrhizal on spruce and fir. The English name reflects its odour, though it is inedible. *Camphoratus*, however, reflects a scent of camphor.

Green webcap *C. scaurus*, scattered, commoner in the Highlands, on soil usually in coniferous woodland, occasionally in deciduous woodland, for instance under beech in the New Forest. *Scaurus* is Latin for 'large ankled'.

Red banded webcap *C. armillatus*, generally common throughout Britain in summer and autumn in woodland soil under birch. *Armillatus* is Latin for 'collared', from the girdles surrounding the stem.

Saffron webcap *C. croceus*, poisonous, fairly common in coniferous and mixed woodland, particularly mycorrhizal on pines, spruces and birches, from August to November. *Croceus* is Latin for 'saffron yellow'.

Stocking webcap *C. torvus*, possibly poisonous, widespread and locally common in late summer and autumn on soil in deciduous and mixed woodland, especially under beech and birch. *Torvus* is Latin for 'savage' or 'wild'.

Sunset webcap *C. limonius*, recorded from a few locations in England and on Anglesey, but mostly noted in upland Scotland on soil in coniferous forest, especially under pine. It is on the Scottish Biodiversity List of Species of Principal Importance for Biodiversity Conservation (2005), and is Near Threatened in the RDL (2006). *Limus* is Latin for 'mud'.

Surprise webcap *C. semisanguineus*, locally common and widespread in late summer and autumn on soil in woodland under conifers, especially spruce, or birch, and often in moss. The blood-red gills are a 'surprise' given the pallid cap and stem, hence the common name. Latin *semisanguineus* = 'half blood-red'.

Variable webcap *C. anomalus*, widespread and generally common in summer and autumn in mixed woodland, especially under birch and pine. *Anomalus* is Latin here meaning 'irregular', from the variable appearance.

White webcap *Leucocortinarius bulbiger* (Tricholomataceae), not a true webcap, with a few scattered records in Britain, Near Threatened in the RDL (2006), found in late summer and autumn in woodland soil and in acid grassland. Greek *leucos* + the genus *Cortinarius*, and Latin *bulbiger* = 'bulb-bearing'.

Woolly webcap *C. laniger*, recorded from Surrey (on heathland soil) and Kent (on soil in deciduous woodland), but mostly noted in upland Scotland on soil in coniferous forest, especially under pine. It is on the Scottish Biodiversity List of Species of Principal Importance for Biodiversity Conservation (2005), and is Vulnerable in the RDL (2006). *Laniger* is Latin for 'wool-bearing'.

WEBWORM Cotoneaster webworm *Athrips rancidella* (Gelechiidae, Gelechiinae), wingspan 12 mm, a small nocturnal North American moth first recorded as an adventive colonist in 1971 at West Wickham, Kent, breeding in suburban gardens, currently restricted to seven 10 km squares in west Kent and Surrey. Adults fly in June–July. Larvae feed on cotoneaster in May–June, inside a silk tube. Greek *athroos* + *ips* = 'in heaps' + 'larva', from the gregariousness of the caterpillars, and Latin *rancidus* = 'stinking', reason unknown.

WEDGE SHELL Banded wedge-shell *Donax vittatus* (Donacidae), a filter-feeding bivalve, shell size up to 4 cm, common and widespread in clean, firm sand in the intertidal and shallow sublittoral (surf zone) on exposed sandy shores and bays. The binomial is Latin for 'scallop' and 'striped' (literally beribboned).

WEEVER FISH Coastal fishes in the family Trachinidae. Venom glands are located on the first dorsal fin and on spines on the gill cover, and if trodden on the sting is acute and intense. 'Weever' dates from the early seventeenth century, perhaps as a transferred use of Old French *wivre* = 'serpent, dragon', from Latin *vipera* = 'viper'.

Greater weever fish *Trachinus draco*, length up to 53 cm, found round much of the coastline buried in clean sandy substrates from the low water mark to the shallow sublittoral down to 50 m. Spawning is in summer. Diet is of small crustaceans and small demersal fish. Greek *trachys* = 'rough', and Latin *draco* = 'dragon'.

Lesser weever fish *Echiichthys vipera*, length 15 cm, found around the British coast, especially in the south and west, buried in clean sandy substrates from the low water mark to the shallow sublittoral down to 50 m. Spawning is in summer. Diet is of small crustaceans and small demersal fish. Greek *echieys* + *ichthys* = 'a little viper' + 'fish'.

WEEVIL Members of the beetle superfamily Curculionoidea, in particular the 'true' weevils (family Curculionidae, see 1 below), but also Anthribidae (fungus weevils, see 2), Apionidae (seed weevils, see 3), Attelabidae (q.v., and see leaf-roller), Dryophthoridae (see 4), Erirhinidae (wetland weevils, see 5), Nanophyidae (see 6), and Rhynchitidae (tooth-nosed snout weevils, see 7), as well as some leaf beetles (Chrysomelidae, see 8). Adults have an elongated head that forms a snout (rostrum). 'Weevil' comes from Old English *wifel* = 'beetle', from a Germanic term meaning 'move briskly'.

1) Members of the 'true' weevil family Curculionidae, mostly in the subfamily Curculioninae.

Rarities include **apple bud weevil** *Anthonomus piri* (Cheshire, Norfolk and Kent), **fern weevil** *Syagrius intrudens* (member of an Australian genus, but this fern-eating species not been found there, so the small, scattered British sites in south Devon and Kent contain the only known world populations), **Gilkicker weevil** *Pachytychius haematocephalus* (Channel Islands, and in a 5 km strip of shingle between Gilkicker Point and Browndown, near Gosport, Hampshire), and **Lundy cabbage leaf weevil** *Ceutorhynchus insularis* (flightless, endemic to Lundy).

Acorn weevil *Curculio glandium*, 4–8 mm length, common and widespread in oak woodland in the southern two-thirds of England and in Wales. The female uses her rostrum to drill into the centre of an acorn to lay an egg. The larva feeds within the acorn then bores its way out. *Curculio* is Latin for 'weevil', *glans* for 'acorn'.

Alder jumping weevil *Orchestes testaceus*, recorded from Ireland Cheshire, the Midlands, Cambridgeshire, East Anglia and Kent, a leaf miner in alder.

Alfalfa weevil Synonym of clover leaf weevil.

Apple blossom weevil *Anthonomus pomorum*, scattered in England and Wales, a major pest in pesticide-free apple orchards. After overwintering mainly in woodland and hedgerows adjacent to orchards, adults (3.5–6 mm length) emerge in early spring to fly into orchards to oviposit in apple buds, which are subsequently damaged by the larvae. *Anthonomos* is Greek for a plant fed on by bees, *pomus* Latin for a fruit tree.

Banded pine weevil *Pissodes pini* and *P. castaneus* (Molytinae), scattered in England and Scotland, feeding on conifers. *Pissodes* is Greek for 'yielding pitch', *pinus* and *castaneus* Latin for, respectively, 'pine' and 'chestnut'.

Bird-cherry weevil *A. rectirostris*, mainly recorded from south-central Wales, with a few sites noted in northern England and Scotland. Latin *rectus* + *rostrum* = 'straight' + 'snout'.

Black marram weevil *Otiorhynchus atroapterus* (Entiminae), common around the British coastline, adults and larvae respectively feeding on leaves and roots of marram on dunes. Greek *otoeis* + *rhynchos* = 'eared' + 'snout', and Latin *ater* + *aptera* = 'black' + 'wingless'.

Bloody cranesbill weevil *Zacladus exiguus* (Ceutorhynchinae), scattered in southern England, evident from July to September in habitats containing small-flowered geraniums such as bloody crane's-bill. Greek *za* + *clados* = 'very' + 'branch', and Latin *exiguus* = 'small'.

Brookline gall weevil *Gymnetron veronicae*, scattered in England and Wales. Greek *gymnos* + *netron* = 'naked' + 'spindle', and Latin for 'of Veronica' (speedwell).

Brown leaf weevil *Phyllobius oblongus* (Entiminae), 3.7–6 mm length, scattered in England and Wales, occasionally recorded in Scotland and Ireland, in woodland and hedgerows, feeding on trees such as hawthorn, willows, apple and elm. *Phyllon* is Greek for 'leaf'.

Cabbage gall weevil *Ceutorhynchus assimilis* (Ceutorhynchinae), a locally common pest of brassicas, widespread in England and Wales. Greek *ceutho* + *rhynchos* = 'to hide' + 'snout', and Latin *assimilis* = 'similar'.

Cabbage leaf weevil *Ce. contractus*, a locally common pest of brassicas, widespread in England and Wales. *Contractus* is Latin for 'contracted'.

Cabbage seed weevil *Ce. obstrictus*, a locally common pest of brassicas widespread in England, especially in the central Midlands and Lincolnshire. *Obstrictus* is Latin for 'inducement'.

Cabbage stem weevil *Ce. pallidactylus*, a locally common pest of brassicas widespread in England and Wales. *Pallidactylus* may be from Latin *pallidus* = 'pale' + Greek *dactylos* = 'foot'.

Clay-coloured weevil *Ot. singularis*, widespread in Britain, occasional in Northern Ireland, adults feeding in spring and summer on buds and flowers, larvae on roots of rhododendron, and locally important as a pest of raspberry in Scotland. *Singularis* is Latin for 'unique'.

Clover leaf weevil *Hypera postica* (Hyperinae), 3–5 mm length, common and widespread in England and Wales feeding on clovers and lucerne. *Hyperos* is Greek for 'pestle', *postica* Latin for 'rear'.

Clover root weevil *Sitona hispidulus* (Entiminae), 2.8–4.6 mm length, common and widespread in England and Wales, occasional in Wales and Ireland. Adults are present all year, feeding on the leaves, larvae from April to August on the roots, of leguminous plants, including clovers and lucerne, habitats therefore including pasture, wasteland, roadsides and scrub. *Siton* is Greek for 'cereal field' or 'granary', *hispidulus* Latin for 'a little hairy'.

Common leaf weevil *Phyllobius pyri*, 5–7 mm length, common and widespread on trees and shrubs in parks, gardens and woodland edge, particularly in May–June. *Pyrus* is Latin for 'pear tree'.

Elm flea weevil or **European elm flea weevil** *Orchestes alni*, scattered in England as far north as Lincolnshire, and in south Wales, as a larval leaf miner of elm (not alder, *Alnus*, as implied by the specific name).

Figwort weevil *Cionus scrophulariae*, widespread and locally abundant in England, Wales and Ireland, with a few records from Scotland, feeding on figwort and mullein. When disturbed the adult drops from the plant and is camouflaged as a crumb of soil. Larvae can cover themselves with a sticky tar-like substance that acts as a deterrent to predation and parasitism. *Cion* is Greek for 'pillar', *Scrophularia* Latin for 'figwort'.

Flea weevil Some members of the family Curculionidae, subfamily Curculioninae, tribe Ramphini, which can leap to avoid disturbance. See elm flea weevil above.

Golden keyhole weevil *Sibinia sodalis*, scarce, mainly recorded from cliff grassland in Cornwall, Devon and south-west Wales, with a few other records from England, feeding on thrift and sea-lavender. *Sibyne* is Greek for 'spear', *sodalis* Latin for 'companion'.

Green leaf weevil *Phyllobius maculicornis*, locally common and widespread in Britain on nettles and a range of other plants. Latin *maculus* + *cornu* = 'spot' + 'horn'.

Green nettle weevil See nettle weevil below.

Ground weevil *Barynotus obscurus* (Entiminae), 8–9.5 mm length, scattered in England and Wales, a ground-dwelling species found year-round under stones, on the soil surface or on low vegetation. Greek *barys* + *notos* = 'heavy' + 'the back', and Latin *obscurus* = 'dark'.

Hairy spider weevil *Barypeithes pellucidus* (Entiminae), 3–4 mm length, widespread in England and Wales, more scattered in Scotland and Ireland, adults found in late spring and early summer in various habitats, often on the ground in moss or on low vegetation, but also under bark. Larvae feed on black medick. Greek *barys* + *peithes* = 'heavy' + 'persuasion', and Latin *pellucidus* = 'transparent'.

Heather weevil *Neliocarus sus* (Entiminae), scattered in Britain on heather in lowland heathland and upland moorland. Greek *neilos* + *care* = 'Nile' + 'head', and Latin *sus* = 'pig', reason unknown.

Hop root weevil *Mitoplinthus caliginosus* (Molytinae), adults 6–9 mm length, often found under stones, larvae feeding in the roots of hops and other woody and herbaceous plants, mainly in Kent and East Sussex, but also recorded in Hampshire and

Gloucestershire. Greek *mitos* + *plinthos* = 'thread' + 'brick', and Latin *caliginosus* = 'dark'.

Iris weevil *Mononychus punctumalbum* (Ceutorhynchinae), recorded from Avon south-eastwards to Hampshire, with a record from the East Riding of Yorkshire, in grassland, recorded on flag iris. Greek *monas* + *onychos* = 'single' + 'claw', and Latin *punctus* + *album* = 'point' + 'white'.

Jumping weevil See alder jumping weevil above.

Large pine weevil See pine weevil below.

Large thistle weevil *Cleonis pigra* (Lixinae), with a few records from the Midlands and eastern England from Kent north to the East Riding of Yorkshire. Cleon was an Athenian general. *Pigra* is Latin for 'lazy' or 'tiresome'.

Long-snout weevil Alternative name for nut weevil, redundant in that all weevils characteristically have long rostra.

Lucerne weevil Synonym of clover leaf weevil.

Marram weevil *Philopedon plagiatum* (Entiminae), locally abundant on marram around the English, Welsh and to an extent Scottish dune coastline. Greek *philos* + *pedon* = 'loving' + 'earth', and Latin *plagiatum* = 'stolen'.

Meadow cranesbill weevil *Zacladus geranii* (Ceutorhynchinae), 2–3 mm length, with a scattered distribution in England and Wales, evident in July–September in habitats containing the food plants meadow and bloody crane's-bills. Greek *za* + *clados* = 'very' + 'branch', and from Latin *geranii* = 'geraniums'.

Nettle weevil *Phyllobius pomaceus*, 9 mm length, widespread in England and Wales, occasionally recorded in Scotland and Ireland, most commonly on nettles Latin *poma* = 'fruit tree'.

Nut leaf weevil *Strophosoma melanogrammum* (Entiminae), 4–6 mm length, scattered and common in woodland, parkland, hedgerows and gardens, feeding on a range of broadleaf and coniferous trees and shrubs, preferring oak and hazel. They can be a pest of conifers, especially of larch. Adults damage leaves and strip rings of bark from young twigs. Larvae feed on the roots of a variety of herbaceous plants. Greek *strophos* + *soma* = 'twisted' + 'body', and *melas* + *gramma* = 'black' + 'mark'.

Nut weevil *Curculio nucum*, 6–8.5 mm length, common and widespread in the southern two-thirds of England and in Wales. Development takes two years: adults emerge in spring from the soil where they have been overwintering, feed on hazel buds and leaves, and the females drills into a hazel nut with her rostrum (which is larger than the male's) to lay an egg. The larva subsequently drills out of the nut. Latin *nux* = 'nut'.

Pea leaf weevil *Sitona lineatus*, common and widespread in England and Wales, adults feeding on the leaves, larvae the roots, of leguminous plants. It is at worst a minor pest of pea and bean crops. *Lineatus* is Latin for 'lined'.

Pear weevil *Magdalis barbicornis* (Magdalidinae), with a scattered distribution in the southern third of England, larvae damaging pear tree buds. Greek *magdalis* = 'charcoal', and Latin *barba* + *cornu* = 'beard' + 'horn'.

Pine weevil *Hylobius abietis* (Molytinae), 10–13 mm length, scattered but locally abundant pest in conifer plantations. Adults, which live for up to four years, mainly breed in stumps and roots of felled conifers. They hibernate over winter, then feed on the bark of, and damaging mainly sapling fir, pine and spruce. Greek *hyle* = 'a wood', and Latin *abies* = 'fir'.

Raspberry weevil Synonym of clay-coloured weevil.

Red-legged weevil *Ot. clavipes*, predominantly found in the southern third of England and Lincolnshire. *Clavipes* is Latin for 'club-foot'.

Rough strawberry-root weevil *Ot. rugostriatus*, scattered in England and Wales, larvae causing damage to cultivated strawberry roots. Latin *rugosus* + *striatus* = 'wrinkled' + 'lined'.

Sea-wormwood weevil *Polydrusus pulchellus* (Entiminae), locally common on the coast from the Solway Firth southwards round to the Humber estuary, and in north-east Ireland, feeding on a variety of saltmarsh plants, adults on leaves, larvae on roots. Greek *poly* + *drysos* = 'many' + 'oak', and Latin *pulchellus* = 'pretty'.

Seed weevil Members of the family Apionidae.

Silver-green leaf weevil *Phyllobius argentatus*, locally common and widespread on both broadleaf trees and low herbaceous vegetation such as nettles. *Argentatus* is Latin for 'silvery'.

Small green nettle weevil *Phyllobius roboretanus*, scattered in England and Wales, occasionally recorded in Scotland and Ireland, associated with a number of tree and shrub species. *Robor* is Latin for 'oak'.

Small heather weevil *Micrelus ericae*, widespread but scattered on heather in both lowland heathland and upland moorland. *Erica* is Latin for 'heather'.

Small nettle weevil *Nedyus quadrimaculatus* (Ceutorhynchinae), 3 mm length, common and widespread, especially in England and Wales, on nettles from spring into July. *Nedys* is Greek for 'winter', *quadrimaculatus* Latin for 'four-spotted'.

Spider weevil *Barypeithes araneiformis* (Entiminae), common and widespread in England and Wales, scattered in Scotland and Ireland, often in woodland litter. *Araneiformis* is Greek for 'spider-shaped'.

Strawberry blossom weevil *An. rubi*, common and widespread in England and Wales, occasional in Scotland and Ireland, an important pest of cultivated strawberry (and to an extent raspberry) with adults (2.5–3 mm length) feeding on leaves, larvae in flower buds. Adults may be attracted to pheromones produced by the strawberries. Latin *rubus* = 'bramble'.

Strawberry-root weevil *Ot. ovatus*, scattered in England and Wales, rare in Scotland, larvae causing damage to the roots of cultivated strawberry. Latin *ovatus* = 'oval'.

Turnip gall weevil Synonym of cabbage gall weevil.

Vine weevil *Ot. sulcatus*, adults 9 mm length, common and widespread, more local in Scotland and Ireland, infesting a range of ornamental plants and fruits, especially when growing in containers, adults (which are all female) feeding on leaves in spring and summer, larvae – the more serious stage as a pest – on roots from summer to the following spring. Listed as second in the Royal Horticultural Society's list of worst garden pests of 2017. *Sulcatus* is Latin for 'furrowed'.

Violet weevil *Orobitis cyaneus* (Orobitinae), widespread and scattered in England and Wales, a small species, similar in appearance to the seed capsules of Violaceae, the larval food source. *Orobitis* is Greek for 'like chickpea', *cyaneus* Latin for 'blue'.

Willow gall weevil *Archarius salicivorus*, common and widespread in England and Wales, plus a few records from Scotland and Ireland, adults (2.5 mm length) found on willow in spring and summer, laying eggs in sawfly galls where the subsequent weevil larvae, as hyperparasites, prey on the sawfly larvae. *Salicivorus* is Latin for 'willow-eating'.

Withy weevil *Cryptorhynchus lapathi* (Cryptorhynchinae), uncommon, mostly scattered in England and Wales, larvae boring into the stems of willow species. Greek *cryptos* + *rhynchos* = 'hidden' + 'snout', and Latin *lapathus* = 'sorrel'.

2) **Fungus weevils** in the family Anthribidae. *Anthribus fasciatus* and *A. nebulosus* are parasitoids of scale insects. *Bruchela rufipes* develops in the fruits of *Reseda lutea*, but most species are found in association with dead wood and fungi.

Cramp ball fungus weevil *Platyrhinus resinosus*, the commonest fungus weevil, found in the south-eastern half of England and in south Wales, associated with the fungus King Alfred's cakes (cramp balls). Greek *platys* + *rhinos* = 'broad' + 'nose', and Latin *resinosus* = 'resinous'.

3) **Seed weevils** in the family Apionidae.

Clover seed weevil *Protapion apricans*, *P. assimile* and *P. trifolii*, females 2 mm in length (males smaller), all common and widespread in England and Wales. Greek *protos* + *apion* = 'first' + 'pear'. *Apricans* is Latin for 'warming in the sun', *assimile* for 'similar', and *trifolium* for 'clover'. See also white clover seed weevil below.

Dog's-mercury seed weevil *Kalcapion pallipes*, with a scattered distribution in England and Wales, rare in Scotland.

Greek *calos* + *capion* = 'beautiful' + 'smoking', and Latin *pallipes* = 'covered foot'.

Gorse seed weevil *Exapion ulicis*, body length 2–3 mm, feeding on gorse, common and widespread in England and Wales, and with a few records from Scotland and Ireland. *Ulicis* is from Latin for 'gorse' (*ulex*).

Petty whin weevil *E. genistae*, scarce, feeding on dyer's-greenweed, with a scattered distribution mainly in south-east England, listed in RDB 3, and a priority species in the UK BAP. *Genistae* is from Latin for 'broom'.

Red clover seed weevil Alternative name for a clover seed weevil (*P. trifolii*).

Sallow guest weevil *Melanapion minimum*, with a scattered distribution in (mostly) eastern England, becoming increasingly scarce, listed in RDB 3, and a priority species in the UK BAP. Larvae are inquilines in the galls of *Pontania* sawflies on willow leaves. *Melas* is Greek for 'black'.

White clover seed weevil *Protapion fulvipes*, 1.5–2 mm length, common and widespread, less so in Scotland and Ireland, larvae feeding on clover flowers. Latin *fulvipes* = 'red-footed'.

4) Members of the family Dryophthoridae, stored-product pests that include rice weevil *Sitophilus oryzae* and maize weevil *S. zeamais*.

Granary weevil grain weevil *Sitophilus granarius*, body length 3–5 mm, common, causing significant damage to stored grain. Larvae feed inside the grain until pupation; they then bore a hole out of the grain and emerge. Greek *sitos* + *philos* = 'grain' + 'loving'.

Wood-boring weevil *Dryopthorus corticalis*, a rarity found in woodland in Windsor Great Park. Greek *drys* + *phthora* = 'oak' + 'destruction', and Latin *cortex* = 'bark'.

5) **Wetland weevils** in the family Erirhinidae.

Azolla weevil *Stenopelmus rufinasus*, introduced from North America in 1921, scattered in England and south Wales, feeding on (and used for biological control of) water fern. It was first reported in Ireland in 2007 from Co. Fermanagh and Co. Cork, associated with infestation of water fern. Greek *stenos* + *pelma* = 'narrow' + 'sole of the foot', and Latin *rufus* + *nasus* = 'red' + 'nose'.

Duckweed weevil *Tanysphyrus lemnae*, 1.4–2 mm length, widespread in England and Wales, rare in Ireland. Adults become active during warm days in early March and by April–May are common, remaining so until at least mid-July. Adults cannot swim, but they and larvae both feed on duckweed. Greek *tanyo* + *sphyron* = 'to stretch' + 'ankle', and Latin *lemna* = 'duckweed'.

Horsetail weevil *Grypus equiseti* (Erirhinidae), 4.1–6.7 mm length, widespread, if scattered and local in wetland and riparian habitats, feeding on horsetails, larvae in the stems. The binomial is from Greek *grypos* = 'hook-nosed', and Latin *equisetum* = 'horsetail'.

6) Members of the family Nanophyidae.

Loosestrife weevil *Nanophyes marmoratus*, widespread in England and Wales, with a few Irish records, feeding on purple-loosestrife in damp habitats. Greek *nano* + *phye* = 'small' + 'growth', and Latin *marmoratus* = 'marbled'.

7) **Tooth-nosed snout weevils** in the family Rhynchitidae. See also apple twig cutter, birch leaf roller under leaf-roller, and rhynchites.

Apple fruit weevil *Tatianaerhynchites aequatus*, 2.5–4.5 mm length, common and widespread in England and Wales, found in May–July in woodland and hedgerows with hawthorn and blackthorn, as well as orchards where they damage developing fruitlets by drilling holes. The binomial is from the name Tatiana + Greek *rhynchos* = 'snout', and Latin *aequa* = 'fair'.

Aspen leaf-rolling weevil Synonym of poplar leaf-rolling weevil.

Poplar leaf-rolling weevil *Byctiscus populi*, 5 mm length, declining in range and now with just a few records from south-east England and the west Midlands, Rare in RDB 3, and a species of conservation concern in the UK BAP. It feeds on aspen leaves in woodland and scrub. The female nips the leaf petiole then rolls the leaf into which she lays her eggs. The leaf falls off and larvae feed in decaying tissue, later to pupate in the soil. Adults are most numerous in June, but there are a number of generations a year. Greek *byctes* = 'swelling', and Latin *populus* = 'poplar'.

Twig-cutting weevil Alternative name for apple twig cutter.

8) Leaf beetles in the family Chrysomelidae (Bruchinae).

Bean weevil *Acanthoscelides obtectus*, 2–5 mm. an introduction with grain, with occasional records from south-east England and East Anglia, feeding on beans, vetches and other leguminous plants. Greek *acantha* + *scelos* = 'thorn' + 'leg', and Latin *obtectus* = 'matted'.

WEEVIL-WASP Species of *Cerceris* (Crabronidae), from Latin meaning to be rough or hard.

Five-banded weevil-wasp *C. quinquefasciata*, Rare in RDB 3, with records from south-east England and East Anglia, on sandy soils in which it excavates its nesting burrows, provisioned with weevils. Latin *quinquefasciata* = 'five-banded'.

Four-banded weevil-wasp *C. quadricincta*, very rare, on light, sandy soils in which it excavates its nesting burrows, with records only from Essex and (recently) Kent, Endangered in RDB 1. Latin *quadricincta* = 'four-banded'.

WEIGELIA *Weigela florida* (Caprifoliaceae), an evergreen Chinese neophyte shrub introduced in 1845, scattered in the mid-Thames Valley and Lancashire, naturalised on railway banks and waste ground. Christian von Weigel (1748–1831) was a German botanist. Latin *florida* = 'with many flowers'.

WELD *Reseda luteola* (Resedaceae), a common and widespread biennial archaeophyte herb (more local in northern Scotland and western Ireland), typically occurring on neutral or base-rich soils on roadsides and waste ground, in gravel-pits, urban demolition sites and, occasionally, arable or grassy areas. 'Weld' dates to late Middle English, perhaps related to *wold* (= 'wooded upland'). Latin *resedo* = 'to calm' and *luteus* = 'yellow'.

WELLINGTONIA *Sequoiadendron giganteum* (Cupressaceae), a coniferous neophyte tree introduced from California in 1853, common in lowland regions as a parkland ornamental, shade-intolerant, and requiring moist soils of poor to medium nutrient status. Wellingtonia was the original but invalid genus name honouring the recently deceased Duke of Wellington. Sequoia was a famous Native American of the Cherokee tribe. An etymological study published in 2012, however, suggested that the name may have originated from Latin *sequi* (= 'follow') since the number of seeds per cone in the newly classified genus *Sequoia* showed a mathematical sequence with the other genera in the suborder. *Dendron* is Greek for 'tree'.

WELS *Silurus glanis* (Siluridae), a fish reaching lengths of 1.3–1.6 m in England (up to 5 m in its native central and eastern Europe), with a number of introductions in the late nineteenth century, but especially imported for sports angling, and escaping into various English and Welsh waterways, since the 1970s, living in deep, slow-flowing rivers and lakes. The common name comes from German. Greek *silouros* refers to a kind of large river fish, perhaps 'catfish', and New Latin *glanis* = 'catfish'.

WELSHMAN'S BUTTON *Sericostoma personatum* (Sericostomatidae), a common and widespread semivoltine caddis fly (one generation taking two years) of streams, rivers and stony lakes, preferably neutral to alkaline, on substrates of sand and gravel. Larvae are omnivorous but feed mostly as shredders, and construct tubular cases using sand grains. Adults fly in May–September. Greek *sericon* + *stoma* = 'silk' + 'mouth', and Latin *personatum* = 'masked'.

WENTLETRAP Common wentletrap *Epitonium clathrus* (Epitoniidae), a sea snail with a shell up to 4 cm length, on the south and west coasts of England, and west Wales, on sand

and mud substrates in the sublittoral, preying on anemones, migrating to the lower shore during spring and summer to spawn. 'Wenteltrap' is Dutch for spiral staircase, reflecting the vertical ridges on the shell. *Clathrus* is Latin for 'latticework'.

WHALE Large marine mammals within the infraorder Cetacea, in the suborders Mysticeti (baleen whales; see also rorqual) and Odontoceti (toothed whales, dolphins and porpoises). Beluga (or white whale), bowhead, fin, Gervais' beaked, humpback, minke, pygmy sperm, sei, Sowerbury's beaked and True's beaked whale are sometimes seen, some not uncommonly, in coastal waters, and are also evident from strandings. **Killer whale** is not a whale; see also orca.

Cuvier's beaked whale *Ziphius cavirostris* (Ziphiidae), body length 5–7 m, weight 2,500 kg, commonly stranded on the Atlantic coastline, for instance five animals dying in the Atlantic then drifting to Islay, Tiree, Mull and Lewis in the Hebrides in 2008, probably from acoustic interference by navy sonar. Also in 2008 came the first record of stranding from the Moray Firth. A northern range expansion in UK waters may explain the increasing number of strandings. Food is mainly squid, plus fish and crustaceans. They are protected under Schedule 5 of the WCA (1981) and the Nature Conservation (Scotland) Act (2004). The species in named after the French anatomist Georges Cuvier (1769–1832). Greek *xiphos* = 'sword', referring to the distinctive beak, and Latin *cavus* + *rostrum* = 'hollow' + 'beak', referring to the indentation on the head in front of the blowhole.

Northern bottlenose whale *Hyperoodon ampullatus* (Ziphiidae), male body length 8.5–9.8 m, females 7.5–8.7 m, weight 5,800–7,500 kg, with a bulbous melon (forehead) and a dolphin-like beak, hence its English name. In British waters it is usually sighted in the Faroe–Shetland Channel and Rockall Trough, but it has been reported from coastal areas. In January 2006 a disorientated individual swam up the Thames to Chelsea, London; it was lifted on to a barge with the aim of taking it back to sea, but it died following a convulsion. Its skeleton is now in the Natural History Museum. The preferred food is squid. The species is protected under Schedule 5 of the WCA (1981) and the Nature Conservation (Scotland) Act (2004). Greek *hyperoe* + *odontos* = 'upper mouth' + 'tooth' (misleading, as the only two teeth are in the lower jaw), and Latin *ampulla* = 'flask' or 'bottle'.

Sperm whale *Physeter macrocephalus* (Physeteridae), largest of the toothed whales, males up to 18 m long, females 11 m, the head representing up to a third of the length. The UK has the highest number of sightings in northern Europe. Five animals were stranded near Skegness, Lincolnshire, and a further specimen at Hunstanton, Norfolk, in January 2016, the highest number recorded since records began in 1913: these deep-sea whales probably became disorientated in shallow water where they were uncharacteristically looking for food. They are protected under Schedule 5 of the WCA (1981) and the Nature Conservation (Scotland) Act (2004). The head contains a liquid wax, spermaceti, from which the species gets its English name. Greek *physeter* = 'blower' or 'bellows' (a reference to the blow spout), and Latin for 'large head'.

WHEAT Cereals in the genus *Triticum* (Poaceae), Latin for 'wheat', which itself is from Old English *hwæte*.

Bread wheat *Triticum aestivum*, one of the commonest crops of lowland arable land this annual also occurs as a casual in cereal fields, roadsides and waste ground as a relic of crops and as a birdseed alien. *Aestivus* is Latin for 'relating to summer'.

Durum wheat, pasta wheat *Triticum durum*, a minor crop in southern England, found as a relic of cultivation or grain alien in a few paces. *Durum* is Latin for 'hard'.

Rivet wheat *Triticum turgidum*, an annual rarely grown as a crop (mainly for animal feed) but scattered and occasional in Britain as a relic of cultivation in fields and on roadsides, and also on waste ground originating as a contaminant of grain and from birdseed. *Turgidum* is Latin for 'swollen'.

WHEATEAR *Oenanthe oenanthe* (Turdidae), wingspan 29 cm, length 15 cm, weight 24 g, an insectivorous migrant breeder, wintering in central Africa, mainly found in March–October on moorland and heathland in western and northern Britain and western Ireland, with smaller numbers in southern and eastern England, with an estimated 230,000 pairs (160,000–300,000 pairs) in Britain in 2009, numbers declining by 11% between 1995 and 2014 (BBS data). Nests, in rock crevices or disused burrows, have 5–6 eggs which hatch after around 14 days, the altricial chicks fledging in 16–17 days. 'Wheatear' is a sixteenth-century corruption of 'white' and 'arse', referring to the bird's white rump. Greek *oenos* + *anthos* = 'wine' + 'flower', and refers to the bird's return to Greece in spring as grape-vines blossom. Other wheatear species are scarce vagrants.

WHELK Predatory marine gastropods, from Old English *wioloc* and *weoloc*, a number of other species found from the low sublittoral to greater depths.

Common whelk *Bittium undatum* (Cerithiidae), a prosobranch, shell length 10 cm, width 6 cm, widespread, occasionally intertidal but mainly subtidal down to 1,200 m, on muddy sand, gravel and rock, a scavenger taking carrion as well a predator on bivalves and polychaetes. Empty egg-masses ('sea wash balls') are often found on the strand-line, sometimes mistaken for sponges. *Undatum* is Latin for 'wavy'.

Dog whelk *Nucella lapillus* (Muricidae), shell height 3–6 cm, width 2 cm, common and widespread on rocky shores from the midshore downwards, rarely in the sublittoral though they may be abundant in areas exposed to strong tidal stress. They are gregarious among barnacles and mussels on which they feed, drilling a hole through the latter's shell to pour in digestive enzymes before sucking up the resultant 'soup'. The binomial is from Latin for 'little nut' and 'stone'.

Needle whelk *B. reticulatum*, shell length 1 cm, on the coasts of south-west England, north-west Wales, west Scotland and south and west Ireland, grazing on microalgae on sandy and muddy shores, and on rocks and stones, from the lower shore to depths down to 250 m. *Reticulatum* is Latin for 'like a net'.

Netted dog whelk *Nassarius reticulatus* (Nassariidae), shell up to 3 cm high by 6 mm wide, common and widespread in the lower rocky shore and sublittorally to 15 m on soft sediments, feeding on carrion. *Nassarius* is Latin for a narrow-necked fish basket, *reticulatus* for 'like a net'.

Thick-lipped dog whelk *Na. incrassatus*, shell up to 12 mm high by 6 mm wide, common and widespread from low tide level in silted areas of rocky shores, feeding on bryozoans, and is most abundant in the shallow sublittoral. *Incrassatus* is Latin for 'thickened'.

WHIMBREL *Numenius phaeopus* (Scolopacidae), a Red List wading bird with a wingspan 82 cm, length 41 cm, weight 430 g, breeding from April to August in north Scotland, Orkney and Shetland, with a decline by >50% between 1995 and 2007, when 400–500 pairs were noted. It is also a scarce passage migrant to British and Irish coastal areas in spring and autumn on its way from and to its wintering areas in South Africa. The nest on a bare scrape usually contains four eggs, incubation taking 27–8 days, and fledging of the precocial young 35–40 days. The breeding grounds diet is of insects, snails and slugs, on passage, of small crabs, shrimps, molluscs and worms, taken from the surface of or just under the tidal or estuarine mud. 'Whimbrel' is supposedly imitative of the call. Greek *neos* + *mene* = 'new' + 'moon', a reference to the crescent-shaped bill, and *phaios* + *pous* = 'dusky' + 'foot'. **Hudsonian whimbrel** *N. hudsonicus* and **little whimbrel** *N. minutus* are both scarce accidentals.

WHIN Scottish name for gorse.

Petty whin *Genista anglica* (Fabaceae), a small spiny shrub, found in Britain in lowlands on damp grass-heath and around the drier fringes of bogs. In upland areas it occurs in heathy, damp, unimproved pastures and moorland. Away from the New Forest, with its long history of grazing, there has been a substantial decline. The binomial is Latin for 'broom' and 'English'.

WHINCHAT *Saxicola rubetra* (Turdidae), wingspan 22 cm, length 12 cm, weight 17 g, an insectivorous migrant breeder in parts of northern and western Britain, with a few also found in Ireland, in upland grassland with scrub, grassy heathland and moorland. Suitable habitat has become sandwiched between intensive agriculture at lower levels and higher land unsuitable for breeding (April–September), and also limited by aspect. It winters in central and southern Africa. There were 47,000 pairs (19,000–75,000) in Britain in 2009. Numbers declined by 52% between 1970 and 2015 and by 53% between 1995 and 2015, but showed a slight increase (85%) in 2010–15. It was moved from the Amber to the Red list in 2015. It is also a passage migrant. The ground nest contains 4–7 eggs which hatch after around 13 days, the altricial young fledging in 14–15 days. 'Whin' is 'gorse'; 'chat' is an onomatopoeic representation of the bird's call. Latin *saxum* + *colere* = 'rock' + 'to dwell', and modern Latin *rubetra* for a type of small bird.

WHIPWORT Foliose liverworts in the genera *Bazzania* (Lepidoziaceae) and *Mastigophora* (Masigophoraceae) (Greek *mastigos* = 'whip').

Arch-leaved whipwort *B. pearsonii*, on humus and peat in moist woodland, moorland and rock ledges in the Highlands and Islands and in western Ireland, occasionally in northern England and west Wales.

Greater whipwort *B. trilobata*, a calcifuge occasional in upland regions, rare in parts of southern England, on rocks, trees and logs in oakwood, peat on moorland. *Trilobata* is Latin for 'three-lobed'.

Lesser whipwort *B. tricrenata*, in upland regions on rocky, heather-covered slopes, and among boulders in woodland, frequent in the Highlands, occasional in northern England, Wales and Ireland. *Tricrenata* is Latin for 'three-notched'.

Wood's whipwort *M. woodsii*, in western Scotland and western Ireland in acidic, oceanic-montane heaths in mixed turf, often among dwarf shrubs and boulders.

WHITE (BUTTERFLY) Species of *Pieris* (one of the Muses in Greek mythology) and *Leptidea* (Greek *leptos* + *eidos* = 'thin' + 'appearance'), both Pieridae, together with one species of *Melanargia* (Greek *melanos* + *arges* = 'black' + 'white') in the Nymphalidae.

Cabbage white Synonym of both the large and small white butterflies. Larvae of the large white prefer to feed on the outer leaves of the food plant, those of small white on leaves closer to the heart.

Cryptic wood white *L. juvernica* (Dismorphiinae), wingspan 42 mm, taxonomically separated from wood white and Réal's wood white in 2011, and widespread in Ireland in open habitats such as scrubby grassland and verges, where larvae and adults both feed on trefoils and vetches, adults adding the nectar of other species such as bramble and buttercups. Juverna is a variant of Hibernia, Latin for Ireland.

Green-veined white *P. napi* (Pierinae), wingspan 40–50 mm, very common, with numbers increasing by 72% between 2005 and 2014, and arguably the most widespread butterfly in the British Isles, with subspecies in Ireland, Scotland, and England and Wales (*britannica*, *thomsoni* and *sabellicae*, respectively), found in grasslands, woodland rides, hedgerows, parks and gardens, usually (but not exclusively) in damp areas. Both larvae and adults feed on a range of plant species, the former favouring crucifers. Adults fly in May–September. *Brassica napi* is rape, a food plant.

Large white *P. brassicae*, wingspan 57–66 mm, common (though numbers declined by 28% over 2005–14) and one of the most widespread butterfly species in the British Isles, in flight in April–October. The female is distinguished from the male by two black spots, together with a black dash on the forewing upper side. Favourite larval foods include crucifers, and since

its wide range of habitats includes gardens where it is found, for example, on cabbage and brussel sprout it is often viewed as a pest. (It is one of the butterflies also called cabbage white.) Adult diet is very varied. *Brassicae* reflects the brassica food plants.

Marbled white *M. galathea* (Satyrinae), wingspan 53–8 mm, found in often sizeable colonies, flying in June–August, south-east of a line between south Wales and north-east Yorkshire, though it is rare in East Anglia. Largest colonies are found on chalk and limestone downland, but other unimproved grass-lands such as field margins and verges are also suitable. Young larvae feed on grasses during the day, more mature caterpillars at night, especially on fescues (red fescue may be essential) and Yorkshire-fog; adults show a preference for purple flowers, such as thistles, marjoram, field scabious and knapweeds, but are also found on clovers and yarrow. The species has extended its range northwards and eastwards in recent years (overall by 26% over 2005–14, with numbers increasing by 25%). Galatea was a nymph in Greek mythology, the statue of whom, sculpted by Pygmalion, came to life.

Réal's wood white Former name of cryptic wood white before taxonomic separation from wood white.

Small white *P. rapae*, wingspan 46–55 mm, very common in a variety of habitats, and one of the most widespread butterflies in the British Isles, though less common in northern Scotland. Numbers nevertheless did decline by 25% over 1976–14, but saw a resurgence by 9% in 2005–14. Favourite larval foods include crucifers and other plants with white flowers, and since its wide range of habitats includes gardens, where it is found, for example, on cabbage and nasturtium this is often viewed as a pest. (It is one of the butterflies also called cabbage white.) Adults fly in April–October. Diet is very varied. *Brassica rapa* is wild turnip, a food plant.

Wood white *L. sinapis*, wingspan 36–48 mm, with a scattered distribution in woodland rides and margins in central and southern England and the Welsh Borders, in meadow and other open habitats in south-west England, and on limestone pavement on the Burren, Ireland. Larval food plants include vetches and trefoils. Nectar sources include a variety of flowers, favourites being bramble, bugle, ragged-robin and bird's-foot-trefoil. In hot weather, males sometimes take mineral salts from puddles. Flight is in May–August. Reduction of coppicing has led to a decline in range and numbers (both decreasing by 89% over 1976–2014, and by 25% and 18%, respectively, over 2005–14), so this is a priority species for conservation. *Sinapis* is a synonym of *Brassica*, a putative food plant.

WHITE (MOTH) **Bordered white** *Bupalus piniaria* (Geom-etridae, Ennominae), wingspan 28–35 mm, common and wide-spread, day-flying (May–June, to August in the north), with a yellow form tending to occur in the south, a white forms in northern England and Scotland, with some gradation. Larvae feed on Scots pine and other conifers. The binomial is from the Greek sculptor Bupalus (fl.540 BCE), and the food plant *pinus* = 'pine'.

Broad-barred white *Hecatera bicolorata* (Noctuidae, Haden-inae), wingspan 28–35 mm, nocturnal, widespread and common in dunes, shingle beaches, rough grassland, chalk downland and gardens, more scattered in Scotland and Ireland. Adults fly in June–August. Larvae feed on such plants as hawk's-beards and mouse-ear-hawkweed. Hecate was the Greek goddess of the Underworld. Latin *bicolorata* = 'two-coloured'.

WHITE-MARKED (MOTH) *Cerastis leucographa* (Noctu-idae, Noctuinae), wingspan 34–8 mm, local in broadleaf woodland, mature hedgerows and scrub, discontinuously in England from Kent to Yorkshire, and westwards to the Welsh Border. Adults fly at night in March–April, visiting sallow flowers. Larvae feed on a variety of plants. Greek *cerastes* = 'horned', from the male antennae, and *leucos* + *graphe* = 'white' + 'marking', from the wing pattern.

WHITE-POINT *Mythimna albipuncta* (Noctuidae, Hadeninae), wingspan 30–5 mm, a migrant nocturnal moth appearing in southern England from June to September, in most years in sufficient numbers to breed. It has increased dramatically since around 2000 and continues to extend its range northwards, probably as a result of increasing temperatures, and can be widespread in East Anglia. Mithimna is a town on the Greek island of Lesbos. *Albipuncta* is Latin for 'white-spotted'.

WHITE-SPECK *Mythimna unipuncta* (Noctuidae, Hadeninae), a migrant nocturnal moth appearing on the south and south-west coasts of England and in Wales in recent years, particularly in August–October, often in sufficient numbers to breed (possibly resident in Cornwall), and around the Irish coastline (last recorded in 2008). Mithimna is a town on the Greek island of Lesbos. *Unipuncta* is Latin for 'one-spotted'.

WHITEBEAM Deciduous trees in the genus *Sorbus* (Rosaceae), Latin for 'rowan'. Also includes the rare endemics **least whitebeam** and **Ley's whitebeam** *S. minima* and *S. leyana* (both limestone crags, Brecon), **Llangollen whitebeam** *S. cuneifolia* (limestone crags, Llangollen, Denbighshire), and **Scannell's whitebeam** *S. scannelliana* (in limestone wood, Ross Island, Co. Kerry).

Arran whitebeam *S. arranensis*, an endemic shrub or small tree, the 400 or so specimens known only from streamsides and granite outcrops on Arran.

Broad-leaved whitebeam *S. latifolia*, a medium-sized neophyte introduced from Europe in 1866, sometimes planted in parks and large gardens, occasionally naturalised in woodland, scrub, rocky slopes and riverbanks. This species has a number of hybrids, and a number of rare and endemic aggregate forms: **Bristol**, **broad-leaved**, **Devon**, **orange**, **sharp-toothed** and **Somerset whitebeams**. *Latifolia* is Latin for 'broad-leaved'.

Common whitebeam *S. aria*, common and widespread in Britain, more scattered in Ireland, in woodland, scrub and rocky habitats, mostly on calcareous soils, and often planted and naturalised. This species has a number of hybrids, and many rare and endemic aggregate forms: **bloody**, **Cheddar**, **Doward**, **Gough's**, **grey-leaved**, **Irish**, **Lancastrian**, **Leigh Woods**, **Llanthony**, **Margaret's**, **round-leaved**, **ship rock**, **Stirton's**, **Symonds Yat**, **thin-leaved**, **twin cliffs**, **Welsh**, **White's** and **Wilmott's whitebeams**. *Aria* is Latin for 'open space' or 'park'.

English whitebeam *S. anglica*, a shrub or small tree, found locally in south-west England, Wales and Co. Kerry in open rocky woods and scrub, and on cliffs and stony slopes, usually on Carboniferous limestone, but occasionally also on other base-rich rocks. Latin *anglica* = 'English'.

Swedish whitebeam *S. intermedia*, a widespread if scattered neophyte from Scandinavia, cultivated by 1789, now often planted as an ornamental, especially in town streets and parks, frequently self-sown in copses and on waste ground.

WHITEFISH Species in the genus *Coregonus*. See gwyniad, powan and schelly (some taxonomists maintaining the view that these are all a single species, common or European whitefish, *C. lavaretus*), pollan and vendace.

WHITEFLY Members of the family Aleyrodidae (Sternorryncha), with 22 species in 17 genera, many of which are widespread and common pests of crops and trees. Their powdery white secretions covering their body give them their common name. Wingspan is <4 mm.

WHITETHROAT *Sylvia communis* (Sylviidae), wingspan 20 cm, length 14 cm, weight 16 cm, a common and widespread insectivorous migrant breeder, wintering in sub-Sahara, with 1.1 million territories in Britain in 2009. A drought in the Sahel region of Africa in 1968 caused a 90% drop in whitethroat numbers in Britain, and they have probably still not fully recovered, though there was a 33% increase between 1995 and 2015 (though including a non-significant –3% over 2010–15). The cup nests built near the

ground in bramble or scrub usually contain 4–5 eggs, incubation lasting 12–13 days, and fledging of the altricial young taking 12–14 days. Diet is insectivorous, with fruit also taken in autumn to boost fat reserves before migration. Latin *Sylvia* = a woodland sprite, in turn from *silva* = 'wood', and *communis* = 'common'.

Lesser whitethroat *S. curruca*, wingspan 18 cm, length 13 cm, weight 12 g, a migrant breeder mainly found in woodland in England and Wales (its scarcity in Ireland making it Amber-listed there), with 74,000 territories in Britain in 2009, numbers increasing by 23% over 1970–2015, if remaining fairly stable (3% increase) over 2010–15. The cup nests built near the ground in bramble or scrub usually contain 3–5 eggs, incubation lasting 11–12 days, and fledging of the altricial young taking 12–13 days. Diet is insectivorous, with fruit also taken in autumn to boost fat reserves before migration. *Curruca* is a bird mentioned by Juvenal.

WHITEWASH Elder whitewash *Hyphodontia sambuci* (Hymenochaetales, Schizoporaceae), a common and widespread pathogenic fungus evident as a chalky-coloured film on dead branches, mainly of elder (*Sambucus*, giving the specific name). Greek *hyphe* + *odontos* = 'web' + 'tooth'.

WHITLOWGRASS Species of *Draba* (Greek *drabe* = 'sharp' or 'acrid', describing the burning taste of the leaves) and *Erophila* (*eros* + *philos* = 'love' + 'loving') in the Brassicaceae. Plants were a supposed cure for a whitlow, an infection of the fingertip.

Common whitlowgrass *E. verna*, a widespread, locally common ephemeral annual of limestone rock, dunes, shingle, quarries, spoil tips, walls, waste ground and pavement cracks. *Verna* is Latin for 'spring'.

Glabrous whitlowgrass *E. glabrescens*, widespread and locally common on limestone, chalk downland, rocky streamsides, river shingle, dunes, sandy grassland, walls, verges and pavement cracks.

Hairy whitlowgrass *E. majuscula*, an ephemeral annual with a scattered distribution, commoner in the south, on limestone rock and calcareous grassland (where it may grow on ant hills), sandy habitat, walls and railway lines. *Majuscula* is a diminutive of Latin *majus* = 'large'.

Hoary whitlowgrass *D. incana*, a biennial or perennial tufted herb, mainly found in the Highlands, local in northern England, north-west Wales, and north and west Ireland, usually on limestone, occasionally in open grassland on thin dry soils. It also occurs on sandy habitats, base-rich mica-schists and igneous rocks, often in upland habitat. *Incana* is Latin for 'hairy' or 'grey'.

Rock whitlowgrass *D. norvegica*, a perennial tufted herb of base-rich rocks, local in the Highlands on cliff ledges, crevices and scree. Latin *norvegica* = 'Norwegian'.

Wall whitlowgrass *D. muralis*, a winter-annual, widespread, found as a native in south-west and northern England, on limestone rocks and screes. Elsewhere it is a colonist on old walls, forest tracks and railways, and as a garden weed where summer-dry, winter-moist conditions are found. *Muralis* is from Latin for 'wall'.

Yellow whitlowgrass *D. aizoides*, a short-lived, cushion-forming, perennial herb restricted to limestone rocks in the Gower Peninsula, south Wales. This is the county flower of Glamorgan as voted by the public in a 2002 survey by Plantlife. *Aizoides* means 'like the genus *Aizoon*'.

WHORL-GRASS *Catabrosa aquatica* (Poaceae), a stoloniferous herb of damp meadow, marsh, pond edges, canals and slow-flowing streams throughout Britain and, as var. *uniflora*, widespread in west and north-west Scotland, perhaps also in west Ireland, on wet coastal sand.

WHORL SNAIL Pulmonates in the family Vertiginidae in the genera *Vertigo* (Latin for 'turning around') and *Truncatellina* (*trunca* = 'footless'), shell lengths 1.4–3 mm, widths 0.7–1.6 mm.

Scattered and uncommon species include **Lilljeborg's whorl snail** *V. lilljeborgi* (Vulnerable in the RDB, in upland Wales, the Lake District, Scotland and Ireland, in calcareous flushes and swampy lake shores), **mountain whorl snail** *V. alpestris* (in England, mainly in the Lake District, and Wales found on old stone walls, in Scotland among moss in limestone rubble), **narrow-mouthed whorl snail** *V. angustior* (with 26 sites recorded in the UK, Endangered in the RDB, with highest populations around estuarine margins in East Anglia; and 20 sites, most declining in viability, in Ireland), **round-mouthed whorl snail** *V. genesii* (first recorded in 1979, found in northern England and the Highlands, Endangered in the RDB, living at the base of sedges and in moss), and **wall whorl snail** *V. pusilla* (sinistral, declining in Britain, local and rare in Ireland, in woodland, scrub and rough pasture, south-facing chalk or limestone slopes, and dunes).

Rarities include **British whorl snail** *T. callicratis* (from Devon to the Isle of Wight), **cross whorl snail** *V. modesta* (a glacial relict discovered in 1993, found at two high mountain sites in the Highlands), **cylindrical whorl snail** *T. cylindrica* (five sites in Bedfordshire, Norfolk and Yorkshire), and **Geyer's whorl snail** *V. geyeri* (discovered in 1978 in the Lake District).

Common whorl snail *V. pygmaea*, common and widespread in open habitats such as damp pasture and dune grassland, as well as gardens and hedgerows. *Pygmaea* is Latin for 'dwarfish'.

Desmoulin's whorl snail *V. moulinsiana*, scattered and locally common (with 150–60 known sites) in England, Wales and Ireland, but listed as Endangered in the RDB, in base-rich swamp and marsh with tall emergent plants. In 1996 presence of this species on the site of the planned Newbury bypass caused construction to be postponed until the snails had been moved to a new habitat two miles away, but by 2006 they had died out there. Alexandre des Moulins (1798–1875) was a French naturalist.

Meadow whorl snail *V. antivertigo*, widespread and locally common in damp meadow, and river and lake margins.

Striated whorl snail *V. substriata*, common and widespread in damp grassland and scrub, especially on calcareous substrates. *Substriata* is Latin for 'near-striped'.

WHORTLEBERRY Southern synonym of bilberry.

WIDOW (FUNGUS) **Weeping widow** *Lacrymaria lacryma-bunda* (Agaricales, Psathyrellaceae), common, widespread, poorly edible, found in summer and autumn in soil under grass in woodland clearings and on verges, and in parkland. The English name comes from the black watery droplets that appear at the cap rim and on the edges of the gills when they are moist. The binomial derives from Latin *lacrimae* = 'tears'.

WIG 1) **Landlady's wig** *Ahnfeltia plicata* (Ahnfeltiaceae), a common and widespread red seaweed with fronds 15 cm long, found as turf on shallow sublittoral bedrock and in rock pools on the lower shore, often partly buried by sand, its bushy appearance prompting the English name. Latin *plicata* = 'folded into pleats or furrows'.

2) **Lawyer's wig** Alternative name for shaggy inkcap fungus.

WIGEON *Anas penelope* (Anatidae), wingspan 80 cm, length 48 cm, male weight 800 g, females 650 g, around 170 pairs breeding in northern England, and central and northern Scotland in 2010. With large numbers of wintering Icelandic, Scandinavian and Russian birds (440,000 individuals) found throughout the British Isles it is an Amber List species. Nests are on the ground, clutch size is 8–9, incubation takes 24–5 days, and the precocial young fledge in 40–5 days. Feeding is on aquatic plants and grasses. Suggestions for the etymology of 'wigeon' are unconvincing. *Anas* and *penelops* are respectively the Latin and Greek for 'duck'.

WILD BOAR *Sus scrofa* (Suidae), head-body length up to 200 cm, tail 15–40 cm, exterminated in Britain in medieval times, but imported from the Continent for meat farming in the 1980s, with escaped animals establishing themselves in the wild in the early 1990s, mainly in Kent and the Forest of Dean (Gloucestershire, spreading into Herefordshire and Wales) but also in East Sussex, Dorset, possibly Devon and parts of Scotland. Estimated numbers in England in 2005 were <500,

but in the Forest of Dean alone an estimate of 600 animals was made in 2012, and a culling programme began, with targets of 100–200 animals p.a. In 2015 DEFRA conservatively estimated a population of around 1,000 animals in the UK. It is a mainly nocturnal omnivore, eating a variety of plants, including roots, seeds and fruits, carrion, young mammals, and the eggs and young of ground-nesting birds. As a farmed animal, wild boar are subject to the Dangerous Wild Animals Act 1976. In 2008 DEFRA published 'Feral wild boar in England: an action plan', setting out the Government's position that free-roaming wild boar are feral wild animals and do not belong to any individual. 'Boar' comes from Old English *bar*, itself of West Germanic origin. The binomial is Latin for 'pig' and 'sow'.

WILDCAT *Felis silvestris* ssp. *silvestris* (Felidae), an isolated island population of the European wildcat, male head-body length 790–950 mm, tail 30 cm, weight 3.8–7.3 kg, females 700–865 mm, tail 30 cm, weight 2.3–4.7 kg, found in the Highlands up to 800 m, but usually no higher than 650 m, rare and endangered, with the scattered populations declining for a number of reasons, critically because of hybridisation with domestic and feral cats, transmission of disease from such cats, and habitat loss and fragmentation. Data from 2012 indicated that 150 breeding pairs remained (though one pessimistic suggestion by the Scottish Wildcat Association was of a mere 35 individuals). Adults are solitary nocturnal or crepuscular hunters. Diet is mainly of rodents, hares, birds and especially rabbits, which can form up to 70% of the diet in eastern Scotland where wildcats favour the margins of moorland, pasture and woodland; in western Scotland they prefer uplands with rough grazing and moorland with limited pasture. Rocky habitat provides den shelters for breeding females. The main mating season is January–March. Gestation lasts 63–8 days. There is normally one litter a year, in April–May, with 1–8 (usually 3–4) kittens which begin to hunt with their mother when 10–12 weeks old, becoming fully weaned by 10–14 weeks. It was formerly listed on Schedule 5 for protection under the WCA (1981), but was removed through an amendment to the Conservation (Natural Habitats etc.) Regulations in Scotland in 2007. It was added to the Scottish Biodiversity List in 2005 as a Species of Principal Importance, and in 2007, it was added to the revised UK (BAP) list of priority species. In 2008 a 650 km² 'Wildcat Haven' was established by the Scottish Wildcat Association on the Ardnamurchan and Sunart peninsula on the west coast, in which feral cats were to be neutered or eliminated to prevent hybridisation, reported in 2014 to have been a success. In March 2016, plans were announced to create a further 18,100 km² of protected land. In 2013 it was reported that, as part of a £2 million conservation project, a programme of neutering and vaccinating feral cats would be undertaken by Scottish National Heritage in Angus Glens; Strathbogie, Aberdeenshire; Strathpeffer, Easter Ross; Strathavon, Moray; Northern Strathspey and Morvern, Lochaber. A further £973,000 was provided in 2014 through National Lottery funding. The binomial is Latin for 'wild cat'.

WILLOW (MOTH) Nocturnal moths in the family Noctuidae (Xyleninae).

Pale mottled willow *Caradrina clavipalpis*, wingspan 26–35 mm, widespread and common in grassland, farmland and gardens. Population levels increased by 275% between 1968 and 2007. Adults mainly fly in July–September. Larvae feed on seeds of grasses and cereals, including those that have been harvested. *Caradrina* is an old name for the Black Drin river in Albania. Latin *clavus* + *palpus* = 'club' + 'labial palp'.

Small mottled willow *Spodoptera exigua*, wingspan 26–32 mm, nocturnal, a common migrant from the continent appearing most frequently in southern coastal counties, less often inland and northwards to southern Scotland and the Inner Hebrides, and around the coastline of much of Ireland. Arrivals have been noted in most months from February to November, but peak arrival is in August–October. In some years several hundred may appear. Larvae are rarely evident. Greek *spodos* + *pteron* = 'wood ashes' + 'wing', from the forewing speckling, and Latin *exigua* = 'small'.

WILLOW (TREE) Deciduous trees or shrubs in the genus *Salix* (Salicaceae), a taxonomically complex genus with much hybridisation, 68 combinations being recognised by Stace (2010), of which 20 are hybrids between just three species. Once the flowers are pollinated by insects, the female catkins transform into woolly white seeds which are wind-dispersed. The slender, flexible stems of some species have been much used in basketry. Many taxa are being grown for biomass production. Larger stems were traditionally used to make small sailing boats. A hybrid of white willow and crack willow is used to make cricket bats. Traditionally willows were used to relieve pain: aspirin is derived from salicin, a compound found in the bark of all *Salix* species. Willows were once seen as trees of celebration, but recently they have become more associated with sadness and mourning. 'Willow' comes from Old English *welig*.

Almond willow *S. triandra*, an archaeophyte shrub or small tree, widespread if commonest in England and south-east Ireland, in damp habitats by streams and ponds and in marsh and osier beds, much planted for basketry. *Triandra* is Greek for 'three stamens'.

Bay willow *S. pentandra*, a large shrub or small tree, widespread but mainly in northern England north to central and north-east Scotland, and north Ireland, in marsh, fen and wet woodland, in winter-flooded dune slacks, and by ponds and streams. Male plants are widely planted as ornamentals. *Pentandra* is Greek for 'five stamens'.

Crack willow *S. fragilis*, a widely planted often pollarded archaeophyte tree growing in hedgerows, marsh, fen and wet woodland, and by ponds and ditches, as well as by streams and rivers to help stabilise their banks. The common name comes from the sound made when branches fall off; because the species often grows beside rivers, broken material is carried downstream, helping the trees propagate since the branches readily take root. This willow can tolerate polluted and salt-laden air. *Fragilis* is Latin for 'fragile' or 'brittle'.

Creeping willow *S. repens*, a common and widespread shrub, though scarce in the Midlands, growing in a range of habitats. The prostrate var. *argentea* is generally found in maritime heaths and heathy grassland, and on inland heaths and moorland. The prostrate var. *repens* is associated with dune slacks. The erect var. *fusca* is found in East Anglian fens. The species becomes more confined to moist or wet habitats in the south and east of its range. *Repens* is Latin for 'creeping'.

Cricket-bat willow See white willow (*Salix alba* var. *caerulea*) below.

Dark-leaved willow *S. myrsinifolia*, a shrub, occasionally a small tree, locally common in northern England, Scotland and northern Ireland, on rocks or gravelly riverbanks and lake shores, less frequently on marshy ground or by wet woodland margins. In Scotland it sometimes grows in wet dune slacks and as a dwarf shrub spreading on wet rock ledges. Latin *myrsinifolia* = 'with myrrh-like leaves'.

Downy willow *S. lapponum*, a shrub of moist or wet, moderately base-enriched sites on rocky mountain slopes and cliffs in the Highlands. It tolerates a wider range of soil conditions than most montane *Salix*, but is now largely confined to cliffs, and probably declining in status. *Lapponum* is Latin for 'from Lapland'.

Dwarf willow *S. herbacea*, a dwarf shrub of open, often bryophyte-rich communities in conditions of extreme exposure, mainly in the Highlands, but also in north Wales, north-west England, and parts of Ireland. It grows on stony ground and scree, and locally on ledges and in montane grass-heath, from near sea level at Fethaland (Shetland) to 1,310 m on Ben Nevis (Westerness), but usually >600 m. *Herbacea* is Latin for 'herbaceous'.

Eared willow *S. aurita*, a widespread and locally common shrub (though scarce in the east Midlands) growing on acidic soils on heathland and moorland, in scrub, and on rocky streamsides and hills. *Aurita* is Latin for 'eared'.

Ehrhart's willow *Salix* x *ehrhartiana* (*S. pentandra* x *S. alba*), a small neophyte tree of European garden origin, scattered in England, planted on riversides, pond edges and ditches, and on verges and in hedges. Only male clones are known in this country, but it can become naturalised through suckering. Jakob Ehrhart (1742–95) was a German botanist.

European violet willow *S. daphnoides*, a widespread and scattered tall Scandinavian neophyte shrub or small tree most frequent as an ornamental, having been planted for amenity in parks and landscaped areas since 1829, but sometimes naturalised. *Daphnoides* means resembling the genus *Daphne*.

Goat willow *S. caprea*, also known as **great sallow**, a very common and widespread shrub or small tree found in damp places in woodland and woodland edge, scrub, hedgerows and waste ground, and around lakes and streams. Traditional uses for its wood included clothes pegs, while the leaves were used as a winter feed for cattle. The wood burns well, making a good fuel. *Capra* is Latin for 'goat' (not from *caprea* = 'roe deer').

Green-leaved willow *Salix* x *rubra* (*S. purpurea* x *S. viminalis*), a widespread if scattered shrub or small tree often found with the parents in osier beds (it is much used in basketry) and other thickets, and on roadsides. *Rubra* is Latin for 'red'.

Grey willow *S. cinerea* (especially ssp. *oleifolia*), also known as **common sallow**, a very common and widespread shrub or small tree found in damp habitats such as woodland, marsh, fen and waste ground, and by streams and bog. *Cinerea* is Laton for 'ash-grey'.

Heart-leaved willow *S. eriocephala*, a North American neophyte shrub naturalised in bog in East Sussex, Warwickshire, Lancashire, Dyfed (Cardiganshire) and Co. Londonderry. Greek *eriocephala* = 'wool-headed'.

Holme willow *Salix* x *calodendron* (*S. viminalis* x *S. caprea* x *S. cinerea*), a locally frequent and widespread shrub or small tree often found in damp habitats, now planted for biomass production. Greek *calodendron* = 'beautiful wood'.

Laurel-leaved willow *Salix* x *laurina* (*S. cinerea* x *S. phylicifolia*), a shrub found in northern England and Scotland, occasionally in Ireland, often growing with its parents in damp habitats or as a garden relic.

Mountain willow *S. arbuscula*, a shrub locally abundant on base-rich substrates on Scottish mountains, in moist or wet habitats, mostly in flushes, on gravelly soil near streams and on damp calcareous rock ledges. *Arbuscula* is Latin for 'small tree'.

Net-leaved willow *S. reticulata*, a creeping dwarf shrub locally found in the Highlands, growing on base-rich montane ledges of limestone or calcareous schist. *Reticulata* is Latin for 'netted'.

Olive willow *S. elaeagnos* (Salicaceae), a deciduous south European neophyte shrub or much-branched tree planted (since 1820) or as a relic of cultivation in damp sites, scattered in Britain, occasionally naturalised. It grows poorly in shade or in competition with other shrubs. Greek *elaia* + *agnos* = 'olive' + 'pure'.

Purple willow *S. purpurea*, a locally common and widespread shrub or small tree on wet ground, woodland edges and damp hillsides, by streams and rivers, on river shingle, in marsh and fen, and much planted as an osier for basketry.

Pussy willow Alternative name for both goat and grey willows, or sallows, so called because of the silky male catkins that resemble cat paws.

Sachalin willow *S. udensis*, an East Asian neophyte shrub or small tree cultivated since the 1950s, scarce and scattered as a garden throw-out in damp woodland and on roadsides, riverbanks and waste ground. Sakhalin is a large island off the Russian Pacific coast, Uden a district in Siberia.

Sharp-stipuled willow *Salix* x *mollissima* (*S. triandra* x *S. viminalis*), an archaeophyte shrub which grows in damp places, much planted for basketry, especially in Northern Ireland. Var. *undulata* has recently locally been planted as a biomass plant. It often occurs in the absence of both parents; in such cases it is introduced. *Mollissima* is Latin for 'very soft'.

Shiny-leaved willow *Salix* x *meyeriana* (*S. pentandra* x *S. fragilis*), a tall shrub or small tree, scattered in England, Wales and eastern Ireland, scarce elsewhere, in moist, often water-logged habitats including river margins and ditches. Plants of this hybrid between native and alien parents are usually of cultivated origin, planted as an ornamental. *Acutifolia* is Latin for 'sharp/pointed leaves'.

Siberian violet-willow *S. acutifolia*, a widespread, occasionally found tall Scandinavian neophyte shrub or small tree, males having been planted for amenity in parks and landscaped areas since 1798, and also on roadsides and riverbanks, rarely found in semi-natural habitats, notably on dunes in south Lancashire. Latin *acutifolia* = 'sharp-leaved'.

Tea-leaved willow *S. phylicifolia*, a shrub or small tree widespread in northern England and Scotland, growing by ponds, streams and rivers, and in damp rocky places, preferring base-rich soils and sometimes associated with Carboniferous limestone. In Co. Leitrim and Co. Sligo it is a montane species. *Phylicifolia* means having leaves like the genus *Phylica*.

Violet willow See European and Siberian violet willows, both above.

Weeping willow *Salix* x *sepulcris* (*S. alba* x *S. babylonica*), a long-lived deciduous neophyte tree, planted in ornamental and amenity sites in parks and on riversides since 1869, superseding the parental *Salix babylonica* by the early twentieth century, now widespread and common in England, less common elsewhere.

White willow *S. alba*, a common and widespread archaeophyte tree growing in marsh and by ponds, ditches, streams and rivers, and much planted. Var. *caerulea* (**cricket-bat willow**) is vulnerable to watermark disease involving the bacterium *Erwinia salicis*. *Alba* is Latin for 'white'.

Whortle-leaved willow *S. myrsinites*, a low, spreading shrub of the Highlands, mainly in moist base-enriched montane sites, restricted to ungrazed or lightly grazed areas. *Myrsinites* means like the genus *Myrsine*.

Woolly willow *S. lanata*, a shrub found in the Highlands on damp base-rich mountain rock, usually calcareous schist, with late snow-lie. Populations have dwindled because of increased grazing by deer and sheep: in places numbers have been reduced to single plants, and other populations have been lost. Fencing is being used to try to arrest the decline. *Lanata* is Latin for 'woolly'.

WILLOW BEETLE Members of the family Chrysomelidae. **Blue willow beetle** *Phratora vulgatissima* and **brassy willow beetle** *P. vitellinae* (Chrysomelinae), 4–5.5 mm length, common and widespread, both adults and larvae feeding on willow leaves. Adults mainly overwinter under bark. Greek *phratoria* = 'clan', and Latin *vulgatissima* = 'very common', *vitellus* = 'yolk' or 'little calf'.

Brown willow beetle *Galerucella lineola* (Galerucinae), 5–6 mm, common and widespread, mainly living on willows but also on alder, hazel, poplars and birches. Adults hibernate in sedge tussocks, under bark on mature willows and dead wood, and in umbellifer stems. Latin *galerum* + *eruca* = 'helmet' + 'caterpillar', and *lineola* = 'small line'.

WILLOW FLY Fisherman's name for the stonefly *Leuctra geniculata*.

WILLOWHERB Herbaceous perennial species of *Epilobium*, from Greek *epi* + *lobos* = 'upon' + 'capsule', the flower and capsule appearing together, the corolla borne on the end of the ovary, and *Chamerion*, from Greek *chamae* + *nerium* = 'low' + 'oleander' (Onagraceae).

Alpine willowherb *E. anagallidifolium*, stoloniferous, mainly found in the Highlands, from 155 m near Inchnadamph (west Sutherland) to 1,190 m on Ben Lawers (mid-Perth), but also the Southern Uplands and north Pennines, in mossy mountain flushes, on steep wet slopes and by streams. Latin *anagallidifolium* = 'pimpernel-leaved'.

American willowherb *E. ciliatum*, a common and wide-spread North American neophyte found on disturbed ground, for example gardens, waste ground, walls and pavement cracks. It spreads readily by wind-borne seed, and has become the commonest willowherb in England and Wales, often colonising newly disturbed sites and continuing to expand its range. Latin *cilia* = 'hair'.

Broad-leaved willowherb *E. montanum*, common and widespread in woods, hedge banks, gardens and waste ground, sometimes on rock ledges, often in shade.

Bronzy willowherb *E. komarovianum*, a New Zealand neophyte occasionally found as a casual garden weed. Komarovo is a settlement near St Petersburg, Russia.

Chickweed willowherb *E. alsinifolium*, found in the Highlands, Cumbria and north-west Wales, by mountain streams or on wet mountain ledges, often in moss carpets. Latin *alsinifolium* = 'chickweed-leaved'.

Great willowherb *E. hirsutum*, common and widespread (except for central and north Scotland) in open habitats, including marsh, stream and pond margins, damp woodland edges and waste ground. It thrives in wet, fertile, neutral to basic habitats, but can also tolerate dry ground. It spreads from wind-blown seed, or from branching rhizomes which may result in dense stands. Latin *hirsutum* = 'hairy'.

Hoary willowherb *E. parviflorum*, frequent throughout lowland Britain in marsh, fen, streamsides and other disturbed wet places, spreading by wind-blown seed, but also growing in dry habitats, including quarries and waste ground, and as a street weed. Latin *parviflorum* = 'small-flowered'.

Marsh willowherb *E. palustre*, widespread and locally common, stoloniferous, on wet acidic sites, in bog, marsh and ditches, and on stream or lake margins. It spreads by wind-borne seed and by turions which form at the ends of the stolons and can be carried away by winter floods. *Palustre* is from Latin for 'marsh'.

New Zealand willowherb *E. brunnescens*, a creeping neophyte, first recorded in the wild in Edinburgh in 1904, now widespread, though common only in south-west and northern England, Wales, Scotland, and (excepting central) Ireland, in moist open habitats, on gravel or stony soils. It grows on streamsides, ditches, scree, quarries, railway sidings and walls, spreading by rooting at the nodes, and by seed, and becoming naturalised in many remote localities. Latin *brunneus* = 'brown'.

Pale willowherb *E. roseum*, widespread and particularly frequent in England and central Scotland in moist disturbed habitats, including damp woodland, shaded banks, street gutters, waste ground and gardens. Latin *roseum* = 'rose-red'.

Rockery willowherb *E. pedunculare*, a New Zealand neophyte first recorded in 1953, found on damp roadsides in a few places in central Scotland and western Ireland, and a rare garden weed elsewhere. *Pedunculare* is Latin for 'foot' (peduncle).

Rosebay willowherb *C. angustifolium*, common and wide-spread, rhizomatous, on moderately fertile soils, forming dense stands on disturbed, often burnt ground, in woodland rides and clearings, on heaths and dunes, along tracksides, roadsides and railways, and in waste places and gardens; also in upland areas on rock ledges and scree. It rapidly colonised derelict bomb sites during World War II, and remains a common plant in many towns and cities. This is the 'county' flower of London as voted by the public in a 2002 survey by Plantlife. Latin *angustifolium* = 'narrow-leaved'.

Short-fruited willowherb *E. obscurum*, on cultivated and waste ground, marsh, streamsides and woodland edges, with a preference for damp habitats, but tolerating dry ones, spreading by wind-borne seed and long stolons. Common and widespread, but with some declines recently noted in eastern England and western Scotland. Latin *obscurum* = 'dusky'.

Spear-leaved willowherb *E. lanceolatum*, in southern England (mainly the south-west) and south Wales, growing in dry habitats such as roadsides, walls, quarries, streets and dunes. Elsewhere, it is more commonly a garden weed. Latin *lanceolatum* = 'lance-shaped'.

Square-stalked willowherb *E. tetragonum*, locally common in England, Wales and eastern Ireland in waste ground, gardens, quarries, streamsides, hedgerows and woodland edges. Greek *tetragonum* = 'four-angled'.

WINEBERRY Japanese wineberry *Rubus phoenicolasius* (Rosaceae), an erect to spreading deciduous neophyte east Asian shrub cultivated from the 1870s for ornament and fruit, scattered mainly in southern England, naturalised on roadsides and in hedges, woodland, scrub and rough grassland. Latin *rubus* = 'bramble', and Greek *phoinix* + *lasios* = 'purple-red' + 'hairy'.

WING (CADDIS FLY) See grouse wing.

WING (MOTH) See scorched wing.

WING-MOSS Bird's-foot wing-moss *Pterogonium gracile* (Leucodontaceae), widespread, more common in western Britain and western Ireland, often abundant on base-rich igneous rocks, and also found on acidic rocks, occasionally on base-rich sedimentary rocks. It also grows on bark of mature trees in open woodland. Greek *pteron* + *gonia* = 'wing' + 'angle', and Latin *gracile* = 'slender'.

Capillary wing-moss *Pterigynandrum filiforme* (Pterigynandraceae), locally frequent in upland Scotland on exposed sandstone, limestone and igneous rocks. It is also occasional on mature trees with base-rich bark, especially ash, as with most of the small number of English and Welsh records. Greek *pteron* + *gyne* + *andros* = 'wing' + 'woman' + 'man', and Latin *filiforme* = 'thread-like'.

Pendulous wing-moss *Antitrichia curtipendula* (Leucodontaceae), scattered and widespread, commoner in upland regions on rock and scree. It also grows on trees in open woodland or scrub. Greek *anti* + *trichos* = 'against' + 'hair', and Latin *curtipendula* = 'short pendulous'.

WING WEED Species of red seaweed.

Branched wing weed *Pterocladiella capillacea* (Pterocladiaceae), in large pools and lagoons in the lower littoral and shallow subtidal from Anglesey to Dorset, the Isle of Man and around Ireland. Greek *pteron* + *clados* = 'wing' + 'branch', and Latin *capillacea* = 'thread-like'.

Feathered wing weed *Ptilota gunneri* (Wrangeliaceae), a subtidal tufted seaweed with fronds up to 30 cm long, mainly found in north-east England, Scotland and Ireland, epiphytic on kelp. Greek *pilotos* = 'winged'.

WINGNUT Caucasian wingnut *Pterocarya fraxinifolia* (Juglandaceae), a fast-growing deciduous neophyte tree introduced by 1782, planted in woodland, on field borders and railway banks, occasionally naturalised, scattered in Britain but mostly found in southern England. Greek *pteron* = 'wing' + the genus *Carya*, and Latin *fraxinifolia* = 'with ash-like leaves'.

WINTER-CRESS *Barbarea vulgaris* (Brassicaceae), a common and widespread biennial or perennial herb favouring damp substrates, widespread by streams, and growing on roadsides, by hedges and in ditches and waste ground. It requires disturbance, for example from seasonal flooding or human activity. *Barbarea* derives from St Barbara. *Vulgaris* is Latin for 'common'.

American winter-cress *B. verna*, a biennial, occasionally annual neophyte which despite its common name is of European origin, grown for centuries as a substitute for water-cress, and now widely cultivated again after a hiatus in the eighteenth and nineteenth centuries, and scattered and locally common

as a garden escape on waste ground, by roads and on railways. *Verna* is Latin for 'spring'.

Medium-flowered winter-cress *B. intermedia*, a biennial European neophyte (sometimes perennial or annual), of waste ground, disturbed habitats, and by roadsides, once common as an arable weed. *Intermedia* is Latin for 'medium'.

Small-flowered winter-cress *B. stricta*, a biennial or perennial European neophyte present by the mid-nineteenth century, found in damp habitats by rivers, ditches, canals and marsh, naturalised in a few places in southern England and south Wales, a rare casual of waste places elsewhere. *Stricta* is Latin for 'upright'.

WINTER MOTH *Operophtera brumata* (Geometridae, Larentiinae), wingspan 22–33 mm, nocturnal, widespread and very common in woodland, orchards, parks and gardens. Females have only vestigial wings and crawl up tree-trunks to await the attention of the males, which are active from late autumn to January–February. Larvae feed on a number of tree species (also heather), sometimes in great numbers to reach pest status, especially in orchards, occasionally completely defoliating small trees. Greek *opora* + *ptheiro* = 'fruit' + 'to destroy', and Latin *bruma* = 'winter'.

Northern winter moth *O. fagata*, wingspan 32–40 mm, nocturnal, common in woodland, heathland, scrub, gardens and orchards throughout Britain, with three post-2000 records from Northern Ireland. Females have vestigial wings, but males fly in October–December. Larvae feed on a number of tree species, including birches, alder, beech and apple. *Fagata* derives from *Fagus* = 'beech'.

WINTERGREEN Rhizomatous, mycorrhizal, herbaceous, evergreen perennials in the Ericaceae. See also chickweed-wintergreen.

Common wintergreen *Pyrola minor*, scattered, most widespread in northern England and Scotland. In southern England it is found in damp woodlands with deep litter; elsewhere it grows in damp places in heathland, plantations and disused railways, and on rock ledges. The binomial is Latin for 'pear-like' and 'smaller'.

Intermediate wintergreen *P. media*, locally frequent in the Highlands, local elsewhere in Scotland, northern England and north Ireland, on well-drained, mildly acidic to slightly basic soils in woodland and heathland, and characteristic of bearberry-heather submontane heath derived from former woodland.

One-flowered wintergreen *Moneses uniflora*, rare, found in north Scotland (this is the county flower of Moray as voted by the public in a 2002 survey by Plantlife), spreading in ericaceous dwarf shrub communities of mature pine plantations and native Scots pine forest. Greek *monos* = 'alone' and Latin *uniflora* = 'one-flowered', both for the solitary flower.

Round-leaved wintergreen *P. rotundifolia*, local, scattered and generally declining. In England it usually grows in damp, calcareous sites including fen and disused chalk-pits. In Scotland it inhabits open pine woodland, riverbanks and gullies in open moorland. Ssp. *maritima* is especially found in south-east England, north and south Wales, and Lancashire on calcareous dune slacks and fixed dunes, usually growing on drier fringes of slacks, and persisting in conifer plantations on dunes. *Rotundifolia* is Latin for 'round-flowered'.

Serrated wintergreen *Orthilia secunda*, mainly found in the Highlands, very local in northern England, central Wales and Ireland, in damp heathy birch and pine woodland, moorland, wet rock ledges and rocky stream banks. Greek *orthos* = 'straight', from the one-sided raceme, as with Latin *secunda* = 'secund', here meaning having the leaves in a row down one side.

WIREPLANT *Muehlenbeckia complexa* (Polygonaceae), a sprawling or weakly climbing deciduous neophyte shrub introduced from New Zealand in 1842, grown in gardens in mild climates in the Channel Islands, Scillies and south-west and southern England, and becoming increasingly established on cliffs, rocky slopes and walls, in hedgerows and quarries, and on waste ground. Heinrich Mühlenbeck (1798–1845) was a physician and plant collector from Alsace. *Complexa* is Latin for 'twining'.

WIREWEED *Sargassum muticum* (Sargassaceae), an invasive brown seaweed, the thallus up to 4 m long, supported by air bladders, native to the Pacific and appearing at Bembridge, Isle of Wight, in 1973, having spread there from France, and now found intertidally and subtidally along the south and south-west English coasts, Wales, Isle if Man, the Firth of Clyde, and north-east and west Ireland. Fast-growing, it competes with native plants such as sea-grasses and is a nuisance in estuaries, harbours, beaches and shallow waters. Reproductive plants detach easily, continuing to reproduce while drifting. *Sargasso* is Portuguese for this kind of seaweed. *Muticum* is Latin for 'blunt'.

WIREWORM Larvae of click beetles in the genera *Agriotes* and *Athous* (Elateridae), many with a widespread distribution, growing to lengths of up to 30 mm, and living for up to four years, often becoming pests of potato, carrot and other root crops.

WITCHES BEARD Alternative name for witches' whiskers lichen.

WITCH'S HAIR Species of fruticose lichen in the genus *Alectoria* (order Lecanorales), alectoria being the name dating from the fourteenth century of a magical stone supposedly found in the crop of a cock. Species are local on acid and often base-rich rock in the Highlands, **green witch's hair** *A. ochroleuca* also on Anglesey.

WITCH'S HAT Alternative name for blackening waxcap.

WITHY Name given to the osier, or the tough flexible branch of an osier or other willow used for binding or basketry, from Old English *withig*.

WOAD *Isatis tinctoria* (Brassicaceae), a biennial or perennial archaeophyte herb, found in ruderal habitats such as quarries, arable fields, docks and waste ground. It has been used as a dye plant since the Iron Age, but in the 1930s the world's last two woad mills, in Lincolnshire, were closed and since then it has generally been a rare casual, though it persists in a few places. 'Woad' derives from Old English *wad*. *Isatis* is the Ancient Greek name of this plant, *tinctoria* Latin for 'related to dyeing'.

WOLF See bee wolf, a solitary wasp.

WOLF SPIDER Members of the family Lycosidae, body lengths 3.5–18 mm, ground-hunting predators. See also fox-spider under spider.

Coastal wolf spider *Pardosa purbeckensis*, widespread around the British coastline and locally common, in saltmarsh and tidal habitats on mudflats and estuaries, running over mud and tidal debris. As the tide comes in, it stays at the base of the plants or walks down the stem and submerges itself, taking a bubble of air trapped on the dense hairs on the abdomen. Adults are found from early to midsummer. *Pardos* is Greek for 'leopard'. *Purbeckensis* refers to the Isle of Purbeck, Dorset.

Common wolf spider *Pa. pullata*, common and widespread, mainly found in summer in open habitats such as grassland, dunes, heathland, moorland, bog, woodland clearings and verges. In lowland agricultural areas it is most often found in tussocky grassland. *Pullata* is Latin for 'clothed in black' (*pulla* = 'dark').

Pirate wolf spider *Pirata piraticus*, female body length up to 9 mm, males 6 mm, widespread and locally common, an ambush or actively hunting predator in wetland habitats, for example pond and stream margins, marsh, fen, upland blanket bog, sphagnum seeps and reedbed, able to move on the water surface because of water-repellent hairs on its legs, found between late spring and autumn, but mainly from early to midsummer. *Pirata* is Latin for 'pirate'.

Rustic wolf spider *Trochosa ruricola*, female body length up to 15 mm, males 10 mm, common and widespread in England and Wales, scattered in Scotland, rare in Ireland, a ground hunter mainly found from late spring to midsummer, often under stones and logs in damp habitats. Greek *trochos* = 'wheel', and Latin *ruricola* = 'rustic' (living in the countryside).

Spotted wolf spider *Pa. amentata*, body length 5–8 mm, common and widespread, found in summer in often damp open habitats, and in gardens is often the commonest in its genus. It actively hunts on the ground. *Amentata* is Latin for 'strapped'.

WOLF'S-BANE *Aconitum lycoctonum* (Ranunculaceae), a perennial herb introduced from central Europe, an occasional garden escape found in woodland and by streams. It is poisonous, the principal alkaloids being aconite and (the key toxin) aconitine. Ingestion of even a small amount results in severe gastrointestinal upset, but it is by slowing the heart rate that it can be fatal. 'Wolfsbane' comes from Latin *lycoctonum*, in turn from Greek *lycotonon*, from *lycos* + *cteino* = 'wolf' + 'to kill'. *Aconiton* is Greek for this plant.

WOOD (BUTTERFLY) See speckled wood.

WOOD-BORING BEETLE See borer (beetle).

WOOD-MOSS Species of *Hylocomium* (Hylocomiaceae) (Greek *hyle* + *come* = 'wood' + 'hair').

Glittering wood-moss *H. splendens*, common and widespread in heathland, moorland and acidic woodland. *Splendens* is Latin for 'bright'.

Oake's wood-moss *H. pyrenaicum*, a rarity in the Highlands, scattered in turf in rocky sites, in scree and at the edge of stony flushes, favouring base-rich habitats. *Pyrenaicum* is Latin for 'Pyrenean'.

Shaded wood-moss *H. umbratum*, locally common in montane western Britain and western Ireland in open woodland, turf, rocks and scree. *Umbratum* is Latin for 'shaded'.

Short-beaked wood-moss *H. brevirostre*, common and widespread, especially in mountainous parts, in woodland on grassy banks, on rocks and on trees and logs, favouring damp, base-rich habitats. *Brevirostre* is Latin for 'short-beaked'.

WOOD-RUSH Herbaceous perennials in the genus *Luzula* (Juncaceae), being the Latinised form of the Italian *lucciola* for these plants. **Curved wood-rush** *L. arcuata* and **spiked wood-rush** *L. spicata* are both local in the Highlands on bare rocky ridges.

Field wood-rush *L. campestris*, common and widespread in short, relatively infertile grassland on moderately acidic to slightly alkaline soil, and on spoil tips. *Campestris* is from Latin for 'field'.

Great wood-rush *L. sylvatica*, common and widespread except in parts of central and eastern England, and central Ireland, in damp, acidic, shaded habitats such as woodland, on peaty heathland and moorland, and on rocky montane streamsides. It is intolerant of grazing. *Sylvatica* is Latin for 'forest'.

Hairy wood-rush *L. pilosa*, common and widespread, more local in Ireland, in woodland and moist but well-drained shaded habitats such as on roadside banks, in hedgerows, and among heather on moorland, generally on fairly acidic soils. *Pilosa* is Latin for 'hairy'.

Heath wood-rush *L. multiflora*, common and widespread, preferring fairly acid soils and shaded places, on heathland, moorland and drier parts of bogs, in meadow, and in open woodland and woodland margins. *Multiflora* is Latin for 'many-flowered'.

Southern wood-rush *L. forsteri*, locally common in the Channel Islands, southern England, the south-west Midlands and south Wales, in damp woodland and other moist, well-drained shaded habitats, often on roadside banks and in hedgerows. It is commonest on mildly acidic soils. Johann Forster (1729–98) was a German-born naturalist.

White wood-rush *L. luzuloides*, a European neophyte introduced before 1800, scattered and naturalised as a garden escape in woodland and other moist shady habitats, often by streams, sometimes in peaty habitats in upland areas, usually on acidic soils.

WOOD-RUST Alternative name for rustwort *Nowellia curvifolia*.

WOOD-SEDGE *Carex sylvatica* (Cyperaceae), a common and widespread perennial found in woodland, scrub and hedges, especially where the soil is base-enriched, moist and clayey, and frequent along the sides of paths and tracks. 'Sedge' comes from Old English *secg*. The binomial is Latin for 'sedge' and 'forest'.

Starved wood-sedge *C. depauperata*, a perennial with Critically Endangered status, protected under Schedule 8 of the WCA (1981), and listed in Ireland on the Flora (Protection) Order (1999). Once reduced to a single plant, populations are now recovering through restocking, appropriate habitat management and chance, but the species is currently found only at Axbridge, Somerset, and Ockford Wood, Godalming, Surrey (both sites in semi-shade on lane-side banks), and Killavullen, Co. Cork (in a wooded limestone area). *Depauperata* is Latin for 'impoverished', probably from its producing few stems each with just one or two flowers.

Thin-spiked wood-sedge *C. strigosa*, a perennial, locally common in England, scattered in Wales and Ireland, on moist, base-rich, sometimes clayey soils in woodland, often near streams or seepages, most frequently in clearings and along tracks, though sometimes in deep shade. *Strigosus* is Latin for 'featureless', but may here derive from *stria* = 'bristle'.

WOOD-SORREL *Oxalis acetosella* (Oxalidaceae), a common and widespread perennial creeping herb of woodland, hedgerows, and other moist, usually shaded, habitats; also in montane grassland and in limestone pavement. It is one of the few species able to survive the deep shade of conifer plantations. Greek *oxys* = 'sharp' or 'sour', and New Latin *acetosella* = 'a little acid' or 'sour'.

WOOD SPARK *Perrotia flammea* (Heliotales, Hyaloscyphaceae), a rare fungus found in the Highlands on dead wood of willows and aspen. Louis Perrot (1785–1865) was a French botanist. *Flammea* is Latin for 'flame-coloured'.

WOODBINE Synonym of honeysuckle.

WOODCOCK *Scolopax rusticola* (Scolopacidae), wingspan 58 cm, length 34 cm, weight 280 g, widespread in damp woodland and heathland, with an estimated 78,000 (62,000–96,000) males in 2003, though numbers fell by 76% over 25 years, and it is Red-listed in Britain (Amber in Ireland). The male performs a courtship flight known as 'roding' at dusk in spring. Nests are lined cups on the ground with a clutch usually of 4 eggs, incubation taking 21–4 days, and fledging of the precocial young in 15–20 days. It is also a summer migrant in the Highlands, and a winter visitor, especially in south-west England. Its eyes are set far back on its head to give it 360° vision. Feeding is mainly on earthworms, but also beetles and other insects, and plant material, by probing in damp ground usually at night. Greek *scolopax* = 'woodcock', and Latin *rusticus* + *colere* = 'rural' + 'to live'.

WOODHOPPER *Arcitalitrus dorrieni* (Talitridae), the only amphipod to occur inland, originally discovered on the Isles of Scilly, this native of east Australia is now well established in south-west England and occurs patchily along the west coasts of Britain as far north as the Western Isles, and is continuing to spread, probably dispersed via plant nurseries and garden centres. It occurs under stones and dead wood and among detritus and in gardens, damp scrub and woodland. Latin *arcus* + *talitrum* = 'bow' + 'a finger rap'.

WOODLARK *Lullula arborea* (Alaudidae), wingspan 28 cm, length 15 cm, weight 30 g, despite its name found in heathland and woodland edge rather than woodland itself, with 3,100 pairs (2,500–3,700 pairs) estimated in 2006, breeding mainly in southern and eastern England, for example in the New Forest, Surrey–Berkshire heathland, Breckland and some Suffolk heaths. Overwintering birds are usually found in Hampshire, west Surrey and Devon, and in recent years in East Anglia. Numbers increased by 126% over 1970–2015, and by 62% in 2010–15. It was moved from the Red list to Amber in 2009, and to Green in 2015. The nest is made of grass, bracken, roots and moss, constructed in a depression on the ground. Clutch size is 3–5, incubation lasts 14 days, and fledging of the altricial young 11–13 days. It feeds on the ground on invertebrates, also taking seeds in winter. *Lullula* is an onomatopoeic representation of the song; *arborea* is from Latin for 'tree'. In Scotland 'woodlark' refers to tree pipit.

WOODLOUSE Isopod crustaceans with a rigid, segmented, long exoskeleton and 14 jointed limbs, most species being nocturnal and feeding on dead plant material.

Ant woodlouse *Platyarthrus hoffmannseggii* (Platyarthridae), 5 mm long and blind, living underground in habitats such as gardens, verges and semi-natural grassland in association with ants such as *Lasius* and *Myrmica*, probably tolerated because they keep the ant nest clear of droppings; they also eat fungi. It is widespread across much of southern Britain and south-east Ireland, becoming rare in northern England, and restricted to coastal sites in Scotland. Greek *platys* + *arthron* = 'flat' + 'joint'. Johann Graf von Hoffmannsegg (1766–1849) was a German naturalist.

Beach pill woodlouse *Armadillidium album* (Armadillidiidae), up to 6 mm long, uncommon, sporadic around the British coastline as far north as the Scottish Borders, and known from several sites on the east coast of Ireland, in undisturbed dune systems, occasionally in saltmarsh, typically associated with storm strand-line debris, or at depths of 20–30 cm in the sand. *Armadillidium* is a diminutive of Spanish *armado* = 'armed man', from Latin *armatus* = 'armed'. *Album* is Latin for 'white'.

Common pill woodlouse *A. vulgare*, up to 18 mm long, locally abundant in south-east England in many habitats, becoming scarcer further north, restricted to lowland sites with calcareous soils and high levels of insolation. Scottish sites are mostly coastal. It is commonly found under stones and dead wood, but also in grass litter and within tussocks. *Vulgare* is Latin for 'common'.

Common pygmy woodlouse 1) *Trichoniscus provisorius* (Trichoniscidae), widespread, less common in Scotland, especially associated with sunny, dry, calcareous sites including semi-natural limestone grassland, woodland and gardens, found under stones, dead wood, etc. Greek *trichos* = 'hair' + *oniscos*, literally 'ass' or 'onion', but used for various arthropods. *Provisorius* is Latin for 'worker'. 2) *T. pusillus*, similar to (and often confused with) *T. provisorius*, but better suited to cooler, damper habitats, and usually found close to the soil surface, typically on the underside of stones and dead wood or among leaf litter and moss. *Pusillus* is Latin for 'small'.

Common shiny woodlouse *Oniscus asellus* ssp. *asellus* (Oniscidae), with rough woodlouse the most abundant woodlouse, widespread in woodland, hedge banks, damp meadow, coastal cliffs, waste ground, parks and gardens. Large numbers are often found under dead wood and stones, in leaf litter and compost heaps, among rubbish and inside damp bathrooms. In south-west England, south Wales and southern Ireland it is partially replaced by a genetically distinct subspecies, western shiny woodlouse *Oniscus asellus* ssp. *occidentalis*, preferring damper and non-synanthropic habitats. *Oniscos*, literally Greek for 'ass' and 'onion', is also used for various arthropods. *Asellus* is Latin for 'little ass'.

Common striped woodlouse *Philoscia muscorum* (Philosciidae), 11 mm in length, abundant in lowland England, Wales and Ireland, becoming less common in northern England, and restricted to coastal sites and low-lying river valleys in Scotland, found in a variety of habitats, gregarious, typically spending the day concealed beneath stones, logs, etc. feeding on dead organic material (including cowpats).

Least pygmy woodlouse *T. pygmaeus*, common and widespread wherever there is suitable friable soil, including semi-natural grassland, woodland and supralittoral coastal habitats, and synanthropic sites such as churchyards, old quarries and railway cuttings.

Painted woodlouse *Porcellio spinicornis* (Porcellionidae), widespread and locally common in dry, relatively exposed, calcareous substrates such as dry limestone and loosely mortared walls, but also quarries, cuttings, cliffs, and occasionally on trees with calcareous bark. It is often associated with buildings. *Porcellio* is Latin for 'woodlouse', *spinicornis* for 'spined horn'.

Pill woodlouse Mainly *Armadillidium*. They can roll themselves into a tight ball when disturbed.

Plum woodlouse *Porcellionides pruinosus* (Porcellionidae), up to 12 mm long, widespread but most common in central and eastern England characteristically in manure heaps, but also associated with outbuildings of dairy farms, stables or riding schools, and compost heaps in gardens and allotments. *Porcellionides* is from the genus *Porcellio* + Greek *eides* = 'appearance (of)', and Latin *pruinosus* = 'frosted'.

Rathke's woodlouse *Trachelipus rathkii* (Trachelipodidae), up to 15 mm, mostly recorded in south-east England, tolerant of flooding, making it a characteristic woodlouse of riverside meadows. It also inhabits poorly drained scrub, gravel-pits, churchyards, etc., under stones and dead wood, beneath bark on logs, among grass litter, within tussocks and in flood debris. Jens Rathke (1769–1855) was a Norwegian zoologist. Greek *trachelos* + *pous* = 'neck' + 'foot'.

Rosy woodlouse *Androniscus dentiger* (Trichoniscidae), common and widespread in lowland regions in semi-natural coastal habitats such as thinly vegetated shingle, boulder beaches, and synanthropic sites such as churchyards, gardens and waste ground, often on the underside of stones or among rubble. Greek *andros* + *oniscos* = 'man' + 'ass', and Latin *dens* + *gero* = 'tooth' + 'to carry'.

Rough woodlouse *Porcellio scaber*, up to 17 mm long, widespread, with common shiny woodlouse the most commonly found species (though more tolerant of drier conditions), found under stones and dead wood, and the woodlouse most commonly encountered inside houses. *Scaber* is Latin for 'rough'.

Rough pygmy woodlouse *Trichoniscoides albidus* (Trichoniscidae), widespread in southern and eastern England, rare in Ireland, in damp soils in wet deciduous woodland, alluvial meadows, stream banks, ditches and farmyards, usually on the underside of stones and dead wood, and in damp rubble. *Trichoniscoides* is from the genus *Trichoniscus* + Greek *eidos* = 'appearance (of)', and Latin *albidus* = 'whitish'.

Western shiny woodlouse A subspecies of common shiny woodlouse.

WOODPECKER Birds in the family Picidae.

Great spotted woodpecker *Dendrocopos major*, wingspan 36 cm, length 22 cm, weight 85 g, found in woodland (and gardens) throughout Britain (with 130,000–150,000 pairs in 2009), and a colonist in eastern Ireland since 2007 or 2008, nesting in oakwood at a few sites. Numbers increased by 350% between 1970 and 2015, and by 136% between 1995 and 2015, though including no change noted in 2010–15. Tree-hole nests contain 4–6 eggs, incubation taking 14–16 days, the altricial young fledging in 20–4 days. Feeding is on insects, seed and nuts. Communication includes drumming on tree-trunks (10–40 strikes per second) making the timber resonate, the force on the bill offset by shock-absorbent tissue at the base of the skull. Greek *dendron* + *copos* = 'tree' + 'striking', and Latin *major* = 'greater'.

Green woodpecker *Picus viridis*, wingspan 41 cm, length 32 cm, weight 190 g, locally common in Britain except for north-west Scotland, Amber-listed, with a population in 2009 of 52,000 (47,000–58,000), with a doubling of numbers between 1970 and 2015, and an increase of 31% between 1995 and 2015, though including a weak decline (–6%) over 2010–15. Nests are in tree holes, containing 4–6 eggs, incubation taking 19–20 days, the altricial young fledging in 21–4 days. Feeding is on ants on short grass. Its loud calls are known as 'yaffling', giving this species its dialect name of yaffle. Greek *picos* = 'woodpecker' and Latin *viridis* = 'green'.

Lesser spotted woodpecker *Dryobates minor*, wingspan 26 cm, length 14 cm, weight 21 g, a Red List species, scarce in England and Wales, totalling 1,000–2,000 pairs in 2009, highest densities being in south-east England. Data suggest a strong decline in numbers, by 83% between 1970 and 2015, and 38% in 2010–15. The tree-hole nest contains 4–6 eggs, incubated for 14–15 days, the altricial young fledging in 19–22 days. Diet is mainly of insects and spiders, generally taken from the tree canopy. Communication includes drumming on tree-trunks making the timber resonate, the force on the bill offset by shock-absorbent tissue at the base of the skull. Greek *dryos* + *bates* = 'woodland' + 'walker', and Latin *minor* = 'smaller'.

WOODPIGEON *Columba palumbus* (Columbidae), wingspan 78 cm, length 41 cm, weight 450 g, common and widespread in woodland and field, and increasingly seen in urban parks and gardens, with 5.3 million pairs estimated in 2009, and numbers in Britain increasing by 36% between 1995 and 2015 (and by 122% since 1970). Winter numbers are augmented by continental birds. The stick nests built in trees usually contain two eggs that hatch after 17 days, the altricial young fledging in 33–4 days. Diet is mainly of plant material: leaves, seeds, berries, buds, flowers and root and brassica crops. *Columba* is Latin for 'dove', *palumbus* for 'woodpigeon'.

WOODRUFF 1) or **sweet woodruff** *Galium odoratum* (Rubiaceae), a common and widespread rhizomatous perennial herb of deciduous woodland, scrub and shaded hedge banks on base-rich or neutral, often damp, soils, and a good indicator of ancient woodland in lowland England. 'Woodruff' dates to Old English *wudurofe*, from *wudu* = 'wood' + an element of unknown meaning. Greek *galion* = 'milk', some species having been used to curdle milk in cheese-making, and Latin *odoratum* = 'fragrant'.

2) Species of *Asperula* (Rubiaceae), from Latin diminutive of *aspera* = 'rough'.

Blue woodruff *A. arvensis*, an annual European neophyte, introduced into cultivation by the sixteenth century, scattered in England and Wales, generally via birdseed, as a casual on waste places and rubbish tips. *Arvensis* is from Latin for 'field'.

Pink woodruff *A. taurina*, a herbaceous perennial south European neophyte introduced by 1739, locally naturalised in damp woodland and on shaded riverbanks in central Scotland. *Taurina* is Latin for 'like a bull' or 'with a tough hide'.

WOODSIA Ferns in the genus *Woodsia* (Woodsiaceae), **alpine woodsia** *W. alpina* and **oblong woodsia** *W. ilvensis* both very rare, recorded from rock crevices in upland Scotland, the latter also from Cumbria. Joseph Woods (1776–1864) was an English botanist.

WOODSIACEAE Fern family named in honour of the English botanist Joseph Woods (1776–1864). See bladder-fern, lady-fern, oak fern under fern and woodsia.

WOODTUFT Sheathed woodtuft *Kuehneromyces mutabilis* (Agaricales, Strophariaceae), a common, widespread and tasty fungus, found from spring to winter on stumps and dead wood of broadleaf trees, especially birch. The binomial honours the American mycologist Calvin Kuehner (1922–2011) + the Greek *myces* = 'fungus', with the Latin *mutabilis* = 'changeable', from the cap colour varying with moisture level.

WOODWART 1) Inedible fungi in the genus *Hypoxylon* (Xylariales, Xylariaceae) (Greek *hypo* + *xylon* = 'beneath' + 'wood'). All are widespread and common, evident year-round as globular fruiting bodies or crusts on dead wood, sporulating in autumn. **Beech woodwart** *H. fragiforme* is found on beech wood, **birch woodwart** *H. multiforme* on birch, **hazel woodwart** *H. fuscum* on hazel and alder, and **rusty woodwart** *H. rubiginosum* on ash.

2) **Woolly woodwart** *Lasiosphaeria ovina* (Sordaliales,

Lasiosphaeriaceae), minute white spherical fungi, widespread and locally common in England, rare elsewhere, found year-round though mainly from late summer to late autumn, spreading as a mat over rotten wood. Greek *lasios* + *sphaera* = 'woolly' + 'sphere', and from Latin *ovis* = 'sheep'.

WOODWASP Giant woodwasp Alternative name for greater horntail wasp.

Sirex woodwasp Alternative name for lesser horntail wasp.

WOODWAX Fungi in the genus *Hygrophorus* (Agaricales, Hygrophoraceae) (Greek *hygros* + *phoros* = 'wet' + 'bearing'). See also herald of winter.

Rarities include **almond woodwax** *H. agathosmus* (mainly Highlands), **arched woodwax** *H. camarophyllus* (12 records between 1982 and 2009 scattered over Britain and Ireland), **oak woodwax** *H. nemoreus*, **rosy woodwax** *H. pudorinus* (one recent record from Hampshire), and **twotone woodwax** *H. unicolor*.

Ashen woodwax *H. mesotephrus*, locally common in southern England, occasional in Wales and Ireland in autumn on broadleaf woodland soil, especially under beech. Greek *mesos* + *tephra* = 'middle' + 'ashes'.

Blistered woodwax *H. pustulatus*, uncommon, on soil in coniferous plantation in northern England and Scotland, generally under spruce. *Pustulatus* is Latin for 'blistered'.

Forest woodwax *H. arbustivus*, with a scattered distribution in southern England and the Welsh Borders, Near Threatened in the RDL (2006), in autumn on soil and litter in broadleaf woodland. *Arbustivus* is Latin for 'tree-planted'.

Gold flecked woodwax *H. chrysodon*, uncommon, scattered in the southern half of England in autumn and early winter on soil in coniferous and mixed woodland, especially under pines, and often beside trackways. Greek *chrysos* + *odontos* = 'gold' + 'teeth'.

Ivory woodwax *H. eburneus*, edible, widespread and locally common, more evident in the southern half of England from summer to late autumn on often calcareous soil in grass in broadleaf woodland, mycorrhizal with oak and beech, and (disturbance tolerant) often found on verges and by trackways. *Eburneus* is Latin for 'ivory'.

Larch woodwax *H. lucorum*, locally common in the south-east quarter of England in soil and litter in coniferous plantations under larch. *Lucorum* is Latin for 'green'.

Matt woodwax *H. penarius*, with a scattered distribution in England, Vulnerable in the RDL (2006), in autumn on broadleaf woodland soil, especially under beech.

Yellowing woodwax *H. discoxanthus*, scattered in the southern half of England, in autumn on soil in broadleaf woodland, often under beech. Greek *discos* + *xanthos* = 'disc' + 'yellow'.

WOODWORM (BEETLE) (Damage caused by) one of a number of wood-boring beetles, for example death-watch, furniture, house longhorn or powderpost beetles.

WOOLLY BEAR 1) Caterpillar of moths in the family Erebidae, usually more specifically that of the garden tiger. 2) Alternative name for the larva of the museum beetle or larder beetle.

WOOLLYFOOT Wood woollyfoot *Gymnopus peronatus* (Marasmiaceae), a common, widespread, poisonous fungus found in summer and autumn in soil and leaf litter in deciduous, less often coniferous woodland, sometimes under bracken in heathland. Its common name comes from the lower half of the stem being covered in fine white hairs. Greek *gymnos* + *pous* = 'naked' + 'foot'; and Latin *peronatus* = 'sheathed', referring to the woolly-booted appearance of the stem.

WOOLLYWORT Handsome woollywort *Trichocolea tomentella* (Trichocoleaceae), a locally common foliose liverwort found in valleys in the lowlands of western Britain, especially on damp stream banks, and the mossy tops of stones in water. Greek *trichos* + *coleos* = 'hair' + 'sheath', and Latin *tomentosus* = 'hairy'.

WORM See earthworm, and bootlace, coral, football jersey, parchment, Ross, sludge and horseshoe worms. See also slow worm and Nematomorpha (horse worm).

WORMWOOD, THE (MOTH) *Cucullia absinthii* (Noctuidae, Cucullinae), wingspan 32–40 mm, on cliffs, quarries and embankments along the coasts of the southern half of England and of Wales. From the middle of the twentieth century it has also been recorded inland on derelict industrial sites in the Midlands, but records on such sites have since become fewer. Adults fly at night in July. Larvae feed on the flowers and seed of wormwood (*Artemisia absinthium*, giving the English and specific names) and mugwort. Latin *cucullus* = 'hood', from the cowl-like thoracic crest.

WORMWOOD (PLANT) *Artemisia absinthium* (Asteraceae), an aromatic archaeophyte perennial herb, cultivated since at least 1200 for flavouring and medicine (aiding digestion, reducing flatulence), common and widespread in England and Wales, scattered elsewhere, on waste and rough ground, waysides, railway sidings, gravel-pits, quarries and other anthropogenic habitats. 'Wormwood' comes from Old English *wermod*. Artemis was the Greek goddess of chastity. *Apsinthion* is Greek for this plant.

Field wormwood *A. campestris*, a herbaceous perennial (ssp. *campestris*) found at three sites in the Breckland in short open grassland, grass-heath, on forest rides and tracks, in abandoned arable fields, and on roadsides. A naturalised population (ssp. *maritima*) on dunes in Glamorgan since 1956 appears to be in decline. A population on dunes at Crosby, Lancashire, has been noted since 2004. *Campestris* is from Latin for 'field'.

Sea wormwood *A. maritima*, an aromatic herbaceous perennial locally common on coasts north to Arran and Aberdeen, and on the west and east sides of Ireland, in the upper, drier parts of saltmarsh; also on shingle, sea-cliffs, waste ground and walls close to the sea, and on the banks of tidal rivers.

WOUNDWORT Species of *Stachys* (Lamiaceae), Greek for 'flower spike'. See also yellow-woundwort.

Downy woundwort *S. germanica*, a biennial or short-lived perennial herb of woodland margins, grassy banks, ancient hedgerows and green lanes overlying oolitic limestone. It has recently been recorded from only four localities in an area of 20 x 7 km in west Oxfordshire, where survival is precarious, though conservation work has helped maintain the populations. It is Endangered in the RDB, and listed in Schedule 8 of the WCA 1981.

Field woundwort *S. arvensis*, an archaeophyte summer- or winter-annual, widespread but scattered, less common in northern England, Scotland and much of Ireland, in arable fields, allotments and gardens, waste ground and verges, usually on non-calcareous soils, though it occurs on limestone outcrops in western Ireland. The English name is that of a *wyrt* (Old English), or herb, that staunches blood and serves as an antiseptic. *Arvensis* is from Latin for 'field'.

Hedge woundwort *S. sylvatica*, a common and widespread rhizomatous perennial of woodland, hedgerows, stream banks, rough grassland and waste ground, and a persistent garden weed, characteristically growing in moist, fertile, mildly acidic to basic soils in disturbed and moderately shaded sites. *Sylvatica* is from Latin for 'woodland'.

Limestone woundwort *S. alpina*, a perennial European neophyte in cultivation by 1597, though possibly native, in open woodland, woodland edge, hedge banks and trackways on thin soils overlying calcareous rock. There are only two known populations: one in Gloucestershire may have spread shortly after its discovery in 1897, but is now apparently stable, maintained by sowing seed. It was also found in Denbighshire in 1927, but these colonies are slowly declining and remaining plants are often reintroductions. This is the county flower of Denbighshire as voted by the public in a 2002 survey by Plantlife.

Marsh woundwort *S. palustris*, a common and widespread

perennial growing by streams, ditches and ponds, in fen, marsh and swamp, on rough ground and in places in arable fields. *Palustris* is Latin for 'marsh'.

WRACK Large brown seaweeds (macroalgae) in the order Fucales.

Bladder wrack *Fucus vesiculosus* (Fucaceae), thallus up to 2 m, common and widespread, on rocky substrates in the midlittoral. *Phycos* is Greek for these macroalgae. Latin *vesiculosus* refers to the small vesicles or 'bladders'.

Bushy berry wrack *Cystoseira baccata* (Sargassaceae), thallus up to 1 m long, in the lower intertidal in large sandy pools or lagoons in south England, west Ireland, and west Scotland. Greek *cystis + seira* = 'bladder' + 'chain' or 'rope', Latin *baccata* = 'berries'.

Channelled wrack *Pelvetia canaliculata* (Fucaceae), thallus 8–15 cm long, common and widespread on hard substrates on the upper littoral, tolerating very sheltered to moderately exposed conditions. It is usually infected by an obligate endophytic fungus *Mycosphaerella acophylli* (Ascomycetes), which may help it survive high in the intertidal. *Canaliculata* is Latin for 'small channel'.

Guiry's wrack *F. guiryi*, scarce in the intertidal in parts of Britain, the Isle of Man and west Ireland. Michael Guiry (b.1949) is an Irish algologist.

Horned wrack *F. ceranoides*, thallus 30 cm long, widespread in estuaries and near freshwater streams on the midshore, attached to stones, rocks or gravel. It does not have air bladders, but the side of the fronds are often inflated. *Ceranoides* comes from Greek *ceraos* = 'horned'.

Knotted wrack *Ascophyllum nodosum* (Fucaceae), common and widespread, bladdered fronds 0.5–2 m long, these attaching to midshore rocks and boulders in a range of habitats, from estuaries to relatively exposed coasts. Var. *mackaii* is found on very sheltered shores in sea lochs, especially common on the west coasts of Scotland and Ireland. In Ireland it is sustainably harvested for the extraction of alginic acid, a polysaccharide used in foods and in biotechnology. Greek *ascos + phyllon* = 'bladder' + 'leaf', and Latin *nodosum* = 'knotted'.

Rainbow wrack *C. nodicaulis* (Latin for 'knotted stalk') and *C. tamariscifolia* ('tamarisk-like leaves'), with thalli up to 1 m long, found in large tidal rock pools and lagoons, and in the shallow subtidal, often together, commoner in south and south-west England, Wales and Ireland, but reaching the Hebrides.

Saw wrack Alternative name for serrated wrack.

Serrated wrack *F. serratus*, common and widespread, thalli up to 30 cm long, on hard substrates on the lower littoral in more sheltered areas.

Spiral wrack *F. spiralis*, common and widespread, thalli up to 40 cm long, in the upper littoral attached to rocky substrates on sheltered to moderately exposed shores.

Toothed wrack Alternative name for serrated wrack.

WRANGELIACEAE Family of red algae in the order Ceramiales. The name may refer to the Russian explorer Ferdinand von Wrangel (1797–1870).

WRASSE Fish in the family Labridae, mostly found on rocky shores. See also goldsinny and rock cook. 'Wrasse' dates from the late seventeenth century, from the Cornish *wrah* (related to the Welsh *gwrach* = 'old woman').

Baillon's wrasse *Symphodus bailloni*, length 18 cm, widespread on rocky shores among seaweed and in rock pools, extending to depths of 50 m, feeding on crustaceans and molluscs. Males build and guard a dish-shaped nest of seaweed in rock crevices or among seaweed or sea-grasses in sediment areas. Greek *syn + odous* = 'together' + 'tooth'.

Ballan wrasse *Labrus bergylta*, length 30–50 cm, common inshore off all coasts among seaweed-covered rocks and in lower rock pools, and in the subtidal on rocky coasts from 5 to 30 m. All are female for their first 4–14 years before a few change into males. Food is mainly of crustaceans and molluscs. Spawning is in June when the female builds a nest of fine seaweed in a crevice. Some stocks are dwindling as wrasse are caught off the English coast for introduction into Scottish salmon fisheries as cleaner fish to control fish lice, although some wrasse are being specifically reared for such activity, for example at Loch Fyne. 'Ballan' may be mid-eighteenth-century, either from Gaelic for a 'ball spot' or for 'pail' or 'tub'. Greek *labros* = 'furious'.

Corkwing wrasse *S. melops*, length 15–25 cm, widespread on rocky shores among seaweed and in rock pools, feeding on crustaceans and molluscs. Males build and guard a nest of mounded seaweed in rock crevices or among seaweed or sea-grasses in sediment areas.

Cuckoo wrasse *L. mixtus*, male length up to 35 cm, females 30 cm, found around most of the coastline (less so on the east) over rocks and hard substrates and in the algal zone between 2 and 200 m, feeding on crustaceans and molluscs. Young fish are all female, a few changing to males at about six years old. The English name comes from Cornish fishermen who associated the male's blue markings with bluebell flowers; in Cornish, a bluebell is *bleujenn an gog* = 'cuckooflower'. *Mixtus* is Latin for 'mixed'.

Small-mouthed wrasse Alternative name for rock cook.

WREN *Troglodytes troglodytes* (Troglodytidae), wingspan 15 cm, length 10 cm, weight 10 g, common and widespread in a range of habitats, especially deciduous woodland, but also gardens, farmland, heathland and moorland, Britain's commonest breeding bird, with 7.7 million territories in 2009, with an increase in numbers by 67% over 1970–2015, 20% between 1995 and 2015, and 31% over 2010–15. The globe-shaped nest of grass, moss, lichens or leaves is built in a hole in a tree-trunk or wall, or a crack in a rock, with 5–6 eggs which hatch after 16–18 days, the altricial young fledging in 15–18 days. Diet is of insects and spiders. The common name has evolved from Old English *wrenna*. Greek *trogle + dytes* = 'hole' + 'diver', from the cave-like nest.

WRYNECK *Jynx torquilla* (Picidae), wingspan 26 cm, length 16 cm, weight 38 g, a rare passage migrant in the woodpecker family, seen in autumn, less so in spring, in south and especially eastern England, the Highlands, Orkney and Shetland, occasionally breeding. In the early 1800s wrynecks bred in almost every county of England and Wales, but numbers then declined to 40–80 pairs in 1966, and only one pair recorded (in Buckinghamshire in 1985) after 1977. Since 1978, single breeding pairs have been confirmed only eight times in the Highland region, and twice in north-east Scotland, the most recent being in Ross-shire (Highland) in 2002. Nests are in tree holes with 7–10 eggs, incubation takes 12–14 days, and the altricial young fledge in 18–22 days. Feeding is mainly on ants. The English name comes from the bird's ability to turn its heads almost 180°. Greek *iunx* = 'wryneck' and Latin *torquere* = 'to twist'.

X

XANTHIDAE Family name of furrowed and Risso's crabs (Greek *xanthos* = 'yellow').

XANTHORIA Cushion xanthoria *Xanthoria polycarpa* (Teloschistaceae), a common and widespread foliose lichen, nitrogen-tolerant and so found on nutrient-enriched branches and twigs, especially of willows and poplars. Greek *xanthos* = 'yellow' and *polycarpa* = 'many-fruited'.

 Leafy xanthoria Synonym of golden shield lichen.

XANTHORRHOEACEAE Family name of asphodel, from Greek *xanthos* + *rheo* = 'yellow' + 'to flow', from the yellow gum that can be extracted from some species.

Y

Y (MOTH) Members of the family Noctuidae (Plusiinae).

Beautiful golden Y *Autographa pulchrina*, wingspan 35–40 mm, widespread and common in a range of habitats, including woodland, downland, hedgerows and gardens, though scarcer in Scotland. Adults fly in June–August. Larvae feed on stinging nettle, dead-nettles, hogweed and honeysuckle. Greek *autographos* = 'autograph', from markings on the forewings, and Latin *pulcher* = 'beautiful'.

Plain golden Y *A. jota*, wingspan 38–46 mm, common and widespread. Adults fly in June–August. Larvae feed on a range of plants, including nettle. *Jota* comes from the Greek letter *iota* ('i'), for the forewing mark.

Scarce silver Y *Syngrapha interrogationis*, wingspan 34–41 mm, day- and night-flying in June–August, on moorland throughout Scotland and northern England, locally in Ireland, Wales and south-west England, and an occasional migrant in south-east England. Larvae feed on bilberry, heather, etc. Greek *syn* = 'with' + contracted form of *Autographa*, and Latin *interrogationis* = 'question mark', from the forewing mark.

Silver Y *A. gamma*, wingspan 35–40 mm, a common and widespread immigrant moth that can number in thousands. In autumn the breeding population from spring migrants is swelled by further migration. Adults are found from spring until late autumn, flying by day and night. Larvae feed on a range of herbaceous plants. *Gamma* is the Greek letter 'g', the capital letter resembling a Y, from the forewing mark.

YAFFLE Alternative name for green woodpecker, from its loud calls, known as yaffling.

YARD-GRASS Species of *Eleusine* (Poaceae), generally annual South American, African and Asian neophytes, introduced with wool shoddy and as grain impurities, occasionally found as casuals on waste ground. The genus is named after Eleusis, a Greek town where the Temple of Ceres (goddess of grain and agriculture) was located.

YARROW *Achillea millefolium* (Asteraceae), a common and widespread herbaceous perennial found in a variety of (usually short) grassland habitats, from lawns to montane communities, and on coastal dunes and stabilised shingle, waysides and waste ground. It tolerates drought, and grows in most soils except the most nutrient-poor, waterlogged or acidic. 'Yarrow' comes from Old English *gearwe*. *Achillea* is named after Achilles, who used plants of this genus to staunch the wounds of his soldiers at the siege of Troy. *Millefolium* is Latin for 'a thousand leaves'.

Fern-leaf yarrow *A. filipendulina*, a west Asian neophyte herb scattered in England and Wales, naturalised in rough ground, and on walls and roadsides. Latin *filipendulina* = '*Filipendula* (meadowsweet)-like'.

Southern yarrow *A. ligustica*, from southern Europe, naturalised on waste ground at Newport Docks, Monmouthshire, since 1953. *Ligustica* is Latin for 'from Liguria', north-west Italy.

Tall yarrow *A. distans*, a European neophyte scattered in England and Wales, naturalised in grassy habitats. *Distans* is Latin for 'separated'.

YEAST A microscopic single-celled fungus. Using fermentation, yeasts such as *Saccharomyces cerevisiae* convert carbohydrates to carbon dioxide (used as a leavening agent in baking) and alcohol (used for beer and wine).

YELLOW (BUTTERFLY) Clouded yellow *Colias croceus* (Pieridae, Coliadinae), male wingspan 52–8 mm, female 54–62 mm, common and widespread on a variety of open habitats, though less so in the northern half of Britain, primarily an immigrant butterfly from North Africa and southern Europe. Overwintering has been noted in southern England in recent years. Distribution increased by 84% over 1976–2014, but 2005–14 saw a decline by 19%, and over these two periods numbers, respectively, increased by 734% and declined by 57%, a reflection of how both range and numbers fluctuate from year to year. Larvae feed on leguminous plants, especially clovers and lucerne. Adults take nectar from a variety of plants, including thistles, vetches, knapweeds and ragwort. Colias is the name of a promontory in Greece where there was a temple dedicated to Aphrodite. *Croceus* is Latin for 'saffron yellow'.

YELLOW (MOTH) Members of the family Geometridae.

Barred yellow *Cidaria fulvata* (Larentiinae), wingspan 20–5 mm, widespread and common in broadleaf woodland, calcareous grassland, hedgerows, scrub and gardens. Adults fly at night in June–July. Larvae feed on leaves of roses. Cidaria is a variant name of Ceres, goddess of grain and agriculture. *Fulvus* is Latin for 'tawny yellow'.

Speckled yellow *Pseudopanthera macularia* (Ennominae), wingspan 23–8 mm, day-flying (May–June), common in southern England, Wales and the Burren, Ireland, more scattered in Scotland, in open woodland and scrub, larvae mainly feeding on wood sage, but also white dead-nettle, woundwort and yellow archangel. Greek *pseudos* + *panther* + 'false' + 'leopard', from the black spots on yellow, and Latin *macula* = 'spot'.

YELLOW-CRESS Herbaceous perennials in the genus *Rorippa* (Brassicaceae), the Latinised form of *rorippen*, a Saxon vernacular name.

Austrian yellow-cress *R. austriaca*, a European neophyte, probably introduced with grain, scattered in England and south Wales, naturalised on waste ground, beside roads and rivers, and in meadows.

Creeping yellow-cress *R. sylvestris*, common and widespread except for north Scotland, on damp bare ground, often in disturbed, winter-flooded habitats, on the margins and banks of streams, canals and ditches, by lakes and ponds, and in depressions in pastures. It is also an arable weed. It is a vigorous pioneer, spreading by seed and by broken pieces of rhizome, but intolerant of competition. *Sylvestris* is Latin for 'woodland' or 'wild'.

Great yellow-cress *R. amphibia*, common and widespread in much of England and Ireland, scarce elsewhere, along the edges of streams, by lakes and ponds, often in sites which are flooded in winter and where some water remains in the summer, and generally where the water is calcareous and eutrophic. Spread is mainly by fragmentation of mature plants. *Amphibia* is Latin for growing both in and out of water.

Hybrid yellow-cress *Rorippa* x *anceps* (*R. sylvestris* x *R. amphibia*), an often fertile hybrid found in damp places, commonly in the absence of one or other parent. Scattered in Britain, where most records are from riverbanks, but in Ireland it is as frequent on lake shores. *Anceps* is Latin for 'doubtful'.

Marsh yellow-cress *R. palustris*, summer-annual, rarely perennial, locally common except in north Scotland as an early colonist of riverbanks, mud exposed above receding lake margins, and summer-dry ponds. It is also an occasional weed in arable fields, railway cinders, waste ground and gardens. *Palustris* is from Latin for 'marsh'.

Northern yellow-cress *R. islandica*, an annual or short-lived perennial distinguished from marsh yellow-cress in 1968, scattered mainly in Cumbria, south-west Wales, Scotland and north and west Ireland, in open, muddy habitats, for example lake, pond and pool margins, ditch banks and turloughs. It has also been recorded from waste ground. *Islandica* is Latin for 'Icelandic'.

Thames yellow-cress *Rorippa* x *erythrocaulis* (*R. palustris* x *R. amphibia*), local by the Thames in Surrey and Middlesex, and the Avon in Warwickshire, all with both parents, and by the Avon in east Gloucestershire, without either. *Erythrocaulis* is Greek for 'red-stemmed'.

YELLOW-EYED-GRASS *Sisyrinchium californicum* (Iridaceae), a semi-evergreen herbaceous North American neophyte perennial introduced in 1796, scattered and naturalised on lake shores, in marshy meadows and in other damp, grassy habitats often near the coast. *Sisyrinchium* is from Greek for 'pig's snout', referring to the roots being eaten by swine.

Pale yellow-eyed-grass *S. striatum*, a herbaceous South American neophyte perennial, scattered in England and Wales, naturalised or casual on waste ground and grassy habitats. *Striatum* is Latin for 'striped'.

YELLOW HORNED (MOTH) *Achlya flavicornis* (Drepanidae, Thyatirinae), wingspan 35–40 mm, widespread and common in woodland, and on heathland and moorland. Adults fly at night in March–April. Larvae feed on birch leaves. Greek *achlys* = 'of darkness', possibly referring to time of flying, and Latin *flavus* + *cornu* = 'yellow' + 'horn', from the colour of the antennae.

YELLOW MAY Angler's name for the mayfly *Heptagenia sulphurea* (Heptageniidae).

YELLOW-RATTLE *Rhinanthus minor* (Orobanchaceae), a common and widespread annual root-hemiparasite of nutrient-poor grasslands, including permanent pasture, hay meadows, drier parts of fens, flushes in lowland and upland grassland, and on roadsides and waste ground. By taking nutrients from grasses it reduces competition for other plants, thereby increasing species diversity in created wild-flower meadow. Greek *rhinos* + *anthos* = 'nose' + 'flower', from the shape of the upper lip of the corolla, and Latin *minor* = 'smaller'.

Greater yellow-rattle *R. angustifolius*, an annual root hemiparasite once a common and widespread arable weed in eastern Britain but now local on the Surrey North Downs in grassland and open scrub on chalk. In Lincolnshire, it grows on peat in an area of cleared bracken and on railway ballast. In Angus, a small colony survives in sandy coastal grassland. It is a rare casual in a few other places. *Angustifolius* is Latin for 'narrow-leaved'.

YELLOW SALLY Fisherman's name for some perlodid stoneflies; small yellow sally for chloroperlid stoneflies.

YELLOW-SEDGE Perennial grass-like members of the genus *Carex* (Cyperaceae), from Old English *secg*, *carex* being Latin for these plants.

Common yellow-sedge *C. demissa*, common and widespread, scarcer in the south and east Midlands and East Anglia, on wet ground on soils ranging from acidic to neutral in open habitats such as sandy shores, tracks, streamsides, wet fields and heathland, and pond and lake margins. *Demissa* is Latin for 'drooping'.

Large yellow-sedge *C. flava*, now found only in Roudsea Wood, Cumbria, and at Malham Tarn, Yorkshire Dales, on peaty fen soil flushed by calcareous water. *Flava* is Latin for 'yellow'.

Long-stalked yellow-sedge *C. lepidocarpa*, growing in fen and calcareous mire, especially at sites with winter flooding, over much of upland Britain and Ireland. It is also frequent base-rich flushes and wet ledges on sea and inland cliffs. Greek *lepis* + *carpos* = 'scale' + 'fruit'.

Small-fruited yellow-sedge *C. oederi*, widespread if scattered, more frequent by the coast in north and west Scotland in dune slacks, upper saltmarsh, lake and pond margins, fen and marsh, on acidic and, locally, on base-rich soils. George Oeder (1729–91) was a Danish botanist.

YELLOW-SORREL Naturalised neophyte garden weeds in the genus *Oxalis* (Oxalidaceae) (Greek *oxys* = 'sharp' or 'sour'), including **Chilean yellow-sorrel** *O. valdiviensis* (found in central and southern England and north Wales), **least yellow-sorrel** *O. exilis* (from New Zealand and Tasmania, on walls and waste ground in England and Wales), **procumbent yellow-sorrel** *O. corniculata* (present by 1656, naturalised on walls and waste ground, common in much of England and Wales, scarce in Scotland and Ireland), and **upright yellow-sorrel** *O. stricta* (introduced from North America in 1658, now scattered over much of Britain).

YELLOW-TAIL (MOTH) *Euproctis similis* (Erebidae, Lymantriinae), wingspan 28–35 mm, common in hedgerows, woodland, scrub and gardens throughout much of England, parts of Wales, the east coast of Ireland, and increasingly in Scotland. Adults fly at night in July–August. Larvae feed on tree and shrub species. Larval hairs are extremely irritating, causing rashes on exposed parts of human skin. Greek *eu* + *proctos* = 'good' + 'anus', from the anal tuft, and Latin *similis* = 'resembling', from the similarity with the brown-tail.

YELLOW-WORT *Blackstonia perfoliata* (Gentianaceae), an annual or biennial herb common and widespread in England, Wales and much of Ireland, on open dry (though commonly winter-wet), shallow basic soils. Its main habitats are calcareous grasslands and fixed dunes, but also disturbed ground such as quarries and railway cuttings, and verges and path sides. John Blackstone (1713–53) was an English botanist. Latin *perfoliata* = 'perfoliate' (the stem penetrating the leaf).

YELLOW-WOUNDWORT Species of *Stachys* (Lamiaceae), from Greek for 'flower spike'. See also woundwort.

Annual yellow-woundwort *S. annua*, an annual European neophyte cultivated by 1713, rare and scattered in southern England, in arable fields, on waste ground and occasionally in gardens, usually casual but sometimes persisting, and originating as a contaminant of grain and oil-seed.

Perennial yellow-woundwort *S. recta*, a perennial herb naturalised on waste ground at Barry Docks, Glamorgan, and casual at four widely separated sites in England. Latin *recta* = 'upright'.

YELLOWHAMMER *Emberiza citrinella* (Emberizidae), wingspan 26 cm, length 16 cm, weight 31 g, a widespread resident seed-eating bunting found in open countryside with scrub and hedges, with an estimated 700,000 territories in 2009, but with numbers declining by 56% between 1970 and 2015 and by 14% over 1995–2015, though only by 2% over 2010–15, a Red List species. The cup nest on or near the ground usually contains 3–4 eggs which hatch after 13–14 days, the altricial young fledging in 13–16 days. The male's song is traditionally rendered as having a rhythm like 'a little bit of bread and no cheese'. The English name dates from the mid-sixteenth century, derived from *yelambre*, from Old English *geolu* = 'yellow' + *amore*, a kind of bird, possibly conflated with *hama* = 'feathers', or from Old High German *amaro* = (a bird that feeds on) emmer wheat. Old German *embritz* = 'bunting', and Latin *citreus* = 'of the citrus tree' (i.e. 'yellow').

YELLOWLEGS (FUNGUS) Alternative name for trumpet chanterelle.

YELLOWSTREAK Dusky yellowstreak *Electrogena lateralis* (Heptageniidae), common though localised in upland Britain. Nymphs live in stream riffle areas, occasionally on wave-affected lake shores, clinging to submerged plants and stones, feeding by scraping periphyton from the substrate or by gathering fine particulate organic detritus from the sediment. Adults emerge in May–September; unusually for mayflies they can do so under water, flying immediately they break the water surface thereby minimising risk of predation. Greek *electron* + *genos* = 'amber' + 'race', Latin *lateralis* = 'lateral'.

YEW *Taxus baccata* (Taxaceae), on chalk (for example as ancient woodland in the North Downs) and limestone, also some acid sandstone. Commonly planted as an individual tree, especially in churchyards, and as a hedging or topiary plant. 'Yew' comes from Old English *iw* or *eow*. *Taxus* is Latin for yew (from Greek *taxon* = 'bow', the tree being used in making longbows), and *baccata* = 'berried'.

YOKE-MOSS Species of *Amphidium* (Rhabdoweisiaceae) and *Zygodon* (Orthotrichaceae) (Greek *zygos* + *odontos* = 'yoke' + 'teeth').

Green yoke-moss *Z. viridissimus*, common and widespread on base-enriched bark such as of ash. It is also common on shaded walls, and occasionally grows on base-rich rocks and concrete. Latin *viridissimus* = 'very green'.

Knothole yoke-moss *Z. forsteri*, uncommon in southern England, growing mainly on pollarded or damaged, veteran beech trees in well-lit sites, favouring bark where water runs down. Benjamin Forster (1764–1829) was an English botanist.

Lapland yoke-moss *A. lapponicum*, an upland species of base-rich crags.

Lesser yoke-moss *Z. conoideus*, common and widespread in humid woodland in valleys in western Britain, growing on the trunks of ash, sycamore and willow. Further east, it is commonest on elder.

Mougeot's yoke-moss *A. mougeotii*, a common and widespread upland species found on siliceous crags with some base enrichment. Jean-Baptiste Mougeot (1776–1858) was a French botanist.

Park yoke-moss *Z. rupestris*, scattered but mainly found in southern and south-west England, Wales and west Scotland, favouring the bark of old trees, especially ash and oak, and in parks. It occasionally grows on shaded, base-rich rocks. *Rupestris* is from Latin for 'rocks'.

Slender yoke-moss *Z. gracilis*, a rarity in the mid-Pennines on old dry-stone walls of Carboniferous limestone and unshaded limestone rock. *Gracilis* is Latin for 'slender'.

YORKSHIRE-FOG *Holcus lanatus* (Poaceae), a common and widespread perennial grass found in hay meadow, pasture, chalk and limestone downland, and in hedge banks, open woodland and moorland. It grows in dry to winter-wet, acidic to calcareous soils, favouring moist habitats. It withstands mowing and grazing, but not heavy trampling. *Holcos* is Greek for this kind of grass, *lanatus* Latin for 'woolly'.

YPONOMEUTIDAE Family name for ermine and some of the ermel moths, from Greek *yponomeuo* = 'to make underground mines', larvae being leaf miners.

YPSOLOPHIDAE Family name (Greek *hypsilophos* = 'high-crested') for some smudge moths, stem-moths and honeysuckle moth.

Z

ZANDER *Stizostedion lucioperca* (Percidae), length up to 130 cm, weight 12 kg, introduced to Woburn (Bedfordshire) from North America in 1878, distributed to other water bodies in the twentieth century, and now found in a number of places in England, particularly in East Anglia, in both deep and shallow, still and slow-flowing water, preferring oxygen-rich and turbid conditions with little vegetation. Adults hunt smaller fish, and they are cannibalistic. Spawning is on shallow sandy or stony bottoms in April–June. 'Zander' is German for this species. Greek *stizo* + (possibly) *stethion* = 'to puncture' + 'small breast' (a further suggestion is to mean 'pungent throat'), and Latin *lucioperca* = 'pike-perch'.

ZIPHIIDAE Family name of toothed whales, from Greek *xiphos* = 'sword', a reference to the distinctive beak.

ZIPPERBACK *Chrysotoxum elegans* (Syrphidae), wing length 9.5–12 mm, a hoverfly found on dry open grassland, mainly in south-west England and south-west Wales, and on chalk downland in south-east England. Adults feed on nectar and pollen from spring to early autumn. Larvae probably feed on root aphids being farmed by ants for their honeydew. In 2010, *The Guardian* launched an annual competition for the public to suggest common names for previously unnamed species, this being a winner in 2011 from the discontinuous, zipper-like bands across the abdomen unlike the continuous bands of true wasps. Greek *chrysos* + *toxon* = 'gold' + 'a bow', and Latin *elegans* = 'elegant'.

ZODARIIDAE Family name for ant-eating spider, with four species of *Zodarion*, Greek for 'bovine', reason unclear.

ZOPHERIDAE See Colydiidae.

ZORIDAE Family name of ghost spiders (Greek *zora* = 'strong' or 'pure').

ZOSTERACEAE Family name for eelgrass (Greek *zoster* = 'belt' or 'girdle', from the ribbon-shaped leaves).

ZYGAENIDAE Family name for burnet and forester moths. The name, used by Fabricius, is Greek for 'hammerhead shark', clearly with no entomological meaning, though it may come from *zygos* = 'yoke'.

ZYGOPTERA Suborder of damselflies (Greek *zygos* + *pteron* = 'yoke' + 'wing') with four families in Britain: Calypterygidae, Lestidae, Platycnemididae and Coenagriidae.

Bibliography

Books

Atherton, I.D.M., Bosanquet, S.D.S. and Lawley, M. (eds) (2010) *Mosses and Liverworts of Britain and Ireland: A Field Guide*. British Bryological Society (privately published).

Barnard, P.C. (2011) *The Royal Entomological Society Book of British Insects*. Oxford: Wiley-Blackwell.

Battersby, J. and Tracking Mammals Partnership (eds) (2005) *UK Mammals: Species Status and Population Trends*. Peterborough: JNCC/Tracking Mammals Partnership. Available as a PDF at http://jncc.defra.gov.uk/pdf/pub05_ukmammals_speciesstatusText_final.pdf (accessed 30 March 2019).

Benton, T. (2006) *Bumblebees* (New Naturalist 98). London: HarperCollins.

Cameron, R. (2016) *Slugs and Snails* (New Naturalist 133). London: HarperCollins.

Chambers, P. (2009) *British Seashells: A Guide for Collectors and Beachcombers*. Barnsley: Pen & Sword Books.

Corbet G.B. and Harris, S. (1991) *The Handbook of British Mammals*, 3rd edition. Oxford: Blackwell Scientific Publications.

Dobson, F. (2011) *Lichens: An Illustrated Guide to the British and Irish Species*, 6th edn. Slough: Richmond Publishing.

Duff, A.D. (2012) *Checklist of Beetles of the British Isles*, 2nd edn. Iver: Pemberley Books.

Edwards, S.R. (2012). *English Names for British Bryophytes*. British Bryological Society, Special Vol 5 (privately published).

Everard, M. (2013) *Britain's Freshwater Fishes*. Princeton, NJ: Princeton University Press.

Fox, R., Asher, J., Brereton, T., Roy, D. and Warren, M. (2006), *The State of Butterflies in Britain and Ireland*. Newbury: Pisces Publications.

Gilbert, O.L. (2000) *Lichens* (New Naturalist 86). London: HarperCollins.

Gledhill, D. (2008) *The Names of Plants*, 4th edn. Cambridge: Cambridge University Press.

Grigson, G. (1974) *A Dictionary of English Plant Names*. London: Allen Lane.

Harris, S. and Yalden, D.W. (eds) (2008) *Mammals of the British Isles*, 4th edn. Southampton: The Mammal Society.

Hayward, P.J. and Ryland, J.S. (2017) *Handbook of the Marine Fauna of North-west Europe*, 2nd edn. Oxford: Oxford University Press.

Jaeger, E.C. (1950) *A Source-book of Biological Names and Terms*, 2nd edn. Springfield, IL: Charles C. Thomas Publishers.

Jobling, J.A. (2010) *The Helm Dictionary of Scientific Bird Names: From Aalge to Zusii*, new edn. London: Christopher Helm.

Jones, R. (2018) *Beetles* (New Naturalist 136). London: HarperCollins.

Lever, C. (2009) *The Naturalized Animals of Britain and Ireland*. London: New Holland Publishers.

Perks, J. (2016) *Freshwater Fishes of Britain*. London: New Holland Publishers.

Porley, R.D. (2013) *England's Rare Mosses and Liverworts: Their History, Ecology, and Conservation*. Princeton, NJ: Princeton University Press.

Preston, C.D., Pearman, D.A. and Dines, T.D. (2002) *New Atlas of the British and Irish flora: An Atlas of the Vascular Plants of Britain, Ireland, the Isle of Man and the Channel Islands*. Oxford: Oxford University Press.

Roberts, M.J. (1993) *The Spiders of Great Britain and Ireland*, compact edn. Vester Skerninge, Denmark: Apollo Books (orig. pub. Harley Books).

Roy, H., Booy, O. and Wade, P.M. (2015) *Field Guide to Invasive Plants and Animals in Britain*. London: Bloomsbury.

Sell, P. and Murrell, G. (1997–2018), *Flora of Great Britain and Ireland*, vols. 1–5. Cambridge: Cambridge University Press.

Smith, A.J.E. and Smith, R. (2004) *The Moss Flora of Britain and Ireland*, 2nd edn. Cambridge: Cambridge University Press.

Smith, C., Aptroot, A., Coppins, B., Fletcher, A., Gilbert, O., James, P. and Wolseley, P. (eds) (2009) *The Lichens of Great Britain and Ireland*. London: British Lichen Society.

Stace, C. (2010) *New Flora of the British Isles*, 3rd edn. Cambridge: Cambridge University Press.

Stace, C.A. and Crawley, M.J. (2015) *Alien Plants* (New Naturalist 129). London: HarperCollins.

Stearn, W.T. (1996) *Stearn's Dictionary of Plant Names for Gardeners*, revised edn. London: Cassell.

Watson, L. and Dallwitz, M.J. (2003–) *The Families of British Spiders*, version 4 January 2012. Available at http://delta-intkey.com (accessed 30 March 2019).

Organisations and web resources

All web sites accessed 30 March 2019

General

British Plant Gall Society: https://www.british-galls.org.uk
First Nature: https://www.first-nature.com
GB Non-native Species Secretariat: http://www.nonnativespecies.org/home/index.cfm
Ireland's Wildlife: https://www.irelandswildlife.com/category/species-profiles
Marine Life Information Network: https://www.marlin.ac.uk/species
National Biodiversity Data Centre (Ireland): http://www.biodiversityireland.ie
National Biodiversity Network: https://nbnatlas.org
National Museums Northern Ireland: http://www.habitas.org.uk
Natural History Museum: http://www.nhm.ac.uk/our-science/data/uk-species.html

Mammals

Bat Conservation Trust: http://www.bats.org.uk/pages/uk_bats.html
Conserve Ireland: http://www.conserveireland.com
Mammal Society: http://www.mammal.org.uk

Birds

BirdWatchIreland: https://www.birdwatchireland.ie/IrelandsBirds/tabid/541/Default.aspx
British Trust for Ornithology: https://www.bto.org/about-birds/birdfacts/british-list
British Trust for Ornithology BirdTrends 2017: https://www.bto.org/about-birds/birdtrends/2017
Royal Society for the Protection of Birds: https://www.rspb.org.uk/birds-and-wildlife/wildlife-guides/bird-a-z

Reptiles and Amphibians

Reptiles & Amphibians of the UK: http://www.herpetofauna.co.uk

Fishes

List of Fish in Ireland: https://en.wikipedia.org/wiki/List_of_fish_of_Ireland
List of Fishes of Great Britain: https://en.wikipedia.org/wiki/List_of_fishes_of_Great_Britain

Insects

Amateur Entomologists' Society: https://www.amentsoc.org/insects/fact-files/orders/

Bees, Wasps and Ants Recording Society: http://www.bwars.com/species_gallery
British Bugs: http://www.britishbugs.org.uk/systematic_het.html
British Dragonfly Society: https://www.british-dragonflies.org.uk/content/uk-species
Bumblebee Conservation Trust: https://www.bumblebeeconservation.org
Butterfly Conservation: https://butterfly-conservation.org
Dipterists Forum: http://www.dipteristsforum.org.uk
Elateridae of the British Isles: http://elateridae.co.uk/species-accounts
Gelechiid Recording Scheme: http://www.gelechiid.co.uk/full-species-list
Leaf and Stem Mines of British Flies and Other Insects: http://www.ukflymines.co.uk
Moths Ireland: http://www.mothsireland.com/specieslist.htm
Orthoptera and Allied Insects: https://www.orthoptera.org.uk/species
Riverfly Partnership (Plecoptera, Ephemeroptera, Trichoptera): http://www.riverflies.
    org/riverflies-1
UK Beetle Recording: http://www.coleoptera.org.uk/beetle-families
UK Butterflies: http://www.ukbutterflies.co.uk/index.php
UK Moths: https://www.ukmoths.org.uk/systematic-list
Urban Butterfly Garden: http://urbanbutterflygarden.co.uk

Arachnids

British Arachnological Society: britishspiders.org.uk
Spider and Harvestman Recording Scheme: http://srs.britishspiders.org.uk/
    portal/p/A-Z+Species+Index

Myriapods and Isopods

British Myriapod and Isopod Group: http://www.bmig.org.uk

Molluscs

Marine Bivalve Shells of the British Isles: https://naturalhistory.museumwales.ac.uk/
    BritishBivalves/Browse_taxa.php
Non-marine Molluscs of Great Britain: https://en.wikipedia.org/wiki/List_of_
    non-marine_molluscs_of_Great_Britain
Non-marine Molluscs of Ireland: https://en.wikipedia.org/wiki/List_of_non-marine_
    molluscs_of_Ireland

Vascular Plants

Botanical Society of Britain & Ireland: http://sppaccounts.bsbi.org
Ecological Flora of the British Isles: http://ecoflora.org.uk
Online Atlas of the British and Irish Flora: https://www.brc.ac.uk/plantatlas
Plantlife: http://www.plantlife.org.uk/uk/discover-wild-plants-nature/plant-fungi
    -species

Mosses

British Bryological Society: http://rbg-web2.rbge.org.uk/bbs/Activities/BBSFGspac.
    htm

Lichens

British Lichen Society: http://www.britishlichensociety.org.uk/resources/species
    -accounts
Lichens of Wales: http://wales-lichens.org.uk/content/species-communities

Fungi

British Mycological Society: https://www.britmycolsoc.org.uk
British Mycological Society GB Checklist of Fungal Names: http://www.nhm.ac.uk/
    our-science/data/uk-species/checklists/BMSSYS0000000001/version2.html
First Nature: https://www.first-nature.com/fungi/index1binom.php
Fungi and Lichens of Great Britain and Ireland: http://fungi.myspecies.info